Biochemistry & Molecular Biology of Plants

Bob B. Buchanan
University of California
Berkeley

Wilhelm Gruissem
Swiss Federal Institute
of Technology
Zürich

Russell L. Jones
University of California
Berkeley

American Society of Plant Physiologists
Rockville, Maryland

Cover: Portion of the ribbon structure of dinitrogenase (MoFe-protein); see Figure 16.7, adapted with permission from Schindelin H., Kisker, C., Schlessman, J. L., Howard, J. B., Rees, D. C. 1987. Structure of ADP•IAF$_4$–stabilized nitrogenase complex and its implications for signal transduction. *Nature* 387: 370–376. © 1987 Macmillan Magazines Ltd.

Part-title pages

Part 1: Iceplant (*Carpobrotus chilensis*) at sunset, coastal bluffs of Marin Headlands, California, 1991; Part 2: Spring wildflowers on Vollmer Peak, Tilden Regional Park, Berkeley, California; Part 3: Sunbeams in morning fog, Berkeley, California, 1996; Part 4: Coast redwood, Big Basin Redwoods State Park, California, 1995; Part 5: The High Sierra as seen from the Central Valley, near Madera, California, on a clear winter day. Used with permission from Galen Rowell/Mountain Light Photography, Emeryville, California.

Illustration concepting and illustrations: J/B Woolsey Associates
Text design and composition: J/B Woolsey Associates

Cover design: Barbara Alonso
Cover production: Debra Naylor, Naylor Design, Inc.

http://www.aspb.org/biotext

Address editorial correspondence to American Society of Plant Biologists, *Biochemistry and Molecular Biology of Plants*, 15501 Monona Drive, Rockville, MD 20855-2768 USA.

For North, Central, and South America
Customer Service
John Wiley & Sons, Inc.
Distribution Center
1 Wiley Drive
Somerset, NJ 08875-1272
Phone: (732) 469-4400 or (800) 225-5945
Fax: (732) 302-2300
E-mail: bookinfo@wiley.com

Rest of World
Customer Service Department
John Wiley & Sons, Ltd.
1 Oldlands Way
Bognor Regis
West Sussex PO22 9SA
United Kingdom
Phone: 44 (0) 1243-843291
Fax: 44 (0) 1243-843303
E-mail: cs-books@wiley.co.uk

Library of Congress Cataloging-in-Publication Data
Biochemistry and molecular biology of plants / edited by Bob B. Buchanan, Wilhelm Gruissem, Russell L. Jones.

     p. cm.
   Includes bibliographical references (p.).
   ISBN 0-943088-37-2 hardcover (alk. paper)
   ISBN 0-943088-39-9 softcover (alk. paper)
   ISBN 0-943088-40-2 CD-ROM of illustrations

   1. Botanical chemistry. 2. Plant molecular biology. I. Buchanan, Bob B. II. Gruissem, Wilhelm. III. Jones, Russell L.

   QK861 .B45 2000
   572.8'2–dc21

                     00-040591

Printed in the United States of America
First impression (split edition) July 2000, Courier Companies, Inc.
Second impression (casebound) September 2000, Courier Companies, Inc.
Third impression (paperback) February 2001, Courier Companies, Inc.
Fourth impression (paperback) July 2002, Courier Companies, Inc.
Fifth impression (paperback) May 2005, Courier Companies, Inc.
Sixth impression (paperback) Oct 2006, Printed in Singapore, Craft Print International Limited.

Joe Varner
1921–1995

Preface

The origin of this book lies with Joe Varner, who in the mid-1990s decided to develop a third edition of *Plant Biochemistry*, the highly successful textbook he had edited with James Bonner 30 years earlier. Unfortunately, Joe died before the wheels could be put in motion. Recognizing the need to maintain this important resource, the American Society of Plant Physiologists asked us to take on the project. We agreed, but soon found the scope growing beyond the bounds of a traditional biochemistry book.

Reflecting on the needs and future of the field, we concluded that for a contemporary biochemistry textbook to be of maximal use, it should present the biochemistry of plants in the context of relevant elements of their physiology and cellular and molecular biology. The ASPP leadership enthusiastically supported this concept, and the plan was thus put in place.

We have organized *Biochemistry and Molecular Biology of Plants* around the elements required for life: membranes, energy and metabolism, and reproduction. The first four of the five sections of the book follow this theme. The fifth section, however, represents a diversion and extends relevant scientific fundamentals to environmental aspects of biochemistry and biotechnology—dynamic areas in which the unique capabilities of plants are applied to solve contemporary societal problems.

The development and production of this book required the talent, expertise, and sustained effort of many individuals. We wish to highlight the efforts of the contributors, who not only admirably integrated information from diverse fields in composing the chapters, but also endured what seemed at times to be an unending series of editorial suggestions and revisions to both text and artwork. Thanks are in order to the able reviewers of the individual chapters; to the ASPP publications staff—both full-time and free-lance; to Kimberly Cline and Liz Burke, who coordinated the project from the University of California at Berkeley; and to the illustration and production staff at J/B Woolsey Associates.

We especially wish to acknowledge the outstanding contribution of Kathleen Vickers, our developmental editor. We are indebted to Kathleen not only for her perseverance and positive attitude in meeting an endless array of deadlines, but also for her unfailing vigilance to scientific accuracy and for her persistent efforts to integrate diverse material to make a whole from many parts. It is difficult to imagine completing this project without her.

Most important, we want to express appreciation to our wives, Melinda, Barbara, and Frances, who during the past four years not only tolerated the textbook, but came to accept it as a family member.

Bob B. Buchanan
Wilhelm Gruissem
Russell L. Jones

March 31, 2000
Berkeley, California

Acknowledgments

The editors wish to acknowledge the capable guidance and assistance given throughout this project by staff associated with the publisher—the American Society of Plant Physiologists—and the illustrator, J/B Woolsey Associates. The contributions of the individuals listed below are especially noteworthy.

ASPP

John Lisack, Jr., Executive Director
Nancy Winchester, Director of Publications
Kenneth Beam, former Executive Director
Melinda (Jody) Moore, former Director of Publications
Susan Wantland, former Director of Publications
Ellen Brennan, Indexer
Elizabeth Burke, Project Coordinator
Kimberly Cline, Project Coordinator
Morna Conway, Publishing Consultant, The Conway Group
Christine Cotting, Project Manager, UpperCase Publication Services, Ltd.
Virginia S. Marcum, Copyeditor, Mark-Em Editorial Services
Caroline Polk, Proofreader
Kathleen Vickers, Developmental Editor

J/B Woolsey Associates, Inc.

John B. Woolsey, President
Patrick Lane, Senior Art Director
Laura Colangelo, Production Manager
Greg Gambino, Compositor
Regina Santoro, Designer/Illustrator

The Editors

Bob B. Buchanan

A native Virginian, Bob B. Buchanan obtained his Ph.D. in microbiology at Duke University and did postdoctoral research at the University of California at Berkeley. In 1963, he joined the Berkeley faculty and is currently a professor in the Department of Plant and Microbial Biology. He has taught general biology and biochemistry to undergraduate students and graduate-level courses in plant biochemistry and photosynthesis. Initially focused on pathways and regulatory mechanisms in photosynthesis, his research has more recently dealt with the regulation of seed germination. This latter work is finding application in several areas.

Bob Buchanan has served as department chair at UC–Berkeley and was president of the American Society of Plant Physiologists from 1995 to 1996. A former Guggenheim Fellow, he is a member of the National Academy of Sciences and a fellow of the American Academy of Arts and Sciences and the American Association for the Advancement of Science. His other honors include the Bessenyei Medal from the Hungarian Ministry of Education, the Kettering Award for Excellence in Photosynthesis from the American Society of Plant Physiologists, and the Distinguished Achievement Award from his undergraduate alma mater, Emory and Henry College.

Wilhelm Gruissem

Wilhelm Gruissem was born in Germany, where he studied biology and chemistry and obtained his Ph.D. from the University of Bonn. He did postdoctoral research at the University of Marburg and the University of Colorado at Boulder, and in 1983 joined the faculty of the University of California at Berkeley. He chaired the Department of Plant and Microbial Biology at UC–Berkeley from 1993 to 1998, and since 1998 has been director of a collaborative research program between the department and the Novartis Agricultural Discovery Institute in San Diego. In July 2000, he will join the Swiss Federal Institute of Technology in Zürich as professor of plant biotechnology. He has taught general biology and plant molecular biology to undergraduate and graduate students. His research focuses on pathways and molecules involved in plant growth control and regulation of chloroplast development.

Willi Gruissem is an elected fellow of the American Association for the Advancement of Science and a member of several learned societies. He serves on the editorial boards of several professional journals and has received a number of honors and awards for his research program.

Russell L. Jones

Russell Jones was born in Wales and completed his B.Sc. and Ph.D. degrees at the University of Wales, Aberystwyth. He spent one year as a postdoctoral fellow at the Michigan State University–Department of Energy Plant Research Laboratory with Anton Lang before being appointed to the faculty of the Department of Botany at the University of California at Berkeley in 1966. He is now a professor of plant biology at UC–Berkeley, where he teaches undergraduate classes in general biology and graduate courses in plant physiology and cell biology. His research focuses on hormonal regulation in plants using the cereal aleurone as a model system, with approaches that exploit the techniques of biochemistry, biophysics, and cell and molecular biology.

Russell Jones was president of the American Society of Plant Physiologists from 1993 to 1994. He was a Guggenheim Fellow at the University of Nottingham in 1972, a Miller Professor at UC–Berkeley in 1976, a Humboldt Prize Winner at the University of Goettingen in 1986, and a RIKEN Eminent Scientist, RIKEN, Japan, in 1996.

The Contributors and Reviews

The Contributors

Julia Bailey-Serres
Department of Botany and
Plant Sciences
University of California, Riverside

Tobias Baskin
Department of Biological Science
University of Missouri, Columbia

Paul Bethke
Department of Plant and
Microbial Biology
University of California at Berkeley

J. Derek Bewley
Department of Botany
University of Guelph
Guelph, Ontario

Gerard Bishop
Institute of Biological Sciences
University of Wales, Aberystwyth

Stephen Blakeley
Department of Biology
Queens University
Kingston, Ontario

Elizabeth A. Bray
Department of Botany and
Plant Sciences
University of California, Riverside

John A. Browse
Institute of Biological Chemistry
Washington State University
Pullman

Nicholas Carpita
Department of Botany and
Plant Pathology
Purdue University
West Lafayette, Indiana

Maarten J. Chrispeels
Department of Biology
University of California San Diego

Gloria Coruzzi
Department of Biology
New York University
New York City

Nigel Crawford
Department of Biology
University of California San Diego

Rodney Croteau
Institute of Biological Chemistry
Washington State University
Pullman

Alan Crozier
Department of Botany
University of Glasgow
Glasgow, Scotland

Jeffrey L. Dangl
Department of Biology
University of North Carolina
Chapel Hill

David A. Day
Division of Biochemistry and
Molecular Biology
Australian National University
Canberra

David T. Dennis
Performance Plants, Inc.
Kingston, Ontario

Robert A. Dietrich
Novartis Crop Protection, Inc.
Research Triangle Park
North Carolina

Peter Doerner
Institute of Cell and
Molecular Biology
University of Edinburgh
Edinburgh, Scotland

Robert Ferl
Department of Horticultural
Sciences
University of Florida, Gainesville

Donald B. Fisher
Department of Botany
Washington State University
Pullman

Kim Hammond-Kosack
Monsanto Company
Cambridge, United Kingdom

Frederick D. Hempel
Department of Plant and
Microbial Biology
University of California at Berkeley

Jan G. Jaworski
Department of Chemistry
Miami University
Miami, Ohio

Jonathan D. G. Jones
The Sainsbury Laboratory
John Innes Centre
Norwich, United Kingdom

Michael Kahn
Institute of Biological Chemistry
Washington State University
Pullman

Yuji Kamiya
RIKEN, Wako-shi, Japan

Leon V. Kochian
U.S. Plant Soil and
Nutrition Laboratory
Cornell University
Ithaca, New York

Toni M. Kutchan
Leibniz Institut für
Pflanzenbiochemie
Universität Halle
Halle, Germany

Robert L. Last
Cereon Genomics LLC
Cambridge, Massachusetts

Thomas Leustek
Center for Agricultural
Molecular Biology
Rutgers University
New Brunswick, New Jersey

Norman G. Lewis
Institute of Biological Chemistry
Washington State University
Pullman

Sharon R. Long
Department of Biological Sciences
Stanford University
Stanford, California

Richard Malkin
Department of Plant and
Microbial Biology
University of California at Berkeley

Maureen McCann
Department of Cell Biology
John Innes Centre
Norwich, United Kingdom

Sheila McCormick
U.S. Department of Agriculture/
Plant Gene Expression Center
Albany, California

Eldon H. Newcomb
Department of Botany
University of Wisconsin, Madison

Krishna Niyogi
Department of Plant and
Microbial Biology
University of California at Berkeley

John B. Ohlrogge
Department of Botany
Michigan State University
East Lansing

Anna Lisa Paul
Department of Horticultural
Sciences
University of Florida, Gainesville

Natasha Raikhel
MSU–DOE Plant Research
Laboratory
Michigan State University
East Lansing

Dale Sanders
Department of Biology
University of York
Heslington, United Kingdom

James N. Siedow
Department of Botany
Duke University
Durham, North Carolina

Chris Somerville
Plant Biology Department
Carnegie Institution of Washington
Stanford, California

Linda Spremulli
Department of Chemistry
University of North Carolina
Chapel Hill

L. Andrew Staehelin
Department of Molecular and
Cell Development Biology
University of Colorado, Boulder

Masahiro Sugiura
Center for Gene Research
Nagoya University
Nagoya, Japan

Yutaka Takeda
Bioscience Center
Nagoya University
Nagoya, Japan

Howard Thomas
Cell Biology Department
Institute of Grassland and
Environmental Research
Aberystwyth, Wales

Anthony Trewavas
Institute of Cell and
Molecular Biology
University of Edinburgh
Edinburgh, Scotland

Elizabeth Weretilnyk
Department of Biology
McMaster University
Hamilton, Ontario

Takao Yokota
Department of Bioscience
Teikyo University
Utsunomiya, Japan

Patricia Zambryski
Department of Plant Biology
University of California at Berkeley

The Reviewers

Richard Amasino
University of Wisconsin, Madison

Nikolaus Amrhein
Swiss Federal Institute of
Technology
Zürich, Switzerland

Sarah Assmann
Pennsylvania State University
University Park

Rebecca Boston
North Carolina State University
Raleigh

Anne Britt
University of California, Davis

Judy Callis
University of California, Davis

H. Maelor Davies
University of Kentucky, Lexington

Deborah Delmer
University of California, Davis

Malcolm Drew
Texas A&M University
College Station

David W. Emerich
University of Missouri, Columbia

Gad Galili
Weizmann Institute of Science
Rehovot, Israel

Charles Gasser
University of California, Davis

Simon Gilroy
Pennsylvania State University
University Park

Richard Hallick
University of Arizona, Tucson

Jeffrey Harborne
University of Reading
Reading, United Kingdom

Rüdiger Hell
Institute for Plant Genetics and
Crop Plant Research (IPK)
Gatersleben, Germany

Steven Huber
North Carolina State University
Raleigh

Dirk Inzé
Universiteit Gent
Gent, Belgium

Anita Klein
University of New Hampshire
Durham

Anthony Moore
University of Sussex
Brighton, United Kingdom

Donald R. Ort
University of Illinois
Champaign–Urbana

M. V. Parthasarathy
Cornell University
Ithaca, New York

John Patrick
University of Newcastle
Newcastle, Australia

Tony Pryor
CSIRO Plant Industry
Canberra, Australia

Daniel Schachtman
CSIRO Plant Industry
Glen Osmond, Australia

Christopher Staiger
Purdue University
West Lafayette, Indiana

William F. Thompson
North Carolina State University
Raleigh

Jan A. D. Zeevaart
Michigan State University
East Lansing

Contents in Brief

Contents

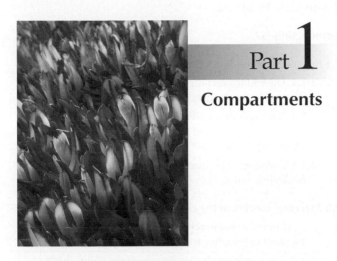

Part 1

Compartments

1 Membrane Structure and Membranous Organelles 2

Part 2

Cell Reproduction

6 Nucleic Acids 260

M Ecell Briology

7 Genome Organization and Expression 312

Part 3

Energy Flow

12 Photosynthesis 568

Part 4

Metabolic and Developmental Integration

17 Biosynthesis of Hormones and Elicitor Molecules 850

20 Senescence and Programmed Cell Death 1044

Part 5

Plant Environment and Agriculture

21 Responses to Plant Pathogens 1102

Contents

PART 1
Compartments

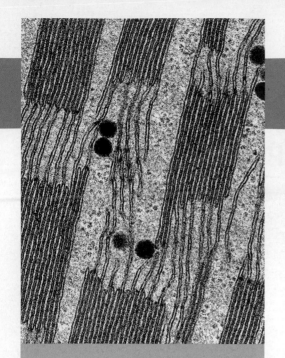

Biochemistry & Molecular Biology of Plants, B. Buchanan, W. Gruissem, R. Jones, Eds.
© 2000, American Society of Plant Physiologists

CHAPTER *1*

Membrane Structure and Membranous Organelles

L. Andrew Staehelin
Eldon H. Newcomb

Introduction

Cells, the basic units of life, require **membranes** for their existence. Foremost among these is the plasma membrane, which defines each cell's boundary and helps create and maintain electrochemically distinct environments within and outside the cell. Other membranes enclose eukaryotic organelles such as the nucleus, chloroplasts, and mitochondria. Membranes also form internal compartments, such as the endoplasmic reticulum (ER) in the cytoplasm and thylakoids in the chloroplast (Fig. 1.1).

The principal function of membranes is to serve as a barrier to the diffusion of most water-soluble molecules. These barriers delimit compartments wherein the chemical composition can differ from the surroundings and can be optimized for a particular activity. Membranes also serve as scaffolding for certain proteins. As membrane components, proteins perform a wide array of functions: transporting molecules and transmitting signals across the membrane, processing lipids enzymatically, assembling glycoproteins and polysaccharides, and providing mechanical links between cytosolic and cell wall compounds.

This chapter is divided into two parts. The first is devoted to the general features and molecular organization of membranes. The second provides an introduction to the architecture and functions of the different membranous organelles of plant cells. Many later chapters of this text focus on metabolic events that involve these organelles.

(A) Mesophyll

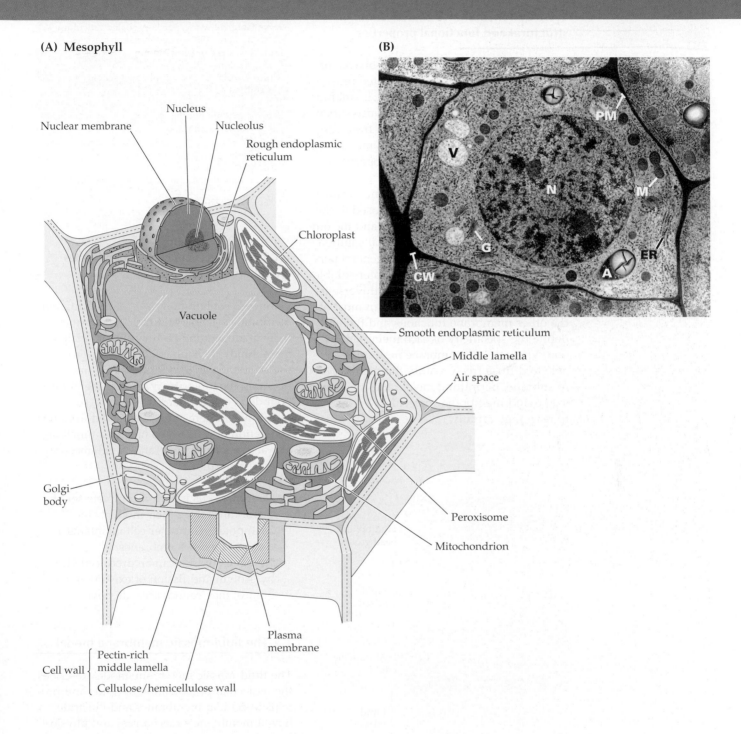

(B)

Figure 1.1

(A) A diagrammatic representation of a mesophyll leaf cell, depicting the principal membrane systems and cell wall domains of a differentiated plant cell. Note the large volume occupied by the vacuole. (B) Thin-section transmission electron micrograph (TEM) through a meristematic root tip cell preserved by rapid freezing. The principal membrane systems shown include amyloplast (A), endoplasmic reticulum (ER), Golgi stack (G), mitochondrion (M), nucleus (N), vacuole (V), and plasma membrane (PM). Cell wall (CW).

1.1 Common properties and inheritance of cell membranes

1.1.1 Cell membranes possess common structural and functional properties.

All cell membranes consist of a bilayer of polar lipid molecules and associated proteins. In an aqueous environment, membrane lipids self-assemble with their hydrocarbon tails clustered together, protected from contact with water (Fig. 1.2). Besides mediating the formation of bilayers, this property causes membranes to form closed compartments. As a result, every membrane is an asymmetrical structure, with one side exposed to the contents inside the compartment and the other side in contact with the external solution.

The lipid bilayer serves as a general permeability barrier because most water-soluble (polar) molecules cannot readily traverse its nonpolar interior. Proteins perform most of the other membrane functions and thereby define the specificity of each membrane system. Virtually all membrane molecules are able to diffuse freely within the plane of the membrane, permitting membranes to change shape and membrane molecules to rearrange rapidly.

Table 1.1 Membrane types found in plant cells

Plasma membrane
Nuclear membrane
Endoplasmic reticulum
Golgi cisternae (cis, medial, trans types)
Trans-Golgi network/partially coated reticulum
Clathrin- and COP[a]-coated transport vesicles
Endocytic vesicle membrane
Endosomal membrane
Multivesicular body/autophagic vacuole
    membranes
Tonoplast
Peroxisomal membrane
Glyoxysomal membrane
Chloroplast envelope membranes (2: inner
    and outer)
Thylakoid membranes
Mitochondrial membranes (2: inner and outer)

[a]COP, "coat protein."

1.1.2 All basic types of cell membranes are inherited.

Plant cells contain no fewer than 17 different membrane systems and may contain more than 20, depending on how sets of related membranes are counted (Table 1.1). From the moment they are formed, cells must maintain the integrity of all their membrane-bounded compartments to survive, so all membrane systems must be passed from one generation of cells to the next in a functionally active form. Membrane inheritance follows certain rules:

- Daughter cells inherit a complete set of membranes from their mother.
- Each potential mother cell maintains a complete set of membranes.
- New membranes can be produced only by growth and fission of existing membranes; they cannot arise de novo.

1.2 The fluid-mosaic membrane model

The fluid-mosaic membrane model describes the molecular organization of lipids and proteins in cellular membranes and illustrates how a membrane's mechanical and physiological traits are defined by the physicochemical characteristics of its various molecular components. The focus of this section will first be the molecular properties of membrane lipids, their assembly into bilayers, and the regulation of membrane fluidity.

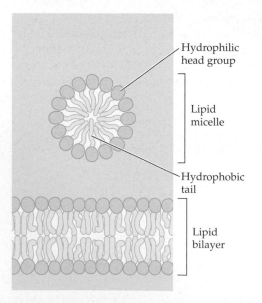

Figure 1.2
Cross-sectional views of a lipid micelle and a lipid bilayer in aqueous solution.

Hydrophilic head group

Lipid micelle

Hydrophobic tail

Lipid bilayer

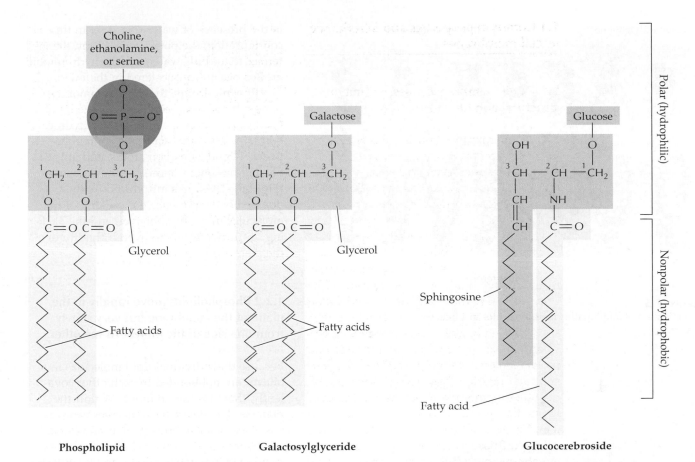

Phospholipid   **Galactosylglyceride**   **Glucocerebroside**

Figure 1.3
Plant membrane lipids.

The different mechanisms by which membrane proteins associate with lipid bilayers will next be described, and finally, the fluid-mosaic membrane model will be shown to integrate these different findings into a hypothesis that predicts many of the diverse properties of cell membranes.

1.2.1 The amphipathic nature of membrane lipids allows for the spontaneous assembly of bilayers.

In most cell membranes, lipids and proteins (glycoproteins) make roughly equal contributions to the membrane's mass. The lipids belong to several classes, including phospholipids, galactosylglycerides, glucocerebrosides, and sterols (Figs. 1.3 and 1.4). These molecules share an important physicochemical property: They are **amphipathic,** containing both **hydrophilic** ("water-loving") and **hydrophobic** ("water-hating") domains. When brought into contact with water, these molecules spontaneously self-assemble into higher-order structures. The hydrophilic head groups maximize their interactions

with water molecules, whereas the hydrophobic tails interact with each other, minimizing their exposure to the aqueous phase (see Fig. 1.2). The geometry of the resulting lipid assemblies is governed by the shape of the amphipathic molecules and the balance between the hydrophilic and hydrophobic domains. For most membrane lipids, the bilayer configuration is the minimum-energy self-assembly structure, that is, the structure that takes the least amount of energy to form

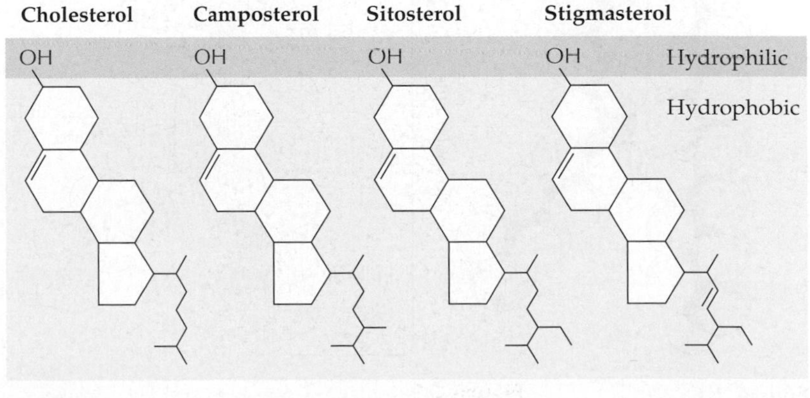

Cholesterol	Camposterol	Sitosterol	Stigmasterol	
OH	OH	OH	OH	Hydrophilic
				Hydrophobic

Figure 1.4
Sterols found in plant plasma membranes.

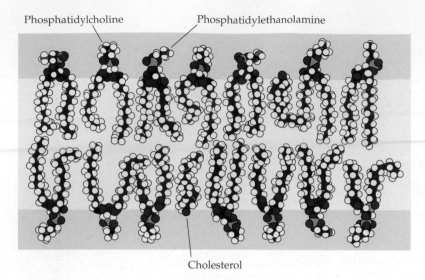

Phosphatidylcholine   Phosphatidylethanolamine

Cholesterol

Figure 1.5
Organization of amphipathic lipid molecules in a bilayer.

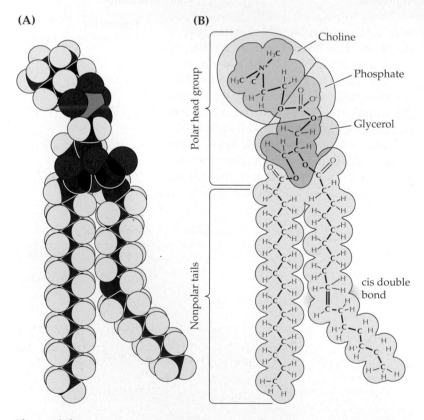

(A)   **(B)**

Polar head group

Choline

Phosphate

Glycerol

Nonpolar tails

cis double bond

Figure 1.6
(A) Space-filling model of a phosphatidylcholine molecule. (B) Diagram defining the functional groups of a phosphatidylcholine molecule.

in the presence of water (Fig. 1.5). In this configuration, the polar groups form the interface to the bulk water, and the hydrophobic groups become sequestered in the interior.

Phospholipids, the most common type of membrane lipid, contain a charged, polar head group and two hydrophobic hydrocarbon tails. The fatty acid tails contain between 14 and 24 carbon atoms, and at least one tail has one or more cis double bonds (Fig. 1.6). The kinks introduced by these double bonds influence the packing of the molecules in the lipid bilayer, and the packing, in turn, affects the overall fluidity of the membrane.

1.2.2 Phospholipids move rapidly in the plane of the membrane but very slowly from one side of the bilayer to the other.

Because the individual lipid molecules in a bilayer are not bonded to each other covalently, they are free to move. Within the plane of the bilayer, the molecules can slide past each other freely. A membrane can assume any shape without disrupting the hydrophobic interactions that stabilize its structure. Aiding this general flexibility is the ability of lipid bilayers to close on themselves to form discrete compartments, a property that also enables them to seal damaged membranes.

Studies of the movement of phospholipids in bilayers have revealed that these molecules can diffuse laterally, rotate, flex their tails, bob up and down, and flip-flop (Fig. 1.7). The exact mechanism of lateral diffusion is unknown. One theory suggests that individual molecules hop into vacancies ("holes") that form transiently as the lipid molecules within each monolayer exhibit thermal motion. Such vacancies arise in a fluid bilayer at high frequencies, and the average molecule hops approximately 10^7 times a second, which translates to a diffusional distance of approximately 1μm traversed in a second. Both rotation of individual molecules around their long axes and up-and-down bobbing are also very rapid events. Superimposed on these motions is a constant flexing of the hydrocarbon tails. Because this flexing increases toward the ends of the tails, the center of the bilayer has the greatest degree of fluidity.

In contrast, the spontaneous transfer of phospholipids across the bilayer, called flipping, rarely occurs. A flip would require the polar head to migrate through the nonpolar interior of the bilayer, an energetically unfavorable event. Some membranes contain "flippase" enzymes, which mediate the movements of newly synthesized lipids across the bilayer (Fig. 1.8). Different flippases specifically catalyze the translocation of particular lipid types and thus can flip their lipid substrates in only one direction. The energy barrier to spontaneous flipping and flippase specificity, together with the specific orientation of the lipid-synthesizing enzymes in the membranes, result in an asymmetrical distribution of lipid types across membrane bilayers.

Membrane sterols in lipid bilayers behave somewhat differently from phospholipids, primarily because the hydrophobic domain of a sterol molecule is much larger than the uncharged polar head group (see Fig. 1.4). Thus, membrane sterols are not only able to diffuse rapidly in the plane of the bilayer; they can also flip-flop without enzymatic assistance at a higher rate than phospholipids.

1.2.3 Cells optimize the fluidity of their membranes by controlling lipid composition.

Like all fatty substances, membrane lipids exist in two different physical states, as a semicrystalline gel and as a fluid. Any given lipid, or mixture of lipids, can be converted from gel to fluid—that is, melted—by a temperature increase. This change in state is known as **phase transition,** and for every lipid this transition occurs at a precise temperature, called the temperature of melting (T_m, see Table 1.2). Gelling brings most membrane activities to a standstill and increases permeability. At high temperatures, on the other hand, lipids can become too fluid to maintain the permeability barrier. Nonetheless, some organisms live happily in the frigid waters of the Arctic, whereas others thrive in boiling hot springs around the world. Many plants survive temperature fluctuations of 30°C on a daily basis. How do organisms adapt the fluidity of their membranes to suit their mutable growth environments?

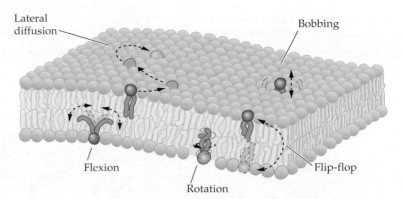

Figure 1.7
Mobility of phospholipid molecules in a lipid bilayer.

To cope successfully with the problem of temperature-dependent changes in membrane fluidity, virtually all poikilothermic organisms—those whose temperatures fluctuate with the environment—can alter the composition of their membranes to optimize fluidity for a given temperature. Mechanisms exploited to compensate for low temperatures include shortening the fatty acid tails, increasing the number of double bonds, and increasing the size or charge of the head groups (see also Section 1.3.2). Changes in sterol composition can also alter membrane responses to temperature. Membrane sterols serve as membrane fluidity "buffers," increasing the

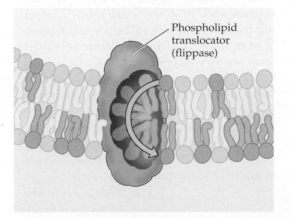

Figure 1.8
Mechanism of action of a "flippase," a phospholipid translocator.

Table 1.2 The effects of fatty acid chain length and double bonds on the temperature of melting (T_m) of some defined phospholipids

	T_m (°C)		
Types of chains[a]	Phosphatidyl-choline	Phosphatidyl-ethanolamine	Phosphatidic acid
Two $C_{14:0}$	24	51	
Two $C_{16:0}$	42	63	67
Two $C_{18:0}$	55	82	
Two $C_{18:1}$ (cis)	−22	15	8

[a]The shorthand nomenclature for the fatty acyl chains denotes how many carbon atoms (first number) and double bonds (second number) they contain.

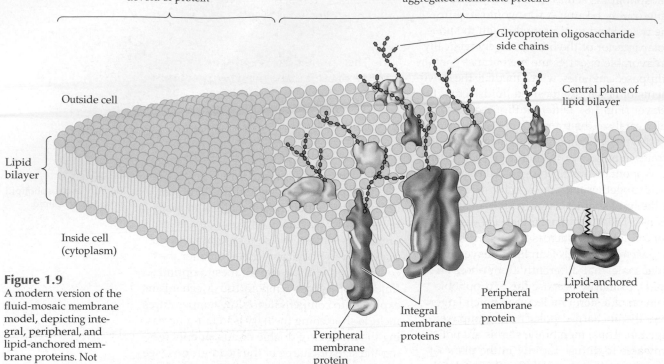

Gelled lipid domain
devoid of protein

Fluid lipid domain with
aggregated membrane proteins

Glycoprotein oligosaccharide
side chains

Central plane of
lipid bilayer

Outside cell

Lipid
bilayer

Inside cell
(cytoplasm)

Peripheral
membrane
protein

Integral
membrane
proteins

Peripheral
membrane
protein

Peripheral
membrane
protein

Lipid-anchored
protein

Figure 1.9
A modern version of the
fluid-mosaic membrane
model, depicting inte-
gral, peripheral, and
lipid-anchored mem-
brane proteins. Not
drawn to scale.

fluidity at lower temperatures by disrupting
the gelling of phospholipids, and decreasing
fluidity at high temperatures by interfering
with the flexing motions of the fatty acid tails.
Because each lipid has a different T_m, lowering the temperature can induce one type of
lipid to undergo a fluid-to-gel transition and
form semicrystalline patches, whereas other
lipids remain in the fluid state. Like all cellular molecules, membrane lipids have a finite
life span and have to be turned over on a regular basis. This turnover also enables plant
cells to adjust the lipid composition of their
membranes in response to seasonal changes
in ambient temperature.

1.2.4 Membrane proteins associate with lipid bilayers in many different ways.

The many different ways in which membrane-
bound proteins associate with lipid bilayers
reflect the diversity of enzymatic and structural functions they perform. The original
fluid-mosaic membrane model included
two basic types of membrane proteins: peripheral and integral (Fig. 1.9). More recent
research has led to the discovery of four
additional classes of membrane proteins—
**fatty acid–linked, prenyl group–linked,
phosphatidylinositol-anchored,** and **choles-
terol-linked**—all of which are attached to
the bilayer by lipid tails (Fig. 1.10).

By definition, **peripheral proteins** are
water-soluble and can be removed by washing membranes in water or in salt or acid solutions that do not disrupt the lipid bilayer.
Peripheral proteins bind either to integral
proteins or to lipids through salt bridges,
electrostatic interactions, hydrogen bonds, or
some combination of these. Some peripheral
proteins also provide links between membranes and cytoskeletal systems. In contrast,
integral proteins are insoluble in water. Because at least one domain lies embedded in
the hydrophobic interior of the bilayer, an
integral protein can be removed and solubilized only with the help of detergents or organic solvents, which degrade the bilayer.

Both the fatty acid–linked and the
prenyl group–linked proteins bind reversibly
to the cytoplasmic surfaces of membranes to
help regulate membrane activities. Cycling
between the membrane-bound and free states
is mediated in most cases by phosphoryla-
tion/dephosphorylation or by GTP/GDP
binding cycles. The fatty acid–linked proteins are attached either to myristic acid

(C_{14}), by way of an amide linkage to an amino terminal glycine, or to one or more palmitic acid (C_{16}) residues, by way of thioester linkages to cysteines near the carboxy terminus. Prenyl lipid–anchored proteins are attached to one or more molecules of farnesyl (C_{15}; 3 isoprene units) or geranylgeranyl (C_{20}; 4 isoprene units), which are also coupled to cysteine residues in carboxy-terminal CXXX, CXC, and XCC motifs (Fig. 1.10).

In contrast to the fatty acid– and the prenyl group–linked proteins, the phosphatidylinositol-anchored proteins are bound to the lumenal/extracellular surfaces of membranes (Fig. 1.10). Interestingly, these proteins are first produced as larger, integral proteins with one transmembrane domain. Enzymatic cleavage between the transmembrane domain and the globular surface domain produces a new C terminus on the globular domain, to which the lipid is coupled by ER-based enzymes. The remaining transmembrane domain is then degraded by proteases. Many arabinogalactan proteins (AGPs) appear to be linked to the plasma membrane via a glycosylphosphatidylinositol (GPI) anchor, which contains a ceramide lipid moiety. Cholesterol-anchored signaling molecules have recently been discovered in animal cells, attached to the external surface of plasma membranes. No sterol-linked proteins are yet known in plants.

1.2.5 The fluid-mosaic membrane model predicts structural and dynamic properties of cell membranes.

Although the original fluid-mosaic membrane model was developed at a time when membrane researchers knew only of peripheral and integral proteins, slight modifications to its basic premises have accommodated more recent discoveries, including lipid-anchored

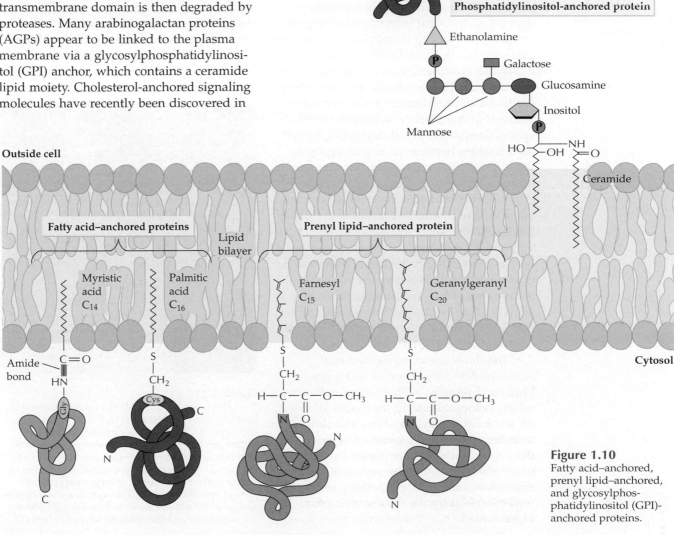

Figure 1.10
Fatty acid–anchored, prenyl lipid–anchored, and glycosylphosphatidylinositol (GPI)-anchored proteins.

proteins and membrane protein–cytoskeletal interactions. Thus, with these minor adjustments, it is still valid to view cell membranes from the perspective of the fluid-mosaic membrane model, formulated over a quarter of a century ago by Jonathan Singer and Garth Nicolson.

Membrane fluidity involves the movement not only of lipid molecules but also of integral proteins that span the bilayer and of the different types of surface-associated membrane proteins. This ability of membrane proteins to diffuse laterally in the plane of the membrane is crucial to the functioning of most membranes: Collisional interactions are essential for the transfer of substrate molecules between many membrane-bound enzymes and of electrons between the electron transfer chain components of chloroplasts and mitochondria (see Chapters 12 and 14). Such movements are also critical for the assembly of multiprotein membrane complexes. In addition, many signaling pathways depend on transient interactions among defined sets of integral membrane proteins and peripheral or lipid-anchored proteins.

Tethering structures regulate and restrict the movement of membrane proteins, often limiting their distribution to defined membrane domains. This tethering can involve connections to the cytoskeleton and the cell wall, bridges between related integral proteins, or junction-type interactions between proteins in adjacent membranes.

A particularly striking example of the latter type of interaction occurs in the grana stacks of chloroplast membranes (see Figs. 1.51 and 1.53). Grana stack formation has been shown to affect the lateral distribution of all major protein complexes in thylakoid membranes and to regulate the functional activity of the photosynthetic reaction centers and other components of the photosynthetic electron transport chain (see Chapter 12).

The fluid-mosaic model can readily accommodate various ratios and types of lipids and proteins without any fundamental alterations. Similarly, the model allows for the local expansion of any membrane domain as well as the turnover of individual proteins and lipids without disruption of membrane function. In sum, the elegance and functionality of cell membranes is well described by the fluid-mosaic membrane model.

1.3 Plasma membrane

The plasma membrane forms the outermost boundary of the living cell and functions as an active interface between the cell and its environment (see Figs. 1.1 and 1.11). In this capacity it controls the transport of molecules into and out of the cell, transmits signals from the environment to the cell interior, participates in the synthesis and assembly of cell wall molecules, and provides physical links between elements of the cytoskeleton and the extracellular matrix. In conjunction with specialized domains of the ER, the plasma membrane produces **plasmodesmata,** membrane tubes that cross cell walls and provide direct channels of communication between adjacent cells (Fig. 1.12). As a result of these plasmodesmal connections, almost all the living plant cells of an individual plant share a physically continuous plasma membrane. This contrasts sharply with the situation in animals, where virtually every cell has a separate plasma membrane, and cell-to-cell communication occurs instead through protein channels known as gap junctions.

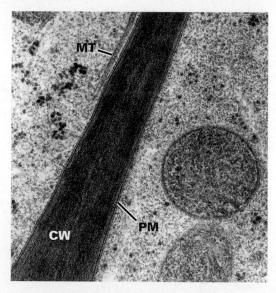

Figure 1.11
The plasma membrane (PM) of a turgid plant cell is pressed tightly against a cell wall (CW). These adjacent cryofixed plant cells have been processed by techniques that preserve the close physical relationship between plasma membrane and cell wall. Cells preserved with chemical fixatives for observation under an electron microscope often demonstrate artifacts of specimen preparation, such as a wavy conformation of the plasma membrane and a gap between the membrane and the cell wall. Microtubule (MT).

Yet another important difference between plants and animals is that plant cells are normally under turgor pressure, whereas animal cells are isoosmotic with their environments. Turgor pressure forces the plasma membrane tightly against the cell wall (see Fig. 1.11).

1.3.1 The lipid composition of plasma membranes is highly variable.

Plasma membranes of plant cells consist of lipids, proteins, and carbohydrates in a molecular ratio of approximately 40:40:20. The lipid mixture contains phospholipids, glycolipids, and sterols, the same classes found in animal plasma membranes. In plant plasma membranes, the ratio of lipid classes varies remarkably among the different organs in a given plant and among identical organs in different plants—in contrast to the far more constant ratios in animal cells. Barley root cell plasma membranes, for example, contain more than twice as many free sterol molecules as phospholipids (Table 1.3). In leaf tissues this ratio is generally reversed, but it varies: In barley leaf plasma membranes, the phospholipid-to-free sterol ratio is 1.3:1, whereas in spinach it is 9:1.

This striking variability, which continues to puzzle researchers, indicates that ubiquitous plasma membrane enzymes can function in widely different lipid environments. These results have led to the suggestion that the lipid composition of plant plasma membranes may have little bearing on their functional properties and that the only important lipid parameter is membrane fluidity. If this were true, it would mean that virtually all lipid classes are interchangeable so long as a given combination of lipids yields a bilayer of desired fluidity at a particular temperature. This provocative idea may well be an overstatement, reflecting our ignorance about the functional roles of specific lipid types; moreover, it seems to be contradicted by the recent finding that the activity of corn root **proton-translocating ATPase** (H^+-ATPase) molecules reconstituted into artificial membranes can be modulated by changes in sterol composition. Clearly, more research is needed to clarify how different lipid classes contribute to plasma membrane function.

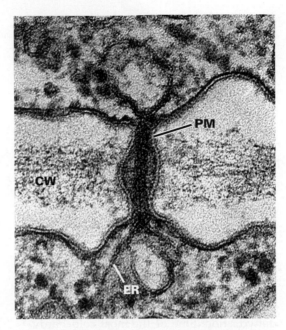

Figure 1.12 Longitudinal section through a plasmodesma. Plasma membrane (PM), endoplasmic reticulum (ER), cell wall (CW).

The most common free **sterols** of plant plasma membranes are campesterol, sitosterol, and stigmasterol (see Fig. 1.4). Cholesterol, the principal free sterol of mammalian plasma membranes, is a minor component in the vast majority of plant species analyzed to date, oat being a notable exception to this trend. Sterol esters, sterol glycosides, and acylated sterol glycosides are more abundant in plants than in animals. Sterol glycosylation, a reaction catalyzed by UDP-glucose:sterol glycosyltransferase, has been exploited as a marker for isolated plant plasma membranes. Sphingomyelin, another major type of lipid formed in mammalian plasma membranes, has yet to be found in plants. Interesting differences in the fatty acid tails of plant and mammalian plasma membrane **glycerolipids** have also been reported. Whereas plants principally utilize palmitic

Table 1.3 Lipid composition of plasma membranes from various non-cold-acclimated species and tissues (mole %)

Lipid type	Barley root	Barley leaf	*Arabidopsis* leaf	Spinach leaf
Phospholipids	26	44	47	64
Free sterols	57	35	38	7
Steryl glucosides	7	–	5	–
Acylated steryl glycosides	–	–	3	13
Glucocerebrosides	9	16	7	14

$(C_{16:0})$, linoleic $(C_{18:2})$, and linolenic $(C_{18:3})$ acids, mammals use palmitic $(C_{16:0})$, steric $(C_{18:0})$, and arachidonic $(C_{20:4})$ acids.

1.3.2 Cold acclimation leads to characteristic changes in plasma membrane lipid composition.

Low temperature is one of the most important factors limiting the productivity and distribution of plants. All plants able to withstand freezing temperatures possess the ability to freeze-proof their cells by a process known as **cold acclimation.** This metabolic process involves altering the composition and physical properties of membranes, cytoplasm, and cell walls so that they can withstand not only freezing temperatures but also freeze-induced dehydration. One of the most cold-hardy woody species is the mulberry tree. After cold acclimation in midwinter, these trees can withstand freezing below −40°C, but in mid-summer, when they are not cold-acclimated, they can be injured by a freeze below −3°C.

Among the most pronounced and critical alterations that occur during cold acclimation are changes in lipid composition of plasma membranes. One might expect cold acclimation–induced lipid changes to vary among species, given the differences in plasma membrane lipid composition already noted (Table 1.3). However, in all cold-hardy herbaceous and woody species studied to date, cold acclimation has been reported to cause an increase in the proportion of phospholipids and a decrease in the proportion of glucocerebrosides. In addition, the mole percent of phospholipids carrying two unsaturated tails increases. Species in which the cold-acclimated plasma membranes contain the highest proportion of diunsaturated phospholipids and the lowest proportion of glucocerebrosides tend to be the most cold hardy.

1.3.3 Plasma membrane proteins serve a variety of functions.

Most plasma membrane proteins involved in transmembrane activities such as transport and signaling, the anchoring of cytoskeletal elements to cell wall molecules, and the assembly of cellulose fibrils from cytosolic substrates are of the integral type. However, these proteins often form larger complexes with peripheral proteins. The extracellular domains of many integral proteins are glycosylated, bearing N- and O-linked oligosaccharides. The recent cloning of several plasma membrane proteins has opened up a new era in plasma membrane studies, but researchers are still far from understanding how these proteins perform their specialized functions at the molecular level. Nevertheless, the stage is now set for rapid progress toward these goals.

1.3.4 The electrochemical gradient produced by H$^+$-ATPase drives many other transport systems.

The plasma membrane H$^+$-ATPase (P-type H$^+$-ATPase) is the principal **primary active transport** system of plant cells. Its function is to couple ATP hydrolysis to the transmembrane transport of protons from the cytosol to the extracellular space. This proton pumping has two effects. First, it acidifies cell walls and alkalizes the cytosol, thereby affecting cell growth and expansion (see Chapter 2) as well as many other cellular activities. Second, it produces an **electrochemical** gradient across the plasma membrane. Electrochemical gradients consist of an electrical potential (inside negative in this case) and a concentration gradient (in this case a pH gradient outside acidic); the gradients can be used by **secondary active transport** systems to drive the transport of ions and solutes against their respective concentration gradients (see Chapter 3). Chemical analysis of the 100-kDa P-type H$^+$-ATPase has shown that this molecule is phosphorylated and dephosphorylated during each cycle of ATP hydrolysis. P-type H$^+$-ATPase genes and gene families (as many as 10 genes in *Arabidopsis* and tobacco) have been isolated from several species. Isoforms of these proteins appear to serve distinct functions and may be expressed differentially in specific tissues and during particular phases of development.

Some plasma membrane transport systems are indirectly coupled to the activity of P-type H$^+$-ATPase-dependent transporters. For example, two plant cell systems regulate the uptake of K$^+$, which is required for growth and osmoregulation. One system corresponds to a low-affinity K$^+$-uptake channel and a second to a high-affinity, H⁺-gradient–dependent K⁺-uptake carrier. Genes for several such secondary active transport systems and ion channels have been identified by functional complementation of yeast mutants that are defective in the uptake of K⁺ and other ions (see Chapters 3 and 23). Similar studies have led to the cloning of various H⁺-coupled sugar and amino acid transporters. Plant cell plasma membranes also contain proteins known as aquaporins, which form water channels.

1.3.5 Some plasma membrane receptors have been identified or cloned.

Although our knowledge of plasma membrane-localized receptor proteins remains limited, great interest in this field of research will probably yield major discoveries in the not-too-distant future. Significant advances have been made in the identification and physiological characterization of several receptors that bind hormones, oligosaccharins (see Section 1.5.3), peptides, and toxins, but the purification of these receptor proteins, which are present in the plasma membrane at low concentrations, has proved very difficult. As a result, many researchers have adopted a genetic approach to identifying and isolating receptor-protein genes. Computer analyses of several putative receptor gene sequences suggest that the plasma membrane receptors of plants possess many of the same structural motifs as are found in mammalian receptors and that they often exhibit protein kinase activities comparable with those reported for human receptor systems.

1.3.6 Several classes of plasma membrane proteins mediate interactions with the cell wall.

Plasma membrane proteins participate in a variety of interactions with the cell wall, including formation of physical links to cell wall molecules, synthesis and assembly of cell wall polymers, and creation of a highly hydrated, tissue-specific interfacial domain.

The presence of physical connections between the plasma membrane and the cell wall was first deduced from the presence of thread-like strands connecting the protoplasts of plasmolyzed cells to the cell wall (Fig. 1.13). The strands are known as **Hechtian strands** in honor of Kurt Hecht, who is credited with their discovery in 1912. During cold acclimation, the number of Hechtian strands increases, suggesting that increasing the strength of the protoplast–cell wall interactions helps protect the protoplasts from the stress of freeze-induced dehydration. Electron microscopic analysis has shown that these strands are thin tubes of cytoplasm delineated by a plasma membrane that retains tight contacts with the cell wall. The strands remain continuous with the plasma membrane. Although the molecules that link the plasma membrane to the cell wall have not yet been identified, indirect studies suggest they may be integrin-type receptors that recognize the amino acid sequence Arg-Gly-Asp (RGD) in cell wall constituents. A protein known as WAK1, a plasma membrane receptor with kinase activity, is another candidate protein.

Cellulose synthase and callose synthase complexes constitute a second major class of plasma membrane proteins. Both of these polymers (cellulose: β-1,4-linked glucose; callose: β-1,3-linked glucose) are secreted directly into the cell walls (see Chapter 2).

(A)

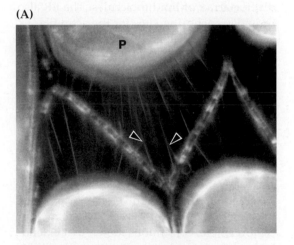

(B)

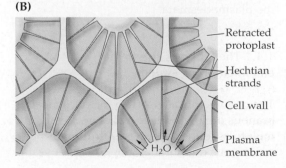

Figure 1.13
(A) A light micrograph of plasmolyzed onion epidermal cells. Hechtian strands (arrowheads) connect the protoplasts (P) to the cell wall. (B) Diagram illustrating features of plasmolyzed plant cells.

AGPs, a third class of cell surface proteins, are highly glycosylated proteoglycans that derive more than 90% of their mass from sugar. The classical-type AGPs appear to be anchored to the external surface of the plasma membrane by means of GPI lipid anchors—thereby providing a carbohydrate-rich interface between the cell wall and the plasma membrane. The fact that AGPs are expressed in a tissue- and developmental stage–specific manner suggests they may play a role in differentiation.

1.4 Endoplasmic reticulum

ER is the most extensive, versatile, and adaptable organelle in eukaryotic cells. It consists of a three-dimensional network of continuous tubules and flattened sacs that underlie the plasma membrane, course through the cytoplasm, and connect to the nuclear envelope but remain distinct from the plasma membrane. In plants, the principal functions of ER include synthesizing, processing, and sorting proteins targeted to membranes, vacuoles, or the secretory pathway as well as adding N-linked glycans to many of these proteins and synthesizing a diverse array of lipid molecules. The ER also provides anchoring sites for the actin filament bundles that drive cytoplasmic streaming and plays a critical role in regulating the cytosolic concentrations of calcium, which influence many other cellular activities.

The classical literature distinguishes three types of ER membranes: **rough ER, smooth ER,** and **nuclear envelope.** However, researchers now recognize many more morphologically distinct subdomains that perform various different functions (Fig. 1.14). Despite this functional diversity, virtually all ER membranes are physically linked and enclose a single, continuous lumen that extends beyond the boundaries of individual cells via the plasmodesmata.

1.4.1 The ER gives rise to the endomembrane system.

The endomembrane system includes membranous organelles that exchange membrane molecules, either by lateral diffusion through continuous membrane or by transport vesicles that bud from one type of membrane and fuse with another (Fig. 1.15). The principal membrane systems connected in this manner include the nuclear envelope, membranes of the secretory pathway (ER, Golgi, *trans*-Golgi network, plasma membrane, vacuole, and different types of transport/secretory vesicles), and membranes associated with the endocytic pathway (plasma membrane, endocytic vesicles, multivesicular bodies, partially coated reticulum, vacuoles, and transport vesicles). Extensive traffic between these compartments not only transports secreted molecules to the cell surface and vacuolar proteins to the vacuoles but also distributes membrane proteins and membrane lipids from their sites of synthesis, the ER and Golgi cisternae, to their sites of action, all of the endomembrane organelles. A plethora of sorting, targeting, and retrieval systems regulate traffic between the different compartments, ensuring delivery of molecules to the correct membranes and the maintenance of organelle identity (see Chapter 4).

All membranes of the endomembrane system are connected by both forward (anterograde) and backward (retrograde) traffic (Fig. 1.15). The anterograde pathway usually delivers newly synthesized molecules to their destination. In the retrograde pathway, membrane molecules dispersed by transport processes are recycled to their sites of origin, and "escaped" molecules are returned to their normal site of action. Because the volume of membrane traffic is large and the accuracy of sorting is less than 100%, a certain percentage of mislocalized proteins remains in all endomembrane systems. This normal "contamination" of endomembranes provides a never-ending challenge for researchers interested in obtaining "pure" membrane fractions.

1.4.2 The ER forms a dynamic network, the organization of which changes during the cell cycle and development.

In living plant cells, the spatial organization and kinetic behavior of ER membranes can be visualized by means of the lipophilic fluorescent stain $DiOC_6$. Light microscopic images of such cells show a lace-like network of lamellar and tubular cisternae that continuously undergoes architectural rearrangements (Fig. 1.16). Electron microscopic studies

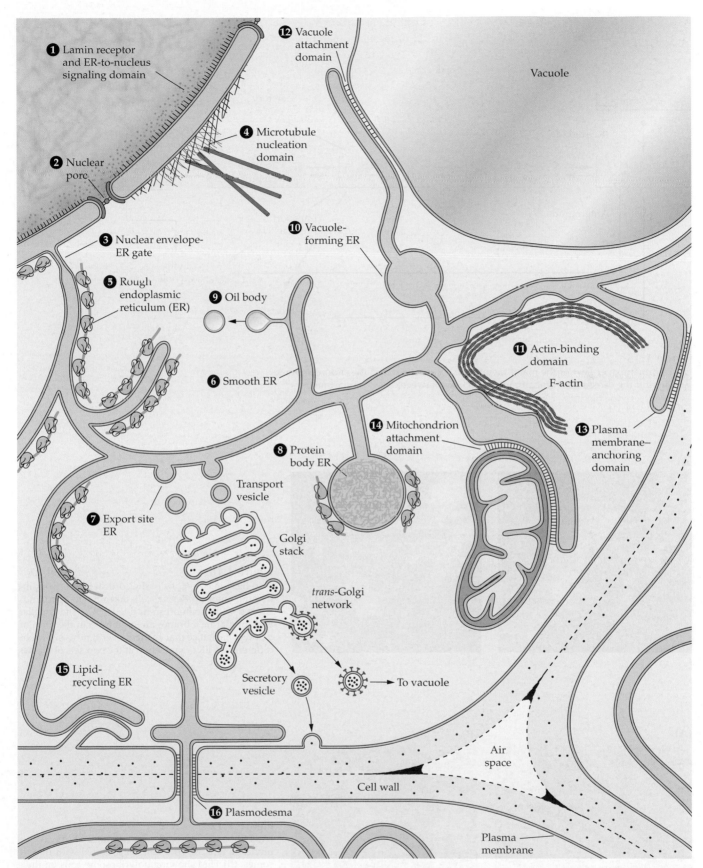

Figure 1.14
Diagram illustrating the functional domains of the plant ER system.

The labeled components are:

1. Lamin receptor and ER-to-nucleus signaling domain
2. Nuclear pore
3. Nuclear envelope-ER gate
4. Microtubule nucleation domain
5. Rough endoplasmic reticulum (ER)
6. Smooth ER
7. Export site ER
8. Protein body ER
9. Oil body
10. Vacuole-forming ER
11. Actin-binding domain
12. Vacuole attachment domain
13. Plasma membrane–anchoring domain
14. Mitochondrion attachment domain
15. Lipid-recycling ER
16. Plasmodesma

Other labels: Vacuole, F-actin, Transport vesicle, Golgi stack, *trans*-Golgi network, Secretory vesicle, To vacuole, Air space, Cell wall, Plasma membrane

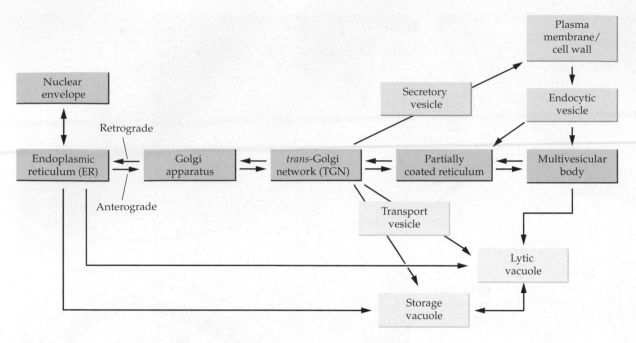

Figure 1.15
Diagrammatic overview of the major membrane compartments of the endomembrane system and the directions of membrane trafficking between these compartments.

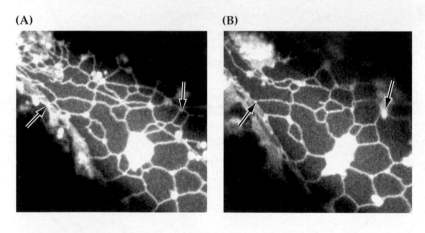

Figure 1.16
Two light micrographs of the cortical ER network in a living onion epidermal cell, taken 5 min apart and stained with the fluorescent dye DiOC$_6$. The lamellar and tubular membranes are organized in the form of a polygonal lattice that changes over time as shown by difference in ER organization at the two sets of arrows.

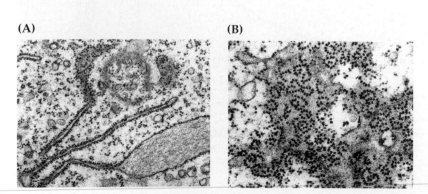

Figure 1.17
(A) TEM cross-sectional view of rough ER membranes in a radish root hair cell studded with ribosomes. (B) TEM showing polyribosome coils on the surface of a rough ER membrane domain in a young root epidermal cell of radish.

have shown that the lamellar regions correspond to polysome-bearing rough ER (Fig. 1.17), the tubular regions to smooth ER (Fig. 1.18). New tubules can grow from existing membranes and then fuse with other ER cisternae to create new network polygons while other tubules rupture and are being reabsorbed into the network.

In interphase cells, the ER underlying the plasma membrane, called the cortical ER, is highly developed and, because of its links to the plasma membrane, is less mobile than the ER cisternae that pass through the cell interior. Indeed, the cortical ER could serve as a semi-immobile platform for the actin filament bundles that drive cytoplasmic streaming (Fig. 1.19). During mitosis, the ER undergoes a series of characteristic rearrangements consistent with the idea that it might regulate spindle activities and cell plate assembly by controlling local calcium concentrations. Specialized aggregates of ER membranes are also present in the gravity-sensing columella cells of root tips; again, these cisternae have been postulated to participate in the gravitropic response by way of calcium fluxes. In seeds, extensive arrays of rough ER develop in storage cells during the accumulation of storage proteins. In contrast, intricate networks of smooth ER tubes are typically seen in cells (e.g., oil glands [see Fig. 1.18]) that synthesize and secrete significant amounts of lipophilic substances such as terpenes, cutins, waxes, and flavonoids (see Chapters 10 and 24).

1.4.3 Oil bodies and some types of protein bodies are formed by specialized ER domains.

Two important agricultural products, vegetable oils and dietary seed proteins, are produced by enzymes associated with the ER. The storage oils of seeds—triacylglycerols—provide energy and membrane lipid building blocks during early stages of embryo germination and growth. During their synthesis by ER enzymes, triacylglycerols partition into the interior of the ER bilayer and then accumulate at sites defined by molecules known as **oleosins.** Oleosins have a thumbtack-like architecture, with the "shaft" portion consisting of hydrophobic amino acids and the head exhibiting an amphipathic

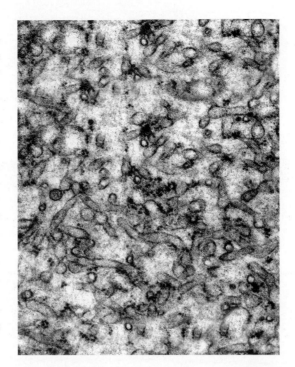

Figure 1.18
TEM of smooth ER tubules in a glandular trichome cell of a mint leaf.

structure. As the newly synthesized triacylglyceride molecules accumulate between the bilayer leaflets at the oleosin sites, they give rise to spherical oil bodies that bud into the cytosol (Fig. 1.20). The surface layer of each oil body consists of a phospholipid monolayer, embedded oleosins, and other proteins.

Seed storage proteins are synthesized during seed development and serve as the principal source of amino acids for germination and seedling growth. Cereal plants produce two classes of storage proteins, the water-soluble **globulin** proteins and the more hydrophobic, alcohol-soluble **prolamins.** Globulin proteins are synthesized by rough

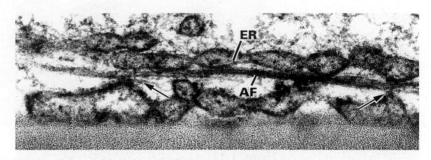

Figure 1.19
TEM showing sites where an actin filament (AF) attaches to cortical ER (arrows). The TEM corresponds to domain 11 in Figure 1.14.

(A) (B)

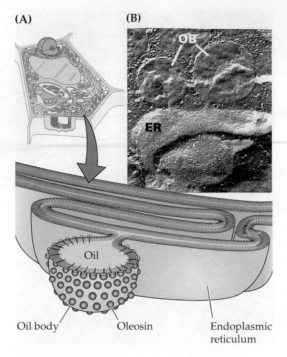

Figure 1.20
(A) Triacylglycerides accumulate between the two lipid monolayers of a smooth ER membrane and bud off as oil bodies at sites defined by oleosin molecules. (B) TEM of oil bodies (OB) budding from an ER membrane into the cytoplasm. See domain 9 in Fig. 1.14.

Oil

Oil body Oleosin Endoplasmic reticulum

1.4.4 Transport vesicles mediate the transfer of newly synthesized secretory/storage/membrane proteins from the ER to the Golgi apparatus.

Transport vesicles that bud from smooth ER mediate the ER-to-Golgi transport of soluble proteins destined for the cell surface and the vacuole and membrane proteins for the different compartments of the endomembrane system (see Fig. 1.15). These vesicles are produced by an ER domain known as **export site ER,** which separates the secretory proteins from the resident ER proteins before the former enter the transport vesicles and are sent to *cis*-Golgi cisternae (see Fig. 1.14, domain 7). In animal cells and in algae such as *Chlamydomonas* that produce glycoprotein walls, export site ER domains with many budding vesicles are often seen close to the cis side of the Golgi stacks. In contrast, such a spatial relationship is rarely observed in plant cells because of the low volume of protein transport between ER and Golgi. Only in seed cells producing large quantities of globulin-type storage proteins and in gland cells of insectivorous plants that actively secrete digestive enzymes have the vesicle-budding, export site ER elements been regularly observed close to the cis side of Golgi stacks. In other types of plant cells, transport vesicles appear to bud most frequently from the edges of ER cisternae (Fig. 1.22); these sites exhibit no defined spatial relationship to any part of nearby Golgi stacks. How these vesicles "find" their way to the *cis*-Golgi cisternae has yet to be determined.

ER polysomes and then in most cases transported to storage vacuoles via the Golgi apparatus. In contrast, prolamins aggregate into protein bodies (Fig. 1.21) where they enter the ER lumen. Nascent prolamin molecules associate with the aggregating proteins, trapping polysomes on the surface of the forming **protein bodies.**

(A) (B)

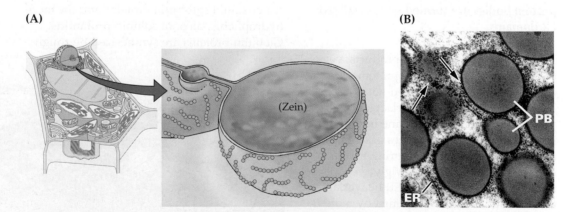

(Zein)

PB

ER

Figure 1.21
(A) Prolamin storage proteins (e.g., zein) aggregate in protein bodies that bud from specialized rough ER. (B) TEM of protein bodies (PB) forming in maize endosperm. Polysomes (arrows) are attached to the delimiting ER membrane. See Figure 1.14, domain 8.

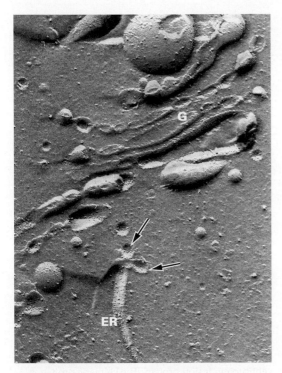

Figure 1.22
TEM of transport vesicles (arrows) budding from ER cisterna in the vicinity of a Golgi stack (G).

1.5 Golgi apparatus

The term Golgi apparatus refers to the complement of **Golgi stacks** and associated *trans*-**Golgi networks** (TGNs) within a given cell. As shown in Figure 1.15, the Golgi apparatus occupies a central position in the secretory pathway, receiving newly synthesized proteins and lipids from the ER and directing them to either the cell surface or the vacuoles (see Chapter 4). In plants, the Golgi apparatus is involved in assembling complex polysaccharides of the cell wall matrix, synthesizing and processing the O- and N-linked oligosaccharide side chains of membrane, cell wall, and vacuolar glycoproteins and producing glycolipids for the plasma membrane and tonoplast. The glycosyltransferases and glycosidases that carry out these reactions are integral proteins with their active sites facing the interior space of the flattened Golgi cisternae. Specific membrane transporters deliver sugar nucleotides from the cytosol to the glycosyltransferases that synthesize the different carbohydrate products.

1.5.1 The plant Golgi apparatus consists of dispersed Golgi stack–TGN units that are carried around by cytoplasmic streaming.

The functional unit of the plant Golgi apparatus is the Golgi stack, its associated TGN, and the **Golgi matrix** that encompasses both structures (Fig. 1.23). Each stack consists of a set of five to eight flattened cisternae that exhibit a distinct morphological polarity and possess fenestrated and bulbous margins (Figs. 1.23 and 1.24). The TGN tends to exhibit a more tubulovesicular structure and is always closely associated with the trans side of the stack. The Golgi matrix is a fine, filamentous, cage-like structure that excludes ribosomes and other cytosolic proteins from

(A)

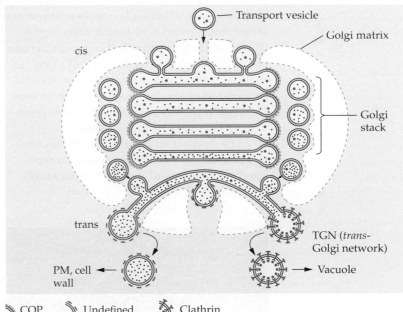

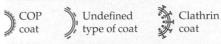

(B)

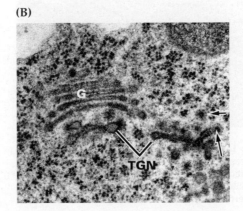

Figure 1.23
(A) Diagram illustrating the spatial relationship of a Golgi stack to its associated *trans*-Golgi network (TGN) and the Golgi matrix. The distribution of COP- and clathrin-coated budding vesicles is also shown. (B) TEM of a Golgi stack (G) and its TGN in a meristematic root tip cell of *Nicotiana*. The Golgi matrix appears as a ribosome-excluding zone around the cisternae. Arrows point to typical clathrin-coated buds associated with the TGN cisterna.

the immediate vicinity of these membrane systems. Its postulated functions include protecting stacks from shearing and preventing the loss of intercisternal transport vesicles from the stack.

Unlike in animal cells, where the Golgi apparatus occupies a position close to the cell center, the Golgi stack–TGN units of plants are always dispersed throughout the cytoplasm, either as individual units or in small clusters (Fig. 1.25). This dispersed organization, together with the fact that the stacks are transported by actin-based cytoplasmic streaming, ensures that even in the large, vacuolated cells common in plants, secretory products reach their destinations. Analysis of Golgi stack dynamics has shown that they move in a stop-and-go manner. This has led to the hypothesis that the stacks may pause at ER export sites to pick up cargo and at growing cell wall sites to secrete their contents. Plant Golgi stacks also remain structurally intact and functionally active during mitosis to provide the membrane and cell wall molecules needed for cell plate formation. New stacks arise by fission, which can start at either the cis or the trans side of the stack (Fig. 1.26) and usually occurs during the G2 phase of the cell cycle.

The number of Golgi stack–TGN units per cell varies widely, depending on the species, the size and developmental stage

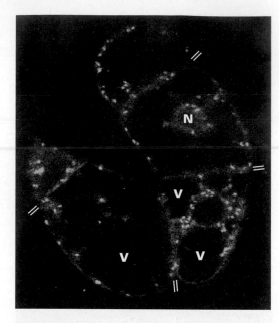

Figure 1.25
Light micrograph of a group of transformed tobacco BY-2 cells expressing a mannosidase I–green fluorescent protein fusion protein. The chimeric protein is localized to the Golgi stacks. Labeled Golgi are seen in the cortical cytoplasm as well as in the transvacuolar cytoplasmic strands that connect the nucleus (N) with the cortical cytoplasm. Cross-walls between cells are marked with double lines. Vacuoles (V).

of the cell, and the volume and type of secretory/storage materials produced. For example, small-shoot apical meristem cells of *Epilobium* (willow-herb) contain approximately 20 Golgi stacks, whereas the giant fiber cells of cotton contain more than 10,000. Onion root apical meristem cells contain approximately 400 stacks.

1.5.2 Golgi stack–TGN units consist of morphologically distinct cisternae and give rise to different types of coated vesicles.

Electron micrographs of tissues preserved by cryofixation demonstrate that the Golgi stacks consist of morphologically distinct classes of cisternae (Fig. 1.27). These morphological disparities arise because each type of cisterna contains a different set of enzymes and performs distinct, specialized functions. Furthermore, Golgi stacks are remodeled in a cell- and tissue-specific manner to meet the functional needs of specific cell types. This functional retailoring of Golgi stacks creates tissue-specific differences in

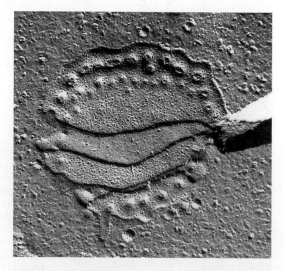

Figure 1.24
TEM presenting a face-on view of a freeze-fractured Golgi stack of a pea root tip cell. Two rows of fenestrae are seen to delineate the vesicle-budding domains of the cisternal margins.

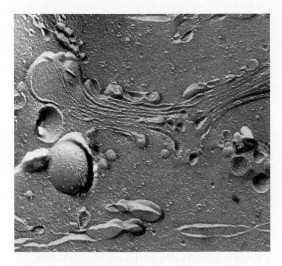

Figure 1.26
TEM of a dividing Golgi stack in a peripheral root cap cell of pea. The stack is seen to be dividing by fission from one side to the other.

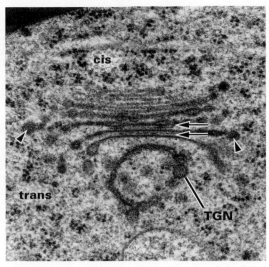

Figure 1.27
TEM of a cross-sectioned Golgi stack from a root cap columella cell. The cisternae exhibit a distinct cis-to-trans polarity with the lumenal contents of the trans side cisternae staining more densely. Arrowheads: COP-coated budding vesicles; arrows: intercisternal elements; *trans*-Golgi network (TGN).

Golgi stack morphology and glycosylation activities. For example, when the transmitting tissue-specific protein of tobacco styles is expressed in other tobacco tissues, the protein is substantially underglycosylated, indicating that only the stylar cells possess all of the enzymes required to glycosylate this protein fully.

Golgi cisternae are usually subdivided into cis, medial, and trans types, based on their position in a stack, their staining patterns, and their functional properties. The lightly staining *cis*-cisternae serve as receiving stations for products transported in vesicles from the ER. This is also the recycling site for escaped ER proteins. After an initial round of processing, ER-derived products as well as newly formed Golgi products are passed on to the *medial-* and *trans*-cisternae and, finally, sent to the TGN. Thus, products move in a cis-to-trans direction through the stacks. *Trans*-Golgi cisternae are the most morphologically conspicuous because of the tight appression of their membranes and darkly staining contents, a result of the *trans*-cisternae and the TGN being the most acidic Golgi compartments (acidified by vacuolar [V-type] H^+-ATPases). Aside from regulating enzyme activities, this low-pH environment appears to cause an osmotic collapse of the cisternal lumen, which squeezes the newly synthesized products into the budding vesicles. **Intercisternal elements** are another

type of *trans*-Golgi cisterna–associated structure (Fig. 1.27). These parallel protein fibers lie sandwiched between cisternae and may serve as anchors for the glycosyltransferases involved in the synthesis of large polysaccharide slime molecules, such as those secreted by the outermost cells of root caps.

Most vesicles that bud from Golgi cisternae and the TGN appear to be assembled with the aid of a protein coat. Such coats serve several different functions, regulating the types of molecules packaged into a given vesicle, controlling the vesicle assembly process, and specifying both the target membrane and the machinery that ultimately enables the vesicle to fuse with that target. Most Golgi-associated vesicles move Golgi enzymes in a retrograde direction to compensate for the cisternal maturation-associated anterograde movement of cisternae and their products. Vesicles involved in ER-to-Golgi, intra-Golgi, and Golgi-to-TGN transport are produced by a cisternal budding mechanism that involves **coat protein (COP) coats** (Fig. 1.27). On completion of the budding process, the coat proteins dissociate from the vesicle and are recycled. COP coats differ in morphology and chemical composition from the better-known **clathrin coats** (see Fig. 1.23) that assemble on TGN buds and participate in sorting vacuolar enzymes and storage proteins into vesicles destined for the vacuoles (see Chapter 4). Those molecules not sorted into clathrin-coated vesicles are subsequently packaged into another type of vesicle, which delivers the molecules to the cell surface in what is known as the default pathway. Clathrin-coated vesicles with different "adapter" proteins are also involved in the uptake of soluble and membrane proteins during endocytosis (see Fig. 1.31).

1.5.3 The sugar-containing molecules produced in Golgi cisternae serve diverse functions.

The Golgi apparatus is involved in assembling the N-linked and O-linked glycans of glycoproteins and proteoglycans and in synthesizing complex polysaccharides (see Chapter 4). One of the principal functions of glycosylation is to protect proteins against proteolysis, thereby increasing their life span. In specific instances, the sugar groups may also specify plasma membrane–cell wall interactions, prevent the premature activation of lectins, or contribute to protein folding or assembly of multiprotein complexes. Most proteins subject to N-linked glycosylation perform enzymatic functions, whereas O-glycosylated proteins often serve structural roles. Both N- and O-linked glycan side chains are present in many highly glycosylated arabinogalactan proteins that, at more than 90% sugars, are classified as proteoglycans. The complex cell wall polysaccharides produced by the Golgi apparatus perform structural functions and can bind water and heavy metals. In addition, these polysaccharides contain cryptic regulatory oligosaccharide domains that can be released by specific enzymes to yield regulatory molecules known as oligosaccharins.

1.5.4 The Golgi apparatus is a carbohydrate factory.

The synthesis of N-linked glycans starts in the ER with the assembly of a 14-sugar oligosaccharide on a molecule of dolichol, a large lipid composed of 14 to 24 isoprene units. Once complete, this oligosaccharide is transferred by oligosaccharyl transferase to selected asparagine residues in nascent polypeptide chains. Most of the subsequent processing of the oligosaccharide occurs in the Golgi, including the systematic removal of mannose residues and the addition of other types of sugars (see Chapter 4). The enzymes involved in these reactions are not randomly distributed among Golgi cisternae but rather appear to be localized to subsets of *cis-*, *medial-*, or *trans-*cisternae, depending on which sequential step they catalyze as a given N-linked glycan moves through the stack from cis to trans.

Oxygen-linked glycans are important components of hydroxyproline-rich glycoproteins and AGPs, many of which serve structural roles. The sugars, mostly arabinose and galactose, are attached to amino acids that contain hydroxyl groups, such as hydroxyproline, serine, and threonine. Very little is known about the synthesis of O-linked glycans. On newly synthesized proteins destined for O-glycosylation, selected proline residues are probably converted to hydroxyproline in the ER, whereas the enzymes that add arabinose sugars to those hydroxyprolines are located in the *cis-*cisternae.

The matrix polysaccharides of plant cell walls are structurally complex molecules that play a central role in determining cell size and shape (see Chapter 2). Defined fragments of such molecules also function as signaling molecules in pathways that control plant growth, organogenesis, and the elicitation of defense responses. All complex (i.e., branched) cell wall polysaccharides are synthesized by enzymes in Golgi and TGN cisternae; note that the linear polymers cellulose and callose, which are synthesized at the cell surface, are not considered complex. The molecular details of the assembly pathways have yet to be elucidated; however, an outline of the spatial organization of the xyloglucan and pectic polysaccharide pathways in the Golgi stacks has been developed by using immunolocalization to track the sites of appearance of specific carbohydrate groups. Probably the most striking result of these studies is that the assembly of pectic polysaccharides appears to involve enzymes in all types of Golgi cisternae, whereas enzymes that produce xyloglucan are largely confined to the *trans-*cisternae and the TGN (Fig. 1.28). Current work focuses on the identification and characterization of enzymes

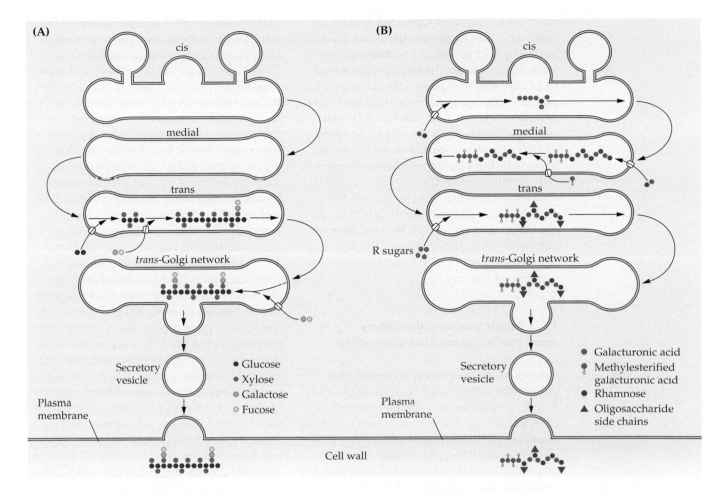

(A) cis — medial — trans — *trans*-Golgi network — Secretory vesicle — Plasma membrane

- Glucose
- Xylose
- Galactose
- Fucose

(B) cis — medial — trans — R sugars — *trans*-Golgi network — Secretory vesicle — Plasma membrane — Cell wall

- Galacturonic acid
- Methylesterified galacturonic acid
- Rhamnose
- Oligosaccharide side chains

Figure 1.28
Spatial organization of the xyloglucan (A) and pectic polysaccharide (B) biosynthetic pathways in plant Golgi stacks, as deduced from immunolabeling studies. The synthesis of xyloglucan, a hemicellulose, is postulated to be initiated in *trans*-cisternae and completed in the *trans*-Golgi network. By contrast, the synthesis of pectic polysaccharides appears to involve all types of cisternae, with the back bone assembled in *cis*- and *medial*-cisternae and the side chains added in the *trans*-Golgi cisternae. How the products are transported from one cisterna to the next has yet to be determined conclusively (see Chapter 4).

that give rise to these and other complex polysaccharides as well as to the many types of glycoproteins and proteoglycans made in the Golgi apparatus.

1.6 Exocytosis and endocytosis

Exocytosis is the process by which secretory vesicles derived from the TGN fuse with the plasma membrane, releasing their contents into the extracellular space. In growing cells, this process delivers the proteins and lipids needed for expansion of the plasma membrane as well as the complex polysaccharides, glycoproteins, and proteoglycans required for cell wall growth. Because of the large surface-to-volume ratio of secretory vesicles, exocytosis delivers more membrane to the cell surface than is needed for the expansion of the plasma membrane. Then **endocytosis** retrieves excess membrane for recycling through the formation of plasma

membrane infoldings that give rise to endocytic vesicles. In addition, some excess lipid appears to be removed by a molecular recycling method in which lipids are transferred between adjacent membranes by means of a "hopping" mechanism. Endocytosis is also used to turn over plasma membrane and cell wall molecules and to remove activated receptors from the cell surface. In animal cells, endocytosis plays a major role in the uptake of nutrient molecules, but little evidence suggests such a role in plants.

1.6.1 In plants, turgor pressure affects membrane events associated with exocytosis and membrane recycling.

When a secretory vesicle in an animal cell fuses with the plasma membrane, its contents are expelled to the extracellular space and the vesicle membrane becomes part of the plasma membrane. As this occurs, the plasma membrane is expanded slightly, an expansion that can be readily accommodated by changes in the surface architecture of the animal cell. **Turgor pressure** prevents this from happening in plant cells. In turgid plant cells, the plasma membrane is pressed tightly against the cell wall (see Fig. 1.11) and cannot expand unless the cell wall expands as well. When a spherical secretory vesicle fuses with the plasma membrane of a turgid cell, the turgor pressure not only forces out the vesicle contents, but also flattens the vesicle into a disc-shaped infolding of the plasma membrane (Fig. 1.29). Because the plasma membrane cannot expand, the infolding remains in place until the excess membrane is removed (see below).

1.6.2 Turgor pressure also affects endocytosis and membrane recycling.

How is excess membrane removed from the plasma membrane? In animal cells, the recycling of excess membrane components from the plasma membrane occurs via endocytosis of clathrin-coated vesicles. In plants, clathrin-coated vesicles are also used for the internalization of plasma membrane proteins and lipids (see Section 1.6.3). However, this process demands much more energy in plants than in animal cells because sizable hydrostatic pressure forces must be overcome to form plasma membrane invaginations.

To circumvent some of these problems, plants have evolved a second pathway for the internalization of plasma membrane lipids, known as **molecular recycling.** In this process, plasma membrane lipids are returned to the ER by a mechanism that involves neither endocytic vesicles nor the transient fusion of ER and plasma membranes. Instead, individual lipid molecules appear to hop directly from the plasma membrane to adjacent ER membranes by a mechanism that has yet to be elucidated.

Evidence for this second pathway has come from both lipid uptake research and electron microscopic studies of ultrarapidly frozen cells. To investigate lipid uptake, researchers applied a fluorescent analog of the membrane lipid phosphatidylcholine to the outer surface of the plasma membrane. This molecular marker was quickly translocated to peripheral ER cisternae with no evidence of vesicular intermediates. However, the uptake required that an extracellular enzyme convert phosphatidylcholine to diacylglycerol before the transfer. Electron micrographs of cryofixed cells suggest that this rapid lipid transfer is mediated by unique ER membrane extensions that form tight caps over the plasma membrane appendages left behind by the fused secretory vesicles (Fig. 1.30; see also domain 15 in Fig. 1.14). During this process the disc-shaped appendages tip over to form characteristic horseshoe-shaped plasma membrane invaginations. Once capped by an ER extension, the invagination shrinks until the excess membrane is gone. The ER then retracts from the plasma membrane.

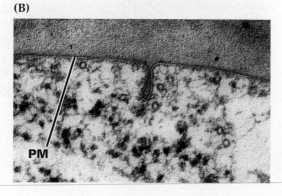

(A)

(B)

Figure 1.29
Cross-sectional images of secretory vesicles. (A) TEM of a secretory vesicle in the process of initiating membrane fusion with the plasma membrane (PM). (B) TEM of a secretory vesicle that has fused with the plasma membrane and discharged its contents. The disc-shaped infolding of the membrane shown here is observed only in turgid cells.

1.6.3 The membrane compartments associated with endocytosis can be identified by following the uptake of tracer molecules.

The process of endocytosis can be visualized by exposing cells to tracer molecules that can be internalized by clathrin-coated endocytic vesicles. Two classes of molecular markers are used in such investigations. Compounds in one group serve as membrane markers and bind to the plasma membrane (e.g., cationized ferritin); members of the other class, known as fluid-phase markers, are internalized with the aqueous phase (e.g., lanthanum nitrate). In contrast to the vast literature on endocytosis in animal systems, the number of comparable plant studies remains remarkably small. The low level of endocytic activity of plant cells and the presence of the cell walls greatly impede the access of tracer molecules to the plasma membrane and thereby make experimentation more difficult. Protoplasts can be used for tracer uptake studies, but the concentrated sucrose medium that protects the protoplasts from bursting also affects endocytosis. Nevertheless, a consensus has emerged among researchers on the following model for the endocytic pathway in plants (see Fig. 1.15):

Plasma membrane/extracellular space
(see Fig. 1.11)
↓
Clathrin-coated pits and vesicles (Fig. 1.31)
↓
Noncoated vesicles
↓
Partially coated reticulum, which seems to be an extension of the TGN (see Fig. 1.23) or of another endosomal compartment
↓
Multivesicular bodies, organelles containing materials that may not be destined for degradation (Fig. 1.32)
↓
Vacuoles (see Fig. 1.33)

Endocytosed molecules have also been observed in Golgi cisternae on occasion.

It is not yet clear which of the above compartments corresponds to the early and late endosomes of animal cells, which recycle plasma membrane and TGN receptor molecules, respectively, before passing the endocytosed molecules to the lysosomes (lytic vacuoles). The partially coated reticulum may provide the sorting functions of an early endosome.

1.7 Vacuoles

Vacuoles, fluid-filled compartments encompassed by a membrane called the **tonoplast,** are conspicuous organelles of most plant cells: They usually occupy more than 30% of the cell volume (Fig. 1.33). Apical meris-

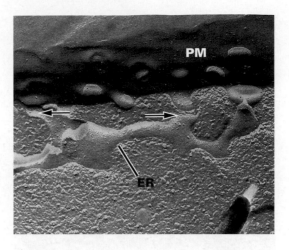

Figure 1.30
TEM showing a lipid-recycling ER domain. The ER cisterna has characteristic extensions that form specialized contact sites (arrows) with freshly fused and collapsed secretory vesicles in the plasma membrane (PM) of a cryofixed pea root tip cell. These structures correspond to domain 15 in Fig. 1.14.

tem cells typically contain numerous small vacuoles (see Fig. 1.1B), which coalesce into one or a few larger vacuoles as the cell matures and expands (see Fig. 1.1A). In large, mature cells, the space occupied by the vacuolar compartment(s) can approach 90% of the cell volume, with most of the cytoplasm confined to a thin peripheral layer connected to the nuclear region by transvacuolar strands of cytoplasm.

Vacuoles store a large variety of molecules, including inorganic ions, organic acids, sugars, enzymes, storage proteins, and many types of secondary metabolites. Proteins in the tonoplast transport all of these molecules—except for storage proteins—into the vacuole

(A)   **(B)**

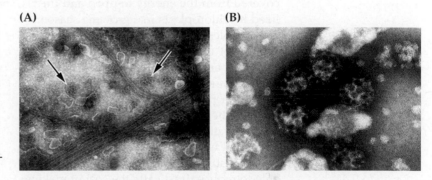

Figure 1.31
(A) TEM of negatively stained clathrin-coated endocytic vesicles budding from a plasma membrane (arrows). The rod-like structures are microtubules. (B) Higher magnification TEM of clathrin-coated vesicles. Note the lattice-like organization of the clathrin triskelion molecules that make up the coat.

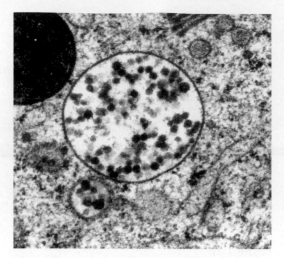

Figure 1.32
TEM of one large and one small multivesicular body in the cytoplasm of a tobacco cell cultured in suspension.

(see Chapter 3). The resulting accumulation of solutes in the vacuole drives the osmotic uptake of water, producing the turgor pressure needed for cell enlargement. Many hydrolytic enzymes found in vacuoles resemble those present in the lysosomes of animal cells, suggesting a role for vacuoles in the turnover of cellular constituents in plant cells. Vacuoles play a wide range of essential roles in the life of a plant. Considering the structural simplicity of vacuoles, this functional diversity is quite remarkable.

1.7.1 Plants use vacuoles to produce large cells cheaply.

One of the major challenges faced by plants during evolution was to develop a structural design capable of producing large solar collectors at a metabolic price that could be recovered from the energy trapped and utilized by chloroplasts in a growing season. This problem was solved by increasing the volume of the vacuolar compartment to drive cell enlargement while keeping the amount of nitrogen-rich cytoplasm constant. This latter point is particularly important for plants, whose growth is often limited by nitrogen availability. By filling a large volume of the cell with "inexpensive" vacuolar contents, mostly water and minerals, plants are able to reduce drastically the cost of making expanded structures such as leaves, which are essentially throw-away solar collectors.

Plant cell expansion is driven by a combination of osmotic uptake of water into the

vacuoles and altered cell wall extensibility. The water taken into the vacuoles generates turgor pressure, which not only expands the primary cell wall, but also creates stiff, load-bearing structures in conjunction with the walls. This exploitation of internal hydrostatic pressure to stiffen thin primary cell walls resembles the use of air pressure in an inner tube to convert a pliable, flat bicycle tire into a stiff circle capable of supporting heavy loads. Wilting and the associated softening of plant organs is caused by the loss of water from the vacuoles and surrounding cytosol.

To maintain the turgor pressure of continuously expanding cells, solutes must be actively transported into the growing vacuole to maintain its osmolarity. An electrochemical gradient across the tonoplast provides the driving force for this uptake of solutes. The gradient, in turn, is produced and maintained by two electrogenic proton pumps: V-type H^+-ATPase and vacuolar H^+-pyrophosphatase (H^+-PPase). The principal solutes in vacuoles include the ions K^+, Na^+, Ca^{2+}, Mg^{2+}, Cl^-, SO_4^{2-}, PO_4^{3-}, and NO^{3-}, and primary metabolites, such as amino acids, organic acids, and sugars. The movement of water across the tonoplast is mediated by aquaporin channels consisting of tonoplast intrinsic proteins (TIPs) (see Chapter 3).

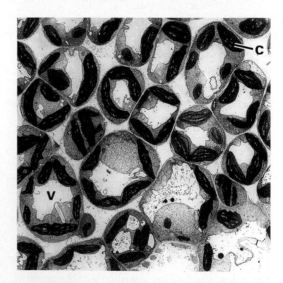

Figure 1.33
TEM showing a cross-sectional view of spongy mesophyll cells in a bean leaf illustrating the large amount of cell volume occupied by the central vacuoles (V). Chloroplasts (C).

1.7.2 Plant vacuoles are multifunctional compartments.

Vacuoles play several metabolic roles in addition to growth.

Storage: Aside from the ions, sugars, polysaccharides, pigments, amino acids, and organic acids mentioned above, plants also store large amounts of proteins in their vacuoles, especially in seeds. All of these primary metabolites can be retrieved from vacuoles and used in metabolic pathways to sustain growth. Interestingly, most of the flavors of fruits and vegetables can be traced to compounds that are stored in the vacuoles.

Digestion: Vacuoles have been shown to contain the same types of acid hydrolases found in animal cell lysosomes. These enzymes, which include proteases, nucleases, glycosidases, and lipases, together allow for the breakdown and recycling of nearly all cellular components. Such recycling is needed not only for the normal turnover of cellular structures but also for the retrieval of valuable nutrients during programmed cell death associated with development and senescence (see Chapter 20).

pH and ionic homeostasis: Large vacuoles serve as reservoirs of protons and metabolically important ions such as calcium. Typically, plant vacuoles have a pH between 5.0 and 5.5, but the range extends from 2.5 or so (lemon fruit vacuoles) to greater than 7.0 in unactivated protein storage vacuoles. By controlling the release of protons and other ions into the cytosol, cells can regulate not only cytosolic pH but also the activity of enzymes, the assembly of cytoskeletal structures, and membrane fusion events.

Defense against microbial pathogens and herbivores: Plant cells accumulate an amazing variety of toxic compounds in their vacuoles, both to reduce feeding by herbivores and to destroy microbial pathogens (see Chapters 21 and 24). These compounds include the following:

- phenolic compounds, alkaloids, cyanogenic glycosides, and protease inhibitors to discourage insect and animal herbivores
- cell wall–degrading enzymes, such as chitinase and glucanase, and defense molecules, such as saponins, to destroy pathogenic fungi and bacteria
- latexes, wound-clogging emulsions of hydrophobic polymers that possess insecticidal and fungicidal properties and also serve as antiherbivory agents.

Sequestration of toxic compounds: Plants cannot escape from toxic sites, nor can they efficiently eliminate by excretion toxic materials such as heavy metals and toxic metabolites such as oxalate. Instead, plants sequester these compounds into vacuoles. For example, to remove oxalate, specific cells develop vacuoles containing an organic matrix within which oxalate is allowed to react with calcium to form calcium oxalate crystals. In other plant cell types, members of the ABC family of transporters are used to transport **xenobiotics** (chemicals of human manufacture) from the cytoplasm into the vacuoles (see Chapter 3). Accumulation of toxic compounds in leaf vacuoles is one of the reasons leaves are shed on a regular basis.

Pigmentation: Vacuoles that contain anthocyanin pigments are found in many types of plant cells. Pigmented flower petals and fruits are used to attract pollinators and seed dispersers, respectively. Some leaf pigments screen out UV and visible light, preventing photooxidative damage to the photosynthetic apparatus (see Chapter 12). This screening appears to be essential for the survival of the leaves of evergreens that grow in climates where freezing conditions during winter months prevent the absorbed light energy from being used in photosynthesis.

1.7.3 Many plant cells contain two different vacuole systems.

For years, plant researchers puzzled about how storage proteins and hydrolytic enzymes might coexist in vacuoles. This problem has now finally been resolved by the discovery that many plant cells contain, at least during some stage of development, two

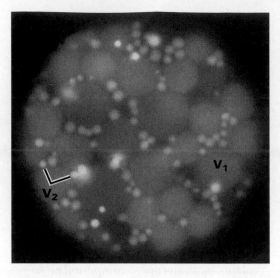

Figure 1.34
Light micrograph of an aleurone protoplast stained with a fluorescent dye. Two types of vacuoles are depicted: large protein storage vacuoles (V_1) and smaller lytic vacuoles (V_2).

functionally different kinds of vacuoles: neutral protein storage vacuoles and acidic, lytic vacuoles. During development, storage products may be mobilized either by fusing the two vacuole types or by delivering specific lytic enzymes to the storage vacuoles.

Evidence in support of this discovery has come from investigations using several types of probes, including heavy metal stains, fluorescent molecules sensitive to pH (Fig. 1.34) or lytic enzyme activity, and antibodies directed against two types of tonoplast proteins: α-TIP and TIP-Ma27. Anti-α-TIP antibodies label only vacuoles that have a pH near 7 and contain storage proteins such as lectins. In contrast, anti-TIP-Ma27 antibodies label only acidified vacuoles, which contain hydrolytic enzymes such as the cysteine protease aleurain.

The discovery of multiple vacuole types leads to other interesting questions. How do proteins produced in response to pathogens survive in the vacuoles of cells that contain only one acidic vacuole system, one that presumably contains lytic enzymes? In which types of vacuoles do calcium, pigments, and toxic compounds accumulate? Finally, will future studies lead to the identification of yet other classes of vacuoles?

1.7.4 Vacuoles may be the only membrane compartment that can be created de novo.

Unlike all other membrane systems, some vacuoles may arise by a process of differentiation from other membrane systems rather than by inheritance and growth. However, the study of vacuole biogenesis has been made difficult by wide variations in the morphology and cytochemical staining properties of the membrane systems that appear to give rise to vacuoles. In addition, difficulties arise in interpreting static electron micrographs of what are clearly very dynamic membrane events associated with vacuole formation. Vacuole biogenesis studies using fluorescence-tagged marker proteins in living

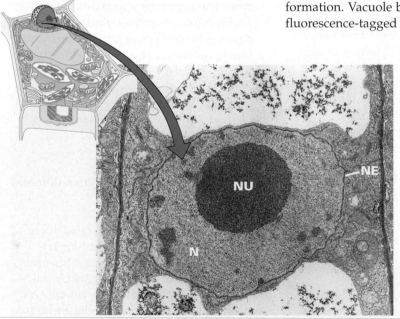

Figure 1.35
TEM showing the nucleus (N) of a bean root tip cell. Note the two membranes of the nuclear envelope (NE) and the large central nucleolus (NU).

cells may resolve some of these questions in the future. Some of the reported differences in vacuole biogenesis probably stem from different cell types producing different kinds of vacuoles, which may be assembled by different mechanisms. In some cells, moreover, vacuolar organization changes during the cell cycle.

Two membrane systems have been postulated to give rise to the initial vacuolar compartments: a TGN-like membrane system and subdomains of smooth ER. The preponderance of evidence supports an ER origin of vacuoles, but variations in the assembly pathways are evident from published micrographs. Most frequently, **provacuoles** appear to arise from regionally differentiated ER domains in which both tonoplast membrane proteins (V-type H^+-ATPase, α-TIP) and storage proteins accumulate. After their separation from the ER, these provacuoles appear to inflate and become full-fledged vacuoles. A very different sequence of events has been described for the seed coat cells of linseed, in which the central vacuole appears to arise from smooth ER tubes that encompass the future vacuole site. As these tubes assemble into a cage-like structure, the engulfed cytoplasm is cleared of organelles and then is either autophagocytosed into the forming vacuole or displaced and collapsed as the vacuolar tubes fuse and inflate. Vacuole biogenesis is clearly a field that offers exciting opportunities for future studies.

1.8 The nucleus

The nucleus contains most of the cell's genetic information and serves as the center of regulatory activity (Fig. 1.35). Although the DNA–protein complexes that make up the chromosomes are evident only as an irregular network of chromatin during interphase, the individual chromosomes nevertheless occupy discrete domains throughout this part of the cell cycle. Interphase is the most important stage of the cell cycle for gene expression because during this period chromosomes are actively transcribed.

A typical interphase plant cell nucleus also contains one to several **nucleoli** lying free in the nuclear matrix, or **nucleoplasm** (Fig. 1.35). These prominent, densely staining, often spherical bodies house the machinery that manufactures cytoplasmic ribosomes. A brief introduction is given below to the ultrastructure and activities of the nucleolus as well as to the structure of the nuclear pores, through which nucleolar products must move to reach the cytoplasm.

1.8.1 The nuclear envelope is a dynamic structure with many functions.

The nucleus is bounded by an envelope consisting of two bilayer membranes separated by a lumen, the **perinuclear space** (Fig. 1.36). The envelope separates the genetic material in the nucleus from protein synthesis in the cytoplasm and controls nucleocytoplasmic exchange by means of highly complex **nuclear pores.** Seen in cross-section, the continuity between the outer and inner envelope membranes at the pore boundaries becomes obvious (Fig. 1.36). Most RNA synthesized in the nucleus passes through nuclear pores to the cytoplasm for use in protein synthesis, whereas those proteins synthesized in the cytoplasm and required in the nucleus are imported through nuclear pores.

The outer membrane of the nuclear envelope is continuous with membranes of the ER through narrow connections and resembles the ER in having functional ribosomes on its cytoplasmic face (see Fig. 1.14). The perinuclear space is therefore continuous with the ER lumen. A meshwork of 10-nm-diameter filaments, called the **nuclear lamina,** underlies the inner envelope membrane. The lamina links the nuclear pore complexes and anchors and organizes the interphase chromatin at the nuclear periphery.

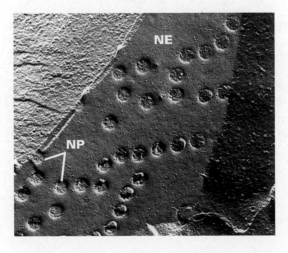

Figure 1.36
TEM of a nuclear envelope (NE) with nuclear pores (NP). The continuity of the inner and outer membranes becomes apparent where the membranes are seen in cross-section.

1.8.2 Nuclear pore complexes function both as molecular sieves and as active transporters.

The density of nuclear pores embedded in the envelope varies considerably, depending on the type of cell. In plant cells pores occupy from 8% to 20% of the envelope surface (Fig. 1.36), at a pore density of 6 to 25/μm². The pattern of pore distribution over the envelope varies in different organisms and cell types.

Each pore consists of an elaborate macromolecular assemblage known as the nuclear pore complex (Fig. 1.37). Nuclear pore complexes appear generally similar in size and architecture throughout the plant and animal kingdoms. A nuclear pore complex measures 120 nm in depth, with a pore diameter of 50 nm. The complex has a total protein mass of 124 MDa and contains approximately 120 different kinds of proteins, called nucleoporins, several of which have been identified and characterized.

A nuclear pore complex has octagonal symmetry about a central axis. Eight peripheral, aqueous channels, each 9 nm in diameter, permit free diffusion of small molecules. A larger, regulated, central channel functions in the active transport of proteins and RNA molecules. This central channel contains a particle 35 nm in diameter called the "plug" (Fig. 1.37A), which is believed to be the actual transporter protein complex that controls nucleo-cytoplasmic exchange.

Proteins in the cytoplasm can gain access to the nucleus only if they contain a **nuclear localization signal** sequence of amino acids. Recognition of this signal by the nuclear pore complex causes the central channel of the pore to open (see Chapter 4). An additional signal is apparently needed for exit from the nucleus. Most imported proteins remain in the nucleus, even if they are in soluble form and can move freely in the nucleoplasm. However, some proteins are known to move back and forth between nucleus and cytoplasm. Energy in the form of ATP is required to transport most proteins or particles through the central channel in either direction. Hydrolysis of ATP probably leads to a conformational change in the transporter that causes the pore to open.

(A)

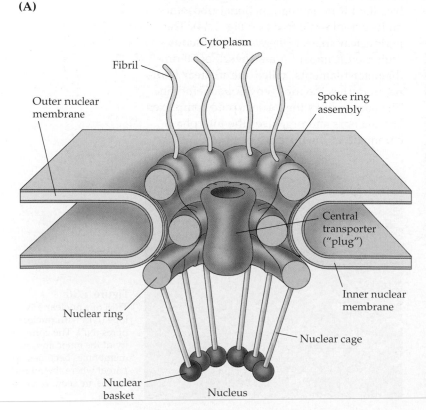

(B)

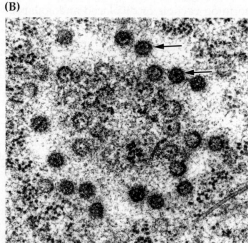

Figure 1.37
(A) Diagram of a nuclear pore complex in a nuclear membrane. (B) TEM showing a tangential thin section through nuclear pore complexes of a tobacco root tip cell. Arrows indicate pores in which the central transporter plug depicted in (A) is clearly seen.

1.8.3 The nucleolus, a prominent organelle in the interphase nucleus, is the ribosome factory of the cell.

Although the **nucleolus** is not membrane-bounded, it is a distinct organelle forming a highly specialized region of the nucleus (Fig. 1.38). A product of active ribosomal genes, the nucleolus, when assembled, contains more than 100 different proteins and nucleic acids. It is responsible for transcribing ribosomal DNA, processing ribosomal RNA transcripts, and assembling ribosomal RNA and imported ribosomal proteins into ribonucleoprotein subunits for transport to the cytoplasm. The nucleolus contains three substructures clearly visible under the electron microscope:

- one or more fibrillar centers, in which ribosomal transcription is believed to take place
- dense fibrillar components surrounding the centers, where the ribosomal RNA transcripts are processed
- a granular component that surrounds the entire fibrillar region.

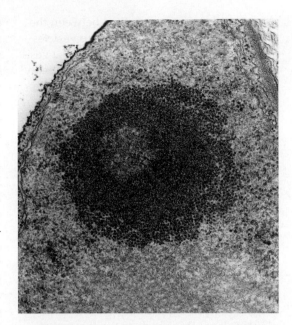

Figure 1.38
TEM of a nucleolus within the nucleus of a carnation cell. Note the three characteristic subdomains: the fibrillar center, where ribosomal transcription takes place; the surrounding dense fibrillar domain, where the rRNA transcripts are processed; and the outermost granular layer that surrounds the whole structure.

The granular component consists of preribosomal particles and ribosomal subunits 15 to 20 nm in diameter. These are in various stages of assembly and are smaller than fully developed ribosomes in the cytoplasm. On maturation they will be exported from the nucleus through the central channel of the nuclear pore, presumably utilizing an export signal on one of the ribosomal proteins. The nucleolus breaks down at the beginning of prophase and reforms when ribosomal genes of the chromosomal nucleolar organizing region become active in the young daughter nuclei during telophase.

1.8.4 During mitosis, the nuclear envelope disassembles into vesicles that participate in the formation of new envelopes around the daughter nuclei.

Electron microscopic studies of mitotic cells reveal that nuclear pore complexes disappear from the nuclear envelope, leaving holes in envelope membranes before the envelope itself breaks down at the start of prometaphase. A protein kinase called mitosis-promoting factor appears to play a central role in disassembly and reassembly of the envelope during mitosis. Phosphorylation of the lamina bound to the inner surface of the envelope is correlated closely with envelope breakdown, and dephosphorylation of the lamina is correlated with reconstitution of the envelope (see Chapter 5). Phosphorylation is also believed to play a part in assembly and disassembly of the pore complex, because phosphorylation of some of the pore proteins during mitosis prevents their assembly into pores.

When the nuclear envelope breaks down, the membranes of the envelope do not completely dissolve but instead disperse into numerous membrane-bounded vesicles that serve later as precursors of the nuclear envelopes surrounding the daughter nuclei. After prometaphase, the vesicular remnants of the old envelope are partitioned into the two halves of the dividing cell, and at early telophase these envelope components bind to the surface of the chromatin, thereby excluding cytoplasmic proteins. During reconstitution of the nuclear envelope, the nuclear pore complexes are assembled between the borders of neighboring membrane vesicles and serve to hold them together. These vesicles eventually fuse, and envelope assembly is completed by further membrane growth and coalescence.

1.9 Peroxisomes

Peroxisomes are structurally simple but functionally diverse organelles present in virtually all eukaryotic cells. They are composed of a single membrane that surrounds the finely granular peroxisome matrix (Fig. 1.39A). In plants, their role depends on the organ or tissue in which they occur. In germinating fat-storing seeds, peroxisomes participate in lipid mobilization; in leaves of C_3 plants (plants that produce the three-carbon compound phosphoglycerate as the first stable photosynthetic intermediate), they play a key role in photorespiration; and in some legume root nodules, they are involved in

the conversion of recently fixed N_2 into nitrogen-rich organic compounds. So-called unspecialized peroxisomes, also found in living cells of roots and many other plant tissues (Fig. 1.39), contain catalase, glycolate oxidase, urate oxidase, and enzymes for β-oxidation, but what functions they perform in the cells are not yet known.

Most commonly, peroxisomes are roughly spherical, with diameters ranging from approximately 0.2 to 1.7 μm, but they may have elongate or irregular shapes or interconnections that form a reticulum. They have a characteristic high density (1.23 g/cm^3, compared with 1.18 g/cm^3 for mitochondria). Depending on the plant species and cell type, their matrices may contain crystalline or fibrillar inclusions or densely amorphous regions (nucleoids), but they lack internal membranes, and all of their proteins are encoded by nuclear genes. The peroxisomes in the leaves of many temperate plants often contain strikingly beautiful catalase crystals (Fig. 1.40A). Within a given cell, peroxisomes can vary widely in number, size, enzyme constitution, and metabolic function, depending on the environment and the developmental stage of the plant. Nearly 50 enzymes have been localized to plant and animal peroxisomes. Catalase, which is always present, is used as a marker enzyme for the organelle.

Peroxisomes were called **microbodies** when first observed with the electron microscope in animal cells in the 1950s and in plant cells in the 1960s. In 1965 Christian de Duve isolated similar organelles from liver cells by centrifugation; he called them *peroxisomes* because they both generated and destroyed hydrogen peroxide. Although organelles fitting the above description are still sometimes called microbodies, they invariably contain catalase and are more properly named peroxisomes.

1.9.1 The toxic H_2O_2 produced by peroxisomal oxidases is destroyed in situ by catalase.

The oxidases of peroxisomes are flavoproteins that produce hydrogen peroxide by transferring electrons (as hydrogen atoms) from a substrate (R) to oxygen:

$$RH_2 + O_2 \longrightarrow R + H_2O_2$$

(A)

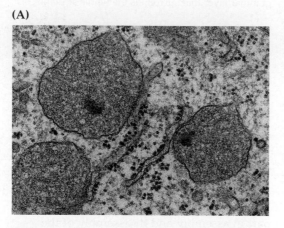

(B)

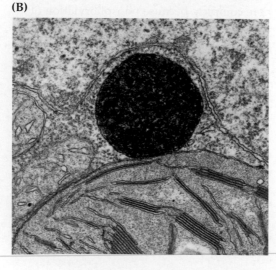

Figure 1.39
(A) TEM showing three roundish, unspecialized peroxisomes closely associated with ER cisternae in a bean root cell. (B) TEM of a tobacco leaf peroxisome stained with diaminobenzidine to demonstrate the presence of catalase in the organelle.

(A)

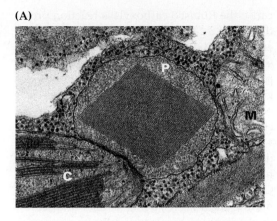

Key

1 Ribulose-1,5-bisphosphate carboxylase/oxygenase
2 Phosphoglycolate phosphatase
3 Glycolate oxidase
4 Glutamate:glyoxylate aminotransferase
5 Glycine decarboxylase and serine hydroxymethyl transferase
6 Serine:glyoxylate aminotransferase
7 Hydroxypyruvate reductase
8 Glycerate kinase
9 Catalase

(B)

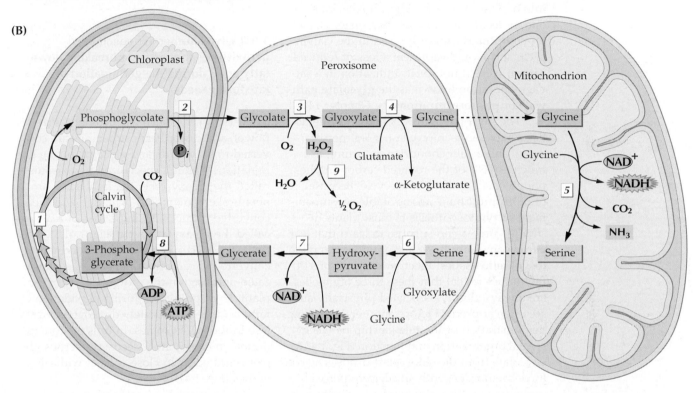

Figure 1.40
(A) TEM of a tobacco leaf peroxisome (P) in close physical contact with both a chloroplast (C) and a mitochondrion (M). Note the large catalase crystal in the peroxisome. (B) Diagram of the glycolate pathway, which involves enzyme systems located in peroxisomes, chloroplasts, and mitochondria.

Because catalase is always present in peroxisomes, the harmful H_2O_2 generated by these flavoprotein oxidases is broken down immediately within the organelle and never reaches the cytoplasm. In destroying hydrogen peroxide, the enzyme usually functions catalatically:

$$2H_2O_2 \longrightarrow O_2 + 2H_2O$$
The **catalatic** reaction

$$R'H_2 + H_2O_2 \longrightarrow R' + 2H_2O$$
The **peroxidatic** reaction

Cytochemistry on the electron microscopic level can demonstrate the presence of catalase in a peroxisome. When leaf tissue is fixed in glutaraldehyde and incubated with the reagent 3,3'-diaminobenzidine (DAB) and hydrogen peroxide, DAB is oxidized to insoluble polymers through the peroxidatic action of catalase. The formation of the polymers, and hence the presence of catalase, is confirmed by incubating the tissue in osmium tetroxide. The DAB polymers reduce OsO_4 and deposit it in place as osmium black, an electron-opaque product (Fig. 1.39B).

1.9.2 Leaf peroxisomes participate with chloroplasts and mitochondria in the glycolate pathway (photorespiration).

Rubisco (ribulose-1,5-bisphosphate carboxylase/oxygenase) is the Calvin-cycle enzyme that catalyzes CO_2 fixation during photosynthesis (see Chapter 12). In this reaction, Rubisco functions as a **carboxylase,** adding atmospheric CO_2 to ribulose-1,5-bisphosphate to form two molecules of 3-phosphoglycerate. However, Rubisco also acts as an **oxygenase,** catalyzing an inseparable, competitive reaction that splits ribulose-1,5-bisphosphate into one molecule of 3-phosphoglycerate and one molecule of 2-phosphoglycolate. The latter compound cannot be used in the Calvin cycle and would represent a loss of reduced carbon were it not for its utilization in a salvage operation known as the **glycolate pathway,** or **photorespiration** (see Chapter 14).

The glycolate pathway involves interactions between a chloroplast, a leaf peroxisome, and a mitochondrion, returning as much as 75% of the reduced carbon in phosphoglycolate to the Calvin cycle. Because photorespiration is an inevitable accompaniment of photosynthesis in most plants, it should come as no surprise to learn that leaf peroxisomes are large and abundant in photosynthetic tissues. Electron micrographs of the tissues reveal that these three organelles are always closely associated physically and are often appressed to one another—a striking demonstration of the relationship between spatial organization and function (Fig. 1.40A). Glycolate from the chloroplast diffuses across three membranes into an adjacent peroxisome, where it is oxidized to glyoxylate in a reaction that yields hydrogen peroxide. The abundant catalase in the peroxisome decomposes the H_2O_2. Several subsequent steps involving the peroxisome and a nearby mitochondrion result finally in the production of a molecule of glycerate, which diffuses back into a chloroplast, is phosphorylated, and reenters the Calvin cycle. The pathway is called photorespiration because the oxygenase reaction of Rubisco results in the light-dependent uptake of O_2 and release of CO_2 (Fig. 1.40B).

This description of leaf peroxisomes applies to plants carrying out the typical Calvin cycle in the mesophyll cells. As mentioned above, these plants are called C_3 because the Rubisco carboxylase reaction produces the 3-carbon compound phosphoglycerate (see Chapter 12). But what of peroxisomes in C_4 plants, for which the first stable photosynthetic intermediate is a 4-carbon organic acid? In these plants, the Calvin cycle occurs not in the mesophyll but in the vascular bundle sheath cells (see Chapter 12), where the oxygenase activity of Rubisco is decreased because CO_2 concentration is much higher than in the ambient atmosphere. In C_4 plants, peroxisomes are concentrated in bundle sheath cells, being smaller and relatively scarce in mesophyll cells.

1.9.3 Glyoxysomes are specialized peroxisomes that assist in breaking down fatty acids during the germination of fat-storing seeds.

In oil-rich seeds, such as cucumber, sunflower, and castor bean, stored oils are converted to fatty acids and glycerol during germination. Within a specialized peroxisome called the glyoxysome, the fatty acids are next broken down by β-oxidation and then converted to succinate in a series of reactions called the **glyoxylate cycle** (Fig. 1.41A). The final steps occur in the cytosol: Succinate is converted to carbohydrate, which is translocated and used in the growing parts of the plant. H_2O_2 produced during β-oxidation of fatty acids is immediately destroyed by catalase in the glyoxysomes. In cells of storage organs in germinating seeds, numerous glyoxysomes occur in close contact with oil bodies (Fig. 1.41B).

In some seeds—castor bean, for example—oil is stored in the endosperm, which shrivels as the fat is metabolized and the carbohydrate products are translocated out of the endosperm. In seeds such as sunflower and watermelon, however, oil is stored in cotyledons. During the first few days of germination, levels of characteristic glyoxylate cycle enzymes increase rapidly in cotyledon glyoxysomes. During this period, the cotyledons are yellow or pale green, the chloroplasts exhibit only rudimentary thylakoid development, and the enzymes characteristic of leaf peroxisomes are still absent. After a few days' exposure to light, the lipid stores are metabolized and the cotyledons undergo greening. During this transition,

(A)

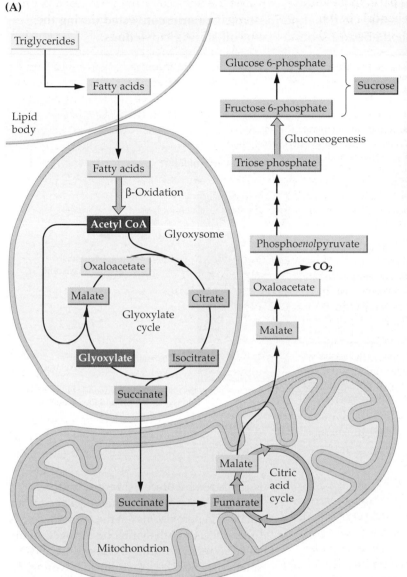

Triglycerides

Fatty acids

Lipid body

Fatty acids

β-Oxidation

Acetyl CoA   Glyoxysome

Oxaloacetate

Malate   Citrate

Glyoxylate cycle

Glyoxylate   Isocitrate

Succinate

Glucose 6-phosphate

Fructose 6-phosphate   } Sucrose

Gluconeogenesis

Triose phosphate

Phospho*enol*pyruvate

→ CO₂

Oxaloacetate

Malate

Malate   Citric acid cycle

Succinate → Fumarate

Mitochondrion

(B)

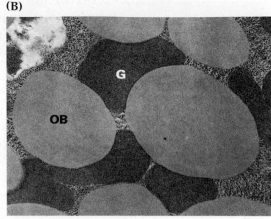

Figure 1.41
(A) Diagram illustrating gluconeogenesis in fat-storing seedlings. (B) TEM of glyoxysomes (G) surrounding oil bodies (OB) in a cotyledon of a germinating tomato seed.

levels of glyoxysomal enzymes decline rapidly, and the enzymes characteristic of leaf peroxisomes become first detectable and finally predominant (see Box 1.1).

1.9.4 In some leguminous root nodules, peroxisomes play an essential role in the conversion of recently fixed N₂ into ureides for nitrogen export.

Several legumes of major economic importance, including bean, soybean, and cowpea, are known as "ureide transporters" because they fix N₂ in organic intermediates that become ureides, nitrogen-rich compounds useful in translocating nitrogen to the growing and storage organs of the plants (Fig. 1.42). Both N₂ fixation and ureide formation take place entirely in root nodules (see Chapter 16). The metabolic sequence leading from N₂ to ureides occurs in two different cell types in the nodule. Early steps take place only in cells infected with Rhizobium or related bacteria, the organisms that actually perform the fixation reactions. The final steps take place only in interstitial cells, which remain uninfected.

One of the later steps in ureide production, the conversion of uric acid to allantoin,

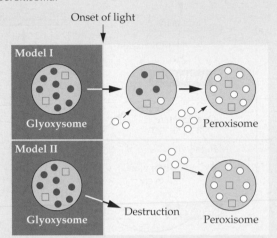
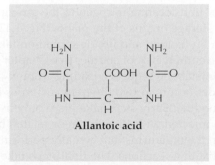

Allantoin

Allantoic acid

Figure 1.42
Ureides are nitrogen-rich compounds formed in the root nodules of some legumes that form symbiotic associations with rhizobial bacteria.

is catalyzed by urate oxidase, a hydrogen peroxide–generating enzyme packaged with catalase in peroxisomes of the interstitial cells. Cells infected with symbiotic nitrogen-fixing bacteria have no urate oxidase and only small, degenerate peroxisomes. In younger nodules of soybean, where N_2 fixation and ureide production remain low, peroxisomes of uninfected cells are small and contain only a scanty matrix. As these nodules mature and fix nitrogen more vigorously, the peroxisomes become larger and more numerous and their matrix becomes much denser.

1.9.5 New peroxisomes arise by division of preexisting peroxisomes and import peroxisomal proteins synthesized on cytosolic ribosomes.

New peroxisomes arise through the division of existing peroxisomes and grow by importing lipids, membrane proteins, and matrix proteins from the cytosol. Specialized peroxisomes, like those of C_3 leaves and oilseeds, grow very rapidly in number and volume as their metabolic role increases in the early stages of organ growth. This requires an enormous movement of metabolites across peroxisomal membranes. Peroxisomes are known to have pore-forming proteins in their membranes that permit the passage of small molecules as large as 800 Da. Protein molecules, however, can be incorporated into the growing membrane or permitted access to the peroxisomal matrix only if they bear a specific targeting signal as part of their structure (see Chapter 4). Unlike plastids and mitochondria, peroxisomes contain neither DNA nor ribosomes, so all peroxisomal proteins must be encoded by nuclear genes. The proteins are then translated on free polysomes in the cytosol and are targeted post-translationally to the organelle.

Nobel laureate Christian de Duve has suggested that peroxisomes represent an ancient class of respiratory particle lost from many cell types as its functions were gradually superseded by those of the more efficient mitochondrion. According to this theory, peroxisomes have been retained only where they perform highly specialized reactions, as they do in photorespiration and the other processes considered above.

1.10 Plastids

Plastids are major organelles found only in plant and algal cells. They are responsible for photosynthesis, for the storage of a wide variety of products, and for the synthesis of key molecules required for the basic architecture and functioning of plant cells. Like mitochondria, plastids are enclosed in an envelope—a pair of concentric membranes. Plastids also resemble mitochondria in being semiautonomous and containing the genetic machinery required to synthesize a few of their own proteins.

As is implied in the name (from the Greek *plastikos:* "molded"), plastids vary in size, shape, content, and function. Interest in this morphological plasticity has been rekindled recently by studies of transgenic tobacco plants expressing green fluorescent protein (GFP) fused to a chloroplast stroma localization signal. As illustrated in Figure 1.43, this GFP probe can be used to follow rapid changes in plastid shape, particularly

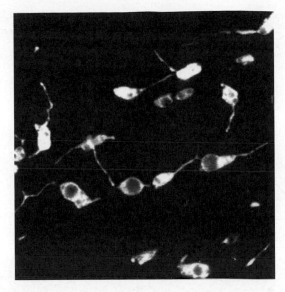

Figure 1.43
Light micrograph of plastids with thin, tubular stromule extensions. The stromules are highlighted by green fluorescent protein (GFP) expressed in the chloroplast stroma.

the formation and retraction of tubular extensions known as stromules. Stromules not only have the ability to grow and contract; they can also fuse with other plastids. This creates stromal bridges between plastids through which genetic material can be exchanged. Plastids also possess a remarkable capacity to differentiate, dedifferentiate, and redifferentiate (Fig. 1.44). They can be colorless in some cell types and pigmented in others. The numbers of any particular type of plastid in a cell can vary greatly among cell types as well as at different stages in the life of the same cell. Plastids reproduce solely by division, independently of cell division.

1.10.1 All types of plastids are developmentally related to proplastids.

Proplastids are the precursors of other plastids and are always present in the young meristematic regions of the plant. For example, angiosperm meristems contain approximately 20 proplastids per cell. Proplastids commonly range in diameter from approximately 0.2 to 1.0 μm and are most frequently spherical or ovoid (Fig. 1.45). The internal milieu of the proplastid, called the **stroma,** appears rather uniformly dense and finely granular. Proplastids contain far fewer

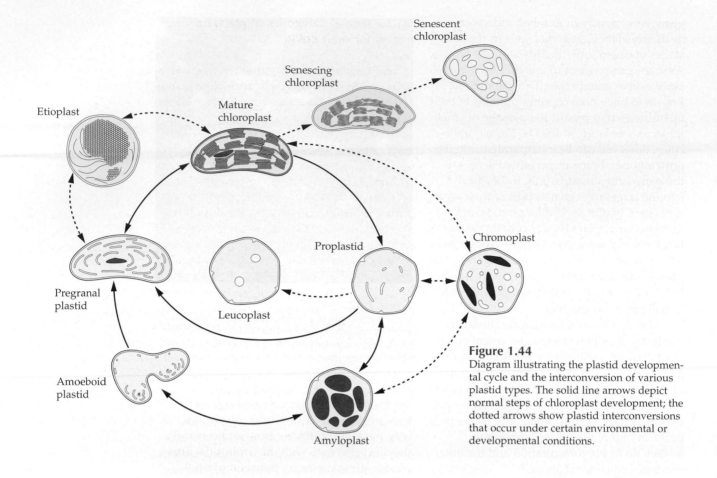

Senescent chloroplast

Senescing chloroplast

Mature chloroplast

Etioplast

Pregranal plastid

Amoeboid plastid

Leucoplast

Proplastid

Chromoplast

Amyloplast

Figure 1.44
Diagram illustrating the plastid developmental cycle and the interconversion of various plastid types. The solid line arrows depict normal steps of chloroplast development; the dotted arrows show plastid interconversions that occur under certain environmental or developmental conditions.

ribosomes than do the more highly differentiated etioplasts and chloroplasts. One or more nucleoids are sometimes identifiable in the stroma as electron-lucent regions containing fine (3-nm-diameter) DNA fibrils. In proplastids, the internal membrane system remains poorly developed, consisting only of a few invaginations of the inner membrane and a small number of flattened sacs called **lamellae.** Occasionally, a vesicle containing

storage proteins is also present. The proplastid stroma sometimes contains deposits of phytoferritin (Fig. 1.46), a storage form of iron similar to the ferritin of animal cells. Phytoferritin occurs almost exclusively in plastids, most abundantly in the plastids of storage organs.

1.10.2 Amyloplasts are starch-storing plastids.

Amyloplasts (Fig. 1.47) are unpigmented plastids that resemble proplastids but contain starch granules. Their name is derived from amylose, a linear, water-soluble polysaccharide composed of $\alpha(1\rightarrow4)$-glucopyranosyl units, which constitutes a major component of starch. Chloroplasts also frequently store starch in quantities that vary according to the balance between carbohydrate synthesis and export.

Amyloplasts are especially common organelles in storage organs. Starch grains, which can be quite massive, occur free in the

Figure 1.45
TEM showing a proplastid (left) adjacent to a mitochondrion in a bean root cell. Note that the ribosomes in the two organelles (small arrows) are smaller than the cytoplasmic ribosomes (large arrows). The large, electron-dense particles in the proplastid are lipid droplets, and the large round structure in the middle is a protein deposit.

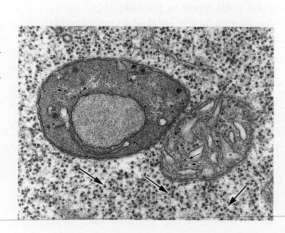

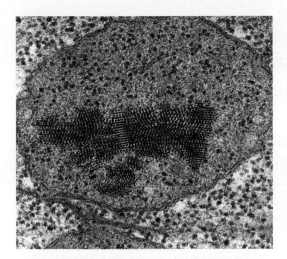

Figure 1.46
TEM illustrating phytoferritin deposits inside a proplastid in a root apical meristem cell of soybean.

stroma. In potato tubers, for example, starch grains commonly fill the entire plastid except for a thin layer of stroma lining the inner membrane of the envelope. In the gravity-sensing columella cells of root caps, amyloplasts serve as statoliths, sedimenting in response to gravity and thereby triggering the gravitropic response. The structure of these statolith amyloplasts differs somewhat from their storage counterparts by having a stroma-excluding matrix that encompasses all of the starch granules (Fig. 1.47).

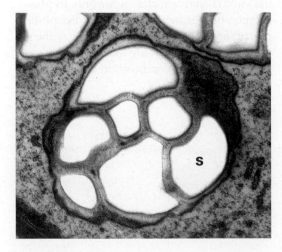

Figure 1.47
TEM of amyloplasts containing many starch granules (S) in a columella root-cap cell of white clover. These amyloplasts serve both as statoliths of the gravity-sensing apparatus of root tips and as starch storage organelles.

1.10.3 Several categories of plastid are named for their color.

Leucoplasts are colorless plastids involved in the synthesis of monoterpenes, the volatile compounds contained in essential oils. Many of these compounds are exploited by humans as flavors or pharmacological agents. Their synthesis is carried out by specialized secretory gland cells associated with leaf and stem trichomes or with the secretory cavities of citrus peel. The literature on leucoplasts is somewhat confusing because the term has also been used as a synonym for an amyloplast. However, monoterpene-producing leucoplasts constitute a unique type of plastid characterized by two envelope membranes, a dense stroma, very few internal membranes and ribosomes, and small plastoglobuli, that is, lipid droplets that stain darkly with osmium (Fig. 1.48). Leucoplasts are typically surrounded by an extensive network of tubular smooth ER membranes, which also participate in the synthesis of lipid molecules.

Etioplasts (Fig. 1.49A) are plastids for which the development from proplastids to chloroplasts has been arrested by the absence of light or by very low light conditions in leaves. They do not represent an intermediate stage in the normal development that proplastids undergo in becoming chloroplasts in the light but rather are a diversion during development into a special kind of plastid found only in white or pale-yellow etiolated leaves and other normally green tissues kept in the dark. Etioplasts lack chlorophyll, but produce a large amount of a colorless chlorophyll precursor called **protochlorophyllide.**

Etioplasts store membrane lipids in the form of a prominent, quasicrystalline membranous structure called the **prolamellar body** (Fig. 1.49A). This structure forms through the continued synthesis of the lipids that normally compose the internal membrane systems of chloroplasts, accompanied by severely reduced synthesis of the necessary membrane proteins. The high proportion of lipids in prolamellar bodies (approximately 75% lipids) results in tubes that branch in three dimensions to form a semi-crystalline lattice (Fig. 1.49B–D). When etioplasts are illuminated, they begin to develop into chloroplasts. Light triggers the synthesis of chlorophyll from protochlorophyllide as well as the assembly of stable chlorophyll–protein

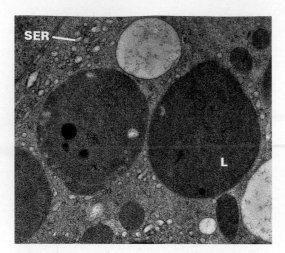

Figure 1.48
TEM showing leucoplasts (L) in an actively secreting glandular trichome of peppermint (*Mentha piperita*). The electron-dense globules correspond to lipid droplets. Note the absence of internal membranes in the leucoplasts and the presence of extensive tubular smooth ER cistenae (SER) in the surrounding cytoplasm.

complexes, resulting in dispersal of the lipid molecules of the prolamellar body into the growing, photosynthetically active chloroplast membranes known as **thylakoids** (Fig. 1.50).

Chloroplasts, the green photosynthetic plastids responsible for energy capture, are commonly hemispherical or lens-shaped in vascular plants (Fig. 1.51) but vary considerably in shape in mosses and algae. Each cell in a filament of the familiar alga *Spirogyra,* for example, contains a single ribbon-shaped chloroplast. Plant chloroplasts commonly measure approximately 5 to 8 μm long and 3 to 4 μm thick.

Different cells contain different numbers of chloroplasts. A palisade mesophyll cell of a castor bean leaf contains approximately 36 chloroplasts; a spongy mesophyll cell contains only approximately 20. A castor bean leaf contains approximately 500,000 chloroplasts per square millimeter, of which 82% occur in the palisade mesophyll cells. A mature leaf mesophyll cell of *Arabidopsis thaliana,* the small crucifer widely used for studies of plant molecular genetics, contains approximately 120 chloroplasts. The photosynthetic apparatus of chloroplasts is contained within the expansive thylakoid membrane system (see Chapter 12).

Chromoplasts (Fig. 1.52) are yellow, orange, or red, depending on the particular combination of carotenes and xanthophylls present. Chromoplasts are responsible for the colors of many fruits (e.g., tomatoes, oranges), flowers (e.g., buttercups, marigolds), and roots (e.g., carrots, sweet potatoes). Chromoplasts can develop directly from proplastids or by the redifferentiation of chloroplasts, as happens in ripening tomato fruits. Occasionally chromoplasts will redifferentiate into chloroplasts, as they do in the light-exposed surface area of carrot roots or in some yellow and orange citrus fruits that regreen under the appropriate conditions.

Chromoplast development is accompanied by a massive induction of enzymes that catalyze carotenoid biosynthesis. Although inactive forms of these enzymes occur in the stroma, active forms occur only in plastid membranes, where the highly lipophilic precursors of the carotenes, as well as the carotenes themselves, are localized. The great variation in composition of the large

(A)

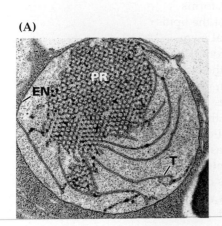

(B)

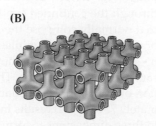

(C)

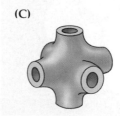

(D)

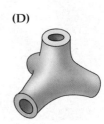

Figure 1.49
(A) TEM of an etioplast with a large prolamellar body (PR) and associated unstacked thylakoids (T) in a maize leaf. Envelope membranes (EN). (B–D) Three-dimensional models of the membrane lattice structures that make up prolamellar bodies.

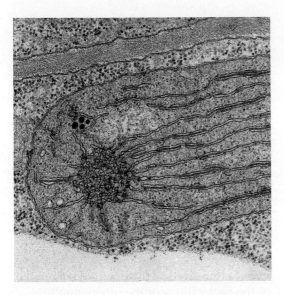

Figure 1.50
TEM of a pea leaf cell, depicting an early stage of grana thylakoid development in a greening etioplast. Remnants of a prolamellar body are seen on the left.

carotenoid deposits results in major differences in the sizes, shapes, and internal structure of chromoplasts.

1.10.4 The outer and inner membranes of the plastid envelope differ in composition, structure, and transport functions.

Most studies of the plastid envelope have used chloroplasts not only because these organelles have great metabolic importance but also because they are readily available in large quantities. However, most of the properties ascribed to chloroplast envelopes are probably applicable to the envelopes of all plastid types. Extensive metabolite traffic flows in both directions across the two membranes that separate the photosynthetically active chloroplast interior from the surrounding cytosol. Unlike other membranes of most eukaryotic cells, the envelope and the thylakoid membranes of the chloroplast are poor in phospholipids and rich in galactolipids, a difference that some investigators believe might reflect the evolution of plants in a marine environment, where phosphorus is a limiting element (see Chapter 10).

The **outer envelope membrane** contains a nonspecific pore protein that freely permits water and a variety of ions and metabolites—of sizes up to approximately 10 kDa—to pass into the **intermembrane space,** the aqueous compartment between the membranes. Difficulties in isolating material from this intermembrane space have limited studies on its chemical composition or physiological properties. The outer membrane is smooth and never becomes confluent with the inner membrane at any point. Electron microscopic studies, however, have revealed contact sites where the proteins connecting the membranes mediate the transfer of nuclear-encoded chloroplast proteins from the cytosol to the chloroplast stroma (see Chapter 4). Among the polypeptides localized to the outer membrane are enzymes involved in galactolipid metabolism.

The **inner envelope membrane** is freely permeable to small uncharged molecules, including O_2 and NH_3, and to low–molecular mass, undissociated monocarboxylic acids. Most metabolites, however, cross the inner

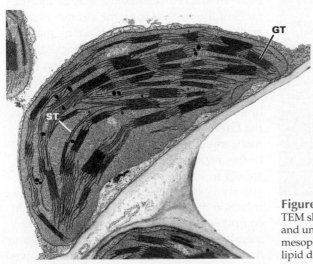

Figure 1.51
TEM showing a chloroplast with stacked grana (GT) and unstacked stroma (ST) thylakoids in a maize leaf mesophyll cell. The dark granules are plastoglobuli, lipid droplets that stain strongly with osmium.

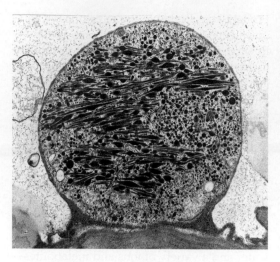

Figure 1.52
TEM of a chromoplast surrounded by a very thin layer of cytoplasm in the wall of a ripe fruit of Jerusalem cherry *(Solanum pseudocapsicum)*. The dense bodies in the plastid contain the carotenoids.

membrane with the aid of specific transporters. The inner envelope membrane also contains enzymes involved in the assembly of thylakoid membrane lipids.

1.10.5 The photosynthetic grana and stroma thylakoid membranes form a physically continuous three-dimensional network.

As illuminated proplastids differentiate into chloroplasts, a highly complex membrane system develops in the stroma. These thylakoid membranes contain the chlorophylls,

carotenoids, proteins, and associated carriers that capture sunlight and store it in the chemical bonds of NADPH and ATP (see Chapter 12).

Within each chloroplast, the thylakoids form a continuous network enclosing a single anastomosing chamber, the thylakoid lumen. The thylakoid network contains two distinct types of membrane domains, the stacked **grana** (singular **granum**) thylakoids and the unstacked **stroma thylakoids,** which interconnect the grana stacks through the chloroplast stroma (Fig. 1.53). In young developing chloroplasts, the initial thylakoids are unstacked and highly perforated. During chloroplast development, stacked grana arise from tongue-like outgrowths in the margins of the perforations of the originating thylakoids. As each outgrowth expands into a flattened sac, it contributes to the stacked grana membrane domain. As this process repeats, the grana stacks increase in height and differentiate further into intricate three-dimensional thylakoid networks (Figs. 1.54 and 1.55). The number of thylakoids in the grana of a mature chloroplast may vary from several to more than 40, depending on the plant species, light conditions, and so forth. For example, plants grown in the shade generally have not only more numerous grana but also thicker grana, that is, more thylakoids per granum, than individuals of the same species growing in bright light.

Membranes of unstacked and stacked thylakoids differ in important ways from one another. For example, photosystem I and ATP synthase complexes reside exclusively in the stroma thylakoids and the unappressed regions of stacked thylakoids, that is, the surfaces of the stacks facing the stroma. Photosystem II, on the other hand, occurs mainly in the appressed regions of the stacked thylakoids (see Fig. 12.1). This separation of components between appressed membranes of the stacked regions and unappressed membranes of the unstacked regions is known as the **lateral heterogeneity** of the photosynthetic membrane system (see Chapter 12). In the freeze-fracture electron micrograph shown in Figure 1.56, this lateral heterogeneity is evidenced by the nonrandom distribution of the membrane particles (protein complexes) between the grana and stroma membranes. Thylakoid stacking is ubiquitous in green plants and is believed to be involved

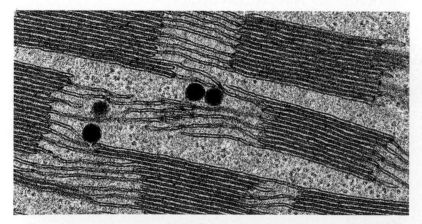

Figure 1.53
High magnification TEM of stacked grana and unstacked stroma thylakoids in a leaf chloroplast of timothy grass. Note that a given membrane can be stacked in one area and not stacked in another.

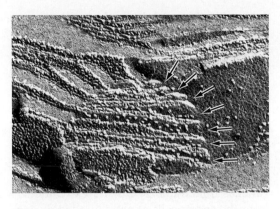

Figure 1.54
TEM depicting a single granum and associated stroma thylakoids of a freeze-fractured pea chloroplast. The arrows point to sites where the two types of thylakoid domains are continuous (compare with the diagram in Fig. 1.55).

Rubisco, the key enzyme in photosynthesis and photorespiration, serves as a good example. Although the small subunit of this enzyme is encoded in nuclear DNA and translated on cytoplasmic ribosomes, the large subunit is encoded in chloroplast DNA and translated on chloroplast ribosomes. Because the enzyme makes up approximately 50% of the total soluble protein in leaves and may even be the most abundant protein in nature, the pattern of its continual breakdown and replacement helps explain the large number of ribosomes present in chloroplasts.

1.10.7 Plastids reproduce by division of existing plastids.

Plastids arise from the binary fission of existing plastids, independently of cell division. Division of proplastids, etioplasts, and young chloroplasts is most common, but even fully developed chloroplasts can divide. First reported in the late 1800s, chloroplast division has since often been observed in algae and young leaves under the light microscope. In root tips, shoots, and other meristems, proplastid division keeps pace with cell division, so the daughter cells possess approximately the same number of plastids as the parent cells. The number of proplastids per cell at this stage is not great, approximately 20. As cell expansion supersedes cell division, the number of plastids per cell increases because of the continued plastid division. As leaf cells expand, for example, the number of chloroplasts per cell may increase severalfold. Synchronous division of plastids appears to be common.

in regulating the distribution of radiant energy between photosystem I and photosystem II during photosynthesis.

1.10.6 Plastids are partially autonomous, encoding and synthesizing some of their own proteins.

Chloroplasts contain double-stranded, circular chromosomes similar in some respects to those of bacteria. These plastids also contain protein-synthesis machinery that accounts for as much as 50% of the total ribosomal complement of photosynthetic cells. Given that most chloroplast polypeptides are synthesized on cytoplasmic ribosomes, clearly the development of chloroplasts requires the integrated activities of both chloroplast and nuclear genomes.

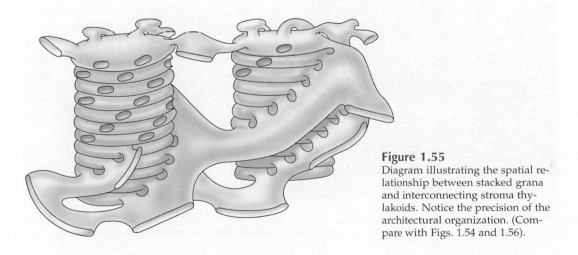

Figure 1.55
Diagram illustrating the spatial relationship between stacked grana and interconnecting stroma thylakoids. Notice the precision of the architectural organization. (Compare with Figs. 1.54 and 1.56).

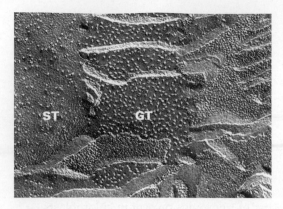

Figure 1.56
Freeze-fracture TEM revealing differences in grana (GT) and stroma (ST) thylakoid composition. The large granules seen on the grana membrane fracture faces in the center correspond to photosystem II complexes.

Division is first made evident by a constriction in the center of the plastid. The constriction deepens and tightens, creating an extremely narrow isthmus before the two daughter plastids separate completely (Fig. 1.57). At the isthmus stage of division, an electron-opaque plastid-dividing ring appears on the cytoplasmic face of the outer envelope membrane. Recent studies have shown that this ring contains a protein that is related to the contractile FtsZ protein of bacteria, which is required for septum formation during cell division.

1.10.8 Plastids are inherited maternally in most flowering plants but paternally in gymnosperms.

Cytoplasmic organelles can be passed from one generation to the next by way of the egg (maternal inheritance), the sperm (paternal inheritance), or both (biparental inheritance). However, genetic evidence indicates that in most angiosperms, the plastids and mitochondria are inherited maternally. These organelles from the male either are excluded from the sperm cells or are degraded during male gametophyte development or double fertilization. Whether from egg or sperm, plastids are in the proplastid stage of development at the time they are inherited.

Biparental inheritance of plastids and mitochondria has been documented both cytologically and genetically in a few flowering plant genera, including *Pelargonium* (geranium), *Plumbago* (leadwort), and *Oenothera* (evening primrose). In *Pelargonium,* male plastids and mitochondria differ ultrastructurally from those of the female, making it possible to establish that organelles from both are present in the developing embryo. *Plumbago,* a model of particular interest, has dimorphic sperm—one of the male gametes is rich in mitochondria, the other in plastids—and exhibits preferential fertilization, the plastid-rich sperm fusing with the

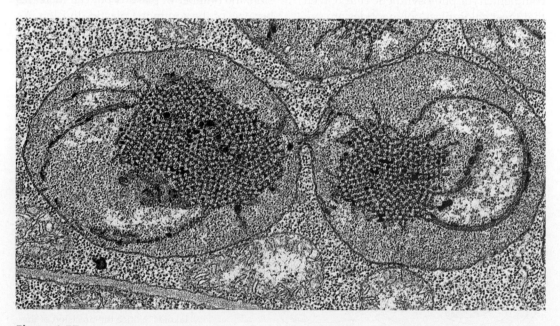

Figure 1.57
TEM of a dividing etioplast in an etiolated leaf of a bean seedling. Note the central constriction.

egg most commonly and the mitochondrion-rich sperm doing so only rarely. Gymnosperms differ strikingly from angiosperms in their mode of plastid inheritance: Plastids are usually passed to the next generation through the sperm.

1.10.9 Plastids synthesize chlorophylls, carotenoids, and fatty acids and reduce some inorganic nutrients.

In addition to their all-important role in energy capture and their function as storage organelles, plastids carry on several other indispensable metabolic processes, including the biosynthesis of chlorophylls, carotenoids, purines, pyrimidines, and fatty acids. Fatty acids are major components of all membranes and are essential participants in many cellular processes. In plants, their de novo biosynthesis takes place exclusively in the stroma of plastids (see Chapter 10). Plastids also reduce the important inorganic ions nitrite (NO_2^-), the product of cytosolic nitrate reduction, and sulfate (SO_4^{2-}) (see Chapter 16).

1.11 Mitochondria

Mitochondria are found in nearly all eukaryotic cells. These essential organelles house the respiratory machinery that generates ATP by way of the citric acid cycle and associated electron transfer chain. Mitochondria also supply various compounds, including organic acids and amino acids, that are used as building blocks in synthetic reactions elsewhere in the cell (see Chapter 14). The basic mitochondrial architecture is the same in both plants and animals and can be immediately recognized in electron micrographs regardless of tissue source. The structure similarities shared by mitochondria from widely divergent eukaryotic organisms suggest the ancient origin of this organelle.

Plant mitochondria in cells fixed for electron microscopy commonly appear spherical to ellipsoid, generally approximately 1 μm thick and 1 to 3 μm long (Fig. 1.58). Viewed in living cells under a light microscope, however, these organelles have a **pleiomorphic** appearance, frequently undergoing rapid changes in shape.

Mitochondria move about the cytoplasm, most likely in association with the actin-myosin system that drives cytoplasmic streaming (see Chapter 5). A plant cell may contain hundreds or thousands of mitochondria, depending on the cell type and the stage of development. The central cells of the maize root cap have approximately 200 mitochondria when young, and 2000 to 3000 when fully enlarged and mature. The ratio of mitochondria per unit volume of cytoplasm, however, probably does not change greatly. In the stem apex of willow-herb *(Epilobium)*, an interphase cell contains approximately 120 mitochondria just before division, whereas newly formed daughter cells have about half this number. In very actively metabolizing nonphotosynthetic plant cells, as much as 20% of the cytoplasmic volume consists of mitochondria (see Chapter 14).

1.11.1 Similarity in the basic architecture of all mitochondria reflects the universality of their mechanism for generating energy.

Like plastids, mitochondria are bounded by two membranes (Fig. 1.59). The **inner membrane,** which is much larger in area than the **outer,** folds into **cristae** that extend deeply into the **mitochondrial matrix.** The aqueous space between the two envelope membranes, called the **intermembrane space,** is continuous

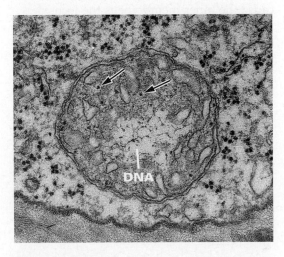

Figure 1.58
Thin-section TEM of a mitochondrion in a bean root tip cell. Note outer and inner membranes, matrix containing ribosomes (arrows), and the fibrous, translucent DNA-containing area.

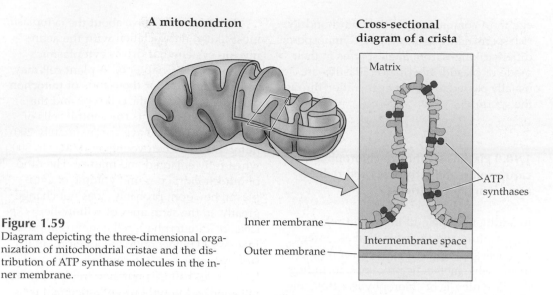

A mitochondrion

Cross-sectional diagram of a crista

Matrix

ATP synthases

Inner membrane

Intermembrane space

Outer membrane

Figure 1.59
Diagram depicting the three-dimensional organization of mitochondrial cristae and the distribution of ATP synthase molecules in the inner membrane.

with and includes the interior region between the two closely opposed membranes of each crista. The outer membrane contains only a small percentage of the total mitochondrial mass and relatively few enzymes. The inner membrane (including cristae) contains 80% to 95% of the total membrane-associated protein and more than 90% of the total mitochondrial lipid, including a small amount of an unusual phospholipid, cardiolipin (diphosphatidylglycerol: Fig. 1.60). The matrix is composed of soluble enzymes, mitochondrial DNA (mtDNA), and ribosomes.

Cristae form flattened sacs, or cisternae, that greatly increase the surface area of the inner membrane. This point is significant because the inner membrane is the locus of multiprotein complexes that include both the respiratory electron transfer chain and ATP synthase. These ATP synthases utilize the proton electrochemical gradient across the inner membrane, which is produced by the electron transfer chain. The chemiosmotic mechanism that drives ATP synthesis is discussed in greater detail in Chapters 3, 12, and 14.

The number of cristae in a cell's mitochondria bears a close relationship to the capacity of the mitochondria to generate ATP. Plant mitochondria, for example, generally have fewer cristae and a lower crista-to-matrix ratio than the mitochondria in animal tissues with high energy demands, such as muscle. However, in specialized plant cells such as transfer cells, where energy demands are great, the mitochondria also possess many cristae (Fig. 1.61).

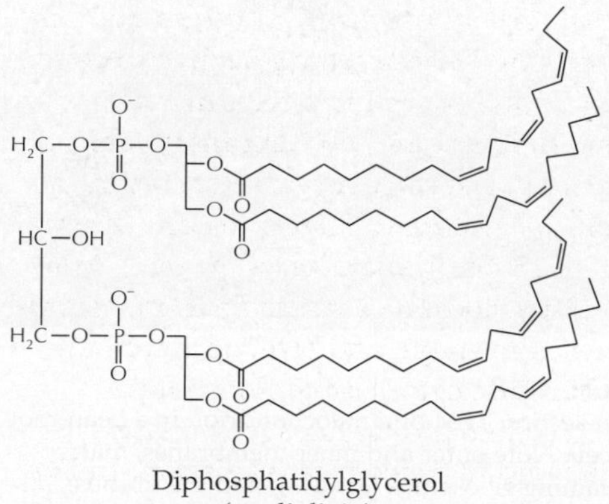

1.11.2 Small solutes cross the outer and inner mitochondrial membranes sequentially, whereas large proteins destined for the matrix cross both membranes simultaneously at sites where the two membranes adhere.

To understand the movement of molecules across the two membranes of a mitochondrion, it is helpful to realize that two different systems contribute to transmembrane transport. Small solutes, metabolites, and ions do not follow the transport pathway used to import larger protein molecules.

The outer mitochondrial membrane is highly permeable to small molecules, whereas the inner membrane has a very low permeability. The transport of small molecules across the outer membrane is mediated by

$H_2C-O-P-O-$

$HC-OH$

$H_2C-O-P-O-$

Figure 1.60
Structure of cardiolipin, an unusual phospholipid component of the inner mitochondrial membrane.

Diphosphatidylglycerol
(cardiolipin)

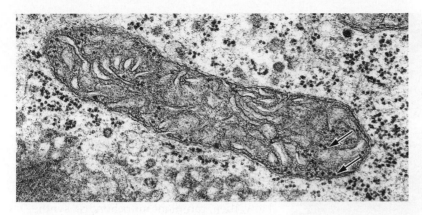

Figure 1.61
TEM depicting a longitudinal section through a transfer cell mitochondrion showing numerous cisternae and mitochondrial ribosomes (arrows), which are smaller than cytoplasmic ribosomes.

protein complexes known as mitochondrial **porins,** which resemble the previously discovered bacterial porins. Each complex creates an aqueous channel through which ions and molecules can move freely. In contrast, solute transport across the inner membrane is controlled by several carrier-type proteins, including, for example, phosphate carriers and exchange transporters such as the ATP/ADP translocator and the α-ketoglutarate transporter (see Chapter 14).

Proteins destined for the mitochondrial matrix must traverse both membranes and the intermembrane space. Some parts of the envelope inner membrane are close to the outer membrane, but other parts (the cristae) are not. In those regions where the inner membrane lies close to the outer, protein import is aided by the existence of numerous membrane adhesion sites. These sites occupy approximately 5% to 10% of the outer membrane surface. At these sites, the membranes are not actually in contact but are closely and uniformly spaced only 4 to 6 nm apart. By stabilizing mitochondrial structure and maintaining a uniform spacing between the membranes, the membrane adhesion sites aid in the formation of translocation contact sites where protein import takes place. Protein targeting to the mitochondrion is described in Chapter 4.

1.11.3 Mitochondria resemble prokaryotes in numerous important properties.

Mitochondria and plastids are now widely thought to have arisen from the "invasion" of early eukaryotic cells by prokaryotes. These endosymbionts became essential cell components during the course of evolution.

Like plastids, mitochondria possess the genetic capacity to make some of their own proteins, but in mitochondria, as in plastids, most of the formerly complete genetic system of the original endosymbiont has been transferred to the nucleus over time in what appears to be an ongoing process. Genes are first translocated from organelle to nucleus, so that they coexist for a time in both, before being gradually lost from the organelle. As a result of this gene loss, mitochondria and plastids now require nuclear gene products for function and replication. As is true of prokaryotes, the DNA of mitochondria is in the form of one or more membrane-associated nucleoids. In prokaryotes, the DNA is aggregated and attached to a specific region of the plasma membrane, but it is not segregated from the other cellular constituents by a membrane. Similarly, in electron micrographs of mitochondria the DNA occupies a relatively clear region of the matrix, in which it is visible as a fine tangle of strands when post-stained with heavy metal salts (see Fig. 1.58).

Mitochondrial (and plastid) ribosomes resemble those of prokaryotes in several important respects, even sharing similarities in RNA and ribosomal protein sequences (see below). The ribosomes of mitochondria, like those of prokaryotes, are sensitive to certain antibiotics that are ineffective against cytoplasmic ribosomes (e.g., chloramphenicol). Mitochondrial ribosomes are visible in the matrix as electron-opaque particles approximately 15 nm in diameter (Fig. 1.61). Mitochondrial ribosomes resemble those of prokaryotes in shape and size, and both are smaller than cytoplasmic ribosomes.

Also like prokaryotes (and plastids), mitochondria reproduce by fission. In Figure 1.62 the narrow neck and presence of

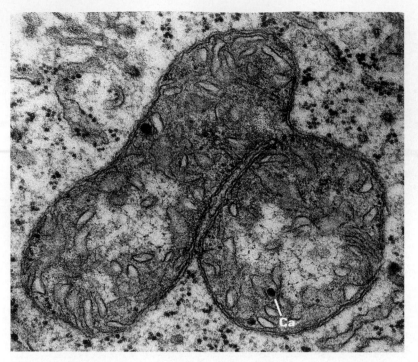

Figure 1.62
TEM of a mitochondrion in a bean root tip cell, showing the final phase of division by constriction. Note the fibrillar DNA in the nucleoid regions in each body. The dark granules (Ca) are calcium phosphate deposits.

are 15,000 to 18,000 nucleotides long, plant mtDNAs range in size from approximately 200,000 nucleotides in some crucifers to a remarkable 2,600,000 nucleotides in muskmelon.

Another notable feature of plant mitochondrial genomes is that they do not contain a complete set of tRNA genes but encode only approximately 16 tRNAs, specific for 12 to 14 amino acids. Twenty tRNAs, at least one for each of 20 common amino acids, are generally necessary for protein synthesis. How, then, can plant mitochondria synthesize any proteins at all? The best evidence to date is that the "missing" tRNAs are encoded by nuclear genes and imported into mitochondria from the cytosol (see Chapter 6).

Still another special feature of plant mitochondria is the presence of some chloroplast DNA sequences in their genomes. These sequences, some of which are tRNA genes, represent transfers from chloroplast DNA and incorporation into the mtDNA during the course of evolution. Most of them are not functional in the mitochondria (see Chapter 6).

Plant mitochondria probably contain several hundred different proteins. Only a very few of these, however—some 15 or 20—are encoded and synthesized in the mitochondria. Most of these few are components of the electron transfer chain or part of the ATPase complex in the mitochondrial inner membrane. All of the remaining mitochondrial proteins must be encoded by nuclear genes, synthesized on ribosomes in the cytosol, and imported into the mitochondria (see Chapter 6).

a nucleoid in both presumptive daughter mitochondria suggest that fission is nearly complete. Note that there is no evidence here of a dividing ring at the isthmus, as there would be in a dividing plastid. A mitochondrion-dividing ring analogous to the dividing rings of plastids has been discovered at the constricted isthmus of mitochondria in dividing cells of *Cyanidium caldarium*, a primitive unicellular red alga, but has not yet been observed in other algae or plants.

1.11.4 Like plastids, mitochondria are semiautonomous and possess the genetic machinery to make some of their own proteins.

Plant mitochondria, like the mitochondria of all eukaryotes, contain DNA genomes that encode ribosomal RNAs (rRNAs), transfer RNAs (tRNAs), and a few proteins. However, the enzymes concerned with DNA replication, RNA transcription and processing, and translation are apparently all encoded in the nucleus. Among the distinctive features of plant mitochondrial genomes are their large size and complexity in comparison with mitochondria of other organisms. For example, whereas mammalian mtDNAs

Summary

Cell membranes are inherited structures that serve as barriers to the diffusion of most water-soluble molecules and enable cells to create compartments in which the chemical composition differs from the surroundings. They are composed of polar lipids that form a bilayer continuum and proteins that are responsible for most membrane functions, including transport, receptor-based signaling, and the generation of ATP. Membrane proteins are classified as integral, peripheral, or lipid-anchored based on the nature of their noncovalent association with the lipid bilayer. Both lipids and proteins can diffuse laterally in the plane of the bilayer.

The plasma membrane serves as the diffusional boundary of individual cells and controls the transport of molecules into and out of cells. Plasma membranes contain receptors that participate in the interactions of cells with their surroundings. The process of endocytosis participates in the recycling of molecules from the plasma membrane.

The endomembrane system comprises membranes that are continuous with the ER (e.g., nuclear envelope) or derived from the ER (e.g., Golgi apparatus, *trans*-Golgi network, plasma membrane, vacuole, transport vesicles). The ER forms a dynamic network that permeates all regions of the cytosol and is differentiated into numerous functional domains. The sheet-like rough ER domains are defined by the presence of bound polysomes that produce membrane and secretory/vacuolar proteins, whereas the tubular smooth ER membranes are mostly involved in the synthesis of lipidic compounds. The Golgi apparatus consists of flattened cisternae organized into stacks. Its principle function is to serve as a carbohydrate factory, adding sugars to glycoproteins and assembling complex polysaccharides such as hemicelluloses and pectins. The *trans*-Golgi network sorts secretory and vacuolar molecules and packages them into separate vesicles. Vacuoles enable plant cells to grow rapidly with expenditure of minimal amounts of energy and proteins. They perform a multitude of functions including, storage, digestion, ionic homeostasis, defense against pathogens and herbivores, and sequestration of toxic compounds.

The nuclear genome-containing nucleus is surrounded by an envelope with pore complexes that regulate trafficking between the nuclear matrix and the cytosol. The envelope breaks down during mitosis and reassembles around the daughter nuclei. Ribosomes are assembled in the nucleolus and are then exported into the cytosol.

Peroxisomes and glyoxysomes are small vesicle-like organelles that contain many enzymes in their lumen. Peroxisomes work together with chloroplasts and mitochondria in the glycolate pathway. In certain tissues, peroxisomes can be transiently converted to glyoxysomes, which participate in the mobilization of fatty acids and in conversion of fixed N_2 to ureides.

The term plastid refers to a family of organelles that are developmentally related to proplastids, reproduce by fission, and are semi-autonomous. The principal types of plastids are the photosynthetic chloroplasts, the starch-storing amyloplasts, the carotenoid-forming and colorful chromoplasts, and the monoterpene-producing leucoplasts. The aqueous stroma contains the plastid's genetic machinery as well as enzymes and variable amounts of internal membranes. The photosynthetic apparatus of chloroplasts is located in thylakoid membranes, which are organized into three-dimensional networks with granal (stacked) and stroma-expose domains.

Mitochondria, like chloroplasts, are semi-autonomous organelles that divide by fission. Mitochondria consist of two membranes and an internal matrix. Their principal function is to generate ATP, which is produced by enzymes in the inner membrane and the matrix.

Further Reading

Assmann, S. M., Haubrick, L. L. (1996) Transport proteins of the plant plasma membrane. *Curr. Opin. Cell Biol.* 8: 458–467.

Battey, N., Carroll, A., Van Kesteren, P., Taylor, A., Brownlee, C. (1996) The measurement of exocytosis in plant cells. *J. Exp. Bot.* 299: 717–728.

Bhatnagar, R. S., Gordon, J. I. (1997) Understanding covalent modifications of proteins by lipids: where cell biology and biophysics mingle. *Trends Cell Biol.* 7: 14–20.

Cheniclet, C., Carde, J.-P. (1985) Presence of leucoplasts in secretory cells and of monoterpenes in the essential oil: a correlative study. *Isr. J. Bot.* 34: 219–238.

Coleman, J.O.D., Blake-Kalff, M.M.A., Davies, T.G.E. (1997) Detoxification of xenobiotics by plants: chemical modification and vacuolar compartmentation. *Trends Plant Sci.* 2: 144–151.

Cowin, P., Burke, B. (1996) Cytoskeleton–membrane interactions. *Curr. Opin. Cell Biol.* 8: 56–65.

de Duve, C. (1996) The birth of complex cells. *Sci. Am.* 274(4): 50–57.

De Mendoza, D., Cronan, J. E. (1983) Thermal regulation of membrane lipid fluidity in bacteria. *Trends Biochem. Sci.* 8: 49–52.

Goldberg, M. W., Allen, T. D. (1995) Structural and functional organization of the nuclear envelope. *Curr. Opin. Cell Biol.* 7: 301–309.

Grabski, S., de Freijter, A. W., Schindler, M. (1993) Endoplasmic reticulum forms a dynamic continuum for lipid diffusion between contiguous soybean root cells. *Plant Cell* 5: 25–38.

Gunning, B.E.S., Steer, M. W. (1996) *Plant Cell Biology: Structure and Function.* Jones and Bartlett, Sudbury, MA.

Hawes, C., Crooks, K., Coleman, J., Satiat-Jeunemaitre, B. (1995) Endocytosis in plants: fact or artifact? *Plant Cell Environ.* 18: 1245–1252.

Hoober, J. K. (1984) *Chloroplasts.* Plenum Press, New York.

Larsson, C., Møller, I. M., Widell, S. (1990) An introduction to the plant plasma membrane: Its molecular composition and organization. In *The Plant Plasma Membrane,* C. Larsson and I. M. Møller, eds. Springer-Verlag, Berlin, pp. 1–15.

Masters, C., Crane, D. (1995) *The Peroxisome: A Vital Organelle.* Cambridge University Press, Cambridge, UK.

Mélèse, T., Xue, Z. (1995) The nucleolus: an organelle formed by the act of building a ribosome. *Curr. Opin. Cell Biol.* 7: 319–324.

Miller, D., Hable, W., Gottwald, J., Ellard-Ivey, M., Demura, T., Lomax, T., Carpita, N. (1997) Connections: The hardwiring of the plant cell for perception, signaling and response. *Plant Cell* 9: 2105–2117.

Neuhaus, J.-M., Rogers, J. C. (1998) Sorting of proteins to vacuoles in plant cells. *Plant Mol. Biol.* 38: 127–144.

Olsen, L., Harada, J. (1995) Peroxisomes and their assembly in higher plants. *Annu. Rev. Plant. Physiol. Plant Mol. Biol.* 46: 123–146.

Panté, N., Aebi, U. (1996) Molecular dissection of the nuclear pore complex. *Crit. Rev. Biochem. Mol. Biol.* 31: 153–199.

Schuler, I., Duportail, G., Glasser, N., Benveniste, P., Hartmann, M. A. (1990) Differential effects of plant sterols on water permeability and on acyl chain ordering of soybean phosphatidylcholine bilayers. *Proc. Natl. Acad. Sci. USA* 88: 6926–6930.

Schultz, C., Gilson, P., Oxley, D., Youl, J., Bacic, A. (1998) GPI-anchors on arabinogalactan proteins: implications for signaling in plants. *Trends Plant Sci.* 3: 426–431.

Singer, S. J., Nicholson G. L. (1972) The fluid mosaic model of the structure of cell membranes. *Science* 175: 720–731.

Staehelin, L. A. (1997) The plant ER: a dynamic organelle composed of a large number of discrete functional domains. *Plant J.* 11: 1151–1165.

Staehelin, L. A., Moore, I. (1995) The plant Golgi apparatus: structure, functional organization and trafficking mechanisms. *Annu. Rev. Plant Physiol. Plant Mol. Biol.* 46: 261–288.

Staehelin, L. A., van der Staay, G.W.M. (1996) Structure, composition, functional organization and dynamic properties of thylakoid membranes. In *Oxygenic Photosynthesis: The Light Reactions,* D. R. Ort and C. F. Yocum, eds. Kluwer Academic Publishers, Dordrecht, The Netherlands, pp. 11–30.

Tobin, A. K., ed. (1992) *Plant Organelles.* Cambridge University Press, Cambridge, UK.

Wink, M. (1993) The plant vacuole: a multifunctional compartment. *J. Exp. Bot.* 44: 231–246.

Yoshida, S., Uemura, M. (1990) Responses of the plasma membrane to cold acclimation and freezing stress. In *The Plant Plasma Membrane,* C. Larsson and I. M. Møller, eds. Springer-Verlag, Berlin, pp. 293–319.

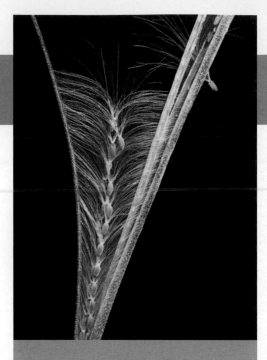

Biochemistry & Molecular Biology of Plants, B. Buchanan, W. Gruissem, R. Jones, Eds.
© 2000, American Society of Plant Physiologists

CHAPTER

2

The Cell Wall

Nicholas Carpita
Maureen McCann

The protoplast is the cell's way of making more wall.

—*Joe Varner*

CHAPTER OUTLINE

Introduction

The shape of a plant cell is dictated largely by its cell wall. When a living plant cell is treated with cell-wall–degrading enzymes to remove the wall, the resulting membrane-bound **protoplast** is invariably spherical (Fig. 2.1). In living cells, the cell wall constrains the rate and direction of cell growth, exerting a profound influence on plant development and morphology. Cell walls contribute to the functional specialization of cell types. Within *Zinnia* leaves, for example, the shape of spongy parenchyma cells maximizes both the volume of the intercellular spaces and the surface area available for gas exchange (Fig. 2.2A), whereas the branched structure of trichomes (Fig. 2.2D) may be adapted for sending mechanical stimuli. In contrast, the papillae-shaped epidermal cells of a snapdragon petal reflect light, attracting the attention of pollinators (Fig. 2.2B). In some cells, including tracheary elements (Fig. 2.2C), the protoplast disintegrates during development, and the mature cell consists entirely of cell wall.

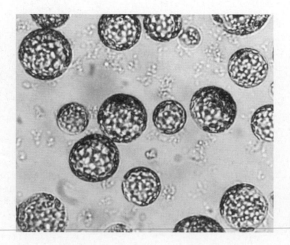

Figure 2.1
Without its wall, a protoplast adopts a spherical form.

(A) **(B)** **(C)** **(D)**

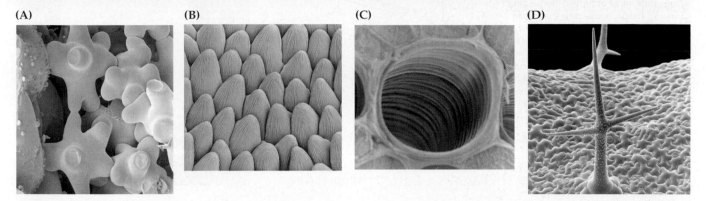

Figure 2.2

A developing cell can change its wall architecture to provide myriad forms. (A) The spongy parenchyma of a *Zinnia* leaf minimizes cell contact and maximizes cell surface for gas exchange. (B) The specialized shapes of these epidermal cells reflect light to enrich the colors of a snapdragon petal. (C) The secondary thickenings of a tracheid prevent collapse of the wall from the tension created by the transpirational pull. (D) An *Arabidopsis* trichome is an exquisitely branched modified epidermal cell.

The plant cell wall is a dynamic compartment that changes throughout the life of the cell. The new **primary cell wall** (Fig. 2.3) is born in the cell plate during cell division (see Chapter 5) and rapidly increases in surface area during cell expansion, in some cases by more than a hundred-fold. The **middle lamella** forms the interface between the primary walls of neighboring cells. Finally, at differentiation, many cells elaborate within the primary wall a **secondary cell wall** (Fig. 2.4), building complex structures uniquely suited to the cell's function.

The plant cell wall is a highly organized composite of many different polysaccharides, proteins, and aromatic substances. Some structural molecules act as fibers, others as a cross-linked matrix, analogous to the glass-fibers-and-plastic matrix in fiberglass. The molecular composition and arrangements of the wall polymers differ among species, among tissues of a single species, among individual cells, and even among regions of the wall around a single protoplast (Fig. 2.5).

Not all specialized functions of cell walls are structural. Some cells walls contain molecules that affect patterns of development and mark a cell's position within the plant. Walls contain signaling molecules that participate in cell–cell and wall–nucleus communication. Fragments of cell wall polysaccharides may elicit the secretion of defense molecules, and the wall may become impregnated with protein and lignin to armor it against invading fungal and bacterial pathogens (Fig. 2.6;

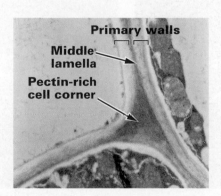

Figure 2.3

The primary walls of cells are capable of expansion. The middle lamella is formed during cell division and grows coordinately during cell expansion. Contact between certain cells is maintained by the middle lamella, and the cell corners are often filled with pectin-rich polysaccharides. In older cells (not shown), the material in the cell corners is sometimes degraded and an air space forms.

(A)

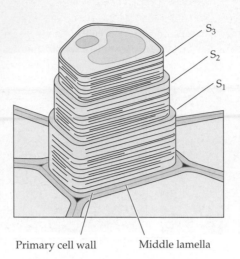

(B)

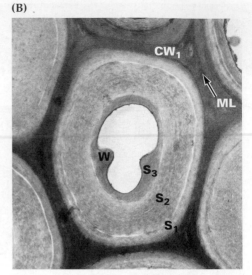

Figure 2.4
When they have achieved their final size and shape, some cells elaborate a multilayered secondary wall within the primary wall. In the diagram (A), the lumen of the cell is sometimes surrounded by several distinct kinds of secondary walls (here, S_1–S_3), with the original primary wall (CW_1) and the middle lamella (ML) constituting the outermost layers. As shown in this example of a fiber cell from a young stem of a locust tree, fibers may also contain "warts" (W), which are a last stage of wall thickening before the protoplast disintegrates (B).

Primary cell wall   Middle lamella

see also Chapters 21 and 24). In other instances, the walls participate in early recognition of symbiotic nitrogen-fixing bacteria (see Chapter 16). Surface molecules on cell walls also allow plants to distinguish their own cells from foreign cells in pollen-style interactions (Fig. 2.7; see also Chapter 19).

2.1 Sugars: building blocks of the cell wall

Polysaccharides, polymers of sugar, are the principal components of the cell wall and form its main structural framework. Polysaccharides are long chains of sugar molecules

(A)

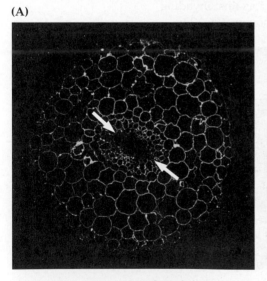

(B)

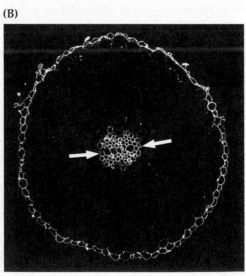

(C)

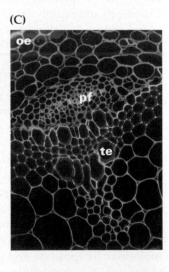

Figure 2.5
The production of complex, chemically diverse molecules within cell walls is developmentally regulated. (A, B) Staining walls of the oat root with two antibodies against arabinogalactan proteins demonstrates that different members of this class of molecules occur in specific tissues. One epitope is present only in cortical cell walls (A), whereas a second epitope is present only in walls of the vascular cylinder (arrows) and epidermis (B). (C) Antibodies against methyl-esterified pectin and unesterified pectins show that these epitopes are enriched not only in different cells but also in different regions of the walls of a single cell. In this cross-section of an *Arabidopsis* stem, domains enriched in methyl-esterified pectin are labeled in green, whereas deesterified pectins, shown in yellow, are found in specialized cells, such as the outer epidermal wall (oe), fiber cells of phloem (pf), some cell corners and cross-walls, and tracheary elements of the xylem (te). The autofluorescence of lignin is shown in red.

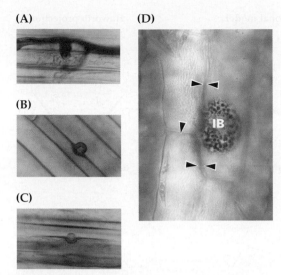

Figure 2.6
Cells often respond to potential pathogens and symbionts through alterations in their cell wall. In response to the attempt of a fungal hypha of *Colletotrichum*, stained with lactophenol-cotton Blue (A), to penetrate a maize cell, the cell produces a wall apposition called a papilla. Largely composed of callose, the papilla also accumulates lignin, as shown by staining with phloroglucinol (B) and by staining with syringaldehyde to detect laccase activity (C). (D) In its response to *Colletotrichum* invasion, an infected cell of sorghum accumulates phytoalexins in inclusion bodies (IB), and the neighboring cells armor their walls with the reddish phenylpropanoids (arrowheads).

configuration (four carbons and an oxygen) are called **furanoses,** whereas those forming six-membered rings (five carbons and an oxygen) are **pyranoses.** The ring conformation a sugar adopts is not defined by the number of carbons: Both pentoses and hexoses can occur in either ring form.

Pyranose and furanose rings can be diagrammed by using either flat **Haworth projections** or conformational models (Fig. 2.8). The pyranose sugars assume the so-called "chair" conformation, whereas the furanose sugars are a "puckered" five-membered ring. The five tetrahedral carbon atoms forming a pyranose ring project the hydrogen and hydroxyl groups either in **equatorial** positions away from the ring or in **axial** positions above and below the ring. Pyranoses adopt one of two possible "chair" forms that will place as many hydroxyl or other bulky groups as possible in the equatorial position. For the common cell wall D-sugars, such as D-glucose, the $^4C_1$ conformation is more favorable, but for the deoxysugars L-fucose and L-rhamnose, the $^1C_4$-chair is the more favorable conformation (Fig. 2.9).

In aldoses, including the hexose glucose and the pentose arabinose, the C-1 is the **anomeric carbon,** the only carbon bound to two oxygen atoms. The other carbons are

covalently linked at various positions, some being decorated with side chains of various lengths. Familiarity with the chemistry and nomenclature of sugars will greatly facilitate understanding of the many biological functions of the cell wall polysaccharides.

Sugars represent a vast spectrum of polyhydroxyl aldehydes **(aldoses)** and ketones **(ketoses)** that can be grouped according to their chemical formula, configuration, and stereochemical conformation. Almost all cell wall sugars are aldoses. Many sugars have the empirical formula $(CH_2O)_n$, from which the term **carbohydrate** is derived.

Numerical prefixes define how many carbons a sugar contains. For example, a triose has three carbon atoms, a pentose five, and a hexose six. All sugar molecules can assume a straight-chain conformation, and those containing four carbons or more can also rearrange into heterocyclic rings (Fig. 2.8). Sugars that adopt a five-membered ring

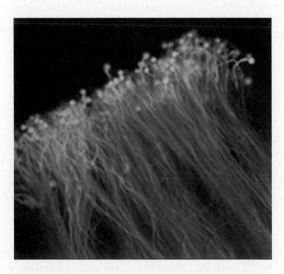

Figure 2.7
In a self-incompatibility response, growth of pollen tubes is terminated coincident with swelling of the pollen tube tips and formation of callose plugs, as observed here by staining with analine blue.

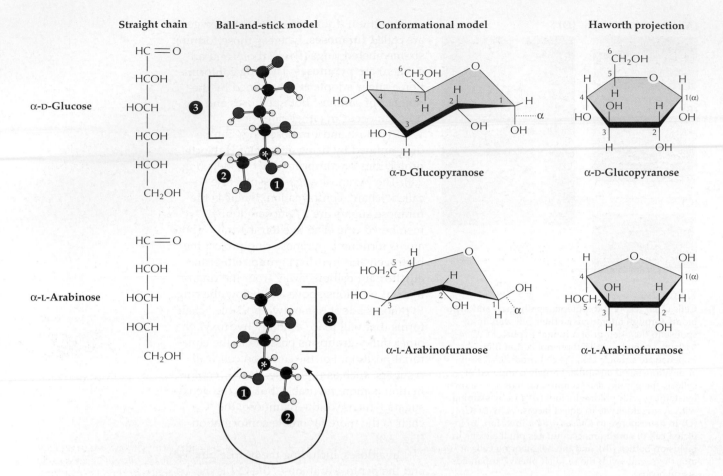

Straight chain	Ball-and-stick model	Conformational model	Haworth projection

α-D-Glucopyranose

α-D-Glucopyranose

α-L-Arabinofuranose

α-L-Arabinofuranose

Figure 2.8

Sugar nomenclature. Both D-glucose and L-arabinose are shown (from left to right) in straight-chain models, ball-and-stick models, conformational models, and Haworth projections. The ball-and-stick model demonstrates the convention whereby the last asymmetric carbon (marked with an asterisk) is oriented with its hydrogen group to the rear. The three asymmetric groups are labeled numerically, 1 being the smallest and 3 the largest. The size of the three asymmetric groups increases clockwise for D-sugars or counterclockwise for L-sugars. The conformational model distinguishes the relative axial and equatorial positions of the hydroxyl groups around the ring structure of pyranoses. α-D-Glucose is the most stable of the hexoses because every hydroxyl group of the ring and the C-6 primary alcohol group are in the equatorial position, which is energetically more favorable than other orientations. By convention, the α configuration of L-arabinofuranose is in the "up" equatorial position.

numbered sequentially around the ring. The hydroxyl group of the anomeric carbon can be in (α) or (β) position. In solution, the hydroxyl group of the anomeric carbon will **mutarotate,** flip-flopping between the α and β configurations as the ring spontaneously opens and closes (see Chapter 13, Fig. 13.11). However, the hydroxyl becomes locked into a specific configuration when the anomeric carbon becomes linked to another molecule.

The designation D or L should always accompany reference to an α or β configuration. The D or L designation refers to the position of the hydroxyl group on the asymmetric **(chiral)** carbon farthest from the C-1 (i.e., the C-5 of hexoses and the C-4 of pentoses). An asymmetric carbon is one in which all four substituent groups are different, so that **enantiomers** (mirror images) of the structures cannot be superimposed. The D-configuration means that when viewed in a **ball and stick model,** the three larger groups of the last asymmetric carbon increase in size clockwise, whereas in the L-configuration they increase in size counterclockwise (see Fig. 2.8). Note that D and L do not correlate to the (+) or (−) designations for **dextrorotatory** or **levorotatory** optical rotation, which define as clockwise or counterclockwise, respectively, the direction in which a chemical species rotates plane-polarized light.

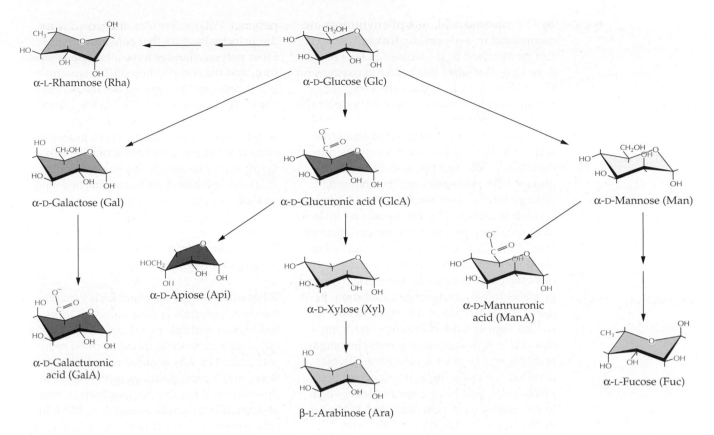

Figure 2.9

The common sugars of plant cell walls and their interconversion. The modifications needed to convert D-Glc into the other sugars are shown in red. L-Rha and L-Fuc adopt an alternative ring form to accommodate as many of the bulkier side groups as possible in the equatorial position. By convention, the "up" axial position is now the α configuration. The abbreviations of each of these sugars are specified by IUPAC-IUBMB convention and should be used exclusively. Note that L-Ara can assume both the α-furanose conformation shown in Fig. 2.8 and the β-pyranose conformation shown here.

2.1.1 The monosaccharides in cell wall polymers are derived from glucose.

D-**Mannose** (Man) and D-**galactose** (Gal) are **epimers** of D-glucose (Glc) made by converting the C-2 and C-4 hydroxyl groups, respectively, from the equatorial to the axial positions (Fig. 2.9). The C-6 primary alcohols of these three sugars can be oxidized to a carboxylic acid group to form D-**mannuronic acid** (ManA), D-**galacturonic acid** (GalA), or D-**glucuronic acid** (GlcA). Enzymatic removal of the carboxyl group from D-glucuronic acid forms the pentapyranose D-**xylose** (Xyl), a sugar in which all of the carbons are part of the heterocyclic ring. A rare branched sugar, D-**apiose** (Api), also is formed from D-GlcA. The C-4 epimer of D-xylose is L-**arabinose** (Ara). The D to L conversion occurs because, in this instance, the epimerization occurs at C-4, the last asymmetric carbon.

Carbon 6 of some hexapyranoses also can be dehydrated to methyl groups, creating deoxysugars. In plants, the two major cell wall deoxysugars are 6-deoxy-L-mannose, called L-**rhamnose** (Rha), and 6-deoxy-L-galactose, called L-**fucose** (Fuc). This spontaneously changes the conformation of L-Rha and L-Fuc from the $^4C_1$-chair conformation to the more favorable $^1C_4$-chair conformation (Fig. 2.9).

2.1.2 Polymers of specific sugars are further defined by their linkage type and the configuration of the anomeric carbon.

Sugars in polymers are always locked in pyranose or furanose rings. During sugar polymerization, the anomeric carbon of one sugar molecule is joined to the hydroxyl group of another sugar, a sugar alcohol, a hydroxylamino acid, or a **phenylpropanoid** compound in a **glycosidic linkage.** A sugar can be attached to D-glucose at O-2, O-3, O-4, or O-6, that is, to the hydroxyl oxygens on C-2, C-3, C-4, or C-6. Only the O-5 position is unavailable, because it constitutes part of the ring structure.

A disaccharide can be described with respect to both linkage and anomeric configuration. For example, cellobiose is β-D-glucosyl-(1→4)-D-glucose: The anomeric linkage forms when the C-1 of one D-glucose residue is replaced by the equatorial hydroxyl at the C-4 position of the other D-glucose (Fig. 2.10). Only one D-glucose is locked in the β configuration; the other D-glucose is undesignated because its anomeric hydroxyl group is free to mutarotate in solution. Because the aldehyde of this sugar is able to reduce copper under alkaline conditions, it classically is described as a **reducing sugar,** and this end of even a very long polymer is called the **reducing end.** Branched polysaccharides will have a **nonreducing sugar** at the end of each side chain and at the terminus of the backbone but only a single reducing end.

Note that in cellobiose the anomeric carbon of one glucose is linked to the hydroxyl group farthest away from the anomeric carbon of the other glucose. For the β-linkage to occur with the C-4 equatorial hydroxyl group, the sugars to be linked must be inverted almost 180° relative to each other; iteration of this linkage produces a nearly linear molecule. In contrast, the units of laminaribiose, or β-D-glucosyl-(1→3)-D-glucose (Fig. 2.10), are linked somewhat askew; iteration of this linkage produces a helical polymer. Polysaccharides are named after the principal sugars that constitute them. Most polysaccharides have a backbone structure, and the composition of this structure is indicated by the last sugar in the polymer's name. For example, xyloglucan has a backbone of C-4–linked glucosyl residues to which xylosyl units are attached, glucuronoarabinoxylan has a backbone of C-4–linked xylosyl residues to which glucosyluronic acid and arabinosyl units are attached, and so forth.

2.1.3 Carbohydrate structures offer great functional flexibility.

What makes sugar subunits such versatile building materials is their ability to form linkages at multiple positions. With 11 different sugars commonly found in plant cell walls (see Fig. 2.9), 4 different linkage positions, and 2 configurations with respect to the oxygen atom, the permutations of pentasaccharide structure exceed 5×10^9! Glucose alone can form almost 15,000 different pentameric structures. Moreover, multiple bonding positions make possible the formation of branched polysaccharides, enormously increasing the number of possible structures. Fortunately for cell wall researchers, the structures of some of the major wall components are relatively conserved among species.

The two major tools that carbohydrate chemists use to elucidate the structure of highly complex polysaccharides are methylation analysis (Box 2.1) and nuclear magnetic resonance (NMR) spectroscopy (Box 2.2).

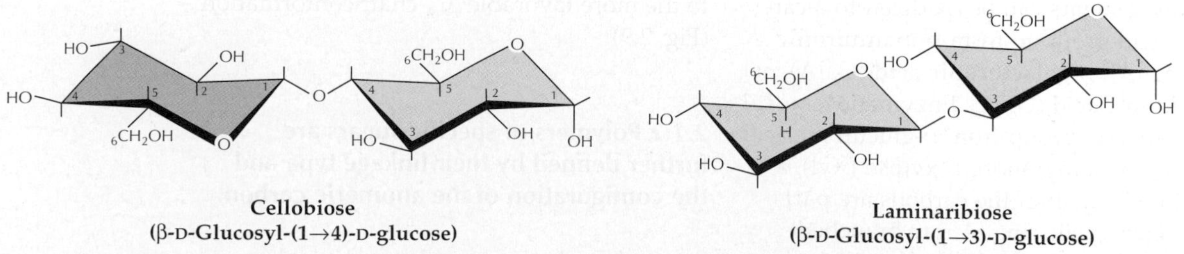

Cellobiose
(β-D-Glucosyl-(1→4)-D-glucose)

Laminaribiose
(β-D-Glucosyl-(1→3)-D-glucose)

Figure 2.10
Linkage structures of cellobiose and laminaribiose. The (1→4)β-D-linkage of cellobiose inverts the glucosyl unit about 180° with respect to each neighbor, whereas the (1→3)β-D-linkage is only slightly askew. (The shading of the glucose units is used to illustrate the 180° inversion).

Strong acids can break the glycosidic linkages of polysaccharides, freeing the individual monosaccharide components. The monosaccharides can then be chemically reduced with borohydride and acetylated with acetic anhydride to make **alditol acetates.** For example, the complex xyloglucan shown in panel A of the figure is hydrolyzed to its four major sugar components. Ketoses (not shown in figure) are reduced to a mixture of C-2 epimers of the equivalent alditols. The alditol acetates derived from different sugars are volatile at different temperatures and can be separated on that basis by **gas–liquid chromatography.** The derivatives are injected onto a thin capillary column coated with a highly polar, waxy, liquid phase to which they adhere. The temperature is then raised to the point at which the interaction with the wax is broken, and the derivatives are swept out of the column by an inert carrier gas to a detector. If a temperature gradient is applied, different sugars will elute at different times (panel B). By comparing these results with the behavior of similarly treated standards, one can determine the molar ratio of each sugar in the original sample. Frequently, the types of sugars present in the sample and their molar ratios will give a good indication of the type of polysaccharide.

However, complex polysaccharides cannot be identified from their derivative content alone. Polymer backbones are built from unit structures of one to a few repeated sugars and linkages but are frequently substituted with appendant sugar units and side chains that may themselves be further branched. Linkage structure of polysaccharides can be deduced indirectly by a method called **methylation analysis.** Polysaccharides are chemically methylated with methyl iodide at every free hydroxyl group. The carbon atoms that participate in glycosidic linkages and in the ring do not have a free hydroxyl group and thus are not methylated. The methylated polysaccharide is hydrolyzed into its components of partly methylated monosaccharides, and then these components are reduced and acetylated as before. However, because linkage analyses of certain derivatives depend on the ability to differentiate the top of the molecule or molecular fragment from its bottom,

(Continued)

(A)

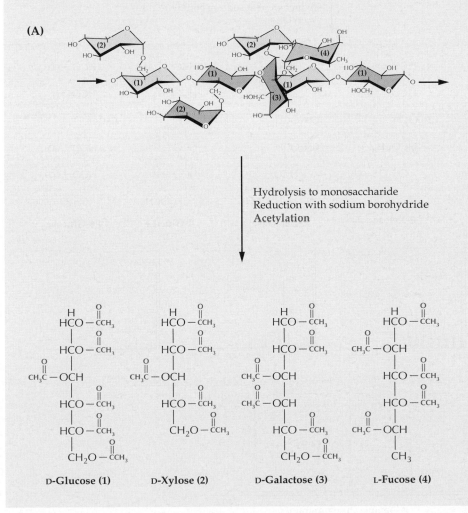

Hydrolysis to monosaccharide
Reduction with sodium borohydride
Acetylation

(B)

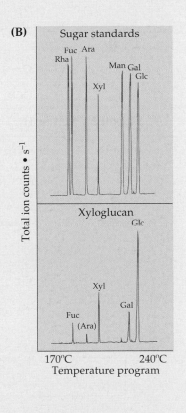

(C)

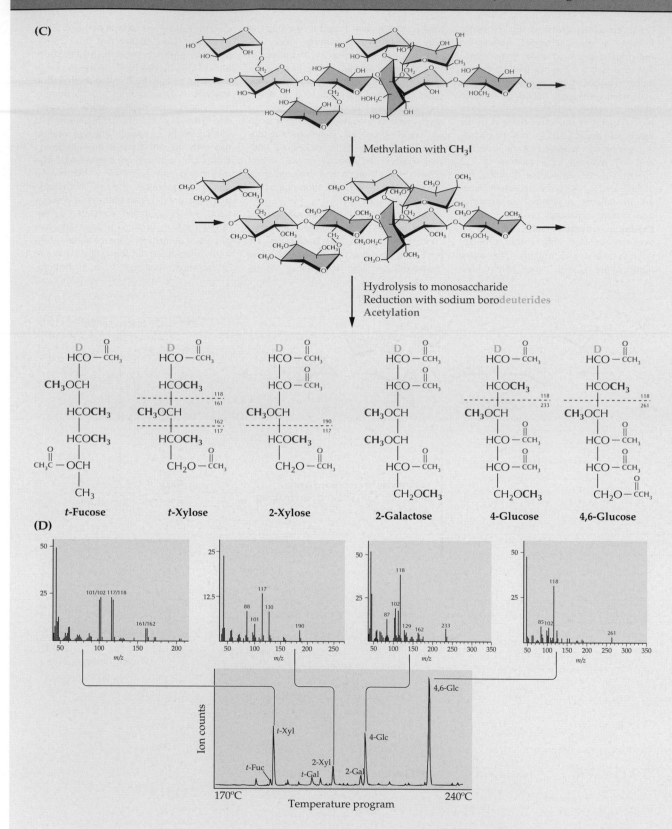

Methylation with **CH₃I**

Hydrolysis to monosaccharide
Reduction with sodium borodeuterides
Acetylation

t-Fucose   *t*-Xylose   2-Xylose   2-Galactose   4-Glucose   4,6-Glucose

(D)

the reducing agent borodeuteride is used to label the C-1 alcohol and thus fix its position in later analyses. These partly methylated alditol acetates are resolved by gas–liquid chromatography. The xyloglucan methylated before hydrolysis yields an array of derivatives (panel C), the nature of which depends on the number and positions of the acetyl groups, which in turn indicate the positions to which another sugar was attached (see arrows). The *t* designation refers to the terminal sugar in a side chain.

Unequivocal determination of the structure is made by **electron-impact mass spectrometry** (EIMS). As the partly methylated alditol acetate derivatives exit the gas chromatograph column, an electron beam breaks each derivative into fragments. The bonds in each derivative have different susceptibilities to breakage, so the molecule breaks into a characteristic set of fragments, which are like the pieces of a jigsaw puzzle. The larger fragments are more diagnostic, but each of these may be broken into still smaller fragments

that give added complexity to the spectrum. The relative abundance of the mass of each fragment, in combination with the spectrum of masses for the fragments from a particular derivative, is diagnostic of its structure. Identification of the one or two extra acetyl groups instead of methyl groups indicates which carbons were participating in glycosidic bonds and ring formation. For example (panel D), a C-4–linked glucose unit in xyloglucan gives characteristic mass/charge ratios (m/z values) of 118 and 233; the latter fragment indicates the acetyl group at the O-4 position, which was protected from methylation by the linkage of a sugar. Likewise, the 4,6-linked glucose branch point contains yet another acetyl group, increasing the signal for this fragment to m/z 261. Note that the presence of the deuterium atom at C-1 is essential for discriminating between 2-Xyl and 4-Xyl, which are symmetrical derivatives. The major fragments from the 2-Xyl derivative are m/z 190 and 117, whereas those from 4-Xyl are m/z 118 and 189; without the

deuterium at the C-1, the results would be m/z 117 and 189 for both derivatives, which would thus be indistinguishable.

EIMS of the partly methylated alditol acetates is the main procedure used by carbohydrate biochemists to establish the linkage structure of unknown polymers. New mass spectrometry (MS) techniques, however, greatly extend the mass range that can be analyzed and can provide linkage and sequence information about underivatized oligosaccharides. One such technique is **matrix-assisted laser desorption ionization–time of flight (MALDI-TOF) MS.** Underivatized polymers are mixed with a material that ionizes them on exposure to brief pulses of a laser. The polymeric ions are differentially accelerated in the mass spectrometer, depending on their size, and molecular masses as great as 200 kDa are detected by a finely tuned calculation of the time interval between laser bombardment and contact with the mass spectrometer anode.

2.2 Macromolecules of the cell wall

2.2.1 Cellulose is the principal scaffolding component of all plant cell walls.

Cellulose is the most abundant plant polysaccharide, accounting for 15% to 30% of the dry mass of all primary cell walls and an even larger percentage of secondary walls. Cellulose exists in the form of **microfibrils,** which are paracrystalline assemblies of several dozen (1→4)β-D-glucan chains hydrogen-bonded to one another along their length (Fig. 2.11). In plants, on average, each microfibril is 36 individual chains thick in cross-section, but microfibrils of algae can form either large, round cables or flattened ribbons of several hundred chains. Microfibrils of angiosperms have been measured to be between 5 and 12 nm wide in the electron microscope. Each (1→4)β-D-glucan chain may be just several thousand units (about 2 to 3 μm long), but individual chains begin and end at different places within the microfibril

to allow a microfibril to reach lengths of hundreds of micrometers and to contain thousands of individual glucan chains. This structure is analogous to a spool of thread that consists of thousands of individual cotton fibers, each about 2 to 3 cm long.

By electron diffraction, the (1→4)β-D-glucan chains of cellulose are arranged parallel to one another; that is, all of the reducing ends of the chains point in the same direction. Recently, bacterial cellulose was shown to be synthesized by the adding of glucose units to the nonreducing ends of glucan chains.

Callose differs from cellulose in consisting of (1→3)β-D-glucan chains, which can form helical duplexes and triplexes. Callose is made by a few cell types at specific stages of wall development, such as in growing pollen tubes and in the cell plates of dividing cells. It is also made in response to wounding (see Section 2.4.3) or to attempted penetration by invading fungal hyphae (see Fig. 2.6).

Box 2.2

Nuclear magnetic resonance spectroscopy can help determine polymer structure.

Nuclear magnetic resonance (NMR) spectroscopy is a nondestructive method that provides many different kinds of information about the polysaccharides that make up the cell wall. Specific NMR experiments have been developed to probe the chemical structure and composition of polysaccharides, their relative orientation, their mobility, and the nature of their mutual interactions.

An NMR spectrometer consists of a powerful magnet and special electronic circuitry that is designed to measure the magnetic properties of the atomic nuclei contained in a sample, such as a purified polysaccharide. Atomic nuclei are charged particles (panel A) that have an intrinsic angular momentum, sometimes called the "nuclear spin." A spinning charge generates a magnetic field, and many common nuclei, such as ^{1}H and ^{13}C, are magnetic. The nucleus most frequently observed by

NMR is the proton (^{1}H), which is the nucleus of virtually all hydrogen atoms. In contrast, the magnetically active carbon isotope ^{13}C represents only 1.1% of all naturally occurring carbon, which makes ^{13}C-NMR much less sensitive than ^{1}H-NMR. Nevertheless, ^{13}C-NMR is sensitive enough to probe the structures of polysaccharides that are available in milligram quantities.

When a magnetic nucleus is placed in the spectrometer's magnetic field, a force is exerted that tends to line up the spin of this nucleus with the applied magnetic field. However, the rules of quantum mechanics do not allow a nucleus to arbitrarily adopt just any orientation. That is, the angular momentum (or spin) of the nucleus is "quantized" (i.e., limited to a few well-defined values). The interplay of the quantized angular momentum and the magnetic force causes the nucleus to "wobble" the way a spinning top does under the influ-

ence of gravity (panel B). This motion, called **precession,** has a characteristic "resonance frequency" that depends on the strength of the magnetic field and the magnetic properties of the nucleus.

NMR experiments observe populations of nuclei, rather than individual nuclei. Under most conditions, the precession of a population of nuclei is not "coherent"; that is, the nuclear spins in the sample are tilted in different directions (panel C, left) so the precession cannot be detected. To be observed by NMR, the nuclei must be made to precess coherently. This is accomplished by subjecting the sample to a brief magnetic pulse, called a radiofrequency (RF) pulse, that oscillates at the nuclear resonance frequency. This pulse redistributes the nuclear spin orientations so that they become partially aligned, and the precessing components of nuclear spin can be observed (panel C, right).

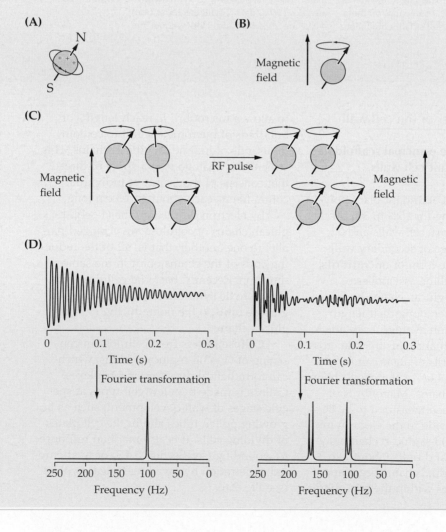

The signal observed as a result of the RF pulse is usually composed of many slightly different resonance frequencies. The RF pulse is thus analogous to ringing a bell to produce a mixture of different audio frequencies. All the frequencies are recorded simultaneously, just as we hear all of the frequencies of a ringing bell simultaneously. A mathematical operation called **Fourier transformation** is used to convert the NMR data into a form (called a frequency-domain spectrum) that makes it easier to distinguish and tabulate signals with different frequencies (panel D).

The different frequencies in the NMR spectrum arise because the electron cloud surrounding each nucleus "shields" it from the applied magnetic field. Neighboring atoms with different degrees of electronegativity surrounding the nucleus being observed produce a change in its resonance frequency called the **chemical shift.** This chemical shift, which is measured in parts per million (ppm) relative to the resonance frequency of a nucleus in a standard compound such as trimethylsilane, provides information about the chemical environment of the nucleus within the molecule. For example, the electron-withdrawing oxygen attached to C6 of glucose decreases the electronic shielding for this nucleus, resulting in a chemical shift of approximately 60 ppm. On the other hand, two electron-withdrawing oxygen atoms are attached to the anomeric carbon (C1) of glucose, which is thus even less shielded than C6 and has a chemical shift of approximately 100 ppm.

Chemical shifts can provide specific structural information, as illustrated in the ^{13}C-NMR spectrum of β-methylcellobioside (panel E). Here, the glycosidic bond between two glucosyl residues causes the chemical shift (80.2 ppm) of C4′ (at the point of attachment) to be substantially greater than the chemical shift (71.0 ppm) of C4. This difference, called a **glycosylation effect,** is often used to establish the specific point at which two sugars are joined by a glycosidic bond.

Nuclei that are connected by three or fewer molecular bonds often exhibit an interaction called **scalar coupling.** This coupling leads to the familiar splitting of signals that is a characteristic feature of ^{1}H-NMR spectra. The magnitude of a scalar coupling interaction depends on molecular geometry. For example, the scalar coupling between two protons that are separated by three bonds depends on a geometric parameter called the dihedral angle, which provides a convenient means of determining whether the anomeric configuration of a sugar is α or β. The anomeric (H1) resonance of a typical sugar is split into a doublet by scalar coupling with H2. For a β-linked glucosyl residue (as in cellobiose), the dihedral angle H1–C1–C2–H2 (i.e., the angle between the plane containing H1, C1, and C2 and the plane containing H2, C2, and C1) is approximately 180°, which results in a relatively large H1–H2 scalar coupling of 8.0 Hz. That is, the H1 resonance of a β-linked glucosyl residue is split into a doublet consisting of two signals separat-

ed by 8.0 Hz. Conversely, the dihedral angle H1–C1–C2–H2 for an α-linked glucosyl residue (as in maltose) is approximately 60°, which results in a relatively small H1–H2 scalar coupling of 3.6 Hz. Thus, the anomeric configuration of a glucosyl residue can be unambiguously determined simply by measuring the distance (in Hertz) between the two components of the anomeric proton doublet.

As one might predict, the ^{1}H- and ^{13}C-NMR spectra of a polysaccharide can be quite complicated, and it is often difficult to determine the complete primary structure of the polymer by this technique. Nevertheless, the structures of oligosaccharides with 10 or more glycosyl residues can be completely characterized by two- and three-dimensional NMR techniques. NMR analysis is not limited to determining the primary structures of biopolymers. For example, ^{13}C-NMR analysis of the products generated when a plant is fed a ^{13}C-labeled substrate can be used to study biosynthetic pathways in vivo. NMR is an important tool for analyzing the conformational and dynamic properties of oligosaccharides in solution. **Solid-state NMR** techniques can distinguish regions of a polysaccharide that differ in their mobility or conformation, thus facilitating the development of models for the assembly of polysaccharides to form complex, dynamic structures within the cell wall. *Will York, Complex Carbohydrate Research Center, University of Georgia, Athens, contributed this section.*

(E)

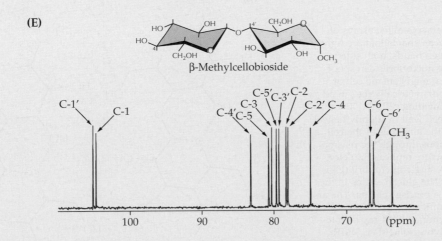

β-Methylcellobioside

2.2.2 Cross-linking glycans interlock the cellulosic scaffold.

Cross-linking glycans are a class of polysaccharides that can hydrogen-bond to cellulose microfibrils: They may coat microfibrils but are also long enough to span the distance between microfibrils and link them together to form a network. Most cross-linking glycans are often called "hemicelluloses," a widely used but archaic term for all materials extracted from the cell wall with molar concentrations of alkali, regardless of structure.

The two major cross-linking glycans of all primary cell walls of flowering plants are **xyloglucans** (XyGs) and **glucuronoarabino-xylans** (GAXs) (Fig. 2.12). The XyGs cross-link the walls of all dicots and about one-half of the monocots, but in the cell walls of the "commelinoid" line of monocots, which includes bromeliads, palms, gingers, cypresses, and grasses, the major cross-linking glycan is GAX (Fig. 2.13).

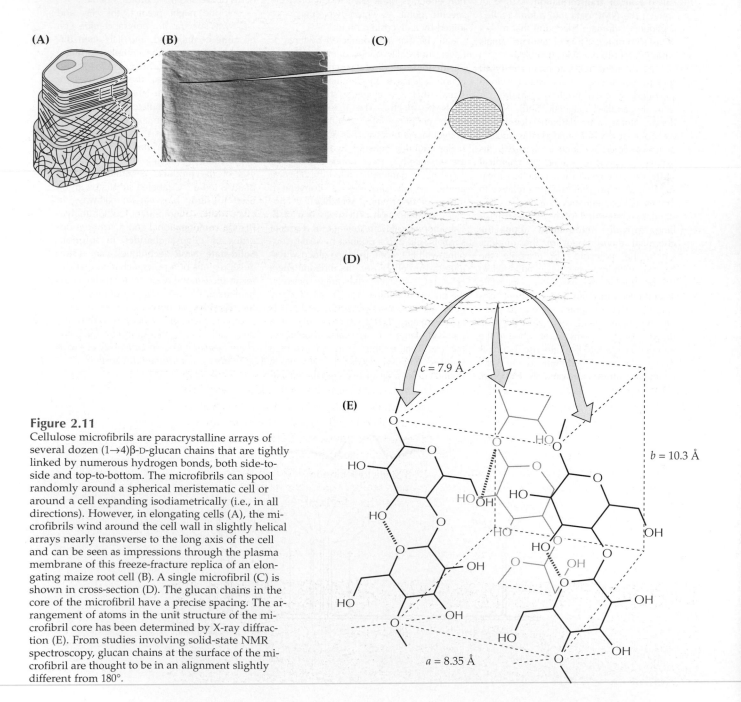

(A) **(B)** **(C)** **(D)** **(E)**

$c = 7.9$ Å
$b = 10.3$ Å
$a = 8.35$ Å

Figure 2.11
Cellulose microfibrils are paracrystalline arrays of several dozen (1→4)β-D-glucan chains that are tightly linked by numerous hydrogen bonds, both side-to-side and top-to-bottom. The microfibrils can spool randomly around a spherical meristematic cell or around a cell expanding isodiametrically (i.e., in all directions). However, in elongating cells (A), the microfibrils wind around the cell wall in slightly helical arrays nearly transverse to the long axis of the cell and can be seen as impressions through the plasma membrane of this freeze-fracture replica of an elongating maize root cell (B). A single microfibril (C) is shown in cross-section (D). The glucan chains in the core of the microfibril have a precise spacing. The arrangement of atoms in the unit structure of the microfibril core has been determined by X-ray diffraction (E). From studies involving solid-state NMR spectroscopy, glucan chains at the surface of the microfibril are thought to be in an alignment slightly different from 180°.

The XyGs consist of linear chains of (1→4)β-D-glucan with numerous α-D-Xyl units linked at regular sites to the O-6 position of the Glc units. Some of the xylosyl units are substituted further with α-L-Ara or β-D-Gal, depending on the species, and sometimes the Gal is substituted further with α-L-Fuc. Sequence-dependent hydrolases are used to elucidate the fine structure of XyGs by cleaving them at specific sites along the glucan backbone into fragments that are small enough to characterize fully (Box 2.3). A convention has been adopted to describe certain ubiquitous side chains of XyG in which the entire subtending side chain along the glucan is designated by a single letter on the basis of its terminal sugar (Table 2.1).

The XyGs are constructed in block-like unit structures containing 6 to 11 sugars, the proportions of which vary among tissues and species. The XyGs can form three major variants of structure. All of the noncommelinoid monocots and most of the dicots are fucogalacto-XyGs (see Fig. 2.12A). The fundamental structure is composed of nearly equal amounts of XXXG and XXFG, but variations can occur, and α-L-Ara is added at some places along the glucan chain. Solanaceous species and peppermint have arabino-XyGs (see Fig. 2.12B), in which only two of every four glucosyl units contains a xylose unit, and the xylosyl units are substituted with either one or two α-L-Ara units to produce a mixture of AXGG, XAGG, and AAGG subunits. Curiously, an acetyl group replaces the third xylosyl unit in the arabino-XyGs. The commelinoid monocots also contain small amounts of XyG, but these include random additions of xylosyl units and rarely any further subtending sugars.

All angiosperms also contain at least small amounts of GAXs, but their structure may vary considerably with respect to the degree of substitution and position of attachment of α-L-Ara residues. In the commelinoid monocots, where GAXs are the major cross-linking polymers, the Ara units are invariably on the O-3 position (Fig. 2.12C). However, in species where XyG is the major cross-linking glycan, the α-L-Ara units are more commonly found at the O-2 position (Fig. 2.12D). In all GAXs, the α-D-GlcA units are attached to the O-2 position.

In the order Poales, which contains the cereals and grasses, a third major cross-linking glycan, called **"mixed-linkage" (1→3),(1→4)β-D-glucans** (β-glucans), distinguishes these species from the other commelinoid species (Fig. 2.14). These unbranched polymers consist of 90% cellotriose and cellotetraose units in a ratio of about 2:1 and connected by (1→3)β-D-linkages. The cellotriosyl and cellotetraosyl units together make up corkscrew-shaped polymers about 50 residues long that are spaced by oligomers of four or more contiguous (1→4)Glc units.

Other, much less abundant noncellulosic polysaccharides, such as **glucomannans, galactoglucomannans,** and **galactomannans,** potentially interlock the microfibrils in some primary walls (Fig. 2.15). These mannans are found in virtually all angiosperms examined.

2.2.3 Pectin matrix polymers are rich in galacturonic acid.

Pectins—a mixture of heterogeneous, branched, and highly hydrated polysaccharides rich in D-galacturonic acid—have been defined classically as material extracted from the cell wall by Ca^{2+}-chelators such as ammonium oxalate, EDTA, EGTA, or cyclohexane diamine tetraacetate. They are thought to perform many functions: determining wall porosity and providing charged surfaces that modulate wall pH and ion

Table 2.1 Single-letter designators of xyloglucan side groups based on the nonreducing terminal sugar

Single-letter designator	Terminal sugar	Side group on the glucan chain
G	D-Glucose	None
X	D-Xylose	α-D-Xyl-(1→6)-
L	D-Galactose	β-D-Gal-(1→2)-α-D-Xyl-(1→6)-
F	L-Fucose	α-L-Fuc-(1→2)-β-D-Gal-(1→2)-α-D-Xyl-(1→6)-
A	L-Arabinose	α-L-Ara-(1→2)-α-D-Xyl-(1→6)-

(A) (Fucogalacto)Xyloglucans

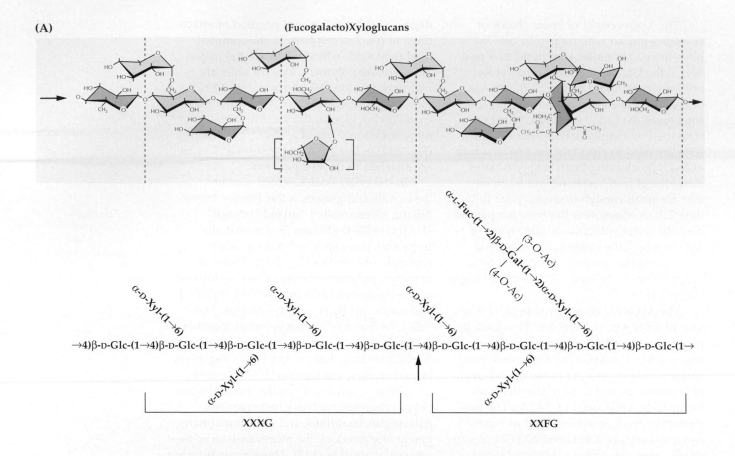

α-L-Fuc-(1→2)β-D-Gal-(1→2)α-D-Xyl-(1→6)

(3-O-Ac)

(4-O-Ac)

α-D-Xyl-(1→6)

α-D-Xyl-(1→6)

α-D-Xyl-(1→6)

α-D-Xyl-(1→6)

→4)β-D-Glc-(1→4)β-D-Glc-(1→4)β-D-Glc-(1→4)β-D-Glc-(1→4)β-D-Glc-(1→4)β-D-Glc-(1→4)β-D-Glc-(1→4)β-D-Glc-(1→

α-D-Xyl-(1→6)

α-D-Xyl-(1→6)

XXXG   **XXFG**

(B) Solanaceous (arabino)xyloglucans

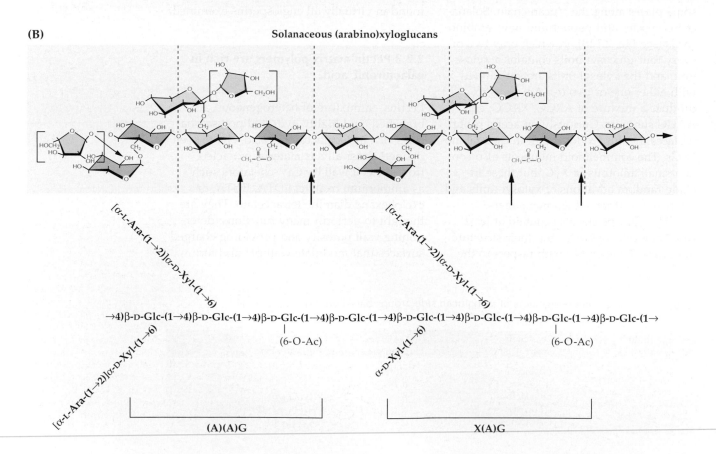

[α-L-Ara-(1→2)][α-D-Xyl-(1→6)]

[α-L-Ara-(1→2)][α-D-Xyl-(1→6)]

[α-L-Ara-(1→2)][α-D-Xyl-(1→6)]

→4)β-D-Glc-(1→4)β-D-Glc-(1→4)β-D-Glc-(1→4)β-D-Glc-(1→4)β-D-Glc-(1→4)β-D-Glc-(1→4)β-D-Glc-(1→4)β-D-Glc-(1→

(6-O-Ac)

(6-O-Ac)

α-D-Xyl-(1→6)

(A)(A)G   **X(A)G**

Commelinoid glucuronoarabinoxylans

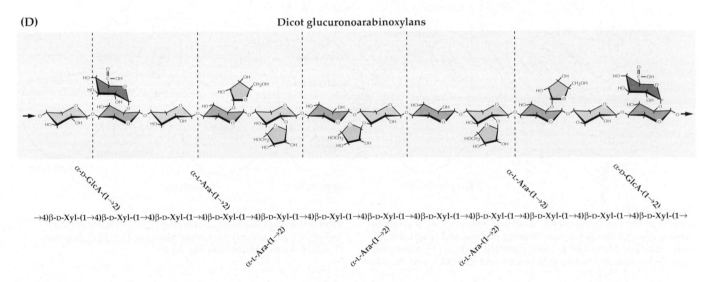

(D)

Dicot glucuronoarabinoxylans

Figure 2.12

The chemical structures of the major cross-linking glycans of the primary walls of flowering plants. The dotted lines delineate the β(1→4) disaccharide units of glucose and xylose inherent in all the cross-linking glycans. Such a linkage requires the sugars to be oriented about 180° from each other. (A) Fucogalactoxyloglucans. In most xyloglucans, the α-D-xylosyl units are added to three contiguous glucosyl units of the backbone to produce a heptasaccharide unit structure. On about one-half of these unit structures, a *t*-α-L-Fuc-(1→2)β-D-Gal- is added to the O-2 of the Xyl side group nearest the reducing end, forming a nonasaccharide unit. Attachment of an α-L-Ara unit to the O-2 position of the backbone glucose unit at a few positions along the backbone blocks hydrogen bonding of the XyG to cellulose at those sites. The arrows denote the only linkages able to be cleaved by the *Trichoderma* endo-β-D-glucanase. According to the single-letter designator convention (Table 2.1), these two oligomers are XXXG and XXFG. (B) Arabinoxyloglucans. In the Solanales and Lamiales, the major repeating unit is a hexamer, rather than a heptamer, with one or two α-L-Ara units added directly to the O-2 position of the Xyl units. The Solanaceae XyG units are separated by two unbranched Glc units rather than one, and the penultimate Glc contains an acetyl group at the O-6 position. The arrows denote the linkages that can be cleaved by the *Trichoderma* endo-β-D-glucanase. In the single-letter designator convention (Table 2.1), these two oligomers are AAG or XAG, if the Ara units are attached; both are XXG if no Ara units are attached. (C) Commelinoid glucuronoarabinoxylans. In the GAX from commelinoid monocot walls, the α-L-Ara units are added strictly to the O-3 position of the xylosyl units of the backbone polymer. Feruloyl groups (and sometimes other hydroxycinnamic acids) are esterified to the O-5 position of the α-L-Ara units and are spaced about every 50 Xyl units of the backbone. The α-D-GlcAs are added to the O-2 position of the xylosyl units. (D) Other glucuronoarabinoxylans. The noncommelinoid monocots and all dicots also contain GAX in addition to the more abundant XyG. However, the α-L-Ara units of these GAXs are attached to the O-2 position as well as to the O-3 position. As with the commelinoid GAX, the α-D-GlcA units are attached only at the O-2 position. XLFG can be a significant decasaccharide in some species.

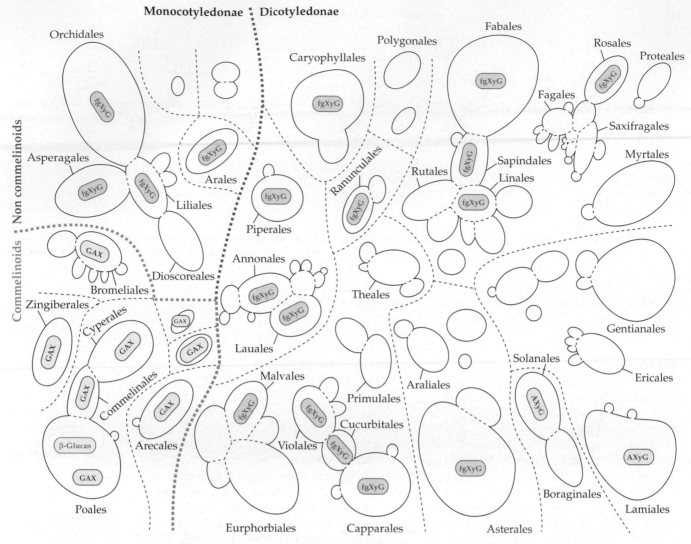

Figure 2.13

Orders of the flowering plants, dicot and monocot, with symbolic descriptions of the major cross-linking glycans and major distinctions illustrated between grasses, commelinoids, and noncommelinoid monocots, and between Solanales and Lamiales and the other dicot orders. fgXyG, fucogalactoxyloglucan; AXyG, arabinoxyloglucan; GAX, glucuronoarabinoxylan (see Fig. 2.12); β-glucan, (1→3),(1→4)β-D-glucan (see Fig. 2.14).

balance; regulating cell–cell adhesion at the middle lamella; and serving as recognition molecules that alert plant cells to the presence of symbiotic organisms, pathogens, and insects. Particular cell wall enzymes may bind to the charged pectin network, constraining their activities to local regions of the wall. By limiting wall porosity, pectins may affect cell growth, regulating the access of wall-loosening enzymes to their glycan substrates (see Section 2.5.7).

Two fundamental constituents of pectins are **homogalacturonan** (HGA; Fig. 2.16A) and **rhamnogalacturonan I** (RG I; Fig. 2.16C).

HGAs are homopolymers of (1→4)α-D-GalA that contain as many as 200 GalA units and are about 100 nm long. There are two kinds of structurally modified HGAs, **xylogalacturonan** (Fig. 2.16B) and **rhamnogalacturonan II** (RG II; Fig. 2.17A). RG II has the richest diversity of sugars and linkage structures known, including apiose, aceric acid (3-C'-carboxy-5-deoxy-L-xylose), 2-O-methyl fucose, 2-O-methyl xylose, Kdo (3-deoxy-D-*manno*-2-octulosonic acid), and Dha (3-deoxy-D-*lyxo*-2-heptulosaric acid). Its very highly conserved structure among flowering plants suggests an important

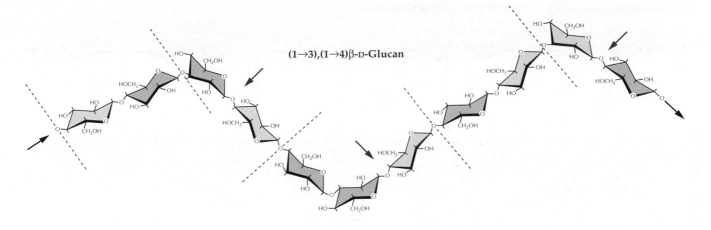

(1→3),(1→4)β-D-Glucan

→4)β-D-Glc-(1→4)β-D-Glc-(1→3)β-D-Glc-(1→4)β-D-Glc-(1→4)β-D-Glc-(1→3)β-D-Glc-(1→4)β-D-Glc-(1→4)β-D-Glc-(1→4)β-D-Glc-(1→3)β-D-Glc-(1→4)β-D-Glc-(1→

Figure 2.14
The mixed-linkage (1→3),(1→4)β-D-glucan unique to the Poales. The red arrows indicate cleavage sites by the *Bacillus subtilis* endoglucanase. The dotted lines demonstrate cellobiose units within the polymer.

function despite its low abundance in cell walls. Dimers of RG II have been found to be cross-linked by two diester bonds per boron atom between apiose units in the complex side groups (Fig. 2.17B).

RG I is a rod-like heteropolymer of repeating (1→2)α-L-Rha-(1→4)α-D-GalA disaccharide units. The best-documented RG I forms are isolated from the cell walls by enzymatic digestion with **polygalacturonase** (PGase), but the length of RG I is unknown because there may be runs of HGA on the ends of the molecule.

Other polysaccharides, composed mostly of neutral sugars—such as **arabinans, galactans,** and highly branched type I **arabinogalactans (AGs)** of various configurations and sizes—are attached to the O-4 of many of the Rha residues of RG I (see Fig. 2.16D). In general, about half of the Rha units of RG I have side chains, but this ratio can vary with cell type and physiological state. There are two types of AG structures. Type I AGs are found associated only with pectins and are composed of (1→4)β-D-galactan chains with mostly *t*-Ara units at the O-3 of the Gal units. Type II AGs constitute a broad group of short (1→3)- and (1→6)β-D-galactan chains connected to each other by (1→3,1→6)-linked branch point residues and are associated with specific proteins, called **arabinogalactan proteins (AGPs)** (see next section).

2.2.4 Structural proteins of the cell wall are encoded by large multigene families.

Although the structural framework of the cell wall is largely carbohydrate, structural proteins may also form networks in the wall. There are four major classes of structural proteins, three of them named for their uniquely enriched amino acid: the **hydroxyproline-rich glycoproteins** (HRGPs), the **proline-rich proteins** (PRPs), and the **glycine-rich proteins** (GRPs) (Fig. 2.18). All of them are developmentally regulated, their relative amounts varying among tissues (Fig. 2.19) and species. Like other secretory proteins destined for the cell wall, the structural proteins are cotranslationally inserted into the endoplasmic reticulum (ER). Thus, all mRNAs for cell wall proteins encode signal peptides that target the proteins to the secretory pathway (see Chapter 4).

Extensin, encoded by a multigene family, is one of the best-studied HRGPs of plants. Extensin consists of repeating Ser-(Hyp)$_4$ and Tyr-Lys-Tyr sequences that are important for secondary and tertiary structure (see Fig. 2.18). The repeating Hyp units predict a "polyproline II" rod-like molecule.

Another large multigene family codes for PRPs. The conformational structure of PRPs is unknown, but their similarity to extensin suggests that they also may be rod-shaped proteins.

(A) **(Galacto)Glucomannan**

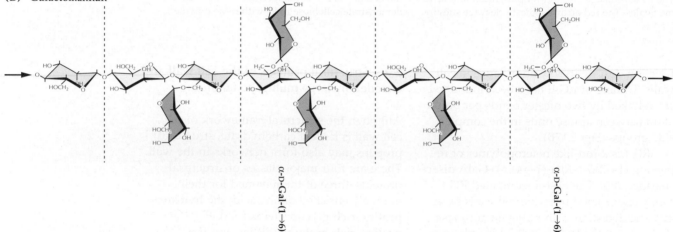

→4)β-ᴅ-Glc-(1→4)β-ᴅ-Glc-(1→4)β-ᴅ-Glc-(1→4)β-ᴅ-Glc-(1→4)β-ᴅ-Man-(1→4)β-ᴅ-Man-(1→4)β-ᴅ-Man-(1→4)β-ᴅ-Man-(1→4)β-ᴅ-Man-(1→

α-ᴅ-Gal-(1→6)

(B) **Galactomannan**

→4)β-ᴅ-Man-(1→4)β-ᴅ-Man-(1→4))β-ᴅ-Man-(1→4)β-ᴅ-Man-(1→4)β-ᴅ-Man-(1→4)β-ᴅ-Man-(1→4)β-ᴅ-Man-(1→4)β-ᴅ-Man-(1→4)β-ᴅ-Man-(1→

α-ᴅ-Gal-(1→6)   α-ᴅ-Gal-(1→6)   α-ᴅ-Gal-(1→6)   α-ᴅ-Gal-(1→6)   α-ᴅ-Gal-(1→6)

(C) **Mannan**

→4)β-ᴅ-Man-(1→4)β-ᴅ-Man-(1→4))β-ᴅ-Man-(1→4)β-ᴅ-Man-(1→4)β-ᴅ-Man-(1→4)β-ᴅ-Man-(1→4)β-ᴅ-Man-(1→4)β-ᴅ-Man-(1→4)β-ᴅ-Man-(1→

Figure 2.15
Cross-linking glycans that contain mannose. The dotted lines mark (1→4)β-disaccharide unit structures (see Fig. 2.12). (A) (Galacto)Glucomannans are roughly equimolar mixtures of (1→4)β-ᴅ-Man and (1→4)β-ᴅ-Glc units, with various amounts of terminal α-ᴅ-Gal units added to the O-6 position of the Man units. (B) Galactoman-nans have backbones composed exclusively of (1→4)β-ᴅ-Man with the α-ᴅ-Gal units added at the O-6 positions. (C) Pure mannans can hydrogen bond into paracrystalline arrays, which are similar in structure to cellulose.

GRPs, some of which contain more than 70% glycine, are predicted to be β-pleated sheets rather than rod-shaped molecules. GRPs are thought to form a plate-like structure at the plasma membrane–cell wall interface. The cell wall face of the pleated sheet contains an arrangement of aromatic amino acids, the function of which is not known (see Fig. 2.18). Like HRGPs, cell wall GRPs are difficult to extract and may become cross-linked into the wall. In one example, bean GRPs are synthesized in xylem parenchyma cells but are targeted and exported to the walls of neighboring cells. GRPs constitute

---

Box 2.3

Polysaccharide sequences can be inferred by analyzing the cleavage products of sequence-specific glycanases.

A direct method for determining the sequence of sugars in a complex carbohydrate is not available. However, much like the restriction endonucleases that recognize and cleave specific sequences of DNA, **sequence-dependent glycanases** that cleave polysaccharides can be used to generate small oligosaccharides, for which the structure can be determined. Glycanases obtained from fungi and bacteria cleave specific glycosidic linkages. For example, the activity of an endo-β-D-glucanase from the fungus *Trichoderma viride* is blocked by appendant groups of the glucan chain; the enzyme can hydrolyze only unsubstituted (1→4)β-D-glucosyl linkages. This feature makes the enzyme useful in determining the frequency of contiguous attachment of xylosyl units onto the (1→4)β-D-glucan chain of xyloglucan.

Sequence-dependent glycanases either require or are restricted by structural features of the polysaccharide. A *Bacillus subtilis* endoglucanase cleaves a (1→4)β-D-glucosyl linkage only if preceded by a (1→3)β-D-linkage, and a *B. subtilis* xylanase cleaves (1→4)β-D-xylosyl linkages only at sites with appendant glucuronic acid units. Several of these enzymes have been used to yield oligomers characteristic of repeating unit structures, which provides a reasonable approximation of the general sequence of very large polymers.

In general, sugars and oligosaccharides are not well resolved by conventional **high-performance liquid chromatography** (HPLC) systems. However, special alkali-resistant HPLC apparatus can support **high-pH anion-exchange HPLC** (HPAE-HPLC), a useful new tool for analyzing the oligosaccharide products of sequence-dependent glycanases. The hydroxyl groups of sugars are weak acids and become negatively charged at high pH. This property may be exploited, given that the relative retention on an anion-exchange column depends on the number, positions, and degrees of freedom of the hydroxyl groups of the oligomer. The oligosaccharides become charged when introduced into a stream of NaOH solution as concentrated as 0.5 M. The anionic oligosaccharides bind to the column and are eluted in a gradient of increasing sodium acetate in NaOH. As the sugars elute, they are detected by an electrochemical cell called a **pulsed amperometric detector** (PAD). A small proportion of the sugars are oxidized as they pass over the gold-plated detector, and when these oxidized sugars bind to the detector plate, the concentration is measured as the relative decrease in the standing potential of the detector. A brief pulse of reversed polarity repels the oxidized sugar from the plate. This pulse-cycle is repeated every 300 ms or so. The measuring pulses are summed to provide a chromatogram that is digitally integrated. The PAD is reasonably specific for sugars, in that few other compounds are oxidized by comparable electrical pulses. Because most of the sample is not oxidized by the PAD, peak fractions can be collected individually, neutralized, and deionized before being subjected to nuclear magnetic resonance, methylation analysis, or other mass spectrometry techniques to determine the linkage structure (see Boxes 2.1 and 2.2).

In the HPAE-HPLC chromatographs shown here, xyloglucans from seed flours of *Tamarindus* and from leaves of *Arabidopsis* and tobacco (a solanaceous plant) are digested to their respective oligosaccharide units by endo-β-D-glucanase and separated by HPAE-HPLC. For *Tamarindus*, peak 7 is XXXG, peaks 8 and 8' are XLXG and XXLG, respectively, and peak 9 is XLLG; for *Arabidopsis*, peak 7 is the same as for *Tamarindus*, peak 9 is XXFG, and peak 10 is XLFG. The single broad peak for tobacco (combined peaks 6 and 7) contains both XAG and AAG (the fourth glucosyl unit at the reducing end is hydrolyzed from the oligomer by the endo-β-D-glucanase).

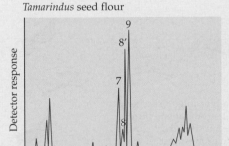

Tamarindus seed flour

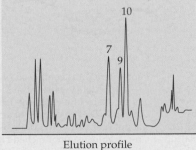

Arabidopsis leaf

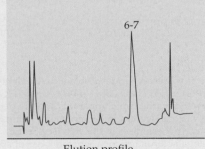

Tobacco cells

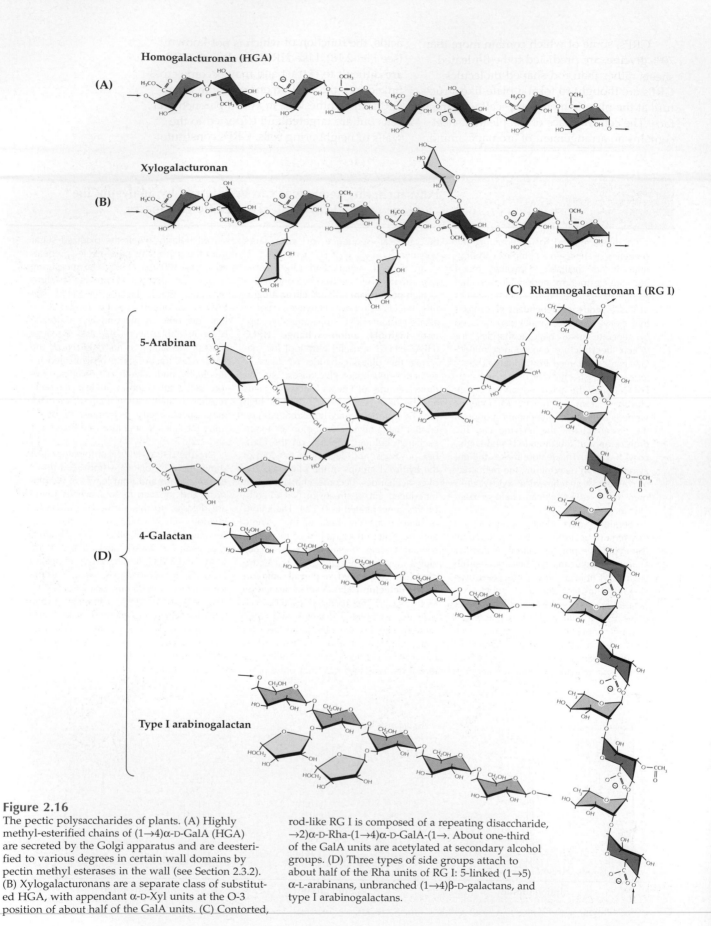

Figure 2.16

The pectic polysaccharides of plants. (A) Highly methyl-esterified chains of (1→4)α-D-GalA (HGA) are secreted by the Golgi apparatus and are deesterified to various degrees in certain wall domains by pectin methyl esterases in the wall (see Section 2.3.2). (B) Xylogalacturonans are a separate class of substituted HGA, with appendant α-D-Xyl units at the O-3 position of about half of the GalA units. (C) Contorted, rod-like RG I is composed of a repeating disaccharide, →2)α-D-Rha-(1→4)α-D-GalA-(1→. About one-third of the GalA units are acetylated at secondary alcohol groups. (D) Three types of side groups attach to about half of the Rha units of RG I: 5-linked (1→5) α-L-arabinans, unbranched (1→4)β-D-galactans, and type I arabinogalactans.

(A) RG II monomer

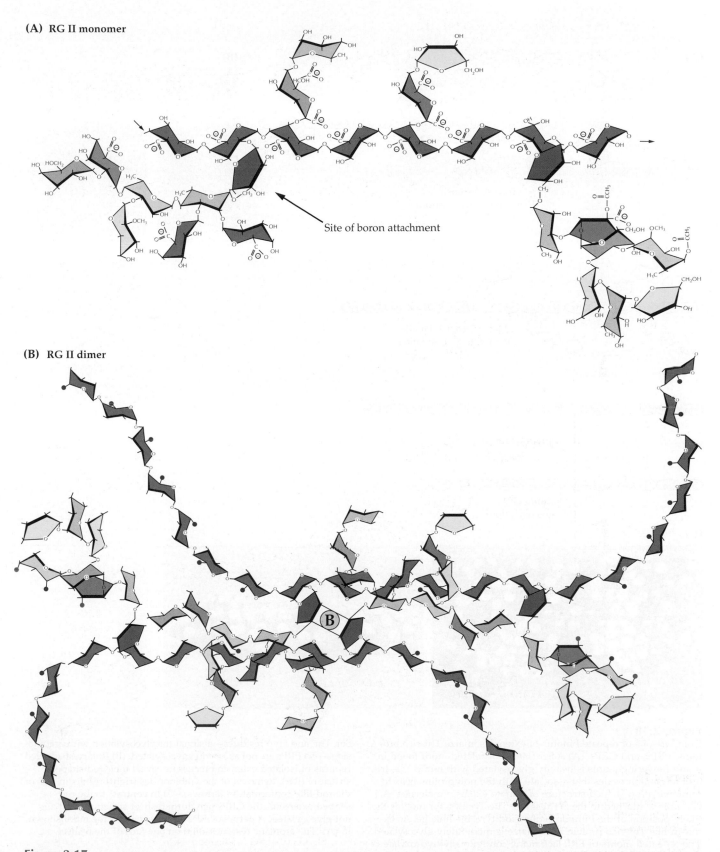

Site of boron attachment

(B) RG II dimer

B

Figure 2.17
(A) RG II is a complex HGA with four distinct side groups containing several different kinds of sugar linkages. (B) RG II monomers of about 4200 kDa can dimerize as boron di-diesters of the apiose residues. The red dots indicate methylation, blue dots are acetylation.

(A)

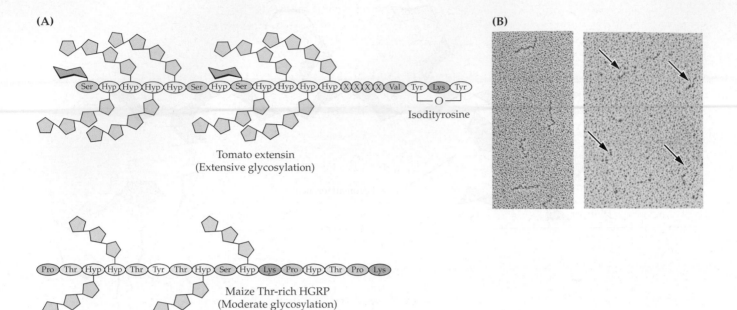

Tomato extensin
(Extensive glycosylation)

Isodityrosine

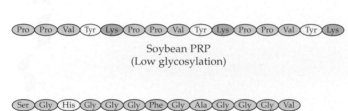

Maize Thr-rich HGRP
(Moderate glycosylation)

Pro Pro Val Tyr Lys Pro Pro Val Tyr Lys Pro Pro Val Tyr Lys

Soybean PRP
(Low glycosylation)

Ser Gly His Gly Gly Gly Phe Gly Ala Gly Gly Gly Val

Petunia GRP
(No glycosylation)

(C)

Figure 2.18

Comparisons of repeated motifs of extensins, maize Thr-rich proteins, PRPs, and GRPs. (A) A Ser-(Hyp)₄ or related motif found in many flowering plants is heavily glycosylated with mono-, di-, tri-, and tetraarabinosides that associate with the polyproline helix to reinforce a rod-shaped structure of tomato extensin molecules. A Gal unit is attached to the Ser residue. The Tyr-Lys-Tyr motif is the likely position of the intramolecular isodityrosine linkage. An extensin-like Thr-rich protein from maize is moderately glycosylated. The repeated motifs of PRP lack many contiguously hydroxylated Ser, Thr, and Hyp residues—a signal for glycosylation with arabinosides—so PRPs are not as heavily glycosylated. (B) Rotary-shadowed replicas of isolated extensin precursors reveal their rod-shaped structure (left). Removal of the arabinosides (right) results in loss of the rod-like conformation (arrows). (C) In contrast to the rod-shaped extensins, the GRPs may form β-sheet structures and are not glycosylated. A petunia GRP has 14 repeats of the motif shown in (A). The aromatic residues align on one face of the β-sheet.

a diverse group of glycoproteins that may function as structural elements inside the cell as well as in the cell wall, and genes for GRPs have now been found in many species.

The fourth major class of structural proteins, AGPs (Fig. 2.20), are more aptly named proteoglycans because they can be more than 95% carbohydrate. AGPs constitute a broad class of molecules located in Golgi-derived vesicles, the plasma membrane, and the cell wall. The site of glycosylation of the AGPs remains unknown but is likely to occur in the Golgi apparatus because it involves the attachment of large, highly branched galactan chains and subsequent decoration with Ara units. Characterization of the polysaccharide contents of the Golgi apparatus and secretory vesicles—including their glycosylated proteins—shows that a majority of the material present is AGP. Another characteristic of all AGPs is their ability to bind the Yariv reagent, a β-D-Glc derivative of phloroglucinol (Fig. 2.20).

Only recently have the genes encoding a few of the AGPs of flowering plants been cloned and the proteins' amino acid sequences deduced. Like GRPs, the AGPs are a diverse family of proteins, many of which are unrelated except for the glycan structures. The few proteins that have been identified can be characterized as enriched in Pro(Hyp), Ala, and Ser/Thr. They possess no distinguishing common motifs but do contain domains with similarity to some PRPs, extensins, and the solanaceous lectins (Fig. 2.20). No clear-cut function has been described for AGPs, or indeed for any of the structural proteins. Some proteins may have more subtle architectural roles (e.g., as nucleation sites for wall assembly) or may directly bind polymers together like the clamps that interlock scaffolding poles.

2.2.5 Aromatic substances are present in the nonlignified walls of commelinoid species.

The primary walls of the commelinoid orders of monocots and the Chenopodiaceae (such as sugar beet and spinach) contain significant amounts of aromatic substances in their nonlignified cell walls—a feature that makes them fluorescent under ultraviolet (UV) light. A large fraction of plant aromatics consists of **hydroxycinnamic acids,** such as ferulic and *p*-coumaric acids (Fig. 2.21; see also Chapter 24). In grasses, these hydroxycinnamates are attached as carboxyl esters to the O-5 position of a few of the Ara units of GAX. A small proportion of the ferulic acid units of neighboring GAXs may cross-link by phenyl–phenyl or phenyl–ether linkages to interconnect the GAX into a large network (Fig. 2.22). In the Chenopodiaceae, ferulic acids are attached to Gal or Ara units on side chains subtending some RG I molecules. Hydroxycinnamic acids are also reduced in the plant to hydroxycinnamoyl alcohols, which form the common precursors for lignin and lignan structures (see Chapter 24).

2.3 Cell wall architecture

2.3.1 The primary wall consists of structural networks.

Buildings come in many styles. So, too, does the architecture of cell walls. Any wall model is generic, and in the following subsections, we establish architectural principles rather than detail a specific cell wall.

The primary cell wall is made up of two, sometimes three, structurally independent but interacting networks. The fundamental framework of cellulose and cross-linking glycans lies embedded in a second network of matrix pectic polysaccharides. The third independent network consists of the structural proteins or a phenylpropanoid network. Evidence for these networks comes partially from direct imaging of walls (Box 2.4). Subsequent sections of the chapter will address two distinct cell wall types, Type I and Type II, which differ in chemical composition and are associated with distinct plant taxa (see Fig. 2.13).

2.3.2 Walls of angiosperms are arranged in two distinct types of architecture.

The walls of most dicots and the noncommelinoid monocots contain about equal amounts of XyGs and cellulose. These kinds of wall we denote as **Type I walls.** XyGs

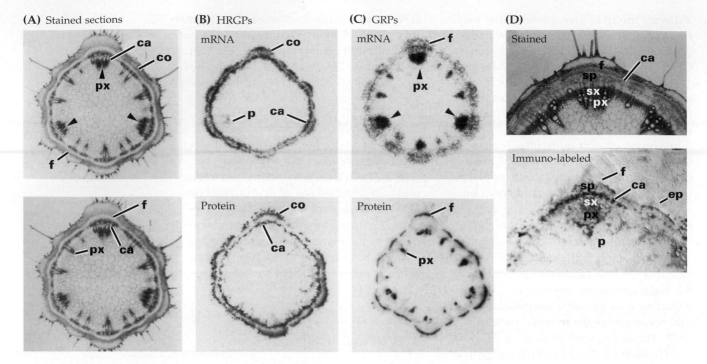

(A) Stained sections   **(B)** HRGPs   **(C)** GRPs   **(D)**

Figure 2.19
When cleanly sliced sections of a plant are pressed firmly for a few seconds to a nitrocellulose sheet, molecules of soluble carbohydrate, protein, and nucleic acids are left behind, imprinted on the nitrocellulose in a nearly cell-specific pattern. Coined **tissue printing** by Joe Varner, this technique has been instrumental in providing biologists with an extremely simple experimental tool of many uses. (A) Stained sections of the elongating second internode of soybean stem are provided for comparison with the other pairs of sections. (B) In situ hybridization with an HRGP cDNA probe and immuno-localization with an HRGP-specific antibody reveal that HRGPs and the mRNA transcripts that encode them are colocalized in cortical (co) and cambial (ca) cells. (C) The same techniques show that GRP mRNA transcripts and their protein products are enriched in the phloem (f) and protoxylem (px). Thus, the GRP proteins and transcripts are found primarily in vascular or lignified cells, whereas the HRGP proteins and transcripts are located in meristematic cells and cells around the cambium. (D) Magnification of the stained section and tissue print shows that HRGP can be immunolocalized to the cellular level. ep, epidermis; p, pith; sp, secondary phloem; sx, secondary xylem.

occur in two distinct locations in the wall. They bind tightly to the exposed faces of glucan chains in the cellulose microfibrils, and they lock the microfibrils into the proper spatial arrangement by spanning the distance between adjacent microfibrils or by linking to other XyGs. With an average length of about 200 nm, XyGs are long enough to span the distance between two microfibrils and bind to both of them. Such spanning XyGs have been observed by electron microscopy (see Box 2.4).

In Type I walls (Fig. 2.23A), the cellulose-XyG framework is embedded in a pectin matrix that controls, among other physiological properties, wall porosity. HGA is thought to be secreted as highly methyl-esterified polymers, and the enzyme **pectin methylesterase** (PME), located in the cell wall, cleaves some of the methyl groups to initiate binding of the carboxylate ions to Ca^{2+}. The helical chains of HGAs can condense by cross-linking with Ca^{2+} to form "junction zones," thereby linking two antiparallel chains (Fig. 2.24A). The strongest junctions occur between two chains of at least seven unesterified GalA units each. If sufficient Ca^{2+} is present, some methyl esters can be tolerated in the junction, and the HGAs can bind in both parallel and antiparallel orientation (Fig. 2.24B). The spacing of the junctions is postulated to create a cell-specific pore size. Rha units of RG I and their side chains interrupt the Ca^{2+} junctions and contribute to the pore definition (Fig. 2.25).

Other properties of the pectin network also regulate porosity. The extent of methyl esterification may remain high in the walls of some cells, and a type of gel may form

(A)

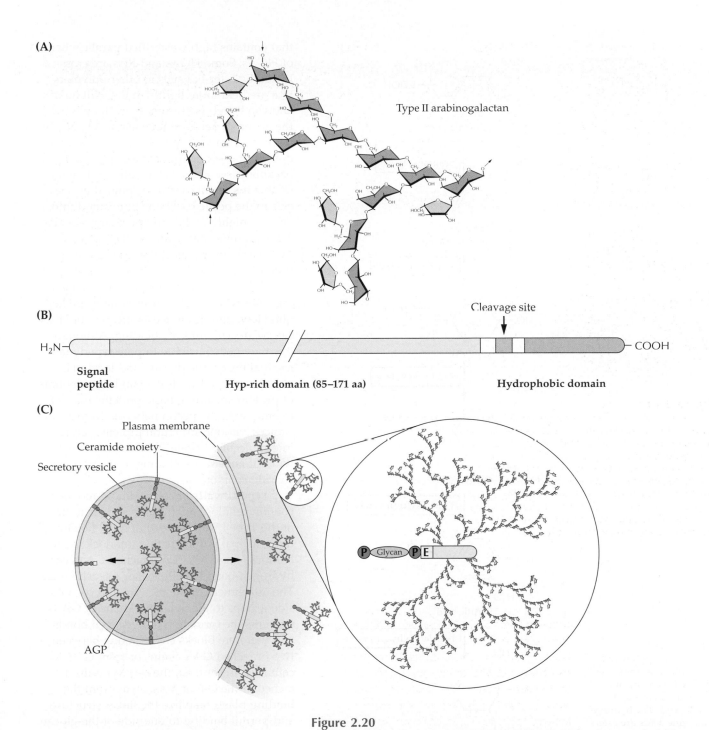

Type II arabinogalactan

(B)

Cleavage site

H₂N— ——COOH

Signal peptide **Hyp-rich domain (85–171 aa)** **Hydrophobic domain**

(C)

Plasma membrane

Ceramide moiety

Secretory vesicle

P Glycan P E

AGP

(D)

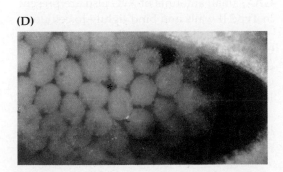

Figure 2.20
(A) The arabinogalactan proteins (AGPs) are proteoglycans, many of which are glycosylated with type II AG structures. (B) The few AGP genes that have been identified encode proteins enriched for Hyp, Ala, and Ser/Thr residues. The domains are thought to contain the AG chains but, as with the GRPs, there are no clear-cut unifying motifs. Some of the AGPs contain domains that are homologous to extensins or PRPs and contain Cys-rich domains. Each of these domains is glycosylated in different ways. (C) Some AGPs, after cleavage of the C terminus, covalently attach to glycosylphosphatidyl inositides (called GPI anchors) during synthesis and secretion in the ER and Golgi apparatus. Note that the GPI anchor and its attached protein are enlarged relative to the AG carbohydrates. Once at the exterior of the cell, the AGP portion may be cleaved from the ceramide moiety of the GPI anchor and serve as a signal molecule. E, ethanolamine. (D) The Yariv reagent, a reddish phloroglucinol derivative with three β-linked galactosyl units, specifically stains AGPs (see asterisk for example). AGPs are particularly enriched in the styles of many flowers, such as this one from *Nicotiana alata*.

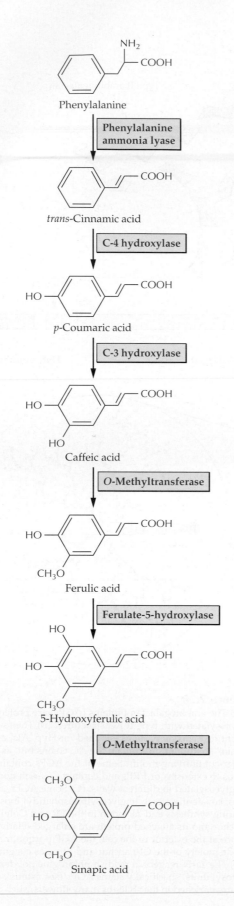

Figure 2.21
Conversion pathway from phenylalanine to the hydroxycinnamic acids of grasses and chenopods. The hydroxycinnamic acids are esterified to the Ara units of Type II arabinoxylans and to Gal and Ara units of some pectic polysaccharides. These hydroxycinnamic acids are the basis for the UV fluorescence observed in nonlignified walls of the commelinoid species. They also may be reduced to their respective cinnamoyl alcohols, which then are precursors of lignin.

that contains highly esterified parallel chains of HGAs. Some HGAs and RGs are cross-linked by ester linkages to other polymers that are held more tightly in the wall matrix and can be released from the wall only by the action of deesterifying agents. Other pectic polymers may separate the sites of borate di-diester cross-linking along the pectic backbone (see Fig. 2.17). Neutral polymers (arabinans or galactans) are pinned at one end to the pectic backbone but extend into, and are highly mobile in, the wall pores (Fig. 2.25). At some stages of cell development, hydrolases are released that trim these neutral polymers, potentially increasing the pore size in the walls. In meristems and elongating cells, where Ca^{2+} concentrations are kept quite low, significant deesterification of HGA can occur without Ca^{2+} binding. Although this may not contribute to pore size dynamics, it alters charge density and local pH.

Some Type I walls contain large amounts of protein, including basic proteins that can interact with the pectin network. In these instances, various structural proteins can form intermolecular bridges with other proteins without necessarily binding to the polysaccharide components.

Type II walls of commelinoid monocots contain cellulose microfibrils similar to those of the Type I wall; instead of XyG, however, the principal polymers that interlock the microfibrils are GAXs. Unbranched GAXs can hydrogen-bond to cellulose or to each other. The attachment of the α-L-Ara and α-L-GlcA side groups to the xylan backbone of GAXs prevents the formation of hydrogen bonds and therefore blocks cross-linking between two branched GAX chains or from GAX to cellulose. In contrast, the α-D-Xyl units attached at the O-6 of XyG, away from the binding plane, stabilize the linear structure and permit binding to one side of the glucan backbone. Despite the predominance of GAX, small amounts of XyG also are present in Type II walls and bind tightly to cellulose.

In general, Type II walls are pectin-poor, but an additional contribution to the charge density of the wall is provided by the α-L-GlcA units on GAX. These walls have very little structural protein compared with dicots and other monocots, but they can accumulate extensive interconnecting networks of phenylpropanoids, particularly as the cells stop expanding (see Fig. 2.23B).

(A)

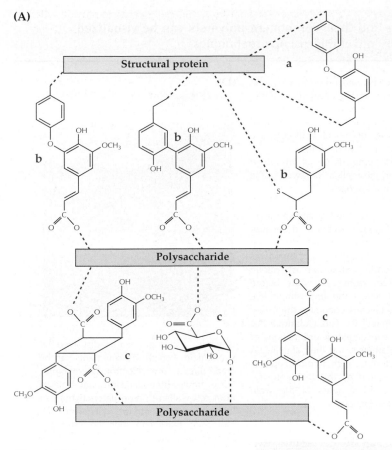

(B)

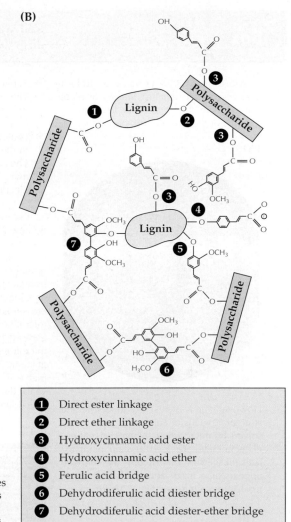

❶	Direct ester linkage
❷	Direct ether linkage
❸	Hydroxycinnamic acid ester
❹	Hydroxycinnamic acid ether
❺	Ferulic acid bridge
❻	Dehydrodiferulic acid diester bridge
❼	Dehydrodiferulic acid diester-ether bridge

Figure 2.22

A representation of the many kinds of lignin interactions in plants. (A) Several kinds of polysaccharide–polysaccharide and polysaccharide–protein cross-bridges are possible, and many of these involve aromatic substances. Isodityrosine forms an intrapeptide linkage needed to stabilize the extensin rods (a). Tyrosine, lysine (not shown), and sulfur-containing amino acids can form ether and aryl linkages with hydroxycinnamic acids esterified to polysaccharides (b). Neighboring polysaccharides may contain cross-bridges esterified directly to sugars (c). (B) A summary of the kinds of aromatic ester and ether cross-links between carbohydrate and lignin.

2.3.3 Polymers remain soluble until they can be cross-linked at the cell surface.

Many polymers are modified by esterification, acetylation, or arabinosylation for solubility during transport. Later, extracellular enzymes deesterify, deacetylate, or dearabinosylate at free sites along the polymers for cross-linking into the cell wall. These sites are determined by the long-range binding order of polysaccharides, permitting their assembly into very precise cell wall architectures. A range of cross-linking possibilities exists, including hydrogen-bonding, ionic bonding with Ca^{2+} ions, covalent ester linkages, ether linkages, and van der Waals interactions. AGPs, because they constitute a majority of the material of secretory vesicles but never accumulate to a comparable extent in the wall, may function as a chaperone-like matrix that prevents premature associations and keeps enzymatic functions in a quiescent state until the secretory materials are assembled in the wall.

Assembly processes occur in an aqueous environment, and one of the major components of cell walls is water. This is of structural importance from the viewpoint of maintaining polymers in their proper conformations. Water is the medium that permits the passage of ions and of signaling molecules through the apoplast; it also

Box 2.4

Cell walls and their component polymers can be visualized directly by novel microscopy techniques.

More than 300 years ago, Robert Hooke recorded the first images of plant cell walls in this *camera lucida* print of sections of the bark of cork oak (panel A).

(A)

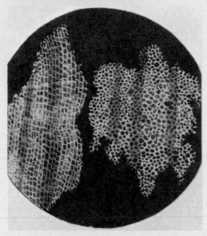

However, many years passed before the structural details of cell walls could be resolved by microscopy. When conventional techniques are used to stain plant tissues for electron microscopy, the cell wall appears as a fuzzy zone with little structural information (see Fig. 2.3B for an example). However, an alternative method of preparing samples, **fast-freeze, deep-etch, rotary-shadowed replica,** allows visualization of cell walls at high resolution and with good preservation of the three-dimensional spatial relationships of the polymers (panel B; scale bar represents 200 nm). This technique requires four steps:

(B)

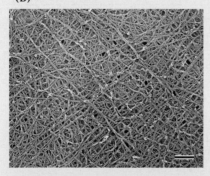

- Wall material is frozen rapidly with liquid nitrogen or helium.

- Surface ice is removed under vacuum.
- The exposed surface is coated with a thin film of platinum and carbon to produce a replica, essentially a three-dimensional contour map of the cell wall polymers in their proper orientation.
- The underlying tissue is dissolved away, and the replica is viewed in the electron microscope.

Because the walls are frozen so quickly, little ice damage occurs, and because no chemical fixatives or dehydrants are used, the normal spacing of components remains. Gentle extraction of pectic polysaccharides from cell walls before freezing can reveal the fine, thread-like cross-linking glycans spanning the larger microfibrils (panel B). This technique has allowed more accurate determinations of microfibril diameters and their spacing. A glancing fracture through an onion epidermal wall prepared by this technique reveals the individual strata of the wall, called lamellae (panel C).

(C)

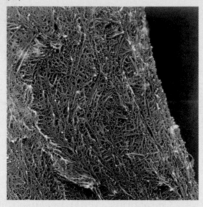

An alternative technique uses acid-etching of material fixed and sectioned for conventional transmission electron microscopy. A glancing section through the thick outer wall of an epidermal cell wall, which was then acid-etched and viewed by conventional transmission electron microscopy, reveals an iterative pattern of microfibril orientation in the individual lamellae (panel D).

Isolated cell wall molecules can also be imaged with the replica technique. The molecules are mixed in glycerol and sprayed onto a freshly cleaved sheet of mica. After the glycerol has been dried under vacuum, a rotary-shadowed replica

(D)

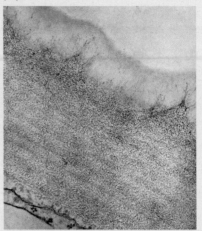

is made, and the mica is dissolved by strong acid. Length measurements can be made directly from electron micrographs. For example, homogalacturonans isolated from cell walls of onion parenchyma can be as long as 400 nm (panel E). Minimum molecular mass can then be estimated from length data.

(E)

Another technique, **atomic force microscopy,** is now being used to image cell walls in even greater detail (panel F). This method is sensitive enough to resolve the spacing of individual glucose molecules in a microfibril.

(F)

(A) Type I wall

(B) Type II wall

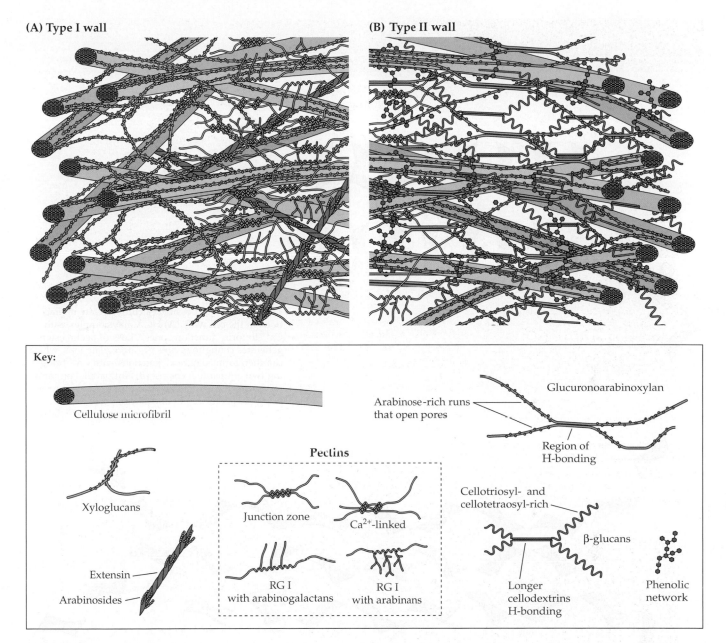

Key:

Cellulose microfibril

Xyloglucans

Extensin

Arabinosides

Pectins

Junction zone

Ca²⁺-linked

RG I
with arabinogalactans

RG I
with arabinans

Arabinose-rich runs
that open pores

Glucuronoarabinoxylan

Region of
H-bonding

Cellotriosyl- and
cellotetraosyl-rich

β-glucans

Longer
cellodextrins
H-bonding

Phenolic
network

Figure 2.23

(A) A three-dimensional molecular model of the Type I wall shows the molecular interactions between cellulose, XyG, pectins, and wall proteins. The framework of cellulose microfibrils and XyG polymers is embedded in a matrix of pectic polysaccharides, HGA, and RG I, the latter being substituted with arabinan, galactan, and arabinogalactan. Because XyGs have only a single face that can hydrogen bond to another glucan chain and can self-associate, we depict several XyGs as woven to interlace the microfibrils. Many other associations are possible, including bridging of two microfibrils by a single XyG. With microfibril diameters of 5 to 10 nm and a spacing of about 20 to 30 nm, a primary wall 80 nm thick can contain only about 5 to 10 strata (only three lamellae of microfibrils are shown for clarity). During growth, cleavage or dissociation of XyGs by enzymes loosens the cellulose–XyG network, allowing microfibrils to separate in the long axis of the page. After growth, extensin molecules may interlock the separated microfibrils to reinforce the architecture. Additional proteins may also be inserted to cross-link extensin, forming a heteropeptide network. Formation of intramolecular covalent bonds among the individual wall proteins may help to terminate cell elongation. (B) A three-dimensional molecular model of the Type II wall shows the molecular interactions between cellulose, GAX, pectins, and aromatic substances. The microfibrils are interlocked by GAXs instead of XyGs. Unlike XyGs, the xylans are substituted with Ara units, which block hydrogen bonding. The xylans are probably synthesized in a highly substituted form, and dearabinosylated in the extracellular space to yield runs of the xylan that can bind on either face to cellulose or to each other. Porosity of the GAX domain could be determined by the extent of removal of the appendant units. Some highly substituted GAX remains intercalated in the small amount of pectins that are also found in the primary wall. Unlike the Type I wall, a substantial portion of the noncellulosic polymers are "wired on" to the microfibrils by alkali-resistant phenolic linkages. Depicted here is the special Type II wall of the grasses (Poales), which synthesize β-glucans during cell enlargement. These corkscrew-shaped molecules contain linear runs of cellodextrins about every 50 glucosyl units or so; these runs are targets for endo-β-D-glucanases, which cleave them.

Figure 2.24
There are several possible calcium–pectate interactions in the cell wall. (A) HGA may complex with Ca^{2+} to form "junction zones." Loss of pectin esters generates contiguous runs of carboxylate ions, which can then form "egg box" junctions with Ca^{2+} bridging two antiparallel chains. (B) Antiparallel or parallel chains of partly methylated HGA can also cross-bridge in less firmly bound calcium complexes.

(B)

provides an environment for enzymes to function. The pH of apoplastic space is thought to be 5.5; whether this pH changes appreciably during growth is not yet resolved. The pore diameter that limits diffusion of molecules through the primary wall is about 4 nm, but some molecules larger than this can pass through the wall to the plasma membrane, perhaps by way of a small number of larger pores or because they are rod-like rather than globular.

2.4 Cell wall biosynthesis and assembly

2.4.1 Cell walls are born in the developing cell plate.

Cell walls originate in the developing cell plate. As plant nuclei complete division during telophase of the mitotic cell cycle, the **phragmosome,** a flattened membranous vesicle containing cell wall components, forms across the cell within a cytoskeletal array

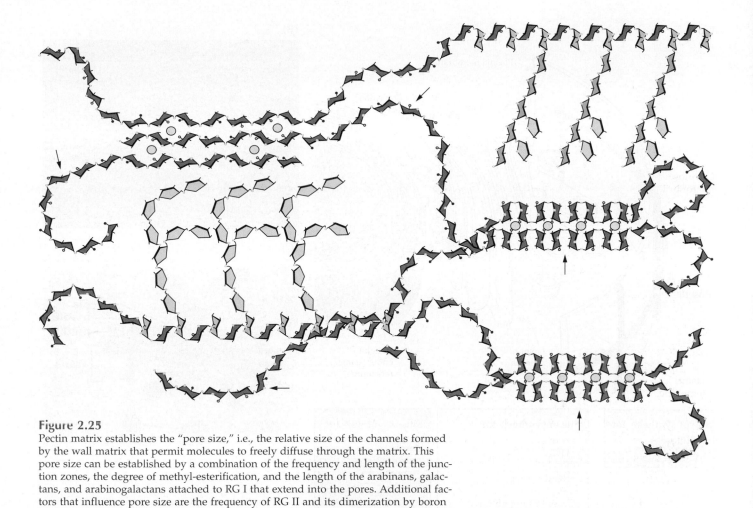

Figure 2.25

Pectin matrix establishes the "pore size," i.e., the relative size of the channels formed by the wall matrix that permit molecules to freely diffuse through the matrix. This pore size can be established by a combination of the frequency and length of the junction zones, the degree of methyl-esterification, and the length of the arabinans, galactans, and arabinogalactans attached to RG I that extend into the pores. Additional factors that influence pore size are the frequency of RG II and its dimerization by boron (not shown). The charge density depends on the extent of deesterification of the HGA to yield carboxyl acids unbound to Ca^{2+}. Arrows denote domains of negative-charge density in HGAs.

called the **phragmoplast.** The noncellulosic cell wall polysaccharides synthesized in the Golgi apparatus and packaged in vesicles fuse with the growing cell plate. The plate grows outward until the edges of the membranous vesicle fuse with the plasma membrane, creating two cells. Finally, the new cell wall fuses with the existing primary wall (see Chapter 5).

The plant Golgi apparatus is a factory for the synthesis, processing, and targeting of glycoproteins (Fig. 2.26). The Golgi apparatus also has been shown by autoradiography to be the site of synthesis of noncellulosic polysaccharides. Thus—except for cellulose—the polysaccharides, the structural proteins, and a broad spectrum of enzymes are coordinately secreted in Golgi-derived vesicles and targeted to the cell wall.

2.4.2 Golgi-localized enzymes interconvert the nucleotide sugars, which serve as substrates for polysaccharide synthesis.

The reactions that synthesize noncellulosic cell wall polysaccharides in the Golgi apparatus utilize several nucleotide sugars as substrates. Beginning with formation of UDP-glucose and GDP-glucose (Fig. 2.27), pathways for **nucleotide sugar interconversion** produce various nucleotide sugars de novo in enzyme-catalyzed reactions (Figs. 2.28 and 2.29). Many of these interconversion enzymes (e.g., epimerases and dehydratases) appear to be membrane-bound and localized to the ER-Golgi apparatus. Guanosine-based nucleotide sugars (Fig. 2.29), such as GDP-Glc and GDP-Man, are used in the synthesis of glucomannan, and GDP-Fuc

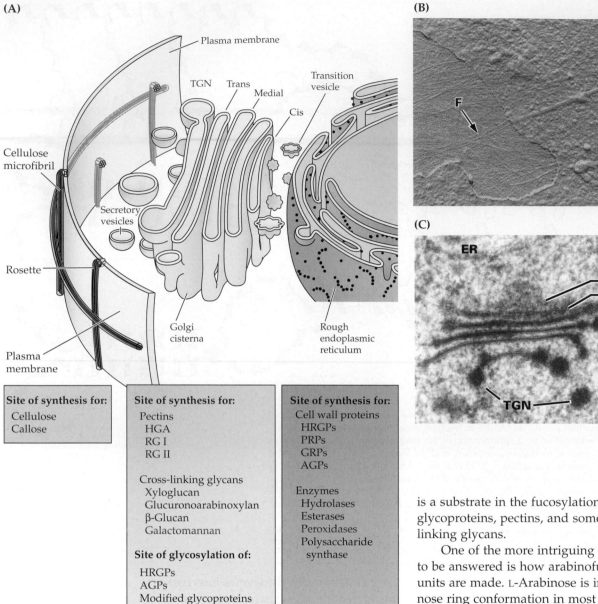

Plasma membrane

TGN Trans Medial Cis

Transition vesicle

Cellulose microfibril

Secretory vesicles

Rosette

Plasma membrane

Golgi cisterna

Rough endoplasmic reticulum

(B)

F

(C)

ER

cis
med(E)
med(L)

trans

TGN

Site of synthesis for:
Cellulose
Callose

Site of synthesis for:
Pectins
  HGA
  RG I
  RG II

Cross-linking glycans
  Xyloglucan
  Glucuronoarabinoxylan
  β-Glucan
  Galactomannan

Site of glycosylation of:
HRGPs
AGPs
Modified glycoproteins

Site of synthesis for:
Cell wall proteins
  HRGPs
  PRPs
  GRPs
  AGPs

Enzymes
  Hydrolases
  Esterases
  Peroxidases
  Polysaccharide
   synthase

Figure 2.26
(A) Biosynthesis of the wall requires a coordination of the synthesis of cellulose microfibrils at the plasma membrane surface, with the synthesis and glycosylation of proteins and wall-modifying enzymes at the rough ER and the synthesis of all noncellulosic polysaccharides at the Golgi apparatus. Material destined for the cell wall is packaged into secretory vesicles, transported to the cell surface, and integrated with the newly synthesized microfibrils. Assembly of the new wall stratum is estimated to begin when no more than 10 glucose residues of a cellulose chain are made. (B) A replica of an E-face (exterior leaf of a fractured membrane bilayer). The numerous vesicles aggregating at the surface resemble small craters. A portion of the membrane has torn away to reveal the underlying microfibrils (F). (C) A cross-section of a single dictyosome body of the Golgi apparatus shows the characteristic development of several membrane sheets, from the cis face nearest the ER to the trans face; these develop into the distinct *trans*-Golgi network (TGN), the principal vesicle-secreting body. E and L, early and late *medial*-Golgi stacks, respectively.

is a substrate in the fucosylation of complex glycoproteins, pectins, and some cross-linking glycans.

One of the more intriguing questions yet to be answered is how arabinofuranosyl units are made. L-Arabinose is in the furanose ring conformation in most plant polymers containing this sugar, including GAX, 5-linked arabinans, AGP, and extensin, whereas UDP-Ara (see Fig. 2.28) is exclusively in the pyranose form. An arabinosyltransferase may differ from other glycosyltransferases in its ability to permit ring rearrangement before the sugar is added to the polymer.

The precise topographic location of the polysaccharide synthase enzymes on or within the Golgi membranes has not been established. For branched polysaccharides, the nucleotide sugar substrate used for synthesis of the backbone may be donated from either the cytosolic or the lumenal side of the Golgi, but the branch units probably are added only from the lumenal side.

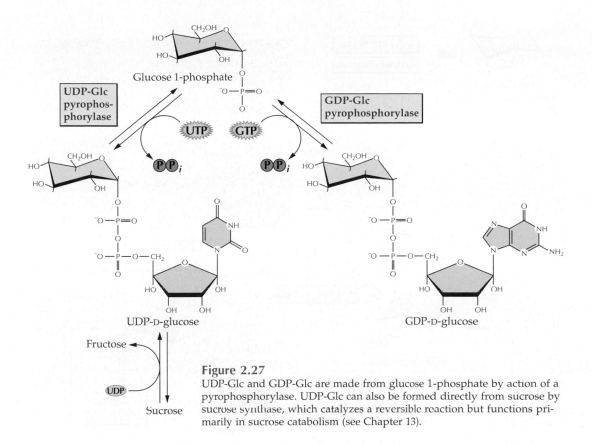

Figure 2.27

UDP-Glc and GDP-Glc are made from glucose 1-phosphate by action of a pyrophosphorylase. UDP-Glc can also be formed directly from sucrose by sucrose synthase, which catalyzes a reversible reaction but functions primarily in sucrose catabolism (see Chapter 13).

Hence, synthesis of complex polysaccharides must be coordinated with transport of some of the nucleotide sugars into the Golgi apparatus.

Formation of nucleotide sugars occurs via two distinct pathways. The de novo synthesis pathways initially produce the full array of nucleotide sugars, which are then used as substrate for the synthesis of polysaccharides, glycoproteins, and several other glycosylation reactions. However, several monosaccharides other than Glc may be incorporated into nucleotide sugars via salvage pathways involving C-1 kinases and nucleotide diphosphate (NDP)–sugar pyrophosphorylases (see Figs. 2.28 and 2.29). The salvage pathways are essential for reuse of these monosaccharides after their hydrolysis from polymers during cell wall assembly and during turnover of the cell wall. Some sugars, such as Rha and Xyl, do not have C-1 kinases (see Fig. 2.28), and their carbons must be reused via other pathways. Xyl carbons are returned by way of the pentose phosphate pathway after isomerization to xylulose.

2.4.3 Membrane fractions enriched for Golgi membranes can synthesize many noncellulosic polysaccharides in vitro.

Many reports of the synthesis of noncellulosic wall polysaccharides in vitro include use of membrane preparations enriched in Golgi membranes. In mixed-membrane preparations containing plasma membrane, Golgi, and UDP-Glc as a substrate, the predominant product is the (1→3)β-D-glucan, callose—a polymer thought to be made as a default product by damaged cellulose synthase. This reaction is activated by calcium ions, the concentrations of which increase markedly in cells that have been damaged. Thus, the other noncellulosic polysaccharides of interest must be detected and quantified in the presence of a huge background of callose.

By treating polysaccharides with sequence-dependent glycanases, researchers can detect the characteristic unit structures despite the presence of the wound-induced callose. This technique, which verifies that isolated membranes are capable of

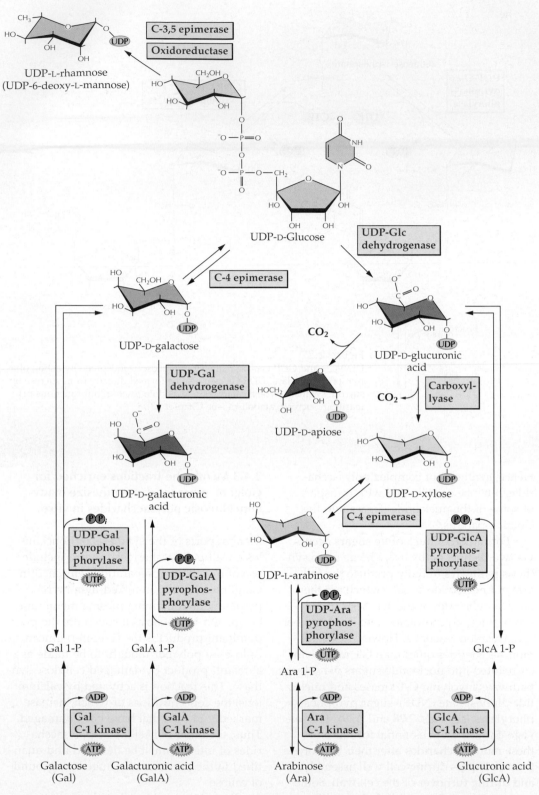

Figure 2.28

Synthesis of UDP-sugars. UDP-Gal is made from UDP-Glc in an equilibrium reaction catalyzed by a C-4 epimerase. UDP-Xyl and UDP-L-Ara are interconverted similarly by a C-4 epimerase. A C-3,5 epimerase and a hypothetical oxidoreductase are thought to convert UDP-Glc to UDP-Rha. Oxidation of UDP-Gal or UDP-Glc by a dehydrogenase yields UDP-GalA or UDP-GlcA, respectively.

Decarboxylation of UDP-GlcA by a carboxyl-lyase constitutes another pathway to UDP-Xyl. Finally, Gal, GalA, Ara, and GlcA can be uridinylated via salvage pathways, in which the sugars are phosphorylated by C-1 kinases to generate sugar 1-phosphates, which in turn act as substrates for the pyrophosphorylases that generate nucleotide sugar end-products.

synthesizing unit structures identical to those made in vivo, has been particularly useful for studying in vitro synthesis of XyG.

The isolated Golgi apparatus of plants with Type I walls can synthesize very short (1→4)β-D-glucan chains from either UDP-Glc or GDP-Glc, but the synthesis of XyG is greatly enhanced by addition of nearly millimolar amounts of both UDP-Glc and UDP-Xyl, with Mn^{2+} or Mg^{2+} as cofactor. Cleavage of the XyG reaction products yields substantial amounts of the diagnostic XXXG heptasaccharide unit, verifying that the cellular machinery needed to make a complete unit structure can be preserved in a cell-free system. Because only very short chains of (1→4)β-D-glucan backbone can be made from UDP-Glc in the absence of UDP-Xyl, the glucosyl- and xylosyltransferases appear to be tightly coupled, catalyzing the formation of the repeating heptasaccharide units cooperatively. The attachment of additional sugars to the Xyl units may occur coordinately as well, but additional elaboration involving galactosyl- and fucosyltransferases continues during later stages of transit through the Golgi apparatus (see Fig. 2.26). These types of coordinated glycosyl transfers are observed in the synthesis of glucomannans with GDP-Glc and GDP-Man, galactomannans with GDP-Man and UDP-Gal, and arabinoxylans with UDP-Xyl and UDP-Ara. In addition, the synthesis of HGA in vitro appears to be coordinated with its methyl-esterification.

UDP-Glc is the substrate for the mixed-linkage β-glucan synthase of grass species, and either Mg^{2+} or Mn^{2+} is required as a cofactor. The preservation of β-glucan synthase activity is demonstrated by digestion of the β-glucan product with a *Bacillus subtilis* endoglucanase into characteristic tri- and tetrasaccharides ($G_4 G_3 G$ and $G_4 G_4 G_4 G$, respectively) in the correct ratio (Fig. 2.30).

None of the genes encoding the polysaccharide synthases that make the backbone chains have been identified, but numerous candidates are potential homologs of cellulose synthases, in particular within sequences encoding the UDP-Glc–binding domains (see Section 2.4.5). In contrast, the genes of several fucosyl- and galactosyltransferases, which add single sugars subtending the backbone as side groups, have now been identified. These genes encode

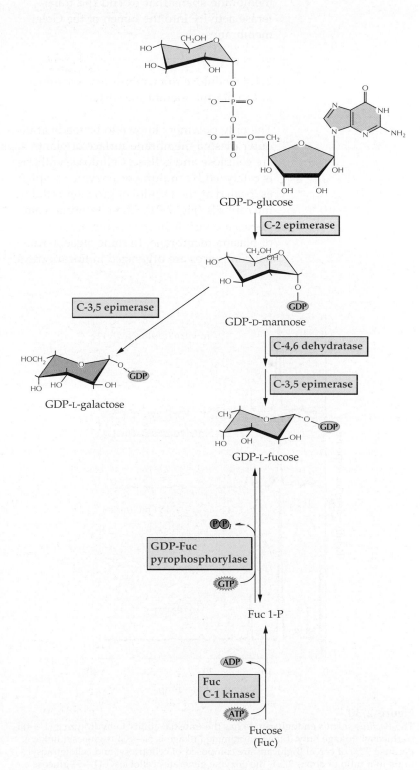

Figure 2.29
Synthesis of GDP-sugars. GDP-Glc is converted to GDP-Man by a C-2 epimerase. GDP-Man yields GDP-L-Gal in a reaction catalyzed by a C-3,5 epimerase. Two enzymes, a C-4,6 dehydratase and the C-3,5 epimerase active in GDP-L-Gal biosynthesis, participate sequentially in the conversion of GDP-Man to GDP-L-Fuc. A salvage pathway generates GDP-L-Fuc from fucose.

a special class of proteins with single-membrane spans that extend the transferase activity into the lumen of the Golgi membranes.

2.4.4 Cellulose microfibrils are assembled at the plasma membrane surface.

The only polymers known to be made at the outer plasma membrane surface of plants are cellulose and callose. Cellulose synthesis is catalyzed by multimeric enzyme complexes located at the termini of growing cellulose microfibrils (Fig. 2.31). These **terminal complexes** are visible in freeze-fracture replicas of plasma membrane. In some algae, terminal complexes are organized in linear arrays,

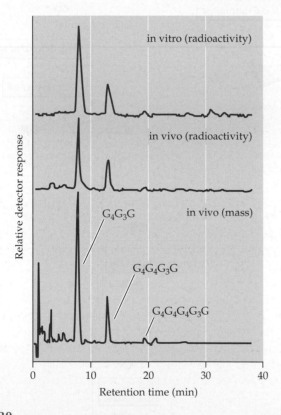

Figure 2.30
The *Bacillus subtilis* endoglucanase has the unusual ability to hydrolyze a (1→4)β-D-glucosyl linkage only if the penultimate linkage is (1→3)β-D-glucosyl linkage. Because 90% of cereal β-glucans are composed of cellotriosyl and cellotetraosyl units in a ratio of about 2.5:1, the enzyme generates cellobiosyl-(1→3)-glucose (G_4G_3G) and cellotriosyl-(1→3)-glucose ($G_4G_4G_3G$) oligosaccharides in this ratio also. When the radioactive products of such a digest are separated by HPAE-HPLC, the ratio of these oligomeric units is the same whether from β-glucan synthesized in vivo by living plants fed [^{14}C]glucose or from isolated Golgi membranes treated with UDP-[^{14}C]glucose. Characterization of polysaccharide synthase reaction products in such a way confirms that synthesis in vitro parallels that in vivo. $G_4G_4G_4G_3G$, cellotetraosyl-(1→3)-glucose.

whereas in others—and in all angiosperms—they form **particle rosettes** (Fig. 2.31). Terminal complexes appear in the plasma membrane coincident with the activity of cellulose synthesis. Kinetic studies monitoring the pathway of glucose to cellulose have established that UDP-Glc is the primary substrate for cellulose synthase. Isoforms of sucrose synthase, an enzyme that produces UDP-Glc directly from sucrose (see Fig. 2.27), also are associated with the plasma membrane; here, in close association with cellulose synthase, these isoforms may contribute substrate directly into the catalytic site of the enzyme.

Although callose synthesis in isolated membranes may have resulted from damage to the cellulose synthase, callose is a natural component of the initial cell plate, the pollen tube wall, and transitional stages of wall development of certain cell types. Unlike the wound polymer, pollen tube callose is synthesized by enzymes that do not require Ca^{2+} as cofactors; moreover, these synthases must be enzymes distinct from cellulose synthases.

Although cellulose synthase generates one of the most abundant biopolymers on earth, this plant enzyme has proven curiously difficult to purify in active form. Cellulose synthase activity from plants disappears even under conditions of gentle plasma membrane isolation, and no terminal complexes have been found in isolated membranes. The complex may require cytoskeletal or cell wall components for it to stabilize in the membrane.

2.4.5 Cellulose synthase genes from plants have been cloned on the basis of sequence identity with bacterial enzymes.

After a search spanning nearly four decades, several genes encoding cellulose synthases of plants have been identified. The molecular groundwork for this research was laid by identifying cellulose synthase genes in the bacteria *Acetobacter xylinum* and *Agrobacterium tumefaciens*, which extrude extracellular ribbons of cellulose. These synthase genes and those of other enzymes requiring nucleotide sugars encode four highly conserved domains thought to be critical for binding and catalysis of UDP-Glc (and certain other UDP-sugars) (Fig. 2.32). So far, all other kinds of synthases able to make contiguous (1→4)β-glycosyl linkages, such as chitin synthases and hyaluronate synthases, contain these four highly homologous sequences of amino acids.

Searching sequences in a cDNA library made from cotton fibers at the onset of secondary wall cellulose formation (see Section 2.6.3), Deborah Delmer and colleagues found two plant sequences homologous to the *Acetobacter* cellulose synthase gene, sharing all four UDP-Glc binding sequences. The plant *Ces*A genes are highly expressed in fibers at the time of active synthesis of secondary wall cellulose. The polypeptides they encode (size about 110 kDa) are predicted to have eight transmembrane domains, to bind the substrate UDP-Glc, and to contain two large domains unique to plants.

Genetic proof of the function of a *Ces*A homolog has been obtained by complementing an *Arabidopsis* mutant unable to make cellulose at a high temperature. Transformation of the temperature-sensitive mutant with the wild-type *Ces*A gene restores the normal phenotype, providing the first evidence that a plant *Ces*A gene functions in the formation of cellulose microfibrils. However, direct proof of the function of the gene product is still lacking, because cellulose cannot at present be synthesized to any great extent in vitro.

The discovery of *Ces*A genes has now opened the door for the identification of many other cell wall polysaccharide synthase genes. The *Arabidopsis* genome contains several homologs of cotton *Ces*A, and other sequences share significant identity with one or more of the suspected UDP-Glc binding domains.

Why do plants have so many different *Ces*A and related genes? Almost all plant cell types can be identified by unique features of their cell wall, so the large family of *Ces*A genes may represent the diversity of cellulose synthases needed to build specialized walls. In addition, some of the related genes may encode other kinds of synthases that make a broad range of noncellulosic polysaccharides. Of all of the polymers built from pyranosyl sugars, those that possess runs of (1→4)β-linkages of either glucose, xylose, or mannose invert the orientation of one sugar unit with respect to its neighbor (see Figs. 2.12,

2.14, and 2.15). The steric problem is the same as for cellulose synthase, and it is tempting to speculate that an ancestral cellulose synthase may have been modified to confine the location of the synthase to the Golgi apparatus and to alter the nucleotide sugar substrate.

2.5 Growth and cell walls

2.5.1 The cell wall is a dynamic structure.

Cell expansion involves extensive changes in the mass and composition of the cell wall. Cell growth, an irreversible increase in cell volume, can occur by expansion (increase in cell size in two or three dimensions) or by elongation (expansion constrained to one dimension). Variety in cell shape may result if either of these two processes occurs at specific regions of the cell surface.

During elongation or expansion, existing cell wall architecture must change to incorporate new material, increasing the surface area of the cell and inducing water uptake by the protoplast. The osmotic pressure (turgor) exerted by the protoplast is necessary to drive cell expansion, and this pressure usually remains a relatively constant driving force for expansion. The regulation of **wall loosening** is considered the primary determinant of rates of cell expansion. The cell wall architecture must be extensible; that is, mechanisms must exist that allow discrete biochemical loosening of the cell wall matrix, permitting microfibril separation and insertion of newly synthesized polymers. Cells may extend to tens, hundreds, or even thousands of times their original length while maintaining a constant wall thickness. Thus, wall loosening and continued deposition of new material into the wall must be tightly integrated events (Fig. 2.33).

2.5.2 Most plant cells grow by uniform deposition of cell wall materials, whereas some demonstrate tip growth.

In the vast majority of plant cells, growth and the deposition of new wall material occur uniformly along the entire expanding wall. However, **tip growth,** the growth and deposition of new wall material strictly at

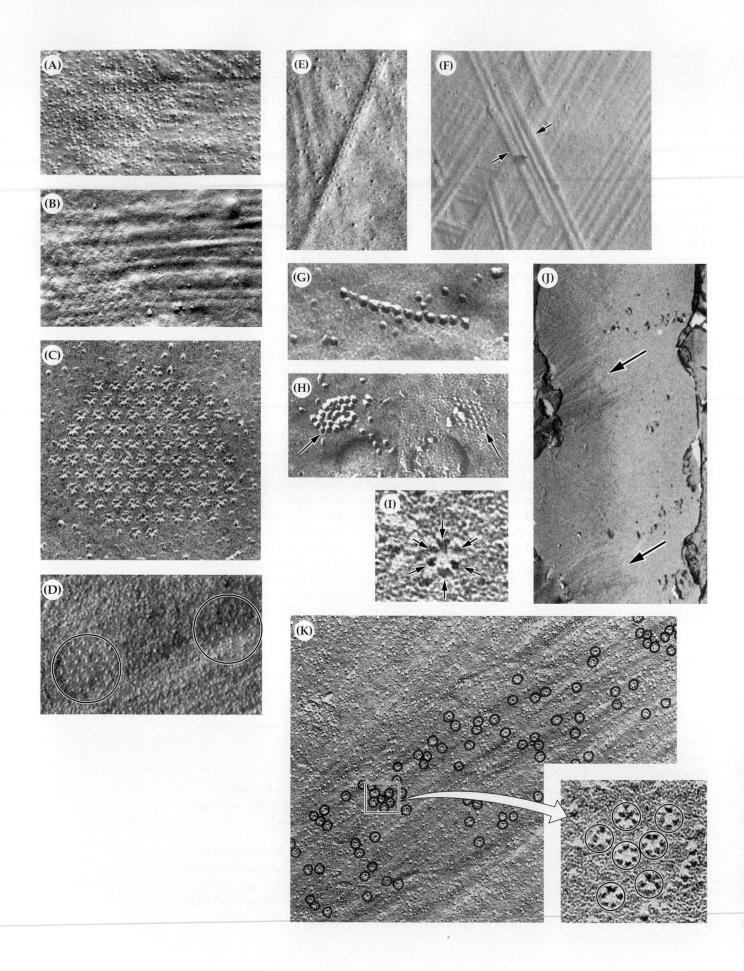

Figure 2.31 *(Facing page)*
The terminal complexes associated with cellulose microfibril biosynthesis are imaged by freeze-fracture and rotary-shadowing of membranes. The six-membered particle rosettes were first seen in the desmid *Micrasterias* (A–C). The rosettes are seen on the "P-face," the inner leaf of the plasma membrane bilayer (A), whereas the impression of the microfibrils and larger particles that apparently fit into the center of each rosette are located in the outer leaf, or "E-face" (B). In this alga, the individual rosettes aggregate in the membrane to form even larger hexagonal arrays that are associated with formation of flat ribbons of cellulose (C). The green alga *Spirogyra* also assembles aggregates of rosettes (D), but some algae, such as *Oocystis*, have linear arrays of particles (E). These terminal complexes often pair up at the plasma membrane and initiate synthesis of microfibrils in opposite directions (see arrows) (F). The cellular slime mold, *Dictyostelium discoideum*, forms two kinds of terminal complexes: A linear array is associated with the extracellular ribbon of cellulose formed by streaming cells (G), whereas the cell wall of the stalk cells is made by aggregate arrays that appear to be rudimentary rosettes (H). In contrast to algae, all flowering plants examined contain single rosette structures (I); arrows point to each of the six particles. Such single rosettes are abundant during the formation of the secondary wall thickenings (arrows) visible in this longitudinal section of a vessel element of cress (*Lepidium sativum* L.) (J). The rosettes (circled) are found in great abundance only in the membrane underlying the developing thickening (K); individual rosettes are magnified in the inset.

the tip of a cell, occurs in some plant cells, notably root hairs and pollen tubes (Fig. 2.34).

Plant cells possess arrays of **cortical microtubules** underlying and connected to the plasma membrane (see Chapter 5). The orientation of the cortical microtubule array often predicts the orientation of cellulose microfibril deposition. The cortical array may align the cellulose synthase complexes, either by direct protein-mediated connection or by defining channels in the membrane in which the synthase can move. In turn, cellulose microfibril orientation controls whether cells expand or elongate and determines the plane of elongation. In cells that grow by uniform expansion, layers of microfibrils lie unaligned in the wall matrix, but in cells that elongate, microfibrils align in transverse or helical orientations to the axis of elongation (see Fig 2.11).

2.5.3 The multinet growth hypothesis has been developed to explain axial displacement of cellulose microfibrils during growth of the cell wall.

From studies of developing cotton fibers, the **multinet growth hypothesis** was developed to explain how cellulose microfibrils that have been deposited in a transverse or slightly helical orientation are displaced axially as elongation proceeds (Fig. 2.35A). New microfibrils deposited in strata on the inner surface of the wall in a generally transverse orientation functionally replace older microfibrils. The older microfibrils are pushed into the outer layers of the wall and are passively reoriented in a longitudinal direction as the cell elongates. Consistent with this hypothesis, the changes in the diffraction patterns of polarized light by algal cells, induced by reorientation of the parallel

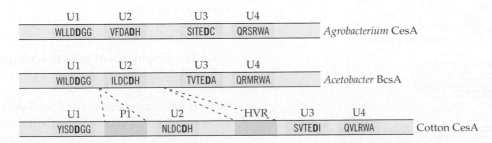

Figure 2.32
Cellulose synthases from *Acetobacter xylinum*, *Agrobacterium tumefaciens*, and *Gossypium hirsutum* (cotton) are characterized by four motifs (called U-motifs) critical for binding the substrate UDP-Glc. Three of the four domains demonstrate absolute conservation of an aspartyl residue (in red) predicted to be necessary for catalysis of glycosidic bond formation. In addition to the three aspartyl residues, the QxxRW motif in U4 also is conserved in every synthase of a polysaccharide with contiguous (1→4)β–linked units in which one sugar is inverted with respect to its neighbors. Two sequences, P1 and HVR, are plant specific.

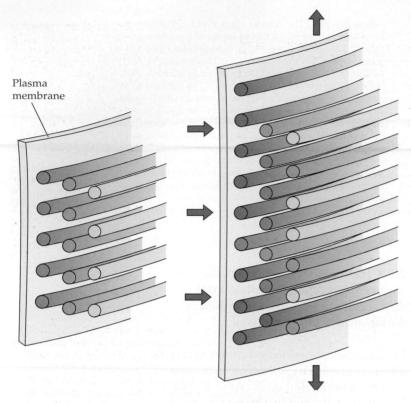

Plasma
membrane

Figure 2.33
Wall loosening and incorporation of new wall polymers are integrated with each other so that wall thickness is maintained during cell expansion. Because the walls are only a few strata thick, loosening of the wall with no insertion of new wall material would very quickly thin the wall during growth and cause rupture. In contrast, deposition without loosening would increase wall thickness, because the walls would not expand. Vertical arrows denote the direction of expansion, and horizontal arrows show the addition of new microfibrils (red) on the inner surface of the wall.

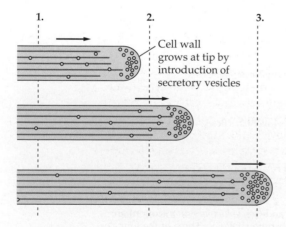

1. 2. 3.

Cell wall
grows at tip by
introduction of
secretory vesicles

Figure 2.34
In tip growth, the rate of vesicle delivery and deposition at the tip determines the rate of growth. Vesicles are delivered to the tip via actin microfilaments (see Chapter 5).

microfibrils, occur primarily in the inner layers of the wall close to the plasma membrane. The driving force for wall extension is generally viewed to be the turgor generated by the protoplast, but it is the tension created on the microfibrils 90° perpendicular to the outward push of the protoplast that leads to separation of the microfibrils (Fig. 2.35B). A turgor pressure of 1 MPa can generate several hundred megapascals of tension because the volume of a relatively large protoplast is resisted by a very thin cell wall.

The essence of the multinet growth hypothesis is descriptive for many kinds of cells, but microfibrils do not necessarily reorient axially, and substantial extension is possible with little reorientation if many wall strata contribute to the direction of expansion. Cellulose microfibrils woven in a shallow helix around the cell prevent the growing cell from becoming spherical. By analogy, a spring-like toy such as Slinky or Flexi stretches easily along its longest axis but resists attempts to increase its diameter (Fig. 2.35C); when stretched, the spring extends substantially with only a small change in the helical angle but a wide separation of the coils. Extension of a cell wall might be viewed as a series of tightly interacting concentric Slinkys in both right-handed and left-handed orientations that reorient at crossed angles during separation (Fig. 2.35A). Given the estimated thickness of the primary wall (80 to 100 nm in meristematic and parenchymatous cells) and the dimensions of the matrix components (see Box 2.4, panel C), only about 5 to 10 strata make up the wall. Moreover, microfibrils from several strata may merge somewhat to fill in gaps as new strata are deposited onto the inner surface (see Fig. 2.33).

2.5.4 The biophysics of growth underpins cell wall dynamics.

Several different classes of cell wall polymers constitute nearly independent determinants of strength in elongating cells. These include (*a*) microfibrils arranged in the transverse axis, which are interconnected to the cross-linking glycans; (*b*) putative networks involving structural proteins or phenylpropanoid compounds; and (*c*) elements of the pectin network. When plant growth

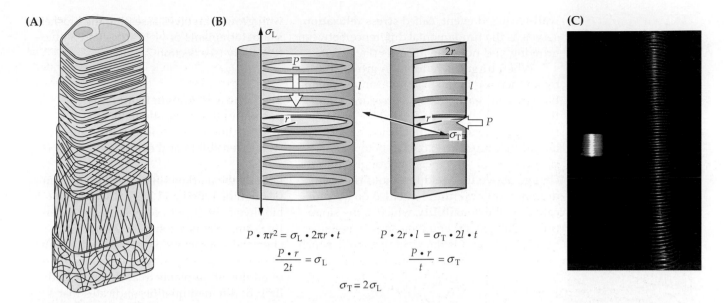

(A)   **(B)**   **(C)**

$$P \cdot \pi r^2 = \sigma_L \cdot 2\pi r \cdot t \qquad P \cdot 2r \cdot l = \sigma_T \cdot 2l \cdot t$$

$$\frac{P \cdot r}{2t} = \sigma_L \qquad\qquad \frac{P \cdot r}{t} = \sigma_T$$

$$\sigma_T = 2\sigma_L$$

Figure 2.35

(A) The original multinet growth hypothesis explains that as walls stretch during growth, the microfibrils reorient passively from a transverse direction on the inner wall to a longitudinal direction at the outer wall. (B) The hydrostatic pressure developed by the protoplast is resisted by a relatively thin cell wall, and the tensile force pulling the microfibrils apart is several orders of magnitude higher than cell turgor pressure. For example, a spherical cell with a radius (r) of 50 μm and 1 MPa of turgor pressure (P), enveloped by a cell wall only 0.1 μm thick (t), develops 250 MPa of tension in the wall. This enormous tension changes as the cell geometry changes. When this cell begins to elongate and become cylindrical, the tension increases to 500 MPa tangentially because of the change in cell dimension. l, length; σ_L, longitudinal stress; σ_T, tangential stress. (C) Although a Slinky is difficult to extend radially because of the orientation of the coils, it is easily extended lengthwise. Cell shape in plants is controlled similarly. Altering the interaction between the tethering cross-linking glycans and cellulose is the principal determinant of the rate of cell expansion.

regulators such as auxin and gibberellin change the *direction* of growth, they do so through changing the orientation of cortical microtubules and cellulose microfibrils. When they change the *rate* of growth, their mechanisms include dissociation or breakage of the tethering molecules from microfibrils. Mathematical formulations that describe the growth rate of plant cells allow us to define the cell wall properties that must be modified to permit growth.

Equation 2.1

$$dl/dt = L_p \left(\Delta \psi_w \right)$$

Equation 2.2

$$dV/dt = A \cdot L_p \left(\Delta \psi_w \right)$$

Equation 2.3

$$\text{Rate} = m \left(\psi_p - Y \right)$$

The rate of cell elongation can be quantified by Equation 2.1, where dl/dt is the change in length per unit time, L_p is hydraulic conductivity (i.e., the rate at which water can flow across the membrane), and $\Delta \psi_w$ is the water potential difference between the cell and the external medium. The difference in water potential is the driving force for water movement and has two components, $\psi_s + \psi_p$, the osmotic potential and pressure potential (turgor), respectively. Revising the equation to include any change in cellular volume (V) yields Equation 2.2, where growth is defined as a change in volume per unit time and depends on the surface area (A) of the plasma membrane available for water uptake. Thus the rate of growth is proportional to membrane surface area, the conductivity of the membrane, and the water potential difference driving water uptake. In nongrowing cells, $\Delta \psi_w$ (and thus dV/dt) = 0, because the rigid cell wall prevents water uptake, and the turgor pressure rises to a value equal to that of the osmotic potential of the cell. By contrast, in growing cells, $\Delta \psi_w$ does not reach zero because the wall tethers have been loosened. As a result, cell volume increases irreversibly. This wall-localized event, called **stress relaxation,** serves as the fundamental difference between growing and nongrowing cells (Fig. 2.36).

When turgor is reduced in growing cells by an increase in the external osmotic potential, growth ceases at some pressure before the turgor reaches zero. This value, called the **yield threshold,** defines the pressure potential that must be exceeded before expansion can occur. The increment of growth rate change above the yield threshold is dependent not only on turgor but also on a factor called wall **extensibility,** which is the slope (m) of the general equation shown as Equation 2.3, where Y is the yield threshold. Much of the work remaining in the study of cell wall growth involves assessing the biochemical determinants of yield threshold and extensibility (see Section 2.5.6).

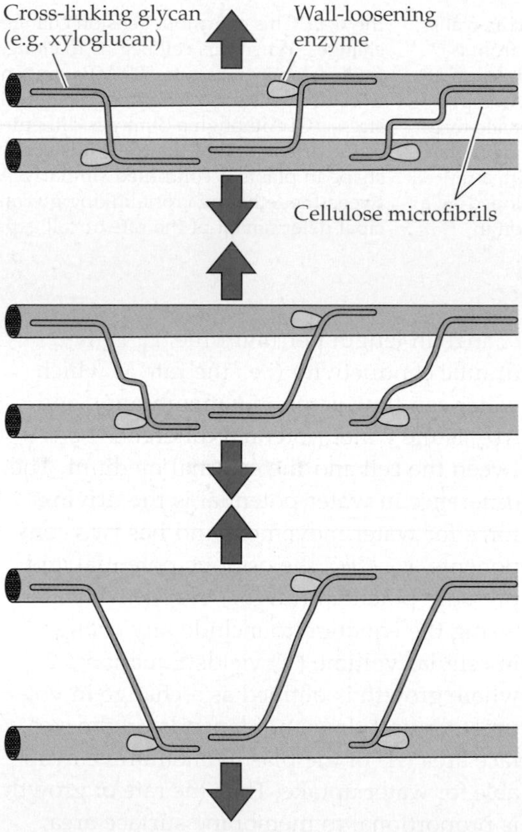

2.5.5 The acid-growth hypothesis postulates that auxin-dependent acidification of the cell wall promotes wall extensibility and cell growth.

Despite the marked differences in the composition of Type I and Type II walls, the biophysics of growth of grasses and other flowering plants are similar. Although the chemical complexity of the wall is daunting, the similarity of the physiological responses of all flowering plants to acid, auxin, and light of different qualities indicates that a few common mechanisms of wall expansion exist, regardless of the kinds of molecules that tether the microfibrils.

The extraction of several kinds of polysaccharide hydrolases from the cell walls of tissues rich in growing cells raised the possibility that the regulation of these enzymes was a mechanism by which auxin could cause wall expansion. A major breakthrough came with the discoveries that auxin caused acidification of the medium in which elongating tissue sections were bathed and that H^+ could substitute effectively for the auxin hormone (Fig. 2.37A). This **acid-growth hypothesis** proposes that auxin activates a plasma-membrane proton pump, which acidifies the cell wall. The low pH, in turn, activates apoplast-localized growth-specific hydrolases, which cleave the load-bearing bonds that tether cellulose microfibrils to other polysaccharides. Cleavage of these bonds results in loosening of the cell wall, and the water potential difference causes uptake of water. Relaxation of the wall (i.e., separation of the microfibrils) passively leads to an increase in cell size.

The basic tenets of the acid-growth hypothesis have stood the test of time, but three problems persist. First, no enzymes have been found that hydrolyze cell wall cross-linking glycans exclusively at pH conditions lower than 5.0. Second, no reasonable explanation exists for how growth is kept in check once the hydrolases are activated. Third, no hydrolases extracted from the wall and added back to the isolated tissue sections, regardless of external pH, cause extension in vitro.

Figure 2.36
Stress-relaxation is considered the underlying basis of cell expansion. When an elongating cell is stretched by turgor, the longitudinal stress (indicated by arrows) is borne more or less equally by the glycans tethering the cellulose microfibrils. If some of the tethers are dislodged from the microfibrils, or hydrolyzed, they temporarily "relax" and the yield threshold is breached because the other tethers are strained. Water uptake results in expansion of the microfibrils to take up the slack of the relaxed glycans, which are once again placed under tensile stress.

2.5.6 At present, two kinds of enzymes are being evaluated as having possible wall-loosening activities.

Two candidate wall-loosening enzymes are currently being studied. One of these, **xyloglucan endotransglycosylase** (XET), carries out a transglycosylation of XyG in which one chain of XyG is cleaved and reattached to the nonreducing terminus of another XyG chain. Given such a mechanism, microfibrils could undergo a transient slippage but the overall tensile strength of the interlocking XyG matrix would not diminish (Fig. 2.38).

XETs may also function in the realignment of XyG chains in different strata during growth and in the assembly of the wall as newly synthesized XyGs are incorporated. In some cases, the correlation between growth and XET activity is not clear, and some XETs appear to function hydrolytically.

Other proteins catalyze wall extension in vitro without any detectable hydrolytic or transglycolytic events. Called **expansins,** these proteins probably catalyze breakage of hydrogen bonds between cellulose and the load-bearing cross-linking glycans. Such an activity could disrupt the tethering of

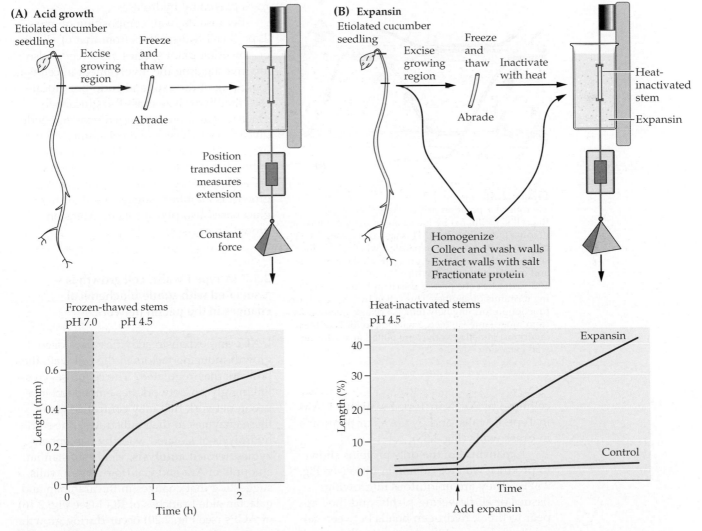

Figure 2.37
Sections of growing epicotyl, hypocotyl, or coleoptile are frozen, thawed, and placed under a constant stress. A position transducer measures elongation continuously. (A) When the bathing solution of the frozen-thawed sections is changed from pH 7 to 4.5, extension occurs almost immediately. (B) If the sections are heat-inactivated, no extension occurs at any pH. However, when expansins are added, extension is restored at acidic pH. Neither plant hydrolases nor xyloglucan endotransglycosylase is able to cause this effect in vitro.

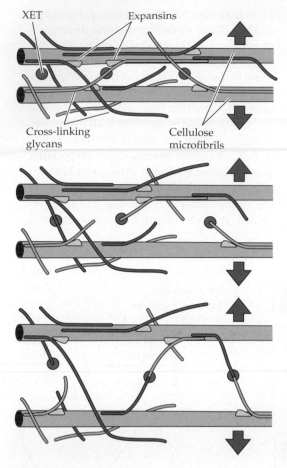

XET   Expansins

Cross-linking glycans   Cellulose microfibrils

Figure 2.38
Microfibril separation driven by osmotic pressure of the cell is facilitated by loosening of the cross-linking glycans that tether them. This may be accomplished by coordinate action of expansins, which break the steric interactions between the cross-linking glycans and cellulose, and XET, which hydrolyzes a glycan and reattaches one part of the chain to the nonreducing terminus of another. This action by XET may also function in forming new tethers because microfibrils from inner lamellae merge with microfibrils of the outermost lamellae as they are pulled apart during wall extension (see Fig. 2.33).

cellulose by XyGs in Type I walls, by GAXs in Type II walls, and by GAXs or β-glucans in grass walls (Fig 2.38).

Expansins are the only proteins shown to produce wall expansion in vitro (see Fig. 2.37B). They are ubiquitous in growing tissues of all flowering plants, and they appear to break hydrogen bonds between cellulose and several kinds of cross-linking glycans regardless of their chemical structure. Also, because grass expansins induce extension of tissues with Type I walls, it is attractive to think that expansins are ubiquitous

enzymes involved in the rapid growth responses of both Type I and Type II walls. However, further studies show that a second multigene family of β-expansins, found predominantly in the grasses, have no appreciable activity on Type I cell walls. Exogenous application of expansins to a meristem induces bulging of the meristem at the sites of application—bulges that develop into leaf primordia.

The genes of several XETs and expansins have been cloned, and experiments designed to reduce the amounts of each in transgenic plants are being tested to determine their relative significance. XET and expansin may not be the only wall-loosening agents, and work continues to determine the roles played by hydrolases.

Because the wall composition of grass Type II cell walls differs from that of all other flowering plant species, researchers also are investigating the involvement in cell wall loosening of the exo- and endo-β-D-glucanases that hydrolyze grass β-D-glucans to glucose. Addition of purified exo- and endo-glucanases to heat-killed coleoptiles cannot induce extension growth. However, when antibodies directed against these enzymes are added to enzyme-active walls, wall growth is inhibited, suggesting that these glucanases also play a role in extension growth in grasses.

2.5.7 In Type I walls, cell growth is associated with subtle biochemical changes in the pectin network.

If XET and expansin activities are indeed growth-inducing factors in the cell wall, then how are they regulated? The cellulose/cross-linking glycan network lies embedded in a pectin network that may control access of these enzymes to their substrates. The self-hydrolysis of isolated walls by nascent enzymes, termed **autolysis,** yields substantial amounts of Ara and Gal from Type I walls, suggesting that changes in the arabinan and galactan side-branches of RG I (see Fig. 2.16) or AGPs (see Fig. 2.20) occur during growth.

The most marked change revealed by biochemical analysis is the increase in the degree of methyl-esterification of the wall HGAs when the newly synthesized pectins are deposited during elongation. The deposition of highly esterified HGAs is succeeded by a deesterification event that can remove a large proportion of methyl ester groups from the wall when growth stops. The ionized carboxyl groups can form Ca^{2+}-HGA junction zones, causing the wall to become more rigid (see Fig. 2.24). The cell walls of meristematic and elongation zones are characteristically low in Ca^{2+}, and Ca^{2+}-HGA junction zones are observed more frequently after cell elongation has stopped.

2.5.8 More obvious biochemical changes occur in growing Type II walls than in Type I walls.

More obvious chemical changes occur with the cross-linking glycans of the Type II walls of commelinoid monocots. A highly substituted GAX (HS-GAX), in which six of every seven Xyl units bears an appendant group, is associated with the maximum growth rate of coleoptiles. The number of side groups of Ara and GlcA along the GAX chains varies markedly, from GAXs for which Xyl units are nearly all branched to those with only 10% or less of the Xyl units bearing side groups. Side groups not only prevent hydrogen bonding but also render the GAX water-soluble. In dividing and elongating cells, highly branched GAXs are abundant, whereas after elongation and differentiation, more and more unbranched GAX polymers accumulate. Cleavage of the Ara and other side groups from contiguously branched Xyl units can yield runs of unbranched xylan capable of binding to other unbranched xylans or to cellulose microfibrils.

Type II walls also have a low pectin content. Chemically, pectins of the Type II wall include both HGA and RG, and HS-GAX is closely associated with these pectins. Interactions between the HS-GAX, HGA, and RG could control wall-loosening activities, as suggested for the Type I wall. The spacing of the appendant Ara and GlcA units of GAXs could determine porosity and surface charge, functionally replacing the predominant pectic substances in the Type I cell wall, as illustrated in Fig. 2.23B. The HGAs of maize pectins, which are methyl-esterified (see Fig. 2.16A), also contain nonmethyl

esters, the formation and disappearance of which coincide with the most rapid rate of cell elongation. We do not yet know the chemical nature of the nonmethyl esters. Some arabinans, particularly the 5-linked arabinans, are found in the walls of dividing cells but are not made during cell expansion.

When grass cells begin to elongate, they accumulate the mixed-linkage β-glucans in addition to GAX (see Fig. 2.14). β-Glucans are unique to the Poales (see Fig. 2.13) and are one of the few known developmental stage–specific polysaccharides. Absent from meristems and dividing cells, β-glucans accumulate to almost 30% of the noncellulosic cell wall material during the peak of cell elongation and then are largely hydrolyzed by the cells during differentiation. Because synthesis and hydrolysis of the β-glucans occur simultaneously throughout elongation, the amount that accumulates is a small fraction of the total synthesized. The appearance of β-glucans during cell expansion and the acceleration of their hydrolysis by growth regulators all implicate direct involvement of the polymer in growth.

2.5.9 Once growth has ceased, cell wall shape must be locked in place by wall components.

Once elongation is complete, the primary wall locks the cell into shape by becoming much less extensible. One component of the locking mechanism for Type I walls may be HRGPs, such as extensin. The flexible regions of these proteins could wrap around cellulose microfibrils, and the rod-like regions could serve as spacers. However, we do not yet know how extensins are cross-linked in the wall. Extensin units, and perhaps other proteins, may covalently bind to each other, forming an independent interlocking network, as illustrated in Figure 2.23A. The loss of solubility of extensins in Type I walls is associated with an increase in the tensile strength of the wall.

Other proteins may be necessary to actually lock the extensins together. One candidate is a 33-kDa PRP. Greater quantities of PRP are synthesized later in cell development and particularly in the same vascular cells as extensin. One hypothesis holds that extensin precursors accumulate in the cell wall during cell division and elongation but cannot cross-link wall components without the presence of PRPs.

Just as in Type I walls, Type II walls are locked into form during differentiation; unlike Type I walls, however, they do not use Hyp-rich extensin. Instead, Type II walls contain a threonine-rich protein with sequences reminiscent of the typical extensin structure; cross-linking phenolic compounds are also present. As with extensin, soluble precursors of the threonine-rich protein accumulate early in the cell cycle and become insoluble during cell elongation and differentiation. This polymer is prevalent in vascular tissue and in special, reinforced wall structures, such as the pericarp. However, much of the cross-linking function in Type II walls probably rests with the esterified and etherified phenolic acids, and formation of these cross-linkages accelerates at the end of the growth phase. Once the phenolic cross-bridges are in place, they may account for a substantial part of the load-bearing role in the fully expanded cells.

2.6 Cell differentiation

2.6.1 The plant extracellular matrix is a coat of many colors.

At the light microscopy level, traditional histochemical stains reveal a broad diversity in the distribution of various polysaccharides in different cells. The complexity of polysaccharide structure is resolved even more effectively when highly specific probes are used. This has been achieved by using the natural specificity of enzymes for their substrates and that of antibodies for particular antigens (Box 2.5). More recently, the advent of microspectroscopy, in which a microscope with modified optics is attached to a spectrometer, has made **chemical imaging** possible. The distribution of a particular functional group of a molecule can be mapped at the single-cell level (Box 2.5).

Even in a single cell, modifications occur that distinguish between transverse and longitudinal walls. For example, a preferential digestion of end walls occurs in sieve tube members. Some substances, such as waxes, are secreted only to a cell's outer epidermal face. Within a single wall there are zones of

different architectures—the middle lamella, plasmodesmata, thickenings, channels, pit-fields, and the cell corners—and there are domains within the thickness of a wall in which the degree of pectin esterification and the abundance of RG I side chains differ (Fig. 2.39).

The size of such microdomains in the wall, compared with the sizes of the polymers that must fit in these domains, implies that mechanisms must exist for packaging and positioning of large molecules. For example, unesterified pectins can be as long as 700 nm and yet, in some cell types, are accommodated in a middle lamella 10 to 20 nm wide and so must be constrained at least to lie parallel to the plasma membrane. This microdiversity within walls is now changing our view of the wall—from that of a homogeneous and uniform building material to

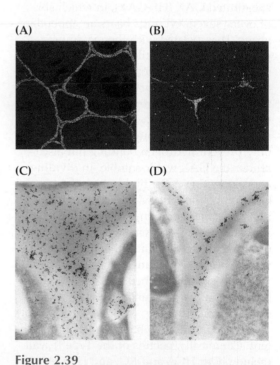

Figure 2.39
Potato cell walls labeled with immunogold and silver-enhanced to reveal the localization of pectin galactan side chains and calcium pectate junctions, with use of the monoclonal antibodies LM5 and 2F4, respectively, in the light microscope. The green color shows where light is reflected from silver-enhanced gold particles in the confocal microscope. The galactan side chains are localized to the primary wall (A), whereas calcium pectate is predominantly in cell corners (B). A methyl-esterified pectic epitope is present throughout the cell walls of vascular cells (C) but restricted to an outer layer of the walls of palisade cells (D) in electron micrographs of a *Zinnia* leaf immunogold-labeled with the JIM7 monoclonal antibody.

one of a mosaic of different wall architectures in which the various components contribute to the multifunctional properties of the apoplast.

2.6.2 Fruit-ripening involves developmentally regulated changes in cell wall architecture.

Most fruits in which the pericarp or endocarp softens during ripening develop thickened primary walls that are markedly enriched in pectic substances, primarily HGA and RG I. The texture of the ripe fruit pulp is governed by the extent of wall degradation and loss of cell–cell adhesion. For example, the walls of the apple cortex undergo little change in rigidity and exhibit little separation, whereas the walls of the peach and tomato pericarp soften considerably through wall swelling and loss of cell adhesion (Fig. 2.40). In tomato, the locules containing the seeds dissolve completely, in a process called **liquefaction**.

Pectins often constitute more than 50% of the fruit wall. The softening process in tomato parenchyma tissue is associated with loss of methyl esters of HGA, a consequence of the activity of PME, which removes methyl ester groups from the GalA residues of pectic polysaccharide backbones. The deesterified HGA backbone then is susceptible to the activity of PGase. PGase I, with a molecular mass of approximately 100 kDa, consists of 46-kDa PGase II tightly complexed with a β-subunit. The β-subunit is a unique **aromatic amino acid-rich protein** thought to function as an anchoring component for the PGase II subunit and is believed to be synthesized early in fruit development. The β-subunit may solubilize pectins from the cell wall, facilitating progressive hydrolysis by PGase II of the glycosidic bonds within the unbranched HGA backbone. Pectin modification within the wall during ripening is a tightly regulated process, the result of such mechanisms as substrate modification, which restricts enzyme access to the substrate, or the presence of enzyme inhibitors, as in the inhibition of PGase II activity by the diffusible products of pectin depolymerization.

However, despite extensive deesterification and depolymerization of the pectin polymers during ripening, fruit softening does not appear to result directly from these modifications to the pectic network. Using antisense inhibition of PGase, researchers were able almost completely to inhibit such activity, essentially preventing pectic depolymerization in transgenic fruit; however, little or no reduction in softening accompanied this achievement. Although the cross-linking glycans do not appear to undergo extensive ripening-related depolymerization, several glycan-modifying enzymes increase in activity, including XET, expansin, and glucan hydrolases. Thus these enzymes may be involved in restructuring the tethering by cross-linking glycans, thereby changing the properties of the network throughout the primary cell wall and contributing to the softening of the pericarp tissue and the whole tomato fruit.

2.6.3 Secondary walls are elaborated after the growth of the primary wall has stopped.

For many cell types, the differentiation process is associated with formation, on the plasma-membrane side of the primary wall, of a distinct secondary wall. Regardless of chemical composition, the primary wall is always defined as the structure that participates in irreversible expansion of the cell. When cells stop growing, the wall is cross-linked into its ultimate shape (see Section 2.5.9). At that point, deposition of the secondary wall begins.

Secondary walls often exhibit elaborate specializations. The cotton fiber, for example, consists of nearly 98% cellulose at maturity (Fig. 2.41A). In some cells, such as sclereids, vascular fibers, and the stone cells of pear, the secondary wall becomes uniformly thick, composed largely of cellulose microfibrils that can sometimes fill the entire lumen of the cell (Fig. 2.41B). The secondary wall may, however, contain additional noncellulosic polysaccharides, proteins, and aromatic substances such as lignin. In tracheids, secondary walls can form special patterns, such as annular or helical coils or reticulate and pitted patterns (Fig. 2.42). These walls typically contain glucuronoxylans or 4-O-methylglucuronoxylans in addition to cellulose. Unlike GAXs of grass walls, these xylans are devoid of Ara but contain a

(A)

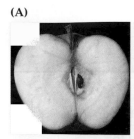

(B)

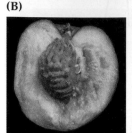

(C)

Figure 2.40
Cell walls are the principal textural elements of fresh fruits and vegetables. The walls of fruit change their architecture in different ways during ripening. (A) Apple walls may stiffen and maintain cell adhesions. The "mealy" texture of an overripe apple is a result of loss of cell–cell adhesion by dissolution of the middle lamella. The middle lamella is enriched in many pectic substances, but the mechanism of the cementing process for cell–cell adhesion is unknown. (B) The walls of the pericarp cells of peach swell and soften during ripening, whereas those of the seed coat become an exceptionally tough protective shield for the seed. (C) Like the peach, the tomato pericarp walls swell and soften during ripening, but some of the walls completely disintegrate, by a process called liquefaction, to create locules for the developing seed.

$(1{\rightarrow}2)\alpha$-D-GlcA residue about every 6 to 12 xylosyl residues. The collenchyma generally confines the secondary wall to thickened corners of these cells (see Fig. 2.41C).

The walls of many cells function long after the cells that produced them are dead and desiccated. For example, the orientation of the polymers assembled in the walls of living cells results in mechanical strains on desiccation, strains that result in abscission of plant parts and the dehiscence of fruit coats along defined planes. The paper-thin wings of a maple samara and the feathery tufts of hair cells of dandelion and willow fruits form only on drying: Each structure aids in the scattering of seeds (Fig. 2.43).

Many structural proteins are cell-specific, occurring only in the secondary wall. For example, some PRPs concentrate in the secondary walls of protoxylem elements of bean, whereas some of the GRP family members are synthesized in xylem parenchyma cells and exported to the primary walls of the neighboring protoxylem elements (see Fig. 2.19). Other GRPs are found in sclereids, associated with both the primary and secondary wall. HRGPs generally are found in the primary walls of all tissues, although in widely variable proportions, but a threonine-rich, extensin-like protein of maize is more abundant in the secondary walls of the firm pericarp of popcorn.

Not all cell wall secondary thickenings represent distinct secondary walls. Some thickened walls have a composition typical of a primary wall but simply contain many more strata. Guard cells and epidermal cells thicken the wall facing the environment to a much greater degree than they do the side walls or the inward-facing wall. Pairs of guard cells contain thickenings of radially arranged cellulose microfibrils, which are needed to withstand the enormous turgor pressures generated by the cell during stomatal opening (see Fig. 2.41D). Epidermal cells form specialized exterior layers of cutin and suberin (see next section) to prevent the loss of water vapor, and the contiguous side walls of endodermal cells are impregnated with suberin to force the water to move symplasmically into the stele.

2.6.4 Secondary deposition of suberin and cutin can render cell walls impermeable to water.

Suberin (see Chapters 10 and 24) material is found in specific tissues and cell types, notably, the root and stem epidermis, cork cells of the periderm, the surfaces of wounded cells, and parts of the endodermis and bundle sheath cells. It is recognized by lipid-specific stains such as Sudan IV, which detects the presence of long-chain fatty acids and alcohols, dicarboxylic acids, and hydroxylated fatty acids. The core of suberin is lignin-like, and the attachment of the long-chain hydrocarbons imparts a strongly hydrophobic characteristic to suberin that prevents water movement.

The polyester cutin and its associated waxes (see Chapter 10) also are found on leaf and stem surfaces, providing a strong barrier to the diffusion of water. Waxes generally are esters of long-chain fatty acids and alcohols but are better described as complex mixtures of these hydrocarbon esters with ketones, phenolic esters, terpenes, and sterols.

2.6.5 Lignin is a major component of some secondary walls.

The most obvious distinguishing feature of secondary walls is the incorporation of lignins, complex networks of aromatic compounds called phenylpropanoids. With a few exceptions, no lignin exists in primary walls. Synthesis of lignin is initiated solely when secondary wall deposition commences. The phenylpropanoids, hydroxycinnamoyl alcohols, and "monolignols"—*p*-coumaryl,

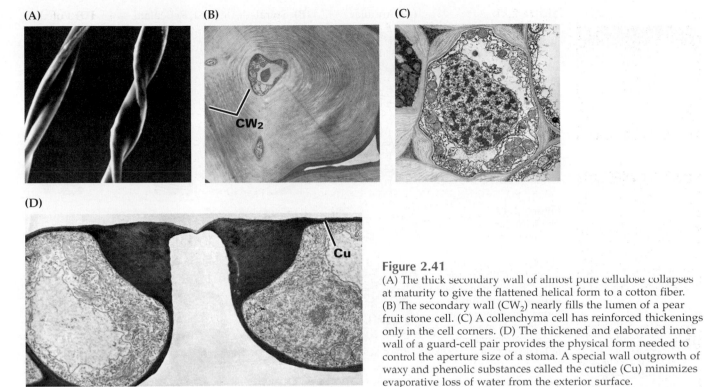

(A)

(B)

CW₂

(C)

(D)

Cu

Figure 2.41
(A) The thick secondary wall of almost pure cellulose collapses at maturity to give the flattened helical form to a cotton fiber. (B) The secondary wall (CW₂) nearly fills the lumen of a pear fruit stone cell. (C) A collenchyma cell has reinforced thickenings only in the cell corners. (D) The thickened and elaborated inner wall of a guard-cell pair provides the physical form needed to control the aperture size of a stoma. A special wall outgrowth of waxy and phenolic substances called the cuticle (Cu) minimizes evaporative loss of water from the exterior surface.

coniferyl, and sinapyl alcohols—account for most of the lignin networks (see Chapter 24). The monolignols are linked by way of ester, ether, or carbon–carbon bonds. The diversity of monolignols and their possible intralignol linkages impart remarkable complexity of structure.

Lignins can be detected in tissue sections by specific stains, such as the Mäule reagent, acid fuchsin (Fig. 2.44A), and the Wiesner reagent (phloroglucinol; Fig. 2.44B and 2.44C). The Mäule reagent, a mixture of $KMnO_4$ and ammonia, is particularly useful because it distinguishes syringyl lignin and guaiacyl lignin, indicating the former by a bright red color and the latter by yellow (Fig. 2.44D and 2.44E). Total lignin is quantified by various methods, either chemically or by NMR and other spectroscopic methods. Degradative acidolysis and thioacidolysis, permanganate, nitrobenzene, and alkali/copper oxide oxidation all are used to detect lignin. The oxidation products from these reactions include *p*-hydroxybenzaldehyde, vanillin, and syringaldehyde from *p*-hydroxylphenyl, guaiacyl, and syringyl lignin, respectively. Because all plants contain *p*-hydroxylphenyl, lignins are classified strictly by their guaiacyl and syringyl

content. Gymnosperms contain predominantly guaiacyl, whereas woody angiosperms and grasses exhibit a broad range of ratios of guaiacyl–syringyl residues.

The synthesis of monolignols is fairly well documented in plants, and all synthetic reactions appear to occur in the cytosol (see Chapter 24). How monolignols form lignin, however, is not clear. Lignols can be glycosylated in reactions associated with the ER and the Golgi apparatus, and this glycosylation may be necessary for membrane transport and targeting. The extent to which monolignols begin to condense and form associations with carbohydrates or other materials during secretion is unclear. Once in the wall, monolignols and their initial condensation products are polymerized by **peroxidases,** which utilize H_2O_2 as a substrate, or by **laccases,** which use O_2 and participate in the formation of dilignols.

As cells differentiate, many other kinds of ester, ether, and phenyl–phenyl bonds can tightly link an aromatic framework to carbohydrate (see Fig. 2.22). Even lignin is covalently linked to cellulose and xylans in ways that indicate the orientations of polysaccharides may serve as a template for the lignin patterning.

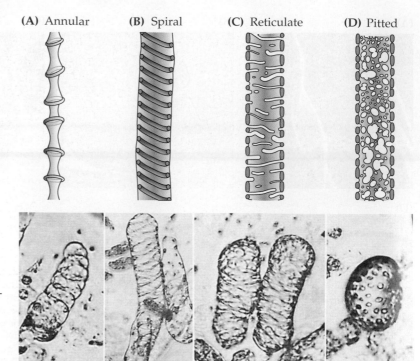

(A) Annular **(B)** Spiral **(C)** Reticulate **(D)** Pitted

Figure 2.42
The secondary wall thickening patterns of vessel elements in a single plant may be annular, spiral, reticulate, or pitted. Each of these patterns is evident during formation of the tracheary elements in vitro by *Zinnia* mesophyll cells in liquid culture.

Figure 2.43
Cells that produce special walls can function long after death. The sycamore-maple uses the paper-thin walls of a samara as wings for seed dispersal (A). The hard fruit coat of a willow cracks open along a dehiscence zone on desiccation (B). The pattern of microfibril deposition creates a physical tension that causes the fruit to snap open. Like the dandelion (C), the feather tufts of willow trichomes are used for long-distance seed dispersal.

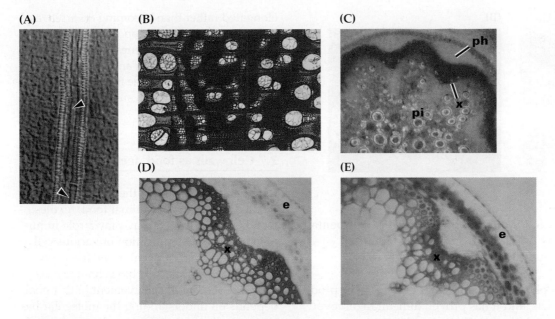

Figure 2.44

Several different procedures can be used to visualize lignin, including (A) acid fuchsin, which produces an orange fluorescence. In this *Arabidopsis* leaf, lignin can be seen in the spiral thickenings of the vessels. Phloroglucinol is a general stain for lignin, as seen in the cross-section of a *Robinia pseudoacacia* stem (B), or the xylem (x) and extraxylary fibers of an *Arabidopsis* floral stem (C). The Mäule reagent produces a red color with syringyl lignin, which is made predominantly from sinapyl alcohols, but forms a yellow color on reacting with guaiacyl lignin, which is made primarily from coniferols (D). This reagent here differentiates a mutant of *Arabidopsis* unable to make syringyl lignin (E). ph, phloem; pi, pith; e, endodermis.

2.6.6 Some secondary walls can serve as storage materials.

Another site of diversity among plant species is in the secondary walls of the cotyledon and the endosperm of developing seeds. These walls contain little or no cellulose but rather consist of a single noncellulosic polysaccharide typically found in the primary wall (see Figs. 2.12, 2.14, and 2.15). These secondary walls serve two functions. First, they provide a strong wall to protect the embryo or impose mechanical dormancy. Second, they contain specialized storage carbohydrates that are digested during germination and converted to sucrose for transport to the growing seedling.

The cotyledon walls of *Tamarindus* and similar legumes, as well as species of Primulaceae (primrose family), Linaceae (flax family), and Ranunculaceae (buttercup family) abound in a Gal-rich XyG. Glucomannans predominate in the cotyledon walls of some lilies and irises. Seeds of date (Fig. 2.45), coconut, and other palms; coffee bean; ivory

nut; and seeds of some Apiaceae all contain a thick cotyledon or endosperm wall of almost pure mannan. The endosperm wall of lettuce seeds, which constitutes the mechanical determinant of dormancy, is more than 70% mannan. All endospermic legumes store galactomannans, but the Man:Gal ratio can vary markedly, yielding a variety of galactomannans with very different physical properties. For example, fenugreek *(Trigonella)* makes an almost fully branched galactomannan, whereas guar *(Cyamopsis)* galactomannans and those of the carob or locust bean *(Ceratonia)* are much less branched, which changes their viscosities. Seeds of yet other species accumulate neutral polysaccharides typically found associated with pectins. For example, lupines contain large amounts of $(1{\rightarrow}4)\beta$-D-galactan, and some arabinan.

All of the grasses accumulate $(1{\rightarrow}3)$, $(1{\rightarrow}4)\beta$-D-glucan in the walls of the endosperm at some stage in embryo development. Oat and barley brans are notably enriched in β-glucans, which make up as much as 30% of the aleurone layer cell walls at maturity.

(A) 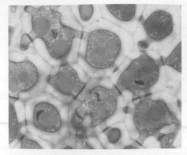 **(B)**

Figure 2.45
Endosperm cells of date (*Phoenix dactylifera*) have extremely hard and thick walls that accumulate mannans. (A) A low-magnification light micrograph of date endosperm shows that these mannans exclude the toluidine blue O dye. (B) Cotyledon walls of the Brazilian legume jatobá (*Hymanaea courbaril*) accumulate "amyloid" xyloglucans that stain lightly with iodine (B).

2.6.7 Walls can be modified experimentally by use of environmental adaptation, mutation, and genetic engineering.

Polysaccharides are not primary gene products, and it has proven difficult to discover precisely the role that each component (and its modifications) plays in the overall mechanical and functional properties of the cell wall or of the tissues that contain them. Cells in liquid culture (Fig. 2.46) have proved useful in a whole range of studies on cell wall metabolism in which wall modifications are induced in response to various environmental stresses. However, given the estimated 1000 gene products involved in cell wall biosynthesis, assembly, and turnover, genetically defined variation offers the most comprehensive approach to addressing function. *Arabidopsis* mutants with altered carbohydrate components in the primary wall have been detected by gas chromatography of alditol acetates of their neutral sugars. More than three dozen mutants have been classified by mapping to 11 different loci in which one or several specific sugars are over- or underrepresented compared with the sugar composition in wild-type plants. Of these, *mur*1 has been identified as a GDP-mannose-4,6-dehydratase, *mur*2 as a fucosyltransferase, and *mur*4 as a C-4 epimerase. However, many cell wall mutants also have been selected on the basis of a growth or developmental phenotype. A mutant in primary wall cellulose synthase, *rsw*1, was selected by a root radial swelling phenotype; a secondary wall cellulose synthase mutant, *irx*3, was selected by its collapsed xylem phenotype. A dwarf hypocotyl mutant, *korrigan*, results from a lesion in a membrane-bound endoglucanase, whereas another dwarf, *acaulis*, is an XET mutant. The *tch*4 mutant is an XET mutant but remains

elongated rather than becoming dwarfed, as wild-type plants do, in response to a breeze or gentle touch. As the genes of many cell-wall–modifying enzymes and structural proteins become available, the function of their products will be tested in suppression and overexpression studies.

2.7 Cell walls as food, feed, and fibers

Cell walls directly affect the raw material quality of human and animal food, textiles, wood, and paper and may play a role in human medicine. Modification of various cell wall constituents is a goal in the food processing, agriculture, and biotechnology industries. Successful achievement of this goal depends on understanding the molecular basis for mechanical and textural properties of plant-derived materials.

At present, the food industry uses isolated AGPs and pectins as gums and gelling agents. Fungal and bacterial wall hydrolases are used to adjust food textures and states. The cell walls of fruits and vegetables are now recognized as important dietary components and may protect against cancer of the colon, coronary heart disease, diabetes, and other ailments. β-Glucans are the causal agents in the ability of oat and barley brans to lower serum cholesterol and reduce the insulin demand of people with diabetes. Some pectins may have antitumor activities, possibly by stimulating the immune system after uptake through particular cells of the gastrointestinal tract.

With the advent of biotechnology, agricultural researchers are investigating particular enzymes involved in cell wall metabolism in hopes of producing crops with desired characteristics by enhancing commercially valuable traits (e.g., fiber production in flax, cotton, ramie, and sisal) or abolishing costly ones (e.g., lignification of some plant tissues). For example, the pulp and paper industry, which processes trees into cellulose, and the livestock industry, which depends on the transformation of cell walls into muscle tissue, are striving to reduce the lignin content in their respective sources of fiber and fodder. Reducing lignin content would reduce organochlorine wastes and cut costs tremendously for the paper industry, which currently uses chemical

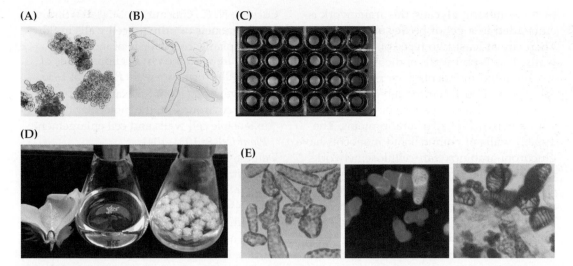

(A) **(B)** **(C)**

(D)

(E)

Figure 2.46
In addition to the many plant systems used in cell wall studies, cells in liquid (or suspension) culture are a very useful resource—abundant, homogeneous, and available at all times. Pioneering work with cells of the sycamore-maple *(Acer pseudoplatanus)* provided the first structures of XyG and RG I and II. Cell cultures are generally started by swirling pieces of tissue in flasks of culture medium until the cells break off. The culture is then propagated by diluting a sample of the culture into a fresh flask of medium every 7 days or so. (A) Most media are devised so that small clumps of dividing cells are produced, as shown with this carrot cell suspension culture. (B) The medium conditions can also be altered to induce cell elongation. (C) Sections of immature tomato pericarp will ripen in vitro if provided the proper developmental cues (e.g., ethylene). This system permits plant physiologists to study empirically the biochemical changes that occur in the cell wall during the swelling and softening events. (D) One of the more remarkable examples of cell development in vitro is the formation of cotton fibers by unfertilized ovules

in liquid culture. The cotton fibers are epidermal hair cells that elongate for about 3 weeks, achieving a length of almost 3 cm. Toward the end of the elongation phase and primary wall formation, a thick secondary wall of almost pure cellulose is deposited, increasing the mass more than 20-fold and nearly filling the lumen of the fiber cell at maturity, about 2 weeks later. At maturity the fibers of the plant are more than 95% cellulose. (E) *Zinnia* mesophyll cells have been used to study the development of tracheary elements, including the deposition of secondary walls, in vitro. Intact single cells are obtained aseptically by gently mashing young leaves of *Zinnia* in a mortar and pestle and incubating them in a medium containing cytokinin and auxin (left panel). Culture conditions have been optimized so that more than 70% of the cells undergo xylogenesis synchronously. Like those in the plant, these tracheary elements develop thickened secondary walls, which yield a bright fluorescence when stained with Calcofluor (middle panel), and become lignified, as seen by staining with phloroglucinol (right panel).

extractions to purify cellulose from the wood. Lignin–carbohydrate interactions exert a great influence on digestibility of forage crops by animals, and the *kind* of lignin present, rather than the total amount, is often the critical factor. Hence, mutants with altered lignin type may yield new forage crops that exhibit greater digestibility without sacrifice of the strengthening function of lignin to the water-conducting cells of the plant.

Particular foods are being genetically engineered to maximize consumer appeal and to improve their storage characteristics. Target enzymes include PGase, PME, and several polysaccharide hydrolases. The results of attempts to change only one parameter of wall metabolism are rarely those predicted, because of the inherent complexity of the wall and the ability of plant cells to adapt to change. Such biotechnological applications will become ever more common as we understand more of the regulation of cell wall

metabolism, particularly regulatory mechanisms for polymer synthesis and wall loosening. As we approach a revolution in the genetic engineering of the crops in our fields, the cell wall is a scientific frontier.

Summary

Cell walls are composed of polysaccharides, proteins, and aromatic substances. The primary wall of the cell is extensible but constrains the final size and shape of every cell. Facing walls of adjacent cells adhere to each other at the middle lamella. In some cells, secondary walls are deposited on the inner surface of the primary wall after growth has stopped. Cell walls become specialized for the function of the approximately 40 cell types that plants comprise.

The cellulose microfibrils form the scaffold of all cell walls and are tethered together by cross-linking glycans; this framework is embedded in a gel of pectic substances. There are at least two types of primary walls. The Type I walls of dicots and non-commelinoid monocots have xyloglucan–cellulose networks embedded in a pectin-rich matrix and can be further cross-linked with a network of structural proteins. The Type II walls of commelinoid monocots have glucuronoarabinoxylan–cellulose networks in a relatively pectin-poor matrix. Ferulate esters and other hydroxycinnamic acids and aromatic substances cross-link the Type II walls.

The cell wall is born at the cell plate. Cellulose microfibrils are synthesized at the surface of the plasma membrane at terminal complexes called particle rosettes, whereas all noncellulosic cross-linking glycans and pectic substances are made at the Golgi apparatus and secreted. All cell wall sugars are synthesized de novo from interconversion of nucleotide sugars, which are the substrates for polysaccharide synthases and glycosyltransferases.

Cell enlargement depends on the activities of endoglycosidase, endotransglycosylase, or expansin, or some combination of these, but cell shape is largely governed by the pattern of cellulose deposition. Cell enlargement also is accompanied by numerous changes in the structure of the wall's cross-linking glycans and pectin matrix. Termination of cell growth is accompanied by cross-linking reactions involving proteins and aromatic substances.

In addition to their use in wood, paper, and textile products, cell walls are the major textural component in fresh fruits and vegetables and constitute important dietary fibers in human nutrition. Transgenic plants with altered cell wall structures will become an important factor in crop and biomass improvement.

Further Reading

Carpita, N. C. (1996) Structure and biogenesis of the cell walls of grasses. *Annu. Rev. Plant Physiol. Plant Mol. Biol.* 47: 445–471.

Carpita, N. C., Gibeaut, D. M. (1993) Structural models of primary cell walls in flowering plants: consistency of molecular structure with the physical properties of the walls during growth. *Plant J.* 3: 1–30.

Cosgrove, D. J. (1997) Relaxation in a high-stress environment: the molecular bases of extensible cell walls and cell enlargement. *Plant Cell* 9: 1031–1041.

Delmer, D. P. (1999) Cellulose biosynthesis: exciting times for a difficult field of study. *Annu. Rev. Plant Physiol. Plant Mol. Biol.* 50: 245–276.

Gibeaut, D. M., Carpita, N. C. (1994) The biosynthesis of plant cell wall polysaccharides. *FASEB J.* 8: 904–915.

Hadfield, K. A., Bennett, A. B. (1998) Polygalacturonases: many genes in search of a function. Plant Physiol. 117: 337–343.

Jarvis, M. C. (1984) Structure and properties of pectin gels in plant cell walls. *Plant Cell Environ.* 7: 153–164.

McCann, M. C. (1997) Tracheary element formation: building up to a dead end. Trends Plant Sci. 2: 333–338.

McCann, M. C., Roberts, K. (1991) Architecture of the primary cell wall. In *The Cytoskeletal Basis of Plant Growth and Form*, C. W. Lloyd, ed. Academic Press, London, pp. 109–129.

Meier, H., Reid, J.S.G. (1982) Reserve polysaccharides other than starch in higher plants. In *Encyclopedia of Plant Physiology*, Vol. 13A, F. A. Loewus and W. Tanner, eds. Springer-Verlag, Berlin, pp. 418–471.

Rees, D. A. (1977) *Polysaccharide Shapes*. Chapman and Hall, London.

Reiter, W.-D. (1998) *Arabidopsis thaliana* as a model system to study synthesis, structure, and function of the plant cell wall. *Plant Physiol. Biochem.* 36: 167–176.

Showalter, A. M. (1993) Structure and function of plant cell wall proteins. *Plant Cell* 5: 9–23.

Taiz, L. (1984) Plant cell expansion: regulation of cell-wall mechanical properties. *Annu. Rev. Plant Physiol.* 35: 585–657.

Biochemistry & Molecular Biology of Plants, B. Buchanan, W. Gruissem, R. Jones, Eds.
© 2000, American Society of Plant Physiologists

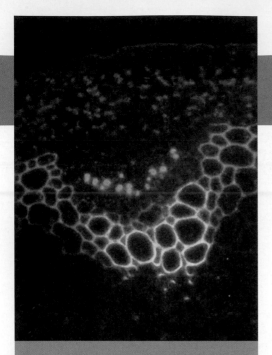

CHAPTER *3*

Membrane Transport

Dale Sanders
Paul Bethke

Introduction

The selective movement and redistribution of ions and small organic molecules is essential for plant growth and cellular homeostasis. Because of this, plants have evolved numerous proteins that facilitate the transport of minerals, sugars, metabolites, and other compounds through the limiting membranes of cells and organelles. These membrane transporters are selective for the compound being transported, and their activities are often tightly regulated. Membrane transporters are usually members of multigene families that show developmental and tissue-specific expression. How the activities of individual transporters are coordinated at the cell or tissue level is only beginning to be understood.

3.1 Overview of membrane transport

3.1.1 Membranes facilitate compartmentalization.

One of the earliest events in the evolution of life was the development of a barrier between the sensitive biochemical machinery of replication and the dispersive forces of the outside world. This barrier is the cell membrane. Delimited by its membrane, the cell can support metabolic, reproductive, and developmental activities that require a stable physicochemical environment. The hydrophobic nature of the lipid bilayer ensures that hydrophilic compounds, including most metabolites, can be sequestered on one side of the membrane or the other. During the 1930s, groundbreaking work by the Finnish plant biologist Runar Collander established that the ability of molecules to permeate plant cell membranes correlates directly with their oil:water partition coefficient (Box 3.1). The same principle applies to all biological membranes.

The evolution of eukaryotes, concomitant with development of an endomembrane system, allowed this homeostatic function of membranes to be carried one step further. Compartmentalization of solutes within membrane-bound organelles not only concentrates reactants and catalysts but also segregates incompatible processes. Plant cells contain many specialized organelles, responsible for diverse biosynthetic, catabolic, and storage functions (see Chapter 1). This division of labor facilitates metabolic flexibility and efficiency.

3.1.2 Selective permeability of biological membranes derives from transport systems that contain integral membrane proteins.

Controlled transport of metabolites across the plasma membrane and intracellular membranes is required for integrated organismal and cellular metabolism. Transport is made possible by membrane-spanning proteins within the lipid bilayer. These so-called transport systems can be regarded as conventional enzymes in almost all respects, with the important exception that transport events are **vectorial** (i.e., defined by a magnitude and a direction in space), whereas enzyme-catalyzed reactions are **scalar** (i.e., defined entirely by magnitude). Like enzymes, all transport systems exhibit some degree of substrate-specificity and work by lowering the activation energy required for transport.

Integral membrane proteins are characterized by stretches of amino acid sequence that are predominantly hydrophobic and can interact favorably with the fatty acyl chains of membrane phospholipids. **Hydropathy analysis** (Box 3.2) identifies integral membrane proteins on the basis of the principle that an α-helix must contain about 20 amino acid residues to span the hydrophobic sector of the lipid bilayer. Although all transport systems have membrane-spanning components, some also require soluble protein subunits for full catalytic activity.

3.1.3 Membrane transport underlies many essential cell biological processes.

The complete sequence of the yeast genome reveals that about 2000 of the 6000 genes encode proteins associated with membranes, of which a large proportion are transport system components. In plant cells, membrane transport underpins a wide range of essential processes, including the following:

- *Turgor generation.* The presence of a cell wall in the vast majority of plant cells enables them to generate turgor (positive pressure). The cell wall provides structural rigidity, allowing the cell to survive in dilute media without bursting. Turgor generation is accomplished by accumulating salts. In the mature cells of most plants, K^+ accumulates in the cytoplasm and in the large central vacuole, whereas

Box 3.1

Early experiments with giant internodal cells of charophyte algae linked lipophilicity and membrane permeability.

Although the hydrophobic nature of the permeability barrier that characterizes cell membranes has been established for about 100 years, it was experiments carried out during the 1930s by Runar Collander that quantified very elegantly the relationship between solute lipophilicity and membrane permeability. Collander's studies revealed that the rapidity with which uncharged compounds permeate the plasma membrane of a plant cell is related to their ability to dissolve in a hydrophobic phase—namely (in his studies), olive oil.

Collander took single, giant internodal cells of the charophyte alga *Chara tomentosa* (then known as *C. ceratophylla*) and incubated them in various bathing solutions, each of which contained a different uncharged organic compound. After specific periods, he removed the cells, washed them and, with careful chemical analysis, determined the content of each tested compound within the cells. Eventually, the concentration of each compound inside the cell was equal to the concentration outside, although the time taken to reach this equilibrium state varied, de-

pending on the compound, from a few minutes to several days. Collander calculated a half-time ($t_{0.5}$) for filling based on the time course for attainment of the equilibrium. From the $t_{0.5}$ value, he was able to estimate a permeability coefficient (P_S) for each solute, based on the relationship shown in Equation 3.B1, where ln 2 is the natural logarithm of 2, and V and A are the volume and area of the cell, respectively. V and A were quite easily measured because the cells are cylindrical in shape, several centimeters long, and about 1 mm in diameter. The formulae for calculating the volume and surface area of a cylinder could then be applied.

Equation 3.B1

$$P_S = \ln 2 \bullet V/t_{0.5}A$$

Because solutes with a low relative molecular mass (M_r) diffuse faster than those with a high M_r, Collander "corrected" for this effect by multiplying the estimates of PS by the square root of the M_r of the solute. Plotting the resulting values as a function of the oil:water partition coefficient for each compound shows a re-

markable linear relationship, apparent over three orders of magnitude. This relationship is shown for 30 compounds in the accompanying figure, which is presented on a log-log scale to encompass the wide range of permeabilities derived. Each point represents a different compound, and some of the more familiar compounds are identified. The regression line to which these data can be fitted indicates a strong correlation between lipophilicity and membrane permeability over a wide range of both parameters. Water, an obvious outlier on the plot, exhibits two interesting features. First, the unique hydrogen-bonding capacity of water endows the molecule with a surprising degree of lipophilicity. Second, the measured membrane permeability is two orders of magnitude greater than would be predicted for a molecule with this degree of lipophilicity—an indication of the presence of a permeability pathway for water that bypasses solubility in the membrane lipid. The identity of this pathway is explored further in Section 3.7.

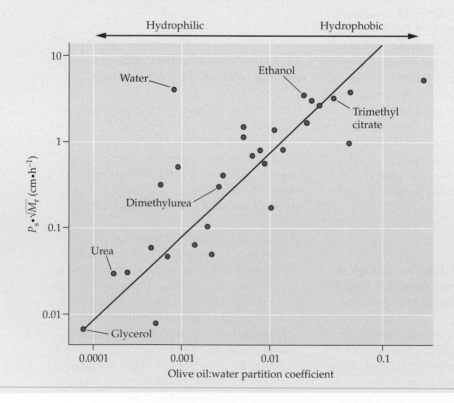

Box 3.2

Hydropathy analysis uses sequence data to identify membrane-spanning polypeptides.

Because all transport systems are intrinsic membrane proteins, an obvious pair of questions is, "Which domains of the protein traverse the membrane, and which are exposed to the aqueous phases?" For proteins with a known protein sequence, a simple test has been devised. This test incorporates the recognition that a polypeptide in a predominantly hydrophobic environment (e.g., a membrane) will adopt an α-helical conformation. Given the well-established dimensions of an α-helix (one amino acid residue for every 0.15 nm rise along the helix) and those of the hydrophobic portion of the membrane lipid (about 3 nm for the fatty acyl side chains of the phospholipid bilayer), about 20 amino acid residues are required to span the hydrocarbon core of a membrane, assuming that the α-helix is perpendicular to the plane of the membrane.

The propensity of a stretch of polypeptide to reside in a bilayer can be predicted on the basis of the mean hydrophilicity of the component amino acid residues. In 1982, Jack Kyte and Russell Doolittle assigned each of the common amino acids a "hydropathy index," reflecting the aqueous solubility of the amino acid; values ranged from +4.5 for the most hydrophobic amino acid (isoleucine) to –4.5 for the most hydrophilic amino acid (arginine). Stretches of a protein sequence can be scanned by computer to determine the mean hydropathy index of that stretch. Kyte and Doolittle determined that a stretch, or "window," of 19 residues and a mean hydropathy index of greater than 1.6 would identify known transmembrane domains of proteins and would distinguish those from the hydrophobic domains buried in the center of globular proteins. Thus, a mean hydropathy index is calculated for residues 1 through 19 of a protein, then residues 2 through 20, and so on. The analysis predicts that highly hydrophilic, even charged, residues can reside in the membrane lipid, provided the number of hydrophobic residues in the vicinity is sufficient to favor stable partitioning in a hydrophobic phase. This prediction is supported by experimental evidence: Mutations that introduce charged residues into transmembrane domains do not alter transport function.

A sample hydropathy plot resulting from such an analysis for the plant K[+] channel protein KAT1 is shown in panel A of the accompanying figure, and the interpretation of the results in terms of transmembrane topology is shown in panel B. Results of hydropathy analyses are theoretical, however, and experimental confirmation of predictions arising from the analysis is desirable (though not often performed). Ideally, transmembrane domains can be identified from a crystal structure; except for some proteins in the thylakoid membrane (see Chapter 12), however, no plant membrane protein has yet been visualized at atomic resolution. An alternative experimental approach is to construct chimeric proteins, inserting a reporter enzyme, such as alkaline phosphatase, at various positions within the transmembrane protein. A transgene is constructed that encodes an alkaline phosphatase inserted at one of two locations within KAT1: glutamate-95 (D95) or lysine-128 (K128). The chimeric gene is expressed in *Escherichia coli*. If the alkaline phosphatase domain of the chimeric protein is located in the cytosol of *E. coli*, the enzyme activity is unstable; if located in an extracellular domain, however, the activity of the enzyme is stable (panel C). Alkaline phosphatase is detected for chimera D95 but not for chimera K128. This finding is consistent with the hydropathy analysis because D95 and K128 are on opposite sides of the membrane and therefore must be separated by a transmembrane domain. In practice, many such constructs must be performed to determine the topology of the complete protein.

More refined theoretical approaches have been devised since Kyte and Doolittle developed hydropathy analysis, including techniques to maximize charge–charge pairings within the membrane. Viewed from above the plane of the membrane, the α-helices of the membrane protein can be visualized as helical wheels (panel D). In this representation of the first 22 amino acids of the S4 transmembrane domain of KAT1, each letter represents an amino acid residue (standard amino acid code). Positively charged arginine residues are shown in yellow.

Charged residues of opposite polarity on different helices tend to associate with each other, contributing to the tertiary structure of the protein.

(A)

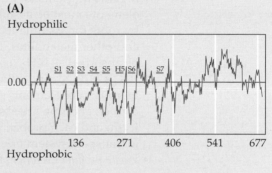

(B)

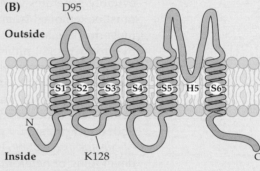

(C)

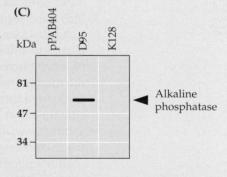

(D)

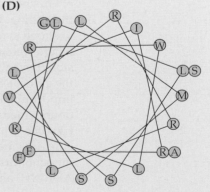

in halophytes the principal cation is usually Na^+. The cations must be balanced by a corresponding concentration of anions to achieve electroneutrality; in the vacuole, the principal anion is usually Cl^- (depending on its extracellular availability) or malate.

- *Nutrient acquisition.* Unlike animals, plants synthesize organic biomolecules from inorganic nutrients. Many of these essential elements must be absorbed from the soil by roots for assimilation into amino acids and other metabolites. Thus, nitrogen can be absorbed as NH_4^+ or NO_3^-, phosphorus as $H_2PO_4^-$, and sulfur as SO_4^{2-}. Numerous trace elements such as boron, zinc, copper, and iron are absorbed in inorganic form, and each probably requires a specific transport system (see Chapter 23).

- *Waste product excretion.* Metabolism inevitably generates waste products that must be removed from the cytosol. In most nutritional conditions, the principal by-product is H^+, although OH^- production can dominate in plants that assimilate HCO_3^- and NO_3^- into organic compounds. To expel H^+ from the cytosol, plants have evolved proton pumps at both the vacuolar and plasma membranes.

- *Metabolite distribution.* In terrestrial plants, developing or heterotrophic tissues require reduced carbon and other metabolites to be supplied by autotrophic tissues. Long-distance transport from source tissues to sinks is accomplished by the phloem. Sucrose and amino acids cross the membranes of companion cells and are loaded into the phloem by way of specific transport systems (see Chapter 15).

- *Compartmentalization of metabolites.* Given the many metabolic pathways present within a plant cell, compartmentalization of enzymes and metabolites prevents futile cycling. One classic example relates to starch synthesis within amyloplasts, where starch can be synthesized and stored even while glycolysis proceeds in the cytosol. Starch synthesis depends on importing glucose 6-phosphate across the double membrane of the amyloplast (see Chapter 13). Compartmentalization also enhances metabolic efficiency. In the mitochondrial matrix, for example, the ADP/ATP and NADH/NAD$^+$ ratios are greater than in the cytosol, which pro-

vides a substrate/product ratio that favors respiratory activity. Correspondingly, specific transport mechanisms are required for the export of ATP and NAD^+ to the cytosol (see Chapter 14).

- *Energy transduction.* Membrane transport lies at the heart of the conversion of free energy into biologically useful forms. Light energy stimulates the photosynthetic electron transport chain to pump H^+ into the thylakoid lumen. Similarly, oxidation of NADH provides the energy for pumping H^+ from the mitochondrial matrix into the intermembrane space. In each case, the spontaneous, exergonic flow of H^+ back across the membrane is used to generate ATP. Understanding the mechanisms by which light and high-energy electrons are harnessed to phosphorylate ADP therefore demands knowledge of membrane transport.

- *Signal transduction.* Many biotic and abiotic signals for plant growth and development trigger their respective responses by transiently increasing the concentration of cytosolic free Ca^{2+}, which is accomplished through the activities of two kinds of transport systems. First, Ca^{2+}-translocating ATPases remove Ca^{2+} from the cytosol by pumping it across the plasma membrane and intracellular membranes. Second, Ca^{2+}-permeable channels open in response to particular stimuli and allow the passive entry of Ca^{2+} into the cytosol, thereby propagating the signal.

3.2 Organization of transport at plant membranes

3.2.1 Metabolically coupled H^+ pumps underlie a proton-based energy economy in plants.

Protons (H^+) constitute one of the major energy currencies of the plant cell, on a par with NAD(P)H and ATP. At the inner mitochondrial membrane and at the thylakoid membrane, a transmembrane H^+ potential is generated and used to energize the synthesis of ATP. At all other membranes in the cell, **H^+ pumps** hydrolyze ATP to power the transport of protons out of the cytosol, establishing electrochemical potentials across these membranes as well. The resulting

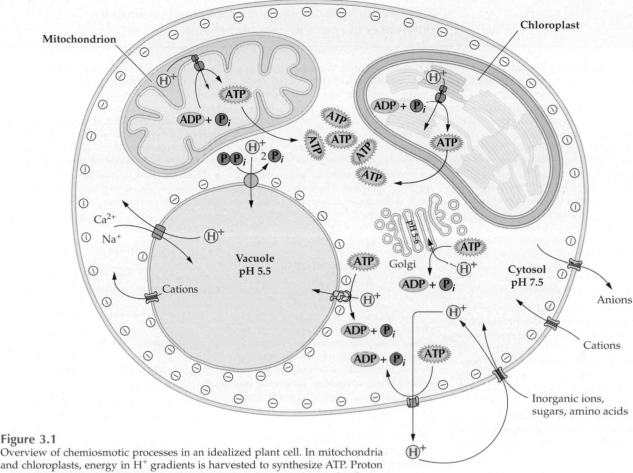

Figure 3.1

Overview of chemiosmotic processes in an idealized plant cell. In mitochondria and chloroplasts, energy in H^+ gradients is harvested to synthesize ATP. Proton gradients also are established across the plasma membrane and tonoplast by pumps that hydrolyze ATP and PP_i. The electrochemical potential established by these pumps is used to move many ions and small metabolites through integral membrane channels and carriers.

transmembrane H^+ potentials are then used to power the transport of other ions and solutes across the membranes (Fig. 3.1).

These general principles were elucidated by Peter Mitchell in the early 1960s, in the context of ATP synthesis and bacterial transport. The chemiosmotic coupling hypothesis is now widely accepted as a universal mechanism of biological energy conservation, and Mitchell was awarded the Nobel Prize for chemistry in 1978.

The net direction of transport across a membrane is dictated by the driving force on the transport system. Driving forces can be quantified in terms of the free energy relationships inherent in transmembrane **solute potentials** (Box 3.3), whether these involve protons, other inorganic ions, or organic solutes such as sugars and amino acids. We can describe the magnitude of a

transmembrane potential for an uncharged solute ($\Delta\mu$) or for an ion ($\Delta\bar{\mu}$) in the units of free energy (kJ/mol) normally associated with scalar reactions. For an ion bearing a net charge, two factors contribute to the overall driving force: the concentration difference across the membrane, and the membrane electrical potential difference (**membrane potential,** or V_m, Boxes 3.3 and 3.4).

The polarity (sign) of the transmembrane (electro)chemical potential for a solute determines the direction of flow. At the plasma membrane, a negative value of $\Delta\mu$ or $\Delta\bar{\mu}$ indicates passive flow *from* the external medium *into* the cytosol. Conversely, a positive value indicates that energy input is required for uptake of the solute or ion because the passive direction of flow is out of the cell.

In the particular case of H^+, a simplified expression based on pH, rather than $[H^+]$, can

The **chemical potential** of an uncharged solute, S, in a given medium can be defined as in Equation 3.B2, where μ^o_S is the **standard chemical potential** of S at 1 M, R is the gas constant (8.314 J•mol^{-1}•K^{-1}), T is absolute temperature (in K), and ln a is the natural logarithm of the activity, a, of the solute. For uncharged solutes, a can be expressed as molar concentration, but for charged solutes, a is used to account for dissociation of the solute. The units of chemical potential are J•mol^{-1}.

Equation 3.B2

$$\mu_S = \mu^o_S + RT•\ln a$$

Suppose we have a membrane separating two aqueous media, one inside the cell (denoted by the subscript c) and the other outside the cell (denoted by the subscript o), as shown in the figure below. We can calculate the chemical potential of an uncharged solute S in each of the two media by using Equation 3.B2. For S in medium c and medium o, we can make the calculations shown in Equations 3.B3a and 3.B3b, respectively.

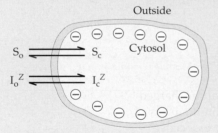

Outside

Cytosol

Equation 3.B3a

$$(\mu_S)_c = \mu^o_S + RT•\ln[S]_c$$

and

Equation 3.B3b

$$(\mu_S)_o = \mu^o_S + RT•\ln[S]_o$$

Now we can calculate the **chemical potential difference** for S, $\Delta\mu_S$, across the membrane simply by subtracting Equation 3.B3b from Equation 3.B3a. The standard chemical potential terms cancel each other out, and we are left with Equation 3.B4, the relationship that defines the energy stored in a transmembrane solute concentration difference. The term $RT•\ln\{[S]_c/[S]_o\}$ is similar to the Mass

Action Ratio term in the equation defining Gibbs free energy, $RT•\ln\{[\text{products}]/[\text{reactants}]\}$, except that here the reactants and products are defined by their position in space, not by their chemical modification. The vectorial reaction for S is shown in the figure. Convention dictates that we relate the cytosolic compartment to the extracytosolic compartment when defining chemical potential differences (thus, Eq. 3.B3b was subtracted from Eq. 3.B3a, and not vice versa).

Equation 3.B4

$$\Delta\mu_S = RT•\ln[S]_c - RT•\ln[S]_o = RT•\ln\{[S]_c/[S]_o\}$$

The polarity of $\Delta\mu_S$ is critical in expressing the energetic relationship of the solute across the membrane. A positive value indicates that free energy must be applied to translocate solute into the cytosol, whereas a negative value indicates that the solute will move passively, given an appropriate transport system. A value of zero indicates equilibrium; i.e., the solute concentration is the same on each side of the membrane. Just as the Gibbs free energy change describes the equilibrium position of a reaction but not the rate at which equilibrium is reached, so too the magnitude of the chemical potential difference describes the energetics of transport but cannot be used to predict rates of transport.

Let us now consider an ion, I, carrying a net charge of z, which is positive for cations and negative for anions. Because the ion bears a charge, the overall energetic potential of the ion will be influenced by the electrical potential of the medium. This involves adding an extra term to Equation 3.B2, so that we can now express the **electrochemical potential** of the ion as in Equation 3.B5, where $\bar{\mu}^o_I$ is the **standard electrochemical potential** of I at 1 M, F is the Faraday constant (96,500 C•mol^{-1} [coulombs per mole of electrons]), and ψ is the electrical potential of the medium (units: volts). The inclusion of F in the final term has the effect of converting the electrical units into J•mol^{-1}, and because z is dimensionless, all terms in the equation have the same units.

Equation 3.B5

$$\bar{\mu}_I = \bar{\mu}^o_I + RT•\ln[I] + zF\psi$$

If we now return to the model cell in which the two media (cytosolic, or c, and outside, or o) are separated by a membrane, as shown in the figure, we can calculate an **electrochemical potential difference** for the ion I across the membrane in a way analogous to that used to calculate the chemical potential difference for S. Thus, for ion I in media c and o, respectively, we can derive Equations 3.B6a and 3.B6b from Equation 3.B5.

Equation 3.B6a

$$(\bar{\mu}_I)_c = \bar{\mu}^o_I + RT•\ln[I]_c + zF•\psi_c$$

and

Equation 3.B6b

$$(\bar{\mu}_I)_o = \bar{\mu}^o_I + RT•\ln[I]_o + zF•\psi_o$$

Equation 3.B7

$$\Delta\bar{\mu}_I = RT•\ln\{[I]_c/[I]_o\} + zF•V_m$$

Subtracting Equation 3.B6b from 3.B6a, we obtain Equation 3.B7, where $\Delta\bar{\mu}_I$ the electrochemical potential difference for I, and V_m (= $\psi_c - \psi_o$) is the electrical potential difference or, more commonly, the membrane potential (see Box 3.4). The electrochemical potential difference for an ion therefore is simply the sum of the chemical and electrical potentials. When the chemical potential is equal and opposite to the electrical potential, the sum is zero and there is no overall driving force on the ion (i.e., the ion is at equilibrium). Each ionic species will have its own chemical potential difference across a membrane, depending on its concentration, but the electrical potential has a unique value at any one time.

A membrane potential is the difference in electrical potential between two aqueous media separated by a membrane. Two common abbreviations for membrane potential are encountered in the literature: V_m (favored by electrophysiologists) and $\Delta\psi$ (favored by bioenergeticists). We have chosen V_m for this chapter. Normally, V_m across the plasma membrane of plant cells is about -150 mV. Other membranes are less polarized: A value of -20 mV is commonly cited for the tonoplast.

Two important features of V_m are essential to understanding its nature and biological relevance. First, membrane potential is a macroscopic parameter. That is, if V_m is measured by comparing the potentials reported by two electrodes placed on each side of the membrane, then no matter how far from the membrane the electrodes are placed, the potential should be the same. Second, a membrane potential results from an imbalance in the number of cations and anions in the aqueous media separated by the membrane.

In terms of concentration, the cation-to-anion imbalance is very small indeed and can be calculated as follows: The charge (Q, in coulombs) stored on a membrane at a particular voltage is a function of the membrane capacitance (C, in coulombs per volt per square meter [$C \cdot V^{-1} \cdot m^{-2}$], i.e., farad per square meter [$F \cdot m^{-2}$]) according to the relationship $Q = CV$, where V is voltage. For most biological membranes, C is of the order 0.01 $F \cdot m^{-2}$, and for the plasma membrane of a plant cell, V_m averages -150 mV. This yields a net charge of 0.0015 $C \cdot m^{-2}$ as an excess of anions over cations inside the cell compared with the situation outside the cell. For a plant cell of radius 30 μm, the area of the plasma membrane is 10^{-8} m^2, assuming the shape of the cell approximates a sphere. Hence, converting coulombs to molar equivalents by using the Faraday constant, the excess negative charge in the cytosol is (1.5×10^{-3} $C \cdot m^{-2}$)(10^{-8} m^2 membrane)/(96,500 $C \cdot mol^{-1}$), roughly 1.5×10^{-16} eq. (Because F refers to the number of ionized species, the results are expressed here in terms of molar equivalents rather than moles.) If the cytosol of this cell occupies 10% of the intracellular volume (with the vacuole occupying most of the rest), the volume of cytosol is 10^{-11} l, so the excess negative charge is (2×10^{-16} eq)/(10^{-11} l) = 20 μeq$\cdot l^{-1}$. Given that the total ionic strength of the cytosol is more than 200 meq$\cdot l^{-1}$, the ionic imbalance does not exceed 1 part in 10,000.

Membrane potentials arise across biological membranes in several ways (see accompanying figure). First, most cytosolic proteins—bacterial, fungal, plant, or animal—bear a net negative charge at a physiological pH of 7.5. This excess of negatively charged amino acid residues is countered largely by K^+ ions. Membranes are impermeable to proteins, yet always exhibit a finite K^+ permeability through K^+-conducting channels. Because the external concentration of K^+ is invariably less than the cytosolic concentration (maintained at about 100 mM), the tendency of K^+ to leak from the cell generates an excess of negative charge inside, yielding a negative value of V_m. This combination of fixed immobile (negative) charges and mobile (positive) charges is known as a **Donnan potential.**

Electrogenic transport of H^+ out of the cytosol, the other major factor exerting a major influence on the resting value of V_m in plant cells, also tends to drive V_m to negative values (see figure). Pumping H^+ from the cytosol is achieved by metabolically coupled pumps (usually ATPases). By driving positive charge (current) from the cytosol, electrogenic pumps have the effect of leaving an uncompensated negative charge, which can result in **hyperpolarization** of the membrane; i.e., V_m can become even more negative than the value defined by high K^+ permeability. At steady state, continuing operation of electrogenic pumps requires the compensating movement of countercharges through other transport systems. This charge movement may consist of a flow of positive charges into the cytosol or a flow of negative charges out of it.

The balance between the inherently high K^+ permeability and electrogenic H^+ pump activity in plant membranes is usually the dominant feature that determines the resting V_m of plant membranes. However, any transport system capable of conducting ions, and therefore current, across the membrane will affect V_m. In addition, in some conditions, membranes can exhibit transiently high permeabilities to anions or to Ca^{2+}. This will ordinarily lead to a transient positive swing in V_m, or membrane **depolarization,** as anions flow out of the cytosol or Ca^{2+} ions flow into it. The role of ion channels in evoking transient changes in V_m is discussed further in Sections 3.6.3 and 3.6.8.

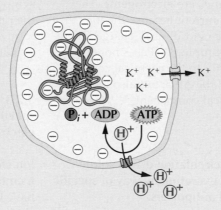

be obtained for the electrochemical potential difference. The resulting parameter, expressed in units of mV rather than kJ mol^{-1}, is known as the **proton motive force,** or simply pmf (Box 3.5). The sign rule given above for the transmembrane electrochemical potential ($\Delta\mu$) also applies to the pmf ($\Delta\mu_H^+/F$), because the conversion factor, the Faraday constant (F), is greater than zero.

The quantitative definition of the pmf (see Box 3.5) can be used to estimate the free energy contained in the proton potential that is generated across the plasma membrane by a H^+-pumping ATPase (sometimes called the plasma membrane proton pump). Typically, the cytosolic pH is 7.5, the apoplastic pH is near 5.5, and V_m is on the order of -150 mV. Under these particular conditions, the

Box 3.5

Proton motive force relates transmembrane pH difference to membrane potential.

The proton motive force (pmf) is a measure of the free energy stored in a transmembrane electrochemical potential difference for H^+ and is expressed in volts. The pmf can therefore be defined as shown in Equation 3.B8, where $\Delta\bar{\mu}_H^+$ can be determined by writing Equation 3.B7 from Box 3.3 explicitly for H^+ (Eq. 3.B9). For H^+, $z = +1$, and so can be ignored. Dividing both sides of Equation 3.B9 by F, and converting the natural logarithm into a $base_{10}$ logarithm by multiplying by 2.303, we obtain Equation 3.B10. Equation 3.B10 can be simplified considerably, because R and F are constants, and the value for T usually falls within a fairly narrow range. Furthermore, because

$log[H^+]_c = -pH_c$ and $log\{1/[H^+]_o\} = pH_o$, we can write, for conditions at 25°C, Equation 3.B11, in which the first term on the right-hand side simplifies to 0.0591 V. However, it is more convenient in biological systems to express the pmf (and V_m) in mV, so after multiplying through by 10^3, Equation 3.B11 becomes pmf = $59.1\{pH_o - pH_c\} + V_m$, which states that each pH unit change is energetically equivalent to 59 mV of membrane potential.

Equation 3.B8

$$pmf = \Delta\bar{\mu}_H^+/F$$

Equation 3.B9

$$\Delta\bar{\mu}_H^+ = RT \cdot \ln\{[H^+]_c/[H^+]_o\} + zF \cdot V_m$$

Equation 3.B10

$$pmf = (RT/F) \cdot (2.303)\log\{[H^+]_c/[H^+]_o\} + V_m$$

Equation 3.B11

$$pmf = [(8.314)(298)(2.303)/(96,500)] [pH_o - pH_c] + V_m$$

resulting pmf would be –268 mV; values ranging from –200 to –300 mV are common in plant cells.

The proton pumps at the vacuolar and plasma membranes are **electrogenic.** They create electrical current because the ions they remove from the cytosol carry charge. Therefore, these H^+-pumping ATPases not only contribute directly to the chemical component of the pmf, the ΔpH, but also tend to make the electrical component, V_m, more negative.

3.2.2 Proton recirculation drives solute absorption and excretion by way of specific symporters and antiporters.

Transport systems that couple the downhill (exergonic) flow of ions such as H^+ or Na^+ to the uphill (endergonic) flow of inorganic ions and solutes are called **carriers.** Carriers that catalyze solute flux in the same direction as H^+ or Na^+ flux are known as **symporters.** Because protons flow passively across most membranes in the direction of the cytosol, symporters typically energize uptake of solute into the cytosol, either from the external medium or from intracellular compartments. Conversely, excretion from

the cytosol can be accomplished by **antiporters,** which exchange solutes for protons. Antiporters are present at the plasma membrane and at endomembranes (Fig. 3.1). Both symporters and antiporters tend to dissipate the pmf, and this energy is conserved in the form of an electrochemical potential for particular solutes.

3.2.3 Channels catalyze the movement of specific ions in the net direction of their electrochemical potential driving forces.

Ion **channels** are ubiquitous in biological membranes. Several classes of ion channels have been described in plant systems, including those specific for K^+, Ca^{2+}, and anions. The net direction of ion flux through a channel is determined solely by the electrochemical driving force acting on that ion (see Box 3.3). Thus, in contrast to the operation of symporters and antiporters, the pmf plays no direct role in driving the passage of ions through channels. However, the V_m is an integral part of the electrochemical driving force on ions, and any membrane potential generated by the action of electrogenic pumps will contribute to the overall force that drives ions through channels.

3.2.4 Turnover rates among classes of transport system vary, affecting protein abundance in membranes.

Although all transport systems are integral membrane proteins by definition, their distinct functional attributes have important consequences for their prevalence in biological membranes. For pumps, long-range and complex conformational transitions that extend across large polypeptides or assemblies of polypeptides serve to couple the metabolic reactions to those involving transport. The turnover rate of pumps is correspondingly slow—on the order of 10^2 molecules transported per second. Carriers also undergo conformational changes during transport as they switch between states in which the transport binding sites are oriented either toward or away from the cytosolic face of the membrane. Nevertheless, there are no long-range interactions with soluble substrates, and turnover rates are greater than those of pumps, typically being about 10^3 s^{-1}. Channels, in contrast to pumps and carriers, do not undergo conformational changes during transport and can catalyze ion fluxes of 10^6 to 10^8 s^{-1}.

These considerations help explain why the concentrations of different transport proteins vary markedly. Pumps not only have a slow turnover number but also generate the pmf for an array of symporters and antiporters. Thus, H$^+$ pumps are fairly abundant and are prominent on sodium dodecyl sulfate (SDS)–containing polyacrylamide gels after electrophoresis of membrane proteins (Fig. 3.2). Although a square micrometer of membrane may include several hundred to several thousand pump proteins, it typically contains only 1 to 10 channel proteins, too few to permit the biochemical approaches that are applicable to the analysis of pumps. However, the rapid turnover rates and diffuse distribution of channels facilitate electrophysiological analysis of single channel molecules (see Section 3.5.2).

Figure 3.2
SDS–polyacrylamide gel electrophoresis (PAGE) of fractions obtained during purification of the plasma membrane H$^+$-ATPase from spinach leaves. $C_{14}E_8$, polyoxyethylene 8 myristal ether; DM, dodecyl-β-D-maltoside; FPLC, fast performance liquid chromatography.

3.3 Pumps

3.3.1 F-type H$^+$-pumping ATPases at the inner mitochondrial and thylakoid membranes are pumps that operate in reverse mode to synthesize ATP.

Proton-pumping ATPases known as F-type ATPases are found in plants at the inner mitochondrial and thylakoid membranes that synthesize ATP. These two kinds of membranes contain proton-pumping electron transport chains driven by redox potential and light energy, respectively (see Chapters 12 and 14). The pmf established by the electron transport chains serves to drive H$^+$ flow back through the F-type ATPases, thereby resulting in ATP synthesis. One sector of an F-type ATPase, called F_0 in plant mitochondria and CF_0 in plant chloroplasts, traverses the membrane and forms an H$^+$-permeable conduit. The other sector of the enzyme, called F_1 in mitochondria and CF_1 in chloroplasts, readily dissociates from the transmembrane sector, contains adenine nucleotide–binding sites, and can hydrolyze ATP in vitro. The

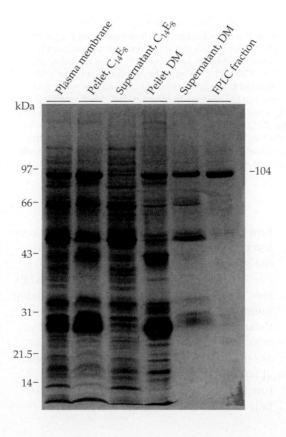

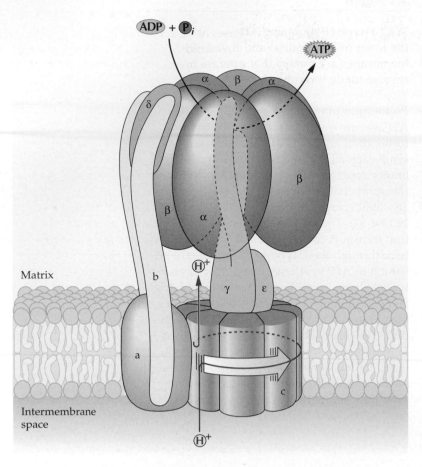

Figure 3.3
Structure of the core subunits of the mitochondrial ATP synthase, an F-type H$^+$-ATPase. The common subunits of the F$_0$ sector are known as a, b, and c. In the complex from *Escherichia coli,* these subunits are present in a stoichiometry of ab$_2$c$_{9-12}$. Subunit c is a small (8 kDa) and extremely hydrophobic subunit with two transmembrane spans. Subunit b possesses one transmembrane span and is thought to project from the membrane surface to interact with the F$_1$ sector. A core of five distinct subunit types are present in the F$_1$ sector in the stoichiometry $\alpha_3\beta_3\gamma\delta\epsilon$. Three nucleotide-binding sites are located primarily on the three β sub-units. The yellow arrow indicates the direction in which the c subunits are proposed to rotate during proton translocation. Rotation of the c subunits is thought to drive rotation of the γ-subunit and thereby alter the conformation of the nu-cleotide-binding sites (see Fig. 3.4).

flow of H$^+$ through F$_0$ drives long-range conformational transitions in F$_1$ that result in the synthesis of ATP.

The subunit composition of F-type ATP-ases varies according to the ATP–synthesizing membrane in which they are found. Nevertheless, members of this class of H$^+$-ATPase contain a "core" of common subunits that are essential for catalysis (Fig. 3.3). Additional subunits probably perform regulatory functions. Regulation is particularly important in chloroplasts at night, when ATP

hydrolysis by F$_1$ must be prevented in the absence of a light-imposed pmf (see Chapter 12).

Crystallographic studies of the F$_1$ sector $\alpha_3\beta_3\gamma$ complex from bovine mitochondria, performed by John Walker, demonstrated that the three adenine nucleotide–binding sites of the complex (located primarily on the three β subunits), show three distinct conformations. At any given time, one binding site is in an open conformation, another is binding a nucleotide loosely, and the third is binding a nucleotide tightly. These studies lend structural support to Paul Boyer's conformational model for proton-driven ATP synthesis, which postulates that ATP is synthesized by F-type ATPases through a process of rotational catalysis. Boyer and Walker received the Nobel Prize for chemistry in 1997.

According to the conformational model, ADP and inorganic phosphate first bind to the open nucleotide–binding sites. Proton flow through the F$_0$ sector of the enzyme causes the central F$_1$ subunit (γ) to rotate, altering the conformation of all three nucleotide–binding sites. The tight binding site opens and releases ATP into the aqueous medium, the open site is converted to a loose site, and the loose binding site forms a tight pocket in which ATP is formed spontaneously (Fig. 3.4). In net terms, a total of three or perhaps four protons are admitted through the F$_0$ sector for each ATP synthesized.

3.3.2 Plasma membrane H$^+$-pumping ATPase performs a variety of physiological functions and is a P-type ATPase.

Despite similarities in the overall transport reactions they catalyze, the plasma membrane and F-type H$^+$-ATPases differ markedly in their protein chemistry, their reaction mechanism and, hence, their evolutionary origins. In contrast to the multipartite mitochondrial and chloroplast ATP synthases, plasma membrane H$^+$-ATPase is a single polypeptide of about 100 kDa that both binds ATP and catalyzes H$^+$ transport.

As we have already seen, the pmf generated by plasma membrane H$^+$-pumping ATPase powers transport through a variety of carriers; it also influences ion channel activity through its impact on V_m. The enzyme therefore constitutes a foundation for

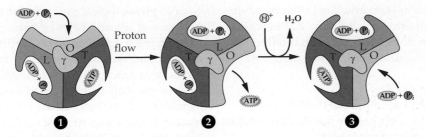

Figure 3.4

Model for the synthesis of ATP by an F-type H⁺-pumping ATPase. At any one time, each of the three nucleotide-binding sites has a distinct conformation: One site is open (O), another binds loosely (L), and the third site binds tightly (T). Proton flow through the ATP synthase causes the γ-subunit to rotate. This rotation changes the conformation of the three nucleotide-binding sites such that the O site becomes an L site, the L site becomes a T site, and the T site becomes an O site. ADP + P$_i$ initially bind to a site in the O conformation and are condensed to form ATP on conversion of the binding site to the T conformation. Release of ATP is from the O site.

transport processes at the plasma membrane. However, in addition to driving fluxes through other transport systems, plasma membrane H⁺-ATPase plays several other essential roles in plant cell biology. Activation of plasma membrane H⁺-ATPase is one of the earliest responses to auxin, occurring within one to two minutes of exposure. The resulting acidification of the cell wall in turn activates expansins, proteins that loosen H-bonds within the wall and allow turgor-generated growth (see Chapter 2). Plasma membrane H⁺-ATPase is also critical in removing excess H⁺ from the cytosol. Many metabolic pathways result in the net production of H⁺, some of which must be removed to prevent acidification of the cytosol. In fact, cytosolic pH is remarkably stable, remaining at 7.3 to 7.5. The low pH

optimum of plasma membrane H⁺-ATPase, 6.6, means that if conditions tend to lower the cytosolic pH, the pump is poised to respond with an increased rate of turnover.

Plasma membrane ATPase pumps a single H⁺ out of the cell for each MgATP hydrolyzed. During hydrolysis, the γ-phosphate of ATP becomes transiently but covalently bound to an aspartyl residue on the enzyme, forming an acyl-phosphate bond. Hydrolysis of this acyl-phosphate bond provides the driving force for the pump reaction cycle, in which different conformations of the enzyme, known as the E1 and E2 states, expose the H⁺–binding site to alternate sides of the membrane (Fig. 3.5). The conformational change that exposes the bound H⁺ to the apoplast is linked to additional conformational transitions that lower binding-site affinity, permitting H⁺ dissociation.

The existence of a covalent enzyme–phosphate (E-P) transition state during the reaction cycle firmly distinguishes plasma membrane H⁺-ATPase from F-type ATPases, placing it in a class of ion-motive ATPases known as P-type ATPases (to reflect the presence of the E-P form). This family of enzymes includes the fungal plasma membrane H⁺-ATPase, the Na⁺/K⁺-exchanging ATPase ubiquitous in animal cell plasma membranes, Ca²⁺-ATPases of the plasma and endomembranes of animals and plants (see Section 3.3.6), and the H⁺/K⁺-exchanging ATPase of mammalian gastric mucosa. (For work on the Na⁺/K⁺-ATPase, J. Skou was a corecipient of the 1997 Nobel Prize for Chemistry.) All P-type ATPases are inhibited by orthovanadate (H$_2$VO$_4^-$), which forms an analog of the

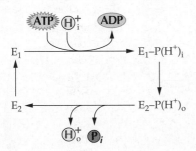

Figure 3.5

Reaction cycle for P-type ATPases. Protons bind to the ATPase in the E$_1$ conformation. Hydrolysis of ATP and phosphorylation of the enzyme cause a change to the E$_2$ conformation. The lower affinity of E$_2$ for H⁺ results in the release of protons on the other side of the membrane from where they were acquired. Hydrolysis of the enzyme–phosphate bond releases P$_i$ and returns the ATPase to the E$_1$ state.

E-P transition state and blocks the reaction cycle. Furthermore, all P-type ATPases share a common domain structure that reflects their functional attributes (Fig. 3.6).

3.3.3 Plasma membrane H⁺-ATPase is encoded by a multigene family that exhibits tissue-specific expression.

In *Arabidopsis*, plasma membrane H⁺-ATPases are encoded by genes of the *AHA* family. At least 10 *AHA* genes have been identified, each encoding a distinct plasma membrane H⁺-ATPase isoform. The plasma membrane H⁺-ATPases of tomato and tobacco also are members of multigene families. Studies using the *GUS* reporter gene fused to the promoter regions of specific *AHA* genes have revealed that isoform expression is tissue-specific. For example, *AHA3* is expressed selectively in phloem companion cells, the micropyle, and the developing seed funiculus (Fig. 3.7A). In contrast, *AHA10* is expressed principally in developing seeds, especially in the integument surrounding the embryo (Fig. 3.7B).

No physiological rationale for the appearance of specific AHA isoforms in given tissue types is yet possible. Nevertheless, distinct biochemical properties of the isoforms have emerged from expression studies in yeast mutants that lack native plasma membrane H⁺-ATPase activity. This work has established that AHA1 and AHA2 have K_m values for ATP of 0.15 mM, whereas the K_m of AHA3 is 10-fold higher. Sensitivities to orthovanadate also differ. Such findings reveal the futility of using crude plant homogenates as the starting material for biochemical analyses of enzymes that occur in multiple isoforms.

3.3.4 Plasma membrane H⁺-ATPase is regulated by an array of mechanisms.

Most enzymes that play pivotal roles in metabolism are subject to intense regulation, and plasma membrane H⁺-ATPase is no exception. Activation in response to lowered cytosolic pH has already been mentioned. More profound, perhaps, is the impact of the C terminus cytosolic region of the enzyme

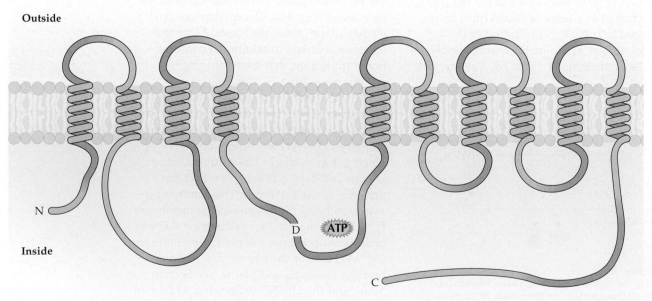

Figure 3.6

Membrane disposition of plasma membrane H⁺-ATPase. The large hydrophilic domain between the fourth and fifth transmembrane spans contains the ATP-binding site and the aspartyl residue (D) that undergoes phosphorylation. This domain of the enzyme is highly conserved among the different ion pumps. Binding sites for H⁺ probably reside in the transmembrane spans, predominantly spans 6 through 10 (at right). The smaller hydrophilic loop between transmembrane spans 2 and 3 may participate in catalyzing dephosphorylation of the Asp-P intermediate.

(see Fig. 3.6), which forms an autoinhibitory domain. Removal of the autoinhibitory domain, by either tryptic cleavage or genetic modification, activates the H^+-ATPase considerably. Point mutations in other cytosolic domains of the enzyme have a similar effect, suggesting that the autoinhibitory domain folds back into the protein to accomplish inhibition.

A striking twist to C-terminal H^+-ATPase inhibition has been observed during studies with fusicoccin, a toxin produced by the fungus *Fusicoccum amygdali*, a pathogen affecting peach and almond trees (Fig. 3.8). Fusicoccin elicits an increase in guard cell turgor, thereby opening stomata and causing leaves to wilt. Studies during the 1970s established that a primary mode of action of fusicoccin is stimulation of plasma membrane H^+-ATPase,

not just in stomatal guard cells but also in a whole range of plant cell types. The fusicoccin receptor, which was not identified until 1994, is a member of a family of soluble signal transduction proteins known as 14-3-3 proteins. These proteins function as dimers and bind to target proteins at a defined consensus sequence that includes a phosphorylated serinyl residue. Fusicoccin- and trypsin-induced activation of plasma membrane H^+-ATPase exhibit similar kinetic characteristics and are nonadditive, indicating that fusicoccin may activate the enzyme by relieving C-terminal autoinhibition. Plausible models for the control of plasma membrane H^+-ATPase by 14-3-3 proteins and the C terminus are shown in Figure 3.9. Activation of the enzyme is proposed to be evoked through 14-3-3 binding to the C terminus, which is subsequently displaced. Binding of the 14-3-3 protein to the H^+-ATPase is facilitated either by phosphorylation of the H^+-ATPase or by fusicoccin, which binds to both the 14-3-3 protein and the C terminus of the protein pump. These models provoke many unanswered questions, including the question of whether a physiological ligand exists for the 14-3-3 protein, because plants do not themselves produce fusicoccin.

A quite separate mechanism of activation of H^+-ATPase is associated with auxin-induced stimulation of proton pumping. In this case, auxin up-regulates expression of the pump. Induction of H^+-ATPase expression correlates with the auxin-responsiveness

(A)

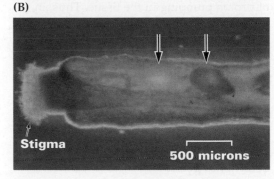

(B)

Figure 3.7
Tissue-specific expression of genes for the plasma membrane H^+-ATPase in *Arabidopsis*. (A) Stem cross-section from *Arabidopsis* expressing an AHA3–c-Myc fusion protein and labeled for immunofluorescence detection of the H^+-ATPase. Cell-specific labeling is seen in a subset of phloem (p) cells but not in cortical (c) or autofluorescent xylem (x) cells. (B) The *AHA10* gene promoter is expressed in developing seeds, as shown by β-glucuronidase (GUS) staining. Two developing seeds (arrows) are shown in an *Arabidopsis* silique. The blue color in the seed to the right indicates expression of an AHA10–GUS fusion protein.

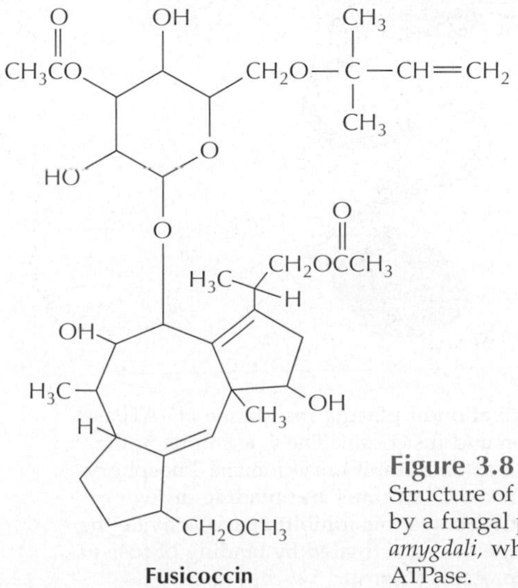

Fusicoccin

Figure 3.8
Structure of fusicoccin, a toxin produced by a fungal pathogen of plants, *Fusicoccum amygdali*, which stimulates the P-type H^+-ATPase.

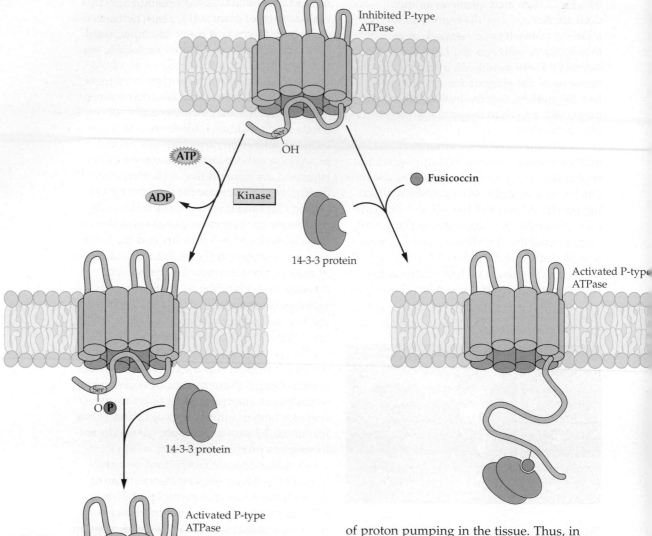

Figure 3.9
Models for the activation of plasma membrane H$^+$-ATPase by phosphorylation and fusicoccin. The C terminus of the enzyme functions as an autoinhibitory domain. Phosphorylation of a serine in the C terminus and binding of two 14-3-3 proteins (left) remove this inhibition and activate the enzyme. The enzyme is also activated by binding of fusicoccin and 14-3-3 proteins (right).

of proton pumping in the tissue. Thus, in maize coleoptiles, although auxin stimulates pump expression and apoplastic acidification in nonvascular tissues, the vascular bundles do not display auxin-enhanced expression of the pump and do not contribute to auxin-induced acidification of the cell wall.

3.3.5 H$^+$:ATP stoichiometry determines poise.

Why do F-type H$^+$-ATPases synthesize ATP at the expense of the pmf, whereas plasma membrane H$^+$-ATPases hydrolyze ATP and generate a pmf? The relative sizes of the driving forces do not answer this question of poise; the ΔG for ATP hydrolysis in the mitochondrial matrix is probably similar to that in the cytosol, being about −50 kJ•mol^{-1}. Likewise, the electrochemical potential for protons (pmf•F) is similar across both the inner mitochondrial and plasma membranes, with a value near +25 kJ•mol^{-1}. To resolve this apparent paradox, we must also take into account the reaction stoichiometry, the number (n) of H$^+$ moved per ATP hydrolyzed or synthesized. In Reaction 3.1, H^+_i and H^+_o refer, respectively, to H$^+$ on the inside or outside of the membrane. Depending on the membrane involved, inside refers to the chloroplast stroma (thylakoid membrane), the mitochodrial matrix (inner mitochondrial membrane), or the cytosol (plasma membrane); outside designates the thylakoid lumen, the intermembrane space, the lumen of endomembrane organelles, or the extracellular medium. The free energy relationship is defined in Equation 3.1. If ΔG_{pump} is negative, the pump reaction proceeds from left to right, as written, whereas a positive value indicates the reverse reaction. Thus, for plasma membrane H$^+$-ATPase, which hydrolyzes an ATP for each proton pumped, n − 1 and ΔG_{pump} = –25 kJ•mol^{-1}. In contrast, for mitochondrial F-type ATPase, which translocates several protons per molecule of ADP phosphorylated, $n = 3$ and ΔG_{pump} = +25 kJ•mol^{-1}. This difference in net direction can be likened to pushing cars uphill. A person might make headway with one car, but additional cars will ultimately overcome the force applied by the person and will roll back down the hill.

Reaction 3.1

$$n\text{H}^+_i + \text{ATP} \rightleftharpoons n\text{H}^+_o + \text{ADP} + \text{P}_i$$

Equation 3.1

$$\Delta G_{pump} = n(\text{pmf}•F) + \Delta G_{ATP}$$

3.3.6 Ca^{2+}-ATPases, another group of P-type ATPases, are distributed among various plant membranes.

Calcium-pumping ATPases are distributed in the plasma membrane, the ER, the chloroplast envelope, and vacuolar membranes. These enzymes pump Ca^{2+} out of the cytosol, thereby maintaining the cytosolic concentration of free Ca^{2+} at about 0.2 μM. This low concentration of free Ca^{2+} is essential in all cells to prevent precipitation of phosphates, but in eukaryotes it has become a base on

which to build stimulus–response coupling pathways (see Chapter 18). Thus, cytosolic Ca^{2+} can be increased manyfold by selected physiological stimuli without exceeding the low micromolar range.

All Ca^{2+}-ATPases are P-type ion-motive ATPases. Despite the common hydrolytic and transport reactions they execute, the molecular structures of Ca^{2+}-ATPases vary considerably and depend on the membrane type in which they occur. They have been best characterized in animals, where they are subdivided into two classes. The plasma membrane (PM) type is activated by calmodulin. The presence of a C-terminal calmodulin-binding domain increases the molecular mass of the catalytic subunit to 130 kDa. Conversely, endoplasmic reticulum (ER)-type Ca^{2+}-ATPases possess no calmodulin-binding domain, and their molecular masses are typical for P-type ATPases (110 kDa).

An intriguing picture is emerging in plants. In *Arabidopsis,* an ER-type Ca^{2+}-ATPase identified by molecular cloning has been localized immunochemically to the ER. In plasma membranes, Ca^{2+}-ATPase activity is enhanced by calmodulin, as in animal cells. Curiously, however, although Ca^{2+}-ATPase homologs of the PM-type have been identified in plant cells, none has thus far been localized to the plasma membrane. One of these enzymes has been associated with the chloroplast inner envelope membrane; another resides in the vacuolar membrane. The vacuolar Ca^{2+}-ATPase is calmodulin-activated and, accordingly, possesses a calmodulin-binding domain. Interestingly, this calmodulin-binding domain is present as an extension at the N terminus rather than the C terminus (Fig. 3.10). The plant endomembrane "PM-type" Ca^{2+}-ATPases are the only examples yet reported of P-type ATPases regulated by their N terminus.

The electrochemical potential against which Ca^{2+} must be pumped by these enzymes is vast, not only because the concentration of free Ca^{2+} is several orders of magnitude less than that on the other side of the membrane, but also because the cytosol-negative membrane potential opposes the export of divalent cations with double the effective force imposed on monovalent cations. The free energy of the Ca^{2+} electrochemical potential difference across the plasma membrane, roughly –60 kJ•mol^{-1}, may actually

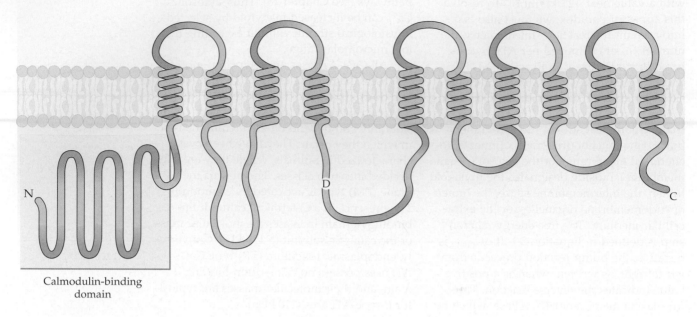

Figure 3.10
Activity of the vacuolar Ca^{2+}-ATPase is regulated by the binding of calmodulin to the N terminus.

Calmodulin-binding
domain

exceed the free energy input available from ATP hydrolysis ($-50\,kJ \bullet mol^{-1}$). In cases where this issue has been rigorously addressed, the enzyme has been found to catalyze Ca^{2+}/H^+ exchange. This exchange reaction can offset the large opposing driving force from Ca^{2+}. Ca^{2+}-ATPases probably use both ATP and the pmf to accomplish an energetically daunting task.

3.3.7 Vacuolar and other membranes are energized through vacuolar H⁺-ATPases.

Plant vacuoles contain a highly acidic solution, with a pH near 5.5, about 2 pH units lower than that of the cytosol. In most unripe fruits and some citrus fruits, the vacuolar pH can drop below 3, imparting a sour taste to the tissue as a whole. Proton pumping into the vacuolar lumen not only energizes the membrane for carrier-mediated transport but also generates the low pH of the vacuole, where proteases, glucosidases, phosphatases, and nucleotidases with acidic pH optima reside.

Proton pumping is catalyzed by vacuolar-type (V-type) H⁺-ATPases. Sequence analysis has demonstrated that these enzymes are distant relatives of F-type H⁺-ATPases, but

V-type ATPases operate solely in the direction of ATP hydrolysis. The ratio of H⁺ translocated per ATP hydrolyzed has been measured as 2, although some evidence suggests that pump stoichiometry depends on the lumenal pH, with a low pH resulting in average ratios less than 2.

At the protein level, the structure–function partitioning of F-type and V-type H⁺-ATPases has been retained. Thus the V-type enzyme can be separated into a soluble V1 sector, analogous to F_1 and including the adenine nucleotide–binding sites, and a membrane-bound V_0 sector, analogous to F_0 and composing the H⁺ pathway through the membrane. As might be anticipated, there is some homology between subunits in the V_1 and F_1 sectors and also between subunits in the V_0 and F_0 sectors. However, V-type H⁺-ATPases possess a more complex subunit composition than that of F-type H⁺-ATPases, and some evidence indicates that subunit composition for the V-type enzyme is variable (Fig. 3.11).

V-type H⁺-ATPases are potently and specifically inhibited by the macrolide antibiotic bafilomycin A_1 (Fig. 3.12). This extremely hydrophobic compound is produced by *Streptomyces* and interacts with the V_0 sector of the enzyme.

Except for mitochondria and chloroplasts, all cellular organelles maintain an acid interior. An acidic lumenal pH is thought to contribute to vesicle sorting and hence to membrane trafficking and protein targeting (see Chapter 4). V-type ATPases contribute to this lumenal acidification, and evidence is strong that these enzymes are present in membranes from the ER, Golgi, and coated vesicles of plant cells, as is the case in yeast and animal cells.

3.3.8 The plant vacuolar membrane also possesses a unique H⁺-pumping inorganic pyrophosphatase (H⁺-PPase).

A supplementary H^+ pump is ubiquitous at the vacuolar membrane of plants. This transport system uses the free energy of hydrolysis of inorganic pyrophosphate (PP_i), rather than ATP, to pump H^+. In contrast to V-type ATPases, the PPase is a comparatively simple enzyme, containing a single class of 80-kDa polypeptide, but possessing as many as 16 membrane-spanning domains. The functional unit of the pump is probably a homodimer. The H^+-PPase is unique to plants and a few species of *Archaebacteria*.

The bona fide substrate of H^+-PPase for hydrolysis is dimagnesium PP_i, which is present in micromolar amounts in the cytosol. Unlike the V-type H^+-ATPase at the same membrane, H^+-PPase activity is potently inhibited by Ca^{2+} and also competitively inhibited by the substrate analog aminomethylenediphosphonate. Potassium ions also are required at the cytosolic face of the pump for both hydrolytic and H^+ pumping activity, although whether K^+ is actually transported is controversial.

It is not absolutely clear why the vacuolar membrane possesses two H^+ pumps. Perhaps the best guess arises from the observation that the H^+-PPase tends to be more active than the V-type H^+-ATPase in immature tissue, whereas the converse holds in mature tissue. Synthetic pathways for biopolymers such as cellulose, proteins, and nucleic acids—pathways that predominate in immature tissue—all form PP_i as a by-product. One way to remove this unwanted PP_i would be to hydrolyze it with a soluble PPase, a well-characterized process in animals and yeast. By coupling hydrolysis to proton translocation,

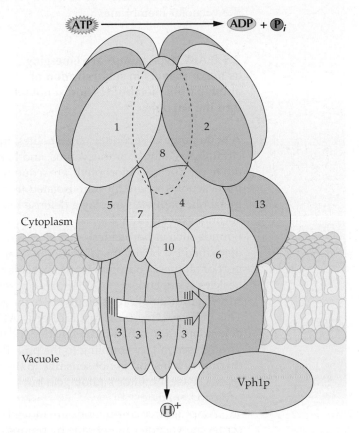

Figure 3.11
Structural model of a V-type H^+-ATPase from *Saccharomyces cerevisiae*. The numbered subunits correspond to the protein products of *VMA1–VMA8, VMA10,* and *VMA13*; an additional subunit, Vph1p, also is shown. A hexameric subunit array in V_1 is generated by three copies each of subunit 1 and subunit 2 (homologous to the β and α subunits of F_1, respectively). Another eight or so subunits also contribute to V_1. The V_0 sector contains multiple copies of subunit 3 (a dominant, highly hydrophobic subunit of 16 kDa that arose from a duplication and fusion of the gene for the 8-kDa c subunit of F_0) and four additional subunits.

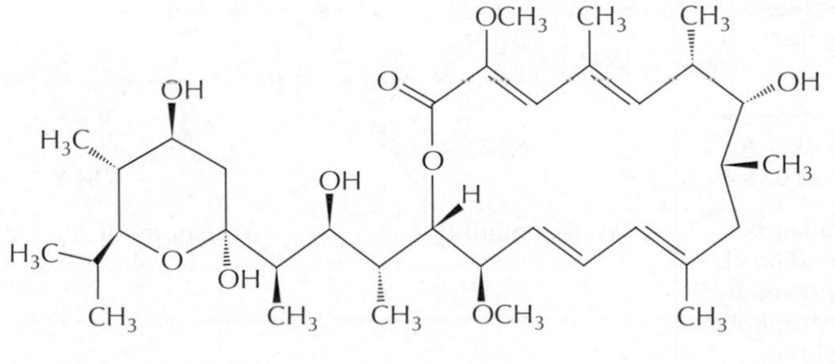

Bafilomycin A₁

Figure 3.12
Structure of bafilomycin A_1, a toxin produced by *Streptomyces* species that specifically inhibits the V-type H^+-ATPase.

some of the free energy that would otherwise be lost as heat is conserved as a pmf across the vacuolar membrane.

3.3.9 ABC-type pumps are emerging as major players in sequestration of amphipathic metabolites and xenobiotics into the vacuole.

A wide range of secondary metabolites, including flavonoids, anthocyanins, and breakdown products of chlorophyll, are sequestered into vacuoles where they are isolated from metabolic pathways or play a defense role. In addition, several xenobiotics—synthetic compounds such as herbicides—can be effectively detoxified by vacuolar sequestration. Transport of all these compounds into vacuolar vesicles is dependent on ATP but is not sensitive to protonophores that dissipate the pmf. This indicates that transport is directly ATP-dependent rather than indirectly dependent through secondary coupling to the pmf. Furthermore, transport is not sensitive to the V-type ATPase inhibitor bafilomycin but is sensitive to vanadate.

Amphipathic compounds are moved across the vacuolar membrane by pumps known as ATP-binding cassette (ABC) transporters. The ATP-binding cassette itself is widely distributed among enzymes that bind ATP (including F-type ATPases) and includes the so-called Walker A and B motifs (Fig. 3.13). Mammalian ABC transporters, which have

been extensively characterized, transport a wide range of compounds, including chemotherapeutic drugs and proteins. In bacteria, some ABC transporters are involved in uptake of some simple nutrients such as phosphate and amino acids, and in yeast an ABC transporter exports mating factor. The defining feature of ABC transporters is their domain structure: a stretch of transmembrane spans, followed by a nucleotide-binding fold that incorporates the Walker motifs. This domain structure is duplicated in both the N- and C-terminal halves of the transporter (Fig. 3.13). Several distinct expressed sequence tags (ESTs) for ABC transporters have been identified in plants, suggesting the presence of roles as diverse as those described for other organisms. For example, many xenobiotics are sequestered in plant vacuoles after glucosylation, and ABC transporters are possibly involved in sequestration.

Other compounds, including flavonoids and xenobiotics, are known to be transported by plant ABC transporters as glutathione conjugates (GS-conjugates). In these cases, the compound must first be linked to glutathione by a glutathione S-transferase reaction. Yet other compounds (e.g., the linear tetrapyrrole catabolites of chlorophyll) are transported without prior conjugation. One *Arabidopsis* ABC transporter, MRP1, appears to translocate GS-conjugates exclusively. Another, MRP2 (Fig. 3.13), translocates not only GS-conjugates but also unmodified chlorophyll catabolites (Fig. 3.14). Several mammalian ABC transporters also demonstrate broad substrate specificities that are atypical of transport systems in general; most non-ABC transporters exhibit strong selectivity for the preferred substrate.

Research into the properties of plant ABC transporters is only just beginning, and many interesting questions are yet unanswered. For example, whether these transporters function as pumps or "flippases" has not been clarified. In the pump model, the transported substrate passes through an aqueous domain formed within the membrane by the ABC transporter, much as is envisaged for inorganic ion pumps. In the flippase model, the transported substrate first partitions into the membrane lipid phase and then is "flipped" to the other side of the membrane in a reaction occurring at the interface of the ABC transporter and the membrane lipid.

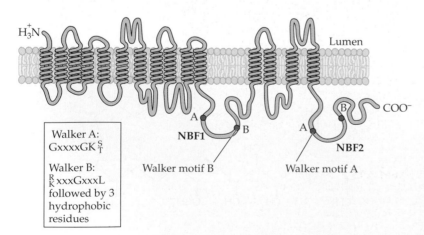

Figure 3.13
Model of MRP2, a vacuolar ATP-binding cassette (ABC) transporter from *Arabidopsis*. Two nucleotide-binding folds, NBF1 and NBF2, each contain Walker motifs A and B. The NBFs are separated by transmembrane domains containing multiple integral membrane helices.

(A)

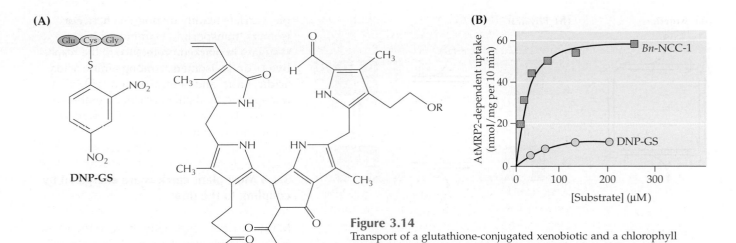

DNP-GS

Bn-NCC-1

(B)

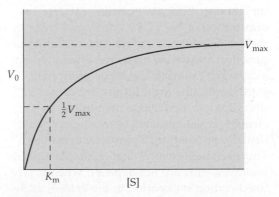

Figure 3.14
Transport of a glutathione-conjugated xenobiotic and a chlorophyll catabolite by AtMRP2, an ABC transporter from *Arabidopsis*. (A) Structures of the glutathione-conjugated xenobiotic dinitrophenol (DNP-GS) and a linear tetrapyrrole from *Brassica napus* (*Bn*-NCC-1). In the *Bn*-NCC-1 structure, R represents a malonyl group. (B) Concentration-dependent uptake of DNP-GS and *Bn*-NCC-1 by AtMRP2.

3.4 Carriers

3.4.1 Carriers exhibit Michaelis–Menten kinetics that indicate conformational changes during transport.

Unlike pumps, carriers do not catalyze scalar reactions such as ATP hydrolysis. In other words, the transport process does not involve chemical modification of any of the compounds bound to the carrier. Rather, carriers catalyze solely vectorial reactions, i.e., the movement of inorganic ions and simple organic solutes across membranes.

One defining feature of carriers is that they display saturability when the kinetics of transport are expressed relative to the substrate concentration (Fig. 3.15). Frequently, these kinetics can be expressed in terms of the Michaelis–Menten equation, yielding a K_m for the substrate and a $\frac{1}{2}V_{max}$ for the transport rate. An important inference can be drawn from these saturation kinetics: Carriers undergo conformational changes during transport (Fig. 3.16).

3.4.2 Carriers translocate a wide variety of inorganic ions and small organic solutes with high specificity.

The array of ions and solutes translocated by carriers is vast. The principal inorganic nutrients, including NH_4^+, NO_3^-, P_i (probably as $H_2PO_4^-$), K^+, and SO_4^{2-}, are all translocated into cells by plasma membrane carriers. Carriers also are responsible for taking up ions

that play less central roles in metabolism, such as Cl^-. Among the organic solutes translocated into cells by carriers are the fundamental building blocks of biopolymers: sugars, amino acids, and purine and pyrimidine bases.

At the plasma membrane, carriers not only are central to nutrient absorption from the soil but also play fundamental roles in the mobilization and storage of metabolites. For example, in plant species in which

Figure 3.15
Carriers exhibit saturation kinetics. At low concentrations of substrate (S), the overall rate of transport is limited by the binding of S to the carrier, and therefore transport rate increases linearly with [S]. At high concentrations of S, transport is rate-limited by the conformational change associated with access of binding sites from one side of the membrane or the other. The substrate concentration at which one-half of the maximal velocity is achieved is the Michaelis–Menten constant (K_m).

(A) Kinetic

(B) Physical

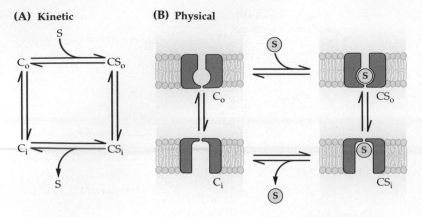

Figure 3.16
Kinetic (A) and cartoon (B) representations of the activity of carrier C, illustrating the transport of solute S from an extracytosolic compartment (outside, o) to the cell interior (inside, i). Carrier binding sites are not actually thought to move from one side of the membrane to the other. Instead, conformational changes associated with the transport reaction may be quite subtle.

reduced carbon is exported from leaves as sucrose, carriers specifically transporting sucrose are responsible for loading the phloem (see Chapter 15). Similarly, many species reduce NO_3^- in the leaves, and carriers must therefore load the resulting organic nitrogen compounds into the phloem to be transported to sink tissue (see Chapters 15 and 16). During germination (see Chapter 19), hydrolysis of storage polymers yields sugars and amino acids that must be mobilized to growing tissue, and carriers facilitate this movement.

Carriers also play crucial roles at endomembranes. Tonoplast carriers catalyze sequestration of Na^+, Ca^{2+}, Mg^{2+}, and NO_3^- as well as sucrose and amino acids. At other organelles, metabolite exchange often predominates. For example, the triose phosphate translocator, the most prominent carrier at the chloroplast envelope membrane, exchanges equal amounts of triose phosphate (usually stromal) for P_i (usually cytosolic). This exchange ensures that chloroplast phosphate contents are not depleted when newly fixed carbon is exported to the cytosol as dihydroxyacetone phosphate (see Chapter 13). At the mitochondrial inner membrane, a carrier executes the stoichiometric exchange of matrix ATP for cytosolic ADP, poising the substrate–product ratio at values conducive to mitochondrial ATP synthesis (see Chapter 14).

Like the vast majority of enzymes, carriers are relatively specific for their substrates. In the case of organic substrates, for example, carriers readily distinguish between isomers, transporting L-amino acids and D-sugars in marked preference to the respective D- and L-isomers. Among amino acids, acidic, basic, and neutral amino acids all are transported by distinct plasma membrane carriers. Thus, a wide range of carrier types are found in all membranes.

3.4.3 Most plant carriers are energized by coupling to the pmf.

Many carriers—especially those at the vacuolar and plasma membranes—behave as H^+-symporters or -antiporters (see Section 3.2.2). As a result, the carrier substrate is translocated against its own electrochemical potential. Experimental evidence for H^+-coupled transport has been derived from two classes of experiments. The first method, which uses microelectrodes to monitor the V_m across the cell membrane, is restricted to assessment of plasma membrane symporter activity (Fig. 3.17A). When a cell is bathed in carrier substrate—even an uncharged one like a sugar—the plasma membrane often depolarizes very rapidly, indicating that positive electrical charge is flowing into the cell. These depolarizations show that the H^+-coupled transport systems are **electrophoretic** (carry electrical charge). By evolving transport systems with H^+:substrate stoichiometries great enough to carry positive charge, plants are able to exploit the highly negative V_m to provide part of the driving force. This might be particularly important for plants growing in alkaline environments, where the chemical component of the pmf is absent or even inverted.

An alternative approach that is applicable to a variety of membranes involves the use of fractionated membrane vesicles (Fig. 3.17B). Vesicular uptake of a radioisotope-labeled substrate is compared in the presence and absence of the pmf. In the presence of a pmf of the appropriate polarity, substrate accumulation inside the vesicles indicates H^+-coupled transport.

The prominence of H^+-coupled transport systems in plant cell plasma membranes illustrates that organisms living in relatively dilute media depend on energized solute transport. In many animal cells, in contrast, transport systems for most nutrients are not energized by ion-coupling, presumably

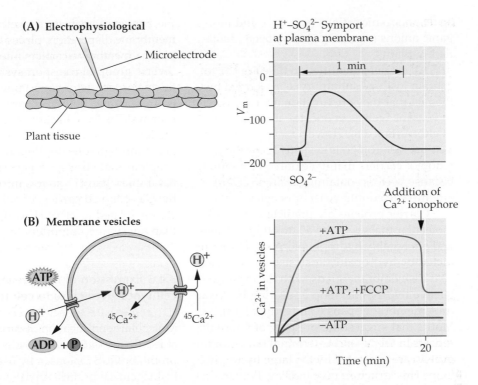

Figure 3.17
Methodologies for demonstrating H^+-coupled solute transport by carriers. (A) The cell is impaled with a glass micropipette containing an electrolyte and an electrode. This microelectrode, in conjunction with another electrode in the bathing solution, is used to measure the voltage across the membrane. An uncharged or anionic substrate, such as SO_4^{2-}, is introduced into the bathing solution. If a cation is cotransported with the substrate, positive charge flows into the cell, the membrane depolarizes, and V_m increases. Ion substitution experiments and alkalization of the medium often identify the cation as H^+. If the transported substrate is anionic, membrane depolarization implies that at least two H^+ are transported per monovalent anion, or three H^+ per divalent anion. (B) Vesicles are incubated with a radiolabeled substrate (e.g., $^{45}Ca^{2+}$), and filtered though narrow-pore nitrocellulose. The filter is then washed and the vesicle-associated radioactivity is quantified. This method allows comparison of vesicular uptake in the presence and the absence of a pmf. If the cytosolic face of a vesicle is exposed to the medium, endogenous H^+-ATPase activity can establish an inside-acid pmf. This gradient is used to drive uptake of the radiolabeled substrate, as shown in the accompanying graph of Ca^{2+} uptake (+ATP, green line). Addition of a Ca^{2+}-selective ionophore (indicated by vertical arrow above graph) results in Ca^{2+} leakage from the vesicle and demonstrates that Ca^{2+} has been accumulated within the vesicle rather than bound to it. In the absence of ATP (–ATP, blue line), or in the presence of compounds that eliminate the proton gradient (FCCP: carbonyl cyanide *p*-[trifluoromethoxy]phenylhydrazone, red line), Ca^{2+} accumulation is much reduced.

because such cells are bathed in nutrient-rich body fluids. However, some plant carriers are not energized by ion-coupling. For example, vacuolar uptake of glucose and some amino acids is probably facilitated by carriers that do not translocate H^+ but are instead coupled to gradients in concentration of the transported solute.

3.4.4 Molecular identification of carriers defines them as members of the major facilitator superfamily.

Carriers are very difficult to identify by biochemical approaches because they are not typically abundant (see Section 3.2.4) and do not execute reactions that can be assayed without reconstitution in membranes. Fortunately, genetic approaches have yielded considerable dividends. The most widely applicable and successful approach applied to plant carriers has been yeast complementation. Yeast transport mutants that are defective in growth on the solute of interest are transformed with cDNA from a plant library. Transformed yeast colonies that are able to grow on the particular solute must accordingly contain functional copies of the relevant plant transport system, which can then be sequenced from the vector. The large number of yeast transport mutants, and the ease and rapidity with which genetic studies can be conducted in yeast, make this an attractive experimental system. As a result, plasma membrane carrier systems for sugars, amino acids, inorganic cations (e.g., K^+), and inorganic anions (e.g., SO_4^{2-}) have been identified in plants.

All carriers identified to date appear to contain just one subunit type, as judged by the fact that they are functionally expressed in yeast as well as in oocytes after injection of cRNA (Box 3.6). However, experiments that use heterologous systems favor the isolation of carriers that have only one subunit, because vectors containing a single cDNA are introduced into yeast or oocytes.

Carrier proteins are predicted to have molecular masses in the 40- to 50-kDa range and are notably hydrophobic. A structure predicted by hydropathy analysis often exhibits 12 transmembrane spans, with the most extensive hydrophilic loop appearing between transmembrane spans 6 and 7 (Fig. 3.18). Motifs that suggest the presence of internal repeats in the N- and C-terminal halves of the protein are separated by the large hydrophilic loop. This structure may indicate that an ancient progenitor of these carrier systems underwent a gene duplication and fusion.

Sequence analysis, as well as the general membrane disposition, places these plant plasma membrane carriers into a large and diverse group of transport systems known as the major facilitator superfamily (MFS). This group includes both H^+-coupled sugar transport systems from bacteria and uniporters, carriers from animal cells that are not energized through ion-coupling. In animals, ion-coupled solute transport generally utilizes Na^+ rather than H^+, so it is intriguing that the Na^+-coupled carriers of animals are less closely related to plant H^+-coupled carriers than are animal uniporters.

3.4.5 Expression of carriers in particular cell types gives clues to cell function.

Molecular identification of carriers has enabled their locations to be identified by promoter-GUS fusions or by use of immunohistochemical probes. A particularly intriguing example of the insights that can emerge from such analysis is seen in studies on the

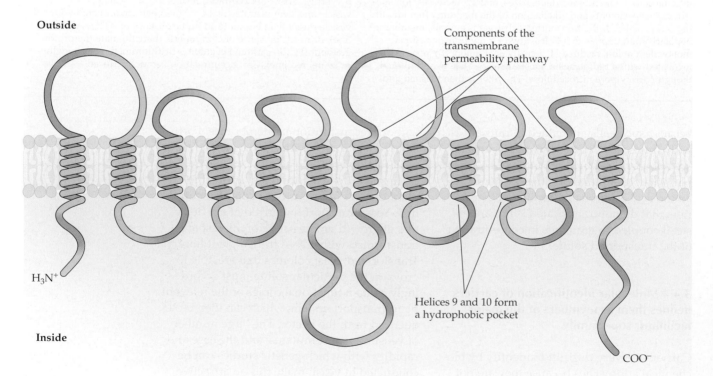

Figure 3.18
Structural model illustrating the orientation of a generalized carrier in a membrane.

Box 3.6

Observation of plant carriers expressed in heterologous systems can provide insight into carrier function.

Cloned plant transporters have been characterized in various heterologous systems. In yeast, a popular and relatively easy system to study, carrier-mediated fluxes can be measured with radioisotopes. Solute specificity of the transport system is most easily examined this way in yeast strains in which the native transporters for the solute of interest have been deleted.

However, because most carriers execute electrophoretic transport reactions, the transport system will also be sensitive to V_m. A full characterization of transport will therefore include the sensitivity of the system to V_m. Yeast is not readily amenable to electrophysiological experiments because the cells are small and are bounded by a tough chitinous cell wall.

A popular alternative heterologous expression system for carrier activity has emerged in the form of oocytes from the South African clawed toad *Xenopus laevis*.

The large size of the oocytes, more than 1 mm in diameter, not only permits the use of radiotracers to measure solute fluxes across the plasma membranes of individual cells but also accommodates impalement by micropipettes and microelec-

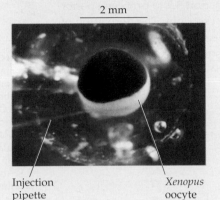

2 mm

Injection pipette   *Xenopus* oocyte

trodes (see figure), which can be used to monitor transporter-mediated membrane currents. Furthermore, the plasma membrane of the oocytes is relatively quiescent with respect to transport activity, so the background activity of endogenous transporters rarely contributes substantially to the fluxes or currents measured in experiments with heterologously expressed transporters.

Expression of heterologous transporters in oocytes is achieved by microinjection of a cRNA-containing vector into single cells. The transporter cDNA is used as a template for the cRNA, which in turn provides the template for translation. Typically, transport activity is monitored between 2 and 4 days after cRNA microinjection, when expression is at a maximum.

distribution of sucrose-H^+ symporters. In *Plantago* and *Arabidopsis* the sucrose transporter SUC2 is expressed only in the companion cells of the phloem and not in the sieve elements (Fig. 3.19). This finding suggests a major role for companion cells in phloem loading, with sucrose accumulating first in the companion cells and then diffusing into the sieve elements through the many plasmodesmatal connections. The driving force for the uptake of sucrose into companion cells would be provided by the pmf generated by the plasma membrane H^+-ATPase AHA3 isoform in *Arabidopsis* (see Section 3.3.3). In contrast, the sucrose transporter SUT1 from solanaceous species locates to the sieve elements but not to the companion cells. *SUT1* mRNA is also found in the mature sieve elements (which possess no nuclei) in greater amounts than in companion cells. These findings imply not only that sieve elements themselves are responsible for phloem loading of sucrose in solanaceous species, but also that the *SUT1* mRNA is targeted through plasmodesmata from the companion cells to the sieve elements, where it is translated (Fig 3.20).

3.4.6 Transcriptional and posttranslational controls regulate carrier activity.

The capacity of H^+-coupled transporters to generate a solute potential gradient can be calculated from the free energy relationship of the transport reaction that the carrier

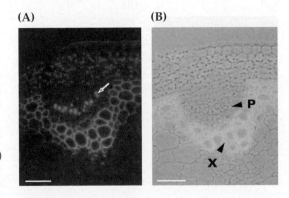

(A)   **(B)**

Figure 3.19
Localization of the sucrose transporter SUC2 to companion cells of the phloem. (A) Immunofluorescent localization of SUC2 in *Arabidopsis* stems (arrow). Fluorescence from the xylem is nonspecific autofluorescence. (B) Same section as in A but viewed with transmitted light. P, phloem; X, xylem.

catalyses. For example, Reaction 3.2 describes a symporter catalyzing transport of H^+ and the substrate S, where n is the stoichiometric ratio of H^+:S translocated in the reaction; the subscripts e and c represent the extracytosolic and cytosolic sides of the membrane, respectively; and z is the valence (charge) of S. The sum of the electrochemical potential differences of the cotransported species, $n\Delta\bar{\mu}_{H^+} + \Delta\bar{\mu}_S$, must be less than $0\ J\ mol^{-1}$ for Reaction 3.2 to proceed spontaneously from left to right. The derivation of $\Delta\bar{\mu}_{H^+}$ and $\Delta\bar{\mu}_S$ (see Box 3.3) gives the maximum potential solute accumulation ratio (Eq. 3.2) which can be plotted with respect to V_m (Fig. 3.21).

Reaction 3.2

$$nH^+_e + S^z_e \rightleftharpoons nH^+_c + S^z_c$$

Equation 3.2

$$[S]_c/[S]_e =$$
$$[H^+]^n_e/[H^+]^n_c \bullet \exp[-(n+z)F \bullet V_m/RT]$$

Biologically, these quantitative relationships yield two important conclusions. First, H^+-coupled transport systems are capable of generating impressive amounts of solute accumulation. Second, and as a corollary, H^+-coupled carriers must be subject to tight regulation if physiological concentrations of solutes are to be maintained in the cytosol.

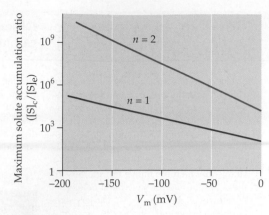

Figure 3.21
Maximum solute accumulation ratio plotted against V_m, for representative conditions across the plasma membrane: Extracellular and cytosolic pH values are 5.5 and 7.5, respectively, and $V_m = -150$ mV. For an electrically neutral solute ($z = 0$) that cotransports with one proton ($n = 1$), the resulting maximum accumulation ratio is very large indeed: 36,500. For $n = 2$, an astonishing solute accumulation ratio of 1.34 $\times 10^9$ can be calculated. $[S]_c$, cytosolic concentration of solute; $[S]_e$, extracellular concentration of solute.

Two forms of control predominate. Transcriptional control is evident for many carrier-mediated transport systems, with gene expression being derepressed during periods of substrate-starvation. A burst of transcriptional activity lasting from minutes to hours is evident when substrate is provided after starvation of the plant. In the continued presence of substrate, expression of the carrier is repressed and transport is evident only at a low rate. An example of probable transcriptional control is shown in Figure 3.22 for uptake of K^+ into roots of *Arabidopsis*: K^+ starvation, which enhances uptake of K^+ into the roots, correlates with expression of the K^+ carrier KUP3. The marked exception to this generalization is carrier-mediated NO_3^- transport, which is induced rather than repressed by its substrate.

In addition, posttranslational control can play a role in preventing unacceptable accumulation ratios. This has been documented in the giant internodal cells of the charophyte alga *Chara*, which can be internally perfused with defined media. There, a high rate of H^+-coupled Cl^- transport into the cell is observed in the absence of cytosolic Cl^-, but the rate declines almost to zero as the cytosolic Cl^- increases to the modest concentration of 10 mM. This phenomenon, known as

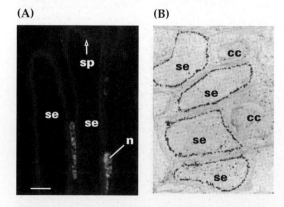

(A)   **(B)**

Figure 3.20
Localization of the sucrose transporter SUT1 to sieve elements (se). (A) Immunofluorescent localization of SUT1 in a longitudinal section of a potato stem. sp, sieve plate; n, nucleus. (B) Silver-enhanced immunogold localization of SUT1 in cross-section of a potato petiole. cc, companion cells.

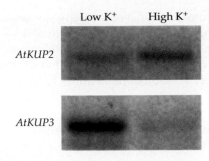

Low K⁺   High K⁺

AtKUP2

AtKUP3

Figure 3.22
Differential transcription of *AtKUP2* and *AtKUP3*, which encode K⁺ uptake transporters. RNA was isolated from roots of *Arabidopsis* plants grown in medium containing low or high concentrations of K⁺ (40 μM or 2 mM, respectively). A greater quantity of *AtKUP3* mRNA accumulated under low-K⁺ conditions.

transinhibition, results from relatively tight binding of cytosolic Cl⁻ to the carrier at the active site. Thus, the greater the cytosolic Cl⁻ concentration, the more the carrier becomes locked up in its substrate-bound form, and the less the carrier is able to take up Cl⁻ from outside the cell.

3.4.7 In some cases, ion-coupled solute transport involves Na⁺ rather than H⁺.

Although H⁺-coupling prevails as a mechanism for energizing carrier-mediated transport in plants, a growing number of examples of Na⁺ symport have been found. In some marine algae, uptake of NO_3^- and some amino acids is Na⁺-dependent. Perhaps this is not too surprising, given the plentiful concentration of Na⁺ in the sea (approximately 480 mM): There will be a large inwardly directed concentration difference for Na⁺ across the plasma membrane (Fig. 3.23). However, uptake of K⁺ at micromolar concentrations is also Na⁺-dependent in some freshwater charophyte algae and in some angiosperms. Although the Na⁺ concentration of freshwater is considerably less than that of sea water, it is nevertheless possible for Na⁺-coupling to energize transport because the V_m component of the electrochemical driving force for Na⁺ is sufficiently large to overcome the modest concentration difference for this ion across the plasma membrane.

Molecular evidence for Na⁺-coupling of K⁺ transport has been provided by the identification of a wheat cDNA clone that, when expressed in yeast or in frog oocytes, catalyzes joint uptake of K⁺ and Na⁺. This putative high-affinity K⁺ transporter, HKT1, is expressed in the root cortex, but physiological studies on wheat roots have not provided evidence for Na⁺-coupling of K⁺ transport in planta. The role of HKT1 in root K⁺ uptake therefore remains to be defined.

3.5 General properties of ion channels

3.5.1 Ion channels are ubiquitous in plant membranes.

During their classic studies on squid axons in the early 1950s, Alan Hodgkin and Andrew Huxley coined the term "channel" to describe the elements in the plasma membrane that responded to electrical stimuli by opening and facilitating selective fluxes of ions during action potentials. Action potentials are generated when the membrane is depolarized to a voltage value more positive than the **threshold voltage.** A substantial further depolarization then follows, but this is

Sea water
480 mM Na⁺

Cytosol
50 mM Na⁺
$V_m = -80$ mV

Figure 3.23
A large electrochemical potential for Na²⁺ exists across the plasma membrane of marine algae, such as *Acetabularia*, shown here.

transient, and V_m returns spontaneously to its negative resting potential, typically after 0.001 to 0.002 s in axons. Such cells are said to be "excitable." In the nervous systems of animals, action potentials have well-known signaling roles. Some types of plant cells are excitable, although the action potentials are about three orders of magnitude slower than those of their animal counterparts. The best-studied action potentials in plants occur in the giant internodal cells of charophyte algae, where their function—if any—remains unknown (Fig. 3.24). A few terrestrial plant cells also are excitable, especially those exhibiting rapid movements. In the sensitive plant (*Mimosa pudica*), leaf stroking evokes an action potential that causes the pulvinus, a hinge region at the base of the leaf, to lose turgor and collapse the leaf (Fig. 3.25). Some insectivorous plants (e.g., sundews, *Drosera*) also use action potentials to couple the sensing of prey to subsequent leaf movements.

These rather specialized examples might well imply that the distribution and roles of plant channels are rather restricted. Indeed, despite extensive speculation, little was known about plant ion channels until the 1980s, when the development of new technologies for neurophysiological investigations and the application of these methods to plant cells demonstrated the presence of ion channels in a nonexcitable plant cell, the guard cell. The guard cell ion channels were determined to play key roles in mediating

the solute fluxes that accompany stomatal opening and closure. Channels are now known to be present in all plant cell types—at the vacuolar and plasma membranes, and probably at all other membranes as well.

3.5.2 Ion channels are studied with electrophysiological techniques.

As so often occurs in biology, the dramatic advances in our understanding of plant ion channels made during the mid-1980s resulted from the application of a novel technique—in this case, the **patch clamp** technique. Originally developed by Erwin Neher and Bert Sakmann for detecting and characterizing channels in excitable animal cells, the technique is also applicable to plant plasma membranes and intact plant vacuoles. In essence, the patch clamp technique allows detection of the tiny electrical currents that ions carry as they flow through channels. The experimenter observing these currents can control all the components of the electrochemical driving force on individual ionic species, including the media on both sides of the membrane and, by using a feedback amplifier, the transmembrane voltage. The outstanding capability of the patch clamp technique is that it can resolve the activity of single protein molecules (channels) as they catalyze ion translocation (Box 3.7). An alternative technique for the measurement of single channels involves the construction of planar lipid bilayers (Fig. 3.26). This technique has applications for recording activities of channels at intracellular membranes (e.g., ER) that, by virtue of their dimensions, are not amenable to conventional patch clamp analysis.

Although single-channel recording techniques have yielded immense insights into the diversity of channel types in plant membranes, a role remains for the classical voltage-clamping approach, in which intact, walled cells are impaled with microelectrodes (Fig. 3.27). However, this approach is applicable to only a limited number of cell types, because intercellular currents spreading through plasmodesmata compromise the capacity for a spatially uniform voltage clamp. In this respect, guard cells are ideal material for investigation because, unlike the vast majority of higher plant cells, they lack intercellular

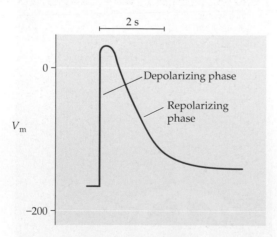

Figure 3.24
An action potential as seen in the large internode cells of some algae. In response to a depolarizing, stimulating pulse, the membrane potential rapidly depolarizes further. This depolarization phase is followed by a much slower repolarizing phase, during which ion fluxes return the V_m to nearly its initial value.

connections. Despite being restricted to whole-cell recordings of membrane currents, the voltage clamp technique has the advantage over patch clamp, in that the cell wall and regulatory machinery within the cell can remain relatively intact during recordings.

3.5.3 Ionic fluxes through channels are driven solely by electrochemical potential differences.

Ion flow through channels is passive. Thus, in contrast to pumps or ion-coupled carrier activity, the direction of flow of a particular ion through a channel is dictated simply by the electrochemical potential gradient for that ion, $\Delta\bar{\mu}_{ion}$ (see Box 3.3). The current passing through a single channel (Fig. 3.28A) can be plotted as a function of the membrane potential to yield a so-called current–voltage (I-V) relationship for the channel (Fig. 3.28B). Ordinarily the single channel I-V relationship is fairly linear on either side of the zero-current potential, so Ohm's law (I = V/R) holds. The slope of this relationship, $1/R$, is the single channel conductance (g), measured in pico-siemens (pS). The value of g is a measure of the permeability of the channel and is characteristic for a given channel type in particular conditions. Typically, g increases with the concentration of the permeant ion, so it is important to specify the recording conditions under which g is measured.

Channel-mediated currents that flow into the cytosol are defined as negative, are shown as downward deflections on current traces, and may be carried either by an influx of cations or an efflux of anions. Conversely, outward currents carried by cation efflux or anion influx are defined as positive and are shown as upward deflections on current traces. When $\Delta\bar{\mu}_{ion}$ = 0, the net flux of that ion (and hence the current carried by it) through the channel also will be zero. Thus the equation that defines $\Delta\bar{\mu}_{ion}$ (see Box 3.3) can be rewritten for this equilibrium condition as Equation 3.3, in which E_{ion} is known as the equilibrium potential for that particular ion. E_{ion} defines the voltage at which the current for a given ion reverses between inward and outward. At this point, in terms of experimental variables, E_{ion} is simply a function of the ion concentration difference across the membrane.

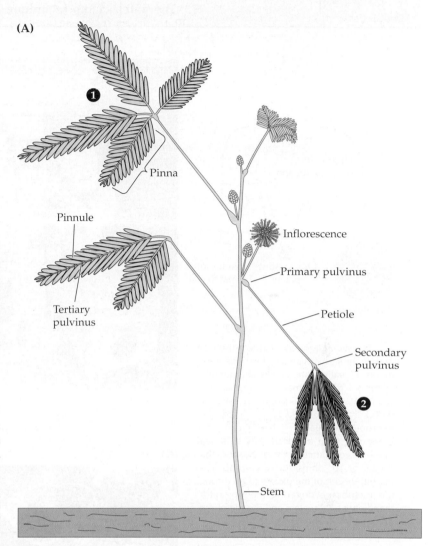

(A)

Pinna

Pinnule

Inflorescence

Primary pulvinus

Tertiary pulvinus

Petiole

Secondary pulvinus

Stem

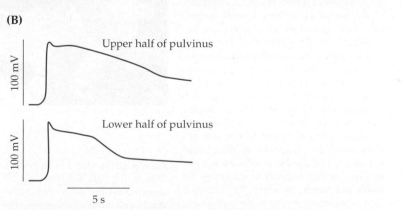

(B)

100 mV — Upper half of pulvinus

100 mV — Lower half of pulvinus

5 s

Figure 3.25
An action potential in the pulvinus of *Mimosa pudica* leaves results in a reversible collapse of the leaf. (A) Each leaf has a pulvinus at the base of its petiole as well as more numerous secondary and tertiary pulvini located at the bases of its leaflets. When turgid, these organs hold the leaf away from the stem with the leaflets expanded. An action potential results in loss of turgor in the pulvini and a rapid lowering of the petiole and folding of the leaflets. (B) Action potential for the upper and lower halves of the *Mimosa pudica* pulvinus.

Box 3.7   **The patch clamp technique is used to measure ionic currents.**

(A)

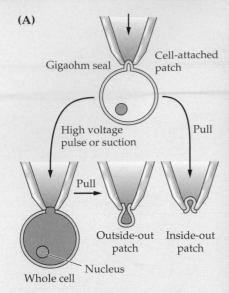

(B)

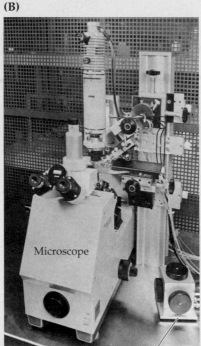

3-axis
micromanipulator

(C)

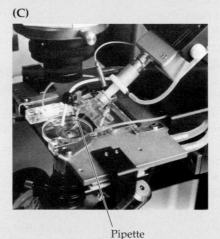

Pipette

The patch clamp technique arose from two technical advances. First, developments in solid-state electronics enabled accurate measurement of picoampere (10^{-12} A) currents. Second, investigators found that pushing a blunt-tipped glass micropipette against a biological membrane can form an electrically tight seal. The high resistance of this seal (more than 1 GΩ, or 10^9 ohms) forces currents flowing into or out of the pipette to go through the membrane covering the tip, rather than between the glass and the membrane lipid (panel A, above).

Typical patch clamp recording apparatus is shown in panels B and C. The patch clamp technique utilizes several recording modes (panel A), all of which are accessible with this apparatus. The initial seal formation attains **cell-attached mode,** which can record the activities of individual ionic channels but does not allow control of medium composition on the cytosolic side of the membrane. Despite its limited usefulness, this recording mode can be applied to channels that lose activity when their contact with the cytosol is broken. Pulling the pipette away from the rest of the membrane generates an **inside-out patch,** in which the cytosolic face is exposed to the bathing medium. As with cell-attached recording, the activities of single ionic channels can be assessed, but the solution composition on both sides of the membrane is defined. Alternatively, the membrane patch that covers the pipette tip can be disrupted with a high-voltage pulse or suction,

thereby giving electrical access to the inside of the cell. Currents can be monitored across the entire membrane in this **whole-cell mode** of recording. The activities of the ensemble of ion channels in the cell membrane can then be monitored. The relatively large volume of medium in the pipette exchanges rapidly with the cell contents, defining the intracellular solution. Finally, if the pipette is pulled away from the cell after attainment

of the whole-cell mode, a membrane bleb is also pulled away and reseals itself across the pipette tip as an **outside-out patch.** This recording mode is particularly useful for testing the effects of putative cytosolic regulators on channel activity.

The cell wall represents a barrier that must be removed for patch clamp studies on plant material. Enzymatic digestion of the wall yields protoplasts for which the membranes are sufficiently clean to form gigaohm seals. Intact vacuoles suitable for patch clamping can be isolated either by gentle lysis of protoplasts, or by slicing tissue in an isoosmotic medium. Because the vacuole is the only organelle in biology large enough to be easily patch-clamped, plant biologists have a more complete description of ion channel activity in native endomembranes than do researchers who work on animal or yeast cells.

Turnover rates of individual channels can be calculated from currents flowing when a channel opens and the recording mode is configured for an inside-out or an outside-out membrane patch. The channel current (in amperes, i.e., coulombs per second) can be converted to an electrical flux (mol charge s^{-1}) by division by F. To convert the electrical flux to a rate of transport (ions s^{-1}), the resulting value is multiplied by Avogadro's number and divided by the valence of the ion. The channel-mediated current of 1 pA for a monovalent cation represents a turnover rate of more than 6×10^6 molecules per second:

$$[(10^{-12}\ \text{C} \bullet \text{s}^{-1})(6.02 \times 10^{23}\ \text{ions} \bullet \text{mol}^{-1})]/[(96{,}500\ \text{C} \bullet \text{mol}^{-1})(\text{valence term})]$$
$$= 6.24 \times 10^6\ \text{ions} \bullet \text{s}^{-1}.$$

The patch clamp technique can be applied to studying the electrical properties of any electrophoretic transport system, including pumps and carriers. However, in these cases, because the turnover rates are markedly lower than those of channels (see Section 3.2.2), the activities of individual pump or carrier transport systems cannot be recorded; instead, experimental investigation is restricted to the whole-cell or whole-vacuole recording modes.

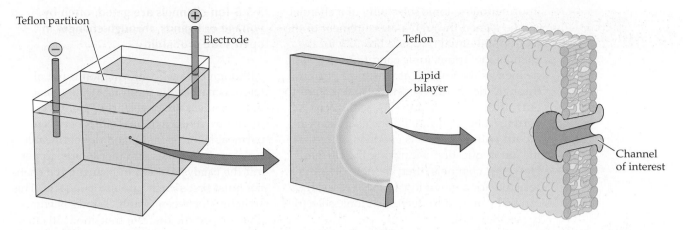

Figure 3.26
Measurement of channel activity in planar lipid bilayers. A bilayer of purified lipids can be formed across a small (0.1-mm-diameter) hole in a Teflon partition that separates two chambers. Electrodes are dipped into solutions on either side of the partition and connected to an amplifier. Fractionated plant membrane vesicles introduced into one of the chambers fuse spontaneously with the artificial membrane, thereby incorporating channels into the bilayer and allowing the currents they carry to be measured.

Equation 3.3

$$V_m = E_{ion} = RT/zF \cdot \ln\{[ion]_o/[ion]_c\}$$

Channels are not thought to require conformational change to catalyze transport. Nevertheless, it would be wrong to think of channels merely as sieve-like pores in a membrane, because they exhibit two additional properties that are essential for their function: **ionic selectivity** and **gating,** which are discussed and defined in the next two sections.

3.5.4 Ion channels exhibit ionic selectivity.

To various degrees, ion channels display selectivity for their ionic substrates. Most classes of ion channels in plant cells usually discriminate in favor of either cations or anions. Cation channels can be further subdivided into those that select K^+ over other monovalent cations, those that are relatively nonselective among monovalent cations, and those that are selective for Ca^{2+}. Most plasma membrane anion channels allow permeation of a wide range of anions, including Cl^-, NO_3^-, and organic acids. Other anion channels in the vacuolar membrane select specifically for malate.

Up to this point, we have emphasized that ion fluxes through channels are monitored as electrical currents. However, currents themselves provide no information on which ions are flowing. In general, this

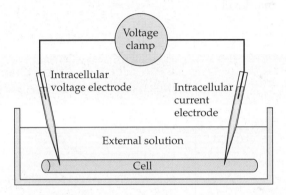

Figure 3.27
Recording whole-cell membrane currents with a two-electrode voltage clamp. The cell is impaled with a voltage microelectrode, which is used to establish the electrical potential across the membrane, and a current electrode, which measures the current flowing in or out of the cell.

Single-channel current

Single-channel current (pA)

Outward current

Membrane potential (mV)

Inward current

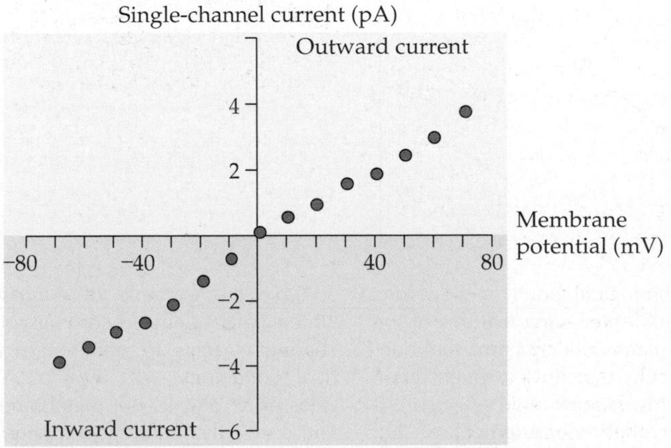

Figure 3.28
Current–voltage (*I-V*) curve for a single channel.

information on ionic selectivity of a channel can be gained through a measurement of the reversal potential (E_{rev}; see Box 3.8) for the channel in various ionic conditions. E_{rev} is defined as the voltage at which the current through the channel reverses from inward to outward; in other words, E_{rev} is the zero-current voltage. If E_{rev} for the channel is in accord with E_{ion}, this is good evidence that the ion in question is carrying the current. However, channels often are not perfectly selective for a given ion; in these cases, E_{rev} can be evaluated to determine ionic selectivity (see Box 3.8).

The fact that channels are selective implies that binding sites capable of distinguishing specific ions must be located within the channel pores. In some cases, these ion-binding sites have been identified at a molecular level.

3.5.5 Ion channels are gated, often by voltage or ligands, through changes in open state probability.

Although channel conformation is not altered as part of the ion translocation reaction, channels are nevertheless tightly controlled by conformational shifts between permeable (or open, O) and nonpermeable (or closed, C) states. In other words, to permit the catalysis of ion movement, the channel must first switch into its open state. This switching between O and C states is what gives rise to the discrete transitions in current shown in single-channel recordings (Fig. 3.29). This alternation between the O and C states, known as gating, is represented by Reaction 3.3. In nearly all channels, gating is controlled by membrane voltage, a ligand, or both. When a gating factor activates a

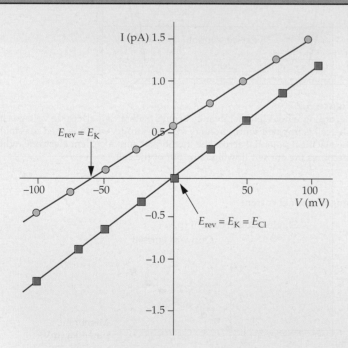

From a biological point of view, one of the principal interesting features of ion channels in plant cells concerns the identities of the ions that flow through them. Frequently, this issue of ionic selectivity is addressed through measurement of E_{rev} for the channel-mediated currents in various ionic conditions.

Take the example of a channel for which patch clamp recordings of single-channel currents are obtained in symmetrical conditions (100 mM KCl on both sides of an outside-out membrane patch). These records of current can be plotted as a function of voltage in the form of a current–voltage (I-V) relationship (see the

figure at left). As seen for the curve with the square data points, the zero-current potential (E_{rev}) is 0 mV under these symmetrical conditions. However, for either K^+ or Cl^-, the natural logarithm term in Equation 3.3 that defines the E_{ion} reduces to 0 because the concentration ratio of either ion is 1 (and ln 1 = 0). Now, let us suppose that the external KCl concentration is decreased to 10 mM in the continued presence of 100 mM KCl in the pipette. Now there is a 10-fold concentration difference across the pipette, which is reflected in the I-V curve with circular data points (figure at left). The value for E_{rev} is now –59 mV. Comparing this value with E_K (–59 mV) and E_{Cl} (+59 mV) shows, by the coinciding values of E_{rev} and E_K, that the channel has perfect selectivity for K^+ over Cl^-. If the channel is permeable to more than one ionic species, its reversal potential is the mean of the individual ionic equilibrium potentials, weighted according to their respective permeabilities.

Most cation channels do indeed discriminate very effectively against anions, although selectivity for a given cation over other cations is rarely perfect (and similarly for anion channels). Selectivity ratios for different permeant ions are often determined in so-called bi-ionic conditions, in which each permeant ion is

channel, the equilibrium of Reaction 3.3 shifts from left to right.

Reaction 3.3

$$\text{Channel}_c \rightleftharpoons \text{Channel}_o$$

The contribution of an individual type of channel to whole-cell or whole-vacuole currents can be defined as in Equation 3.4, where I is the ionic current across the whole cell or vacuole, N is the number of channels, i is the current that flows through the channel when open, and P_o is the probability that the channel is in its open state. Activation of a channel-mediated current usually involves an increase in P_o through gating (Fig. 3.30). For example, voltage-dependent gating can transform a linear I-V relationship measured at the single-channel level (see Fig. 3.28)

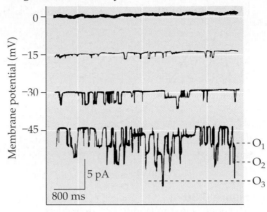

Figure 3.29
Single channel activities in a plant vacuolar membrane patch. The prevalence of the conductive (O state) increases with imposed membrane voltage in the negative range. O_1, one channel open; O_2, two channels open; O_3, three channels open.

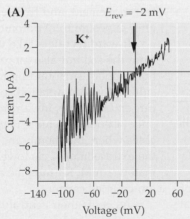

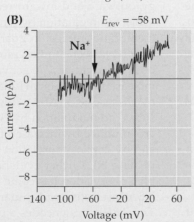

selectively present on just one side of the membrane. In the pair of graphs at left, for example, the same outside-out patch was bathed with either 100 mM K^+ (A) or 100 mM Na^+ (B). In both cases, the pipette contained 100 mM K^+. As expected, E_{rev} was close to 0 mV with equal concentrations of K^+ across the membrane. With K^+ inside and Na^+ outside, however, E_{rev} shifted to −58 mV. This value can be used to compute a permeability ratio for Na^+ and K^+. For the example of a cation channel with permeability both to K^+ and Na^+, then $E_{rev} = (RT/zF)\ln\{P_{Na}[Na^+]_o/P_K[K^+]_i\}$, where P_{Na}/P_K is the permeability ratio for Na^+ with respect to K^+.

A word of caution: This approach to the determination of selectivity ratios assumes a constant electrical field through the membrane and independent movement of ions through the channel. Examples of nonindependent ion movement through some channels have been noted, but to date no better approach for estimating selectivity ratios has been devised.

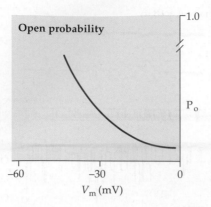

Figure 3.30
The open state probability (P_o), defined as the proportion of total recording time that the channel resides in its open state, is plotted as a function of the holding voltage. Here, P_o shows a progressive increase as V_m decreases.

to a distinctly nonlinear *I-V* relationship across the whole plasma membrane (Fig. 3.31). In addition, effective changes in *N* can be achieved through covalent modification (e.g., phosphorylation).

Equation 3.4

$$I = N \bullet i \bullet P_o$$

3.6 Ion channels in action

3.6.1 Voltage-dependent K$^+$ channels at the plasma membrane stabilize V_m and allow controlled K$^+$ uptake and loss.

Whole-cell recordings made over a range of holding voltages (see Fig. 3.31) indicate that currents flow across the plasma membrane in both directions, into and out of the cytosol. These inward and outward currents have two components: an instantaneous current, and a time-dependent current that increases over a period of several hundred milliseconds. Plotting the time-dependent currents as a function of membrane voltage makes it apparent that the currents are voltage-activated. Activation occurs only beyond a threshold voltage, which is less than E_K for the inward current and greater than E_K for the outward current. The time-dependent currents are widely distributed in different types of plant cells and represent the activi-

ties of channels that are highly selective for K$^+$ (Fig. 3.32). Molecular evidence (see Section 3.6.2) demonstrates definitively that the time-dependent inward and outward currents are carried by separate classes of ion channel. The channels carrying these currents are said to **rectify.** Like valves, rectifying channels carry current in one direction but not the other. For this reason, the channels are known as K$^+$ inward rectifiers and K$^+$ outward rectifiers. Both are inhibited by millimolar concentrations of tetraethylammonium, a diagnostic blocker of K$^+$ channels.

The functions of rectifying K$^+$ channels are now well established. Outward rectifiers stabilize V_m at relatively negative values in the region of E_K. For a cell bathed in 1 mM K$^+$ and having 100 mM K$^+$ in the cytosol, E_K is –120 mV. Thus, if other channels or electrophoretic carriers depolarize the membrane to values more positive than –120 mV, outward rectifying channels open; the resulting increased membrane permeability to K$^+$ tends to prevent further depolarization. In addition, outward rectifiers contribute to the net release of salt from guard cells during stomatal closure, from pulvinar cells during leaf movement, and from euryhaline algae after exposure to hypotonic medium.

Inward rectifiers take up K$^+$ not only from the soil but also more generally from the apoplast surrounding most cells; the resulting K$^+$ accumulation contributes to cell turgor. Of course, inward rectifiers work only at potentials more negative than E_K. In cases where the membrane is depolarized or the extracellular K$^+$ is lower than approximately 1 mM, K$^+$ must be accumulated by ion-coupled carriers.

As befits their central role in cell biology, outward and inward rectifiers also are subject to regulation by factors other than membrane voltage. In guard cells, where they have been studied most intensively, outward rectifiers are activated by the small increase in cytosolic pH that can be elicited by abscisic acid (ABA). Inward rectifiers are subject to modulation by G-proteins and are inhibited by increased free cytosolic Ca$^+$ and by dephosphorylation by protein phosphatase 2B. Further evidence for control by phosphorylation has come from a study of ABA-insensitive *abi1* mutants of *Arabidopsis*. The *ABI1* locus encodes a protein phosphatase 2C. Mutants display much less outward

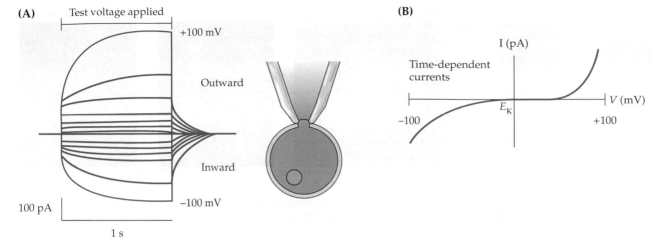

(A) Test voltage applied

+100 mV

Outward

Inward

−100 mV

100 pA

1 s

(B)

I (pA)

Time-dependent
currents

E_K

−100

+100

V (mV)

Figure 3.31

Whole-cell currents in a plant cell as measured by the patch clamp technique. (A) A pipette is sealed to the plasma membrane in the whole-cell configuration. In response to a series of 1-s voltage pulses, current flows through channels in the plasma membrane. The 13 current traces, corresponding to 13 different voltages between −100 and 100 mV, are superimposed in this plot. (B) Current–voltage relationship for the time-dependent component of currents in A. Note that this whole-cell I-V curve is decidedly nonlinear.

rectifier activity and even abolished rectifier response to ABA, which up-regulates outward rectifiers and down-regulates inward rectifiers in wild-type plants. These regulatory factors are all components of guard cell signal transduction networks, and their coordinated action tightly controls K^+ uptake and loss.

3.6.2 Plant cell inward rectifiers are members of the *Shaker* family of voltage-gated channels.

Inward rectifiers have been cloned by complementation of yeast mutants defective in K^+ uptake. Electrophysiological analyses of plant inward rectifiers expressed in frog oocytes, insect cells, and yeast have confirmed their presumed function.

Plant inward rectifier subunits are products of a multigene family, members of which exhibit tissue-specific expression. One member, KAT1 (see Box 3.2), is expressed selectively in guard cells (Fig. 3.33A); another, AKT1 (Fig. 3.34), is found in roots and hyadothodes (Fig. 3.33B). The *AKT1* gene of *Arabidopsis* has been disrupted by using T-DNA insertional mutagenesis (see Chapter 6). The resulting mutant, referred to as *akt1-1*, has dramatically reduced K^+ uptake (Table 3.1) and substantially less growth in media

(A)

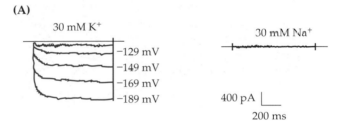

30 mM K^+

−129 mV
−149 mV
−169 mV
−189 mV

30 mM Na^+

400 pA

200 ms

(B)

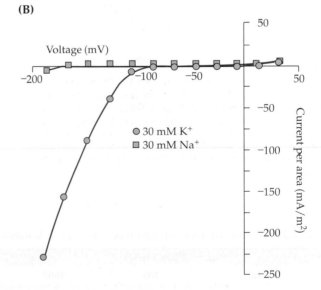

50

Voltage (mV)

−200   −100   −50   50

● 30 mM K^+
■ 30 mM Na^+

−50

−100

−150

−200

−250

Current per area (mA/m²)

Figure 3.32

(A) Time-dependent currents through K^+-selective inward rectifiers in maize root cortical protoplasts. K^+ flows freely through these channels, but sodium is much less permeable. (B) The *I-V* relationship for the currents in A demonstrates the channels' selectivity for K^+.

Figure 3.37
Structure of the *Shaker*
potassium channel,
viewed from the side
(left) and top (right). In
the side view, the fore-
most of the four sub-
units that contribute to
the pore is omitted to
show the center of the
channel. Positively
charged regions are
shown in blue, nega-
tively charged regions
in red.

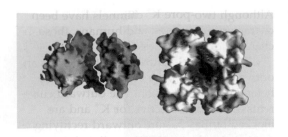

3.6.4 Voltage-insensitive cation channels may be a major pathway for Na⁺ uptake across the plasma membrane and for salt release to the xylem.

Unlike time-dependent currents, currents that arise instantaneously are effectively present at all values of V_m (Fig. 3.40). Analysis of these currents and the underlying channels demonstrates that although these are cation-selective, they do not select as effectively for K⁺ as the inward rectifiers do. Such channels probably constitute a major pathway for Na⁺ uptake into plant cells and could have important implications for salinity tolerance. These channels are partially blocked by external Ca²⁺.

An even less selective channel has been observed in the plasma membrane of xylem parenchyma cells. This channel is almost as permeable to anions as to cations; such low selectivity is unusual in plasma membrane ion channels. Perhaps this channel could provide a pathway for release of salts from the symplasm into the xylem.

3.6.5 Monovalent cation channels at the vacuolar membrane are Ca²⁺-sensitive and mediate vacuolar K⁺ mobilization.

Vacuoles are usually thought of as organelles for accumulation and storage of solutes, but in some circumstances massive solute loss from vacuoles is required. One important example is during osmotic adjustment, which can involve massive net salt loss from the cell and hence from the vacuole. Thus, hypoosmotic stress involves mobilization of ions from vacuoles to restore normal cell turgor, and during stomatal closure, ions must be lost from guard cell vacuoles. Ion channels facilitate this mobilization of ions. In the case of guard cells, two different classes of K⁺-permeable channel are competent in releasing K⁺ from the vacuole. The interesting features of these channels relate to their complementary modes of regulation.

Both types of channels activate instantaneously in response to an imposed voltage. The fast vacuolar (FV) channel, which exhibits little selectivity among monovalent cations, is inhibited when cytosolic Ca²⁺ concentrations exceed 1 μM and is activated when cytosolic pH increases (Fig. 3.41). Conversely, the vacuolar K⁺ (VK) channel, which is highly selective for K⁺ over other monovalent cations, is activated by cytosolic Ca²⁺ in the nanomolar to low micromolar concentration range and is inhibited by increasing cytosolic pH.

Stomatal closure is often but not always preceded by an increase in the concentration of free Ca²⁺ in guard cell cytosol, so VK channels are well poised to respond by opening and releasing K⁺. However, stomata also can close in the absence of a change in free Ca²⁺, in which case an increase in guard cell cytosolic pH is likely to play a role in responding to the primary closing stimulus. In these cases, FV channels will open. The presence of complementary channel types, each fulfilling the same function of vacuolar K⁺ release, probably indicates a convergence point for different signaling pathways.

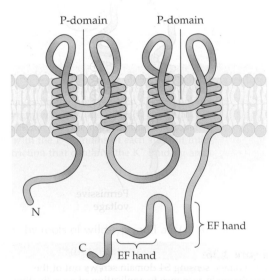

P-domain   P-domain

N

C

EF hand

EF hand

Figure 3.38
An outward rectifying K⁺ channel. KCO1 has just four transmembrane spans but two P-domains. Toward the C terminus are two high-affinity calcium binding motifs known as EF hands.

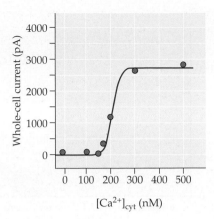

Figure 3.39
Calcium dependence of the outward rectifier KCO1. Ion-channel activity of insect cells expressing KCO1 was measured by the patch clamp technique. Shown is the mean current amplitude in the presence of different concentrations of free Ca^{2+}. Activation of the channel is seen to occur between approximately 150 nM and 300 nM free Ca^{2+}.

3.6.6 Calcium-permeable channels in the plasma membrane provide potential routes for entry of Ca^{2+} to the cytosol during signal transduction.

A wide range of metabolic and developmental signals in plants trigger an increase in cytosolic free Ca^{2+}. This Ca^{2+} increase is thought to be central to signal transduction (see Chapter 18). Thus, the increase in the concentration of free Ca^{2+} activates downstream targets that further transduce the initial signal into the end response (Fig. 3.42). Immediate targets of the increase in free Ca^{2+} include not only ion channels (see Section 3.6.5) but also calmodulin, calmodulin-domain protein kinases (CDPKs), phosphatases similar to protein phosphatase 2B (e.g., calcineurin), and calmodulin-binding enzymes.

Because the increase in cytosolic Ca^{2+} occurs through activation of Ca^{2+}-permeable channels, these channels can be considered upstream elements in Ca^{2+}-based signal transduction pathways, and the variety of ways in which the channels are activated is a critical feature of the early stages of the signaling. A wide array of Ca^{2+}-permeable channel types has been reported in several plant cell membranes, including the tonoplast, the plasma membrane, and the ER. A large electrochemical potential difference, directed into the cytosol, is maintained across

all these membranes by Ca^{2+}-pumping ATPases, H^+-coupled carriers, or both.

Voltage-gated Ca^{2+}-permeable channels reside in the plasma membranes of a range of cell types and have been characterized by both patch clamp and planar lipid approaches. The channels exhibit various degrees of selectivity for Ca^{2+} over K^+—from about 2:1 to 20:1. The channels are activated by membrane depolarization (Fig. 3.43). Thus, although membrane depolarization decreases the considerable electrochemical driving force on Ca^{2+} into the cytosol, the influx of Ca^{2+} is increased because the probability of being in the open state, P_o, is increased by membrane depolarization. By implication, signals that

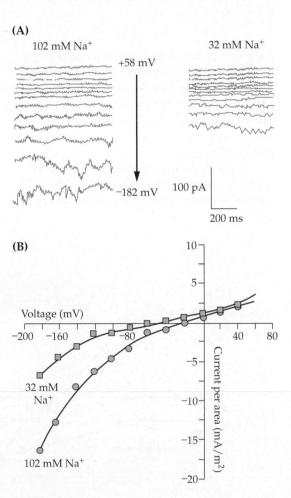

(A)

102 mM Na$^+$ 32 mM Na$^+$

+58 mV

−182 mV   100 pA

200 ms

(B)

Voltage (mV)

32 mM Na$^+$

102 mM Na$^+$

Current per area (mA/m^2)

Figure 3.40
(A) Na$^+$-dependent currents across the plasma membrane of protoplasts prepared from maize root cortical cells. Currents are present at both positive and negative voltages and are almost instantaneous, showing no lag after application of the voltage pulse. (B) I-V relationship for the currents in A showing substantial influx of Na$^+$.

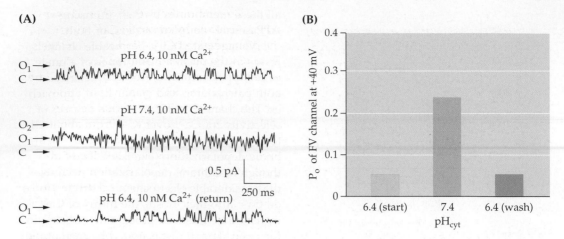

(A)

pH 6.4, 10 nM Ca^{2+}

O_1 →
C →

pH 7.4, 10 nM Ca^{2+}

O_2 →
O_1 →
C →

0.5 pA
250 ms

pH 6.4, 10 nM Ca^{2+} (return)

O_1 →
C →

(B)

P_o of FV channel at +40 mV

6.4 (start)   7.4   6.4 (wash)
pH$_{cyt}$

Figure 3.41
Activity of vacuolar FV K$^+$ channels depends on cytosolic pH and Ca^{2+}. (A) The activity of a single FV channel from a *Vicia faba* guard cell tonoplast is strongly influenced by pH. O_1, one channel open; O_2, two channels open; C, closed. (B) Increasing cytosolic pH from pH 6.4 to 7.4 increases the open-state probability (P_o) of the channel.

elicit membrane depolarization could trigger entry of Ca^{2+} from the external medium. As discussed earlier (see Section 3.6.3), the depolarizing signal can be switched off—and uncontrolled Ca^{2+} entry into the cytosol prevented—by the Ca^{2+}-induced activation of outward rectifiers.

In patch clamp experiments, applying gentle suction to the patch pipette when recording from membrane patches has re-

vealed the existence of additional Ca^{2+}-permeable channels. These stretch-activated channels, so named because they become active when the membrane is stretched by the applied suction, may be involved in signaling during the early stages of mechanosensory transduction. A touch-responsive rapid increase in cytosolic free Ca^{2+} is thought to play a role in inhibiting growth and producing the stunted habit exhibited by plants in

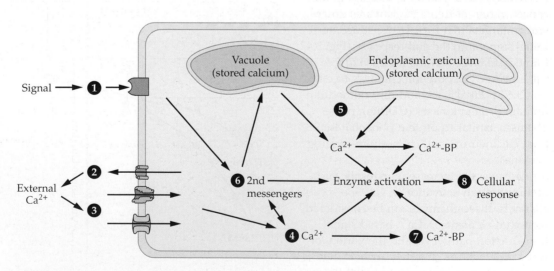

Figure 3.42
Calcium-based signal transduction in a typical plant cell. An extracellular signal (1) is perceived by a receptor that may be a protein integral to the plasma membrane. Signal perception may change the activity of Ca^{2+}-ATPases (2) or plasma membrane Ca^{2+} channels (3), resulting in a change in cytosolic free calcium concentration (4). The signal might also result in the release of Ca^{2+} from internal stores (5) by way of additional second messengers (6) such as IP$_3$. Cross talk between Ca^{2+} and other second messengers is likely. Increased free calcium changes the activity of Ca^{2+}-binding proteins (Ca^{2+}-BP) (7), which activate enzymes that initiate the events leading to the final cellular response (8).

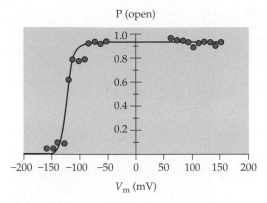

P (open)

Figure 3.43
Activation of a wheat root plasma membrane Ca^{2+} channel by voltage. The channel activates strongly as V_m shifts to above –150 mV and is fully activated at –100 mV.

windy environments. However, no direct evidence for a role of any given class of plasma membrane Ca^{2+} channel in a particular signal transduction pathway has yet been established.

3.6.7 Calcium-permeable channels in endomembranes are activated by both voltage and ligands.

No fewer than four different classes of Ca^{2+}-permeable channels are present in intracellular membranes of plants. Two of these channel types, which reside in the vacuolar membrane, are activated by ligands—inositol 1,4,5-trisphosphate (IP_3) and cyclic ADP-ribose (cADPR)—known to have potent Ca^{2+}-mobilizing properties in animal cells (Fig. 3.44). The latter compound releases Ca^{2+} through ryanodine receptors, a class of channels named for the plant alkaloid they bind.

In almost all respects, IP_3- and cADPR-activated plant channels resemble their animal counterparts, including dose–response relationships to activating ligands, ionic selectivity (the IP_3 receptor is very highly selective for Ca^{2+} over K^+, the ryanodine receptor much less so), and pharmacology (the IP_3 receptor is inhibited by heparin, the ryanodine receptor by ruthenium red). The two respects in which they differ are membrane location—the plant channels reside in the vacuolar membrane, the animal channels in the ER—and the failure of the plant channels to be activated by cytosolic free Ca^{2+}.

Activation of the animal IP_3 and ryanodine receptors by cytosolic Ca^{2+} underlies the phenomenon of Ca^{2+}-induced Ca^{2+} release (CICR), in which a small Ca^{2+} signal is amplified. In plants, CICR might be generated through the activity of one of the two voltage-activated Ca^{2+}-permeable channels at the vacuolar membrane. Although one of these channels is activated by membrane hyperpolarization and is not sensitive to cytosolic Ca^{2+}, the other exhibits rather slow, time-dependent activation over several hundred milliseconds in response to membrane depolarization and is strongly activated by Ca^{2+}-calmodulin (Fig. 3.45). This latter channel is known as the slowly activating vacuolar (SV) channel. Protein phosphatase 2B, which is Ca^{2+}-activated, provides negative feedback control of SV channel activity.

Why are there so many types of Ca^{2+}-permeable channels in plant cells? There is as yet no definitive answer to this question, but we can speculate. The activation of different channel types might result in different dynamic patterns of Ca^{2+} signal, thereby endowing the Ca^{2+} signal with additional

(A) The guard cell vacuolar membrane during ABA-induced stomatal closure

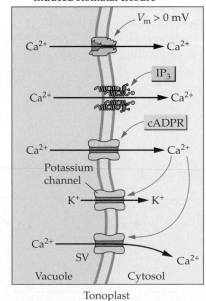

Tonoplast

(B) During plasma membrane-based signal transduction

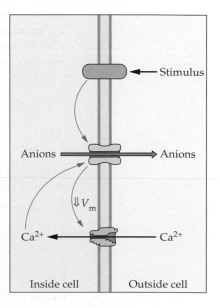

Figure 3.44
Diagram illustrating channel activities at the guard cell vacuolar membrane during stomatal closure (A) and during plasma membrane–based signal transduction (B). SV, slowly activating vacuolar channel; IP_3, inositol 1,4,5-trisphosphate; cADPR, cyclic ADP-ribose.

(A)

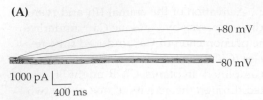

+80 mV

−80 mV

1000 pA

400 ms

(B)

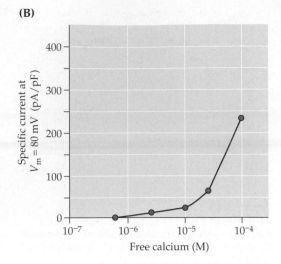

Free calcium (M)

Figure 3.45
Activity of the SV channel increases with increasing cytosolic concentration of Ca^{2+}. (A) Slow activation of the SV channel in barley aleurone vacuoles in response to positive voltages. (B) Ca^{2+}-dependence of whole-vacuole channel activity. Increasing free calcium above approximately 1 µM increases the activity of the channel. Ca^{2+} is thought to interact with calmodulin associated with the channel or a channel regulatory protein. pF, picofarads.

information that might render the downstream responses stimulus-specific. This notion is supported by stimulus-specific Ca^{2+} signal kinetics observed in plants and by Ca^{2+}-based signaling pathways that rely on different Ca^{2+} pools to various degrees.

3.6.8 Plasma membrane anion channels facilitate salt release during turgor adjustment and elicit membrane depolarization after stimulus perception.

Anion channels are ubiquitous in plant plasma membranes, where they are thought to play several essential roles. Their most obvious function is controlling salt release dur-

ing turgor adjustment, either in response to hypotonic stress or to a signal that changes the set-point around which turgor is controlled, as seen in guard cells during stomatal closure. Cellular loss of anions, principally Cl^-, is effected by Ca^{2+}-activated anion channels, of which at least two kinetically distinct classes, the rapidly activating R-type and the slowly activating S-type, have been identified in guard cells. These Ca^{2+}-activated anion channels are also gated by voltage (Fig. 3.46). E_{Cl} ordinarily attains a positive value because cytosolic Cl^- concentrations exceed the extracellular Cl^- concentration. Therefore, opening these channels results in not only loss of Cl^- but also membrane depolarization. This depolarization probably

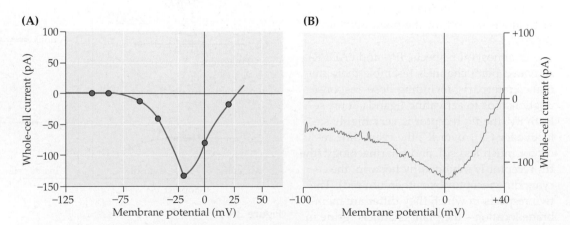

Figure 3.46
Anion channels in guard cells. (A) Current–voltage relationship for rapidly activating (R-type) anion channels. (B) Current–voltage relationship for slowly activating (S-type) channels.

activates outward rectifying K^+ channels during salt loss. Anion channel activity is therefore the pacemaker of turgor control.

A second function of anion channels relates simply to their ability to depolarize the plasma membrane. Anion channel opening appears to be an early event in many signal transduction pathways and probably leads to the activation of voltage-gated Ca^{2+} channels (see Section 3.6.6).

Third, some classes of anion channel are activated by extreme membrane hyperpolarization. These channels are probably able to participate in control of V_m, a function that could achieve prominence if inward rectifier activity is limited (e.g., through near-absence of extracellular K^+).

Calcium-activated anion channels also are subject to additional types of V_m- and Ca^{2+}-independent regulation. In a range of cell types, including guard cells, ATP has an activating effect. Anion channel activity is not supported by nonhydrolyzable ATP analogs, and the inhibitory effects of ATP removal are prevented by protein phosphatase inhibitors. All this evidence points to control of channel activity by a protein kinase.

Chloride channels have been extensively described in animal cells. Use of the polymerase chain reaction on regions conserved in animal and yeast anion channels has yielded a clone from tobacco that carries anion currents when expressed in frog oocytes. Hydropathy plots of the derived protein sequence predict as many as 13 transmembrane spans. The membrane location of this channel in planta has yet to be confirmed.

3.6.9 Vacuolar malate channels participate in malate sequestration.

In most glycophytic plants, the principal anionic constituent of the vacuole is malate. Plants exhibiting crassulacean acid metabolism (CAM) take up and release malate from their vacuoles on a diurnal basis. Ion channels known as VMAL channels catalyze the uptake of malate into the vacuole. Although they have been observed in numerous plant species, the VMAL channels are most easily studied in CAM plants such as *Kalanchoë*, where they dominate the electrical characteristics of the membrane. Unlike the other channels described in this section,

VMAL channels have no known counterpart in animal cells.

Malate uptake into the vacuolar lumen is driven by the cytosol-negative V_m (Fig. 3.47). Currents are very strongly inward rectifying, so for an anion such as malate, reverse flow into the cytosol is highly unfavorable, even in the presence of a positive membrane potential. Malate is therefore likely to exit the vacuoles of CAM plants through a completely independent pathway, perhaps by way of a carrier. Selectivity studies, in which the ionic forms of malate are titrated between the unprotonated (malate^{2-}) and the singly protonated (H•malate$^-$) forms, strongly suggest that the permeant form is malate^{2-} (Fig. 3.48), which would effectively maximize the use of V_m as a driving force. Because the vacuolar pH is notably lower than that of the cytosol (by about 2 pH units in most plants, but by as much as 4 pH units at the end of the night in CAM plants), the malate^{2-} that enters the lumen will be rapidly protonated and its concentration effectively lowered in favor of the H•malate$^-$ and H_2•malate forms. The pH difference across the tonoplast helps sustain the concentration difference that helps drive malate^{2-} into the vacuole.

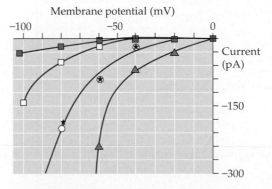

Figure 3.47
Current–voltage relationship for vacuolar uptake of malate through time-dependent anion channels in the tonoplast. Malate uptake by anion channels is strongly promoted by negative membrane potentials and increases with cytosolic malate concentration. In this figure, cytosolic malate concentrations were 10 mM (filled squares), 20 mM (open squares), 50 mM (open circles), and 100 mM (filled triangles)— all with a vacuolar malate concentration of 10 mM. Malate uptake with equal concentrations of malate (50 mM) present on both sides of the membrane is indicated by stars.

3.6.10 Integrated channel activity at the vacuolar and plasma membranes yields sophisticated signaling systems.

Channels can interact in two fundamentally different ways. First, passage of an ion (usually Ca^{2+}) through one channel can result in concentration changes that gate a different channel. Second, channel opening can change V_m, leading to activation of voltage-gated channels. Both aspects of channel interaction are combined in membrane-based signaling pathways that transduce and amplify incoming signals.

One well-studied example relates to Ca^{2+} signaling during stomatal closure (Fig. 3.49). In guard cells, the cytosolic concentration of free Ca^{2+} can be increased by opening hyperpolarization- or ligand-gated Ca^{2+} channels at the vacuolar membrane or depolarization-activated channels at the plasma membrane. The greater concentration of cytosolic free Ca^{2+} in turn will activate VK channels, driving the V_m towards E_K, which is probably positive. This brings V_m into a range in which it can activate SV channels, which additionally will be activated by the increase of cytosolic Ca^{2+}. The effective result is CICR. In this way, at least three separate classes of channels participate in generating a sustained Ca^{2+} signal.

Channel interactions in signaling also occur at the plasma membrane (Fig. 3.49). Signals including blue light, fungal elicitors, and red light initiate rapid but transient changes in membrane potential, in a sequence thought to be initiated by activation of anion channels, which provide the main depolarizing thrust. Voltage-activated Ca^{2+} channels are then activated, increasing the concentration of cytosolic Ca^{2+}. The increase in Ca^{2+} then activates the outward rectifying K^+ channels, returning the membrane potential to a relatively negative value close to E_K.

3.7 Water transport through aquaporins

3.7.1 Directionality of water flow is determined by osmotic and hydraulic forces.

Most biologists have, at some time or another, observed the simple effects of hyperosmotic treatment of plant tissue: The tissue becomes flaccid, and at the cellular level, plasmolysis occurs. The rapidity of plasmolysis bears testimony to the high degree of permeability biological membranes exhibit toward water.

Two forces acting on water determine the directionality of water flow through the plasma membrane: the hydrostatic pressure difference (ΔP) across the plasma membrane/cell wall complex, and the osmotic pressure difference ($\Delta \pi$) between the inside and the outside of the cell. Formally, the flux of water across the membrane (J_v) is proportional to the driving force (Eq. 3.5), where L_p, the constant of proportionality, is the permeability of the membrane to water, expressed in terms of surface area, $m \cdot s^{-1} \, MPa^{-1}$ (i.e., $m \cdot s^{-1} \, m^2 \, N^{-1}$, where N is the force in Newtons, $kg \cdot m \cdot s^{-2}$); σ, known as the reflection coefficient, expresses the ability of the osmotically relevant solutes to permeate the membrane relative to water. For completely impermeant solutes, $\sigma = 1$, whereas for solutes with permeability equal to that of water, $\sigma = 0$. In practice, for most osmotically active solutes encountered in physiological conditions (ions, sugars, and so forth),

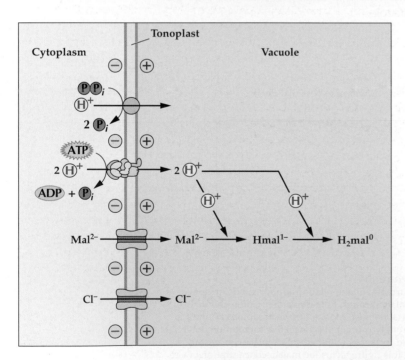

Figure 3.48
Accumulation of malate in the roots of CAM plants. Malate^{2-} is thought to enter the vacuole through malate-selective channels. These channels are strongly inward rectifying and do not allow substantial malate^{2-} efflux. Once inside the vacuole, malate^{2-} is protonated to H•malate^{1-} and H$_2$•malate^0. This maintains the effective concentration difference for malate^{2-} across the membrane.

the permeability of water is so high that σ can be considered to equal 1. In conditions of osmotic equilibrium, $J_v = 0$ by definition, so ΔP and $\sigma\Delta\pi$ are equal and opposite.

Equation 3.5

$$J_v = L_p(\Delta P - \sigma\Delta\pi)$$

3.7.2 Membrane permeability to water can be defined with either an osmotic coefficient (P_f) or a diffusional coefficient (P_d).

Membrane permeability to water can be measured essentially in either of two ways. In the first approach, imposition of an osmotic or a hydrostatic pressure difference across the membrane can be used to generate a water flow. Techniques used for such measurements are discussed in Box 3.9. L_p is calculated from the water flux J_v (see Eq. 3.5) and can be converted to units of a permeability coefficient, P_f (Eq. 3.6), where V_w is the partial molar volume of water. The units of P_f are m s^{-1}.

Equation 3.6

$$P_f = L_p\, RT/V_w$$

The second approach to assessing water permeability relies on measuring the diffusional permeability (P_d) with isotopic water (i.e., enriched with either ^{2}H or ^{3}H). Unlike the estimate for P_f, which relates to the net flux of water through the membrane, P_d reflects the unidirectional flux. Nevertheless, if water molecules move independently through the membrane—as they would, for example, if they permeated the lipid in the bilayer—then P_d and P_f should be equal. Estimating P_d is fraught with difficulties, however, principally because the very high permeability of membranes to water means that artifacts related to lack of isotope equilibration in the unstirred layer surrounding the membrane require careful correction. Nevertheless, when estimates of P_f and P_d have been made critically on the same cell type, a difference between them becomes evident. For example, in *Chara* internodal cells, P_f has a value of about 250 μm•s^{-1}, whereas P_d is only 8 μm•s^{-1}.

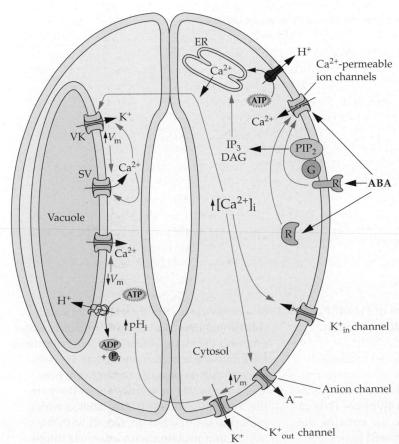

Figure 3.49
Ca^{2+} signaling coordinates the activities of multiple ion channels and H$^+$-pumps during stomatal closure. In this model, perception of ABA by a receptor (R) results in an increase in cytosolic free Ca^{2+} through Ca^{2+} influx or Ca^{2+} release from internal stores. Increased cytosolic Ca^{2+} promotes opening of plasma membrane anion and K$^+_{out}$ channels and inhibits opening of K$^+_{in}$ channels. As more ions leave the cell than enter it, water follows, turgor is lost, and the stomatal pore is closed.

Box 3.9

The osmotic coefficient for water permeability (P_f) of plant membranes can be determined.

Several techniques are available for measuring the water permeability of plant membranes. The simplest is that of **deplasmolysis,** which measures the rate of volume change of a plasmolyzed cell in response to a hypotonic solution. Although simple, this method is inaccurate because unstirred layers around the cell wall prevent rapid mixing of solutions.

A second method, applicable only to giant-celled algae, involves **transcellular osmosis** (see panel A). A cell is positioned in a grooved barrier that separates two initially identical solutions. The solution in the open chamber is then changed to a solution of higher osmolarity, and the rate of change of volume in the closed chamber is recorded as water flows through the cell from the medium of lower osmolarity. Although unstirred layers are easier to con-

trol in this method, the range of cell types for which determinations can be made obviously is restricted.

The favored method for most cells involves use of the **pressure probe** (panel B). A glass micropipette, usually filled with silicone oil, is inserted into a cell. The oil makes contact through a pressure-tight seal with a pressure transducer, which records the cell turgor. A micrometer screw is also in contact with the oil. Rapid changes in cell turgor can be elicited by turning the screw, which moves the plunger a fixed distance. The volume change in the system can be calculated, and the flux of water across the cell is calculated from the rate of relaxation of turgor. The measurements must be interpreted carefully if there are plasmodesmatal connections between cells in a tissue, be-

cause these will effectively increase the area of membrane through which water can flow.

All of these cellular methods for measuring water permeability involve, in effect, measurement of the permeabilities of the plasma membrane and tonoplast in series. Measurements of **light scattering** by purified membrane vesicles circumvent this problem. Rapid mixing devices can be used to obtain a change in solution osmotic pressure in less than 5 ms. The scattered light intensity, which is a function of vesicle size, can be monitored as the vesicles swell or shrink, and a value for the water permeability can be calculated from the rate of change in light scattering and from the vesicle size in steady-state conditions (determined by electron microscopy).

(A)

(B)

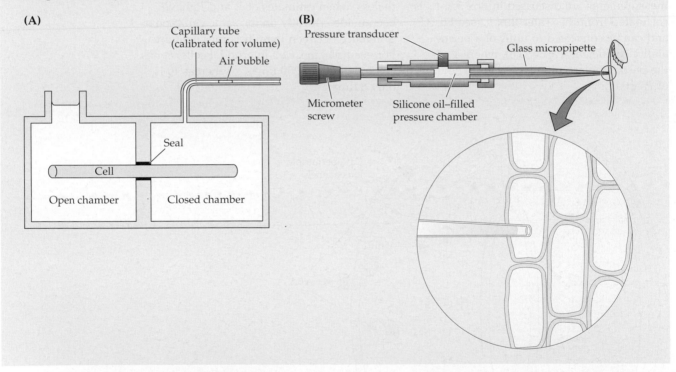

3.7.3 The nonequivalence of P_f and P_d provides evidence for water channels.

Why might P_f and P_d differ so radically? Figure 3.50 depicts one possible answer: a single-file, multiple-occupancy water channel. The estimate of P_f involves net flow of water, so each water molecule entering the channel, say, from the left, will knock out

one molecule on the right. By contrast, for diffusional flow, a molecule of labeled water entering the channel from the left can diffuse back into the solution on the left as well as continue its passage through the pore. The more water molecules are in the pore, the more the passage of the labeled water molecule is impeded. In accord with the notion that proteinaceous channels might reside in the lipid bilayer and facilitate water flow, it can be demonstrated that mercury ions (Hg^{2+}) reversibly decrease L_p.

To conclude that water flow through the lipid bilayer is not significant, however, would be wrong. Water passes relatively freely through the lipid component of biological membranes, despite the fundamentally hydrophobic characteristics of the fatty acyl chains in the phospholipid bilayer. The high permeability of bilayer membranes to water can be demonstrated in artificial systems—either liposomes or planar lipid bilayers—and results from the capacity of water to form ice-like hydrogen-bonded structures in hydrophobic environments. This lipid-mediated water flow probably accounts for most of the membrane permeability estimated as P_d.

Thus there are two parallel pathways of water movement across those membranes (Fig. 3.51). The first pathway is simple movement of water across the membrane lipid; the second involves water channel proteins.

3.7.4 Aquaporins are members of the major intrinsic protein family, which can form water channels when expressed in heterologous systems.

Molecular biological and biochemical studies confirm the presence of water channels in plants. Water channels, or aquaporins, are highly expressed in plant membranes such that they are easily visualized in membrane preparations examined by SDS-PAGE. Members of a family of transmembrane channels known as the major intrinsic protein (MIP) family, these proteins are relatively small (only 25 to 30 kDa), probably span the bilayer six times, and contain an internal repeat sequence indicative of origins from a gene duplication and fusion (Fig. 3.52). Aquaporins also are characterized by the highly conserved NPA (Asn-Pro-Ala) residues present in both the N- and C-terminal halves of the protein. In plants, aquaporins have been identified in both the vacuolar and plasma membranes, forming discrete subfamilies known as tonoplant intrinsic protein (TIP) and plasma membrane intrinsic protein (PIP), respectively.

Aquaporin function can be confirmed by expression of the cRNA in *Xenopus* oocytes. When the oocytes are then subjected to hypo-osmotic shock, the rate of swelling is considerably faster in the aquaporin-expressing oocytes. The enhanced swelling can be blocked by Hg^{2+} for some plant aquaporins, thereby strongly suggesting that aquaporin activity underlies the L_p estimates in intact plant cells.

Intriguingly, all aquaporins identified thus far are highly selective for water transport. Thus, after expression in oocytes, no increase in membrane electrical conductance

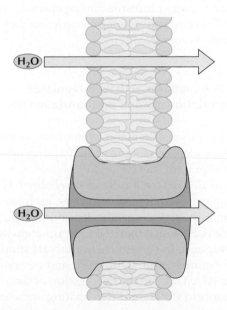

Figure 3.51
Water movement across biological membranes occurs through both the lipid bilayer and the pores formed by water channels.

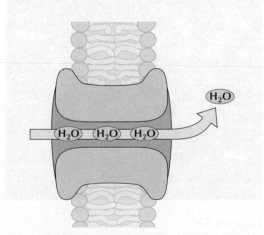

Figure 3.50
Model for water flow through a single-file, multiple-occupancy aquaporin.

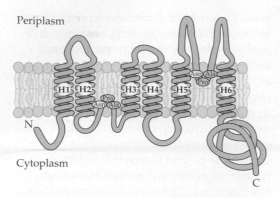

Periplasm

H1 H2 H3 H4 H5 H6

Asn Ala Pro

Pro Asn Ala

N

Cytoplasm

C

Figure 3.52
Structure of an aquaporin showing the six transmembrane helices and two conserved NPA (Asn-Pro-Ala) residues.

is observed—such as would be anticipated if water channels conducted ions. Some very limited permeation by small hydrophilic molecules such as formamide, however, can be inferred from the effects of Hg^{2+} on the estimates of σ for these molecules.

What do water channels look like in three dimensions? Modeling studies, coupled with biochemical and structural analysis of animal aquaporins (also MIP family members), have produced the predictions shown in Figure 3.53. Although aquaporins are thought to exist as tetramers, a pore is believed to be formed within each subunit, the NPA boxes being folded into the membrane between transmembrane spans 2 to 3 and 5 to 6 to provide the constriction that confers water selectivity.

3.7.5 Aquaporin activity is regulated transcriptionally and posttranslationally.

Arabidopsis thaliana, and by implication other species also, possess many aquaporin isoforms in both the TIP and PIP families, with the result that each cell type probably has aquaporin activity. Nevertheless, each isoform has a tissue-specific distribution, and there is evidence that some are up-regulated in response to certain environmental stimuli. For example, blue light, ABA, and gibberellic acid (GA_3) all enhance expression of the aquaporin PIP1b in differentiating and elongating tissues as well as in guard cells.

In addition to control of expression, aquaporin activity can be regulated by phosphorylation. For both the TIP and PIP fami-

lies, phosphorylation is catalyzed by a Ca^{2+}-dependent protein kinase that may form a component of a signaling pathway linking water stress to channel activity.

3.7.6 Plasma membrane aquaporins may play a role in facilitating transcellular water flow.

The prominence of aquaporins at a protein level alone provokes the question of why the vacuolar and plasma membranes of plants possess water channels, given that water can permeate the membrane lipid at relatively high rates. One answer might lie in the requirement for a low-resistance pathway for water flow through cells in conditions of high transpirational throughput. Thus, although the xylem presents an effective low-resistance pathway for water transport from root to shoot, the path of water to the xylem is likely to involve at least one transcellular step (i.e., through the endodermis), and facilitating the symplasmic transfer of water through other cell types in the root and leaf has the potential to increase transpiration rates beyond those possible simply through the apoplastic route. Furthermore, if water transport can be directed principally through aquaporins, then the plant has more control

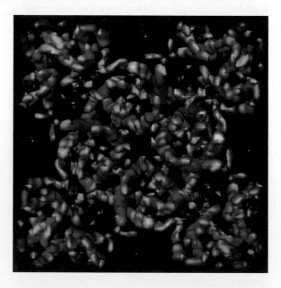

Figure 3.53
Three-dimensional structure of aquaporin-1 from human erythrocytes. Extracellular view of eight asymmetrical subunits that form two tetramers. One of the monomers of the central tetramer is colored gold.

over resistance to root-to-leaf water movement than would be possible if the principal pathways were solely the apoplast and the lipid bilayer component of membranes.

3.7.7 Differential water permeabilities of the vacuolar and plasma membranes can prevent large changes in cytoplasmic volume during water stress.

Measurements on membrane vesicles demonstrate that the water permeability (P_f) of the vacuolar membrane is about 100-fold greater than that of the plasma membrane. What could be the physiological significance of this observation? One likely possibility relates to the requirement for maintenance of cytosolic volume during osmotic stress. For a cell undergoing modest volume shrinkage in dehydrating conditions, potentially catastrophic effects on cytosolic volume could ensue if the large central vacuole failed to achieve rapid osmotic equilibrium across the tonoplast: Water would be withdrawn rapidly from the cytosol, and metabolite concentrations would be significantly perturbed. In effect, the high permeability of the tonoplast to water enables the vacuole to act as a "water buffer," enabling ready mobilization of water reserves in the vacuole, which minimizes cytosolic volume change.

Summary

Membrane transport plays a fundamental role in many biological processes in plant cells, including generation of cell turgor, energy and signal transduction, nutrient acquisition, waste product excretion, and metabolite distribution and compartmentalization. Four fundamental classes of transport system are present at all membranes. *Pumps* catalyze transport of ions or complex organic molecules against their thermodynamic gradients. At membranes other than the ATP-synthesizing membranes of mitochondria and chloroplasts, pumps are generally driven by ATP hydrolysis. At all membranes, H^+ pumps dominate the transport characteristics, removing H^+ from the cytosol and generating a pmf across each membrane. *Carriers* translocate a wide range of simple solutes, including ions, sugars, and amino acids. Carriers are distinguished from pumps by not executing scalar reactions such as ATP hydrolysis. Solute transport through carriers is generally energized through coupling of pmf-driven H^+ transport to the uphill flow of the solute. *Ion channels* are purely dissipative with respect to catalysis of transport and operate at very high turnover rates. Channels gate between open and closed states, and gating is frequently controlled either by membrane voltage or by a ligand, depending on the function of the channel. Channels that are highly selective for K^+ and Ca^{2+} reside in the vacuolar and plasma membranes. Less selective cation and anion channels also are present in both membranes, as are malate-selective channels in the tonoplast. *Aquaporins* facilitate rapid transport of water across the plasma membrane and tonoplast, bypassing an alternative pathway for water transport that involves permeation through the lipid bilayer. All classes of transport system have been identified at a molecular level; in many cases, structural aspects of the transport system can be related to solute permeation and to control of transport system activity.

Further Reading

Pumps

Coleman, J.O.D., Blake-Kalff, M. M., Davies, T.G.E. (1997) Detoxification of xenobiotics by plants: chemical modification and vacuolar compartmentation. *Trends Plant Sci.* 2: 144–151.

De Buer, B. (1997) Fusioccin—a key to multiple 14-3-3 locks? *Trends Plant Sci.* 2: 60–66.

Harper, J. F., Hong, B., Hwang, I., Guo, H. Q., Stoddard, R., Huang, J. F., Palmgren, M. G., Sze, H. (1998) A novel calmodulin-regulated Ca^{2+}-ATPase (ACA2) from *Arabidopsis* with an N-terminal autoinhibitory domain. *J. Biol. Chem.* 273: 1099–1106.

Lüttge, U., Ratajczak, R. (1997) The physiology, biochemistry and molecular biology of the plant vacuolar ATPase. *Adv. Bot. Res.* 25: 253–296.

Michelet, B., Boutry, M. (1995) The plasma membrane H^+-ATPase: a highly regulated enzyme with multiple physiological functions. *Plant Physiol.* 108: 1–6.

Palmgren, M. G. (1997) Plant pumps turned on by yeast. *Trends Plant Sci.* 2: 43–45.

Palmgren, M. G. (1998) Proton gradients and plant growth: role of the plasma membrane H^+-ATPase. In *Advances in Botanical Research,* J. A. Callow, ed. Academic Press, San Diego, CA, pp. 1–70.

Rea, P. A. (1998) From vacuolar GS-X pumps to multispecific ABC transporters. *Annu. Rev. Plant Physiol. Plant Mol. Biol.* 49: 727–760.

Zhen, R. G., Kim, E. J., Rea, P. A. (1997) The molecular and biochemical basis of pyrophosphate-energized proton translocation at the vacuolar membrane. *Adv. Bot. Res.* 25: 297–337.

Carriers

Blumwald, E., Gelli, A. (1997) Secondary inorganic ion transport at the tonoplast. *Adv. Bot. Res.* 25: 401–417.

Bush, D. R. (1993) Proton-coupled sugar and amino acid transporters in plants. *Annu. Rev. Plant Physiol. Plant Mol. Biol.* 44: 513–542.

Rentsch, D., Boorer, K. J., Frommer, W. B. (1998) Structure and function of plasma membrane amino acid, oligopeptide and sucrose transporters from higher plants. *J. Membrane Biol.* 162: 177–190.

Channels

Allen, G. J., Sanders, D. (1997) Vacuolar ion channels of higher plants. *Adv. Bot. Res.* 25: 217–252.

Maathuis, F.J.M., Ichida, A. M., Sanders, D., Schroeder, J. I. (1997) Roles of higher plant K^+ channels. *Plant Physiol.* 114: 1141–1149.

Tester, M. (1990) Tansley review no. 21. Plant ion channels whole-cell and single-channel studies. *New Phytol.* 114: 305–340.

Tyerman, S. D. (1992) Anion channels in plants. *Annu. Rev. Plant Physiol. Plant Mol. Biol.* 43: 351–373.

Ward, J. M. (1997) Patch-clamping and other molecular approaches for the study of plasma membrane transporters demystified. *Plant Physiol.* 114: 1151–1159.

Ward, J. M., Pei, Z.-M., Schroeder, J. I. (1995) Roles of ion channels in initiation of signal transduction in higher plants. *Plant Cell* 7: 833–844.

White, P. J. (1998) Calcium channels in the plasma membrane of root cells. *Ann. Bot.* 81: 173–183.

Aquaporins

Maurel, C. (1997) Aquaporins and water permeability of plant membranes. *Annu. Rev. Plant Physiol. Plant Mol. Biol.* 48: 399–42.

Biochemistry & Molecular Biology of Plants, B. Buchanan, W. Gruissem, R. Jones, Eds.
© 2000, American Society of Plant Physiologists

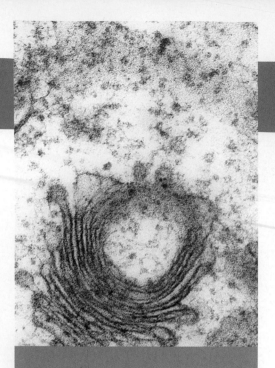

CHAPTER *4*

Protein Sorting and Vesicle Traffic

Natasha Raikhel
Maarten J. Chrispeels

Introduction

A typical plant cell contains 5000 to 10,000 different polypeptide sequences and billions of individual protein molecules. If such a cell is to function properly, it must direct these proteins to specific metabolic compartments, cytoplasmic structures, and membrane systems. Accurate protein sorting is required at all times, both when cellular structures are formed in dividing and differentiating cells, and when proteins in mature structures are degraded and replaced. Examples of the proteins that must be sorted include soluble enzymes, intrinsic membrane proteins, and structural proteins in the cell wall matrix.

Soluble enzymes are present in all subcellular compartments, including the cytosol, vacuole, cell wall, mitochondrial matrix, chloroplast stroma, thylakoid lumen, peroxisome/glyoxysome, lumen of the endoplasmic reticulum (ER), cisternae of the Golgi apparatus, and nuclear sap. Membrane-bound proteins occur in more than a dozen different lipid bilayers that delimit these compartments (e.g., the vacuolar membrane or tonoplast, the plasma membrane, ER, Golgi membranes, outer membrane of the chloroplast envelope), including membranes that lie within organelles (e.g., thylakoids). Some proteins are unique to a particular structure, compartment, or membrane. Alternatively, very similar proteins with comparable amino acid sequences, structures, and functions can occur in more than one compartment. For example, acid invertases occur in the vacuole and cell wall (see Chapter 13), and water-channel proteins (aquaporins) are found in the tonoplast and plasma membrane (see Chapter 3). Cells therefore require the necessary machinery to sort each protein and direct it to its proper destination.

4.1 The machinery of protein sorting

4.1.1 Protein sorting requires peptide address labels and protein-sorting machinery.

How do these thousands of proteins each find their way to the correct subcellular location? All proteins, except those that remain in the compartment where they are translated, include one or more **targeting domains** that act as an address label. Targeting domains are usually short peptides or amino acid motifs but can also be glycans (oligosaccharides), as in the case of mammalian lysosomal hydrolases. Targeting domains are often located at the amino-(N-)terminal end of a protein but may be present in the carboxyl (C) terminus or elsewhere in the amino acid sequence. Two isoforms of invertase with different targeting domains are shown in Figure 4.1. Specific cellular machinery interacts with this information to translocate the protein into, or retain it in, the proper compartment. Different names apply to targeting domains, depending on the organelle to which a protein is being targeted (Table 4.1).

Each compartment and membrane system requires a different targeting domain and sorting machinery. In addition, some proteins contain domains required for interaction rather than localization. Certain cytosolic structures, such as microtubules and microfilaments result from self-assembly of specific monomeric proteins. No protein sorting occurs in this process, but informational polypeptide domains that allow the monomers to form higher-order structures are required.

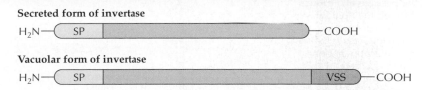

Secreted form of invertase

H_2N—(SP)——————————COOH

Vacuolar form of invertase

H_2N—(SP)—————————(VSS)—COOH

Figure 4.1
Sorting domains direct proteins to specific compartments. Certain enzymes such as acid invertase, chitinase, and glucanase exist in two forms: One resides in the vacuole and the other is secreted. The secreted form carries a single targeting domain at its N terminus, a signal peptide (SP) for entry into the ER. The vacuolar form has two targeting domains, a signal peptide at the N terminus and a vacuolar sorting signal (VSS) at the C terminus.

Although the targeting domain is essential for protein transport, it may not be part of the active protein, and proteases in the target location often remove the targeting domain to create a functional, mature polypeptide.

4.1.2 To reach its destination, a protein often has to cross at least one membrane.

Most proteins have hydrophilic surfaces and therefore do not readily pass through the hydrophobic core of the lipid bilayer. Translocation through a membrane involves a proteinaceous pore or channel through which a protein passes, not in its globular form but in an extended or unfolded configuration (Fig. 4.2).

As a polypeptide passes through a pore, it is assisted by molecular **chaperones,** proteins that bind the polypeptide during the folding and assembly process (see Chapter 9). By inhibiting molecular interactions that may cause polypeptides to fold incorrectly and aggregate, chaperones increase the yield of correct tertiary structures but not the rate of protein folding. Some cytosolic chaperones interact with newly synthesized proteins, keeping them unfolded so they can pass through a protein pore to an appropriate compartment or membrane. Other

Table 4.1 Peptide targeting domains for transport to different organelles

Organelle	Address label (targeting domain)
ER	Signal peptide (SP)
Chloroplast	Transit peptide
Mitochondrion	Presequence
Nucleus	Nuclear localization signal (NLS)
Peroxisome	Peroxisome targeting signal (PTS)
Vacuole	Vacuolar sorting signal (VSS)

Figure 4.2
Chaperones facilitate the passage of a protein through the lipid bilayer. Chaperones bind a polypeptide as it makes its way through a proteinaceous pore in a lipid membrane. The chaperones keep the polypeptide in an unfolded state on one side of the membrane and help it fold correctly on the other side.

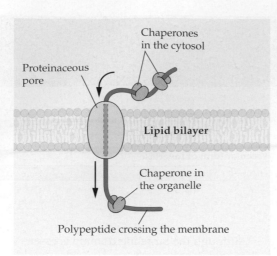

chaperones bind to the amino acid chain as it emerges from the membrane and facilitate folding. Still others function as repair stations to correct minor misfolding. The many members of one family of chaperones—the 70-kDa heat shock proteins (Hsp70)—fill all these roles, interacting with a wide spectrum of proteins. Hsp70 chaperones are present in all cells, and the synthesis of many Hsps is up-regulated in response to heat stress, perhaps to prevent misfolding and to repair misfolded proteins. As will be discussed later in this chapter, proteins such as Hsp70 perform several different functions in protein transport across membranes, in addition to translocation and folding.

Nearly all of a cell's proteins are encoded by nuclear DNA and synthesized by cytosolic ribosomes, which may or may not be attached to the ER. However, proteins encoded by mitochondrial and chloroplast DNA (approximately 100 or so) are synthesized by ribosomes within these organelles and incorporated directly into the organelle compartments.

4.1.3 Protein sorting can be a multistep process requiring more than one targeting domain.

The first sorting event for all proteins made in the cytosol separates them into two groups. Proteins in the first group are released in the cytosol and may either remain in that compartment or be targeted to a variety of destinations, including plastids, mitochondria, peroxisomes, and nuclei (Fig. 4.3A).

By contrast, proteins in the second group are targeted to the ER by **signal peptides** located in the N terminus. Translation of a signal peptide causes the ribosome to bind to the ER during protein synthesis. ER studded with such ribosomes is referred to as rough ER. Proteins synthesized on the rough ER enter the **secretory pathway,** an intracellular system of vesicles and cisternae (flattened sacs) that includes the ER, Golgi complex, tonoplast, and plasma membrane. These proteins can then be secreted or targeted to the various compartments of the secretory system (Fig. 4.3B). Integral membrane proteins with signal peptides do not enter the ER lumen but are integrated into the ER membrane, from which they can translocate to other membranes of the secretory system. Translocation to a final destination or retention in a compartment usually requires a second targeting domain (see Fig. 4.1).

The transport of both soluble and membrane proteins is mediated by vesicles that carry proteins from one compartment to the next. Thus, vesicles are constantly being formed, transported, and fused with a compartment different from the one in which they originated. Such transport implies that vesicles can recognize their destination according to the cargo they carry. This specificity of recognition is part of the sorting machinery. As polypeptide chains move through the secretory system, they first can be modified by the attachment of N-linked glycans, then folded correctly, and assembled into oligomers—dimers, trimers, tetramers, and so forth. These events are catalyzed by proteins that reside in the ER.

Ribosomes, when synthesizing proteins that lack a signal peptide, do not associate with the ER. These cytosol-localized ribosomes are described as "free," although they are often attached to the cytoskeleton. Both soluble proteins and proteins that later become incorporated into membranes can be translated by free ribosomes.

Although details vary, depending on the protein being transported, the general features of protein sorting appear to apply to any organelle. Most transported proteins carry a targeting domain that is recognized by a cytosolic factor, often a chaperone. The chaperone delivers the unfolded protein to a specific receptor on the target membrane. The targeting domain typically binds to the

Box 4.1

Targeting signals have been analyzed using isolated organelles, permeabilized cells, and transgenic plants.

Different approaches have been employed to study protein targeting signals. Some targeting signals, such as those for targeting proteins to chloroplasts, mitochondria, and peroxisomes, can be studied in vitro by incubating the isolated and purified organelles with precursor proteins and determining the conditions necessary for protein uptake into the organelle. This approach makes it possible to determine which cytosolic proteins and small molecules (e.g., ATP or GTP) are necessary for import. Instead of using isolated organelles, it is also possible to incubate precursor proteins with permeabilized protoplasts. For such experiments, protoplasts are first centrifuged to remove vacuoles and then given a mild osmotic shock, which causes them to leak. The small molecules and most of the cytosolic proteins diffuse out, but the organelles remain inside and intact. This approach has been used successfully to study nuclear import (see Section 4.4.3).

The secretory system is not a single organelle, but a complex system of membranes that cannot be reconstituted in vitro with isolated organelles. Although studies using permeabilized protoplasts would be feasible, researchers have relied on transient expression systems or stable transformants to study protein targeting in the secretory system. Transgenic plants are constructed by using the natural gene transfer system of *Agrobacterium tumefaciens,* a pathogenic bacterium that can move a DNA segment from its Ti plasmid into the genome of a host plant (see Chapter 6). By replacing these transferred bacterial genes with the coding region of a gene of interest, flanked by appropriate regulatory sequences, it is possible to transform plants or cultured cells with a chimeric gene suitable for protein targeting studies. Immunocytochemistry or organelle fractionation can then be used to determine the subcellular compartment(s) in which the protein accumulates when cells or tissues express the transgene (see Box 4.3). Investigators determined early on that if a protein accumulates in a given organelle of one plant species, it will accumulate in

the same organelle in another species. Since targeting signals are conserved between plant species, it is possible to delete or mutate a putative targeting signal and then examine the effect of this mutation on the subcellular location of the protein in the transgenic plant. Such experiments have led to the conclusion that the glycans of vacuolar proteins contain no targeting information (see Section 4.6.2), and that the addition of a signal peptide (SP) to a cytosolic protein leads to the secretion of that protein (see Section 4.5.3). Removing a vacuolar domain from a vacuole-sorting protein also leads to its secretion.

In one targeting experiment, the coding sequence for phytohemagglutinin, a vacuolar protein of bean seeds, was elongated

by the addition of nucleotides that encode the amino acids Lys-Asp-Glu-Leu (KDEL) at the C terminus (see Section 4.5.7). Tobacco was transformed with a construct consisting of this chimeric gene and a strong seed-specific promoter, and immunocytochemical techniques were used to visualize phytohemagglutinin in the seeds of the transformed tobacco plants. In the absence of the KDEL targeting domain, all the phytohemagglutinin is transported to the protein storage vacuoles (this control is not shown). However, when the KDEL tetrapeptide is present, phytohemagglutinin accumulates in the ER and the nuclear envelope, as shown by the labeling with electron-dense gold particles (see figure below).

Phytohemagglutinin

H_2N—(SP)————————————————————)—COOH

Phytohemagglutinin modified with C-terminal KDEL

H_2N—(SP)——————————————(KDEL)—COOH

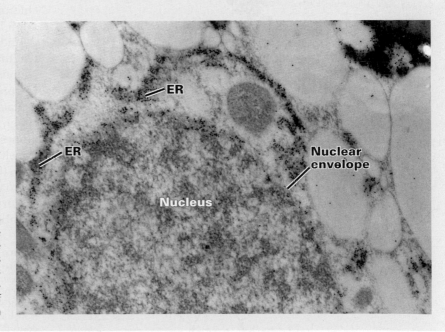

receptor, opening a transmembrane channel within the membrane. Such channels usually consist of several polypeptides and are coupled to a peripheral protein translocation motor that hydrolyzes nucleoside triphosphates to drive protein transport. After passing through the membrane, the transported protein folds with the help of chaperones and ATP.

4.2 Targeting proteins to the plastids

4.2.1 Transport of proteins into chloroplasts involves a removable transit peptide.

Chloroplasts, found in leaves and other green organs of plants and in algae, belong to a family of closely related organelles called plastids (see Chapter 1). Although plastids contain their own DNA and ribosomes, most of their proteins are encoded in nuclear DNA and imported from the cytoplasm. The contents of a plastid are separated from the cytoplasm by two distinct membranes, the outer and inner membranes of the chloroplast envelope. In addition, chloroplasts have an internal membrane system consisting of closed flattened sacs (thylakoids) that lie stacked on one another. The three membranes define three aqueous compartments: the intermembrane space in the envelope, the stroma within the inner membrane, and the thylakoid lumen (Fig. 4.4). Each aqueous compartment and each membrane contain unique proteins. Thus, protein targeting into the chloroplast operates at two levels of complexity: Proteins must be directed not only into the plastid, but also to the proper location within the organelle.

Chloroplast proteins encoded in nuclear DNA are synthesized in the cytosol by free ribosomes. These proteins are translated as precursors with an N-terminal **transit peptide** of 40–50 amino acids that targets the polypeptides to the chloroplast and further enables their translocation across the chloroplast envelope into the stroma. After their translocation through the chloroplast envelope, a peptidase removes the transit peptides of stromal precursor proteins. Transit peptides are both necessary and sufficient for chloroplast import: Proteins that lack a transit peptide cannot be imported, and if the transit peptide is added to the N terminus of a protein that is foreign to the chloroplast, this chimeric precursor protein is imported. The import of chloroplast precursors occurs at contact sites (proteinaceous channels) between the outer and inner envelope membranes, and precursor proteins that are in transit span both envelope membranes. To date, there is no consensus as to whether the transit peptide interacts initially with the receptor proteins or with the lipids of the outer membrane; possibly, both interactions occur.

(A) Free ribosomes in cytosol

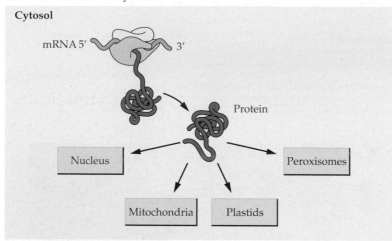

(B) Membrane-bound ribosomes

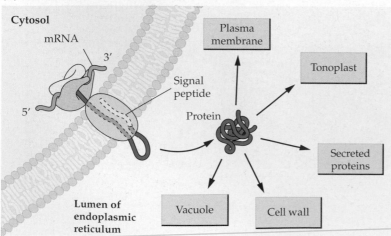

Figure 4.3
(A) Proteins synthesized on free ribosomes either remain in the cytosol or are targeted to the nucleus, mitochondria, plastids, or peroxisomes. (B) Proteins synthesized on membrane-bound ribosomes are first translocated into the lumen of the ER and transported to the Golgi. These proteins may subsequently be targeted to the plasma membrane or the tonoplast, secreted or sent to the vacuole. Some proteins remain in the ER or the Golgi because they have special retention signals.

4.2.2 To enter chloroplasts, proteins pass through a proteinaceous channel with the aid of molecular chaperones.

Import into chloroplasts can be studied by reconstituting the process in a cell-free system (Fig. 4.5A) or in transgenic plants. In the cell-free system approach, chloroplasts are first purified from a tissue homogenate and then incubated with radioactively labeled precursor proteins (obtained by in vitro translation of mRNA in the presence of radioactive amino acids). The precursor proteins first bind to the chloroplasts, which can then take up the labeled polypeptides efficiently. By varying conditions during the incubation, we can determine the efficiency (Fig. 4.5B) and biochemical requirements for protein import into chloroplasts.

Researchers studying this cell-free system have proposed a model for chloroplast protein import that entails two functional steps and requires several more proteins in addition to the transit peptide of the transported species. The process requires chaperones on both sides of the chloroplast envelope and a group of proteins, collectively called the protein import apparatus, that span both the outer and inner membranes and come into contact at the point of entry. Both steps require energy in the form of nucleoside triphosphates.

In step one, cytosolic chaperones (Hsp70 homologs) hold proteins in an unfolded or partially folded state. Transit peptides on these proteins interact with the lipids of the outer membrane or with the proteins of the import apparatus, which form the proteinaceous channels through which the imported polypeptides pass (Fig. 4.6). During the late 1990s, researchers identified several of these import apparatus proteins and obtained the

cDNAs encoding them. These components have been named Toc (translocon of the outer envelope membrane of the chloroplast) and Tic (translocon of the inner envelope membrane). Two of these proteins, Toc34 and Toc159, are specific GTP-binding proteins tightly anchored in the outer membrane with their GTP-binding domains exposed to the cytosol. A third protein, Toc75, shows no obvious similarity to the proteins for which the functions have been identified; however, this component is also deeply embedded in the outer membrane. A fourth protein, Hsp70 IAP (import intermediate-associated protein), a homolog of the Hsp70 molecular chaperone, exhibits the biochemical characteristics of an integral membrane protein. The precise interactions of these proteins and their involvement in the transport process are not yet understood. Only one component of the inner envelope translocation machinery, Tic110, has been identified to date.

In step two, the protein enters the chloroplast stroma, where a protease removes the transit peptide. Once inside the chloroplast, the polypeptide has several possible fates. If

Figure 4.4
Biosynthesis of chloroplast proteins and their targeting to five different compartments within the chloroplast. Chloroplast proteins may be encoded by nuclear DNA (nDNA) or chloroplast DNA (ctDNA); the respective mRNAs are translated by ribosomes in the cytosol (80S ribosomes) or in the chloroplast stroma (70S ribosomes). Proteins made as precursor polypeptides in the cytosol may be targeted to the outer envelope membranes or may enter the chloroplast stroma. Once across the envelope membranes, proteins may remain in the stroma compartment or may be targeted to the thylakoid membrane, thylakoid lumen, or inner envelope membrane.

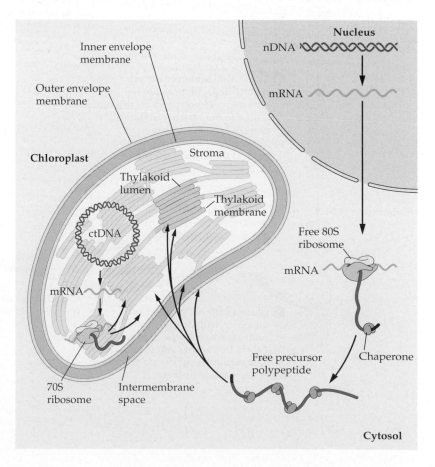

(A) A chloroplast import experiment

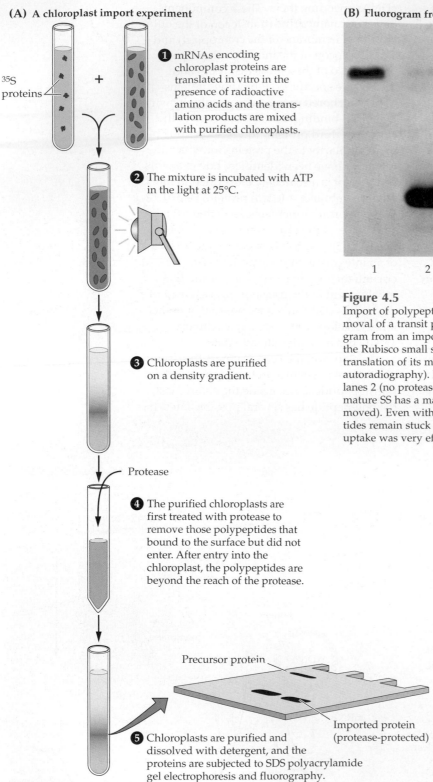

^{35}S proteins

1 mRNAs encoding chloroplast proteins are translated in vitro in the presence of radioactive amino acids and the translation products are mixed with purified chloroplasts.

2 The mixture is incubated with ATP in the light at 25°C.

3 Chloroplasts are purified on a density gradient.

Protease

4 The purified chloroplasts are first treated with protease to remove those polypeptides that bound to the surface but did not enter. After entry into the chloroplast, the polypeptides are beyond the reach of the protease.

Precursor protein

Imported protein (protease-protected)

5 Chloroplasts are purified and dissolved with detergent, and the proteins are subjected to SDS polyacrylamide gel electrophoresis and fluorography. This method separates polypeptides on the basis of their size.

(B) Fluorogram from an import experiment

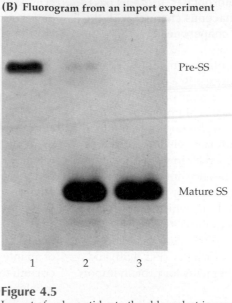

Pre-SS

Mature SS

1   2   3

Figure 4.5

Import of polypeptides to the chloroplast is accompanied by removal of a transit peptide. (A) An import experiment. (B) Fluorogram from an import experiment. The radioactive polypeptide of the Rubisco small subunit polypeptide (pre-SS), made by in vitro translation of its mRNA, is visualized by fluorography (a type of autoradiography). In lane 1, the pre-SS has a mass of 23 kDa. In lanes 2 (no protease treatment) and 3 (after protease treatment), mature SS has a mass of 18 kDa (the transit peptide has been removed). Even without protease treatment, no full-length polypeptides remain stuck to the chloroplasts (lane 2), which suggests that uptake was very efficient under these conditions.

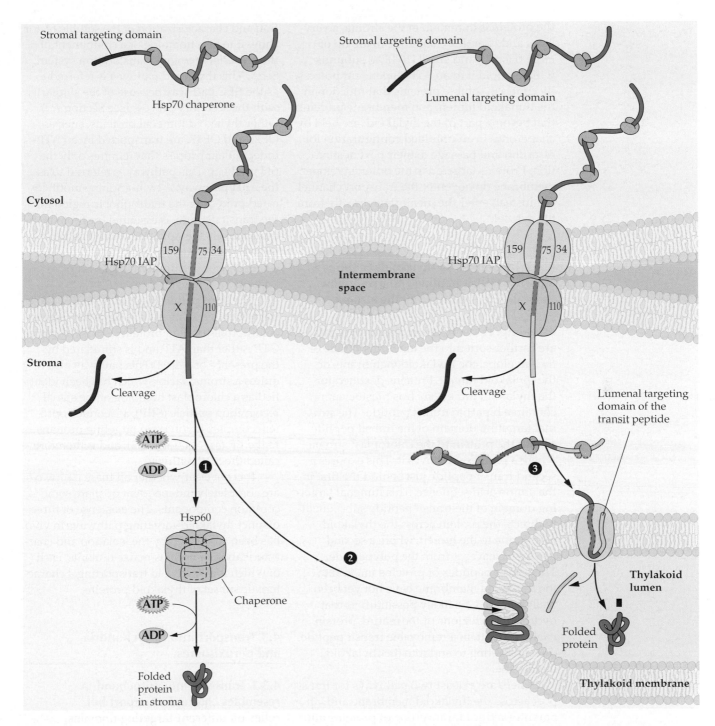

Figure 4.6

Mechanism of protein import into chloroplasts. At top of figure, chaperones in the cytosol hold a protein with its transit peptide exposed in an unfolded configuration. The protein on the left has a stromal targeting domain only (red), whereas the protein on the right has a bipartite transit peptide with both stromal and lumenal (yellow) domains. Whether these two types of proteins enter through distinct types of pores or through the same pore is not known. The transit peptide binds first to the lipids and proteins of the outer membrane of the chloroplast envelope and is then engaged by the import apparatus that forms the proteinaceous pore. The apparatus consists of outer membrane proteins (Toc159, Toc75, and Toc34) and inner membrane proteins (X and Tic110). The composition of this complex may be dynamic, changing during the translocation process. After translocation to the inside, the protein can have three different destinations. Proteins that remain in the stroma (e.g., Rubisco) lose their transit peptides and are folded and assembled in a process that requires ATP and a large chaperone containing Hsp60 (Path 1). Proteins that function in the thylakoid membrane are engaged by a chloroplast signal recognition particle (SRP, not shown); insertion into the thylakoid membrane requires GTP (Path 2). Proteins destined for the lumen of the thylakoid interact with one of two different import complexes in the thylakoid membrane; one is shown in the diagram. These proteins lose their lumenal transit domain after passage into the thylakoid (Path 3). Depending on the protein being transported, this translocation process requires only a pH differential or a pH differential and ATP.

the protein is to remain in the stroma, a very large and complex chaperone made up of large (Hsp60) and small (10 kDa) subunits will help fold it in an ATP-dependent process. By contrast, soluble proteins that function in the thylakoid lumen, and membrane proteins that become part of the thylakoid, are held by chaperones in an unfolded configuration for insertion and passage to their next destination. Proteins targeted to the outer envelope membrane do not enter the transport channel but instead enter the membrane directly from the cytosol.

4.2.3 Targeting into thylakoids requires a bipartite transit peptide and may follow three different paths from the stroma.

After import, many chloroplast polypeptides are further sorted to the inner membrane of the envelope, the thylakoid membrane, or the thylakoid lumen. Proteins destined for the thylakoid lumen, such as plastocyanin, contain a bipartite transit peptide. The stromal targeting domain of the transit peptide guides the protein to the chloroplast stroma, where a protease removes it. This exposes a second transit peptide just behind the first in the amino acid sequence. This lumenal targeting domain of the transit peptide subsequently directs the protein across the thylakoid membrane to the lumen, where a second protease cleaves it from the polypeptide. The transit peptides of proteins integral to the thylakoid membrane have not yet been well defined. One likely possibility is that hydrophobic regions of the mature protein itself, rather than a removable transit peptide, direct targeting to and into the thylakoid membrane.

There are at least two pathways for transport across the thylakoid membrane and, apparently, a third for insertion of proteins into the thylakoid membrane. Evidence for the existence of these pathways was obtained by analysis of energy requirements for import, by precursor competition studies, and by identification of translocation components.

The translocation of lumenal proteins, such as OE33 (a component of the oxygen-evolving complex) and plastocyanin, requires ATP and a soluble protein and is stimulated by a pH difference (ΔpH) between the chloroplast stroma and the thylakoid lumen. Isola-

tion and characterization of the soluble factor show it to be a homolog of a component of the bacterial secretory translocation system, SecA. This thylakoid pathway is referred to as the **SEC pathway** because of the similarity with the secretory process (see Section 4.5). Other thylakoid lumenal proteins, such as OE23 and OE17, are transported by an ATP-independent process that requires only the pH gradient. This pathway is referred to as the Δ**pH pathway**. A twin-arginine motif, located preceding the hydrophobic region in the transit peptide, is essential for protein transport via the ΔpH pathway. Whether these two pathways use the same proteinaceous pore or use different pores is not known.

The third pathway, for integral thylakoid membrane proteins such as light-harvesting chlorophyll-binding protein (LHCP), requires GTP rather than ATP and is stimulated by the presence of a ΔpH. This pathway also requires a stromal factor, which has been identified as a chloroplast homolog of the signal recognition particle (SRP), a ribonucleoprotein complex involved in targeting proteins to the ER (see Section 4.5.3), and is therefore called the **SRP pathway.**

It is not clear whether all these pathways are completely independent or share some common components. The existence of three distinct thylakoid-targeting pathways in vivo has been confirmed by the isolation and characterization of various maize mutants, each of which is defective in transporting a characteristic subset of thylakoid proteins.

4.3 Transport into mitochondria and peroxisomes

4.3.1 Transport into mitochondria resembles chloroplast import but relies on different targeting domains, called presequences, and a different import apparatus.

Mitochondria are present in nearly all eukaryotic cells and house the machinery of cellular respiration (see Chapter 14). Like chloroplasts, they have both an outer membrane (OM) and an inner membrane (IM). These two membranous compartments define two aqueous compartments: the intermembrane space and the mitochondrial matrix (Fig. 4.7A). The inner membrane is highly convoluted but continuous, and the two aqueous spaces are not in contact with each other. As in other organelles, each membrane and each aqueous compartment contain unique proteins. The enzymes of the electron transport chain lie in the inner membrane, whereas glycine decarboxylase and most of the enzymes of the citric acid cycle are found in the matrix. Like chloroplasts, mitochondria have their own DNA, RNA, and protein synthesis machinery. Nevertheless, the majority of mitochondrial proteins (hundreds) must be imported from the cytosol because they are encoded in nuclear DNA and translated in the cytosol.

Protein transport into mitochondria has been studied most extensively in the yeast *Saccharomyces cerevisiae* and in the bread mold *Neurospora crassa*. Less work has been done in plants, but mitochondrial import is probably very similar in plants and fungi as well as other eukaryotes. Most nuclear-encoded mitochondrial proteins are synthesized as precursors with an N-terminal targeting domain, called a **presequence,** that facilitates protein translocation across both mitochondrial membranes. Despite the superficial resemblance between chloroplast transit peptides and mitochondrial presequences, each is quite specific in its function. Presequences do not target chloroplast proteins to chloroplasts, and transit peptides do not target proteins to mitochondria. Analysis of mitochondrial presequences indicates that many can form positively charged amphipathic α-helices. An amphipathic helix has hydrophobic amino acid residues on one face of the helix and charged residues on the other (Fig. 4.7B and C). After a protein enters a mitochondrion, the presequence is cleaved by an endopeptidase, just as in the chloroplast.

As in chloroplasts, proteins are transported to the mitochondria in an unfolded state with the help of a chaperone, cytosolic Hsp70. The import process is mediated by a protein import apparatus that spans both inner and outer membranes at a point of contact. Once the protein is in the matrix, folding is catalyzed by another chaperone, Hsp60. Many genes that encode proteins necessary for import into mitochondria have been cloned. These proteins are termed TOMs and TIMs (translocases of outer and inner membranes, respectively).

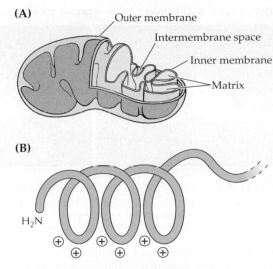

(A)
Outer membrane
Intermembrane space
Inner membrane
Matrix

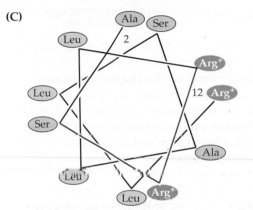

(B)

(C)

Figure 4.7
(A) The four compartments of a mitochondrion. (B) Schematic representation of an amphipathic polypeptide helix. (C) Two-dimensional projection of amino acids 2 through 12 of the presequence of the β-subunit of ATPase from *Nicotiana plumbaginifolia*. The hydrophobic amino acids (Ala, Leu) are almost entirely on one side of the helix, the charged amino acids (Arg) on the other.

Several features distinguish mitochondrial from chloroplast protein transport. Mitochondrial import machinery proteins do not share sequence identity with those of the chloroplast. Furthermore, transport into the mitochondrial matrix requires an electrochemical potential across the inner membrane, in addition to ATP. These differences have prompted researchers to speculate on the existence of mechanisms at the surface of these organelles that contribute to the fidelity of protein sorting between mitochondria and chloroplasts.

Again, as in chloroplasts, transport of some proteins into the mitochondrial inner membrane and the intermembrane space requires two signals. Such proteins possess a hydrophobic amino acid sequence located immediately after the amino-terminal presequence. When the presequence is cleaved by a matrix peptidase, the newly exposed hydrophobic sequence functions as a presequence to guide the protein from the matrix into or across the inner membrane. After

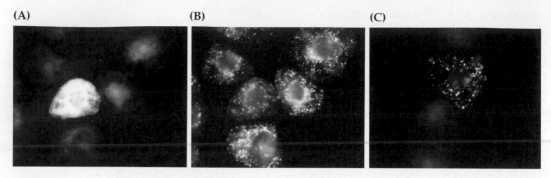

Figure 4.8

The peroxisome targeting signal Ser-Lys-Leu (SKL) directs a bacterial enzyme to peroxisomes. The tobacco cells are viewed with a fluorescence microscope. Proteins are flagged with antibodies coupled to a fluorescent dye. By using specific antibodies, different proteins can be visualized. The cells have been transformed with a construct that includes (A) the bacterial gene for chloramphenicol acetyltransferase (CAT) and (C) CAT with the signal SKL added to the C terminus; (B) antibodies against the endogenous protein catalase have been used. In panel (A), fluorescence is distributed throughout the cell, indicating that the CAT enzyme is cytoplasmic, whereas the staining pattern for CAT-SKL is punctate in panel (C). A similar punctate pattern, visible in panel (B), shows the presence of catalase, a peroxisomal enzyme.

proteins destined for the intermembrane space have been inserted into the inner membrane, this second presequence is cleaved in the intermembrane space by another peptidase, releasing the mature polypeptide as a soluble protein.

4.3.2 Uptake of proteins by peroxisomes involves removable or intrinsic peroxisome targeting signals.

Peroxisomes are specialized organelles that participate in many essential metabolic pathways and are found in all eukaryotes (see Chapter 1). Plants contain at least three types of peroxisomes. Young seedlings and senescent leaves contain glyoxysomes, which function in the mobilization of stored triglycerides (see Chapters 1, 10, and 14). Second, peroxisomes in leaves play a major role in the photorespiration reactions that accompany carbon dioxide fixation in C_3 plants (see Chapters 1 and 14). Third, root nodules of some tropical legumes that export ureides contain peroxisomes with unique enzymes involved in nitrogen metabolism (see Chapters 1 and 16).

In contrast to chloroplasts and mitochondria, peroxisomes are surrounded by a single membrane and do not contain DNA or ribosomes; as a consequence, all peroxisomal proteins must be synthesized by cytosolic ribosomes and imported from the cytoplasm. At least two distinct peroxisome targeting signals, PTS1 and PTS2, are involved in the import of proteins into the peroxisomal matrix.

The tripeptide PTS1 (Ser-Lys-Leu, SKL) or a closely related variant targets most proteins destined for the peroxisomal matrix. Differing from other targeting domains, this short sequence is located at the C terminus of many peroxisomal proteins and is not removed after translocation into the peroxisome. Deletion of this C-terminal signal abolishes import of nearly all peroxisomal proteins that carry it. In a few proteins, import is not abolished, suggesting that additional peroxisome targeting signals may be present within the protein. PTS2 consists of a cleavable N-terminal targeting domain and is used by a subset of peroxisomal matrix proteins. Passenger proteins fused to PTS-like sequences are directed correctly into peroxisomes (Fig. 4.8).

Protein import to peroxisomes is known to require not only a peroxisomal targeting signal but also energy in the form of ATP. Although the transport of some PTS1-containing proteins requires cytosolic factors, such as members of the Hsp70 family, it is not known whether chaperones are needed for import of all unfolded proteins and oligomers, and the relationship between chaperones and oligomer import is an active area of research. A candidate for a PTS1 receptor was first identified in yeast through a screen for peroxisome assembly mutants. The yeast mutant

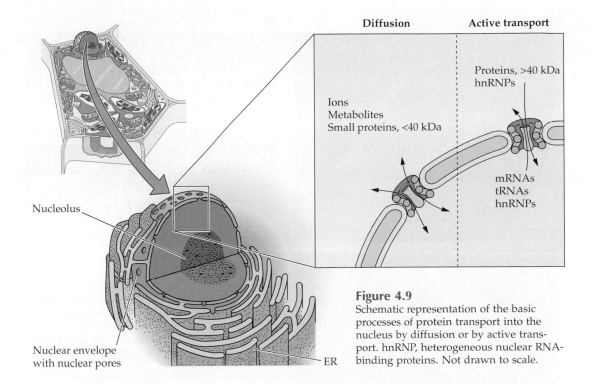

Diffusion

Active transport

Ions
Metabolites
Small proteins, <40 kDa

Proteins, >40 kDa
hnRNPs

mRNAs
tRNAs
hnRNPs

Nucleolus

Nuclear envelope
with nuclear pores

ER

Figure 4.9
Schematic representation of the basic processes of protein transport into the nucleus by diffusion or by active transport. hnRNP, heterogeneous nuclear RNA-binding proteins. Not drawn to scale.

pas8 is unable to import a PTS1-containing polypeptide into its peroxisomes but is fully competent to import a PTS2-containing polypeptide. A functional homolog of the *PAS8* gene, one that can complement the yeast mutant, has been cloned from human cells, which suggests that transport has been evolutionarily conserved. Another yeast mutant is deficient only in the PTS2 import pathway. Genes representing mutations in both pathways have been cloned from yeast, but we do not yet know the function of the proteins they encode.

4.4 Transport in and out of the nucleus

4.4.1 The nuclear pore is the site for macromolecular movement into and out of the nucleus.

A wide variety of macromolecules, including regulatory proteins, histones, RNA polymerases, and heterogeneous nuclear RNA-binding proteins, are synthesized in the cytoplasm and then selectively transported into the nucleus (Fig. 4.9). In some cases, DNA from plant pathogenic bacteria and viruses is also selectively imported from the cytoplasm into the nucleus. Concurrently, tRNAs and mRNAs are synthesized, modified, and pro-

cessed in the nucleus and exported to the cytoplasm. Sometimes the transport process can be even more complex: Ribosomal proteins are produced in the cytoplasm, imported into the nucleus, assembled with RNA into ribosomal particles, and then exported to the cytoplasm as part of a ribosomal subunit. The nuclear envelope, which regulates these transport activities, comprises three structural elements: the perinuclear compartment, consisting of the outer and inner membranes and the intermembrane space; nuclear pore complexes; and the nuclear lamina. The outer nuclear membrane is continuous with the ER and, like rough ER, is studded with ribosomes.

Pores in the nuclear envelope are formed by nuclear pore complexes, located where the inner and outer nuclear membranes meet. Each nucleus has hundreds of nuclear pore complexes, which can easily be seen in grazing sections that provide a surface view of the nuclear envelope (Fig. 4.10A). Proteins and RNA move in and out of the nucleus through these pores. The morphology of nuclear pore complexes has been determined in detail by electron microscopy. These massive (124 MDa) protein complexes are composed of 100 or more polypeptides and possess octagonal radial symmetry (Fig. 4.10B). Aside from certain repeat motifs, few similarities shared by these proteins have been identified.

(A)

(B)

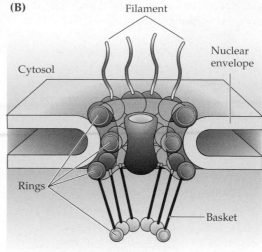

Filament

Cytosol

Nuclear envelope

Rings

Basket

Figure 4.10
(A) Two images of the surface of the nuclear envelope obtained by freeze-fracture and freeze-etching of a specimen viewed with the electron microscope. A few of the numerous nuclear pores are indicated by slanted lines. Bars are 0.5 μm. (B) Schematic representation of a nuclear pore complex. The structure extends approximately 100 nm into the nucleoplasm.

Box 4.2   **Nuclear pore complexes contain glycoproteins.**

The best evidence to date for the role of nuclear pore complex proteins in nuclear import derives from studies with wheat germ agglutinin, a lectin that blocks protein import into vertebrate nuclei. Some vertebrate nuclear pore proteins are glycoproteins containing single O-linked *N*-acetylglucosamine (O-GlcNAc) residues, the site of wheat germ agglutinin binding. The GlcNAc moiety itself is not essential for import, but wheat germ agglutinin is thought to block import by simply occluding the pore, as shown in the figure to the right. Nuclear pore complexes can be isolated from animal cells, dissolved into their individual components and then reconstituted. If GlcNAc-containing proteins are first removed by passing the protein mixture over a wheat germ agglutinin affinity column, the pore complexes reconstituted from the remaining proteins are inactive, suggesting that the glycoproteins are essential for transport.

Import of nuclear proteins into plant nuclei is not blocked by wheat germ agglutinin, although this lectin blocks the uptake of large protein complexes (e.g.,

antibody-antigen complexes) or proteinsingle stranded DNA complexes. Electron microscopic studies of pore complexes in plants have shown them to resemble those described in other organisms. However, analysis of the carbohydrate structure of plant pore complex glycoproteins reveals that unlike vertebrate proteins, which contain single O-linked GlcNAc residues, plant proteins contain an O-linked oligosaccharide composed of five or more sugar residues, including a terminal GlcNAc. Perhaps the wheat germ agglutinin attached to this longer oligosaccharide chain does not occlude the pore sufficiently to prevent the entry of nuclear proteins (see figure below).

Mammalian nuclear pore complex

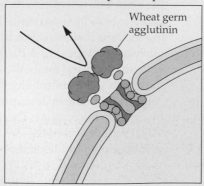

Wheat germ agglutinin

Plant nuclear pore complex

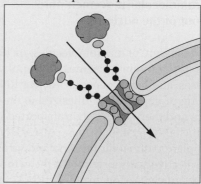

4.4.2 Nuclear localization signals target proteins to the nucleus.

The nuclear pore complex constitutes a passive diffusion channel approximately 9 nm in diameter. Small proteins of less than 40 kDa can diffuse passively through this pore, whereas larger proteins enter the nucleus through active transport. However, even small nuclear proteins such as the 20-kDa histones usually enter the nucleus by active transport rather than by diffusion.

Most polypeptides destined for the nucleus have address labels, called **nuclear localization signals** (NLSs). NLSs are functionally similar to the other targeting domains discussed so far but consist of one or more short, internal sequences with basic amino acids. The classical NLSs share certain characteristics, but there is no strict consensus sequence (Table 4.2). They generally contain several residues of arginine and lysine and may also have residues such as proline that disrupt helical domains. The position of the NLS within the amino acid sequence of a protein varies from one protein to another. Unlike targeting domains that are removed during the translocation of soluble proteins into other organelles, NLSs are not cleaved from nuclear proteins. As a result the proteins can reenter the nucleus if they are exported into the cytosol or released when the nuclear envelope breaks down during mitosis. Many nuclear proteins require more than one NLS for efficient nuclear targeting in vivo, which suggests that protein structure may play an important role in NLS presentation and in the ability of independent NLSs to function in a cooperative manner.

Virtually all classical NLSs may be categorized into one of three classes, all of which have been described in plants. Many viral proteins have NLSs, and the NLS of the large T-antigen of the simian virus SV40 has been studied in greatest detail. This NLS has been functionally defined as Pro-Lys-Lys-Lys-Arg-Lys-Val. A mutation that causes a single amino acid change at the third position (Pro-Lys-Thr-Lys-Arg-Lys-Val) abolishes NLS function and prevents nuclear import. A second class of NLS, designated bipartite, is typified by nucleoplasmin, a protein of the African clawed toad *Xenopus laevis*. The nucleoplasmin signal consists of two peptide regions that contain basic residues and are separated by a spacer of 10 or more residues. Many bipartite NLSs permit a spacer of variable length. Mutations in either basic region alone do not affect function, whereas mutations in both basic regions significantly impair nuclear targeting. The third class of NLS is typified by the N-terminal signal of the yeast protein Matα2, Lys-Ile-Pro-Ile-Lys. Here, the basic residues are separated by three hydrophobic residues. Examples from each of the three classes of NLSs are present in plant nuclear proteins (Table 4.2).

4.4.3 Nuclear import can be studied both in vivo and in vitro.

Nuclear import is studied in vivo by using DNA constructs that encode a reporter enzyme fused to an NLS. The construct is introduced into living tissues, and the reporter gene product is visualized. Onion epidermis is particularly well suited for such work (Fig. 4.11 A and B). SV40-like, bipartite, and Matα2-like signals are sufficient to independently direct the reporter enzyme β-glucuronidase (GUS) to the nucleus of onion epidermal cells.

The import of nuclear proteins through the nuclear pore complex requires two steps: binding and translocation. Using an in vitro system consisting of either permeabilized protoplasts or purified nuclei, researchers can distinguish between these two steps (Fig. 4.11 C and D). Both steps require several cytosolic factors. The first step, binding of NLS-containing protein domains to the nuclear pore complex, requires soluble cytosolic factors and energy in the form of GTP. Typical of

Table 4.2 Nuclear localization signals in two maize transcription factors

Protein	NLS[a]	Class
Opaque-2	RKRKESNRESARRSRYRK	Bipartite
R	MSERKRREKL	SV40-like
R	MISEAIRKAIGKR	Matα2-like

[a]Basic residues are underlined. A hydrophobic sequence in the Matα2-like NLS is boxed.

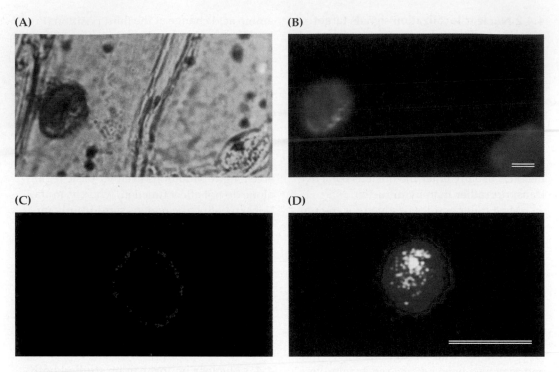

(A)

(B)

(C)

(D)

Figure 4.11
(A, B) Import of a chimeric protein consisting of β-glucuronidase (GUS) and the nuclear localization signal (NLS) of the maize regulatory protein Opaque-2 (O2). O2, which localizes to the nucleus, possesses two independent NLSs, one of which is bipartite. The image shows that this bipartite NLS is able to direct the reporter protein GUS to the nucleus of an onion epidermis cell. First, the nucleotide sequence that encodes GUS is fused with the NLS of O2. This chimeric gene is inserted into a plant expression vector with a strong promoter, and the expression construct is introduced into onion epidermis cells by using biolistic ("biological ballistic") particle bombardment. GUS activity can then be localized with a histochemical stain that the enzyme converts to an insoluble blue product. In (A) this stain appears in the nucleus; in (B) the identity of the organelle is confirmed by staining with the DNA-binding dye 4′,6′-diamidino-2-phenylindole (DAPI). Bar = 10 μm. (C, D) Binding to the nuclear envelope can be distinguished from nuclear import by confocal microscopy. The chimeric protein O2-human serum albumin (HSA) is shown binding to a purified nucleus (C) and after import into the nucleus of a permeabilized cell (D). The protein is visualized with antibodies to HSA that are tagged with a fluorescent dye. In an optical section through the nucleus, binding is seen as a fluorescent ring around the nucleus (C), whereas import stains the interior of the nucleus (D). Bar = 10 μm.

a facilitated process, the rate of import depends on temperature. Import into plant nuclei slows at temperatures below 23°C but continues even at 0°C. Import into nuclei of mammalian cells, on the other hand, stops at 4°C. Several soluble cytosolic factors that bind NLSs and stimulate import have been characterized: importin-α and importin-β. Protein import occurs when heterodimers of importin-α and importin-β bind NLS-containing proteins in the cytoplasm, forming a trimeric complex. Docking of the trimeric complex at the nuclear pore complex is mediated by the importin-β subunit. Translocation of the trimeric complex through the pore requires free GTP and a small GTPase, Ran.

In addition to the classical NLSs defined in Table 4.2, several other nuclear import and export targeting motifs have been iden-

tified and characterized. Each type of signal is linked to an import pathway, an export pathway, or both. Signal-mediated import and export are facilitated by members of the importin-β protein family, in conjunction with Ran.

4.4.4 Nuclear import is controlled by several mechanisms, providing an additional level of regulation.

Nuclear proteins often shuttle between the nucleus, where they exert their normal function, and the cytoplasm, where they may be stored. This traffic can be regulated by complexing with a cytoplasmic protein that masks the NLSs (Fig. 4.12), by phosphorylation and dephosphorylation, or by environmental stimuli such as light. Some nuclear proteins show tissue-specific nuclear localization: They are present in the nucleus in some tissues but excluded in others. Information about this regulation of nuclear protein shuttling is limited, but this aspect of gene regulation will undoubtedly be investigated more closely as we begin to understand the mechanistic details of nuclear import.

Light regulation of nuclear import was demonstrated for COP1 (constitutive photomorphogenesis), a protein that negatively regulates photomorphogenesis (see Chapter 18). (Note: COP1 bears no relationship to the COPI and COPII discussed in Section 4.5.8.) When expressed as a GUS fusion protein either transiently in onion epidermal cells or stably in transgenic *Arabidopsis* plants, COP1 can be detected in the nucleus and in the cytoplasm. In dark-grown plants, the protein is present only in the nucleus; when these plants are exposed to continuous light, some GUS activity appears in the cytosol after a period of hours. Consistent with the role of COP1 as a negative regulator of photomorphogenesis, the COP1 fusion protein is present in both the cytoplasm and nucleus in leaves of plants grown in the light. However, in roots, which do not display light-induced development, the COP1 fusion protein appears only in the nucleus. In the dark, therefore, COP1 probably acts as a negative nuclear regulator that turns off the photomorphogenic program. Light-induced development apparently requires COP1 to be released from its binding sites in the nucleus, so that it becomes less able to suppress photomorphogenesis. By regulating the relative nuclear abundance of COP1, this mechanism could achieve different levels of suppression of photomorphogenic development under different intensities of light.

4.5 The role of ER in protein sorting and assembly

Eukaryotic cells contain an extensive membrane system consisting of a series of cisternae, small vesicles, and other membrane-bound compartments, including vacuoles and endosomes—all of which are collectively referred to as the secretory system or pathway (see Chapter 1). Although first discov-

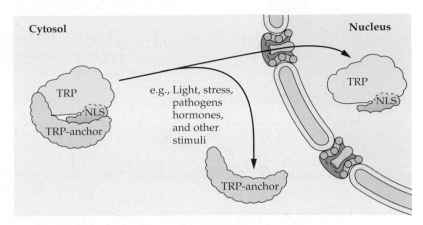

ered in highly specialized mammalian cells that secrete soluble and extracellular matrix proteins, the secretory system has other functions as well. After synthesis, proteins that enter the system are modified, assembled, and sorted to various destinations, including the vacuoles. Membrane proteins that enter the system can be retained at different points or delivered to the tonoplast or plasma membrane. The secretory system also stores calcium, synthesizes lipids for delivery elsewhere, and is involved in the biosynthesis and secretion of noncellulosic polysaccharides (see Chapter 2).

4.5.1 The first sorting decision takes place as ribosomes attach to ER.

For nearly all proteins, the first sorting decision is made when ribosomes that are synthesizing proteins destined to enter the secretory pathway become attached to the ER, creating rough ER. The rough ER is particularly abundant in cells that specialize in protein secretion (e.g., cereal aleurone cells triggered by gibberellic acid) or storage of vacuolar proteins (e.g., storage parenchyma cells of developing seeds). In most cell types, relatively little protein enters the secretory system, and the rough ER is not abundant. The rough ER consists of an extensive system of interconnected cisternae that often lie parallel to each other (Fig. 4.13). The lumenal space of these cisternae is topologically distinct from the cytosol and equivalent to the vacuolar and extracellular spaces. Once a protein is in the lumen of the ER, no further change in topology is necessary; that is, the protein need not pass through another

Figure 4.12
Regulation of nuclear import. Proteins that function in the nucleus, such as transcription factors (TRP = transcription regulatory protein), are usually found both in the cytoplasm and in the nucleus. In the cytoplasm, they may be complexed with other proteins, here called anchors, that prevent them from entering the nucleus. A variety of signals may cause this complex to be disrupted, possibly by phosphorylation or dephosphorylation, leaving the nuclear protein free to enter the nucleus.

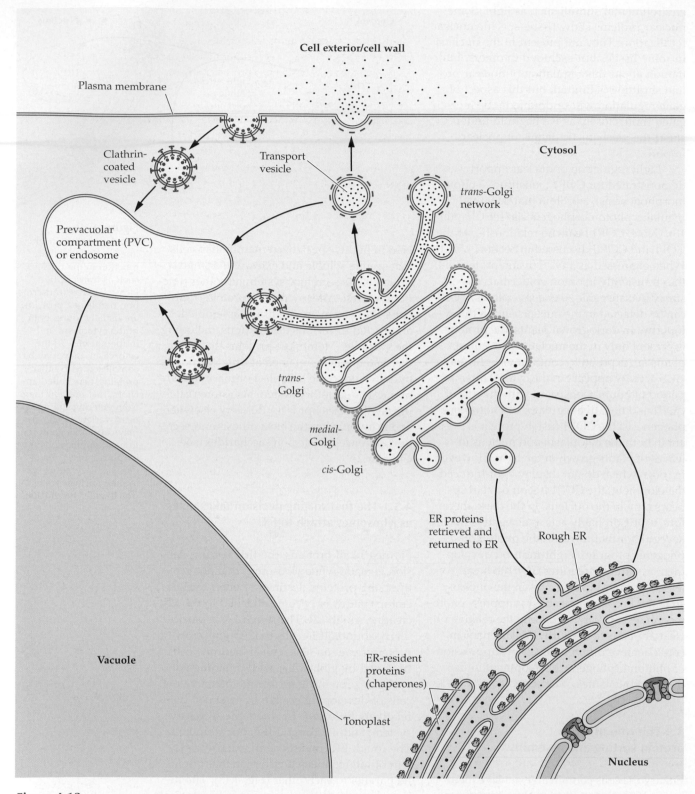

Cell exterior/cell wall

Plasma membrane

Cytosol

Clathrin-
coated
vesicle

Transport
vesicle

trans-Golgi
network

Prevacuolar
compartment (PVC)
or endosome

trans-
Golgi

medial-
Golgi

cis-Golgi

ER proteins
retrieved and
returned to ER

Rough ER

ER-resident
proteins
(chaperones)

Vacuole

Tonoplast

Nucleus

Figure 4.13

The secretory pathway for protein synthesis and sorting. Proteins are synthesized by ribosomes bound to the ER membrane. They pass into the lumen of the ER cisterna, where they are properly folded—a process aided by ER-resident chaperone proteins. The newly synthesized proteins are then sequestered in vesicles that move to the *cis*-Golgi reticulum and subsequently pass through the Golgi to the *trans*-Golgi network (TGN). In the TGN, secreted proteins are sorted from vacuolar proteins. From the TGN, proteins destined for the vacuole are first delivered to the prevacuolar compartment, or endosome, by clathrin-coated vesicles (CCVs). Endocytosis is carried out by CCVs, and the proteins taken up in this manner end up in the vacuole.

membrane to become an extracellular or a vacuolar protein. When vesicles that contain cargo protein fuse with the tonoplast or plasma membrane, the cargo protein is released into the vacuole or extracellular space.

When plant cells are homogenized in a buffered iso-osmotic medium, membranes break up and form small vesicles called **microsomes.** Some microsomes originate from the Golgi complex, plasma membrane, and tonoplast, but the bulk are formed from the ER, the most massive component of the endomembrane system. Newly synthesized proteins released in the lumen of the ER end up in the lumen of ER-derived microsomes. Sucrose density gradients can separate the microsomal vesicles derived from different membrane systems (see Box 4.3).

Most proteins imported into the ER lumen or membrane move to the Golgi complex (Fig. 4.13). This organelle consists of a series of slightly curved cisternae referred to as the *cis-, medial-,* and *trans-*cisternae, depending on their location with respect to the ER. The trans face is connected to a system of vesicles and tubules called the *trans-*Golgi network (TGN). Mammalian cells, but not plant cells, contain a similar system of vesicles and tubules found at the cis face and called the *cis-*Golgi network, or intermediate compartment. In some mammalian cells, the various Golgi compartments have been shown to contain specialized enzymes for the modification of glycoproteins (see Sections 4.7.1 and 4.7.2), but there is as yet little evidence for functional subcompartmentation of the Golgi complex in plants (see Chapter 1).

4.5.2 Proteins travel through the secretory system like cargo in containers.

Proteins traverse the secretory system in vesicles that bud off one cisterna or compartment and first dock and then fuse with the next. Within the secretory pathway, proteins travel from the ER to the Golgi, through the Golgi from cis to trans locations, into the TGN, and ultimately to the plasma membrane for secretion or to the tonoplast for delivery to the vacuole (Fig. 4.13). Vesicle formation requires specific cytosolic proteins that self-assemble to coat the vesicles. Both vesicle budding and fusion at the target membrane consume energy in the form of GTP (see Box 4.6). To dock

and fuse, a vesicle must shed its protein coat, thereby exposing integral membrane proteins that interact with proteins integral to the target membrane. These integral membrane proteins and the soluble proteins recruited from the cytosol during vesicle formation are all part of the protein-sorting machinery and determine the specificity of the sorting process. Vesicles that bud from the rough ER differ from Golgi-derived vesicles in their protein coat composition and in their complement of integral membrane docking proteins.

Protein-filled vesicles move continuously between compartments of the secretory system. This forward (anterograde) traffic must be matched by reverse (retrograde) traffic of nearly equal magnitude if the different compartments are to maintain their characteristic complement of proteins. Furthermore, soluble and integral membrane proteins unique to each compartment must remain behind when vesicles form to carry cargo forward (Fig. 4.14). Like docking, the sorting process that separates cargo from resident proteins requires integral membrane proteins—receptors that interact with cargo proteins inside the vesicle and with vesicle coat proteins on the cytosolic side. The interaction between the cargo protein and the receptor is mediated by targeting domains on the cargo protein, usually short peptides that may be proteolytically removed after delivery. Not all protein sorting at the TGN has been proved to require receptors, and cargo proteins can perhaps also be segregated by other mechanisms (e.g., protein aggregation).

After vesicles bud from the TGN, some go on to fuse with the plasma membrane, others with the prevacuolar compartment (PVC) (see Fig. 4.13). Vesicles that fuse with the plasma membrane carry integral plasma membrane proteins, components of the extracellular

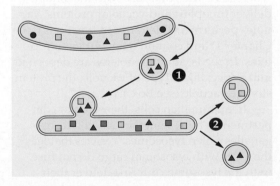

Figure 4.14
Proteins travel as cargo in vesicles after they enter the secretory system. In step 1, two types of cargo protein (yellow ■ and purple ▲) are sorted from resident proteins (red ●), which stay behind. In step 2, the two types of cargo proteins are sorted into different vesicles and the resident proteins (green ■) stay behind.

In cell biology, a question that arises often is, In which subcellular compartment is a particular protein located? One way to answer this question is to homogenize the tissue and then to separate the various organelles on sucrose gradients either on the basis of their size (rate zonal centrifugation) or on the basis of their density (isopycnic or equilibrium density centrifugation). This method does not completely separate organelles, but the position in the gradient is diagnostic of each organelle fraction.

First, the tissue must be homogenized gently in a buffered medium that is more or less iso-osmotic with the cytosol and will keep the organelles intact (e.g., 0.4 M sucrose). Most methods of homogenization will cause the vacuoles, Golgi apparatus, plasma membrane and ER to fragment and form vesicles. Removal of cell wall fragments, starch grains, and nuclei by a brief centrifugation at 2000g results in a cleared homogenate that can be loaded on top of a sucrose gradient and centrifuged to separate the components.

Gradients with a sucrose concentration of 15% at the top and 25% at the bottom can be used for rate zonal separations of organelles. During a brief (15 to 30 minutes) centrifugation, the largest organelles (nuclei and chloroplasts) quickly migrate to the bottom of the centrifuge tube, the smaller organelles (mitochondria, peroxisomes) travel shorter distances, and the microsomal vesicles (see Section 4.5.1) remain near the top.

Organelles or vesicles that have the same size and shape cannot be separated on rate zonal gradients. To separate organelles by density, isopycnic gradients with a sucrose concentration of 16% at the top and 48% or 60% at the bottom of the tubes are centrifuged for longer periods of time (3 to 18 hours), because each organelle must travel to the region of the

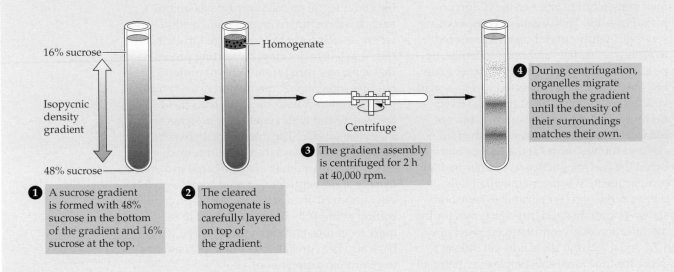

16% sucrose

Isopycnic density gradient

48% sucrose

1 A sucrose gradient is formed with 48% sucrose in the bottom of the gradient and 16% sucrose at the top.

Homogenate

2 The cleared homogenate is carefully layered on top of the gradient.

Centrifuge

3 The gradient assembly is centrifuged for 2 h at 40,000 rpm.

4 During centrifugation, organelles migrate through the gradient until the density of their surroundings matches their own.

matrix (e.g., hydroxyproline-rich glycoproteins; see Chapter 2) and secreted enzymes, whereas vesicles that fuse with the tonoplast deliver tonoplast and vacuolar proteins. Vacuoles perform a wide range of functions (see Chapter 1) and contain many different proteins. In seeds, storage proteins are deposited into a specialized type of vacuole, the protein storage vacuole (see Box 4.4).

In mammalian cells, the vacuolar compartment is represented by small lytic organelles called lysosomes. Vesicles that leave the TGN with lysosomal cargo do not fuse with the lysosomes but first deliver their proteins to **endosomes,** membrane-bound organelles in which the secretory and endocytic pathways converge. The **endocytic pathway** imports proteins from outside the cell. This transport is mediated by vesicles that originate from the plasma membrane. Plant cells also engage in endocytosis, and molecules that enter the cells are delivered first to endosomes. Cargo carried in vesicles derived from the TGN is first delivered to a prevacuolar compartment and then is moved. Whether the prevacuolar compartment and the endosomes overlap in plant cells, as they do in mammalian cells, is not yet known.

gradient whose density matches its own (see step 4 in the figure in this box). Such gradients can be used to separate organelles that have the same size, such as mitochondria and peroxisomes, or the microsomal vesicles derived from the different membrane systems. The density of these vesicles reflects the protein-to-lipid ratio of the different organelles from which they are derived.

Both rate zonal and isopycnic gradients can be fractionated after centrifugation, and the resulting fractions assayed to determine which organelle or organelles house a particular protein of interest. This method requires that each organelle (whether intact or fragmented into vesicles) be identified by means of a well characterized marker enzyme or protein, preferably one unique to that organelle. The subcellular locations of various marker proteins have been established over the years using a variety of techniques. Among the enzymes utilized as markers are cytochrome oxidase (associated with the mitochondria), catalase (peroxisomes), callose synthase (plasma membrane), NADH-cytochrome c reductase (ER membrane), vacuolar pyrophosphatase (tonoplast), and sugar nucleotide transferases (Golgi apparatus–derived vesicles) (see step 6 in the figure below). Clathrin-coated vesicles (CCVs) and other organelles that lack characteristic enzyme activities can be located with antibodies raised against a characteristic protein (e.g., clathrin). An enzyme assay or immunoblot can be used to determine which gradient fractions contain the protein of interest. By comparing the distribution of the protein of interest with the distribution of marker proteins, investigators can identify which organelle(s) contain the protein of interest.

5 The centrifuge tube is punctured at the bottom and the gradient fractions are collected (fractions can also be collected from the top of the gradient by aspiration).

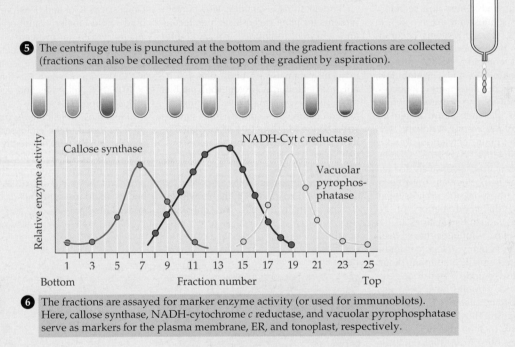

6 The fractions are assayed for marker enzyme activity (or used for immunoblots). Here, callose synthase, NADH-cytochrome c reductase, and vacuolar pyrophosphatase serve as markers for the plasma membrane, ER, and tonoplast, respectively.

4.5.3 Signal peptides allow proteins to enter the secretory pathway.

As previously noted, ribosomes that are synthesizing secretory pathway proteins associate with the rough ER. As they are being synthesized, soluble proteins destined for the vacuole or cell exterior cross the membrane completely, whereas integral membrane proteins become part of the ER membrane and can remain in the ER or proceed to other destinations. Why do the ribosomes synthesizing these classes of proteins bind to the ER, and how do proteins cross the lipid bilayer? The nascent polypeptide chains of these proteins have at their N terminus a **signal peptide** of 16 to 30 amino acids that directs the polypeptide chain to the ER membrane and begins the process of translocating the entire protein into the lumen of the ER (Fig. 4.15). Such signal peptides are found on nearly all secreted and vacuolar proteins as well as on proteins that reside in the lumen of the ER, the Golgi complex, and the endosomes. In addition, these peptides are present on many integral membrane proteins of the secretory system. Signal peptides of proteins that enter the secretory pathway differ from one another but share important structural features. A signal peptide typically consists of one or more positively charged amino acids followed by a stretch of 6 to 12 hydrophobic amino acids and then several additional amino acids (Table 4.3). The signal peptides of different secretory proteins are interchangeable, not only among plant proteins, but also among plant, animal, and yeast proteins. Furthermore, the presence of a signal peptide is sufficient to

Box 4.4 — Globulin storage proteins accumulate in the protein storage vacuoles of developing seeds.

Many basic features of the secretory system of plant cells were discovered by studying protein storage deposition in developing seeds. The seeds of most plants contain a limited number of very abundant proteins, so-called storage proteins, that accumulate in protein storage vacuoles. Half of all the protein synthesized in developing seeds enters the secretory system and is sequestered in vacuoles. In a few species, storage proteins accumulate in protein bodies derived from the ER.

The most common storage proteins are the salt-soluble 7S and 11S **globulins,** found in all dicots and many monocots, and the alcohol-soluble **prolamins,** which are unique to the endosperm of cereals. Because globulins are so abundant, and because developing cotyledons readily take up radioactive precursors, cotyledons have been used extensively for experiments that help elucidate the role of organelles in transport, the processing of polypeptides, the modification of N-linked glycans, and other features of the secretory system. Globulins have been studied most thoroughly in legumes and a few other commercially important dicots such as sunflower and oilseed rape. Unfortunately, these proteins have been given species-specific names that confuse the nonspecialist. Thus, the 7S (vicilin-type) globulins include the homologous proteins phaseolin from the common bean *(Phaseolus vulgaris),* vicilin from the garden pea *(Pisum sativum),* and conglycinin from soybean *(Glycine max).*

The 7S globulins are synthesized as 45–50-kDa polypeptides, and 5 to 10 different polypeptides, each encoded by a different gene, may be present in the seeds of a single species. After removal of the signal peptide and glycosylation in the ER, these peptides form trimers and proceed to the Golgi, where the glycans are modified. The 11S (legumin-type) are synthesized as polypeptides of 60–65 kDa. After removal of the signal peptide, the polypeptides form trimers in the ER, proceed to the Golgi, and then to the protein storage vacuoles where they form hexamers. This process depends on a single event of proteolytic processing that creates polypeptides of 40 and 20 kDa. Processing is carried out by a specific vacuolar protease. As with the 7S globulins, a number of different genes encode the 11S globulins, and purified 11S proteins can contain complex mixtures of these polypeptides.

Protein storage vacuoles also contain proteins for plant defense such as lectins and enzyme inhibitors. For example, seeds of the common bean contain about 5% phytohemagglutinin, a lectin that is toxic to mammals and birds, and 1% α-amylase inhibitor, a protein that inhibits the α-amylases of mammals and insects. Like globulins, these proteins are synthesized on the rough ER and glycosylated in the ER lumen, traverse the secretory system, and accumulate in protein storage vacuoles.

Protein storage vacuoles arise de novo during the expansion of the cotyledon storage parenchyma cells. The contents of protein storage vacuoles (PSV) are electron dense, presumably because of the high protein concentration. Other vacuoles, assumed to be lytic vacuoles (LV), are without protein aggregates (A, magnification ×2,760). Specific storage proteins can be visualized using antibodies labeled with colloidal gold. In (B) and (C), Golgi stacks (Go) and dense vesicles (DV) bind gold-labeled antibodies raised against vicilin (arrowheads). The dense vesicle in (D) appears to contain vicilin, but a neighboring clathrin-coated vesicle (CCV) is not labeled (B, magnification ×40,000; C, magnification ×65,000, D, magnification ×245,000).

Prolamins, the alcohol-soluble storage proteins found in the endosperm of many cereals, behave very differently at the cellular level: They generally form aggregates within the lumen of the ER and pinch off from the ER to form protein bodies that are surrounded by an ER-derived membrane.

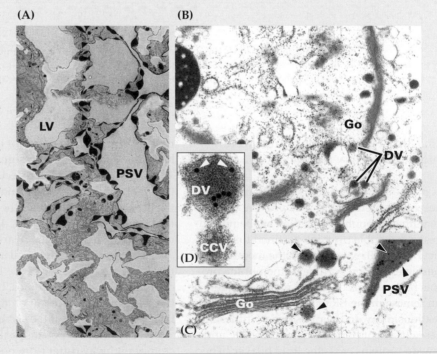

(A) (B) (C) (D)

Chapter 4 Protein Sorting and Vesicle Traffic

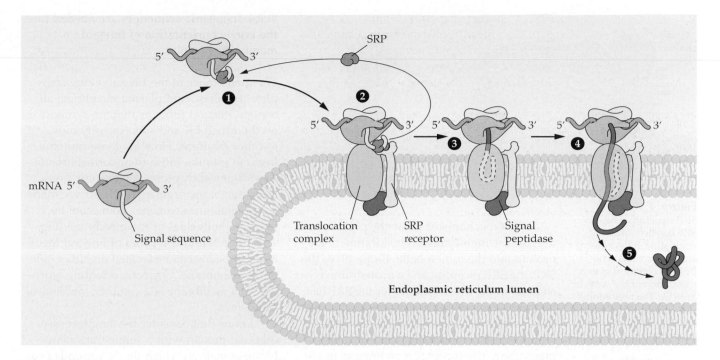

Figure 4.15

Entry of proteins into the lumen of the ER. Step 1: When the nascent polypeptide contains approximately 70 amino acids, the signal peptide protrudes far enough from the ribosome to bind the signal recognition particle (SRP). Step 2: The SRP docks the entire complex onto the SRP receptor. Step 3: The signal peptide releases both the SRP and one of the two polypeptides of the SRP receptor and is inserted into the membrane channel of the translocation complex. Step 4: Signal peptidase cleaves the signal peptide, and continued translation causes the nascent chain to protrude into the lumen of the ER, where it binds chaperones (not shown). Step 5: Cycles of chaperone binding and removal complete the protein-folding process.

direct the secretion of any stable and correctly folded protein. Recombinant DNA techniques have been used to produce transgenes that encode chimeric proteins that consist of a signal peptide from any secretory protein and a cytosolic or bacterial polypeptide that is not ordinarily secreted. When such a nucleic acid construct is introduced into a plant cell, the resulting protein is synthesized, enters the ER, and is secreted.

Before it can bind to the ER, a signal peptide must be recognized by the **signal recognition particle** (SRP), a ribonucleoprotein complex (Fig. 4.16). The SRP contains a small RNA molecule of 300 nucleotides that binds six different proteins, each with specific functions. SRPs ordinarily reside in the cytosol, where they await the protrusion of a signal peptide on a nascent protein that is being made on a polysome. Recognition between the signal peptide and the SRP must occur when the nascent polypeptide is about 70 amino acids long, with approximately 40 amino acids protruding beyond the surface of the ribosome. One SRP component, the P54 protein, is a GTPase containing clustered

methionine residues, the side chains of which point outward and bind to the hydrophobic core of the signal peptide. When the SRP binds to the signal peptide, the SRP polypeptides P9 and P14 act to arrest translation of the nascent polypeptide chain. This is an important control step in the production of proteins. If synthesis were to continue, the protein might begin to fold in the cytosol, possibly misfolding or aggregating. Protein folding is assisted by chaperones, and the chaperones needed to help fold secretory proteins are found in the lumen of the ER, not in the cytosol. Furthermore, many secretory proteins are glycosylated at specific asparagine residues, a process that occurs on the lumenal side of the ER membrane and is usually necessary for correct folding.

Table 4.3 Examples of signal peptide sequences

Protein	Signal peptide sequence
Barley lectin	MKMMSTRALALGAAAVLAFAAATAHAN
Sporamin	MFILTRTTLSKWQRCIATMQPHN
Phytohemagglutinin	MASSNFSTVLSLALFLPLLTHANS

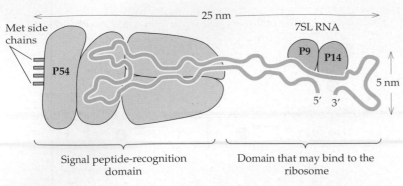

Met side chains

25 nm

7SL RNA

P54

P9 P14

5 nm

5' 3'

Signal peptide-recognition domain

Domain that may bind to the ribosome

Figure 4.16
SRP structure. The SRP consists of a 300-nucleotide RNA (green) and six proteins. The P9/P14 complex arrests translation of the nascent protein. The methionine side chains of P54 make contact with the signal peptide.

What mechanisms allow the polypeptide to cross the membrane? Translocation of the protein into the lumen of the ER involves the SRP, the SRP receptor, and a protein-lined channel in the membrane. After the SRP has bound to a signal peptide protruding from a ribosome, the entire assembly makes contact with the SRP receptor and docks on an ER membrane. The receptor is an integral membrane protein composed of two different GTP-binding polypeptides. Once docking has occurred, the SRP and one of the two polypeptides of the receptor dissociate from the signal peptide, utilizing the energy obtained by GTP hydrolysis. After the SRP is released into the cytosol, the ribosome and its nascent polypeptide dock onto the translocation complex, a transmembrane protein channel in the ER membrane (see Fig. 4.15). This translocation complex consists of three integral membrane proteins. Thus, the SRP and SRP receptor start the process of inserting the nascent polypeptide chain into the protein channel, but translocation of the entire polypeptide requires both the further hydrolysis of GTP as a source of energy and the activity of molecular chaperones in the lumen of the ER. Translation and translocation into the ER occur simultaneously. As the polypeptide enters the lumen of the ER, the signal peptide is cleaved by signal peptidase, an enzyme complex that resides in the lumen of the ER. As a result, the signal peptide is usually not found on mature vacuolar or secreted proteins.

Some proteins can be targeted to the ER by an SRP-independent pathway. These proteins are synthesized on free ribosomes and then transferred to the ER posttranslationally. This pathway has been found only in yeast, but it probably also exists in other eukaryotic cells.

4.5.4 Topogenic sequences are needed for the correct orientation of integral membrane proteins.

The membranes of the ER and Golgi complex, tonoplast, and plasma membrane all contain integral proteins that are synthesized on the rough ER and subsequently move to other locations. How are these proteins inserted into the ER in their correct orientation? Many such proteins have only one transmembrane segment. Others have many transmembrane segments, connected by loops on both sides of the membrane (Fig. 4.17). The N and C termini of integral membrane proteins can be located on either side of the membrane. Correctly orienting a protein in a membrane is a complex topological problem.

Let us first consider the simplest case, that of a protein with a single transmembrane domain in which the N terminus protrudes into the lumen of the ER. Such proteins have N-terminal signal peptides that function exactly as described in the previous section. The signal peptide is cleaved as the protein begins to traverse the membrane. Then, partway through the membrane, protein translocation stops. A hydrophobic transmembrane sequence in the protein acts as a "stop transfer," or **membrane anchor sequence.** The size of the polypeptide domain that remains in the cytosol depends on the location of the transmembrane domain within the polypeptide. This type of integral membrane protein, with a cytoplasmic C terminus, is called a type I membrane protein. Proteins that exhibit the reverse orientation, called type II, do not contain a removable signal peptide. Rather, the internal hydrophobic transmembrane domain acts as a signal peptide and directs integration of the protein into the ER membrane, leaving a large portion of the N terminus in the cytosol. The C-terminal segment of the protein translocates into the lumen of the ER. Enzymes in the Golgi apparatus of mammalian cells have this orientation. The orientation of a protein in the membrane is also influenced by charged amino acids in the two hydrophilic domains. Usually, but not always, the domain that carries more positive charges remains in the cytosol, which is negatively charged relative to the apoplast and subcellular compartments.

Box 4.5

sec mutants and vps mutants of Saccharomyces cerevisiae have provided important insights into the secretory process.

Genetic and biochemical approaches initiated in the laboratory of Randy Schekman have proven extremely useful for studying the secretory processes in the yeast *Saccharomyces cerevisiae*. A number of elegant selection and screening strategies have been employed to isolate secretory *(sec)* mutants that are temperature-sensitive for cell growth, division, and secretion. They are called **conditional mutants** because they normally grow at 27°C (permissive temperature), but their growth is arrested at 37°C (nonpermissive temperature). Many of these mutants accumulate secretory proteins in intracellular pools at the nonpermissive temperature. Using this approach, at least 30 genes have been implicated in the processes of delivering membrane and secretory proteins to the cell surface. These genes have been cloned and the proteins identified. The figure at right shows the numerical names of these Sec proteins and indicates the process that is interrupted when a particular protein is disabled by a mutation in the corresponding gene.

Other investigators have become interested in isolating vacuolar protein sorting *(vps)* mutants. Carboxypeptidase Y (CPY), a yeast vacuolar hydrolase, has been shown to be sorted to the vacuole from the *trans*-Golgi compartment. The precursor form of CPY carries positive sorting information that directs the protein to the vacuole, and in the absence of this sorting signal the protein is secreted. To identify

genes whose products participate in the sorting process, researchers devised a scheme of positive selection for mutants that secrete CPY. Complementation testing among a large collection of such *vps* mutants revealed the existence of greater than 40 *VPS* genes that influence CPY sorting, suggesting that the process of transport to the vacuole is very complex. A large number of these genes have been cloned and the proteins identified.

Several laboratories have also set up in vitro reconstitution systems that allow the complementary biochemical dissection

of secretory machinery. Subcellular organelles are isolated and then incubated with cytosolic factors. This approach has revealed that a complex of five integral ER membrane proteins, three of which are encoded by known *SEC* genes, seems to participate directly in the translocation of proteins into the lumen of the ER. The most hydrophobic member of this complex, Sec61p, makes direct contact with a secretory polypeptide during the translocation event. Translocation-competent vesicles have been reconstituted from detergent-solubilized membranes.

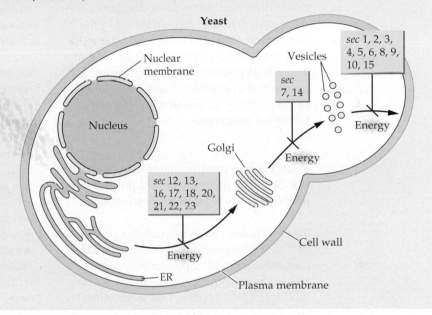

Yeast

Nuclear membrane

Nucleus

Golgi

Vesicles

sec 7, 14

sec 1, 2, 3, 4, 5, 6, 8, 9, 10, 15

Energy

Energy

sec 12, 13, 16, 17, 18, 20, 21, 22, 23

Energy

Cell wall

ER

Plasma membrane

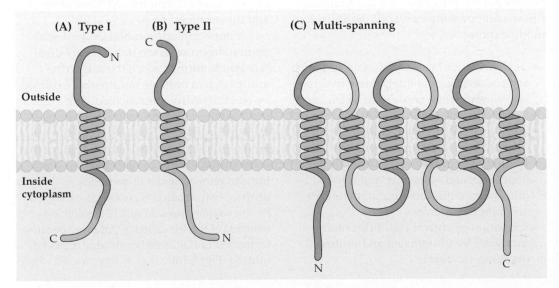

(A) Type I   **(B) Type II**   **(C) Multi-spanning**

Outside

Inside cytoplasm

N   C

C   N

N   C

Figure 4.17
Orientation of membrane proteins. Integral membrane proteins have one or more α-helical domains of 20–25 amino acids. These proteins can be oriented in the membrane in different ways. The type I and type II proteins (A and B) each have only one membrane-spanning domain but differ in the location of the N and C termini. The larger protein (C) has six membrane-spanning domains. Its N and C termini are in the cytosol.

Insertion of proteins with multiple hydrophobic membrane-spanning domains is even more complex. The loops between the transmembrane domains are usually hydrophilic and will remain at the cytosolic side of the ER membrane or within the lumen of the ER. The first transmembrane domain acts as a signal peptide (the N terminus remaining in the cytosol), and the second domain functions as a membrane anchor sequence; the third domain acts again as a signal peptide, the fourth as a membrane anchor sequence, and so on. Although both SRP and SRP receptors are necessary for the integration of the first transmembrane domain, they are not needed for the third, fifth, and subsequent domains. Once these proteins have been inserted in the ER membrane, they can be retained there (ER residents) or targeted to other destinations. There is at present no information on the domains that target membrane proteins to other destinations or retain them along the transport pathway.

4.5.5 Protein modifications that occur in the ER allow proteins to fold properly and proceed to their targeted destinations.

The ER lumen contains a number of enzymes and chaperones that ensure correct folding and may degrade imperfect polypeptides rather than export them, forming a sort of quality-control system. This mechanism ensures that only properly folded and assembled proteins are transported along the secretory pathway to their final destinations. ER-resident proteins carry out the following modifications:

- modification of certain amino acids, such as proline (e.g., peptidylproline hydroxylase catalyzes the conversion of proline to hydroxyproline)
- addition of N-linked glycans and removal of glucose residues from these glycans
- folding of proteins correctly, catalyzed by peptidylprolyl isomerase and assisted by immunoglobulin binding protein (BiP) and other chaperones
- formation of correct disulfide bonds, catalyzed by glutathione and protein disulfide-isomerase

- assembly of oligomeric proteins, also assisted by chaperones
- degradation of proteins or possible retranslocation to the cytosol for degradation

Not all proteins undergo all of these modifications, and other modifications may occur after proteins leave the ER. For example, the modification of asparagine-linked glycans (see Section 4.7.1) continues in the Golgi complex, and O-linked glycans are added there. Proteins that are not correctly folded or assembled may be retained in the ER, where they may be broken down by proteases.

BiP, a chaperone of approximately 70 kDa that resides in the lumen of the ER, was discovered in mammalian cells that secrete immunoglobulin G. BiP and other ER-resident chaperones bind intermediates of protein folding and assembly. Such binding appears to contribute directly to ER retention of intermediates as part of a quality-control system, ensuring that only properly folded and functional proteins are transported along the secretory pathway to their final destinations. BiP functions by binding to nascent or partially folded proteins, specifically to polypeptide segments rich in hydrophobic amino acids, which are not exposed on the surface of a properly folded protein. By binding, BiP prevents the nonspecific interaction of such domains during folding, thereby preventing proteins from denaturing or aggregating. BiP is an ATPase, and the hydrolysis of ATP causes it to release the partially folded protein, affording additional chances to fold correctly. Correct folding of a protein may involve several cycles of BiP binding, ATP hydrolysis, and dissociation.

Many secreted, vacuolar, and integral membrane proteins are stabilized by disulfide bonds, formed when the sulfhydryl groups of two cysteine residues are oxidized by reacting with a suitable oxidizing agent (e.g., oxidized glutathione). However, because polypeptide chains often contain more than two cysteines, the disulfide bridges that form are not always the correct ones. Breaking and reforming disulfide bonds so that the proper conformation is created is catalyzed by the enzyme protein disulfide-isomerase, a resident of the ER called a "foldase" because formation of disulfide bonds aids correct folding (Fig. 4.18).

4.5.6 Oligomerization and attachment of N-linked glycans occur in the ER.

Many proteins do not consist of a single polypeptide but are oligomers of two, three, four, or more subunits. Oligomerization occurs in the ER, and the monomeric subunits of oligomeric proteins cannot leave the ER before oligomerization. For example, the seed storage proteins vicilin and phaseolin accu-mulate in protein storage vacuoles as trimers of 45-kDa polypeptides. These trimers are formed in the ER. When recombinant DNA techniques are used to produce a truncated phaseolin that cannot form trimers, this monomer has prolonged interactions with BiP and fails to be transported out of the ER. Rather than accumulating in the ER as a monomer, the truncated phaseolin is slowly degraded (Fig. 4.19).

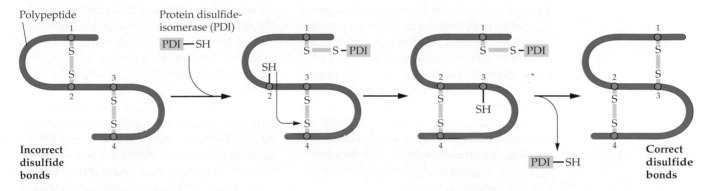

Figure 4.18
Protein disulfide-isomerase (PDI) rearranges disulfide bonds. PDI catalyzes the breakage and reformation of disulfide bonds and there-by accelerates protein folding. Disulfide bonds form easily in the oxidizing environment of the ER lumen, but these bonds are often incorrect. PDI breaks an incorrect bond, generating a free sulfhydryl group that can react with an existing disulfide bond to form a correct disulfide and a free sulfhydryl. Finally, with the creation of the second correct disulfide bond, PDI is released.

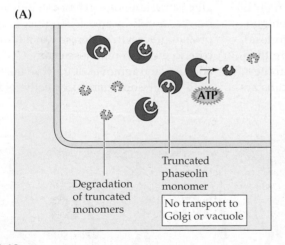

Figure 4.19
The 7S seed storage globulins, such as vicilin and phaseolin, are transported out of the ER as trimers. (A) A phaseolin mutant that lacks the C-terminal 59 amino acids is unable to form trimers when it is expressed in tobacco leaf protoplasts; (B) wild-type phaseolin, however, trimerizes efficiently and rapidly in this heterologous system. Proteins from pulse-labeled protoplasts expressing wild-type phaseolin and truncated phaseolin were studied with immunopre-cipitation and rate zonal centrifugation. Antibodies raised against immunoglobulin-binding protein (BiP) immunoprecipitated far more truncated phaseolin than wild-type phaseolin. Analysis of the wild-type phaseolin on rate zonal gradients showed that BiP was as-sociated only with the not-yet-assembled monomeric form, not with the much more abundant trimeric phaseolin. The BiP–phaseolin complex could be dissociated by addition of ATP, indicating that the association is specific. Thus, BiP binds to monomeric phaseolin that has not yet formed, or is incapable of forming, oligomers. Monomers probably undergo cycles of binding and release from BiP until the BiP–binding sites are concealed by trimer formation. Whereas wild-type phaseolin trimerizes and progresses through the secretory path-way to the vacuole, truncated phaseolin is degraded.

As noted above, quality-control mechanisms act to retain proteins in the ER during folding and assembly and to degrade structurally defective polypeptides that cannot mature properly. Because BiP and other ER-resident chaperones (such as calnexin, calreticulin, and endoplasmin) bind partially folded and assembled proteins, they might contribute directly to the retention of secretory proteins in the ER as part of the quality-control system. Disposal of some defective secretory proteins involves translocation from the ER lumen into the cytosol, ubiquitination, and degradation by a proteasome (see Chapter 9). However, we still do not know whether other proteins are, instead, degraded in the ER itself and, in general, which features of defective proteins determine their sorting for degradation by the quality-control system. Normal subunits of heteromultimeric proteins synthesized in the absence of partner subunits, and therefore unable to oligomerize, are treated as defective and usually degraded.

Many secretory and integral membrane proteins of the secretory systems are glycoproteins, containing glycans (oligosaccharides) that are N-linked (to asparagine) or O-linked (to serine, threonine, or hydroxyproline). Attachment of N-linked glycans occurs as the nascent polypeptide chain is translocated into the ER lumen. Such glycans, which exhibit a (glucose)$_3$(mannose)$_9$ (N-acetylglucosamine)$_2$ structure, are then modified, first in the ER and later in the Golgi (see Section 4.7.1). These glycans are built up, one sugar residue at a time, on dolichol pyrophosphate, a polyisoprenoid lipid of the ER, and are then transferred as a unit by the ER enzyme oligosaccharide protein transferase to certain asparagine residues (Fig. 4.20). The residues that receive these glycans are invariably in an Asn-X-Ser or Asn-X-Thr configuration, where X can be any amino acid except proline. However, not all Asn-X-Ser/Thr configurations in a given protein are necessarily N-glycosylated. After a glycan is transferred to a nascent polypeptide chain, the three glucose residues are quickly removed by two α-glucosidases that reside in the ER. The resulting glycan, with the structure (Man)$_9$(GlcNAc)$_2$, is referred to as a high-mannose glycan.

N-linked glycans have a major role in protein folding and quality control in the ER. Calnexin and calreticulin, two ER-resident chaperones, help glycoproteins fold by binding to high-mannose chains from which two of the three glucose residues have already been removed. Such binding may result in the release of a correctly folded glycoprotein. In cases where these chaperones release an incorrectly folded glycoprotein, the protein may need to be reglycosylated before the chaperones can bind again. After the last glucose residue of a glycoprotein is removed, an ER-located UDP-glucose:glycoprotein glucosyltransferase can add one glucose to the high-mannose chain. Cycles of addition and removal of glucose take place until the glycoprotein is correctly folded. A

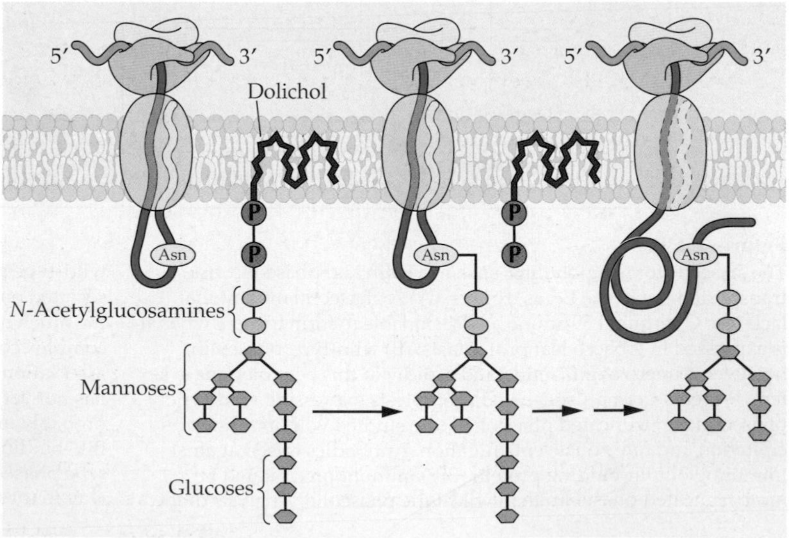

Figure 4.20
Cotranslational transfer of a glycan from a lipid carrier to a nascent polypeptide chain. Glycans consisting of (Glc)$_3$(Man)$_9$(GlcNAc)$_2$ are first assembled on a dolichol carrier in the ER membrane and then transferred to the nascent polypeptide chain by an ER-resident enzyme. As the protein begins to fold, aided by chaperones such as BiP and protein disulfide-isomerase, the three glucose residues are removed by glucosidases.

folded glycoprotein is not a substrate for readdition of glucose residues and thus cannot interact with calnexin or calreticulin. This system monitors the process of protein folding and may participate with other chaperones in retaining unfolded polypeptides in the ER lumen.

4.5.7 The C-terminal tetrapeptide Lys-Asp-Glu-Leu ensures that soluble ER-resident proteins are returned to the ER if they escape.

Proteins that have been released into the lumen of the ER and properly assembled can proceed to other destinations. If they have no other targeting or retention domains, they are secreted. For this reason, the pathway leading to secretion from the cell is often called the default pathway. Why then do some proteins, such as BiP and protein disulfide-isomerase, remain in the ER? Two mechanisms appear to retain ER residents. First, these proteins seem to have high affinity for each other. Because of the relatively high concentration of calcium in the ER (see Chapter 3), perhaps resident proteins—many of which contain calcium-binding domains—may form a bulky network that is unable to enter vesicles budding off from this compartment. However, experiments show that a certain fraction of the ER residents escape, travel as far as the *cis*-Golgi, and then are returned to the ER by retrograde transport vesicles. Such ER residents have a C-terminal sequence, Lys-Asp-Glu-Leu, called the KDEL tail after the single-letter names for the amino acids. The amino acid sequence may also be HDEL or a similar variant. A KDEL receptor, ERD2p, is an integral membrane protein found primarily in the *cis*-Golgi and in vesicles that shuttle between the ER and the Golgi (see Fig. 4.13). The receptors package proteins bearing the KDEL tail into vesicles bound for the ER. Thus, retention of ER-resident proteins may result from the combination of network formation in the ER and retrieval from the *cis*-Golgi by retrograde vesicles. Experimentally adding a KDEL tail to proteins that ordinarily are secreted usually causes their retention in the ER (see Box 4.1), and removing KDEL from BiP allows its secretion.

One interesting protein with a KDEL motif is the 22-kDa auxin-binding protein (ABP), usually found in the ER. How ABP functions in the auxin signal transduction pathway is not yet clearly understood (see Chapter 18), but recent evidence indicates that some ABP is always present at the cell surface, putative site of the auxin's primary action. Exogenously added ABP has been shown to induce changes in the membrane potential of tobacco protoplasts, a response ordinarily associated with auxin action. The observation that ABP may actually function at the plasma membrane and outside of the cell has led to the idea that when necessary, the cell may be able to allow the secretion of ER residents. A conformational change in the protein could alter the way the KDEL tail is displayed and block its interaction with the KDEL receptor, ERD2p.

4.5.8 Transport from the ER to the Golgi involves forward (anterograde) and backward (retrograde) transport of vesicles.

The transport of proteins from one compartment to the next is mediated by vesicles that bud off from a donor membrane and, after transport, fuse with an acceptor membrane. When formed, these vesicles contain cargo proteins as well as integral membrane proteins that are part of the protein-sorting and vesicle formation/targeting machinery. After these vesicles fuse with their target compartment, the proteins intrinsic to the donor membrane must be returned to it. Thus, a forward, or anterograde, movement of vesicles must be matched by a return, or retrograde, movement of equal magnitude. In yeast cells, two types of transport vesicles shuttle materials between the ER and the Golgi (Fig. 4.21A). COPII vesicles function in anterograde transport, moving cargo from the ER to the Golgi, whereas COPI vesicles participate in retrograde transport, returning escaped ER residents and integral membrane components of the vesicle targeting system. (COP stands for coat protein here, and is not the same as COP1 discussed in Section 4.4.4.) The formation of vesicles requires recruitment of proteins from the cytosol to the surface of the membrane where the proteins can form a coat on the vesicle. These

(A)

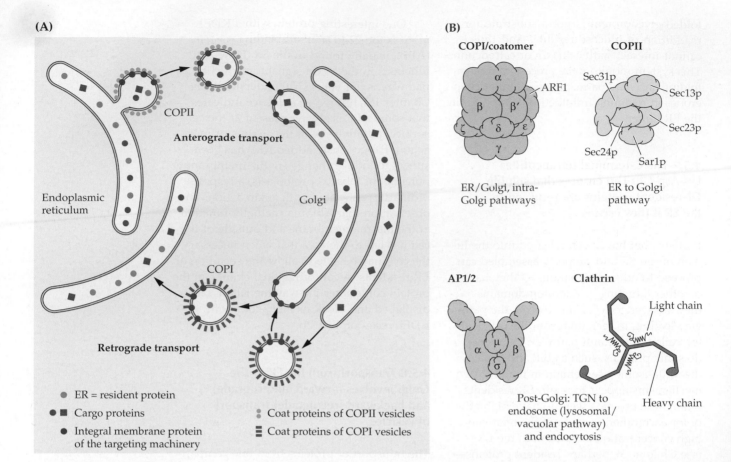

(B)

COPI/coatomer

α
β β′ ARF1
ζ δ ε
γ

ER/Golgi, intra-
Golgi pathways

COPII

Sec31p
Sec13p
Sec23p
Sec24p Sar1p

ER to Golgi
pathway

AP1/2

α μ β
σ

Post-Golgi: TGN to
endosome (lysosomal/
vacuolar pathway)
and endocytosis

Clathrin

Light chain
Heavy chain

ER = resident protein
Cargo proteins
Integral membrane protein
of the targeting machinery
Coat proteins of COPII vesicles
Coat proteins of COPI vesicles

Figure 4.21
(A) Proposed roles for COPI and COPII in vesicle traffic between
the ER and *cis*-Golgi. COPII vesicles are shown participating in anterograde transport, moving cargo molecules and intrinsic membrane proteins from the ER to the Golgi. COPI vesicles are shown
mediating a cycle of retrograde transport that returns membrane
proteins and escaped ER residents (not shown) to the ER. (B) Components of the protein coats of different types of transport vesicles.

CCVs have coats of clathrin triskelions and adapter proteins (AP)
that form the bridge between the clathrin coat and the membrane.
COPI/coatomer vesicles have coats consisting of seven different
proteins. The coats of COPII vesicles contain five different proteins.
The names shown are those used for yeast and correspond to various *sec* mutants.

---

coats can be distinguished on the basis of
their appearance in electron micrographs.
A third class of vesicles, which has a distinctive spiked appearance, has protein
coats of clathrin and adaptins (Fig. 4.21B).
These CCVs are involved in endocytosis
(see Section 4.8) and also function in transport from the *trans*-Golgi to other compartments. Three heavy chains (180 kDa) and
three light chains (35 to 40 kDa) form a
clathrin triskelion (three-legged structure).
Other types of transport vesicles, such as
COPI and COPII, are collectively referred
to as non-CCVs. Their surfaces appear
smooth in electron micrographs, and their
coats contain numerous different proteins
(Fig. 4.21B).

4.5.9 Vesicle budding can be studied in vitro.

In yeast, COPII vesicles bud from the ER in
a process that requires at least six proteins
and GTP as a source of energy. The proteins
involved include an ER integral membrane
protein, Sec12p, and five cytosolic proteins:
the small GTP-binding protein Sar1p (a
member of the Rab family; see Box 4.6) and
two heterodimer complexes made up of
Sec23p-Sec24p and Sec13p-Sec31p. Sec12p,
which functions as a guanine nucleotide exchange factor (see Box 4.6) lies in the membrane but has its N-terminal domain in the
cytosol. This domain recruits Sar1p to the
membrane and catalyzes a nucleotide

Box 4.6

GTP-binding proteins participate in vesicular transport.

Cells use GTP-binding proteins and GTP hydrolysis as molecular switches in a variety of important processes that must be regulated precisely. A number of different families of GTP-binding proteins have been identified in all higher organisms, and one such family, the Rab family, functions in vesicle traffic. The different Rab proteins share a similar size (about 220 amino acids) and structure, but each functions at a different step in the secretory and endocytotic pathways.

All Rab proteins are GTPases: They bind GTP and hydrolyze it to GDP and inorganic phosphate. Rab proteins exist in two states: an active form, which binds GTP, and an inactive form, which binds GDP. Although they demonstrate a low level of intrinsic GTPase activity, Rab proteins must interact with another protein, GAP (GTPase activating protein), to hydrolyze GTP effectively. To return to the active state following GTP hydrolysis, Rab proteins exchange GDP for GTP; this reaction requires yet another protein, the guanine nucleotide exchange factor (GEF). Alternatively, Rab can be maintained in the inactive state by yet another protein, GDP dissociation inhibitor (GDI). Thus, the cycle of Rab activation and inactivation involves at least three proteins in addition to Rab itself.

Some active Rab-GTP is associated with membranes, while a pool of inactive Rab-GDP is present in the cytosol. The exchange of GDP for GTP results in a conformational change that exposes an isoprenoid lipid anchor attached at the C-terminal cysteine residue. This hydrophobic tail allows Rab to bind to membranes. An additional component, Rab escort protein (REP), binds the GDP-bound form of both unprenylated and prenylated Rab proteins, inhibits GDP release, and delivers Rab protein directly to the membrane following prenylation.

Rab and its associated proteins function in vesicle targeting. As a vesicle is budding from a donor compartment, the exposed lipid anchor of Rab-GTP can bind to a receptor protein on the vesicle membrane. The presence of the Rab-GTP on the membrane allows other proteins to be recruited from the cytosol and form a complex with v-SNARE, an integral membrane protein of the vesicle.

After budding is complete, the vesicle is able to dock at a target membrane through an interaction between the v-SNARE and a t-SNARE, an integral membrane protein of the target membrane. During vesicle docking, a Sec1p homolog is displaced from the t-SNARE, possibly through the action of Rab-GTPase. Docking is followed by the release of the cytosolic proteins, including Rab. GTP hydrolysis takes place just before Rab is released and the resulting conformational change facilitates the efficient dissociation of Rab-GDP from the membrane. Once docked, the coiled-coil motifs of the v- and t-SNAREs interact in a head-to-head manner, drawing the vesicle and target membranes into close contact and possibly driving the fusion of the lipid bilayers. Cargo is delivered when the membranes fuse, while the v-SNARE/t-SNARE complex remains together in the target membrane.

Two general factors that act ubiquitously to activate all SNAREs throughout the cell are the ATPase, NSF (N-ethylmaleimide-sensitive factor) and α-SNAP (soluble NSF attachment factor). The name SNARE, meaning SNAP receptor, is derived from one of the functions of this protein complex. α-SNAP binds the v-SNARE/t-SNARE complex and recruits NSF to it. The ATPase activity of NSF then breaks up the SNARE complex, freeing the v-SNARE and t-SNARE for subsequent fusion events.

The various proteins that participate in the Rab cycle and the proteins that form the vesicle coats have different names depending on the organism (mammals or yeast) and on the particular vesicle transport process. In general, plant biologists who study protein sorting follow the yeast nomenclature.

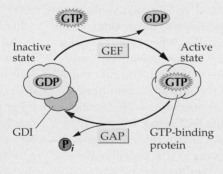

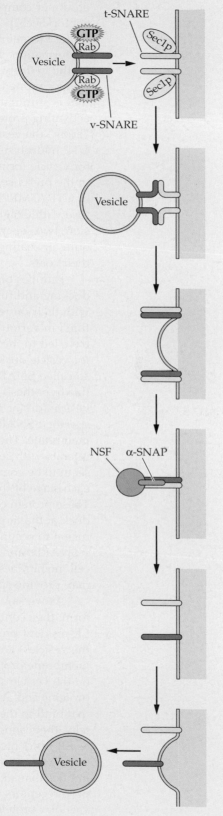

exchange reaction in which Sar1p releases GDP and binds GTP, forming the active Sar1p configuration. Next, the two heterodimer complexes (Sec23p-Sec24p and Sec13p-Sec31p; see Fig. 4.21B) are recruited to the ER membrane surface, initiating vesicle formation. This process is accompanied by the hydrolysis of GTP, either during budding or immediately thereafter, hydrolysis that is stimulated by Sec23p, a GTPase-activating protein (see Box 4.6). This last step "arms" the vesicle for docking and fusion. GTP hydrolysis is not absolutely required for vesicle formation, but vesicles that form in the presence of a nonhydrolyzable analog of GTP (GMPPNP) cannot be targeted to or fuse with Golgi cisternae. Docking and fusion, two separate processes, appear to require uncoating; unarmed vesicles retain their coats.

Families of proteins involved in the docking and fusion of transport vesicles with their target membranes have been identified in various organisms. These studies have led to the proposal of a general model for vesicle docking on target membranes, the so-called SNARE hypothesis. According to this hypothesis, a specific v-SNARE in the vesicle (donor membrane) interacts with a specific t-SNARE in the target or acceptor membrane. The SNARE complex may also interact with certain soluble factors that seem to be common components found at each stage of the pathway (see Box 4.6). These protein complexes enable a vesicle to dock at the target membrane and permit fusion to occur. Different isoforms of the v-SNAREs and t-SNAREs reside in various cell membranes and vesicles, where they may provide specificity to the docking event.

As we saw earlier, when COPII vesicles form, they contain cargo proteins but leave ER-resident proteins behind. The ER (and the vesicles) are thought to contain integral membrane protein receptors with domains on the vesicle side that recognize the cargo proteins and also contain cytosolic domains that bind to the vesicle coat proteins. Proteins that cannot bind to the receptor, such as ER residents, cannot be selected for transport to the next compartment. This model implies that, at least for ER-to-Golgi transport, receptors must exist for secreted and vacuolar proteins and that transport at this stage is not by bulk flow.

The model discussed above has been developed from studies on yeast and mammalian cells. A convenient way to hunt for plant homologs of targeting machinery components already identified in other organisms is to search the Expressed Sequence Tag (EST) databases for sequences that share identity or homology with genes known to encode targeting machinery proteins. A promising EST can identify or be used to isolate a full-length plant cDNA that can then be tested to determine whether the protein it encodes will complement a yeast mutant in which the gene targeting a known protein has been knocked out. Such experiments led to the successful identification of several components of the protein-targeting machinery in plants, including AtERD2p, the KDEL receptor (see Section 4.5.7).

Alternatively, functional complementation of available yeast mutants in a plant expression library can be used directly to isolate the gene of interest. With this approach, several other components of the secretory machinery, such as AtPEP12p, AtSEC12p, and AtSAR1p, have been isolated from *Arabidopsis*.

4.6 Vacuolar targeting and secretion

4.6.1 Transport to vacuoles can occur by at least two different pathways: vesicle fusion and vesicle autophagy.

The secretory system delivers proteins to the plasma membrane for secretion as well as to the tonoplast for incorporation into the vacuole. Much of the early evidence about the role of the Golgi and Golgi-derived vesicles in the delivery of vacuolar proteins has come from the study of storage parenchyma cells in developing legume seeds (see Box 4.4). In developing legume seeds and a few other plant tissues, a large proportion of the newly synthesized proteins enters the secretory system, to be delivered to protein storage vacuoles. These vacuoles appear to arise de novo, possibly by the fusion of Golgi-derived vesicles, as the storage parenchyma cells enlarge. The transport of protein from either the Golgi cisternae or the TGN to the vacuoles occurs by electron-dense protein-filled vesicles, which can be readily observed under the electron microscope (see Box 4.4).

These vesicles do not have a clathrin coat and do not contain the cargo receptor BP-80 (Fig. 4.22A).

Transport to the larger lytic vacuoles of plant cells is also mediated by vesicles. These CCVs, which contain BP-80, leave the TGN and transport their cargo to the prevacuolar compartment. From there the cargo moves to the vacuoles. How this transport occurs is not yet known, but it is probably mediated by vesicles. Most of the recent discoveries about vacuolar protein targeting (see Sections 4.6.2 and following) pertain to this transport route (Fig. 4.22B).

A second mechanism of vacuolar delivery, one that bypasses the Golgi, has been demonstrated for nonglycoprotein seed storage proteins in developing cucumber seeds, and for prolamins in the endosperm of certain cereals (see Box 4.4). Endosperm cells of cereal grains have modified ER regions, called protein bodies, that contain prolamin, a storage protein, surrounded by an ER membrane. These cells thus have two morphologically distinct rough ER membranes: cisternal ER and protein body ER. Endosperm cells also contain a globulin storage protein, glutelin, which accumulates in vacuoles. Evidence suggests that glutelin and prolamin transcripts are not randomly associated with particular ER types. Hybridization analysis using thin sections of endosperm tissue and subcellular fractions enriched in the two ER types has confirmed that the prolamin mRNAs are preferentially associated with protein body ER and the glutelin mRNAs with cisternal ER. The molecular basis for the segregation of prolamin transcripts to the surface of the protein body ER is not yet known.

The two types of storage proteins seem to utilize different vacuolar delivery mechanisms. Globulins are synthesized on rough ER and transported to the vacuole by way of the Golgi. By contrast, wheat and rice prolamins are usually sequestered in the ER-derived protein bodies but also appear in the vacuoles. Is this protein delivery to the vacuole also Golgi-mediated, or is there a transport route that bypasses the Golgi? Recent work demonstrates that the latter is likely, as prolamin and BiP, ordinarily an ER resident, are present both in the protein bodies linked to the ER and in the vacuole. Perhaps ER-derived protein bodies are taken up directly

into vacuoles, either by membrane fusion or, more likely, by engulfment akin to autophagy (Figs. 4.22C and 4.23). This novel pathway to vacuoles may be confined to developing seeds, in which rates of protein synthesis are high, but the mechanism of this Golgi bypass is not yet well understood. In other species (e.g., maize and sorghum), the ER-derived protein bodies persist in the cytoplasm and do not enter the vacuoles by autophagy.

(A) Transport to PSV in dense vesicles

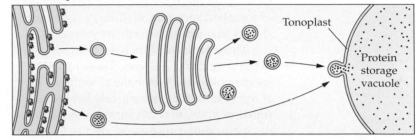

(B) Transport in CCV to lytic vacuoles

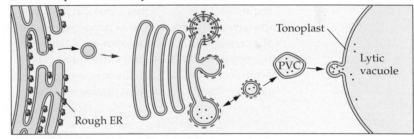

(C) Transport of prolamins by autophagy

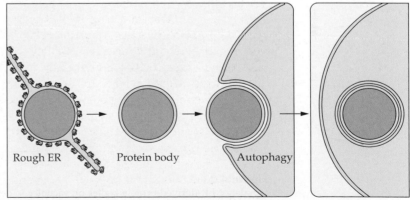

Figure 4.22
Comparison of different pathways for the delivery of storage proteins to vacuoles in seeds. (A) Golgi-mediated pathway for the delivery of storage proteins to protein storage vacuoles (PSVs). Dense protein-filled vesicles bud off the rough ER and the cisternae of the Golgi. (B) Golgi-mediated pathway in which CCVs bud off the TGN and transfer proteins to the prevacuolar compartment (PVC) before transport to the lytic vacuole. (C) ER-derived protein bodies filled with prolamins are autophaged by vacuoles in the endosperm cells of some cereals.

4.6.2 Targeting to the vacuole depends on a short vacuolar sorting signal.

Protein transport to the vacuole usually occurs by way of the Golgi although, as discussed above, this is not always the case. Delivery of proteins to the vacuole requires specific vacuolar sorting signals (VSSs) as well as targeting machinery. Proteins lacking these signals follow the default pathway, which leads to secretion at the cell surface. There are many similarities between plants and yeast with respect to vacuolar targeting, but plant VSSs do not function properly in yeast, even though some other components of the plant secretory machinery have been shown to operate correctly in yeast cells.

Three types of VSSs have been identified in plant vacuolar proteins. These include short peptides, either at the C terminus or at the N terminus immediately behind the signal peptide, that are removed as part of the maturation process and surface loops of undetermined size that remain as part of the mature protein (Fig. 4.24A). Polypeptide domains that are lost when a protein matures are generally referred to as propeptides, and the terminal VSSs are known as N-terminal propeptides (NTPPs) and C-terminal propeptides (CTPPs). Proteolytic enzymes remove these propeptides during or after transport to the vacuole. Although some common sequence elements occur in NTPPs, the three types of VSSs do not appear to share any specific amino acid sequence motif responsible for vacuolar targeting (Fig. 4.24B). Extensive site-directed mutagenesis of CTPP sequences has revealed which features are essential for correct function. Because these studies indicate that CTPPs interact with components of the sorting machinery, the sorting machinery presumably recognizes some structural feature or features of the CTPP.

NTPPs and CTPPs are both necessary and sufficient for vacuolar targeting. If a gene construct is introduced into a plant cell and the CTPP or NTPP is absent from the coding sequence, then the resulting protein is secreted (default pathway; Fig. 4.25). Conversely, a CTPP or NTPP added to a nonvacuolar protein is sufficient to direct that protein to the vacuole. These properties have not been demonstrated for the surface loops appearing to function as VSSs in proteins that lack NTPPs or CTPPs.

It is interesting to compare the location of VSSs in plant proteins with the location of similar targeting information in yeast and mammalian proteins. In yeast, only two vacuolar proteins have been analyzed and in both, carboxypeptidase Y and proteinase A,

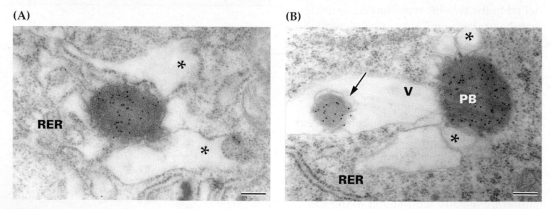

(A)

(B)

Figure 4.23

Uptake of ER-derived protein bodies by vacuoles in developing wheat endosperm. Electron micrographs of developing wheat endosperm cells at 16 days after flowering show two progressive stages of the internalization of protein bodies into vacuoles by an autophagy-like mechanism. (A) Electron lucent vesicles (indicated by asterisks), which are apparently small vacuoles (also called provacuoles), attach to the surface of protein body surrounded by rough ER (RER). The provacuole membrane is probably formed directly from the ER membrane. (B) Two different protein bodies in different stages of entering the provacuoles. The right protein body (PB) is surrounded partially by rough ER-enriched cytoplasm and partially by the electron lucent provacuoles (indicated by asterisks). The left protein body is already inside the small vacuole and is also surrounded by a loosely attached membrane (arrow). Sections were immunogold-labeled with anti-wheat γ-gliadin serum. Bars: 0.25 μm.

(A)

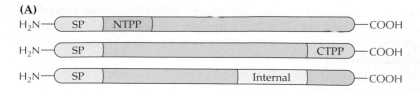

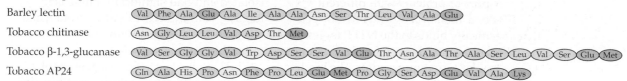

Figure 4.24
Vacuolar sorting signals.
(A) Location of signals
along the polypeptide.
(B) Amino acid se-
quences of vacuolar
sorting signals.

(B)

N-terminal propeptides (NTPP)

Sweet potato sporamin — His Ser Arg Phe Asn Pro Ile Arg Leu Pro Thr Thr His Glu Pro Ala

Barley aleurain — Ser Ser Ser Ser Phe Ala Asp Ser Asn Pro Ile Arg Pro Val Thr Asp Arg Ala Ala Ser Thr Leu Glu

C-terminal propeptides (CTPP)

Barley lectin — Val Phe Ala Glu Ala Ile Ala Ala Asn Ser Thr Leu Val Ala Glu

Tobacco chitinase — Asn Gly Leu Leu Val Asp Thr Met

Tobacco β-1,3-glucanase — Val Ser Gly Gly Val Trp Asp Ser Ser Val Glu Thr Asn Ala Thr Ala Ser Leu Val Ser Glu Met

Tobacco AP24 — Gln Ala His Pro Asn Phe Pro Leu Glu Met Pro Gly Ser Asp Glu Val Ala Lys

---

the vacuolar targeting information is in an N-terminal domain immediately behind the signal peptide. In mammalian lysosomal hydrolases, lysosomal targeting information is found in special phosphorylated glycans (mannose 6-phosphate residues) as well as in polypeptide domains (possibly similar to the surface loops found in some plant proteins). The knowledge that glycans can carry important targeting information has led investigators to study the role of glycans in the transport and targeting of plant vacuolar glycoproteins. When glycosylation of plant vacuolar glycoproteins is prevented, either by specific inhibitors or by abolishing glycosylation sites through site-directed mutagenesis, the unglycosylated proteins are still delivered to the vacuoles. Thus, glycans are not required for correct vacuolar targeting.

4.6.3 Homologs of yeast and mammalian vacuolar sorting sequences and SNAREs are present in plants.

Proteins that end up in vacuoles or lysosomes are sorted from secreted proteins in the TGN because they bind to transmembrane receptors that direct them to the appropriate compartment. These receptors bind specifically to the peptide domains (or glycans, in mammalian hydrolases) that function as sorting signals. Receptors of this type have been identified in yeast and in

	Protoplasts		Media	
Chase time (h)	0	12	0	12

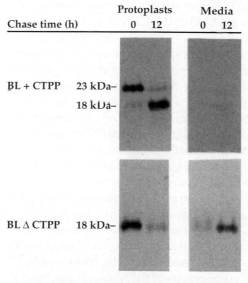

BL + CTPP   23 kDa–
             18 kDa–

BL Δ CTPP   18 kDa–

Barley lectin (BL) is an 18-kDa protein found in the vacuoles of barley seedlings. It is synthesized as a 23-kDa precursor, here called BL + CTPP, that is proteolytically processed with the loss of a 5-kDa C-terminal propeptide (CTPP).

By introducing a stop codon at the end of the sequence that encodes the 18-kDa polypeptide, here called BL Δ CTPP, it is possible to allow cells to synthesize a protein that lacks the CTPP.

Figure 4.25
Pulse–chase labeling experiment showing the fate of a protein with and without its vacuolar sorting signal. Protoplasts were obtained from the leaves of transgenic *Arabidopsis* plants expressing either native barley lectin containing a CTPP (BL + CTPP) or a truncated barley lectin lacking the vacuolar sorting sequence (BL Δ CTPP). The protoplasts were labeled for 6 hours with radioactive methionine and cysteine (pulse), the label was removed, and the incubation was continued for 12 hours (chase). During the chase period, no new radioactive proteins are synthesized, and the previously synthesized radioactive proteins move to their final destination. Samples of the cells and the incubation medium were taken at the beginning of the chase (0 hours) and at the end (12 hours). The BL was purified with antibodies and analyzed by SDS-PAGE and fluorography. At the start of the chase, the majority of the BL + CTPP had the correct size (23 kDa) and a small amount had been processed to the 18-kDa form. At the end of the chase, this situation was reversed. Relatively little BL + CTPP was accumulated in the culture medium. The situation was different for the BL Δ CTPP protein. It also had the correct size at the end of the chase (18 kDa), but during the chase it was secreted into the medium and did not accumulate in the cells. This experiment shows that the CTPP is necessary to keep BL in the cells; other experiments not described here show that CTPP causes BL to accumulate in vacuoles.

mammals. These receptors move with their cargo in anterograde vesicles from the TGN to the endosomes and are returned empty in retrograde vesicles. By analogy with these systems and by means of experiments showing that the vacuolar sorting process is saturable, plant cells are hypothesized also to have vacuolar sorting receptors that recognize VSSs. We do not yet know whether proteins with different VSSs are transported to the vacuole in the same vesicle or whether various classes of vesicles are responsible for the transport of proteins with different VSSs.

BP-80, a membrane protein from pea that specifically binds to the NTPP targeting signal, has recently been identified as a possible vacuolar targeting receptor. The binding specificity of BP-80 has been examined by using affinity columns that contain peptides corresponding to vacuolar sorting domains of several proteins. Although BP-80 binds to NTPPs from various vacuolar proteins, it does not bind to the CTPP signal peptide. BP-80 may thus mediate the sorting of a subset of vacuolar proteins, but it is probably not involved in the transport of all proteins to the vacuole. BP-80 is encoded by a multigene family in pea and *Arabidopsis*. Several BP-80 homologs are localized to Golgi and are found in highly purified CCVs. It is not yet known whether the protein cycles between two compartments, as would be expected for a transport receptor.

In addition to sorting receptors that recognize VSSs, correct sorting also depends on other membrane-associated proteins that function in vesicle budding and fusion, such as v-SNAREs and t-SNAREs. Interestingly, such components of the transport machinery are conserved among yeast, plants, and mammals, even though there is no similarity in the VSSs of the transported proteins. Proteins implicated in vesicle docking and fusion at the presynaptic membranes of neuronal cells have homologs in yeast that function in vesicle trafficking through the secretory pathway. The *pep12* mutant of yeast, a *vps* mutant (see Box 4.5), is defective in sorting proteins to the vacuole. The *PEP12* gene of yeast encodes a protein that belongs to the t-SNARE family. Based on this finding, the protein (Pep12p) was thought probably to function in vesicle transport between the TGN and the vacuole of yeast. In a mutant with a defective *pep12* gene, the vacuolar protease car-

boxypeptidase Y (CPY) is secreted instead of being transported to the vacuole. Because CPY ordinarily is activated by proteolysis in the vacuole, the secreted CPY is inactive. This phenotype of the mutant yeast has been utilized in a screen for an *Arabidopsis thaliana* homolog of PEP12p. The yeast *pep12* mutant was transformed with material from an *Arabidopsis* cDNA library, and colonies were screened for the restoration of vacuolar CPY activity. A cDNA *(AtPEP12)* that complements the yeast mutant (Fig. 4.26) was isolated and had significant amino acid identity with the yeast Pep12p protein and with other t-SNAREs. With use of biochemical and immunocytochemical approaches, AtPEP12p was localized to a prevacuolar compartment through which some proteins move en route to the vacuole. AtPEP12p is thus a component of the plant vacuolar transport machinery.

4.6.4 Phospholipid modification may play a part in vesicle budding.

Vesicle budding may require not only recruitment of cytosolic proteins to the membrane (see above for ER-to-Golgi transport) but also modification of phospholipids. In both mammals and yeast, phosphatidylinositol-3-kinase (PI-3-kinase) plays an important role in many membrane-associated activities. Yeast mutants that lack PI-3-kinase are unable to deliver proteins to the vacuole. The activity of this enzyme can be inhibited by micromolar concentrations of the fungal metabolite wortmannin, thus providing a tool to study the importance of PI-3-kinase for transport. The cDNA for this enzyme has been cloned from *Arabidopsis*, and the polypeptide has considerable amino acid identity with the yeast enzyme. To determine whether PI-3-kinase is required for vacuolar protein delivery in plants, investigators treated with wortmannin suspension cultures of tobacco cells that express recombinant vacuolar proteins with either a CTPP or an NTPP. The inhibitor caused missorting of a CTPP vacuolar protein but not of an NTPP vacuolar protein. This finding indicates that CTPP- and NTPP-mediated transports occur by different mechanisms. Surprisingly, missorting of this CTPP vacuolar protein was not caused by the inhibition

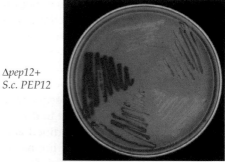

Δpep12

Δpep12+
S.c. PEP12

Δpep12+
A.t. cDNA2

Δpep12+ unrelated
A.t. cDNA

Δpep12+
A.t. cDNA1

Figure 4.26
Complementation of a yeast *pep12* mutant by *Arabidopsis thaliana* cDNAs. Mutations in the yeast *PEP12* gene cause the secretion of protease precursors such as carboxypeptidase Y (CPY), which ordinarily are located in the vacuole. Because secreted CPY is inactive, restoration of CPY activity provides a convenient assay for complementation of the *pep12* mutant. The *pep12* mutant was transformed with the yeast *(Saccharomyces cerevisiae; S.c.) PEP12* gene or *A. thaliana (A.t.)* cDNA yeast expression library. The resulting colonies were transferred to plates containing galactose to induce GAL1 promoter-driven expression of the cDNAs. CPY activity was assayed with the *N*-acetyl-DL-phenylalanine naphthyl ester (APE) overlay test. In this assay, colonies are overlaid with agar containing dimethylformamide, which permeabilizes the cells, and APE, a substrate for CPY. The α-naphthol produced by CPY reacts with Fast Garnet, producing an insoluble red dye. Colonies containing active CPY are therefore red, whereas colonies lacking CPY activity are yellow.

of PI-3-kinase but resulted from inhibition of the synthesis of PI-4-phosphate and major phospholipids. Therefore, this experiment does not provide evidence for the involvement of PI-3-kinase in protein sorting. It does, however, support the conclusion that the continued synthesis of specific phospholipids is required for the vesicular delivery of certain types of proteins.

4.6.5 Details of tonoplast targeting remain unclear.

How are membrane proteins transported to the tonoplast? Despite successful identification of VSSs in several soluble vacuolar proteins, few plant biologists are studying the sorting of vacuolar membrane proteins, and the signals and mechanisms by which targeting to the tonoplast is achieved are not well understood. α-Tonoplast intrinsic protein (α-TIP), a tonoplast protein with six membrane-spanning domains, travels through the secretory pathway to the vacuole. Fusion of the sixth transmembrane domain and the cytoplasmic tail of α-TIP to the reporter protein phosphinothricin acetyltransferase is sufficient to direct the reporter to the tonoplast of lytic vacuoles in transgenic tobacco plants. Because an α-TIP deletion mutant lacking the cytoplasmic tail is correctly targeted to the tonoplast, the sixth transmembrane domain is apparently sufficient for tonoplast targeting of the reporter protein. However, it is not clear whether this domain contains specific information or whether vacuolar targeting is the default

pathway for membrane proteins entering the secretory system in plant cells and hence requires no information.

The mechanisms of transport of a vacuolar membrane protein (α-TIP) and a soluble vacuolar protein (phytohemagglutinin) have been compared in transgenic tobacco protoplasts through use of the vesicle transport inhibitors monensin and brefeldin A. Both of these inhibitors prevent the transport of phytohemagglutinin to the vacuole. However, transport of α-TIP is not affected by either monensin or brefeldin A; therefore, some vesicle transport must continue in the presence of these compounds. These data suggest that phytohemagglutinin and α-TIP transport occur by different mechanisms and that these two proteins are probably carried in different types of vesicles.

4.6.6 The secretory system transports cargo proteins from the *trans*-Golgi to the plasma membrane—the default destination—by way of vesicles.

The sorting of proteins destined for the vacuole or for the plasma membrane occurs in the TGN. For soluble proteins, secretion is a default destination: Proteins lacking a VSS or the KDEL tail are transported to the cell surface in vesicles. Vesicles carrying proteins targeted for secretion bud from the TGN and travel to the cell surface, where they fuse with the plasma membrane. In yeast, components of the transport machinery acting between the Golgi apparatus and the plasma membrane have been isolated by genetic and biochemical means. A screen for secretion-defective (*sec*) mutants (see Box 4.5) resulted in the identification of several genes that encode proteins required specifically in this late stage of the secretory pathway. Some of these proteins form part of a large, multisubunit protein complex located in the cytoplasm and associated with the plasma membrane; however, the precise function of these proteins is not yet known.

As discussed earlier, studies in various organisms, particularly in mammalian neuronal cells and yeast, have led to the proposal of a SNARE-receptor hypothesis. The exact role of the SNAREs is still being debated, and the precise mechanism of membrane fusion remains unknown. Vesicle fusion with the plasma membrane is finely regulated in animal neurons, whereas in yeast, docking and fusion seem to occur continuously and appear to be unregulated.

We do not know whether transport of membrane proteins to the plasma membrane requires targeting information. Several plant proteins have been expressed in yeast and are found to be targeted to the plasma membrane correctly. Although most proteins found at the plasma membrane arrive there by way of the ER and Golgi, there are some exceptions to this rule. For example, a sucrose-binding protein found on the outer face of the plasma membrane in soybean cotyledons has no signal peptide, appears to be synthesized on free ribosomes in vivo, and is not transported into microsomes in vitro. Thus, this protein may be transported to the plasma membrane by a yet-to-be-discovered pathway.

4.7 Protein modification in the Golgi

4.7.1 Complex N-linked glycans are derived from high-mannose N-linked glycans during processing in the Golgi.

Many secretory and integral membrane proteins are glycoproteins with short glycans attached to specific amino acid residues. Glycans attached to the amide nitrogen of asparagine (Asn) are referred to as N-linked glycans, and those attached to the hydroxyl of threonine, serine, or hydroxyproline are called O-linked glycans. Two major types of N-linked glycans occur on plant glycoproteins: high-mannose glycans, with the structure $(Man)_{6-9}(GlcNAc)_2$ (see Section 4.5.6), and complex glycans, with the core structure $Xylose(Man)_3Fucose(GlcNAc)_2$. The latter are derived from the former by modifications that take place in the Golgi and involve glycosidases and glycosyltransferases. Glycans initially attached to Asn residues in the ER are in the high-mannose form. Subsequently, some of these glycans are modified in the Golgi. Why certain glycans are modified and others are not remains unclear, but important factors include the manner in which a glycan is displayed on the surface of the protein and the enzymatic environment in the Golgi apparatus. Mammalian glycoproteins also have high-mannose and complex glycans, but mammalian complex glycans are larger and very different from those found in plants (Fig. 4.27). Typically, plant complex glycans have α-1,3 fucose and β-1,2 xylose residues attached to the proximal GlcNAc and the core mannose residue, respectively. Because residues with these linkages are not found on mammalian glycoproteins, plant glycoproteins may be unusually immunogenic when injected into mammals.

Conversion of a high-mannose glycan to a complex glycan starts with the removal of four mannose residues by mannosidase I (Fig. 4.28). This is followed by the addition of a single terminal GlcNAc by GlcNAc transferase I and the removal of two more mannose residues by mannosidase II. Subsequently, fucose, xylose, and another terminal GlcNAc are added to create a complex glycan with two terminal GlcNAc residues. One galactose and one fucose residue are then usually attached to the two terminal GlcNAc residues by Golgi enzymes, and proteins with these large complex glycans are abundant in the plasma membrane and the cell wall. Vacuolar glycoproteins contain small complex glycans consisting only of $Xylose(Man)_3Fucose(GlcNAc)_2$. It is likely that the peripheral sugars are removed by vacuolar hydrolases, but this remains to be shown.

Mutant plants that lack GlcNAc transferase I are unable to convert high-mannose glycans to complex glycans. These plants can still complete their life cycles when raised under normal growth conditions, which suggests that complex glycans are not essential for growth and development.

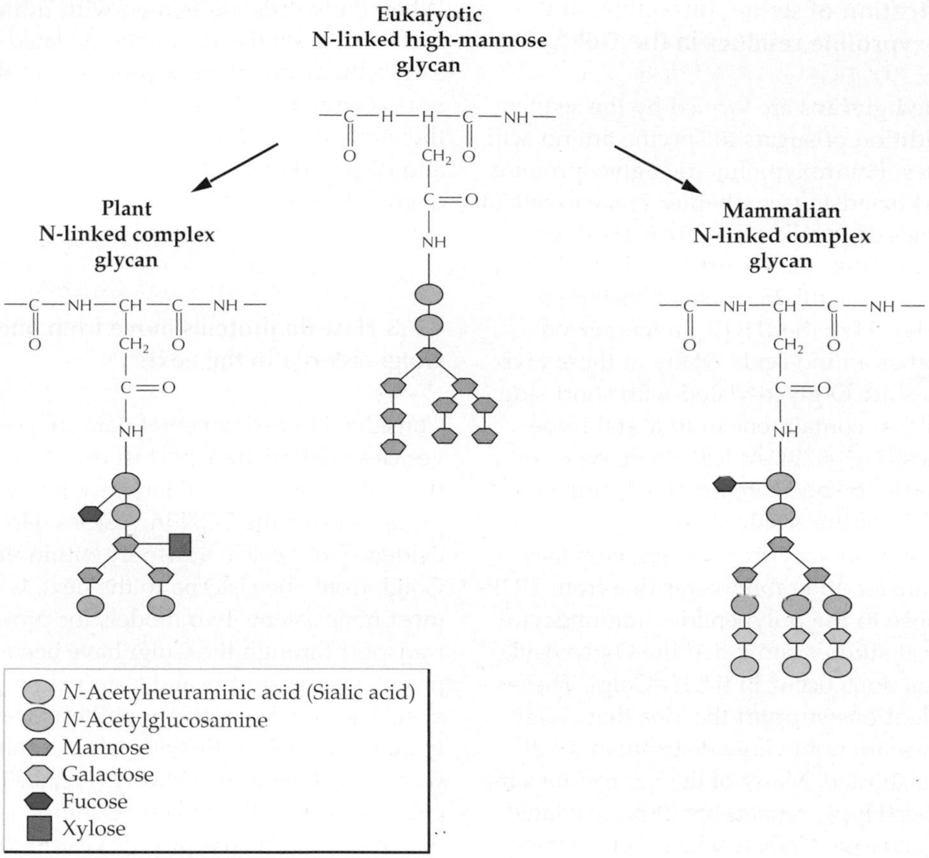

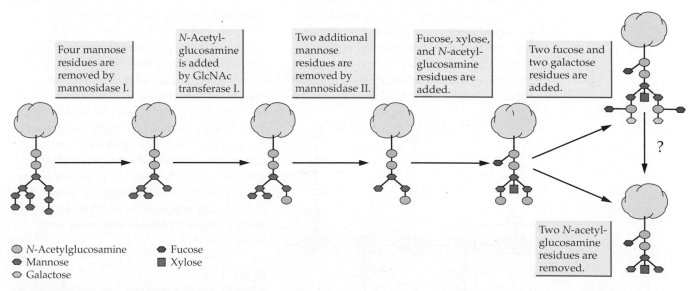

Eukaryotic N-linked high-mannose glycan

Plant N-linked complex glycan

Mammalian N-linked complex glycan

- N-Acetylneuraminic acid (Sialic acid)
- N-Acetylglucosamine
- Mannose
- Galactose
- Fucose
- Xylose

Figure 4.27
Structures of typical asparagine-linked glycans found on plant and mammalian glycoproteins. In complex glycans of plant glycoproteins, fucose is attached by an α-1,3 linkage, and in mammalian complex glycans fucose is attached by an α-1,6 linkage. The xylose residue, attached by a β-1,2 linkage to the core mannose, is unusually antigenic and is also found in the glycoproteins of invertebrates.

Four mannose residues are removed by mannosidase I.

N-Acetyl-glucosamine is added by GlcNAc transferase I.

Two additional mannose residues are removed by mannosidase II.

Fucose, xylose, and N-acetyl-glucosamine residues are added.

Two fucose and two galactose residues are added.

?

Two N-acetyl-glucosamine residues are removed.

- N-Acetylglucosamine
- Mannose
- Galactose
- Fucose
- Xylose

Figure 4.28
Processing of N-linked glycans by enzymes in the Golgi apparatus. As glycoproteins with N-linked glycans arrive in the Golgi and are transported through the Golgi stack, the glycans are modified by an ordered sequence of reactions. Many glycans are modified by the attachment of galactose and fucose residues to the terminal N-acetylglucosamine residues. Such glycans are found on extracellular glycoproteins. Glycans without N-acetylglucosamine residues are found in vacuoles.

4.7.2 Polypeptides undergo O-linked modification of serine, threonine, and hydroxyproline residues in the Golgi.

O-linked glycans are formed by the sequential addition of sugars to specific amino acid residues. Hydroxyproline-rich glycoproteins such as extensin (see Chapter 2) are excellent examples of proteins that are extensively O-glycosylated. Carrot extensin is a 33-kDa polypeptide with 25 repeats of Ser-Hyp-Hyp-Hyp-Hyp [Ser(Hyp)$_4$] interspersed with other amino acids. Many of these Hyp residues are O-glycosylated with short side chains that contain one to four arabinose residues (Fig. 4.29). At least three types of glycosidic linkages are involved; thus, the synthesis of these side chains requires three different arabinosyltransferases, enzymes that transfer an arabinose residue from UDP-arabinose to the polypeptide. Immunocytochemical studies show that the O-glycosylation reactions occur in the *cis*-Golgi. These investigations support the idea that Golgi enzymes are not evenly distributed in all Golgi cisternae. Many of the Ser residues in these Ser(Hyp)$_4$ repeats are O-glycosylated with galactose. Galactosylation of Ser residues has not been studied in plants but, by analogy to mammalian cells, this process is thought to occur in the Golgi complex.

One of the major functions of glycans on glycoproteins is to aid in protein folding and stabilization against proteolytic breakdown. Glycosylation can be prevented by certain drugs; for example, tunicamycin is an effective inhibitor of N-linked glycosylation. Suspension-cultured tobacco cells ordinarily synthesize large amounts of an extracellular invertase that accumulates in the cell walls. When these cells are treated with tunicamycin, they synthesize unglycosylated extracellular invertase, a protein that does not accumulate. The invertase polypeptide, therefore, must either be improperly folded and disposed of by the ER-based protein degradation system or secreted and immediately degraded after arrival in the cell wall.

4.7.3 How do proteins move from one Golgi cisterna to the next?

Considerable experimental data suggest that vesicles mediate transport from the ER to the Golgi complex. Similarly, we know that proteins leave the TGN in vesicles. However, evidence for vesicle transport within the Golgi, from one cisterna to the next, is almost nonexistent. Two models for protein transport through the Golgi have been proposed: vesicle shuttle and cisternal progression. The vesicle shuttle hypothesis was widely accepted until 1999, when doubts were raised about its validity (Fig. 4.30). One problem with the vesicle shuttle model is that no specific receptor or vesicle coat proteins have been identified for this portion of the pathway. Alternatively, cisternal progression postulates new cisternae forming at the cis face and old cisternae breaking up at the trans face—a mechanism known to function in Chrysophycean algae that secrete cell-wall scales. In these cells, a single polysaccharide/glycoprotein scale fills an entire cisterna, and the scale gradually moves through the Golgi complex and matures. There is good evidence that in certain mammalian cells, the *cis*-, *medial*-, and *trans*-Golgi cisternae have different glycan-modifying enzymes. To make such an observation compatible with the cisternal progression model, we have to postulate either efficient retrieval and retrograde movement of the Golgi enzymes as the cisterna matures, or rapid synthesis and breakdown of the enzymes.

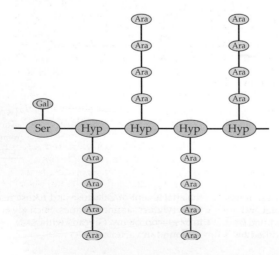

Figure 4.29
Schematic representation of a glycosylated region of extensin, a glycoprotein rich in hydroxyproline (Hyp). Carrot extensin has 25 such repeats. Gal, D-galactose; Ara, arabinose; Ser, serine. Linkages between arabinose residues and between arabinose and hydroxyproline are of four different types (not shown).

4.8 Endocytosis

Just as cells have a vesicle-mediated pathway for protein secretion, there is also a pathway for vesicle-mediated uptake of extracellular macromolecules, the endocytic pathway. Here, the plasma membrane invaginates and small vesicles bud off inside the cell and move to endosomes (see Fig. 4.13). The process is analogous to the formation of coated vesicles described above, and the protein coats of endocytic vesicles consist of clathrin (see Fig. 4.21B). In electron micrographs, these clathrin-coated vesicles appear to have spikes (Fig. 4.31). Recruitment of clathrin from the cytosol requires dynamin, a small GTP-binding protein, and adaptins, protein complexes that contain four different polypeptides (Fig. 4.21B). Adaptins are situated between the membrane and the clathrin coat. They bind clathrin on one side and the cytoplasmic domains of transmembrane receptors on the other. The extracellular domains of these transmembrane receptors interact with the proteins being internalized and determine the specificity of the endocytic process, a process referred to as receptor-mediated endocytosis. After the cargo-laden clathrin-coated vesicle has formed, a 70-kDa chaperone-like protein causes the clathrin-coated vesicle to uncoat in a process that requires ATP. The uncoated vesicle can then fuse with an endosome.

In plant cells, endocytosis was first investigated with electron-dense cationized ferritin particles, because these are readily taken up in an endocytic process and easily visualized with the electron microscope. Recently, immunocytochemical techniques have been used to localize specific proteins, but such studies lack a dynamic dimension. Cells that are incubated with ferritin can be fixed and examined at different intervals to follow the route of ferritin internalization. When ferritin is taken up, it is seen first in small endocytic vesicles and then in organelles that have been identified as multivesicular bodies. These organelles contain acid phosphatase, a typical vacuolar enzyme. After prolonged incubation with ferritin (30 minutes to 3 hours), ferritin is seen in Golgi cisternae and in vacuoles. Such observations suggest that the multivesicular bodies are part of the endosomal

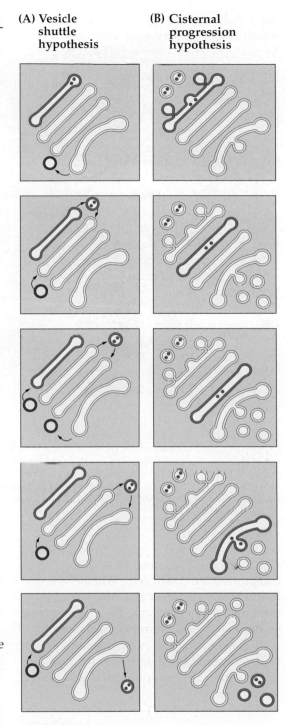

(A) Vesicle shuttle hypothesis

(B) Cisternal progression hypothesis

Figure 4.30
Two hypotheses of how proteins move through the Golgi stack. According to the vesicle shuttle hypothesis (A), the Golgi cisternae are static and proteins move from one cisterna to the next by vesicles, being sorted from resident proteins at each step. Red vesicles move in the anterograde direction, dark blue vesicles in the retrograde direction. According to the cisternal progression hypothesis (B), the entire cisterna moves from cis to trans. New cisternae are added at the cis side, pass through the Golgi stack, and are degraded at the trans side. Retrograde movement of vesicles is not shown.

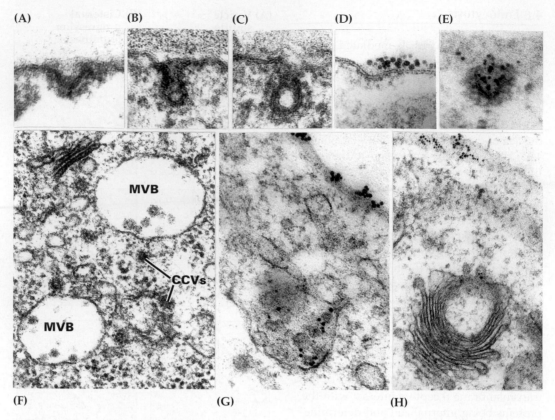

(A)   (B)   (C)   (D)   (E)

(F)   (G)   (H)

Figure 4.31
CCVs participate in endocytosis. (A–C) A coated vesicle is formed from a coated pit. Invagination of the plasma membrane of a cauliflower cell results in the formation of a vesicle (magnification ×200,000). (D, E) *Vicia faba* protoplasts, incubated with concanavalin A linked to colloidal gold. The gold can be seen at the cell surface (D) and in a CCV (E) (×200,000). (F) CCVs in the cytoplasm and in multivesicular bodies (MVBs) of cauliflower cells (×100,000). (G, H) Uptake of concanavalin A-gold into MVB (G) and Golgi stack (H) of a *V. faba* protoplast.

compartment in plant cells. Whether the endosomal compartment has morphological components in addition to the multivesicular bodies is not clear. Immunocytochemistry studies using antibodies to vacuolar sorting receptors should help elucidate the structure and function of the endosomal compartment in plant cells.

Summary

Cells contain numerous metabolic compartments and membranes that have unique sets of proteins. Nearly all of these proteins are synthesized in the cytosol. How do they reach their correct destinations? Proteins have sorting domains that act as address labels; there is a unique label for each destination. These sorting domains are usually short segments of polypeptide that may or may not be removed after the proteins reach their destination.

Many sorting domains interact with proteinaceous pores in membranes. Such pores consist of numerous different polypeptides and are unique to each organelle. Passage of a polypeptide through a pore requires the participation of chaperones and energy in the form of ATP, GTP, or both. Chaperones help keep proteins unfolded in the cytosol before passage through the pore and are required to fold the protein correctly once it has crossed the membrane. All living cells maintain an active secretory pathway having several functions besides protein secretion. Proteins move between the various compartments of this pathway as cargo in vesicles. All proteins that have a signal peptide are translated on rough ER and enter the secretory pathway via translocation into the ER lumen. Correct folding is necessary not only for proper protein function but also for transport. Proteins that are not correctly folded are degraded in the ER. Vesicle formation and sorting are complex processes that involve integral membrane proteins (e.g., receptors) and cytosolic proteins (e.g., vesicle coat proteins). Coat formation is necessary for vesicle budding. Vesicles always uncoat before fusing with their target compartments. Cells make extensive use of highly regulated GTP-binding proteins/GTPases to regulate and provide energy for secretory pathway activity. The secretory pathway targets proteins to the ER, Golgi, TGN, prevacuolar compartment, vacuole, and plasma membrane. Cells also have a vesicle-mediated mechanism for taking up proteins from their surroundings and delivering them to various compartments.

Further Reading

Nuclear Targeting

Gorlich, D. (1997) Nuclear protein import *Curr. Opin. Cell Biol.* 9: 412–419.

Nigg, E. A. (1997) Nucleocytoplasmic transport: signals, mechanism, and regulation. *Nature* 386: 779–787.

Smith, H.M.S., Raikhel, N. V. (1999) Protein targeting to the nuclear pore—what can we learn from plants? *Plant Physiology* 119: 1157–1163.

Protein Translocation

Schatz, G., Dobberstein, B. (1996) Common principles of protein translocation across membranes. *Science* 271: 1519–1526.

Glycosylation

Lerouge, P., Cabaness-Macheteau, M., Rayon, C., Fitchette-Lainé, A.-C., Gomord, V., Faye, L. (1998) N-glycosylation biosynthesis: recent development and future trends. *Plant Mol. Biol.* 38: 31–48.

Sorting in the Secretory System

Battey, N. H., James, N. C., Greenland, A. J., Brownlee, C. (1999) Exocytosis and endocytosis. *Plant Cell* 11: 643–659 [Special Issue].

Herrmann, J., Malkus, P., Schekman, R. (1999) Out of the ER—outfitters, escorts, and guides. *Trends Cell. Biol.* 9: 5–7.

Kirchhausen, T., Bonifacino, J. S., Reizman, H. (1997) Linking cargo to vesicle formation: receptor tail interactions with coat proteins. *Curr. Opin. Cell Biol.* 9: 488–495.

Marty, F. (1999) Plant vacuoles. *Plant Cell* 11: 587–599 [Special Issue].

Pelham, H.R.B. (1998) Getting through the Golgi complex. *Trends Cell. Biol.* 8: 45–49.

Sanderfoot, A. A., Raikhel, N. V. (1999) The specificity of vesicle trafficking: coat proteins and SNAREs. *Plant Cell* 11: 629–641 [Special Issue].

Traub, L. M., Kornfeld, S. (1997) The *trans*-Golgi network: a late secretory sorting station. *Curr. Opin. Cell Biol.* 9: 527–533.

Vitale, A., Denecke, J. (1999) The endoplasmic reticulum: gateway of the secretory system. *Plant Cell* 11: 615–628 [Special Issue].

Chloroplast

Keegstra, K., Cline, K. (1999) Protein import and routing systems of chloroplasts. *Plant Cell* 11: 557–570 [Special Issue].

Mitochondria

Neupert, W. (1997) Protein import into mitochondria. *Annu. Rev. Biochem.* 66: 863–917.

Peroxisome

Subramani, S. (1998) Components involved in peroxisime import, biogenesis, proliferation, turnover, and movement. *Physiol. Rev.* 78: 171–188.

Biochemistry & Molecular Biology of Plants, B. Buchanan, W. Gruissem, R. Jones, Eds.
© 2000, American Society of Plant Physiologists

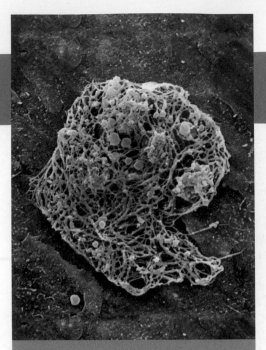

C H A P T E R 5

The Cytoskeleton

Tobias I. Baskin

Introduction

Despite the importance of membrane-bound organelles, as illustrated in the preceding chapters, the eukaryotic cell requires more than a set of defined compartments to function. Eukaryotic cells organize their components spatially, fixing some at defined locations in the cell, moving others to attain optimal positions. The contents of eukaryotic cells (and in some cases, the cells themselves) are mobile. Poking a cell with a fine microneedle stimulates its contents to move vigorously. In the first half of this century, these directed movements were considered *the* unmistakable indicator of life. Amusingly, the term coined to describe this essential cellular property was "irritability."

Spatial organization within the eukaryotic cell and directed movements of the cell or its contents are mediated by the **cytoskeleton,** a network of filamentous protein polymers that permeates the cytosol. The cytoskeleton comprises three major families of proteins: intermediate filaments, actin, and tubulin. In this chapter, each family and its principal functions will be described. Mitosis and cytokinesis will also be discussed, given the paramount role of the cytoskeleton in these processes. The cytoskeleton evolved before plants diverged from animals, and the main features of the cytoskeleton have been conserved in both. Thus, much of the information presented applies to both animals and plants. However, the plant cytoskeleton has evolved unique functions that differ from those in animals, as will also be highlighted in this chapter.

5.1 Introduction to the cytoskeleton

5.1.1 Cells contain a dynamic, filamentous network called the cytoskeleton.

When Robert Hooke viewed thin slices of cork through his microscope, he called the large empty spaces he saw "cells," because they reminded him of the bare cells inhabited by monks. We now know that Hooke saw only the walls of dead cells. Living cells, by contrast, are far from bare, empty spaces. Even the image of a cell with nucleus, chloroplasts, mitochondria, and the extensive endoplasmic reticulum (ER) network all floating free in a more or less clear cytoplasm ignores the true complexity of the cell. Modern microscopes and staining techniques reveal that cells are also packed with a dynamic filamentous network that anchors, guides, and transports myriad macromolecules, supramolecular complexes, and organelles (Fig. 5.1).

5.1.2 The cytoskeleton provides structure and motility and facilitates information flow.

At any one time, a typical cell contains millions of protein molecules engaged in thousands of activities, ranging from biosynthesis to degradation. Few of these proteins function as single polypeptide chains. Instead, they form complexes comprising anywhere from several to many thousand subunits. Moreover, the complexes themselves do not function in isolation. The products of a reaction sequence must find their way from one complex to another, for example, in biochemical pathways or as information flows from the exterior of the cell to the nucleus. All of this miniaturized machinery must withstand intense molecular bombardment from **Brownian motion**—thermal noise—that occurs at physiological temperatures (see Box 5.1). At 25°C, a typical temperature for a plant, a globular protein 5 nm in diameter will be hit by solutes approximately 10^9 times a second.

It should thus come as no surprise that cells have evolved means to organize their teeming internal contents, not only to dampen the effects of thermal noise but also to facilitate reactions of increasing duration and complexity. One such adaptation uses membrane-bound compartments to sequester and concentrate certain components (see Chapter 1). A second such invention is the cytoskeleton.

(A)
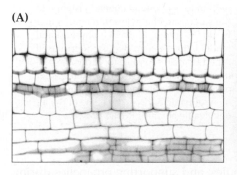

(B)

(C)
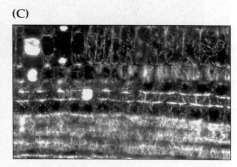

Figure 5.1
Different views of the same cells. A root of *Arabidopsis thaliana* was fixed, embedded, and sectioned at a thickness of 2 μm. (A) The section was stained with a reagent specific for cell walls. This image resembles what Hooke saw when he coined the term "cell." (B) An adjacent section imaged with Nomarksi optics, a technique that makes visible many organelles, such as nuclei, mitochondria, or plastids. These and other organelles were discovered in the 19th and 20th centuries as the light microscope improved. (C) The same section seen in (B), imaged with fluorescence microscopy. The bright fibrous elements are microtubules, a major component of the cytoskeleton. The section was treated with an antibody that recognizes microtubules and then was stained with a fluorochrome-tagged second antibody that binds the first. The cell cytoskeleton was discovered only after further refinements in microscopy and the development of specific molecular reagents.

Robert Brown (1773–1858), pictured here (A) in an 1835 portrait by Pickersgill, discovered the motion that bears his name while pursuing his passion, botany. Even though Brown studied medicine at university and was serving in the army, he had amassed so great a knowledge of botany that Sir Joseph Banks hired him to sail as a ship's naturalist aboard *The Investigator* on a voyage of discovery to Australia. The voyage lasted for years, circumnavigating Australia twice. During that time, Brown collected more than 4,000 species, most of them new to science and many in previously unknown genera. The difficulties he faced were physical and intellectual: *The Investigator* was cramped, damp, and unseaworthy, and the botany of his day had only just begun to classify plants "naturally" (that is, evolutionarily). Nonetheless, Brown's treatment of the Australian flora remains essentially intact today.

Brown's zeal for botany spurred him to become an excellent light microscopist. He realized that traits needed to support a natural classification of plants could be found in the early stages of development, which are microscopic. In Brown's day, the best microscopes had only a single lens, because the theory for properly correcting compound lenses had yet to be developed. Despite their simplicity, single-lens microscopes can reveal even subcellular detail. The micrograph below (B) shows a peel of onion epidermis viewed through a microscope used by Brown. Using dark-field illumination with the substage mirror shows the nuclei clearly against a dark background. From observations such as this, Brown coined the term "nucleus," and before the advent of cell theory he cor-

rectly surmised that it was a feature of every plant cell. He discovered cytoplasmic streaming in the stamen hairs of *Tradescantia* and delighted in showing this vivid cellular action to his friends, including such luminaries as Charles Darwin and William Hyde Wollaston (British chemist and inventor of the prism used in Nomarski optics, which today provides the best images of streaming in the stamen hairs—see Box 5.2). Brown's major microscopical interest was the fertilization of plants. He was the first to correctly outline the anatomy of seeds and their embryos, and he discovered the naked ovule of gymnosperms, which allowed him to rationalize the classification of this group.

Important though these discoveries were, Brown's enduring fame rests on an 1827 discovery, made while examining pollen. He saw particles inside the grain moving randomly—rapidly and without cessation. (It was not the pollen itself he saw move, as often erroneously stated. Recent skepticism about his observations has been refuted by Brian J. Ford, who recreated Brown's 1827 observations of Brownian motion with the very microscope Brown himself used, pictured here ready for use [C]). Brown was not the first person to observe this incessant movement; others had seen it before and hoped they were seeing the essence of life. But Brown observed the motion in pollen grains preserved for months in alcohol as well as in suspensions made by grinding a variety of rocks and minerals. He correctly concluded that the motion has a physical, not an organic, explanation. Later scientists quantified the motion to derive fundamental insight into the stochastic character of the molecular world. For example, French physicist and Nobel Laureate Jean-Baptiste Perrin used Albert Einstein's formula for Brownian motion to calculate the size of the water molecule. It is apt that the incessant movement of particles bears the name of the indefatigable botanist, Robert Brown.

(A)

(B)

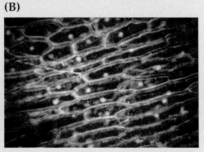

(C)

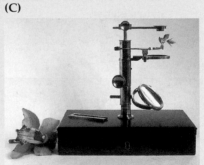

5.1.3 The cytoskeleton consists of a network of fibrous polymers.

The cytoskeleton is a network of interconnected fibrous polymers that run throughout the cell within the cytosol (Fig. 5.2). This network provides structural stability to cytoplasm, anchoring proteins and other macro-molecules, and supporting organelles during and after their synthesis. Besides structural stability, the cytoskeleton also gives cells the property of motility, both internal and external. Cellular components can move within the cell, such as during cytoplasmic streaming, and many cell types change shape and move within their environment.

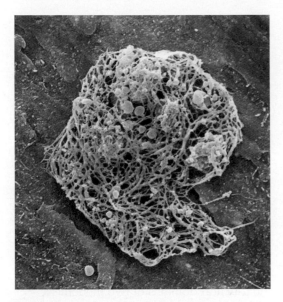

Figure 5.2
Electron micrograph of cytoskeletal meshwork, showing detergent-resistant fibrous elements from a suspension-cultured carrot cell.

The cytoskeleton also participates in processing cellular information. For example, elements of the cytoskeleton are known to converge on the plasma membrane at sites associated with transmembrane receptors. Finally, many elements of the cytoskeleton consist of asymmetric subunits so that the polymers themselves are like an arrow, providing directional cues within the cell (Fig. 5.3).

5.2 Intermediate filaments

In animal cell cytoskeletons, **intermediate filaments** function similarly to the bones of the vertebrate skeleton. Intermediate filaments were first discovered as a fibrous network that remained when the soluble components of a cell were washed out by incubation in a concentrated salt solution containing detergent. These filaments are named for their thickness; at 10 to 15 nm in diameter, they are thinner than microtubules and thicker than actin filaments.

5.2.1 Intermediate filaments are complex quaternary protein structures that lend strength and resiliency to the cell.

The different classes of intermediate filaments are defined by their subunits,

(A) Anchorage

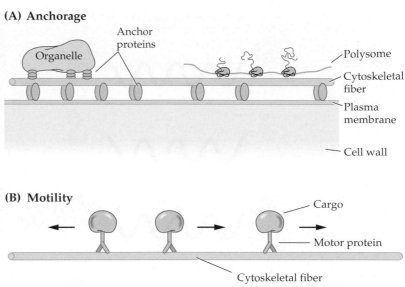

(B) Motility

(C) Information

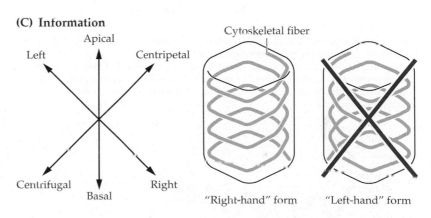

(D) Polarity

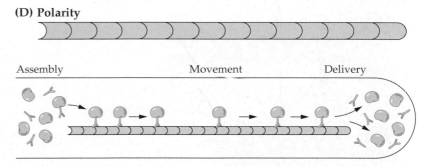

Figure 5.3
Diagrams showing different functions of the cytoskeleton. (A) Anchorage. The cytoskeleton provides support to the plasma membrane and anchors organelles and other macromolecular assemblies (e.g., polysomes). (B) Motility. The cytoskeleton supports the directed intracellular movement of cellular components. (C) Information. The cytoskeleton can provide spatial cues to cellular geometry. In this example, the cell builds cytoskeletal fibers into a right-handed helical array, not a left-handed one. The fibers define a third axis of symmetry, in addition to the apical/basal and centripetal (toward the center)/centrifugal (away from the center) axes. (D) Polarity. The informational content of cytoskeletal fibers depends on the polarity of the fibers, which are formed from asymmetric subunits that define a direction along the polymer. In this example, cargo is assembled at one site, moved through the cell via the cytoskeleton, and delivered to a second location.

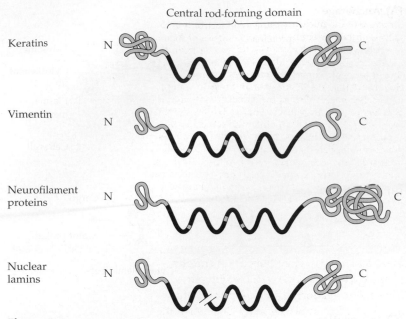

Central rod-forming domain

Keratins N   C

Vimentin N   C

Neurofilament proteins N   C

Nuclear lamins N   C

Figure 5.4
Monomers of intermediate filaments consist of a conserved, rod-forming domain, flanked by diverged globular domains at each end. The rod-forming domain is approximately 300 amino acids long and folds into an extended α-helix. The specific characteristics of the different classes of intermediate filament proteins are determined by the N- and C-terminal domains at each end, which differ greatly in size and sequence.

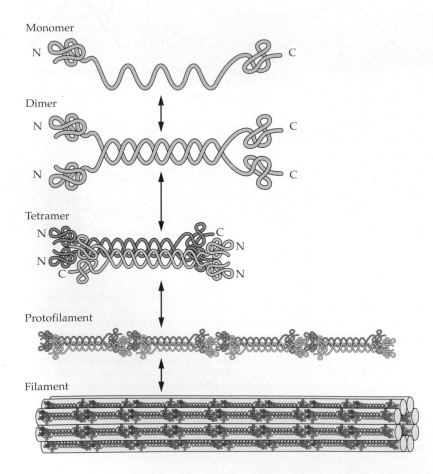

Monomer N   C

Dimer N   C
       N   C

Tetramer N   C N
          N   C N

Protofilament

Filament

including lamin, keratin, vimentin, and neuronal filament proteins. The subunit monomers are rod-shaped, with long α-helical domains that are highly conserved among the different classes (Fig. 5.4) and support a similar mode of polymerization. In addition to the conserved rod-forming domains, intermediate filament proteins have other functional domains, which are generally small and may be located at either terminus or in the middle of the protein. These domains give each class of filament a distinct character. In assembled filaments, these domains project from the fiber, providing ligands for interaction with other cellular components.

During intermediate filament assembly, the α-helical domains of two monomeric subunits associate through coiled-coil interactions to form a bipolar dimer. These dimers associate laterally to form tetramers, which pack together in long, rope-like fibers (Fig. 5.5). All animal cells are thought to make at least one class of intermediate filament, and many cells make more than one type; however, any particular fiber contains proteins from only a single class of monomer. Unlike microtubules or actin filaments, intermediate filaments do not have an intrinsic polarity.

Intermediate filaments appear to add mechanical strength and resiliency to the cell. For example, keratin fibers are abundant in skin cells, imparting to that organ its characteristic toughness. Similarly, the long, thin axons of nerve cells are filled with neurofilaments, which presumably prevent the axons from being sheared apart by every bend and bounce of the animal.

The intermediate filaments with the most conserved function are the lamins. These form the **nuclear lamina,** a sheet-like structure lying just inside the nuclear envelope (Fig. 5.6A). The lamina is thought to stabilize the nuclear envelope, to maintain the shape of the organelle, to ensure that

Figure 5.5
Assembly of intermediate filaments. Identical monomers pair through coiled-coil interactions to form a dimer. Dimers associate laterally to form tetramers. In the tetramer, the dimers are antiparallel (i.e., head to tail) and staggered, so that further lateral associations can take place. Tetramers polymerize to form protofilaments. Finally, protofilaments associate laterally to form a rope-like filament.

(A)

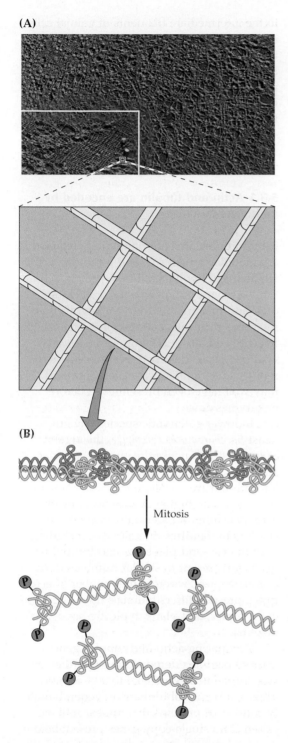

(B)

Mitosis

Figure 5.6
Structure of the nuclear lamina. (A) Electron micrographs of the nuclear lamina from oocytes of *Xenopus laevis*. The overview shows the nuclear lamina meshwork. The inset highlights a region from which the nuclear pore complexes have been removed mechanically. Note the intersection of fibers at roughly 90°. (B) Schematic view of the association of lamin subunits to form the nuclear lamina. Lamin subunits are always dimeric. During interphase, the lamin dimers associate "head to head" to form a tetramer, which polymerizes into the filaments that form the nuclear lamina meshwork. During mitosis, the lamin dimers are phosphorylated, which prevents tetramer formation and thus causes the lamin filaments to disassemble and the nuclear lamina to break down.

maximally active near the end of mitotic prophase (see Chapter 11). Conversely, after the transition from metaphase into anaphase, this kinase activity declines and lamin-bound phosphates are removed. During telophase, the nuclear lamina and nuclear envelope reform around the decondensing chromosomes.

5.2.2 Evidence for intermediate filaments in plant cells remains inconclusive.

Are homologs of the intermediate filament proteins of animals present in plant cells? Plants are known to contain homologs of nuclear lamins, but for the other intermediate filament proteins, the evidence is no more than suggestive. Filamentous structures, reminiscent of animal intermediate filaments, have been seen in plant cells and in some cases are resistant to extraction by detergent and high salt (Fig. 5.7). These plant "cytoskeletons" contain proteins that are immunologically related to certain intermediate filament proteins in animal cells. Furthermore, when isolated, these plant proteins can undergo assembly and disassembly reactions typical of animal intermediate filament proteins (see Fig. 5.5).

However, the above evidence may demonstrate only a functional convergence among filament-forming proteins of different evolutionary origins. In animal intermediate filament proteins, much of the structure is specified by the coiled-coil interactions of the rod-forming domains of the monomers. Rod-like proteins formed from coiled-coil interactions are known to have evolved

chromatin can be successfully anchored and, possibly, to reinforce the structure of the large nuclear pore complexes (see Chapters 1, 4). When the nuclear envelope breaks down during mitosis, the nuclear lamina dissolves after reversible protein phosphorylation (Fig. 5.6B). Lamin proteins contain several phosphorylation sites, substrates for a cyclin-dependent protein kinase that is

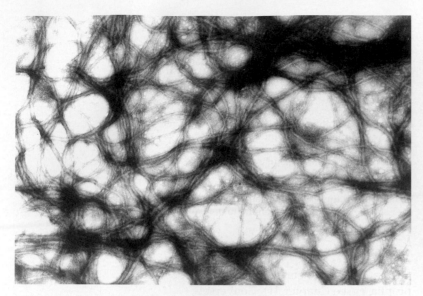

Figure 5.7
Electron micrograph of intermediate-like filaments formed from a purified plant protein. The purified protein forms these filaments under biochemical conditions that are similar to those required to form bona fide animal intermediate filaments.

independently many times (Fig. 5.8). A ciliate-derived protein that fulfilled the same immunological and biochemical criteria as the putative intermediate filament proteins of plants was discovered to be, in fact, a glycolytic enzyme. Thus, until gene sequences are available, it is prudent to use the term "intermediate filament-like" proteins for the structures found in plants.

Little can be said about the function of the intermediate filament-like structures in plants. Interestingly, studies have found these filaments colocalizing with and even stabilizing microtubules (Fig. 5.9). Many of the mechanical functions typically ascribed

Every fourth amino acid is hydrophobic.

The resulting stripe-shaped hydrophobic domains can interact, forming a coiled-coil.

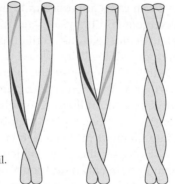

Figure 5.8
The structure of a coiled-coil. Rod-like proteins are commonly formed from coiled-coils. Two long α-helices can wrap around one another when each has hydrophobic amino-acid side chains at every fourth position. Coiled-coil associations hold together intermediate filament subunits, myosin II, kinesin, and many others.

to the intermediate filaments of animal cells are probably carried out by the plant cell wall. Therefore, the evolution of durable filamentous structures in plant cells may relate to mechanical problems that are purely intracellular, for example, anchoring the nucleus.

5.3 Actin and tubulin gene families

5.3.1 Actin and tubulin are encoded by multigene families.

Microtubules are heterodimeric polymers of the globular proteins α- and β-tubulin; **actin filaments** are polymers of the protein actin. Both actin and tubulin occur in all eukaryotes, and presumably they both evolved at the time eukaryotic cells became established. Recently, the product of the *Escherichia coli FtsZ* gene was identified as a tubulin homolog, so the major cytoskeletal proteins may well have been present even before eukaryotes arose.

In lower eukaryotic species (e.g., the yeast *Saccharomyces cerevisiae*, the alga *Chlamydomonas reinhardtii*, and the cellular slime mold *Dictyostelium discoideum*), actin and tubulin are usually encoded by single-copy genes. In higher eukaryotes, however, actin and tubulin are usually encoded by small **gene families.** In animals—including vertebrates—and plants, α- and β-tubulin genes are present in similar numbers (four to nine copies), whereas the number of actin genes present differs (animals have about eight copies, and plants typically have dozens).

Comparing actin and tubulin genes across kingdoms demonstrates that they encode proteins that are 80% to 90% identical. Thus, actin and tubulin are, in general, highly conserved proteins that appear to have arisen from single-copy genes present before multicellular eukaryotes diverged (Fig. 5.10).

5.3.2 Several models have been proposed to explain gene family evolution.

Why are these cytoskeletal proteins encoded by gene families? Nearly all members of each family are expressed, so these families do not result from the accumulation of

(A)   **(B)**

(C)   **(D)**

(E)   **(F)**

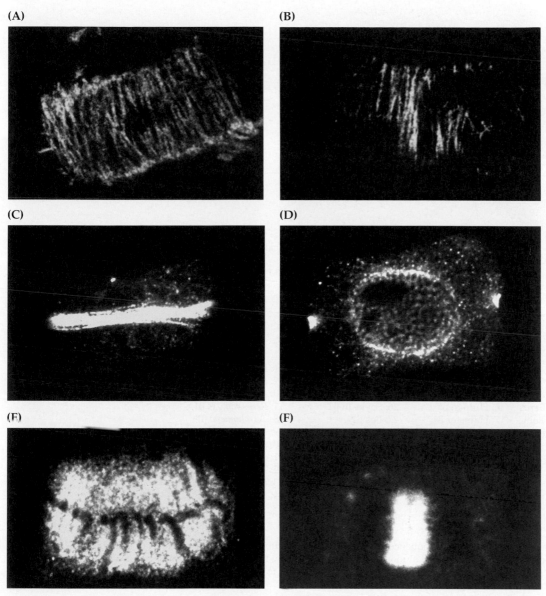

Figure 5.9
Fluorescence micrographs of wheat root-tip cells stained with an antibody that recognizes an intermediate filament-like protein in plant cells and a bona fide intermediate filament in animal cells (one of the keratins). At all stages of the cell cycle, the antibody stains fibrous elements that resemble arrays formed by microtubules (see Figs. 5.28, 5.43). (A, B) Interphase cells demonstrate fibrous staining in the cortex, similar to that expected for microtubules. (C, D) A prophase cell, brightly stained at the so-called preprophase band at the cell periphery (C) and the polar regions of the nucleus (D). Both of these locations have abundant microtubules. (E, F) Dividing cells show fibrous staining more or less congruent with microtubules in the mitotic spindle (E) and the phragmoplast (F).

pseudogenes (see Chapter 7). The elaboration of gene families often reflects an evolutionary pattern initiated by gene duplication. Having two functional copies of a gene in the genome frees one copy to evolve and perhaps acquire a new function. A family may grow from repeated rounds of duplication and selection. In one view, each member of a gene family, or **isotype,** diverges in response to selection for protein function. Because the cytoskeleton participates in many processes, and because tubulin and actin proteins must interact with numerous accessory proteins that are probably cell type-specific, this hypothesis—of functional divergence among isotypes—is plausible (Fig. 5.11).

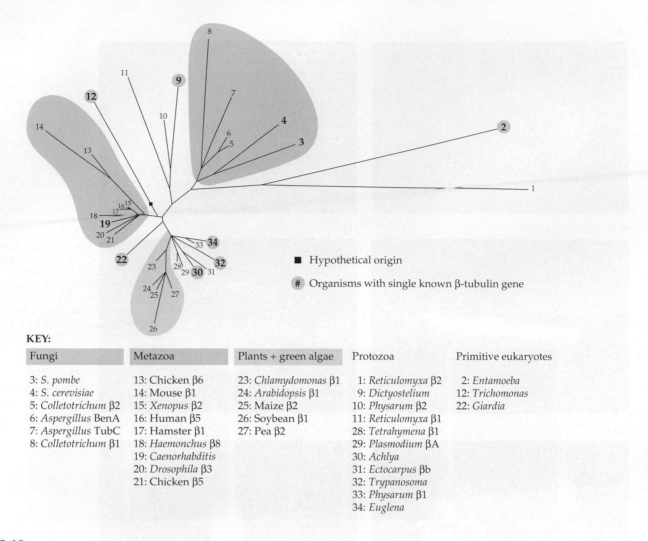

Figure 5.10
Overview of evolutionary divergence of β-tubulin sequences. The length of each branch is proportional to the rate of evolution of the sequence. The tree has been constructed to emphasize rates of divergence, not phylogenetic relationships. Note the clustering of tubulins within the major groups of eukaryotes and the predicted origin of β-tubulin before the radiation of the eukaryotes. As shown by the long branches, only two β-tubulin sequences are thought to have undergone comparatively rapid evolutionary change.

■ Hypothetical origin

Organisms with single known β-tubulin gene

KEY:

Fungi	Metazoa	Plants + green algae	Protozoa	Primitive eukaryotes
3: *S. pombe*	13: Chicken β6	23: *Chlamydomonas* β1	1: *Reticulomyxa* β2	2: *Entamoeba*
4: *S. cerevisiae*	14: Mouse β1	24: *Arabidopsis* β1	9: *Dictyostelium*	12: *Trichomonas*
5: *Colletotrichum* β2	15: *Xenopus* β2	25: Maize β2	10: *Physarum* β2	22: *Giardia*
6: *Aspergillus* BenA	16: Human β5	26: Soybean β1	11: *Reticulomyxa* β1	
7: *Aspergillus* TubC	17: Hamster β1	27: Pea β2	28: *Tetrahymena* β1	
8: *Colletotrichum* β1	18: *Haemonchus* β8		29: *Plasmodium* βA	
	19: *Caenorhabditis*		30: *Achlya*	
	20: *Drosophila* β3		31: *Ectocarpus* βb	
	21: Chicken β5		32: *Trypanosoma*	
			33: *Physarum* β1	
			34: *Euglena*	

A contrasting view is that the extra gene copies provide regulatory flexibility, which may be required in the diverse cell types of higher organisms. In this view, gene duplication accommodates disparate, noncoding sequences that regulate expression. Regulation of the expression of actin and tubulin genes is likely to be complex: The expression level of these proteins must be high, because the cytoskeleton requires large amounts of them. In addition, the expression level must be controlled accurately because the concentrations of actin and tubulin partially determine the cytoskeleton's characteristics (see Section 5.4.1).

Finally, the elaboration of gene families has been said to reflect historical accident. In this view, the different isotypes are neither adapted functionally nor advantageous for regulation but instead result from a series of accidental and irreversible duplications, followed by random mutation of copies. Consistent with this premise, the number of genes in a gene family can vary tremendously among species. For example, *Arabidopsis thaliana* contains only 10 actin genes, whereas petunia has more than 100. It is difficult to envision that being a petunia imposes more than 90 unique demands on the function or regulation of actin.

(A)

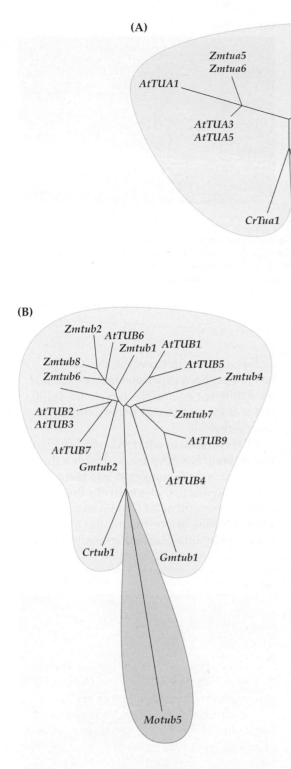

Figure 5.11
Phylogenetic trees showing putative evolutionary relationships for the coding sequences of α- and β-tubulins in plants. Trees have been rooted with sequences from mouse *(Mo)* as the outgroup. For both proteins, diverged taxa—*Arabidopsis thaliana* *(At,* a dicot) and *Zea mays (Zm,* a grass)—have closely related isotypes, indicating that a gene family was already present in the common ancestor and suggesting that the continued evolution of the family members has been shaped, at least in part, by selection. (A) α-Tubulin sequences. A sequence from *Chlamydomonas reinhardtii (Cr)* is also included. (B) β-Tubulin sequences. Sequences from *C. reinhardtii* and soybean *(Gm)* are also included.

5.3.3 Some evidence supports and some opposes the functional specialization of tubulin isotypes in higher plants.

Determining whether specific isotypes of either tubulin or actin have unique functions presents a formidable challenge. Isotype function may differ in different cell types or in the different cytoskeletal arrays made by a single cell. Delineating the functions of specific isotypes is made even more complex because these proteins are modified posttranslationally in several ways (including phosphorylation, acetylation, glutamylation, and detyrosination). Thus, each genetic isotype may give rise to several biochemical isotypes.

To date, the best example of functional specialization of isotypes comes from the β-tubulins of the fruit fly, *Drosophila melanogaster.* In males, a single isotype of this protein is made by the germ cell lineage. During the production of spermatozoa, this cell lineage builds several different microtubule arrays, including the apparatus needed to separate chromosomes in mitosis and also meiosis, several types of cytoplasmic arrays required for cell motility and, finally, the axoneme of the sperm's flagellum. During the 1990s, transgenic techniques were used to

Contrasting explanations for the multigene families of cytoskeletal proteins, including functional selection, regulatory flexibility, and historical accident, are not mutually exclusive, and each probably describes part of the evolutionary process.

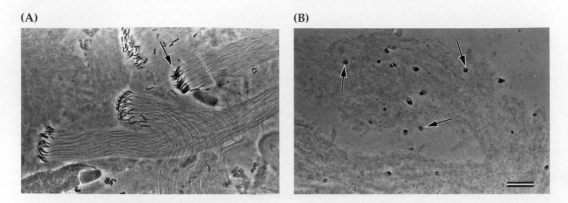

(A)   (B)

Figure 5.12
Light micrographs of excised testes from the fruit fly, *Drosophila melanogaster*. In *D. melanogaster*, male germline cells predominantly express a single isotype of β-tubulin (β2). When this isotype is replaced by another fruit fly β-tubulin isotype, β3, most microtubules fail to form. As a result, meiosis, mitosis, and formation of the sperm's flagellum are disrupted, and the flies are rendered sterile. (A) Testes of fertile male that correctly expresses the β2 isotype. Spermatids are bundled (bracket) and have elongated nuclei (arrow). (B) Testes of sterile male in which β3 is expressed in place of β2. Nuclei are round (arrows) and elongated flagella have not formed.

make male flies in which expression of the normal β-tubulin isotype (β2) was completely replaced by the expression of a different isotype (β3). These flies were sterile, and their germline cells made few of the customary microtubule arrays, thereby showing that the isotypes are not functionally interchangeable (Fig. 5.12). In plants, although most organs and tissues synthesize various mixtures of both actin and tubulin isotypes, the male gametophyte and its precursors predominantly express a single α-tubulin isotype (Fig. 5.13). However, this isotype has not yet been experimentally exchanged for a different one, so the functional significance of this expression pattern cannot be ascertained.

Experiments in which tubulin from bovine brain has been marked with a fluorochrome and microinjected into plant cells demonstrate that some cells do not require an exact blend of tubulin isotypes. Tagged bovine or porcine tubulin incorporates rapidly into plant microtubule arrays

(A)   (B)

Figure 5.13
The pollen of higher plants predominantly expresses a single isotype of α-tubulin. The photographs show histochemical localization of glucuronidase (GUS) activity in *Arabidopsis thaliana* plants transformed with a construct consisting of the promoter of the α-tubulin isotype fused to the coding sequence of a gene for GUS from a bacterium. GUS activity and hence transcription of the transgene is indicated by the presence of a bright blue precipitate. (A) Mature flower showing intense GUS staining in the anthers. (B) A fragment of the stylar end of a pistil with germinated pollen. Note the blue staining in the exposed pollen tubes.

without impairing their function (see Figs. 5.34 and 5.35A for results of this type of experiment). Although the injected tubulin probably amounts to no more than 20% of the host cell's tubulin, the absence of any detrimental effect indicates that these cells do not require an exact blend of isotypes.

5.4 Polymerization of actin and tubulin

Actin protein self-assembles into long polymers called actin filaments or microfilaments, and tubulin protein polymerizes into structures called microtubules (see Figs. 5.16 and 5.17 in Section 5.5). Actin filaments and microtubules assemble by a similar process and share several characteristics.

Like the synthetic polymer nylon, many biological polymers consist of molecular subunits linked by covalent bonds (e.g., polypeptide chains, single-stranded nucleic acids). In contrast, cytoskeletal polymers have macromolecular subunits linked by many noncovalent bonds, typical of the quaternary level of protein structure. Therefore, microtubules and actin filaments are dynamic, assembling or disassembling in response to factors that govern protein–protein interactions, such as ionic strength or temperature.

5.4.1 Spontaneous assembly of cytoskeletal polymers allows detailed laboratory analysis of polymerization.

The ability of actin filaments and microtubules to assemble spontaneously has the practical consequence that researchers can isolate and handle the protein in the form of subunits and can initiate polymerization in vitro by one of several simple manipulations. The time course of the polymerization reaction can be conveniently assayed by measuring the amount of light scattered by the solution, because light scattering increases as the polymer lengthens.

Polymerization occurs in three stages (Fig. 5.14). First, a lag occurs between the onset of conditions favoring polymerization and the first detected increase in polymer. The lag reflects the need for an initial group of several subunits to associate correctly, forming a template for further assembly.

This first stage of polymer assembly, the formation of a template, is called **nucleation.** The requirement for nucleation can be bypassed by adding small pieces of preformed polymer, often referred to as "seeds."

During the second stage, **elongation,** templates formed during nucleation grow by endwise addition of subunits. The rate of elongation represents the difference between the rate of subunit addition and the rate of subunit loss. The rate of subunit addition is the product of the association constant (k_{on}) and the concentration of free subunits in solution; the rate of subunit loss equals the dissociation constant (k_{off}). Consequently, the rate of elongation is proportional to the prevailing subunit concentration. Elongation continues until a plateau concentration of polymer is reached.

The third stage is a **steady state** in which a constant amount of polymer is maintained over time. The steady state is dynamic, the polymer gaining and losing subunits constantly. However, at steady state, the rate of subunit addition is balanced exactly by the rate of subunit loss (i.e., k_{on}[subunit] = k_{off}). Thus, steady state occurs when the concentration of subunits equals the ratio of the dissociation and association rate constants (i.e., [subunit] = k_{off}/k_{on}), the

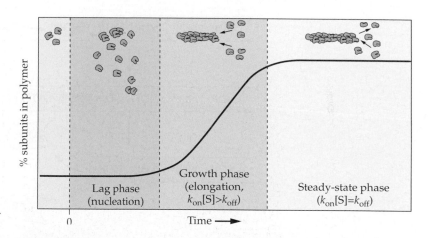

Figure 5.14
Diagram illustrating process and kinetics of polymerization. A solution of individual subunits is induced to polymerize at time zero. During nucleation (pink panel), subunits must associate to create a stable template that serves as a platform for further elongation. In this example, using actin filaments, templates are formed when three subunits associate, generating a trimer. Because several subunits must interact simultaneously to form a template, this process is kinetically slower than that of adding subunits to the ends of an extant polymer and results in a lag phase. During elongation (growth phase, blue panel), subunits are rapidly added to the ends of growing polymers. A steady-state phase (yellow panel) is reached when the rate of subunit addition (k_{on}[S]) is balanced exactly by the rate of subunit loss (k_{off}).

concentration known as the **critical concentration.** At concentrations less than this, polymer will not form spontaneously; at greater concentrations, the polymer will grow until the subunits are depleted to the critical concentration.

5.4.2 Cytoskeletal polymers have an intrinsic polarity.

Actin filaments and microtubules are polar structures because their protein subunits are asymmetrical. In polymer assembly, the asymmetric subunits line up end-to-end with a uniform orientation (Fig. 5.15A). The polarity thus conferred to the polymer means that each end has a different bio-

chemical character. Therefore, each end of the polymer may have different rate constants for the assembly and disassembly reactions. In addition, each end of the polymer can be recognized specifically, so that cells may build polymer arrays with uniform polarity. The more dynamic end of the polymer is designated "plus" and the less active end "minus." Note that in this context these terms do not indicate electrical charge.

5.4.3 Actin and tubulin bind and hydrolyze nucleotides.

Actin binds and hydrolyzes ATP, whereas tubulin binds and hydrolyzes GTP. These nucleotides play an important role in

(A)

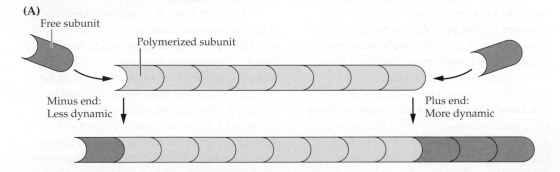

Free subunit

Polymerized subunit

Minus end:
Less dynamic

Plus end:
More dynamic

(B)

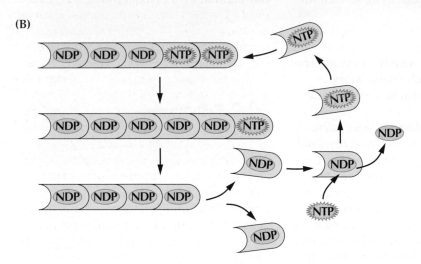

Figure 5.15
(A) Because actin filaments and microtubules form by head-to-head addition of intrinsically polar subunits, these cytoskeletal polymers also have a defined polarity, and each end has distinct biochemical properties. One end of the filament is more dynamic than the other end, having increased rates of both subunit addition and loss. The more dynamic end is called the "plus" end, and the less dynamic end is called the "minus" end (these terms do not refer to electric charge). (B) Actin and tubulin each bind nucleotides. The soluble proteins have a high affinity for the triphosphate form (NTP), and NTP-bound subunits have a high affinity for the assembled polymer (that is, k_{on} S-NTP>>k_{on} S-NDP). After the subunit is incorporated into the polymer, the subunit's nucleotide is hydrolyzed. This lowers the affinity of the subunit for the polymer and may promote depolymerization. Subunits usually disassemble in the NDP-bound form and must exchange this nucleotide for NTP before again binding to the polymer.

determining the properties of the polymers. Soluble subunits bind nucleotide triphosphates (NTPs), which greatly increase their assembly rate. This enhanced rate of assembly does not depend on energy from the γ-phosphate bond, because assembly is promoted equally by nonhydrolyzable nucleotide analogs. The γ-phosphate bond is hydrolyzed only after a subunit has bound to the polymer. The energy released by binding is stored in the polymer and used subsequently to enhance polymer dynamics, as described in Section 5.5.2. After a subunit dissociates, the nucleotide diphosphate (NDP) is rapidly exchanged for NTP, and the subunit is ready for another round (Fig. 5.15B). The presence of nucleotides means that, in addition to the intrinsic difference between plus and minus polymer ends, each end may exist in at least two states, with subunits containing NDP or NTP. These different states extend the range of behavior possible for cytoskeletal polymers.

5.5 Characteristics of actin and tubulin

5.5.1 Actin filaments are slender, tightly helical polymers, and microtubules are, literally, little tubes.

Soluble actin is a globular protein (sometimes termed "G-actin," in contrast to the polymeric or filamentous form called "F-actin") of 375 amino acids. Subunits polymerize into a tightly helical filament, approximately 8 nm wide (Fig. 5.16). For historical reasons having to do with the appearance of actin filaments when decorated with the protein myosin, the minus end is also sometimes called the "pointed" end, and the plus end is the "barbed" end.

The subunits that assemble into microtubules are heterodimers of the globular proteins α- and β-tubulin (Fig. 5.17). The dimers associate to form a hollow structure 25 nm in diameter. Each member of the dimer binds a guanine nucleotide, but only the β-subunit engages in GTP hydrolysis and GDP–GTP exchange.

In the microtubule, tubulin dimers bind together both at their sides and at their ends. The dimers lie in straight columns called protofilaments. Most microtubules consist of 13 protofilaments, but the number can vary

from 11 to 16. The lateral bonds between tubulin dimers displace dimers in adjacent protofilaments by several nanometers toward one end. Thus, the lattice of subunits that makes up the microtubule wall has a helical character. Although its identification is still controversial, the best evidence to date indicates that the plus end of the microtubule corresponds to the β-tubulin end of the dimer.

5.5.2 Differences in the biochemical properties of actin and tubulin give the polymers distinct dynamic behaviors.

Several features besides nucleotide specificity distinguish actin filament and microtubule polymerization. First, for actin, the lag time for nucleation is proportional to the third power of concentration, which suggests that nucleation requires the assembly of a trimer. For microtubules, the nucleation lag is proportional to a much higher power of concentration, as would be expected if as many as 13 subunits must assemble to form a template. Second, the critical concentration for assembly of actin is approximately 0.2 μM, very much lower than the concentration of actin typically found in cells (0.1 to 1 mM). Therefore, actin concentration does not present a barrier to nucleation of actin filaments; instead, cells must prevent unwanted polymerization. Cells have evolved accessory proteins to sequester soluble actin and to control its polymerization into filaments (see Box 5.2). By contrast, the critical concentration for microtubule assembly (for animal tubulin) is approximately 8 μM, only modestly below usual total tubulin concentrations (approximately 20 μM). Not surprisingly, cells have evolved structures for nucleating microtubules but apparently not for sequestering soluble tubulin.

Finally, in the two types of polymer, nucleotide hydrolysis has different consequences. For actin, the presence of ATP at the plus (or barbed) end results in an assembly rate far exceeding that at the minus (or pointed) end, where the actin subunits contain ADP. This difference underlies a type of dynamic behavior called **treadmilling,** which occurs when the prevalent concentration of free subunits supports growth at the plus end but results in shrinkage at the

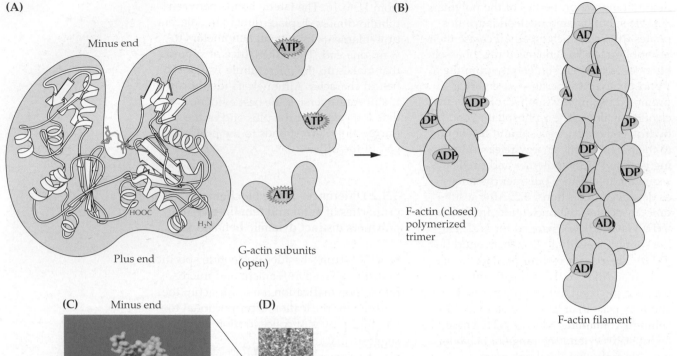

(A)

Minus end

Plus end

G-actin subunits
(open)

(B)

F-actin (closed)
polymerized
trimer

F-actin filament

(C) Minus end

Plus end

(D)

Figure 5.16
The structure of actin and actin filaments. (A) Ribbon model of a single actin molecule. The yellow structure in the middle is the bound ATP. (B) Diagrams of G-actin monomers, a nucleated F-actin trimer, and a short actin filament, emphasizing the "open" and "closed" conformations. In the polymerized form, the conformation changes to enclose the active site, preventing the loss or exchange of nucleotide. Staggered subunit associations give the polymer its helical character. (C) Three-dimensional model of an actin filament, in which each ball represents an amino acid. Note the different shapes of the plus and minus ends. (D) Actin filaments assembled from purified actin and imaged via cryoelectron microscopy, without fixation or heavy metal staining.

minus end. The net rate of change in polymer length can be exactly zero, during which subunits incorporated at the plus end will "treadmill" through the polymer and eventually be released from the minus end (Fig. 5.18). Because the difference in assembly rate at the plus and minus ends is much larger for actin than for microtubules, treadmilling is presumed to be more common among actin assemblies. However, both polymers demonstrate this behavior.

Nucleotide hydrolysis gives rise to a different kind of behavior in microtubules, called **dynamic instability.** Here, energy released by GTP hydrolysis accelerates the rate of depolymerization. Growing microtubules add GTP-bound subunits. Because hydrolysis lags slightly behind assembly, the growing ends of a microtubule will have a cap of subunits with bound GTP. This cap stabilizes the structure and favors further growth. However, when the supply of tubulin dimers runs low, or when the rate of GTP hydrolysis accelerates, the end of the microtubule may contain a majority of GDP-bound subunits. The lateral bonds between protofilaments are strained by this arrangement. When freed of the GTP cap, the protofilaments peel apart and initiate a catastrophic disassembly, hundreds of times faster than the rate of growth (Fig. 5.19). A process exists that can rescue these rapidly shrinking microtubules, halting

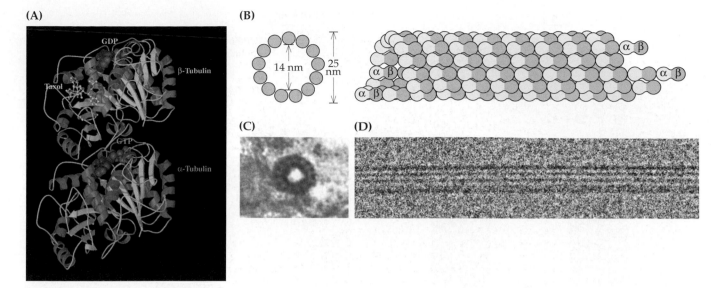

Figure 5.17
Structure of microtubules. (A) Ribbon diagram of the tubulin dimer within a microtubule, resolved at 0.37 nm. The dimer is viewed from the inside of the microtubule, looking out. Regions of the β-sheet structure are shown in green and the α-helix in blue. Also shown are the positions of bound nucleotides and of taxol (a compound that stabilizes microtubules that was used in the structural determination). The plus end is at the top. Note that GTP is hydrolyzed by β-tubulin but not usually by α-tubulin; note also the overall similarity of structure between α- and β-tubulin. (B) Diagrams of a microtubule in cross-section and side view. The subunits of the microtubule, dimers of α- and β-tubulin, align head to tail in long parallel columns called protofilaments. The protofilaments are displaced longitudinally from one another, giving a helical character to the lattice of subunits. The dimers assemble head to tail, giving microtubules a structural polarity; α-tubulin is exposed at one end of the polymer, β-tubulin at the other. (C) Electron micrograph of a microtubule in cross-section from a plant cell, showing thirteen protofilaments. (D) Electron micrograph of a microtubule assembled from purified tubulin and viewed from the side. Unlike (C), this side view was prepared by using cryotechniques, without fixation or staining with heavy metals.

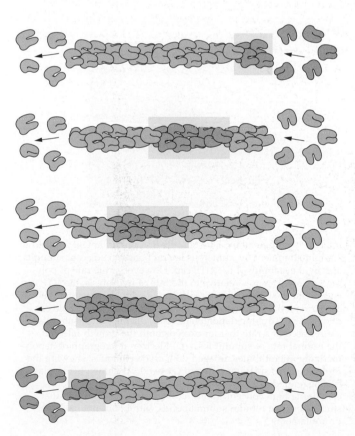

Figure 5.18
Treadmilling, a dynamic behavior observed in cytoskeletal polymers, is supported by nucleotide hydrolysis. Because the bound nucleotide is hydrolyzed after subunit polymerization, the substrates of the assembly reaction differ from the products of the disassembly reaction. Therefore, each end of the polymer may have different values of k_{on} and k_{off} and hence exhibit different critical concentrations. As a result, subunits may add preferentially at the plus end and be lost preferentially at the minus end without changing the length of the polymer. As shown in the diagram, subunits (purple) added at one end "treadmill" through the polymer and are released from the other end. Although the diagram illustrates actin, treadmilling also occurs in microtubules.

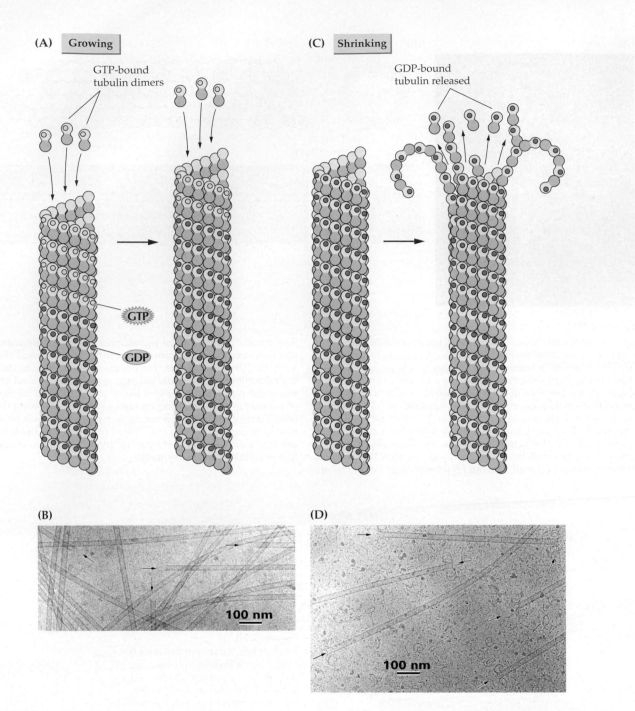

(A) Growing

GTP-bound
tubulin dimers

GTP

GDP

(C) Shrinking

GDP-bound
tubulin released

(B)

100 nm

(D)

100 nm

Figure 5.19

Model for microtubule dynamic instability. (A) Diagram of grow-ing microtubules. Since hydrolysis of GTP usually lags behind polymerization of new subunits, the growing ends of microtubules are rich in subunits in which β-tubulin monomers bind GTP. Such microtubules are said to have a "GTP-cap." In cells, the minus end of the microtubule is usually embedded in some nucleating ma-terial, so GTP-caps are commonly considered only for the plus end. Note that the inert GTPs bound to α-tubulin are not shown. (B) Electron micrograph of growing microtubules. GTP-tubulin subunits pack into the microtubule lattice with minimal distortion, yielding straight ends for the growing microtubules (arrows). The image was obtained by cryofixing a population of rapidly elongat-ing microtubules. (C) Hydrolysis of GTP causes a conformational change that tends to bend the protofilaments outward, weakening lateral contacts between dimers in adjacent protofilaments. In an

intact microtubule, the GDP-subunits in a protofilament are held in straight alignment by numerous lateral bonds to adjacent subunits and by the stability of the GTP cap. However, if the rate of poly-merization decreases relative to the rate of hydrolysis, the GTP cap is lost and the GDP-subunits can more readily assume their bent conformation. In this situation, the destabilized microtubule can undergo catastrophic depolymerization, during which subunits are released at a rate that far exceeds both the growth rate and the normal rate of subunit loss. (D) Electron micrograph of depoly-merizing microtubules, imaged with cryotechniques, showing the protofilaments splaying outward (arrows). As shown here, splaying is exaggerated when microtubules depolymerize in solutions containing high concentrations of magnesium, a condition that strengthens the tubulin subunit bonds within, but not between, protofilaments.

their disassembly and allowing them to resume growth, but its nature is not yet well understood.

As a result of treadmilling or dynamic instability, a population of polymers may display prolonged growth and shrinkage simultaneously. The dynamic properties of actin filaments and microtubules are controlled to modify the behavior of the cytoskeleton as required for distinct cell functions.

5.6 Cytoskeletal accessory proteins

If actin filaments and microtubules are the cell's scaffolding, then accessory proteins are the joints, motors, and tools that link, move, and modify that scaffolding. The different functions performed by polymers of the highly conserved proteins actin and tubulin are determined largely by a suite of less highly conserved accessory proteins. Defined operationally as proteins that are copurified with cytoskeletal polymers, accessory proteins can be further divided by function.

5.6.1 Mechanochemical enzymes convert chemical energy into work.

Mechanochemical enzymes are colloquially referred to as "motor proteins." The three superfamilies of known motor proteins—myosin, dynein, and kinesin—all use ATP as their energy source. Furthermore, all contain a globular, force-producing "head" domain that binds a cytoskeletal polymer and a rod-shaped "tail" domain that binds cargo (Fig. 5.20). The families differ in which type of polymer they recognize, what type of cargo they will carry, and which biochemical pathways they use to convert chemical energy into work.

The first such protein to be characterized was myosin, the force-producing enzyme of muscle. We now know that members of the myosin family are present in all eukaryotic cell types and that they power many types of cellular motility involving actin filaments (Fig. 5.21). Of the 13 different classes of myosin described to date, only 2 have been demonstrated unequivocally in plants.

The other two motor proteins, dynein and kinesin, bind microtubules; like myosin, they are represented by large gene families, with members present throughout the eukaryotes (Fig. 5.22). Dynein transports cargo to the minus end of the microtubule, whereas most, but not all, kinesins move toward the plus. Dynein has two major forms, axonemal and cytoplasmic. The axonemal form, the force-producing protein of eukaryotic cilia, has presumably been lost from seed plants, which lack ciliated sperm. The cytoplasmic form, which was only recently identified in plants, supports a variety of microtubule-mediated motility in animal cells, including vesicular traffic toward the cell center and microtubule organization. Similarly, kinesins

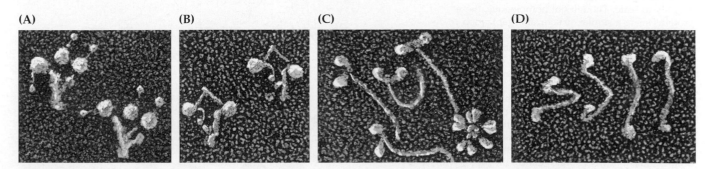

(A)   **(B)**   **(C)**   **(D)**

Figure 5.20

Electron micrographs of mechanochemical motor proteins. The proteins have been highly purified, deposited on a clean surface, frozen to fix their structure, etched to remove water, and shadowed with metal to provide contrast. This technique, called "freeze-etch, rotary shadowing," images macromolecules in three dimensions (see Box 2.4, Chapter 2, for additional images generated with this technique). (A) The flagellar form of dynein has three spherical head domains. (B) The cytoplasmic form of dynein has two head domains. (C) Myosin II. The asterisk-shaped structure in the lower right of the image is an IgM antibody used to purify the myosin. (D) Kinesin. In this image, the two small motor domains at one end of the rod-shaped protein and the light chains at the other end are not distinguishable from each other.

Figure 5.21

Myosin proteins interact with actin filaments, moving from the minus (pointed) end to the plus (barbed) end of the polymer. Myosin II proteins, such as those in muscle, have long rod-like domains that promote assembly into bipolar (head-to-tail) filaments. Myosin II filaments pull actin filaments with opposite polarity past one another, mediating local contractions. Myosin I motors have a short tail domain that can bind an actin filament or a membrane. Force transduction by the myosin I head may then (1) move an actin filament relative to another, (2) move a vesicle along an actin filament, or (3) move an actin filament along a membrane. Myosin proteins are found in plants, but the equivalent forms of myosins I and II have not yet been described for plants.

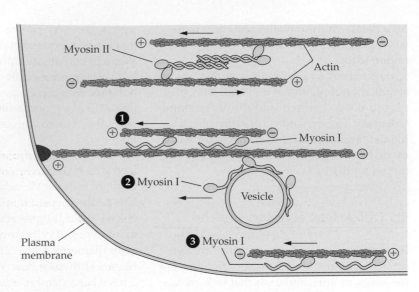

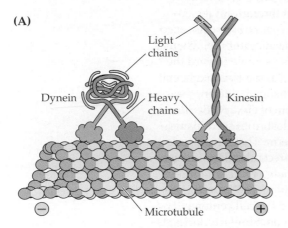

Figure 5.22

Diagrams of two microtubule-associated motor proteins. (A) Cytoplasmic dynein and kinesin, drawn roughly to scale. Though not homologous, these proteins share a similar structural organization, each containing a pair of heavy chains and several accessory light chains. The heavy chains form globular head domains and elongated tail domains. The head binds to the microtubule, contains the ATP-binding site, and produces most if not all of the force. The tail domain is responsible for binding the specific cargo. The light chains may regulate the activity of the motor or may modify its binding properties. (B) Schematic view of kinesin in action. Alternate binding by the two heads ensures that a head is almost always bound; hence, the motor with cargo can move along the microtubule for long distances before dissociating. Although this action is often likened to a biped walking, the two heavy chains are the same, not mirror images (as left and right legs are), and so the head rotates by 180° with each step to be in proper orientation for binding the microtubule. (C) Dyneins move cargo toward the minus end of a microtubule, whereas most kinesins move cargo toward the plus end.

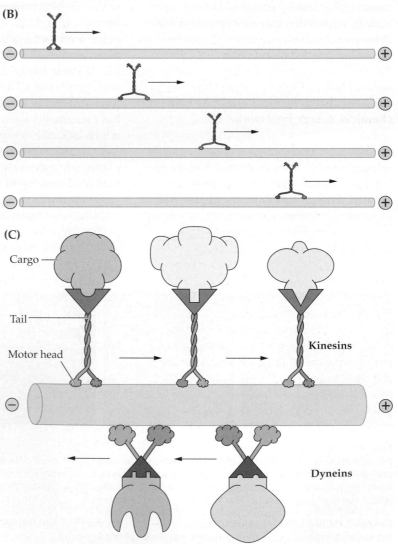

in both animals and plants are implicated in vesicle traffic and formation of **mitotic spindles** (see Section 5.11.1); plant members of the kinesin superfamily have been cloned and are being studied.

5.6.2 Other accessory proteins bind, sever, or cap cytoskeletal polymers.

In addition to motor proteins, several other types of accessory proteins affect cytoskeleton function (Fig. 5.23). Some, called "cross-linking" or "bundling" proteins, form bonds between cytoskeletal polymers of the same type. Many such proteins that act on actin filaments or microtubules have been well characterized, and a few examples for actin will now be described.

Cytoskeletal polymers in cells commonly are arranged in bundles (i.e., side-by-side). Several proteins localize to these bundles in vivo and can even bundle polymers in vitro. Among the members of this class are two actin-binding proteins, fimbrin and α-actinin. Fimbrin links actin filaments having the same polarity to form a parallel array in which all plus ends face the same direction. In contrast, α-actinin links actin filaments of opposite polarities, creating an antiparallel array in which half the plus ends face one direction and half the other. A parallel array can support movement in a uniform direction, as in cytoplasmic streaming, whereas an antiparallel array can be pushed apart or drawn together by a motor protein, as occurs in the **contractile ring** during animal cytokinesis (see Section 5.12.3). Other proteins (e.g., filamin) cross-link actin filaments at an angle, promoting the formation of a gel-like network that is known to influence the motility and cytoplasmic structure of animal cells. A gel-like network may also provide a path for transmitting a signal through the cell rapidly, because a mechanical disturbance initiated at one location in the gel will propagate through the network much faster than a chemical messenger can diffuse or be translocated.

Some cytoskeletal accessory proteins stabilize polymers; others sever or weaken them. Some proteins bind specifically to one of the polymer ends. Others, including the pollen allergen profilin, lower the concentration of subunits available for polymerization by binding to soluble actin (Box 5.2). Finally, some proteins cross-link the cytoskeleton to other components of the cell, such as membranes, biosynthetic enzymes, or signal transduction components (see Section 5.10).

5.7 Role of actin filaments in directed intracellular movement

5.7.1 Cytoplasmic streaming in algae and higher plants requires actin.

Actin filaments are known to be important in many kinds of intracellular motility. The most thoroughly studied example is cytoplasmic streaming in the giant internodal cells of algae in the family Characeae. The large size of these internodes has facilitated development of a cell model in which the streaming process can be stopped and then reactivated with purified components, allowing researchers to characterize the molecular machinery (Fig. 5.24).

In these internodal cells, the cytoplasm is partitioned into a stationary layer in the cell cortex and a streaming layer internal to the cortex, called the endoplasm. At or near the boundary between layers lie long parallel bundles of actin filaments with uniform polarity. These bundles are firmly anchored to rows of chloroplasts, themselves immobilized in the cortex by ramifying connections to the cell wall that to date remain uncharacterized (Section 5.10). In the motile endoplasm, the ER and other large organelles have myosin bound to their membranes. Moving along the fixed actin filaments, the myosin drives organelle movement in a uniform direction, as specified by the polarity of the actin filaments. The motion of the large organelles drags the entire endoplasm into the flow.

Streaming in cells of higher plants is likewise thought to involve myosin-bound organelles moving along actin filaments. Unlike the algal internodes, most plant cells have less coherent streaming; beyond the requirement for actin, little is known about the biochemical regulation of streaming in such cells. Dynamic actin filaments, like the static bundles in the giant internodes, are also able to support the streaming movements of cytoplasm (see Box 5.2).

Box 5.2   On the trail of plant profilin—from sneezes to signals

If you cry and sneeze around plants in flower, your allergy may be triggered by a fascinating, actin-binding protein present in pollen, called profilin. Medical research and plant cell biology, usually worlds apart, were united in 1991 by the discovery that many patients with pollen allergies synthesize antibodies against profilin. The antibodies recognized profilin from a variety of plants, a finding that offered the first explanation for the nonselectivity of typical pollen allergies. Even more intriguing, the patients' antibodies also recognized human profilin, suggesting the hypothesis that the allergy is aggravated through autosensitization. Although this finding has yet to dry a nose, it has simplified diagnoses and helped allergists evaluate preventive measures.

Beyond the etiology of allergy, the discovery of large quantities of profilin in pollen has proved a boon for students of the plant cytoskeleton. Profilin has now been cloned from several plant species. A small (10- to 15-kDa) protein, it occurs in all organs, although its abundance increases 10- to 100-fold as pollen matures. The predicted amino acid sequences are typically 70% to 80% similar among plants, but when profilin sequences are compared across widely separated taxa, the similarity drops to 20% to 40%. Despite this degree of sequence divergence, the function of profilin has been widely conserved. If profilin is overexpressed in British Hedgehog Kidney (BHK) fibroblasts, the effects are identical, whether the encoding gene came from a plant or a mammal. Furthermore, a profilin gene from maize completely rescues a *D. discoideum* mutant that lacks profilin. The similar properties of profilins from different sources may result from shared residues at both N and C termini and a similar three-dimensional structure. The ribbon diagrams above compare profilins from birch and acanthamoeba, a fora-

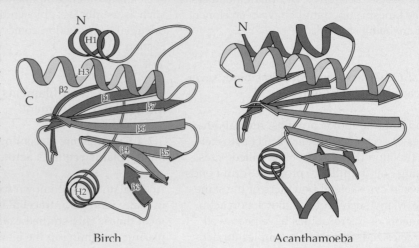

Birch   Acanthamoeba

minifer. The conserved tertiary structure includes a core, composed of a six-stranded antiparallel β-sheet, and the actin-binding domain (blue), composed predominantly of the C-terminal α-helix (H3) and the bottom three strands of the central sheet (β4, β5, and β6).

The first property of profilin to be characterized was its high affinity for G-actin but not F-actin, which suggests that profilin sequesters actin monomers, releasing them locally when and where the cell needs to polymerize actin. Indeed, the name profilin was derived from "profilamentous" actin, a term for sequestered actin that is competent to polymerize. The discovery of profilin in mature pollen fits this conception: Mature pollen has little F-actin, but abundant actin filaments form within minutes of hydration. A dramatic confirmation of the ability of profilin to sequester actin has been found in plants, where high concentrations of profilin microinjected into living *Tradescantia* stamen hair cells inhibited cytoplasmic streaming (A) and cytokinesis (B) by destroying the

actin network in a few minutes. Presumably, profilin bound up all of the G-actin and prevented any new polymerization from balancing the normal depolymerization.

The profilin story has become complicated by subsequent experiments. Profilin seems to do more than sequester actin, as shown by a recent demonstration of the localization of profilin in living pollen tubes. Because actin filaments in pollen tubes are largely absent from the 20 to 30 μm closest to the tip and plentiful elsewhere, as shown by microinjecting a fluorescent reporter that labels filamentous actin (C, upper panel), profilin content was expected to be high at the apex and low elsewhere. Surprisingly, profilin was found to permeate the cytosol uniformly. The pollen tube was microinjected with fluorescently tagged profilin (C, middle panel) as well as with a tagged dextran fluorescing at a different wavelength (C, lower panel). Dextran, which is inert and about the size of profilin, reports the total space in the cytosol accessible to diffusion. The amount of profilin injected amounted to

5.7.2 Movement and anchoring of some organelles depend on actin filaments.

Cytoplasmic streaming involves the bulk movement of cytoplasm, which tends to occur in large and specialized cell types. However, the cytoskeleton interacts with organelles in all cells to move them specifically or to anchor them at defined positions. In the cell cortex, the ER spreads out into a network of fine tubules connected by polygonal cister-

nae (see Fig. 1.16). In animal cells, spreading of the ER depends on microtubules; in plant cells, however, evidence suggests that the ER has myosin bound to its membrane and moves along actin filaments. The Golgi apparatus in animal cells is clustered beside the nucleus by means of minus-end–directed microtubule motor proteins that move the Golgi membranes toward the **centrosome**. In plant cells, the Golgi apparatus is randomly distributed throughout the cell. Recent

(A)

Cytoplasmic streaming along transvacuolar strands is evident in an uninjected cell.

Ten minutes after profilin is microinjected, streaming has stopped and most of the transvacuolar strands have broken down.

Streaming continues unabated 10 minutes after bovine serum albumin (BSA) is microinjected into a control cell.

no more than 4% of the estimated endogenous amount, much less than was used to disrupt actin in (A) and (B), and did not affect pollen tube growth or structure.

Besides sequestering actin, profilin is now known to have other biochemical activities. It can also stimulate actin polymerization and enhance the stability of actin filaments. Profilin has been shown to up-regulate the exchange of ADP for ATP on G-actin, an activity expected to stimulate polymerization. Also, the protein interacts with a regulatory subunit of adenyl cyclase and binds phosphoinositidyl lipids with a greater affinity than it binds actin. Therefore, profilin seems to have no single mode of action on the actin cytoskeleton. Sneezing and watery eyes remind us there is much yet to learn about how this protein acts in pollen and throughout the plant.

(B)

Anaphase cell undergoing chromosomal separation in an uninjected cell.

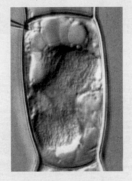

Fifty minutes after profilin is microinjected, chromosomal separation is complete, but a cell plate has not formed.

Twenty-five minutes after BSA is microinjected, a cell plate separates the control cell into two daughter cells.

(C)

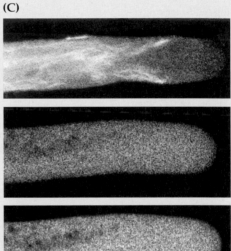

10 µm

experiments, imaging the Golgi apparatus in living plant cells, have discovered that these organelles move rapidly along actin filaments.

Mitochondria are known to move bidirectionally on microtubules in the cells of animals and fungi, but the cytoskeletal basis for mitochondrial movements in plants has not been studied thoroughly. However, plant plastids undergo complex movements known to involve actin filaments. In photosynthetic cells, as the light environment changes, chloroplasts move to absorb light optimally (Fig. 5.25). Plastid movements have been reported frequently to depend on actin filaments, but few mechanistic details are known (Fig. 5.26).

Finally, the nucleus is anchored and relocated in cells by the cytoskeleton. Again, the mechanisms that mediate nuclear positioning have not been documented, but evidence suggests that a collaboration between actin and microtubules is responsible.

Nucleating protein

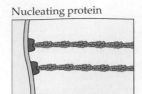

Severing protein

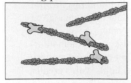

Cross-linking protein

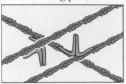

Capping (end-blocking) protein

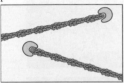

Side-binding protein

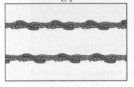

Motor protein

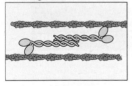

Bundling protein

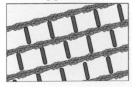

Figure 5.23
Proteins that associate with the cytoskeleton have various functions, as illustrated here with actin filaments. Although these proteins have been studied mainly in animal cells, some similar and possibly homologous proteins have been identified in plant cells, and many such are probably present.

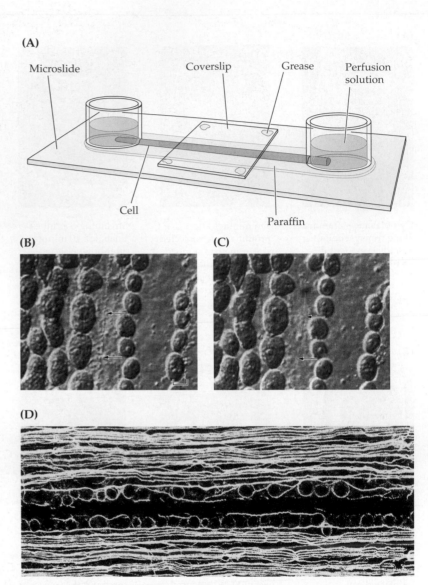

Figure 5.24
The giant internodes of Characean algae were used in the first functional studies of cytoplasmic streaming. (A) Perfusion chamber designed by Richard Williamson. An internode is placed on a microscope slide, and each end of the cell is placed under a glass ring with a notch cut to accommodate the cell. The perfusion solution is placed in the rings, and the portion of the cell between the rings is covered with liquid paraffin. A coverslip is then placed atop the cell, supported by filets of grease. Under these conditions, intact cells continue to stream for more than a week. To begin an experiment, one severs the ends of the cell and allows the solution to flow through the cell. This removes the central vacuole and much of the endoplasm; however, the cortical cytoplasm containing the chloroplasts and associated cytoskeletal elements remains. (B, C) Light micrographs of a perfused cell in the chamber. Arrows point to a bundle of actin filaments that has been dislodged from a nearby row of chloroplasts. In (B), the bundle is covered with small particles. In (C), a solution containing ATP was perfused through, which reactivates streaming, moving the particles along the bundle and out of the field of view. This movement is now known to be powered by myosin and to underlie cytoplasmic streaming in many plant cells. (D) Fluorescence micrograph showing a region of the cortical cytoplasm from a giant algal internode stained with fluorescent phalloidin to localize actin filaments (yellow). Files of chloroplasts are also visible beneath the bundles.

(A)

(B)

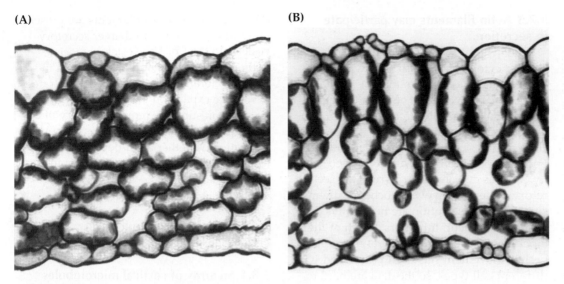

Figure 5.25
As shown in cross-sections through the leaf of *Arabidopsis thaliana*, chloroplasts move to maximize or minimize their absorption of light. (A) Under dim light, chloroplasts move to walls parallel to the leaf surface, thus maximizing light absorption for photosynthesis. (B) In bright light, chloroplasts migrate to cell walls perpendicular to the leaf surface, thus minimizing light absorption and photodamage.

(A)

(B)

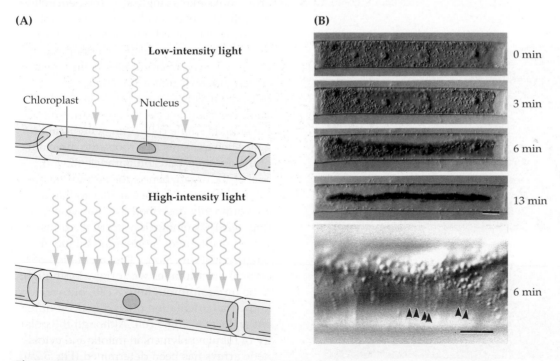

Figure 5.26
Studies of chloroplast movement often take advantage of species in the green algal genus *Mougeotia*, which are large, cylindrical cells with a single, plate-shaped chloroplast that spans the entire cell. (A) In dim light, the flat side of the chloroplast faces the light, whereas in bright light, the edge faces the light. (B) A chloroplast rotates in response to high-fluence white light. The light micrographs were imaged with Nomarski optics and taken at defined intevals. The final image is at higher magnification and shows the edge of the chloroplast moving along cytoplasmic filaments demonstrated to contain actin (arrowheads). Scale bar = 10 μm.

5.7.3 Actin filaments may participate in secretion.

During cytokinesis, it has been widely assumed but never proven that vesicles move to the forming cell wall on actin filaments, microtubules, or both. In exocytosis in nondividing cells, one can observe a close association of vesicles with actin filaments. Furthermore, inhibitors of actin-based motility, such as cytochalasin, disrupt the secretory system. However, biochemically measured rates of secretion are not always diminished by actin inhibitors. Much of the biochemical and ultrastructural research on vesicle movement has been performed with different cell types, so the data are hard to correlate. Thus, the general significance of actin in exocytotic vesicle transport requires further study.

One example in which actin filaments do play a pivotal role in exocytosis occurs in cells that undergo "tip growth," that is, those that confine exocytosis to the extending tip. In tip-growing cells (e.g., pollen tubes), polarized actin filaments are present in the tube, arranged to deliver secretory vesicles to the tip. The vesicles associate with actin filaments and have myosin in their membranes (Fig. 5.27). The high rates of relative expansion sustained by tip-growing cells presumably require delivery of a large quantity of wall precursors without interruption. Whether actin-dependent exocytosis is used by other cell types remains to be determined.

5.8 Cortical microtubules and cell expansion

5.8.1 An array of cortical microtubules helps orient cell expansion.

An array of microtubules usually lies just inside the plasma membrane of plant cells, in the cortical cytosol. The microtubules are found within 200 nm of the plasma membrane and sometimes appear to contact it. In immunofluorescence light micrographs, the array looks like a single structure, encircling the cell. However, in electron micrographs, the array is resolved into a series of short (less than 10 μm), partially overlapping microtubules. The microtubules within the array are oriented uniformly. In elongating cells, the majority of microtubules lie perpendicular to the direction of most rapid expansion. In single cells or in cylindrical organs such as stems and roots, the direction of maximum expansion parallels the long axis of the organ; hence, the cortical microtubules, oriented perpendicular to that axis, are commonly described as "transverse" (Fig. 5.28). As elongation stops, the transverse microtubules reorient to oblique or longitudinal directions. The orientation of cortical microtubules was originally thought to be uniform throughout a cell, but we now know that orientation often differs on different walls of a single cell. Although the polarity of tubulin polymers in mitotic and cytokinetic arrays has been determined (Fig. 5.29), the polarity of microtubules in the cortical array has not.

The mechanism that organizes the cortical array remains an outstanding problem in plant cell biology. The cortical array is not associated with any detectable microtubule-organizing structure, such as the basal body

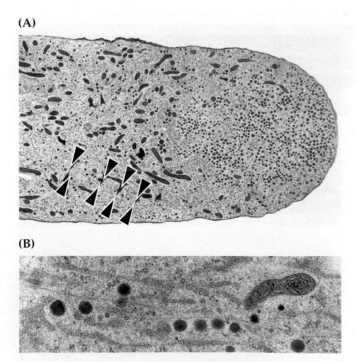

(A)

(B)

Figure 5.27
Actin filaments provide tracks for the movement of vesicles to the tip in the growing pollen tube. Secretory vesicles undergo localized exocytosis at the tube tip to restrict growth to that region. (A) Electron micrograph of the tip region of a pollen tube. Note that the extreme tip is filled with secretory vesicles and that other organelles are excluded. Bundles of actin filaments are shown by the paired arrowheads. (B) High-magnification view from the body of the tube, showing secretory vesicles lined up along actin filaments, presumably frozen on their way to the tip.

of an axoneme or the centrosome of an animal cell. A divergent member of the tubulin family, γ-tubulin, found in animal centrosomes, confers the ability to nucleate microtubules (see Section 5.11.1). Recently, γ-tubulin was discovered in the plant cortical array (Fig. 5.30). However, other than in developing guard cells, the protein has not been found in any detectable pattern, and its role in organizing the array remains uncertain.

5.8.2 Mechanisms that control microtubule orientation resemble a signal transduction pathway.

It is useful to view microtubule organization in the context of signal transduction, where three events can be delineated: signaling, perception, and response (see Chapter 18). To produce a group of uniformly oriented microtubules, *(a)* a signal must specify a direction, *(b)* that signal must be perceived by some cellular components and transmitted to the microtubules, and *(c)* the microtubules themselves must line up accordingly.

Models have been proposed to explain each of these steps. The signal providing a polarity cue could be mechanical, electrical, or chemical. The transduction mechanism (i.e., perception) is widely thought to involve some kind of transmembrane protein or protein complex, in which the extracellular domain "senses" the signal, and the cytosolic domain constrains microtubule behavior. Microtubule behavior (i.e., the response) can be modified by altering the interactions between microtubules, by rotating microtubules into correct alignment with a putative rotary motor protein, or by selectively stabilizing those microtubules that happen by chance to polymerize at the correct angle. Cortical microtubule behavior is known to be affected by the components of cellular signal transduction, such as calmodulin and protein kinases (Fig. 5.31).

5.8.3 The cortical array helps align cellulose microfibrils.

The principal documented function of the cortical array is to influence the direction of cellulose microfibril deposition, which in

(A) **(B)**

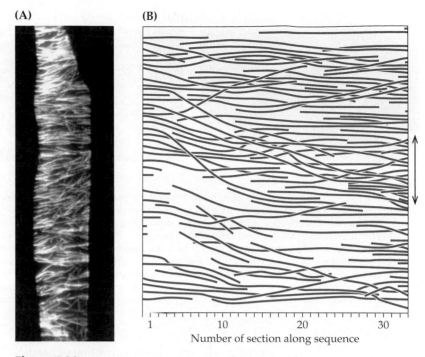

Figure 5.28
(A) Fluorescence micrograph of a cell from the root of *Arabidopsis thaliana*, showing the cortical array of microtubules. The microtubules are oriented mainly with their axes transverse to the long axis of the cell (and root). (B) Drawing of a region of the cortical array from a cell in the root of *Azolla pinnata*, based on reconstructions from serial sections examined with electron microscopy. The double-headed arrow represents the direction of maximum expansion rate.

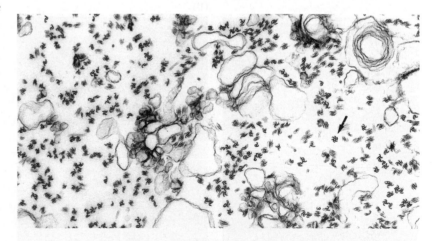

Figure 5.29
Electron micrograph through the phragmoplast of a higher plant illustrates the experimental determination of microtubule polarity. When microtubules are exposed to exogenous tubulin, new tubulin assembles onto existing microtubules along the walls, forming curved, ribbon-like extensions. Often called "hooks," they appear in cross-section as arcs protruding from the circular microtubule. The direction of hook curvature is specified by the polarity of the microtubule. Here, most of the hooks bend in the same direction (counterclockwise, marked with arrow), showing that the polarity of the microtubules is highly uniform.

(A)   (B)   (C)

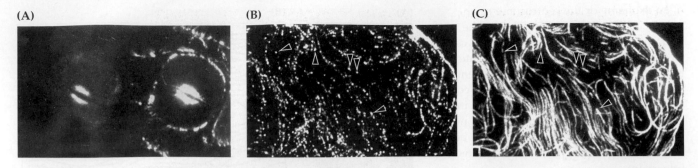

Figure 5.30

The recently discovered third class of tubulin, γ-tubulin, is present in plant cells and may participate in organizing cortical microtubule arrays. (A) Fluorescence micrographs of a developing pair of guard cells in an onion leaf. In these cells, the cortical array is tightly focused at the middle of the ventral wall, where the future pore will open, and fans out from there to reach the rest of the cortex. Experimental studies suggest that this ventral focus is the site of microtubule organization. The guard cells were double-labeled with one antibody that recognizes γ-tubulin (left) and with another that recognizes β-tubulin (right); imaging with confocal optics delineated a single optical section. The γ-tubulin is present mainly at the putative focus of microtubule organization. (B, C) Fluorescence micrographs of cortical microtubules from a suspension-cultured tobacco cell. Microtubules have been double-labeled with antibodies raised against γ-tubulin (B) and β-tubulin (C). Arrowheads point to punctate γ-tubulin staining, which appears to correspond to the ends of microtubules, suggesting a role for γ-tubulin at the minus ends of cortical microtubules.

growing cells specifies the direction of maximal cell expansion. In nearly all plant cells (except tip-growing cells such as pollen tubes and root hairs), the alignment of the cellulose microfibrils parallels that of the microtubules (Fig. 5.32). Cellulose microfibrils closely parallel the underlying cortical microtubules when protoplasts regenerate a cell wall. Both cellulose and microtubules become oriented in helices as cells mature. Cellulose microfibrils also parallel the microtubules in developing secondary walls. When thickenings form in tracheids, the banded deposition of cellulose is preceded by a bundling of cortical microtubules. Finally, when microtubules are lost, either by experimental treatment with inhibitors or naturally, as in the leaf sheath cells of certain bulb-forming monocots, cellulose microfibrils are deposited randomly and cell expansion is no longer directional.

Although such evidence makes clear that cortical microtubules influence the deposition of cellulose, the mechanism behind this influence has not been discovered. Motor proteins may drive cellulose-synthesizing enzymes along cortical microtubules. Or the enzymes may lie trapped between barriers of parallel cortical microtubules. Refutation of either hypothesis awaits further experiments.

(A)   (B)

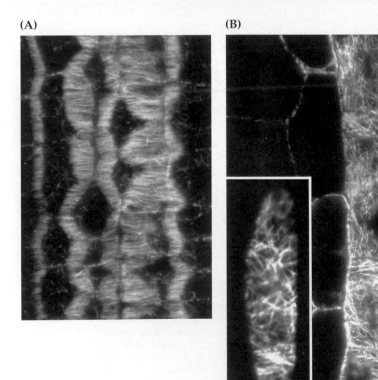

Figure 5.31

Protein phosphorylation plays a role in microtubule organization. (A) Cortical microtubule arrays in a growing region of the *Arabidopsis thaliana* root possess the typical transverse orientation. (B) Cortical arrays in a similar region of a root treated with staurosporine, an inhibitor of protein kinases, are organized aberrantly. Inset is another cell at higher magnification. Certain inhibitors of protein phosphatases cause similar disorganization (not shown). The protein kinase inhibitor could be interfering with the signal, its perception, or the response.

5.9 Observing cytoskeletal dynamics

The name "cytoskeleton" conjures up images of a static network of struts or scaffolding, and these metaphors dominated early discussions of cytoskeletal function. Although clues were available to cytoskeletal dynamics—for example, from polarized light microscopy (Box 5.3)—not until scientists imaged actin filaments and microtubules directly in living cells did the extraordinary dynamics of these polymers become widely appreciated.

Imaging cytoskeletal polymers directly in living cells requires introducing a detectable reporter molecule into the cell. The reporter molecule must not interfere with the function of native structures. Researchers found that a small fluorochrome (e.g., fluorescein or rhodamine) could be conjugated to tubulin or actin without appreciably altering the biochemical properties of the protein. The flagged subunits could then be microinjected into living cells and incorporated into the native cytoskeletal polymers, thus rendering the cytoskeleton visible in fluorescence microscopy. The general name for this method is fluorescent-analog cytochemistry (Fig. 5.33).

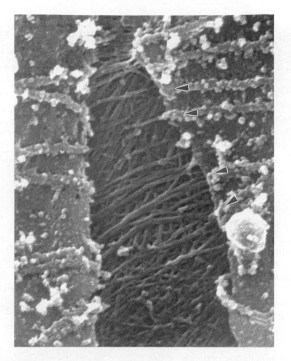

Figure 5.32
Field-emission scanning electron micrograph of the interior of an onion root cell, looking toward the cell wall. Shown is the cortical cytoplasm on either side of a tear in the plasma membrane, through which the cell wall is visible. The orientation of cortical microtubules (noted with arrowheads) can be seen to parallel the direction of many of the microfibrils in the cell wall.

5.9.1 In plants, cytoskeletal polymers are imaged by confocal laser-scanning fluorescence microscopy.

Although the cytoskeleton was observed in living animal cells in the late 1970s, plant cells were more difficult to microinject because of their thick cell walls and large turgor pressures. These obstacles were overcome in 1990 by Peter Hepler and coworkers who made microneedles of optimal size and used hydrostatic pressure to introduce fluorescence-tagged mammalian tubulin into stamen hair cells. Once the reporter molecule was successfully introduced, conventional epifluorescence microscopy was found to damage the cell. The solution was to use a confocal laser-scanning fluorescence microscope. Although the laser excitation is intense, its dwell time at any one location on the sample is extremely short, minimizing overall the amount of incident light energy. In fact, plant cells have proved superior to animal cells for long-term imaging through the confocal microscope, perhaps because plants have evolved mechanisms to resist damage by high light energy.

Fluorescent tubulin from brain was found to be rapidly incorporated into all microtubule arrays of stamen hair cells without disturbing the cell or altering the kinetics of mitosis (Fig. 5.34).

Besides stamen hair cells, microtubules in living cells have been imaged in pea epicotyls (Fig. 5.35A) and in the giant internodes of the Characean algae. In all cases, microtubule arrays have been found to be highly dynamic. Actin filaments have also been imaged in living plant cells. Although fluorescence-tagged mammalian muscle actin damages plant cells and therefore cannot be used to image plant actin filaments in vivo, fluorescent conjugates of the small bicyclic peptide phalloidin are usable though imperfect markers for actin filaments in plants (Fig. 5.35B). Not only can phalloidin affect actin filament behavior, it can become sequestered in vacuoles and lost from the cell through plasmodesmata, preventing long-term observations. Nevertheless, use of this probe has generated important knowledge about the role of actin in plant cells.

Box 5.3

The development of polarized light microscopy provided the first images of the cytoskeleton in living cells.

Well over a century ago, microscopists identified a fibrous structure associated with the chromosomes of dividing cells. They called the structure the mitotic spindle. To image the spindle, they fixed and stained cells with complex chemical mixtures of unspecified activity. As knowledge of the chemical properties of organic matter became more sophisticated in the 1920s and 1930s, chemists seeking physicochemical explanations for cell activities dismissed these mitotic spindle fibers as artifacts of chemical preparation. Efforts to visualize spindle fibers in living cells failed. Furthermore, the chemists synthesized fibrous structures that looked like mitotic spindles from colloidal suspensions and even from cytoplasm itself.

The situation changed in the early 1950s, when Shinya Inoué convincingly demonstrated spindle fibers in a variety of living mitotic cells. Inoué succeeded by enhancing polarized light microscopy.

To produce contrast between the image of an object and its background, different types of microscopy depend on different interactions between light and matter. Bright-field microscopy depends on absorption, and so cannot image the transparent structures of living cells. Polarized light microscopy depends on molecular order and images objects such as crystals, in which most component molecules are aligned. However, the conventional polarized light microscope, which is subject to specific optical aberrations, is not sensitive enough to detect the small degree of molecular order typical of most biological objects. Inoué invented rectified optics, an optical system that removed these aberrations and allowed the polarized light mi-

(A)

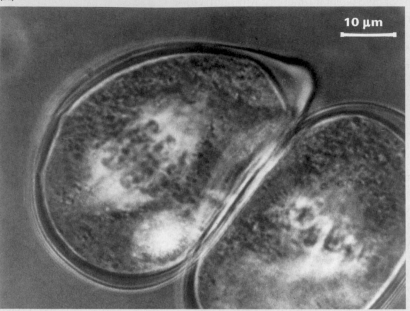

10 µm

croscope to image the ordered contents of living cells at high magnification.

Polarized light microscopy is powerful because of its ability not only to reveal but also to quantify the extent of molecular order in the image. A sample containing aligned molecules will have different refractive indices for light waves polarized in different directions. The difference between the maximal and minimal refractive index is termed birefringence. The signal in a polarized light microscope, termed retardation, is proportional to the product of the sample's birefringence multiplied by its thickness. Retardation is quantified with

respect to a crystal of known retardation, called a compensator.

Inoué used polarized light microscopy first to demonstrate the reality of the mitotic spindle, as shown in the image of living *Lilium longiflorum* microsporocytes in anaphase (left cell) and metaphase (right cell) of meiosis (A). The bright, fibrous material is the mitotic spindle, imaged by virtue of the birefringence of aligned microtubules. With the reality of the spindle thus proven, Inoué then delineated how its structure changes during mitosis. The sequence of micrographs in (B) shows the spindle in a living endosperm cell of

5.9.2 Green fluorescent protein provides an alternative to microinjection.

Despite the success of fluorescent-analog cytochemistry in imaging the plant cytoskeleton, the technical difficulties of microinjection have slowed widespread adoption of this method. A potential alternative is the use of green fluorescent protein (GFP). Originally isolated from a luminescent jellyfish *(Aequora victoria)*, this small protein contains amino acids with side chains that form a brightly fluorescent fluorophore. Unlike

fluorophores that must be conjugated to polypeptides and microinjected, the GFP is introduced as a label at the gene level, the GFP coding sequence having been fused to the genes for a variety of proteins. When these chimeric genes are expressed in cells, the fluorescent protein frequently retains full activity. GFP is fast becoming the reporter of choice for localizing proteins in living cells (see Fig. 1.25). This approach promises exciting opportunities for observing cytoskeletal dynamics in living plant cells (Fig. 5.36).

(B) Subtractive contrast · Additive contrast **(C)**

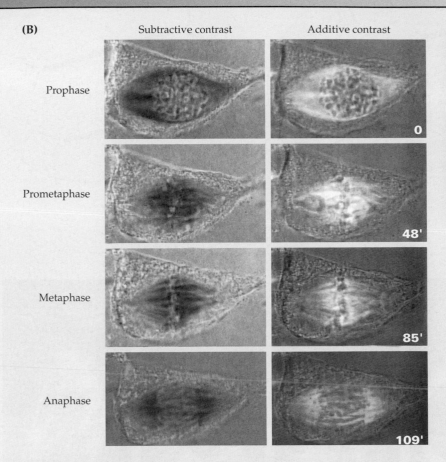

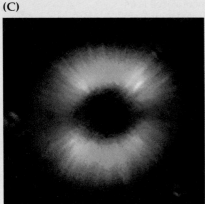

Prophase

Prometaphase

Metaphase

Anaphase

Haemanthus katherinii as it traverses prophase, prometaphase, metaphase, and anaphase. The micrographs are paired, with the compensator providing subtractive contrast (left) and additive contrast (right). Because the spindle fibers are birefringent, they reverse in contrast, appearing darker than background at one setting and lighter than background in the other. Images of the mitotic spindle obtained with polarized light are comparable with those obtained 30 years later with microinjected fluorescent tubulin (see Fig. 5.34). Inoué also manipulated the spindle and quantified the resulting changes. By so doing, he developed a theory for the dynamic behavior of the spindle components that correctly anticipated many of the properties of microtubules discovered years later, after tubulin had been identified and purified.

Imaging techniques for polarized light microscopy, pioneered by Inoué, continue to advance. Panel C shows a sea urchin aster, a radial array of birefringent microtubules (see Fig. 5.41). The intensity of each pixel is proportional to the amount of retardation. In addition, the color of each pixel represents the orientation of the optical axis of each birefringent element of the structure (red is horizontal, green is vertical). The micrograph was taken on a polarized light microscope equipped with Cambridge Research Instruments' "LC PolScope" system, invented by Rudolph Oldenbourg and collaborators, which improves the speed and accuracy of quantitative polarized light microscopy.

5.10 The cytoskeleton and signal transduction

5.10.1 Connections between plant cell wall, plasma membrane, and cytoskeleton may participate in processing information.

Cytoskeletal anchorage to the cell wall may play an informational role. The structural state of the cell wall reflects both internal parameters, such as turgor pressure or expansion rate, and external parameters, such as mechanical vibration, water status, or gravity. To sense these parameters, the plant cell might report the mechanical status of the wall to the cell interior through localized regions where the cell wall associates intimately with the plasma membrane (Fig. 5.37). The importance of such specialized attachment sites was first suggested by analogy to animal cells, in which transmembrane receptor proteins, called integrins, physically link the cytoskeleton to the extracellular matrix and mediate signal transduction based on the status of that linkage.

Figure 5.33
Schematic representation of fluorescent analog cytochemistry as applied to cytoskeletal polymers. A cell is microinjected with a small volume of a solution containing the protein of interest conjugated to a bright fluorescent molecule (a fluorochrome). A short time later, fluorescent subunits as well as native subunits are incorporated into cytoskeletal structures.

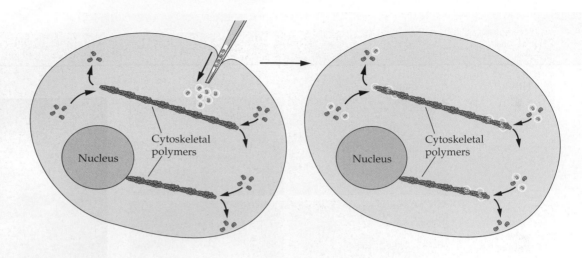

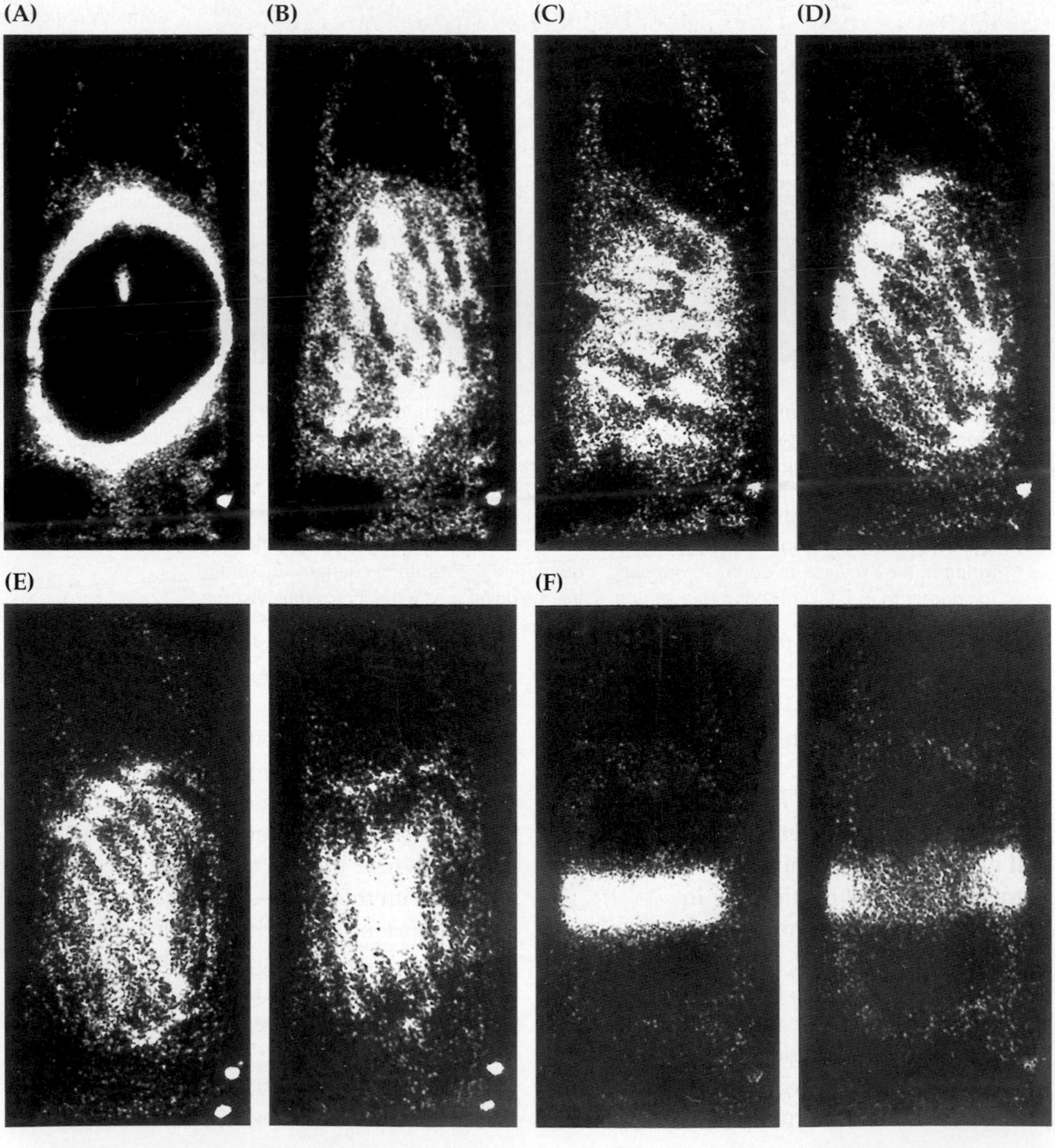

Nucleus

Cytoskeletal polymers

Nucleus

Cytoskeletal polymers

(A)   **(B)**   **(C)**   **(D)**

(E)   **(F)**

Figure 5.34
Confocal fluorescence micrographs of living stamen hair cells that were microinjected with tubulin purified from cow brain and derivatized with a flurochrome. These images represent the first use of fluorescent-analog cytochemistry to label the cytoskeleton in living plant cells. The disposition of microtubules is shown as the cell moved from (A) prophase through (B) prometaphase, (C) metaphase, (D) anaphase, (E) telophase, and (F) cytokinesis.

(A)

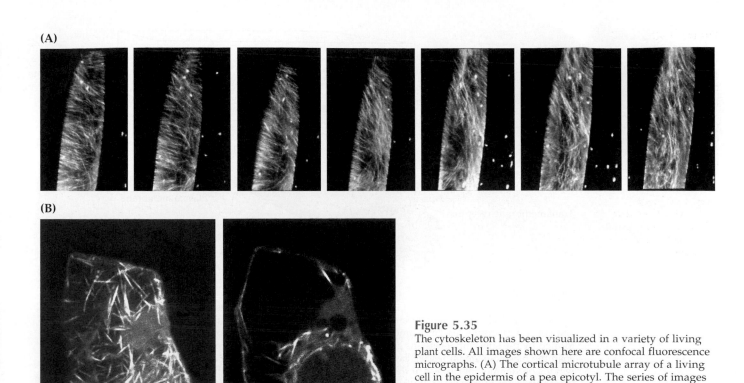

(B)

Figure 5.35
The cytoskeleton has been visualized in a variety of living plant cells. All images shown here are confocal fluorescence micrographs. (A) The cortical microtubule array of a living cell in the epidermis of a pea epicotyl. The series of images shows the same cell at approximately 10-minute intervals, as the orientation of the array changes from being mainly transverse to longitudinal. (B) Actin filaments of a living leaf epidermal cell of *Tradescantia virginiana* at two different focal planes.

Plant cell biologists are painting an analogous picture, although to date only a few brush strokes have been laid down. Evidence suggests that the cytoskeleton is linked to the plasma membrane (Fig. 5.38). For example, when fragments of the plasma membrane are made by adhering protoplasts to a substrate and then lysing them, cortical microtubules remain bound to the membrane. This association is lost when the protoplasts are first treated briefly with a protease, presumably because the protease inactivates a transmembrane protein that binds microtubules. Likewise, convincing evidence shows that the plasma membrane is anchored to the wall at localized attachment sites. These sites are demonstrated by placing cells in a solution that is sufficiently hypertonic to cause them to shrink to a volume smaller than the space enclosed by the cell wall, a process called **plasmolysis**. In plasmolyzed cells, the protoplast does not completely shrivel inward but remains stuck to the wall in many regions. Often, these sites are small in area, and upon

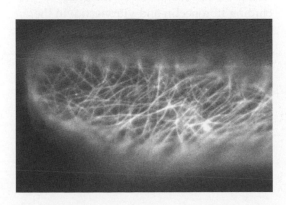

Figure 5.36
Green fluorescent protein (GFP) provides a powerful alternative to microinjection for labeling cytoskeletal structures. This example shows cortical microtubules in a bean leaf epidermal cell expressing GFP fused to the microtubule-binding domain of a microtubule-binding protein. The GFP was fused to an accessory protein because GFP-tubulin might disrupt assembly or disassembly. These microtubules are brightly labeled and can be observed for as long as the cell continues to express the fusion protein.

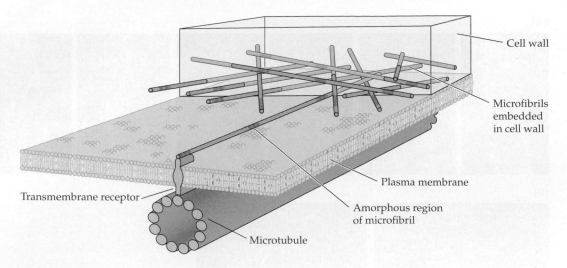

Cell wall

Microfibrils embedded in cell wall

Plasma membrane

Amorphous region of microfibril

Microtubule

Transmembrane receptor

Figure 5.37
Schematic view of how mechanical information may be sensed at the plasma membrane. This cutaway view of the cell wall illustrates the relationship between cellulose microfibrils, receptors in the plasma membrane, and cortical microtubules. Some amorphous regions on the microfibrils (or on other wall polymers) are deformed by the substantial tensions generated by turgor pressure or other mechanical loads. Receptors that span the plasma membrane have two binding sites: At their extracellular domain they bind the deformed regions of the microfibrils, and at their cytosolic domain they bind microtubules (or other cytoskeletal polymers). Changes in the mechanical loading will change the extent of deformation of the microfibril, which in turn could increase or decrease its binding affinity to the receptor. Occupancy of the receptor's extracellular binding site determines the activity of its cytosolic site. The cytosolic site could have kinase activity and participate in signal transduction directly, or it could affect the behavior of cytoskeletal polymers.

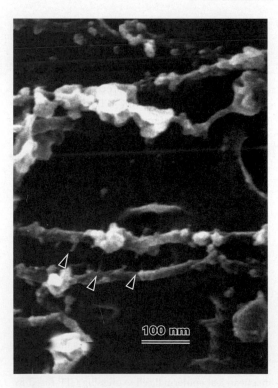

100 nm

Figure 5.38
Field-emission scanning electron micrograph of an onion root cell cortex showing connections (arrowheads) between cortical microtubules and the plasma membrane.

plasmolysis, the wall and the shriveled protoplast are linked by a large number of thin, membranous strands. These strands are called Hechtian strands, after the botanist who described them in the 1910s (see Fig. 1.13).

5.10.2 Evidence links adhesion proteins of plant cell walls and animal integrins.

In addition to data that indicate physical links between wall, cytoskeleton, and membrane, immunological evidence suggests that plants contain proteins related to animal integrins and to the extracellular matrix adhesion proteins recognized by integrins, including fibronectin and vitronectin. In the plant cell, these proteins have been implicated in adhesion, mediating wall-to-membrane attachment during diverse activities, such as pollen tube growth through the pistil style, gravity sensing, acclimation to salt stress, and even invasion by pathogens. In the walls of onion epidermis, these proteins have been localized to small, discrete patches, being of a size and density plausibly corresponding to the attachment sites of the Hechtian strands. Despite the immunological cross-reactivity, adhesion proteins in plant cell walls may be analogous to rather than homologous to their counterparts in animal cells. Recently, a gene for a plant cell wall protein related immunologically and functionally to vitronectin was sequenced and found to be nearly identical to a translation factor.

Whether they are homologs or analogs, we have only just begun to elucidate the activities of the proteins that join cell wall to cytoskeleton. Although the cytoskeleton seems certain to be anchored to the cell wall, the details of this linkage are unknown. In addition to providing physical connections, these attachment sites probably constitute conduits through which information about the cell's surroundings flows into the cell. A great challenge for the future will be to detect this information flow and to decipher its meaning.

5.11 The cytoskeleton and mitosis

Every cell on the planet was formed by cell division. The regulation of cell division is discussed in Chapter 11; here we describe the mechanics of the division process. **Mitosis** defines that stage in the cell cycle when the duplicated chromosomes are physically separated into two new nuclei. **Cytokinesis** describes the separation of one cell into two. Both mitosis and cytokinesis are accomplished by means of cytoskeletal structures: the mitotic spindle and the phragmoplast, respectively. In each process, key questions include how the cytoskeletal structures are assembled and how forces are exerted to move chromosomes or separate the newly formed cells.

5.11.1 Mitotic spindles in plant and animal cells differ in spindle pole structure.

The mitotic spindle consists of hundreds to thousands of microtubules as well as their associated proteins. Other components are actin filaments and membrane-bound organelles derived from the endoplasmic reticulum. The mitotic spindle has two poles, which define not only the sites where the separated chromosomes will finally arrive but also the spindle's bipolar organization.

Microtubules radiate away from each pole with their minus ends toward the nearest pole and their plus ends away (Fig 5.39). Some microtubules run from the pole to specialized sites of attachment on the chromosomes, called kinetochores (see Section 5.11.3), whereas others extend from the pole for various lengths. Some of the latter meet and interact at the spindle equator with microtubules emanating from the opposite pole, forming a zone of overlap that stabilizes the spindle structure. Microtubules in the spindle are highly dynamic, with half-lives on the order of 1 min. Kinetochore microtubules are longer lived but still turn over many times during mitosis. Entire microtubules may turn over, depolymerizing down to the pole, while new ones are nucleated at the pole and grow outward. Many microtubules also undergo treadmilling (Section 5.5.2), in which microtubule subunits are added at the kinetochore and are lost at the pole. Treadmilling creates a coherent flux of tubulin dimers moving through the spindle, from kinetochore to pole.

The major difference between the spindles of animals and plants is the structure of the pole. In animal cells, the spindle pole is tightly focused; in plant cells, the pole is usually broad (Fig. 5.39). The focused pole of an animal mitotic spindle is usually built around a conspicuous structure called a **centrosome** (Fig. 5.40). At interphase, the centrosome is adjacent to the nucleus and nucleates a radial array of microtubules; during S phase it replicates, and the replicated centrosomes separate during prophase to form two radial arrays of microtubules (each often called an aster) that become the spindle poles (Fig. 5.41). Plant cells have no centrosomes. Instead, the plant spindle has diffuse poles, containing numerous microtubule foci linked by interdigitating microtubules (see Fig. 5.39). As in animal cells, the assembly of the plant mitotic spindle begins during prophase. In interphase plant cells, most microtubules are part of the cortical array, but some radiate from the nuclear envelope, nucleated there by proteins that are possibly related to constituents of the animal centrosome. In prophase, cortical microtubules disappear, except for those in the preprophase band (see Section 5.12.2). Myriad microtubules come to lie parallel to the nuclear envelope surface, radiating from two foci on opposite sides of the nucleus that define the future poles of the mitotic spindle. This structure is often called the prophase spindle (Figs. 5.42 and 5.43). What determines the placement of the opposed microtubule foci of the prophase spindle remains unknown, as is the reason why the two focused poles of the prophase spindle become diffuse after nuclear envelope breakdown.

5.11.2 Despite similarities, plant mitotic spindles and oocyte meiotic spindles do not share an organizational mechanism.

Lacking centrosomes, the poles of the mitotic spindle in plant cells have been compared to the poles of the meiotic spindles of oocytes, which in many animal species also lack centrosomes. However, the meiotic spindles of

(A) Animal spindle

(B) Plant spindle

(C)

(D)

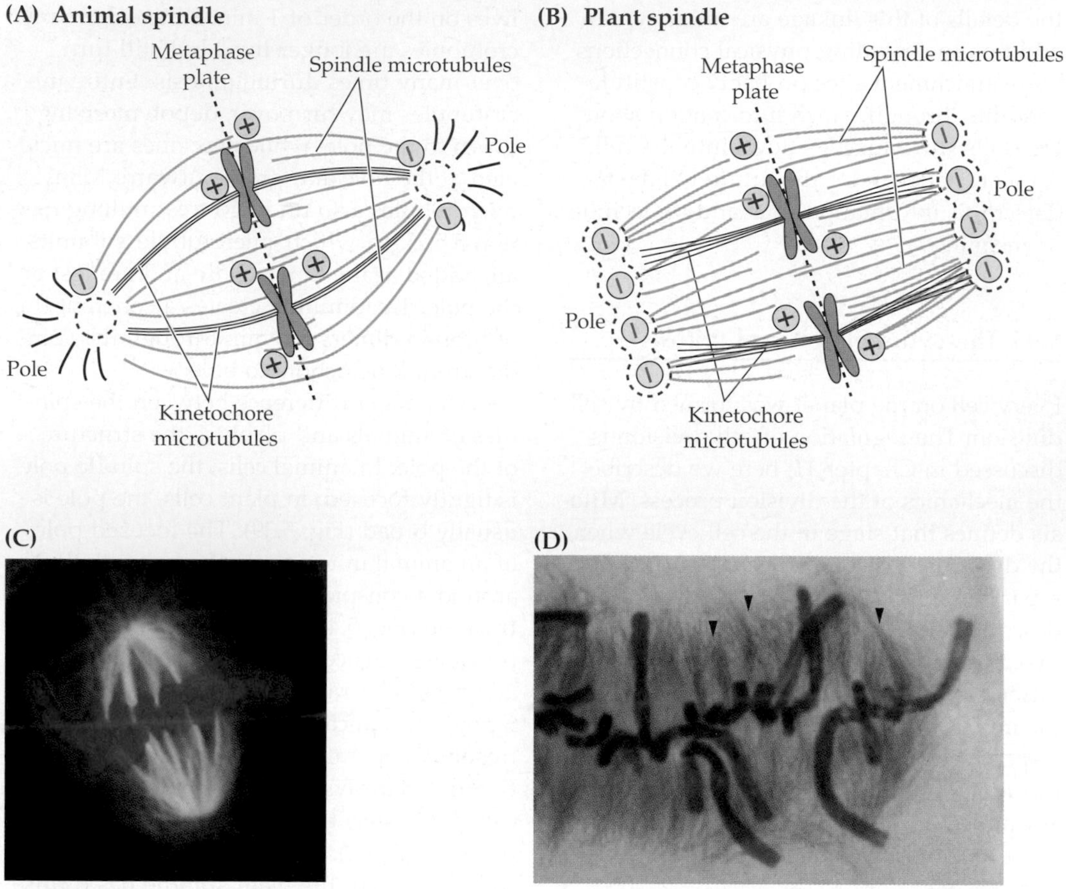

Figure 5.39
(A, B) Schematic comparison of the organization of the mitotic spindle in cells of animals and plants. Both spindles are at metaphase (only two pairs of chromosomes are shown for clarity). The plane occupied by the chromosomes (the equator, or metaphase plate) is a plane of symmetry for the spindle. Each half-spindle contains a pole, kinetochore microtubules (prominent bundles of microtubules that run from the kinetochore up to or near the pole), and many spindle microtubules that may enter the area occupied by the other half-spindle but do not reach the opposite pole. The minus ends of most of the microtubules are near or facing the poles, whereas the plus ends grow away from the poles. In animal cells, the poles are tightly focused at the centrosome, whereas in plant cells, the poles are diffuse, containing many subsidiary foci and probably centrosomal components spread out among the foci. (C) Fluorescence micrograph of a metaphase spindle from a vertebrate tissue culture cell. The chromosomes are blue and the microtubules are green. This spindle has the focused poles and prominent kinetochore bundles typical of animal spindles. (D) Light micrograph of a region of the metaphase spindle of an endosperm cell from *Haemanthus katherinii*. These wall-less cells spread out on the microscope slide so that the usual barrel-shaped spindle is flattened into a highly linear structure. Although lacking an overall polar organizing center, numerous subsidiary foci of microtubules are present (arrowheads). In this image, microtubules were made visible with a primary anti-tubulin antibody followed by a gold-conjugated secondary antibody, which in transmitted light appears red brown.

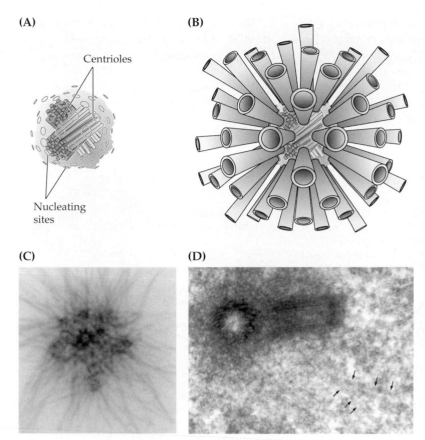

(A) Centrioles

Nucleating sites

(B)

(C)

(D)

Figure 5.40
(A) Schematic view of an animal centrosome, composed of a pair of centrioles oriented at right angles and surrounded by a protein-rich matrix of indeterminate structure, termed pericentriolar material. One component of this material is γ-tubulin, which along with other proteins forms ring-shaped templates that play a role in nucleating microtubules at the centrosome. (B) Microtubules grow out from the centrosome with their minus ends bound to the γ-tubulin rings buried in the pericentriolar material. The structural interaction between a ring and a microtubule is not yet known with certainty. (C) Electron micrograph of an isolated centrosome and microtubules from *Drosophila melanogaster.* One of the centrioles is just visible as a dark ring near the center of the pericentriolar material (arrow). (D) Electron micrograph through a centrosome from a mouse embryo. A pair of dark centrioles is seen with paler radiating microtubules (arrows). The protein-rich pericentriolar material—often termed "osmiophilic fuzz" because of its high affinity for osmium tetroxide (a heavy metal reagent used routinely in transmission electron microscopy).

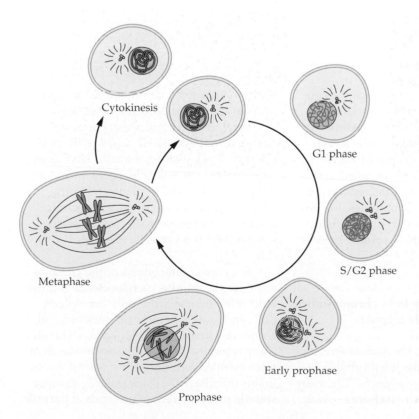

Cytokinesis

G1 phase

S/G2 phase

Metaphase

Early prophase

Prophase

Figure 5.41
The centrosomal cycle of an animal cell. During G1 phase of the cell cycle, the centrosome consists of pericentriolar material surrounding a single pair of centrioles. The pair of centrioles duplicates during S phase, but the two pairs remain closely associated in a single complex. In early prophase, the complex splits, forming two separate centrosomes, each nucleating a radial array of microtubules, called an aster. As prophase continues, the centrosomes separate further and define the two poles of the mitotic spindle, which forms between them. By the end of mitosis, when the chromosomes have reached the poles, each daughter nucleus reforms near the centrosome. Each new cell generated by cytokinesis inherits a single centrosome.

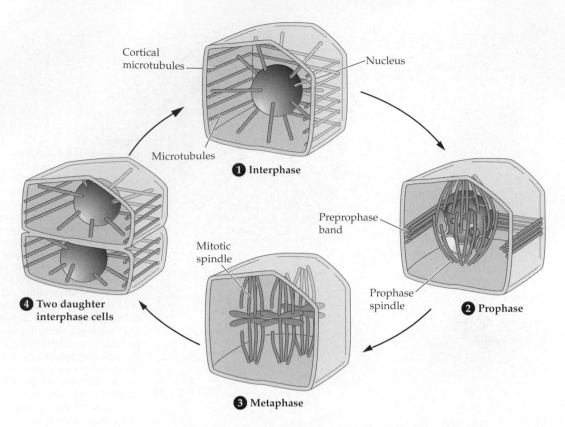

Figure 5.42
Steps in the assembly of the mitotic spindle of higher plants. (1) In an interphase cell, microtubules are present in the cortical array and radiate from the nuclear envelope. (2) During prophase, the cortical microtubules coalesce into a band at the plane of the future new cell wall, called the preprophase band. The nuclear envelope gains microtubules, primarily at two foci on opposite sides of the nucleus. This formation is called the prophase spindle, and these foci define the axis of the future mitotic spindle. (3) At metaphase, the preprophase band has been lost and microtubules form the mitotic spindle. (4) After cytokinesis, the new cell wall meets the parent cell wall at the location previously occupied by the preprophase band, and the cortical and nuclear microtubule arrays reform.

oocytes form in a process in which the condensed chromosomes nucleate randomly oriented microtubule bundles. The bipolar organization of the mature spindle subsequently emerges as the divergent microtubules coalesce, apparently bundled and focused by motor proteins directed toward the minus ends of the microtubules. In contrast, bipolar organization in plant cells is established by the prophase spindle before nuclear envelope breakdown.

5.11.3 The kinetochore links chromosomes and spindle in a flexible manner.

At the end of prophase, the nuclear envelope breaks down, and spindle microtubules begin to interact with chromosomes. The main sites of interaction are **kinetochores,** which form on each sister chromatid at opposite sides of the paired centromeric region. The kinetochore is a cup-shaped, protein-rich structure associated closely with centromeric DNA. Various proteins have been localized to animal kinetochores, including dynein and kinesin. In plants, the ultrastructural appearance of the kinetochore resembles that in animals, but the protein composition has yet to be elucidated (Fig. 5.44).

For correct attachment of the chromosomes to the spindle, each kinetochore must bind microtubules from only one spindle pole, and sister kinetochores must bind microtubules from opposite poles. Each kinetochore typically binds dozens to hundreds of microtubules, forming a fiber. Microtubules in this fiber have the same polarity as other spindle microtubules: minus ends at the pole

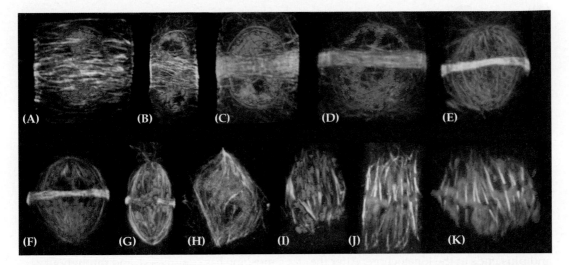

Figure 5.43
Confocal fluorescence micrographs of fixed wheat root cells, showing the assembly of the mitotic spindle. Microtubules are green and yellow; DNA is blue. (A–D) The preprophase band develops from the coalescence of cortical microtubules before or during prophase. (E–H) The prophase spindle forms from foci of microtubules at either end of the nucleus. (G, H) In late prophase, the preprophase band disassembles. (I–K) The mitotic spindle forms after breakdown of the nuclear envelope as the poles spread laterally and kinetochores capture bundles of spindle microtubules.

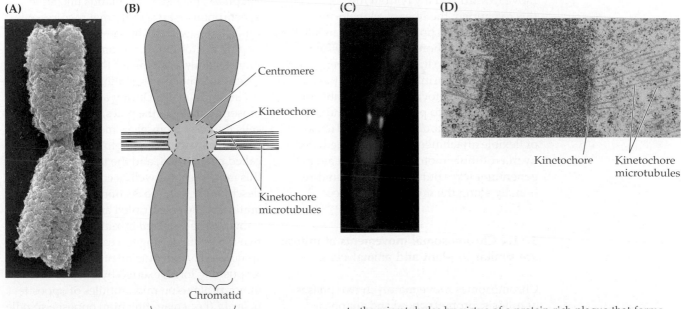

Figure 5.44
The kinetochore. (A) Classical studies of genetics and cytology identified a chromosomal region that lacked genes and recombined infrequently as the location where sister chromatids attach to each other. This region, termed the centromere, can often be discerned in light micrographs as a constriction of the chromosome and is shown clearly in this scanning electron micrograph of a plant metaphase chromosome. (B) The centromeric region of the chromosome is now known to mediate attachment of the mitotic spindle to the microtubules by virtue of a protein-rich plaque that forms at each sister centromere, called the kinetochore. The kinetochore binds to the plus ends of microtubules. (C) The kinetochore is a flat, plate-like structure. An Indian muntjac chromosome in metaphase is shown double-labeled for DNA (red) and for several kinetochore proteins (green). Two distinct kinetochores are visible, one for each chromatid. (D) Electron micrograph through the centromeric region of a chromosome at metaphase, showing both kinetochores. Note the layered structure of the kinetochore and the microtubules terminating there. A kinetochore may typically bind 30 to 50 microtubules but, depending on its size, may bind only a few or more than 100.

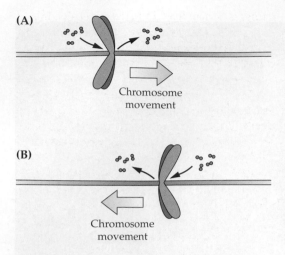

(A)

Chromosome
movement

(B)

Chromosome
movement

Figure 5.45
Kinetochore binding of microtubules is flexible. During prometaphase, chromosomes become bipolarly oriented and usually move back and forth toward each pole, eventually reaching a stable position at the metaphase plate. While the chromosomes are moving, microtubule subunits are gained and lost, mostly at the kinetochores; the kinetochores thus must be able to hold onto the microtubules while still allowing them to grow or shrink. (A) A bipolarly oriented chromosome is moving toward the right-hand pole: The microtubules attached to the leading kinetochore must lose subunits, whereas the microtubules attached to the trailing kinetochore must gain them. (B) A few moments later, the direction of a chromosome's movement has reversed, and so too must the microtubule assembly/disassembly reactions at each kinetochore.

and plus ends at the kinetochore. After the breakdown of the nuclear envelope, the kinetochore pairs face random directions and may form various inappropriate attachments to microtubules. However, such attachments are unstable, and during the first 10 to 30 minutes of prometaphase all the chromosomes usually attain their correct bipolar attachment to the spindle. Once a chromosome is bipolarly oriented, it still must be able to move toward and away from the poles. For this to happen, the fiber of a kinetochore moving toward a pole must shrink, while the fiber of a kinetochore moving away from a pole must grow. All kinetochore fiber growth and most fiber shrinkage occurs at the kinetochore. Thus, kinetochores must be able to remain attached to polymerizing and depolymerizing microtubules (Fig. 5.45). This kind of flexible attachment possibly is mediated by microtubule motor proteins that are not generating force but are diffusing unidirectionally along the microtubule lattice.

5.11.4 Chromosomal movements at mitosis are similar in plant and animal cells.

Chromosomes move mainly in two phases of mitosis, prometaphase and anaphase (Fig. 5.46). At prometaphase, the chromosomes "**congress,**" moving from the random positions they held when first captured by the mitotic spindle to gather at the metaphase plate. Congression is not direct but usually consists of successive, alternating trajectories toward and away from the nearest pole; the trajectories are unequal so that movement toward the equator predominates. Once a chro-

mosome has congressed, and throughout metaphase, it may sporadically oscillate, moving briefly toward one pole and then back to the equator. At anaphase, the replicated sister chromatids, previously held together at the centromeres, uncouple and move apart to each pole. The signal to begin anaphase can be blocked by even a single chromosome pair that has not yet congressed at the metaphase plate. At the start of anaphase, the sister chromatids uncouple synchronously, an activity that requires a topoisomerase (see Chapter 6) but does not require microtubules. Chromosomes move more steadily in anaphase than during congression, although oscillations in anaphase occur occasionally. During anaphase, chromosomes move to the poles, and in many cells the poles themselves move farther apart. Chromosomal movement to the poles is called anaphase A, and the forces powering this movement (as well as congression) are described below (see Section 5.11.5). The separation of the poles, called anaphase B, is known to be powered by either (or both) of two forces: (*a*) pulling forces exerted on each spindle pole from outside of the spindle, and (*b*) pushing forces exerted by microtubule motor proteins on microtubules of opposite polarity (i.e., emanating from opposite spindle poles) located in the zone of overlap in the middle of the spindle.

The behavior and kinetics of chromosomal movements are similar in plants and animals, except that plant chromosomes rarely oscillate during prometaphase and metaphase. Instead, chromosomes in prometaphase plant cells usually move directly toward the equator.

5.11.5 Chromosomes move in response to forces exerted by (or through) the kinetochore, as well as forces external to the kinetochore.

When a chromosome congresses or moves to the pole in anaphase A, force is exerted so as to pull the chromosome toward the pole. How this tension is exerted is controversial, with four possibilities being commonly considered (Fig. 5.47). First, because the force moves the chromosome toward the minus ends of microtubules, minus-end–directed motor proteins such as dynein, which do occur at kinetochores, could exert the required force. Second, because the kinetochore forms a cap over the ends of the bundled microtubules, the only way the kinetochore can actually move toward the pole is if the microtubules depolymerize. Microtubule depolymerization does occur at kinetochores, and energy from GTP hydrolysis, stored in the microtubule lattice, is capable of doing work. Third, kinetochore microtubules also depolymerize at the poles, which could transmit a tension through the microtubules to move the chromosomes poleward. Fourth, kinetochore microtubules themselves could be pulled poleward by motor proteins bound to other spindle microtubules or even to a nonmicrotubular matrix of some kind permeating the spindle. This model also requires kinetochore microtubules to depolymerize at the pole. In both the third and fourth models, the kinetochore binds the microtubules statically.

These four explanations of the mechanism for force generation at mitosis are not necessarily exclusive, and one or all of them could apply to chromosomal movements during different phases of mitosis or in different cell types. Although most evidence supporting each hypothesis has been obtained in animal cells, some comparable evidence has been gleaned from plant cells, and nothing has been found to suggest that tension generation differs between animals and plants.

5.11.6 In animal cells, congression may involve kinetochore cooperation.

Although force at the kinetochore is almost certainly sufficient to explain the chromosomal movements of anaphase A, the same force may not entirely explain metaphase congression, at least in vertebrate cells. In these cells, kinetochores are known to exist in two states: an active state during which they produce (or transmit) tension and move toward the pole, and an inactive state where no tension is produced (or transmitted) and the kinetochores coast. The oscillatory movements during prometaphase involve a pair of sister kinetochores switching cooperatively between the two states.

During one phase of an oscillation, one kinetochore pulls while the other coasts; they then switch, reversing the direction in which the chromosome moves. This can certainly produce oscillations but does not explain how the chromosomal position becomes stabilized at the metaphase plate. Recently, a model for congression has been developed based on the discovery that forces are also exerted on chromosomes by nonkinetochore microtubules emanating from the spindle poles. These microtubules have been likened to a "**polar wind,**" pushing on chromosome arms throughout their length. The force from the polar wind can reasonably be postulated to be proportional to microtubule density, which diminishes with distance from the pole. Thus, the metaphase plate, which lies halfway between the poles, is an equilibrium position, where the nonkinetochore forces balance. The model hypothesizes that the paired kinetochores are "smart"—preferring to coast when pulling would "fight" the polar wind (i.e., when near to and facing a pole) and preferring to pull when running before the wind. In this view, the metaphase plate is an equilibrium between opposing forces originating at the poles and sensed at the kinetochores (Fig. 5.48).

5.11.7 Congression in plant cells remains unexplained.

Although the polar wind model fits congression in vertebrates, it does not apply to plant cells. Among the strongest evidence for the polar wind is the fate of **acentric fragments,** portions of chromosomes that lack centromeres. Acentric fragments occur spontaneously but rarely in cells. They can also be generated experimentally by cutting chromosome arms with high-energy light. Although artificial, the fragments' behavior is mimicked by long arms of intact chromosomes

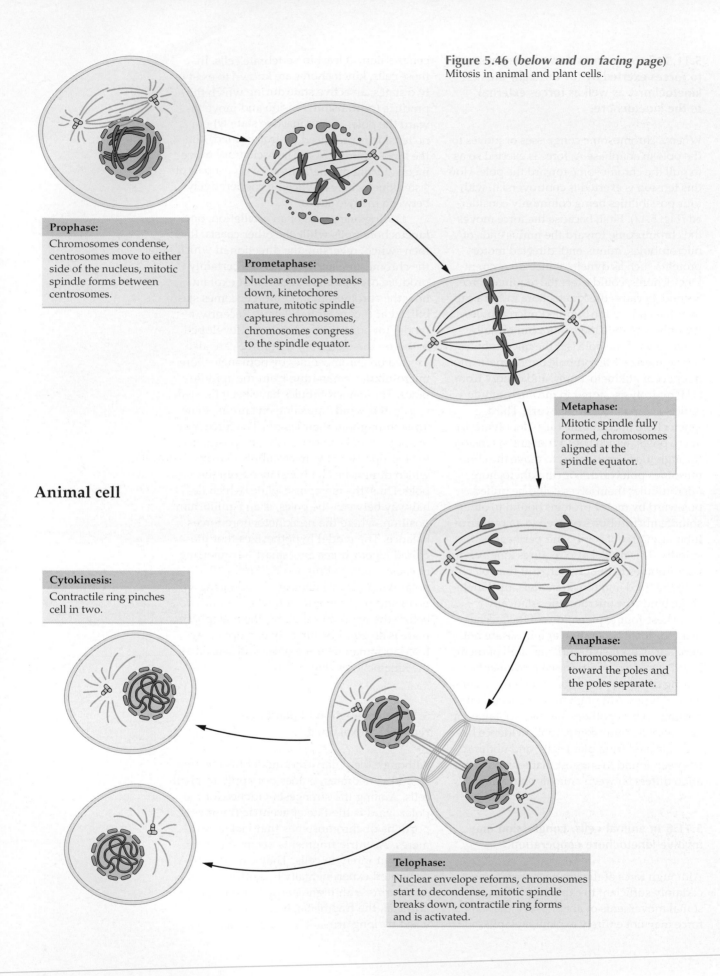

Figure 5.46 (*below and on facing page*)
Mitosis in animal and plant cells.

Prophase:
Chromosomes condense, centrosomes move to either side of the nucleus, mitotic spindle forms between centrosomes.

Prometaphase:
Nuclear envelope breaks down, kinetochores mature, mitotic spindle captures chromosomes, chromosomes congress to the spindle equator.

Metaphase:
Mitotic spindle fully formed, chromosomes aligned at the spindle equator.

Animal cell

Cytokinesis:
Contractile ring pinches cell in two.

Anaphase:
Chromosomes move toward the poles and the poles separate.

Telophase:
Nuclear envelope reforms, chromosomes start to decondense, mitotic spindle breaks down, contractile ring forms and is activated.

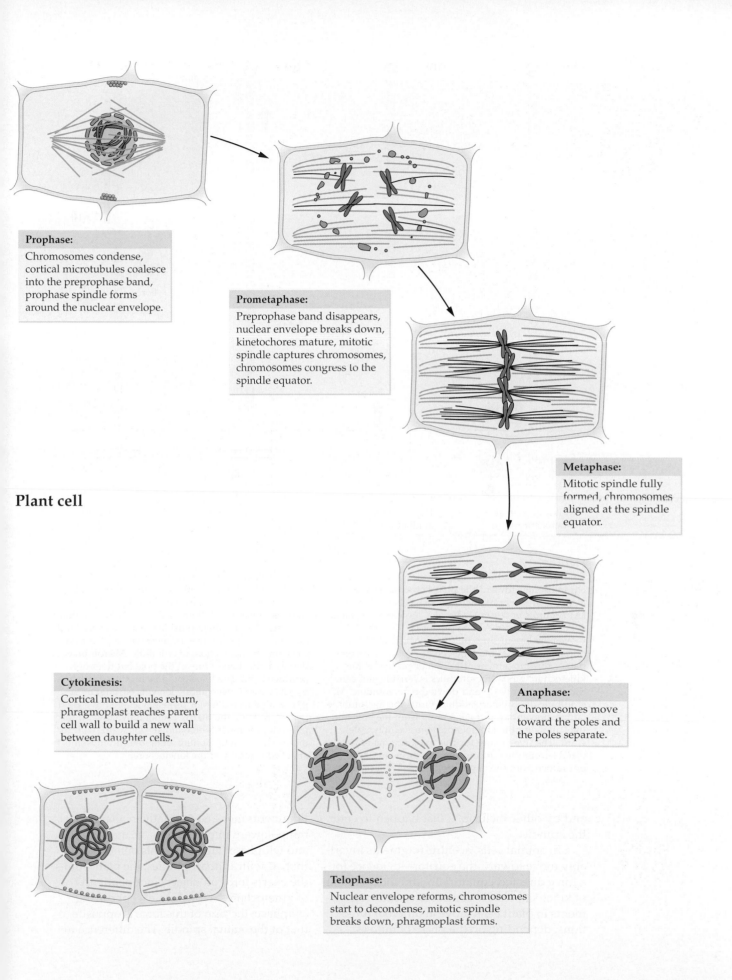

Prophase:
Chromosomes condense, cortical microtubules coalesce into the preprophase band, prophase spindle forms around the nuclear envelope.

Prometaphase:
Preprophase band disappears, nuclear envelope breaks down, kinetochores mature, mitotic spindle captures chromosomes, chromosomes congress to the spindle equator.

Plant cell

Metaphase:
Mitotic spindle fully formed, chromosomes aligned at the spindle equator.

Cytokinesis:
Cortical microtubules return, phragmoplast reaches parent cell wall to build a new wall between daughter cells.

Anaphase:
Chromosomes move toward the poles and the poles separate.

Telophase:
Nuclear envelope reforms, chromosomes start to decondense, mitotic spindle breaks down, phragmoplast forms.

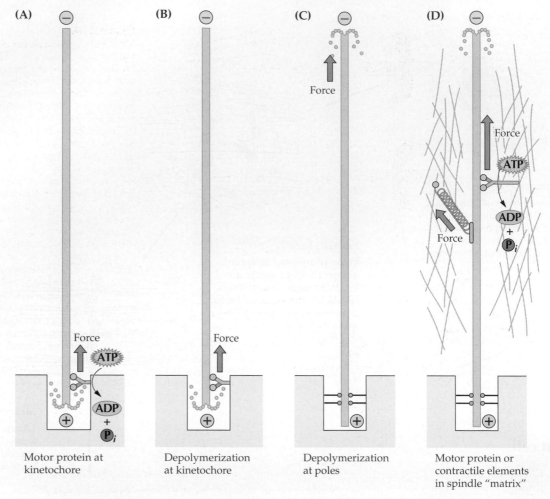

(A)   **(B)**   **(C)**   **(D)**

Motor protein at kinetochore   Depolymerization at kinetochore   Depolymerization at poles   Motor protein or contractile elements in spindle "matrix"

Figure 5.47

Four possible methods for producing force to move chromosomes toward the pole. For simplicity, only a single kinetochore microtubule has been drawn. (A) A minus-end–directed motor protein at the kinetochore hydrolyzes ATP to produce a force that moves the kinetochore poleward. Microtubule depolymerization at the kinetochore passively accompanies this movement. (B) A motor protein at the kinetochore binds microtubules reversibly and can diffuse randomly up and down the microtubule. Microtubule depolymerization deforms the microtubule lattice, preventing diffusion away from the pole and thus allowing the kinetochore to move only poleward. The free energy for this work originates not as usual—from the motor protein hydrolyzing ATP—but rather from the newly polymerized subunits of the microtubule lattice hydrolyzing GTP. (C) Similar to (B), except the microtubules depolymerize at the poles, and a force is transmitted through the microtubules to the kinetochore, which binds microtubules statically. (D) A protein outside the kinetochore binds an immobile spindle matrix and exerts a poleward force on the kinetochore microtubule. Microtubule depolymerization occurs at the pole but does not produce force. The force could be produced either by prestressed mechanical elements (spring) or by a plus-end–directed motor protein anchored to a rigid matrix. Because the matrix cannot move toward the plus end, the microtubule moves poleward, with its minus end leading, pulling along the chromosome attached statically at the kinetochore.

and by other inclusions that happen to enter the spindle.

In animal cells, acentric fragments invariably move toward the equator—evidence for a force directed equatorially and independent of kinetochores. By contrast, acentric fragments in plant cells move in different directions, depending on the phase of mitosis: Fragments move equatorially in anaphase but poleward in prometaphase, metaphase, and telophase (Fig. 5.49). This demonstrates that, as with animal spindles, the plant spindle exerts force on objects independently of kinetochores. However, except during anaphase, the sign of this force is opposite that of the animal spindle. The difference in

(A) Polar wind—kinetochore force opposed

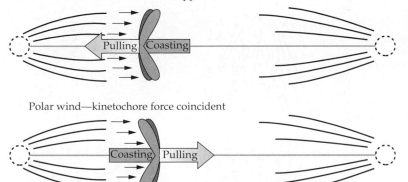

Polar wind—kinetochore force coincident

(B) Force directly proportional to length of microtubule

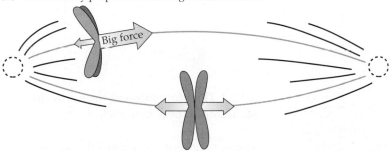

Figure 5.48
Models for metaphase congression. (A) Model invoking the polar wind and smart kinetochores. For a bipolarly oriented chromosome moving toward a near pole, force is exerted to pull the near kinetochore poleward while the other kinetochore coasts. In this situation, the force at the pulling kinetochore is opposed by a force that increases as distance to the pole decreases. This latter force, which may arise from the polymerization of growing microtubules pushing against any protruding region of the advancing chromosome, has been called the "polar wind." When the chromosome moves in the opposite direction, the paired kinetochores have changed states so that force is now exerted to pull the far kinetochore toward the pole it faces while the near one now coasts. In this situation the force at the kinetochore parallels the force from the polar wind. The model postulates that paired kinetochores are "smart" and prefer to move with the polar wind, rather than against it, and thus wind up halfway between the two poles, where the polar forces are balanced. (B) An alternative model for metaphase congression postulates that the poleward force exerted at the kinetochore is directly proportional to the length of the kinetochore microtubules. Such a force would be equal and opposite at a pair of kinetochores when the chromosomes are midway between poles. Although neither of these models excludes the other, there is no direct evidence that the force acting at kinetochores is length-dependent, as predicted in (B).

sign of these polar forces may result from the different pole structures. To understand the origin and significance of the kinetochore-independent poleward-directed forces generated by the plant mitotic spindle, we must learn much more about the behavior of the spindle pole and congressing chromosomes.

5.11.8 Microtubule dynamics regulate the rate of chromosome movements during anaphase.

Although it remains unproved whether microtubule depolymerization generates the force that drives chromosome movement, considerable evidence indicates that microtubule depolymerization regulates the rate at which chromosomes move during anaphase. Inoué and collaborators subjected the spindles of various cell types (including plant endosperm) to treatments that enhanced the rate of microtubule depolymerization and observed the spindle by using polarized light microscopy (see Box 5.3). In the earliest stages of anaphase, low temperatures or high

concentrations of colchicine depolymerized the spindle microtubules extensively, halting chromosome movement. However, less extreme cold or lower colchicine concentrations only modestly depolymerized spindle microtubules and accelerated chromosome movement.

Studies with calcium, another agent that stimulates microtubule depolymerization, have yielded similar results. In stamen hair cells in anaphase, microinjection of high doses of calcium depolymerized kinetochore microtubules almost completely and stopped chromosomal movement, whereas lower doses accelerated chromosome velocity and modestly increased kinetochore microtubule depolymerization (Fig. 5.50). These experiments may be taken as evidence that microtubule depolymerization itself provides the force to move chromosomes. They clearly show that the kinetochore microtubules limit the velocity chromosomes can attain. Limiting chromosomal velocity may enhance faithful segregation by minimizing opportunities for chromosomal breakage.

Figure 5.49

The mitotic spindle moves objects as well as kinetochores. During specific phases of mitosis in plant cells, chromosome arms as well as acentric chromosome fragments and unidentified inclusions move within the spindle in directions that can differ from kinetochore movements. Moving particles are shown as small dots. (A) In the prometaphase and metaphase, chromosome arms, acentric fragments, and other inclusions move poleward. (B) In the anaphase, movement is predominantly toward the equator. (C) In the telophase, movement is once again poleward.

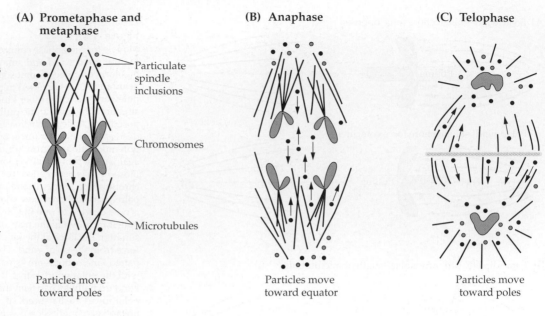

(A) Prometaphase and metaphase

Particulate spindle inclusions

Chromosomes

Microtubules

Particles move toward poles

(B) Anaphase

Particles move toward equator

(C) Telophase

Particles move toward poles

5.12 The cytoskeleton and cytokinesis

Cytokinesis occurs in plants with the construction of a new wall between newly formed cells. Because the new wall is permanent, the plant cell has several mechanisms to ensure that the wall gets built in the right location. Two conspicuous structures, the **phragmosome** and the **preprophase band,** aid in orienting cell division.

5.12.1 The phragmosome forms in the plane of the new cell wall.

The phragmosome is a thin layer of cytoplasm, more or less reticulated, that forms in the plane of the future cell wall. This structure may form early in the cell cycle, possibly along with the initiation of S phase. As the first visible sign of division, the nucleus migrates to a position where it will undergo mitosis, after which the phragmosome coalesces around it (Fig. 5.51). The phragmosome contains actin filaments and microtubules, and both cytoskeletal elements participate in nuclear migration and phragmosome formation. Experiments show that once a phragmosome has formed, reprogramming the plane of division becomes difficult.

Phragmosomes are visible in vacuolated dividing cells, such as cambial cells, and in mature cells that have been stimulated to divide by wounding. Whether phragmosomes are also present in meristematic cells, which have no large vacuoles, is not clear. Recently, microinjection of the actin filament tag rhodamine phalloidin into the densely cytoplasmic cells of developing leaf epidermis did not reveal phragmosome-like organization. The function of the phragmosome is not known; perhaps it fixes the future plane of division or establishes a cytoplasmic avenue through the vacuole for the expanding cytokinetic apparatus.

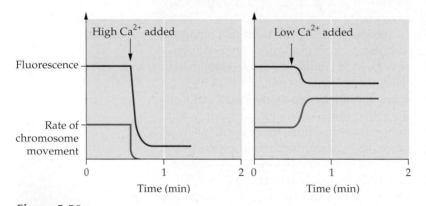

High Ca^{2+} added

Low Ca^{2+} added

Fluorescence

Rate of chromosome movement

Time (min)

Time (min)

Figure 5.50

At anaphase, modest increases in cytosolic concentrations of calcium enhance microtubule depolymerization and accelerate chromosome movement. When living stamen hair cells are injected with fluorescently labeled tubulin, the fluorescence of the kinetochore microtubules and the velocity of chromosome movement can be compared quantitatively. On the left, injecting a calcium-rich solution (cytosolic calcium approximately 10 μM) removes nearly all kinetochore microtubules and stops chromosomal movement; on the right, injecting a more dilute calcium solution (cytosolic calcium approximately 1 μM) removes some kinetochore microtubules and accelerates the chromosomes.

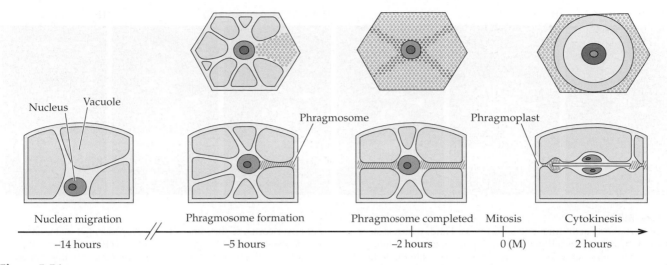

Nucleus   Vacuole

Phragmosome

Phragmoplast

| Nuclear migration | Phragmosome formation | Phragmosome completed | Mitosis | Cytokinesis |

−14 hours   −5 hours   −2 hours   0 (M)   2 hours

Figure 5.51
Formation of the phragmosome. When cell division is to happen in a highly vacuolated cell, as in the pith parenchyma cell sketched here, first the nucleus migrates to the center of the cell and the cytoplasm coalesces in the plane of the future cell wall. This sheet-like accretion of cytoplasm is called the phragmosome. Cytoplasmic strands progressively accumulate and fuse to form the phragmosome, which may provide a conduit for the growing cell plate. Construction of the cell plate begins at the cell center and must move outward through the vacuole to reach the parent cell wall. Time 0 (M) indicates mitosis.

5.12.2 The preprophase band predicts the site of the new cell wall.

The other cytoskeletal structure involved in orienting the cell plate is the preprophase band (Fig. 5.52), which despite its name usually forms during prophase. The preprophase band is a ring of actin filaments and microtubules that encircles the cell just inside the plasma membrane at the site where the future cell wall will join the parent wall. Initially, the band is wide and

(A)   **(B)**   **(C)**

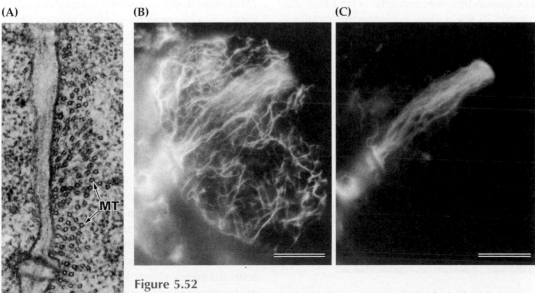

MT

Figure 5.52
Structure of the preprophase band. (A) Electron micrograph of a leaf epidermal cell, showing a cross-section through a preprophase band with abundant microtubules (Mt). (B, C) Fluorescence micrographs of a suspension-cultured tobacco cell double-labeled to show actin (B) and microtubules (C). The microtubule band is narrow and the rest of the cortex lacks cortical microtubules, whereas the actin band is wider and actin filaments are present throughout the cortex. Bars indicate 10 μm.

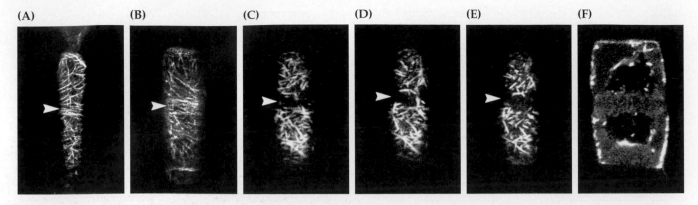

(A)   **(B)**   **(C)**   **(D)**   **(E)**   **(F)**

Figure 5.53
Before prophase, actin filaments also form a band, which is selectively removed as the cell goes through prometaphase. A living stamen hair cell microinjected with fluorescent phalloidin to label actin filaments was observed during mitosis. (A, B) In prophase, a loosely organized transverse band of actin filaments forms at the division site (arrowheads). Images shown are from the upper and lower cortex of the same cell. (C) Prophase, (D) metaphase, (E) anaphase, and (F) telophase of another cell. Actin filaments are lost from the region of the preprophase band late in prophase (arrowheads) but remain elsewhere in the cortex. (C–E) A focal plane at the upper edge of the cell; (F) a midplane.

rather loosely organized; as prophase continues, however, the band narrows and nearly all of the actin filaments and microtubules become closely parallel (Fig. 5.52; also see Fig. 5.43). Around the time of nuclear envelope breakdown, the polymers of the preprophase band all depolymerize. This removes all microtubules from the cortical cytoplasm but, interestingly, actin filaments remain in the cortex, except at the location previously occupied by the preprophase band (Fig. 5.53).

Although the preprophase band faithfully predicts the site at which the cell plate joins the parent walls, its function is not known. Some have suggested that the preprophase band modifies the structure of the parent wall to enable subsequent insertion of the daughter wall (Fig. 5.54). Others have suggested that the band helps guide the cytokinetic apparatus to the correct site. Both views may be correct.

5.12.3 The cytokinetic organelle in plant cells is called the phragmoplast.

Although mitosis in plants resembles mitosis in animals, cytokinesis differs. An animal cell physically pinches itself in two, whereas a plant cell erects a partition, in the form of a new cell wall. To divide, animal cells have evolved a cytokinetic organelle called the contractile ring, a band of actin filaments encircling the cell in the future plane of division. The contractile ring contains myosin, which pulls together adjacent, antiparallel actin filaments, shortening the circumference of the ring and squeezing the cell in two (Fig. 5.55A). In contrast, to build a new wall between cells, plants have evolved a cytokinetic organelle called the **phragmoplast**. The phragmoplast contains actin and myosin as well as microtubules, and all three are

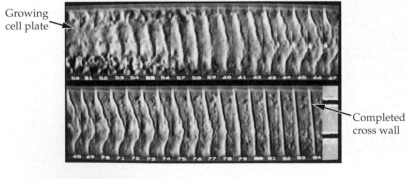

Growing cell plate

Completed cross wall

Figure 5.54
Time-lapse light micrographs at 1-minute intervals demonstrating cell plate formation in a living stamen hair cell. The cell plate undulates as it forms; a short time after the plate reaches the parent wall (77 to 81 minutes), the undulations vanish abruptly. The preprophase band formed at the site where the cell plate reaches the parent walls may condition the wall for subsequent events involving cell plate fusion.

required for its function (Fig. 5.55B). Although the phragmoplast and the contractile ring presumably evolved from some common ancestral apparatus for division, by now the divergence is so great that any meaningful similarity between the two structures is removed.

5.12.4 The phragmoplast initially forms between the separating chromosomes at late anaphase and then grows toward the cell wall.

As the phragmoplast begins to form, residual microtubules in the spindle midzone are joined by abundant, newly polymerized microtubules emanating from the poles and by newly polymerized actin filaments (Fig. 5.56). Both microtubules and actin filaments run perpendicular to the plane of division (Fig. 5.57). The polarity of the microtubules is the same as for the spindle: The minus ends are toward the poles, which are now occupied by the reforming nuclei. The actin filaments also form with uniform polarity, the minus (or pointed) end being directed toward the pole (Fig. 5.58). The zone in which microtubules overlap at the spindle equator defines the midplane of the phragmoplast. Newly recruited microtubules retain a narrow region of overlap at that

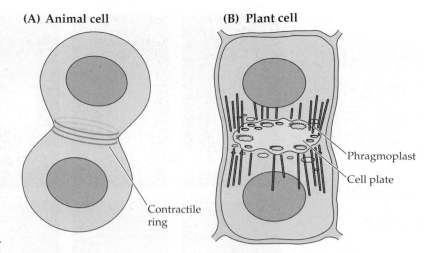

(A) Animal cell

(B) Plant cell

Phragmoplast

Cell plate

Contractile ring

Figure 5.55
Comparison of cytokinesis in animal and plant cells. (A) In many animal cell types, the contractile ring is responsible for cytokinesis. Within the cortex of an animal cell with two new nuclei, the contractile ring encircles the cell at a position midway between the nuclei. The contractile ring, made of actin filaments and myosin, pulls adjacent filaments together, thus pinching the cell in two. Alternatively, in highly motile cells such as fibroblasts, daughter cells can use their attachments to the substrate to crawl in opposite directions and thus essentially pull themselves apart; this mechanism also relies on actin and myosin. (B) The phragmoplast of the plant cell. Once new nuclei have formed, the phragmoplast builds a new cell wall between them, thus creating two cells. The phragmoplast contains both actin filaments and microtubules, which are oriented perpendicularly to the forming cell wall. In many fungi and algae, construction of the new wall begins at the edges and moves to the center, analogous to the animal contractile ring; in plants, in contrast, the wall forms in the center of the cell and grows toward the edges.

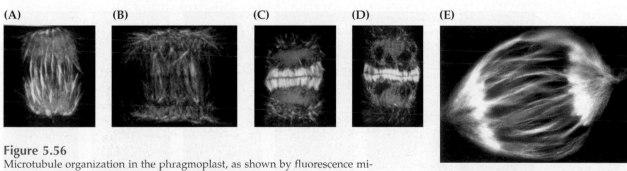

(A) **(B)** **(C)** **(D)** **(E)**

(F)

Figure 5.56
Microtubule organization in the phragmoplast, as shown by fluorescence micrographs of anti-tubulin–labeled phragmoplast microtubules from different plant cell types. (A–D) Wheat root cells, with microtubules in green or yellow and DNA in blue. (E, F) Endosperm cells of *Haemanthus katherinii*. The phragmoplast begins to form at late anaphase (A, E), as interzonal microtubules of the spindle are joined by newly polymerized microtubules at the equator. By telophase (B), the bipolar organization of the phragmoplast is well established: The plane of the forming cell wall is defined by a band of microtubule overlap, and microtubules extend away on either side. At the zone of overlap, a dark band appears in these images because the high density of membranous vesicles and proteins excludes the antibodies. As telophase progresses, phragmoplast microtubules tend to shorten and disappear from the central region, forming an annulus of overlapping microtubules that expands centrifugally, eventually reaching the parental wall (C, D, F).

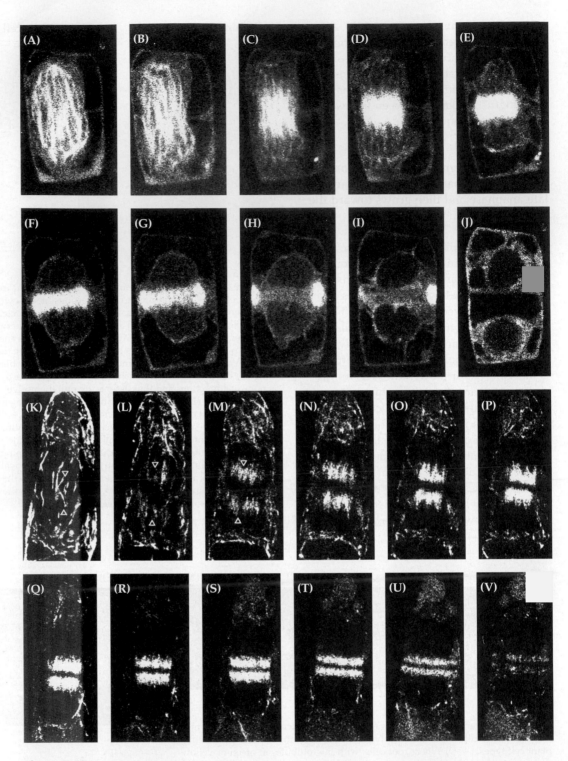

Figure 5.57
Comparison of microtubules and actin filaments in the phragmoplast of living stamen hair cells. (A–J) Fluorescent tubulin illuminates the phragmoplast as it condenses out of the spindle mid-zone (C–E), becomes an annular structure, and grows to reach the parent walls. The absence of a dark band at the equator indicates that microtubules containing fluorescent subunits do penetrate that region. (K–V) Actin filaments labeled with fluorescent phalloidin are recruited to the phragmoplast along with microtubules. Actin filaments do not form as obvious an annular structure but instead tend to remain present throughout the forming cell plate. The dark band at the equator indicates that actin filaments do not overlap significantly across the equator.

position, whereas the actin filaments form in a band on either side of the overlap zone (see Fig. 5.57).

The phragmoplast forms in the center of the cell between the reforming nuclei. It then grows centrifugally toward the cell edges. As it grows, new cytoskeletal polymers are recruited to the periphery and lost from the interior, so that soon the phragmoplast attains a ring-like structure (Fig. 5.59). In large, dividing cells, such as cambial initials, the phragmoplast may take half a day to reach the cell edges. Microtubules in the phragmoplast are highly dynamic, with half-lives on the order of 1 minute. The dynamic state of these polymers presumably facilitates phragmoplast growth, although it is not known whether microtubules are translocated laterally or are turned over differentially (with a net loss at the interior of the phragmoplast and a net gain at the exterior).

Recently, a microtubule motor protein has been purified from isolated phragmoplasts, and the gene for the protein has been cloned. This protein, which resembles kinesin in sequence and biochemistry, may be responsible for the microtubule translocation observed in the phragmoplast or for stabilizing the phragmoplast's bipolar arrangement of microtubules (Fig. 5.60).

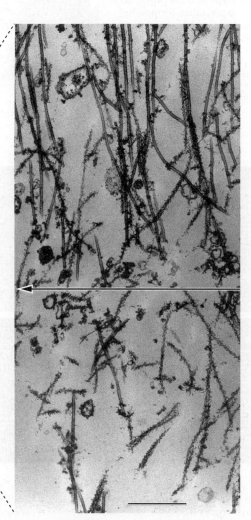

Figure 5.58
An electron micrograph of an isolated phragmoplast illustrates the polarity of actin filaments in that structure. The inset provides an overview, the labeled box indicating the region used for the main figure. Before embedding, the actin filaments were decorated with a myosin fragment that binds asymmetrically on the filaments, forming "arrowheads" that reveal the filament's polarity. Most of the actin filaments are perpendicular to the equator (double-headed arrow) and most have arrowheads that point away from the equator, indicating that most minus ends face the nuclei.

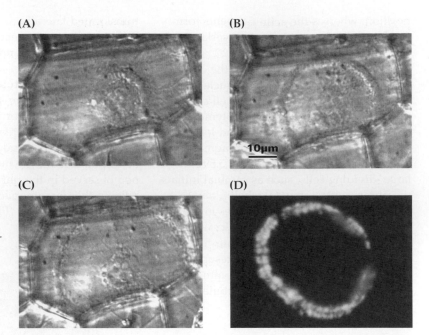

Figure 5.59
The annular structure of the phragmoplast. (A–C) A series of light micrographs of a dividing leaf epidermal cell at 30-minute intervals. The plane of the forming cell wall is parallel to the page, and the phragmoplast is viewed from above. An annulus, formed by the phragmoplast microtubules viewed end on, grows outward and eventually reaches the cell periphery. If fixed and stained to show microtubules, the annular structure in (C) would resemble the phragmoplast shown in (D).

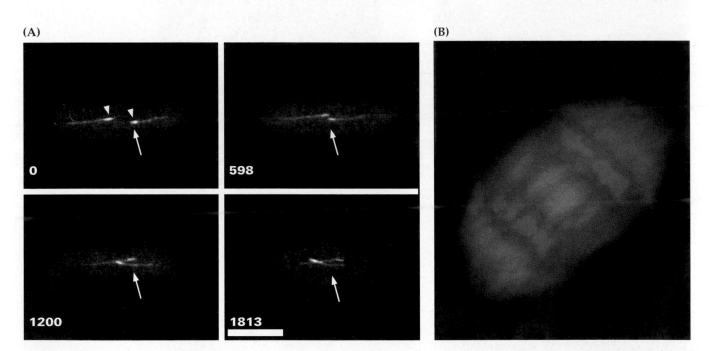

Figure 5.60
The phragmoplast contains the first microtubule motor protein to be purified and characterized from plants. The gene has been cloned for this protein, which belongs to the kinesin superfamily and has been named TKRP125 (for Tobacco Kinesin-Related Protein of 125 kDa). (A) TKRP125 purified from the tobacco phragmoplast moves microtubules toward their minus ends in an ATP-dependent manner. The purified protein was used to coat one surface of a microscopic observation chamber and was allowed to bind fluorescently labeled microtubules, prepared with a small region of heightened fluorescence marking their minus ends (arrowheads). ATP was added to the chamber and images captured at successive intervals (time is given in seconds; the arrows mark a fixed reference location). The microtubules move minus end first, which is equivalent to a protein moving cargo toward the plus end of an anchored microtubule, typical of kinesin motors. In vivo, the motor may extrude microtubules from the inner regions of the phragmoplast as it expands as an annulus toward the parent cell wall. (B) The motor protein is abundant in the spindle midzone, suggesting that it also participates in phragmoplast formation. Here, an anaphase mitotic spindle from a tobacco cell has been triple-labeled to show DNA (blue), microtubules (green), and TKRP125 (red). The orange color in the spindle midzone where the phragmoplast is forming results from the colocalization of the microtubules and TKRP125.

5.12.5 The mechanism of cell plate construction by the phragmoplast remains unknown.

The function of the cytoskeletal elements in the phragmoplast is not known with certainty. The phragmoplast probably guides the expanding new wall to the edges of the cell (see above) and aids in the construction of the new wall. In addition to cytoskeletal filaments, the phragmoplast harbors swarms of vesicles (Fig. 5.61A). Vesicles deliver proteins and polysaccharides to the nascent cell wall and deliver plasma membrane to the growing cell plate. Additionally, vesicles return from the cell plate to the Golgi apparatus, removing excess membrane and recycling cell wall components. Intense secretory activity is required to build the cell plate, and the cytoskeleton presumably plays a major role in directing this secretory traffic. However, very little is known about the mechanisms responsible.

(A)

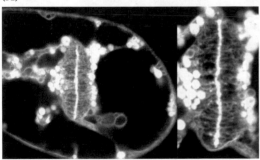

(B)

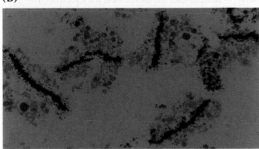

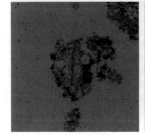 

Summary

The cytoskeleton, a fundamental component of all eukaryotic cells, comprises filamentous protein polymers of three types: intermediate filaments, actin filaments, and microtubules. Intermediate filaments are the most "skeletal" in nature, giving the cytoplasm mechanical strength. Actin filaments and microtubules not only are skeletal but also "muscular"; they are responsible for the movements of cytoplasm and organelles. Actin filaments and microtubules are dynamic, turning over many times during their functional lifetimes. These polymers also are polar, in the sense of having different biochemical properties at each end. Their dynamic behavior and polarity are enhanced by free energy from the hydrolysis of nucleoside triphosphates. The biochemical properties of cytoskeletal polymers themselves appear to be widely conserved among eukaryotes. The different functions of the cytoskeleton are mediated by various accessory proteins, which assemble, disassemble, bundle, sever, cap, cross-link, or move the polymers. Actin filaments play starring roles in organelle movement. Microtubules are important for cell polarity and for controlling the direction of cell expansion.

The cytoskeleton provides the machinery that separates the replicated chromosomes at mitosis and that partitions the

Figure 5.61
(A) Fluorescence micrograph of a vacuolated suspension-cultured tobacco cell at telophase, stained with a membrane marker to indicate the large amounts of membranous material throughout the phragmoplast and the forming cell plate. The right-hand panel is an enlargement of the forming cell plate. (B) A promising system for studying the function of the phragmoplast. Certain lines of suspension-cultured tobacco cells can be treated to divide nearly synchronously; these can hence serve as starting material from which phragmoplasts can be isolated with relatively high yield and purity. Here, isolated phragmoplasts were pulsed with radiolabeled glucose or xylose and then fixed, embedded, sectioned, and monitored by autoradiography. Radioactivity from glucose is found extensively at the cell plate (upper panel), whereas that from xylose is less abundant and found mainly in material around the nucleus (lower panels). Systems such as this can provide tools for dissecting the biochemical and cell biological pathways of cell wall formation.

daughter cells at cytokinesis. Mitosis depends on the mitotic spindle, a structure built largely from microtubules. Spindle microtubules interact with chromosomes at the kinetochore, a specialized protein-rich region that forms at the centromere. The dynamic properties of the microtubules, the architecture of the spindle, and the force-generating proteins at the kinetochore participate jointly in chromosome movement. The importance of each may vary at different phases of mitosis and in different cell types. Cytokinesis depends on cytoskeletal structures to ensure the appropriate orientation of the cell's division and to divide the cell. During interphase, the phragmosome, a raft of cytoskeletal-rich cytoplasmic strands, often forms in the plane in which the cell division will take place. In prophase, the preprophase band, a ring of microtubules and actin filaments, usually marks the location where the new cell plate will later fuse with the parental cell wall. During cytokinesis itself, the phragmoplast, a structure built from microtubules and actin filaments, orchestrates the intense secretory activity needed to build the forming cell plate and guides the growing cell plate to the parent cell wall.

Further Reading

Alberts, B., Bray, D., Lewis, J., Raff, M., Roberts, K., Watson, J. D. (1994) *Molecular Biology of the Cell.* Garland, New York.

Asada, T., Collings, D. (1997) Molecular motors in plants. *Trends Plant Sci.* 2: 29–37.

Asada, T., Kuriyama, R., Shibaoka, H. (1997) TKRP125, a kinesin-related protein involved in the centrosome-independent organization of the cytokinetic apparatus in tobacco BY-2 cells. *J. Cell Sci.* 110: 179–189.

Bajer, A. S., Molè-Bajer, J. (1972) Spindle dynamics and chromosome movements. *Int. Rev. Cytol. Suppl.* 3: 1–273.

Baskin, T. I., Cande, W. Z. (1990) The structure and function of the mitotic spindle in flowering plants. *Annu. Rev. Plant Physiol. Plant Mol. Biol.* 41: 277–315.

Baskin, T. I., Miller, D. D., Vos, J. W., Wilson, J. E., Hepler, P. K. (1996) Cryofixing single cells and multicellular specimens enhances structure and immunocytochemistry for light microscopy. *J. Microsc.* 182: 149–161.

Cleary, A. L. F-actin redistributions at the division site in living *Tradescantia* stomatal complexes as revealed by microinjection of rhodamine-phalloidin. *Protoplasma* 185: 152–165.

Cyr, R. J. (1994) Microtubules in plant morphogenesis: role of the cortical array. *Annu. Rev. Cell Biol.* 10: 153–180.

Davies, E. (1993) Intercellular and intracellular signals and their transduction via the plasma membrane-cytoskeleton interface. *Semin. Cell Biol.* 4: 139–147.

Ford, B. J. (1992) Brownian movement in *Clarkia* pollen: a reprise of the first observations. *Microscope* 40: 235–241.

Fosket, D. E., Morejohn, L. C. (1992) Structural and functional organization of tubulin. *Annu. Rev. Plant Physiol. Plant Mol. Biol.* 43: 201–240.

Frankel, S., Mooseker, M. S. (1996) The actin-related proteins. *Curr. Opin. Cell Biol.* 8: 30–37.

Gens, J. S., Reuzeau, C., Doolittle, K. W., McNally, J. G., Pickard, B. G. (1996) Co-visualization by computational optical-sectioning microscopy of integrin and associated proteins at the cell membrane of living onion protoplasts. *Protoplasma* 194: 215–230.

Gunning, B.E.S. (1982) The cytokinetic apparatus: its development and spatial regulation. In *The Cytoskeleton in Plant Growth and Development*, C. W. Lloyd, ed. Academic Press, London, pp. 229–292.

Gunning, B.E.S., Steer, M. W. (1996) *Plant Cell Biology: Structure and Function.* Jones and Bartlett, Sudbury, MA.

Hepler, P. K., Cleary, A. L., Gunning, B.E.S., Wadsworth, P., Wasteneys, G. O., Zhang, D. H. (1993) Cytoskeletal dynamics in living plant cells. *Cell Biol. Int.* 17: 127–142.

Hush, J. M., Overall, R. L. (1996) Cortical microtubule reorientation in plants: dynamics and regulation. *J. Microsc.* 181: 129–139.

Hush, J. M., Wadsworth, P., Callaham, D. A., Hepler, P. K. (1994) Quantification of microtubule dynamics in living plant cells using fluorescence redistribution after photobleaching. *J. Cell Sci.* 107: 775–784.

Hyams, J. S., Lloyd, C. W., eds. (1994) *Microtubules.* Wiley-Liss, New York.

Hyman, A. A., Karsenti, E. (1996) Morphogenetic properties of microtubules and mitotic spindle assembly. *Cell* 84: 401–410.

Inoué, S. (1953) Polarization optical studies of the mitotic spindle. I. The demonstration of spindle fibers in living cells. *Chromosoma* 5: 487–500.

Inoué, S. (1986) *Video Microscopy.* Plenum Press, New York.

Inoué, S., Salmon, E. D. (1995) Force generation by microtubule assembly/disassembly in mitosis and related movements. *Mol. Biol. Cell* 6: 1619–1640.

Khodjakov, A., Cole, R. W., Bajer, A. S., Rieder, C. L. (1996) The force for poleward chromosome motion in *Haemanthus* cells acts along the length of the chromosome during metaphase but only at the kinetochore during anaphase. *J. Cell Biol.* 132: 1093–1104.

Khodjakov, A., Rieder, C. L. (1996) Kinetochores moving away from their associated pole do not exert a significant pushing force on the chromosome. *J. Cell Biol.* 135: 315–327.

Lloyd, C. W., ed. (1991) *The Cytoskeletal Basis of Plant Growth and Form.* Academic Press, London.

Lloyd, C. W., Shaw, P. J., Warn, R. M., Yuan, M. (1996) Gibberellic-acid–induced reorientation of cortical microtubules in living plant cells. *J. Microsc.* 181: 140–144.

McDowell, J. M., Huang, S., McKinney, E. C., An, Y.-Q., Meagher, R. B. (1996) Structure and evolution of the actin gene family in *Arabidopsis thaliana. Genetics* 142: 587–602.

Meagher, R. B. (1995) The impact of historical contingency on gene phylogeny: plant actin diversity. *Evol. Biol.* 28: 195–215.

Meagher, R. B., Williamson, R. F. (1994) The plant cytoskeleton. In *Arabidopsis,* E. M. Meyerowitz and C. R. Somerville, eds. Cold Spring Harbor Laboratory Press, Cold Spring Harbor, NY, pp. 1049–1084.

Menzel, D. (1993) Chasing coiled coils: intermediate filaments in plants. *Bot. Acta* 106: 294–300.

Mineyuki, Y., Gunning, B.E.S. (1990) A role for preprophase bands of microtubules in maturation of new cell walls, and a general proposal on the function of preprophase band sites in cell division in plants. *J. Cell Sci.* 97: 527–537.

Mizuno, K. (1995) A cytoskeletal 50 kDa protein in plants that forms intermediate-sized filaments and stabilizes microtubules. *Protoplasma* 186: 99–112.

Moreno Diaz de la Espina, S. M. (1995) Nuclear matrix isolated from plant cells. *Int. Rev. Cytol.* 162B: 75–139.

Palevitz, B. A. (1993) Morphological plasticity of the mitotic apparatus in plants and its developmental consequences. *Plant Cell* 5: 1001–1009.

Pluta, A. F., Mackay, A. M., Ainsztein, A. M., Goldberg, I. G., Earnshaw, W. C. (1995) The centromere: hub of chromosomal activities. *Science* 270: 1591–1594.

Pollard, T. D., Cooper, J. A. (1985) Actin and actin-binding proteins: a critical evaluation of mechanisms and functions. *Annu. Rev. Biochem.* 55: 987–1035.

Raff, E. C., Fackenthal, J. D., Hutchens, J. A., Hoyle, H. D., Turner, F. R. (1997) Microtubule architecture specified by a β-tubulin isotype. *Science* 275: 70–73.

Rothkegel, M., Mayboroda, O., Rohde, M., Wucherpfennig, C., Valenta, R., Jockusch, B. M. (1996) Plant and animal profilins are functionally equivalent and stabilize actin filaments in living cells. *J. Cell Sci.* 109: 83–90.

Samuels, A. L., Giddings, T. H., Staehelin, L. A. (1995) Cytokinesis in tobacco BY-2 and root tip cells: a new model of cell plate formation in plants. *J. Cell Biol.* 130: 1345–1357.

Sheterline, P., Clayton, J., Sparrow, J. (1995) Actin. *Protein Profile* 2: 1–103.

Shibaoka, H. (1994) Plant hormone-induced changes in the orientation of cortical microtubules: alterations in the cross-linking between microtubules and the plasma membrane. *Annu. Rev. Plant Physiol. Plant Mol. Biol.* 45: 527–544.

Sohn, R. H., Goldschmidt-Clermont, P. J. (1994) Profilin: at the crossroads of signal transduction and the actin cytoskeleton. *BioEssays* 16: 465–472.

Sonobe, S., Takahashi, S. (1994) Association of microtubules with the plasma membrane of tobacco BY-2 cells in vitro. *Plant Cell Physiol.* 35: 451–460.

Spurck, T., Forer, A., Pickett-Heaps, J. (1997) Ultraviolet microbeam irradiations of epithelial and spermatocyte spindles suggest that forces act on the kinetochore fibre and are not generated by its disassembly. *Cell Motil. Cytoskeleton* 36: 136–148.

Staehelin, L. A., Hepler, P. K. (1996) Cytokinesis in plants. *Cell* 84: 821–824.

Biochemistry & Molecular Biology of Plants, B. Buchanan, W. Gruissem, R. Jones, Eds.
© 2000, American Society of Plant Physiologists

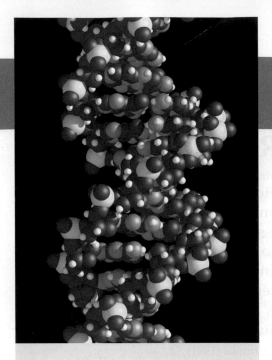

CHAPTER 6

Nucleic Acids

Masahiro Sugiura
Yutaka Takeda

CHAPTER OUTLINE

Introduction

The nucleic acids—deoxyribonucleic acid (DNA) and ribonucleic acid (RNA)—are polymers that can store and transmit genetic information. The blueprints for the biochemical machines that manufacture living organisms are encoded in the DNA molecules that make up the **genome** of the cell. During **transcription,** sequences of DNA serve as templates for the synthesis of RNA. Certain RNAs, called messenger RNAs (mRNAs), are subsequently decoded by ribosomes during **translation** (see Chapter 9). The information that is stored in the translated sequence of mRNA specifies the amino acid sequence of one or more proteins, which ultimately determine the phenotypic characteristics of the organism. When cells divide, DNA **replication** generates a duplicate set of genetic instructions for the new cell (Fig. 6.1). DNA replication and repair are important processes because the survival of the individual organism depends on the stability of its genome. However, long-term survival of a population can be promoted by the genetic variation that results from changes in the DNA blueprints of its individual members.

Living cells store genetic information in the form of double-stranded DNA. In contrast, viral genomes consist of either double-stranded or single-stranded nucleic acids and contain either DNA or RNA. The genomes of viruses are generally small and encode only a few of the proteins that are required for viral propagation. To replicate their nucleic acids and multiply, viruses must therefore exploit the biochemical machinery of a host cell. For example, when an RNA virus infects a cell, the cellular machinery can translate the viral RNA directly into protein, or can use the viral genome as a template and synthesize complimentary RNAs for subsequent translation. Some RNA viruses encode reverse transcriptase, an enzyme that uses RNA as a template for DNA synthesis. Once this enzyme has catalyzed the **reverse transcription** of a DNA copy of the viral RNA genome, the transcriptional and translational machinery of the host cell produces the other components necessary for virus multiplication (Fig. 6.1).

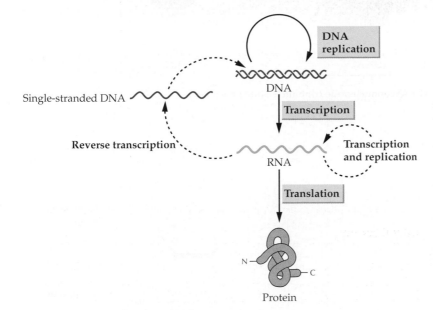

Figure 6.1
In living organisms, genetic information is stored in genomes of double-stranded DNA. Single-stranded RNA is transcribed from this DNA template. Some RNA molecules encode instructions for the synthesis of specific proteins. Others participate in RNA processing or translation of RNA sequences into proteins. In contrast, viral genomes are varied; different viruses encode the genetic information required for their amplification in either single-stranded or double-stranded genomes that contain either DNA or RNA. The dotted lines indicate pathways involved in the reproduction of single-stranded RNA and single-stranded DNA viruses.

6.1 Composition of nucleic acids and synthesis of nucleotides

6.1.1 DNA and RNA are polymers of purine and pyrimidine nucleotides.

DNA and RNA are long, unbranched polymers each composed of four types of building blocks, called **nucleotides.** Each nucleotide comprises a purine or pyrimidine base, a pentose sugar, and a phosphate group (Fig. 6.2). The ribonucleotides in RNA contain ribose, a sugar with the chemical formula $(CH_2O)_5$. The deoxyribonucleotides that make up DNA contain 2-deoxyribose, from which the hydroxyl group linked to C-2 has been removed. The presence of reactive 2′-hydroxyl groups makes RNA much less stable than DNA, particularly in alkaline solutions.

Each nucleic acid includes four nitrogenous bases, two purines and two pyrimidines. DNA contains the purines adenine and guanine and the pyrimidines cytosine and thymine. RNA also contains adenine, guanine, and cytosine, but thymine is replaced with another pyrimidine, uracil. As we will learn later, unusual or modified bases also are found in some ribonucleic acids, particularly transfer RNAs (tRNAs, see Section 6.7.3).

6.1.2 Plant cells synthesize pyrimidine and purine nucleotides de novo as well as by way of salvage pathways.

The de novo synthesis of pyrimidine nucleotides initiates with a series of six reactions that make up the orotic acid pathway. The pyrimidine orotate is produced from simple molecules: CO_2, aspartate, and the amide group of glutamine (Fig. 6.3). Once synthesized, orotate is joined to 5-phosphoribosyl-1-pyrophosphate (PRPP), and subsequently modified to form the ribonucleotides uridine triphosphate (UTP) and cytidine triphosphate (CTP). In plants, all enzymes required for pyrimidine nucleotide biosynthesis are found in plastids, which are thought to be the primary site of cellular pyrimidine biosynthesis.

Purine nucleotide synthesis in plants is not well understood but is thought to proceed by a pathway defined in other organisms. Whereas the pyrimidine orotate is first

(A) Bases

Purines

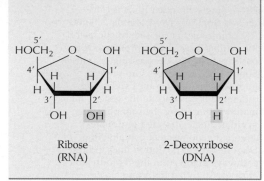

Adenine (A)   Guanine (G)

Pyrimidines

Thymine (T)
(DNA)

Cytosine (C)

Uracil (U)
(RNA)

(B) Pentose sugars

Ribose
(RNA)

2-Deoxyribose
(DNA)

(C) A ribonucleotide (ribonucleoside phosphate)

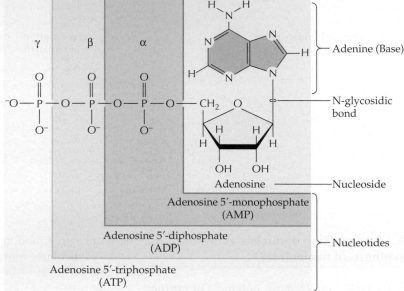

Adenine (Base)

N-glycosidic
bond

Adenosine — Nucleoside

Adenosine 5'-monophosphate
(AMP)

Adenosine 5'-diphosphate
(ADP)

Nucleotides

Adenosine 5'-triphosphate
(ATP)

Figure 6.2
The chemical components of nucleic acids.
(A) The purine bases adenine and guanine
are present in both DNA and RNA, as is the
pyrimidine base cytosine. The pyrimidine
base thymine is present in DNA. Uracil replaces thymine in RNA. (B) RNA nucleotides
contain the pentose sugar ribose. DNA contains a less reactive pentose, 2-deoxyribose.
In nucleosides and nucleotides, the carbons
of the pentose sugar are numbered 1' to 5',
as shown here. (C) A nucleoside comprises
a purine or pyrimidine base and a pentose
sugar. A nucleotide is a nucleoside in which
C-5 is linked to one, two, or three phosphate
groups by a phosphoester bond.

synthesized as a free base and then linked to
PRPP by an N-glycosidic bond, purine nucleotides are synthesized directly from PRPP
by sequential addition of purine moiety precursors: glycine, CO_2, amide groups from aspartate and glutamine, and methenyl and
formyl tetrahydrofolates (Fig. 6.4). Unlike
pyrimidine biosynthesis, purine nucleotide
synthesis is thought to occur in the cytosol.

Deoxyribonucleotides are derived from
the corresponding ribonucleotides. The ribose moiety of the ribonucleotide is reduced
by ribonucleotide reductase. The enzyme
thymidylate synthase catalyzes the formation of deoxythymidine 5'-monophosphate
(dTMP) by transferring a methyl group from
methenyl tetrahydrofolate to the C-5 of deoxyuridine 5'-monophosphate (dUMP). Because the thymidylate synthase reaction is
the only intracellular source of de novo
dTMP, this enzyme is essential for maintaining a balanced supply of the four deoxynucleotide triphosphates required for DNA
replication. The enzyme also has an important role in regulating the concentration of
deoxyuridine nucleotides, which must be
kept low to prevent incorporation of these
nucleotides into DNA. Indirect evidence
suggests that certain plants (e.g., *Lemna
major*) may contain an alternative pathway
that methylates deoxycytidine 5'-diphos-
phate and deaminates the resulting product
to yield deoxythymidine 5'-diphosphate.

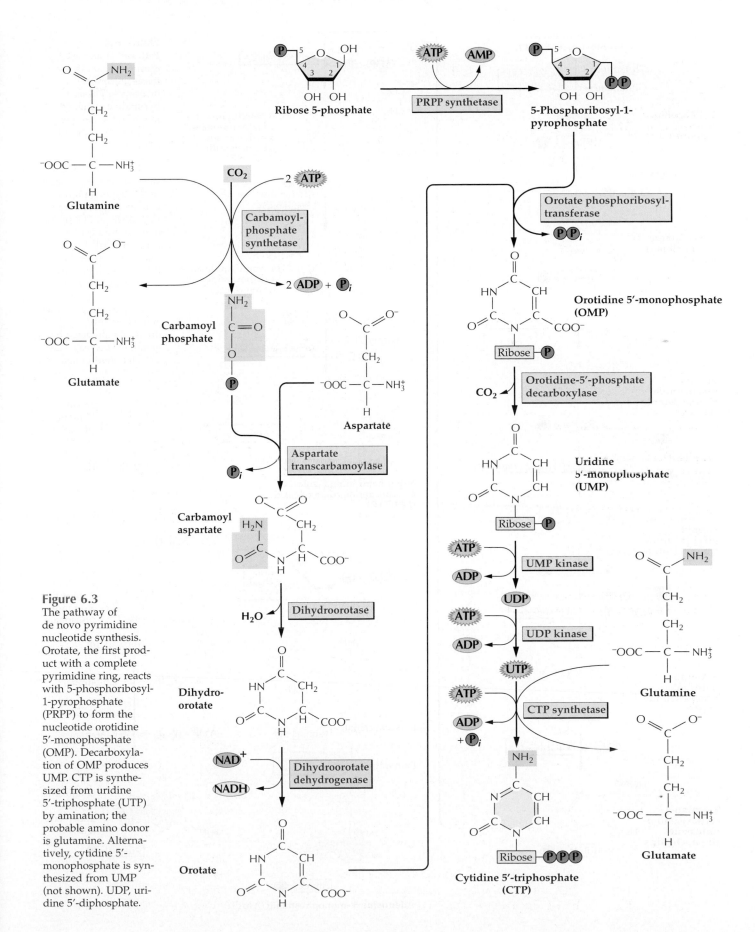

Figure 6.3
The pathway of de novo pyrimidine nucleotide synthesis. Orotate, the first product with a complete pyrimidine ring, reacts with 5-phosphoribosyl-1-pyrophosphate (PRPP) to form the nucleotide orotidine 5'-monophosphate (OMP). Decarboxylation of OMP produces UMP. CTP is synthesized from uridine 5'-triphosphate (UTP) by amination; the probable amino donor is glutamine. Alternatively, cytidine 5'-monophosphate is synthesized from UMP (not shown). UDP, uridine 5'-diphosphate.

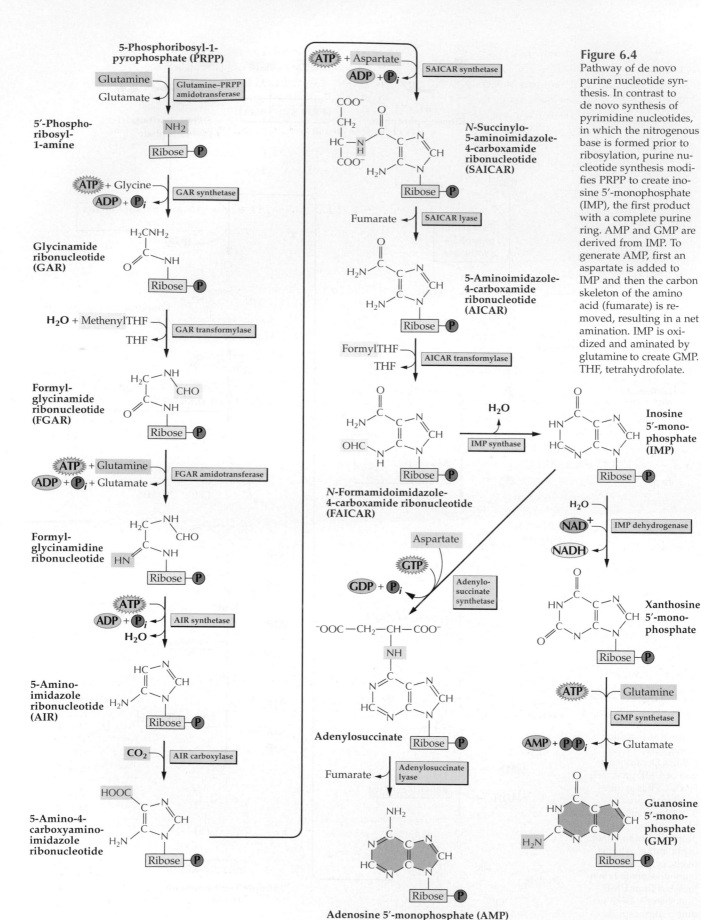

Figure 6.4
Pathway of de novo purine nucleotide synthesis. In contrast to de novo synthesis of pyrimidine nucleotides, in which the nitrogenous base is formed prior to ribosylation, purine nucleotide synthesis modifies PRPP to create inosine 5'-monophosphate (IMP), the first product with a complete purine ring. AMP and GMP are derived from IMP. To generate AMP, first an aspartate is added to IMP and then the carbon skeleton of the amino acid (fumarate) is removed, resulting in a net amination. IMP is oxidized and aminated by glutamine to create GMP. THF, tetrahydrofolate.

5-Phosphoribosyl-1-pyrophosphate (PRPP)
Glutamine
Glutamine–PRPP amidotransferase
Glutamate
5'-Phospho-ribosyl-1-amine
NH₂ Ribose Ⓟ
ATP + Glycine
GAR synthetase
ADP + Ⓟᵢ
Glycinamide ribonucleotide (GAR)
H₂O + MethenylTHF
GAR transformylase
THF
Formyl-glycinamide ribonucleotide (FGAR)
ATP + Glutamine
FGAR amidotransferase
ADP + Ⓟᵢ + Glutamate
Formyl-glycinamidine ribonucleotide
ATP
AIR synthetase
ADP + Ⓟᵢ
H₂O
5-Amino-imidazole ribonucleotide (AIR)
CO₂
AIR carboxylase
5-Amino-4-carboxyamino-imidazole ribonucleotide

ATP + Aspartate
SAICAR synthetase
ADP + Ⓟᵢ
N-Succinylo-5-aminoimidazole-4-carboxamide ribonucleotide (SAICAR)
Fumarate
SAICAR lyase
5-Aminoimidazole-4-carboxamide ribonucleotide (AICAR)
FormylTHF
AICAR transformylase
THF
N-Formamidoimidazole-4-carboxamide ribonucleotide (FAICAR)
H₂O
IMP synthase
Inosine 5'-monophosphate (IMP)
H₂O
NAD⁺
IMP dehydrogenase
NADH
Xanthosine 5'-monophosphate
ATP + Glutamine
GMP synthetase
AMP + ⓅⓅᵢ + Glutamate
Guanosine 5'-monophosphate (GMP)

Aspartate
GTP
GDP + Ⓟᵢ
Adenylo-succinate synthetase
Adenylosuccinate
Fumarate
Adenylosuccinate lyase
Adenosine 5'-monophosphate (AMP)

264

The catabolism of nucleotides in plants is not well understood. DNA or RNA is first hydrolyzed to oligonucleotides by deoxyribonucleases or ribonucleases, respectively. The oligonucleotides are further hydrolyzed by phosphodiesterases to yield mononucleotides, which are then hydrolyzed by nucleotidases and phosphatases to yield nucleosides. The free nucleosides and bases that are formed during the breakdown of nucleotides can be converted back to nucleotides by way of salvage pathways. Unlike the de novo pathways of purine and pyrimidine synthesis, which are well established and thought to be similar in animals and plants, the salvage pathways of nucleotide biosynthesis are less understood and more diverse. These pathways have received considerable attention following the discovery that some diseases in animals and humans result from deficiencies in salvage pathway enzymes (Box 6.1).

Most salvage pathways of nucleotide biosynthesis fall into one of two principal categories. The first type is a one-step pathway catalyzed by phosphoribosyltransferase. This reaction is reversible in the presence of pyrophosphatase, which is found in the plastids but not the cytosol of plant cells (see Chapter 13).

Reaction 6.1: Phosphoribosyltransferase

$$\text{Base} + \text{PRPP} \rightleftharpoons \text{Ribonucleotide} + \text{PP}_i$$

The second type of salvage pathway requires two steps that are catalyzed by nucleoside phosphorylase and nucleoside kinase, respectively. The latter reaction is irreversible.

Reaction 6.2: Nucleoside phosphorylase

$$\text{Base} + \text{(deoxy)Ribose 1-P} \rightleftharpoons \text{(deoxy)Nucleoside} + \text{P}_i$$

Reaction 6.3: Nucleoside kinase

$$\text{(deoxy)Nucleoside} + \text{ATP} \longrightarrow \text{(deoxy)Nucleotide} + \text{ADP}$$

Not all of the possible reactions suggested by the two pathways above are thought to occur in vivo. For example, although the nucleoside cytidine is present in plants, the free base cytosine has not been observed. Although certain deoxynucleosides can be salvaged by comparable reactions, experimental analysis has focused primarily on thymidine metabolism.

6.1.3 Nucleic acids are composed of nucleotides linked by phosphodiester bonds.

In a nucleic acid, nucleotides are linked together in a polynucleotide chain. Covalent phosphodiester bonds join the 5′ carbon of one nucleotide to the 3′ carbon of the next, forming a sugar–phosphate backbone to

---

Box 6.1

Defects in the salvage pathways of nucleotide biosynthesis are associated with diseases in humans.

The importance of the salvage pathways of nucleotide biosynthesis is illustrated by the severity of the diseases that result from deficiencies in salvage pathway enzymes. For example, **Lesch–Nyhan syndrome**, a severe genetic disorder in human children, is caused by lack of functional hypoxanthine-guanine phosphoribosyltransferase (HGPRTase), which catalyzes the condensation of hypoxanthine and PRPP. The inability to use PRPP results in its accumulation, which in turn stimulates glutamine-PRPP amidotransferase, thereby causing the overproduction of purine nucleotides. Oxidation of the excess purines results in the production of high concentrations of uric acid and affects especially the central nervous system, which may reflect the brain's particular dependence on the salvage pathways. Symptoms of the disease include mental retardation, poor coordination, and self-mutilating behavior.

Reduced activities of purine-nucleoside phosphorylase and adenosine deaminase affect the development and function of both T cells and B cells, causing a severe human immunodeficiency disease. Purine nucleoside phosphorylase deficiency results in a large accumulation of dGTP in T cells, which most likely is toxic during T cell development. Adenosine deaminase deficiency leads to an accumulation of deoxyadenosine in T cells, which has a negative effect on the activity of ribonucleotide reductase and thereby lowers the concentration of other deoxynucleotide triphosphates. The basis for B cell toxicity is less well understood.

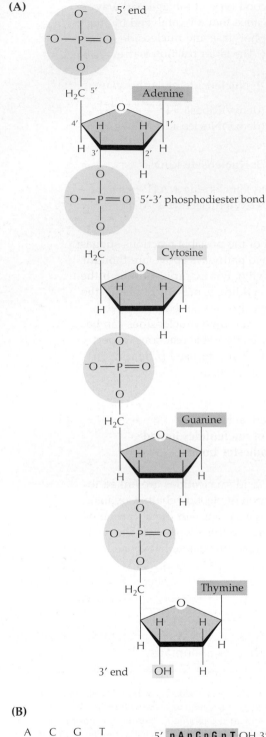

(A)

5′ end

5′-3′ phosphodiester bond

Adenine

Cytosine

Guanine

Thymine

3′ end   OH   H

(B)

A   C   G   T

5′ **pApCpGpT** OH 3′

5′ **ACGT** 3′

Figure 6.5
(A) A polynucleotide chain of DNA. The alternating sugar–phosphate polymer forms the backbone of the polynucleotide chain. In RNA, a 2′-hydroxyl group replaces the 2′-hydrogen. (B) Representations of nucleic acids are drawn 5′ to 3′, from left to right. In most cases, nucleic acid sequences are presented in a format that does not include symbols for the sugar–phosphate backbone (lower right).

which the four bases are attached (Fig. 6.5A). The directional nature of these bonds means that a linear nucleotide chain has two distinct ends with different biochemical characteristics. The 5′ end of a nucleic acid usually has a free phosphate group, while the 3′ end is typically hydroxylated. By convention, polynucleotide sequences are written 5′ to 3′, from left to right (Fig. 6.5B).

The predominant form of DNA is a double-stranded, double helical structure formed from two separate polynucleotide chains by specific pairing of bases. Hydrogen bonds form between a larger, purine base on one chain (A or G, respectively) and a smaller, pyrimidine base on the other chain (T or C, respectively). A–T pairing involves two hydrogen bonds; G–C pairing involves three hydrogen bonds (Fig. 6.6). The free energy that drives the hybridization of two single strands to form a duplex DNA results from the overall increase in entropy that occurs when the hydrophobic bases are moved from an aqueous environment into the center of the duplex. (Although duplex formation decreases the entropy of the nucleotide chains, it increases the entropy of the surrounding water molecules.) The two chains are antiparallel, such that the sugar–phosphate backbones of the two strands are oriented in opposite directions. The 5′ and 3′ ends of the genomes of single-stranded DNA viruses are typically linked by a phosphodiester bond, resulting in a circular molecule.

Most cellular RNA is single-stranded, although the nucleotide chain often folds locally. The resulting domains generally contain stems, double-stranded regions that contain both "Watson–Crick" basepairs (i.e., G–C and A–U) and many types of atypical basepairs (e.g., G–U, U–U, G–A, etc.), and terminal, single-stranded loops. These domains then associate with each other through various types of specific tertiary interactions. As a result, RNAs can assume very compact structures.

6.2 Replication of nuclear DNA

Before a cell can divide, it must produce a new copy of each of its chromosomes. In eukaryotic cells, chromosomes are made up of **chromatin,** a filamentous complex of DNA and proteins (see Chapter 7). Considering

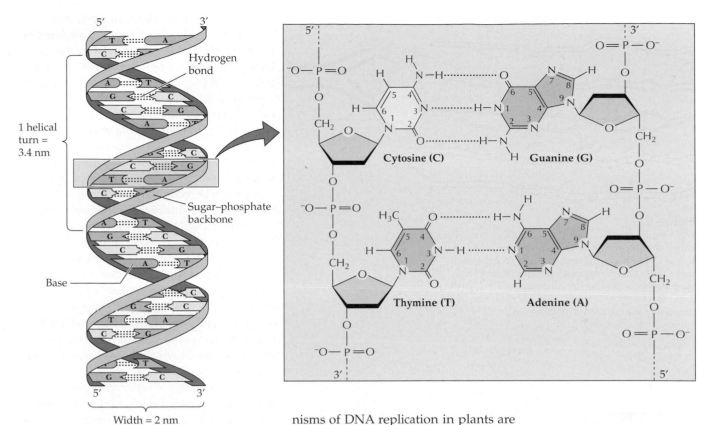

Figure 6.6

The DNA double helix. Two DNA strands associate when their complementary bases pair. Three hydrogen bonds link cytosine and guanine, whereas only two hydrogen bonds form between thymine and adenine. The two DNA strands are antiparallel, running from 5' to 3' in opposite directions.

that the long DNA molecules are tightly packaged into DNA–protein complexes, it is not surprising that the replication of DNA involves a large number of enzymes and regulatory proteins. Replication of nuclear DNA has been studied extensively in bacteria, yeast, and mammalian cells. Regulation of this process during the eukaryotic cell cycle is discussed in more detail in Chapter 11.

In all organisms, DNA replication involves three stages: initiation, elongation, and termination. In eukaryotes, the ends of the linear chromosomes also are modified to prevent DNA loss during subsequent rounds of replication. In this section we will focus primarily on the general enzymology of nuclear DNA replication in eukaryotes. The prevailing view is that processes are very similar in all eukaryotes, although it can be expected that some of the regulatory mecha-

nisms of DNA replication in plants are tailored specifically to their unique developmental and physiological functions.

6.2.1 Nuclear DNA synthesis begins at discrete origins of replication and requires complex cellular machinery composed of many proteins.

Nuclear DNA replication occurs during the S phase of the cell cycle (see Chapter 11) and is a highly regulated process. Replication initiates simultaneously at a number of specific sites, termed **origins of replication** (Fig. 6.7). The identification of eukaryotic chromosomal replication origins has been most successful in the budding yeast *Saccharomyces cerevisiae*, where a specific sequence of 200 basepairs (bp)—called the autonomously replicating sequence (ARS)—is the minimal sequence requirement for the initiation of chromosomal DNA replication. In mammalian cells, initiation of DNA replication occurs in specific regions that are more than 10,000 bp (i.e., 10 kilobases [kb]) apart. The minimal sequence that constitutes the DNA replication origin in these regions is still unknown. Specific replication origins have not been identified in plants.

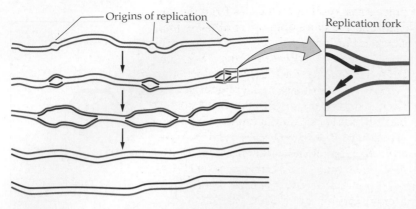

Figure 6.7
DNA replication initiates at specific DNA sequences, called origins of replication, which are spaced apart in the DNA. Each origin produces two replication forks that move in opposite directions until they meet replication forks initiated at neighboring origins. The parental DNA strands are shown in blue, the newly synthesized strands in red.

Many possible origins of replication exist in the chromosome, but not all function in every cell. The segment of DNA that is replicated from a single origin is termed a replicon. In plant cells, for example, a typical replicon is 50 to 70 kb in length. *Triticale*, a hybrid allohexaploid grain, has replicons of about 22 kb. In contrast, the parent plants (rye and wheat) have replicons of 50 to 60 kb. Apparently, replication origins that remain silent in the cells of the parental species can become active in *Triticale*. Experiments with synchronized cell populations also have shown that not all of the replication origins on a given chromosome are activated at the same time when DNA replication is initiated during S phase. Origins of replication are activated ("fire") in a reproducible order. The timing of firing at a given origin is tightly regulated by the cell and may depend on the state of condensation of the chromatin in which the replication origin resides (see Section 6.2.4).

The activation of replication origins during S phase is now understood best in yeast and mammalian cells (see Chapter 11). DNA replication initiates as the chromatin structure unfolds. The two DNA strands separate at an AT-rich region and replication-specific proteins bind to the DNA (Fig. 6.8). Because the DNA double helix is very stable under normal conditions, specific ATP-dependent enzymes, called DNA helicases, must catalyze strand separation at the origin and ahead of the replication fork. Following DNA helicase activity, replication protein A (RP-A) binds to single-stranded DNA and

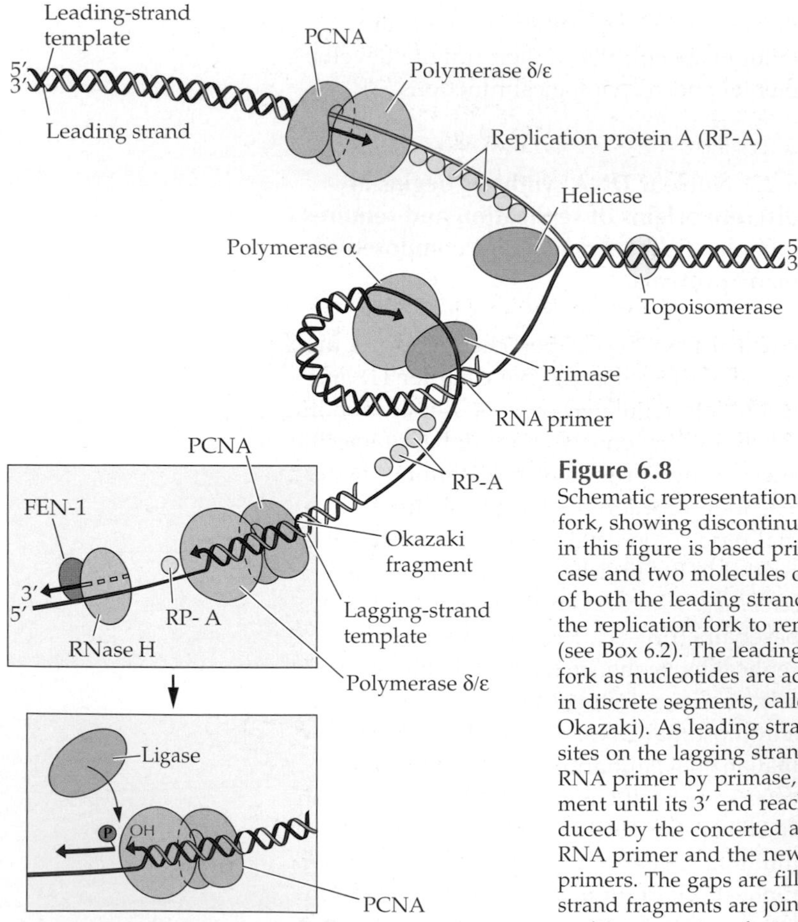

Figure 6.8
Schematic representation of the organization of the eukaryotic DNA replication fork, showing discontinuous replication of the lagging strand. The model shown in this figure is based primarily on research in yeast. A complex containing helicase and two molecules of DNA polymerase carries out the coordinated synthesis of both the leading strand and the lagging strand. Topoisomerase acts ahead of the replication fork to remove the supercoils generated by unwinding the DNA (see Box 6.2). The leading DNA strand is lengthened continuously toward the fork as nucleotides are added. In contrast, the lagging DNA strand is synthesized in discrete segments, called Okazaki fragments (after their discoverer, Reiji Okazaki). As leading strand synthesis progresses and the fork opens, priming sites on the lagging strand template are exposed. Following synthesis of a short RNA primer by primase, DNA polymerase α adds nucleotides to the new fragment until its 3′ end reaches the primer of the adjacent fragment. Gaps are produced by the concerted action of FEN-1 nuclease, which cleaves between the RNA primer and the new DNA strand, and RNase H, which degrades the RNA primers. The gaps are filled by DNA polymerases δ/ε, and the lagging DNA strand fragments are joined by DNA ligase. PCNA (proliferating cell nuclear antigen) is a protein cofactor of the DNA polymerase complex. RP-A (replication protein A) binds to and stabilizes single-stranded DNA.

stabilizes its structure. As DNA replication proceeds, DNA topoisomerases remove supercoils that form in front of the replication fork (Box 6.2).

6.2.2 Nuclear DNA replicates semiconservatively and semidiscontinuously.

As DNA replication proceeds from the replication origin, a multienzyme complex that contains **DNA polymerase** catalyzes the addition of deoxyribonucleotides to the 3' end of a template-bound nucleic acid. Thus, DNA synthesis proceeds in the 5' to 3' direction, while the template strand is read 3' to 5', creating an asymmetric structure at the replication fork (Fig. 6.8). DNA polymerases can only add nucleotides to a preexisting chain **(primer)** and are not active unless a primer with a free 3' hydroxyl is hydrogen-bonded to the DNA template being replicated. The primers are synthesized by specific ribonucleotide polymerases, called DNA primases, which initiate primer synthesis with a purine ribonucleotide.

As the result of the asymmetric structure of the replication fork, DNA synthesis is **semidiscontinuous.** Addition of nucleotides to one strand of newly synthesized DNA is continuous. This strand, known as the **leading strand,** begins at a single primer. The other strand, known as **lagging strand,** is synthesized in short, discontinuous segments of DNA called Okazaki fragments, each of which requires its own primer (Fig. 6.8). To produce a continuous DNA strand from the many fragments of the lagging strand, a special DNA repair system that includes a specific ribonuclease, RNase H, removes the RNA primer and replaces it with DNA. Another enzyme, DNA ligase, joins the 3' end of the new DNA fragment to the 5' end of the downstream DNA fragment. Since both strands of nuclear DNA serve as templates for DNA synthesis, the replication products each contain one original and one newly synthesized complimentary strand. This type of DNA replication is termed **semiconservative** (see Fig. 6.7).

Three principal DNA polymerase holoenzymes that replicate nuclear DNA have been identified in the nuclei of eukaryotes, including plants: α, δ, and ε. DNA polymerase α often dissociates from the template and therefore does not synthesize long chains of nucleotides, but two of its four subunits have DNA primase activity. Therefore, DNA polymerase α is thought to function primarily as a primase in lagging strand DNA synthesis (Fig. 6.8). After DNA polymerase α dissociates from the template, a primer terminus is exposed. A multisubunit replication factor complex binds the terminus, facilitating the assembly of a functional DNA polymerase δ or ε complex to complete the synthesis of the Okazaki fragment. In contrast to DNA polymerase α, DNA polymerases δ and ε remain tethered to the DNA template by PCNA (proliferating cell nuclear antigen), an essential protein cofactor of the DNA polymerase complex that forms a ring-like structure around the DNA. For pedagogical reasons, most of the components involved in DNA replication are shown as individual complexes in Figure 6.8. However, we now know that these proteins probably cooperate to form a large, multisubunit complex of replication machinery at the replication fork. In addition, DNA replication occurs at discrete foci in the nuclei of mammalian cells. These foci, called replication centers, may contain as many as 100 replication forks. Biochemical studies with *Xenopus* egg extracts have revealed that replication centers form during nuclear assembly, which precedes DNA synthesis (see Chapter 11).

DNA replication proceeds with remarkably high fidelity, producing on average only one error for every 10^9 basepairs replicated. Mismatch repair and the 3'→5' exonuclease function of DNA polymerases δ and ε maintain this high fidelity. DNA polymerases are self-correcting enzymes that remove their own polymerization errors during replication. Separate, intrinsic 3'→5' exonuclease activities enhance fidelity by a process called **proofreading.** These enzymatic activities can clip mismatched nucleotides from the 3' end of the newly synthesized strand. DNA polymerase α lacks this 3'→5' exonuclease activity, but demonstrates high fidelity nonetheless. The lack of proofreading by DNA polymerase α may be compensated for by the removal of the RNA primers and the few deoxyribonucleotides polymerized by the enzyme, followed by DNA mismatch repair (see Section 6.3.4).

Box 6.2

DNA topoisomerases perform a number of functions during DNA replication.

In double-stranded DNA, approximately 10 basepairs correspond to one complete turn about the axis of the double helix structure. As DNA replication begins, the DNA double helix must unwind to expose the leading and lagging strand templates. When one section of the molecule is underwound, the regions proximal and distal to it become overwound, thereby creating supercoils (see left panel of the diagram). The DNA ahead of the replication fork would have to rotate rapidly to remove these supercoiled regions, allowing DNA replication to proceed. However, DNA in circular molecules or chromosomes is not free to rotate. Instead, DNA topoisomerases remove (or in some cases add) supercoils to maintain favorable DNA topologies and to allow DNA replication to continue. The electron micrograph shows a supercoiled circular DNA molecule (A) and a relaxed circular DNA molecule (B) for comparison.

During the replication of circular DNA molecules such as the chloroplast genome, discussed in Section 6.5.4, the oppositely oriented replication forks eventually collide to yield two catenated (linked) circles of double-stranded DNA (see middle panel of diagram and structure C in the electron micrograph). In addition to removing supercoils, DNA topoisomerases also are required to separate the interlocked circular replication products (see right panel of diagram).

Two classes of DNA topoisomerase that can attach covalently to a DNA phosphate, type I and type II, are found in both prokaryotes and eukaryotes. Topoisomerase I causes a single-strand break (or nick) in a DNA duplex by breaking a phosphodiester bond. Topoisomerase II creates transient double-strand breaks. Cleavage of the phosphodiester bonds is reversible and rapid, and does not require additional energy. Topoisomerase I can only remove supercoils from DNA, whereas topoisomerase II (also called DNA gyrase) can either add or remove supercoils and can either catenate or decatenate circular duplex DNA molecules. Genetic analysis of topoisomerase function in yeast has shown that topoisomerase II is essential for cell growth and division. In addition to decatenating the products of replication, this class of enzymes participates in chromosome condensation during mitosis and

meiosis, constitutes a major component of the chromosome scaffold and nuclear matrix, and may also play a role in the structure of the DNA–matrix complex.

The topoisomerases found in plant nuclei resemble their yeast and animal counterparts. A type I enzyme isolated from chloroplasts, however, is more similar to prokaryotic topoisomerase I, which is consistent with the prokaryotic origin of this organelle (see Section 6.5.1). Topoisomerases of types I and II also have been found in the mitochondria of mammals and trypanosome, but not yet in plant mitochondria.

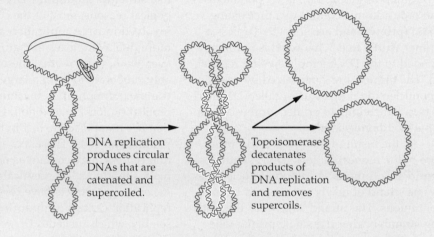

DNA replication produces circular DNAs that are catenated and supercoiled.

Topoisomerase decatenates products of DNA replication and removes supercoils.

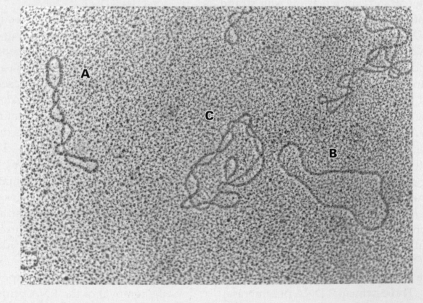

Termination occurs in regions where newly synthesized DNA strands from two replication origins meet. In the *Escherichia coli* chromosome, specific sequences near the termination region, called *Ter* sites, arrest new strand synthesis in one direction. The newly synthesized strand from the other direction passes through the *Ter* site and homologous recombination (see Section 6.4.2) occurs between the two newly synthesized strands. Although there is no direct evidence, a similar mechanism may operate in eukaryotes, as DNA sequences resembling *Ter* sites have been found in yeast.

The fidelity of DNA replication is exploited by viruses, which have evolved specific mechanisms that direct the host cell DNA replication machinery to multiply and propagate the viral DNA. In turn, scientists have harnessed the ability of DNA viruses to invade cells and replicate: Viruses have been used as vectors to deliver foreign genes of interest into cells, where the genes can be replicated to high copy numbers and be overexpressed to yield substantial concentrations of the desired gene product. A unique group of single-strand plant DNA viruses, called geminiviruses, provide good examples of the strategies used by viruses to replicate their DNA and the potential use of viruses in plant genetic engineering techniques (Box 6.3).

6.2.3 Unlike prokaryotic DNA, eukaryotic chromosomes have ends that are protected by telomeres.

As described above, DNA polymerases require nucleotide primers to initiate DNA replication. If a priming site was located at the end of a linear chromosome, this short DNA segment would remain unreplicated after the primer is degraded and would not be available during the next round of DNA replication. As a result, a short segment of the chromosome would be deleted during each round of DNA replication. Normally, however, such deletions do not occur and chromosome length remains stable because chromosomes terminate in special sequences of DNA called **telomeres.** These sequences, which are similar in all eukaryotic cells, consist of multiple tandem repeats of a short DNA sequence. The repeated telomeric

sequence is TTAGGG in humans and TTTAGGG in *Arabidopsis* (Fig. 6.9).

An enzyme called telomerase recognizes the G-rich strand of the telomere and extends the 3′ end of chromosomal DNA by adding new repetitions of the telomeric sequence. All known telomerases contain specific RNA molecules that are part of the enzyme complex and serve as the complementary template for the telomeric repeat unit (Fig. 6.10, p. 274). After telomerase has copied several telomere repeats from its integral RNA template, the complementary C-rich DNA strand is synthesized. The telomeric repeats may serve as a primer site for chain elongation, allowing DNA polymerase to complete the unfinished strand.

Telomerase was discovered in the protozoan *Tetrahymena,* and has been isolated from yeast and animals. Plant chromosomes have highly conserved telomeric sequences and telomerase activity has been found in plants, suggesting that plant telomeres are maintained by mechanisms similar to those identified in other eukaryotes.

6.2.4 The timing of nuclear DNA replication is tightly regulated, although the mechanisms involved are not well understood.

All chromosomal DNA replicates during the S phase of the cell cycle, but the timing of replication of a DNA segment appears to depend in part on the structure of the chromatin region in which the DNA segment resides. The chromatin that remains in a highly condensed conformation during the interphase of the cell cycle is called **heterochromatin;** replication of DNA in segments

Human	TTAGGGTTAGGGTTAGGGTTAGGG
Paramecium	TTGGGGTTGGGGTTGGGGTTGGGG
Trypanosoma	TTAGGGTTAGGGTTAGGGTTAGGG
Arabidopsis	TTTAGGGTTTAGGGTTTAGGG

Figure 6.9
Telomeres are simple, repeated DNA sequences found at the ends of eukaryotic chromosomes. Despite the evolutionary divergence of eukaryotic organisms, the repeated sequences of telomeric DNA are remarkably conserved.

Geminiviruses are transmitted by white flies or grasshoppers and cause disease in several important crops, including maize, bean, squash, and tomato. These viruses have attracted research interest as plant pests, as models of viral replication, and as potential vectors for the genetic engineering of plants.

The name of this group of viruses refers to the unusual twin icosahedral capsid structure. Depending on the viral species, each paired capsid encloses either one or two molecules of covalently closed, circular, single-stranded DNA. In species with two DNA molecules, the two genomic components are of similar size but encode different genes. One component, designated A, encodes all of the viral genes necessary for DNA replication and encapsidation (packaging of the nucleic acid within a protein coat). Among these is the Rep protein, which is required

for viral DNA replication. The other component, designated B, contains genes required for cell-to-cell and systematic spread of virus in plants. Both components are thus required for infectivity.

As shown in the figure, after infection the single-stranded viral DNA (+ strand, blue) is transported to the host cell nucleus and converted to double-stranded DNA by host-directed synthesis of a complementary strand (– strand, red). The resulting double-stranded DNA then serves as a template for transcription of the *Rep* gene, which is required for further replication. Geminivirus DNA replicates using a rolling-circle mechanism (see Section 6.5.4). The virus-encoded Rep protein introduces a nick into the DNA at a specific inverted repeat site on the (+) strand and covalently binds to the free 5′ terminus. The 3′ terminus is then used as a primer for the synthesis of a new (+) strand,

which displaces the parental (+) strand from the intact (–) strand. Host cell enzymes catalyze these steps in the viral DNA replication. After completion of one round of DNA synthesis, the Rep protein cleaves the new (+) strand and ligates the ends to produce a new (+) strand circular viral DNA. This new (+) strand can serve as another template during the early infection cycle. Later in the infection cycle the (+) strand DNA is encapsidated.

Geminivirus genomes achieve high copy numbers in inoculated cells and their genes can be expressed strongly. These features make geminiviruses attractive as potential vector systems for the expression of specific gene products in plants. In such a vector, a foreign gene of interest would typically replace the gene that encodes the viral coat protein. The engineered virus can be transmitted to the plant by mechanical inoculation of

---

of heterochromatin occurs very late in S phase. Studies in the budding yeast *S. cerevisiae* have also revealed that the DNA located near telomeres is replicated late in S phase, whereas the DNA associated with centromeres (see Chapter 7) is replicated earlier. A telomere-associated ARS will replicate sooner if relocated to another site on the chromosome. Similarly, if small segments of telomeric DNA sequence are inserted into a chromosome near an origin of replication that initiates DNA replication early, activation of the origin is delayed. In contrast, DNA segments that are actively transcribed are located in **euchromatin,** regions of the chromosome in which the chromatin is less condensed. These DNA segments are replicated early during S phase, suggesting that gene expression can affect the timing of DNA replication. For example, genes that are transcribed constitutively in all cells replicate early, whereas genes that encode proteins with tissue-specific functions are replicated early in cells that express the genes but late in cells where the genes are inactive. Together, these observations suggest that the DNA sequence of the replication origin does not necessarily dictate replication timing, and that other regulatory elements probably control initiation of DNA replication.

6.3 DNA repair

6.3.1 Damage to DNA can result in mutations.

DNA molecules are constantly subjected to physical and chemical stresses in vivo. Oxygen, ultraviolet (UV) light, alkylating agents, and radicals all can cause random changes in the DNA sequence. These changes can result from strand breaks, chemical modification of bases, and incorporation of mismatched bases during replication. Common types of DNA changes include the formation of pyrimidine dimers, alkylation of bases, and deamination of bases (Fig. 6.11, p. 275). If left uncorrected, spontaneous and induced DNA changes would rapidly alter the DNA sequence. For example, spontaneous deamination of cytosine to uracil is estimated to occur at a rate of 100 bases per genome per day. During replication, uracil would permanently change the DNA sequence at this site by pairing with adenine instead of guanine. Most DNA changes occur in sequences that do not encode genes; such events are therefore inconsequential. However, the possibility that DNA damage may result in mutations that disrupt the function of essential enzymes or structural proteins

leaves. The foreign gene would be amplified during viral DNA replication. Because the engineered virus DNA lacks the gene for the capsid protein it would not assemble into viral particles, so its ability to spread would be limited.

In the presence of the Rep protein, part of the geminivirus genome can also replicate in the host cell as a plasmid, an extrachromosomal, autonomously replicating molecule of DNA. This particular feature may provide a basis for developing plasmid DNA vectors to transform plant cells. Transgenic plants grown from callus that has been inoculated with this plasmid might produce seeds carrying the desired gene. Plasmid DNA vectors, which are used to transform bacteria (see Chapter 7) and to transfect mammalian cells, can be expressed strongly to yield high concentrations of protein.

poses a significant risk to the cell. We now know that several efficient DNA repair mechanisms have evolved to protect organisms against the potential deleterious effects of mutations.

6.3.2 Pyrimidine dimers, which are caused by UV-B, are repaired using visible light or UV-A.

UV light can fuse pyrimidines to create cyclobutane dimers (Fig. 6.11) and pyrimidine(6→4)pyrimidinone dimers. The UV-A (320 to 400 nm) and UV-B (280 to 320 nm) wavelengths of sunlight cause approximately equal amounts of DNA damage. Although each photon of UV-B radiation contains more energy than a photon of UV-A radiation and is therefore more likely to induce damage, far fewer photons of UV-B reach the earth's surface than do photons of UV-A. In bacteria and plants, some UV-induced mutations can be reversed by high-energy wavelengths of visible light or by UV-A. This conversion, called **photoreactivation,** is catalyzed by photolyases, which absorb radiant energy of wavelengths 300 to 600 nm and convert cyclobutane dimers to pyrimidine monomers (Fig. 6.12, p. 276).

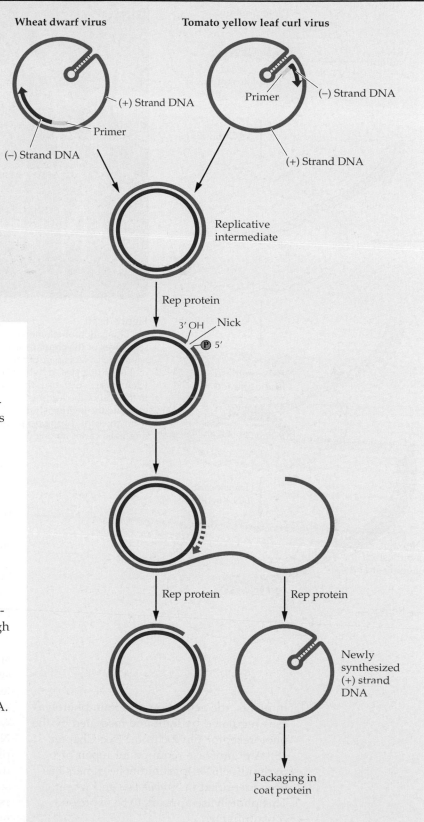

(A)

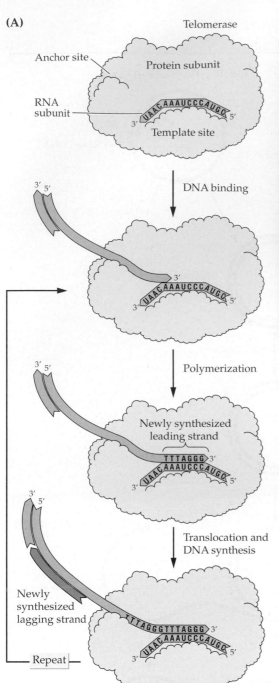

Telomerase

Anchor site

Protein subunit

RNA subunit

3′ UAAC AAAUCCC AUGC 5′

Template site

DNA binding

3′ 5′

3′ UAAC AAAUCCC AUGC 5′

Polymerization

3′ 5′

Newly synthesized leading strand

TTTAGGG 3′
UAAC AAAUCCC AUGC 5′

Translocation and DNA synthesis

3′ 5′

Newly synthesized lagging strand

TTTAGGGTTTAGGG 3′
UAAC AAAUCCC AUGC 5′

Repeat

(B)

Older embryo, 30 kb
Immature embryo, 80 kb
Mature inflorescence spike, 20 kb

Young inflorescence spike, 45 kb

Leaves, 23 kb

Figure 6.10
(A) Elongation of telomeres. Telomerase recognizes the overhanging 3′ telomere sequences of the chromosome. An RNA molecule that is complementary to the telomeric DNA is part of the telomerase enzyme complex. The overhanging DNA of the leading strand hybridizes with the telomerase RNA and is elongated by telomerase, which uses the RNA as a template. Telomerase then translocates to the end of the leading strand. Primase and DNA polymerase subsequently elongate the lagging strand of telomere DNA. (B) Telomere length does not remain constant during development. The length of the telomeres of barley chromosomes varies among different organs and developmental stages.

6.3.3 Excision repair mechanisms can remove either individual bases or longer nucleotide chains.

Two major pathways have been identified for DNA excision repair: **base excision repair** and **nucleotide excision repair.** In base excision repair, the chemically modified base (e.g., uracil, 5-methylcytosine, or 3-methyladenine) is first removed by a DNA glycosylase that is specific for that type of DNA lesion. The action of DNA glycosylase results in an apurinic or apyrimidinic (AP) site with an intact sugar–phosphate backbone. Next, an AP endonuclease cleaves a phosphodiester bond at the AP site. A pathway that includes several enzymes repairs the resulting nick. First, deoxyribose phosphodiesterase removes the deoxyribose-phosphate molecule. Next, repair polymerase (DNA

In plants, the activation of certain photolyases is regulated by light and mediated by the photoreceptor phytochrome (see Chapter 18). A photolyase required for repair of pyrimidine(6→4)pyrimidinone dimers has been identified in *Arabidopsis* and wheat. This photolyase appears to be expressed constitutively.

Spontaneous

Deamination

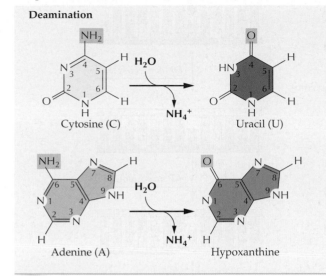

Cytosine (C) → Uracil (U)

Adenine (A) → Hypoxanthine

Induced

Exposure to UV light

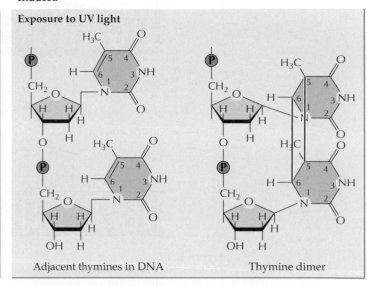

Adjacent thymines in DNA   Thymine dimer

Depurination

Guanine

DNA chain   DNA chain

AP site

Alkylation

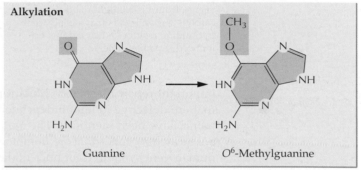

Guanine O^6-Methylguanine

Figure 6.11
Chemical reactions that cause DNA mutations either can occur spontaneously or be induced by radiation or chemicals. One major type of spontaneous DNA damage results from deamination of the bases cytosine or adenine. Depurination, another type of spontaneous event, forms an apurinic (AP) site in the DNA. UV light induces the formation of pyrimidine dimers between adjacent pyrimidine bases. Alkylation of DNA bases results in the addition of methyl groups (or sometimes ethyl groups; not shown) at various positions in the ring structure.

polymerase β in animal and yeast cells) adds a single nucleotide to the 3' end of the DNA strand at the gap. Finally, DNA ligase seals the nick (Fig. 6.13). Genes homologous to the yeast genes that encode proteins involved in base excision repair have been isolated from plants. The plant gene products can excise specific DNA damage products in vitro.

The second pathway, nucleotide excision repair, involves the DNA replication machinery. This repair pathway is capable of removing DNA damage that distorts the DNA double helix, including various pyrimidine dimers created by UV light. In eukaryotic cells, nucleotide excision repair removes and replaces a DNA fragment approximately 30 bases long. A multienzyme complex that recognizes the distortion in the DNA cleaves the damaged strand on both sides of the lesion. The oligonucleotide is then removed by DNA helicase. The resulting gap is repaired by DNA polymerase δ or ε and closed by DNA ligase (Fig. 6.14). With few exceptions, these activities have been identified in plants, suggesting that plants also utilize this

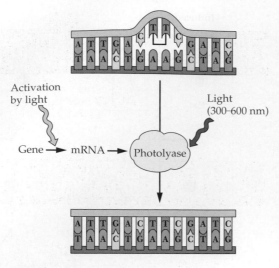

Figure 6.12
Photolyases catalyze the repair of pyrimidine dimers. The enzyme uses energy from visible light to break the carbon–carbon bonds that join adjacent pyrimidine residues, such as UV-induced thymine dimers.

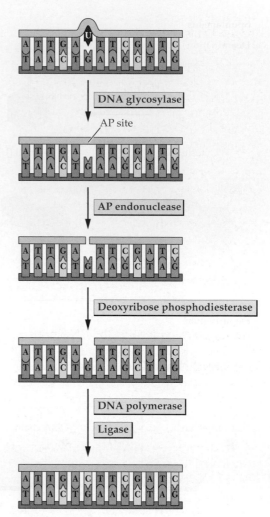

Figure 6.13
The base excision repair pathway. In the pathway shown in this figure, a uracil formed by deamination of cytosine is removed from the sugar–phosphate backbone of the DNA strand by DNA glycosylase. AP endonuclease recognizes the resulting AP site and cleaves the DNA strand at that site. Deoxyribose phosphodiesterase then removes the remaining deoxyribose molecule. DNA polymerase and ligase repair the gap, restoring the cytosine/guanine basepair.

important repair machinery. Mutant plants that are defective in light-independent DNA repair have been identified. The mutations are located in genes that are homologous to human genes required for nucleotide excision repair, indicating that plants utilize the same repair pathways identified in other eukaryotes.

6.3.4 Mismatch repair corrects errors made during DNA replication.

In *E. coli*, adenine residues in the palindromic sequence GATC are methylated shortly after DNA replication. If DNA polymerase introduces an error during DNA replication that results in a mismatched basepair, repair occurs before methylation, when the newly synthesized DNA strand is unmethylated and the two strands can be distinguished (Fig. 6.15). This repair requires the function of three proteins that form a complex. MutS recognizes and binds to the base mismatch. MutH binds to DNA at the hemimethylated GATC sites and cleaves the unmethylated (newly synthesized) strand. MutL binds to MutS and MutH and is required for the final repair steps. The combined action of additional proteins, namely the UvrD helicase and exonucleases, results

in the removal of the damaged DNA strand beginning at the MutH cleavage site. The resulting gap is filled by DNA polymerases and sealed by DNA ligase.

Mismatch repair is a highly conserved mechanism that appears to function similarly in eukaryotes and bacteria. Several homologs of the bacterial *mutS* and *mutL* genes have been found in the genomes of yeast and mammals, and a *mutS* homolog is also present in human mitochondrial DNA. The mechanism that allows the bacterial MutS–MutL–MutH complex to discriminate between the parental and newly synthesized strands of DNA is not known in eukaryotes, in which GATC methylation has not been detected and *mutH* homologs have not been identified. Families of eukaryotic genes that are homologous to *mutS* and *mutL* arose before the evolutionary divergence of animals, plants, and fungi. Homologs of these two genes have been cloned from several plants, suggesting that this type of mismatch repair system is similar in all eukaryotic cells.

6.3.5 Error-prone repair allows DNA polymerase to read through damaged sites on the template.

If during DNA replication the DNA polymerase complex encounters damage in the template strand, the enzyme complex bypasses the damaged site and continues DNA synthesis. This process is potentially mutagenic because the DNA polymerase does not repair the damage but instead inserts adenine nucleotides at the gap regardless of the original DNA sequence that would otherwise result at the damaged site.

In bacteria subjected to potentially lethal conditions, such as extensive UV irradiation, this **error-prone repair** mechanism, also called **translesion replication,** is activated. In *E. coli,* the proteins UmuC and UmuD bind to DNA polymerase and alter the stringency of nucleotide incorporation. Adenines are inserted into the newly synthesized strand at the lesion site irrespective of the sequence of the parental DNA strand. If the damaged sites are thymidine dimers, the newly synthesized DNA is not mutated. However, cytosine dimers result in mutations at the lesion sites. Although this type of repair can generate a large number of mutations, it may allow the cells to survive extensive DNA damage. A similar bypass mechanism has been identified in yeast. The products of the *umuC* gene family are required for dimer bypass and include proteins active in both error-prone and high fidelity repair mechanisms. A gene homologous to one member of the *umuC* gene family, the human XP-V gene, has been identified in plants.

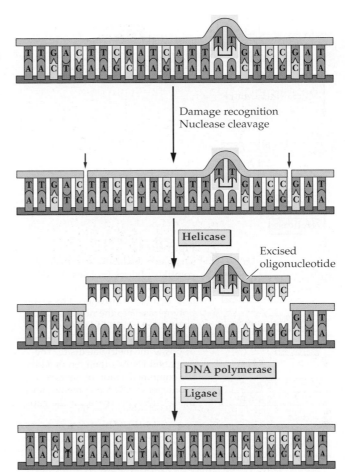

Figure 6.14
Nucleotide excision repair of a pyrimidine dimer. The nucleotide excision repair pathway utilizes the DNA replication machinery to remove large distortions in the DNA double helix, such as pyrimidine dimers induced by UV light. Damaged DNA is first recognized by endonucleases and cleaved (red arrows). The oligonucleotide that contains the damaged bases is then excised after helicase unwinds the DNA at this site. DNA polymerase and ligase repair the resulting gap.

6.3.6 Extensive DNA damage also can be repaired by homologous recombination.

When DNA damage is extensive or affects a longer stretch of nucleotides, an alternate repair pathway can be induced as well. In recombination repair (Fig. 6.16), DNA polymerase stops replication when a lesion is encountered in the parental (template) strand. DNA replication resumes at the next priming site on the template, which may be several hundred nucleotides downstream from the damaged site. The discontinuity, or daughter strand gap, in the newly synthesized strand is then repaired by homologous DNA recombination with the complementary parental strand using specific

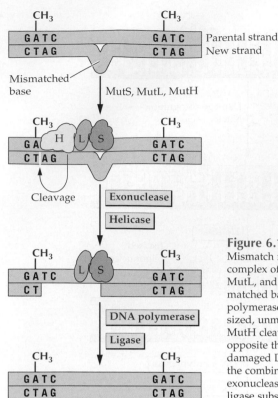

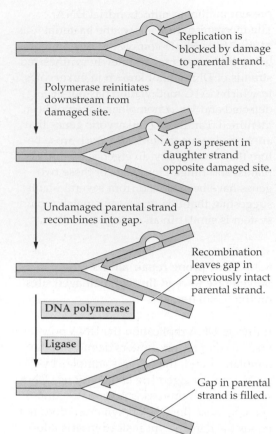

Figure 6.15
Mismatch repair in *Escherichia coli*. A complex of three proteins, MutH, MutL, and MutS, recognizes mismatched bases introduced by DNA polymerase into the newly synthesized, unmethylated DNA strand. MutH cleaves the new DNA strand opposite the methylated site. The damaged DNA is then removed by the combined action of helicase and exonuclease. DNA polymerase and ligase subsequently repair the gap.

Figure 6.16
Post-replication repair. When the DNA replication machinery encounters an unrepaired lesion (e.g., a pyrimidine dimer), DNA polymerase can bypass the block and reinitiate at a downstream site. The resulting gap in the newly synthesized DNA strand is then repaired by recombination with the undamaged parental strand. DNA polymerase and ligase fill the resulting gap in the previously intact parental DNA. The pyrimidine dimer can later be removed by excision repair (not shown) to produce two intact, double-stranded DNA molecules.

recombination enzymes (see Section 6.4.2). DNA polymerases and DNA ligase repair the resulting gap in the complementary parental strand.

6.4 DNA recombination

6.4.1 DNA recombination plays an important role in both meiotic cell division and evolution.

Although accurate DNA replication and repair of DNA damage are crucial for genetic stability in individuals, the long-term survival of a species is also influenced by genetic variation that allows its members to adapt to changing environments. **Genetic recombination** of DNA has an important role in evolution because it rearranges DNA sequences to generate new combinations of DNA molecules. The creation of novel genes during DNA recombination can result in new types of RNAs and proteins and therefore can yield new phenotypes. Furthermore, during meiotic cell division, DNA recombination can result in genetically distinct gametes and thereby promotes the production

of varied genotypes, which are acted on by environmental factors in the process of natural selection. The mechanisms involved in genetic recombination are related to DNA replication and repair. DNA recombination events are not rare, and have been observed in all classes of living organisms and in viruses.

6.4.2 Homologous recombination occurs between long nucleotide sequences that are similar.

Three types of DNA recombination mechanisms have been identified: homologous recombination, site-specific recombination, and illegitimate recombination. **Homologous recombination** requires that the DNA sequences involved have regions that are substantially similar (that is, homologous). The frequency of recombination generally increases with the length of the homologous DNA segments. The mechanisms used in homologous recombination can be inferred from the structure of the products (Fig. 6.17). When no information encoded in the two DNA duplexes is lost during the exchange of sequences by **crossover,** the recombination event is termed reciprocal. Nonreciprocal recombination, or **gene conversion,** results when one DNA duplex donates sequence information rather then exchanging it. Three molecular models—single-strand annealing, double-strand break repair, and one-sided invasion—have been proposed to explain the particular products of specific DNA recombi-

nation events as well as the general finding that double-strand DNA breaks enhance homologous recombination rates (Fig. 6.18).

Humans have exploited homologous recombination for thousands of years in the production of important agricultural crop plants. Recent advances in molecular breeding techniques and transgenic plant technologies (see Box 6.4) are now providing new insights into the processes that underlie meiotic and somatic DNA recombination events in plants. The **double-strand break repair model** (Fig. 6.18) seems to explain the recombination events that occur during meiosis in plants. The **single-strand annealing model** (Fig. 6.18) best describes the extrachromosomal recombination events that occur in somatic plant cells, for example, when plasmid DNA molecules that contain overlapping parts of a specific reporter gene are transferred into plant cells and recombine

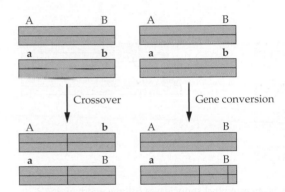

Figure 6.17
Two types of homologous DNA recombination produce different results. Crossover between two molecules of double-stranded DNA results in the reciprocal exchange of DNA sequences (left panel). Gene conversion involves a nonreciprocal transfer of a nucleotide sequence from one double-stranded DNA molecule to another. The donor sequence (purple) remains unchanged while the recipient sequence (pink) loses genetic information (right panel).

Box 6.4

Gene targeting is a powerful technology for investigating gene function.

Homologous recombination presents many possible applications for manipulating the plant genome. One of these techniques, gene targeting, has been developed to achieve integration of transfected DNA molecules into homologous sequences of genomic DNA. When the transfected DNA includes both sequences homologous to the target gene of interest and one or more genes for selectable markers (e.g., antibiotic resistance), its insertion into the genome can efficiently inactivate the target gene, creating a mutant with a readily selectable phenotype. Such mutants can reveal the function of the target gene. This technique has been used extensively to analyze the function of

genes in yeast and mouse cells. Although experiments that employ gene targeting have been performed in *Arabidopsis* and tobacco, use of this technique in plants has proved to be very inefficient and impractical for large-scale investigations into gene function.

One reason for this low efficiency is that, in plant cells, homologous recombination occurs less frequently than does illegitimate recombination (see Section 6.4.5). In the haploid genome of the moss *Phycomitrella* (see figure), homologous recombination is the normal pathway of DNA integration. Gene targeting experiments in this organism therefore promise new insights into gene function. Genetic

engineering of the plant genome to express components of the recombination machinery from other organisms may also help to boost the recombination efficiency and allow the targeted knocking-out of genes in angiosperms.

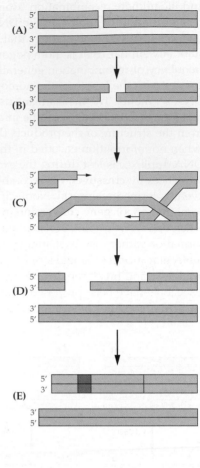

(1) Single-strand annealing model

(A)

(B)

(C)

(D)

(2) Double-strand break repair model

(A)

(B)

(C)

(D)

(E)   or

(3) One-sided invasion model

(A)

(B)

(C)

(D)

(E)

Figure 6.18

Three models of homologous recombination. The donor DNA is purple, the recipient DNA pink. In the single-strand annealing model (1), a double-strand break must occur in both DNA duplexes (A). Exonucleases remove nucleotides beginning at the double-strand breaks and expose single-stranded homologous regions (B). The complementary single strands of DNAs anneal to each other (C), followed by the removal of the non-homologous DNA overhangs and gap repair (D). This recombination mechanism is non-conservative, because only one chimeric DNA duplex can survive. In the double-strand break repair (DSBR) model (2), the recombination event initiates with a double-strand break in one of two DNA duplexes (A). After the double-strand break is enlarged by exonucleases (B), the overhanging 3' ends of the DNA bind to complementary sequences on the other duplex and act as primers for DNA repair synthesis (C). The remaining gaps are repaired using the intact donor DNA as a template (D). The branched DNA molecules resulting from this process resolve into one of two possible combinations (E). DSBR is a conservative recombination mechanism because both DNA duplexes participating in the recombination event are restored. In the one-sided invasion model (3), a double-strand break in the acceptor DNA (A) is enlarged by exonucleases (B). One of the two resulting 3' ends invades a complementary DNA sequence on an intact DNA duplex (C). DNA synthesis lengthens the invading strand (D). Illegitimate recombination occurs upstream from the double-strand break, resulting in gene conversion (E).

with each other to restore an intact reporter gene. Naturally occurring extrachromosomal DNA molecules, such as the transferred DNA (T-DNA) of *Agrobacterium tumefaciens* (see Chapter 21) or viral DNAs, show a similar recombination behavior.

6.4.3 Some proteins involved in homologous recombination have been identified in plants.

Homologous recombination in both meiotic and mitotic cells requires the function of many gene products. Of these, the yeast proteins RAD51 and DMC1 are well characterized. RAD51 is required for both mitotic and meiotic recombination, while DMC1 is expressed and active only during meiosis. Both are homologous to the *E. coli* protein RecA, which plays a central role during homologous recombination in bacteria. RecA catalyzes the strand transfer reaction, allowing single-stranded DNA to invade a DNA duplex at a homologous region (Fig. 6.19). In the presence of RP-A, which binds to single-stranded DNA, RAD51 and DMC1 also catalyze strand transfer. Strand transfer activity

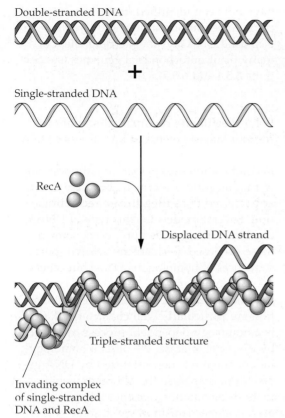

Double-stranded DNA

+

Single-stranded DNA

RecA

Displaced DNA strand

Triple-stranded structure

Invading complex
of single-stranded
DNA and RecA

Figure 6.19
The role of RecA in strand transfer. The *Escherichia coli* protein RecA is a 38-kDa monomer that binds to single-stranded DNA by a cooperative and stoichiometric mechanism. The resulting nucleoprotein complex aggregates with double-stranded DNA in a triple-stranded DNA complex in which the nucleotide bases do not pair. This complex facilitates invasion of the single-stranded DNA and complementary pairing of bases. Strands are subsequently exchanged. A stable heteroduplex region forms and functions as a template for branch migration (not shown).

is required by all three mechanisms of homologous recombination (see Fig. 6.18).

A number of plant genes that share homology with *recA* have been identified, some of which may encode proteins that function in the chloroplast. Homologs of *RAD51* have been cloned from several plants, including *Lilium, Arabidopsis,* rice, wheat, and maize. Interestingly, expression of this gene is activated after treatment of plants with X-rays, suggesting that somatic DNA recombination is induced by DNA damage. Plant proteins homologous to RecA or RAD51 most likely function in homologous recombination. A cDNA homologous to *DMC1* has been isolated from *Arabidopsis.* Proteins similar to DCM1 and RAD51 colocalize during some stages of meiosis in *Lilium*; these proteins may act cooperatively in the alignment and pairing of complementary DNA strands during homologous recombination. They are attached to chromatin loops during the leptotene and zygotene stages of meiosis, and are associated with the synaptonemal complex during pachytene (Fig. 6.20).

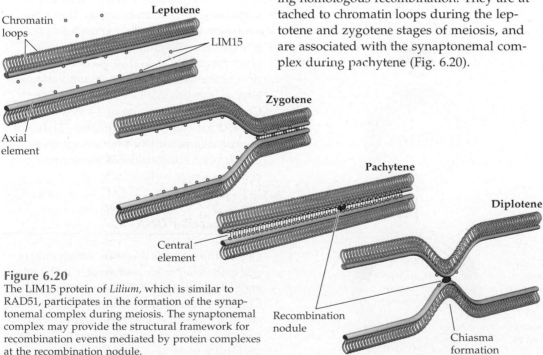

Leptotene

Chromatin
loops

LIM15

Axial
element

Zygotene

Pachytene

Diplotene

Central
element

Recombination
nodule

Chiasma
formation

Figure 6.20
The LIM15 protein of *Lilium,* which is similar to RAD51, participates in the formation of the synaptonemal complex during meiosis. The synaptonemal complex may provide the structural framework for recombination events mediated by protein complexes at the recombination nodule.

6.4.4 Site-specific recombination involves enzymatic activities and defined DNA loci.

Another mechanism of DNA exchange that has been identified in *E. coli*, called **site-specific recombination,** occurs at specific loci, does not require long regions of homologous DNA, and is independent of RecA. The experimental system in which site-specific recombination is best characterized is the integration of bacteriophage λ into the *E. coli* chromosome. Two proteins and a shared DNA sequence are necessary and sufficient for λ integration. The proteins are a phage-encoded recombinase activity (integrase) and a host-encoded DNA-binding protein (IHF). The sequence requirement is a 31-bp palindrome that is present in both phage DNA and the host chromosome (Fig. 6.21).

Site-specific recombination also occurs in eukaryotic organisms. For example, the 2μ plasmid of budding yeast encodes a recombinase (FLP) that is required for site-specific recombination at inverted repeats during DNA replication (see Section 6.5.4). In mammalian B and T cells, the rearrangement of immunoglobulin genes results from site-specific recombination. Although site-specific recombination events in the plant nucleus have not been identified, this process appears to be important during the replication of plastid DNA and the rearrangement of many plant mitochondrial genomes (see Sections 6.5.4 and 6.5.5).

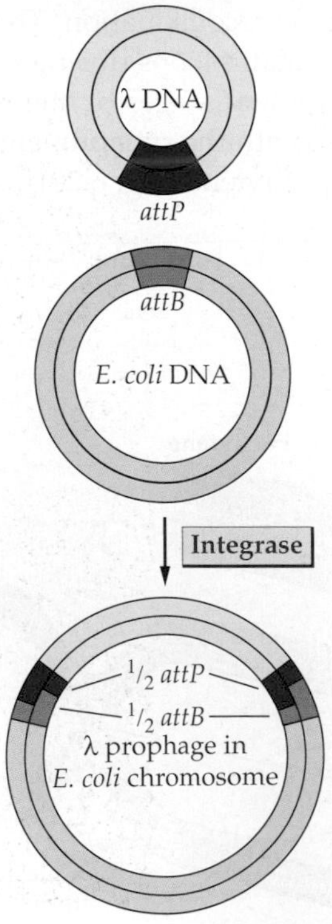

6.4.5 Illegitimate recombination does not require long segments of homologous DNA.

Recombination events that do fall under any of the categories described above are generally referred to as **illegitimate recombination.** Several models for this type of DNA recombination have been proposed; some of these proposed mechanisms involve short sequences of homologous DNA, but others do not require any DNA homology. In *E. coli,* two mechanisms of illegitimate recombination are particularly well characterized: One is a homology-dependent process catalyzed by RNA polymerase, the other a homology-independent reaction catalyzed by DNA gyrase (topoisomerase II). Although illegitimate recombination does not appear to be a general mechanism in yeast, some gene products that participate in this type of DNA recombination have been identified.

Recombination products that cannot be explained by typical meiotic or somatic homologous DNA recombination mechanisms often are observed in plants. Illegitimate recombination models often are used to explain these novel recombinants. If this assumption is correct, plant genomes appear to experience illegitimate recombination more frequently than they do homologous recombination. Because integration and excision of transposable DNA elements (see Chapter 7) and insertion of T-DNA into plant chromosomes (see Chapter 21) do not require significant DNA homology, these events often are considered illegitimate recombination as well.

6.5 Organellar DNA

One of the major features by which eukaryotic cells differ from prokaryotes is the presence of various subcellular organelles (see Chapter 1), including the mitochondria (present in almost all eukaryotes) and plastids (present only in plants and algae). Plastids perform a wide variety of anabolic reactions

Figure 6.21
Integration of λ phage DNA into the *Escherichia coli* chromosome involves site-specific recombination between the *attP* sequence of the phage genome and the *attB* sequence of the bacterial genome. The recombination is catalyzed by an integrase.

that include photosynthetic carbon reduction (see Chapter 12), amino acids biosynthesis (see Chapter 8), and fatty acid production (see Chapter 10). Plant mitochondria engage in respiration (see Chapter 14) and act in concert with other organelles to perform photorespiration (see Chapter 14) and gluconeogenesis (see Chapters 10 and 14).

Among the organelles of eukaryotic cells, plastids and mitochondria are unique because they possess their own genetic systems and protein synthesis machinery. In different taxa of algae and plants, transmission of these organelles (and thus of their genomes) varies, but is typically uniparental. Since the discovery almost 40 years ago that plastids and mitochondria contain DNA, much of the molecular research has focused on the structure and function of the organellar genomes. We now know that the genomes in plastids and mitochondria encode a small number of genes for proteins that function in the organelles, as well as components for the maintenance of their own genetic system. The expression of organellar genomes is under tight control of the nuclear genome, but regulatory mechanisms also have evolved in plants that coordinate the expression of nuclear, plastid, and mitochondrial genes for proteins that function in the cytoplasmic organelles.

6.5.1 During evolution, chloroplasts and mitochondria originated from endosymbiotic bacteria.

Plastids and mitochondria share many traits with free-living prokaryotic organisms. For example, the organization of organellar genomes and bacterial genomes is similar. Likewise, chloroplasts carry out photosynthesis in much the same way as cyanobacteria. Comparison of ribosomal RNA (rRNA) sequences from organelles and certain free-living prokaryotes suggests that plastids share a common ancestor with modern cyanobacteria and mitochondria with modern proteobacteria. This RNA evidence supports the **endosymbiont hypothesis,** that chloroplasts and mitochondria probably evolved from prokaryotes that were engulfed by protoeukaryotic cells during the origin of the eukaryotic lineage (Fig. 6.22). Extant symbioses between photosynthetic organisms and nonphotosynthetic hosts lend additional support to the endosymbiont hypothesis. For example, the biflagellate protist *Cyanophora paradoxa* acquires an endosymbiotic cyanobacterium, called a cyanelle, which functions as a chloroplast and provides the host cell with photosynthetically reduced carbon. Other examples are marine nudibranchs (sea slugs) that feed on algae. These

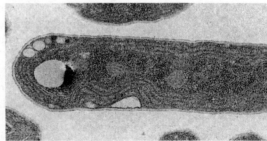

Synechococcus lividus

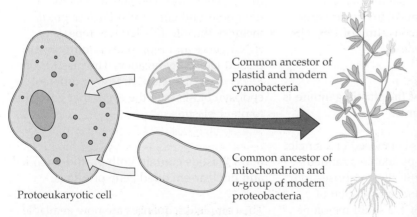

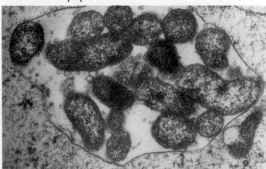

Rickettsiella popilliae

Figure 6.22
Eukaryotes probably evolved from endosymbiotic associations between different prokaryotic organisms. A comparison of ribosomal RNA sequences suggests that plant plastids share a common ancestor with modern cyanobacteria, such as the free-living organism *Synechococcus lividus*. Mitochondrial rRNA sequences of extant eukaryotes share greatest homology with members of the α-group of the proteobacteria, which includes several genera of intracellular parasites (e.g., *Agrobacterium, Rhizobium,* and *Rickettsia*). The electron micrograph at lower right shows cells of *Rickettsiella popilliae* within a blood cell of its insect host.

Common ancestor of plastid and modern cyanobacteria

Common ancestor of mitochondrion and α-group of modern proteobacteria

Protoeukaryotic cell

animals can incorporate the algal chloroplasts into their own tissue and maintain the foreign organelles for nearly two months, receiving nourishment from plastid photosynthesis. Considering these contemporary examples, it is not difficult to envision that endosymbiotic events gave rise to oxygen-tolerant, photosynthetic eukaryotes.

Neither the chloroplasts nor the mitochondria of contemporary organisms can exist autonomously outside of the eukaryotic cell. During evolution, most of the DNA once present in the organelles was transferred to the nuclear genome. This process of gene transfer appears to be ongoing at the present time. Similar to the genomes of bacteria and unlike the nuclear genomes of eukaryotes, the DNA that remains within the organelles is predominantly circular and does not form extensive supramolecular complexes with proteins. Some organellar DNA sequences resemble the **operons** of prokaryotes, in which genes encoding proteins that function in a common pathway or assemble into a complex are clustered together. For example, many sequences in the plastid genome are similar to the corresponding operons in cyanobacteria, including regulatory elements (e.g., promoters and transcription terminators), gene clusters for a number of ribosomal proteins (e.g., the *rpl23* operon), and ATPase subunit genes *(atp)*. For example, the genes for plastid rRNAs of most plants, ranging from the bryophyte *Marchantia* to the monocot maize, are organized into an operon consisting of 16S, 23S, 4.5S, and 5S rRNA genes.

6.5.2 The structure of the plastid genome is conserved among plants.

The plastid genome is composed of a single, circular chromosome of double-stranded DNA. The genome typically comprises four segments (Fig. 6.23A): a large region of single-copy genes (LSC), a small region of single-copy genes (SSC), and two copies of an inverted repeat that separate the single copy regions (IR$_A$ and IR$_B$, respectively). However, other modes of plastid DNA organization also have been found. Plastid DNA is generally homogeneous within a given plant species although, in genomes that contain IR regions, plastid DNA often consists

of two groups of molecules differing in the relative orientation of the single-copy regions. In the same way that all of the somatic cells in an individual organism possess the same set of nuclear genes but may express those genes in different combinations, all of the plastids in a single plant contain identical genetic material but may differ in developmental fate and metabolic activity.

The plastid DNAs of different plant and algal species vary greatly in length. The siphonous green alga *Codium fragile* has the smallest known plastid genome of any photosynthetic eukaryote, 89 kb. In contrast, plastids of the giant green alga *Acetabularia* contain the largest known genome, estimated at 400 kb. Most plastid genomes, including those of tobacco, maize, rice, and *Marchantia*, range in size from 120 to 160 kb and share similar distribution and organization of genes (Fig. 6.23B). Much of the variation in size observed among plant plastid genomes can be accounted for by changes in the length of the IR region, which ranges from 0.5 to 76 kb in length. The presence or absence of IR regions in plastid DNA has been used to categorize plastid genomes. Most plants contain IR regions but the plastid genomes of certain legumes, conifers, and algae lack IR regions. It is probable that IR regions were present in the common ancestor of plastids and that one IR region was lost in some legumes and conifers during evolution. The plastid genome of the photosynthetic protist *Euglena gracilis* includes three to five tandem repeats of rRNA genes and represents a third type of plastid DNA organization. The plastids of *Euglena* may have arisen from a different endosymbiosis than the event that gave rise to plant plastids.

6.5.3 Plastids contain both plastid-encoded and nuclear-encoded gene products.

The complete sequences are now available for plastid genomes from more than a dozen plants, including tobacco, maize, rice, and pine. Most plastid DNAs contain all of the genes for rRNAs and a full complement of genes for tRNAs. However, they include only about 100 single-copy genes, most encoding proteins that are required for photosynthetic functions (Table 6.1).

(A)

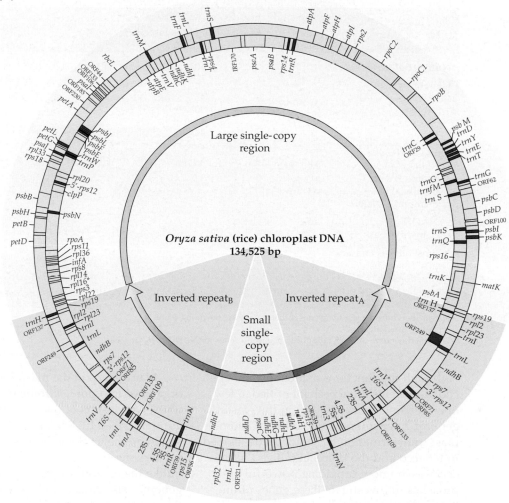

Large single-copy region

Inverted repeat_B

Small single-copy region

Inverted repeat_A

Oryza sativa (rice) chloroplast DNA
134,525 bp

(B)

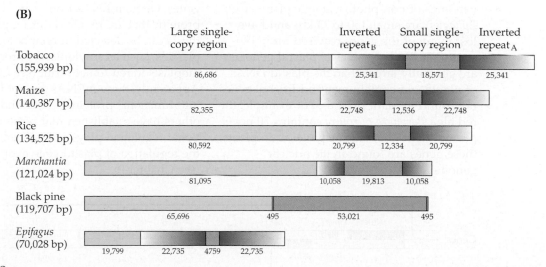

	Large single-copy region			Inverted repeat_B	Small single-copy region	Inverted repeat_A
Tobacco (155,939 bp)	86,686			25,341	18,571	25,341
Maize (140,387 bp)	82,355			22,748	12,536	22,748
Rice (134,525 bp)	80,592			20,799	12,334	20,799
Marchantia (121,024 bp)	81,095			10,058	19,813	10,058
Black pine (119,707 bp)	65,696	495	53,021			495
Epifagus (70,028 bp)	19,799	22,735	4759	22,735		

Figure 6.23

(A) The chloroplast genome map of rice is representative of the organization of plastid genes in most flowering plants. IRF, intron reading frame; ORF, open reading frame. (B) Schematic representations of several plant plastid genomes, showing the conserved features: two regions of single-copy genes and two inverted repeats (IRs). The length of the IRs can vary widely among plastid genomes of different species. The single-copy regions in the plastid genome of the nonphotosynthetic parasitic plant *Epifagus* are quite small because most of the genes encoding photosynthetic proteins have been eliminated. The number beneath each DNA segment refers to its length in basepairs.

Table 6.1   Genes identified in complete plastid genome sequences

Gene products	Gene acronym	Plants		Algae	
		Photosynthetic plants	Epifagus[a]	Euglena	Porphyra[b]
Number of genes		101–150	40	82	182
Genetic system					
rRNA	*rrn*	4	4	3	3
tRNA	*trn*	30–32	17	27	35
Ribosomal protein	*rps, rpl*	20–21	15	21	46
Other		5–6	2	4	18
Photosynthesis					
Rubisco and complexes of the thylakoid membrane system	e.g., *rbcL, psa, psb, pet, atp*	29–30	0	26	40
NADH dehydrogenase[c]	*ndh*	11	0	0	0
Biosynthesis and miscellaneous functions		1–5	2	1	40
Number of introns		18–21	6	155	0

[a]*Epifagus* (beechdrops) is a nonphotosynthetic, parasitic flowering plant.
[b]*Porphyra* is a red alga.
[c]The plastid genome of black pine does not encode genes for NADH dehydrogenase.

Although the functions of most plastid-encoded proteins have been identified, a few open reading frames (ORFs) in the plastid genome have unknown functions. Chloroplast transformation technologies developed in tobacco and the unicellular alga *Chlamydomonas* are now providing powerful approaches to investigate these genes of unknown function. Interestingly, the plastid genomes of non-photosynthetic plants (e.g., *Epifagus*) are small (50 to 73 kb) and have lost many of the genes required for photosynthesis. In contrast, algal plastid genomes are generally larger than the plastid DNAs of plants and contain many additional genes not found in plant plastids. For example, the red alga *Porphyra purpurea* contains 70 novel plastid genes that encode proteins; in plants, these genes are found in the nuclear genome.

Many genes in the plastid genome are organized into **polycistronic** transcription units, i.e., clusters of two or more genes that are transcribed by RNA polymerase from a single promoter. Some of these resemble prokaryotic operons (Fig. 6.24). However, unlike the operons of prokaryotes, plastid genomes also contain polycistronic transcription units composed of functionally distinct genes. Often, mRNAs from genes encoding proteins that are required during photosynthesis can be detected in nonphotosynthetic plastids, such as amyloplasts in root or chromoplasts in red tomato fruit (see Chapter 1). Apparently, these mRNAs are not translated into functional proteins. These findings and others have established that posttranscriptional regulation plays a significant role in the regulation of plastid gene expression (see Chapter 9).

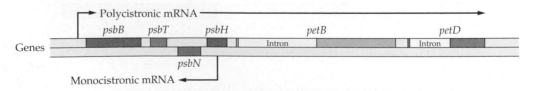

Figure 6.24
The *psbB* operon of plant plastids exemplifies the typical organization of plastid genes into operons similar to those of prokaryotic genomes. Operons are transcribed into polycistronic mRNAs that often encode several related gene products. In the case of the *psbB* operon, the *psbN* gene encoded by the opposite strand appears to be transcribed as a monocistronic mRNA that encodes only one gene product.

While the plastid genome encodes several of the RNAs and proteins involved in the translation of organellar mRNA, most of the genes that encode proteins required for photosynthesis are found in the nuclear genome. As discussed above, many genes originally present in the progenitor organelle were probably transferred to the nucleus during evolution. For example, the enzyme that is required for CO_2 fixation, ribulose-1,5-bisphosphate carboxylase/oxygenase (Rubisco), is a multiprotein complex with two types of subunits. The gene for the smaller subunit (*rbcS*) is found in the nucleus, whereas the gene for the larger subunit (*rbcL*) is present in the plastid genome (see Chapters 9 and 12). This distribution of genes between the nuclear genome and the plastid genome is typical of proteins that participate in the assembly of photosynthetic protein complexes. For example, photosystems I and II contain proteins encoded in both the nuclear and plastid genomes, as does the cytochrome $b_6 f$ complex and the plastid ATP synthase (see Chapter 12). Chloroplasts and other types of plastids contain many important biochemical pathways for which all of the enzymes are encoded in the nuclear genome, translated in the cytoplasm, and transported into the plastids (see Chapters 4 and 9).

genomes can vary widely with the developmental stage and the type of plastid (see Chapter 11). The high number of DNA molecules present in chloroplasts probably reflects the high demand for expression of proteins active in photosynthesis, because substantially fewer DNA molecules are found in root amyloplasts and other non-photosynthetic plastids. Our understanding of the regulation of plastid DNA replication is still very limited, but it is now established that DNA replication in organelles is mostly independent of DNA replication in the nucleus. Plastid DNA amplification appears to be regulated by the frequency of DNA replication initiation, either in synchrony with the cell cycle in rapidly dividing cells, or by a cell-cycle–independent mechanism during cellular differentiation. The process of plastid DNA replication is probably similar to the double rolling-circle model that describes the replication of the 2μ plasmids of budding yeast (Fig. 6.25). In these plasmids, as in the plastid genome, replication origins and site-specific recombination sites are located within long inverted repeats.

Electron microscopic analysis of DNA replication intermediates has been used to map the replication origins of plastid DNAs

6.5.4 The mechanism of plastid DNA replication is not well understood.

Usually each chloroplast contains many copies of its circular genome, often as many as 150. However, the copy number of plastid

Figure 6.25
The double rolling-circle model was first developed to describe the amplification of the yeast 2μ plasmid, but also serves as a model for the replication of chloroplast DNA. The small red arrows indicate the direction of DNA synthesis. During double rolling-circle replication, a single initiation event produces two replication forks with opposite orientations. A site-specific recombination catalyzed by FLP recombinase reorients the forks so that they move in the same orientation. A second recombination event returns the replication forks to their initial, opposing orientations, allowing the forks to meet and replication to terminate. The resulting replication product is a long concatameric DNA from which monomers can be generated by FLP-mediated recombination or other recombination mechanisms. These concatameric DNA molecules have been observed in plastids.

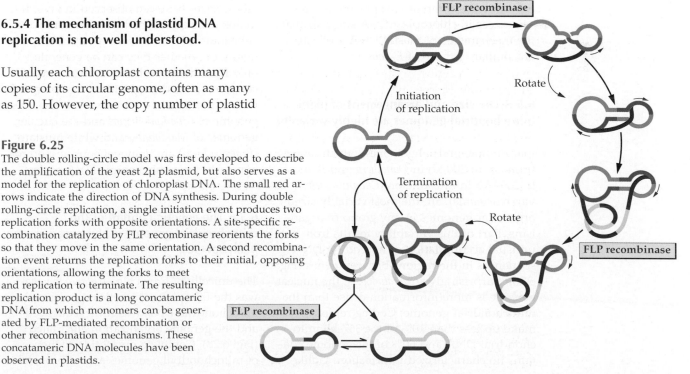

in some plants. In *Oenothera* and tobacco, DNA replication begins at specific origins in the IR regions, near the gene for the 16S rRNA. Although pea and *Euglena* do not have IR regions, the replication origin still appears to be close to the rRNA genes, suggesting that the rRNA operon has retained DNA sequences that interact with the DNA replication machinery to initiate replication. However, in the chloroplast DNAs of *Oenothera*, pea, maize, and tobacco, displacement or D-loops have been observed at multiple sites. The presence of more than two D-loops, replicative intermediates in which the unidirectional synthesis of a single new strand displaces one parental strand, suggests that plastid genomes may contain DNA replication origins in addition to those associated with the rRNA operon.

DNA sequences available from several different plants indicate that all of the enzymes and regulatory proteins essential for plastid DNA replication are encoded in the nuclear genome. Specific enzymatic activities, including DNA polymerases, primase, topoisomerases, and helicases, have been detected in plant plastids; their function in the replication process is under investigation. DNA sequences in the chloroplast genomes of *Chlamydomonas* and *Porphyra* share homology with the *E. coli* genes *dnaA* and *dnaB*, which encode proteins that function in DNA replication and repair. The presence of these sequences in chloroplast DNA suggests that the mechanisms of plastid DNA replication are similar to those of *E. coli*.

6.5.5 The size and arrangement of plant mitochondrial genomes are highly variable.

Plant mitochondrial genomes, which range from about 200 kb in *Oenothera* and *Brassica* to 2600 kb in muskmelon (*Cucumis melo* var. *reticulatus*), are the most variably sized organellar genomes in any group of organisms. Part of this variability results from an unusual accumulation of noncoding DNA sequences in the regions between genes (Fig. 6.26). Surprisingly, in *Arabidopsis*, the nuclear genome is more information-dense than the mitochondrial genome: Coding regions make up less than 10% of the 367-kb mitochondrial DNA and 60% of the genome contains no characterized information. Unlike much of the noncoding intergenic DNA in the nuclear genome, plant mitochondrial intergenic regions are not made up of extensive repetitive DNA sequences. In stark contrast, known animal mitochondrial genomes are very compact (about 16 kb) and essentially lack noncoding intergenic sequences.

Plant mitochondrial genomes sometimes exist as circular DNA molecules of variable size. Sequence analysis suggests that the smaller DNA circles, called **subgenomic circles**, are derived from a larger circle and that the combined DNA sequence content of the subgenomic circles can account for the entire mitochondrial genome. The subgenomic circles of maize mitochondrial DNA are typical of this type of mitochondrial genome organization (Fig. 6.27). The largest possible maize mitochondrial circular DNA molecule, called the **master circle,** would encode the complete set of mitochondrial genes. However, the master circle DNA has never been isolated. In fact, it is conceivable that this large circular DNA molecule does not exist. Several short regions of different repeated DNA sequences that are distributed throughout the maize mitochondrial genome participate in the recombination events that form subgenomic DNA circles. These direct and inverted DNA repeat sequences probably shield the functional genes from deleterious rearrangements. The formation of subgenomic DNA circles has been observed in vivo. It remains to be established whether some or all subgenomic DNA circles replicate independently or whether they can be generated only from the hypothetical master circle. Recombination and fragmentation of plant mitochondrial DNA is not universal: The linear genome of *Chlamydomonas* and the circular genomes of *Marchantia* and white mustard (*Brassica hirta*) are homogeneous and do not appear to form subgenomic DNA circles.

6.5.6 The genetic content of the mitochondrial genome is conserved among plant species.

The unicellular green alga *Chlamydomonas* was the first photosynthetic organism whose mitochondrial DNA was fully sequenced and the genome organization elucidated (Fig. 6.28). Complete nucleotide sequences of mitochondrial genomes are now available

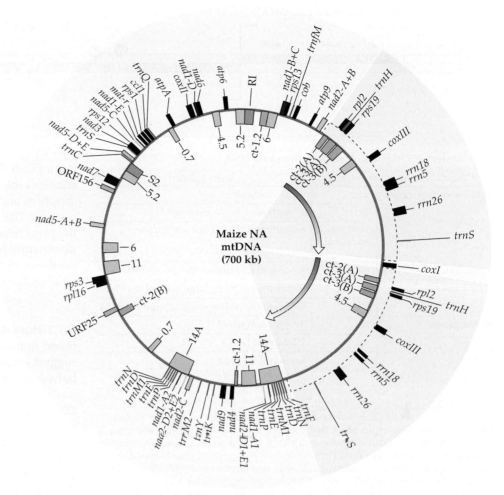

Figure 6.26
The map of the mitochondrial genome of maize is based on a hypothetical master circle DNA molecule that contains all of the mitochondrial genes. Although the mitochondrial DNA in maize is considerably larger than the chloroplast DNA, it contains fewer genes. Several inverted and direct repeat DNA sequences (shown as blue, green, and magenta boxes on the inner circle) have been identified that participate in recombination events to produce small subgenomic circular DNA molecules shown in Figure 6.27. URF, unidentified reading frame.

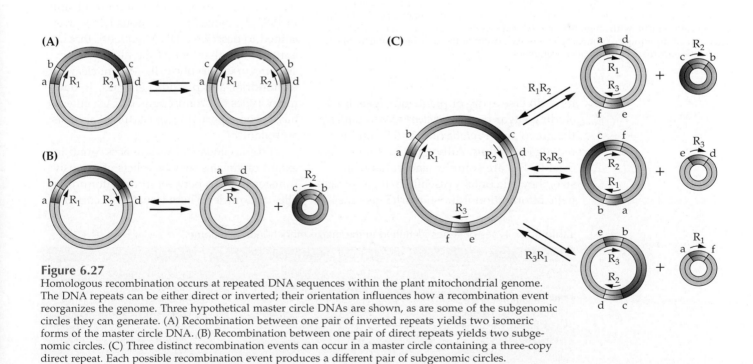

Figure 6.27
Homologous recombination occurs at repeated DNA sequences within the plant mitochondrial genome. The DNA repeats can be either direct or inverted; their orientation influences how a recombination event reorganizes the genome. Three hypothetical master circle DNAs are shown, as are some of the subgenomic circles they can generate. (A) Recombination between one pair of inverted repeats yields two isomeric forms of the master circle DNA. (B) Recombination between one pair of direct repeats yields two subgenomic circles. (C) Three distinct recombination events can occur in a master circle containing a three-copy direct repeat. Each possible recombination event produces a different pair of subgenomic circles.

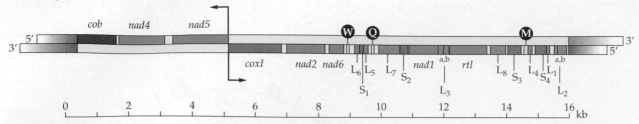

Chlamydomonas mitochondrial DNA

Gene content of *Chlamydomonas* mitochondrial DNA

	Gene	Gene product
Protein-coding genes	*cob*	Apocytochrome
	coxI	Cytochrome oxidase subunit I
	nad1, nad2, nad4, nad5, nad6	Subunits of NADH dehydrogenase
	rtl	Reverse transcriptase-like reading frame
rRNA genes	$L_1 - L_8$	Segments encoding large subunit of rRNA
	$S_1 - S_4$	Segments encoding small subunit of rRNA
tRNA genes	W	Tryptophan tRNA
	Q	Glutamine tRNA
	M	Methionine tRNA (elongator, not initiator)

Figure 6.28
The mitochondrion of *Chlamydomonas* contains a linear DNA duplex of 15.8 kb. The genome contains eight genes that code for proteins. One striking and unique feature is the scrambled arrangement of rRNA gene segments. Both LSU and SSU rRNA genes are discontinuous. Throughout a 6-kb region, segments encoding specific rRNA domains are interspersed with one another and with intact genes that encode protein and tRNA. The rRNA-coding segments are numbered S1 through S4 and L1 through L8, according to the 5' to 3' order in which the homologous segments occur in the 16S and 23S rRNAs of *Escherichia coli*, respectively. The mtDNA encodes only three tRNAs, necessitating the import of numerous tRNAs from the cytoplasm. The ends of the genome are characterized by terminal inverted repeats with single-stranded 3' extensions. These extensions, which are identical in sequence (not complementary), exclude the possibility of genome cyclization or the formation of concatamers.

for a diverse array of plant and algae, including sugar beet, *Arabidopsis*, *Marchantia*, the green alga *Prototheca* (Box 6.5), and the red alga *Chondrus*. Although plant mitochondrial DNAs are complex and variable in size, they all contain essentially the same genetic information. Mitochondrial genomes do not code for many genes, as most of the enzymes required for DNA replication, transcription, and translation are encoded by the nucleus. The products of genes encoded in the mitochondrial genome participate predominantly in oxidative respiration, ATP synthesis, or mitochondrial translation (Table 6.2).

6.5.7 Homologous DNA sequences found in more than one plant genome suggest extensive movement of DNA between genomes.

DNA sequencing projects revealed that the mitochondrial genomes of plants contain nucleotide sequences homologous to sequences in chloroplast DNA (Fig. 6.29). In addition, sequences of mitochondrial and chloroplast DNA were found in nuclear genomes. These discoveries suggested the possibility of DNA transfer between organelles and nuclei, and in 1982 the name "promiscuous DNA" was coined to describe a DNA sequence that occurs in more than one of the three genetic systems of eukaryotic cells—the nuclear, mitochondrial, and plastids genomes. To date, many types of promiscuous DNA segments have been reported in yeast, fungi, animals, and plants.

Promiscuous DNA sequences found in extant organisms may be relics of extensive transfer of DNA between progenitor organelles and nuclei. These transfers occurred

Table 6.2 Types of genes identified in the maize mitochondrial genome

Gene products	Gene abbreviations	Function
rRNAs	*rrn18, rrn26, rrn5*	Protein synthesis
tRNAs	*trn*	Protein synthesis
Ribosomal proteins	*rps, rpl*	Protein synthesis
NADH dehydrogenase	*nad*	Respiratory electron transport
Cytochrome *c* oxidase	*cox*	Respiratory electron transport
Apocytochrome	*cob*	Respiratory electron transport
F_0F_1-ATPase proteins	*atp*	ATP synthesis

over a long time span and provide insight into the evolution of plants. For example, the *tufA* gene, which encodes plastid protein synthesis factor EF-Tu, is a plastid gene in algae but a nuclear gene in plants. The transfer of *tufA* from the plastid to the nuclear genome was probably an early event during the evolution of the Charophyceae, more than 500 million years ago. Another example is *rpl21*, a gene that encodes the plastid ribosomal protein L21. This gene is present in the plastid genome of liverwort and the nuclear genomes of angiosperms. Thus, *rpl21* was probably transferred to the nuclear genome some time after the divergence of nonvascular plants and flowering plants. The *rpl22* gene for the plastid ribosomal protein L22 may represent an even later transfer event because it is present in the plastid DNA of all plants except legumes, in which it is a nuclear gene.

Transfer of genes between genomes is probably not a unidirectional process: Migration of nuclear genes to organellar genomes may also have taken place during evolution. The *rpl23* gene that encodes plastid ribosomal protein L23 may be such an example. Unlike most plastid ribosomal proteins, the L23 ribosomal protein is more similar to proteins in the cytoplasmic ribosomes of eukaryotes than to prokaryotic L23 proteins. However, the gene is present in the plastid genome of most plants, except in spinach and related species in which the gene is located in the nuclear genome. Thus, it is likely that a nuclear gene for L23 evolved and was transferred to the plastid, after which the prokaryote-like *rpl23* gene was inactivated and deleted from the plastid genome.

Interestingly, spinach plastid DNA contains a pseudogene of *rpl23*, probably a remnant of an intact ancestral prokaryotic gene, suggesting that transfer of the nuclear gene for L23 to the plastid genome may not have been completed in this plant species.

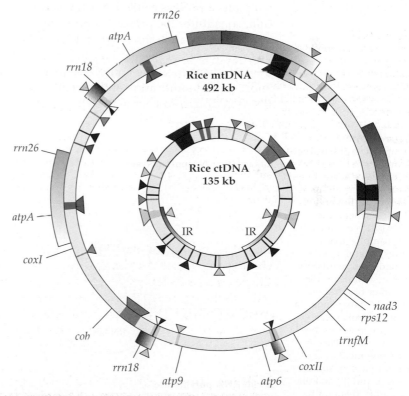

Figure 6.29
Comparison of mitochondrial and chloroplast DNA sequences from rice reveals the extent of promiscuous DNA. The colored triangles and boxes shown on the rice chloroplast genome (inner circle) represent DNA sequences that have been transferred to the mitochondria and inserted at different sites in the mitochondrial DNA (outer circle). It is interesting to note that mitochondrial DNA sequences have not been found inserted into the chloroplast genome, suggesting that DNA transfer may be unidirectional or that the chloroplast may have an effective mechanism to protect itself against invading DNA molecules.

Table 6.3 The nuclear RNA polymerases of plant cells

Polymerase	Location	Products
I	Nucleolus	25S, 17S, and 5.8S rRNAs
II	Nucleoplasm	mRNAs, U1, U2, U4, U5
III class 1	Nucleoplasm	5S rRNA
III class 2	Nucleoplasm	tRNAs
III class 3	Nucleoplasm	U3; U6; other small, stable RNAs

6.6 DNA transcription

The nuclear and organellar genes discussed in the previous sections contribute to the function of individual cells and the development of the multicellular organism. Expression of these genes must be regulated for cells to respond to changes in their environment and coordinate their activities within the tissues that make up complex organisms. Particularly in the nucleus, the organization of genes into chromatin and chromosomes is an important aspect of their regulation. However, the synthesis of RNA on the DNA template, a process called **DNA transcription,** constitutes the primary control step.

All RNAs are transcribed by RNA polymerases, a class of specific enzymes that copy the DNA sequence into an RNA sequence. Subsequent modification of the resulting RNAs is discussed later in this chapter. This section will review briefly the types of RNA polymerases found in the nuclei and organelles of plants. The control of transcription by RNA polymerases and associated regulatory proteins is described in Chapter 7.

6.6.1 Three nuclear RNA polymerases each transcribe different types of RNAs.

Eukaryotic nuclei contain three classes of DNA-dependent RNA polymerases, called RNA polymerases I, II, and III (Table 6.3). RNA polymerases bind to and initiate transcription at specific sites on the chromosome, termed **promoters** (see Chapter 7). Although bacterial RNA polymerases can begin transcribing at promoters in the absence of additional proteins, eukaryotic RNA polymerases must interact with specific proteins called transcription factors to initiate transcription (Fig. 6.30).

The nuclear RNA polymerases are complex enzymes that contain two large subunits (125 to 220 kDa) and several smaller subunits. Five of the smaller subunits are common to all three RNA polymerases. As a result of these structural similarities, the three nuclear RNA polymerases share several functional properties, including the requirement for accessory proteins to initiate transcription correctly at cognate promoter regions. The largest subunit of RNA polymerase II has a C-terminal domain (CTD) that contains multiple repeats of the consensus sequence Tyr–Ser–Pro–Thr–Ser–Pro–Ser. A regulatory protein complex called transcription factor II H (TFIIH), which interacts with RNA polymerase II following recruitment of the enzyme to the promoter, contains a protein kinase that can phosphorylate Ser and Thr residues in the CTD. This phosphorylation is thought to release RNA polymerase II from its association with the promoter-initiation complex and allow the enzyme to begin synthesis of the mRNA from the template strand of the DNA. We now understand many details of the mechanisms by which the three nuclear RNA polymerases initiate transcription in yeast and mammalian cells, although additional regulatory proteins remain to be discovered. The types of nuclear RNA polymerases present in plants are similar to those observed in other eukaryotes. Moreover, the genetic and

Figure 6.30
Comparison of eukaryotic and prokaryotic RNA polymerases reveals that these enzymes differ in complexity. Eukaryotic RNA polymerase II (RNAPII) is a multisubunit complex that requires several accessory proteins and regulatory protein complexes for promoter recognition at the TATA-box region. In contrast, prokaryotic RNA polymerases, typified by *Escherichia coli* RNA polymerase, consist of four core subunits (α, α, β, and β′) and one regulatory σ-subunit. The increased complexity of the eukaryotic RNA polymerase II reflects the regulation of nuclear transcription by numerous signals (see Chapters 7 and 18). TFII, transcription factors for RNA polymerase II.

Eukaryotic RNA polymerase

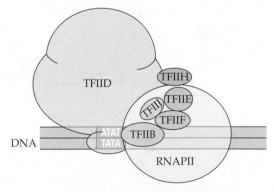

E. coli **RNA polymerase**

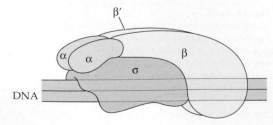

biochemical information available to date suggests that the initiation of transcription in plants is controlled by regulatory mechanisms that resemble those in yeast and mammals.

6.6.2 Plastids contain multiple RNA polymerases.

The principal RNA polymerase of plastids resembles that of *E. coli*. The *E. coli* RNA polymerase holoenzyme consists of four core subunits—α, α, β, and β'—and one of several regulatory σ-subunits. Chloroplast DNA contains genes that encode proteins similar to the *E. coli* subunits α, β, and β' that can form an enzyme complex in the plastid (Fig. 6.31). Several genes for σ-like subunits have also been identified in the nuclear genomes of plants. In a current model, plastid and nuclear gene products combine to form an *E. coli*-like plastid RNA polymerase required for the transcription of plastid genes.

As would be expected for genes transcribed by the plastid-encoded, prokaryote-type RNA polymerase, most chloroplast genes have regions proximal to their transcription initiation sites that often contain DNA sequences similar to the *E. coli* "–10" and "–35" consensus promoter motifs. Analysis of these DNA sequences by mutagenesis has shown that RNA polymerase recognizes this *E. coli*-type chloroplast promoter during the initiation of transcription. However, subsequent investigations of chloroplast transcription in the presence of inhibitors of plastid protein synthesis, combined with analysis of mutants defective in chloroplast-localized protein synthesis, revealed that transcriptional activity was reduced but not abolished when the *E. coli*-type RNA polymerase was impaired. Plastids must therefore import a nuclear-encoded RNA polymerase that functions in addition to the *E. coli*-type RNA polymerase. This novel RNA polymerase, which resembles the single-peptide RNA polymerases of bacteriophage T3 and T7, recognizes promoter regions that differ in DNA sequence from the conserved "–10" and "–35" consensus promoter motifs. The nuclear-encoded enzyme appears active in all plastid types (e.g., etioplasts, chromoplasts, and amyloplasts). Results suggest that constitutive gene transcription of the plastid genome maintains chloroplasts and nonphotosynthetic plastids.

Several nuclear genes have been cloned that encode proteins similar in sequence to the single-peptide RNA polymerases of bacteriophage T3 and T7. The gene products can be imported into either plastids (Fig. 6.31) or mitochondria, suggesting that a nuclear-encoded RNA polymerase may be common to both organelles. Although it is likely that these proteins have a function in transcribing the organellar genomes, RNA polymerases that can account for the constitutive transcription of the plastid genome

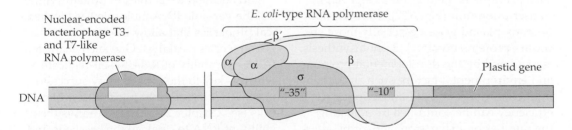

Figure 6.31
Plant chloroplasts have two types of RNA polymerases. The existence of plastid genes encoding proteins similar to subunits α, β, and β' of the *Escherichia coli* RNA polymerase, combined with biochemical analyses, suggest the presence of a prokaryotic-type chloroplast RNA polymerase (right panel). Genes for σ-like subunits have been identified in the nuclear genomes of many plants, indicating that both plastid and nuclear genomes cooperate in the regulation of chloroplast transcription. The plastid-encoded RNA polymerase recognizes chloroplast promoters that contain typical "–10" and "–35" prokaryotic consensus sequences. Recent results suggest that chloroplast genes can also be transcribed by a nuclear-encoded, single-subunit RNA polymerase that resembles the RNA polymerases of bacteriophage T3 and T7 (left panel). When specific inhibitors of translation are used to block in the chloroplast, the nuclear-encoded RNA polymerase activity continues to transcribe chloroplast genes. The promoter sequences recognized by this novel type of RNA polymerase are not well understood.

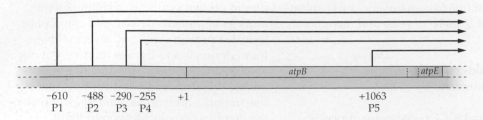

Figure 6.32

Multiple promoters direct transcription of the *atpB–atpE* gene cluster in tobacco chloroplasts. The *atpB* and *atpE* genes encode the β- and ε-subunits of ATP synthase, and their coding regions overlap each other by a few nucleotides. There are five transcription initiation sites (P1 through P5), indicated by red arrows. Four of the transcription initiation sites contain "–10" and "–35" consensus sequence motifs, but the sequence around the –290 site (P3) has no such motifs.

have not yet been isolated. T3 and T7 RNA polymerases have strict promoter requirements and recognize only one specific sequence. If the nuclear-encoded T3- and T7-like proteins have polymerase functions, they must be able to recognize multiple plastid promoters with similar, but divergent DNA sequences.

6.6.3 The genes of plastids and mitochondria can have multiple promoters.

Unlike nuclear genes, which typically have one promoter that is recognized by RNA polymerase, transcription of many plastid genes and operons is initiated either at multiple promoters or at multiple sites. One such example is the dicistronic *atpB–atpE* transcription unit (Fig. 6.32). Transcription of many plastid genes, especially those encoding proteins involved in photosynthesis, is regulated by the developmental program and environmental factors, such as changing light conditions. Overall transcriptional efficiency can be controlled by adjusting the number of active promoters or by altering promoter strength. Plastids appear to utilize both of these mechanisms. Certain mitochondrial genes also have multiple promoters that may control their differential expression. It is not known at present whether the different types of plastid RNA polymerases or the cofactors for these enzymes are involved in either promoter selection or control of transcription initiation at multiple sites.

6.7 Characteristics and functions of RNA

RNA is the primary product of gene expression and has essential roles in protein synthesis and other cellular functions. Most central to cell function and development is the decoding of information in DNA into proteins. rRNAs form complex three-dimensional structures that combine with polypeptides to create ribosomes, the organelles responsible for protein synthesis (see Chapter 9). Ribosomes serve as a platform for reading mRNAs, which carry the instructions that dictate the amino acid sequences of proteins. tRNAs act as adapters to translate the codons of mRNA (specific three-nucleotide sequences) into particular amino acids (see Chapter 9).

In addition to its roles in protein synthesis, the versatile RNA molecule has biochemical properties that allow it to function as a genome or as a catalyst. One such characteristic is the ability of RNA to self-replicate, which is key to the evolutionary success of many RNA viruses that infect eukaryotic cells (see Chapter 21). Another trait is the ability of RNA to catalyze the breakage and formation of covalent bonds between nucleotides (e.g., during RNA splicing, see Section 6.8). Some RNA molecules can assume specific structures that cause particular chemical groups in one or more of its constituent nucleotides to become highly reactive. These RNAs, often referred to as **ribozymes,** can catalyze the cleavage and rejoining of polynucleotide chains. Some RNAs, such as the small nuclear RNAs

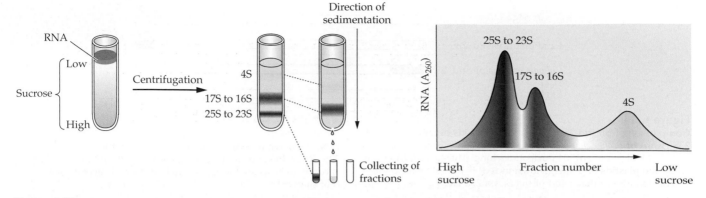

Figure 6.33

The separation of RNA species by zone centrifugation. A solution containing a mixture of RNA molecules is loaded onto the top of a sucrose gradient (typically ranging from 2.5% to 15% sucrose) and centrifuged at high speed. When subjected to strong gravitational field, large RNA molecules sediment more rapidly than small RNA molecules. After centrifugation, sequential fractions are collected from the bottom of the centrifugation tube and the amount of RNA in each fraction is determined by measuring UV absorption at 260 nm (A_{260}).

(snRNAs) and the RNA component of telomerase (see Section 6.2.3), are found in the eukaryotic nucleus and participate in RNA processing and DNA replication. The following sections will discuss the different types of RNAs found in eukaryotic cells as well as what we know about how these molecules function in plants.

6.7.1 RNAs are classified according to their function and size.

When bulk RNA is extracted from cells, it consists of a complex mixture of polynucleotide chains that vary in lengths. RNA molecules are commonly designated according to the sedimentation coefficient, measured in Svedberg units (S, or 10^{-13} seconds). Three main RNA fractions—23S to 25S, 16S to 17S, and 4S RNA—can be separated by zone sedimentation (Fig. 6.33). The 23S to 25S and 16S to 17S fractions contain rRNAs, while the 4S fraction includes tRNAs and other small RNAs. Most of the RNAs in these three classes are relatively stable molecules and can be detected easily using agar gel electrophoresis and ethidium bromide staining. In contrast, mRNAs are far less abundant in the cell and typically make up only 1% to 2% of the total cellular RNA. Molecules in this class of RNA, which vary considerably in size and stability, do not separate as a discrete fraction when subjected to zone sedimentation or gel electrophoresis.

6.7.2 The bulk of cellular RNA is ribosomal.

All ribosomes consist of a large subunit and a small subunit. The small ribosomal subunit contains only one type of rRNA molecule (small-subunit rRNA), whereas the large subunit contains one long rRNA molecule (large-subunit rRNA) and one or more short rRNA molecules (see Chapter 9, Table 9.1). Plants, algae, and photosynthetic protists each contain three classes of ribosomes: cytoplasmic, plastid, and mitochondrial. The short 5.8S rRNA is unique to cytoplasmic ribosomes and associates with 25S rRNA. Interestingly, the 5.8S rRNA is homologous to sequences in the 5′ end of *E. coli* 23S rRNA. Plastid 23S rRNAs resemble those of *E. coli* and cyanobacteria both in sequence and length (Fig. 6.34), although it is broken into shorter RNA chains in some plants or algae. Plant mitochondrial ribosomes also contain a short 5S rRNA molecule, which is absent in animal mitochondria. Prokaryotic rRNAs typically contain methylated nucleotides, up to 10 in 16S rRNA and up to 20 in 23S rRNA, but little is known about the methylation state of plant and animal rRNAs because their sequences have been deduced from the genes.

Both small and large rRNAs fold into complex secondary structures (Fig. 6.35). Comparisons of rRNA sequences from many different organisms and organelles suggest that the secondary structures of rRNAs are highly conserved despite considerable

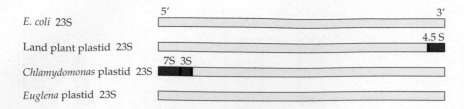

E. coli 23S

Land plant plastid 23S   4.5 S

Chlamydomonas plastid 23S   7S 3S

Euglena plastid 23S

Figure 6.34

Comparison of plastid 23S rRNAs from various organisms with the 23S rRNA of *Escherichia coli* reveals differences in organization. The 23S rRNA of *E. coli* is 1904 nucleotides long. The 4.5S rRNA encoded by the plastids of plants (shown in red) is homologous to the 100 nucleotides at the 3′ end of the *E. coli* 23S rRNA. The

Chlamydomonas reinhardtii chloroplast genome contains 7S and 3S rRNAs (shown in red) that are homologous to the 5′ end of *E. coli* 23S rRNA. In contrast, the plastid-encoded 23S rRNA of *Euglena*, is continuous, as are those of cyanobacteria and several algae (not shown).

sequence divergence among organisms. The proposed structures receive additional support from experimental analysis, because many regions that are predicted by comparative analysis to form intramolecular basepairs also resist digestion by RNases and chemical reagents that degrade single-stranded RNA.

6.7.3 Plant cells contain three distinct sets of transfer RNAs.

The second type of abundant RNA, transfer RNA (tRNA), makes up most of the 4S RNA fraction in a whole-cell rRNA extract (see Fig. 6.33). Acting as adapters, tRNAs bind specific amino acids and basepair with defined mRNA codons, facilitating the translation of nucleic acid sequences into polypeptides (see Chapter 9). A particular tRNA can accept only one of the 20 standard amino acids, but in most cases several different tRNA species, called isoaccepting tRNAs, bind the same amino acid. All tRNAs sequenced to date fit the cloverleaf model (Fig. 6.36). The conserved primary and secondary structures that characterize this class of molecules make purification of individual tRNA species difficult. In plants, the cytoplasm, plastids, and mitochondria each contain a unique population of tRNAs.

While mammalian mitochondrial DNA encodes all the tRNA genes necessary for translation, mitochondria from plants and many other organisms lack a full set of tRNA genes. Mitochondrial import of tRNAs from the plant cytoplasm was first suggested

(A) Tobacco plastid 16S rRNA

II. Central domain

III. 3′ major domain

IV. 3′ minor domain

I. 5′ domain

(B) Tobacco plastid 23S and 4.5S rRNAs

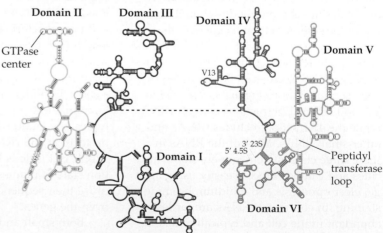

Domain II   Domain III   Domain IV   Domain V

GTPase center

V13

Domain I

3′ 23S   5′ 4.5S

Peptidyl transferase loop

Domain VI

Figure 6.35

Ribosomal RNAs can fold into complex secondary structures by pairing of complementary sequences. (A) The predicted secondary structure of tobacco plastid 16S rRNA reveals four major domains (I to IV). The structure of the tobacco 16S rRNA is based on the model for the structure of *Escherichia coli* 16S rRNA.

(B) Secondary structure predictions for tobacco plastid 23S and 4.5S rRNAs also reveal a conserved organization with six major domains (I to VI). Domain V is the principal site of tRNA binding to the 50S subunit.

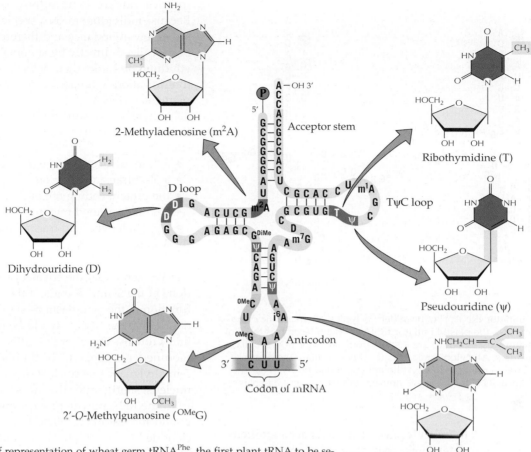

Figure 6.36

The cloverleaf representation of wheat germ tRNA^{Phe}, the first plant tRNA to be sequenced. The structures of the modified nucleotides that occur at specific positions in plant tRNAs are shown. The four arms are named for their structure and function. The acceptor stem consists of nucleotide basepairs that extend to an unpaired CCA sequence at the 3' end. The TψC loop is named for its pseudouridine residue (ψ). The anticodon, which varies among different tRNAs, constitutes the three nucleotides that interact with mRNA during translation (see Chapter 9). The D loop is named for its dihydrouridine residue (D). A notable feature of all tRNA molecules is the presence of modified (or unusual) bases.

Labels in figure: 2-Methyladenosine (m²A); Dihydrouridine (D); D loop; Acceptor stem; Ribothymidine (T); TψC loop; Pseudouridine (ψ); Anticodon; Codon of mRNA; 2'-O-Methylguanosine (^{OMe}G); N⁶-Isopentenyladenosine (i⁶A)

by the findings of hybridization experiments in which genes encoding some mitochondrial tRNAs were detected in nuclear DNA. Direct evidence was subsequently obtained from transgenic potato plants that expressed a nuclear tRNA gene from bean. The bean tRNA could be detected in both the cytoplasm and the mitochondria of the transgenic potato plants. The mechanism of mitochondrial tRNA import remains unknown. In *Arabidopsis,* a single base mutation of tRNA^{Ala} blocks both its aminoacylation and import, suggesting that aminoacyl-tRNA synthases (see Chapter 9) participate in tRNA import into plant mitochondria. In contrast to mitochondrial DNA, the plastid genome of most plants encodes approximately 30 species of tRNAs (see Table 6.1),

which are generally assumed to be sufficient for translation of proteins within the organelle. However, the chloroplast DNA of the nonphotosynthetic parasitic plant *Epifagus* contains only 17 tRNA genes (see Table 6.1). Translation in *Epifagus* plastids must therefore involve tRNAs of nuclear origin, suggesting that plastids may also be able to import tRNAs.

Plant cytoplasmic tRNAs are similar in sequence to those found in the cytoplasm of yeast and other eukaryotes, but many plastid tRNAs have attributes that are typically prokaryotic (Fig. 6.37). Consistent with their mixed origin, some plant mitochondrial tRNAs resemble those of prokaryotes, while others resemble those found in plant cell cytoplasm.

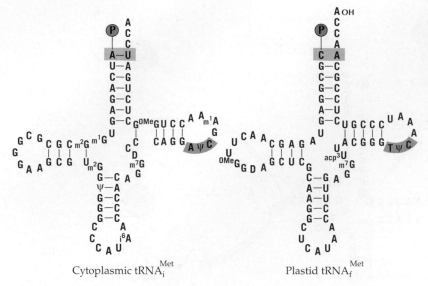

Cytoplasmic tRNA$_i^{Met}$

Plastid tRNA$_f^{Met}$

Figure 6.37
The cytoplasmic initiator tRNA^{Met} of bean (left) is typical of eukaryotic cytoplasmic tRNAs. In contrast, the plastid initiator tRNA^{Met} (right) resembles those of prokaryotes. Initiator tRNAs, which are distinct from elongator tRNA^{Met}, have certain unique features. All cytoplasmic initiator tRNAs contain the sequence AψC instead of TψC in the TψC loop, Although the plastid tRNA^{Met} has the typical TψC sequence, its 5′ nucleotide is unusual in that it cannot basepair to the acceptor stem.

6.7.4 Cytoplasmic mRNAs are modified extensively after transcription.

Messenger RNAs encode the amino acid sequences of proteins. The number of distinct mRNA species in a given cell, which can exceed several thousand, is comparable to the number of different proteins present in the cell. As discussed above, mRNAs constitute no more than 1% to 2% of total cellular RNA. Consistent with the range of molecular masses observed for polypeptides, mRNAs vary in length from a few hundred nucleotides to several thousand nucleotides. In addition, mRNAs generally are more labile than rRNAs and tRNAs, although the

turnover rate for some mRNAs is quite slow. Because individual mRNA species usually are of low abundance and therefore cannot be visualized directly by staining a gel with ethidium bromide, their detection requires hybridization with labeled probes of complementary DNA.

A typical cytosolic mRNA consists of five distinct regions (Fig. 6.38):

- a "cap" structure at the 5′ end
- a 5′-untranslated region
- the protein coding region
- a 3′-untranslated region
- a polyadenylic acid tract added to the 3′ end

We know that the covalent modifications at the 5′ and 3′ ends of the mRNA function in the cytoplasm as signals designating that the RNAs should be translated into proteins. The 5′ of the mRNA is first modified, or "capped," by addition of a 7-methylguanosine nucleotide (m^{7}G). In the nucleus, shortly after the mRNA emerges from the RNA polymerase complex, a guanosine residue is joined to the 5′ nucleotide by way of a triphosphate bridge. The N-7 of the guanine is subsequently methylated (Fig. 6.39). The 5′ cap plays an important role during the initiation of translation (see Chapter 9) and also seems to protect the mRNA from degradation during its synthesis.

The coding region, which is the nucleotide sequence translated into the amino acid sequence, begins with an initiation codon (usually AUG), and ends with a termination codon (UAG, UAA, or UGA). The sequence between start and stop codons constitutes an ORF (see Chapter 9). It is flanked by 5′ and 3′ untranslated regions (UTRs) that are highly variable in length. Consequently, the length of a mRNA does not always correlate with the length of the amino acid sequence it encodes. The precise function of the 5′ and 3′ UTRs is still poorly understood for many mRNAs, but often these regions contain RNA sequences that can form secondary structure and interact with proteins, which regulate transport, translation, and stability of the mRNA.

Most cytoplasmic mRNAs contain a 3′ sequence of polyadenylic acid. This poly(A) tail is not encoded in the DNA, but is added

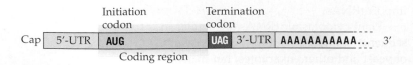

Figure 6.38
Structure of a typical mature, nuclear-encoded, eukaryotic mRNA. Most mature mRNAs have 5′ and 3′ untranslated regions (UTRs) that often contain cis-regulatory sequences involved in the regulation of translation or mRNA stability. Specific regulatory sequences in the 3′-UTR are recognized by poly(A) polymerase, which synthesizes the addition of a polyadenylic acid tail at the 3′ end.

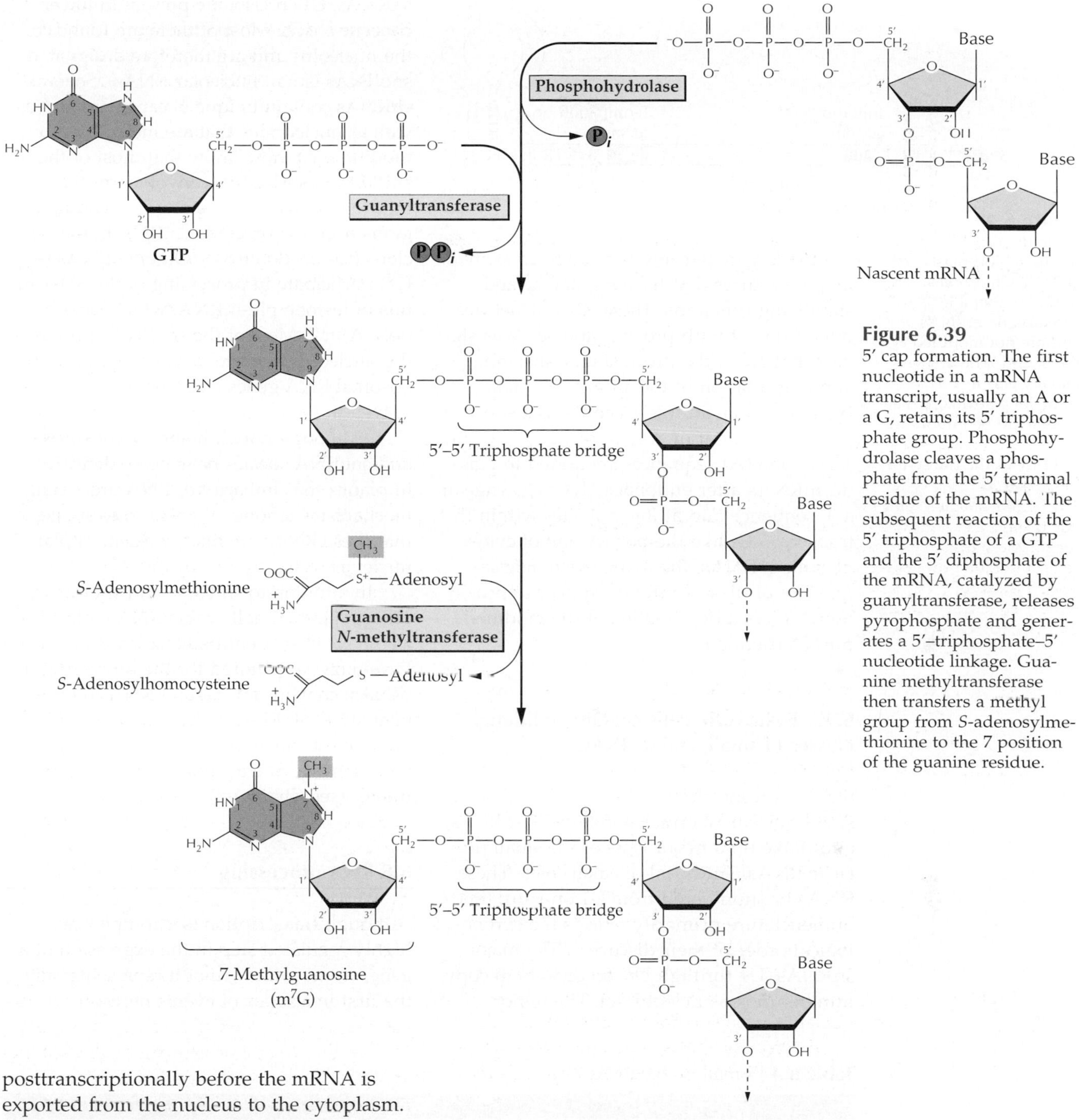

Phosphohydrolase

Guanyltransferase

GTP

P_i

PP_i

5'–5' Triphosphate bridge

Nascent mRNA

S-Adenosylmethionine

Guanosine N-methyltransferase

S-Adenosylhomocysteine

CH_3

S$^+$—Adenosyl

S—Adenosyl

7-Methylguanosine
(m^{7}G)

CH_3

5'–5' Triphosphate bridge

Base

Base

Base

Figure 6.39
5' cap formation. The first nucleotide in a mRNA transcript, usually an A or a G, retains its 5' triphosphate group. Phosphohydrolase cleaves a phosphate from the 5' terminal residue of the mRNA. The subsequent reaction of the 5' triphosphate of a GTP and the 5' diphosphate of the mRNA, catalyzed by guanyltransferase, releases pyrophosphate and generates a 5'–triphosphate–5' nucleotide linkage. Guanine methyltransferase then transfers a methyl group from *S*-adenosylmethionine to the 7 position of the guanine residue.

posttranscriptionally before the mRNA is exported from the nucleus to the cytoplasm. An enzyme complex that includes an endonuclease and poly(A) polymerase recognizes a signal sequence (5'–AAUAAA–3') near the 3' end of the transcript, cleaves the pre-mRNA downstream from the signal, and adds 20 to 250 adenosine residues to the 3' end. The poly(A) tail facilitates export of the mRNA from the nucleus, stabilizes the mRNA against exonucleolytic degradation, and appears to have a role in the initiation of translation.

The structure of plastid and mitochondrial mRNAs differs from that of cytoplasmic mRNA. The organellar mRNAs lack both the 5' cap and 3' poly(A) tail (Fig. 6.40). As a result of differences in posttranscriptional RNA processing, mitochondrial mRNAs commonly retain the 5' triphosphate group, while the 5' ends of most plastid mRNAs are monophosphorylated. Similar to prokaryotic mRNAs, the 5'- and 3'-UTRs

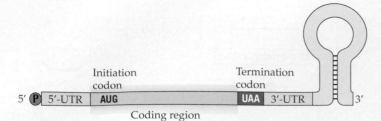

Initiation codon · Termination codon

5′ Ⓟ 5′-UTR | AUG | UAA | 3′-UTR | 3′

Coding region

Figure 6.40
Structure of a typical mature chloroplast mRNA. Chloroplast mRNAs resemble prokaryotic mRNAs and are not modified at the 5′ end. The mRNAs usually end in a 3′ stem–loop structure that appears to be important for the processing and stability of the transcript.

of most organellar mRNAs can form stem-loop structures that have regulatory and stabilizing functions. These RNA structures often interact with proteins and serve as signals that affect the processing, translation, and degradation of the mRNA. A small fraction of plastid mRNA contains a short poly(A) segment or A-rich tract at the 3′ end. These poly(A) sequences are added to plastid mRNAs after endonucleolytic cleavage at a 3′ sequence and at internal sites within the transcript. Unlike the poly(A) tail of cytoplasmic mRNAs, the 3′ polyadenylate sequences of plastid mRNAs appear to promote efficient degradation of the modified mRNA fragment.

6.7.5 Eukaryotic cells contain additional classes of small, stable RNAs.

Stable, uridine-rich RNAs are components of small nuclear ribonucleoproteins (UsnRNPs) and make up a major class of the small nuclear RNAs (snRNAs) of eukaryotes. These RNAs are numbered from U1 onward; this nomenclature primarily reflects the chronological order of their discovery. The major snRNAs, U1 through U6, occur in high copy number in the nucleoplasm. The minor

snRNAs, U7 to U40, are present in lower concentrations. Most of them are found in the nucleolus and are therefore designated snoRNAs (small nucleolar RNAs). Several snRNAs contain unique 5′ cap structures as well as nucleotides that are methylated or modified. We now know that most of the snRNAs in snRNPs are involved in the splicing (removal) of intervening non-coding regions from the precursor mRNA in the nucleus (see Section 6.8.2). Other snRNAs (e.g., U7) participate in processing of the 3′ terminus of histone pre-mRNA, which lacks a poly(A) tail. Most of the snoRNAs present in the nucleolus, the site of transcription of ribosomal RNA genes (see Chapter 7), are involved in the processing of rRNA precursors. Although not all homologs of snRNA and snoRNA species have been identified in plants, the similarity of RNA processing mechanisms among all eukaryotes suggests that these RNAs are likely present in plant nuclei as well.

In addition to snRNAs and snoRNAs, other types of small, stable RNA molecules also have been identified (Table 6.4). RNase P, an enzyme required for the processing of tRNAs, contains an integral RNA component. The 7SL RNA of the signal recognition particle functions in conjunction with proteins during the transport of proteins into the ER (see Chapter 4).

6.8 RNA processing

Although transcription is a complex and highly regulated step in the expression of a gene, we now know that it represents only the first in a series of events necessary to

Table 6.4 Small, stable RNAs

Type	Names	Location	Proposed function
Small nuclear RNAs (snRNAs)	Ul, U2, U4 through U9, U11, U12	Nucleoplasm	Pre-mRNA splicing
	U7	Nucleoplasm	Histone pre-mRNA processing, poly(A) addition
Small nucleolar RNAs (snoRNAs)	U3, U8, U13 through U40	Nucleolus	Pre-rRNA processing
Other small RNAs	RNase P	Nucleoplasm, plastids, mitochondria	Pre-tRNA processing
	7SL	Cytoplasm	Protein transport

Box 6.6 Introns can sometimes yield functional gene products.

In mammalian cells, the genes that encode U14 through U40 are located in the introns of other genes. Once these introns are excised from the transcript, further processing releases the snoRNAs. The genes that code for U14 through U24 (except U22) and U32 through U40 lie within the introns of protein-coding genes, and are co-transcribed with their host pre-mRNAs. In contrast, the introns that contain genes for U22 and U25 through U31 are found in a gene whose spliced RNA does not encode a protein and is short lived. In this unusual gene, the introns carry genetic information but the exons do not.

produce a functional RNA. From the previous sections we have learned that RNA can be modified extensively. However, most primary transcripts in eukaryotic cells must also be processed to yield mature RNA products. The coding regions of many eukaryotic genes are interrupted by sequences that do not appear in the mature RNA. The primary transcripts synthesized by RNA polymerases therefore include both stretches of RNA sequence that contribute to the final gene product **(exons)** and noncoding, intervening sequences that must be removed before the RNA can function **(introns).** In the case of pre-mRNAs transcribed from genes containing introns, the intron sequences must be removed to produce a mRNA molecule that codes directly for the protein. In some cases, introns can give rise to functional gene products (Box 6.6). The process by which the introns are excised from a pre-RNA and the exons are rejoined to yield the mature RNA sequence is called **RNA splicing.**

6.8.1 Introns are found in RNAs encoded in all three genomes of the plant cell.

Introns have been found in most classes of genes and occur in the nuclear, chloroplast, and mitochondrial genomes. However, the number of introns and frequency of intron-containing genes varies among plant species and organelles. For example, 155 introns account for 40% of the plastid genome of *Euglena,* while the chloroplast DNA of the red alga *Porphyra* lacks introns entirely. Most protein-coding genes in the plant cell nucleus contain one or more introns, but some, such as the maize zein gene, do not. It is not unusual for some members of a gene family to have multiple introns while other members contain none. No introns have been reported in 5.8S and 5S rRNA genes or in any of the plant snRNA genes.

Most introns can be classified into four general categories based on their structural features and the mechanisms by which they are spliced (Table 6.5). Nuclear pre-mRNA introns are the most prevalent type. In mammalian genes, these introns are highly variable in length, ranging from less than 100 nucleotides to 100 kb, but they are typically shorter in plant genes. Nuclear pre-tRNA introns, which are found only in a subset of nuclear tRNA genes, are usually short and distinct from mRNA introns. Group I and Group II introns, which are classified according to their primary and secondary structures, occur in the mitochondrial genomes of fungi,

Table 6.5 The four major types of introns

Type	Structural features: Length (nucleotides)	Structural features: Splice sites	Occurrence
Nuclear pre-mRNA	> 70	G$^{\downarrow}$GU. . . . AG$^{\downarrow}$N[a]	Nuclear mRNAs
Nuclear pre-tRNA	11–13	not conserved	Nuclear tRNAs
Group I-type	> 200	U$^{\downarrow}$N . . . . G$^{\downarrow}$N	Plastid tRNA, rRNA and mRNAs
Group II-type	< 200 to >400	N$^{\downarrow}$GYGCG. . . . AY$^{\downarrow}$N[b]	

[a]Minor classes of nuclear pre-mRNA introns contain the splice site sequence N$^{\downarrow}$GC . . . . AG$^{\downarrow}$U and N$^{\downarrow}$AU. . . . AC$^{\downarrow}$N, in which N represents any nucleotide.
[b]Y represents a pyrimidine nucleotide.

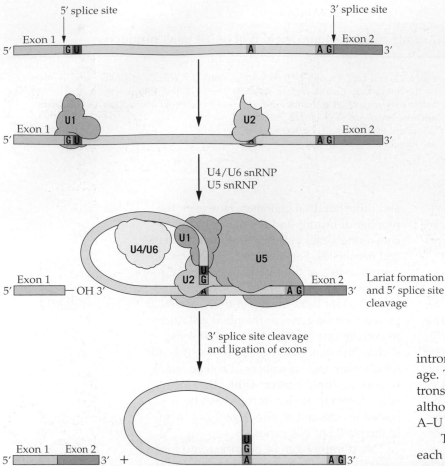

5' splice site

3' splice site

Exon 1 | G U | A | A G | Exon 2
5'————————————————————3'

Exon 1 | U1 | G U | U2 | A | A G | Exon 2
5'————————————————————3'

U4/U6 snRNP
U5 snRNP

Exon 1
5'——— OH 3'

U4/U6 | U1 | U5 | U2 | U G | A | A G | Exon 2 | 3'

Lariat formation and 5' splice site cleavage

3' splice site cleavage and ligation of exons

Exon 1 | Exon 2
5'——————3' +

U G | A | A G 3'

Figure 6.41
Steps in the intron splicing of nuclear pre-mRNA. Pre-mRNA intron splicing occurs within a large ribonucleoprotein complex, called the spliceosome, which assembles on pre-mRNA. The spliceosome contains the snRNAs U1, U2, the U4/U6 complex, and U5, which are associated with small ribonuclear proteins (snRNPs), and non-snRNP protein factors. The splicing reaction is initiated by the nucleophilic attack of the 2'-hydroxyl group of the branch site adenosine on the 3'→5'-phosphodiester bond at the 5' splice site. The 5' guanosine of the exon–intron boundary and the branch site adenosine together form a 2'→5' phosphodiester bond, generating the lariat intermediate and releasing the 5' exon. Splicing is completed with a second nucleophilic attack by the 3'-hydroxyl group of the 5' exon on the phosphodiester bond at the 3' splice site, which ligates the exons and releases the intron lariat.

as well as in plant mitochondria and plastids. A fifth category of introns, Group III, has recently been identified in the genus *Euglena,* in which more than 100 examples of this type of intron are now known.

6.8.2 The introns of plant nuclear pre-mRNA tend to be AU-rich and have conserved sequences at their splice junctions.

Introns in plant pre-mRNAs vary considerably in length—from 70 nucleotides to well over 7 kb—but most range from 80 to 140 nucleotides. There appears to be a minimal length required for efficient splicing of these introns, because introns of less than 70 nucleotides are rarely found in protein-coding genes. In contrast to the introns in the nuclear genes of animals, plant nuclear pre-mRNA introns tend to be enriched for adenosine and uridine ribonucleotides. In dicots, these

introns contain 70% A–U basepairs on average. The composition of the pre-mRNA introns of monocots is 60% A–U on average, although some contain as few as 30% A–U basepairs.

The nucleotide sequences surrounding each exon–intron junction are highly conserved. In almost all of the nuclear mRNA introns of plants, the boundary sequences at the 5' splice site (donor site) and the 3' splice site (acceptor site) consist of 5' GU and AG 3'; this characteristic is known as the GU–AG rule. In rare instances, the intron splice site sequences are 5' GC. . . . AG 3' or 5' AU. . . . AC 3'. An additional structural element, the branch site, is usually located 20 to 40 nucleotides upstream from the 3' splice site. During intron excision, the 5' end of the intron binds to an adenine residue at the branch site, forming a lariat (Fig. 6.41). A conserved branch site sequence, UACUAAC is found in yeast introns, but this sequence is not strictly conserved in plant or mammalian introns. The branch site adenine in the conserved sequence has been underlined.

Our understanding of pre-mRNA splicing in plants is very limited because it has been difficult to obtain from plant tissues an in vitro splicing system that faithfully reproduces the in vivo splicing reaction. The fundamental mechanism of pre-mRNA splicing in plants probably resembles that of mammals and yeast, because all of the conserved

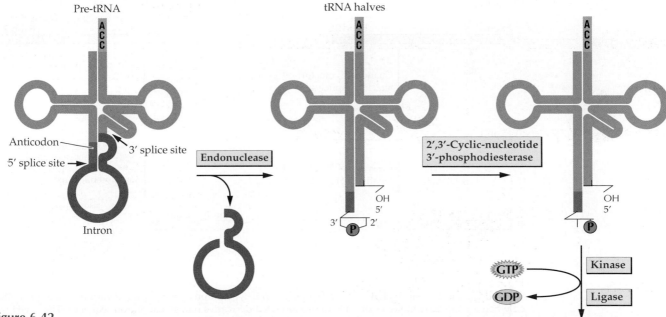

Pre-tRNA

Anticodon
5' splice site →
3' splice site
Intron

Endonuclease

tRNA halves

2',3'-Cyclic-nucleotide
3'-phosphodiesterase

OH
5'
3' 2'
P

OH
5'
P

GTP
GDP

Kinase
Ligase

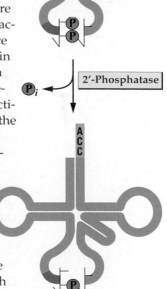

2'-Phosphatase

Mature tRNA

Figure 6.42
The splicing pathway of pre-tRNA. An endonuclease cleaves the pre-tRNA at both ends of the intron (red arrows). This results in the formation of a cyclic 2',3'-phosphate group at the 3' end of the 5' tRNA segment, as well as a free 5'-hydroxyl group at the 5' end of the 3' tRNA segment. The cyclic phosphate group is cleaved to form a 2'-phosphate group. After the free 5' hydroxyl of the 3' tRNA segment is phosphorylated, both halves of the tRNA are joined by an RNA ligase. A 2'-phosphatase removes the 2'-phosphate group to yield the mature spliced tRNA.

elements present in plant introns, i.e., the 5' and 3' splice sites and the branch site, are similar to those found in the introns of mammals and yeast.

6.8.3 The position of nuclear tRNA introns is conserved, but their sequences are not.

Introns in eukaryotic nuclear tRNA genes are short (11 to 60 nucleotides). The 5' splice site is located one nucleotide distal to the anticodon. There is no conserved sequence at the splicing junctions but the introns contain a sequence that is complementary to the anticodon of the tRNA. Nuclear tRNA splicing is best understood in yeast. An in vitro pre-tRNA splicing system from wheat germ has provided a model for nuclear pre-tRNA splicing in plants, which is similar to that in yeast cells (Fig. 6.42).

6.8.4 Group I introns can self-splice and act as mobile genetic elements.

Group I introns are distributed in the mitochondria of yeasts and other fungi, the nu-

clear rRNA genes of certain unicellular eukaryotes (e.g., *Tetrahymena*), and the organellar genes of plants. Group I introns are ribozymes (RNA molecules with catalytic activity) and reflect the presumed importance of RNA-catalyzed splicing reactions early in evolution. Group I intron sequences, which fold into complex secondary structures, begin the splicing reaction by binding and activating a G nucleotide. The 3'-hydroxyl of the activated G attacks the 5' splice site group and catalyzes the cleavage of the phosphodiester bond at that site, releasing an exon (Fig. 6.43A). The 3'-hydroxyl of the free exon then attacks the 3' splice site to release the intron and ligate the exons. Some self-splicing Group I introns remain today, for example, in the nuclear rRNA genes of *Tetrahymena* (where they were first discovered by Thomas Cech and colleagues) and in certain chloroplast and mitochondrial genes.

Several Group I introns, including the intron of the plastid 23S rRNA gene from *Chlamydomonas reinhardtii*, are mobile genetic elements. In a process known as "intron homing," these mobile introns are transmitted with high frequency to the progeny of

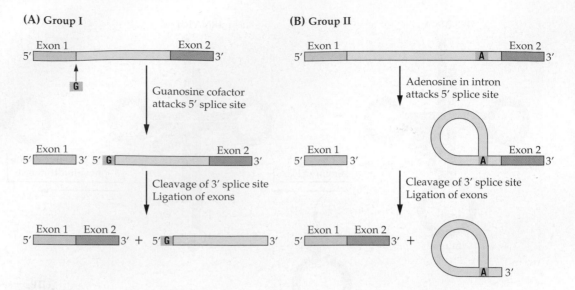

(A) Group I

Exon 1   Exon 2

G

Guanosine cofactor
attacks 5′ splice site

Cleavage of 3′ splice site
Ligation of exons

Exon 1 Exon 2   +   5′ **G**

(B) Group II

Exon 1   A   Exon 2

Adenosine in intron
attacks 5′ splice site

Cleavage of 3′ splice site
Ligation of exons

Exon 1 Exon 2   +

Figure 6.43
Self-splicing mechanisms remove Group I- and Group II-type introns. (A) Splicing of Group I introns is initiated when a guanosine molecule or its 5′ phosphorylated derivative binds to an intron sequence. The 3′ OH group of the bound guanosine attacks the phosphate group at the 5′ splice site, cleaving the phosphodiester bond at that site. The 3′ OH at the end of exon 1 reacts with the phosphate at the 3′ splice site, ligating the two exons and releasing a linear intron that is subsequently circularized (not shown). (B) In contrast, the mechanism of Group II introns self-splicing resembles splicing of nuclear pre-mRNAs, but does not involve the formation of a spliceosome (see Fig. 6.41).

genetic crosses between parents that have the intron and parents that do not. All mobile introns encode a DNA endonuclease that specifically recognizes the intron-less allele and generates a double-strand break in the DNA near the insertion or "homing" site. This break is thought to initiate the replicative integration of the intron into the intron-less allele.

6.8.5 Group II introns and nuclear pre-mRNA introns share the same splicing mechanism.

Group II introns are found in fungal mitochondrial genes, as well as in plant mitochondrial and plastid genes. Although Group II introns differ significantly from nuclear pre-mRNA introns in their structure, they do share certain common features. Group II introns follow the GU–AG rule: The sequences at their 5′ and 3′ splice sites resemble those of nuclear pre-mRNA introns and they contain a conserved branch site adenine. This finding indicates that the Group II introns and the nuclear pre-mRNA introns have a common evolutionary origin and share the same splicing mechanism. Some of the Group II introns in yeast mito-

chondria self-splice in vitro, using a particularly reactive A residue in the intron sequence as the attacking group for the formation of the lariat structure (Fig. 6.43B). Autocatalytic splicing has not been observed in Group II introns from plant plastids or mitochondria, suggesting that these introns require additional factors for efficient splicing.

6.8.6 Precursor RNAs are processed extensively to create functional RNA molecules.

Most eukaryotic genes are transcribed as precursor RNA (pre-RNA) molecules, which then are processed into shorter RNA species (mature RNAs). Extensive processing and modification are highly regulated and critical aspects in the production of functional RNA forms. Only certain species of small, stable RNAs that are transcribed by RNA polymerase III do not undergo extensive processing.

The transcription units in the nuclear rRNA gene cluster that encode the 17S, 5.8S, and 25S rRNA molecules are transcribed by RNA polymerase I into a single long precursor molecule. The rRNA precursor molecule

Box 6.7

Certain Group II introns of plants are removed by a novel trans-splicing mechanism.

Splicing reactions are typically intramolecular and join exons from a single transcript (cis-splicing). However, splicing events discovered in plant chloroplasts and mitochondria involve two or more RNA molecules (trans-splicing). The tobacco chloroplast *rps12* gene (see figure) is organized as two separate loci in the genome. Each locus produces an independent transcript: one encodes exon 1 and the other encodes the RNA product exon 2–intron–exon 3. Following transcription,

trans-splicing joins exons 1 and 2; exons 2 and 3 are spliced in cis after removal of the intron. The *Chlamydomonas* chloroplast *psaA* gene (see figure) is another example of a discontinuous gene in which three exons are encoded by three separate loci in the genome. In this case, two trans-splicing events generate the mature mRNA from three different transcripts.

In angiosperm mitochondria, the *nad1*, *nad2*, and *nad5* genes are also discontinuous, and their transcripts require trans-

splicing to produce the functional mRNAs. In contrast, the *nad2* and *nad5* genes in the mitochondrial genomes of ferns are continuous and their introns are spliced in cis. These continuous genes are thought to be the ancestral genes that gave rise to the discontinuous genes of flowering plants. The unique trans-splicing mechanism may have co-evolved with the angiosperm genes as the latter were cleaved into separate transcription units.

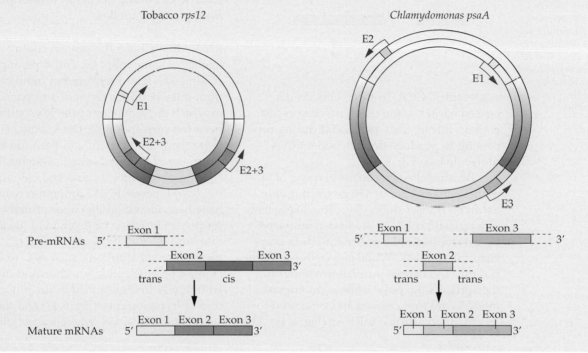

then undergoes a series of cleavages and methylation steps to yield mature the 17S, 5.8S, and 25S rRNA molecules (Fig. 6.44). The 5.8S rRNA is derived from the same RNA intermediate as the 25S rRNA; the bases of these two RNA products remain paired even after the intergenic RNA spacer sequences have been removed. Transcription of the 5S rRNA from the corresponding nuclear genes is independent and is catalyzed by RNA polymerase III. This cytoplasmic 5S rRNA is unusual in that it requires no processing.

All four plastid rRNA molecules are encoded as a polycistronic transcription unit that also includes the two tRNA genes encoding tRNA^{Ile} and tRNA^{Ala}, each of which contains an unusually long intron. Following

a complex processing pathway involving specific nucleases, the precursor RNA is cleaved into 16S, 23S, 4.5S, and 5S rRNAs (Fig. 6.45). Once removed from the polycistronic RNA, the cleaved tRNA precursors are processed and their introns are spliced to produce functional tRNA^{Ile} and tRNA^{Ala} molecules. Ribosomal RNAs and tRNAs in plant mitochondria also are extensively processed from precursor molecules.

Unlike 5S rRNA, the tRNAs encoded by clusters of nuclear genes are transcribed by RNA polymerase III into precursor molecules that contain additional sequences at both the 5' and 3' ends. These extra sequences are then removed by specific nucleases, followed by the addition of three

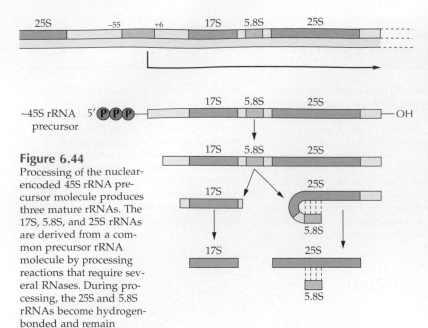

Figure 6.44
Processing of the nuclear-encoded 45S rRNA precursor molecule produces three mature rRNAs. The 17S, 5.8S, and 25S rRNAs are derived from a common precursor rRNA molecule by processing reactions that require several RNases. During processing, the 25S and 5.8S rRNAs become hydrogen-bonded and remain paired after the processing is complete.

nucleotides, CCA, to the 3' end. As discussed earlier, some tRNA precursors also contain introns that are spliced during processing to produce the functional tRNA molecule (see Fig. 6.42).

Because the chloroplast genome of most plants contains about 150 genes, but only about 60 transcription units, it is likely that most plastid mRNAs are first synthesized by chloroplast RNA polymerase into polycistronic mRNAs. While polycistronic bacterial mRNAs can be translated without further processing, polycistronic chloroplast mRNAs are first cleaved and processed to create monocistronic mRNAs (Fig. 6.46). The

significance of this difference between bacterial and chloroplast polycistronic mRNAs is not well understood, although analysis of certain mutants that lack chloroplast functions have shown that in some cases processing is required for plastid mRNA translation. This suggests that details of the mRNA translation process in plastids differ from those in prokaryotes. With the exception of some closely linked genes, plant mitochondrial protein-coding genes are transcribed into monocistronic mRNAs.

6.8.7 RNA editing occurs in transcripts from plant organelles.

RNA editing alters the protein-coding sequence of some mRNAs. This type of processing was first discovered in transcripts from mitochondrial genes of a trypanosome, in which the sequence of the RNA differs from the corresponding DNA sequence. Since then, RNA editing has been found in a number of different genetic systems, including plant mitochondria and chloroplasts.

Two types of RNA editing mechanisms have been identified that alter primary transcripts: (*a*) insertion or deletion of nucleotides and (*b*) nucleotide conversion (i.e., modification or substitution), such as C to U (Table 6.6). Nucleotide insertion and deletion editing was first examined in the mitochondria of trypanosomes (Fig. 6.47) and the slime mold *Physarum*, where transcripts

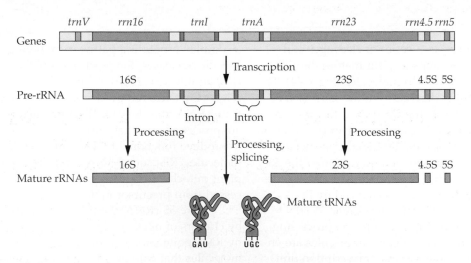

Figure 6.45
Processing of chloroplast pre-rRNA from plants. Unlike nuclear-encoded rRNAs, the chloroplast rRNA operon also encodes two tRNAs in the spacer regions that separate the 16S and 23S rRNA. These tRNAs are interrupted by long introns (approximately 1 kb) and require both processing and splicing to produce the mature tRNAs.

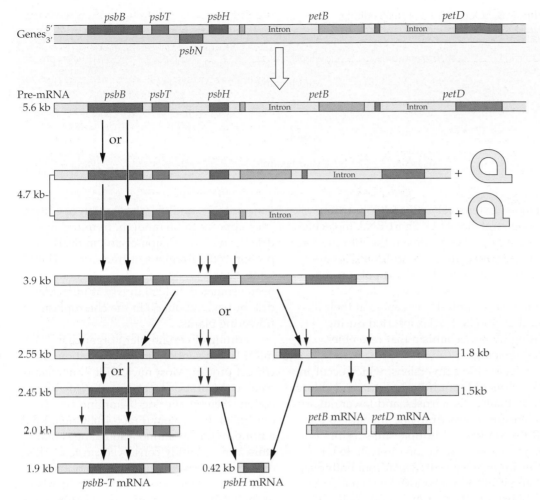

Figure 6.46

The transcript from the *psbB* operon in plant plastids is a typical example of the complex processing and splicing reactions that are required to cleave the polycistronic mRNA precursor into mature monocistronic mRNAs that can be efficiently translated. Black arrows indicate possible alternative processing steps, and red arrows indicate cleavage sites that have been experimentally confirmed. The *psbN* gene is encoded by the opposite DNA strand and therefore is not part of the polycistronic transcript.

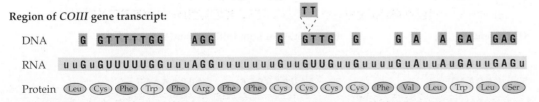

Figure 6.47

RNA editing in protein-coding regions of mitochondrial transcripts from the protozoan *Trypanosoma brucei*. Comparison of the DNA sequence (purple) and RNA sequence (green) for the mitochondrial cytochrome oxidase III reveals striking differences. The RNA sequence is made considerably longer than the DNA sequence as the result of RNA editing, in this case the insertion of long stretches of Us. The protein encoded by the edited, mature mRNA has an amino acid sequence resembling the consensus sequence of cytochrome oxidase III proteins. The information required for RNA editing is provided by guide RNAs (not shown), which are complementary to portions of the edited mRNA. Note that UGA, a stop codon in the universal genetic code (see Chapter 9), functions as a Trp codon in some mitochondrial genes. *COIII*, gene encoding cytochrome oxidase III.

Table 6.6 Occurrence of RNA editing

Type	Organism	Organelle	Transcript	Modification
Insertion/ deletion	Trypanosome	Mitochondria	mRNAs	U insertion/ deletion
	Physarum	Mitochondria	mRNAs	C insertion
Conversion	Plants	Mitochondria	mRNAs	C → U, U → C
			tRNAs	C → U
			rRNAs	C → U
		Plastids	mRNAs	C → U, U → C
	Mammals	Nuclei	mRNAs	C → U
	Metazoans	Mitochondria	tRNAs	Various

often are extensively edited by insertion of long stretches of U. Short RNA molecules, called guide RNAs, contain the information required for editing. The guide RNAs are complementary to the edited portions of the mature mRNA. These guide RNAs contain short poly-U sequences at their 3′ end that donate the Us inserted during editing by a mechanism that resembles RNA splicing.

Nucleotide conversions, which occur in transcripts of plant mitochondria and chloroplasts, mammals, viruses, and lower eukaryotes, result in less dramatic changes to the mRNA sequence. Multiple site-specific conversions from C to U (or, rarely, U to C) occur in primary transcripts from both plastid and plant mitochondrial genomes. The frequency of RNA editing is organelle- and gene-specific, and the distribution of editing sites appears to be random. In mitochondria, C to U conversion occurs in nearly all protein-coding regions of transcripts. The total number of editing sites per genome has been estimated at 1200 in wheat mitochondria but only about 30 in the chloroplasts of flowering plants.

Editing of certain plant organellar mRNAs is essential for the translation of a correct protein. Most nucleotide conversions change the first or second nucleotide in a codon, thereby altering the amino acid sequence of the protein (Fig. 6.48). Often, RNA editing results in amino acid changes that restore a more highly conserved protein. This conclusion, drawn using data from many different organisms, is based on comparison of the DNA sequence of a gene with the

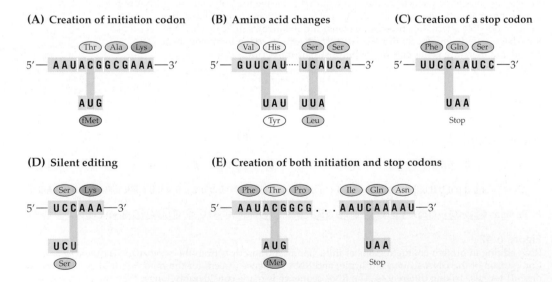

Figure 6.48
Several types of mRNA editing involving base substitutions can significantly alter the polypeptide products encoded by the RNA sequences. Base substitutions can produce translation initiation codons (A), result in amino acid changes (B), or create a stop codon (C), possibly leading to premature termination of translation. Certain base substitutions within coding regions may not alter the amino acid sequence (D), whereas others may introduce initiation and stop codons that result in a new open reading frame (E).

sequence of its edited mRNA transcript. In rare cases, RNA editing occurs at the third codon position. Often, as a result of redundancy in the genetic code, such editing is silent and does not modify the predicted protein (see Chapter 9).

Besides amino acid changes, editing can shorten ORFs predicted from DNA sequences or produce new ORFs. The editing of glutamine codons (CAA and CAG) and arginine codon (CGA) yields termination codons (UAA, UAG, and UGA), thereby introducing a premature termination of the ORF in both mitochondria and plastids. These edited stop codons generally convert the predicted C-terminal extension of a protein to the conserved protein size. Likewise, a threonine codon (ACG) can be converted to AUG. C to U editing can thus create both an initiation and a termination codon within the same transcript, producing a new ORF. These examples clearly demonstrate that genomic DNA sequencing alone is not always sufficient to predict the products of organellar genes.

Editing processes in plant mitochondria and chloroplasts share certain features: C to U conversion, and strong preferences for second codon positions and certain codon transitions. Furthermore, common sequence motifs around the editing sites are found in the transcripts of both organelles. These similarities suggest editing systems in chloroplast and mitochondria share components and perhaps mechanisms, although it is likely that organelle-specific factors also are required. RNA editing in chloroplasts does not require translation, indicating that all protein components necessary for editing are encoded by the nuclear genome and imported into the chloroplast.

Summary

Genetic information is encoded in the nucleotide polymer of DNA. The DNA molecule is a double-stranded helix of two complementary strands of nucleotides that form hydrogen bonds between A–T and G–C basepairs. The genetic information is duplicated during DNA replication when DNA polymerases synthesize a new complementary nucleotide polymer, using each of the two parental strands in the double helix as templates. Viruses, which are simple complexes of short DNA (or RNA) molecules encapsulated in protein coats, exploit the host cell DNA replication machinery for their own amplification and propagation. Although DNA is a stable molecule, it sometimes suffers damage from UV light or chemicals, or errors introduced by DNA polymerase during replication. If not repaired, this damage can cause mutations. Several repair mechanisms function to remove damaged nucleotides or incorrectly paired bases. Exchange of genetic information between DNA molecules in chromosomes is an important mechanism that drives evolution and establishes diversity among organisms. Genetic information exchange occurs during recombination when double-stranded DNA molecules are broken and rejoined with different DNA molecules. Before the genetic information stored in DNA can be expressed into a co-linear sequence of amino acids in proteins, it must be transcribed into RNAs by multiprotein enzyme complexes called RNA polymerases. Plants, like other eukaryotes, contain three types of nuclear RNA polymerase. Chloroplast (and probably mitochondrial) genes are transcribed by prokaryote-type RNA polymerases, which reflects their evolutionary origin. Most RNAs transcribed by RNA polymerases must undergo extensive processing before they can function in the translation of proteins or in the formation of ribonucleoprotein complexes such as ribosomes. The protein-coding regions of many genes are also interrupted by non-coding sequences, called introns, which must be spliced out to produce functional RNAs. In some cases the genetic information contained in genes differs from the information found in their corresponding RNAs, which have nucleotides added, deleted, or substituted. This RNA editing usually restores cryptic DNA information to produce proteins with evolutionary conserved amino acid sequences.

Further Reading

Britt, A. B. (1996) DNA damage and repair in plants. *Annu. Rev. Plant Physiol. Plant Mol. Biol.* 47: 75–100.

Busby, S., Ebright, R. H. (1994) Promoter structure, promoter recognition, and transcription activation in prokaryotes. *Cell* 79: 743–746.

DePamphilis, M. L., ed. (1996) *DNA Replication in Eukaryotic Cells.* Cold Spring Harbor Laboratory Press, Plainview, NY.

Fauron, C., Casper, M., Gao, Y., Moore, B. (1995) The maize mitochondrial genome: dynamic, yet functional. *Trends Genet.* 11: 228–235.

Gray, M. W. (1989) Origin and evolution of mitochondrial DNA. *Annu. Rev. Cell. Biol.* 5: 25–50.

Gruissem, W. (1989) Chloroplast gene expression: how plants turn their plastids on. *Cell* 56: 161–170.

Koleske, A. J., Young, R. A. (1995) The RNA polymerase holoenzyme and its implications for gene regulation. *Trends Biochem. Sci.* 20: 113–116.

Kornberg, A., Baker, T. A. (1992) *DNA Replication,* 2nd ed. W. H. Freeman, New York.

Maier, L.R.M., Neckermann, K., Igloi, G. L., Kössel, H. (1995) Complete sequence of the maize chloroplast genome: gene content, hotspots of divergence and fine tuning of genetic information by transcription editing. *J. Mol. Biol.* 251: 614–628.

Puchta, H., Hohn, B. (1996) From centimorgans to base pairs: homologous recombination in plants. *Trends Plant Sci.* 1: 340–348.

Rochaix, J. D. (1996) Post-transcriptional regulation of chloroplast gene expression in *Chlamydomonas reinhardtii. Plant Mol. Biol.* 32: 327–341.

Schuster, W., Brennicke, A. (1994) The plant mitochondrial genome: physical structure, information content, and gene migration to the nucleus. *Annu. Rev. Plant Physiol. Plant Mol. Biol.* 45: 61–78.

Simpson, L., Thiemann, O. H. (1995) Sense from nonsense: RNA editing in mitochondria of kinetoplastid protozoa and slime molds. *Cell* 81: 837–840.

Sugiura, M. (1992) The chloroplast genome. *Plant Mol. Biol.* 19: 149–168.

Thosness, P. E., Weber, E. R. (1996) Escape and migration of nucleic acids between chloroplasts, mitochondria, and the nucleus. *Int. Rev. Cytol.* 165: 207–234.

Timmermans, M.C.P., Prem Das, O., Messing, J. (1994) Geminiviruses and their uses as extrachromosomal replicons. *Annu. Rev. Plant Physiol. Plant Mol. Biol.* 45: 79–112.

Tjian, R. (1995) Molecular machines that control genes. *Scientific Am.* 272(2): 54–61.

Wang, J. C. (1996) DNA topoisomerases. *Annu. Rev. Biochem.* 65: 635–692.

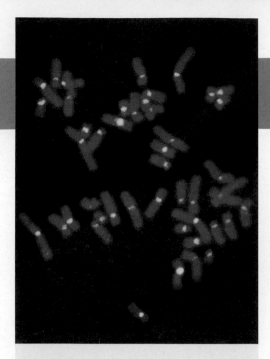

Biochemistry & Molecular Biology of Plants, B. Buchanan, W. Gruissem, R. Jones, Eds.
© 2000, American Society of Plant Physiologists

CHAPTER 7

Genome Organization and Expression

Robert Ferl
Anna-Lisa Paul

CHAPTER OUTLINE

Introduction

The essence of every characteristic of a species, from morphology to timing of senescence, resides in information stored in the species' DNA: its genome. All of that information, encoded in thousands of genes in even simple organisms, requires some sort of organization. In eukaryotes, the first level of organization is the chromosome (Fig. 7.1). With so many genes and so few chromosomes (five, for example, in *Arabidopsis*), a vast array of genes must reside on any one chromosome. How all of these genes are organized is an important question of molecular genetic research. This chapter will focus on the informational organization of the genes within chromosomes. Our current understanding of genome organization and expression reflects the capabilities and limitations of widely varied current technologies. Merging the data from these technologies into a complete picture of genome structure is a major research goal. Most of an organism's genes reside on nuclear DNA. However, plastids and mitochondria also contain DNA that encodes some of the gene products required for organelle function and reproduction. Several traits encoded by organelle genes are in commercial use in plant crops, such as cytoplasmic male sterility in sorghum and maize. Traits arising

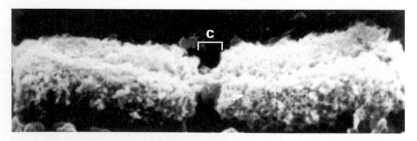

Figure 7.1
Electron micrograph of a metaphase chromosome, showing the structure of a chromosome in its most compact state. C, centromere.

from organelle genes can be clearly recognized by their uniparental inheritance patterns, because in many species mitochondria and chloroplasts are inherited only through the maternal contribution to the seed.

In the 1950s, plant geneticists found evidence that pieces of genetic material could move from place to place in the DNA. These mobile genetic elements were subsequently found to be ubiquitous components of genomes, providing the possibility for genomic changes on a larger scale than the accumulation of single-point mutations and thereby speeding evolutionary change.

Our examination of genome organization and genetic regulation begins with a historical perspective and proceeds to a state-of-the-art view of the function of single genes (Fig. 7.2). The historical perspective conveys the development of genetic theory and illustrates the continued value of "classic" approaches. However, our ability to characterize the chromosome and the genes contained therein with increasingly finer resolution has developed tremendously since the time of Mendel. Elucidating the workings of genetics has been, and continues to be, one of the most fruitful of biological disciplines, changing the way we view our world and think about ourselves as biological organisms. In 1958 George Beadle, Joshua Lederberg, and Edward Tatum were awarded the Nobel Prize for their work in gene regulation. Since then, almost half of all the Nobel Prizes awarded for Chemistry and Medicine or Physiology have been for innovations in our understanding of genetics and gene regulation.

7.1 Genes and chromosomes

7.1.1 An elegant series of experiments confirmed that nucleic acids function as genetic material.

In the 1850s Gregor Johann Mendel conducted a series of experiments with peas *(Pisum sativum)* that laid the foundation for modern

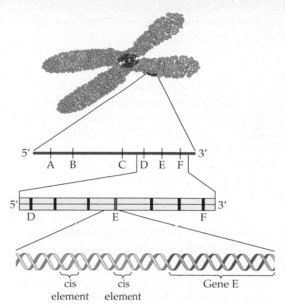

Figure 7.2
Levels of genetic and molecular organization: Chromosomes (especially those in metaphase) can be characterized by structural features, such as the position of the centromere, and by banding patterns seen in the presence of certain cytological dyes. A genetic map can be created for a chromosome by assigning genes to their relative positions on that chromosome. Smaller sections of the genome can be precisely ordered into a physical map by using molecular tools such as restriction fragment length polymorphism (RFLP) analyses (see Fig. 7.18). Individual genes within the physical map can be analyzed as to their specific sequence of nucleotides by using DNA sequencing (not shown). Functional analyses evaluate the importance of certain sequence features to the transcriptional regulation of that gene (see Box 7.5).

genetics. As a result of these experiments, he proposed that "particular factors" gave rise to specific traits.

Almost 100 years later, the question still raged as to the composition of these factors, which we now know as **genes.** (The term "gene" was not used until 1909, when it was coined by W. L. Johannsen.) By the 1940s chromosomes were understood to consist mostly of two types of molecules: protein and deoxyribonucleic acid (DNA). Researchers had also deduced that any molecule responsible for transmitting genetic information must be able to accomplish three tasks. First, it must encode all of the information needed for cell growth, development, structure, and reproduction. Second, it must replicate itself accurately to ensure that progeny cells contain the same information as the parent. Finally, it must be capable of variation to accommodate the changes and adaptations evidenced by evolution. Although DNA seemed too simple to carry so much information, other features made a compelling argument for its being the genetic molecule. DNA is very stable and does not undergo the metabolic turnover observed for many proteins. Moreover, the amount of DNA is

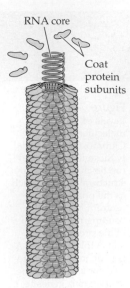

Figure 7.3
The tobacco mosaic virus (TMV) consists of a helical RNA core surrounded by a helical array of identical protein subunits.

roughly associated with the complexity of the organism (e.g., bacteria have less DNA than humans).

Between the late 1920s and the mid-1950s, researchers combined the sciences of biochemistry and genetics in a series of experiments that confirmed DNA was the genetic material. Frederick Griffith in 1928 showed that a component of heat-killed virulent bacteria could transform avirulent bacteria into the virulent form. Oswald T. Avery, Colin M. MacLeod, and Maclyn McCarty utilized the same biological system in 1944 to demonstrate that Griffith's "transforming principle" was, in fact, DNA. In 1952, Alfred Hershey and Martha Chase used a virus that infects bacteria (a **bacteriophage**) to show that the infectious component of the phage was the DNA, not the protein. Using X-ray crystallography data generated by Rosalind Franklin and Maurice Wilkins, as well as quantitative analysis of DNA composition performed by Erwin Chargoff, Francis Crick and James Watson postulated the structure of DNA—the now-familiar double helix—in 1953.

Not long after, a second nucleic acid—RNA—was identified as the genetic material of tobacco mosaic virus (TMV). In 1956, A. Gierer and G. Schramm discovered that purified RNA alone could initiate an infection of TMV in tobacco leaves, which suggested that the RNA carried all of the genetic information necessary for the synthesis of new viruses (Fig. 7.3). The next year, Heinz Fraenkel-Conrat and Bea Singer confirmed this hypothesis, using hybrids of two distinct strains of TMV—each hybrid carrying the RNA of one strain and the protein of the other. The progeny of the hybrid viruses in infected tobacco leaves always matched the type represented by the RNA component (Fig. 7.4).

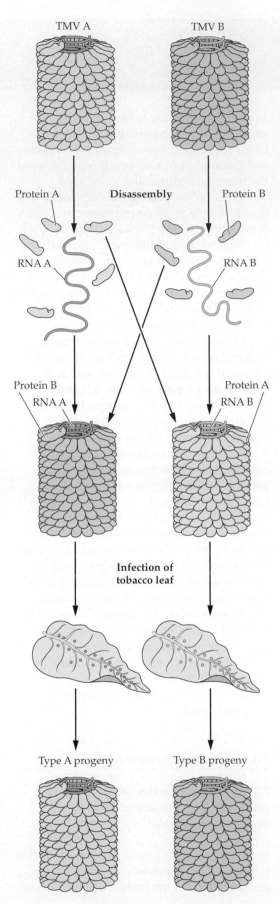

Figure 7.4
The Fraenkel-Conrat and Singer experiment, demonstrating that the RNA component of TMV particles carried the genetic information. Hybrid virus particles were made from two different strains of TMV (TMV-A and TMV-B). When particles consisting of TMV-A protein + TMV-B RNA were used to infect tobacco plants, all members of the next generation of virus particles recovered from infected plants were of the TMV-B type. Conversely, infecting plants with the TMV-A RNA + TMV-B protein hybrids resulted in type TMV-A type progeny. Thus, the RNA contained the genetic information to make new virus particles.

Figure 7.5
Gregor Johann Mendel, often referred to as the father of the science of genetics.

7.1.2 Genes encode heritable traits.

Mendel's pea experiments, the first experiments to correlate physical traits **(phenotype)** with heritable components **(genotype),** represent only one example of how plants have played a central role in the foundations of genetics. Mendel (Fig. 7.5) studied eight pairs of characteristics in the garden pea: flower color (which is also reflected in seed coat color), seed color, seed shape, seed pod color, petal color, seed pod shape, plant height, and the position of the flower in the plant (Fig. 7.6). Several key factors contributed to his success. First, the garden pea self-fertilizes, so Mendel could control the matings (Fig. 7.7). Had Mendel selected a less tractable model plant, his results might have been much more difficult to interpret (Box 7.1). Second, the traits he chose to follow exhibit simple **dominance.** This meant that (*a*) each

Figure 7.6
Mendel used the common garden pea *(Pisum sativum)* for genetic experiments to follow the inheritance of eight different traits: seed shape (round or wrinkled), seed color (yellow or green), seed coat color (gray or white), petal color (purple or white), pod shape (inflated or constricted), pod color (green or yellow), plant height (tall or short), and flower position on the plant (axial or terminal). We know now these traits are unlinked and display simple dominance—features that contributed to Mendel's success.

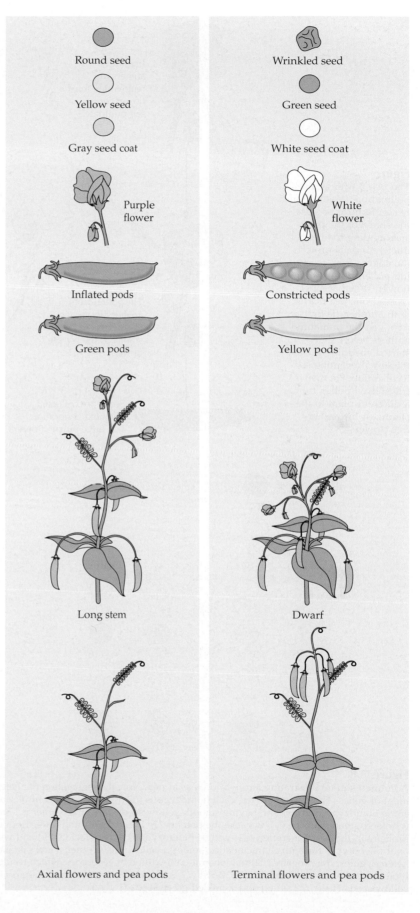

Round seed

Yellow seed

Gray seed coat

Purple flower

Inflated pods

Green pods

Long stem

Axial flowers and pea pods

Wrinkled seed

Green seed

White seed coat

White flower

Constricted pods

Yellow pods

Dwarf

Terminal flowers and pea pods

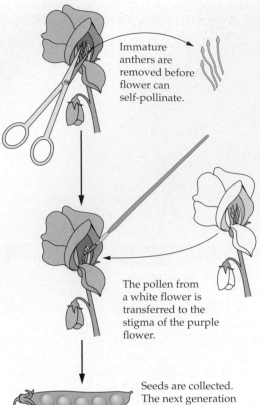

Figure 7.7

The pea flower and self-pollination. Peas typically self-fertilize because the shape of the petals surrounding the reproductive organs protects the stigma from receiving pollen from another flower. When Mendel crossed one pea flower with another, he first removed the immature anthers before they started shedding pollen, thereby preventing self-fertilization. He then transferred pollen from a different plant to the stigma of the anther-free flower.

Immature anthers are removed before flower can self-pollinate.

The pollen from a white flower is transferred to the stigma of the purple flower.

Seeds are collected. The next generation will have purple flowers.

trait exhibited two distinct forms and (b) when plants that bred true for one form were crossed with plants that bred true for the other, only one form (the dominant) was visible in the offspring. Third, Mendel chose to study no more than two traits at a time, which enabled him to follow the transmission of each trait in a controlled genetic background. Fourth, he observed traits that were not linked (another fortunate choice). Finally, Mendel diligently carried out hundreds of crosses and screened thousands of progeny. His large sample size and the fact that he kept meticulous records enabled him to compile statistical data and thus make predictions for the outcomes of new crosses.

On the basis of his findings, Mendel proposed two laws of genetics, which hold true for all unlinked genes (Fig. 7.8).

- **The principle of segregation:** The two alleles of a gene segregate during the formation of gametes; i.e., one gamete gets one allele, the other gets the other allele.
- **The principle of independent assortment:** The factors (genes) for different traits assort independently from one another.

The wrinkled-seed phenotype—the most frequently discussed of Mendel's pea traits—is now understood at the molecular level, i.e., the gene responsible for the phenotype has been cloned. (See Box 7.2 for a description of one approach to cloning.) The phenotype is caused by a mutation in the gene which encodes a starch-branching enzyme called SBEI (see Chapter 13). The defect reduces the amount of starch in the seed and increases the amount of sucrose. Increasing the sucrose content causes the developing seed to take up more water, which in turn increases the cell size and fresh weight. When the mature seeds dry, the mutants lose the extra water, resulting in wrinkled coats.

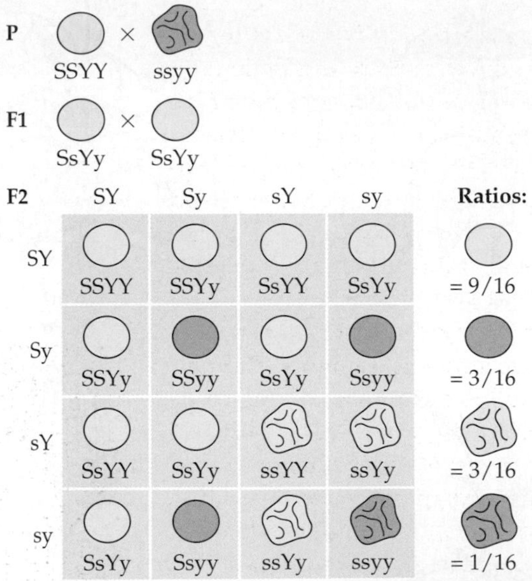

Figure 7.8

A Punnett square illustrating a dihybrid cross in pea. Note the segregation of alleles and independent assortment of genes. The parents (P) are each homozygous for two genes that control seed characteristics. One parent is homozygous dominant, having smooth yellow seeds (SSYY); the other is homozygous recessive, having wrinkled green seeds (ssyy). Because the parents are homozygous and differ from one another, all of the progeny in the first generation (F1) will be heterozygous at both loci (SsYy) and will demonstrate the dominant phenotypes for both characteristics: smooth, yellow seeds. When a cross is made between two of the F1 heterozygotes, the progeny (F2) demonstrate nine different genotypes, which result in four different phenotypes: smooth yellow seeds (1 SSYY, 2 SSYy, 2 SsYY, 4 SsYy), smooth green seeds (1 SSyy, 2 Ssyy), wrinkled yellow seeds (1 ssYY, 2 ssYy), and wrinkled green seeds (1 ssyy).

Box 7.1

Some results of Mendel's genetic studies did not conform to his principles of segregation and independent assortment.

Mendel's data came to light in 1900 with the discovery of a series of letters in the estate of Carl von Nägeli, a famous German botanist of Mendel's time. In addition to the classic pea experiments, these letters referred to experiments with other plants. Notably, Mendel failed miserably in his attempts to apply his theories of inheritance to *Hieracium,* a dandelion-like plant that commonly reproduces asexually and often exhibits uniform, nonsegregating progenies. Mendel's initial conclusion from his work with *Hieracium* was that the laws developed for garden pea had only limited application. The letters go on to state, however, that Mendel, undaunted, later conducted crosses with maize, and these data *did* corroborate the results from the garden pea. Just think what would have happened if Mendel had tried *Hieracium* first!

7.1.3 Genes reside on chromosomes.

Walter Sutton and Theodor Boveri in 1903 postulated the **chromosome theory of heredity,** based on the cytological observation that chromosomes—visible and individually distinguishable in the light microscope—were consistently transmitted from one generation to the next, as were certain traits. They concluded that Mendelian factors (the term "gene" was still not used) were found on chromosomes. In 1916, Calvin B. Bridges correlated the **nondisjunction** of the X chromosome in *Drosophila* with specific heritable traits. Nondisjunction occurs when homologous chromosomes fail to separate during meiosis; thus, some daughter cells are left with two copies of the chromosome, while others have no copies.

Some excellent examples of the correlation between genes and chromosomes in plants were published shortly thereafter, during study of the cytogenetics of maize (Fig. 7.9). Maize has 10 chromosomes—varying in size from chromosome 1, the longest, to chromosome 10, the shortest (Fig. 7.10). In the early 1930s Barbara McClintock studied primary trisomics in maize (individuals with $2n + 1$ chromosomes), correlating the "extra" chromosome with specific groups of genes that yielded trisomic phenotype ratios in crosses (Fig. 7.11).

7.1.4 Genes can be recombined into novel combinations through chromosomal exchange.

The first evidence that portions of chromosomes could exchange, leading to novel genetic combinations, came in 1905. In experiments with sweet pea (*Lathyrus odoratus;* Fig. 7.12), W. Bateson, E. R. Saunders, and P. C. Punnett noted that independent assortment of gene pairs did not always occur, a violation of Mendel's principle of independent assortment. When two parents differing in two distinct traits are bred (a **dihybrid cross**), the resulting F1 (first filial) generation displays the dominant phenotype for each trait. However, if F1 plants are crossed together, independent assortment yields a 9:3:3:1 ratio of phenotypes in the F2 (second filial) generation (see Fig. 7.8). Bateson and colleagues crossed two parental phenotypes: plants having long pollen and purple flowers, and plants having red flowers and round pollen. While the F1 plants exhibited the expected dominant phenotype—purple flowers with long pollen—most of the F2 progeny had either purple flowers with long pollen or red flowers with round pollen (Fig. 7.13). In other words, members of the F2 generation resembled one parental plant or the other. However, these crosses also yielded a small proportion of nonparental phenotypes (i.e., purple flowers with round pollen, and red flowers with long pollen). What did this mean?

At the same time that Bateson and coworkers were studying *Lathyrus,* Thomas H. Morgan and his colleagues conducted similar dihybrid crosses with fruit flies (*Drosophila melanogaster*) that differed in eye color and wing size; these too gave results that were inconsistent with independent assortment. Morgan and his colleagues proposed that traits might be inherited together because their Mendelian factors are located in close proximity to one another on a single chromosome. Genes that are usually inherited together are described as **linked,** and a set of linked genes constitutes a **linkage group.** The small proportion of nonparental phenotypes could be explained if the linkage between the two factors were

Figure 7.9
Corn (*Zea mays*), or maize, is an excellent system in which to study genetics.

Figure 7.10
The chromosomes of
maize. (A) A micro-
graph of a maize micro-
sporocyte nucleus at
the pachytene stage,
showing the 10 chromo-
somes and the nucleo-
lus. (B) Schematic draw-
ing identifying the
chromosomes. In the
pachytene stage, homol-
ogous chromosomes re-
side side-by-side and
appear thick enough to
enable the identification
of individual chromo-
somes by their struc-
tural features (e.g., po-
sitions of centromere
and knobs).

(A)

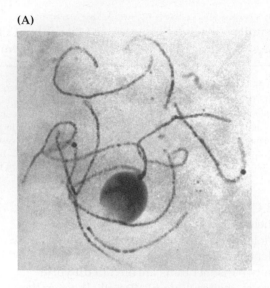

(B)

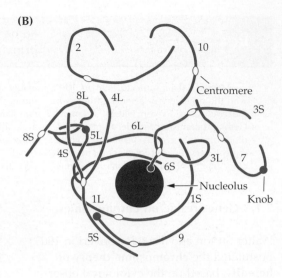

occasionally disrupted at some point during
the formation of gametes, such that new
combinations of Mendelian factors were produced. This process is called **recombination.**
Morgan proposed that the more closely the
factors were linked, the less likely a recombination event would occur to separate them.
Further work led to the suggestion that
recombination is the result of a physical exchange of genetic material between homologous chromosomes in a process called
crossing-over, and that the frequency of recombination might be used as an index
of the distance between two genes on the
chromosome.

The first direct evidence that chromosomes physically exchange material through

crossing-over came from maize cytogenetics.
Harriet B. Creighton and Barbara McClintock
used a strain of maize with an unusual chromosome. In this strain, the short arm of chromosome 9 possessed a large, darkly staining knob (see Fig. 7.10). Also, the long arm
was longer than usual because of an "interchange" with chromosome 8 (a translocation). This aberrant chromosome 9 had been
previously shown to contain several genes,
including seed aleurone color (*C*—dominant
to colorless and located close to the knob),
and waxy endosperm (*wx*—recessive to non-
waxy and located close to the translocated
piece of chromosome 8).

Thus, Creighton and McClintock were
armed with both **cytological markers** (the

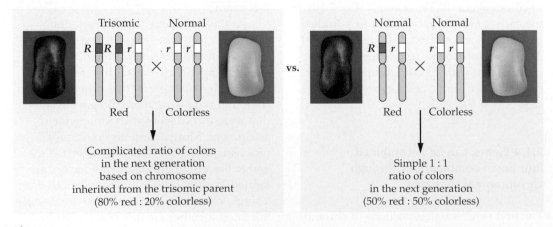

Figure 7.11
Assigning a gene to a chromosome by utilizing trisomy and a phenotypic trait. Barbara McClintock was able
to correlate a gene affecting aleurone color (*R*, red) with a trisomic chromosome of maize. The ratio of progeny
phenotypes from a backcross (*Rr* × *rr*) differs depending on whether the gene is carried on a chromosome
exhibiting trisomy (*RRr*; left panel) or a normal chromosome pair (*Rr*; right panel).

Figure 7.12
Sweet pea *(Lathyrus odoratus).*

knob and the extra-long arm) and **genetic markers** (the two genes and their traits). In crosses of maize plants containing a normal chromosome 9 with those having the aberrant chromosome 9, Creighton and McClintock could correlate the physical features of the chromosomes with the inheritance patterns of the genes contained on them. In all cases where they observed a recombinant phenotype, the cytological markers indicated that a physical exchange of chromosome material had occurred by crossing over at a site between the knob/*C* position and the translocated 8/*wx* position (Fig. 7.14).

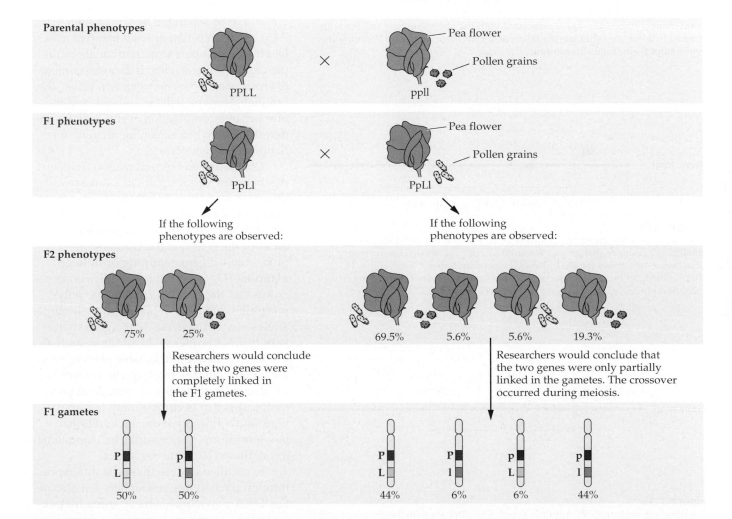

Parental phenotypes

Pea flower

Pollen grains

PPLL × ppll

F1 phenotypes

Pea flower

Pollen grains

PpLl × PpLl

If the following phenotypes are observed: If the following phenotypes are observed:

F2 phenotypes

75%   25%   69.5%   5.6%   5.6%   19.3%

Researchers would conclude that the two genes were completely linked in the F1 gametes.

Researchers would conclude that the two genes were only partially linked in the gametes. The crossover occurred during meiosis.

F1 gametes

P   P   P   P   P   P
L   l   L   l   L   l
50%   50%   44%   6%   6%   44%

Figure 7.13
Gene linkage in sweet pea. The genetics of sweet pea observed by Bateson, Saunders, and Punnett seemed to defy Mendel's law of independent assortment of traits. In a series of experimental crosses, the genes governing the traits for pollen and flower color were shown to be partially linked. PPLL × ppll parents gave all F1 PpLl progeny, exhibiting the dominant phenotypes: purple flowers and long pollen. If the genes are completely linked, crossing the F1 generation (PpLl × PpLl) should yield 25% PPLL, 50% PpLl, and 25% ppll in the F2 generation. However, the observed result was that the F1 plants generated 6% Pl (purple, round) and 6% pL (red, long) gametes. The only explanation for these data is that although the genes for the two traits reside near one another on the same chromosome, some mechanism must allow homologous chromosomes to exchange information to create a new arrangement of genes.

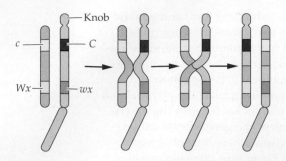

Figure 7.14
Physical evidence that chromosomes exchange material through crossing-over. Two of the genes on maize chromosome 9 are located close to cytological markers: *C* is located close to a knob, and *wx* is located close to a translocated piece of chromosome 8. A cross between a maize line with a normal chromosome 9 containing *c* and *Wx* and a line with an aberrant chromosome 9 containing *C* and *wx* results in a chromosome pair that is heterozygous both for alleles and for chromosomal morphology. When crossover occurs, an exchange of chromosomal segments results in rearrangement of both genetic and cytological markers. One recombinant chromosome has *C* associated with the knob structure and *Wx* with normal 9, while the other has *wx* associated with the translocated chromosome 8 and *c* from the normal chromosome 9.

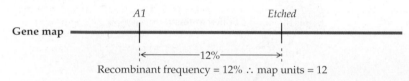

Figure 7.15
Creating a unit for genetic maps. Thomas H. Morgan defined a map unit as the linkage distance that gives 1% recombination. As shown here for maize (*Zea mays*), two genes for seed phenotype on chromosome 3 are said to be 12 map units apart because the recombination frequency between the gene *A1* (anthocyanin color) and the gene *Etched* (etched seed) is 12%.

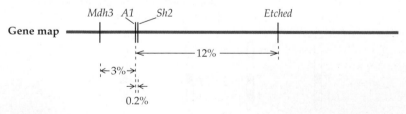

Figure 7.16
Creating a genetic map for multiple genes. Experiments with the seed-trait genes described in Figure 7.15 were used to quantify the distance between linked genes. Four linked characters—etched seed (*Etched*), purple seeds (*A1*), shrunken kernels (*Sh2*), and a gene for malate dehydrogenase (*Mdh3*)—were observed to deviate occasionally from complete linkage as a result of a recombination event. Recombination occurred between *Etched* and *A1* in 12% of meioses, between *A1* and *Mdh1* in 3% of meioses, and between *A1* and *Sh2* in only 0.2% of meioses. A genetic map showing the relative positions of each gene on the chromosome can be created by assigning a map unit for every percentage point of recombination frequency.

7.1.5 Genes can be mapped to relative positions within chromosomes.

As Morgan and colleagues suggested, the greater the distance between linked genes, the greater the frequency of recombination. By definition, genes are said to be linked whenever more than 50% of the gametes produced are of the parental combination. Other factors besides physical distance influence frequency of recombination, however, so we speak of *relative* units of mapping distance. The linkage distance that gives 1% recombination is defined as 1 **map unit.** Thus, a linked pair of genes for which 7% of progeny show nonparental gene combinations—a recombination frequency of 7%—are said to be separated by 7 map units (Fig. 7.15). A complication arises when two genes lie far enough apart to accommodate multiple crossing-over events. If an even number of crossovers occurs between two genes, the original genotype will be restored, and the number of recombination events may be underestimated when screening progeny for nonparental phenotypes.

Comparing the recombination frequencies between various pairs in a linkage group—a set of genes, each having been linked to at least one other member of the set—enables researchers to determine the order of the genes on the chromosome, a process called **gene mapping** or **linkage mapping** (Fig 7.16). An informative cross requires that the genes in question be **polymorphic**—that is, be represented by multiple **alleles,** distinct versions of the gene that vary in DNA sequence. In addition, each allele must produce a detectable phenotype in the parent. Linkage maps of chromosomes are generated by taking advantage of polymorphisms that occur naturally in populations and artificially inducing additional polymorphisms, for example, by using ionizing radiation to create mutations.

In addition to detecting gene differences through phenotypes, researchers can also detect **DNA polymorphisms** by using molecular probes. Highly polymorphic regions are usually located between genes, where small variations in DNA sequence have no effect on gene expression. In 1980, researchers created a DNA polymorphism linkage map in humans, utilizing **restriction endonucleases,** enzymes that cut DNA at specific nucleotide

(A)

Enzyme	Restriction Site
*Bam*HI	G|G A T C C C C T A G|G
*Eco*RI	G|A A T T C C T T A A|G
*Hind*III	A|A G C T T T T C G A|A
*Hpa*II	C|C G G G G C|C
*Mbo*I	G|A T C C T A|G
*Hae*III	G G|C C C C|G G
*Eco*RV	G A T|A T C C T A|T A G

(B)

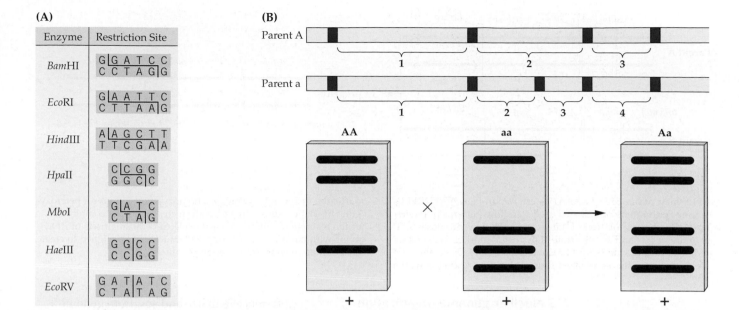

Figure 7.17

Restriction endonucleases cut DNA at specific sites. (A) Restriction enzymes recognize particular basepair sequences in DNA and cleave the molecule at a precise location. When DNA is cleaved with restriction enzymes, the resulting fragments typically have either cohesive (overlapping) ends or blunt ends. (B) DNA fragments cleaved by restriction enzymes can be resolved on the basis of their size through use of gel electrophoresis. The DNA fragments are loaded into wells in a gel matrix, after which an electric current is applied. DNA, which is negatively charged, will move toward the positive electrode. Smaller fragments move more quickly through the gel than larger ones, creating the appearance of a "ladder," with the larger fragments being closer to the negative electrode and the smaller ones closer to the positive electrode. Restriction sites in DNA are inherited just like any other trait, so restriction site polymorphisms (variability) among individuals (e.g., parents A and a) can also be used to follow patterns of inheritance. Variations in the distribution of restriction sites (restriction fragment length polymorphisms, RFLP) can be visualized by hybridization with a labeled DNA probe and used as phenotypic trait markers in a dihybrid cross the same way as seed or flower color.

sequences. A sequence-specific endonuclease cuts a particular DNA reproducibly into fragments of various lengths (Fig. 7.17). The lengths of the fragments depend on the location of the restriction site (the DNA sequence recognized by a specific endonuclease) in that individual's genome. Because the endonuclease recognizes specific sequences, the pattern of restriction fragment lengths will differ among DNA molecules that contain these sequences at different sites. Such polymorphisms could result from the loss of a restriction site (through alteration or deletion of nucleotides within the site) or from an insertion or deletion in the stretch of DNA between two restriction sites. Thus, **restriction fragment length polymorphism** (RFLP) analysis allows for genetic characterization of DNA at the molecular level (Fig. 7.18).

Many differences in the genome can be seen in phenotypes—such as the difference between a round and a wrinkled pea—but how does one *see* a RFLP? The technique of Southern blotting provides a means of visualizing the pattern of fragment lengths. DNA fragments that have been resolved electrophoretically on an agarose gel matrix and subsequently denatured are transferred to a membrane. The membrane is incubated with small, radioactively labeled fragments of known DNA sequence. These smaller fragments act as probes that hybridize to specific stretches of sequence found within the electrophoretically resolved molecules fixed to the membrane.

A probe that is useful for RFLP analysis and mapping will highlight a restriction fragment that often varies in length among different individuals, thereby generating clearly discernible patterns. Because RFLP markers are inherited just like genes that code for detectable phenotypes, the markers can be mapped relative to those genes. Using RFLPs for mapping often creates a "bridge" of reference points between genes that lie far apart.

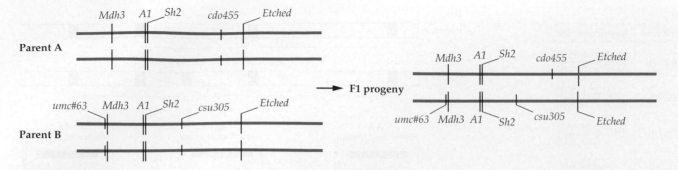

Figure 7.18
RFLP map of maize *Sh2* and *A1*. As shown in Figure 7.17, RFLPs among individuals can also be used to follow patterns of inheritance. Further annotation of the section of maize chromosome 3 containing *Sh2* and *A1* illustrates how RFLP markers (labeled in red, e.g., *cdo455*) can be used as phenotypic traits. Each allele has a characteristic set of sized fragments when restricted with a particular enzyme. Crossing two homozygotes will give a heterozygotic F1 that exhibits a hybrid pattern created by the combination of the restriction patterns of each allele. Several small sets of RFLP maps with patterns that overlap and match can be aligned to create a large map over an entire region of interest.

7.2 Nuclear genome organization

The previous section discussed how genes are arranged on chromosomes and how this arrangement facilitates recombination and genetic diversity. However, there is much more to the organization of the genome than aligning genes on a chromosome. Although we have been discussing genes as if they were attached to a chromosome at discrete sites, genes are part of a complex organization. We think of the role of the genome as one of providing gene products, but in many genomes only 1% or so of the DNA is transcribed and translated during normal cellular activities. Striking evidence that the actual coding capacity is likely to be relatively constant among plants is seen when comparing the genomes of *Arabidopsis* and maize. Sequence information obtained from cDNAs (produced by reverse transcription of mRNA) indicates that both genomes code for essentially the same number of genes, although the genome sizes differ by two orders of magnitude. Similarly, maize and sorghum are closely related plants that both have 10 chromosomes, but the maize genome is more than three times the size of that of sorghum. When DNA fragments from maize were used in hybridization analyses with sorghum sequences, homology was shared predominantly by low-copy-number sequences and unique sequences. In fact, several of the genes in sorghum show the same chromosomal arrangement as their counterparts in maize (see Section 7.2.5). From these and similar analyses, the "extra" DNA that accounts for the difference in maize and sorghum genome size apparently comprises mostly noncoding repetitive sequences between genes. This finding supports the conclusion that the majority of nuclear DNA may play a supporting role in the structure and organization of the genome but does not contribute directly to its protein-coding capacity.

7.2.1 Genome size is highly variable among flowering plants.

The size of the nuclear genome varies among organisms. The DNA content of haploid eukaryotic cells (**C value**) ranges from 10^7 to 10^{11} basepairs (bp). The human genome lies in the middle of this range, at 3×10^9 bp. Although it has been assumed that organism complexity correlates roughly with genome size—humans have larger genomes than most insects, and insects have larger genomes than fungi—this correlation is by no means universal. For example, some amphibians have genomes almost 50 times larger than that of humans, and cartilaginous fish generally have larger genomes than bony fish.

Interestingly, genomes of plants are represented throughout the size range (Fig. 7.19). The smallest known plant genome of 7×10^7 bp belongs to *Arabidopsis thaliana* and one of the largest is a member of the lily family, *Fritillaria assyriaca,* with 1×10^{11} bp. Rice, maize, and wheat fall in between those two, having genome sizes of 5×10^8, 6.6×10^9, and 1.6×10^{10} bp, respectively. The lack of a direct relationship between genome size and

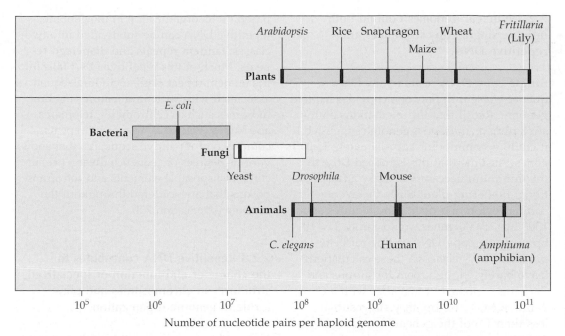

Figure 7.19
C values (haploid genome size in base-pairs) from various organisms. Most eukaryotes have haploid genome sizes between 10^7 and 10^{11} bp of DNA.

Number of nucleotide pairs per haploid genome

organism complexity is called the **C-value paradox.** We have no satisfactory explanation yet for the C-value paradox, but in plants, at least, we know that genome size can to some degree be attributed to repetitive DNA and duplicated genomes (polyploidy).

Given that *Arabidopsis* and *Fritillaria* are similar in terms of the basic functions that characterize a plant (Fig. 7.20), it follows that *Arabidopsis* probably has about the same number of genes as most plants. The average gene takes up approximately 1300 DNA bases (1.3 kb). However, those bases may not be contiguous; most genes also contain flanking sequences and intervening sequences that do not code for the gene product. The typical gene may actually cover approximately 4 kb. On that basis, there is just enough capacity in the *Arabidopsis* genome to accommodate 15,000 genes; among plants, therefore, *Arabidopsis* may be an extreme example of streamlining in genome organization. This characteristic makes *Arabidopsis* an excellent model system for plant genetics—similar to the animal model systems *D. melanogaster* or *Caenorhabditis elegans*—and provides counterpoint to genetic analyses with more complex plant genomes.

The intervening and flanking noncoding sequences that contribute to the diversity in genome size are integral parts of a typical protein-encoding gene. The intervening sections of noncoding DNA are called **introns.** Introns subdivide the coding sequence into

sections referred to as **exons.** The noncoding sequence that flanks the 5′ end of the gene usually contains DNA sequence elements that function in the regulation of gene transcription. The 3′ flanking region contains sequence elements for modifying mRNA and a transcription stop site, and it may contain additional regulatory elements (Fig. 7.21). Determining where the 3′ flanking region of one gene "ends" and the 5′ flanking region of the next gene "begins" is not always straightforward. Much of the "extra" DNA that expands the genomes of plants to 10^{10} bp lies within this region, and we are still learning about the functional role of noncoding DNA sequences.

(A)   **(B)**

Figure 7.20
The two extremes in genome size in plants: (A) *Arabidopsis* and (B) *Fritillaria. Fritillaria* is a genus in the lily family. Like many members of the lily family, *Fritillaria* species have very large genomes (10^{11} bp). *Arabidopsis,* a member of the mustard family, has a genome size of 10^7 bp. Although the genome of *Arabidopsis* is 10,000 times smaller than that of *Fritillaria,* both plants have roughly the same requirements for living and reproducing in a temperate habitat.

7.2.2 Nuclear genomes contain both unique, single-copy sequences and repetitive DNA.

Single-copy DNA refers to any DNA sequence present in only one copy per haploid genome. Recall that the calculations on the streamlined *Arabidopsis* genome suggested it could accommodate approximately 15,000 genes. Assume that this is indeed close to the minimum necessary to make a fully functional flowering plant. Since the average coding region of a plant gene requires about 1300 bp, 15,000 genes would imply 2×10^7 bp of single-copy DNA coding for gene products. In *Arabidopsis*, these calculations leave barely enough room for introns and flanking regions, but in maize, with a C value of 6.6×10^9 bp, they suggest that less than 1% of the genome is coding DNA. This rough approximation is actually very close to the number of basepairs that are transcribed to mRNA in a similarly sized species. In tobacco (*Nicotiana*, C = 1.7×10^9), only 2% of the genome is transcribed into mRNA. Biochemical analyses, however, indicate that as much as 40% of the tobacco genome consists of single-copy DNA. Apparently, therefore, the genome contains many single-copy sequences that are not transcribed.

Repetitive DNA consists of groups or families of similar but not necessarily identical repeated sequences. The sequence and length of repeats vary among repetitive families within a species, and a species may have from 1000 to 40,000 repetitive families.

Based on its organization in the genome, repetitive DNA can be subdivided into two classes: **tandem repeats** and **dispersed repeats.** Much of this repetitive DNA falls into the tandem repeat class, including sequences associated with structural features of chromosomes such as centromeres, telomeres, and knobs (Fig. 7.22). Dispersed repeats, composed of repetitive sequences that show various degrees of sequence divergence, include transposable elements and retrotransposons that are scattered throughout the genome (see Section 7.3).

7.2.3 Repetitive DNA contributes to the character and function of specialized structures in chromosomes and plays a role in genome organization.

Noncoding, tandem repetitive DNA is referred to as **satellite DNA.** First identified on cesium chloride density gradients, these bands had separated from the majority of the DNA because of their unique basepair composition; accordingly, they were thought of as a "satellite" of the bulk of the genomic DNA. In most animals and yeast, satellite DNA is AT-rich, but examples from plants tend to be GC-rich. Satellite DNA is primarily associated with either the centromere or the telomeres in plants and is usually heterochromatic; that is, it remains condensed and inactive.

The **centromere** is a region of the chromosome to which the spindle fibers attach for the separation of the replicated chromatids

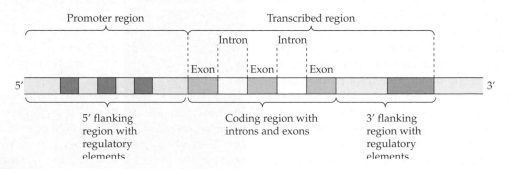

Figure 7.21

The structure of a gene. Eukaryotic genes are organized into a regulatory region (promoter) and a region that contains the protein-coding sequence (transcribed region). Within both these regions are discrete sequence motifs that confer regulatory information. The coding region includes intervening sequences (introns) that do not contribute to the final message to be translated into a protein. Sequences within the coding region that do encode for the protein are called exons.

(A)   **(B)**   **(C)**

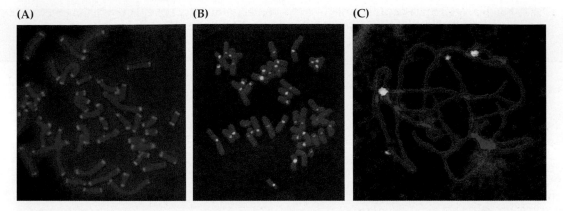

Figure 7.22
Localization of repetitive sequence motifs in chromosomes. In situ hybridization localizes repetitive sequences to discrete regions of the chromosomes. Fluorescence in situ hybridization (FISH) with various repetitive sequence motifs shows the localizations of those motifs within chromosomes. The micrographs of the chromosomes are shown with the sequence motifs used for the FISH analysis: (A) telomeres, (B) centromeres, (C) knobs.

in mitosis and meiosis. The centromere contains certain repetitive elements in short tandem sequences that can be repeated millions of times. These sequences appear to be essential for the recognition of the centromere by the spindle apparatus, because mutations in this region disrupt centromere function in yeast. Although the organization and arrangement of repetitive DNA in centromere regions is highly conserved (i.e., contained within the same chromosome structures) among organisms, the actual sequence composition of these regions is extremely variable, even among related species.

The **telomere** is the structure that defines the end of a chromosome (see Chapter 6). This specialized "chromosomal cap" offsets the tendency for DNA to shorten with each round of replication. Telomeres also contain repetitive elements, but unlike those of centromeres, these sequences are highly conserved among eukaryotes in both sequence and arrangement (see Fig. 6.9). For example, telomeres in humans and trypanosomes (a group of flagellated protozoa) have the same sequence, TTAGGG, repeated thousands of times. The repeated telomere sequence of *Arabidopsis* differs from this by only one base: TTTAGGG. As with centromeres, telomeres play a crucial role in the replication of the genome. At the end of the chromosome, after the several thousand copies of the telomere sequence, lies a section of single-stranded DNA composed of only two

or three copies of the sequence. In eukaryotes, a specialized enzyme called **telomerase** maintains the single-stranded overhang, to keep it from being shortened with each round of replication (see Fig. 6.10). Telomerase is a reverse transcriptase with a segment of RNA integrated into its three-dimensional structure. The RNA functions as the template for the new telomere sequence. Telomeres also seem to function in maintaining a "nonsticky" end on the chromosome. By contrast, according to cytogenetic analyses, the ends of broken chromosomes are very sticky, readily joining with any other available DNA fragment. Telomeres also appear to play a role in the organization of chromosomes in the nucleus, attaching the chromosome to the fibrous nuclear lamina on the inner surface of the nuclear envelope (see Chapter 5).

Dispersed repeat sequences make up a significant portion of the genome. These sequences differ from the tandem repeats described above in that copies are dispersed throughout the genome rather than adjacent to one another. Many dispersed repeat sequences are related to transposable elements (see Section 7.3). In some plants, transposable elements make up the most abundant class of dispersed repetitive sequences. For example, more than a dozen dispersed repeat sequence elements derived from inactive transposon variants occupy the regions surrounding the *Adh1* gene of maize (Fig. 7.23).

Box 7.2

Once a difficult procedure, cloning of genes or
DNA fragments is now routine.

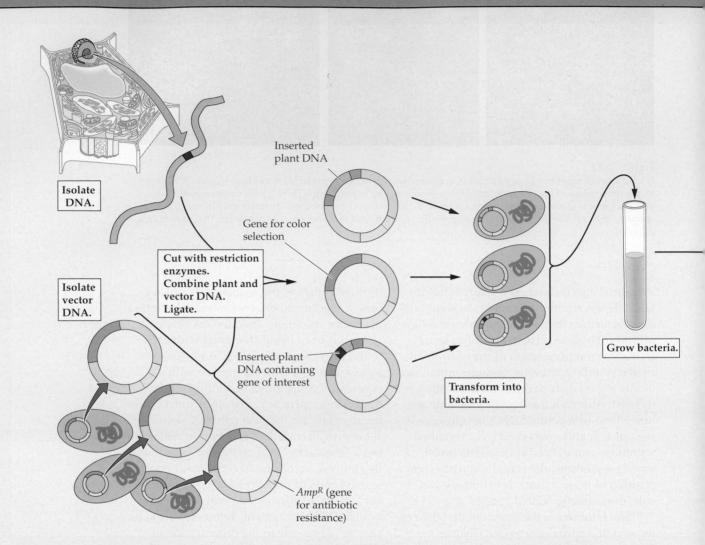

Isolate DNA.

Isolate vector DNA.

Cut with restriction enzymes. Combine plant and vector DNA. Ligate.

Inserted plant DNA

Gene for color selection

Inserted plant DNA containing gene of interest

Amp^R (gene for antibiotic resistance)

Transform into bacteria.

Grow bacteria.

Technically, the word cloning simply means to make multiple identical copies of something. In the context of recombinant DNA technology, however, cloning is a multistep process as illustrated in the figure above:

1. Create a recombinant molecule in a bacterial plasmid DNA vector.

DNA is isolated from an organism, either directly from the genome or by using reverse transcriptase to transcribe complementary DNA (cDNA) from a messenger RNA (mRNA) template.

The DNA is then cleaved with a restriction endonuclease. The same endonuclease is also used to cut a population of bacterial **plasmids**—typically small, circular molecules of DNA that have been modified to encode genes for selectable markers, such as an enzyme that inactivates an antibiotic chemical and conveys antibiotic resistance to bacteria harboring the plasmid. The endonuclease digestions create complementary cohesive ends that allow the sample DNA and the plasmids to hybridize and be ligated (connected) together with the help of the enzyme T4 ligase.

2. Transform a host with the recombinant plasmid.

Host bacteria (usually specialized strains of *Escherichia coli*) are first chemically treated with a solution (such as $CaCl_2$) to permeabilize the bacterial wall and membrane and make them more receptive to taking up foreign DNA. The permeabilized bacteria are then incubated with a dilute solution of the recombinant plasmid and plated out to grow on selective agar medium.

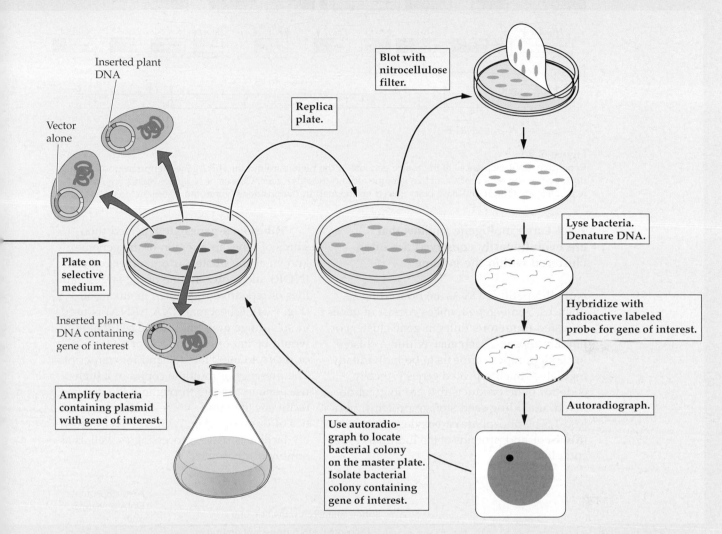

Inserted plant DNA

Vector alone

Plate on selective medium.

Inserted plant DNA containing gene of interest

Amplify bacteria containing plasmid with gene of interest.

Replica plate.

Blot with nitrocellulose filter.

Lyse bacteria. Denature DNA.

Hybridize with radioactive labeled probe for gene of interest.

Autoradiograph.

Use autoradiograph to locate bacterial colony on the master plate. Isolate bacterial colony containing gene of interest.

3. Screen the transformed bacteria.

As mentioned above, for commercial plasmid vectors that come engineered with genes conferring antibiotic resistance, only transformed bacteria containing the plasmid can grow on medium containing the antibiotic. In addition, many plasmids are engineered with genes that create colored colonies. When DNA is inserted into the color-producing genes, color is no longer produced. Locating colorless colonies therefore allows selection for plasmids carrying cloned DNA.

4. Select transformed colonies for amplification.

The colonies that grew on the antibiotic medium are duplicated in the same spatial arrangement by picking up a replica of the colonies on a piece of sterilized material (e.g., velvet) and then applying this to another plate of medium. After growth of the colonies on the replica plate, the plate surface is transferred to a piece of nitrocellulose membrane. The bacteria stuck to the membrane are lysed in place, and the DNA contained in them is chemically denatured and fixed to the membrane. The nylon membrane is hybridized with a labeled DNA probe homologous to the gene of interest—either a section of the gene itself or a fragment of the gene from another organism. If the probe hybridizes to DNA from a colony, the autoradiograph can be aligned with the original plate and the colony containing the DNA of interest can be selected and grown.

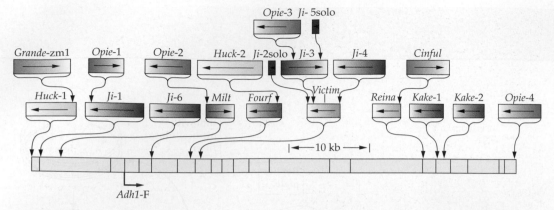

Figure 7.23

Repetitive DNA sequences of the maize genome in the region around the *Adh1* gene. Retrotransposons make up much of the region of the maize genome around the *Adh1* gene. Classes of sequence motifs are named [e.g., *Opie* for (blue)] and have been placed on the map in the positions where the insertions occur.

7.2.4 Large multigene families that are evolutionarily conserved are often clustered within the genome.

Not all repetitive DNAs are noncoding sequences. Some gene families consist of genes repeated numerous times in gene clusters or tandem repeats. Each gene within a cluster or tandem repeat appears to be individually regulated. Such repeated genes typically code for gene products that are in great demand, including seed storage proteins, ribulose-1,5-bisphosphate carboxylase/oxygenase (Rubisco), and proteins of the light-harvesting complex.

Ribosomal genes are repeated thousands of times in a region of the genome known as the **nucleolar organizer region (NOR)** and represent one of the largest families of repetitive sequences in eukaryotes (Fig. 7.24). Ribosomal RNA (rRNA) is used in such huge quantities in the cell that it would be impossible for one or a few copies of rDNA to meet the demand for transcripts. The presence of multiple copies of a high-use gene in clusters throughout the genome facilitates the rapid synthesis of large quantities of transcript (Fig. 7.25). The NOR probably facilitates rRNA processing as well as assembly of ribosomes.

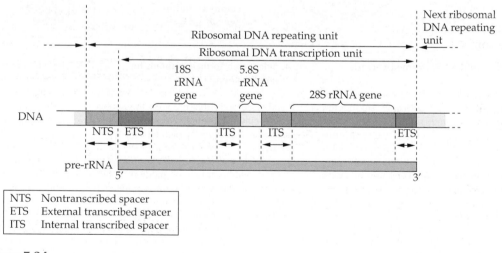

Figure 7.24

The eukaryotic ribosomal DNA repeating unit. In plants, most ribosomal genes lie within a repeating unit. Ranging from 7800 to 185,000 bp long, the repeating unit is composed of the highly conserved rRNA genes separated by short stretches of spacer sequences (NTS, ETS, and ITS). There are four rRNA genes, of sizes 18S, 5.8S, 28S, and 5S. The first three are contained on a rDNA repeating unit, whereas the 5S gene is found elsewhere in the genome. Each gene codes for a different rRNA molecule that contributes to the structure of the ribosome.

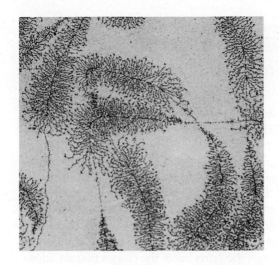

Figure 7.25
Electron micrograph of rRNA being transcribed. The rDNA genes are transcribed at a very high rate, and the resulting rRNA transcripts are readily visible with electron microscopy (if the rate were slow, fixing an image in which transcription was proceeding would be difficult). This photo shows multiple RNA polymerase molecules transcribing ribosomal RNA along the same stretch of DNA that contains the rDNA gene units.

Histone proteins, which make up a major component of the chromatin proteins, are also needed in great abundance within the cell. The five histone genes are arranged in a cluster, and the entire cluster is repeated 10 to 600 times. The coding regions of the histone genes are highly conserved, even among diverse species, but their organization within the cluster varies (Fig. 7.26).

Some genes occur in families containing 20 to 25 repeats. Although these genes may be clustered or linked, they are not tandemly repeated. One example is the maize zein family, which encodes seed storage proteins. The family contains two subsets, each with approximately 25 members. The coding region homology among members of each group is very high (greater than 90% identity in derived amino acid sequence), but the homology between the two groups is lower (Fig. 7.27).

7.2.5 Related plant species display evolutionarily conserved organization and arrangement of single-copy genes.

Genome mapping projects have revealed that large segments of chromosomes are often conserved among related species (Fig. 7.28). As mentioned earlier, maize and sorghum contain many of the same genes and linkage groups residing at the same loci. This colinearity of loci is called **synteny.** Species in the Solanaceae family also contain conserved sections of genomes. The genomes of tomato and potato are very similar, even to the degree that the organiza-

tion of many linkage groups is conserved. The grasses barley and wheat also show a great deal of similarity in the organization of genes within their genomes.

7.2.6 Genes can be mapped to specific physical locations on chromosomes.

Whereas a genetic map defines a chromosome in terms of recombination among loci to give a relative genetic distance between them, a **physical map** defines specific distances in terms of the number of bases between genes. High-resolution physical mapping uses cloned sections of DNA to create maps of contiguous, overlapping fragments of DNA, which can span very long distances. The region defined by such a map is called a **contig,** since it is constructed from contiguous clones. In this way, a specific, physical distance in nucleotide basepairs can be assigned between reference points.

The biggest challenge of creating a physical map is assembling all the pieces correctly. It is easy to state that one simply creates a contiguous map of overlapping fragments, but the only way to do this is if

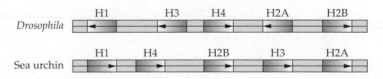

Figure 7.26
Organization of histone genes in the animal genome. The histone genes are another set of highly conserved genes that are required in great abundance. They, too, are clustered into transcription units in the genome.

Chromosome 7

Short arm   Centromere   Long arm

Zp6-h Zp22 Zp28 Zp30 Zp27 FI2   Zp14   Su1 Zp12   GI4   Zp15   Zp10

Figure 7.27
Maize zein genes code for abundant, functionally related storage proteins in seeds. The zein genes are clustered on chromosome 7 in maize, but are not arranged as tandem repeats or transcription units.

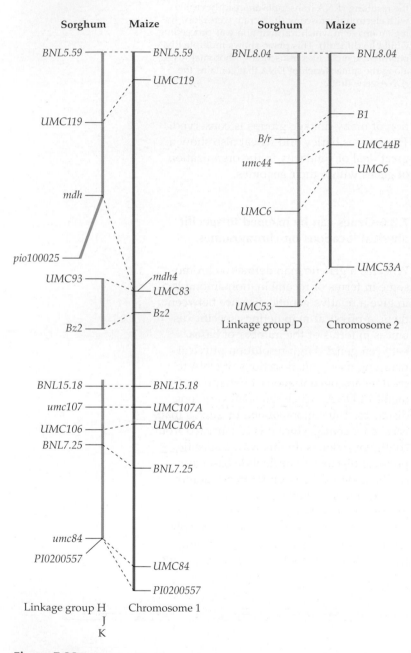

Figure 7.28
Many of the genes common to maize and sorghum are organized into similar positions in the genomes of these two plants. A comparison of genetic maps of chromosomes 1 and 2 from maize with the maps of several linkage groups from sorghum show the two grass genomes similarly organized with respect to the relative positions of the genes.

unique landmarks are available to determine *where* the overlaps should be aligned. One strategy that has been used in the *Arabidopsis* genome project is to generate sets of overlapping clones 10 to 25 kb long that have been "fingerprinted" by the restriction endonuclease *Hind*III. *Hind*III cleaves the clones into a distinctive pattern of fragments. Computer analyses of the patterns enable researchers to align the portions of the clones that are identical, leading to a map of contiguous, overlapping fragments (Fig. 7.29).

Another mapping strategy uses a eukaryotic vector developed from yeast, the **yeast artificial chromosome (YAC).** YACs contain the structures necessary for replication in yeast cells—at least one origin of replication, telomeres, a centromere, and a selectable marker (Fig. 7.30). YACs can accommodate as much as 1 million bp (1000 kb) of inserted DNA, but YACs of 100–300 kb are more common. Thus a physical map of overlapping YAC clones can span an entire chromosome, integrating a linkage map with a physical map of a chromosome (Figs. 7.31 and 7.32). More recently, similar techniques for cloning large DNA fragments have been developed, using **bacterial artificial chromosomes (BACs).** These techniques are becoming increasingly popular in large-scale DNA mapping and sequencing projects.

7.3 Transposable elements

7.3.1 Transposable elements are mobile DNA sequences that can make up a significant portion of the nuclear genome.

Transposable elements are sections of DNA (sequence elements) that move, or **transpose,** from one site in the genome to another. These mobile DNA elements carry genetic information with them as they transpose, making them important features of genome organization. Transposable elements from organisms as diverse as *Drosophila*, yeast, and maize show a substantial conservation in organization and the mode of transposition.

Once a gene is identified, cloned, and sequenced, most research journals require that the sequence be made available to the public by way of an electronic sequence database like GenBank. Collections of the sequences of many of the genes that are expressed as mRNAs (called ESTs, for Expressed Sequence Tags) are available on-line for several plant species, and additional ESTs are being placed in databases daily. *Arabidopsis* is the most sequenced of all plants, with rice and maize at second and third, respectively.

There are two major reasons to access these sequence databases. First, an investigator who has cloned and sequenced a gene can compare its sequence with that of numerous related genes from other organisms. For example, researchers interested in phylogeny and molecular evolution access these data to create evolutionary models.

Similarly, having identified a sequence motif that looks interesting from a biochemical point of view (e.g., a potential regulatory element), one can scan the database for similar sequences to see if the motif is used in another gene in a similar capacity.

The second major use is to "clone by phone." Researchers can find out who cloned a gene of particular interest and contact the investigator, who is usually glad to share the clone with others. (Increasingly, moreover, clearinghouses such as the *Arabidopsis* Biological Resource Center maintain and freely distribute clones and ESTs.) The clones can then be either used directly in experiments, or included in a heterologous hybridization assay to help identify the comparable gene in another research organism.

Database information for *Arabidopsis* is available at www.tigr.org (see figure

above). The TIGR database contains information pertaining to the genome sequencing project and EST assembly databases and can lead a researcher to other *Arabidopsis* resources.

Chromosome

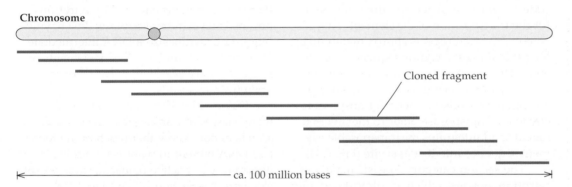

Cloned fragment

ca. 100 million bases

Figure 7.29
A contig map consists of a set of contiguous overlapping fragments of cloned DNA that can span an entire chromosome, creating a physical map.

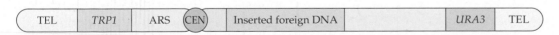

| TEL | *TRP1* | ARS | CEN | Inserted foreign DNA | | *URA3* | TEL |

Figure 7.30
Structure of a Yeast Artificial Chromosome (YAC). A YAC contains sequences from a bacterial plasmid, which facilitate the insertion of foreign DNA sequence, as well as yeast chromosomal sequences that confer the ability of the artificial molecule to replicate in a yeast host cell. The CEN motif represents centromeric sequences, the ARS (autonomous replicating sequence) encodes an origin of replication (see Chapter 6), and the TEL sequences function as telomeres. Metabolic yeast genes (e.g., *TRP1, URA3*) also are included in a YAC to facilitate screening on selective media. The sequences contained in the artificial chromosome are sufficient for replication by the yeast cell as though the chromosome were natural. As much as 1 megabase (1 Mb; 10^6 bp) of DNA can be inserted into a YAC.

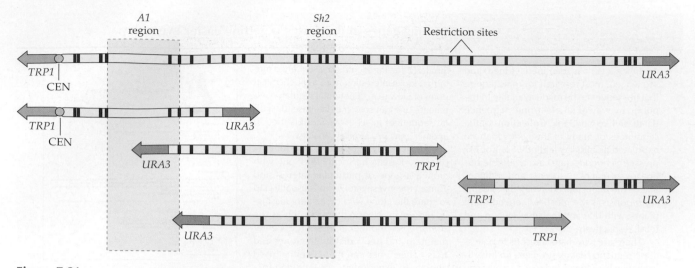

Figure 7.31
YAC clones of the *A1–Sh2* region of the maize genome. The restriction maps of four YAC clones containing portions of two adjacent maize genes, *A1* and *Sh2*, are shown with their map position aligned. Overlapping YAC clones can be used to construct physical maps for whole genes by aligning the restriction sites among the clones. The segments labeled *TRP1* and *URA3* are selectable markers that form part of the YAC vector.

There are two basic types of transposable elements. The first type of transposable elements was described by Barbara McClintock in the 1940s. The elements in this category code for one or two gene products necessary for the transposition of the element. They also contain inverted repeats approximately 10 bp long that flank the coding sequence at both ends. The inverted repeats on either end of the element are recognized by **transposase,** an enzyme encoded by certain transposable elements. Transposase binds to the inverted repeats and integrates the transposable element sequence into its target site (Fig. 7.33).

The second category consists mainly of **retrotransposons,** which are almost certainly viral in origin. They resemble the structure left by the integrated form of RNA tumor viruses and accomplish transposition by way of an RNA intermediate (Fig. 7.34). One of the proteins encoded by retrotransposons is reverse transcriptase, the enzyme required to synthesize a DNA intermediate from an RNA template. The *Ty* elements of yeast and the *copia* elements of *Drosophila* are retrotransposons.

Having identified and characterized many classes of transposable elements, researchers now know that much of the repetitive DNA present in plant genomes is actually composed of active or inactive transposable elements (see Section 7.2 and Fig. 7.23).

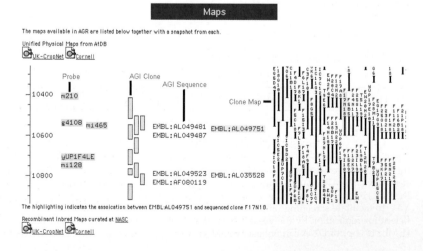

Figure 7.32
Part of the physical map of *Arabidopsis* chromosome 2. The various YAC clones have been placed in order on the chromosome, based on their hybridization to various probes (see//weeds.mgh.harvard.edu/goodman/c2_a.html, an example of which is shown here). The status of the *Arabidopsis* physical map can be viewed at //humgen.upenn.edu/~atgc/physical-mapping/physmaps.html.

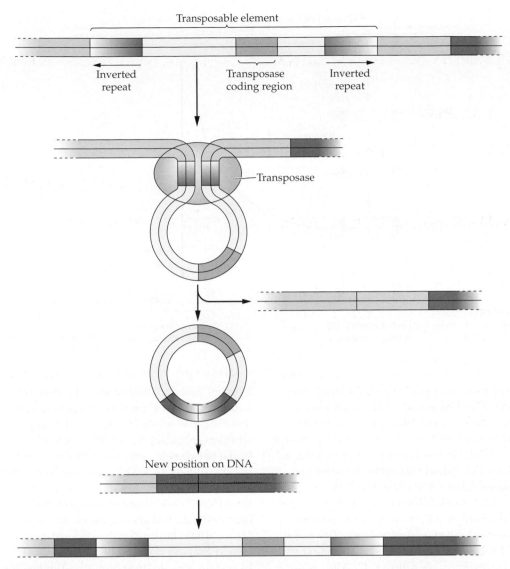

Figure 7.33
Structure and transposition of a transposable element. A common feature of transposable elements is the flanking of the element by short inverted repeat sequences. A model for transposition suggests that the enzyme transposase recognizes these sequences, creates a stem/loop structure, and then excises the loop from that region of the genome. The excised loop can then be inserted into another region of the genome.

7.3.2 The *Ac/Ds* system is the first example of transposable elements described for eukaryotes.

Barbara McClintock was the first person to envision the idea of transposable elements (Box 7.4). As mentioned in Section 7.1.4, McClintock had assigned a series of cytological and genetic markers to the short arm of chromosome 9 in maize (see Fig. 7.14). During her experiments with chromosome 9, she devised a system to initiate a break in the short arm during meiosis. Broken ends of chromosomes are "sticky" (cohesive), tending to fuse with any other broken end. As a result, during synapsis and crossing-over in these experiments, two abnormal chromosomes were generated: one with two centromeres, the other with none. The acentric chromosome is lost during anaphase since it does not have a centromere to attach the spindle apparatus. The bridged dicentric chromosome, meanwhile, is pulled apart as the spindle apparatus attempts to move each

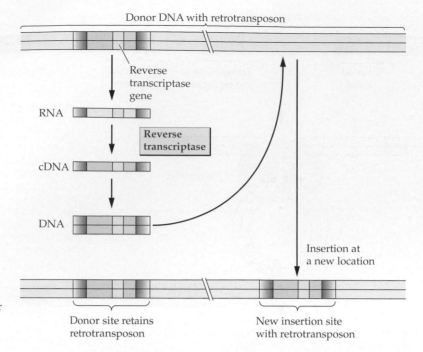

Figure 7.34
Retrotransposon transposition, an alternative means of transferring a section of DNA from one point to another. In this case, the element to be transposed is copied in place through creation of an RNA intermediate, after which the duplicate copy is inserted into a new location.

Donor DNA with retrotransposon

Reverse transcriptase gene

RNA

Reverse transcriptase

cDNA

DNA

Insertion at a new location

Donor site retains retrotransposon

New insertion site with retrotransposon

centromere to opposite poles during anaphase. This break occurs at a position between the two centromeres, but sister chromatids of the broken chromosome fuse again into a bridged chromosome after prophase. McClintock called this cycle the Breakage–Fusion–Bridge (BFB) cycle (Fig 7.35).

McClintock initiated these experiments to generate deletions in certain genetic markers along the chromosome. The genes for three seed traits were located next to the centromere as follows: *C*, color; *sh*, shrunken endosperm; *wx*, waxy endosperm; centromere. McClintock expected (and usually observed) deletions that abolished the activity of the marker farthest from the cen-

tromere first. In other words, the progeny kernels had lost *C* and were typically colorless. She was working with a particular strain of maize, however, that typically broke between *wx* and the centromere during the BFB cycle, resulting in the loss of all three marker traits. Reasoning that there was something weak about this region, McClintock assigned it a genetic marker, calling it **Ds** for **dissociation**. Further observation revealed that the position of *Ds* was not stable. In certain genetic lines of maize, the position of *Ds* seemed to "jump" from a site between *wx* and the centromere to a position within the *C* gene, disrupting the function of this gene and rendering kernels colorless. In

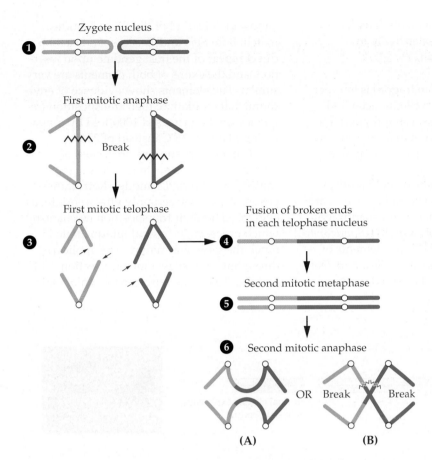

Zygote nucleus

1

First mitotic anaphase

2 Break

First mitotic telophase

3

Fusion of broken ends
in each telophase nucleus

4

Second mitotic metaphase

5

6 Second mitotic anaphase

OR Break   Break

(A)   **(B)**

Figure 7.35
Breakage–Fusion–Bridge model in a pair of chromosomes. If each gamete contributes a chromosome
that has been broken in the anaphase of the division
preceding gamete formation, the zygote nucleus (1)
will then contain two such chromosomes, each composed of two sister chromatids fused at the position
of the previous anaphase break. In the first anaphase
of the zygotic division (2), these two chromosomes
form bridge configurations as the centromeres of the
sister chromatids move to opposite poles. Breaks
occur in each bridge at some position between the
centromeres. The chromosome complement of a
telophase nucleus (3) includes two chromosomes
(one yellow, one blue), each with a newly broken
end (see red arrows). Fusion of broken ends of the
chromosomes occurs in each telophase nucleus, establishing a dicentric chromosome (4). Following
metaphase (5), several types of configurations may
result from separation of the sister centromeres, two
of which are shown (6A and B). In the anaphase
bridge configuration (6B), breaks occur in each
bridge at some position between the centromeres.
The subsequent behavior of the broken ends, from
telophase to telophase, follows the six steps diagrammed here.

rare cases, *Ds* jumped back out of *C* during
development of the seed. These mutable
kernels were largely colorless but exhibited
small sectors of color where *Ds* has apparently left the *C* locus, restoring gene activity.
Because only some plants were unstable
with regard to *Ds* position, McClintock proposed that the nonstable lines also contained
some type of **"activator"** *(Ac)* and that both
Ds and *Ac* were required for *Ds* transposition (Fig. 7.36).

McClintock's genetic analyses and subsequent molecular studies have shown that
the *Ac/Ds* transposable element system is a
complex set of structures that consists of two
types of elements: the **autonomous element**
(Ac) and the **nonautonomous element** *(Ds)*.
Activator *(Ac)* element is "autonomous" because it encodes all products required for its
transposition. *Ac* is 4.6 kb long. The central
region of the element codes for a 3.5-kb transcript of the transposase that excises both
Ac and all members of the *Ds* family. Inverted repeats of 11 bp on each end of the element function in transposase recognition. In

addition, when the element inserts into a
new position in the genome, a pair of 8-bp
direct repeats generated from the host genome is added to flank the element. These
direct repeats, which remain fixed in the
genome even after the element transposes
out again, are referred to as a **"footprint"** of
the transposition event.

The dissociation *(Ds)* element is actually
a family of related elements recognized by the
transposase encoded by *Ac*. The *Ds* elements
are referred to as nonautonomous because
they do not encode their own transposase.
Rather, *Ds* transposition depends on the presence of *Ac* and its gene products. The *Ds* elements fall into three structurally diverse classes: *Ds*, *Ds2*, and *Ds1*. *Ds* elements are simple
deletion derivatives of *Ac*. *Ds2* elements contain very few *Ac* sequences in general, and
their internal sequences are completely unrelated to *Ac*. *Ds1* elements are homologous to
Ac only in their terminal repeats, which contain the sequences recognized by the *Ac* transposase. So far, about 20 members of the *Ds*
family of elements have been characterized.

7.3.3 Transposable elements have been characterized extensively in maize and snapdragon.

Antirrhinum majus (snapdragon) is another plant species that has well-characterized transposable elements. In snapdragon, the *Tam3* elements are very similar in structure and function to the *Ac* element of maize, and they code for a very similar transposase (approximately 60% amino acid identity). Other sets of transposable elements that are very similar in both maize and snapdragon include the Suppressor/mutator *(Spm/dSpm)* family in maize and the *Tam1* elements of snapdragon. A comparison of *Spm* and *Tam1* shows that the terminal inverted repeats are almost identical: 12 of the 13 nucleotides match. Both *Spm* and *Tam1* generate a 3-bp direct repeat of the host genome upon insertion, and the exons of both elements are very similar. *Tam* elements show evidence of environmental regulation: their transposition frequency can be increased 1000-fold by propagating plants at 15°C instead of 25°C.

The retrotransposons are a class of transposable elements that differs significantly from the *Ac/Ds* model. Retrotransposons are almost certainly viral in origin, as evidenced by their transposition mechanism, which proceeds via a viral intermediate in much the same way a retrovirus replicates throughout a host genome (see Section 7.3.1 and Fig. 7.34). One retrotransposon of maize,

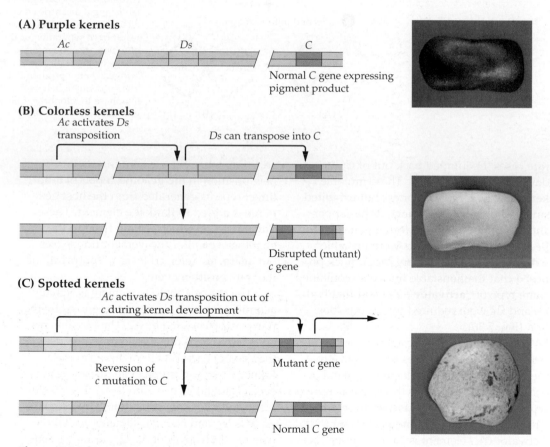

(A) Purple kernels

Ac *Ds* *C*

Normal *C* gene expressing pigment product

(B) Colorless kernels

Ac activates *Ds* transposition

Ds can transpose into *C*

Disrupted (mutant) *c* gene

(C) Spotted kernels

Ac activates *Ds* transposition out of *c* during kernel development

Reversion of *c* mutation to *C*

Mutant *c* gene

Normal *C* gene

Figure 7.36
Variegated corn and effects of transposition of the color patterns in corn kernels. Barbara McClintock was the first scientist to recognize the possibility of transposable elements. She proposed the theory that genes, or sections of genes, could "jump" from one position in the genome to another, based on her observations of the genetics of certain maize kernel phenotypes. The following example illustrates the progression of events: (A) The *C* gene encodes an enzyme that catalyzes the formation of maize pigments, so when *C* is active, kernels are purple. (B) *Ac* encodes a transposase that mediates transposition of the *Ds* element. *C* is inactivated when the transposable element *Ds* is inserted into it, resulting in a colorless mutant phenotype *(c)*. (C) If *Ds* is transposed ("jumps") out of the *c* gene during the course of development, *c* is restored to activity *(C)*, and purple sectors are seen within a colorless background.

the mutator *(Mu)* element, transposes so frequently that a genome containing this element will produce mutated progeny at a rate 50 times higher than an identical genome without it. At least eight *Mu* elements have been identified, which vary in the composition of their internal sequences; upon insertion, however, all *Mu* elements generate a 9-bp repeat from the host genome. First characterized in the late 1970s, *Mu* has been used to create and characterize numerous mutants in maize.

7.3.4 Transposable elements can function in diverse species.

Transposable elements from one species can be active in another. The maize transposable element *Ac* was first introduced into tobacco. Transposition was demonstrated by the presence of *Ac* in new positions in the genome and by the existence of the characteristic sets of direct repeats that *Ac* leaves as a footprint upon insertion and excision in the genome. *Ac* has been introduced into many plant species, including *Arabidopsis* (Fig. 7.37). Apparently *Ac*—and perhaps many other transposons—can function independently of the host genetic background.

The ability to follow a transposition, even in a heterologous species, has given rise to the technology of **transposon tagging,** in which a phenotypic trait is screened for a putative insertion event and hybridization techniques are used to identify the mutated gene containing the transposon. Once this has been done, sequences from the mutated gene can be used to generate additional probes and characterize additional sections of the gene.

7.3.5 The impact of transposons on genome organization is complex.

The remnants of transposition events and the similarity of transposons in diverse species suggest that these mobile genetic elements have been associated with the eukaryotic genome for a long time. Data indicate that transpositions are influenced by developmental and perhaps environmental signals and may play a role in the temporal and spatial patterns of gene expression. The possibility that they exist as simply extraneous sequences is unlikely. Instead, they may act as a complement of the genome, increasing its diversity and adaptability. Perhaps genomes with active transposons are better able to cope with evolutionary pressures and therefore have an adaptive advantage. Another possibility is that transposons represent "selfish DNA" and have no other function than to replicate.

7.4 Gene expression

7.4.1 Cell differentiation is a function of regulated gene expression and does not involve loss of genetic material.

Early on, researchers thought that cells differentiated by retaining only those chromosomes or parts of chromosomes needed for the particular cell type they were destined to become. In the 1950s, Carlos Miller, Folke Skoog, and their colleagues refuted this idea by performing a key experiment in plant science. They showed that isolated root cells from a mature plant could be induced to regenerate into a complete plant that was identical to the root cell donor. This result was possible only if each fully differentiated root cell were **totipotent,** that is, retained the full complement of the plant's genes and the potential to regenerate into a fully differentiated plant. Similarly, the cytoplasm of an

(A) **(B)**

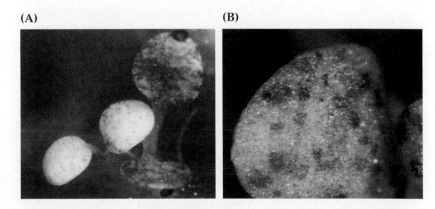

Figure 7.37
The maize *Ac* element also functions in *Arabidopsis*. Explants were transformed with a streptomycin-resistant gene interrupted by an *Ac* element and cultured to regenerate plantlets. Somatic excision of *Ac* yields streptomycin-resistant green sectors (i.e., restores chloroplast function) in a streptomycin-sensitive (white) background, shown in (A) and (B).

enucleated frog egg is capable of development when guided by the genome of a nucleus transferred from a differentiated frog cell. The animal that develops will be identical to the one from which the differentiated nucleus was taken.

Given that each nucleus of a multicellular organism contains a full complement of genes, how do the multitudes of differences among cell types occur? The answer is that not all genes are active in all cells at all times. Gene expression is regulated by both developmental and environmental signals. Regulatory regions of genes can detect these signals and respond by initiating or suppressing gene expression. Gene expression is controlled on several levels. This section focuses on the DNA elements that regulate when, or if, a gene is transcribed into RNA and discusses how genes detect and respond to signals that promote transcription.

7.4.2 Some genes are strictly regulated by a developmental program and are active only in certain tissues or organs.

As a plant grows and develops, various suites of genes become active or quiescent with the changing demands of the cellular processes. Transitions from juvenile to mature forms, and from vegetative to reproductive structures, are controlled by distinct sets of genes that govern cell fate. In most annual plants, the developmental program that directs the transition from juvenile seedling to reproductive maturity is a linear process tied to shoot production. Furthermore, the linear progress along the main shoot is recapitulated along the lateral shoots and even within individual leaves. In all cases, the tissues closest to the base of the plant or to the base of the stalk exhibit the most juvenile state of development, and those furthest from the base are the most mature developmentally. In maize, individual leaf blades can be partitioned into distinct developmental stages, the juvenile stages being at the tip of the leaf.

A well-characterized gene that plays a role in cell fate determination in maize is *Knotted1 (Kn1)*. *Knotted1* is required to maintain shoot meristem, but when expressed incorrectly in leaves where the gene is normally inactive, it stimulates the proliferation of vascular cells that form "knots" of tissue as cells divide out of the plane of the leaf (Fig. 7.38). These abnormal cell divisions occur after cell proliferation should have stopped in the blade of the fully differentiated leaf. *Knotted1* is referred to as a **dominant** or **"gain of function"** mutation—one in which the phenotype results from expression of a gene rather than from loss of expression. *Knotted*-like genes have been identified in several plant systems (e.g., *KNAT1* in *Arabidopsis*), and in each case the gene plays a role in leaf morphology and development (Fig. 7.39).

7.4.3 Some genes are environmentally regulated, becoming active only after responding to certain environmental cues.

Environmentally regulated genes are governed by cues from the external environment that initiate signal transduction for transcription. Major sources of environmental stimuli include light, pathogens, wounding, and other stresses, although each can be

Figure 7.38
Expression of *Knotted1 (Kn1)* in maize leaf. When *Kn1* is expressed in leaf tissues at an inappropriate stage of development, the vascular cells running through the ligule will overgrow and form "knots" of tissue as cells divide outside the plane of the leaf, resulting in this aberrant phenotype. *Kn1* is an example of a homeobox gene, a class of proteins that often play a role in gene regulation.

(A)   **(B)**

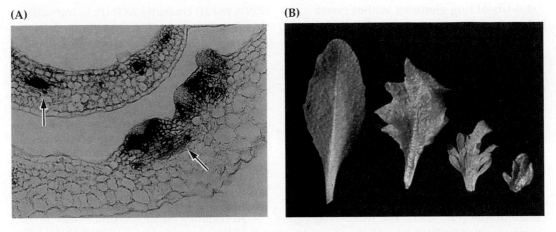

Figure 7.39
Localization of gene expression of *KNAT1*, a knotted-like gene from *Arabidopsis*. *KNAT1* codes for a protein that plays several roles in the developmental regulation of *Arabidopsis* and is ordinarily expressed only in specific groups of cells. When *KNAT1* is expressed ectopically in transgenic *Arabidopsis* at inappropriate times during development, the protein causes aberrant leaf morphology in the form of ectopic meristems arising from the vegetative leaf surface. (A) In situ localization of *KNAT1* in leaf cross-sections. *KNAT1* expression is seen only in the developing ectopic meristems (arrows point to dark blue stain that indicates presence of *KNAT1* mRNA). (B) The morphology of a normal leaf (left) differs from that of transgenic leaves expressing *KNAT1*. The degree of leaf lobing, however, did not correlate with the amount of *KNAT1* transcript present.

linked. For instance, drought imposes water stress on the roots of a plant, inducing the activation of genes encoding enzymes that synthesize the plant hormone abscisic acid (ABA). ABA then initiates a cascade of events that eventually results in the efflux of ions from the guard cells, thereby causing stomatal closure to minimize transpiration and conserve water (Fig. 7.40). However, it is sometimes difficult to determine which genes are induced directly by an environmental stress such as drought, and which are induced as a secondary or more long-term response to the initial signal.

(A)   **(B)**

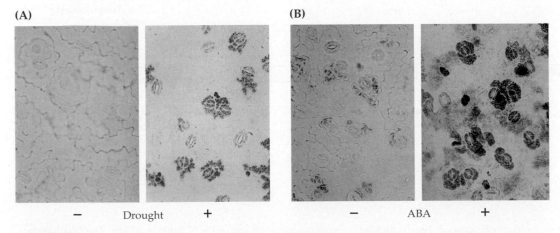

Figure 7.40
Stomatal guard cells express a drought stress gene in response to dehydration and ABA. (A) Transgenic *Arabidopsis* plants containing the *CDeT6-19-GUS* gene (the promoter of a drought-responsive gene from the resurrection plant fused to the *GUS* reporter gene) show expression of the drought-responsive gene in the stomata when exposed to drought. (B) Expression of this transgene is also induced by ABA, suggesting that ABA is part of the signal transduction pathway for relaying drought stress to the nucleus of a stomatal guard cell.

7.4.4 Cis-acting elements within genes help coordinate gene expression.

Much of the regulatory portion of plant genes is located primarily upstream, or 5′, from the transcription start site and can be generally referred to as the gene **promoter.** The promoter contains various sequence elements that function in the recruitment of protein factors that facilitate transcription of the protein-coding region of the gene. Regulatory elements located on the same strand as the coding region of the gene are called **cis-elements** (Fig. 7.41). The most basic cis-element is the **TATA box,** which is found in most eukaryotic genes, located around position −30 (i.e., 30 nucleotides upstream from the transcription initiation site, which corresponds to the first nucleotide of the RNA). The TATA box is often juxtaposed to another basic cis-element, the **CAAT box.** Each element is named on the basis of its nucleotide sequence. TATA is responsible for positioning RNA polymerase II

(RNA pol II) correctly to initiate transcription. A gene may still be able to recruit RNA pol II to a promoter with an altered TATA box, but the efficiency of transcription initiation may be greatly compromised. Inducible genes almost always contain a TATA box and at least two other cis-elements that play a role in the final stages of environmental signal transduction. Housekeeping genes, on the other hand, have less diversity in their cis-elements and may not even contain a recognizable TATA box.

Regulatory elements can be also be found far upstream from the TATA box, downstream in the 3′ flanking regions, in the untranslated leader sequence (noncoding sequence contained at the 5′ end of the mRNA), and even within introns. These elements typically act as **enhancers,** contributing to the efficiency of RNA pol II in initiating transcription of the gene (Fig. 7.42). Enhancer sequences may span hundreds of basepairs and can contain cassettes of repeated

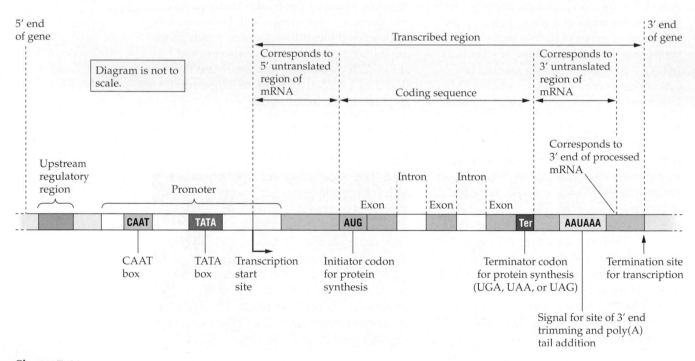

Figure 7.41

Structure and organization of a eukaryotic gene. A gene is divided into several sections. The transcribed region functions as a template for synthesis of RNA, which is then edited and translated into the protein product of the gene. The transcribed region is interspersed with noncoding sequences that partition the region into coding sections (exons) and noncoding sections (introns). The transcribed region is flanked on either side by noncoding sequences that play a role in regulation of the gene. Most of the regulatory sequence elements are in the 5′ flanking region. The first 1000 bp or so of the 5′ flanking region is referred to as the gene promoter, as it contains sequence motifs important for the "promotion" of transcription. These sequence motifs are called "cis-acting elements." The most highly conserved cis-element is the TATA box, which is usually found within the first 50 bp of the transcription start site. The TATA box coordinates the recruitment of RNA polymerase to the gene.

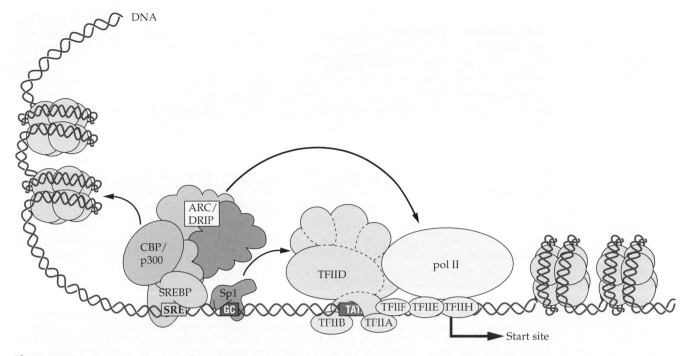

Figure 7.42

A gene enhancer element can function in gene regulation at a distance. One way an enhancer can function in promoting gene transcription is by facilitating a secondary structure such as a bend or loop. Transcription factors (TFs) that have both DNA-binding and transcription activating domains can make a bridge between a distal element and the TATA box, creating a favorable structure for RNA polymerase II (pol II) binding and initiation of transcription. ARC, activator recruited complex; CBP, CREB binding protein; DRIP, vitamin D receptor interaction protein; Sp1, specificity protein 1; SREBP, sterol response element binding protein.

sequences, each of which may function independently as cis-elements. Enhancers can function in either orientation in the chromosome and can be located a considerable distance from the coding region of the gene. However, as cis-elements, they must by definition reside on the same strand as the genes they influence. Enhancers may also function in regulating the specificity of expression, dictating whether a gene is expressed in a particular tissue or organ. This role is often retained when an enhancer is removed from its normal context and placed in another gene, where it imposes tissue-specific expression on the new host gene. In contrast to enhancers, structurally similar **silencers** act to downregulate gene expression.

Enhancers can be identified by using an **enhancer trap** system, which takes advantage of transposon activity (see Section 7.3). In this procedure, an *Ac* element is modified so that it cannot transpose itself, and the coding region for the transposase is put under the control of the CaMV 35S promoter. The enhancer trap element is a *Ds* element containing a promoterless β-glucuronidase gene (*GUS*). If the *Ds/GUS* construct transposes downstream from a promoter in the vicinity of appropriate enhancers, that cell will express the enzyme GUS and can be identified colorimetrically.

7.4.5 The organization of DNA elements within the gene promoter and in enhancer regions is complex.

Promoter elements often appear similar for genes with a related function (Fig. 7.43). For example, a number of environmentally induced genes in plants contain a conserved cis-element called the **G-box** (5'-CCACGTGG-3'). The G-box is required for recognition of many environmental stimuli (see Chapter 22). Deletion and mutational analyses in promoters responsive to light, ultraviolet radiation, ABA, coumaric acid (a phenolic compound), cold, and dehydration all have shown that disruption of the G-box compromises the ability of the promoter to respond to its respective stimulus. However, a salient feature of all these promoters is the requirement of at least

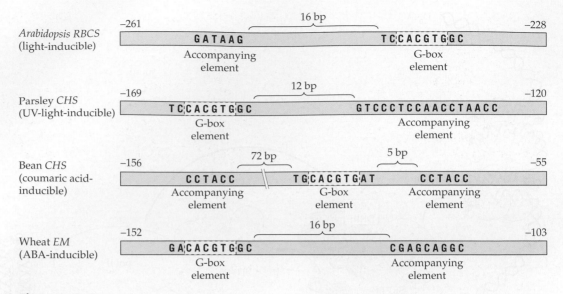

Figure 7.43
Comparison of the promoters of several stress-response genes containing the G-box. G-box DNA sequence elements are often found in association with other regulatory elements. Together, these element pairs confer specific types of inducible responses to genes. Examples of familiar genes containing G-boxes include the small subunit of Rubisco *(RBCS)*, which is induced by light; chalcone synthase *(CHS)*, induced by ultraviolet light and other stresses (see Chapters 20, 21, and 24); and *EM*, a drought-responsive late-embryogenesis gene induced by the plant hormone ABA (see Chapter 22).

one cis-element in addition to the G-box for appropriate transcriptional activation. One of these, the **abscisic acid-responsive element** (ABRE), is found in the promoters of several genes that are induced by ABA and is an example of an extended cis-element that contains the G-box core sequence. Mutational and deletion studies with promoters containing the ABRE show that although the G-box is necessary for the function of ABRE-containing genes, the presence of the G-box is not sufficient for detection of the ABA signal. Rather, the G-box acts in conjunction with other cis-elements to confer ABA responsiveness. Thus, the G-box appears to have a role in the recruitment of a more general transcription factor that functions in conjunction with specific regulatory proteins to activate RNA pol II (see Fig. 7.42).

7.4.6 Transcription factors interact with promoter elements to facilitate transcription.

Promoter cis-elements function in concert to recruit trans-acting, DNA-binding proteins that will interact with RNA pol II at the precise time and location needed for the gene to become active. The trans-acting proteins that bind to specific cis-elements are called **transcription factors.** Some molecular techniques for identifying functional promoter elements are based on the binding of transcription factors to DNA (Box 7.5)

Transcription factors typically have at least two domains—one that functions in the recognition and binding of the cis-element target sequence, and one that functions in organizing additional proteins involved in activating transcription. Transcription factors are categorized on the basis of certain structural motifs, which are conserved among species. These can be either in the DNA-binding domain or in the functional domain of the protein and fall into four major categories: helix-turn-helix motifs, basic leucine zippers, zinc fingers, and high-mobility group (HMG) box motifs.

Helix-turn-helix proteins contain two α-helical segments separated by an intervening loop. This type of factor always binds to DNA as a dimer. One helix of each monomer functions to keep the dimer together by interacting with its counterpart, whereas the other pair of helices creates a scissors-like structure in which the two "blades" of the scissors fit into the adjacent major grooves in the DNA molecule (Fig. 7.44A).

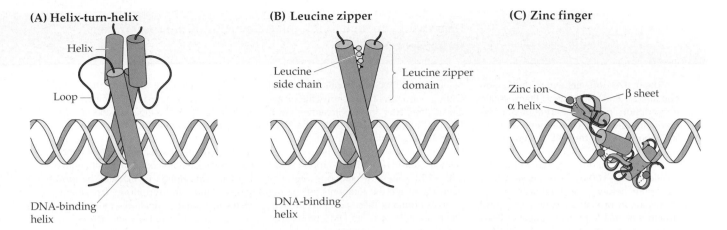

(A) Helix-turn-helix

Helix

Loop

DNA-binding
helix

(B) Leucine zipper

Leucine
side chain

Leucine zipper
domain

DNA-binding
helix

(C) Zinc finger

Zinc ion

α helix

β sheet

Figure 7.44
Three major categories of transcription factor proteins: (A) helix-turn-helix motif, (B) basic leucine zipper, and (C) zinc finger.
Each example is illustrated by a model showing the DNA/protein complex.

Basic leucine zipper proteins also create a scissors-like dimer that binds in the major groove of the DNA molecule. Each monomer contains an α helix in which every seventh amino acid is a leucine. The leucines face each other in the dimer, and their structural arrangement "zips" the helices together into a coiled coil. The coil holds the dimer together and holds the basic region in place within the major groove (Fig. 7.44B).

Zinc finger proteins do not necessarily form dimers; rather, they create their own duplicated set of projections to insert into the major groove of the DNA molecule. The projections are created by coordinating a zinc ion with four amino acids (a combination of histidines and cysteines, or just four cysteines). A zinc finger protein can have from two to nine of these projections, which insert into successive turns of the major grooves along the DNA molecule (Fig. 7.44C).

HMG box motif proteins get their name from the high-mobility-group chromatin proteins from which they were first identified. HMG box proteins also typically bind to DNA as a dimer. Each monomer contains three α helices, two of which bind to DNA. The structure created when these factors bind to the promoter distorts the DNA, bringing other regulatory sequences that are outside of the HMG recognition motif in close enough to allow their joint interaction with additional transcription factors.

Transcription factors are often encoded by families of genes that collectively code for a population of related proteins. Hetero-

dimers—created from two different members of a transcription factor family—allow for variation in the DNA binding specificity. For example, many components of cis-elements have a dyad nucleotide sequence symmetry recognized by a symmetric protein dimer. A good example of this type of interaction is the G-box. The G-box is recognized by **G-box binding factors,** a family of basic leucine zipper proteins that freely form heterodimers. Thus, if one family member typically recognizes CCACGTGG as its binding motif and another recognizes GCACGTGC, a heterodimer of the two could also recognize GCACGTGG. Because the G-box is so widespread in its distribution, heterodimers with slightly different binding affinities may confer specificity among the wide variety of promoters that contain the G-box.

7.4.7 Some transcription factors do not bind DNA directly.

Some transcription factors function indirectly, interacting with proteins that then bind to promoter elements. Although these proteins do not themselves bind to a DNA template, they participate in generating a transcription factor complex that in turn promotes transcription. The most basic, but important, example, the **transcription initiation complex,** includes RNA pol II and many associated transcription factors (Fig. 7.45). RNA pol II cannot bind to the DNA template alone; rather, it requires the presence of several

Box 7.5

Identifying functional promoter elements—from footprinting to transgenes

The maize *Adh1* gene promoter region contains several cis-elements that regulate transcription in response to hypoxia. Hypoxia initiates the recruitment of transcription factors to the ARE (anaerobic response element), which is centered on position −100, and to two other elements positioned at −180 and −90 (as shown in the figure below). The binding of trans-acting factors probably creates a promoter structure favorable for recruitment of RNA polymerase to the TATA box. Without the bound factors, the gene remains quiescent. In vivo footprinting is a technique that uses specialized chemical probes to reveal the position of proteins bound to a DNA molecule. Dimethyl sulfate (DMS) is a DNA-modifying chemical. DNA with a bound protein molecule will be either more or less accessible to the effects of DMS than protein-free ("naked") DNA. Thus, a protected or enhanced sensitivity to DMS in vivo compared with naked DNA exposed to DMS in vitro, reveals the "footprint" of a protein bound to that section of DNA in the nucleus of the living cell. The cartoon below shows the gene promoter in uninduced (U) and induced (I, transcriptionally active) configurations. Note the proteins bound to the footprinted positions. Lane U is the cleavage pattern

facilitated by DMS modification of naked DNA in vitro, and lane I shows the cleavage pattern after in vivo treatment with DMS of hypoxia-stressed cells. The DNA is protected from DMS modification in vivo around positions −130 and −110 and is more sensitive to DMS around −180 and −100. These protections and enhancements are the footprints left by the proteins bound at these positions in vivo. Open triangles indicate DMS protection, and filled triangles indicate DMS hypersensitivity.

In vivo footprinting has also identified four putative *cis* elements of *Adh2* that interact with protein factors within the DNase I hypersensitive domains of the 5′ flanking region (as shown in panel A of the figure on the facing page). The power of in vivo footprinting to identify functionally significant sites within a gene promoter was tested by biochemical and transgenic analyses of the putative element at position −160. Biochemical analyses show that proteins isolated from maize cell suspensions will bind to the *Adh2* promoter in vitro to generate a footprint at −160 that is identical to that seen in vivo (see panel A of figure on the facing page). DNA constructs were used to assess the impact of the −160 element on expression

of the reporter gene *GUS* (as shown in panel B of the figure on the facing page) in transgenic *Arabodopsis* plants. *NPTII* encodes neomycin phosphotransferase, which confers resistance to kanamycin (Kan^R) and allows selection of plant transformants on media containing that antibiotic. Its transcription is regulated by promoter and transcription terminator elements from the nopaline synthase gene of *Agrobacterium tumefaciens* (*NOS*). In the experiment construct, *GUS* expression is driven by the *Adh2* −160 DNA sequence element fused to a minimal CaMV 35S promoter that lacked cis elements. *GUS* expression was driven by the minimal 35S promoter in the negative control, and by the strong, full-length CaMV 35S promoter in the positive control. Transgenic analyses of *GUS* expression patterns indicate that the −160 element of the maize *Adh2* promoter acts as an activator in the meristem and vascular tissue of roots and in the vascular tissue of stems and leaves (see panel C of the figure on the facing page). The photographs show the *GUS* expression pattern associated with the −160 element, while the cartoons describe the *GUS* expression for each of the three constructs.

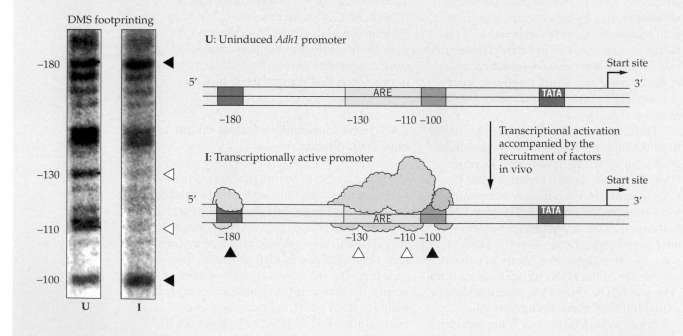

(A)

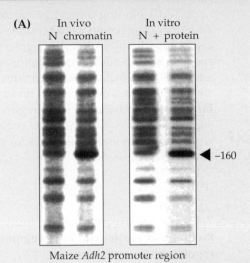

In vivo
N chromatin

In vitro
N + protein

◄ −160

Maize *Adh2* promoter region

(B)

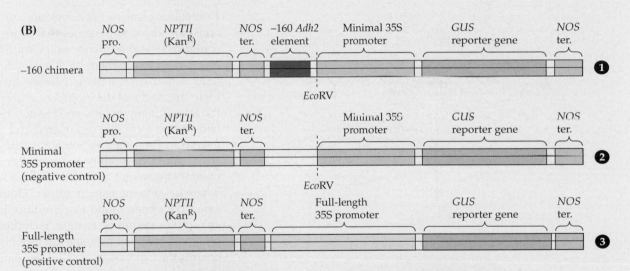

−160 chimera

NOS pro. | *NPTII* (Kan^R) | *NOS* ter. | −160 *Adh2* element | Minimal 35S promoter | *GUS* reporter gene | *NOS* ter. | ❶

*Eco*RV

Minimal 35S promoter (negative control)

NOS pro. | *NPTII* (Kan^R) | *NOS* ter. | Minimal 35S promoter | *GUS* reporter gene | *NOS* ter. | ❷

*Eco*RV

Full-length 35S promoter (positive control)

NOS pro. | *NPTII* (Kan^R) | *NOS* ter. | Full-length 35S promoter | *GUS* reporter gene | *NOS* ter. | ❸

(C)

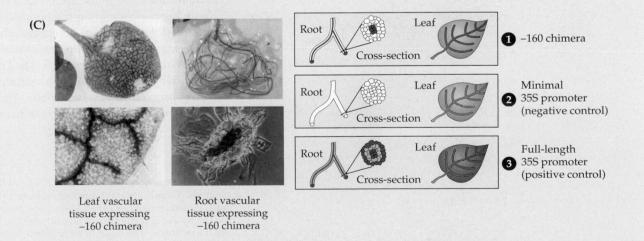

Leaf vascular tissue expressing −160 chimera

Root vascular tissue expressing −160 chimera

Root | Leaf | Cross-section | ❶ −160 chimera

Root | Leaf | Cross-section | ❷ Minimal 35S promoter (negative control)

Root | Leaf | Cross-section | ❸ Full-length 35S promoter (positive control)

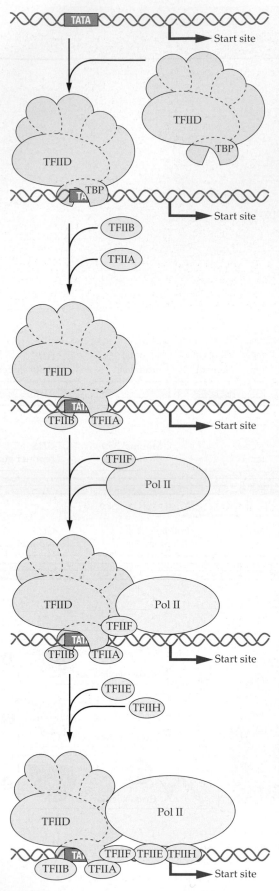

core transcription factors (e.g., TFII-A, B, D, E, F). TFIID, itself a multiprotein complex, constitutes the key set of proteins in the transcription initiation complex. A TFIID component, TATA-binding protein, recognizes the TATA box. The bound TFIID complex creates a structure that RNA pol II recognizes. TFIID can also bind to additional proteins, such as a transcription factor associated with an upstream (5′) enhancer element. This capacity introduces additional levels of regulation, such that transcription initiation occurs only when the additional enhancer factor is also bound to the promoter.

7.4.8 Homeobox proteins are transcription factors that participate in development by regulating gene activity.

Homeobox proteins (or homeodomain proteins) are encoded by genes that share a common 180-bp sequence called the "homeobox" (Fig. 7.46). These have been referred to as master control genes because they regulate the stages of development and specify the fate of many organs and tissues. Homeobox genes were first discovered and characterized in *Drosophila* as homeotic genes that, when mutated, produce dramatic developmental effects such as formation of a leg where an antenna usually grows. Homeobox genes have been found in all multicellular eukaryotes examined to date, including plants. Homeobox proteins are thought to control the expression of other regulatory proteins, including transcription factors.

In all cases, the transcription factors encoded by genes containing the homeobox motif are of the helix-turn-helix class; they regulate target genes in a precise spatial and temporal, or ontological, pattern. The DNA-binding domain of the protein is encoded by the conserved 180-bp homeobox. The regions outside of the homeobox domain

Figure 7.45
Structure of the RNA polymerase II (RNA pol II) complex. Various general transcription factors (TF) must be sequentially recruited to the gene promoter for the RNA pol II complex to be competent to initiate transcription. The order of recruitment assembly is TFIID, TFIIB, TFIIA, TFIIF, and RNA pol II, followed by TFIIE and TFIIH.

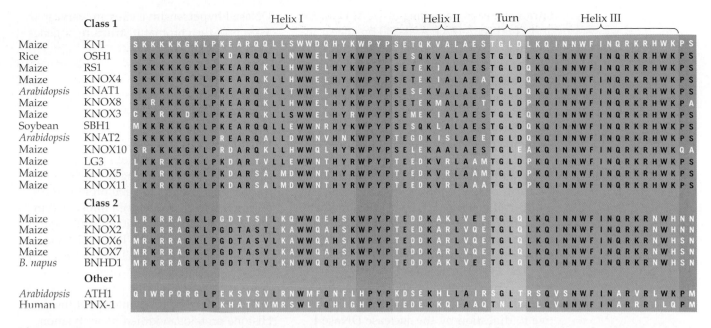

	Class 1	Helix I	Helix II	Turn	Helix III

Figure 7.46
Plants have three classes of homeobox proteins. Class 1 are Knotted1-like proteins, class 2 are KNOX-like, and all others form a third class. Among themselves, the class 1 and class 2 homeobox proteins have their helix III motif almost completely conserved.

are not conserved among species, or even within species. At present, two classes of homeobox genes, *Knotted1 (Kn1)* and *ZMH1/ZMH2*, have been identified in maize. The amino acid sequence of the homeodomains is conserved between class 1 and class 2 proteins (57 of 64 residues being identical), but the remaining sequences are very dissimilar.

The maize *Knotted1* gene was the first homeobox gene to be identified in plants. Before it was cloned, *Kn1* was defined phenotypically by several dominant gain-of-function mutations that altered leaf morphology by causing vascular cells in the leaf to proliferate outside of their normal developmental context (see Fig. 7.38). Since the discovery of *Kn1*, additional homeotic genes have been isolated from maize, *Arabidopsis*, and other plants. Most plant homeobox proteins seem to fall into two classes, based on sequence similarity and expression patterns. Within each class, the amino acid content of the homeobox protein is highly conserved, the sequences of the third helix being virtually identical for all members (Fig. 7.46). By definition, the homeodomains of these proteins resemble those of proteins from outside the plant kingdom as well. A comparison of the homeodomains from plants and humans reveals a profound similarity, suggesting

that the homeobox genes originated before the divergence of the three kingdoms represented by these organisms.

7.5 Role of chromatin in chromosome organization and gene expression

7.5.1 Histones organize DNA into nucleosomes and chromatin and influence DNA sensitivity to nucleolytic enzymes.

A diploid maize cell contains 10^{10} bp, or 10 m, of DNA, which is a highly negatively charged and rather rigid molecule until the charges along its length are neutralized. This DNA must fit inside a nucleus less than 10 μm wide without knotting or breaking. In addition, once this DNA is packaged, it must be accessible to transcriptional signals and replication enzymes. Cells accomplish this task by utilizing **chromatin,** a highly organized complex of DNA and protein.

The most localized level of chromatin structure is the **nucleosome** array, called first-order condensation to distinguish it from subsequent higher-order condensation. A nucleosome consists of DNA wrapped two full turns (166 bp) around a globular octamer of histone proteins made up of two tetramers, each consisting of H2A, H2B, H3, and H4 (Fig. 7.47A). A 20–200-bp spacer intervenes between adjacent nucleosomes. Under the electron microscope, the nucleosome array looks like a string of beads and is approximately 10 nm in diameter (Fig. 7.47B). A single nucleosome bead may be excised from chromatin by using the bacterial enzyme micrococcal nuclease, which preferentially cuts the more exposed spacer DNA between nucleosomes. This free bead contains 1.8 turns of DNA and the histone core. An additional histone, H1, binds outside the nucleosome core; one of its functions is to stabilize both the nucleosome array and higher-order chromatin structures (Fig. 7.47C).

The initial correlation between gene expression and chromatin structure established that active genes are generally more sensitive to digestion by the nuclease **DNase I** than are inactive genes. This differential sensitivity arises because the chromatin surrounding a transcriptionally active gene is less condensed than genomic chromatin as a whole and is thereby more accessible to nuclease digestion. **DNase I hypersensitivity** defines yet another conformational level of chromatin in active genes. These hypersensitive regions are thought to be nucleosome-free, allowing transcription factors greater access to promoter cis-elements (Fig. 7.48).

DNase I hypersensitive sites are pervasive aspects of the chromatin structure of genes undergoing transcription.

7.5.2 Modification of histones in chromatin affects DNA accessibility.

Histones help regulate transcription by manipulating the degree of DNA condensation, thereby influencing the access of transcription factors to gene promoters. Histones have recently been implicated in a larger role in transcriptional regulation. H1 and the core histones H2A, H2B, H3, and H4 influence the level of DNA condensation and accessibility through selective acetylation of the histone protein "tails" that face away from the interior of the core. Histone proteins modified by acetylation create nucleosomes that are less compact than those without modification. Thus, histone modification by specific enzymes such as histone acetyl-transferases can inhibit condensation and thereby render that region of the genome more accessible to transcription factors (Fig. 7.49). Histone deacetylase can remove acetyl groups, which increases the compactness of chromatin and renders the genes less accessible to RNA polymerase.

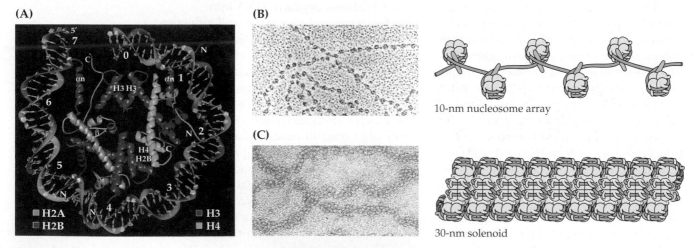

Figure 7.47

Nucleosomes in chromatin structure. (A) The core nucleosome particle contains 146 bp of DNA wrapped around an octamer of histone proteins. The octamer contains two tetramers of H2A, H2B, H3, and H4. (B) The 10-nm–diameter fiber, or nucleosome array, is often referred to as "beads on a string" because of its appearance under transmission electron microscopy. The core of the "bead" is an octamer of histone proteins H2A, H2A, H3, and H4, while another histone, H1, resides just outside of the core. (C) The 10-nm–diameter fiber is further condensed into a 30-nm–diameter fiber called the "solenoid" structure. This structure is stabilized by H1, which links coils consisting of six nucleosomes. An easy way to visualize this structure is to think of H1 molecules as paper clips holding together the coils of a "Slinky" (a helical toy for children).

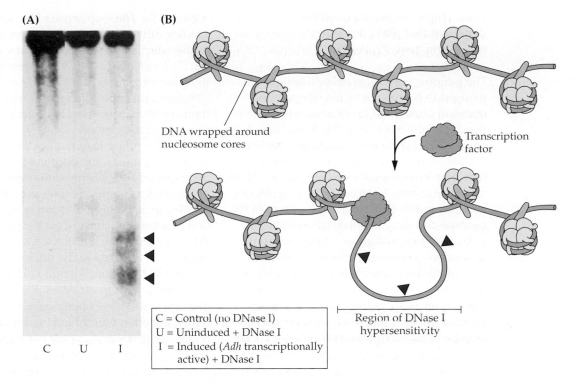

(A)

C   U   I

C = Control (no DNase I)
U = Uninduced + DNase I
I = Induced (*Adh* transcriptionally active) + DNase I

(B)

DNA wrapped around nucleosome cores

Transcription factor

Region of DNase I hypersensitivity

Figure 7.48

DNase I hypersensitive sites in the maize *Adh1* promoter. As a gene becomes transcriptionally active, regions of the promoter become less condensed to accommodate transcription factors. This opening of the promoter chromatin can be visualized as an increase in accessibility to DNase I digestion. DNase I is an endonuclease with very little sequence specificity. DNA held in a condensed chromatin configuration is difficult for DNase I to access, but when the chromatin of a gene promoter becomes less condensed—to afford access to the transcriptional machinery—it also becomes more accessible to DNase I. Thus, hypersensitivity to DNase I is a hallmark of the less-condensed chromatin configuration necessary for a gene to become transcriptionally active. (A) The *Adh1* gene promoter is less accessible to DNase I in cells where the gene is "off" (uninduced, U) than in cells where it is transcriptionally active (induced, I). The transcriptionally active promoters from induced cells show more regions of DNase I hypersensitivity (indicated with red triangles) than do those where the gene is not being transcribed. (C is the control.) (B) The hypersensitive region may reflect a nucleosome-free region, a state potentiated by the binding of one or more transcription factors.

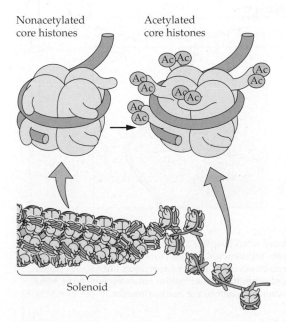

Nonacetylated core histones

Acetylated core histones

Solenoid

Figure 7.49

Histones can be modified by acetyltransferases and deacetylases to influence the degree of chromatin condensation. Both histones within the nucleosome and the linker histone H1 can be modified at the N-terminal region by acetyltransferases. An acetyl group on a histone creates steric hindrance for surrounding structures, thereby generating a local region of decondensation. A decondensation of chromatin can facilitate recruitment of transcription factors to the surrounding sequence. Histone deacetylases (not shown) remove acetyl groups to allow the N termini of histones to interact more tightly with DNA.

7.5.3 Higher-order chromatin structure also plays a role in regulating gene expression.

The genomes of higher eukaryotes are thought to be organized into looped domains of chromatin by attachments to a nuclear scaffold (Fig. 7.50). The loops, which are anchored at **matrix attachment regions** (MARs), vary widely in size from 5 to 200 kbp. The MARs themselves are AT-rich DNA sequence motifs 200 to 1000 bp long. MARs function in the structural organization of the genome and may facilitate transcription of a gene or group of genes by promoting the formation of less-condensed chromatin structures. This latter role is thought to involve the topology created when the chromatin fiber is anchored to a fixed protein scaffold. In addition, some proteins that participate in transcriptional processes associate with MARs. The loops created by MARs are topologically independent from one another, as they display various degrees of supercoiling. The topology of DNA has been shown to influence gene expression in several eukaryotes, and supercoiling in particular influences gene expression in bacteria and in the chloroplast genome.

MAR binding may require the presence of distinct proteins in addition to the proteins that create the structural framework of the nuclear matrix. Several specialized MAR-associated proteins have been identified in several organisms; such proteins may act as the tether by which the MAR DNA sequence is attached to the nuclear matrix.

MARs have been identified in association with several plant genes, for which they appear to have operational importance. MARs are often found flanking the coding regions of genes and are often associated

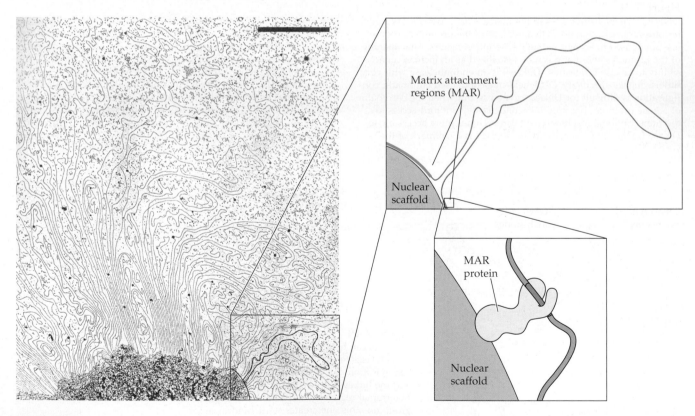

Figure 7.50
Matrix attachment regions (MARs) play a role in the organization of the genome, creating topologically independent, supercoiled structures in the genome through their attachment to the nuclear matrix scaffold. MARs may play a regulatory role by allowing differential levels of condensation of the genome in discrete regions. The electron micrograph (left) and cartoon blow-ups (right) show histone-depleted DNA loops in a nucleus. In addition to the proteins that create the structural framework of the nuclear matrix, there appear to be several classes of specialized MAR-associated proteins. These proteins may act as the tether by which the MAR DNA sequence is attached to the nuclear matrix.

with regulatory elements. The concerted function of MARs and regulatory elements has been documented for such genes as chicken lysozyme, human apolipoprotein, and several *Drosophila* and yeast genes. In each of these cases, MARs also coincide with the boundaries of general DNase I sensitivity. Similar trends are seen in the MARs that have been identified in plant genes. The 5′ flanking region of the β-phaseolin gene from bean (*Phaseolus vulgarus*) contains a MAR that acts in concert with a transcriptional enhancer sequence; moreover, this region of the β-phaseolin gene displays increased DNase I sensitivity. MARs have also been characterized for genes in soybean, tobacco, pea, and maize and in all cases appear to influence gene structure or function. Expression of transgenic constructs that combine MARs with the gene of interest shows that MARs can confer an expression pattern similar to that of the endogenous gene. Various models predict that a transgenic construct containing MARs creates its own chromatin domain favorable for transcription when the transgene becomes incorporated into the genome. This phenomenon further illustrates how chromatin structure can influence gene regulation.

7.6 Epigenetic mechanisms of gene regulation

Gene expression can be influenced by changes in the genome that do not involve mutations in DNA sequence. These heritable yet readily reversible changes are termed **epigenetic.** The molecular mechanisms responsible for these epigenetic changes are an active area of research, and much of our understanding about these mechanisms derives from the study of plant genes and transposable elements.

Plant genes or transposable elements that influence the production of colorful pigments have been especially useful for identifying and investigating epigenetic regulatory mechanisms. The intensity and patterning of pigmentation is typically under strict genetic and developmental control. Epigenetic changes that lead to altered pigmentation properties are easy to identify, and the inheritance of these changes can be followed through somatic development and into the

next generation simply by visual inspection. These model systems show that some epigenetic changes are reversed before meiosis, whereas others are meiotically transmitted. Epigenetic changes that occur as a normal process of plant development are typically erased, or reset, at or before meiosis.

7.6.1 Imprinting involves gamete-specific epigenetic changes in gene expression.

Imprinting is an example of epigenetic change that is reversed during somatic development. Imprinting describes the observation that the expression of certain alleles differs, depending on the gamete of origin. This is probably a universal phenomenon, with examples now known in mammals and fungi as well as plants.

Imprinted loci can be inferred or directly observed in the triploid endosperm of flowering plants. In most flowering plants, the standard ratio of two maternal genomes to one paternal genome is absolutely essential for endosperm development. In maize, manipulations that alter dosages of chromosome arms reveal the existence of loci that require paternal transmission to ensure proper endosperm development. Imprinting of the *r (red)* gene that affects endosperm pigmentation can be observed directly. Certain *r* alleles *(R)* are strongly expressed (solid seed color) when transmitted through the female ovule but weakly expressed (variegated seed color) when transmitted through the male pollen (Fig. 7.51). This result is not due to the difference in *r* gene dosage but rather reflects a male-specific epigenetic change in *r* gene activity.

7.6.2 Paramutation can be transmitted through meiosis.

Paramutation is an example of an epigenetic change that is meiotically heritable. Paramutation describes an allelic interaction whereby the expression of one allele in a heterozygote is altered by the presence of the other. Several examples have been described for loci that control pigmentation in *Zea mays* (Fig. 7.52). The strongly expressed *b (booster)* allele *B–I* confers robust plant pigmentation—in contrast to the weakly

Figure 7.51
Imprinting in the *R* gene of maize. Imprinting is an epigenetic effect observed when the expression of a gene is influenced by its parent of origin. The distinct phenotypes shown here are not the result of differences in dosage but rather reflect an epigenetic change in the *r* gene of the male parent.

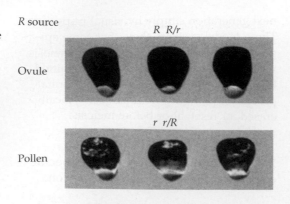

R source
R R/r
Ovule

r r/R
Pollen

expressed *B'* allele. The *B–I/B'* heterozygote is weakly colored, and only *B'* alleles are transmitted sexually from the heterozygote. The two *b* alleles, *B'* and *B–I*, have identical DNA sequences, yet their rates of transcription differ by 10- to 20-fold. Because *B'* can spontaneously arise from *B–I*, the *B'* and *B–I* designations serve only to discriminate different epigenetic states of a single allele.

Other examples of paramutation occur at the *r* and *pl* (purple plant) loci (Fig. 7.52). The strongly expressed *Pl* allele that normally confers solid pigmentation of the anthers can spontaneously change to weaker activity states *(Pl')*, which result in variegated pat-

terns of anther pigment. Similarly to *B–I*, *Pl* invariably changes to *Pl'* in the *Pl/Pl'* heterozygote; only *Pl'* epigenetic states are sexually transmitted. The requirement for allelic interactions is underscored by the observation that *Pl'* can change back to the strongly expressed *Pl* state when *Pl'* is heterozygous with other *pl* alleles that do not display epigenetic changes. A key question is whether the interaction between *Pl* and *Pl'* requires physical contact, or pairing, between the interacting alleles or chromosomal regions.

7.6.3 Epigenetic changes in gene expression can be induced by repeated DNA sequences or chromosomal location.

A correlation between genes repeated at specific loci and silencing has emerged from several transgenic studies in *Arabidopsis*, petunia *(Petunia hybrida)*, and tobacco *(Nicotiana tobaccum)*: The more copies of the transgene that are present at a given locus, the greater the chance that expression from those cis-linked transgenes is silenced, and the greater their ability to act in trans to

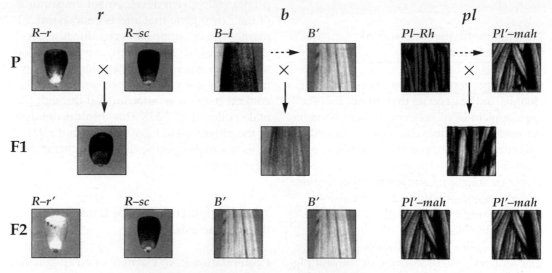

r   *b*   *pl*

P   *R–r*   ×   *R–sc*   *B–I*   ×   *B'*   *Pl–Rh*   ×   *Pl'–mah*

F1

F2   *R–r'*   *R–sc*   *B'*   *B'*   *Pl'–mah*   *Pl'–mah*

Figure 7.52
Paramutation in *Zea mays*. Three of the maize anthocyanin regulatory loci (*r*, *b*, and *pl*) have alleles that exhibit paramutation. The examples shown are those affecting coloration of seed (*r* expression), husk (*b* expression), and anther (*pl* expression). Parental, F1, and the resulting segregant phenotypes are represented in successive rows of the figure. Activity of the paramutable *R–r* allele (left) is heritably reduced after exposure to the paramutagenic *R–sc* allele in the F1 heterozygote. The *R–r* and *R–sc* alleles are structurally distinct from one another and the *R–sc* allele is always strongly expressed. Weakly expressed paramutagenic *B'* states (center) can arise spontaneously from strongly expressed paramutable *B–I* alleles (dashed arrow). *B–I* alleles change exclusively to *B'* when exposed to *B'* in the F1 heterozygote. Weakly expressed paramutagenic *Pl'–mah* (*Pl'*) states (right) can arise spontaneously from strongly expressed paramutable *Pl–Rh* (*Pl*) alleles (dashed arrow). *Pl–Rh* alleles change exclusively to *Pl'–mah* when exposed to *Pl'–mah* in the F1 heterozygote.

silence related transgenes elsewhere in the genome. Endogenous loci with repeated gene sequences in maize, *Arabidopsis*, snapdragon, and soybean *(Glycine max)* also show similar cis- and trans-silencing behaviors. In both cases, decreasing the number of copies of the transgene or of the endogenous gene typically results in reactivation of previously cis-silenced regions and the inability to cause trans-silencing. Whether the ability to cause gene silencing, either in cis or in trans, depends on gene dosage or repetitive sequences per se is still unknown.

Specific chromosome structures or organization may also play a causative role in certain examples of epigenetic gene silencing. In petunia, certain varieties have colorless flowers because of mutations in a gene required to synthesize pigment. Pigmentation can be restored by introducing transgenes that contain the maize *A1* gene. These transgenes can reside at various sites in the genome and in multiple copies. Most plants with single copies of the transgene have pigmented flowers. However, in some petunia plants expressing a single copy of *A1*, the transcription of the transgene can be epigenetically silenced, resulting in loss of flower pigment (Fig. 7.53). This silenced state is somatically and meiotically heritable. The silenced *A1* transgene can induce epigenetic silencing of transcription of another strongly expressed *A1* transgene only when the new transgene inserts near the chromosomal location of the original silenced *A1* transgene; other *A1* transgenes found elsewhere in the genome are unaffected.

7.6.4 Transgenes can induce epigenetic silencing of endogenous homologous genes.

Other examples of epigenetic gene silencing do not result in meiotically heritable changes of gene activity. In **cosuppression,** transgenes cause silencing of endogenous genes that share sequence identity. Early efforts to increase pigment production in petunia by introduction of a transgene encoding a rate-limiting enzyme led to the efficient silencing of both transgene and endogenous gene activity and resulted in colorless flowers as well as striking patterns of pigmentation (see Fig. 7.54). Cosuppression does not lead to a meiotically heritable change, because endogenous gene activity is restored once the silencing transgene is segregated away by sexual crosses. A possibly related phenomenon involves the ability of cytoplasmically replicating RNA viroids and plant viruses to cause silencing of homologous sequences found in the nuclear genome. Grafting experiments demonstrate that, once triggered by viral infection, such silencing can occur systemically throughout the plant. Given that both cosuppression and viral-induced gene silencing affect the expression of genes post-transcriptionally, such silencing behaviors are probably mediated by RNA.

The biochemical mechanisms responsible for these epigenetic phenomena are not well understood. Some mechanisms operate to prohibit transcription, whereas others act at a post-transcriptional level. Even within these two broad categories, several different molecular events are likely to contribute to these epigenetic processes. Because epigenetic changes are, by definition, not caused by direct DNA mutations, the underlying mechanisms may well lie in the realm of chromatin and the influence of heritable changes in chromatin structure.

Figure 7.53
Gene silencing in petunia. Shown are floral phenotypes of trangenic petunias that carry a *Zea mays A1*-expression construct. Pigmentation of the flowers is a visual indicator of transgene activity. The various epigenetic states of transgene activity shown represent uniform expression (red flowers at top), variegated expression (mottled flower at lower right), and complete inactivation (white flower at lower left).

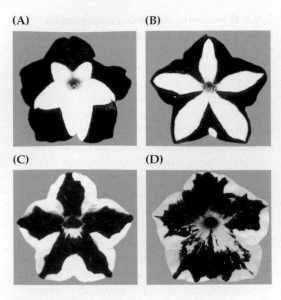

(A)   **(B)**

(C)   **(D)**

Figure 7.54
Sense cosuppression of chalcone synthase genes in pigmented petunia flowers can produce a variety of attractive flower color patterns in addition to pure white flowers (not shown). The spatial organization of the white sectors is determined principally by corolla morphology. Patterns may be determined by cells at the junctions of the five fused petals (A), by the midveins of each petal (B), by both (C), or by combinations of morphological and non-morphological determinants that yield complex, irregular patterns (D).

7.6.5 DNA methylation affects gene expression and developmental regulation.

Although models of heritable changes in nucleosome positioning, packing, and covalent modifications have been proposed to explain epigenetic changes in gene expression, the best-characterized modification is DNA methylation. DNA methylation is an attractive model for epigenetic changes because mechanisms exist to ensure both mitotically and meiotically inheritance of specific DNA methylation patterns. In many eukaryotes, cytosine residues in genomic DNA can be methylated at the 5′ position of the cytidine moiety (Fig. 7.55). The DNA-methyltransferase enzymes that carry out this reaction in mammals have preference for cytosines found in a CpG dinucleotide context (the p represents a phosphodiester bond). In plants, cytosines found in a CpNpG context are also strong substrates. In either case, newly replicated DNA that contains hemimethylated CpG or CpNpG sites is a strong substrate for DNA-methyltransferase activity. This maintenance methylase activity ensures the maintenance of preexisting methylation patterns in both daughter chromosomes (see Chapter 6).

In mice, a specific DNA-methyltransferase has been shown to be essential for

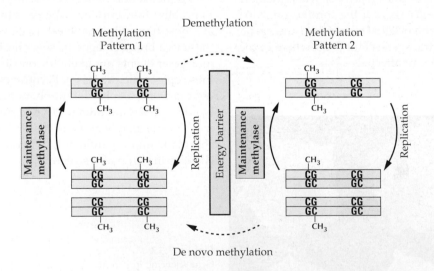

Figure 7.55
Cytosine methylation and inheritance of methylation patterns. Methylation at the 5′ position of cytosine can occur at CpG and CpA/TpG residues in plants and is often correlated with the activity of the gene. The pattern of methylation of a gene is a combination of the action of maintenance methylases and de novo methylase activity on one hand, and the demethylation and replication activities on the other. Inheritance of the methylation pattern of a gene may be responsible for epigenetic effects.

normal embryonic development and necessary for the continued silencing of certain imprinted genes. Interference with DNA-methyltransferase activity in *Arabidopsis* also leads to pleiotropic effects on growth and development (Fig. 7.56), implicating a central role for DNA methylation in the epigenetic regulation of plant development. Multiple DNA methyltransferases are found in many eukaryotes; the challenges now are to understand how their activities are regulated and identify the specific roles each one plays in de novo methylation and maintenance methylation reactions.

Many eukaryotic genes and transposable elements exhibit a strong inverse correlation between density of DNA methylation and transcriptional or transpositional activity. Whether cytosine methylation is causal in this effect or is simply reflective of an altered chromatin structure that directly affects gene function remains unclear. Research in plants shows clear correlations between increased transgene copy numbers, increased methylation, and decreased transcriptional activity. In the example of petunias that carry an *A1* transgene susceptible to gene silencing and paramutation (see Fig. 7.53), changes in the pigmenting activity of the transgene mirror changes in cytosine methylation found in the 5′ flanking sequences. However, such correlations are not universal for all examples of gene silencing in plants. Despite the strong reduction in transcriptional activity that occurs during *b* paramutation (see Fig. 7.52), there is no evidence for changes in DNA methylation patterns.

The fact that similar gene silencing behaviors occur in both yeast and *Drosophila*, organisms that do not utilize DNA methylation, indicates that other elements of chromatin structure can also be efficiently replicated to new daughter chromosomes. The examples discussed suggest that diverse epigenetic mechanisms are in place to affect the heritable activity of plant genes and, perhaps, of larger chromosomal regions. Indeed, the recognition of these dynamic regulatory behaviors illustrates the central role of genome organization and chromosome function in the control of gene expression.

Figure 7.56
Interference with DNA-methyltransferase activity. Phenotypic comparisons of wild-type (left) and a *MET1* antisense line of *Arabidopsis thaliana*. Interference with cytosine methylation activity encoded by *MET1*, using an antisense-RNA knock-out approach, is associated with abnormal phenotypes during (A) vegetative, (B) inflorescence, and (C) floral development.

Summary

The characteristics of an organism are encoded in its DNA. In the eukaryotic nucleus, most of this information resides within thousands of genes and is organized into linear chromosomes. A few genes are also encoded by DNA found in plant mitochondria and plastids. The entirety of this genetic material is referred to as an organism's genome. The study of the organization and regulation of the nuclear genome is the fundamental basis for genetic and molecular genetic research.

In the 1850s Gregor Mendel developed the science of genetics by proposing that particular factors (now called genes) gave rise to specific traits. Mendel's pea experiments were the first to correlate physical traits (phenotype) with heritable components (genotype). About 50 years after Mendel's work, genes were postulated to reside on chromosomes in the nucleus; it took another 50 years for scientists to identify the molecule responsible for transmitting genetic information, deoxyribonucleic acid (DNA). During the century after Mendel, elucidation of the organization of the nuclear genome included the discovery that genes are aligned on chromosomes and can be inherited in linkage groups and, moreover, that genetic recombination within a linkage group plays a role in the generation of new phenotype combinations. The analysis of genetic recombination also gave rise to the ability to map genes to their relative positions in a chromosome.

The role of a gene is to encode a product that contributes to the phenotype of that organism; often, however, only a small percentage of an organism's genome actually encodes gene products. The intervening and flanking noncoding sequences are integral parts of a typical gene and contribute to some of the diversity in genome size observed among similar organisms. The intervening sections of noncoding DNA are called introns, and the sections of coding sequence are referred to as exons. In addition, repetitive noncoding DNA sequences contribute to the character and function of specialized structures in chromosomes such as the centromere and telomere. Another specialized class of DNA sequence that can make up a significant portion of the genome is the transposable elements. Transposable elements of sections of DNA move, or transpose, from one site in the genome to another, carrying genetic information with them as they transpose.

Not all genes of a nuclear genome are expressed all the time. Selective expression gives rise to cellular differentiation during development and enables cells to respond to environmental signals accordingly. Genes are divided into two fundamental regions. The coding region is transcribed into the gene product, and the regulatory region contains a variety of sequence elements that function in the recruitment of protein transcription factors, which facilitate transcription. Another feature of the genome that plays a role in the selective expression and regulation of genes is the chromatin structure associated with a given gene. Chromatin refers to the complex of DNA and protein that makes up the chromosome and organizes the genome within the nucleus. In addition to these large-scale roles, the fine-scale features of chromatin influence gene regulation by manipulating the accessibility of a gene promoter to the factors required for initiating transcription.

Further Reading

Dean, C., Schmidt, R. (1995) Plant genomes: a current description. *Annu. Rev. Plant Physiol. Plant Mol. Biol.* 46: 395–418.

Doring, H. P., Starlinger, P. (1986) Molecular genetics of transposable elements in plants. *Annu. Rev. Genet.* 20: 175–200.

Fedoroff, N. V. (1989) About maize transposable elements and development. *Cell* 56: 181–191.

Fedoroff, N., Botstein, D., eds. (1992) *The Dynamic Genome.* Cold Spring Harbor Laboratory Press, Plainview, NY.

Fosket, D. E. (1994) *Plant Growth and Development.* Academic Press, San Diego, CA.

Franklin, A. E., Cande, W. Z. (1999) Nuclear organization and chromosome segregation. *Plant Cell* 11: 523–534.

Freeling, M. (1984) Plant transposable elements and insertion sequences. *Annu. Rev. Plant Physiol.* 35: 277–298.

Gehring, W. J, Affolter, M., Bürglin, T. (1994) Homeodomain proteins. *Annu. Rev. Biochem.* 63: 487–526.

Glick, B. R., Paternak, J. J. (1998) *Molecular Biotechnology: Principles and Applications of Recombinant DNA.* ASM Press, Washington, DC.

Goodrich, J. A., Cutler, G., Tjian, R. (1996) Contacts in context: promoter specificity and macromolecular interactions in transcription. *Cell* 84: 825–830.

Holmes-Davis, R., Comai, L. (1998) Nuclear matrix attachment regions and plant gene expression. *Trends Plant Sci.* 3: 91–97.

Chapter 7 Genome Organization and Expression

Karp, G. (1995) *Cell and Molecular Biology.* John Wiley & Sons, New York.

Pikaard, C. S. (1998) Chromosome topology—organizing genes by loops and bounds. *Plant Cell* 10: 1229–1232.

Rhoades, M. M. The early years of maize genetics. (1984) *Annu. Rev. Genet.* 18: 1–30.

Russell, P. J. (1992) *Genetics,* 3rd ed. Harper-Collins Publishers, New York.

Russo, V.E.A., Martienssen, R. A., Riggs, A. D. (eds.) (1996) *Epigenetic Mechanisms of Gene Regulation.* Cold Spring Harbor Laboratory Press, Plainview, NY.

Sherratt, D., ed. (1995) *Mobile Genetic Elements.* IRL Press, Oxford and New York.

Smith, J., Hill, R., Baldwin, J. (1995) Plant chromatin structure and posttranslational modifications. *Crit. Rev. Plant Sci.* 14: 299–328.

Somerville, C., Meyerowitz, E. M., eds. (1994) *Arabidopsis.* Cold Spring Harbor Laboratory Press, Plainview, NY.

Struhl, K. (1998) Histone acetylation and transcriptional regulatory mechanisms. *Genes Dev.* 12: 599–606.

Weil, C. F., Wessler, S. R. (1990) The effects of plant transposable element insertion on transcription initiation and RNA processing. *Annu. Rev. Plant Physiol. Plant Mol. Biol.* 41: 527–552.

Biochemistry & Molecular Biology of Plants, B. Buchanan, W. Gruissem, R. Jones, Eds.
© 2000, American Society of Plant Physiologists

CHAPTER 8

Amino Acids

Gloria Coruzzi
Robert Last

CHAPTER OUTLINE

Introduction

In addition to their obvious role in protein synthesis (see Chapter 9), **amino acids** perform essential functions in both primary and secondary plant metabolism. Some amino acids serve to assimilate nitrogen and transport it from sources to sinks; others serve as precursors to secondary products such as hormones and compounds involved in plant defense. Thus, the synthesis of amino acids directly or indirectly controls various aspects of plant growth and development. Recent investigations of genes involved in amino acid biosynthesis reveal that this is a dynamic process controlled by metabolic, environmental, and developmental factors. This chapter will highlight examples in which combined molecular, biochemical, and genetic approaches have helped define the pathways and uncover regulatory mechanisms of amino acid biosynthesis in plants. These studies have implications for both basic and applied research because amino acid biosynthesis genes are targets for herbicide action and metabolic engineering of transgenic crop plants. A comprehensive review of the biochemistry of amino acid synthesis can be found elsewhere (see Further Reading).

8.1 Amino acid biosynthesis in plants: research and prospects

8.1.1 Amino acid biosynthesis pathways in plants have been inferred largely from microbial pathways.

The carbon skeleton backbones used for amino acid biosynthesis in plants are derived from glycolysis, photosynthetic carbon reduction, the oxidative pentose phosphate pathway, and the citric acid cycle (Fig. 8.1; see also Chapters 12 through 14). The pathways proposed for amino acid biosynthesis in plants (Fig. 8.2) are inferred in large part from those defined in *Escherichia coli* and yeast, where the steps and regulatory mechanisms have been identified through using a combination of genetics, biochemistry, and molecular biology (Box 8.1). At present, we do not know whether plants use all the enzymatic pathways and control points observed in bacteria or yeast.

Amino acid biosynthesis pathways in plants demonstrate levels of complexity not found in microbes. Whereas bacteria possess small genomes and lack organelles, plants may contain multiple genes for each step in a pathway, and sequential steps may occur in distinct subcellular compartments. Given that the mechanisms controlling the subcellular and intercellular transport of amino acids and their intermediates are at present largely unknown, it is impossible to predict the in vivo function of an **isoenzyme** based solely on in vitro biochemistry. Plant extracts often contain mixtures of isoenzymes that do not coexist in the same organelle or cell type within the plant. Therefore, the in vitro flux measurements used to define rate-limiting steps in the biosynthetic pathways of unicellular organisms have limited significance for study of multicellular plants.

8.1.2 *Arabidopsis* mutants reveal aspects of amino acid biosynthesis and its regulation in plants.

Our knowledge of amino acid biosynthetic pathways in microbes resulted largely from the biochemical characterization of **auxotrophic** mutants that require amino acid supplementation. Historically, the isolation of comparable amino acid biosynthesis mutants in plants was hampered by several factors, including gene redundancy and problems associated with supplementing auxotrophs. In recent years, however, several different types of **genetic screens** have identified plant mutants in amino acid biosynthetic enzymes. These include **positive selections** for tryptophan synthesis mutants (see Box 8.5), screens for the accumulation of metabolic intermediates (see Box 8.5), screens for photorespiratory mutants (see Fig. 8.11), and extensive screens for loss of isoenzymatic activity in glutamate dehydrogenase (see Fig. 8.16) and aspartate aminotransferase mutants. The phenotypes of these whole-plant mutants have begun to reveal the in vivo role of particular genes involved in amino acid biosynthesis and to define the key or rate-limiting steps in multistep pathways. Moreover, molecular studies on the regulation of amino acid biosynthesis genes in plants have revealed that in some cases transcriptional control plays a major role in regulating pathways. For example, the genes controlling glutamine and asparagine synthesis are not constitutively expressed "housekeeping" genes; rather, they are expressed differentially by cell type and are regulated by factors such as light and metabolic control. In plants, investigation of amino acid biosynthesis mutants has centered on developing a molecular-genetic blueprint of the amino acid biosynthesis

Carbon metabolite   **Amino acid derivative(s)**

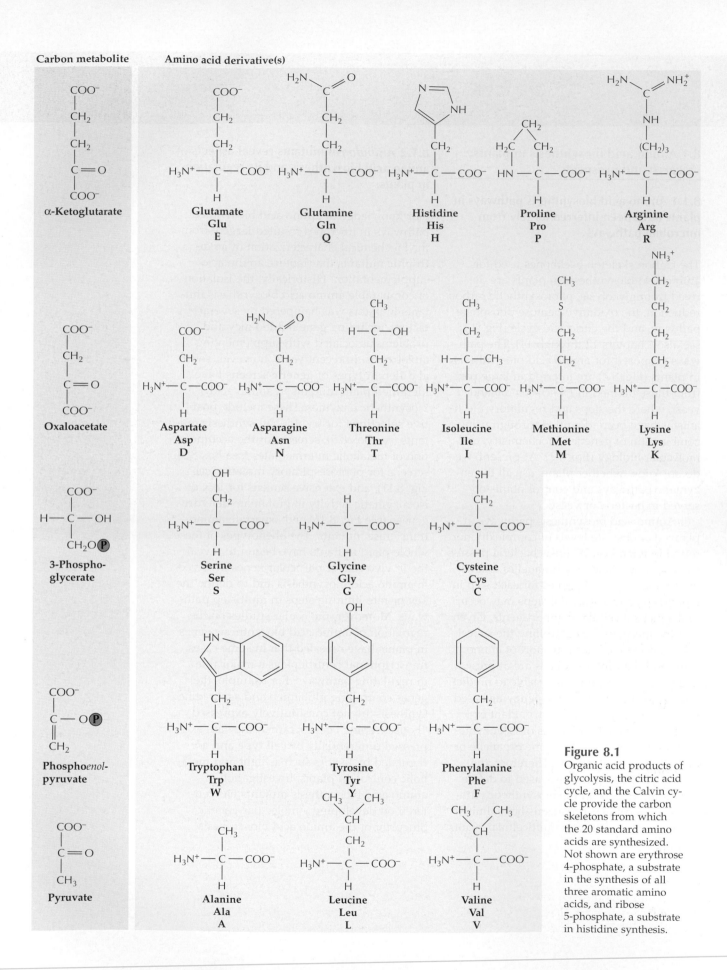

Figure 8.1
Organic acid products of glycolysis, the citric acid cycle, and the Calvin cycle provide the carbon skeletons from which the 20 standard amino acids are synthesized. Not shown are erythrose 4-phosphate, a substrate in the synthesis of all three aromatic amino acids, and ribose 5-phosphate, a substrate in histidine synthesis.

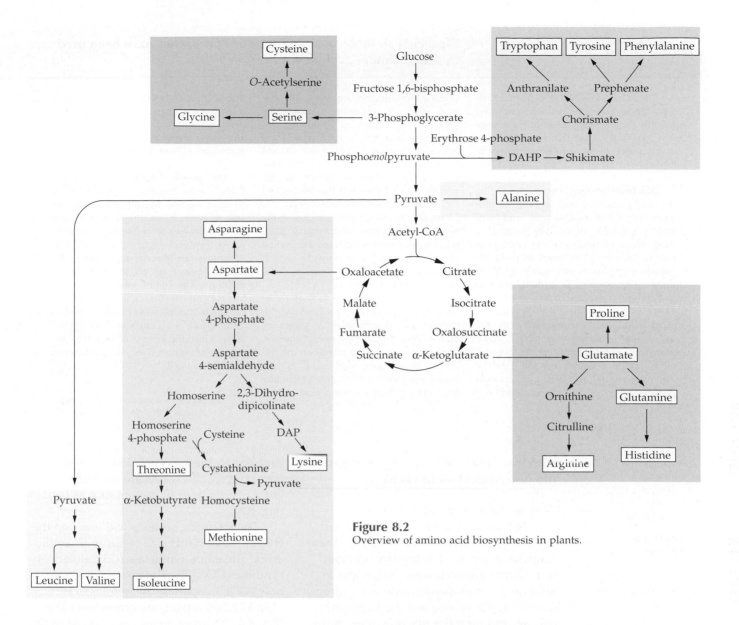

Figure 8.2
Overview of amino acid biosynthesis in plants.

pathways in a single model species, the dicot *Arabidopsis thaliana,* analogous to that generated in *E. coli* or yeast.

8.1.3 Amino acid pathways in plants are targets for basic and applied research.

Uncovering the features of amino acid biosynthesis unique to plants should stimulate both fundamental and applied research. One goal is to understand the genes that control growth-limiting processes (e.g., the assimilation of inorganic nitrogen into amino acids) and the factors that regulate the synthesis of plant secondary compounds from amino acid

precursors. Such studies should also provide a framework for manipulating amino acid biosynthesis pathways in transgenic plants. For example, enzymes in several pathways have been identified as targets for herbicides. In some cases, the genes encoding these enzymes have been used for engineering herbicide resistance (see Box 8.4). Eventually, future progress in amino acid biosynthesis research may provide enhanced crop resistance to osmotic stress (see Chapter 22) and improved food protein composition. Therefore, the structural and regulatory genes controlling amino acid biosynthesis in plants are of interest not only to biochemists but also to agricultural biotechnologists.

Box 8.1

Biochemical, molecular, and genetic approaches have been used to dissect amino acid biosynthesis in plants.

The methodologies that plant biologists use to dissect biosynthetic pathways fall into three general catgories: biochemical, molecular, and genetic. These frequently used terms are defined below in the context of amino acid biosynthesis research.

Biochemistry typically focuses on the structure, function, and activity of proteins. Traditional biochemical studies of amino acid biosynthesis have involved in vitro assays to determine the concentration or activity of enzymes in plant organelle preparations or organ extracts. In some cases, purified or partially purified enzyme preparations have enabled researchers to determine the biochemical properties of an enzyme, such as K_m, V_{max}, and substrate specificity. Major drawbacks to biochemical approaches include problems associated with enzyme instability and the presence of nonprotein enzyme inhibitors. In addition, conditions under which enzymes are assayed in vitro (e.g., substrate availability, pH, and the presence or absence of additional proteins) may bear little resemblance to the microenvironment of the enzyme in vivo. Purified enzymes have been used to produce specific antibodies, which are important tools for detecting the amount and the subcellular site of the enzyme in plants and, in some cases, for cloning the corresponding genes. Biochemical pathways have also been defined by feeding plants radiolabeled precursors (which contain isotopes such as ^{14}C or ^{15}N) or inhibitors of enzyme function.

A variety of **molecular biology** methods, including heterologous DNA hybridization, use of antibodies to screen for cDNAs encoding the enzyme of interest, and functional complementation of microbial auxotrophs with plant cDNA, have been used to clone genes encoding enzymes involved in amino acid biosynthesis. Once cloned, the cDNA encoding a biosynthetic enzyme may be used to examine gene regulation or to produce large amounts of the enzyme in a heterologous expression system for biochemical characterization.

The **genetic** approach involves isolation of plant mutants deficient in the enzyme of interest. Mutants missing a specific enzyme in an amino acid biosynthetic pathway can be used to assess the role of the missing enzyme in the plant, to determine rate-limiting steps, and to determine effects on related downstream pathways. The combination of molecular biology and genetics (molecular genetics) has made it possible to clone the mutant gene encoding a particular defective enzyme. Once the gene is cloned, the critical residues required for enzyme function can be identified.

8.2 Assimilation of inorganic nitrogen into N-transport amino acids

Plants assimilate inorganic nitrogen into these **N-transport amino acids:** glutamate, glutamine, aspartate, and asparagine. These compounds are used to transfer nitrogen from source organs to sink tissues and to build up reserves during periods of nitrogen availability for subsequent use in growth, defense, and reproductive processes. Nitrogen assimilated into glutamate and glutamine is readily disseminated into plant metabolism, because these amino acids donate nitrogen in the biosynthesis of amino acids, nucleic acids, and other N-containing compounds. Alternatively, nitrogen assimilated into glutamate and glutamine may be incorporated into aspartate and asparagine. Aspartate is a metabolically reactive amino acid that serves as the nitrogen donor in numerous aminotransferase reactions; asparagine is relatively inert and serves primarily as a nitrogen transport and storage compound. Glutamate, glutamine, aspartate, and asparagine are the major amino acids translocated in the phloem of most species, including corn, pea, and *Arabidopsis.* The concentrations of these transported amino acids are not static but are modulated by factors such as light (Fig. 8.3).

The following sections will focus on the enzymes that synthesize N-transport amino acids: glutamine synthetase (GS), glutamate synthase (GOGAT), glutamate dehydrogenase (GDH), aspartate aminotransferase (AspAT), and asparagine synthetase (AS) (Fig. 8.4). These enzymes are involved in the **primary assimilation** of inorganic nitrogen from the soil as well as in the reassimilation **(secondary assimilation)** of free ammonium within the plant. In plants, ammonium is released from organic compounds by means of several metabolic processes, including the deamination of amino acids during seed germination, the synthesis of specific amino acids or lignin, and photorespiration. Unlike animals, plants do not excrete nitrogenous wastes. Instead, the liberated ammonium must be reassimilated to support plant growth. For example, photorespiratory ammonium release may exceed primary nitrogen assimilation by 10-fold, so a plant unable to recycle this ammonium would quickly deplete its nitrogen stores.

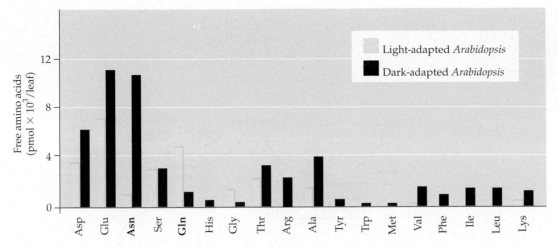

Figure 8.3

Concentrations of free amino acids in light- and dark-treated *Arabidopsis*, determined by HPLC. Aspartate, glutamate, asparagine, and glutamine constitute 70% of total free amino acids. Asparagine concentrations are induced dramatically in dark-adapted plants, whereas glutamine concentrations are increased in the light. These light-induced reciprocal changes in concentrations of asparagine and glutamine reflect the distinct natures of these amino acids. Glutamine, a metabolically reactive amino acid, is preferentially synthesized in the light. Asparagine, which is relatively inert, is preferentially synthesized in the dark. Asparagine, which carries more nitrogen atoms per carbon atom than does glutamine, is therefore a more economical compound to transport nitrogen when carbon skeletons are limiting (in the dark). Increased concentrations of glycine in light-adapted plants result from photorespiration, which produces glycine as a byproduct.

8.2.1 The GS/GOGAT cycle is the principal nitrogen assimilation pathway in plants.

Glutamine synthetase catalyzes the ATP-dependent assimilation of ammonium into glutamine, using glutamate as a substrate. GS functions in a cycle with glutamate synthase (glutamine-2-oxoglutarate aminotransferase), which catalyzes the reductive transfer of the amide group from glutamine to α-ketoglutarate, forming two molecules of glutamate (Fig. 8.5). Distinct isoenzymes of both GS and GOGAT have been identified in all higher plant species examined. Isoenzymes of GS found in the cytosol (GS1) and chloroplast (GS2) are the products of homologous but distinct nuclear genes (Fig. 8.6). The two major classes of GOGAT enzymes in higher plants are a ferredoxin-dependent GOGAT (Fdx-GOGAT) found exclusively in photosynthetic organisms, and an NAD(P)H-dependent GOGAT (NAD(P)H-GOGAT) found in plants and bacteria. Subcellular fractionation and identification of plastid-targeting sequences have shown that both Fdx-GOGAT and NAD(P)H-GOGAT are plastid-localized (Fig. 8.6). Some proteins in the latter class are more active in the presence of NADH than NADPH and are therefore termed NADH-GOGAT.

The GS/GOGAT cycle is most likely the principal route of ammonium assimilation in plants. GS, which has a very high affinity for ammonium, can operate at the low ammonium concentrations present in living cells (K_m 3 to 5 μM). Labeling studies that trace the fate of $^{15}NH_4^+$ confirm that the label is incorporated primarily into the amide group of glutamine, subsequently appearing in the amino groups of glutamate (see Fig. 8.5) and other amino compounds, including glutamine. The addition of GS inhibitors, e.g., the glutamate analogs methionine sulfoximine or L-phosphinothricin (MSO or L-PPT, respectively; Fig. 8.7), inhibits but does not completely block labeling of the amido group of glutamine and the amino group of glutamate. The GOGAT inhibitor azaserine (a glutamine analog) blocks incorporation of the radiolabel into glutamate. These results, among others, support the hypothesis that the majority of inorganic nitrogen is assimilated through the GS/GOGAT pathway in plants.

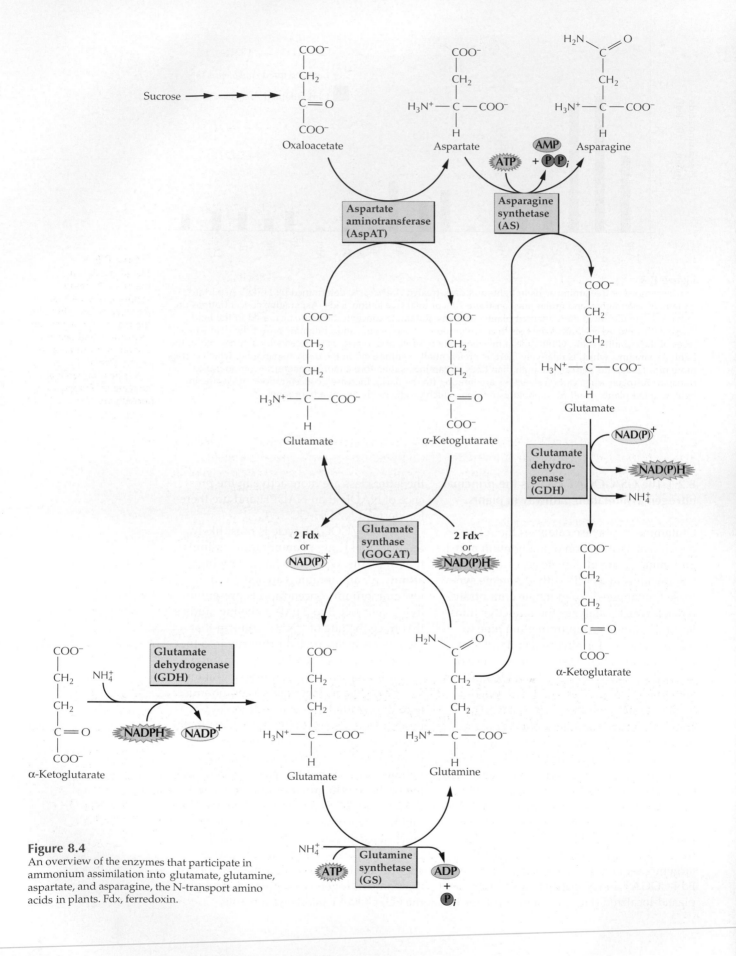

Figure 8.4
An overview of the enzymes that participate in ammonium assimilation into glutamate, glutamine, aspartate, and asparagine, the N-transport amino acids in plants. Fdx, ferredoxin.

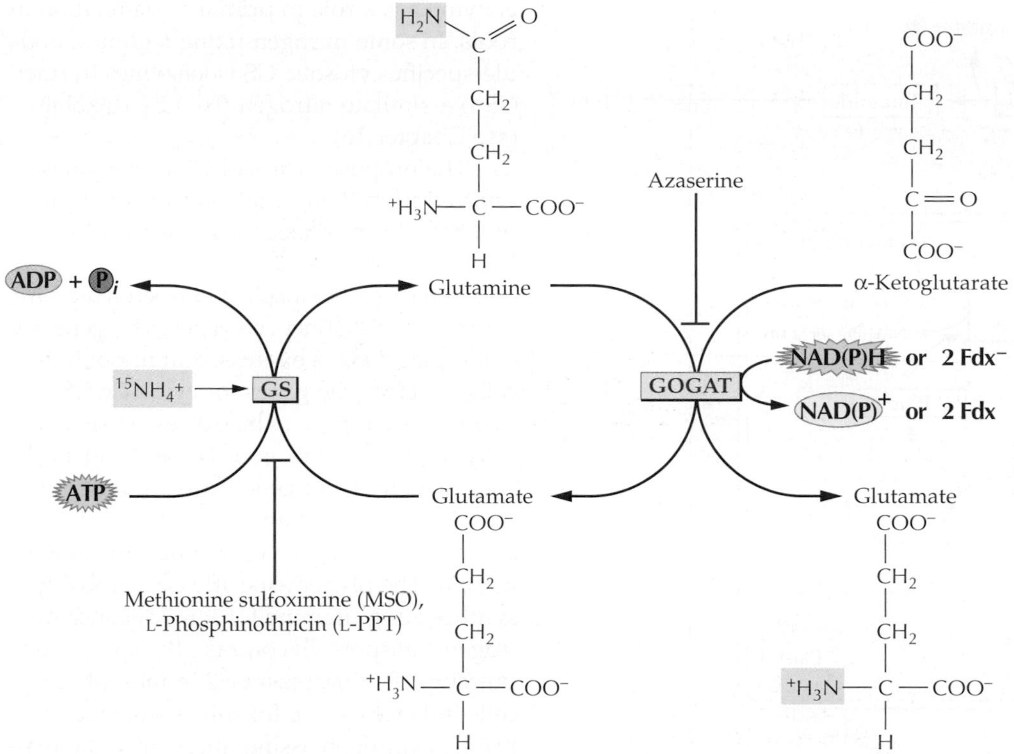

Figure 8.5
The glutamine synthetase–glutamate synthase (GS/GOGAT) pathway is thought to be the principal mechanism of primary and secondary ammonium assimilation. Sites of action are shown for several enzyme inhibitors. Fdx, ferredoxin.

8.2.2 Molecular and genetic studies demonstrate that cytosolic and chloroplast GS isoenzymes perform nonoverlapping roles in vivo.

GS isoenzymes can be separated into two classes by ion-exchange chromatography—one localized in the cytosol (GS1), the other in the chloroplast (GS2). Genes encoding each GS isoenzyme have been cloned by several different methods, including cross-hybridization to animal *GS* cDNAs and complementation of microbial *GS* mutants. Plant *GS* cDNAs are able to complement bacterial *GS* mutants, despite the fact that their respective holoenzymes are distinct in subunit structure and sequence. Plants examined thus far appear to possess a single nuclear gene encoding GS2 and multiple (two to four) nuclear genes encoding GS1 subunits. The GS holoenzymes in plants function as octamers, and GS1 polypeptides can assemble into homo- or heterooctamers. Although the chloroplast and cytosolic GS holoenzymes do not appear to differ significantly in their biochemical properties when assayed

---

Box 8.2

Not all screens for photorespiration mutants have identified GS mutants.

A screen for photorespiratory mutants of barley in 1987 identified plants specifically deficient in chloroplast GS2 and pointed to a key role for that isoenzyme in reassimilation of photorespiratory ammonia. Curiously, however, no *GS* mutants were recovered in a 1980 screen for photorespiratory mutants in *Arabidopsis*. This finding has several possible explanations.

Perhaps the *Arabidopsis* photorespiratory screen was not saturating, i.e., not extensive enough to identify all possible loci. This is unlikely, however, because multiple alleles for many enzymes in the photorespiratory pathway were isolated, including 58 mutations affecting Fdx-GOGAT (see Section 8.2.3).

If cytosolic GS1 and chloroplast GS2 are both expressed in mesophyll cells of *Arabidopsis,* a mutation in one gene could be masked by a functional copy of the other. Similarly, if *Arabidopsis* contains more than one gene for chloroplast GS2, a mutation in only one gene might be masked. Finally, if a mutation in either chloroplast or cytosolic GS is lethal in *Arabidopsis,* failure of the mutant plants to survive would prevent their isolation.

Ongoing investigations on the *GS* genes and isoenzymes in *Arabidopsis* may answer these questions and clarify the role of GS in photorespiration and primary assimilation.

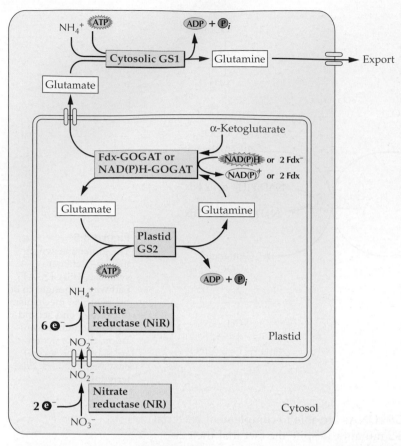

Figure 8.6
Isoenzymes of glutamine synthetase (GS) are present in both the plastids (GS2) and the cytoplasm (GS1). Fdx, ferredoxin.

enzyme has a role in primary assimilation in roots. In some nitrogen-fixing legumes, nodule-specific cytosolic GS isoenzymes (termed GSn) assimilate nitrogen fixed by rhizobia (see Chapter 16).

The proposed roles of GS isoenzymes inferred from their organ-specific distribution have been refined by more recent molecular and genetic analysis. Genes encoding chloroplast and cytosolic GS isoenzymes are expressed in distinct cell types. The gene for chloroplast GS2 is expressed in mesophyll cells, whereas the genes for cytosolic GS1 isoenzymes appear to be expressed specifically in phloem (Fig. 8.9). These distinct cell-specific patterns of gene expression suggest that the chloroplast and cytosolic GS isoenzymes perform nonoverlapping functions in vivo. The phloem-specific GS1 probably synthesizes glutamine for long-distance nitrogen transport. In contrast, the specific expression of chloroplast GS2 in mesophyll cells indicates a role for this isoenzyme in primary nitrogen assimilation, or in the reassimilation of photorespiratory ammonium. Genetic evidence supports the latter conclusion. First, mutants in chloroplast GS2 display a conditional lethality: They die in air but grow in a 1% CO_2 atmosphere that suppresses photorespiration. Thus, the chloroplast GS2 enzyme is responsible for reassimilating photorespiratory ammonium released in the mitochondria. This example demonstrates how a mutant can be used to define the flux through a pathway in vivo when the parameters regulating intra- and intercellular transport are unknown. Second, the GS2 mutants have normal amounts of cytosolic GS1 (see Fig. 8.8), confirming that the GS isoenzymes play nonoverlapping roles. Moreover, the phloem-specific expression of

in vitro, they have distinct in vivo functions. The chloroplast GS2 holoenzyme is the predominant GS isoenzyme in leaves (Fig. 8.8), where it is thought to function both in primary ammonia assimilation and in the reassimilation of photorespiratory ammonia. Cytosolic GS1 isoenzymes are present at low concentrations in leaves and at higher concentrations in roots, suggesting that this iso-

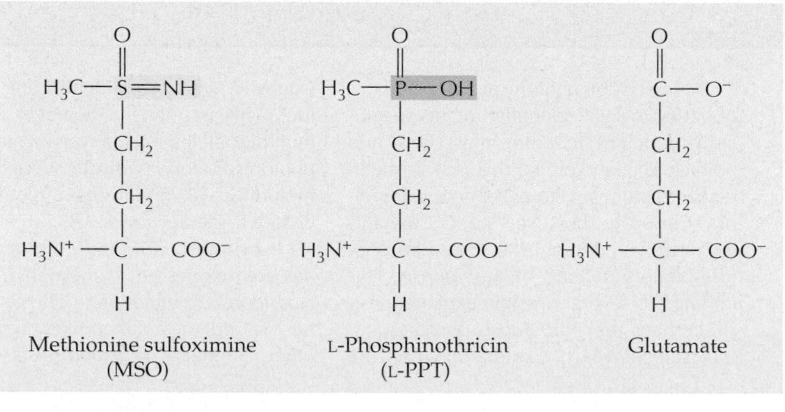

Figure 8.7
Methionine sulfoximine and phosphinothricin, competitive inhibitors of glutamine synthetase.

Methionine sulfoximine (MSO)

L-Phosphinothricin (L-PPT)

Glutamate

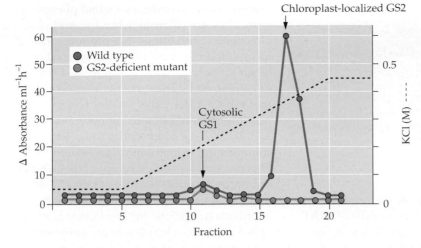

Chloroplast-localized GS2

Figure 8.8
The plastid-localized isoenzyme GS2 is the principal source of GS activity in leaves. Shown are the results of separating GS2 and cytosolic GS1 by ion-exchange chromatography. Desalted extracts from young, expanding leaves were applied to an ion-exchange column and eluted with a linear KCl gradient (dotted line).

cytosolic GS1 may explain why this isoenzyme is unable to compensate for the loss of the chloroplast GS2 isoenzyme in leaf mesophyll cells of the barley GS2 photorespiratory mutants (Box 8.2).

8.2.3 Mutants indicate a major role for Fdx-GOGAT in photorespiration.

Despite extensive biochemical characterizations of Fdx-GOGAT and NADH-GOGAT in plants, it is not currently resolved whether these isoenzymes perform overlapping or distinct functions in vivo. Quantitative analyses of each isoenzyme in various tissues have been used as circumstantial evidence to propose an in vivo role. For example, Fdx-GOGAT is the predominant GOGAT isoenzyme in leaves and can account for as much as 95% to 97% of total leaf GOGAT activity, as determined in *Arabidopsis* and barley (Table 8.1). In contrast, the NADH-GOGAT isoenzyme is present in low amounts in leaves but constitutes the predominant isoenzyme in nonphotosynthetic tissues such as

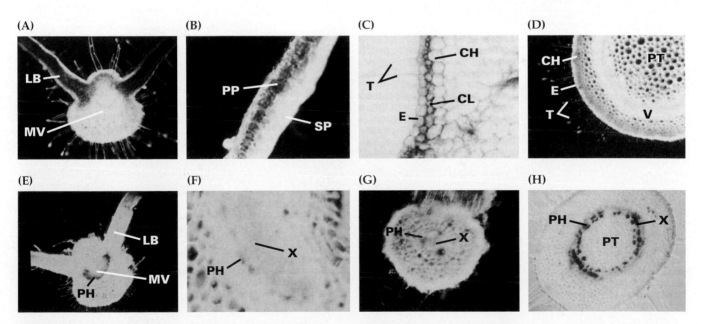

Figure 8.9
Photographs and light micrographs of mature transgenic tobacco plants show that GS1 and GS2 promoters are expressed in different tissue types. Use of β-glucuronidase staining to analyze promoter-specific expression patterns is described in Chapter 7, Box 7.5. Here, the promoter for chloroplast-localized GS2 is expressed in photo-synthetic cells (A–D), whereas the promoter for cytosolic GS1 is expressed in phloem. CH, chlorenchyma; CL, collenchyma; E, epidermis; LB, leaf blade; MV, midvein; PH, phloem; PP, palisade parenchyma; PT, pith parenchyma; R, root; SP, spongy parenchyma; T, trichome; V, vasculature; X, xylem.

Table 8.1 Levels of GOGAT isoenzymes in wild-type and mutant plants

	Percentage of total wild-type GOGAT activity		
	Fdx-GOGAT	NADH-GOGAT	Total GOGAT
Wild-type			
Arabidopsis	95	5	100
Barley	97	3	100
Fdx-GOGAT mutants			
Arabidopsis gluS alleles	5	5	10
Barley Rpr 82/9	<1	3	<5

roots. These organ-specific distribution patterns suggest a major role for Fdx-GOGAT in primary nitrogen assimilation and photorespiration in leaves, whereas NADH-GOGAT may function predominantly in primary assimilation in the roots (Fig. 8.10).

The in vivo role of Fdx-GOGAT has been further elucidated by the isolation and characterization of photorespiratory mutants. In the case of the *Arabidopsis* mutants *(gluS)*, Fdx-GOGAT activity in leaves is reduced to less than 5% of wild-type levels, whereas the low amounts of NADH-GOGAT activity detected in wild-type plants remains unaffected. Similar results were obtained for barley Fdx-GOGAT mutants (Table 8.1). All of the Fdx-GOGAT–deficient mutants have a conditional lethal phenotype: They are chlorotic when grown in atmospheric conditions that promote photorespiration (air), and they are rescued when photorespiration is suppressed (1% CO_2) (Fig. 8.11). These results suggest both that Fdx-GOGAT plays a major role in the reassimilation of photorespiratory ammonia and that Fdx-GOGAT–deficient mutants are competent in primary nitrogen assimilation. This presents a paradox because Fdx-GOGAT ordinarily accounts for 95% of GOGAT activity in leaves, the major site of primary assimilation. Perhaps the low concentrations of NADH-GOGAT in leaves are sufficient for primary assimilation. Alternatively, all *gluS* mutants may be leaky and may generate sufficient amounts of Fdx-GOGAT to support primary assimilation. The latter assertion is supported by the fact that *Arabidopsis* has two genes for Fdx-GOGAT, which are differentially expressed in leaves and roots and are differentially regulated by light (Fig. 8.12). Although *GLU1* maps to the location of the *gluS* mutation, the presence of a second Fdx-GOGAT gene *(GLU2)* may allow primary assimilation to proceed in leaves of *gluS* mutants at rates sufficient to maintain primary assimilation and growth.

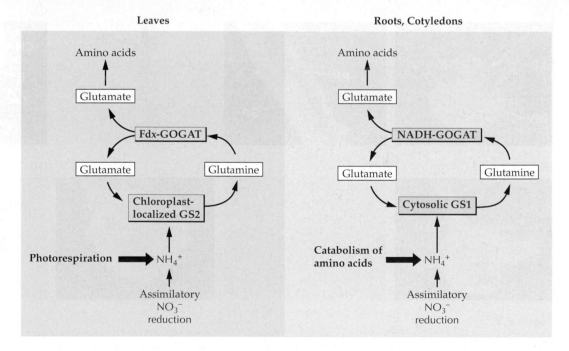

Figure 8.10
NADH-dependent and Fdx-dependent isoenzymes of GOGAT (NADH-GOGAT and Fdx-GOGAT, respectively) play different physiological roles in plant metabolism.

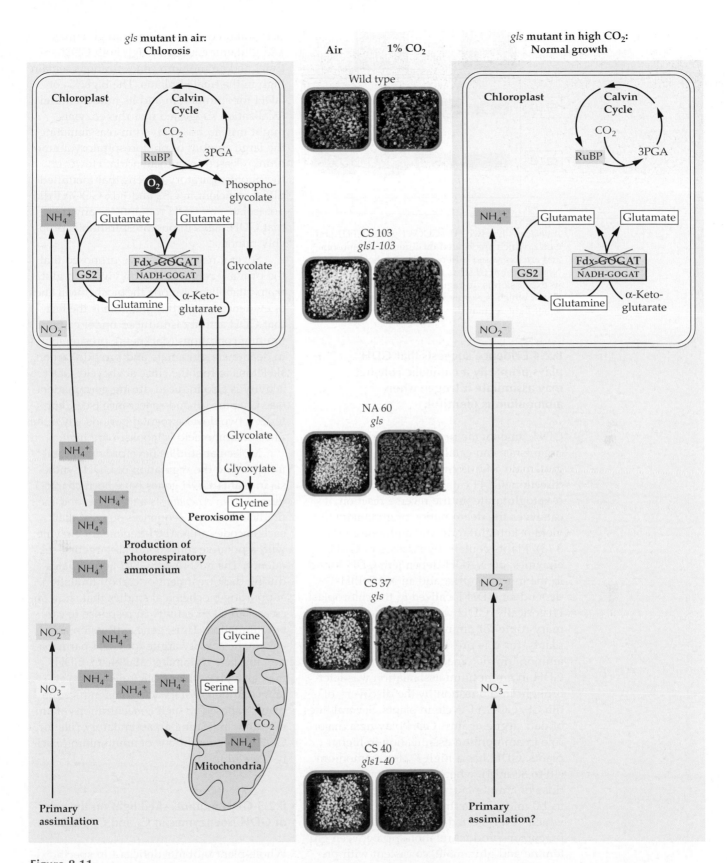

Figure 8.11
Proposed role of Fdx-GOGAT in primary nitrogen assimilation and in assimilation of ammonium produced as a by-product of photorespiration. *gls* (Fdx-GOGAT) mutants of *Arabidopsis* demonstrate a chlorotic phenotype when grown in air (approximately 0.03% CO_2) but can tolerate growth in a 1% CO_2 atmosphere that suppresses photorespiration.

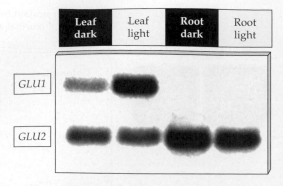

| Leaf dark | Leaf light | Root dark | Root light |

GLU1

GLU2

Figure 8.12
Arabidopsis has two Fdx-GOGAT genes, *GLU1* and *GLU2*, which are located on different chromosomes and are expressed differentially. As shown in this Northern blot, *GLU1* is expressed only in leaf tissues; its expression is up-regulated in response to light. *GLU2*, which is expressed in both leaves and roots, is not light-regulated.

8.2.4 Evidence suggests that GDH plays primarily a catabolic role but may assimilate nitrogen when ammonium is plentiful.

GDH, an enzyme present in nearly all living organisms, can catalyze both the synthesis of glutamate and its catabolism. In the forward direction, GDH catalyzes the amination of α-ketoglutarate; in the reverse reaction, it catalyzes the deamination of glutamate to yield α-ketoglutarate and ammonium (Fig. 8.13). Plants contain two classes of GDH enzymes: an NADH-dependent GDH found in the mitochondria, and an NAD(P)H-dependent GDH localized to the chloroplast. Historically, GDH was thought to be the primary route for ammonium assimilation in plants, for it is present in microbes grown in ammonium-rich media. However, the role of GDH in ammonium assimilation was later brought into question by the discovery of the GS/GOGAT cycle in plants. Several lines of data argue against GDH playing a major role in ammonium assimilation in higher plants. GDH has a high K_m for ammonium (10 to 80 mM), whereas tissue concentrations of ammonia typically range from 0.2 to 1.0 mM. Furthermore, experiments in which $^{15}NH_4^+$ is fed to plants indicate a precursor–product relationship between glutamine and glutamate, consistent with primary nitrogen assimilation by GS/GOGAT. Additionally, treatment of plants with the GS inhibitor MSO (see Fig. 8.7) prevents in-corporation of ammonium into glutamate and glutamine, even though both GDH activity and ammonium concentrations remain high in the treated plants. The high K_m of GDH for ammonium and its mitochondrial localization suggested that this enzyme might instead be involved in reassimilating the large amount of photorespiratory ammonium released in mitochondria. However, the photorespiratory screens that identified plants deficient in GS2 and Fdx-GOGAT did not isolate any GDH mutants, so any role that GDH plays in photorespiration is probably minor.

Several researchers have proposed that the primary role for GDH in vivo is in glutamate catabolism. Among the biochemical lines of evidence supporting this role is the fact that GDH activity is induced under carbon-limiting conditions (darkness), presumably to deaminate glutamate and provide carbon skeletons to fuel the citric acid cycle. GDH activity is also induced during germination (see Chapter 19) and senescence (see Chapter 20), two developmental periods when the rates of amino acid catabolism are high.

Molecular studies have provided some insights into the regulation of GDH synthesis in plants. *GDH* genes have been characterized from *Arabidopsis* and maize, and their predicted protein sequences suggest that each encodes an NADH-dependent enzyme with a putative mitochondrial-targeting sequence. The mRNA for GDH accumulates during dark-treatment or carbon limitation, supporting biochemical studies that show increases in GDH activity in response to carbon limitation. Thus, gene expression and enzymatic activity argue for GDH participation in plant glutamate catabolism. GDH may also play an anabolic role under certain conditions. GDH activity is induced in plants exposed to high concentrations of ammonium, indicating an assimilatory role for GDH under conditions of ammonium toxicity (Fig. 8.13).

8.2.5 GDH mutants shed light on the role of GDH isoenzymes in C₃ and C₄ plants.

Whole-plant mutants deficient in specific GDH isoenzymes have been identified both in a C₄ plant (maize) and in a C₃ plant (*Arabidopsis*). Mutants were confirmed by

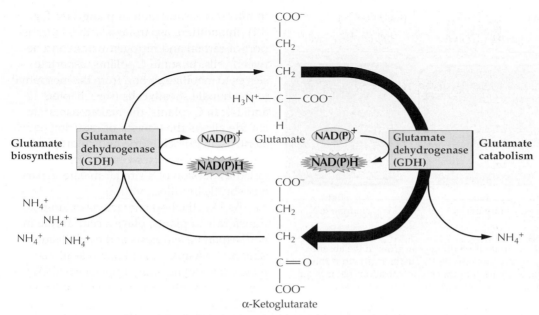

COO⁻
|
CH₂
|
CH₂
|
H₃N⁺—C—COO⁻
|
H

Glutamate

NAD(P)⁺

NAD(P)H

Glutamate
biosynthesis

Glutamate
dehydrogenase
(GDH)

NAD(P)⁺

NAD(P)H

Glutamate
dehydrogenase
(GDH)

Glutamate
catabolism

NH₄⁺
NH₄⁺
NH₄⁺   NH₄⁺

NH₄⁺

COO⁻
|
CH₂
|
CH₂
|
C=O
|
COO⁻

α-Ketoglutarate

Figure 8.13
GDH is thought to function primarily in glutamate catabolism (deamination) but can also assimilate inorganic nitrogen into glutamate when ammonium concentrations are high.

assaying crude leaf extracts for loss of GDH isoenzymes, using a native gel assay in which the GDH holoenzymes of wild-type plants are resolved as seven bands on the gel (Fig. 8.14). These seven GDH isoenzymes are formed by the association of two types of GDH subunits (GDH1 and GDH2) into hexameric complexes—two distinct homohexamers and five heterohexamers formed by the association of GDH1 and GDH2 monomers (Fig. 8.14). GDH isoenzyme analyses predict that the seven GDH isoenzymes are the products of two non-allelic genes. Subsequent molecular studies have uncovered two GDH genes in *Arabidopsis*, one of which comaps with a mutation (*gdh1-1*) that causes specific deficiencies in the GDH1-containing homomeric and heteromeric holoenzymes. A similar mutant lacking the GDH1-containing holoenzymes has also been described in maize. Neither the maize nor the *Arabidopsis gdh1* mutants are null for GDH activity, because each possesses normal amounts of the GDH2 homohexamer. Phenotypic analysis of the *Arabidopsis* and maize *gdh1* mutants has clarified the in vivo roles of the GDH1 holoenzyme in C₃ and C₄ plants. Compared with wild-type *Arabidopsis*, the *gdh1-1* mutant displays an impaired growth phenotype when grown under conditions of high inorganic nitrogen (Fig. 8.15). This conditional phenotype offers additional evidence that GDH1 may play a role in ammonia assimilation under conditions of inorganic nitrogen excess. The *gdh1* mutant of *Arabidopsis* will be valuable for defining the function of this enzyme in photorespiration in C₃ plants. In contrast, a maize *gdh* mutant displays a growth phenotype only when grown under low night temperatures. The photorespiratory rate is very low or not detectable in maize, suggesting that GDH must be playing some other role in C₄ plants, possibly involvement in a stress response.

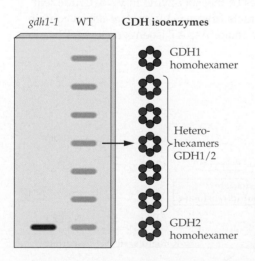

gdh1-1   WT   **GDH isoenzymes**

GDH1 homohexamer

Hetero-hexamers GDH1/2

GDH2 homohexamer

Figure 8.14
The seven isoenzymes of GDH in *Arabidopsis* are derived from the products of two genes, *GDH1* and *GDH2*. *gdh1* mutants contain only one GDH isoenzyme, a homohexamer of GDH2 protein. WT, wild type.

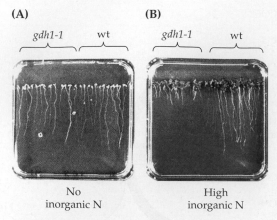

(A)
gdh1-1   wt

(B)
gdh1-1   wt

No
inorganic N

High
inorganic N

Figure 8.15
gdh1-1 mutant seedlings can grow in the absence of inorganic nitrogen (left), but their growth is inhibited (right) by high nitrogen concentrations (20 mM ammonium plus 40 mM nitrate). wt, wild type.

8.2.6 Studies of mutant plants have defined the isoenzymes that control nitrogen assimilation into the N-transport amino acid aspartate.

After its initial assimilation into glutamine and glutamate, nitrogen can be distributed to many other compounds by the action of enzymes called **transaminases**. Because glutamate-utilizing transaminations regenerate α-ketoglutarate, the carbon precursor needed for ammonium assimilation, these transaminations permit primary assimilation to continue in the absence of net α-ketoglutarate synthesis. In particular, synthesis of aspartate regenerates the carbon skeletons required for further nitrogen assimilation by transferring an amino group from glutamate to oxaloacetate. Thus, AspAT plays a key role

in nitrogen assimilation in plants (see Fig. 8.4). In addition, aspartate is active in transport of carbon and nitrogen within and between cells. In some C_4 plants, aspartate serves to mobilize carbon from the mesophyll to the bundle sheath cells (see Chapters 12 and 14). In C_3 plants, the malate–aspartate shuttle allows the transfer of reducing equivalents from mitochondria and chloroplasts into the cytoplasm (see Chapter 14 for detailed discussion of a similar malate–aspartate shuttle in mitochondria).

AspAT, the best characterized aminotransferase in plants, plays a central role in both aspartate synthesis and catabolism (Fig. 8.16). AspAT, also known as glutamate:oxaloacetate aminotransferase (GOT), is a pyridoxal phosphate-dependent enzyme (Fig. 8.17)

The ability of AspAT to interconvert these important carbon and nitrogen compounds places it in a key position to regulate plant metabolism. In line with the fact that aspartate is used to transfer carbon, nitrogen, and reducing equivalents between intracellular compartments, isoenzymes of AspAT have been localized to the cytosol, mitochondria, chloroplasts, and peroxisomes.

In *Arabidopsis*, the entire gene family of AspAT isoenzymes has been characterized. Five different *AspAT* cDNA clones have been identified, including those encoding the plastid, mitochondrial, peroxisomal, and cytosolic forms. In vitro studies on the partially purified AspAT isoenzymes indicate a broad range of K_m values and substrate specificities.

Recent molecular genetic studies using *AspAT* mutants have helped discern the roles of the isoenzyme in vivo. Crude leaf extracts from wild-type *Arabidopsis* contain two major AspAT isoenzymes, AAT2

Figure 8.16
Aspartate aminotransferase (AspAT) catalyzes the reversible transamination of oxaloacetate by glutamate to yield α-ketoglutarate and aspartate.

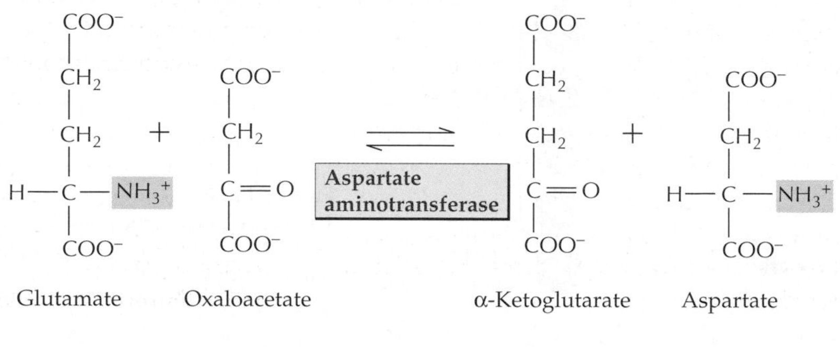

Glutamate   Oxaloacetate   α-Ketoglutarate   Aspartate

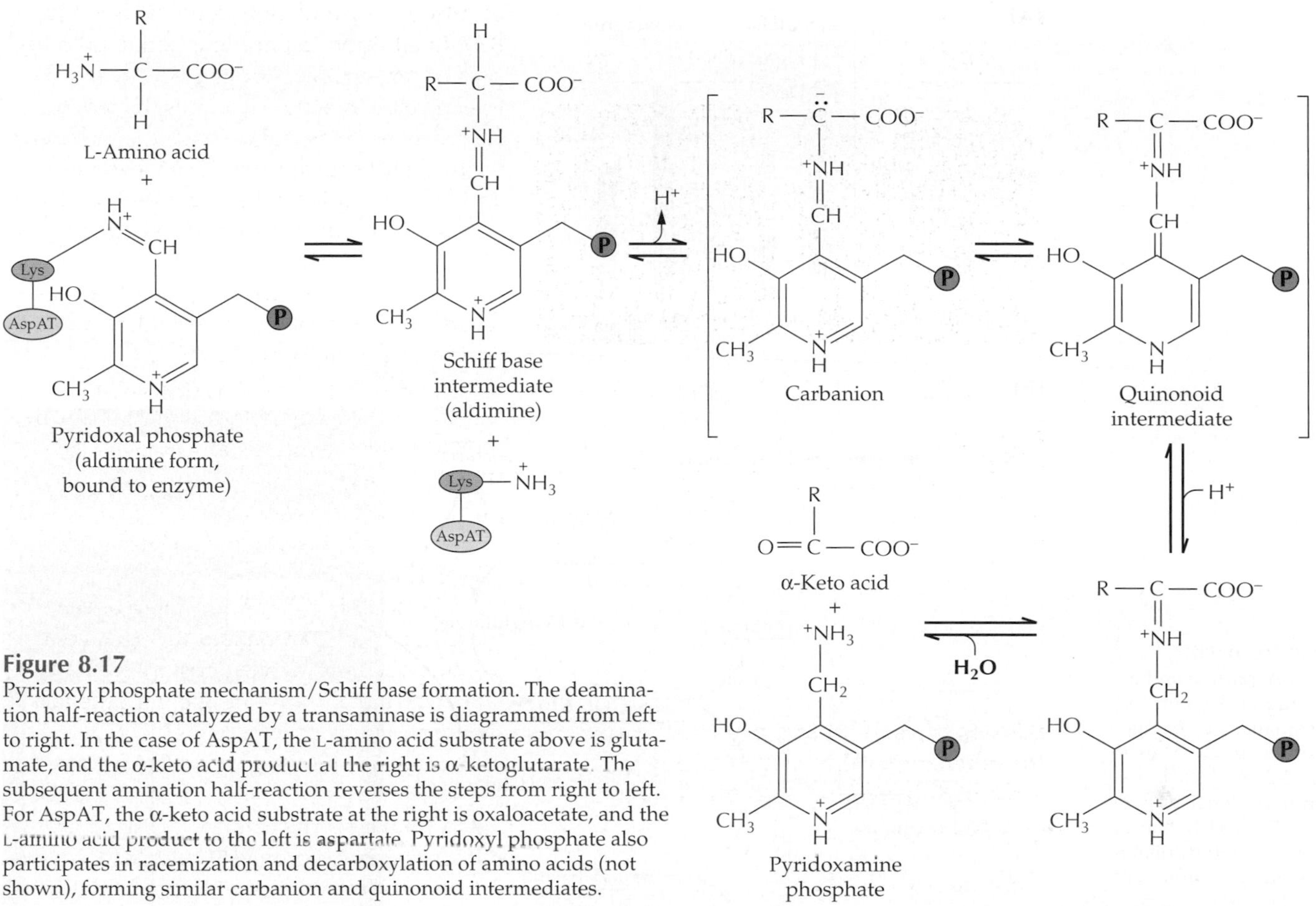

Figure 8.17
Pyridoxyl phosphate mechanism/Schiff base formation. The deamination half-reaction catalyzed by a transaminase is diagrammed from left to right. In the case of AspAT, the L-amino acid substrate above is glutamate, and the α-keto acid product at the right is α-ketoglutarate. The subsequent amination half-reaction reverses the steps from right to left. For AspAT, the α-keto acid substrate at the right is oxaloacetate, and the L-amino acid product to the left is aspartate. Pyridoxyl phosphate also participates in racemization and decarboxylation of amino acids (not shown), forming similar carbanion and quinonoid intermediates.

(cytosol) and AAT3 (chloroplast), as detected by activity staining on native gels. Mutant plants specifically deficient in either AAT2 or AAT3 have been identified with this assay. Amino acid analysis of leaf extracts taken from light-grown plants reveals that the free aspartate concentrations are diminished by 80% in *aat2*, a mutant of AAT2 (Fig. 8.18A). Thus, the cytosolic form, AAT2, controls the bulk of aspartate synthesis in light-grown plants. The *aat2* mutants also show a specific decrease in the concentrations of asparagine in dark-adapted plants, suggesting that aspartate synthesized in the light by cytosolic AAT2 is the precursor for asparagine synthesized in the dark (Fig. 8.18B). Although three mutants deficient in chloroplast AAT3 had no obvious phenotype, no conclusions can be drawn regarding the function of this isoenzyme, because we do not know whether these were null alleles, totally deficient in AAT3 activity.

8.2.7 Light represses the biosynthesis of asparagine, an amino acid used for N-transport and storage.

Amidation of aspartate by glutamine or ammonium yields asparagine, an inert amino acid used to store nitrogen and transport it from sources to sinks. Asparagine is the major nitrogenous compound detected in the phloem of several legumes, with concentrations as great as 30 mM reported. Crystallized out of asparagus extracts almost 200 years ago, asparagine was the first amino acid discovered.

Despite its historical, biochemical, and gastronomic importance (Box 8.3), the enzymology of asparagine biosynthesis in plants has been plagued by problems of enzymatic instability and remains enigmatic. Three possible routes for asparagine synthesis have been proposed, of which the one utilizing the glutamine-dependent enzyme AS is now

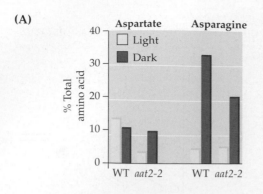

(A)

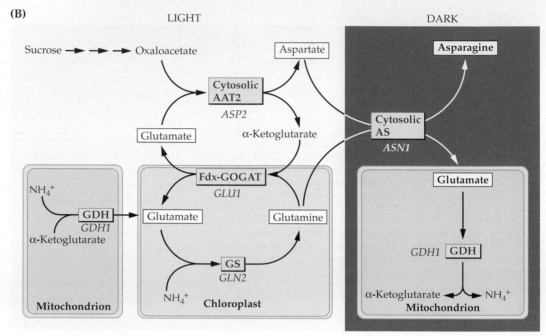

LIGHT **DARK**

Figure 8.18
(A) Aspartate and asparagine account for a smaller percentage of free amino acids in *aat2-2*, an *Arabidopsis* mutant deficient in cytosolic AAT2, than in wild-type (wt) plants. (B) AAT2 synthesizes aspartate during periods of illumination. This aspartate is subsequently converted to asparagine in the dark.

accepted as the major route for asparagine biosynthesis in plants. In an ATP-dependent reaction, AS catalyzes the transfer of the amido group from glutamine to aspartate, generating glutamate and asparagine (Fig. 8.19A). Biochemical studies on partially purified plant AS enzymes have been hindered by the copurification of a heat-stable, dialyzable inhibitor; the instability of AS in vitro; and the presence of contaminating asparaginase activity. Although glutamine is the preferred substrate for nearly all of the AS enzymes studied in higher plants, some evidence indicates that ammonium-dependent asparagine synthesis may also occur in plants, as it does in *E. coli* (Fig. 8.19B). Partially purified plant AS enzymes from soybean can also use ammonium as a substrate, albeit at a much higher K_m (3.0 mM) than

that for glutamine (0.18 mM). Maize roots contain an AS enzyme that has a less stringent requirement for glutamine ($K_m = 0.5$ mM) than for ammonium ($K_m = 2.0$ mM). Tracer studies with $^{15}NH_4^+$ have shown that asparagine is labeled after glutamine is, so the direct amination of aspartate does not appear to occur at significant rates in vivo. However, production of asparagine by direct assimilation, when plants encounter toxic concentrations of ammonium, has also been proposed.

The cloning of plant *AS* genes has provided insights into the properties of the AS isoenzymes in plants. Initially, cDNA clones encoding plant AS were obtained from pea, with use of a human *AS* cDNA clone as a heterologous probe. Additional *AS* genes have been isolated from a variety of plants,

Box 8.3   Asparagine research has a long and flavorful history.

Asparagine, the first amino acid discovered, was initially recognized as a substance that crystallized out of asparagus extracts and was named accordingly. Early studies in the 1800s also showed that the quantities of free asparagine were increased in plants exposed to dark treatment. These early biochemical studies explain why spargel (white asparagus; see photograph), a culinary delicacy grown under mounds of soil, has an intense asparagine flavor. This etiolation treatment results in the accumulation of high concentrations of asparagine and a reduction in the formation of unsavory woody tissue.

Recent molecular studies have shown that genes involved in asparagine synthesis are preferentially transcribed in the dark in several species (including *Arabidopsis*), confirming the initial discovery of light repression of asparagine synthesis reported as early as the 1800s.

including legumes, nonlegumes, and monocots. The AS polypeptides encoded in each of these cDNAs contain a purF-type glutamine-binding domain that was previously described for glutamine amidotransferases involved in purine biosynthesis in *E. coli*, another indication that glutamine is the substrate of the encoded plant AS enzyme. Whether any of the encoded AS enzymes can also utilize ammonium as a substrate in vivo is an open question. In humans, the *AS* gene encoding a glutamine-dependent AS enzyme can also catalyze an ammonium-dependent reaction in vitro, albeit at a higher K_m. In bacteria, glutamine-dependent and ammonium-dependent AS enzymes are encoded by separate genes. One AS enzyme can utilize only ammonium

(A) Glutamine-dependent asparagine synthesis

Glutamine + Aspartate $\xrightarrow[\text{Asparagine synthetase}]{\text{ATP} \quad \text{AMP} + \text{P P}_i}$ Glutamate + Asparagine

(B) Ammonium-dependent asparagine synthesis

Aspartate + NH_4^+ $\xrightarrow[\text{Asparagine synthetase}]{\text{ATP} \quad \text{AMP} + \text{P P}_i}$ Asparagine

Figure 8.19
Asparagine synthetases from plants preferentially use glutamine as a nitrogen donor (A) but can catalyze the assimilation of inorganic nitrogen when ammonium is plentiful (B).

as a substrate, whereas the other can use either glutamine or ammonium as the nitrogen donor. The biochemical characterization of plant AS enzymes expressed in microbes should further define the ammonium- versus glutamine-dependent activities of each isoenzyme.

8.2.8 Light and carbon metabolism regulate the assimilation of nitrogen into amino acids.

The notion that amino acid biosynthesis is a static or "housekeeping" process is inconsistent with current data. The incorporation of inorganic nitrogen into the assimilatory amino acids glutamate, glutamine, aspartate, and asparagine is a dynamic process that is regulated by external factors such as light and by internal stores of carbon and nitrogen metabolites.

As early as the 1850s, studies showed that asparagine concentrations were dramatically influenced by light in several plant species. Subsequent investigations revealed that the asparagine content of phloem exudates and the amount of AS enzyme activity are negatively regulated by light or by sucrose in numerous plant species, including *Arabidopsis* and asparagus. Asparagine accumulation and AS activities are induced when light-grown plants are dark-adapted. Conversely, glutamine concentrations are increased in light-grown plants relative to dark-treated plants (see Fig. 8.3).

Expression of the genes controlling assimilation of inorganic nitrogen into amino acids is subject to control by both light and metabolic factors. For example, genes for nitrate reductase and nitrite reductase are induced by light and by sucrose (see Chapter 16). Similarly, light up-regulates the expression of genes involved in assimilation of ammonium into glutamine and glutamate, particularly that of chloroplast GS2 and Fdx-GOGAT. By contrast, light represses the expression of genes for AS. In all three cases, this transcriptional regulation is mediated, at least in part, by the plant photoreceptor phytochrome (see Chapter 18). Additionally, the light-induced synthesis of carbohydrates also appears to affect the expression of genes involved in nitrogen assimilation. Sucrose or glucose can at least partially substitute for light in inducing the expression of genes for GS2 or Fdx-GOGAT, as demonstrated in tobacco and in *Arabidopsis*. In contrast, sucrose represses the expression of AS genes when supplied to dark-adapted maize explants or to whole *Arabidopsis* plants. The effects of sucrose on GS/AS gene expression can be antagonized by amino acids. Sucrose induction of GS expression is repressed by amino acids, whereas amino acids relieve sucrose repression of AS gene expression.

Based on the reciprocal effects of carbon and organic nitrogen on the expression of GS2 and ASN1, a metabolic control model has been proposed (Fig. 8.20). In this model, GS2 expression is induced by light *or* when carbon skeletons required for ammonia assimilation are abundant. Thus, in the light, nitrogen is assimilated and transported as metabolically reactive glutamine, a substrate in numerous anabolic reactions. In contrast, ASN1 expression is induced when darkness prevents photosynthetic carbon reduction, or when concentrations of organic nitrogen are high relative to those of carbon. Thus, under conditions of low carbon availability or high organic nitrogen, plants direct assimilated nitrogen into inert asparagine, which has a higher N:C ratio than glutamine and therefore can transport and store nitrogen more efficiently when carbon skeletons are limiting.

8.2.9 The mechanisms by which plants receive and transduce signals relating to their carbon and nitrogen status remain unresolved.

What are the mechanisms involved in metabolite sensing and signal transduction in plants? Several studies have focused on the metabolic control of photosynthesis and carbon metabolism genes by sugars. Hexokinase is proposed to be the switching enzyme that can sense carbon availability inside the cell. Recent work using hexokinase antisense plants has shown that the sucrose induction of nitrate reductase is mediated at least in part by hexokinase. Although sugar-responsive cis-elements have been identified in some promoters, the consensus DNA sequences appear to be fairly

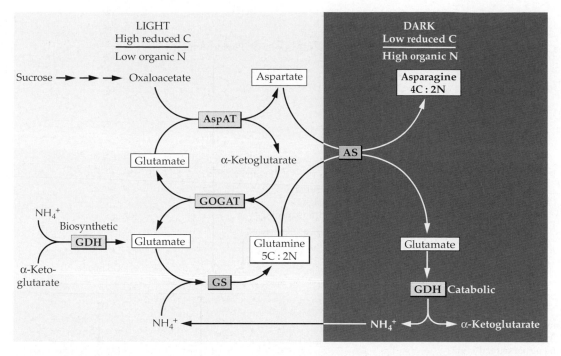

Figure 8.20
Synthesis of glutamine and asparagine is sensitive to light and to the availability of reduced carbon. Expression of plastid localized GS is up-regulated by light or sucrose, promoting formation of the metabolically reactive nitrogen donor, glutamine. AS is inhibited by light or sucrose. This inhibition can be released by increasing the concentration of amino acids present in plant tissues. Darkness promotes AS expression and enhances synthesis of the inert nitrogen storage compound, asparagine. Note that asparagine represents a more efficient use of reduced carbon than glutamine does: Both amino acids bind two nitrogen atoms, but glutamine contains five carbon atoms, asparagine four.

degenerate. Metabolic regulation of transcription in plants by sucrose may also involve posttranslational modification processes such as protein–protein interactions or allosteric modification of transcription factors (Fig. 8.21).

While research into the mechanisms of sugar-dependent metabolic control in plants is at an early stage, the effectors of metabolic regulation in prokaryotes and unicellular eukaryotes have been thoroughly studied. In yeast, sucrose nonfermenting (SNF) mutants led to the identification of a protein kinase. SNF1 is a trans-acting catabolic repressor believed to protect cells against nutritional stresses, particularly those that compromise cellular energy status, by regulating both metabolism and gene expression. The results of these studies in microorganisms have led to identification of a plant homolog of SNF1 in rye, *Arabidopsis*, and barley. It will be important to isolate mutants in these plant SNF1 homologs and test them for altered carbon signaling.

The proteins with which plants sense the supply of organic nitrogen are not known. However, a series of potential candidates have been identified by homology to components of metabolic signaling systems in bacteria and fungi. The nitrogen-regulatory protein NIT2, identified as a positive-acting regulatory protein controlling expression of genes that encode enzymes of nitrogen metabolism in fungi, has been used to identify a plant homolog in tobacco (NTL1). Tobacco NTL1 has the features of a Zn-finger transcription factor; its role in regulating plant nitrogen assimilation remains to be elucidated. Plants may also possess mechanisms to regulate general control of amino acid biosynthesis, a global metabolic control mechanism previously described in yeast (see Box 8.9). Finally, proteins homologous to components of the cellular machinery that senses C:N ratios and regulates GS in *E. coli* have also been uncovered in plants. The function of these putative carbon and nitrogen sensors in plants is currently under study.

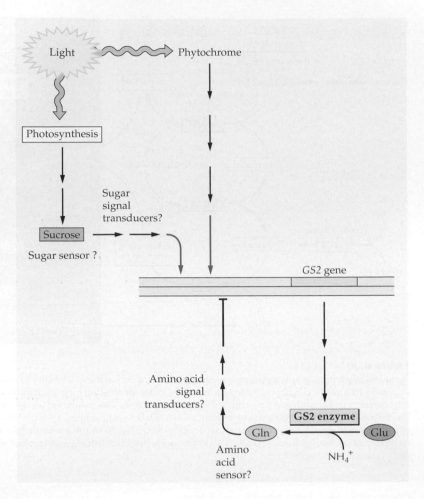

Figure 8.21
Model for regulation of transcription of GS2 gene by light and metabolites. Light can activate the transcription of the GS2 gene either directly, via phytochrome, or indirectly, via photosynthate production. The ability of sucrose to induce GS2 gene expression can be antagonized by amino acids. Studies suggest that plants have mechanisms that sense carbon and amino acids and, in turn, regulate the assimilation of inorganic nitrogen into amino acids. Elucidation of the components involved in metabolite sensing and signal transduction is underway in *Arabidopsis* with utilization of molecular genetic approaches.

8.2.10 The metabolism of N-transport amino acids has biotechnological implications.

The assimilation of inorganic nitrogen (ammonium) into organic form (e.g., glutamine and glutamate) is a key process regulating plant growth. As such, this topic is the focus of two current biotechnological pursuits: the use of biosynthetic enzymes as potential herbicide targets, and the creation of transgenic plants with increased capacity to assimilate inorganic nitrogen into amino acids. Research on GS genes has an impact on both of these goals.

Transgenic tobacco plants that overexpress a cytosolic GS isoenzyme (Fig. 8.22) have increased tolerance to the GS-inhibiting herbicide L-PPT (see Fig. 8.7) and greater GS activity than wild type. Athough these GS-overproducing plants have reduced amounts of free ammonium, they do not demonstrate higher cellular concentrations of glutamine or altered growth phenotype.

In contrast, transgenic tobacco plants that overexpress the chloroplast GS2 isoenzyme have increased photorespiratory capacity, confirming the notion that the chloroplast GS2 isoenzyme is a rate-limiting enzyme in photorespiration. Additionally, these plants are more resistant to photoinhibition (Fig. 8.23). Taken together, these studies suggest that it may be possible to affect primary nitrogen assimilation or the reassimilation of photorespiratory ammonium, or both, by manipulating individual *GS* genes.

The ability of ectopic gene expression to alter the production of specific amino acids is supported by work on AS. In transgenic tobacco, the constitutive overexpression of a pea *AS1* gene led to a 10- to 100-fold increase in the concentrations of free asparagine. Thus, researchers can change the flux through the nitrogen assimilatory pathway by altering the expression pattern of specific key genes. Moreover, because glutamate, glutamine, aspartate, and asparagine all serve to

Figure 8.22
Transgenic tobacco plants that overexpress a *GS1* gene from alfalfa (upper left corner) tolerate the GS inhibitor L-phosphinothricin (L-PPT) at 20 μM, a concentration that is lethal to wild-type plants (lower right corner).

transport nitrogen within a plant, manipulation of the concentrations of these amino acids may aid in the synthesis of downstream amino acids or other products. For example, aspartate is the substrate for aspartate kinase (AK), the first committed step toward the biosynthesis of threonine, lysine, and methionine. Increased transport of aspartate to developing seeds may be required to increase the concentrations of these essential amino

acids in seeds of transgenic plants (see Section 8.4.3).

8.3 Aromatic amino acid synthesis

Bacteria expend most of their metabolic energy in protein synthesis; it is not surprising, therefore, that their aromatic amino acid biosynthetic pathways are almost exclusively

(A)

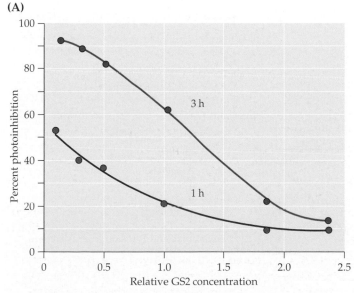

(B)

Figure 8.23
(A) In transgenic tobacco leaves overexpressing a *GS2* gene from rice, tolerance to high-intensity light correlates positively with abundance of GS2 protein. Rates of photoinhibition were estimated from decreases in chlorophyll fluorescence. (B) A wild-type control leaf (left) exposed to high-intensity light for 24 hours is severely damaged, but a transgenic tobacco leaf enriched in GS2 (right) appears unaffected. (Analysis of the removed leaf disks yielded the data plotted in A).

devoted to the synthesis of the aromatic amino acids: phenylalanine, tryptophan, and tyrosine. In contrast, higher plants use the pathways of aromatic amino acid synthesis also for the production of numerous aromatic secondary metabolites, such as the plant hormone auxin (indole-3-acetic acid), pigments (anthocyanins), defensive phytoalexins (see Chapters 21 and 24), bioactive alkaloids, and structural lignin (Fig. 8.24; see also Chapter 24). The importance of these pathways in plants is underscored by the fact that about 20% of the carbon fixed by plants flows through the common aromatic amino acid pathway, largely to make lignin (see Chapters 2 and 24).

8.3.1 Synthesis of chorismate constitutes the common aromatic amino acid pathway.

Chorismic acid is the final precursor common to the synthesis of the three aromatic amino acids (phenylalanine, tryptophan, and tyrosine) as well as the vitamin p-aminobenzoic acid (a component of the C_1 carrier tetrahydrofolate) and quinone electron transport cofactors (see Chapters 12 and 14). Except for one isoenzyme of chorismate mutase (see Section 8.3.2), to date all the enzymes of aromatic amino acid synthesis localized with certainty are found in plastids (see Section 8.3.8).

The seven-step synthesis of chorismate (Fig. 8.25) begins with the condensation of two intermediates of carbohydrate metabolism, phospho*enol*pyruvate (PEP) from glycolysis and erythrose 4-phosphate from the pentose phosphate pathway (see Chapter 13). This reaction is catalyzed by 3-deoxy-D-*arabino*-heptulosonate-7-phosphate (DAHP) synthase. Although plant and bacterial enzymes share similar kinetic properties, plant DAHP synthases are distinct from bacterial enzymes by not being feedback-inhibited by any of the aromatic amino acids. In fact, the homodimeric DAHP synthases purified from plant cells are activated by tryptophan as

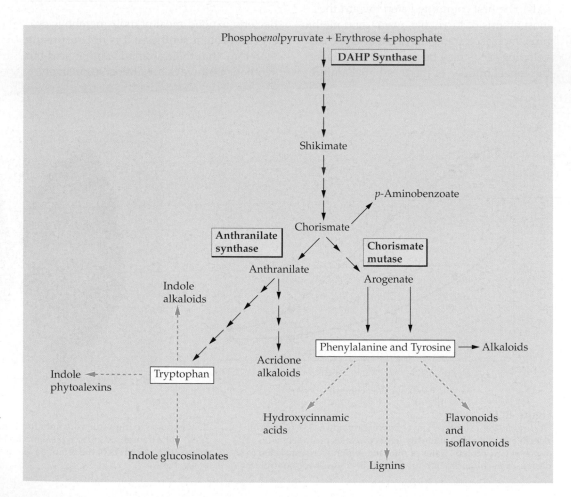

Figure 8.24
Synthesis of aromatic amino acids in plants. In addition to their functions in proteins, phenylalanine, tyrosine, and tryptophan serve as precursors for the synthesis of numerous primary and secondary metabolites.

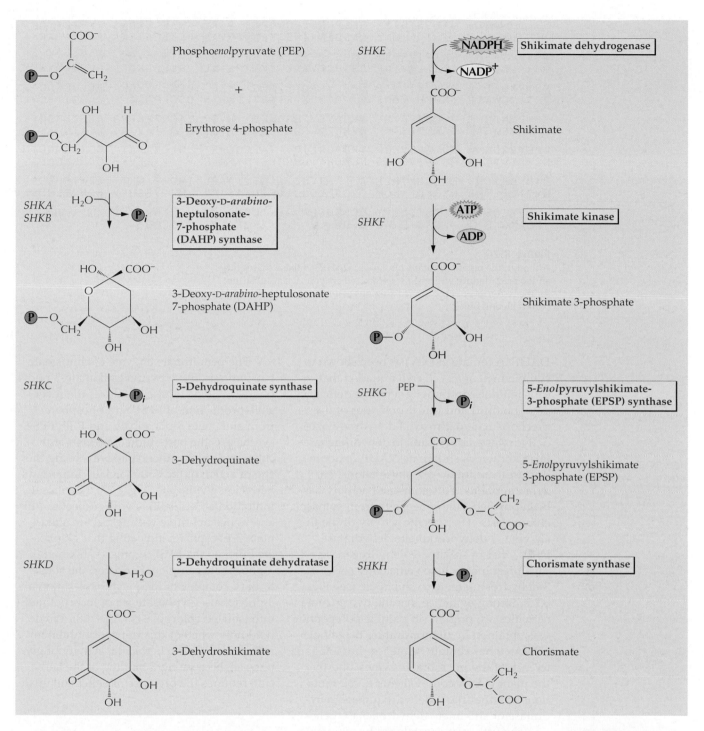

Figure 8.25
Chorismate biosynthesis from phospho*enol*pyruvate and erythrose 4-phosphate. The abbreviations *SHKA* through *SHKH* are notations proposed for the genes encoding shikimate pathway enzymes in higher plants.

well as Mn^{2+}. Consistent with this functional difference, the plant and bacterial enzymes share only about 20% amino acid identity (Fig. 8.26). Despite the primary sequence divergence and differences in regulatory properties, the plant enzyme can substitute for a defective *E. coli* or yeast enzyme, relieving the requirement for amino acid supplementation and restoring **prototrophy.**

The second enzyme of aromatic amino acid biosynthesis, 3-dehydroquinate synthase, catalyzes the first cyclization step (see Fig. 8.25). Although in *E. coli* this enzyme is thought to catalyze a complex redox and cyclization reaction, little has been published characterizing this enzyme from plants. The activity requires Co^{2+} and NAD^+ as cofactors, and the plant enzyme has a K_m of 25 μM for

Solanum tuberosum	MALSSTSTTN	SLLPNRSLVQNQPLL-PSPL	KNAF-FSNNS	TKTVRFVQP-
Escherichia coli aroG	MNY	QNDDLR-IKEIKELLPPVAL	LEKFPATENA	ANTVAHARKA
	ISAVHS-SDS	NKIPIVSDKP	SKSSPPAATATTAP APAV	TKTEWAVDS-	W KSKKALQL
	IHKILKGND-	-DRLLVVIGP	CSIHDPVAAK-EYATRLLAL	-REEL-KDEL	ELVMRVYFEK
	PEYPNQEE-L	RSVLKTID E	FPPIVFAG-E-ARS--LEE	RLGEAAMGRA	FLLQGGDCAE
	PRTTVGWKGL	INDPH-MDNS	F--QINDGLRIARKLLLDIN	DSGLPAAG-E	FLDMITPQYL
	SFKEFNANNI	RDTFRILL	QMGAVLMFGGQMPV-IKVGR	MAGQFAKPRS	DSFEEKDGVK
	A-DLMSWGAI	GARTTES---	QVHRELASGLSCPVGFKNGT	D-GTIKV-AI	D-AINAAGAP
	LPSYRGDNVN	GD-AFDVKSR	TPDPQRLIR---AYCQSAAT	LNLLRAFAT-	GGYAAMQRIN
	HC-FLSVTKW	GHSAIVNTSG	NGDCHIILRGGKEPNYSAKH	V----AEVKE	GLNKAGLPA-
	QWNLDFTEHS	E-QGDRYREL	A---SRVDE-ALGFMTAAGL	TMDHPIMKTT	EFWTSHECLL
	QVMIDFS-HA	NSSKQFKKOM	DVCADVCQQIAGGEKAIIGV	-M--VESHLV	EGNQSLESSE
	--LPYEQSLT	RRDSTSGLYY	DCSAHFLWVGERTRQLDGAH	VEFLRGIA	- Plus 141 residues
	-PLAYGKSIT	--DACIG-WE	DTDA-LL------RQLANA	VKARRG	

Figure 8.26
Although the potato (*Solanum tuberosum*) DAHP synthase shares little amino acid identity with the *Escherichia coli* enzyme, it can substitute for the bacterial enzyme in vivo. Regions of amino acid identity are highlighted in blue.

DAHP. A tomato cDNA has been shown to complement an *E. coli aroB* mutation that causes 3-dehydroquinate synthase deficiency.

The third and fourth activities of the prechorismate pathway, 3-dehydroquinate dehydratase and shikimate dehydrogenase (also known as shikimate:NADP⁺ oxidoreductase), are found on a bifunctional enzyme in plants. This combined activity has been purified to homogeneity from spinach chloroplasts. The specific activity of the first enzyme, 3-dehydroquinate dehydratase, is 10% that of shikimate dehydrogenase. This presumably helps efficiently convert 3-dehydroshikimate to shikimate (see Fig. 8.25). Strong evidence that the two enzyme activities are present on a single polypeptide was obtained by demonstrating that single cDNA clones contain sequences homologous to each of the microbial enzymes, and that the plant cDNA complements *E. coli* mutations in either the 3-dehydroquinate dehydratase or shikimate dehydrogenase genes.

The fifth step is catalyzed by shikimate kinase, which phosphorylates shikimate at the 3-hydroxyl (see Fig. 8.25). Biochemical regulation of this chloroplast activity has been reported, including inhibition by the reaction products shikimate 3-phosphate and ADP, as has light-dependent modulation in isolated organelles. Catalytically active shikimate kinase produced from tomato cDNAs should facilitate detailed investigation of the biochemical regulation of this enzyme.

The penultimate enzyme of chorismate biosynthesis, 5-*enol*pyruvylshikimate-3-phosphate (EPSP) synthase, catalyzes the reversible production of EPSP and phosphate from shikimate 3-phosphate and PEP. EPSP synthase is the best-studied enzyme of the chorismate biosynthetic pathway, being the site of action of the commercially important herbicide glyphosate (Fig. 8.27). Glyphosate inhibition is competitive with respect to PEP and noncompetitive with respect to shikimate 3-phosphate, indicating that glyphosate binds to the EPSP synthase–shikimate 3-phosphate complex. However, the stable ternary complex of EPSP synthase–shikimate 3-phosphate–glyphosate is not incorporated into isolated chloroplasts (Fig. 8.28). We do not know whether this transport inhibition is important to the herbicidal activity of glyphosate, because tight binding of the herbicide requires the presence of either shikimate

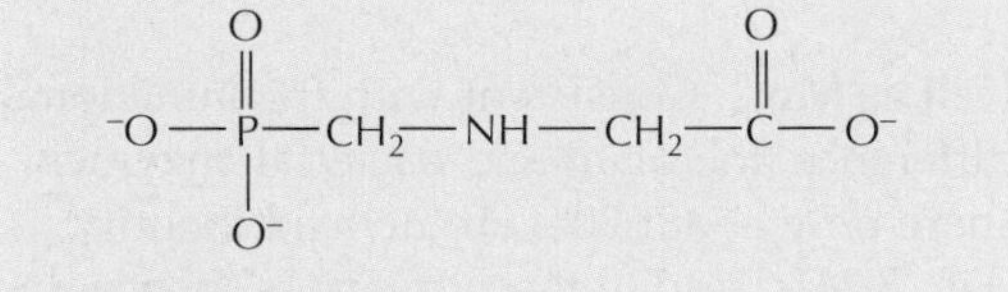

Glyphosate
(*N*–[Phosphonomethyl]glycine)

Figure 8.27
Structure of the commercial herbicide glyphosate, a competitive inhibitor of EPSP synthase.

3-phosphate or EPSP, both of which may be present only in the plastid. EPSP synthase genes have been cloned from many plant species, and mutations that lead to glyphosate tolerance have been identified (Box 8.4).

The final enzyme of the pathway, chorismate synthase, catalyzes a 1,4-trans elimination of phosphate from EPSP to produce chorismate (see Fig. 8.25). The reaction is unusual in that it requires a reduced flavin nucleotide cofactor ($FMNH_2$; see Fig. 14.6) typically associated with redox reactions, but does not appear to involve a net change in the oxidation state of the reactants.

8.3.2 Chorismate mutase is the committing enzyme in phenylalanine and tyrosine synthesis.

The phenylalanine and tyrosine pathways in plants differ from those in enteric bacteria and fungi by having the amino-transfer step as the penultimate, rather than the final, reaction. This means that arogenate is the final intermediate in the pathway (Fig. 8.29).

Chorismate mutase (CM), which catalyzes the intramolecular rearrangement of the enolpyruvyl side chain of chorismate to produce prephenate, is the **committing enzyme** for phenylalanine and tyrosine

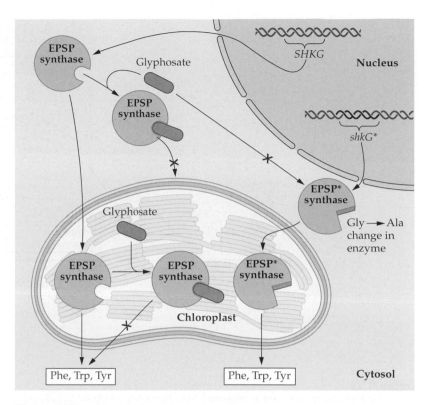

Figure 8.28
Mutations in EPSP synthase render the enzyme insensitive to the inhibitor glyphosate. Binding of glyphosate to the wild-type enzyme blocks enzyme activity and inhibits transport of EPSP synthase–shikimate 3-phosphate into the chloroplast. (The substrate molecule in this ternary complex is not shown.) Mutant EPSP* synthase (encoded by *shkG**) has a lower affinity for glyphosate and is catalytically active in the presence of herbicide.

<table>
<tr><td>**Box 8.4**</td><td>**Detoxification genes and resistant enzymes can provide crop plants with herbicide tolerance.**</td></tr>
</table>

Several enzymes of amino acid biosynthesis are targets for commercially significant herbicides. GS is inhibited by L-PPT (Basta), acetohydroxy acid synthase (AHAS) by three classes of herbicides (imidazolinones, sulfonylureas, and triazolopyrimidines), and EPSP synthase by glyphosate (Roundup).

Herbicide tolerance is considered a desirable agricultural trait because it allows farmers to control weeds without affecting crop plant growth. Both classical breeding and molecular genetic strategies have been used to introduce these traits into economically important plants. One successful approach for enhancing tolerance to herbicides is to express enzymes that detoxify or degrade the inhibitor. For example, expression of the *Streptomyces hygroscopicus* phosphinothricin acetyltransferase *bar* gene makes plants resistant to L-PPT. Another proven technique is to introduce mutant target enzymes that are less sensitive to the inhibitor. A naturally occurring form of EPSP synthase from an *Agrobacterium* strain provides tolerance to high concentrations of glyphosate.

Mutant EPSP synthase has also been characterized in *E. coli*. Here, a backbone plot of the *E. coli* enzyme highlights active site residues (blue, numbered in red) and a conserved region (red). Conversion of the yellow Gly[96] residue to Ala makes the protein resistant to glyphosate.

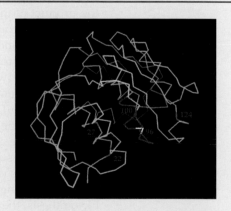

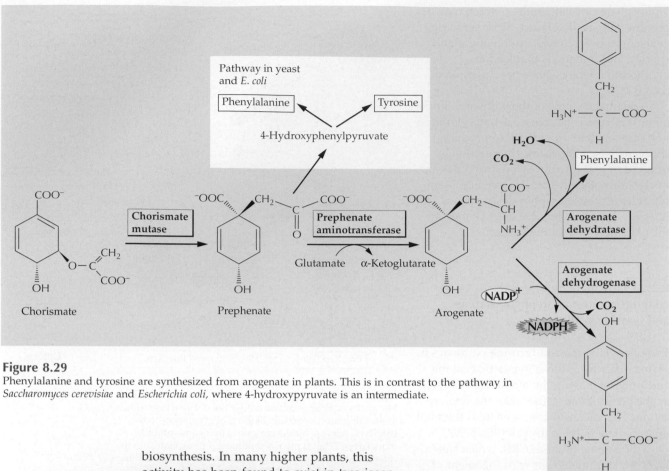

Figure 8.29
Phenylalanine and tyrosine are synthesized from arogenate in plants. This is in contrast to the pathway in *Saccharomyces cerevisiae* and *Escherichia coli,* where 4-hydroxypyruvate is an intermediate.

biosynthesis. In many higher plants, this activity has been found to exist in two isoenzyme forms, CM1 and CM2, which are regulated quite differently from one another. The plastid localization and regulatory behavior of CM1 are consistent with a role for this activity as a committing enzyme in an amino acid biosynthetic pathway.

CM1 is feedback-inhibited by each of the end products, phenylalanine and tyrosine. CM1 is also activated by tryptophan, the product of the other branch of the pathway, and tryptophan reverses the inhibition by phenylalanine or tyrosine (Fig. 8.30). This scheme serves to regulate flux into the two competing pathways by increasing synthesis of phenylalanine and tyrosine when tryptophan is plentiful, and suppressing synthesis of phenylalanine and tyrosine when the supply of these amino acids is adequate.

The contrasting features of the CM2 isoform make this enzyme enigmatic. Unlike most committing enzymes of amino acid biosynthesis, the activity of CM2 is not regulated by any of the aromatic amino acids. Furthermore, although every other aromatic amino acid biosynthetic enzyme for which

compelling biochemical or molecular biological data exist has been localized to the plastid, CM2 is cytosolic, and the proteins encoded by *CM2* cDNAs from tomato and *Arabidopsis* lack any obvious N-terminal plastid target sequences (Fig. 8.31). Taken together, these results suggest that CM2 may not be involved in amino acid biosynthesis. CM2 does, however, seem likely to function in chorismate metabolism, because in several plants the K_m of CM2 for chorismate is lower than that of CM1. For example, in sorghum the CM2 enzyme has a K_m of 9 μM for chorismate, compared with 150 μM for CM1. Backing up these biochemical data is the observation that products of plant *CM2* genes can rescue CM^- mutants from *E. coli* and yeast. The role of CM2 in plants remains to be determined, probably by a combined genetic and biochemical approach.

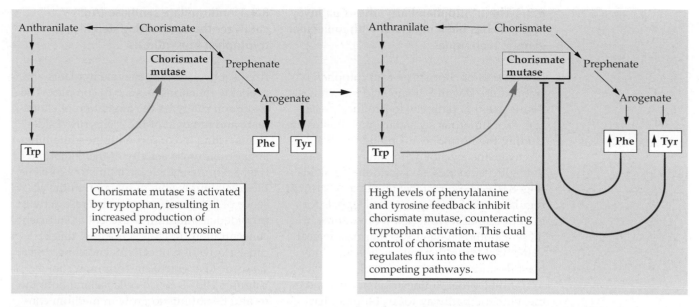

Figure 8.30
Allosteric regulation of chorismate mutase controls flux from chorismate into phenylalanine and tyrosine.

The final step common to phenylalanine and tyrosine biosynthesis is catalyzed by prephenate aminotransferase. This heat-stable enzyme can utilize glutamate and aspartate as amino donors and has a fairly strong preference for prephenate. It is not inhibited by end products, consistent with the enzyme not acting at a branch point in the pathway.

8.3.3 In plants, the pathway that synthesizes phenylalanine and tyrosine is regulated by its final reactions.

The biosynthesis of phenylalanine and tyrosine in plants represents an unusual case, wherein the committing reactions are also the last steps of the pathways. Observations suggest that the enzymes that catalyze these steps are regulated by their amino acid end products, although this hypothesis has not been confirmed with genetic data. Arogenate dehydrogenase catalyzes the conversion of arogenic acid to tyrosine. This $NADP^+$-dependent enzyme is strongly feedback-inhibited by tyrosine, as would be expected for a committing enzyme in the pathway. The final enzyme of phenylalanine biosynthesis, arogenate dehydratase, is inhibited by phenylalanine and activated by tyrosine.

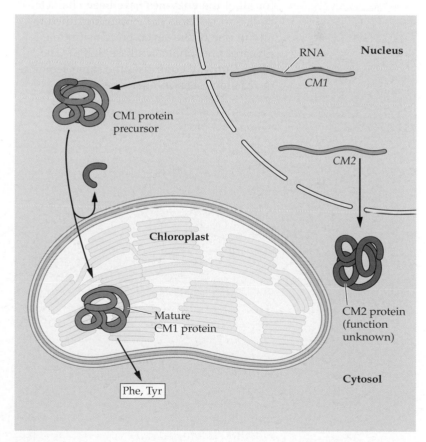

Figure 8.31
CM1 and CM2 isoforms have distinct subcellular localization and appear to have different physiological roles. CM1 is plastid-localized, whereas CM2 is cytosolic. Because the substrate for CM is produced in the plastid, the function of the cytosolic isoenzyme remains unknown.

8.3.4 The tryptophan biosynthesis pathway in plants has been dissected with molecular genetic techniques.

Conversion of chorismate to tryptophan has significance beyond amino acid biosynthesis. Plants use this pathway to produce precursors for numerous secondary metabolites, including the hormone auxin, indole alkaloids, phytoalexins, cyclic hydroxamic acids, indole glucosinolates, and acridone alkaloids (Fig. 8.32). These metabolites serve as growth regulators, defense agents, and signals for insect pollinators and herbivores. Some of these alkaloids have great pharmacological value, including the anticancer drugs vinblastine and vincristine.

Unlike other amino acids in plants, the biosynthetic pathway for tryptophan has proven to be amenable to detailed molecular genetic analysis (Box 8.5); as a result of these studies, a generally accepted in vivo pathway has been constructed (Fig. 8.33). Genes for all of the enzymes have been characterized, and mutants have been identified for all but one of the seven proteins (indole-3-glycerol-phosphate synthase; IGPS). The plant pathway comprises the same sequence of reactions seen in microorganisms.

8.3.5 Anthranilate synthase (AnS) catalyzes the committing step in tryptophan biosynthesis.

AnS is a two-subunit enzyme in plants and probably functions as an $\alpha_2\beta_2$ complex. The α-subunit catalyzes the amination of chorismate and removal of the *enol*pyruvyl side chain, acting in concert with the glutamine aminotransferase activity of the β-subunit (Fig. 8.34). As is the case in microorganisms, the α-subunit is able to function in the absence of β-subunit aminotransferase activity, provided ammonium is present at sufficient concentration (Rx. 8.1). We know this because expression of cDNAs encoding *Arabidopsis* or *Ruta graveolens* (common rue) α-subunits allows *E. coli* cells lacking both the α- and β-subunits to grow in medium containing high concentrations of ammonium. However, the glutamine aminotransferase activity is likely to be important for AnS function in plants, as illustrated by the isolation of *trp4* mutants defective in the AnS β-subunit gene 1 *(ASB1)*. These mutations suppress accumulation of the blue fluorescent intermediate anthranilate in the *trp1-100* mutant by reducing the rate of conversion of chorismate to anthranilate (Box 8.5).

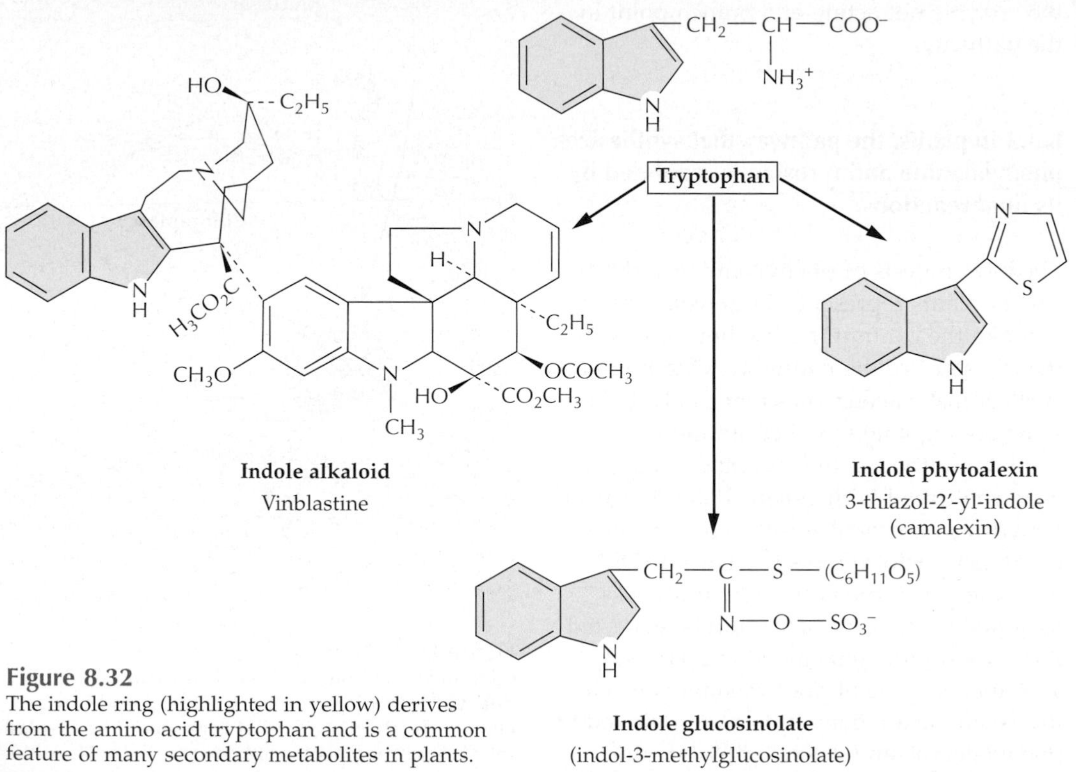

Figure 8.32
The indole ring (highlighted in yellow) derives from the amino acid tryptophan and is a common feature of many secondary metabolites in plants.

Indole alkaloid
Vinblastine

Indole phytoalexin
3-thiazol-2′-yl-indole
(camalexin)

Indole glucosinolate
(indol-3-methylglucosinolate)

Although the *trp4* and *trp1-100* single mutants are able to grow normally without tryptophan supplementation, the double-mutant plants are auxotrophs, indicating that the AnS β-subunit plays an essential role in tryptophan biosynthesis in vivo.

Reaction 8.1: Ammonium assimilation by α-subunit of AnS

$$\text{Chorismate} + NH_4^+ \longrightarrow \text{anthranilate} + \text{pyruvate}$$

As expected for the committing enzyme in the tryptophan pathway, AnS activities from plant species are feedback-inhibited by micromolar concentrations of tryptophan in vitro, and toxic tryptophan analogs can be substituted as false-feedback inhibitors (Box 8.5). This has allowed the identification of mutants with altered AnS regulation in several species. Dominant *trp5-1*^{D} *Arabidopsis* mutants that are resistant to tryptophan analogs were shown to have feedback-insensitive AnS activity (with an apparent K_I, the dissociation constant for the enzyme-inhibitor complex, approximately three times greater than that of wild type). All four of these mutants resulted from the same aspartate-to-asparagine change in the AnS α-subunit gene *ASA1*, a mutation close to a region of microbial AnS enzymes known to be important for feedback regulation (Fig. 8.35). Therefore, as in bacteria, the plant AnS α-subunit is involved in tryptophan binding. The *trp5-1*^{D} mutant has a threefold increase in free tryptophan. Both *trp5-1*^{D} mutants and feedback-insensitive wild-type isoforms (Box 8.6) illustrate the importance of allosteric control in regulating tryptophan pools.

8.3.6 Biochemical characterizations of PAT, PAI, and IGPS lag behind molecular and genetic analyses.

Although relatively little is known about the enzymology of plant phosphoribosylanthranilate transferase (PAT, Fig 8.36), the *Arabidopsis* enzyme has been the subject of detailed genetic and molecular analysis. PAT mutants fall into two general classes: auxotrophic plants, which have such low PAT enzyme activity that seedlings cannot grow on sterile medium without addition of

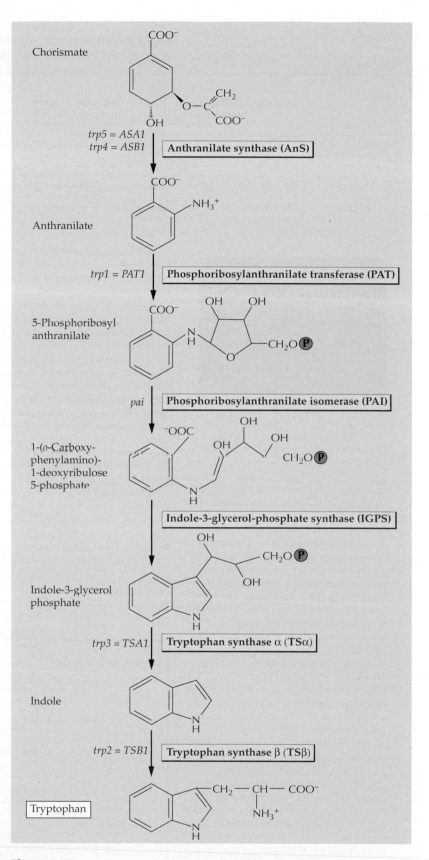

Figure 8.33
Tryptophan biosynthetic pathway of *Arabidopsis*. The wild-type gene names (e.g., *ASA1*) and mutant designations (e.g., *trp5*) are indicated to the left of the arrows.

(A)

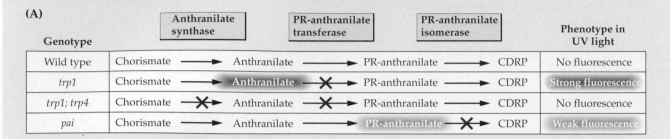

Genotype	Anthranilate synthase		PR-anthranilate transferase		PR-anthranilate isomerase		Phenotype in UV light
Wild type	Chorismate → Anthranilate		→ PR-anthranilate		→ CDRP		No fluorescence
trp1	Chorismate → Anthranilate		✗→ PR-anthranilate		→ CDRP		Strong fluorescence
trp1; trp4	Chorismate ✗→ Anthranilate		✗→ PR-anthranilate		→ CDRP		No fluorescence
pai	Chorismate → Anthranilate		→ PR-anthranilate		✗→ CDRP		Weak fluorescence

(B)

In microorganisms, it is very easy to identify mutants auxotrophic for amino acids. This is done by replica-plating colonies (see Chapter 7, Box 7.2) and comparing their growth on minimal and supplemented media. Auxotrophs are identified as those colonies that are unable to grow without the addition of amino acids. Because replica plating is not an option for plant geneticists, the search for mutants that require medium supplementation is much more difficult, and plant biosynthetic pathways have been less amenable to genetic analysis. This problem can be overcome by devising rapid screens for mutants that accumulate an intermediate in the biosynthetic pathway, or that continue to grow in the presence of compounds that would be converted into toxic products if all the enzymes in the pathway were present and active. There are several such examples in the tryptophan pathway.

The strong blue fluorescence of anthranilate, the first intermediate in the tryptophan pathway, provides a phenotype for identifying Arabidopsis mutants with reduced activities for the first three pathway enzymes by screening for seedlings that glow blue under ultraviolet (UV) light. Mutants with loss of function of phosphoribosylanthranilate (PR-anthranilate) transferase (trp1) or PR-anthranilate isomerase (pai; see Box 8.7) have striking blue fluorescent leaves that are filled with anthranilate compounds because they are unable to effectively convert this intermediate to downstream products. This screen also provides a method for identifying trp4 mutants with reduced anthranilate synthase (AnS) β-subunit activity by screening trp1 mutants for second-site suppressor mutations (see Fig. 8.35) that decrease the ability to convert chorismate to anthranilate. Here, a diagram (A) of the phenotypic effects of mutations that abolish the activity in the first three enzymes of the pathway is accompanied by a photograph (B) taken under UV light, which reveals a single blue fluorescent trp1 mutant plant growing among many wild-type Arabidopsis plants.

Toxic analogs of tryptophan and pathway intermediates have been used to identify mutants with both loss of function and gain of function. Analogs of pathway intermediates are toxic to plants because the enzymes of tryptophan biosynthesis convert them to tryptophan analogs (C). For example, 6-methylanthranilate is metabolized to 4-methyltryptophan, and 5-fluoroindole metabolizes to 5-fluorotryptophan—compounds that starve the plant for tryptophan because they inhibit AnS. Anthranilate analog-resistant plants fall into two categories: those with reduced activity of the enzymes that convert anthranilate to tryptophan (trp1, pai, trp3, and trp2 mutants), and gain-of-function feedback-insensitive AnS trp5D mutants. The latter class, relaxed **allosteric regulation** mutants, are also obtained by directly selecting for resistance to tryptophan analogs (α-methyltryptophan, for example). Just as mutants of the first three steps of the pathway were targeted by fluorescence screening, tryptophan synthase (TS)-deficient mutants have been identified by screening for resistance to exogenous 5-fluoroindole, which is converted by the TS β-subunit to toxic 5-fluorotryptophan.

(C)

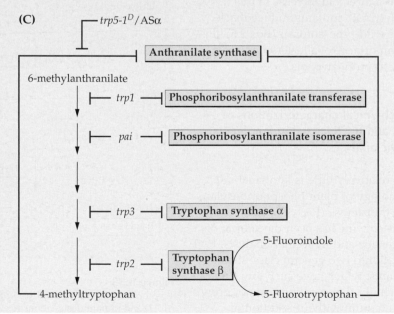

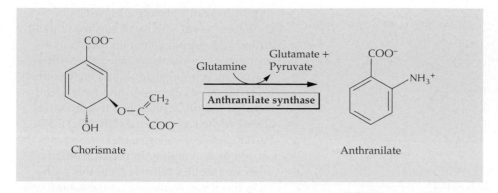

Figure 8.34
Anthranilate synthase catalyzes the first step of tryptophan biosynthesis.

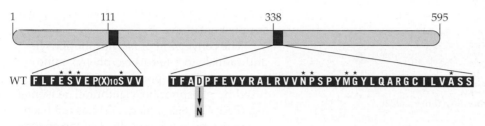

Figure 8.35
A single aspartate-to-asparagine substitution confers feedback insensitivity to the *Arabidopsis* AnS α-subunit. Asterisks indicate positions of amino acids that can strongly affect the tryptophan sensitivity of the *Escherichia coli* and *Saccharomyces cerevisiae* AnSα.

Box 8.6   An alkaloid-producing plant has a feedback-insensitive AnS.

While feedback regulation of AnS is a logical mechanism for controlling the accumulation of tryptophan, it could reduce the plant's ability to rapidly synthesize secondary metabolites in response to environmental stimuli.

The existence of naturally occurring tryptophan-insensitive AnS isoforms is one mechanism by which plants appear to avoid this potential difficulty. After treatment with fungal cell wall elicitor, *Ruta graveolens* cell cultures rapidly accumulate anthranilate-derived antimicrobial acridone alkaloids such as rutagravin (A); during this period, both AnS activity and AnS α-subunit mRNA increase dramatically. Both control and elicitor-induced cultures have similar amounts of tryptophan-sensitive AnS activity, but elicited cultures accumulate a large amount of feedback-resistant AnS. These biochemical data correlate well with results of gene expression studies. *R. grave-*

olens has two characterized AnSα genes: AnSα1 mRNA accumulation is stimulated after elicitation, whereas the amount of AnSα2 mRNA does not change (B). Further characterization of the two AnS α-isoforms expressed in *Escherichia coli* has shown that the inducible AnSα1 activity is barely affected by the presence of unusually high concentrations of tryptophan (up to 100 μM). In contrast, the unin-

ducible AnSα2 product has an apparent K_I of less than 3 μM tryptophan, very similar to that observed for other characterized plant AnS activities. These results suggest that this alkaloid-producing plant makes an inducible AnS that is not subject to tryptophan inhibition, which allows synthesis of anthranilate for secondary metabolism even under conditions where free tryptophan is plentiful.

(A)

Rutagravin

(B)

	Elicitor					Sterile water (control)				
Hours	0	3	6	9	12	0	3	6	9	12
AnSα1										
AnSα2										

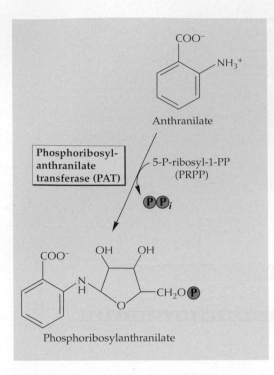

Figure 8.36
PAT catalyzes the second step of tryptophan biosynthesis.

tryptophan, and prototrophs, which have enough residual activity to survive without amino acid supplementation. Unlike the prototrophic mutants, which grow with a normal morphology, adult *trp1* auxotrophic mutants are small, bushy, and have greatly reduced fertility, even when tryptophan is added to their growth medium (Fig. 8.37). Perhaps these developmental defects are caused by an inability to produce auxin or other metabolites derived from this pathway. Biochemical analysis of the prototrophic *trp1* mutants indicates that the PAT enzyme activity is present in vast excess of the concentration required for maintaining pathway flux. Five of the prototrophic mutants have 1% of wild-type enzyme activity or less, yet grow normally without tryptophan supplementation.

The conversion of *N*-phosphoribosyl-anthranilate to 1-(*o*-carboxyphenylamino)-1-deoxyribulose 5-phosphate (CDRP) is catalyzed by phosphoribosylanthranilate isomerase (PAI; Fig. 8.38). Plant PAI has not been characterized biochemically, but the enzyme has been the subject of molecular genetic analysis in *Arabidopsis.* Originally, no PAI-deficient plants were uncovered from extensive screens for blue fluorescent mutants in *Arabidopsis.* The discovery of three genetically unlinked and transcriptionally active *PAI* genes in the wild-type Columbia ecotype used for these screens explains the lack of *pai* mutants: Loss of any one gene would not reduce enzyme activity sufficiently to produce a mutant phenotype. In the single PAI-deficient *Arabidopsis* mutant for which data have been published, deletion of one locus and DNA methylation-associated epigenetic silencing of another gene were required to reduce the enzyme activity enough to observe a phenotype (Box 8.7; see also Chapter 7).

Even transgenic antisense plants with 10% to 15% of wild-type PAI enzyme activity do not show reduced growth rate or require tryptophan for growth. These observations are consistent with a model in

trp1   WT

Figure 8.37
Morphology of the tryptophan-requiring *trp1* mutants is dramatically altered from that of wild-type (WT) plants and includes small stature and reduced apical dominance.

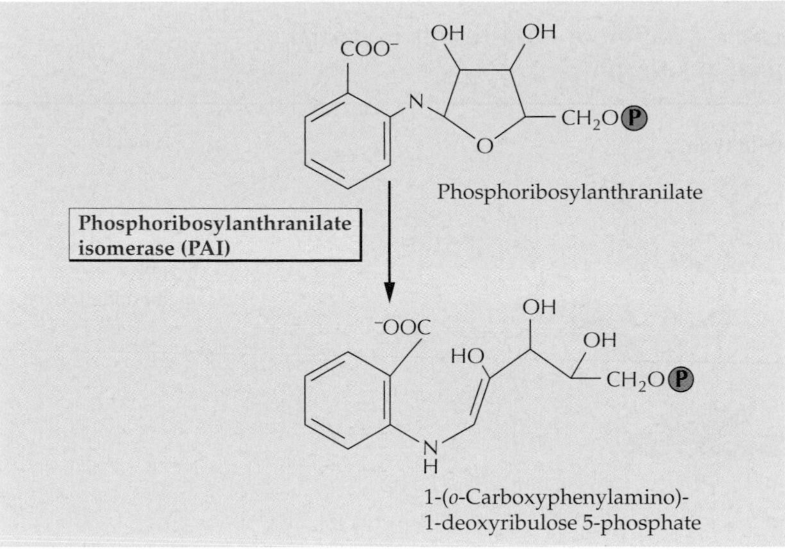

Figure 8.38
PAI catalyzes the third step of tryptophan biosynthesis.

Phosphoribosylanthranilate

Phosphoribosylanthranilate isomerase (PAI)

1-(o-Carboxyphenylamino)-
1-deoxyribulose 5-phosphate

which PAI activity is present in excess of that required for normal growth. The three *PAI* genes in Columbia are very similar to one another; *PAI1* and *PAI2* are 99% identical, including their nontranslated regions. This is very different from the low degree of conservation observed for prokaryotic PAI proteins and suggests a recent gene duplication event or a mechanism that actively maintains the similarity of these genes.

The penultimate enzyme of tryptophan biosynthesis, indole-3-glycerol-phosphate synthase (IGPS), catalyzes the decarboxylation and ring closure of CDRP (Fig. 8.39). This is the least well-studied step of the plant tryptophan pathway. IGPS cDNA clones from *Arabidopsis* can complement an *E. coli trpC⁻* mutation despite low amino acid identity (22% to 28%) with known microbial enzymes. Considering that the reaction produces the indole ring, a component of many secondary metabolites, this enzyme deserves further study.

8.3.7 Tryptophan synthase (TS) catalyzes the final step in tryptophan synthesis.

TS catalyzes the conversion of indole-3-glycerol phosphate (IGP) and serine to tryptophan (Fig. 8.40). The best characterized of the microbial tryptophan biosynthetic enzymes, TS has been unusually amenable to molecular genetic analysis in plants, in part because of the wealth of information available from microbes. The bacterial TS enzyme exists as an $\alpha_2\beta_2$ heterotetramer, and the biochemically purified subunits are capable of catalyzing independent reactions:

Reaction 8.2: TS α-subunit

$$IGP \longrightarrow indole + glyceraldehyde\ 3\text{-phosphate}$$

Reaction 8.3: TS β-subunit

$$Indole + serine \longrightarrow tryptophan + H_2O$$

Biochemical and X-ray crystallographic studies of the *Salmonella typhimurium* TS enzyme have revealed that the indole intermediate is not ordinarily released from the enzyme but rather travels through a hydrophobic tunnel connecting the active sites of α-subunit (TSA) and β-subunit (TSB) (Fig. 8.41). The efficient function of each activity depends on a successful collaboration between the two proteins, and each subunit can cause conformational changes that greatly affect the catalytic activity of the other. For example, α-subunit binding stimulates β activity (see Rx. 8.3) 30-fold compared with uncomplexed β. Conversely, the rate of cleavage of IGP at the α-subunit active site (see Rx. 8.2) is increased 20-fold by binding of serine to the pyridoxal phosphate at the β-subunit active site.

Box 8.7

Epigenetic regulation of *PAI2* results in partial tryptophan auxotrophy.

Although blue fluorescence screening failed to reveal *pai* mutants in the *Arabidopsis* Columbia ecotype, an unusual blue fluorescent mutant with diminished PAI activity was identified in the Wassilewskija ecotype. Unlike the large number of *trp1* mutants and antisense-*PAI* lines that are uniformly fluorescent, the phenotype of this mutant is unstable, showing a high frequency of somatic reversion to wild type. Unlike the three genes in Columbia, wild-type Wassilewskija (A) has four *PAI* genes, with *PAI4* found in an inverted repeat with *PAI1* at the same genetic site as the Columbia *PAI1* locus. The blue fluorescent *pai* mutant contains a deletion that simultaneously inactivates the linked *PAI1* and *PAI4* genes, leaving only *PAI2* and *PAI3* intact and making the plant dependent on expression of the *PAI2* gene. The Wassilewskija *PAI2* gene is subject to epigenetic regulation and is normally hypermethylated (shown as Me-*PAI2** in panel B), which leads to low mRNA expression. The instability of the mutant phenotype can be accounted for by random reduction in methylation of Me-*PAI2** (C), which reactivates the gene and creates epi-revertant regions of the plant that are phenotypically wild type (i.e., not blue fluorescent). These revertant tissues have enough PAI enzyme activity to prevent accumulation of anthranilate and thus allow normal growth of the plant. Because the blue fluorescence is cell-autonomous, the phenotype of each part of the plant accurately reflects its epi-genotype.

This is an example of how genetic tools that become available from studies of biochemical genetics can open up new approaches to studying seemingly unrelated areas of biology. In fact, because of the powerful blue fluorescence phenotype and well-characterized *PAI* gene family, the epigenetically silenced *PAI2** gene has led to new approaches for studying the interesting areas of gene silencing and epigenetics (see Chapter 7).

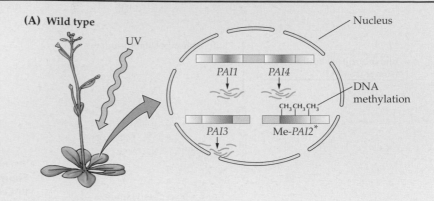

(A) Wild type

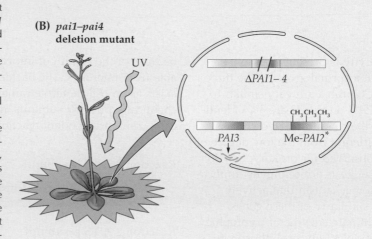

(B) pai1–pai4 deletion mutant

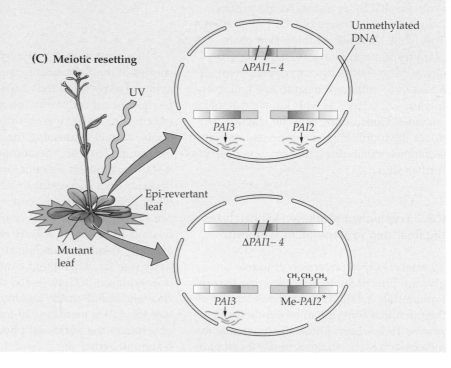

(C) Meiotic resetting

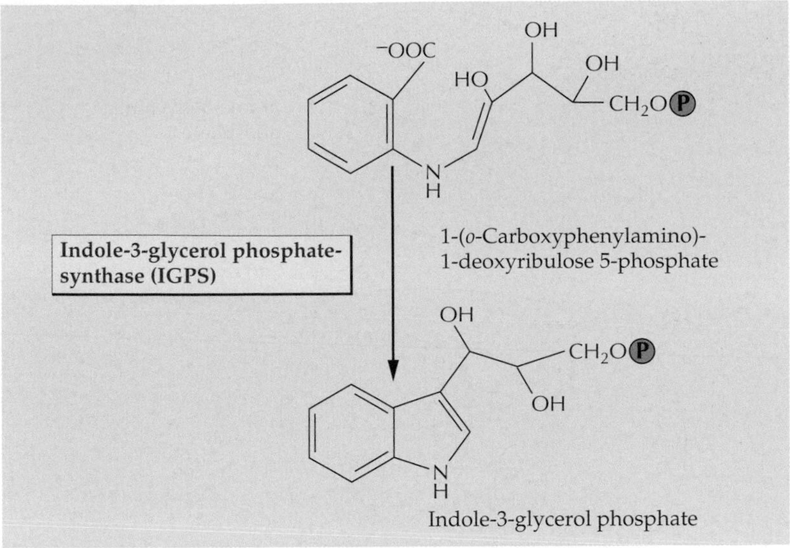

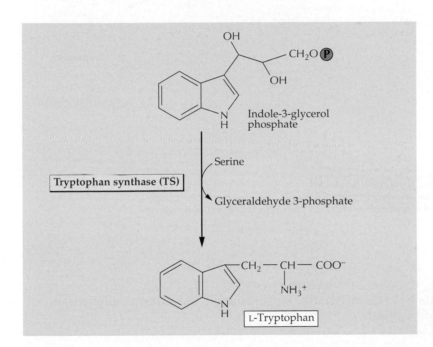

Figure 8.39
The IGPS reaction produces the indole ring found in tryptophan and secondary metabolites.

1-(*o*-Carboxyphenylamino)-1-deoxyribulose 5-phosphate

Indole-3-glycerol phosphate-synthase (IGPS)

Indole-3-glycerol phosphate

Indole-3-glycerol phosphate

Serine

Tryptophan synthase (TS)

Glyceraldehyde 3-phosphate

L-Tryptophan

Figure 8.40
Tryptophan synthase catalyzes the final step in tryptophan biosynthesis.

In contrast to fungal TS enzymes, which consist of a single protein containing both α and β domains, independent α- and β-subunit activities have been resolved by biochemical fractionation of plant extracts. This suggests that the two activities are present on separate protein subunits, as in prokaryotes. In fact, *TSA* and *TSB* clones from *Arabidopsis* or maize encode single subunit proteins, with no sign of protein fusion.

Several observations suggest that plant TS also functions as a stable αβ heteroenzyme complex. Immunoaffinity chromatography of whole *Arabidopsis* leaf extracts shows that the α- and β-subunit proteins copurify. Also, successful functional complementation of an *E. coli trpA⁻* mutation with the *Arabidopsis TSA1* cDNA requires simultaneous expression of an *Arabidopsis TSB1* cDNA. This suggests that TSA1 subunit activity is stimulated by interaction with the plant β-subunit. However, a TSA-like protein from maize does not require TSB for activity (Box 8.8).

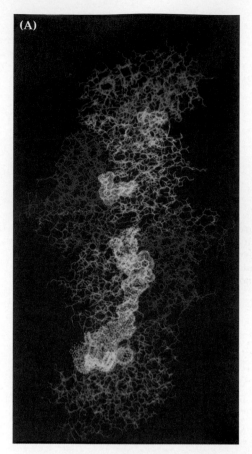

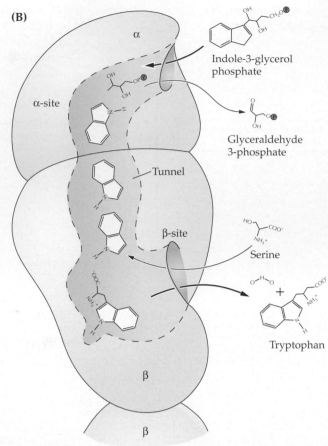

Figure 8.41
The tryptophan synthase (TS) complex consists of α- and β-subunits that join to form two active sites with a hydrophobic tunnel between them. (A) The structure of the TS complex from *Salmonella typhimurium*. Alpha-subunits are blue, β-subunit N-terminal residues are yellow, and C-terminal residues are shown as red. A red molecule of indole-3-glycerol phosphate is bound to α-subunit active site.

Box 8.8

Plants have an alternative single-subunit TSA enzyme efficient in indole synthesis.

Indole has been implicated as an intermediate in synthesis of some secondary metabolites, including auxin and cyclic hydroxamic acids such as 2,4-dihydroxy-7-methoxy-1,4-benzoxazin-3-one (DIMBOA), the structure of which is shown in the box figure. This would appear to create a dilemma, because in the well-characterized bacterial TS enzyme, indole moves through a channel from the TSA active site to TSB and does not escape from the TS heteroenzyme. A novel variation on the TS theme is found in maize, where a mutant (*bx1*, for benzoxazin-less) that fails to accumulate the normally abundant DIMBOA has been shown to have a defect in a *TSA*-like gene *(BX1)*.

Unlike the *Arabidopsis* TSA1 protein, expression of the BX1 protein complements a *trpA⁻* mutation even in the absence of a plant β-subunit. This result argues that the BX1 protein can efficiently cleave IGP to form free indole without being activated by a TSB subunit. Kinetic analysis of the BX1 protein expressed in *Escherichia coli* has confirmed that the homomeric BX1 enzyme is 30-fold more efficient in catalyzing IGP cleavage than the *E. coli* $\alpha_2\beta_2$ heterotetramer. This result suggests that plants have modified an enzyme of tryptophan biosynthesis to serve as the committing enzyme of indole-derived secondary metabolism while retaining the αβ heteroenzyme for trypto-

DIMBOA
(2,4-Dihydroxy-7-methoxy-1,
4-benzoxazin-3-one)

phan biosynthesis. Because of its unique catalytic properties, the BX1 protein may be renamed indole synthase.

8.3.8 Aromatic amino acid biosynthesis occurs in the plastid.

Aromatic amino acids are synthesized in the chloroplast, and evidence shows that the enzymes for biosynthesis of phenylalanine, tryptophan, and tyrosine are found in plastids. Isolated chloroplasts incorporate $^{14}CO_2$ into aromatic amino acids, and enzymes of chorismate biosynthesis can be assayed in chloroplast extracts (Fig. 8.42). Consistent with this view, genes containing plastid target sequences have been cloned for all of the enzymes of chorismate and tryptophan biosynthesis and for chorismate mutase-1. DAHP synthase, shikimate kinase, EPSP synthase, PAI, and TSA are imported into isolated chloroplasts.

Cytosolic isoforms of chorismate mutase of aromatic amino acid biosynthesis exist in plants (see Section 8.3.2). In addition, several cytosolic enzymes are described as catalyzing the same reaction as DAHP synthase and are activated by Co^{2+} rather than Mn^{2+}. However, it seems unlikely that these are true enzymes of chorismate biosynthesis, because (a) they can utilize a wide variety of aldehyde substrates in place of erythrose 4-phosphate, and (b) their K_m values for erythrose 4-phosphate are 10-fold higher than that of the Mn^{2+}-activated enzyme. Thus, these enzymes probably have metabolic roles unrelated to aromatic amino acid biosynthesis. It is more difficult to dismiss the importance of cytosolic CM2 in amino acid biosynthesis because its kinetic characteristics and ability to complement CM-mutations in E. coli argue that this enzyme efficiently converts chorismate to prephenate.

8.3.9 Aromatic amino acid biosynthesis is stress inducible.

Unlike well-studied microbes such as E. coli and yeast, the enzymes of plant aromatic amino acid biosynthesis are regulated by environmental signals. This is logical considering that in contrast to the low abundance of the amino acids themselves, plants produce abundant and, in many cases, highly inducible secondary metabolites from these pathways. In fact, these enzymes of intermediary metabolism appear to be coordinately regulated with the pathways of aromatic secondary metabolism (see Box 8.9).

The first evidence for this was obtained in studies of the committing enzyme of aromatic amino acid biosynthesis. Mn^{2+}-dependent DAHP synthase activity and mRNA accumulation increase in response to several treatments known to induce secondary metabolite accumulation: wounding of plants, and treatment of suspension cells with fungal elicitor or bacterial pathogens. The kinetics of wound-inducible DAHP synthase mRNA induction (Fig. 8.43A) are similar to that of phenylalanine ammonia lyase (PAL), the committing enzyme of aromatic secondary metabolism. This coinduction of gene expression in response to the environment is not limited to the first enzyme of the prechorismate pathway, because shikimate kinase, EPSP synthase, chorismate synthase, and PAL are all maximally induced within several hours after fungal elicitor treatment of tomato suspension cells. Similarly, CM1 activity is induced by wounding in potato tubers, and CM1 and PAL mRNA are coordinately induced in elicited Arabidopsis suspension culture cells.

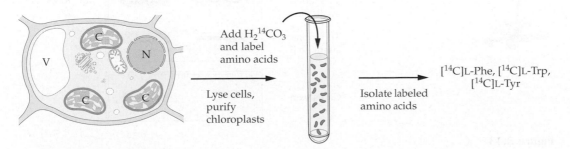

Add H$_2$^{14}CO$_3$ and label amino acids

Lyse cells, purify chloroplasts

Isolate labeled amino acids

[^{14}C]L-Phe, [^{14}C]L-Trp, [^{14}C]L-Tyr

Figure 8.42
Isolated chloroplasts (C) can convert labeled CO_2 into labeled aromatic amino acids. This is evidence that the chloroplast has a full complement of enzymes for biosynthesis of phenylalanine, tryptophan, and tyrosine. V, vacuole; N, nucleus.

There is also a strong correlation between up-regulation of the genes of *Arabidopsis* tryptophan biosynthesis and accumulation of the indolic secondary metabolite camalexin (see Fig. 16.51) in response to a variety of stressful environmental conditions. Both mRNAs and proteins for all of the enzymes of the tryptophan pathway are induced after treatment with bacterial pathogens, and the rates of induction are similar to that for camalexin accumulation. Not only is the timing of the responses coordinated, but the amount of camalexin that accumulates also is tightly correlated with the degree of tryptophan pathway enzyme induction after inoculation with various different bacterial pathogens (Fig. 8.43B).

8.4 Aspartate-derived amino acid biosynthesis

Three pathways lead directly from aspartic acid to the amino acids lysine, threonine, and methionine (Fig. 8.44). Aspartate is also needed for synthesis of other compounds: Threonine is a precursor to isoleucine, and methionine is an intermediate in the biosynthesis of *S*-adenosylmethionine (SAM), ethylene, and polyamines.

Aspartate-derived amino acids are required in the diets of nonruminant animals, including humans. Some diets that rely on plant foods are deficient in one or more of these protein building blocks and can lead to health problems if the amino acid

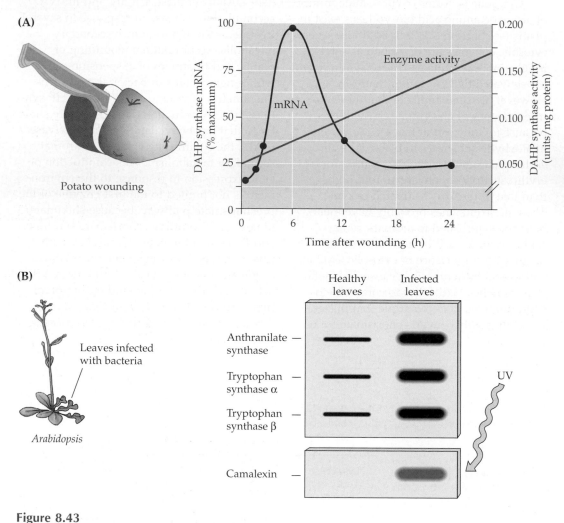

Figure 8.43
Plants induce the enzymes of aromatic amino acid biosynthesis under conditions that cause increases in aromatic secondary metabolism. (A) Wounding of potato tubers causes increased DAHP synthase mRNA concentrations and enzyme activity. (B) Infection with a bacterial pathogen increases mRNAs for *Arabidopsis* tryptophan pathway enzymes and increases production of the antimicrobial indolic secondary metabolite camalexin, which fluoresces blue under ultraviolet light.

deficiencies of grains and legume seeds are not complemented by eating both in combination. Human vegetarians who do not eat any animal products must balance their diets with care to provide all essential amino acids in the proportions that are needed for protein synthesis.

Like humans, farm animals cannot subsist on a diet composed entirely of grain. For example, corn is commonly used in animal feed because it provides calories at low cost, but it has a poor amino acid content, being especially deficient in lysine. On the other hand, soybean is lysine-rich but methionine-poor. Supplementing the diets of soybean- or grain-fed animals with lysine, threonine, or methionine increases the growth rate of these animals but is expensive. Not surprisingly, researchers are interested in using conventional breeding and transgenic metabolic engineering approaches to increase the amounts of nutritionally important amino acids in crop plants.

8.4.1 Threonine, lysine, and methionine are products of a branched pathway with complex biochemical regulation.

Aspartate forms the entire carbon skeleton for threonine, which makes it useful for us to consider threonine biosynthesis as the main pathway (Fig. 8.45), with branches to

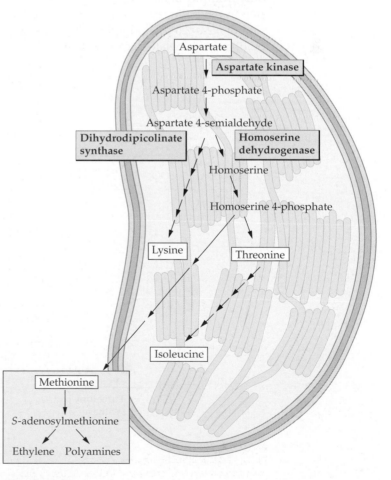

Figure 8.44
The aspartate-derived amino acids lysine, threonine, and isoleucine are produced in the plastid.

lysine and methionine (see Fig. 8.44). Aspartate is activated by phosphorylation by aspartate kinase (AK; also called aspartokinase), the committing enzyme for all aspartate-derived amino acid biosynthesis. Aspartate 4-phosphate (β-aspartyl phosphate) then undergoes two NADPH-mediated reductions, which are catalyzed sequentially by aspartate-semialdehyde dehydrogenase and homoserine dehydrogenase (HSDH), to yield homoserine. Homoserine is then phosphorylated at the 4-position by homoserine kinase, and threonine synthase catalyzes hydrolysis of the phosphate and rearrangement of the hydroxyl group to produce threonine.

Despite its importance in human and agricultural animal nutrition, except for the first and last steps, the pathway of lysine synthesis has not been fully identified. The committing enzyme, dihydrodipicolinate synthase (DHDPS), converts aspartate-4-semialdehyde plus pyruvate to 2,3-dihydrodipicolinate (Fig. 8.46). The final step in the pathway appears to be conversion of *meso*-2,6-diamino-pimelate to lysine by diaminopimelate decarboxylase.

8.4.2 Of two possible methionine biosynthesis pathways, only one appears to be important.

There are two ways that plants can make methionine, yet only the transsulfuration pathway is considered biosynthetically important in plants. In this pathway (Fig. 8.47), the thiol group of cysteine is transferred to homoserine to produce homocysteine, through a cystathionine intermediate. Cystathionine γ-synthase catalyzes the sulfur-linked joining of cysteine and homoserine 4-phosphate to form cystathionine and orthophosphate. Next, the C_3 skeleton of cysteine is cleaved by cystathionine β-lyase, leaving sulfur attached to the homoserine carbon skeleton; this produces homocysteine, pyruvate, and ammonium, making the reaction essentially irreversible. Homocysteine is then converted to methionine by tetrahydrofolate-mediated methylation of homocysteine, catalyzed by methionine synthase (homocysteine methylase). This activity not only is involved in de novo methionine synthesis but also functions in recycling

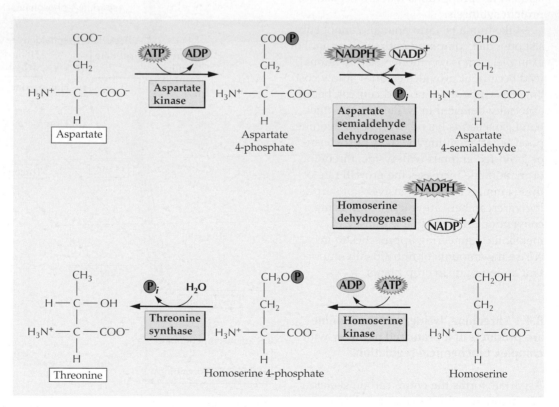

Figure 8.45
The threonine biosynthetic pathway.

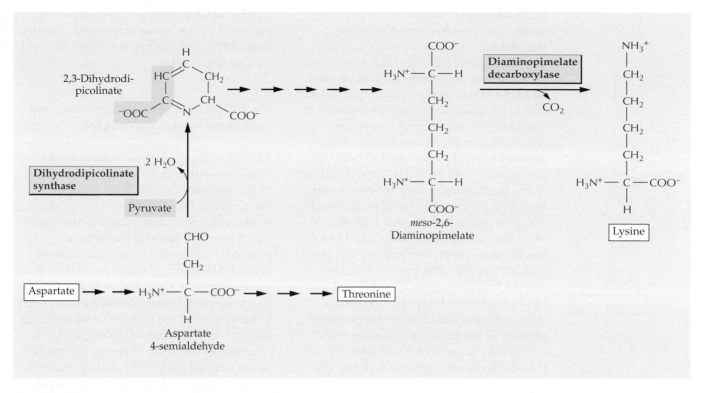

Figure 8.46
Lysine biosynthesis branches from the threonine pathway.

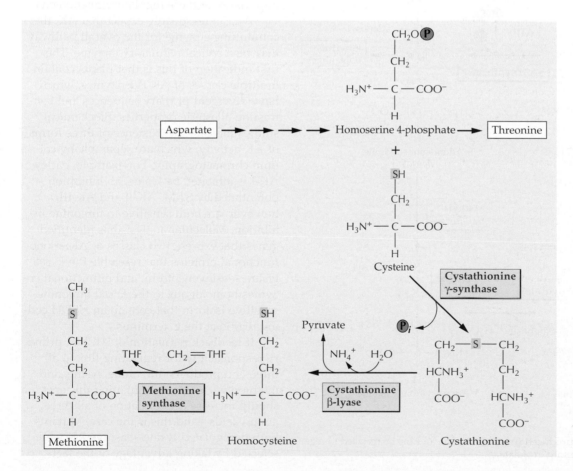

Figure 8.47
The methionine biosynthetic pathway. THF, tetrahydrofolate.

the S-adenosyl-L-homocysteine produced when SAM donates its methyl group (Fig. 8.48).

Methionine has two major fates: incorporation into proteins or conversion into SAM. SAM is a methyl donor used in DNA and RNA modification and in synthesis of abundant plant structural components, including lignin precursors, choline and its derivatives, and pectin (methyl esters of polygalacturonic acid). The carbon skeleton of the methionine moiety of SAM is also used as a precursor to the plant hormone ethylene and to polyamines. Radiotracer experiments with the aquatic plant *Lemna paucicostata* (duckweed) indicate that more than 80% of the label from ^{14}C-methyl-labeled methionine was incorporated into lipids, pectins, chlorophyll, and nucleic acids, whereas less than 20% appeared in protein. Thus, apparently the majority of methionine is converted into SAM for transmethylation reactions in plants. Pro-

duction of SAM is an energy-demanding process, all three phosphates being liberated from ATP during its synthesis by SAM synthetase (Fig. 8.48).

8.4.3 Regulation of threonine, lysine, and methionine synthesis is complex.

Not surprisingly, the biochemical mechanisms regulating flux through the branches of the aspartate-derived amino acid pathway are complex. First, aspartate-derived metabolism can follow three major routes, creating at least five branch point enzymes, four of which—aspartate kinase, threonine synthase, HSDH, and cystathionine γ-synthase—are allosterically regulated in vitro. Second, the products of these pathways are expected to be needed by the plant in vastly different amounts at each stage of development. An overview of the types of regulatory mechanisms inferred by in vitro studies is shown in Figure 8.49. As described below, experiments with feedback-insensitive mutants and transgenic plants are clarifying the in vivo importance of these regulatory phenomena.

We can reasonably expect that AK, the committing enzyme for the overall pathway, may be a critical regulatory enzyme. The first indication of this is that plants contain multiple classes of AK isoenzymes, which have divergent primary sequences and contrasting allosteric properties. Biochemical studies indicate the existence of three forms of AK activity, which are separable by column chromatography. For example, barley AK-I is inhibited by lysine, an inhibition potentiated by SAM. AK-II and AK-III, however, are both sensitive to threonine inhibition. Molecular analysis has identified genes that express two classes of AK: monofunctional proteins that resemble the *E. coli* lysine-sensitive isoform, and bifunctional enzymes homologous to the *E. coli* threonine-sensitive isoform that contain an HSDH coding region at the C terminus.

If feedback regulation of AK had primary responsibility for regulating flux to all three end products (threonine, lysine, and methionine/SAM), deregulating this enzyme should increase accumulation of all three amino acids—and this is the case. Mutants with deregulated lysine-sensitive AK were selected by taking advantage of the fact

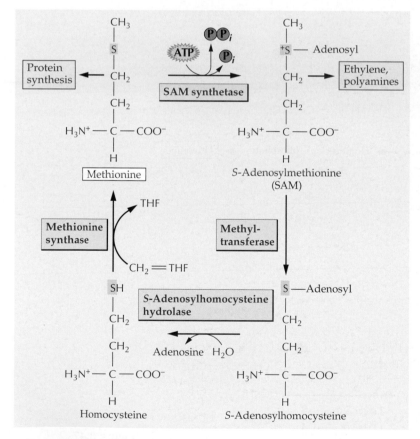

Figure 8.48
S-Adenosylmethonine is produced from methionine and can be recycled to regenerate methionine. THF, tetrahydrofolate.

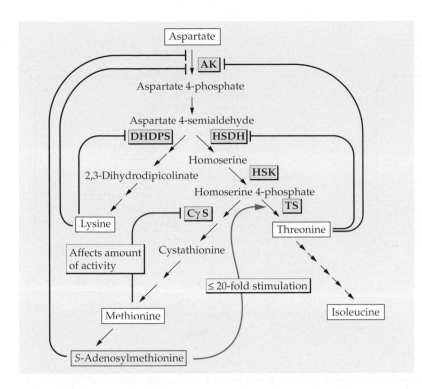

Figure 8.49
Biochemical regulatory mechanisms proposed to regulate the synthesis of amino acids derived from aspartate. Red lines indicate end-product inhibition, the purple line refers to a reduction in detectable enzyme, and the green arrow indicates stimulation of activity. AK, aspartate kinase; DHDPS, dihydrodipicolinate synthase; HSDH, homoserine dehydrogenase; HSK, homoserine kinase; CγS, cystathionine gamma-synthase; TS, threonine synthase.

that administration of threonine plus lysine causes starvation for methionine by reducing total AK activity through feedback inhibition. Mutants were selected with altered AK activities that were less sensitive to lysine inhibition (Fig. 8.50). In an analogous approach, transgenic tobacco or potato plants were constructed to express a *lysC* gene encoding a feedback-desensitized AK protein from *E. coli.* The observation that these mutants and transgenic plants over-accumulated threonine shows both that AK plays an important role in modulating threonine accumulation in plants and that lysine and methionine accumulation are influenced at later steps in the pathway.

In plants, DHDPS activity is very sensitive to lysine inhibition (the concentration to achieve 50% inhibition, or I_{50}, is 10 to 50 μM), making it a likely candidate for a key regulator of lysine accumulation. Plant mutants with desensitized DHDPS were identified in several species by selection for resistance to the toxic lysine analog *S*-aminoethyl-L-cysteine (AEC). This molecule competes with lysine for incorporation into proteins and allows the identification of lysine-overproducer plants. AEC resistance in *Nicotiana sylvestris,* caused by production of a lysine-insensitive mutant DHDPS enzyme with a single amino acid change, led

to these plants accumulating about 10-fold more lysine than did the wild type (Fig. 8.51 and Box 8.10). Similar results were obtained with transgenic plants that expressed bacterial DHDPS proteins, which are naturally about 20-fold less sensitive to lysine inhibition than plant enzymes are. In agreement with the mutant results, tobacco and oilseed rape transformants accumulated greater amounts of lysine than controls did. These results demonstrate the importance of allosteric regulation of DHDPS in controlling lysine accumulation in plants.

Transgenic plants and mutants that simultaneously express both feedback-insensitive AK and DHDPS accumulate free lysine concentrations that exceed those in plants expressing the mutant DHDPS alone and have less threonine than plants with the desensitized AK alone. These results provide two important lessons about regulation of flux through this pathway. First, the relative accumulation of threonine and lysine seems to be affected by the ability of DHDPS and HSDH to compete for limiting amounts of aspartate 4-semialdehyde. Second, the very high sensitivity of DHDPS to feedback inhibition ordinarily restricts flux into lysine biosynthesis, explaining why AK single-mutant plants preferentially accumulate threonine rather than lysine.

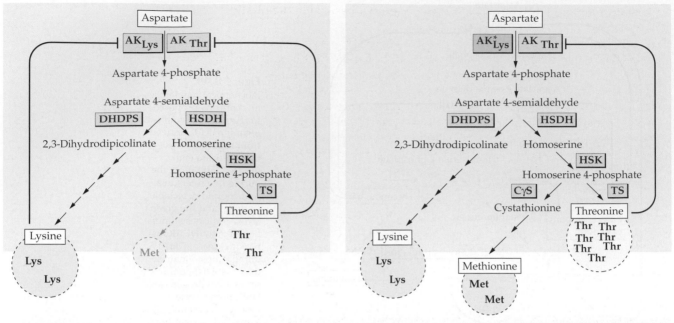

Figure 8.50

Treatment of plants with lysine plus threonine allows selection for feedback-insensitive aspartate kinase mutants. AK_{Lys} and AK_{Thr} refer to AK isoforms that are inhibited by lysine or threonine, respectively. (A) Growth of wild-type plants or plant cell cultures on medium containing a mixture of lysine and threonine causes death by methionine starvation because of inhibition of AK activity. (B) Mutants arise that are insensitive to this normally toxic amino acid mixture. These plants (AK^*_{Lys}) contain amino acid changes in their lysine-sensitive aspartate kinase activities that make them insensitive to feedback inhibition. This restores the plant's ability to synthesize methionine. AK, aspartate kinase; DHDPS, dihydrodipicolinate synthase; HSDH, homoserine dehydrogenase; HSK, homoserine kinase; CγS, cystathionine gamma-synthase; TS, threonine synthase.

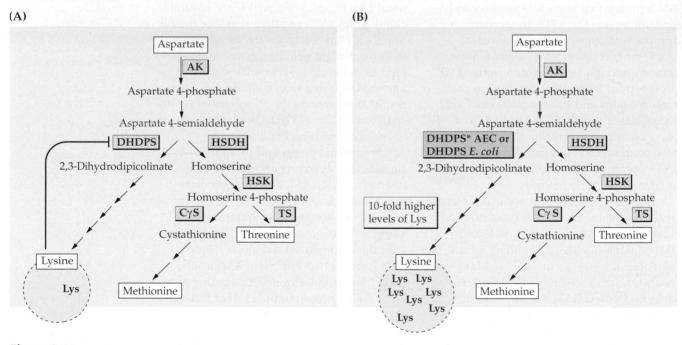

Figure 8.51

Dihydrodipicolinate synthase (DHDPS) regulates the accumulation of lysine in plants. (A) Feedback inhibition of DHDPS acts to control the flux from aspartate 4-semialdehyde into lysine. (B) Dominant feedback-insensitive DHDPS mutants can be selected by use of classical genetics (DHDPS* AEC) or by expression of feedback-insensitive *Escherichia coli* enzyme in transgenic plants (DHDPS *E. coli*). These plants overproduce lysine in comparison with the plant that expresses the wild-type DHDPS activity. AK, aspartate kinase; DHDPS, dihydrodipicolinate synthase; HSDH, homoserine dehydrogenase; HSK, homoserine kinase; CγS, cystathionine gamma-synthase; TS, threonine synthase.

Homoserine 4-phosphate is the final intermediate common to synthesis of threonine/isoleucine and methionine/SAM. Thus, the relative ability of threonine synthase and cystathionine γ-synthase to compete for this intermediate is likely to be important for regulating metabolic flux through these pathways. Two mechanisms have been proposed to influence this partitioning: (a) stimulation of threonine synthase activity by the product of the competing pathway, SAM; and (b) modulation of cystathionine γ-synthase activity by methionine availability. Thus, flux from homoserine 4-phosphate into threonine or methionine biosynthesis may be primarily regulated by the amounts of methionine and SAM present.

8.4.4 Most aspartate-derived amino acids are synthesized in the plastid.

Many of the enzymes of lysine, threonine, and methionine biosynthesis have been localized to the chloroplast. In addition, comparison of the DNA sequences of cloned plant genes for pathway enzymes with those of microbes has revealed N-terminal extensions that resemble characterized plastid-target sequences. The final steps of methionine and SAM biosynthesis appear to be the exception to this rule. Enzyme localization and DNA sequence analysis argue that methionine synthase and SAM synthase are cytosolic. The localization of cystathionine β-synthase is less clear. Although a cloned

---

Box 8.10 — **High-lysine plants exhibit developmental defects and induced lysine catabolism.**

Very high quantities of lysine accumulate in plants expressing mutant DHDPS. In fact, foliar lysine accounts for as much as half the mole fraction of total soluble amino acids in transgenic *Arabidopsis* plants. This dramatic perturbation of metabolism causes some striking effects on development. *N. sylvestris* plants that accumulate very high concentrations of lysine demonstrate decreased chlorophyll content and altered vegetative development, including lack of apical dominance and alteration of leaf size and morphology. (A) compares a lysine-overproducing

plant (bottom) with the parental wild-type control (top). These lysine-overproducing plants also fail to make the transition into reproductive phase. Similar but less severe effects are observed in high-lysine transgenic tobacco plants, which have a mosaic green color and a partial loss of apical dominance. Transgenic soybean seeds that accumulate very high amounts of lysine (15 mol% of total amino acids) produce wrinkled seeds that germinate poorly. These abnormalities may make it difficult to commercialize plants with increased free lysine content.

The goal of increasing the nutritional value of crops requires engineering an increased lysine content of seeds, but simply increasing lysine biosynthesis may not be enough to achieve this goal. Targeting high-level expression of deregulated enzyme activities to tobacco seeds causes induction of synthesis of the first enzyme of lysine catabolism, lysine–ketoglutarate reductase (LKR), leading to increased lysine breakdown (B). This induction of LKR activity occurs directly in response to lysine accumulation.

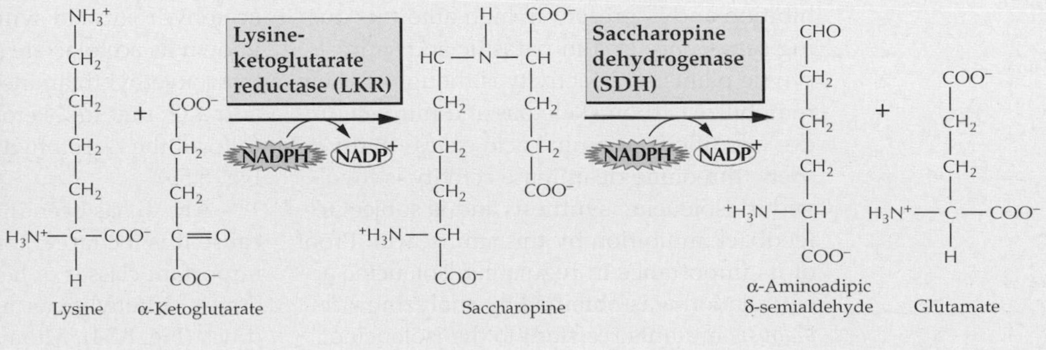

gene from *Arabidopsis* encodes an N-terminal extension that presumably targets the enzyme to the chloroplast, barley cell fractionation studies indicate that only 60% of the enzyme activity is in the chloroplast. Perhaps there are multiple isoforms, targeted to different organelles.

8.5 Branched-chain amino acids

The synthesis of isoleucine, leucine, and valine by plants commands considerable interest for various reasons. First, as 3 of the 10 essential amino acids that animals must obtain from their diet, they are nutritionally important. Second, this pathway is a commercially significant target for a variety of low-use-rate herbicides. Finally, the amino acid products serve as precursors for secondary metabolism (Fig. 8.52).

8.5.1 Threonine deaminase participates in the synthesis of isoleucine but not of valine.

Isoleucine and valine are synthesized in chloroplasts from two parallel pathways with four common enzymes, each of which has dual substrate specificity. The exception is the committing enzyme of the isoleucine pathway, threonine deaminase (also known as threonine dehydratase), which catalyzes the synthesis of the oxoacid 2-ketobutyrate from threonine but has no role in valine metabolism (Fig. 8.53). Two general classes of threonine deaminase are found in plants: biodegradative (catabolic) and biosynthetic (anabolic) forms. The degradative activity, which can use both threonine and serine as substrates, is not sensitive to end-product inhibition and is present in high amounts during senescence and in parasitic and saprophytic plants. This activity is thought to help remobilize carbon skeletons and nitrogen under conditions of amino acid excess. Biosynthetic threonine deaminase activity is involved in isoleucine synthesis and is subject to feedback inhibition by this amino acid. Proof of its importance in regulating isoleucine accumulation was obtained by analyzing *Arabidopsis* mutants resistant to the isoleucine analog L-*O*-methylthreonine. These threonine deaminase mutants accumulated 20-fold more isoleucine than did the wild type.

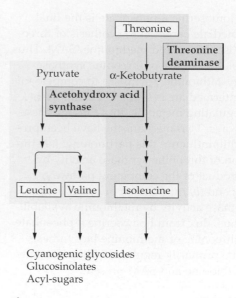

Figure 8.52
The branched-chain amino acids serve as precursors to plant secondary metabolites.

Developmentally regulated, the biosynthetic enzyme threonine deaminase is one of the most abundant proteins in mature tomato and potato flowers. In fact, the mRNA is found in flowers at concentrations more than 500-fold greater than in leaf; it can be induced in both organs by wounding and by treatment with abscisic acid (ABA) or methyl jasmonate. Perhaps this regulation reflects the need for precursors for stress-inducible secondary metabolites derived from the isoleucine pathway.

8.5.2 The isoleucine and valine biosynthetic pathways share four enzymes in common.

The first parallel reaction is catalyzed by acetohydroxy acid synthase (AHAS; also known as acetolactate synthase), which uses hydroxyethyl thiamine pyrophosphate to add a C_2 unit to 2-ketobutyrate or pyruvate, to form the 2-acetohydroxy derivatives Fig. 8.53).

AHAS has been intensively studied because it is the target for three commercially important classes of herbicide: imidazolinones, sulfonylureas, and triazolopyrimidines (Fig. 8.54). Mutants resistant to each of these herbicide types express AHAS proteins with altered amino acid sequences that cause increased inhibitor resistance.

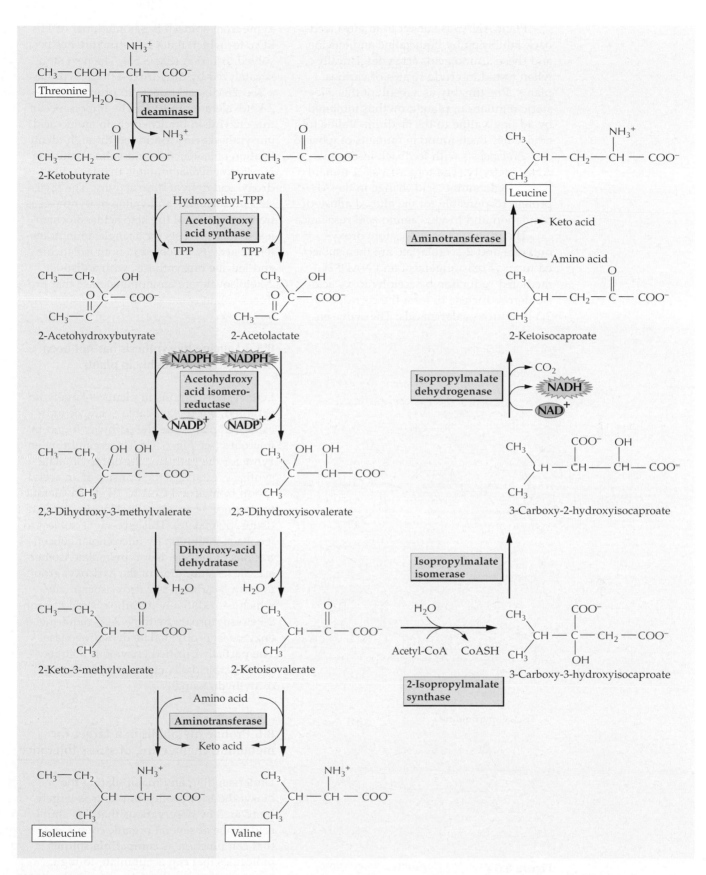

Figure 8.53
Biosynthesis of isoleucine, leucine, and valine.

Plant AHAS is subject to in vitro feedback inhibition by both valine and leucine, and these amino acids act synergistically when tested in crude lysates of various plants. Presumably as a result of this allosteric regulation, plant growth is inhibited by adding valine to the medium. Valine tolerance has been found in mutants of tobacco and *Arabidopsis* with feedback-insensitive AHAS activity. The tobacco *Val^R-1* mutant has a single amino acid change in the AHAS protein responsible for the altered allosteric regulation and in vivo amino acid resistance.

The intermediates 2-acetohydroxybutyrate and 2-acetolactate are then subjected to alkyl rearrangement and NADPH-mediated reduction by acetohydroxy acid isomeroreductase to form the corresponding 2,3-dihydroxyvaleric acids. The active en-

zyme from spinach is a homodimer of 114 kDa, for which the X-ray structure has been solved to 1.65 Å (Fig. 8.55). The next step is catalyzed by dihydroxy-acid dehydratase, a 2Fe–2S cluster protein, to produce the 2-ketovalerate products. The importance of this enzyme in branched-chain amino acid biosynthesis was confirmed through identification of isoleucine- and valine-requiring *N. plumbaginifolia* mutants that lack dihydroxy-acid dehydratase activity. The final step in isoleucine and valine biosynthesis is transamination of the keto acids. Biochemical evidence exists for a single aminotransferase activity involved in both isoleucine and leucine biosynthesis, with a distinct 2-ketoisovalerate aminotransferase that produces valine.

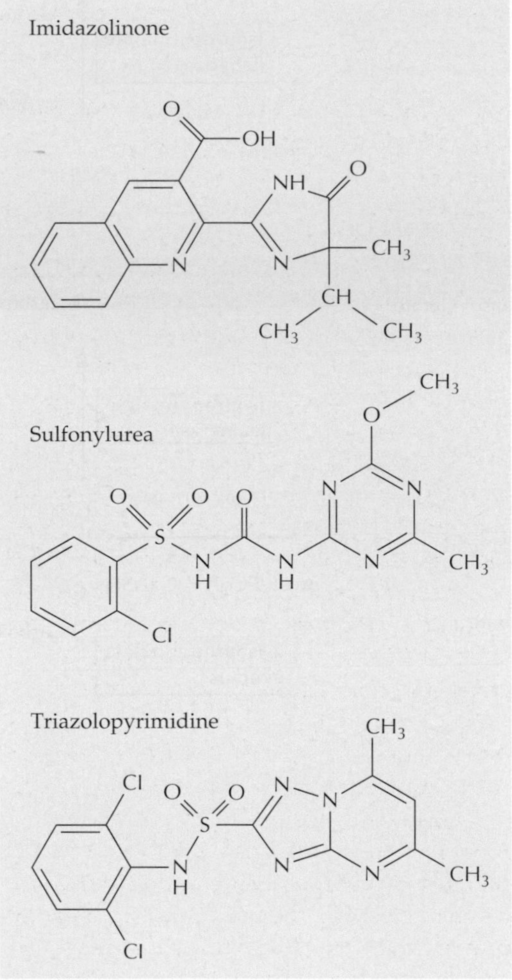

8.5.3 Leucine biosynthesis has not been investigated thoroughly in plants.

Leucine biosynthesis in plants is largely unexplored, but the available evidence indicates that plants use the pathway found in microbes (see Fig. 8.53). The committing enzyme for the pathway, 2-isopropylmalate synthase, catalyzes the transfer of an acetyl group from acetyl-CoA to 2-ketoisovalerate to produce 3-carboxy-3-hydroxyisocaproate (isopropylmalate). This enzyme is subject to feedback inhibition by micromolar concentrations of leucine. Isopropylmalate isomerase causes migration of the hydroxyl group to form 3-carboxy-2-hydroxyisocaproate, which is oxidatively decarboxylated to form 2-ketoisocaproate by the NAD^+-requiring enzyme isopropylmalate dehydrogenase. The partially purified enzyme is a target of the herbicidal compound *O*-isobutenyl oxalylhydroxamate.

8.6 Proline metabolism: a target for metabolic engineering of stress tolerance

Understanding and manipulating the biosynthesis of proline in plants is largely motivated by observations that this amino acid is one of several organic compounds that can function as **compatible solutes,** molecules that can accumulate to high concentrations in the cytoplasm without disrupting cellular activity. Compatible

Figure 8.54
Examples of the three classes of commercial herbicides that are AHAS inhibitors.

solutes allow plants to lower their water potential and maintain turgor during dry or saline conditions (see Chapter 22). Water stresses cause considerable loss of agricultural productivity, making it desirable to develop drought-tolerant plants by conventional or transgenic strategies. Studies of proline-overproducing mutants provide clear evidence for the ability of proline to confer increased osmotic tolerance in *E. coli*; they also suggest the potential for genetic engineering of proline biosynthesis and catabolism to increase stress resistance in plants.

8.6.1 In plants, proline is produced by two distinct pathways.

Proline apparently is synthesized by two different pathways in plants. One originates from glutamate, the other from ornithine, although the details of the ornithine pathway are less clear (Fig. 8.56). The pathway from glutamate is identical to that in *E. coli*. The committing enzyme of this pathway,

Figure 8.55
Ribbon structure of the spinach acetohydroxy acid isomeroreductase homodimer. Each monomer comprises two domains: the α/β N-terminal domain, which contains the NADPH-binding site, and the α C-terminal domain, which binds the inhibitor by means of two Mg^{2+} ions. The inhibitor and the NADPH are held stacked together at the domain junction. Here, the N- and C-terminal domains are represented in light and dark orange, respectively, for the first monomer and in light and dark green, respectively, for the second monomer. NADPH, the inhibitor, and the Mg ions are shown in purple, grey, and yellow, respectively.

Δ^1-pyrroline-5-carboxylate synthetase, is a bifunctional enzyme in plants (Box 8.11). Its first activity, γ-glutamyl kinase, catalyzes the ATP-dependent phosphorylation of L-glutamate, and the γ-glutamyl phosphate produced is then converted to glutamic γ-semialdehyde (GSA) by the NADPH-dependent GSA reductase. This intermediate spontaneously cyclizes to form Δ^1-pyrroline-5-carboxylate, which is converted to proline by the NADPH-dependent enzyme Δ^1-pyrroline-5-carboxylate reductase.

Two routes proposed for the synthesis of proline from ornithine (Fig. 8.56) differ as to whether Δ^1-pyrroline-5-carboxylate or Δ^1-pyrroline-2-carboxylate is the final intermediate in the pathway. A moth bean (*Vigna aconitifolia*) cDNA encoding a protein with ornithine-δ-aminotransferase activity has been identified, suggesting that Δ^1-pyrroline-5-carboxylate may be an intermediate in the synthesis of proline from ornithine in plants.

8.6.2 Proline synthesis and breakdown are environmentally regulated in plants.

Regulation of proline accumulation in plants occurs both at the enzyme level and through changes in gene expression. The committing enzyme Δ^1-pyrroline-5-carboxylate synthetase appears to be rate limiting in plants; for example, overexpression of the moth bean enzyme in transgenic tobacco led to a 14-fold increase in the soluble proline content of unstressed plants. These transgenic plants were relatively insensitive to osmotic stress, showing both negligible changes in leaf osmotic potential under drought conditions and less inhibition of shoot and root development in response to NaCl treatment. Consistent with the notion that Δ^1-pyrroline-5-carboxylate reductase is not rate limiting, 50-fold overexpression of moth bean reductase activity had no effect on free proline in transgenic tobacco.

More evidence for the importance of Δ^1-pyrroline-5-carboxylate synthetase in regulating proline accumulation was found by characterizing the moth bean enzyme expressed in *E. coli*: The γ-glutamyl kinase activity is feedback-inhibited by proline ($K_I = 1$ mM). This allosteric regulation is an obvious target for engineering further increases in proline accumulation, given that

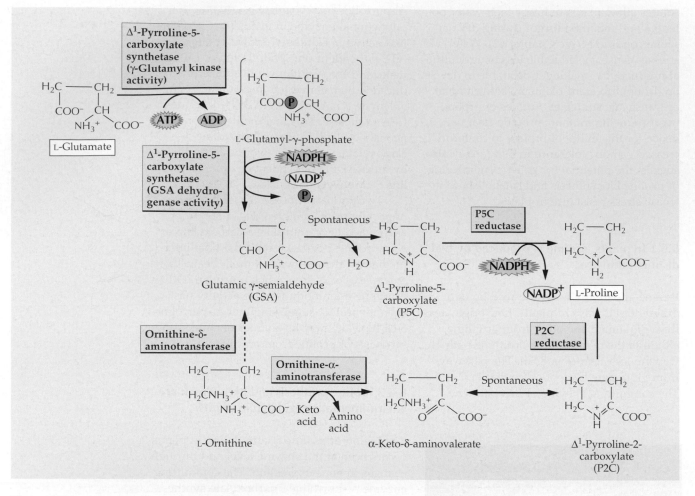

Figure 8.56

Multiple pathways are proposed for proline biosynthesis in plants. A glutamate-derived pathway is well established (black arrows), and increasing evidence suggests the importance of ornithine as a precursor (red arrows). The relative importance of the pathways through ornithine-δ-aminotransferase (dotted red arrow) or ornithine-α-aminotransferase (pathway with all red arrows) remains to be determined. In plants, the γ-glutamyl kinase (proB in *Escherichia coli*) and GSA dehydrogenase (proA in *E. coli*) are synthesized in a protein fusion. This activity is referred to as Δ^1-pyrroline-5-carboxylate synthetase.

an enzyme with as much as 200-fold reduction in sensitivity to feedback inhibition was created by using in vitro mutagenesis.

The genetic regulation of enzymes of proline metabolism is sensitive to environmental conditions that affect the concentrations of free proline. In *Arabidopsis*, mRNA for the Δ^1-pyrroline-5-carboxylate synthetase accumulates rapidly in response to desiccation, NaCl stress, and treatment with ABA. This induction correlates well with free proline concentrations, which increase four- to eightfold within 24 hours of desiccation or ABA treatment. This correlation is strengthened by the observation that both mRNA and free proline decrease coordinately when dehydrated plants are watered.

Is there evidence for regulation of the putative ornithine pathway enzymes in response to changes in proline accumulation? Yes, but moth bean ornithine-δ-aminotransferase and the glutamate-derived pathway enzyme pyrroline-5-carboxylate synthetase message appear to be differentially regulated at the level of transcription. Moth bean ornithine-δ-aminotransferase mRNA decreased slightly after salt stress and nitrogen starvation, conditions that induced accumulation of pyrroline-5-carboxylate synthetase. In contrast, ornithine-δ-aminotransferase expression was induced after treatment with high concentrations of nitrogen fertilizer, a condition that repressed accumulation of pyrroline-5-carboxylate synthetase. These

Box 8.11 — Plant amino acid biosynthetic pathways exhibit numerous gene duplications but few protein fusions.

The genetic organization of the enzymes of amino acid biosynthesis in plants is in some ways quite different from what is found in microorganisms. In prokaryotes and fungi it is relatively unusual to discover enzymes with multiple isoforms encoded by separate genes. In contrast, many plant biosynthetic enzymes are encoded by multiple genes. In cases such as AK or the AnS α-subunit from *Ruta graveolens,* these isoforms have dissimilar biochemical properties. In other instances, the genes are regulated differently, or they encode proteins with distinct subcellular locations (e.g., GS). However, for the vast majority of proteins encoded by multigene families, we do not know what, if any, functional significance can be ascribed to the existence of multiple isoforms.

Another clear difference between plants and microbes is that multifunctional proteins are unusual in plants, and most enzymes of amino acid metabolism catalyze a single pathway step. Exceptions to this rule include the bifunctional enzymes AK-HSDH, Δ^1-pyrroline-5-carboxylate synthetase, 3-dehydroquinate dehydratase–shikimate dehydrogenase, and the lysine catabolic enzyme lysine–ketoglutarate reductase–saccharopine dehydrogenase. In prokaryotes and fungi, fusion proteins are extremely common and sometimes include three or more different activities. For example, the ARO1 protein of yeast is a pentafunctional protein that catalyzes all but the first and final steps of chorismic acid biosynthesis!

results suggest that pyrroline-5-carboxylate may be synthesized from either glutamate or ornithine, and the relative importance of these two pathways may be modulated in response to environmental factors.

Analysis of *Arabidopsis* proline dehydrogenase, which catalyzes conversion of proline back to pyrroline-5-carboxylate, the first degradative step in the yeast pathway, suggests that catabolism plays a role in determining the concentration of free proline in plants. Proline dehydrogenase mRNA is regulated in a manner reciprocal to the biosynthetic enzyme pyrroline-5-carboxylate synthetase during both dehydration/rehydration and NaCl stress and recovery. These results suggest that the degradative pathway is suppressed during periods of rapid de novo synthesis, to prevent a futile cycle, and then is activated to bring proline concentrations back to prestress conditions.

Summary

Amino acids, the building blocks of proteins in all organisms, play additional roles in plants. For example, amino acids serve as precursors to plant hormones and serve to transport nitrogen from sources to sinks. As such, the control of amino acid synthesis in plants affects many aspects of growth and development. Additionally, the synthesis of essential amino acids in plants and amino acid composition of seeds relate indirectly to animal nutrition. Thus, understanding the pathways that control amino acid synthesis in plants has significance with regard to basic research on the control of metabolic pathways as well as practical implications. Although amino acid biosynthetic pathways have been well worked out in microbes, the situation in plants is less well defined, in part because of additional unique complexities. For example, in many instances, plants have multiple isoenzymes that catalyze biosynthetic reactions. These isoenzymes may be localized in distinct organelles or distinct cell types. Defining each step in an amino acid biosynthetic pathway and determining how each step is regulated are some of the key aspects of current research in amino acid biosynthesis in plants. This chapter highlights examples in which molecular, genetic, and biochemical approaches have been combined to elucidate the steps of these pathways in plants and to understand the regulation of these pathways at the level of gene regulation and beyond. Plant mutants in amino acid biosynthetic enzymes have shown that the synthesis of amino acids in vivo affects numerous diverse processes, including photorespiration, hormone biosynthesis, and plant development. Use of transgenic approaches has revealed the feasibility of manipulating amino acid biosynthesis pathways in plants, with applications to engineering herbicide resistance, altering amino acid composition in seed, and altering resistance to photoinhibition. Thus, although they are products of primary metabolism, amino acids also control many diverse aspects of plant growth and development.

Further Reading

Anderson, J. W. (1990) Sulfur metabolism in plants. In *The Biochemistry of Plants,* Vol. 16, *Intermediary Nitrogen Metabolism,* B. J. Miflin and P. J. Lea, eds. Academic Press, New York, pp. 328–381.

Bryan, J. K. (1990) Advances in the biochemistry of amino acids. In *The Biochemistry of Plants,* Vol. 16, *Intermediary Nitrogen Metabolism,* B. J. Miflin and P. J. Lea, eds. Academic Press, New York, pp. 161–195.

Coschigano, K. T., Melo-Oliveira, R., Lim, J., et al. (1998) *Arabidopsis gls* mutants and distinct Fdx-GOGAT genes: implications for photorespiration and primary nitrogen assimilation. *Plant Cell* 10: 741–752.

Delauney, A. J., Verma, D. P. (1993) Proline biosynthesis and osmoregulation in plants. *Plant J.* 4: 215–223.

Galili, G. (1995) Regulation of lysine and threonine biosynthesis. *Plant Cell* 7: 899–906.

Halford, N. G., Hardie, G. (1998) SNF1-related protein kinases: global regulators of carbon metabolism in plants? *Plant Mol. Biol.* 37: 735–748.

Hermann, K. M. (1995) The shikimate pathway: early steps in the biosynthesis of aromatic compounds. *Plant Cell* 7: 907–919.

Jang, J.-C., Sheen, J. (1994) Sugar sensing in higher plants. *Plant Cell* 6: 1665–1679.

Kozaki, A., Takeba, G. (1996) Photorespiration protects C_3 plants from photooxidation. *Nature* 384: 557–560.

Lea, P. J. (1993) Nitrogen metabolism. In *Plant Biochemistry and Molecular Biology,* P. J. Lea and R. C. Leegood, eds. John Wiley & Sons, New York, pp. 155–180.

Lea, P. J., Robinson, S. A., Stewart, G. R. (1990) The enzymology and metabolism of glutamine, glutamate, and asparagine. In *The Biochemistry of Plants,* Vol. 16, *Intermediary Nitrogen Metabolism,* B. J. Miflin and P. J. Lea, eds. Academic Press, New York, pp. 121–159.

Miflin, B. J., ed. (1980) *The Biochemistry of Plants,* Vol. 5, *Amino Acids and Derivatives.* Academic Press, New York.

Oaks, A., Hirel, B. (1985) Nitrogen metabolism in roots. *Annu. Rev. Plant Physiol.* 36: 345–365.

Ogren, W. L. (1984) Photorespiration: pathways, regulation, and modification. *Annu. Rev. Plant Physiol.* 35: 415–442.

Radwanski, E. R., Last, R. L. (1995) Tryptophan biosynthesis and metabolism: biochemical and molecular genetics. *Plant Cell* 7: 921–934.

Sieciechowicz, K. A., Joy, K. W., Ireland, R. J. (1988) The metabolism of asparagine in plants. *Phytochemistry* 27: 663–671.

Siehl, D. L. (1999) The biosynthesis of tryptophan, tyrosine and phenylalanine from chorismate. In *Plant Amino Acids: Biochemistry and Biotechnology,* B. K. Singh, ed. Marcel Dekker, New York, pp. 171–204.

Singh, B. K., Shaner, D. L. (1995). Biosynthesis of branched chain amino acids: from test tube to field. *Plant Cell* 7: 935–944.

Temple, S. J., Vance, C. P., Gantt, J. S. (1998) Glutamate synthase and nitrogen assimilation. *Trends Plant Sci.* 3: 51–56.

Biochemistry & Molecular Biology of Plants, B. Buchanan, W. Gruissem, R. Jones, Eds.
© 2000, American Society of Plant Physiologists

CHAPTER 9

Protein Synthesis, Assembly, and Degradation

Linda Spremulli

CHAPTER OUTLINE

Introduction

Protein synthesis is essential to cell function.

Proteins constitute a large percentage of the plant cell and carry out many different cell functions. It is therefore not surprising that protein synthesis is central to cell growth, differentiation, and reproduction. The processes of transcription and processing of messenger RNA (mRNA), discussed in Chapters 6 and 7, yield a template for the process of translation, which is described in detail in this chapter. **Translation** is the mechanism by which specialized riboprotein complexes "read" the information in the mRNA sequence and "write" a corresponding sequence of amino acids linked by peptide bonds to form a polypeptide chain. This chapter also presents how the polypeptide chain folds to form a precise three-dimensional structure that can carry out one or more specific biological functions (Fig. 9.1). Many proteins must assemble into large supramolecular complexes, called quaternary structures, to perform their specific functions. Photosynthetic complexes serve as examples of such multisubunit structures. To function properly, cells must regulate protein synthesis and degradation, responding to internal and external signals by adjusting the amounts of specific proteins present to suit cellular requirements. Although plants share many features of protein synthesis and metabolism in common with other eukaryotic organisms, certain aspects, such as protein synthesis in plastids and regulation of translation by light, are unique to plant cells.

Protein synthesis occurs in three distinct sites in plant cells.

In plants, protein synthesis occurs in three subcellular compartments. The cytoplasm, plastids, and mitochondria each contain different protein synthetic machinery (Fig. 9.2). About 75% of cell protein is made in the cytoplasm, where the mRNAs transcribed from the nuclear genome are translated. About 20% of the protein in a photosynthetically active cell (e.g., a young mesophyll cell) is synthesized in the chloroplast by means of mRNA templates transcribed from the chloroplast genome. A small amount of protein synthesis (approximately 2% to 5% of the total protein) occurs in mitochondria. This system translates mRNAs transcribed from mitochondrial DNA.

The variety of proteins synthesized also differs among the three compartments. In the cytoplasm, more than 20,000 different proteins may be synthesized. Fifty to 100 proteins are synthesized in chloroplasts, and the number synthesized in mitochondria varies widely among species. About 30 to 40 proteins appear to be synthesized in the mitochondria of the liverwort *Marchantia polymorpha,* for example, whereas the mitochondrial genomes of seed plants typically encode even fewer proteins.

The mechanisms responsible for protein synthesis in the cytoplasm, plastids, and mitochondria are distinct from each other and share few components, if any. Thus, plant cells contain three different types of ribosomes, three groups of transfer RNAs (tRNAs), and three sets of auxiliary factors for protein synthesis. Plastids and mitochondria presumably arose through the endosymbiosis of ancient prokaryotic organisms (see Chapters 1 and 6). Consistent with this theory, the protein synthetic machinery in plastids and mitochondria is more closely related to bacterial systems than to the translation apparatus in the surrounding plant cell cytoplasm. For reasons unknown, chloroplasts and mitochondria have retained a small amount of DNA and have preserved

their capacity to synthesize proteins. In contrast to protein synthesis in the cytoplasm and chloroplast, very little is known about mitochondrial protein synthesis.

9.1 From RNA to protein

9.1.1 During protein biosynthesis, the nucleotide sequence of mRNA is translated into the amino acid sequence of protein.

Much of the genetic information stored in the cell is expressed through the synthesis of protein products. During this process, DNA is first transcribed into pre-mRNA, which often undergoes specific processing reactions required to produce a mature mRNA for translation (see Chapter 6). The mRNA is then translated to yield a protein with a specific amino acid sequence (Fig. 9.3). The term translation is used to describe the process in which the mRNA sequence is read and used to synthesize a corresponding protein chain. The **genetic code** represents the relationship between the nucleotide sequence in the mRNA and the sequence of amino acids in the protein (see Fig. 9.4 and Chapter 8). The mRNA is translated three nucleotides at a time, with each trinucleotide sequence, or **codon,** encoding one amino acid.

Note that for a given mRNA sequence, the codons do not overlap and are not separated by natural divisions. If the protein synthesis machinery mistakenly initiates translation at the U in the AUG in the example above, an entirely different—and probably nonfunctional—amino acid sequence will result.

Crucial research in the 1960s permitted scientists to "break" the genetic code by demonstrating that each codon, or sequence of three bases in mRNA, specifies a

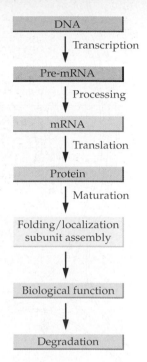

Figure 9.1
The life cycle of a protein.

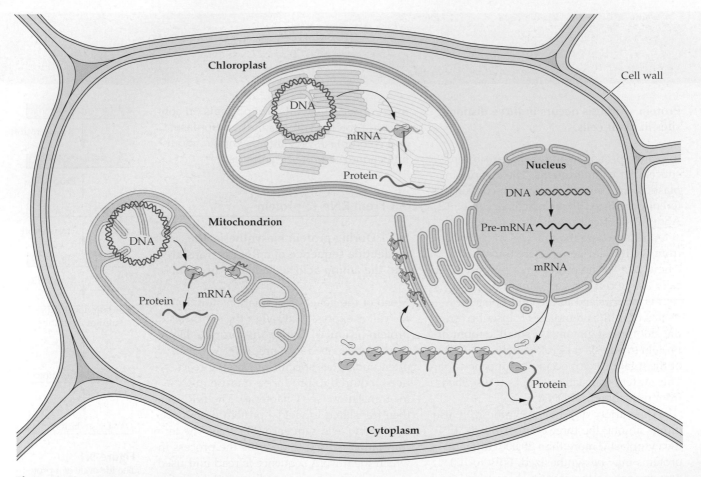

Figure 9.2

Sites for protein synthesis in a plant cell. A typical plant cell synthesizes proteins in three distinct compartments—the cytosol, plastids, and mitochondria. Translation of mRNAs transcribed in the nucleus occurs in the cytosol. In contrast, both transcription and translation of plastid and mitochondrial mRNA take place within those organelles.

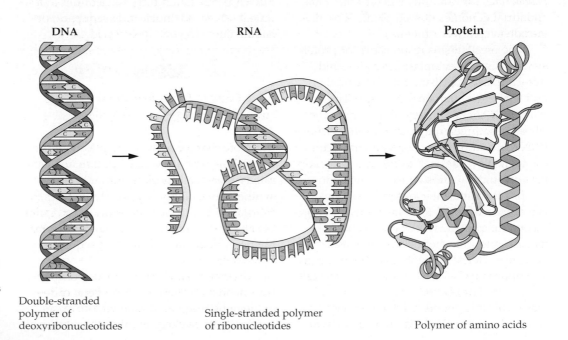

DNA   **RNA**   **Protein**

Figure 9.3

For eukaryotic genes that yield protein products, genetic information flows from double-stranded DNA to single-stranded RNA to protein.

Double-stranded polymer of deoxyribonucleotides

Single-stranded polymer of ribonucleotides

Polymer of amino acids

414

Amino acid	3-Letter code	1-Letter code	Codons
Alanine	Ala	A	GCC, GCU, GCG, GCA
Arginine	Arg	R	CGC, CGG, CGU, CGA, AGA, AGG
Asparagine	Asn	N	AAU, AAC
Aspartic acid	Asp	D	GAU, GAC
Cysteine	Cys	C	UGU, UGC
Glutamic acid	Glu	E	GAA, GAG
Glutamine	Gln	Q	CAA, CAG
Glycine	Gly	G	GGU, GGC, GGA, GGG
Histidine	His	H	CAU, CAC
Isoleucine	Ile	I	AUU, AUC, AUA
Leucine	Leu	L	UUA, UUG, CUA, CUG, CUU, CUC
Lysine	Lys	K	AAA, AAG
Methionine	Met	M	AUG
Phenylalanine	Phe	F	UUC, UUU
Proline	Pro	P	CCU, CCC, CCA, CCG
Serine	Ser	S	UCU, UCC, UCA, UCG, AGU, AGC
Threonine	Thr	T	ACU, ACC, ACA, ACG
Tyrosine	Tyr	Y	UAU, UAC
Tryptophan	Trp	W	UGG
Valine	Val	V	GUU, GUC, GUA, GUG
"Stop"	—	—	UAA, UAG, UGA

Figure 9.4
The name of each amino acid (e.g., alanine) can be abbreviated by using a three-letter code (Ala) or a one-letter code (A). The genetic code relates codons, that is, three-nucleotide sequences of mRNA, to the amino acids they specify. For every three nucleotides of mRNA read during translation, one amino acid is inserted into the growing polypeptide chain.

particular amino acid in a protein. Francis Crick, who codiscovered the double-helix structure of DNA with James Watson, suggested the existence of an adapter molecule to link each mRNA codon to the appropriate amino acid and thereby allow the correct interpretation of the genetic code. This molecule was subsequently shown to be a small RNA referred to as transfer RNA (see Chapter 6). To function properly, a tRNA must participate in two specific reactions. Its 3′ end must bind one and only one of the 20 common amino acids, and its three-nucleotide **anticodon** sequence must pair bases only with mRNA codons that encode the bound amino acid.

Each ribonucleotide of mRNA contains one of four nitrogenous bases: adenine (A), uracil (U), cytosine (C), or guanine (G). At four possible bases per nucleotide and three nucleotides per codon, the number of possible codons is 64 (i.e., 4^3). As a result, the genetic code is **degenerate,** because most of the 20 common amino acids are encoded by more than one codon (Fig. 9.4). Two amino

acids, however, methionine and tryptophan, are each specified by only a single codon. The codon AUG, which encodes the amino acid methionine, also signals the initiation of protein synthesis. Finally, three of the codons do not specify amino acids but instead signal the termination of protein synthesis. These are called stop, or nonsense, codons.

9.1.2 Transfer RNAs link amino acids to mRNA codons.

Translation of the degenerate genetic code links 64 mRNA codons to 20 amino acids by means of an intermediate number of tRNA molecules. This precise process is mediated by two mechanisms. First, some amino acids bind to multiple tRNAs (termed isoacceptor tRNAs), which are able to recognize different codons. Second, certain tRNAs can recognize more than one codon and can tolerate a mismatch at the third position of the codon, a bonding arrangement known as **wobble base pairing.** For example, both

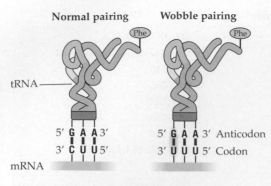

Figure 9.5
Wobble in the translation of the genetic code. Wobble pairing of bases can allow a tRNA to read more than one codon in a mRNA. For example, tRNA^Phe, which has the anticodon 5'-GAA-3' can pair bases with the codon UUU as well as with the complementary codon, UUC. Modified residues in the anticodons of tRNAs also affect the reading of the genetic code.

UUU and UUC specify the amino acid phenylalanine. As shown in Figure 9.5, the 5'-GAA-3' anticodon of the phenylalanine tRNA pairs with both phenylalanine codons. The presence of wobble pairing, predicted

by Crick in the late 1950s, allows 31 tRNAs to read the 61 codons specifying amino acids in the universal code.

Transfer RNAs are generally 70 to 90 residues long. Pairing bases within the tRNA sequence results in an L-shaped three-dimensional conformation. As shown in Figure 9.6, the anticodon is exposed to facilitate hydrogen bonding to the appropriate codon. Individual tRNAs are designated by the specific amino acid they bind and recruit to the growing polypeptide chain. For example, the tRNA that binds the amino acid phenylalanine is referred to as tRNA^Phe.

Enzymes known as **aminoacyl-tRNA synthetases** catalyze the attachment of specific amino acids to the 3' ends of tRNA molecules (see Fig. 9.7 and Chapter 6). Each aminoacyl-tRNA synthetase recognizes a single amino acid and generates a covalent bond between the α-carboxyl group of the amino acid and the ribose moiety at the 3' end of the tRNA. This process, referred to as amino acid activation, involves the initial formation of a high-energy phosphoacid

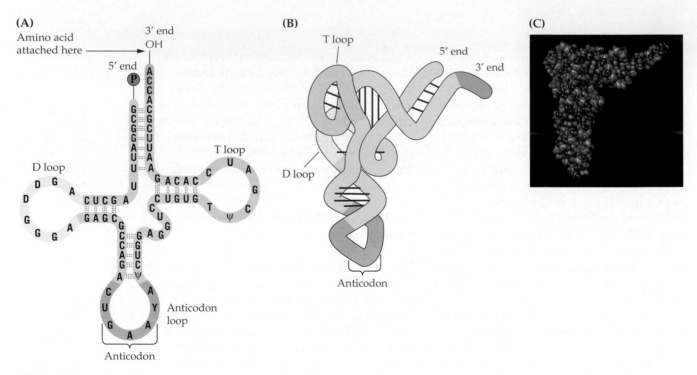

Figure 9.6
Structure of tRNA. Plant tRNAs fold into the same overall conformation as that observed in most systems. (A) The "cloverleaf" structure of tRNA illustrates the pattern of Watson–Crick pairing of bases that gives rise to the secondary structure of the tRNA, a series of stems and loops. Note that the three loops often contain unusual ribonucleotides, such as ψ (pseudouridine), D (dihydrouridine), and Y (N₆-isopentenyladenosine) shown here. (B) The three-dimensional conformation of the tRNA molecule resembles an upside-down L. (C) The compact shape of a tRNA can be seen in this space-filling model.

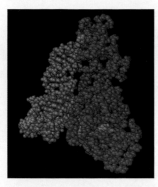

Figure 9.7
Aminoacyl-tRNA synthetases in the cytoplasm, plastids, and mitochondria of plant cells form specific complexes with their cognate tRNAs. Each enzyme recognizes a specific amino acid and only those tRNAs that should have that particular amino acid covalently attached. The enzyme (blue) is shown interacting with a tRNA (orange and green).

as aa-tRNA^{aa} (e.g., Phe-tRNA^{Phe}). It is the aa-tRNAs that participate directly in decoding the information carried in the nucleotide sequence of the mRNA.

9.1.3 Protein biosynthesis occurs on large macromolecular structures called ribosomes.

Protein synthesis occurs on specialized macromolecular complexes called **ribosomes,** which are composed of proteins and ribosomal RNA (rRNA, see Chapter 6). Ribosomes hold charged tRNAs and mRNA in position and catalyze the formation of peptide bonds between amino acid residues. Each subcellular compartment in which protein synthesis takes place contains its own ribosomes. Thus, plant cells contain three different types of ribosomes: cytoplasmic, plastid, and mitochondrial. All ribosomes consist of two subunits, referred to as the large and small subunits (Fig. 9.9). The two subunits reversibly associate and dissociate

anhydride bond between the amino acid and AMP. This step activates the amino acid and facilitates its transfer to the tRNA (Fig. 9.8). Once bound to an amino acid, the tRNA is said to be charged and is represented

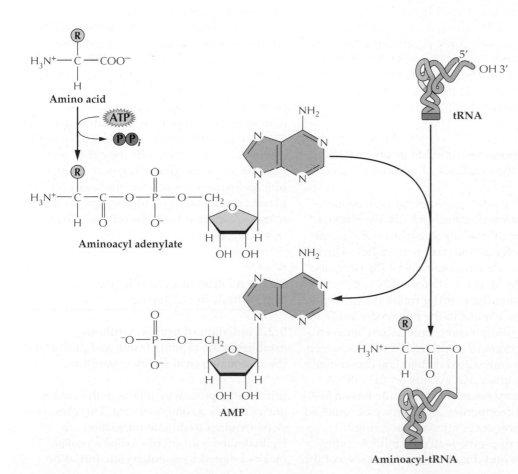

Figure 9.8
Activation of amino acids by aminoacyl-tRNA synthetases. Activation of amino acids is a two-step process. In the first step, shown on the left, an amino acid with side chain R is activated by a specific aminoacyl-tRNA synthetase. In this step, the enzyme hydrolyzes ATP, covalently attaching the α-carboxyl group of the amino acid to the 5′ phosphate of AMP and releasing the β- and γ-phosphates from ATP in the form of pyrophosphate (PP$_i$). The activated amino acid is referred to as the aminoacyl-adenylate. In the second step, shown on the right, the amino acid is transferred to a free hydroxyl group of the terminal ribose at the 3′ end of the tRNA.

Large subunit

Small subunit

Figure 9.9
Structure of the bacterial ribosome, showing the large and small subunits. The structure shown is a reasonable model for the structure of the plastid ribosome. The small subunit is visible in front, partially obscuring the large subunit behind it. The major events of protein biosynthesis occur primarily at the interface between the two subunits.

Table 9.1 Summary of the composition and properties of various ribosome types

| | Svedberg units, S | | | |
	Ribosome	Subunits	RNAs	Number of proteins
Plant cystol	80	40	18	~35
		60	28, 5.8, 5	~50
Plant plastids	70	30	16	22–31
		50	23, 5, 4.5	32–36
Plant mitochondria	~70	30	18	>25
		50	26, 5	>30
Prokaryotic	70	30	16	21
		50	23, 5	31

during the process of protein synthesis. Each subunit contains one or more rRNAs, which fold into highly ordered structures and associate with a variety of proteins (see Table 9.1 and Chapter 6). Formation of peptide bonds on the ribosome is now thought to be catalyzed primarily by the rRNA rather than by a protein enzyme.

Ribosomes are generally designated by their tendency to sediment in a sucrose gradient. The higher the S value (Svedberg value), the faster (and hence farther) the ribosome will sediment when subjected to sucrose density-gradient centrifugation. The two classes of ribosomal subunits differ in their S value. The large subunit has a higher S value, reflecting its greater mass and faster sedimentation rate.

9.1.4 Ribosomes function as an assembly line for the synthesis of proteins.

Like many other polymerization reactions (e.g., DNA replication and RNA synthesis), protein synthesis involves three phases: **initiation, elongation,** and **termination.** During initiation, the small subunit of the ribosome selects the start site on the mRNA, thereby establishing the **reading frame** (the phase in which the triplets in the mRNA are read). This subunit promotes interaction between the anticodon of a specialized tRNA charged with the amino acid methionine (the initiator tRNA) and an AUG codon on the mRNA. Thus, every nascent polypeptide has an N-terminal methionine, which may be removed during post-translational processing.

The ribosome reads the mRNA coding sequence and directs the incorporation of the amino acid corresponding to each codon. During this process, the mRNA is read in a 5′ to 3′ direction and the protein is synthesized from the amino (or N) terminus to the carboxy (or C) terminus (Fig. 9.10). Several ribosomes may read an mRNA simultaneously, forming a structure known as a polyribosome or **polysome** (Fig. 9.11). Targeting domains in the growing polypeptide chains direct some polysomes to membranes, while other polysomes remain free in aqueous solution (see Chapter 4).

After synthesis, many proteins undergo various processing reactions. These post-translational modifications expand the range of amino acids found in proteins beyond the 20 incorporated on the ribosome. Some polypeptides are also targeted to the subcellular compartment where they function (see Chapter 4). In the plant cell, major sites of protein localization include the cytoplasm, plastids, mitochondria, endoplasmic reticulum, peroxisomes, vacuoles, nucleus, and plasma membrane. In addition, certain proteins are exported from the cell (see Chapters 1 and 4).

9.2 Regulation of cytosolic protein biosynthesis in eukaryotes

9.2.1 Initiation of protein synthesis establishes the reading frame and positions the first amino acid for incorporation.

Initiation of protein synthesis in the eukaryotic cytosol is a complex event. The various steps required to initiate translation are facilitated by a group of auxiliary protein factors referred to as **eukaryotic initiation**

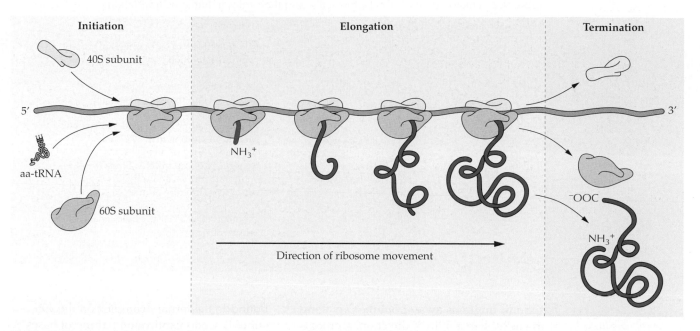

Figure 9.10
Overview of the process of protein synthesis. Translation can be divided into three stages: initiation, elongation, and termination. The mRNA is read from 5′ to 3′, and the polypeptide chain is synthesized from the amino terminus to the carboxy terminus by the sequential addition of amino acids.

factors (eIFs), which are classified on the basis of the general reaction they promote (Table 9.2). Many mechanisms that regulate cytoplasmic protein synthesis in eukaryotes affect the activities of one or more of these factors.

Initiation (Fig. 9.12) begins when eIF2 interacts with an initiator Met-tRNA in the presence of an energy source, GTP, generating the so-called ternary complex. Once formed, the ternary complex binds to a free 40S ribosomal subunit, a process facilitated by several other eIFs. A large and complex factor, eIF3, binds to the 40S subunit.

The small subunit of the ribosome, bound to the Met-tRNA and several eIFs, then interacts with the 5′ cap structure on the mRNA. This step requires the eIF4 family of initiation factors, which recognize the cap structure and facilitate the interaction between the 40S subunit and the mRNA. The subunit must then identify the correct AUG codon at which to begin reading the mRNA. Typically, the Met-tRNA–bound 40S

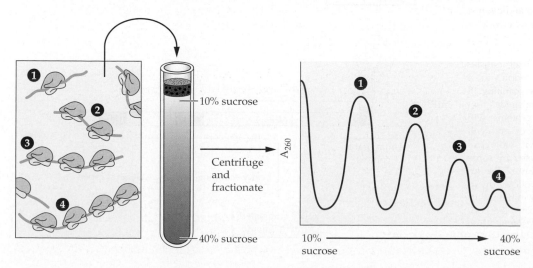

Figure 9.11
Most mRNAs are translated by more than one ribosome at a time. The term polysome refers to a mRNA that binds multiple ribosomes. Sucrose density-gradient centrifugation can be used to separate mRNAs that bind different numbers of ribosomes. The more ribosomes bound, the more massive the polysome, and the greater the distance traveled during sedimentation.

Table 9.2 Eukaryotic initiation factors (eIFs) and their roles in initiating translation

Name of class	Members	General role
eIF1	eIF1, eIF1A	Multiple effects in enhancing initiation of complex formation.
eIF2	eIF2, eIF2B, eIF2C	GTP-dependent recognition of Met-tRNA and nucleotide exchange.
eIF3	eIF3	Ribosomal subunit dissociation. Promotes Met-tRNA binding to 40S subunits.
eIF4	eIF4A, eIF4B, eIF4E, eIF4F, eIF(iso) 4F, eIF4G, eIF4H	Recognizes 5′ cap mRNA. Binds the 40S subunit to mRNA and unwinds the secondary structure of mRNA.
eIF5	eIF5	Promotes eIF2 GTPase activity and release of factors from ribosome.
eIF6	eIF6	Interacts with 60S subunits. Role unknown.

subunit migrates away from the cap along the mRNA in a 5′ to 3′ direction, a process referred to as **scanning.** The ribosome generally selects the first AUG codon it encounters. This selection process is facilitated by codon:anticodon hydrogen bonding between the AUG codon and the Met-tRNA bound to the ternary complex on the 40S subunit. Codon:anticodon pairing of bases fixes the start site on the mRNA and establishes the reading frame. Ribosomal selection of the initiation AUG codon is not strongly dependent on the nucleotide sequence surrounding the codon. In general, a

Box 9.1
Leaky scanning allows plant viruses to translate multiple polypeptides from a single mRNA.

Most eukaryotic mRNAs are monocistronic, encoding a single protein, which is synthesized when translation initiates at the AUG codon closest to the 5′ end. However, some mRNAs encode more than one **open reading frame** (ORF). Several features of a particular mRNA—including the nucleotide sequence surrounding the first AUG, the secondary structure of the mRNA, and the distance between the cap and the first AUG codon—can affect the efficiency of initiation at the first AUG. If the first AUG is in an unfavorable context, some ribosomes may bypass it and initiate translation at a second AUG codon, in a process referred to as **leaky scanning.** If the two AUG codons establish different reading frames, the two proteins will have very different sequences and will probably carry out different biological functions. This strategy is used by some plant viruses, including many of the luteoviruses (e.g., barley yellow dwarf virus). It allows the virus to minimize its genome size while maximizing the coding capacity of the genome (see figure to the right).

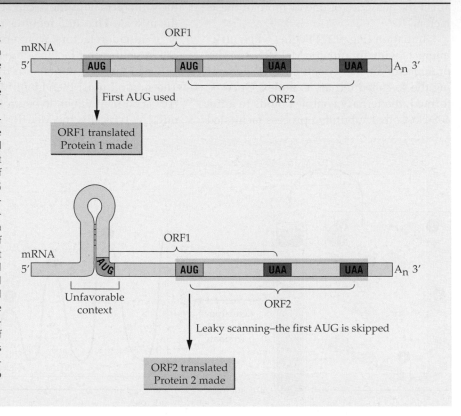

Chapter 9 Protein Synthesis, Assembly, and Degradation

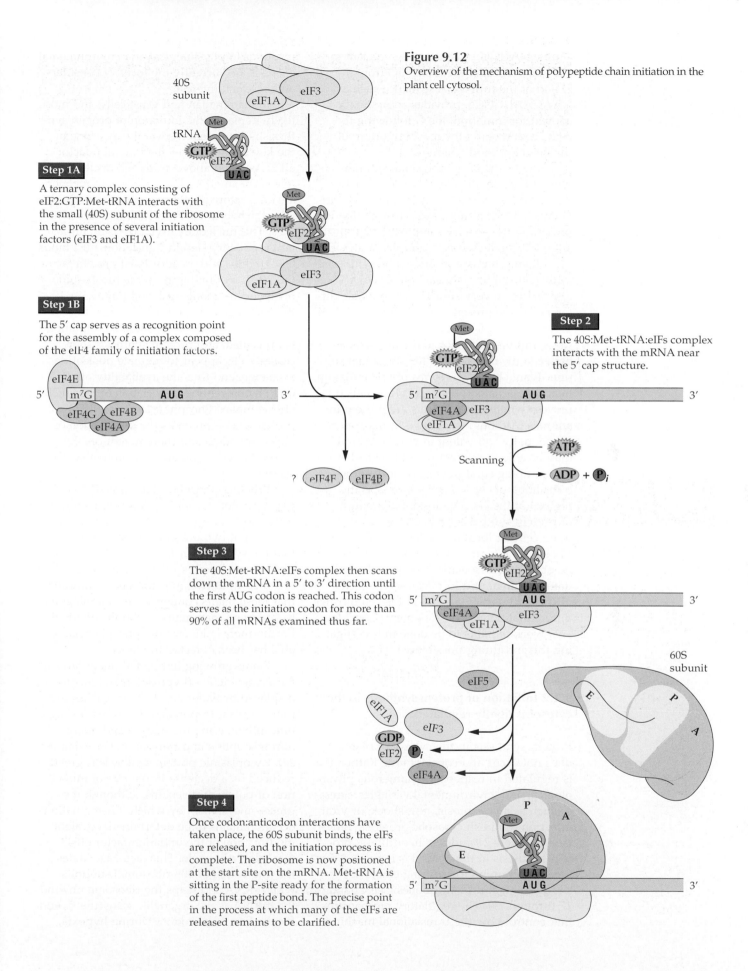

Figure 9.12
Overview of the mechanism of polypeptide chain initiation in the plant cell cytosol.

40S subunit

tRNA

Step 1A

A ternary complex consisting of eIF2:GTP:Met-tRNA interacts with the small (40S) subunit of the ribosome in the presence of several initiation factors (eIF3 and eIF1A).

Step 1B

The 5′ cap serves as a recognition point for the assembly of a complex composed of the eIF4 family of initiation factors.

Step 2

The 40S:Met-tRNA:eIFs complex interacts with the mRNA near the 5′ cap structure.

Scanning

Step 3

The 40S:Met-tRNA:eIFs complex then scans down the mRNA in a 5′ to 3′ direction until the first AUG codon is reached. This codon serves as the initiation codon for more than 90% of all mRNAs examined thus far.

60S subunit

Step 4

Once codon:anticodon interactions have taken place, the 60S subunit binds, the eIFs are released, and the initiation process is complete. The ribosome is now positioned at the start site on the mRNA. Met-tRNA is sitting in the P-site ready for the formation of the first peptide bond. The precise point in the process at which many of the eIFs are released remains to be clarified.

simple ...AUGG... sequence is sufficient to allow initiation at the first AUG from the 5′ end of the mRNA. A sequence such as ...AACA*AUG*GC... provides an optimal context for initiation, the G following the AUG codon being the most important of the neighborhood residues.

Finally, the large ribosomal subunit binds to the small subunit while holding the mRNA and Met-tRNA in the correct orientation. During this stage, the GTP that was present in the ternary complex is hydrolyzed and eIF2 is released as a complex with GDP.

The mechanism of protein biosynthesis in the cytosol of plants and other eukaryotes appears to be very similar. One possible difference is the presence of two forms of eIF4F, a protein complex involved in recognizing the 5′ mRNA cap prior to the interaction between the mRNA and 40S ribosomal subunit. Plants have two forms of this initiation factor, designated eIF4F and eIF(iso)4F. It is unclear whether both forms are present in animals. Although these two forms appear functionally equivalent in vitro, eIF(iso)4F may be able to interact with the cytoskeleton; in doing so, it may help localize the translation of certain mRNAs to specific regions of the cell. A substantial amount of protein synthesis probably occurs at defined cellular locations where mRNAs and ribosomes interact with cytoskeletal elements. These associations may well involve the movement and localization of specific mRNAs to particular sites in the cell during differentiation and development. Considerable work remains to be done in investigating this intriguing possibility.

9.2.2 Initiation of protein synthesis in the cytosol is tightly regulated.

Protein synthesis in the cytosol of plants does not occur at a constant rate. Rather, it is regulated in response to numerous physiological and environmental variables. Stresses such as anaerobiosis, heat shock, or viral infection diminish cytosolic protein synthesis (see Chapters 21 and 22). In addition, protein synthesis in the cytosol is regulated during seed germination, during embryogenesis, by light, and by hormones (see Chapters 17 through 19). The regulatory mechanisms that control how the translational machinery responds to physiological and environmental changes are currently the focus of considerable research interest.

In mammalian and yeast cells, and most likely in plants, the initiation of protein synthesis is regulated in several ways. One of the key steps involves the initiation factor eIF2. As noted above, initiation results in the release of an eIF2:GDP complex. To regenerate eIF2 before another round of chain initiation, the bound GDP must be replaced with GTP. This nucleotide exchange is an important regulatory step in translation. eIF2 binds GDP tightly, and an additional protein factor (eIF2B) is required to promote the guanine nucleotide exchange reaction (Fig. 9.13). If the exchange of GDP for GTP cannot occur, active eIF2 cannot be regenerated, initiation is prevented, and protein synthesis soon ceases. Cells appear to regulate nucleotide exchange (and thus the availability of eIF2–GTP) by preventing dissociation of the eIF2B–eIF2 complex. Phosphorylation of eIF2 by a protein kinase prevents the dissociation of eIF2 from eIF2B and thereby reduces the amount of eIF2 available for protein synthesis (Fig. 9.13).

Whether phosphorylation of eIF2 is a key regulatory event in plant cells as it is in mammalian and yeast cells is not currently known, but two lines of evidence suggest that the systems function similarly. An examination of **expressed sequence tags** (ESTs) obtained from rice and *Arabidopsis* indicates that these plants contain genes homologous to yeast and mammalian subunits of eIF2B. Furthermore, a kinase that phosphorylates eIF2 has been detected in plants.

Plants growing in flooded fields can suffer oxygen deficit (hypoxia) as soil gasses are displaced by water and less air reaches the roots. A plant responds to this stress by significantly altering the pattern and quantity of transcription and translation (see Chapter 22). Cytoplasmic protein synthesis is greatly reduced by a decrease in the rate of initiation of polypeptide chains. Although the precise mechanism by which this regulation occurs remains to be determined, covalent modification of the initiation factor eIF4A may play a key role. This factor facilitates the interaction of 40S ribosomal subunits with mRNA and helps the ribosome unwind the mRNA as the organelle scans the 5′ end for the first AUG codon. During hypoxia,

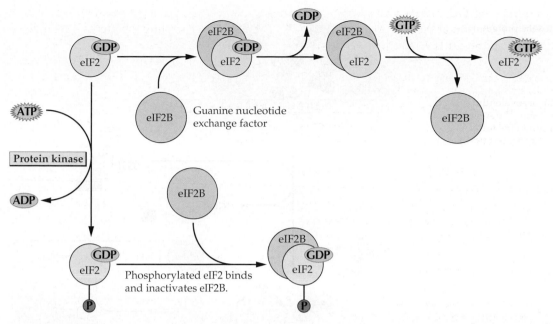

Figure 9.13
Guanine nucleotide exchange reaction in eukaryotic initiation and recycling of eIF2 during protein synthesis. eIF2 is released from the ribosome as an eIF2:GDP complex. Before the protein factor can participate in a new round of initiation, the GDP must be replaced by GTP. The release of GDP and the binding of GTP is promoted by guanine nucleotide exchange factor, eIF2B. When phosphorylated, eIF2 binds eIF2B very tightly. The rate of initiation decreases dramatically when the eIF2B molecules are sequestered in complexes with the phosphorylated eIF2.

oxidative phosphorylation can no longer provide energy for the cell, and fermentative metabolism dominates (see Chapter 22). The cell responds by decreasing the synthesis of most of the normal cellular proteins while increasing the synthesis of enzymes required for glycolysis and ethanolic fermentation. Typically, lactate synthesis acidifies the cytosol during the early stages of anaerobiosis. The timing of eIF4A phosphorylation roughly corresponds to a decrease in intracellular pH (approximately 0.5 unit). The drop in pH may serve as a second messenger to regulate activities of the protein kinases and phosphatases that determine the phosphorylation state of eIF4A.

9.2.3 Elongation involves sequential addition of amino acid residues to the growing polypeptide chain.

The sequential addition of amino acids to the growing polypeptide chain involves the use of three sites on the fully assembled 80S ribosome, designated the P, A, and E sites. The P-site (peptidyl-tRNA-binding site) participates in chain initiation and donates the

growing polypeptide chain to the incoming aa-tRNA. The A-site (aminoacyl-tRNA binding site, or decoding site) exposes the next codon to be read on the mRNA, where the incoming charged tRNA binds. The E-site (exit site) is occupied by tRNA after it has released its amino acid, just before it leaves the ribosome. These three sites are used sequentially as the polypeptide chain is synthesized. A complete cycle requires as little as 0.05 seconds (Fig. 9.14).

As a ribosome starts to read an mRNA, it moves downstream, exposing the start codon and allowing a second ribosome to initiate translation of the mRNA. The resulting polysomes contain ribosomes that are often spaced as close as 80 to 100 nucleotides apart on the mRNA.

9.2.4 Termination of protein synthesis occurs at specific signals in the mRNA.

Polypeptide elongation ceases when a ribosome reaches one of three stop codons on the mRNA: UAA, UAG, or UGA (Fig. 9.15). Specific proteins known as release factors bind to the ribosome and trigger a series of

Figure 9.14

The elongation phase of cytosolic protein synthesis in plants.

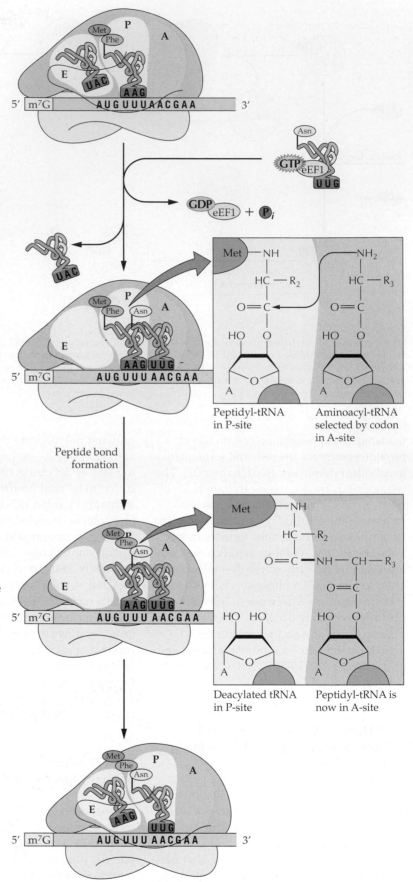

Step 1

A growing polypeptide chain is covalently attached to the tRNA in the P-site. The A-site is empty, exposing the next codon in the mRNA. The E-site is occupied by the uncharged tRNA from the previous cycle.

Step 2

A charged tRNA binds to the ribosomal A-site only if its anticodon matches the exposed codon on the mRNA. When this occurs, the tRNA in the E-site is ejected from the ribosome. A required elongation factor, eEF1, forms a ternary complex with GTP and the charged tRNA and promotes the binding of the charged tRNA to the A-site of the ribosome.

Peptidyl-tRNA in P-site

Aminoacyl-tRNA selected by codon in A-site

Peptide bond formation

Step 3

The ribosome uses the peptidyl transferase center to catalyze the formation of a peptide bond between the growing polypeptide chain and the new amino acid. Recent experiments suggest that rRNA plays a particularly important role as a ribozyme in this catalytic step. The net result of this process is that the nascent polypeptide has been transferred from the tRNA in the P-site to the new amino acid attached to its tRNA in the A-site. The polypeptide is one residue longer, the peptidyl-tRNA now occupies the A-site, and the tRNA in the P-site is free of its amino acid. This tRNA is said to be deacylated.

Deacylated tRNA in P-site

Peptidyl-tRNA is now in A-site

Step 4

The complex must rearrange, exposing the next triplet. This process, called translocation, involves three rearrangements: The deacylated tRNA in the P-site moves into the vacant E-site; the peptidyl-tRNA in the A-site moves into the P-site; and the ribosome moves relative to the mRNA by exactly three nucleotides (one codon), exposing a new triplet in the A-site. Moving one or two nucleotides too little or too much would initiate a new reading frame, probably resulting in an inactive polypeptide.

Figure 9.15

Events during the termination of protein synthesis. Polypeptide chain termination occurs when one of three stop codons appears in the A-site of the ribosome.

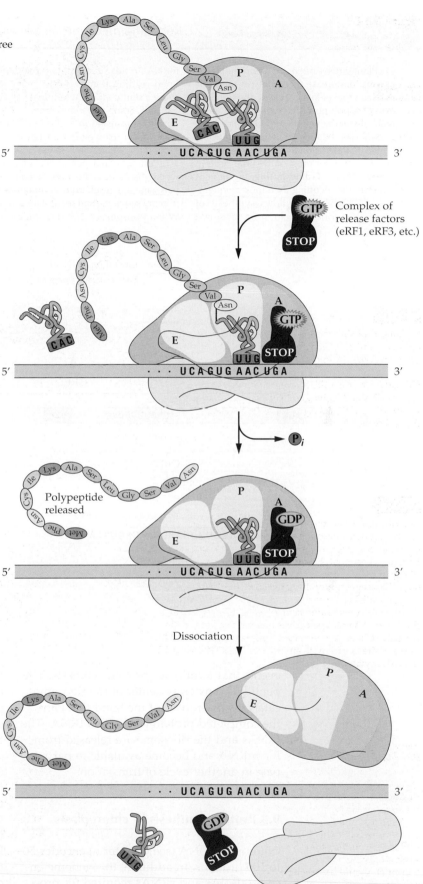

Step 1

A complex of release factors (including eRF1 and GTP-bound eRF3) binds to the stop codon exposed in the A-site. Most likely, although it is not yet proven, the deacylated tRNA in the E-site is released at this point.

Step 2

The peptidyl transferase center on the ribosome catalyzes the hydrolysis of the ester bond that links the completed polypeptide to the final tRNA.

Step 3

The ribosome is dissociated from the mRNA and becomes available for another round of protein synthesis. (Note: the precise timing and function of GTP hydrolysis remain to be investigated more fully.)

Plant viruses often use recoding mechanisms to increase the variety of products they can make. In most cases, the termination of a polypeptide chain occurs automatically at the first stop codon once elongation has begun. In some situations, however, translational recoding can occur, altering some or most of the final protein product made. **Frameshifting,** the type of translational recoding used most frequently by plant viruses, allows synthesis of more than one protein from a single mRNA. In this situation, two overlapping open reading frames (ORF1 and ORF2) share a start codon but terminate at different sites. Some ribosomes read ORF1 and terminate translation at the first in-frame stop codon. However, translation of the second reading frame is made possible by a recoding signal positioned at the site where the frameshift occurs. Recoding signals generally result from a combination of primary sequence and secondary structure. When ribosomes reach this signal, some shift into another reading frame by slipping back one nucleotide (–1 frame) or by sliding forward one nucleotide (+1 frame). This repositioning shifts the ribosomes into a new reading frame, ORF2, which is read until the next in-frame stop codon is reached. Thus, two polypeptide products that differ in size and C-terminal amino acid sequence are translated from a single mRNA (see illustration below).

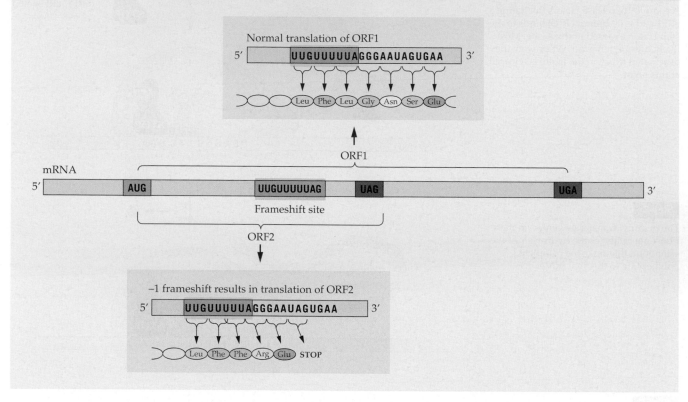

events that terminates protein synthesis. The peptidyl transferase center of the ribosome catalyzes hydrolysis of the bond that joins the completed protein to the final tRNA. The tRNAs and the ribosome are released from the mRNA and become available to participate in another cycle of translation.

9.3 Protein synthesis in chloroplasts

Chloroplast DNA (see Chapter 6) encodes 50 to 100 proteins. In addition, this genome encodes tRNAs and rRNAs required for pro-tein biosynthesis in the chloroplast. Most proteins synthesized in the chloroplast participate either in photosynthesis or in chloroplast-localized transcription and translation.

The chloroplast genome is clearly not autonomous. Most of the proteins required for chloroplast function are encoded within the nuclear genome. These proteins are synthesized in the cytoplasm and subsequently imported into the plastid (see Chapter 4). Indeed, most plastid-encoded proteins function in oligomeric protein complexes containing subunits from both nuclear and

Ricin, a toxic protein from the seeds of the castor bean plant (*Ricinus communis*, pictured at right), is the best-known member of a class of plant polypeptides known as ribosome-inactivating proteins (RIPs). Ricin was made famous as the poison used by the Bulgarian secret police to assassinate the defector Georgi Markov in London in 1978. In a plot that reads like a popular spy novel, a dart-tipped umbrella was used to deliver the toxin to its unfortunate victim.

A potent inhibitor of eukaryotic protein synthesis, ricin acts by irreversibly inactivating eukaryotic ribosomes. To do this, it depurinates a specific adenine nucleotide on the 28S ribosomal RNA. Loss of this adenine inactivates the ribosome. RIPs are being investigated as potential therapeutic agents that could be targeted to kill malignant cells.

chloroplast genomes. A well-studied example is the enzyme ribulose-1,5-bisphosphate carboxylase/oxygenase (Rubisco, Fig. 9.16; see also Section 9.4.9 and Chapter 12). Assembling such multisubunit complexes requires close coordination of transcription and translation in the cytosol and chloroplast.

9.3.1 Chloroplast protein synthesis shows many similarities to bacterial protein synthesis.

Translation in the plastid shares many of the features of bacterial protein synthesis. This

observation is consistent with the proposed endosymbiotic origin of plastids from ancestral free-living cyanobacteria.

Plastid protein synthesis differs from cytoplasmic protein synthesis in several details. In general, the ribosomes, mRNAs, and auxiliary factors required at each step are not interchangeable between the cytosol and the plastid. The numbers and functions of specific initiation factors differ considerably between the two compartments.

Perhaps one of the most fundamental differences between these translation systems can be seen in the process by which the initiation signal on the mRNA is selected in

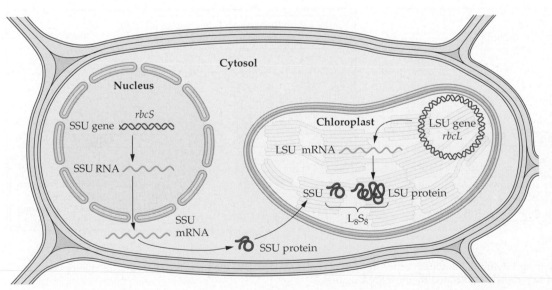

Figure 9.16
In plants, Rubisco consists of 16 protein subunits: 8 copies of a small subunit (SSU) and 8 copies of a large subunit (LSU). *rbcS*, the gene for the SSU, is found in the nucleus, whereas the LSU is encoded by the *rbcL* gene in the plastid. The synthesis of this enzyme, perhaps the most abundant protein on earth, requires the expression of two genomes and the coordination of three subcellular compartments: nucleus, cytosol, and chloroplast.

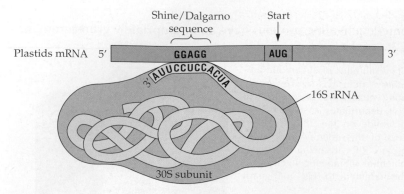

Shine/Dalgarno sequence

Start

Plastids mRNA 5' **GGAGG** **AUG** 3'

3' AUUCCUCCACUA

16S rRNA

30S subunit

Figure 9.17
The Shine/Dalgarno interaction for selection of the start codon. Initiation at the correct AUG codon is facilitated by hydrogen bonding between a polypurine sequence in the mRNA (the Shine/Dalgarno sequence) and a polypyrimidine sequence near the 3' end of the small subunit 16S rRNA.

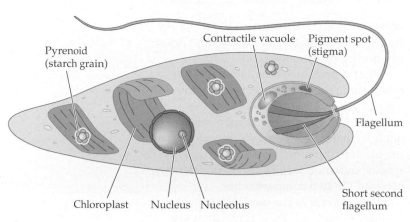

Pyrenoid (starch grain)

Contractile vacuole   Pigment spot (stigma)

Flagellum

Short second flagellum

Chloroplast   Nucleus   Nucleolus

Figure 9.18
Euglena gracilis has been used extensively for studying the mechanisms of protein synthesis in chloroplasts.

the cytoplasm and plastid. As discussed earlier, cytoplasmic mRNAs have a 5' cap, which flags the 5' end for initiation factors (see Fig. 9.12). By contrast, chloroplast mRNAs are not capped and are often transcribed as **polycistronic** messages that contain several reading frames, similar to bacterial mRNAs.

In bacterial systems, the small (30S) subunit of the ribosome selects the correct AUG codon for initiation. The 16S rRNA in this subunit can pair bases with a short mRNA sequence found just upstream of the initiation AUG, called the Shine/Dalgarno sequence after its discoverers (Fig. 9.17). This Shine/Dalgarno sequence on the mRNA facilitates the selection of the start codon by the 30S subunit. In addition, the secondary structure of the mRNA plays an important role in determining the efficiency of translational initiation.

The mechanism by which chloroplast ribosomes recognize the correct AUG for initiation has been studied most extensively in the chloroplasts of the protist *Euglena gracilis* (Fig. 9.18). Chloroplast mRNAs from *Eu. gracilis* fall into two classes (Fig. 9.19). One group contains no conserved sequence signal in the 5' untranslated region (5' UTR). The translational start site in these mRNAs is specified by the presence of an AUG codon in an unstructured or weakly structured region of the mRNA. The chloroplast mRNAs in the second group from *Eu. gracilis* contain

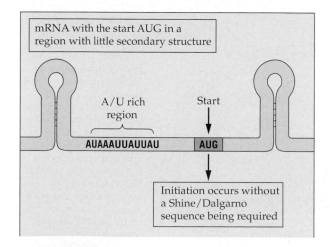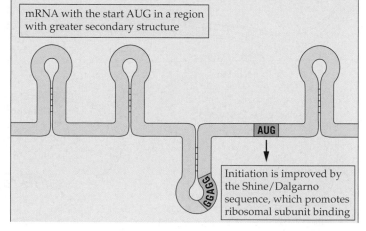

mRNA with the start AUG in a region with little secondary structure

A/U rich region

Start

AUAAAUUAUUAU   **AUG**

Initiation occurs without a Shine/Dalgarno sequence being required

mRNA with the start AUG in a region with greater secondary structure

AUG

GGAGG

Initiation is improved by the Shine/Dalgarno sequence, which promotes ribosomal subunit binding

Figure 9.19
Role of secondary structure in specifying the initiation codon in *Euglena gracilis*. Selection of the start site on *Eu. gracilis* chloroplast mRNAs proceeds by one of two routes, as dictated by the mRNA sequence.

a Shine/Dalgarno sequence upstream from the start codon. In this class of mRNAs, the sequences surrounding the start codons demonstrate a greater degree of secondary structure, and the Shine/Dalgarno sequences typically play a role in initiation. Like the euglenoid transcripts, the mRNAs of plant chloroplasts can probably also be divided into two distinct families.

9.3.2 Thylakoid membrane proteins encoded in chloroplast DNA are translated on membrane-bound ribosomes.

In organelles as well as in the cytosol, polysomes fall into two classes—free and membrane-bound. These two classes translate different sets of proteins. The proteins synthesized by free polysomes are released into the cytosol, plastid stroma, or mitochondrial matrix after translation and are directed to their final destination by protein sequences that act as targeting signals. By contrast, the protein products of membrane-bound polysomes are inserted directly into the membrane during translation. This mechanism is referred to as **cotranslational translocation.**

The light reactions of photosynthesis occur in the thylakoid membranes. Several of the proteins synthesized by the chloroplast are destined for incorporation into enzyme complexes found in these membranes (Fig. 9.20). Protein synthesis in the chloroplast is believed to be initiated on free ribosomes. Once the nascent protein chain has emerged from the ribosome, the amino acid sequence near the N terminus acts as a sig-

nal sequence, directing the ribosome to the thylakoid membrane. The mechanistic details of this process remain to be clarified. Once the ribosome and its nascent chain are associated with the membrane, translation continues and the polypeptide is inserted into the membrane as it is synthesized.

9.3.3 Chloroplast protein synthesis is regulated by light.

As a leaf develops, photosynthetically active chloroplasts arise from proplastids, or from etioplasts in dark-grown tissues. During this process, dramatic morphological changes transform the organelle. In the thylakoids, major protein complexes must be assembled, including photosystems I and II (PSI, PSII) and ATP synthase. Synthesis of many chloroplast proteins is greatly enhanced during the light-induced greening of the plastid. Translation of certain chloroplast mRNAs increases as much as 100-fold, sometimes without significant changes in the amount of mRNA present. In addition to translational control, other mechanisms apparently participate in regulating the synthesis of chloroplast proteins as well: transcription, mRNA stability, cofactor insertion into photosynthetic proteins, and protein turnover (Fig. 9.21). The synthesis of some proteins appears to be controlled to a large extent by transcription, whereas the quantities of other proteins present are regulated by translation or by protein degradation.

The mRNAs for photosynthetic proteins can accumulate to quite high amounts in

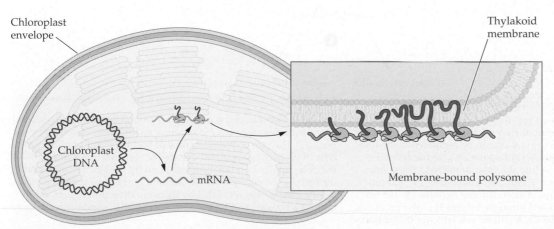

Chloroplast envelope

Thylakoid membrane

Chloroplast DNA

mRNA

Membrane-bound polysome

Figure 9.20
Chloroplast-encoded proteins targeted to multimeric complexes in the thylakoid membrane are synthesized on membrane-bound ribosomes. Translation of these mRNAs is initiated in the soluble portion of the chloroplast (stroma). Signals in the nascent peptides then direct the polysomes to the thylakoid membranes, where protein synthesis is completed.

plastids, even when the plants are in the dark. However, the polypeptide products from these mRNAs are barely detectable. For example, the mRNA for the large subunit of Rubisco is present in dark-grown amaranth (*Amaranthus* sp.), but these transcripts do not associate with polysomes when cotyledons are kept in the dark. When the cotyledons are illuminated, mRNA quantities remain constant, but synthesis of the large subunit protein can be detected within 3 to 5 hours.

A growing body of genetic and biochemical evidence suggests that some nuclear gene products act as regulators of chloroplast protein synthesis, often affecting the translation of a single species of chloroplast mRNA by interacting with its 5' UTR. These protein–mRNA complexes increase in the light, correlating with enhanced translation of the mRNA component (Fig. 9.22). These protein–mRNA interactions may help control the translational activation of

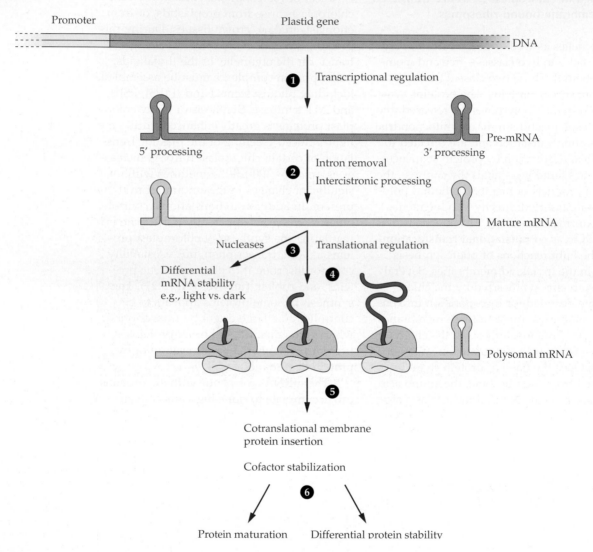

Figure 9.21

Many regulatory processes act together to control the quantities of chloroplast proteins present during different phases of growth and development. (1) Gene expression can be regulated at the transcription of the DNA into RNA. (2) Regulation also occurs in the maturation of the pre-mRNA into the mature mRNA. These steps include processing the 3' and 5' ends, removing introns, and processing polycistronic mRNAs in intercistronic regions. (3) The stability of the mRNA is regulated through the action of the nucleases that determine how long the RNA will be available for translation.

(4) The translation of the mRNA into protein is regulated in several ways, including through the action of specific mRNA-binding proteins. (5) The rate of protein synthesis and the stability of the nascent chain may be regulated during the cotranslational insertion of the polypeptide into the membrane and by the binding of a cofactor such as chlorophyll. (6) The amount of the final protein product from a gene is also regulated by its maturation (folding, modification, localization) and by sequence signals that regulate its lifespan.

certain mRNAs, with the nuclear protein products acting as translational activators to facilitate the synthesis of various polypeptides in the light.

9.3.4 mRNA-binding proteins can be regulated by redox potential.

During light-activated protein synthesis, binding of the translational activator proteins must be coordinated with the photosynthetic activity of the chloroplast. At least two pathways seem to link these processes. The first pathway involves the redox potential of the chloroplast, whereas the second couples translation to the availability of energy in the organelle. Within the chloroplast, photosynthetic electron transfer generates reducing power and drives synthesis of the energy-rich nucleotide ATP (see Chapter 12). Both of these processes appear to act to regulate translation of *psbA* transcripts in the chloroplast of *Chlamydomonas*. The *psbA* gene product is a component of PSII. The redox environment of the chloroplast and the availability of ATP are thought to influence the formation of a multiprotein complex that binds to the 5' UTR of the *psbA* mRNA. Formation of this mRNA–protein complex is promoted by light and correlates with a 50- to 100-fold enhancement of *psbA* translation.

During the light reactions of photosynthesis, water is split and its electrons are transferred to NADP$^+$ by means of a series of electron carriers. One of these redox-active carriers, the small iron–sulfur protein ferredoxin (Fdx), can donate electrons to several proteins, including the enzyme

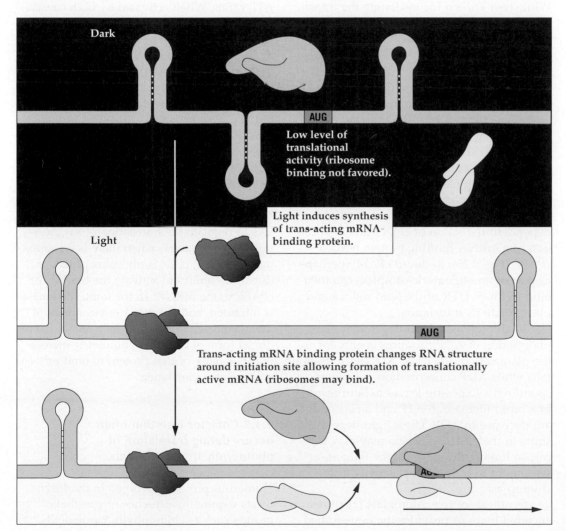

Dark

AUG

Low level of translational activity (ribosome binding not favored).

Light induces synthesis of trans-acting mRNA-binding protein.

Light

AUG

Trans-acting mRNA binding protein changes RNA structure around initiation site allowing formation of translationally active mRNA (ribosomes may bind).

AUG

Figure 9.22
Trans-acting factors that bind to the 5' UTRs of chloroplast mRNAs are thought to play a role in light-dependent activation of translation.

Box 9.4

Protein binding to RNA can often be detected by a gel-mobility shift assay. The RNA is labeled with radioactivity, and its migration during polyacrylamide gel electrophoresis is detected by autoradiography. If a protein binds to the RNA, the protein–RNA complex will migrate more slowly in the gel than will the unbound RNA molecules. The trans-acting factors that bind to chloroplast mRNAs can be detected by this method.

In a specific example (see figure at right), suppose a portion of *psbA* mRNA is subjected to electrophoresis alone (lane 1) or is incubated with an extract of light-grown cells that contain an mRNA-binding protein (lane 2). A protein–mRNA complex forms that migrates at a slower rate during gel electrophoresis. Much less of the mRNA–protein complex is observed when the mRNA is incubated with extracts of dark-grown cells (lane 3).

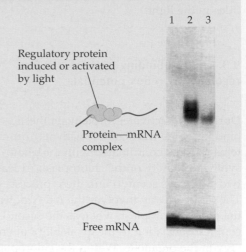

Regulatory protein induced or activated by light

Protein—mRNA complex

Free mRNA

ferredoxin-thioredoxin reductase (FTR). While best known for mediating the transfer of electrons from reduced ferredoxin to the regulatory disulfide protein thioredoxin (Trx) (see Chapters 12 and 13), FTR can also reduce chloroplast protein disulfide isomerase (cPDI), an enzyme that appears to play a direct role in coupling the translation of certain chloroplast mRNAs to the photosynthetic activity of the organelle.

cPDI participates in the formation and isomerization of disulfide bonds during protein folding (see Section 9.4.7). In the current working model for regulation of chloroplast protein synthesis by light, the light-dependent reduction of cPDI leads to reduction of a mRNA-binding protein termed cPABP (Fig. 9.23). Reduced cPABP (perhaps as part of an oligomeric complex) can then bind to the 5′ UTR of the *psbA* mRNA and up-regulate its translation.

In chloroplasts, light-dependent phosphorylation of ADP produces energy for the plant cell. In the dark, ATP concentration falls while ADP concentration rises correspondingly. Exposing leaves to light results in a rapid increase in ATP and a concomitant decrease in ADP. These light-dependent shifts in the ADP/ATP ratio may play a role in linking the status of the light environment to translational activity in the chloroplast.

Synthesis of the chloroplast D1 protein encoded by *psbA* provides the best-studied case for regulation of translation by ADP/ATP ratios. When activated by high concentrations of ADP, a serine-threonine protein kinase is thought to phosphorylate cPDI in an unusual reaction that transfers the β-phosphate from ADP to the protein (Fig. 9.23). The phosphorylated form of cPDI not only fails to promote the reduction of cPABP but also may actively oxidize this protein. This leads to an accumulation of the oxidized form of cPABP. Because the reduced form of cPABP must bind to the *psbA* mRNA to activate translation, the net result of this process is a decrease in the synthesis of D1. Thus, in the current model, a dual system acts to regulate the translation of the *psbA* mRNA. In the dark, when there is an abundance of ADP, cPDI is phosphorylated, reducing its ability to activate the binding of cPABP to the mRNA. In the light, the kinase is inhibited, and the redox environment of the plastid activates cPDI. As a result, the reduced form of cPABP accumulates, increasing the ability of that protein to bind mRNA and activate translation.

9.3.5 Cofactor insertion often occurs during translation of photosynthetic components.

Numerous proteins involved in photosynthesis require the presence of prosthetic groups such as chlorophylls, carotenoids,

Chapter 9 Protein Synthesis, Assembly, and Degradation

quinones, or hemes to carry out their biological functions. Often, a protein must bind its cofactors before it can accumulate in a stable conformation. Perhaps the best-studied proteins in this category are those that bind chlorophyll. Chloroplast thylakoid membranes include several proteins that bind chlorophyll, six of which are synthesized within the plastid itself. These proteins are essential components of PSI, PSII, and the light-harvesting complexes. In addition to chlorophyll, they also contain several other prosthetic groups, such as quinones (see Chapter 12).

These proteins cannot be detected in cells from emerging leaves and cotyledons before the exposure of these tissues to light. However, in most plants, mRNA transcripts for these chlorophyll-binding proteins are present whether the plants are kept in the light or the dark. Surprisingly, some of these mRNAs associate with thylakoid membrane-bound polysomes, even in the dark. This indicates that the synthesis of chlorophyll-binding proteins is initiated in the dark, but some mechanism prevents the accumulation of mature protein products under those conditions.

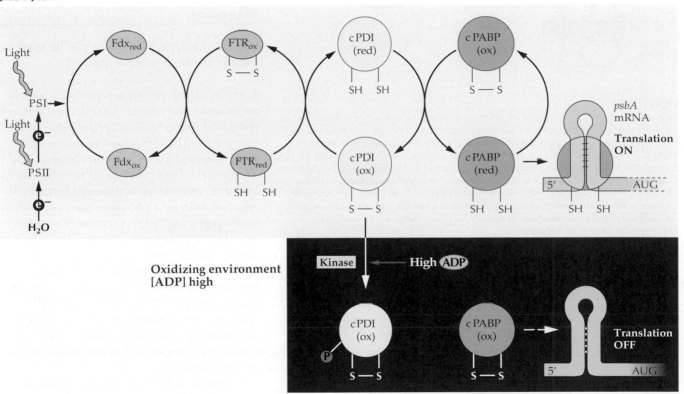

Figure 9.23

Model for how the light can affect synthesis of the chloroplast protein D1, the product of the *psbA* mRNA. In *Chlamydomonas,* a unicellular alga, translation of *psbA* mRNA is enhanced dramatically when a protein complex consisting of at least four distinct polypeptides binds to the 5' UTR (not shown). One of these polypeptides, a chloroplast polyadenylate-binding protein (cPABP), has an affinity for A-rich regions in RNA. When cysteine residues in cPABP are reduced, the binding affinity of cPABP for mRNA is substantially greater. This reduction may be mediated by a second component of the protein complex, chloroplast protein disulfide isomerase (cPDI). The current model postulates that in the dark, concentrations of reduced ferredoxin (Fdx_{red}) are low and cPDI is oxidized. As a consequence, cPABP is also oxidized and binds poorly to the 5' UTR of the *psbA* mRNA. In the absence of reduced

cPABP, there is little translation of the *psbA* mRNA. In the light, electrons flow from water to ferredoxin via photosystems (PS) II and I. Reduced ferredoxin donates electrons to ferredoxin-thioredoxin reductase (FTR), which in turn reduces cPDI. The reduced cPDI subsequently donates electrons to cPABP, which can then bind *psbA* mRNA, greatly enhancing the mRNA translation. A second level of control is exerted by the concentrations of ADP in the chloroplast. In the dark, high concentrations of ADP activate a protein kinase that phosphorylates cPDI. The phosphorylated form of cPDI not only fails to promote the accumulation of reduced cPABP but also appears to actively promote its oxidation. Because the oxidized form of cPABP does not bind well to the *psbA* mRNA, translation is not activated in the dark.

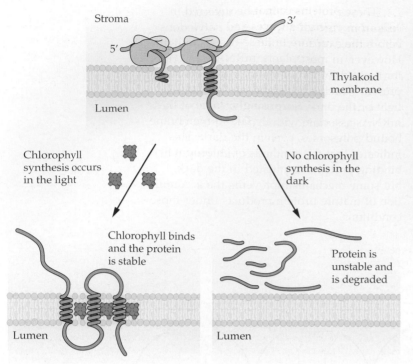

Figure 9.24
Regulation of protein synthesis by cofactor availability: model for stabilization of a chloroplast protein by cofactor binding. Proteins in the thylakoid membranes bind cofactors such as chlorophyll, quinones, or iron–sulfur clusters. Many proteins that bind chlorophyll do not accumulate in dark-grown chloroplasts, although the mRNA for these proteins is present and is often found in polysomes. Plants require light to complete the synthesis of chlorophyll. Although chlorophyll-binding proteins are believed to be synthesized in the dark, the apoprotein (protein without the bound cofactors) formed is unstable and is rapidly degraded. In the light, the plants make chlorophyll, which binds to the growing polypeptide chain and stabilizes it.

In the figure:
- Stroma, 5′, 3′, Thylakoid membrane, Lumen
- Chlorophyll synthesis occurs in the light
- No chlorophyll synthesis in the dark
- Chlorophyll binds and the protein is stable
- Protein is unstable and is degraded
- Lumen, Lumen

Plants must have light to synthesize chlorophyll, since light is required for the conversion of protochlorophyllide to chlorophyll (see Chapter 12). The accumulation of chlorophyll-binding proteins appears to depend on the ability of the chloroplast to complete the process of chlorophyll synthesis. Chlorophyll molecules, and presumably other protein cofactors, are inserted cotranslationally into the growing polypeptide chain (Fig. 9.24). If the cofactors are not present, the apoprotein is degraded—presumably because it fails to fold properly. In this way, the accumulation of chlorophyll-binding proteins is coupled to the light-dependent synthesis of their cofactors.

To allow time for the chloroplast to coordinate the synthesis of a protein, its insertion into a membrane, and the binding of its required cofactors, the ribosomes pause at discrete sites during the synthesis of the nascent polypeptide chain. The best-studied example of the cotranslational insertion of cofactors is D1 protein, the product of the *psbA* gene. This protein includes five transmembrane helices. During the synthesis of D1, the ribosome pauses at discrete positions. These pauses facilitate the interaction of the nascent chain with the thylakoid membrane, the insertion of segments of the growing polypeptide into the membrane, and the binding of cofactors to the protein (Fig. 9.25). One or more nuclear gene products may play an important role in stabilizing intermediates during translation.

9.4 Post-translational modification of proteins

9.4.1 Proteolytic processing can be used to modify the final protein product.

Initially, all polypeptides begin with formyl-methionine (in bacteria, plastids, and mitochondria) or methionine (in the eukaryotic cytoplasm). The formyl group is almost always removed by a ribosome-associated deformylase. In about half of all proteins the initiating methionine is removed from the nascent chain by a ribosome-associated Met-aminopeptidase. The fate of the N-terminal methionine is determined largely by which amino acid occupies the second position; small neighboring residues favor removal of the Met. Occasionally the second amino acid is also removed, although the protease responsible for this step has not been clearly identified. In addition, residues at the C terminus may be removed post-translationally on rare occasions.

After proteins reach their appropriate subcellular location in the cell, proteolytic processing also removes signal sequences (see Chapter 4). Other proteolytic processing events sometimes remove segments of the original polypeptide. One example of proteolytic processing in plant chloroplasts occurs during D1 maturation. A C-terminal domain of this protein is removed on incorporation into PSII. This proteolytic event is essential for subsequent photosystem function.

Proteolytic processing is also involved in generating the response hormone for plant wounds, systemin (Fig. 9.26). This 18-amino acid peptide hormone is secreted by wounded plant cells, after which it is transported to other regions of the affected plant to participate in inducing the synthesis of proteins involved in plant defense. Like many peptide hormones in animals, systemin is initially synthesized as a much larger precursor. Proteolytic processing is thought to be involved in the production of the active peptide hormone after wounding of the plant.

9.4.2 Proteins must fold into a precise three-dimensional structure to carry out their biological function.

During or after the completion of translation, the polypeptide chain must fold into the correct three-dimensional structure. In this process, the linear sequence of amino acids rearranges to give rise to the well-defined three-dimensional conformation of

Pause 1
The first hydrophobic helical segment has been synthesized. However, because the ribosome covers approximately 40 to 60 amino acids of the nascent chain, this region of the polypeptide is not yet exposed. At this pause site, the N terminus of D1 is probably interacting with the thylakoid membrane.

Pause 2
The first helical segment has emerged from the ribosome but has not yet been inserted into the membrane. Note that all of helix 1 must emerge from the ribosome before insertion into the membrane because the C-terminal end of the helix is on the lumenal side of the thylakoid in this protein.

Pause 3
The first two helices have now been completed and have been inserted into the membrane.

Pause 4
Helix 3 has now been synthesized. Pheophytin is thought to bind to helices 2 and 3 and is probably inserted into the growing polypeptide chain at this point.

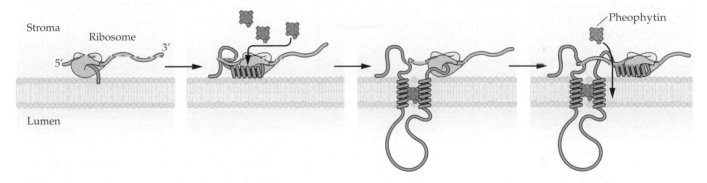

Pause 5
Helices 1, 2, 3, and 4 have now transversed the membrane. Helix 4 can now bind two chlorophyll molecules.

Pause 6
Helix 5 has now been synthesized and is ready to be inserted into the membrane. Quinone and Fe^{2+} cofactors are added at this point (not shown).

Pause 7
The D1 protein has been fully synthesized; it can now be assembled into PSII.

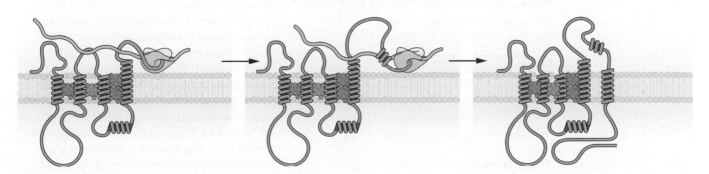

Figure 9.25
Model for cotranslational insertion of cofactors into the D1 protein. Ribosomes synthesizing D1 pause during translation to allow the protein to interact with the membrane with the correct topology and to allow the insertion of cofactors such as chlorophyll, pheophytin, and quinones.

The positions where ribosomes pause on a mRNA can be determined by mapping the edge of the ribosome on the 3′ side of the region of the mRNA covered by the ribosome. In this procedure, known as **toeprinting,** mRNAs with ribosomes bound at the stalled position are isolated and a short DNA primer is hybridized to the mRNA downstream from the predicted pause-site. Reverse transcriptase is used to synthesize a cDNA from the mRNA–primer complex. When no ribosomes are present on the mRNA, the reverse transcriptase can potentially copy the mRNA all the way to the 5′ end. However, if a ribosome is present between the site where the primer is annealed and the 5′ end of the mRNA, the progress of the reverse transcriptase is blocked. When it reaches the edge of the ribosome, the enzyme stops and dissociates from the mRNA, creating a shorter cDNA that locates the leading edge of the ribosome on the mRNA. The products from this procedure are analyzed by gel electrophoresis, which shows investigators the length of the cDNA synthesized (see figure at right).

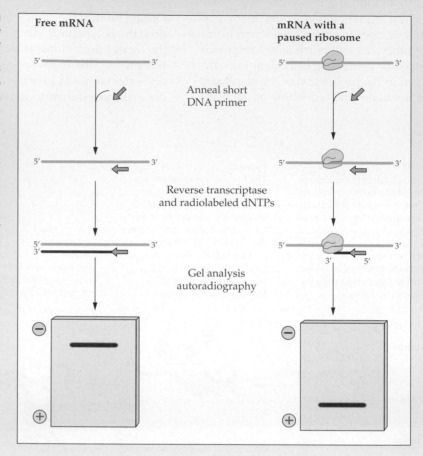

the protein. Much remains to be learned about the process. According to the current model, a newly synthesized protein first folds into a structure in which most of the elements of secondary structure (e.g., α-helices and β-sheets) are present (Fig. 9.27). These structures align with each other so that the orientation approximates the final structure the folded protein will have. This partially folded molecule, referred to as a **molten globule,** is the starting material from which the final three-dimensional structure of the protein will emerge after numerous interactions among the amino acid side chains.

In vitro, many proteins can unfold and then refold correctly within a few microseconds, reestablishing their native

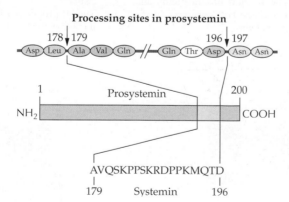

Figure 9.26
Systemin, a peptide hormone, is formed by plant cells in response to wounding. Systemin (18 amino acids) is produced by proteolytic processing of prosystemin, a much larger precursor polypeptide (200 amino acids).

conformations in the absence of other cellular components. This protein behavior has led to the idea that the primary sequence of the polypeptide contains all the information necessary to confer the correct structure on the folded protein. Similar assumptions have been made regarding the formation of oligomeric complexes. However, high concentrations of cellular proteins and the presence of many surfaces with which a folding protein can interact may interfere with proper folding. In the cell cytosol, most protein folding occurs cotranslationally. In chloroplasts, at least a portion of protein folding occurs post-translationally. The folding of proteins in vivo is regulated by a group of proteins referred to as molecular **chaperones,** which promote proper folding of polypeptides and suppress folding patterns that would result in protein aggregation or nonfunctional protein products (Fig. 9.28).

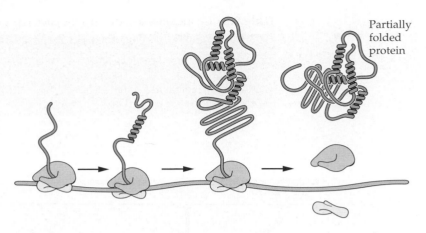

Partially folded protein

Figure 9.27
Steps in protein folding. Protein folding in the cell cytoplasm is primarily cotranslational. Many proteins, especially multidomain proteins, fold as they are being synthesized. Each domain folds as it emerges fully from the ribosome. Proteins in bacteria, and at least some proteins in chloroplasts and mitochondria, fold after the completion of their synthesis (i.e., post-translationally).

9.4.3 Protein-assisted folding occurs in the cell.

Molecular chaperones are proteins that bind and stabilize otherwise unstable protein conformations. Through coordinated binding and releasing of partially folded polypeptides, chaperones facilitate several processes, including protein folding, oligomer assembly, subcellular localization, and protein degradation. Chaperones do not direct specific folding patterns; rather, they prevent the formation of incorrect interactions within a polypeptide, between polypeptides, or between polypeptides and other macromolecules. Thus, chaperones increase the yield of fully functional proteins.

Chaperones fall into two major groups (Table 9.3 and Fig. 9.29). The first group, the Hsp70 family, is found in bacteria and in most compartments of eukaryotic cells. Chaperones in this family maintain an unfolded polypeptide in a soluble form ready for folding. Hsp70 chaperones can also transfer folding-competent polypeptides to members of the second family, the chaperonins, which promote proper polypeptide folding. Chaperonins are also found in all cells, from bacteria to eukaryotic cytoplasm to subcellular organelles such as chloroplasts and mitochondria.

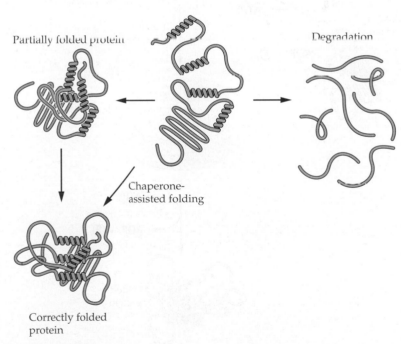

Partially folded protein

Degradation

Chaperone-assisted folding

Correctly folded protein

Figure 9.28
Current view of the process of protein folding. The newly synthesized protein first enters a molten globule state in which various secondary structural interactions have taken place. The partially folded protein then undergoes many tertiary interactions until the most stable conformation is reached. Many of these interactions are facilitated by molecular chaperones. Proteins that fail to fold correctly are generally degraded.

Table 9.3 Chaperones found in plants, bacteria, and organelles

Hsp70 family		Chaperonins	
Bacteria	dnaK	Bacteria	GroEL
Eukaryotic cytosol	Hsp70	Mitochondria	Hsp60
Endoplasmic reticulum	BiP	Chloroplasts	Rubisco-binding protein
Mitochondria	Grp25	Cytosol	TRiC

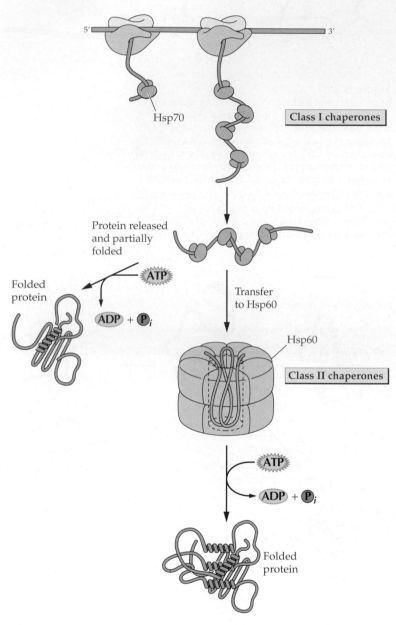

9.4.4 The Hsp70 family of molecular chaperones maintains polypeptides in an unfolded state.

During protein synthesis, stable folding cannot occur until at least one complete domain (i.e., 100 to 200 amino acids) of a nascent protein chain emerges from the ribosome. These nascent chains contain hydrophobic amino acids that tend to aggregate so as to minimize their interaction with the aqueous environment. A cell must maintain these aggregation-prone nascent chains in a nonaggregated, folding-competent state.

Molecular chaperones in the Hsp70 family bind to amino acid chains emerging from the ribosome during protein synthesis (Fig. 9.30). These proteins, named for their size (approximately 70 kDa), bind and release hydrophobic segments of unfolded proteins in an ATP-dependent manner. Hsp70 binding stabilizes unfolded proteins and prevents their aggregation. The controlled release of the unfolded protein may permit it to proceed successfully along the folding pathway (Fig. 9.31).

9.4.5 Chaperonins play a crucial role in facilitating the folding of many proteins.

In the current model for protein folding in bacteria, chloroplasts, and mitochondria, nascent polypeptide chains interact with Hsp70 and its cochaperone Hsp40 on the ribosome. For some cellular proteins, particularly in chloroplasts and mitochondria, subsequent successful folding requires transfer to a second folding complex, the chaperonin system. Chaperonins are the most varied and structurally complex group of the molecular chaperones. They are divided into two families: the GroEL group (i.e., the Hsp60, chaperonin 60, or cpn60 group), and the TRiC (TCP-1 ring) family (see Table 9.3).

Figure 9.29
Molecular chaperones are divided into two classes. The Hsp70 chaperones function in the early stages of the protein-folding process, often during translation, by recognizing small hydrophobic patches on the surface of the nascent polypeptide. Many polypeptides can fold with the aid of the Hsp70 pathway alone. The chaperonins act later in the process, after translation, and provide a central cavity in which protein folding is facilitated. Both Hsp70 and chaperonin pathways require energy in the form of ATP hydrolysis for proper functioning.

(A) No Hsp70 present

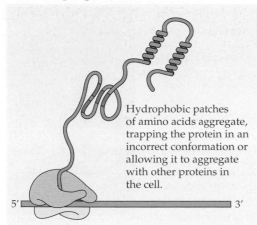

Hydrophobic patches of amino acids aggregate, trapping the protein in an incorrect conformation or allowing it to aggregate with other proteins in the cell.

(B) Hsp70 present

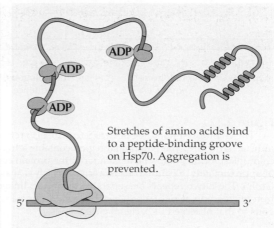

Stretches of amino acids bind to a peptide-binding groove on Hsp70. Aggregation is prevented.

Figure 9.30
Hsp70 is believed to play a role in preventing improper folding and aggregation of nascent polypeptide chains and newly released polypeptide chains. (A) The three-dimensional structure of a protein often involves interactions between distant segments of the polypeptide chain. As the nascent peptide emerges from the ribosome during synthesis, the side chains of hydrophobic amino acids cluster together to avoid contact with water. As a result, the protein may fail to fold correctly, or the peptide chain may aggregate with other cellular proteins. (B) Hsp70 and its cochaperone Hsp40 bind to the nascent chain, impeding unfavorable interactions within the nascent chain and between the unfolded polypeptide and other proteins in the cell. Hsp70 thus prevents the protein from becoming trapped in an incorrectly folded conformation and prevents the aggregation of the nascent polypeptide with other proteins in the cell.

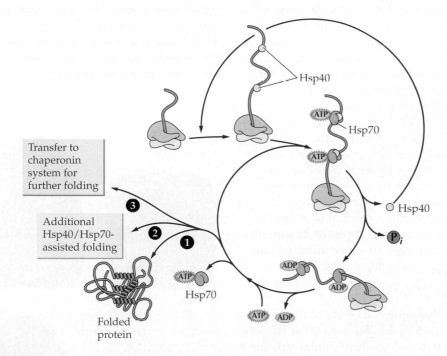

Figure 9.31
Model for the role of Hsp70 in protein folding. The unfolded polypeptide is bound by Hsp40, a cochaperone that facilitates the interaction of an Hsp70–ATP complex with the unfolded polypeptide. The Hsp70 protein contains two domains: The N-terminal domain contains the ATP-binding site, whereas the C-terminal domain folds into a conformation that can bind short peptide sequences exposed on an unfolded protein. After ATP hydrolysis, a stable ternary complex of the unfolded polypeptide, Hsp70, and ADP forms. A nucleotide exchange factor, often designated GrpE, promotes the exchange of ADP for ATP. In the presence of ATP, the interaction of Hsp70 with the polypeptide is reduced, and Hsp70 dissociates from the folding protein. The protein can then (1) complete its folding and enter the native state, (2) interact with Hsp40 and Hsp70 again to undergo additional folding steps, or (3) be transferred to an additional folding system, the chaperonin system.

(A)

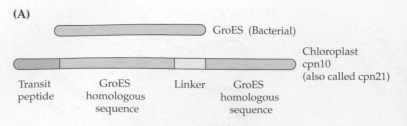

GroES (Bacterial)

Transit peptide | GroES homologous sequence | Linker | GroES homologous sequence

Chloroplast cpn10 (also called cpn21)

(B)

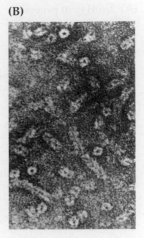

Figure 9.32
(A) The chloroplast homolog of GroES (cpn10, also referred to as cpn21) has an unusual structure. The N-terminal region contains a transit peptide specifying chloroplast localization. The remainder of the protein consists of two sequences of approximately 100 amino acids, each of which is analogous to the bacterial GroES. These two repeats are separated by a short linker region. (B) In the electron micrograph at right, cpn10 can be seen forming a ring-like structure analogous to that observed in bacterial GroES.

The GroEL group is found in bacteria, mitochondria, and chloroplasts; the TRiC family functions in the eukaryotic cytosol. All are oligomeric proteins. In mitochondria and chloroplasts, as in bacteria, GroEL chaperonins consist of two stacked seven-membered rings arranged in the shape of a barrel. Each subunit has a molecular mass of 60 kDa. Bacteria and mitochondria appear to contain one chemical species of GroEL subunit, whereas chloroplasts have two. In plants, plastid and mitochondrial GroEL are encoded in the nuclear genome and are imported into the organelle. These proteins have primary sequences very similar to the bacterial factor and work by the same mechanism. GroEL works in cooperation with a smaller protein, GroES. In *Escherichia coli*, GroES consists of 10-kDa molecular mass subunits that form heptameric rings. This protein is sometimes referred to as cpn10, based on the molecular mass of its subunit in *E. coli*. However, this nomenclature can be misleading because the chloroplast homolog has a molecular mass of 24 kDa (Fig. 9.32).

The best-studied member of the GroEL family is from *E. coli*, but organellar chaperonins are believed to share similar structures and functions (Fig. 9.33). Electron microscopy and X-ray crystallography have shown that GroEL is a cylindrical complex, approximately 150 Å in height and 140 Å in diameter. The cylinder encloses a central cavity approximately 50 Å wide. Unfolded polypeptides bind in this cavity in a collapsed molten globule-like conformation. GroES plays a crucial role in promoting the binding and release of the unfolded polypeptide and induces significant conformational changes in GroEL (Fig. 9.34).

Figure 9.33
Structures of GroEL and GroES. These space-filling representations illustrate the central cavity in which folding takes place. The upper GroEL ring is lavender, the lower GroEL ring is blue, and GroES is green.

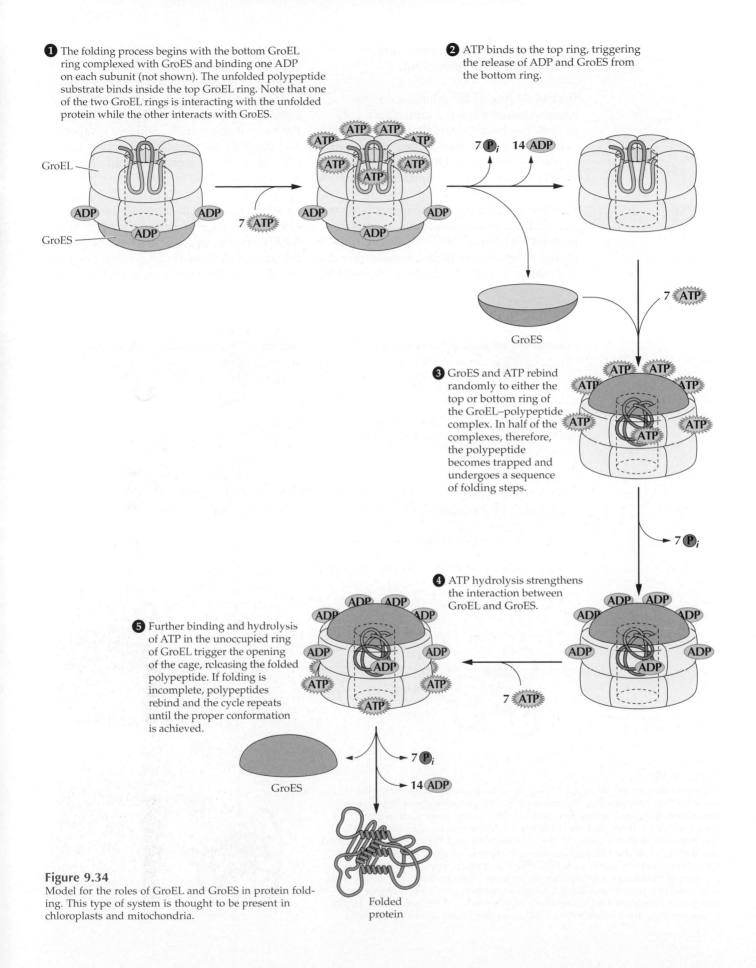

❶ The folding process begins with the bottom GroEL ring complexed with GroES and binding one ADP on each subunit (not shown). The unfolded polypeptide substrate binds inside the top GroEL ring. Note that one of the two GroEL rings is interacting with the unfolded protein while the other interacts with GroES.

GroEL

GroES

❷ ATP binds to the top ring, triggering the release of ADP and GroES from the bottom ring.

7 P_i 14 ADP

7 ATP

GroES

7 ATP

❸ GroES and ATP rebind randomly to either the top or bottom ring of the GroEL–polypeptide complex. In half of the complexes, therefore, the polypeptide becomes trapped and undergoes a sequence of folding steps.

7 P_i

❹ ATP hydrolysis strengthens the interaction between GroEL and GroES.

7 ATP

❺ Further binding and hydrolysis of ATP in the unoccupied ring of GroEL trigger the opening of the cage, releasing the folded polypeptide. If folding is incomplete, polypeptides rebind and the cycle repeats until the proper conformation is achieved.

GroES

7 P_i

14 ADP

Folded protein

Figure 9.34
Model for the roles of GroEL and GroES in protein folding. This type of system is thought to be present in chloroplasts and mitochondria.

9.4.6 Protein folding in eukaryotic cytoplasm is a complex event.

Protein folding in the cytoplasm occurs cotranslationally but not until after a polypeptide has been "targeted" to remain in the cytoplasm or to be translocated into the endoplasmic reticulum (ER; see Chapter 4). In the cytoplasm, polysomes attached to the ER translocate nascent polypeptides into the ER membrane or lumen as protein synthesis occurs. Soluble or free polysomes synthesize proteins that are to remain in the cell cytoplasm. Whether a polysome remains free or is localized to the ER membrane depends on the interaction of the nascent chain with one of two different protein complexes as it emerges from the ribosome. When a sequence signal for secretion is present, the signal recognition particle (SRP) binds to the nascent chain, halting protein synthesis and targeting the translation complex to the ER (Fig. 9.35). If the nascent chain does not contain a signal sequence, the ribosome complex performing the synthesis binds an alternative protein complex, the nascent chain-associated complex (NAC). The NAC competes with SRP for binding to the nascent chain as the chain emerges from the ribosome. SRP has a greater affinity for

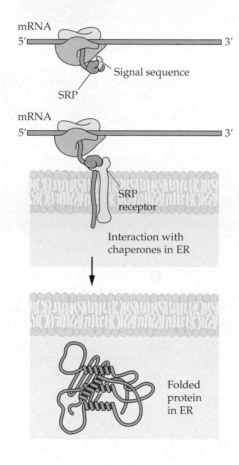

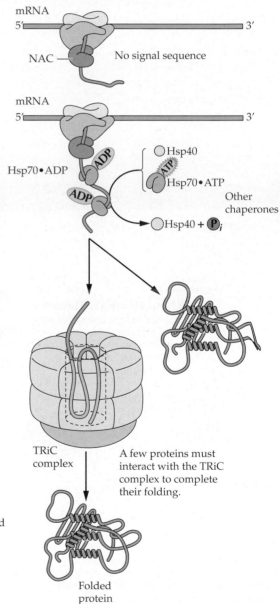

Figure 9.35
Protein folding in the plant cell cytoplasm is a complex process. When the nascent chain emerges from the ribosome it interacts with one of two protein complexes, the signal recognition particle (SRP) or the nascent chain-associated complex (NAC). If the nascent chain has a signal sequence, it interacts with SRP. The ribosome is then directed to the ER membrane and the protein is synthesized into the lumen of the ER (see Chapter 4). If there is no signal sequence present, the nascent chain interacts with NAC. Binding to NAC appears to play a role in preventing the incorrect localization of proteins. Nascent chains interacting with NAC fold in a process mediated through interaction with several chaperones. One major chaperone that is important in this process is the cytoplasmic Hsp70. A few proteins must interact with TRiC, the cytoplasmic equivalent of GroEL.

nascent chains that carry signal sequences for targeting to the ER, whereas NAC has a lesser affinity for these sequences. If NAC binds to the nascent protein, SRP cannot bind and the growing chain is protected from accidental targeting to the ER.

If a nascent chain is targeted into the ER, the chain is not folded until substantial portions of the polypeptide have been translocated into the lumen of the ER. In the lumen, specific chaperones facilitate the folding of the polypeptide chains. Polypeptides that are to remain in the cytosol are believed to fold cotranslationally as each domain emerges from the ribosome. Folding in the cytosol involves several classes of chaperones that are not yet well understood. Nascent chains in the plant cytosol interact with Hsp70 and its cochaperone Hsp40 (Fig. 9.35). These interactions probably occur analogously to those described above for the Hsp70 family (Section 9.4.4) and are believed to play a crucial role in preventing aggregation of the unfolded polypeptides. A few polypeptides in the cytoplasm are thought

to use a chaperonin referred to as the TRiC complex. This complex is not as well understood as the GroEL–GroES complex, but it may play a similar role in folding. In plants, the TRiC complex contains at least six distinct polypeptides, many of which appear to be related to each other in sequence. As with GroEL, TRiC subunits are arranged in two stacked rings with a hole in the center. Many of the general principles discussed above for protein folding in chloroplasts and mitochondria probably apply to plant cytoplasm as well. However, much work remains to be done to develop a detailed understanding of this process.

9.4.7 Protein folding is also catalyzed by isomerases that promote correct disulfide bond formation and proline isomerization.

Several proteins, particularly those that are secreted, contain disulfide bonds that link cysteine residues (Fig. 9.36). The formation of disulfide bonds between the correct

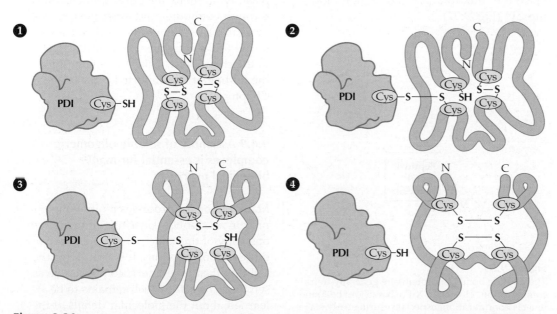

Figure 9.36
Protein disulfide isomerase (PDI) promotes the formation of correct disulfide bonds in proteins. Different forms of this enzyme are localized in the chloroplast (see Section 9.3.4) and the ER. In the ER, PDI is thought to play a crucial role in the formation of disulfide bonds in proteins destined for secretion from the cell or targeted to organelles of the secretory pathway. As proteins are folding, incorrect disulfide bonds often form (1). PDI has several highly reactive Cys residues that can attack S–S bonds in folding proteins, creating a transient disulfide bond between the folding protein and PDI (2). These intermediates rearrange as the many combinations of disulfide bonds occur during folding (3). When the folding protein establishes the correct combination of disulfide bonds and is in its thermodynamically most stable conformation, the disulfide bond with PDI breaks and the completed protein is released (4).

two cysteine residues is facilitated by an ER-localized enzyme called protein disulfide isomerase (PDI), a homolog of the chloroplast PDI discussed in Section 9.3.4. PDI is directly involved in the synthesis of disulfide bonds in proteins; it also speeds the folding of proteins rich in cysteine residues by repeatedly breaking disulfide bonds and promoting the formation of new disulfide bonds. These interactions proceed until the new protein adopts the most thermodynamically stable conformation supported by cysteine pairing.

Peptide bonds in proteins generally have a trans configuration. However, approximately 6% of the bonds involving proline residues in fully folded proteins are in a cis conformation. During folding, X-Pro bonds (where X can be any amino acid) must isomerize to the final cis or trans configuration. Because spontaneous isomerization is very slow, enzymatic catalysis of this step is required for protein folding to occur within a biologically useful time. A group of enzymes called peptidyl prolyl isomerases (PPIases) catalyze the cis–trans isomerization of X-Pro bonds, which helps the protein reach the final folded conformation more rapidly (Fig. 9.37).

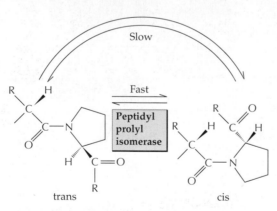

Figure 9.37
The peptide bonds in proteins are generally trans in configuration. However, X-Pro bonds can be found in a cis configuration, especially during early stages of folding. Isomerization of the peptide bonds involving proline is an important part of protein folding; indeed, proline isomerization can limit the rate at which protein folding occurs. Spontaneous isomerization is slow because the partial double-bond character of the peptide bond is a significant energy barrier to the isomerization process. PPIase increases the rate of the isomerization reaction approximately 300-fold by twisting the plane of the peptide bond, thereby reducing the energy barrier for isomerization.

9.4.8 Protein folding and localization are coupled processes.

Many proteins are secreted from cells or are localized in cellular compartments other than their sites of synthesis (see Chapter 4). Protein synthesis, transport, and folding events vary according to where the protein is localized. Proteins destined for secretion from the cell or localized to membrane-bound organelles of the secretory pathway are simultaneously translated on rough ER and translocated into the ER lumen, where molecular chaperones facilitate their folding. Similarly, nuclear-encoded proteins targeted to the plastids, mitochondria, or peroxisomes remain unfolded in the cytosol until they are imported by their target organelles. In both of these examples, protein transport and folding must be tightly coordinated.

This coordination between protein folding and transport is well illustrated by the import of nuclear-encoded proteins into chloroplasts. These proteins associate with chaperones in the cytoplasm until they interact with components of the import machinery in the outer membrane of the chloroplast. From there, the polypeptide is moved through both outer and inner membranes. In many cases, the protein is then transferred to the equivalent of GroEL (cpn60) in the chloroplast. This interaction promotes the folding of the imported protein (see Chapter 4).

9.4.9 Assembly of soluble oligomeric complexes is essential for many biological processes.

Many biological processes require oligomeric protein assemblies consisting of more than one copy of the same polypeptide or many different polypeptides. It is crucial for the cell that these oligomeric complexes be assembled correctly. Much remains to be learned about the molecular details of the assembly processes.

The chloroplast enzyme Rubisco has been studied as an example of the assembly of a soluble oligomeric protein complex. In plants, Rubisco consists of eight copies of a large subunit (53 kDa) and eight copies of a small subunit (approximately 14 kDa) (Fig. 9.38). The large subunits form an octameric

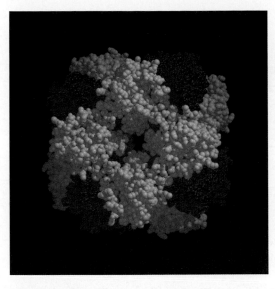

Figure 9.38
Three-dimensional structure of Rubisco, consisting of eight copies of the large subunit and eight copies of the small subunit.

core, whereas the small subunits form two layers of four subunits each, on opposite sides of the core. As shown in Figure 9.16, the large subunit is encoded by chloroplast DNA and synthesized in the chloroplast, whereas the small subunit is transcribed in the nucleus, translated in the cytoplasm, and imported into the chloroplast. Rubisco assembly requires the active participation of the chloroplast chaperonin (cpn60), another plastid-localized product of a nuclear gene (Fig. 9.39). The cpn60 forms a high-molecular-mass complex with the large subunit of Rubisco shortly after the subunit is released from the ribosome. In an ATP-dependent step, the folding of the large subunit is facilitated as described in Section 9.4.5. Once folded, the large subunit is released from the chaperonin and can assemble with large and small subunits to form a functional holoenzyme.

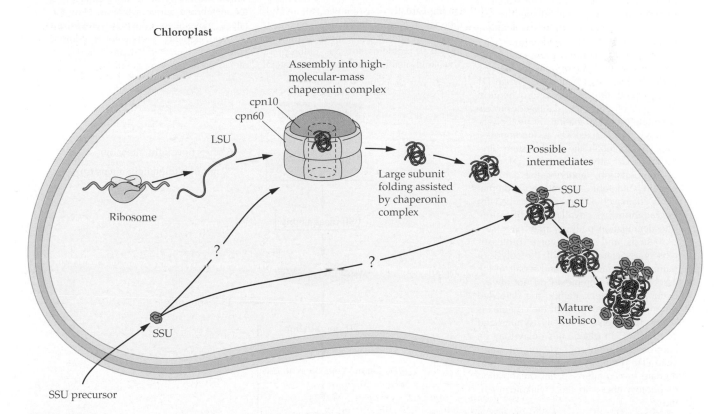

Figure 9.39
Synthesis and assembly of Rubisco requires the participation of two genomes and two translational systems as well as chloroplast chaperones. The large subunit of Rubisco is encoded in the chloroplast genome. The correct folding of the large subunit requires the chaperonin system consisting of cpn60 (the analog of GroEL) and its cochaperonin cpn10 (the analog of GroES). Both cpn60 and cpn10 are synthesized in the cell cytosol as precursors. After they are imported into the chloroplast, their transit peptides are removed, and the mature proteins assemble into the chaperonin complex. The small subunit of Rubisco is synthesized in the cytosol as a precursor protein. After transport across the chloroplast membranes, the precursor is processed into its mature form. Eight copies of the folded large subunit must then assemble with eight copies of the small subunit to form the active oligomeric enzyme complex.

Box 9.6   **The D1 repair cycle operates to repair photodamage to PSII.**

Although light is the source of energy for photosynthesis, it can also cause extensive damage to the photosynthetic apparatus, especially PSII. Light can lead to the inactivation of electron transport and can promote oxidative damage to the reaction centers, particularly to proteins D1 and D2. Plants have evolved a complex and energetically expensive mechanism to repair photoinduced damage, which occurs at a rate directly proportional to the intensity of the incident light. When damage to PSII outpaces the plant's ability to repair itself, **photoinhibition** results. This stress condition can lead to reduced growth of the plant (see the figure at top right and Chapter 22). The severity of photoinhibition varies with the plant species, the physiological state of the plant, and its life history. The extent of photoinhibition is greatly affected by other environmental conditions, including temperature and water availability.

D1 and D2 constitute the core polypeptides of the photosynthetic reaction center of PSII (see Chapter 12). When D1 is damaged during the photochemical reactions of PSII, the damaged protein is targeted for degradation by proteolysis. A newly synthesized D1 is incorporated into PSII to repair the photosystem. Presumably, oxidative damage to D1 alters its conformation, rendering the damaged protein vulnerable to proteases. Alternatively, damaged D1 may be marked for degradation by covalent modification. The degradation pathway may have several steps and may involve proteases closely associated with PSII. Proteolysis of damaged D1 molecules and repair of the associated PSII complexes do not always occur immediately. Rather, the damaged protein is held in PSII until the complex is ready to receive a newly synthesized copy of D1. In plants, this delay may be controlled by phosphorylation of damaged D1. The delay is required to accommodate the complex spatial organization of photosynthesis in the thylakoid membrane.

Once D1 has been damaged, PSII is thought to undergo a partial disassembly. The photodamaged D1 is degraded, new D1 protein is synthesized de novo on chloroplast ribosomes and inserted into

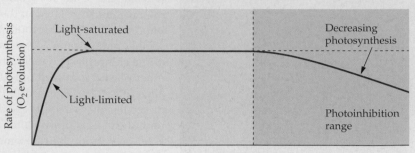

the thylakoid membrane, and the PSII complex is reassembled (see figure below). Direct D1 degradation and de novo D1 biosynthesis can become rate-limiting under high light intensities, resulting in the accumulation of 160-kDa complexes containing partially disassembled PSII and additional proteins of an unknown nature.

The D1 repair cycle poses significant logistical problems for the chloroplast. Thylakoid membranes demonstrate lateral heterogeneity. PSII is located primarily in closely adjacent, or appressed, membranes. By contrast, D1 synthesis occurs in nonappressed membranes exposed to the stroma, where ribosomes can bind and insert the growing polypeptide into the membrane during synthesis. How the lateral movements of newly synthesized D1 and partially disassembled components of the damaged PSII are orchestrated remains an interesting question.

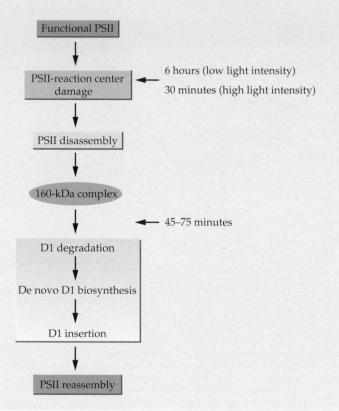

9.5 Protein degradation

9.5.1 Protein degradation plays many important physiological roles in the cell.

Degradation of cellular proteins is a constant and ongoing process. Protein degradation removes abnormal proteins, facilitates the recycling of amino acids, and regulates protein activity by eliminating molecules that are no longer needed. One of the most important roles of protein degradation might be called "cellular housekeeping." Abnormal proteins arise in cells as a result of mutations, errors in protein synthesis or folding, spontaneous denaturation, disease, stress, or oxidative damage. If not removed, these damaged proteins eventually poison the cell, often by forming large insoluble aggregates. Plant cells do not divide rapidly enough to keep concentrations of damaged proteins dilute, so eliminating these proteins is especially crucial for plants. Photoinhibited D1 protein in PSII is a good example of a damaged protein that is proteolyzed during the repair process (Box 9.6).

A second role of protein degradation is to promote the accumulation of the subunits of oligomeric protein complexes in the correct stoichiometry and to maintain the correct ratio of enzymes and their cofactors. One example can be seen in the coordinated accumulation of the large and small subunits of Rubisco. When the concentration of the large subunit in the chloroplast decreases—for example, if synthesis is experimentally prevented—small subunits synthesized in the cytoplasm are rapidly degraded when imported into chloroplasts. Likewise, when chlorophyll synthesis is prevented, the chlorophyll a/b-binding protein is rapidly degraded (see Fig. 9.24).

Protein degradation plays a particularly important role in regulating numerous biological processes. For example, consider enzymes catalyzing the first step or the rate-limiting step of a metabolic cascade. These enzymes generally are short-lived because of proteolytic degradation. The plant can alter the activity of a particular metabolic pathway simply by reducing the synthesis of specific enzymes. Likewise, many key regulatory proteins are subject to proteolytic regulation, including effectors of the cell cycle, development, and differentiation. In plants,

phytochrome A (PhyA) provides a good example (see Chapter 18). Phytochromes exist in two forms that can be interconverted by light. The Pr form of PhyA absorbs red light, has little biological activity, and resists proteolysis. The Pfr form of PhyA absorbs far-red light, acts as a photoreceptor to initiate many light-regulated developmental processes, and is rapidly degraded. When PhyA Pr absorbs red light, it is converted to the Pfr form and becomes active. The rapid degradation of the biologically active Pfr form, coupled to the continual synthesis of Pr, allows the plant to respond to its continually changing light environment (Fig. 9.40).

Proteolysis in plants also supplies amino acids necessary for maintaining cellular homeostasis and for new growth. About half of the complete set of proteins in a plant is

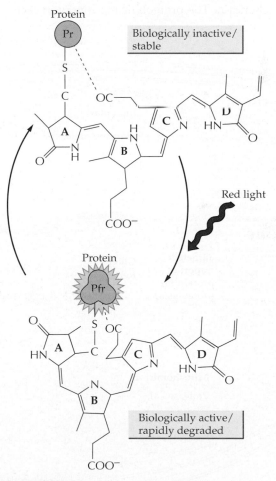

Figure 9.40
Light-dependent conversion of phytochrome and protein turnover. The biologically inactive form of phytochrome (P_r) resists proteolytic degradation, but the biologically active form (P_{fr}) is rapidly degraded.

replaced every 4 to 7 days. In many cases, new proteins are synthesized from recycled amino acids. A good example of the role of proteolysis in supplying amino acids is seed germination (see Chapter 19). On germination, most of the amino acids required for the growth of the seedling are supplied by the degradation of seed storage proteins.

9.5.2 Proteases that catalyze protein degradation are found at several sites in the cell.

The degradation of proteins in the cell is carried out by several complex proteolytic pathways. Plants probably have specific proteolytic pathways in each cellular compartment (Fig. 9.41). Thus, distinct systems are probably present in vacuoles, in the cytosol, in the nucleus, in chloroplasts, and in mitochondria. The proteolytic machinery in each cellular compartment reflects the evolutionary origin of that compartment. Thus, the proteolytic mechanisms observed in chloroplasts and mitochondria are more closely related to bacterial systems than to those localized in the plant cell cytosol.

Vacuoles are very conspicuous organelles and may occupy as much as 95% of the plant cell volume. Rich in hydrolytic enzymes, vacuoles perform a variety of functions. The proteases they contain may play a role in protein degradation similar to that of lysosomes in animal cells. Some researchers suggest that during nutrient starvation, plants transport a series of cytosolic proteins into the vacuole, where the proteins are degraded to generate needed amino acids. In seed-reserve tissues, specialized vacuoles called **protein bodies** have been observed (see Chapters 4 and 19). Proteins are deposited in the protein body during seed maturation and remain there in a stable

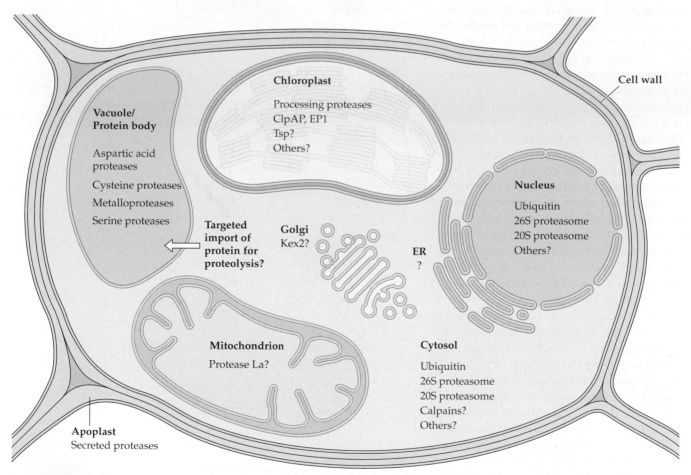

Figure 9.41
Many of the subcellular compartments in a plant cell—including the cytosol, nucleus, chloroplasts, and vacuoles—contain proteases. Mitochondria, the ER, and the Golgi are also thought to have proteolytic enzymes. In addition, many plant cells secrete proteases.

state until the seed germinates. During germination, a specific protease is imported and the storage proteins are degraded, supplying amino acids to the new seedling.

Chloroplasts contain approximately 50% of the protein found in photosynthetic tissues, so protein degradation must occur in chloroplasts. One of the proteases observed in chloroplasts has homology to the ClpAP protease of *E. coli* (Fig. 9.42). This protease, a multimeric enzyme complex, is believed to play an important role in degrading chloroplast proteins. The activity of this protein is particularly important in preventing the accumulation of abnormal proteins as well as during leaf senescence, when much of the protein loss observed occurs in chloroplasts. Other plant organelles, including mitochondria, peroxisomes, and the ER, also have systems for the degradation of proteins. Little is currently known about these systems, and the proteases remain to be purified and characterized. The better-characterized cytosolic systems are described below.

9.5.3 Protease activity must be regulated tightly to prevent degradation of essential cellular proteins.

A plant cell may contain more than 10,000 different proteins at any given time. It is imperative that the cell have precise mechanisms for controlling the activity of proteases and for targeting individual proteins for degradation at the correct time. Without such regulation, proteolytic enzymes would degrade essential cellular proteins indiscriminately. Proteolysis in plants generally requires ATP. Because the hydrolysis of peptide bonds is energetically favorable, ATP

hydrolysis is thought to act in the regulation of proteolytic activity by accessory proteins, which are involved in the recognition of proteins targeted for degradation. These protease systems are complex, and progress in identifying their regulatory mechanisms has been slow.

Cytoplasmic proteins destined for short lives are marked by the N-terminal amino acid residue (Section 9.5.5) and perhaps by short sequence elements. For example, polypeptide segments rich in Pro (P), Glu (E), Ser (S), and Thr (T), which are called PEST sequences, seem to serve as proteolytic signals. In many cases, conditions that promote protein denaturation (e.g., heat, desiccation, heavy metal exposure, and pathogen infection) also activate proteolytic pathways. Recognition of damaged or incorrectly folded proteins probably involves a perceived increase in the number of hydrophobic residues present on the surface of the protein. These residues ordinarily are buried in the protein interior to prevent their interaction with water. However, when a protein is damaged or incorrectly folded, hydrophobic side chains become exposed. These hydrophobic patches probably provide binding sites for the proteins that regulate the activity of various proteolytic enzymes.

9.5.4 Proteins in cytoplasm may be tagged with ubiquitin to target them for degradation.

Most protein degradation in the cytoplasm and nucleus is carried out by the **proteasome,** an extremely large oligomeric protein complex with a molecular mass of more than 1.5 megadaltons (MDa). The complex pathway catalyzed by the proteasome involves products of more than 100 genes. Proteins destined for degradation are delivered to the proteasome after they are covalently

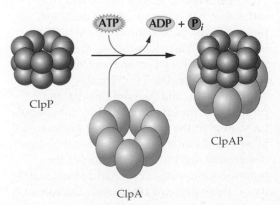

Figure 9.42
The ClpAP protease system found in chloroplasts is composed of two types of subunits: ClpP, a 21-kDa serine protease, and ClpA, an 81-kDa ATPase that activates the protease activity of ClpP and unfolds protein substrates for degradation. The degradation of proteins by this system is carried out by an oligomeric complex in which ClpP forms two heptameric complexes that assemble with a single (probably heptameric) ring of ClpA.

(A)

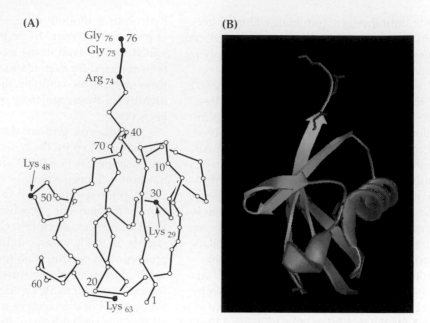

(B)

Figure 9.43
(A) The three-dimensional structure of ubiquitin. Ubiquitin consists of a compact globular domain with a flexible protruding C-terminal segment. The Lys residues at positions 29, 48, and 63 often participate in forming multiple ubiquitin chains. The protein to be degraded is covalently attached to the C-terminal residue, Gly76. (B) A ribbon diagram of ubiquitin.

modified by conjugation to the small protein ubiquitin (Fig. 9.43). Ubiquitin is a highly conserved protein with an identical amino acid sequence in all plants. The conjugation of ubiquitin to a protein requires ATP and is carried out by a multienzyme pathway (Fig. 9.44). This pathway catalyzes the formation of a peptide bond between the ε-amino group on the side chain of a lysine residue and the C terminus of ubiquitin. Additional ubiquitin molecules can also be added, forming a multiubiquitin chain. Multiple lysine residues on the target protein can be conjugated in this manner.

Once ubiquitinated, the protein is delivered to the proteasome for degradation. The proteasome occurs in two forms: the 20S core proteasome and the 26S proteasome. The core consists of four stacked rings, each made up of seven subunits (note the similarity to the GroEL chaperone). A channel through the center of the stacked rings contains the active sites for proteolysis (Fig. 9.45). Placement of these catalytic centers inside the channel helps protect other cellular proteins from degradation. Access to the channel is governed by another protein complex, which associates with the core proteasome to form the 26S structure. This second protein complex serves as the "mouth" of the proteasome, engulfing substrates, probably helping unfold the target proteins, and injecting them into the channel of the stacked ring structure, where degradation occurs.

Degradation of most short-lived and abnormal proteins in plants is thought to be carried out by the ubiquitin-mediated pathway. One important example of this is phytochrome, the first protein that was discovered to be ubiquitinated in vivo. When the Pr form is converted to the Pfr (active) form in etiolated tissues, the phytochrome protein becomes ubiquitinated and is rapidly degraded. This system also plays a crucial role in degrading proteins that have been damaged by heat shock or that encountered errors in synthesis.

9.5.5 Amino acid sequence at the N terminus of a protein can affect protein lifespan.

The N-terminal sequence of a protein influences the protein's stability in the cell. This observation, termed the N-end rule, states that when certain amino acids are present at the N terminus (e.g., Met, Thr, Ser, Gly, and Val), the proteins are relatively stable. However, when amino acids such as Lys, Arg, His, Phe, Tyr, Trp, Ile, Leu, Asp, Glu, Gln, and Asn are present, the proteins are rapidly degraded by the ubiquitin system in the cytoplasm and nucleus. Most cytoplasmic proteins have stabilizing N-terminal residues. Acetylation of the α-amino group may confer additional stability on these proteins (Fig. 9.46). The role of

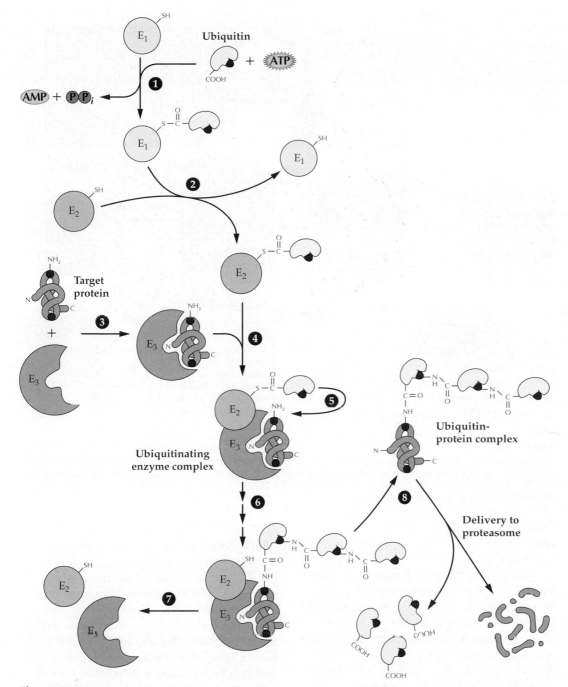

Figure 9.44
Tagging a protein with ubiquitin marks it for degradation. This process is carried out by a ubiquinating enzyme complex. This enzyme complex carries out a complicated series of reactions that are shown in simplified form here. The complex consists of a ubiquitin-activating enzyme (E1), a ubiquitin-conjugating enzyme (E2), and a ubiquitin-ligating enzyme (E3). Step 1: The C-terminal carboxyl group of a ubiquitin chain is covalently attached to a Cys residue sulfhydryl in E1. Step 2: Ubiquitin is transferred to a Cys residue sulfhydryl of E2. Step 3: The target protein carrying the degradation signal is recognized by E3, which binds it in a noncovalent complex. Step 4: The (E3:target protein) complex binds E2, forming the ubiquinating enzyme complex. Step 5: Ubiquitin is transferred from E2 to the target protein through an isopeptide bond between the C-terminal carboxyl group of ubiquitin and the side chain amino group of a lysine residue (shown as a red dot) in the target protein. Step 6: The transfer of ubiquitin is repeated a number of times, forming a multi-ubiquitin chain. The C-terminal carboxyl group of each additional ubiquitin is attached to an internal lysine residue (red dot) of the preceding ubiquitin chain. In general, a series of about four ubiquitins must be added to the target protein before it can be delivered to the proteasome effectively. Steps 7 and 8: E2 and E3 dissociate and the ubiquitin-target protein complex is delivered to the proteasome for degradation. Ubiquitin is released from the target protein in this step so that it can be reused.

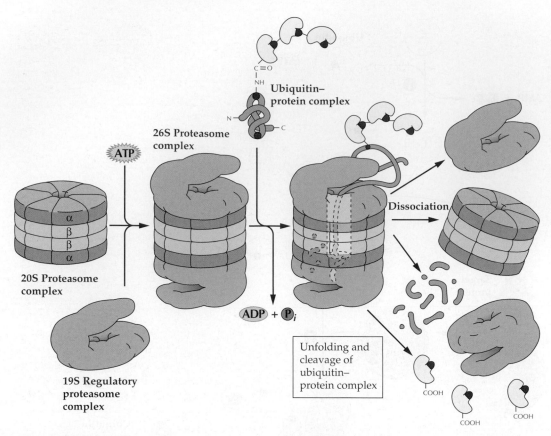

Figure 9.45
Structure of the proteasome. The core proteasome (20S proteasome) has a distinctive hollow cylindrical shape created by the assembly of four stacked rings. Each ring contains seven polypeptides. A 19S regulatory subunit binds to the ends of the 20S proteasome, forming the 26S proteasome. *Ubiquitinated* target proteins are delivered to the proteasome for degradation.

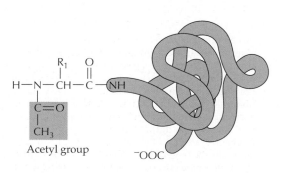

Figure 9.46
Acetylation of the N terminus of a protein. Acetylation of the terminus is thought to increase the life of the protein by protecting it from proteolysis.

N-acetylation in the turnover of plant proteins is not yet understood.

In contrast to the stabilizing amino acids, destabilizing residues are often present at the N termini of proteins that are transported to cellular compartments such as the ER, the Golgi, and other membrane-bound organelles. Thus, one postulated function of the N-end rule pathway is to degrade proteins that were incorrectly compartmentalized or that were previously compartmentalized in other parts of the cell, such as the ER, but have "leaked" back into the cytoplasm. Such leakage has been observed between the ER and cytoplasm.

As noted earlier, all proteins are initiated with methionine, normally a stabilizing residue. However, the N-terminal Met can be removed by an aminopeptidase to expose the second amino acid, which becomes the new N terminus of the protein. Moreover, the N terminus of a protein can also be

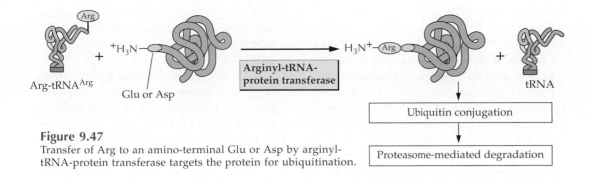

Figure 9.47

Transfer of Arg to an amino-terminal Glu or Asp by arginyl-tRNA-protein transferase targets the protein for ubiquitination.

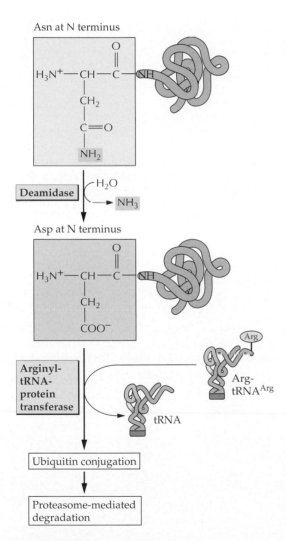

Figure 9.48

Deamidation of Gln or Asn leads to the formation of Glu or Asp at the N terminus. The new N terminus can then be modified by the addition of Arg and thus targeted for ubiquitin-mediated degradation.

modified by the addition of an Arg residue through the action of the enzyme arginyl-tRNA-protein transferase (Fig. 9.47). This enzyme transfers the amino acid from Arg-tRNA to the N terminus of the target proteins, which typically have N-terminal Glu or Asp residues. Proteins with these N-terminal dicarboxylate amino acids are not recognized directly by the ubiquitin pathway. Attachment of Arg, however, renders these proteins subject to ubiquitin-mediated degradation. Thus, Arg can be described as a primary destabilizing residue, whereas Glu and Asp are considered secondary destabilizing residues. The tertiary destabilizing residues, Gln and Asn, are acted on by a specific amidase that converts them to the secondary destabilizing residues, Glu and Asp (Fig. 9.48).

Summary

Plants have three compartments in which proteins are synthesized. Protein synthesis mechanisms differ among these compartments, probably reflecting their distinct evolutionary origins. Protein synthesis is carefully regulated in response to the physiological needs of the plant. This regulation is particularly apparent in chloroplasts, where protein synthesis is coordinated with the photosynthetic activity of the organelle. After synthesis, proteins must be localized to the appropriate cellular location and may also be modified. Chaperones are required for the proper folding of proteins in cells. Protein folding and localization are coupled. Proteins are degraded to modulate the quantities of key metabolites present, to provide amino acids, and to remove improperly folded proteins.

Further Reading

Aro, E.-M., Virgin, I., Andersson, B. (1993) Photoinhibition of photosystem II. Inactivation, protein damage and turnover. *Biochim. Biophys. Acta* 1143: 113–134.

Boston, R., Viitanen, P., Vierling, E. (1996) Molecular chaperones and protein folding in plants. *Plant Mol. Biol.* 32: 191–222.

Browning, K. (1996) The plant translational apparatus. *Plant Mol. Biol.* 32: 107–144.

Futterer, J., Hohn, T. (1996) Translation in plants—rules and exceptions. *Plant Mol. Biol.* 32: 159–189.

Kim, J., Mayfield, S. (1997) Protein disulfide isomerase as a regulator of chloroplast translational activation. *Science* 278: 1954–1957.

Mayfield, S., Yohn, C., Cohen, A., Danon, A. (1995) Regulation of chloroplast gene expression. *Annu. Rev. Plant Physiol. Plant Mol. Biol.* 46: 147–166.

Vierstra, R. (1996) Proteolysis in plants. *Plant Mol. Biol.* 32: 275–302.

Biochemistry & Molecular Biology of Plants, B. Buchanan, W. Gruissem, R. Jones, Eds.
© 2000, American Society of Plant Physiologists

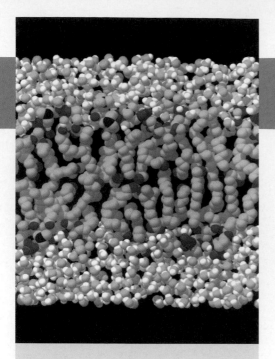

CHAPTER 10

Lipids

Chris Somerville
John Browse
Jan G. Jaworski
John B. Ohlrogge

CHAPTER OUTLINE

Introduction

The term **lipid** refers to a structurally diverse group of molecules that are preferentially soluble in a nonaqueous solvent such as chloroform. Lipids include a wide variety of fatty acid–derived compounds, as well as many pigments and secondary compounds that are metabolically unrelated to fatty acid metabolism. Although we will limit our discussion of lipids to those compounds with origins in fatty acid synthesis, this limitation still provides a broad group of compounds to explore, many of which are vital to the normal functioning of a cell. Each plant cell contains a diverse range of lipids, often located in specific structures. Furthermore, different plant species may contain different lipids.

Although the metabolism of fatty acids and lipids in plants has many features in common with other organisms, the lipid pathways in plants are complex and not well understood. The complexity arises primarily from cellular compartmentalization of the pathways and the extensive intermixing of lipid pools between these compartments (Fig. 10.1). In addition, higher plants collectively accumulate more than 200 different fatty acids, so there are many open questions about the nature of the enzymes involved in the synthesis of these compounds. Among the many challenges facing plant biochemists today is the complete elucidation of these pathways and the mechanisms that regulate them.

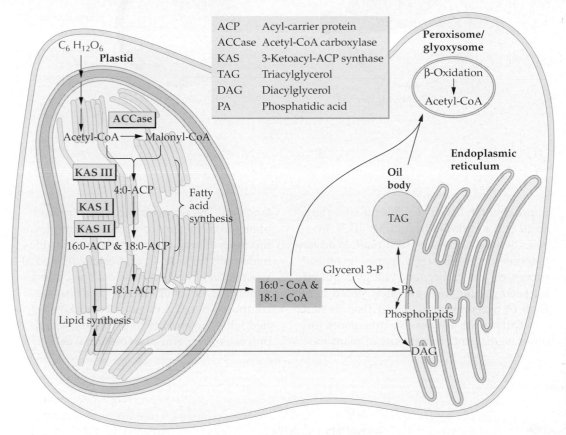

ACP	Acyl-carrier protein
ACCase	Acetyl-CoA carboxylase
KAS	3-Ketoacyl-ACP synthase
TAG	Triacylglycerol
DAG	Diacylglycerol
PA	Phosphatidic acid

Figure 10.1
Lipid synthesis and metabolism take place in various organelles and in some cases involve movement of lipids from one cellular compartment to another.

10.1 Structure and function of lipids

10.1.1 Lipids have diverse roles in plants.

Lipids serve many functions in plants (Table 10.1). As the major components of biological membranes, they form a hydrophobic barrier that is critical to life (see Chapter 1). Membranes not only separate cells from their surroundings; they also separate the contents of organelles, such as chloroplasts and mitochondria, from the cytoplasm. Cellular compartmentalization depends on polar lipids forming a bilayer that prevents free diffusion of hydrophilic molecules between the cellular organelles and prevents diffusion in and out of the cells. The membranes of chloroplasts, in which the light reactions of photosynthesis take place, primarily contain **galactolipids.** Membranes external to plastids are composed mainly of mixtures of **phospholipids.** Although a single gram of leaf tissue may contain as much as 1 m^2 of membrane, lipids make up a relatively small proportion of the total mass of plant tissue (Fig. 10.2).

Lipids also represent a substantial chemical reserve of free energy. Because fatty acids are substantially more reduced organic molecules than carbohydrates, fatty acid oxidation has a higher potential for producing energy. Furthermore, **triacylglycerols** are largely hydrophobic and exist in an essentially anhydrous environment. Carbohydrates, however, are hydrophilic, and the water of hydration adds substantially to their mass. Thus, on a mass basis, the ATP yield from catabolism to CO_2 and H_2O is

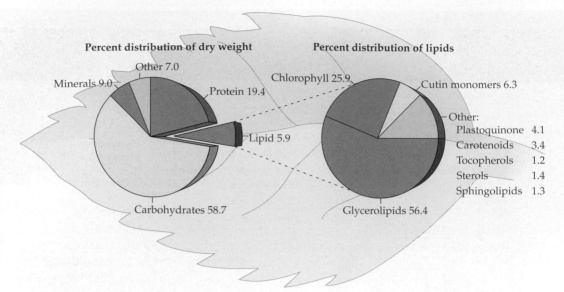

Figure 10.2
Approximate distributions of cellular constituents (as a percentage of total dry weight) and lipid types (as a percentage of total lipids by weight) in leaf tissues of *Arabidopsis*. Some of the values were extrapolated from results obtained with other species.

Percent distribution of dry weight

Other 7.0
Minerals 9.0
Protein 19.4
Lipid 5.9
Carbohydrates 58.7

Percent distribution of lipids

Chlorophyll 25.9
Cutin monomers 6.3
Other:
Plastoquinone 4.1
Carotenoids 3.4
Tocopherols 1.2
Sterols 1.4
Sphingolipids 1.3
Glycerolipids 56.4

approximately twice as high for triacylglycerols as for carbohydrates (Fig. 10.3). In cases where a compact seed mass is advantageous for facilitating dispersal or other processes, the carbon and energy required for seed germination are often stored in the form of triacylglycerols rather than as starch.

Fatty acids are also the precursors for other significant components of plant metabolism. The **waxes** that coat and protect plants from the environment are complex mixtures of long-chain hydrocarbons, aldehydes, alcohols, acids, and esters derived almost entirely from fatty acids. The **cutin** and **suberin** layers of epidermal cells also are composed of oxygenated fatty acids esterified with one another to produce a tough, polyester skin. Thus, in epidermal cells of

ATP/g from sugar

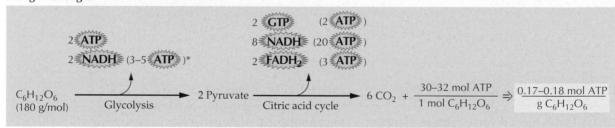

2 ATP
2 NADH (3–5 ATP)*

2 GTP (2 ATP)
8 NADH (20 ATP)
2 FADH$_2$ (3 ATP)

$C_6H_{12}O_6$ (180 g/mol) $\xrightarrow{\text{Glycolysis}}$ 2 Pyruvate $\xrightarrow{\text{Citric acid cycle}}$ 6 CO_2 + $\dfrac{\text{30–32 mol ATP}}{\text{1 mol } C_6H_{12}O_6}$ $\Rightarrow$ $\dfrac{\text{0.17–0.18 mol ATP}}{\text{g } C_6H_{12}O_6}$

ATP/g from fatty acid

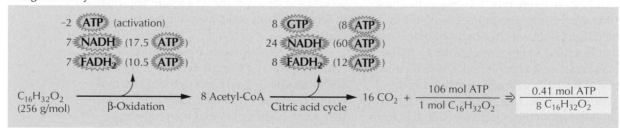

–2 ATP (activation)
7 NADH (17.5 ATP)
7 FADH$_2$ (10.5 ATP)

8 GTP (8 ATP)
24 NADH (60 ATP)
8 FADH$_2$ (12 ATP)

$C_{16}H_{32}O_2$ (256 g/mol) $\xrightarrow{\beta\text{-Oxidation}}$ 8 Acetyl-CoA $\xrightarrow{\text{Citric acid cycle}}$ 16 CO_2 + $\dfrac{\text{106 mol ATP}}{\text{1 mol } C_{16}H_{32}O_2}$ $\Rightarrow$ $\dfrac{\text{0.41 mol ATP}}{\text{g } C_{16}H_{32}O_2}$

Figure 10.3
Comparison of energy yield in animals from metabolism of fatty acids and carbohydrates to CO_2 and H_2O. Metabolism of fatty acids produces 0.41 mol of ATP per gram of fatty acid, whereas metabolism of carbohydrate yields 0.17 to 0.18 mol/g. Oxidation of one NADH by the mitochondrial electron transport chain is assumed to yield 2.5 ATP, whereas oxidation of one FADH$_2$ is assumed to yield 1.5 ATP. *The number of ATPs derived from one NADH oxidized in the cytosol or peroxisome varies according to the mechanism by which the reducing equivalents (electrons) are transferred into the mitochondrion (see Chapter 14).

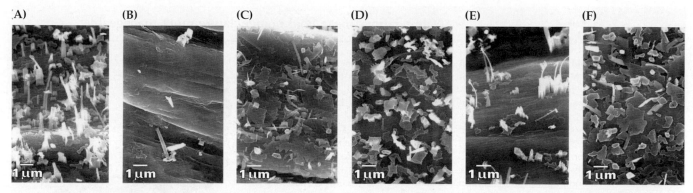

(A) 1 μm **(B)** 1 μm **(C)** 1 μm **(D)** 1 μm **(E)** 1 μm **(F)** 1 μm

Figure 10.4
Scanning electron micrographs of wild-type and mutant *Arabidopsis* stems (A–C) and siliques (D–F). Wild-type surfaces (A,D) are covered with various tubes and lobed plates of wax. Surfaces of the *cer7* mutant (B,E) have both fewer wax crystals and altered crystalline structures. Surfaces of *cer17* mutants (C,F) have fewer tubes than the wild-type plants.

aerial organs, the bulk of fatty acid synthesis is devoted to production of wax and cutin for protection (Fig. 10.4).

Some fatty acids may play major roles in certain signal transduction pathways. The synthesis of the growth regulator **jasmonic acid** from linolenic acid and the activities of jasmonates as plant hormones and second messengers are widely studied (see Section 10.8.5 and Chapters 17 and 21). Similarly, phosphatidylinositol and its derivatives may be messengers with regulatory roles, similar to messenger compounds found in other eukaryotes, which play important roles in signal transduction pathways (see Chapter 18). Fatty acids may also be involved in regulating various cellular processes via acylation of proteins.

Table 10.1 Functions of lipid molecules in higher plants

Function	Lipid types involved[a]
Membrane structural components	Glycerolipids
	Sphingolipids
	Sterols
Storage compounds	Triacylglycerols
	Waxes
Compounds active in electron transfer reactions	Chlorophyll and other pigments
	Ubiquinone, plastoquinone
Photoprotection	Carotenoids (xanthophyll cycle)
Protection of membranes against damage from free radicals	Tocopherols
Waterproofing and surface protection	Long-chain and very-long-chain fatty acids and their derivatives (cutin, suberin, surface waxes)
	Triterpenes
Protein modification	
Addition of membrane anchors	
Acylation	Mainly 14:0 and 16:0 fatty acids
Prenylation	Farnesyl and geranylgeranyl pyrophosphate
Other membrane anchor components	Phosphatidylinositol, ceramide
Glycosylation	Dolichol
Signaling	
Internal	Abscisic acid, gibberellins, brassinosteroids
	18:3 Fatty acid precursors of jasmonate
	Inositol phosphates
	Diacylglycerols
External	Jasmonate
	Volatile insect attractants
Defense and antifeeding compounds	Essential oils
	Latex components (rubber, etc.)
	Resin components (terpenes)

[a] The isoprenoids and related lipids are described in Chapter 24.

Box 10.1
Abbreviations make lipid nomenclature more manageable.

A simple shorthand notation based on molecule length and the number and position of double bonds has been developed to designate fatty acids. For example, the saturated C_{16} fatty acid, palmitic acid (hexadecanoic acid), is designated 16:0. The first value, 16, represents the number of carbon atoms. The second value, 0, indicates the number of double bonds. The monounsaturated 18-carbon fatty acid, oleic acid (cis-9-octadecenoic acid), is designated $18:1^{\Delta 9}$. The $\Delta 9$ superscript designates the position of the single double bond, counting the carboxyl group as carbon atom number 1. Because the double bonds in fatty acids are almost exclusively cis isomers, no designation for the configuration of double bond is used unless it is a trans isomer, as in $16:1^{\Delta 3t}$. As shown in the illustration of oleic and elaidic acids, introduction of a cis unsaturation creates a bend in the acyl chain, whereas a trans unsaturation does not.

Some authors also designate the position of the double bonds relative to the terminal methyl group (the ω carbon). Thus,

an $\omega 3$, or n-3, fatty acid contains a double bond three carbons from the methyl end of the fatty acid (e.g., the polyunsaturated α-linolenic acid, $18:3^{\Delta 9,12,15}$ is an $\omega 3$ fatty acid). A limitation of this nomenclature is that $18:1^{\Delta 15}$ would also be referred to as an $\omega 3$ or n-3 fatty acid.

Abbreviations are also used to designate the position at which a fatty acid is esterified to the glycerol backbone of glycerolipids. sn-3 (stereospecific nomenclature-3) denotes the terminal hydroxyl that is phosphorylated in glycerol 3-phosphate, sn-2 refers to the central hydroxyl, and sn-1 is the terminal hydroxyl that is not phosphorylated.

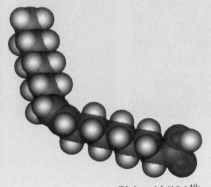

Oleic acid $(18:1^{\Delta 9})$

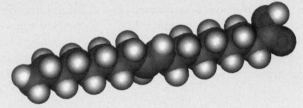

Elaidic acid $(18:1^{\Delta 9t})$

10.1.2 Most, but not all, lipids contain fatty acids esterified to glycerol.

Fatty acids are carboxylic acids of highly reduced hydrocarbon chains. The typical fatty acids found in the membranes of plants contain 16 or 18 carbons and are listed in Table 10.2, along with some unusual fatty acids that typically accumulate only in the storage triacylglycerols of seeds. Some of the nomenclature used in abbreviations for fatty acids and lipids is described in Box 10.1.

A major fraction of the fatty acids in plants are the polyunsaturated fatty acids linoleic acid $(18:2^{\Delta 9,12})$ and α-linolenic acid $(18:3^{\Delta 9,12,15})$. Only a few plants accumulate fatty acids with double bonds closer to the carboxyl group than the $\Delta 9$ position. In addition to the C_{16} and C_{18} common fatty acids, some plants also produce fatty acids of 8 to 32 carbons in length. These are usually accumulated in storage lipids or epicuticular wax. The fatty acid composition of lipids can be determined by using gas chromatography to separate the methylated derivatives of the fatty acids (Box 10.2).

Glycerolipids consist of fatty acids esterified to derivatives of glycerol. Four principal types are found in plants: triacylglycerols, phospholipids, galactolipids, and a sulfolipid. In addition, plants contain small amounts of sphingolipids. In most cases, purification of a particular type of lipid from a plant extract yields a complex mixture. For example, seven different classes of phospholipid (Table 10.3) are defined by the structure of the head group, and each class is composed of distinct molecular species defined by the fatty acids attached to the sn-1 and sn-2 positions of the glycerol backbone. Thus, the phosphatidylcholine molecule depicted in Figure 10.5A has a saturated fatty acid esterified to the sn-1 position and a diunsaturated fatty acid esterified to the sn-2 position. Some of the factors that control the fatty acid composition of lipids are discussed in Sections 10.5.3, 10.7.1, and 10.7.4.

Lipids are usually stored as triacylglycerols, three fatty acids esterified to glycerol (Fig. 10.5B). Triacylglycerols are frequently referred to as neutral lipids because of their nonpolar nature. Found primarily in seeds

Table 10.2 Selected fatty acids present in plants

Common name	Systematic name	Structure	Abbreviation[a]
Saturated fatty acids			
Lauric acid	*n*-Dodecanoic acid	$CH_3(CH_2)_{10}COOH$	12:0
Palmitic acid[b]	*n*-Hexadecanoic acid	$CH_3(CH_2)_{12}CH_2CH_2COOH$	16:0
Stearic acid[b]	*n*-Octadecanoic acid	$CH_3(CH_2)_{12}CH_2CH_2CH_2CH_2COOH$	18:0
Arachidic acid	*n*-Eicosanoic acid	$CH_3(CH_2)_{12}CH_2CH_2CH_2CH_2CH_2CH_2COOH$	20:0
Behenic acid	*n*-Docosanoic acid	$CH_3(CH_2)_{12}CH_2CH_2CH_2CH_2CH_2CH_2CH_2CH_2COOH$	22:0
Lignoceric acid	*n*-Tetracosanoic acid	$CH_3(CH_2)_{12}CH_2CH_2CH_2CH_2CH_2CH_2CH_2CH_2CH_2CH_2COOH$	24:0
Unsaturated fatty acids			
Oleic acid[b]	*cis*-9-Octadecenoic acid	$CH_3(CH_2)_7CH=CH(CH_2)_7COOH$	$18:1^{\Delta 9}$
Petroselenic acid	*cis*-6-Octadecenoic acid	$CH_3(CH_2)_{10}CH=CH(CH_2)_4COOH$	$18:1^{\Delta 6}$
Linoleic acid[b]	*cis,cis*-9,12-Octadecadienoic acid	$CH_3(CH_2)_4CH=CH-CH_2-CH=CH(CH_2)_7COOH$	$18:2^{\Delta 9,12}$
α-Linolenic acid[b]	*all-cis*-9,12,15-Octadecatrienoic acid	$CH_3CH_2CH=CH-CH_2-CH=CH-CH_2-CH=CH(CH_2)_7COOH$	$18:3^{\Delta 9,12,15}$
γ-Linolenic acid	*all-cis*-6,9,12-Octadecatrienoic acid	$CH_3(CH_2)_4CH=CH-CH_2-CH=CH-CH_2-CH=CH(CH_2)_4COOH$	$18:3^{\Delta 6,9,12}$
Roughanic acid	*all-cis*-7,10,13-Hexadecatrienoic acid	$CH_3CH_2CH=CH-CH_2-CH=CH-CH_2-CH=CH(CH_2)_5COOH$	$16:3^{\Delta 7,10,13}$
Erucic acid	*cis*-13-Docosenoic acid	$CH_3(CH_2)_7CH=CH(CH_2)_{11}COOH$	$22:1^{\Delta 13}$
Some unusual fatty acids			
Ricinoleic acid	12-Hydroxyoctadec-9-enoic acid	$CH_3(CH_2)_5\overset{OH}{\underset{H}{C}}-CH_2-CH=CH(CH_2)_7COOH$	$12\text{-}OH\text{-}18:1^{\Delta 9}$
Vernolic acid	12,13-Epoxyoctadec-9-enoic acid	$CH_3(CH_2)_4CH-CH-CH_2-CH=CH(CH_2)_7COOH$ (12,13 epoxide)	

a See Box 10.1 for an explanation of abbreviation nomenclature.

b These five fatty acids are commonly found as the principal constituents of membrane lipids. The others are found principally in storage lipids.

Table 10.3   Major classes of membrane lipids

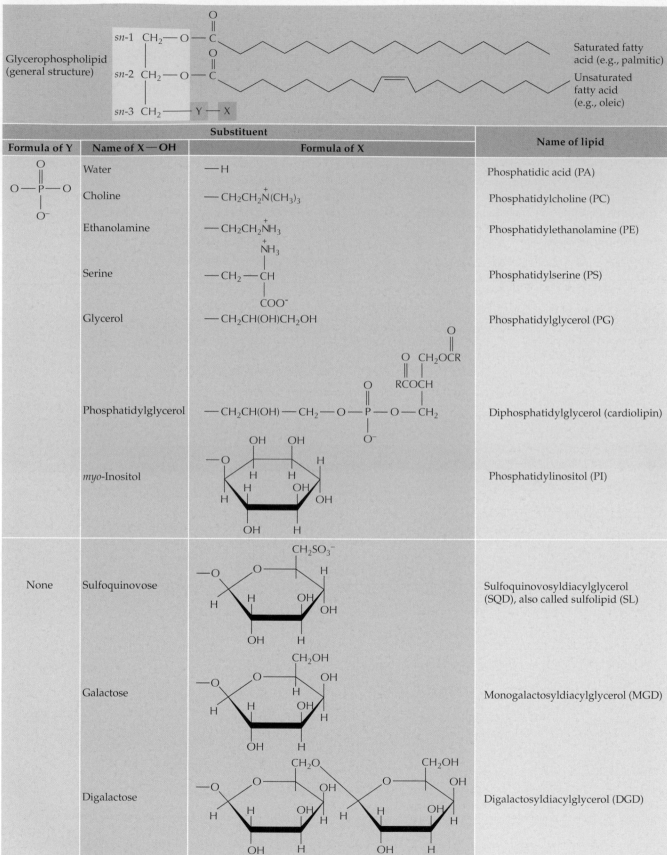

Formula of Y	Name of X—OH	Formula of X	Name of lipid
$O=P(O^-)$ group	Water	—H	Phosphatidic acid (PA)
	Choline	—$CH_2CH_2\overset{+}{N}(CH_3)_3$	Phosphatidylcholine (PC)
	Ethanolamine	—$CH_2CH_2\overset{+}{N}H_3$	Phosphatidylethanolamine (PE)
	Serine	—CH_2—$CH(\overset{+}{N}H_3)$—COO^-	Phosphatidylserine (PS)
	Glycerol	—$CH_2CH(OH)CH_2OH$	Phosphatidylglycerol (PG)
	Phosphatidylglycerol	—$CH_2CH(OH)$—CH_2—O—$P(O^-)$—O—CH_2	Diphosphatidylglycerol (cardiolipin)
	myo-Inositol	(ring structure)	Phosphatidylinositol (PI)
None	Sulfoquinovose	(ring structure)	Sulfoquinovosyldiacylglycerol (SQD), also called sulfolipid (SL)
	Galactose	(ring structure)	Monogalactosyldiacylglycerol (MGD)
	Digalactose	(ring structure)	Digalactosyldiacylglycerol (DGD)

Major classes of membrane lipids. The basic structure of a glycero-lipid is shown at the top. The C_3 backbone (highlighted in dark yellow) is usually esterified to two fatty acids at the carbons labeled *sn*-1 and *sn*-2. The modifications of the *sn*-3 carbon can be described by the substituents *X* and *Y*, which correspond to the compounds shown in the lower part of the table. *sn* numbers refer to the stereochemical nomenclature system, which is, by convention, based on the structures of D- and L-glyceraldehyde. The convention with respect to glycerol is that, in a Fisher projection of L-glycerol (not shown), the central hydroxyl is shown to the left. By definition, the carbon above the *sn*-2 carbon is the *sn*-1 position and the position below is the *sn*-3 position.

and pollen, triacylglycerols serve as energy and carbon stores. Because neutral lipids are not soluble in the aqueous phase of cells, they do not contribute to the osmotic potential of the cell. This is important to their role as storage materials because they accumulate in amounts that would otherwise disrupt the maintenance of normal cellular osmolality.

Phospholipids are synthesized by esterification of fatty acids to the two hydroxyl groups of *sn*-glycerol 3-phosphate to produce phosphatidic acid. All other phospholipids are derived from phosphatidic acid by esterification of a polar "head group" to the phosphoryl group (Table 10.3). Phospholipids are **amphipathic,** containing both hydrophobic (noncharged, nonpolar) fatty acids and a hydrophilic (charged, polar) head group. This property allows phospholipids—and other amphipathic glyerolipids—to form a bilayer in which the hydrophilic heads are in contact with an aqueous environment such as the cytosol, while the hydrophobic tails remain in contact with other hydrophobic tails (Fig. 10.6; see also Chapter 1).

Galactolipids, another major class of glycerolipids, are localized in the plastid membranes. These lipids have a galactosyl or sulfoquinovosyl group replacing the phosphoryl head group of the phospholipids (Table 10.3). The three major lipids belonging to this class of lipids are the galactolipids monogalactosyldiacylglycerol and digalactosyldiacylglycerol and the plant sulfolipid sulfoquinovosyldiacylglycerol. These **glycolipids** contain high concentrations

(A) Phosphatidylcholine

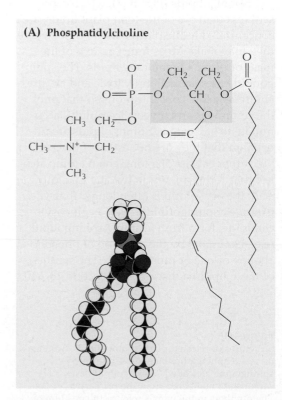

(B) Triacylglycerol

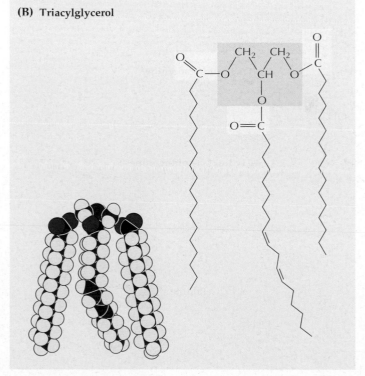

Figure 10.5
Space-filling and conformational models of (A) the phospholipid phosphatidylcholine and (B) triacylglycerol. The ester linkages are highlighted in yellow, and the glycerol backbone is in orange.

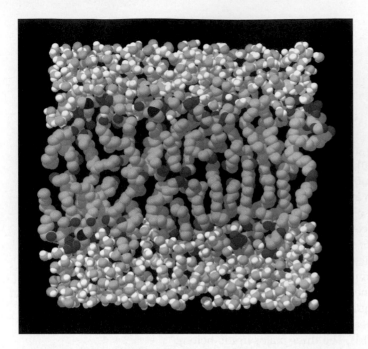

Figure 10.6
Computer simulation of a cross-section of a membrane. The coloring is as follows: phosphorus, dark green; nitrogen, dark blue; lipid oxygens, red; terminal chain methyl groups, magenta; other carbons, gray; water oxygens, yellow; water hydrogens, white. For clarity, the radii of heavy atoms are reduced slightly from their van der Waals values, the radii of water hydrogens are increased, and carbon-bound hydrogens are omitted.

of polyunsaturated fatty acids. In the photosynthetic tissue of some plants, α-linolenic acid ($18:3^{\Delta9,12,15}$) can constitute as much as 90% of the fatty acids in glycolipids. Some plants, such as peas, in which glycolipids contain exclusively C_{18} polyunsaturated fatty acids, are sometimes referred to as "18:3 plants." Others, such as spinach, contain glycolipids having appreciable amounts of a C_{16} polyunsaturated fatty acid, $16:3^{\Delta7,10,13}$, localized exclusively in the *sn*-2 position, and are called "16:3 plants." The underlying difference between 18:3 plants and 16:3 plants is now understood: 18:3 plants synthesize most or all of their lipids in the endoplasmic reticulum (ER), whereas 16:3 plants utilize biosynthetic pathways in the plastid as well (see Sections 10.7.2 and 10.7.3).

Sphingolipids (Fig. 10.7), which represent less than 5% of the total plant lipids, are concentrated in the plasma membrane, where they may make up as much as 26% of the mass of plasma membrane lipids. The sphingolipid bases are generally toxic and are present in very low concentrations so that only the ceramides and glycosylceramides accumulate to any extent. Sphingolipids are unusual in that they are not esters of glycerol, but rather consist of a long-chain amino alcohol that forms an amide linkage to a fatty acid; the acyl group is often longer than C_{18}. Complex sphingolipids, such as glucosylceramide, form from a simple sphingolipid (e.g., ceramide) by the addition of phosphocholine or one or more sugars. The synthesis of these lipids is discussed in Section 10.7.10.

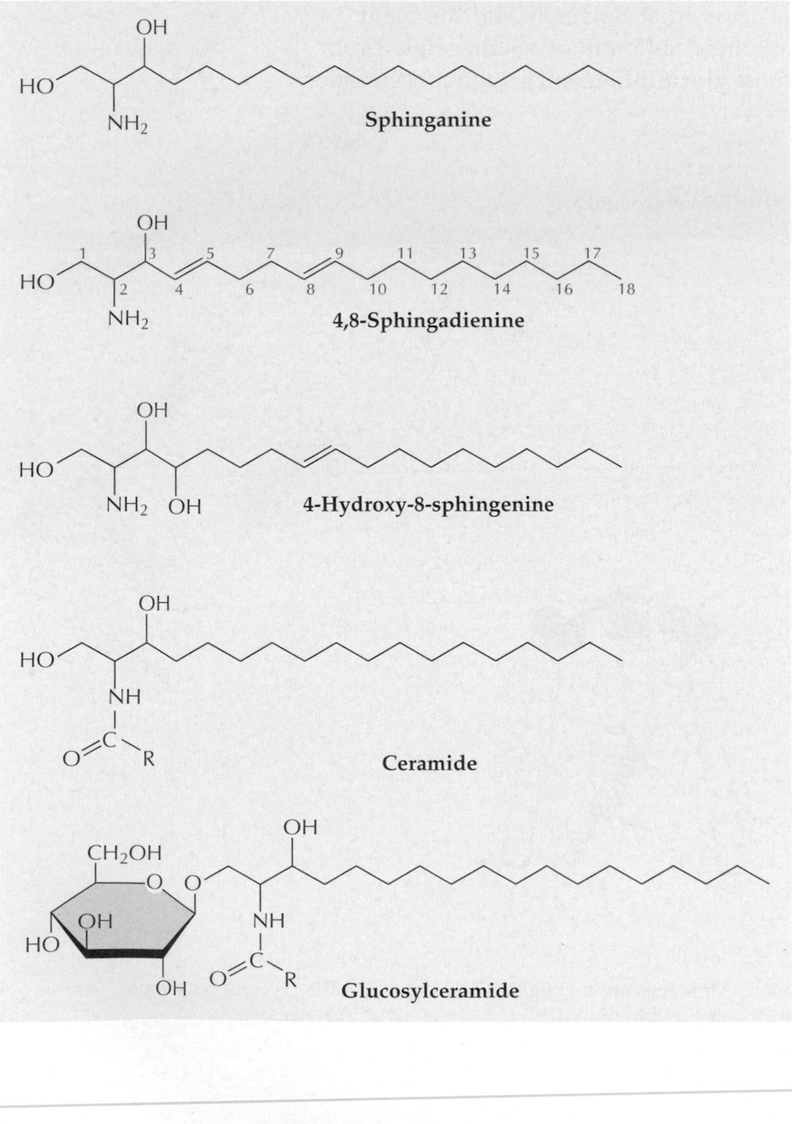

Sphinganine

4,8-Sphingadienine

4-Hydroxy-8-sphingenine

Ceramide

Glucosylceramide

Figure 10.7
Structures of selected plant sphingolipids. Sphinganine, 4,8-sphingadienine, and 4-hydroxy-8-sphingenine are called "sphingolipid bases." The carbons are numbered from the primary hydroxyl. The numbers in the names refer to the position of double bonds or of other functional groups as indicated. The fatty acids (R) are usually saturated or diunsaturated $C_{16}-C_{24}$ hydroxy fatty acids.

Box 10.2

Methyl esters of fatty acids can be quantified by gas chromatography.

The fatty acid composition of samples is commonly measured by gas chromatography of the methyl esters. The esterification renders the fatty acids more volatile so they can be analyzed at temperatures that do not alter the chemical structure of most fatty acids. The esters are typically formed by Fisher ester synthesis: The fatty acid is protonated by a strong acid such as HCl, which facilitates its reaction with methanol:

$$RCOOH + CH_3OH \xrightleftharpoons{H^+} RCOOCH_3 + H_2O$$

10.2 Fatty acid biosynthesis

10.2.1 Fatty acid biosynthesis in plants is similar to that in bacteria.

Fatty acid biosynthesis in plants takes place within plastids, organelles widely thought to have originated from a photosynthetic bacterial symbiont. Thus it is perhaps not surprising that fatty acid metabolism in plants closely resembles that of bacteria.

During fatty acid biosynthesis, a repeated series of reactions incorporates acetyl moieties of acetyl-CoA into an acyl group 16 or 18 carbons long. The enzymes involved in this synthesis are acetyl-CoA carboxylase (ACCase) and fatty acid synthase (FAS) (Fig. 10.8). The name fatty acid synthase refers to a complex of several individual enzymes that catalyze the conversion of acetyl-CoA and malonyl-CoA to 16:0 and 18:0 fatty acids (see Section 10.4). **Acyl-carrier protein** (ACP), an essential protein cofactor, is generally considered a component of FAS.

Fatty acid biosynthesis is initiated by the ATP-dependent carboxylation of acetyl-CoA to form malonyl-CoA. The malonyl group is transferred next to ACP. Subsequent decarboxylation of the malonyl moiety acts to drive a condensation reaction, in which a carbon–carbon bond forms between C-1 of an acetate "primer" and C-2 of the malonyl group on ACP. This two-carbon chain length extension results initially in the formation of acetoacetyl-ACP. Subsequently, a sequence of three reactions—reduction, dehydration, and reduction again—leads to the formation of the fully reduced acyl-ACP. This sequence, progressing in three steps from a 3-ketoacyl group to a saturated acyl group, is a common reaction series found in biochemical pathways. For example, both β-oxidation and the citric acid cycle use the same series of reactions, but in reverse order.

Fatty acyl chains and their derivatives are among the most reduced molecules found in cells. Producing these molecules from their more-oxidized precursors requires a large investment of reducing power. As indicated above, each cycle of two-carbon addition involves two reduction steps. Thus, for a typical C_{18} fatty acid, 16 molecules of NAD(P)H are consumed. In illuminated chloroplasts, abundant reducing power is available from Photosystem I. In the dark and in tissues lacking chloroplasts, the oxidative pentose phosphate pathway is the most likely origin of reduced NADPH (Chapter 13).

10.2.2 Carbon precursors for fatty acid synthesis can be provided by reactions inside or outside the plastids.

Acetyl-CoA is the initial substrate for synthesis of the carbon backbone of all fatty acids. Also a central intermediate in many aspects of cellular metabolism, it is produced and consumed by dozens of reactions in the cell. Until recently, there has been some uncertainty as to which of these reactions produce the acetyl-CoA used in fatty acid biosynthesis. Most likely, acetyl-CoA from more than one reaction finds its way into fatty acid synthesis (Fig. 10.9). Contributions of individual reactions depend on developmental and other factors.

Experimental results have shown that within plastids, pyruvate dehydrogenase can directly produce acetyl-CoA from pyruvate generated during glycolysis (see Chapter 13). The activity of pyruvate dehydrogenase in isolated oilseed plastids is sufficient to

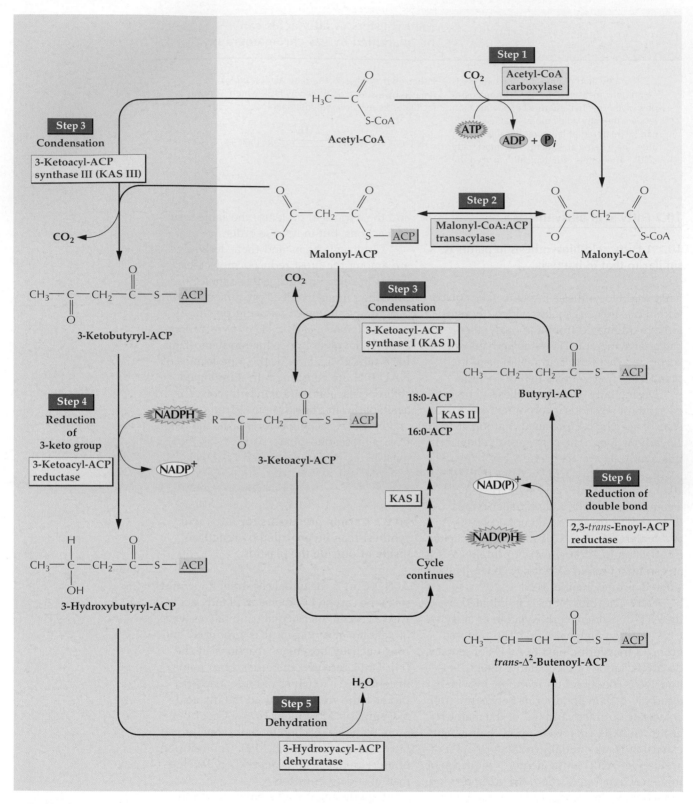

Figure 10.8

Overview of fatty acid synthesis. Fatty acids grow by addition of two-carbon (C_2) units. The reactions highlighted in yellow show how malonyl-CoA enters the cycle; those highlighted in orange represent the cyclic reactions. Synthesis of a C_{16} fatty acid requires that the cycle be repeated seven times. During the first turn of the cycle,

the condensation reaction (step 3) is catalyzed by ketoacyl-ACP synthase (KAS) III. For the next six turns of the cycle, the condensation reaction is catalyzed by isoform I of KAS. Finally, KAS II is used during the conversion of 16:0 to 18:0.

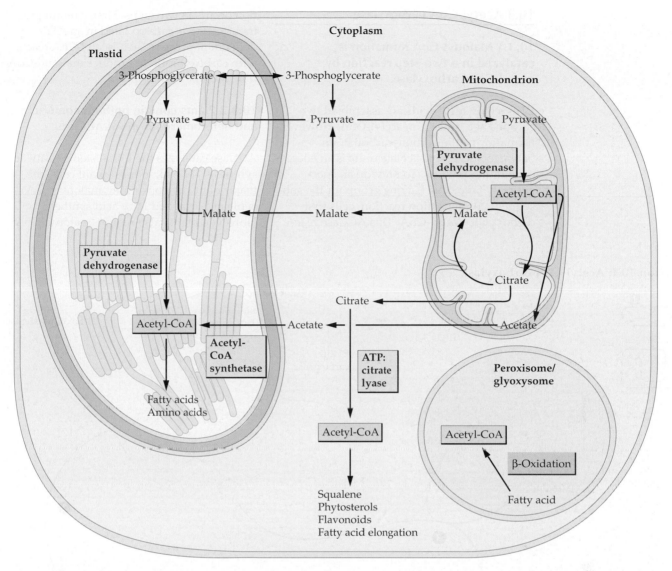

Figure 10.9

The central role of acetyl-CoA in metabolism. Acetyl CoA may be the most central intermediate in cellular metabolism, providing a link between many pathways. The major pathways involved in its production include glycolysis (via pyruvate dehydrogenase) and fatty acid oxidation. Acetyl-CoA is the starting material for biosynthesis of fatty acids, several amino acids, flavonoids (via chal-cone synthase), sterols, and many isoprenoid derivatives synthesized in the cytosol. During respiration, acetyl-CoA is the source of carbon input into the citric acid cycle in the mitochondria. Despite this central role, acetyl-CoA is not believed to cross membranes and must be produced in the compartment in which it is utilized.

account for in vivo rates of fatty acid synthesis. Furthermore, pyruvate is one of the most efficient precursors for synthesis of fatty acids by isolated oilseed plastids. However, some chloroplasts with low pyruvate dehydrogenase activity can sustain in vivo rates of fatty acid synthesis when supplied with exogenous free acetate but not with exogenous pyruvate. These results suggest that the acetate for chloroplast fatty acid synthesis might be generated outside the plastid. Candidate reactions include mitochondrial pyruvate dehydrogenase and cytosolic ATP citrate-lyase, which converts citrate, ATP, and coenzyme A (CoASH) to acetyl-CoA, oxaloacetate, ADP, and inorganic phosphate. Free acetate is probably taken into plastids and activated by acetyl-CoA synthetase in the stroma (Fig. 10.9) because acetyl-CoA does not cross membranes by diffusion and no transporters are known in plants.

10.3 Acetyl-CoA carboxylase

10.3.1 Malonyl-CoA formation is catalyzed in a two-step reaction by acetyl-CoA carboxylase.

Long-chain fatty acids are assembled two carbons at a time from acetyl-CoA. However, formation of the carbon–carbon bond between successive acetate units is an energy-consuming process. To provide an energetically favorable leaving group for the subsequent condensation reaction, cells first carboxylate acetyl-CoA. This ACCase reac-

tion occurs in two steps. First, a biotin prosthetic group is carboxylated in an ATP-dependent process; then, acetyl-CoA reacts with carboxybiotin to produce malonyl-CoA (Fig. 10.10).

10.3.2 Plants contain both homomeric and heteromeric forms of ACCase.

ACCase catalyzes the initial step in fatty acid synthesis; indeed, in most plant cells, the major pathway consuming malonyl-CoA is plastid-localized fatty acid synthesis. However, plants also require malonyl-CoA

Reaction 10.1: Acetyl-CoA carboxylase

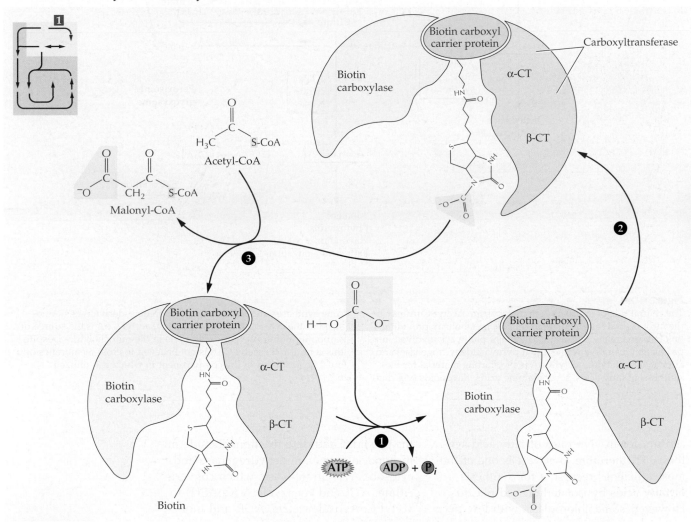

Figure 10.10

Schematic diagram of the acetyl-CoA carboxylase (ACCase) reaction. ACCase catalyzes step 1 in the reaction sequence shown in Figure 10.8. ACCase has three functional components that form malonyl-CoA from acetyl-CoA. (1) In an ATP-dependent reaction, biotin carboxylase activates CO_2 (as HCO_3^-) by attaching it to a nitrogen in the biotin ring of biotin carboxyl carrier protein (BCCP). (2) The flexible biotin arm of BCCP carries the activated CO_2 from the biotin carboxylase active site to the carboxyltransferase site (α-CT and β-CT). (3) The transcarboxylase transfers activated CO_2 from biotin to acetyl-CoA, producing malonyl-CoA.

outside the plastid, where it serves as a substrate for the flavonoid biosynthetic pathway, for fatty acid elongation reactions at the ER, for malonylation of some amino acids, and for the ethylene precursor, aminocyclopropanecarboxylic acid (Fig. 10.11). Because ACCase activity is believed to be highly regulated and a major determinant of the overall rate of fatty acid synthesis, many research groups have concentrated efforts toward understanding this enzyme. Until 1994, however, there was much confusion about its structure because different research groups could not agree on whether plant ACCase had a structure similar to that described for bacteria, with four separate subunits, or to that of fungi and animals, with a single polypeptide. It is now established that most plants utilize a prokaryotic-type ACCase in plastids and a multifunctional ACCase in the cytosol (Box 10.3).

The plastid form of ACCase has four subunits: biotin carboxyl carrier protein (BCCP), biotin carboxylase (BC), and the α- and β-subunits of carboxyltransferase (CT). These four subunits form a heteromeric complex of more than 650 kDa, which may be membrane-associated (Fig. 10.12). Three of the subunits are encoded by distinct nuclear genes, whereas the fourth (β-CT) is encoded in the chloroplast genome.

The cytosolic form of ACCase is a single large (greater than 500 kDa) homodimeric protein. The four subunits described above are integrated into domains of a single polypeptide, two of which associate to make up

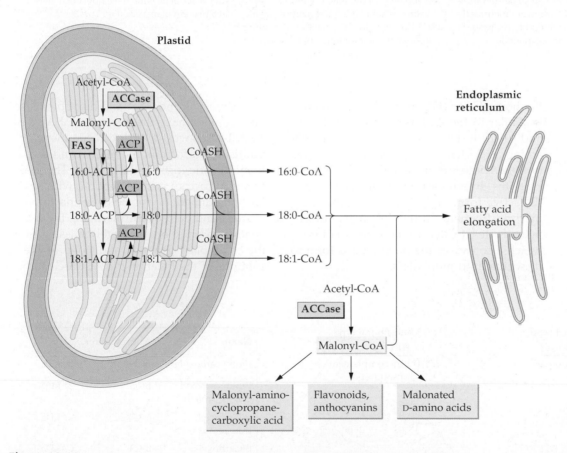

Figure 10.11
Multiple fates of malonyl-CoA. Malonyl-CoA produced by the ACCase reaction enters several pathways in plants. Within the plastid, malonyl-CoA is used exclusively for the production of fatty acids. In the cytosol, malonyl-CoA is the carbon donor for fatty acid elongation, producing the precursors for surface waxes and certain seed lipids. Condensation of three molecules of malonyl-CoA produces a diverse range of flavonoids and their derivatives (see Chapter 24). In many plant tissues, the major form of the ethylene precursor 1-aminocyclopropane-1-carboxylic acid is the inactive malonylated derivative, produced by reaction with malonyl-CoA. Finally, many D-amino acids and other secondary metabolites react with malonyl-CoA to form malonylated derivatives.

In 1972, Gamini Kannangara and Paul Stumpf reported that spinach chloroplasts have a prokaryotic type ACCase with four separate subunits for biotin carboxyl carrier protein (BCCP), biotin carboxylase (BC), and carboxyltransferase (CT). Although some later work also supported this interpretation, for the next 20 years almost all efforts to purify the enzyme or to obtain clones came to a different conclusion. Several laboratories purified ACCase and obtained a structure similar to that of the animal enzyme. Furthermore, the first cDNA clones for ACCase also clearly encoded a single, large, multifunctional polypeptide with high similarity to fungal and animal sequences. In light of these results, many early experiments that had suggested multiple subunits for ACCase were incorrectly reinterpreted as reflecting proteolytic breakdown of a larger polypeptide.

A clearer understanding of ACCase emerged when Yukiko Sasaki and coworkers in Japan followed up on information originating from the sequencing of the chloroplast genome. One open reading frame of unknown function had sequence similarity to carboxylases, including acetyl-CoA and propionyl-CoA carboxylases. Sasaki raised antibodies to a part of this protein and found that the antibodies both inhibited ACCase activity from chloroplasts and coprecipitated a biotin-containing protein. Recent results from several other laboratories have now shown that both prokaryotic and eukaryotic ACCase enzymes exist in plants; recognition of this fact has explained much of the apparently contradictory results.

After Sasaki's characterization of the β-CT subunit, identifying the remaining components of the multisubunit structure might have proceeded by purification of the chloroplast enzyme. However, the prokaryotic-type ACCase complex readily dissociates and loses activity, making conventional approaches to protein purification difficult. Fortunately, the large amount of DNA sequence data available for plants has led to identification of three additional subunits, based on their similarity to bacterial ACCase gene sequences. The BC and BCCP were identified as a result of projects that sequenced large numbers of anonymous cDNA clones. The α-CT subunit was first identified as a clone encoding a chloroplast inner membrane protein and was only later shown to be part of ACCase. All four subunits known to participate in the bacterial enzyme have now been characterized in plants (see Fig. 10.12).

the homodimer (Fig. 10.12). This structure is similar to the ACCase found in animals and fungi. Indeed, there is as much as 50% amino acid identity between the sequences of the enzymes from organisms in these kindoms. In animals and fungi, however, the active form of ACCase is a highly polymerized filament. In these systems, the enzyme is deactivated by phosphorylation, which also causes the polymerized form to dissociate into monomers.

Almost all monocot and dicot plants so far examined have the two types of ACCase described above—with the heteromeric, multisubunit form in the plastid and the homodimeric form in the cytosol. The grass family (Poaceae) is an exception, with similar homodimeric forms being found in both plastids and cytosol. Furthermore, the β-CT gene is partially or completely missing from the chloroplast genome, and none of the subunits of the heteromeric ACCase can be

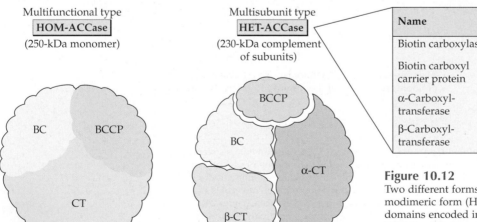

Multifunctional type
HOM-ACCase
(250-kDa monomer)

Multisubunit type
HET-ACCase
(230-kDa complement of subunits)

Name	Abbreviation	Size
Biotin carboxylase	BC	50 kDa
Biotin carboxyl carrier protein	BCCP	21 kDa
α-Carboxyltransferase	α-CT	91 kDa
β-Carboxyltransferase	β-CT	67 kDa

Figure 10.12
Two different forms of ACCase occur in plants. A homodimeric form (HOM-ACCase) has three functional domains encoded in a single polypeptide of about 250 kDa. A heteromeric form (HET-ACCase) consists of four subunits that together form a plastid-localized complex of 650 to 700 kDa.

detected. Although the reason for the different ACCase organization in grasses remains unknown, it has substantial practical significance in agriculture. Several widely used grass-specific herbicides are now known to kill grasses by specifically inhibiting the homodimeric form of ACCase in the plastid, thereby blocking plastid fatty acid biosynthesis.

10.3.3 Malonyl-CoA formation is the first committed step in fatty acid synthesis.

In many primary metabolic pathways, biochemical regulation occurs at the first committed step. Although malonyl-CoA is used in several pathways in the cell, fatty acid biosynthesis is the only known fate for malonyl-CoA within the plastid. Therefore, the ACCase reaction in plastids is the first committed step for fatty acid synthesis. Several lines of evidence indicate that plastid ACCase activity is tightly regulated and that this regulation determines, in large part, the overall rate of fatty acid synthesis. First, the concentration of malonyl-CoA in chloroplasts changes quickly during light–dark transitions and remains proportional to the rate of fatty acid synthesis. Second, for herbicides that specifically target the plastid ACCase, the same concentrations inhibit foliar fatty acid synthesis both in vivo and in vitro. Finally, addition of exogenous lipids to suspension cultures slows production of new fatty acids. Analysis of the substrates and products of plastid fatty acid metabolism indicates this regulation occurs at the ACCase reaction. Thus, plastid-localized ACCase appears to be a highly regulated enzyme subject to feedback and other biochemical controls. Recently, modification of ACCase by thioredoxin and by phosphorylation has been discovered.

10.4 Fatty acid synthase

10.4.1 Different types of FAS exist in different kingdoms.

FAS refers to all enzyme activities in fatty acid biosynthesis except ACCase. Although the reactions catalyzed by FAS are essentially the same for all organisms, two distinctly different types of FAS are found in nature. Animals and yeast use a Type I FAS, a single multifunctional enzyme complex characterized by large subunits (250 kDa). Each subunit is capable of catalyzing several different reactions. By contrast, plants and most bacteria have a Type II FAS, in which each enzyme activity resides on an individual protein that can be readily separated from the other activities participating in fatty acid synthesis. Type II FAS also includes ACP. The Type II FAS functions much like a metabolic pathway, whereas Type I FAS functions like a large protein complex (e.g., pyruvate dehydrogenase).

The assembly of a C_{18} fatty acid from acetyl-CoA in Type II fatty acid synthesis requires 48 reactions involving at least 12 different proteins. How is this complex pathway organized? Although to date no direct evidence has been established, some type of supramolecular organization seems very likely. The estimated concentrations of many acyl-ACP intermediates of the pathway are in the nanomolar range, far below the values predicted by kinetic analyses of the enzymes involved. Calculations suggest the enzyme activities available at these low substrate concentrations are not sufficient to support observed in vivo rates of fatty acid synthesis. Accordingly, some form of substrate channeling seems essential. Furthermore, in osmotically lysed chloroplasts, neither acetyl-CoA nor malonyl-CoA competes with radiolabeled free acetate for incorporation into fatty acids, suggesting that acetate is channeled directly into fatty acid synthesis. Thus, Type II fatty acid synthesis is catalyzed by separate enzymes that appear to be complexed in a tightly coupled pathway.

10.4.2 ACP transports intermediates of fatty acid synthesis through the pathway.

After the formation of malonyl-CoA in the ACCase reaction, the assembly of fatty acids involves a central cofactor, ACP (Fig. 10.13). This small protein is about 80 amino acids long and contains a phosphopantetheine prosthetic group covalently linked to a serine residue near the middle of the polypeptide chain. The phosphopantetheine group, also found in CoASH, contains a terminal

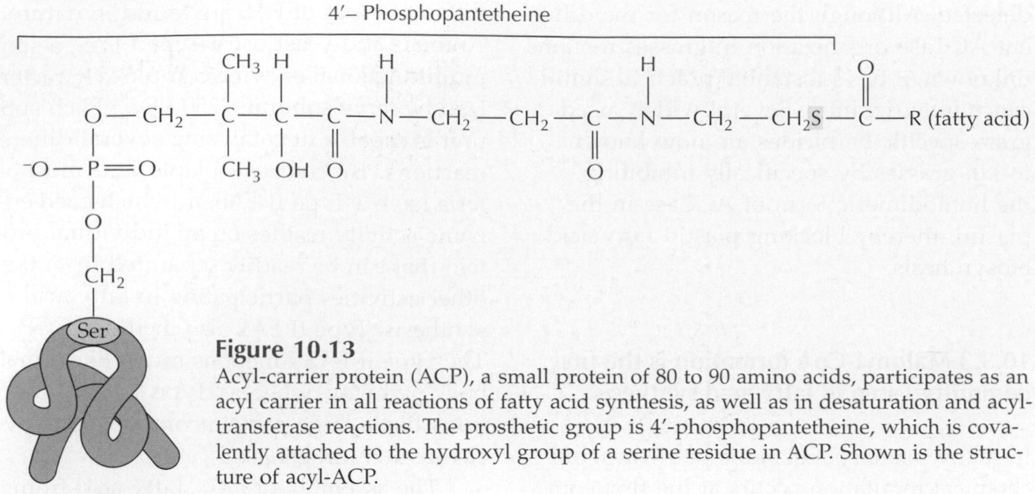

$$^-O-CH_2-\underset{\underset{CH_3}{|}}{\overset{\overset{CH_3}{|}}{C}}-\underset{\underset{OH}{|}}{\overset{\overset{H}{|}}{C}}-\underset{\underset{O}{\|}}{\overset{H}{C}}-N-CH_2-CH_2-\underset{\underset{O}{\|}}{C}-N-CH_2-CH_2S-\overset{\overset{O}{\|}}{C}-R \text{ (fatty acid)}$$

Figure 10.13

Acyl-carrier protein (ACP), a small protein of 80 to 90 amino acids, participates as an acyl carrier in all reactions of fatty acid synthesis, as well as in desaturation and acyl-transferase reactions. The prosthetic group is 4'-phosphopantetheine, which is covalently attached to the hydroxyl group of a serine residue in ACP. Shown is the structure of acyl-ACP.

sulfhydryl. The thioester linkage that forms between a fatty acid and this sulfur is a high-energy bond with a free energy of hydrolysis similar to that for ATP.

10.4.3 Malonyl-CoA:ACP transacylase transfers a malonyl moiety from CoASH to ACP.

ACP first becomes involved in the fatty acid synthesis pathway when the malonyl group produced by ACCase is transferred from CoASH to the sulfhydryl of ACP by the reactions catalyzed by malonyl-CoA:ACP transacylase (Rx. 10.2; see also step 2, Fig. 10.8).

The reaction mechanism involves a covalent malonyl–enzyme intermediate (Fig. 10.14). Analysis of the malonyl transacylase (note: the terms transacylase and acyltransferase are equivalent) from *E. coli* has demon-

strated that this intermediate is a serine ester. Two isoforms of this enzyme have been found in both soybean and leek, although no functional difference between the two forms is known. After the malonyl-transacylation, all subsequent reactions of fatty acid synthesis involve ACP.

10.4.4 The three plant isoforms of 3-ketoacyl-ACP synthase demonstrate different substrate specificities.

The defining reaction of fatty acid synthesis is the elongation of a "primer" acyl chain by two carbons donated from malonyl-ACP (step 3, Fig. 10.8). The condensation reaction to form a new carbon–carbon bond is catalyzed by 3-ketoacyl-ACP synthase (KAS), commonly called condensing enzyme (Fig. 10.15). All plants examined to date contain

Reaction 10.2: Malonyl-CoA:ACP transacylase

Chapter 10 Lipids

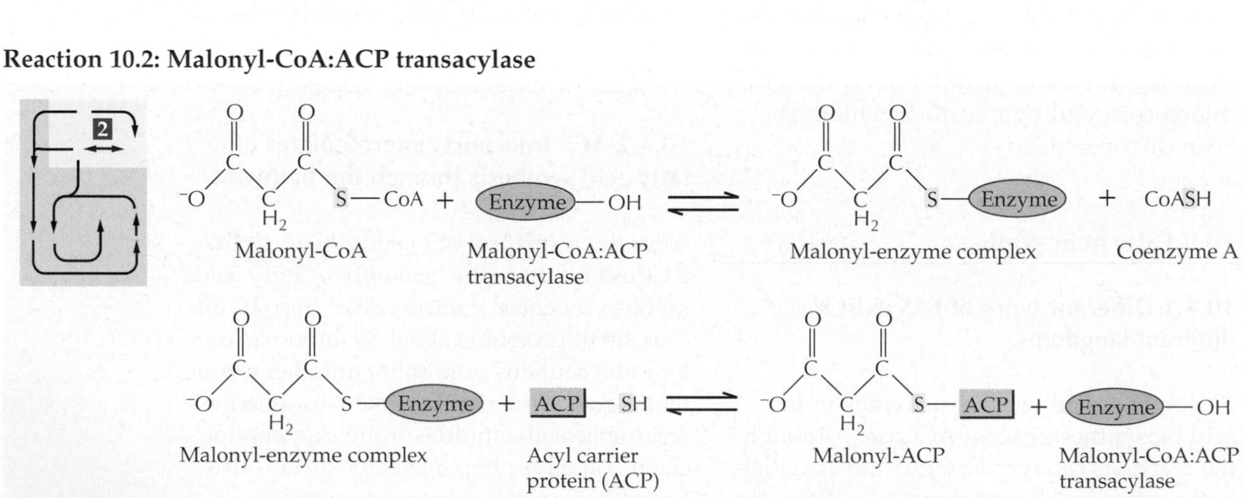

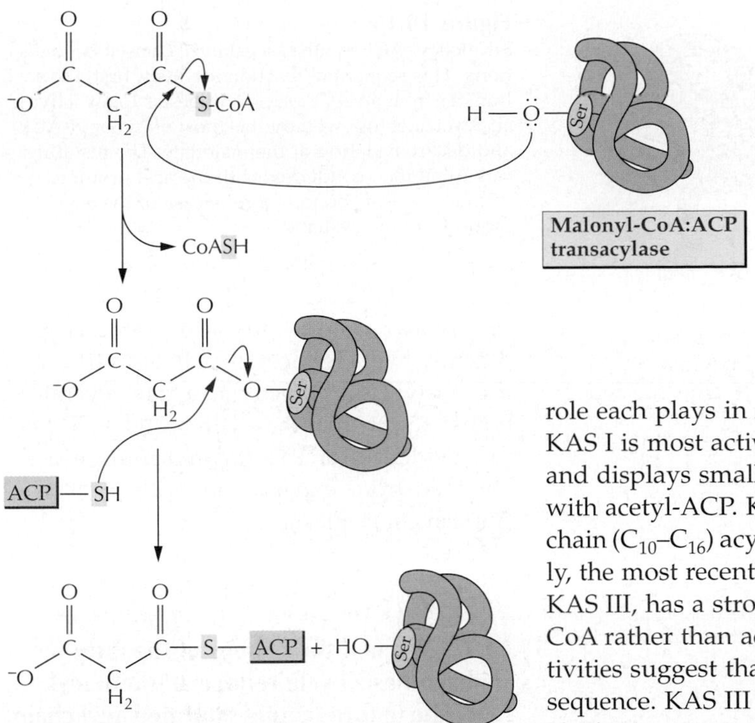

Figure 10.14
Mechanism of the malonyl-CoA:ACP transacylase reaction. In the first step of the reaction, the malonyl group is transferred to a serine residue on the enzyme and CoASH is released. In the second step, the acyl group is transferred to the phosphopantetheine sulfhydryl group of ACP to form the malonyl-ACP thioester.

three KAS isoenzymes (I, II, and III), each distinguished by its substrate specificity. The general reaction for KAS occurs in two steps (Rx. 10.3). The in vitro substrate specificity of each of the KAS isoenzymes suggests the role each plays in fatty acid biosynthesis. KAS I is most active with C_4–C_{14} acyl-ACPs and displays small but significant activity with acetyl-ACP. KAS II accepts only longer-chain (C_{10}–C_{16}) acyl-ACPs as substrates. Finally, the most recently discovered isoenzyme, KAS III, has a strong preference for acetyl-CoA rather than acyl-ACP. These in vitro activities suggest that KAS I isoenzymes act in sequence. KAS III initiates fatty acid biosynthesis, using acetyl-CoA as a primer. KAS I then extends the acyl chain to C_{12}–C_{16}. Finally, KAS II completes the synthesis to C_{18}.

The characterization of KAS III raises the question of the role of acetyl-ACP in fatty acid biosynthesis. Acetyl-ACP had long been considered the first substrate in bacterial

Reaction 10.3: 3-Ketoacyl-ACP synthase (KAS)

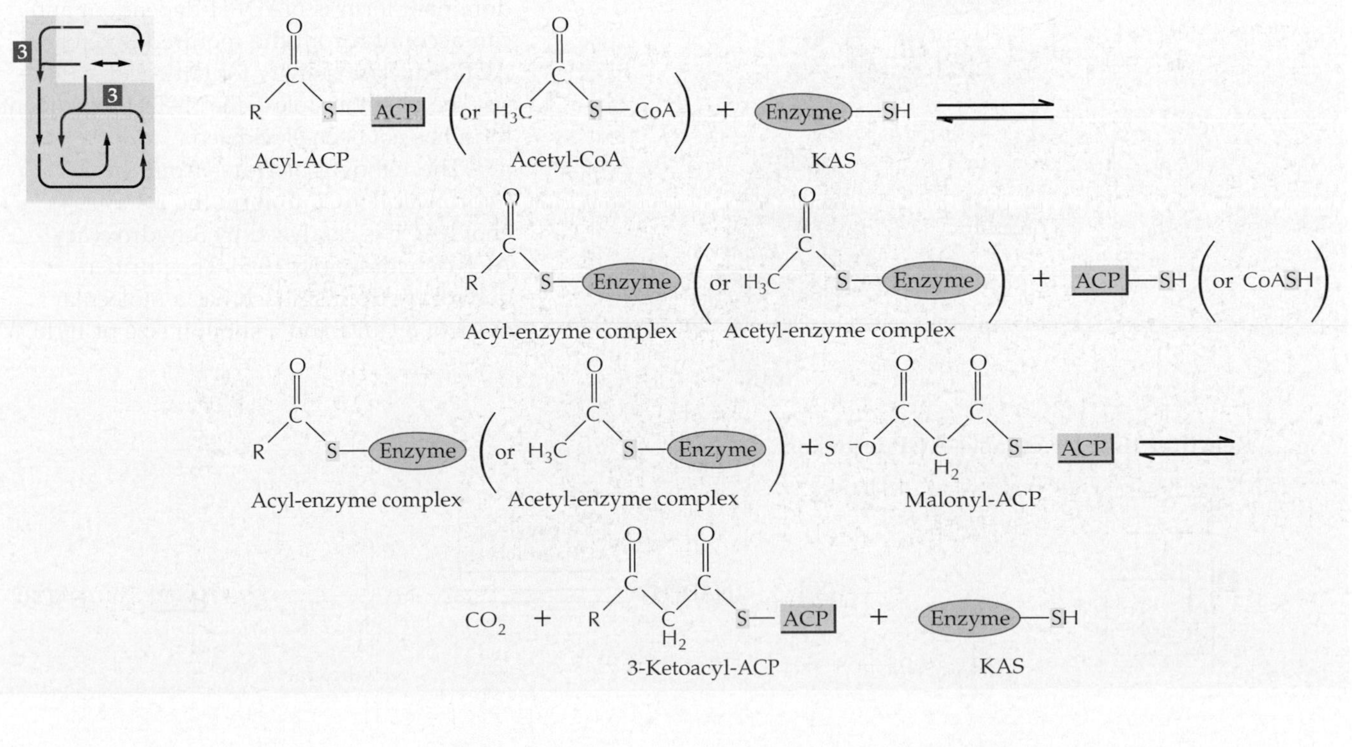

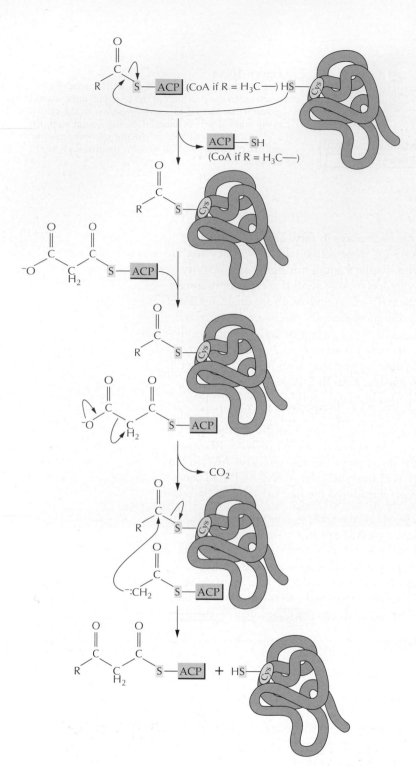

Figure 10.15
3-Ketoacyl-ACP synthases catalyze Claisen condensations. This sequential reaction involves, first, the acyl transfer from an ACP or CoA thioester to an active-site cysteine, followed by the entry of malonyl-ACP and decarboxylation of the malonate. The resulting carbanion then condenses with the acyl group to form a new C–C bond before release of the acyl group from the cysteine.

and plant fatty acid synthesis. However, the favored route of fatty acid synthesis apparently bypasses acetyl-ACP: KAS III can use acetyl-CoA directly at a rate several times greater than that with acetyl-ACP. Direct evidence for this favored route was obtained from experiments with isolated spinach chloroplasts.

10.4.5 The last three steps of the fatty acid synthesis cycle reduce a 3-ketoacyl substrate to form a fully saturated acyl chain.

The initial reductive step of fatty acid biosynthesis is the conversion of a 3-ketoacyl-ACP to a 3-hydroxyacyl-ACP (Rx. 10.4). Native 3-ketoacyl-ACP reductase from avocado has a molecular mass of 130 kDa and a subunit mass of 28 kDa, suggesting that it is a tetramer. At least two isoforms of this reductase occur in avocado, an NADPH-dependent form and an NADH-dependent form. The predominant form is NADPH-dependent and can account for all the required 3-ketoacyl-ACP reductase activity for fatty acid synthesis. The metabolic role of the NADH-dependent form has not been elucidated.

The removal of water from the 3-hydroxyacyl-ACP to form the 2,3-*trans*-enoyl-ACP is catalyzed by 3-hydroxyacyl-ACP dehydrase (Rx 10.5). The purified dehydrase from spinach has a molecular mass of 85 kDa and a subunit size of 19 kDa.

Reaction 10.4: 3-Ketoacyl-ACP reductase

Reaction 10.5: 3-Ketoacyl-ACP dehydratase

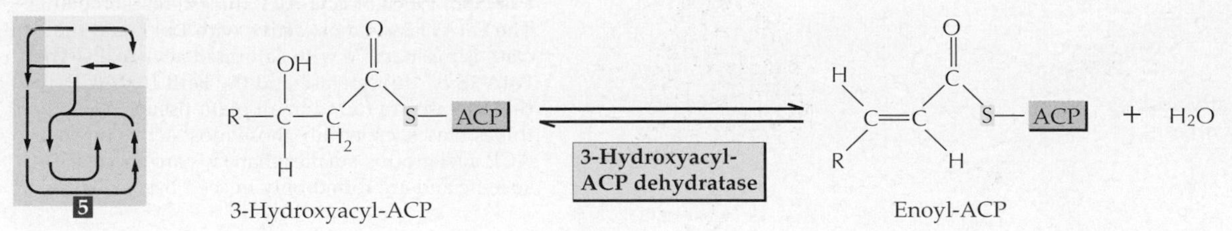

3-Hydroxyacyl-ACP ⇌ Enoyl-ACP + H_2O (3-Hydroxyacyl-ACP dehydratase)

Thus it appears to be a homotetramer. Its substrate specificity is very broad, with high activity demonstrated for acyl groups ranging from C_4 to C_{16}.

In the final reduction step, enoyl-ACP reductase converts the 2,3-*trans*-enoyl-ACP to the corresponding saturated acyl-ACP (Rx. 10.6). Enoyl-ACP reductase exists in two isoforms. The major form is specific for NADH; a homotetramer with a native molecular mass of 115 to 140 kDa and a subunit molecular mass of 32.5 to 34.8 kDa, this isomer has been purified from several plants. A second isoform uses either NADPH or NADH and is specific for longer-chain (C_{10}) enoyl-ACPs.

10.4.6 Thioesterase reactions terminate the fatty acid biosynthesis cycle.

Each cycle of fatty acid synthesis adds two carbons to the acyl chain. Typically, fatty acid synthesis ends at 16:0 or 18:0, when one of several reactions stops the process.

The most common reactions are hydrolysis of the acyl moiety from ACP by a thioesterase, transfer of the acyl moiety from ACP directly onto a glycerolipid by an acyl transferase, or double-bond formation on the acyl moiety by an acyl-ACP desaturase. The thioesterase reaction yields a sulfhydryl ACP (Rx. 10.7).

Two principal types of acyl-ACP thioesterases occur in plants (Fig. 10.16). The major class, designated FatA, is most active with $18:1^{\Delta 9}$-ACP. A second class (FatB), typified by 16:0-ACP thioesterase, is most active with shorter-chain, saturated acyl-ACPs. In both cases, however, the metabolic consequence of these reactions is the same. The cleavage of the acyl moiety from ACP prevents extension and targets the acyl group for export out of the plastid by an unknown mechanism. Thioesterases play an important role in plants that have unusually short fatty acids, such as coconut, many species of *Cuphea,* and California bay *(Umbellularia californica).* These plants have thioesterases that are especially active with C_{10} to C_{12} acyl-ACPs; by prematurely

Reaction 10.6: Enoyl-ACP reductase

Enoyl-ACP + NAD(P)H ⇌ Acyl-ACP + NAD(P)$^+$ (Enoyl-ACP reductase)

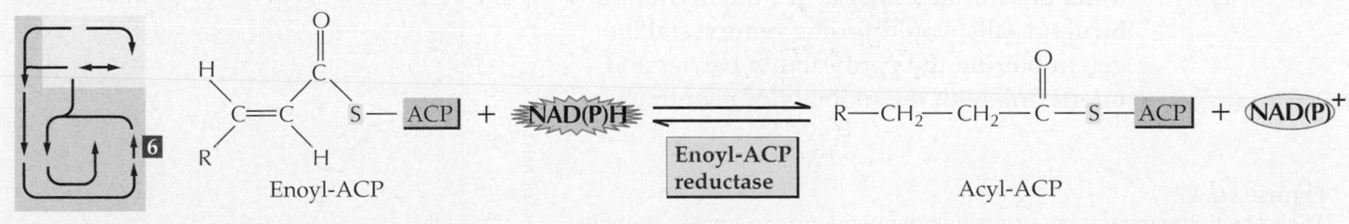

Reaction 10.7: Thioesterase

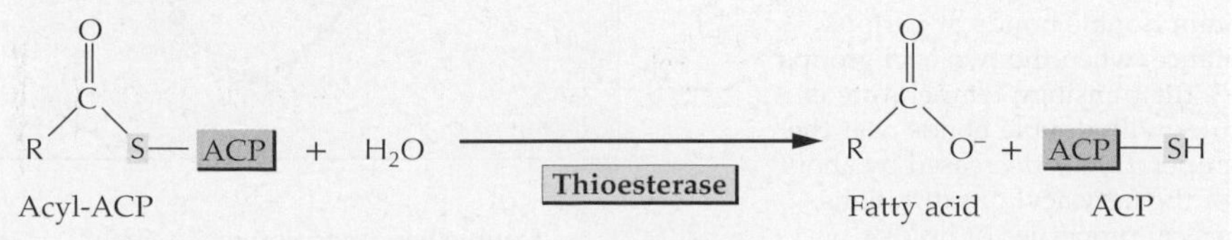

Acyl-ACP + H_2O → Fatty acid + ACP—SH (Thioesterase) ACP

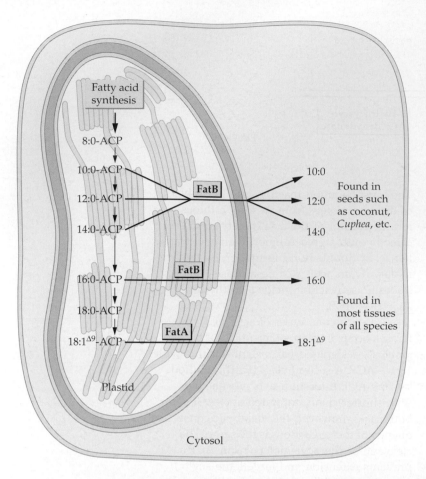

Figure 10.16
Principal types of acyl-ACP thioesterases in plants. The FatA class is most active with 18:1$^{\Delta9}$ and the FatB class is most active with saturated acyl-ACPs. The FatA 18:1$^{\Delta9}$ thioesterase and the FatB 16:0-ACP thioesterase are found in all plant tissues. Some FatB thioesterases, especially those most active on acyl-ACP acyl groups smaller than C$_{16}$, are species-specific and are found only in seed tissue.

components. By contrast, cis-double bonds, which introduce "kinks" into a fatty acid chain, can enhance membrane fluidity by dramatically lowering the temperature at which these ordered matrices melt. For example, the desaturation step that converts stearic acid (18:0) to oleic acid (18:1$^{\Delta9}$) lowers the melting point of the fatty acid from 69°C to 13.4°C. Double bonds at other positions in the chain also exert large effects on the melting temperature of fatty acids and of the lipids that contain them (Fig. 10.17; see also Chapter 1, Table 1.2). Formation of such double bonds is catalyzed by various desaturase enzymes, which generate a diverse array of unsaturated lipids found in membranes, storage reserves, and extracellular waxes.

10.5.1 Plants contain a soluble, plastid-localized stearoyl-ACP desaturase.

Unsaturated C$_{18}$ acyl chains found in membrane lipids throughout the plant cell are downstream products of a soluble chloroplast enzyme, stearoyl-ACP Δ9-desaturase (Fig. 10.18). All eukaryotes and some prokaryotes have desaturases that catalyze

terminating fatty acid biosynthesis, thioesterase activity results in the accumulation of 10:0 and 12:0 in the seed triacylglycerols.

10.5 Desaturation and elongation of C$_{16}$ and C$_{18}$ fatty acids

If membranes contained only saturated or trans-unsaturated fatty acids, the hydrophobic lipid tails would form a semicrystalline gel, impairing the permeability barrier and interfering with the mobility of membrane

Figure 10.17
The transition temperature, or melting temperature, of a lipid is strongly influenced by the presence and position of double bonds in the acyl groups. In this example, the transition temperature has been determined for several molecular species of phosphatidylcholine in which the two acyl groups are C$_{18}$ fatty acids that contain double bonds at various positions along the chains. Thus, for instance, when the two acyl groups have double bonds between C-2 and C-3, the transition temperature is approximately 40°C, whereas for molecules with double bonds near the center of the acyl groups, the melting temperature is decreased by about 60°C. Thus, by controlling the position of the fatty acyl unsaturations, organisms can exert control over the physical properties of lipids.

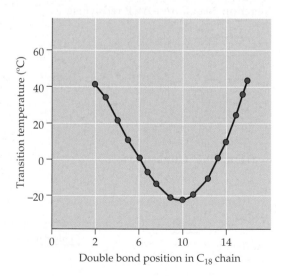

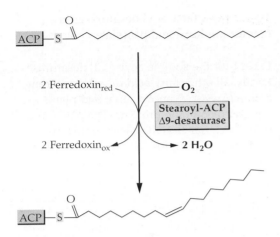

Figure 10.18
Desaturation of stearic acid is catalyzed by stearoyl-ACP Δ9-desaturase.

similar reactions, but in most cases these enzymes utilize acyl-CoA substrates and are integral membrane proteins. Some bacteria such as *Escherichia coli* lack such desaturases and instead introduce double bonds during the synthesis of the fatty acids. Recently, stearoyl-ACP Δ9-desaturase has come to be recognized as the prototype of a distinct family of structurally similar enzymes that introduce double bonds at various loca-

tions along the acyl chain. Species-specific genes for two isoforms, which catalyze desaturations of palmitic acid at the Δ4 or Δ6 position, have recently been isolated from coriander (*Coriandrum sativum*) and black-eyed susan (*Thunbergia alata*), respectively.

Unlike the stearoyl-CoA desaturases of fungi and animals, which are located in the ER membranes, the members of the stearoyl-ACP desaturase enzyme family are soluble, a fact that has greatly facilitated structural and mechanistic studies. The tertiary structure of castor bean (*Ricinus communis*) stearoyl-ACP Δ9-desaturase has been solved by X-ray crystallography (Fig. 10.19A). As the crystal structure shows, the protein contains a cavity that can bind the 18:0 substrate in the correct orientation with respect to the active site (Fig. 10.19B). The amino acid sequences of known soluble desaturases from several plants are sufficiently homologous that we can deduce their active sites by mapping the sequences of the Δ4- and Δ6-desaturases onto the structure of the Δ9-desaturase. When the gene encoding the castor bean enzyme was expressed in *E. coli*, it produced a functional enzyme that could be purified in large amounts. Mössbauer spectrometry of the recombinant castor bean enzyme from cultures of *E. coli*

(A)

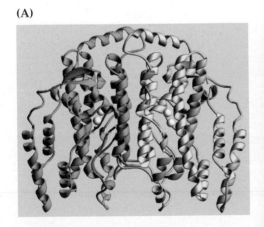

(B)

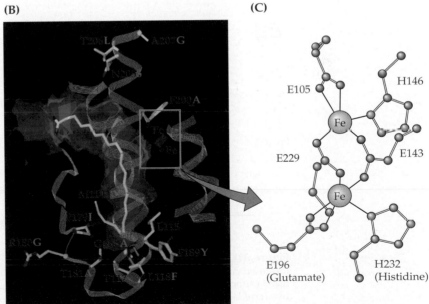

(C)

Figure 10.19
(A) Tertiary structure of stearoyl-ACP Δ9-desaturase dimer from *Ricinus communis*. The four white spheres near the center of the enzyme are two pairs of iron ions that catalyze the desaturation reaction. (B) Schematic view of the substrate channel of a stearoyl-ACP Δ9-desaturase monomer. A model of the stearoyl substrate moiety is fitted in the binding pocket of the desaturase. (C) Details of the residues that coordinate the diiron-oxo group at the active site.

grown in ^{57}Fe revealed the presence of a non-heme iron center in the desaturases—a structure now known to be an Fe-O-Fe (diiron) center of the type found in bacterial methane monooxygenase (Fig. 10.19C). This provides an explanation for many properties of the desaturases, such as the observation that the overall desaturation reaction requires transfer of two electrons from a donor such as ferredoxin or cytochrome b_5 (Fig. 10.20).

10.5.2 Most fatty acyl desaturases are membrane-localized proteins.

Except for the soluble acyl-ACP desaturase family, all other fatty acid desaturases from animals, yeast, cyanobacteria, and plants are integral membrane proteins. The plant and cyanobacterial enzymes desaturate fatty acids of glycerolipids, whereas some yeast and animal desaturases act on acyl-CoAs.

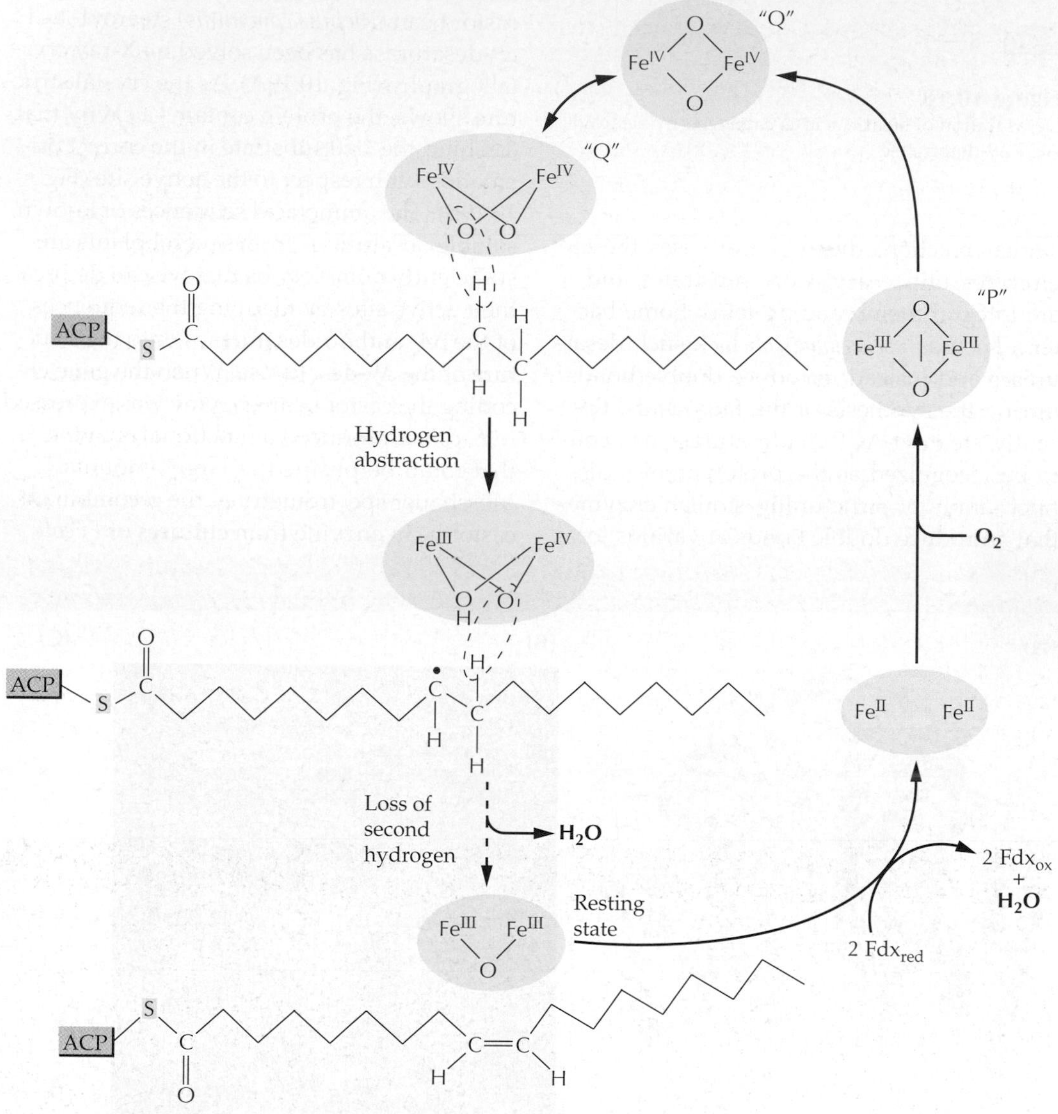

Figure 10.20

Proposed catalytic mechanism for fatty acid desaturation. In the resting state, the diiron center is in the oxidized (diferric, or Fe^{III}–Fe^{III}) form with a μ-oxo bridge. Reduction of both iron ions by electron transfer from two ferredoxins (Fdx) results in the reduced (diferrous, or Fe^{II}–Fe^{II}) form. The reduced enzyme binds molecular oxygen, resulting in the formation of a peroxo intermediate, "P." Scission of the O–O bond results in the formation of an activated form of the diiron center, "Q" (diferryl, or Fe^{IV}–Fe^{IV}). By analogy with the methane monooxygenase reaction, Q has been proposed to perform an energy-demanding hydrogen abstraction from the methylene group of the unactivated fatty acid to yield a radical intermediate. Loss of the second hydrogen results in formation of the double bond along with the loss of H_2O and regeneration of the oxidized active site and the μ-oxo bridge.

Solubilizing and purifying the plant enzymes have proven very difficult, limiting investigation by traditional biochemical techniques. Genetic analysis in *Arabidopsis* has provided an alternative approach to biochemical methods for assessing the function of the enzymes.

The number and properties of different desaturases in plants are known from the isolation of a comprehensive collection of *Arabidopsis* mutants with defects in each of eight desaturase genes. The enzymes encoded by these genes differ in substrate specificity, subcellular location, mode of regulation, or some combination of these (Table 10.4). The mutants were identified by analyzing lipid samples from leaves or seeds of individual plants from heavily mutagenized populations of plants. The biochemical defect of each class of mutants is shown by breaks in the pathway in Figure 10.21. The mutations that disrupt the activity of specific fatty acid desaturases are designated *fab2* and *fad2* through *fad8*. Two of the desaturases are ER-localized enzymes, oleate desaturase (FAD2) and linoleate desaturase (FAD3). Three structurally related enzymes are located in the plastids: one oleate desaturase (FAD6) and two functionally similar linoleate desaturases (FAD7 and FAD8). The identities of these enzymes have been confirmed by cloning the corresponding genes and using the cloned genes to complement the corresponding mutations in transgenic *Arabidopsis* plants, thereby restoring enzyme activity.

The ER-localized enzymes act on fatty acids esterified to phosphatidylcholine, and possibly other phospholipids, and utilize cytochrome b_5 as an intermediate electron donor. Cytochrome b_5 is reduced by another membrane protein, cytochrome b_5 reductase. Thus, three proteins are required for the overall desaturation reaction catalyzed by the ER-localized desaturases. In contrast, the membrane-localized chloroplast enzymes use soluble ferredoxin as an electron donor and act on fatty acids esterified to galactolipids, sulfolipids, and phosphatidylglycerol. Analysis of the mutants suggests the FAD4 enzyme is completely specific for phosphatidylglycerol, and the FAD5 enzyme appears to be specific for monogalactosyldiacylglycerol (Fig. 10.21). Despite differences in the glycerolipid substrates and electron donors used, and the position in which they insert the double bond, all of the membrane-localized plant desaturases cloned thus far are structurally related to each other and to the membrane-localized desaturases from animals and yeast. They contain a pair of histidine-rich sequences (HXXHH) that have been proposed to bind the two iron ions required for catalysis (see Fig. 10.19). Although the tertiary structure of the membrane-bound desaturases is not known, the iron-binding site may resemble that found in the diiron protein hemerythrin (Fig. 10.22).

From analysis of the effects of the mutations on glycerolipid fatty acid composition,

Table 10.4 Fatty acid desaturases of *Arabidopsis*

Name	Subcellular location	Fatty acid substrates	Site of double-bond insertion	Notes
FAD2	ER	$18:1^{\Delta 9}$	$\Delta 12$	Preferred substrate is phosphatidylcholine
FAD3	ER	$18:2^{\Delta 9,12}$	$\omega 3$	Preferred substrate is phosphatidylcholine
FAD4	Chloroplast	16:0	$\Delta 3$	Produces 16:1-*trans* at *sn*-2 of phosphatidylglycerol
FAD5	Chloroplast	16:0	$\Delta 7$	Desaturates 16:0 at *sn*-2 of monogalactosyldiacylglycerol
FAD6	Chloroplast	$16:1^{\Delta 7}$ $18:1^{\Delta 9}$	$\omega 6$	Acts on all chloroplast glycerolipids
FAD7	Chloroplast	$16:2^{\Delta 7,11}$ $18:2^{\Delta 9,12}$	$\omega 3$	Acts on all chloroplast glycerolipids
FAD8	Chloroplast	$16:2^{\Delta 7,11}$ $18:2^{\Delta 9,12}$	$\omega 3$	Isoenzyme of FAD7 induced by low temperature
FAB2	Chloroplast	18:0	$\Delta 9$	Stromal stearoyl-ACP desaturase

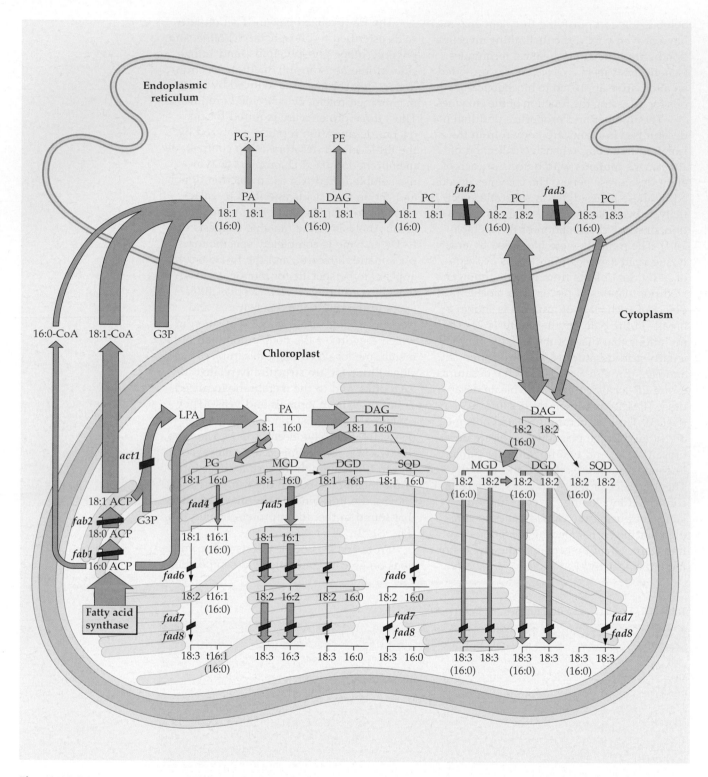

Figure 10.21

Abbreviated scheme for lipid synthesis in leaves of *Arabidopsis*. The set of reactions occurring solely within the chloroplast are termed the prokaryotic pathway; those that involve glycerolipid synthesis in the ER and subsequent transfer to the chloroplast constitute the eukaryotic pathway. The widths of the arrows represent the relative fluxes through the various steps of the pathways. The breaks in the pathway (red) represent some of the sites at which mutations have been obtained in *Arabidopsis* (see Table 10.4). DAG, diacylglycerol; DGD, digalactosyldiacylglycerol; G3P, glycerol 3-phosphate; LPA, lysophosphatidic acid; PA, phosphatidic acid; PC, phosphatidyl-choline; PE, phosphatidylethanolamine; PG, phosphatidylglycerol; PI, phosphatidylinositol; MGD, monogalactosyldiacylglycerol; SQD, sulfoquinovosyldiacylglycerol.

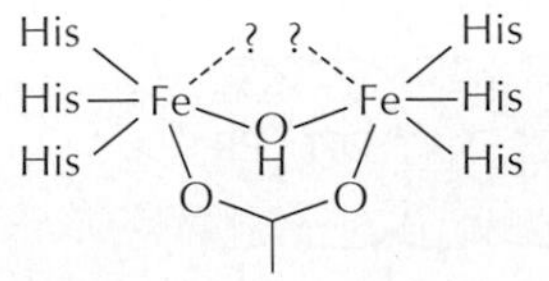

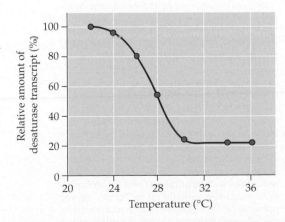

Figure 10.22

Proposed ligation sphere of the integral membrane desaturases and related enzymes possessing two HXXHH motifs, in which all of the histidine residues have been shown to be essential for catalysis. The model is based on the structure of hemerythrin and shows a hydroxo bridge and a single carboxylate ligand that has yet to be defined. Question marks indicate the presence of two undefined ligands.

a loss-of-function mutant appears to be available for every desaturase in *Arabidopsis* except stearoyl-ACP Δ9-desaturase, which is encoded by a family of genes. Thus, the *fab2* mutation, thought to inactivate one of the stearoyl-ACP desaturase genes, causes a significant increase in stearate concentrations but does not completely eliminate unsaturated fatty acids.

10.5.3 What factors determine glycerolipid desaturation?

A central question concerning the mechanisms that control desaturation is, What factors determine the extent of desaturation of a specific glycerolipid in a particular membrane? In cyanobacteria, the steady-state concentration of an oleate Δ12-desaturase mRNA is inversely correlated with temperature (Fig. 10.23). The stimulatory effect of low temperature on desaturase gene expression can be mimicked by using nonlethal catalytic hydrogenation to reduce the number of double bonds in membranes of living cells, thereby decreasing the fluidity of the membranes. Thus, cyanobacteria appear to have a mechanism that senses the fluidity of the membrane and regulates the level of desaturase mRNA accordingly.

Most plants also increase the extent of membrane glycerolipid desaturation when grown at low temperature. However, with one exception, the amount of mRNA for the desaturases genes in *Arabidopsis* does not appear to change significantly in response to

changes in growth temperature. The one exception is the *FAD8* gene, which exhibits a strong increase in mRNA abundance at low temperature. If a regulatory mechanism controlled desaturase gene expression in response to changes in the physical properties of membranes, we would expect this mechanism to detect large changes in membrane composition and to signal an increase in desaturase gene expression. However, expression of the cold-inducible *fad8* gene is not altered in *fad7 fad8* double mutants, which are highly deficient in $16:3^{Δ7,10,13}$ and $18:3^{Δ9,12,15}$ fatty acids. Indeed, none of the loss-of-function *fad* mutations alters *FAD* gene expression, indicating that if such a sensing mechanism exists in plants, it must act at the posttranscriptional level. Presumably, the complexities associated with regulating the composition of many intracellular membranes in plants may require posttranscriptional mechanisms that can adjust the composition of each membrane independently.

10.5.4 Specialized elongase systems produce long-chain fatty acids.

Fatty acid synthesis generally results in C_{16} and C_{18} fatty acids, yet plants have numerous requirements for longer fatty acids. All plants produce waxes, usually derived from C_{26} to C_{32} fatty acids. Often, sphingolipids contain C_{22} and C_{24} fatty acids. And, in some plants, triacylglycerols contain large

Figure 10.23

In cyanobacteria, the amount of desaturase mRNA is regulated by growth temperature.

amounts of C_{20} and C_{22} fatty acids. Where and how are these very-long-chain fatty acids synthesized?

Plants and most other eukaryotic organisms have a specialized elongase system for extension of fatty acids beyond C_{18}. These elongase reactions have several important features in common with FAS reactions (see Fig. 10.8). Each uses a reaction series that condenses two carbons at a time from malonyl-CoA to an acyl primer, followed by reduction, dehydration, and a final reduction. The result is that the same acyl intermediates are used for the two processes. Although the enzymology of the elongase is not well understood, several important differences between the two systems are known (Fig. 10.24):

- Fatty acid elongases are localized in the cytosol and are membrane-bound.
- ACP is not involved in this process.
- The elongase 3-ketoacyl-CoA synthase (elongase KCS) catalyzes the condensation of malonyl-CoA with an acyl primer.

Recent cloning of elongase KCS genes from *Arabidopsis* and jojoba (*Simmondsia chinensis*) has revealed that these 60-kDa enzymes bear very little sequence similarity to any other condensing enzymes. The identification of additional clones suggests that the elongase KCSs in *Arabidopsis* belong to a

Figure 10.24
Sequence of events during a single cycle of fatty acid elongation by the membrane-bound fatty acid elongase system, which consists of at least four components: 3-ketoacyl-CoA synthase (KCS), 3-ketoacyl-CoA reductase (KR), 3-hydroxyacyl-CoA dehydrase (DH), and enoyl-CoA reductase (ER). The initial cycle of elongation starts with a C_{18} acyl-CoA, either stearoyl-CoA or oleoyl-CoA. Steps 1–3 are catalyzed by KCS. First, the acyl-CoA acylates an active-site cysteine. Next, the malonyl-CoA binds to the active site. Finally, a concerted reaction occurs in which the malonyl moiety is decarboxylated and a Claisen condensation with the acyl group results. Then Steps 4–6, a reduction, a dehydration, and a second reduction, occur sequentially on the remaining components of the elongase system and result in the release of an acyl-CoA that is two carbons longer than the starting acyl-CoA. This acyl-CoA may then undergo additional cycles of elongation or be used in other pathways of lipid metabolism.

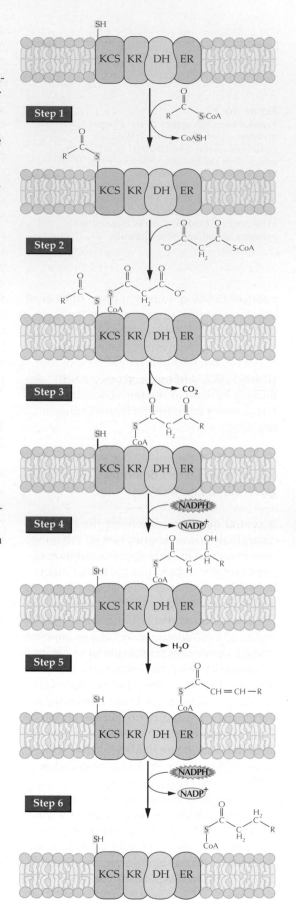

complex gene family with at least 15 members and possibly more than 25. The requirement for multiple genes may be related to the relatively narrow acyl substrate specificity of elongase KCSs, so that several different elongase KCSs would be required to synthesize a C_{30} fatty acid. Elongase KCSs may also be specific for a particular physiological function, such as wax biosynthesis or sphingolipid biosynthesis.

The additional components of the elongase system remain largely uncharacterized. The reductase and dehydrase activities are membrane-localized, and the reductases prefer NADPH. To date, none has been purified or characterized. However, *glossy8*, a gene in maize that is required for normal wax composition, has been cloned and may encode a 3-ketoacyl reductase.

10.6 Synthesis of unusual fatty acids

10.6.1 More than 200 fatty acids occur in plants.

Extensive surveys of the fatty acid composition of seed oils from different plants species have resulted in the identification of more than 200 naturally occurring fatty acids, which can be broadly classified into 18 structural classes. The classes are defined by the number and arrangement of double or triple bonds and various functional groups, such as hydroxyls, epoxys, cyclopentenyl or cyclopropyl groups, or furans (Fig. 10.25).

The most common fatty acids, which often occur in both membrane and storage lipids, belong to a small family of C_{16} and C_{18} fatty acids that may contain as few as zero, or as many as three, cis-double bonds. All members of the family are descended from the fully saturated species as the result of a series of sequential desaturations that begin at C-9 and progress in the direction of the methyl carbon (Fig. 10.26). Fatty acids that cannot be described by this simple algorithm are generally considered "unusual" because each is found almost exclusively in the seed oil of only a few plant species. Several, however, such as lauric (12:0), erucic ($20:1^{\Delta 14}$), and ricinoleic (12-OH, $18:1^{\Delta 9}$) acids, are of substantial commercial importance (see Section 10.11.4).

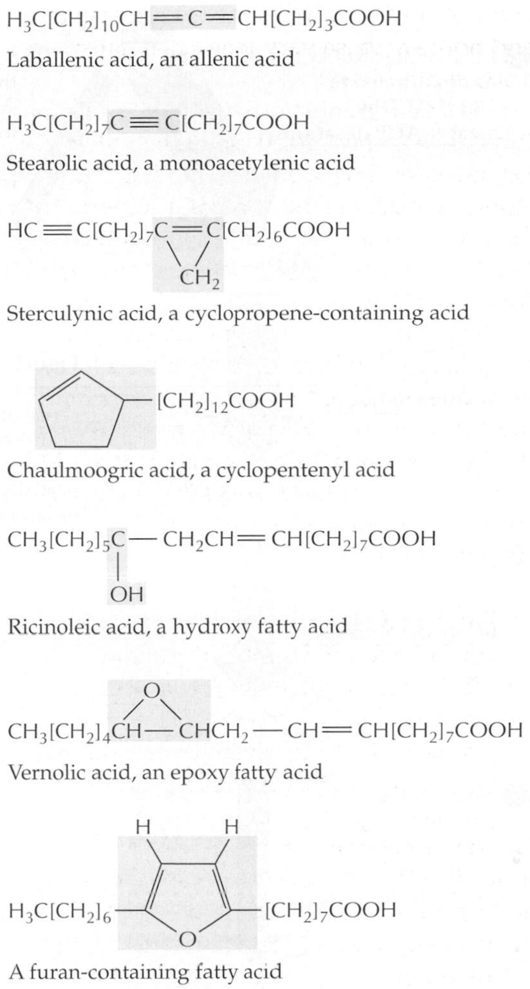

$$H_3C[CH_2]_{10}CH{=}C{=}CH[CH_2]_3COOH$$

Laballenic acid, an allenic acid

$$H_3C[CH_2]_7C{\equiv}C[CH_2]_7COOH$$

Stearolic acid, a monoacetylenic acid

$$HC{\equiv}C[CH_2]_7C{=}C[CH_2]_6COOH$$

Sterculynic acid, a cyclopropene-containing acid

Chaulmoogric acid, a cyclopentenyl acid

$$CH_3[CH_2]_5C{-}CH_2CH{=}CH[CH_2]_7COOH$$
with OH below

Ricinoleic acid, a hydroxy fatty acid

$$CH_3[CH_2]_4CH{-}CHCH_2{-}CH{=}CH[CH_2]_7COOH$$
(epoxide O above)

Vernolic acid, an epoxy fatty acid

$$H_3C[CH_2]_6{-}\text{(furan)}{-}[CH_2]_7COOH$$

A furan-containing fatty acid

Figure 10.25
Some of the functional groups found in unusual plant fatty acids.

10.6.2 Some enzymes that synthesize unusual fatty acids resemble enzymes involved in the biosynthesis of common fatty acids.

Much of the research concerning unusual plant fatty acids has been focused on the identification of new structures or cataloging the composition of fatty acids found in various plant species. Less is known about the mechanisms responsible for the synthesis and accumulation of unusual fatty acids, or of their significance to the fitness of the plants that accumulate them. Recently, however, a gene for the oleate Δ12-hydroxylase that catalyzes the synthesis of ricinoleic and other hydroxylated fatty acids was cloned from castor bean and *Lesquerella fenderli*. This enzyme exhibits about 70% amino acid sequence

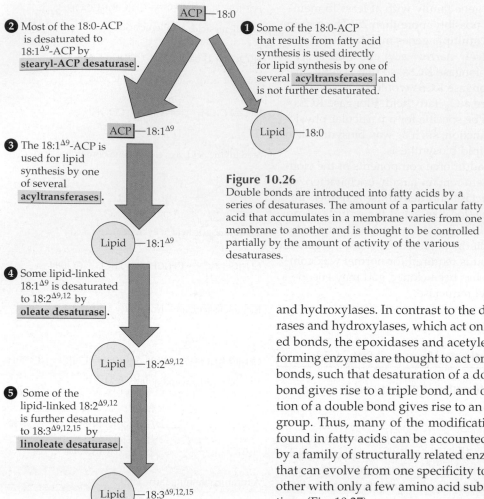

2 Most of the 18:0-ACP is desaturated to $18:1^{\Delta 9}$-ACP by **stearyl-ACP desaturase**.

ACP—18:0

1 Some of the 18:0-ACP that results from fatty acid synthesis is used directly for lipid synthesis by one of several **acyltransferases** and is not further desaturated.

ACP—$18:1^{\Delta 9}$

Lipid—18:0

3 The $18:1^{\Delta 9}$-ACP is used for lipid synthesis by one of several **acyltransferases**.

Figure 10.26
Double bonds are introduced into fatty acids by a series of desaturases. The amount of a particular fatty acid that accumulates in a membrane varies from one membrane to another and is thought to be controlled partially by the amount of activity of the various desaturases.

Lipid—$18:1^{\Delta 9}$

4 Some lipid-linked $18:1^{\Delta 9}$ is desaturated to $18:2^{\Delta 9,12}$ by **oleate desaturase**.

Lipid—$18:2^{\Delta 9,12}$

5 Some of the lipid-linked $18:2^{\Delta 9,12}$ is further desaturated to $18:3^{\Delta 9,12,15}$ by **linoleate desaturase**.

Lipid—$18:3^{\Delta 9,12,15}$

and hydroxylases. In contrast to the desaturases and hydroxylases, which act on saturated bonds, the epoxidases and acetylene-forming enzymes are thought to act on double bonds, such that desaturation of a double bond gives rise to a triple bond, and oxidation of a double bond gives rise to an epoxy group. Thus, many of the modifications found in fatty acids can be accounted for by a family of structurally related enzymes that can evolve from one specificity to another with only a few amino acid substitutions (Fig. 10.27).

similarity to the microsomal oleate Δ12-desaturase from the same species. To identify the amino acid differences responsible for causing the enzyme to be a hydroxylase instead of a desaturase, investigators have used site-directed mutagenesis to test the importance of the amino acids conserved in one class of enzymes but not in the other. On the basis of these studies, it appears that the hydroxylase can be converted to a desaturase by as few as seven amino acid substitutions that alter the geometry of the active site. The fact that such a small number of amino acid substitutions can alter the outcome of the enzymatic reaction indicates that the diiron center used by these enzymes is capable of catalyzing different reactions, depending on the precise geometry of the active site. Indeed, recent evidence also indicates that the enzymes that introduce epoxy groups and triple bonds are also structurally similar to the desaturases

10.6.3 Taxonomic relationships between plants having similar or identical kinds of unusual fatty acids are not predictable.

In some cases, particular fatty acids occur mostly or solely in related taxa. For example, the cyclopentenyl fatty acids (e.g., chaulmoogric acid, Fig. 10.25) have been found only in the family Flacourtiaceae, although the presence of cyclopentenylglycine, the biosynthetic precursor of the cyclopentenyl fatty acids, in the Passifloraceae (passionflower family) and Turneraceae (e.g., yellow alder) suggests that these fatty acids may also be found in these other families of the order Violales. Petroselinic acid is most commonly found in the related families Apiaceae (carrot and parsley family), Araliaceae (ginseng family), and Garryaceae (silk tassel family), but it has also been observed in more divergent families.

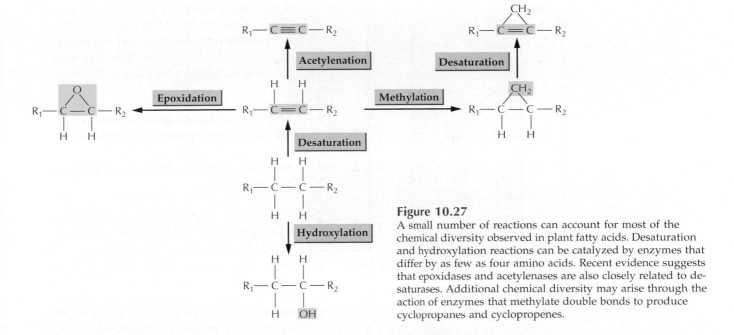

Figure 10.27
A small number of reactions can account for most of the chemical diversity observed in plant fatty acids. Desaturation and hydroxylation reactions can be catalyzed by enzymes that differ by as few as four amino acids. Recent evidence suggests that epoxidases and acetylenases are also closely related to desaturases. Additional chemical diversity may arise through the action of enzymes that methylate double bonds to produce cyclopropanes and cyclopropenes.

In other cases there does not appear to be a direct link between taxonomic relationships and the occurrence of unusual fatty acids. For example, lauric acid is prominent in two unrelated families, the Lauraceae (laurel family) and Arecaceae (palm family). Similarly, ricinoleic acid (see Fig. 10.25) has now been identified in 12 genera from 10 families. If any conclusion can be drawn from the taxonomic distribution of unusual fatty acids, it would be that the ability to synthesize some unusual fatty acids appears to have evolved several times independently, whereas for others it may have evolved only once. The studies of the desaturases and related enzymes described above provide an insight into how this may have happened, in at least some cases, by the accumulation of a small number of amino acid substitutions.

10.6.4 Unusual fatty acids occur almost exclusively in seed oils and may serve a defense function.

Given that the ability to synthesize various unusual fatty acids must have evolved independently, the common confinement of triacylglycerols to seed indicates some selective constraint or functional significance. One possible function of unusual fatty acids is that by being toxic or indigestible they protect the seed against herbivory. Some unusual fatty acids may be inherently toxic, such as the acetylenic fatty acids or some of their metabolites. Other unusual fatty acids are toxic upon catabolism by the herbivore, such as the 4-fluoro fatty acids of *Dichapetalum toxicarium*. Cyclopentenyl fatty acids were long used in the treatment of leprosy, and the activity of hydnocarpic acid against many *Mycobacterium* species has been demonstrated. The cyclopropenoid fatty acids also appear to have biological activities, possibly because of the accumulation in animal tissues of partial catabolites containing the cyclopropene ring, which inhibits β-oxidation of fatty acids. Malvalic and sterculic acids, produced by cotton plants, inhibit the growth of seed-eating lepidopteran larvae and may be part of a defense mechanism against these insects. These fatty acids may also be effective antifungal agents, inhibiting the growth of some plant pathogenic fungi at concentrations that appear biologically relevant.

From a dietary viewpoint, the most intensely studied of the unusual fatty acids is erucic acid (see Table 10.2), because of fears that the consumption of rapeseed oil might be detrimental to human health. Chronic feeding of erucic acid to experimental animals has a range of deleterious effects (Fig. 10.28), but whether these are sufficiently severe to propose a herbivore-defense role for erucic acid in seeds remains questionable.

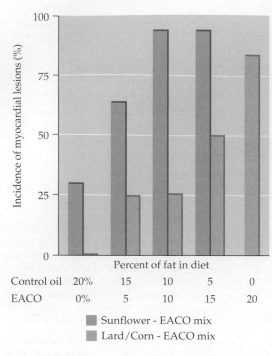

Figure 10.28
Relationship between the incidence of myocardial lesions in rats and the percentage of dietary fat derived from high erucic acid–containing oil (EACO). Shown are the ratios of EACO to control oil (either sunflower oil or a 3:1 [by wt.] lard:corn oil mixture). In both experiments, myocardial lesions increase dramatically with the proportion of erucic acid in the diet.

10.7 Synthesis of membrane lipids

In 1979, Grattan Roughan and Roger Slack first proposed that there are two distinct pathways for membrane synthesis in higher plants and named these the "prokaryotic pathway" and the "eukaryotic pathway." An abbreviated schematic diagram of the two pathways of glycerolipid synthesis in *Arabidopsis* leaves is shown in Figure 10.29. The prokaryotic pathway refers to the synthesis of lipids within the plastid. The eukaryotic pathway refers to the sequence of reactions involved in synthesis of lipids in the ER, transfer of some lipids between the ER and the plastid, and further modification of the lipids within the plastid.

Superimposed on the main pathways for glycerolipid synthesis are several additional pathways for the synthesis of other lipids, such as waxes, cutin, sphingolipids, and sterols. Relatively little is known about the cellular localization of these pathways.

Sphingolipid synthesis is likely to take place in the ER. Similarly, elongation of fatty acids for wax synthesis and production of oxygenated fatty acids for cutin synthesis are also thought to take place in the ER.

10.7.1 Phosphatidic acids formed in the plastids via the "prokaryotic pathway" and in the ER via the "eukaryotic pathway" differ in fatty acyl composition and position.

Glycerolipid synthesis involves a complex and highly regulated interaction between the chloroplast, where fatty acids are synthesized, and other membrane systems of the cell. The 16:0-, 18:0-, and $18:1^{\Delta 9}$-ACP products of plastid fatty acid synthesis may be either incorporated directly into chloroplast lipids by the plastid-localized prokaryotic pathway or exported to the cytoplasm as CoA esters, which are then incorporated into ER lipids by an independent set of eukaryotic pathway acyltransferases (see Fig. 10.29).

Figure 10.29 *(Facing page)*
Prokaryotic and eukaryotic pathways of glycerolipid synthesis. The prokaryotic pathway takes place in plastids and predominantly esterifies palmitate to the *sn*-2 position of lysophosphatidate (LPA). The eukaryotic pathway occurs outside the plastid, primarily in the ER, and results in C_{18} fatty acids being esterified to the *sn*-2 position of glycerolipids. In the prokaryotic pathway, acyl-ACP is condensed with glycerol 3-phosphate (G3P) by a soluble enzyme, G3P acyltransferase (1). The product, LPA, rapidly partitions into the membranes, where it is converted to phosphatidate (PA) by a membrane-localized LPA acyltransferase (2). PA is then converted to the other lipids found in chloroplasts. Most of the reactions of lipid synthesis by the prokaryotic pathway are thought to take place in the inner envelope of the plastids. The initial reactions of the eukaryotic pathway are similar except that acyl-CoA substrates are used and the G3P acyltransferase is thought to be associated with the ER. Lipids move from the ER to the other organelles, including the outer envelope of plastids. Eukaryotic lipids in the outer envelope are transferred into the inner membranes by an unknown mechanism and are there modified by the replacement of head groups and the action of desaturases. CDP-DAG, cytidine diphosphate-diacylglycerol; DAG, diacylglycerol; DGD, digalactosyldiacylglycerol; PC, phosphatidylcholine; PE, phosphatidylethanolamine; PG, phosphatidylglycerol; PI, phosphatidylinositol; PS, phosphatidylserine; MDG, monogalactosyldiacylglycerol; SQD, sulfoquinovosyldiacylglycerol.

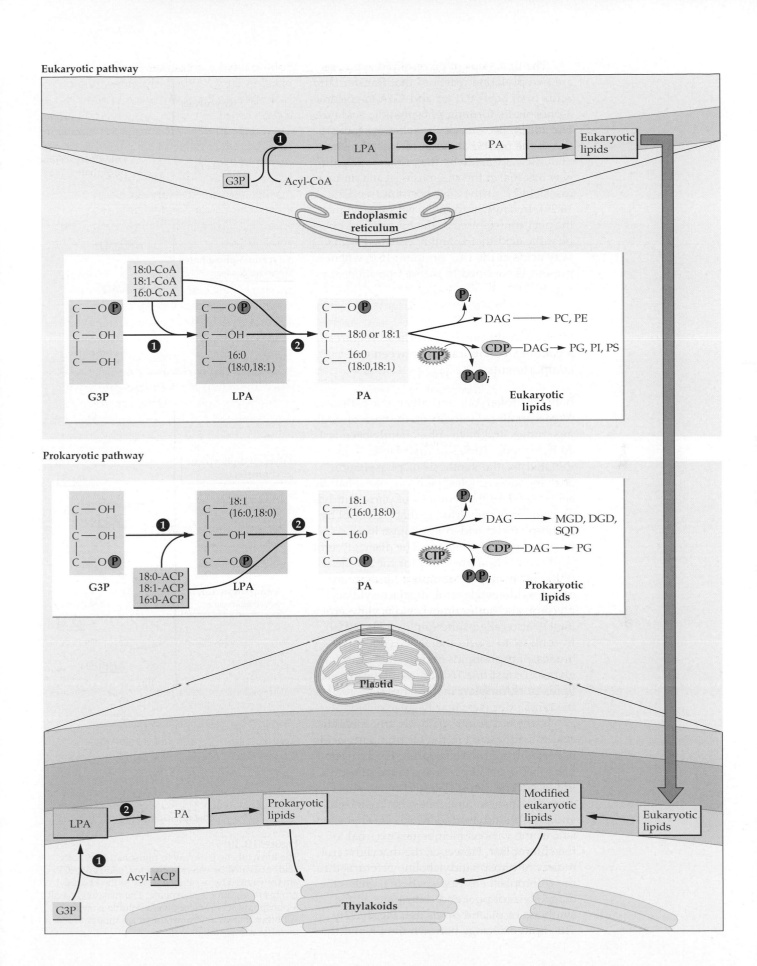

The first steps of glycerolipid synthesis are two acylation reactions that transfer fatty acids from acyl-ACP or acyl-CoA to glycerol 3-phosphate, forming phosphatidic acid (see Fig. 10.30). Because of the substrate specificities of the plastid acyltransferases, the phosphatidic acid made by the prokaryotic pathway has 16:0 at the *sn*-2 position and, in most cases, $18:1^{\Delta 9}$ at the *sn*-1 position (see Fig. 10.29). In contrast to the plastid isoenzymes, the acyl-transferases of the ER produce phosphatidic acid that is highly enriched with C_{18} fatty acids at the *sn*-2 position; 16:0, when present, is confined to the *sn*-1 position (see Fig. 10.29).

10.7.2 Membrane lipid synthesis requires a complex collaboration between cell compartments.

Both the prokaryotic and eukaryotic pathways were formulated initially on the basis of labeling studies and have subsequently been tested by genetic studies in *Arabidopsis*. Both are initiated by the synthesis of phosphatidic acid. In the chloroplast pathway, phosphatidic acid is used for the synthesis of phosphatidylglycerol or is converted to diacylglycerol by a phosphatidic acid–phosphatase located in the inner plastid envelope. The diacylglycerol pool can act as a precursor for the synthesis of the other major chloroplast lipids: monogalactosyldiacylglycerol; digalactosyldiacylglycerol; and sulfoquinovosyldiacylglycerol (SQD), also called sulfolipid (see Fig. 10.29). In contrast to most other eukaryotic membranes, phospholipids (mostly phosphatidylglycerol) constitute 16% or less of the glycerolipids in chloroplast membranes (Fig. 10.31); the remainder is primarily galactolipid.

In the eukaryotic pathway, phosphatidic acid is synthesized in the ER from fatty acids exported from the chloroplast. ER-derived phosphatidic acid gives rise to the phospholipids, such as phosphatidylcholine, phosphatidylethanolamine, phosphatidylinositol, and phosphatidylserine, which are characteristic of the various membranes external to the chloroplast. However, the diacylglycerol moiety of phosphatidylcholine is returned to the chloroplast envelope, where it enters the diacylglycerol pool and contributes to the synthesis of plastid lipids (see Fig. 10.21). This lipid exchange between the ER and the

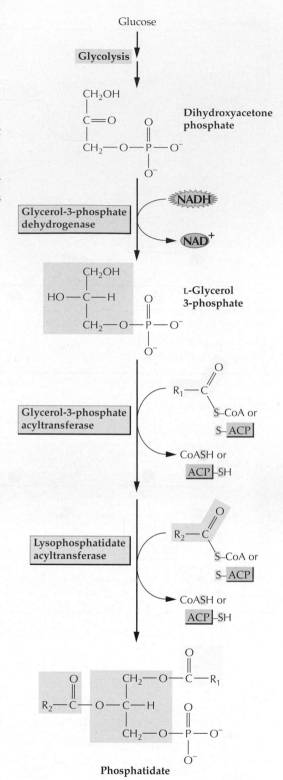

Figure 10.30
The biosynthetic pathway to phosphatidate. Fatty acids activated by esterification to CoASH or ACP are transferred by acyltransferases to the hydroxyl groups of glycerol 3-phosphate. The plastid glycerol-3-phosphate acyltransferase is a soluble enzyme. All other known acyltransferases are thought to be membrane-associated.

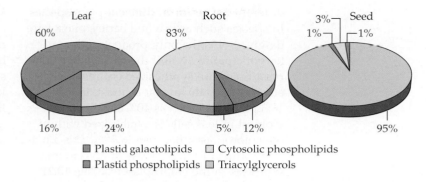

Figure 10.31
Glycerolipid compositions differ in different cell types. In leaves, where the most abundant membranes are chloroplast lamellae, galactolipids are the most abundant class. In roots, by contrast, most of the membranes are present in extraplastidic systems such as the endoplasmic reticulum and Golgi apparatus. Seeds of oilseed species, such as *Arabidopsis*, contain mostly triacylglycerols.

■ Plastid galactolipids □ Cytosolic phospholipids
■ Plastid phospholipids ▨ Triacylglycerols

chloroplast is reversible to some extent (see Section 10.7.4). The mechanism by which lipids move between the ER and chloroplast is not known. Some steps that have been hypothesized are shown in Figure 10.32. The eukaryotic pathway is the principal route of glycerolipid synthesis in all nonphotosynthetic tissues as well as in the photosynthetic tissues of many higher plants. Among the angiosperms, only so-called 16:3 plants, such as spinach and *Arabidopsis*, produce more than 10% of their glycerolipids by the prokaryotic pathway (see Section 10.1.2 and below).

10.7.3 The fatty acid composition of lipids can reveal their pathway of origin.

The relative amount of glycerolipid synthesized in plastids and the ER may vary in

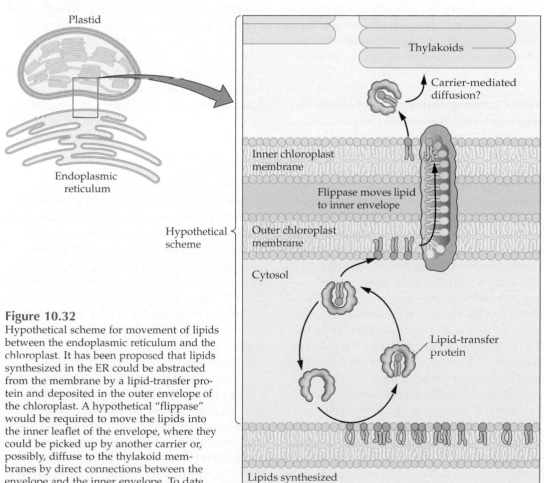

Figure 10.32
Hypothetical scheme for movement of lipids between the endoplasmic reticulum and the chloroplast. It has been proposed that lipids synthesized in the ER could be abstracted from the membrane by a lipid-transfer protein and deposited in the outer envelope of the chloroplast. A hypothetical "flippase" would be required to move the lipids into the inner leaflet of the envelope, where they could be picked up by another carrier or, possibly, diffuse to the thylakoid membranes by direct connections between the envelope and the inner envelope. To date, there is no direct evidence for any of the steps shown.

different tissues or in different plant species. In species such as pea and barley, phosphatidylglycerol is the only product of the prokaryotic pathway, and the remaining chloroplast lipids are synthesized entirely by the eukaryotic pathway. By contrast, in leaves of 16:3 plants as much as 40% of cellular glycerolipid in leaf cells is synthesized within the chloroplasts. Because $16:3^{\Delta7,10,13}$ is a major product of the pathway of glycerolipid synthesis in the chloroplast (see Fig. 10.21), the relative flux through the chloroplast prokaryotic pathway can be determined readily by the presence of this fatty acid (hence the name, 16:3 plants). Plants with very small amounts of $16:3^{\Delta7,10,13}$ are called 18:3 plants. However, this terminology has probably outlived its usefulness because at least some plants appear to vary the relative flux between the two pathways, depending on the growth temperature. The contribution of the eukaryotic pathway to monogalactosyldiacylglycerol, digalactosyldiacylglycerol, and sulfolipid synthesis is diminished in nonvascular plants, and in many green algae the chloroplast is almost entirely autonomous with respect to membrane lipid synthesis.

10.7.4 Large quantities of lipid appear to move between the ER and chloroplasts.

Each plant cell is autonomous with respect to glycerolipid synthesis, and there is no extensive transport of fatty acids or glycerolipids between cells. However, membrane biogenesis in plants involves a massive movement of fatty acids or lipids from one organelle to another (see Fig. 10.21). The quantitatively most significant movement of lipids is between the ER and the chloroplasts, which contain as much as 80% of leaf-cell glycerolipids. In epidermal cells, there is also substantial movement of wax and cutin monomers from the ER to the extracellular space. The mechanisms by which lipids move between the ER and the chloroplasts, or from the ER to the outside of the cell, are not known.

Several lines of evidence indicate the existence of regulatory mechanisms that coordinate the activity of the two pathways for glycerolipid synthesis in plants. The *act1* mutants of *Arabidopsis* are deficient in activity of chloroplast acyl-ACP:*sn*-glycerol-3-phosphate acyltransferase, the first enzyme of the prokaryotic pathway (see Fig. 10.21). This deficiency severely diminishes the amount of movement through the prokaryotic pathway but is compensated for by increased synthesis of chloroplast glycerolipids via the eukaryotic pathway (Fig. 10.33). These and related results indicate that even in the face of a major disruption of one of the pathways of glycerolipid synthesis, mechanisms exist to ensure the synthesis and transfer of enough glycerolipids to support normal rates of membrane biogenesis. This striking example raises many unanswered questions. In particular, how is the demand for increased glycerolipid synthesis communicated to the lipid biosynthetic pathways during membrane expansion?

A partial answer to this question lies in the recognition that, at least for the ER and the plastid, lipid moves in both directions between the membranes. Detailed analysis of the effects of various *Arabidopsis fad* mutations on the fatty acid composition of membranes indicates that even though each enzyme is located exclusively in one compartment or the other, most of the mutations affect the composition of both chloroplast and extrachloroplast membranes (see Fig. 10.21). Thus, lipids must travel to and from chloroplast membranes. Because most of the chloroplast glycerolipids are not found in other membranes, these lipids must be converted to some other form (e.g., phosphatidate, diacylglycerol, or phosphatidylcholine) before or immediately after transfer. Similarly, phosphatidylcholine, the principal glycerolipid in the extrachloroplast membranes, occurs only in the outer leaflet of the chloroplast envelope. Furthermore, chloroplasts completely lack phosphatidylethanolamine, the other major component of the ER. Thus, phosphatidylcholine or another lipid is probably converted to one of the common chloroplast lipids in the chloroplast envelope before being transferred to the inner membranes of the chloroplast.

A confounding problem is that no conclusive evidence exists for vesicular traffic between the ER and the plastids. Thus, there has been sustained interest in the possible role of lipid-transfer proteins as carriers of lipids between the various membranes (see Fig. 10.32). These small proteins are defined by their ability to catalyze the in vitro exchange of lipids between membranes.

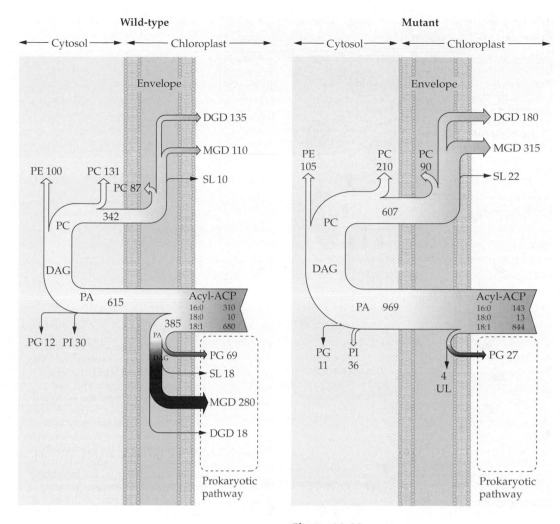

Wild-type

← Cytosol → ← Chloroplast →

Envelope

DGD 135
MGD 110
SL 10

PE 100   PC 131
          PC 87
          342

PC

DAG

PA   615

Acyl-ACP
16:0   310
18:0   10
18:1   680

385
PA

PG 12 PI 30

DAG

PG 69
SL 18
MGD 280
DGD 18

Prokaryotic pathway

Mutant

← Cytosol → ← Chloroplast →

Envelope

PE   PC   PC
105   210   90

DGD 180
MGD 315
SL 22

607

PC

DAG

PA   969

Acyl-ACP
16:0   143
18:0   13
18:1   844

PG   PI
11   36

4
UL

PG 27

Prokaryotic pathway

Figure 10.33

Flux diagrams for wild-type *Arabidopsis* and *act1* mutant, showing how fatty acids synthesized in chloroplasts are directed to different lipids in the chloroplasts and extraplastidic membranes. The diagram shows the average fate of 1000 molecules of fatty acid. In wild-type *Arabidopsis* approximately 39% of the fatty acids are used for lipid synthesis within the chloroplast. Of the remaining fatty acids, about 27% are used for synthesis of lipids for extrachloroplast membranes such as the ER, Golgi, and plasma membrane, and roughly 34% are transferred to the chloroplast for lipid synthesis. In a mutant of *Arabidopsis* deficient in plastid glycerol-3-phosphate acyltransferase activity, most of the fatty acids are exported to the cytoplasm and then approximately 61% are reimported into the chloroplast. This suggests a mechanism that adjusts the flow of lipid between the membranes when necessary. SL, sulfolipid; UL, unknown lipid; other abbreviations as in Figure 10.29.

However, the plant lipid-transfer proteins that were originally identified by in vitro assays apparently are located outside the plasma membrane in the cell wall space of epidermal cells. The function of these proteins is not known but, based on their location, they may play a role in transferring wax and cutin monomers from the plasma membrane to the cell surface (Fig. 10.34).

10.7.5 During de novo glycerolipid synthesis, the change in free energy that drives attachment of the polar head group is provided by nucleotide activation of either diacylglycerol or the head group itself.

A general reaction common to all phospholipid biosynthesis pathways involves the attachment of the phospholipid head group. In one case, a hydroxyl located on the head group carries out a nucleophilic attack on the β-phosphate group of CDP-diacylglycerol. In the other, the *sn*-3 hydroxyl of diacylglycerol attacks the β-phosphate of a CDP-activated

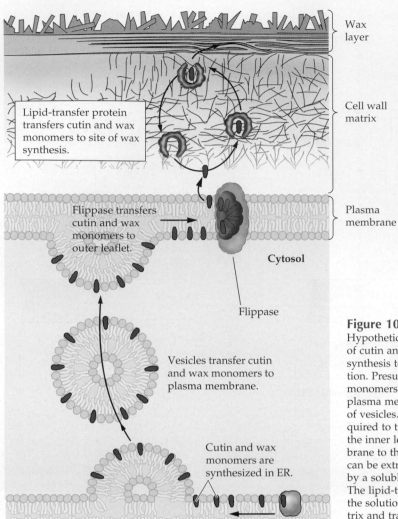

Wax layer

Cell wall matrix

Plasma membrane

Lipid-transfer protein transfers cutin and wax monomers to site of wax synthesis.

Flippase transfers cutin and wax monomers to outer leaflet.

Cytosol

Flippase

Vesicles transfer cutin and wax monomers to plasma membrane.

Cutin and wax monomers are synthesized in ER.

Figure 10.34
Hypothetical scheme for the transport of cutin and wax monomers for cutin synthesis to the site of polymerization. Presumably, cutin and wax monomers move from the ER to the plasma membrane in the membranes of vesicles. A "flippase" may be required to transfer the monomers from the inner leaflet of the plasma membrane to the outer leaflet, where they can be extracted from the membrane by a soluble lipid-transfer protein. The lipid-transfer protein diffuses in the solution phase of the cell wall matrix and transfers the monomers to enzymes that polymerize them near the surface of epidermal cells.

head group (Fig. 10.35). Thus, phospholipid biosynthesis can be divided into two general types of pathways, referred to as the CDP-diacylglycerol and the diacylglycerol pathways, distinguished by whether the activated substrate is CDP-diacylglycerol or a CDP-head group.

Unlike prokaryotes, in which only the CDP-diacylglycerol pathway is present, plants, yeast, and animals synthesize phospholipids by using both the CDP-diacylglycerol and the diacylglycerol pathways, but the contributions of each pathway vary. For example, yeasts can synthesize all their phospholipids by using the CDP-diacylglycerol pathways, but under some conditions, they can repress a CDP-diacylglycerol pathway and use diacylglycerol instead. Animals make their major phospholipids—phosphatidylcholine and

phosphatidylethanolamine—by using diacylglycerol pathways. The details of the synthesis of phospholipids in plants are not well characterized, and much of what is known has been learned by analogy to other organisms. Many of the pathways present in animals and yeast, for example, also are present in plants, although the extent to which each pathway contributes to the final phospholipid components of a cell may vary among organisms.

10.7.6 Phosphatidate is a substrate for both the CDP-diacylglycerol and the diacylglycerol pathways.

CDP-diacylglycerol is synthesized from phosphatidate and CTP by CDP:diacylglycerol

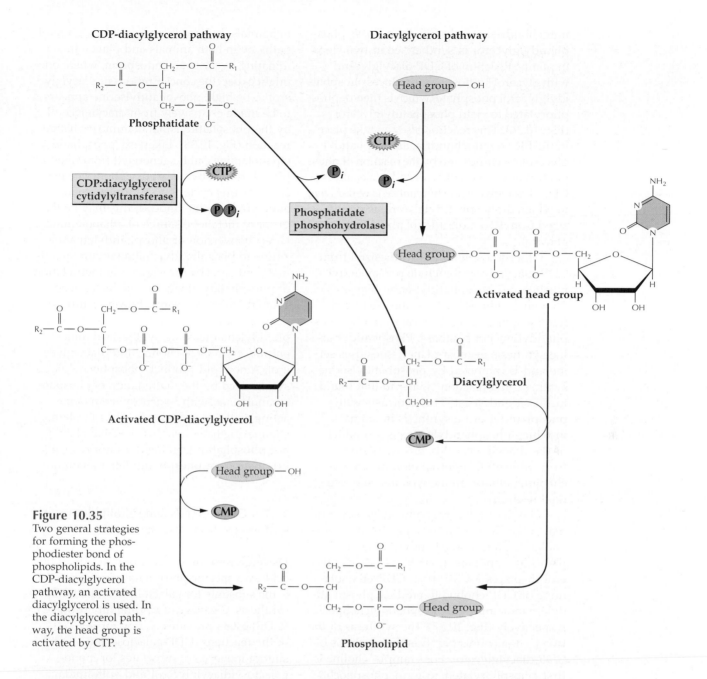

Figure 10.35
Two general strategies for forming the phosphodiester bond of phospholipids. In the CDP-diacylglycerol pathway, an activated diacylglycerol is used. In the diacylglycerol pathway, the head group is activated by CTP.

cytidylyltransferase (Fig. 10.35). In plants, this membrane-bound enzyme is associated with several different organelles, being localized in the inner chloroplast envelope and also found in the inner mitochondrial envelope. Two CDP:diacylglycerol cytidylyltransferases occur in castor bean endosperm: one localized in the mitochondria, the other in a microsomal fraction. Diacylglycerol is produced by the dephosphorylation of phosphatidate by phosphatidate phosphohydrolase, an enzyme in the inner chloroplast envelope membranes, microsomes, and soluble fractions. Analogy to yeast phospholipid

synthesis makes it likely that a mitochondrial form of this enzyme also exists, although it has not yet been detected in plants.

10.7.7 CDP-diacylglycerol and diacylglycerol pathways generate distinct types of lipids.

Phospholipids derived from CDP-diacylglycerol include phosphatidylglycerol, the only phospholipid found in chloroplast thylakoids, and diphosphatidylglycerol (cardiolipin), which occurs exclusively in the inner mitochondrial membrane. As in *E. coli*, phosphatidylglycerol is synthesized in two steps in plants. Reaction of CDP-diacylglycerol with glycerol 3-phosphate produces phosphatidylglycerol phosphate, which is then dephosphorylated to yield phosphatidylglycerol (Fig. 10.36). This reaction also can take place in the ER. In mitochondria, diphosphatidylglycerol is synthesized by the reaction of phosphatidylglycerol with a second molecule of CDP-diacylglycerol. This contrasts with *E. coli*, in which diphosphatidylglycerol is synthesized from two molecules of phosphatidylglycerol.

Other phospholipids synthesized from CDP-diacylglycerol include phosphatidylinositol and phosphatidylserine (Fig. 10.36). Phosphatidylinositol is synthesized from free inositol in a reaction catalyzed by phosphatidylinositol synthase. Phosphatidylserine synthesis in most plant tissues uses serine and is catalyzed by phosphatidylserine synthase in a reaction similar to that found in *E. coli* and yeast. This contrasts with phosphatidylserine synthesis in animals, in which phosphatidylserine is a product of the diacylglycerol pathway, arising from phosphatidylethanolamine by exchanging ethanolamine with a serine (see next section).

The diacylglycerol pathway is used primarily for the synthesis of phosphatidylethanolamine and phosphatidylcholine in both plants and animals. In each case, diacylglycerol displaces a CMP from CDP-ethanolamine or CDP-choline to produce phosphatidylethanolamine or phosphatidylcholine, respectively (Fig. 10.37). The synthesis of the two CDP-alcohols parallels the synthesis of CDP-diacylglycerol. For example, choline is first phosphorylated to form phosphocholine before reacting with CTP to produce CDP-choline.

10.7.8 Pools of CDP-diacylglycerol–derived phospholipids and diacylglycerol-derived phospholipids interact in plants and in animals.

Tracing the origin of specific lipids is often more difficult than the preceding section might suggest. For example, phosphatidylethanolamine, a phosphatidylserine precursor in animals, is a product of the diacylglycerol pathway in both animals and plants. In germinating castor bean endosperm, where one might expect the concentrations of diacylglycerol to be high, phosphatidylserine appears to be made exclusively from diacylglycerol by the phosphatidylethanolamine exchange reaction (Fig. 10.38). Likewise, phosphatidylethanolamine can be generated from phosphatidylserine by decarboxylation. The net effect of this cycle is to produce ethanolamine from serine. Indeed, this may be the primary metabolic source of ethanolamine. Direct conversion of phosphatidylethanolamine to phosphatidylcholine occurs in animals and yeast by the sequential methylation of phosphatidylethanolamine by *S*-adenosylmethionine. In plants, however, this appears to be only a minor pathway, phosphatidylcholine being synthesized almost entirely from CDP-choline and diacylglycerol. A nonlipid product, phosphocholine, is synthesized by the methylation of phosphoethanolamine, with *S*-adenosylmethionine acting as the methyl donor. Thus in plants, choline originates mainly from ethanolamine, but phosphatidylcholine does not typically originate from phosphatidylethanolamine.

10.7.9 Galactolipids and sulfolipids are synthesized from diacylglycerol.

Diacylglycerol produced from phosphatidic acid by a specific phosphatase in the plastid is the substrate for galactolipid and sulfolipid synthesis. Cleavage of uridine diphosphate (UDP) esters provides energy for several of the reactions. UDP-galactose and UDP-sulfoquinovose are substrates for monogalactosyldiacylglycerol and sulfolipid synthesis, respectively. The synthesis of digalactosyldiacylglycerol involves the transfer of the galactose head group from one monogalactosyldiacylglycerol molecule to a second by the action of a monogalactosyldiacylglycerol dismutase or galactolipid galactosyltransferase (Fig. 10.39).

Sulfolipid is of interest because relatively little is known about how the carbon–sulfur bond forms. From in vitro assays and from analyzing mutants of the photosynthetic purple bacterium *Rhodobacter sphaeroides,* the final step in the synthesis

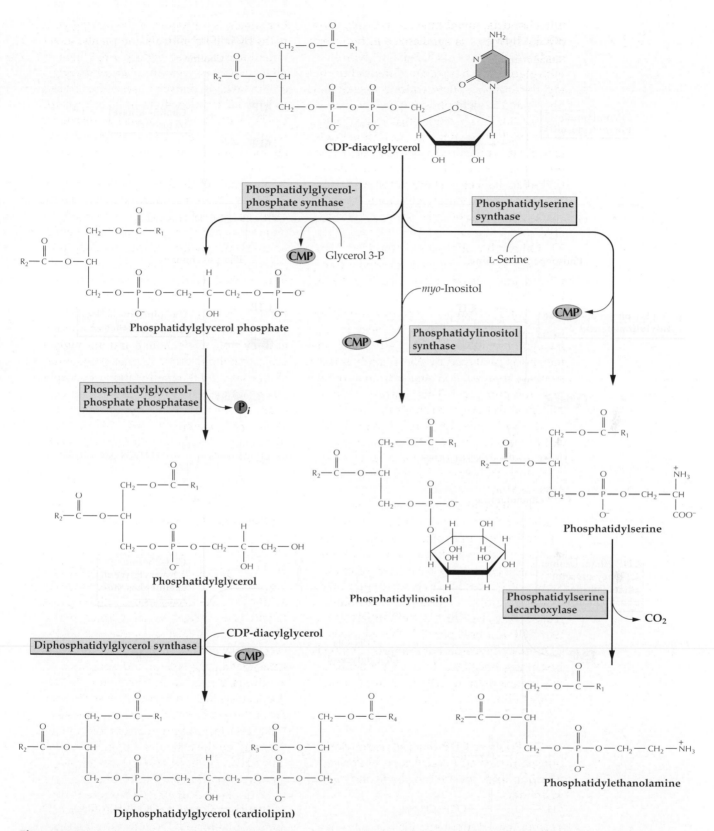

Figure 10.36
Synthesis of phospholipids by the CDP-diacylglycerol pathway. Phosphatidylserine, phosphatidylinositol, and phosphatidylglycerol phosphate are synthesized when the head group displaces CMP from CDP-diacylglycerol. Phosphatidylglycerol phosphate is dephosphorylated to phosphatidylglycerol. Diphosphatidylglycerol (cardiolipin) is formed from two molecules of phosphatidylglycerol.

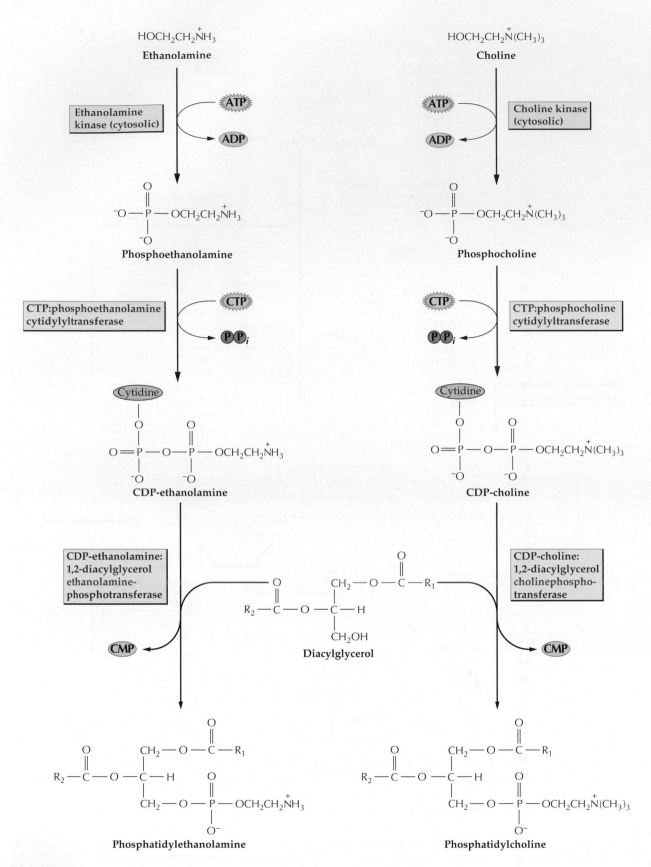

Figure 10.37
The diacylglycerol pathway is the principal route for synthesis of phosphatidylcholine and phosphatidylethanolamine in plants.

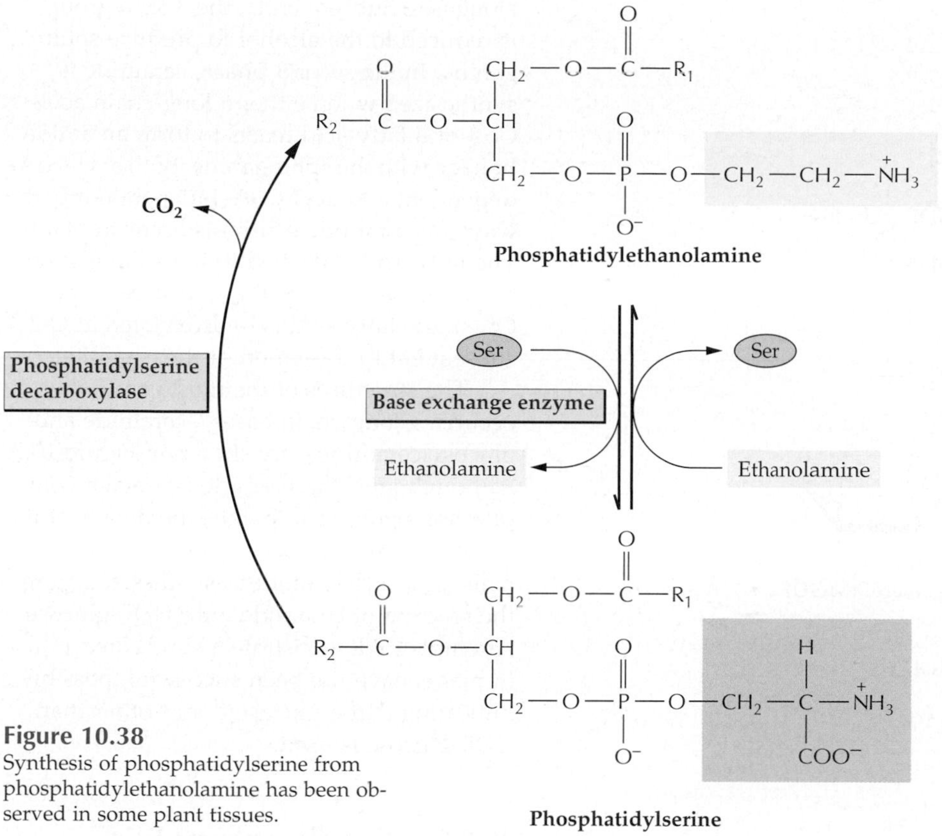

O
‖
CH₂—O—C—R₁

O
‖
R₂—C—O—CH   O
                   ‖
CO₂ ←   CH₂—O—P—O—CH₂—CH₂—$\overset{+}{N}H_3$
                   |
                   O⁻

Phosphatidylethanolamine

Phosphatidylserine decarboxylase

Ser → Ser

Base exchange enzyme

Ethanolamine ← Ethanolamine

O
‖
CH₂—O—C—R₁

O
‖
R₂—C—O—CH   O   H
                   ‖ |
                   CH₂—O—P—O—CH₂—C—$\overset{+}{N}H_3$
                   | |
                   O⁻   COO⁻

Phosphatidylserine

Figure 10.38
Synthesis of phosphatidylserine from phosphatidylethanolamine has been observed in some plant tissues.

of sulfoquinovosyldiacylglycerol, the plant sulfolipid, is the transfer of the sulfoquinovosyl group from UDP-sulfoquinovose to diacylglycerol (Fig. 10.40). UDP-sulfoquinovose could be formed by addition of sulfite to either UDP-4-ketoglucose-5-ene or UDP-glucose-5-ene, followed by reduction. These intermediates can, in principle, be formed by enzymes such as deoxythymidine diphosphate–glucose-4,6-dehydratase. Four genes required for sulfolipid synthesis in *R. sphaeroides* have been cloned and should be useful in testing hypotheses about the early steps of the pathway. Recently, sequence information about the *R. sphaeroides* genes has been used to identify homologs in cyanobacteria and higher plants. Thus, the pathway is probably similar or identical in bacteria and higher plants.

All organisms that carry out oxygenic photosynthesis contain sulfolipid as a major component of their photosynthetic lamellae—a fact stimulating sustained interest in the possibility that this lipid plays some essential role in supporting the light reactions of photosynthesis. However, mutants of

R. sphaeroides or cyanobacteria that are unable to synthesize sulfolipid have normal rates of photosynthesis. Apparently, the main reason photosynthetic organisms have sulfolipid is to minimize the phosphate required for synthesizing the large amounts of membrane required to support high rates of light capture (Box 10.4).

10.7.10 The limited information on sphingolipid biosynthesis in plants is derived mostly by analogy to animal systems.

Just as diacylglycerol is the basic component of all glycerolipids, ceramide is the basic component of all sphingolipids. Complex sphingolipids, such as the glucosylsphingolipids that predominate in plants, are formed by modification of ceramide.

The initial phase of ceramide synthesis involves first condensation of a palmitoyl-CoA with serine, resulting in a 3-ketosphinganine (Fig. 10.41). This reaction is analogous to the condensation reactions catalyzed by FASs and

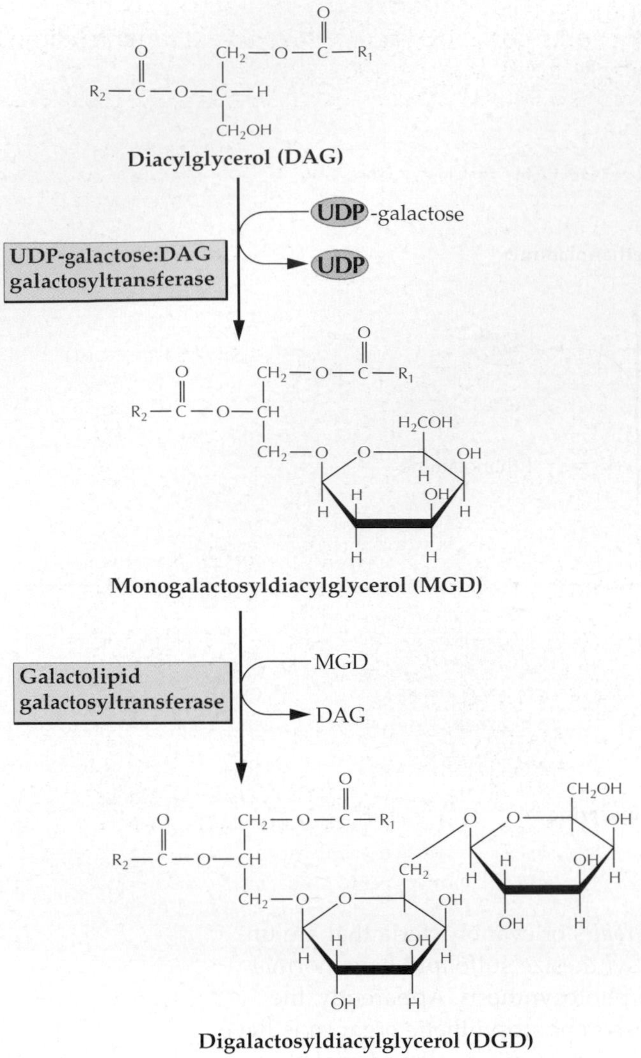

Figure 10.39
Pathway of synthesis of galactolipids. The synthesis of monogalactosyldiacylglycerol is catalyzed by what appears to be a typical glycosyltransferase that utilized a nucleotide sugar-activated substrate. In contrast, the synthesis of digalactosyldiacylglycerol is thought to utilize a dismutation reaction in which the galactosyl moiety of one monogalactosyldiacylglycerol is transferred to another.

elongases. Subsequently, the 3-keto group is reduced to the alcohol to produce sphinganine. In the second phase, ceramide is synthesized when either a long-chain acyl-CoA or a fatty acid reacts to form an amide linkage with the sphinganine. Both acyl-CoA–dependent and acyl-CoA–independent pathways for ceramide synthesis occur in plants. The fatty acids attached to the sphinganine can range from C_{16} to C_{26} in different species. Often, the fatty acid is hydroxylated at C-2 subsequent to formation of the ceramide.

The structures of the most commonly occurring long-chain bases—ceramide and glucosylceramide—are shown in Figure 10.7. Although glucosylceramide is a major complex sphingolipid in plants, the details of its synthesis have not been elucidated. In animals, glucosylceramide is synthesized from the reaction of ceramide with UDP-glucose. Attempts to demonstrate a similar reaction in plants have not been successful, possibly indicating that a glucose donor other than UDP-glucose is used.

10.8 Function of membrane lipids

10.8.1 Membrane lipid composition affects plant form and function.

Each membrane in a plant cell has a characteristic and distinct complement of lipid types (Fig. 10.42), and within a single membrane each class of lipids has a distinct fatty

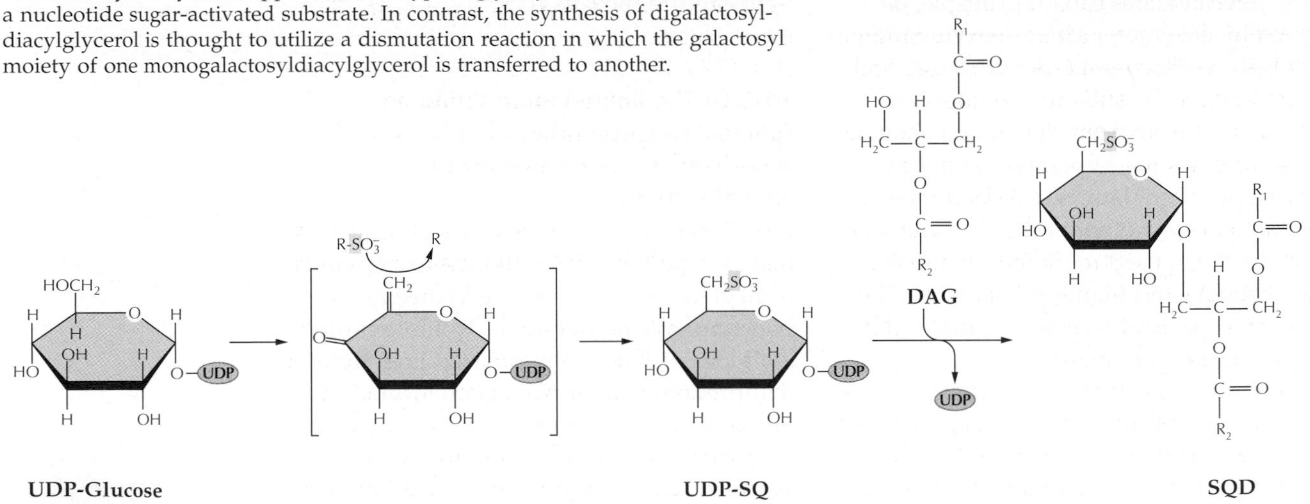

Figure 10.40
Hypothetical scheme for the synthesis of sulfoquinovosyldiacylglycerol (SQD). The identity of the sulfur donor is not known. Only the final step of the pathway, the condensation of UDP-sulfoquinovose (UDP-SQ) with DAG, is known with certainty.

Box 10.4

Plants conserve phosphate by using sulfolipids and galactolipids for chloroplast membrane synthesis.

Each gram of leaf material contains approximately 1 m² of chloroplast membrane. If this membrane were composed primarily of phospholipids, as the membranes from animals and fungi are, plants would require much more phosphate for growth than they currently do. Given that phosphate is a limiting nutrient in many natural ecosystems, it is advantageous to plants to minimize the need for phosphate in membrane synthesis. In what appears to be an evolutionary adaptation to the problem of phosphate limitation, plants and other photosynthetic organisms use monogalactosyldiacylglycerol (MGD), digalactosyldiacylglycerol (DGD), and sulfoquinovosyldiacylglycerol (SQD, or sulfolipid) as substitutes for phospholipids such as phosphatidylglycerol (PG). For example, in the chloroplast membranes of spinach (a 16:3 plant), less than 15% of the bilayer lipids are phospholipids (see graph).

Evidence in support of this concept was obtained by the isolation of sulfolipid-lacking mutants of the photosynthetic purple bacterium *Rhodobacter sphaeroides* and the cyanobacterium *Synechococcus* PCC7942. When the mutants were grown in the presence of high concentrations of phosphate, their growth rate and photosynthetic characteristics were indistinguishable from those of the wild type. Under conditions of limiting phosphate, however, the growth of the mutants was inhibited. In addition, as the amount of phosphate available for membrane synthesis became limiting, several novel glycolipids accumulated to high concentrations.

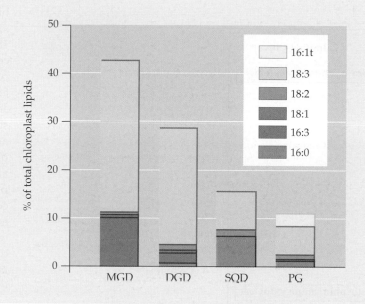

acid composition. Until recently, we knew almost nothing about the reasons for this remarkable diversity. Our knowledge remains limited, but the examples discussed in this section demonstrate some of the important developmental processes and responses to environmental factors that are affected by membrane lipid composition.

One of the most visually dramatic phenotypes produced by a change in lipid composition is that of the *fab2* mutant of *Arabidopsis*. The *fab2* plants are miniatures, a result of the accumulation of 18:0 in membrane lipids (Fig. 10.43A). The decrease in leaf size results from decreased size of several specific cell types. The failure of mesophyll and epidermal cells to enlarge produces a dramatic "brick-wall" appearance to the cross-section of a mutant leaf, contrasting with the characteristic less-compact leaf anatomy evident in the wild type (Fig. 10.43B,C). We do not know the mechanism by which increased 18:0 produces the miniature phenotype, but biophysical principles lead to the expectation that increased temperature would ameliorate the effects of increased saturation on the bilayer. Accordingly, the morphology of *fab2* more closely resembles that of the wild type when plants are grown at 35°C (Fig. 10.43D,E). Furthermore, at this temperature, *fab2* leaves develop typical palisade and spongy mesophyll layers.

10.8.2 Photosynthesis is impaired in plants that lack polyunsaturated membrane lipids.

Highly unsaturated $18:3^{\Delta 9,12,15}$ and $16:3^{\Delta 7,10,13}$ fatty acids account for approximately 70% of all the thylakoid membrane fatty acids in plants and more than 90% of the fatty acids in monogalactosyldiacylglycerol, the most abundant chloroplast lipid (see Fig. 10.42). These very high amounts are noteworthy because free radical byproducts of the photosynthetic light reactions stimulate oxidation of polyunsaturated fatty acids. If plants maintain high levels of unsaturation in thylakoids despite the risk of oxidation, one might reason that photosynthesis is critically dependent on membrane unsaturation. Surprisingly, an *Arabidopsis* triple mutant (*fad3 fad7 fad8*), completely lacking $18:3^{\Delta 9,12,15}$ and $16:3^{\Delta 7,10,13}$, exhibits normal rates of vegetative growth and photosynthesis at 22°C (Fig. 10.44). These results clearly demonstrate that $18:3^{\Delta 9,12,15}$ and $16:3^{\Delta 7,10,13}$ are not essential for photosynthesis in plants. These two fatty acids are not irrelevant, however: Conservation of thylakoid composition attests to their importance, but their role is more subtle than expected. In fact, the loss of $18:3^{\Delta 9,12,15}$ and $16:3^{\Delta 7,10,13}$ has noticeable effects on photosynthesis only at low (below 10°C) and high (above 30°C) temperatures.

Although eliminating triunsaturated $18:3^{\Delta 9,12,15}$ and $16:3^{\Delta 7,10,13}$ has only minor consequences for photosynthesis, the process is greatly affected in an *Arabidopsis fad2 fad6* mutant lacking the diunsaturated fatty acids $18:2^{\Delta 9,12}$ and $16:2^{\Delta 7,10}$ and their downstream triunsaturated derivatives $18:3^{\Delta 9,12,15}$ and $16:3^{\Delta 7,10,13}$. These mutants lose nearly all photosynthetic capacity and cannot grow autotrophically. Nevertheless, *fad2 fad6* plants grow on sucrose media, under which conditions growth and organ development are remarkably normal (Fig. 10.44). These observations indicate that the vast majority of receptor-mediated and transport-related membrane functions required to sustain

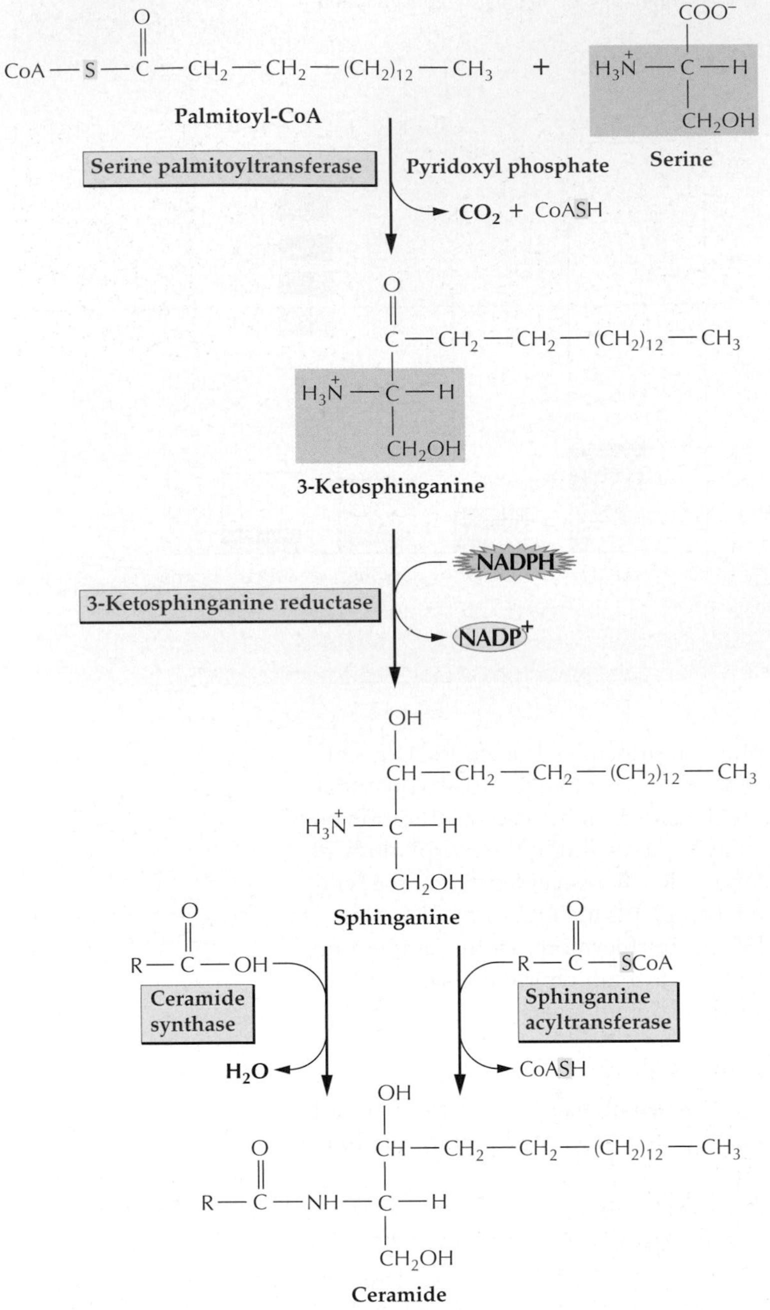

Figure 10.41
Tentative pathway for ceramide biosynthesis in plants, including the two mechanisms for amide bond formation found in plant membrane preparations. The acyl chain (designated R in the figure) may contain a hydroxyl group at the C-2 and the sphingosine base may contain hydroxyl groups or double bonds at C-4 and C-8.

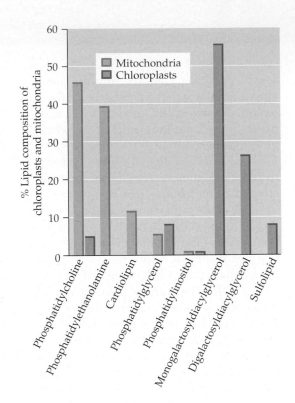

Figure 10.42
Comparison of the lipid compositions of chloroplasts and mitochondria.

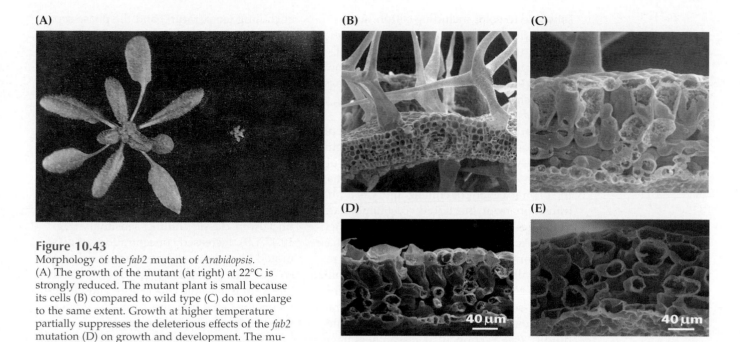

(A)

(B)

(C)

(D)

(E)

Figure 10.43
Morphology of the *fab2* mutant of *Arabidopsis*.
(A) The growth of the mutant (at right) at 22°C is
strongly reduced. The mutant plant is small because
its cells (B) compared to wild type (C) do not enlarge
to the same extent. Growth at higher temperature
partially suppresses the deleterious effects of the *fab2*
mutation (D) on growth and development. The mutant grown at 36°C more closely resembles the wild-
type control (E).

the organism and to induce proper development are well supported in this double
mutant, which contains almost no polyunsaturated lipids. Apparently, therefore, photosynthesis is the only vegetative cell function that requires high levels of membrane
polyunsaturation.

The work with *Arabidopsis* mutants indicates that plants require polyunsaturated
lipids to maintain the photosynthetic machinery, but this is not true of cyanobacteria.
Mutants of *Synechocystis* PCC6803 that lack
polyunsaturated fatty acids can photosynthesize normally except at low temperature.

10.8.3 Does lipid composition affect chilling sensitivity?

One of the most extensively studied issues
in membrane biology is the relationship between lipid composition and the ability of organisms to adjust to temperature changes.
Chilling-sensitive plants undergo sharp reductions in growth rate and development at
temperatures between 0°C and 12°C (Fig.
10.45). **Chilling injury** includes physical and
physiological changes induced by low temperatures as well as subsequent symptoms of
stress. Many economically important crops are

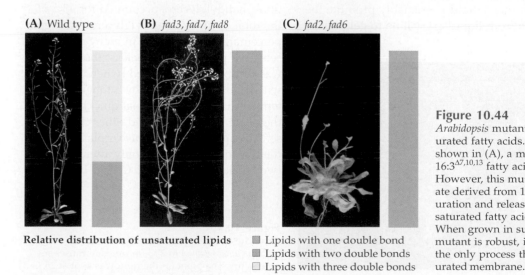

(A) Wild type   **(B)** *fad3, fad7, fad8*   **(C)** *fad2, fad6*

Relative distribution of unsaturated lipids

☐ Lipids with one double bond
☐ Lipids with two double bonds
☐ Lipids with three double bonds

Figure 10.44
Arabidopsis mutants reveal the roles of polyunsaturated fatty acids. Compared with a wild-type plant
shown in (A), a mutant lacking $18:3^{\Delta9,12,15}$ and
$16:3^{\Delta7,10,13}$ fatty acids (B) grows normally at 22°C.
However, this mutant is male-sterile because jasmonate derived from $18:3^{\Delta9,12,15}$ is required for pollen maturation and release. A mutant deficient in all polyunsaturated fatty acids cannot grow autotrophically.
When grown in sucrose-enriched medium (C), this
mutant is robust, indicating that photosynthesis is
the only process that absolutely requires a polyunsaturated membrane.

sensitive to cold, including cotton, soybean, maize, rice, and many tropical and subtropical fruits. In contrast, most plants of temperate origin, which continue to grow and develop at low temperatures, are classified as chilling-resistant plants, including *Arabidopsis*.

In attempts to link the biochemical and physiological changes that characterize chilling injury with a single "trigger" or site of damage, investigators have suggested that the primary event of chilling injury is a phase transition from the liquid crystalline state to the gel state in cellular membranes (Fig. 10.46). According to this proposal, the phase transition from liquid crystalline to gel results in alterations in the metabolism of chilled cells and leads to injury and death of the chilling-sensitive plants. Because desaturation of membrane lipids favors membrane fluidity (see Section 10.5), researchers have sought to define the relationship between membrane composition and chilling sensitivity. A related hypothesis specific to chloroplast membranes has been proposed, in which molecular species of chloroplast phosphatidylglycerol containing a combination of saturated fatty acid (16:0 and 18:0) at the *sn*-1 position of the glycerol backbone and either a saturated fatty acid or $16:1^{\Delta 3}$-*trans* fatty acid at the *sn*-2 position are suggested to confer chilling sensitivity on plants. Because the trans-double bond leaves the $16:1^{\Delta 3}$-*trans* fatty acid with a structure similar to 16:0 (see illustration in Box 10.1), these molecules are termed disaturated phosphatidylglycerol. The presence of significant quantities of disaturated phosphatidylglycerol in the chloroplast membranes would presumably promote the change from liquid crystalline to gel phase at chilling temperature, and the phase separation within the membranes would cause chilling sensitivity.

As shown in studies with five different *Arabidopsis* mutants, diminished unsaturation resulted in plants that grew well at 22°C but were less robust than wild type when grown at 2°C to 5°C. These results were observed even though the lipid changes in most of the mutants were insufficient to cause a lipid-phase transition. In addition, the low-temperature symptoms that developed in these lines appeared to be quite distinct from classic chilling sensitivity (Fig. 10.47A,B). Increased concentrations of disaturated phosphatidylglycerol (to as much as 60% of total phosphatidylglycerol) were obtained in transgenic *Arabidopsis*, and in these plants the damaging effects of low temperature became evident more quickly.

A complementary series of experiments was carried out in tobacco, a chilling-sensitive plant. Transgenic expression of exogenous genes was used to specifically decrease the concentrations of disaturated phosphatidylglycerol or to bring about a general increase in membrane unsaturation, and the damage caused by chilling was alleviated to some extent. These findings indicate that the extent of membrane unsaturation or the presence of particular lipids such as disaturated phosphatidylglycerol can affect the low-temperature responses of plants.

However, some recent research results indicate that the relationship between membrane unsaturation and plant temperature responses is subtle and complex. In one mutant, *fab1*, disaturated molecular species of phosphatidylglycerol accounted for 43% of the total leaf phosphatidylglycerol—a higher percentage than is found in many chilling-sensitive plants. Nevertheless, the mutant was completely unaffected (when compared with wild-type controls) by a range of low-temperature treatments that quickly led to the death of cucumber and other chilling-sensitive plants. Growth of *fab1* plants slowed, relative to the wild type, only after more than two weeks' exposure to 2°C (Fig. 10.47C,D).

Figure 10.45
Exposure of chilling-sensitive plants such as cucumber *(Cucumis sativus)* to 2°C for one day causes severe injury. The plant on the right was kept at 25°C.

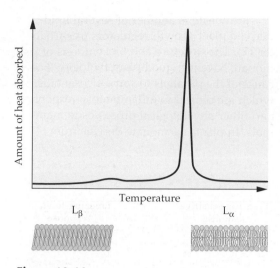

Figure 10.46
Thermal transition from the gel phase (L_β) to liquid crystalline phase (L_α) in a pure phosphatidylcholine bilayer (lower panel). At low temperatures, motion of the fatty acid chains is limited by Van der Waals forces. As the temperature is raised through the phase transition, heat is absorbed (upper panel) and Van der Waals forces are disrupted to form a bilayer with melted fatty acid chains (L_α).

10.8.4 Membrane lipid composition can influence plant cell responses to freezing.

Freezing stress in plants differs from chilling stress, and tolerance of freezing requires a specialized set of biological mechanisms. During initial freezing of plant tissue (down to –5°C), ice forms outside the plasma membrane. Because solutes are excluded from the ice, the solute concentration in the remaining extracellular aqueous phase increases, forcing water out through the plasma membrane by osmosis, and plasmolyzing the cell. When the temperature later increases and the ice melts, cellular damage may result.

If plants are allowed to acclimate to cold by growth at low, but not freezing, temperatures for a few days before exposure to freezing, many can withstand freezing to temperatures that would otherwise cause extensive damage or death (Fig. 10.48). Protoplasts from rye leaves have been used as a model system to investigate the molecular basis for this acclimation response (Fig. 10.49).

When nonacclimated protoplasts are placed in a hyperosmotic medium, the plasma membrane buds off endocytotic vesicles, as shown in Figure 10.49. However, if the protoplasts are first preincubated with monounsaturated or diunsaturated species of phosphatidylcholine, so that the phospholipid is incorporated into the plasma membrane, hyperosomotic treatment results in the formation of exocytotic extrusions. Disaturated species of phosphatidylcholine do not induce this change. These differences in plasma membrane behavior correlate with protoplast survival during freezing: Preincubation of nonacclimated protoplasts with monounsaturated or diunsaturated phosphatidylcholine is as effective as cold acclimation in promoting protoplast survival, whereas preincubation with disaturated phosphatidylcholine has no beneficial effect. These observations suggest that one

Figure 10.47
Three different chilling responses in lipid mutants of *Arabidopsis*. (A) Compared with wild-type plants (left), the *Arabidopsis fad6* mutant (right) becomes chlorotic after three weeks at 5°C. (B) *fad2 Arabidopsis* plants die after seven weeks at 6°C. (C) Compared with wild-type plants (left), the *fab1* mutant (right) is unaffected by up to one week of exposure to chilling at 2°C. (D) After four weeks at 2°C, however, *fab1* plants (right) show clear symptoms of chlorosis and reduced growth.

aspect of cold acclimation is an increase in unsaturated phospholipids in the plasma membrane. Physicochemical considerations indicate that such changes could mediate the shift from formation of endocytotic vesicles to formation of exocytotic extrusions.

10.8.5 Membrane lipids function in signaling and in defensive processes.

Plants, animals, and microbes all use membrane lipids as precursors for the synthesis of compounds that have intracellular or long-range signaling activities. Phosphatidylinositols, frequently referred to as phosphoinositides, have important functions in signal transduction pathways. Their role in regulation of metabolic pathways has been most extensively studied in animal systems, but recent studies suggest that they may serve a similar function in plant systems (see Chapter 18). In the principal active molecule, the phosphatidylinositol 4,5-bisphosphate (PIP$_2$), the inositol moiety is phosphorylated at C-4 and C-5. In animals, phosphoinositides serve as second messengers to extracellular signals. The signal activates a phosphatidylinositol-specific phospholipase C, which cleaves PIP$_2$ to produce inositol 1,4,5-triphosphate (IP$_3$) and diacylglycerol, each of which acts as a second messenger (Fig. 10.50).

Jasmonate is another of several lipid-derived plant growth regulators (see Chapter 17). The structure and biosynthesis of jasmonate have intrigued plant biologists because of the parallels to some eicosanoids, which are central to inflammatory responses and other physiological processes in mammals. In plants, jasmonate derives from

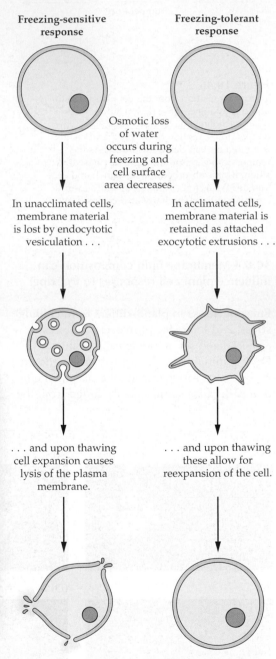

Freezing-sensitive response

Freezing-tolerant response

Osmotic loss of water occurs during freezing and cell surface area decreases.

In unacclimated cells, membrane material is lost by endocytotic vesiculation . . .

In acclimated cells, membrane material is retained as attached exocytotic extrusions . . .

. . . and upon thawing cell expansion causes lysis of the plasma membrane.

. . . and upon thawing these allow for reexpansion of the cell.

Figure 10.48
Cold acclimation allows plants to survive freezing. The *Arabidopsis* plants on the right were incubated at 4°C for 4 days to acclimate the plants, after which both pots were kept at –5°C for 4 days, and then transferred to growth conditions at 23°C for 10 days.

Figure 10.49
During freezing stress, changes in plasma membrane morphology determine death or survival of the cell.

$18:3\Delta^{9,12,15}$, presumably released from membrane lipids by phospholipase A_2. The linolenic acid is oxidized by lipoxygenase, and the resulting product, 9-hydroperoxylinolenic acid or 13-hydroperoxylinolenic acid, may be further metabolized by one of three routes to produce a wide variety of oxylipins, a diverse group of fatty acid derivatives that includes jasmonate (Fig. 10.51).

The pathways by which 13-hydroperoxylinoleic acid may be metabolized are shown in Figure 10.52. The enzyme hydroperoxide lyase catalyzes α-scission of the trans-11,12-double bond to produce a C_6 aldehyde, cis-3-hexenal, and a C_{12} compound, 12-oxo-cis-9-dodecenoic acid. The acid is subsequently metabolized to 12-oxo-trans-10-dodecenoic acid, also known as the wound hormone traumatin. The enzyme hydroperoxide dehydratase (allene-oxide synthase) catalyzes the dehydration of hydroperoxides to unstable allene oxides, which readily decompose to form 9,12-ketols or 12,13-ketols. The allene oxide of 13-hydroperoxylinolenic acid may also be converted by allene-oxide cyclase to 12-oxo-phytodienoic acid, which can be further metabolized to 7-isojasmonic acid. Recently, $16:3^{\Delta7,10,13}$ also has been found to metabolize, presumably by a similar route, to a biologically active compound called dinoroxo-phytodienoic acid.

The actions of jasmonate are dramatic and wide-ranging. In the last few years, it has become clear that jasmonate is a key component of a wound-signaling pathway that allows plants to protect themselves against insect attack. When experimentally applied to plants at low concentrations, jasmonate induces expression of genes that lead to the production of proteinase inhibitors and other defense compounds. Furthermore, mutants of tomato and *Arabidopsis* that are deficient in jasmonate synthesis are much more susceptible to insect damage (Fig. 10.53). More recently, a very different role for jasmonate was revealed by the *fad3 fad7 fad8* triple mutant of *Arabidopsis*, which cannot synthesize jasmonate because it lacks the precursor $18:3^{\Delta9,12,15}$. The plants are male-sterile because pollen does not mature properly and is not released from the anthers. Application of jasmonate or linolenic acid to the anthers restores fertility, demonstrating that jasmonate is a key signal in pollen development. The same mutant has

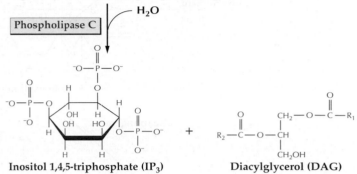

Phosphatidylinositol 4,5-bisphosphate (PIP$_2$)

Inositol 1,4,5-triphosphate (IP$_3$)   **Diacylglycerol (DAG)**

Figure 10.50
Phospholipase C hydrolyzes phosphatidylinositol 4,5-bisphosphate to inositol 1,4,5-triphosphate and diacylglycerol.

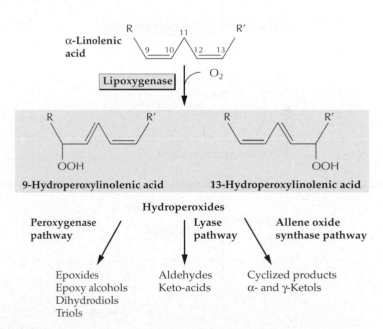

Figure 10.51
The lipoxygenase pathway. In the dioxygenase reaction catalyzed by lipoxygenase, there is no net oxidation or reduction. A cis–trans conjugated diene, 13-hydroperoxylinolenic acid, forms in the reaction. The hydroperoxy acid can be metabolized by three separate pathways to a wide variety of products, including jasmonate. The pathway starts with the action of lipoxygenase on $18:3^{\Delta9,12,15}$ to produce 9-hydroperoxylinolenic acid or 13-hydroperoxylinolenic acid, depending on the source of the enzyme.

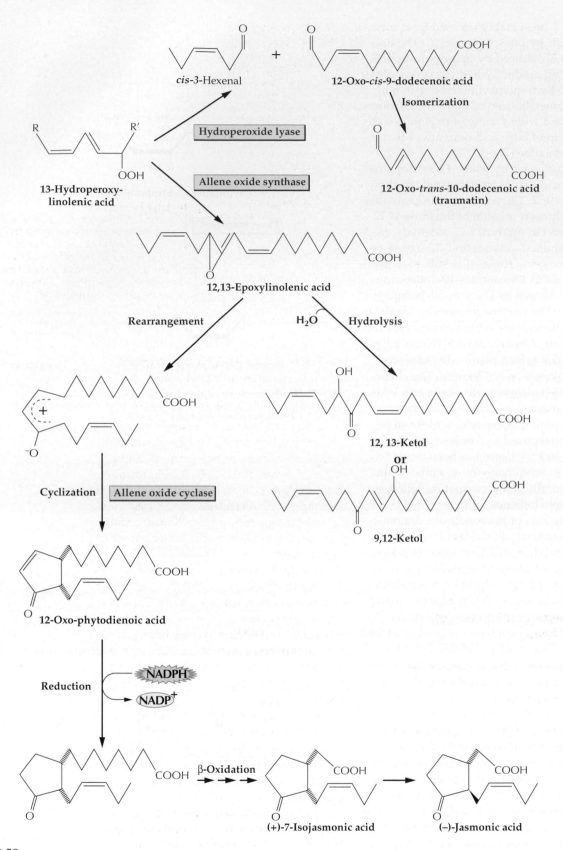

Figure 10.52

Metabolism of 13-hydroperoxylinolenic acid. Additional reactions not shown include conversion of 11,12-epoxylinolenic acid to 13-hydroxylinolenic acid by peroxygenase and rearrangement of the 13-hydroxylinolenic acid to form 15,16-epoxy-13-hydroxyoctadecenoic acid. Also *cis*-3-hexenal and its corresponding alcohol are thought to give rise to several isomers.

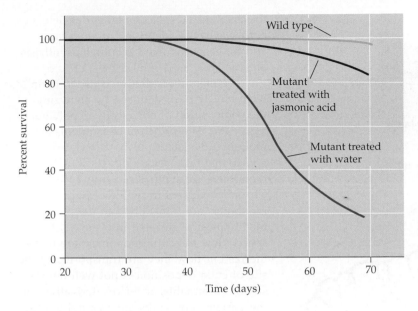

Figure 10.53
Jasmonic acid signaling protects plants from insect predation. Wild-type *Arabidopsis* and a *fad3 fad7 fad8* mutant were enclosed with adult flies of the fungal gnat *Bradysia impatiens*. The mutant plants, which were unable to synthesize jasmonic acid because of the *fad* mutations, were sprayed with either water or low concentrations of jasmonic acid.

been used to demonstrate that jasmonate (along with ethylene) is an important chemical signal in nonhost resistance against fungal pathogens.

In addition to jasmonate, several of the other oxylipins also have been reported to function as signal molecules. In particular, the oxylipin traumatin has been suggested to trigger cell division at the site of wounds, leading to the development of a protective callus. The lipoxygenase product 13-hydroxylinolenic acid triggers phytoalexin production. Similarly, C_6–C_{10} alkenals act as volatile elicitors of a defense response in cotton.

10.9 Synthesis and function of structural lipids

10.9.1 Cutin and suberin provide an epidermal barrier to water loss and pathogen infection.

Most epidermal cells of the aerial parts of vascular plants and some bryophytes are covered by a film of soluble and polymerized lipids, collectively called the cuticle (Fig. 10.54). The primary role of this layer is to provide a permeability barrier against water loss; it also provides some degree of resistance to pathogens and insects. Plant roots contain a group of cells called the endodermis to control the entry of materials and organisms from the soil. The walls of

these cells contain a highly polymerized lipophilic material called suberin.

The fine structure of the cuticle is not completely understood. In part, this is because the microscopic appearance and chemical composition may vary among different species, or within one species at different stages of development (Fig. 10.55). In spite of the apparent differences, however, the underlying processes that give rise to the cuticle and the general elements of its overall structure are thought to be similar among higher plants.

A major component of the cuticle is a lipid polyester called cutin, which in turn is covered by surface waxes (Fig. 10.55). Cutin is a polymeric network of oxygenated C_{16} and C_{18} fatty acids cross-linked by ester bonds, such that the carboxyl group of one fatty acid is linked·to a primary or secondary hydroxyl group of another (Fig. 10.56). Thus, the interesterified acyl chains form a highly cross-linked and relatively inelastic meshwork with a strongly hydrophobic character. However, because of the relatively large "pore size" of the cutin network, it is unlikely that cutin provides a significant barrier to water loss. Rather, it may act as a relatively inelastic outer skin that assists in providing rigidity to turgid plant tissues. Because of its physical strength, cutin may provide some defense against penetration of pathogens. To penetrate the cutin layer, pathogens probably need to secrete cutinases, enzymes that hydrolyze

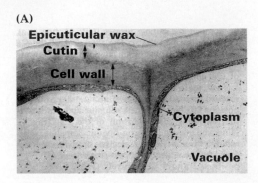

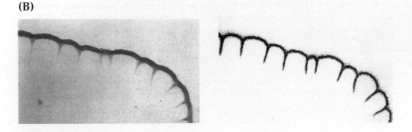

Figure 10.54
(A) Transmission electron micrograph of a transverse section through the cuticle of a leaf of *Clivia miniata*. (B) Cutin is deposited on the outer surface of epidermal cells. In the upper panel, a section of a leaf has been stained with Sudan III, which stains lipophilic materials. The lower panel is an autoradiograph of a leaf after incubation with radioactive fatty acids for 24 h. Subsequent extraction with methanol and methanol:chloroform (1:1 by vol.) removed the soluble fatty acids, leaving only cutin in place.

the ester linkages. No mutants deficient in cutin have been identified in any plant, and no genes involved in cutin biosynthesis are known.

The monomers that give rise to cutin have largely been deduced by analyzing the composition of chemically depolymerized cutin. The principal constituents are monohydroxylated, polyhydroxylated, and epoxidated fatty acids (Fig. 10.57). Cutin monomers are synthesized from fatty acyl-CoAs by oxidases in the ER. Apparently, the monomers are transported to the plasma membrane and secreted into the cell wall space. How the cutin monomers are polymerized or how they are transported to the site of cutin deposition is not well understood. Presumably, acyl-CoA derivatives of the various monomers are secreted into the cell wall space (see Fig. 10.34), where acyltransferases catalyze synthesis of the esters.

Epidermal cells of root tissues do not appear to be covered with cutin to the same extent as the aerial tissues. This may reflect the fact that root epidermal tissues function to take up water and solutes and must therefore maintain hydrophilic contact with soil moisture. However, roots do have an internal hydrophobic layer composed of suberin,

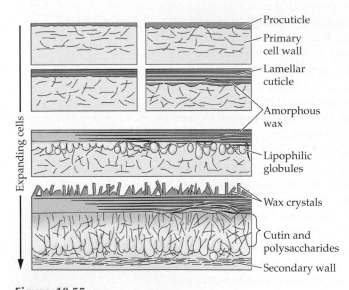

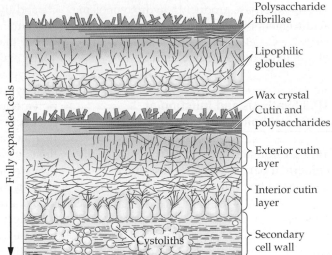

Figure 10.55
Stages in the development of a plant cuticle. At early stages, the primary cell wall is covered by a thin amorphous layer of wax. As the leaf expands, the amount of wax increases by agglomeration of secreted globules. Near the time that leaf expansion stops, wax crystals start to appear on the surface and the deposition of cutin begins. After leaf expansion, cutin deposition increases and secondary cell wall deposition starts. The cutin layer may take on a fibrillar appearance, thought to reflect codeposition of cutin and

secondary cell wall materials such as hemicellulose. In the fully expanded mature leaf, distinct zones—the exterior cutin layer and the interior cutin layer—may be visible. Some studies indicate that these layers differ in chemical composition. In some species, lipophilic globules called cystoliths are observed at late stages of cuticle development and are thought to contain cutin precursors that have been secreted from the epidermal cell.

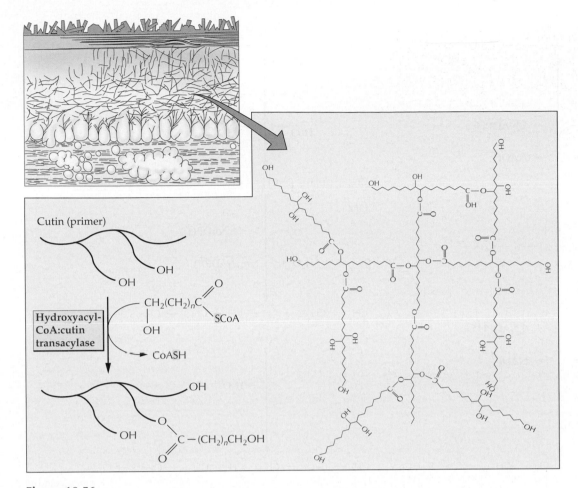

Figure 10.56
Hypothetical structure of a region of a cutin network. For the sake of clarity, this example was constructed by randomly linking carboxyl groups of each of 11 identical cutin monomers (9,10,18-trihydroxyoctadecanoic acid) to the hydroxyl group of the other 10 fatty acids. In a leaf, the structure would reflect the availability of fatty acids produced by the plant. For instance, in leaves of *Clivia*, the substrates are approximately 23% 9,16- or 10,16-dihydroxyhexanoic acid, 47% 9,10-epoxy-18-hydroxyoctadecanoic acid, 18% 18-hydroxy-9-octadecenoic acid, and 7% 9,10,18-trihydroxyoctadecanoic acid, the remainder being composed of roughly equal amounts of 23 other fatty acids. The esters are thought to be synthesized by the condensation of a fatty acyl-CoA and an alcohol group of another cutin precursor or of partially esterified cutin by the enzyme hydroxyacyl-CoA:cutin transacylase.

the components of which are similar to those of cutin. The principal constituents are very-long-chain fatty acids, fatty alcohols, diacids, and phenolics such as *p*-coumaric acid (Fig. 10.58). The main differences between cutin and suberin are thought to be that the fatty acids in suberin do not have secondary alcohols or epoxy groups and are usually longer than 18 carbons. In addition, suberin has a relatively high phenolic content (Chapter 24). These components are thought to be interesterified in much the same way that cutin is. However, because of the relatively low content of hydrophilic groups, suberin is more hydrophobic than cutin; as a result,

the endodermis appears highly impermeable to aqueous solutions. Thus, materials must move across the endodermis and into the vascular tissues by symplasmic transport. Suberin is also found surrounding bundle sheath cells in C_4 species. The function in this case may be to prevent CO_2 produced by decarboxylation reactions from diffusing out of the bundle sheath cells.

10.9.2 Epicuticular wax reduces water loss.

The aerial surfaces of plants are covered with a layer of chloroform-soluble nonvolatile

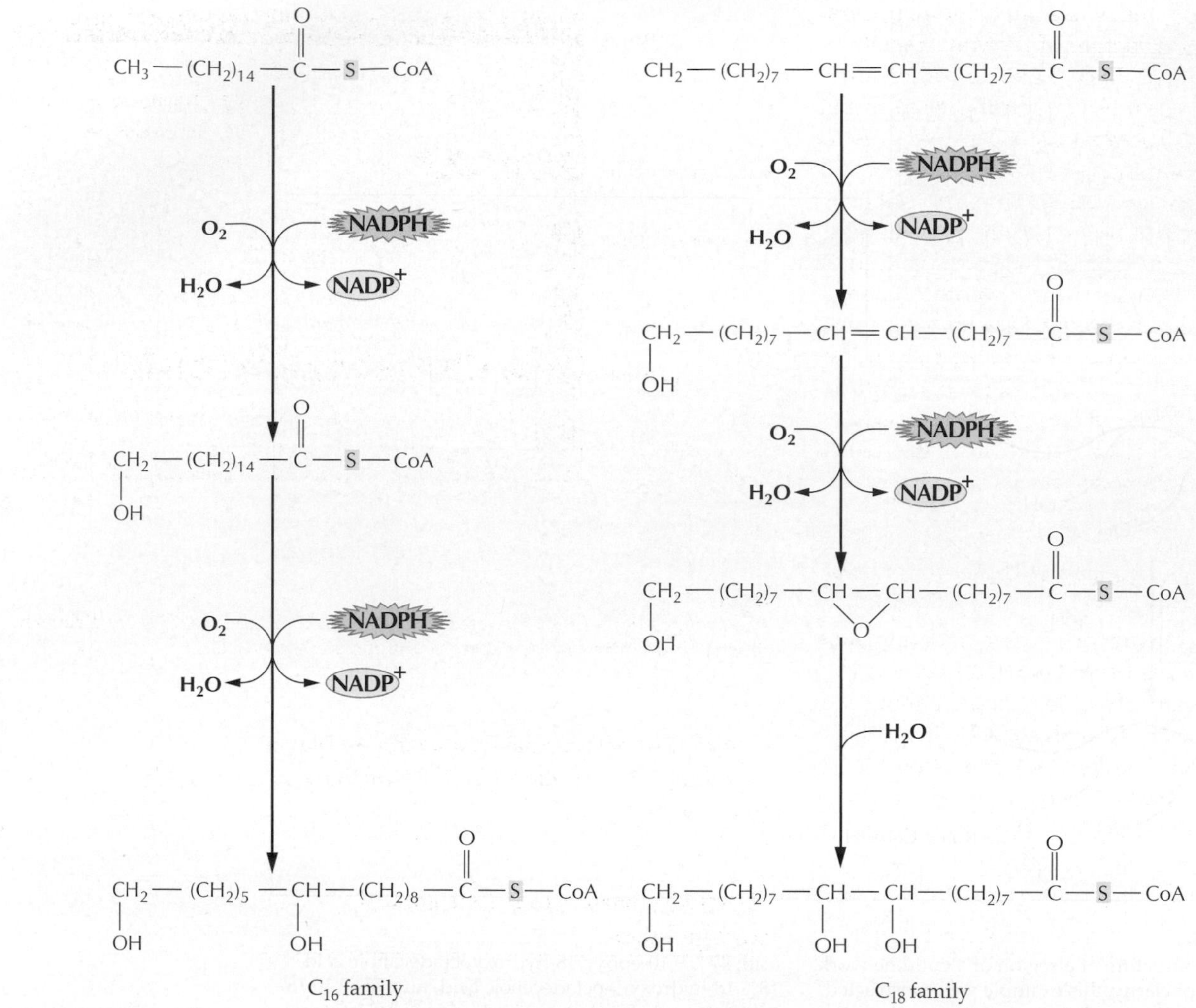

Figure 10.57
The principal components of cutin are synthesized from 16:0 and 18:1 fatty acyl-CoAs by oxidations thought to be catalyzed by cytochrome P450 enzymes located in the endoplasmic reticulum. In most cases the enzymes have not been purified or characterized in detail and the reactions shown here are inferred from assays using crude extracts. All of the fatty acid moieties of the acyl-CoAs shown here are found as constituents of cutin but the relative amounts of each may vary from one plant species to another. It is not known if the order in which the hydroxylations take place only occurs as shown or whether the sequence of reactions can vary.

$CH_3(CH_2)_mCOOH$
Long-chain fatty acids

$HOCH_2(CH_2)_nCOOH$
Hydroxy fatty acids

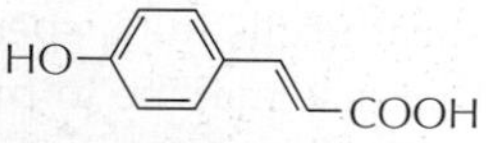

p-Coumaric acid

(m = 18–30; n = 14–20)

$CH_3(CH_2)_mCH_2OH$
Fatty alcohols

$HOOC(CH_2)_nCOOH$
Fatty diacids

Figure 10.58
Principal components of suberin. The first four compounds are derived from fatty acid metabolism.

lipids, collectively called wax (see Fig. 10.54). The wax layer reduces water loss by orders of magnitude, thereby making terrestrial plant life possible. The amounts and composition of wax deposited are controlled by the plant in response to environmental factors such as relative humidity, soil moisture, and light intensity (Fig. 10.59). Wax composition varies from one plant species to another, but wax generally contains a mixture of long-chain hydrocarbons, acids, alcohols, ketones, aldehydes, and esters. The functional significance of interspecies differences in wax composition is not known. However, from the study of mutants with altered wax composition, the wax composition appears to affect the structure of the wax crystals. Some plants produce filaments, others produce plates or tubes or spiral-shaped forms (Fig. 10.60). The mechanisms responsible for the shape of the crystals are not known. Different crystal structures may vary in their ability to reflect light—a property that may be useful in adaptation to growth in different light intensities. More important, perhaps, some pathogens and herbivorous insects are attracted to or repelled by specific wax compositions. Thus, the wax composition of particular plant species may reflect the balance of selective pressures exerted on a particular plant species.

Wax monomers are thought to be synthesized from C_{16} and C_{18} fatty acids by an elongase complex located in the ER (Fig. 10.61). Although the reactions involved in elongation, reduction, decarbonylation, and oxidation have been demonstrated in cell extracts, relatively little is known about the specific enzymes involved. As with cutin biosynthesis, it is not well understood how the wax monomers move from the site of synthesis to the surface of epidermal cells. Because the principal components are completely insoluble in aqueous solutions, the monomers are likely transported by lipid-transfer proteins (see Fig. 10.34).

10.9.3 Wax is required for productive pollen–pistil interactions.

Pollen grains are covered with a waxy layer called the exine, which is composed of a complex polymer called sporopollenin. Embedded in the exine is a layer of lipophilic material called tryphine or pollenkitt (Fig. 10.62; see also Chapter 19). In *Arabidopsis*, the tryphine layer includes small lipid bodies that appear to contain primarily C_{28} and C_{30} wax monomers. Probably an important function of the tryphine layer is to reduce water loss from the pollen grain. However, experiments using *Arabidopsis cer* mutants with altered wax composition hint at a more complex role for wax. Some *cer* mutants are male-sterile. However, the fertility of *cer* pollen can be rescued by mixing irradiation-killed wild-type pollen with *cer* pollen; accordingly, some factor in the wild-type pollen appears to complement the defect in the *cer* pollen. These experiments

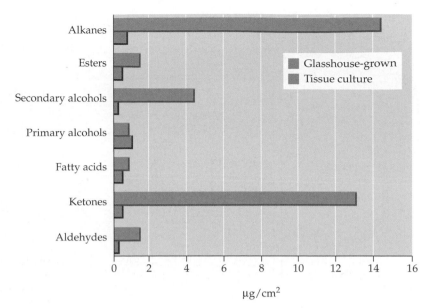

Figure 10.59
Wax composition of *Brassica oleracea* plant grown in high-humidity conditions in tissue culture or in low-humidity conditions in a glasshouse.

suggest that the normal wax coat may play a signaling role in pollen–pistil interactions (see Chapter 19).

10.10 Synthesis and catabolism of storage lipids

Lipids in the form of triacylglycerols are widely found as a major carbon and chemical energy reserve in seeds, fruits, and pollen grains (Fig. 10.63). One of the few plants that stores lipids in a form other

Figure 10.60
Scanning electron micrograph of the surface of a sorghum leaf sheath. In this species, the wax crystallizes as filaments that appear to emerge from defined regions of the cuticle. Scale bar (lower left) = 10 μm.

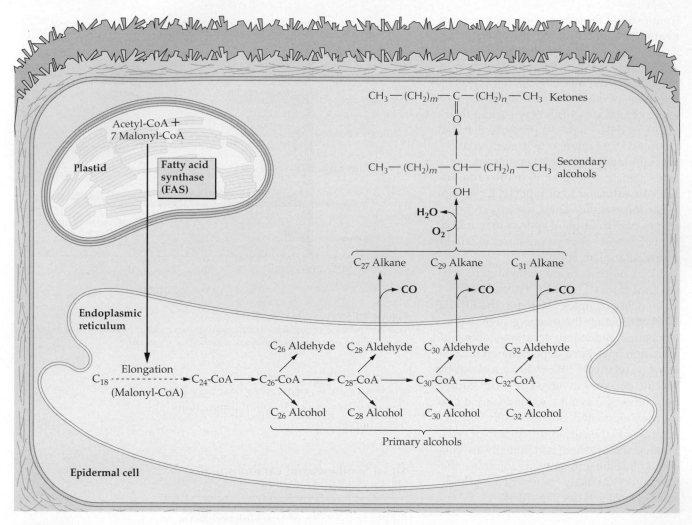

Figure 10.61

Proposed pathway for synthesis of the principal components of wax. C_{16} or C_{18} precursors for wax are produced by the de novo fatty acid synthesis pathway in the plastids. After export from the plastid, these acyl chains are elongated to C_{26} to C_{32} fatty acyl chains in the ER. The fatty acids (attached to coenzyme A) can be reduced to alcohols or to aldehydes, and the aldehydes can be decarbonylated to form alkanes with an odd number of carbons. The reactions forming secondary alcohols and ketones are not yet well understood. Together, the numerous long-chain substrates, including the primary alcohols, aldehydes, alkanes, secondary alcohols, and ketones diagrammed here, contribute to wax formation.

than triacylglycerols is jojoba, a perennial shrub that stores fatty acids as wax esters in seeds. Plant storage lipids are also an important source of dietary fats for humans and other animals. Approximately 40% of the daily energy requirement of humans in industrialized countries is supplied by dietary triacylglycerols, of which about half is from plant sources. Furthermore, triacylglycerols find use in manufacturing industries, particularly in the production of detergents, coatings, plastics, and specialty lubricants. For both food and industrial applications, the fatty acid composition of the oil determines its usefulness and, therefore, its commercial value.

10.10.1 Triacylglycerol synthesis involves acyltransferase and acyl-exchange reactions that move fatty acids between pools of membrane and storage lipids.

Many of the biochemical reactions and some aspects of subcellular compartmentalization of triacylglycerol synthesis in developing oilseeds are the same as for membrane lipids. However, because of interest in the factors that control the precise fatty acid composition of triacylglycerols, it is useful to consider seed lipid biosynthesis in terms of a scheme that emphasizes the acyltransferase and acyl-exchange reactions (Fig. 10.64). This

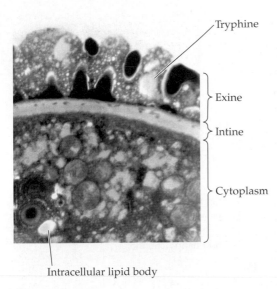

Figure 10.62
Transmission electron micrograph through an *Arabidopsis* pollen grain.

scheme was developed from the results from many oilseed species but is drawn to describe the metabolism of developing *Arabidopsis* seeds. *Arabidopsis* seed lipids contain substantial proportions of both unsaturated C_{18} fatty acids (30% $18:2^{\Delta9,12}$, 20% $18:3^{\Delta9,12,15}$) and long-chain fatty acids (22% $20:1^{\Delta11}$) de-

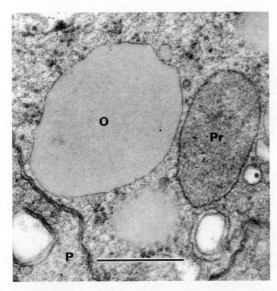

Figure 10.63
Electron micrograph of a thin section of an oil body (O), peroxisome/glyoxysome (Pr), and a small portion of a plastid (P) in the subapical zone of a one-day-old shoot apex of maize seedling. The bar represents 0.5 μm.

rived from $18:1^{\Delta9}$. As a result, *Arabidopsis* is a good model for the biochemistry of both $18:2^{\Delta9,12}/18:3^{\Delta9,12,15}$-rich oilseeds and those species containing longer fatty acids.

As in other tissues, 16:0-ACP and $18:1^{\Delta9}$-ACP are usually the major products of plastid fatty acid synthesis and 18:0-ACP desaturase activity in oilseeds. These products either are utilized for lipid synthesis within the plastid or are hydrolyzed to free fatty acids by thioesterases in the plastid stroma and transferred through the plastid envelope by an unknown mechanism. The free fatty acids are converted to acyl-CoAs in the outer plastid envelope, forming the substrates for subsequent reactions in other cellular compartments. Newly produced $18:1^{\Delta9}$-CoA, 18:0-CoA, and 16:0-CoA can be used for the synthesis of phosphatidylcholine, as shown in Figure 10.64. Phosphatidylcholine is the main substrate for the sequential desaturation of $18:1^{\Delta9}$ to $18:2^{\Delta9,12}$ and $18:3^{\Delta9,12,15}$. Cholinephosphotransferase is freely reversible, so in many oilseeds phosphatidylcholine is a direct precursor of highly unsaturated species of diacylglycerol used for triacylglycerol synthesis. However, the acyl-CoA pool does not contain only 16:0 and $18:1^{\Delta9}$: Exchange of $18:1^{\Delta9}$ from $18:1^{\Delta9}$-CoA with the fatty acid at position *sn*-2 of phosphatidylcholine provides inputs of $18:2^{\Delta9,12}$ and $18:3^{\Delta9,12,15}$ back into the cellular acyl-CoA pool. In some oilseeds, including *Arabidopsis* and rapeseed, $18:1^{\Delta9}$-CoA can be modified by elongation to $20:1^{\Delta11}$-CoA and $22:1^{\Delta13}$-CoA. Synthesis of diacylglycerol may also involve these components of the acyl-CoA pool, as does the final acylation of diacylglycerol to form triacylglycerol by the enzyme acyl-CoA:1,2-diacylglycerol *O*-acyltransferase.

10.10.2 Triacylglycerols accumulate in discrete subcellular organelles called oil bodies.

Electron microscopic studies show that in oilseeds, oil bodies are surrounded by what appears to be a "half-unit" membrane instead of the more usual bilayer. The oil body is thus a droplet of triacylglycerol surrounded by a monolayer of phospholipids, with the hydrophobic acyl moieties of the phospholipids interacting with the triacylglycerols and the hydrophilic head groups

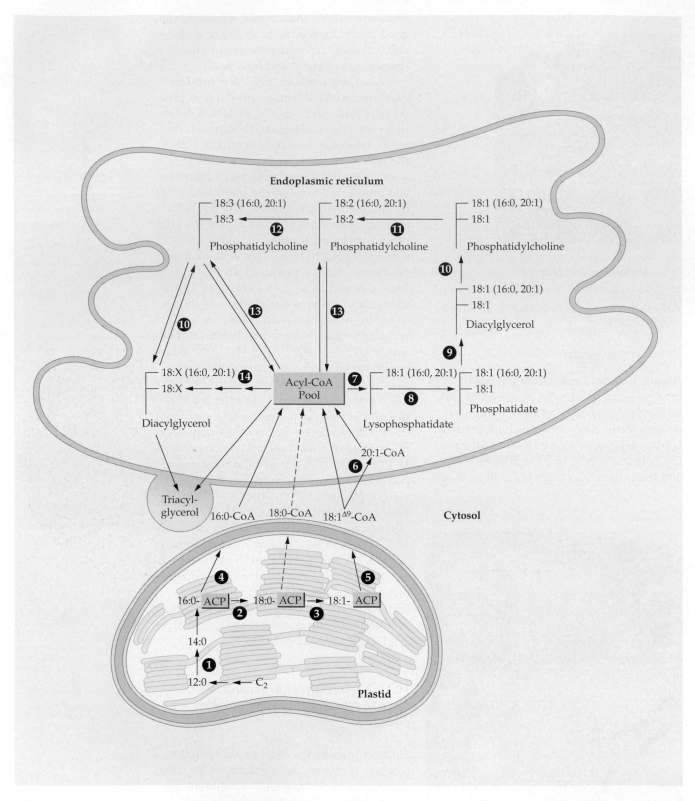

Figure 10.64

Abbreviated scheme for the reactions of triacylglycerol synthesis in seeds of *Arabidopsis* and other oilseeds. The enzyme-catalyzed steps are indicated by numbers and involve the following enzymes: (1) KAS I– and KAS III–dependent FAS; (2) KAS II–dependent FAS; (3) stearoyl-ACP desaturase; (4) palmitoyl-ACP thioesterase; (5) oleoyl-ACP thioesterase; (6) oleate elongase; (7) acyl-CoA:glycerol-3-phosphate acyltransferase; (8) acyl-CoA:lysophosphatidate acyltransferase; (9) phosphatidate phosphatase; (10) CDP-choline: diacylglycerol cholinephosphotransferase; (11) oleate desaturase, *FAD2*; (12) linoleate desaturase, *FAD3*; (13) acyl-CoA: *sn*-1 acyllysophosphatidylcholine acyltransferase; (14) same as in (7), (8), and (9), but any fatty acids used are from the acyl-CoA pool.

facing the cytosol. Oil bodies also contain major protein components called oleosins, which are not found in significant amounts in any other cellular location (Fig. 10.65). Oleosins are low-molecular-mass proteins (15 to 25 kDa), of which the defining feature is a sequence of 70 to 80 hydrophobic amino acids toward the middle of the protein. The sequence of this hydrophobic domain is conserved in oleosins from different plant species, but these proteins are not found in animals, bacteria, or fungi. Although some question of the protein secondary structure in the hydrophobic domain (β-strand or β-sheet) remains, there is general agreement that it protrudes into the triacylglycerol core of the oil body. Very possibly, the more hydrophilic N- and C-terminal domains form amphipathic helices at the oil body surface.

Although it is tempting to suppose that oleosins help stabilize the interface between the oil body and the aqueous cytoplasm, this is probably not their function. Oil bodies in fruit tissues (such as those of avocado and olive) do not contain oleosin homologs. Oleosins are found only in oil bodies of seeds and pollen, both of which undergo dehydration during maturation. Thus, oleosins may function to stabilize oil bodies at low water potential when hydration of the surface phospholipids is not sufficient to prevent the oil bodies from coalescing and fusing. Oleosins also may regulate the size of oil bodies by imparting a defined curvature to the surface—which could be important for regulating the surface-to-volume ratio to facilitate rapid breakdown of oil bodies during germination. In this respect, the mesocarp lipids in avocado and olive are not thought to contribute to the germination or growth of the seedling but are probably made by the plant to facilitate seed dispersal by animals.

The ontogeny of oil bodies is not absolutely clear, but one popular view suggests that they arise by deposition of triacylglycerols between the two leaflets of the ER and then develop into discrete organelles that may or may not remain attached to the ER membranes (Fig. 10.66). This model accounts for the half-unit membrane of oil bodies and is consistent with biochemical results that demonstrate high rates of triacylglycerol synthesis by microsomal membrane preparations. The exact nature of the relationship between the oil bodies and ER takes on extra significance because diacylglycerols are precursors for the synthesis of both triacylglycerols and the major membrane phospholipids. Because many oilseeds contain high amounts of unusual fatty acids that are largely excluded from the membrane phospholipids, it follows that diacylglycerols must occupy a critical branch point in oilseed lipid metabolism.

10.10.3 Membrane and storage lipids often have distinct compositions.

Unusual fatty acids (see Table 10.2 and Fig. 10.25) often constitute 90% or more of all fatty

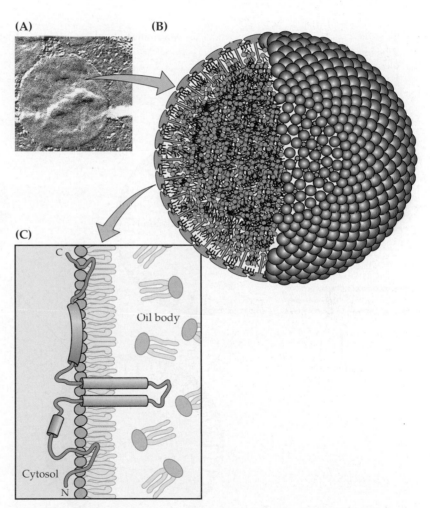

Figure 10.65
(A) Transmission electron micrograph close-up of oil body. (B) Scale-model of a maize oil body. The oleosin molecule is depicted in the shape of an 11-nm-long hydrophobic stalk attached to an amphipathic and hydrophilic globular structure that forms the outer surface of the oil body. (C) Proposed model of the conformation of a maize 18-kDa oleosin. The cylinders depict helices.

acids produced in the seed, but in almost all cases these fatty acids are present only in triacylglycerols and are excluded from membrane lipids. This probably occurs because fatty acids with unusual structures could create undesirable physical or chemical properties in the membrane or perturb membrane fluidity. Therefore, plants that produce such unusual fatty acids must also have mechanisms to prevent the accumulation of these compounds in membranes. Furthermore, hydroxy-fatty acids, epoxy-fatty acids, and some other unusual fatty acids are first synthesized on phosphatidylcholine, a major membrane lipid. What

mechanisms ensure first the removal of these modified fatty acids from phospholipids and then their targeting to storage oils? At present, no clear and complete answer is available. However, developing seeds from species that produce unusual fatty acids often have specific phospholipases that can remove the fatty acids from polar lipids. Furthermore, these seeds usually have a suite of specialized acyltransferases and other enzymes specific for metabolizing unusual fatty acids.

One possible method for keeping membrane and storage fatty acids distinct may be subcellular compartmentation. Although

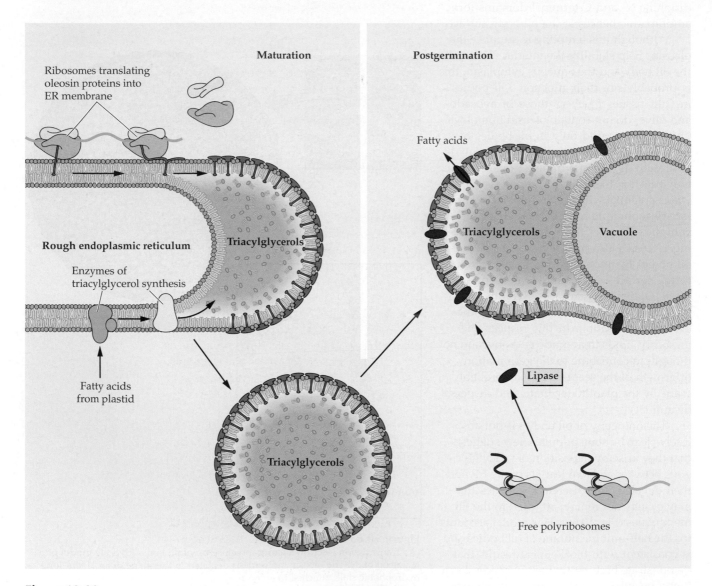

Figure 10.66
Model of the synthesis and degradation of an oil body in a maize embryo during seed maturation and postgermination.

membrane and storage lipid synthesis proceed by similar pathways and share common chemical intermediates (phosphatidate, diacylglycerol, and phosphatidylcholine; see Fig. 10.64), the specialized enzymes involved in storage oil biosynthesis may have subcellular locations distinct from enzymes that process the "common" fatty acids for incorporation into membranes.

10.10.4 Mobilization of storage lipids provides carbon and chemical energy for germination and pollination.

Triacylglycerols function as an efficient source of carbon and energy for germination of seeds and pollen because lipids are a far more compact form of storage than is carbohydrate or protein (see Fig. 10.3). Relatively few species accumulate significant quantities of storage lipids in roots, tubers, or other storage organs where compact size is not important. Because of their hydrophobicity and insolubility in water, triacylglycerols are segregated into lipid droplets, which do not raise the osmolarity of the cytosol. Furthermore, unlike polysaccharides, triacylglycerols do not contain extra weight as water of solvation. The relative chemical inertness of triacylglycerols allows their intracellular storage in large quantity without risking undesired chemical reactions with other cellular constituents.

The same properties that make triacylglycerols good storage compounds present problems in their utilization. Because of their insolubility in water, triacylglycerols must be hydrolyzed to fatty acids before they are available for metabolism. The relative stability of the C–C bonds in the alkyl moiety of a fatty acid is overcome by activation of the fatty acid carboxyl group at C-1 by attachment to CoASH, followed by oxidative attack at the C-3 position. This latter carbon atom is also called the β-carbon in common nomenclature, from which the oxidation of fatty acids gets its common name: β-oxidation. The oxidation of long-chain fatty acids to acetyl-CoA by a four-step cycle is a central energy-yielding pathway during the germination of seeds with high lipid contents and during the germination and growth of pollen tubes during fertilization. The acetyl-CoA formed during β-oxidation may be converted, via the glyoxylate cycle and gluconeogenesis, to carbohydrate (Fig. 10.67; see also Chapters 1 and 14).

Although the biological role of fatty acid oxidation differs from organism to organism, the enzyme reactions are essentially the same in plants and animals. Details of the enzymatic mechanisms can be found in biochemistry textbooks that focus primarily on animal metabolism. Therefore, the discussion below highlights those features of the pathway that differ between plants and animals.

10.10.5 β-Oxidation takes place in peroxisomes and glyoxysomes.

The major site of fatty acid oxidation in animal cells is the mitochondrial matrix. In contrast, fatty acid oxidation in plants occurs primarily in the peroxisomes of leaf tissue and the glyoxysomes of germinating seeds (Fig. 10.67). Plant peroxisomes and glyoxysomes are similar in structure and function. Glyoxysomes may be considered specialized peroxisomes (see Chapter 1).

Unlike in animals, the β-oxidation pathway is not used solely for production of metabolic energy in plants. The biological role of β-oxidation in peroxisomes and glyoxysomes is to provide biosynthetic precursors from stored lipids. During germination, triacylglycerols stored in seeds are converted into glucose and a wide variety of essential metabolites. Fatty acids released from triacylglycerols are activated to form their CoA derivatives and are oxidized in glyoxysomes by the same four-step process that occurs in peroxisomes. The acetyl-CoA produced is converted by way of the glyoxylate cycle to C_4 precursors for gluconeogenesis (Fig. 10.67; see also Chapter 14, Fig. 14.41). Glyoxysomes, like peroxisomes, contain high concentrations of catalase, which converts the H_2O_2 produced by β-oxidation to H_2O and O_2 (see Chapter 1).

In animal mitochondria, each of the four enzymes of β-oxidation is a separate, soluble protein, similar in structure to the analogous enzyme in Gram-positive bacteria. In contrast, the enzymes of peroxisomes and glyoxysomes form a complex of proteins. As in the oxidation of fatty acids in mitochondria, the intermediates are CoA derivatives, and the process consists of four steps (Fig. 10.68):

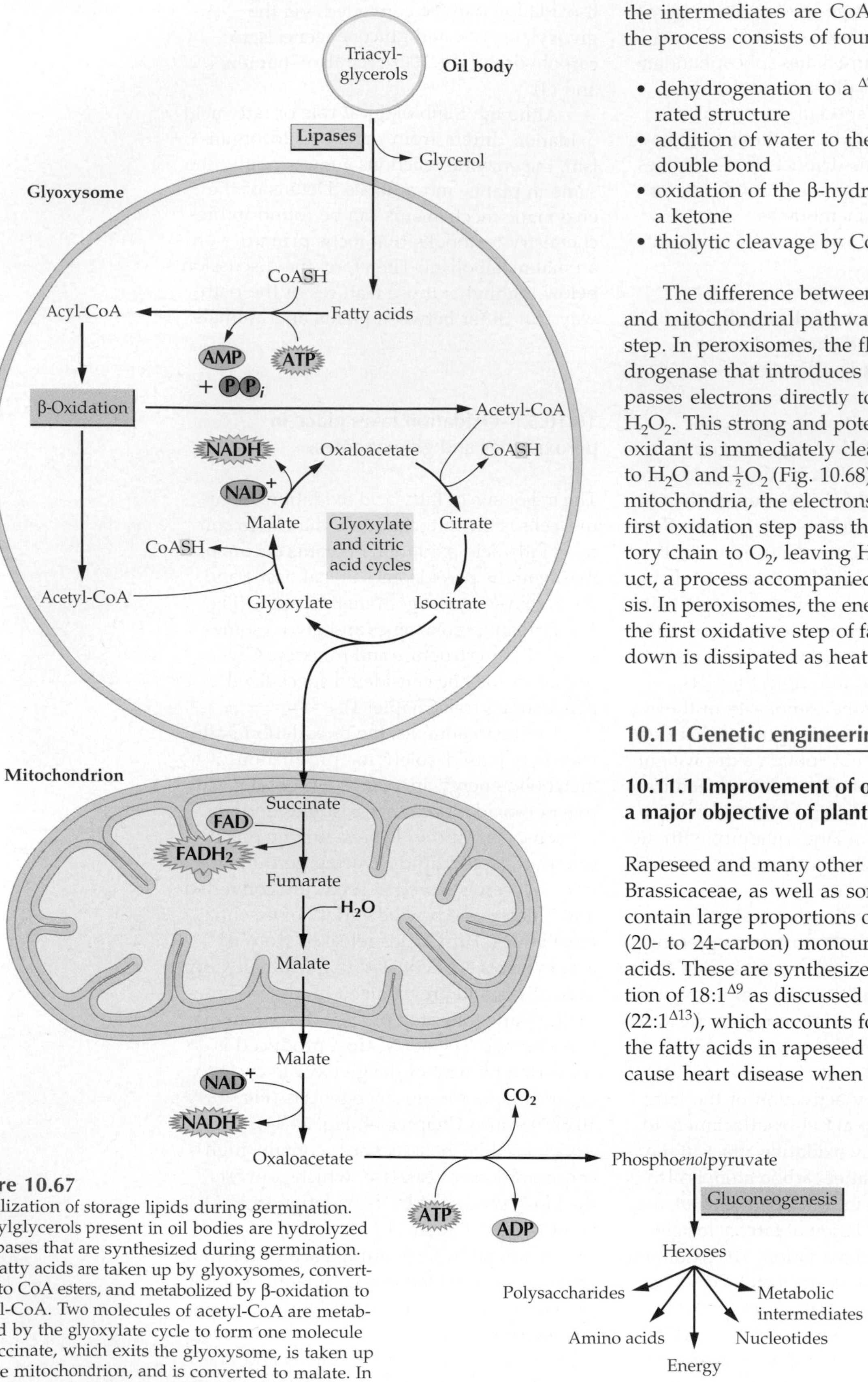

- dehydrogenation to a $^{\Delta2}$-*trans* unsaturated structure
- addition of water to the resulting double bond
- oxidation of the β-hydroxyacyl-CoA to a ketone
- thiolytic cleavage by CoASH

The difference between the peroxisomal and mitochondrial pathways lies in the first step. In peroxisomes, the flavoprotein dehydrogenase that introduces the double bond passes electrons directly to O_2, producing H_2O_2. This strong and potentially damaging oxidant is immediately cleaved by catalase to H_2O and $\frac{1}{2}O_2$ (Fig. 10.68). In contrast, in mitochondria, the electrons removed in the first oxidation step pass through the respiratory chain to O_2, leaving H_2O as the product, a process accompanied by ATP synthesis. In peroxisomes, the energy released in the first oxidative step of fatty acid breakdown is dissipated as heat.

10.11 Genetic engineering of lipids

10.11.1 Improvement of oil quality is a major objective of plant breeders.

Rapeseed and many other members of the Brassicaceae, as well as some other species, contain large proportions of very-long-chain (20- to 24-carbon) monounsaturated fatty acids. These are synthesized by chain elongation of 18:1$^{\Delta9}$ as discussed earlier. Erucic acid (22:1$^{\Delta13}$), which accounts for about 50% of the fatty acids in rapeseed oil, was shown to cause heart disease when included in the

Figure 10.67
Mobilization of storage lipids during germination. Triacylglycerols present in oil bodies are hydrolyzed by lipases that are synthesized during germination. The fatty acids are taken up by glyoxysomes, converted into CoA esters, and metabolized by β-oxidation to acetyl-CoA. Two molecules of acetyl-CoA are metabolized by the glyoxylate cycle to form one molecule of succinate, which exits the glyoxysome, is taken up by the mitochondrion, and is converted to malate. In the cytosol, malate is oxidized and the resulting oxaloacetate is converted into hexose by gluconeogenesis.

diet of laboratory animals (see Fig. 10.28). Although the high erucic acid content made rapeseed oil useful for certain industrial applications, it prevented the widespread use of rapeseed (Brassica napus) as an edible oil crop. In the 1950s, an extensive search was made for varieties of B. napus with low contents of erucic acid and glucosinolates, another antinutritional component of oil from Brassica species (Fig. 10.69; see also Chapter 16). During a period of about 20 years, several natural isolates containing reduced amounts of erucic acid were identified, and the two loci responsible for the phenotype were introgressed into cultivars of B. napus by many rounds of back-crossing. To distinguish the cultivars that produced oil with low erucic acid (LEAR) from those with high erucic acid (HEAR) content, the LEAR cultivars are now called Canola (Fig. 10.69). Expression of a jojoba β-ketoacyl-CoA synthase in transgenic Canola plants restored the high erucic acid trait. Thus, the two genes introgressed into B. napus to produce Canola apparently are naturally occurring mutations in two genes for β-ketoacyl-CoA synthase. The presence of two functionally homologous genes in Canola is probably related to the fact that the plant is a tetraploid. This is one of the few instances in which the biochemical basis for an important agronomic trait has been discovered.

10.11.2 Edible oils can be improved by metabolic engineering.

Approximately 20% of the calories consumed in developed countries are derived from plant oils. The fatty acid composition of dietary oils, particularly the saturated fatty acid content, is believed to influence the etiology of major diseases such as atherosclerosis and cancer. Many attempts have been made to alter the fatty acid compositions of food oils and reduce the proportions of 16:0, 18:0, $18:2^{\Delta9,12}$, and $18:3^{\Delta9,12,15}$ in favor of $18:1^{\Delta9}$. The move to develop monounsaturated oils is based on epidemiological data and laboratory experiments that suggest that these oils may reduce atherosclerosis (and associated heart attacks and strokes) by increasing the ratio of high-density lipoproteins to low-density lipoproteins in the blood. Furthermore, oils low in $18:2^{\Delta9,12}$ and $18:3^{\Delta9,12,15}$ are

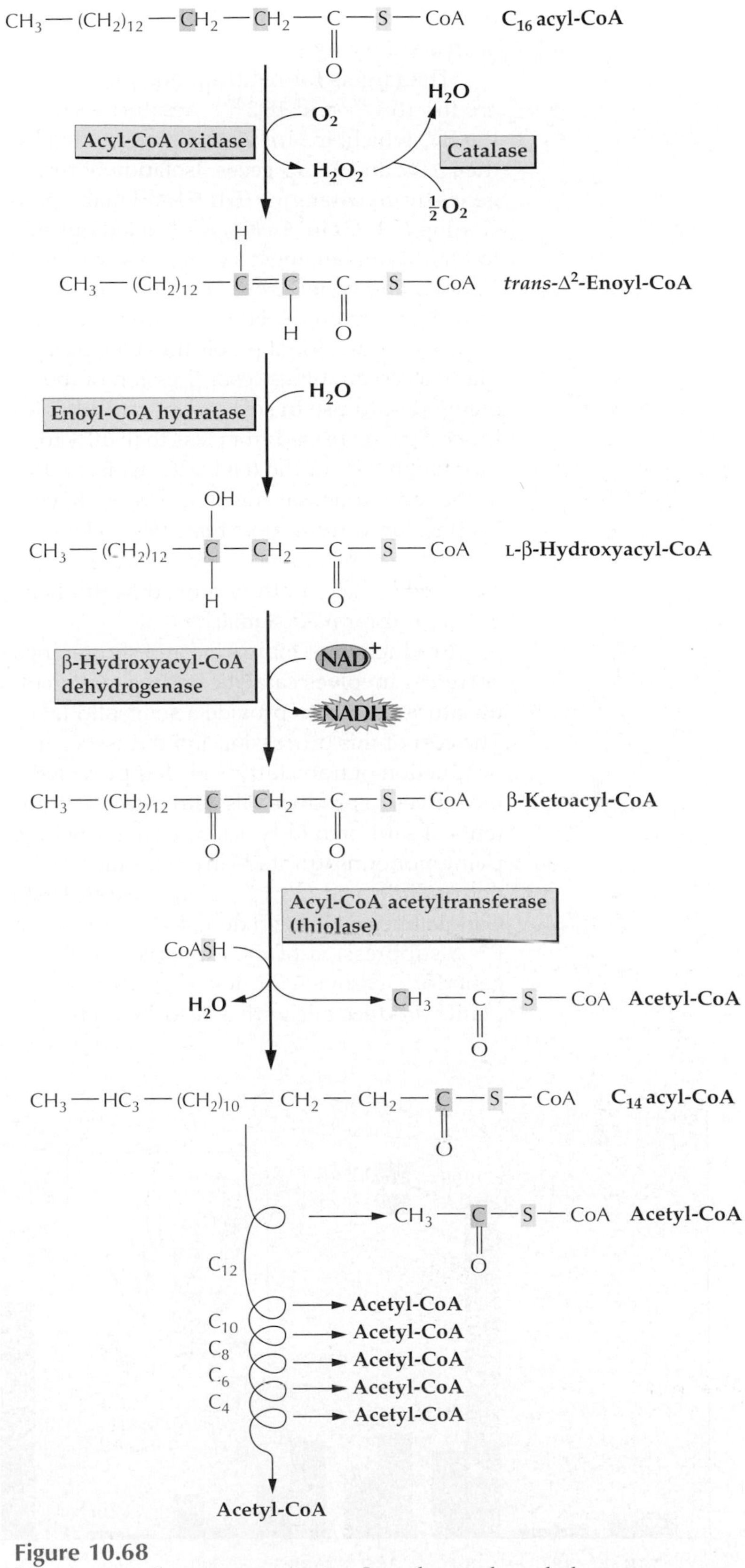

Figure 10.68

β-Oxidation pathway in peroxisomes. In each pass through the sequence, one acetyl residue is removed in the form of acetyl-CoA from the carboxyl end of an acyl-CoA. Seven passes through the cycle are required to oxidize a C_{16} fatty acid to eight molecules of acetyl-CoA.

much more stable, particular in high-temperature food frying.

The targets for reducing polyunsaturates are the $18:1^{\Delta 9}$- and $18:2^{\Delta 9,12}$ desaturases of the ER, which, in *Arabidopsis,* are encoded by the *FAD2* and *FAD3* genes. Isolation of these genes by gene-tagging (*FAD2*) and map-based cloning (*FAD3*) in *Arabidopsis* has led quickly to identifying analogous genes in soybean, Canola, and other crops. This has now allowed genetic engineering of edible oils with improved nutritional properties. One particular success involves cosuppression of the oleoyl desaturase in soybean; as a result, oleic acid has increased from less than 10% to more than 85% of the total fatty acids, and at the same time, saturated fatty acids have been reduced from more than 15% to less than 5%. Oil from these soybean varieties is expected to have both improved health benefits and improved stability.

Production of margarine and shortenings currently involves catalytic hydrogenation of unsaturated oils to provide a semisolid fat. The cost of this processing and the associated production of trans fatty acids has provided incentives to produce oils with increased contents of saturated fatty acids or a high-melting point monounsaturated fatty acid, such as petroselinic acid ($18:1^{\Delta 6}$). High stearate (18:0) Canola lines have been designed by antisense RNA suppression of the expression of the gene for stearoyl-ACP desaturase. These plants produce oils with 30% to 40% saturated fatty acids, a content that causes the oil to be semisolid at room temperature and thus suitable for margarine or shortening manufacture. Because petroselinic acid is a cis-unsaturated fatty acid but has a melting point above room temperature, oils containing this fatty acid could provide physical properties suitable for margarine manufacture but without the high saturated fatty acid content associated with health problems. A novel acyl-ACP desaturase responsible for petroselinic acid biosynthesis has been cloned from coriander and can function in transgenic plants. However, at least two other genes are required for high-level petroselinic acid biosynthesis, and efforts are underway to combine these in oilseed crops such as Canola.

10.11.3 Molecular genetic approaches have been used to increase oil yields.

In addition to efforts to modify the types of fatty acids produced in plant seeds, there is considerable interest in increasing the yield of oil obtainable from oilseed crops. But a major unanswered question in plant lipid metabolism remains: What determines the quantity of oil stored in a seed? In addition to its interest to basic researchers, this question is of considerable practical importance if chemicals derived from oilseeds are to compete economically with petrochemical alternatives.

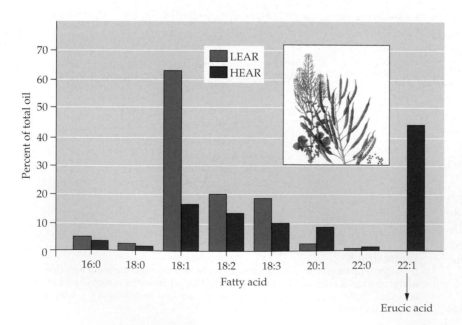

Figure 10.69
Fatty acid composition of seed oil from high- and low-erucic acid rapeseed (HEAR and LEAR, respectively) varieties. Additional improvements by plant breeding have reduced the 18:3 content of the most recent cultivars to a small percentage of the total fatty acids, with corresponding increases in the 18:1 content. Rapeseed is the major oilseed crop of Europe, Canada, and many countries with short growing seasons. After soybean and oil palm, it is the world's third largest source of vegetable oils. Varieties that have been bred to contain low amounts of erucic acid are referred to as Canola. Because of the high oil content (45% of seed weight) and the relative ease of transformation, Canola has become the first crop genetically engineered to produce new oils.

More than 30 reactions are required to convert acetyl-CoA to triacylglycerol, so many genes could control the yield of the end-product storage oil. Recently, two approaches have resulted in increased oil content in seeds. In the first, the gene for the cytosolic ACCase of *Arabidopsis* was fused to a plastid transit peptide sequence and the promoter of the gene for the *B. napus* seed storage protein napin. This chimeric gene was used to transform rapeseed. Because plastid ACCase is a highly regulated enzyme, its overexpression might result in its down-regulation and therefore might fail to increase net movement through the pathway. By targeting the cytosolic enzyme to the plastid, researchers hoped to circumvent feedback or other mechanisms that control the plastid enzyme. This approach was at least partially successful: The resulting plants had seed oil contents 5% higher than control transformants (Fig. 10.70). The second attempt to modify oil yields required manipulation near the end of the triacylglycerol biosynthesis pathway. Expression of a yeast *sn*-2 acyltransferase in *Arabidopsis* and rapeseed increased seed oil content by more than 25%. Both the ACCase and acyltransferase results have so far been demonstrated only in growth chambers and must be confirmed in field trials to establish whether increased oil yields per hectare can be achieved.

10.11.4 Fatty acids have numerous industrial applications.

Soybean oil accounts for approximately 68% of the US production of vegetable oil, with approximately 6.5 million tons of oil produced annually. Most plant-derived oil is currently used for food. However, nonfood industrial uses of plant lipids include the manufacture of soaps and detergents, paints, varnishes, lubricants, adhesives, and plastics (Table 10.5). The largest nonfood use, by volume, is lauric acid (C_{12}) from coconut and palm kernel for the production of detergents. Another large use is the production of ricinoleic acid (12-hydroxyoctadecenoic acid) for the production of a wide variety of compounds. Ricinoleic acid can be pyrolyzed to sebacic acid, which is used to produce certain types of nylon. The lithium salts of sebacic acid are also used as high-temperature greases for jet engines. Erucic acid is used to make erucamide, a slip agent in the plastics industry that makes plastic films and other products easier to handle.

At present, genetic engineering of oilseeds is largely concerned with altering the quality of oil produced in temperate crops. An immediate goal is to expand the range of fatty acids available from crop species so that the uses of plant fatty acids can be expanded. In the long term, perhaps plant

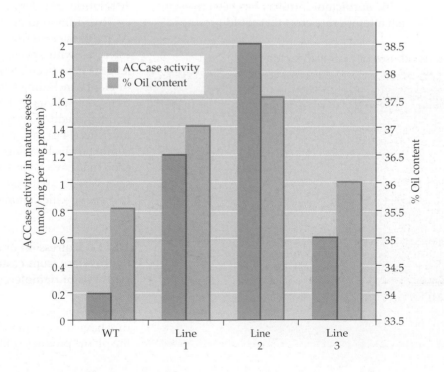

Figure 10.70
Expression of a cDNA that encodes cytosolic ACCase in *Brassica* plastids results in increased oil content of seeds. WT, wild type.

lipids will also be used as a source of fuel such as "biodiesel." Methyl esters of plant fatty acids have virtually the same performance characteristics in diesel engines as petroleum-derived diesel fuel but lack the pollutants produced by burning diesel fuel. Perennial tropical species such as oil palm are capable of producing as much as 100 barrels of oil per hectare per year with very low inputs of agrochemicals. Thus, when it becomes possible to genetically engineer oil palm, we may be able to produce a wide range of chemicals from a renewable source at a cost that is competitive with petroleum.

10.11.5 High-lauric-content rapeseed: a case study in successful oilseed engineering.

The acyl-ACP thioesterases have become the first enzymes of lipid biosynthesis for which genes were engineered and introduced into transgenic plants to produce a commercial product. A major ingredient of soaps, shampoos, detergents, and related products is the surfactants derived from medium-chain fatty acids and obtained from coconut or palm kernel oils grown in the tropics. Worldwide, approximately $1 billion worth of such oils are used for production of surfactants. In part because of periodic price instability in these raw material supplies, a long-term goal of the surfactant industry has been to establish a temperate crop that could produce

medium-chain fatty acids. Such a crop could provide alternative supplies and thereby lower or stabilize prices.

In the early 1980s, the advent of plant gene transfer techniques raised the question of whether temperate oilseed crops could be engineered to produce medium-chain fatty acids and other novel fatty acid compositions with the use of foreign genes. For example, it was unclear whether addition of new, unusual fatty acids might disrupt lipid metabolism or some other process in oilseed cells. More specifically, would a new fatty acid such as lauric acid be correctly targeted to triacylglycerol and excluded from membranes? This question arose because oilseeds usually segregate unusual fatty acids into the storage triacylglycerols and exclude them from membrane glycerolipids.

Work at the biotechnology company Calgene led first to biochemical demonstration of the existence of medium-chain acyl-ACP thioesterases and then to the cloning of 12:0-ACP thioesterase from the California bay (*U. californica*). Introduction of a single gene encoding the medium-chain acyl-ACP thioesterase into transgenic plants dramatically altered the chain length of fatty acids stored in the seed oils. In 1995, the first commercial production of a genetically engineered oil was obtained when about 1 million pounds of oil was extracted from rapeseed plants engineered to produce 40% to 50% lauric acid (Fig. 10.71).

In addition to the laurate-specific acyl-ACP thioesterase from California bay, thioesterases with specificity for other chain lengths have been identified from a variety of plants, including *Cuphea*. In several cases, transgenic plants that produce oils enriched in other medium-chain fatty acids, such as myristic (14:0) and decanoic (10:0), are becoming available. This type of research may soon make available a range of specialty rapeseed varieties with oil compositions tailored to meet specific commercial applications.

10.11.6 Expression of a glycerolipid hydroxylase from castor bean can drive synthesis of ricinoleic acid in tobacco.

Many of the unusual fatty acids found in nature have important industrial uses. However, the plants producing them are often poorly

Table 10.5 Some nonfood uses of plant fatty acids

Lipid type	Example	Major sources	Major uses	Approximate U.S. market 1989 ($ 10^6)
Medium-chain	Lauric acid (12:0)	Coconut, palm kernel	Soaps, detergents, surfactants	350
Long-chain	Erucic acid (22:1)	Rapeseed	Lubricants, slip agents	100
Epoxy	Vernolic acid	Epoxidized soybean oil, *Vernonia*	Plasticizers, coatings, paints	70
Hydroxy	Ricinoleic acid	Castor bean	Coatings, lubricants, polymers	50
Trienoic	Linolenic acid (18:3)	Flax	Paints, varnishes, coatings	45
Wax esters	Jojoba oil	Jojoba	Lubricants, cosmetics	10

suited for high-production agriculture. As an alternative, the isolation of key genes directing the synthesis of a particular fatty acid can provide the means to genetically engineer agronomically suitable oilseed crops to produce the desired oil more easily and cheaply. For example, ricinoleic acid (12-OH-18:1$^{\Delta 9}$) produced by castor bean is an extremely versatile natural product with industrial applications that include the synthesis of nylon-11, lubricants, hydraulic fluids, plastics, cosmetics, and other materials. However, castor bean contains ricin, an extremely toxic lectin (see Chapter 9, Box 9.3), as well as other poisons and allergens. In addition, agronomic problems result in poorer yields than other crops. As a result of these and other factors, castor bean is a minor crop grown mainly in nonindustrialized countries.

The successful cloning of a gene that encodes the hydroxylase responsible for 12-OH-18:1$^{\Delta 9}$ synthesis from 18:1$^{\Delta 9}$ relied on a detailed understanding of the biochemistry involved. For some time, the conversion of 18:1$^{\Delta 9}$ to 12-OH-18:1$^{\Delta 9}$ was thought probably to involve a single enzyme inserting a hydroxyl group at the same position as the double bond introduced by FAD2 desaturase (Fig. 10.72). Consideration of the probable reaction cycle of the desaturases suggested that 12-OH-18:1$^{\Delta 9}$ might be produced by a stalled desaturase. Partial sequencing of randomly chosen cDNAs from a library derived from developing castor bean endosperm

identified a relatively abundant cDNA that encodes a protein with homology to the glycerolipid desaturases of other oilseeds. Expression of the castor bean cDNA in transgenic tobacco plants led to the synthesis of 12-OH-18:1$^{\Delta 9}$ in seed tissues.

10.11.7 A Δ6-desaturase from borage, identified by homology to sequences of conserved membrane-bound desaturases, can catalyze synthesis of γ-linolenic acid in transgenic plants.

The successful use of random sequencing projects to clone genes for enzymes involved in the synthesis of unusual fatty acids does not require high levels of sequence homology. Because all the membrane-bound desaturases described from animals, microbes, and plants contain three highly conserved histidine-rich sequences, new members of the gene family can be identified easily, even when overall homology is low. For example, production of γ-linolenate (18:3$^{\Delta 6,9,12}$) in seeds of borage, black currant, and evening primrose is known to involve a Δ6-desaturase active on 18:2$^{\Delta 9,12}$ at the sn-2 position of phosphatidylcholine. To clone the glycerolipid Δ6-desaturase gene, researchers partially sequenced 600 cDNAs from a subtracted library prepared from developing borage seeds. One class of cDNAs showed low overall homology to other desaturases—about 30% identity at the amino acid

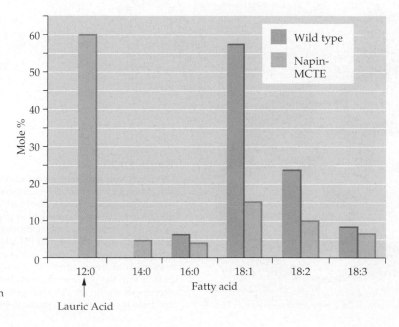

Figure 10.71
Expression of a cDNA from California bay that encodes a medium-chain thioesterase (MCTE) under control of the seed-specific napin promoter diverts carbon into medium-chain fatty acids to produce an oil with high laurate content.

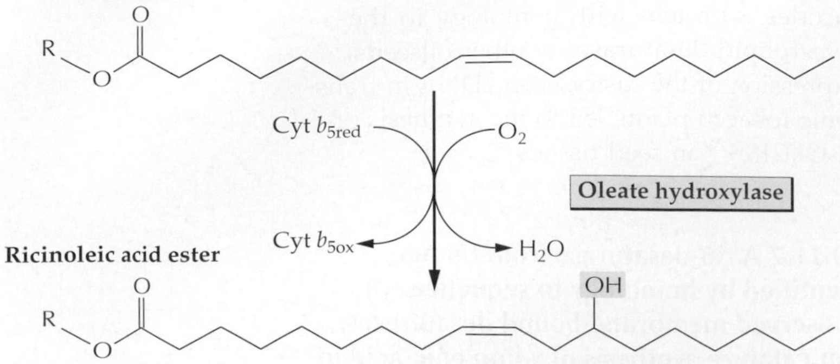

Oleic acid ester

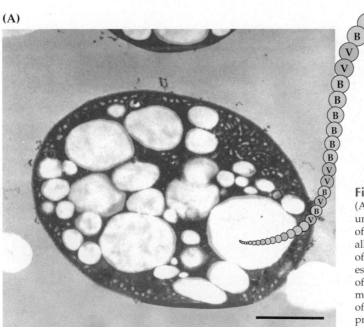

Ricinoleic acid ester

Figure 10.72
Reaction catalyzed by oleate hydroxylase. The R group is thought to be a lysolipid such as lysophosphatidylcholine.

level—but did contain characteristic histidine boxes with spacings similar to those in the Δ6-desaturase of *Synechocystis*. Expression of the borage cDNA in transgenic tobacco and carrot suspension culture protoplasts led to the production of γ-linolenate and 18:4$^{\Delta6,9,12,15}$.

10.11.8 Biodegradable plastics can be produced in plants.

Many species of bacteria synthesize and accumulate granules of biodegradable plastics called polyhydroxyalkanoates (PHAs). The majority of PHAs are composed of *R*-(–)-3-hydroxyalkanoic acid monomers ranging from 3 to 14 carbons long. More than 40 PHAs have been characterized, with some polymers containing unsaturated bonds or other functional groups. One such PHA produced commercially by fermentation of the bacterium *Alcaligenes eutrophus* is Biopol, a random copolymer of 3-hydroxybutyrate and 3-hydroxyvalerate units (Fig. 10.73). The copolymer combines biodegradability with water resistance and good physical properties, making it a suitable polymer for a wide variety of uses. The major drawback of Biopol and other bacterial PHAs is their high production cost, which makes them substantially more expensive than synthetic plastics and thereby restricts their large-scale use in consumer products.

Polyhydroxybutyrate (PHB), a PHA that is a linear polyester of 3-hydroxybutyrate, is synthesized from acetyl-CoA by the sequential action of three enzymes (Fig. 10.74). The first enzyme of the pathway, 3-ketothiolase, catalyzes the reversible condensation of two acetyl-CoA moieties to form acetoacetyl-CoA. Acetoacetyl-CoA reductase subsequently reduces acetoacetyl-CoA to *R*-(–)-3-hydroxybutyryl-CoA, which is then polymerized by the action of PHA synthase to form PHB. PHB is typically produced as a polymer of 10^3 to 10^4 monomers and accumulates as inclusions of 0.2 to 0.5 μm in diameter. In *A. eutrophus*,

(A)

(B)

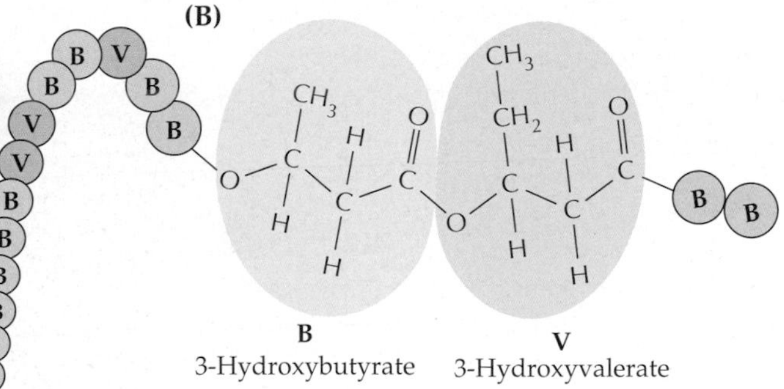

B
3-Hydroxybutyrate

V
3-Hydroxyvalerate

Figure 10.73
(A) Electron micrograph of a section through a cell of the bacterium *Azobacter vinelandii*. The light-colored granules are composed of polyhydroxybutyryrate. The inset bar is 1 μm. (B) Polyhydroxyalkanoates (PHAs) accumulate in granules composed of thousands of linear polymers containing thousands of monomers linked by ester bonds. One commercial PHA is Biopol, a random copolymer of hydroxybutyrate (B) and hydroxyvalerate (V), in which a B or V may occur at any position along the length of a polymer. The ratio of B to V in the polymer depends both on the concentration of the precursors within the cell and on the relative activity of the PHA synthase toward the monomeric substrates.

PHB inclusions can typically accumulate to 80% of the dry weight of bacteria grown in media containing excess carbon, such as glucose, but limited in one essential nutrient, such as nitrogen or phosphate. Under these conditions, PHB synthesis acts as a carbon reserve and an electron sink. When growth-limiting conditions are alleviated (by addition of phosphorus or nitrogen), PHB is catabolized to acetyl-CoA. PHAs are biodegradable because PHB and related copolymers are readily degraded by enzymes (PHA depolymerases) secreted by bacteria in a wide range of environments, including anaerobic sewage, compost, and landfill.

Recently, the possibility of producing PHA in plants has been investigated by expressing the PHB biosynthetic genes of *A. eutrophus* in *Arabidopsis*. Because the major movement of acetyl-CoA occurs in plastids, the pathway was targeted to this organelle. Plastids are also the site of starch accumulation (starch, like PHB, is synthesized as osmotically inert inclusions in plastids without disruption of organelle function). The *A. eutrophus* enzymes 3-ketothiolase, acetoacetyl-CoA reductase, and PHB synthase were targeted to the plastid by modifying the genes to encode the transit peptide of the small subunit of ribulose bisphosphate carboxylase at the N termini of the proteins. Genes for the modified bacterial enzymes were expressed in *Arabidopsis* under the control of a constitutive promoter. Transgenic plants expressing the genes produced PHB inclusions exclusively in the plastids (Fig. 10.75). The maximal amount of PHB detected in fully expanded leaves of transgenic plants was 10 mg/g fresh weight (about 14% dry weight). No significant deleterious effect on growth or seed yield was detected in plants that accumulated high amounts of PHB in plastids. These results demonstrated the possibility of producing PHA in plants to industrially significant amounts without detrimental effects on plant growth and viability.

Oilseed crops are regarded as the most amenable targets for seed-specific PHA production. Because both oil and PHB are derived from acetyl-CoA, metabolic engineering of plants for the diversion of acetyl-CoA toward PHB accumulation can be more directly achieved in the seeds of crops having a naturally high movement of carbon through acetyl-CoA.

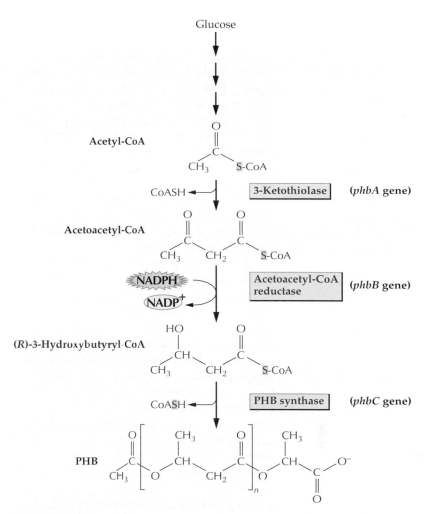

Figure 10.74
Pathway of polyhydroxybutyrate (PHB) synthesis in *Alcaligenes eutrophus*.

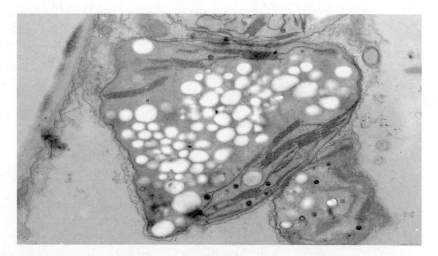

Figure 10.75
Accumulation of PHA inclusions in a chloroplast of a transgenic *Arabidopsis* plant. The oval inclusions are PHB grains. The normal morphology of the chloroplast has been distorted by the accumulation of the granules.

Summary

Lipids have diverse and essential roles in plants. As the hydrophobic barrier of membranes, they are essential for integrity of cells and organelles. In addition, they are a major form of chemical energy storage in seeds and are now recognized as a key component of some signal transduction pathways. Most lipids, but not all, contain fatty acids esterified to glycerol, and consideration of this area of metabolism involves, first, the synthesis of the fatty acid and, second, the synthesis of lipid after esterification of the fatty acid to form phosphatidic acid.

Fatty acid biosynthesis in plants is very similar to that in bacteria and is carried out in the plastid. Fatty acid synthesis begins with the carboxylation of acetyl-CoA to malonyl-CoA by acetyl-CoA carboxylase. This is the first committed step of fatty acid synthesis and is a likely site for regulation of the whole pathway. Acetyl-CoA and malonyl-CoA are subsequently converted into fatty acids by a series of reactions that add two carbons at a time to a growing chain. Acyl-carrier protein is a 9-kDa protein that transports the intermediates of fatty acid synthesis through the pathway. In plants, each reaction is catalyzed by a separate gene product—in contrast to fatty acid synthesis in animals, which depends on a multifunctional protein.

Fatty acid biosynthesis can be terminated by several different reactions, including hydrolysis of the thioester bond, transfer of the acyl group to a glycerolipid, and acyl desaturation. Plants contain an unusual stearoyl-ACP desaturase that is soluble and plastid-localized. Most fatty acyl desaturases are membrane proteins localized in the endoplasmic reticulum or the plastid.

Plants are capable of synthesizing unusual fatty acids, and more than 200 different fatty acids having been found in plants. Some of the enzymes that synthesize unusual fatty acids bear close resemblances to common fatty acid enzymes, such as membrane-bound desaturases. These unusual fatty acids are found almost exclusively in seed oils, and it is thought that they may serve a defense function.

There are two distinct pathways for the synthesis of membrane glycerolipids. The prokaryotic pathway, located in the chloroplast inner envelope, uses 18:1-ACP and 16:0-ACP for the sequential acylation of glycerol 3-phosphate and synthesis of glycerolipid components of the chloroplast membranes. The eukaryotic pathway involves (a) export of 16:0 and 18:1 fatty acids from the chloroplast to the endoplasmic reticulum as acyl-CoAs and (b) their incorporation into phosphatidylcholine and other phospholipids that are the principal structural lipids of all the membranes of the cell except the chloroplast. In addition, the diacylglycerol moiety of phosphatidylcholine can be returned to the chloroplast envelope and used as a second source of precursors for the synthesis of chloroplast lipids.

Membrane lipids serve as a hydrophobic barrier, delimiting the cell and dividing it into functional compartments. The membrane lipid composition also affects plant form as well as many cellular functions. For example, photosynthesis is impaired in plants lacking polyunsaturated membrane lipids, and lipid composition can affect chilling sensitivity and influence plant cell responses to freezing. In addition, membrane lipids function in signal transduction pathways and defensive processes.

Storage lipids play a distinctly different role from membrane lipids. Storage lipids are almost exclusively triacylglycerols and accumulate in discrete subcellular organelles called oil bodies. The mobilization and catabolism of triacylglycerols provides energy for germination and pollination. The catabolism of the released fatty acids occurs via β-oxidation in peroxisomes and glyoxysomes.

Further Reading

Browse, J., Somerville, C. R. (1994) Glycerolipids. In *Arabidopsis*, E. Meyerowitz and C. R. Somerville, eds. Cold Spring Harbor Laboratory Press, Cold Spring Harbor, NY, pp. 881–912.

Harwood, J. (1996) Recent advances in the biosynthesis of plant fatty-acids. *Biochim. Biophys. Acta* 301: 7–56.

Huang, A.H.C. (1993) Oil bodies in maize and other species. In *Biochemistry and Molecular Biology of Membrane and Storage Lipids of Plants,* N. Murata and C. Somerville, eds. American Society of Plant Physiologists, Rockville, MD, pp. 215–227.

Jeffree, C. E. (1996) Structure and ontogeny of plant cuticles. In *Plant Cuticle*, G. Kerstiens, ed. Biosis Scientific, Oxford, pp. 33–82.

Kerstiens, G. (1996) Signalling across the divide: a wider perspective of cuticular structure–function relationships. *Trends Plant Sci.* 1: 125–129.

Kolattukudy, P. E. (1996) Biosynthetic pathways of cutin and waxes, and their sensitivity to environmental stresses. In *Plant Cuticle*, G. Kerstiens, ed. Biosis Scientific, Oxford, pp. 83–108.

Kolattukudy, P. E. (1998) Biopolyester membranes of plants: cutin and suberin. *Science* 208: 990–1000.

Lynch, D. V., Spence, R. A., Theiling, K. M., Thomas, K. W., Lee, M. T. (1993) Enzymatic reactions involved in ceramide metabolism. In *Biochemistry and Molecular Biology of Membrane and Storage Lipids of Plants*, N. Murata and C. Somerville, eds. American Society of Plant Physiologists, Rockville, MD, pp. 183–190.

Moore, T. Jr., ed. (1993) *Lipid Metabolism in Plants*. CRC Press, Boca Raton, FL.

Murphy, D. J. (1996). Engineering oil production in rapeseed and other oil crops. *Trends Biotechnol.* 14: 206–213.

Ohlrogge, J. B. (1994) Design of new plant products: engineering of fatty acid metabolism. *Plant Physiol.* 104: 821–826.

Ohlrogge, J., Jaworski, J. (1997) Regulation of plant fatty acid biosynthesis. *Annu. Rev. Plant Phys. Plant Mol. Biol.* 48: 109–136.

Poirier, Y., Nawrath, C., Somerville, C. R. (1995) Production of polyhydroxyalkanoates, a family of biodegradable plastics and elastomers in bacteria and plants. *Biotechnology* 13: 142–150.

Post-Beittenmiller, D. (1996) Biochemistry and molecular biology of wax production in plants. *Annu. Rev. Plant Biochem. Mol. Biol.* 47: 405–430.

Robbelen, G., Downey, R. K., Ashri, A. (1989) *Oil Crops of the World: Their Breeding and Utilization.* McGraw-Hill, New York.

Schmid, K., Ohlrogge, J. B. (1996) Lipid metabolism in plants. In *Biochemistry of Lipids, Lipoproteins and Membranes*, D. E. Vance and J. Vance, eds. Elsevier Press, Amsterdam, pp. 363–390.

Somerville, C. R., Browse, J. (1996) Dissecting desaturation: plants prove advantageous. *Trends Cell Biol.* 6: 148–153.

Vick, B. A., Zimmerman, D. C. (1987) Pathways of fatty acid hydroperoxide metabolism in spinach leaf chloroplasts. *Plant Physiol.* 85: 1073–1078.

Biochemistry & Molecular Biology of Plants, B. Buchanan, W. Gruissem, R. Jones, Eds.
© 2000, American Society of Plant Physiologists

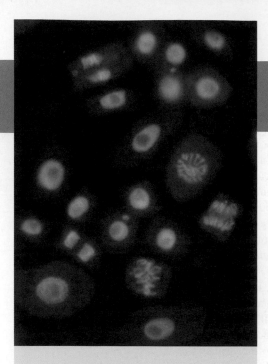

CHAPTER 11

Cell Division Regulation

Peter Doerner

CHAPTER OUTLINE

Introduction

Every cell is the product of a cell cycle. Cell proliferation is controlled precisely by the cell division control machinery, which ensures that cells divide only at appropriate times and that crucial components are replicated with adequate fidelity. The basic mechanisms governing cell division arose early in eukaryotic evolution and are highly conserved.

The cell cycle comprises two alternating events: **cell division,** during which a cell replicates its genome and partitions it to daughter cells, and **cell growth,** during which cells increase their mass by synthesizing proteins, membrane lipids, and other essential components. Cell growth and cell division serve different purposes. Cell growth increases the fitness of the individual cell and increases the likelihood of its survival. Controlled cell division perpetuates and spreads genetic information, thereby increasing the likelihood that a species will be successful and survive. To reconcile these divergent purposes, cell division and growth are flexibly coupled such that, in a given environment, most cells divide when reaching a particular size (Fig. 11.1). However, growth alone is generally insufficient to initiate division; cells must be stimulated by growth factor signals to divide.

In addition to regulating the timing of cell division, cell cycle control mechanisms perform a quality-control function to prevent transmission of incompletely replicated or damaged genomes to daughter cells. Incomplete replication of the genome, rereplication of DNA without division, or division before DNA replication is complete can all wreak havoc on the cell's progeny. To prevent such errors from occurring, all organisms utilize molecular mechanisms to monitor cell division progress at specific steps in the cell cycle, called **checkpoints.**

Cell growth and most biochemical processes are continuous, but the cell cycle proceeds in discrete, incremental steps. Eukaryotic cells have evolved specialized protein kinases, phosphatases, and proteases that function as switches to impose this stepwise mode of progression

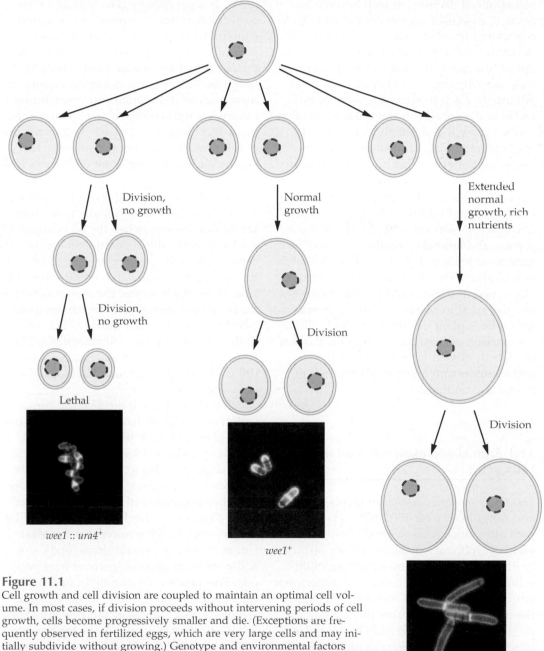

**Division,
no growth**

**Normal
growth**

**Extended
normal
growth, rich
nutrients**

**Division,
no growth**

Division

Division

Lethal

wee1 :: ura4⁺

wee1⁺

wee1⁺(5×)

Figure 11.1

Cell growth and cell division are coupled to maintain an optimal cell volume. In most cases, if division proceeds without intervening periods of cell
growth, cells become progressively smaller and die. (Exceptions are frequently observed in fertilized eggs, which are very large cells and may initially subdivide without growing.) Genotype and environmental factors
(e.g., nutrient availability) can also influence cell size. The division-without-
growth phenotype is observed in the fission yeast (*Schizosaccharomyces
pombe*) mutant for the Wee1 protein kinase, which ordinarily prevents premature mitosis (see Section 11.5.4). The left panel shows cells in which *wee1*
is disrupted by another gene (*ura4*). With adequate nutrients available,
a specific maximum cell volume is set and cells oscillate between this unit volume (immediately before division) and half of this volume (immediately after division). This homeostasis of cell volume is partly controlled
by the *wee1⁺* gene (middle panel). In above-optimal conditions, the set maximum cell volume increases.
This situation can be mimicked by increasing the amount of the wee1 protein kinase (right panel).

on the processes of DNA replication and cytokinesis. This network of regulatory proteins involved in monitoring cell cycle progression is also well suited to coupling cell cycle stages with the environmental conditions and facilitating quality control.

This chapter discusses the basic mechanisms of cell division, as well as some important aspects of growth control and DNA replication in plants. The discussion of cell division control mechanisms will initially focus on the molecules fundamentally responsible for cell cycle control in all eukaryotes. Although much progress has recently been made in dissecting the molecular events that control growth and cell division in plants, our understanding of these processes remains far better in other model systems. In particular, much of the experimental evidence discussed in this chapter is the result of research in budding yeast (*Saccharomyces cerevisiae*), fission yeast (*Schizosaccharomyces pombe*), and animal cell cultures, as well as in genetic systems such as the fruitfly *Drosophila melanogaster*. The molecular mechanisms elucidated in these model organisms are often also used in plants to regulate growth throughout plant development. To facilitate comparison of information from the various models discussed later in the chapter, we will compare and contrast cell division in plants and animals.

11.1 Animal and plant cells and their cell cycles

The last common ancestor of plants, animals, and fungi lived over 1500 million years ago, well after eukaryotes evolved. Despite this great evolutionary distance and the distinctly different appearances of modern multicellular organisms, the cellular machinery that performs fundamental functions, including the molecules that regulate the division cycle, is highly conserved in all extant eukaryotes. Among these conserved molecules are the protein kinases (see Box 11.1) that control major cell cycle transitions and their regulatory subunits; the enzymes responsible for DNA replication (see Chapter 6); the cytoskeletal structures necessary for chromosome movements during mitosis (see Chapter 5); and components of the ubiquitin-dependent pathway for degradation of protein (see Chapter 9), which performs several essential roles during the cell cycle. Somatic cells also share the fundamental four-phase subdivision of the cell division cycle (see next section).

The most distinctive features of plant cells with respect to control of cell division are their separation by cell walls and their replication of three genomes (nuclear, mitochondrial, and plastid). Although these structures require specialized cell division processes that are unique to the production of multicellular photoautotrophic organisms, most nuclear and cellular processes during mitosis are indistinguishable in plants, animals, and fungi (e.g., yeasts). A comparison of cell division in plants and animals, however, reveals some important differences (see Chapter 5). One notable distinction is that in animal cells, unlike plant cells, chromosomes move toward a discrete spindle pole. Also, late in cell division, when the two daughter nuclei separate, this separation in animal cells involves the progressive constriction of the plasma membrane at a central contractile ring. In plants, however, the two daughter nuclei are partitioned to form two separate cells by a cell plate that grows at the equator of the mother cell. The cell plate ultimately fuses with the plasma membrane and the cell wall that now surrounds both daughter cells (Fig. 11.2).

The structural and biochemical conservation of regulatory molecules over long periods of evolutionary time does not mean that these molecules are regulated similarly in all organisms. For example, multicellular plants and animals use entirely different approaches to control cell homeostasis. In animals, damaged or dead cells are replaced by cell division, which ensures that the organism maintains its general shape, body size, and number of appendages over long periods. This process requires dedicated stem cells to define many different cell lineages. A balance of cell division and programmed cell death ultimately maintains appropriate steady-state cell numbers. In contrast, plants grow continuously by repeated organ formation, and damaged or dead plant organs are replaced entirely. Therefore, animals have evolved very elaborate regulatory mechanisms to prevent inappropriate cell proliferation and can suffer catastrophic illness when these regulatory mechanisms fail, such as

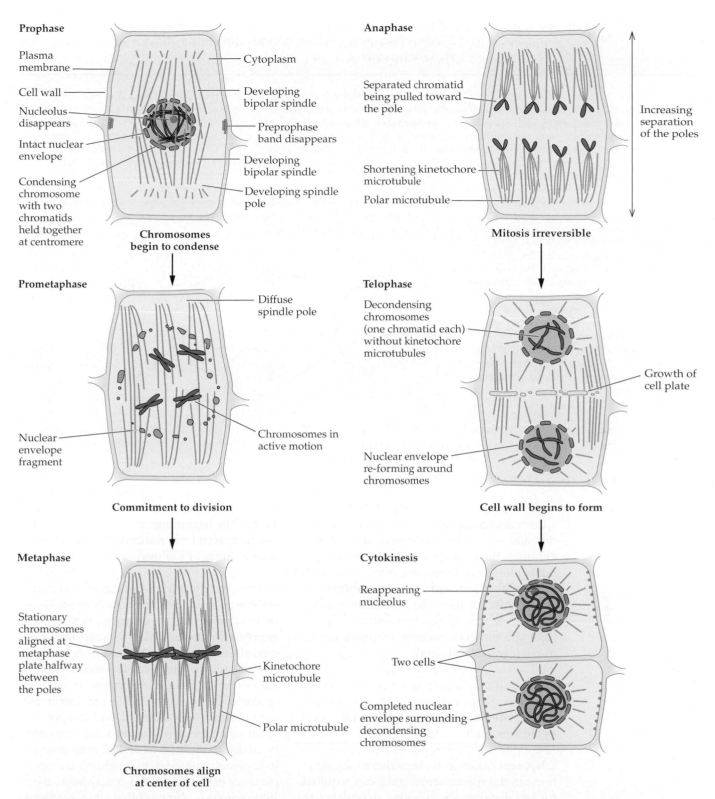

Prophase

- Plasma membrane
- Cell wall
- Nucleolus disappears
- Intact nuclear envelope
- Condensing chromosome with two chromatids held together at centromere
- Cytoplasm
- Developing bipolar spindle
- Preprophase band disappears
- Developing bipolar spindle
- Developing spindle pole

Chromosomes begin to condense

Prometaphase

- Diffuse spindle pole
- Nuclear envelope fragment
- Chromosomes in active motion

Commitment to division

Metaphase

- Stationary chromosomes aligned at metaphase plate halfway between the poles
- Kinetochore microtubule
- Polar microtubule

Chromosomes align at center of cell

Anaphase

- Separated chromatid being pulled toward the pole
- Shortening kinetochore microtubule
- Polar microtubule
- Increasing separation of the poles

Mitosis irreversible

Telophase

- Decondensing chromosomes (one chromatid each) without kinetochore microtubules
- Growth of cell plate
- Nuclear envelope re-forming around chromosomes

Cell wall begins to form

Cytokinesis

- Reappearing nucleolus
- Two cells
- Completed nuclear envelope surrounding decondensing chromosomes

Figure 11.2

Mitosis in plants. By the end of the 19th century, microscopy of plant cells had shown that steady increases of cell volume in proliferating cells were punctuated by the disappearance of the nucleus and the appearance of condensed chromosomes (prophase), followed by their alignment along the cell's equator (metaphase). Chromosomes appeared to separate and move toward the cell poles (anaphase), which was followed by the reappearance of two nuclei (telophase), and the subsequent formation of separate daughter cells, completing cytokinesis. Most aspects of mitosis in plant cells are similar to mitosis in animal cells, but the two types of cells differ in their spindle poles (which are more diffuse in plant cells than in animal cells) and in the mechanisms by which daughter nuclei are separated (see Chapter 5, Fig. 5.46 for a comparison).

Protein kinases and phosphatases are important components of the signaling networks that control cell division, growth, cytoskeleton, and metabolism. They function as signal amplifiers and modify the cellular response relative to the magnitude of input signals.

Most protein kinases modify proteins on specific serine, threonine, or tyrosine residues by catalyzing the reversible covalent addition of a phosphate group ($-PO_3^{2-}$) released by hydrolysis of ATP or GTP. After phosphorylation, the polar but uncharged side chains of these amino acids are transformed into charged residues. This usually leads to marked changes in the three-dimensional structure of the kinase substrate. Such changes generally affect protein activity, but they can also affect a protein's stability or its subcellular localization. Protein phosphatases hydrolytically cleave phosphate groups from their substrate proteins. In combination, the activities of protein kinases and phosphatases determine what fraction of the total number of molecules of a specific protein is phosphorylated.

Phosphorylation/dephosphorylation can represent a binary mechanism of regulation, in which the substrate protein demonstrates one of two activity states, depending on whether it is phosphorylated (see diagram). However, many highly regulated proteins are phosphorylated at several different sites. In such cases, covalent modification does not usually render the protein active or inactive, like a toggle switch, but rather leads to more subtle, graded changes of activity, like a dimmer switch. The activities of protein kinases and phosphatases can themselves be regulated by phosphorylation and dephosphorylation, often being modulated by alterations in access of the requisite substrate to the active site of the kinase or phosphatase.

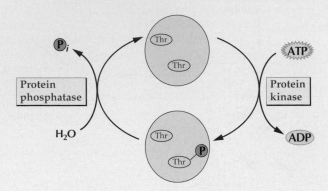

when cancers develop. Uncontrolled cell division is typically not observed in plants, although the reasons for this difference are not understood. Moreover, plant growth and cell proliferation are very responsive to environmental conditions because plant organs must grow toward nutrient sources (e.g., light and minerals), whereas animals can move toward their food.

11.2 Historical perspective on cell cycle research

DNA replication and chromosome segregation are the fundamental processes required for cell division. Proliferating cells alternate between these two mutually incompatible states. Transitions between these states are regulated by cell cycle control mechanisms. Elucidation of these mechanisms has been a very active area of research during the final decades of the 20th century.

11.2.1 The beginnings of cell cycle research can be traced to important discoveries in several areas of biology.

In 1869, Friedrich Miescher discovered that DNA was a major component of the nucleus. By the end of the 19th century, microscopic examination had revealed that proliferating cells alternate between two states, **interphase** and **mitosis.** In interphase, no substantial nuclear structures are discernible. In mitosis, chromosomes become visible, are distributed to opposite poles of the cell, and disappear again when the newly formed cells separate by division. By 1944, DNA had been shown to be genetic material, highlighting the importance of the nucleus in cell division. In 1951, scientists observed that DNA synthesis in bean root tip cells occurred during interphase. Thus, the cell cycle was determined to consist of four phases rather than two: Interphase is subdivided into a synthesis phase (S) and two "gap" phases (G1 and G2) that separate DNA synthesis from mitosis

(M). In sequence—G1, S, G2, M—the four phases constitute one complete round of the cell cycle (Fig. 11.3). The discovery in 1953 by James Watson and Francis Crick that the nucleotide bases of DNA are arranged as a complementary double helix immediately suggested a likely mechanism for how the encoded genetic information is replicated during S phase. The stage was set for the remarkable advances in cell cycle research that have characterized the past three decades.

11.2.2 During the past 30 years, genetics, biochemistry, and cell biology have been instrumental in elucidating molecular details of the cell cycle.

Progress in several fields of biological research has combined to rapidly advance our understanding of how cells divide. Genetic analysis, biochemical complementation, and cell fusion have provided important pieces of evidence that together reveal a great deal about the cell cycle puzzle. First, in 1970 Lee Hartwell and colleagues showed that cell cycle regulation could be dissected genetically in the single-celled yeast *S. cerevisiae*. Genetic analysis was used extensively to identify the genes encoding the essential components of the cell division machinery and to clarify the functional relationships of the gene products. Conditional cell cycle mutants were especially useful because they could grow at permissive temperatures, usually around 20°C to 25°C, but not at a restrictive temperature, usually 36°C. Fortunately, the budding yeast cell cycle also has a readily visible marker that facilitates screening for mutants—bud formation. The nascent bud, which appears at around the time that DNA synthesis is initiated, grows throughout the cell cycle, so that a mutant phenotype can be related to a specific ratio of the volume of the mother cell to that of the daughter cell. The logic behind screening for **cell division control** (*cdc*) mutants was to identify colonies in which growth was arrested at a particular stage when the culture was transferred to the nonpermissive temperature (Fig. 11.4A). In contrast, mutations in non–cell cycle-related genes (e.g., genes encoding metabolic enzymes) were predicted to result in an arrest of proliferation at random time points of the cell cycle, that is,

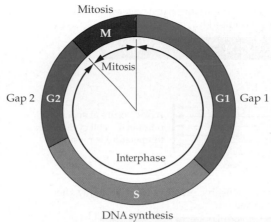

Figure 11.3
Phases of the cell cycle. Microscopic investigation of the cell cycle reveals two distinct states: mitosis, characterized by chromosome condensation and nuclear division; and interphase, during which the chromosomes appear diffuse and cells do not divide. The discovery that DNA synthesis occurs during interphase led to the identification of four cell cycle phases: mitosis (M), DNA synthesis (S), and two gap phases (G1 and G2) that separate these events.

times at which the cell was deprived of a particular metabolite (Fig. 11.4B). Time-lapse photography was used to identify the *cdc* mutants. The arrest phenotype of the *cdc* mutant identified how far the cell cycle could progress without the presence of the wild-type gene product—the **termination** or **arrest point.** However, it did not reveal when the wild-type gene product was normally required. Accomplishing this task required the use of another technique, **synchronization.**

In a proliferating population, individual cells are in different phases of the cell cycle, making it extremely difficult to study proteins or RNA molecules that are specific to individual phases of the cell cycle. However, cultures can be synchronized by blocking the progression of cells at a specific point in the cell cycle. Synchronization of yeast cells can be achieved by using drugs that inhibit crucial processes, such as DNA synthesis. Once all the cells in a culture have been arrested at the same stage of the cell cycle, the cell cycle block is relieved by removing the inhibitor, and the cells subsequently proceed through the cell cycle synchronously for one or two rounds. By transferring synchronized cells at successive stages of the cycle to a restrictive temperature and subsequently monitoring their fate by time-lapse photography, investigators can identify the point in the cycle at which the wild-type gene product is required—the **execution point** (Fig. 11.4C).

Biochemical analysis, particularly in combination with cell biology, also provides powerful tools that researchers have used to examine the function of specific proteins. Two techniques often used together are analysis of phase-specific protein extracts

(A) Cell division cycle *(cdc)* mutants arrest uniformly after temperature is raised.

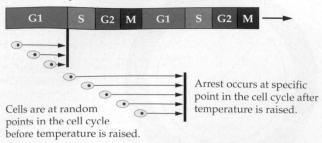

Cells are at random points in the cell cycle before temperature is raised.

Arrest occurs at specific point in the cell cycle after temperature is raised.

(B) Other temperature-sensitive mutants arrest randomly after temperature is raised.

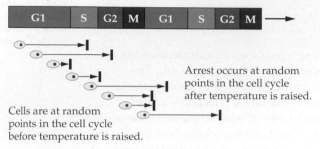

Cells are at random points in the cell cycle before temperature is raised.

Arrest occurs at random points in the cell cycle after temperature is raised.

(C) Mapping the point when *CDC* gene function is required

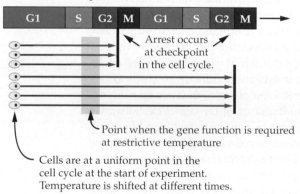

Arrest occurs at checkpoint in the cell cycle.

Point when the gene function is required at restrictive temperature

Cells are at a uniform point in the cell cycle at the start of experiment. Temperature is shifted at different times.

Figure 11.4

A genetic screen to identify conditional cell division cycle *(cdc)* mutants of yeast. Unlike conditional *cdc* mutants (A), which arrest at the same phase in the cell cycle after the temperature is increased to the restrictive temperature, other conditional mutants (B) arrest at random points in the cell cycle after exposure to the restrictive temperature. Such experiments allow the identification of the termination, or arrest, point—the point in the cell cycle beyond which mutant cells cannot progress without the gene product. Once a *cdc* mutant is identified, synchronized cell cultures can be used to determine the execution point—the point in the cell cycle at which the gene product is required (C). The temperature is shifted at different times for different synchronized cell cultures, as denoted by the transition of the arrow color from red to blue. Those cultures exposed to the restrictive temperature before they reach the execution point will arrest at the next termination point. Cultures exposed to the restrictive temperature after they reach the execution point can proceed through another round of the cell cycle before arresting.

and biochemical complementation (Fig. 11.5). In one experiment, clam oocytes were stimulated to divide; then protein extracts from the cells were prepared at different times after stimulation and the pattern of proteins was analyzed. This approach identified a set of unstable proteins (**cyclins;** see Section 11.5.3) that appear and disappear at specific phases of the cell cycle. In other experiments, protein extracts from cells undergoing mitosis were fractionated, and the fractions were combined with extracts from arrested oocytes or injected into arrested oocytes. By identifying which fractions released the arrested extracts or oocytes, investigators were able to purify a **mitosis-promoting factor** (MPF). The peak cyclin abundance was later found to correlate well with the maximum biochemical activity of MPF.

Finally, a series of cell fusion experiments were performed, first with human cell lines by Potu Rao and Robert Johnson (Fig. 11.6) and later also with plant cells by Denes Dudits. In these experiments, cells in different stages of the cell cycle were fused to observe how the nuclei responded. When cells in G1 were fused with S-phase cells, the G1 nuclei rapidly advanced into DNA replication. This indicated that although G1 cells were competent to replicate their DNA, this process required activation by a factor already present in S-phase cells. However, when G2- and S-phase cells were fused, only the S-phase nucleus continued to replicate its DNA. Moreover, the G2 nucleus did not progress into M phase after this fusion, as it ordinarily would have. This indicated that G2 nuclei were not competent to rereplicate and that a factor present in S-phase cells prevented G2 nuclei from entering mitosis. When G2 cells and G1 cells were fused, the G1 nucleus replicated DNA, but on a schedule similar to that of nonfused G1 cells, whereas the G2 nucleus did not replicate. Together, these observations demonstrated that S-phase cells contain a labile activator that can stimulate G1 nuclei (but not G2 nuclei) to enter S phase. Cells also contain a repressor of mitosis, which is active in S-phase cells and inhibits DNA rereplication before mitosis.

Genetic analysis, biochemical complementation, and cell fusion experiments allowed scientists to identify *cdc* mutants and

(A)

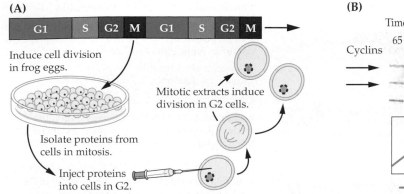

Induce cell division in frog eggs.

Mitotic extracts induce division in G2 cells.

Isolate proteins from cells in mitosis.

Inject proteins into cells in G2.

(B)

Time after fertilization (min)

Cyclins   65 70 75 80 85 90 95 100 105 110 115 120 125

—— Cyclin abundance   ▲ MPF activity

Figure 11.5

(A) Extracts taken from mitotic *Xenopus* oocytes contain proteins that induce G2 cells to undergo mitosis. This and similar experiments were used to isolate a mitosis-promoting factor (MPF) capable of inducing G2 cells to divide. (B) Biochemical analyses of fractions of cell extract taken at various stages of the cell cycle reveal the presence of cyclins, soluble protein factors that demonstrate cyclic changes in abundance during the cell cycle. Cyclin abundance correlates with the activity of MPF.

indicated that regulatory molecules were involved in coordinating the complex events required for cell division. Subsequent research efforts that focused on identifying these factors led to a convergence in the 1980s of the previously separate genetic, molecular, and biochemical approaches to cell cycle analysis. Technical advances continue to assist researchers in their efforts to elucidate cell cycle control mechanisms (Box 11.2).

Subsequent sections of the chapter will discuss the regulation of DNA replication and mitosis and will identify molecules that function during these processes. The molecular nomenclature of these cell cycle proteins, which can be very confusing, has been simplified whenever possible. For historical reasons, a protein conserved in many organisms may have many different names in the literature. In subsequent discussions here, however, only one name will be used for homologous genes or proteins. Where necessary to avoid confusion, a prefix will be added to the general name to distinguish between species (e.g., *AtCDC2* for *Arabidopsis thaliana CDC2*).

11.3 DNA replication

11.3.1 DNA replication is strictly controlled during the cell cycle.

As stated earlier, dividing cells alternate between two mutually incompatible states; that is, initiation of DNA synthesis is inhibited in wild-type cells during G2, M, and G1.

This inhibition suppresses two different types of cell division errors. DNA synthesis in G2 and early M would lead to a change in **ploidy** (DNA content and genome copy number) and interfere with chromosome segregation, whereas DNA synthesis in anaphase, telophase, and G1 would not be coupled to cell growth.

Although DNA synthesis remains inhibited during late M and G1, the assembly of the protein complexes that mediate initiation of DNA synthesis is promoted. As a result, the cell becomes competent to initiate DNA synthesis. In contrast, DNA synthesis is promoted during S phase, while assembly of the protein complexes that facilitate the initiation of DNA synthesis is inhibited. Thus, although replication of nuclear DNA in S phase is the critical event of the chromosomal cycle, this event is entirely dependent on the cell reacquiring competence during M and G1 phases to initiate replication later in S phase.

DNA synthesis in S phase is initiated at discrete origins of replication distributed at regular intervals throughout the genome, occurring on average every 36 kb in yeast, 66 kb in dicotyledonous plants, and 47 kb in monocotyledonous plants (see also Chapter 6). Origins of replication are important in the regulation of DNA synthesis, and a large number of proteins interact directly and indirectly with the origins to control progression through the chromosomal cycle. Not all origins initiate DNA synthesis simultaneously at the beginning of S phase; rather, DNA synthesis is initiated throughout

(A)

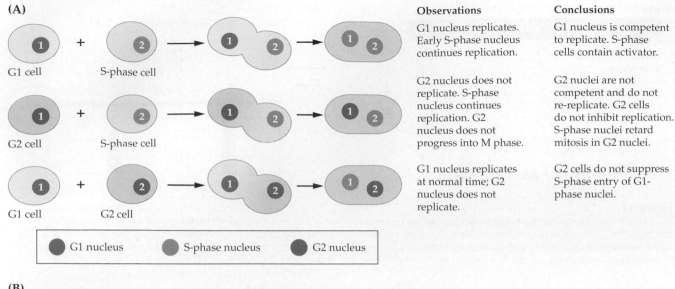

Observations	Conclusions
G1 nucleus replicates. Early S-phase nucleus continues replication.	G1 nucleus is competent to replicate. S-phase cells contain activator.
G2 nucleus does not replicate. S-phase nucleus continues replication. G2 nucleus does not progress into M phase.	G2 nuclei are not competent and do not re-replicate. G2 cells do not inhibit replication. S-phase nuclei retard mitosis in G2 nuclei.
G1 nucleus replicates at normal time; G2 nucleus does not replicate.	G2 cells do not suppress S-phase entry of G1-phase nuclei.

(B)

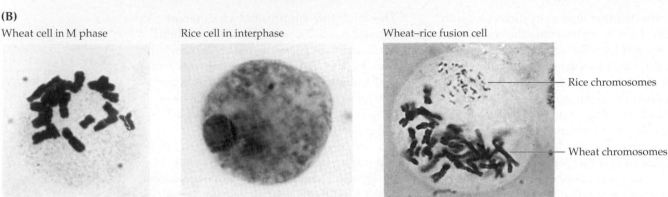

Figure 11.6
(A) Comprehensive fusion experiments conducted with animal cells revealed that diffusible factors regulated cell division progress, whereas nondiffusible factors associated with chromosomes determined whether the chromosomes were competent to respond to the diffusible factors. (B) Cell fusions of plant cells at different stages of the cell cycle. Plant cell protoplasts, generated by removing the cell walls by enzymatic digestion, can be fused together. To readily distinguish the origin of chromosomes in the fused protoplasts, investigators can use cells from different species with distinct chromosome morphologies. In this experiment, mitotic wheat protoplasts (condensed chromosomes, left panel) were fused with rice protoplasts in interphase (chromosomes are not condensed and therefore not visible, middle panel). After fusion, the rice chromosomes rapidly condensed and became visible (right panel). This suggests that the mitotic wheat cells contain diffusible factors sufficient to initiate chromosome condensation in interphase cells.

S phase. Therefore, origins of DNA replication are classified into early, intermediate, and late types (Fig. 11.7). At present, how these differences in timing are regulated is not well understood.

11.3.2 Many molecules are involved in controlling S-phase progression.

Origins of DNA replication are bound throughout the cell cycle by an **origin recognition complex** (ORC; Fig. 11.7 and Table 11.1). In budding yeast, this complex consists of six proteins, designated Orc1 to 6 in order of increasing size. The ORC serves as a docking point for additional proteins that recognize the pre- and postreplicative states of origins. ORC also interacts with the DNA polymerases that catalyze DNA synthesis. Additional proteins—e.g., Cdc6p, MCM proteins, Cdc45p, and the Cdc7p/Dbf4p protein kinase—are required in yeast to organize the proper function of origins of replication (Fig. 11.8 and Table 11.1). The association of ScCdc6p with ORC is required to establish

Box 11.2

New techniques continue to advance molecular analysis of the cell cycle.

Recent technological advances have allowed the development of automated methods to study cell cycle events. These techniques include the use of **flow cytometry** and **DNA microarray** or **gene chip** analysis.

Flow cytometry allows the relative DNA content of individual cells to be measured as shown here in panel A. A dilute cell suspension (or, in the case of plants, a suspension of isolated nuclei) is treated with a fluorescent DNA-specific dye and expelled from a small nozzle, generating small droplets. These drops are irradiated with a laser; the resulting fluorescence of each cell is proportional to its amount of DNA. By analyzing a sufficient number of cells, one can obtain a representative distribution of the cell cycle phases within a population. In a nonsynchronized proliferating population, most of the cells at any given time are in G1

phase, with a smaller number in S phase and G2/M. This is reflected in the flow diagram: The largest fraction of cells contains the relative amount of DNA corresponding to the nonreplicated state (green), followed by a population in S phase with an intermediate DNA content (orange) and a fraction of cells in G2 (blue) or M (red) with a DNA content corresponding to the replicated state. Flow cytometry is often used to monitor DNA synthesis and mitosis in a synchronized population. By examining the population at different times after release of the cells from cell cycle arrest, one can determine when the population commences DNA synthesis or initiates mitosis. Flow cytometry is also an excellent tool for determining the execution point of cell cycle mutants. Because the nuclei of plant cells undergoing mitosis lack envelope membranes (see Fig.

11.2), however, they cannot be visualized with this technique.

DNA microarray or gene chip analysis is a new technique used to monitor the expression of genes at the mRNA level. Once an organism's genome has been sequenced, DNA fragments corresponding to all the transcriptional units of the genome can be immobilized in a grid on a glass microscope slide (microarray) or silicon chip (gene chip). The DNAs can then be hybridized with fluorescently labeled transcripts isolated from cells at different phases of the cycle. After excitation by laser light, the amount of fluorescence emitted is proportional to the number of hybridized transcripts. This permits the quantitative, parallel analysis of essentially all transcripts in a cell as a function of cell cycle phase, as shown in panel B.

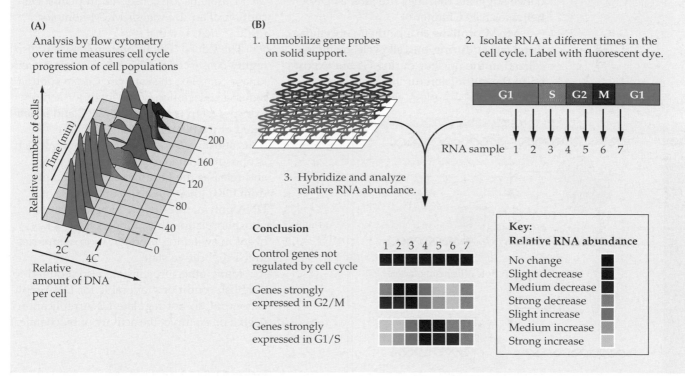

(A)

Analysis by flow cytometry over time measures cell cycle progression of cell populations

(B)

1. Immobilize gene probes on solid support.

2. Isolate RNA at different times in the cell cycle. Label with fluorescent dye.

3. Hybridize and analyze relative RNA abundance.

Conclusion

Control genes not regulated by cell cycle

Genes strongly expressed in G2/M

Genes strongly expressed in G1/S

Key:
Relative RNA abundance

No change
Slight decrease
Medium decrease
Strong decrease
Slight increase
Medium increase
Strong increase

the prereplication complex. ScCdc6p is a very unstable protein that accumulates only briefly at the end of M phase. After the ScCdc6p–ORC complex is formed in telophase, an additional 50 bp of DNA flanking the origin of replication becomes complexed with proteins, indicating that ScCdc6p

facilitates formation of a larger DNA–protein complex.

MCM proteins are a family of six related essential proteins required for initiating DNA replication. Experiments to reveal protein–protein interactions suggest that MCM proteins assemble into a large

Figure 11.7
A multiprotein complex binds to origins of replication. The origin recognition complex (ORC) is a landmark that instructs the cell where to initiate DNA replication. The ORC also serves as a docking point for replication proteins. Individual origins initiate replication at different times during S phase (early, middle, and late origins). This model for replication of discrete origins was developed from electron microscopy data. The dotted line represents the lagging DNA strand.

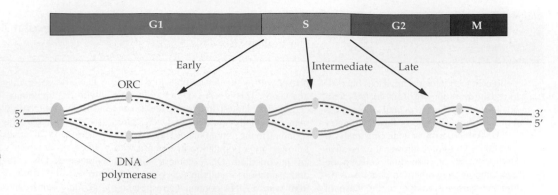

doughnut-shaped complex around the DNA helix. Stabilization of this protein complex, which binds weakly to the ORC, requires Cdc45p. MCM proteins are much more abundant than other proteins associated with ORC and may also associate with the intervening chromatin that separates the replication origins. MCM proteins bind and hydrolyze ATP, and certain biochemical evidence suggests that they function as DNA helicases (see Chapter 6).

MCM proteins change their subcellular localization during the cell cycle, being localized in the nucleus during G1 but exported to the cytoplasm during S phase and

remaining there in G2. This redistribution and their interaction with the Cdc6p–ORC complex suggest that MCM proteins are involved in ensuring that the replication origins fire (initiate replication) only once during the chromosomal cycle. This activity has been called **licensing factor,** to indicate that nuclear-localized MCM proteins give the cell a license for one round of DNA replication. MCM proteins are conserved in plants. Disruption of an *Arabidopsis* MCM homolog, *PROLIFERA*, is lethal.

The Cdc7p/Dbf4p protein kinase is also highly conserved. In this heteromeric protein kinase complex, the catalytic Cdc7p subunit requires association with Dbf4p for activity. Cdc7p/Dbf4p is tethered to ORC and is activated at the beginning of the S phase. One of its substrates is the MCM5 protein, which is phosphorylated at this stage of S phase and releases the hexameric MCM complex from ORC, thereby facilitating access of DNA polymerases to the template. MCM phosphorylation thus appears to be a key event in switching prereplication complexes to the replicative state.

Many other proteins are also required to establish conditions favorable for DNA replication but do not regulate DNA replication itself. For example, the activity of biochemical

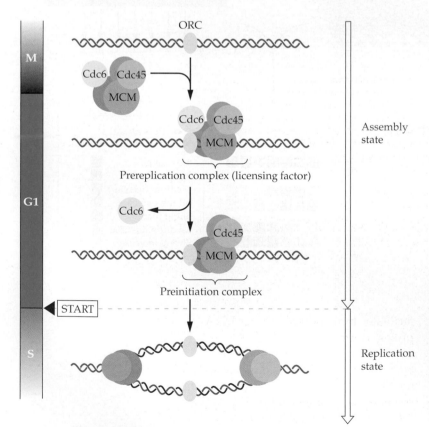

Figure 11.8
Stepwise assembly of the prereplication complex. To prevent premature DNA replication, the replication proteins that associate with ORC are assembled in steps. Cdc6 protein binds first, followed by MCM proteins and Cdc45 protein. The prereplication complex is activated by protein phosphorylation at the restriction point (START), the point in the cell cycle at which a cell is committed to a round of cell division, which defines the beginning of S phase. Phosphorylation by the Cdc2p/Dbf4p protein kinase (discussed in text) is not shown.

pathways required to synthesize the nucleotide triphosphate substrates used during DNA synthesis (see Chapter 6) is greatly stimulated just before the onset of S phase. During replication, chromosomal DNA remains organized in **chromatin,** a complex composed largely of **histone** proteins (see Chapter 7). Therefore, a substantial amount of new histones must be synthesized in S phase. Histone gene expression and protein accumulation are strongly stimulated in early S phase (Fig. 11.9).

11.3.3 Establishing the prereplicative state is a multistep process.

The chromosome cycle initiates at the transition from metaphase to anaphase (see Fig. 11.2). At this time, the sister chromatids are irreversibly separated and the block to DNA rereplication is lifted. The metaphase-to-anaphase transition requires the activation of the anaphase-promoting complex (APC), which facilitates the destruction of specific proteins (see Section 11.5.5). APC activation subsequently also results in the degradation of mitotic cyclins, which marks the end of mitosis and permits the cell to exit from M phase.

In yeast, the first step of a new chromosome cycle occurs when Cdc6p interacts with ORC (see Fig. 11.8). The intrinsically very labile Cdc6p associates with ORC proteins only in the absence of mitotic kinase activity. Thus, Cdc6p–ORC complexes can form only after M phase is complete and before S phase begins. During this period, the ORC–Cdc6p prereplication complex serves

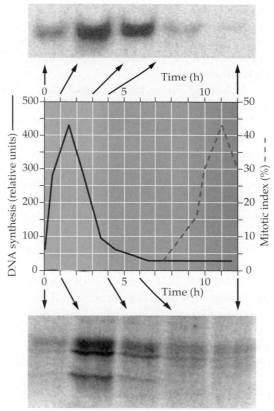

Figure 11.9
Histone expression is regulated by the cell cycle. Histones are basic proteins involved in organizing chromatin structure. During S phase, the cellular content of histones must be doubled to assemble with the newly replicated DNA. Therefore, the synthesis of histone-specific RNA (top panel) and the accumulation of histones (bottom panel) are strongly stimulated in S phase (0 to 4 hours in this experiment) but not in M phase (8 to 12 hours). The activity of cell cycle phase-specific transcription factors is responsible for this peak of histone mRNA accumulation.

Table 11.1 Molecules involved in progression through S and M phases

Name	Function
ORC	Origin recognition complex. Contains many proteins and binds to replication origins.
Cdc6p	Small, labile protein that interacts with the ORC to signal the ability to prepare for S phase.
MCM	Six related proteins that interact with ORC. May facilitate DNA replication by DNA polymerases, involved in restricting DNA synthesis to once per cycle.
Cdc45p	Stabilizes the binding of MCM proteins to the ORC.
Cdc7p	Forms heteromeric complex with Dbf4p. Complex triggers onset of S phase by activating DNA replication at the ORC.
Dbf4p	Forms heteromeric complex with Cdc7p. Complex triggers onset of S phase by activating DNA replication at the ORC.
APC	Anaphase-promoting complex. Ubiquitinates target proteins for subsequent destruction.
Scc1p	A cohesin that maintains linkage between sister chromatids during the period separating S phase and metaphase.
Pds1p	Inhibitor of anaphase onset. Inhibits Cut1p activity.
Cut1p	Promotes the separation of sister chromatids.

as a docking center and platform for the assembly of MCM proteins. The association of MCM proteins with the ORC–Cdc6p pre-replication complex remodels the replication origin, which then assumes the primed pre-replication state.

In late G1 phase, the cell makes the decision whether to proceed with proliferation at the **restriction point** (or START in the budding yeast cell cycle). Once the cell has passed this point in the cell cycle, it is committed to S phase. Although replication origins are competent to initiate DNA synthesis at this time, cells are not able to enter S phase before the expression of S-phase genes (e.g., histone genes) has been activated. S phase is then initiated when primed prereplication complexes are activated by phosphorylation of Cdc46p (MCM5), which is catalyzed by the Dbf4p-dependent kinase, Cdc7p. High activities of S-phase kinase, and later of M-phase kinase, are required to maintain the postreplicative state of DNA until telophase, at which time new Cdc6p–ORC complexes can assemble. If mitotic kinase activity is experimentally suppressed in S or G2 phase, rereplication will occur without intervening mitosis.

To maintain genome integrity, mitosis must be suppressed in G1 to prevent chromosome segregation before DNA replication, a process that would lead to chromosome loss **(aneuploidy).** Mitosis must also be delayed in S phase to prevent segregation before replication is complete (mitotic catastrophe). Mitosis in G1 is prevented by Cdc6p, which is inactivated after initiation of DNA replication. Therefore, a different signal is required to prevent premature mitosis before completion of S phase. This signal is probably the continued activity of DNA polymerase ε (POL2) during DNA synthesis, because mutation of a noncatalytic C-terminal domain of this enzyme relieves the inhibition of mitosis during S phase.

11.4 Mitosis

Most of our recent understanding of the molecules involved in organizing mitosis is drawn from genetic and physiological experiments with animal cells and yeast. Although far less is known about the nature and regulation of the molecules involved in mitosis in plants, current experimental evidence indicates that regulation of mitosis in plants resembles that in animals and yeast.

11.4.1 After S phase, cells acquire competence for chromosome segregation.

How does a cell become competent for mitosis? As we learned earlier, mitosis is suppressed during phases G1, S, and G2. The stepwise process that promotes mitosis is initiated in S phase. Mitosis commences with the initiation of chromosome condensation and the disassembly of the nuclear envelope that separates the nuclear matrix from the cytoplasm (see Fig. 11.2). This is often referred to as the onset of M phase, but the cells are not yet competent for chromosome segregation. Full competence for separation of the duplicated DNA is achieved only when condensed chromosomes are aligned along a plane in the center of the cell, with each chromosome comprising two chromatids that, although attached to each other, are connected by microtubules to opposite ends of the cell. Chromosome segregation is then initiated by severing the link between sister chromatids.

Cells have evolved similar principles for regulating DNA replication and mitosis. Each process requires a sequence of multiple preparatory steps that takes place during a period in the cell cycle when DNA replication or mitosis is prevented. How are these two different cellular states defined at the molecular level?

11.4.2 Structural and regulatory molecules are involved in controlling the initiation of mitosis.

Several different classes of molecules are involved in establishing full competence for mitosis. Proliferating cells synthesize proteins collectively called condensins and cohesins, which are essential for assembling the long chromatin fibers into chromosomes and thereby permit the replicated DNA to be segregated without damage (Fig. 11.10).

The molecular details of how chromosome condensation and cohesion are achieved are still poorly understood. Cohesion proteins such as Scc1p and Smc1p, which have been identified in yeast and animals, are synthesized during S phase. These proteins interact with both sister chromatids (daughter strands) of a chromosome to establish and maintain their tight linkage until anaphase (see Chapter 6). Experiments in which *SCC1* gene expression was induced in G2 phase but not in S demonstrated the requirement for sister chromatids to be joined in S phase; otherwise, the chromosomes will separate prematurely. How this linkage occurs and how it is regulated in cells so that only sister chromatids are joined are not known.

Recent experiments suggest that this linkage includes centromeres and extends further into the chromosome arms.

Sister chromatid condensation and cohesion are tightly coupled. The high degree of DNA packing density in early M phase involves a complete remodeling of chromatin structure. Histone H1, which in interphase binds to linker DNA between two nucleosomes (see Chapter 7), is phosphorylated and subsequently is unable to bind DNA. Condensing chromosomes are assembled around a central axis of scaffold proteins,

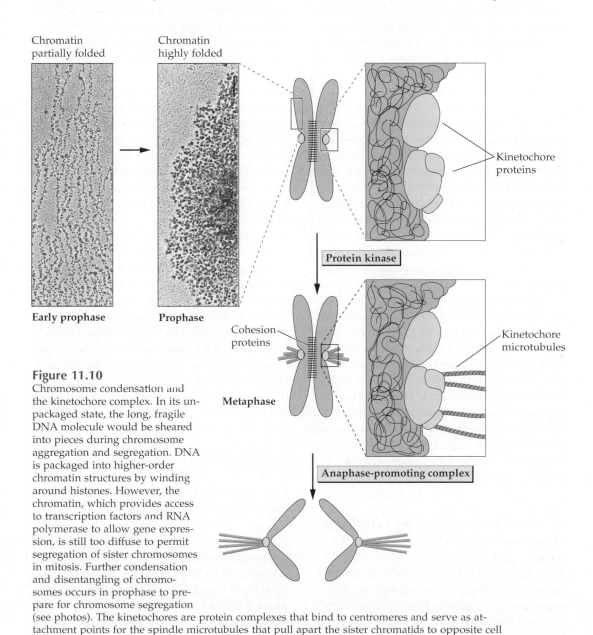

Figure 11.10
Chromosome condensation and the kinetochore complex. In its unpackaged state, the long, fragile DNA molecule would be sheared into pieces during chromosome aggregation and segregation. DNA is packaged into higher-order chromatin structures by winding around histones. However, the chromatin, which provides access to transcription factors and RNA polymerase to allow gene expression, is still too diffuse to permit segregation of sister chromosomes in mitosis. Further condensation and disentangling of chromosomes occurs in prophase to prepare for chromosome segregation (see photos). The kinetochores are protein complexes that bind to centromeres and serve as attachment points for the spindle microtubules that pull apart the sister chromatids to opposite cell poles. The anaphase-promoting complex, which is activated by protein phosphorylation, tags the mitosis inhibitor Pds1p for proteolysis. This ultimately leads to the destruction of the cohesion proteins that join sister chromatids and results in chromatid separation during anaphase.

presumably arrays of cohesin and condensin molecules, which bind to long loops of DNA. During the early steps of chromosome condensation, the two sister chromatids condense together as one large cylinder. On further condensation in late prophase and early metaphase, topoisomerases introduce supercoils into the DNA helix (see Chapter 6). Supercoiling increases the packing density of DNA even further, and at this point the sister chromatids become distinguishable in the light microscope. The more-organized chromosome structure at this phase of the cell cycle exposes the cohesins to regulatory factors that break the linkage between sister chromatids in anaphase. Throughout this process, inhibitory proteins such as Pds1p begin to accumulate, which act to safeguard against precocious entry into mitosis. Pds1p protein binds to and antagonizes the Cut1p protein, which breaks the linkage between sister chromatids and therefore triggers chromosome separation at the onset of anaphase. Genetic analysis in yeast indicates that Pds1p must bind to Cut1p to target the Cut1p correctly to the site where it acts in late metaphase. Thus, Pds1p binding simultaneously suppresses premature activation of Cut1p and ensures its correct localization.

APC, which targets proteins that must be degraded during M phase, will be discussed in more detail in Section 11.5.5. A second large complex, the kinetochore, attaches to the centromeres of chromosomes. The microtubule network, which is ultimately responsible for congregation of the chromosomes along the metaphase plate and their subsequent movement toward the cell poles, attaches to the kinetochores (Fig. 11.10). Kinetochore assembly is coupled to the completion of DNA replication. The molecular composition of the kinetochore complex is still poorly understood (see also Chapter 5).

11.4.3 Proteases regulate the initiation of chromosome separation.

The capacity of the cell to undergo ordered mitosis is established during S phase and continues until mitosis is complete. Ordered mitosis begins with the insertion of cohesins into the newly replicated helices and the removal of catenations that would preclude chromosome separation. Once chromosomes are disentangled and condensed, they are fully competent for segregation. However, the cell is not yet competent for mitosis until the sister chromatids are aligned along its equator. Cells at this time also monitor whether the kinetochores of paired chromatids are attached to the opposite cell poles. To prevent premature sister chromatid separation while these processes are in progress, segregation is suppressed during metaphase by Pds1p, an inhibitor of Cut1p (see previous section). At the end of metaphase, the protein complexes required for chromosome separation are fully poised, in a state analogous to the primed prereplication complex at the transition from G1 to S. What trigger activates segregation of chromosomes? As in DNA synthesis, mitosis is initiated by protein phosphorylation. The metaphase-to-anaphase transition is activated by phosphorylation of APC, which is catalyzed by cyclin-activated kinases (see next section) and by the kinase encoded by *CDC5*. The role of kinases as biochemical switches that coordinate cell growth with the stepwise progression of the cell cycle is described in Box 11.3.

Activation of APC results in the ubiquitination of Pds1p. Ubiquitinated Pds1p is recognized by the 26S proteasome as a substrate and is degraded (see Chapter 9). Destruction of Pds1p activates Cut1p, which breaks the cohesin-established linkages between sister chromatids. Once these bonds are broken, the mechanical forces generated by motor proteins along the kinetochore microtubules draw each "sister" of the formerly paired chromatids to opposite ends of the cell (Fig. 11.10).

11.5 Mechanisms of cell cycle control

11.5.1 Specific kinase complexes advance the cell through the cell cycle.

The molecular mechanisms of cell cycle control in eukaryotes are remarkably well conserved. Cell cycle progression is controlled by activity changes of **cyclin-dependent kinases** (CDKs). Cyclin-dependent protein kinase complexes are composed of two different subunits—one functioning as a catalyst, the other activating catalysis. The catalytic subunit alone lacks activity; association

Cell growth and many biochemical reactions in the cell proceed continuously. In contrast, cell division proceeds stepwise. How do cells process continuous (analog) information into discrete, on–off (digital) information to punctuate a gradual process such as growth? One important mechanism involves protein kinases that function in series, known as **protein kinase cascades.** Protein kinases can act as signal transducers and amplifiers; they respond to a change in input signal by a proportional change in kinase activity—their output signal. Signal processing by kinases involves two events: amplification of the magnitude of an output in relation to the input signal, and a change in sensitivity toward the perceived input signal. At low input, **magnitude amplification** increases the output linearly with respect to the input signal (see figure, green plot). **Sensitivity amplification** increases the percent change in response to a given percent change in stimulus, usually over a limited concentration range (see red plot). Responses in which the output signal dampens the input signal are also known (see yellow plot). Sensitivity amplification, particularly in ultrasensitive systems, has two important consequences for signal transduction pathways: First, it restricts most of the response to a limited concentration range of the stimulus, thus minimizing interference from the biochemical background of the reactions (i.e., noise). Second, a steep stimulus–response curve

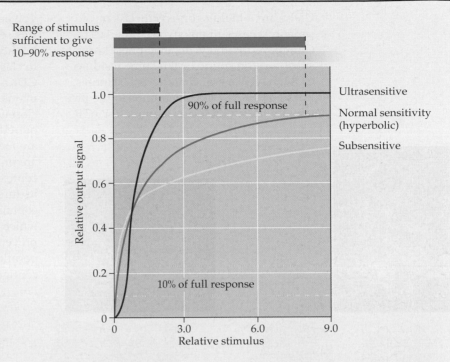

makes for a strong, robust response output, thereby ensuring that small changes in a continuous input signal are translated into an "on–off" output.

Mitogen-activated protein kinase (MAPK) cascades (see Section 11.7.4 and Chapter 18) couple environmental signal perception to cellular responses and act as an ultrasensitive signal transducer. Other kinase cascades directly target the

activation of cyclin-dependent protein kinases (see Section 11.5). The unique ability of kinase cascades to generate very strong responses over a small range of stimuli provides one explanation for the widespread occurrence of signaling pathways made up of several kinases functioning in series, including those controlling cell division.

---

with a cyclin is the first step in activation of the kinase complex. Many single-celled eukaryotes such as yeast have a single CDK catalytic subunit, whereas all multicellular eukaryotes appear to have multiple CDKs. All eukaryotic cells have multiple classes of cyclins, each of which is required for specific regulatory steps during the cell cycle. In yeast, these different cyclins interact sequentially with the CDK catalytic subunit, thereby changing the substrate specificity of the enzyme complex during the cell cycle. In multicellular organisms, certain CDKs can interact with certain classes of cyclins. These interactions determine the activity and specificity of the enzyme complexes at particular points in the cell cycle. Thus, the

association of CDKs with specific cyclins is generally thought to be a key regulatory mechanism, advancing the cell through the various stages of the cell cycle. Several cell cycle–specific phosphatases counteract CDK activity, resetting proteins to their nonphosphorylated ground state, but most dephosphorylation during the cell cycle is mediated by phosphatases that are not exclusively dedicated to cell cycle regulation. In addition, proteases function in cell cycle regulation at two critical, committing steps. At the start of the cycle, proteases are required to initiate DNA replication, and at the metaphase-to-anaphase transition, proteases degrade the cohesion proteins that connect sister chromatids.

11.5.2 Multicellular eukaryotes have a complex pathway of CDKs.

CDKs are a highly conserved class of protein kinases in eukaryotes (Fig. 11.11). Complexed with cyclin, they phosphorylate substrates on serine or threonine residues. CDKs have several characteristic features. Located close to the N terminus of the protein are the amino acids required for ATP binding, followed by a small PSTAIRE domain (named for the single-letter codes of the amino acids it comprises), which is involved in cyclin binding. Several variants of this sequence motif are known, and the differences among the variants have been used to classify CDKs into different subgroups. CDKs with similar or identical PSTAIRE domains are likely to interact with the same or closely related cyclins. The amino acids required for catalytic activity of the enzyme are located throughout the molecule; they are organized into their appropriate spatial context only after the protein has folded into its three-dimensional structure. CDKs also contain a flexible domain called the T-loop, which can either mask the catalytic site to prevent substrate binding or swing open to permit substrate phosphorylation. The open configuration is possible only after a specific

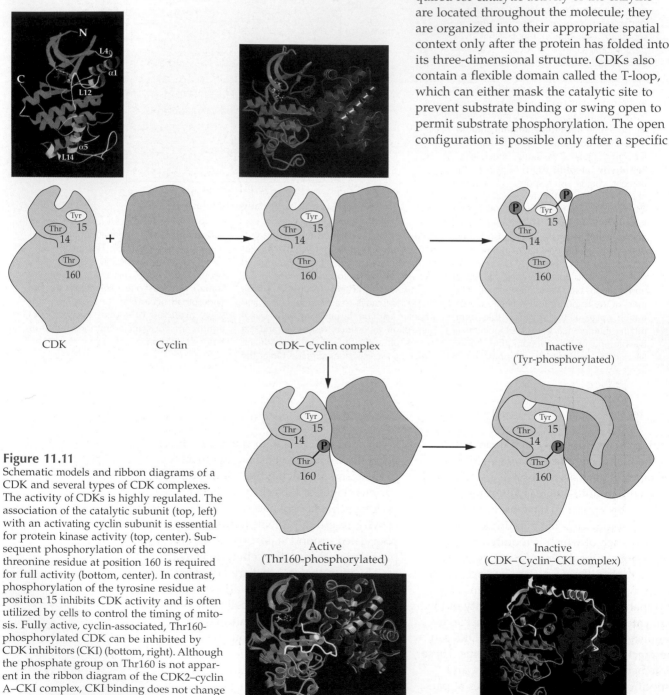

Figure 11.11
Schematic models and ribbon diagrams of a CDK and several types of CDK complexes. The activity of CDKs is highly regulated. The association of the catalytic subunit (top, left) with an activating cyclin subunit is essential for protein kinase activity (top, center). Subsequent phosphorylation of the conserved threonine residue at position 160 is required for full activity (bottom, center). In contrast, phosphorylation of the tyrosine residue at position 15 inhibits CDK activity and is often utilized by cells to control the timing of mitosis. Fully active, cyclin-associated, Thr160-phosphorylated CDK can be inhibited by CDK inhibitors (CKI) (bottom, right). Although the phosphate group on Thr160 is not apparent in the ribbon diagram of the CDK2–cyclin A–CKI complex, CKI binding does not change the phosphorylation state of the CDK.

threonine residue in the T-loop has been phosphorylated (Fig. 11.11).

Complex CDK gene families have evolved in multicellular eukaryotes. Multiple CDK genes have been identified in all plant species studied. In most plants, several CDK genes encode proteins with the canonical N-terminal PSTAIRE motif, and additional CDK genes encode variant PSTAIRE motifs. However, the specific function of these proteins in any individual cell cycle transition is still very poorly understood. In animals, distinct CDKs are required at different phases of the cycle. CDK1 is essential for the transition from G2 to M. CDK2 and CDK3 are required for the G1-to-S transition, with CDK2 also being required during S phase. CDK4 and CDK6 function during G1 to control the decision about whether to enter a new cell cycle.

In addition to CDKs, at least two other protein kinases are essential for the chromosomal cycle and mitosis: In yeast, these are encoded by the genes *CDC5* and *CDC7*. As described in Section 11.3.2, the Cdc7 protein associates with a regulatory subunit encoded by *DBF4* to activate DNA synthesis at replication origins. Although homologous genes have been identified in animals, their function is not understood in much detail. A *CDC7* homolog has also been identified in plants during genome sequencing efforts. Cdc5 protein kinase activity is required in M phase to activate APC and to regulate the motor proteins required for chromosome segregation. Genes homologous to *CDC5* have been identified in animals; as in yeast, they are necessary for M-phase progression.

11.5.3 Cyclins determine the specificity and subcellular localization of CDKs.

Association with cyclins is required for CDK protein kinase activity. Cyclins also confer substrate specificity to the cyclin–CDK complex and are involved in targeting CDKs to specific subcellular compartments during the cell cycle. Cyclins were discovered as a small group of proteins in fertilized and proliferating clam eggs that accumulated and disappeared periodically during the cell cycle (see Fig. 11.5).

Cyclins can be classified into two groups. The first group, called **mitotic cyclins,**

includes **B-type cyclins** (also known as M cyclins) and **A-type cyclins** (also known as S cyclins). Proteins in the first group are characterized by a large conserved central domain, called the **cyclin box,** that interacts with the kinase subunit. They also carry a protein domain called the **mitotic destruction box,** which mediates cyclin degradation late in mitosis. B-type cyclins commit the cells to mitosis at the G2-to-M transition. A-type cyclins, which have essential functions in S phase, also contain destruction boxes. Animal and yeast genomes encode approximately 10 different types of A-type and B-type cyclins. Plants, however, appear to have a much larger number of genes for these types of cyclins. The significance of this is not well understood, and only very few functional studies of plant cyclins have been done to date.

The second group of cyclins, called **G1 cyclins,** is composed primarily of **D-type cyclins,** but also includes **E-type cyclins** that are synthesized in late G1 and early S phase. D-type cyclins contain a distinct central cyclin box, but it is less well conserved than that of B-type cyclins. D-type cyclins are generally unstable proteins, but destruction boxes do not mediate their degradation. Some D-type cyclins accumulate only in G1 and are degraded rapidly at the beginning of S phase, whereas others are present at low concentrations throughout the cell cycle. In animals, the accumulation of D-type cyclins is stimulated by growth hormones, indicating that this class of cyclins is important for the advancement of cells into G1 and S phases. Similarly, in *Arabidopsis* the G1 cyclin D3 is regulated by cytokinins and probably couples the control of plant growth to the cell cycle. In yeast, the D-type cyclins are encoded by CLN1, CLN2, and CLN3. Cln3p activates the SBFs (SC box-binding factors) and MBFs (MluI cell cycle box-binding factors) required for cell cycle–regulated gene expression in G1 and S phases, whereas Cln1p and Cln2p are required to initiate S phase.

11.5.4 CDK activity is regulated by kinases, phosphatases, and specific inhibitors.

The activities of CDK–cyclin complexes are controlled by specialized kinases and phosphatases, which have important functions in regulating cell cycle progression (Fig. 11.12). Wee1-type kinases, which are named for the small-cell phenotype of fission yeast mutant for this protein (see Fig. 11.1), phosphorylate adjacent threonine and tyrosine residues near the N terminus of the CDK molecule. This phosphorylation of Thr14 and Tyr15 inhibits CDK activity. A homolog of *wee1*+ was recently identified in plants. The yeast *CDC25* gene encodes a multifunctional phosphatase that removes phosphate groups from tyrosine, threonine, and serine residues. The most important reaction catalyzed by Cdc25p is the removal from the N terminus of the CDK phosphates added by Wee1p (Fig. 11.12). In fission of yeast and animals this dephosphorylation results in the abrupt increase in CDK activity observed at the onset of the breakdown of the nuclear envelope. Cdc25p activity is controlled in part by pathways that prevent damaged cells from replicating. In the event of DNA damage, Cdc25p is not activated, which allows the cell to pause and repair the damage before initiating mitosis.

CDK inhibitors (CKIs) have also been identified. Under certain conditions, CKI proteins associate with the activated CDK–cyclin complex to prevent it from phosphorylating substrates (see Fig. 11.11). This inhibition of CDK activity is reversible and does not alter the composition or phosphorylation state of the CDK–cyclin complex. CKIs are used by the cell to control CDK–cyclin complex activity before undergoing cell cycle transitions and to arrest the cell cycle temporarily in

response to DNA damage or to other signaling pathways. In yeast, the *SIC1* gene product ScSic1p is essential to prevent premature activation of S-phase CDK activity during G1.

In humans, at least three classes of CKIs are known. The *INK4* gene product, which serves an important function in controlling the G1-to-S transition, is frequently mutated in neoplastic cells that have lost the ability to control proliferation. The *CIP1* gene product p21 is an important effector of cellular damage control pathways activated by other key regulatory proteins, and the product of the *KIP1* gene, p27, functions similarly to SIC1 to inhibit premature activation of S-phase CDK activity late in G1.

The *Arabidopsis ICK1* gene encodes a KIP1-like CKI and is induced by treatment with abscisic acid (ABA). This protein is probably involved in mediating the cell cycle arrest observed in plants after exposure to such growth regulators.

11.5.5 Ubiquitin-dependent proteolysis occurs at key transitions in the cell cycle.

Proteolysis of CKIs is necessary to trigger the G1-to-S transition. Similarly, degradation of the cohesion proteins that connect sister chromatids is required for the irreversible transition from metaphase to anaphase (Fig. 11.13). How are these and other proteins, which are degraded at key transition points in cell cycles, selected and marked for

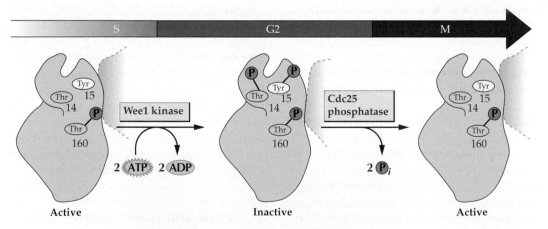

Figure 11.12
Phosphorylation of a yeast CDK inhibits the protein kinase activity of the enzyme. The Wee1 kinase phosphorylates two conserved adjacent threonine and tyrosine residues (Thr14 and Tyr15) near the N terminus of the CDK catalytic subunit, inhibiting protein kinase activity, whereas the Cdc25 phosphatase removes these inhibitory phosphate residues. The Cdc25 phosphatase therefore regulates the timing of entry into mitosis in many organisms, but whether a similar enzyme functions in plants is still unknown.

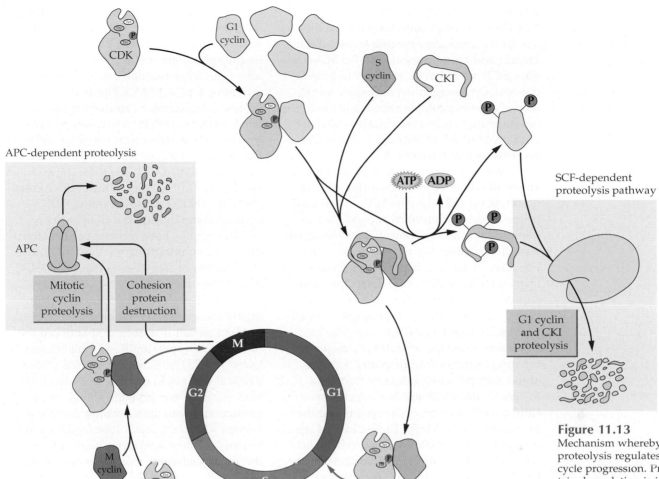

APC

Mitotic cyclin proteolysis

Cohesion protein destruction

M cyclin

CDK

M

G2

G1

S

G1 cyclin

S cyclin

CKI

ATP ADP

SCF-dependent proteolysis pathway

G1 cyclin and CKI proteolysis

Figure 11.13
Mechanism whereby proteolysis regulates cell cycle progression. Protein degradation is instrumental in controlling cell cycle progress at two major transitions. At the G1-to-S transition, two types of cell cycle proteins are substrates for regulatory proteolysis: CKIs, which bind to S-phase cyclin–CDK complexes and prevent CDK activity, and G1 cyclins, which must be degraded to commit the cell irreversibly to the start of a new cycle. At the metaphase-to-anaphase transition, proteolysis is required to sever the links between sister chromatids to permit chromosome segregation and to degrade mitotic cyclins. The APC ubiquitinates the chromatid cohesion proteins, targeting them for degradation. The APC also participates in degradation of the mitotic cyclins.

destruction? A common feature of all such proteins is their modification by covalent attachment of polyubiquitin to the side chains of lysine residues. Ubiquitinated proteins are recognized by the 26S proteasome (see Chapter 9), which functions throughout the cell cycle and also in quiescent cells to degrade proteins. The SCF (Skp1–Cullin–F-box protein) complex and the APC regulate ubiquitination of cell cycle regulatory molecules at the G1-to-S transition and in M phase, respectively.

The core SCF complex is composed of three proteins: Skp1p, Cdc53p, and Cdc34p. Cdc34p is the ubiquitin ligase (E3) that attaches ubiquitin to target proteins. Cdc53p forms a scaffold to which both Cdc34p and Skp1p attach. Skp1p contains a domain

called the F-box that can interact with individual selector proteins, which themselves recognize specific substrates. By binding to Skp1p, selector proteins present these substrates to the SCF complex in the appropriate orientation for ubiquitination by Cdc34p. The selector protein encoded by the yeast *CDC4* gene, for example, mediates the recognition of the CKI Sic1p and the ORC assembly factor Cdc6p. A different selector protein is required for degrading G1 cyclins. The SCF complex recognizes only phosphorylated proteins as substrates for ubiquitination. Therefore, SCF activity complements G1 CDK activity to mediate the G1-to-S phase transition: Proteins that are phosphorylated by CDKs can be permanently removed by the proteasome after binding to the selector protein and being presented to the SCF complex. SCF may also be involved in other, non–cell cycle–signaling networks. For example, several proteins involved in auxin signal transduction in plants share homology with proteins involved in SCF-dependent proteolysis (see Chapter 18).

Anaphase onset is triggered by the activation of APC, a protein complex with ubiquitin ligase (E3) activity. APC, which comprises at least 12 proteins, recognizes and ubiquitinates several proteins, including proteins that contain the mitotic destruction box. Like SCF, APC may exist in several forms that have different selector proteins. At least two different APC selector proteins have been identified. The proteins encoded by *CDC20* in yeast and the *fizzy* gene in *Drosophila* associate with APC to mediate the ubiquitination of Pds1p and S cyclins at the metaphase-to-anaphase transition. In conjunction with another selector protein, Cdh1p, APC recognizes M cyclins and the M-phase kinase Cdc5p. These proteins are degraded only in late M phase. APC is activated by CDKs and Cdc5 in late metaphase. Unlike SCF, however, APC recognizes proteins that are not phosphorylated.

The basic components of the ubiquitin-mediated degradation machinery are conserved in plants. Stimulating proliferation in quiescent cells activates the coordinate expression of some plant proteasome components, and adding a proteasome inhibitor prevents cells from entering anaphase. In addition, identification of the genes that encode proteins to subunits of the SCF and APC complexes suggests that the G1-to-S and metaphase-to-anaphase transitions in plants are mediated by mechanisms similar to those of animals and yeasts.

11.5.6 The structural basis of CDK–cyclin complex regulation has been elucidated recently.

CDKs, currently recognized as the most highly regulated group of serine–threonine protein kinases, are at the center of the cellular network that coordinates and controls the progression of cell division. A stepwise process is required for full activation of CDKs. After association of the catalytic subunit with a cognate cyclin, the CDK is activated by phosphorylation of a conserved threonine (Thr160), a reaction catalyzed by CDK-activating kinase (CAK). The activity of the CDK–cyclin complex can then be modulated further by inhibitory threonine/tyrosine phosphorylation near the N terminus (see Section 11.5.4) or reversibly inhibited by association with a CKI. The three-dimensional structures of CDK2, the CDK2–cyclin A complex, the Thr160-phosphorylated CDK2–cyclin A complex, and the CDK2–cyclin A–CKI complex have recently been resolved, providing an intriguing window into the mechanisms of stepwise regulation of CDK activity (see Fig. 11.11).

CDK2 has a two-lobed structure with highly conserved ATP-binding and catalytic residues nested in a deep cleft between the N-terminal (upper) and C-terminal (lower) lobes. The PSTAIRE region, located within a helical domain in the smaller N-terminal lobe, is highly conserved in CDKs. In the monomeric state, the activation domain (or T-loop) within the lower lobe (residues 146 to 166) folds like a lid over the cleft to block the site for substrate recognition and to interact with the PSTAIRE helix. In the absence of cyclin, the relative positioning of these two structural elements of the CDK protein is not favorable for catalysis. Thus, regulation of the CDK catalytic subunit includes structural changes to prevent premature activation of the kinase.

Monomeric cyclin A is a globular protein consisting of 12 α-helices that organize into two domains of five helices each. Cyclin A does not undergo substantial conformational changes when binding CDK2. However, association of cyclin A with CDK does cause major structural changes in the CDK polypeptide, priming it for full activation (see Fig. 11.11). Cyclin A binds to both lobes of the kinase subunit toward one side of the catalytic cleft, thus changing the relative orientation of the lobes and disrupting the tight interaction of the PSTAIRE helix and the T-loop. The highly conserved cyclin box interacts with the PSTAIRE helix, reorienting into the catalytic site the side chain that is involved in ATP binding and thereby bringing all the catalytic residues into the conformation essential for protein kinase activity.

Additionally, cyclin A binding causes the T-loop to move further away from the catalytic cleft, which exposes the Thr160 residue.

Activation of the CDK2–cyclin A complex is not complete, however, because the catalytic cleft remains partially covered by the T-loop. Still needed is phosphorylation of Thr160 by CAK, which increases CDK activity by 100- to 300-fold, a result of major conformational changes in the CDK2–cyclin A complex (see Fig. 11.11). The phosphate group interacts with three arginine side chain residues, thereby moving the T-loop away from the catalytic cleft to fully expose the active site. Simultaneously, additional contacts are made between CDK2 and cyclin A, further stabilizing their interaction. Thus, phosphorylated Thr160 becomes a long-range structural organizing center for the active conformation of the kinase complex.

Inhibition of CDK activity by CKIs is also an important mechanism of CDK regulation. Structural analysis has revealed that one CKI, Kip1p (see Section 11.5.4), is an extended, nonglobular protein made up of various coiled, helical, and sheet domains. None of these domains interacts noticeably with one another, but each one interacts with a specific surface of the CDK2–cyclin A complex (see Fig. 11.11). On cyclin A, Kip1p binds strongly to highly conserved amino acid residues within the cyclin box, which perturbs substrate recognition. On CDK2, Kip1p rearranges the N-terminal upper lobe, which destabilizes ATP binding. In addition, inserting a Kip1p helical domain into the catalytic cleft of CDK further blocks kinase activity. Thus, Kip1p suppresses CDK activity by a combination of three distinct, complementary mechanisms.

11.6 The logic of cell cycle control

11.6.1 Cell cycle progression is regulated by intrinsic and external signals.

The discovery of conditional cell division control mutants in yeast raised the question of whether cell cycle progression was governed by the activity state of the protein substrates, which execute specific steps in the cell cycle, or by the sequential activation of cell cycle regulators. In the first model, the "domino" model, progression through each step of the cell cycle depends on completion of the preceding step. The second model, the "clock" model, suggests the existence of an independent timer mechanism that entrains the separate events of the cycle. Initial analysis of *cdc* mutants at restrictive temperatures (see Fig. 11.4) indicated that the domino model was generally valid. However, in later observations with fertilized frog eggs, drugs that blocked spindle assembly were unable to inhibit the continued oscillation of a previously identified MPF activity, now known as the CDK–cyclin complex. Because the domino model would not have predicted continued MPF activity in the absence of spindle formation, these findings provided support for an independent mechanism. Additional observations have reinforced the notion that the regulatory pathways controlling the biochemical reactions of the cell cycle represent a combination of both domino and clock models. Such a combination would have important advantages. Regulatory mechanisms based on the state of the execution substrate can prevent cell cycle steps from occurring in an inappropriate sequence, and the existence of additional independent mechanisms can make the decision to divide conditional on environmental factors.

In living systems, biochemical reactions do not always proceed to completion, and a cell undergoing division can experience adverse conditions that could damage its DNA or spindle apparatus. To account for these factors, the cell cycle contains dedicated checkpoints at which the cell can monitor the completion of specific reactions (e.g., DNA replication, chromosome condensation) or the integrity of complex structures (e.g., spindle apparatus, sister chromatid adhesion complex). Additional monitoring is required to link entry into the cell cycle and cell cycle progression with internal and external cues that control growth, development, and the social context of cells. The next sections will consider the logic of the cell cycle as it relates to checkpoint controls and other signals.

11.6.2 Regulation of cell cycle progression depends on CDK and protease activities.

Cells have evolved elaborate mechanisms to safeguard that S phase and M phase occur only once per cell cycle and in the correct order. However, these mechanisms do not instruct the cell when to initiate a new cycle or when to initiate S phase and M phase. Such intrinsic mechanisms are unable to couple cell cycle control to the cell's environment, growth, and metabolism. The overriding control of cell cycle progression is established through the interplay of CDKs and ubiquitin-dependent proteases. To best illustrate the basic principles of this control, let us consider the following simplified models of how the cell cycle operates in budding yeast (Fig. 11.14) and animal cells (Fig. 11.15).

CDK activity requires association with either G1, S, or M cyclins. Experiments with yeast show that the activity of one type of complex (e.g., S-phase CDK–cyclin complex) suppresses the synthesis or activity of the other type (e.g., M-phase CDK–cyclin complex), while simultaneously activating the mechanism for its own destruction. This ensures temporal alternation of the different types of CDK–cyclin complexes during the

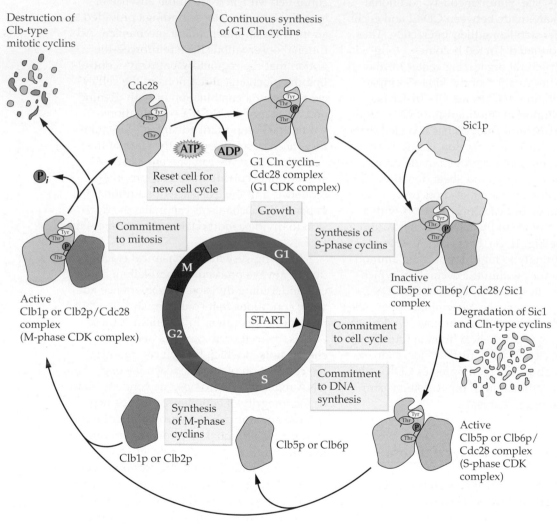

Figure 11.14

Cell cycle regulation in budding yeast. In early interphase (G1), yeast cells rapidly increase in volume. This cell growth is accompanied by a gradual increase in the abundance of G1 (Cln) cyclins but not other cyclins (top). Phosphorylation of Cln cyclins and of the CKI protein Sic1p (not shown) targets them for degradation (right). Proteolysis of these two classes of proteins enables budding yeast cells to accumulate the S-phase cyclins Clb5p and Clb6p and, subsequently, to enter S phase (right). Synthesis of the mitotic cyclins Clb1p and Clb2p is initiated in late G2. These cyclins then associate with CDC28, stimulating entry into mitosis. Progression from metaphase to anaphase is regulated by the ubiquitination and subsequent degradation of negative regulators of mitosis such as Pds1p (not shown). Destruction of Pds1p is followed shortly by the proteolysis of the mitotic cyclins (i.e., S and M cyclins, upper left). This resets the cell cycle to its ground state in early G1. Note that although the names Clb5p and Clb6p suggest that they are B-type (M-phase) cyclins, these proteins function as A-type (S-phase) cyclins.

cell cycle. The sequential transitions between the CDK–cyclin complexes are enforced by proteolysis of the components that were required for the preceding phase (see Fig. 11.13).

Once cell division is completed, how does the cell regulate its entry into a new cell cycle? In somatic cells, G1 cyclins (D-type and E-type cyclins) play an important role in this process. The abundance of G1 cyclins is rate-limiting for G1 progression, and their synthesis is coupled to cell growth. This establishes that cells proceed into S phase only when they have achieved adequate mass. Cyclin D-dependent CDK activity abruptly increases, by a yet unknown process, when cell size exceeds a certain threshold. The CDK/cyclin D complex phosphorylates substrates to activate four important processes. First, the phosphorylation turns off APC-mediated proteolysis of cyclins that contain destruction boxes (i.e., S and M cyclins), which permits a buildup of S cyclin/CDK complexes. Second, it activates

the transcription of genes required for S phase, leading to enhanced expression of essential biosynthetic and structural gene products such as histones (see Fig. 11.7). Third, phosphorylated inhibitors of the S cyclin/CDK complex (CKIs) are recognized as substrate by the SCF-dependent proteolysis machinery and subsequently degraded; this results in a strong and abrupt activation of preaccumulated CDK activity in S phase, which is required to activate DNA synthesis at the origins of replication. Fourth, the phosphorylated G1 cyclins are targeted for destruction by proteolysis, thereby completing the irreversible transition from G1 CDK activity to S-phase CDK activity and thus from G1 to S phase.

Once DNA replication is complete, how does the cell complete mitosis and return to G1? The synthesis of M-phase cyclins in late G2 prepares the cell for mitosis. The rapid increase of mitotic CDK activity at the G2-to-M transition initiates mitosis and

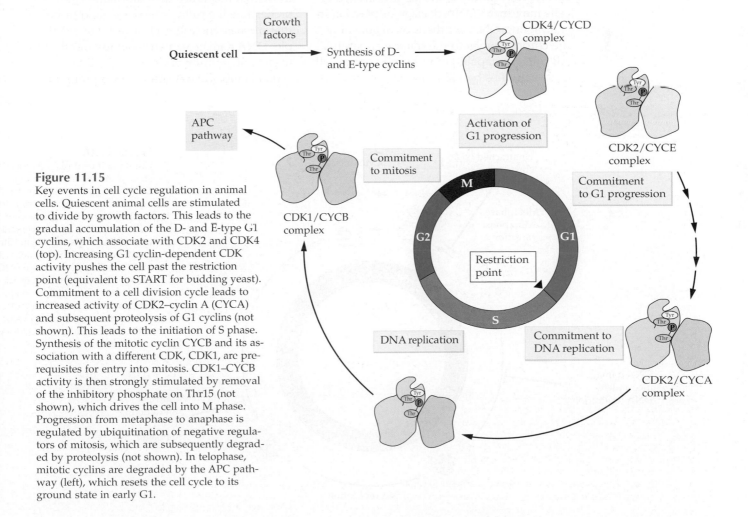

Figure 11.15
Key events in cell cycle regulation in animal cells. Quiescent animal cells are stimulated to divide by growth factors. This leads to the gradual accumulation of the D- and E-type G1 cyclins, which associate with CDK2 and CDK4 (top). Increasing G1 cyclin-dependent CDK activity pushes the cell past the restriction point (equivalent to START for budding yeast). Commitment to a cell division cycle leads to increased activity of CDK2–cyclin A (CYCA) and subsequent proteolysis of G1 cyclins (not shown). This leads to the initiation of S phase. Synthesis of the mitotic cyclin CYCB and its association with a different CDK, CDK1, are prerequisites for entry into mitosis. CDK1–CYCB activity is then strongly stimulated by removal of the inhibitory phosphate on Thr15 (not shown), which drives the cell into M phase. Progression from metaphase to anaphase is regulated by ubiquitination of negative regulators of mitosis, which are subsequently degraded by proteolysis (not shown). In telophase, mitotic cyclins are degraded by the APC pathway (left), which resets the cell cycle to its ground state in early G1.

cytokinesis, beginning with chromosome condensation and alignment of the chromosomes at the metaphase plate. The activation of APC, for which both mitotic CDK activity and Cdc5p protein kinase are required, initiates separation of the sister chromosomes and destruction of the mitotic cyclins. The decline in mitotic CDK activity relieves the repression of G1 cyclin synthesis, allowing accumulation of D-type cyclins, which had been suppressed from S phase through telophase. Thus, G1 CDK activity can gradually accumulate, which permits the assembly of new replication complexes and maintains APC activity to prevent any residual mitotic CDK activity.

11.6.3 Checkpoint controls are activated by DNA damage or incomplete cell cycle events.

Cell proliferation often occurs under unfavorable conditions. Frequently, therefore, cells may incur DNA damage or may lag in completing DNA synthesis or in attaching the chromosomes to the spindle apparatus. All cells are capable of restoring normal cell cycle activities by using special repair mech-

anisms, but cell cycle progression must be arrested to prevent the damage from progressing to a mitotic catastrophe. When damage occurs, cell cycle progression is arrested at a checkpoint, of which three major ones have been discovered. Arrest at a checkpoint can occur prior to the restriction point at the G1-to-S transition, prior to the S-to-M transition that couples mitosis to the completion of DNA synthesis, or prior to sister chromatid separation at the metaphase-to-anaphase transition. Each of the three checkpoints immediately precedes a committing event that is irreversible (Fig. 11.16). Checkpoint controls have not been studied extensively in plants, but the basic mechanisms discovered in other eukaryotes are likely to operate in plants as well.

The highly conserved proteins involved in checkpoint control and the regulatory network in which they function can be grouped into three classes: those involved in perceiving damage or incomplete cell cycle events, those that transduce this information and orchestrate the cellular response, and those that mediate the cell cycle arrest. For example, DNA damage is perceived (by mechanisms not yet fully understood) and this signal is initially transmitted by a group of

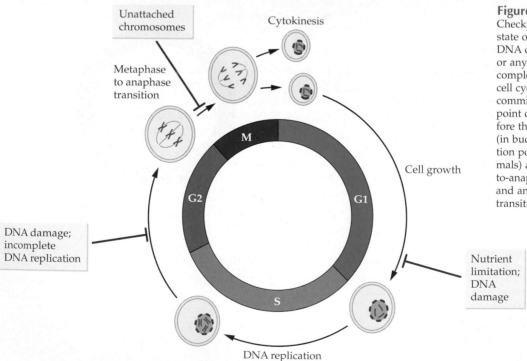

Figure 11.16
Checkpoint controls monitor the state of the cell and, if serious DNA damage has been incurred or any biosynthetic process is incomplete, arrest progression of the cell cycle before any irreversible, committing steps are taken. Checkpoint controls are highly active before the cell cycle reaches START (in budding yeast) or the restriction point (its equivalent in animals) as well as at the metaphase-to-anaphase transition (in yeast and animals) and at the G2-to-M transition in animals.

proteins encoded by *RAD* genes. These genes were originally identified in yeast and are required for resistance to ionizing radiation. RAD3, a DNA-dependent protein kinase, plays a central role in the initial steps of the pathway; it is activated by binding to single-stranded DNA—a hallmark of damaged DNA. A human homolog of *RAD3* is encoded by the *ATM* (ataxia telangiectasia mutated) gene, which, when defective, predisposes the cell to cancer. In most eukaryotes, the signals resulting from incomplete DNA replication (generated by continued engagement of DNA polymerases; see Section 11.3.3) and DNA damage (generated by RAD proteins) converge on the DNA-dependent protein kinase (RAD3).

Activated RAD3 stimulates further protein kinases, notably the Chk1 (checkpoint kinase 1) protein. These kinases couple the perception of genome distress to cell cycle arrest and also activate repair mechanisms. Chk1p targets the genes that control CDK activity at the G2-to-M transition, whereas a different protein kinase, RAD53, mediates arrest at START. Chk1p inhibits the activity of CDK1 by phosphorylating the tyrosine residue at position 14, near the N terminus of the protein. Active Chk1p also inhibits the CDC25 phosphatase required to remove the inhibitory Tyr14 phosphate, thus maintaining the CDK–cyclin complex in an inactive state. Active Chk1p also stimulates the tyrosine protein kinases (such as Wee1p; see Fig. 11.12) responsible for the inhibitory Tyr14 phosphorylation of CDK1, thereby reinforcing the inactivation of the CDK–cyclin complex.

11.6.4 Accessory proteins are required to enforce CDK control of cell cycle progression.

In animals, members of the E2F transcription factor family, particularly E2F1, E2F2, and E2F3, are required for the transcriptional activation of genes for proteins that are essential during DNA replication. Thus, E2F transcription factors are critical effectors of the decision to pass the G1-to-S restriction point and allow the cell to proceed into S phase. Certain E2F proteins bind to their own promoters, thereby creating an autocatalytic transcription loop that results in a very

rapid and decisive commitment to entering S phase (Fig. 11.17). However, such autocatalytic control mechanisms operate efficiently only if a sufficient amount of preformed (but inactive) E2F is present in the cell. Moreover, without a mechanism to dampen and reverse such signal amplification, E2F autoactivation can disrupt normal cell cycle regulation. Thus, there is a strong requirement for an inhibitor that can inactivate E2F transcription factors, analogous to the CDK inhibitors. Such proteins are encoded by members of the retinoblastoma (Rb) gene family. The Rb protein, which was first identified because of its ability to cause oncogenic cell growth when mutated, binds to E2F proteins and suppresses their activity. In mammalian cells, pRb and pRb-related proteins control the transition from G1 to S, but their function in other cell cycle stages is still not well understood. Homologs of Rb and E2F proteins have been identified in monocots and dicots, suggesting that a similar G1-to-S regulatory mechanism is functioning in plants.

pRb can interact with members of the E2F transcription factor family and suppress their activity only when it is not phosphorylated. During G1, pRb becomes progressively phosphorylated on multiple sites by CDK4–cyclin D complexes. When pRb is fully phosphorylated, E2F factors dissociate from it and are active. Later in S phase, the CDK2–cyclin A complex phosphorylates E2F proteins, abolishing their ability to bind DNA. By this mechanism, Rb proteins enforce the decision to proliferate, activating cell cycle–dependent transcription to produce the DNA synthesis machinery at the transition from G1 to S.

Rb proteins are targets in additional pathways for control of cellular growth and therefore are often mutated when proliferation spins out of control in cancer cells. Cellular parasites, such as DNA viruses that depend on the host replication machinery for multiplication, may target Rb for inactivation. Rb inactivation is sufficient to activate DNA synthesis pathways and thereby allow viral replication. The recently discovered plant Rb homologs have also been shown to interact with proteins encoded by geminiviruses, which suggests that these plant viruses exploit a mechanism for initiating their replication in ways similar to animal oncoviruses.

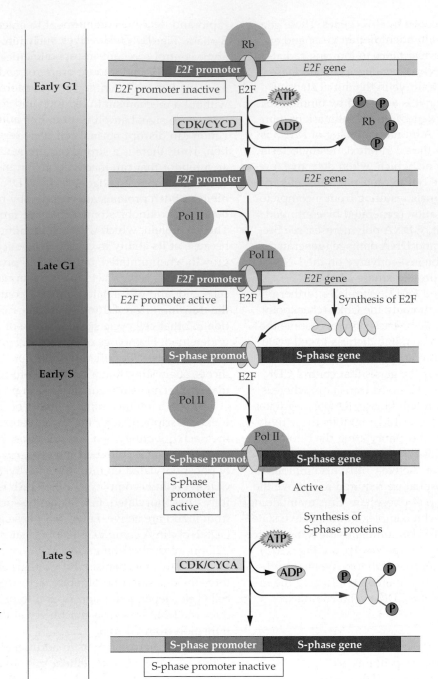

Figure 11.17
E2F transcription factor activation at G1-to-S enforces the commitment to S phase. The E2F transcription factor activates its own transcription by binding to a site in its own promoter. In G1, E2F activity is suppressed by binding of the retinoblastoma protein (Rb). As a consequence of passage through the restriction point, the CDK4–cyclin D (CYCD) complex phosphorylates Rb, resulting in its dissociation from E2F and relieving the inhibition of E2F. This activates E2F as a transcription factor, resulting in a positive-feedback loop that leads to accumulation of large amounts of E2F protein as well as other S-phase–specific proteins. Phosphorylation of E2F by the CDK–cyclin A (CYCA) complex then inactivates E2F-dependent S-phase transcription. Pol II, RNA polymerase II.

11.7 Cell cycle control in multicellular organisms

11.7.1 Intercellular communication controls the cell cycle during growth and development.

In single-celled organisms, nutrient limitation is the single most important factor restricting cell growth; therefore, cell size is the major cue for division. In multicellular eukaryotes with specialized cell types and functions, cell division control necessarily becomes more complex. In plants, only a few stem cells organized into meristems divide to produce the plant body. These cells depend completely on nutrients provided by the rest of the plant; although cell size signals are also required, they are insufficient to direct the complex patterns and rates of cell division in multicellular organisms. Furthermore, control of cell proliferation is only one aspect of development: Cell cycle control is integrated with the regulation of cell expansion, differentiation, and cell death. Other types of growth control, called social controls, evolved to regulate development and proliferation within complex organisms. In most multicellular organisms, social control prevails over nutritional control of the cell cycle.

The emergence of social controls resulted in the loss of cellular autonomy over proliferation control. This had important consequences for dividing and differentiating cells. First, cells in multicellular organisms are normally quiescent (a state often termed G0), and quiescent cells do not proliferate without stimulation (Fig. 11.18). Second, the stimuli that mediate cell proliferation, for example, growth factors, provide the signaled cell with information about the status of the whole organism rather than just its individual cells. Third, if only a few cells permanently retain the capacity to divide, the majority of cells that have lost this capacity must somehow be instructed to do so.

11.7.2 Cell division is tightly controlled in shoot meristems and during organ formation.

Cell division activity during vegetative plant development is restricted to **meristems,** organ primordia, and cambial tissues. Shoot

and root apical meristems have two functions: to maintain themselves and to initiate plant organs (Fig. 11.19A). Cambial tissues are a specialized form of meristem that generates only vascular cells during secondary growth of tissues; they are not involved in organogenesis. Organ formation in shoots occurs on the flanks of meristems, where cell division activity is generally increased to produce the large number of new cells required. Less cell division activity at the meristem center is sufficient for its maintenance. How social controls regulate cell proliferation in distinct zones of plant meristems is currently a topic of extensive investigation. A slowly dividing population of self-renewing stem cells at the center of the meristem not only is important for its maintenance, but also may be critical for ensuring low cumulative mutation rates in long-lived plants such as bristlecone pines, in which meristems can function for millennia.

Although we still do not understand how meristems are organized, and which genes set up meristematic domains, several genes have been identified in *Arabidopsis* that are required for its proper function. Genetic analysis of the shoot apical meristem has allowed researchers to group these genes into three classes (Fig 11.19B). One class of genes is required to establish and maintain the indeterminate central zone required for self-renewal of the meristem. A second class of genes directs differentiation in organ primordia. A third class regulates localized cell

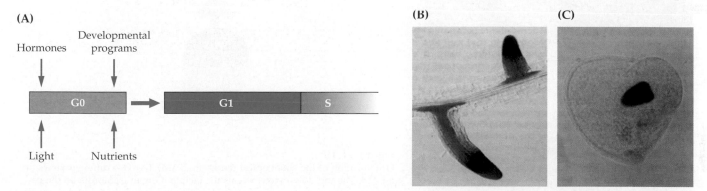

(A)

Hormones   Developmental programs

G0 → G1   S

Light   Nutrients

(B)   **(C)**

Figure 11.18
All cells require stimulation to proceed with proliferation. (A) Signaling pathways couple the perception of environmental cues to the control of proliferation, allowing quiescent (G0) cells to progress into G1. (B) The plant hormone auxin stimulates formation of lateral roots, which is initiated with stimulation of cell division activity in the root pericycle tissue. The plant cyclin gene, *AtCYCB1,* is a

good marker for cell division activity. When the *AtCYCB1* promoter is fused to the gene for β-glucuronidase (GUS), the enzyme reports the expression of the cell cycle gene during auxin-stimulated lateral root formation. (C) *AtCYCB1* also is expressed during embryogenesis, as shown here for a heart-stage embryo.

division in organ primordia. Mutations in genes of each of these three classes affect all aspects of normal shoot meristem activity, indicating that the individual activities of domains in the meristem in which these genes are expressed are not autonomous.

The first class of genes is defined by several different homeodomain-type putative transcription factors required to establish and maintain meristems. The first gene of this type to be discovered was the maize *Knotted* gene, which was identified as a mutant that develops ectopic meristematic growth on the vascular tissues of leaves (see Chapter 7). When *Knotted* or the related *Arabidopsis* gene *KNAT1* is activated inappropriately, shoot meristems appear ectopically, for example, on leaves. Another *Arabidopsis* gene, *SHOOT MERISTEM-LESS* (*STM*), is necessary to maintain the capacity for self-renewal or indeterminate growth in the center of the meristem. In the absence of a functional *STM* gene, the meristem disappears and the cells at its position differentiate. During embryogenesis, the *Arabidopsis* *WUSCHEL* gene is required to specify stem cells.

The second class of genes promotes differentiation in organ primordia. When these are disrupted by mutation, the meristem increases in size because cells fail to differentiate. In *Arabidopsis*, loss of *CLAVATA* (*CLV*) gene function results in extremely enlarged meristems and increased organ numbers. *CLV* encodes a receptor-like protein kinase (see Chapter 18) and might function to process intercellular signals. *clv3* mutants have phenotypes very similar to *clv1* plants. *CLV3* encodes a short polypeptide that might be the ligand for *CLV1*.

Leaf organ primordia initially resemble needles but soon broaden in the plane of the future leaf blade. The third class of genes required for meristem function controls proliferation locally in the developing organ. The *PHANTASTICA* (*PHAN*) gene in *Antirrhinum* regulates cell proliferation along the upper (dorsal) lamina to generate the leaf blade. The phenotype associated with recessive mutations in this gene is a ventralized, radially symmetrical, needle-like leaf consisting of just the midrib. Furthermore, impaired leaf development in *phan* mutants suppresses proliferation and further organogenesis in the apex, indicating that retrograde signaling from nascent organs is important for maintaining shoot meristem function.

(A)

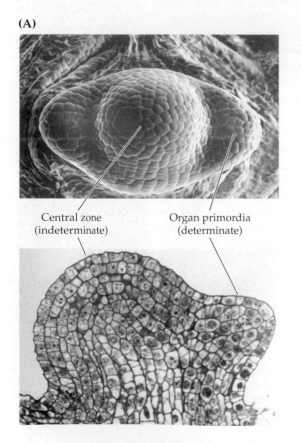

Central zone (indeterminate)

Organ primordia (determinate)

(B)

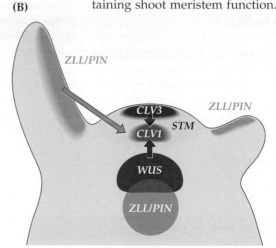

Figure 11.19
Organization of the shoot apical meristem (SAM). (A) Two different views of a SAM. The top-down view reveals the incipient organ primordia on the periphery of the meristem. The longitudinal section shows the layered tissue organization of the SAM as well as a developing new organ. (B) Regulatory genes required for the establishment and maintenance of the SAM are expressed in specific domains. The *WUS* gene is expressed at the base of the meristem and is required for expression of the *STM* gene in a domain above it. The products of the *CLV1* and *CLV3* genes presumably interact to determine the size of the indeterminate central zone of the meristem. The *ZLL/PIN* genes are required for maintaining SAM activity.

Meristems efficiently organize organogenesis but are not a prerequisite for cell production within the plant body. Loss of shoot meristem function does not lead to a complete loss of the ability of cells in shoot tissues to divide, and root meristems are not affected. For example, *stm* mutant plants can form, at a low frequency, single new leaf organs at ectopic positions. Moreover, embryo development and the initial stages of lateral root development proceed without meristems. Thus, meristems are not required for activation of cell division per se, but they are needed to organize and maintain specific types of stem cells within the plant body. Individual social control pathways are tissue- and organ-specific, not universal regulators of proliferation.

11.7.3 Specific patterns of cell division may not be required for appropriate root cell speciation.

In contrast to the apparently random arrangement of cells in shoot organ primordia, the root apex is highly structured, and tissues derived from the root apical meristem are often organized into long, contiguous cell files. This organization results from an underlying pattern of cell division. The cell files appear to converge onto a central zone within the root apex, the **quiescent center** (QC), in which cells divide only infrequently (Fig. 11.20). The QC is surrounded by rings of initial cells that are located at the apical end of specific cell files. Microsurgical experiments with maize root tips, in which the QC was removed together with the root cap, have demonstrated that cell division and tissue differentiation could resume from such root stumps but only after a new QC had formed. Root apical segments were able to sustain root growth in culture only when they included the QC. Based on these observations, a model was developed that the QC directs the regular pattern of cell division,

which in turn determines appropriate tissue differentiation. The model suggests that control over rates and patterns of cell division is at the top of a regulatory hierarchy coordinating the development of the root.

More recent experiments and theoretical considerations now indicate that regular patterns of cell division may not be responsible for correct tissue differentiation (Fig. 11.21). First, laser ablation of the four QC cells in *Arabidopsis* results in their displacement by vascular initial cells. These cells rapidly assume QC cell properties, indicating that the QC can be replaced without massive cell redifferentiation. Second, laser ablation of only two QC cells caused the differentiation of abutting initial cells, suggesting that the QC maintains the self-renewing properties of the initial cells. Third, laser ablation of the immediate progeny of the initial cells activated the production in neighboring cell files of additional cells that subsequently acquired the identity of the cells they displaced. This process occurs only when the next older tier of cells is intact, which suggests that differentiated cells, rather than cell lineage or the QC, provide the information that directs the maturation of younger cells. Fourth, root mutations, such as *transparent testa glabrous* (*ttg*) or *fass*, or the partial suppression of cell division activity by the ectopic expression of a dominant-negative (inactive) CDK perturbs the otherwise regular division patterns in the root apices of *Arabidopsis* and tobacco. However, this does not prevent cells from adopting their appropriate identity or the root from developing a phenotypically normal morphology. Theoretical modeling of cell wall patterning in relation to root apical growth also predicts the existence of cell files and their convergence into a QC, where stem cell division rates are low relative to division rates in the first few derivative cells. Thus, the pattern of cell division observed in many roots appears to facilitate, rather than dictate, the pattern of root development.

(A)

(B)

Figure 11.20
(A) A longitudinal section through the root apex of *Arabidopsis,* as viewed by light microscopy. Patterns of mitotic activity in the *Arabidopsis* root apex are visualized by the activity of a cyclin–GUS chimeric protein, which is expressed exclusively in mitotic cells. (B) Patterns of DNA synthesis reveal low amounts of proliferation in the quiescent center, the dark void in the upper half of the panel. This DNA synthesis was visualized by growing roots in a solution of radioactive thymidine, which was incorporated into the newly synthesized DNA. The longitudinal section was then exposed to a photographic emulsion. The white spots represent cells with newly synthesized DNA.

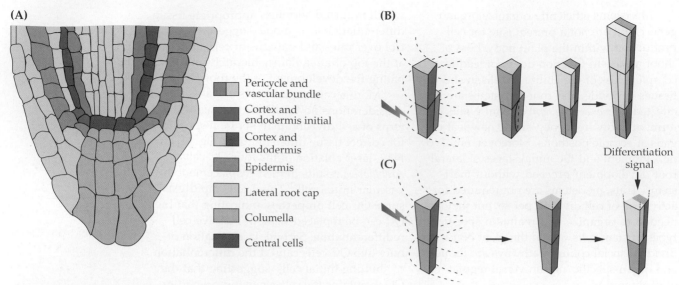

(A)

Pericycle and vascular bundle

Cortex and endodermis initial

Cortex and endodermis

Epidermis

Lateral root cap

Columella

Central cells

(B)

(C)

Differentiation signal

Figure 11.21

Model for the specification of cell identity in the root apex. (A) Schematic diagram of the tissue organization of the root apex. (B) Ablation of cortex initial cells shown in (A) leads to their replacement by pericycle cells, which subsequently give rise to the appropriate derivatives: the cortical and endodermal cell files. (C) Death of individual daughter cells of the cortex initial prior to the formative division that yields cortical and endodermal cell files has no consequence for subsequent cell differentiation. However, cell differentiation requires that the cells next to the ablated cell be in contact with more-mature cells. This implies that the latter (i.e., further differentiated) cells provide cues to immature cells for differentiation.

11.7.4 Plant growth regulators affect the activity of specific cell cycle regulators.

Like other multicellular organisms, plants use long-range signaling to communicate between distant organs (e.g., shoots and roots) and to coordinate their growth. Small molecules play a pivotal role in such mitogenic signaling. Plant growth regulators, including nonpeptide hormones (e.g., auxins, cytokinins, ethylene, ABA, and brassinosteroids; see Chapter 17), lipooligosaccharides (e.g., Nod factors; see Chapter 16), and peptides (e.g., systemin; see Chapter 9), have been implicated in various aspects of growth regulation. Mutants that are defective in hormone synthesis, perception, or response often show drastic changes in their appearance and growth habits (Fig. 11.22).

Auxins and cytokinins are two plant growth regulators likely to have a direct role in cell division regulation (Fig. 11.23). Freshly established plant cell cultures require auxins and cytokinins for continued proliferation; withdrawal of auxin or cytokinin from the medium of fresh tobacco cell cultures, for example, results in their cell cycle arrest in G1 or G2, respectively. The lack of cytokinin in tobacco cell cultures is correlated with increased tyrosine phosphorylation of a tobacco CDK, which probably inhibits kinase activity (see Fig. 11.12). If such arrested cultures are

Figure 11.22

Morphologies of plant hormone mutants. Abnormalities in hormone synthesis, perception, or response strongly affect control of plant growth. (A) The inability to synthesize brassinosteroids compromises the ability of plant cells to expand appropriately and affects their ability to respond to light (see pea plant on left; plant at right is wild type) (B) Enhanced synthesis of auxin stimulates excessive lateral root formation. In the *Arabidopsis superroot* mutant, this phenotype is ultimately lethal. (C) The inability of the maize *vp1* mutant to maintain seed dormancy in response to ABA results in premature germination on the ear (vivipary) and jeopardizes the ability of the seed to survive desiccation.

(A)　　**(B)**　　**(C)**

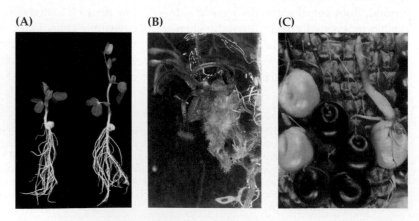

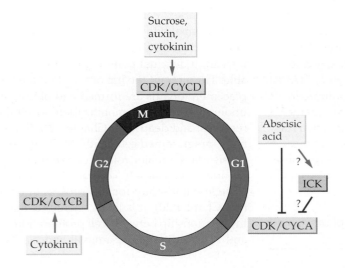

Figure 11.23
Plant hormones and cell cycle control. Plant growth regulators affect cell cycle progression at specific stages of the cell cycle. Sucrose, cytokinin, and auxin promote accumulation of G1 cyclins (shown here as CYCD) and therefore support entry into a new cell cycle. ABA suppresses CDK activity in *Arabidopsis*, presumably by promoting accumulation of CKI. The CKI may then bind to a CDK–cylin complex (shown here as CDK/CYCA) to prevent entry into S phase. Cytokinin promotes entry into M phase, presumably by activating a phosphatase that removes inhibitory CDK tyrosine phosphorylation (CDK shown here in association with mitotic cyclin CYCB).

supplied with cytokinins, tyrosine phosphorylation is reduced and cell division activity resumes. In *Arabidopsis*, cytokinin is also required to stimulate the expression of a D-type (G1) cyclin (CYCD3). Constitutive expression of this cyclin relieves the requirement for cytokinin in tissue explants.

Treatment of intact plants with plant hormones does not stimulate growth in all tissues. For example, application of auxins to roots induces branch root formation but inhibits apical meristem activity. Changes in the expression of cell division regulators, such as cyclin (Fig. 11.24), and overall cell division activity are early responses to such hormone treatments. Although changing the concentrations of auxin clearly affects cell division activity, the nature of the response is not uniform and depends on the tissue.

ABA is primarily involved in signaling drought stress responses (see Chapters 18 and 22). ABA concentrations in the root increase during water stress, and the root apical meristem is arrested under such conditions. Expression of the CDK–cyclin complex inhibitor ICK is induced, suggesting that the inhibitor regulates CDK activity in roots during stress conditions. Brassinosteroids and gibberellins mostly enhance cell growth and expansion. In the dark, however, brassinosteroids suppress the activation of leaf organogenesis, presumably by controlling cell division activity. Gibberellins have a stimulatory effect of cyclin expression in deep-water rice shoot internodes, where cells divide and expand rapidly in response to flooding.

Bacteria of the genus *Rhizobium* and related genera stimulate formation of the root nodules involved in symbiotic nitrogen fixation in legumes (see Chapter 16). This symbiotic interaction, which includes the rapid activation of cell division in root cortical cells, is elicited by Nod factors, lipooligosaccharides that are active at very low concentrations. Consequently, this signaling cascade might modulate the endogenous ratio of auxins to cytokinins, because treatment of legume roots with auxin transport inhibitors or with rhizobia that produce cytokinins will also generate nodule-like structures.

In animals and yeast, mitogenic information is often transduced by components of the MAPK cascade (see Chapter 18). Genes that encode proteins homologous to the MAPK pathway have been cloned from alfalfa, *Arabidopsis*, pea, and tobacco. Expression of *Arabidopsis* MAPK1 can be induced by auxin, whereas expression of several other MAPKs is stimulated by ABA. One important negative regulator of growth

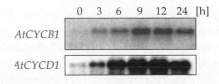

Figure 11.24
Plant growth stimulators affect the expression of cell cycle regulators. The accumulation of mRNAs for mitotic (*AtCYCB1*) and G1 (*AtCYCD1*) cyclins increases after stimulation of *Arabidopsis* roots with auxin. The accumulation is greatest 12 hours after stimulation and then decreases.

responses to ethylene in *Arabidopsis*, CTR1, shares homology with a MAPK kinase kinase (i.e., a kinase that phosphorylates a second enzyme, MAPK kinase, which in turn phosphorylates a third enzyme, MAPK). Ongoing sequencing projects continue to reveal that plant genomes contain many more MAPK genes than do other eukaryotic genomes, suggesting that these enzymes might be involved in the stimulus–response coupling of many signals perceived by plants.

11.8 Cell cycle regulation in plant growth and development

11.8.1 The plant lifestyle requires specific controls for cell division.

Terrestrial plants are **sessile,** attached to an immobile substrate. As plants deplete nearby reserves, they must grow continuously toward new points of access to soil nutrients and light. To exploit their environment optimally, plants require flexible growth rates and patterns. Although growth is not the only plant response to a changing environment, cell division and expansion are essential to bringing the plant body into contact with untapped resources. Therefore, the pathways regulating cell proliferation in plants are likely to respond to environmental and metabolic signals differently from those in animals.

Postmitotic cells in plants do not permanently lose the ability to resume cell division activity or even to regenerate an entire plant. After their displacement from meristems, plant cells are maintained in a differentiated and nonproliferative state by continuous cell–cell interactions with neighboring cells. In other words, plant cell identity is not dependent on cell lineage. The relationship of cell proliferation to cell differentiation therefore differs in plants and animals.

After cell expansion, differentiated plant cells become very large compared with cells in the meristem or animal cells. In dividing cells, cell volume is correlated to DNA content (ploidy). The ploidy of many quiescent plant cells also increases concomitantly with their differentiation, expansion, and maturation. Consequently, plants must have specific mechanisms for uncoupling DNA synthesis from mitosis.

In marked contrast to the maintenance of the animal body, plants do not replace dead or damaged cells within mature organs; instead, individual plant organs are dispensable. Throughout the life of the plant, new replacement organs are formed and old organs undergo senescence, mobilization of cellular reserves, and death (see Chapter 20). In animals, programmed cell death (PCD) by apoptosis plays a major role in determining cell numbers in tissues. In contrast, PCD appears to be less important for this process in plants (see Chapter 20). It is likely, therefore, that the relationship between cell proliferation and PCD is different in animals and plants.

11.8.2 Under some circumstances, cell division activity can limit plant growth.

The responsiveness of plant organ production and growth to environmental conditions raises the question of whether cell division activity is directly regulated by environmental signals. The cyclin subunit of the CDK–cyclin complex is a good candidate target for such a signal pathway. First, cyclin abundance correlates well with growth activity in many organisms. Second, cyclins are labile proteins and therefore well suited to mediate changing signals. In transgenic *Arabidopsis* plants that expressed an increased concentration of a mitotic cyclin, root growth was accelerated and the growth response to auxin was altered (Fig. 11.25A). This result indicates that cyclin concentrations limit cell production and suggests that the amount of cyclin expression couples proliferation with growth control. However, the magnitude of growth acceleration in these plants did not span the whole range of growth rates observed in *Arabidopsis* in response to environmental changes, which suggests that the activity of other cell cycle regulators also might be targets of growth control pathways.

If accelerated cell division enhances growth in an indeterminate organ such as the root, do cell division rates then determine the size of plant organs or whole plants? This question has been investigated experimentally. Expression in transgenic tobacco plants of a dominant-negative plant CDK mutant lowered the overall rates of cell production. Surprisingly, most organs in these tobacco plants were indistinguishable in size

Figure 11.25
Plant growth and cell division are not tightly coupled. (A) Increased expression of a mitotic cyclin in transformed *Arabidopsis* plants results in accelerated root growth and overall increased plant body mass. (B) Alterations in the amount of CDK activity do not significantly affect plant morphology and stature. In transgenic tobacco plants, reduction of CDK activity by expression of a dominant-negative mutant (CDK) gene that encodes an inactive CDC2A protein inhibits proliferation. The resulting plant (right) has fewer but larger cells than wild type, as shown in cross-sections of leaves (bottom panels). The overall stature of the plant is remarkably similar to that of the wild type (top panel), indicating that increased cell expansion compensates for decreased proliferation.

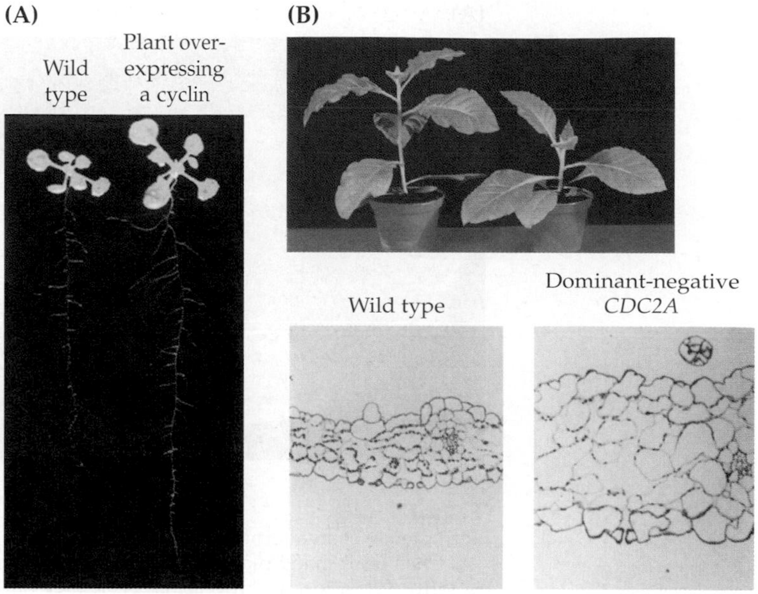

(A)
Wild type   Plant over-expressing a cyclin

(B)

Wild type

Dominant-negative *CDC2A*

and shape from the wild type because the cells they contained, although far fewer, were larger than the wild-type cells (Fig. 11.25B). This indicates that plants can sense the size of developing organs and compensate for insufficient cell division activity with increased cell expansion. Whether the opposite is true, however, is not known, because the final sizes of leaves and leaf cells in the *Arabidopsis* plants with greater cyclin abundance were not examined closely. Precise control of overall proliferation rates is probably not crucial for appropriate plant organ morphogenesis, but localized activation of cell division appears essential for some aspects of morphogenesis, (e.g., in leaf blade development).

11.8.3 Totipotency is a rarely used alternative developmental path.

Most quiescent plant cells and many animal cells resume proliferation when removed from a tissue containing neighbor cells and cultured in a nutrient medium supplemented with plant hormones. This suggests that continuous cell–cell communication between adjacent cells is required to suppress uncontrolled proliferation within tissues.

Most animal cells are replaced many times over the lifetime of an organism. Cell replacement is mediated by different types of stem cells, each of which can generate a limited number of cell types. In contrast, plant cells within tissues of mature organs are not appreciably replaced, and plant growth is mediated by stem cells that can generate every type of cell. Moreover, as shown in genetic and grafting experiments, plant cell differentiation is not determined by lineage but by cell position within the plant. When isolated from mature, quiescent tissues, plant cells lose both their identity and their contact-mediated inhibition of growth, reverting to a ground state. This reactivation of cell division has not been studied in molecular detail but probably involves profound changes of regulatory programs, given that neither CDK1 nor cyclins are expressed in quiescent cells.

In many plant cell cultures, single cells are able to regenerate an entire plant under the appropriate conditions. This can occur by two alternative pathways: organogenesis and embryogenesis. In the former, morphogenesis proceeds by forming shoot organs first, followed by root organs (Fig. 11.26A). Many plants also proliferate asexually in this manner; in many species, roots can generate new shoots, or vice versa. During somatic embryogenesis, embryos are formed directly in culture. The ability of single cells to recapitulate embryogenesis and pattern formation somatically, without an intervening reproductive phase (flowering), is called **totipotency**. In some genera (e.g., *Kalanchoe*), somatic embryogenesis can occur in intact plants and is not dependent on loss of tissue integrity (Fig. 11.26B). Thus, organogenesis and somatic embryogenesis are alternative reproductive strategies to flowering. These asexual mechanisms may be developmental relics that predate the evolution of flowering plants and perhaps have been preserved in some plant species to allow clonal propagation. The signaling pathways responsible for

(A)

(B)

Clonal plants

Figure 11.26
Totipotency is an alternative developmental path.
(A) Cells from many plant species can be cultured
in vitro and induced to regenerate vegetative organs
by application of suitable combinations of plant hormone. The photograph shows plantlets growing from
disorganized tissues called calli (singular: callus) of
petunia leaf tissue. (B) For some plant species (e.g.,
Kalanchoe), the formation of clonal offspring is a competitive advantage. Small plantlets formed in the ax-
ils of serrated leaves are viable and can root if the
parent leaf is torn off or when it senesces.

somatic embryogenesis and the associated
changes in the regulation of cell division
have not been studied in detail. Cell division
appears to be only a late downstream consequence of the activation of the developmental pathways that regulate embryogenesis or
organ formation.

11.8.4 Developmental or phylogenetic ploidy increases are common in plants.

Increased genome copy number—
polyploidy—and genome amplification
during the development of the individual—
endoreduplication—are widespread phenomena in plants. In the latter process,
genome copy number increases concomitantly with cell maturation. These increases can
be dramatic: The suspensor cells of *Phaseolus
coccineus* go through 12 additional cycles of
DNA replication without intervening mitosis, resulting in approximately 8000-fold
greater DNA contents per cell. Endoreduplication occurs preferentially in individuals of
species with low total DNA content such as
Arabidopsis (Fig. 11.27). Specific cell cycle
controls are probably required to inhibit

mitosis during repeated cycles of DNA replication and to reset replication origins to their
prereplication state, but the mechanisms that
control this process are not known.

Why endoreduplication or polyploidy
occurs and how they are regulated are not
well understood. Polyploid plants generally
have a greater capacity to adapt to a broad
range of environmental conditions. Increased
genome copy number may constitute a
means to accelerate rates of gene evolution,
genome reorganization, and ultimately speciation, by providing increased material for
genetic recombination. Moreover, polyploidy
might modify epigenetic regulation of gene
activity in plants, which can lead to selective gene silencing when multiple copies of
closely related genes are present in a genome.
Endoreduplication is developmentally regulated and correlates with cell expansion and
cell maturation (Fig. 11.28). As postmitotic
plant cells expand, their volumes increase
100- to 1000-fold. Perhaps endoreduplication
provides the increased gene dosage needed
to accommodate this increased volume. Endoreduplication may also be an evolutionary strategy to allow small genome sizes in
meristematic cells. By generating additional

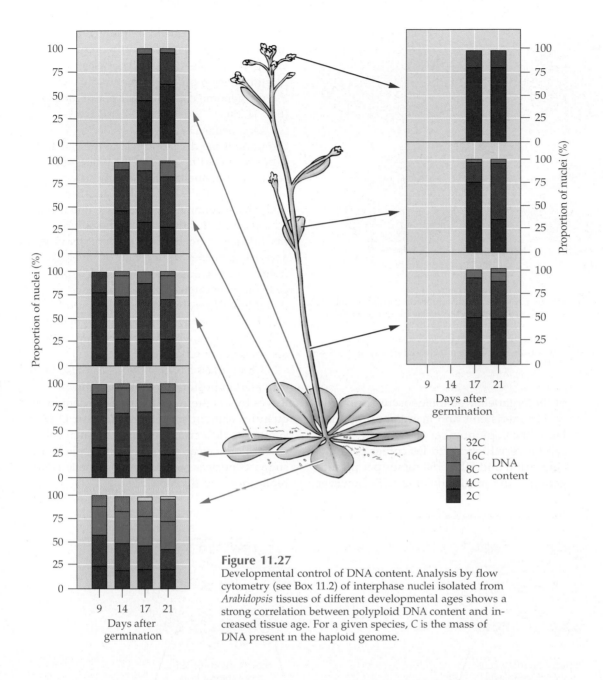

Figure 11.27
Developmental control of DNA content. Analysis by flow cytometry (see Box 11.2) of interphase nuclei isolated from *Arabidopsis* tissues of different developmental ages shows a strong correlation between polyploid DNA content and increased tissue age. For a given species, C is the mass of DNA present in the haploid genome.

copies of wild-type genes, it might also protect against the effects of recessive mutations. The function of endoreduplication should become better understood once mutants are available that lack this form of gene amplification.

11.8.5 Plant cells must replicate and maintain three genomes.

Unlike cells of animals or fungi, plant cells have three different genomes in three cellular compartments: the nucleus, the plastids,

and the mitochondria. Plant mitochondrial genomes are generally much larger than their animal or yeast counterparts and often range from 200 to 2400 kb (see Chapter 6). The plastid genome is approximately 130 to 150 kb (see Chapter 6). Genomes in mitochondria and plastids generally encode only a fraction of the proteins required for the functioning of these organelles, which therefore depend on the import of proteins encoded by nuclear genes.

Within an individual plant, the copy number of mitochondrial and plastid genomes is surprisingly flexible. Replication

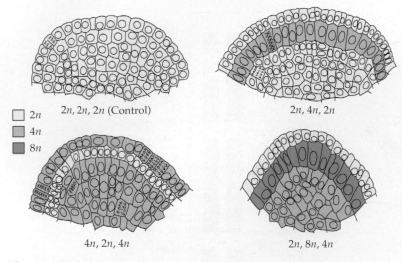

2n □

4n ▨

8n ■

2n, 2n, 2n (Control)

2n, 4n, 2n

4n, 2n, 4n

2n, 8n, 4n

Figure 11.28
Correlation of ploidy with cell size. Chimeras (individuals composed of cells with different genotypes) can be readily generated in plants by use of grafting techniques. The analysis of plants with chimeric meristems demonstrates that cell volume is determined not by position but by genome copy number (ploidy, n). A strong positive correlation was observed between increased ploidy and cell size.

the number of organelles remains relatively low. As postmitotic cells mature, replication of organellar DNA gradually ceases and plastid division proceeds, eventually reducing the genome copy number per organelle (Fig. 11.29).

Organellar DNA synthesis occurs throughout the division cycle of the cell and is not restricted to any specific phase of the cell cycle. Organellar DNA synthesis and division are spatially uncoupled, and because of the high genome copy number, one is not a prerequisite for the other. Segregation of plastids and mitochondria to daughter cells appears to be stochastic when the average number of plastids and mitochondria per cell is high enough to ensure that each daughter cell receives its complement of organelles. However, the mechanisms that control partitioning when only a few organelles are present are unknown.

Chloroplast DNA synthesis can also be activated in postmitotic cells in mature leaves by long-term changes of incident light quantity without concomitant nuclear DNA synthesis. However, the general spatial coincidence of nuclear and organellar replication suggests nuclear control over organelle DNA replication and division.

of the organellar genomes occurs primarily within meristems and in organ primordia. Here, the copy number of organelle genomes can be very high: 20 to 100 copies per mitochondrion and 50 to 150 copies per plastid genome. In rapidly dividing cells, however,

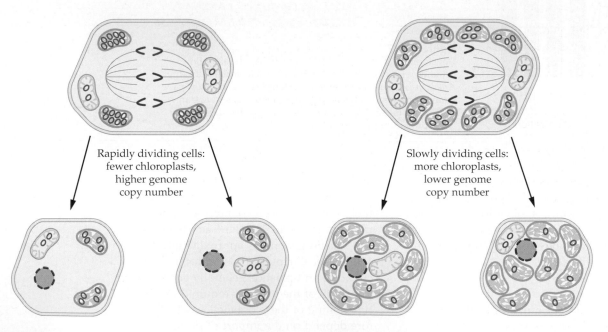

Rapidly dividing cells: fewer chloroplasts, higher genome copy number

Slowly dividing cells: more chloroplasts, lower genome copy number

Figure 11.29
Replication of chloroplast DNA is not coupled to nuclear replication. Rapidly dividing cells contain relatively few chloroplasts but have high copy numbers of the circular plastid genome. DNA synthesis in these plastids continues even when nuclear DNA is not

being replicated. When cells gradually cease to proliferate as they exit the meristem, the genome copy number of individual plastids decreases because the plastids continue to divide one or two more times even in the absence of plastid DNA synthesis.

Summary

In proliferating cells, cell growth is coupled with cell division. The most fundamental functions of the cell division process are the precise replication of chromosomes and their subsequent distribution to daughter cells. These events are controlled by protein kinases and proteases, which have been identified in plants as well as in better-studied models (e.g., animals and yeast). Although functional characterization of these regulatory proteins is not complete, initial observations indicate that they function in similar ways in plants and other organisms but may be regulated differently.

Further Reading

DNA replication

Dutta, A., Bell, S. P. (1997) Initiation of DNA replication in eukaryotic cells. *Annu. Rev. Cell Dev. Biol.* 13: 293–332.

Toone, W. M., Aerne, B. L., Morgan, B. A., Johnston, L. H. (1997) Getting STARTed: regulating the initiation of DNA replication in yeast. *Annu. Rev. Microbiol.* 51: 125–129.

Tye, B. K. (1999) MCM proteins in DNA replication. *Annu. Rev. Biochem.* 68: 649–686.

Waga, S., Stillman, B. (1998) The DNA replication fork in eukaryotic cells. *Annu. Rev. Biochem.* 67: 721–751.

Proteolysis in the cell cycle

Page, A. M., Hieter, P. (1999) The anaphase-promoting complex: new subunits and regulators. *Annu. Rev. Biochem.* 68: 583–609.

Zachariae, W., Nasmyth, K. (1999) Whose end is destruction: cell division and the anaphase-promoting complex. *Genes Dev.* 13: 2039–2058.

Chromosome dynamics

Amon, A. (1999) The spindle checkpoint. *Curr. Opin. Genet. Dev.* 9: 69–75.

Biggins, S., Murray, A. W. (1998) Sister chromatid cohesion in mitosis. *Curr. Opin. Cell Biol.* 10: 769–775.

Skibbens, R. V., Hieter, P. (1998) Kinetochores and the checkpoint mechanism that monitors for defects in the chromosome segregation machinery. *Annu. Rev. Genet.* 32: 307–337.

Uhlmann, F., Lottspeich, F., Nasmyth, K. (1999) Sister-chromatid separation at anaphase onset is promoted by cleavage of the cohesin subunit Scc1. *Nature* 400: 37–42.

CDKs

Morgan, D. O. (1997) Cyclin-dependent kinases: engines, clocks, and microprocessors. *Annu. Rev. Cell Dev. Biol.* 13: 261–291.

Checkpoints

Longhese, M. P., Foiani, M., Muzi-Falconi, M., Lucchini, G., Plevani, P. (1998) DNA damage checkpoint in budding yeast. *EMBO J.* 17: 5525–5528.

Plant cell cycle

Jacobs, T. W. (1995) Cell cycle control. *Annu. Rev. Plant Physiol. Plant Mol. Biol.* 46: 317–339.

Mironov, V., Veylder, L. D., Van Montagu, M., Inzé, D. (1999) Cyclin-dependent kinases and cell division in plants— the nexus. *Plant Cell* 11: 509–522.

PART 3

Energy Flow

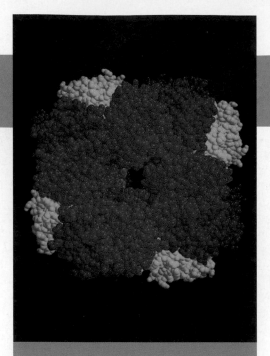

Biochemistry & Molecular Biology of Plants, B. Buchanan, W. Gruissem, R. Jones, Eds.
© 2000, American Society of Plant Physiologists

CHAPTER *12*

Photosynthesis

Richard Malkin
Krishna Niyogi

CHAPTER OUTLINE

Introduction

The synthesis of organic compounds from inorganic precursors requires energy and reducing power (low-potential electrons). For **chemoautotrophic** bacteria and the living communities dependent on their activity (e.g., in the fauna of deep sea vents), the ultimate source of this energy is chemical bonds. Such organisms are, however, a distinct minority. In nearly all biological systems, the synthesis of organic molecules is driven directly or indirectly by energy from the sun.

The overall process whereby plants, algae, and prokaryotes directly use light energy to synthesize organic compounds is called **photosynthesis.** This process supports most autotrophic producers of organic material as well as the heterotrophic consumers they support. In addition to providing food, biomass, and fossil fuels, photosynthesis in plants produces as a by-product the oxygen required for respiratory activity by all multicellular and many unicellular organisms.

Photosynthesis encompasses both a complex series of reactions that involve light absorption, energy conversion, electron transfer, and a multistep enzymatic pathway that converts CO_2 and water into carbohydrates. This chapter will explore these life-sustaining processes in detail. Other topics frequently covered in discussions of photosynthesis are discussed elsewhere, such as sucrose and starch metabolism in Chapter 13 and photorespiration in Chapter 14.

12.1 Overview of photosynthesis

12.1.1 Photosynthesis is a biological oxidation–reduction process.

Most photosynthetic organisms convert light energy into stable chemical products, according to Reaction 12.1. As shown in the equation, photosynthesis is a biological **oxidation–reduction** (or **redox**) process. CO_2 is the electron acceptor, and H_2A is any reduced compound that can serve as the electron donor. (CH_2O) represents the carbohydrate generated by the reduction, and A represents the product formed by oxidation of H_2A:

Reaction 12.1: Photosynthesis

$$CO_2 + 2H_2A \xrightarrow{h\upsilon} (CH_2O) + 2A + H_2O$$

In the case of oxygen-evolving **(oxygenic)** photosynthesis, the focus of this chapter, water serves as the reductant. Water is oxidized and the electrons released are energized and ultimately transferred to CO_2, yielding oxygen and carbohydrate. This light-driven, **endergonic** (energy-requiring) reaction has a free energy change ($\Delta G^{\circ\prime}$) of +2840 kJ per mol of hexose formed. Plants, algae, and prokaryotic cyanobacteria utilize a water-cleaving photosynthetic reaction:

Reaction 12.2: Oxygenic photosynthesis

$$CO_2 + 2H_2O \xrightarrow{h\upsilon} (CH_2O) + O_2 + H_2O$$

Many prokaryotes perform **anoxygenic** photosynthetic reactions that are consistent with Reaction 12.1 but utilize electron donors other than water. For example, as shown in Reaction 12.3, purple sulfur bacteria use electron donors such as H_2S, producing elemental sulfur rather than oxygen as a photosynthetic product. These bacteria retain the capacity to reduce CO_2 to carbohydrate, and

in some cases the mechanism of the carbon-fixation reactions is almost identical to those found in oxygenic organisms. Anoxygenic photosynthetic bacteria are essential components of many terrestrial and aquatic ecosystems. In addition, these prokaryotes have served as important models for studying various aspects of photosynthesis. The advent of molecular genetics has made this group of organisms even more attractive as experimental systems (see Box 12.1 below).

Reaction 12.3: Photosynthetic sulfur reduction

$$CO_2 + 2H_2S \xrightarrow{h\upsilon} (CH_2O) + 2S + H_2O$$

12.1.2 In photosynthetic eukaryotes, photosynthesis takes place in the chloroplast, a specialized organelle.

In eukaryotes, the biophysical and biochemical reactions of photosynthesis occur in a specialized plastid, the **chloroplast** (see Chapter 1). All the reactions required for photosynthesis take place in this organelle. The plant chloroplast is thought to have arisen from the endosymbiotic association of a protoeukaryotic cell and a photosynthetic bacterium related to modern cyanobacteria (see Chapters 1 and 6). The complex structure of the chloroplast reflects its diverse biochemical functions. Chloroplasts from higher plants (Fig. 12.1) are surrounded by a double-membrane system consisting of an **outer and inner envelope** and also contain a complex internal membrane system. The internal membrane system, known as the **thylakoid** membrane, contains distinct regions. Some thylakoids **(granal thylakoids)** are organized into **grana,** stacks of appressed membranes, whereas others **(stromal thylakoids)** are unstacked and thus are exposed to the surrounding fluid medium, the chloroplast **stroma.** The thylakoid membranes are all interconnected and enclose an internal space known as the **lumen.**

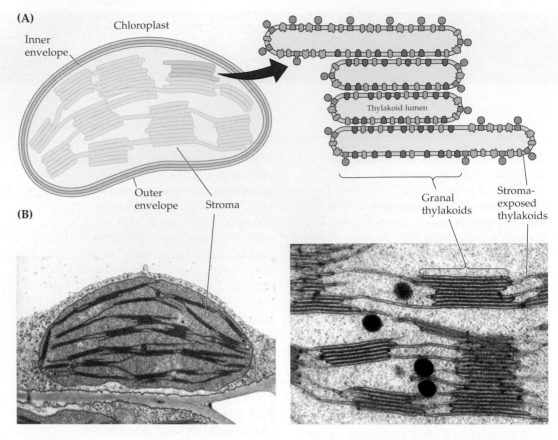

(A)

Chloroplast

Inner
envelope

Thylakoid lumen

Outer
envelope   Stroma

Granal
thylakoids

Stroma-
exposed
thylakoids

(B)

Figure 12.1
(A) Schematic diagram of plant chloroplast, showing compartmentation of the organelle. In a typical plant chloroplast, the internal membranes (thylakoids) include stacked membrane regions (granal thylakoids) and unstacked membrane regions (stromal thylakoids). (B) Transmission electron micrographs of plant chloroplast reveal its ultrastructure. The higher magnification presentation emphasizes the membrane stacking and also includes electron-dense lipid bodies known as plastoglobuli.

12.1.3 Photosynthesis requires the coordination of two phases: light reactions and carbon-linked reactions.

During the 1950s and 1960s, Danial I. Arnon and colleagues showed that isolated chloroplasts can convert CO_2 to carbohydrate in the light (see Rx. 12.2). These experiments also documented that the photosynthetic process involves two phases: the so-called light reactions, which produce O_2, ATP, and NADPH; and carbon-linked reactions (the carbon reduction cycle, also called the Calvin cycle, after one of its discoverers, Melvin Calvin), which reduce CO_2 to carbohydrate and consume the ATP and NADPH produced in the light reactions.

The concept that water oxidation and CO_2 reduction were not obligately linked was first advanced in the late 1930s by Robert Hill. Hill found that various redox-active compounds, including nonphysiological iron-containing compounds, could serve as electron acceptors (A) and could support the evolution of oxygen by chloroplasts in the light, even in the absence of CO_2. This generalized reaction has become a standard measure of activity for the light phase of photosynthesis by way of the so-called Hill reaction:

Reaction 12.4: Hill reaction

$$H_2O + A \xrightarrow{h\upsilon} \tfrac{1}{2}O_2 + H_2A$$

The two phases of photosynthesis occur in different regions of the chloroplast (Fig. 12.2). The thylakoid membranes contain the multiprotein photosynthetic complexes Photosystems I and II (PSI and PSII), which include the reaction centers responsible for converting light energy into chemical bond energy. These reaction centers are part of a photosynthetic electron transfer chain that also contains a transmembrane cytochrome complex (cytochrome b_6f), a water-soluble copper protein (plastocyanin), and a lipid-soluble quinone (plastoquinone). Located primarily in the thylakoids, the photosynthetic electron transfer chain moves electrons from water in the thylakoid lumen to soluble redox-active compounds in the stroma (e.g., $NADP^+$). ADP is phosphorylated on the surface of chloroplast ATP synthase, a large stroma-exposed protein complex located in thylakoid membranes. In contrast, the carbon-linked Calvin cycle reactions occur in the stroma.

In the past, the Calvin cycle has been referred to as the "dark" reactions of photosynthesis. This terminology may be misleading,

---

Box 12.1 Molecular genetic approaches offer new methods for studying the photosynthetic apparatus.

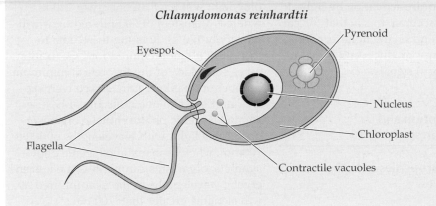

Chlamydomonas reinhardtii

Recent research in photosynthesis has benefited from new techniques of molecular genetics and molecular biology. This is particularly true for studies of the light reactions in photosynthetic prokaryotes, model systems amenable to targeted mutagenesis to yield altered proteins that affect activity. This approach is based on the cloning and sequencing of genes for photosynthetic proteins. Comparison of sequences for similar proteins from a variety of species has also resulted in an understanding of critical protein domains. Sequence analyses also have provided clues to the tertiary structure of these proteins.

An even more powerful technique uses cloned genes to prepare deletion mutants, in which specific protein subunits have been eliminated. These studies are followed by experiments in which the proteins and their functions are restored by transforming cells with strategically modified DNA. Such techniques have been used extensively in studies of photochemical reaction center complexes and electron transfer proteins in cyanobacteria and anoxygenic photosynthetic bacteria. Because these prokaryotic organisms can grow under heterotrophic conditions, they can be used to study mutations

of the photosynthetic apparatus that would be lethal in most photosynthetic eukaryotes, most of which are obligately dependent on photosynthesis. Similar research has shown the feasibility of assembling in prokaryotic organisms photosynthetic membrane complexes that lack a single component. This approach has proven particularly useful in studies of PSI and PSII in various cyanobacteria, providing insight into the specific functions of many individual subunits.

The pace of discovery has not been as rapid in studies of eukaryotic organisms, where cooperation of the nuclear and chloroplast genomes (see Chapters 6 and 9) presents new and somewhat daunting challenges. *Chlamydomonas reinhardtii* (see figure), a unicellular green alga, has emerged as the best eukaryote for molecular studies of chloroplast formation. Like the photosynthetic bacteria described above, *C. reinhardtii* can grow heterotrophically in the dark if supplied with a carbon substrate such as acetate, thus allowing researchers to study mutations that abolish photosynthesis. Another major advantage of *C. reinhardtii* is its amenability to transformation of the chloroplast genome by biolistic transfor-

mation, in which DNA is shot into the cell. Although studies of the nuclear genome in this organism are less advanced, the technology for nuclear transformation is rapidly emerging and should greatly accelerate work on eukaryotic photosynthesis. However, assembly of photosynthetic complexes lacking a single component appears to be difficult in *C. reinhardtii*. Deletion mutants in which a single gene has been mutated yield a phenotype that usually lacks the entire photosynthetic complex, presumably because of destabilization and degradation of the remaining components of the complex (see Chapter 9). Nonetheless, *C. reinhardtii* continues to be the photosynthetic eukaryote of choice for studies of PSI, PSII, and the cytochrome b_6f complex.

The use of molecular manipulation to study photosynthesis in higher plants has been limited because photosynthetic mutations are generally lethal in such systems. However, the use of antisense technology to regulate the quantities of specific proteins in the cell appears promising. In this technique, a cDNA of a target gene is inserted into an appropriate vector but in an orientation opposite to that of the native gene, after which the DNA construct is transferred into the plant genome. The "antisense" mRNA derived from the cDNA is then complementary to the "sense" transcript of the endogenous gene and thus interferes with the translation of the sense transcript, suppressing expression of the native protein. Concentrations of specific proteins active in the Calvin cycle have been diminished through the use of antisense technology. In some cases, this technique has yielded plants having reduced rates of photosynthesis. In other cases, antisense plants have revealed unknown complexities in the regulation of photosynthesis.

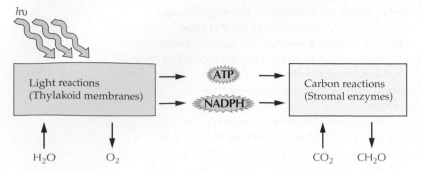

Figure 12.2
The light and carbon (formerly "dark") reactions of photosynthesis occur in distinct chloroplast compartments. Light is required for the synthesis of ATP and NADPH substrates in a series of reactions that occur in thylakoid membranes of the chloroplast. These products of the light reactions are then used by a series of stromal enzymes that fix CO_2 into carbohydrates during the carbon reactions.

because these reactions occur during the day. Furthermore, some of the enzymes involved in CO_2 fixation require activation by light. It is therefore more accurate to say that the carbon reactions depend on the light reactions both for high-energy products and for coordinating (regulatory) signals.

12.2 Light absorption and energy conversion

12.2.1 Light has properties of both particles and waves.

One of the major advances of 20th-century physics was the finding that light, a form of electromagnetic radiation, has properties of both waves and particles. In the case of the quantum mechanical description, radiant energy is visualized as being a stream of energy-carrying particles called quanta (singular: quantum). Each quantum of light, or **photon,** contains a discrete quantity of energy. The energy of a single photon is equal to Planck's constant, h (6.626×10^{-34} Joule•s), multiplied by the frequency of the radiation, υ, in cycles per second, s^{-1} (see Eq. 12.1).

Equation 12.1: Energy is directly proportional to frequency

$$E = h\upsilon$$

Thus, not all colors of light (wavelengths) have equal energy. The energy of a photon of a particular wavelength can be described by Equation 12.2, where c is the velocity of light (3.0×10^8 m s^{-1}) and λ is wavelength (m). As this relationship indicates, the energy content of light is inversely proportional to its wavelength. For example, 1 mol of photons (an einstein) of 490-nm blue light will have an energy of 240 kJ, whereas 1 einstein of 700-nm red light will have only 170 kJ. Relating these data to the +2840 kJ/mol required for CO_2 fixation gives some idea of the theoretical minimum number of quanta required to convert six CO_2 molecules to one molecule of hexose sugar. Oxygenic photosynthetic organisms use visible light, with wavelengths of 400 to 700 nm, whereas many anoxygenic photosynthetic organisms can harness the less energetic wavelengths in the near infrared at wavelengths greater than 700 nm.

Equation 12.2: Energy is inversely proportional to wavelength

$$E = hc/\lambda$$

12.2.2 Light is absorbed by pigment molecules.

For light energy to be used by any system, the light must first be absorbed. This is a significant problem for photosynthetic organisms, because shading and reflection can result in large losses of available light. Molecules that absorb light are called **pigments.** The absorption of a photon by a pigment molecule results in the conversion of the pigment from its lowest-energy (ground) state to an excited state (pigment*):

Reaction 12.5: Pigment excitation

$$\text{Pigment} \xrightarrow{h\upsilon} \text{Pigment*}$$

A pigment molecule becomes excited when absorption of light energy causes one of its electrons to shift from a lower-energy molecular orbital, which is closer to the pigment's atomic nuclei, to either of two more-distant, higher-energy orbitals. The laws of quantum mechanics (Eq. 12.3) dictate that transitions to an excited state will occur only when the energy exactly matches the energy gap between the ground state energy (E_g) and the excited state energy (E_e)—in this case, the gap between the ground state and the excited state energies of the pigment molecule. That is, not all transitions are allowed.

Equation 12.3: Quantitative requirements for pigment excitation

$$E_e - E_g = hc/\lambda$$

In the case of molecules, in contrast to atoms, the ground state and the excited states have many closely spaced substates, the result of molecular vibration and rotations. These are shown in Figure 12.3, along with the first and second excited **singlet states** for a molecule that has two major absorption bands, such as chlorophyll, which has one absorption band in the blue spectral region and one in the red. Excitation causes transitions from the lowest substate of the ground state to any one of these higher-energy substates, depending on the energy relationship shown in Equation 12.3. The existence of this series of substates in molecules yields broad absorption bands rather than the sharp absorption bands found with atoms.

In molecules, two types of excited states can exist. The so-called singlet state is relatively short-lived and contains electrons with opposite (antiparallel) spins; the more long-lived **triplet state** has electron spins that are aligned (parallel). Triplet states generally have much longer lives (take longer to deexcite) than singlet states and are at a

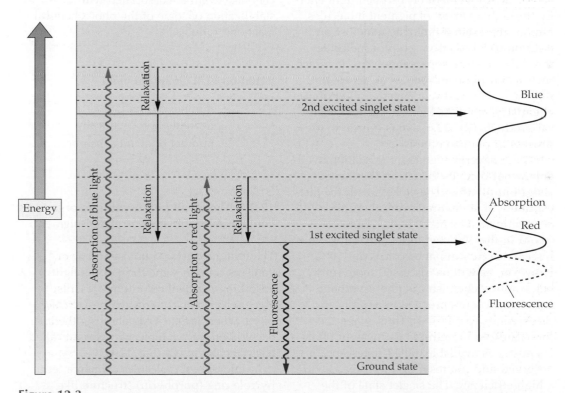

Figure 12.3
Energy levels in the chlorophyll molecule. Absorption of blue or red light causes the chlorophyll molecule to convert into an excited state, with blue light absorption resulting in a higher excited state because of the greater energy of blue light relative to red light. Internal conversions or relaxations result in the conversion of higher excited states to the lowest excited state, regardless of whether excitation was from a red or a blue photon, with a concomitant loss of energy as heat. Light may be reemitted from the lowest excited state through the process of fluorescence. The spectra for fluorescence and absorption are shown at the right of the figure. The short-wavelength absorption band corresponds to a transition to the higher excited state, and the long-wavelength absorption band corresponds to a transition to the lower excited state.

lower energy level. Transitions from the singlet state to the triplet state can occur, but these are low-probability events.

Once excited, an electron can return to the more-stable ground state by one of several paths. In each, the electron returns to the ground state, but the energy released may take several forms. In the simplest case, the energy is released as heat during a nonradiative decay **(relaxation)** (Fig. 12.3). A second mechanism involves the emission of a photon in a process known as **fluorescence** (Fig. 12.3). The emitted light has a slightly longer wavelength than the absorbed light because fluorescence always arises from a decay from the first excited state to the ground state and is preceded by deexcitation from higher substates to the first excited state via vibrational relaxations. A third mechanism involves the transfer of energy to another molecule, usually one in close proximity to the excited molecule. This process, called **energy transfer,** is an important vehicle for the movement of absorbed light energy through an array of pigment molecules. Finally, the excited molecule may lose an electron to an electron-acceptor molecule through a **charge separation** event, in which the excited pigment reduces an acceptor molecule. This last mechanism, called **photochemistry,** converts light energy into chemical products (Rx. 12.6) and is central to the process of photosynthesis.

With several avenues of relaxation available, what dictates the fate of the excited state? Simply this: The process with the fastest rate will be favored over others and will predominate. For many of the pigments found in the photosynthetic apparatus, fluorescence occurs in nanoseconds (ns, 10^{-9}s). However, as will be discussed more fully below, photochemistry in photosynthetic organisms occurs much more rapidly, in picoseconds (ps, 10^{-12} s). Thus, when the thousandfold more rapid pathway of photochemistry is available, little fluorescence is observed and photosynthesis proceeds with a high efficiency. The singlet state of the photosynthetic pigment chlorophyll (see below) participates in energy transfer photo-

chemistry. The relatively long life of triplet-state chlorophyll (as evidenced by how long it takes the molecules in the triplet state to diminish to one-half their original number) suggests that it is not an intermediate in the charge separation events in photosynthesis.

Using Equation 12.4, we can estimate the efficiency of photochemistry in any system by determining the **quantum yield,** phi (ϕ), of the photochemical event. According to this equation, a quantum yield of 1.0 would indicate that every absorbed photon is converted into chemical product. Lower values would indicate that other decay pathways diminish the efficiency of the photochemical reaction. In photosynthetic systems under optimal conditions, the measured quantum yield of photochemistry is approximately 1, indicating that other decay routes do not occur to any substantial extent and that almost all absorbed photons are utilized for photochemical charge separation. Losses in efficiency in other steps of the primary photochemical event are associated with the stabilization of some of the photochemical reaction products.

Equation 12.4: Quantum yield

$$\phi = \frac{\text{number of products formed photochemically}}{\text{number of quanta absorbed}}$$

12.2.3 Almost all photosynthetic organisms contain chlorophyll or a related pigment.

The vast majority of photosynthetic organisms contain some form of the light-absorbing pigment chlorophyll. Plants, algae, and cyanobacteria synthesize chlorophyll, whereas anaerobic photosynthetic bacteria produce a molecular variant called bacteriochlorophyll (Fig. 12.4).

Chlorophyll molecules contain a tetrapyrrole ring (porphyrin) structure like that found in the heme prosthetic group of hemoglobin and the cytochromes (see

Reaction 12.6: Photochemistry

$$\text{Pigment} + \text{acceptor} \xrightarrow{h\upsilon} \text{pigment}^* + \text{acceptor} \longrightarrow \text{pigment}^+ + \text{acceptor}^-$$

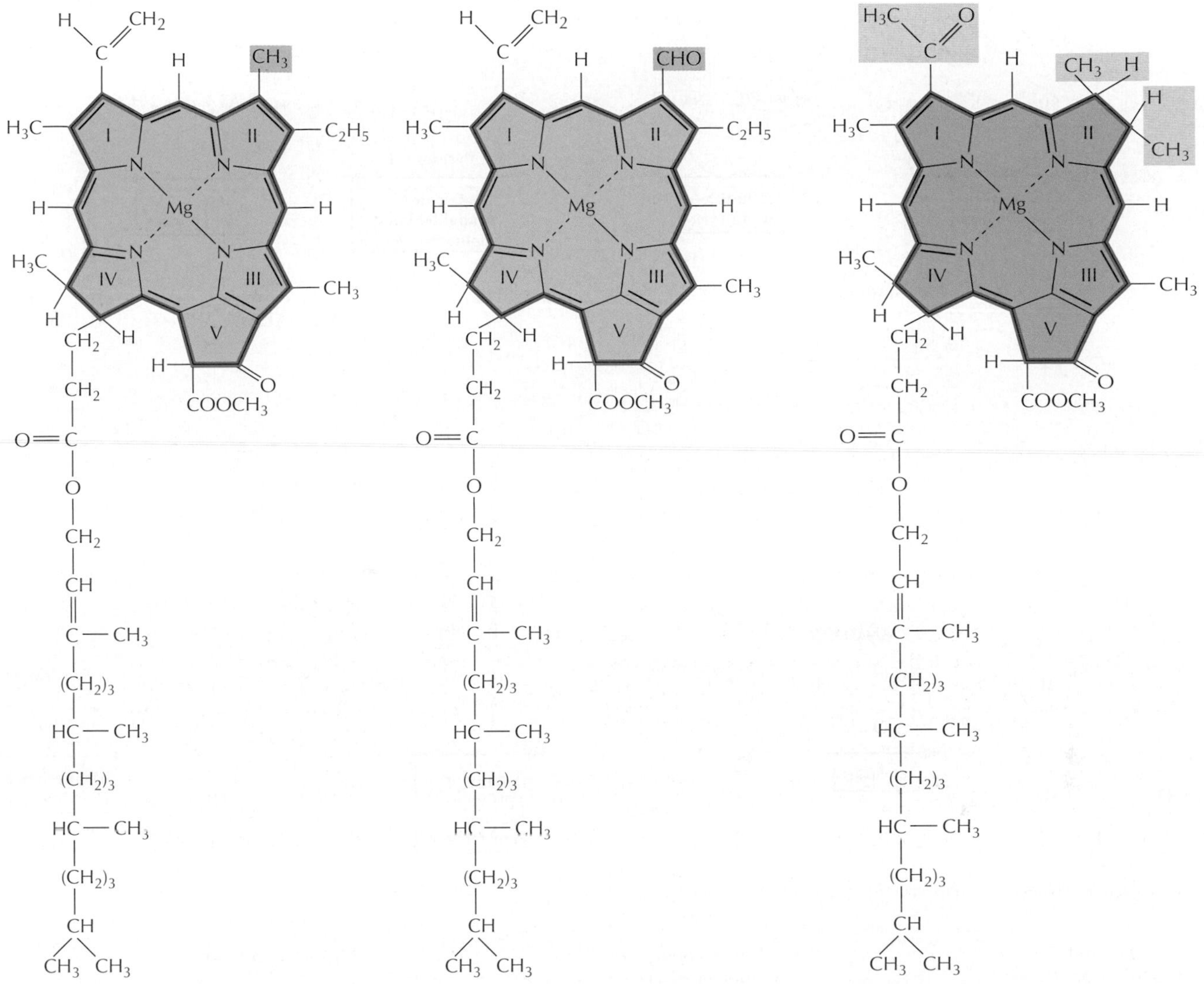

Chlorophyll *a* Chlorophyll *b* Bacteriochlorophyll *a*

Figure 12.4
Structures of chlorophylls. Chlorophyll molecules have a porphyrin-like ring structure that contains a central Mg atom coordinated to the four modified pyrrole rings. Chlorophylls also contain a long hydrocarbon tail that makes the molecules hydrophobic. Various chlorophylls differ in their substituents around the ring structure. In the case of chlorophylls *a* and *b*, a methyl group is present in the former, whereas a formyl group is present at the same position in the latter. Bacteriochlorophyll, found in prokaryotic photosynthetic bacteria, also shows side-chain modifications when compared with the chlorophyll *a* molecule.

Chapter 14, Fig. 14.17). Several steps in the biosynthesis of chlorophyll and heme are shared. However, chlorophyll binds a magnesium atom in the center of its tetrapyrrole ring, whereas heme binds an iron atom. In addition, a long (C_{20}) hydrophobic side chain, known as a phytol tail, is attached to the tetrapyrrole ring structure of chlorophyll and renders the molecule extremely nonpolar.

The biosynthetic pathway for chlorophyll has been elucidated through a combination of labeling studies, enzyme biochemistry, and mutant analysis. The first committed precursor in the biosynthesis of chlorophyll and heme is δ-aminolevulinic acid (ALA). In plants and cyanobacteria, ALA is produced from glutamate in a series of reactions involving $tRNA^{Glu}$ (Fig. 12.5). Although the synthesis of ALA differs in plants and animals, the reactions that convert ALA to the branch point intermediate, protoporphyrin IX, are common to both groups of

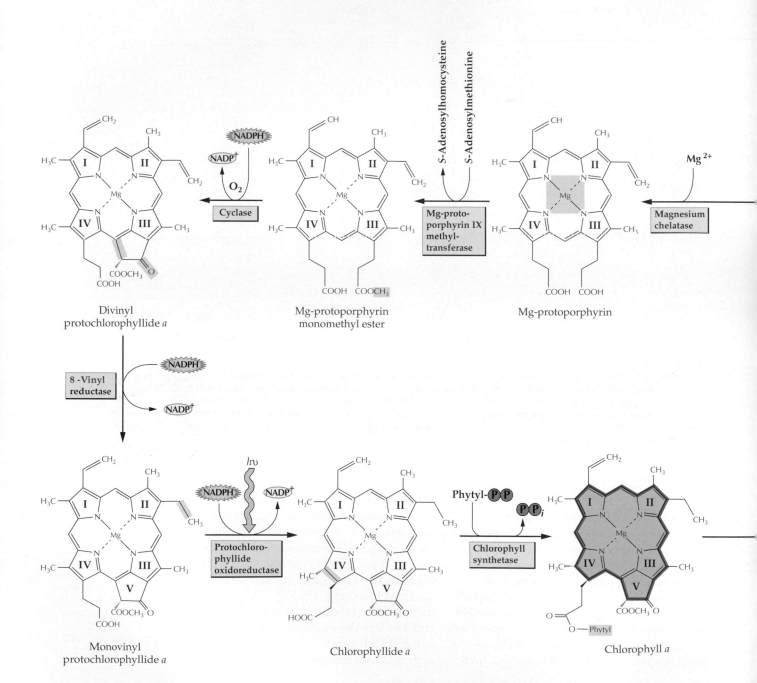

Glutamic acid

Glutamyl-tRNA

Glutamate-1-semialdehyde

δ-Aminolevulinic acid (δ-ALA)

Porphobilinogen (PBG)

Divinyl protochlorophyllide *a*

Mg-protoporphyrin monomethyl ester

Mg-protoporphyrin

Monovinyl protochlorophyllide *a*

Chlorophyllide *a*

Chlorophyll *a*

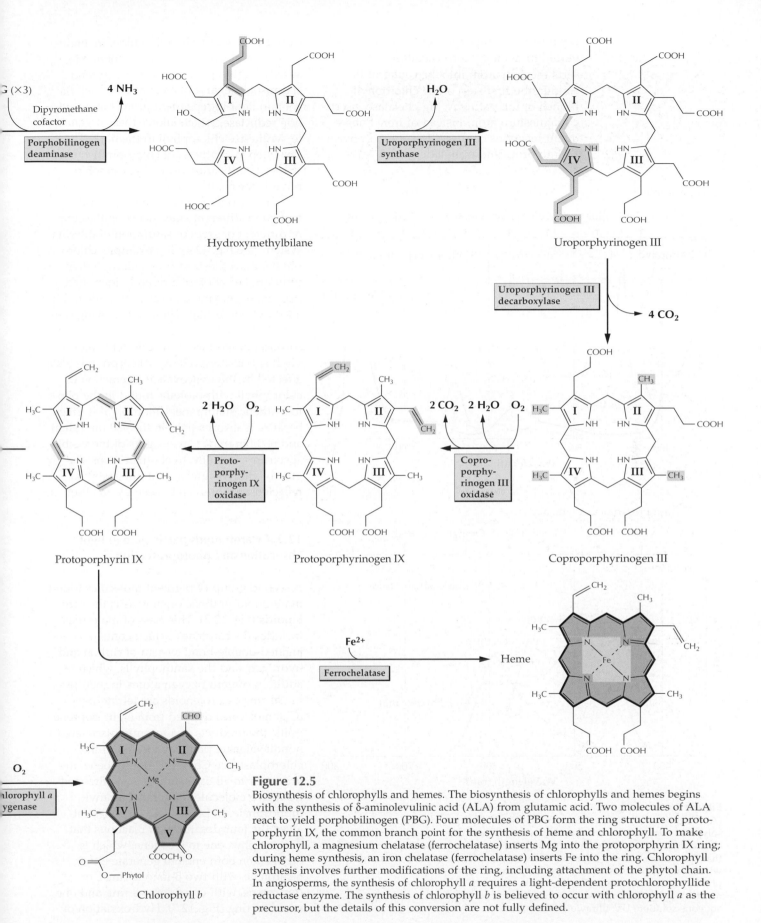

Figure 12.5

Biosynthesis of chlorophylls and hemes. The biosynthesis of chlorophylls and hemes begins with the synthesis of δ-aminolevulinic acid (ALA) from glutamic acid. Two molecules of ALA react to yield porphobilinogen (PBG). Four molecules of PBG form the ring structure of protoporphyrin IX, the common branch point for the synthesis of heme and chlorophyll. To make chlorophyll, a magnesium chelatase (ferrochelatase) inserts Mg into the protoporphyrin IX ring; during heme synthesis, an iron chelatase (ferrochelatase) inserts Fe into the ring. Chlorophyll synthesis involves further modifications of the ring, including attachment of the phytol chain. In angiosperms, the synthesis of chlorophyll *a* requires a light-dependent protochlorophyllide reductase enzyme. The synthesis of chlorophyll *b* is believed to occur with chlorophyll *a* as the precursor, but the details of this conversion are not fully defined.

organisms. Protoporphyrin IX can be converted to heme by ferrochelatase, which inserts the iron atom into the center of the ring. In the first step of the chlorophyll branch of the pathway, Mg-chelatase inserts a magnesium atom instead of iron. Subsequent reactions convert Mg-protoporphyrin IX into protochlorophyllide, which is reduced to generate chlorophyllide. In angiosperms, protochlorophyllide reduction is strictly light-dependent, whereas gymnosperms, algae, and photosynthetic bacteria contain light-independent protochlorophyllide reductase, which allows these organisms to synthesize chlorophyll in the dark. The final step in synthesis of chlorophyll *a* involves esterification of the hydrophobic phytol side chain.

The distinct forms of chlorophyll demonstrate different side chains on the ring or different degrees of saturation of the ring system (see Fig. 12.4). For example, chlorophyll *b* is synthesized from chlorophyll *a* through the action of a recently identified oxygenase enzyme that converts a methyl to a formyl side group. These small changes in chemical structure substantially alter the absorption properties of the different chlorophyll species (Fig. 12.6A). Absorption is also affected by the noncovalent interaction of chlorophyll with proteins found in the photosynthetic membrane. Chlorophyll is green because it absorbs light in the 430 nm (blue) and 680 nm (red) wavelengths of the visible spectrum more effectively than it absorbs green light. The green light not absorbed is reflected and thus can be seen by an observer.

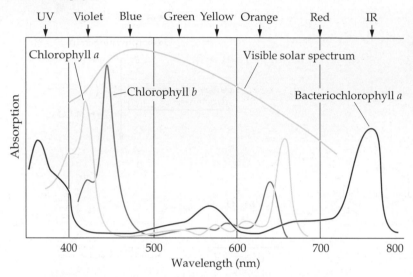

(A) Chlorophylls

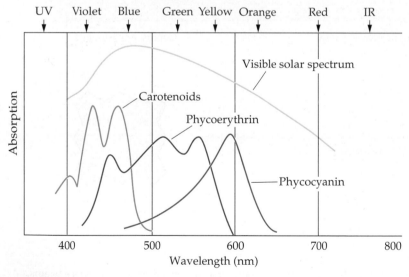

(B) Other photosynthetic pigments

Figure 12.6
(A) Absorption spectra of chlorophylls. The absorption spectra of pigments dissolved in nonpolar solvents are shown for chlorophylls *a* and *b* and bacteriochlorophyll *a*. The visible region of the solar spectrum is also diagrammed. Note that the spectra of these pigments show substantial shifts in absorbance in vivo, where they are associated with specific proteins. (B) Absorption spectra of other photosynthetic pigments. The absorption spectrum of the carotenoids is for pigments dissolved in nonpolar solvents; the remaining spectra are for pigments in aqueous solution. UV, ultraviolet; IR, infrared.

12.2.4 Carotenoids participate in light absorption and photoprotection.

A second group of pigment molecules found in all photosynthetic organisms is the **carotenoids** (Fig. 12.7). This class of molecules includes the carotenes, which contain a conjugated double-bond system of carbon and hydrogen, and the xanthophylls, which in addition contain oxygen atoms in their terminal rings. Carotenoids are tetraterpene (C_{40}) molecules derived from eight isoprene units, the products of a recently discovered nonmevalonate pathway located in the chloroplast (see Chapter 24). Phytoene, the precursor to all carotenoids, is synthesized from two molecules of geranylgeranyl diphosphate. In plants, two enzymes catalyze the four desaturation reactions that convert phytoene to lycopene, which is cyclized on both ends to generate either β-carotene, with two β-ionone rings, or α-carotene, with one β-ionone ring and one ε-ionone ring (Fig. 12.7). Hydroxylation of

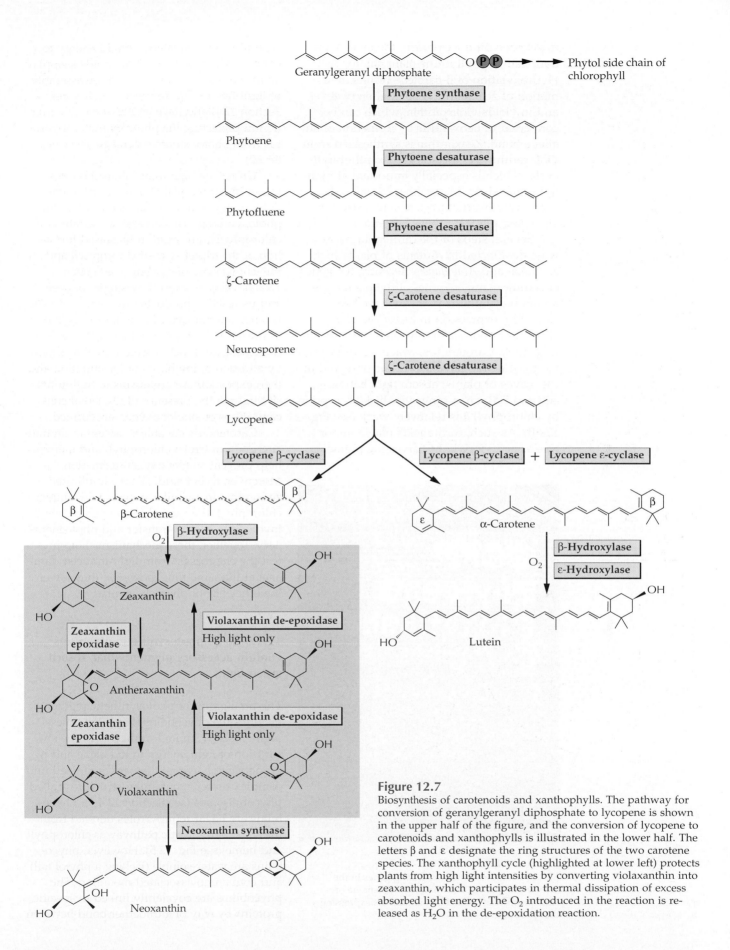

Figure 12.7

Biosynthesis of carotenoids and xanthophylls. The pathway for conversion of geranylgeranyl diphosphate to lycopene is shown in the upper half of the figure, and the conversion of lycopene to carotenoids and xanthophylls is illustrated in the lower half. The letters β and ε designate the ring structures of the two carotene species. The xanthophyll cycle (highlighted at lower left) protects plants from high light intensities by converting violaxanthin into zeaxanthin, which participates in thermal dissipation of excess absorbed light energy. The O_2 introduced in the reaction is released as H_2O in the de-epoxidation reaction.

α-carotene produces lutein, the most abundant xanthophyll in the plant chloroplast. Hydroxylation of β-carotene results in formation of zeaxanthin. Epoxidation of zeaxanthin yields violaxanthin, which can be converted to neoxanthin by formation of an allene bond. Zeaxanthin is synthesized from violaxanthin via the so-called **xanthophyll cycle,** which is especially important at high light intensities. Many other carotenoids with diverse structures are synthesized by algae and photosynthetic bacteria.

Several steps of the carotenoid pathway are affected in mutants of plants such as maize and *Arabidopsis*. Recently, the genes encoding many enzymes of the plant carotenoid biosynthesis pathway have been cloned by expression in *Escherichia coli,* which does not ordinarily make carotenoids (Fig. 12.8). Carotenoids, which are responsible for the orange-yellow colors observed in the leaves of plants, absorb light between 400 and 500 nm, a range in which absorption by chlorophyll *a* is relatively weak (see Fig. 12.6B). As such, carotenoids play a minor role as accessory light-harvesting pigments, absorbing and transferring light energy to chlorophyll molecules. Carotenoids also play an important structural role in the assembly of light-harvesting complexes (LHCs, see Section 12.4) and have an indispensable function in protecting the photosynthetic apparatus from photooxidative damage (see Chapter 22).

Under the high light intensities often found in nature, plants may absorb more light energy than they can actually use for photosynthesis. This excessive excitation of chlorophylls can result in increased formation of the triplet state of chlorophyll and the singlet state of oxygen (see Section 12.2.2). Damage caused by singlet oxygen and its reactive products can decrease the efficiency of photosynthesis in a process known as **photoinhibition** (see Chapter 9, Box 9.6). If a researcher blocks carotenoid biosynthesis by addition of inhibitors or by mutation and then exposes these organisms to high-intensity light in the presence of O_2, lethal concentrations of singlet oxygen are formed.

Carotenoids are able to accept excitation energy from triplet chlorophyll and thereby help prevent singlet oxygen formation. Moreover, recent studies have implicated zeaxanthin in a similar process with singlet chlorophyll, the excited form of chlorophyll involved in energy transfer and photochemistry. Together, these mechanisms help protect the chloroplast from light-induced damage; at the same time, however, they lower the efficiency of photochemistry.

12.2.5 Some photosynthetic organisms contain accessory pigments that absorb green light.

The final group of photosynthetic pigments includes the phycobilins (Fig. 12.9). These pigments play an important role in the absorption of light energy in red algae and in cyanobacteria, where they are organized into complex protein-containing structures called phycobilisomes (see Section 12.4.4). Phycobilins are linear tetrapyrroles derived from the same biosynthetic pathway as chlorophyll and heme (see Fig. 12.5). However, phycobilins are water-soluble (lacking a phytol tail) and contain no associated metal ion. The phycobilins are covalently linked to specific proteins by way of a thioether bond between

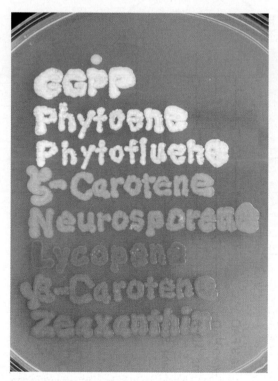

Figure 12.8
Escherichia coli colonies expressing genes in the carotenoid biosynthesis pathway. The name of each product is spelled out in colonies expressing that product.

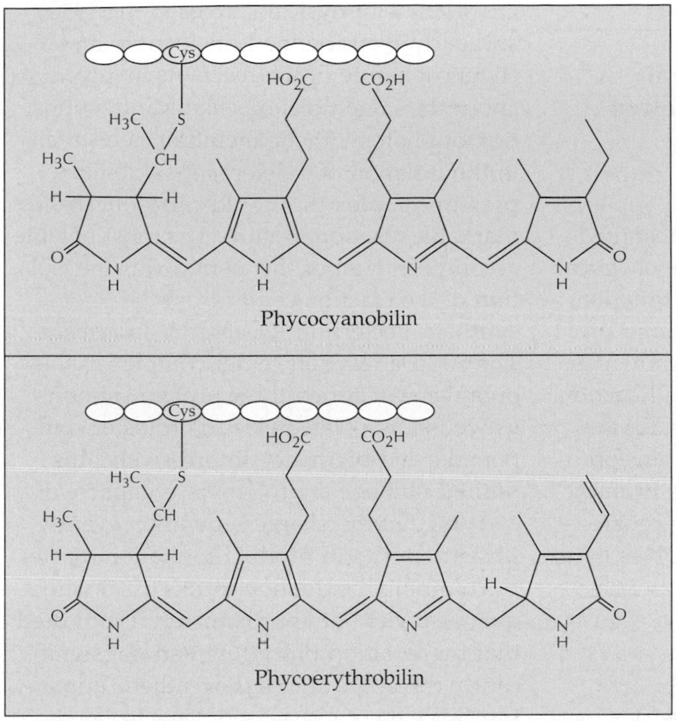

Figure 12.9
Structure of phycobilins. Shown are the structures of two chromophores, phycocyanobilin and phycoerythrobilin, which bind to phycocyanin and phycoerythrin proteins, respectively, by way of thioether linkages involving cysteine residues. The complete sequences of the proteins are not shown.

a cysteine residue in the protein and a vinyl side chain of the phycobilin. The protein phycoerythrin binds the chromophore phycoerythrobilin, whereas the proteins phycocyanin and allophycocyanin bind the chromophore phycocyanobilin. Red algae and cyanobacteria generally contain several different phycobilin pigments, which absorb light in the 500- to 650-nm range (see Fig. 12.6B).

As summarized in Table 12.1, all oxygen-evolving photosynthetic organisms contain chlorophyll *a.* Photosynthetic eukaryotes contain a second form of chlorophyll, chlorophyll *b, c,* or *d.* Most prokaryotic cyanobacteria are unusual in that they contain only one form of chlorophyll, chlorophyll *a.* In addition, all photosynthetic organisms contain

several different carotenoid molecules and some contain phycobilins as well. In organisms that contain chlorophyll or bacteriochlorophyll, the presence of multiple pigments broadens the range of wavelengths that can be absorbed, resulting in a more effective utilization of visible light energy than could be achieved with only one pigment. This arrangement is particularly important in some submerged aquatic niches, where penetration of red light is limited (see Section 12.4.4). (An unusual group of salt-loving bacteria can synthesize ATP with light as the energy source, but these *Halobacterium* species do not contain chlorophyll and, though fascinating, are tangential to the chlorophyll-based light reactions discussed here.)

Table 12.1 Pigment composition of oxygen-evolving photosynthetic organisms

| Organism | Chlorophyll | | | | Carotenoids | Phycobilins |
	a	*b*	*c*	*d*		
Plants	+	+	−	−	+	−
Green algae	+	+	−	−	+	−
Diatoms	+	−	+	−	+	−
Dinoflagellates	+	−	+	−	+	−
Brown and yellow algae	+	−	+	−	+	−
Red algae	+	−	−	+	+	+
Cyanobacteria	+	−	−	−	+	+

12.3 The reaction center complex

12.3.1 Reaction centers are integral membrane protein complexes involved in conversion of light energy into chemical products.

In photosynthesis, light energy is absorbed and converted into relatively stable chemical products. The initial energy transformations in photosynthesis take place at specific sites in the photosynthetic membrane, known as **reaction centers.** Here, as shown in Reaction 12.7, specially bound chlorophyll (Chl) absorbs a photon and subsequently transfers an electron to an acceptor (A). This primary charge separation event is the only photosynthetic reaction that directly involves light.

Reaction 12.7: Photochemistry in a reaction center

$$\text{Chl A} \xrightarrow{h\upsilon} \text{Chl* A} \longrightarrow \text{Chl}^+\text{A}^-$$

Many biophysical studies in the 1950s and early 1960s defined the identity and chemical nature of the reactants involved. At about the same time, biochemical fractionation of photosynthetic membranes resulted in the isolation of a detergent-solubilized protein complex that could carry out this remarkable reaction in vitro. The most notable accomplishment of this period was the isolation of the reaction center from the photosynthetic bacterium *Rhodobacter sphaeroides*. The 80-kDa reaction center complex in this organism contained three protein subunits as well as associated electron transfer components and pigments. Impressively, this limited number of proteins was capable of carrying out the charge separation event. Measurements of product formation (bacteriochlorophyll oxidation in this case) with a quantum yield of approximately 1 indicated that the isolation procedure had not significantly disrupted the photosynthetic apparatus. Since then, researchers have isolated

---

Box 12.2 — Biophysical techniques have been used to investigate photosynthetic electron transfer events.

(A)

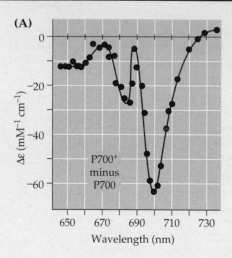

Studies of photosynthesis, particularly those related to the light reactions, have benefited from a vast array of experimental techniques used to examine specific components of the photosynthetic apparatus and their function. Various spectroscopic techniques have been used in these studies, taking advantage of the many oxidation–reduction reactions that occur during photo-synthesis as a result of light conversion and subsequent electron transfer events. These approaches rely on characteristic "signatures" that distinguish the oxidized and reduced species of a given molecule. When these molecules engage in electron transfer reactions, changes in their redox state can be monitored. Two techniques that have been widely applied to these systems are optical absorbance spectroscopy, based on the interaction of light with the molecules, and electron paramagnetic resonance (EPR) spectroscopy, based on the magnetic properties of the molecules.

Absorbance spectroscopy has been used extensively to detect the state of pigment molecules. Such measurements have provided important details on the kinds of pigments found in photosynthetic membranes and their organizational features. An extension of this technique is absorbance difference spectroscopy, which has been applied to the study of pigments that show absorbance changes under differing conditions, for example, when they undergo oxidation–reduction reactions in the light. Because the concentration of these molecules in the photosynthetic membrane is very low and is masked by the large ab-sorbance of pigmented molecules that do not undergo redox changes, these small absorbance changes are difficult to detect. By measuring the difference in the redox state of the molecule in the light versus that in the dark, one can detect these small absorbance changes without needing to obtain the absolute spectrum of the component. For example, the reaction center chlorophyll P700 takes its name from such an absorbance change, a decrease centered at approximately 700 nm when this pigment undergoes oxidation in the light (see panel A). Other electron carriers also have characteristic absorbance changes associated with their oxidation or reduction. This trait allows the monitoring of any individual component in a complex mixture of components, such as are found in a photosynthetic membrane. To obtain these spectra, one can activate the samples by using either steady-state illumination lasting 1 or more seconds or ultrashort flashes, in the range of picoseconds, nanoseconds, or microseconds. The latter technique allows determination of the time-resolved kinetics of a particular redox process. In the case of submicrosecond measurements, the technology is rather sophisticated, using high-energy lasers as

reaction center complexes from a large number of photosynthetic organisms, including plants. Characterization of these complexes has strengthened the concept that the reaction center exists as a discrete protein complex within the photosynthetic membrane and contains all of the essential elements for carrying out the primary charge separation reaction of photosynthesis.

12.3.2 Reaction centers contain both special chlorophyll and electron acceptor molecules involved in energy conversion.

Spectroscopic studies have indicated that the reaction center chlorophyll participating in the primary charge separation event shown in Reaction 12.7 is actually a chlorophyll or bacteriochlorophyll dimer called the special pair. In the case of the bacterial reaction center complex from *Rb. sphaeroides*, this pigment has been named P865 after the wavelength of light for which oxidation-linked changes in absorbance are greatest (Box 12.2). This absorbance change associated with the primary charge separation provides a means to detect active fractions during isolation of reaction centers from intact membranes. In the case of the reaction centers from plants, the pigments that undergo oxidation in the light have been named P700 and P680, again referring to absorbance bands in the difference spectra (see Box 12.2) for these pigments.

In addition to chlorophyll, reaction center complexes contain several electron acceptors. The chemical nature of these acceptor molecules depends on the type of reaction center complex. A general feature of electron transfer events in all reaction centers is the transfer of an electron from the special pair, such as P865, to another pigment molecule, such as a bacteriopheophytin (pheophytin is a chlorophyll derivative lacking magnesium), or even to another chlorophyll molecule. This reaction is followed by subsequent electron

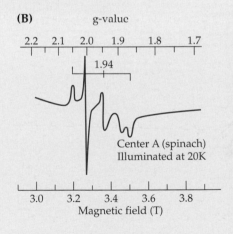

(B)

g-value

Center A (spinach)
Illuminated at 20K

Magnetic field (T)

the light source. Such measurements have been used to identify early intermediates in the charge separation events in photosynthesis as well as to monitor the overall kinetics of specific electron transfer events.

Another useful spectroscopic technique applied to photosynthetic systems is EPR spectroscopy, also known as electron spin resonance (ESR). Instead of measuring changes related to the optical properties of molecules as occur during optical absorbance spectroscopy, EPR is based on the magnetic properties of paramagnetic molecules, those that contain unpaired electrons. For example, in all photosynthetic reaction centers, both the oxidized and the reduced products of the primary charge separation event are paramagnetic (having been generated by the transfer of a single electron), and both can be detected by the EPR technique. In addition, the process of oxidation and reduction of electron carriers in the electron transfer chain alters the number of unpaired electrons in these molecules. EPR detects transitions that occur between electron spin states that have been established by placing the paramagnetic molecule in a magnetic field. Researchers can use this technique not only to follow the redox state of the electron carrier but also to gather information on the chemical nature of the paramagnetic species. Different paramagnetic species are characterized by different values for g (a unitless number associated with paramagnets). For example, the free electron (a theoretical construct) has a g-value of 2.0023, whereas the unpaired electron in a photooxidized molecule of the reaction center of chlorophyll has a g-value of 2.0026 and the plas- tosemiquinone free radical has a g-value of 2.0045. These slight differences are related to the environment of the unpaired electron in each molecule, relative to a free electron not subject to environmental influences. Metal ion–containing centers, such as Fe-S centers found in ferredoxin and PSI, have more complex EPR spectra because of the involvement of different orbitals for their unpaired electrons compared with those for free radicals. The first-derivative EPR spectrum of F_A (iron–sulfur center A) in PSI in liquid helium is shown in panel B on this page. This Fe-S center is characterized by three g-values: 2.04, 1.94, and 1.86. F_B and F_X, the other Fe-S centers of PSI, have slightly different g-values, allowing for their spectral resolution. Copper-containing proteins in the Cu^{2+} state, such as the oxidized form of the copper-containing soluble protein plastocyanin, have g-values of 2.22 and 2.05, substantially different from the g-values for Fe-S centers. Cu^+ is diamagnetic, so reduction of plastocyanin results in a loss of its EPR signal.

transfer to nonpigment molecules, such as quinones or Fe-S centers. The generalized sequence of events is described schematically in Reaction 12.8.

12.3.3 The structure of a reaction center from a photosynthetic bacterium has been determined.

A major advance in our understanding of the structure and function of reaction centers came about in the 1980s, when Hartmut Michel, Johann Deisenhofer, and Robert Huber reported the high-resolution structure of a reaction center complex isolated from the photosynthetic bacterium, *Rhodopseudomonas viridis*, determined by X-ray crystallography (Fig. 12.10A). Because this organism is very similar to *Rb. sphaeroides*

discussed above, the structure of the complex from *Rb. sphaeroides* has also since been solved. A major feature of the structure is the two axes of symmetry around a central axis perpendicular to the plane of the membrane. The central axis bisects the reaction center bacteriochlorophyll dimer and intersects a single iron atom (Fig. 12.10B). The structure has two branches, each of which contains a monomeric bacteriochlorophyll molecule, a monomeric bacteriopheophytin molecule, a ubiquinone, and one partner of the bacteriochlorophyll dimer (P865). This detailed structure confirmed the dimeric nature of the reaction center chlorophyll, as had been predicted by spectroscopic analysis. The complete solution of this structure, a major achievement in the field of photosynthesis and membrane biology, earned these workers the 1988 Nobel Prize in chemistry.

Reaction 12.8: Primary change separation in the reaction center

$$\text{Chl A}_0\text{A}_1\text{A}_2 \xrightarrow{h\upsilon} \text{Chl}^*\text{A}_0\text{A}_1\text{A}_2 \longrightarrow \text{Chl}^+\text{A}_0^-\text{A}_1\text{A}_2 \longrightarrow \text{Chl}^+\text{A}_0\text{A}_1^-\text{A}_2 \longrightarrow \text{Chl}^+\text{A}_0\text{A}_1\text{A}_2^-$$

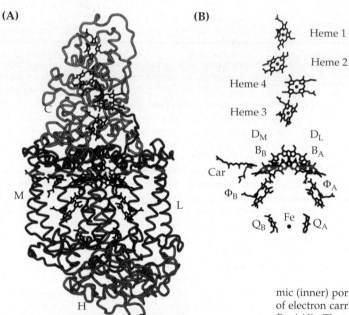

Figure 12.10
(A) Structure of the protein subunits of the reaction center complex of the purple bacterium *Rhodopseudomonas viridis*, resolved by high-resolution X-ray crystallography. Eleven α-helical regions of the L (dark red) and M (purple) subunits form the major transmembrane portion of the structure, with the C (cytochrome, green) subunit being attached to the periplasmic (outer) portion of the structure, and the H subunit (magenta) being attached to the cytoplas-

mic (inner) portion of the structure. (B) Organization of electron carriers in the reaction center complex of *R. viridis*. The series of electron carriers is organized around a two-fold axis of symmetry, a line bisecting the P870 reaction center special pair (D_M/D_L) and intersecting the single Fe atom. The four heme groups identified in the structure are part of the *c*-type cytochrome on the periplasmic face of the structure. B_A, accessory bacteriochlorophyll in the active branch; B_B, accessory bacteriochlorophyll in the inactive branch; Car, carotenoid; Φ_A, bacteriopheophytin in the active branch; Φ_B, bacteriopheophytin in the inactive branch; Q_A, primary electron acceptor quinone; Q_B, secondary electron acceptor quinone.

Resolution of the bacterial reaction center structure, a notable accomplishment in itself, provided a model for the structure of PSII reaction center complexes in plants. For example, similarities among the amino acid sequences of subunits in the bacterial complex and proteins in PSII of oxygen-evolving organisms led to proposals of similar structure. In addition, some of the early electron acceptors of the bacterial charge transfer pathway are similar to those found in the PSII complex from plants, further strengthening this analogy. More direct evidence is emerging that the PSII reaction center has a highly symmetrical structure, similar to that in bacteria (see Section 12.3.5).

12.3.4 The kinetics of primary charge separation events are understood in great detail.

Within several picoseconds of the initial charge separation event of photosynthesis, the light energy conversion reaction is complete and chemical products have been formed. The technique of rapid kinetic absorbance spectrophotometry, in which ultrashort laser flashes activate electron transfer components in the reaction center complex, has made it possible to follow the electron transfer steps in the bacterial reaction center, from the special pair to the sequential electron acceptors. The initial charge separation event, resulting in the formation of oxidized chlorophyll (P865) and the first reduced electron acceptor (BPha⁻), takes place in approximately 1 ps. After an additional 100 to 200 ps, the electron has been transferred to a bound quinone (Q_A). The terminal quinone electron acceptor (Q_B) is reduced on a slower—microsecond—time scale (Fig. 12.11). Only one of the two highly symmetrical arms or pathways available to the electron is actually used. The second arm remains inactive by processes that are not yet fully understood.

Electron transfer events within the reaction center complex are independent of temperature. Thus, it is possible to measure the charge separation at temperatures as low as 4 K (the temperature of liquid helium). That these photophysical events are independent of temperature is not surprising, since they do not require molecular collisions. Mea-

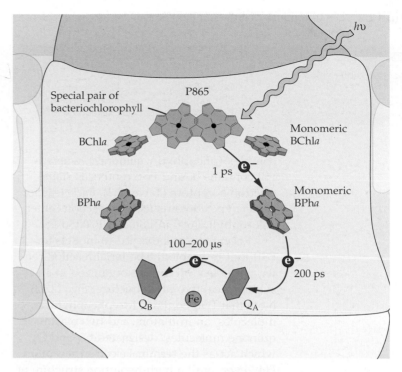

Figure 12.11
Kinetics of electron transfer in the bacterial reaction center. The special pair of bacteriochlorophylls in the reaction center undergoes oxidation in the light. The electron lost from P865 is rapidly transferred to a monomeric bacteriochlorophyll (BChla) and then on to bacteriopheophytin (BPha). These reactions occur in approximately 1 ps. In approximately 200 ps, this electron is transferred to one of two ubiquinone molecules, Q_A, and then on to a second ubiquinone, Q_B, on a slower time scale. The electron transfer proceeds down one of the structure's two arms, although the mechanism by which the single arm is selected is not understood.

surements of charge separation at cryogenic temperatures have been used extensively in the characterization of primary reactants in both bacterial and plant reaction center complexes.

12.3.5 Oxygenic photosynthetic organisms contain two photochemical reaction centers, PSI and PSII.

The photosynthetic membranes of anoxygenic photosynthetic bacteria contain only one type of reaction center complex, which varies among organisms. In contrast, all oxygen-evolving organisms, including cyanobacteria, algae, and plants, contain two different reaction center complexes. As mentioned above, these complexes have been designated Photosystem I (PSI) and Photosystem II (PSII). PSII contains electron carriers very similar to those in the R. viridis

Table 12.2 Electron transfer carriers in the PSI and PSII reaction center complexes

Carrier	PSI	PSII
Reaction center chlorophyll	P700	P680
A_0	Chlorophyll *a*	Pheophytin *a*
A_1	Phylloquinone	Plastoquinone (Q_A)
A_2	F_X (Fe-S center)	Plastoquinone (Q_B)

complex (pheophytin, quinones), whereas PSI contains bound Fe-S centers as stable electron acceptors (Table 12.2). Each reaction center type contains its own reaction center chlorophyll, P680 in PSII and P700 in PSI.

Recent research has led to models for PSII that resemble the bacterial reaction center, with electron carriers organized in a symmetrical two-arm structure (Fig. 12.12). Note that PSII contains two pheophytin molecules, an iron atom, and two plastoquinone molecules, designated Q_A and Q_B, which act as the terminal electron acceptors. However, until a high-resolution structure of PSII is available, this organization model for PSII must be considered only tentative.

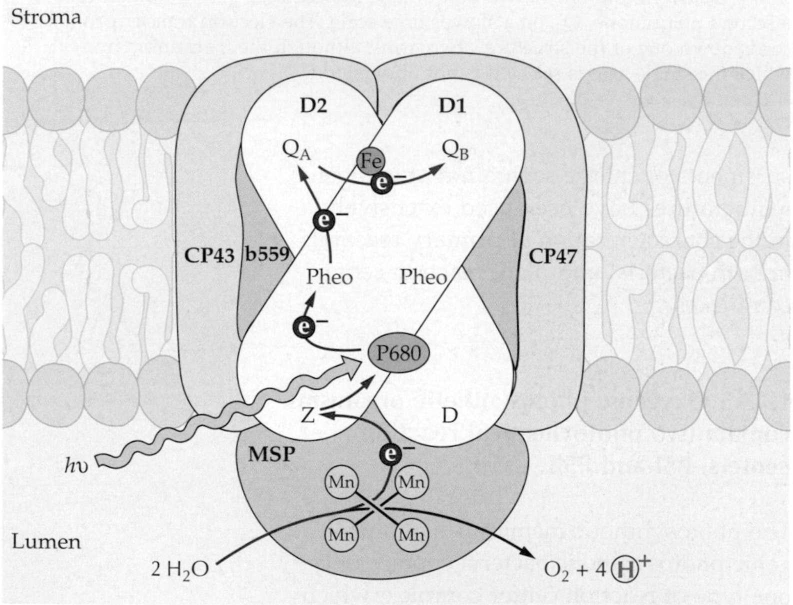

Figure 12.12
Structural model of the PSII reaction center, a schematic representation showing the structure dominated by the two PSII reaction center proteins D1 and D2. The model is based on analogies with the bacterial reaction center complex (see Fig. 12.10). Electrons are transferred from P680 to pheophytin (Pheo) and subsequently to two plastoquinone molecules, Q_A and Q_B. P680$^+$ is reduced by Z, a tyrosine residue in the D1 subunit. The oxidation of water by the Mn cluster is also indicated. CP43 and CP47, chlorophyll *a*–binding proteins. D1 is susceptible to photochemical damage and undergoes active turnover (see Chapter 14; Box 9.6).

A comparable tentative model for the organization of the electron carriers in PSI is shown in Figure 12.13. Ongoing structural analysis of a cyanobacterial PSI complex, currently at a resolution of 4.0 Å, reveals the organization of its many protein subunits, including several associated with extrinsic subunits localized to the stromal face of the membrane (Fig. 12.14). The two-arm symmetical structure previously noted with the *R. viridis* complex is also apparent in this PSI complex, particularly in the organization of the chlorophylls that function as early electron acceptors (Fig. 12.15). Although current resolution of the PSI complex is insufficient to identify all the electron acceptors and define their interaction with the protein subunits, this structure has provided preliminary evidence for the organization of pigments and electron transfer components that resembles those of the *R. viridis* complex in several important respects.

12.4 The photosystem

12.4.1 A photosystem contains a photochemical reaction center and multiple antennae (auxiliary light-harvesting pigment–protein complexes).

All photosystems contain a reaction center with its full complement of electron transfer components as well as an array of light-harvesting, or **antenna,** pigments. These antennae function to absorb light energy, transferring it to the reaction center, where the energy is then converted into stable chemical products.

Analyses of the antenna sizes of PSI and PSII show that in most plants approximately 250 chlorophyll molecules are associated with each reaction center. The transfer of excitation energy from one pigment molecule to another, by a mechanism known as Förster energy transfer, or resonance, does not require emission and reabsorption of photons. Proximity of the donor and acceptor molecules within the antennae is critical because the efficiency of energy transfer is inversely proportional to the sixth power of the distance separating the two molecules. For two pigments separated by approximately 1.5 Å, an energy transfer time of less than 1 ps has been observed. The relative

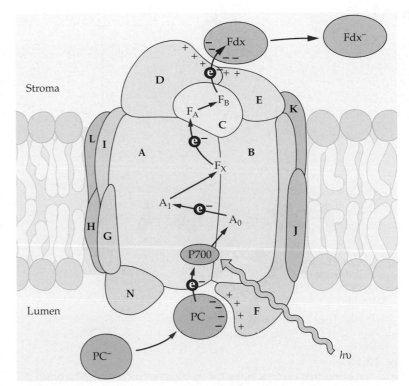

Figure 12.13
Structural model of the PSI reaction center, a schematic representation showing the organization of the two major proteins in this complex, the psaA and psaB subunits, designated here as A and B. Electrons are transferred from P700 to a chlorophyll molecule, A_0, then on to the A_1 electron acceptor, phylloquinone. Electron transfer then proceeds through a series of Fe-S centers, designated F_X, F_A, and F_B, and ultimately to the soluble iron–sulfur protein, ferredoxin (Fdx). $P700^+$ received electrons from reduced plastocyanin (PC). Several PSI subunits, such as psaF, psaD, and psaE are involved in the binding of soluble electron transfer substrates to the PSI complex.

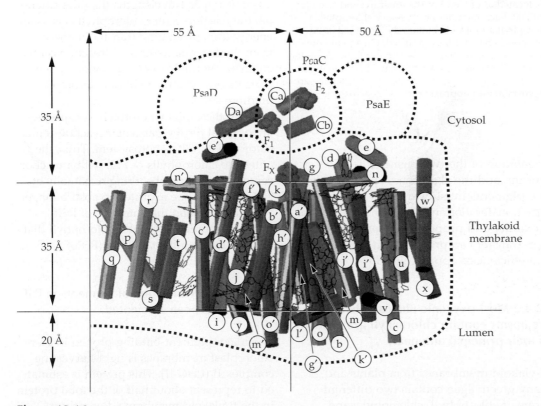

Figure 12.14
Structure of the PSI reaction center complex. This 4.0-Å resolution structure of a cyanobacterial PSI complex indicates a large number of transmembrane helices (designated with lowercase letters). The cytosolic side of the complex is at the top of the figure, and the lumenal side is at the bottom. α–Helices associated with the cytosol-exposed subunits PsaD and PsaC are labeled Da and Ca or Cb, respectively. The three Fe-S centers that function in the electron acceptor complex of PSI (here labeled F_X, F_1, and F_2) are visible on the cytosolic side of the complex.

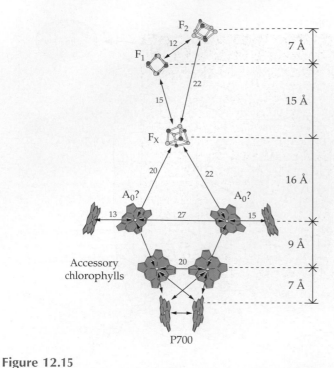

Figure 12.15
Organization of electron carriers in the PSI reaction center complex. The organization shown is based on the 4.0-Å resolution structure of the cyanobacterial PSI complex. The P700 dimer is located at one end of the structure, and two symmetrical arms radiate out from P700. Each includes an accessory chlorophyll a molecule and a monomeric chlorophyll a molecule tentatively identified as A_0. The location of acceptor A_1 (not shown) has not been resolved. On a plane running through the middle of the structure is located the Fe-S center F_X; at the outer side of the complex, two additional Fe-S centers (F_1 and F_2) have been identified. Distances between the electron carriers are shown at the right. In transferring an electron from P700 to F_X, the electron moves approximately 30 Å across the presumed membrane.

orientation of the two pigments is also significant, and the absorbance spectra of the two pigments must overlap for efficient energy transfer. For antenna pigments in close proximity and proper spatial orientation, energy transfer can proceed with high speed and efficiencies approaching 99%.

12.4.2 Most oxygenic photosynthetic organisms contain chlorophyll *a/b* proteins as their principal antennae.

Thylakoid membranes from plants and many green algae contain two different forms of chlorophyll, chlorophyl a and chlorophyll b (see Fig. 12.4). Chlorophyll a is found in all reaction center complexes as well as in antennae, whereas chlorophyll b is found only in antenna complexes. Fractionation of chloroplast membranes with

nondenaturing detergents, followed by electrophoretic analysis, shows that all of the chlorophyll in a chloroplast membrane is associated with specific proteins. Approximately 15 different chlorophyll-binding proteins have been identified, some associated with PSI and others with PSII (Table 12.3). All are encoded in the nucleus and must therefore be imported into the chloroplast before binding with chlorophyll and associating with their proper photosystem. In addition to the chlorophyll pigments, carotenoids are also commonly found in antenna. Generally, a carotenoid/total chlorophyll ratio approaching 0.5 is typical for antennae complexes in most plants. As indicated in Section 12.2.4, carotenoids are capable of absorbing light energy in the 450- to 500-nm region and transferring this energy to chlorophyll molecules.

When chlorophyll associates with specific proteins, its peak absorption wavelength shifts toward the red (lower energy) spectral region. Because the reaction center complex absorbs longer wavelengths than the antenna, the reaction center chlorophyll acts as an energy trap, promoting transfer of energy from the antenna toward the reaction center complex. Not only does each photosystem have its own complement of chlorophyll-binding antenna proteins, but also the properties of these pigment–protein complexes are optimal for the particular reaction center chlorophyll in the photosystem. Thus, the antennae chlorophylls of PSI, with a reaction center chlorophyll absorbing maximally at 700 nm, would be expected to absorb longer wavelengths than the antennae of PSII, which has a reaction center chlorophyll that absorbs maximally at 680 nm (P680).

12.4.3 The organization of pigments in PSI and PSII has been elucidated.

The major pigment-binding protein in plant chloroplast membranes is light-harvesting complex II (LHC-II). This protein is estimated to represent about half of the total protein in the thylakoid membrane. Recently, researchers reported the structure of LHC-II at a resolution of 3.4 Å. The three transmembrane helices of this protein bind about 12 chlorophyll a and b molecules. In addition, two carotenoid molecules serve as scaffolding

Table 12.3 Properties of light-harvesting chlorophyll protein complexes

Complex	Chl *a/b* ratio	Gene[a]	Mol. mass (kDa)
Light-harvesting complexes (LHC) associated with PSI			
LHC-Ia	2.0–3.1	*lhca3*	20.5
(LHC-I 680)		*lhca2*	18
LHC-Ib	2.2–4.4	*lhca1*	20
(LHC-I 730)		*lhca4*	20
Light-harvesting complexes associated with PSII			
LHC-IIa	4.0	*lhcb4*	29
(CP29)			
LHC-IIb	1.35	*lhcb1*	27–28
		lhcb2	25–27
		lhcb3	25
LHC-IIc	2.9	*lhcb5*	26.5
(CP26)			
LHC-IId	1.51	*lhcb6*	24
(CP24)			

[a] Genes are nuclear-encoded.

for helices A and B (Fig. 12.16A). LHC proteins are often organized into trimeric structures (Fig. 12.16B).

Detergent fractionation of chloroplast membranes yields intact PSI and PSII complexes, complete with full arrays of pigment molecules. Analysis of pigment–protein complexes by mildly denaturing gel electrophoresis methods results in separation of individual chlorophyll–protein complexes in which each protein retains its native chlorophyll array, allowing researchers to construct a detailed model for the chlorophyll organization in the respective photosystems (Fig. 12.17).

12.4.4 The light-harvesting antennae of organisms that contain phycobilins are structurally distinct.

Red algae and cyanobacteria, both oxygen-evolving organisms, contain light-harvesting antennae different from those of plants and anoxygenic photosynthetic bacteria. These organisms contain phycobilisomes, water-soluble light-harvesting units. **Phycobilisomes** are complex structures that lie on the surface of the thylakoid membrane, associating only with PSII. They contain phycobiliproteins as well as linker polypeptides that do not bind pigments and may be involved in the binding of phycobilisomes to PSII.

Phycobiliproteins themselves are actually complexes containing arrays of phycobilins associated with specific proteins (see Fig. 12.9). Two protein subunits (α and β) are each bound covalently to molecules of bilin pigment (phycoerythrobilin, phycocyanobilin, or allophycocyanobilin). Phycobiliproteins can be differentiated on the basis of their absorbance properties: Phycoerythrin absorbs in the 450- to 570-nm region, phycocyanin in the 590- to 610-nm region, and allophycocyanin in the 650- to 670-nm region (see Fig. 12.6B). These phycobiliprotein arrays are organized into rod-like phycobilisomes radiating out from the surface of the membrane like a fan (Fig. 12.18). Ordinarily, five or six rod-like elements form these fan structures. Within each rod, phycoerythrin (PE) occupies the outermost position, phycocyanin (PC) the intermediate region, and allophycocyanin (APC) the region nearest to the thylakoid membrane. The pigments are arranged in order of decreasing energy of the excited state, ensuring an efficient transfer of energy from the outermost region of the rod to the photosystem complex in the membrane (Rx. 12.9).

Reaction 12.9: Energy transfer in a phycobilisome

PE $\longrightarrow$ PC $\longrightarrow$ APC $\longrightarrow$ chlorophyll *a*

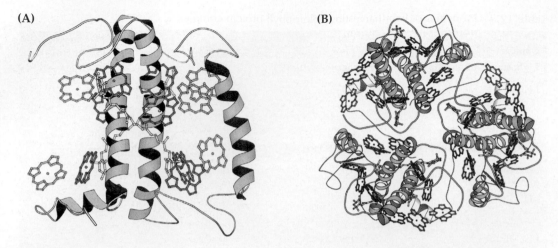

(A) **(B)**

Figure 12.16
(A) Monomeric structure of LHC-II determined by electron microscopy. The complex contains three membrane-spanning helices and binds approximately 12 molecules of chlorophyll *a* (dark green) and *b* (light green), as well as several carotenoid molecules (yellow). The relative positions of the bound chlorophyll molecules are shown. (B) Trimeric structure of LHC-II. In the photosynthetic membranes of plants, LHC-II is present as a trimeric structure organized around the perimeter of the PSII reaction center complex (see Fig. 12.17).

The presence of green-absorbing phycobiliproteins (phycoerythrin) in cyanobacteria and red algae allows these organisms to absorb green light efficiently in aquatic environments. Red light does not transmit through water as effectively as shorter wavelengths do, resulting in a narrowing of the available spectrum of light to blue-green wavelengths with increasing depth. Therefore, cyanobacteria and red algae might be expected to flourish in deep water, although other algae also have pigment adaptations, such as certain carotenoids, that allow them to thrive in green light.

12.5 Organization of the thylakoid membrane

12.5.1 Protein complexes of the thylakoid membrane exhibit lateral heterogeneity.

As shown in Figure 12.1, two types of internal membranes are present in the chloroplast: stacked or appressed membranes, also known as granal membranes, and exposed or unstacked membranes, known as stromal membranes. An important organizational feature of thylakoids

Figure 12.17
Organization of chlorophyll in PSI and PSII, showing the association of specific chlorophyll-binding complexes with PSI and PSII. In PSII, an internal core of chlorophyll *a*–binding proteins, designated CP43 and CP47, are closely associated with the D1/D2 reaction center complex; several peripheral chlorophyll *a/b* binding proteins, or LHC-II complexes, are also present. In PSI, the core complex contains approximately 90 chlorophyll *a* molecules; additional chlorophyll is present in LHC-I complexes, which contain both chlorophyll *a* and *b*. A trimeric organization of the LHCs is shown in both photosystems.

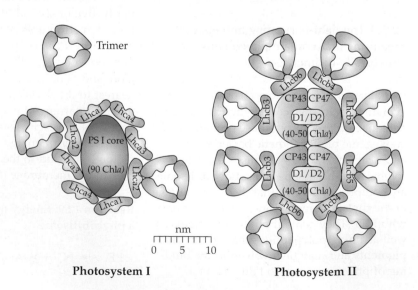

is that PSI and PSII are not distributed randomly throughout the membrane system. PSI lies primarily in the unstacked and stroma-exposed membranes, whereas PSII is found mostly in the appressed (stacked or granal) membranes. This organization, along with the distribution of other complexes, is shown schematically in Figure 12.19.

Other protein complexes integral to the thylakoid membrane are also distributed unequally between these two membrane types (Table 12.4). For example, the ATP synthase is located almost entirely in the stroma-exposed membrane. Even soluble electron transfer proteins such as plastocyanin are distributed unequally within granal and stromal regions of the continuous thylakoid lumen. This phenomenon, called **lateral heterogeneity,** indicates that the two photosystems, which cooperate in transferring electrons from H_2O to $NADP^+$ in the electron transport chain, are spatially separated from each other. According to this model, mechanisms for long-distance electron transfer must exist. Note that not all membrane-bound proteins are distributed unequally in the thylakoid. The cytochrome b_6f complex, which transfers electrons between the two photosystems, is distributed quite evenly (Fig. 12.19).

12.5.2 Phosphorylation of LHC-II may influence the distribution of energy between PSI and PSII.

LHC-II, the light-harvesting protein complex that functions as an auxiliary antenna for PSII, is found almost exclusively in granal membranes (see Table 12.4). Extensive stacking of membranes correlates with the presence of LHC-II. Cyanobacterial membranes, although capable of oxygenic electron transport, are unstacked and contain no LHC-II. Chloroplasts from some mutant plants lacking LHC-II also lack stacked membranes but remain photosynthetically competent. The function of membrane stacking in chloroplasts may be related to the efficient distribution of light energy between PSI and PSII complexes. Because balanced excitation of both photosystems is required for maximum electron transport efficiency, it is critical that one photosystem not receive preferential

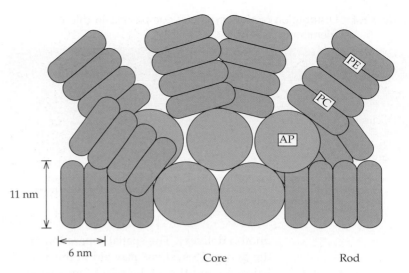

Figure 12.18
Phycobilisome structure. Phycobilins, nonchlorophyll light-harvesting complexes found in cyanobacteria and red algae, contain phycobiliproteins, which are covalently bound phycobilin pigments and proteins. The phycobiliproteins are organized into rod-like structures. Light of shorter wavelengths is absorbed at the periphery of the rods by phycoerythrin (PE), and its energy is transferred to PSII reaction centers in the thylakoid membrane via pigments that absorb longer wavelengths—first phycocyanin (PC) and then allophycocyanin (AP) localized in the core.

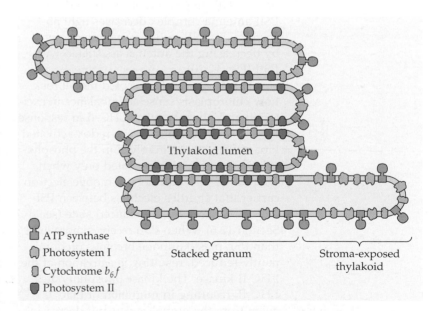

Figure 12.19
Lateral heterogeneity of the chloroplast membrane complexes. PSII is localized primarily in the stacked membrane regions of the thylakoid membrane, whereas PSI and ATP synthase are almost exclusively localized in the unstacked membrane regions. The cytochrome b_6f complex is distributed evenly throughout the membrane regions. The separation of the photosystems necessitates mobile electron carriers such as plastoquinone and plastocyanin, which shuttle electrons between the spatially separated membrane complexes.

Table 12.4 Distribution of photosynthetic components in chloroplast membrane regions

| Component | Thylakoids (%) | |
	Appressed	Stroma-exposed
PSII	85	15
PSI	10	90
Cytochrome b_6f complex	50	50
LHC-II	90	10
ATP synthase	0	100
Plastocyanin[a]	40	60

[a]The percentages indicate relative distribution of each component in either appressed or stroma-exposed membranes, except in the case of plastocyanin, where localization refers to the lumen of the respective membrane region.

photon delivery. The spatial separation of the two photosystems may aid in this regulation. In addition, chloroplasts are able to adjust LHC-II association with PSII to regulate distribution of quanta between the photosystems.

A small pool of the total LHC-II undergoes reversible phosphorylation in the light. Phosphorylation of LHC-II changes the surface charge on the protein. After phosphorylation, LHC-II molecules become negatively charged and are displaced from the hydrophobic core of the granal stacks to the less hydrophobic, exposed stromal membrane region. This migration of a portion of the PSII antenna complex decreases light absorption by PSII in the granal membrane by decreasing the antenna associated with PSII (Fig. 12.20).

The most widely accepted model for how chloroplasts sense an imbalance in excitation and phosphorylate LHC-II in response involves the crucial role of a redox-activated kinase. The kinase involved in the phosphorylation of LHC-II is activated only when the pool of plastoquinone, a mobile electron carrier that shuttles electrons between PSI and PSII, is in a highly reduced state (see Section 12.6). When PSII receives more light than PSI, plastoquinone becomes predominantly reduced, resulting in activation of the LHC-II kinase. The kinase phosphorylates LHC-II, resulting in migration of LHC-II away from the granal region and decreasing the light absorption by PSII. As plastoquinone becomes more oxidized by PSI activation, the kinase becomes less active and a phosphatase acts to dephosphorylate LHC-II, allowing it to migrate back to the granal membrane and thereby increasing the amount of light absorbed by PSII. This feedback system

allows for a subtle control of light distribution between the two photosystems, even though they lie in different parts of the thylakoid.

12.6 Electron transport pathways in chloroplast membranes

12.6.1 The chloroplast noncyclic electron transport chain produces O_2, NADPH, and ATP and involves the cooperation of PSI and PSII.

Between the 1940s and the early 1960s, three seminal experimental findings formed the basis for our current understanding of the coordinated functioning of the two photosystems in the chloroplast electron transport chain. The first was discovery of the so-called "red drop." The efficiency of photosynthesis was found to decrease dramatically at wavelengths greater than approximately 680 nm, even though there was still substantial absorption of light by chlorophyll. In effect, light of wavelengths longer than 680 nm appeared relatively ineffective in photosynthesis, regardless of absorption by the photosynthetic pigments. Questions raised by the "red drop" were answered by the second observation, the "enhancement effect." When red and far-red light were used in combination, the rate of photosynthesis was greater than the sum of the rates achieved by using only red or only far-red light. The third experiment revealed that red and far-red wavelengths had an antagonistic effect on the oxidation and reduction of cytochromes in the electron transport chain—red light causing reduction of cytochromes, and far-red causing their oxidation.

These three experimental findings form the basis for the **Z-scheme** (Fig. 12.21), a model of the chloroplast electron transfer chain that takes into account the differing pigment and reaction center compositions of the two photosystems. PSI, with its longer-wavelength reaction center P700, is more efficient in far-red light. PSII, however, operates more efficiently in red light because of its relatively shorter wavelength reaction center, P680. For photosynthesis to occur with optimal efficiency, the two photosystems must cooperate.

This cooperation of two photosystems that absorb optimally in different spectral regions explains the red drop phenomenon and the enhancement effect, given that the efficiency of photosynthesis would be expected to diminish if one of the two photosystems were preferentially activated, as would happen under either red or far-red light alone, but not under the combination of the two. The Z-scheme also explains the antagonistic effects of red and far-red light. The far-red light that activates PSI photochemistry oxidizes the photosystem, which would in turn oxidize electron carriers located between the photosystems (e.g., cytochromes). By contrast, the red light that activates PSII photochemistry and oxidizes PSII generates reduced compounds, which would reduce subsequent electron carriers in the chain.

The Z-scheme remains the preeminent model of **noncyclic photosynthetic electron transfer** and has been elaborated in great detail. In the light, the primary charge separation in PSII produces a strong oxidant ($P680^+$) and a relatively stable reductant, a plastosemiquinone (Q_A^-); in PSI the charge separation produces a strong, stable reductant (a reduced Fe-S center, F_X^-) and a weak oxidant ($P700^+$). $P680^+$ produced by PSII provides the oxidizing power to remove electrons from water, whereas Q_A^- provides the reducing power that ultimately donates electrons to $P700^+$ via a series of energetically "downhill" electron transfers involving additional electron carriers, including the transmembrane protein complex, cytochrome b_6f. The energy released by this **exergonic** electron transfer contributes to the formation of a proton gradient that can be used for the synthesis of ATP. An intermediate product of P700 oxidation, reduced ferredoxin, serves as an electron donor in many important reactions, including reduction of $NADP^+$, assimilation of nitrogen (see Chapter 8), and reduction of the regulatory disulfide protein thioredoxin (see Box 12.3).

An up-to-date depiction of the Z-scheme, describing the electron carriers in reference to their midpoint oxidation–reduction potentials, is shown in Figure 12.22A. A corresponding schematic, Figure 12.22B, illustrates the movement of electrons and protons during noncyclic electron transfer in oxygenic photosynthetic organisms. This electron transfer pathway generates three

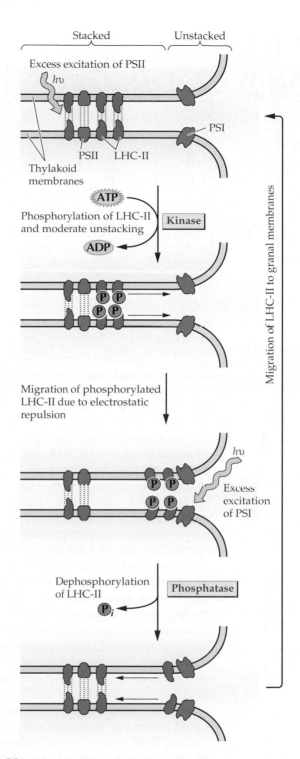

Figure 12.20

Phosphorylation of LHC-II controls energy distribution. During electron transport, phosphorylation of LHC-II occurs upon activation of a kinase by reduced plastoquinone. This results in an unstacking of the membranes because of electrostatic repulsion of negatively charged LHC-II molecules (not shown) and a migration of some phosphorylated LHC-II from the stacked membrane region to the unstacked membrane region. This effectively reduces the PSII antenna size and favors absorption of quanta by PSI. Excessive PSI activation results in plastoquinol oxidation and phosphatase activation, the latter being able to hydrolyze the phosphate group of LHC-II, allowing it to migrate back to the more hydrophobic environment of the stacked membrane region. This mechanism allows for regulation of quantal distribution between PSI and PSII as dictated by the rate of noncyclic electron transport.

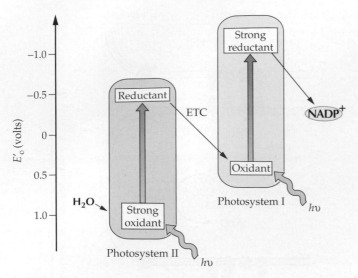

Figure 12.21
Conceptual diagram of the Z-scheme, showing the cooperation of PSII and PSI in the transfer of electrons from water to NADP$^+$. In the light, PSII generates a strong oxidant, capable of oxidizing water, and a reductant. Illuminated PSI, in contrast, generates a strong reductant, capable of reducing NADP$^+$, and a weak oxidant. The two photosystems are linked by an electron transfer chain (ETC) that allows the PSI oxidant to receive electrons from the PSII reductant.

principal products, O_2, ATP, and NADPH. The two photosystems are connected by a series of electron carriers that include plastoquinone, the cytochrome b_6f complex, and plastocyanin. Oxidation of water and exergonic electron transport produce a proton electrochemical gradient that drives synthesis of ATP by the transmembrane ATP synthase.

As described above (see Table 12.4), PSI and PSII are found in different regions of the thylakoid membrane, whereas the cytochrome b_6f complex is distributed equally between appressed and stroma-exposed membranes. This distribution of complexes dictates that mobile electron carriers connect the membrane complexes, and both plastocyanin and plastoquinone have been proposed to serve as these diffusible electron carriers. Both of these electron carriers would be expected to be reduced in the granal region, which contains the major proportion of PSII. Plastoquinol could be oxidized by cytochrome b_6f complexes in either the granal or stromal membranes, but plastocyanin would have to diffuse to the stromal membranes because of the almost exclusive localization of PSI complexes in this membrane region. The presence of the cytochrome complex

in PSI-enriched stromal membranes is also consistent with its postulated role in a PSII-independent cyclic electron transport pathway (see Section 12.6.12).

12.6.2 Photosystem stoichiometry varies by species and is influenced by light environment.

The concept that PSII and PSI cooperate in the noncyclic transport of electrons from water to NADP$^+$ led to the expectation that the two photosystems would be present in equal amounts in thylakoid membranes. However, recent evidence has not supported this prediction. The ratio of PSII to PSI (photosystem stoichiometry) can vary between 0.4 and 1.7 among various wild-type photosynthetic organisms such as green and red algae, cyanobacteria, and plants (Table 12.5). The ratio in mutants may vary even more.

Photosystem stoichiometry can also be regulated by light quality. In the case of both peas (*Pisum sativum*) and cyanobacteria (*Synechococcus* 6301), substantial changes in the ratio of PSII to PSI are observed during growth under different light conditions. These results demonstrate the ability of the photosynthetic organisms to alter their membrane composition in response to gradients of light quality, such as those observed in a forest canopy or in aquatic environments.

12.6.3 PSII functions as a light-dependent water–plastoquinone oxidoreductase.

PSII is an integral membrane complex containing the P680 reaction center. In addition to the electron transport components involved in the primary charge separation, the complex contains more than 20 proteins. Two proteins, D1 and D2, bind electron transfer prosthetic groups, such as P680, pheophytin, and plastoquinone. Other proteins, such as CP43 and CP47, bind chlorophyll *a* antenna pigments. Still other PSII proteins (33, 23, and 17 kDa) are involved specifically in water oxidation. The function of many of the remaining low-molecular-mass subunits, including cytochrome b_{559}, remains unknown. Table 12.6 summarizes the properties of the polypeptides known to be components of the PSII core complex (the more peripheral LHC-II subunits are not included). As with other photosynthetic membrane complexes, PSII contains both nuclear and chloroplast-encoded gene products, indicating a level of cooperation between organelle and nucleus in the final assembly of the complex. Molecular genetic studies have revealed that the low-molecular-mass

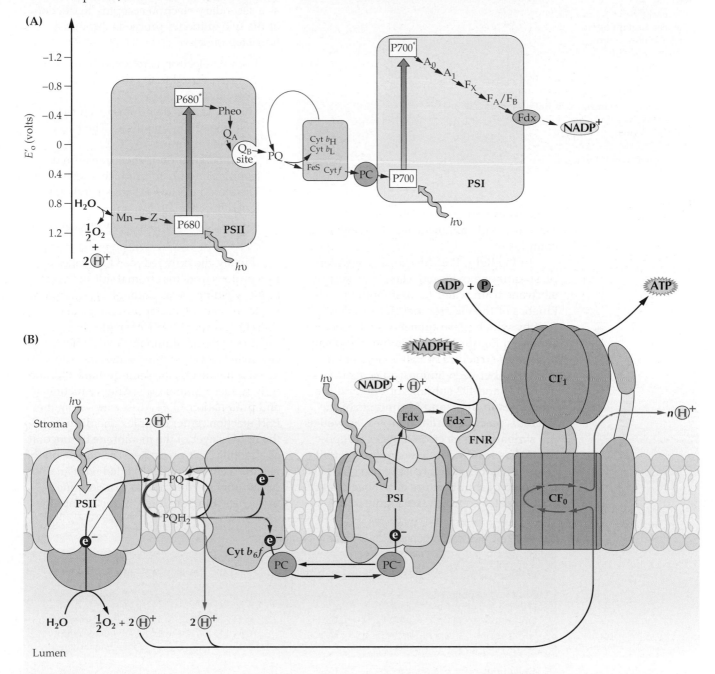

Figure 12.22
(A) The current Z-scheme, showing E_m values of electron carriers. The vertical placement of each electron carrier of the noncyclic electron transfer chain corresponds to the midpoint of its redox potential. These voltage values have been verified experimentally. (B) Membrane organization of the Z-scheme. The components of the chloroplast electron transport chain and the ATP-synthesizing apparatus are illustrated in the thylakoid membrane. Four membrane complexes—PSII, PSI, the cytochrome b_6f complex, and the ATP synthase ($CF_1 – CF_0$)—are shown. Electrons are transferred from water to $NADP^+$; accompanying this electron transfer, a proton gradient is established across the membrane. This electrochemical gradient is ultimately utilized for the synthesis of ATP by the ATP synthase. The transfer of electrons is illustrated with red lines; the translocation of protons is illustrated with blue lines. Fdx, ferredoxin; FNR, ferredoxin-NADP$^+$ reductase.

Table 12.5 Stoichiometry of the photosystems in different oxygen-evolving systems and in response to changes in light quality

System	PSII/PSI ratio
Cyanobacteria	0.4
Red algae	0.4
Green algae	1.4
Tobacco mutant *Su/Su*	2.7
Barley chlorina mutant	3.0
Intermittent-light developing plastids	4.1
Pea chloroplasts	
From plants grown in PSII light (550-660 nm)	1.2
From plants grown in sunlight	1.8
From plants grown in PSI light (>660 nm)	2.3
Synechococcus 6301	
Grown in PSII light	0.3
Grown in sunlight	0.5
Grown in PSI light	0.7

proteins in PSII are required for the assembly and stabilization of active PSII complexes.

In the light, PSII functions as a water-plastoquinone oxidoreductase, transferring electrons from water to plastoquinone. The kinetics of the electron transfer reactions in PSII by which plastoquinone is reduced are shown in Figure 12.23. Although the detailed molecular structure of PSII is not yet known, similarity between amino acid sequences in the D1 and D2 subunits of PSII and the sequences in the L and M subunits of the *R. viridis* reaction center complex, as well as the similarity in electron transfer components, has resulted in a model for the PSII complex that is strongly influenced by what is known of the structure of the bacterial reaction center complex (see Figure 12.10).

The PSII reaction center complex binds two quinones, denoted Q_A and Q_B. Q_A is tightly bound to the PSII reaction center and functions as the first stable electron acceptor. Q_B, which is bound more loosely, functions as a secondary electron acceptor. Reduction of the two quinones proceeds through a four-step process:

1. The first electron is released from P680 and transferred to Q_A to produce a plastosemiquinone, Q_A^-.
2. This electron is then transferred to Q_B to yield the semiquinone Q_B^-. The loss of the electron returns Q_A^- to Q_A.
3. A second electron is then transferred from P680 to Q_A to produce a second Q_A^-.
4. The second electron is subsequently transferred from Q_A^- to Q_B^- to produce a fully reduced Q_B^{2-} molecule. Again, Q_A^- reverts to Q_A.

Finally, the fully reduced Q_B^{2-} takes up two protons from the stromal side of the membrane, yielding a plastoquinol, Q_BH_2. According to this model, under physiological conditions Q_A is capable only of single electron reduction to the semiquinone level, whereas Q_B can switch between three states: the fully oxidized quinone Q_B, the semi-quinone Q_B^-, and fully reduced quinol Q_B^{2-}. After reduction and protonation, Q_BH_2 dissociates from the PSII reaction center complex and diffuses into the lipid bilayer of the membrane to function as a mobile electron carrier. The Q_B site on the reaction center complex is filled with another plastoquinone from the pool of quinone molecules diffusing freely in the membrane.

Table 12.6 Protein subunits of the PSII core complex

Protein	Gene	Location of gene[a]	Mol. mass (kDa)	Function
Hydrophobic subunits				
D1	*psbA*	C	32	Reaction center protein
D2	*psbD*	C	34	Reaction center protein
CP47	*psbB*	C	51	Antenna binding
CP43	*psbC*	C	43	Antenna binding
Cyt b_{559}				
α subunit	*psbE*	C	9	Unknown
β subunit	*psbF*	C	4	Unknown
PsbH–PsbN	*psbH–psbN*	C	3.8–10	Unknown
22 kDa	*psbS*	N	22	Photoprotection
Hydrophilic subunits				
33 kDa	*psbO*	N	33	Oxygen evolution
23 kDa	*psbP*	N	23	Oxygen evolution
17 kDa	*psbQ*	N	17	Oxygen evolution
10 kDa	*psbR*	N	10	Unknown

[a]Tables 12.6 through 12.9: C, chloroplast; N, nucleus.

The reduction of Q_A occurs in about 200 ps because this reaction is directly linked to the primary charge separation. The reduction of Q_B is considerably slower, requiring 100 μs. This indicates that Q_B^- must bind firmly to the PSII reaction center complex while waiting to receive a second electron. In contrast, the binding of Q_BH_2 is relatively weak: This quinol is easily displaced from its binding site by fully oxidized quinones.

12.6.4 The cytochrome b_6f complex transfers electrons from reduced plastoquinone to oxidized plastocyanin.

The oxidation of plastoquinol occurs by an integral membrane protein complex known as the cytochrome b_6f complex. This complex is similar in structure and function to the cytochrome bc_1 complex (Complex III) of the mitochondrial respiratory chain (Chapter 14). The cytochrome b_6f complex contains three electron carriers: a high-potential c-type cytochrome (cytochrome f), a high-potential 2Fe–2S Fe-S protein (the Rieske Fe-S protein), and a b-type cytochrome (cytochrome b_6) with two b-type hemes (see Chapter 14, Figs. 14.17 and 14.18). The cytochrome complex contains both nuclear and chloroplast gene products, including several low-molecular-mass subunits of unknown function (Table 12.7). The cytochrome b_6f complex functions as a plastoquinol-plastocyanin oxidoreductase, transferring electrons from plastoquinol to plastocyanin, a copper-containing protein resident in the lumen of the thylakoid membrane. This electron transfer is accompanied by the translocation of protons across the membrane, from stroma to lumen. An unusual feature of the cytochrome complex mechanism permits the translocation of two protons for every electron transferred to plastocyanin and thereby facilitates formation of the proton gradient that drives ATP synthesis.

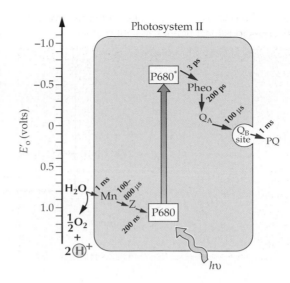

Figure 12.23
PSII electron carriers and their kinetic behavior. Pheophytin, the first electron acceptor, receives electrons from reaction center P680 and transfers them to two plastoquinones, the first of which (Q_A) is bound tightly to the complex; the second, being mobile, is able to bind to the Q_B site when oxidized (PQ), but not when fully reduced (PQH_2). Transfer of a single electron to Q_A occurs in approximately 200 ps, whereas the two-electron reduction of PQ bound in the Q_B site occurs in approximately 100 μs.

12.6.5 Proton translocation via cytochrome b_6f is thought to involve a Q-cycle.

The most widely accepted mechanism to describe the reactions of the cytochrome complex is the Q-cycle (Fig. 12.24). According to this model, the cytochrome complex contains one quinol-binding site (Q_p) and one quinone-binding site (Q_n), present on opposite sides of the membrane. Quinol oxidation proceeds in two discrete steps at the Q_p site, located at the lumenal side of the membrane. At Q_p, quinol is oxidized to the semiquinone by the Rieske Fe-S center and the released electron passes from the Fe-S center to cytochrome f, then on to plastocyanin. The plastosemiquinone is then oxidized by one

Table 12.7 Polypeptide subunits of the cytochrome b_6f complex

Protein	Gene	Location of gene	Mol. mass (kDa)	Function
Cyt f	petA	C	32	Cyt f apoprotein
Cyt b_6	petB	C	24	Cyt b_6 apoprotein
RFeS	petC	N	19	Rieske Fe-S apoprotein
Subunit IV	petD	C	17	Quinone binding at Q_p
PetG, PetM	petG, petM	N	4.0	Unknown
PetL	petL	C	3.4	Unknown

(A) First turnover

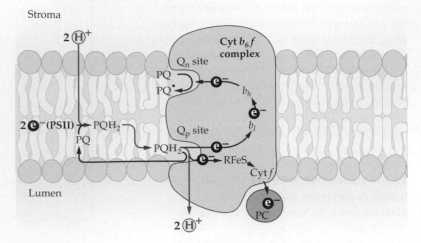

Figure 12.24
The Q-cycle. The transmembrane organization of the cytochrome b_6f complex is shown. At the quinol-binding site, Q_p, a plastoquinol molecule (PQH_2) is oxidized and two protons are released to the lumen. During the first turnover of the complex (A), one electron released from plastoquinol is passed through the high-potential electron carriers, the Rieske Fe-S protein (RFeS) and cytochrome f, to the electron acceptor, plastocyanin (PC). The other electron passes through the two cytochrome b_6 hemes (b_l and b_h) to a quinone-binding site on the stromal side of the membrane, Q_n, where it reduces a quinone molecule to form a plastosemiquinone ($PQ^•$). During the oxidation of a second plastoquinol molecule (B), the pathways of electron transfer are identical except that the second electron reduces the plastosemiquinone at the Q_n site to a fully reduced plastoquinol ($PQ^{••}$), which takes up protons from the stroma, is released from the complex, and enters the plastoquinone pool.

(B) Second turnover

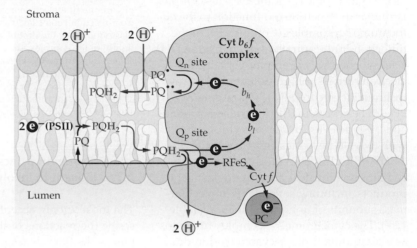

of the two hemes, known as b_l, (b-low potential), located in cytochrome b_6 near the lumenal side of the membrane. Accompanying this oxidation, protons are released from the quinol into the lumen. The electron in the b_l heme is transferred across the membrane to the second b-type heme, known as b_h (b-high potential). This electron is then transferred to a quinone molecule that has bound at the Q_n site on the stromal side of the membrane, producing a semiquinone. This cycle repeats itself once again to oxidize a second plasto-quinol, with one electron being passed to plastocyanin and the second being transferred to the Q_n site to produce a fully reduced quinol molecule (PQH_2); the fully reduced quinol picks up two protons from the stroma and dissociates from the Q_n site. The net result of this cycle is that a plastoquinol molecule is oxidized to a quinone (PQ; at the Q_p site), two electrons are transferred to plastocyanin (PC), and four protons are transferred from the stroma to the chloroplast lumen (Rx. 12.10).

Reaction 12.10: The Q-cycle in the chloroplast

$$PQH_2 + 2PC_{ox} + 2H^+_{stroma} \longrightarrow PQ + 2PC_{red} + 4H^+_{lumen}$$

The reactions of the cytochrome b_6f complex are among the most important in limiting the rate of electron flux. Whereas Q_BH_2 is formed in approximately 100 μs, quinol oxidation occurs in 10 to 20 ms. Direct measurements of the rate of reactions through the high-potential carriers, including oxidation of cytochrome f by plastocyanin, have shown half-reaction times no longer than 1 to 2 ms. However, the reactions of cytochrome b_6, particularly the kinetics of the oxidation of the high-potential heme, b_h, are considerably slower and appear to be the rate-limiting steps in the complex.

Copious kinetic data support the functioning of the Q-cycle, but detailed information on the structure of the complex is not yet available. For example, the molecular details of the Q_p and Q_n sites are not known. A model for the organization of the subunits, based on hydropathy profiles of the individual subunits, indicates the lumenal location of the Fe-S cluster in the Rieske protein and of the heme group in cytochrome f (Fig. 12.25). Additional information on the structure of cytochrome b_6f is suggested by recent advances in elucidating the structure of a similar complex associated with a mitochondrial Q-cycle, cytochrome bc_1. The crystal structure of the cytochrome bc_1 complex, resolved to 2.9 Å (see Chapter 14, Fig. 14.22), has defined the distances between the prosthetic groups of the complex and has revealed features of inhibitor binding sites. The structure also shows that the cytochrome bc_1 complex is a functional dimer. Researchers have been unable to obtain crystals of the chloroplast cytochrome b_6f complex for X-ray analysis, but strong biochemical evidence indicates that the functional form is a dimer as well.

Although a complete structure of the cytochrome b_6f complex has not yet been obtained, water-soluble portions of two of the electron carriers in the complex, cytochrome f and the Rieske Fe-S protein, have been crystallized. The structure of the lumenal portion of cytochrome f from *Chlamydomonas reinhardtii* (Fig. 12.26), which is similar to a structure for turnip cytochrome f, has provided useful information on the nature of the heme binding and the possible

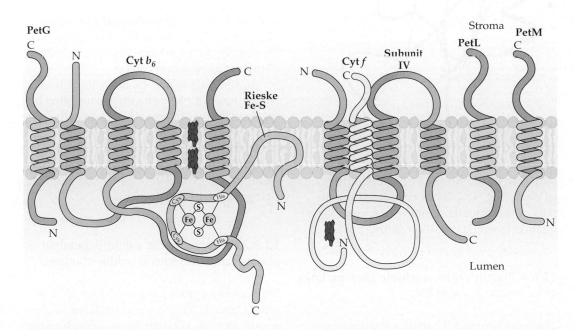

Figure 12.25

Organization of the protein subunits of the cytochrome b_6f complex. The proposed structures of the individual protein subunits of the cytochrome complex are based on analyses of amino acid sequences and hydropathy profiles. Several of the subunits, such as PetG, PetA (cyt f), PetL, and PetM are thought to contain only one membrane-spanning domain, whereas PetB (cyt b_6) and Subunit IV contain several transmembrane helices. The cyt b_6 heme groups are shown on opposite sides of the membrane and are part of the quinol oxidation and reduction sites, respectively; the heme group of cyt f and the Fe-S center of the Rieske protein are localized within the lumen of the chloroplast. The orientations of all proteins except PetL have been determined through proteolysis studies with isolated thylakoid membranes.

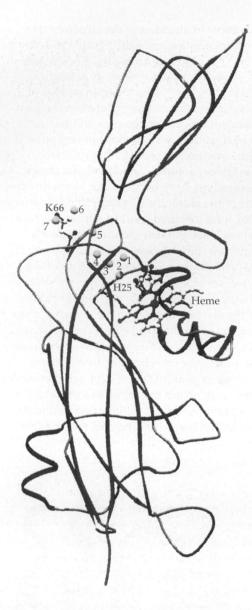

Figure 12.26
Structure of a truncated form of cytochrome *f* from *Chlamydomonas*. α-Helices are shown in blue; β-sheets are shown in red. The molecule consists of two protein domains linked by the heme group. Histidine 25 ligates the iron of the heme group. A chain of seven water molecules (yellow spheres), extending from the heme group to the upper surface of the molecule, is also shown. Lysine 66 is one of a conserved patch of lysines speculated to be the docking site for plastocyanin.

K66 6
7   5
4 2 1
3
H25
Heme

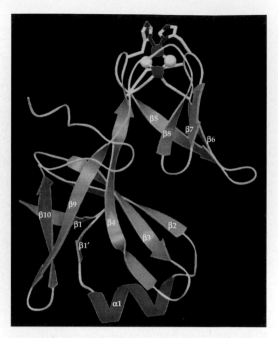

Figure 12.27
Structure of the Rieske Fe-S protein from spinach. Shown is a soluble fragment of the spinach chloroplast Rieske protein at 1.8 Å resolution. Two cysteine ligands coordinate one iron of the 2Fe–2S center, and two histidine ligands bind the other iron atom.

mode of interaction with plastocyanin. The structure of the Rieske Fe-S protein from spinach (Fig. 12.27) has elucidated the details of the unusual binding of Cys and His ligands to the Fe-S cluster in this protein.

12.6.6 Plastocyanin, a soluble protein, links cytochrome $b_6 f$ and PSI.

The cytochrome $b_6 f$ complex reduces the mobile electron carrier plastocyanin, a low-molecular-mass (11-kDa), copper-containing protein. Electron transfers involving plastocyanin are very rapid: Its reduction by the cytochrome $b_6 f$ complex takes about 100 to 200 μs, and its oxidation by PSI requires about 10 μs. This has led to some proposals

that the three components might be organized into a "supercomplex" that eliminates the requirement for long-distance movement of the mobile electron carrier.

In some algae and cyanobacteria, the biosynthesis of plastocyanin is controlled by the availability of copper in the growth medium. In the absence of copper, these cells are unable to accumulate plastocyanin and instead synthesize a *c*-type cytochrome, cytochrome *c*-553, which is functionally interchangeable with plastocyanin.

12.6.7 PSI functions as a light-dependent plastocyanin–ferredoxin oxidoreductase.

PSI contains approximately 15 protein subunits (Table 12.8). PsaA and PsaB (each approximately 80 kDa) are involved in binding the major electron transfer carriers, such as P700, the chlorophyll *a* acceptor molecule (A_0), phylloquinone (vitamin K1, the A_1 acceptor), and the bound Fe-S center F_X. The low-molecular-mass PsaC subunit (9 kDa) binds Fe-S centers, F_A and F_B. Like PSII, PSI contains a large number of protein subunits

Table 12.8 Polypeptide subunits of the PSI core complex

Protein	Gene	Location of gene	Mol. mass (kDa)	Function
Hydrophobic subunits				
PsaA	*psaA*	C	83	Reaction center protein
PsaB	*psaB*	C	82	Reaction center protein
PsaG	*psaG*	N	11	Unknown
PsaI	*psaI*	C	4	Unknown
PsaJ, PsaM	*psaJ, psaM*	C	5, 3.5	Unknown
PsaK, PsaL	*psaK, psaL*	N	8.4, 18	Unknown
Hydrophilic subunits				
Stromal orientation				
PsaC	*psaC*	C	9	Fe-S apoprotein
PsaD	*psaD*	N	18	Ferredoxin docking
PsaE	*psaE*	N	10	Cyclic electron transport
PsaH	*psaH*	N	10	LHC-I linker
Lumenal orientation				
PsaF	*psaF*	N	17	PC docking
PsaN	*psaN*	N	10	Unknown

that do not bind prosthetic groups, many of which have as yet undefined function. Two subunits, PsaD and PsaF, have been implicated in the binding of ferredoxin and plastocyanin, respectively, to the complex.

The path and kinetics of electron transfer through PSI are shown in Figure 12.28. Plastocyanin is the electron donor to PSI; the terminal electron acceptor is the soluble Fe-S protein, ferredoxin. Thus, the PSI complex functions as a light-dependent plastocyanin–ferredoxin oxidoreductase. Ferredoxin is a strong reductant with a redox midpoint potential of –420 mV. The reduced ferredoxin generated by PSI is capable of reducing $NADP^+$ in a thermodynamically favorable reaction in the physiological pH range.

12.6.8 Electrons from PSI are transferred to $NADP^+$ in the stroma in a reaction requiring ferredoxin and ferredoxin-$NADP^+$ reductase.

Electrons from PSI are transferred to the 2Fe–2S Fe-S protein ferredoxin, located in the chloroplast stroma. This electron carrier does not transfer electrons directly to $NADP^+$, but rather by way of an intermediate enzyme called ferredoxin-$NADP^+$ reductase (FNR; see Fig 12.22). Strong evidence indicates that ferredoxin and FNR form a complex through electrostatic interactions of the two proteins. FNR is an FAD-containing enzyme that can be reduced in two single-

electron steps. The first electron reduces FNR to the flavin semiquinone state; the second, to the fully reduced state, $FADH_2$ (see Chapter 14, Fig. 14.6). FNR then transfers the two electrons to $NADP^+$. FNR is loosely associated with the thylakoid membrane and is easily dissociated. High-resolution structures of both ferredoxin and FNR are available from X-ray crystallography, and models for the interaction of the two proteins have been developed from the structural data.

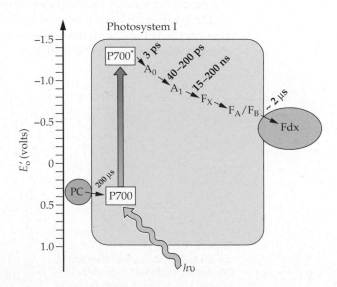

Figure 12.28
PSI electron carriers and their kinetic behavior. The pathway of electron transfer through the PSI complex is shown to involve P700, a monomeric chlorophyll *a* (A_0), a phylloquinone (A_1), and a series of additional electron carriers that include three different Fe-S centers (F_X, F_A, and F_B). Electron transfers through these carriers occur in the time range of picoseconds to nanoseconds, with the terminal electron acceptor, ferredoxin (Fdx), being reduced in approximately 2 μs.

12.6.9 Oxidation of water produces O_2 and releases electrons required by PSII.

The reaction center chlorophyll of PSII, P680, undergoes light-induced oxidation to produce the strong oxidant, P680$^+$, which can oxidize water, releasing O_2. This implies that the oxidation–reduction midpoint potential of the P680/P680$^+$ couple is more electropositive than +820 mV at pH 7, because splitting water requires at least this much oxidizing potential. The oxidation of water is not a direct process but involves a complex series of reactions on the oxidizing (lumenal) side of PSII. The oxidation of water involves the transfer of four electrons:

Reaction 12.11: Oxidation of water

$$2H_2O \longrightarrow O_2 + 4H^+ + 4e^-$$

Analysis of the thermodynamics of water oxidation has been used to predict that this reaction proceeds either in two steps involving two electrons each or in one step involving four electrons. Single-electron transfers would form highly energetic oxidizing intermediates that would presumably destroy components of the photosynthetic membrane system. This inference implies the existence of a charge-storage apparatus on the oxidizing side of PSII that can couple the single positive charges produced in the reaction center at P680 with the multiple positive charges required for water oxidation.

The mechanism of O_2 evolution has been investigated by measuring the amount of O_2 evolved during a series of short flashes of light. With dark-adapted chloroplasts, little O_2 appears after the first two flashes, but maximal amounts of O_2 appear after the third flash and after every subsequent fourth flash. The evolution finally reaches a steady-state level after a large number of flashes (Fig. 12.29).

The S-state model (Fig. 12.30) was proposed to explain these observations. This model postulates a light-driven charge-accumulation mechanism by which the oxygen-evolving machinery of PSII progresses through five successive states of increasing oxidation, S_0 through S_4, with S_4 being a strong oxidant capable of oxidizing water. Each charge separation event in the P680 reaction center yields P680$^+$, which ultimately oxidizes the charge accumulator, advancing it to the next S-state and increasing its charge by +1. The only state from which O_2 is evolved is S_4.

In dark-adapted chloroplasts, the charge accumulator is predominantly in the S_1 state, resulting in a maximal yield of O_2 on the third flash, as observed experimentally (see Fig. 12.29). The first flash advances the oxidation state from the dark-adapted S_1 to S_2. The second flash oxidizes S_2 to S_3, and the third flash yields the strong oxidant, S_4, triggering the evolution of oxygen and returning the charge accumulator to the S_0 state. Each succeeding cycle requires four

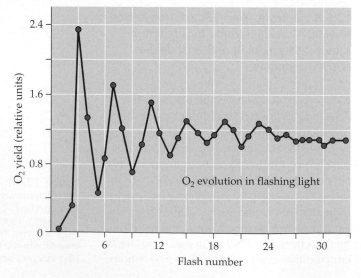

Figure 12.29
O_2 evolution in flashes. The pattern of oxygen evolution is given in response to a series of short flashes of light. The maximum amount of oxygen is produced on the third flash, after which a periodic spike is observed after every fourth flash. The flash yield damps out after approximately 20 flashes to a steady-state value.

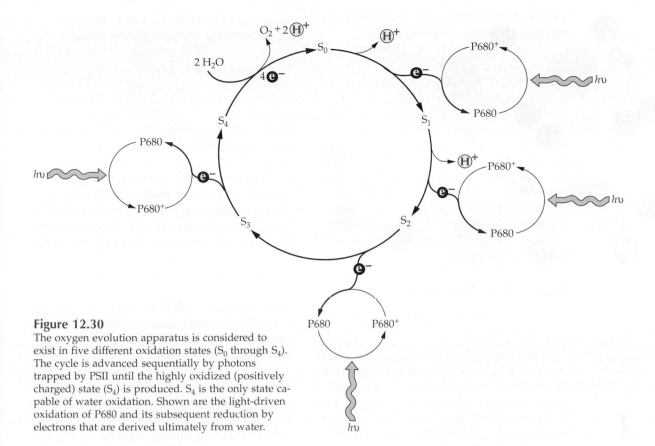

Figure 12.30
The oxygen evolution apparatus is considered to exist in five different oxidation states (S_0 through S_4). The cycle is advanced sequentially by photons trapped by PSII until the highly oxidized (positively charged) state (S_4) is produced. S_4 is the only state capable of water oxidation. Shown are the light-driven oxidation of P680 and its subsequent reduction by electrons that are derived ultimately from water.

flashes because four photons are required per O_2 evolved. The reactions are complicated by "misses" (a flash that fails to oxidize P680), "double hits" (one flash oxidizes P680 twice), and "relaxations" (S_2 or S_3 decays to S_1 in the dark). Such events cause the oxygen yield to stabilize over a series of many flashes. In general terms, however, the S-state scheme has provided an excellent explanation for many of the basic observations on O_2 evolution.

12.6.10 The water-splitting reaction requires manganese and other cofactors.

The S-state model does not provide a direct biochemical framework for understanding O_2 evolution, so an important question remains: What is the chemical nature of the S-state charge accumulator? Early findings that manganese deficiency in algae or chloroplasts specifically affects O_2 evolution, as well as findings in many additional recent studies, have led to wide acceptance that manganese is the main charge accumulator in the S-state model. Manganese atoms

appear to undergo successive oxidations to yield a strongly oxidizing complex in the S_4 state that is capable of water oxidation. A large number of biophysical measurements have helped define the oxidation state and structure of four manganese atoms associated with the oxygen-evolving complex of PSII. One model that shows a probable arrangement of these atoms in the oxygen-evolving complex is a tetranuclear cluster bridged by oxygen atoms (Fig. 12.31).

In addition to the specific requirement for manganese, two additional cofactors for O_2 evolution have been identified: chloride ions and calcium ions. The specific roles of these ions in O_2 evolution remain unclear, although some research suggests that Cl^- may be bound to one of the manganese atoms of the oxygen-evolving cluster (Fig. 12.31).

The link between P680$^+$ and the S-state cluster involving manganese is not direct. An additional electron carrier intermediate, called Z, has been identified (see Rx. 12.12). Z is a specific tyrosine residue in the D1 subunit of the reaction center complex of PSII (Tyr-160). The identity of this residue has

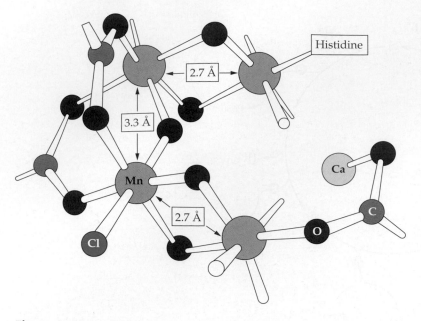

Figure 12.31
Structure of the manganese cluster. The four Mn atoms in the cluster are ligated to amino acid residues of the PSII protein D1, as well as to oxygen and chloride. This model, which also includes calcium, has been developed from several spectroscopic analyses of the oxygen-evolving complex.

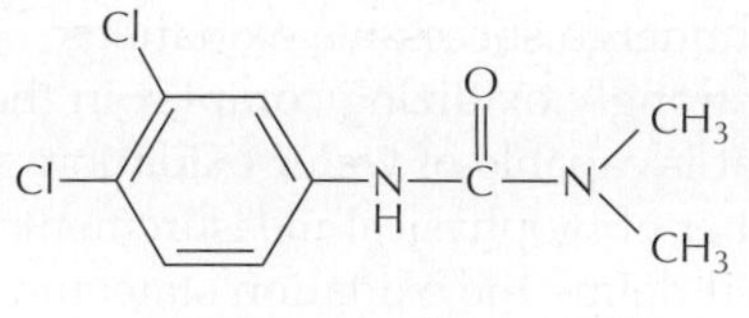

DCMU, an inhibitor of PSII

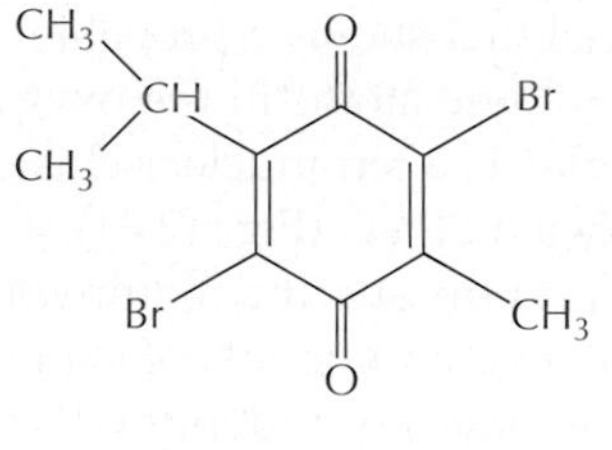

DBMIB, an inhibitor of the Cyt b_6f complex

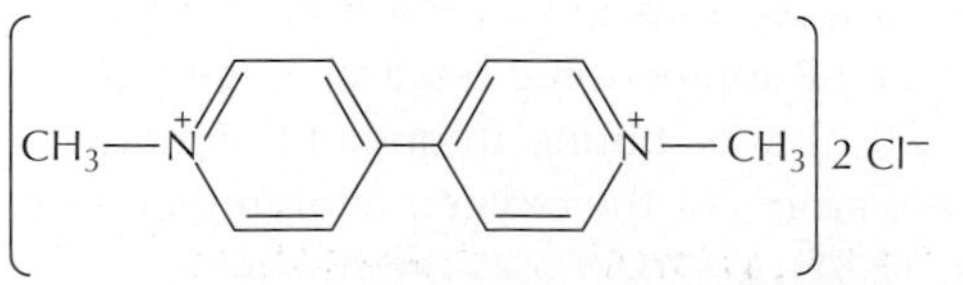

Paraquat (methyl viologen), an inhibitor of PSI

Figure 12.32
Chemical structures of three inhibitors of the photosynthetic electron transport chain: DCMU, 3-(3,4-dichlorophenyl)-1,1-dimethylurea; DBMIB, 2,5-dibromo-3-methyl-6-isopropyl-*p*-benzoquinone; and paraquat.

Reaction 12.12: Oxidation and reduction of P680 during oxygenic photosynthesis

$$S\ Z\ P680 \xrightarrow{h\upsilon} S\ Z\ P680^+ \longrightarrow S\ Z^+\ P680$$
$$\longrightarrow S^+\ Z\ P680$$

been confirmed by site-directed mutagenesis studies with PSII complexes from cyanobacteria. To function as an intermediate between P680 and the S-state, the tyrosine radical formed in the above reaction must be a strong oxidant, capable of advancing the S-state. Measured kinetics of Z^+ formation and decay are consistent with this proposed role.

12.6.11 Specific inhibitors and artificial electron acceptors have been used to study the chloroplast electron transport chain.

Several compounds specifically inhibit the chloroplast electron transport chain and, in so doing, act as herbicides (Fig. 12.32). One general class of inhibitors binds at the Q_B site on the D1 protein in PSII and prevents the reduction of Q_B. The most widely studied of these compounds is the inhibitor Diuron, 3-(3,4-dichlorophenyl)-1,1-dimethylurea (DCMU). The herbicide atrazine also acts at this site. A second class of inhibitors acts at the reducing end of PSI, inhibiting the reduction of ferredoxin. The herbicide paraquat (methyl viologen) is a member of this class. Paraquat's mode of inhibition results from its autoxidizability, which results in the formation of a superoxide radical that is highly reactive and damaging to the photosynthetic apparatus. Paraquat has also been used experimentally as a nonphysiological electron acceptor, substituting for ferredoxin.

Another set of inhibitors includes plastoquinone analogs, such as the compound 2,5-dibromo-3-methyl-6-isopropyl-*p*-benzoquinone (DBMIB; or dibromothymoquinone), which block electron transfer by competing with plastoquinol for binding at the quinol oxidizing (Q_p) site in the cytochrome b_6f complex. The site of action of these inhibitors in relation to the electron carriers of the chloroplast noncyclic electron transfer chain is shown in Figure 12.33. In addition to their commercial value as herbicides, these

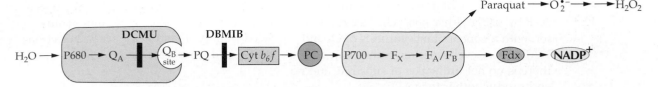

Figure 12.33

Sites of action for the inhibitors of the chloroplast electron transport chain diagrammed in Figure 12.32. DCMU and DBMIB block electron transfer reactions, whereas reduced paraquat autooxidizes to a radical, resulting in the formation of superoxide and other reactive oxygen species.

inhibitors have been useful in defining the electron transfer events that take place in the photosynthetic membrane.

12.6.12 Chloroplasts also contain a cyclic electron transport chain.

As described above, the noncyclic electron transport chain in chloroplasts links the oxidation of water (O_2 evolution) to the reduction of $NADP^+$ and production of ATP. Chloroplasts also can carry out a cyclic electron transport process that involves only PSI and produces only ATP as a product. The cyclic pathway requires a cofactor or catalyst, and there is good evidence that ferredoxin serves as the native cofactor. In a model that describes this pathway, PSI reduces ferredoxin in the light but instead of transferring an electron to $NADP^+$, reduced ferredoxin (Fdx_{red}) is proposed to interact with a Fdx-plastoquinone oxidoreductase that allows for the transfer of electrons into the quinone pool. Plastoquinol can then be oxidized by the cytochrome b_6f complex, allowing for proton translocation across the membrane, possibly via a Q-cycle (Fig. 12.34). According to this model, one would not expect cyclic electron transport to be inhibited by PSII inhibitors such as DCMU but would expect it to be inhibited by inhibitors of the cytochrome b_6f complex such as DBMIB. These expectations have been verified experimentally.

Although the cyclic electron transport pathway and concomitant ATP synthesis can be demonstrated in vitro, the role of this pathway in vivo has always been controversial. Experiments measuring CO_2 fixation in intact chloroplasts under red and far-red light and assessing the response of this system to inhibitors have provided some evidence that cyclic electron transport and

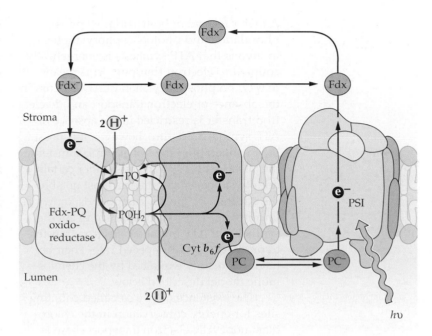

Figure 12.34

Mechanism of cyclic electron transport in chloroplasts. The cyclic electron transport pathway involves PSI, a putative ferredoxin-plastoquinone oxidoreductase, and the cytochrome b_6f complex. The only net product of the pathway is ATP, which is synthesized by using the proton gradient generated through plastoquinol oxidation. Electron transfer is shown in red; proton translocation is shown in blue.

phosphorylation do provide ATP for CO_2 fixation. However, the biochemical documentation for the cycle is still fragmentary. In particular, the key enzyme in the process, Fdx-plastoquinone oxidoreductase, has not been characterized.

12.7 ATP synthesis in chloroplasts

Light-dependent synthesis of ATP, known as **photophosphorylation,** was discovered in isolated chloroplasts during the 1950s by Arnon and collaborators. Photophosphorylation is mechanistically very similar to the oxidative phosphorylation processes in mitochondria. Chloroplasts can synthesize ATP by using either noncyclic electron transfer, with a concomitant synthesis of O_2 and NADPH, or cyclic electron transfer, in which there is no net utilization of substrate and no end product other than ATP.

12.7.1 Electron transport and ATP synthesis are coupled in vivo.

A major feature of both oxidative phosphorylation and photophosphorylation in vivo is that ATP synthesis is energetically coupled to electron transport. This means that (a) no phosphorylation of ADP occurs in the absence of electron transport and (b) electron transfer is restricted in the absence of ATP synthesis. In vitro, however, it is possible to obtain high rates of electron transfer in the absence of ATP synthesis under certain nonphysiological conditions, most notably the addition of specific compounds known as uncouplers. These compounds prevent the synthesis of ATP but allow electron transfer to proceed at an unimpeded rate. Their behavior is readily explained by the chemiosmotic model described below.

The existence of two so-called coupling sites for energy conservation in the chloroplast noncyclic electron transport chain is now generally accepted. These sites involve regions of the electron transport chain where proton accumulation in the thylakoid lumen is coupled to electron transport. One site involves protons liberated when water is oxidized, given that the protons released are localized in the lumenal space of the chloroplast thylakoid membrane system. The second site involves protons released when plastoquinol is oxidized by the cytochrome $b_6 f$ complex; in this case, the quinol oxidation results in the release of protons into the lumenal space.

The amount of ATP synthesized during noncyclic electron transport, expressed as the $P/2e^-$ or $ATP/2e^-$ ratio, has long been a subject of controversy. To some extent, this is because the isolation of well-coupled chloroplasts is an experimentally challenging task. An accurate measurement of the maximum amount of ATP synthesized during electron transfer has been difficult to obtain. Values for noncyclic photophospho-

rylation in chloroplasts are generally in the range of 1.0 to 1.5 ATP molecules synthesized from ADP and P_i for every 2 electrons transferred from water to $NADP^+$.

12.7.2 Chloroplasts synthesize ATP by a chemiosmotic mechanism driven by a proton gradient.

In the 1960s, Peter Mitchell proposed the chemiosmotic model to explain ATP synthesis in chloroplasts as well as in mitochondria. The experimental verification of this hypothesis in the 1960s and 1970s resulted in Mitchell's receiving the 1978 Nobel Prize in chemistry. According to the chemiosmotic model, the main energetic driving force for the synthesis of ATP is an ion gradient across a selectively permeable biological membrane; in chloroplasts and mitochondria, this gradient is known to be a proton gradient that is generated as a consequence of electron transfer. This gradient establishes a concentration difference of protons and can also result in a difference in electric potential across the membrane. These sources of potential energy can be used for the phosphorylation of ADP by ATP synthase, an enzyme that can couple this energetically favorable proton flow to the synthesis of ATP (see Fig. 12.22B and Chapters 3 and 14).

Several important features of the chemiosmotic model have been verified experimentally (see Section 12.7.3). First, this mechanism requires an intact membrane system with low intrinsic permeability to protons. Second, the electron transport components in the membrane must be arranged vectorially to allow electrons to cross the membrane at defined sites. As electrons pass through this series of electron carriers, protons are also translocated across the membrane at these coupling sites. In chloroplasts, protons move from the stroma to the thylakoid lumen as electrons are transferred along the chain. In the noncyclic electron transport pathway, protons are "deposited" in the lumen at the oxygen-evolving complex of PSII and at the Q_p site near the lumenal face of the cytochrome $b_6 f$ complex, which act to oxidize water and plastoquinol, respectively. This unidirectional proton accumulation results in a high concentration of protons on the lumenal side of the membrane and a lower concentration on the stromal side. The resulting gradient supplies the **proton motive force** (pmf).

As summarized by Equations 12.5 and 12.6, there is a linkage between the pH component, arising from a concentration difference of protons across the membrane (ΔpH), and the electrical potential component, arising from a difference in charge across the membrane ($\Delta\psi$). According to the chemiosmotic model, the $\Delta\bar{\mu}_{H^+}$ (in kilojoules per mole) or pmf (in volts) is the driving force for ATP synthesis. In the chloroplast, ΔpH is the major contributor to pmf, whereas in other bioenergetic membrane systems (i.e., the mitochondrion), pmf is determined largely by $\Delta\psi$.

Equation 12.5: Contribution of pH and membrane potential to the proton motive force

$$\Delta\bar{\mu}_{H^+} = \Delta pH + \Delta\psi$$

Equation 12.6: The proton motive force

$$pmf = \Delta\bar{\mu}_{H^+} / 96.5 \text{ kJ V}^{-1} \text{ mol}^{-1}$$

The final element in the chemiosmotic model is the presence in the membrane of an ATP synthase, which is capable of using the pmf to convert ADP into ATP. This enzyme complex is oriented in the thylakoid membrane with a hydrophilic head group on the stromal side of the membrane and a hydrophobic channel (Fig. 12.35). The channel allows H$^+$ ions to diffuse spontaneously from the lumen to the stroma. The chemiosmotic model predicts a fixed stoichiometry between the number of protons transported across the membrane and the number of ATP molecules synthesized, the so-called H$^+$/ATP ratio. Most recent measurements indicate a value of 4 H$^+$/ATP. According to our present model of the noncyclic electron transport chain, the net transfer of two electrons through the chain results in the accumulation of six protons in the lumen. These numbers are clearly critical in understanding proton translocation sites as well as the energetics of ATP synthesis in chloroplasts.

The chemiosmotic model explains many observations of coupled ATP synthesis in

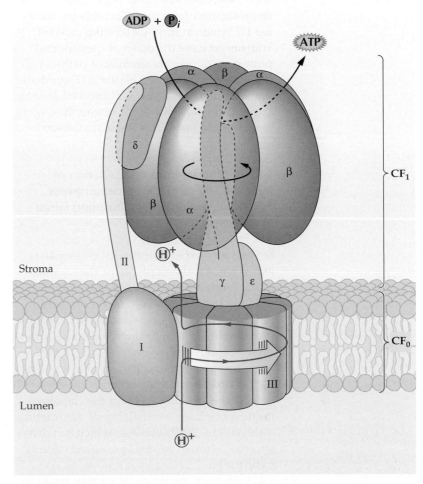

Figure 12.35

Model for the ATP synthase complex. The subunit structure of the ATP synthase indicates two major regions in the protein: an integral membrane protein portion (CF$_0$), which functions as a channel for protons passing through the membrane, and an extrinsic portion (CF$_1$), which contains the catalytic sites involved in ATP synthesis. CF$_1$ consists of five different subunits (α, β, γ, δ, and ϵ), whereas CF$_0$ contains at least four different subunits (I, II, III, and IV), multiple copies of subunit III being present in the membrane.

both chloroplasts and mitochondria. For example, the coupling of phosphorylation and electron transfer can be understood as a feedback system. As the electron transport chain moves protons into the lumen, the lumenal concentration of protons would be expected to increase, generating a back pressure that slows further proton pumping. As the ATP synthase transports protons out of the lumen, the lumenal concentration of protons decreases, relieving the back pressure and facilitating the accumulation of more protons in this space. The ability of nonphysiological uncoupling agents to overcome this effect is easily understood: Most uncouplers are weak acids that enter the lumen in their unprotonated form and diminish the lumenal H^+ concentration by binding protons. Transmembrane diffusion of these bound protons provides an alternative pathway to the reactions catalyzed by the ATP synthase. Thus electron transport will proceed at high rates in the presence of the uncoupling agent, even though no ATP is being synthesized.

12.7.3 Experimental manipulation of lumenal and stromal pH can promote light-independent ADP phosphorylation in chloroplasts.

Several key experiments with chloroplasts led to a general acceptance of the chemiosmotic model. First, measurements of light-induced proton uptake showed that large pH gradients could be established across the thylakoid membrane. Under optimum conditions, gradients as large as 3 to 3.5 pH units have been measured, sufficient to drive ATP synthesis. Further experiments confirmed another prediction of the chemiosmotic model, the reversibility of the reaction catalyzed by ATP synthase, which can drive stroma-to-lumen protons pumping via ATP hydrolysis.

However, the most convincing result in support of the model was "acid–base phosphorylation." In this experiment, chloroplasts were first incubated at about pH 4, allowing them to establish this pH internally. The chloroplast suspension was then rapidly adjusted to pH 8, establishing an artificial pH gradient across the membrane. This artificial gradient promoted ATP synthesis until it was discharged and the pH difference was insufficient to drive ADP phosphorylation. Acid–base ATP synthesis required no light and was insensitive to electron transport inhibitors, such as DCMU, indicating that a pH gradient alone was sufficient to drive ATP synthesis. However, the acid–base ATP synthesis was sensitive to the addition of uncoupling agents, which affect the proton gradient directly. Subsequent research has confirmed and extended the original observations, and the chemiosmotic model is now widely accepted as the mechanism for ATP synthesis by both chloroplasts and mitochondria.

12.7.4 The thylakoid ATP synthase complex contains numerous subunits.

The chemiosmotic model proposed a key role for a vectorial ATP synthase that would use the energy stored in a proton electrochemical potential for the synthesis of ATP. This enzyme was first identified on the basis of its ability to hydrolyze ATP (see F-type ATPase in Chapter 3). In native membranes the ATPase activity of this enzyme is totally inactive, allowing for a unidirectional synthesis of ATP in the light. The pmf, established across the thylakoid membrane in the light, functions to activate the ATP synthase and inhibit the wasteful ATPase activity of this enzyme. Under some in vitro conditions, such as when a disulfide bond in one of the subunits of the ATP synthase is reduced, the ATPase activity of the enzyme can be activated.

The thylakoid ATP synthase consists of two segments: a transmembrane segment, called CF_0, and a hydrophilic segment on the stromal surface, called CF_1. CF_0 participates in translocating protons across the membrane to the catalytic portion of the enzyme, CF_1; CF_1 is involved in the actual conversion of ADP and P_i to yield ATP, using energy stored in the proton gradient. The enzyme is often referred to as the CF_0–CF_1 complex (Fig. 12.35).

Chloroplast ATP synthase is a 400-kDa enzyme that contains nine different subunits (Table 12.9) and includes both chloroplast and nuclear-encoded gene products. CF_1 consists of two relatively large subunits, α and β, each of which has a molecular mass of approximately 50 kDa, plus three smaller subunits called γ, δ, and ε (Fig. 12.35). CF_1 contains three copies of each of the large subunits and one copy each of the other subunits to give an overall stoichiometry of $\alpha_3\beta_3\gamma\delta\varepsilon$. The α and β subunits bind ADP and phosphate and catalyze the phosphorylation of ADP into ATP. The δ subunit links CF_0 to CF_1, and the γ subunit appears to control proton-gating through the enzyme. The ε subunit blocks catalysis in the dark, preventing the breakdown of ATP, and may also be involved in proton gating through interactions with the γ subunit. The γ subunit also participates in a regulatory mechanism mediated by the ferredoxin/thioredoxin system, which enhances activation of ATP

Table 12.9 Polypeptide subunits of the ATP synthase complex

Protein	Gene	Location of gene	Mol. mass (kDa)	Function
CF$_1$				
α subunit	*atpA*	C	55	Catalytic
β subunit	*atpB*	C	54	Catalytic
γ subunit	*atpC*	N	36	Proton gating
δ subunit	*atpD*	N	20	Binding of CF$_1$ to CF$_0$, regulation
ε subunit	*atpE*	C	15	ATPase inhibition
CF$_0$				
I	*atpF*	C	17	Binding of CF$_0$ to CF$_1$
II	*atpG*	N	16	Binding of CF$_0$ to CF$_1$
III	*atpH*	C	8	Proton translocation
IV	*atpI*	C	27	Binding of CF$_0$ to CF$_1$

synthase in the light and deactivation of the enzyme complex in the dark, thereby blocking wasteful hydrolysis of ATP at night (see Section 12.8.4).

Although the subunit composition of the CF$_1$ portion is known with some certainty, the composition of the hydrophobic CF$_0$ portion is less well characterized, although CF$_0$ is now generally thought to contain an additional four subunits, I, II, III, and IV. These subunits are probably present as single copies, except for subunit III, which is present in about 12 copies per complex. The subunits of the CF$_0$ complex are generally considered to be involved in binding the membrane portion of the enzyme to the catalytic portion. Some evidence suggests that subunit III may form a pathway for proton translocation from the lumenal space to the stroma.

12.7.5 Structural resolution of the F$_1$–F$_0$ complex from mitochondrial ATP synthase provides insight into the coupling of proton transport and ATP synthesis.

Homologous enzymes involved in the synthesis of ATP have been identified in other energy-transducing membranes, such as the inner mitochondrial membrane (F$_1$–F$_0$; Chapter 14) and bacterial membranes. Recently, a high-resolution structure of the F$_1$ portion of the mitochondrial enzyme has provided new and important structural information related to the mechanism of ATP synthesis. This work resulted in John Walker's being a co-recipient of the 1997 Nobel Prize in chemistry. An important feature of the structure is the description of the alternating α and β subunits, with the γ subunit

extending up through a hydrophobic cavity of the α$_3$β$_3$ hexamer of F$_1$ down into the stalk region. Other low-molecular-mass subunits (δ and ε) were not resolved in the structure. The orientation of the γ subunit is consistent with its proposed role, wherein this subunit transmits the energy of the proton gradient to the catalytic sites for ATP synthesis that are known to be localized on the α and β subunits.

The most widely accepted mechanism of ATP synthesis is the so-called **binding change mechanism** proposed originally by Paul Boyer (a co-recipient of the 1997 Nobel Prize in chemistry) (Fig. 12.36). The key feature in this mechanism for the synthesis of ATP is that conformational changes in the protein are critical. According to this mechanism, the energy stored in the proton gradient is not directly used to drive the synthesis of ATP in a classical sense but rather is used to release a tightly bound form of ATP from its catalytic binding site on the enzyme. The CF$_1$ portion of the enzyme has three distinct nucleotide-binding sites, each of which can exist in one of three distinct conformational states—loose nucleotide binding (L), tight nucleotide binding (T), and a so-called open binding site (O), which is free of nucleotides. At any one time, all three states are present in the CF$_1$ complex, each being associated with one of the three catalytic centers present in the enzyme. According to the binding change mechanism, ADP and P$_i$ initially bind to an unoccupied site in the open state. As protons move from the lumen to the stromal region through the CF$_0$ channel, energy is released that results in a rotation of the γ subunit of CF$_1$, and this rotation causes conformational changes in the three nucleotide binding sites. The T site, which contains

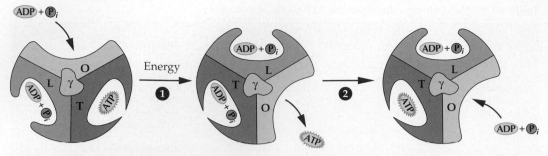

Figure 12.36

Binding change mechanism of ATP synthesis by the CF_0–CF_1 complex. Three nucleotide-binding sites are shown on the enzyme—the O-site (open), available to bind ADP and P_i; the L-site (loose), in which ADP and P_i are loosely bound; and the T-site, a tight nucleotide-binding site. Conformational changes driven by the movement of protons across the membrane in step 1, resulting in a rotation of the γ-subunit in the enzyme, cause interconversion of these sites and change the affinity of the sites for the nucleotides. The formation of ATP in the T-site is shown in step 2, but this condensation of ADP and P_i does not require additional energy.

bound ATP, is converted to an O site as the ATP is released, while the L site (containing the bound ADP and P_i) is converted into a T site; this facilitates the synthesis of ATP although there is no requirement for additional energy for this conversion. Recent evidence from several laboratories has led to strong support for the binding change mechanism, based on observations of the rotation of the γ subunit within the complex. The structure of F_1 that has recently been reported allows for a detailed analysis that should greatly facilitate further studies of this mechanism.

12.8 Carbon reactions in C_3 plants

12.8.1 In C_3 plants, photosynthetic carbon fixation is catalyzed by a single enzyme, Rubisco.

Most plants produce a three-carbon compound, 3-phosphoglycerate (3-PGA), as the first stable product in the multistep conversion of CO_2 into carbohydrate. This functionally defined group, which includes most crop plants, is referred to as **C_3 plants**. The C_3 carbon fixation pathway (Calvin cycle) was first elucidated in the 1950s by Melvin Calvin, Andrew Benson, and James A. Bassham. The researchers used a kinetic approach, applying $^{14}CO_2$ to cell suspensions of the green algae *Chlorella* and *Scenedesmus* and taking samples at short intervals to identify the radiolabeled compounds produced over time. The exact position of the

^{14}C label could be determined for each compound by degrading isolated intermediates. Experiments were carried out in a unique reaction vessel, a "lollipop," designed for the rapid removal of samples at short time intervals (Fig. 12.37).

Although 3-PGA is the first product formed in the fixation of CO_2, it does not form directly from three CO_2 molecules. Instead, 3-PGA forms in two concerted steps from the reaction of CO_2 with a five-carbon sugar, ribulose 1,5-bisphosphate (RuBP). The carboxylation of the C_5 sugar produces a C_6 intermediate that is immediately cleaved into two molecules of 3-PGA (Fig. 12.38). The enzyme that catalyzes this reaction is ribulose bisphosphate carboxylase/

Figure 12.37

"Lollipop" containing illuminated algal suspension with flask of hot ethanol for collecting samples and stopping cellular reactions. This apparatus was used by Calvin and coworkers to determine the first stable products of C_3 photosynthesis.

oxygenase, or Rubisco, one of three enzymes unique to the Calvin cycle (see Section 12.8.2).

As the first enzyme involved in the conversion of CO_2 into carbohydrate, Rubisco plays a critical role in the biochemistry of the chloroplast. Consistent with its role, Rubisco is the most abundant soluble protein in the chloroplast and is possibly one of the most abundant in the biosphere.

In plants, Rubisco consists of eight large (56-kDa) and eight small (14-kDa) subunits called L and S, respectively. Some photosynthetic bacterial enzymes contain only the L subunit, which contains the catalytic domain. The role of the S subunit is less clear. Structural details of plant Rubisco have recently become available, supplementing an earlier structural model for a bacterial Rubisco. These data have allowed site-directed mutagenesis studies of structure and function. In most eukaryotes, L subunits are encoded by the chloroplast genome, S subunits by the nuclear genome (see chapter 9, Fig. 9.39). The latter are transported into the chloroplast, where they combine with the large subunits in the chloroplast stroma to yield an active holoenzyme (Fig. 12.39).

12.8.2 Reduction and regeneration of intermediates follow the first CO_2 fixation reaction.

The Calvin cycle proceeds through 13 steps in three phases: carboxylation, reduction, and regeneration (Fig. 12.40). The carboxylation phase consists of one reaction: the carboxylation of RuBP to produce two molecules of 3-PGA. The two-step reductive phase converts 3-PGA into the triose phosphate, glyceraldehyde 3-phosphate (GAP). ATP and NADPH are used in this phase of the cycle. The last and largest set of reactions regenerates RuBP; in this process, an additional ATP is consumed during the conversion of ribulose 5-phosphate to RuBP. Overall, three molecules of ATP and two of NADPH supply energy for specific steps for each cycle. All of the 13 enzymes required in the Calvin cycle are located in the stroma, and 10 of the 13 enzymes are involved in the regeneration phase of the cycle. In addition to Rubisco, the enzymes that are unique to the cycle are sedoheptulose-1,7-bisphosphatase, which dephosphorylates a

diphosphosugar to yield a monophosphosugar, and phosphoribulokinase, which phosphorylates ribulose 5-phosphate to form RuBP, thus regenerating the initial CO_2 acceptor.

Addition of three molecules of CO_2 to three molecules of the C_5 sugar RuBP yields six molecules of 3-PGA, each of which is phosphorylated and reduced to generate the 3-carbon sugar GAP. Five GAP and three ATP molecules are used to regenerate the three RuBP molecules. The remaining GAP molecule, the net product of carbon fixation, can be used to build carbohydrates or other cellular constituents (Fig. 12.41).

The energetic requirements for the synthesis of one hexose from six CO_2 molecules are nine molecules of ATP and six molecules of NADPH. Using standard free energy changes for the hydrolysis of ATP and the oxidation of NADPH by O_2 (−30.5 and −220 kJ/mol, respectively), one can calculate that the utilization of ATP and NADPH to form GAP is approximately 90% efficient. This calculation assumes nonphysiological concentrations of reactants, however. Measurements of the efficiency of the Calvin cycle under physiological conditions also give relatively high efficiencies (greater than 80%).

12.8.3 The Calvin cycle is regulated by light via changes in pH and Mg^{2+} concentration.

Many of the Calvin cycle enzymes that catalyze reversible reactions (e.g., aldolase, glyceraldehyde-3-phosphate dehydrogenase) are common to the glycolytic pathway for carbohydrate degradation (see Chapter 13). Because enzymes involved in both the synthesis and degradation of carbohydrates are present within the chloroplast, it is essential that the synthetic apparatus be "on" and the degradative apparatus be "off" in the light. Specific regulatory mechanisms are required to prevent futile cycling and ensure optimal activity.

Changes in stromal pH and Mg^{2+} concentration are important regulators of enzymes such as Rubisco, fructose-1,6-bisphosphatase, and phosphoribulokinase. Rubisco activation involves the formation of a carbamate–Mg^{2+} complex on a specific

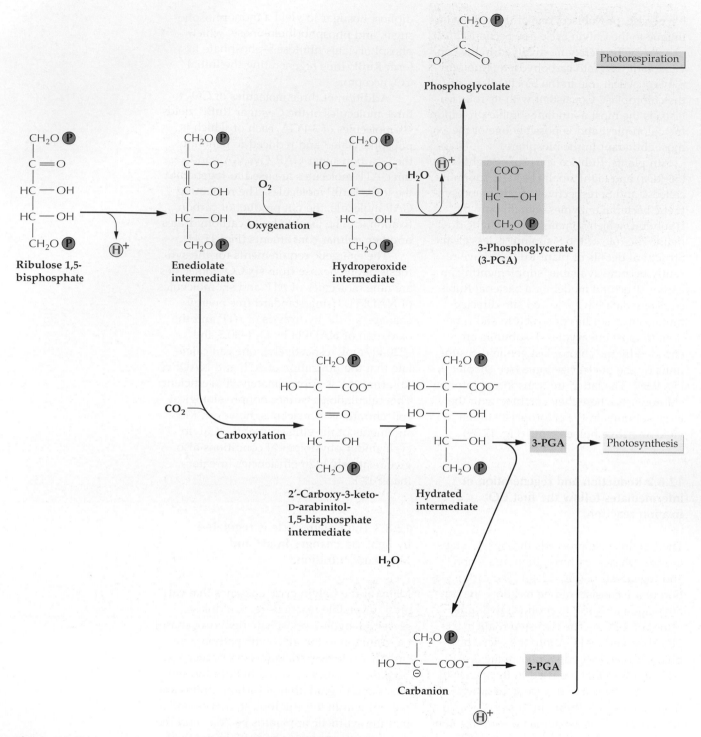

Figure 12.38

Rubisco catalyzes two types of reactions, carboxylation and oxygenation. Carboxylation of ribulose 1,5-bisphosphate (RuBP) yields two molecules of 3-phosphoglycerate (3-PGA), the first stable intermediate in C_3 photosynthesis. Oxygenation of RuBP yields one molecule of 3-PGA and one of 2-phosphoglycolate. This C_2 phosphoacid is converted to 3-PGA by the photorespiratory cycle (see Chapter 14). Both reaction sequences are initiated by the binding of RuBP to the Rubisco active site, followed by proton extraction and rearrangement to form a 2,3-enediolate intermediate. Reaction of CO_2 with the enediolate (carboxylation) generates a C_6 intermediate,

2'-carboxy-3-keto-D-arabinitol 1,5-bisphosphate. Hydration of this intermediate leads to cleavage between C-2 and C-3, producing 3-phosphoglycerate from the lower three carbons and the carbanion of 3-phosphoglycerate from the upper three carbons. Protonation of the carbanion completes the carboxylation catalytic cycle. The reaction of molecular oxygen with the enediolate gives rise to a C_5 hydroperoxide intermediate. A hydration/cleavage/protonation sequence analogous to the carboxylation reaction series gives rise to 3-phosphoglycerate from the lower three carbons and 2-phosphoglycolate from the upper two carbons.

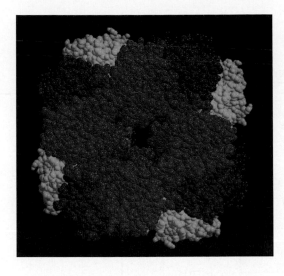

Figure 12.39
Structure of Rubisco (L_8S_8). In plant chloroplasts, Rubisco consists of eight large (56-kDa) and eight small (14-kDa) subunits. The four lobes visible in the structure each contain large and small subunits. Small subunits (of which four are visible) are colored red, and large subunits are colored blue and green in such a way as to show boundaries of dimers. Rubisco, often said to be the most prevalent protein on earth, constitutes up to half the protein of the chloroplast stroma.

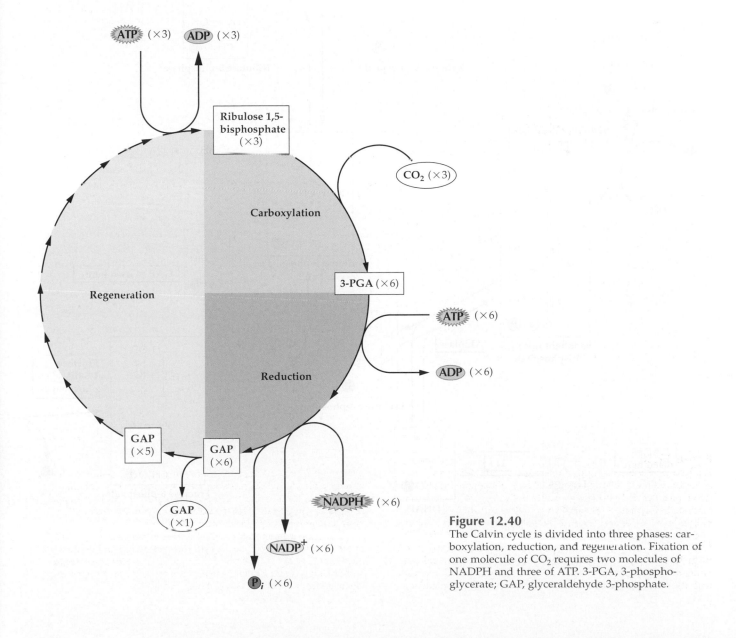

Figure 12.40
The Calvin cycle is divided into three phases: carboxylation, reduction, and regeneration. Fixation of one molecule of CO_2 requires two molecules of NADPH and three of ATP. 3-PGA, 3-phosphoglycerate; GAP, glyceraldehyde 3-phosphate.

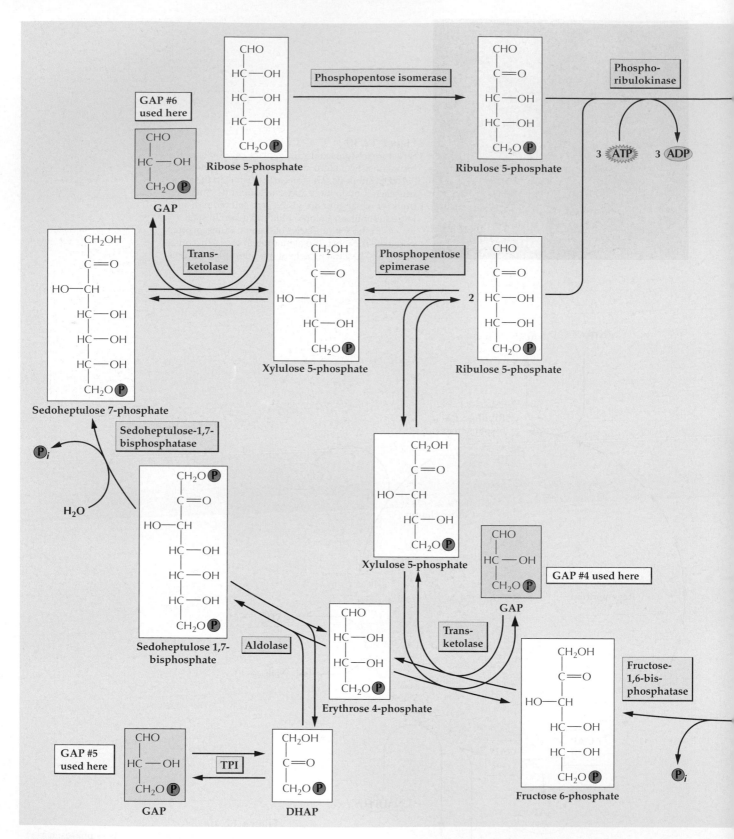

Figure 12.41
Individual steps of the Calvin cycle.

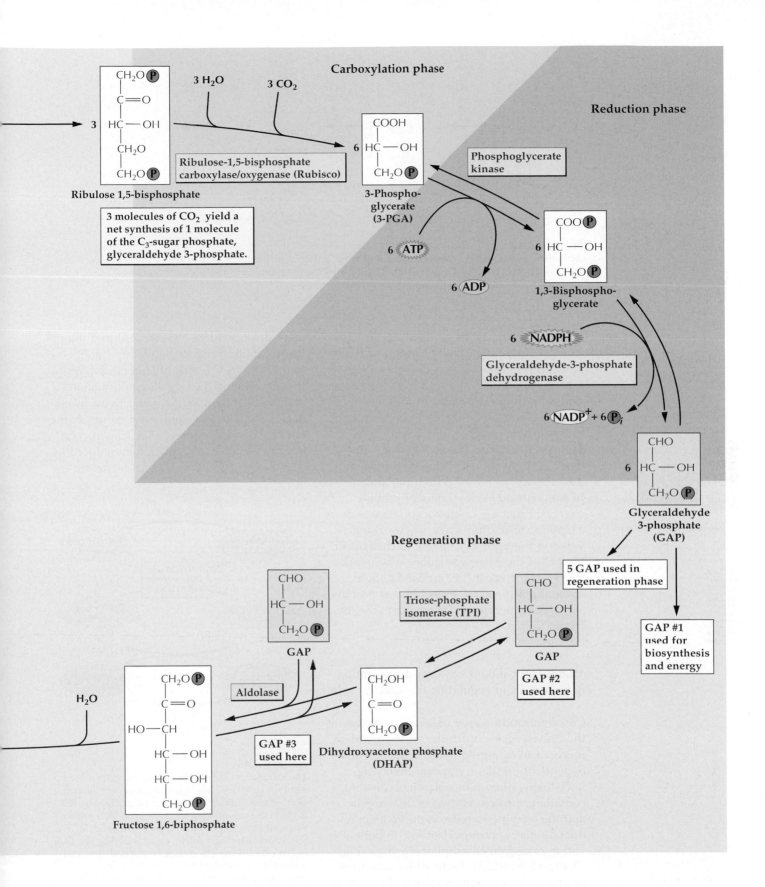

Carboxylation phase

Ribulose 1,5-bisphosphate

Ribulose-1,5-bisphosphate carboxylase/oxygenase (Rubisco)

3 molecules of CO_2 yield a net synthesis of 1 molecule of the C_3-sugar phosphate, glyceraldehyde 3-phosphate.

3 H_2O

3 CO_2

3-Phospho-glycerate (3-PGA)

Reduction phase

Phosphoglycerate kinase

6 ATP

6 ADP

1,3-Bisphospho-glycerate

6 NADPH

Glyceraldehyde-3-phosphate dehydrogenase

6 NADP$^+$ + 6 P$_i$

Glyceraldehyde 3-phosphate (GAP)

Regeneration phase

GAP

Triose-phosphate isomerase (TPI)

5 GAP used in regeneration phase

GAP

GAP #2 used here

GAP #1 used for biosynthesis and energy

Aldolase

H_2O

GAP #3 used here

Dihydroxyacetone phosphate (DHAP)

Fructose 1,6-biphosphate

615

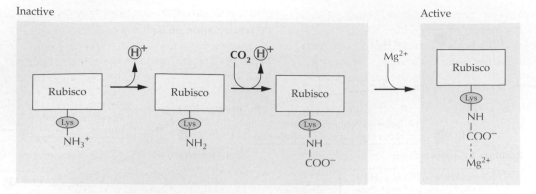

Inactive   Active

Figure 12.42

Activation of Rubisco by carbamylation. Rubisco is activated by binding CO_2 and Mg^{2+}. The activating reaction, which releases protons, is promoted by the increases in stromal pH and Mg^{2+} concentration associated with chloroplast illumination.

lysine (Lys-201) at the active site (Fig. 12.42). The CO_2 molecule involved in this activation reaction differs from the CO_2 substrate required for catalysis. Stromal Mg^{2+} concentration is an important factor in this activation. Evidence indicates that Mg^{2+} concentrations in the stroma increase from approximately 1–3 mM to approximately 3–6 mM in the light. As light drives H^+ into the thylakoid lumen, Mg^{2+} is released from the lumen to compensate for the influx of positive charges. Fructose-1,6-bisphosphatase and, to a lesser extent, other enzymes are also activated in the light in response to changes in the pH of the stroma solution. When dark-treated chloroplasts are exposed to light, stromal pH increases from 7 to 8 as protons are translocated from the stroma to the thylakoid lumen.

12.8.4 Light-linked covalent modification is important in regulating the Calvin cycle.

Covalent modification of several enzymes in the Calvin cycle constitutes an independent mechanism of regulation by light. A redox-dependent regulatory system involving ferredoxin, a low-molecular-mass disulfide-containing protein known as thioredoxin, and the enzyme ferredoxin-thioredoxin reductase plays an important role in light activation of certain chloroplast enzymes (Fig. 12.43 and Box 12.3). Reduced ferredoxin is produced by the light reactions of photosynthesis and reacts with oxidized thioredoxin f or m in a reaction catalyzed by ferredoxin-

thioredoxin reductase (FTR). Reduced thioredoxin can itself reduce regulatory intramolecular disulfide bonds of target enzymes, thereby altering the target's conformation and modulating its enzymatic activity. In the dark, the regulatory sulfhydryl groups of the enzyme become oxidized. The Calvin cycle enzymes subject to regulation by reduced

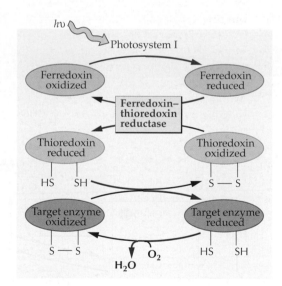

Figure 12.43

The ferredoxin-thioredoxin system provides a mechanism for light-dependent activation or deactivation of enzymes. Photosynthetic electron transfer by PSI leads to the reduction of ferredoxin, an Fe-S protein. Reduced ferredoxin in turn reduces thioredoxin, a regulatory disulfide protein, as catalyzed by the Fe-S enzyme ferredoxin-thioredoxin reductase. Reduced thioredoxin can reduce the disulfide bonds of numerous target proteins, modulating their activity. The path of electron transfer is shown in red. In the dark, O_2 oxidizes both the target enzyme and the thioredoxin (not shown).

thioredoxin are activated by reduction in the light and deactivated by oxidation in the dark. This simple sequence of reactions links the light reactions to CO_2 fixation, ensuring that carbohydrate synthesis will proceed in the light. This mechanism has been demonstrated for several chloroplast enzymes, including fructose-1,6-bisphosphatase, sedoheptulose-1,7-bisphosphatase, phosphoribulokinase, $NADP^+$-glyceraldehyde-3-phosphate dehydrogenase, and Rubisco activase, as well as the ATP synthase. Research has demonstrated that the thioredoxin system up-regulates additional light-dependent processes in the chloroplast, including translation and fatty acid biosynthesis. At the same time, the reduced thioredoxin formed in the light inhibits catabolic processes, which occur primarily in the dark (e.g., the oxidative pentose phosphate pathway; see Chapter 13).

Several unique mechanisms for the regulation of Rubisco are undoubtedly related to its central role in CO_2 fixation. Although researchers have long known that Rubisco must be activated to function fully in CO_2 fixation, and that this activation requires CO_2 binding at a specific site in the enzyme (see Fig. 12.42), several problems have emerged in elucidating Rubisco activation. For example, the in vitro carbamylation reaction requires millimolar concentrations of CO_2, whereas the in vivo concentration of CO_2 is in the micromolar concentration range. Moreover, sugar phosphates also bind to Rubisco, and this binding affects the activation reaction with CO_2. The enzyme Rubisco activase is specifically involved in the activation of Rubisco through

carbamylation. Although details of its activity remain unclear, Rubisco activase appears to promote the dissociation of sugar phosphate molecules bound at the active site of Rubisco, thereby allowing for carbamylation. Apparently, ATP is hydrolyzed as part of this activation reaction, although this requirement is not fully understood (Fig. 12.44). Finally, in some plants Rubisco activity is very low in dark-adapted leaves and cannot be reactivated by the addition of CO_2 and Mg^{2+}; in contrast, Rubisco from leaves from the same plant obtained in the middle of the day is activated by this treatment. This difference in activation is related to the presence of a tightly bound inhibitor, 2-carboxyarabinitol-1-phosphate (CA1P), which is an analog of the C_6 reaction intermediate in the Rubisco catalytic mechanism. Two different mechanisms appear to exist for the removal of CA1P from Rubisco, one requiring Rubisco activase and the other involving light-dependent degradation of the inhibitor—which explains the differing degrees of Rubisco inactivation observed in dark-adapted and light-grown plants. These multiple mechanisms of Rubisco regulation point to the key role this enzyme plays in general aspects of chloroplast metabolism.

12.8.5 Rubisco also functions as an oxygenase.

In addition to its role as a carboxylase, Rubisco also has an oxygenase activity that uses molecular oxygen as a substrate instead of CO_2. Rubisco catalyzes the reaction

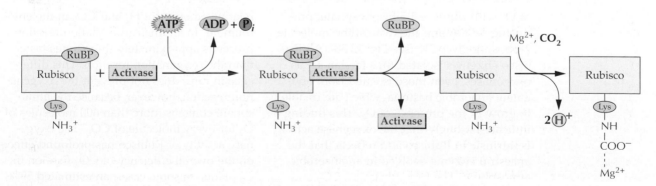

Figure 12.44
Rubisco activase removes bound RuBP from inactive, decarbamylated Rubisco in an ATP-dependent reaction. The free Rubisco can then be activated by carbamylation, binding CO_2 and Mg^{2+} as shown in Figure 12.42. Rubisco activase also can be activated by light via thioredoxin.

Box 12.3

Thioredoxins with contrasting evolutionary histories share in the redox regulation of chloroplast enzymes by the structurally defined ferredoxin/thioredoxin system.

Thioredoxins are ubiquitous 12-kDa proteins, the primary function of which lies in regulation of protein activity. The highly conserved active site of thioredoxin—Trp-Cys-Gly-Pro-Cys—contains a disulfide group that undergoes reversible redox changes:

$$S-S \rightleftharpoons 2\ SH$$

The chloroplast ferredoxin/thioredoxin system is made up of ferredoxin, the Fe-S enzyme, ferredoxin-thioredoxin re-ductase (FTR), and *f*- and *m*-type thioredoxins. Thioredoxin *f* was named for the target enzyme, fructose-1,6-bisphosphatase (FBPase), and thioredoxin *m* for NADP$^+$-malate dehydrogenase (NADP-MDH), an enzyme functional in C_4 photosynthesis (see Section 12.9.4). The three-dimensional structures of the components of the system and the original target enzymes have now been determined (see figure, below). The structures of most of the other target enzymes are still unknown.

Despite sharing a similar function, the thioredoxins participating in enzyme regulation in oxygenic photosynthesis differ in their phylogenetic history. Gene and protein sequencing, together with bio-chemical activity measurements, indicate that thioredoxin *m* resembles prokaryotic thioredoxins, whereas thioredoxin *f* is more similar to thioredoxins from eukaryotes. According to this view, the *m*-type thioredoxins are derived from the endosymbiotic prokaryotic ancestor of the

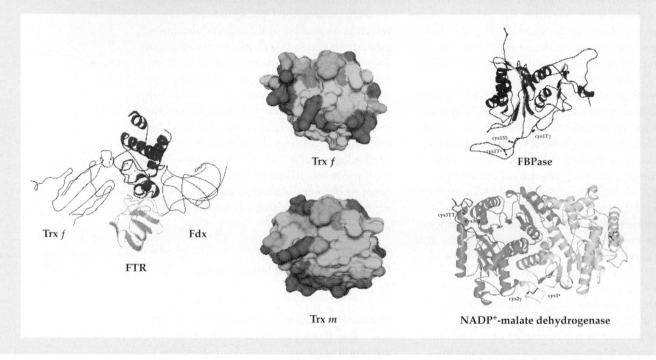

Trx *f*   Fdx

FTR

Trx *f*

Trx *m*

FBPase

NADP$^+$-malate dehydrogenase

of O_2 with ribulose 1,5-bisphosphate, producing 3-PGA and the two-carbon molecule, 2-phosphoglycolate (see Fig. 12.38). All Rubisco enzymes isolated thus far can act as oxygenases, even those isolated from obligately anaerobic bacteria, which are unable to grow in the presence of O_2. This finding indicates strongly that the oxygenase activity is intrinsic to Rubisco and reflects that the ancestral enzyme evolved in an anaerobic atmosphere. The two substrates, CO_2 and O_2, compete for the same active site on the enzyme, and the activity of the enzyme toward these substrates is dictated by the relative amounts of O_2 and CO_2 in the environment. In air, the carboxylation reaction proceeds approximately three times faster than the oxygenation reaction. This difference in rates does not prevent the oxygenation reaction, however, because the atmosphere contains more than 600 molecules of O_2 for every molecule of CO_2. The oxygenase activity of Rubisco has profound effects on the overall efficiency of CO_2 fixation in C_3 plants: In some cases, an estimated 50% of the CO_2 fixed via photosynthesis is lost through the process of **photorespiration** (see Chapter 14).

chloroplast, where they evolved to function in the regulation of oxygenic photosynthesis (via phosphoribulokinase) and the oxidative pentose phosphate pathway (via glucose-6-phosphate dehydrogenase).

Recent sequence analysis of the chloroplast genome supports this series of events. Most photosynthetic eukaryotes encode thioredoxin *m* in the nucleus. However, the nucleotide sequences for red algae, such as *Porphyra purpurea*, revealed a plastid-encoded thioredoxin *m* gene. This pattern of gene distribution is consistent with thioredoxin *m* having originated from an engulfed prokaryotic endosymbiont. Thioredoxin *f*, on the other hand, apparently originated in the nucleus of the eukaryotic host and was subsequently modified for transport to the chloroplast (see figure below).

A comparison of the protein and gene sequences, as well as the location of introns, indicates that thioredoxin *f* is most closely related to the nuclear-encoded thioredoxin of animals. Over time, thioredoxin *f* became the principal regulator of photosynthetic enzymes. The research leading from the discovery of the ferredoxin/thioredoxin system to the elucidation of its function, structure, and phylogenetic history took place over three decades.

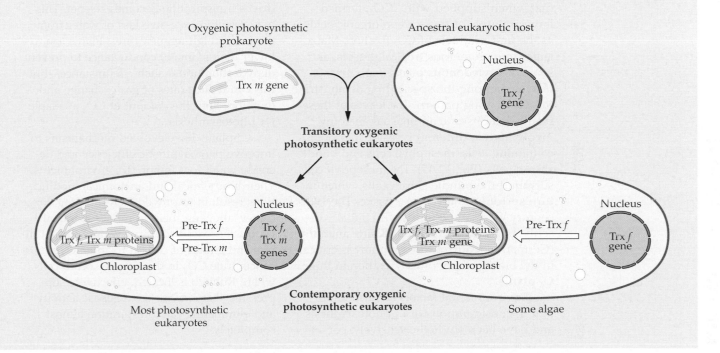

12.9 Variations in mechanisms of CO₂ fixation

12.9.1 Some photosynthetic bacteria do not fix carbon via the Calvin cycle.

Although the Calvin cycle is generally viewed as an almost ubiquitous mechanism for the conversion of CO_2 into carbohydrates in photosynthetic organisms, it is not universal. In some anaerobic photosynthetic bacteria, CO_2 fixation proceeds by very different processes.

Organisms such as the photosynthetic green sulfur bacteria fix carbon via the **reductive carboxylic acid cycle.** Essentially a reversal of the citric acid cycle (see Chapter 14), this pathway synthesizes organic acids from acetyl-CoA and CO_2, with reduced ferredoxin serving as the main reductant for CO_2 fixation. Another Rubisco-independent photosynthetic mechanism is the **hydroxypropionate pathway,** present in the green nonsulfur bacterium *Chloroflexus*. As with the reductive carboxylic acid cycle, CO_2 fixation involves a reaction with acetyl-CoA, but the product is 3-hydroxypropionate and the net fixation product of the cycle is glyoxalate. These interesting variants of photosynthesis have survived eons of evolutionary history. Other variants, found in higher plants, are discussed more fully below.

12.9.2 C₄ plants contain two distinct CO₂-fixing enzymes and have specialized foliar anatomy.

Although Rubisco is present in all plants, not all plants generate 3-PGA as the first stable photosynthetic intermediate. In the 1960s, several plant species were identified that, when supplied with $^{14}CO_2$, formed large amounts of four-carbon organic acids as the first products of CO_2 fixation. Maize, sugarcane, numerous tropical grasses, and some dicotyledonous plants (e.g., *Amaranthus*) are among the species that demonstrate this C₄ labeling pattern. The leaves of these plants demonstrate an unusual anatomy involving two different types of chloroplast-containing cells: mesophyll cells and bundle sheath cells (Fig. 12.45). The mesophyll cells surround the bundle sheath cells, which in turn surround the vascular tissue. The 19th-century German botanists who initially described this feature called it **Kranz anatomy** (German: wreath). Kranz anatomy is critical in the biochemistry of CO_2 fixation in these **C₄ plants.**

Environmental factors are also important in considerations of C₄ photosynthesis and have been studied extensively. For ex-

ample, at higher temperatures, the oxygenase activity of Rubisco is favored over the carboxylation activity, thus decreasing the efficiency of photosynthesis. This effect results in part from temperature-dependent changes in the solubilities of CO_2 and O_2, which decrease the ratio of available CO_2/O_2 as the temperature is increased. Shifts in temperature also alter Rubisco's affinity for each of its substrates. As a result, the ratio of carboxylation to oxygenation of Rubisco declines as the temperature warms, and the ratio of photorespiration to photosynthesis increases. In addition, as a leaf warms, the amount of water vapor in its air spaces increases and the concentration gradient that drives transpiration becomes steeper. This may result in excessive loss of water from the leaf at high temperatures. Plants can lower their stomatal conductance to prevent dehydration under such circumstances, but this can reduce rates of gas exchange and therefore limit the amount of CO_2 available for photosynthesis.

C₄ plants have evolved mechanisms to improve photosynthetic efficiency and decrease water loss in hot, dry environments. Their biochemical and anatomical modifications result in improved CO_2-trapping efficiency, allowing them to reduce stomatal conductance and conserve water without diminishing rates of carbon fixation. C₄ plants concentrate CO_2 in the bundle sheath cells, where Rubisco is located, effectively suppressing that enzyme's oxygenase activity and eliminating photorespiration almost completely.

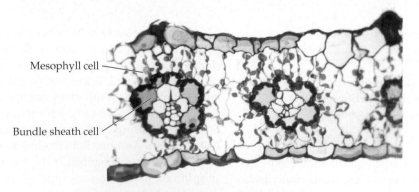

Figure 12.45
Micrograph showing Kranz anatomy in maize, a C₄ plant. The closely spaced vascular bundles are surrounded by large bundle sheath cells. In this species, large chloroplasts are located at the periphery of the bundle sheath cells which are surrounded by cells of the mesophyll.

12.9.3 The C$_4$ pathway increases the concentration of CO$_2$ in bundle sheath cells.

The C$_4$ pathway of CO$_2$ fixation is based on a complex interaction between mesophyll and bundle sheath cells. The pathway involves an initial HCO$_3^-$ fixation in the outer mesophyll cells, transfer of an organic acid from mesophyll to bundle sheath cells, and release of CO$_2$ for subsequent refixation in the Calvin cycle (Fig. 12.46). Biochemical fractionation analyses have shown that en-zymes unique to the Calvin cycle are located only in the bundle sheath chloroplasts, whereas oxaloacetate is generated from HCO$_3^-$ and phospho*enol*pyruvate (PEP) in the cytosol of mesophyll cells. Electron micrographs of bundle sheath and mesophyll chloroplasts (Fig. 12.47) show structural differences that reflect distinct function. The bundle sheath chloroplasts have no stacked membranes and demonstrate little PSII activity, whereas mesophyll chloroplasts have retained both PSII and PSI activities.

All C$_4$ carbon-fixation pathways begin in the mesophyll, where PEP carboxylase converts PEP to oxaloacetate. The C$_1$ substrate for this enzyme is not CO$_2$ but HCO$_3^-$. The aqueous equilibrium of these two species strongly favors HCO$_3^-$ ion over gaseous CO$_2$, allowing for a more efficient initial fixation step. Moreover, PEP carboxylase cannot fix oxygen, which has a three-dimensional structure similar to that of CO$_2$ but not HCO$_3^-$.

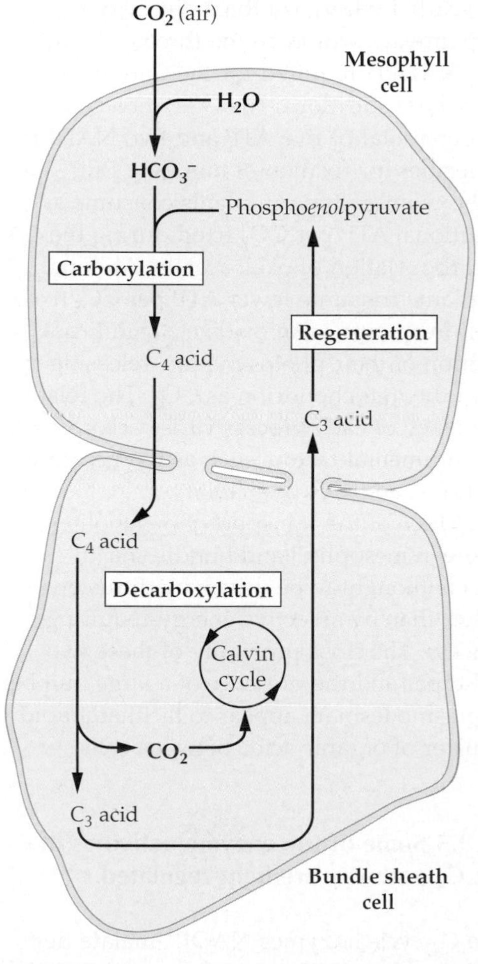

Figure 12.46
General aspects of the C$_4$ pathway. CO$_2$ enters the mesophyll cell and is converted to HCO$_3^-$ in the aqueous environment of the cytosol. This bicarbonate ion reacts with PEP to form a C$_4$ acid (oxaloacetate), which is converted to a second C$_4$ acid (malate or aspartate) and then transported to a neighboring bundle sheath cell. There, the C$_4$ acid is decarboxylated, and the CO$_2$ released is fixed by Rubisco and converted to carbohydrate by the Calvin cycle. The C$_3$ acid product of decarboxylation is transported back to the mesophyll cell to regenerate PEP.

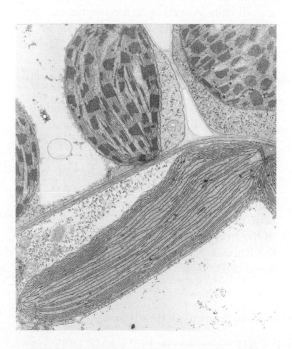

Figure 12.47
Electron micrograph comparing the chloroplasts of a bundle sheath cell (bottom) and a mesophyll cell (top) in a C$_4$ plant (sorghum). The chloroplast morphologies reflect their biochemical functions. The bundle sheath chloroplast lacks stacked thylakoids and contains little PSII. In contrast, the mesophyll chloroplasts contain all the transmembrane complexes required for the light reactions of photosynthesis but little or no Rubisco.

Reaction 12.13: NADP⁺-malic enzyme

$$\text{Malate} + \text{NADP}^+ \longrightarrow \text{Pyruvate} + CO_2 + \text{NADPH}$$

Reaction 12.14: NAD⁺-malic enzyme

$$\text{Malate} + \text{NAD}^+ \longrightarrow \text{Pyruvate} + CO_2 + \text{NADH}$$

Reaction 12.15: PEP carboxykinase

$$\text{Oxaloacetate} + \text{ATP} \longrightarrow \text{PEP} + \text{ADP} + CO_2$$

The subsequent metabolism of oxaloacetate differs among species, but in all cases, a C_4 acid is transported to the bundle sheath cells, where it is decarboxylated. The CO_2 released is refixed by Rubisco. The resulting C_3 acid is then transported back to the mesophyll cells for regeneration of PEP.

Three variations of C_4 photosynthesis are known, differing in the C_4 acids transported between mesophyll and bundle sheath cells as well as in the mechanism of decarboxylation (see Rxs. 12.13–12.15) in the bundle sheath cells (Table 12.10). The variants are named on the basis of the decarboxylase enzymes in the bundle sheath cells (Fig. 12.48).

The intracellular compartmentation and energetic requirements for each of these three systems differ slightly (see Chapter 14, Fig. 14.42 and Table 14.2). This diversity reflects that C_4 photosynthesis is a product of convergent evolution, having developed on many separate occasions in very different taxa. In all cases, however, CO_2 is released in the bundle sheath cell and is refixed by the Rubisco localized in the chloroplasts of these cells.

Reaction 12.16: Pyruvate orthophosphate dikinase

$$\text{Pyruvate} + P_i + \text{ATP} \longrightarrow \text{PEP} + \text{AMP} + PP_i$$

12.9.4 The C_3 and C_4 pathways have different energy costs.

One of the key reactions in C_4 photosynthesis is the synthesis of PEP (Rx. 12.16), catalyzed by pyruvate-orthophosphate dikinase (PPDK). The AMP subsequently reacts with ATP to form two ADP, and the pyrophosphate is subsequently hydrolyzed to two P_i, driving the reaction to the right. The C_4 cycle thus requires a net conversion of two ATP molecules to two ADP and two P_i per CO_2 molecule released into bundle sheath cells. The total energy required to fix CO_2 into carbohydrate via the malic enzyme C_4 pathways would be the three ATP and two NADPH required for the Calvin cycle plus these additional two ATP molecules, giving a total of five ATP and two NADPH molecules for fixation of one CO_2. PEP carboxykinase-type C_4 plants consume an additional ATP per CO_2 fixed during the decarboxylation of oxaloacetate. Although C_3 plants consume fewer ATP per CO_2 fixed, photorespiration can oxidize a significant fraction of their photosynthate, releasing it from the mitochondrion as CO_2. The relative efficiency of each process varies according to environmental factors such as temperature, humdity, and CO_2 availability.

Much of the transport of metabolites between mesophyll and bundle sheath cells is thought to be driven by diffusion rather than by an active energy-requiring process. The close proximity of these two cell types and the presence of a large number of plasmodesmata appear to facilitate rapid transfer of organic acids between cells.

12.9.5 Some of the enzyme activities of the C_4 pathway are light-regulated.

The C_4 cycle enzymes NADP⁺-malate dehydrogenase, PEP carboxylase, and PPDK are regulated by light. This regulation is critical to maintain coordination between the

Table 12.10 Variations in C_4 photosynthesis

C_4 acid transported to bundle sheath cells	C_3 acid transported to mesophyll cells	Decarboxylase	Plant examples
Malate	Pyruvate	NADP⁺-malic acid	Maize, sugarcane
Aspartate	Alanine	NAD⁺-malic enzyme	Millet
Aspartate	Alanine, PEP, or pyruvate	Phospho*enol*pyruvate carboxykinase	*Panicum maximum*

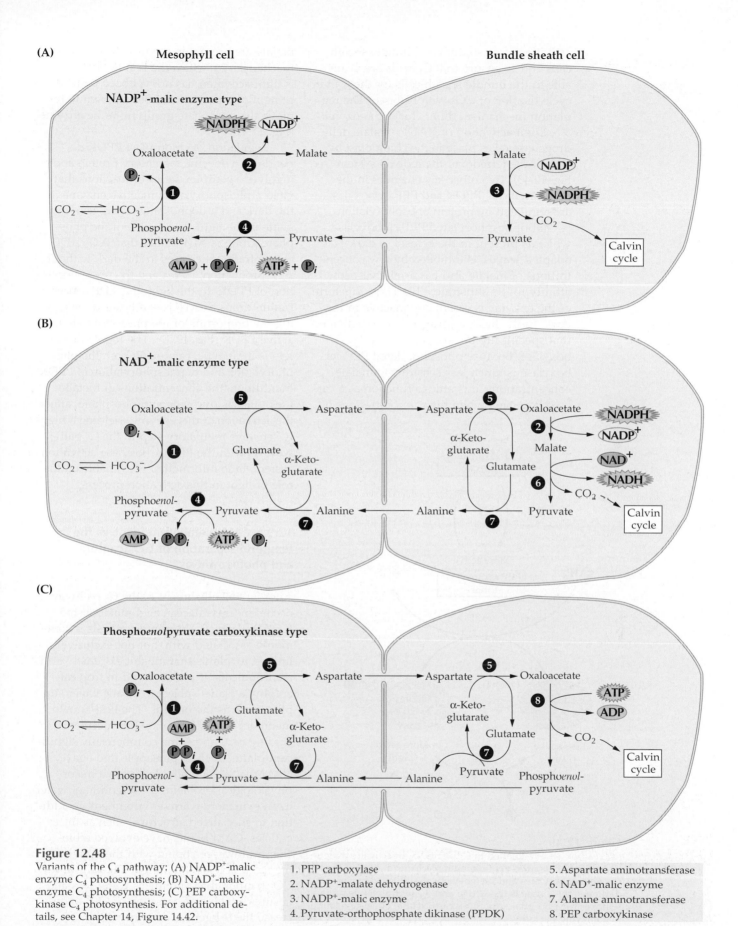

(A) Mesophyll cell   Bundle sheath cell

NADP$^+$-malic enzyme type

(B) NAD$^+$-malic enzyme type

(C) Phospho*enol*pyruvate carboxykinase type

Figure 12.48
Variants of the C$_4$ pathway: (A) NADP$^+$-malic enzyme C$_4$ photosynthesis; (B) NAD$^+$-malic enzyme C$_4$ photosynthesis; (C) PEP carboxykinase C$_4$ photosynthesis. For additional details, see Chapter 14, Figure 14.42.

1. PEP carboxylase
2. NADP$^+$-malate dehydrogenase
3. NADP$^+$-malic enzyme
4. Pyruvate-orthophosphate dikinase (PPDK)

5. Aspartate aminotransferase
6. NAD$^+$-malic enzyme
7. Alanine aminotransferase
8. PEP carboxykinase

623

activities of mesophyll and bundle sheath cells and to ensure that C_4 acids are available to the bundle sheath cells for the Calvin cycle fixation of CO_2 with Rubisco. The regulation mechanism differs for each enzyme.

Light activation of $NADP^+$-malate dehydrogenase takes place in the chloroplast by way of the ferredoxin-thioredoxin system (see Fig. 12.43). Two other key enzymes in the mesophyll cell, PPDK and PEP carboxylase, are regulated by protein phosphorylation.

Nonphosphorylated PEP carboxylase, which is present in the cytosol of dark-adapted leaves, is inhibited by low concentrations of malate and has a relatively low affinity for its substrate, PEP. Thus, this form of the enzyme is essentially inactive in the dark. In the light, a serine kinase is activated that specifically phosphorylates PEP carboxylase. Because phosphorylated PEP carboxylase is much less sensitive to malate concentrations, it is able to function as a carboxylase even when high concentrations of malate are present in the mesophyll cell. The mechanism by which the regulatory kinase is light-activated has not yet been defined. A protein phosphatase has been shown to remove the phosphate group from the enzyme (Fig. 12.49).

Regulation of chloroplast PPDK depends on a specific regulatory protein and involves a complex series of reactions that yield a phosphorylated, inactive enzyme in the dark and a dephosphorylated, active enzyme in the light. The source of the phosphate group is ADP rather than ATP. ADP concentration increases in the dark without ATP production, increasing the phosphorylation of PPDK. In the light, the ADP concentration drops as ATP is synthesized, increasing the proportion of nonphosphorylated, active PPDK (Fig. 12.50). The regulatory protein is bicatalytic, catalyzing both the phosphorylation and dephosphorylation of PPDK. In addition, the concentrations of metabolites, such as pyruvate and phosphate, affect the activation of this enzyme, yielding a highly sensitive regulatory step in the C_4 pathway. As with the PEP carboxylase activation system, little information is available on the role of light in this activation process.

12.9.6 CAM metabolism involves the temporal separation of CO_2 capture and photosynthesis.

Another embellishment to the C_3 pathway occurs in Crassulacean acid metabolism (CAM) plants. CAM photosynthesis is commonly associated with, but not exclusively limited to, plants that inhabit extremely arid environments. It is also found in tropical epiphytic plants subject to erratic water supplies, such as the orchids. The best known members of this group are the succulents, including cacti and some commercially significant plants, such as pineapple and agave. CAM plants generally deal with a major physiological problem, the retention of water in an extremely warm environment. In addition to specialized structures such as thick cuticles, CAM plants have evolved a biochemical way of coping with their environmental problems, utilizing a mechanism for CO_2 fixation that minimizes water loss and ensures a high concentration of CO_2 to decrease the oxygenase activity of Rubisco.

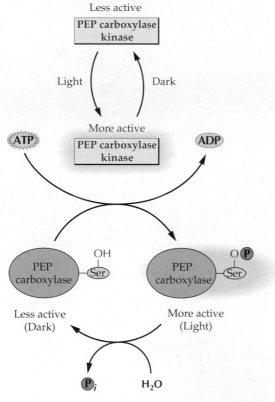

Figure 12.49
PEP carboxylase regulation in C_4 plants. Light activates a regulatory kinase by an as yet unknown mechanism. In turn, the kinase phosphorylates and thereby activates PEP carboxylase. In darkness, the kinase is less active and the hydrolytic cleavage removes the phosphate from PEP carboxylase, down-regulating that enzyme's activity.

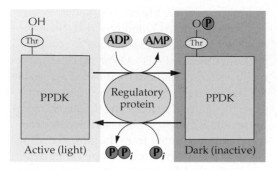

Figure 12.50
Regulation of pyruvate-orthophosphate dikinase (PPDK), an enzyme active in C_4 photosynthesis. PPDK activity is modulated by a regulatory protein. This protein promotes phosphorylation of PPDK in the dark, which renders the enzyme inactive. Concentrations of the phosphate donor, ADP, are increased in darkness, i.e., in the absence of photophosphorylation. PPDK is dephosphorylated, and thereby activated, in the light.

CO_2 fixation in CAM plants bears some similarity to the C_4 pathway, but there are also some very important differences. The initial fixation reaction involves PEP carboxylase, which uses HCO_3^- in a manner similar to the initial fixation in a C_4 plant. However, instead of separating the two CO_2 fixation events spatially in bundle sheath and mesophyll cells, CAM plants utilize temporal separation (Fig. 12.51). CO_2 is fixed by PEP carboxylase at night, forming oxaloacetate, which is subsequently converted to malate. During the night, the malate is stored in vacuoles, where it can reach very high concentrations. As the light period commences, this malate is transported out of the vacuole and decarboxylated to produce CO_2 and pyruvate. The CO_2 released is then refixed by Rubisco and the conventional Calvin cycle converts this into carbohydrate.

A notable feature of CAM plants is their regulation of stomatal opening and closing. Stomata remain open during the cool and relatively humid night, allowing CO_2 entry while sustaining minimal water loss. During the heat and dryness of the day, the stomata close, preventing water loss. Closed stomata also prevent CO_2 entry. However, the CO_2 is supplied by malate, the internal CO_2 source. The CO_2 concentration in CAM leaves rises to very high levels because CO_2 cannot escape through the closed stomata. The high concentrations of CO_2 further increase the efficiency of Rubisco as a carboxylase. Under very extreme drought conditions, CAM plants can close their stomata at night and recycle CO_2 within their cells. While maintaining the health of the plant, this mechanism does not support plant growth.

Note that the mechanisms that inhibit PEP carboxylase activity in dark-treated C_4 plants do not operate here. In CAM plants, the activation of PEP carboxylase at night and its deactivation during the day are mediated by endogenous circadian rhythms rather than by exogenous light and dark signals. The day form is inhibited by malate, which is present at high concentrations as it leaves the vacuole, whereas the night form is insensitive to malate, thereby allowing this metabolite to accumulate. The regulation of these two forms involves protein phosphorylation and dephosphorylation and prevents the useless carboxylation/decarboxylation cycle that

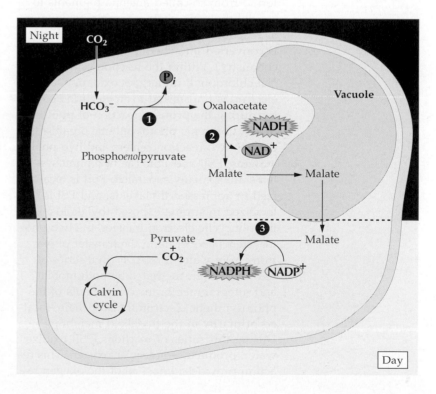

Figure 12.51
Crassulacean acid metabolism (CAM) is an evolutionary adaptation to photosynthesis in an arid environment. At night, CAM plants open their stomata, allowing CO_2 to enter. PEP carboxylase (1) incorporates this CO_2 (as HCO_3^-) into the C_4 organic acid oxaloacetate, which is reduced to malate by malate dehydrogenase (2). The malate is stored in the vacuole overnight. In the light, CAM plants close their stomata, preventing water loss. The stored malate is decarboxylated by $NADP^+$-malic enzyme (3), and the resulting CO_2 is converted to carbohydrate via the Calvin cycle.

would occur during the day if both PEP carboxylation and malate decarboxylation were fully active.

Summary

Photosynthesis produces organic compounds from inorganic carbon by using the energy of sunlight. These processes are carried out in plants, algae, and various bacteria. In all cases, the photosynthetic reactions may conveniently be divided into two phases, the light reactions and the carbon-fixing reactions.

In the case of eukaryotic organisms, photosynthesis takes place in the chloroplast. This organelle is surrounded by a double membrane and contains a complex internal membrane system, the thylakoid membranes. The two phases of photosynthesis take place in different regions of the chloroplast, with the light reactions being localized to the thylakoid membranes and the carbon-fixing reactions occurring in the stroma.

The light reactions of photosynthesis involve the photosynthetic pigments, the photosynthetic electron transport chain, and the ATP synthesis machinery. Light is absorbed by pigments that are localized into pigment–protein complexes within the thylakoid membrane. This light energy can be transferred from so-called antenna pigments to special pigment–protein complexes, known as reaction centers, where the light energy is converted into chemical products (photochemistry). Within the reaction center, a special chlorophyll undergoes oxidation, and an electron is transferred to an electron acceptor during the primary reaction of photosynthesis. Oxygenic photosynthetic organisms contain two reaction centers and two photosystems, PSII and PSI. The two photosystems are spatially separated: PSII is localized in appressed thylakoids, and PSI is localized in stroma-exposed thylakoids. During noncyclic electron transfer, the two photosystems cooperate in the transfer of electrons from water to $NADP^+$ in a series of redox reactions mediated by both mobile and integral membrane components of the photosynthetic electron transfer chain, located primarily in the thylakoid membranes. During this series of reactions, PSII oxidizes water, producing molecular oxygen. This reaction provides almost all of the oxygen required for aerobic life on our planet.

In addition to O_2 and NADPH, the noncyclic electron transfer reactions are coupled to the synthesis of ATP. ATP synthesis is driven by a proton gradient established across the thylakoid membrane. During electron transport, the interior space of the thylakoid membrane, the lumen, becomes acidified as protons are translocated from the stroma during electron transport or released by water oxidation in the thylakoid lumen. Use of this electrochemical gradient as an energy source requires a large protein complex located in the membrane, the ATP synthase. This enzyme is composed of two portions—one intrinsic to the membrane and involved in transporting protons through the membrane, and the second, which is extrinsic, involved in the actual conversion of ADP and P_i into ATP. Estimates of the stoichiometry for ATP synthesis have indicated that four H^+ are required per ATP molecule. The actual synthesis of ATP is believed to involve complicated conformational changes in ATP synthase, driven by the movement of protons through the enzyme.

In addition to the noncyclic pathway for ATP synthesis, chloroplasts also synthesize ATP via a cyclic pathway that involves only PSI. This pathway generates a proton gradient that can also be used for the synthesis of ATP, but does not yield O_2 or NADPH.

The reduction of CO_2 to carbohydrates requires NADPH and ATP, which are synthesized during the light reactions of photosynthesis. Plants employ the C_3 photosynthetic pathway (Calvin cycle) to fix CO_2, using the key enzyme Rubisco to convert CO_2 and RuBP into the C_3 product, 3-PGA. The multienzyme Calvin cycle involves three phases—carboxylation, reduction, and regeneration—and requires three ATP and two NADPH molecules per molecule of CO_2 fixed. All Calvin cycle reactions are catalyzed by soluble enzymes localized in the chloroplast stroma. Regulation of the Calvin cycle occurs by multiple mechanisms, including changes in ionic strength and pH as well as protein-mediated reactions.

Variants of C_3 photosynthesis exist in many plants. In one variation, known as the C_4 pathway, plants fix CO_2 into a C_4 acid in mesophyll cells, and transport this fixed CO_2 to anatomically distinct bundle sheath cells, where the CO_2 is released and refixed by Rubisco. This sequence of reactions provides a higher concentration of CO_2 for Rubisco in the bundle sheath cell and aids in inhibiting the oxygenase activity of Rubisco, a wasteful side-reaction that reduces the efficiency of CO_2 fixation in the C_3 chloroplasts, where O_2 and CO_2 compete for the Rubisco active site. In the case of CAM plants, another variant, CO_2 is fixed at night into organic acids that are subsequently decarboxylated during the day to provide CO_2 for Rubisco. CAM photosynthesis aids in the retention of water in arid environments. Key photosynthetic enzymes in C_4 and CAM plants are regulated to ensure the efficient interaction of these CO_2-concentrating mechanisms and the Calvin cycle for which they provide substrate.

Further Reading

Books

Clayton, R. K. (1980) *Photosynthesis: Physical Mechanisms and Chemical Patterns.* Cambridge University Press, Cambridge.

Cramer, W. A., Knaff, D. B. (1989) *Energy Transduction in Biological Membranes—A Textbook of Bioenergetics.* Springer-Verlag, Berlin.

Ort, D. R., Yocum, C. F., eds. (1996) *Oxygenic Photosynthesis: The Light Reactions.* Kluwer Academic Publishers, Dordrecht, The Netherlands.

Rochaix, J.-D., Goldschmidt-Clermont, M., Merchant, S. (1998) *The Molecular Biology of Chloroplasts and Mitochondria in* Chlamydomonas. Kluwer Academic Publishers, Dordrecht, The Netherlands.

Reviews

Bassham, J. A. (1979) The reductive pentose phosphate cycle and its regulation. In *Photosynthesis II—Photosynthetic Carbon Metabolism and Related Processes*, M. Gibbs and E. Latzko, eds. Springer-Verlag, Berlin, pp. 9–30.

Chitnis, P. R. (1996) Photosystem I. *Plant Physiol.* 111: 661–669.

Deisenhofer, J., Michel, H. (1989) The photosynthetic reaction centre from the purple bacterium *Rhodopseudomonas viridis. EMBO J.* 8. 2149–2170.

Gutteridge, S., Gatensby, A. A. (1995) Rubisco synthesis, assembly, mechanism and regulation. *Plant Cell* 7: 809–819.

Hartman, F. C., Harpel, M. R. (1994) Structure, function, regulation and assembly of D-ribulose-1,5-bisphosphate carboxylase/oxygenase. *Annu. Rev. Biochem.* 63: 197–234.

Hatch, M. C. (1987) C_4 photosynthesis: a unique blend of modified biochemistry, anatomy and ultrastructure. *Biochim. Biophys. Acta* 895: 81–106.

Jansson, S. (1994) The light-harvesting chlorophyll *a/b*-binding proteins. *Biochim. Biophys. Acta* 1184: 1–19.

Kuhlbrandt, W. (1994) Structure and function of the plant light-harvesting complex, LHC-II. *Curr. Opin. Struct. Biol.* 4: 519–528.

Melis, A. (1991) Dynamics of photosynthetic membrane composition and function. *Biochim. Biophys. Acta* 1058: 87–106.

Pakrasi, H. B. (1995) Genetic analysis of the form and function of Photosystem I and Photosystem II. *Annu. Rev. Genet.* 29: 755–776.

Ray, T. B., Black, C. C. (1979) The C_4 pathway and its regulation. In *Photosynthesis II—Photosynthetic Carbon Metabolism and Related Processes*, M. Gibbs and E. Latzko, eds. Springer-Verlag, Berlin, pp. 77–101.

Schürmann, P., Buchanan, B. B. (in press) The structure and function of the ferredoxin/thioredoxin system. In *The Regulatory Aspects of Photosynthesis,* B. Andersson and E.-M. Aro, eds. Kluwer Academic Publishers, Dordrecht, The Netherlands.

Research articles

Allen, J. F., Bennett, J., Steinback, K. E., Arntzen, C. J. (1981) Chloroplast protein phosphorylation couples plastoquinone redox state to distribution of excitation energy between photosystems. *Nature* 291: 21–25.

Krauss, N., Schubert, W.-D., Klukas, O., Fromme, P., Witt, H. T., Saenger, W. (1999) Photosystem I at 4 Å resolution: a joint photosynthetic reaction center and core antenna system. *Nat. Struct. Biol.* 3: 965–973.

Markwell, J. P., Thornber, J. P., Boggs, R. T. (1979) Evidence that in plant chloroplasts all the chlorophyll exists in chlorophyll–protein complexes *Proc. Natl. Acad. Sci. USA* 76: 1233–1235.

Nanba, O., Satoh, K. (1987) Isolation of a photosystem II reaction center consisting

of D-1 and D-2 polypeptides and cytochrome *b*-559. *Proc. Natl. Acad. Sci. USA*
84: 109–112.

Schreuder, H. A., Knight, S., Curmi, P.M.G.,
Andersson, I., Cascio, D., Sweet, R. M.,
Brändén, C.-I., Eisenberg, D. (1993) Crystal
structure of activated tobacco Rubisco complexed with the reaction-intermediate analog 2-carboxy-arabinitol 1,5-bisphosphate.
Protein Sci. 2: 1136–1146.

Biochemistry & Molecular Biology of Plants, B. Buchanan, W. Gruissem, R. Jones, Eds.
© 2000, American Society of Plant Physiologists

C H A P T E R *13*

Carbohydrate Metabolism

David T. Dennis
Stephen D. Blakeley

CHAPTER OUTLINE

Introduction

Carbohydrate metabolism has been studied extensively in animals and microorganisms but has received far less attention in plants. Researchers, as well as the authors of most general biochemistry texts, have assumed that metabolic data from any model system could be directly applied to plants because the individual reactions involved in carbohydrate metabolism appear to be very similar in all organisms. As our knowledge of metabolism has increased, however, we have discovered many aspects of plant carbohydrate metabolism that are unique.

Several important attributes distinguish plants from other organisms and help explain the singular nature of plant carbohydrate metabolism. These factors result in a unique carbohydrate metabolism that only superficially resembles carbohydrate metabolism in other organisms.

- Plants are autotrophic.
- Plants are photosynthetic.
- Plants are sessile.
- Plants contain plastids.
- Plants have cell walls.

Plants are autotrophic. Almost all plants convert simple nutrients such as carbon dioxide, water, and inorganic ions into all the intermediates required for the biosynthesis of nucleic acids, proteins, lipids, and polysaccharides as well as coenzymes and numerous secondary metabolites. Thus, in addition to fulfilling the energy requirements of the cell, plant carbohydrate metabolism must feed numerous anabolic pathways (Fig. 13.1). This biosynthetic investment affects the regulation of carbohydrate metabolism.

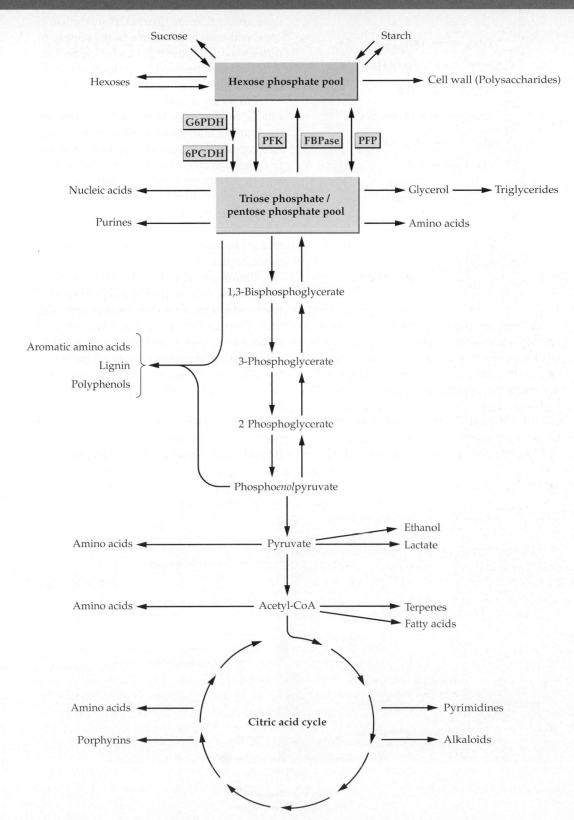

Figure 13.1
Central role of carbohydrate metabolism in the supply of carbon skeletons for biosynthetic reactions. G6PDH, glucose-6-phosphate dehydrogenase; 6PGDH, 6-phosphogluconate dehydrogenase; PFK, ATP-dependent phosphofructokinase; FBPase, fructose-1,6-bisphosphatase; PFP, pyrophosphate-dependent phosphofructokinase.

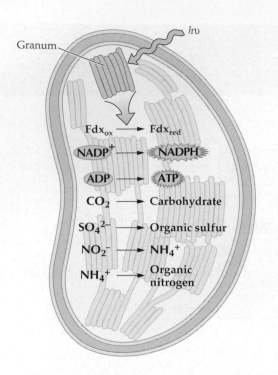

Figure 13.2
Photosynthetic electron transfer generates ATP and energy-rich reductants used to assimilate carbon, nitrogen, and sulfur. Fdx, ferredoxin.

reduction and ammonia assimilation (see Fig. 13.2 and Chapters 8 and 16). In addition, photosynthesis supplies many biosynthetic pathways with reducing equivalents and ATP. Because the processes associated with light absorption are diurnal, plants must cope with major variations in the supply of nutrients during the light and dark periods. This imposes the need for a flexibility in metabolism that is not seen in other organisms.

Plants are sessile. Terrestrial plants are unable to relocate when faced with physical or chemical stress and must tolerate changes in temperature; fluctuations in the availability of light, water, and nutrients; and herbivory. To ensure survival in unforgiving circumstances, plant metabolism is highly flexible (see Box 13.4 for discussion). In addition, plants expend considerable resources in the production of defense compounds (see Chapter 24).

Plants contain plastids. These organelles, found only in plants, perform most of the biosynthetic reactions that occur in plant cells. Duplication of some pathways in both cytosol and plastid, and the requirement for coordination of metabolism in all parts of the cell, have resulted in the evolution of several plastid envelope transporters (Fig. 13.3). The properties and functions of these transporters differ among various tissues and plastid types. Besides linking the cytosol and plastid compartments, these transporters

Plants are photosynthetic. Apart from a very few parasitic species, plants harness light and use its energy to fix and reduce carbon dioxide (see Chapter 12). The resulting triose phosphates produced can supply carbon to leaf cells or be converted to sucrose for export to other parts of the plant. Other activities associated with the capture of light energy include nitrite and sulfate

Figure 13.3
Transporters located in the inner envelope membrane of plastids exchange inorganic phosphate (P_i) for carbon metabolites. In the transport processes illustrated here, photosynthetic chloroplasts export C_3 intermediates, whereas the nonphotosynthetic plastids import a more diverse group of sugars and organic acids. All the transport reactions are reversible, and the direction of transport is dependent on the concentrations of P_i and metabolites on each side of the envelope. 3-PGA, 3-phosphoglycerate; 2-PGA, 2-phosphoglycerate.

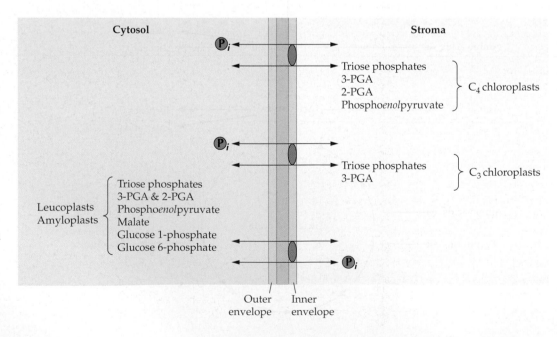

(A)

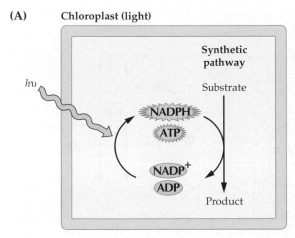

Chloroplast (light)

Synthetic pathway

Substrate

$h\upsilon$

NADPH

ATP

NADP$^+$

ADP

Product

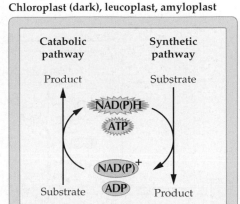

Chloroplast (dark), leucoplast, amyloplast

Catabolic pathway

Synthetic pathway

Product

Substrate

NAD(P)H

ATP

NAD(P)$^+$

ADP

Substrate

Product

(B)

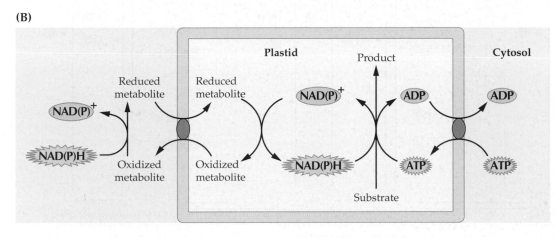

Plastid

Product

Cytosol

Reduced metabolite

Reduced metabolite

NAD(P)$^+$

ADP

ADP

NAD(P)$^+$

NAD(P)H

Oxidized metabolite

Oxidized metabolite

NAD(P)H

ATP

ATP

Substrate

Figure 13.4
Mechanism for supplying cofactors to plastids. Cofactors such as NAD(P)H or ATP are required for biosynthetic pathways. (A) Cofactors can be generated within the plastid by catabolic pathways located in the organelle, or by light energy ($h\upsilon$) in the case of the chloroplast. (B) A simple shuttle system for the transfer of reducing equivalents into a plastid. Cofactors may be imported from the cytosol either by direct transport (e.g., exchange of ADP for ATP) or via shuttles, as shown here.

also play important roles in metabolic regulation. The segregation of synthesis reactions into plastids presents another problem for the plant cell. Syntheses require an abundant supply of ATP and reducing power in the form of NAD(P)H, both of which must be generated within the plastid. Various mechanisms have evolved to satisfy this need (Fig. 13.4).

Plants have cell walls. Cell walls are composed of cellulose, hemicellulose, and lignin (see Chapters 2 and 24). Synthesis of these wall components represents a major drain on carbohydrates, because the pathways for the synthesis of wall components can account for 30% or more of cellular carbohydrate metabolism.

The topic of carbohydrate metabolism is typically discussed in terms of discrete anabolic and catabolic pathways. However, these pathways can be envisioned as a series of "pools" of metabolic intermediates linked by reversible enzyme reactions (Fig. 13.5).

Metabolites can be added to or withdrawn from these pools to serve the needs of the various pathways.

Two pools feed plant carbohydrate metabolism. One is composed of hexose phosphates, the other of pentose phosphate pathway intermediates and the triose

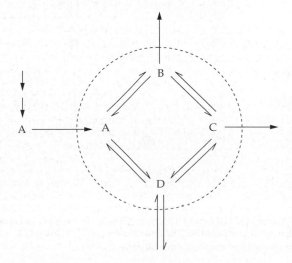

A

A

B

C

D

Figure 13.5
A metabolic pool, showing four metabolites at equilibrium. Flow through this pool will be dictated by the addition of specific metabolites (A), or withdrawal of intermediates (B and C). In many cases, the metabolic status of the cell will determine whether a particular compound (D) enters or leaves the pool.

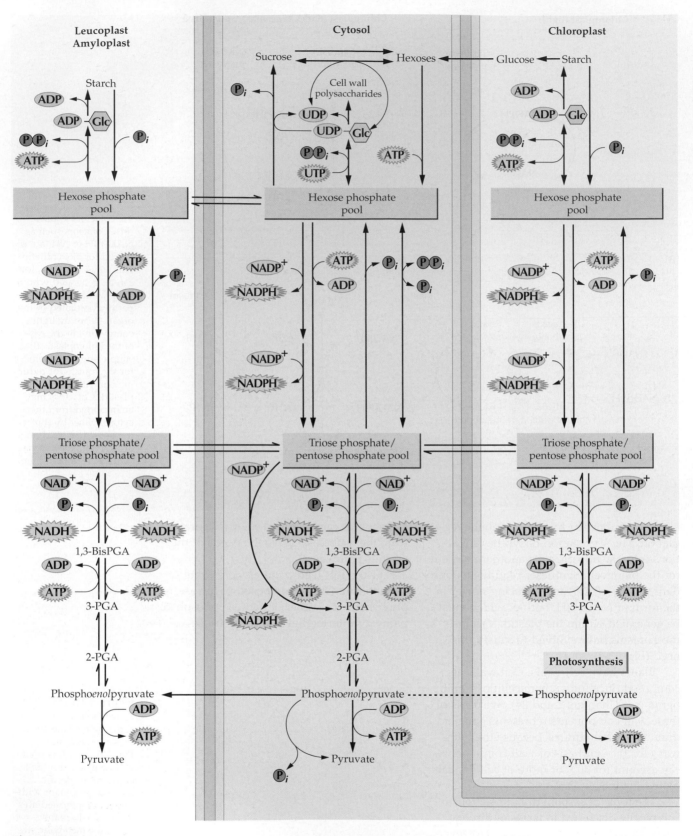

Figure 13.6

The use and synthesis of cofactors (NADH, NADPH, and ATP) in photosynthetic and nonphotosynthetic plastids. The contribution of photosynthetic electron transfer and photophosphorylation to the supply of cofactors in chloroplasts has not been included. 1,3-BisPGA, 1,3-bisphosphoglycerate.

phosphates glyceraldehyde 3-phosphate and dihydroxyacetone phosphate. The direction of flow through these pools depends on the requirements of the cell (Fig. 13.6). In photosynthetic tissues, carbon flux through pathways also is governed by whether cells are in light or darkness. Metabolite pools in the cytosolic and plastid compartments communicate by way of highly specific carrier proteins (see Fig. 13.3 and Chapter 3).

To highlight the function of metabolite pools, in this chapter we will frame some pathways differently from most biochemistry texts. For example, glycolysis in animals is typically described as the catabolism of either glucose or the storage polysaccharide glycogen to pyruvate. Starch metabolism in plants, however, is compartmentalized in plastids, and the complements of glycolytic enzymes present are different in the cytosol, chloroplasts, and heterotrophic plastids. When viewed as only one of many pathways linking the hexose phosphate and triose phosphate/pentose phosphate pools, glycolysis is revealed as a set of reactions contributing to an integrated whole plant metabolism rather than a discrete pathway to be memorized and quickly forgotten.

13.1 The hexose phosphate pool

13.1.1 Three interconvertible hexose phosphates make up the hexose phosphate pool.

The hexose phosphate pool consists of three metabolic intermediates: glucose 6-phosphate, glucose 1-phosphate, and fructose 6-phosphate (Fig. 13.7). The three metabolites are kept in equilibrium through the action of phosphoglucomutase and glucose-6-phosphate isomerase (hexose-phosphate isomerase). These reactions are readily reversible and apparently remain close to equilibrium in vivo.

Reaction 13.1: Phosphoglucomutase

$$\text{Glucose 1-P} \rightleftharpoons \text{Glucose 6-P}$$

Reaction 13.2: Glucose-6-P isomerase

$$\text{Glucose 6-P} \rightleftharpoons \text{Fructose 6-P}$$

Fructose 1,6-bisphosphate, another hexose phosphate present in plant cells, is not ordinarily considered part of the hexose phosphate pool, because most organisms interconvert fructose 1,6-bisphosphate and fructose 6-phosphate by means of irreversible regulatory reactions catalyzed by ATP-dependent phosphofructokinase (PFK) and fructose-1,6-bisphosphatase. In plants, however, these metabolites are linked by a reversible reaction catalyzed by pyrophosphate-dependent phosphofructokinase (PFP), an enzyme whose role in the hexose phosphate metabolism is not yet known (see Box 13.1 and Section 13.5.3).

Carbon can enter the hexose phosphate pool through gluconeogenesis (see Chapter 14), by phosphorylation of free hexoses, as the product of starch and sucrose degradation, and by way of the glycolytic pathway acting in reverse on the triose phosphate products of photosynthesis. Carbon can leave this pool through starch and sucrose synthesis, cell wall formation, and the oxidative reactions of the pentose phosphate pathway. A further major drain on this pool is the glycolytic pathway (Fig. 13.8).

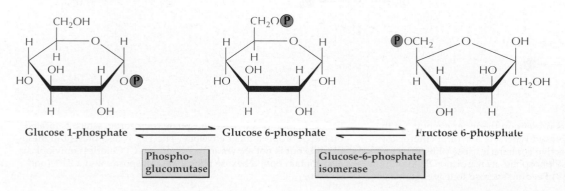

Figure 13.7
Intermediates of the hexose phosphate pool are interconverted by two enzymes, phosphoglucomutase and glucose-6-phosphate isomerase.

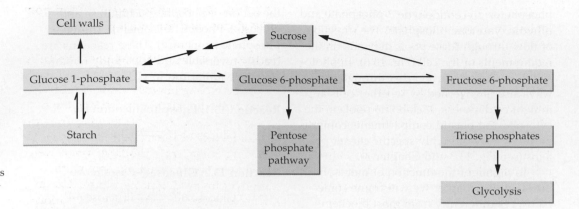

Figure 13.8
The hexose phosphate pool contributes intermediates to glycolysis as well as to many biosynthetic processes.

13.1.2 Most chloroplasts cannot transport hexose phosphates directly, but colorless plastids can.

Both the cytosol and plastids contain hexose phosphate pools. Leaf chloroplasts typically do not contain a hexose phosphate carrier. Instead, the chloroplast and cytosolic hexose phosphate pools normally communicate by way of the C_3 intermediates dihydroxyacetone phosphate and 3-phosphoglycerate. The chloroplast inner membrane contains a **triose phosphate translocator** (TPT), an antiporter (see Chapter 3) that exchanges these C_3 intermediates for inorganic phosphate (Fig. 13.9A). This counterexchange is strictly coupled because there is little transport of phosphate in the absence of an exchangeable substrate. Chloroplasts in some specialized tissues (e.g., developing fruit) can import hexose phosphates directly, but the major function of these chloroplasts is biosynthesis rather than photosynthesis. Expression of a chloroplast hexose phosphate translocator also can be induced by feeding leaves high concentrations of sucrose (Fig. 13.9B).

In some chloroplasts, for example, in specialized tissues and in C_4 plants, as well as in some nonphotosynthetic plastids, other translocators can be found that are similar to the TPT but are able to exchange phosphate for a wider array of substrates (see Fig. 13.3). The chloroplast envelope also contains a free hexose transporter, which has a much lower activity and higher K_m value than does the TPT. This carrier appears to be active in exporting the free hexoses generated by starch degradation in the dark.

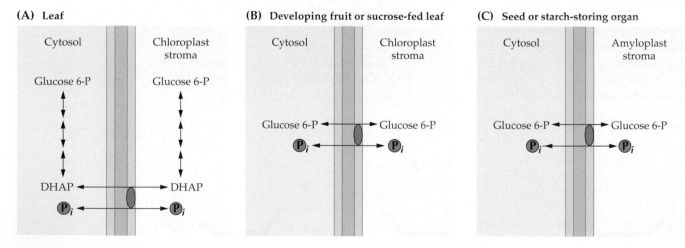

Figure 13.9
The metabolite transporters present on the inner envelope of the plastid vary according to plastid type. (A) A typical chloroplast that is photosynthetically active contains a triose phosphate translocator (TPT) but no hexose phosphate transporter. DHAP, dihydroxyacetone phosphate. (B) Feeding sucrose to leaves stimulates the expression of a hexose phosphate transporter, which is located on the chloroplast inner envelope. The TPT is also present in these plastids but is not shown in the diagram. (C) Colorless amyloplasts also contain both a hexose phosphate transporter and a TPT (not shown).

Unlike chloroplasts, the colorless amyloplasts involved in long-term starch storage can transport glucose 6-phosphate or glucose 1-phosphate, facilitating a direct interaction between the cytosolic and plastid hexose phosphate pools. In pea roots and cotyledons, the transported species is glucose 6-phosphate, whereas potato tuber amyloplasts appear to transport glucose 1-phosphate. In contrast, amyloplasts from monocotyledonous endosperm appear to transport both of these hexose phosphates as well as adenine diphosphate glucose (ADP-glucose) in some cases. This functional difference between chloroplasts and amyloplasts probably reflects the fact that chloroplasts serve as only a temporary storage compartment for recently fixed carbon (Fig. 13.9A).

13.2 Biosynthetic pathways that consume hexose phosphates: synthesis of sucrose and starch

13.2.1 Glucose 1-phosphate can be reversibly converted to UDP-glucose.

Uridine diphosphate glucose (UDP-glucose; Fig. 13.10) is a substrate in several anabolic reactions, including sucrose synthesis (see below) and cell wall synthesis (see Section 13.3.2). Glucose 1-phosphate can enter or leave the hexose phosphate pool through the action of UDP-glucose pyrophosphorylase:

Reaction 13.3: UDP-glucose pyrophosphorylase

$$\text{Glucose 1-P} + \text{UTP} \rightleftharpoons \text{UDP-glucose} + \text{PP}_i$$

The reaction catalyzed by UDP-glucose pyrophosphorylase is readily reversible, and the equilibrium in vivo depends on the concentration of pyrophosphate. The latter remains relatively constant in plant cells. In animals, pyrophosphatase activity is considered essential to drive biosynthetic reactions in the direction of synthesis through the removal of pyrophosphate. Plants do not contain a cytosolic pyrophosphatase and the pyrophosphate concentration can be as great as 0.3 mM in that compartment (Box 13.1).

13.2.2 Sucrose is synthesized in cytosol from UDP-glucose and fructose 6-phosphate.

Sucrose is a major product of photosynthesis in green leaves, accounting for much of the CO_2 fixed during photosynthesis. It also serves as the principal long-distance transport compound in most plants and as a storage compound in some (including sugar beet, sugar cane, and carrot). Sucrose is well suited to its transport role. By joining the carbonyl carbons of glucose and fructose in a stable glycosidic bond, sucrose formation protects these potentially reactive groups from oxidation. For this reason, sucrose is described as a nonreducing sugar (Fig. 13.11).

Because sucrose is a principal end product of photosynthesis, researchers assumed that sucrose synthesis took place in the chloroplast. However, numerous attempts to demonstrate plastid-localized sucrose synthesis proved unsuccessful. Instead, triose phosphates exported from the chloroplast by way of the TPT supply materials for hexose production and subsequent sucrose synthesis in the cytosol. In oil-rich tissues that lack vascular connectors to photosynthetic tissues (e.g., germinating oil seeds, pollen grains), the conversion of lipid to sugar provides precursors for sucrose synthesis (see Chapters 10 and 14). The principal route

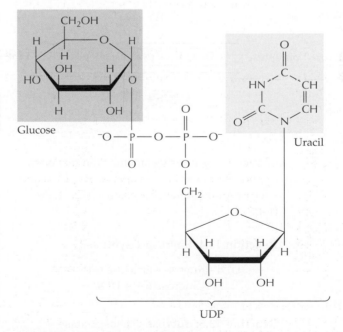

Figure 13.10
Structure of uridine diphosphate-glucose (UDP-glucose).

Box 13.1   **Pyrophosphate is a mysterious player in plant metabolism.**

Many students of biochemistry will recognize the following dogma about pyrophosphate: "Pyrophosphate never accumulates in cells because an active pyrophosphatase breaks it down as quickly as it forms." This assumption has a thermodynamic rationale: Pyrophosphate hydrolysis has a $\Delta G^{\circ\prime}$ of -33.4 kJ/mol and can therefore drive reactions that are otherwise energetically unfavorable, including many involved in biosynthesis.

Pyrophosphate is, indeed, generated in biosynthetic reactions, such as the synthesis of UDP-glucose, the substrate for glycogen synthesis in animals and cellulose synthesis in plants (see Rx.13.3). Other anabolic reactions, such as amino acid activation for protein synthesis, also release pyrophosphate. In animals and many microorganisms, hydrolysis of pyrophosphate by pyrophosphatase removes one of the products of these reactions, driving the equilibrium in the direction of synthesis. This is also the case in plant plastids, which contain an active pyrophosphatase. Removal of pyrophosphate drives the reversible ADP-glucose pyrophosphorylase

reaction in the direction of ADP-glucose synthesis; see (A) below.

The investigators who discovered pyrophosphate in plant cells traveled a circuitous route. No one initially looked for it because it was not expected to be there. However, isolation of an abundant pyrophosphate-dependent phosphofructokinase from plants convinced researchers to look for pyrophosphate. To general surprise they found substantial concentrations (up to 0.3 mM) in the plant cytosol, which contains no pyrophosphatase. Cytosol-localized synthesis of sucrose and cellulose consumes UDP-glucose. UDP-glucose pyrophosphorylase equilibrates UDP-glucose and its precursor, glucose 1-phosphate; see (B) below.

Overexpression of a cytosol-localized bacterial pyrophosphatase in transgenic tobacco inhibits growth, demonstrating that plants require cytosolic pyrophosphate. In all plant cells, the cytosolic pyrophosphate concentration remains remarkably stable under a variety of conditions. Levels are maintained even in tissues where cytosolic biosynthesis does not account for adequate

pyrophosphate production, so some elusive mechanism must regulate pyrophosphate synthesis and degradation. It has been proposed that pyrophosphate-dependent phosphofructokinase, acting in the reverse direction (fructose 6-phosphate synthesis), could supply pyrophosphate. But transgenic plants with very low levels of this enzyme can sustain steady levels of cytosolic pyrophosphate, suggesting that other mechanisms are involved.

Another player in pyrophosphate metabolism is a proton-pumping pyrophosphatase located in the tonoplast (see Chapter 3). Energy released by pyrophosphate hydrolysis drives the proton pump, acidifying the vacuole and generating an electrochemical gradient used to power membrane transport. This H^+-pyrophosphatase may participate in controlling cytoplasmic pyrophosphate levels, but its specific role has not been defined.

For now, pyrophosphate metabolism in plants is enigmatic. We know neither why nor how plant cells maintain a stable pool of cytosolic pyrophosphate.

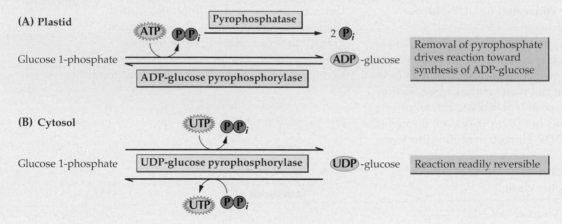

(A) Plastid

(B) Cytosol

of sucrose synthesis combines the reactions of sucrose-phosphate synthase (Rx. 13.4) and sucrose-phosphate phosphatase (Rx. 13.5) (Fig. 13.12).

Reaction 13.4: Sucrose-P synthase

$$\text{UDP-glucose} + \text{fructose 6-P} \rightleftharpoons \text{Sucrose 6-P} + \text{UDP}$$

Reaction 13.5: Sucrose-P phosphatase

$$\text{Sucrose 6-P} + H_2O \longrightarrow \text{Sucrose} + P_i$$

The use of fructose 6-phosphate as substrate results in the synthesis of sucrose 6-phosphate. The hydrolysis of sucrose 6-phosphate to sucrose plus phosphate has a large negative free energy change ($\Delta G^{\circ\prime} = -16.5$ kJ/mol). The formation of UDP-glucose from glucose 1-phosphate and UTP has a small negative free energy change ($\Delta G^{\circ\prime} = -2.88$ kJ/mol), as has the formation of sucrose 6-phosphate by sucrose-phosphate synthase ($\Delta G^{\circ\prime} = -5.7$ kJ/mol). However, overall, the formation of sucrose from glucose 1-phosphate by this route has a large negative free energy change, strongly favoring the formation of sucrose ($\Delta G^{o\prime}$ = –25 kJ/mol). Consequently, this sequence of reactions is essentially irreversible.

Another enzyme, sucrose synthase (see Rx. 13.9), is capable of catalyzing both sucrose synthesis and degradation. However, the concentration of sucrose-phosphate synthase is high in tissues that synthesize sucrose, indicating that it is the dominant enzyme for sucrose biosynthesis. In contrast, sucrose synthase is present in higher concentrations in sucrose-utilizing tissues, such as developing seeds, than in sucrose-exporting tissues, such as photosynthetic leaves and germinating oil seeds. The principal role of sucrose synthase is thus thought to be sucrose catabolism (see Sections 13.3.1 and 13.3.2).

Sucrose synthesis is regulated by the activity of sucrose-phosphate synthase. Covalent modification and allosteric modulation of this enzyme combine to provide a very sensitive control of sucrose synthesis that is related to the status of the hexose phosphate pool (Fig. 13.13). When hexose phosphates are abundant, sucrose-phosphate synthase is directly activated by glucose 6-phosphate. Additionally, a kinase that phosphorylates and thereby down-regulates sucrose-phosphate synthase is itself inhibited by glucose 6-phosphate. Recent evidence indicates that the phosphorylation-dependent down-regulation of sucrose-phosphate synthase is enhanced by 14-3-3 proteins in a manner similar to regulation of nitrate reductase (see Chapter 16). High concentrations of cytosolic phosphate inhibit both sucrose-phosphate synthase and a phosphatase that up-regulates the biosynthetic enzyme.

Figure 13.11
Some sugars can act as reductants. (A) Glucose undergoes an interconversion at the anomeric carbon (C-1) that yields an aldehyde (D-glucose) and two hemiacetal isomers (α-D-glucopyranose and β-D-glucopyranose). The aldehyde group can reduce many inorganic substrates and is therefore described as the reducing end of the molecule. (B) In the disaccharide maltose, one of the anomeric carbons of glucose remains unbound and reactive. For this reason, maltose is a reducing sugar. (C) During sucrose synthesis, the anomeric carbon of fructose (C-2) is joined to the anomeric carbon of glucose by a hydrolytic bond. This bond protects the reducing ends of both monomers and defines sucrose as a nonreducing sugar.

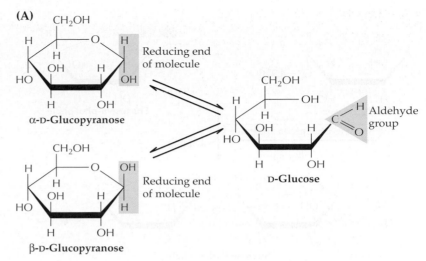

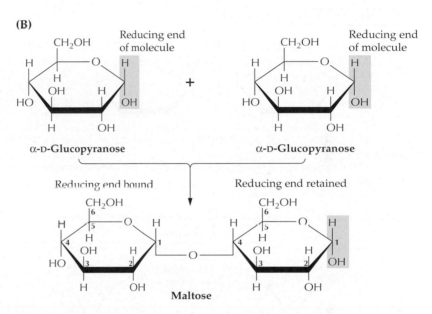

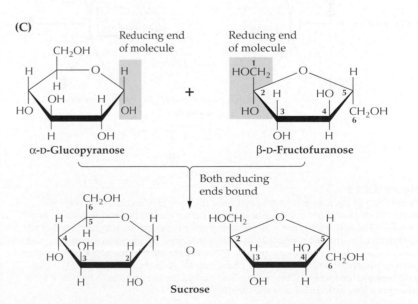

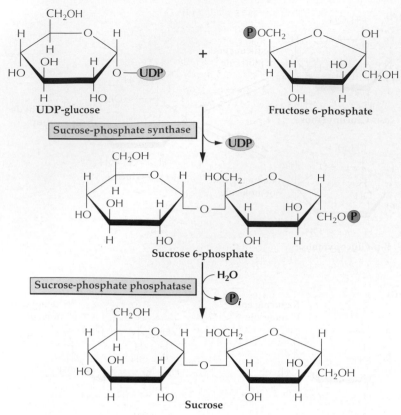

Figure 13.12
The synthesis of sucrose by sucrose-phosphate synthase and sucrose-phosphate phosphatase.

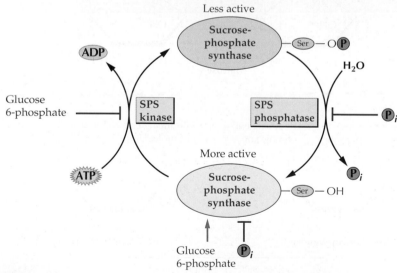

Figure 13.13
Regulation of sucrose-phosphate synthase (SPS). SPS is regulated by both allosteric effectors and phosphorylation/dephosphorylation of a serine residue in the enzyme. The active enzyme is dephosphorylated and can be activated further, allosterically, by glucose 6-phosphate. SPS is inhibited by inorganic phosphate and, possibly, sucrose. In addition, glucose 6-phosphate inhibits the SPS kinase, and inorganic phosphate inhibits the SPS phosphatase. As a result, a high ratio of glucose 6-phosphate to inorganic phosphate maintains the enzyme in its active form, whereas a low ratio promotes the inactive form.

13.2.3 Starch synthesis occurs in plastids.

Starch, a polymer of glucose, is synthesized and stored in plastids—temporarily in chloroplasts or for longer periods in the amyloplasts of storage tissues such as tubers or seeds. By linking thousands of glucose monomers into a small number of molecules, starch synthesis protects plastids from osmotic disruption. If a comparable number of hexose units were stored in the plastid as sucrose, the stromal solution would contain too many solute particles, and water from the cytosol would flood the plastid, causing it to swell and burst.

The pathways for sucrose and starch synthesis are segregated in the cytosol and the plastids, respectively, and are fed by different hexose phosphate pools. Starch synthesis in plants begins with ADP-glucose. In animals, the storage compound analogous to starch is glycogen, which is synthesized in the cytosol from UDP-glucose. In bacteria, however, glycogen is synthesized from ADP-glucose. The use of ADP-glucose for the synthesis of storage polysaccharides in both bacteria and plastids supports the endosymbiont hypothesis for the origin of these organelles (see Chapter 6).

In the chloroplast, the Calvin cycle (see Chapter 12) can feed carbon into the starch synthesis pathway (Fig. 13.14). This occurs when the export of sucrose from the cell cannot keep pace with photosynthesis. The excess fixed carbon enters the hexose phosphate pool and is finally stored in the chloroplast as starch.

In amyloplasts, carbon for starch synthesis must be imported from the cytosol. At first, this carbon was assumed to enter through the triose phosphate transporter, but elegant labeling experiments showed that this could not be the case. These experiments demonstrated that in wheat seeds incubated with glucose radiolabeled at a single carbon (C-1 or C-6), the starch that was formed retained most of the label in the same position. If the carbon had entered the amyloplast by way of the TPT, the label would have been more extensively randomized between these two positions (Fig. 13.15). Instead, the results indicated that intact hexose molecules were taken into the plastid directly.

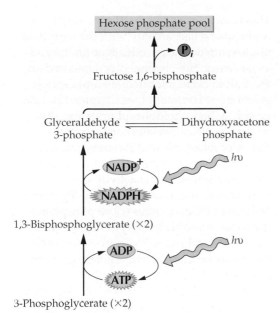

Figure 13.14
Relationship between photosynthesis and the chloroplast hexose phosphate pool. In chloroplasts, the hexose phosphate pool has a source of carbon not available to colorless plastids. The light reactions of photosynthesis can convert 3-phosphoglycerate, the first product of CO_2 fixation, into triose phosphates. When these are not exported to the cytosol, they can be converted to fructose 1,6-bisphosphate and enter the hexose phosphate pool through the action of fructose-1,6-bisphosphatase.

Another assumption was that ADP-glucose was synthesized only in plastids, but a recent report has indicated that in maize endosperm, the major activity of ADP-glucose pyrophosphorylase (see next section) is in the cytosol. If correct, this means that ADP-glucose could be synthesized in the cytosol and imported into plastids. A transporter that exchanges ADP-glucose for ADP has been found in envelopes of amyloplasts from barley. This process may be a general feature of starch synthesis in cereals.

13.2.4 Starch synthesis is regulated by ADP-glucose pyrophosphorylase.

The enzyme UDP-glucose pyrophosphorylase would be poorly suited for regulating sucrose synthesis; such regulation would interfere with the other metabolic pathways for which UDP-glucose is a substrate. By contrast, all ADP-glucose is directed into starch synthesis, and ADP-glucose pyrophosphorylase (Rx. 13.6) is the major regulatory enzyme in the starch biosynthetic pathway.

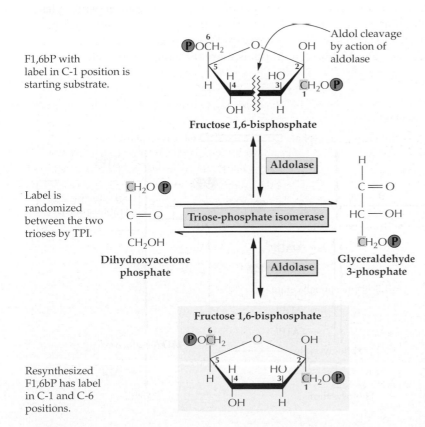

F1,6bP with label in C-1 position is starting substrate.

Label is randomized between the two trioses by TPI.

Resynthesized F1,6bP has label in C-1 and C-6 positions.

Figure 13.15
Randomization of label by triose-phosphate isomerase (TPI). Aldolase cleaves fructose 1,6-bisphosphate (F1,6bP) labeled at the C-1 position, yielding labeled dihydroxyacetone phosphate (DHAP) and unlabeled glyceraldehyde 3-phosphate (GAP). However, label is immediately randomized between GAP and DHAP by triose-phosphate isomerase. Resynthesis of fructose 1,6-bisphosphate after label randomization yields products labeled at both C-1 and C-6. The starch synthesized by wheat grains fed C-1-labeled glucose is labeled predominantly at the C-1 position, which suggests that this carbon probably enters the amyloplast as hexose phosphate. If the carbon entered the plastid as triose phosphate, the label would be randomized by TPI to yield starch containing glucose residues labeled at both C-1 and C-6.

Reaction 13.6: ADP-glucose pyrophosphorylase

$$\text{Glucose 1-P} + \text{ATP} \rightleftharpoons \text{ADP-glucose} + \text{PP}_i$$

ADP-glucose pyrophosphorylase is a heterotetramer of two large and two small subunits. As revealed by in vitro expression of these subunits, the small subunit is sufficient for catalytic activity, but it has a lower affinity than the native enzyme for the activator 3-phosphoglycerate (see below). Multiple isoenzymes of both subunits of ADP-glucose pyrophosphorylase exist, which may be expressed on a tissue-specific basis or under different environmental conditions. At present, however, we do not understand the function of these isoenzymes. In barley and maize, which appear to have both plastid and cytosolic forms of the enzyme, ADP-glucose is now thought to be synthesized in the cytosol at later stages of seed development and transported into the amyloplast in exchange for ADP.

In both photosynthetic and heterotrophic tissues, ADP-glucose pyrophosphorylase is activated by 3-phosphoglycerate and inhibited by inorganic phosphate (P_i). However, the availability of these compounds varies according to metabolic status and plastid type. The P_i/3-phosphoglycerate ratio is the principal regulator of starch synthesis, at least in green tissues (Fig. 13.16).

This mechanism links ADP-glucose pyrophosphorylase activity with the supply of photosynthate (i.e., products of photosynthesis), because the enzyme is activated when the Calvin cycle intermediate 3-phosphoglycerate is abundant (see Chapter 12). Conversely, when photophosphorylation slows or ceases, the phosphate concentrations in the chloroplast rise and starch synthesis is repressed. The concentration of plastid P_i is also determined by the activity of the TPT. In the dark, and sometimes in the light as well, the TPT exchanges triose phosphates from the chloroplast for cytosolic P_i, increasing the P_i/3-phosphoglycerate ratio within the organelle and thereby inhibiting starch synthesis.

In amyloplasts and other colorless plastids, the rationale for ADP-glucose pyrophosphorylase regulation is less intuitive. The inhibition of this enzyme by phosphate is understandable, because nonphotosynthetic plastids synthesize starch from glucose phosphate, which is imported into the plastid by a hexose phosphate/P_i antiporter. High concentrations of inorganic phosphate in the plastid, therefore, indicate that cytosolic hexose phosphates are scarce (Fig. 13.17). Given these substrate-limiting conditions, inhibition of ADP-glucose pyrophosphorylase by P_i seems appropriate. Activation of this enzyme by 3-phosphoglycerate in amyloplasts is more difficult to comprehend.

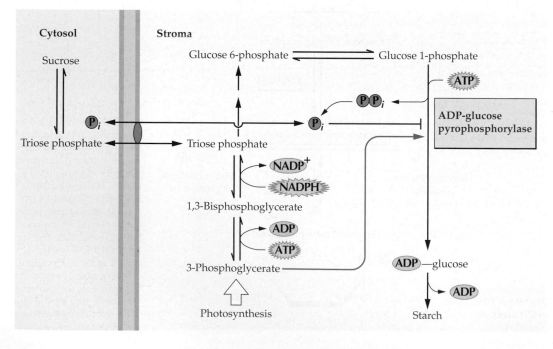

Figure 13.16
Regulation of starch synthesis in chloroplasts. When 3-phosphoglycerate is abundant, starch synthesis is activated. Inorganic phosphate, an indicator of the status of the triose phosphate pool, inhibits starch synthesis.

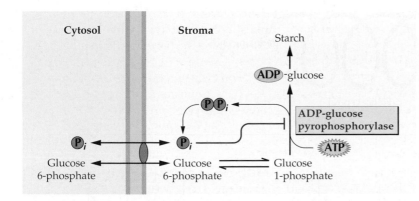

Figure 13.17

Starch synthesis in amyloplasts. In contrast to chloroplasts (see Fig. 13.16), amyloplasts can import hexose phosphates and incorporate them directly into starch. The hexose phosphate/P_i antiporter is shown. As in chloroplasts, ADP-glucose pyrophosphorylase is inhibited by phosphate. Starch breakdown is stimulated by high phosphate concentrations in part because it serves as a substrate for starch-degrading enzymes (see Fig. 13.25).

The importance of ADP-glucose pyrophosphorylase in the regulation of starch biosynthesis has been demonstrated by the expression of an unregulated bacterial ADP-glucose pyrophosphorylase in potato tubers. The presence of this enzyme boosted the amount of starch by as much as 60%, suggesting that in wild-type plants this enzyme catalyzes the major regulatory step in the pathway and that the bacterial enzyme is able to bypass this regulation. In storage tissues, enzymes other than ADP-glucose pyrophosphorylase also may limit starch formation.

13.2.5 Amylose and amylopectin, two distinct types of starch molecules, have different branching patterns.

Starch consists of two molecules: amylose and amylopectin (Fig. 13.18). In many cases, amylose accounts for approximately 30% of the total starch. However, the percentage of amylose depends on the species and the organ used for starch storage. The proportion of amylose to amylopectin and the size and structure of the starch grain give distinctive properties to different extracted starches, properties important in food and industrial uses (Box 13.2). Because glucose residues are not symmetrical, the ends of starch molecules differ in reactivity (see Fig 13.11).

Starch is organized into grains which range in size from less than 1 µm in diameter to greater than 100 µm and contain both amylose and amylopectin. Starch grain size varies by plant and tissue type, with the grains in the leaves tending toward the small end of the range. Starch grains grow by adding layers, and growth rings within the grain represent areas of faster and slower growth.

Three enzymes are responsible for the synthesis of starch from hexose phosphates: ADP-glucose pyrophosphorylase (see Rx. 13.6), starch synthase (Rx. 13.7), and starch-branching enzyme (Rx. 13.8). In addition, a plastid-specific isoenzyme of pyrophosphatase cleaves the pyrophosphate generated by the first of these reactions, driving ADP-glucose synthesis (Fig. 13.19; see also Box 13.1).

Reaction 13.7: Starch synthase

$$\text{ADP-glucose} + \alpha\text{-glucan}_{(n)} \rightleftharpoons$$
$$\alpha\text{-Glucan}_{(n+1)} + \text{ADP}$$

Reaction 13.8: Starch-branching enzyme

$$\text{Linear } (1{\rightarrow}4)\alpha\text{-glucan} \rightleftharpoons$$
$$\text{Branched } (1{\rightarrow}6),(1{\rightarrow}4)\alpha\text{-glucan}$$

Starch synthase adds individual glucose molecules from ADP-glucose to the nonreducing end of a preexisting amylose or amylopectin chain, forming a $(1{\rightarrow}4)\alpha$-bond (see Fig. 13.18). Starch synthase has a number of isoenzymes. Some are found in the soluble stromal phase of the plastid; others are bound within the growing starch granules. Researchers initially thought that only one starch synthase existed and that some of the enzyme molecules became engulfed in growing starch grains. However, we now know that single isoenzyme mutants can synthesize an altered form of starch, indicating that the isoenzymes have distinct biochemical functions. For example, the waxy mutants found in several plant species are deficient in granule-bound starch synthase

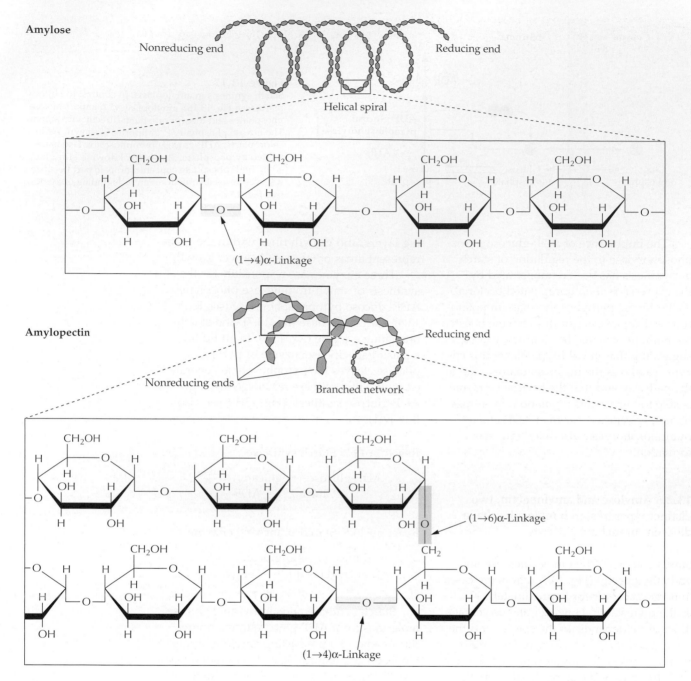

Figure 13.18

Structures of amylose and amylopectin. Starch contains α-amylose, long, unbranched chains of (1→4)α-linked glucose units, and amylopectin, which is branched by (1→6)α-linkages. Branch points in amylopectin are separated by an average of 20–30 glucose residues. Diagrammatic representations of the two molecules are shown.

and produce starch that does not contain amylose (Box 13.2). This indicates that amylose formation is catalyzed by bound starch synthase, whereas the (1→4)α-glucans generated by soluble starch synthase are subsequently converted by the branching enzyme to amylopectin. The embedding of granule-bound starch synthase may be a prerequisite for the production of amylose as an end product. Multiple isoenzymes of soluble starch synthase exist, but their specific functions are not known. One granule-bound starch synthase is prominent; the functions of the other minor isoenzymes found have not been elucidated.

The (1→6)α-linkages in starch are produced by starch-branching enzyme (see Fig. 13.19). This enzyme, also called amylo-(1→4),

The structure of starch affects the texture and baking properties of the food we eat. Starches from various plants react differently when they are heated and therefore can be used for different purposes.

Amylopectin molecules in starch grains are crystalline rather than amorphous, primarily because of hydrogen-bonding between the hydroxyl groups of the glucose residues of adjacent chains in the amylopectin. When mixed with water, starch grains form a suspension. Upon heating, the grains swell as they absorb water, thickening the suspension. As the temperature increases, amylose separates from amylopectin; on subsequent cooling, the amylose molecules form hydrogen bonds with other amylose molecules and with amylopectin. Water becomes trapped in a network of amylose and amylopectin, resulting in a gelatinous mass.

This coagulated starch can taste better than its description sounds; mixed with milk, sugar, and flavorings, it becomes pudding. When pudding is frozen or cooled for long periods, the amylose molecules associate more closely, forming dense crystals. The resulting watery, unappetizing mess can be rescued by reheating and cooling, which reestablishes the amylose–amylopectin gel. To make instant pudding, a manufacturer spray-dries the hot starch gel before the amylose molecules can associate. When milk is added to the resulting powder, the amylose and amylopectin molecules congeal.

Ordinarily, the same disastrous changes would occur in the starchy filling of frozen pies. To overcome this, the pie manufacturer uses a substance called waxy starch, found in the waxy mutant of maize (see Chapter 7). This mutant lacks granule-bound starch synthase and so produces only amylopectin. With no amylose to crystallize out of the gel, the pie can be frozen and thawed without loss of texture. A similar change in amylose structure occurs in day-old bread. Bread loses its texture and becomes stale with age, not because it loses moisture or is moldy, but because the amylose molecules—which were evenly dispersed when the bread was baked—crystallize. Stale bread can be revived by dampening and heating to redisperse the amylose, which may in part explain why restaurants often serve bread warm.

(1→6)-transglycosylase, cleaves a (1→4)α-linkage, then joins the reducing end of the severed fragment to the C-6 of a glucose approximately 20 residues downstream from the end of the parent chain, forming a (1→6)α-linkage (see Fig 13.18). As a result, side chains in starch molecules are not randomly attached, but are usually separated by approximately 20 glucose residues. Two isoforms of branching enzyme, I and II (formerly termed B and A, respectively), differ in specificity. The I form has a higher affinity for unbranched starch (amylose), whereas the II form preferentially branches amylopectin. Isoform II also transfers shorter glucan chains than does I and, in this way, produces a more branched form of amylopectin. Keep in mind, however, that most of these data come from in vitro studies; the role of these enzymes in vivo is not yet fully understood. Developmentally regulated expression of starch-branching enzymes I and II may result in distinct starch grain morphologies. In developing pea embryos, the II form is expressed at early stages of development and the I form later. In concert with this expression pattern, the branch length of the amylopectin in the starch grains of the seed increases during development. Probably both isoforms are required for complete development of the starch grain, given the inhibition of starch synthesis in mutants lacking the II isoform (Box 13.3).

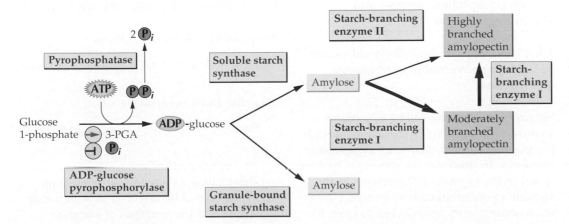

Figure 13.19
Synthesis of amylose and amylopectin. Regulation by 3-phosphoglycerate and inorganic phosphate is indicated. The two branching enzymes have been shown to catalyze independent reactions but probably act in concert to produce branched amylopectin. 3-PGA, 3-phosphoglycerate.

Box 13.3 Mendel's round and wrinkled peas reflect different alleles of a starch-branching enzyme.

A century before the structure of DNA was discovered, an Austrian monk named Gregor Mendel bred peas in the monastery garden in the town of Brno. Mendel crossed round peas with wrinkled peas and followed these traits through a series of generations, eventually coming to the conclusion that a pair of factors governed the appearance of the seeds and that these factors, now called genes, were passed from parent to offspring. The results of these crosses were published by Mendel in 1865 and were the initiation of the discipline that came to be known as classical genetics.

In 1990, modern molecular biology revealed the basis of the round and wrinkled pea phenotypes. The mutation that Mendel had followed was described as a transposon-like insertion into the gene encoding the II-isoform of starch-branching enzyme. In these mutants, this loss of function is not fully compensated for by the presence of isoform I, and overall starch production falls by about 30%. Carbon that would have been stored as starch accumulates as sucrose that sweetens the flavor of the pea seeds. The excess sucrose also lowers the osmotic potential of the developing seed, causing it to swell with fluid. On drying, the loss of the additional water results in the seed having a wrinkled appearance. The pod illustrated at right contains both round and wrinkled peas. Note the 3:1 ratio of dominant (round) to recessive (wrinkled) phenotype.

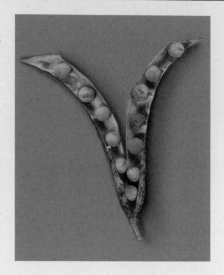

At first glance, starch biosynthesis appears to require just three enzymes: ADP-glucose pyrophosphorylase, starch synthase, and starch-branching enzyme. However, the various isoforms of these enzymes can combine to create starches with very different properties, some of which prove useful to industry—for example, starches used in paper, fiber boards, paint, packaging, bioplastics, and various foods. The manipulation of these isoenzymes by genetic modification may lead to the production of specific starch molecules designed for distinct industrial uses.

13.3 Catabolic pathways that generate hexose phosphates: sucrose and starch degradation

13.3.1 Sucrose can be hydrolyzed to free hexoses or converted to UDP-glucose and fructose.

Sucrose can enter the cell by two routes. In most cells, it enters by diffusion through the plasmodesmata, that is, by symplasmic transport. In this case, the sucrose is broken down in the cytosol or sequestered in the vacuole to maintain passive import. However, in some tissues, sucrose has to cross the plasma membrane to enter the cell, often against a concentration gradient. This apoplastic pathway is particularly important to developing embryos, which have no direct connection to tissues of the maternal plant. The sucrose can be either transported into these cells unchanged or converted to glucose and fructose by an invertase located in the cell wall. The significance of the various means of sucrose transport and degradation was recently underscored when researchers found that carbohydrates play a role in the expression of many genes and that the specific form of carbohydrate is important in this regulation.

Sucrose can be degraded by both sucrose synthase (Fig. 13.20) and invertase (Fig. 13.21), as in the following reactions:

Reaction 13.9: Sucrose synthase

$$\text{Sucrose} + \text{UDP} \rightleftharpoons \text{UDP-glucose} + \text{fructose}$$

Reaction 13.10: Invertase

$$\text{Sucrose} + H_2O \longrightarrow \text{Glucose} + \text{fructose}$$

As mentioned in Section 13.2.2, sucrose synthase catalyzes a readily reversible reaction and could be involved in both the degradation and the synthesis of sucrose.

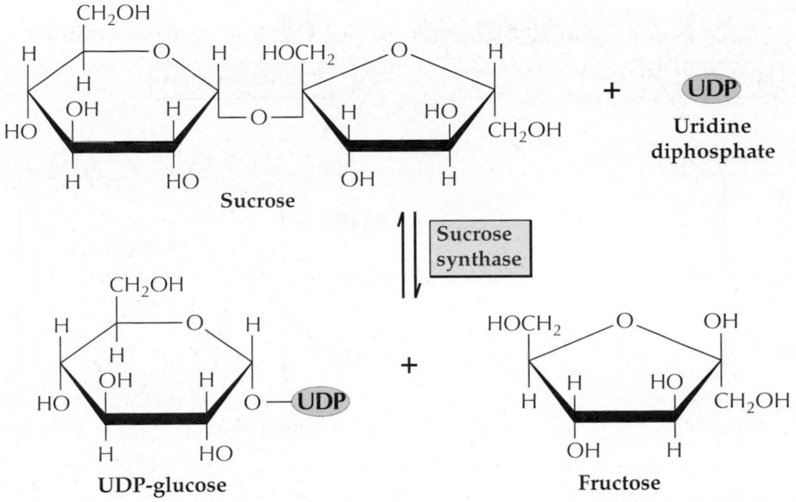

Figure 13.20
Sucrose synthase catalyzes a reversible reaction that can synthesize or degrade sucrose, but in plant cells this enzyme is associated primarily with sucrose degradation.

However, higher concentrations of this enzyme are present in tissues that degrade sucrose than in tissues that synthesize it, suggesting a largely catabolic function.

By contrast, the reaction catalyzed by invertase is irreversible, leading only to sucrose degradation. There are two forms of invertase, one with an acid pH optimum, the other being most active under alkaline conditions. These forms function in different cellular compartments: alkaline invertase in the cytosol and acid invertase in both the vacuole and the cell wall. Sucrose stored in the vacuole is probably hydrolyzed within this compartment, the resulting free hexoses being subsequently transported to the cytosol. The cell-wall form of invertase is firmly bound to the matrix components of the wall.

The precise roles of invertase and sucrose synthase in sucrose breakdown are not fully understood, since both often appear in the same tissue. In potato tubers, the amount of invertase is too low to account for the required rate of sucrose breakdown, so utilization of sucrose synthase can be inferred. However, such evidence is not available for most tissues. In addition, both sucrose synthase and invertase occur as multiple isoenzymes, each of which probably functions in specific tissues. Expression of the various genes encoding these isoenzymes is tightly controlled.

A significant difference between the two sucrose-degrading enzymes lies in the energy status of the products and their relation to the cytosolic hexose phosphate pool. The invertase reaction produces free hexoses that can be phosphorylated only at the expense of ATP. By contrast, sucrose synthase generates UDP-glucose residues, which can react with pyrophosphate to produce glucose

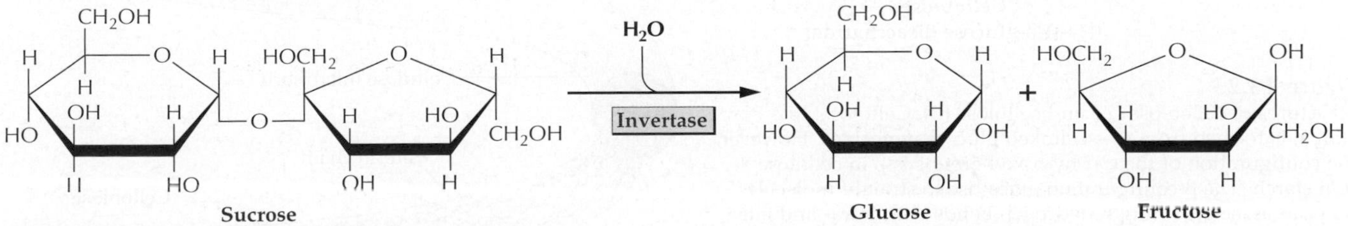

Figure 13.21
Degradation of sucrose by invertase.

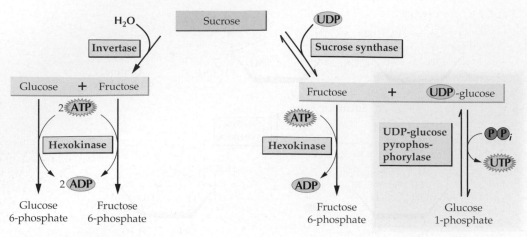

Figure 13.22
Hexokinase uses ATP to phosphorylate glucose and fructose, the products of invertase. Sucrose synthase and UDP-glucose pyrophosphorylase act together to generate hexose phosphate by an ATP-independent pathway, here highlighted in yellow.

1-phosphate and uridine triphosphate (UTP). Thus, sucrose synthase can combine with UDP-glucose pyrophosphorylase (see Rx. 13.3) to provide an ATP-independent pathway for hexose phosphorylation (Fig. 13.22).

13.3.2 Sucrose degradation can generate substrate for cell wall biosynthesis.

Cell walls make up a major part of the dry weight of plants, and at times of rapid growth, their synthesis may be a major drain on the resources of the cell. Cell walls contain two types of carbohydrate-derived components, polysaccharides and lignins. The biosynthesis of polysaccharides is closely linked to the hexose phosphate pool,

whereas the lignins derive from erythrose 4-phosphate and phospho*enol*pyruvate (see Chapter 24; see also Fig. 13.1).

Cellulose is a (1→4)β-linked polyglucan with a chain length of 2,000 to 20,000 glucose units (Fig. 13.23). Whereas the (1→4)α-linked glucose polymer amylose has a helical secondary structure, the β-linkage rotates each residue 180° relative to its neighbors. The alternating chain assumes a linear configuration. Approximately 36 parallel cellulose molecules can associate to form a crystalline fibril held together by intra- and interchain hydrogen bonds between the hydroxyl groups of the glucose residues.

Cellulose is synthesized by an enzyme complex associated with the plasma membrane. The substrate for cellulose synthesis,

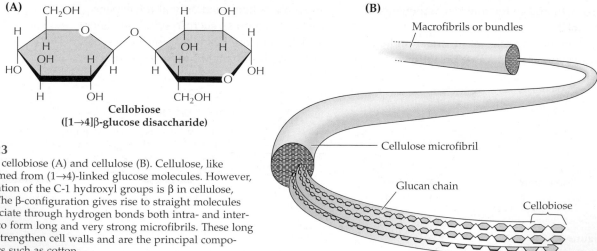

Figure 13.23
Structures of cellobiose (A) and cellulose (B). Cellulose, like starch, is formed from (1→4)-linked glucose molecules. However, the configuration of the C-1 hydroxyl groups is β in cellulose, α in starch. The β-configuration gives rise to straight molecules that can associate through hydrogen bonds both intra- and inter-molecularly to form long and very strong microfibrils. These long microfibrils strengthen cell walls and are the principal components of fibers such as cotton.

UDP-glucose, can be formed from hexose phosphates by UDP-glucose pyrophosphorylase (see Rx. 13.3) and from sucrose by sucrose synthase (see Rx. 13.9). In one proposed model, cellulose synthase spans the plasma membrane, using cytosolic UDP-glucose as a precursor for extracellular cellulose synthesis. In another, a plasma-membrane–bound isoenzyme of sucrose synthase forms a complex with cellulose synthase, channeling UDP-glucose from sucrose catabolism directly into cell wall synthesis (Fig. 13.24). If the latter hypothesis is correct, rapidly growing young tissues would require imported sucrose for cellulose biosynthesis. The amount of sucrose synthase associated with cellulose synthase in vivo is not known.

The cell wall also contains noncellulose polysaccharides. These polymers consist of hexoses, pentoses, and uronic acids. The precursor of all these polysaccharides is, again, UDP-glucose. Unlike cellulose, these polysaccharides are formed in the Golgi apparatus and are exported to the external surface of the plasma membrane in Golgi vesicles. Thus, cell wall synthesis requires that both the cytosol and the Golgi lumen contain supplies of UDP-glucose and its derivatives as well as a system for transporting sugar nucleotides into the Golgi apparatus.

13.3.3 Phosphorylytic starch degradation may be controlled by availability of inorganic phosphate.

The mechanisms that execute and regulate starch degradation remain unclear. The enzymes that degrade starch participate in either phosphorolytic or hydrolytic cleavage reactions. One hypothesis, which is not universally accepted, holds that starch phosphorylase cannot degrade intact starch grains and that the initial degradation depends on hydrolytic endoamylase activity.

At least three enzymes—starch phosphorylase, debranching enzyme, and glucosyltransferase—contribute to phosphorolytic starch degradation:

Reaction 13.11: Starch phosphorylase

$$\alpha\text{-Glucan}_{(n)} + P_i \rightleftharpoons$$
$$\alpha\text{-Glucan}_{(n-1)} + \text{glucose 1-P}$$

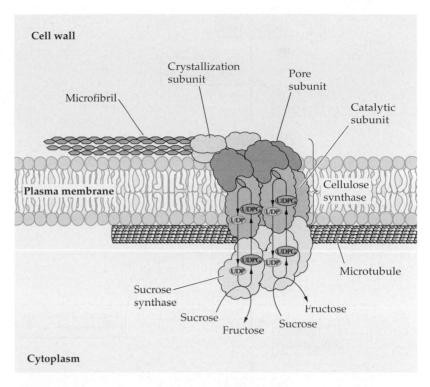

Figure 13.24
Model of cellulose synthase complex. Cellulose is formed outside the cell by the addition of glucose units to the growing cellulose molecule. The glucose units for this synthesis come from UDP-glucose (UDPG), which might be supplied by sucrose synthase associated with the complex that synthesizes cellulose. This would solve the problem of transferring UDP-glucose across the plasma membrane of the cell.

Reaction 13.12: Debranching enzyme (R-enzyme)

$$\text{Branched } (1{\rightarrow}6),(1{\rightarrow}4)\alpha\text{-glucan} \longrightarrow$$
$$\text{Linear } (1{\rightarrow}4)\alpha\text{-glucans}$$

Reaction 13.13: Glucosyltransferase (D-enzyme)

$$\alpha\text{-Glucan}_{(m)} + \alpha\text{-glucan}_{(n)} \rightleftharpoons$$
$$\alpha\text{-Glucan}_{(m+n-1)} + \text{glucose}$$

Starch phosphorylase cleaves individual glucose residues from the nonreducing end of the starch molecule, generating glucose 1-phosphate. Starch phosphorylase can attack only bonds located at least four glucose residues from a branch point. Continued phosphorolytic attack on the starch molecule is made possible by a debranching enzyme (also called R-enzyme or pullulanase) that cleaves the $(1{\rightarrow}6)\alpha$-bonds, releasing linear chains on which the starch phosphorylase can act. In addition, the D-enzyme, or

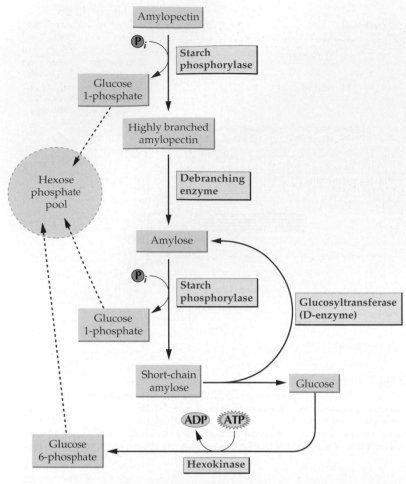

Figure 13.25
Phosphorolytic starch degradation. Starch phosphorylase cannot completely degrade amylopectin to glucose 1-phosphate but requires debranching enzyme to remove the (1→6)-linkages and glucosyltransferase to form longer-chain glucan polymers from small glucans that are too short for breakdown by phosphorylase.

glucosyltransferase, can condense short glucan polymers to produce new substrate for the phosphorylase (Fig. 13.25).

Isoenzymes of starch phosphorylase occur predominantly in plastids, although phosphorylase activities also have been found in the cytosol. In some seeds, degradation of starch coincides with disintegration of plastids, so that the starch can be broken down by cytosolic starch phosphorylase. Interestingly, in the early stages of potato tuber development, starch phosphorylase occurs only in the amyloplasts; in more mature tubers, the phosphorylase is predominantly cytosolic. Both plastid and cytosolic phosphorylase activities appear to be the product of a single gene, so the localization of this enzyme may be regulated by amyloplast protein-uptake mechanisms.

In animals, the activity of glycogen phosphorylase—the analog of starch phosphorylase—is regulated in a complex manner involving hormones. The enzyme is regulated allosterically, and the allosteric properties can be altered by phosphorylation/dephosphorylation brought about by a cascade mechanism involving the secondary messenger cAMP. No regulatory properties of starch phosphorylase have been described in plants, which is surprising, given the central role played by this enzyme in carbohydrate metabolism. The activity of the plant enzyme may well be determined by the availability of inorganic phosphate, which in turn is dependent on the predominant direction of phosphate transport by TPT.

13.3.4 Starch hydrolysis during cereal germination has been studied extensively.

Starch also can be cleaved by a series of hydrolytic enzymes known as amylases. Many isoforms of amylase are known, but their role in starch mobilization has not been fully elucidated. α-Amylase catalyzes the internal cleavage of glucosyl bonds, giving rise to small-molecular-mass glucans called limit dextrins and some glucose and maltose (Rx. 13.14). The main product of α-amylase, β-amylase (Rx. 13.15), cleaves maltose residues from the nonreducing end of the starch molecule. Maltose, short-chain glucans, and maltosaccharides or limit dextrins can be further degraded to glucose molecules by the action of α-glucosidase (Rx. 13.16). These three enzymes, combined with debranching enzyme, can effectively degrade starch to glucose (Fig. 13.26).

Reaction 13.14: α-Amylase

$$(1{\rightarrow}4)\alpha\text{-D-Glucan}_{(n)} \longrightarrow (1{\rightarrow}4)\alpha\text{-D-Glucan}_{(x)}$$
$$+ (1{\rightarrow}4)\alpha\text{-D-glucan}_{(y)}\ (n \geq 3;\ x + y = n)$$

Reaction 13.15: β-Amylase

$$(1{\rightarrow}4)\alpha\text{-D-Glucan}_{(n)} \longrightarrow$$
$$(1{\rightarrow}4)\alpha\text{-D-Glucan}_{(n-2)} + \text{maltose}$$

Reaction 13.16: α-Glucosidase

$$(1{\rightarrow}4)\alpha\text{-D-Glucose}_{(n)} \longrightarrow$$
$$(1{\rightarrow}4)\alpha\text{-D-Glucose}_{(n-1)} + \text{D-glucose}\ (n \geq 2)$$

(A)

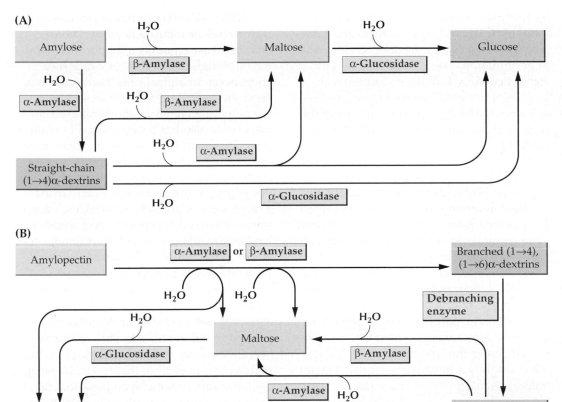

Figure 13.26
Hydrolytic starch degradation. Complete hydrolytic cleavage of amylose (A) requires the cooperative action of several hydrolytic enzymes. In the case of amylopectin (B), debranching enzyme is also required.

(B)

The enzymes involved in hydrolytic starch cleavage are particularly active during seed germination, when their expression is up-regulated by gibberellins secreted by the embryo (see Chapter 20). During the germination of cereal seeds, there is a rapid breakdown of starch in the endosperm. The most studied of the starch-hydrolyzing enzymes is α-amylase, which is synthesized in the aleurone layer surrounding the starchy endosperm (Fig. 13.27). This de novo synthesis is induced by increasing concentrations of gibberellins. A precursor of β-amylase is present in seeds before germination and is activated during germination by removal of a small peptide from the C terminus of the enzyme. Short-chain glucans and maltosaccharides are hydrolyzed by α-glucosidase. Some of the latter enzyme is present in the dried seed, but large amounts are synthesized during germination in the aleurone in response to the increase in gibberellic acid. Seed-localized debranching enzyme and α-amylase activities are further regulated

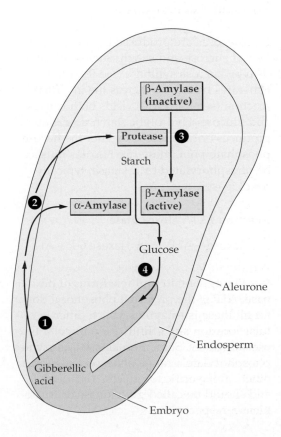

Figure 13.27
Role of gibberellic acid (GA) in mobilizing the carbohydrate reserves of germinating cereal seeds. The breakdown of starch in the endosperm of monocots such as barley is triggered by the release of gibberellins from the embryo (1), which activates expression of α-amylase in the aleurone layer. α-Amylase is released into the endosperm. GA also promotes expression of a protease (2), which is released from the aleurone and activates β-amylase by cleaving a propeptide from an inactive form of the starch-degrading enzyme (3). Together with starch-debranching enzyme (not shown), α-amylase and proteolytically activated β-amylase hydrolyze the starch to form glucose, which is exported from the endosperm to feed the developing embryo through the scutellum (4).

by specific disulfide proteins that act as inhibitors. Thioredoxin h reduces and thereby inactivates these inhibitor proteins early in germination. The glucose liberated by starch breakdown in the endosperm can be phosphorylated by hexokinase (see Section 13.3.5) and converted to sucrose in the scutellum for transport to the developing embryo.

The hydrolytic enzymes associated with starch breakdown in storage tissues also appear to play a role in the mobilization of starch stored temporarily in the chloroplast. Transgenic tobacco leaves that contain negligible amounts of chloroplast starch phosphorylase can mobilize starch almost as rapidly as wild-type leaves can. This suggests that the hydrolytic starch cleavage can account for all of the required degradation reactions. The hydrolytic pathway results in the formation of free hexoses, which may be transported across the chloroplast envelope to the cytosol by a hexose transporter. Mutant lines of *Arabidopsis* that lack a chloroplast hexose transporter cannot effectively degrade starch at night.

13.3.5 Free hexoses are phosphorylated by isoenzymes of hexokinase.

Free hexoses are produced by the degradation of sucrose and starch. Sucrose catabolism by sucrose synthase (see Rx. 13.9) generates fructose, whereas the invertase reaction (see Rx. 13.10) yields both fructose and glucose. Hydrolytic starch degradation produces only glucose. To enter the hexose phosphate pool, these free hexoses must be phosphorylated by a kinase, typically hexokinase:

Reaction 13.17: Hexokinase

$$\text{Hexose} + \text{ATP} \longrightarrow \text{Hexose 6-P} + \text{ADP}$$

Plants contain multiple forms of hexokinase. ATP is the preferred phosphoryl donor for all these isoenzymes, which differ in cellular location and affinity for particular hexose substrates. Some hexokinase isoenzymes phosphorylate a range of hexoses, whereas others are specific for glucose or fructose and should be called glucokinase or fructokinase, respectively.

Although certain types of plastids appear to lack hexokinase activity, a nonspecific hexokinase has been found in the stroma of plastids in developing castor seed endosperm. In animal cells, hexokinase has been shown to bind to porin molecules in the outer mitochondrial membrane; a similar association also has been reported in plant cells. This bound hexokinase may use newly synthesized ATP exported from the mitochondrion and produce ADP that can be imported by the organelle to maintain oxidative phosphorylation. In animals, the energy status of the cell appears to affect whether hexokinase is bound to the mitochondrion or is free in the cytosol; a similar situation probably occurs in plants.

As with other enzymes of carbohydrate metabolism, the various genes for the isoenzymes of hexokinase are differentially expressed in different tissues. Expression of hexokinase isoenzymes may depend on the free hexoses present in that tissue. Growing evidence supports the hypothesis that the expression of genes involved in carbohydrate metabolism in plants can be regulated by hexoses, and differential expression of hexokinase isoenzyme genes may contribute to this regulation.

13.4 The triose phosphate/pentose phosphate metabolite pool

Whereas the hexose phosphate pool contains only three metabolic intermediates, the **triose phosphate/pentose phosphate pool** includes a diverse set of sugar intermediates: ribulose 5-phosphate, ribose 5-phosphate, xylulose 5-phosphate, dihydroxyacetone phosphate, glyceraldehyde 3-phosphate, sedoheptulose 7-phosphate, erythrose 4-phosphate, and fructose 1,6-bisphosphate. Fructose 6-phosphate is usually considered part of the hexose phosphate pool, but as a substrate for the enzymes transaldolase and transketolase, this compound functions as a component of both metabolite pools (Fig. 13.28).

Considering these sugar phosphates as components of metabolite pools that participate in several different pathways provides an opportunity to investigate how those pathways interrelate. Glycolysis is usually depicted as a linear sequence of reactions in which glucose (or in animals, glycogen) is

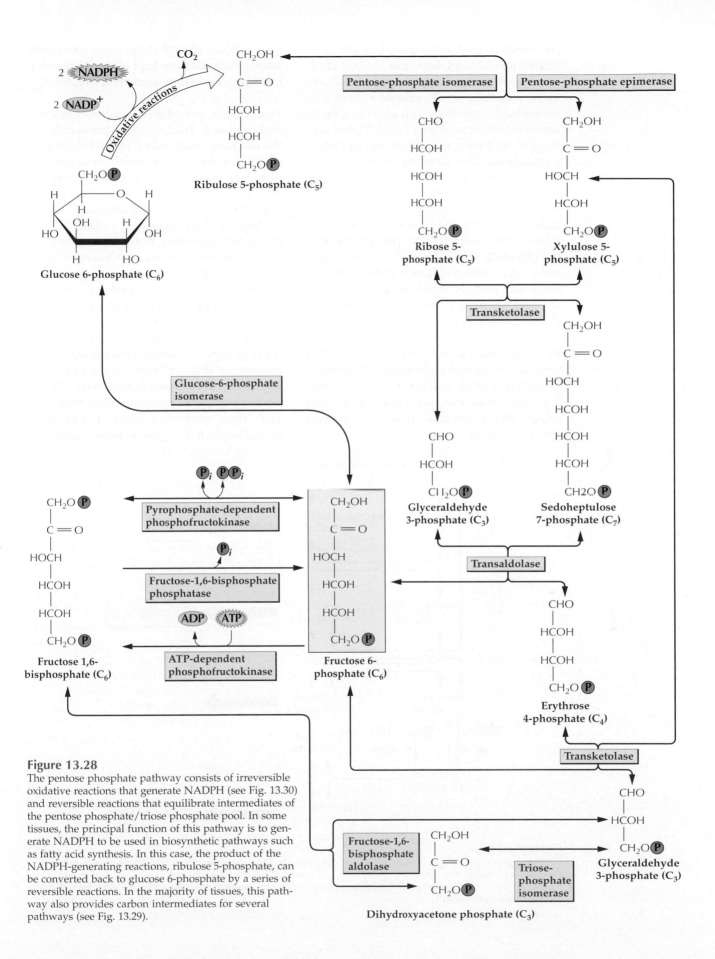

Figure 13.28
The pentose phosphate pathway consists of irreversible oxidative reactions that generate NADPH (see Fig. 13.30) and reversible reactions that equilibrate intermediates of the pentose phosphate/triose phosphate pool. In some tissues, the principal function of this pathway is to generate NADPH to be used in biosynthetic pathways such as fatty acid synthesis. In this case, the product of the NADPH-generating reactions, ribulose 5-phosphate, can be converted back to glucose 6-phosphate by a series of reversible reactions. In the majority of tissues, this pathway also provides carbon intermediates for several pathways (see Fig. 13.29).

converted to pyruvate and a small net gain in ATP and NADH. Similarly, the oxidative pentose phosphate pathway is depicted as a cyclic process in which glucose 6-phosphate is oxidized to ribulose 5-phosphate, with the concomitant formation of NADPH and subsequent multistep regeneration of glucose 6-phosphate. This viewpoint, however, effectively isolates two pathways that are actually linked by several shared intermediates. Although these two pathways can perform independently in plant cells, they also can interact. The various sugar phosphates of the glycolytic and pentose phosphate pathways occur both in plastids and in the cytosol, and there is communication between the pathways across the plastid envelope. Plants use both pathways not only to produce energy-rich cofactors, but also to generate carbon skeletons required for biosynthetic reactions. In short, glycolysis, the pentose phosphate pathway, and various biosynthetic pathways are integrated in plants by their relationship to the triose phosphate/pentose phosphate pool.

13.4.1 Triose phosphate/pentose phosphate pool metabolites are kept at equilibrium by numerous reversible enzymatic reactions.

In glycolysis, after the conversion of fructose 6-phosphate to fructose 1,6-bisphosphate, the reactions catalyzed by fructose-1,6-bisphosphate aldolase and triose-phosphate isomerase are reversible. Similarly, apart from the oxidative reactions, the pentose phosphate pathway is also reversible, including the steps catalyzed by ribulose-5-phosphate epimerase, ribose-5-phosphate isomerase, transketolase, and transaldolase. These reversible reactions serve to keep the cytosolic and plastid-localized pools of triose phosphate/pentose phosphate metabolites in equilibrium (Fig. 13.28).

There are three major drains on the triose phosphate/pentose phosphate of metabolites (Fig. 13.29). First, the triose phosphates are utilized in the energy-conserving reactions of glycolysis (see Section 13.8). These compounds also can be transported across the envelope of the plastid

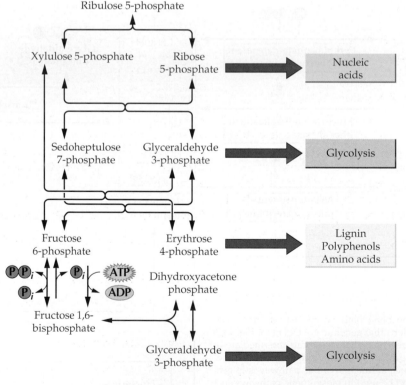

Figure 13.29
Three metabolic pathways, both anabolic and catabolic, drain the pentose phosphate/triose phosphate pool. Glycolysis partially oxidizes triose phosphates to organic acids. Ribose is utilized in nucleotide synthesis. Erythrose 4-phosphate enters the shikimate pathway and contributes to the production of lignin, polyphenols, and aromatic amino acids.

by the TPT. Second, the shikimate pathway uses erythrose 4-phosphate for the formation of lignin, amino acids, and polyphenols (see Chapters 8 and 24). Third, ribose 5-phosphate is required in the synthesis of nucleic acids. As metabolites are drained from this pool, the reversible reactions adjust to allow production of the metabolite that is being depleted. Because the reactions are reversible, metabolites can flow in either direction, toward product or substrate formation. The oxidative reactions, which are not reversible, operate only when reducing power (i.e., NADPH) is needed.

13.4.2 Fructose-1,6-bisphosphate aldolase and triose-phosphate isomerase participate in the interconversion of fructose 1,6-bisphosphate, glyceraldehyde 3-phosphate, and dihydroxyacetone phosphate.

Fructose-1,6-bisphosphate aldolase (Rx. 13.18) catalyzes the aldol cleavage of fructose 1,6-bisphosphate to glyceraldehyde 3-phosphate and dihydroxyacetone phosphate:

Reaction 13.18: Fructose-1,6-bisphosphate aldolase

$$\text{Fructose 1,6-P}_2 \rightleftharpoons$$
$$\text{DHAP + glyceraldehyde 3-P}$$

The standard free energy change of this reaction indicates that the equilibrium favors fructose 1,6-bisphosphate. However, the actual equilibrium position is determined by the concentrations of all three compounds involved, so fructose 1,6-bisphosphate can accumulate only when its cleavage products achieve a threshold concentration.

Aldolases exist in two quite distinct forms: Class I and Class II. Despite their similar functions, there is no sequence similarity between the two isoenzymes, and their common functions appear to have resulted from convergent evolution. Animals and plants have only Class I aldolases, yeast and pro-karyotes have Class II aldolases, and the photosynthetic protist *Euglena* has both. Plants have isoenzymes of aldolase in the cytosolic and plastid compartments, both belonging to Class I.

Triose-phosphate isomerase, which occurs in both the cytosol and plastids, catalyzes the interconversion of glyceraldehyde 3-phosphate and dihydroxyacetone phosphate and is one of the most active enzymes known.

Reaction 13.19: Triose-phosphate isomerase

$$\text{Glyceraldehyde 3-P} \rightleftharpoons \text{DHAP}$$

The reaction is readily reversible, although the equilibrium in the cell favors dihydroxyacetone phosphate by approximately 14:1, making this compound the principal triose phosphate in the cell. The effect of both the aldolase and triose-phosphate isomerase equilibria is to keep the concentration of glyceraldehyde 3-phosphate low. However, the lower part of the glycolytic pathway can proceed because its energetically favorable reactions consume glyceraldehyde 3-phosphate, shifting aldolase and triose-phosphate isomerase equilibria toward glyceraldehyde 3-phosphate production (see Section 13.8).

13.4.3 Enzymes that catalyze reversible reactions of the pentose phosphate pathway may not be present in cytosol of mesophyll cells.

The reactions catalyzed by ribulose-5-phosphate epimerase (Rx. 13.20), ribose-5-phosphate isomerase (Rx. 13.21), transketolase (Rx. 13.22), and transaldolase (Rx. 13.23) are all reversible.

Reaction 13.20: Ribulose-5-P epimerase

$$\text{Xylulose 5-P} \rightleftharpoons \text{Ribulose 5-P}$$

Reaction 13.21: Ribulose-5-P isomerase

$$\text{Ribulose 5-P} \rightleftharpoons \text{Ribose 5-P}$$

Reaction 13.22: Transketolase

(1) Sedoheptulose 7-P + glyceraldehyde 3-P
$$\rightleftharpoons \text{Xylulose 5-P + ribose 5-P}$$

(2) Xylulose 5-P + erythrose 4-P $\rightleftharpoons$
Fructose 6-P + glyceraldehyde 3-P

Reaction 13.23: Transaldolase

Glyceraldehyde 3-P + sedoheptulose 7-P $\rightleftharpoons$
Erythrose 4-P + fructose 6-P

All the above enzymes occur in various types of plastids but are reportedly absent from the cytosol of mesophyll cells. This finding may explain why sedoheptulose 1,7-bisphosphate does not accumulate in mesophyll cytoplasm. This compound, the condensation product of erythrose 4-phosphate and dihydroxyacetone phosphate, is broken down by sedoheptulose-1,7-bisphosphatase, an enzyme unique to plastids. In nonphotosynthetic tissues, the cytosolic enzymes for the reversible reactions of the oxidative pentose phosphate cycle are present in detectable amounts, but we lack detailed studies of these isoenzymes.

The absence of cytosolic enzymes for the reversible reactions of the pentose phosphate pathway raises a dilemma: What is the fate of the ribulose 5-phosphate formed by the oxidative reactions? Under some circumstances the hexose phosphate transporter can also transport pentose phosphates; this may be the mechanism by which ribulose 5-phosphate is removed.

13.5 Interactions between the hexose phosphate and pentose phosphate/triose phosphate pools

13.5.1 Two pentose phosphate pathway enzymes oxidize glucose 6-phosphate to ribulose 5-phosphate and NADPH.

Glucose-6-phosphate dehydrogenase (Rx. 13.24) catalyzes the oxidation of glucose 6-phosphate to 6-phosphoglucono-δ-lactone with the concomitant formation of NADPH from $NADP^+$ (Fig. 13.30). The plant enzymes cannot use NAD^+ as an electron acceptor.

Reaction 13.24: Glucose-6-P dehydrogenase

Glucose 6-P + $NADP^+$ $\rightleftharpoons$
6-Phosphoglucono-δ-lactone + NADPH + H^+

High concentrations of both cytosolic and plastid-localized glucose-6-phosphate dehydrogenase isoenzymes have been found in all plant materials studied, but tissue-specific isoenzymes have not been reported. The plastid isoenzyme is less stable than the cytosolic isoenzyme, which makes study of the former more difficult. Genes for both isoenzymes have been cloned and indicate approximately 75% identity of amino acid sequences. Although reduction of the product 6-phosphoglucono-δ-lactone back to glucose-6-phosphate is thermodynamically feasible, the lactone is unstable and spontaneously decomposes to form 6-phosphogluconate, making the oxidative reaction essentially irreversible. A lactonase further increases the rate of 6-phosphogluconate formation.

The cytosolic isoenzyme of glucose-6-phosphate dehydrogenase has no allosteric regulatory properties, which is surprising, considering its important role in supplying

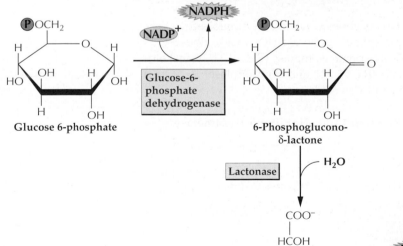

Figure 13.30
The oxidative reactions of the pentose phosphate pathway convert glucose 6-phosphate to ribulose 5-phosphate, generating NADPH in the process.

NADPH. However, the enzyme is strongly inhibited by one of its products, NADPH; accordingly, the $NADP^+/NADPH$ ratio appears to determine the activity of the enzyme. This control mechanism indicates that the function of the enzyme is to provide NADPH for biosynthetic reactions.

In contrast, the plastid isoenzyme shows complex regulation. This isoenzyme can occur in reduced and oxidized forms, the latter being the active form. Reduction of the enzyme is related to the redox state of thioredoxin m and therefore to the availability of light. During photosynthesis, $NADP^+$ can be reduced by the photosynthetic electron transfer chain (see Chapter 12), which makes the oxidative steps of the pentose phosphate pathway superfluous. Under these conditions, plastid-localized glucose-6-phosphate dehydrogenase is deactivated by photoreduced thioredoxin m.

In plastids of the green alga *Selenastrum*, the reduction of glucose-6-phosphate dehydrogenase and photosynthesis are linked. Adding nitrate to a cell suspension greatly increases the demand for reducing power, such that the photosynthetic electron transport chain cannot supply sufficient NADPH. As a result, the thioredoxin m pool becomes oxidized and, in the absence of reduced thioredoxin m, glucose-6-phosphate dehydrogenase becomes oxidized and active (Fig. 13.31). Apparently, supplies of reducing power from photosynthesis and the oxidative pentose phosphate pathway are interchangeable in the chloroplast of this alga.

Phosphogluconate dehydrogenase (Rx. 13.25) catalyzes the irreversible oxidative decarboxylation of 6-phosphogluconate to ribulose 5-phosphate and carbon dioxide, with the concomitant reduction of $NADP^+$ (see Fig. 13.30).

Reaction 13.25: 6-Phosphogluconate dehydrogenase

$$6\text{-Phosphogluconate} + NADP^+ \longrightarrow$$
$$\text{Ribulose 5-P} + CO_2 + NADPH + H^+$$

All plant tissues appear to have distinct cytosolic and plastid isoforms of this enzyme. The cytosol of some plants contains several different isoenzymes. No regulatory properties of the enzyme have been described. Its activity is probably determined by the availability of 6-phosphogluconate generated by glucose-6-phosphate dehydrogenase.

13.5.2 In plants, fructose 6-phosphate and fructose 1,6-bisphosphate are interconverted freely by the action of three enzymes.

In plants, the interconversion of fructose 6-phosphate and fructose 1,6-bisphosphate (Fig. 13.32) involves the regulatory metabolite fructose 2,6-bisphosphate and three enzymes: PFK (Rx. 13.26), PFP (Rx. 13.27), and fructose-1,6-bisphosphatase (Rx. 13.28). Animals and most other eukaryotes do not contain PFP but instead use the irreversible reactions catalyzed by PFK and fructose-1,6-bisphosphatase to phosphorylate the C-1 of fructose 6-phosphate and to dephosphorylate the C-1 of fructose 1,6-bisphosphate, respectively.

Reaction 13.26: ATP-dependent PFK

$$\text{Fructose 6-P} + ATP \longrightarrow$$
$$\text{Fructose 1,6-}P_2 + ADP$$

Reaction 13.27: PP_i-dependent PFP

$$\text{Fructose 6-P} + PP_i \rightleftharpoons \text{Fructose 1,6-}P_2 + P_i$$

Reaction 13.28: Fructose-1,6-bisphosphatase

$$\text{Fructose 1,6-}P_2 + H_2O \longrightarrow$$
$$\text{Fructose 6-P} + P_i$$

Although the relationship between fructose 6-phosphate and fructose 1,6-bisphosphate has been studied for many years, numerous questions remain unanswered about the roles played by PFK and fructose-1,6-bisphosphatase, which are found in both the cytosolic and plastid compartments, and by PFP, which occurs exclusively in the cytosol.

13.5.3 The physiological role of PFP remains unknown.

PFP was discovered in propionic acid bacteria and later found in anaerobic intestinal parasites such as *Entamoeba* and *Giardia*.

Photosynthesis alone
(Thioredoxin mostly reduced)

Photosynthesis plus nitrate and sulfate reduction
(Thioredoxin mostly oxidized)

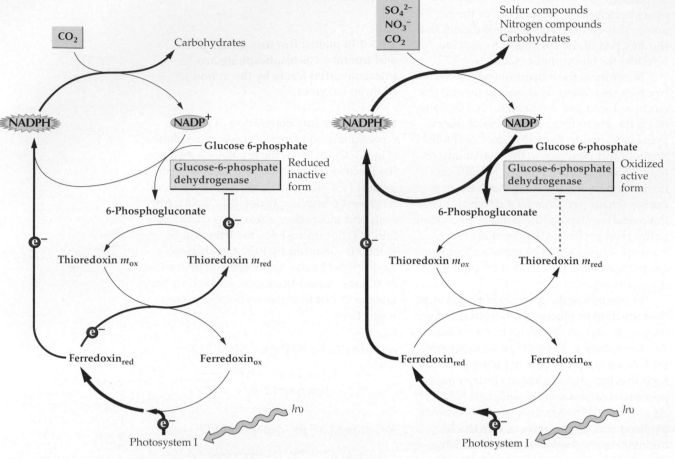

Figure 13.31

Relationship between photosynthesis and the activity of glucose-6-phosphate dehydrogenase in the green alga *Selenastrum*. Chloroplasts can reduce $NADP^+$ to NADPH using electrons from either the light reactions of photosynthesis or the oxidative reactions of the pentose phosphate pathway. During the light reactions of photosynthesis, Photosystem I reduces ferredoxin, which in turn reduces both $NADP^+$ and thioredoxin *m*, a regulatory protein. Reduced thioredoxin *m* transfers hydrogens to glucose-6-phosphate dehydrogenase, inhibiting the enzyme. However, when other metabolites such as nitrate are being reduced in the chloroplast, the demand for NADPH increases and electrons are preferentially diverted to the reduction of $NADP^+$. Under these conditions, the thioredoxin pool becomes oxidized and ceases to inhibit glucose-6-phosphate dehydrogenase. Increased activity of this enzyme further enhances production of NADPH, which is needed for CO_2 fixation, SO_4^{2-} reduction, and NO_2^- reduction. The thickness of the arrows reflects electron transport activity.

In plants, the enzyme was first observed in 1979, but the amount found appeared low and inconsequential. Not until the discovery that PFP was powerfully activated in the direction of fructose 1,6-bisphosphate formation by fructose 2,6-bisphosphate (Fig. 13.33) was the enzyme considered an important component of plant carbohydrate metabolism.

PFP catalyzes a readily reversible reaction, unlike the irreversible action of PFK. In many tissues PFP is present at higher concentrations than PFK.

In most plant tissues, PFP is a heterotetramer composed of two α- and two β-subunits, although some researchers have reported finding the enzyme as a β-subunit homodimer. The genes for the subunits have been cloned. The amino acid sequences of the two subunits show many differences, suggesting that they are not the product of a recent gene duplication. In addition to sequence dissimilarity, the genes for the two subunits differ structurally. A comparison of the sequence of the subunits of PFP with the sequence of PFK suggests the β-subunit

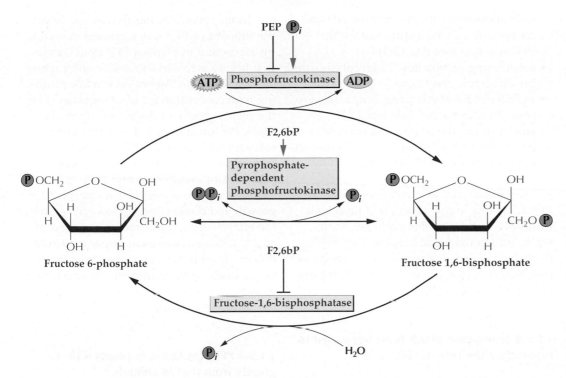

Figure 13.32
Interconversion in the cytosol of fructose 6-phosphate and fructose 1,6-bisphosphate by ATP-dependent phosphofructokinase, PP_i-dependent phosphofructokinase, and fructose-1,6-bisphosphatase. PEP, phospho*enol*pyruvate; F2,6bP, fructose 2,6-bisphosphate.

binds fructose 6-phosphate. The α-subunit appears to play a regulatory role, perhaps binding the activator fructose 2,6-bisphosphate (Fig 13.34). The affinity of PFP for the activator is very high. At observed cellular concentrations of fructose 2,6-bisphosphate, the enzyme should be fully activated. However, because the actual concentrations of enzyme and activator are estimated to be similar, the ratio of enzyme to activator may be more important than the absolute concentrations of either.

Many roles have been postulated for PFP, but none has been verified. In all organisms, the conversion of fructose 6-phosphate to fructose 1,6-bisphosphate has been assumed to be the principal regulatory step in glycolysis. Regulation, however, usually occurs at an irreversible reaction, such as that catalyzed by PFK, whereas the reaction PFP catalyzes is readily reversible. It is therefore somewhat difficult to imagine that PFP might have a role in the regulation of carbon flow through glycolysis. Other roles proposed for PFP include the following:

- generating pyrophosphate required for the breakdown of sucrose by way of the sucrose synthase pathway
- bypassing PFK during times of phosphate

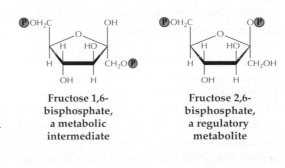

Fructose 1,6-bisphosphate, a metabolic intermediate

Fructose 2,6-bisphosphate, a regulatory metabolite

Figure 13.33
Structures of fructose 1,6-bisphosphate and fructose 2,6-bisphosphate. The regulatory metabolite and the metabolic intermediate differ in the position of one phosphate group.

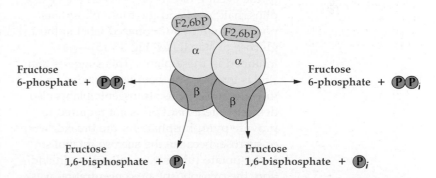

Figure 13.34
Schematic representation of PFP. The β-subunit binds substrates and catalyzes the interconversion of fructose 6-phosphate and fructose 1,6-bisphosphate. The α-subunit binds the activator fructose 2,6-bisphosphate (F2,6bP). It is possible that the active site is formed cooperatively between the α- and β-subunits. In the PFK from animal tissues, the active site is a cooperative interaction between two identical subunits. The binding of an activator to the animal enzyme probably causes a conformational change that enhances the catalytic efficiency of the active site. A similar mechanism may exist in the PP_i-dependent enzyme in plants.

starvation when the amounts of ATP decrease and may be insufficient for the PFK reaction (see Box 13.4)

- stimulating carbon flow in glycolysis during anaerobic conditions
- maintaining a stable pyrophosphate concentration in the cell (see Box 13.1)
- equilibrating the hexose and pentose phosphate/triose phosphate pools during times of increased biosynthetic activity.

Circumstantial evidence indicates that PFP functions in the direction of glycolysis to supply carbon for that pathway. Meanwhile, strong evidence suggests that PFK is the principal enzyme regulating the entry of metabolites into glycolysis to supply the energy needs of the cell.

13.5.4 Transgenic plants have been used to investigate the role of PFP.

The use of transgenic plants has allowed some of the hypotheses for the function of PFP to be tested. By inserting a clone for PFP into plants in the antisense orientation, the expression of the resident gene for PFP can be reduced to less than 2% of that found in wild-type plants. Tobacco plants and potato tubers with these very low amounts of PFP appeared to grow and develop normally, except that the tubers showed a minor reduction in the concentration of starch. Glycolysis was inhibited slightly but not enough to affect the growth of the plant. Interestingly, the cycling of carbon between the hexose phosphate pool and the triose phosphate pool—shown by randomized label in the glucose of starch (see Fig. 13.15)—was inhibited in these plants. This suggests that PFP is involved in the interaction between the two pools. These transgenic plants also demonstrated that PFP is not required to provide pyrophosphate for the breakdown of sucrose, because the sucrose catabolism in the potato tubers was still active. In addition, the pyrophosphate concentration was not affected in these plants, suggesting that the amount of this compound present was maintained by other mechanisms. The transgenic plants were also able to withstand stress conditions to the same extent as wild-type plants.

In the reverse of the above experiment, the amount of PFP was increased manyfold by expressing in plants a PFP from *Giardia* that is very active in the absence of fructose 2,6-bisphosphate. Again, even while producing high concentrations of unregulated PFP, the plants grew and developed normally. Thus, the function of PFP remains unresolved. One reason for this may be that we have endeavored to place PFP in a central role in plant metabolism and have assumed that the enzyme is essential for normal growth and development. However, perhaps the enzyme merely enhances the fitness of plants in the environment under certain conditions. Even if this enhancement in fitness is too small for us to measure its effect, the enzyme still would likely be retained during the course of evolution.

13.5.5 PFK regulation in plants differs greatly from that in animals.

In organisms other than plants, PFK has been studied in great detail. The enzyme is the principal regulator of the flux of carbon through glycolysis. In animals, but not plants, it is activated by fructose 2,6-bisphosphate and is indirectly under hormonal control. In addition, the product of the reaction, fructose 1,6-bisphosphate, activates pyruvate kinase in animals and other organisms, so that activation of PFK switches on the entire glycolytic pathway by a feed-forward mechanism. In contrast, plant pyruvate kinase is not affected by fructose 1,6-bisphosphate, so activation of PFK cannot have the same overall impact on the pathway (Fig. 13.35).

The situation in plants is made more complex by the presence of plastid and cytosolic isoenzymes of PFK. In most of the research on this enzyme in plants, the isoenzymes have not been separated, which makes it difficult to interpret the results. The plastid enzyme specifically feeds carbon from the hexose phosphate pool into the pentose phosphate/triose phosphate pool within the organelle, whereas the cytosolic enzyme performs this function in the cytoplasm. Of the two isoenzymes, the plastid form is more highly regulated and is similar to the bacterial enzyme. Both of the

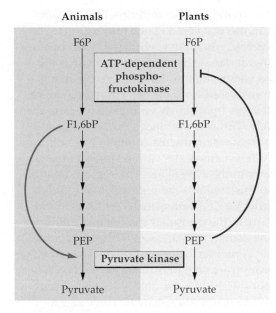

Animals   **Plants**

F6P   F6P

ATP-dependent phospho-fructokinase

F1,6bP   F1,6bP

PEP   PEP

Pyruvate kinase

Pyruvate   Pyruvate

Figure 13.35
Regulation of glycolysis in plant and animal cells. In animals there is a "top down" regulation through the activation of the pyruvate kinase by fructose 1,6-bisphosphate (F1,6bP). This activation does not occur in plants. Instead there is a "bottom up" regulation through the inhibition of PFK by phospho*enol*pyruvate (PEP).

plant isoenzymes are strongly inhibited by phospho*enol*pyruvate, and to a lesser extent by other metabolites at the lower end of the glycolytic pathway, suggesting that the pathway is regulated by a negative-feedback mechanism (Fig. 13.35).

In plants, this negative feedback may involve the TPT that links plastid metabolism with that in the cytosol. In green cells, the source of carbon for metabolism is triose phosphate from photosynthesis. If dihydroxyacetone phosphate from the chloroplast stroma can supply the cytosol with adequate substrate for the energy-conserving reactions of glycolysis, there is no need to expend ATP to generate the glycolytic substrate fructose 1,6-bisphosphate.

In contrast to the effect of phospho*enol*pyruvate, inorganic phosphate is a powerful activator of both the plastid and cytosolic isoenzymes of PFK. This again illustrates the central role of the TPT in plant metabolism. High concentrations of phosphate in a compartment indicate that triose phosphate is being translocated out of that compartment, signaling a need to resupply

the triose phosphate pool. In illuminated chloroplasts, the plastid triose phosphate pool can be replenished through photosynthesis. In colorless plastids and in the dark, this function is performed by way of starch breakdown and glycolysis. Hence, the principal variable controlling the activity of plant PFK is the phospho*enol*pyruvate/phosphate ratio, which links the activity of the enzyme to the status of the lower (energy-conserving) end of the glycolytic pathway (Fig. 13.36).

13.5.6 Fructose-1,6-bisphosphatase is differentially regulated in plastids and cytosol and is absent from some colorless plastids.

Fructose-1,6-bisphosphatase catalyzes the hydrolytic cleavage of phosphate from the C-1 position of fructose 1,6-bisphosphate, forming fructose 6-phosphate and inorganic phosphate. Generally, the enzyme is present in both the cytosol and chloroplasts, although it is typically absent from nonphotosynthetic plastids. Although this enzyme is usually shown to be part of the pentose phosphate pathway when it is drawn as a cycle (see Fig. 13.28), the absence of the enzyme in some tissues, coupled with the absence of other enzymes of the pathway, casts doubt on the truly cyclic nature of this pathway.

Cytosolic fructose-1,6-bisphosphatase is involved in the interconversion of starch and sucrose in green leaves (see below). As with the animal enzyme, fructose 2,6-bisphosphate strongly inhibits cytosolic fructose-1,6-bisphosphatase. The concentration of fructose 2,6-bisphosphate is controlled by the status of the triose phosphate pool, so the enzyme is active only when carbon is supplied by the chloroplast (see below).

The plastid isoenzyme of fructose-1,6-bisphosphatase is not affected by fructose 2,6-bisphosphate. Fructose-1,6-bisphosphatase is not present in plastids from heterotrophic tissues such as roots, developing cereal grains, and tubers, so carbon entering these organelles by way of the TPT cannot be converted to hexose phosphate for starch synthesis by gluconeogenesis. Hence, hexose phosphate must be transported across the envelope of these plastids.

13.6 Starch used as an overflow when the synthesis of sucrose exceeds the capacity of the leaves to export it: an example of the integrated control of metabolism in two cell compartments

13.6.1 Fructose 2,6-bisphosphate plays a key role in controlling diurnal carbohydrate cycling.

Our knowledge of the kinetics and regulation of the enzymes involved in carbohydrate metabolism allows us to postulate how starch might be used as an overflow mechanism to store carbohydrate when the synthesis of sucrose exceeds the capacity of the leaf to export it. It also indicates how the leaf can use the stored starch and convert it to sucrose when photosynthesis is inactive during darkness. In some plants, such as tobacco, almost half of the total sucrose that is transported from the leaf during the night derives from starch synthesized during the day. This is not a binary, on/off, mechanism but rather represents continuous monitoring of the status of the various pools of metabolites in the cytosol and chloroplast. Central to this model is the regulatory compound fructose 2,6-bisphosphate, which is found only in the cytosol of the cell and inhibits the activity of fructose-1,6-bisphosphatase.

Fructose 2,6-bisphosphate is synthesized from fructose 6-phosphate by fructose-6-phosphate 2-kinase and is converted back to fructose 6-phosphate by fructose-2,6-bisphosphatase (Fig. 13.37). Fructose 2,6-bisphosphate is present in the cytosol of plants, but not in the plastid, and is a powerful inhibitor of fructose-1,6-bisphosphatase. Hence the concentration of fructose 2,6-bisphosphate exerts a profound effect on the activity of cytosolic fructose-1,6-bisphosphatase, which in turn regulates the flow of carbon from triose phosphates into the hexose phosphate pool.

Fructose-6-phosphate 2-kinase is activated by inorganic phosphate and strongly inhibited by triose phosphates. In contrast, fructose-2,6-bisphosphatase is inhibited by inorganic phosphate. Hence, the concentration of fructose 2,6-bisphosphate is related to the activity of the TPT, which determines the concentrations of triose phosphate and P_i in the cytosol and is dictated by the supply of photosynthate from the chloroplast. In addition, the concentration of fructose

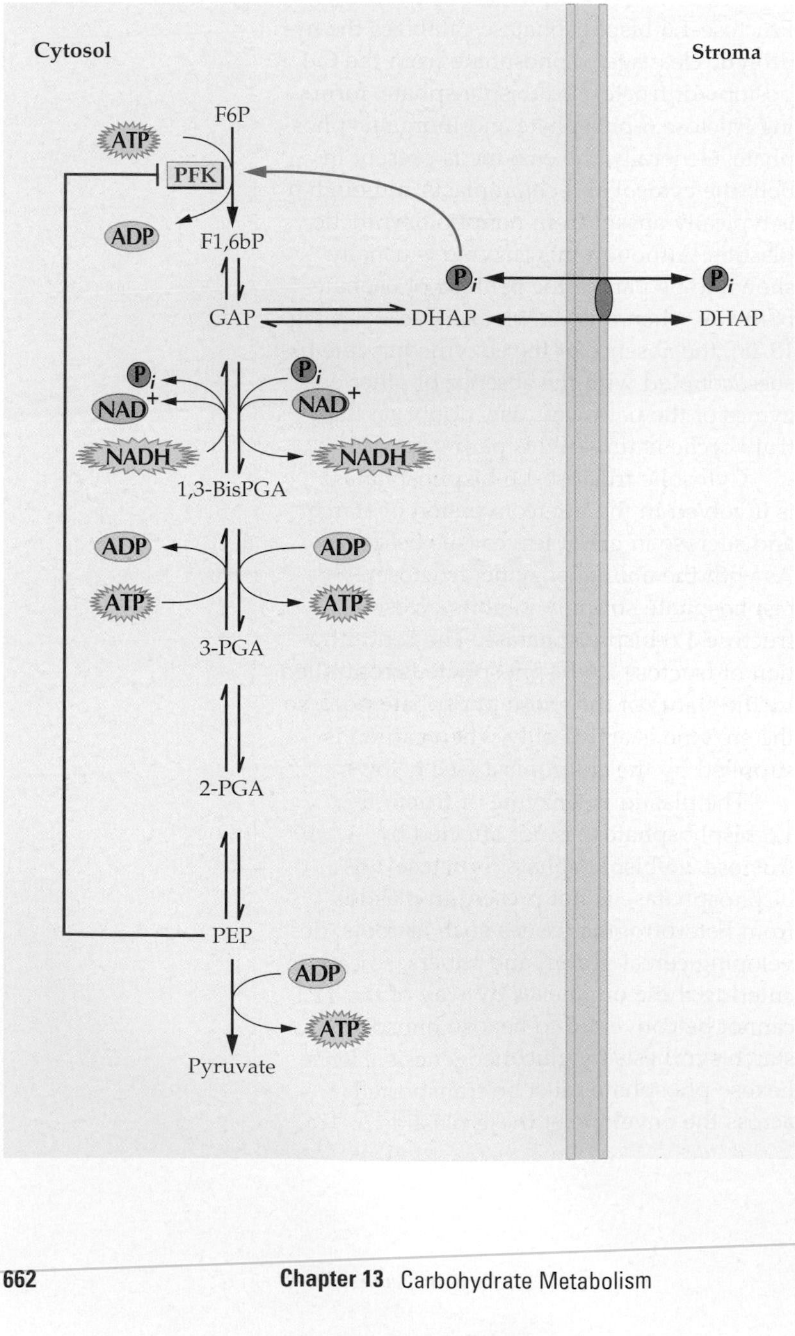

Figure 13.36
PFK regulation and relationship to the triose phosphate translocator (TPT). Inorganic phosphate is both an activator of PFK and an indicator of the status of the TPT. Phospho*enol*pyruvate (PEP) is a powerful inhibitor of PFK and indicates the status of the lower half of the glycolysis process, which in turn is linked to the TPT. PFK is therefore finely regulated by the PEP:P_i ratio, which in turn reflects the status of the passage of carbon and phosphate into and out of the plastid compartment. F6P, fructose 6-phosphate; F1,6bP, fructose 1,6-bisphosphate; GAP, glyceraldehyde 3-phosphate; 1,3-BisPGA, 1,3-bisphosphoglycerate; 3-PGA, 3-phosphoglycerate; 2-PGA, 2-phosphoglycerate.

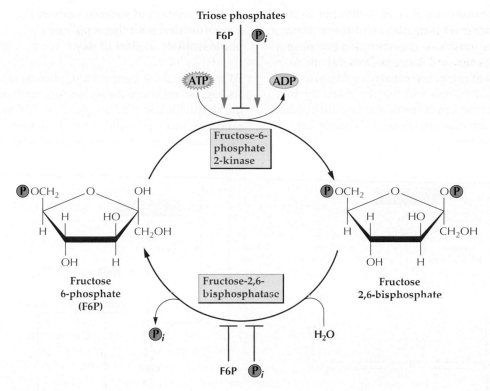

Figure 13.37
Formation and degradation of fructose 2,6-bisphosphate by fructose-6-phosphate 2-kinase and fructose-2,6-bisphosphatase. Regulation by inorganic phosphate, triose phosphate, and fructose 6-phosphate is indicated.

2,6-bisphosphate also is affected by the status of sugar export from the leaf. If sucrose synthesis exceeds export, its synthesis is inhibited and the content of the hexose phosphate pool increases. One component of this pool, fructose 6-phosphate, activates fructose-6-phosphate 2-kinase and inhibits fructose-2,6-bisphosphatase, thereby increasing the fructose 2,6-bisphosphate concentrations (Fig. 13.37). Hence, the concentration of fructose 2,6-bisphosphate reflects both the supply of photosynthate from the chloroplast and the rate at which sucrose is exported from the leaf.

13.6.2 The equilibrium position of the aldolase reaction also contributes to the effectiveness of this control.

Under standard conditions, at pH 7.0, the equilibrium position of the aldolase reaction favors the formation of fructose 1,6-bisphosphate. The equilibrium position is defined by Equation 13.1:

Equation 13.1

$$K_{eq} = \frac{[\text{Glyceraldehyde 3-P}]\,[\text{DHAP}]}{[\text{Fructose 1,6-P}_2]}$$

Because the two triose phosphates are rapidly interconverted by triose-phosphate isomerase, this equation shows that the concentration of fructose 1,6-bisphosphate is determined by the square of the triose phosphate concentration ($[\text{triose phosphate}]^2$), making the aldolase reaction very sensitive to the triose phosphate concentration. That is, the concentration of fructose 1,6-bisphosphate remains low at low triose phosphate concentrations but increases exponentially as triose phosphate concentrations rise. Thus, triose phosphate concentrations determine the concentration of fructose 1,6-bisphosphate available as substrate for the reaction catalyzed by fructose-1,6-bisphosphatase. In addition, increased triose phosphate concentrations effectively inhibit the formation of fructose 2,6-bisphosphate. The equilibrium of the aldolase reaction acts in concert with regulation of the fructose 2,6-bisphosphate concentration by triose phosphates and inorganic phosphate, to regulate finely the activity of fructose-1,6-bisphosphatase and thereby relate the rate of sucrose synthesis to the amount of photosynthate exported from the chloroplast (Fig 13.38).

13.6.3 Initiation of sucrose synthesis is coordinated with the supply of photosynthate at start of day.

At the start of the light period, photosynthesis begins and triose phosphate accumulates in the chloroplast. This triggers the TPT to exchange triose phosphate for phosphate so that triose phosphate starts to accumulate in

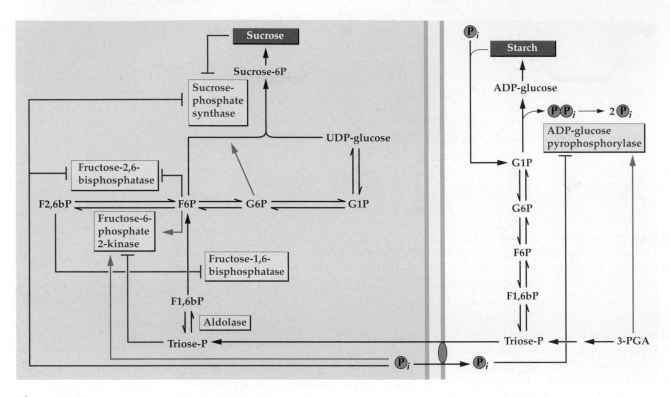

Figure 13.38

Regulation of the interconversion of starch and sucrose. F26,bP, fructose 2,6-bisphosphate; F6P, fructose 6-phosphate; G6P, glucose 6-phosphate; G1P, glucose 1-phosphate; F1,6bP, fructose 1,6-bisphosphate; 3-PGA, 3-phosphoglycerate. This interconversion provides an example of how metabolite levels are constantly monitored in the leaf cell and pathways activated and inhibited to keep them balanced.

Early daylight: The onset of photosynthesis causes the accumulation of triose phosphates in the chloroplast. These triose phosphates are exported to the cytosol in exchange for inorganic phosphate. The triose phosphates inhibit fructose-6-phosphate 2-kinase, while the removal of cytosolic phosphate weakens the activation of this enzyme and relieves inhibition of fructose-2,6-bisphosphatase and sucrose-phosphate synthase. As a result F2,6bP concentration drops, inhibition of cytosolic fructose-1,6-bisphosphatase is relieved, and hexose phosphates accumulate. The increased concentration of G6P activates sucrose-phosphate synthase, resulting in sucrose synthesis.

Midday: The model postulates that sucrose-phosphate synthase is inactivated by negative feedback from sucrose accumulation, but this has not been confirmed. In any event, diminished sucrose-phosphate synthase activity results in accumulation of its substrate, F6P, which inhibits fructose-2,6-bisphosphatase and activates fructose-6-phosphate 2-kinase. The resulting increase in the concentration of F2,6bP leads to inhibition of fructose-1,6-bisphosphatase. Because the aldolase reaction is reversible, triose phosphates

accumulate in the cytosol, preventing export of triose phosphates from the chloroplast. This results in accumulation of 3-PGA, which activates ADP-glucose pyrophosphorylase, promoting starch synthesis. At the same time, the phosphate concentration in the cytosol increases as a result of decreased export of plastid triose phosphates and this contributes to the inhibition of sucrose synthesis through inhibition of sucrose-phosphate synthase and fructose-2,6-bisphosphatase, and activation of fructose-6-phosphate 2-kinase.

The onset of darkness: As the light intensity falls during the evening, photosynthesis progressively declines. The synthesis of 3-PGA stops, the activation of ADP-glucose pyrophosphorylase is lost, and starch is no longer synthesized. Phosphate accumulates in the chloroplast, further inhibiting ADP-glucose pyrophosphorylase. It also is used as a substrate for starch phosphorylase. The triose phosphates from starch breakdown are transported to the cytosol and inhibit the formation of fructose 2,6-bisphosphate, allowing carbon to flow into the hexose phosphate pool and thence into sucrose. However, it must be remembered that much of the starch is broken down by the hydrolytic pathway and the glucose so formed is exported directly to the cytosol.

Although the legend to this diagram presents the situation in the leaf as a series of distinct events, it must be remembered that metabolite levels are constantly monitored and starch and sucrose syntheses are kept in step with the rate of photosynthesis at all times.

the cytosol and the cytosolic concentration of phosphate declines; as a result, fructose-6-phosphate 2-kinase is progressively inhibited and fructose-2,6-bisphosphatase is activated. Furthermore, the concentrations of fructose 2,6-bisphosphate diminish, relieving the inhibition of fructose-2,6-bisphosphatase. This, combined with the impact of the equilibrium effect of the aldolase reaction, allows the flow of carbon into the hexose phosphate pool, once the concentration of triose phosphate reaches a threshold value. In turn, the increased concentration of glucose 6-phosphate activates sucrose-phosphate synthase, allowing the synthesis of sucrose (see Fig. 13.13). Enhancing this effect is the stabilization of sucrose-phosphate synthase in the active nonphosphorylated state by the low concentration of phosphate and the high concentration of glucose 6-phosphate (see Fig. 13.13). Hence, this mechanism activates the flow of photosynthate to sucrose once the level of triose phosphate is reached.

13.6.4 Synthesis of starch is an overflow mechanism to store photosynthate when the rate of sucrose synthesis exceeds the rate of sucrose export from the cell.

At midday, under high light conditions, the rate at which the chloroplasts supply photosynthate may exceed the ability of the cell to export the sucrose synthesized from that photosynthate. Under these conditions, starch is synthesized from the excess carbon. When sucrose starts to accumulate in the leaf cell, it reduces the rate of its own synthesis, probably by inhibiting sucrose-phosphate synthase. This increases the size of the hexose phosphate pool, and the resulting increase in fructose 6-phosphate activates fructose-6-phosphate 2-kinase and inhibits fructose-2,6-bisphosphatase. The increase in concentration of fructose 2,6-bisphosphate will inhibit fructose-1,6-bisphosphatase, slowing the flow of carbon into the hexose phosphate pool. Triose phosphate will increase and its export from the chloroplast will be inhibited. The increase of triose phosphate in the chloroplast causes an increase in the plastid concentrations of 3-phosphoglycerate, which in turn activates ADP-glucose pyrophosphorylase (see

Fig. 13.16). Hence, carbon flow is diverted from sucrose to starch.

Clearly, inorganic phosphate plays a major role in the fine regulation of this overflow mechanism, for this molecule affects a number of enzymes. The TPT also plays a central role in controlling carbon flow between the chloroplast and the cytosol, since the direction of phosphate transport reflects whether carbon is entering or leaving the chloroplast.

13.6.5 At night the reserve of starch is broken down and converted to sucrose.

As the light intensity declines and night approaches, the supply of photosynthate decreases and, in turn, the rate of sucrose synthesis is reduced; as a result, sucrose export exceeds the rate of sucrose synthesis. This decreases the concentration of the hexose phosphates, which in turn results in a decreased concentration of fructose 2,6-bisphosphate, relieving the inhibition of fructose-1,6-bisphosphatase and allowing carbon to flow from triose phosphate to the hexose phosphate pool. Triose phosphate is exported from the chloroplast in exchange for inorganic phosphate, and the activation of ADP-glucose pyrophosphorylase by phosphoglycerate is lost. In addition, the influx of inorganic phosphate into the chloroplast provides a substrate for phosphorolytic starch degradation. However, much of the starch in the chloroplast is broken down hydrolytically during the night, possibly as a response to the shortage of ATP in dark-treated chloroplasts, which may limit the phosphofructokinase reaction. It is not known what triggers the hydrolytic pathway.

Hence, the leaf presents an elegant mechanism for controlling the flow of carbon into sucrose to reflect the supply of photosynthate from the chloroplast. Central to the control are the TPT, the regulation of various enzymes by phosphate, and the indicators of the status of the hexose phosphate pool and of the triose phosphate/pentose phosphate pool. This mechanism has developed in the cell to monitor constantly the status of the various metabolic pools so that the flow to sucrose or starch is optimized for the particular environmental conditions of the plant.

13.6.6 The diurnal cycling model does not explain how some mutant plants may interconvert starch and sucrose.

Plants have a way of making even the most elegant model seem inadequate. For example, naturally occurring mutants of the Florida weed, *Flaveria linearis*, have no detectable cytosolic fructose-1,6-bisphosphatase, although the chloroplast isoenzyme is present. Without this enzyme, the plants cannot convert fructose 1,6-bisphosphate to fructose 6-phosphate in the cytosol. These plants grow more or less normally and can clearly interconvert sucrose and starch in their leaves with near-normal efficiency. How can these plants survive without this enzyme? Could PFP perform this role, using inorganic phosphate and fructose 1,6-bisphosphate to generate fructose 6-phosphate? This is unlikely because the reaction would require high concentrations of fructose 2,6-bisphosphate (which do not occur) and would result in the production of large amounts of pyrophosphate (which would inhibit UDP-glucose pyrophosphorylase). More likely, the chloroplasts of these plants export glucose, which can be phosphorylated in the cytosol to form hexose phosphate. (Transgenic plants expressing the TPT in the antisense orientation appear to use this strategy as well.) Plants of this type demonstrate the remarkable flexibility of plant metabolism and the ability of plants to use alternative routes without any apparent impact on their survival or vigor (see Box 13.4).

13.7 Modulation of gene expression by carbohydrates

Carbohydrates play an important role in bacterial gene regulation, the classic example being the *lac* operon of *Escherichia coli*. In animal cells, however, hormones usually act as intermediates between the sensing of sugar levels and the cellular response. As has become increasingly apparent, in some cases, plant genes follow the bacterial pattern and are regulated directly by carbohydrates. This process is still a very active area of research, but details of its workings are emerging. As in microorganisms, some plant genes appear to be part of an ancient system that responds to nutrient availability.

One class of plant genes ("famine genes") is affected by sugar depletion. Under this condition, the expression of genes involved in photosynthesis (for example, ribulose-1,5-bisphosphate carboxylase/oxygenase, ATP synthase, and the TPT) is enhanced, as is that of genes that encode enzymes involved in the breakdown of starch, lipids, and proteins and in sucrose mobilization. The expression of a second class of genes ("feast genes") is enhanced by an abundance of sugars—genes involved in accumulation of storage protein and starch, such as ADP-glucose pyrophosphorylase, starch synthase, and branching enzyme.

The integration of putative mechanisms for sensing carbohydrate status with the expression of genes involved in energy metabolism and synthesis of storage compounds brings a coherence to plant metabolism that is only beginning to be appreciated.

13.8 Energy-conserving reactions of glycolysis

The energy-conserving reactions of the lower half of glycolysis provide an excellent example of evolutionary engineering. The cell is presented with a dilemma: how to generate a phosphorylated compound for which hydrolytic removal of the phosphate results in a large, negative free energy change. Such molecules would be able to transfer a phosphate moiety to ADP, generating ATP through the formation of an acid anhydride bond (Fig. 13.39). Neither inorganic phosphate nor the phosphate groups of triose phosphates can be transferred directly to ADP.

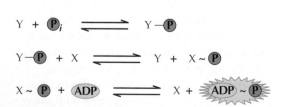

Figure 13.39
Classical scheme for generating a phosphorylated compound with a high free energy of hydrolysis. The bond with the high free energy of hydrolysis is shown as a tilde (~). The nature of the high-energy bond was debated at length before Peter Mitchell's chemiosmotic hypothesis was accepted.

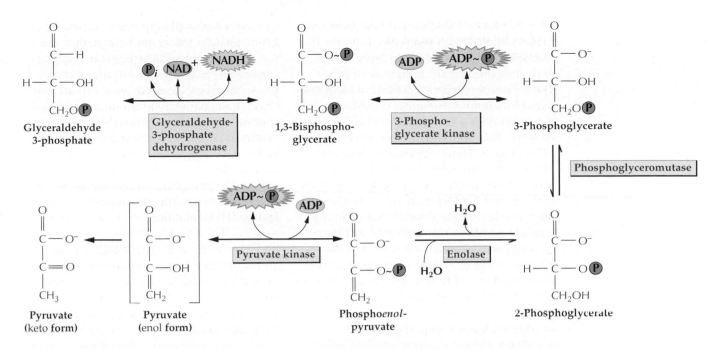

Figure 13.40
Oxidative and ATP-forming reactions of the glycolytic pathway, showing the relationship between the oxidation of the three-carbon molecules of the lower half of glycolysis and the creation of phosphate bonds with high negative free energies of hydrolysis (~).

The problem is solved by oxidation of triose phosphate molecules to pyruvate, during which energy is conserved in high-energy phosphate compounds that can be used to synthesize ATP from ADP. These high-energy compounds are generated by two enzymes, glyceraldehyde-3-phosphate dehydrogenase, which forms the mixed acid anhydride 1,3-bisphosphoglycerate, and by enolase, which generates a phosphate ester linked to the enol form of pyruvic acid, phospho*enol*pyruvate (Fig. 13.40). Because the instability of the enol form of pyruvate gives the phosphate ester a high negative free energy of hydrolysis, the phosphate group can be transferred to ADP, yielding ATP.

13.8.1 Plants contain three distinct glyceraldehyde-3-phosphate dehydrogenase activities.

Glyceraldehyde-3-phosphate dehydrogenase (Rx. 13.29) catalyzes the only reaction in glycolysis in which the oxidation of the substrate (glyceraldehyde 3-phosphate) is linked directly to the reduction of NAD$^+$ to NADH. This example of substrate-level phosphorylation has been studied extensively.

Reaction 13.29: Glyceraldehyde-3-P dehydrogenase

Glyceraldehyde 3-P + NAD$^+$ + P$_i$ $\rightleftharpoons$ 1,3-bisphosphoglycerate + NADH + H$^+$

The mixed acid anhydride product, 1,3-bisphosphoglycerate, is a high-energy compound capable of donating its C-1 phosphate to ADP. This example of substrate-level phosphorylation was considered the model for all cellular ATP formation until the Mitchell chemiosmotic scheme for ATP synthesis was accepted (see Chapters 3, 12, and 14).

The glyceraldehyde-3-phosphate dehydrogenase reaction is readily reversible, although the equilibrium position favors glyceraldehyde 3-phosphate over 1,3-bisphosphoglycerate by approximately 10:1. It is the subsequent phosphoglycerate kinase reaction that pulls the glyceraldehyde-3-phosphate dehydrogenase reaction in favor of the products. In the cell, the equilibrium position is also determined by the NAD$^+$/NADH ratio, because these molecules bind competitively to the enzyme. The concentration of NAD$^+$ in the cell is ordinarily maintained at approximately 10 times the concentration of NADH, again favoring 1,3-bisphosphoglycerate formation.

The reaction mechanism of this enzyme, first elucidated with pea seeds, involves the oxidation of glyceraldehyde 3-phosphate with the concomitant reduction of NAD$^+$. The 3-phosphoglyceric acid formed remains covalently bound through a thioester bond to a cysteine residue at the active site of the enzyme. This bond has a high free energy of hydrolysis. Hence, the energy made available by the oxidation of glyceraldehyde 3-phosphate is conserved not only through the formation of NADH from NAD$^+$ but also by the formation of the thioester bond. In the second half of the reaction catalyzed by the enzyme, phosphate also is bound to the active site of the enzyme and then transferred to the 3-phosphoglyceric acid moiety. The reaction is driven in the direction of 1,3-bis-phosphoglycerate synthesis by phosphorolysis of the thioester bond (Fig. 13.41).

Three distinct forms of glyceraldehyde-3-phosphate dehydrogenase exist in plants. In addition to the cytosolic NAD$^+$-dependent enzyme, an NADP$^+$-dependent enzyme in the chloroplast is involved in photosynthesis. A quite different enzyme, a nonphosphorylating NADP$^+$-dependent glyceraldehyde-3-phosphate dehydrogenase in the cytosol, generates 3-phosphoglyceric acid instead of 1,3-bisphosphoglycerate. This enzyme does not conserve energy. The nonphosphorylating enzyme may become important during phosphate starvation, allowing carbon flow through the pathway to continue when the concentrations of ADP are too low for the synthesis of ATP to proceed (Box 13.4).

13.8.2 The phosphoglycerate kinase reaction is reversible but strongly favors ATP formation.

Phosphoglycerate kinase (Rx. 13.30) catalyzes the formation of 3-phosphoglycerate from 1,3-bisphosphoglycerate. In the process, the phosphate group from the 1-position of 1,3-bisphosphoglycerate is transferred to ADP, forming ATP.

Reaction 13.30: 3-Phosphoglycerate kinase

$$\text{1,3-Bisphosphoglycerate} + \text{ADP} \rightleftharpoons$$
$$\text{3-Phosphoglycerate} + \text{ATP}$$

The equilibrium strongly favors the products of the reaction, because the mixed acid anhydride bond of 1,3-bisphosphoglycerate is very unstable and has a high free energy of hydrolysis. At a high ratio of ATP to ADP, however, the reverse reaction can occur. Both cytosolic and plastid forms of this enzyme exist, and both have properties similar to those from other organisms. In contrast to other pairs of plastid and cytosolic isoenzymes, however, the two isoenzymes of phosphoglycerate kinase are more similar than anticipated, suggesting their recent evolution from a common eukaryotic precursor.

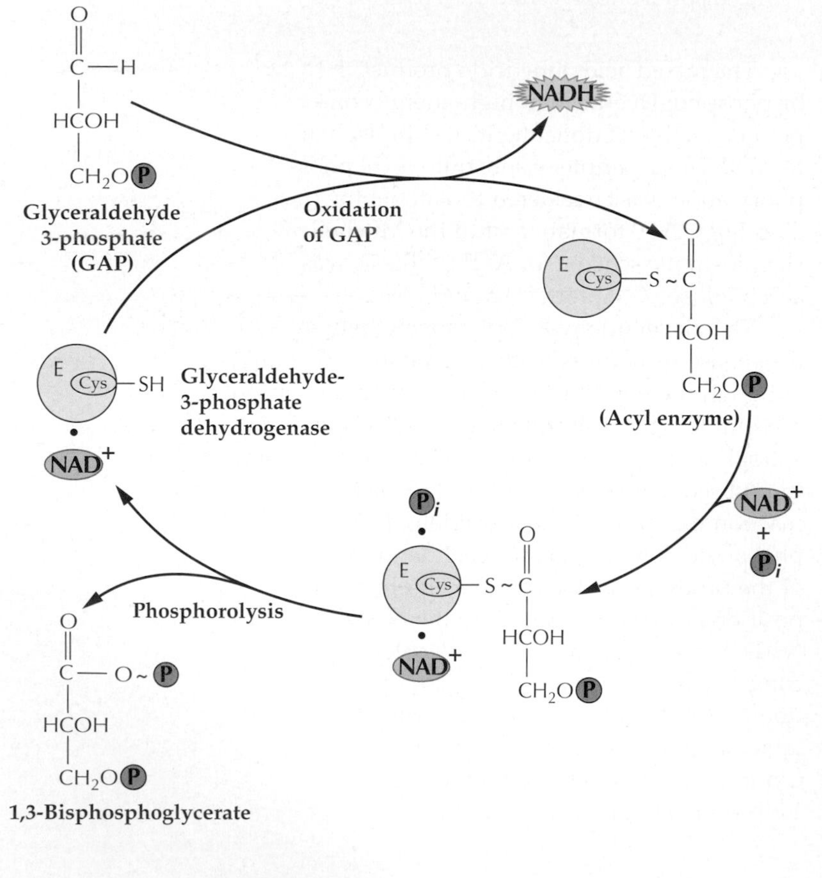

Figure 13.41
Substrate-level phosphorylation by glyceraldehyde-3-phosphate dehydrogenase. In this reaction a thioester bond is formed between phosphoglyceric acid (derived from GAP) and a cysteine residue in the enzyme. This bond, which has a high negative free energy of hydrolysis, is broken by inorganic phosphate via a phosphorolysis reaction in which the phosphate forms a mixed acid anhydride with phosphoglyceric acid (i.e., 1,3-bisphosphoglyceric acid). Like the thioester, the mixed acid anhydride bond has a high negative free energy of hydrolysis and is designated by a tilde (~). In the subsequent reaction catalyzed by phosphoglycerate kinase (not shown), this phosphate group is transferred from 1,3-bisphosphoglyceric acid to ADP to form ATP.

Plants are anchored to the ground and subjected to whatever the environment has to offer them; they cannot move to a more hospitable location. To overcome this vulnerability, plants have developed a highly flexible metabolism that makes it possible for them to accommodate environmental changes metabolically. Plants often have several different ways of accomplishing the same step in a metabolic pathway. This metabolic flexibility in plants is perhaps best illustrated when genetic engineering is used to knock out supposedly critical enzyme activities and the resulting transgenic plants are able to grow and develop more or less normally.

One factor contributing to this plasticity is that plants metabolize carbohydrate in two cellular locations, the cytosolic and plastid compartments. Depending on the transporters available in the plastid envelope, the presence of these compartments and their distinct pathways affords flexibility to the cell.

A second attribute of plant metabolism is the presence of more than one enzyme capable of catalyzing a specific step. For example, three reactions of glycolysis require adenine nucleotides. Plants contain three enzymes that bypass these glycolytic reactions and do not require adenine nucleotides (see the figure at right). These bypass enzymes—PFP, a nonphosphorylating glyceraldehyde-3-phosphate dehydrogenase that produces only NADPH and 3-phosphoglycerate, and a vacuolar phospho*enol*pyruvate phosphatase—are not found in animals or in most other eukaryotic organisms. The reaction catalyzed by pyruvate kinase also can be bypassed by a combination of phospho*enol*pyruvate carboxylase, malate dehydrogenase (see Chapter 14), and malic enzyme (see Rx. 13.33). The alternative oxidase pathway of plant mitochondria (see Chapter 14) is another way plants bypass enzymes that are essential in animal cells.

The function of these bypasses is not fully understood. They may provide the plant with an ability to withstand stresses that occur during phosphate starvation, a common condition. In some plants, the amount of PFP increases when plants are starved of phosphate. Under these conditions, concentrations of adenylate nucleotides are severely reduced, and reactions requiring ADP or ATP may become inhibited. Activation of PFP would allow the use of pyrophosphate as an alternative

energy source, thereby conserving ATP. The nonphosphorylating glyceraldehyde-3-phosphate dehydrogenase and phospho*enol*pyruvate phosphatase also would bypass reactions that require ADP as a substrate, thereby allowing carbon to pass through glycolysis without the need for ATP synthesis. Hence, the presence of bypass enzymes may allow carbon flow to continue under adverse conditions. We note, however, that transgenic tobacco plants lacking PFP do not appear less tolerant to phosphate stress than normal plants are. Sensitivity to phosphate starvation could be species-specific.

A surprising example of the flexibility of plant metabolism is the ability of transgenic plants to grow normally in the absence of cytosolic pyruvate kinase in their leaves. In humans, the lack of this enzyme

causes fatal hemolytic anemia. Similarly, yeast and bacteria cannot grow on hexoses in the absence of this enzyme. However, tobacco mutants lacking cytosolic pyruvate kinase appear nearly normal. Apparently, the reactions that bypass pyruvate kinase can convert sufficient phospho*enol*pyruvate to pyruvate to maintain growth. However, these plants do exhibit an unexpected phenotype. Although root cytosolic pyruvate kinase activity is not affected in the transgenic plants, their root-to-shoot ratios are smaller than those of wild-type plants, an indication that the supply of nutrients for transport may be altered by the absence of cytosolic pyruvate kinase in leaves. Thus, a mutation in one part of a plant may influence growth and development in another.

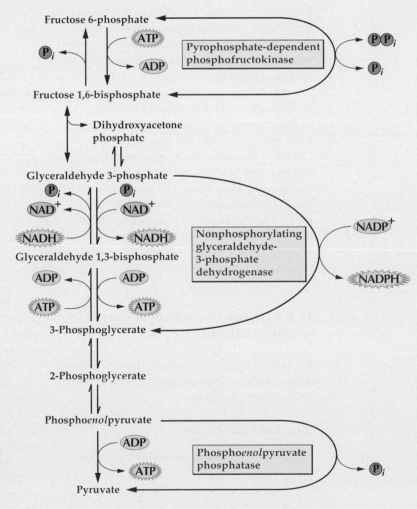

13.8.3 Phosphoglyceromutase is present in cytosol and in heterotrophic plastids.

Phosphoglyceromutase (Rx. 13.31) catalyzes the readily reversible interconversion of 3-phosphoglycerate and 2-phosphoglycerate. The enzyme moves the phosphate group from the 3-position to the 2-position, such that subsequent dehydration by enolase produces a high-energy compound.

Reaction 13.31: Phosphoglyceromutase

3-Phosphoglycerate $\rightleftharpoons$ 2-Phosphoglycerate

Of the two forms of phosphoglyceromutase, one requires a cofactor (2,3-bisphosphoglycerate) and one is cofactor independent. All animals have the cofactor-dependent type, whereas in plants all cytosolic and plastid isoenzymes belong to the cofactor-independent class; both isoenzyme types are found in the eubacterial kingdom. These two functionally similar types of phosphoglyceromutase share no sequence homology and must therefore have evolved from different ancestral proteins.

Phosphoglyceromutase is not present in all plastids. The absence of this enzyme in chloroplasts may be essential, because 3-phosphoglycerate is a product of the Calvin cycle and phosphoglyceromutase activity would draw carbon from this cycle into plastid glycolysis. The enzyme is present in colorless plastids from seeds and roots. The cytosolic and plastid isoenzymes are difficult to separate and are very similar, although some distinct properties have been reported.

13.8.4 Enolase concentrations can rise in response to abiotic stresses.

Enolase (Rx. 13.32) catalyzes the reversible dehydration of 2-phosphoglycerate to form phospho*enol*pyruvate.

Reaction 13.32: Enolase

2-Phosphoglycerate $\rightleftharpoons$ PEP + H_2O

The dehydration of 2-phosphoglycerate generates phosphopyruvate in the enol configuration. Because this configuration is unstable, the phosphate attached to the 2-position of pyruvate has a high negative free energy of hydrolysis, allowing transfer of the phosphate group to ADP in the subsequent reaction catalyzed by pyruvate kinase. Cloning of the plant enzyme shows it to be similar to the enzyme from other organisms.

Plastid and cytosolic isoenzymes occur, but the amount of the plastid form present varies considerably; it appears to be absent in leaves but abundant in plastids from seeds. In general, the amount of plastid isoenzyme and the level of biosynthetic activity of the tissue in nonphotosynthetic tissue apparently are correlated positively.

In some plants, enolase is recognized as a heat-shock protein; its concentration increases when the plant is exposed to heat stress. In maize, enzyme activities are enhanced fivefold when the tissue is stressed through anaerobiosis, possibly to facilitate an increased ATP production in the absence of oxidative phosphorylation (see Chapter 22).

13.8.5 Pyruvate kinase regulation in plants is complex but does not resemble mechanisms regulating that enzyme in mammalian liver.

Pyruvate kinase (Rx. 13.33) catalyzes the transfer of the phosphate group from phospho*enol*pyruvate to ADP with the formation of ATP and pyruvate.

Reaction 13.33: Pyruvate kinase

PEP + ADP $\longrightarrow$ Pyruvate + ATP

The reaction is virtually irreversible because the product of the reaction is the enol form of pyruvate (see Fig. 13.40). In phospho*enol*pyruvate, the enol form is stabilized by the phosphoester bond. Removal of the phosphate group allows the resulting enol form of pyruvate to convert to the more stable keto form, removing the product from the reaction and driving the equilibrium toward product formation.

Several plastid and cytosolic isoenzymes of pyruvate kinase exist. Unlike other enzymes of glycolysis, the plastid form is present in all plant tissues studied thus far. Examples of both the plastid and cytosolic

Box 13.5

The mutable nature of plant carbohydrate metabolism presents a challenge to plant scientists.

This chapter would have been much easier to write, and perhaps to read, if glycolysis, the pentose phosphate pathway, and starch and sucrose degradation and synthesis had been presented as independent pathways. Most textbooks take this approach, but plants themselves do not. Pools of metabolic intermediates are shared by several pathways, with the direction of metabolite flow being determined by the needs of the cell. These needs change according to the availability of light and other environmental factors as well as the tissue type and developmental stage involved. Because plants have evolved a highly flexible metabolism, plant research can be difficult and often perplexing.

For example, to collect consistent data on the carbohydrate content of transgenic and wild-type tobacco leaves, leaves must be harvested at the same time of day, at exactly the same developmental stage, from plants grown in the same chamber. Furthermore, the seeds from which these experimental plants are grown must have been harvested from plants that were themselves grown under identical conditions because the quality of the seed can affect the metabolism throughout the life of the plant. Given this extreme sensitivity to environmental and developmental variables, it is often difficult to reconcile the data collected from plants grown on different continents.

Similarly, each plant tissue contains many distinct cell types, all of which are different. Homogenized leaves, the source material for many biochemical analyses, contain an unresolvable mixture of the various enzymes and intermediates present in palisade and spongy mesophyll cells, companion cells, vascular sheath cells, and many others. Each of these cell types performs distinct biochemical roles that change during development.

Likewise, the generalization that an enzyme is or is not present in chloroplasts belies the enormous functional diversity of those organelles. A chloroplast from a mature leaf fixes CO_2 for conversion to sucrose and transport to other parts of the plant. A chloroplast from a young leaf is basically a biosynthetic organelle manufacturing compounds for cell growth and development. A green plastid from developing cotyledons in developing rapeseed uses energy from photosynthesis to provide cofactors for fatty acid biosynthesis within the organelle, for synthesis of storage oils. In contrast, a green plastid from pea cotyledons is basically an amyloplast, synthesizing and storing starch. All use the metabolite pools in different ways, and perform distinct tasks, but all are referred to as chloroplasts.

In short, understanding and researching plant metabolism is a challenge but offers unique opportunities unlike any to be found investigating other organisms.

isoenzymes have been cloned, and the sequences of these enzymes suggest that they all employ the same reaction mechanism, similar to the reaction mechanism observed in pyruvate kinases from nonplant sources. In most cases, the enzyme is a homotetramer, although other multimeric forms of the enzyme have been reported.

Recently, regulatory kinetics have been described for some pyruvate kinases of photosynthetic organisms. Cytosolic pyruvate kinase from the green alga *Selenastrum* is regulated by amino acids such as glutamate and glutamine (Fig. 13.42). This couples the activity of the enzyme to the supply of carbon for amino acid biosynthesis and nitrogen assimilation. A similar regulation has now been found for the enzyme from biosynthetically active tissues from higher plants. This form of regulation of the enzyme is related to the autotrophic nature of plants, which requires that a large amount of carbon be directed into biosynthetic activity.

Plastids have at least two distinct isoenzymes of pyruvate kinase. The sequences of these isoenzymes differ, which suggests that they have evolved independently. The genes for these two isoenzymes also differ—not only from one another but also from the gene for the cytosolic enzyme. In addition, there are distinct differences in the way these forms of pyruvate kinase are transported into the plastid. Finally, one plastid form has a high affinity for substrates and is present in large amounts in biosynthetically active tissues, where it may provide ATP for these activities (see Section 13.9). This form is inhibited by amino acids and is activated by metabolites of carbohydrate metabolism such as dihydroxyacetone phosphate—regulatory kinetics similar to those of some cytosolic isoenzymes (Fig. 13.42).

Pyruvate kinase from mammalian liver and from many other sources is highly regulated. Specific sites on these enzymes are subject to phosphorylation and dephosphorylation by a protein kinase and phosphatase; the phosphorylated enzyme is inactive. None of the plant enzymes, however, appear to undergo phosphorylation. Furthermore, sequence data from the plant enzyme show no sites similar to the phosphorylation sites of the mammalian enzyme.

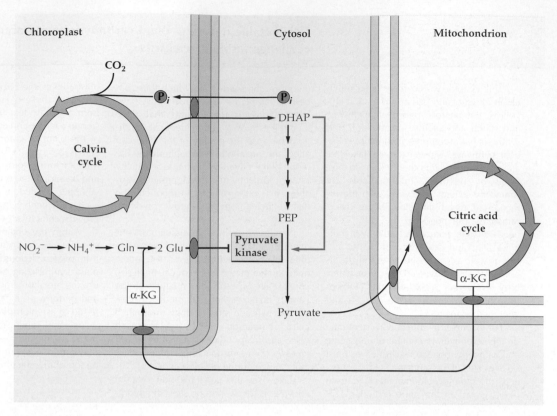

Figure 13.42

Regulation of cytosolic pyruvate kinase and ammonia assimilation in the chloroplast of the green alga *Selenastrum minutum*. The regulation of pyruvate kinase in the cytosol is related to the synthesis of amino acids in the chloroplast. Gln, glutamine; Glu, glutamate; DHAP, dihydroxyacetone phosphate; PEP, phospho-*enol*pyruvate; α-KG, α-ketoglutarate.

As described above, pyruvate kinase is also allosterically activated by fructose 1,6-bisphosphate in animals and most other organisms (see Fig. 13.35). Thus, the activity of pyruvate kinase is linked to that of PFK by a feed-forward loop. Although sequences similar to the fructose 1,6-bisphosphate activation site do occur in the plant enzyme, no evidence has been found for any feed-forward regulation.

13.9 Supply of energy and reducing power for biosynthetic reactions

Clearly, carbohydrate metabolism occurs in both the cytosolic and plastid compartments of the plant cell. Synthesis of compounds derived from hexoses would be expected to occur in the plastids, because these organelles are responsible for photosynthesis and starch storage. Thus, the discovery of plastid-localized mechanisms for the conver-

sion of hexoses to other metabolites did not surprise researchers. But why should the enzymes of the lower part of glycolysis also occur in both the cytosolic and plastid compartments?

In most organisms, carbohydrate metabolism occurs in the cytosol, as a means of funneling carbon into the mitochondrion for oxidative phosphorylation and for the production of biosynthetic precursors. Glycolysis serves this latter function in certain plastids as well, supplying carbon precursors for biosynthetic pathways that branch off from the energy-conserving glycolytic reactions.

An important additional function of glycolytic enzymes, especially in nonphotosynthetic plastids, may be the supply of cofactors (see Fig. 13.6). Anabolic pathways (e.g., fatty acid biosynthesis) can consume large amounts of energy, in the form of ATP, and reducing power, in the form of NAD(P)H. In animals, fatty acid biosynthesis occurs in the cytosol, where there is a supply of cofactors. In chloroplasts, photosynthesis can provide the needed cofactors. However, in nonphotosynthetic plastids, the cofactors must be generated from the activity of metabolic pathways.

Plastids have an adenylate transport system that exchanges ATP for ADP. No known transport system exists for the direct import of NADH or NADPH, although reducing equivalents can be imported or exported by one or more shuttle systems (see Fig. 13.4). In most plastids, cofactors appear to be generated within the organelle by use of sequestered metabolic pathways, such as glycolysis, or the oxidative reactions of the pentose phosphate pathway. In isolated colorless plastids of castor seed endosperm, an active tissue for fatty acid synthesis, the most effective substrate for fatty acid biosynthesis is malate. A transporter exchanges malate for inorganic phosphate. Within the plastid, malic enzyme (Rx. 13.34) oxidatively decarboxylates malate, generating NADPH and pyruvate.

Reaction 13.34: Malic enzyme

$$\text{Malate} + \text{NADP}^+ \longrightarrow$$
$$\text{Pyruvate} + \text{NADPH} + \text{CO}_2$$

The oxidative decarboxylation of pyruvate in the plastid generates acetyl-CoA for fatty acid biosynthesis and the reducing agent NADPH. The operation of pyruvate kinase in this organelle can supply the required ATP. Hence, both the cofactors and the acetyl-CoA substrate consumed in fatty acid biosynthesis are generated within the organelle (Fig. 13.43).

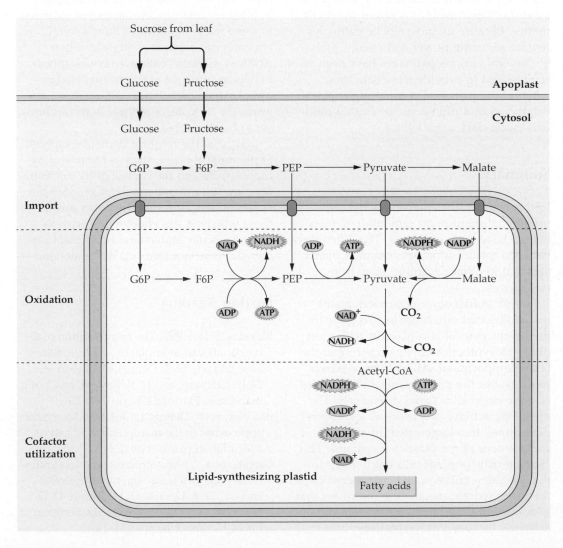

Figure 13.43
Use of photoassimilate by lipid-storing seeds to generate cofactors for fatty acid synthesis in a lipid-synthesizing plastid. In colorless plastids, sequestration of glycolytic enzymes, NADPH-dependent malic enzyme, and the pyruvate dehydrogenase complex provides not only carbon for fatty acid biosynthesis but also cofactors such as ATP, NADH, and NADPH required for this biosynthetic pathway. G6P, glucose 6-phosphate; F6P, fructose 6-phosphate; PEP, phospho*enol*pyruvate.

In other oilseeds, such as rapeseed (canola), isolated plastids at an early stage of seed development appear to use hexose phosphates as the preferred carbon source. These plastids have all the glycolytic enzymes necessary to convert the hexose phosphate to pyruvate, which can be further converted to acetyl-CoA by a plastid-localized pyruvate dehydrogenase complex. Glycolysis generates both NADH and ATP as well as the carbon precursor. The NADPH required for fatty acid biosynthesis is supplied by the oxidative reactions of the pentose phosphate pathway. At later stages of seed development, pyruvate becomes the preferred substrate, and plastid-localized glycolysis probably provides the ATP. Under these circumstances, NADH is generated by the pyruvate dehydrogenase complex and NADPH by the pentose phosphate pathway.

The source of cofactors for biosynthetic reactions is clearly a major issue within the plastid compartment in nonphotosynthetic tissues. Because all biosynthetic pathways require reducing power and energy, probably several various pathways have been sequestered to provide these cofactors, whether the principal product of the plastid is fatty acids, amino acids, or natural products (secondary metabolites).

Summary

In this chapter, the complexity and the unique nature of plant carbohydrate metabolism have been presented. The differences between this form of metabolism in plants and that in other organisms stem from two factors.

First, unlike other organisms, plants metabolize carbohydrates in two compartments, the cytosol and plastids, with most of the biosynthetic activity occurring in the latter compartment. All plant tissues have plastids, but the primary metabolic activity of these organelles varies extensively. The metabolic activity of a tissue as a whole is determined to a large extent by the enzyme complement of the plastids it contains. This characteristic of plant cells has a major impact on the regulation and characteristics of the carbohydrate metabolism pathways that generate building blocks for the biosynthetic activities localized in the plastids.

Carbohydrate metabolism in the cytosol is connected with plastid metabolism by a series of carriers in the plastid envelope. Each plastid type has a different group of carriers, which usually reflects the nature of the biosynthetic activity within the organelle. Universally present, however, is the TPT, which exchanges triose phosphate for inorganic phosphate. The overall regulation of carbohydrate metabolism is linked to the activity of this and the other transporters. The TPT connects the cytosol with the plastid stroma at about the midpoint of the glycolytic pathway. For this reason, the regulation of glycolysis differs from that in other organisms, in which metabolism is regulated by the supply of nutrients.

Second, all cells in a plant have to withstand changes in the external environment without the homeostatic mechanisms that stabilize the milieu surrounding animal cells. This has led to an inherent flexibility in plant metabolism reminiscent of that seen in some microorganisms. Often, several enzymes can catalyze a single step in a pathway. Certain of these enzymes appear to bypass principal reactions involved in carbohydrate metabolism, although mechanisms for the control of these bypasses have yet to be elucidated.

Some of the most fundamental aspects of plant metabolism, such as the role of pyrophosphate and the control of its concentration, are still not understood. Considerable opportunities for future discovery await those who read this chapter and wish to pursue a better understanding of plant carbohydrate metabolism and its regulation.

Further Reading

Blakeley, S. D. (1997) The manipulation of resource allocation in plants. In *Plant Metabolism*, 2nd ed., D. T. Dennis, D. H. Turpin, D. D. Lefebvre, and D. B. Layzell, eds. Longman Press, Harlow, UK, pp. 580–591.

Blakeley, S. D., Dennis, D. T. (1993) Molecular approaches to the manipulation of carbon allocation in plants. *Can. J. Bot.* 71: 765–778.

Carpita, N. C. (1996) Structure and biosynthesis of plant cell walls. In *Plant Metabolism*, 2nd ed., D. T. Dennis, D. H. Turpin, D. D. Lefebvre, and D. B. Layzell, eds. Longman Press, Harlow, UK, pp. 124–147.

Carpita, N. C. (1997) Structure and biogenesis of the cell walls of grasses. *Annu. Rev. Plant Physiol. Plant Mol. Biol.* 47: 445–476.

Coultate, T. P. (1996) *Food: The Chemistry of Its Components*, 3rd ed. RSC Paperbacks, Cambridge, UK.

Delmer, D. P., Amor, Y. (1995) Cellulose biosynthesis. *Plant Cell* 7: 987–1000.

Dennis, D. T., Blakeley, S. D. (1995) The regulation of carbon partitioning in plants. In *Carbon Partitioning and Source Sink Interactions in Plants*, M. A. Madore and W. J. Lucas, eds. *Curr. Topics Plant Physiol.* 13: 258–267.

Dennis, D. T., Greyson, M. F. (1987) Fructose 6-phosphate metabolism in plants. *Physiol. Plant* 69: 395–404.

Dennis, D. T., Huang, Y., Negm, F. B. (1996) Glycolysis, the pentose phosphate pathway and anaerobic respiration. In *Plant Metabolism*, 2nd ed., D. T. Dennis, D. H. Turpin, D. D. Lefebvre, and D. B. Layzell, eds. Longman Press, Harlow, UK, pp. 105–123.

Dennis, D. T., Miernyk, J. A. (1982) Compartmentation of non-photosynthetic carbohydrate metabolism. *Annu. Rev. Plant Physiol.* 33: 27–50.

Duff, S.M.G, Moorhead, G. B., Lefebvre, D. D., Plaxton, W. C. (1989) Phosphate starvation inducible "bypasses" of adenylate and phosphate-dependent glycolytic enzymes in *Brassica nigra* suspension cells. *Plant Physiol.* 90: 1275–1278.

Emes, M. J., Dennis, D. T. (1997) Regulation by compartmentation. In *Plant Metabolism*, 2nd ed., D. T. Dennis, D. H. Turpin, D. D. Lefebvre, and D. B. Layzell, eds. Longman Press, Harlow, UK, pp. 69–80.

Emes, M. J., Tobin, A. K. (1993) Control of metabolism and development in plant plastids. *Int. Rev. Cytol.* 145: 149–216.

Hajirezaei, M., Sonnewald, U., Viola, R., Carlisle, S., Dennis, D. T., Stitt, M. (1994) Transgenic potato plants with strongly decreased expression of pyrophosphate:fructose 6-phosphate phosphotransferase show no visible phenotype and only minor changes in tuber metabolism. *Planta* 192: 16–30.

Huber, S. C., Huber, J. L. (1996) Role and regulation of sucrose-phosphate synthase in plants. *Annu. Rev. Plant Physiol. Plant Mol. Biol.* 47: 431–444.

Kennedy, R. A., Rumpho, M. E., Fox, T. C. (1992) Anaerobic metabolism in plants. *Plant Physiol.* 100: 1–6.

Koch, K. E. (1996) Carbohydrate-modulated gene expression in plants. *Annu. Rev. Plant Physiol. Plant Mol. Biol.* 47: 509–540.

Kruger, N. J. (1997) Carbohydrate synthesis and degradation. In *Plant Metabolism*, 2nd ed., D. T. Dennis, D. H. Turpin, D. D. Lefebvre, and D. B. Layzell, eds. Longman Press, Harlow, UK, pp. 83–104.

Lin, M., Turpin, D. H., Plaxton, W. C. (1989) Pyruvate kinase isozymes from the green alga, *Selenastrum minutum*. II. Kinetic and regulatory properties. *Arch. Biochem. Biophys.* 269: 228–238.

Marsh, J. J., Lebherz, H. G. (1992) Fructose-bisphosphate aldolases: an evolutionary history. *Trends Biochem. Sci.* 195: 110–113.

Martin, C., Smith, A. M. (1995) Starch biosynthesis. *Plant Cell* 7: 971–985.

Paul, M., Sonnewald, U., Dennis, D. T., Stitt, M. (1995) Transgenic tobacco plants with strongly decreased expression of pyrophosphate:fructose-6-phosphate 1-phosphotransferase do not differ significantly from wild type in photosynthate partitioning, plant growth or their ability to cope with limiting phosphate, limiting nitrogen and suboptimal temperatures. *Planta* 196: 277–283.

Perata, P., Alpi, A. (1993) Plant responses to anaerobiosis. *Plant Sci.* 93: 1–17.

Plaxton, W. C. (1996) The organization and regulation of plant glycolysis. *Annu. Rev. Plant Physiol. Plant Mol. Biol.* 47: 185–214.

Plaxton, W.C.P. (1997) Metabolic regulation. In *Plant Metabolism*, 2nd ed., D. T. Dennis, D. H. Turpin, D. D. Lefebvre, and D. B. Layzell, eds. Longman Press, Harlow, UK, pp. 50–68.

Preiss, J., Ball, K., Smith-White, B., Inglesias, A., Kakefuda, G., Li, L. (1991) Starch biosynthesis and its regulation. *Biochem. Soc. Trans.* 19: 539–547.

Smith, A. M., Martin, C. (1993) Starch biosynthesis and the potential for its manipulation. In *Biosynthesis and Manipulation of Plant Products* (Plant Biotechnol. Ser., Vol. 3), D. Grierson, ed. Blackie Academic and Professional Publishers, Glasgow, pp. 1–54.

Stitt, M. (1990) Fructose 2,6-bisphosphate as a regulatory molecule in plants. *Annu. Rev. Plant Physiol. Plant Mol. Biol.* 41: 153–185.

Stitt, M. (1997) The flux of carbon between the chloroplast and cytoplasm. In *Plant Metabolism*, 2nd ed., D. T. Dennis, D. H. Turpin, D. D. Lefebvre, and D. B. Layzell, eds. Longman Press, Harlow, UK, pp. 382–400.

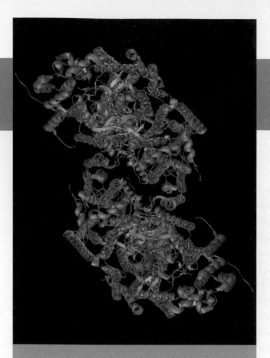

Biochemistry & Molecular Biology of Plants, B. Buchanan, W. Gruissem, R. Jones, Eds.
© 2000, American Society of Plant Physiologists

CHAPTER 14

Respiration and Photorespiration

James N. Siedow
David A. Day

Introduction

Chapters 12 and 13 have described how plants use light energy to assimilate carbon into sugars and starch and how these molecules are subsequently broken down and converted into organic acids and other compounds. In this chapter we examine aerobic respiration—the further oxidation of these compounds to CO_2 and H_2O in the mitochondrion—and review the mechanisms by which the energy released during respiration is conserved as ATP, a process called oxidative phosphorylation. We examine how plants can minimize ATP production while maintaining respiration rates, a mitochondrial attribute that may affect plant responses to environmental stress. We also describe how mitochondria and other cellular compartments interact by means of substrate shuttles across the inner mitochondrial membrane.

In addition to aerobic respiration, another respiratory process takes place in the leaves of many plants. This novel process, photorespiration, is the light-dependent release of CO_2 and uptake of O_2. The O_2 uptake occurs in the chloroplast as a result of the oxygenase reaction of Rubisco, which leads to production of phosphoglycolate. Metabolism of this compound involves a complex interaction among chloroplasts, peroxisomes, and mitochondria and leads to release of CO_2. Although photorespiration in effect drains carbon from plants and adversely affects growth, the process seems to be an unavoidable side reaction of CO_2 fixation in most plants. Some plants, however, have evolved special anatomical and biochemical features that minimize the oxygenase reaction. These plants, known as C_4 plants and Crassulacean acid metabolism (CAM) plants (see Chapter 12), demonstrate very low rates of photorespiration.

14.1 Overview of respiration

Aerobic respiration, a process common to almost all eukaryotic organisms, involves the controlled oxidation of reduced organic substrates to CO_2 and H_2O. Numerous compounds can serve as substrates for respiration, including carbohydrates, lipids, proteins, amino acids, and organic acids. Respiration releases a large amount of free energy, which is conserved in the acid anhydride linkages of ATP molecules. This chemical bond energy can be used to drive metabolic reactions involved in the growth, development, and maintenance of the plant. In addition, the primary pathways of respiration provide metabolic intermediates that serve as substrates for the synthesis of nucleic acids, amino acids, fatty acids, and many secondary metabolites. Although the general process of respiration in plants is the same as in other eukaryotes, several features are unique to plants. These modifications apparently evolved to cope with the unique environmental and metabolic circumstances commonly faced by plants.

14.1.1 Plant mitochondria contain an outer and an inner membrane that separate the organelle into four functional compartments.

The **mitochondrion** is the principal organelle of eukaryotic respiration. Plant mitochondria appear as spherical or rod-shaped entities 0.5 to 1.0 $\bar{\mu}$m in diameter and 1 to 3 $\bar{\mu}$m long. The number of mitochondria per plant cell varies and is related primarily to the metabolic activity of the particular tissue. However, the concentration of mitochondria per unit volume of cytoplasm usually remains similar during different developmental stages of the same cell type. For example, the small cells of the root cap of maize seedlings contain approximately 200 mitochondria each, whereas the much larger mature root tip cells may have approximately 2000 mitochondria each. In very active cells, such

as phloem companion cells, secretory cells, and transfer cells, a large fraction (up to 20%) of the volume of the cytoplasm may be occupied by mitochondria. On the other hand, some unicellular algae (such as *Chlamydomonas*) contain only a few mitochondria per cell. With a few exceptions (e.g., the very active cell types mentioned above), the density of mitochondria observed in most plant cells is less than that found in a typical animal cell. However, the respiratory rates of mitochondria isolated from plants are generally higher than in those isolated from animals.

Mitochondria, including those from plants, contain two sets of membranes that divide the organelle into four compartments: the **outer membrane;** the region between the two membranes **(intermembrane space);** a highly invaginated **inner membrane;** and the aqueous phase contained within the inner membrane (mitochondrial **matrix**) (see Fig. 14.1A). Invaginations of the inner membrane give rise to **cristae,** which are seen as sac-like structures under the electron microscope (Fig. 14.1B). The marked impermeability of the inner membrane necessitates

(A)

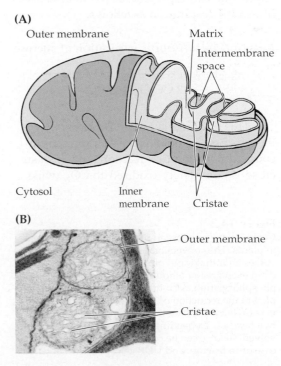

(B)

Figure 14.1
Structural organization of the mitochondrion. (A) Diagram of the membrane organization of a plant mitochondrion. The outer membrane is permeable to molecules of 10 kDa or less because it contains a pore-forming protein called porin. The highly invaginated inner membrane, which contains the components of the mitochondrial electron transfer chain and ATP synthase, serves as the primary permeability barrier for the organelle. The space between the two membranes is called the intermembrane space. The mitochondrial matrix, the region bounded by the inner membrane, contains the mitochondrial genome, the enzymes associated with the citric acid cycle, and the machinery required for mitochondrial protein synthesis. (B) Thin-section electron micrograph of a plant mitochondrion, showing the highly folded inner membrane structures (cristae) and the surrounding outer membrane.

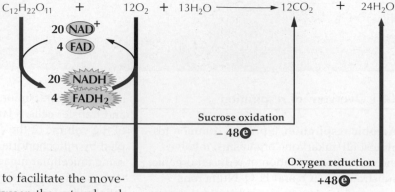

Figure 14.2
Redox reactions of respiration. The oxidation of carbohydrate to carbon dioxide is coupled to the reduction of oxygen to water. The free energy released during this process is linked to the synthesis of ATP.

the presence of carriers to facilitate the movement of metabolites between the cytosol and the mitochondrial matrix. By contrast, the outer membrane is permeable to solutes of molecular mass less than 10 kDa. This permeability is associated with the presence of pore-forming proteins known as **porins.** In some nonplant species, porin channels are known to be gated by voltage (see Chapter 3 for a discussion of voltage gating).

14.1.2 The principal products of respiration are CO_2, H_2O, and free energy conserved as ATP.

Although respiration can oxidize many compounds, the principal substrates provided by photosynthesis (see Chapter 12) for plant respiration are sucrose and starch. Using sucrose as an example of a respiratory substrate, one can represent the overall process of aerobic respiration as follows:

Reaction 14.1: Aerobic respiration of sucrose

$$C_{12}H_{22}O_{11} + 12\ O_2 + 13\ H_2O \longrightarrow 12\ CO_2 + 24\ H_2O$$

This reaction scheme, the reverse of the process of photosynthesis, represents a coupled pair of redox reactions in which sucrose is completely oxidized to CO_2, while

O_2, the terminal electron acceptor, is reduced to water (Fig. 14.2). The standard free energy change ($\Delta G^{o'}$) for this exergonic reaction, -5764 kJ mol^{-1}, provides the thermodynamic driving force for the production of ATP.

The three stages of respiration—**glycolysis,** the **citric acid cycle,** and **electron transfer/oxidative phosphorylation**—take place in different subcellular locations. Glycolysis involves a series of soluble cytosolic enzymes that oxidize sugars to organic acids. In the reaction above, one sucrose molecule is broken into two hexoses, which are then converted by glycolysis to four molecules of the three-carbon (C_3) compound pyruvate (Fig. 14.3) (see Chapter 13). This partial oxidation of sucrose produces the reduced cofactor NADH (Fig. 14.4), as well as ATP. The reactions of the citric acid

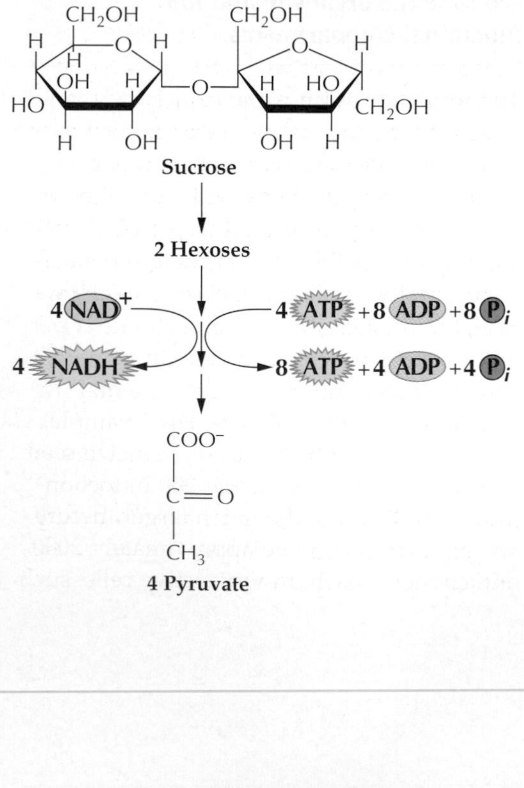

Figure 14.3
Conversion of sucrose to pyruvate during glycolysis in plants. The conversion of sucrose to pyruvate involves an initial input of ATP but leads to formation of an even greater amount of ATP via substrate-level phosphorylation. Also involved is an oxidation coupled to the reduction of NAD^+ to NADH. The products of sucrose catabolism are described here as two hexoses. Depending on the enzyme activity involved, sucrose can be converted to fructose and glucose or to fructose and UDP-glucose (see Chapter 13).

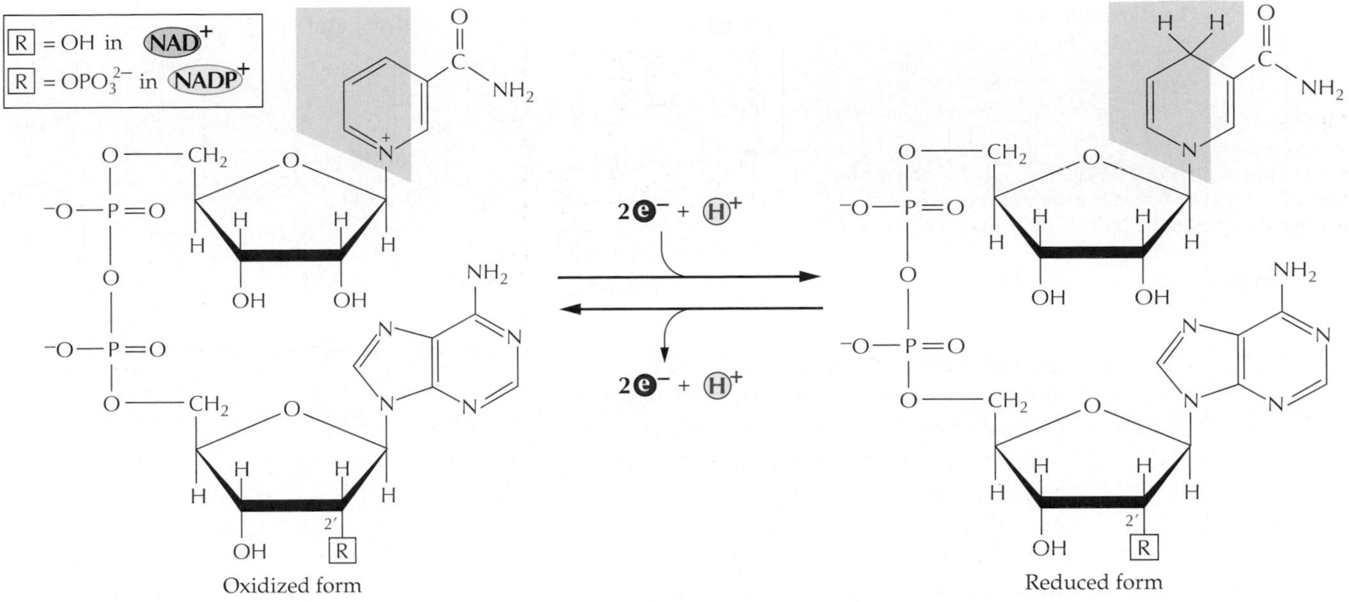

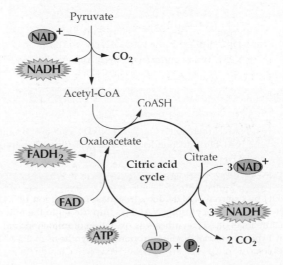

Figure 14.4
Structure and redox chemistry of nicotinamide adenine dinucleotide (NAD^+) and nicotinamide adenine dinucleotide phosphate ($NADP^+$). The nicotinamide moiety of the oxidized species NAD^+ gains two electrons and a proton to produce the reduced species

NADH. The related coenzyme $NADP^+$ differs from NAD^+ by the addition of a phosphate group to the 2'-hydroxyl of the ribose sugar moiety (see variable group R in the diagram).

cycle occur within the mitochondrial matrix and completely oxidize pyruvate to CO_2 (Fig. 14.5); the electrons are transferred to NAD^+ and another cofactor, FAD (Fig. 14.6), yielding NADH and $FADH_2$. The citric acid cycle also phosphorylates ADP directly (see Fig. 14.5). Finally, in the inner mitochondrial membrane, the reduced cofactors generated during glycolysis and the citric acid cycle are oxidized by a set of electron transfer proteins that ultimately donate the electrons to oxygen, producing H_2O. The large amount of free energy released during this electron transfer is used to drive the formation of a proton electrochemical gradient ($\Delta\bar{\mu}_H{}^+$) across the inner mitochondrial membrane, and the energy stored in this gradient subsequently drives the conversion of ADP and P_i to ATP by a process known as oxidative phosphorylation (Fig. 14.7; see also Table 14.1). It is important to recall that not every molecule of sucrose or C_6 unit of starch metabolized via glycolysis and the citric acid cycle is completely oxidized in this manner. Many intermediates from these two pathways are diverted along the way to provide the primary carbon building blocks for important cellular compounds (Fig. 14.8).

14.2 Citric acid cycle

Our detailed discussion of respiration will focus on mitochondrial reactions (glycolysis is described in Chapter 13). Known by several names, including the tricarboxylic acid cycle and the Krebs cycle (after its discoverer, Hans Krebs), the citric acid cycle moves electrons from organic acids to the oxidized redox cofactors NAD^+ and FAD, forming NADH, $FADH_2$, and carbon dioxide.

Figure 14.5
The citric acid cycle leads to the complete oxidation of pyruvate. As pyruvate is progressively oxidized to three molecules of CO_2, its electrons are transferred to redox cofactors, reducing four molecules of NAD^+ to NADH and one molecule of FAD to $FADH_2$. This series of reactions takes place entirely within the mitochondrial matrix. In addition, one molecule of ATP is synthesized directly from ADP and P_i.

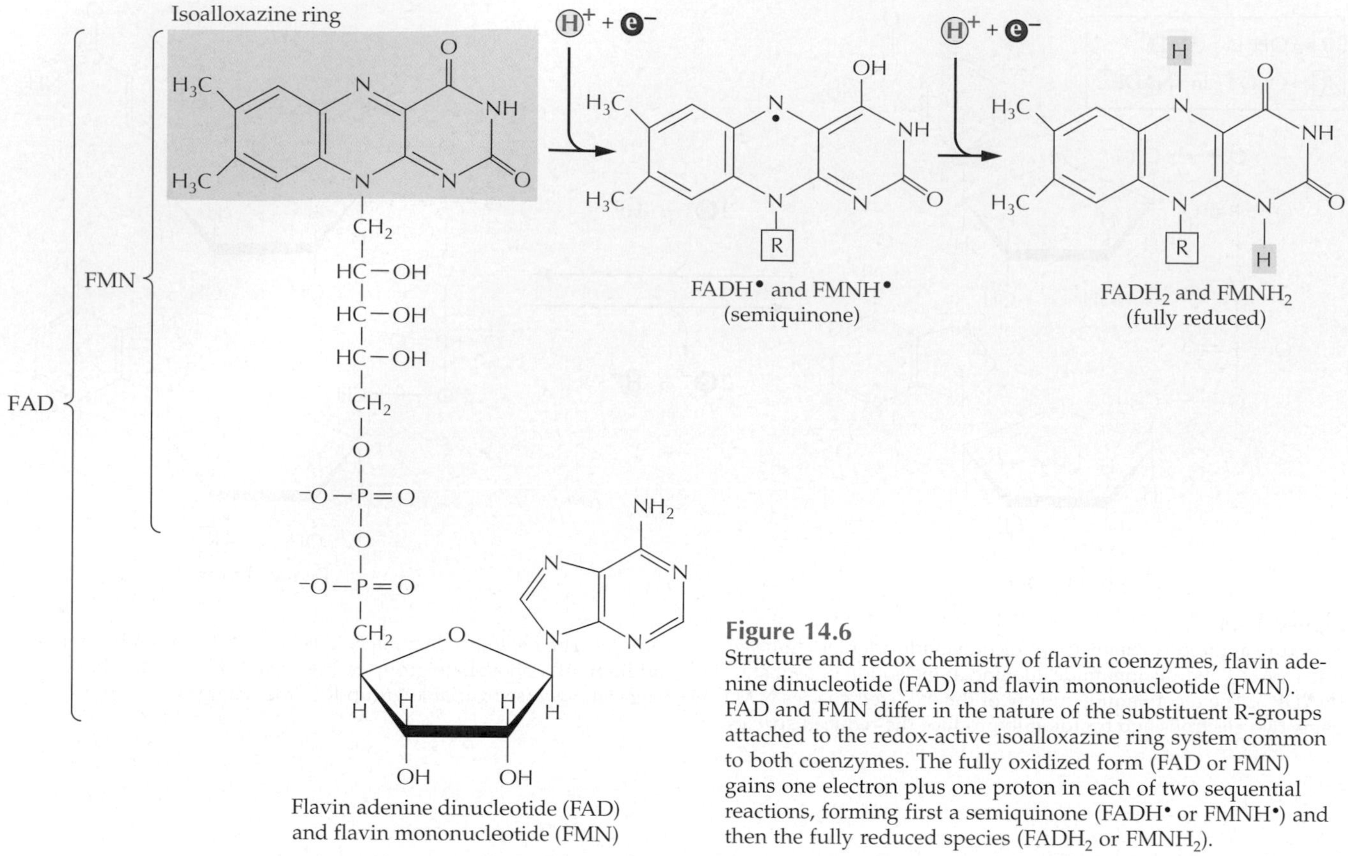

Figure 14.6
Structure and redox chemistry of flavin coenzymes, flavin adenine dinucleotide (FAD) and flavin mononucleotide (FMN). FAD and FMN differ in the nature of the substituent R-groups attached to the redox-active isoalloxazine ring system common to both coenzymes. The fully oxidized form (FAD or FMN) gains one electron plus one proton in each of two sequential reactions, forming first a semiquinone (FADH• or FMNH•) and then the fully reduced species (FADH$_2$ or FMNH$_2$).

Table 14.1 Net stoichiometry of sucrose oxidation

Metabolic pathway	Substrates	Products	ATP yield, no. of molecules
Glycolysis			
	1 Sucrose 4 ADP + 4 P$_i$ 4 NAD$^+$ (cytosolic)	4 Pyruvate 4 ATP 4 NADH (cytosolic)	4
Citric acid cycle			
	4 Pyruvate 4 ADP + 4 P$_i$ 16 NAD$^+$ (mitochondrial) 4 FAD	12 CO$_2$ 4 ATP 16 NADH (mitochondrial) 4 FADH$_2$	4
Oxidative phosphorylation			
	12 O$_2$ 4 NADH (cytosolic) 16 NADH (mitochondrial) 4 FADH$_2$	24 H$_2$O 4 NAD$^+$ (cytosolic or mitochondrial)[a] 16 NAD$^+$ (mitochondrial) 4 FAD	6–10[a,b] 40[b] 6[b]
Cumulative ATP yield			60–64[a,b]

[a] Oxidation of cytosolic NADH by external NADH dehydrogenase (see Section 14.3.4) supports the same level of ATP synthesis as oxidation of mitochondrial FADH$_2$. However, if cytosolic NADH is imported into the mitochondrion via the malate–aspartate shuttle (see Fig. 14.39A), it can support the same level of ATP synthesis as oxidation of mitochondrial NADH.
[b] Assuming that oxidation of 1 mitochondrial NADH supports the synthesis of 2.5 ATP and that oxidation of FADH$_2$ supports the synthesis of 1.5 ATP (see Section 14.4.3).

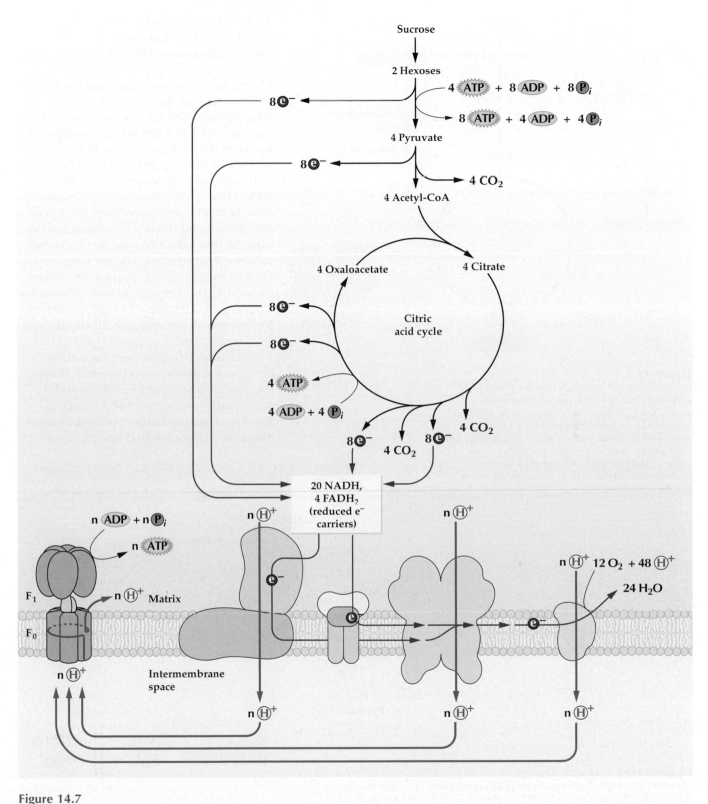

Figure 14.7

The general mechanism of oxidative phosphorylation in mitochondria. Electrons released during the oxidative steps of glycolysis and the citric acid cycle produce 20 molecules of NADH and 4 molecules of FADH$_2$. These reduced coenzymes are subsequently oxidized by the mitochondrial electron transfer chain. The free energy released during mitochondrial electron transfer is coupled to the translocation of protons across the inner mitochondrial membrane from the matrix into the intermembrane space, generating an electrochemical proton gradient ($\Delta\bar{\mu}_H{}^+$) across the inner membrane. The free energy subsequently released by the movement of protons back across the inner membrane through the F$_0$ proton channel of the ATP synthase complex is used by the catalytic site on the F$_1$ component of the complex to convert ADP and P$_i$ to ATP in the mitochondrial matrix. (See also legend to Fig. 14.3.)

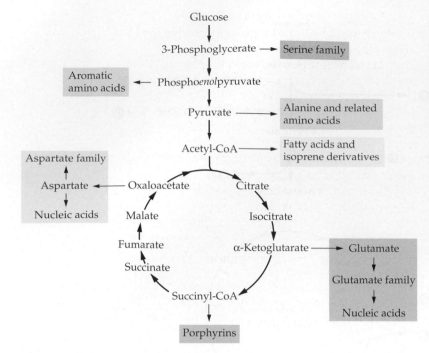

Figure 14.8
Intermediates produced during the reactions of glycolysis and the citric acid cycle serve as substrates for numerous plant biosynthetic pathways. The primary intermediates for the production of amino acids, lipids, nucleic acids, porphyrins, cell wall sugars, and many other essential components of the plant cell are derived from compounds that originate from either glycolysis or the citric acid cycle.

14.2.1 Cytosolic reactions generate products that are transported into the mitochondria to feed the citric acid cycle.

The nature of the end product of the glycolytic reactions in the cytosol of plants is determined by the relative activities of the three enzymes that can utilize phospho*enol*pyruvate (PEP) as a substrate. Both pyruvate kinase and PEP phosphatase form pyruvate; PEP carboxylase generates oxaloacetate (OAA) (Fig. 14.9). Pyruvate is transported directly into the mitochondrion. OAA is either transported directly into the mitochondrion or first reduced to malate by cytosolic malate dehydrogenase. The reduction of OAA in the cytosol can provide an extramitochondrial, nonfermentative mechanism for oxidizing NADH formed by glyceraldehyde-3-phosphate dehydrogenase during glycolysis (see Chapter 13).

The mitochondrial inner membrane contains separate carriers by which malate and pyruvate are taken into the mitochondrial matrix (see Section 14.6.1). Once in the matrix, malate can be oxidized by two enzymes: a mitochondrial isoenzyme of

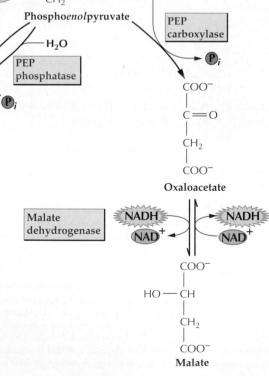

Figure 14.9
Conversion of phospho*enol*pyruvate (PEP) to pyruvate or malate during plant respiration. The conversion of PEP to pyruvate by the enzyme pyruvate kinase is coupled to phosphorylation of ADP. PEP phosphatase can bypass this ATP synthesis step, releasing inorganic phosphate. Alternatively, PEP can react with HCO_3^- via the enzyme PEP carboxylase, releasing the phosphate and producing the C_4 product, oxaloacetate (OAA). OAA is commonly reduced to malate by NADH through the action of malate dehydrogenase. Both pyruvate and malate can be readily transported into mitochondria via carriers located in the inner mitochondrial membrane (see Fig. 14.38).

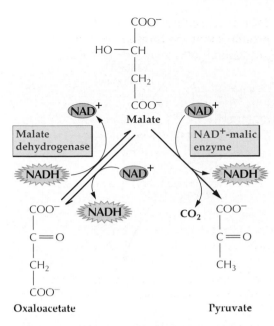

Figure 14.10
Alternative routes for the oxidation of malate in plant mitochondria. Once transported into the mitochondrion, malate can be acted on either by malate dehydrogenase, which couples the oxidation of malate to the reduction of NAD^+, producing OAA and NADH, or by NAD-malic enzyme, which catalyzes the oxidative decarboxylation of malate to pyruvate and CO_2 and the concomitant reduction of NAD^+ to NADH. Both enzymes are located in the mitochondrial matrix.

malate dehydrogenase that yields OAA and NADH, and an NAD^+-linked malic enzyme, which generates pyruvate, CO_2, and NADH (Fig. 14.10). Therefore, importing mitochondrial malate can transfer reducing equivalents from the cytosol into the mitochondrion.

14.2.2 Pyruvate enters the citric acid cycle through the action of the pyruvate dehydrogenase enzyme complex.

Before entering the citric acid cycle proper, pyruvate is oxidized and decarboxylated by the pyruvate dehydrogenase enzyme complex to form CO_2, acetyl-CoA, and NADH. The pyruvate dehydrogenase enzyme complex, which requires the bound cofactors thiamine pyrophosphate, lipoic acid, and FAD as well as free coenzyme A (CoASH) and NAD^+, links the citric acid cycle to glycolysis. The complex contains three separate enzymes: pyruvate dehydrogenase, dihydrolipoyl transacetylase, and dihydrolipoyl dehydrogenase (Fig. 14.11).

The pyruvate dehydrogenase complex is subject to sophisticated regulatory mechanisms, including phosphorylation/dephosphorylation by a protein kinase/phosphatase couple; ATP-dependent phosphorylation inhibits the dehydrogenase enzyme (Fig. 14.12). In turn, various metabolites modulate the kinase activity. Pyruvate inhibits the protein kinase, ensuring that the dehydrogenase is active when substrate is plentiful. Ammonia, on the other hand, stimulates the kinase and therefore inhibits pyruvate oxidation. This latter effect may underlie the observed inhibition of the citric acid cycle in illuminated leaves (see Section 14.6.6) because ammonia is produced within the mitochondrion during operation of the photorespiratory cycle. Pyruvate dehydrogenase is also subject to feedback inhibition by its products, acetyl-CoA and NADH.

14.2.3 The citric acid cycle generates reducing equivalents, CO_2, and ATP.

As the citric acid cycle proper begins, acetyl-CoA and OAA condense to form the C_6 molecule citrate and free CoASH (Fig. 14.13), in a reaction catalyzed by the enzyme citrate synthase. Aconitase then isomerizes citrate to isocitrate. Next, NAD-linked isocitrate dehydrogenase oxidatively decarboxylates isocitrate to form CO_2, α-ketoglutarate, and NADH. The α-ketoglutarate thus formed is oxidized further to succinyl-CoA, CO_2, and NADH in a reaction catalyzed by the α-ketoglutarate dehydrogenase enzyme complex. The structure of this enzyme complex is similar to that of the pyruvate dehydrogenase complex, and the reaction catalyzed is chemically analogous to the formation of acetyl-CoA from pyruvate (see Fig. 14.11). The reaction mechanisms are very similar: For example, lipoamide participates in the α-ketoglutarate dehydrogenase reaction, and dihydrolipoyl dehydrogenase (L-protein) is the same in both complexes. However, α-ketoglutarate dehydrogenase activity does not appear to be regulated by reversible phosphorylation.

Succinyl-CoA synthetase catalyzes the conversion of succinyl-CoA to succinate, with the concomitant phosphorylation of ADP to ATP; this is the only citric acid cycle reaction that conserves energy directly by

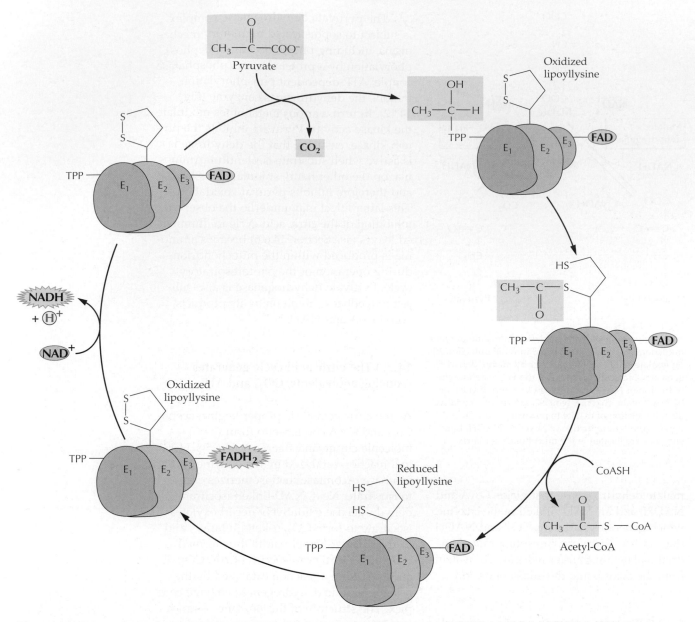

Figure 14.11

Reaction sequence catalyzed by the pyruvate dehydrogenase complex. The overall reaction catalyzed by this complex involves the oxidative decarboxylation of pyruvate and leads to the formation of acetyl-CoA, CO_2, and NADH. The enzyme pyruvate dehydrogenase (E_1) initially catalyzes the decarboxylation of pyruvate to produce acetaldehyde, a C_2 product that reacts with bound thiamine pyrophosphate (TPP) to form hydroxyethyl-TPP. The hydroxyethyl derivative is transferred to the oxidized lipoamide group of dihydrolipoyl transacetylase (E_2), forming the acetyl thioester of reduced lipoic acid. E_2 next transfers the acetate to the thiol moiety of coenzyme A (CoASH) to give acetyl-CoA and a fully reduced lipoamide group on the dihydrolipoyl transacetylase. Dihydrolipoyl dehydrogenase (E_3) then catalyzes the transfer of two electrons and two protons from the reduced lipoamide group of E_2 to the bound FAD of E_3, oxidizing the lipoamide and producing $FADH_2$. In the final reaction of the series, E_3 transfers two electrons and a proton to NAD^+, yielding NADH.

forming ATP. In plants, this enzyme differs from its animal counterpart in being specific for ADP rather than GDP. Succinate dehydrogenase, which catalyzes the oxidation of succinate to fumarate, is the only membrane-associated enzyme of the citric acid cycle and represents the catalytic component of Complex II of the respiratory electron transfer chain (see Section 14.3.2).

Fumarase reversibly hydrates fumarate to form malate. Like succinate dehydrogenase, fumarase is unique to the

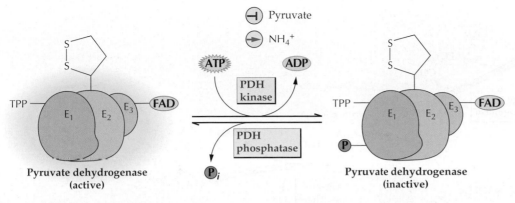

Figure 14.12

Regulation of the pyruvate dehydrogenase (PDH) complex by phosphorylation/dephosphorylation. PDH kinase, which is activated by ammonia and inhibited by pyruvate, inactivates the PDH complex by catalyzing the phosphorylation of E_1. Reactivation of PDH involves subsequent dephosphorylation by a PDH phosphatase.

mitochondrion and therefore is a convenient marker for the mitochondrial matrix. Malate dehydrogenase catalyzes the final step of the citric acid cycle, oxidizing malate to OAA and producing NADH. The reaction is freely reversible, but the equilibrium in vitro strongly favors the reduction of OAA. In vivo, however, the products are consumed rapidly; OAA condenses with acetyl-CoA to produce citrate, and NADH is oxidized by the respiratory electron transfer chain. This shifts the malate dehydrogenase equilibrium toward product formation. Malate dehydrogenase is inhibited by its product, NADH, and by acetyl-CoA.

Whereas most citric acid cycle enzymes are NAD^+-linked, $NADP^+$-dependent isoforms of isocitrate dehydrogenase and malate dehydrogenase do exist. The NADPH produced by these enzymes has many possible fates. The electron transfer chains of plant mitochondria can oxidize matrix NADPH (see Section 14.3.4). In addition, several mitochondrial reactions use NADPH as an electron donor. These include reduction of dihydrofolate to tetrahydrofolate, a substrate for the C_2 photorespiratory cycle (see Section 14.8); production of reduced glutathione, which protects against reactive oxygen species generated during mitochondrial electron transport (see Chapter 22); and reduction of mitochondrial thioredoxin, which may regulate enzymatic activities (e.g., the alternative oxidase) by reducing critical disulfide bonds (see Section 14.5.3).

Overall, during the pyruvate dehydrogenase reaction and one subsequent turn of the citric acid cycle proper, the three carbon atoms of pyruvate are released as CO_2, one molecule of ATP is formed directly, and four NADH and one $FADH_2$ molecules are produced (Fig. 14.13; see also Fig. 14.5). This stoichiometry assumes that the NADP-dependent isoforms of isocitrate and malate dehydrogenase do not participate in the reaction.

14.2.4 Regulation of carbon flux through the citric acid cycle is not well understood in plants.

The mechanisms of citric acid cycle regulation in plants have not yet been resolved, but current data suggest that cycle activity is controlled by the energy status of the cell, which affects pyruvate dehydrogenase phosphorylation, and by mitochondrial matrix concentrations of NADH and acetyl-CoA, which inhibit various dehydrogenases through negative feedback. Therefore, the citric acid cycle turnover rate in vivo depends on the rate of NADH reoxidation by the respiratory electron transfer chain and on the cellular rate of ATP utilization. The rate at which plastid and cytosolic pathways provide citric acid cycle substrate may also be very important. For example, in pea leaves, which undergo large diurnal shifts in carbohydrate content, respiratory rates are

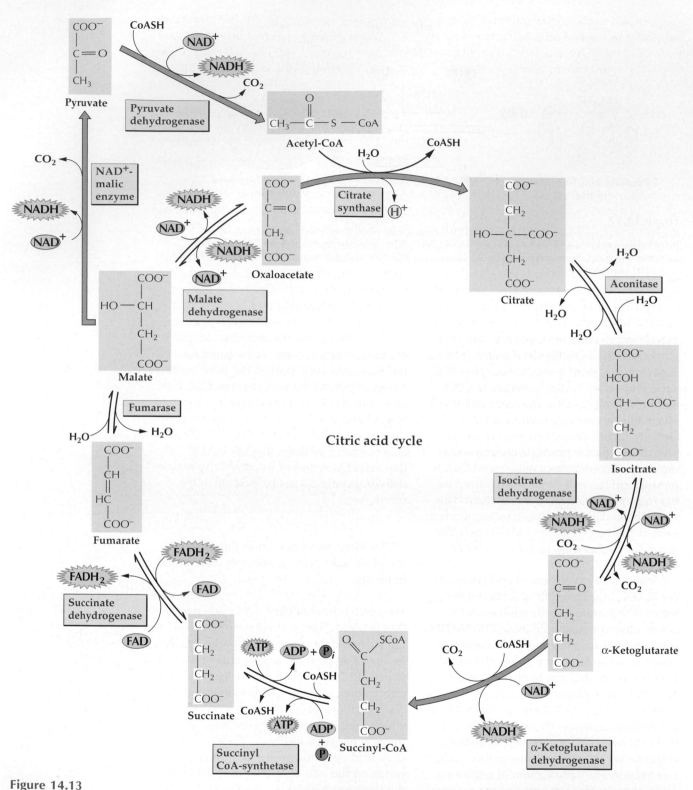

Figure 14.13

Reaction sequence of the citric acid cycle. In a reaction catalyzed by citrate synthase, acetyl-CoA formed by the pyruvate dehydrogenase complex combines with OAA to produce the C_6 tricarboxylic acid, citrate. In the overall cycle, the citrate is oxidized to produce two molecules of CO_2 in a series of reactions that leads to the formation of one OAA, three NADH, one $FADH_2$, and one ATP. The resulting OAA reacts with another molecule of acetyl-CoA to continue the cycle. The oxidative decarboxylation of pyruvate yields an additional CO_2 and NADH. Thus the citric acid cycle brings about the complete oxidation of pyruvate to 3 CO_2 plus 10 e^-, which are stored temporarily as 4 NADH and 1 $FADH_2$. Malate, which can be an alternative product of glycolysis in plants (see Fig. 14.9), is converted to pyruvate, CO_2, and NADH by the action of NAD-malic enzyme. The irreversible reactions of the cycle are identified with solid brown arrows. The two carbons donated by acetyl-CoA are highlighted through the first symmetrical intermediate succinate, at which point they become randomized and can no longer be traced.

proportional to the concentration of carbon substrates in the leaf cells, being very low at the end of the night and increasing substantially during the day as photosynthesis replenishes carbon reserves.

14.2.5 The citric acid cycle can oxidize amino acids and fatty acids.

Pyruvate and malate from glycolysis are typically the most prevalent mitochondrial substrates in vivo, especially in the dark or in nonphotosynthetic tissues. However, other compounds are also metabolized in the mitochondrial matrix and interact with the citric acid cycle. Glycine produced during photorespiration is a special case, discussed later in detail (see Section 14.8.2), but other amino acids may also serve as substrates for mitochondrial respiration in some tissues, particularly in seeds rich in stored protein. Amino acid oxidation may be preceded by a transamination reaction that generates a citric acid cycle intermediate, or the oxidation may occur directly. One example of direct oxidation within the mitochondrion involves the reversible reaction catalyzed by glutamate dehydrogenase:

Reaction 14.2: Glutamate dehydrogenase

$$\text{Glutamate} + \text{NAD}^+ + \text{H}_2\text{O} \rightleftharpoons$$
$$\alpha\text{-ketoglutarate} + \text{NADH} + \text{NH}_4^+$$

The role of this enzyme is not apparent in all plant tissues. In cotyledons of some legumes (e.g., pea and soybean), glutamate dehydrogenase and mitochondrial aspartate aminotransferase participate in the degradation of stored protein, oxidizing glutamate as a source of energy. The ammonia released by this catabolic reaction is typically reassimilated (see Chapter 8). In leaves, glutamate dehydrogenase has been proposed to operate in the reverse direction as an ancillary enzyme in the reassimilation of ammonia released during photorespiration, although recent screens for photorespiratory mutants of barley and *Arabidopsis* did not identify plants deficient in glutamate dehydrogenase (see Chapter 8).

There is also evidence that β-oxidation of fatty acids to acetyl-CoA can occur in plant mitochondria. However, most fatty acid oxidation in plants seems to occur in

peroxisomes (see Chapter 10). A specialized interaction through which glyoxysomes and mitochondria convert lipids to sugars is discussed in Section 14.6.4 and Chapter 10.

14.3 Plant mitochondrial electron transport

In all mitochondria, the principal respiratory electron transfer chain consists of a series of membrane-bound redox centers that catalyze the multistep transfer of electrons from NADH and FADH_2 to oxygen, forming water and translocating protons from the matrix to the intermembrane space. This **endergonic** (energy-consuming) proton pumping is driven by the **exergonic** (energy-releasing) transfer of electrons from strong reducing agents to a strong oxidant. The 2 e^- reduction of $\frac{1}{2}\text{O}_2$ by NADH involves a reduction potential difference ($\Delta E^{\circ\prime}$) of 1.14 V (Fig. 14.14). This translates to 219.2 kJ of free energy released for every mole of NADH

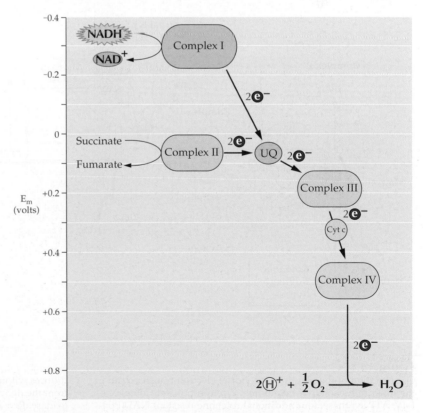

Figure 14.14
Approximate positions of the components of the respiratory chain on the redox potential scale. Release of energy drives proton translocation at three sites on the chain: between Complex I and ubiquinone (UQ); between UQ and cytochrome *c*; and between cytochrome *c* and O_2.

oxidized. If NADH or $FADH_2$ reduced oxygen directly, the heat released would not be biologically useful. The electron transfer chain facilitates several modestly exergonic redox reactions, rather than a single explosive one, and conserves energy through proton translocation. The F_0F_1–ATP synthase provides a path for proton diffusion into the matrix and uses the free energy released by this spontaneous diffusion to drive the phosphorylation of ADP.

In addition to the electron transfer chain shared by all aerobic eukaryotes, plant mitochondria also possess novel pathways for both the oxidation of NAD(P)H and the reduction of oxygen. These unique bypass pathways do not pump protons. The free energy released as electrons flow through them is lost as heat and cannot be used for ATP synthesis.

14.3.1 The standard mitochondrial electron transfer chain contains both peripheral and integral membrane proteins and a lipid-soluble quinone.

The mitochondrial electron transfer chain conserved among eukaryotes, which we will call the standard mitochondrial electron transfer chain, consists of four multiple-subunit protein complexes, commonly referred to as Complexes I through IV. F_0F_1– ATP synthase, which does not have electron transfer activity, is occasionally referred to as Complex V (Fig. 14.15). Complex I is an NADH dehydrogenase that oxidizes the NADH generated in the mitochondrial matrix by citric acid cycle activity, regenerating NAD^+ and reducing ubiquinone. Complex II, which contains a component enzyme of the citric acid cycle (succinate

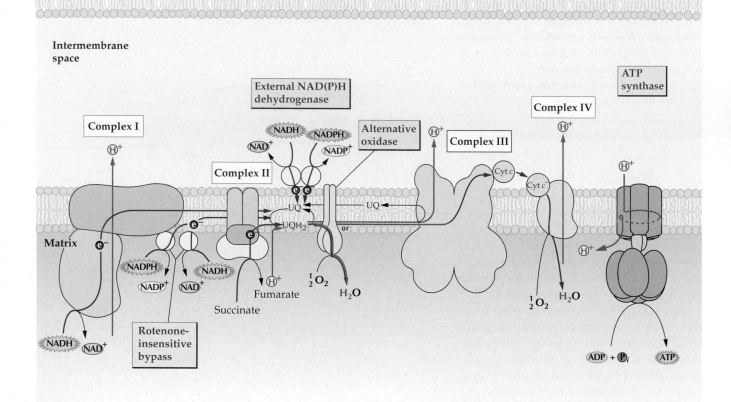

Figure 14.15
Organization of the plant mitochondrial electron transfer chain in the inner mitochondrial membrane. Electron transfer Complexes I–IV, ATP synthase, four additional rotenone-resistant NAD(P)H dehydrogenases, and the alternative oxidase are shown in diagrammatic fashion, incorporating what is currently known about their topological orientation in the inner membrane. A large pool of oxidized ubiquinone (UQ) and reduced ubiquinol (UQH_2) freely diffuses within the inner membrane and transfers electrons derived from the dehydrogenases to either Complex III or to the alternative oxidase. Red arrows indicate electron transfer; blue arrows indicate proton translocation. Transmembrane proton movement generates a proton electrochemical gradient that drives the synthesis of ATP from ADP and P_i via the ATP synthase.

dehydrogenase), oxidizes succinate to fumarate. Like Complex I, Complex II transfers electrons to ubiquinone, an abundant mobile electron transfer cofactor that consists of a substituted *p*-benzoquinone head group and a C_{45} to C_{50} prenyl side chain (Fig.14.16). Ubiquinone can carry one or two electrons. Both fully oxidized ubiquinone and fully reduced ubiquinol molecules are highly hydrophobic and capable of transverse and lateral movement within the inner membrane. A free pool of ubiquinone, present at concentrations far in excess of the proteins with which it interacts, shuttles electrons from Complexes I and II to Complex III.

Complex III transfers electrons from ubiquinol to cytochrome *c*, the only protein in the electron transfer chain not tightly associated with an integral membrane protein complex. A small (12.5 kDa), peripheral membrane protein located on the outer surface of the inner membrane and exposed to the intermembrane space, cytochrome *c* carries one electron at a time from Complex III to Complex IV (cytochrome *c* oxidase), the terminal electron carrier in the electron transfer chain. For every four molecules of cytochrome *c* oxidized, one molecule of oxygen is reduced to two molecules of water.

Numerous electron transport cofactors participate in respiration, including bound FAD and FMN moieties (see Fig. 14.6), hemes (Fig. 14.17), iron–sulfur (Fe–S) centers (Fig. 14.18), and copper atoms.

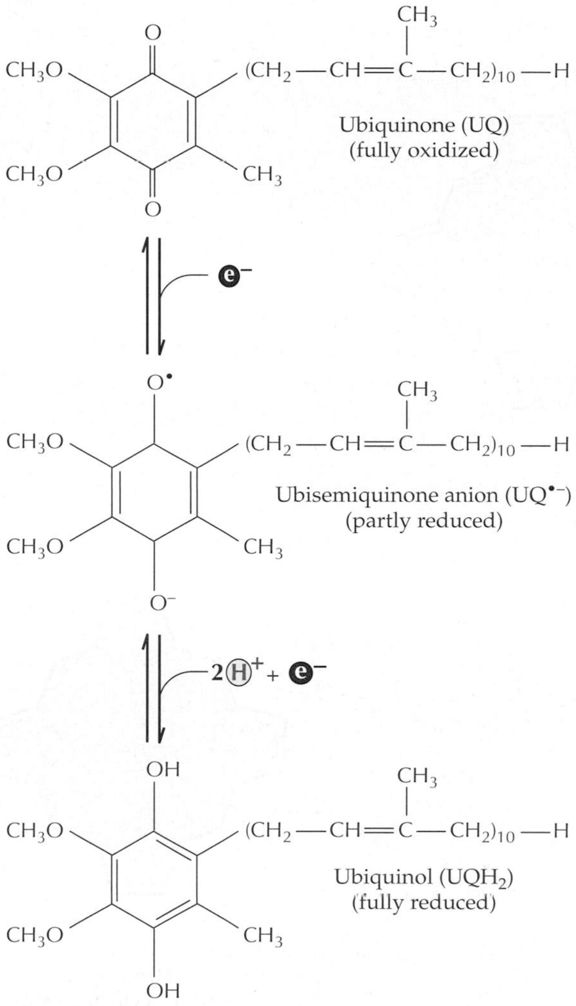

Figure 14.16
Structure and redox reactions of ubiquinone. Reduction of oxidized ubiquinone (UQ) with one electron forms the ubisemiquinone anion (UQ•⁻). The second reduction step involves the addition of an electron and two protons to give the fully reduced ubiquinol (UQH₂). The hydrophobic prenyl side chain, which in plants commonly contains 45 or 50 carbons, retains UQ near the center of the inner membrane, between adjacent leaflets of the phospholipid bilayer.

| Box 14.1 | **Some electron transfer chain components in the mitochondria resemble comparable complexes in the chloroplast.** |

The mitochondrial electron transfer chain shares several striking structural similarities with the photosynthetic electron transfer chain. In the thylakoid membranes of the chloroplast, three large multisubunit protein complexes (photosystem II, cytochrome b_6f complex, and photosystem I) are linked by a membrane-associated quinone (plastoquinone) and a small mobile protein (plastocyanin). The photosystems perform a light-harvesting function that has no mitochondrial parallel, whereas the electron transport cofactors of the cytochrome b_6f complex mirror those of Complex III: two *b*-type cytochromes, a *c*-type cytochrome (cytochrome *f*), and a Rieske Fe–S protein. The cytochrome b_6f complex interacts with plastoquinone in a Q cycle that translocates protons from the stroma to the thylakoid lumen. Furthermore, plastocyanin resembles cytochrome *c* in some key characteristics. Both are membrane-associated soluble metalloproteins that can be reduced by a cytochrome complex and carry one electron. However, the redox active metal that gives plastocyanin its name and blue color is copper, not iron. The similarity of the electron transfer chains and ATP synthases present in these organelles may derive from a common prokaryotic heritage.

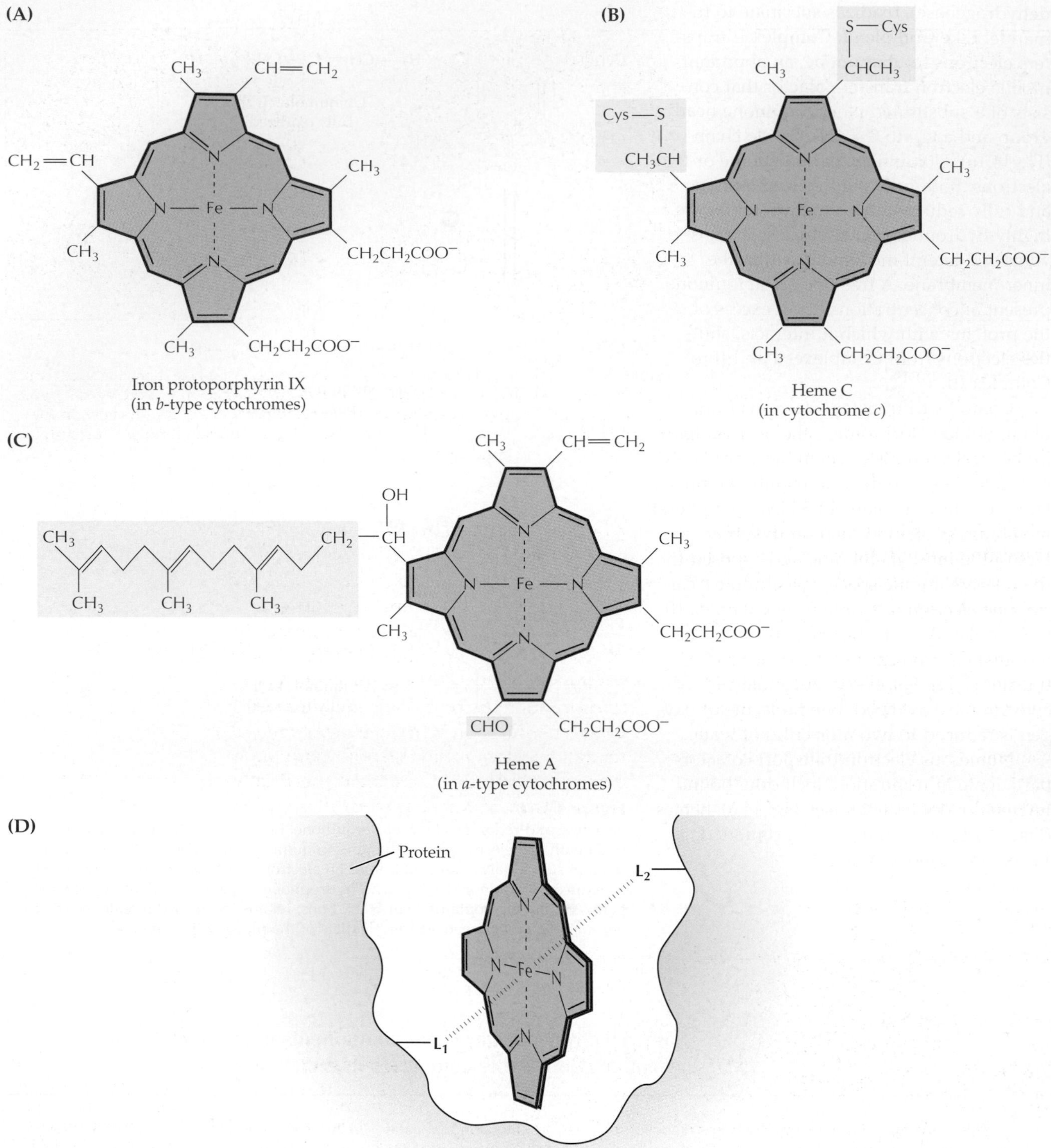

(A)

Iron protoporphyrin IX
(in *b*-type cytochromes)

(B)

Heme C
(in cytochrome *c*)

(C)

Heme A
(in *a*-type cytochromes)

(D)

Figure 14.17

Structure of the heme prosthetic groups found in cytochromes. These porphyrin prosthetic groups contain four pyrrole rings linked in a macrocyclic ring system surrounding a central iron atom coordinated to the four pyrrole nitrogen atoms. Both *b*-type (A) and *c*-type (B) cytochromes contain iron protoporphyrin IX, with the iron-porphyrin complex bound to the protein through iron-binding "axial" ligands (L_1 and L_2) on either side of the plane of the por-phyrin ring system (D). In *c*-type cytochromes, in addition to the axial ligands, the porphyrin is covalently bound to the protein via thioether bonds between the vinyl side chains of the porphyrin and one or two cysteine residues of the protein (see yellow highlighting in B). The *a*-type cytochromes contain heme A (C), which has a C_{15} isoprenoid tail attached to one of the pyrrole rings and is noncovalently bound to the protein through one or two axial ligands.

(A)   (B)   (C)

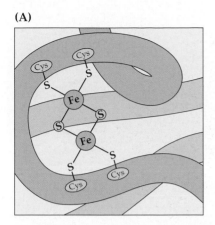   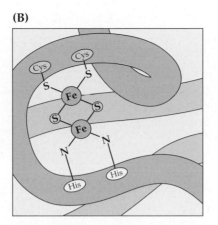   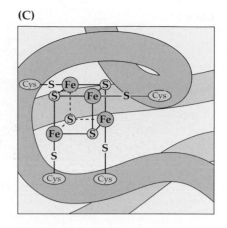

Figure 14.18

Structure of the metal clusters found in iron–sulfur proteins. Iron–sulfur proteins associated with Complexes I and II of the mitochondrial electron transfer chain contain both 2Fe–2S clusters and 4Fe–4S clusters. A Rieske iron–sulfur protein is located in Complex III. Four cysteine residues serve as ligands to the iron atoms of the low-redox potential 2Fe–2S centers (A), whereas the high-redox potential Rieske 2Fe–2S centers are attached to the protein by two cysteine residues and two histidine residues (B). The irons in the 4Fe–4S centers also are bound to the protein through four cysteine ligands (C).

14.3.2 Three of the four multiprotein respiratory complexes located in the inner mitochondrial membrane participate in proton translocation.

Complex I is a large multisubunit complex of 30 to 40 polypeptides. Depending on the species, the plant mitochondrial genome may encode as many as nine Complex I components. The plant mitochondrial Complex I is similar to the better-characterized mammalian and fungal NADH dehydrogenase complexes. Electron microscopic analysis of purified Complex I from *Neurospora* indicates that the complex has a pronounced *L* shape (Fig. 14.19). Complex I oxidizes NADH produced by the reactions of the citric acid cycle and by other NAD-linked enzymes in the matrix. The passage of electrons through the complex is accompanied by H$^+$ translocation across the membrane, although the mechanism remains poorly understood. Complex I is inhibited specifically by the flavonoid rotenone (a rat poison) and

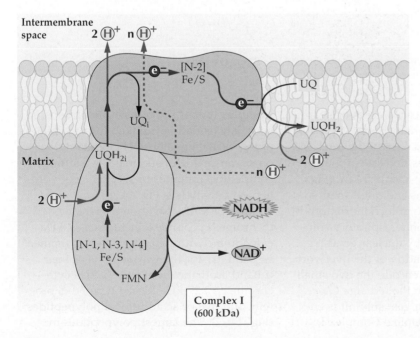

Figure 14.19

Proposed structure and membrane topography of mitochondrial complex I (NADH:UQ oxidoreductase). The complex acts as an NADH-ubiquinone oxidoreductase, transferring electrons from matrix NADH to ubiquinone. This transfer involves FMN, four iron–sulfur centers (N1–N4) and an internal quinone (UQ$_i$). The H$^+$/2e$^-$ ratio is thought to be four, with two H$^+$ taken up at the internal quinone site and another two translocated by a poorly defined mechanism (dotted line). The complex is readily broken into two smaller complexes, one of which (blue) is very hydrophobic and contains all of the subunits encoded in the mitochondrion. The other arm of the complex (purple) protrudes into the matrix and is composed of nucleus-encoded subunits.

Complex

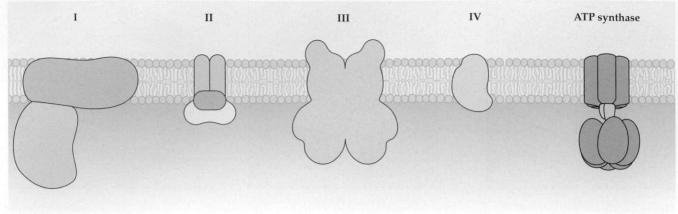

Inhibitor

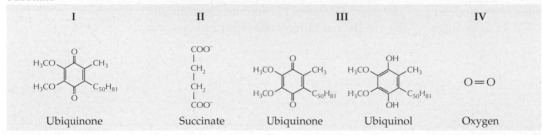

I

Rotenone

II

Malonate
COO⁻ — CH₂ — COO⁻

III

Antimycin A

Myxothiazol

IV

C≡N⁻ Cyanide

N=N=N⁻ Azide

C≡O Carbon monoxide

ATP synthase

Oligomycin B

Substrate

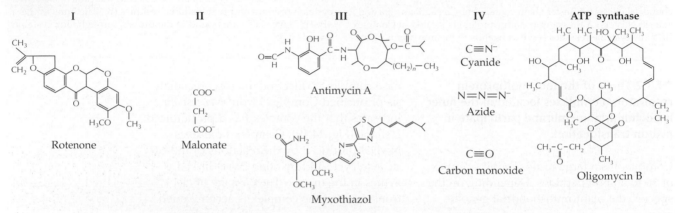

I	II	III	III	IV
Ubiquinone	Succinate	Ubiquinone	Ubiquinol	Oxygen

Figure 14.20
Inhibitors of the mitochondrial electron transfer chain. Specific inhibitors of each of the mitochondrial electron transfer complexes are shown, as are the substrates with which the inhibitors are thought to compete.

its analogs, which appear to act at or near the site of ubiquinone reduction (Fig. 14.20).

Complex II, the smallest of the four electron transfer complexes, consists of four proteins of sizes 70, 27, 15, and 13.5 kDa (Fig. 14.21). In gymnosperms and angiosperms, Complex II contains no mitochondrial products, making this the only respiratory complex encoded entirely by nuclear genes. However, the mitochondria of the liverwort *Marchantia polymorpha* encode the two smaller polypeptides, and in the red alga *Chondrus crispus*, only the largest subunit is encoded in the nucleus. Unlike Complex I,

Complex II does not translocate protons, so succinate oxidation is linked to the synthesis of less ATP than is NADH oxidation. Malonate, an analog of succinate, is a strong competitive inhibitor of succinate dehydrogenase activity (see Fig. 14.20).

Complex III, also known as the cytochrome bc_1 complex, includes a 42-kDa cytochrome with two b-type hemes (named b_{566} and b_{560} for the wavelengths of their α-band asorbance peaks), a 31-kDa cytochrome c_1, a 27-kDa Rieske-type Fe–S protein, and five or so additional polypeptides (Fig. 14.22). The largest polypeptides in

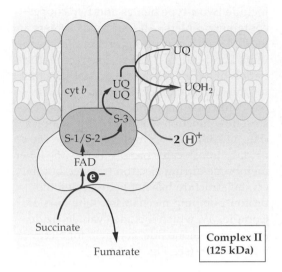

Figure 14.21
Proposed structure and membrane topography of mitochondrial Complex II (succinate:ubiquinone oxidoreductase). Two large hydrophilic peripheral subunits that make up succinate dehydrogenase are bound to the inner membrane through two small, hydrophobic membrane-spanning subunits. Covalently bound flavin adenine dinucleotide (FAD) is associated with the subunit F_p ("flavin protein"; yellow), while three iron–sulfur clusters (S-1, S-2, and S-3) are bound to I_p ("iron protein"; brown). The two hydrophobic subunits contain one b-type cytochrome and a ubiquinone (UQ) pair, the exact locations of which are uncertain. The proposed path of electron flow is succinate→FAD→S-1/S-2→S-3→the UQ pair→a UQ molecule from the mitochondrial UQ pool.

(A)

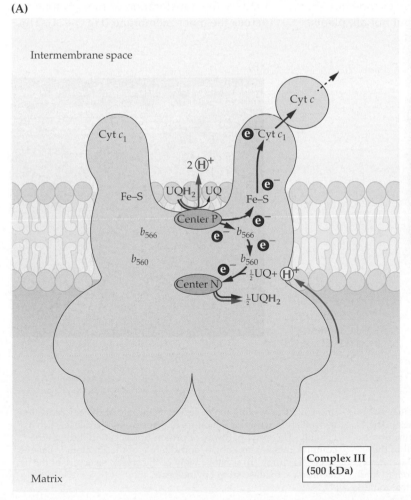

(B)

Figure 14.22
(A) Diagram illustrating proposed structure and membrane topography of mitochondrial Complex III (ubiquinone:cytochrome-c oxidoreductase, also known as cytochrome bc_1). The complex is a dimer, with each monomer containing multiple subunits. Ubiquinol (UQH_2) is oxidized at Center P, while ubiquinone (UQ) is reduced at Center N. The two electrons from UQH_2 take divergent paths, one being transferred to mobile cytochrome c via a Rieske iron–sulfur center and cytochrome c_1, the other reaching Center N via two b-type cytochromes. The inhibitory sites of antimycin and myxothiazol are at Centers N and P, respectively.
(B) Crystal structure of a dimeric mammalian cytochrome bc_1 complex.

693

Complex III are known as core proteins; the plant complex contains three of these, ranging in size from 51 to 55 kDa. Complex III has two quinone-binding sites, one that oxidizes ubiquinol (Center P) and one that reduces ubiquinone (Center N). These binding sites and the electron transport cofactors described above participate in a proton translocation mechanism known as the Q cycle (see Section 14.3.3). The widely utilized inhibitor antimycin A binds to Center N and thus blocks reduction of ubiquinone. Myxothiazol, another commonly used inhibitor of Complex III in mitochondria, blocks the oxidation of ubiquinol at Center P (see Fig. 14.20).

The final complex of the standard respiratory chain is Complex IV, cytochrome c oxidase. As its name implies, this complex accepts an electron from reduced cytochrome c on the cytosolic side of the inner membrane and eventually donates it to oxygen on the matrix side of the membrane. The complex contains seven to nine polypeptides, the largest three of which are encoded by the mitochondria in most, but not all, plants.

The redox-active species of Complex IV include two a-type hemes and two copper-containing centers. All four redox centers are located within mitochondrion-encoded subunits I and II (Fig. 14.23A). The recent elucidation of a high-resolution crystal structure for the mammalian cytochrome c oxidase has confirmed many predictions of the structure–function features of this complex (Fig. 14.23B). Like Complex I, cytochrome c oxidase translocates protons across the inner membrane during electron transfer, but the crystal structure has not yet revealed the proton-pumping mechanism. Inhibitors of Complex IV, which include cyanide, azide, and carbon monoxide, compete for electrons with oxygen (see Fig. 14.20).

14.3.3 Proton pumping at Complex III is accomplished by the Q cycle.

At Complex III, transfer of one electron pair from ubiquinol to cytochrome c is accompanied by the translocation of four protons across the inner membrane (Fig. 14.24). This

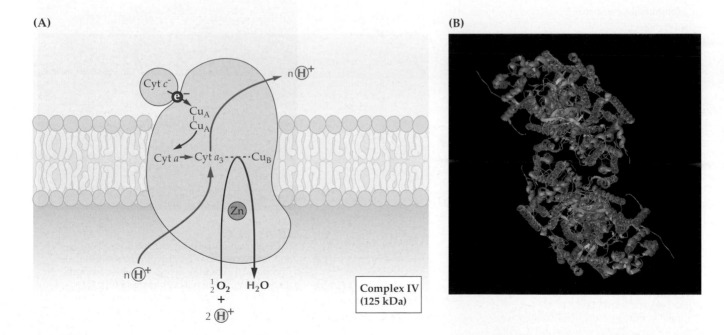

(A)

(B)

Complex IV (125 kDa)

Figure 14.23

(A) Diagram illustrating proposed structure and membrane topography of mitochondrial Complex IV (cytochrome c oxidase). The electron transfer center that initially oxidizes reduced cytochrome c, Cu_A, contains two copper atoms and resides in subunit II of the complex. All other electron carriers in cytochrome c oxidase are found on subunit I. From Cu_A, electrons flow to cytochrome a and then to the coupled binuclear metal center formed by cytochrome a_3

and Cu_B. The a_3–Cu_B center serves as the site of oxygen reduction as well as of action of inhibitors of cytochrome c oxidase, such as cyanide. The exact pathway of proton translocation through the complex remains unknown, and the role of the Zn atom is believed to be structural. (B) Cystolic view of the crystal structure of the cytochrome c oxidase from bovine heart.

mechanism, called the Q-cycle, presents the only case in which proton translocation by the electron transport chain is largely understood. During the Q-cycle, ubiquinone is reduced to ubiquinol on the matrix side of the inner membrane. Three electron transfer chain sites can reduce ubiquinone: one each at Complexes I and II, and Center N on Complex III. The reduction of ubiquinone to ubiquinol is accompanied by the uptake of two protons from the mitochondrial matrix, regardless of which ubiquinone reduction site is utilized. Ubiquinol then diffuses to the outer face of the membrane and binds to Center P, a ubiquinol oxidation site on Complex III. Once bound to Center P, the ubiquinol is oxidized; one electron reduces the Rieske Fe–S center, and the other reduces cytochrome b_{566}. The oxidation of ubiquinol to ubiquinone also releases two protons into the intermembrane space. The electron on the Rieske Fe–S center is subsequently transferred to cytochrome c_1 and then to cytochrome c. The electron residing on cytochrome b_{566} is transferred to cytochrome b_{560}, which donates electrons to ubiquinone at Center N.

(A) First UQH₂ oxidized

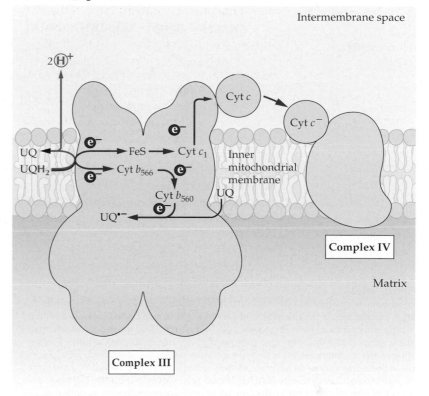

(B) Second UQH₂ oxidized

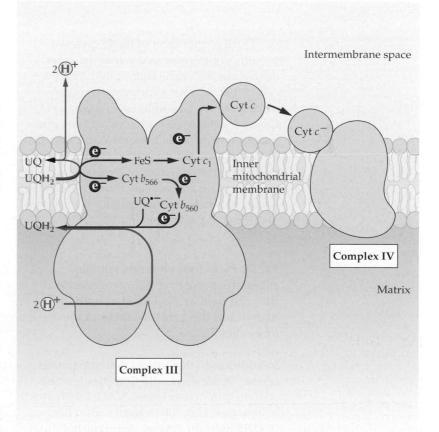

Figure 14.24
Operation of the proton motive Q-cycle during electron transfer through Complex III. (A) Ubiquinol (UQH_2) binds at Center P (not shown) in a pocket formed by cytochrome b and the Rieske iron–sulfur protein. Initially, one electron is transferred to the Rieske protein, and two protons are released into the intermembrane space. The resulting ubisemiquinone ($UQ^{\bullet-}$) transfers the remaining electron to the low-potential cytochrome b_{566}. The electron transferred to the Rieske protein is subsequently transferred to cytochrome c_1 and then onto cytochrome c. The electron on cytochrome b_{566} is transferred to the high-potential cytochrome b_{560}, which is oriented toward the matrix side of the inner membrane and is associated with a second UQ binding site in Complex III, Center N (not shown). Oxidized ubiquinone (UQ) binds to Center N and is reduced by one electron, forming a relatively stable ubisemiquinone anion ($UQ^{\bullet-}$). (B) After a second round of electron transfer from cytochrome b_{566} to cytochrome b_{560}, the $UQ^{\bullet-}$ is further reduced to UQH_2 in a reaction that includes the uptake of two protons from the mitochondrial matrix. This UQH_2 dissociates from Center N, enters the UQ pool in the inner mitochondrial membrane, and diffuses across the membrane to Center P to initiate the reaction sequence again. UQH_2 is also formed from UQ by the action of Complexes I and II and the rotenone-insensitive NAD(P)H dehydrogenases. As a result of operation of the Q-cycle, four protons are translocated from the matrix to the intermembrane space for every two electrons transferred from Complex III to cytochrome c.

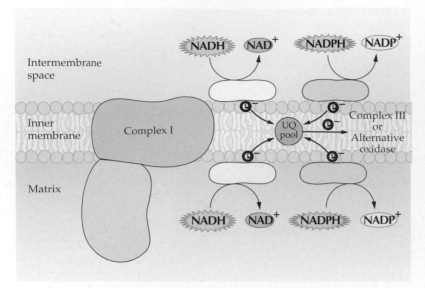

Figure 14.25

Rotenone-insensitive NADH and NADPH dehydrogenases of the inner mitochondrial membrane. In addition to Complex I, plant mitochondria possess simpler (single polypeptide) dehydrogenases on both surfaces of the mitochondrial inner membrane. These do not pump protons and are insensitive to Complex I inhibitors such as rotenone. Four dehydrogenases have been described, although not all of these may occur in all plant tissues. The two external dehydrogenases are thought to oxidize cytosolic NAD(P)H and feed electrons into the UQ pool. The two enzymes on the inner surface provide additional routes for oxidation of the NADH and NADPH formed in the matrix. The proteins involved in these processes have not yet been firmly identified.

During operation of the Q cycle, two protons are translocated across the inner membrane for each electron that reduces cytochrome *c*. This stoichiometry occurs because every two ubiquinols oxidized at Center P release four protons into the intermembrane space, but only two of the four electrons proceed down the electron transfer chain to ultimately reduce oxygen. The other two electrons are recycled through the *b*-type hemes to re-reduce one ubiquinone to ubiquinol at Center N (Fig. 14.24).

14.3.4 Plant mitochondria contain rotenone-insensitive dehydrogenases that can oxidize NAD(P)H at both the matrix and the cytosolic faces of the inner membrane.

In addition to the NADH dehydrogenase activity associated with Complex I, plants contain a matrix-oriented NADH dehydrogenase (see Figs. 14.15 and 14.25). This NADH dehydrogenase, distinguished from

Complex I by its insensitivity to the inhibitor rotenone, is often referred to as the rotenone-insensitive bypass. As a result of the bypass, oxidation of NAD-linked substrates by isolated plant mitochondria is not fully inhibited by adding rotenone, but the oxygen uptake that persists is coupled to synthesis of less ATP (Fig. 14.26). Like Complex I, the rotenone-insensitive bypass transfers electrons from NADH to the ubiquinone pool, but it does not translocate protons across the inner membrane. The rotenone-insensitive bypass has substantially less affinity for NADH than does Complex I and consequently operates only when matrix concentrations of NADH are high.

A separate dehydrogenase that can oxidize matrix NADPH in a rotenone-insensitive and non-proton-translocating fashion has also been identified in some plant mitochondria (see 14.25). The role of this latter enzyme is not clear, but it may participate in interactions between NAD^+ and $NADP^+$ pools in the mitochondrion.

Plant mitochondria can also oxidize both NADH and NADPH via dehydrogenases localized on the outer face of the inner membrane (see Figs. 14.15 and 14.25). These external NAD(P)H dehydrogenases are not inhibited by rotenone and are not linked to proton translocation. In vivo, these dehydrogenases presumably oxidize NAD(P)H produced in the cytosol. Regulation of the external dehydrogenases has not been elucidated, except that both activities depend strongly on the presence of calcium. However, these dehydrogenases have the potential to influence the redox balance of the cytosolic pyridine nucleotide pools—that is, the ratio of $NAD(P)^+$ to NAD(P)H—and therefore may influence many cytosolic reactions.

14.3.5 Plant mitochondria have an alternative, cyanide-resistant oxidase that transfers electrons to oxygen.

In addition to optional pathways for the oxidation of NAD(P)H, plant mitochondria contain a pathway for the transfer of electrons from ubiquinol to oxygen that bypasses cytochrome *c* oxidase. This nuclear-encoded alternative oxidase is found in plants, many algae and fungi, and some protozoa. Electron flow through the alternative oxidase is

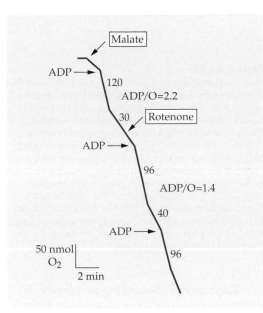

Malate

ADP →

120

ADP/O=2.2

30   Rotenone

ADP →

96

ADP/O=1.4

40

ADP →

50 nmol
O₂

96

2 min

Figure 14.26
Operation of the rotenone-insensitive bypass of Complex I in plant mitochondria. The diagram shows an oxygen electrode recording obtained with isolated mitochondria oxidizing malate in a reaction medium containing an excess of inorganic phosphate. Oxidation of the malate within the matrix generates NADH, which in turn is oxidized by the respiratory chain, leading to oxygen consumption. Addition of a small amount of ADP stimulates the rate of oxygen uptake by allowing the ATP synthase to operate, dissipating the proton motive force across the inner mitochondrial membrane. When all of the added ADP has been phosphorylated, oxygen uptake reverts to a slow resting rate. These changes in rate allow researchers to calculate an **ADP/O ratio** (the number of ADP molecules phosphorylated per molecule of oxygen reduced), which is an indication of the number of proton translocating sites. The ratio before addition of rotenone was greater than 2, indicating the operation of three translocation sites. Addition of rotenone blocks electron flow through Complex I and redirects electrons to the alternative NADH dehydrogenase on the inner surface of the inner membrane; this enzyme does not pump protons, and the ADP/O ratio decreases accordingly. Note that the resting rate of oxygen consumption is a little higher when electrons bypass proton-pumping Complex I.

insensitive to classic inhibitors of cytochrome c oxidase (cyanide, azide, and carbon monoxide) and those of Complex III (antimycin A, myxothiazol). However, the alternative oxidase can be specifically inhibited by salicylhydroxamic acid (SHAM) and n-propylgallate (Fig. 14.27). Tissues of most higher plants examined to date demonstrate some amount of cyanide-resistant, SHAM-sensitive oxygen uptake that is diagnostic for the alternative oxidase, but the total amount of this activity can vary widely. The alternative oxidase is tightly bound to the inner membrane and diverts electrons from the standard electron transfer chain (commonly referred to as the **cytochrome pathway**) at the level of the ubiquinone pool, transferring electrons from ubiquinol to oxygen and generating water as the product. Because no proton electrochemical gradient ($\Delta\bar{\mu}_{H^+}$) is generated during the electron transfer between ubiquinol and oxygen, all the free energy released during electron flow through the alternative oxidase is lost as heat and cannot be used for ATP synthesis. Thus, when electrons from NADH oxidation flow through the alternative oxidase, at least two of the three sites of proton translocation are bypassed, the resulting proton motive force is diminished, and fewer molecules of ATP are synthesized (see Section 14.4.1).

The alternative oxidase is associated with a single polypeptide of approximately 32 kDa. A putative structure has been in-

ferred by examining cDNA sequences from numerous plants, fungi, and a protozoan (Fig. 14.28). The oxidase exists in the membrane as a homodimer, in which the two reduced (–SH) monomers are apparently held together by noncovalent interactions. Its oxygen-reducing activity suggests that the alternative oxidase active site probably contains one or more transition metal centers, and studies on the fungal oxidase indicate that expression of an active oxidase requires iron. No overall amino acid sequence similarities have been observed between the alternative oxidase and any other known protein, but conserved iron-binding motifs associated with a large family of proteins containing coupled di-iron centers, such as ribonucleotide reductase, have been identified in the C-terminal domain.

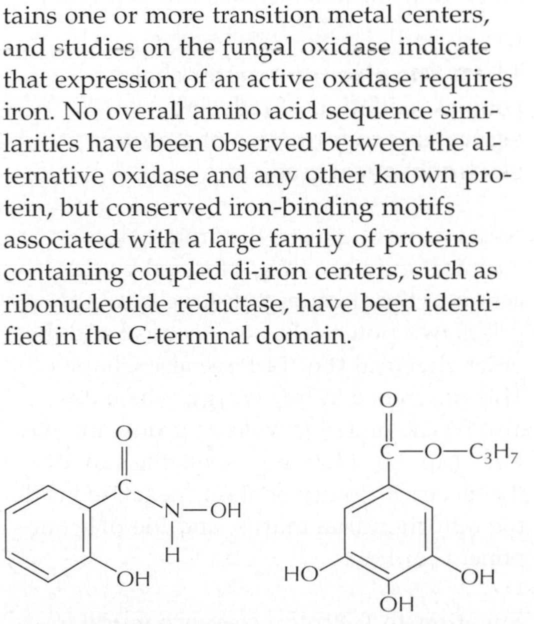

Salicylhydroxamic acid (SHAM)   **n-Propylgallate**

Figure 14.27
Structures of salicylhydroxamic acid (SHAM) and n-propylgallate, inhibitors of the alternative oxidase. SHAM is thought to act as a competitive inhibitor of ubiquinol (UQH_2).

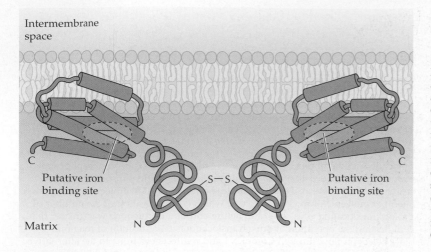

Intermembrane space

Matrix

C

Putative iron binding site

S—S

Putative iron binding site

C

N

N

Figure 14.28

Proposed structure and membrane topography of the alternative oxidase dimer, as deduced from derived amino acid sequences. In plants, the alternative oxidase exists in the membrane as a dimer. Although the nature of the protein–protein interactions that maintain the dimeric structure when the intermolecular disulfide is reduced remain unknown, the proximity of the N-terminal hydrophilic domains required to form the intermolecular disulfide bond suggests their possible involvement. Two large hydrophilic domains extend into the mitochondrial matrix. The amino acids that make up the iron-binding motif of the putative di-iron center are located in the C terminal hydrophilic domain.

14.4 Plant mitochondrial ATP synthesis

The mitochondrial electron transfer chain translocates protons from the matrix side of the inner membrane to the cytosolic side, concentrating H^+ in the intermembrane space and thereby creating an electrochemical gradient ($\Delta\bar{\mu}_{H^+}$). Protons diffuse back into the matrix through the F_0F_1–ATP synthase complex, which catalyzes the conversion of ADP and P_i to ATP. Numerous inner membrane transporters facilitate this process.

14.4.1 The electron transfer chain couples oxidation of reducing equivalents to formation of a proton electrochemical gradient.

As its name suggests, the proton electrochemical gradient ($\Delta\bar{\mu}_{H^+}$) formed by the action of the electron transfer chain comprises two potentials, one electrical and the other chemical (Eq. 14.1; see also Chapter 3). This difference in free energy, which can also be calculated in volts as proton motive force (Δp, Eq. 14.2), represents the sum of the electrical component $\Delta\psi$ (negative inside the mitochondrial matrix) and the pH component (ΔpH).

Equation 14.1: Proton electrochemical gradient

$$\Delta\bar{\mu}_{H^+} = \Delta\psi + \Delta pH$$

Equation 14.2: Proton motive force

$$\Delta p = \Delta\bar{\mu}_{H^+}/96.5 \text{ kJ V}^{-1} \text{ mol}^{-1}$$

The concept of coupling two reactions through a proton electrochemical gradient forms the basis of the chemiosmotic theory, proposed in 1960 by Nobel laureate Peter Mitchell. In mitochondria, including those from plants, the principal contributor to $\Delta\bar{\mu}_{H^+}$ is $\Delta\psi$ (–150 mV to –200 mV); the intermembrane space pH is only 0.2 to 0.5 units greater than that of the matrix. Typical $\Delta\bar{\mu}_{H^+}$ values for plant mitochondria are –200 to –240 mV in the resting state (i.e., in the absence of ADP). In chloroplasts, by contrast, ΔpH represents most of the total $\Delta\bar{\mu}_{H^+}$ because chloroplasts are more permeable to some ions and contain gated ion channels that are not present in mitochondria.

The exact number of protons translocated per pair of electrons transferred from NADH to oxygen remains controversial, since the mechanisms that couple proton translocation to electron flow have not been clearly elucidated for Complexes I and IV. However, measurements with isolated mitochondria from different sources give a value of 10 H^+ translocated per NADH oxidized by the cytochrome pathway.

14.4.2 The F_0F_1–ATP synthase complex in mitochondrial inner membrane couples dissipation of the proton electrochemical gradient with ATP formation.

F_0F_1–ATP synthase, a multisubunit complex that spans the inner membrane, is occasionally referred to as Complex V of the respiratory chain. Although mitochondrial function

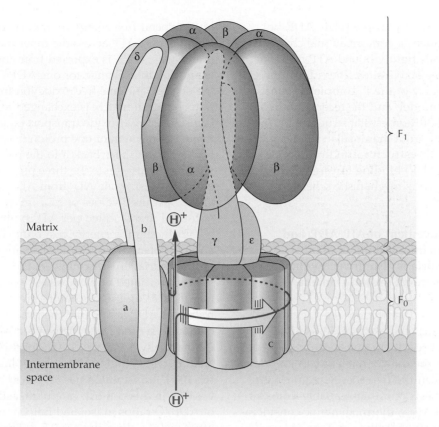

Matrix

Intermembrane
space

Figure 14.29

Proposed structure and membrane topography of F_0F_1–ATP synthase. The peripheral F_1 complex, which extends into the mitochondrial matrix, consists of at least five subunits ($\alpha_3\beta_3\gamma\delta\varepsilon$). The catalytic sites for conversion of ADP and P_i to ATP are located principally on the β-subunits. The α- and β-subunits are arranged in an alternating hexagonal array surrounding the γ-subunit, which consists of two long, coiled α-helices that are believed to connect with the c-subunits of the F_0 complex. The exact positions of the δ- and ε-subunits are uncertain. The F_0 complex is an integral membrane-spanning complex that acts as a proton channel, providing the pathway for movement of protons across the inner membrane into the matrix. The F_0 complex consists of at least three proteins ($a_1b_2c_{10-12}$). The c-subunits are thought to form the proton channel, and the a- and b-subunits are believed to act as stators, holding the α/β hexamer in place while the γ-subunit, which is bound directly to the c-subunits, rotates during proton translocation.

is commonly associated with proton diffusion coupled to ATP synthesis, in some circumstances the F_0F_1 complex can hydrolyze ATP to ADP and P_i and drive proton translocation from the matrix into the intermembrane space.

F_0F_1–ATP synthase is composed of two major components (Fig. 14.29). The integral membrane protein complex, F_0, functions as a proton channel through the inner membrane and consists of as many as eight or nine separate polypeptides. The peripheral membrane protein complex, F_1, extends into the matrix space on a stalk anchored at its base to the F_0 complex. The F_1 complex contains at least five separate polypeptides, α through ε, in a stoichiometry of $\alpha_3\beta_3\gamma\delta\varepsilon$. The catalytic sites for ATP synthesis are localized primarily on the β-subunits. Three of the F_0

polypeptides and one of the F_1 polypeptides (α) are commonly encoded by plant mitochondrial DNA. The F_1 stalk includes the γ-polypeptide and several other subunits, one of which is known as the oligomycin-sensitivity–conferring protein (OSCP) because it binds the antibiotic oligomycin. When oligomycin binds OSCP, this prevents proton translocation through the F_0 complex, inhibits ATP synthesis, and thereby limits mitochondrial oxygen uptake (see Section 14.5.1).

The postulated reaction mechanism of the ATP synthase involves the "conformational" model proposed by 1997 Nobel laureate Paul Boyer and supported by the crystal structure of the F_1 complex obtained by a fellow co-laureate John Walker (Fig. 14.30). According to this hypothesis, the free energy input from proton diffusion is not required to phosphorylate ADP, but rather induces a conformational change that releases tightly bound ATP from one of the three active sites. The $\alpha_3\beta_3$ subunit stoichiometry of the F_1 complex is consistent with this model, and the recent solution of a high-resolution crystal structure for the F_1 complex from mammalian mitochondria demonstrates structural and nucleotide (i.e., ATP and ADP) binding features that generally support this mechanistic scheme.

14.4.3 Movement of ATP, ADP, and phosphate into and out of plant mitochondria is also driven by the electrochemical proton gradient.

As mentioned in Section 14.4.1, evidence indicates that the oxidation of NADH by Complex I results in translocation of 10 protons from the matrix to the intermembrane space. Given the free energy required for ATP synthesis and that available from the measured $\Delta\bar{\mu}_{H^+}$, thermodynamic considerations indicate that three protons are needed per ATP molecule synthesized. However, when the energy cost of transporting adenine nucleotide and phosphate across the inner membrane is also taken into account, the H^+/ATP ratio increases (Fig. 14.31). Unlike the chloroplast, where ATP synthesis and consumption both take place in the same compartment (stroma), the mitochondria must expend free energy to import ADP and P_i and export ATP across the inner membrane. One ATP^{4-} is exported from the mitochondrion in exchange for one ADP^{3-} from the cytosol, and one hydroxide ion from the mitochondrial matrix is exchanged for one cytosolic P_i. Hydroxide transport equates to the net movement of one proton from the intermembrane space back into the matrix. Adding this proton to the three protons needed to synthesize one ATP from ADP and P_i at the measured values of $\Delta\bar{\mu}_{H^+}$ gives a value of four H^+ translocated per ATP synthesized.

Including the energy cost of nucleotide and phosphate transport in calculations of mitochondrial efficiency lowers the ADP/O values associated with substrate oxidation (see Fig. 14.26). If four protons are needed to transport ADP and P_i into the mitochondrion and to synthesize ATP, then the ADP/O ratio actually is 2.5, i.e., 1 ADP phosphorylated per 4 H^+ translocated, multiplied by the 10 H^+ translocated per $\frac{1}{2}O_2$ reduced by NADH. In this context, careful measurement of ADP/O ratios in isolated mitochondria often yields values close to 2.5 during oxidation of NAD-linked substrates. Oxidation of succinate by the bound FAD of Complex II, or of NAD(P)H by the rotenone-insensitive dehydrogenases, drives the synthesis of only about 1.5 molecules of ATP per oxygen atom reduced, because these reactions utilize only two of the three proton translocation sites.

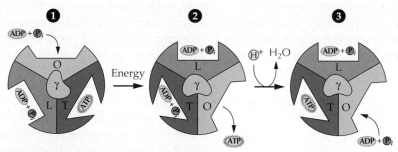

Figure 14.30
The conformational (binding) change model of ATP synthesis. The F_1 complex has three nucleotide-binding sites. Each of these active sites can exist in one of three distinct conformational states: loose nucleotide binding (L), tight nucleotide binding (T), and nucleotide-free, or open (O). At any one time, all three states are present in the F_1 complex, one being associated with each of the three catalytic centers present in the enzyme complex. ADP and P_i initially bind to an unoccupied site in the open state (1). Energy released by proton movement through the F_0 channel results in rotation of the γ-subunit. This rotation changes the conformations of each of the three nucleotide-binding sites simultaneously. The tight site containing bound ATP is converted to the open state, and the ATP is released. At the same time, the loose site containing the bound ADP and P_i is converted to a tight-binding, hydrophobic pocket that facilitates ATP synthesis. The open site that bound ADP and P_i in step 1 is converted to the loose conformation (2). The tightly bound ADP and P_i are subsequently converted to ATP in a step that does not require additional energy input or conformational change (3).

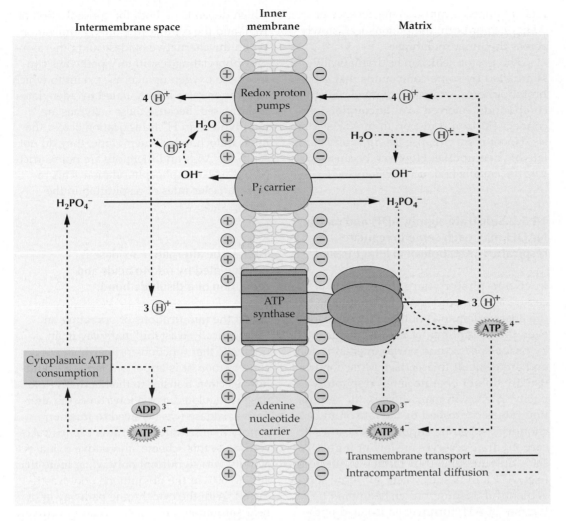

Figure 14.31
The adenine nucleotide carrier and the phosphate carrier expend proton motive force to supply ATP synthase with substrate. The adenine nucleotide carrier catalyzes the electrogenic exchange of cytosolic ADP^{3-} for mitochondrial ATP^{4-}. The net negative charge on the matrix side of the membrane ($\Delta\psi$) drives ATP export against a large concentration gradient and supports a high ATP/ADP ratio in the cytosol. In addition, the outer face of the carrier has a high affinity for ADP and therefore can import ADP to maintain oxidative phosphorylation even at low cytosolic ADP concentrations. The P_i carrier catalyzes an electroneutral exchange of negatively charged phosphate for hydroxide ion and thus balances the electrogenic adenine nucleotide exchange by dissipating the equivalent amount of ΔpH.

14.5 Regulation of mitochondrial respiration

14.5.1 Regulation of respiratory activity in isolated mitochondria depends on availability of ADP and P_i.

In isolated mitochondria, the rate of electron transfer, and therefore the rate of oxygen uptake, is controlled primarily by the availability of ADP and P_i, a phenomenon described as **respiratory control.** In the absence of ADP or P_i, the F_0 proton channel of the ATP synthase is blocked, and $\Delta\bar{\mu}_{H^+}$ builds until it exerts a back pressure that restricts further proton translocation across the inner membrane. Because electron transport is obligately linked to proton translocation, a large $\Delta\bar{\mu}_{H^+}$ will also restrict the rate of oxygen uptake (Fig. 14.32). In steady state, the rate of electron transfer is determined by the rate at which protons in the intermembrane space flow back into the matrix. When ADP and P_i are available, the backflow of protons via the

ATP synthase is rapid; in the absence of ADP or P_i or both, the protons leak slowly across the inner membrane.

The proton leak can be dramatically stimulated by some compounds that act as protonophores, or proton channels. Such compounds, referred to as **uncouplers,** disconnect electron transport from ATP synthesis. Uncouplers collapse the $\Delta\bar{\mu}_{H^+}$ and stimulate oxygen uptake. However, because no $\Delta\bar{\mu}_{H^+}$ is established, no ATP is formed.

14.5.2 Substrate supply, ADP, and matrix NADH may each serve to regulate respiratory metabolism in intact tissues.

In vivo respiratory rates are generally accepted as being at least somewhat linked to the energy demands of the plant cell. However, the controlling factors may be dictated by tissue-specific and environmental factors and are difficult to ascertain. Mitochondria usually do not operate at full respiratory capacity in vivo. In many tissues, the respiration rate is controlled by the rate of mitochondrial substrate supply, which slows if carbohydrate reserves are low or if glycolysis is downregulated by metabolic effectors (e.g., ATP). In tissues where substrate supply is plentiful, respiration can be limited by the rate of ATP turnover. In isolated mitochondria, the respiration rate is controlled by adding ADP. In vivo, respiration is also influenced by the cytosolic ATP/ADP ratio,

which depends on both the concentration of ADP and the cellular rate of ATP utilization. When the alternative oxidase and other non-phosphorylating respiratory pathways contribute to oxygen uptake, the extent to which respiration rates are regulated by adenylates is decreased. Because these enzymes are not involved in H^+ translocation across the inner mitochondrial membrane, they do not sense the $\Delta\bar{\mu}_{H^+}$ and therefore are not restricted by oxidative phosphorylation. This results in faster rates of respiration in the resting state.

14.5.3 The alternative oxidase is up-regulated by α-keto acids and reduction of a disulfide bond.

Given the ramifications of operating an "energetically wasteful" pathway in an organelle that functions primarily in energy conservation, it is important to understand how electron transfer to the alternative oxidase is regulated. For a long time, the alternative oxidase was believed to function simply in an electron overflow capacity. According to this scheme, alternative oxidase activity was significant only when more than 40% to 60% of the ubiquinone pool was reduced. With the cytochrome pathway at or near saturation, ubiquinol would thus divert electrons to the alternative oxidase. As a consequence of this paradigm, alternative oxidase activity was measured by the extent

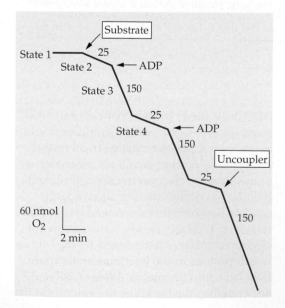

Figure 14.32
Respiratory control and the impact of uncouplers on the ADP/O ratio. The rate of oxygen uptake is described by four respiratory states. State 1 occurs when no oxidizable substrate is present; oxygen consumption (O_2 nmol min^{-1} mg^{-1} of protein) in state 1 is insignificant. Some oxygen is reduced (consumed) when substrate alone is added (state 2), but oxygen uptake is stimulated dramatically by adding ADP (state 3). When all ADP has been consumed, the oxygen uptake rate again decreases (state 4). The state 3 to state 4 ratio for rates of electron transfer, termed the **respiratory control ratio,** indicates how tightly ADP phosphorylation and electron transport are coupled. Chemical uncouplers collapse the proton gradient and release respiratory control, increasing oxygen consumption to the state 3 rate.

(A) Electron overflow model (considered out-of-date)

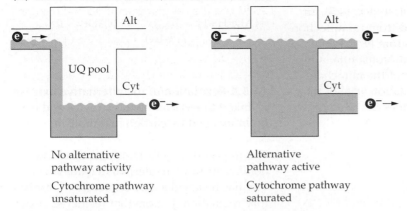

No alternative
pathway activity

Cytochrome pathway
unsaturated

Alternative
pathway active

Cytochrome pathway
saturated

(B) Electron distribution model (reflects current thinking)

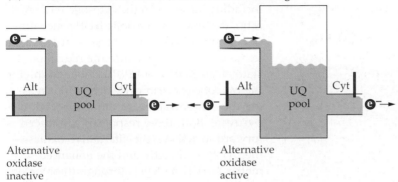

Alternative
oxidase
inactive

Alternative
oxidase
active

Figure 14.33
Two models for regulation of electron flow through the alternative oxidase. (A) In the electron overflow model, no appreciable electron transfer through the alternative pathway takes place until electron flow through the cytochrome pathway is at or near saturation. This could result from the effects of respiratory control, if the rate of mitochondrial ATP production exceeds its rate of utilization in the cytosol, or from some externally imposed stress, such as low temperature. Under such circumstances, the UQ pool becomes sufficiently reduced to allow electrons to flow through the alternative oxidase, the latter requiring that the UQ pool be 40% to 60% reduced to attain significant activity. (B) In the electron distribution model, the alternative and cytochrome pathways both show significant activity at low levels of UQ pool reduction, and electrons are distributed between the two pathways on the basis of the relative activities of each pathway. The activity of the alternative oxidase under these circumstances is thought to be regulated by the action of α-keto acids and by reduction/oxidation of the intermolecular disulfide bond, as well as by additional regulatory mechanisms not yet characterized.

to which alternative oxidase inhibitors such as SHAM (see Fig. 14.27) inhibited respiration. This approach assumed that the cytochrome pathway was already saturated and could not compensate for the inhibited alternative pathway. Thus, any observed decrease in the respiration rate would reflect how active the alternative oxidase had been in the absence of the inhibitor. This model, now considered out of date, is diagrammed in Figure 14.33A.

The alternative oxidase is currently known to be regulated by both carbon metabolites and covalent modification. The activity of the alternative oxidase is markedly enhanced in the presence of α-keto acids, such as pyruvate and glyoxylate (Fig. 14.34). This activation allows the alternative oxidase to compete for electrons with the cytochrome pathway even when ubiquinol concentrations are low (Fig. 14.33B). Direct measurements of electron distribution between the cytochrome and alternative pathways in the absence of inhibitors have verified this effect. If the cytochrome pathway is unsaturated, increased electron flow through

it can mask inhibition of the alternative pathway. Therefore, respiration rates measured in the presence and absence of SHAM do not quantify alternative oxidase activity.

Electron flow through the alternative pathway is also influenced by the oxidation state of an intermolecular disulfide bond that can covalently link the two monomers of the alternative oxidase dimer. Enzyme activity increases four- to fivefold when the disulfide bond is reduced to form two sulfhydryl groups (Fig. 14.35). The oxidized state is also markedly less sensitive to stimulation by α-keto acids. Reduction of this regulatory disulfide bond can be induced by applying citrate or isocitrate to isolated mitochondria and may involve NADPH generated in the matrix by NADP-linked isocitrate dehydrogenase (see Section 14.2.3).

The ability of this sulfhydryl/disulfide system to regulate alternative oxidase activity requires the presence of a redox-sensing mechanism in the mitochondrial matrix. One possible candidate is the thioredoxin system. Thioredoxin, a 12-kDa disulfide protein present in mitochondria, is known to reduce

$$R-\overset{\overset{\displaystyle O}{\|}}{C}-\overset{\overset{\displaystyle O}{\|}}{C}-O^-$$

α-Keto acid

Figure 14.34
Structure of an α-keto acid.

regulatory disulfide bonds of numerous proteins. Thioredoxin reductase (Rx. 14.3), an enzyme recently purified from rat mitochondria, transfers electrons from NADPH to thioredoxin (Trx) and may constitute a link between the redox poise of the mitochondrial matrix and the oxidation state of the regulatory cysteine residues in the alternative oxidase.

Reaction 14.3: Thioredoxin reductase

$$\text{NADPH} + \text{H}^+ + \text{Trx}_{ox} \rightleftharpoons \text{NADP}^+ + \text{Trx}_{red}$$
$$(-\text{S-S-}) \qquad\qquad\qquad (-\text{SH HS-})$$

14.5.4 Regulation of the alternative oxidase is linked to environmental stresses and may be influenced by carbon metabolism.

The impact of α-keto acids and disulfide bond reduction on alternative oxidase activity has revealed a clearer picture of respiratory regulation. Factors that restrict electron flow though the cytochrome pathway—including increases in the cytosolic ATP/ADP ratio, decreases in cytosolic ADP concentration, and environmental stresses such as low temperature—will enhance the reduction state of the matrix NAD(P)^+ pool and inhibit the rate of citric acid cycle activity, increasing the concentrations of mitochondrial pyruvate. Both these responses could feed forward to activate the alternative oxidase—the pyruvate directly and the enhanced matrix NAD(P)H levels through thioredoxin-mediated disulfide reduction. Once activated, the alternative oxidase could compete effectively with the cytochrome pathway for electrons from the ubiquinone pool, allowing the plant to maintain a greater rate of aerobic respiration than if the alternative pathway were not present. This respiratory activity may prevent damage associated with fermentative metabolism and the formation of active oxygen species (see next section).

The alternative oxidase is also regulated at the level of gene expression. In many plant tissues, two or more alternative oxidase proteins are detectable, and these have been shown to be encoded by separate genetic loci that may be regulated in a tissue-specific manner. In soybean, for example, one isoform appears to be expressed mainly in photosynthetic tissues. In some plants, the alternative oxidase is present in low amounts until the tissue is stressed in some manner, whereupon increased concentrations of the alternative oxidase protein are rapidly synthesized. Environmental conditions that can induce alternative oxidase synthesis include low temperatures, drought, generation of superoxide or hydrogen peroxide, certain herbicides, inhibitors of mitochondrial protein synthesis or electron transfer, and ripening

Inactive form (oxidized)

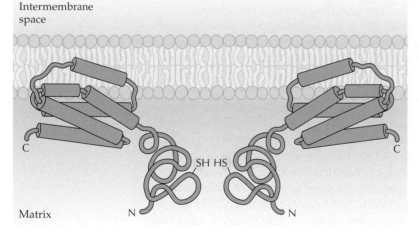

Active form (reduced)

Figure 14.35
Proposed regulation of the alternative oxidase through action of a redox-active regulatory sulfhydryl/disulfide system. Reduction of an intermolecular disulfide bond increases the sensitivity of the alternative oxidase dimer to activation by α-keto acids, possibly by reaction of the α-keto acid with the sulfhydryl. In both the oxidized and reduced states, the alternative oxidase remains associated as a homodimeric species. The in vivo source of the reducing agent has yet to be established, but plant mitochondria are known to contain the disulfide-reducing protein, thioredoxin.

in some fruits. In most cases, induction involves increased amounts of mRNA. In cultured tobacco cells, addition of the citric acid cycle intermediate citrate to the growth medium also triggers alternative oxidase synthesis, suggesting a link between carbon metabolism and regulation of alternative oxidase gene transcription. Because citrate can also generate NADPH for the regulatory sulfhydryl/disulfide system, both the activity and the synthesis are integrated with the carbon and energy status of the cell (Fig. 14.36).

14.5.5 Nonphosphorylating bypasses represent a unique aspect of plant respiratory metabolism, but their role is not well understood.

The existence of the rotenone-insensitive NADH dehydrogenase and the alternative oxidase suggests that respiring plant mitochondria might be able to shunt electrons solely through these nonphosphorylating bypasses and completely eliminate ATP synthesis, the function most commonly associated with mitochondria. It is unlikely that a plant would experience this extreme state, but the presence of these non–energy-conserving bypasses raises obvious questions about the role they play in plant metabolism.

With one unusual exception (see Box 14.2), putative bypass functions are not supported by direct evidence, but several hypotheses regarding their roles have been proposed—based on the principle that electron transfer through both bypasses is not constrained by respiratory control. When ADP concentrations are low, the nonphosphorylating bypasses may support higher respiration rates than the cytochrome pathway. If true, this would ensure a stable supply of metabolites needed for biosynthetic reactions, or in some cases this might be required for C_4 or CAM photosynthesis (see Section 14.6.5). An alternative, somewhat related view suggests that the nonphosphorylating bypasses act in an "energy overflow" capacity. This would constitute a coarse control on carbohydrate metabolism by which substrates could be oxidized if they accumulated to quantities greater than those required for growth, storage, and ATP synthesis. Both of these concepts are derived

from the now-defunct notion that electrons flowed to the alternative oxidase only when the standard cytochrome pathway was saturated. Although recently identified regulatory features do not rule out the roles envisioned above for the nonphosphorylating bypasses, evidence for these ideas is lacking.

A different hypothesis postulates enhanced bypass performance under unfavorable circumstances. Many environmental conditions (e.g., chilling, drought, osmotic stress) can inhibit plant respiration at the mitochondrial level. Since both the rotenone-insensitive NADH dehydrogenase and the alternative oxidase are far less complex than their proton-translocating counterparts, perhaps these simpler proteins function under stress more effectively than does the standard electron transfer chain. If so, the diversion of electrons through these bypasses in

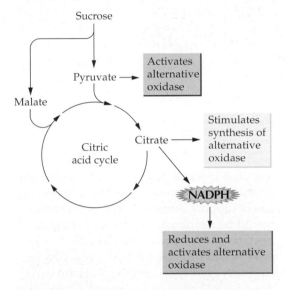

Figure 14.36
Impact of sugar metabolism on alternative oxidase activity. Activation and synthesis of the alternative oxidase are linked to the carbon status of the cell. When carbon flux to the mitochondrion exceeds the capacity of the electron transport chain to accept electrons, carbon intermediates such as pyruvate and citrate accumulate and the matrix pyridine nucleotide pool becomes highly reduced. Accumulation of citrate leads to synthesis of more alternative oxidase protein, whereas accumulated pyruvate and NADPH activate the enzyme. This feed-forward control ensures that the potentially wasteful alternative oxidase is active only when the carbon supply in the cell is plentiful. Activation of alternative oxidase prevents overreduction of other respiratory chain components, thereby decreasing the production of harmful active oxygen intermediates.

The alternative oxidase is active during thermogenesis in some flowers.

The alternative oxidase has one documented role: thermogenesis during flowering in a few plants, mostly members of the family Araceae. These plants, which include skunk cabbage and voodoo lily, produce a club-like structure (called an appendix) on the developing floral apex (see illustration). The outer layers of the appendix contain many more mitochondria than most plant tissues do. During **anthesis,** mitochondria in the appendix use the alternative oxidase to respire at very high rates. The free energy is released as heat raises tissue temperatures to 10°C to 25°C above ambient conditions and volatilizes odorous compounds that attract pollinators. In some cases, this mechanism can mimic the scent of rotting flesh and is used to deceive insects that lay their eggs on carrion.

The alternative oxidase is unlikely to promote thermogenesis in other plant tissues, for plants are not generally thermogenic. Observed rates of plant respiration are simply too low to generate significant amounts of heat, even if all electron flow were to be shunted through the alternative oxidase. Moreover, the high respiration rates observed in aroid appendices would release considerable heat even if the electrons were to travel through the cytochrome pathway; if thermogenesis were linked to ATP synthesis, however, the tissue would need to utilize large amounts of ATP very rapidly, or insufficient ADP would soon limit respiration.

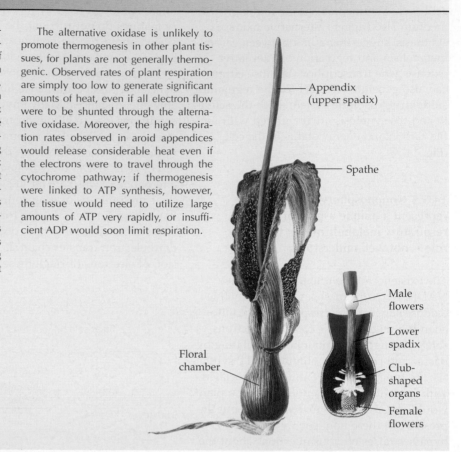

Appendix (upper spadix)

Spathe

Male flowers

Lower spadix

Club-shaped organs

Female flowers

Floral chamber

response to stress might maintain higher respiration rates than the cytochrome pathway could support. Although ATP production would be restricted, oxidation of cytosolic NADH by the external NADH dehydrogenase would provide the NAD$^+$ required to maintain glycolysis, and the alternative oxidase would continue to support aerobic respiration. Thus, the plant could avoid shifting to fermentative metabolism, which can have potentially deleterious consequences (Chapter 22). This concept is somewhat favored by the fact that the alternative oxidase diverts electrons from the cytochrome pathway at the level of the ubiquinone pool. Overreduction of the electron transfer chain can lead to formation of superoxide anion, hydrogen peroxide, and hydroxyl radical. When the mitochondrial quinone pool attains a high extent of reduction, these reactive oxygen species are generated by the autoxidation of ubiquinol. Alternative oxidase

activity may prevent the formation of such damaging oxygen compounds.

If a stress inhibits the plant mitochondrial electron transfer chain at a site downstream from the ubiquinone pool, then pyruvate, citric acid cycle intermediates, and reduced pyridine nucleotides will accumulate and activate the alternative oxidase. This activation would attenuate the inhibitory effects of the stress on respiration and, in the process, prevent overreduction of the ubiquinone pool. Interestingly, superoxide anion induces synthesis of alternative oxidase in fungal cells, as does hydrogen peroxide in cultured plant cells. On the other hand, a similar logic cannot be applied to activation of the rotenone-insensitive NADH dehydrogenase, because stress-induced activation of this protein would lead to further reduction of the ubiquinone pool. Nevertheless, the rotenone-insensitive bypass may play a role in maintaining respiration rates in response to a given stress by modulating matrix concentrations of NADH and consequently influencing carbon flow through the citric acid cycle (see Section 14.2.4).

At present, the role of the nonphosphorylating bypasses remains speculative. Recent years have seen the development of transgenic plants that lack alternative oxidase activity, so the phenotypes demonstrated by such plants should shed further light on how this protein functions in plant metabolism.

14.6 Interactions between mitochondria and other cellular compartments

In addition to synthesizing ATP, plant mitochondria generate biosynthetic precursors and carry out reactions that are linked to other metabolic pathways, including photorespiration, C_4 photosynthesis and Crassulacean acid metabolism, and gluconeogenesis during oil seed germination. These processes are facilitated by an extensive array of membrane transport proteins.

14.6.1 Movement of metabolites into and out of plant mitochondria is regulated by a series of specific transporters.

The inner membrane of the mitochondrion is selectively permeable. Gases and water move rapidly across the membrane, as do a few other uncharged molecules, such as the protonated species of small or lipophilic organic acids (e.g., acetic acid). However, since mitochondrial chemiosmotic energy conservation is supported primarily by $\Delta\bar{\mu}_{H^+}$, movement of charged molecules across the inner membrane must be controlled so as to not dissipate the $\Delta\bar{\mu}_{H^+}$.

Charged compounds enter or leave the mitochondrion via selective carriers that are regulated by the $\Delta\bar{\mu}_{H^+}$. The types of movements that can occur are summarized in Figure 14.37. Transport can be linked to the membrane potential ($\Delta\psi$), the pH gradient (ΔpH), or both ($\Delta\bar{\mu}_{H^+}$); this last case occurs when proton cotransport in one direction creates a charge imbalance in an exchange reaction. The $\Delta\bar{\mu}_{H^+}$ can drive transport directly or can be harnessed indirectly by coupling the transport of one species to another.

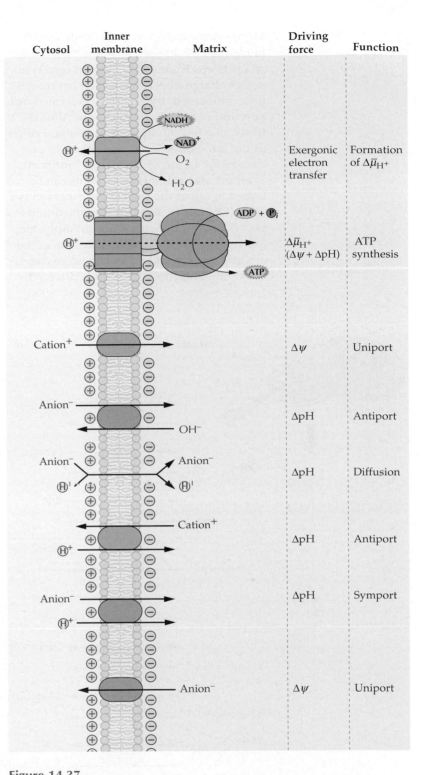

Figure 14.37

Mitochondrial transport is energized by $\Delta\bar{\mu}_{H^+}$. The proton motive force established by proton pumping of the respiratory chain can be used to drive ion transport across the inner mitochondrial membrane. Electrogenic transport involves the net movement of charge across the membrane; transport without net charge transfer is termed electroneutral. Inner membrane carriers that move only one ion (uniporters) are linked directly to $\Delta\psi$. Two chemical species may be cotransported in the same direction by symporters, or in opposite directions by antiporters. In the mitochondrial membrane, such coupled transport may be driven by the ΔpH, if one of the ions is a H^+ or OH^-; by $\Delta\psi$, if the combined transport is electrogenic; or by a combination of ΔpH and $\Delta\psi$ ($\Delta\bar{\mu}_{H^+}$).

Electrogenic transport involves the net movement of charge across the membrane, whereas **electroneutral transport** results in no net charge transfer. When an ion moves unaccompanied, the transport is termed **uniport** and is linked directly to $\Delta\psi$. Alternatively, one ion can move with another, either in the same direction **(symport),** or in exchange for an ion of like charge **(antiport).** Symport and antiport movements can be linked to either the ΔpH, if the other ion is a H^+ or OH^-, or to $\Delta\psi$ if the combined transport is not electroneutral. For example, the exchange of $ATP^{4-}_{out}/ADP^{3-}_{in}$ results in the electrogenic movement of a negative charge out of the mitochondrion and is driven directly by the $\Delta\psi$. The P_i and pyruvate carriers, on the other hand, catalyze the electroneutral exchange of an anion for OH^-, utilizing ΔpH. Because transporters such as the dicarboxylate carrier exchange a dicarboxylate anion for P_i, they too are dependent on maintenance of a ΔpH across the inner membrane.

Plant mitochondria interact in many ways with a variety of cellular metabolic processes and therefore require numerous transporters that exchange metabolites across the inner mitochondrial membrane. These transporters are summarized in Figure 14.38. Transport proteins provide numerous paths by which carbon and cofactors can

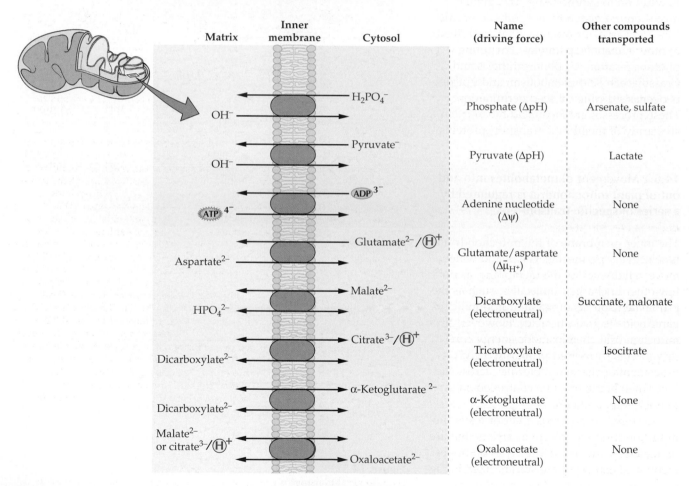

Figure 14.38
Carriers (substrate transporters) of the inner mitochondrial membrane. All of these secondary active transporters are indirectly linked to the proton motive force across the membrane and are capable of accumulating substrates inside the mitochondrial matrix. Uptake of pyruvate and P_i are linked to the ΔpH and exchange for hydroxide ions. The P_i gradient so generated can then be used to drive uptake of dicarboxylate anions via exchange on the dicarboxylate carrier. Dicarboxylates in turn can exchange for α-ketoglutarate or tricarboxylates; in the latter exchange, electroneutrality is maintained by cotransport of a proton with the citrate. This is also true for the exchange catalyzed by the OAA transporter. The exchange of ADP for ATP and glutamate for aspartate are electrogenic and driven by the $\Delta\psi$ across the inner membrane. Note that glutamate exchange is coupled to cotransport of a proton to cause a charge imbalance that is thought to be important for the operation of malate/aspartate shuttle (see Fig. 14.39).

be shipped to and from the mitochondria. The dicarboxylate and pyruvate carriers probably import substrates to the mitochondria, whereas the α-ketoglutarate and tricarboxylate carriers may export citric acid cycle carbon (see next section).

Plant mitochondria also contain specific transporters for the net uptake of NAD^+, CoASH, and thiamine pyrophosphate, all of which are essential cofactors for operation of the citric acid cycle. These transporters were discovered because isolated plant mitochondria lose these cofactors during prolonged storage and become unable to oxidize substrates such as malate and pyruvate. Adding back these cofactors restores respiratory activity. Although slower than those involved in transport of other metabolites across the mitochondrial inner membrane, the action of these transporters can facilitate accumulation of substantial cofactor concentrations. In vivo, these transporters are presumably involved in mitochondrial biogenesis; organelle division would dilute and eventually deplete these cofactors if they were not imported.

14.6.2 Metabolite shuttles transfer carbon and reducing equivalents between mitochondria and other cellular compartments.

One of the major challenges in plant mitochondrial research is developing an integrated view of mitochondria in the context of cell metabolism. Understanding how various carriers of the inner membrane interact to supply substrates for both mitochondrial and cytosolic processes is a crucial step toward this goal.

The α-ketoglutarate carrier is thought to operate with the glutamate/aspartate carrier in a substrate cycle known as the malate/aspartate shuttle (Fig. 14.39A). This shuttle, which can be demonstrated in isolated plant mitochondria, imports cytosolic reducing equivalents, i.e., NAD(P)H, into the mitochondrion in the form of malate. NADH imported via the malate/aspartate shuttle can reduce Complex I, so oxidation of imported NADH is associated with greater ATP synthesis than is oxidation of cytosolic NAD(P)H by the externally facing dehydrogenases of the respiratory chain. The

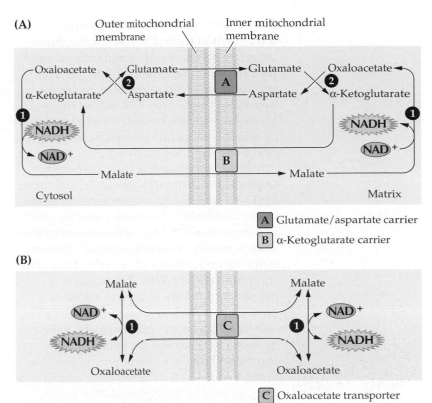

A Glutamate/aspartate carrier

B α-Ketoglutarate carrier

C Oxaloacetate transporter

Figure 14.39

Examples of mitochondrial metabolite shuttles: substrate shuttles in plant mitochondria. Plants can transfer reducing equivalents from cytoplasmic NAD(P)H to the mitochondrial matrix by means of substrate shuttles across the inner membrane. Two such shuttles, which have been demonstrated in isolated mitochondria, are shown. Whether either of these operates in vivo is not known. (A) Malate/aspartate shuttle. Isoenzymes of malate dehydrogenase (1) and aspartate aminotransferase (2) isoenzymes are found on both sides of the mitochondrial membrane. OAA reduction in the cytosol oxidizes NADH (or NADPH) and produces malate, which moves into the mitochondrion, where the reverse reaction occurs, liberating NADH in the matrix. Because this NADH can be oxidized by the respiratory chain, the reducing equivalents of the NADH in the cytosol have effectively been transferred into the mitochondrion. This allows formation of more ATP from NADH oxidation. Transamination of the OAA formed in the matrix prevents product inhibition of MDH. The aspartate and α-ketoglutarate formed are transported out of the mitochondrion in exchange for glutamate (carrier A) and malate (carrier B), respectively. Electrogenic glutamate/aspartate exchange is driven by the proton motive force across the inner membrane, thereby providing directionality to the shuttle (i.e., reducing power is transferred into the mitochondrion only). (B) Malate/OAA shuttle. Here MDH isoforms on either side of the mitochondrial membrane are linked by malate/OAA exchange on the OAA carrier (carrier C). This shuttle is readily reversible in isolated mitochondria when sufficient MDH is added to the reaction medium.

glutamate/aspartate exchange is electrogenic in nature: glutamate and one proton are imported from the cytosol, and aspartate is exported from the matrix. Transporting a net negative charge out of the mitochondrion ensures that this shuttle is unidirectional in respiring mitochondria, because the $\Delta \psi$ across the inner membrane (negative inside matrix) drives aspartate out of the mitochondrion. However, the malate/aspartate shuttle can function only when the $NADH/NAD^+$ ratio is higher in the cytosol than in the matrix. The malate/aspartate shuttle may be involved in glyoxysome–mitochondria interactions in oil seeds (see discussion later in Section 14.6.4 and Fig. 14.41).

Plant mitochondria also possess an OAA carrier that exchanges malate or citrate for OAA (the dicarboxylate carrier does not transport OAA). This simpler shuttle, which has been demonstrated in isolated mitochondria, is readily reversible and can transfer reducing equivalents into or out of the mitochondrion (Fig 14.39B). The direction of the shuttle is likely to depend on the concentration of the participant metabolites on either side of the mitochondrial inner membrane. In leaves, the malate/OAA shuttle is thought to participate in the photorespiratory cycle. The reducing equivalents formed by glycine oxidation in the mitochondrial matrix are exported to the peroxisome, where they are used to reduce hydroxypyruvate (see Section 14.8.1).

14.6.3 The citric acid cycle provides carbon skeletons for ammonia assimilation and amino acid synthesis.

An important ancillary function of mitochondria in all tissues is the production and export of α-ketoglutarate for ammonia assimilation or transamination reactions in the cytosol and plastids. Ammonia is assimilated via the GS/GOGAT (glutamine synthetase/glutamine:2-oxoglutarate aminotransferase) enzyme couple (see Chapter 8); the GOGAT reaction transfers the γ-amino group of glutamine to α-ketoglutarate. The only net source of this α-keto acid is the citric acid cycle. In addition, mitochondria in some roots may export citric acid cycle intermediates such as malate and citrate, which are secreted into the rhizosphere to solubilize cations and facilitate plant uptake of Fe^{3+} and other mineral nutrients (see Chapter 23).

Export of citric acid cycle intermediates requires the mitochondrial import of substrates that can generate both acetyl-CoA and OAA. If pyruvate were provided as the only carbon source, export of the citric acid

cycle intermediates would prevent regeneration of OAA and bring the citric acid cycle to a halt. This need for **anaplerotic** reactions that replenish pools of citric acid cycle intermediates may explain why plant mitochondria possess relatively large amounts of NAD-linked malic enzyme.

The α-ketoglutarate, OAA, and citrate carriers of the inner mitochondrial membrane may function in the export of α-ketoglutarate and citrate. Various schemes can be drawn for this carbon export, but it has not yet been determined which pathway operates in vivo. For example, acting in concert with the pyruvate carrier, the OAA carrier may promote the uptake of OAA and pyruvate into the mitochondrial matrix, where these can be used to synthesize citrate for subsequent export (Fig. 14.40A). Because there is also a cytosolic isoform of isocitrate dehydrogenase, the exported citrate can be used to generate α-ketoglutarate for ammonia assimilation or transamination. Alternatively, malate/α-ketoglutarate exchange on the α-ketoglutarate carrier, or malate/citrate exchange on the tricarboxylate carrier, could be involved in a more complex scheme, in which NADH is transferred into the mitochondrion and more energy is generated from matrix NADH via the respiratory chain (Fig. 14.40B). When provided with the necessary extra mitochondrial enzymes, isolated plant mitochondria are able to catalyze all of the exchanges required for both schemes.

14.6.4 Through gluconeogenesis, some plant tissues can convert lipids to sugars.

Most plant tissues store and respire carbohydrates. An important exception to this occurs in oil seeds, such as castor and soybean, which store and respire lipids. The storage tissues of these seeds, namely, the endosperm in castor bean and the cotyledons in soybean, contain substantial lipid reserves stored as triglycerides in specialized organelles called oil bodies (see Chapters 1 and 10). Utilization of these triglycerides involves their conversion to sucrose, which is then translocated to other organs in the growing seedling. This interconversion involves three different processes: β-oxidation of fatty acids, the glyoxylate cycle, and gluconeogenesis. Mitochondria interact closely with glyoxysomes during operation of the glyoxylate cycle.

Triglycerides are hydrolyzed to free fatty acids and glycerol via lipases in the oil bodies. The glycerol is converted to triose phosphates, which in turn are used to synthesize sucrose in the cytosol (see Chapter 13). β-Oxidation of fatty acids occurs in the glyoxysomes and produces NADH and acetyl-CoA (see Chapter 10). The acetyl-CoA is then converted to succinate through the action of the glyoxylate cycle (Fig. 14.41). This cycle utilizes two different enzymes, isocitrate lyase and malate synthase, which bypass the oxidative phase of the citric acid cycle. By preventing the complete oxidation of organic acids to CO_2, the glyoxylate cycle diverts carbon skeletons toward sugar biosynthesis. By contrast, animals lack isocitrate lyase and malate synthase and are therefore unable to convert lipids to carbohydrates.

Acetyl-CoA produced from oxidation of the fatty acids condenses with OAA to form citrate; this citrate is converted to glyoxylate and succinate by isocitrate lyase. The glyoxylate combines with another molecule of acetyl-CoA to form malate in a reaction catalyzed by malate synthase. This malate enters the cytosol and is converted first to OAA via malate dehydrogenase and then to PEP by the enzyme PEP carboxykinase. PEP is converted to fructose 1,6-bisphosphate by a reversal of glycolysis. The fructose 1,6-bisphosphate is subsequently converted to sucrose (see Chapter 13). The conversion of OAA to sugar is called **gluconeogenesis.**

Mitochondria participate in gluconeogenesis because glyoxysomes lack the enzymes to process succinate. Succinate, which is formed in the glyoxysomes, is transported to the mitochondria, where it is converted to OAA via the citric acid cycle. Furthermore, glyoxysomes lack the mitochondrial electron transfer chain, so NADH formed during fatty acid oxidation is transported to the mitochondria via the malate/aspartate shuttle. Thus, aspartate aminotransferase in the mitochondrion interconverts OAA and glutamate to aspartate and α-ketoglutarate, which are transported to the glyoxysome, where the reverse reactions occur to form OAA. Conversion of this OAA to malate oxidizes the excess NADH in the glyoxysome. The malate is recycled to the mitochondria

to complete the shuttle, and the NADH formed in the mitochondrion is oxidized via the respiratory chain. Operation of a malate/aspartate shuttle, rather than the simpler malate/OAA shuttle, has been proposed because isolated castor endosperm mitochondria do not transport OAA readily and because the equilibrium of the malate dehydrogenase reaction does not favor OAA formation. Amination of OAA therefore helps drive the oxidation of malate in the mitochondria. During respiration, the driving force for malate oxidation would be provided by citric acid cycle demand for OAA, but during gluconeogenesis, carbon must be returned to the glyoxysome to provide substrate for further sugar production. The ATP needed for gluconeogenesis is provided by mitochondrial respiration.

The overall stoichiometry of these reactions is shown in Fig. 14.41. The net result is conversion of fats to sucrose, which can be translocated to other parts of the seedling, and synthesis of ATP in the mitochondria.

14.6.5 Some C_4 and Crassulacean acid metabolism plants use mitochondrial reactions to concentrate carbon dioxide for photosynthesis.

The photosynthetic pathways of some C_4 plants and many plants that carry out CAM include mitochondria-associated reactions. These reactions are unique to plant mitochondria and reflect the extent to which the central organelle of heterotrophic metabolism has evolved to support autotrophic carbon fixation.

C_4 photosynthesis (Chapter 12) takes advantage of an aqueous equilibrium that favors HCO_3^- ion over CO_2 gas. In mesophyll cells, PEP carboxylase incorporates bicarbonate into a C_4 organic acid, which is then transported to bundle sheath cells, the site of CO_2 fixation by ribulose-1,5-bisphosphate carboxylase/oxygenase (Rubisco). The C_4 acid is decarboxylated in the bundle sheath cells, concentrating CO_2 and preventing oxygen fixation by Rubisco.

There are three C_4 photosynthesis pathways, each named for the bundle sheath enzyme that releases CO_2: $NADP^+$-malic enzyme (Fig. 14.42A), NAD^+-malic enzyme

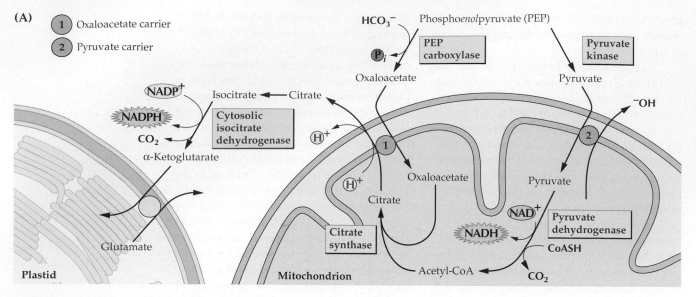

(A)

① Oxaloacetate carrier
② Pyruvate carrier

Figure 14.40 *(above and on facing page)*

Export of organic acids from the mitochondrion for support of ammonia assimilation in the plastid. Two possible substrate shuttles in the mitochondrion have been proposed. The $NADH/NAD^+$ and $NADPH/NADP^+$ ratios in the cytosol may determine which scheme operates by regulating the activity of MDH and $NADP^+$–isocitrate dehydrogenase, respectively. (A) OAA and pyruvate enter the mitochondrion and are used to generate citrate, which effluxes from the mitochondrion (in exchange for OAA) and is converted to α-ketoglutarate in the cytosol by $NADP^+$-linked isocitrate dehydrogenase. (B) Two molecules of malate enter the mitochondrion and are oxidized via MDH and NAD^+-linked malic enzyme to produce OAA and pyruvate in the matrix, which are subsequently converted to either citrate or α-ketoglutarate. The latter compounds exchange for malate via either the α-ketoglutarate or tricarboxylate carriers. Note that mitochondria and chloroplasts are not drawn to size.

(Fig. 14.42B), and PEP carboxykinase (Fig. 14.42C). Whereas the $NADP^+$-malic enzyme pathway does not require mitochondrial activity, the NAD^+-malic enzyme and PEP carboxykinase C_4 pathways both utilize mitochondrial NAD^+-malic enzyme to decarboxylate malate. In NAD^+-malic enzyme plants, this enzyme supplies Rubisco with CO_2 for photosynthesis (Fig. 14.42B), and the bundle sheath mitochondria process an enormous amount of carbon to supply the chloroplast. The photosynthetic rates attained by NAD^+-malic enzyme plants suggest that carbon flux through the bundle sheath mitochondria is 10- to 20-fold greater than the standard respiratory carbon flux, and severalfold greater than the flux of glycine through the mitochondria of C_3 plants during photorespiration.

In PEP carboxykinase plants, the situation is more complex. Although mitochondrial NAD^+-malic enzyme is known to be active in this pathway, cytosolic PEP carboxykinase is thought to provide most of the CO_2 fixed by Rubisco (Fig. 14.42C). Why are both enzymes required? One possible role for NAD^+-malic enzyme might involve oxidative phosphorylation. PEP carboxykinase converts OAA to PEP and CO_2 in a reaction that consumes one ATP. If the NADH generated by NAD^+-malic enzyme is utilized for ATP synthesis, the predicted stoichiometry would be roughly two malate molecules oxidized per five molecules of PEP produced.

CAM, a variation on the C_4 theme, allows plants to conserve water while still carrying out photosynthesis (Chapter 12). In CAM plants, HCO_3^- uptake by PEP carboxylase takes place at night, while stomata are open. The resulting C_4 acid, typically malate, is stored in the vacuole. Malate decarboxylation and CO_2 fixation by Rubisco take place in the same cell during the following light period, with the stomata closed to prevent water vapor from escaping (see Fig. 12.51). As with C_4 plants, subtypes of CAM photosynthesis are named for the decarboxylating enzyme. Malic enzyme CAM plants use both cytosolic NADP-malic enzyme and mitochondrial NAD-malic enzyme to decarboxylate malate. By contrast, PEP carboxykinase CAM plants contain very low concentrations of malic enzyme. This does not preclude mitochondrial

(B)

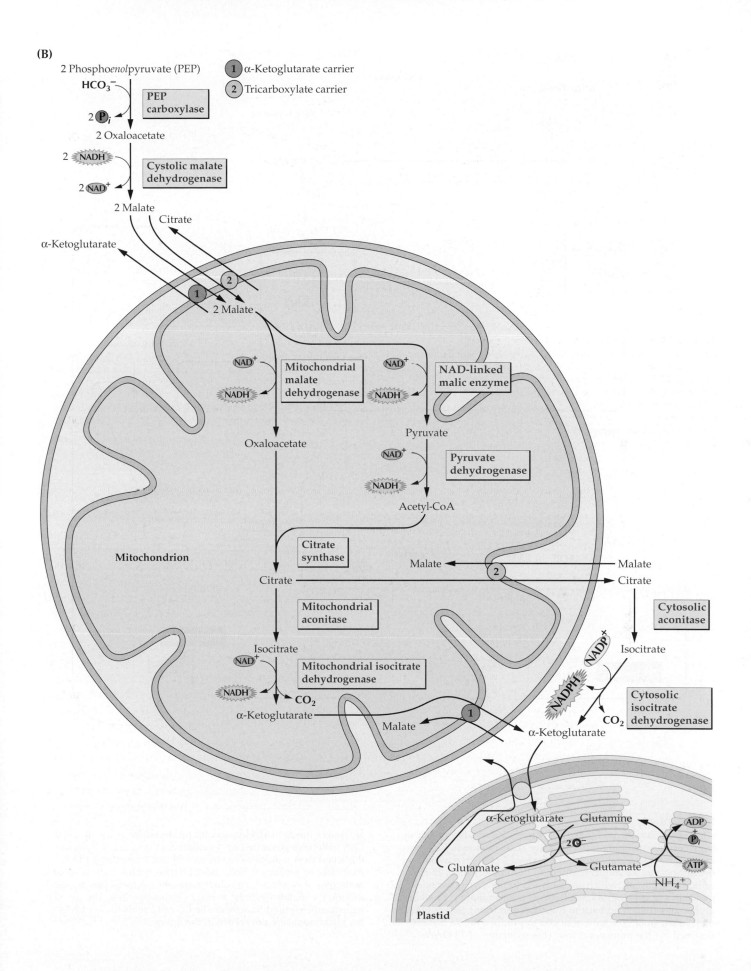

2 Phospho*enol*pyruvate (PEP)

(1) α-Ketoglutarate carrier

(2) Tricarboxylate carrier

HCO₃⁻

PEP carboxylase

2 Pᵢ

2 Oxaloacetate

2 NADH

Cystolic malate dehydrogenase

2 NAD⁺

2 Malate

Citrate

α-Ketoglutarate

2 Malate

Mitochondrion

NAD⁺

Mitochondrial malate dehydrogenase

NADH

NAD⁺

NAD-linked malic enzyme

NADH

Oxaloacetate

Pyruvate

NAD⁺

Pyruvate dehydrogenase

NADH

Acetyl-CoA

Citrate synthase

Malate

Malate

Citrate

Citrate

Cytosolic aconitase

Mitochondrial aconitase

Isocitrate

NADP⁺

Isocitrate

NAD⁺

Mitochondrial isocitrate dehydrogenase

NADH

CO_2

NADPH

Cytosolic isocitrate dehydrogenase

CO_2

α-Ketoglutarate

Malate

α-Ketoglutarate

α-Ketoglutarate

Glutamine

ADP + Pᵢ

2 e⁻

Glutamate

Glutamate

ATP

NH_4^+

Plastid

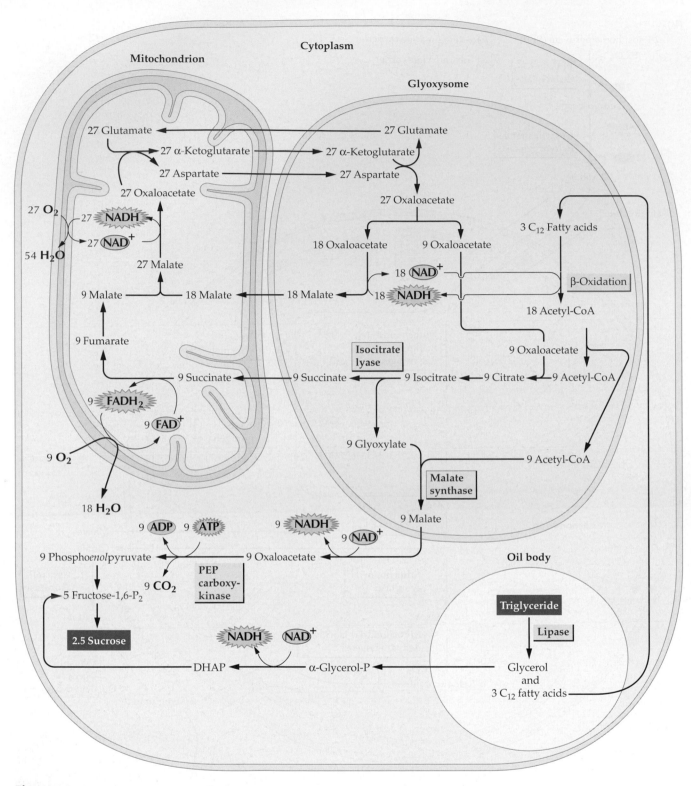

Figure 14.41

Gluconeogenesis from lipids. The conversion of a hypothetical triglyceride containing three C_{12} fatty acids is shown. Eighteen malate and 9 succinate molecules formed in the glyoxysome are transported to the mitochondrion and ultimately form 27 molecules of aspartate, which are returned to the glyoxysome with 27 molecules of α-ketoglutarate. Reverse transamination produces 27 glutamates (which go back to the mitochondrion) and 27 OAA molecules, 18 of which are reduced to produce the 18 malates destined for the mitochondrion. This reduction of 18 OAAs reoxidizes the 18 NADH molecules produced for every 18 acetyl-CoA molecules generated by β-oxidation of fatty acids. Nine of the acetyl-CoA molecules condense with the 9 remaining OAA molecules to produce 9 molecules of citrate, which, in turn, yield 9 molecules of succinate. The other 9 acetyl-CoA molecules are used to form the malate destined to enter gluconeogenesis. The six ATP equivalents required to activate the C_{12} fatty acids to acyl-CoAs in the glyoxysome are not shown in the figure.

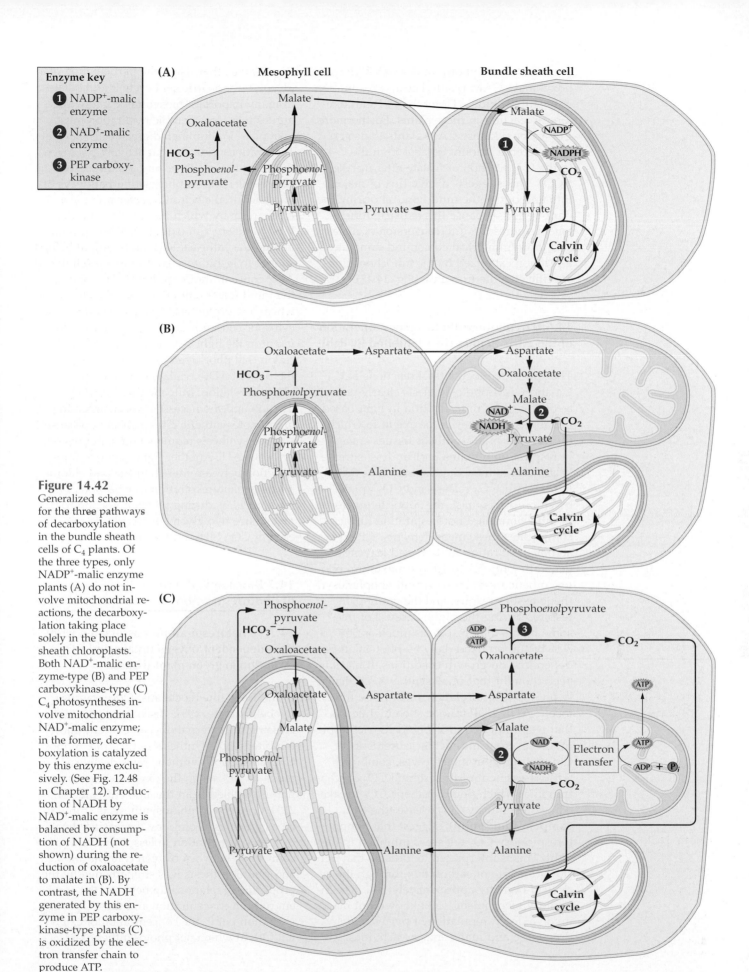

Figure 14.42
Generalized scheme for the three pathways of decarboxylation in the bundle sheath cells of C_4 plants. Of the three types, only NADP⁺-malic enzyme plants (A) do not involve mitochondrial reactions, the decarboxylation taking place solely in the bundle sheath chloroplasts. Both NAD⁺-malic enzyme-type (B) and PEP carboxykinase-type (C) C_4 photosyntheses involve mitochondrial NAD⁺-malic enzyme; in the former, decarboxylation is catalyzed by this enzyme exclusively. (See Fig. 12.48 in Chapter 12). Production of NADH by NAD⁺-malic enzyme is balanced by consumption of NADH (not shown) during the reduction of oxaloacetate to malate in (B). By contrast, the NADH generated by this enzyme in PEP carboxykinase-type plants (C) is oxidized by the electron transfer chain to produce ATP.

participation in photosynthesis by PEP carboxykinase CAM plants, because respiration may still provide ATP for the PEP carboxykinase reaction in these plants. Furthermore, PEP carboxykinase CAM plants transfer a considerable amount of label from the C-4 to the C-1 position of malate after uptake of $^{13}CO_2$. This suggests a large flux of malate into and out of the mitochondria during the dark period, because this redistribution of label requires the interconversion of malate and the symmetrical compound fumarate, a reaction catalyzed by the mitochondrion-specific enzyme fumarase (Fig. 14.43).

14.6.6 In photosynthetic tissues, operation of the citric acid cycle is inhibited by light.

The specific mitochondrial reactions of C_3, C_4, and CAM photosynthesis clearly point to a role for plant mitochondria in light-based metabolism. However, the extent to which illuminated photosynthetic tissues engage in respiration remains unclear. It is extremely difficult to measure respiration in the light, because O_2 release and CO_2 uptake by chloroplasts mask the opposite processes in mitochondria. Photorespiration further complicates these measurements. Rates of aerobic respiration in darkened leaves are generally only 5% to 10% of maximum photosynthetic rates. However, all nonphotosynthetic tissues respire, and the entire plant respires during the night. Because of this, some plants may respire as much as 50% to 70% of carbon fixed daily by photosynthesis, depending on growth conditions. It has long been thought that photosynthesis, particularly photophosphorylation, must poise the cytosolic ATP/ADP ratio at such high levels that respiratory control would restrict respiration severely. Indirect measurements of CO_2 output from intact leaves, combined with assessment of the external CO_2 concentration at which photosynthetic CO_2 uptake equals respiratory CO_2 release (the **CO_2 compensation point**), suggest that nonphotorespiratory CO_2 release is inhibited in the light. This is interpreted to indicate that light slows the rate of carbon flux through the citric acid cycle by approximately 50%. The inhibitory mechanism is not known but may involve down-regulation of pyruvate dehydrogenase and malic enzyme activities.

On the other hand, the participation of glycolytic and citric acid cycle intermediates in many important biosynthetic pathways argues that some aerobic respiration must take place in illuminated green cells. Several observations indicate that actively photosynthesizing tissues do respire. As discussed above, the nonphosphorylating pathways of mitochondrial electron transfer provide a mechanism by which plant mitochondria, at least in theory, can oxidize NADH without having to synthesize ATP and without being subject to respiratory control when cellular ADP concentrations are low. Furthermore, the rapid fractionation of organelles from wheat leaf protoplasts has demonstrated that cytosolic concentrations of ADP are no different in the light than in the dark, an indication that photosynthesis does not deplete the pool of ADP available for oxidative phosphorylation. Mitochondrial ATP production appears necessary for maintaining high photosynthetic rates, probably because sucrose synthesis requires high amounts of cytosolic ATP. Another light-dependent interaction between mitochondria and chloroplasts is photorespiration (see Sections 14.7 through 14.9), during which mitochondria metabolize two glycine molecules to one each of CO_2, NH_3, and serine.

14.7 Biochemical basis of photorespiration

14.7.1 Photorespiration is associated with light-dependent oxygen uptake and CO_2 evolution in green plant tissues.

In the 1920s, the biochemist Otto Warburg (Nobel laureate, 1931) discovered that increasing the external oxygen concentration inhibited photosynthesis in *Chlorella*. In most C_3 plants, photosynthetic rates decline as much as 50% when the oxygen concentration is doubled from the ambient value of 21%. Conversely, photosynthesis is stimulated as much as twofold by decreasing the oxygen concentration to less than 2%. The connection between oxygen concentration and photosynthesis was strengthened by the proportional relationship between the ambient oxygen concentration and the CO_2 compensation point, the concentration of external CO_2 at which net photosynthetic CO_2

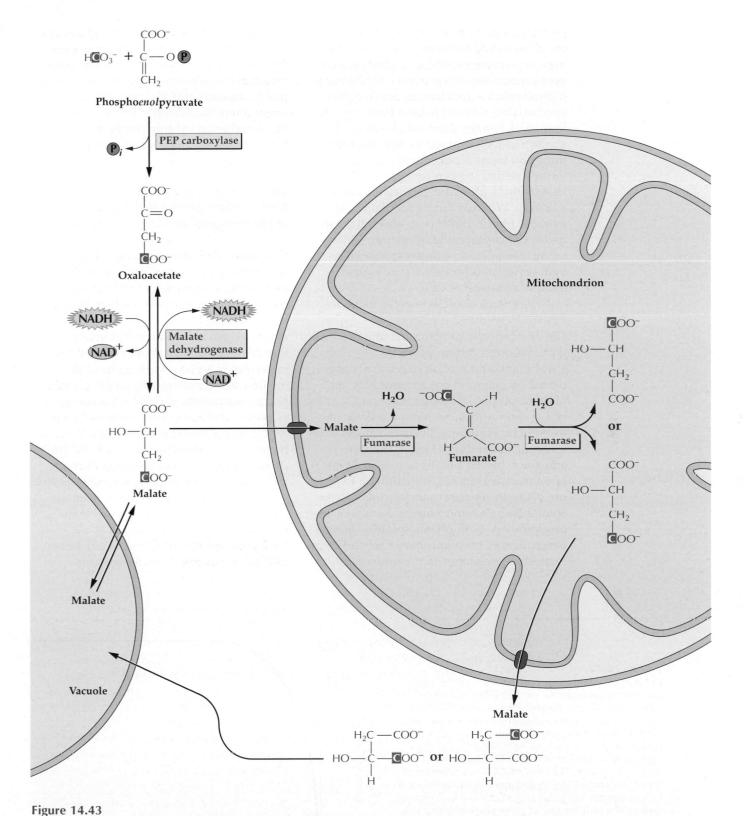

Figure 14.43
Randomization of label in PEP carboxykinase-type CAM plants suggests a considerable amount of flux through the mitochondria takes place at night after CO_2 is taken up in the cytosol to form malate. The reaction of $H^{13}CO_3^-$ with PEP catalyzed by PEP carboxylase results in malate labeled exclusively at the C-4 position. The transfer of large amounts of label to the C-1 position of malate stored in the vacuole suggests the movement of malate through the mitochondria. The carbon atoms at the C-1 and C-4 positions of malate are scrambled by the action of the citric acid cycle enzyme fumarase, which converts malate to the symmetrical molecule fumarate. When the resulting fumarate is converted back to malate, fumarase is equally likely to hydroxylate C-2 or C-3; therefore, the labeled carbon (previously C-4) can be at either the C-1 or C-4 position in the resynthesized malate molecule.

exchange equals zero (Fig. 14.44). The CO_2 compensation point increases with increasing oxygen concentration, suggesting a competition between oxygen and CO_2 during photosynthesis. In addition, actively photosynthesizing tissues release a burst of CO_2 immediately after illumination ceases. The amount of CO_2 evolved during this **postillumination burst** is also directly proportional to the external oxygen concentration. These observations ultimately revealed a novel pathway. Whereas almost all plants are capable of photosynthesis, in which light stimulates CO_2 uptake and oxygen evolution, many also perform **photorespiration**, a light-stimulated process that consumes oxygen and evolves CO_2.

High rates of photorespiration are restricted to so-called C_3 plants (see Chapter 12). The majority of plants fall into this category, although certain plants (i.e., C_4 and CAM plants) have evolved relatively sophisticated biochemical mechanisms that limit photorespiration by concentrating CO_2 at the site of carbon fixation. Many oxygenic photosynthetic organisms living in aquatic environments, including algae, cyanobacteria, and some higher plants, have also developed mechanisms for concentrating CO_2 or HCO_3^- in cells and transporting it to the chloroplast, thus minimizing the rate of photorespiration. Both photorespiration and its consequences for plants have been linked inextricably to changes in the ratio of CO_2 to

O_2 in the earth's atmosphere that have taken place since life appeared. During the 20th century, large-scale anthropogenic CO_2 emissions have accompanied an increase in global atmospheric CO_2 concentrations that might affect and inhibit photorespiration, thereby altering existing competitive relationships among many plant species.

14.7.2 Oxygenase activity of Rubisco catalyzes the initial step of photorespiration.

The origin of photorespiration is found in the kinetic properties of Rubisco, the enzyme that catalyzes the uptake of CO_2 in the initial reaction of photosynthetic carbon metabolism (see Chapter 12). In photosynthesis, this enzyme catalyzes the carboxylation of ribulose 1,5-bisphosphate (RuBP) to form two molecules of 3-phosphoglycerate (3-PGA), the first stable intermediate in the C_3-reductive photosynthetic carbon or Calvin cycle. However, Rubisco is also capable of catalyzing an oxygenase reaction, in which a molecule of oxygen reacts with RuBP to produce one molecule of 3-PGA plus one molecule of 2-phosphoglycolate (Fig. 14.45).

The Rubisco reaction mechanism involves the initial binding of RuBP to the enzyme, followed by formation of a 2,3-enediolate intermediate (Fig. 14.45). Either CO_2 or O_2 can react directly with this

Figure 14.44

Effect of carbon dioxide concentration on the carbon dioxide fixation rate. As the ambient concentration of CO_2 is increased in the presence of saturating light, net CO_2 uptake during photosynthesis increases in a roughly linear fashion, saturating at a low CO_2 concentration in C_4 plants and only beginning to show signs of saturation at ambient CO_2 levels (360–375 μl/l) in C_3 plants. The external CO_2 concentration at which net change in CO_2 is 0 μmol m^{-2} s^{-1} defines the CO_2 compensation point and reflects the CO_2 concentration at which the rate of gross photosynthetic CO_2 uptake exactly equals the respiration rate. The total respiration rate comprises both mitochondrial-linked (dark) respiration and photorespiration. That the CO_2 compensation point for C_3 plants (20 to 100 μl/l) is greater than for C_4 plants (0 to 5 μl/l) is associated with the presence of photorespiration in C_3 plants and its virtual absence in C_4 plants.

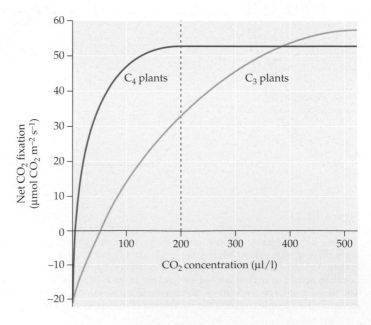

enediolate species, producing, respectively, an unstable C_6 or C_5 intermediate that breaks down to the final products rapidly and irreversibly. As a result of this mechanism, CO_2 and O_2 act as competitive substrates for Rubisco, with O_2 inhibiting RuBP carboxylation and CO_2 inhibiting RuBP oxygenation.

14.7.3 The relative carboxylase and oxygenase activities of Rubisco depend on kinetic properties of the enzyme.

The ratio of RuBP carboxylase to RuBP oxygenase activity (v_c/v_o, Eq. 14.3) is an important determinant of photosynthetic efficiency and can be expressed as a function of the kinetic parameters of the two competing reactions:

Equation 14.3: Ratio of carboxylase activity to oxygenase activity

$$\frac{v_c}{v_o} = \left(\frac{V_c}{K_c}\right)\left(\frac{K_o}{V_o}\right)\left(\frac{[CO_2]}{[O_2]}\right)$$

where V_c and V_o are the maximal velocities (V_m), and K_c and K_o are the Michaelis–Menten constants (K_m), for the carboxylase and oxygenase reactions, respectively. This formula assumes that CO_2 and O_2 are each present at a concentration lower than its apparent K_m. The ratio V_m/K_m defines the pseudo second-order rate constant for each reaction, so the set of kinetic constants $\left(\frac{V_c}{K_c}\right)\left(\frac{K_o}{V_o}\right)$ constitutes a **specificity factor** that represents the ratio of the carboxylation and oxygenation rates when the two gaseous substrates are present in equal amounts. Among land plants, Rubisco demonstrates an average specificity factor of 100, with values ranging between 80 and 130. In today's atmosphere, the air-equilibrated concentrations of the two substrates are approximately 8 μM CO_2 (0.036%) and 250 μM O_2 (21%) at 25°C. Therefore, a specificity factor of 100 gives a v_c/v_o ratio of 3.2.

14.8 Photorespiratory pathway

Given a specificity factor of 100 for Rubisco, the ratio of carboxylase to oxygenase activity is approximately 3 to 1 under present atmospheric conditions. Therefore, an appreciable amount of RuBP is converted to 2-phosphoglycolate, a compound that cannot be utilized by the C_3 reductive photosynthetic carbon cycle. A second cycle, known as the **C_2 oxidative photosynthetic carbon cycle** or photorespiratory carbon oxidation cycle, salvages this carbon so that it is not lost to photosynthetic metabolism. During the C_2

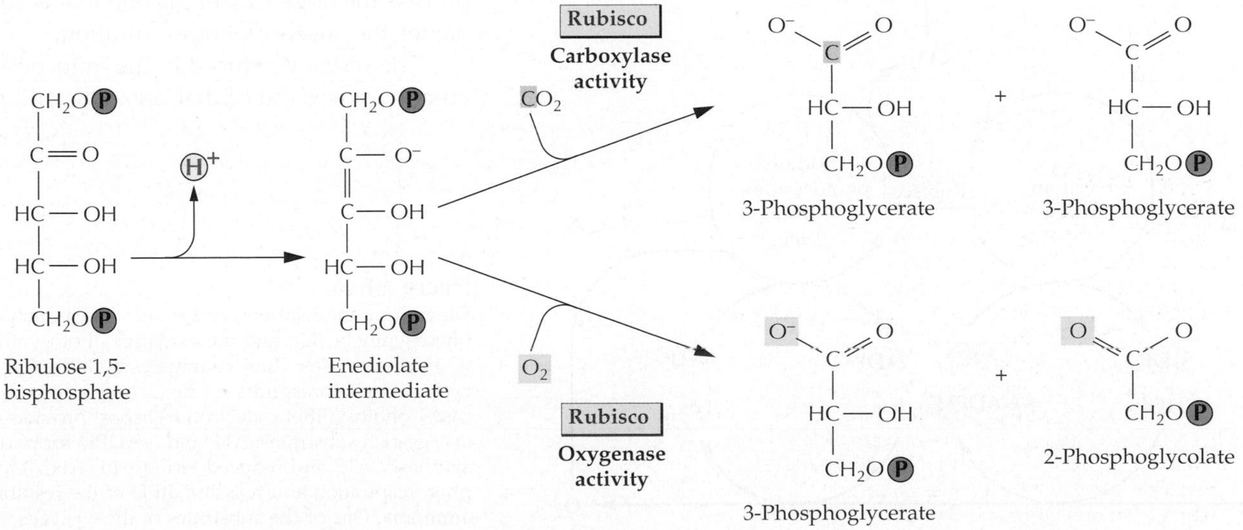

Figure 14.45
Rubisco catalyzes two types of reactions, carboxylation and oxygenation. In both cases the substrate is an enediolate intermediate formed by deprotonation of RuBP. Carboxylation yields two molecules of 3-phosphoglycerate (3-PGA), the first stable intermediate in C_3 photosynthesis. The oxygenation reaction yields one molecule of 3-PGA and one of 2-phosphoglycolate.

cycle, two molecules of phosphoglycolate are converted to one CO_2 and one molecule of 3-PGA, which can return to the C_3 cycle. When thinking of photosynthetic carbon metabolism in C_3 plants, remember that the C_3 and C_2 cycles operate together in an integrated fashion and not as separate, independent pathways (Fig. 14.46).

14.8.1 Photorespiratory reactions occur in three organelles: chloroplast, peroxisome, and mitochondrion.

The C_2 cycle (Fig. 14.47) is initiated in the chloroplast with the formation of 2-phosphoglycolate, which is subsequently converted to glycolate and P_i by the action of phosphoglycolate phosphatase. The glycolate exits the chloroplast through a glycolate transporter on the inner membrane of the chloroplast envelope and then enters a peroxisome, possibly by diffusion. In the peroxisome, glycolate reacts with O_2 to produce glyoxylate and H_2O_2. This reaction is catalyzed by a peroxisomal FAD-containing enzyme, glycolate oxidase. Two molecules of glyoxylate are aminated to form glycine in a reaction

catalyzed by two aminotransferases found in the peroxisome, serine:glyoxylate and glutamate:glyoxylate aminotransferases. Both enzymes must function for the continued operation of the C_2 cycle (see Section 14.9.2). The H_2O_2 is removed by the abundant catalase present in the peroxisome; two molecules of H_2O and one of O_2 are formed for every two molecules of phosphoglycolate that enter the C_2 cycle.

The glycine produced in the peroxisome then moves into the mitochondrion. For every two molecules of glycine that enter the mitochondrion, one molecule each of serine, CO_2, and NH_3 is produced, and one NAD^+ is reduced to NADH. This takes place through the action of two enzymes, glycine decarboxylase and serine hydroxymethyltransferase. One CO_2 is produced for every two molecules of O_2 taken up by the action of RuBP oxygenase.

Glycine decarboxylase is a very abundant and complex enzyme that catalyzes the oxidative decarboxylation of glycine to produce CO_2, NH_3, NADH, and methylene tetrahydrofolate (Fig. 14.48). Serine hydroxymethyltransferase joins this methylene tetrahydrofolate to a second molecule of glycine, producing the C_3 amino acid serine in a reaction that requires pyridoxal phosphate as a cofactor. Leaves maintained in the dark demonstrate low activities of these enzymes, but expression of the enzymes is stimulated on exposure to light, probably to process the large flux of glycine that accompanies the onset of photorespiration.

The serine produced in the mitochondrion now moves out of that organelle and into

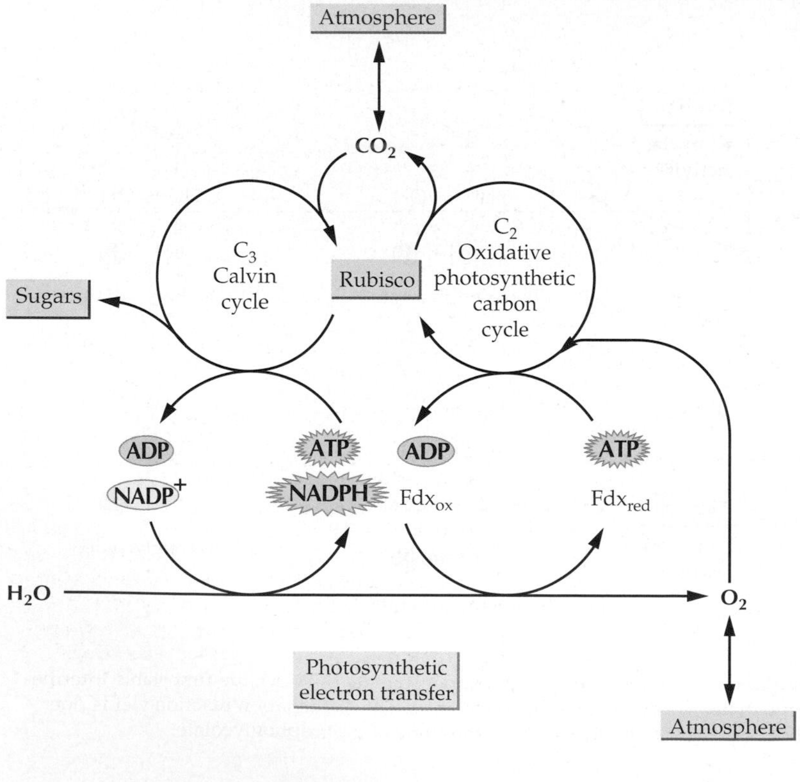

Figure 14.46
Diagram of the relationship between the reductive photosynthetic (C_3) and the oxidative photosynthetic (C_2) carbon cycles. Rubisco initiates both the Calvin cycle and photorespiration (the C_2 cycle). In both cases, photosynthetic electron transport provides energy-rich substrates: ATP and NADPH for photosynthesis, ATP and reduced ferredoxin (Fdx$_{red}$) for photorespiration and reassimilation of the resulting ammonia. One of the substrates of the C_3 cycle, CO_2, is a product of the C_2 cycle; in turn, the substrate of the C_2 cycle, O_2, is a product of C_3 photosynthesis. The gaseous substrates/products of both processes are derived from and contribute to the same atmospheric pools.

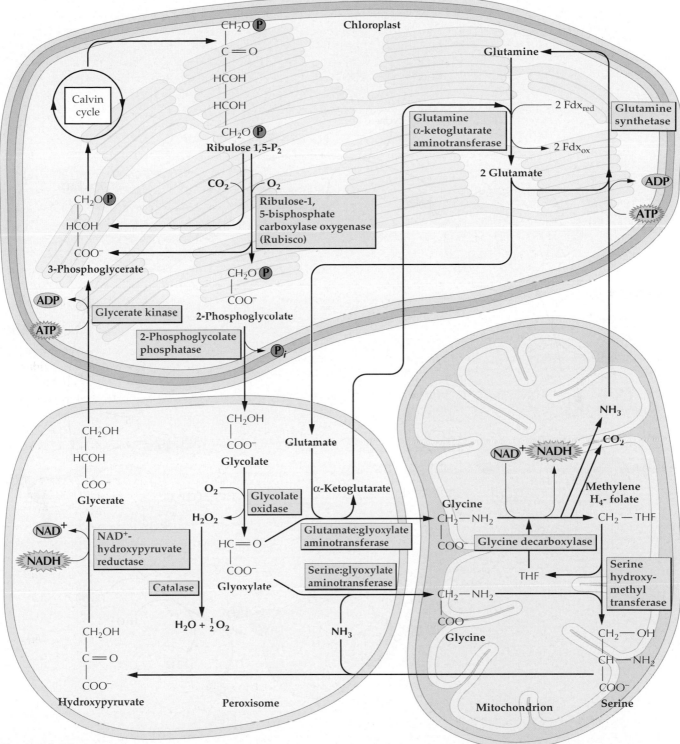

Figure 14.47

Reactions of the oxidative photosynthetic carbon (C₂) pathway. The operation of the C₂ pathway involves a cooperative interaction among three separate subcellular organelles: chloroplasts, peroxisomes, and mitochondria. The pathway is initiated in the chloroplast when the enzyme Rubisco catalyzes the oxygenation of ribulose 1,5-bisphosphate to produce one molecule of 3-phosphoglycerate and one molecule of 2-phosphoglycolate. The reactions of the C₂ pathway bring about the metabolic conversion of two molecules of 2-phosphoglycolate to one molecule of 3-phospho-glycerate, which can be used in the operation of the C₃ cycle, and one molecule of CO₂. Additionally, operation of the C₂ pathway leads to the uptake of a third molecule of oxygen and the release of one molecule of ammonia (NH₃). Reassimilation of the NH₃ requires the net input of two reducing equivalents (e.g., one molecule of NAD(P)H or two molecules of Fdx_red) plus one molecule of ATP. The final step in the formation of 3-phosphoglycolate also requires a molecule of ATP.

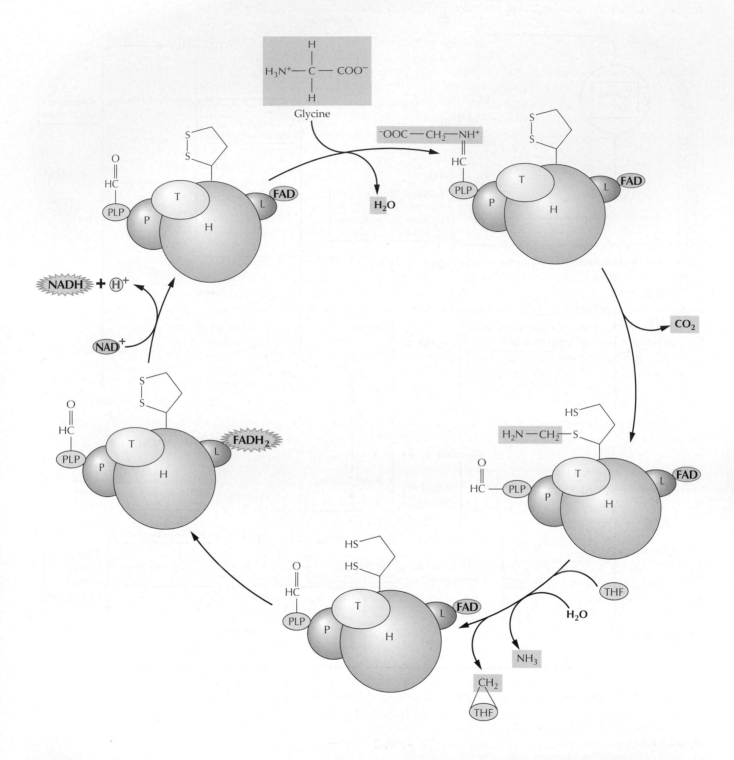

Figure 14.48

Reaction mechanism of the glycine decarboxylase complex. The complex consists of four enzymes, including a pyridoxal phosphate (PLP)–containing protein (P-protein), a lipoamide-containing protein (H-protein), a protein that interacts with tetrahydrofolate (T-protein), and a FAD-containing lipoamide dehydrogenase (L-protein). The reaction sequence is initiated by formation of a Schiff's base between glycine and the PLP on the P-protein. The P-protein catalyzes an oxidative decarboxylation of the glycine and transfers the reducing equivalents and the remaining methylamine to the lipoamide moiety on the H-protein. While not entirely the same, the remaining mechanistic features of the glycine decarboxylase resemble those found in the pyruvate dehydrogenase (see Fig. 14.11) and α-ketoglutarate dehydrogenase reactions of the citric acid cycle. The same L-protein is used in all three complexes.

the peroxisome, where it is deaminated by serine:glyoxylate aminotransferase, forming hydroxypyruvate. Another peroxisomal enzyme, hydroxypyruvate reductase, catalyzes the reduction of hydroxypyruvate to glycerate, using NADH as an electron donor. The glycerate exits the peroxisome and is imported into the chloroplast by the same transporter that exports glycolate. Once inside the chloroplast stroma, glycerate is phosphorylated by ATP, yielding 3-PGA and ADP. This reaction, catalyzed by the enzyme glycerate kinase, completes the conversion of two molecules of phosphoglycolate to one of 3-PGA, which can now enter the C_3 cycle.

The phenotypes associated with mutations in genes that encode the C_2 cycle enzymes support the cycle's proposed function in recycling 2-phosphoglycolate produced by the oxygenase reaction. Such mutants die under normal atmospheric conditions that promote photorespiration (21% O_2, 0.036% CO_2) but grow normally when placed in a high-CO_2 (greater than 0.2%) atmosphere, where the oxygenase activity of Rubisco is minimal. An *Arabidopsis* mutant lacking phosphoglycolate phosphatase was among the first plants reported to demonstrate this phenotype, but additional mutations have now been found in peroxisomal catalase, serine:glyoxylate aminotransferase, glycine decarboxylase, serine hydroxymethyltransferase, and chloroplast-localized isoenzymes of GS and ferredoxin-dependent GOGAT.

The reducing equivalents for the GOGAT reaction are provided by the reduced ferredoxin formed during photosynthetic electron transfer, whereas the ATP consumed by the GS reaction is produced by photophosphorylation (see Chapter 12).

The need to reassimilate the ammonia produced during the operation of the C_2 cycle helps explain why two different aminotransferases are required in the peroxisome for the conversion of glyoxylate to glycine (see Fig. 14.47). For every two glyoxylates that pass through the C_2 cycle, only one serine is produced in the mitochondria. The conversion of this serine to 3-PGA requires that serine:glyoxylate aminotransferase catalyze deamination of the serine to produce hydroxypyruvate. This reaction aminates one of the two glyoxylates, generating glycine. Amination of the second glyoxylate is associated with reassimilation of photorespiratory ammonia. Ammonia assimilation forms a molecule of glutamate that must be deaminated to regenerate the GOGAT substrate α-ketoglutarate. If glutamate is allowed to accumulate, ammonia assimilation will be inhibited by excess product and insufficient substrate. Glutamate:glyoxylate aminotransferase provides a mechanism for converting glutamate back to α-ketoglutarate. The photorespiratory nitrogen cycle and the two peroxisomal glyoxylate aminotransferases ensure that movement of nitrogen into and out of the C_2 cycle remains in balance.

14.8.2 Production of ammonia during photorespiration requires an ancillary cycle for its efficient reassimilation.

During operation of the C_2 cycle, glycine decarboxylation releases NH_3 inside the mitochondrion. To avoid toxic ammonia accumulation while preventing nitrogen loss through volatilization, photorespiring plant cells must reassimilate this ammonia efficiently. Reassimilation takes place in the chloroplast through the sequential action of two enzymes, GS and GOGAT (Rx. 14.4; see also Chapter 8).

Reaction 14.4: Net reaction of GS/GOGAT

$$\alpha\text{-Ketoglutarate} + NH_3 + ATP + 2e^- \rightleftharpoons$$
$$\text{Glutamate} + ADP + P_i$$

14.8.3 Photorespiration increases the energy costs associated with photosynthesis.

The stoichiometry associated with the C_2 cycle is not straightforward. Oxygenation of two RuBPs generates two molecules of 3-PGA and two molecules of 2-phosphoglycolate. Reaction 14.5 depicts the net conversion of the latter to one molecule of 3-PGA:

Reaction 14.5: The C_2 cycle

$$2\,RuBP + 3\,O_2 + 2\,Fdx_{red} + 2\,ATP \rightleftharpoons$$
$$3\,\text{3-PGA} + CO_2 + 2\,Fdx_{ox} + 2\,ADP + 2\,P_i$$

In photosynthesis, the relationship of O_2 evolved to CO_2 taken up is 1:1. In photorespiration, by contrast, 3 O_2 are taken up for every CO_2 evolved. Salvaging phosphoglycolate by converting it to phosphoglycerate requires a net input of two reduced ferredoxin (Fdx_{red})—i.e., one NAD(P)H equivalent—and two ATPs. The reducing equivalents and one ATP molecule are consumed during photorespiratory ammonia assimilation. The remaining ATP is used to phosphorylate glycerate. Whether this exact stoichiometry operates in vivo is less certain. No endogenous source of NADH exists in the peroxisome, so the reducing equivalents needed for the hydroxypyruvate reductase reaction must be imported. This probably involves malate, which can generate NADH via the peroxisomal malate dehydrogenase, forming OAA as a product. As discussed in Section 14.6.2, malate and OAA can be exchanged between the mitochondrial matrix and the cytosol through the operation of the OAA carrier. This malate/OAA shuttle provides a convenient mechanism for using the NADH produced by the mitochondrial glycine decarboxylase reaction in the peroxisome, where it can provide electrons for the reduction of hydroxypyruvate. If this mitochondrial shuttle operates, there is no need for the mitochondrial electron transfer chain to function in conjunction with the C_2 cycle, and the stoichiometry cited above would hold. However, chloroplasts also possess an OAA transporter and could utilize a malate/OAA shuttle to transfer reducing equivalents to the peroxisome. If the chloroplast-localized malate/OAA shuttle is used and all the NADH produced during glycine decarboxylation is oxidized through the mitochondrial electron transfer chain, operation of the C_2 cycle would consume two NAD(P)H equivalents and actually produce one ATP for every two CO_2 evolved.

If each NADH is capable of producing 2.5 ATPs during oxidative phosphorylation, the net energy cost of the operation of the C_2 cycle per CO_2 evolved is 4.5 ATP equivalents, regardless of which pathway or combination thereof operates in vivo (2 ATP, 2 Fdx_{red}). The cost of converting the resulting 3-PGA to 0.6 RuBP via the C_3 reductive photosynthetic carbon cycle adds an additional 4 ATP equivalents (i.e., 1 NADPH + 1.5 ATP) to the total. However, the operation of the C_2 cycle also leads to the loss of one CO_2, and the cost of refixing this CO_2 is another 8 ATP equivalents (i.e., 2 NADPH + 3 ATP). Assuming that C_3 plants fix three molecules of CO_2 for every molecule of oxygen fixed by the C_2 cycle, then the cost of fixing three molecules of carbon dioxide is 32.25 ATP equivalents in a C_3 plant—but only 30 ATP equivalents in many C_4 plants (Table 14.2). The C_2 cycle clearly constitutes a major energy drain that lowers the efficiency of CO_2 uptake during C_3 photosynthesis. This point is emphasized when we consider that C_4 plants can make up for using additional ATP during photosynthesis by minimizing photorespiration. Plants that have evolved mechanisms to minimize photorespiration are thus at a definite advantage under the atmospheric CO_2 concentrations extant in more recent times.

Table 14.2 Energy costs of photorespiration and photosynthesis

1 O_2 fixed by Rubisco	
C_2 cycle	2.25 ATP equivalents
Calvin cycle (regenerating RuBP from 3-PGA)	2 ATP equivalents
Recapturing $\frac{1}{2} CO_2$ released by C_2 cycle	4 ATP equivalents
Total	**8.25 ATP equivalents**
1 CO_2 fixed by Rubisco (C_3 plant)	
Calvin cycle	8 ATP equivalents
Total	**8 ATP equivalents**
1 HCO_3^- incorporated into C_4 acid (C_4 plant)	
C_4 cycle	2 or 3 ATP equivalents[a]
Calvin cycle	8 ATP equivalents
Total	**10 or 11 ATP equivalents[a]**

Note: Assuming a cost of 2.5 ATP per NAD(P)H oxidized.
[a] 2 ATP equivalents for NAD-malic enzyme or NADP-malic enzyme plants, 3 ATP equivalents for PEP-carboxykinase plants.

14.9 Role of photorespiration in plants

14.9.1 The rate of photorespiration can represent an appreciable percentage of the photosynthetic rate.

Under current atmospheric conditions, and given a specificity factor of 100, the rate of carboxylation to oxygenation should be approximately 3 to 1. Because each oxygenation reaction produces only 0.5 CO_2, the rate of photorespiratory CO_2 evolution should be equivalent to 16% of the rate of gross photosynthetic CO_2 uptake (i.e., 0.5 CO_2 released per oxygenation × 1 oxygenation per 3 carboxylations). If the number of carbon atoms metabolized through the C_2 cycle is considered, one can postulate that the rate of carbon flux through the C_2 cycle at any given time will represent more than 65% of the photosynthetic rate (4 carbon atoms metabolized through the C_2 cycle per molecule of photorespiratory CO_2 evolved × 16% of photosynthetic CO_2 uptake). Experimental measurements of photorespiratory carbon flux, which use oxygen isotopes to discriminate between photosynthetic CO_2 uptake and photorespiratory CO_2 efflux, indicate that the rate of photorespiratory CO_2 release ranges from 18% to 27% of the photosynthetic carbon fixation rate. These data conform to the theoretical calculations above. In fact, the latter value suggests that the flux of carbon through the C_2 cycle can exceed the rate of photosynthetic CO_2 fixation.

The values given above clearly illustrate the inhibitory effects of photorespiration on photosynthesis. However, the inhibition of photosynthesis by photorespiration is not associated solely with the evolution of CO_2 during the operation of the C_2 cycle. For every CO_2 released during the operation of the C_2 cycle, two molecules of oxygen react at the Rubisco active site and prevent CO_2 from being fixed. We can hypothesize, therefore, that if photorespiratory CO_2 flux equals 20% of the photosynthetic CO_2 uptake, then replacing these two oxygen molecules with carbon dioxide would increase the rate of photosynthesis by an additional 40% (see Fig. 14.47 and Rx. 14.5). Thus, abolishing RuBP oxygenation and the resulting C_2 cycle would enhance photosynthetic CO_2 fixation not by 20%, but by 60%. This premise is supported by experimental evidence: When net photosynthetic rates in C_3 plants are measured under 1% to 2% oxygen, increases of 50% to 70% are commonly observed. Likewise, growth under high-CO_2 or low-O_2 (relative to that found in the normal atmosphere) enhances production of dry biomass, further demonstrating the energy costs of the C_2 cycle.

14.9.2 The oxygenase activity of Rubisco is consistent with the enzyme's anaerobic origins.

The initial appearance of the enzyme Rubisco in ancient photoautotrophic bacteria is thought to have preceded the buildup of free oxygen in the earth's atmosphere by at least 10^9 years. For a long time, therefore, the environment apparently did not select against the oxygenase activity inherent in Rubisco's catalytic mechanism. After the appearance of free atmospheric oxygen, the selective pressure to minimize oxygenase activity must have been great, even though the early aerobic atmosphere contained less O_2 and more CO_2 than the air we breathe today. As recently as 100 million years ago, during the Cretaceous period, CO_2 accounted for approximately 0.3% of the atmosphere, almost an order of magnitude greater than the current CO_2 content.

The consequence of this building selective pressure can be seen in the range of kinetic specificity factors (see Section 14.7.3) displayed by different photosynthetic organisms. Specificity factor values as low as 15 are seen in bacteria that photosynthesize in anaerobic environments. These prokaryotes could not grow under current aerobic conditions; even if able to tolerate oxygen, they would fix more O_2 than CO_2. Cyanobacteria have specificity factors of 50 to 60, but can augment this with transport systems that concentrate CO_2 at the site of fixation to minimize oxygenase activity. A specificity factor of 100, average for terrestrial plants, appears to reflect the best that evolutionary selective pressure can accomplish with Rubisco alone, given that its oxygenase activity has been selected against for a very long time. The Ice Ages in particular must have impeded RuBP carboxylation—CO_2 concentrations dropped below current levels (0.036%) to as low as 0.030%.

The deleterious impact of photorespiration apparently led to the evolution of C_4 plants, which developed a complex biochemical pathway for concentrating CO_2 near Rubisco, in lieu of evolving a more efficient enzyme. The intricate biochemical, physiological, and structural relationships that make up C_4 photosynthesis suggest that photorespiration will not be overcome by manipulating one or even several genes. Whether genetic engineering of Rubisco can accomplish what several billion years of intense selection pressure have not remains to be seen.

14.9.3 Photorespiration may influence response of C_3 plants to future climatic events.

As temperature rises, changes in Rubisco's kinetic constants decrease the value of the selectivity factor, increasing oxygenase activity relative to carboxylation. Increasing temperature also decreases the aqueous solubility of dissolved CO_2 more than it does that of dissolved O_2, lowering the ratio of CO_2 to O_2 in air-equilibrated solution. Both of these effects enhance the rate of photorespiration relative to photosynthesis as the ambient temperature increases. As a result, in C_3 plants the quantum yield (see Chapter 12) for photosynthesis decreases continually as temperature increases, and the light-saturated optimum for net photosynthesis generally lies somewhere between 25°C and 35°C. In C_4 plants, which have a CO_2-concentrating

mechanism that effectively eliminates photorespiration, the relative quantum yield for photosynthesis remains roughly constant with increasing temperature, and temperature optima for photosynthesis are higher (40°C to 50°C) than those of most C_3 plants. When the atmospheric fraction of O_2 is low (less than 2%), the photosynthetic quantum yield for a typical C_3 plant is greater than that of a C_4 plant and remains constant with increasing temperature (Figs. 14.49 and 14.50).

The projected doubling of the atmospheric CO_2 concentration sometime in the 21st century could have significant consequences for photosynthesis. Theoretical ratios of oxygenase to carboxylase activity will decrease from present values of approximately 0.33 to less than 0.1, which should lead to an increase in net photosynthesis in C_3 plants. Given the direct competition between CO_2 and O_2 for Rubisco-bound RuBP, higher concentrations of CO_2 would enhance carboxylation over oxygenation, even when the overall rate of photosynthesis is not controlled by Rubisco or RuBP availability. C_4 plants that lack measurable rates of photorespiration would not experience this photosynthetic enhancement and might be put at a selective disadvantage relative to C_3 neighbors. Such a change might assist agriculture because many of the world's most pernicious weeds are C_4 plants.

Although the efficiency of photosynthesis will increase with greater CO_2 concentrations, the absolute rate of photosynthesis may not. Numerous studies indicate that

Figure 14.49

Effect of leaf temperature on quantum yield in C_3 and C_4 plants. In C_3 plants, the rate of photorespiration increases with increasing temperature to a greater extent than does the rate of gross photosynthesis. The enhanced relative operation of the oxidative photosynthetic carbon pathway at higher temperatures leads to a greater energy cost per net CO_2 fixed, which is reflected in a decrease in the quantum yield (moles of CO_2 fixed per photon of light absorbed). When C_3 photosynthesis is carried out in the presence of decreased (1%) O_2, photorespiration is effectively eliminated and the resulting quantum yield remains constant with increasing temperature. Plants having the C_4 pathway show a constant quantum yield with increasing temperature, thus reflecting the absence of the operation of photorespiration; instead, C_4 plants concentrate CO_2 at the site of CO_2 fixation in the bundle sheath cells. The lower quantum yield for C_4 plants than for C_3 plants in 1% O_2 reflects the fact that, although both lack photorespiration, C_4 plants retain the added energy cost (2 ATP per CO_2 fixed) needed to operate the C_4 pathway. At lower temperatures, C_3 plants have a higher quantum yield than C_4 plants even in normal air, reflecting the energy tradeoffs of having a lower rate of photorespiration but not incurring the cost of operating the C_4 pathway.

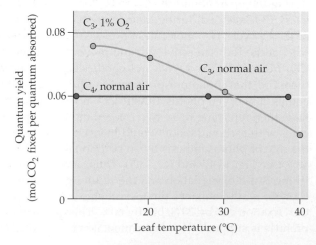

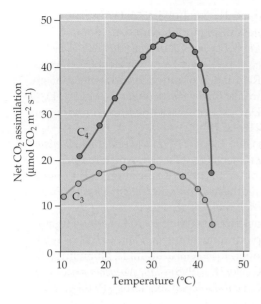

Figure 14.50
Effect of ambient temperature on carbon dioxide assimilation in C_3 and C_4 plants. In C_3 plants the increase in the relative contribution of photorespiration to net photosynthetic CO_2 uptake as temperature increases leads to a lower optimum temperature than that for C_4 plants, where the absence of photorespiration shifts the optimum temperature for net CO_2 fixation to higher temperatures. The steep decline in photosynthetic rates at temperatures above 40°C is common to both C_3 and C_4 plants and reflects an irreversible thermal denaturation of components of the photosynthetic apparatus at high temperatures. When photosynthesis in C_3 plants is measured under conditions of low (1%) O_2, which effectively eliminates photorespiration, the resulting temperature response curve approaches that seen with C_4 plants.

many plants acclimate to continuous exposure to high CO_2, such that their photosynthetic rates eventually resemble those of plants grown in air. Reasons for this response include diminished amounts of Rubisco in plants grown in high-CO_2 and feedback inhibition of photosynthesis as a result of the inability to utilize all the additional carbohydrate initially produced in high-CO_2 conditions. Regardless of this acclimation effect, increased atmospheric CO_2 may affect the balance of photosynthetic and photorespiratory reactions in many plants, with consequences that are at present unknown.

Summary

Aerobic respiration involves the controlled oxidation of reduced organic compounds to CO_2 and H_2O. Much of the free energy released during respiration is conserved in the form of ATP. In glycolysis, the first stage of respiration, carbohydrates are oxidized to organic acids in the cytosol. The organic acids produced during glycolysis are completely oxidized to CO_2 in the mitochondrial matrix via the citric acid cycle. The electrons released during the operation of the citric acid cycle are transferred through a series of multiprotein complexes located in the inner mitochondrial membrane, ultimately reducing O_2 to H_2O. The free energy released during mitochondrial electron transfer is used to generate a proton electrochemical gradient across the inner membrane. The energy available in the proton gradient is subsequently used by another protein complex, ATP synthase, to synthesize ATP from ADP and P_i. Mitochondrial respiration is regulated by the availability of ADP and P_i and by the presence of additional electron transfer complexes that allow respiration to proceed without forming a proton gradient.

Plant mitochondria participate in several metabolic processes besides respiration, including providing reducing equivalents to other cellular compartments and carbon skeletons for amino acid biosynthesis. Plant mitochondria also participate in the biosynthesis of sugars from lipids in some germinating seeds and in the decarboxylation reactions associated with photosynthesis in some plants having C_4 and CAM pathways. The movement of metabolites into and out of mitochondria requires specific transporters in the inner mitochondrial membrane, some of which are regulated by the proton gradient.

Photorespiration involves the light-dependent uptake of O_2 and evolution of CO_2 during photosynthesis in green plant tissues. The first step in photorespiration is associated with the oxygenase activity of the photosynthetic enzyme Rubisco. Phosphoglycolate formed during the oxygenase reaction is metabolized through the photorespiratory carbon cycle to save 75% of the carbon in the form of phosphoglycerate; the remaining 25% is lost as CO_2. The reactions of the photorespiratory carbon cycle occur in three organelles: chloroplasts, peroxisomes, and mitochondria. The loss of CO_2 during photorespiration can represent an appreciable percentage of the carbon fixed during photosynthesis, decreasing the overall efficiency of photosynthesis. Photorespiration reflects the evolutionary origin of Rubisco in an anaerobic environment and may influence the competitiveness of some plants in response to future changes in atmospheric CO_2 concentrations.

Further Reading

Calhoun, M. W., Thomas, J., Gennis, R. B. (1994) The cytochrome oxidase superfamily of redox driven proton pumps. *Trends Biol. Sci.* 19: 325–330.

Day, D. A., Whelan, J., Millar, A. H., Siedow, J. N., Wiskich, J. T. (1995) Regulation of the alternative oxidase in plants and fungi. *Aust. J. Plant Physiol.* 22: 497–509.

Douce, R., Neuburger, M. (1989) The uniqueness of plant mitochondria. *Annu. Rev. Plant Physiol. Plant Mol. Biol.* 40: 371–414.

Dry, I. B., Bryce, J. H., Wiskich, J. T. (1987) Regulation of mitochondrial metabolism. In *The Biochemistry of Plants*, Vol. 11, D. D. Davies, ed., Academic Press, London, pp. 213–352.

Ernster, L., ed. (1992) *Molecular Mechanisms in Bioenergetics*. Elsevier, Amsterdam.

Krömer, S. (1995) Respiration during photosynthesis. *Annu. Rev. Plant Physiol. Plant Mol. Biol.* 46: 45–70.

Mackenzie, S., McIntosh, L. (1999) Higher plant mitochondria. *Plant Cell* 11:571–586.

Møller, I. M., Gardeström, P., Glimelius, K., Glaser, E. (1998) *Plant Mitochondria: From Gene to Function*. Backhuys Publishers, Leiden.

Nicholls, D. G., Ferguson, S. J. (1992) *Bioenergetics*, 2nd ed. Academic Press, London.

Siedow, J. N., Umbach, A. L. (1995) Plant mitochondrial electron transfer and molecular biology. *Plant Cell* 7: 821–831.

PART 4

Metabolic and Developmental Integration

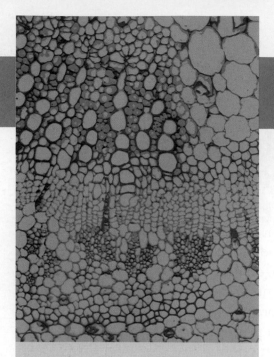

Biochemistry & Molecular Biology of Plants, B. Buchanan, W. Gruissem, R. Jones, Eds.
© 2000, American Society of Plant Physiologists

C H A P T E R *15*

Long-Distance Transport

Donald B. Fisher

Introduction

The vascular plant body is strongly differentiated at all levels of organization and contains many distinct cell types, tissues, and organs. To survive in a terrestrial habitat, plants must coordinate the assimilatory activities of two spatially distant but nutritionally interdependent regions: aerial shoots that harvest light energy and CO_2, and subterranean roots that take up water and mineral nutrients. Although mechanical and spatial factors are also involved, the physiological requirements of this interdependence are met primarily by transport of materials throughout the plant body.

The distinction between short-distance transport and long-distance transport is somewhat arbitrary. Long-distance transport via bulk flow in the vascular tissues is the basis of functional interaction between plant organs, and it is the principal topic of this chapter. However, I will also address short-distance, nonvascular transport in the tissues that surround the conducting cells because these tissues play a central role in controlling the movement of solutes and water into and out of the vasculature. At these shorter distances, diffusive transport becomes increasingly important.

Many specialized tissues participate in the transport of nutrients and metabolites throughout the plant body. Some of the basic features of vegetative plant anatomy, with emphasis on the vascular system, are illustrated in Figure 15.1.

15.1 Overview of diffusive and convective transport in plants

15.1.1 Diffusion and convection are the basis for solute transport.

In general, total solute flux, J_{total} (mol m^{-2} s^{-1}), is the sum of **convective transport,** bulk flow of solution driven by a pressure gradient, and **diffusive transport,** independent movement of molecules driven by a concentration gradient as shown in Equation 15.1, where V is velocity, C is concentration, and D is the diffusion coefficient (m^2 s^{-1}). Diffusion is an effective basis for solute movement only over short distances. This limitation is apparent in the equation that describes the diffusion of a solute into a cylinder after an increase in solute concentration at the cylinder surface: Solute exchange is complete (i.e., net solute movement is zero) when $t = R^2/D$, where t is time, R is the cylinder radius, and D is the diffusion coefficient (m^2 s^{-1}). For metabolites in water, $D \approx 5 \times 10^{-10}$ m^2 s^{-1}. For $R = 10$ μm, the radius of a small cell, diffusive flux is complete in 0.2 s. If R is increased to 10 mm, a radius typical for some branches or berries, solute exchange does not reach equilibrium for 50 hours. These calculated values do not take into account the heterogeneous nature of biological cylinders; nevertheless, they illustrate the "distance-squared" relationship limiting the maximum dimensions that can be served by diffusive transport. For dissolved metabolites, this effective distance is well under 1 mm, for which $R^2/D \approx 30$ min.

Equation 15.1

$$J_{total} = J_{convection} + J_{diffusion} = VC - D\frac{\Delta C}{\Delta x}$$

15.1.2 Water and solutes can move from cell to cell along either of two parallel aqueous pathways.

Plants contain two parallel aqueous pathways for intercellular solute movement, separated by the plasma membrane (Fig. 15.2).

The aqueous phase of the **apoplast** lies outside the plasma membrane and consists almost entirely of the cell walls and conducting cells of the xylem. The **symplasm,** which lies within the plasma membrane and is connected from cell to cell by plasma membrane–bound tubules (the **plasmodesmata**) consists of the cytoplasm of living nucleate cells and of the enucleate conducting cells of the phloem.

Two significant volumes of the plant body, the intercellular spaces and vacuoles, are left in limbo by the apoplast/symplasm concept. Intercellular spaces are generally included in the apoplast, but they are almost always gas-filled, even in submerged tissues. As such, they have little to do with solute movement. However, being extensively interconnected, they provide a third parallel pathway for long-distance movement of gases. The vacuole is not usually regarded as part of the symplasm because the vacuolar membrane is an effective barrier to most solutes.

15.1.3 Convective transport in the symplasm is usually driven by osmotically generated pressure flow.

Convective transport is usually driven by quite different mechanisms in the apoplast and symplasm. In the former, the motive pressure gradients usually result from transpirational water loss, a topic to be discussed more extensively in Section 15.4. In contrast, convection in the symplasm is driven by **osmotically generated pressure flow** (OGPF).

Turgor pressure in plant cells is a consequence of their high solute content and, at constant water potential (Box 15.1), varies directly as the solute concentration changes. As a cell accumulates more solute, it absorbs water and its turgor increases. Conversely, solute loss or utilization leads to water loss and a decline in turgor. If cells are hydraulically isolated, as with stomatal guard cells, these changes in turgor proceed independently in each cell (Fig. 15.3A). In contrast, if the cells are hydraulically linked, a pressure

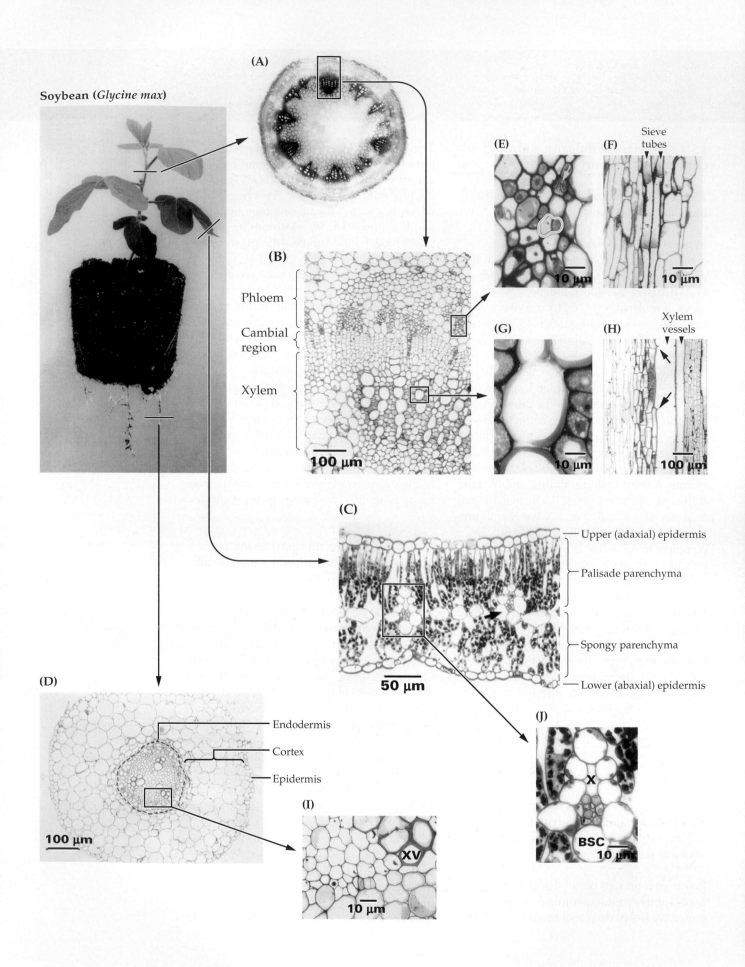

Soybean (*Glycine max*)

(A)

(B)

Phloem
Cambial region
Xylem
100 μm

(E) 10 μm

(F) Sieve tubes 10 μm

(G) 10 μm

(H) Xylem vessels 100 μm

(C)
Upper (adaxial) epidermis
Palisade parenchyma
Spongy parenchyma
Lower (abaxial) epidermis
50 μm

(D)
Endodermis
Cortex
Epidermis
100 μm

(I) XV 10 μm

(J) X BSC 10 μm

Figure 15.1 *(Facing page)*
Anatomical overview of the vascular system of soybean (*Glycine max*), a dicot. Except for (A), the illustrations (B–J) are bright-field micrographs of semithin plastic sections, 1 μm thick, stained first with the periodic acid–Schiff reaction for carbohydrates (red), followed by aniline blue black staining for proteins (blue). The tissues were fixed with glutaraldehyde. (A) A hand-cut cross-section of an unfixed stem internode. The shoot had been cut off and the cut stem placed in a solution of methylene blue for 1 hour, followed by 1 hour in water, to stain the water-conducting elements of the xylem. Dye has begun to diffuse from the xylem vessels into the walls of surrounding parenchyma cells. (B) A somewhat higher magnification of the stem vascular tissues, showing the arrangement of xylem, phloem, and meristematic tissues (cambium) in a vascular bundle. (C) Principal tissues of a leaf, as seen in cross-section. Two minor veins are present (box and arrow). (D) Cross-section of a root, about 2 cm from the tip, illustrating the principal tissues. In the endodermis (dotted line), the innermost layer of cortical cells, the radial and transverse cell walls contain suberin, a hydrophobic waxy compound. Water, therefore, must cross the plasma membrane of an endodermal cell to enter the vascular tissues. (E) Cross-section, at high magnification, of a cluster of sieve elements and companion cells in the phloem. The larger, empty-appearing sieve elements (SE) are paired with highly cytoplasmic, smaller companion cells (CC). Approximately 10 sieve element/companion cell (SE/CC) pairs are shown here, of which 1 pair is outlined. SE and CC are formed from a common mother cell, but this relationship is not always apparent in the fully differentiated state. A sieve plate is present (yellow arrow). As a result of wounding, the pores (white dots) of the sieve plate are filled with callose, which, though a carbohydrate,

is not stained by the periodic acid–Schiff reaction. (F) A longitudinal section of the phloem at high magnification. Two sieve tubes are shown (arrowheads at top), each consisting of a file of sieve tube members. (Sieve elements, a more generic term for conductive phloem cells, include both the sieve tube members of angiosperms and the sieve cells of gymnosperms; the latter do not form sieve tubes.) The accumulation of proteinaceous material on the sieve plates (crosswalls between successive sieve elements) is part of an efficient sealing mechanism in wounded phloem. Sudden pressure release causes the sieve tube contents to surge toward the wound, filtering out proteinaceous material on the sieve plate. Surging is difficult to avoid and has been a source of controversy in determining the structure of functioning sieve elements (see Section 15.6.1). (G) Cross-section of the xylem at high magnification, showing a xylem vessel and surrounding parenchyma cells. (H) Longitudinal section showing several xylem vessels (arrowheads at top). The ends of individual vessel elements are marked by cusps (arrows) in the vessel walls. Another class of xylem element, tracheids, are not labeled in the figure. Tracheids are present in both gymnosperms and angiosperms and can be distinguished from vessels by the absence of open-ended cell walls. (I) Cross-section of vascular tissues of a root at high magnification. A mature xylem vessel is indicated (XV); other vessels retain some cytoplasm and are not yet mature. One sieve element is marked by an asterisk (*); most of the other empty-appearing cells are also sieve elements. Companion cells are not present. (J) Cross-section of a minor vein, showing vacuolate bundle sheath cells (BSC) surrounding the vein, a xylem vessel (X), and the phloem (immediately below the xylem). Two sieve elements (small, empty-appearing cells) are present, each flanked by two much larger companion cells.

Cytosol   Plasmodesmata   Vacuole   Cell wall   Intercellular space

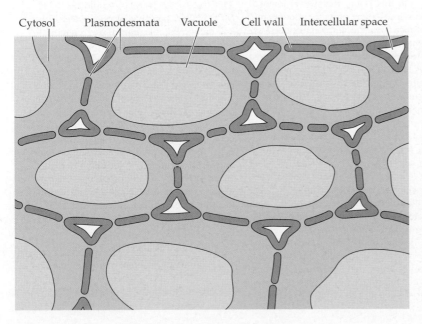

Figure 15.2
Diagram of the apoplast/symplasm concept in its simplest form. Solutes may move through a tissue via an apoplastic pathway outside the plasma membranes (see all cell walls in red) or via the symplasmic pathway that connects the cytoplasmic contents of cells (blue). Intercellular spaces (in white, part of the apoplast) form a third parallel pathway for gaseous movement. Because most solutes cannot readily cross the tonoplast, vacuoles (gray) are not usually regarded as part of the symplasm.

The water potential equation states that the total **water potential,** ψ_w, is the sum of contributions from the **pressure potential, solute** (or, equivalently, osmotic) **potential, gravitational potential,** and **matric potential:**

EQUATION 15.B1

$$\psi_w = \psi_p + \psi_s + \psi_g + \psi_m$$

Although quantitative work in plant water relations is not usually described in thermodynamic terms, the definition of water potential is based directly on the thermodynamic concept of chemical potential, in this case the chemical potential of water, μ_w:

EQUATION 15.B2

$$\psi_w \equiv \frac{\mu_w - \mu_{w,0}}{\overline{V}_w}$$

EQUATION 15.B3

$$\mu_w = \mu_{w,0} + \overline{V}_w P + RT(\ln a_w) + m_w g h$$

μ_w, given in J mol^{-1} (i.e., kg m^2 s^{-2} mol^{-1}), is defined in Equation 15.B3 in terms of the following:

$\mu_{w,0}$ = the chemical potential of water under standard conditions

$\overline{V}_w$ = the (partial) molar volume of water (l mol^{-1}, i.e., 0.001 m^3 mol^{-1})

P = pressure relative to atmospheric pressure (i.e., absolute pressure minus atmospheric pressure; "gauge pressure") (MPa, i.e., 1×10^6 N m^{-2})

R = gas constant (m^3 MPa mol^{-1} K^{-1})

T = absolute temperature (K)

a_w = activity of water (dimensionless)

m_w = mass per mole of water (kg mol^{-1})

g = gravitational acceleration (m s^{-2})

h = height relative to a reference point (m)

For pure water, $a_w = 1.0$, $P = 0$ MPa, and $h = 0$ m, so $\mu_{w,0} = 0$ J mol^{-1}. Water potential is a construct limited almost entirely to the field of plant water relations. At the time of its introduction, the concept of water potential represented a compromise between defining the activity of water in terms of chemical potential (energy per mole) and continuing to use pressure-based descriptions of plant water relations, some of which had not been rigorously defined. Equation 15.B2 provides a clear thermodynamic basis for plant water relations as described in units of pressure, a more useful parameter for conceptualizing water movement.

We assume that a_w is affected only by solutes, so that $RT (\ln a_w) = -\pi \overline{V}_w$, where π is **osmotic pressure** (RTC_s, in which C_s is molality, moles of solute per kilogram of solvent). Given this assumption, Equation 15.B4 can be derived from the definition of ψ_w (see Eq. 15.B2). The density of water, ρ_w, is equal to $m_w/\overline{V}_w$.

EQUATION 15.B4

$$\psi_w = P - \pi + \rho_w g h$$

On comparing Equations 15.B1 and 15.B4, we can conclude the following: $\psi_s = -\pi$ (i.e., solute potential is the negative of osmotic pressure) and $\psi_g = \rho_w g h$. Note that there are not separate terms for the pressure potential, ψ_p, and the matric potential, ψ_m.

P, however, is evidently the sum of $\psi_p = \psi_m$. On this point, the water potential equation loses some of its thermodynamic rigor, since ψ_m is a term introduced for convenience when finely porous structures open to the atmosphere (almost always the cell walls or soil) are involved. Pressure within these structures arises from surface tension (see Fig. 15.8), and is designated ψ_m. Because they are open to the atmosphere, no other pressure source is involved, so $P = \psi_m$. In the absence of a matrix (e.g., a vacuole or the lumen of a xylem vessel), $\psi_m = 0$ and $P = \psi_p$. In effect then, $P = \psi_m$ or ψ_p, depending on the presence or absence of a matrix.

Although the introduction of ψ_m as a separate pressure term may seem disconcerting at first, the practice is convenient when dealing with water status and flow in the soil and cell walls. However, the arbitrary nature of matric potential must be kept in mind, because ψ_m has also been used to signify surface effects involving a_w,

rather than surface tension (P), especially in colloid **imbibition.** For example, in germinating seeds, fixed surface charges in the seed storage materials are balanced by counterions that make a strong osmotic contribution and consequently generate a balancing pressure, evident as colloid swelling ("colloid pressure").

In dealing with water flow, water potential gradients are important but are not the only factor involved. In general, $J_w = -L_p(\Delta P - \sigma \Delta \pi)$, where J_w = volume flux of water (m^3 m^{-2} s^{-1}, i.e., m s^{-1}); L_p = hydraulic conductivity, usually of a membrane (m s^{-1} MPa^{-1}); and σ = the solute reflection coefficient (dimensionless), which varies from 0 (no barrier present, as in cell walls, sieve tubes, and xylem vessels) to 1.0 (perfectly permeable to water but impermeable to solutes).

If $\sigma = 1.0$, as is true for most plant membranes and endogenous solutes, then $J_w = -L_p \Delta \psi_w$. The significance of σ may be appreciated by setting $J_w = 0$ m s^{-1} with water on one side of the membrane, in which case the pressure developed will only be a fraction (σ) of the osmotic pressure, π. Or, if $\Delta P = 0$ MPa, then J_w would be reduced to a fraction (σ) of the maximum possible rate.

$\Delta \psi_w$ is not a meaningful driving force if there is a temperature gradient across the diffusion pathway. Temperature has a strong effect on the vapor pressure of water. At 20°C, pure water ($\psi_w = 0$) has a vapor pressure of 0.234 MPa. The vapor pressure over a 0.5 M NaCl solution at 21°C ($\psi_w = -2.24$ MPa) would be 0.245 MPa, higher than that of pure water only 1°C cooler. For this reason, $\Delta \psi_w$ cannot be applied to the diffusion of water vapor away from a leaf, because leaf and air temperatures are rarely the same. Thus, the proper measure of the driving force for gaseous water transport is water vapor pressure or water vapor concentration.

Because ψ_w is a thermodynamic concept, it is often unsuited as an indicator of water stress, a physiological condition that may differ substantially for different plants at the same ψ_w. Another measure, **relative water content** (RWC), is a more useful measure of physiological water status:

EQUATION 15.B5

$$RWC = \frac{\text{Actual water content}}{\text{Water content at full turgidity}}$$
(i. e., when $\psi_w = 0$)

differential generated by localized solute accumulation and utilization will result in OGPF: Solution will flow from regions of high pressure to regions of low pressure (Fig. 15.3B). Moreover, flow will occur more efficiently between these regions if they are linked by specialized conducting cells (Fig. 15.3C).

This mechanism of pressure flow in the symplasm was proposed by Ernst Münch in 1929. Ingeniously simple, the model depends only on the presumption that cells are interconnected hydraulically via plasmodesmata, with some cells in the plant body being engaged in solute accumulation, acting as **sources,** whereas others are engaged in solute utilization, acting as **sinks.** As will be noted later, some amendment is necessary, especially when discontinuities are present in the symplasm or apoplast (see Section 15.6.10). However, the underlying biophysical mechanism envisioned by Münch still applies.

15.1.4 Concentration gradients over short distances tend to drive both diffusion and OGPF.

OGPF is almost always invoked in the context of flow along sieve tubes. Clearly, however, concentration gradients will also tend to cause bulk flow in other parts of the symplasm. Also, the gradient will drive a diffusive flux over the same pathway; that is, a concentration gradient will affect both terms of Equation 15.1. The relative importance of the two modes of transport will depend on the concentration (as is evident from the convective term of Eq. 15.1) and on the channel dimensions of the pathway. With respect to the latter, a diffusive flux is independent of distance from the channel wall (Fig. 15.4), whereas viscous flow depends almost entirely on distance from the wall. Thus, diffusive transport rapidly increases in relative importance with decreasing channel radius.

The question of relative movement of solutes and water is particularly relevant to postphloem transport in sinks, where osmotic effects are an important factor in the control of sink physiology (see Section 15.6.11 and Table 15.4). Because water and solutes are delivered to many sinks almost entirely via the phloem, their relative movement along the postphloem pathway (i.e., together or independently) has important osmotic implications.

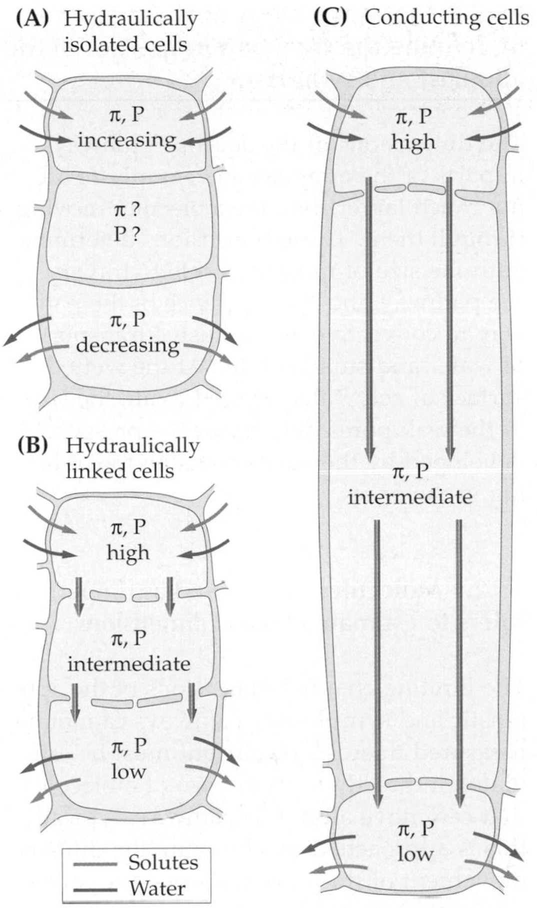

(A) Hydraulically isolated cells

π, P increasing

π? P?

π, P decreasing

(B) Hydraulically linked cells

π, P high

π, P intermediate

π, P low

— Solutes
— Water

(C) Conducting cells

π, P high

π, P intermediate

π, P low

Figure 15.3
Consequences of localized solute fluxes for flow in the symplasm. Solute fluxes may result from solute import or export (as shown here) or from solute production or utilization within cells (e.g., starch mobilization). (A) In the absence of intercellular connections, solute fluxes cause concentrations to change independently in each cell. An example would be the solute and turgor changes in guard cells during stomatal movements. (B) In cells that are interconnected by plasmodesmata, turgor differences cause solution to flow from regions of solute accumulation (sources) to regions of solute utilization (sinks). (C) Solution flow along the symplasm is more efficient if sources and sinks are connected by cells with low resistance to flow.

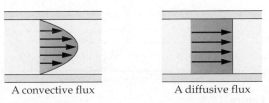

A convective flux A diffusive flux

Figure 15.4
Comparison of diffusive and convective fluxes in a pore. Because a convective flux increases with the square of the radius, pore dimensions are a critical factor in determining the relative contributions of the two modes of transport to the total flux.

15.2 Importance of channel dimensions in defining the transport properties of the apoplast and symplasm

The dimensions of the aqueous pathways in parts of the apoplast and symplasm are not much larger than the molecules moving through them. These dimensions determine both the size of molecules able to traverse the pathway and the resistance of the pathway to convective and diffusive transport of water and small solutes. At the wetted surface of cell walls exposed to air, the size of the wall pores determines the pressure developed by the surface tension forces acting in the pores.

15.2.1 Molecules can be used as (imperfect) rulers to estimate channel dimensions.

The limiting channel dimensions of the apoplastic and symplasmic pathways cannot be measured microscopically but must be estimated indirectly from the size of molecules that can move along the pathway. Typically, this is approached by observing the diffusive movement of fluorescent tracers; fluorescein-conjugated dextrans (F-dextrans) are often used. The **size exclusion limit** (SEL) of a pathway is typically given as the molecular mass of the smallest tracer excluded from movement. The smallest dimensions of a pathway, those that impose the SEL, also restrict the rate at which the pathway can conduct water and small solutes.

The process of inferring channel dimensions from the diffusion of tracer molecules is indirect and imprecise, especially when the results are expressed as molecular mass. The reasons for this not only are relevant to the accuracy of the estimated pore size, but also illustrate several noteworthy aspects of molecular diffusion in small pores.

The effective pore area available for diffusion, A_{eff}, is only a fraction of the actual pore area, A_o. Two factors render A_{eff} smaller than A_o. One is steric hindrance, which prevents a molecule of radius a from occupying the peripheral volume of the pore, reducing the area that can be occupied to $\pi(R - a)^2$, where R is the pore radius (Fig. 15.5). The other factor is increased viscous drag, caused by the steepness of velocity gradients between the pore wall and molecule as water is forced to flow in the reduced area between the two.

Although the SEL is usually given in terms of relative molecular mass (M_r), there is no universal relationship between M_r and molecular size, the actual basis of sieving. Defining the dimensions of a given molecule requires additional information. In particular, the Stokes radius (R_s), the radius of a sphere having the same viscous drag as the molecule in question, is the most relevant to the role of viscous drag. Often, R_s is obtained from the diffusion coefficient and the Stokes–Einstein equation (Eq. 15.2), where R is the gas constant (m^3 MPa mol^{-1} K^{-1}), T is temperature (K), N_A is Avogadro's number (molecules mol^{-1}), and η is viscosity (Pa s). The Stokes radius may also be obtained from viscometry and sedimentation velocity.

Equation 15.2

$$D = \frac{RT}{6\pi\eta N_A R_s}$$

The scaled drawing in Figure 15.6 compares the Stokes radii of some biologically important molecules with pore diameters of 4 and 6 nm, values that approximate physical pore sizes in plasmodesmata. Pores in the cell wall fall in a somewhat larger range (5 to 7 nm).

The M_r/size relationship is further complicated by the substantial dispersion of sizes of molecules present in commercial sources of the F-dextrans commonly used for SEL estimates. Several groups have fractionated their F-dextrans to provide more closely defined sizes.

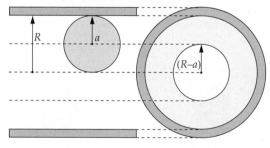

Figure 15.5
Because a molecule cannot be closer to the pore wall than the distance defined by the molecular radius, the effective area available for movement is only a fraction, $(1 - a/R)^2$, of the pore area. Movement is hindered further by steeper velocity gradients between the molecule and pore wall (not shown).

An important shortcoming of using M_r as a size indicator is illustrated by comparing the mass/size relationship for dextrans and globular proteins (Fig. 15.7). Note that the Stokes radius for dextrans increases much more rapidly than for proteins. Especially for larger molecules, the difference can be considerable. A 27-kDa dextran, for example, has the same Stokes radius (3.5 nm) as bovine serum albumin (67 kDa). This is primarily a reflection of the almost complete absence of secondary and tertiary structure in dextrans, for which "size" is a statistically averaged radius of an unbranched random coil. In contrast, the secondary and tertiary structures of proteins impose tighter, more closely defined, molecular dimensions.

Because of the substantial ambiguities between molecular mass and molecular size, and the importance of size, not mass, in exclusion from a pore, this chapter will use the term **mass exclusion limit** (MEL) to designate the exclusion limits stated as molecular mass. When important, a subscript will indicate the probe used (e.g., MEL$_{dex}$ for dextrans). "SEL" will be used only where "size" is given as the molecular radius or diameter.

15.2.2 Surface tension plays a crucial role in excluding air from the apoplast.

Within a plant, the interface between liquid water and air lies at the surface of cell walls. Because of the microporous structure of the wall, surface tension at this interface can balance the considerable tension, or negative pressure, ordinarily present in the liquid water phase. This balance of forces can be readily quantified (Fig. 15.8). Increasing tension in the apoplast is accompanied by a greater sag of the meniscus into the pores (i.e., a decreased contact angle, α). When α reaches 0°, any further increase in tension will empty the pore of fluid, which is replaced by air. The critical tension required to empty a pore of radius R is a fundamental parameter of water status in soils, xylem vessels, tracheids, and cell walls.

Pore diameters in parenchyma cell walls are too small to measure directly and cannot be accurately assessed with electron microscopy because observations of hydrated walls are not possible. Several indirect approaches have been taken to over-

come this problem. In one method, cells are plasmolyzed, and small amounts of a graded series of F-dextrans are added to the plasmolyzing solution. If the F-dextran can penetrate the wall, it will be seen in the space between the wall and protoplast; if not, this space will be nonfluorescent. Such experiments indicate that pores in parenchyma cell

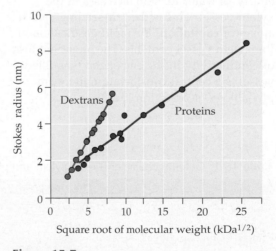

	MW (Da)	R_s (nm)
Water	18	0.15
Glycerol	88	0.26
Glucose	180	0.35
Sucrose	342	0.47
Raffinose	504	0.57
Carboxyfluorescein	374	0.61
Lucifer Yellow CH	443	0.68
Dextran	3,000	1.11
Cytochrome c	12,400	1.65
Dextran	10,000	2.2
Bovine serum albumin	67,000	3.55

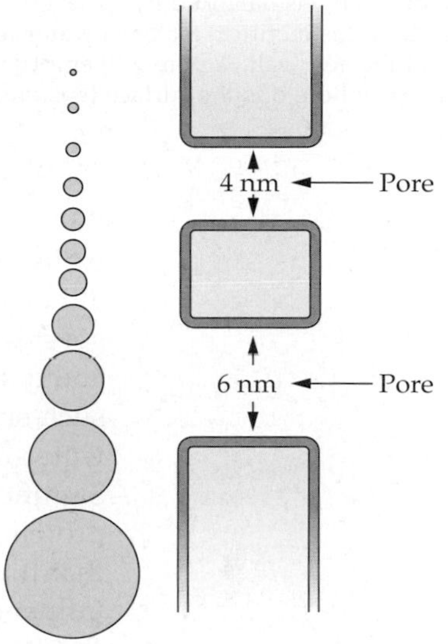

Figure 15.6
Stokes radii of selected molecules, covering a range of molecular weights. These are compared to pores with diameters of 4 nm and 6 nm, representative of limiting pore dimensions in the symplasm. Pores in the cell wall range from about 5 nm to 7 nm.

Figure 15.7
The relationship between molecular mass and molecular size differs markedly for dextrans (blue), which are commonly used to probe pore dimensions, and proteins (red).

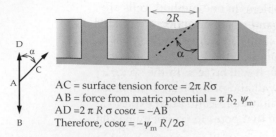

AC = surface tension force = $2\pi R\sigma$
AB = force from matric potential = $\pi R_2\, \psi_m$
$AD = 2\pi R \sigma \cos\alpha = -AB$
Therefore, $\cos\alpha = -\psi_m\, R/2\sigma$

Figure 15.8
Balance of forces in a water-filled pore. The tension exerted by the matric potential (ψ_m) is balanced by surface tension. The ψ_m required to empty a pore of radius R is a critical factor in water relations of xylem vessels, tracheids, the soil, and the cell wall. A pore will empty when α is reduced to zero, i.e., when $\psi_m = 2\sigma/R$, where σ is the surface tension of water (N m^{-1}).

walls are about 5 to 7 nm in diameter, although the exact size varies for different species and tissues. Given that a tension of about 40 MPa would be required to empty such small pores (Fig. 15.8), the pores remain water-filled at physiological water potentials. Significantly, pectinase, but not cellulase or proteinase, enlarges the pores, indicating that the arrangement of pectins has a major influence on wall porosity (see Chapter 2).

Wall pore sizes differ between parenchyma cells and xylem, where water flow between conducting cells is limited to thin spots (**"pits"**) in the otherwise thick walls of vessels and tracheids. To accommodate this flow, the wall across the pit (the **"pit membrane"**) has much larger pores, 0.1 to 0.4 μm in diameter. If air is present, these pores can empty as water tension increases in the xylem vessel. This is the "air-seeding" mechanism of **cavitation,** the rapid gasification of the xylem vessel contents. The attendant loss of conductivity in the embolized (gas-filled) vessel is a central consideration in xylem physiology (see Section 15.4.3).

15.3 Comparison of xylem and phloem transport

15.3.1 Some generalizations and useful comparisons can be made about the composition of xylem and phloem exudates.

Analyses of xylem and phloem exudates reveal that their composition and concentration vary considerably. Major differences occur not only among plant species but also within the same species or even the same plant, in which exudate composition and concentration can vary with changing physiological conditions and time of day. Nonetheless, some useful generalizations and comparisons may be made (Table 15.1). Methods for sampling xylem and phloem exudates are described in Box 15.2.

Phloem and xylem exudates represent samples of the symplasm and apoplast, respectively. Thus, xylem exudates are slightly acidic, typically with a pH of about 6, whereas phloem exudates are slightly alkaline, around pH 8. Phloem exudates contain low concentrations of a wide range of solutes, including reducing sugars, nucleotide phosphates, sugar phosphates, and various enzymes and enzyme cofactors.

Few examples are available in which xylem and phloem exudates have been compared in plants of the same species grown under the same conditions. Two, from lupine and tree tobacco, are given in Table 15.2. Concentrations of almost all solutes, even iron, which is generally characterized as "phloem immobile," are substantially higher in exudate from the phloem than that from the xylem. Calcium, also a phloem-immobile nutrient, is present at similar concentrations in the xylem and phloem.

Table 15.1 General characteristics of xylem and phloem exudates

	Phloem exudates	Xylem exudates
Sugars	100–300 g l^{-1}	0 g l^{-1}
Amino acids (mostly Glu, Asp, Gln, Asn)	5–40 g l^{-1}	0.1–2 g l^{-1}
Inorganics	1–5 g l^{-1}	0.2–4 g l^{-1}
Total solutes	250–1200 mmol kg^{-1} ($\pi \approx 0.6$–3 MPa)	10–100 mmol kg^{-1} ($\pi \approx 0.02$–0.2 MPa)
pH	7.3–8.0	5.0–6.5

Table 15.2 Comparison of xylem and phloem exudates from lupine and tree tobacco

| | *Lupinus angustifolius* | | *Nicotiana glauca* | |
	Xylem (mM)	Phloem (mM)	Xylem (mM)	Phloem (mM)
Sucrose	ND	490.0	ND	460.0
Amino acids	20.0	115.0	2.2	83.0
Potassium	4.6	47.0	5.2	94.0
Sodium	2.2	4.4	2.0	5.0
Phosphorus	NA	NA	2.2	14.0
Magnesium	0.33	5.8	1.4	4.3
Calcium	1.8	1.6	4.7	2.1
Iron	0.02	0.13	0.01	0.17
Zinc	0.01	0.08	0.02	0.24
Nitrate	0.50	Tr	NA	ND
pH	5.9	8.0	5.7	7.9

ND, not detectable; NA, not available; Tr, trace.

15.3.2 Xylem-borne solutes follow the transpiration stream from roots to mature leaves, whereas phloem transports solutes from sources to sinks.

The pressure gradients responsible for flow in the xylem and phloem are generated by independent mechanisms. As a result, the overall patterns of solute movement in the two tissues are very different (Fig. 15.9), despite the fact that the conducting pathways run parallel to each other throughout the plant body. Flow in the xylem is essentially unidirectional, from roots to sites of water evaporation. Reverse movement can occur in principle and is easy to cause experimentally, but it is infrequent and is limited in duration and volume. Reversal can occur, for example, when plant surfaces, especially wounded or woody surfaces, are moistened or when a plant part is cut under water. Movement of a solute to the roots is almost always firm evidence against xylem transport.

The pattern of solute movement in the phloem is driven by the sites of solute production (sources) and utilization (sinks) (Fig. 15.9). These centers of active metabolism shift as the plant grows. As a result, in contrast to the xylem, the pattern of solute movement via the phloem is under

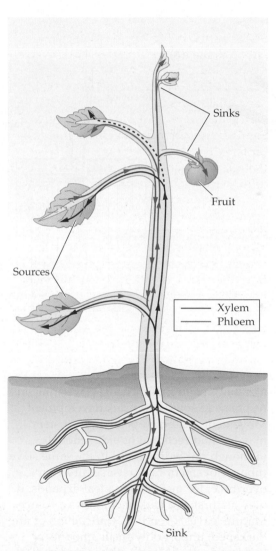

Figure 15.9
Diagram showing patterns of xylem and phloem transport. Movement in the xylem is upwards from the root to the mature leaves, primary sites of transpiration and photosynthesis. A very small amount of xylem transport supplies expanding tissues, and even less is directed to reproductive sinks, both of which have low transpiration rates. Movement in the phloem proceeds from sites of assimilate production, primarily mature leaves, to sites of utilization in expanding tissues and reproductive or storage sinks. Phloem movement can be bidirectional within a single internode but is unidirectional within a given vascular bundle. In the lower stem, movement is downward, because roots are sinks unless they are storage roots in the process of remobilizing their reserves.

Various methods—each with its advantages and shortcomings—have been used to sample the contents of conducting cells in the xylem and phloem. Most require destructive sampling and therefore provide data for only a single point in time.

For the xylem, the easiest and most frequently used method is detopping the plant and collecting the flow resulting from root pressure. Because transpiration has stopped, solute concentrations will be higher than in the intact plant, although this will decline as phloem-supplied energy reserves are depleted. Solutes recirculating from the shoot (see Section 15.3.3) will be especially affected by detopping.

A sample more directly representative of the transpiration stream can be obtained by using a gentle vacuum to suck the xylem vessel contents from stem segments (panel A). Because a vacuum cannot pull air through the pit fields at the ends of the xylem vessels, successive segments are cut from the stem to release the vessel contents. This approach is rarely suited for herbaceous species.

Finally, a pressure bomb (see Box 15.3) may be used to force xylem exudate from excised leaves or shoots or, less frequently, from intact plants. The latter method comes closest to the ideal for sampling the transpiration stream, because it can be used to sample continuously from actively transpiring plants (panel B). However, the plants must be grown in a specially constructed pressure bomb. Xylem exudate collected from excised shoots by standard pressure bomb techniques varies in composition, depending on the volume collected. This reflects that various distinct solutions are being sampled, first from the xylem and then from minor veins; finally, if enough pressure is applied, xylem sap is displaced by water from the symplasm, rapidly depressing the solute concentration of the exudate.

(A)

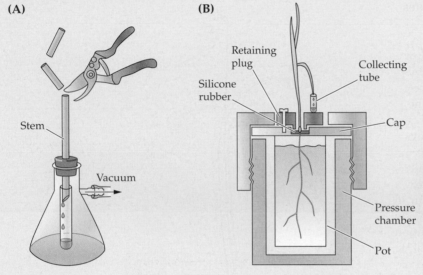

Stem

Vacuum

(B)

Retaining plug

Silicone rubber

Collecting tube

Cap

Pressure chamber

Pot

(C)

metabolic control and changes over time in response to source and sink development. Movement at a given point along a stem may be unidirectional or bidirectional (although not bidirectional in the same sieve tube). Movement may reverse direction over time; for instance, as a leaf expands, it may act first as a sink and later as a source. Mature leaves, even if they are albino or are darkened, import only small amounts of solutes via the phloem—although symplasmic loaders, species in which phloem loading occurs via an entirely symplasmic pathway from the mesophyll cells to the sieve tubes (see Section 15.6.6), may be an exception. These patterns of transport can be visualized by using whole-plant autoradiography (Fig. 15.10).

Phloem transport can be selectively blocked by removing a ring of bark from a woody plant (**girdling**) or by killing the vascular tissues with heat ("heat girdling"; Fig. 15.11). Because the conducting cells of the xylem are nonliving, water flow there is not affected unless embolisms are introduced by the girdling process. Girdling provides a useful experimental means of partly or fully blocking phloem transport to a sink, or of diverting movement along another pathway, without interrupting movement via the xylem. It is also useful in distinguishing xylem transport from phloem transport.

Most phloem exudates have been collected from a shallow incision into the stem. However, plants have very effective mechanisms for sealing damaged phloem, and except for cucurbits (panel C), most herbaceous species do not yield usable amounts of exudate from a cut. Sealing mechanisms can be circumvented by treatment with EDTA. Subtle artifacts in composition may occur, however, and concentrations cannot be determined accurately with this approach.

Collecting exudate from severed aphid stylets comes closest to being an ideal method for sampling sieve tube contents. Aphids and other phloem-feeding insects feed by inserting their mouth parts (stylets)

into a sieve tube (panel D). The stylet bundle may be cut by radiofrequency microcautery, as illustrated in panel E. Several seconds elapsed between each of the first four frames. In the first, the microcautery probe tip has not yet touched the insect's labium (a sheath around the stylet bundle). In the second, the stylet has just been cut. A drop of hemolymph appears in the third, and the stylet begins to exude after the aphid pulls away in the fourth. After a minute has elapsed (fifth frame), the drop is much larger, although most of the water evaporates rapidly.

Phloem exudation rates are usually similar to normal translocation rates, so the impact of aphid stylet exudation on sieve

tube turgor is minimal or nil. In some cases, exudation may continue for long periods (days), allowing continuous measurements on intact plants. Solute concentrations can be measured accurately by collecting exudate under oil to avoid evaporation. Exuding stylets have also been used to measure sieve tube turgor and membrane potential. However, only small exudate volumes can be collected from each stylet, and the duration of exudation is brief (minutes) for most plant species.

(D)

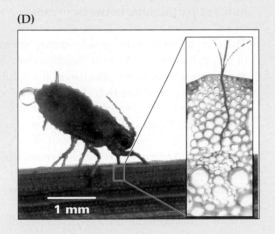

(E)

(A) **(B)** **(C)** **(D)**

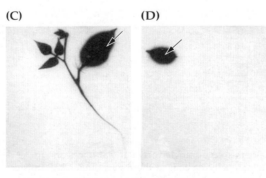

Figure 15.10
Whole-plant autoradiographs showing the pattern of solute distribution via the phloem. Phloem-mobile radioactive phosphate (as $K_2H^{32}PO_4$) was applied to successively younger leaves (arrows) of identical bean plants, which were autoradiographed 24 hours later. Movement of solute is to growing parts of the plant, bypassing mature, transpiring (photosynthesizing) leaves. (The light labeling of mature leaves in A and B probably represents ^{32}P recirculated from the roots.)

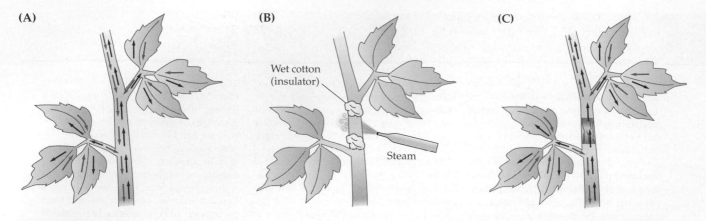

(A) **(B)** **(C)**

Wet cotton
(insulator)

Steam

Figure 15.11
Effect of steam girdling on xylem and phloem transport. (A) Before girdling, phloem transport is downwards at both nodes. (B) Steam girdling of the internode. Wet cotton restricts heating to a short segment of the stem. (C) After girdling, xylem transport is unaffected, as is phloem transport from the lower leaf. Phloem transport from the upper leaf node, however, must be upward, because the internode phloem below has been killed.

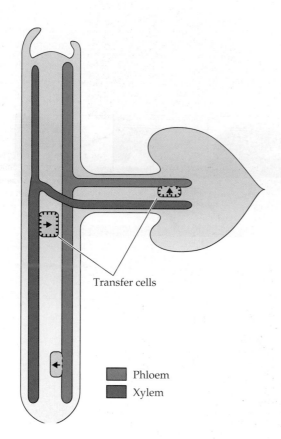

Transfer cells

■ Phloem
■ Xylem

Figure 15.12
Important sites of xylem-to-phloem solute transfer occur at leaf traces and in minor veins of leaves. Vascular parenchyma cells at these sites often have transfer cell morphology, i.e., wall invaginations that amplify the plasma membrane surface area. By the transfer of solutes from the xylem to the phloem, nutrients are delivered more effectively to slowly transpiring, but rapidly growing, sinks such as meristems, fruits, and seeds. Roots are an important site of phloem-to-xylem solute transfer, recirculating a substantial proportion of some phloem-delivered solutes (e.g., amino acids and K$^+$) back to the shoot. The cells that participate in phloem-to-xylem solute transfer do not have transfer cell morphology.

15.3.3 Solute transfer between the xylem and the phloem plays an important role in nutrient partitioning between organs.

The characteristic side-by-side arrangement of the xylem and phloem throughout the plant provides ample opportunity for solute exchange between the two (Fig. 15.12), especially in the highly branched vascular system of source and sink regions. Xylem-to-phloem transfer can be demonstrated by introducing radioactively labeled tracers of endogenous solutes into the transpiration stream, whereupon phloem exudate often becomes labeled. The proportion of the solute removed from the xylem, its possible metabolic conversion, and the extent of subsequent transfer to the phloem depend on the solute involved. A substantial proportion of xylem-borne solutes may be absorbed, distinctly altering the composition of the xylem and phloem transport streams. For example, in comparing **guttation** fluid (xylem sap exuded from leaves in response to positive pressure in root xylem; see Section 15.4.4) with xylem exudate from the roots, concentrations of amino acids, phosphate, and potassium are typically much lower in guttation fluid, but calcium content in the two is similar.

The minor vein endings in transpiring leaves are an important site for xylem-to-phloem solute transfer (see Section 15.4.7), but considerable transfer may also take place in the stem, especially at nodes. In some species, **xylem transfer cells** surround xylem vessels in the stem node and departing leaf vascular traces (Fig. 15.13). Like transfer cells in other tissues (e.g., companion cells in the minor veins of some apoplastic loaders; see Section 15.6.5), these transfer cells are characterized by wall invaginations that substantially increase the surface area of the plasma membrane. Microautoradiographs have shown that these cells actively absorb amino acids from the transpiration stream.

The overall effect of xylem-to-phloem solute transfer is to enrich the nutritional content of the sieve tube sap at the expense of the transpiration stream. This diversion is important because many rapidly growing sinks transpire very slowly and must therefore receive virtually all of their mineral nutrients from the phloem.

Solute transfer in the opposite direction, from the phloem to the xylem, occurs largely in the roots. Again, the process is solute-specific, but a substantial proportion of some solutes delivered to the roots by the phloem may be recirculated back to the shoots via the xylem. An experimental approach to recirculation of nitrogen in wheat seedlings, in which $^{15}NO_3^-$ was applied to half of a split-root system, is shown in Figure 15.14. In this instance, phloem-delivered amino acids accounted for more than 60% of the amino acid-N in the xylem and more than half the total xylem N. In similar experiments, 30% to 40% of xylem K in wheat and lupine came from the phloem. Although such cycling may appear futile, for at least some phloem-mobile nutrients (e.g., N and K) the nutrient status of the shoot affects the rate of nutrient uptake by the roots. Perhaps nutrient cycling provides a feedback mechanism for signaling shoot nutrient status to the roots.

15.4 Transpirational water movement in the xylem

15.4.1 The cohesion–tension mechanism is the leading theory for transpirational water movement.

During transpiration, water evaporates from the moistened cell walls lining the gas-filled intercellular spaces, primarily in leaves. Energetically, the process is driven almost entirely by the net absorption of radiation,

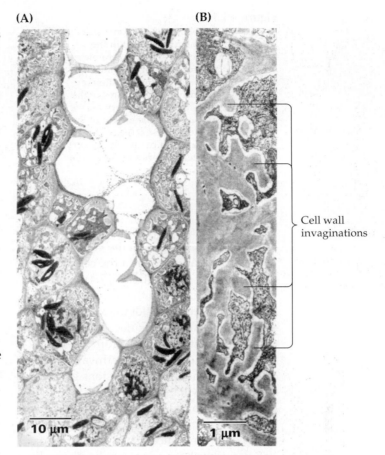

Figure 15.13
Xylem transfer cells surround the xylem vessels in a *Polemonium* cotyledonary trace. (A) A low-magnification electron micrograph shows the complete encirclement of the xylem vessels by transfer cells. Except for the xylem vessels, all the cells shown are xylem transfer cells. Densely staining cytoplasm is a frequent characteristic of transfer cells. (B) A higher-magnification micrograph shows the wall ingrowths more clearly.

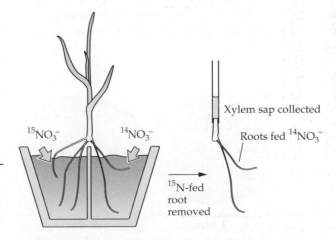

Figure 15.14
Recirculation of phloem-delivered nitrogen in the root system can be evaluated by allowing half of the root system to absorb $^{15}NO_3^-$. Afterwards, the labeled roots are excised, and xylem exudate is collected from the unlabeled root system. ^{15}N appearing in the exudate must be derived from assimilates delivered from the shoot via phloem.

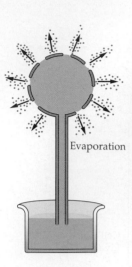

Figure 15.15
A working model of the cohesion–tension mechanism of water movement in the xylem, represented as a porous ceramic cup atop a capillary tube. Water flow through the capillary is driven by the evaporation of water from the menisci in the pores of the cup. Water in the cup and capillary are under tension caused by the weight of the water column and the resistance to flow. This tension is balanced by the pull of surface tension in the capillaries of the ceramic cup (see Fig. 15.8).

Evaporation

for which the sun is the most important source. Surface tension in the cell wall micropores places water in the cell walls under tension (see Section 15.2.4), an effect that is transmitted via the apoplast to water in the conducting cells of the xylem and eventually to the root surface and soil solution. Thus, water is pulled from the soil along a continuous liquid pathway to the cell walls in the leaf. The tensions generated can exceed 5 MPa. The cohesive strength of liquid water, provided by almost four hydrogen bonds per water molecule, is essential to the mechanism's successful operation. Box 15.3 describes how a pressure bomb can be used to quantify the resulting xylem tensions, which can exceed 5 MPa.

Transpiration can be illustrated by a porous ceramic cup attached to the end of a glass capillary (Fig. 15.15). Once the cup and capillary are filled, liquid water is drawn up the capillary by the same forces operating in a vascular plant. Attempts to simulate a plant more closely, however, would quickly

reveal this model's shortcomings. Because pore dimensions in a ceramic cup are much larger than those in a cell wall, the capillary height cannot be greatly increased before the weight of the water column pulls air into the cup. A cup with smaller pores could be used if its walls were sufficiently strong to withstand tensions of 5 MPa or more. However, the risk of pulling gas into the system would still remain.

15.4.2 Cavitation is the Achilles' heel of the cohesion–tension mechanism.

To function, the conducting cells of the xylem (tracheids or vessels) must remain filled with liquid water. In a transpiring plant, however, liquid water is in a metastable state: Given the tensions typically present in the xylem during transpiration, water under equilibrium conditions should be a gas, rather than a liquid. As a result, the xylem must operate under the constant risk that cavitation can fill the conducting cells with gas and thus block water movement.

Cavitation is a fairly regular occurrence in the xylem, which becomes progressively less conductive as the number of embolized vessels increases. In some woody species (not all; see below), this progression takes place over the course of the growing season, with as much as half of the xylem water being lost and replaced with gas. In temperate woody species, gas exclusion by repeated freeze–thaw cycles can leave most of the xylem embolized by winter's end.

The extent of embolized xylem can be visualized by forcing a dye solution through a stem segment (Fig. 15.16). The effect on conductance can be quantified by first measuring the flow rate through a stem segment under a standard pressure differential, followed by a high-pressure flush of water to remove gases, and then remeasuring the stem conductance. Actual cavitation events (very violent, on a microscale) can be monitored by detecting the acoustic vibrations that accompany each event. This approach was a particularly important development because it provided a means for following cavitation in intact plants. It has since been significantly improved through the use of ultrasonic acoustic emissions detectors (Fig. 15.17).

(A)   **(B)**

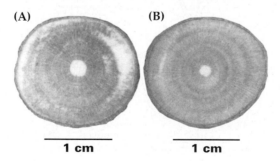

1 cm   1 cm

Figure 15.16
Xylem dysfunction resulting from embolisms can be detected by passing a dye solution through a stem segment. (A) Nonfunctional xylem is unstained. (B) A high-pressure flush with water can restore the xylem to full conductance.

Box 15.3

Xylem tension can be measured with the pressure bomb.

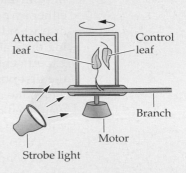

The pressure bomb, developed to measure tensions in the xylem, is perhaps the most widely used instrument in studying plant water relations. In the xylem, where solutes rarely make a significant contribution to ψ_w, the water potential equation (see Box 15.1) simplifies to $\psi_w = \psi_p + \psi_g$ for transport over vertical distances exceeding about 10 m (ψ_g contribution greater than 0.1 MPa) or to $\psi_w = \psi_p$ for transport over shorter vertical distances. (Note that $\psi_m = 0$, because there is no matrix present in the xylem vessels, the site of ψ_w measurement.) By applying a positive pressure equal to $-\psi_p$, the xylem tension and water potential may be measured. That is, $\psi_w + (-\psi_p) = \psi_p + (-\psi_p) = 0$ MPa.

The basic steps in a pressure bomb measurement are illustrated below. On cutting into the stem, air is pulled into the xylem, filling the severed xylem vessels or tracheids. After sealing a detached leaf or leafy stem into the pressure chamber, pressure is increased slowly, raising the water potential of the tissues, causing xylem sap to displace the air. The end point in the measurement is reached when sufficient pressure has been applied to raise the pressure in the xylem to zero, forcing the xylem sap back to the cut surface. In instances where xylem solutes might make a significant contribution to total water potential, overpressurizing the tissue forces xylem sap from the cut stem. The sap can be collected and its osmotic content measured.

The accuracy of pressure bomb measurements of xylem tension has been confirmed experimentally. For example, tensions measured with the pressure bomb are as effective in causing cavitation as equal but externally applied positive pressure (see Fig. 15.17). In addition, pressure bomb measurements correspond to known xylem tensions imposed by subjecting a stem segment to centrifugation. In the illustration above right, a fully hydrated stem segment with a leaf at its center was centrifuged for 15 minutes at a rotational speed measured with a strobe light. Pressure bomb readings (right) made afterwards on the attached leaf were in good agreement with the calculated xylem tension. In a control leaf not attached to a stem segment, centrifugation had only a slight effect on xylem tensions.

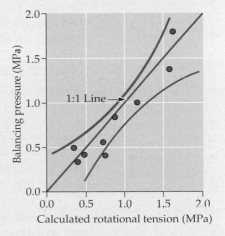

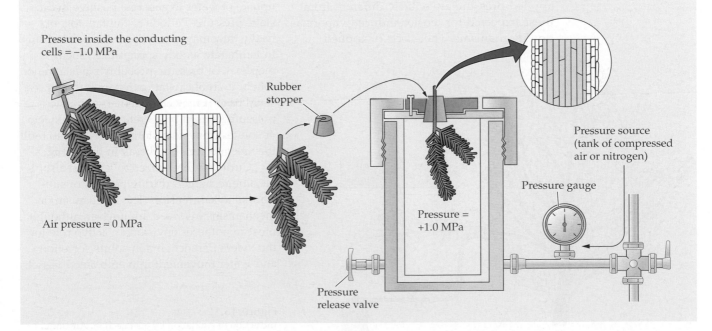

15.4.3 Cavitation occurs by "air seeding."

In principle, cavitation ought to be initiated either by pulling air into the xylem vessel or tracheid through the wall or by vaporizing the water within the vascular element. Experimental evidence decisively supports the first possibility, termed "air seeding," in which the surface tension in the largest cell wall openings is overcome by the tension within the conducting cells.

Pores in most of the wall area of the xylem conducting cells are far too small to allow the aspiration of air through them. However, as mentioned in Section 15.2.4, the size of the pores in the pit membranes through which water moves between vessels ranges from about 0.1 to 0.4 μm in diameter, depending on plant species. Pit membrane pores will withstand considerable tension before emptying (see Fig. 15.8), thereby preventing an embolism from expanding beyond a single tracheid or xylem vessel. As tension increases, more cavitations occur over a range of xylem pressures that correlate with the tensions required to empty the size range of the pit membrane pores involved.

Cavitations also occur in response to applied external pressure, the response pattern matching the effect of an equivalent xylem tension (Fig. 15.17). These effects of applied pressure are entirely different from those expected for tension-induced vaporization mechanisms. The effects of applied pressure and xylem tension can be compared in **xylem vulnerability curves** (Fig. 15.18), originally developed to compare in different species the impact on water conduction of embolisms caused by xylem tension. As predicted by the air-seeding mechanism, externally applied pressure has the same effect on embolism formation as the equivalent xylem tension.

15.4.4 Embolized xylem can be repaired.

Not uncommonly, embolized xylem refills with water during periods of high water potential, restoring its conductive function. Several mechanisms can raise the pressure within gas-filled xylem vessels to values ranging from slightly negative to positive. This increased pressure, augmented by the additional pressure exerted by surface tension in small bubbles, can be sufficient to drive the gas back into solution. In herbaceous species, nightly **root pressure** can play an essential role in reversing daily cavitations, which could otherwise block all xylem vessels after only a few days of even mild water stress. Root pressure results when roots import solutes into the xylem during periods of little transpiration, increasing the xylem's osmotic pressure (π) and driving uptake of water to generate positive hydrostatic pressure. Periods of rainfall, fog, or heavy dew may also relieve some embolisms. In temperate woody species (e.g., sugar maple), root pressure probably plays a minor role in diurnal cavitation repair, but on a seasonal basis it may act together with xylem pressure (caused by secretion into the xylem of solutes from stem storage reserves) to refill the xylem before bud break in the spring.

Curiously, embolized xylem can also become recharged during periods of substantial xylem tension. The mechanism or mechanisms involved are unclear, although most hypotheses envision an active role for the xylem parenchyma in solute secretion and water movement into embolized vessels.

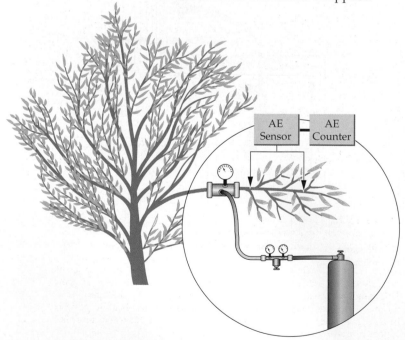

Figure 15.17
Induction of cavitation by air injection. Air under positive pressure, applied via a sealed collar around the branch, causes cavitation in the same manner as tension in the xylem. Acoustic emission (AE) sensors and counters monitor the progress of cavitation.

15.4.5 Cavitation can protect against excessive water loss.

Although cavitation inevitably leads to a decrease in xylem conductivity, it is not necessarily deleterious to overall plant water balance. In large measure, this is because cavitation tends to occur in the extremities of the xylem pathway (i.e., in leaves and small branches), reducing transpirational water loss and protecting against dehydration of larger stems and the root system.

To some extent, a greater tendency toward cavitation in smaller branches might be anticipated, because the ends of the pathway generally have the lowest water potentials. However, this tendency is magnified substantially by the fact that xylem conductivity declines disproportionately toward the ends of the transpiration pathway. As a result, the gradient in xylem tension increases sharply toward the leaves, causing the peripheral parts of the xylem to be at greatest risk of cavitation.

The root system also can be protected from water loss by cavitation in smaller branches and branching junctures. This is particularly important when the soil water potential drops markedly below the water potential of the roots. Although root-to-soil water movement can and does occur at higher root water potentials, embolisms in the root xylem at low water potential minimize such movement under drought conditions.

15.4.6 Xylem structure is only one aspect of a plant's adaptive strategies for maintaining water balance.

Susceptibility of the xylem to cavitation is only one of several factors that allow plants to maintain their water balance. Others include stomatal response, leaf drop, root–soil interactions, and seasonal growth patterns. Strategies may differ in different environments and, even in the same environment, different strategies may achieve the same goal.

An example of some of the tradeoffs that may occur is evident in the xylem vulnerability curves shown for maple (*Acer*) and birch (*Betula*) (Fig. 15.18). Both species grow in moist environments, but maple can also grow in drier conditions. As xylem tension increases in the birch, the stomata must close promptly after cavitations begin if rapid loss of xylem function is to be avoided. This stomatal behavior has been observed in action, and birch xylem remains more than 90% functional in the field. In maples, stomata remain open at much lower water potentials without causing full loss of xylem function; in drier sites, 50% of maple xylem may be embolized. Indeed, the few herbaceous species investigated to date have xylem vulnerability curves qualitatively similar to that illustrated here for maple. Progressive embolism during water stress would presumably occur in these herbaceous species as well but would probably be relieved by root pressure.

Clearly, this is a complex topic that cannot be pursued in detail here. However, largely inspired by recent progress on understanding the relationship between xylem structure and cavitation, this stimulating area of research seeks to integrate several important areas of plant physiology, physiological ecology, and evolution.

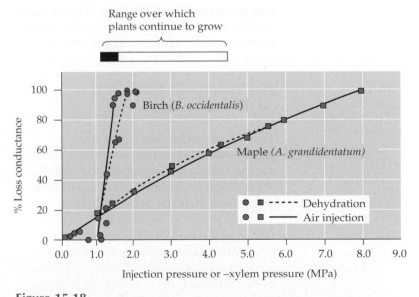

Figure 15.18
Xylem vulnerability curves for birch (*Betula*) and maple (*Acer*). As branches dehydrate during drought and xylem tension increases, cavitation causes a progressive loss of xylem function. The relationship between xylem tension and loss of xylem function can be quite different for different species, affecting the conditions under which they can grow. This range is indicated by the solid bar for birch, and by the open bar for maple. Xylem vulnerability curves developed by application of positive external pressures (see Fig. 15.17) are closely similar to vulnerability curves caused by tension within the xylem.

15.4.7 What happens to solutes in the transpiration stream?

On first consideration, a solution arriving in a leaf would appear simply to flow from the xylem via the bundle sheath cell walls, moving to evaporation sites in the mesophyll, epidermis, or both. Solutes thus would tend to accumulate in walls of these sites as water evaporated. However, attempts to visualize this process with polar dyes have shown that their movement is blocked as soon as they leave the xylem vessels in the minor veins (Fig. 15.19). Observations on fluorescent tracers introduced into the leaf apoplast show that movement there also is slow, even slower than free diffusion. Moreover, the movement of fluorescent tracers is not affected by transpiration rate, as it should be if water were flowing through the cell walls. Thus it appears that both water and solutes largely follow a symplasmic pathway after leaving the xylem. If a membrane is impermeable to a particular solute, the solute is left behind in the minor veins as water moves into the symplasm.

The fate of physiological solutes is more difficult to trace. However, their concentrations, too, would presumably increase in the minor vein endings. This should make them more available for absorption by the phloem and recirculation to sink regions.

15.5 Symplasmic transport via plasmodesmata

15.5.1 In vascular plants, the basic plasmodesmal structure is a tube of plasma membrane surrounding a strand of modified endoplasmic reticulum, with particulate material between them.

Figure 15.20 interprets the ultrastructure of vascular plant plasmodesmata in cross-sections and longitudinal sections. Among its essential features is a cell-to-cell tubule of the plasma membrane that surrounds a cylindrical strand of tightly furled endoplasmic reticulum (ER), the **desmotubule** (or **appressed ER**). A thin, darkly stained **central rod** occupies the center of the desmotubule. A **cytoplasmic sleeve** or, in cross-section, the **cytoplasmic annulus,** lies between the desmotubule and plasma membrane.

Plasmodesmal features that are commonly observed but not universal include the **collar** (a slightly raised annulus of cell wall material at either end of the plasmodesma), restricted **neck regions** just inside the wall, and a widened **central cavity** within the wall. Because plasmodesmata are often constricted at either or both ends, the neck and collar regions have received special scrutiny. Putative **sphincters** are most consistently visualized when tannic acid is included during fixation, appearing as electron-dense material encircling the neck region (Fig. 15.21). However, the actual function of these structures as sphincters has not been fully verified. Callose deposition often occurs rapidly (within seconds) in response to a variety of treatments, including cutting and chemical fixatives. Such rapid deposition raises the issue of its possible role in artifactually restricting the plasmodesma during fixation for electron microscopy. The validity of this concern has been demonstrated by preincubating tissue with 2-deoxy-D-glucose, an inhibitor of callose synthesis, before chemical fixation. Plasmodesmata in material so treated are funnel-shaped at the ends (Fig. 15.22), in contrast to their more constricted appearance in untreated tissue.

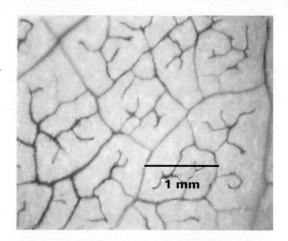

Figure 15.19
As the transpiration stream exits the minor veins in a leaf, xylem-borne solutes that are unable to move with water into the symplasm are left behind in the minor vein apoplast. A detached castor bean leaf was illuminated and allowed to absorb a dilute solution of Evans Blue for several hours, after which the leaf was left for several hours in water. Leaf discs were freeze-substituted in acetone to retain water-soluble compounds and then were cleared in toluene.

(A)

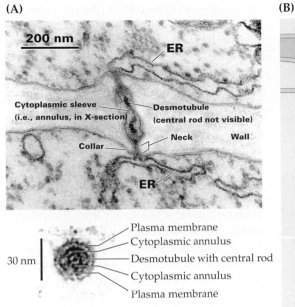

(B)

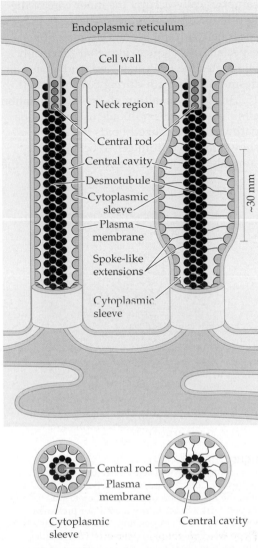

Figure 15.20
(A) Electron micrographs of plasmodesmata in longitudinal and cross-sections, accompanied by the terminology applied to frequently observed structural features. (B) A model of plasmodesmal substructure based on observations of freeze-substituted tobacco leaves. The plasma membrane and endoplasmic reticulum are continuous cell to cell, with the latter tightly furled to form the desmotubule. A sleeve of cytoplasm (the cytoplasmic annulus, in cross-section) occupies the gap between the plasma membrane and desmotubule; in the right plasmodesma, the sleeve has enlarged to form a central cavity. Much of the cytoplasmic annulus is occluded by proteinaceous particles partially embedded in the plasma membrane and the desmotubule. Gaps between these particles evidently represent the physical basis for the molecular sieving properties of symplasmic transport. Within the central cavity, spoke-like structures extend across the cytoplasmic sleeve; their presence in the neck region is unclear.

Both the cytoplasmic sleeve and the desmotubule have been considered as possible pathways for cell-to-cell water and solute movement. For the desmotubule, this was based on the possibility that the ER lumen might be continuous from cell to cell and that an absence of staining indicated an absence of structural material. However, the electron-lucent region of the desmotubule has been shown to be continuous at each end with the unstained lipid interior of the ER membranes, whereas the central rod is continuous with the inner dark surface of the ER lumen. Thus, instead of being hollow, the desmotubule is essentially a solid strand of lipid; the central rod is composed of the

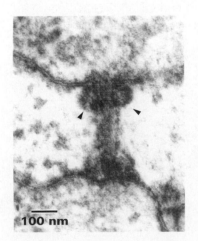

Figure 15.21
Inclusion of tannic acid during fixation often results in the visualization of a ring of darkly staining wall material in the neck region of plasmodesmata. Less frequently, other structural features have been noted in the neck and collar region. Because of their position, they are often referred to as sphincters. However, the relationship between these structures and plasmodesmal gating remains circumstantial.

lipid polar groups and a few proteins that can physically occupy the inner core of the tightly furled lipid bilayer. These relationships between the ER and desmotubule, especially the "squeezing" that practically eliminates the ER lumen, are evident as desmotubules form during cell division (Fig. 15.23). However, although this interpretation eliminates the desmotubule as an aqueous pathway, ER-localized lipids have been shown to move from cell to cell, still presumably via the desmotubule. Curiously, the seemingly less-constrained plasma membrane does not accommodate movement of plasma membrane lipids.

Much of the cytoplasmic annulus is occupied by particulate proteinaceous material (see the black and blue spheres in Fig. 15.20B). These particles are associated with (perhaps embedded in) both the outer surface of the desmotubule and the inner surface of the plasma membrane. In cross-section, seven to nine gaps occur between the particles. The distance across the gaps, about 2 to 3 nm, is comparable with the 4-nm channel diameter estimated from the movement of fluorescent probes (see Section 15.5.5), suggesting that these gaps are the physical basis for the molecular sieving characteristic of cell-to-cell transport. In the central cavity, spoke-like extensions (Fig. 15.20B) radiate between the plasma membrane and desmotubule and may be present in the neck region as well. The close association between the desmotubule and surrounding plasma membrane confers considerable structural stability.

Detergents, proteases, and other agents have been used to probe plasmodesmal composition and to selectively release proteinaceous components (see Section 15.5.9). Detergent removes the plasma membrane and, in corn roots, the desmotubule as well. In fern gametophytes, the desmotubule was removed by protease treatment but was stable to detergent. The experiments confirm the general importance of membranes in plasmodesmal structure but indicate variability in the relative importance of their major components.

A wide range of plasmodesmal morphologies have been observed, occurring at characteristic locations within the same plant. These include differences in length, branching, and size of the central cavity (Fig. 15.24). The functional significance of the variations is largely unknown, although a large MEL has been associated with the simple unbranched form of several postphloem pathways (see Section 15.6.8).

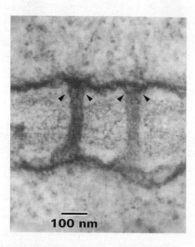

Figure 15.22

In material pretreated with 2-deoxy-D-glucose (DDG) to prevent callose deposition during chemical fixation, the neck region of plasmodesmata was funnel-shaped (arrowheads). In material not pretreated with DDG, callose deposition occurred and the neck region was straight in appearance. Thus the degree of restriction observed in the neck region of many micrographs (e.g., Fig. 15.21) may be less than previously supposed.

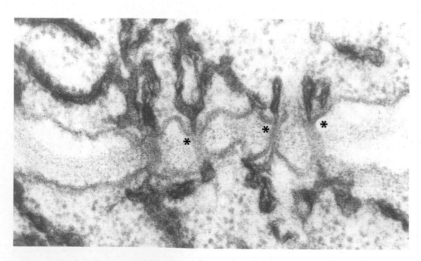

Figure 15.23

"Squeezing" of the ER lumen by the cell plate during formation of primary plasmodesmata. The densely stained ER lumen disappears where ER traverses the cell plate (*).

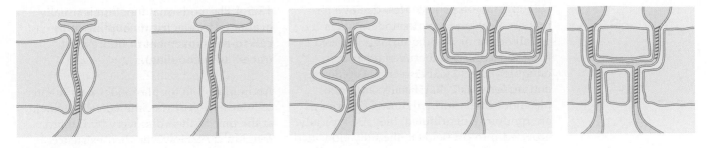

Figure 15.24
Plasmodesmal morphology takes on several forms, but the functional significance of this variation is unknown. The hatched regions indicate ER in its appressed, or desmotubule, form.

15.5.2 Plasmodesmata occur between most living cells of the plant.

Because plasmodesmata are ubiquitous in the plant body, it is simpler to note where they are absent. The primary locations where the plasmodesmata are absent are the interfaces between sporophyte and gametophyte, maternal tissues and embryos, and guard cells and epidermal cells. Between the sieve element/companion cell (SE/CC) complex and surrounding cells along the path from source to sink, plasmodesmata are very infrequent, though not absent. The SE/CC complex in source regions is also symplasmically isolated in **apoplastic loaders,** species in which phloem loading occurs by solute uptake from the apoplast (see Section 15.6.4).

The frequency of plasmodesmata varies substantially, typically ranging from about 0.1 to 10 μm^{-2} of cell wall area. A useful visual impression of symplasmic coupling between cell types can be provided by **plasmodesmograms** (Fig. 15.25). However, because the MEL for plasmodesmata varies

from less than 0.4 to more than 10 kDa, the transport implications of plasmodesmograms—and other data based only on plasmodesmal frequency—must be viewed with reservation.

15.5.3 Primary plasmodesmata form during cell division, whereas secondary plasmodesmata may be added across existing walls.

Primary plasmodesmata form between daughter cells as the growing cell plate constricts strands of ER running through it (see Fig. 15.24). This process is predictable in time and place, making it possible to follow the steps involved. However, **secondary plasmodesmata** can form through preexisting walls as well. This is most apparent when plasmodesmata are found between cells that were initially separate, as in graft unions, host–parasite connections, organs fused postgenitally (e.g., some carpels), and the genetically distinct cells of plant

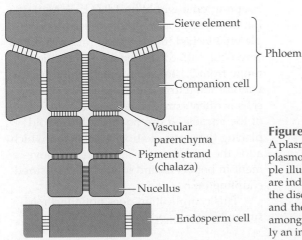

Sieve element
Phloem
Companion cell
Vascular parenchyma
Pigment strand (chalaza)
Nucellus
Endosperm cell

Figure 15.25
A plasmodesmogram shows the relative frequency of plasmodesmata between different cell types. The example illustrated is a rice grain. Plasmodesmal frequencies are indicated by the striping density between cells. Note the discontinuity between the maternal tissues (blue) and the endosperm (red). Because conductance varies among plasmodesmata, their frequency is not necessarily an indication of symplasmic conductivity.

chimeras. In other instances, de novo formation of secondary plasmodesmata is equally clear or may be strongly inferred from their presence between other non-daughter cells (e.g., between the epidermis and underlying cells). Finally, although plasmodesmata formed during cytokinesis are randomly distributed, they are typically grouped into pit fields in the fully elongated cell, in total numbers far exceeding those initially present.

For secondary plasmodesmata, the exact site and timing of formation are usually less evident than for primary plasmodesmata. However, this process has been observed in the graft union between *Vicia* and *Helianthus*, a pairing so disparate that cytological differences between the two make it possible to identify the boundary where plasmodesmata must be forming. The stages of secondary plasmodesma formation in this system are illustrated in Figure 15.26. The final connections are strongly branched, a frequent characteristic of secondary plasmodesmata in other plant systems.

Plasmodesmata may also be occluded or removed after they are formed. This process isolates guard cells from epidermal cells, and sporogenous cells (microspore and megaspore mother cells) from neighboring parental tissues. Plasmodesmata are also occluded or removed when one of the neighboring cells is nonliving at maturity, as between xylem vessel elements and adjacent parenchyma cells. Immunochemical localization of ubiquitin in plasmodesmata destined for removal suggests that the removal process is accomplished at least in part by degradation of plasmodesmal proteins (see Chapter 9).

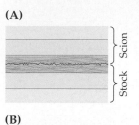
(A)

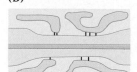
(B)

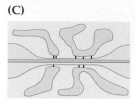
(C)

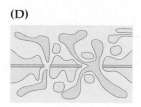
(D)

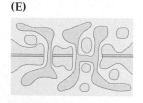
(E)

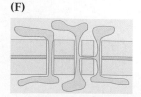
(F)

Figure 15.26

Formation of secondary plasmodesmata during a graft union. Initially, the cells of stock and scion are separated by debris from the cut surfaces (A). The first evidence of the fusion process appears in regions where the ER becomes adnated to the plasma membrane (B). This is followed by localized thinning of the walls (C) to the point where the plasma membrane and ER of adjoining cells can fuse (D). Subsequent wall redeposition (E and F) occurs much as it does during thickening of the cell plate after cell division.

15.5.4 Plasmodesmal function is usually evaluated by electrical coupling or by cell-to-cell movement of fluorescent tracers (dye coupling).

An implied role for plasmodesmata in intercellular transport was immediately evident at the time of their discovery by Eduard Tangl in 1879. However, their existence remained a matter of debate until the advent of electron microscopy. Evidence for symplasmic transport remained indirect for many years and was based largely on comparisons between measured rates of water or solute movement across a cell-to-cell interface and maximum known rates of transmembrane movement. In some cases, observed rates seemed clearly too high to be accounted for by transport across two membranes in series (Fig. 15.27), strongly suggesting that some other pathway must be operative.

Microelectrode measurements of membrane potential have led to the development of more-direct approaches for monitoring symplasmic movement. Because an electrical current in solution is carried by solutes (ions), this methodology can be adapted fairly directly to support or oppose inferences of intercellular solute movement. For this purpose, microelectrodes are inserted into adjacent cells (Fig. 15.28). Membrane potential is monitored in each cell while the current is injected (briefly) into one of the cells to change its membrane potential. The extent of electrical coupling between the two cells is expressed as the **coupling ratio,** $\Delta E_B / \Delta E_A$, where ΔE_A and ΔE_B represent the change in membrane potentials in the injected and adjacent cells, respectively. If both electrodes were in the same cell, the coupling ratio would be 1.0 ($\Delta E_B = \Delta E_A$). If ion movement between the cells were completely blocked, the coupling ratio would be zero ($\Delta E_B = 0$). Measured coupling ratios range from 0.1 to 0.8. In practice, the pathway for ion (current) movement in plant cells is often complicated by the presence of the vacuole. It can be difficult to avoid placing the electrode tips in vacuoles, which adds the tonoplast resistance to ion movement in both cells and lowers the measured coupling ratio.

This complication does not affect measurements made on cells lacking a large

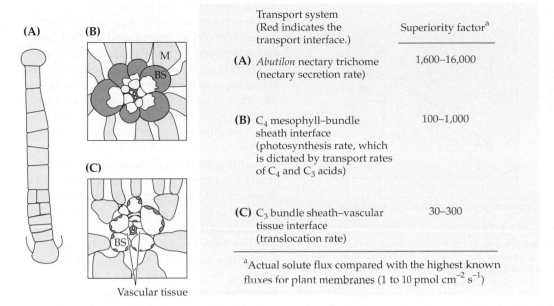

Transport system (Red indicates the transport interface.)	Superiority factor[a]
(A) *Abutilon* nectary trichome (nectary secretion rate)	1,600–16,000
(B) C_4 mesophyll–bundle sheath interface (photosynthesis rate, which is dictated by transport rates of C_4 and C_3 acids)	100–1,000
(C) C_3 bundle sheath–vascular tissue interface (translocation rate)	30–300

[a]Actual solute flux compared with the highest known fluxes for plant membranes (1 to 10 pmol cm^{-2} s^{-1})

Vascular tissue

Figure 15.27
Rates of cell-to-cell solute movement much higher than expected for transmembrane transport implicate plasmodesmata as the pathway for movement. In these three examples, the transport interface for the measured fluxes is indicated in red.

vacuole, as in the tip of the *Azolla* rhizoid (Fig. 15.29). The *Azolla* rhizoid is a particularly useful experimental system because the number of plasmodesmata between the apical cell and its adjacent cells declines systematically during rhizoid growth. Clearly, electrical coupling is strongly correlated with the number of plasmodesmata.

Most often, symplasmic transport is evaluated visually by monitoring the cell-to-cell movement of fluorescent tracers. By analogy with electrical measurements, these are referred to as **dye coupling** experiments. Fluorescent tracer can be pressure-injected or, if charged, can be injected as a current **(iontophoresis).** One of the most important applications of dye coupling experiments has been the use of a series of fluorescent tracers, graded in molecular mass, to determine the MEL (see Section 15.2.1). Examples of a mobile probe and an immobile probe are shown in Figure 15.30. Size gradation is

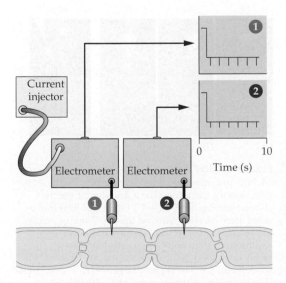

Figure 15.28
Measurement of electrical coupling. While microelectrodes measure the membrane potentials of adjacent cells, periodic short pulses of current are injected into one cell, momentarily changing its membrane potential. If current (ions) can flow to an adjacent cell, its membrane potential will also change, although not as much as in the injected cell. In the example recordings, the initial downward deflection of the trace occurs as the microelectrode enters the cell. On the time scale of the recording, the change in membrane potential appears only as a spike, given the 0.1-second pulse duration. In the example shown here, the coupling ratio (given by the relative spike heights) would be about 2/3. In practice, an oscilloscope is used to follow the time course of the changes with far greater accuracy.

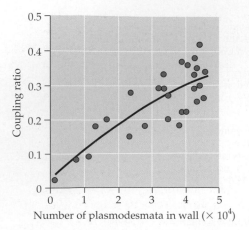

Figure 15.29
Measurements of electrical coupling in the *Azolla* rhizoid demonstrate a strong correlation between electrical coupling and the number of plasmodesmata connecting the cells. The measurements take advantage of the rhizoid's development, which is characterized by a progressive decline in the number of plasmodesmata between the apical cell and adjacent cells.

usually achieved by using fluorescent dextrans of various molecular masses or by preparing fluorescein isothiocyanate derivatives of various-sized peptides.

15.5.5 Most estimates of the plasmodesmal MEL are around 800 Da, but the magnitude may be larger or smaller at some interfaces.

Most estimates of the MEL for free diffusion through plasmodesmata are close to 800 Da, corresponding to an SEL of about 2.0 nm in molecular diameter. However, as dye coupling experiments were applied to an ever-widening range of plant tissues, exceptions accumulated to the MEL "rule of 800." In the context of long-distance transport, larger MELs in the postphloem transport pathway may be necessary to accommodate high solute fluxes there (see Section 15.6.8). A MEL_{dex} of 7 kDa has been observed in tobacco leaf trichomes, but its functional significance is unclear.

In some instances, plasmodesmata are present but apparently not functional. Somewhat surprisingly, root hair cells become symplasmically isolated as they mature (Fig. 15.31). A systematic study of dye coupling between tissues of an aquatic tracheophyte,

Egeria densa, showed strong coupling between most cells and tissues. Exceptions were observed at several well-defined interfaces: between root and root cap, between epidermis and cortex, between older internodes, and between stem and petiole—all cases in which plasmodesmata are present. The effect of these boundaries, and of those established by the absence of plasmodesmata, is to establish symplasmic domains within which solutes can move freely but cannot pass into bordering domains. Additional examples include embryonic tissues, much of the SE/CC complex, and some glands.

For most small molecules, size appears to be the principal factor in symplasmic mobility. In some cases, however, size does not appear to be the sole factor involved. In particular, the presence of aromatic amino acid side chains in a fluorescent probe markedly reduces permeability. Charge-based effects have also been noted. Given that plasmodesmal channels are aqueous pathways presumably lined with polar, charged, and hydrogen-bonding groups, these structural components can be anticipated to interact with transported solutes, especially those at sizes near the SEL. Such selectivity in solute movement could be important but is difficult to assess, particularly for endogenous solutes.

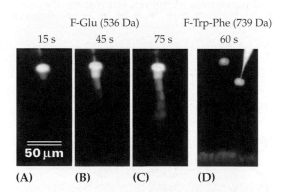

Figure 15.30
Mobile and immobile fluorescent probes in the *Abutilon* nectary trichome. (For a drawing of the trichome, see Fig. 15.27.) (A–C) F-Glu (536 Da) movement after injection into the tip cell. The frequently observed "stair-step" distribution of fluorescence results from diffusion within cells being relatively rapid compared with the much slower cell-to-cell movement. (D) The slightly larger F-Trp-Phe (739 Da) shows no signs of movement after 60 seconds.

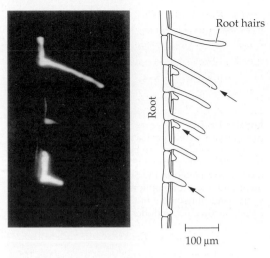

Figure 15.31
Cell-to-cell solute movement may not occur even when plasmodesmata are present. In this example, carboxyfluorescein (CF) introduced into the tips of *Arabidopsis* root hairs (arrows) did not move into adjacent cells, even though plasmodesmata were present. Other studies revealed that these cells are not electrically coupled.

15.5.6 Tissue damage usually causes a decrease in plasmodesmal permeability.

Wounding is almost always accompanied by rapid callose deposition that restricts, but rarely obliterates, plasmodesmata in intact cells bordering the wound. Callose synthase shows a sigmoidal response to Ca^{2+}, a signal often released by wounding. The enzyme is strongly activated at Ca^{2+} concentrations greater than 10 µM. Wound callose forms in response to various treatments and can disappear if no lasting damage occurs, as in some plasmolysis/deplasmolysis treatments (see below). Whether callose is involved in the fine control of plasmodesmal conductance is not clear.

Wounding also causes large turgor differentials between wounded and intact cells. These pressure differentials may play a role in the wound response. In unwounded tobacco leaf trichomes, even small differentials (0.02 to 0.03 MPa) between intact cells can block or sharply reduce symplasmic movement at the site of the altered cell-to-cell pressure difference. Plasmodesmata linking the healthy cell to equally turgid cells were unaffected. However, these responses to wounding are not universal. Exceptions include the leaf bundle sheath cells of C_4

species, glandular trichomes in mint leaves, vascular parenchyma cells in wheat grains, and several meristematic tissues. In these cases, neither the pressure differential nor the increased $[Ca^{2+}]$ that accompanies wounding leads to plasmodesmal sealing.

Cell-to-cell solute movement is sharply reduced in plasmolyzed tissues. Plasmolysis treatments were once presumed to disrupt symplasmic continuity permanently, and this appears to be true for some tissues (e.g., tomato trichomes). In other tissues, however, transport can be restored after deplasmolysis, with recovery time varying from hours (coleoptiles) to minutes (*Egeria* leaves). In the latter case, movement in tissues that were first plasmolyzed and then deplasmolyzed was even more rapid than in untreated controls.

15.5.7 Several experimental treatments have been shown to alter the plasmodesmal MEL.

Some experimental protocols, which might be broadly categorized as stress treatments, can increase the MEL_{dex} to 5 kDa to 10 kDa in some tissues. In wheat roots, anoxia or azide can induce this effect, as can azide in *Setcresea* stamen hairs, which suggests that some ATP-dependent process may be responsible for maintaining a more-restricted pathway in healthy conditions. Likewise, the uncoupler CCCP (though not azide) induces the lateral symplasmic movement of carboxyfluorescein (CF) from the SE/CC complex in *Arabidopsis* roots (Fig. 15.32), although the resulting MEL could not be quantified. Of particular interest is the finding that cytochalasin D, an inhibitor of actin polymerization, can increase the MEL from 0.8 to 20 kDa within 3 to 5 min, strongly implicating plasmodesma-associated actin filaments in the regulation of symplasmic conductance (see also Section 15.5.9). By contrast, metabolic inhibitors decreased symplasmic conductance in oat coleoptiles (cyanide, azide), in *Egeria* leaves (FCCP), and in the large-celled alga, *Chara* (CCCP). However, the contrasting results from metabolic inhibitors may in part reflect differences in experimental technique. The decreased MELs observed in *Chara* and *Egeria* were obtained by using current-assisted

(A)   **(B)**

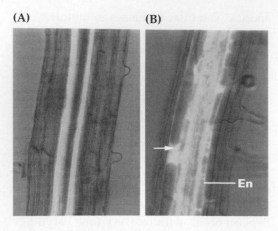

Figure 15.32
In some cases, treatment with metabolic inhibitors causes plasmodesmata to become more conductive. (A) Phloem-transported CF in *Arabidopsis* roots is ordinarily retained in the SE/CC complex. (B) Treatment with the protonophore CCCP induced lateral symplasmic movement of CF all the way into the cortex (arrow). En, endodermis.

injection to measure electrical coupling or to inject dye, whereas the increased MELs seen in *Arabidopsis* involved dyes injected under pressure injection or taken up passively (see Section 15.7.7)

Increased cytosolic concentrations of Ca^{2+}, at less than 1 μM, or inositol triphosphate, which raises the cytosolic Ca^{2+} concentration, reduce symplasmic conductance in *Setcresea* stamen hairs. Here, Ca^{2+} appears to be acting as a second messenger (see Chapter 18). Ca^{2+}-dependent phosphorylation of plasmodesmal proteins may participate in this mechanism.

15.5.8 Symplasmic transport in higher plants is limited by cell-to-cell transport, not by intracellular movement.

Movement within cells (intracellular movement) represents by far the greatest distance traversed along the symplasmic pathway. However, cell-to-cell movement is sufficiently restricted that it limits the overall rate of symplasmic transport unless exceptionally large cells are involved. To a reasonable approximation, solutes within the cell lumen are well-mixed, and concentration gradients occur only between cells. This is evident in the frequently observed "stair-step" distribution of fluorescent tracers in successive cells along the symplasm (see Fig. 15.30). As a consequence, when the path is composed of differing cell sizes, the number of cells traversed is a more relevant measure of movement than is the actual physical distance involved (Fig. 15.33). Finally, in all but very large cells, diffusive mixing of cell con-

tents is sufficiently effective that symplasmic movement is unaffected by the cessation of protoplasmic streaming.

15.5.9 Work is ongoing to characterize the molecular architecture of plasmodesmata.

Substantial progress is being made toward characterizing plasmodesmata at the molecular level by isolating plasmodesmal proteins and by using immunocytochemistry to localize specific proteins in plasmodesmata.

Plasmodesmata are firmly fixed in the cell wall. Although isolating and purifying their components poses unique problems, researchers can "purify" whole plasmodesmata by isolating cell walls. After rigorous treatments to remove membranes and organelle fragments attached to the wall surface, plasmodesmata that appear largely intact in electron micrographs remain embedded in the cell wall. Some results from this approach, using corn mesocotyls, are shown in Figure 15.34. About 12 to 20 proteins remain associated with the purified wall preparation. One of these cross-reacts with a connexin antibody in a Western blot and is localized to plasmodesmata by immunogold labeling. (Connexins are protein subunits of gap junctions, animal cell-to-cell connections with a MEL of about 800 Da.) Antibodies raised against several of the "wall proteins" have been used to confirm the association of those proteins with plasmodesmata. However, the wall proteins have not yet been sequenced, and the possible functions of almost all of these proteins remain unknown.

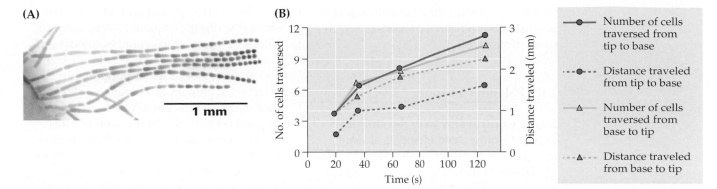

(A)

1 mm

(B)

Legend:
- Number of cells traversed from tip to base
- Distance traveled from tip to base
- Number of cells traversed from base to tip
- Distance traveled from base to tip

Figure 15.33
Cell-to-cell movement limits rates of symplasmic transport. (A) In *Tradescantia* stamen hairs, cell length decreases from the base toward the tip. (B) A fluorescent tracer applied to one end of the hair moves the same number of cells in a given amount of time (solid lines), whether movement is from tip to base or from base to tip. However, because the basal cells are larger, the distance traversed (dashed lines) is greater for movement from base to tip.

In one approach to establishing functional roles for putative plasmodesmal proteins, rigorously purified cell wall proteins from tobacco leaves were used for coinjection experiments in tobacco leaf mesophyll. Cell-to-cell movement of 11-kDa F-dextran was markedly accelerated by the cell wall proteins, whereas the movement of proteins known to traffic from cell to cell (see Section 15.7.6) was reduced. These experiments indicate the presence of at least two functional classes of plasmodesmal proteins: one capable of up-regulating the plasmodesmal SEL, the other interacting with symplasmically mobile proteins during their movement.

The involvement of ATP-dependent processes in the gating of plasmodesmata (see Section 15.5.7) suggested the possibility that protein phosphorylation might regulate or drive gating. To investigate this hypothesis, Ca^{2+}-dependent protein kinase was isolated from purified cell wall proteins from maize mesocotyls and was found to be capable of phosphorylating itself and several other wall-associated proteins. The ability of cytochalasin D, an inhibitor of actin polymerization, to up-regulate the MEL was noted earlier (see Section 15.5.7). Cytochalasin also has been shown by electron microscopy to dilate the neck region of plasmodesmata, while at the same time disorganizing the putative sphincters in the surrounding collar. Actin and myosin have been localized in plasmodesmata by immunocytochemical methods, and both

may be components of the "spokes" between the desmotubule and plasma membrane (see Fig. 15.20).

Although no specific plasmodesmal proteins have yet been isolated, the virus movement proteins (Fig. 15.35) and native transcription factors (e.g., KNOTTED1; see Section 15.7.6) have been shown to bind to plasmodesmata. Quite probably, virus movement proteins and native transcription

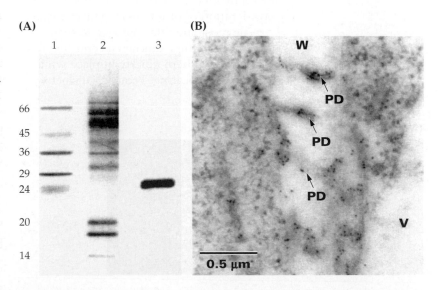

(A)

1   2   3

66
45
36
29
24
20
14

(B)

W
PD
PD
PD
V

0.5 µm

Figure 15.34
Because plasmodesmata are firmly embedded in the cell wall, the most successful approach to the isolation of plasmodesmal proteins has been to purify a cell wall fraction rigorously. (A) About 12 to 20 proteins remain associated with purified cell walls from corn mesocotyl (lane 2). One of the proteins reacts strongly with a connexin antibody (lane 3). Lane 1 contains size markers (in kDa). (B) Immunochemical localization in thin sections confirms that the protein labeled in lane 3 is associated with plasmodesmata (PD). W, wall; V, vacuole.

factors can be exploited to identify and isolate the plasmodesmal proteins to which these proteins bind.

15.6 Phloem transport

15.6.1 Controversy over the structure of functioning sieve elements slowed acceptance of OGPF as the mechanism of phloem transport.

The phloem-conducting pathway consists of interconnected low-resistance elastic tubes under high pressure. Damage results in a sudden pressure release and a rapid surging of the contents toward the site of injury. Sieve element organelles are dislodged from their normal positions and swept along with the surging contents until they are filtered out on the nearest sieve plate (see Fig. 15.1F). This process provides an efficient sealing mechanism. The disruptive effect of surging on microscopic analysis of sieve element structure has proved difficult to avoid completely. Even with considerable caution in tissue preparation, the sieve plate pores fill with proteinaceous material. This occluded configuration is incompatible with the simple pressure-flow mechanism envisioned by OGPF, prompting skepticism and consideration of other possible mechanisms.

The most suitable approach to fixing sieve tubes in their functional condition has been to freeze them quickly in place while they are translocating. Because satisfactory

freezing is critically dependent on dimensions (the distance-squared relationship of diffusive transport also applies to heat conduction), one must first isolate by dissection a short length of vascular bundle, still attached at both ends. Phloem function is confirmed by movement of a radiolabel applied before freezing the sample. Once frozen, the tissue can be prepared for electron microscopy by freeze-substitution (see Box 5.4 in Chapter 5 for a description of this method). Sieve tubes in material prepared by freeze-substitution have sieve plate pores that are largely free of obstruction (Fig. 15.36A). Organelles such as plastids, mitochondria, and ER are located almost exclusively around the periphery of the sieve elements (Fig. 15.36B). This configuration is consistent with OGPF as the mechanism of phloem transport.

An obvious approach to investigating structure–function relationships in sieve tubes would be to make observations on fresh tissue. However, this presents technical difficulties. The optical properties of fresh tissue are poor and sieve tubes sufficiently large for satisfactory cytological observation occur only in relatively thick structures. As in freeze-substitution, dissection is necessary and the functional condition of the observed sieve tubes must be confirmed. However, these difficulties have recently been overcome by using fluorescence staining and confocal microscopy to observe functioning sieve tubes in *Vicia faba* (Fig. 15.37). As in freeze-substituted material, virtually all of the cell contents are distributed around the sieve element periphery. Within the limits of resolution imposed by light microscopy, sieve plate pores were determined not to be occluded.

15.6.2 Physiological evidence strongly supports OGPF as the mechanism of phloem transport.

Several alternatives to OGPF have been suggested as possible mechanisms for solute transport along the phloem. The near-universal occurrence of cytoplasmic streaming in vacuolate plant cells inspired several conceptualizations of streaming-based transport in sieve tubes. Solution flow by electro-osmosis, a process easily observed during

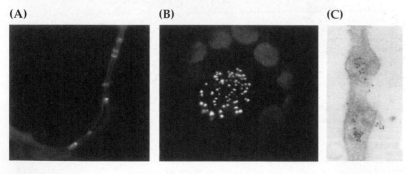

(A) **(B)** **(C)**

Figure 15.35
Several viral movement proteins (MP) have been shown to target plasmodesmata. By fusing green fluorescent protein (GFP) with MP, the binding may be visualized by fluorescence microscopy (A and B). (A) Focal plane through a wall between epidermal cells, revealing a fluorescent chimera of GFP with MP from cauliflower mosaic virus (CMV). (B) Face view of the plasmodesmal pit field between mesophyll cells. (C) A gold-labeled antibody to the CMV-MP binds to plasmodesmata.

(A)

(B)

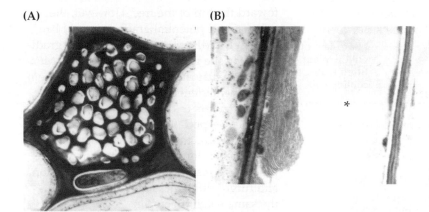

Figure 15.36
Quick-freezing of intact, functioning sieve tubes has been the most successful approach to the elimination of surging. Sieve plate pores are largely free of particulate material (A), and the lumen (∗) is also clear (B).

electrophoresis in a solid support having fixed charges, was suggested as a mechanism that could cause solution flow across the typically "plugged" configuration of sieve plate pores and, as a result, along the entire sieve tube. Here, fixed negative charges on the proteinaceous plug were assumed to be balanced by mobile potassium ions, which would be pulled by an electric potential difference across the sieve plate, in turn pulling along water and other solutes.

Specific objections can be raised to each of these proposals. Each reflects an interpretation of sieve element structure that most workers no longer view as accurate. In addition, in the case of a streaming-based mechanism, maximum streaming rates observed in plant cells are less than 1 mm min^{-1}, far slower than rates typical of phloem transport (often 1 cm min^{-1} or more). In the case of electroosmosis, the efficiency of water movement (the number of water molecules moved per ion) would have to be orders of magnitude higher than has been observed during electroosmosis in nonliving systems.

However, there is a broader energetic distinction between osmotically generated pressure flow and alternative proposals. Without exception, the latter envision some form of metabolically assisted transport along the pathway itself. In contrast, the only energy requirement for OGPF is the maintenance of high solute concentration at the source and low solute concentrations at the sink. Movement along the path is passive; energy is required only to maintain structural integrity. By evaluating whether inhibition of metabolism affects phloem transport along the path, OGPF can be differentiated from alternative mechanisms.

Cold treatment is particularly suited for this purpose because its effect is confined to the point of application. Phloem transport in several species has been shown to be unaffected by chilling a segment of the path to 0°C or even lower (Fig. 15.38). Willow sieve tubes have been observed to function at −12°C. However, cold insensitivity is not characteristic of all species. In chilling-sensitive species, translocation is sharply inhibited below 10 to 15°C, but this is symptomatic of the chilling injury syndrome and does not indicate a fundamentally different mechanism of phloem transport.

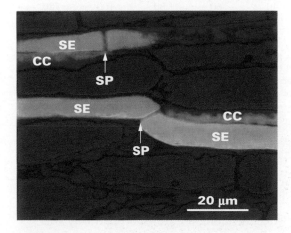

Figure 15.37
Confocal fluorescence microscopy has recently made it possible to examine functioning sieve tubes at high optical resolution. The fluorescence micrograph illustrates a doubly stained, translocating sieve tube in a *Vicia faba* leaf. The phloem was exposed by slicing away shallow paradermal sections with a razor. A locally applied fluorochrome visualized the membranes (red), whereas translocation was revealed by CF (green), which had been introduced several centimeters upstream from the observation point. SE, sieve element; CC, companion cell; SP, sieve plate.

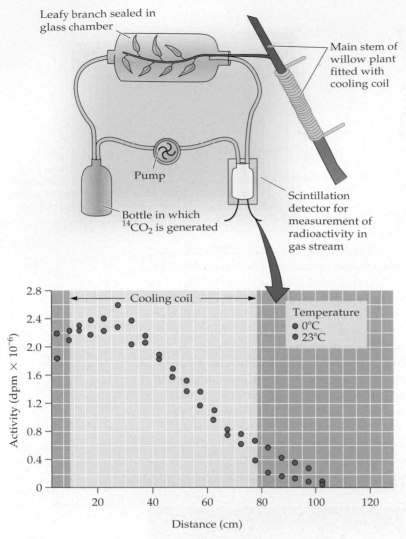

Figure 15.38
Effect on phloem transport of cooling a 65-cm length of the translocation pathway in willow to 0°C. A leafy branch was labeled with $^{14}CO_2$ and the distribution of exported ^{14}C with a control plant labeled at 23°C indicates that cooling to 0°C had only a slight effect on movement along the phloem. The distance referred to on the *x*-axis is the length (cm) of stem along which ^{14}C-labeled assimilates were translocated from the source.

Since the time of Münch's original proposal, evidence that pressure gradients decline along the transport pathway was sought to confirm OGPF. Collection of phloem exudate from trees quickly confirmed that solute concentrations were typically greater in the canopy and decreased down the trunk. A more recent and particularly detailed example is illustrated for tree tobacco (Fig. 15.39). Here, the osmotic gradient is about 0.2 MPa m⁻¹. The turgor gradient in the sieve tubes would be less, given that water potential in the xylem decreases toward the top of the tree. However, the xylem water potential gradient would be at most 0.1 MPa m⁻¹, leaving a turgor gradient along the sieve tubes of at least 0.1 MPa m⁻¹. Note that the complexity of the phloem plumbing system can be a confounding factor in attempts to demonstrate such pressure gradients, especially over short distances. Phloem conductance is high along the sieve tube axis but low between adjacent sieve tubes. As a result, substantial concentration and turgor differences are often found at the same height in a stem.

15.6.3 The SE/CC complex of the transport phloem is symplasmically isolated.

As is the case throughout the phloem, sieve elements and companion cells of the transport phloem readily exchange solutes. Plasmodesmata between the two are frequent and have a distinctive appearance (Fig. 15.40). A single pore in the sieve element wall is linked via a central cavity to branched plasmodesmata on the companion cell side, a structure referred to as a **pore–plasmodesma complex.** The branches of the complex on the companion cell side are occupied by desmotubules, but consensus is lacking on the presence of a desmotubule and associated ER on the sieve element side. Aside from its appearance, the pore–plasmodesma complex also appears to have a distinctively large MEL_{dex} (10 to 20 kDa).

The effective long-distance transport of a solution requires a pipe that is leak-free, or nearly so. In contrast to its strong internal connectedness, the SE/CC complex of the transport phloem is almost completely isolated symplasmically from phloem parenchyma or other adjoining parenchyma cells. This is evident structurally in the very low frequency of plasmodesmata between the SE/CC complex and the surrounding cells and is confirmed experimentally by the near absence of electrical or dye coupling. Microautoradiographs and fluorescence microscopy demonstrate that labeled assimilates and polar fluorochromes, respectively, are confined to sieve tubes and companion cells during transport. Some leaking occurs, but in general, the transport phloem is very active in the absorption of solutes from the apoplast. Judging from the amount of tracer remaining in stems after a pulse of radioactivity has passed through, net loss of solutes per centimeter of SE/CC complex often amounts to 0.5% or less in herbaceous species.

15.6.4 Depending on the species, phloem loading in leaves may occur by transmembrane movement (apoplastic loading) or via plasmodesmata (symplasmic loading).

The process whereby assimilate moves into the collection phloem has been studied almost exclusively in leaves. In some species, the SE/CC complex appears to be as symplasmically isolated in leaves as it is in the transport phloem. This, combined with the frequently high osmotic concentration of phloem exudates (in comparison with that of leaf mesophyll) and the ability of minor vein phloem to absorb sucrose directly from the apoplast, has provided convincing evidence for phloem loading by active solute absorption from the apoplast.

Most crop species translocate sucrose and are apoplastic phloem loaders. Virtually all early work on phloem loading was done with this group of plants, and the consis-

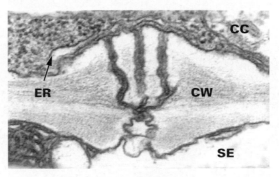

Figure 15.40
Companion cell (CC) and a sieve element (SE) are connected by a pore–plasmodesma complex consisting of a pore in the sieve element wall linked via a central cavity to multiple plasmodesmata in the companion cell wall (CW). ER, endoplasmic reticulum.

tency of the results made it tempting to generalize to all species. However, early experiments with cucurbits, which translocate the **raffinose series of oligosaccharides** (RSOs), indicated that their minor veins were incapable of absorbing RSOs from the apoplast. In addition, when ultrastructural observations were systematically expanded to a wide range of plant groups (700 species from 140 families), many, including RSO translocators, were found to exhibit high plasmodesmal frequencies between the minor vein companion cells and adjacent bundle sheath cells (Fig. 15.41A). These cells, in turn, were linked by plasmodesmata to the leaf mesophyll, providing an uninterrupted symplasmic pathway from the mesophyll to the sieve tubes.

Minor vein configurations can be separated into two broad types, based on the apparent symplasmic continuity between the companion cells and bundle sheath cells. With rare exception, all species within a family are the same type. **Type 1 species,** putative symplasmic loaders, have an "open" configuration with numerous plasmodesmal connections (Fig. 15.41A) between companion cells and bundle sheath cells. **Type 2 species,** putative apoplastic loaders, have a "closed" minor vein configuration, in which the SE/CC complex appears to be symplasmically isolated (Fig. 15.41B). Type 2 species can be divided into two groups: **Type 2b** (closed advanced) species have companion cells with transfer cell morphology (see Fig. 15.13 for an example of a transfer cell found in the xylem), whereas **Type 2a**

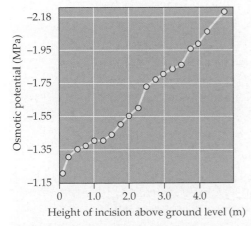

Figure 15.39
Phloem exudate from trees typically decreases in concentration from the canopy to ground level. In this particularly thorough study of tree tobacco (*Nicotiana glauca*), the osmotic gradient is about 0.2 MPa m^{-1}. Water potential (not shown) decreases in the opposite direction. At most, however, this decrease would be no more than 0.1 MPa m^{-1}, so the sieve tube turgor gradient would be at least 0.1 MPa m^{-1}.

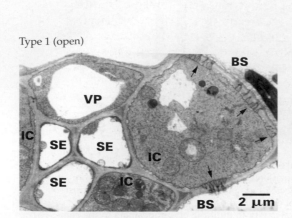

Type 1 (open)

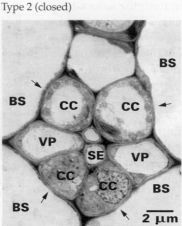

Type 2 (closed)

Figure 15.41

Minor vein anatomy of Type 1 species, putative symplasmic loaders with an "open" configuration (A), and of Type 2 species, putative apoplastic loaders with a "closed" configuration (B). In the open type (A), numerous plasmodesmata (arrows) are present between the companion cell and the bundle sheath cell (BS), which are in turn symplasmically connected to mesophyll cells. In the closed type (B), plasmodesmata are essentially absent from the same interface (arrows), leaving the sieve element/companion cell (SE/CC) complex isolated from the rest of the leaf. IC, intermediary cell (a specialized companion cell characteristic of RSO translocators); VP, vascular parenchyma.

Table 15.3 Minor vein configuration and transport sugars in selected dicotyledons

Family[a]	Species	Common name	Plasmodesmal frequency[b]	Sugar[c]
Type 1 (open)[d]				
Oleaceae	*Fraxinus ornus*	Ash	61	RSOs + mannitol
	Syringa vulgaris	Lilac	55	RSOs + mannitol
Cucurbitaceae	*Cucurbita pepo*	Pumpkin	48	RSOs
Lamiaceae	*Coleus blumei*	Coleus	45	RSOs
Vitaceae	*Vitis vinifera*	Grape	31	Sucrose
Magnoliaceae	*Liriodendron tulipfera*	Tulip tree	24	Sucrose
Salicaceae	*Salix babylonica*	Weeping willow	14	Sucrose
Type 1–2a (intermediate)				
Ericaceae	*Rhododendron caucasicum*	Rhododendron	8.1	RSOs
Malvaceae	*Gossypium hirsutum*	Cotton	6.2	Sucrose
Euphorbiaceae	*Ricinus communis*	Castor bean	4.1	Sucrose
Type 2 (closed)[e]				
Type 2a (closed primitive)[f]				
Solanaceae	*Solanum tuberosum*	Potato	0.12	Sucrose
	Nicotiana tabacum	Tobacco	0.08	Sucrose
Chenopodiaceae	*Beta vulgaris*	Beet	0.06	Sucrose
Type 2b (closed advanced)[g]				
Fabaceae	*Pisum sativum*	Pea	0.08	Sucrose
Asteraceae	*Xanthium strumarium*	Cocklebur	0.06	Sucrose
	Helianthus annuus	Sunflower	0.06	Sucrose
	Lactuca sativa	Lettuce	0.03	Sucrose

[a]With rare exceptions, all species within a family are of the same type.
[b]The number of plasmodesmata per square micrometer between the companion cells and bundle sheath cells.
[c]All species translocate significant amounts of sucrose.
[d]Numerous plasmodesmata between companion cells and minor vein bundle sheath cells. About 80% of the Type 1 species are woody species. [e]Very few plasmodesmata between companion cells and minor vein bundle sheath cells. Herbaceous species make up about 80% of Type 2 species.
[f]Companion cell walls are not invaginated.
[g]Companion cell walls are invaginated.

(closed primitive) species companion cells lack wall invaginations. Table 15.3 lists some examples of this ultrastructural survey, along with the major sugars translocated. The examples were chosen on the basis of familiarity and to illustrate the range and gradation of symplasmic continuity between the minor vein SE/CC complex and bundle sheath cells.

The SE/CC complex never appears to be entirely isolated and, given the gradation in plasmodesmal frequencies, some species are quantitative intermediates (Type 1–2a) with regard to this character. Because this classification system is based solely on plasmodesmal frequency, the distinction between types is somewhat arbitrary, especially between open Type 1 and intermediate Type 1–2a. Leaves of some species contain both open and closed configurations in the same minor vein.

Virtually all species translocate sucrose. To date, verified apoplastic loaders translocate only sucrose. However, exclusive sucrose translocators are also found among Type 1 families (e.g., Vitaceae, Populaceae; see also Section 15.6.6). Woody plants, rarely used in phloem-loading investigations, account for about 80% of putative symplasmic loaders, whereas herbaceous species make up about 80% of putative apoplastic loaders. For almost all plant families, however, the breadth of structural observations to date far outstrips the physiological evidence available on symplasmic versus apoplastic loading.

15.6.5 When the minor vein SE/CC complex is symplasmically isolated (closed configuration), it absorbs solutes from the apoplast by membrane transport.

Studies of apoplastic phloem loading have focused almost exclusively on sugars, specifically sucrose. Sucrose is envisioned to move symplasmically from the mesophyll to the minor veins, where it is released from vascular parenchyma cells into the apoplast for subsequent absorption into the SE/CC complex by H^+-sucrose cotransport. Other transported solutes presumably follow the same pathway, although little is known regarding their mode of uptake into the SE/CC complex.

The feasibility of symplasmic movement to the minor veins is supported by the observed frequencies of plasmodesmata that link the cell types and by the results of dye coupling experiments. Evidence that sucrose itself actually follows this pathway is more circumstantial. Conceivably, sucrose could be released from mesophyll cells into the apoplast and follow an apoplastic pathway to the minor veins. Various washing experiments demonstrate that sucrose and other sugars can, in fact, be leached from leaf tissues. However, the tissue must necessarily be wounded, raising the possibility that some features of sugar release determined this way might be artifacts. Of particular concern is loss from cut vein endings, which is difficult to distinguish from release from the mesophyll.

The approach illustrated in Figure 15.42A avoids the problem of leakage from cut veins. Here, only the epidermis of an otherwise intact, attached leaf was abraded. When the sucrose pool was labeled by supplying $^{14}CO_2$ to the lower leaf surface, no labeled sugars appeared in a buffer solution circulating over the abraded upper leaf surface, even when export of [^{14}C]sucrose had reached its maximum rate. Thus, no [^{14}C]sucrose was released from the mesophyll cells into the apoplast, even though sugar beet is an apoplastic loader. [^{14}C]Sucrose appeared only when unlabeled sucrose was included in the circulating solution, the rate of release varying with the translocation rate (Fig. 15.42B). This is interpreted to mean that sucrose release occurred not from the mesophyll, but from minor vein parenchyma cells, whereupon it was immediately reabsorbed by the SE/CC complex (Fig. 15.41C) unless externally supplied sucrose was present. When unlabeled sucrose competed with photosynthetically labeled [^{14}C]sucrose for uptake, some of the latter escaped to the circulating solution.

That minor veins, not the mesophyll, are the site of sucrose release is supported by additional evidence. In soybean leaves, the kinetics of [^{14}C]sucrose export differ markedly from the labeling kinetics of [^{14}C]sucrose in the photosynthetic cells, arguing against the mesophyll as an immediate source of translocated sucrose. Finally, in a corn mutant unable to export sucrose from the distal portion of its leaves, plasmodesmata in the leaf tip were normal except those between the bundle sheath and vascular parenchyma, which were occluded. Thus, export appeared to be blocked at the last symplasmic step before sucrose release into the apoplast.

In contrast to sucrose release into the apoplast, sucrose absorption from the apoplast by the SE/CC complex is strongly supported by experimental evidence. In sugar beet leaves, 20 mM sucrose supplied via the apoplast (see Fig. 15.42) was sufficient to sustain control rates of translocation, even when photosynthesis and translocation of endogenous sucrose were severely disrupted by plasmolysis of the mesophyll. The in vivo involvement of an intermediate apoplastic step is strongly supported by the effects of the sulfhydryl reagent, PCMBS, which is unable to permeate the symplasm. Translocation in apoplastic loaders is strongly inhibited by PCMBS application under conditions in which photosynthesis remains unaffected, showing that PCMBS has been excluded from the symplasm (Fig. 15.43A). That sucrose is taken up directly from the apoplast into minor veins is shown by tissue autoradiographs of leaf discs exposed briefly to [^{14}C]sucrose. This uptake is eliminated by PCMBS (Fig. 15.43B).

(A)

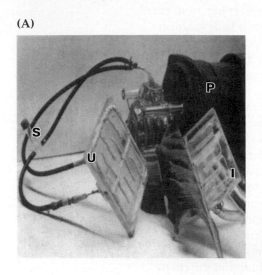

(B)

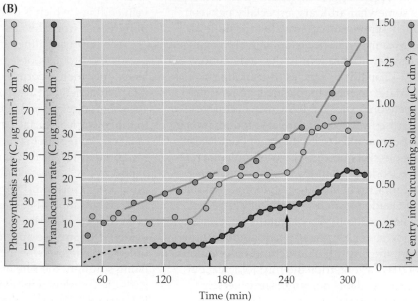

(C)

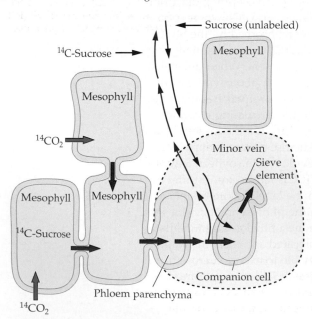

Figure 15.42

(A) Experimental setup for monitoring sugars in the apoplast of a translocating sugar beet leaf. Half of the leaf is visible; the other half has been cut away to show the lower chamber (l) through which air containing $^{14}CO_2$ was circulated. After the upper epidermis was lightly abraded, the upper (u) and lower chambers were sealed and various solutions were circulated over the upper leaf surface by a pump (p). The translocation rate was monitored continually by the arrival rate of ^{14}C-containing photosynthate in a sink leaf. A sampling tube (s) was used to remove aliquots of circulating solution for assay. (B) When sucrose was included in the circulating solution, as in this experiment, a small amount of ^{14}C-containing sucrose was released continually from the leaf into the solution. Increases in the translocation rate, caused by increasing the light intensity (arrows), accelerated the release of [^{14}C]sucrose into the leaf apoplast. In the absence of unlabeled sucrose, no [^{14}C]sucrose appeared in the circulation solution. (C) Diagrammatic interpretation of the observations in (B). Photosynthetically labeled [^{14}C]sucrose moves symplasmically to the minor vein phloem parenchyma, where it is released into the apoplast for immediate reabsorption by the SE/CC complex. Only when external sucrose is present to compete with internally synthesized sucrose for uptake can some of the latter escape reabsorption and subsequently appear in the circulating solution. (For this reason, the circulating sucrose solution is referred to as a "trapping solution.")

The probable involvement of H⁺-sucrose cotransport in supplying the energy for sucrose accumulation received experimental support in the 1970s. Sucrose uptake was shown to be accompanied by alkalinization of the medium, and the response to several experimental treatments (K⁺, fusicoccin, pH) was consistent with H⁺ cotransport. Measurements of sieve tube membrane potential exhibited the expected depolarization in response to exogenous sucrose solutions (Fig. 15.44; see also Chapter 3). As expected for carrier-mediated transport, the depolarization response became saturated at higher sucrose concentrations.

Operation of the apoplastic loading pathway receives strong support from molecular biological investigations. The H⁺-pumping plasma membrane ATPase and several H⁺-sucrose cotransporters have been shown to be expressed strongly in the phloem. Antisense expression of a H⁺-sucrose cotransporter in potato leaves resulted in symptoms of sharply reduced phloem loading, including marked accumulation of soluble sugars and starch in mature leaves, diminished rates of photosynthesis, greatly reduced root growth, and very low tuber yield. Apoplastic expression of invertase in tobacco leaves was accompanied by similar symptoms.

Clearly, there is abundant evidence that sieve tube loading in Type 2 species occurs by energized sucrose uptake from the apoplast. Far less progress has been made concerning sucrose release to the apoplast before uptake.

15.6.6 Symplasmic continuity from the mesophyll to the SE/CC complex allows assimilates to move directly into the translocation stream.

The same experimental approaches used to demonstrate apoplastic loading have provided strong physiological evidence for a continuous membrane-bound pathway from the mesophyll to the SE/CC complex in Type 1 species. The inability of cucurbit leaf minor veins to absorb apoplastically supplied RSOs was noted earlier (see Section 15.6.4). Brief exposure of leaf tissues to [¹⁴C]sucrose or ¹⁴CO₂ results in general labeling of the mesophyll, after which the label accumulates in the veins (Fig. 15.45). This

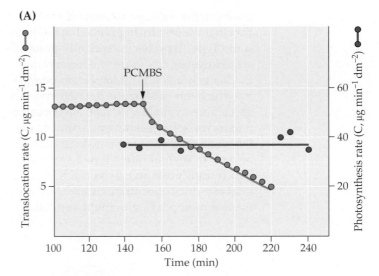

(A)

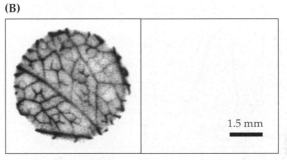

(B)

1.5 mm

Figure 15.43
(A) Effect of PCMBS on phloem transport from a sugar beet leaf. With the same apparatus shown in Figure 15.42, photosynthesis and translocation rates were monitored before and after the introduction of PCMBS into the solution circulating over the upper leaf surface. Export of labeled photosynthate from the leaf decreased promptly upon addition. Photosynthesis was unaffected, indicating that PCMBS had not entered the leaf cells but reacted only with sulfhydryl groups on the plasma membrane surface. (B) Autoradiography confirmed the inhibitory effect of PCMBS on apoplastic sucrose uptake by the minor veins. In leaves of pea, also a Type 2 species, the minor veins can absorb [¹⁴C]sucrose directly from the apoplast. After only a brief uptake period (2 minutes), almost all of the absorbed sucrose was located in the minor veins, rather than the mesophyll (left panel). Uptake was completely eliminated by PCMBS (right panel). Black areas on the autoradiograph correspond to ¹⁴C in the leaf.

contrasts with the results from similar experiments with apoplastic loaders (compare Fig. 15.45A with the left panel of Fig. 15.43B, which shows direct uptake by the veins). The redistribution of ¹⁴C from the mesophyll to the veins is not inhibited by PCMBS, nor does PCMBS inhibit export of assimilate from a Type 1 leaf.

Although this evidence convincingly establishes the fact of symplasmic loading, major features of its operation remain unclear. Because there is no known mechanism for active solute transport through plasmodesmata against a concentration gradient, the increase in osmotic concentration that occurs during phloem loading in some Type 1 species is especially problematic. However, the issue of concentration gradients is a centrally important point for which few systematic data on Type I species are available. To date, a concentration increase has been demonstrated convincingly only for RSO translocators (see below). In several other Type I species, which translocate mostly non-RSO sugars, total osmotic concentrations are similar in the leaf mesophyll and minor vein SE/CC

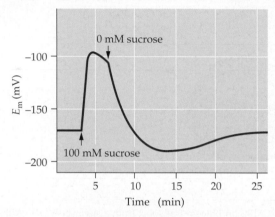

Figure 15.44
Supplying sucrose via the apoplast causes a sharp depolarization of the sieve tube membrane potential, suggesting a H⁺-sucrose cotransport mechanism. An exuding aphid stylet was used to monitor the sieve tube membrane potential in a strip of bark from a willow branch. Introduction of 100 mM sucrose into the solution bathing the cambial surface was followed several minutes later by sucrose-free solution.

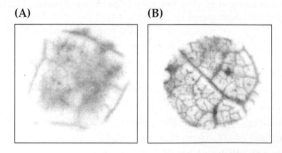

Figure 15.45
In leaves of *Coleus*, a Type 1 species, [¹⁴C]sucrose supplied via the apoplast is first taken up by the mesophyll and then redistributed to the minor veins. (A) Autoradiograph taken after a 2-minute uptake period followed by a 10-minute wash at 1°C to remove nonabsorbed ¹⁴C while minimizing postuptake redistribution. Blackening, which indicates the presence of radioactivity, shows that most of the ¹⁴C is in the mesophyll. (B) Autoradiograph taken after a 2-minute uptake period followed by a 10-minute wash at room temperature. Here, the ¹⁴C initially in the mesophyll has moved to the minor veins.

complex. In *Populus* (Fig. 15.46) and *Salix*, which are both sucrose translocators, mesophyll and phloem cells plasmolyze at similar concentrations. *Salix* plants showed no microautoradiographic evidence of [¹⁴C] sucrose accumulation in the minor veins—results that contrast sharply with observations on Type 2 species. Work with peach, another species with the Type 1 minor vein configuration, showed that concentrations of sorbitol and sucrose in the leaf mesophyll were similar to those in phloem exudate. However, evidence in this case was mixed because assimilate export may have been inhibited by PCMBS.

Although evidence is still meager, these observations suggest that Münch's original perception of an uninterrupted symplasm from photosynthetic cells to the sieve tubes may be applicable to a considerable number of species. The term "phloem loading," if interpreted to imply active solute accumulation by the SE/CC complex, would not apply to such species.

The greatest mechanistic challenge for symplasmic loading lies in those cases in which movement into the SE/CC complex occurs against a concentration gradient, as is reported for RSO translocators. For these species, a "polymer trapping" concept has been proposed to account for the major characteristics of phloem loading. In this model (Fig. 15.47), sucrose is synthesized in the mesophyll and moves via plasmodesmata to intermediary cells (specialized companion cells) in the minor veins. There, sucrose is used to synthesize the larger RSO molecules. The larger size of the RSO sugars prevents them from moving symplasmically from the intermediary cells to the mesophyll, but they can pass through the much larger channels to the sieve element. Instead of energized membrane transport or active transport through plasmodesmata (a proposed mechanism for which there is no known evidence), translocated sugars are concentrated in the sieve tubes by the exergonic reactions used to synthesize the RSOs. This model receives strong support from the finding that RSOs are synthesized in the intermediary cells, where their measured concentrations are much higher than in mesophyll cells, and from the demonstration of dye coupling between intermediary cells and the mesophyll.

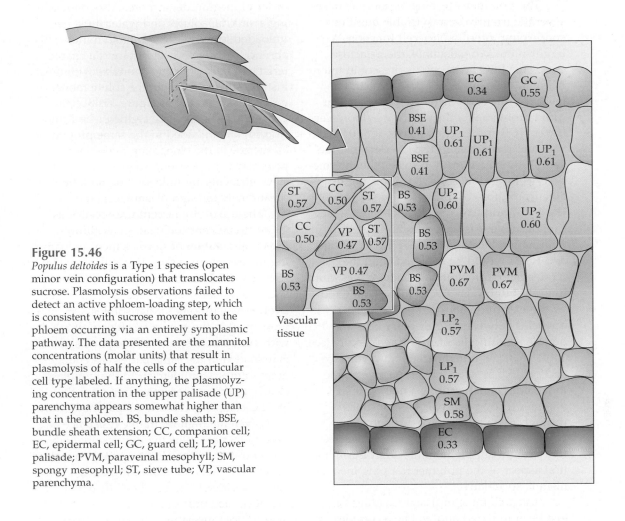

Figure 15.46

Populus deltoides is a Type 1 species (open minor vein configuration) that translocates sucrose. Plasmolysis observations failed to detect an active phloem-loading step, which is consistent with sucrose movement to the phloem occurring via an entirely symplasmic pathway. The data presented are the mannitol concentrations (molar units) that result in plasmolysis of half the cells of the particular cell type labeled. If anything, the plasmolyzing concentration in the upper palisade (UP) parenchyma appears somewhat higher than that in the phloem. BS, bundle sheath; BSE, bundle sheath extension; CC, companion cell; EC, epidermal cell; GC, guard cell; LP, lower palisade; PVM, paraveinal mesophyll; SM, spongy mesophyll; ST, sieve tube; VP, vascular parenchyma.

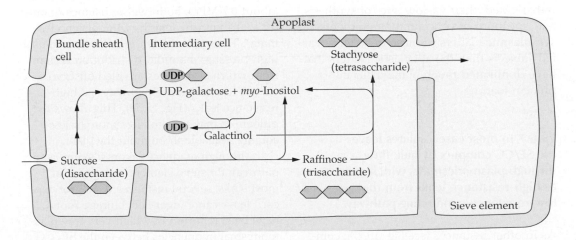

Figure 15.47

The "polymer trapping" model proposes that raffinose and stachyose accumulation in the SE/CC complex results from a symplasmic loading process. According to the model, sucrose (and in some plants, galactinol) moves symplasmically into intermediary cells (a specialized type of companion cell) and is used in the synthesis of raffinose and stachyose. These compounds can move into the sieve element but are too large to diffuse back into the mesophyll.

The polymer trapping hypothesis raises several currently unanswerable questions concerning intermediary cell function. A large amount of galactinol, the galactose donor for RSO synthesis, is present in intermediary cells but is not translocated; the basis for this compartmentation is unclear. More problematic, the size of the fluorescent tracers (CF and Lucifer Yellow, $R_s = 0.61$ and 0.68 nm, respectively) used to demonstrate symplasmic continuity between the intermediary cells and mesophyll is indistinguishable from those of raffinose and stachyose ($R_s = 0.57$ and 0.67 nm, respectively), which cannot move. Also, no substantial size discrimination between sucrose ($R_s = 0.47$ nm) and RSOs would be achieved until the plasmodesmal permeability to sucrose was quite low, although this effect may be countered by the exceptionally high plasmodesmal frequencies between intermediary cells and bundle sheath cells of RSO translocators (see Table 15.3). As noted earlier, however, (see Section 15.5.5) selectivity in passive solute movement is not entirely a matter of size, whether during symplasmic transport, through gap junctions, or through some porins. The extent to which chemical selectivity might supplement the size discrimination aspect of the polymer trapping mechanism is speculative.

Research on symplasmic loading has just been initiated and will be a complex undertaking. Given the mix of sugars translocated, their various sites of synthesis, and questions of concentration gradients, symplasmic loaders promise to be more varied subjects than the apoplastic loaders that have dominated much of the focus in phloem loading.

15.6.7 In most cases, solutes leave the SE/CC complex by bulk flow through plasmodesmata, which serve as high-resistance leaks from the low-resistance conducting pathway.

As in other instances (see Fig. 15.27), comparison between known rates of plant membrane transport and actual efflux rates of water and solutes from sieve tubes suggests a symplasmic pathway for sieve tube unloading, because the efflux rates typically exceed membrane transport rates by an order of magnitude or more. Also, on reaching a sink, all solutes and water must be unloaded from the sieve tube terminus. This absence of transport selectivity is a characteristic of bulk flow and contrasts with most instances of transmembrane solute movement. As illustrated in Figure 15.48, even exogenous solutes (here L-glucose and fluorescein) are unloaded in the same proportion as sucrose in the elongating region of the pea meristem—a strong suggestion that the sieve elements are unloaded by bulk flow, presumably through plasmodesmata.

The use of fluorescent tracers such as CF or fluorescent conjugates has allowed direct observation of sieve element unloading in several sinks (Fig. 15.49; also see Figs. 15.51 and 15.59 for unloading of dextrans and green fluorescent protein [GFP], respectively). As with other polar solutes, CF movement in the transport phloem proceeds with very little loss; on arriving in a sink, it is unloaded in the same pattern as endogenous solutes. Confinement to the symplasm clearly indicates that unloading from the SE/CC complex occurred via plasmodesmata.

Given the high hydraulic resistance of plasmodesmata, plus the universally high concentration of the SE/CC contents compared with that of surrounding cells, a large pressure differential may be anticipated for the sieve tube unloading step. This differential has been measured for wheat grains (approximately 1.0 MPa) and barley root tips (about 0.7 MPa), both over a distance of only 0.2 to 0.4 μm (the length of the plasmodesmata). Thus, the phloem may be viewed as a high-pressure manifold distribution system from which assimilates are bled off from a central low-resistance pathway via high-resistance leaks (Fig. 15.50). This allows the phloem to make assimilates more or less equally available throughout the plant.

Although evidence favors a symplasmic pathway for sieve element unloading in most sinks, several instances (e.g., growing corn leaves and sugar beet storage roots) have been reported in which very low plasmodesmal frequencies between the SE/CC complex and surrounding cells of an active sink argue strongly against symplasmic unloading. Also, some parasitic plants such as dodder (*Cuscuta*) and some fungi (e.g., endomycorrhizae, powdery mildew) are intense sinks but do not establish symplasmic

(A)

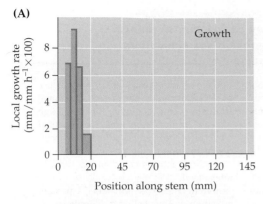

(C)

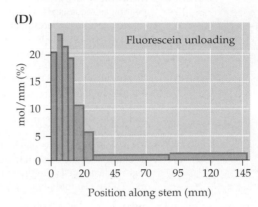

(B)

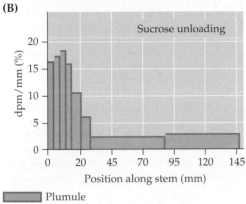

Plumule
Stem

(D)

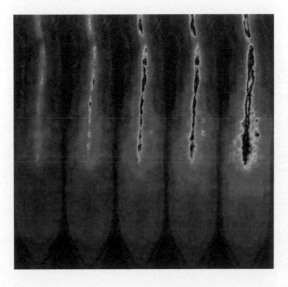

Figure 15.48
Experiments with elongating pea seedlings demonstrate that phloem unloading of sucrose (B) is tightly coupled to the local growth rate (A). The unloading of L-glucose (C) and fluorescein (D), compounds not normally present in the phloem, follows a similar pattern, indicating that sieve tube unloading is entirely nonspecific, as would be expected for bulk flow. The compounds were introduced as solutions applied to the cotyledons.

connections with their host. These observations suggest that the rates of transmembrane solute and water movement are exceptionally high for these sieve elements. Either unknown plant membrane transport carriers or, perhaps more likely, porin-like aqueous channels might be capable of accommodating substantially higher fluxes. Incidentally, because apoplastic solutes will lower cell turgor, the issue of symplasmic versus apoplastic unloading has important implications for the control of turgor pressure within a sink.

15.6.8 Plasmodesmata involved in sieve element unloading and postphloem transport appear to be substantially more conductive than most plasmodesmata.

Especially in the sieve element unloading step and in the immediate vicinity of the sieve tubes, expected assimilate fluxes along the postphloem pathway may be intense. In several instances, the expected flux is sufficiently high to cast doubt on the capacity of the symplasm to accommodate it. However,

most estimates of symplasmic conductance have been extrapolated from systems in which the plasmodesmal MEL is the conventional 800 Da. Estimates of MELs for sieve element unloading and postphloem transport are substantially higher, suggesting a correspondingly greater symplasmic conductance. When F-dextrans are injected into the

Figure 15.49
In favorable material, the use of fluorescent tracers allows real-time observation of phloem unloading. Confocal fluorescence images, taken in false color at 6-minute intervals, illustrate the unloading of CF in an *Arabidopsis* root tip.

Apical meristems are particularly active sinks. Assimilates must be delivered to tissues engaged in rapid elongation. To meet this demand, protophloem sieve elements, capable of elongating in pace with surrounding cells, differentiate to within a

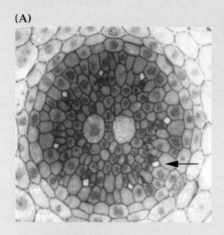

(A)

few millimeters or less of shoot and root tips. In cross-section (panel A), they appear as symmetrically placed empty cells (arrow), often contrasting sharply with the surrounding active, densely cytoplasmic cells of the meristem. Frequently, they lack companion cells. In fully elongated regions, protophloem sieve elements are usually soon crushed, disappearing as recognizable entities.

It is easy to underestimate the importance of such featureless and ephemeral cells. However, they merit special attention as a focal point for intense unloading of assimilates in apical meristems. They are equally intriguing for reasons of basic cell biology: Despite losing their nuclei and protein synthetic machinery as they start to conduct (see red shading in panel B), protophloem sieve elements continue to elongate at the same rate as surrounding nucleate cells, frequently reaching more than 10-fold their original length

(see red shading in panel C). Neither use of stored materials nor growth by simple stretching can account for the large increase in membrane area, for the initial contents of the enucleated cells are insufficient to support the observed extent of elongation, and the cell diameter remains unchanged.

What accounts for sieve element growth in its enucleate state? Similar questions are posed by the necessity of cell maintenance in other sieve elements, especially those that are long lived (to several decades). There, the answer seems to lie in the ongoing exchange of proteins between sieve elements and surrounding cells (see Section 15.7.1). Presumably, such exchange could provide the basis for macromolecular maintenance of the sieve element function, including replacement of plasma membrane components. However, the demands for these materials must be much greater in the protophloem.

(B)

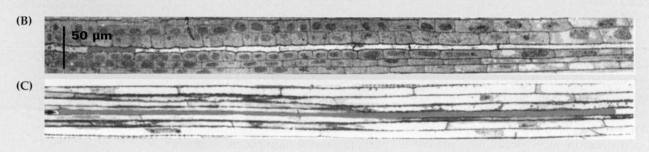

50 µm

(C)

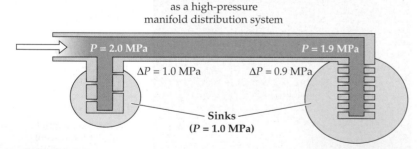

Figure 15.50
In most sinks, the sieve tube sap concentration is substantially higher than that of the surrounding parenchyma cells into which assimilates are unloaded. Consequently, the pressure drop at this step is far greater than that along the sieve tubes. In effect, this makes the phloem a high-pressure manifold distribution system in which the pressure differential for unloading should be fairly similar even at widely separated sinks.

sieve tubes of wheat grains via aphid stylets, 3-kDa dextran ($R_s = 1.1$ nm) unloads from the sieve tubes and moves promptly along the postphloem pathway of the maternal tissues (Fig. 15.51A). Also unloaded is 10-kDa dextran ($R_s = 2.2$ nm) (Fig. 15.51B), which shows extensive but slower subsequent mobility (Fig. 15.51C). Thus, the channel diameter for sieve tube unloading and the postphloem pathway in this system is about 6 to 7 nm. Small proteins, too, have been shown to be mobile in this and other sinks (see Section 15.7.3 and Fig. 15.58). Finally, the MEL_{dex} for SE/CC exchange in *Cucurbita* and *Vicia* phloem is also unusually high, at about 10 to 20 kDa.

Given such large MELs and the generally higher assimilate concentration in and near the sieve tubes, diffusive and

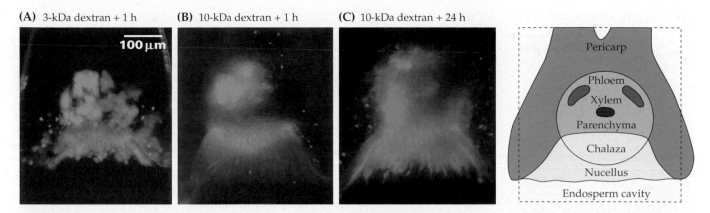

(A) 3-kDa dextran + 1 h   **(B)** 10-kDa dextran + 1 h   **(C)** 10-kDa dextran + 24 h

100 µm

Pericarp
Phloem
Xylem
Parenchyma
Chalaza
Nucellus
Endosperm cavity

Figure 15.51

Fluorescence micrographs showing the unloading of F-dextrans from sieve tubes and subsequent postphloem movement in the maternal tissues of a wheat grain. The dextrans were pressure-injected into sieve tubes via severed aphid stylets and the grains were fresh-sectioned within 1 hour after injection (A and B) or 24 hours later (C). Both dextrans were unloaded from the phloem, but only the smaller one (3 kDa; R_s = 1.1 nm) moved readily along the postphloem pathway (A). Immediately after injection, most of the 10-kDa dextran (R_s = 2.2 nm) remained close to the phloem (B), although there was limited movement to the nucellus. Considerably more movement of 10-kDa dextran to the nucellus occurred in the grains sectioned 24 hours after injection (C).

(especially) convective solute transport will be much greater than for an otherwise equivalent pathway with a MEL of only 800 Da.

15.6.9 After unloading from the SE/CC complex via plasmodesmata, assimilates follow a symplasmic pathway in most sinks, although a later apoplastic step may intervene.

Several approaches have been used to identify the pathway taken by assimilates after leaving the SE/CC complex. Plasmodesmal frequencies in a wide variety of sinks are consistent with postphloem movement via the symplasm. Fluorescent tracers unloaded symplasmically from the sieve tubes allow direct observation of subsequent movement along the postphloem pathway (see Figs. 15.49 and 15.51). Similarly, microautoradiographs have demonstrated strong labeling of the symplasm during the postphloem movement of [^{14}C]sucrose into tissues of the developing wheat grain (Fig. 15.52).

Import into several sinks (e.g., expanding leaves) is insensitive to apoplastically applied PCMBS, arguing against a role for solute absorption across the plasma membrane, a required step in movement of assimilate via the apoplast. In shoot meristems, the rate at which elongating cells can absorb sugars from external solutions is only a small percentage of the rate at which sugar is used for growth, and external sugars do not reduce the rate of endogenous sucrose import via the phloem. Finally, with many sink tissues, washout experiments (i.e., immersion of the tissue in a solution) fail to detect unusual amounts of apoplastic sugars.

In far fewer cases, the sink apoplast may be a significant pathway for cell-to-cell movement of assimilates. Washout experiments show that substantial amounts of sugars are typically present in the apoplast of high-sugar sinks, such as some fruits (e.g., tomato, grape), sugar cane, and sugar beet roots. Although the site of movement into the apoplast has not been clearly established (i.e., whether movement from the symplasm occurs directly from sieve elements or at some later step is not known), the sugar-uptake rates of sugar beet, sugar cane, tomato, and strawberries are sufficient to account for the accumulation rates seen in intact plants.

15.6.10 Discontinuities in the symplasm or apoplast affect solute movement and water relations within sink tissues.

One of the most important modifications to Münch's concept of symplasmic transport is the presence in many sinks, especially fruits and seeds, of discontinuities in the symplasm or apoplast. Both types of discontinuities substantially affect solute and water movement within the sink tissues.

(A) (B)

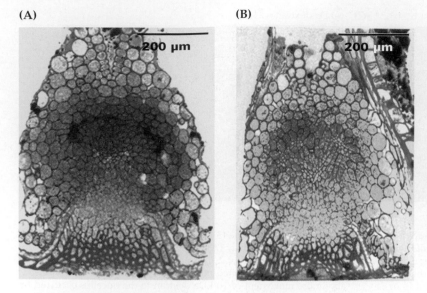

Figure 15.52
Postphloem movement of assimilates via the symplasm can also be demonstrated by microautoradiography. In this experiment, the flag leaf of a wheat plant was pulse-labeled with $^{14}CO_2$, and the maternal tissues of the grain were sampled shortly after arrival of [^{14}C]sucrose in the ear (A) and 90 minutes later, at the end of the pulse (B). Note the near absence of labeling in the phloem in (B), indicating the importance of bulk flow in sieve tube unloading in comparison with diffusion. Also, note the absence of labeled sucrose in the pericarp (top and peripheral cells).

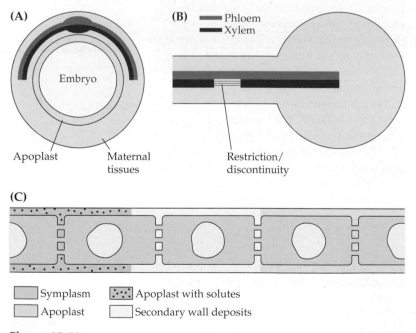

Figure 15.53
Discontinuities in the symplasm or apoplast have important consequences for water and solute movement within a sink. (A) The symplasm is always discontinuous between embryonic and maternal tissues. (B) The xylem can be discontinuous, or nearly so, in the fruiting stalk (e.g., tomato, wheat and barley grains) or just within the fruit (e.g., grape, apple). (C) By limiting apoplastic solute movement, secondary wall deposits can result in locally sharp water potential gradients (e.g., wheat, barley, secretory glands).

The universal absence of plasmodesmata between maternal and embryonic tissues necessitates solute release, followed by uptake, across two membranes in series (Fig. 15.53A). The solution in the intervening apoplastic compartment is close to atmospheric pressure and serves as the nutrient medium for embryo growth.

Apoplastic discontinuities may occur in xylem vessels or in cell walls. Typically, water potentials differ sharply in the tissues separated by the discontinuity, with important implications for sieve tube and symplasmic turgor gradients across the discontinuity. Xylem vessels (Fig. 15.53B) may be morphologically discontinuous, as in the pedicel of some cereal grains, such as wheat and barley, or the funiculus of legume ovules. In other cases, they may be present but offer a high resistance to flow, as in the tomato pedicel or the stretched and partially collapsed vessels that develop in some fruits as they enlarge (e.g., grapes and apples). A restricted xylem connection provides a considerable degree of hydraulic isolation from the transpirationally induced fluctuations in water potential experienced by other parts of the plant. Diurnal variations in water potential are nearly absent in tomato fruit, compared with the 1 MPa variation in leaves on the same plant (Fig. 15.54A). Potato tubers, which have no xylem restriction, track leaf water potential more closely (Fig. 15.54B), although the effect is distinctly dampened by the water storage capacity of the tuber. Within different parts of the same fruit, the degree of hydraulic isolation may differ, as in legumes and citrus, where the pod and rind, respectively, have a xylem supply but the ovules and juice sacs do not.

Discontinuities in the cell wall pathway result from secondary wall deposits of various compositions (see Fig. 15.53C). By restricting solute movement, differences in apoplast solute and matric potentials, as well as water potential, may be sustained across the discontinuity. In sinks with high apoplastic solute concentrations, an apoplastic barrier may prevent solutes from diffusing into the xylem, as in sugar cane stems. More important, in the context of assimilate movement within a sink, water potential gradients in the apoplast will affect turgor gradients in the symplasm. Figure 15.55 illustrates the apoplastic discontinuity in the

(A)

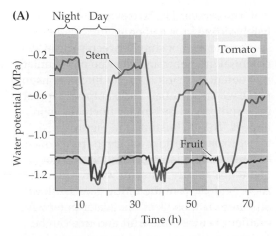

(B)

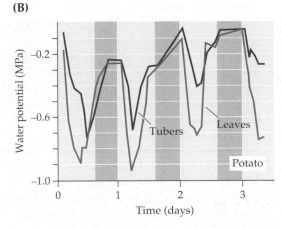

Figure 15.54
In the absence of a strong xylem connection, the water potential of a sink can be relatively independent of changes in plant water potential. (A) Tomato fruits have a very restricted xylem connection in the fruit pedicel. As a result, changes in tomato fruit water potential are only about one-tenth of that in the vegetative portions of the plant. (B) With no restriction to the xylem pathway, changes in potato tuber water potential almost match those in the leaves.

chalazal cell walls of wheat grains between the vascular tissues (with a water potential of approximately –1.1 MPa) and endosperm cavity (water potential approximately –0.7 MPa). Because the symplasmic solute concentration decreases from the phloem toward the endosperm cavity, the difference in water potential mitigates the turgor gradient that would otherwise occur along the postphloem pathway, decreasing the relative importance of bulk flow as the basis for transport. This is an important consideration in the tightly regulated water relations of grains, for which phloem import is the only source of water.

15.6.11 Control of assimilate import by a sink must coordinate sieve element unloading with the characteristics of the particular sink type.

Given the diversity of sink types and the present state of uncertainty about possible linkages between transport and solute utilization within a sink, few generalizations are possible about the control of assimilate import. Just as in a puzzle, some pieces provide stronger organizational focus than others, and the apparent fit of yet other pieces may be misleading.

Mass action principles, which presume a direct causal relationship between concentrations and reaction rates (e.g., between a

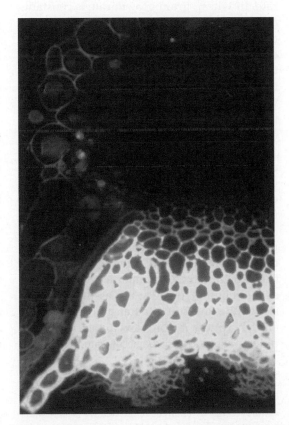

Figure 15.55
Cell wall deposits may divide sink tissues into apoplastic "domains" with markedly different water relations. Here, movement of a fluorescent tracer (Lucifer Yellow) introduced into the endosperm cavity of a wheat grain is blocked at the chalaza. Because the apoplast is blocked, assimilates moving from the phloem to the endosperm cavity must follow a symplasmic pathway. More important, the discontinuity in the apoplastic pathway allows the maintenance of quite different water potentials in the maternal and embryonic tissues (typically about –1.1 and –0.7 MPa, respectively). Clearly, this has important implications for turgor gradients and, consequently, for bulk flow between the two.

direct casual relationship between concentrations and reaction rates (e.g., between concentrations of photosynthate and rates of growth), appear to fall into the latter category. For the most part, the supposed relationships have not been quantified, and the few measurements available do not support the operation of mass action principles. In one of the few direct tests of the idea, soybeans were grown in a range of light intensities and CO_2 concentrations to vary the rates of seed growth. Sugar concentrations in the seed coat and cotyledons were constant over most of the range of seed growth rates; as seed growth rate increased, the flux of sugars through the pools also increased, despite the constant concentration of these intermediate sugar pools. Only at the slowest growth rates was there an apparent relationship between sugar concentration and growth rate. In wheat, grain growth rate is constant, despite substantial variations in sucrose concentrations in sieve tube and endosperm cavity sap. Nonetheless, more critical evaluations of possible mass action effects in partitioning are needed. The growth rate and water relations of seeds are strongly regulated; vegetative sinks may respond differently, as may ovules during differentiation or fertilization. However, some genes are responsive to variations in sugar concentrations. Simple mass action effects must be distinguished from responses to sugar-induced gene expression.

Maintenance of low sink water potential, presumed to steepen the turgor gradient toward the sink end of sieve tubes, has also been suggested as an important factor in sink strength. However, this assumes that sieve tube unloading, a transport step entirely different from flow along the phloem, should also be accelerated by low water potential. The two steps are not obligately linked, nor is there evidence of any systematic relationship between plant water potential gradients and the rate of assimilate movement to a sink. Diurnal fluctuations in plant water potential, for example, may not be accompanied by similar fluctuations in import. This is illustrated in Figure 15.56 for phloem import by an apple fruit, which, because of its relative hydraulic isolation, maintains a fairly constant water potential. As is evident from the inflow and outflow of water via the xylem, the water potential gradient between the fruit and tree can be reversed without apparent effect on phloem import by the fruit. Similar measurements have been made with tomatoes and grapes. In general, there are many instances of rapid phloem transport from low to high water potential, especially for transport to reproductive structures in salt- or water-stressed plants. This is not unexpected because the direction of flow in the phloem will be toward the site of a leak (sink).

In attempting to analyze the possible controls involved, it is useful to view assimilate import by a sink as three basically sequential steps: sieve element unloading, postphloem transport, and utilization as a substrate or osmoticum. In most cases, sieve element unloading proceeds via bulk flow through plasmodesmata. Because of the large pressure differential involved (see Section 15.6.7), this first step is essentially irreversible; subsequent steps of assimilate import can influence partitioning only by their effect on sieve element unloading. Clearly, the number and hydraulic conductance of the aqueous channels involved and their location within the plant body must be crucial determinants of assimilate partitioning.

Figure 15.56

Continual monitoring of apple fruit growth shows that the rate of phloem import was unaffected by changes in the relative water status of the fruit and tree. Sensitive displacement transducers were used to monitor fruit diameter, from which fruit volume changes could be calculated. Total volume change, or growth (G), is the sum of volume changes due to xylem flow (X), phloem import (P), and transpiration (T); that is, $G = (X + P + T)$. T was determined from detached fruit, and steam girdling of fruit pedicels eliminated phloem import, providing for determination of (X + T). Then P and X could be determined: $P = G - (X + T)$, and $X = (X + T) - T$. Over a 24-hour period, fruit water potential was greater than tree water potential (X < 0) during much of the day and lower (X > 0) during the night, but phloem import was unaffected by the change in gradient.

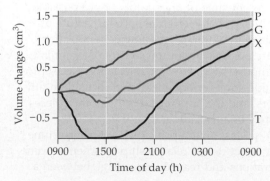

The anatomy and biochemistry of post-phloem transport and assimilate utilization are as diverse as the different types of sinks and are linked intimately to developmental regulation of the particular sink type. Perhaps the most useful clues to the control of assimilate fluxes in sinks come from the frequent invocation of osmotic or turgor regulation, or both, to explain various aspects of sink physiology (Table 15.4). Especially in seeds and meristems, most of the solutes involved are imported assimilates undergoing rapid turnover. In these sinks, osmotic or turgor control is necessarily a dynamic process with important implications for assimilate import. For effective osmotic or turgor regulation, rates of import must balance rates of solute utilization; changes in one must be coordinated with changes in the other.

Variation of import rates in space and time provides some insight into the degree of coordination between sieve tube unloading and assimilate utilization but does not reveal the mechanism. Figure 15.57 illustrates the local elongation rate, turgor pressure, and osmotic pressure in unstressed corn roots and in roots 24 hours after transfer to a solution of 400 mM mannitol. Because local assimilate unloading from sieve elements must closely track the local expansion rate, unloading in the zone of root elongation is greater than in the zones of differentiation or cell division. Because turgor in the expanding cells is constant throughout the zone, as sieve tube turgor would be over such a short distance, these changes occur without any measurable differences in the turgor differential between the sieve tubes and expanding cells. After transfer to

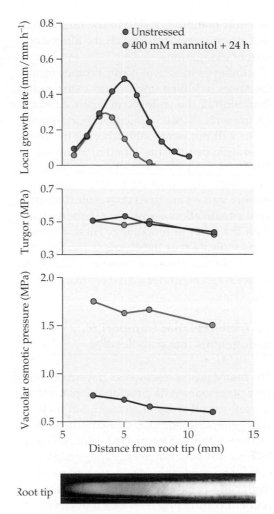

Figure 15.57
As a root grows, sieve tube unloading must be coordinated with the local expansion rate, which changes as a root segment elongates and differentiates. Turgor is constant throughout the root tip. Thus, when root growth is slowed by osmotic stress (400 nM mannitol), flow through the phloem adjusts to match the lower rate of growth without any evident change in turgor differential between the sink cells and sieve tube.

Table 15.4 Osmotic or turgor regulation is an important characteristic of many sinks

Sink	Osmotic or turgor regulation
Developing embryos (many species)	Normal development requires constant apoplast osmotic concentration.
Legume seed coat	Sucrose release rate is turgor-dependent.
Corn root tip, leaf meristem	Turgor is constant throughout the meristem and is unaffected by changing growth rates, water stress.
Sugar beet tap root	Turgor remains constant during sugar accumulation.
Sugar cane stem	Turgor remains constant during sugar accumulation.
Potato tuber	Starch synthesis and sugar uptake are turgor-dependent.

400 mM mannitol, a shift in the unloading pattern must occur to match the altered pattern of elongation.

Other evidence of control comes from instances in which import into a sink ceases. Perfusion of the endosperm cavity of wheat grains with PCMBS inhibits import into the grain without perceptibly altering the concentration or pressure differential (about 1 MPa) between the sieve tubes and surrounding cells. In general, cessation of import into a sink does not result from pressure equalization across the plasmodesmal path involved in unloading. These observations indicate that the channels for sieve tube unloading are not passive conduits but rather must be under active control.

15.7 Intercellular transport of endogenous macromolecules

For many years, researchers have recognized that plasmodesmata provide the pathway for cell-to-cell virus movement, but this was generally regarded as a pathological condition limited to infected cells. Nonetheless, some observations suggested that endogenous macromolecules could move from cell to cell in healthy plants. Although phloem conducting cells lack a nucleus and ribosomes, phloem exudates consistently contain at least a small amount of protein, even when pressure-release artifacts are minimized by collecting exudate

from aphid stylets. Furthermore, it seemed increasingly likely that viruses moving from cell to cell must usurp already-existing mechanisms for macromolecule transport through plasmodesmata rather than relying entirely on viral-encoded information. Trafficking through nuclear pores (see Chapter 4) presents an intriguing model of how plasmodesmata might function in the selective transport of molecules larger than passively retained by the MEL.

15.7.1 The ongoing presence of soluble proteins in the sieve tube requires continual entry of proteins in the source and their removal at the sink; additional turnover occurs by SE/CC exchange along the path.

Except for the cucurbits (see below), all phloem exudates contain low concentrations (0.2 to 2 g l^{-1}) of protein. To date, **sieve tube exudate proteins** (STEPs) from only a few species have been examined in any detail. Despite their low concentrations, a surprisingly large number of proteins (100 to 200) are present, many of which become labeled on application of [^{3}H] amino acids or [^{35}S] amino acids to a source leaf (Fig. 15.58). Because the concentration and composition of exudate change but little between source and sink, evidently the entire protein complement must be loaded at the source and unloaded at the sink. In addition, some exudate proteins are labeled when [^{35}S]methionine is

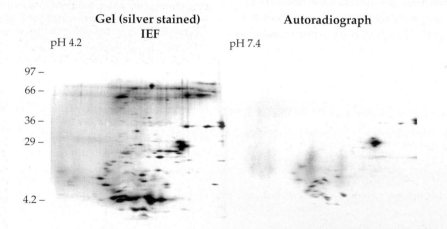

Gel (silver stained)
IEF

pH 4.2

Autoradiograph

pH 7.4

97 –
66 –
36 –
29 –

4.2 –

Figure 15.58
Two-dimensional separation of soluble proteins in aphid stylet exudate collected from wheat plants. Exudate was collected for 5 days; [^{35}S]methionine was fed to the plant on the second day to label the proteins. Many of the exudate proteins have an isoelectric point more basic than 7.4 and so are not retained in the focusing gel used here for the first dimension. Molecular masses (kDa) are listed at left. IEF, isoelectric focusing.

applied only to the phloem path. Thus, some protein exchange between sieve elements and companion cells occurs along the path, with concomitant degradation and resynthesis in companion cells. The unequal participation of different STEPs in this process indicates substantial selectivity in SE/CC exchange or in turnover within CCs.

At 40 to 100 g l^{-1}, the protein concentration in cucurbit phloem exudate is exceptionally high and is clearly an artifact of cutting, because cucurbit fruits are relatively low in protein. Such a high concentration of proteins raises the question of their presence as soluble proteins in undisturbed sieve elements. However, interspecific grafting experiments between cucurbit partners that synthesize different structural phloem proteins **(P-proteins)** have shown that these proteins are phloem-mobile. P-proteins characteristic of one partner were present in exudate from the other, thus circumventing interpretive problems arising from their high concentration in phloem exudate.

15.7.2 STEPs can increase the plasmodesmal MFI and mediate their own cell-to-cell movement.

Coinjection of STEPs and F-dextrans into mesophyll cells has shown that most STEPs are capable of gating plasmodesmata, increasing the MEL_{dex} from a basal value of 1 kDa to values in the range of 20 to 40 kDa. Fluorescein-STEP conjugates (F-STEPs) are themselves also mobile when injected. STEPs covering a 10-fold range of M_r are about equally effective in gating plasmodesmata. These effects of STEPs occurred at concentrations well below that in sieve tube exudate and were not species-dependent: Selected STEPs from *Ricinus* and *Cucurbita* mediated movement in the leaf mesophyll of four plant species. Thus, STEPs appear to have a broad ability to interact with plasmodesmata, either directly or in combination with an endogenous cofactor.

15.7.3 Some non-STEP proteins can also be unloaded from sieve tubes and move along the postphloem pathway.

The full significance of the above findings for intercellular protein movement is not yet clear, for some non-STEP proteins have also been observed to enter and exit sieve tubes and to move along the postphloem pathway. Phloem-specific expression of snow drop lectin (a storage protein in *Galanthus* bulbs; mass of monomer, 13 kDa) in tobacco resulted in the appearance of the lectin in the honeydew of aphids feeding on the tobacco. Similarly, green fluorescent protein (27 kDa) expressed in the phloem of mature *Arabidopsis* leaves was exported to sink tissues (young leaves, root tips, and ovules), unloaded from the sieve tubes, and transported along the postphloem pathway (Fig. 15.59). GFP injected into sieve tubes via severed aphid stylets on wheat grains also was unloaded and moved along the postphloem pathway. Because neither the lectin nor the GFP would be expected to contain a plasmo-desmal recognition factor, the cell-to-cell movement of these proteins presumably results from the higher MELs that are apparently characteristic of the phloem unloading

Figure 15.59
Green fluorescent protein (GFP) expressed in mature *Arabidopsis* leaves under a phloem-specific promoter (A) is exported to sinks, where it is unloaded from the sieve tubes and moves along the postphloem pathway in young leaves (B) and root tips (C, D).

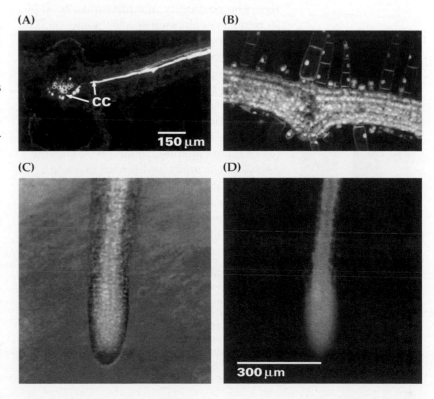

(A)   (B)
150 µm

(C)   (D)
300 µm

pathway (see Section 15.6.8). Thus, there appears to be significant potential for non-specific cell-to-cell protein movement within sink tissues.

15.7.4 STEPs identified to date include enzymes related to carbohydrate metabolism, structural proteins ("P-proteins"), and "maintenance proteins."

Because significant quantities of phloem exudate can usually be collected only by cutting the sieve tubes, reported STEP compositions must be viewed with caution. In *Cucurbita*, the drastically steepened pressure gradient caused by cutting has been shown to sweep otherwise nonmoving companion cell dehydrogenases into the exudate. Also, surging may elicit wound responses involving protein release from companion cells. This problem is minimized with aphid stylet exudate, the flow rates of which in most cases are comparable with the normal rates of flow along sieve tubes. Comparison of exudate collected from wheat plants by wounding and from aphid stylets reveals a substantial number of differences in STEP composition, especially for larger proteins (Fig. 15.60)—although most differences appear to be quantitative rather than qualitative. The possibility of artifact must be kept in mind in evaluating specific points of STEP compositions; more confidence can be placed in stylet exudates, as well as in exudates for which compositions are steady over longer exudation times.

Because they are present in such low concentrations, STEPs do not appear to play a nutritive role, either directly or indirectly (as chelating agents). In the 1970s, enzyme assays of *Cucurbita* and *Robinia* exudate demonstrated the presence of most enzymes of the glycolytic and gluconeogenic pathways. Experiments with aphid stylet exudate

from willows demonstrated the incorporation of $^{32}PO_4^{3-}$ into organic form, including ATP and fructose 1,6-bisphosphate. Similar results were obtained with *Yucca* exudate. Thus, sieve tube contents are apparently capable of directing a small flux of sugar through these pathways—although the significance of this action is unclear, especially since these metabolites should exchange freely with companion cell solutes.

Some proteins appear as either crystalline or amorphous accumulations ("P-protein bodies") during sieve element differentiation. Though common, these structures are absent in some species, including many monocots. The amorphous, fibrillar forms appear to be in equilibrium with soluble monomeric protein, which may be fairly abundant in exudate (see Section 15.7.1). Because of their abundance in cucurbit exudate, these forms have been characterized fairly extensively. Several forms are present in exudate from a given species, and exudates collected from different organs on a plant demonstrate similar protein patterns. However, both gel patterns and antibody cross-reactivity data suggest the existence of significant differences among species. A particularly interesting aspect is the hemagglutinating (lectin) activity of some P-proteins, which has provided another useful diagnostic tool for probing taxonomic relationships; its biological or physiological significance, however, is unclear.

Recent efforts at characterizing STEPs have focused on their possible maintenance role in SE/CC interactions. Functions identified in this category include glutaredoxin, cystatin, ubiquitin, and chaperones from *Ricinus*, and thioredoxin *h* and protein kinase activity from rice.

15.7.5 Some nonviral RNAs are phloem-mobile.

The possibility that nonviral RNAs might be transported in the phloem has received convincing support only recently. Using PCR and primers based on sequences for three proteins present in rice phloem exudate, investigators demonstrated the presence of mRNAs for these proteins in the exudate. Ribonucleases were undetectable, indicating

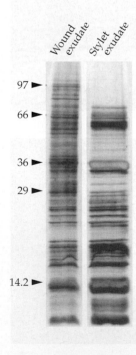

Figure 15.60
The protein composition of phloem exudates from the same plant may differ substantially, depending on the collection method. Here, exudate collected from aphid stylets on a wheat peduncle is compared with exudate collected from the stumps of grain pedicels after the grains were broken off. The values at left are molecular weights (kDa).

that free RNAs should be sufficiently stable to move long distances in the phloem.

In a less direct approach, antibodies to the movement protein of red clover necrotic mosaic virus were used to identify a cross-reacting endogenous protein in pumpkin phloem exudate. The protein was sequenced and shown to mediate both its own cell-to-cell movement and that of its mRNA in leaf mesophyll. Long-distance movement within the phloem was demonstrated by grafting cucumber (scion) onto pumpkin (stock) and collecting exudate from the cucumber phloem (Fig. 15.61). Both the protein and its mRNA were present in exudate from the cucumber scion and in phloem exudate from pumpkin plants. Neither was detected in phloem exudate from nongrafted cucumbers. Thus, their presence in exudate from the heterografted cucumber can be attributed to phloem transport from the pumpkin stock.

A strong candidate for the systemic movement of a nonviral RNA comes from investigation of the cosuppression (see Chapter 7) of the nitrate reductase gene in tobacco (i.e., posttranscriptional silencing of both host and transgene nitrate reductase loci). The cosuppressed state could be monitored visually, because silencing of the gene resulted in chlorosis of the leaves. In a phenomenon termed "systemic acquired silencing," a nonsilenced scion grafted onto a silenced stock became silenced itself as phloem continuity was established between stock and scion (Fig. 15.62). The signal could be transmitted through 30 cm of a nontransgenic interstock segment, clearly indicating movement in the vascular system. Although the identity of the signal involved has not been established unequivocally, previous work on cosuppression strongly suggests that it is some form of RNA produced by the transgene.

15.7.6 Cell-to-cell trafficking of endogenous macromolecules in nonphloem tissues has been demonstrated only recently and holds fundamental implications for plant development.

Although the cell-to-cell movement of STEPs within the phloem holds intriguing implications for developmental control, only limited evidence to date (see Section 15.7.4) suggests functions beyond those entailed by the maintenance of enucleate sieve elements. However, short-distance intercellular macromolecule movement between cell layers of meristems has been shown to influence organ

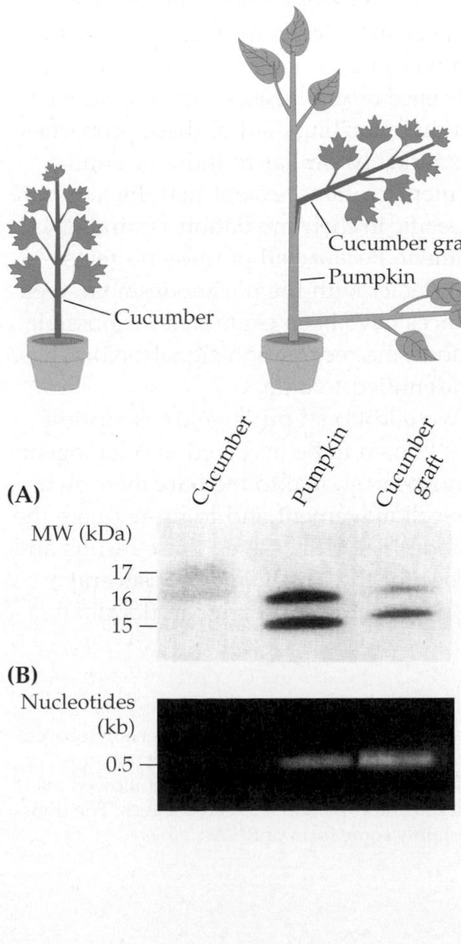

Figure 15.61
Grafting experiments between pumpkin and cucumber demonstrate that a pumpkin-synthesized phloem protein (CmPP16; *Cucurbita maxima* phloem protein, 16 kDa) and its mRNA are transported via the phloem from a pumpkin stock to a cucumber (*Cucumis sativus*) scion unable to synthesize either. Exudate collection sites are indicated in the diagrams. Exudate proteins were separated by sodium dodecyl sulfate–polyacrylamide gel electrophoresis (A) and probed with antiserum to a recombinant CmPP16 from *E. coli* that reacts with two closely related forms of CmPP16. Both forms were present in phloem exudate from pumpkin and from the cucumber grafted onto the pumpkin stock but were absent from cucumber exudate. Similarly, *CmPP16* mRNA, analyzed by the reverse transcriptase polymerase chain reaction (B), was present in exudate from pumpkin and from cucumber grafted onto pumpkin stock but absent from cucumber exudate.

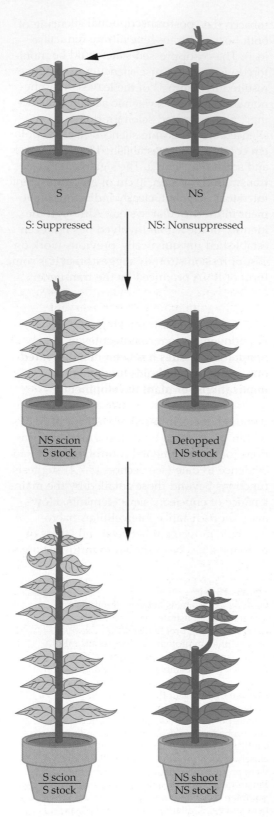

development in periclinal chimeras (plants in which the tissue layers are genetically different). Chimeras demonstrating abnormal development were in several instances controlled by only one cell layer of the developing organ, indicating the transmission of some signal(s) between layers. Although the signal could be a small molecule, several of the genes involved encode transcription factors, which themselves could be mobile. This possibility was confirmed for a maize mutant encoding the protein transcription factor KNOTTED1 (KN1), which is responsible for the knotted appearance of the leaf surface. Immunochemical localization of KN1 showed it was present in the nuclei of all cell layers, including the epidermal (L1) layer (Fig. 15.63A). In situ hybridization, however, failed to detect *kn1* mRNA in the L1 layer (Fig. 15.63B), strongly suggesting that KN1 must have been synthesized in the underlying cell layers and then transported to the dermal layer. Cell injection experiments in tobacco mesophyll with fluorescein-labeled KN1 showed that this factor was capable of mediating its own intercellular movement, with attendant up-regulation of the plasmodesmal MEL_{dex} to 20 to 40 kDa. The fluorescein-labeled factor was also capable of specifically mediating the cell-to-cell movement of its own mRNA (making the absence of *kn1* mRNA from the L1 layer somewhat puzzling). All of these properties are remarkably similar to those of viral movement proteins, except that the latter are nonspecific in their mediation of viral RNA movement. Because all of these proteins must interact with the plasmodesmata, their sequences have been examined for possible plasmodesma recognition signals. None has been identified to date.

Two additional putative transcription factors known to be involved in organogenesis have been shown to mediate their own cell-to-cell movement and to up-regulate the plasmodesmal MEL. Given these results and evidence for the involvement of several other DNA-binding proteins in floral

Figure 15.62
In a phenomenon termed "systemic acquired silencing," cosuppression of the nitrate reductase gene in tobacco is transmitted from a suppressed stock to a nonsuppressed scion. In this experiment, the apex of a nonsuppressed plant is grafted onto a suppressed stock. Transmission of the gene silencing effect can be followed visually, because suppression of the nitrate reductase gene causes marked chlorosis of the affected leaves. The transmitted signal is gene specific and, from earlier work, is almost certainly some form of RNA.

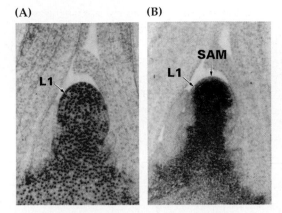

(A) **(B)**

Figure 15.63
(A) Immunolocalization of KN1 in a serial section from a maize vegetative shoot apex. KN1 is present in the epidermal layer (L1). (B) In situ hybridization for *kn1* mRNA in a serial section from a maize vegetative shoot apex. No *kn1* mRNA is detected in L1 (arrows). SAM, shoot apical meristem.

organogenesis, a class of mobile proteins, "supracellular control proteins," has been proposed as a key factor in the control of organogenesis. These exciting developments hold much promise for elucidating the molecular basis of positional information that is central to plant cell development.

Figures 15.64 and 15.65 illustrate models proposed for the trafficking of nucleic acids and proteins, respectively, through plasmodesmata (see below).

15.7.7 Many aspects of viral movement have implications for possible underlying mechanisms of endogenous macromolecule trafficking, including RNA movement.

Experiments showing that the plasmodesmal MEL_{dex} in transgenic tobacco plants expressing the tobacco mosaic virus movement protein (TMV MP) up-regulated to 10 kDa have set in motion a chain of investigations, some noted in the previous sections, that have

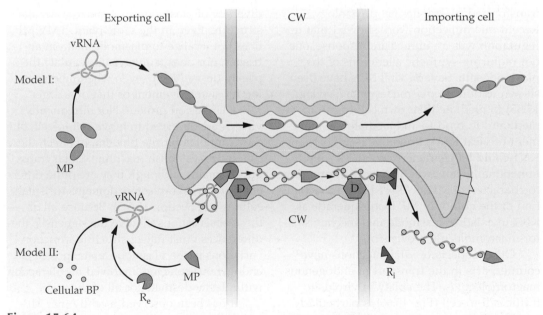

Figure 15.64
Conceptual models for the cell-to-cell trafficking of viral RNA (vRNA). Presumably, a similar process applies to the trafficking of native RNA in uninfected plants. Both models involve unfolding of the RNA as an essential step in transport. Model I: Movement protein (MP) binds to vRNA, and the MP–vRNA complex moves through the plasmodesma. Model II: In addition to MP, this model postulates the involvement of endogenous plant cell proteins in vRNA trafficking. Both endogenous binding protein (BP) and MP are included in the ribonucleoprotein complex, which may also include a cytosolic receptor protein (R_e). At the plasmodesmal orifice, a stationary docking protein (D), perhaps also involved in gating of the MEL or the unfolding of the RNA (or both), initiates the export process and the receptor is recycled. On the importing side, another cytosolic receptor (R_i) recognizes and imports the complex. CW, cell wall.

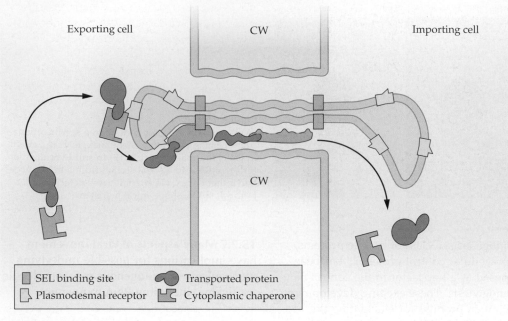

Exporting cell   CW   Importing cell

CW

▨ SEL binding site   ◕ Transported protein
◿ Plasmodesmal receptor   ⊏⊐ Cytoplasmic chaperone

Figure 15.65
A conceptual model for cell-to-cell trafficking of specific proteins of a size larger than the passive SEL. After binding to a chaperone, the protein–chaperone complex interacts with a plasmodesmal receptor site and moves to the plasmodesmal orifice. There, the complex binds to a site that up-regulates the plasmodesmal SEL. Before or upon reaching the SEL binding site, the protein unfolds to a more linear form for passage through the plasmodesmal microchannel. (Proteins smaller than the plasmodesmal SEL may move without unfolding or other specific interaction with the plasmodesma.) CW, cell wall.

transformed plasmodesmal physiology. Recent microinjection work showed that up-regulation was an immediate response, one not requiring synthetic alterations of the plasmodesma. Several viral MPs have been shown to bind to plasmodesmata (see Fig. 15.35) to mediate their own cell-to-cell movement and to bind to and mediate the movement of viral RNA. From this point of view, KN1 could be characterized as "MP-like." Functionally, however, it is more appropriate to characterize MP as "KN1-like," in recognition of the ability of MP to masquerade as KN1 in existing plasmodesmal mechanisms for macromolecule movement.

Other aspects of viral infections have counterparts in the transport of endogenous macromolecules. The ability of viroids to traffic cell-to-cell (Fig. 15.66) is particularly significant because these small, infectious RNA molecules are nontranslatable and must rely on endogenous mechanisms for cell-to-cell movement. That they can do so not only suggests that such a mechanism is present but also implies that cell-to-cell RNA movement is not unusual. Other observations on viral movement emphasize the

diversity of plasmodesmal macromolecular structure. Even in the same plant, TMV MP does not localize to plasmodesmata in all tissues, nor does it always up-regulate the plasmodesmal MEL, even when mediating its own movement or that of a larger MP–GUS fusion protein. Not infrequently, viruses are observed to move cell-to-cell, but they cannot enter the phloem. In effect, these differences establish macromolecular transport domains, although they are quite different from the symplasmic domains for small solutes (see Section 15.5.5). Because all of these observations involve the same MP, the differences must reflect structure/function variations in the plasmodesmata, possible endogenous cofactors involved in interaction with plasmodesmata, or all of these.

It has been observed recently that MP-induced up-regulation did not occur when test F-dextrans were injected by iontophoresis (current-assisted injection), casting some doubt on earlier observations made after using pressure injection. In many of the latter experiments, however, the F-dextrans moved through several cells, indicating that up-regulation was not an artifact of

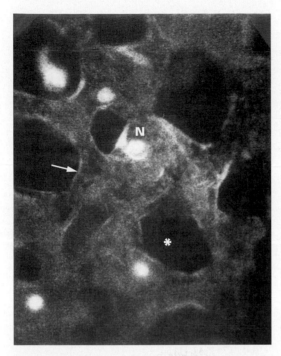

Figure 15.66
Injection of fluorescent-labeled potato spindle tuber viroid (PSTVd) into a tobacco mesophyll cell (arrow) results in prompt movement of the viroid to adjacent cells. The asterisk indicates intercellular space. Note the accumulation of PSTVd in the cell nuclei (N). Plant viroids are nontranslatable circular RNAs, and must depend on endogenous transport mechanisms for their cell-to-cell movement and replication.

pressure-induced changes in the injected cell. Other discrepancies involving current-assisted injection (iontophoresis or electrical coupling) have been noted (see Section 15.5.7). Clearly, additional systematic comparisons of injection methods are needed.

Summary

The long-distance transport of solutes in vascular plants follows two functionally distinct pathways, the xylem and the phloem, which closely parallel each other throughout most of the plant body. Volume flow in the xylem is much greater than in the phloem and is driven by the tension resulting from transpirational water loss. Solute concentration in the phloem is high; movement is driven by a turgor differential between regions of phloem loading ("sources") and unloading ("sinks"). Thus, solutes in the xylem move upward to sites of photosynthesis, whereas movement in the phloem is metabolically directed and variable in direction, depending on the relative positions of sources and sinks.

The principal translocated sugar in most crop species is sucrose. In some species (mostly herbaceous), sucrose is released into the minor vein apoplast and is accumulated in the phloem by H^+-sucrose cotransport across the plasma membranes of the sieve element/companion cell complex ("apoplastic loading"), resulting in a substantially higher concentration of sucrose in the conducting cells than in the leaf mesophyll. In other species (mostly woody), sugars follow a fully symplasmic pathway to the SE/CC complex ("symplasmic loading," which has no active accumulation step). In symplasmic loaders translocating the raffinose series of oligosaccharides, however, the RSOs are synthesized in companion cells, where they reach much higher concentrations than in the mesophyll.

In most sinks, assimilates exit the sieve tube as a bulk flow of solution via plasmodesmata. This step has a high hydraulic resistance and is accompanied by a large pressure drop, making it effectively irreversible. Controls on the location, number, and resistance of these plasmodesmal "leaks" presumably play an important role in assimilate partitioning.

Transport of solutes to and from the xylem and phloem occurs largely by cell-to-cell movement via plasmodesmata. Until recently, these structures seemed essentially static, acting as molecular sieves to prevent the movement of proteins and RNA while allowing passage of solutes smaller than 800 Da. This view continues to have some validity, but requires certain fundamental elaborations. Some proteins and RNAs can move into and out of the phloem conducting cells, which are enucleate and lack protein synthetic machinery. Plasmodesmata in the postphloem symplasmic pathway of sink tissues have MELs of at least 10 kDa, presumably an important factor for accommodating high-solute fluxes there. Several endogenous proteins, including some in phloem exudate, can up-regulate the symplasmic MEL in leaf mesophyll to about 20 kDa. Developmental coordination in meristems has been shown to involve proteins that mediate not only their own transport between cell layers but also that of their own mRNA. Observations on plasmodesmal gating implicate ATP in the regulation of the MEL, and both actin and myosin have been localized to plasmodesmata. Several approaches are yielding insight into the identity of other plasmodesmal proteins. Although their role as molecular sieves remains intact, plasmodesmata must also be viewed as dynamic structures capable of altering their MEL and of interacting with and transporting specific proteins and RNAs.

The author thanks Cora E. Cash-Clark for preparing and organizing the figures for this chapter.

Further Reading

Biophysical aspects

Curry, F.R.E. (1984) Mechanics and thermodynamics of transcapillary exchange. In *Handbook of Physiology. Section 2: The Cardiovascular System, Volume IV: Microcirculation, Part 1*, E. M. Renkin and C. C. Michel, eds. American Physiology Society, Bethesda, MD, pp. 309–374.

Finkelstein, A. (1987) *Water Movement Through Lipid Bilayers, Pores, and Plasma Membranes.* J. Wiley and Sons, New York.

Nobel, P. S. (1991) *Physicochemical and Environmental Plant Physiology.* Academic Press, San Diego.

Water movement in the xylem

Sperry, J. S. (1995) Limitations on stem water transport and their consequences. In *Plant Stems: Physiological and Functional Morphology*, B. L. Gartner, ed. Academic Press, San Diego, pp. 105–125.

Plasmodesmata

Ding, B., Turgeon, R., Parthasarathy, M. V. (1992) Substructure of freeze-substituted plasmodesmata. *Protoplasma* 169: 28–41.

Overall, R. L., Blackman, L. M. (1996) A model of the macromolecular structure of plasmodesmata. *Trends Plant Sci.* 1: 307–311.

Phloem loading

Komor, E., Orlich, G., Weig, A., Kockenberger, W. (1996) Phloem loading—not metaphysical, only complex: towards a unified model of phloem loading. *J. Exp. Bot.* 47: 1155–1164.

Turgeon, R. (1996) Phloem loading and plasmodesmata. *Trends Plant Sci.* 1: 418–423.

Ward, J. M., Kuhn, C., Tegeder, M., Frommer, W. B. (1998) Sucrose transport in higher plants. *Int. Rev. Cytol.* 178: 41–71.

Phloem unloading

Fisher, D. B., Oparka, K. J. (1996) Post-phloem transport: principles and problems. *J. Exp. Bot.* 47: 1141–1154.

Patrick, J. W. (1997) Phloem unloading: sieve element unloading and post-sieve element transport. *Annu. Rev. Plant Physiol. Plant Mol. Biol.* 48: 191–222.

Gamalei, Y. (1991) Phloem loading and its development related to plant evolution from trees to herbs. *Trees* 5: 50–64.

Macromolecule movement

Ding, B. (1998) Intercellular protein trafficking through plasmodesmata. *Plant Mol. Biol.* 38: 279–310.

Ghoshroy, S., Lartey, R., Sheng, J., Citovsky, V. (1997) Transport of proteins and nucleic acids through plasmodesmata. *Annu. Rev. Plant Physiol. Plant Mol. Biol.* 48: 27–50.

Mezitt, L. A., Lucas, W. J. (1996) Plasmodesmal cell-to-cell transport of proteins and nucleic acids. *Plant Mol. Biol.* 32: 251–273.

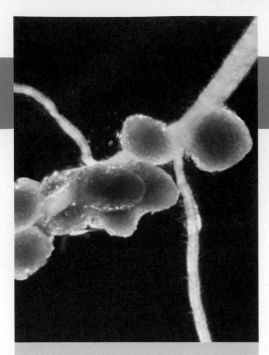

Biochemistry & Molecular Biology of Plants, B. Buchanan, W. Gruissem, R. Jones, Eds.
© 2000, American Society of Plant Physiologists

CHAPTER *16*

Nitrogen and Sulfur

Nigel M. Crawford
Michael L. Kahn
Thomas Leustek
Sharon R. Long

CHAPTER OUTLINE

Introduction

The elements **nitrogen** and **sulfur** are acquired by plants primarily through interaction with the soil solution. As with other mineral nutrients, the acquisition of nitrogen- and sulfur-containing ions is mediated by highly evolved morphological, physiological, and biochemical mechanisms. Unlike other mineral nutrients, however, the inorganic forms of nitrogen and sulfur are often present in soil in oxidized forms, which must be reduced for the element to be used in metabolism. These conversions take place in highly reducing environments (characterized by low $E°$ values) and link nitrogen and sulfur assimilation with pathways that generate reducing potential. Nitrate and sulfate reduction are compartmentalized and regulated to facilitate integration with other cellular metabolism. A combination of biochemical and molecular-genetic approaches is further elucidating these pathways. Although we know more about nitrogen metabolism in plants than we do about sulfur metabolism, our understanding of sulfur is increasing impressively with the recent renewed interest in the subject and the advent of new tools with which to study it.

16.1 Overview of nitrogen in the biosphere and in plants

Nitrogen is the fourth most abundant element in living organisms. It constitutes less than 0.1% of the earth's crust but makes up about 80% of the atmosphere, mostly as dinitrogen (N_2). Despite the apparent disparity in content, the large mass of the earth's crust means that the number of N atoms below ground is about 50 times greater than in the entire atmosphere. Most of this mineral nitrogen is in igneous rocks, but their weathering is not a significant source of entry for inorganic nitrogen into the living world. The cycle of transformation of mineral and organic nitrogen is shown in Figure 16.1. The cycle of N between geochemical and biochemical states is quite complex, given the many oxidation states the N atom can occupy in inorganic and organic compounds (Table 16.1). Most of the nitrogen taken up by organisms is recycled from a pool of nitrogen compounds that have previously been used by other organisms. New inputs to this pool are generated during chemical reactions that result from natural events (e.g., fire and lightning) or from human activity (e.g., use of internal combustion engines and application of chemical fertilizers). The main input to the pool of accessible inorganic N in natural ecosystems, however, is from atmospheric and dissolved N_2 through the biological process of **nitrogen fixation,** a process that converts N_2 to NH_3 (Table 16.2). Nitrogen fixation is carried out only by prokaryotic species. The NH_3 produced by N fixation may be assimilated into amino acids and thence to protein and other N compounds, or it may be converted by nitrifying bacteria to NO_2^- and NO_3^-. In turn, NO_3^- may enter metabolism through reduction to NH_4^+ and subsequent assimilation to amino acids by bacteria, fungi, and plants or can serve as an electron acceptor for denitrifying bacteria when oxygen is limiting. Losses from the nitrogen pool occur physically, when nitrogen (especially nitrate) is leached into inaccessible domains in the soil, and chemically, when denitrification releases N_2.

Plants biosynthesize both the suite of nitrogenous compounds that are present in all living organisms and the many chemical species unique to plants (Fig. 16.2). The most important step in assimilation of inorganic N into organic metabolites is the reaction, catalyzed by glutamine synthetase, a high-flux process subject to extensive regulation in all plants (see Chapter 8). The largest requirement for N is the synthesis of amino acids (see Chapter 8), which function as the building blocks of proteins as well as precursors to many other compounds. Nitrogen is also

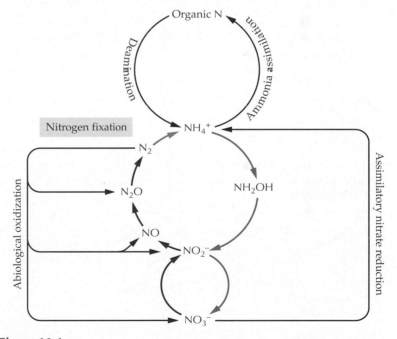

Figure 16.1

The nitrogen cycle. Organic nitrogen compounds, which are constituents of all living organisms, are released into the environment by death and decay and are excreted as waste by some animals. Microorganisms deaminate organic nitrogen, utilizing the carbon as a food source and liberating ammonium in the process. Plants and microorganisms can take up nitrate and reduce it to ammonium for subsequent assimilation into organic nitrogen-containing compounds. Many biological processes that change the oxidation state of nitrogen are catalyzed exclusively by prokaryotes. These include nitrification (in which ammonium or nitrite is oxidized and the energy released is used to fix inorganic carbon), denitrification (in which nitrogen serves as a terminal electron acceptor and is reduced during anaerobic respiration), and nitrogen fixation (in which dinitrogen gas is reduced to ammonium).

Table 16.1 Inorganic nitrogen compounds

Compound	Oxidation state of N	Name
N_2	0	Dinitrogen (nitrogen gas)
NH_3	−3	Ammonia
NH_4^+	−3	Ammonium ion
N_2O	+1	Nitrous oxide
NO	+2	Nitric oxide
NO_2^-	+3	Nitrite
NO_2	+4	Nitrogen dioxide
NO_3^-	+5	Nitrate

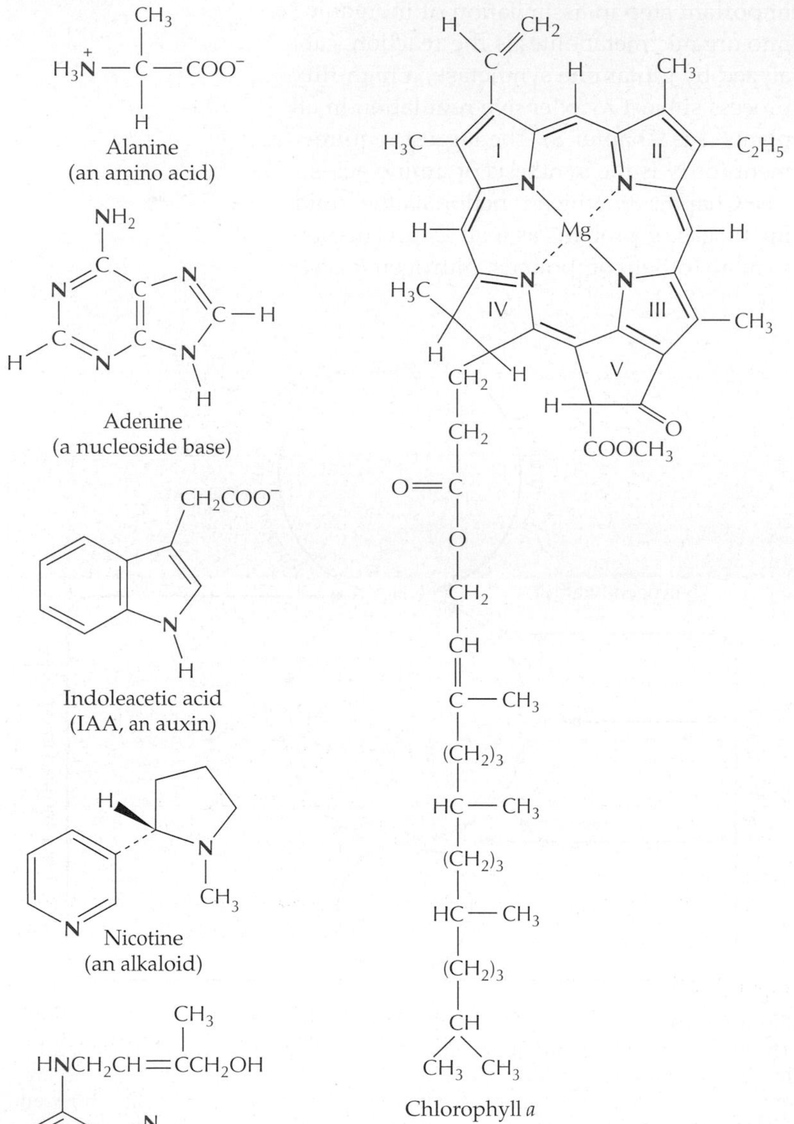

Figure 16.2
Selected nitrogenous compounds of biological importance. Some (e.g., amino acids, nucleoside bases) are found in all living organisms, whereas others are unique to plants.

Alanine (an amino acid)

Adenine (a nucleoside base)

Indoleacetic acid (IAA, an auxin)

Nicotine (an alkaloid)

Zeatin (*Zea mays*) (a cytokinin)

Chlorophyll *a*

an essential component of nucleic acids (see Chapter 6), cofactors, and other common metabolites and is a major component of chlorophyll (see Chapter 12). The characteristic yellow color of nitrogen-starved plants (chlorosis) reflects their inability to synthesize adequate amounts of green chlorophyll under N-limited conditions (Fig. 16.3). In addition, several plant hormones contain N or are derived from nitrogenous precursors (see Chapter 17). Plants synthesize diverse nitrogenous secondary compounds, most prominently the alkaloids (see Chapter 24). Although flavonoids and other plant phenolics do not contain N, their derivation from phenylalanine (see Chapter 24) means their synthesis is linked with amino acid metabolism.

Plants may acquire N through NH_4^+ uptake and incorporation into organic compounds; NO_3^- uptake and reduction to NH_4^+; or, in the case of plant hosts for nitrogen-fixing bacteria, acquisition of fixed N from bacterial **endosymbionts** (from the Greek: "living together within"). Plants display quite diverse strategies in acquiring nitrogen, as is evident both in comparisons of different species with one another and in comparison of individuals of a given species that are grown in different environments. For example, nitrate taken up by roots may be reduced and assimilated in the root ($NO_3^- \rightarrow NH_4^+ \rightarrow$ glutamine), or it may be transported as NO_3^- to the shoot. NO_3^- may also be stored

Figure 16.3
Nitrogen deficiency phenotype in leaves of sugar beet. Nitrogen deficiency is often associated with uniform yellowing (chlorosis) of older leaves. This feature is obvious for the plants grown in nitrogen-deficient soil (left side of photo) relative to counterparts grown with sufficient nitrogen (right side of photo).

in vacuoles in cells of the root or the shoot. Plants able to establish symbioses exert developmental and physiological control over the formation and function of symbiotic structures, according to their nutritional environment: Legume plants grown in the presence of NO_3^-, for example, will use this as a source of N nutrition and will not form symbiotic nodules or allow *Rhizobium* invasion (Fig. 16.4).

Acquisition of nitrogen occurs in a context that is physiological, developmental, and environmental. A comprehensive investigation of nitrogen assimilation requires the study of genes and gene expression, of protein structure and activity, and of root development and physiology. As photoautotrophs, plants are typically limited in their growth by availability of nutrients other than carbon, and nitrogen is frequently the limiting factor for plant productivity. Widespread application of nitrogenous fertilizer has been a key factor in improving agricultural yields during the 19th and 20th centuries (Box 16.1). As the human population grows and the demand for agricultural products increases, it becomes all the more important to understand the mechanisms of nitrogen acquisition for plants.

16.2 Overview of nitrogen fixation

16.2.1 Nitrogen fixation reduces nitrogen gas to ammonia, at the cost of ATP and reducing equivalents.

Dinitrogen constitutes about 80% of the earth's atmosphere, but the stability of its triple bond renders it inert and metabolically inaccessible to most organisms. Eukaryotes cannot utilize dinitrogen, but some prokaryotes are able to catalyze the enzymatic reduction of this compound to ammonia (Rx. 16.1). Unlike the **Haber–Bosch process** (see Box 16.1), the biological nitrogen fixation catalyzed by the enzyme nitrogenase (see Section 16.3) occurs at ambient temperatures and atmospheric pressure.

Reaction 16.1: Nitrogenase

$$N_2 + 16\ ATP + 8e^- + 8H^+ \longrightarrow$$
$$2NH_3 + H_2 + 16ADP + 16P_i$$

Table 16.2 Rates of natural and anthropogenic nitrogen fixation

Source	Amount of N fixed[a]
Lightning	<10 Tg/year
Biological N-fixation in terrestrial systems[b]	90–140 Tg/year
Biological N-fixation in marine systems	30–300 Tg/year[c]
N fertilizer synthesis	80 Tg/year
Fossil fuel combustion	>20 Tg/year

[a]The standard unit of measure is the teragram (Tg), 10^{12} g, equal to 10^6 metric tons.
[b]This estimate includes both natural ecosystems and agricultural nitrogen fixation.
[c]Estimates differ because of variable data.

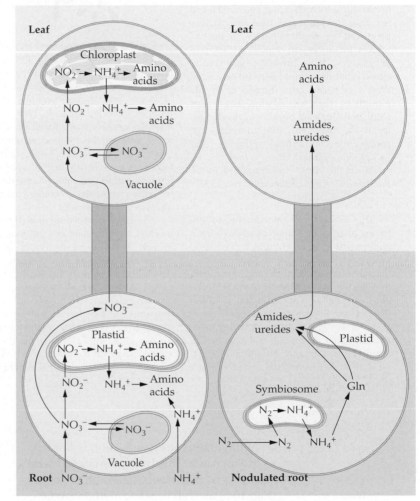

Figure 16.4
Overview of N uptake by a nonnodulated plant (left), and by a nodulated plant with N-fixing symbionts (right). There is considerable variation in the details of nitrogen assimilation in different plants. Plant roots can import nitrate, ammonium, and other nitrogenous compounds from the soil. For use in synthesis of amines and amides, nitrate must be reduced to nitrite and then to ammonium. Nitrate reduction in the cytosol and storage in the vacuoles are processes that can occur in either the roots or the leaves. Nodulated plants are able to take up fixed nitrogen from the soil (not shown) but, through the action of symbiotic bacteria, can generate ammonium also by reducing N_2. The ammonium from nitrogen fixation is assimilated into amino acids and ultimately incorporated into amide amino acids (glutamine or asparagine) or ureides (see Section 16.4.9) for export to the leaves. These conversions are shown in more detail in Figures 16.32 and 16.33.

Box 16.1   **Nitrogenous soil amendments have a long history.**

Varro praeceptis adicit equino quod sit levissimum segetes alendi, prata vero graviore quod ex hordeo fiat multasque gignat herbas . . . inter omnes autem constat nihil esse utilius lupini segete priusquam siliquetur aratro vel bidentibus versa manipulisve desectae circa radices arborum ac vitium obrutis.

[Varro (an earlier Roman writer) adds the employment of the lightest kind of horse dung for manuring cornfields, but for meadowland the heavier manure produced by feeding barley to horses, which produces an abundant growth of grass.... It is however universally agreed that no manure is more beneficial than a crop of lupine turned in by the plough or with forks before the plants form pods, or else bundles of lupine after it has been cut, dug in round the roots of trees and vines.]—*Pliny the Elder (~A.D. 80), Natural History, from vol. 17, vi 54–vii 56, edited and translated by H. Rackham, Harvard University Press, 1971.*

The use of soil amendments to enhance plant growth appears to have been widespread in many cultures according to unwritten tradition and later written records. Agriculture in the Mediterranean, as reflected in the writings of Pliny the Elder in the first century A.D. and of earlier Roman and Greek writers, mentioned use of manure, crop rotation with legumes, and ash, lime, or marl as soil additives to improve plant growth. The three elongated dots in the lower half of the Chinese pictographic character for soybean, *shu* (see calligraphic figure below), have been postulated to signify the relationship between root nodules and plentiful growth of the plants. This character was recorded by Chou scholars in 1000 B.C.

Farmers rotated crops and applied manure or simple minerals with little quantitative rigor until the 18th to 19th century, when chemical analysis of soils and study of plant growth requirements became more defined. By the early 19th century, nitrogen fertilizer was applied on a larger scale, both in the form of manures, including guano, and as of mineral nitrates. However, the use of nitrogen supplies for fertilizer had to compete with the demand for these chemicals as components of explosives and gunpowder.

Ammonia production, initially a byproduct of the coking of coal, was transformed in 1913 by the work of Fritz Haber and Carl Bosch in Germany, who developed a complete practical synthesis of NH_3 from N_2 and H_2 in the presence of an iron catalyst at high temperature and pressure. The reaction of dinitrogen with hydrogen gas is actually exothermic (ΔH_f° for $NH_3(g)$ is -46.1 kJ mol^{-1}), but the activation energy for the reaction is very high because of the triple bond in dinitrogen:

$$N_2(g) + 3H_2(g) \rightarrow 2NH_3(g)$$

At room temperature the reaction is favored but very slow, and at very high temperature the equilibrium favors reactants, not products. K_{eq} is 6.0×10^5 at 298 K, 2.1 ($\times 10^0$) at 450 K, and 4.4×10^{-5} at 800 K. The Haber–Bosch process that produces ammonia industrially is a compromise, using pressures ranging from 150 to 400 atmospheres, temperatures ranging from 400°C to 650°C, metal catalysts, and removal of the ammonia as it is formed (see figure on facing page). The growing dependence of agriculture on NH_3-based nitrogen fertilizers (see Table 16.2) requires high use of fossil fuels, and this may become a serious limitation as energy becomes limiting or more expensive. Industrial nitrogen fixation currently accounts for 80×10^{12} g/year, more than half of the amount of nitrogen fixed biologically by terrestrial systems each year (see Table 16.2). The microbiology of nitrogen-fixing symbiosis also dates to the 19th century, with the discovery by Martinus Beijerinck in 1888 that a microorganism, later confirmed to be a bacterium, is responsible for the formation of root nodules in legumes. In the same year, Hermann Hellriegel and Hermann Wilforth provided careful inferential evidence that nodulated legumes were able to synthesize organic N compounds from nitrogen from the atmosphere. Nonnodulated legumes and nonleguminous species were able to use only mineral nitrogen in solution.

Enzymatic nitrogen fixation is limited to prokaryotes. This trait is associated with members of many eubacterial phylogenies (Fig. 16.5) as well as some methanogenic archaea. A few nitrogen-fixing bacteria from diverse taxa (i.e., cyanobacteria, actinomycetes, and the α-proteobacteria) are able to establish symbiotic associations with plants. In such symbioses, nitrogen fixed by bacteria is exchanged for carbon fixed by the plant. Nitrogen fixation by symbiotic bacteria can be highly productive because interaction with the plant allows fixation to occur under optimized physiological conditions, overcoming constraints that often limit nitrogen fixation by nonsymbiotic bacteria. Much of what we know about the genetics and biochemistry of nitrogen fixation has been determined in studies of free-living eubacteria, such as *Clostridium, Klebsiella, Azotobacter,* and *Anabaena*. Results from these investigations have provided detailed information on factors that constrain all nitrogen-fixing systems, including symbiotic bacteria.

16.2.2 Nitrogen fixation is sensitive to oxygen.

Nitrogen fixation is a unique biochemical reaction that consumes energy-rich compounds

An industry for the production of bacterial inoculant emerged by the beginning of the 20th century. Improved understanding of host specificity and new techniques for bacterial strain determination contributed to better-defined inoculants. However, competition from native rhizobial strains resident in soil sometimes limits nodule colonization by strains having better physiological traits. Improved understanding of plant and bacterial genes used in symbiosis may allow plant breeders and microbiologists to obtain better-functioning host bacteria in agricultural systems.

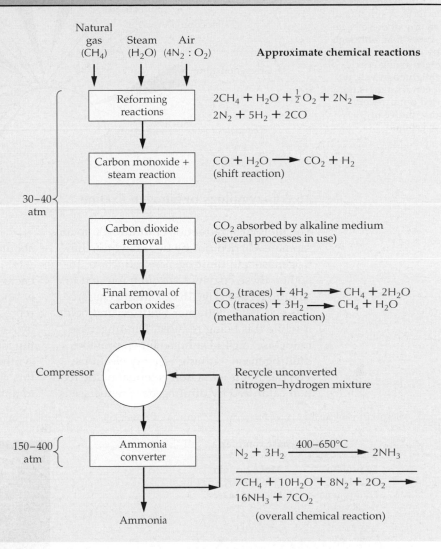

while requiring strong biological reductants. Because nitrogenase and some of the proteins that supply it with reductant are sensitive to oxygen, many nitrogen-fixing bacteria are anaerobes. Neither fermentation nor anaerobic respiration oxidizes reduced carbon compounds as efficiently as aerobic respiration, so anaerobic bacteria must process large quantities of substrate to generate the ATP required for dinitrogen fixation. In contrast, aerobes have the advantage of high ATP production from aerobic metabolism but must contend with the oxygen sensitivity of nitrogenase. In some cases, free-living nitrogen-fixing organisms use mechanical or biochemical barriers to keep oxygen away

from the biological catalysts of nitrogen fixation. In other cases, the nitrogen fixation machinery is segregated spatially in specialized structures. For example, some filamentous cyanobacteria generate **heterocysts,** thick-walled cells that fix nitrogen but cannot complete all the reactions of oxygenic photosynthesis. Heterocysts produce the ATP needed for nitrogen fixation by way of cyclic photophosphorylation, a light-dependent process that does not create oxygen gas (see Chapter 12). Some nonfilamentous cyanobacteria segregate photosynthesis from nitrogen fixation temporally, performing oxygenic photosynthesis in the light and nitrogen fixation in the dark.

Figure 16.5
Phylogenetic distribution of nitrogen-fixing eubacteria. A simplified taxonomy of these bacteria shows that, although many groups contain nitrogen-fixing species (highlighted in yellow), nitrogen fixation is not carried out by every representative of these groups.

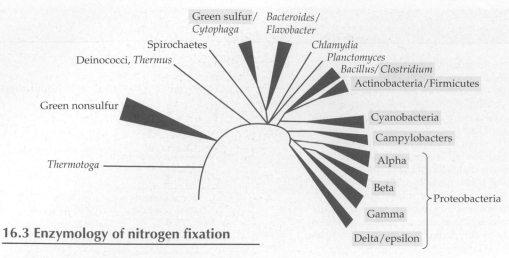

16.3 Enzymology of nitrogen fixation

The biological reduction of dinitrogen to ammonia is carried out by two enzymes, dinitrogenase and dinitrogenase reductase. Together, these enzymes are often referred to as nitrogenase (Fig. 16.6). Dinitrogenase, a 240-kDa heterotetramer, binds N_2 and holds it while it is being reduced. Dinitrogenase reductase, a 64-kDa homodimer, provides dinitrogenase with high-energy electrons: The midpoint reduction potential of the reaction catalyzed by dinitrogenase reductase is about -400 mV when the enzyme is binding ATP. Both proteins have been purified from bacterial species and characterized extensively, and the three-dimensional structures of representative proteins have been determined by X-ray crystallography. For nitrogenase to function, additional proteins are needed to synthesize its unique metal-containing cofactors and to transfer low-potential electrons to dinitrogenase reductase.

(A)

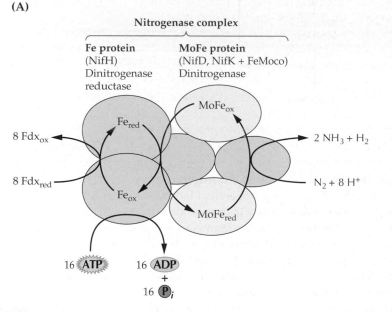

(B)

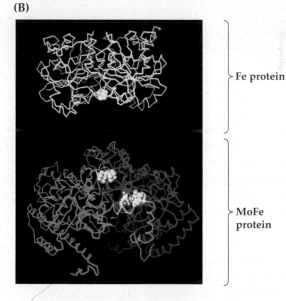

Figure 16.6
(A) Schematic diagram of the nitrogenase complex, showing the flow of reducing power and substrates in enzymatic nitrogen fixation. The Fe-protein, encoded by *nifH*, accepts electrons from a carrier, e.g., ferredoxin, flavodoxin, or another redox-active species of similar potential. The identity of the carrier varies, depending on the biological system involved. The Fe-protein transfers single electrons at very low potential to the MoFe-protein, accompanied by net hydrolysis of ATP. The MoFe-protein, an $\alpha_2\beta_2$ heterotetramer of subunits encoded by *nifD* and *nifK*, accepts electrons and binds H^+ ions and N_2 gas in a stepwise cycle, ultimately leading to the production of H_2 and ammonia. (B) Docking of the nitrogenase FE protein dimer (yellow) with half of the nitrogenase MoFe protein (red, nifD; purple, nifH). A 4Fe:4S cluster is associated with the Fe protein. The P cluster is near the nifD/nifH interface. FeMoCo (green) is mostly associated with nifD.

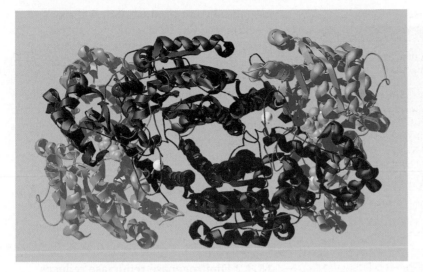

Figure 16.7
Ribbon structure of dinitrogenase (MoFe-protein). The two NifK protein subunits (green, blue) in dinitrogenase associate with each other through the interaction of the NifD protein subunits (red, purple). The FeMoCo (upper) and P-cluster (lower) complexes in the right half of the protein can be seen as yellow groups held within the structure. FeMoCo also is identified by red and turquoise. The two dinitrogenase reductase proteins bind near the interfaces of the NifD and NifK proteins at the left and right of the figure. The Protein Data Bank identifier for the complete structure of nitrogenase is 1N2C.

16.3.1 Dinitrogenase includes unique metal components.

Dinitrogenase is an $\alpha_2\beta_2$ heterotetramer containing two pairs of unusual metal clusters (Fig. 16.7). All dinitrogenases have two identical **P-clusters**, 8Fe–7S iron–sulfur complexes that are integral components of the enzyme, there being six covalent bonds between protein cysteines and the Fe atoms in each P-cluster. In its reduced state (P^N), a P-cluster resembles two distorted 4Fe–4S cubes that share one sulfur at the corner. In the oxidized state (P^{OX}), one of the cubes opens, breaking two of the bonds with this corner sulfur (Fig. 16.8).

A second pair of metal clusters is held in dinitrogenase by only a cysteinyl sulfur and a histidine nitrogen; this cofactor can actually be separated from the enzyme and then reconstituted with the apoenzyme to restore activity. Three forms of this cofactor exist, each named for its component metals. FeMoCo and FeVCo contain one atom of molybdenum or vanadium, respectively, whereas the only metal present in FeFeCo is iron. FeMoCo is the most efficient of the cofactors, followed by FeVCo and FeFeCo. Under conditions of molybdenum deficiency, free-living bacteria that fix nitrogen can synthesize nitrogenases that use FeVCo or FeFeCo. The FeMoCo form of nitrogenase is the only one found thus far in the bacteria that form tight symbiotic relationships with plants. Molybdenum is therefore an essential element for plant-associated nitrogen fixation.

FeMoCo consists of one 4Fe–3S cluster and one 1Mo–3Fe–3S cluster bridged by three atoms of inorganic sulfur. The molybdenum atom is linked to homocitrate, a C_7 analog of citrate (Fig. 16.9). Although FeMoCo is a small complex, its instability hindered the determination of its structure until the crystal structure of the holoenzyme was solved.

Molybdenum dinitrogenase will reduce several compounds with multiple bonds, including dinitrogen and acetylene (Table 16.3). Acetylene is converted to ethylene, providing a convenient assay for nitrogenase activity both in vivo and in vitro. Another convenient assay is based on detecting hydrogen

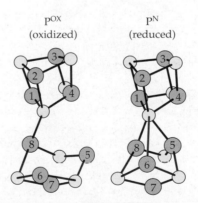

Figure 16.8
Structure of the P-clusters of the MoFe-protein. Oxidized and reduced conformations are shown. Iron atoms are shown in brown, sulfur atoms in yellow. The Fe–S bond of the P^N structure (shown in red) is about 2.92 Å long, whereas the other Fe bonds with this sulfur average 2.34 Å, indicating that the structure is under considerable stress.

Table 16.3 Substrates and products of nitrogenase

Substrate	Common name	Product(s)
N_2	Dinitrogen	NH_3 (ammonia)
H^+	Hydrogen ion	H_2 (hydrogen gas)
N_2O	Nitrous oxide	N_2, H_2O
CN^-	Cyanide	NH_3, CH_4 (methane)
CH_3NC	Methyl isocyanide	CH_3NH_2 (methylamine), CH_4
N_3^-	Azide	N_2, NH_3
C_2H_2	Acetylene	C_2H_4 (ethylene), C_2H_6 (ethane)
H_2NCN	Cyanamide	NH_3, CH_3NH_2
C_3H_4	Cyclopropene	C_3H_6 (cyclopropane)
CH_2N_2	Dazirine	NH_3, CH_3NH_2

gas, an inherent product of the nitrogenase reaction (see Rx. 16.1). However, many nitrogen-fixing organisms have hydrogenases that oxidize hydrogen gas to recover its reducing power and energy; the activity of the hydrogenases prevents accurate inference of nitrogen fixation from hydrogen evolution rates alone. Hydrogenase activity can also reduce oxygen concentrations and can relieve nitrogenase inhibition caused by hydrogen accumulation (see next section). Quantitative physiological assays of nitrogenase function in vivo require correction for possible hydrogenase activity.

16.3.2 Dinitrogenase reductase reduces dinitrogenase and hydrolyzes MgATP.

Dinitrogenase reductase (often called the Fe-protein) is a homodimer that contains a single 4Fe–4S cluster held cooperatively by the two monomers at the protein's surface. This structure is unusual for an iron–sulfur protein because this type of metal cluster is usually buried within the folded protein and is ligated to only one polypeptide chain. Reduced dinitrogenase reductase can bind two molecules of either MgATP or MgADP, but the conformation of the structures differs in such a way that the MgATP complex binds more closely to dinitrogenase (Fig. 16.10). In the reaction cycle, the oxidized Fe-protein accepts an electron from reduced ferredoxin or flavodoxin, binds two MgATP, and transfers the electron to dinitrogenase. The ATP is then hydrolyzed to produce two molecules of MgADP and two of P_i, a reaction that requires both dinitrogenase and the Fe-protein. Subsequent conformational changes allow the now oxidized Fe-protein to be reduced again, acquire two bound MgATP molecules, and continue in the cycle. The Fe-protein can exchange nucleotides after it dissociates from dinitrogenase or while remaining in the complete nitrogenase complex. ATP binding leads to conformational changes in dinitrogenase reductase that lower the potential of the metal clusters. ATP binding and hydrolysis are also involved in the electron transfer process and in the regeneration of a complex that can accept another electron. These electron transfers and the associated binding of protons are designated with solid arrows in Figure 16.11.

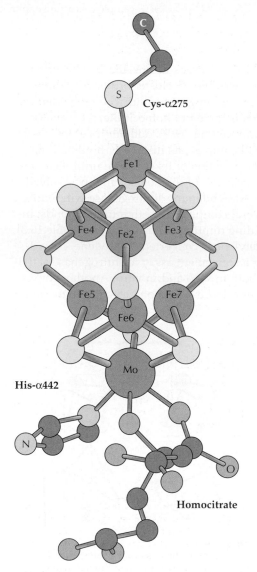

Figure 16.9
Molecular model of molybdenum iron cofactor (FeMoCo). The MoFe type of nitrogenase is present in all symbiotic bacteria, including *Rhizobium* and *Bradyrhizobium*.

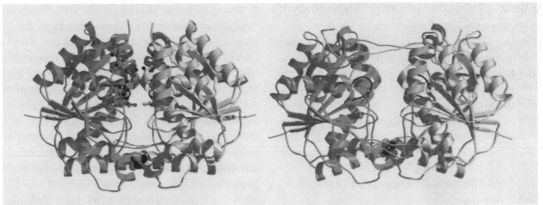

+ATP   −ATP

Figure 16.10
Binding of ATP changes the conformation of the nitrogenase Fe protein from *Azotobacter vinelandii*. The two NifH subunits of the Fe protein hold the purple 4Fe–4S cluster in the region at the bottom of the structure where the polypeptides interact. Comparison of the structure of the free protein (right) with the MgATP-complexed protein (left) shows that the addition of MgATP causes the polypeptides to move toward each other and the 4Fe–4S cluster to move down about 4 Å. This moves the 4Fe–4S cluster toward the P-cluster held in dinitrogenase.

Dinitrogenase reductase transfers single electrons sequentially to dinitrogenase, where they first accumulate in the P-clusters and then are transferred to the FeMoCo-type cluster, at which point the substrate dinitrogen is bound. As dinitrogenase is reduced, the complex acquires as many as four hydrogens before dinitrogen binding and reduction are initiated (Fig. 16.11). As dinitrogen binds, hydrogen gas is released as a byproduct of the overall reaction, but dinitrogenase can also release H_2 nonproductively, without binding N_2 (see dotted arrows in Fig. 16.11).

For example, form E_3H_3 may revert to E_1H_1 if ambient concentrations of N_2 are low. The stoichiometry shown in Reaction 16.1 is not always achieved in vivo because reductant and ATP may be consumed by incomplete cycles, during which protons are reduced and released as hydrogen gas. In addition, H_2 can inhibit dinitrogen reduction by competing with N_2 for the active site of the reduced nitrogenase complex.

Reduced ferredoxin (or, in some organisms, flavodoxin) is thought to donate electrons to dinitrogenase reductase in many

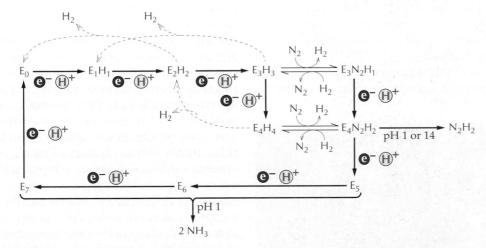

Figure 16.11
The MoFe enzyme cycle. E_n represents the redox state of the nitrogenase enzyme after the enzyme has accepted n electrons. The solid arrows illustrate the transfer of a single electron and association of a proton. In the initial steps of the reaction, hydrogen can be evolved without productive binding of nitrogen, essentially wasting the ATP and reductant invested in creating these intermediates (see dotted arrows). The molecular events in the conversion of N_2 to NH_3 are not precisely known, including where and how N_2 binds to the enzyme, how the partially reduced intermediates coordinate to the metal centers, and when the two molecules of NH_3 are released from the native enzyme. Dinitrogen (N_2) is believed to bind to the enzyme by displacing H_2 from either the E_3H_2 or E_4H_4 forms. Thus, formation of molecular hydrogen is an obligatory step in the enzymatic reduction of dinitrogen. Treatment of the $E_4N_2H_2$ complex with strong acid or base releases hydrazine (N_2H_2). Treatment of later complexes with acid releases ammonia, which suggests that the N atoms in these complexes are associated with more than one H.

free-living heterotrophic nitrogen-fixing bacteria. The enzyme pyruvate ferredoxin (flavodoxin) oxidoreductase can transfer electrons to oxidized ferredoxin (flavodoxin) from α-keto acids such as pyruvate and α-ketoglutarate. Some organisms, including rhizobial symbionts, may have alternative, electron transport-coupled methods of generating the low-potential electrons required to reduce these electron carrier proteins.

16.4 Symbiotic nitrogen fixation

In natural terrestrial ecosystems, 80% to 90% of the nitrogen available to plants is estimated to originate from biological nitrogen fixation. Of that total, approximately 80% is generated in symbiotic associations. Symbiotically fixed nitrogen can increase plant growth and yield considerably. As seen in Figure 16.12, the root nodules that make this growth possible occupy a significant portion of the roots. Thus, the benefits of nitrogen fixation are not without cost. If the plant resources required to establish nodules, fix nitrogen, and transport the resulting ammonia throughout the plant are taken into account, obtaining nitrogen through symbiosis consumes 12 to 17 g of carbohydrate per gram of N fixed. Legumes have mechanisms to suppress nodule formation and function if nitrate or ammonia is available as N sources.

Quantitative isotope ratio measurements are used to assess the contribution of nitrogen fixation to total N in plants. Atmospheric dinitrogen is essentially all ^{14}N. Soil, however, often contains greater amounts of the stable isotope ^{15}N. A plant that obtains fixed N from the atmospheric pool through nitrogen-fixing symbiosis will have less ^{15}N in its total N than if it obtains N only from soil sources. The best controls for this measurement are plants of the same species grown in the same soil but without symbiotic bacteria. **Rhizosphere** (root zone) associations with some free-living nitrogen-fixing bacteria (e.g., *Azospirillum*) have been shown to increase plant yield in crops such as sorghum, but demonstrating that this growth increase results from nitrogen transfer to the plant has been difficult.

16.4.1 Some vascular plants establish nitrogen-fixing symbioses.

There are three major types of nitrogen-fixing symbioses. The first involves a group of Gram-negative bacteria, the rhizobia, that form associations with numerous legume host plants (Fabaceae) and at least one non-legume, *Parasponia* (Ulmaceae). Rhizobial endosymbionts contribute nitrogen to commercially important seed legumes such as peanut, soybean, lentil, bean, and pea and to forage crops such as clover and alfalfa. This best-studied type of nitrogen-fixing symbiosis is discussed below in detail.

A second type of symbiosis occurs between members of a Gram-positive actinomycete genus (*Frankia*) and a diverse group of dicots, generally trees or woody shrubs from about 60 genera in 9 families, including alder (*Alnus*), myrtle (*Myrica*), *Casuarina*, and *Ceanothus*. These associations can play a significant role in the nitrogen economies of forests and other natural ecosystems. Although classical taxonomies do not indicate a close relationship among the plants that associate with actinomycetes, a more recent classification based on the plastid-encoded *rbcL* gene sequences places the plants in a subgroup of the subclass Rosidae and suggests that plants capable of actinomycete symbioses may share characteristics. These molecular phylogenies also suggest that the rhizobial hosts, both legumes and the nonlegume *Parasponia*, may be placed with the *Frankia* hosts in a single clade (Fig. 16.13).

Figure 16.12
Photo of legume grown in N-deficient soil (green pea [*Pisum sativum*]). Note the associated nodules containing the rhizobial symbiont (magnification × 3).

A third group of symbioses exist between cyanobacteria and a diverse array of plants: dicots (e.g., *Gunnera*), cycads, ferns, liverworts, and hornworts. *Azolla*, a water fern that associates symbiotically with the cyanobacterium *Anabaena*, is used as a cocrop in rice paddies and produces sufficient fixed nitrogen to allow sustainable rice cultivation.

16.4.2 Legume–rhizobial symbioses are both diverse and specific.

In the presence of specific rhizobial species, legume hosts form unique structures (the root nodule) in which nitrogen fixation occurs (Fig. 16.14). Plant and bacterial activities create a nodular environment conducive to both nitrogen fixation and microaerobic synthesis of ATP by the bacteria. Plant metabolism in the nodule generates organic acids that both feed the bacteria and provide carbon skeletons for the N-transport compounds used to transfer fixed (reduced) nitrogen to the rest of the plant. In exchange for the carbon substrates, the bacterial symbionts fix nitrogen and release the resulting ammonia to the plant.

The rhizobia were originally classified as a single genus, but more recent taxonomic analyses have grouped them in distinct major clusters within the α-proteobacteria. Several genera are now recognized, including *Rhizobium*, *Sinorhizobium*, *Azorhizobium*, *Mesorhizobium*, and *Bradyrhizobium*. However, these genera do not form a single clade within the alpha subdivision (Fig. 16.15).

Symbioses between groups of plant hosts and particular bacterial species, or biovars, display various levels of specificity (Table 16.4). The host specificity does not

Figure 16.13
Phylogenetic analysis of 99 *rbcL* genes (encoding the large subunit of Rubisco) from genera representing the subclass Rosidae and additional angiosperm taxa. Taxa that engage in nodular nitrogen-fixing symbioses are highlighted in yellow. (*Gunnera*, which hosts cyanobacterial endosymbionts, is highlighted in green.) Note that a single clade includes all the nodule-forming genera included in the tree. Subclades A, C, and D include actinorhizal symbiotic hosts; the legume family (Fabaceae) falls into subclade B. The subfamily Celtoideae, which contains the nitrogen-fixing genus *Paraspoina*, is represented here by the genera *Celtis* and *Trema*.

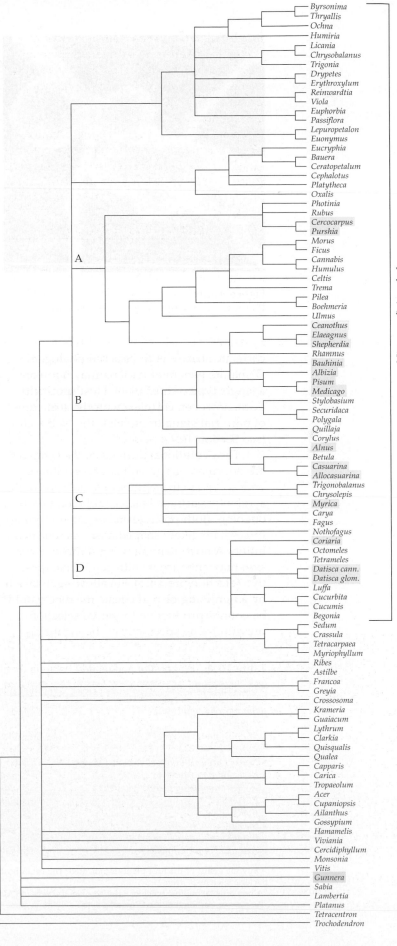

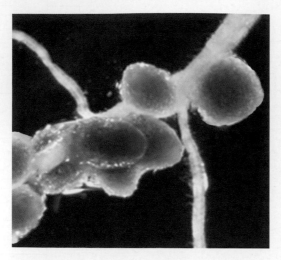

Figure 16.14
Photo of root nodules on pea (magnification × 7.3).

may reflect the balance of these pressures. Nitrogen fixation appears to be most useful to rhizobia when they are symbiotic. Few rhizobia fix nitrogen outside the plant host; of these, only bacteria in the genus *Azorhizobium* can grow using the ammonium they produce. In turn, bacterial nitrogen fixation is most useful to the plant when it has no other nitrogen sources, and both nodule formation and function are subject to downregulation by host plants that are provided with nitrate or ammonia.

16.4.3 Legumes create root nodules to house their rhizobial symbionts.

Legume–rhizobia symbiosis takes place in a unique structure, the root nodule, constructed by the plant in response to bacterial signals (Fig. 16.16). Within the plant, the cellular and organ-level changes associated with root nodule formation help provide the biochemical context required for efficient nitrogen fixation.

Host–symbiont recognition initially occurs in the rhizosphere. In many species, the plant is usually infected through root hairs, which respond to bacterial signals by altering their growth patterns to trap potential symbionts. The invasion structure itself, termed the infection thread, appears to be formed by the plant and is composed of plant cell wall material, although its direction is inward and its growth orientation is from the outside in (Fig. 16.17A). It is plausible that the bacteria reorient plant cell wall growth, perhaps by redirecting

correlate strictly with bacterial phylogeny: Disparate genera of bacteria may nodulate a single type of host plant. Host specificity does, however, correlate with the structures of bacterial signal molecules, the Nod factors (see Section 16.4.4 below).

The evolutionary origins of the symbiosis are not clear, but given that the symbiosis is widespread in the legumes, coevolution of the symbiotic partners has probably long been ongoing. Both positive and negative driving forces have likely shaped host–microbe evolution. A host plant allowing *Rhizobium* invasion may enhance its nitrogen intake (positive selection pressure), but allowing invasion by a nonfixing or pathogenic microbe would be counterproductive (negative selection pressure). The specificity of the symbiosis

Table 16.4 *Rhizobium* and related bacteria that form symbioses with legumes

Bacterial species	Host plants[a]
Sinorhizobium meliloti[b]	*Medicago* (alfalfa), *Trigonella* (fenugreek), *Melilotus* (sweetclover)
S. fredii[b]	*Glycine* (soybean), *Vigna* (cowpea)
Sinorhizobium sp. NGR234	Broad host range, many genera: *Vigna, Leucaena, Macroptilium* (siratro), *Parasponia*[c]
Rhizobium leguminosarum biovar *viciae*	*Vicia* (vetch), *Pisum* (pea), *Cicer* (chickpea), *Lathyrus* (sweet pea)
R. leguminosarum biovar *trifolii*	*Trifolium* (clover)
R. leguminosarum biovar *phaseoli*	*Phaseolus* (bean)
R. tropici	*Phaseolus, Leucaena, Medicago, Macroptilium*
R. etli	*Phaseolus*
Mesorhizobium loti	*Lotus* (trefoil), *Anthyllis* (kidney vetch), *Lupinus* (lupine)
Bradyrhizobium japonicum	*Glycine, Macroptilium, Vigna*
Azorhizobium caulinodans	*Sesbania rostrata*

[a]Only the genus of the host plants is indicated. The list of host plants is not comprehensive, particularly for broad-host-range bacteria such as *S. fredii*.
[b]A previous taxonomic scheme classified *Sinorhizobium* and *Mesorhizobium* as *Rhizobium*.
[c]*Parasponia*, classified in the Ulmaceae, is the only nonlegume known to associate with *Rhizobium*.

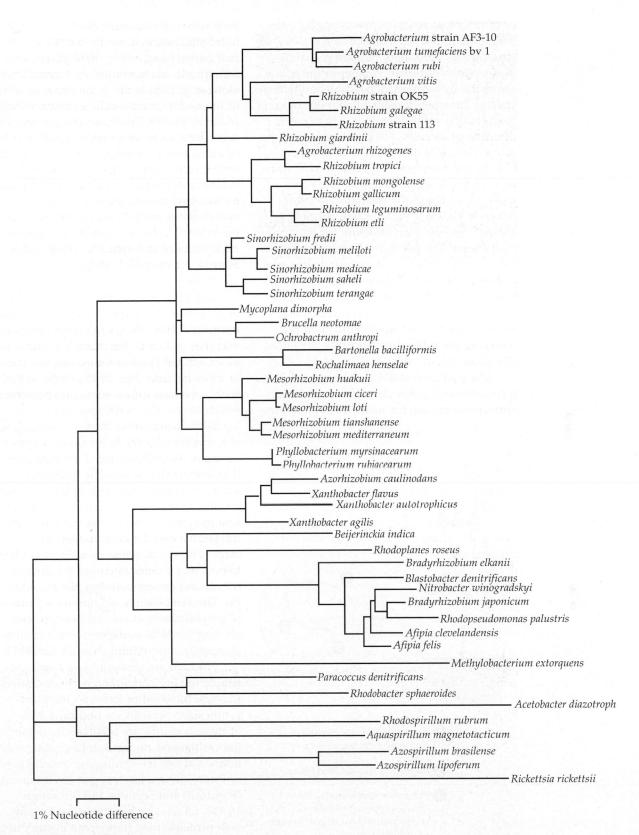

Figure 16.15
Phylogenetic tree of the alpha-proteobacteria, showing the dispersed groups of rhizobia as well as free-living organisms (e.g., *Rhodobacter*) and animal pathogens (e.g., *Brucella*, *Bartonella*, *Rickettsia*). Note that *Rhizobium* and *Sinorhizobium* are very closely related to *Agrobacterium* and more distantly related to *Bradyrhizobium*. Thus the symbiotic habit and host range do not correlate easily with relatedness, as estimated from 16S rRNA sequence comparisons.

enzymes or secretion pathways or by influencing plant cytoskeletal activity. The nodule structure is elaborated by plant cell divisions, which originate in inner or outer layers of the root cortex, depending on the host species. Infection threads filled with dividing bacteria penetrate the root layers to meet dividing plant cells. The plant cells display active reorganization of cytoskeleton, cytoplasm, and membrane systems during invasion. In some species such as *Pisum* (pea), there is prominent reorientation of cytoplasm in cells not yet invaded, possibly providing a preinfection structure for invasion by bacteria. The plant and bacteria both contribute to the matrix that surrounds bacterial cells in the lumen of the infection thread. Specific plant glycoproteins and other proteins are present in infection threads, and some nodulins (nodule-specific plant proteins) are probably directed to the infection thread matrix or wall also.

There are two major developmental plans for root nodules, determinate (or spherical) and indeterminate (or meristematic). In inde-

terminate nodules, such as those generated by alfalfa, pea, and clover, the nodule meristem persists and newly divided (i.e., uninfected) cells are continuously formed at the elongating nodule tip. A longitudinal section of the nodule reveals cells at different stages of development. The base contains senescent cells, the apex is composed of small uninfected cells, and in between lie the mature cells, both infected and uninfected (Fig. 16.17B). In determinate nodules, such as those produced by soybean, trefoil, and bean *(Phaseolus)*, the nodule tissue matures synchronously, with all the infected cells being at the same approximate stage of differentiation (Fig. 16.17C). Mature determinate nodules lack meristematic tissues.

After infection thread penetration of one or more host cell layers, bacteria are released from the walled thread in membrane-bound vacuoles and enter the plant cell. These vacuoles, termed **symbiosomes,** may contain one or more bacteria (Fig. 16.18). In the symbiosomes, bacteria differentiate into **bacteroids,** which in some bacterial species are morphologically distinct from the free-living bacterial forms (Fig. 16.19). In some species such as soybean, the bacteria and their symbiosomes divide extensively; in other cases such as alfalfa, bacterial division is limited to a few rounds. Extensive synthesis and differentiation of symbiosome membrane (also called the peribacteroid membrane) occur at this stage. The symbiosome membrane is likely to be a major determinant of the flow of nutrients and energy between plants and bacteria. The symbiosome membrane is the target of several newly expressed host proteins, including the N-26 aquaporin and a symbiosis-specific ammonium channel, GmSAT1.

Nodules are thought to be relatively simple organs, containing perhaps dozens to a few hundred nodule-specific plant-synthesized proteins, or **nodulins.** Many of these nodulins are involved in metabolizing carbon and nitrogen, transporting compounds across the symbiosome membrane, or establishing a low-oxygen environment (Fig. 16.20 and Sections 16.4.5 through 16.4.9). Others, such as the putative cell wall proteins, may participate in structure, infection, or defense; still other nodulins may mediate signal transduction and cellular response pathways required for nodule development.

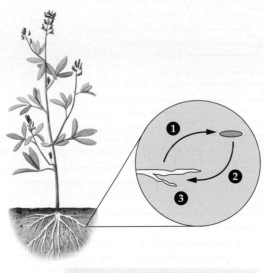

❶ Plant root releases elicitors of *Nod* gene expression.

❷ Bacterium releases Nod factor.

❸ Plant root demonstrates ion fluxes, expresses nodulin proteins, is infected, and undergoes nodule morphogenesis.

Figure 16.16
Overview of events leading to formation of legume–rhizobium symbiosis.

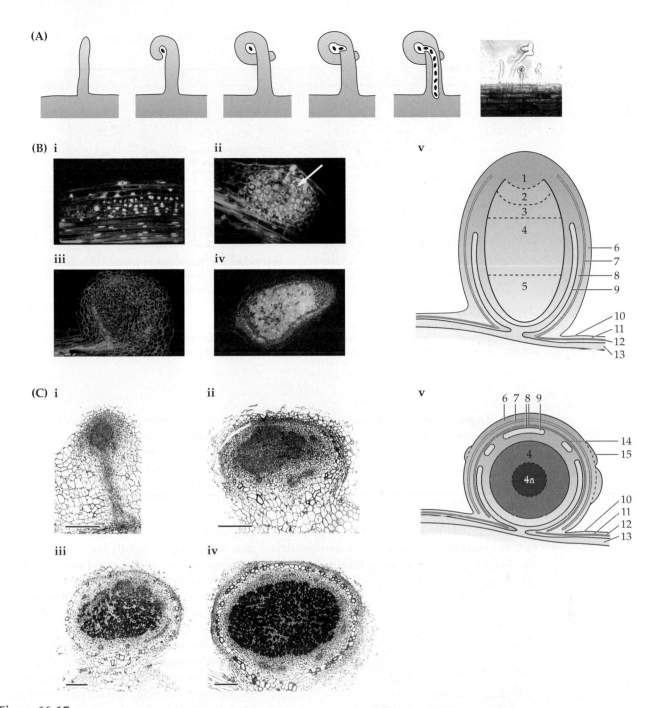

Figure 16.17

Formation of nitrogen-fixing nodules. (A) Diagrammatic representation of root hair invasion (left) and photograph of an infected root hair (right) from alfalfa *(Medicago sativa)*. Root hair growth is perturbed so that the hair deforms and curls as it elongates; bacteria trapped in the curl form an infection thread and proliferate as they invade. (B) Nodule morphogenesis for an indeterminate (cylindrical, meristematic) nodule. Formation of this nodule type begins with cell divisions in the inner cortex (i) (magnification × 100). Invasion occurs as an infection thread (see arrow) penetrates from the epidermis to the inner layers (ii) (magnification × 100). A persistent meristem forms (iii) (magnification × 42), giving rise to an elongated nodule with zones of uninfected, infected, and senescent cells along a distal to proximal axis of the nodule (iv) (magnification × 12.5). Numbers in the accompanying diagram (v) correspond to the following structures: 1, nodule meristem; 2,

zone of infection thread growth and cell penetration; 3, zone of expanding infected cells; 4, mature bacteroid-containing tissue; 5, senesecent bacteroid-containing tissue; 6, outer cortex; 7, nodule endodermis; 8, inner cortex; 9, nodule vascular bundle; 10, root epidermis; 11, root cortex; 12, root endodermis; 13, root xylem and phloem elements. (C) Nodule morphogenesis in determinate (spherical) nodules. Early cell divisions in the outer cortex of the host are followed by invasion of bacteria into plant cells (i); many rounds of host and bacterial cell division give rise to zones of infected and uninfected cells in a spherical, stable nodule (ii–iv). Spacebars = 250 μm. The accompanying diagram (v) shows in schematic form the zones of the fully formed determinate nodule. Numbers are as in (B), plus 4(a), zone where senescence starts in determinate nodules; 14, sclerenchyma; and 15, phellogen.

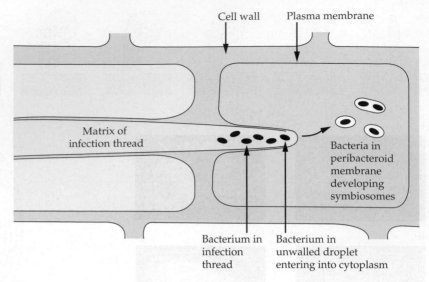

Figure 16.18

Release of bacteria from a walled infection thread into a target cell. The bacteria are taken up into the host plant cell enveloped in host plasma membrane. Bacteria undergo limited cell division (indeterminate nodules) or extensive rounds of cell division accompanying host cell division (determinate nodules).

16.4.4 Legume roots exude inducers of bacterial symbiosis genes.

Accurate mutual recognition of plant host and bacterial symbiont is an essential component of any nitrogen-fixing symbiosis. In legume–rhizobium systems, an exchange of biochemical signals appears to be a prerequisite for establishing symbiosis. Compounds released by the host plant induce bacterial gene expression, which leads to production

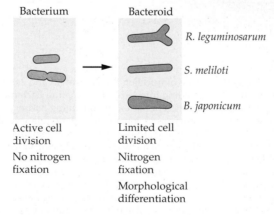

Figure 16.19

After release, bacteria differentiate to form morphologically distinct bacteroids. Different plant–bacterial combinations yield distinct bacteroid forms.

by bacteria of signals that modify plant metabolism and development.

Genetic analysis of the prokaryotic symbionts has revealed specific bacterial genes involved in the sequential stages of symbiosis (Table 16.5 and Fig 16.21). The rhizobial genes responsible for early nodule formation are termed *nod* genes. In some rhizobial species, *nod* genes are tightly clustered; in others, these genes are dispersed over hundreds of kilobases on chromosomes or plasmids. Some genes are common to all rhizobial species (e.g., *nodABC*); others are specific to a single bacterial species or strain. Expression of these bacterial genes depends on a chemical signal from the plant and an endogenous bacterial transcription activator, NodD (Fig. 16.22). Some bacterial species produce a single form NodD, whereas others synthesize multiple forms that may confer diverse regulatory responses. Additional negative and positive regulators of *nod* gene expression include bacterial proteins that respond to nutritional and other environmental signals.

The plant signals associated with *nod* gene expression are commonly flavonoids (Fig. 16.23; see also Chapter 24). For example, two compounds that occur naturally in alfalfa, luteolin (a flavone) and 4-methoxy-chalcone (a chalcone), act as inducers of *nod* gene expression in *Sinorhizobium meliloti*.

Vetch exudates contain diverse flavanones that activate transcription in *Rhizobium leguminosarum* bv. *viciae*. Soybean synthesizes the isoflavones genistein and daidzein, which induce *nod* gene expression in its symbionts, *Bradyrhizobium japonicum* and *S. fredii*. Other compounds that may be used as signals include betaines, such as trigonelline and stachydrine found in alfalfa root exudates, and aldonic acids, found in lupine. Plant phenylpropanoid and flavonoid pathways for synthesis of these inducers probably respond to bacterial presence, because the profile of flavonoids present in exudates changes in response to inoculation with *Rhizobium*. The precise points of this regulation are not completely defined but may affect flavonoid synthesis, processing, and secretion. The pattern of inducer and inhibitor production by the plant also implies both developmental and environmental regulation. For example, inducers appear in greatest concentrations close to the root tip, and flavonoid pathway enhancement in response to *Rhizobium* is similarly localized.

Bacterial responses to plant inducers may be modulated by signals such as nitrogen status and other environmental factors. It has been inferred from genetic data that NodD acts directly as a receptor for the plant signal, although this has not been demonstrated biochemically. The bacterial transcription factor NodD has probably co-evolved with the host plant pathways that generate flavonoids.

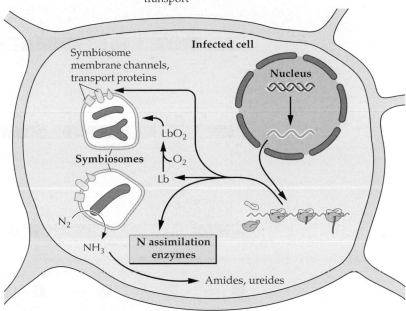

Figure 16.20
The plant nucleus encodes a series of genes that are expressed only during later stages of nodule development, in response to bacterial differentiation. As a general group, these plant genes are termed late nodulins. Some of these transcripts are specific to infected or uninfected cells. The plant-encoded products include the oxygen-binding protein leghemoglobin, specialized membrane proteins targeted to the symbiosome membrane, and enzymes that catalyze ammonia assimilation and synthesis of molecules used for transporting N to the rest of the plant.

Table 16.5 Some bacterial genes used in symbiosis with legumes

Stage of symbiosis	*Rhizobium* genes	Known or proposed functions
Gene regulation in response to host plant signal	*nodD, nolR, nodVW*, others	Activate or repress transcription at *nod* box promoters
Nodule formation, host recognition	*nod, nol, noe*	Enzymatic synthesis of Nod factors (*nodABC, nodH, nodEF, nodSU, nodZ*, and others)
Infection thread growth	*exo, lps, ndv*	Synthesis of extracellular polysaccharides
	Other genes?	Functions not known
Differentiation, bacteroid metabolism	*bacA*, possibly others	Signal import or export?
	dct genes	Import of dicarboxylic acids
Regulation of bacterial nitrogen-fixation genes	*fixL, fixJ, nifA, fixK*	Response to oxygen, transcriptional control at *nif* promoters
Nitrogen fixation	*nifHDK*, other *nif, fix*	Nitrogenase enzyme and cofactors, electron transport

Note: More than 100 genes have been identified in diverse symbiotic bacteria; only a subset are listed here.

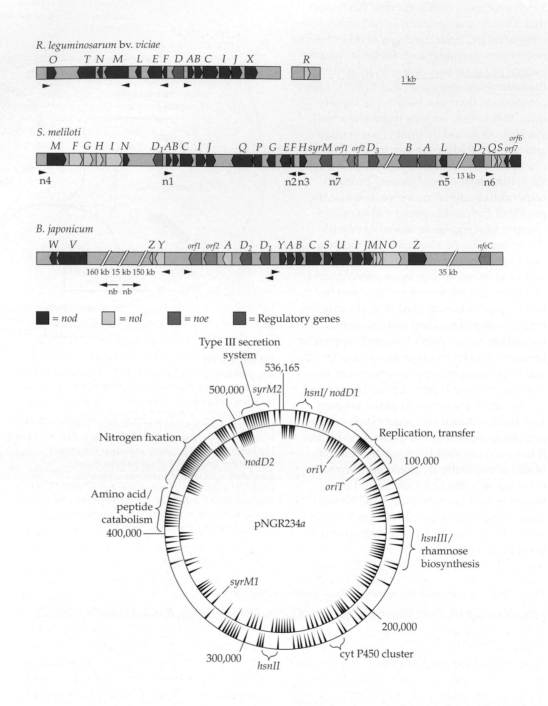

Figure 16.21

Maps of *nod* genes from several different bacterial strains. *nod* genes were originally defined as bacterial genes required for early steps in the formation of nodules; the definition has been expanded to include genes required for host range and genes that are coregulated by plant flavonoids and the bacterial activator, NodD. There are so many nodulation genes that the original letter designation, *nod*, was used up after naming 26 genes (*nodA–nodZ*); succeeding new genes have been designated *nol* and *noe*. Regulatory genes, including *nodD*, are shown in blue in the upper three maps. In some species, for example, *Rhizobium leguminosarum* bv. *viciae*, nodulation genes are clustered tightly on a symbiosis plasmid. In others, such as the *Sinorhizobium* broad-host-range species NGR234, functional analysis and complete DNA sequence determination have revealed that many host-specificity genes are dispersed widely on the symbiosis plasmid. Other bacterial species, such as *Bradyrhizobium japonicum*, carry their nodulation genes on the chromosome.

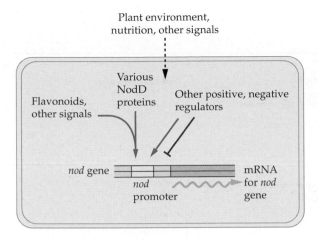

Figure 16.22
Regulatory circuit of NodD and plant inducers. In most *Rhizobium* species, a transcriptional activator, NodD, interacts with a plant inducer, typically a flavonoid, and with the transcriptional machinery of the bacterium to cause expression of the bacterial *nod* genes. Some species, notably *Rhizobium leguminosarum* bv. *viciae*, generate a single form of NodD that activates some genes and represses others. Other rhizobial species have multiple *nodD* genes that encode alternative forms of the transcription factor. Some of these alternative NodD proteins are able to activate *nod* gene expression in the absence of an inducer; other isoforms of NodD activate *nod* gene expression in the presence of nonflavonoid inducers such as betaines (e.g., stachydrine, trigonelline). The circuits for *nod* gene activation may also respond to environmental cues and to thus far unknown signals from the plant.

16.4.5 Rhizobial bacteria produce oligosaccharide and polysaccharide signals.

Several operons, comprising at least 20 genes in most *Rhizobium* species, are subject to NodD-flavonoid or other early symbiotic regulation. Many of these *nod* genes encode enzymes that direct the synthesis of a second signal, a bacterial product that acts as a morphogen to induce nodulation in the plant. These signal molecules are defined operationally as **Nod factors**. The known Nod factors are lipooligosaccharides, derivatized oligomers of chitin (β-1,4-linked *N*-acetylglucosamine [GlcNAc]) (Fig. 16.24). The core structure of Nod factor, three to five GlcNAc residues, is formed by NodC, a gene product common to all rhizobial species. Modifications of the core structure include N-acylation of the nonreducing end residue with diverse acyl groups by NodA and NodB, various C-6 modifications on the reducing and nonreducing ends, and N-methylation or O-carbamoylation of nonreducing end residues. Modifications on the inner GlcNAc residues include fucosyl and acetyl substitutions. Many of the enzymes that catalyze these derivatizations are encoded by so-called host-range genes, and the particular mix of decorations varies among bacteria that infect different plant hosts. Each rhizobial species has a unique array of host-range genes; such genes are found in some rhizobial species and not in others. Direct biochemical assays indicate that the lipooligosaccharide Nod factors are products of the known nodulation genes. However, the Nod enzymes may also form or modify other compounds that play a role in symbiosis, such as lipids and carbohydrates.

Specific extracellular polysaccharide structures may be required for invasion. In *S. meliloti*, mutants with alterations in the structure or processing of extracellular polysaccharides (EPSs) are unable to infect—the growth of their infection threads is feeble and aborts early. Important rhizobial extracellular carbohydrates include loosely associated EPSs (e.g., the succinoglycan illustrated in Fig. 16.25); lipopolysaccharides that extend from the lipid-A anchor in the bacterial outer membrane; K-antigens, unusual EPSs that contain keto-deoxyoctanoic acid; and cellulose, formed by rhizobial cells and important for colony morphology and host adhesion. The roles played by various EPSs in symbiosis appear to vary from one host–microbe pair to another.

Nod factors are sufficient to provoke root hair deformation and nodule formation in host plants but have no known effect on nonhost species. Nod factors may be required for symbiosis to progress beyond its early stages, but if so, they do not act alone. Genetic data indicate that bacterial exopolysaccharides and lipopolysaccharides, alone or in combination, are critical for successful invasion in certain plant hosts. Additional bacterial genes are specifically up-regulated during invasion and may play a role in bacterial growth or differentiation. In the broad-host-range *Sinorhizobium* sp. NGR234, a Type III bacterial secretion system is required for normal infection: By analogy with animal pathogens, the bacterium presumably transfers one or more proteins directly to the host

Host compounds that actively induce *nod* gene expression

Flavonoid inducers

Category	**Generic structure**	**Name/activity**	**Specific structure**
Flavone	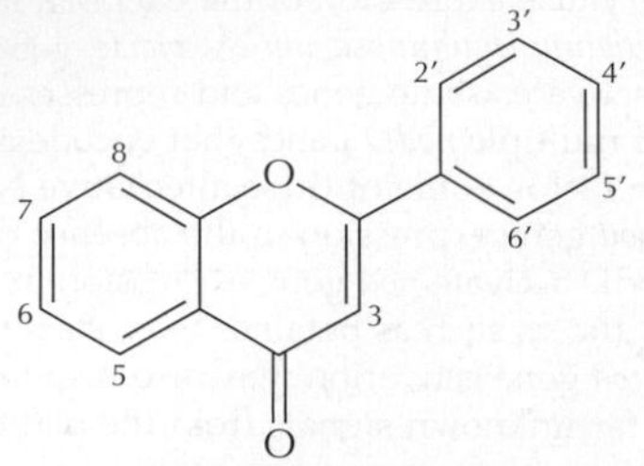	Luteolin, a flavone inducer from *Medicago* spp., active on *S. meliloti*	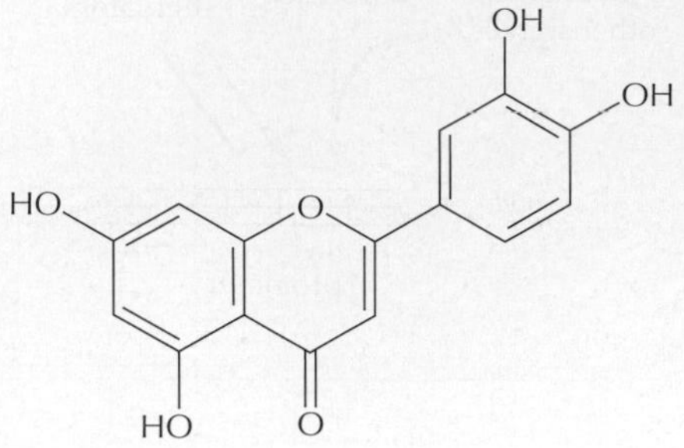
Chalcone	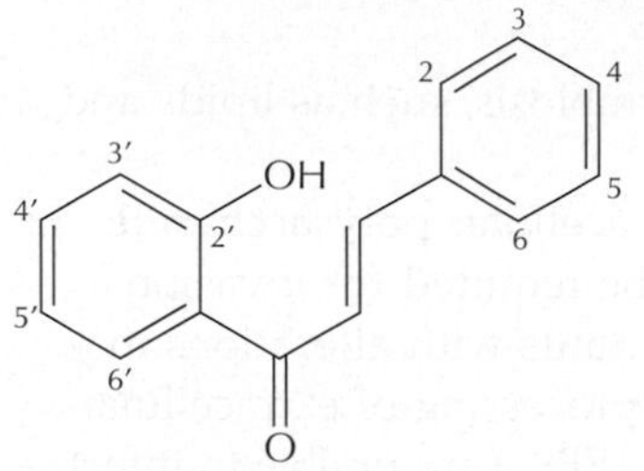	4,4′-Dihydroxy-2′-methoxychalcone, a chalcone inducer from *Medicago*	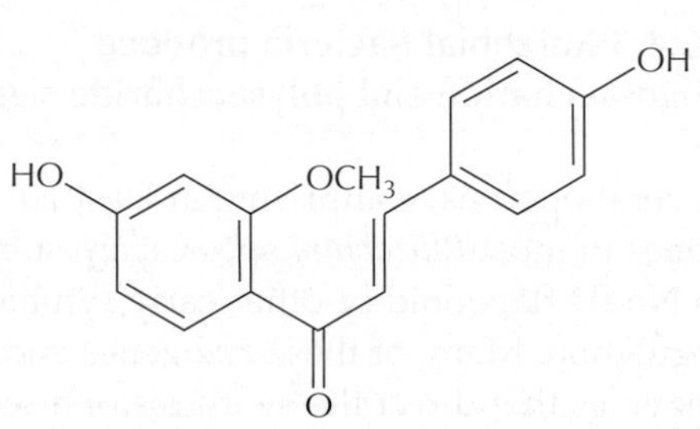
Isoflavone	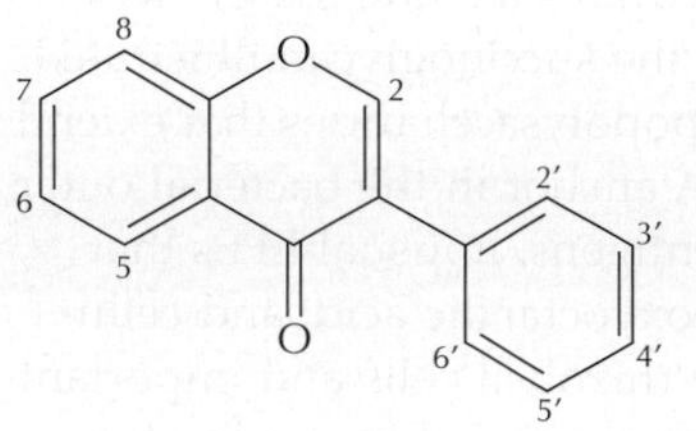	Daidzein, an isoflavone active on *B. japonicum*	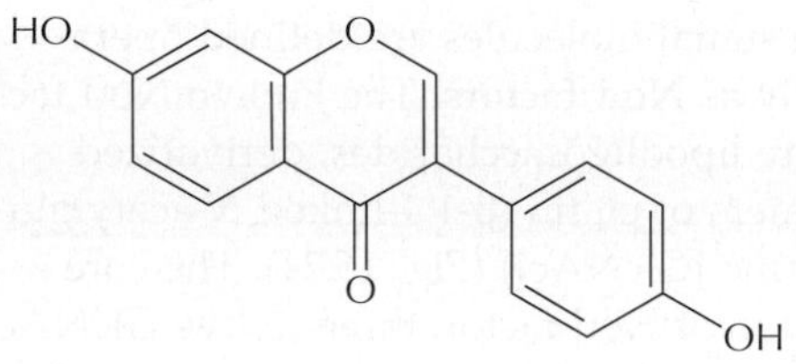
Flavanone	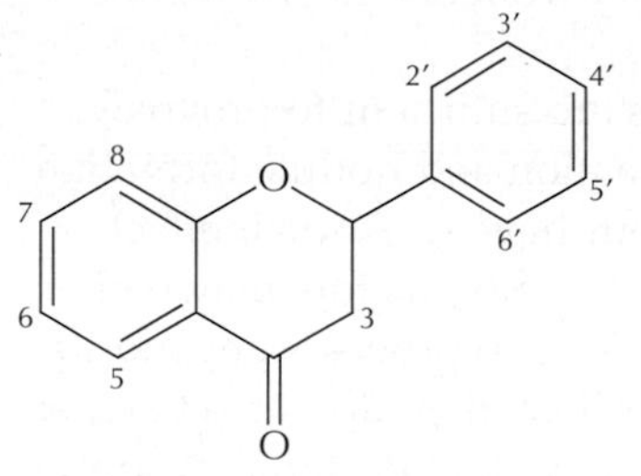	Naringenin, a flavanone active on *R. leguminosarum* bv. *viciae*	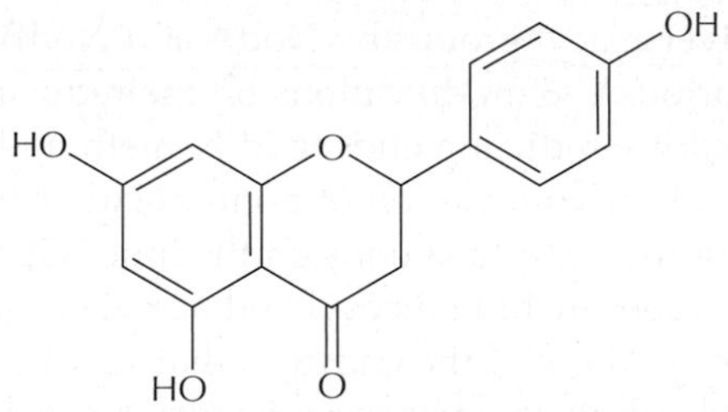

Nonflavonoid inducers

Trigonelline (a betaine)

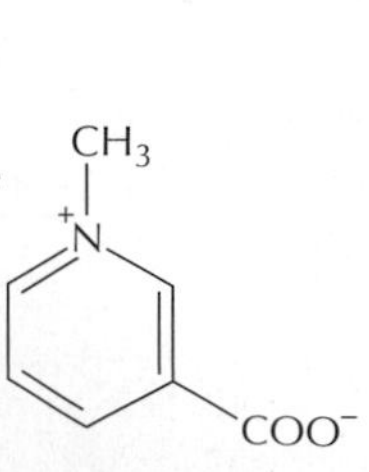

Tetronic acid (an aldonic acid)

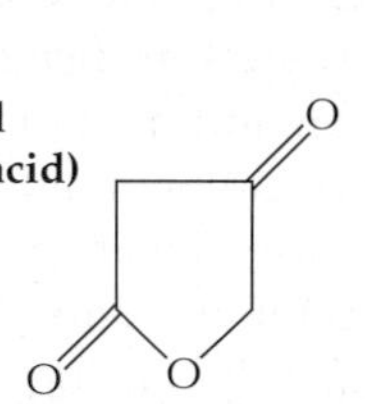

Figure 16.23
Structures of host compounds that induce *nod* gene expression in *Rhizobium*. The plant compounds that most actively induce the expression of nodulation genes vary among different plant–symbiont systems. A bacterium's *nodD* genotype is a primary determinant of the preferred inducer structure. Many plants have been found to secrete several different inducers, which may permit interaction with multiple NodD proteins from a single bacterial strain, or with diverse species or strains of bacterial symbionts.

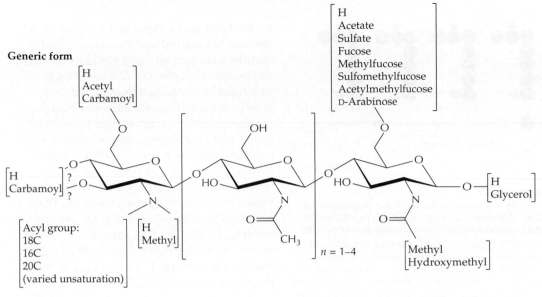

Generic form

$$\begin{bmatrix} H \\ Acetyl \\ Carbamoyl \end{bmatrix}$$

$$\begin{bmatrix} H \\ Acetate \\ Sulfate \\ Fucose \\ Methylfucose \\ Sulfomethylfucose \\ Acetylmethylfucose \\ \text{D-Arabinose} \end{bmatrix}$$

$$\begin{bmatrix} H \\ Carbamoyl \end{bmatrix} \begin{array}{c} ? \\ ? \end{array}$$

$$\begin{bmatrix} H \\ Glycerol \end{bmatrix}$$

$$\begin{bmatrix} Acyl\ group: \\ 18C \\ 16C \\ 20C \\ (varied\ unsaturation) \end{bmatrix}$$

$$\begin{bmatrix} H \\ Methyl \end{bmatrix}$$

$$n = 1\text{–}4$$

$$\begin{bmatrix} Methyl \\ Hydroxymethyl \end{bmatrix}$$

S. meliloti

R. leguminosarum bv. *viciae*

Figure 16.24
Structures of Nod factors, N-acylated chitooligosaccharides. All known Nod factors have a linear backbone of β-1,4-linked *N*-acetylglucosamine. Modifications to the reducing and nonreducing ends differ according to bacterial strain or species. Some bacteria make relatively simple mixtures of Nod factors, whereas others exude a rich and diverse mixture. The structure of a Nod factor, in particular the nature of the modifying groups, determines which host plant(s) will display nodulation-like reactions to the factor.

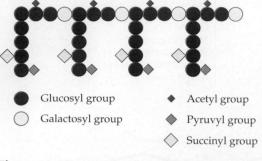

Glucosyl group   Acetyl group

Galactosyl group   Pyruvyl group

Succinyl group

Figure 16.25
Structure of a rhizobial extracellular polysaccharide (EPS). This example is the succinoglycan (EPS-I) of *Sinorhizobium meliloti*. The repeating unit of the EPS is an eight-sugar branched oligomer, consisting of a backbone of four residues [→4)β-Glc-(1→4)β-Glc-(1→4)β-Glc-(1→3)β-Gal-(1→] and a four-residue mixed β1→3 and β1→6 Glc side chain. The backbone and side chain are modified by acidic derivatization. The EPS contains hundred to thousands of these eight-residue repeats, of which four are shown here.

by this means. The processes of invasion and release, which require coordinated activity of bacterial and plant cells, may be characterized by further signal exchange.

16.4.6 Plants exhibit multiple responses to *Rhizobium* cells and signals.

The cellular basis for plant recognition and early response to rhizobial cells and their Nod factors is not known. Probably one or more plant receptors are responsible for tissue-specific responses to Nod factors and for signal amplification. Low- and high-affinity binding sites have been described in roots and cell cultures from several legume species.

Very low concentrations of Nod factor (10^{-9} M or less) can provoke profound morphological changes in the plant host. The earliest responses are in root hairs. In the first few minutes after the correct Nod factor signal is presented, the plasma membranes of alfalfa root hairs depolarize, which is accompanied by fluxes of specific ions (Fig. 16.26). Within 10 minutes, cells display periodic spikes in cytoplasmic calcium (Fig. 16.27). Cytoskeletal rearrangements and further reorientation of cell calcium gradients are seen in root hairs treated with either *Rhizobium* or Nod factors.

Plants express genes specifically in response to rhizobial inoculation. Nod factor alone induces expression of many of these genes. Plant genes expressed early, or in the absence of bacterial infection (as in empty nodules), are termed early nodulins (ENOD sequences). Induction of *ENOD* genes in root hairs has been detected as soon as 6 hours after addition of bacteria or Nod factor and probably occurs earlier. Newly infected root hairs express genes inferred to encode proline-rich proteins (PRPs) and a peroxidase homolog. In emerging nodules, some nodulins are expressed only in cells invaded by infection threads, whereas others are more broadly expressed in the primordium (Fig. 16.28). The functions of the ENOD products and the significance of their expression patterns are likely to be revealed only through combined genetic and biochemical analysis. In any case, they serve as useful markers for the successive stages of plant response to bacteria and as reporters for dissecting the effects of distinct microbial signals in early symbiosis.

Cell division in the cortex is detected within 18 to 30 hours and follows a pattern that is characteristic for each plant. The earliest divisions are sometimes strikingly located in the cortical region on a radius extending outward from each xylem pole. This implies a role for an endogenous host factor that comes from the stele. The signals that coordinate morphogenesis are not known, but molecular and developmental studies imply roles for the plant hormones auxin, cytokinin, and ethylene and for products of nodulin genes. Ethylene is specifically linked to nodule formation in some legumes by data from inhibitor studies and plant mutants and by the pattern of expression of ethylene biosynthesis genes in nodule tissues (Fig. 16.29).

16.4.7 Rhizobia and their plant hosts interact to create a microaerobic nodule environment conducive to nitrogen fixation.

Because nitrogenase is sensitive to oxygen, it is difficult to supply the enzyme with ATP and reductant simultaneously. Within the plant, the signaling and developmental pathways responsible for nodule development generate a microaerobic reducing environment that can stably support aerobic ATP synthesis.

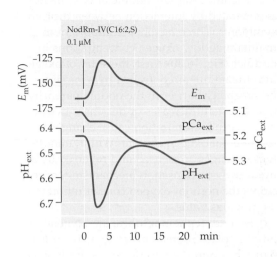

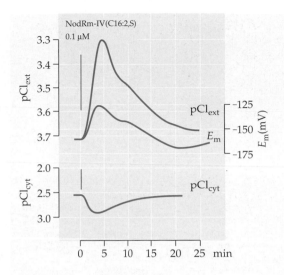

Figure 16.26

Membrane depolarization and ion fluxes occur within a few minutes of alfalfa root hair response to *Rhizobium* Nod factors. At time zero, root hairs were exposed to Nod factor produced by *Sinorhizobium meliloti* (structure NodRm-IV[C16:2,S], indicating that the structure has a backbone of four GlcNAc residues, a 16:2 carbon acyl group, and a sulfate modification at C-6). The root hairs respond rapidly with a change in plasma membrane potential, E_m (both left and right panels), which is accompanied by changes in extracellular calcium ion and pH (left) and in both extracellular and cytoplasmic concentrations of chloride ion (right). pCa and pCl refer to the negative logs of the concentration (M) of Ca and Cl ions, respectively.

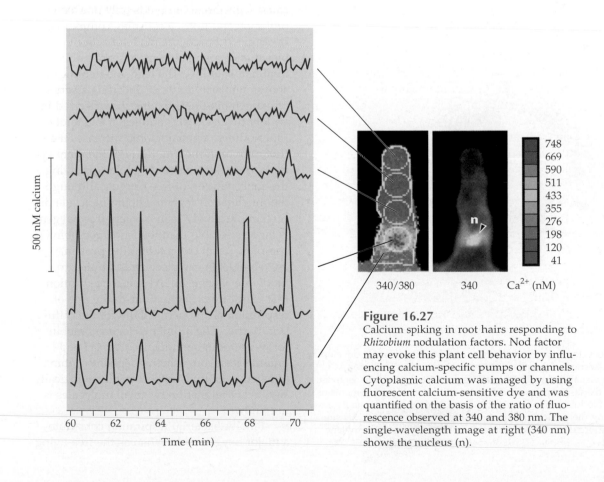

Figure 16.27

Calcium spiking in root hairs responding to *Rhizobium* nodulation factors. Nod factor may evoke this plant cell behavior by influencing calcium-specific pumps or channels. Cytoplasmic calcium was imaged by using fluorescent calcium-sensitive dye and was quantified on the basis of the ratio of fluorescence observed at 340 and 380 nm. The single-wavelength image at right (340 nm) shows the nucleus (n).

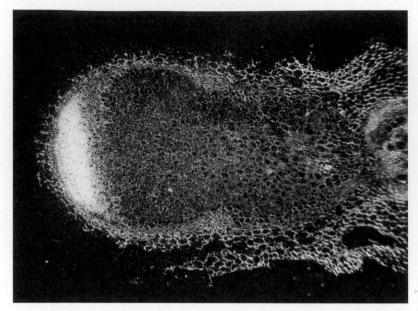

Figure 16.28
Patterns of early nodulin (*ENOD*) gene expression in developing nodules vary according to each gene. For example, *ENOD12* is expressed in the early infection zone of pea nodules, as show in this in situ hybridization photographed in dark-field illumination. Silver grains from probe hybridization scatter light and therefore appear bright against the dark background.

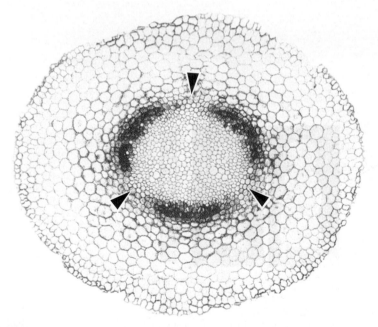

Figure 16.29
Expression of 1-aminocyclopropane-1-carboxylic acid (ACC) oxidase in developing nodules. In this cross-section of the pea root, hybridization of a probe for ACC oxidase transcripts is visualized as dark signal against the cleared root segment. The position of ACC oxidase expression correlates with the phloem elements of the root vasculature. Nodule primordia tend to form in the zones shown by arrowheads, representing the cortical cells just outside the xylem poles. This suggests that nodule primordia are unlikely to form in zones adjacent to ethylene production.

The microaerobic nodule environment is generated by interaction between the two symbionts. Three factors are important in maintaining low oxygen concentrations in nodules (Fig. 16.30). First, the entry of oxygen into the nodule is controlled by a variable-permeability barrier in the nodule parenchyma. Transient increases or decreases in external oxygen concentration cause changes in the internal oxygen content. Within minutes, these changes are sensed and reversed by an unknown compensation mechanism that restores the nodular oxygen concentration to its previous value.

Second, **leghemoglobin,** an oxygen-binding plant protein, plays an active role in regulating and delivering oxygen in the infected cells. Although results of older experiments were interpreted to mean that the heme component of leghemoglobin was synthesized by the rhizobia, that is not true; the entire leghemoglobin protein is made by the plant. A monomer with a single heme moiety, leghemoglobin is transcribed primarily in infected nodule cells and is produced in sufficient abundance to reach millimolar concentrations in the plant cytoplasm. Because leghemoglobin binds only one oxygen molecule and does not display complex cooperative behavior, its mode of action resembles that of myoglobin more than it does animal hemoglobins. The leghemoglobin gene shows similarity to those encoding a series of oxygen-binding proteins characterized in widely diverse plant families. In symbiotic root nodules, leghemoglobin increases the flux of oxygen moving through the plant cytoplasm to the bacteroids while controlling the concentration of free oxygen. Although the much larger leghemoglobin molecule diffuses more slowly than oxygen, free oxygen is far less soluble than the leghemoglobin:O_2 complex. As an oxygen-binding protein, leghemoglobin can also act as a buffer to moderate changes in oxygen concentration that result from fluctuations in the rate of respiration or the permeability of the diffusion barrier. The affinity of leghemoglobin for oxygen can be influenced by pH and organic acids, but whether leghemoglobin, like hemoglobin, is subject to physiologically significant regulation by small molecules is not known.

Finally, bacterial respiration constitutes a major oxygen sink. Whereas the free-living rhizobia typically have a cytochrome oxidase with a K_m for oxygen of around 50 nM, the bacteroid cytochrome oxidase has a very low K_m for oxygen, about 8 nM. Expressed only in nodules, this latter cytochrome oxidase is required for nitrogen fixation. In contrast, K_m values for plant mitochondrial cytochrome oxidases are near 100 nM O_2. Because bacterial respiration is less limited than plant respiration by low oxygen concentrations, most nodule respiration is thought to occur in the bacteroids, although the more peripheral location of the plant mitochondria in infected cells may give them better access to oxygen.

The low oxygen concentration in nodule cells limits oxidative metabolism. If the external oxygen concentration is increased moderately and quickly, bacterial respiration and nitrogen fixation also increase, but these return to previous rates as nodule permeability declines. Thus, the nodular oxygen concentration maintained by the plant supports submaximal nitrogenase activity, a factor that may contribute to the stability of symbiotic nitrogen fixation. Bacterial respiration consumes oxygen, and the ATP generated by this respiration is used by nitrogenase. If the external oxygen concentration is raised to high values, however, inactivation of nitrogenase decreases the rate of catalysis, diminishing ATP consumption. The increased ATP/ADP ratio down-regulates respiration, so that less O_2 is consumed and the oxygen concentration increases, inactivating more nitrogenase and eventually leading to metabolic collapse. Maintaining low concentrations of oxygen might provide a margin of safety for the nitrogen-fixing system.

The low oxygen concentration within the nodule is a key element in regulating bacteroid expression of nitrogenase. In some rhizobial species, an oxygen-sensitive hemoprotein kinase, FixL, controls a regulatory cascade that activates transcription of nitrogen fixation genes (Fig. 16.31). FixL, part of a two-component regulatory system, phosphorylates its partner, FixJ. Once phosphorylated, FixJ activates transcription of other regulatory proteins. Two of these, NifA and FixK, control expression of diverse *nif* and *fix* genes. Only in the correct context of a developed, low-oxygen nodule will the bacteroids express nitrogenase and the associated proteins required for reduction of dinitrogen to ammonia.

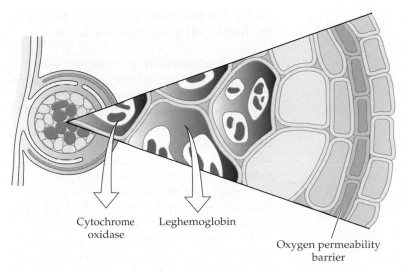

Cytochrome oxidase   Leghemoglobin

Oxygen permeability barrier

Figure 16.30
In nitrogen-fixing nodules, mechanisms that maintain ATP production in an appropriately low oxygen environment include a high-affinity cytochrome oxidase, the oxygen-binding protein leghemoglobin, and a variable-permeability barrier that controls gas exchange at the nodule periphery. The "empty" cells around the central region are thought to contain the permeability barrier that limits the flow of oxygen to the leghemoglobin-containing central region, which contains the infected cells. A photograph showing these cells can be seen in Figure 16.17(C, iv).

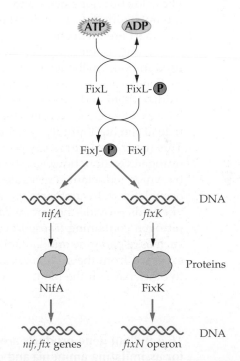

Figure 16.31
In response to a low-oxygen environment, the FixL/FixJ signal cascade activates expression of bacterial genes required for nitrogen fixation. The FixL heme-containing protein is anchored in the bacterial plasma membrane. In the absence of oxygen, FixL is phosphorylated and in turn phosphorylates the soluble FixJ response regulator, which can bind to DNA and activate transcription of the genes needed for nitrogen fixation. Binding of oxygen to the FixL heme inhibits the protein's histidine kinase activity, repressing the expression of these genes under conditions where nitrogenase would be unstable.

16.4.8 The host plant provides carbon to the bacteroids as dicarboxylic acids.

Although photosynthate enters the nodule as sucrose, genetic and biochemical evidence shows that the bacteroids do not need to catabolize mono- or disaccharides to fix nitrogen effectively. Instead, sugars are converted to organic acids before the carbon is made available to the bacteria, probably by using pathways more often associated with fermentation than oxidation. Some evidence indicates that phospho*enol*pyruvate (PEP) is converted to oxaloacetate by PEP carboxylase, and NAD^+ is regenerated by the reduction of oxaloacetate to malate (see Chapters 13 and 14). Intact nodules rapidly incorporate $^{14}CO_2$ label into malate and aspartate, and tissue concentrations of PEP carboxylase and malate dehydrogenase are increased by the synthesis of nodule-specific isoforms of these enzymes. Transport activities further support an important role for organic acids in symbiotic carbon metabolism. The symbiosome membrane can transport dicarboxylic acids but not sugars, and most bacterial mutants with defects in dicarboxylic acid transport are ineffective.

In the nodule, dicarboxylic acids are used in both catabolic and anabolic processes. Some acids are fed to bacteroids, which oxidize them and transfer the electrons to O_2 by way of electron transfer chains that terminate at the high-affinity cytochrome oxidase. This respiration ultimately provides ATP for nitrogen fixation; however, the mechanism by which reductant is generated has not yet been defined. In another major role, dicarboxylic acids provide the carbon backbones for nitrogen-containing transport compounds, such as asparagine and glutamine, which are exported from the nodule to carry nitrogen to other parts of the plant.

16.4.9 Plant gene products are responsible for assimilating ammonia and exporting nitrogen from the nodules.

Unlike free-living bacteria, which assimilate the nitrogen they fix, bacteroids release to the plant the ammonia they produce. Ammonia assimilation occurs in the plant nodule cytosol and organelles. Plant glutamine synthetase (GS) and NADH-dependent glutamate synthase (NADH-GOGAT) are responsible for the initial assimilation of ammonia into organic compounds (Fig. 16.32A; see also Chapter 8). In some plants, a nodule-specific GS is expressed; in others, synthesis of a vegetatively expressed GS is up-regulated.

After its assimilation into glutamine, the fate of ammonia depends largely on the nitrogenous transport compounds used by the plant host. In one major class of legumes that includes alfalfa and pea, nodules export ammonia as **amides,** namely, glutamine and asparagine (Fig. 16.32B). Nodules contain active aspartate aminotransferases. In alfalfa, a nodule-induced aspartate aminotransferase is located in plastids found at the periphery of infected cells, especially near the air spaces between cells. A nodule-induced asparagine synthetase, also present in infected tissues, synthesizes asparagine from aspartate, using glutamine as an amide donor (Fig. 16.32A). In trefoil (*Lotus*), an asparaginase that is usually active in roots is repressed during nodule development. This repression might facilitate export of intact asparagine from the nodules. Amides are easily generated from citric cycle intermediates (see Chapter 14) and can be metabolized in leaves by transamination and transamidation reactions similar to those used in their synthesis (see Chapter 8).

A second major group of legumes, including soybean (*Glycine*) and cowpea (*Vigna*), exports nitrogen from the nodules as the **ureides** allantoin and allantoic acid—compounds produced by synthesizing purines in the infected cells and then oxidizing these purines in neighboring uninfected cells (Fig. 16.33A). Regulation of this pathway is complex. Plants grown in an argon:oxygen atmosphere contain lower concentrations of purine oxidation enzymes, but the enzymes of purine synthesis are almost completely absent. Thus, purine synthesis appears to be driven by ammonia availability, whereas the purine oxidation enzymes are induced as part of a developmental program. The enzymes of the ureide synthetic pathway display compartmentalization, and at least some of the enzymes are expressed selectively in noninfected cells of the nodules (Fig. 16.33B). Both amides and ureides are exported from the nodule primarily through the xylem.

(A)

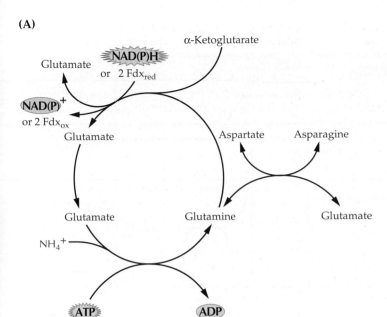

Figure 16.32

Primary nitrogen assimilation pathways in nodules of pea, alfalfa, and other species that export amides. (A) The GS-GOGAT cycle. Ammonium is combined with glutamate to yield glutamine in an ATP-dependent reaction. Reaction of glutamine with α-ketoglutarate and a reductant yields two molecules of glutamate—one of which is used to regenerate the cycle, the other to deliver amino groups to general metabolism. (B) Plant carbon metabolism produces dicarboxylic acids, which are used to generate the ATP and reductant needed for nitrogen fixation in the bacteroids. The reduced nitrogen released is assimilated in the plant cytoplasm and plastids to produce glutamine and asparagine, which are transported through the xylem to the rest of the plant.

(B)

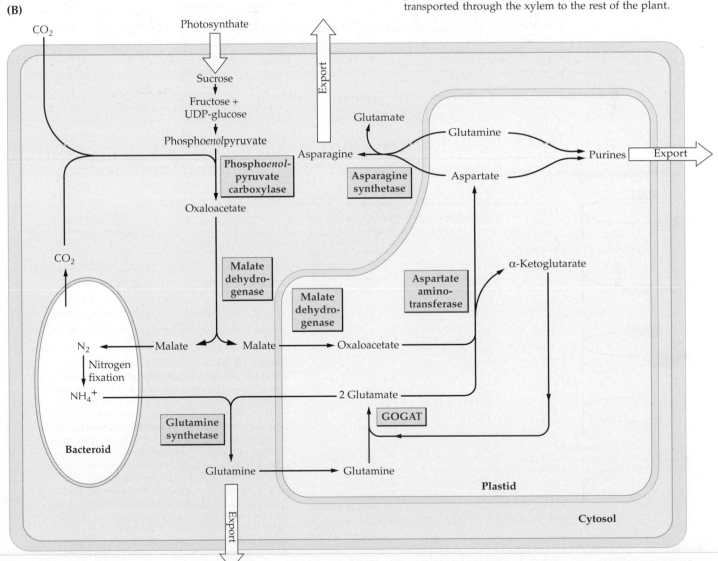

(A)

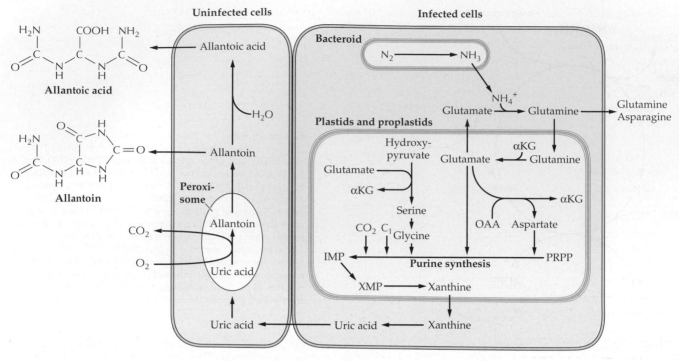

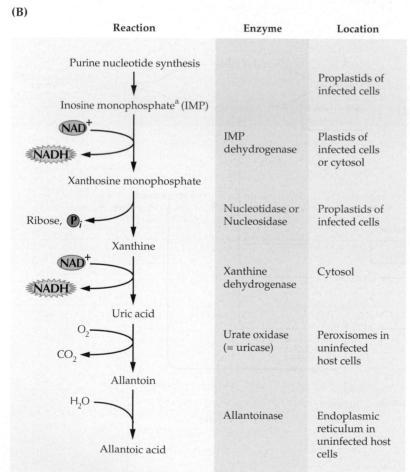

(B)

Reaction	Enzyme	Location
Purine nucleotide synthesis		Proplastids of infected cells
Inosine monophosphate[a] (IMP)	IMP dehydrogenase	Plastids of infected cells or cytosol
Xanthosine monophosphate	Nucleotidase or Nucleosidase	Proplastids of infected cells
Xanthine	Xanthine dehydrogenase	Cytosol
Uric acid	Urate oxidase (= uricase)	Peroxisomes in uninfected host cells
Allantoin	Allantoinase	Endoplasmic reticulum in uninfected host cells
Allantoic acid		

[a]An alternative possibility is conversion of inosine monophosphate to hypoxanthine, which is then oxidized to xanthine by xanthine dehydrogenase.

Figure 16.33
Complex cellular and subcellular compartmentation of ureide biosynthesis enzymes in plants such as soybean and cowpea. (A) In tropical legumes such as cowpea and soybean, nitrogenous compounds in addition to glutamine and asparagine are exported from the infected cells. A substantial fraction of the fixed nitrogen is exported to adjacent uninfected cells as uric acid, a decomposition product of purines. In the uninfected cells, uric acid is degraded to allantoin and allantoic acid, which are then transported through the xylem to the rest of the plant. The uricase reaction is especially interesting because it depends on oxygen, which is in limited supply. (B) The purine degradation pathway in nodules, showing the enzymes involved and their locations.

In the leaves of nodulated plants, ureides are degraded to release ammonia, which is then assimilated by glutamine synthetase and ferredoxin-dependent GOGAT. Ureide catabolism in the leaves releases the nitrogen directly from allantoin by using allantoin amidohydrolase, rather than first generating urea and degrading the urea by way of urease activity.

16.5 Ammonia uptake and transport

Plant cells have the capacity for active transport of ammonium ion (NH_4^+). Plants may encounter substantial concentrations of NH_4^+ ion in acidic soils, where the rates of nitrification and thus the availability of nitrate are low. Physiological studies have revealed multiphasic NH_4^+ uptake in diverse plant species, implying multiple transport systems (see Section 16.6 and Chapter 23). The K_m values for the transporters range from 10 to 70 μM for NH_4^+. Direct genetic studies of *Chlamydomonas* suggest at least two high-affinity systems.

Plant genes coding for ammonium transport have been characterized on the basis of their ability to restore growth to a yeast mutant defective in NH_4^+ uptake or by the results of in vivo functional tests. One set of such genes identified in tomato (*Lycopersicon*) and *Arabidopsis* encodes AMT1 transporters. The genes are homologous to each other and to a yeast transporter of NH_4^+. Inferred peptide sequences for the AMT1 transporters display features expected for an active transport system, such as multiple putative membrane-spanning regions and homology to other transporters. Sequence inspection and physiology studies in heterologous systems suggest that NH_4^+ transport depends on electrochemical potential rather than direct ATP synthesis. K_m values for AMT1 transport of the analog methylamine are about 65 μM. Tissue patterns of expression for cloned AMT type genes have shown wide expression in some cases and root hair–specific expression in others. More complete identification of all transporters will provide a picture of their respective functions in different developmental and physiological contexts.

A specialized ammonium exchange situation may exist in plant cells that host nitrogen-fixing bacteria. The symbiosome membrane acts as the conduit for plant compounds transported to the bacteroids and for bacterial compounds transported to the plant. The latter probably include ammonium, amino acids, or other N-containing compounds. Plants, in contrast, express nodulin proteins targeted to the symbiosome membrane, including a specific ammonium transporter, SAT1, identified in soybean (*Glycine max*). The GmSAT1 protein has unusual features and may have only one transmembrane domain. Direct electrophysiological studies show a K_m for methylamine of about 5 mM, consistent with physiological studies of whole symbiosome membranes. Because the expression of this transporter is specific to nodulated roots, it is unlikely to participate in transporting ammonia from the environment.

16.6 Overview of nitrate uptake and reduction

Nitrate is a major source of nitrogen for plants; indeed, most plants devote a significant portion of their carbon and energy reserves to its uptake and assimilation. Nitrate serves both as nutrient and signal and has profound effects on plant metabolism and growth. Plants have evolved intricate mechanisms to detect nitrate and to integrate its assimilation with photosynthesis and the overall metabolism of nitrogen and carbon. These mechanisms allow plants to control growth rates, root architecture, carbon/nitrogen ratios, concentrations of reductants, and ionic and pH balances under diverse environmental conditions.

The assimilation of nitrate begins with its uptake into the cell (Fig. 16.34). Ordinarily, nitrate is taken up from the soil solution by epidermal and cortical cells of the root. Primary uptake can also occur in leaves, a route that is important for epiphytes and for incorporation of foliar applications of fertilizer. Once within the symplasm, nitrate can be transported into the vacuole, where storage of high concentrations (more than 20 mM) is possible. Major storage organs for nitrate include the roots, stems, and leaf midribs. Nitrate can also be loaded into the xylem for long-distance transport to the shoot. Diverse physiological, genetic, and

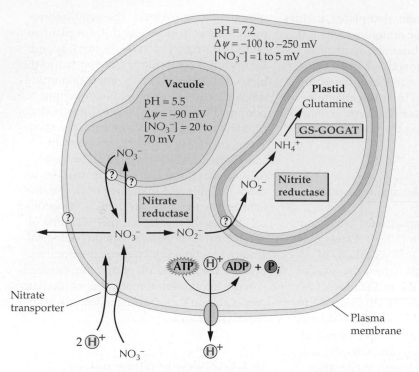

pH = 7.2
$\Delta\psi = -100$ to -250 mV
$[NO_3^-] = 1$ to 5 mV

Vacuole
pH = 5.5
$\Delta\psi = -90$ mV
$[NO_3^-] = 20$ to 70 mV

Plastid
Glutamine

NO_3^-

(?) (?)

(?)

NH_4^+

GS-GOGAT

NO_2^-

Nitrate reductase

Nitrite reductase

$NO_3^- \rightarrow NO_2^-$

(?)

ATP (H)$^+$ ADP + (P)$_i$

Nitrate transporter

2 (H)$^+$

NO_3^-

(H)$^+$

Plasma membrane

Figure 16.34
Nitrate assimilation by plant cells involves transport of nitrate across the plasma membrane and then reduction to ammonia in a two-step process. A proton-pumping ATPase maintains the electrochemical gradient that drives cellular uptake of nitrate. The values shown for electrical potentials and intracellular nitrate concentrations are typical but can vary significantly.

environmental factors determine how nitrate is allocated and stored throughout the plant.

Nitrate itself is not incorporated into organic compounds but is first reduced to ammonium in a two-step process (Fig. 16.34). Nitrate is reduced to nitrite by nitrate reductase (NR), and nitrite is reduced to ammonium by nitrite reductase (NiR). During this eight-electron transfer, the oxidation state of nitrogen drops from +5 to –3. The subsequent reactions that assimilate ammonium into amino acids also consume organic carbon (see Chapter 8). Plants reduce nitrate and nitrite in both root and shoot tissues. Significant quantities of nitrogen are assimilated through this pathway, which consumes large amounts of energy, carbon, and protons.

16.6.1 Nitrate uptake is carrier-mediated.

As nitrate diffuses from the soil solution into the apoplasm of the root, it is taken up by the epidermal and cortical cells. Once in the symplasm, it can be reduced or it can be mobilized across the casparian strip and into the xylem for transport to the shoots. Initial uptake across the plasma membrane is a regulated, active transport process.

Plants utilize both high-affinity and low-affinity transport mechanisms to import nitrate. The high-affinity transport system

(also called HATS, or mechanism I) displays Michaelis–Menten kinetics, saturating at 0.2 to 0.5 mM nitrate with a K_m typically between 10 and 100 µM (Fig. 16.35A). The high-affinity system is further divided into a constitutive component, which is expressed in the absence of nitrate, and an inducible component, which is induced by nitrate treatment. The low-affinity transport system (LATS, or mechanism II) is most observable at nitrate concentrations above 0.5 mM and usually displays nonsaturating uptake kinetics (Fig. 16.35B). These import pathways allow plants to accommodate a wide range of external nitrate concentrations (5 µM to 50 mM) without experiencing severe deficiency or toxicity.

16.6.2 Gene products associated with both high- and low-affinity nitrate uptake have been identified.

Plants have multiple nitrate carriers with distinct kinetic properties and regulation. Two nitrate transporter gene families, *NRT1* and *NRT2*, have been discovered so far. The *NRT2* family encodes transporters that contribute to the inducible high-affinity uptake system. The *NRT1* family is more complex, including nitrate transporters with dual affinity (both low and high K_m) or low affinity.

(A)

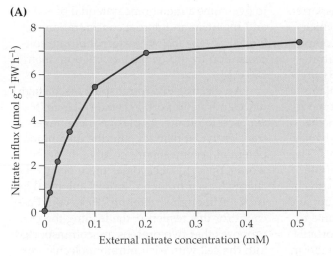

(B)

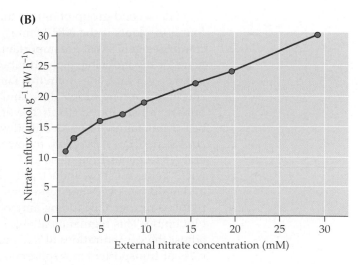

Figure 16.35
Kinetics of nitrate uptake. Nitrate influx was measured as a function of external nitrate concentration in barley roots that had been pretreated with 0.1 mM nitrate to induce the uptake system.

(A) Uptake by the high-affinity system demonstrates Michaelis–Menten kinetics. (B) Uptake by the low-affinity system demonstrates nonsaturating kinetics.

A member of the *NRT1* family was first identified during a study of mutants resistant to chlorate (ClO_3^-), a chlorine analog of nitrate that is toxic when taken up and reduced to chlorite by NR. Mutants that are impaired in either the uptake or reduction of nitrate or chlorate are resistant to chlorate treatment (Fig. 16.36). The *chl1* mutant of *Arabidopsis* exemplifies a class of mutants that has diminished nitrate-uptake activity but wild-type quantities of NR activity. The *CHL1 (AtNRT1)* gene is expressed in the outer cell layers of the root (epidermal, cortical, and endodermal cells), and concentrations of its mRNA are increased by treating plants with nitrate or solutions at slightly acidic pH. The CHL1 protein is hydrophobic, containing 12 membrane-spanning regions, a membrane topology found in almost all cotransporters. The CHL1 protein can be functionally expressed in *Xenopus* oocytes, where it shows electrogenic nitrate-uptake activity. More detailed analysis has shown that CHL1 is a dual-affinity transporter displaying two K_m values for nitrate (approximately 35 µM and 8 mM), properties that are consistent with the *chl1* mutant phenotype: reduced nitrate uptake at both low and high nitrate concentrations. These findings indicate that CHL1 is a component of both the high- and low-affinity nitrate-uptake systems in plants.

AtNRT1:2, a gene closely related to *CHL1,* has been found in *Arabidopsis* and is thought to be another component of the low-affinity uptake system. Expression of this gene is constitutive (i.e., does not require nitrate treatment) and results in nitrate uptake in oocytes. Two closely related genes in the *CHL1* family, one nitrate-inducible and the other constitutive, have been identified in tomato, indicating that, in many dicots, the low-affinity nitrate-uptake system is made up of at least two distinct transporters. *CHL1* is also related to a larger gene family of cotransporters that includes proton/amino acid and proton/dipeptide cotransporters found in yeast, mammals, and plants.

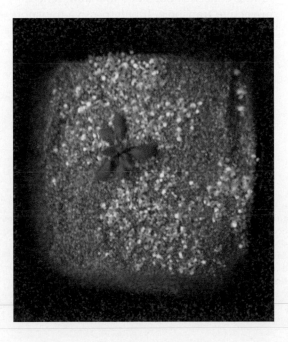

Figure 16.36
Nitrate-uptake mutants can be selected for by treating plants with chlorate. Wild-type plants take up chlorate and reduce it to the toxic product chlorite, which results in chlorosis. Nitrate-uptake mutants are unable to import chlorate and so remain green.

The second group of nitrate transporters found in plants is the *NRT2* family, which comprises genes that are important components of the inducible high-affinity nitrate-uptake system. Genes from this family are conserved among fungi, algae, and plants but are not similar to *CHL1*. The first member of this family to be identified was the *crnA* gene from *Aspergillus*, followed by several *NRT2* genes from algae and plants. The *NRT2* genes all show induction by nitrate and are down-regulated by several forms of reduced nitrogen, including ammonium and glutamine. This regulation allows plants to adjust the concentrations of this important class of transporters in response to the form of nitrogen in the soil solution and the nitrogen needs of the plant.

16.6.3 Nitrate uptake is driven by the proton gradient across the plasma membrane.

Plant cells import and accumulate nitrate against an electrochemical gradient (see Fig. 16.34). The principal electrochemical gradient of the plant cell, a proton gradient established by the plasma membrane proton-pumping ATPase, can generate potentials of –100 to –250 mV (inside negative). When roots are immersed in nitrate solutions at concentrations of 0.1 mM to 10 mM, root epidermal and cortical cells achieve cytosolic nitrate concentrations of 3 to 5 mM, reflecting active transport. If only passive transport were involved, an external nitrate concentration of more than 50 mM would be required

to overcome a membrane potential of –60 mV and maintain an internal concentration of 5 mM; at –120 mV, the external nitrate concentration would be have to be more than 500 mM.

An electrogenic cotransport mechanism drives nitrate import. When a cell is first exposed to nitrate, its plasma membrane transiently depolarizes, and the membrane potential becomes more positive. In other words, the cytosol becomes more positively charged, even though nitrate is anionic (Fig. 16.37). Although several processes could account for this response, the leading theory proposes that two protons are cotransported into the cell with each nitrate molecule, resulting in the import of a net positive charge. This model provides a mechanism for active nitrate transport by both the high- and low-affinity systems, with the proton motive force (pmf) across the plasma membrane energizing nitrate uptake. Within one or two minutes after the onset of depolarization, the plasma membrane repolarizes to its initial potential by increasing pumping activity of the proton ATPase (see Chapter 3).

16.7 Nitrate reduction

The first committed step in the nitrate assimilation pathway is the reduction of nitrate to nitrite (Rx. 16.2). NADH or NADPH serves as the reductant, and a proton is consumed in the reaction. This reaction is catalyzed by NR, a complex metalloenzyme that forms homodimers (Fig. 16.38) and homotetramers. NR has binding sites for NAD(P)H and for

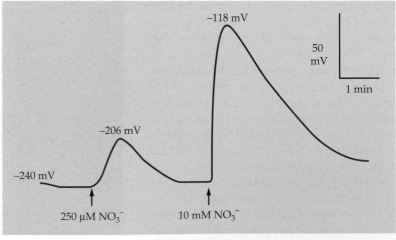

Figure 16.37

Membrane depolarization in response to nitrate. Potentials across the plasma membranes of root epidermal cells of *Arabidopsis* become more positive when plants are first exposed to nitrate. The first depolarization results primarily from uptake by the high-affinity system (measured at 250 μM nitrate), whereas the second depolarization reflects uptake by both the high- and low-affinity systems (measured at 10 mM nitrate). Plants were grown in the absence of nitrate before the measurements were taken.

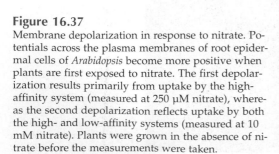

nitrate. Three cofactors—FAD, heme-Fe, and molybdenum cofactor MoCo—provide the redox centers that facilitate the chain of electron transfer reactions diagrammed below. MoCo is a molybdenum ion complexed with molybdopterin (Fig. 16.39). In the native enzyme, the midpoint potentials ($E'°$) for FAD, heme-Fe, and MoCo are –272, –160, and –10 mV, respectively.

Reaction 16.2: Nitrate reductase

$$NO_3^- + NAD(P)H + H^+ \longrightarrow$$
$$NO_2^- + NAD(P)^+ + H_2O$$

Each NR subunit is almost 1000 amino acids long and contains all three cofactors. Most forms of plant NR use NADH, but some are bispecific and use either NADPH or NADH.

16.7.1 NR subunits contain three distinct regions, each associated with a specific electron transport cofactor.

Within the NR holoenzyme, each cofactor is a redox center associated with a distinct functional and structural region of the protein. Partial proteolysis of the enzyme produces discrete fragments that display partial enzymatic activity. One fragment binds FAD and can use NADH to reduce ferricyanide, an artificial electron acceptor. A second fragment contains MoCo and heme-Fe and can reduce nitrate in the presence of methyl viologen (MV), an artificial electron donor. The partial activities associated with NR fragments are consistent with the structural evidence for discrete functional regions. The spatial arrangement of functional regions, inferred from cDNA sequences (see below), are as follows: the MoCo domain is near the N terminus, the heme-Fe domain is in the middle, and the FAD domain is at the C terminus (see Fig. 16.38). The three functional regions are connected by two hinges, one of which contains a regulatory site that binds phosphate and 14-3-3 proteins (see Chapters

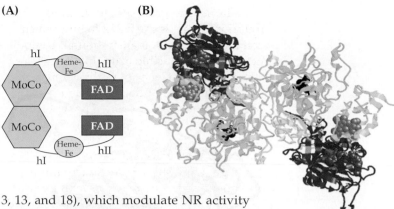

3, 13, and 18), which modulate NR activity (see Section 16.7.4). A proposed model for the structure of NR showing each functional region is provided in Figure 16.38B.

Each functional region of NR can be considered an independent unit, belonging to a distinct protein family. The FAD-binding region (260 to 265 amino acids long) is similar to the ferredoxin-NADP$^+$ reductase (FNR) family of flavin oxidoreductases. This family includes FNR, cytochrome P450 reductase, nitric oxide synthase, and cytochrome b_5 reductase. The crystal structure of the FAD region reveals two domains, each forming a lobe, separated by a cleft (Fig. 16.40). The N-terminal lobe binds the FAD; the C-terminal lobe binds the substrate NAD(P)H. In the C-terminal lobe, a cysteine provides a thiol group that interacts with the NAD(P)H and is thought to position the NAD(P)H to facilitate electron transfer to FAD. Although this

Figure 16.38
(A) Domain structure of nitrate reductase. An NR monomer has three major domains, which bind molybdenum cofactor, heme, and FAD, respectively. The FAD-binding region receives electrons from NAD(P)H; the heme domain shuttles electrons to the MoCo binding region, which transfers electrons to nitrate. hI and hII refer to hinge 1 and hinge 2, which separate the functional domains.
(B) Ribbon diagram of nitrate reductase. The heme prosthetic group is shown in purple, FAD in blue, and MoCo in black. The interface between the two monomers is shown in yellow.

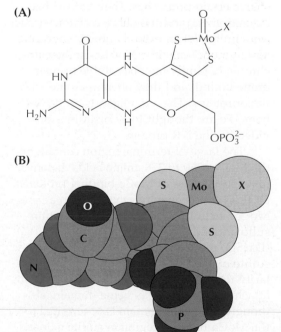

Figure 16.39
Molybdenum cofactor (MoCo) of NR.
(A) Chemical structure.
(B) Space-filling model.

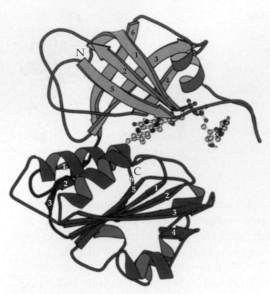

Figure 16.40
Crystal structure of FAD domain of NR. The C-terminal fragment of NR, showing the N-terminal lobe of the fragment (green), a linker domain (blue), and the C-terminal lobe (red). The N-terminal lobe is bound to FAD, whereas the C-terminal lobe is thought to bind NADH (not shown).

cysteine is invariant in the FNR family of flavoproteins, it is not required by NR; a serine substitution for the cysteine renders NR less active but still able to bind NADH and reduce FAD.

The central heme domain of 75 to 80 residues is similar to that of the cytochrome b_5 family of heme proteins. These proteins share certain properties: They can oxidize cytochrome c, and they have a characteristic cytochrome b-type fold that binds the heme moiety noncovalently. Proteins in the cytochrome b_5 family may be soluble or membrane-bound, and they often associate with flavoproteins. The axial ligands of the NR heme-Fe are thought to be histidines provided by the NR protein.

The large N-terminal region containing the MoCo (360 to 370 amino acids) belongs to a special class of MoCo-binding proteins. Of the various enzymes that contain a MoCo, including xanthine oxidase, biotin sulfoxide reductase, dimethyl sulfoxide (DMSO) reductase, and sulfite oxidase, only sulfite oxidase has significant sequence similarity to NR. Like NR, sulfite oxidase has an attached cytochrome b_5 heme domain, although this heme domain is N-terminal to the MoCo-binding region in sulfite oxidase

and C-terminal to the MoCo region in NR. These alternative arrangements of functional protein units suggest that domains have been combined in different configurations to form new enzymes during the course of evolution.

16.7.2 Nitrate is reduced in the cytosol of root and shoot cells.

NR and nitrate reduction are localized in the cytosol of cells throughout the vegetative organs of the plant. In most species, NR is found in both shoots and roots, its distribution depending on environmental conditions. However, some species (e.g., cranberry, white clover, and young chicory) localize almost all their NR in roots, whereas others (e.g., cocklebur) express NR almost exclusively in leaves. Within a specific organ, NR shows cell type–specific localization. At low external concentrations of NO_3^-, NR is found primarily in epidermal cells and cortical cells close to the root surface. At higher external concentrations of NO_3^-, activity is also detected in cells of the cortex and vascular system. In maize, a C_4 plant, NR is located in mesophyll but not in bundle sheath cells—a finding consistent with the far-greater capacity of mesophyll chloroplasts to generate reductant by way of photosynthetic noncyclic electron transfer (see Chapter 12).

16.7.3 Nitrate and other compounds serve as signals to regulate NR gene expression.

The regulation of NR plays a key role in nitrate assimilation. Green tissues have much more NiR activity than NR activity, ensuring that nitrite does not accumulate to toxic amounts. Thus, NR is thought to catalyze the rate-limiting step in the conversion of nitrate to ammonium. Plants use several mechanisms to adjust the concentration and activity of NR in response to such diverse signals as nitrate abundance, nitrogen metabolites (especially glutamine), CO_2, carbon metabolites (especially sucrose), cytokinins, and light. Control of NR gene transcription facilitates long-term responses to these signals (hours to days), whereas posttranslational regulation allows for rapid changes in NR activity (minutes to hours).

(A)

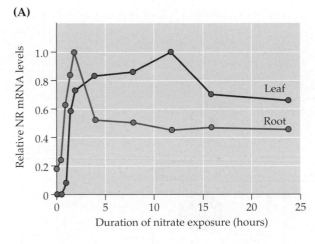

(B)

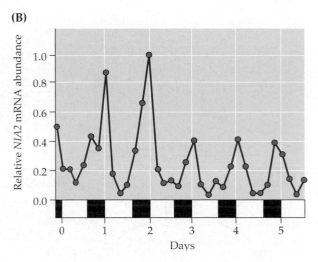

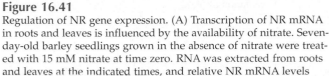

Figure 16.41

Regulation of NR gene expression. (A) Transcription of NR mRNA in roots and leaves is influenced by the availability of nitrate. Seven-day-old barley seedlings grown in the absence of nitrate were treated with 15 mM nitrate at time zero. RNA was extracted from roots and leaves at the indicated times, and relative NR mRNA levels were determined. (B) In plants grown in the presence of nitrate, NR mRNA concentrations demonstrate a diurnal cycle. RNA was extracted from leaves of tomato plants at the indicated times, and relative NR mRNA levels were determined.

The NR gene is substrate-inducible in many plants, although exceptions such as soybean show constitutive and inducible forms of NR. When induced, NR mRNA increases 5- to 100-fold within minutes of nitrate treatment (Fig. 16.41A). This primary transcriptional response is insensitive to protein synthesis inhibitors and results in increased NR mRNA synthesis. The sensitive regulatory system can be triggered by as little as 50 µM nitrate, given as a 10-minute pulse. In leaves, NR induction also requires functional, genetically intact chloroplasts. This coordinated regulation of a cytosolic enzyme by plastid signals probably provides a safeguard by preventing nitrite production if NiR activity is deficient because of a chloroplast defect.

In addition to the primary signal, nitrate, NR mRNA quantities respond to other signals that link nitrate reduction to photosynthesis, carbon metabolism, and diurnal cycles. For maximal induction of NR mRNA, plants require light or a source of reduced carbon. Genetic and physiological data imply that phytochrome (see Chapter 18) mediates this light-induced enhancement of NR mRNA concentrations. Light regulation is most dramatic in etiolated seedlings but also occurs in green tissues. Nearly full induction of NR in plants grown in the dark can also be achieved by providing high concentrations of reduced carbon, e.g., as sucrose. Once the NR gene is induced, NR mRNA quantities cycle in a diurnal rhythm driven by circadian clocks (Fig. 16.41B). A downstream metabolite may act as a signaling molecule that controls NR activity. Glutamine is a likely candidate for this, given the inverse relationship between foliar glutamine concentrations and NR activities in leaves. During diurnal cycling, glutamine concentrations increase as NR transcript concentrations decrease in wild-type plants; moreover, in NR mutant plants, glutamine concentrations are always high and NR mRNA concentrations low. In summary, key regulatory mechanisms control NR gene expression so that nitrate reduction is coordinated with the demand of the shoots for nitrogen, the availability of nitrate and light or reduced carbon, and the proper functioning of the chloroplast.

16.7.4 Phosphorylation-dependent binding of 14-3-3 proteins posttranslationally modulates NR activity in response to endogenous and environmental signals.

Posttranslational mechanisms control NR protein concentration and activity in response to certain physiological conditions. For example, NR proteins decrease when plants are deprived of nitrogen or light for several days even while amounts of NR mRNA remain high. A more rapid and

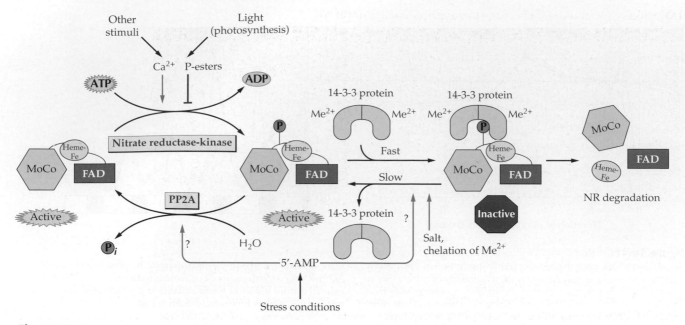

Figure 16.42

Proposed model for regulation of nitrate reductase activity by phosphorylation/dephosphorylation and reversible binding of 14-3-3 protein. NR is phosphorylated by protein kinases on a serine residue in hinge 1. Phosphorylated NR is still active but is bound by 14-3-3 dimers, which inactivate NR. NR in 14-3-3 complexes is more rapidly degraded than free NR and is not an available substrate for phosphatases. If NR is released from 14-3-3 proteins, the regulatory phosphate can be removed by protein phosphatase PP2A.

reversible response that inhibits NR activity occurs when plants are exposed to darkness or low concentrations of CO_2. Under these conditions, NR inhibition occurs within minutes, resulting from the phosphorylation of a conserved serine in the hinge 1 region (see Fig. 16.38) by calcium-dependent protein kinases, followed by Mg^{2+} or Ca^{2+}-dependent binding of 14-3-3 proteins (Fig. 16.42). A protein phosphatase of the type 2A family reactivates NR by dephosphorylating the hinge 1 serine, which prevents 14-3-3 proteins from binding. The reactions that inhibit and reactivate NR are modulated by Ca^{2+}, 5'-AMP, and P_i, which may act as secondary messengers. The kinase/14-3-3 inhibition mechanism allows for rapid and reversible inhibition of NR activity when conditions do not favor nitrate assimilation (i.e., when light or carbon dioxide is limiting).

The regulation of both NR transcription and activity allows plants to fine-tune the amount of nitrate reduction. Light (or its photosynthetic product, sucrose) induces NR transcription, particularly in etiolated plants. Light also enhances the concentrations of glutamine, which suppresses NR transcription (see Chapter 8). However, once induced, NR transcription in photosynthetic tissues demonstrates a cyclic pattern, so that maximal transcript values are achieved just before the appearance of light (see Fig. 16.40B). Nonetheless, even though the NR gene is most actively transcribed predawn, NR protein in photosynthetic tissues is reversibly inhibited by conditions that prevent photosynthetic carbon reduction (e.g., darkness or CO_2 deprivation). Up-regulating NR transcription has the potential to enhance nitrite production. Thus, when NiR activity is low (e.g., in darkness), NR activity must also be down-regulated to prevent accumulation of toxic concentrations of nitrite. A list of signals influencing NR transcription and activity is provided in Table 16.6.

16.8 Nitrite reduction

After nitrate reduction, the next step in the nitrate assimilation pathway is the reduction of nitrite to ammonia, which is catalyzed by NiR (Rx. 16.3). Six electrons are transferred in this step, in contrast to the two that reduce nitrate. The source of electrons is reduced ferredoxin (Fdx_{red}), produced in chloroplasts

Table 16.6 Signals that influence the transcription and activity of NR

Signal	Effect on NR
Glutamine	Down-regulates transcription
Nitrogen starvation	Down-regulates transcription
Circadian rhythm	Modulates transcription depending on time of day
Nitrate	Up-regulates transcription
Cytokinin	Up-regulates transcription
Sucrose	Up-regulates transcription
Light	Up-regulates transcription and activity
Darkness	Down-regulates transcription and activity
High [CO_2]	Up-regulates activity
Low [CO_2]	Down-regulates activity
Oxygen	Down-regulates activity
Anoxia	Up-regulates activity

by photosynthetic noncyclic electron transfer. Nitrite reduction also utilizes Fdx_{red} in the plastids of nonphotosynthetic tissues such as the root. In such colorless plastids, NADPH from the oxidative pentose phosphate pathway reduces ferredoxin in a reaction catalyzed by ferredoxin-$NADP^+$ reductase (Rx. 16.4).

Reaction 16.3: Nitrite reductase

$$NO_2^- + 6\ Fdx_{red} + 8\ H^+ \longrightarrow$$
$$NH_4^+ + 6\ Fdx_{ox} + 2\ H_2O$$

Reaction 16.4: Ferredoxin-$NADP^+$ reductase

$$NADPH + 2\ Fdx_{ox}\ (Fe^{3+}) \longrightarrow$$
$$NADP^+ + 2\ Fdx_{red}\ (Fe^{2+}) + H^+$$

NiR is a nuclear-encoded protein with an N-terminal transit peptide that is cleaved from the mature enzyme. The enzyme, a monomer of 60 to 70 kDa, has two functional domains and cofactors that shuttle electrons from Fdx_{red} to nitrite (Fig. 16.43). The N-terminal half of the enzyme is thought to bind ferredoxin. The C-terminal half, which shares sequence homology with bacterial NADPH-sulfite reductases, contains the binding site for nitrite, as well as two redox centers, a 4Fe–4S center, and a siroheme (Fig. 16.44). These two prosthetic groups are in close proximity, bridged by a sulfur

ligand. Four cysteines located in two clusters provide both the bridging ligand and sulfur ligands for the 4Fe–4S cluster. Mutations that place bulky side chains next to two of these cysteine residues result in diminished NiR activity and can change the spectra of the siroheme in NiR protein, supporting a role for these cysteines in cofactor binding.

NiR is regulated transcriptionally, usually in coordination with NR. Because nitrite is toxic, cells must contain enough NiR to reduce all the nitrite produced by NR. Thus, plants maintain an excess of NiR activity whenever NR is present by inducing NiR gene expression in response to light and nitrate. If NiR concentrations are diminished, either by mutation or antisense expression, plants accumulate nitrite and display chlorosis. In wild-type plants, the regulatory mechanisms that control NR activity are thought to assist in preventing nitrite accumulation.

16.9 Interaction between nitrate assimilation and carbon metabolism

When plants are exposed to nitrate, they redirect carbon from starch synthesis to the production of amino acids and organic acids such as malate by controlling the synthesis and activity of key enzymes. For example, PEP carboxylase, an enzyme that catalyzes

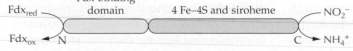

Plant nitrite reductase

Figure 16.43
Structure of nitrite reductase from plants. The N-terminal region oxidizes ferredoxin. The C-terminal region, which binds a 4Fe–4S center and a siroheme group, reduces nitrite to ammonium.

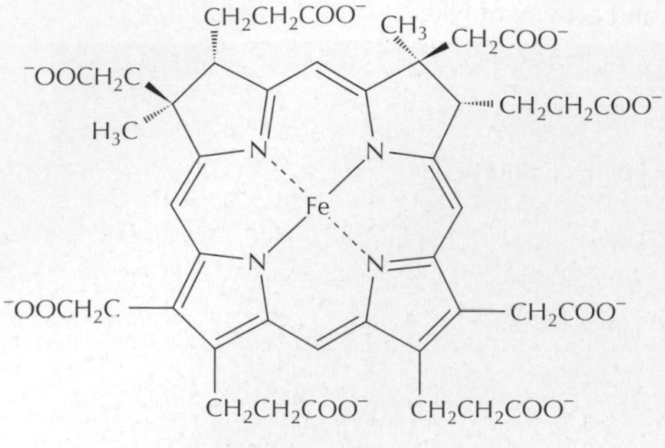

Figure 16.44
Structure of siroheme prosthetic group.

Siroheme

the synthesis of oxaloacetate, a citric cycle intermediate that is readily converted to α-ketoglutarate, is up-regulated by high nitrate. In contrast, ADP glucose pyrophosphorylase, which is required for starch biosynthesis, is down-regulated in response to treatment with high concentrations of nitrate. These regulatory events appear to be in direct response to nitrate because they are expressed most dramatically in NR⁻ mutants, which cannot reduce, but instead accumulate, nitrate. Thus, nitrate acts as a signal that directs plants to redirect carbon flow into compounds that will support nitrogen assimilation into amino acids.

16.10 Overview of sulfate assimilation

Sulfur is an essential macronutrient required for plant growth. It is primarily used to synthesize cysteine, methionine, and numerous essential and secondary metabolites derived from these amino acids. Although of key importance in the life of plants, sulfur is a relatively minor component in comparison with nitrogen. For example, the abundance of sulfur is about 7% that of nitrogen in shoot tissues. The oxidized anion, sulfate, is relatively abundant in the environment and generally is not a growth-limiting nutrient. However, plants have evolved mechanisms for regulating sulfate assimilation in response to the availability of sulfur and for coordinating sulfate assimilation with growth and nitrogen assimilation.

Sulfate enters a plant primarily through the roots by way of an active uptake mechanism (Fig. 16.45). Gaseous sulfur dioxide readily enters the leaves, where it is assimilated, but this source of sulfur is significant only in areas with air pollution. Sulfate is the major transported form of sulfur. To reach the chloroplasts, where most of the reduction to sulfide takes place, a sulfate molecule must traverse at least three membrane systems: the plasma membrane of a root cell at the soil–plant interface, the plasma membranes of internal cells involved in transport, and the chloroplast membranes. Because sulfate is also a major anionic component of vacuole sap, it is also transported across the tonoplast.

Sulfate is assimilated into organic molecules in one of two oxidation states. Most sulfur is reduced to sulfide by the multistep process shown in Reaction 16.5. Sulfide is then incorporated into cysteine, becoming the thiol group. Further metabolism produces thioethers, methylsulfonium compounds, and sulfoxides (Table 16.7 and Fig. 16.46).

Reaction 16.5: Reduction of sulfate to sulfide

$$SO_4^{2-} + ATP + 8\ e^- + 8\ H^+ \longrightarrow$$
$$S^{2-} + 4\ H_2O + AMP + PP_i$$

Another sulfur assimilation pathway directly incorporates sulfate from 3'-phosphoadenosine 5'-phosphosulfate (PAPS) into organic molecules, forming a sulfuryl group. Sulfate can be linked to an oxygen atom to form a sulfate ester, to a nitrogen atom to form a sulfamate, or to a carbon atom to form sulfonic acid (Table 16.7 and Fig. 16.46).

16.11 Sulfur chemistry and function

16.11.1 Cysteine, the sulfur donor in methionine synthesis, is a critical component of proteins, glutathione, and phytochelatins.

Methionine is a sulfur-containing amino acid synthesized from cysteine in a reaction series known as transsulfuration (see Chapter 8). Both cysteine and methionine play pivotal roles in the structural and catalytic functions of proteins. Cysteines are particularly important because oxidizing the thiol groups of two cysteine residues can form a covalent **disulfide bond.** Disulfide bonds are the most important covalent linkages involved in establishing tertiary and, in some cases, quaternary protein structures. Disulfide bonds can also be broken by reduction—a property essential for controlling the redox potential of thiol metabolites and proteins. The redox potential of cellular thiols is ultimately controlled by cellular processes and reactions that generate biological reducing agents (e.g., noncyclic electron transfer and the oxidative pentose phosphate pathway).

Certain proteins (e.g., thioredoxin, glutaredoxin) and a nonprotein tripeptide, **glutathione** (Fig. 16.47A), utilize the chemistry of the dithiol ↔ disulfide interchange to mediate redox reactions. The activities of some chloroplast enzymes involved in carbon metabolism and other photosynthetic processes are regulated by reversible disulfide bond formation (see Chapters 12 and 13).

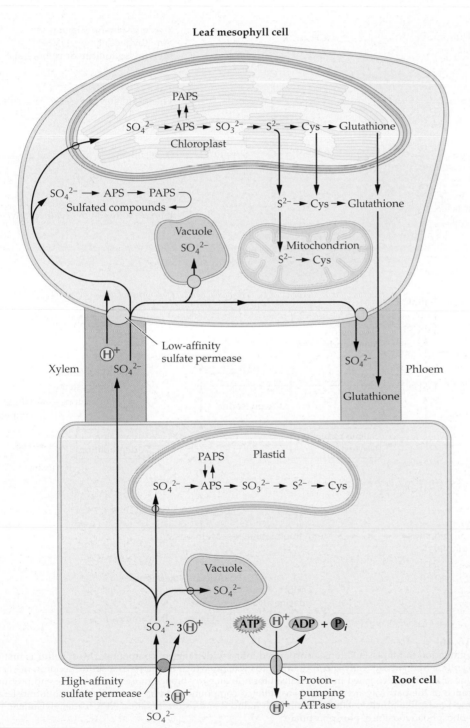

Figure 16.45
Overview of sulfur uptake, reduction, and transport in plants. Like nitrate, sulfate uptake across the plasma membrane is energized by an electrochemical gradient that is maintained by a proton-pumping ATPase. Sulfate is stored in vacuoles. Reduction of sulfate and its assimilation into cysteine take place in the plastids of root and leaf cells. APS, 5-adenylsulfate; PAPS, 3-phosphoadenosine-5'-phosphosulfate.

Table 16.7 Structures and examples of sulfur-containing compounds formed by plants

Compound	Generic structure	Examples
Thiols (mercaptans)	RSH	L-Cysteine, coenzyme A
Sulfides or thioethers	RSR_1	Hydrogen sulfide (H_2S), L-methionine
Sulfoxides	$RSOR_1$	Allicin
Methylsulfonium compounds	$(CH_3)_2S^+R$	S-Adenosyl-L-methionine, S-methylmethionine, DMSP, dimethylsulfonic hydroxybutyrate
Sulfate esters	$R-O-\overset{\overset{O}{\|}}{\underset{\underset{O}{\|}}{S}}-O^-$	Phenol sulfates, polysaccharide sulfates
Sulfamates	$R=N-O-\overset{\overset{O}{\|}}{\underset{\underset{O}{\|}}{S}}-O^-$	Aryl sulfamates, mustard oil glycosides
Sulfonic acids	$R-\overset{\|}{\underset{\|}{C}}-\overset{\overset{O}{\|}}{\underset{\underset{O}{\|}}{S}}-O^-$	Glucose 6-sulfonate, cysteic acid, taurine, sulfoquinovosyl diacylglycerol

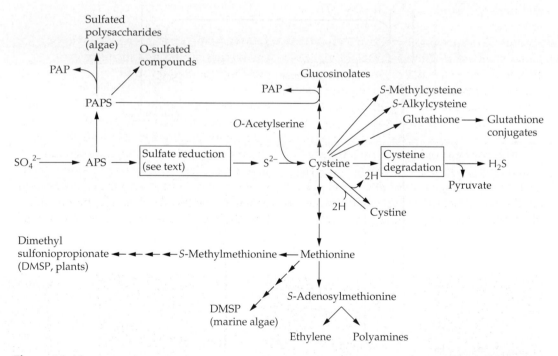

Figure 16.46

Sulfur assimilation in plants. Sulfate is incorporated into a wide range of compounds. Most sulfur is first reduced to sulfide and subsequently incorporated into cysteine, which functions as a sulfur donor in many reactions. Some sulfur is incorporated in the oxidized form in an early branch in the pathway. The end products of this branch of the pathway are referred to as sulfated compounds and are synthesized by sulfotransferases. Some compounds, like glucosinolates, contain both reduced and oxidized sulfur. APS, 5-adenylylsulfate; PAPS, 3'-phosphoadenosine-5'-phosphosulfate.

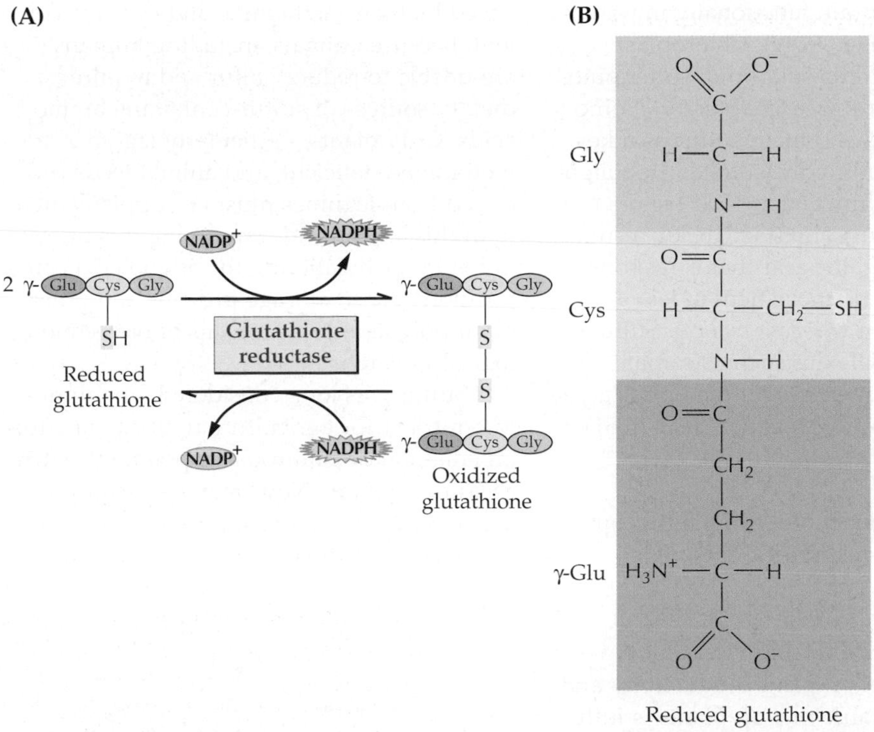

(A)

(B)

Reduced glutathione

Figure 16.47
(A) The tripeptide glutathione functions as a buffer of cellular redox potential, continually being interconverted between the reduced form, in which each glutathione molecule contains a single thiol group, and the oxidized form, in which two glutathione molecules join to form an intermolecular disulfide bond. Usually, glutathione is maintained mostly in the reduced form by glutathione reductase by using electrons derived from NADPH. Under stress conditions, the proportion of oxidized glutathione in the cell increases. (B) Structure of reduced glutathione. Note the unusual peptide linkage that joins the γ-carbonyl of glutamate and the amino group of cysteine.

Glutathione and related compounds are involved in many aspects of growth and development, including redox control, storage and transport of reduced sulfur, metabolism of herbicides, and response to environmental stresses. The three amino acids making up glutathione are glutamate, cysteine, and glycine, with the γ-carboxyl group of glutamate and the amino group of cysteine participating in an unconventional peptide linkage (Fig. 16.47B). Glutathione is not translated; instead, its synthesis is catalyzed by enzymes.

Another class of low-molecular-mass (commonly referred to as LMW, low molecular weight) thiols produced by plants are the **phytochelatins** (Fig. 16.48). These glutathione-derived peptides play a role

in protection against toxic heavy metals. Phytochelatins bind metal ions by using the thiol group as the ligand.

16.11.2 Many plant products, including coenzymes, lipids, and secondary metabolites, contain sulfur.

Plants contain a wide variety of sulfur compounds, some with important functions and others, particularly natural products or secondary metabolites, with no known function. Several group-transfer coenzymes and vitamins, including coenzyme A, S-adenosyl-L-methionine (SAM), thiamine, biotin, and S-methylmethionine (sometimes referred to

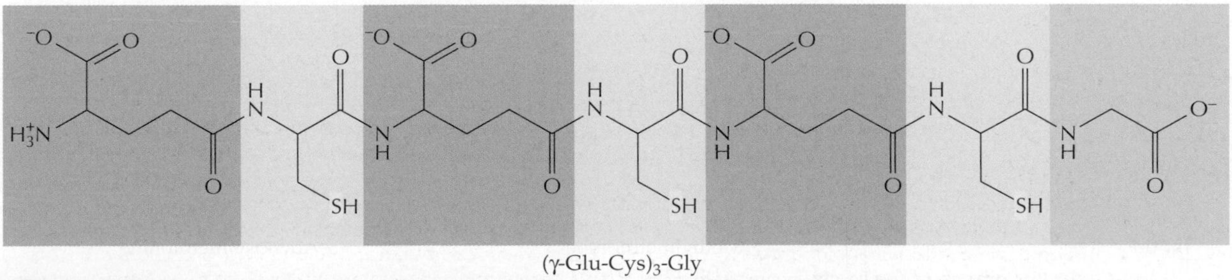

(γ-Glu-Cys)$_3$-Gly

Figure 16.48
Phytochelatin molecule with the structure (γ-Glu-Cys)$_3$Gly.

as vitamin U) contain functionally important sulfur moieties (Fig. 16.49). Chloroplast membranes contain a sulfolipid, sulfoquinovosyldiacylglycerol (see Chapter 10). Some signaling molecules contain sulfur as a key component, including the sulfated lipooligosaccharides that function as Nod factors (see Fig. 16.24), and turgorin (Fig. 16.50), a sulfated derivative of gallic acid that is responsible for thigmotactic movement in leaves of the sensitive plant (*Mimosa pudica*). Sulfur-containing phytoalexins such as camalexin are produced by members of the Brassicaceae in response to pathogen attack (Fig. 16.51).

16.11.3 The sulfur content of plant crops and soils influences nutritional and agricultural practices.

Sulfur availability and sulfur content have a significant impact on plant productivity and utilization. The value of crop plants is influenced by their methionine and cysteine content, because animals, including humans, are unable to reduce sulfur and require dietary sources of sulfur-containing amino acids. Crop plants, particularly legumes, are methionine-deficient, and animal feeds produced from legumes must be supplemented with this amino acid. Improving this nutritional index by altering the amino acid composition of seed storage proteins and other plant polypeptides is a focus of biotechnological research.

Sulfur was long considered a nonlimiting nutrient for agriculture until recent curbs on emission of sulfurous air pollutants such as sulfur dioxide. Now, many agricultural areas, especially in northern Europe, must be fertilized with sulfate if they are to maintain

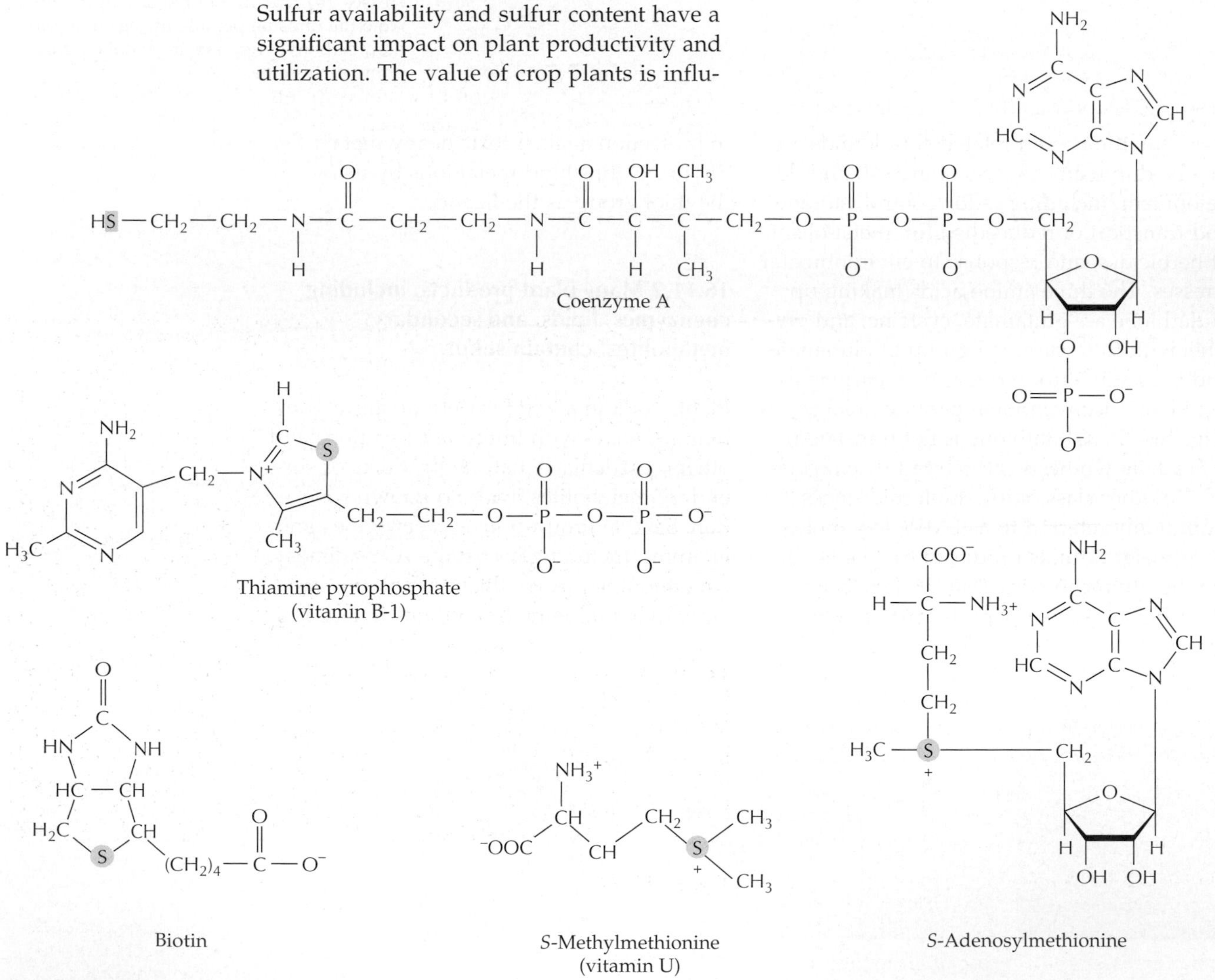

Figure 16.49

Structures of several group-transfer coenzymes, vitamins, and substrates that contain sulfur.

Chapter 16 Nitrogen and Sulfur

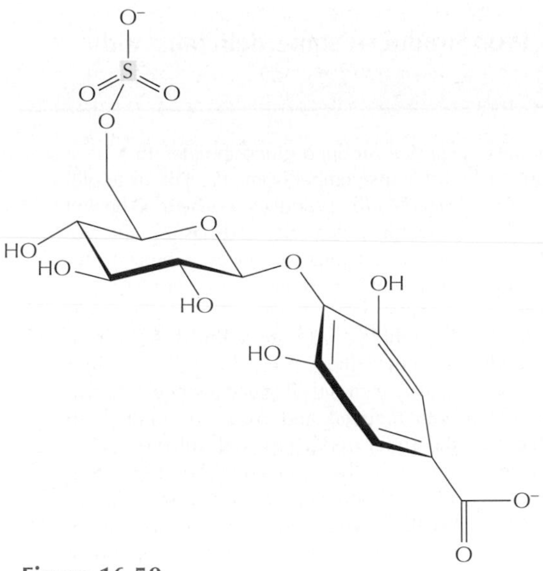

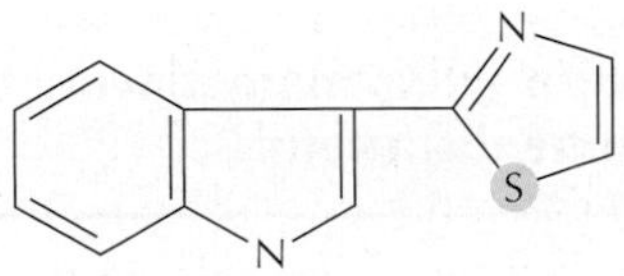

Camalexin

Figure 16.51
Structure of a sulfur-containing phytoalexin, camalexin.

Figure 16.50
Structure of a gallic acid glucoside, a sulfur-containing thigmotactic factor, turgorin.

crop yields and quality (Fig. 16.52). Sulfur deficiency adversely affects the baking quality of wheat flour (Fig. 16.53) and the yield of oil-seed *Brassica* species.

16.11.4 Plants play a major role in the global sulfur cycle.

The interconversion of oxidized and reduced sulfur states on Earth is known as the biogeochemical sulfur cycle (Fig. 16.54). Both plants and microorganisms reduce sulfate to sulfide for assimilation into cysteine. Some anaerobic bacteria use sulfate as a terminal electron acceptor analogous to the way that aerobic organisms use oxygen. This process, known as dissimilation, generates copious amounts of H_2S. Oxidation of reduced sulfur to sulfate completes the cycle. Oxidation occurs in three ways: (*a*) through aerobic catabolism of sulfur compounds, carried out by animals, microorganisms, and plants; (*b*) by bacteria that use reduced sulfur compounds as electron donors for chemosynthetic or photosynthetic reactions; and (*c*) through geochemical mechanisms when organic sulfur compounds are volatilized into the atmosphere. Although the burning of fossil fuels is the greatest contributor to atmospheric sulfur, most of the biogenic component is contributed by marine algae, many of which

produce in abundance the compound dimethylsulfoniopropionate (DMSP), a tertiary sulfur analog of the quaternary nitrogen compounds known as betaines (see Fig. 16.23). DMSP has many functions, including roles as an osmoprotectant, a cryoprotectant, and a repellant against planktonic herbivores. When released from algae, DMSP is degraded to dimethyl sulfide (DMS), which is volatilized into the atmosphere and subsequently oxidized to DMSO (Fig. 16.55A), sulfite, and sulfate. Atmospheric sulfate acts as a nucleus for formation of water droplets and is associated with the formation of clouds. The link between oceanic algae and cloud formation has been proposed as a climate-regulating mechanism (Fig. 16.55B).

Figure 16.52
Aerial photograph of two *Brassica napus* fields, one fertilized with sulfate (foreground), the other not (background). *B. napus*, an oilseed variety, must be fertilized with sulfur to maximize yield. The difference in abundance of blooms between the sulfur-fertilized and the unfertilized fields is clearly evident. *Brassica* species provide clear examples of the requirement for sulfur because their seeds are rich in sulfur-containing compounds.

Sulfur-containing compounds are responsible for numerous distinctive flavors and odors in plants. For example, pungent diallyl compounds are present in alliaceous plants such as onion and garlic. The flavor of onions and garlic is produced by the action of alliinase on the odorless precursor alliin and other S-substituted cysteine conjugates (panel A). Alliinase is localized in the vacuole, whereas alliin is in the cytosol, so the reaction occurs when the onion tissue is crushed or damaged, thereby mixing the enzyme and substrate. The reaction products include allicin, pyruvate or another keto acid, and ammonia.

Allicin has antimicrobial and feeding-deterrent activities and so acts as a chemical defense against pathogens and herbi-

vores. Human cultures and individuals vary greatly in their appreciation of allicin.

Glucosinolates, sulfur-containing defense compounds that are produced by a wide variety of plants, are responsible for the sought-after taste and antioxidant properties of some vegetables (e.g., broccoli) and condiments (mustards and horseradish, among others). The same compounds, however, also result in flavors and odors that diminish the value of canola oil produced from oil-seed *Brassica* crops (see Chapter 10). Glucosinolates are hydrolyzed to release isothiocyanate in response to plant tissue damage (panel B). A diverse group of compounds, glucosinolates have a general 1-β-thioglucopyranoside structure; different plant

species produce glucosinolates that have distinctive embellishments. For example, *Sinapis alba* produces a 4-hydroxybenzyl derivative, sinapine, and nasturtium (*Tropacolum majus*) produces a benzyl glucosinolate derivative (panel C). The biosynthetic pathway of these compounds is complex. All are derived from proteogenic and nonproteogenic amino acids. Because all glucosinolates contain both reduced and oxidized sulfur, both the sulfur reduction and sulfation pathways must be involved in their biosynthesis. It is somewhat ironic that the glucosinolate compounds, which apparently evolved to deter animal consumption of the vegetative tissues of crucifers and other plants, are now one of the reasons why humans eat them.

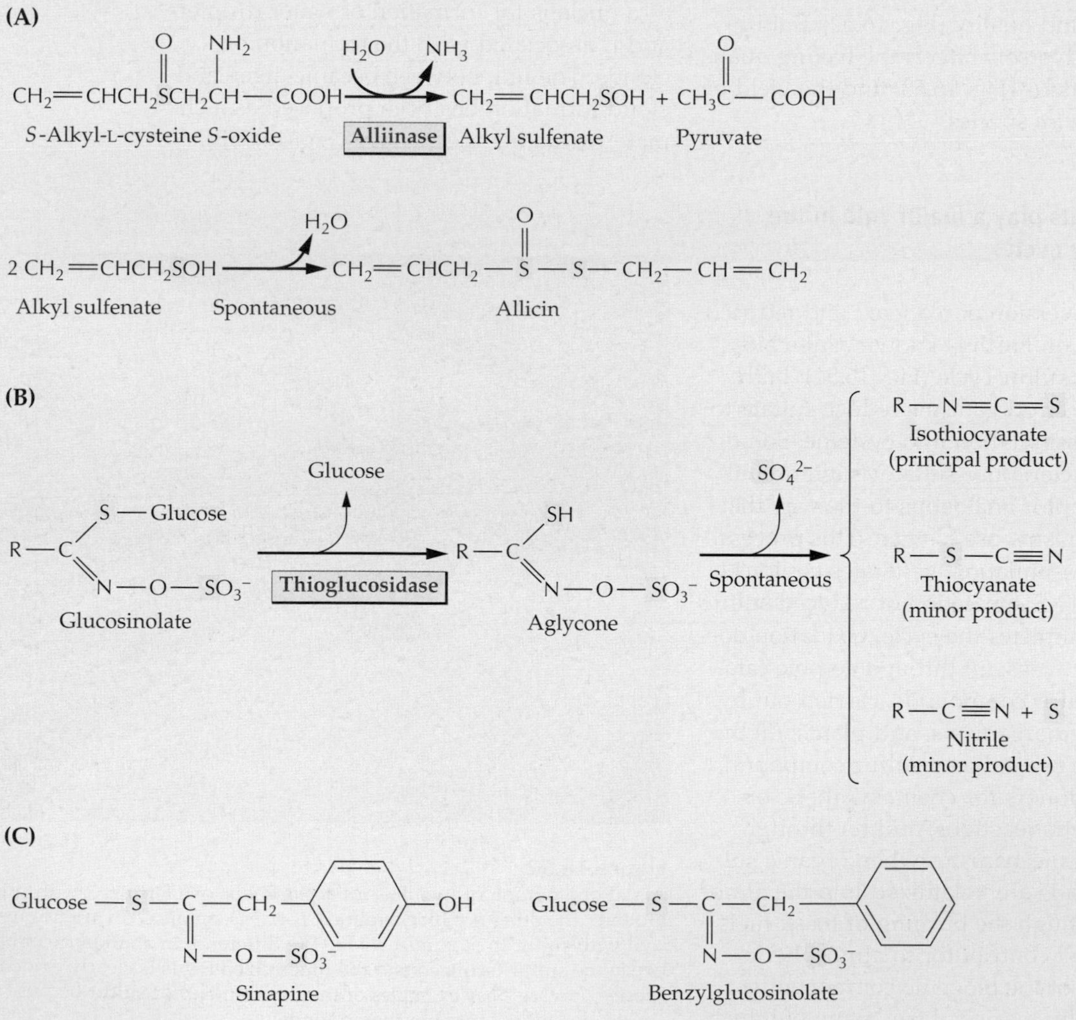

(A)

S-Alkyl-L-cysteine S-oxide ⟶ Alliinase ⟶ Alkyl sulfenate + Pyruvate

2 CH_2=$CHCH_2SOH$ ⟶ (Spontaneous) ⟶ Allicin

Alkyl sulfenate, Spontaneous, Allicin

(B)

Glucosinolate ⟶ Thioglucosidase ⟶ Aglycone ⟶ Spontaneous ⟶ R—N=C=S Isothiocyanate (principal product); R—S—C≡N Thiocyanate (minor product); R—C≡N + S Nitrile (minor product)

(C)

Sinapine   Benzylglucosinolate

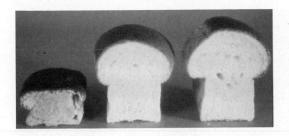

Figure 16.53
Loaves of bread baked from (left to right) low-sulfur wheat, high-sulfur wheat, and excess-sulfur wheat. Limitations in sulfur availability cause a shift in the expression profile of seed storage proteins in wheat. In grain from wheat plants grown under S-limiting conditions, proteins with fewer cysteines predominate. The low-cysteine proteins are unable to form abundant disulfide bonds and adversely affect the baking quality of wheat flour. By contrast, an excess of sulfur leads to a less desirable product, probably because an overabundance of cysteine in the seed storage proteins makes the formation of disulfide bonds more difficult.

16.12 Sulfate uptake and transport

16.12.1 Sulfate uptake is mediated by a transporter that is powered by a proton gradient across root cell plasma membranes.

Sulfate is actively cotransported into plant root cells along with protons at a stoichiometry of one sulfate to three protons. The transport activity is driven by a proton gradient maintained across the plasma membrane by a proton-pumping ATPase (Fig. 16.56). If the pH of the external medium increases or if the proton pump activity decreases, the proton gradient is diminished and sulfate uptake is inhibited. The localization of sulfate transporters to the plasma membrane has been established by characterization of transport activity of isolated plasma membrane vesicles.

The rate of sulfate uptake into roots has been determined to be multiphasic by measurement over a range of sulfate concentrations (Fig. 16.57). This finding can be interpreted to mean that plant roots contain multiple transporters with differing affinities for sulfate, or that there is a single transporter for which the affinity for sulfate is regulated. The sulfur status of the plant determines sulfate uptake activity, which is induced by sulfur starvation (see next section) and repressed by treatment with reduced sulfur. The fact that protein synthesis is necessary for induction of transport activity suggests that transporter expression is regulated. Recent molecular findings have revealed that plants contain multiple transporters with different affinities for sulfate.

The rate of sulfate transport across the tonoplast depends on an electrochemical gradient, but this differs from the process in plasma membranes in that vacuolar transport is mediated by a uniporter. The vacuolar sap is highly acidic, owing to the activity of the tonoplast proton ATPase. The sulfate uniporter utilizes the steep electrical gradient between the cytoplasm and vacuole to transport only sulfate ions. Sulfate transport into plastids is not well understood, but there are several possible transporters. One is the triose phosphate translocator (see Chapter 13), which functions in an antiport mechanism with phosphate. A plasma

Figure 16.54
The biogeochemical sulfur cycle. Sulfate is reduced by sulfur-assimilating organisms, which use it for synthesis of cysteine and other organic sulfur compounds, and by dissimilators, anerobic bacteria that use sulfate as a respiratory electron acceptor in place of oxygen. Many organisms oxidize reduced sulfur to sulfate. Chemoautotrophic bacteria extract electrons for energy, whereas phototrophic bacteria use the electrons for photosynthesis. Reduced sulfur is also oxidized geochemically when oxygen is present.

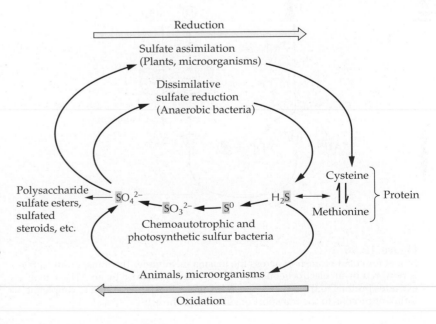

(A)

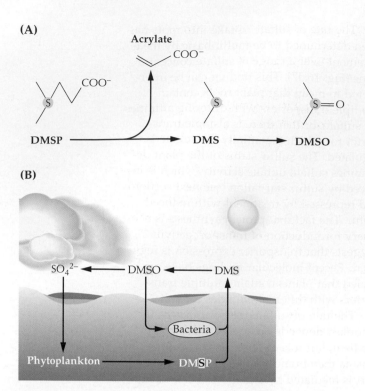

(B)

Figure 16.55
The phytoplankton–climate connection. Phytoplankton-produced DMSP is broken down by bacteria to DMS and acrylate. (A) DMS volatilizes and is oxidized to DMSO and to sulfate, which nucleates water droplets, leading to cloud formation. (B) Sulfate is returned to the sea dissolved in rain. Because cloud cover reduces the growth of phytoplankton and is accompanied by atmospheric cooling, phytoplankton have been proposed to serve as a homeostatic climate regulation mechanism. The extent to which phytoplankton regulate climate is still being keenly debated.

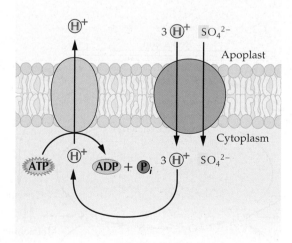

Figure 16.56
Model for sulfate transport across the plasma membrane. The transport of sulfate is powered by an electrochemical proton gradient generated by an ATPase that extrudes protons to the cell exterior. The sulfate transporter is able to couple the influx of protons to the transport of sulfate into the cell.

membrane-like proton–sulfate cotransporter localized in plastids has been identified. In eukaryotic algae, genes for ATP-dependent sulfate transporters resembling those in bacteria have been identified on the basis of homology of their amino acid sequences with those of the bacterial transporters. In plants, as in other organisms, sulfate uptake is inhibited by sulfite, selenate, molybdate, and chromate. These anions compete with sulfate for binding to the transporter.

16.12.2 A gene family encodes high- and low-affinity sulfate transport proteins in plants.

In plants, plasma membrane sulfate transporters are encoded by a gene family. The individual transporters have widely different sulfate affinities and distinct expression patterns, indicating specialized functions. A high-affinity isoform is expressed exclusively in roots. Steady-state levels of the mRNA for this isoform increase rapidly after sulfur starvation, which is consistent with the idea that the high-affinity transporter mediates uptake from soil water, in which the sulfate concentration is relatively low and variable. A second isoform with a lower affinity for sulfate is expressed predominantly, but not exclusively, in leaves and is less responsive than the high-affinity transporter to the external sulfur supply. The lower-affinity type may mediate transport within the plant. Internal cells obtain sulfate from the xylem sap or from subcellular compartments, both of which contain sulfate concentrations that exceed that in soil solution and are less subject to variation.

The cloning and sequencing of plant sulfate transporters have shed light on the evolutionary origin of these proteins. Sulfate-uptake systems can be classified as either facilitated transport systems or sulfate permeases. Facilitated transporters, which occur in heterotrophic bacteria and cyanobacteria, contain two components: a soluble, extracellular sulfate-binding protein, and a membrane-localized transport protein, the transport activity of which is powered by ATP hydrolysis. Sulfate permeases, present in plants and many other eukaryotes, consist of a single polypeptide chain with the 12 membrane-spanning regions characteristic

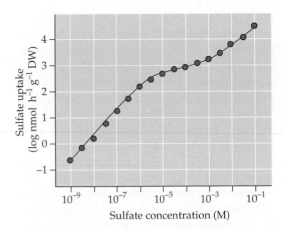

Figure 16.57
Uptake of sulfate by barley roots. The graph shows the multiphasic rate of sulfate uptake by barley roots incubated with a range of sulfate concentrations. DW, dry weight.

of cation/solute cotransporters (Fig. 16.58). Permease-type sulfate transporters from plants share significant amino acid sequence homology with H^+/SO_4^{2-} cotransporters from fungi and mammals. This evolutionary conservation is supported by the finding

that the plant transporters can functionally complement a mutant strain of yeast lacking sulfate permease.

16.13 The reductive sulfate assimilation pathway

The reduction of sulfate to sulfide is an energy-intensive process that consumes 732 kJ mol^{-1}. By contrast, nitrate and carbon assimilation require less energy (347 and 478 kJ mol^{-1}, respectively). In plants, the energy for sulfate assimilation is largely met by the ATP and reductant derived from photosynthesis. In nonphotosynthetic tissues, the energy for sulfate assimilation has not been specifically investigated, but it may be similar to nitrate assimilation in nonphotosynthetic tissues; that is, the reductant is probably generated by the oxidative pentose phosphate pathway and the energy by respiration. Sulfate assimilation is divided into three steps: activation, reduction to sulfide, and incorporation of sulfide into cysteine. Plastids are known to contain the entire pathway for cysteine

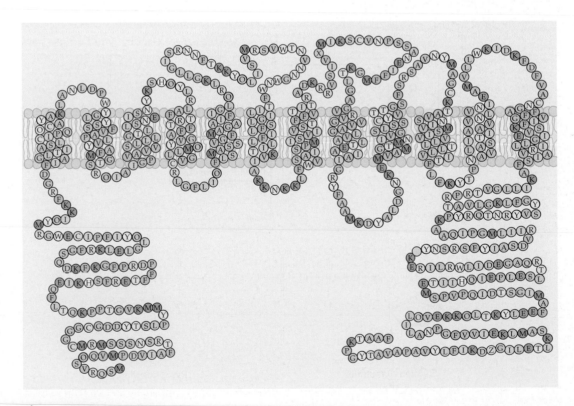

Figure 16.58
Model of a plant sulfate transporter. The protein has 12 regions that span the plasma membrane.

biosynthesis from inorganic sulfate. Chloroplasts are thought to be the primary site for cysteine synthesis—although root plastids are also highly active (see Fig. 16.45). Some but not all sulfur assimilation enzymes exist in cellular compartments outside of plastids, but the exact function of the nonplastid isoenzymes is not understood.

16.13.1 Sulfate activation is catalyzed by ATP sulfurylase.

To enter the assimilatory pathway, sulfate must be activated by the enzyme ATP sulfurylase, as depicted in Figure 16.59 and Reaction 16.6.

Reaction 16.6: ATP sulfurylase

$$SO_4^{2-} + MgATP \rightleftharpoons MgPP_i + \text{5'-adenylylsulfate (APS)}$$

The resulting compound, 5'-adenylylsulfate (sometimes referred to as 5'-adenosinephosphosulfate, hence the acronym APS),

contains a high-energy phosphoric acid–sulfuric acid anhydride bond that potentiates the sulfuryl moiety for subsequent metabolic reactions. APS is a central branch-point intermediate that feeds two pathways: sulfate reduction and sulfation. The free energy of APS formation ($\Delta G^{\circ\prime}$), estimated at +41.8 kJ mol^{-1}, favors ATP formation. Therefore, at equilibrium, APS can accumulate only to approximately 10^{-7} M. The forward reaction is driven by efficient removal of APS and PP$_i$. At least two plant enzymes, APS kinase and APS reductase (now known to be the same as APS sulfotransferase; see Section 16.13.5) metabolize APS. Inorganic pyrophosphatase hydrolyzes PP$_i$ (see Chapter 13).

The properties of plant ATP sulfurylase are known primarily from studies on two isoforms of the enzyme purified from spinach leaf. The major isoform, which is chloroplast-localized, accounts for 85% to 90% of total ATP sulfurylase. The minor isoform is cytoplasmic. Both enzymes are tetramers of 49- to 50-kDa subunits and have similar kinetic constants. The substrates bind in an ordered

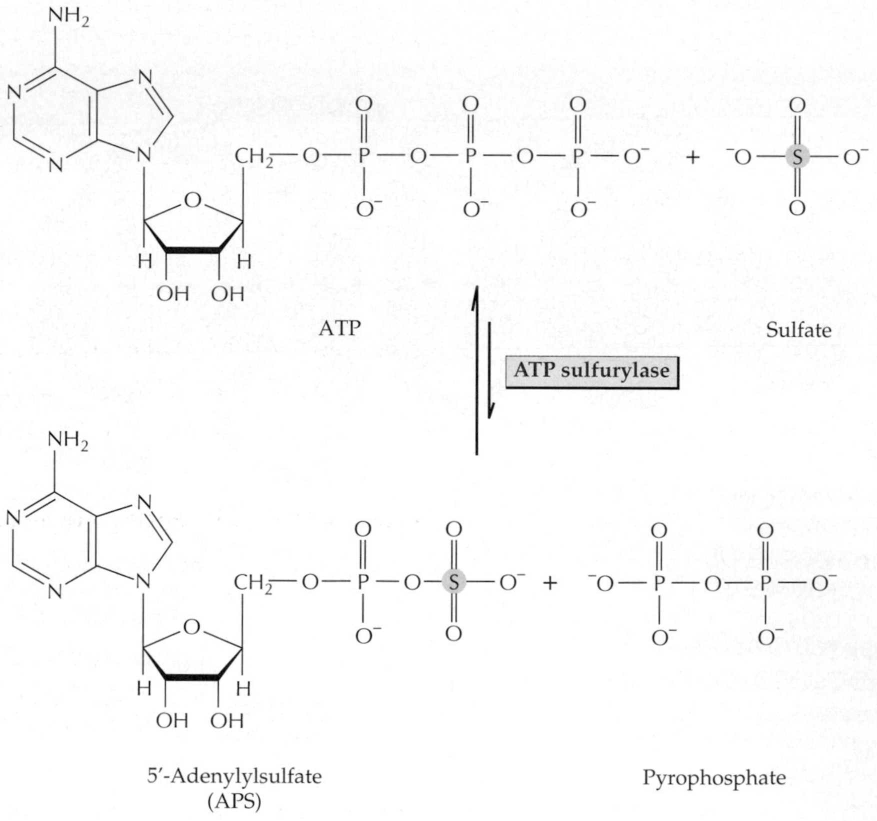

ATP

Sulfate

ATP sulfurylase

5'-Adenylylsulfate
(APS)

Pyrophosphate

Figure 16.59
The reaction catalyzed by ATP sulfurylase.

and synergistic manner with the MgATP binding before the sulfate. Both spinach isoenzymes can utilize molybdate and selenate anions instead of sulfate; however, the resulting adenylyl products are unstable and spontaneously degrade to the anion and AMP. ATP sulfurylase is also present in the plastids of root cells. Whether a cytosolic form also exists in roots has not been determined.

The plastid isoenzyme forms the APS used for sulfate assimilation because plastids contain all the enzymes needed to synthesize cysteine from sulfate. The function of the cytosolic isoform is an enigma because the concentration of pyrophosphate in the cytoplasm is much too high to permit substantial net synthesis of APS.

16.13.2 A gene family encodes ATP sulfurylase isoenzymes.

ATP sulfurylases from a variety of organisms can be classified into two structural types that do not share significant amino acid homology. One type, from sulfate-assimilating prokaryotes, is a heteromeric enzyme that is regulated by an intrinsic GTPase activity. The second enzyme type, from eukaryotes and an endosymbiotic chemolithotrophic bacterium, is homomeric and lacks GTPase activity. The homomeric class to which plant ATP sulfurylase belongs can be further divided into (a) monofunctional enzymes that occur in fungi, algae, higher plants, and the chemolithotroph and (b) bifunctional enzymes with both ATP sulfurylase and APS kinase activities, which are present in animals. The homomeric ATP sulfurylase enzymes vary widely in overall sequence homology, but they include two extended regions that are highly homologous and may constitute the substrate-binding sites. Surprisingly, the plant ATP sulfurylase sequences are most similar to the bifunctional enzymes from animals. Despite the diversity of ATP sulfurylase forms, the plant enzymes can complement auxotrophic ATP sulfurylase mutants of yeast and *Escherichia coli*, indicating that the activities are functionally equivalent.

Investigations have not yet resolved the role that ATP sulfurylase plays in regulating sulfate assimilation in plants. In some but not all plants the activity of ATP sulfurylase

and the steady-state mRNA content have been shown to increase in response to sulfate starvation and to be repressed by feeding of reduced sulfur compounds such as cysteine and glutathione. Thus, perhaps ATP sulfurylase gene expression is regulated by the demand for pathway end products.

16.13.3 Two hypotheses attempt to define the sulfate reduction pathway in plants.

The most contested topic in sulfate assimilation is the question of how plants reduce sulfate. Two leading hypotheses are illustrated in Figure 16.60 and Reactions 16.7 through 16.11.

Hypothesis 1: The carrier-bound pathway

Hypothesis 1 (Fig. 16.60A) is based on the finding that higher plants and eukaryotic algae appear to use APS as a substrate for sulfate reduction. According to this hypothesis, sulfate is transferred from APS to a reduced thiol compound by APS sulfotransferase (Rx. 16.7). The thiol compound has not been identified, but reduced glutathione has been suggested as the most likely candidate. The organic thiosulfate reaction product, termed a thiosulfonate, is acted on by thiosulfonate reductase to form thiosulfide (Rx. 16.8). This hypothesis has been termed the carrier-bound pathway because sulfite or sulfide remains covalently bound to the thiol compound. How the reduced sulfur is incorporated into cysteine is still unclear.

Reaction 16.7: APS sulfotransferase

$$APS + thiol\ compound_{red} \longrightarrow thiosulfonate + 5'\text{-}AMP$$

Reaction 16.8: Thiosulfonate reductase

$$Thiosulfonate + 6\ ferredoxin_{red} \longrightarrow thiosulfide + 6\ ferredoxin_{ox}$$

Hypothesis 2: The microbial pathway

The second hypothesis (Fig. 16.60B) is similar to the sulfate reduction pathway in some sulfate-assimilating microorganisms, including cyanobacteria. First, APS kinase phosphorylates APS, yielding PAPS (Rx. 16.9), which is then reduced to free sulfite by the thioredoxin-dependent enzyme PAPS reductase (Rx. 16.10). Finally, ferredoxin-dependent sulfite reductase reduces sulfite to sulfide (Rx. 16.11).

Reaction 16.9: APS kinase

$$APS + ATP \longrightarrow PAPS + ADP$$

Reaction 16.10: PAPS reductase

$$PAPS + thioredoxin_{red} \longrightarrow SO_3^{2-} + thioredoxin_{ox} + PAP$$

Reaction 16.11: Sulfite reductase

$$SO_3^{2-} + 6\ ferredoxin_{red} \longrightarrow S^{2-} + 6\ ferredoxin_{ox}$$

The primary deficiency of the second hypothesis is the lack of evidence for PAPS reductase in plants, but the hypothesis was proposed to explain certain biochemical observations. Both APS kinase and sulfite reductase are known to exist in plants. In contrast, the APS sulfotransferase and thiosulfonate reductase of Hypothesis 1 have proved difficult to characterize. Moreover, the intermediates of the sulfate reduction pathway are notorious for undergoing side

reactions, yet the evidence for the carrier-bound pathway has been obtained primarily from in vitro biochemical studies with unpurified enzymes. This longstanding controversy appears to have been resolved by recent work. Cloning experiments indicate that plants contain a novel enzyme with thiol-dependent APS reductase activity that yields sulfite as the product (see next section).

16.13.4 Cloned APS reductase cDNAs provide evidence for a novel pathway in plants.

A family of cDNAs recently isolated from *Arabidopsis* encodes plastid-localized enzymes with thiol-dependent APS reductase activity. Like APS sulfotransferase, purified APS reductase produced from the cloned genes uses APS as a substrate and reduced glutathione as a reductant. Along with in vitro enzymological studies, the use of *E. coli* genetics has proved invaluable in demonstrating the properties of plant APS reductase. The cDNAs are able to complement APS kinase (see Rx. 16.9) and PAPS reductase (see Rx. 16.10) mutants of *E. coli*, indicating that the enzyme uses APS as a substrate, as shown in Figure 16.61. The complementation depends on the ability of the *E. coli* strain to synthesize glutathione, supporting the notion

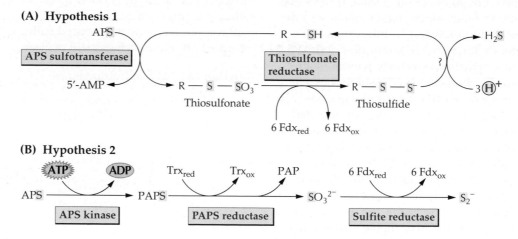

Figure 16.60

Two hypotheses for sulfate reduction in plants. (A) Hypothesis 1. APS sulfotransferase, using APS as a sulfuryl donor, transfers sulfate to an acceptor thiol compound, possibly glutathione. The resulting organic thiosulfonate is reduced to thiosulfide by a ferredoxin-dependent thiosulfonate reductase. (B) Hypothesis 2. APS is first phosphorylated to PAPS by APS kinase. PAPS reductase then produces sulfite by using electrons donated from thioredoxin. Sulfite reductase then completes the reduction to sulfide by using electrons from ferredoxin.

that glutathione is the in vivo reductant in plants. *E. coli* sulfite reductase requires free sulfite as a substrate, indicating that free sulfite must be produced during complementation by APS reductase. These lines of evidence have led to a hypothesis for the reaction carried out by APS reductase (Rx. 16.12).

Reaction 16.12: APS reductase

$$\text{APS} + 2 \text{ glutathione}_{red} \longrightarrow$$
$$SO_3^{2-} + \text{glutathione}_{ox} + \text{AMP} + 2\,H^+$$

The structure of APS reductase reveals how reduced glutathione may serve as a source of electrons (Fig. 16.62). The N-terminal domain of the mature enzyme constitutes the reductase and is moderately homologous to the PAPS reductase of microorganisms. The C-terminal domain is related to thioredoxin and shares a disulfide motif with the redox-active protein glutaredoxin. The C-terminal domain of APS reductase contains two important cysteine residues spaced two amino acids apart. In glutaredoxin and thioredoxin, a similar motif contains the redox active residues, which undergo dithiol ↔ disulfide interchanges as electrons are shuttled to target enzymes. The source of electrons for reduction of the redox proteins differs. Glutaredoxin is specifically reduced by glutathione. In comparison, cytosolic thioredoxin species are reduced by NADPH, whereas chloroplast forms of the protein receive electrons from ferredoxin. APS reductase resembles glutaredoxin in its preference for glutathione. Experimental evidence suggests that the glutaredoxin function of the enzyme probably resides in the C-domain separately from the N-domain. A truncated protein lacking the N-terminal domain catalyzes reactions typical for glutaredoxin but not thioredoxin. In contrast, a protein containing only the N-domain lacks glutaredoxin activity. Because NADPH-dependent glutathione reductase controls the reduction state of glutathione in plants (see Section 16.14.1), the source of electrons for reduction of sulfate to sulfite can be traced to the NADPH generated by the photosynthetic light reactions or by the oxidative pentose phosphate pathway, which reduces NADP$^+$ in heterotrophic tissues and in photosynthetic tissues in the dark. One potentially significant

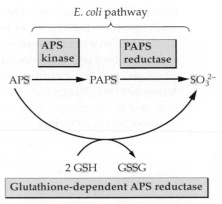

Figure 16.61
Complementation of *Escherichia coli* sulfur assimilation mutants reveals the properties of plant APS reductase. The well-characterized sulfate assimilation system of *E. coli*, which requires the enzymes APS kinase and PAPS reductase, has been used to identify the putative function of an unknown plant enzyme. Mutation of the gene encoding either enzyme produces an *E. coli* cysteine auxotroph. The plant APS reductase cDNA can complement both the APS kinase and PAPS reductase mutants. Complementation requires that the bacterial strain contain glutathione.

aspect of channeling electron flow through glutathione, rather than through thioredoxin, may involve the unique regulatory role thioredoxin plays in regulating photosynthetic processes in plant chloroplasts (see Chapters 12 and 13). Reducing sulfate with electrons from glutathione would separate sulfite formation from reactions regulated by thioredoxin.

Recent biochemical analyses of APS sulfotransferase purified from a marine macroalga and of a cDNA encoding APS sulfotransferase cloned from *Lemna*, an aquatic angiosperm, have shown that this protein is identical to the APS reductase from *Arabidopsis* described

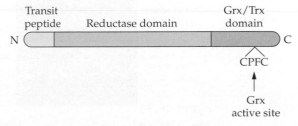

Figure 16.62
Domains of APS reductase. The N-terminal transit peptide, which directs the protein to chloroplasts, is proteolytically removed to form the mature enzyme. The mature enzyme contains a reductase domain (blue shading) and a C-terminal domain (purple shading) that functions as a glutaredoxin.

above. The two also are linked by their association with plastids, by their decreasing abundance during aging, and by their induction and repression properties as a function of sulfur nutrition. APS sulfotransferase activity is increased by sulfur starvation and is repressed by treating plants with reduced forms of sulfur. The steady-state concentration of APS reductase mRNA responds similarly to changes in sulfur nutrition (Fig. 16.63). Further work is needed to clarify the catalytic mechanism(s) of this protein. In discussing the thiol-dependent APS reductase of plants, we must note that a very different type of APS reductase exists in dissimilatory sulfate-reducing bacteria. Although they share a name, the assimilatory plant enzyme and the dissimilatory bacterial enzyme do not share significant amino acid homology. The dissimilatory enzyme is characterized by Fe–S and flavin prosthetic groups.

(A)

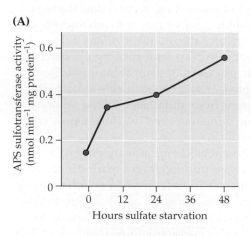

(B)

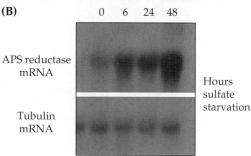

Figure 16.63
Hydroponically grown *Brassica juncea* plants were transferred to sulfate-free nutrient medium and samples taken for measurement of enzyme activity (A) and mRNA (B). The tubulin mRNA was measured as a control because its concentration is unaffected by sulfate starvation.

16.13.5 Sulfite reduction is catalyzed by siroheme 4Fe–4S sulfite reductase.

Although there is still uncertainty about whether sulfate reduction proceeds predominantly via thiosulfonate or free sulfite, it is quite certain that plants contain sulfite reductase, an enzyme that adds six electrons to free sulfite, forming sulfide (see Rx. 16.11). Reduced ferredoxin is the source of electrons. Sulfite reductase is found in the plastids, where it relies on ferredoxin (or flavodoxin) reduced directly by the photosynthetic light reactions in leaves or indirectly by NADPH generated by way of the oxidative pentose phosphate pathway in roots.

Plant sulfite reductase is a homooligomeric hemoprotein composed of two or four identical 64- to 71-kDa subunits, depending on the source. Like the C-terminal half of nitrite reductase (see Section 16.8 and Rx. 16.3), each subunit of sulfite reductase contains one siroheme (see Fig. 16.44) and one iron–sulfur cluster (4Fe–4S). The sequence of the plant enzyme is also homologous to nitrite reductases, especially to the C-terminal region of NiR where the prosthetic groups bind. The sequence relationship indicates that sulfite and nitrite reductases belong to a superfamily of anion redox enzymes. Indeed, the purified sulfite reductase from spinach can reduce nitrite, although its affinity for sulfite is two orders of magnitude greater than for nitrite. In contrast, the NADPH-dependent sulfite reductases characterized from bacteria and yeast are heteromeric enzymes that contain one flavoprotein and one hemoprotein subunit. The amino acid sequence of *Arabidopsis* sulfite reductase, deduced from a cDNA clone, shows that the plant enzyme is homologous with the hemoprotein subunit of bacterial and yeast sulfite reductases.

Unlike the APS reductase discussed above, the steady-state concentration of the mRNA for sulfite reductase in plants remains constant in response to sulfur starvation or to feeding of reduced sulfur compounds, and the activity of sulfite reductase is not affected appreciably by sulfur nutrition. Because sulfite is a toxic anion that would damage cells if it accumulated, presumably sulfite reductase activity is maintained in excess of the preceding enzyme in the pathway, APS reductase.

16.13.6 The combined activities of serine acetyltransferase and *O*-acetylserine(thiol)lyase convert serine to cysteine.

Synthesis of cysteine by condensation of *O*-acetylserine (OAS) and sulfide represents the final step in the reductive sulfate assimilation pathway. OAS is formed from serine and acetyl-CoA. The reaction, catalyzed by serine acetyltransferase (Rx. 16.13 and Fig. 16.64), is specific to cysteine biosynthesis.

Reaction 16.13: Serine acetyltransferase

$$\text{Serine + acetyl-CoA} \longrightarrow$$
$$\text{*O*-acetylserine + CoA}$$

There are several possible sources of acetyl-CoA in plants (see Chapters 10, 13, and 14). Although which source predominates in plastids is not clearly understood, a pyruvate dehydrogenase that generates acetyl-CoA and CO_2 is active in these organelles. OAS and sulfide ion then react to form cysteine. *O*-Acetylserine(thiol)lyase, an enzyme that contains pyridoxal phosphate as a prosthetic group, catalyzes cysteine synthesis (Rx. 16.14 and Fig. 16.64). If the carrier-bound pathway does exist in plants, the substrate for cysteine synthesis would be thiosulfide, however, rather than sulfide. So far, only sulfide has been used to measure OAS(thiol)lyase activity.

Reaction 16.14: *O*-Acetylserine(thiol)lyase

$$\text{*O*-Acetylserine} + S^{2-} \longrightarrow$$
$$\text{cysteine + acetate}$$

Serine acetyltransferase and OAS(thiol)-lyase exist in an enzyme complex that may function to regulate serine acetyltransferase activity. The complex forms through specific protein–protein interactions, but in vivo only a small fraction of the total OAS(thiol)lyase actually associates with serine acetyltransferase; moreover, the enzymes are fully active when isolated free of the complex. Recent evidence shows that the complex synthesizes OAS rather than cysteine; uncomplexed OAS(thiol)lyase is required for cysteine synthesis. In the enzyme complex, OAS(thiol)lyase appears to function as a regulatory subunit, influencing the kinetic behavior of the bound serine acetyltransferase. Because complex formation is promoted by sulfide and is disrupted by OAS, these metabolites therefore act to regulate the activity of serine acetyltransferase, coordinating its activity with that of sulfate reduction pathway (Fig. 16.65).

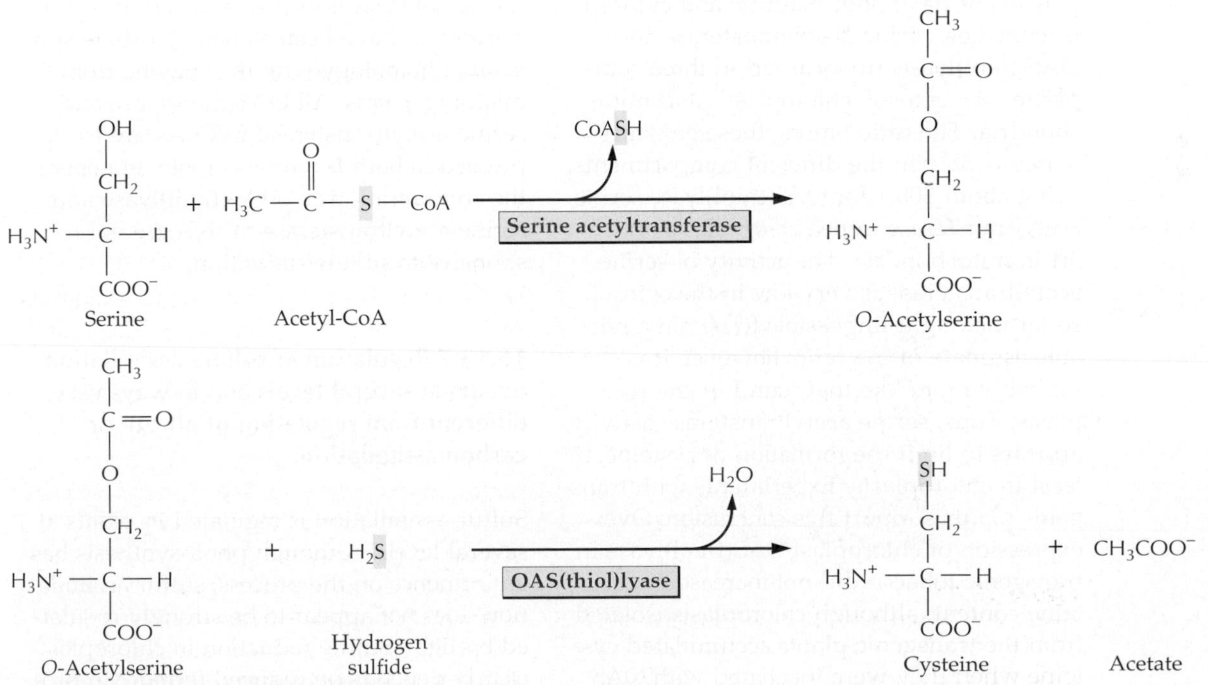

Figure 16.64
Reactions responsible for assimilation of reduced sulfur into cysteine.

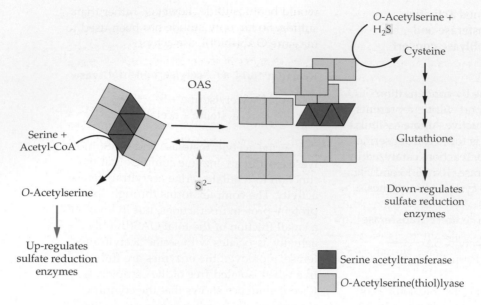

Figure 16.65

Regulation of cysteine synthesis by OAS and thiol compounds. Yellow boxes indicate OAS(thiol)lyase dimers, which are present in excess over the serine acetyltransferase tetramers, indicated as blue triangles. The enzymes associate through specific interaction domains. Sulfide promotes formation of the complex, thereby stimulating OAS formation. OAS positively regulates expression of proteins for sulfate assimilation. If OAS accumulates because of an insufficiency of sulfide, the complex is destabilized, thereby reducing OAS synthesis. OAS also reacts with sulfide to form cysteine catalyzed by free OAS(thiol)lyase dimers. The resulting increased concentrations of cysteine and glutathione repress the expression of sulfate assimilation proteins. The system serves to coordinate the synthesis of OAS with the activity of the sulfate reduction pathway.

Unlike other sulfate assimilation enzymes, which either localize exclusively to plastids or have both plastidic and cytosolic isoenzymes, serine acetyltransferase and OAS(thiol)lyase are localized in three compartments: cytosol, chloroplasts, and mitochondria. The ratio among these enzymes varies widely in the different compartments, being about 300:1 for OAS(thiol)lyase:serine acetyltransferase in pea chloroplasts and 3:1 in mitochondria. The activity of serine acetyltransferase is very low in the cytosol, so far making it impossible to obtain a reliable estimate of the ratio; however, it is probably more like that found in chloroplasts. Thus, serine acetyltransferase activity appears to limit the formation of cysteine, at least in chloroplasts. Experiments with transgenic plants support this conclusion. Overexpression of chloroplast OAS(thiol)lyase in transgenic tobacco did not increase the cysteine content, although chloroplasts isolated from the transgenic plants accumulated cysteine when they were incubated with OAS plus sulfite or sulfate. Addition of sulfate or sulfite alone had no effect.

Plant cDNAs that encode chloroplast-, mitochondrion-, and cytosol-localized isoforms of OAS(thiol)lyase and serine acetyltransferase have been cloned. All share substantial homology with the enzyme from microorganisms. All OAS(thiol)lyase and serine acetyltransferase mRNAs are expressed in both leaves and roots. In general, the concentrations of OAS(thiol)lyase and serine acetyltransferase mRNA are not responsive to sulfate starvation.

16.13.7 Regulation of sulfate assimilation occurs at several levels and in ways very different from regulation of nitrate or carbon assimilation.

Sulfur assimilation is regulated in plants at several levels. Although photosynthesis has an influence on the process, sulfur assimilation does not appear to be strongly regulated by light. Sulfite reduction in chloroplasts clearly depends on reduced ferredoxin produced from the photosynthetic light reactions, but sulfite reduction also occurs in nonphotosynthetic plastids, where the reductant source is most likely generated by the oxidative pentose phosphate pathway. Because the pool of reduced glutathione is essentially constant in plastids, there is no influence of light on the ability of APS reductase to form sulfite. Moreover, although the activities of the sulfur assimilation enzymes increase severalfold when dark-grown plants are illuminated, these enzymes are also active in etiolated plants. In contrast, nitrate reductase and Rubisco are not expressed in etiolated plants. Finally, unlike the enzymes that participate in nitrate and carbon assimilation, the sulfur assimilation enzymes do not demonstrate substantially diurnal oscillations in activity (see Fig. 16.41B).

A more significant factor regulating sulfur assimilation is developmental stage. All the sulfur assimilation enzymes are highly active in young leaves and root tips and decline markedly in older tissues. For example, the activity of sulfate assimilation enzymes is comparatively low in fully expanded leaves that have attained their maximum rate of carbon dioxide fixation. This developmental pattern of expression suggests that sulfur assimilation occurs primarily in actively growing tissues, in which there is a high demand for cysteine and methionine for protein synthesis.

Just as N-assimilating enzymes are regulated by nitrogen availability, plants regulate sulfur assimilation in response to the availability of sulfur. However, the two classes of enzymes respond differently. The key enzymes are up-regulated when nitrate is available and down-regulated when nitrate is unavailable. In contrast, sulfur starvation results in the up-regulation of several activities that function in sulfate assimilation. The response varies, depending on the species and the precise conditions involved, but two activities have consistently been observed to increase substantially in response to sulfur starvation: sulfate transport, and APS reductase. With the recent availability of DNA probes, investigators have found that the steady-state concentrations of mRNA for these enzymes increase in parallel with their activity, which may point to a transcriptional or posttranscriptional mechanism of regulating gene expression. Importantly, activities of sulfate permease and APS reductase and amounts of mRNA are strongly up-regulated in roots, less so in leaves. A response in leaves is

usually observed in seedlings and not in mature plants, which suggests that the leaves of mature plants are buffered against short-term changes in the external sulfate concentration, perhaps by accumulated sulfate in the vacuoles.

Controlling the sulfur reduction pathway by regulation of APS reductase makes intuitive sense. The reduced inorganic sulfur ions, sulfite and sulfide, are toxic and must not be allowed to accumulate. ATP sulfurylase, which produces the substrate for the sulfation pathway, would be a poorer target for regulation than APS reductase because the ability of the former to function in the forward direction is strongly dependent on the removal of APS by APS reductase or APS kinase.

The content of reduced sulfur and nitrogen in plants is strictly maintained at a ratio of 1:20. This indicates that sulfur assimilation must be coordinated with the rate of nitrogen assimilation and with the growth rate. APS reductase activity declines in response to nitrogen starvation, but little else is known about the coordination of sulfur assimilation and nitrogen assimilation. OAS is a prime candidate to be a regulator molecule. Its abundance is a function of synthesis from serine (a nitrogenous compound) and of further metabolism to cysteine (formation of which is a function of sulfide availability). OAS strongly up-regulates sulfate permease and APS reductase mRNA, serving to coordinate sulfate reduction with cysteine synthesis (Fig. 16.65).

16.14 Synthesis and function of glutathione and its derivatives

16.14.1 Glutathione is not gene-encoded but rather is synthesized enzymatically.

As a major end product of the reductive sulfate assimilation pathway, glutathione is the major nonprotein thiol in plants, present in millimolar concentrations that far exceed the low micromolar concentrations of cysteine. Glutathione and its derivatives are involved in the storage and long-distance transport of reduced sulfur, in signal transduction pathways, in scavenging hydrogen peroxide and other reactive oxygen species (see Chapter 22), in detoxifying xenobiotics (e.g., herbicides; see Chapter 3), in activating and conjugating phenylpropanoids and hormones, and, as seen below, in serving as a substrate for phytochelatin synthesis.

Two enzymes make up the glutathione biosynthetic pathway (Fig. 16.66). First, γ-glutamylcysteine synthetase (Rx. 16.15) catalyzes the formation of a peptide bond between the γ-carboxyl group of glutamate and the α-amino group of cysteine. Next, glutathione synthetase (Rx. 16.16) catalyzes the formation of a peptide bond between the cysteinyl carboxyl group of γ-glutamyl-cysteine and the α-amino group of glycine.

ATP is hydrolyzed in each reaction. Both enzymes are localized to plastids and the cytosol but not to mitochondria.

Reaction 16.15: γ-Glutamylcysteine synthetase

$$\text{Glutamate + cysteine + ATP} \longrightarrow \text{γ-glutamylcysteine + ADP + P}_i$$

Reaction 16.16: Glutathione synthetase

$$\text{γ-Glutamylcysteine + glycine + ATP} \longrightarrow \text{glutathione + ADP + P}_i$$

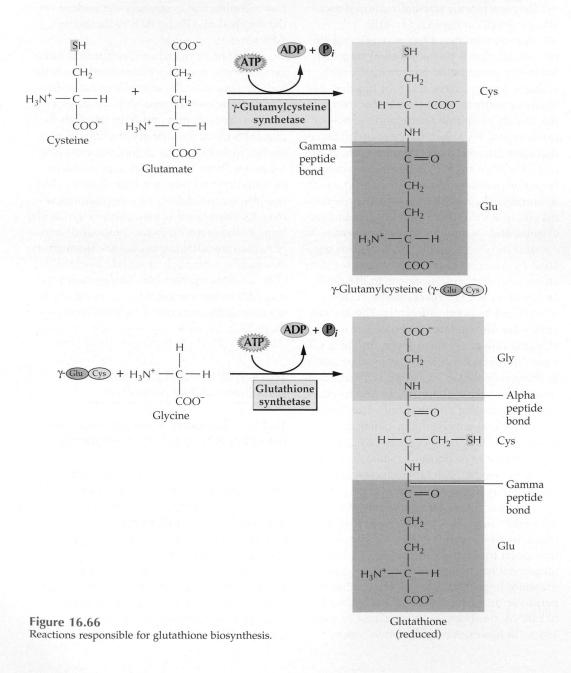

Figure 16.66
Reactions responsible for glutathione biosynthesis.

An *Arabidopsis* cDNA encoding γ-glutamylcysteine synthetase was recently analyzed. The protein product shares little amino acid homology with γ-glutamylcysteine synthetases from *E. coli*, fungi, and mammals; nonetheless, it is able to complement a γ-glutamylcysteine synthetase mutant of *E. coli*. The protein is probably chloroplast-localized, as indicated by the presence of an N-terminal sequence resembling a plastid transit peptide. The origin of the cytoplasmic isoform is not known.

Experiments with transgenic plants have shown that γ-glutamylcysteine synthetase, which catalyzes the first step in glutathione synthesis, is subject to feedback inhibition by physiological concentrations of glutathione. The rate of synthesis also appears to depend on the availability of cysteine (synthesized from OAS) and glycine (a product of photorespiration; see Chapter 14). The feedback inhibition mechanism may be important as a safety valve to prevent the accumulation of high (nonphysiological) concentrations of glutathione if the supply of precursor amino acids becomes too great.

Some plants produce tripeptides related to glutathione. Species in the order Fabales synthesize homoglutathione, in which β-alanine replaces glycine, and species in the Poaceae (grass family) generate hydroxymethylglutathione, in which glycine is replaced by serine. A third class is found in maize, in which glutamate replaces glycine (Fig. 16.67). These plant species also form glutathione. Other than synthesis of phytochelatin-related compounds (see next section), the function of glutathione homologs in these taxa has not been established.

16.14.2 Glutathione and glutathione-related tripeptides are the precursors for phytochelatins.

Exposure to toxic heavy metals such as Cd^{2+}, or to high concentrations of micronutrients such as Cu^{2+}, induces the synthesis of phytochelatins, polymers of various sizes with the general structure $(\gamma\text{-Glu-Cys})_n\text{Gly}$ ($n = 2$ to 11). The relationship of glutathione and phytochelatins was definitively revealed through the isolation of a cadmium-sensitive γ-glutamylcysteine synthetase mutant in *Arabidopsis* termed *cad2*. The mutation maps within the γ-glutamylcysteine synthetase gene, and the γ-glutamylcysteine synthetase DNA clone can complement *cad2*. The glutathione concentration of *cad2* plants is reduced by 70% to 85%, yet the plants appear normal under nonstress conditions. Cadmium sensitivity results from an inability to accumulate phytochelatins. Cadmium-treated *cad2* plants are able to produce phytochelatins to only 10% of the concentrations seen in wild-type plants exposed to cadmium. Buthionine sulfoximine (BSO), an inhibitor of γ-glutamylcysteine synthetase, has a similar effect on plants, diminishing cadmium tolerance by inhibiting synthesis of glutathione and phytochelatin.

An enzyme that is activated by metal ions such as cadmium has been identified and recently cloned, γ-glutamylcysteine dipeptidyl transpeptidase (phytochelatin synthase). This enzyme catalyzes the transpeptidation of a glutathione to the carboxyl group of the cysteine in another glutathione or to the analogous position in a phytochelatin (Fig. 16.68). In this way chains of as

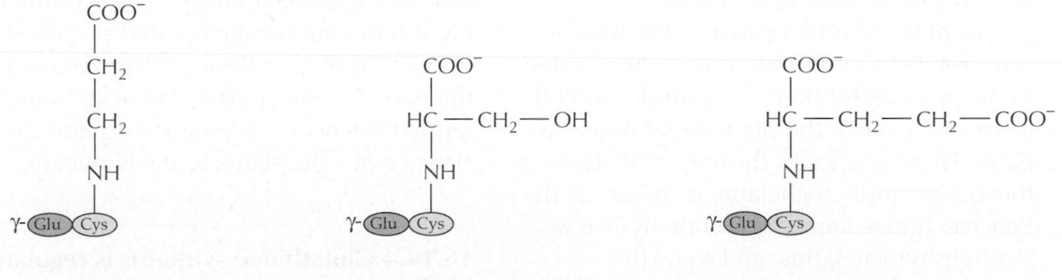

Homoglutathione
(β-alanine replaces Gly; found in some legumes)

Hydroxymethylglutathione
(Ser replaces Gly; found in some grasses)

γ-Glutamylcysteinylglutamate
(Glu replaces Gly; found in maize)

Figure 16.67
Some plant taxa synthesize the glutathione-related compounds homoglutathione, hydroxylmethylglutathione, and γ-glutamylcysteinylglutamate.

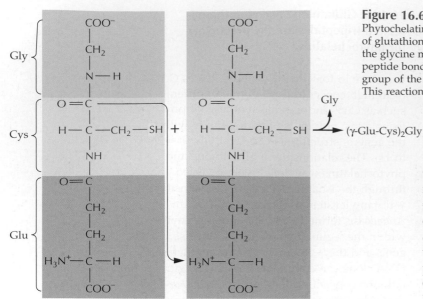

Gly { COO⁻ ... }
Cys { ... }
Glu { ... }

$$\text{Gly} \quad (\gamma\text{-Glu-Cys})_2\text{Gly}$$

Figure 16.68
Phytochelatin synthase catalyzes the transpeptidization reaction of glutathiones to produce $(\gamma\text{-Glu-Cys})_n\text{Gly}$ (n = 2 to 11). Initially, the glycine moiety of one glutathione is removed and a new peptide bond is formed with the residual cysteine and the amino group of the glutamate on a second molecule of glutathione. This reaction is repeated to form the phytochelatin polymers.

the ATP-binding cassette (ABC) type. Within the vacuole a larger complex is formed, termed the high-molecular-weight (HMW) complex, which includes the LMW complex, additional cadmium (transported into the vacuole independently by a Cd^{2+}/H^+ anti-porter), and sulfide. The end product is a CdS crystallite. In essence, the phytochelatins hold cadmium in an ordered structure, thereby allowing the ion to complex with sulfide to form a crystal.

many as 11 γ-glutamylcysteine units are formed. The cadmium-sensitive mutant of *Arabidopisis* termed *cad1* lacks a functional phytochelatin synthase and is completely unable to synthesize phytochelatins, although glutathione synthesis is unaffected in the mutant. This indicates the importance of phytochelatins for resistance to heavy metals. The phenotype of *cad1* does not appear to be adversely affected by growth on Cu^{2+} and Zn^{2+} at micronutrient amounts, bringing into question the hypothesis that phytochelatins play a physiological role in homeostasis of metal nutrients under nonstress conditions. The gene encoding phytochelatin synthase has been cloned from *Arabidopsis* and rice by either positional cloning of the *cad2* allele or screening for cDNAs that confer cadmium tolerance to yeast.

In plant taxa that produce the tripeptides related to glutathione (e.g., homo- and hydroxymethylglutathione), similar modifications appear in the phytochelatin derivatives. Thus, species in the order Fabales synthesize homophytochelatin, members of the Poaceae (grass family) generate hydroxymethylphytochelatins, and so forth.

Phytochelatins serve as a mechanism for sequestration of heavy metals in the plant vacuole and present an example of biomineralization (Fig. 16.69). The LMW phytochelatin–cadmium complexes are transported into the vacuole by a transporter of

16.14.3 Glutathione plays an important role in detoxification of xenobiotics.

Plants are able to detoxify or inactivate many substances, including endogenously produced toxins, hormones, and xenobiotic chemicals such as herbicides by forming conjugates with glutathione. This reaction is catalyzed by glutathione *S*-transferase, which links the reactive thiol group of cysteine to the xenobiotic (Fig. 16.70; see also Chapter 3, Fig. 3.14). The glutathione conjugate is transported actively into the vacuole by a glutathione translocator of the ABC type. Within the vacuole, the glutathione conjugates are hydrolyzed to cysteine conjugates. This mechanism for herbicide detoxification has been applied usefully to agriculture by identifying the substances that stimulate the expression of glutathione *S*-transferase and the vacuolar transporter. These substances, termed safeners, are used to increase the resistance of crop plants to the herbicide.

16.14.4 Glutathione synthesis is regulated by several mechanisms.

As the three preceding sections make clear, glutathione is intimately involved with plant response to environmental stresses. Its versatility derives from the sulfhydryl group,

(A)

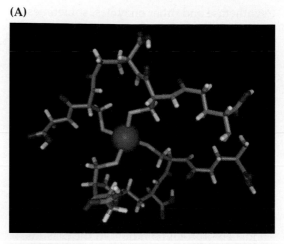

(B)

Figure 16.69
(A) Two phytochelatin molecules [(γ-Glu-Cys)$_n$Gly] coordinated with an ion of cadmium. By binding toxic heavy metal ions, phytochelatins remove them from the cellular machinery. It is thought that the thiol groups of phytochelatins serve as the ligand for heavy metal ions. (B) A model for the mechanism of CdS mineralization and sequestration in plant cells mediated by phytochelatins. Both phytochelatins and low-molecular-weight (LMW) phytochelatin–Cd complexes formed in the cytosol enter the vacuole by way of an ABC transporter (1). Cd ions enter the vacuole by way of an antiporter in exchange for protons (2). HMW, high molecular weight. The electrochemical gradient across the tonoplast is maintained by a proton-pumping ATPase (3).

which gives glutathione the ability to act as a reductant, thereby protecting the cell against electrophiles, such as active oxygen species (see Chapter 22) and xenobiotics. Part of the stress response involves increasing the rate of glutathione synthesis. Under nonstress conditions, the formation of γ-glutamylcysteine is a rate-limiting step in glutathione biosynthesis, and the enzyme

γ-glutamylcysteine synthetase is subject to both positive and negative regulation. The K_m[Cys] of this enzyme is close to the cellular cysteine concentration, which is low. Therefore, in the physiological state, γ-glutamylcysteine synthetase is limited by cysteine and functions below its V_{max}. Accordingly, the concentrations of cysteine and the rate of sulfate assimilation strongly determine the rate of the γ-glutamylcysteine synthetase reaction. Indeed, feeding cysteine to leaves results in glutathione accumulation, as does overexpression of γ-glutamylcysteine synthetase in transgenic plants.

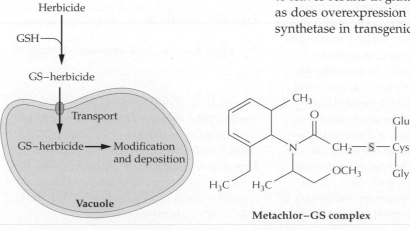

Figure 16.70
Detoxification of an herbicide (e.g., Metachlor) by conjugation to glutathione. The conjugate is then transported into the vacuole.

Summary

Nitrogen and sulfur are used to make diverse plant constituents. These essential macronutrients are involved in biogeochemical cycles and are often obtained from the environment in forms that must be reduced before they can be used. The reduction and assimilation reactions are integrated into general plant metabolism. They are often compartmentalized in the cell and are sometimes located in certain plant cell types or organs. Plants most commonly acquire nitrogen as NH_4^+, which can be assimilated immediately into amino acids, or as NO_3^-, which must be reduced to NH_4^+ for further metabolism. Some bacteria can reduce (fix) atmospheric N_2 into NH_4^+. The key catalyst of nitrogen fixation, nitrogenase, has two functional components. Dinitrogenase is the enzyme that actually reduces N_2. It contains unusual metal clusters, including one type that generally includes an atom of Mo or V. Dinitrogenase reductase uses ATP to lower the potential of reductants obtained from metabolism to a level where they can reduce the dinitrogenase clusters.

A few groups of plants can obtain NH_4 efficiently from N_2 by participating in a symbiosis with nitrogen-fixing bacteria. The best characterized nitrogen-fixing symbioses are the interactions between legumes and rhizobia. Bacteria infect the plant roots and trigger the formation of root nodules. Infection requires communication between the bacteria and the plant. Specific plant compounds, such as flavonoids, are released by developing roots. These are detected by the bacteria and induce the bacteria to make lipooligosaccharide Nod factors, which alter plant gene expression and cell division in roots of specific host plants. As the nodule grows, bacteria invade some plant cells and develop into nitrogen-fixing forms called bacteroids. Bacteroids are surrounded by a plant-derived membrane which controls the exchange of nutrients between the bacteroid and the plant cytoplasm. A mature nodule is organized to support the energy intensive nitrogen fixation reaction by delivering oxygen and carbon sources to the bacteroids at low concentrations of free oxygen in order to protect nitrogenase, a notoriously sensitive enzyme. Ammonia released by bacteroids is first assimilated in the plant by glutamine synthetase and other enzymes, sometimes employing nodule-specific isozymes. Assimilated nitrogen leaves the nodule as either amides or ureides, depending on the host species, which leads to distinct patterns of gene expression, compartmentation, and regulation of these final steps in different hosts.

Nitrate uptake is mediated by high- and low-affinity proton symporters that are encoded by genes belonging to two gene families, NRT1 and NRT2. In the cytosol, nitrate is reduced to nitrite by NAD(P)H-dependent nitrate reductase, a metallo-flavoenzyme that can be inhibited by phosphorylation and 14-3-3 protein binding. In the plastids, nitrite is reduced to ammonium by nitrite reductase, which contains a 4Fe–4S center and a siroheme and uses ferredoxin as the reductant. Most of the nitrate assimilatory genes are regulated at the transcriptional level, being induced by nitrate, light, or sucrose and being repressed by ammonium or glutamine.

Sulfur is essential for life. Its oxidation state is in constant flux as it circulates through the global sulfur cycle. As the primary producers of organic sulfur compounds, plants play a key role in the cycle by coupling photosynthesis to the reduction of sulfate, assimilating sulfur into cysteine, and metabolizing this amino acid to form methionine, glutathione, and many other compounds. The activity of the sulfur assimilation pathway responds dynamically to changes in the sulfur supply and to environmental conditions that alter the need for reduced sulfur. Through molecular-genetic analysis, many of the enzymes involved in the process have been defined, and the mechanisms for regulation are being revealed.

Further Reading

Nitrogen fixation and ammonia transport

Books

Elmerich, C., Kondorosi, A., Newton, W. E., eds. (1997) *Biological Nitrogen Fixation for the 21st Century.* Kluwer Academic Publishers, Dordrecht, The Netherlands.

Legocki, A., Bothe, H., Puhler, A. (1997) Biological fixation of nitrogen for ecology and sustainable agriculture. *NATO ASI Ser. G: Ecol. Sci.* 39 (entire volume).

Spaink, H. P., Kondorosi, A., Hooykaas, P.J.J., eds. (1998) *The Rhizobiaceae.* Kluwer Academic Publishers, Dordrecht, The Netherlands.

Sprent, J. I., Sprent, P. (1990) *Nitrogen Fixing Organisms: Pure and Applied Aspects.* Chapman and Hall, New York.

Stacey, G., Burris, R. H., Evans, H. J., eds. (1992) *Biological Nitrogen Fixation.* Chapman and Hall, New York.

Articles

Appleby, C. A. (1992) The origin and function of leghemoglobin in plants. *Sci. Progr. (Oxford)* 76: 365–398.

Brewin, N. J. (1991) Development of the legume root nodule. *Annu. Rev. Cell Biol.* 7: 191–226.

Downie, J. A. (1998) Functions of rhizobial nodulation genes. In *The Rhizobiaceae,* H. P. Spaink, A. Kondorosi, and P.J.J. Hooykaas, eds. Kluwer Academic Publishers, Dordrecht, The Netherlands, pp. 387–402.

Ferguson, S. J. (1998) Nitrogen cycle enzymology. *Curr. Opin. Chem. Biol.* 2: 182–193.

Freiberg, C., Fellay, R., Bairoch, A., Broughton, W. J., Rosenthal, A., Perret, X. (1997) Molecular basis of symbiosis between *Rhizobium* and legumes. *Nature* 387: 394–401.

Haynes, R. J. (1990) Active ion uptake and maintenance of cation–anion balance: a critical examination of their role in regulating rhizosphere pH. *Plant Soil* 126: 247–264.

Hirsch, A. M. 1992. Developmental biology of legume nodulation. *N. Phytol.* 122: 211–237.

Howard, J. B., Rees, D. C. (1994) Nitrogenase: a nucleotide dependent molecular switch. *Annu. Rev. Biochem.* 63: 235–264.

Hunt, S., Hayzell, D. B. (1993) Gas exchange of legume nodules and the regulation of nitrogenase activity. *Annu. Rev. Plant Physiol. Plant Mol. Biol.* 44: 483–512.

Kahn, M. L., McDermott, T. R., Udvardi, M. K. (1998) Carbon and nitrogen metabolism in rhizobia. In *The Rhizobiaceae,* H. P. Spaink, A. Kondorosi, and P.J.J. Hooykaas, eds. Kluwer Academic Publishers, Dordrecht, The Netherlands, pp. 461–485.

Kaiser, B. N., Finnegan, P. M., Tyerman, S. D., Whitehead, L. F., Bergersen, F. J., Day, D. A., Udvardi, M. K. (1998) Characterization of an ammonium transport protein from the peribacteroid membrane of soybean nodules. *Science* 281: 1202–1206.

Long, S. R. (1996) *Rhizobium* symbiosis: Nod factors in perspective. *Plant Cell* 8: 1885–1898.

Peters, J. W., Stowell, M. H., Soltis, S. M., Finnegan, M. G., Johnson, M. K., Rees, D. C. (1997) Redox-dependent structural changes in the nitrogenase P-cluster. *Biochemistry* 36: 1181–1187.

Phillips, D. A. (1992) Flavonoids: plant signals to microbes. *Rec. Adv. Phytochem.* 26: 201–231.

Schindelin, H., Kisker, C., Schlessman, J. L., Howard, J. B., Rees, D. C. (1997) Structure of ADP·AlF4⁻-stabilized nitrogenase complex and implications for signal transduction. *Nature* 387: 370–376.

Schlaman, H., Phillips, D. A., Kondorosi, E. (1998) Genetic organization and transcriptional regulation of rhizobial nodulation genes. In *The Rhizobiaceae,* H. P. Spaink, A. Kondorosi, and P.J.J. Hooykaas, eds. Kluwer Academic Publishers, Dordrecht, The Netherlands, pp. 361–386.

Soltis, D. E., Soltis, P. S., Morgan, D. R., Swensen, S. M., Mullin, B. C., Dowd, J. M., Martin, P. G. (1995) Chloroplast gene sequence data suggest a single origin of the predisposition for symbiotic nitrogen-fixation in angiosperms. *Proc. Natl. Acad. Sci. USA* 92: 2647–2651.

Udvardi, M. K., Day, D. A. (1997) Metabolite transport across symbiotic membranes of legume nodules. *Annu. Rev. Plant Physiol. Plant Mol. Biol.* 48: 493–523.

Vitousek, P. M., Aber, J. D., Howarth, R. W., Likens, G. E., Matson, P. A., Schindler, D. W., Schlesinger, W. H., Tilman, D. G. (1997) Human alteration of the global nitrogen cycle: sources and consequences. *Ecol. Appl.* 7: 737–750.

von Wiren, N., Gazzarrini, S., Frommer, W. B. (1997) Regulation of mineral nitrogen uptake in plants. *Plant Soil* 196: 191–199.

Nitrate assimilation

Campbell, W. H. (1999) Nitrate reductase structure, function and regulation: bridging the gap between biochemistry and physiology. *Annu. Rev. Plant Physiol. Plant Mol. Biol.* 50: 277–303.

Chrispeels, M. J., Crawford, N. M., Schroeder, J. I. (1999) Proteins for transport of water and mineral nutrients across the membranes of plant cells. *Plant Cell* 11: 661–675.

Crawford, N. M. (1995) Nitrate: nutrient and signal for plant growth. *Plant Cell* 7: 859–868.

Crawford, N. M., Arst, H.N.J. (1993) The molecular genetics of nitrate assimilation in fungi and plants. *Annu. Rev. Genet.* 27: 115–146.

Crawford, N. M., Glass, A.D.M. (1998) Molecular and physiological aspects of nitrate uptake in plants. *Trends Plant Sci.* 3: 389–395.

Daniel-Vedele, F., Filleur, S., Caboche, M. (1998) Nitrate transport: a key step in nitrate assimilation. *Curr. Opin. Plant Biol.* 1: 235–239.

Epstein, E. (1972) *Mineral Nutrition of Plants: Principles and Perspectives.* Wiley, New York.

Huber, S. C., Bachmann, M., Huber, J. L. (1996) Post-translational regulation of nitrate reductase activity: a role for Ca^{2+} and 14-3-3 proteins. *Trends Plant Sci.* 1: 432–438.

Liu, K.-H., Huang, C.-Y., Tsay, Y.-F. (1999) CHL1 is a dual-affinity nitrate transporter of *Arabidopsis* involving multiple phases of nitrate uptake. *Plant Cell* 11: 865–874.

Ourry, A., Gordon, A. J., MacDuff, J. H. (1997) Nitrogen uptake and assimilation in roots and root nodules. In *A Molecular Approach to Primary Metabolism in Higher Plants,* C. H. Foyer and W. P. Quick, eds. Taylor & Francis, Bristol, PA, pp. 237–253.

Rufty, T. W. (1997) Probing the carbon and nitrogen interaction: a whole plant perspective. In *A Molecular Approach to Primary Metabolism in Higher Plants,* C. H. Foyer and W. P. Quick, eds., Taylor & Francis, Bristol, PA, pp. 255–273.

Stitt, M. (1999) Nitrate regulation of metabolism and growth. *Curr. Opin. Plant Biol.* 2: 178–186.

Tsay, Y.-F., Schroeder, J. I., Feldmann, K. A., Crawford, N. M. (1993) A herbicide sensitivity gene *CHL1* of *Arabidopsis* encodes a nitrate-inducible nitrate transporter. *Cell* 72: 705–713.

Wang, R., Liu, D., Crawford, N. M. (1998) The *Arabidopsis* CHL1 protein plays a major role in high affinity nitrate uptake. *Proc. Natl. Acad. Sci. USA* 95: 15134–15139.

Wray, J. L. (1993) Molecular biology, genetics and regulation of nitrite reduction in higher plants. *Physiol. Plant* 89: 607–612.

Zhang, H. M., Jennings, A., Barlow, P. W., Forde, B. G. (1999) Dual pathways for regulation of root branching by nitrate. *Proc. Natl. Acad. Sci. USA* 96: 6529–6534.

Sulfate assimilation

Bick, J. A., Leustek, T. (1998) Plant sulfur metabolism—the reduction of sulfate to sulfite. *Curr. Opin. Plant Biol.* 1: 240–244.

Bogdanova, N., Hell, R. (1997) Cysteine synthesis in plants: protein–protein interactions of serine acetyltransferase from *Arabidopsis thaliana. Plant J.* 11: 251–262.

Byers, M., Franklin, J., Smith, S. J. (1987) The nitrogen and sulphur nutrition of wheat and its effect on the composition and baking quality of the grain. *Aspects Appl. Biol.* 15: 337–344.

Cobbett, C. S. (1999) A family of phytochelatin synthase genes from plant, fungal and animal species. *Trends Plant Sci.* 4: 335–337.

De Kok, L. J., Stulen, I., Rennenberg, H., Brunold, C., Rauser, W. E., eds. (1993) *Sulfur Nutrition and Assimilation in Higher Plants. Regulatory, Agricultural and Environmental Aspects.* SBP Academic Publishing, The Hague, The Netherlands.

Droux, M., Ruffet, M. L., Douce, R., Job, D. (1998) Interactions between serine acetyltransferase and *O*-acetylserine(thiol)lyase in higher plants—structural and kinetic properties of the free and bound enzymes. *Eur. J. Biochem.* 255: 235–245.

Gage, D. A., Nolte, K. D., Rhodes, D., Leustek, T., Cooper, A.J.L., Hanson, A. D. (1997) Biogenic dimethylsulfide: synthesis of its precursor 3-dimethylsulfoniopropionate in marine algae. *Nature* 387: 891–894.

Howden, R., Andersen, C. R., Goldsbrough, P. B., Cobbett, C. S. (1995) A cadmium-sensitive, glutathione-deficient mutant of *Arabidopsis thaliana. Plant Physiol.* 107: 1067–1073.

Leustek, T., Martin, M. N., Bick, J. A., Davies, J. P. (2000) Pathways and regulation of sulfur metabolism revealed through molecular and genetic studies. *Annu. Rev. Plant Physiol. Plant Mol. Biol.* 51: 141–166.

May, M. J., Vernoux, T., Leaver, C., Van Montagu, M., Inzé, D. (1998) Glutathione homeostasis in plants: implications for environmental sensing and plant development. *J. Exp. Bot.* 49: 649–667.

Noctor, G., Foyer, C. H. (1998) Ascorbate and glutathione: keeping active oxygen under control. *Annu. Rev. Plant Physiol. Plant Mol. Biol.* 49: 249–279.

Rauser, W. E. (1995) Phytochelatins and related peptides. Structure, biosynthesis and function. *Plant Physiol.* 109: 1141–1149.

Rea, P. A., Li, Z.-S., Lu, Y.-P., Drozdowicz, Y., Martinoia, E. (1998) From vacuolar GS-X pumps to multispecific ABC transporters. *Annu. Rev. Plant Physiol. Plant Mol. Biol.* 49: 727–760.

Ruffet, M.-L., Droux, M., Douce, R. (1994) Purification and kinetic properties of serine acetyltransferase free of *O*-acetylserine(thiol)lyase from spinach chloroplasts. *Plant Physiol.* 104: 597–604.

Ruffet, M.-L., Lebrun, M., Droux, M., Douce, R. (1995) Subcellular distribution of serine acetyltransferase from *Pisum sativum* and characterization of an *Arabidopsis thaliana* putative cytosolic isoform. *Eur. J. Biochem.* 227: 500–509.

Saito, K., Kurosawa, M., Tatsuguchi, K., Takagi, Y., Murakoshi, I. (1994) Modulation of cysteine biosynthesis in chloroplasts of transgenic tobacco overexpressing cysteine synthase [O-acetylserine(thiol)lyase]. *Plant Physiol.* 106: 887–895.

Schmidt, A., Trebst, A. (1969) The mechanism of photosynthetic sulfate reduction by isolated chloroplasts. *Biochim. Biophys. Acta* 180: 529–535.

Schnug, E., Haneklaus, S. (1994) Sulphur deficiency in *Brassica napus*. *Landbauforsch. Völkenrode* 144: 149–156.

Schürmann, P., Brunold, C. (1980) Formation of cysteine from adenosine 5-phosphosulfate (APS) in extracts from spinach chloroplasts. *Z. Pflanzenphysiol.* 100: 257–268.

Smith, F. W., Ealing, P. M., Hawkesford, M. J., Clarkson, D. T. (1995) Plant members of a family of sulfate transporters reveal functional sub-types. *Proc. Natl. Acad. Sci. USA* 92: 9373–9377.

Suter, M., von Ballmoos, P., Kopriva, S., den Camp, R. O., Schaller, J., Kuhlemeier, C., Schürmann, P., Brunold, C. (2000) Adenosine 5′-phosphosulfate sulfotransferase and adenosine 5′-phosphosulphate reductase are identical enzymes. *J. Biol. Chem.* 275: 930–936.

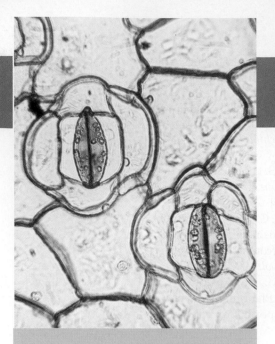

Biochemistry & Molecular Biology of Plants, B. Buchanan, W. Gruissem, R. Jones, Eds.
© 2000, American Society of Plant Physiologists

C H A P T E R *17*

Biosynthesis of Hormones and Elicitor Molecules

Alan Crozier
Yuji Kamiya
Gerard Bishop
Takao Yokota

Introduction

Plant hormones are signal molecules, present in trace quantities. Changes in hormone concentration and tissue sensitivity mediate a whole range of developmental processes in plants, many of which involve interactions with environmental factors. This chapter is concerned with the biosynthetic, catabolic, and conjugation pathways that together control the homeostasis of plant hormone pools. Attention will be focused on the first five plant hormones discovered— gibberellins, abscisic acid, cytokinins, indole-3-acetic acid, and ethylene—as well as on compounds shown more recently to have a regulatory role in plant development, namely, brassinosteroids, polyamines, jasmonic acid, and salicylic acid (Fig. 17.1). Each of these compounds has its own particular properties, so the pathways regulating their production and degradation are quite diverse and have been elucidated by synergistic use of many disciplines, including chemistry, biochemistry, plant physiology, genetics and, more recently, molecular genetics. The most powerful tool has been the use of mutant plants that are unable to catalyze an enzymatic step leading to the formation or degradation of a bioactive hormone. Such mutants have been invaluable in hormone analysis studies and in phenotypic rescue experiments using intermediates in the biosynthetic pathway. They also provide the essential material for cloning the genes that encode the biosynthetic enzymes. In addition, inhibitors of plant hormone biosynthesis have been used to dissect biosynthetic reactions for which relevant mutants have not yet been identified. More recently, transgenic technology has provided insights into pathways by altering the expression of genes involved in the biosynthesis or catabolism of hormones. Furthermore, plants expressing transgenic constructs in situations in which the promoters of the biosynthetic genes are linked to reporter genes have been used to visualize the specific tissues where bioactive hormones are produced. In summary, each hormone biosynthetic pathway has been elucidated by using similar techniques but with biases toward certain methodologies according to the nature of the biosynthetic pathway and the availability of mutants.

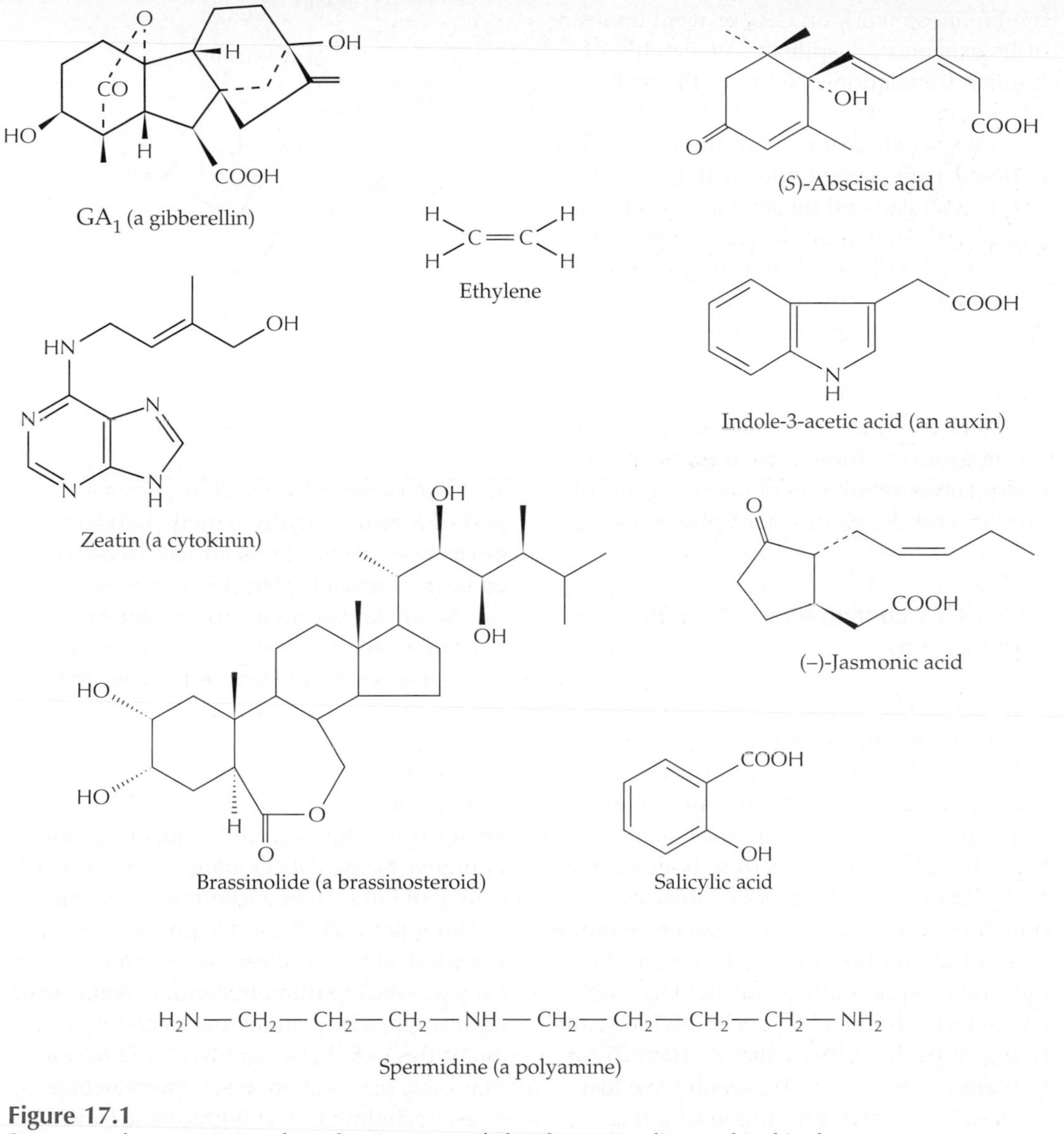

GA₁ (a gibberellin)

Ethylene

(S)-Abscisic acid

Zeatin (a cytokinin)

Indole-3-acetic acid (an auxin)

(–)-Jasmonic acid

Brassinolide (a brassinosteroid)

Salicylic acid

$$H_2N—CH_2—CH_2—CH_2—NH—CH_2—CH_2—CH_2—CH_2—NH_2$$

Spermidine (a polyamine)

Figure 17.1
Structures of representatives from the nine types of plant hormones discussed in this chapter.

17.1 Gibberellins

Gibberellins (GAs) were first isolated from the fungus *Gibberella fujikuroi* in 1926 by the Japanese scientist Eiichi Kurosawa, who was investigating the causative agent of *bakanae*, the "foolish seedling" disease of rice, which was frequently responsible for major reductions in grain yield. Diseased rice seedlings exhibited excessive shoot elongation and yellowish-green leaves. Those that survived to maturity were taller than their healthy counterparts and their grains were either absent or poorly developed. Kurosawa demonstrated that diseased rice plants were infected with a fungal pathogen, *G. fujikuroi*, which secreted a factor that increased the rate of shoot elongation. He also noted that the active factor promoted the growth of maize, sesame, millet, and oat seedlings. In the 1930s, Teijiro Yabuta and Takeshi Hayashi successfully crystallized the fungal growth-inducing factor, which they named gibberellin. At about the same time, scientists in the West were actively investigating hormonal regulation of plant growth by auxin (see Section 17.4). Unfortunately, they either failed to appreciate the significance of the Japanese work on GAs or were unaware of its existence, despite the availability of English translations. Not until the early 1950s did GA research became international in scope, when groups from the US and UK initiated their own studies with *G. fujikuroi*. The British isolated gibberellic acid and the Americans gibberellin X. The compounds proved to be identical, and the structure of gibberellic acid, now known as gibberellin A$_3$ (GA$_3$), was elucidated in 1956 (Fig. 17.2). Shortly thereafter, GAs were shown to be natural components of plants, and it became apparent that they were not merely an interesting group of fungal metabolites but also endogenous regulators of many aspects of growth and development of plants.

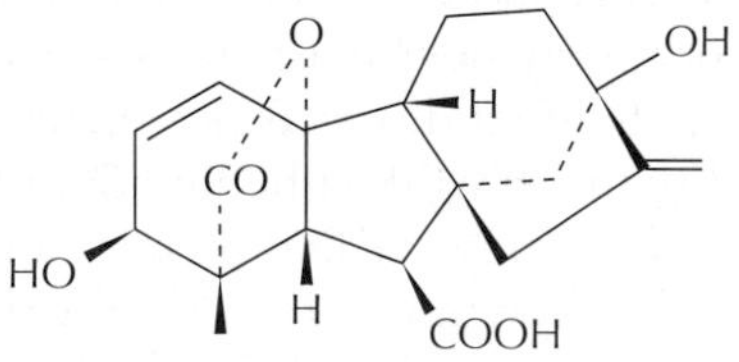

17.1.1 GAs contain either 19 or 20 carbon atoms.

The GAs are a group of tetracyclic diterpenes. At the time of writing, 125 different GAs had been characterized. Twelve GAs occur only in *G. fujikuroi*, 100 are exclusive to plants, and 13 are ubiquitous, having been detected in extracts from both fungal and plant tissues. To avoid confusion, the nomenclature GA$_1$ to GA$_{125}$ has been adopted, and the numbers GA$_{126}$ to GA$_n$ will be allocated sequentially as further GAs are characterized. The GAs can be divided into two groups: C$_{20}$-GAs, which contain 20 carbon atoms, and C$_{19}$-GAs, which have lost the C-20 and carry a γ-lactone ring (Fig. 17.3). The C$_{20}$-GAs are either δ-lactones or

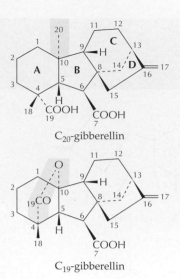

Figure 17.3
Structures of C$_{20}$-GAs and C$_{19}$-GAs.

are characterized by a C-20 that exists as part of a methyl, hydroxymethyl, aldehyde, or carboxyl group; these structures are related biosynthetically (Fig. 17.4).

Most plants contain 10 or more GAs. Variations in the types of GAs present in different species result from differences in the substituent groups or in the sequence in which groups are added to the GA skeleton. The fungus *G. fujikuroi* produces GAs, primarily GA$_3$, as secondary metabolites in amounts that far exceed the quantities found in higher plants. Developing seeds are a rich source of GAs, and concentrations as high as 50 μg per gram fresh weight have been recorded, although these concentrations decline as seeds mature. In contrast, vegetative tissues typically contain low nanogram quantities of GAs per gram. In addition to free GAs, plants contain several GA conjugates, including GA-*O*-β-glucosides and β-glucosyl ethers.

17.1.2 GAs affect many aspects of plant growth and development.

Although best known for their influence on stem elongation (Fig. 17.5), GAs also affect reproductive processes in a wide range of plants. For example, GAs can induce formation of cones in conifers and bring about flowering in plants that ordinarily require a photoperiodic signal or cold treatment for

Figure 17.2
Structure of gibberellic acid, GA$_3$

GA$_3$ (Gibberellic acid)

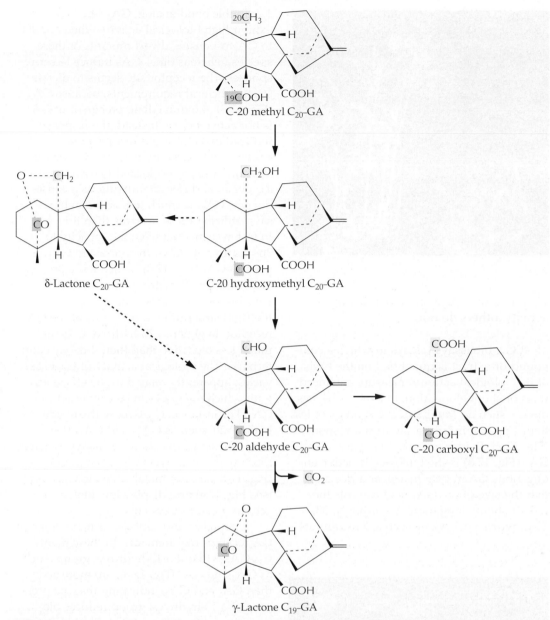

Figure 17.4

Oxidation at C-20 and the biosynthetic relationship between C_{20}- and C_{19}-GAs. On extraction from plant tissues, the C-20 hydroxymethyl C_{20}-GAs form a δ-lactone structure, but the possibility that such structures are also native products cannot be ruled out. In planta, the C-20 hydroxymethyl C_{20}-GAs act as the immediate precursors of C-20 aldehyde C_{20}-GAs, although in some test systems δ-lactones can be similarly converted (see Section 17.1.8). The C-20 aldehyde represents an important branch point in the GA biosynthesis pathway: It can either be converted to C-20 carboxyl C_{20}-GAs or can lose C-20 as CO_2, yielding biologically important C_{19}-GAs with a 19,10-γ-lactone ring.

flowering. In flowering plants that require one or more long days to flower or that require a cold treatment (see Section 17.1.13), a GA application often can substitute for the environmental signal (Fig. 17.6). GAs retard leaf and fruit senescence and promote seed germination (see Chapter 20). GA-induced de novo synthesis of α-amylase and other enzymes in the aleurone layer of barley has been used extensively in studies on the mode of action of GAs (see Chapter 13). Because of this effect, GA_3 is used widely in the malting industry to speed up and regularize the production of malt.

Figure 17.5
Effect of GA_3 on stem elongation of Progress No. 9 dwarf pea seedlings: (left) control plants, (right) plants seven days after treatment with 5 μg of GA_3.

17.1.3 Some GAs exhibit high biological activity, others do not.

Most GA **bioassays** (assays in which a compound or extract is quantified on the basis of its biological activity) measure enhanced rates of shoot elongation. In dwarf varieties of rice, such as Tanginbozu, GA doses of less than 1 ng can induce a significant response (Fig. 17.7). Evidence strongly suggests that GA_1 (Fig. 17.8) is the biologically active endogenous GA in pea, maize, and rice and that the size of the GA_1 pool controls the rate of shoot elongation. Accordingly, 3β, 13-dihydroxy C_{19}-GAs, such as GA_1 and its

Figure 17.6
Effect of GA treatment on flowering in a cold-requiring carrot variety. (Left) Control: no GA and no cold treatment; (center) no cold treatment but 10 μg GA_3 daily for four weeks; (right) six weeks of cold treatment.

1,2-double bond analog, GA_3 (see Fig. 17.8), exhibit high biological activity when applied to GA-responsive dwarf mutants of these species, whereas other GAs induce lesser responses. The receptor site seems to dictate strict structural requirements, because GA_{20} (Fig. 17.8), the immediate precursor of GA_1, is not active per se. Instead, the response induced by GA_{20} is a direct result of its conversion to GA_1 (by way of 3β-hydroxylation). This step is blocked in dwarf maize *d1*, pea *le*, and rice *dy* mutants; as a consequence, GA_{20} is much less active in bioassays utilizing these mutants than it is in test systems that have the capacity to 3β-hydroxylate GAs. In contrast to GA_1, pseudo-GA_1 (Fig. 17.8), which has a 3α- rather than a 3β-hydroxyl group, exhibits very low activity. This is further evidence of the rigid structural requirements of the GA receptor. In general, 2β-hydroxy GAs are much less bioactive than their 2-deoxy counterparts. The biological activity of C_{20}-GAs varies, apparently related to the efficiency with which the GAs can be converted to GA_1. For instance, C_{20}-GAs with an aldehyde C-20, such as GA_{19} and GA_{24} (Fig. 17.8), are active in certain bioassays, whereas those with a carboxyl C-20, such as GA_{28} (Fig. 17.8), are not metabolized to C_{19}-GAs (see Fig. 17.4) and display low biological activity in all test systems.

Cucumber and *Arabidopsis* have different GA receptor requirements. In these plants, C_{19}-GAs without a 13-hydroxyl group, such as GA_4 and GA_7 (Fig. 17.8), are more active than GA_1 and GA_3, indicating that the presence of a 13-hydroxyl group reduces biological activity. Although GA_1 is also present in cucumber and *Arabidopsis*, GA_4 is the endogenous GA that actively promotes stem growth. GA_9 also is highly active when applied to these species, but it induces a much lower response when applied with prohexadione, which inhibits its conversion to GA_4.

17.1.4 GAs are products of the terpenoid pathway.

The initial steps in the biosynthesis of GAs involve the formation of isopentenyl diphosphate and its conversion to geranylgeranyl diphosphate by way of the terpenoid pathway (Fig. 17.9). This pathway is a feature of all higher plants and leads to the production of not only GAs but also a wide range of terpenes and terpene-derived compounds as well as steroids (see Chapters 12 and 24).

The enzymes involved in the synthesis of geranylgeranyl diphosphate and its conversion to *ent*-kaurene, which represents the first committed step in GA biosynthesis, are terpene cyclases that are located in plastids. *ent*-Kaurene is oxidized to bioactive C_{19}-GAs by way of GA_{12}-aldehyde, in a series of steps that are catalyzed by cytochrome P450 monooxygenases situated on endoplasmic reticulum membranes and soluble α-ketoglutarate–dependent dioxygenases. The recent molecular cloning of the genes encoding GA biosynthesis enzymes has provided insights into the regulation of GA homeostasis by different environmental signals.

Until quite recently, the isopentenyl diphosphate used for GA biosynthesis was thought to be derived from mevalonic acid. However, as described in Chapter 24, it has now been established that plant plastids produce isoprenoids by a nonmevalonate, pyruvate/glyceraldehyde 3-phosphate pathway. Cytosolic mevalonate is not involved in GA biosynthesis in planta but is converted, by way of isopentenyl diphosphate, to geranyl diphosphate and then to farnesyl diphosphate, the substrate for synthesis of triterpenoids (e.g., sterols) (Fig. 17.9). Incorporation of [^{14}C]mevalonate into *ent*-kaurene, in yields as high as 70%, in cell-free preparations from liquid endosperm of immature seed of pumpkin (*Cucurbita maxima*) is most probably a consequence of cellular compartmentation being broken such that the preparations contain a mixture of cytosol and plastid enzymes.

17.1.5 In plants, the synthesis of *ent*-kaurene from geranylgeranyl diphosphate is catalyzed by two distinct enzymes.

ent-Kaurene is synthesized by the two-step cyclization of geranylgeranyl diphosphate by way of the intermediate, copalyl diphosphate (Fig. 17.10). The first step is catalyzed by copalyl diphosphate synthase (CPS), and the second by *ent*-kaurene synthase (KS). A CPS gene has been cloned from *Arabidopsis*

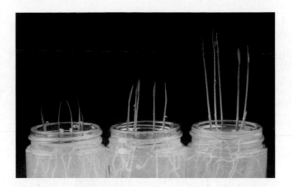

Figure 17.7
Promotion of leaf sheath elongation of Tanginbozu dwarf rice three days after treatment with GA_3: (left) control; (center) 100 pg of GA_3 per seedling; (right) 1 ng of GA_3 per seedling.

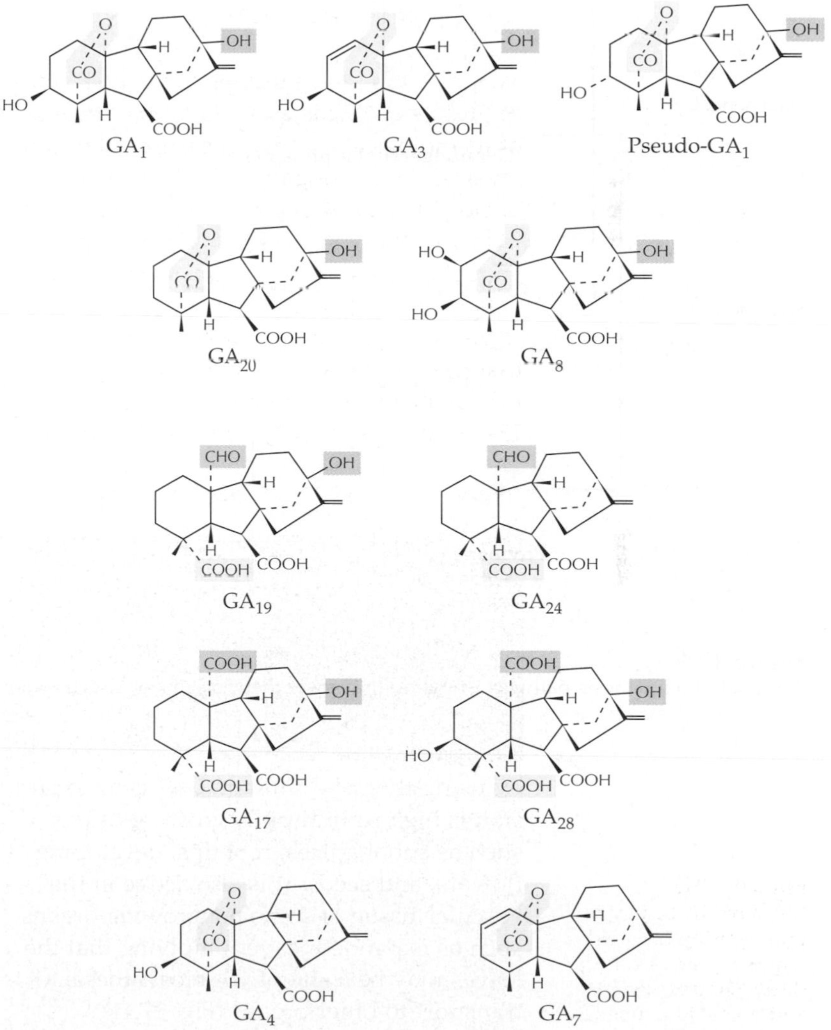

Figure 17.8
Structures of some of the GAs that have been tested in various GA bioassays and that have yielded data providing key information on the strict structural requirements of GA receptor site.

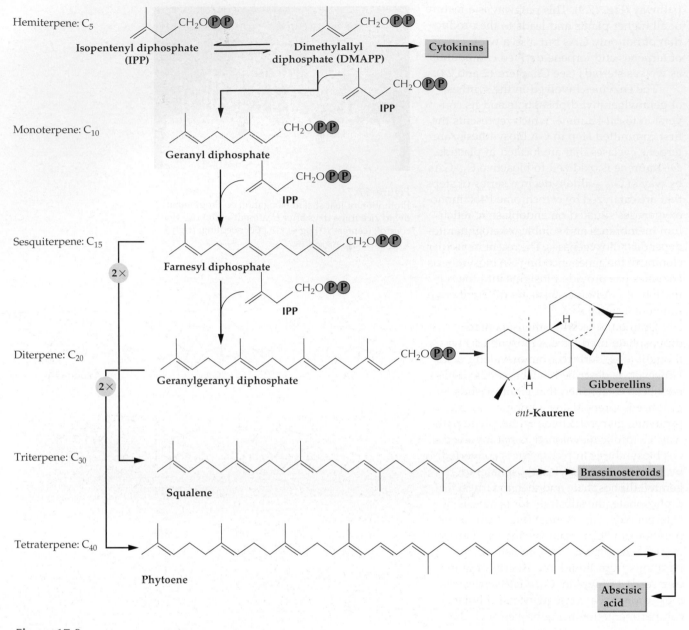

Figure 17.9
Terpenoid biosynthesis pathway, showing the biosynthetic origins of GAs as well as cytokinins, brassinosteroids, and abscisic acid.

by using the *ga1-3* mutant. *GA1* gene expression is highest in rapidly growing tissue, such as shoot apices, root tips, developing flowers, and seeds. It is also active in the vascular tissue of some nongrowing organs such as expanded leaves, implying that the leaves may be a site of GA biosynthesis for transport to other organs (Fig. 17.11).

KS was first purified from crude enzyme extracts of immature pumpkin seeds, a rich source of GA biosynthetic enzymes. Molecular cloning of KS quickly followed its

purification. The KS transcript is abundant in growing tissues, such as apices and developing cotyledons, and is present in all organs of pumpkin seedlings. In contrast, CPS is expressed at a much lower level. The low ratio of CPS transcripts to KS transcripts in pumpkin, *Arabidopsis*, and other species is consistent with a regulatory scheme that controls the conversion of abundant geranylgeranyl diphosphate to copalyl diphosphate more tightly than the subsequent conversion of copalyl diphosphate to *ent*-kaurene.

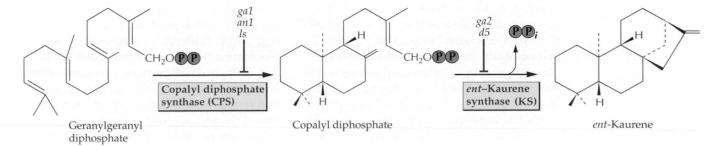

Figure 17.10
Two-step conversion of geranylgeranyl diphosphate to *ent*-kaurene catalyzed by the terpene cyclases CPS and KS. The steps blocked in *Arabidopsis ga1* and *ga2*, maize *an1* and *d5*, and pea *ls* mutants are indicated.

In *Arabidopsis*, both CPS and KS have transit peptides to facilitate their transport into plastids. Whether these peptides function independently or form a complex in vivo is not yet known. The deduced amino acid sequence of KS shares significant homology with other terpene cyclases, most notably CPS from *Arabidopsis* and maize. KS contains the DDxxD motif, which is conserved in casbene synthase, 5-*epi*-aristolochene synthase, and limonene synthase (see Chapter 24). Some researchers have proposed that the motif functions as a binding site for a divalent metal ion–diphosphate complex that is an integral part of diphosphate cleavage. Unlike KS, CPS is not involved in diphosphate cleavage and lacks the DDxxD motif.

Synthesis of *ent*-kaurene in the GA-producing fungus *Phaeospheria* sp. L487 is not identical to *ent*-kaurene production in higher plants. Expression of a cDNA encoding putative fungal CPS in *E. coli* produced a fusion protein with both CPS and KS activity. *Phaeospheria*, therefore, has a single bifunctional enzyme that converts geranyl diphosphate → copalyl diphosphate → *ent*-kaurene.

17.1.6 Cytochrome P450 oxygenases convert *ent*-kaurene to GA₁₂-aldehyde.

The highly hydrophobic *ent*-kaurene is oxidized to GA₁₂-aldehyde by NADPH-dependent, membrane-bound cytochrome P450 monooxygenases, namely, *ent*-kaurene oxidase, *ent*-kaurenoic acid hydroxylase, and GA₁₂-aldehyde synthase (Fig. 17.12). The nitrogen-containing heterocyclic compounds paclobutrazol, uniconazole, tetcyclasis, and ancymidol retard plant growth by inhibiting all three steps catalyzed by *ent*-kaurene oxidase during the conversion of *ent*-kaurene to *ent*-kaurenoic acid. However, these compounds are not specific inhibitors of GA biosynthesis because they also affect, to various degrees, other cytochrome P450 enzymes, including those involved in the biosynthesis of brassinosteroids (see Section 17.6.4 and Fig. 17.64). Several GA-deficient dwarf mutants, including *lh-2* (pea), *dx*

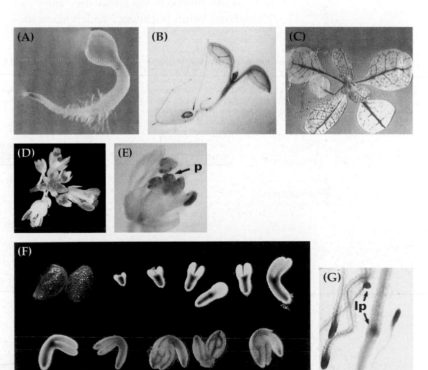

Figure 17.11
Histochemical localization of *GA1* promoter activity indicating CPS expression during the development of transgenic *Arabidopsis* containing the *GA1* promotor–GUS gene fusion *pGA1-103*. (A) One-day-old seedling; (B) five-day-old seedling; (C) three-week-old rosette plant; (D) flower cluster; (E) flower after opening, note the stained pollen (p) grains visible on the stigmatic surface; (F) developing embryos; and (G) root tips (lp, lateral primordia).

(rice), and *ga3* (*Arabidopsis*), have defective *ent*-kaurene oxidase activity. The *Arabidopsis* *GA3* gene has been cloned, and the available data indicate that a single *ent*-kaurene oxidase in *Arabidopsis* catalyzes the three oxidation steps. *GA3* gene is expressed in all tissues examined, with the abundance of mRNA being highest in inflorescence tissue.

The *Zea mays d3* dwarf mutant is GA-responsive, and the *D3* gene encodes a new type of cytochrome P450 monooxygenase that differs from the *GA3* gene and has closest homology to sterol hydroxylases. The *D3* gene perhaps encodes either *ent*-kaurenoic acid hydroxylase or GA_{12}-aldehyde synthase.

17.1.7 Conversion of GA_{12}-aldehyde to C_{19}-GAs involves two pathways: one in which C-13 is hydroxylated early, and one in which C-13 is not hydroxylated.

The multistep conversion of GA_{12}-aldehyde to C_{19}-GAs proceeds via either the "early 13-hydroxylation pathway," which leads to GA_{20} and GA_1, or the "non-13-hydroxylation pathway," which produces GA_9 and GA_4 (Fig. 17.13). Both sets of reactions include 7-oxidation, 13-hydroxylation, 20-oxidation, and 3β-hydroxylation steps, catalyzed either by membrane-bound P450 monooxygenases that require NADPH and oxygen or by soluble dioxygenases from a family of non-heme enzymes that contain iron and utilize α-ketoglutarate as a cosubstrate. Oxidation at C-7, which converts GA_{12}-aldehyde to GA_{12}, can be catalyzed by either dioxygenases or monooxygenases. A dioxygenase that converts GA_{12}-aldehyde to GA_{12} has been cloned from pumpkin endosperm. Although a soluble, α-ketoglutarate–dependent 13-hydroxylase has been detected in cell-free extracts from spinach leaves, the 13-hydroxylases identified in pumpkin endosperm, developing barley, and pea embryos are all monooxygenases.

Cell-free extracts from embryos of *Phaseolus vulgaris*, barley, and pea all utilize the early 13-hydroxylation pathway to produce the C_{19}-GA, GA_{20}, which is 3β-hydroxylated to yield GA_1 (Fig. 17.13). In most species,

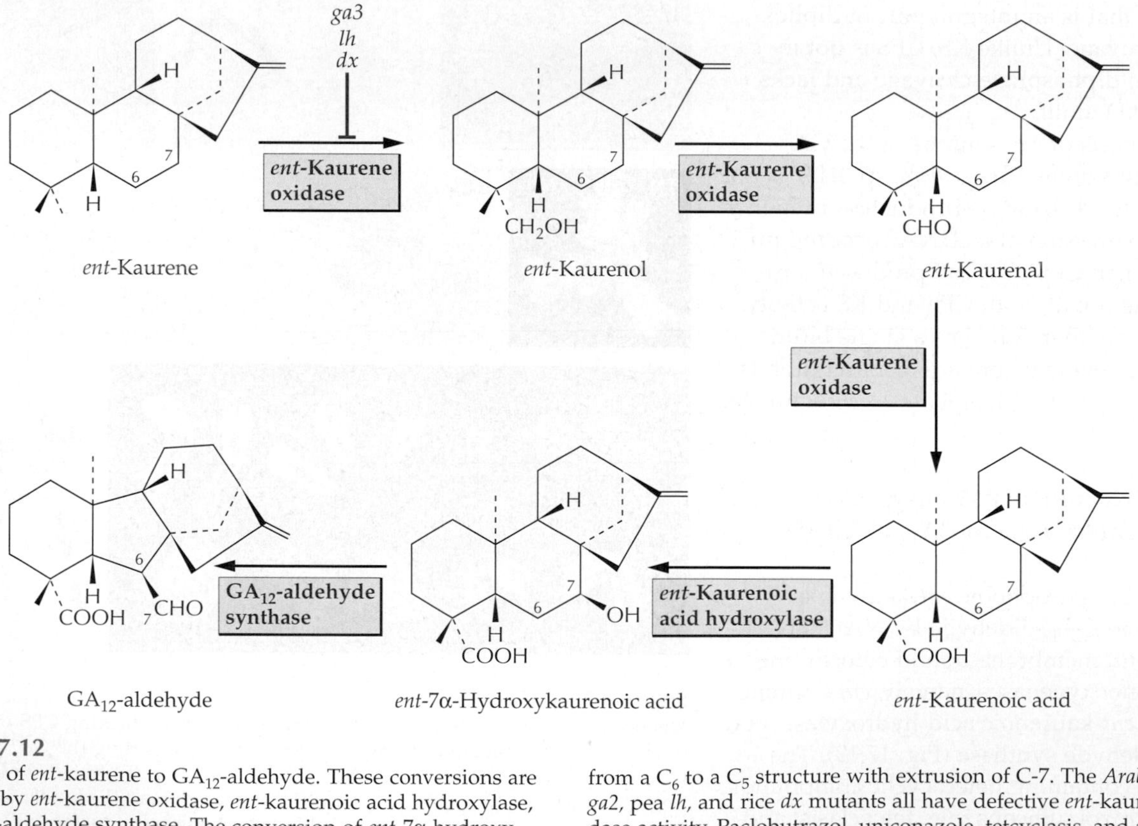

Figure 17.12
Oxidation of *ent*-kaurene to GA_{12}-aldehyde. These conversions are catalyzed by *ent*-kaurene oxidase, *ent*-kaurenoic acid hydroxylase, and GA_{12}-aldehyde synthase. The conversion of *ent*-7α-hydroxy-kaurenoic acid to GA_{12}-aldehyde involves contraction of the B ring from a C_6 to a C_5 structure with extrusion of C-7. The *Arabidopsis ga2*, pea *lh*, and rice *dx* mutants all have defective *ent*-kaurene oxidase activity. Paclobutrazol, uniconazole, tetcyclasis, and ancymidol inhibit the three *ent*-kaurene oxidase–catalyzed steps (not shown).

the preferred substrate for 13-hydroxylation appears to be GA_{12}. However, in embryo cell–free systems from *P. coccineus* and *P. vulgaris*, 13-hydroxylation occurs before 7-oxidation, and the initial steps are GA_{12}-aldehyde → GA_{53}-aldehyde → GA_{53} rather than GA_{12}-aldehyde → GA_{12} → GA_{53}.

GA_9 is produced via the non-13-hydroxylation pathway, and some species, including *Picea abies*, use this route as the major pathway from C_{20}- to C_{19}-GAs. In *P. coccineus*, both the early-13-hydroxylation and non-13-hydroxylation pathways operate. This legume can also catalyze the 13-hydroxylation of GA_4, thereby linking the early 13-hydroxylation and nonhydroxylation pathways and providing an alternative route to GA_1 (Fig. 17.13).

17.1.8 GA 20-oxidase activities catalyze the conversion of GA_{12} to C_{19}-GAs.

C_{20}-GAs undergo successive oxidations at C-20 that result in the loss of CO_2 and the formation of C_{19}-GAs (see Figs. 17.4 and 17.13). The enzymes that catalyze these oxidations are referred to as GA 20-oxidases, which are multifunctional 2-oxoglutarate–dependent dioxygenases. Numerous GA 20-oxidase cDNAs have been cloned, with the predicted amino acid sequences sharing 50% to 75% identity. The enzymes perform similar functions, converting 20-methyl GAs to the corresponding C_{19}-lactones. For example, the three GA 20-oxidase cDNAs cloned from *Arabidopsis* seedlings encode proteins that primarily convert GA_{12} to GA_9, while producing GA_{15} and traces of the biologically inactive tricarboxylic acid, GA_{25}. These cloned genes are expressed differentially in stems, flowers, and siliques (Fig. 17.14). The same *Arabidopsis* enzymes can also slowly metabolize GA_{53} to GA_{20} and to a small amount of another inactive GA, GA_{17}. A GA 20-oxidase from pumpkin seed tends to generate C_{20}-GA products, catalyzing the sequential C-20 oxidations that convert GA_{12} to GA_{25} and GA_{53} to GA_{17}; a small amount of the C_{19}-GA GA_{20} is produced at high enzyme concentrations. Except for the enzyme cloned from rice shoots, where 13-hydroxy

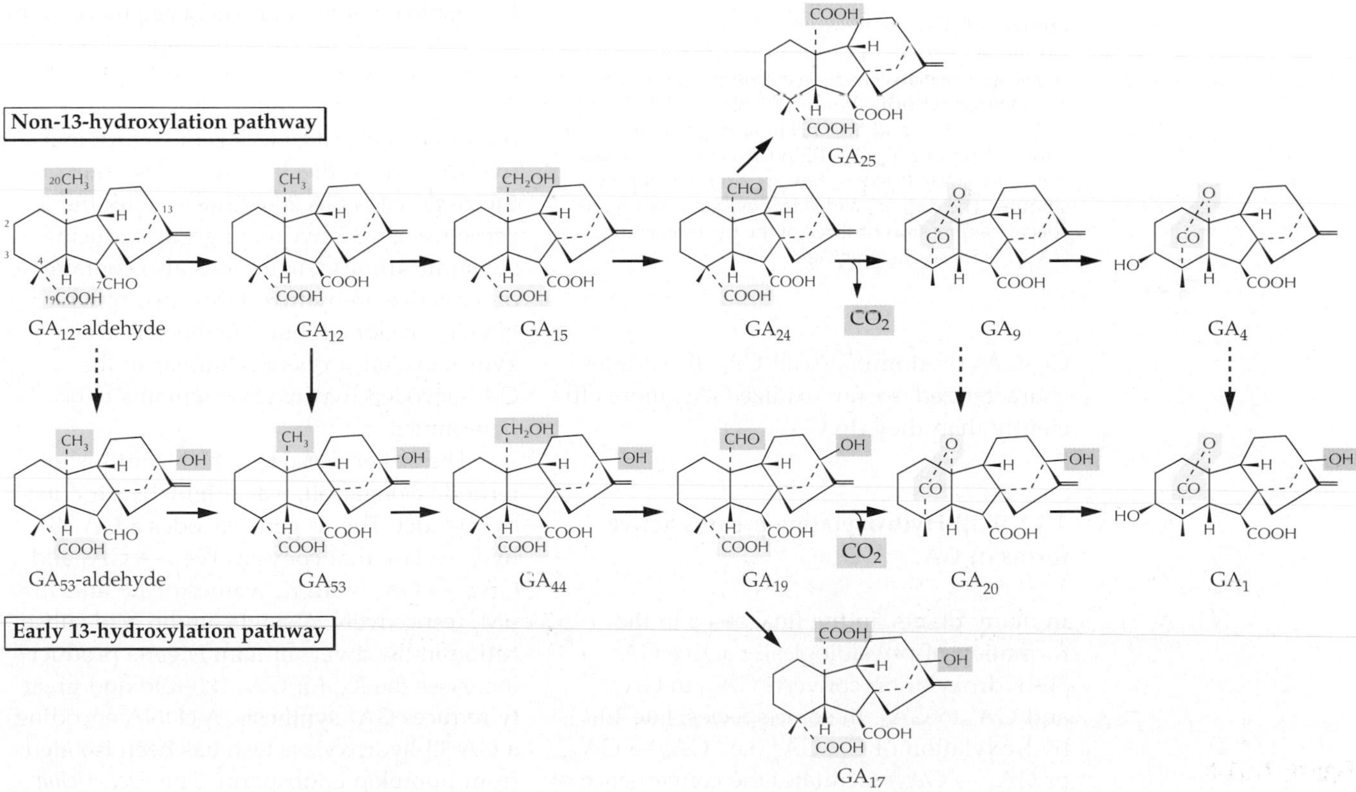

Figure 17.13
Metabolism of GA_{12}-aldehyde to C_{19}-GAs through the early 13-hydroxylation and the non-13-hydroxylation pathways. Dotted arrows indicate minor routes. The conversions involved are 7-oxidation, 13-hydroxylation, C-20 oxidation, and 3β-hydroxylation.

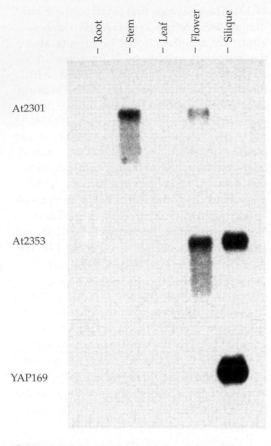

Root | Stem | Leaf | Flower | Silique

At2301

At2353

YAP169

Figure 17.14

Expression pattern of GA 20-oxidases in wild-type *Arabidopsis thaliana* seedlings (ecotype Landsberg *erecta*) after hybridization of 20-oxidase cDNA clones At2301, At2353, and YAP169 to northern blots containing 1 μg of poly(A)$^+$ RNA extracted from roots, stems, expanded leaves, flowers, and developing siliques. The data provide clear evidence of organ-specific expression of three of the genes encoding GA 20-oxidases in *Arabidopsis*.

C_{20}-GAs predominate, all GA 20-oxidases characterized so far oxidize GA_{12} more efficiently than they do GA_{53}.

17.1.9 3β-Hydroxylation creates active forms of GA.

In many plants, in the final steps in the formation of physiologically active GAs, 3β-hydroxylation converts GA_{20} to GA_1 and GA_9 to GA_4. In some species, late 13-hydroxylation of C_{19}-GAs (i.e., $GA_9 \rightarrow GA_{20}$ or $GA_4 \rightarrow GA_1$) facilitates the convergence of the early 13-hydroxylation and nonhydroxylation pathways, opening up a second route to GA_1 (see Fig. 17.13).

Some GA 3β-hydroxylases are multifunctional 2-oxoglutarate–dependent dioxygenases. 3β-Hydroxylase purified from developing embryos of *P. vulgaris* also has 2,3-desaturase and 2,3-epoxidase activities, together with weak 2β-hydroxylase activity, and GA_{20} and GA_9 are equally reactive substrates (Fig. 17.15). The 2,3-desaturase activity of 3β-hydroxylase provides the first step in the production of GA_3 in higher plants. In *Z. mays* shoots, GA_{20} is converted primarily to GA_1, but a $GA_{20} \rightarrow GA_5 \rightarrow GA_3$ branch pathway also exists (Fig. 17.16). The synthesis of GA_1 and GA_3 in maize appears to be catalyzed by one enzyme, because the *d1* lesion not only suppresses the conversion of GA_{20} to GA_1 but also reduces the formation of GA_5 and the metabolism of GA_5 to GA_3.

There is an increasing body of information on the expression of 3β-hydroxylase genes. The *ga4* mutation of *Arabidopsis* results in reduced amounts of GA_1 and GA_4 and a concomitant accumulation of GA_{20} and GA_9 in flowering shoots, indicating that the lesion affects 3β-hydroxylase activity. The *GA4* gene encodes a dioxygenase that has relatively low amino acid sequence identity with the *Arabidopsis* stem-specific GA 20-oxidase (GA5) discussed in Section 17.1.8. The K_m of the recombinant enzyme for the preferred substrate, GA_9, is one-tenth that for GA_{20}. Thus, the 3β-hydroxylase mirrors the *Arabidopsis* GA 20-oxidases, with the presence of a 13-hydroxyl group reducing substrate affinity. *Arabidopsis* also contains other active 3β-hydroxylases, two of which have been cloned, but whether these enzymes exhibit properties similar to the *GA4*-encoded hydroxylase remains to be determined.

The *Le* gene of pea, which controls internode elongation, was originally described by Mendel. The *Le* gene encodes a GA 3β-hydroxylase that converts $GA_9 \rightarrow GA_4$ and $GA_{20} \rightarrow GA_1$, with K_m values of 1.5 and 13 μM, respectively. A single amino acid substitution in the dwarf mutant *le* gene product increases the K_m for GA_9 100-fold and greatly reduces GA_1 synthesis. A cDNA encoding a GA 3β-hydroxylase also has been isolated from pumpkin endosperm. The *Escherichia coli* fusion protein expresses not only 3β-hydroxylase but also 2β-hydroxylase activity, albeit in relatively low amounts.

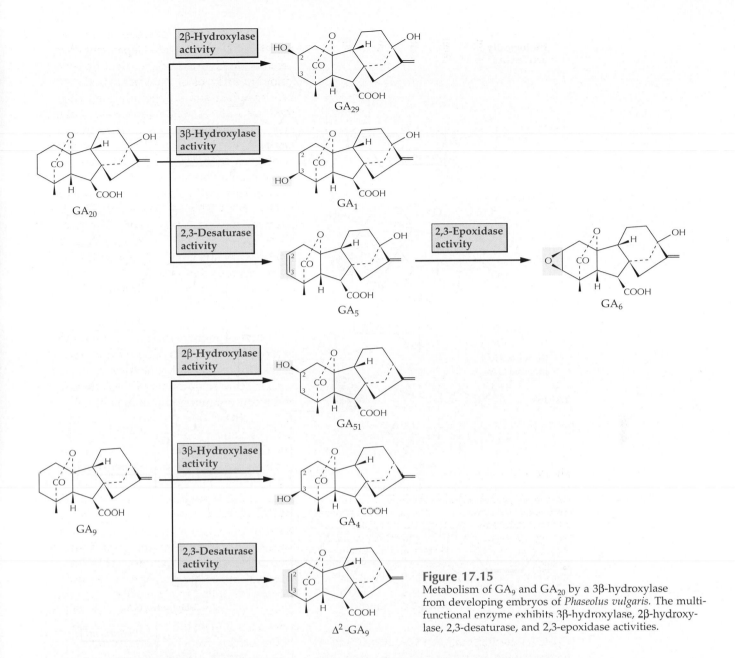

Figure 17.15
Metabolism of GA$_9$ and GA$_{20}$ by a 3β-hydroxylase from developing embryos of *Phaseolus vulgaris*. The multifunctional enzyme exhibits 3β-hydroxylase, 2β-hydroxylase, 2,3-desaturase, and 2,3-epoxidase activities.

17.1.10 Bioactive endogenous GAs are rendered inactive by 2β-hydroxylation.

2β-Hydroxylation steps, such as GA$_1$ → GA$_8$, occur in maize and many other species. These reactions, which form biologically inactive products, are important for the turnover of active GAs. As an extension of this process, GA$_8$ is glycosylated in maize, yielding GA$_8$-2-*O*-β-glucoside; in peas, however, GA$_8$ is oxidized to the 2-keto derivative GA$_8$-catabolite. A different deactivation process occurs in seed of *Pharbitis nil* (Japanese morning glory) (Fig. 17.16). Genes encoding

GA 2β-hydroxylase have been cloned from *P. coccineus* and *Arabidopsis*. Dwarf plants are generated when these GA 2β-hydroxylase clones are expressed in various species. Interestingly, GA 2β-hydroxylase activity also catalyzes the formation of GA$_8$-catabolite.

17.1.11 Some enzymes that participate in GA biosynthesis are feedback-regulated.

Recent studies have shown that GA biosynthesis can be modified by high concentrations of certain GAs in a type of feedback

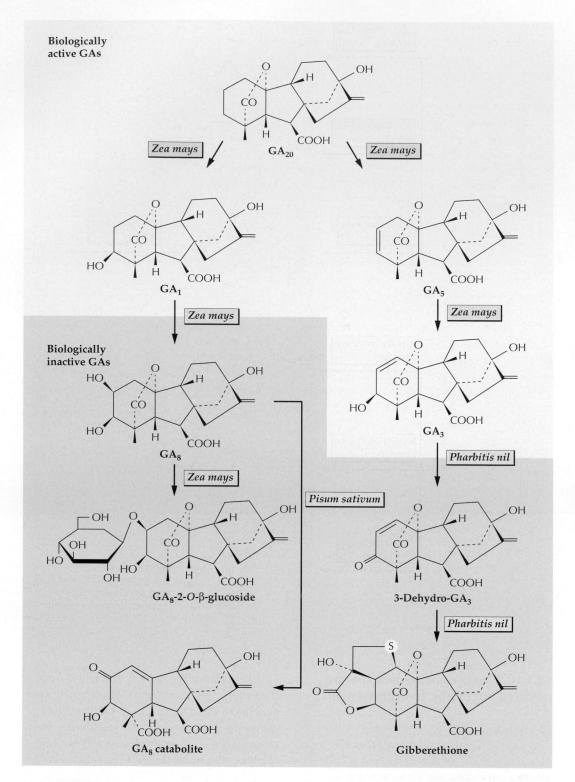

Figure 17.16

Metabolism of GA_{20} leads to formation of biologically active GAs that are then deactivated in different species by various routes. 3β-Hydroxylation of GA_{20} in *Zea mays* leads to the formation of GA_1, which, unlike its immediate precursor, is biologically active per se. GA_1 is then deactivated by the addition of a 2β-hydroxyl group to form GA_8, which is further converted to GA_8-2-*O*-β-glucoside (also inactive). In *Pisum sativum*, GA_8 is converted to the inactive GA_8 catabolite. *Z. mays* also converts GA_{20} to GA_3 by way of GA_5. GA_3 is also a bioactive structure but in *Pharbitis nil* is deactivated by metabolism to gibberethione by way of 3-dehydro-GA_3. Both 2β- and 3β-hydroxylations are blocked by prohexadione, an inhibitor of 2-oxoglutarate–dependent dioxygenases.

In the figure:
- **Biologically active GAs**
- GA_{20}, with *Zea mays* arrows to both sides
- GA_1 and GA_5
- **Biologically inactive GAs**
- GA_8, GA_3
- GA_8-2-*O*-β-glucoside, 3-Dehydro-GA_3
- *Pisum sativum*, *Pharbitis nil*
- **GA_8 catabolite**, **Gibberethione**

regulation. The presence of abnormally high concentrations of C_{19}-GAs in certain GA-insensitive dwarf mutants, such as *Rht3* wheat, *D8* maize, and *gai Arabidopsis,* indicates a link between GA action and biosynthesis. GA action in normal plants has been proposed to result in production of a transcriptional repressor that limits the expression of GA-biosynthetic enzymes. Mutants with an impaired response to GA lack this repressor and therefore have an increased rate of GA production. In addition to increased concentrations of C_{19}-GAs, GA-insensitive dwarfs often contain lower amounts of C_{20}-GAs than do the corresponding wild-type plants. This is probably a consequence of increased GA 20-oxidase activity, such that GA 20-oxidase might be a primary target for feedback regulation (Fig. 17.17). Transcript expression for each of the three *Arabidopsis* GA 20-oxidase genes is much higher in the GA-deficient *ga1-3* CPS mutant than in wild-type plants. GA₃ treatment of the mutants results in a substantial decrease in GA 20-oxidase mRNA (Fig. 17.18) within one to three hours of GA₃ application, long before a growth response is discernible. Strong GA-induced down-regulation of GA 20-oxidase transcript

Figure 17.18
Feedback regulation of GA 20-oxidase expression in floral shoots of the GA-deficient *Arabidopsis ga1-2* mutant. The 20-oxidase cDNA clones At2301, At2353, and YAP169 were hybridized to northern blots containing 1 µg of poly(A)⁺ RNA extracted from control plants and plants sprayed with 100 µM GA₃ 16 hours earlier. Note the extent of expression in the control plants, which is reduced markedly by GA₃ treatment.

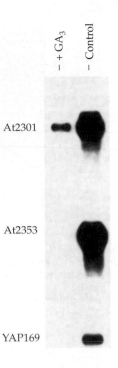

expression has also been observed in pea and rice. Similarly, in wild-type plants treated with paclobutrazol or uniconazole, which inhibit GA biosynthesis (see Fig. 17.12), low concentrations of endogenous GA are associated with increases in GA 20-oxidase mRNA, increases that drop markedly after GA₃ application.

Evidence suggests that, like GA 20-oxidases, 3β-hydroxylase activities are feedback regulated. In pea seedlings, this involves close control of the 3β-hydroxylase that catalyzes the conversion of GA_{20} to GA_1 (a key step that regulates the size of the biologically active GA_1 pool). Evidence for feedback regulation of other steps in the GA biosynthesis pathway is likely to emerge when more genes are cloned. The extent to which a given gene is sensitive to such regulation may well depend on how the enzyme activity affects the rate of turnover of the bioactive GA—which may vary among species.

17.1.12 In some cases, phytochrome appears to mediate light-induced changes in GA metabolism.

The involvement of GAs in the photoperiod-induced bolting of long-day rosette plants is well documented. Transfer of *Silene armeria* plants from short days to long days results in decreased GA_{53} content, consistent with increased GA_{53} metabolism, and a several-fold increase in the amount of GA_1, particularly in the subapical elongation zone of the stem. In spinach, transfer of plants from short to long days initiates changes in both the concentrations of endogenous GA and the activities of foliar enzymes, which suggests enhanced oxidation of GA_{53} and GA_{19} in long days (Fig. 17.19). A spinach leaf GA 20-oxidase that can oxidize both GA_{53} and GA_{19} exhibits increased activity during long days and decreased activity during short days. Furthermore, spinach plants grown

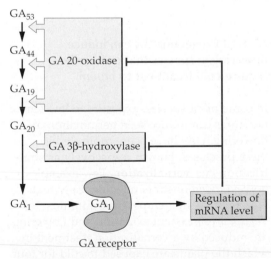

Figure 17.17
Model illustrating regulation of GA_1 synthesis by feedback mechanisms. The three-step conversion of GA_{53} to GA_{20} is catalyzed by GA 20-oxidase. GA_{20} is activated by GA 3β-hydroxylase, yielding GA_1. The GA_1 is recognized by the putative GA receptor, probably located in the cytosol. The signal recognized by the receptor provides feedback to regulate transcription of both the GA 20-oxidase and the GA 3β-hydroxylase.

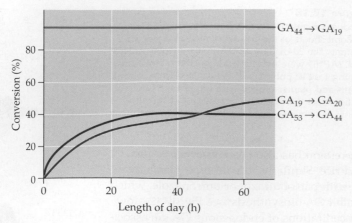

Figure 17.19
Photoperiodic regulation of C_{20}-GAs in spinach leaves, as shown by time courses for conversions of $GA_{53} \rightarrow GA_{44}$, $GA_{44} \rightarrow GA_{19}$, and $GA_{19} \rightarrow GA_{20}$ in cell-free preparations from leaves obtained after transferring spinach plants from a short to a long photoperiod. In contrast to the other two steps, the conversion of GA_{44} to GA_{19} is not affected by photoperiod.

in long days contain more GA 20-oxidase mRNA than do short-day plants. During long days, presumably the GA 20-oxidase activity is sufficient to increase the GA_1 concentration above the threshold required to induce stem extension. In fact, light appears to increase the total flux through the pathway, because *ent*-kaurene synthesis in spinach also is enhanced in long days. Although both GA_{53} oxidase and GA_{19} oxidase activities are regulated by photoperiod, oxidation of GA_{44} remains high irrespective of the prevailing light regime (Fig. 17.19). This latter activity is probably encoded by a separate, specific 20-oxidase gene, the expression of which is not regulated by light.

Many plants demonstrate reduced growth rates under red light and enhanced stem elongation in darkness or far-red light. Despite numerous attempts to correlate these phytochrome-mediated responses with altered GA metabolism, unequivocal data are rare. In cowpea (*Vigna sinensis*) epicotyls, far-red light increases GA_1 concentrations by enhancing 3β-hydroxylation of GA_{20} and reducing 2β-hydroxylation of GA_1. These changes in GA metabolism are accompanied by increased rates of shoot elongation. In peas, enhanced shoot elongation in darkness and far-red–rich light is not associated with increases in GA_1, possibly indicating that light-induced changes in stem elongation are mediated primarily by changes in responsiveness to GA.

A clear link between GA metabolism and phytochrome has been established in the case of red light–induced germination of photodormant Grand Rapids lettuce seed. In a typical phytochrome-mediated response, red light induces germination, but this effect is reversed by subsequent exposure to far-red light (see Chapter 18). Seeds germinate in darkness only after treatment with GA. After a pulse of red light, germination is preceded by increased expression of GA 3β-hydroxylase and increased GA_1 content in the seed. When the red light treatment is followed by a pulse of far-red light, germination is not induced, the GA_1 pool size does not increase, and the 3β-hydroxylase mRNA concentration remains unaltered (Fig. 17.20).

The amounts of the GA_1 precursors GA_{19} and GA_{20} are not affected by red and far-red light treatments. Whether red and far-red light control the turnover rates for these compounds is unclear. Of the two GA 20-oxidase genes cloned from lettuce, one is light-regulated, its expression repressed by red light and enhanced by far-red light, whereas the other is light-independent. Under all light regimes, however, lettuce seeds contain far more GA_{20} than GA_1, so 20-oxidase activity and the rate of GA_{20} synthesis do not limit the amount of substrate available for GA_1 production (Fig. 17.20).

17.1.13 Exogenous GA can induce flowering in species that ordinarily require cold treatment to bloom.

In some plant species, exposure to low temperatures can induce seed germination or flowering. GAs have been implicated in these processes, known respectively as **stratification** and **vernalization**. Few examples, however, demonstrate unequivocally that GAs mediate the temperature stimulus. In *Thlaspi arvense*, stem extension and flowering are induced by a vernalization treatment in which the plants are exposed to cold for four weeks and subsequently returned to higher temperatures. The same effect can be obtained without cold induction by applying GAs; the most active GA tested is GA_9. In non-induced plants, very high concentrations of *ent*-kaurenoic acid accumulate in the shoot tip, the site of perception of the cold stimulus. When plants are returned to warmer temperatures after vernalization, the amounts of *ent*-kaurenoic acid fall to relatively low values within a matter of days. Also, [^{14}C]*ent*-kaurenoic acid is metabolized to GA_9 in thermo-induced shoot tips of *T. arvense* but not in noninduced tips. Microsomes from induced shoot tips convert both *ent*-kaurene and *ent*-kaurenoic acid to GA_9, whereas microsomes from noninduced shoot tips are much less active. In contrast, low temperature does not alter *ent*-kaurenoic acid metabolism in leaves or microsomes extracted from leaves. These results are consistent with low temperature mediating an enhancement of *ent*-kaurene oxidase and *ent*-kaurenoic acid hydroxylase activities (see Fig. 17.12) in shoot tips of *Thlaspi*, although additional effects on downstream GA biosynthetic enzymes cannot be ruled out.

17.2 Abscisic acid

Growth-inhibiting compounds were isolated in the early 1950s by Thomas Bennett-Clark and Ned Kefford, who were studying endogenous auxins (see Section 17.4) in plant extracts. Acidic compounds separated by paper chromatography were tested for their ability to promote growth in oat coleoptiles. These experiments detected a compound that inhibited coleoptile elongation, referred to as the β-inhibitor complex. Subsequent investigations in a number of laboratories correlated high β-inhibitor levels with the suppression of sprouting of potato tubers, abortion of yellow lupine pods, and bud dormancy in deciduous trees such as sycamore and birch. Two groups working in the US and the UK in the early 1960s isolated respectively an abscission-accelerating compound ("abscisin II") from young cotton fruits and a dormancy-inducing factor ("dormin") from sycamore leaves. In 1965 Frederick Addicott's group in the US reported the structure of abscisin II, later called **abscisic acid** (ABA), which turned out to be identical to dormin.

ABA has a single chiral center at its 1'-carbon. Only the (*S*)-enantiomer occurs in nature (Fig. 17.21). In contrast, commercially available ABA is a racemic mixture, although

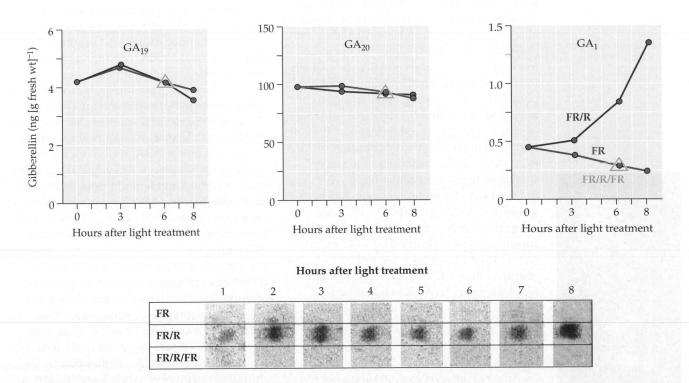

Figure 17.20

Red and far-red light and regulation of GA contents in Grand Rapids lettuce. (Upper panels) Effects of red and far-red light on the levels of GA_{19}, GA_{20}, and GA_1. Note that the phytochrome-mediated, red light–induced germination of the seeds is associated with increased GA_1 content. (Lower panels) Hybridization of a 3β-hydroxylase cDNA clone (*Ls3βOH1*) to northern blots containing 50 µg of total RNA extracted from lettuce seed after treatment with red (R) and far-red (FR) light.

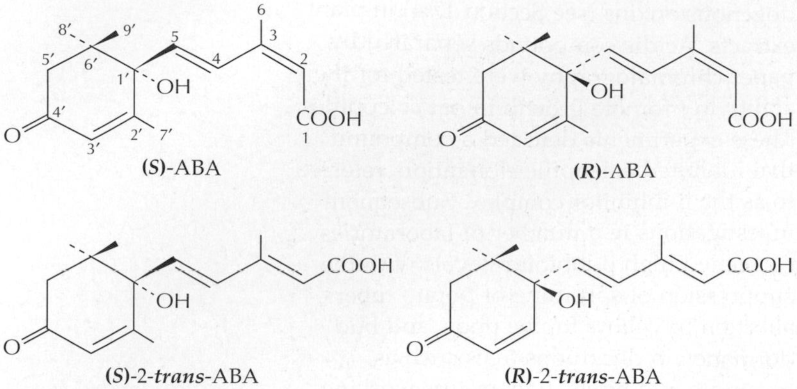

Figure 17.21
Structures of ABA enantiomers. *(S)*-ABA is the naturally occurring form.

17.2.2 Many fungi, including two fungal pathogens of the genus *Cercospora*, synthesize ABA from farnesyl diphosphate.

Although ABA is a relatively simple C_{15} compound, its biosynthetic pathway proved difficult to define—in part, because of poor incorporation of label from putative precursors, even in water-stressed leaves that were synthesizing ABA rapidly. By the early 1970s, two pathways leading to ABA had been proposed. The "direct C_{15}" route envisaged ABA production from the C_{15} precursor farnesyl diphosphate, whereas the "indirect C_{40}" route postulated oxidative cleavage of a putative C_{40} intermediate, such as violaxanthin, to yield a C_{15} intermediate of ABA (Fig. 17.24; also see Fig. 17.9).

Although the first pathway was initially favored, evidence for its operation was scant. Suggestions that ABA might originate from a xanthophyll were based on reports that exposure to light or lipoxygenase could convert violaxanthin to xanthoxin, a C_{15} compound similar in structure to ABA. Xanthoxin is present in plant tissues in trace quantities, and metabolism of [14C]xanthoxin to ABA was demonstrated in tomato and bean plants. Nevertheless, conclusive evidence that endogenous ABA originated from xanthoxin was lacking, and it remained to be established whether xanthoxin was formed in vivo from a C_{40} precursor or more directly from farnesyl diphosphate (Fig. 17.24). In 1977, investigators discovered that the plant pathogen *Cercospora rosicola* produced very large amounts of ABA, well in excess of those found in plants. They quickly appreciated that the fungus might provide the key to elucidating the ABA biosynthesis pathway in much the same way that *G. fujikuroi* had proved invaluable in early studies on GA biosynthesis.

In a simple synthetic liquid medium, *C. rosicola* cultures grown for 20 days can accumulate as much as 30 mg of ABA per liter. This figure can be increased 10-fold by adding corn-steep liquor, $CaCO_3$, and greater proportions of potato broth to the culture medium. Unlike higher plants, the fungus incorporated labeled acetate, farnesol, and other intermediates into ABA and 1'-deoxyABA. When fed to *C. cercospora*, [14C]1'-deoxyABA was metabolized to ABA, indicating that 1'-hydroxylation is the final

stereospecific syntheses of both *(S)*-ABA and *(R)*-ABA have been reported. The 2-*cis* double bond of both enantiomers is isomerized by light to the 2-*trans* isomers, *(S)*-2-*trans*-ABA and *(R)*-*trans*-ABA (Fig. 17.21). For convenience, the term ABA hereafter will refer to the naturally occurring *(S)*-enantiomer.

17.2.1 Contrary to its name, ABA does not induce abscission.

As its name indicates, ABA was originally thought to induce abscission. Now, however, abscission is known to be regulated by ethylene rather than ABA. Moreover, increased internal concentrations of ABA are unlikely to be responsible for the dormancy of either potato tubers or resting buds on deciduous trees. Studies with ABA-deficient mutants of maize have provided no support for conjecture that asymmetric distribution of ABA is involved in the negative geotropic response of roots. Accumulation of ABA during seed development, however, has been associated with maturation of the seed, development of desiccation tolerance, and suppression of vivipary (Fig. 17.22). Certain types of seed dormancy are linked to high concentrations of ABA, and many data link rapid increases in ABA to water stress–induced closure of stomata (Fig. 17.23).

Figure 17.22
Precocious germination (vivipary) of immature seed of ABA-insensitive *vp1* mutant of maize. The colorless *vp1* kernels lack anthocyanins in the aleurone layer, and embryos germinate before maturity because of the insensitivity to ABA. The seed is viable if transplanted directly from the immature ear.

(A)   **(B)**

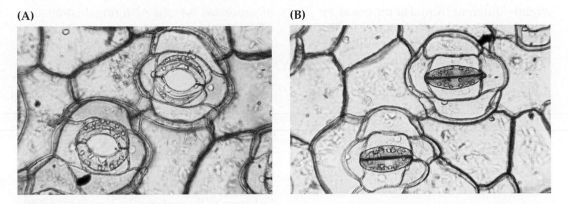

Figure 17.23
ABA-induced stomatal closure. Epidermal strips of *Commelina communis* L. incubated in buffer (10 mM Pipes, pH 6.8) containing 50 mM KCl and supplied with CO_2-free air. The stomata are open wide after two to three hours (A). When transferred to the same solution plus 10 µM ABA, the pores close completely within 10 to 30 minutes (B).

step in the fungal ABA biosynthesis pathway. Studies in *C. rosicola* and a related species, *C. cruenta,* identified α-ionylidene ethanol, α-ionylidene acetic acid, and 4′-hydroxy-α-ionylidene acetic acid as naturally occurring fungal products that were converted to 1′-deoxyABA and ABA; this was demonstrated by isotopically labeling these compounds and feeding them to cultures. Figure 17.25 illustrates one possible pathway for ABA biosynthesis in *C. rosicola*. Note that the sequence of oxidations may vary from that shown, and some evidence suggests that the pathway in *C. cruenta* may be

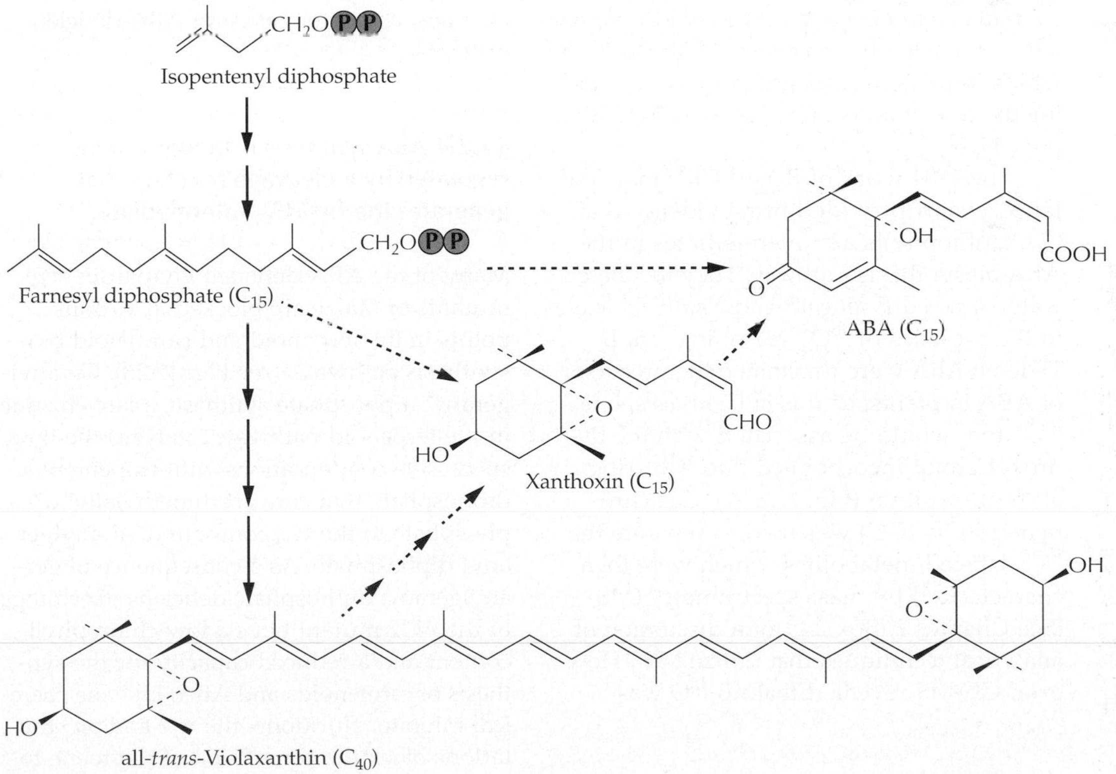

Isopentenyl diphosphate

Farnesyl diphosphate (C_{15})

Xanthoxin (C_{15})

ABA (C_{15})

all-*trans*-Violaxanthin (C_{40})

Figure 17.24
Summary of two possible biosynthetic routes to ABA. In the direct C_{15} pathway, farnesyl diphosphate is modified to yield ABA. Alternatively, in the indirect C_{40} pathway, a carotenoid, 9-*cis*-violaxanthin, is cleaved to form a C_{15} ABA precursor, xanthoxin.

slightly different from that proposed for *C. rosicola*. Nonetheless, it is evident that a direct C_{15} ABA biosynthesis pathway operates in *C. rosicola* and *C. cruenta*. The noncyclic intermediates between farnesyl diphosphate and the α-ionylidene derivatives have not been identified.

17.2.3 Gas chromatography–mass spectrometry shows that plants synthesize ABA from a C_{40} precursor.

After the discovery that 1'-deoxyABA was the immediate precursor of fungal ABA, that compound and α-ionylidene acetic acid were investigated to determine their roles in plant ABA biosynthesis. Although *Vicia faba* cuttings converted both compounds to ABA, neither was metabolized to ABA by *P. vulgaris* leaves, avocado fruit, or barley shoots, suggesting that fungal and plant pathways differed. Data obtained in other plant studies began to provide evidence for an indirect C_{40} ABA biosynthesis pathway from carotenoids. Carotenoid-deficient *Z. mays* mutants, for instance, had reduced ABA content. ABA contents were also depleted in wild-type barley and maize after treatment with fluridone, which inhibits β-carotene synthesis by blocking the conversion of phytoene to phytofluene (Fig. 17.26).

The 1984 work of Robert Creelman and Jan Zeevaart provided firm evidence that C_{40} xanthophylls are intermediates in the ABA biosynthesis pathway. They incubated water-stressed *P. vulgaris* and *Xanthium* leaves in the presence of $^{18}O_2$, reasoning that if 1'-deoxyABA were the immediate precursor of ABA in plants, as it is in *C. rosicola*, one ^{18}O atom would be associated with the hydroxyl group incorporated into ABA ring at the 1'-position (Fig. 17.27A). Gas chromatography (GC) was used to separate the $^{18}O_2$-labeled metabolites, which were then characterized by mass spectrometry (MS). (See Chapter 2, Box 2.1, for a discussion of analytical techniques that utilize MS.) However, GC-MS revealed that no ^{18}O was

incorporated into the ABA ring. Instead, one labeled atom appeared in the side chain carboxyl group, demonstrating that 1'-deoxyABA was not the immediate ABA precursor in water-stressed leaves. These results indicated that ABA was synthesized from a preformed precursor with oxygen atoms at the 1' and 4' positions. This was consistent with a C_{40} pathway to ABA, with $^{18}O_2$ cleaving violaxanthin to form xanthoxin labeled at the 1-CHO group, followed by oxidation of the labeled xanthoxin to yield ABA with one ^{18}O atom in the carboxyl group (Fig. 17.27B).

Xanthophylls (see Chapters 12 and 22) are present in tissues of most higher plants in far higher concentrations than ABA. The large size of the xanthophyll pools impeded efforts to define the ABA biosynthesis pathway with a traditional metabolic approach. Even when a plant or tissue rapidly incorporated ^{14}C-labeled precursors, such as $^{14}CO_2$ or $[^{14}C]$farnesyl diphosphate, only a minor portion of the pools of C_{40} intermediates became radiolabeled and, consequently, very little ^{14}C was incorporated into the C_{15} cleavage products and ABA. Further elucidation of the pathway required the genetic and biochemical characterization of ABA-deficient mutants.

17.2.4 ABA synthesis is thought to be regulated by a cleavage reaction that generates the first C_{15} intermediate.

Many of the ABA-deficient viviparous *(vp)* mutants of maize are blocked at various points in the terpenoid and carotenoid biosynthesis pathways (see Fig. 17.26). Geranylgeranyl diphosphate synthase, a key enzyme in the terpenoid pathway, catalyzes the three successive condensations with isopentenyl diphosphate that convert dimethylallyl diphosphate to the C_{20} compound geranylgeranyl diphosphate. As a consequence of geranylgeranyl diphosphate deficiency, seedlings of the *vp12* mutant have a low chlorophyll content and a reduced capacity for the synthesis of carotenoids and ABA. Like the chemical inhibitor fluridone, the *vp2* and *vp5* mutations block the conversion of phytoene to phytofluene. Available evidence implies that the *vp9* mutant cannot synthesize neurosporene from ζ-carotene; the *vp7* lesion restricts the conversion of lycopene to β-carotene.

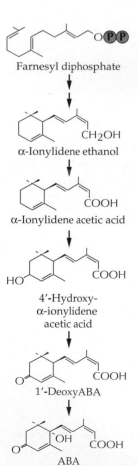

Farnesyl diphosphate

α-Ionylidene ethanol

α-Ionylidene acetic acid

4'-Hydroxy-α-ionylidene acetic acid

1'-DeoxyABA

ABA

Figure 17.25
Possible direct C_{15} pathway for ABA biosynthesis from farnesyl diphosphate in *Cercospora rosicola*.

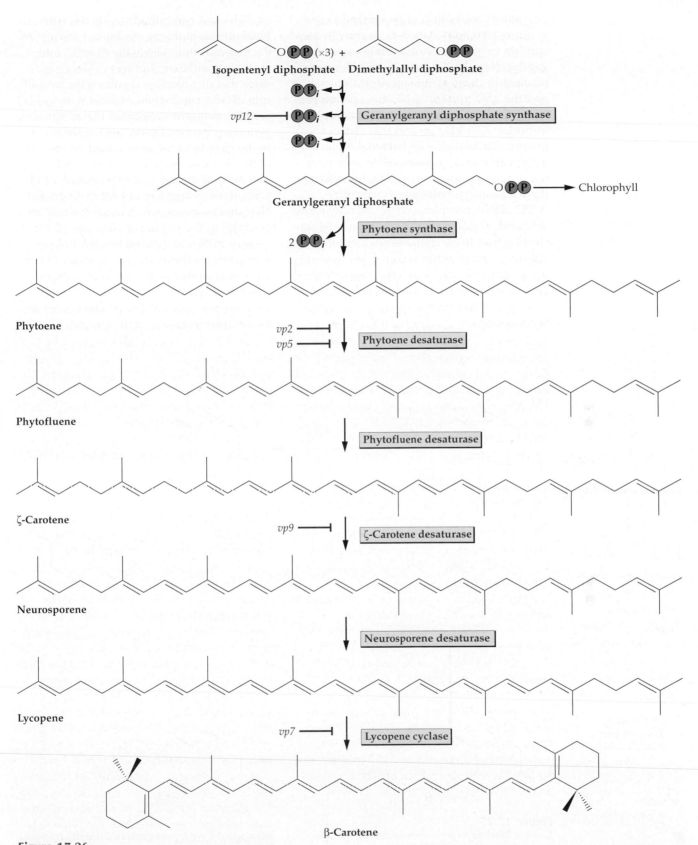

Figure 17.26
Early stages in the indirect C_{40} ABA biosynthesis pathway: production of geranylgeranyl diphosphate and synthesis of β-carotene. Enzymes deficient in maize *vp* mutants are indicated. The chemical inhibitors fluridone and norflorazon block conversion of phytoene to phytofluene (not shown).

After metabolism of β-carotene to zea-xanthin, two epoxidation steps convert zea-xanthin to all-*trans*-violaxanthin via anthera-xanthin (Fig. 17.28). Both epoxidations are blocked in the *aba1* mutant of *Arabidopsis* and the *aba2* mutant of *Nicotiana plumbagini-folia*. The *ABA2* gene encodes a chloroplast-imported 72.5-kDa protein that shares se-quence similarities with bacterial mono-oxygenases and oxidases. ABA2 catalyzes the in vitro conversion of zeaxanthin to an-theraxanthin and all-*trans*-violaxanthin. *ABA2* cDNA complements *N. plumbaginifolia aba2* and *Arabidopsis aba1* mutations, demon-strating that these mutants are homologous. All-*trans*-violaxanthin is converted to 9'-*cis*-neoxanthin via all-*trans*-neoxanthin, 9-*cis*-violaxanthin, or both. To date, no mutant defective in this section of the pathway has been isolated.

The first committed step in the ABA biosynthesis pathway, oxidative cleavage of 9'-*cis*-neoxanthin, yields the first C_{15} inter-mediate, xanthoxin. Indirect evidence indi-cates that this reaction regulates the overall rate of ABA production. Studies with *vp14*, an ABA-deficient viviparous maize mutant generated by transposon tagging, resulted in the cloning of the gene encoding the en-zyme that catalyzes this key conversion. The derived amino acid sequence of VP14 is similar to sequences of bacterial lignostil-bene dioxygenases, which catalyze a reaction resembling the oxidative cleavage of 9'-*cis*-neoxanthin. The recombinant VP14 pro-tein catalyzes the in vitro conversion of both 9'-*cis*-violaxanthin and 9'-*cis*-neoxanthin to xanthoxin, but the corresponding trans iso-mers are not cleaved. The production of xan-thoxin from 9'-*cis*-neoxanthin is also blocked

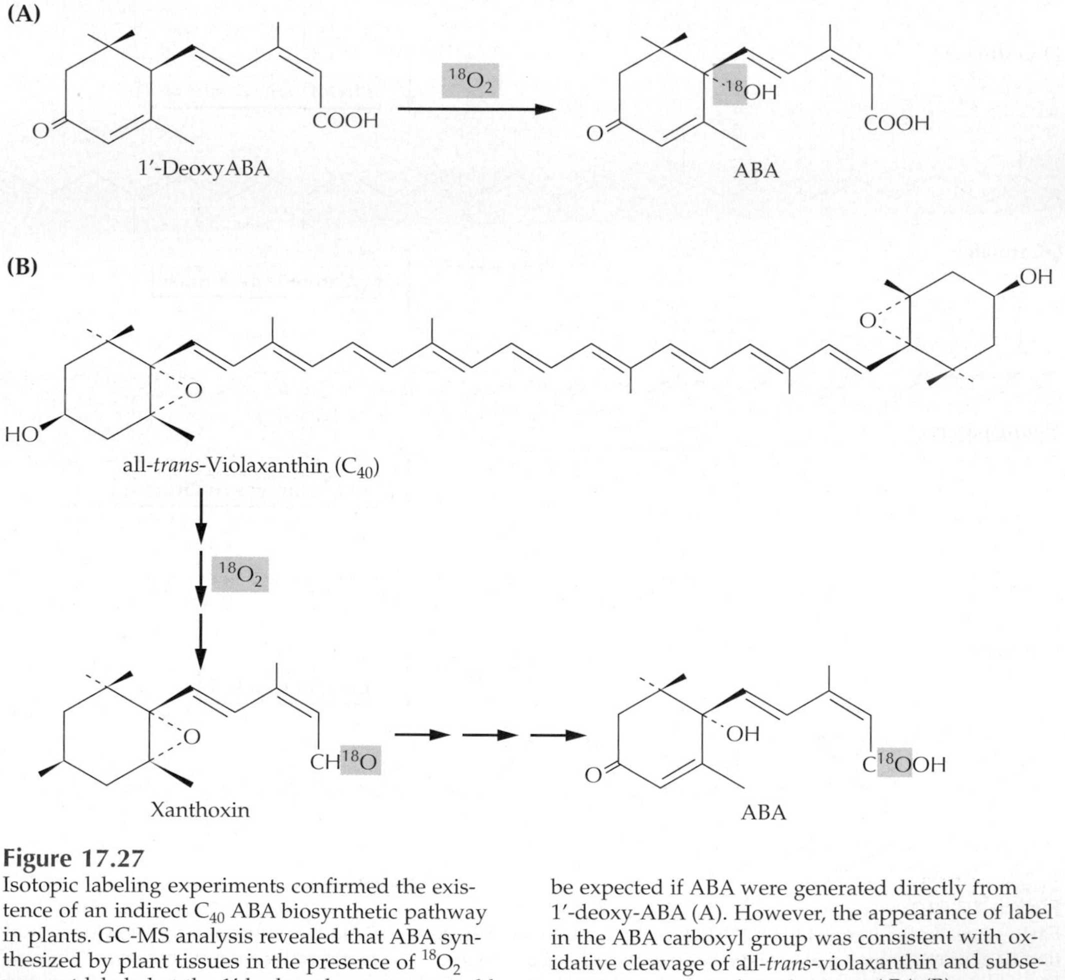

Figure 17.27
Isotopic labeling experiments confirmed the exis-tence of an indirect C_{40} ABA biosynthetic pathway in plants. GC-MS analysis revealed that ABA syn-thesized by plant tissues in the presence of $^{18}O_2$ was not labeled at the 1'-hydroxyl group, as would be expected if ABA were generated directly from 1'-deoxy-ABA (A). However, the appearance of label in the ABA carboxyl group was consistent with ox-idative cleavage of all-*trans*-violaxanthin and subse-quent conversion of xanthoxin to ABA (B).

Chapter 17 Biosynthesis of Hormones and Elicitor Molecules

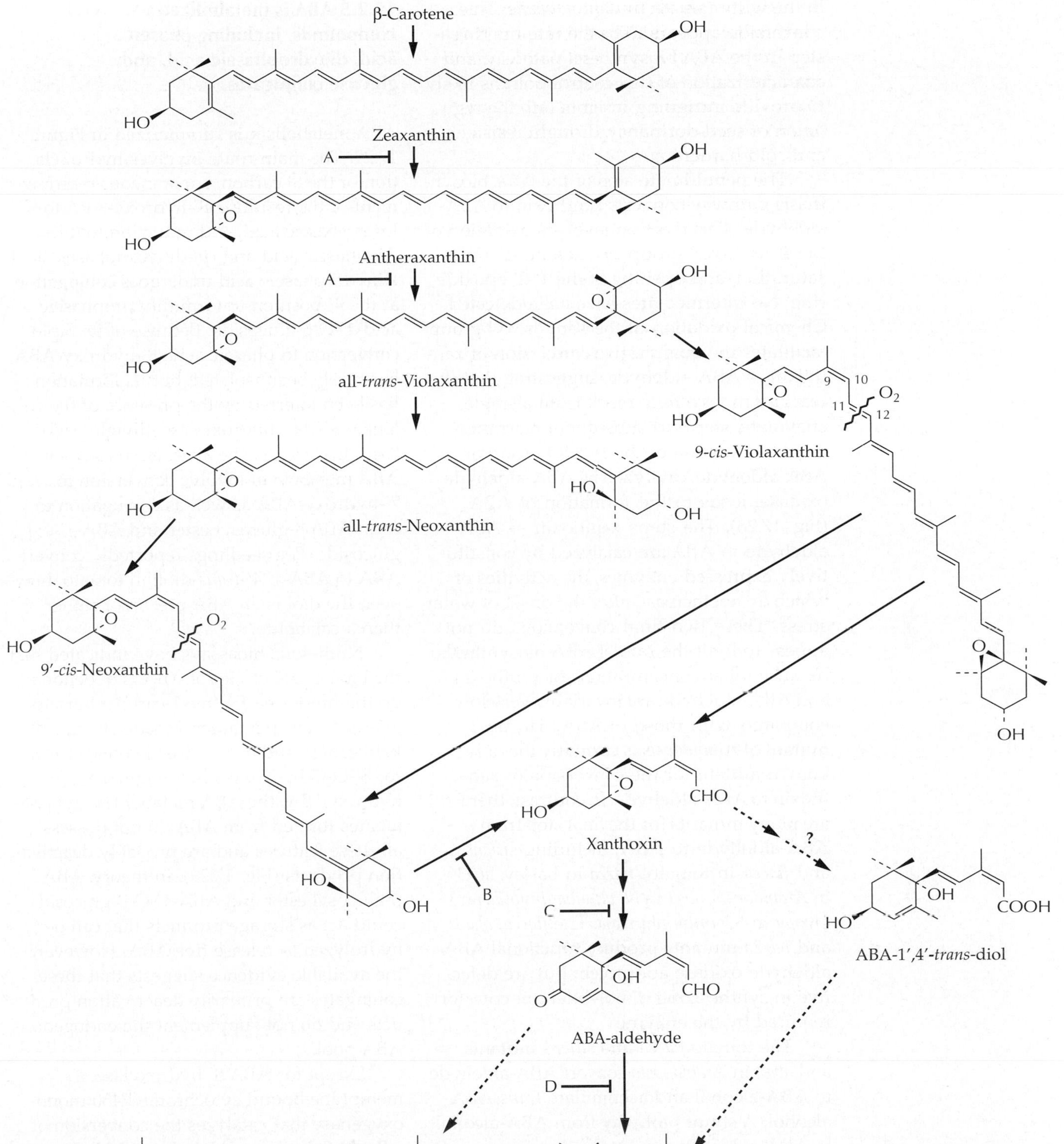

Figure 17.28
Later stages in the indirect C_{40} ABA biosynthesis pathway: β-carotene to ABA. Step A is blocked in *Arabidopsis aba1* and *Nicotiana plumbaginifolia aba2* mutants; step B is blocked in *Zea mays vp14* and *Lycopersicon esculentum* Mill *notabilis* mutants; step C is blocked in *Arabidopsis thaliana aba2* mutant; step D is blocked in *Lycopersicon esculentum* Mill *sitiens* and *flacca*, *Hordeum vulgare nar2a*, *N. plumbaginifolia aba1*, *Arabidopsis aba3*, and *Solanum phureaja droopy* mutants.

in the wilty tomato mutant *notabilis*. This conversion appears to be the rate-limiting step in the ABA biosynthesis pathway, and characterization of this control point is likely to provide intriguing insights into the regulation of seed dormancy, drought resistance, and cold hardening.

The penultimate step of the ABA biosynthesis pathway converts xanthoxin to ABA-aldehyde. This reaction involves oxidation of the 4'-hydroxyl group to a ketone, 2'-3' desaturation, and opening of the 1'-2' epoxide ring. No intermediates have been detected. Chemical oxidation of the 4'-hydroxyl group facilitates the quantitative conversion of xanthoxin to ABA-aldehyde, suggesting that the reaction in vivo may result from a single enzymatic step and subsequent rearrangements. Oxidation of the 1-CHO group of ABA-aldehyde, catalyzed by ABA-aldehyde oxidase, leads to the formation of ABA (Fig. 17.28). The steps xanthoxin → ABA-aldehyde → ABA are catalyzed by constitutively expressed enzymes, the activities of which do not increase after the onset of water stress. These two final conversions do not appear to limit the rate of ABA biosynthesis, because foliar concentrations of xanthoxin and ABA-aldehyde are invariably very low compared with those of ABA. The *aba2* mutant of *Arabidopsis* is currently the only known mutant for the conversion of xanthoxin to ABA-aldehyde. In contrast, there are many mutants for the final step from ABA-aldehyde to ABA, including *sitiens* and *flacca* in tomato, *nar2a* in barley, *aba3* in *Arabidopsis*, *aba1* in *N. plumbaginifolia*, and *droopy* in *Solanum phureaja*. The *flacca*, *aba3*, and *nar2a* mutants produce functional ABA-aldehyde oxidase apoprotein but are defective in synthesizing a molybdenum cofactor required by the enzyme.

The tomato *flacca* and *sitiens* mutants and *aba3* in *Arabidopsis* convert ABA-aldehyde to ABA-alcohol and accumulate *trans*-ABA alcohol. A shunt pathway from ABA-alcohol to ABA, which represents a minor source of ABA in most plants, allows these mutants to synthesize low amounts of ABA. Wild-type tomato and the *flacca* and *sitiens* mutants all contain ABA-1',4'-*trans*-diol and convert [²H]ABA-1',4'-*trans*-diol to [²H]ABA (Fig. 17.28). However, the origins of the diol have not been established, and its relationship to ABA in vivo remains to be clarified.

17.2.5 ABA is metabolized to several compounds, including phaseic acid, dihydrophaseic acid, and glucose conjugates.

ABA metabolism is summarized in Figure 17.29. The main route involves hydroxylation of the 8' carbon, spontaneous rearrangement of the resulting 8'-hydroxy-ABA to form phaseic acid, and reduction to dihydrophaseic acid and *epi*-dihydrophaseic acid. Dihydrophaseic acid undergoes conjugation at the 4'-position to form dihydrophaseic acid-4'-*O*-β-glucoside. Because of its rapid conversion to phaseic acid, 8'-hydroxy-ABA has rarely been isolated, but its formation has been inferred by the presence of the conjugate 8'-*O*-(3-hydroxy-3-methylglutaryl)-8'-hydroxyABA. The alternative routes in ABA metabolism involve conversion to 7'-hydroxy-ABA as well as conjugation to form ABA β-glucosyl ester and ABA-1'-*O*-β-glucoside. Pea seedlings reportedly convert ABA to ABA-1',4'-*trans*-diol; in tomato, however, the diol is an ABA precursor rather than a catabolite.

Numerous bioassays have indicated that the biological activity of ABA is dependent on the presence of a free 1-carboxyl group, a 2-*cis*-4-*trans*-pentadienoic side chain, a 4'-keto group, and a 2'-3' double bond. Except for 8'-*O*-(3-hydroxy-3-methylglutaryl)-8'-hydroxyABA, the ABA metabolites and conjugates formed from ABA do not possess all these features and are probably deactivation products (Fig. 17.29). In theory, ABA β-glucosyl ester and ABA-1'-*O*-β-glucoside could act as storage products that can be hydrolyzed to release free ABA. However, the available evidence suggests that these conjugates are primarily deactivation products and do not supplement the endogenous ABA pool.

Except for ABA 8'-hydroxylase, a membrane-bound cytochrome P450 monooxygenase that catalyzes the conversion of ABA to 8'-hydroxy-ABA (Fig. 17.29), little is known about the enzymes involved in ABA metabolism and conjugation, and none has been characterized at the molecular level. Developments in this area are eagerly awaited, both for their fundamental importance and their biotechnological potential. For example, once ABA 8'-hydroxylase has been cloned, antisense procedures may be able to

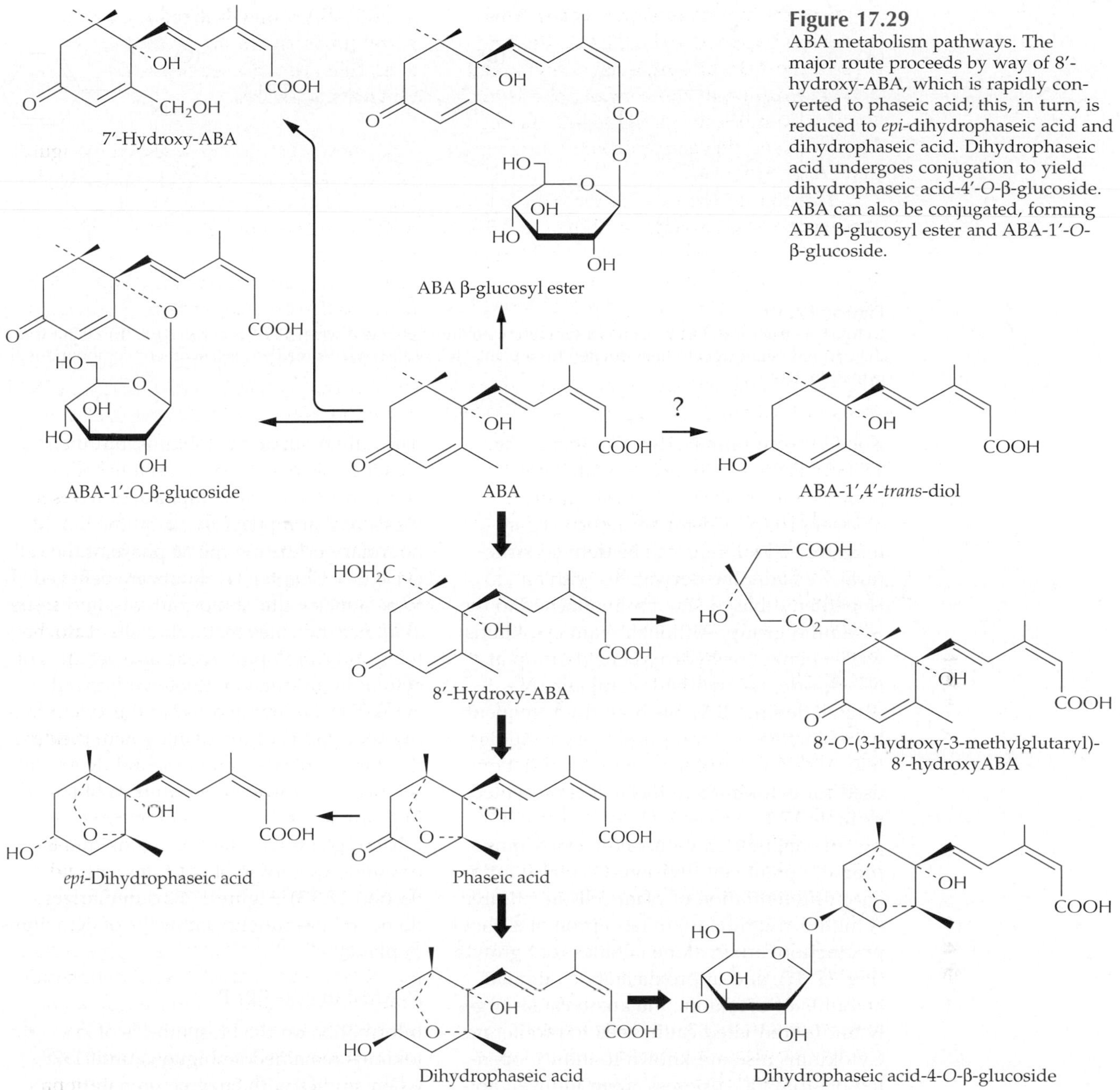

Figure 17.29
ABA metabolism pathways. The major route proceeds by way of 8'-hydroxy-ABA, which is rapidly converted to phaseic acid; this, in turn, is reduced to *epi*-dihydrophaseic acid and dihydrophaseic acid. Dihydrophaseic acid undergoes conjugation to yield dihydrophaseic acid-4'-*O*-β-glucoside. ABA can also be conjugated, forming ABA β-glucosyl ester and ABA-1'-*O*-β-glucoside.

produce transgenic plants in which ABA degradation is blocked. This could have profound implications for agriculture, because increasing the endogenous ABA pool may enhance stress tolerance and other ABA-mediated processes in crop plants.

17.3 Cytokinins

In the 1950s, coconut milk was found to stimulate the growth of immature *Datura* embryos and enhance the proliferation of carrot cells in vitro. At about the same time, Folke Skoog demonstrated that yeast extract, in combination with IAA, promoted new and continued division of tobacco cells in culture. Skoog subsequently showed that these undefined natural factors could be replaced by kinetin, N^6-furfuryladenine, a breakdown product formed during autoclaving of DNA (Fig. 17.30A). Although kinetin is an artifact derived from 2-deoxyadenylate, its biological activity resembles that of zeatin (Z), a native inducer of plant cell division that in 1963 was

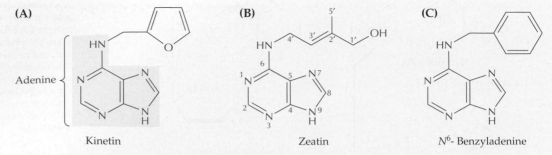

(A) Kinetin

(B) Zeatin

(C) N^6- Benzyladenine

Adenine {

Figure 17.30
Cytokinin structures. (A) Kinetin is a synthetic cytokinin generated when DNA is autoclaved. (B) Zeatin was the first endogenous cytokinin isolated from plants. (C) N^6-Benzyladenine (BA) is a synthetic compound with cytokinin activity.

isolated from immature maize seeds (Fig. 17.30B). Z, biosynthetically related compounds and their metabolites constitute a novel group of plant hormones, the **cytokinins.** Cytokinins can be defined structurally as adenine derivatives with an isopentenyl-based side chain attached to the N^6 amino group. Although plant cytokinins with a benzyl or hydroxybenzyl group at N^6 are rare, the synthetic cytokinin, N^6-benzyladenine (BA) has been used frequently in biochemical and physiological studies (Fig. 17.30C). A full list of the abbreviations used for cytokinins in this text is presented in Table 17.1.

In conjunction with auxin, cytokinins promote plant cell division. They also influence differentiation of plant cells in culture: A high cytokinin/auxin ratio promotes shoot production; auxin alone initiates root growth (Fig. 17.31); and approximately equimolar amounts of cytokinin and auxin cause largely undifferentiated callus cells to proliferate. Cytokinins also are known to induce opening of stomata, suppress auxin-induced apical dominance, and inhibit senescence of plant organs, especially leaves.

17.3.1 Cytokinins are synthesized de novo from 5'-AMP and a diphosphorylated hemiterpene.

Z and its riboside [9R]Z are thought to be cytokinins that are biologically active per se. The principal mechanism by which these cytokinins are produced in plants is de novo biosynthesis. Cis isomers of Z, including *cis*-Z, *cis*-[9R-5'P]Z, and *cis*-[9R-5'P]Z, are also

naturally occurring cytokinins, but their biosynthesis has not been established. In cultured tobacco cells, rapid de novo biosynthesis of Z and [9R]Z occurs at the G2/M boundary or during the M phase of the cell cycle (see Chapter 11). Roots are believed to be a major site of biosynthesis, and some evidence indicates that apical shoot meristems and developing seeds also are sites of cytokinin production. Root-synthesized cytokinins are transported in the xylem to the aerial parts of the plant, where they participate in apical dominance and delay leaf senescence. However, recent data obtained in grafting experiments with transgenic tobacco plants indicate that cytokinins do not undergo root-to-shoot transport (see Section 17.3.3). Figure 17.32 summarizes de novo biosynthetic pathways of cytokinins in plants.

5'-AMP to [9R-5'P]iP

Information on the biosynthesis of cytokinins remained ambiguous until 1978, when studies with enzyme preparations from the slime mold *Dictyostelium discoideum* revealed that 5'-AMP was a direct precursor of [9R-5'P]iP. The enzyme catalyzing this conversion, dimethylallyl diphosphate:5'-AMP transferase (DMAPP:AMP transferase, or isopentenyl transferase), was also found in cell-free extracts from cytokinin-autonomous tobacco callus tissue and maize kernels. The tobacco enzyme has very strict substrate specificities. Neither adenine nor adenosine can substitute for 5'-AMP, and DMAPP is the only acceptable donor for the side chain moiety (Fig. 17.32). This enzyme has not yet been purified to homogeneity

Chapter 17 Biosynthesis of Hormones and Elicitor Molecules

from plant sources; however, the corresponding enzyme from the bacterium *Agrobacterium tumefaciens*, encoded by the *ipt* gene, has been studied in depth at the molecular level (see Section 17.3.2).

[9R-5′P]iP to [9R-5′P]Z

The next step in cytokinin biosynthesis appears to be a trans hydroxylation of the side chain; evidence suggests that this step occurs at the ribotide level in crown gall cells of *Catharanthus roseus* (formerly *Vinca rosea*) and tobacco, with [9R-5′P]iP being metabolized to [9R-5′P]Z. Typically, the [9R-5′P]iP content of plant tissues is very low, implying that it is converted rapidly to [9R-5′P]Z. To date, the *trans*-hydroxylase catalyzing this conversion has not been isolated.

[9R-5′P]Z to Z via [9R]Z

[9R-5′P]Z is dephosphorylated to [9R]Z by 5′-nucleotidase. In turn, [9R]Z is cleaved to form Z by adenosine nucleosidase (Fig. 17.32). Partially purified preparations containing both of these enzymes have been obtained from wheat germ. In general, Z is the most bioactive cytokinin. [9R]Z also exhibits high biological activity and may delay leaf senescence more effectively than Z when introduced through the transpiration stream. Both Z and [9R]Z are probably biologically active per se, but it is difficult to eliminate the possibility that other cytokinins may also possess intrinsic biological activity, because cytokinins with free base, nucleoside, and nucleotide structures are rapidly interconverted when applied to plant tissues (see Section 17.3.5).

[9R-5′P]iP to [9R]Z via [9R]iP and to Z via iP

[9R-5′P]iP is dephosphorylated along an alternative pathway to yield [9R]iP, which is converted to iP by removal of ribose. Metabolism of [9R]iP and iP can then yield [9R]Z and Z, respectively (Fig. 17.33). During both in vivo and in vitro feeding experiments, [9R]Z is often obtained from [9R]iP

Figure 17.31

Arabidopsis callus production is induced by placing tissue on medium containing auxin (IBA) and cytokinin. When callus is subcultured on a medium containing only auxin, roots are produced (left); when subcultured on a medium containing a high cytokinin (zeatin) to auxin ratio, shoots proliferate (right).

Table 17.1 Cytokinin abbreviations

Abbreviation	Compound
iP	N^6-(Δ^2-Isopentenyl)adenine
[7G]iP	7-β-Glucosyl-N^6-(Δ^2-isopentenyl)adenine
[9G]iP	9-β-Glucosyl-N^6-(Δ^2-isopentenyl)adenine
[9R]iP	N^6-(Δ^2-Isopentenyl)adenosine
[9R-5′P]iP	N^6-(Δ^2-Isopentenyl)adenosine 5′-phosphate
[2MeS 9R]iP	2-Methylthio-N^6-(Δ^2-isopentenyl)adenosine
Z	Zeatin
(OG)Z	*O*-β-Glucosylzeatin
(OG)[9R]Z	*O*-β-Glucosyl-9-ribosylzeatin
[7G]Z	7-β-Glucosylzeatin
[9G]Z	9-β-Glucosylzeatin
[9Ala]Z	9-Alanylzeatin
[9R]Z	9-Ribosylzeatin
[9R-5′P]Z	9-Ribosylzeatin 5′-phosphate
[2MeS 9R]Z	2-Methylthio-9-ribosylzeatin
cis-Z	*cis*-Zeatin
cis-(OG)Z	*O*-β-Glucosyl-*cis*-zeatin
cis-(OG)[9R]Z	*O*-β-Glucosyl-9-ribosyl-*cis*-zeatin
cis-[7G]Z	7-β-Glucosyl-*cis*-zeatin
cis-[9R]Z	9-Ribosyl-*cis*-zeatin
cis-[9R-5′P]Z	9-Ribosyl-*cis*-zeatin 5′-phosphate
cis-[2MeS 9R]Z	2-Methylthio-9-ribosyl-*cis*-zeatin
(diH)Z	Dihydrozeatin
(diH OG)Z	*O*-β-Glucosyldihydrozeatin
(diH OG)[9R]Z	*O*-β-Glucosyl-9-ribosyldihydrozeatin
(diH)[3G]Z	3-β-Glucosyldihydrozeatin
(diH)[7G]Z	7-β-Glucosyldihydrozeatin
(diH)[9G]Z	9-β-Glucosyldihydrozeatin
(diH)[9Ala]Z	9-Alanyldihydrozeatin
(diH)[9R]Z	9-Ribosyldihydrozeatin
(diH)[9R-5′P]Z	9-Ribosyldihydrozeatin 5′-phosphate
BA	N^6-Benzyladenine

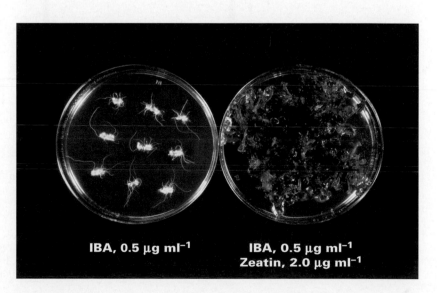

IBA, 0.5 μg ml⁻¹   IBA, 0.5 μg ml⁻¹
Zeatin, 2.0 μg ml⁻¹

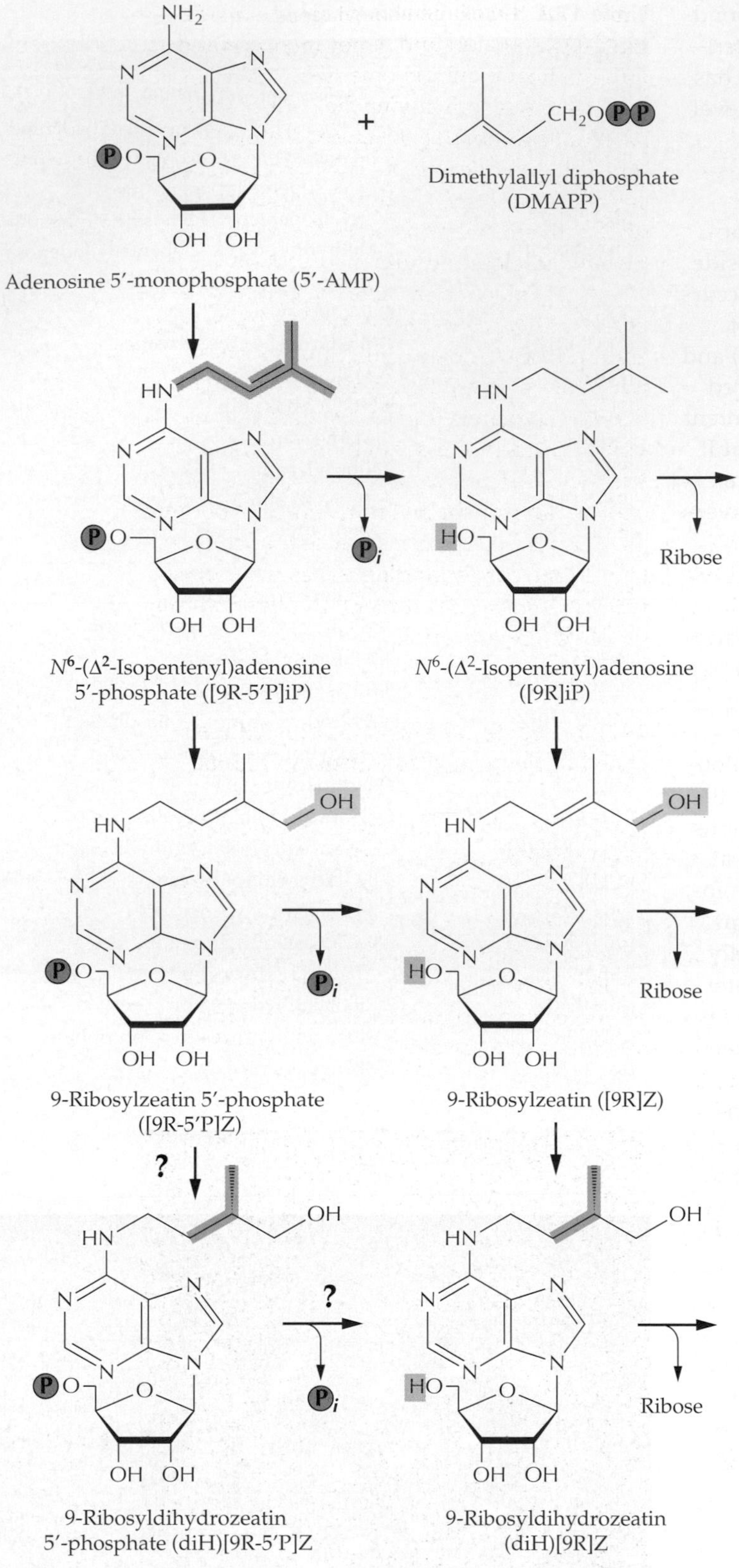

Figure 17.32

Cytokinin de novo biosynthetic pathways in plants. The first committed step is the addition of an isopentenyl side chain to adenosine 5′-monophosphate to produce [9R-5′P]iP. Cross-linked pathways connect [9R-5′P]iP to the putative bioactive cytokinins Z and [9R]Z. However, the major pathway appears to be [9R-5′P]iP → [9R-5′P]Z → [9R]Z → Z. Hydrogenation products, (diH)Z and its derivatives, have reduced biological activity but are major endogenous cytokinins in some legumes.

and Z is produced from iP. In cauliflower, this trans hydroxylation is catalyzed by a microsomal cytochrome P450 oxidase. Free *cis*-Z derivatives have been identified in extracts from many plants, but to date no *cis*-hydroxylase that produces *cis*-Z derivatives has been isolated.

[9R]Z and Z to (diH)[9R]Z and (diH)Z

Stereospecific reduction of the side chain of [9R]Z and Z gives rise to (diH)[9R]Z and (diH)Z, respectively. These dihydro-compounds are major endogenous cytokinins in some legumes. Side chain reduction usually diminishes the biological activity of Z derivatives, but these compounds are unexpectedly active in some test systems, possibly because they resist catabolism by cytokinin oxidase (see Section 17.3.6). A hydrogenase extracted from *P. coccineus* embryos specifically converts Z to (diH)Z but does not modify the side chain of iP, [9R]Z, or *cis*-Z.

17.3.2 Some bacteria encode or express genes for cytokinin biosynthesis.

Some plant pathogenic bacteria utilize cytokinins to influence plant growth. In various dicotyledonous plants and some monocots, *A. tumefaciens* infection results in the production of tumors, a condition referred to as crown gall disease (see Chapter 21). Bacterial pathogenicity and virulence are linked to a tumor-inducing (Ti) plasmid. During the infection process, a segment of the bacterial Ti plasmid, the T-DNA, is transferred into a higher-plant cell, where it integrates into the host genome. Several T-DNA genes—*ipt*, *iaaH*, and *iaaM*—encode enzymes that catalyze the overproduction of cytokinins and IAA. These genes are expressed by the plant under the control of eukaryotic-type promoters, resulting in characteristic tumors, or galls (Fig. 17.33). The *ipt* gene encodes an isopentenyl transferase, and *ipt* expression is associated with the production of [9R-5'P]iP, which is converted rapidly to [9R-5'P]Z by the trans-hydroxylase of the plant host. The *iaaM* and *iaaH* gene products synthesize IAA by way of a unique two-step bacterial pathway (see Section 17.4.7). The Ti-plasmid of octopine-synthesizing strains of *A. tumefaciens* contains the *tzs* gene, which shares considerable homology with *ipt*.

Unlike *ipt*, the *tzs* gene is situated outside the T-DNA region and is not incorporated into the host plant genome. Expression of this gene, which is under the control of a bacterial promoter, is triggered by plant-derived phenolic compounds, such as aceto-syringone, and the resulting cytokinins possibly activate cell division at the wound site, thereby accelerating tumor formation.

After infection of olive, oleander, and privet, *Pseudomonas savastanoi* produces large amounts of IAA and cytokinins, most of which are Z and [9R]Z, leading to the formation of unorganized galls. Plasmids of *Ps. savastanoi* carry the auxin biosynthetic genes *iaaM* and *iaaH*, as well as the *ptz* gene, which encodes an isopentenyl transferase. Although these genes have substantial sequence homology to the corresponding genes in *A. tumefaciens*, *Pseudomonas* tumors involve no transfer of genetic material.

17.3.3 A bacterial cytokinin biosynthesis gene has been used to transform plants.

Investigators have attempted to elucidate physiological functions of cytokinins by studying plants transformed with the *ipt* gene. Transgenic plants expressing the *ipt* gene under the control of the strong cauliflower mosaic virus 35S (CaMV35S) promoter contain increased concentrations of endogenous cytokinins and exhibit

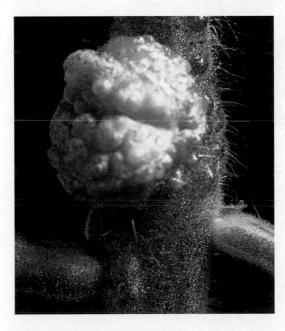

Figure 17.33
Crown gall tumor on a tomato plant. The stem of a one-month-old tomato seedling was wounded with a needle carrying a culture of wild-type *Agrobacterium tumefaciens*. The crown gall tumor was photographed one month later.

phenotypic changes analogous to the well-known effects of exogenous cytokinins, such as retarded leaf senescence and the early release of lateral buds from inhibition by apical buds (Fig. 17.34). In transgenic tobacco plants that express the *ipt* gene or *iaaH-iaaM* genes, IAA overproduction suppresses the rate of cytokinin synthesis and reduces the cytokinin content of plant tissues, whereas cytokinin overproduction has a similar impact on IAA levels and synthesis.

Transgenic tobacco expressing the *ipt* gene under the control of a tetracycline-dependent 35S promoter has been used to study the effects of locally enhanced cytokinin biosynthesis induced by applying tetracycline to individual tissues and organs. Activation of the *ipt* gene can increase local cytokinin content, primarily Z and [9R]Z, as much as 50-fold. The morphological consequences of *ipt* expression are largely restricted to the tetracycline-treated area. For example, *ipt* gene transcription in lateral buds induces the growth of single buds but only

at the site of tetracycline application (Fig. 17.35A). In contrast, in reciprocal grafts of wild-type shoots with cytokinin-rich transgenic root stock, the lateral buds remain inhibited in the wild-type scion. Thus, overloading the roots with cytokinins does not counteract apical dominance in the shoot. Likewise, local production of cytokinins in leaves retards senescence, whereas *ipt* expression in transgenic root stock does not prevent sequential senescence in a wild-type scion (Fig. 17.35B). These observations both lend support to the view that cytokinins act in the region where they are synthesized and challenge the role of cytokinins as root-to-shoot signals that control apical dominance and sequential leaf senescence from a distance.

Tomato has been transformed with an *ipt* gene driven by the fruit-specific 2A11 promoter. The plants appear normal during vegetative and floral development, but fruit ripening and maturation are modified dramatically. The ripening transgenic fruit contain sectors of green pericarp on an otherwise deep red fruit (Fig. 17.36). Z and [9R]Z contents are 10- to 100-fold higher in the transformed fruit tissues than in control fruit. Indeed, cytokinin concentrations in the red regions of the fruit are sixfold higher than in the green sectors—an unexpected finding, given that cytokinins retard senescence.

17.3.4 Base modification of tRNA nucleotides can generate cytokinin moieties.

Some tRNAs from plants, microorganisms, and animals contain nucleotides in which bases have been modified to form cytokinin molecules. These cytokinins are predominantly [9R]iP, *cis*-[9R]Z, and their 2-methylthio derivatives, although [9R]Z and its 2-methylthio derivative are occasionally found as minor constituents (Fig. 17.37). Cultures of *A. tumefaciens* and *Rhodococcus fascians* (formerly *Corynebacterium fascians*), a causative agent of witches' broom disease, frequently contain tRNA cytokinins. The *miaA* gene from *A. tumefaciens* encodes a DMAPP:tRNA transferase responsible for the isopentenylation of tRNAs and shares significant homology with the *ipt* gene.

Figure 17.34
Transgenic tobacco (*Nicotiana tabacum* cv. Petite Havana SR1) expressing the *Agrobacterium tumefaciens ipt* gene under the control of the strong CaMV 35S promoter. The plant contains increased amounts of Z, [9R]Z, and [9R-5'P]Z, and the phenotypic changes are analogous to the effects induced by treatment with exogenous cytokinins, such as retarded leaf senescence and early release of lateral buds from inhibition by apical buds.

(A)

(B)

Figure 17.35
Phenotype of transgenic tobacco (*Nicotiana tabacum* clone IPT5) carrying the *ipt*-gene under the control of a tetracycline-dependent 35S promoter. (A) Formation of lateral shoots after localized application of tetracycline to inhibited lateral buds of clone IPT5. (B) Leaf senescence on grafted plants: (left) a scion of clone IPT5 grafted onto a wild-type root stock; (right) a wild-type scion grafted onto a cytokinin-overproducing root stock of clone IPT5.

As outlined above, [9R]iP and [9R]Z are known to be synthesized de novo in plants. De novo biosynthesis of *cis*-[9R]Z and 2-methylthiocytokinins has not been demonstrated, and cytokinin tRNAs remain a potential source of these compounds. The presence of a cis:trans isomerase in *P. vulgaris* implies that at least one plant species can convert *cis*-cytokinins to their trans isomers, which exhibit far greater biological activity. Overall, however, the contribution of tRNA-derived cytokinins to endogenous cytokinin concentrations is thought to be much less important than de novo synthesis.

17.3.5 Free base, nucleoside, and nucleotide forms of cytokinins may be interconverted in vivo.

Exogenous free base, nucleoside, and nucleotide cytokinins are metabolized rapidly (Fig. 17.38); whether all of these conversions occur in planta, however, is not known. When free base cytokinins are applied to plant tissues, the concentrations of cytokinin ribotides initially increase but then decline as additional metabolites are formed. Thus, nucleotide formation is thought to play a key role in the uptake and subsequent metabolism of exogenous cytokinins. Conversion of free base cytokinins to nucleotide cytokinins (5′-mono-, di-, or triphosphates) is catalyzed by adenine phosphoribosyl transferase. Alternatively, adenosine phosphorylase can convert free base cytokinins to nucleoside cytokinins which, in turn, are converted by adenosine kinase to nucleotides. Nucleotide cytokinins are dephosphorylated to nucleoside cytokinins, from which ribose is removed to yield free base

cytokinins. The enzymes responsible for these conversions are not specific for cytokinins and have higher affinities for adenine, adenosine, and AMP than for the corresponding N^6-substituted compounds. The enzyme systems involved have not yet been well defined.

17.3.6 Cytokinin oxidases remove the isopentenyl side chains of many cytokinins.

The side chain of cytokinins is an essential structural requirement for biological activity. Removal of the side chain is a common metabolic fate of exogenous cytokinins and

Figure 17.36
Fruits of tomato plants transformed with *Agrobacterium tumefaciens ipt* gene under the control of the fruit-specific 2A11 promoter. Paradoxically, given that cytokinins retard fruit senescence, the red regions of the tomatoes contain six times more cytokinins than the green areas.

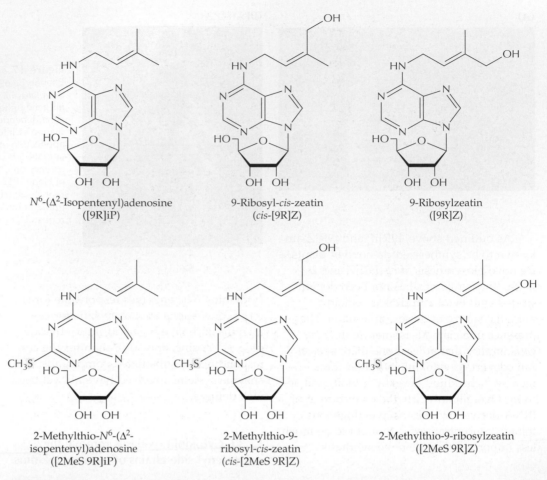

Figure 17.37
Cytokinins found in tRNAs. [9R]iP, *cis*-[9R]Z, and their 2-methylthio derivatives are major tRNA cytokinins.

has a significant role in controlling the pool size of bioactive cytokinins. The enzymes responsible for side chain cleavage, cytokinin oxidases, occur in a variety of plants. Among the naturally occurring cytokinins, iP, [9R]iP, Z, [9R]Z, *N*-glucosides, and *N*-alanyl conjugates are good in vitro substrates for cytokinin oxidases. Like its trans isomer Z, *cis*-Z is also a suitable substrate. The free base cytokinin iP is degraded to form adenine and 3-methyl-2-butenal. The nucleoside [9R]iP is similarly degraded but releases adenosine instead of adenine (Fig. 17.39). (DiH)Z derivatives, ribotide cytokinins, and *O*-glucosides do not serve as substrates and have a reduced affinity for the oxidases. Although kinetin and the synthetic cytokinin BA are generally resistant to cytokinin oxidase, some evidence supports the existence of alternative enzymes that catalyze the oxidative degradation of these compounds.

Cytokinin oxidases are inducible enzymes, the concentrations of which increase rapidly in plant tissues after treatment with exogenous cytokinins, irrespective of whether the applied cytokinin is susceptible or resistant to oxidase-catalyzed degradation. The available evidence suggests that endogenous cytokinins may be protected from the action of cytokinin oxidases through compartmentation. Cytokinin oxidases require molecular oxygen and were once thought to be copper-containing amine oxidases. However, this now appears to be incorrect and recently they have been reclassified as FAD-dependent amine oxidases. The activity of cytokinin oxidase from wheat is strongly inhibited by diphenylurea. Because diphenylurea and related compounds inhibit cytokinin oxidase, they prevent cytokinin catabolism and elicit responses similar to those induced by application of cytokinins. However, these compounds do not

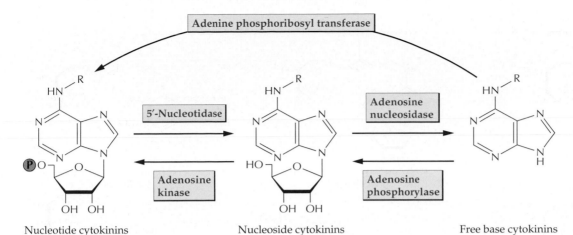

Adenine phosphoribosyl transferase

5′-Nucleotidase

Adenosine kinase

Adenosine nucleosidase

Adenosine phosphorylase

Nucleotide cytokinins

Nucleoside cytokinins

Free base cytokinins

have bona fide cytokinin activity. All the cytokinin oxidases studied thus far have the same substrate specificities but exhibit marked differences in their molecular masses, pH optima, and K_m values. A 63-kDa cytokinin oxidase from *Z. mays* kernels has recently been cloned. The deduced amino acid sequence shares homology domains with FAD-dependent oxidases. When a cytokinin oxidase–encoding gene is overexpressed in plants, it will be of interest to observe if and how this affects their phenotype and to determine the effects of overexpression on the concentrations of endogenous cytokinins.

17.3.7 Glucosylation of the side chain hydroxyl group may reversibly deactivate cytokinins.

O-Glucoside cytokinins, in which the side chain hydroxyl group is glucosylated, are abundant in plants (Fig. 17.40). (OG)Z, (OG)[9R]Z, (diH OG)Z, and (diH OG)[9R]Z are endogenous cytokinins that are also formed frequently in metabolism studies. Cis isomers of (OG)Z, (OG)[9R]Z are present in high concentrations in developing rice grains. Although *O*-glucoside cytokinins may be biologically inert per se, they are readily hydrolyzed by glucosidases to yield active cytokinins. Thus, *O*-glucosylation may have a dual role, deactivating cytokinins while forming storage products that serve as potential reservoirs of biologically active hormone. This proposed function assumes that suitable glycosidases are available to cleave endogenous cytokinin *O*-glucosides in vivo.

As noted in the previous section, *O*-glucoside cytokinins are not susceptible to cytokinin oxidase activity. Curiously, in some bioassays, *O*-glucoside cytokinins are more active than the corresponding free bases. Resistance to cytokinin oxidases may protect the conjugates during transport to target tissues.

O-Glucosyl transferases present in lima bean (*P. lunatus*) embryos recognize Z, but not (diH)Z, *cis*-Z, or [9R]Z. These enzymes utilize UDP-glucose as the favored substrate.

N^6-(Δ^2-Isopentenyl)adenine
(iP)

Adenine

3-Methyl-2-butenal

N^6-(Δ^2-Isopentenyl)adenosine
([9R]iP)

Adenosine

3-Methyl-2-butenal

Figure 17.39
Cytokinin oxidases remove the side chains from cytokinin molecules.

O-Glucosylated cytokinins

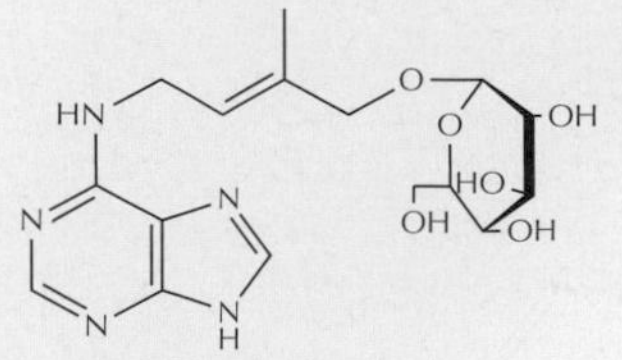

O-β-Glucosylzeatin
(OG)Z*

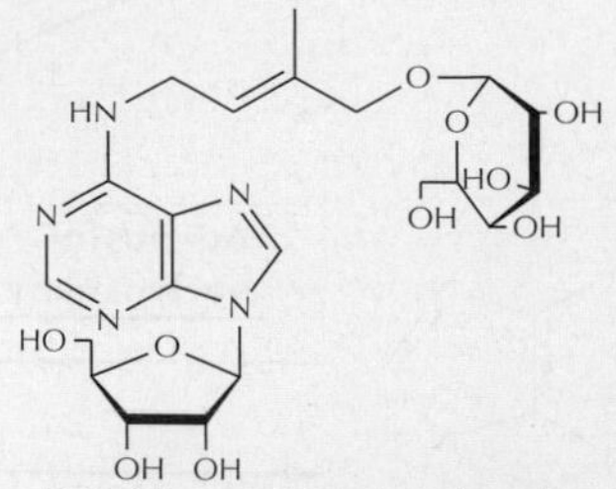

O-β-Glucosyl-9-ribosylzeatin
(OG)[9R]Z*

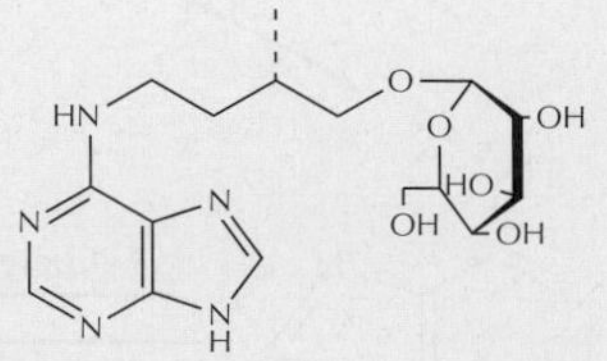

O-β-Glucosyldihydrozeatin
(diH OG)Z*

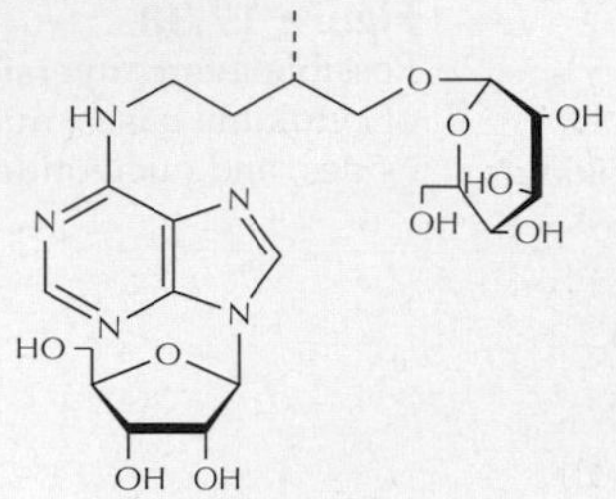

O-β-Glucosyl-9-
ribosyldihydrozeatin
(diH OG)[9R]Z*

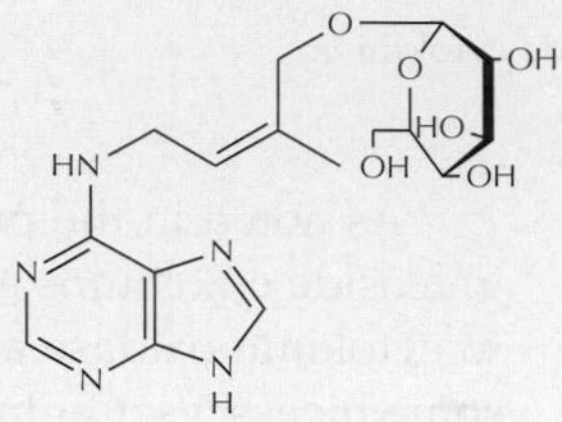

O-β-Glucosyl-cis-zeatin
cis-(OG)Z*

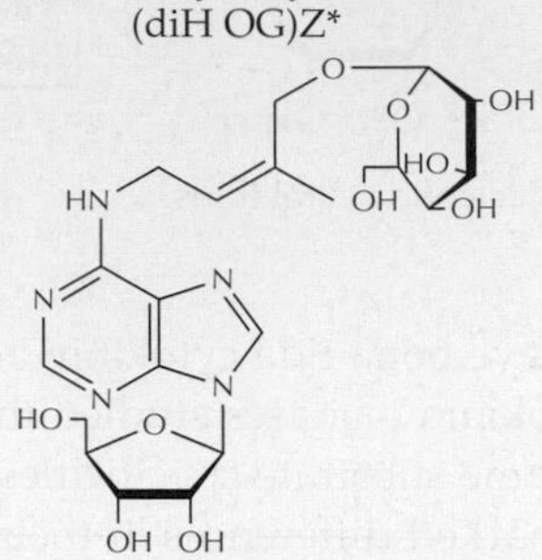

O-β-Glucosyl-9-ribosyl-cis-zeatin
cis-(OG)[9R]Z*

N-Glucosylated cytokinins

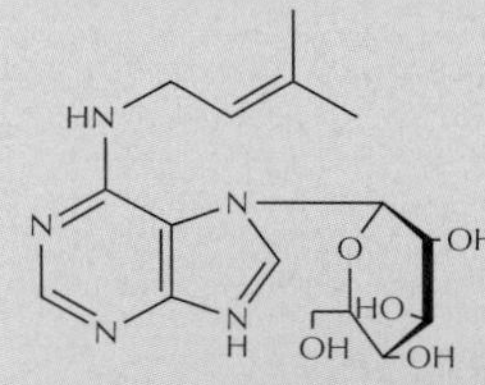

7-β-Glucosyl-N⁶-
(Δ²-isopentenyl)adenine [7G]iP

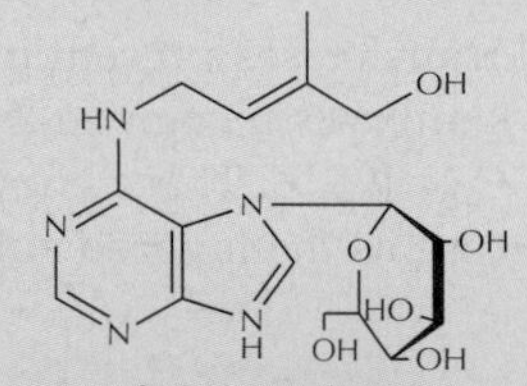

7-β-Glucosylzeatin
[7G]Z*

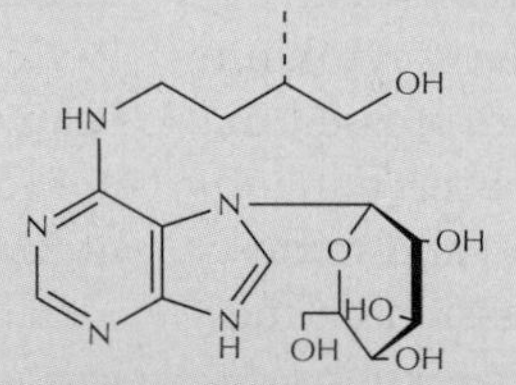

7-β-Glucosyldihydrozeatin
(diH)[7G]Z*

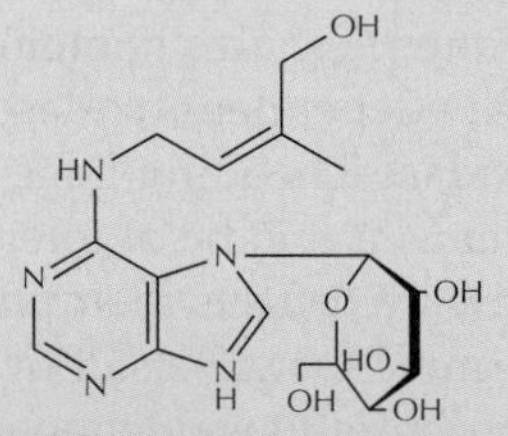

7-β-Glucosyl-cis-zeatin
cis-[7G]Z

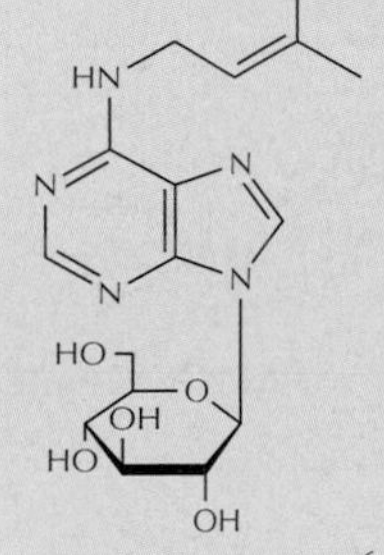

9-β-Glucosyl-N⁶-
(Δ²-isopentenyl)adenine
[9G]iP*

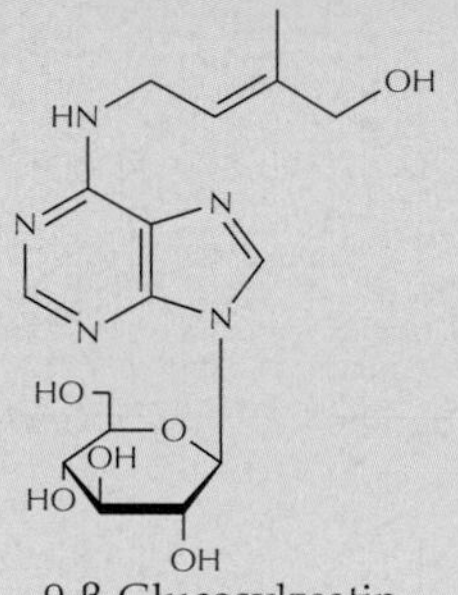

9-β-Glucosylzeatin
[9G]Z*

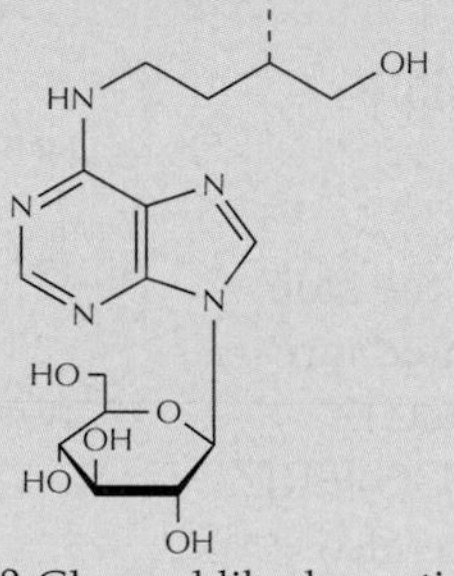

9-β-Glucosyldihydrozeatin
(diH)[9G]Z*

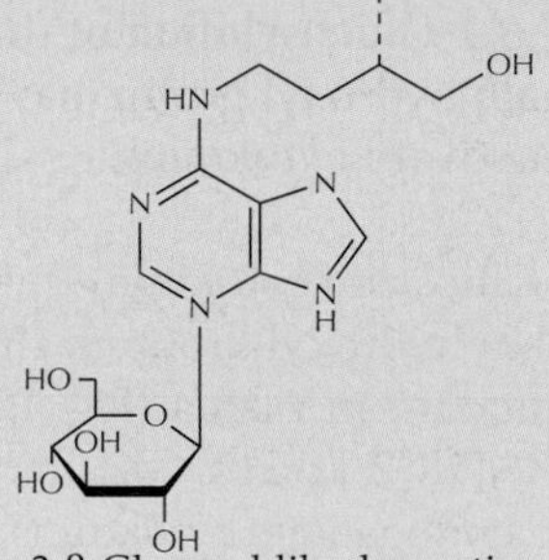

3-β-Glucosyldihydrozeatin
(diH)[3G]Z

Amino acid conjugates

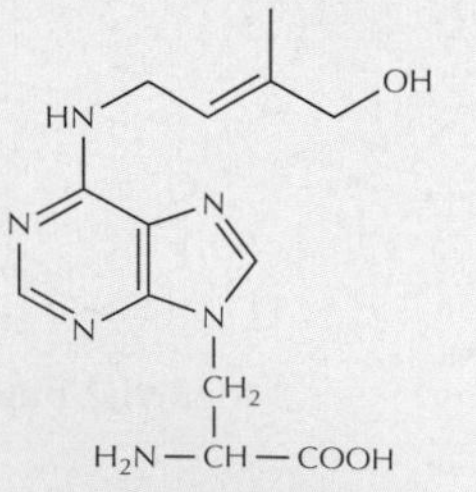

9-Alanylzeatin
[9Ala]Z*

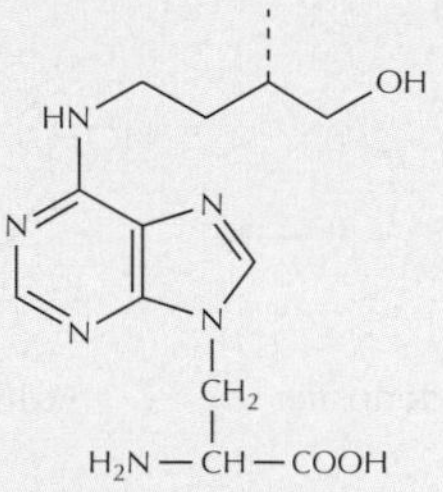

9-Alanyldihydrozeatin
(diH)[9Ala]Z*

Figure 17.40
Conjugates of iP, Z, cis-Z, and [9R]Z found as endogenous compounds or metabolites. Asterisk-marked conjugates are known to be naturally occurring.

The *amp1* mutant of *Arabidopsis*, which accumulates free cytokinins, may be impaired in some aspect of cytokinin glycosylation. *O*-Xylosyltransferase activity that utilizes Z as a substrate has been detected in *P. vulgaris* embryos, and two cDNAs encoding Z-*O*-xylosyltransferases have been cloned. Transgenic tobacco plants expressing one of these genes, under the control of the CaMV35S promoter, have a normal morphological phenotype but are more sensitive to synthetic auxin than are control plants, and in the presence of naphthalene-1-acetic acid (Fig. 17.41) these transgenic plants display symptoms that include leaf chlorosis, restriction of root elongation, and eventual cessation of growth.

17.3.8 N-Glucosylation of the purine moiety is thought to irreversibly deactivate free base cytokinins.

In another class of cytokinin conjugates, the glucose moiety is attached to a purine ring nitrogen at the 3, 7, or 9 position. In the model systems that have been studied most extensively, N-glucosylation is restricted to free base cytokinins. In contrast to *O*-glucoside cytokinins, the *N*-glucosides are distributed in a narrow range of plants, including radish, maize, sweet potato, tobacco, and

Ca. roseus crown gall tissue. Endogenous *N*-glycosylated cytokinins include [7G]Z, (diH)[7G]Z, [9G]iP, [9G]Z, and (diH)[9G]Z (Fig. 17.41). [7G]Z is the only detectable cytokinin in radish seeds. [9G]Z is a major cytokinin in *Ca. roseus* crown gall cells and a minor constituent in *Z. mays* while [9G]iP is one of several endogenous cytokinins in sweet potato. 3-Glucosides only occur as minor metabolites of (diH)Z and BA.

When fed to derooted radish seedlings, iP and Z are converted exclusively to their 7-glucosides, whereas (diH)Z and BA are converted to their 7-glucosides, 9-glucosides, and 3-glucosides, in decreasing order of yield. A purified enzyme system from radish cotyledons catalyzes both 7- and 9-glucosylations of Z, *cis*-Z, (diH)Z, and BA, the 7-glucosides being produced in the largest amounts. Thus, the *N*-glucosyltransferases, which utilize UDP-glucose and TDP-glucose as the glucose donor, have a broad affinity for a variety of side chain structures. Indeed, several enzymes may exist with different glucosylation specificities.

In contrast to *O*-glucoside cytokinins, *N*-glucoside cytokinins are not biologically active, presumably because they are resistant to enzymatic cleavage. Thus, the role of *N*-glucosylation is probably to reduce the quantities of biologically active cytokinins in response to physiological conditions.

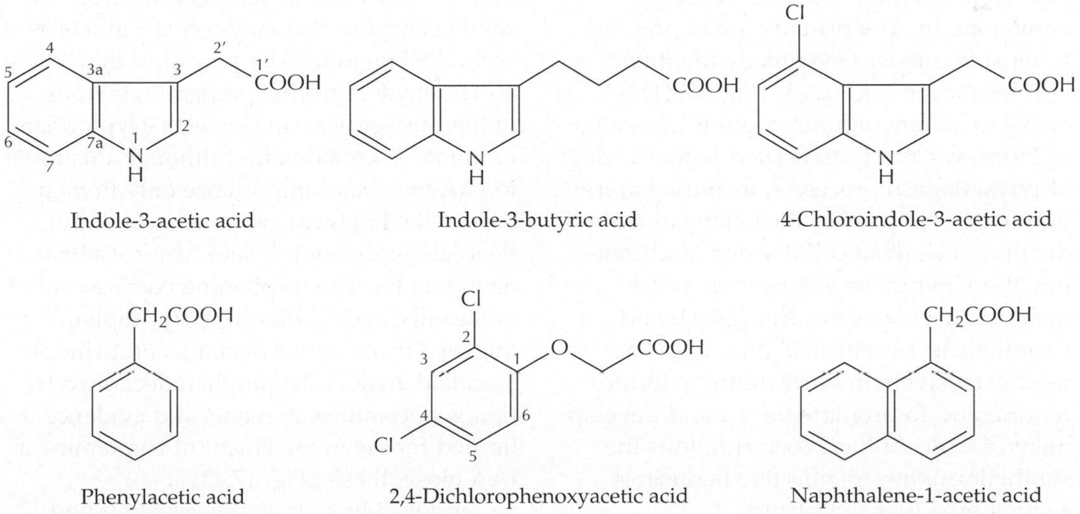

Figure 17.41
Indole-3-acetic acid (IAA), the auxin most widely distributed among plants, and related compounds. Less prevalent naturally occurring compounds that exhibit some degree of auxin-like activity include indole-3-butyric acid, 4-chloroindole-3-acetic acid, and phenylacetic acid. The synthetic auxins 2,4-dichlorophenoxyacetic acid (2,4-D) and naphthalene-1-acetic acid are used commercially, as plant growth regulators in low doses and as herbicides at higher concentrations.

17.3.9 Conjugation with alanine produces extremely stable and inactive metabolites.

The alanyl conjugates [9Ala]Z and (diH)[9Ala]Z were first identified as metabolites of exogenous Z fed to derooted lupine seedlings (Fig. 17.40). This type of conjugation was later observed when Z derivatives and BA were fed to other plant materials, including immature seeds of apple and lupine and seedlings of *P. vulgaris* and soybean. The endogenous occurrence of [9Ala]Z and (diH)[9Ala]Z has been demonstrated in the seeds and pod walls of lupine. A partially purified enzyme responsible for this conjugation utilizes free base cytokinins as a substrate and requires *O*-acetylserine as the amino acid donor. Like cytokinin 7- and 9-glucosides, the alanyl conjugates are stable and not hydrolyzed in vivo. Alanyl conjugation may act to reduce the amounts of biologically active cytokinins.

17.4 Indole-3-acetic acid

During the 19th century, Theophili Ciesielski studied the geotropic responses of plants, and Charles Darwin and his son Francis Darwin investigated phototropism as well as geotropism. These investigations laid the groundwork for Frits Went, who in 1926 obtained from oat coleoptiles a diffusible growth-promoting factor subsequently named **auxin**. The primary auxin present in most plants was eventually identified as indole-3-acetic acid (IAA) (Fig. 17.41). IAA is active in submicrogram amounts in a range of bioassays and is associated with a variety of physiological processes, including apical dominance, tropisms, shoot elongation, induction of cambial cell division, and root initiation. Synthetic auxins such as 2,4-dichlorophenoxyacetic acid (2,4-D) and naphthalene-1-acetic acid (Fig. 17.41) are used extensively in horticulture to induce rooting and to promote the set and development of fruit. At high concentrations the synthetic auxins are effective herbicides against broad-leaved plants.

The IAA content of plant tissues is regulated by several processes. The IAA pool is fed by de novo synthesis from L-tryptophan and nontryptophan precursors and by hydrolysis of IAA conjugates. IAA is rendered inactive by various conjugation and catabolic pathways. IAA contents in individual tissues can also be influenced by a basipetal polar transport system that results in the downward movement of IAA from apical meristems and young leaves towards the root system.

17.4.1 IAA can be synthesized from L-tryptophan.

Evidence accumulated over the past 50 years indicates that plants can synthesize IAA from L-tryptophan by three different routes: the indole-3-pyruvic acid, the indole-3-acetaldoxime, and the tryptamine pathways (Fig. 17.42). The major pathway from L-tryptophan to IAA appears to proceed via indole-3-pyruvic acid and indole-3-acetaldehyde. Indole-3-ethanol and its conjugates are produced in a side shunt from indole-3-acetaldehyde. These compounds may have a storage role, given that they can be rapidly reconverted to indole-3-acetaldehyde and used as a substrate for IAA biosynthesis.

In cabbage, plasma membrane-bound enzymes convert L-tryptophan to IAA via indole-3-acetaldoxime and indole-3-acetonitrile. Indole-3-acetaldoxime is also a precursor of indole-3-methylglucosinolate, which can be metabolized to indole-3-acetonitrile by myrosinase. Indole-3-acetonitrile is converted to IAA by the action of nitrilases. A soluble enzyme also may convert indole-3-acetaldoxime to IAA by way of indole-3-acetaldehyde. Enzyme preparations from numerous species can convert L-tryptophan to indole-3-acetaldoxime, although this route to IAA may be of importance only in members of the Brassicaceae (mustard family), Poaceae (grass family), and Musaceae (banana family). The tryptamine pathway involves decarboxylation of L-tryptophan, followed by oxidative deamination to indole-3-acetaldehyde. L-Tryptophan decarboxylation is uncommon in plants and evidence is limited for the involvement of tryptamine in IAA biosynthesis (Fig. 17.42).

Indole-3-butyric acid has been found in a number of higher plants. It has auxin activity and is used to induce root formation on cuttings. IAA is converted to indole-3-butyric acid in maize and *Arabidopsis* (Fig. 17.42). In vitro studies with maize have demonstrated that indole-3-butyric acid synthase uses acetyl-CoA and ATP as cofactors, and exogenous indole-3-butyric acid is conjugated rapidly by plants. However, metabolism of indole-3-butyric acid to IAA has also been reported. Therefore, whether indole-3-butyric acid is an auxin per se or its biological activity results from conversion to IAA is not clear.

Although the conversion of L-tryptophan → indole-3-pyruvic acid → indole-3-acetaldehyde → IAA was the first plant hormone biosynthesis pathway elucidated, it has not proved especially amenable to detailed study. Enzymatic studies on this three-step IAA pathway have provided far less information than was revealed by comparable research on the longer, more complex pathway that

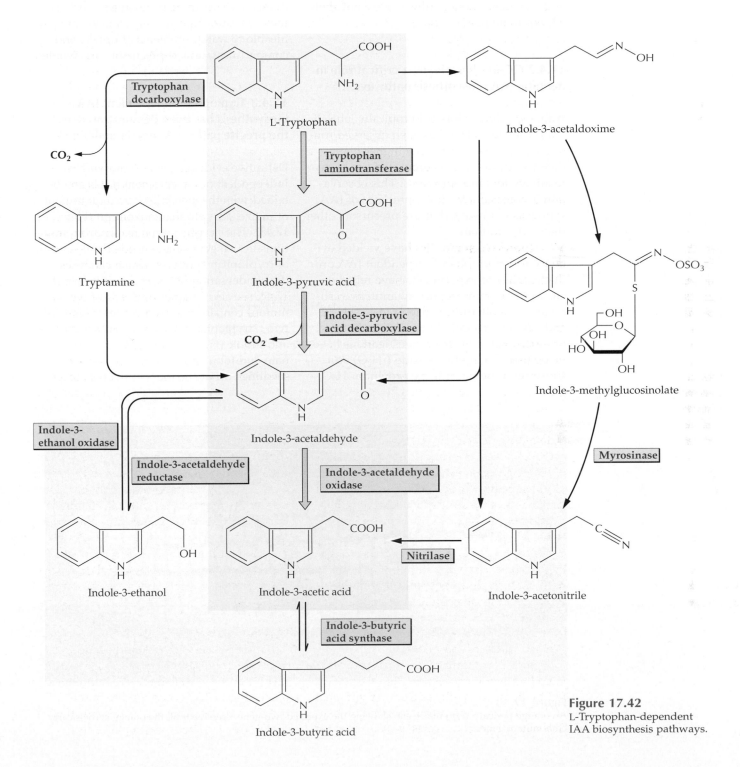

Figure 17.42
L-Tryptophan-dependent IAA biosynthesis pathways.

generates gibberellins. The enzymes involved in IAA biosynthesis have generally exhibited only low activity in vitro; furthermore, indole-3-pyruvic acid is extremely unstable and is not an ideal substrate for metabolic studies. No IAA-deficient mutant has been identified to date, perhaps because IAA deficiency is lethal. Alternatively, IAA may be synthesized by more than one route, and so blocking one pathway may not abolish hormone production.

17.4.2 GC-MS has played a critical role in defining IAA biosynthesis pathways.

Although conversion of isotopically labeled L-tryptophan to IAA has been observed routinely in studies with many higher plants, the amount of label incorporated has been relatively low in some cases. This observation is consistent with the presence of IAA synthesis pathways that use precursors other than L-tryptophan.

Several recent studies have yielded evidence for tryptophan-independent IAA production. These experiments have relied on GC-MS to obtain accurate quantitative estimates of stable isotopic labels (e.g., ^{2}H, ^{13}C, and ^{15}N) incorporated into IAA and other products. Some of these investigations have made use of deuterium oxide (^{2}H$_2$O) as a substrate. When plants are grown in ^{2}H$_2$O,

de novo biosynthesis of aromatic compounds is accompanied by incorporation of deuterium atoms into nonexchangeable positions in the newly formed molecules. The extent of de novo synthesis can be gauged accurately from the amount of deuterium incorporated into the compounds of interest. The special advantage of this technique is that it does not require exact knowledge of the intermediates or the pathways involved. Furthermore, because all cell compartments are permeable to water, problems of uptake and compartmentation of precursors are avoided.

17.4.3 Tryptophan-independent IAA biosynthesis has been demonstrated, but the precise pathways remain undefined.

Definitive evidence for IAA biosynthesis independent of L-tryptophan has been obtained with the *orange pericarp (orp)* mutant of maize, a tryptophan auxotroph (Fig. 17.43). The *orp* phenotype results from mutations in two genes that encode the β-subunit of tryptophan synthase, which catalyzes the condensation of L-serine and indole (Fig. 17.44; see also Chapter 8). Seeds of the *orp* mutant contain increased concentrations of two L-tryptophan precursors, anthranilate and indole (Fig. 17.44). The seeds germinate and form shoots normally, but the seedlings die at the four- to five-leaf stage

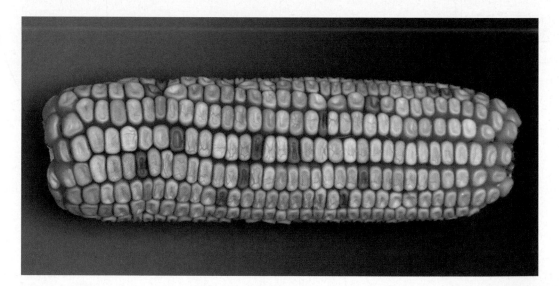

Figure 17.43
An orange pericarp *(orp)* maize cob showing the expected two-gene recessive trait; the orange kernels carry both mutant genes.

unless supplemented with tryptophan. Despite their diminished capacity for tryptophan synthesis, *orp* mutants are rich in IAA. Although *orp* seedlings contain roughly one-seventh the L-tryptophan present in wild-type maize, their IAA contents are increased 50-fold. Furthermore, labeling with ^2H_2O has indicated that isotopic enrichment of the *orp* IAA pool is much greater than in the L-tryptophan pool. [^{15}N]L-Tryptophan does not label the IAA pool significantly in either wild-type or *orp* seedlings, but [^{15}N]anthranilate is incorporated into IAA to a similar extent in both tissues. These findings demonstrate that L-tryptophan is not an intermediate in the major IAA biosynthesis pathway operating in maize seedlings. The IAA precursor is probably an indolic intermediate of tryptophan biosynthesis, downstream from anthranilate and upstream from L-tryptophan. The conversion of indole-3-glycerol phosphate to indole is reversible, and the available data suggest that either of these compounds could be the branch point for IAA biosynthesis (Fig 17.44). Although wild-type maize seedlings convert [^{15}N]indole to IAA as well as L-tryptophan, they contain negligible quantities of endogenous indole. Plant tryptophan synthase may resemble the bacterial enzyme, in which indole produced by the α-subunits is channeled directly to the β-subunits without being released from the protein surface (see Chapter 8).

Similar studies have been carried out with *Arabidopsis trp1, trp2,* and *trp3* mutants, which are conditional tryptophan auxotrophs that require tryptophan for growth under high light but not low light conditions. The *trp2* mutant is defective in tryptophan synthase β activity, whereas the *trp3* lesion affects tryptophan synthase α activity (Fig. 17.44). Unlike *orp,* these mutations do not result in increased concentrations of free IAA. However, the *trp2* and *trp3* mutants accumulate IAA conjugates (19- to 36-fold greater than in wild-type plants) and indole-3-acetonitrile (6- to 11-fold greater than in wild-types). Their normal IAA content, therefore, appears to result from conjugation of excess free IAA (see Section 17.4.5), and the *trp2* and *trp3* phenotypes result from tryptophan deficiency. When *trp2* seedlings are incubated with [^{15}N]anthranilate, three-fold more label is incorporated into IAA than into L-tryptophan. These findings indicate that a tryptophan-independent IAA biosynthesis pathway operates in *Arabidopsis.*

The *Arabidopsis trp1* mutant has defective phosphoribosyl anthranilate transferase and accumulates anthranilate (Fig. 17.44). Despite their reduced L-tryptophan content, *trp1* plants contain wild-type plant amounts of free and conjugated IAA. The most plausible explanation is that the mutation is leaky and that the *trp1* plants are able to preferentially convert the limited amounts of indole or indole-3-glycerol phosphate they synthesize to IAA.

The production of increased concentrations of indole-3-acetonitrile (see Fig. 17.42) by the *trp2* and *trp3* mutants, in which L-tryptophan synthesis is impeded, raises the possibility that indole-3-acetonitrile may be an intermediate in the tryptophan-independent IAA biosynthesis pathway. Studies with the soil bacterium *Azospirillum brasilense* indicate that this pathway may also generate indole-3-pyruvic acid (see Fig. 17.42). In the absence of tryptophan, *Az. brasilense* produces 90% of its IAA by a tryptophan-independent pathway. Disruption of the bacterial gene encoding indole-3-pyruvic acid decarboxylase results in a 95% decrease in IAA production when cultured in a tryptophan-free medium. This indicates that indole-3-pyruvic acid is a potential intermediate in the tryptophan-independent IAA biosynthesis pathway.

The importance of the tryptophan-dependent and tryptophan-independent IAA biosynthetic pathways varies among plant species and even among different developmental stages in the same life cycle. In *P. vulgaris* seedlings germinated for two days, de novo IAA biosynthesis originates primarily from L-tryptophan. In carrot cell suspension cultures, which are incubated in the presence of the synthetic auxin 2,4-D to promote cell division, large amounts of IAA are produced by the tryptophan-dependent pathway. When 2,4-D is withdrawn from the culture medium to induce somatic embryogenesis, the endogenous IAA pool decreases, and the lesser amounts of IAA synthesized are produced by way of a tryptophan-independent pathway. About 50% of the free IAA present in 7-day-old *Arabidopsis* seedlings is formed from L-tryptophan, but 14-day-old plants generate more than 90% of their IAA by a tryptophan-independent route.

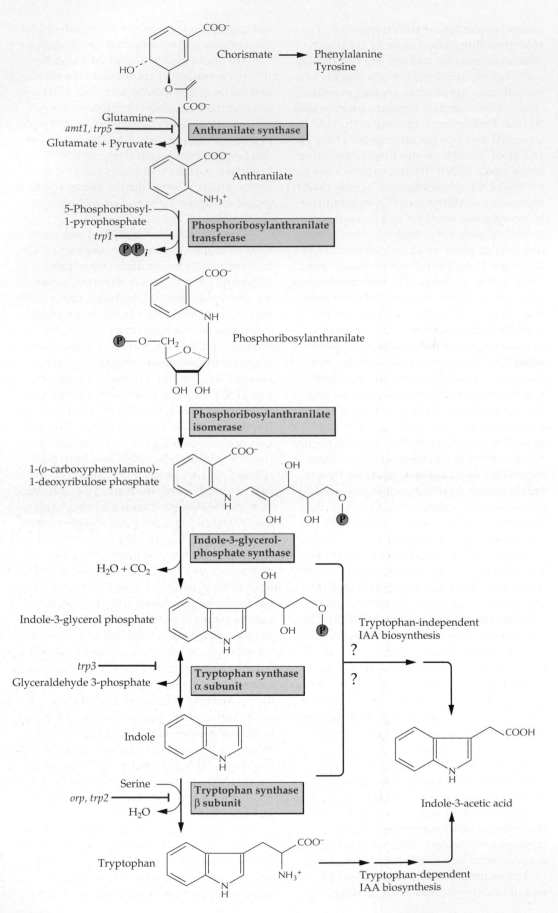

Figure 17.44
Biosynthesis of L-tryptophan and tryptophan-independent biosynthesis of IAA. The enzymes affected by the maize *orp* and *Arabidopsis amt-1*, *trp1*, *trp2*, *trp3*, and *trp5* mutations are indicated.

Chorismate → Phenylalanine / Tyrosine

Glutamine
amt1, trp5 ⊣ **Anthranilate synthase**
Glutamate + Pyruvate

Anthranilate

5-Phosphoribosyl-1-pyrophosphate
trp1 ⊣ **Phosphoribosylanthranilate transferase**
PP_i

Phosphoribosylanthranilate

Phosphoribosylanthranilate isomerase

1-(*o*-carboxyphenylamino)-1-deoxyribulose phosphate

Indole-3-glycerol-phosphate synthase
$H_2O + CO_2$

Indole-3-glycerol phosphate

Tryptophan-independent IAA biosynthesis
?
?

trp3 ⊣ **Tryptophan synthase α subunit**
Glyceraldehyde 3-phosphate

Indole

Serine
orp, trp2 ⊣ **Tryptophan synthase β subunit**
H_2O

Tryptophan

Tryptophan-dependent IAA biosynthesis

Indole-3-acetic acid

17.4.4 Several pathways for IAA conjugation and catabolism have been elucidated.

IAA catabolism results in a loss of auxin activity and irreversibly decreases the size of the IAA pool. Catabolism can proceed by decarboxylative or nondecarboxylative pathways, and either way can involve oxidation of the indole ring. As with cytokinins, catabolism of IAA sometimes involves conjugation reactions. Some investigators speculate that these IAA conjugates can be stored in vacuoles and may represent an important step in preparing the carbon skeleton for degradation and reentry into the general metabolism of the plant. However, although some conjugated IAA species appear to be permanently inactive, others can be cleaved to yield active free hormone (see Section 17.4.5).

Peroxidase-catalyzed decarboxylation of IAA has been studied with a variety of plant materials. In particular, the in vitro activity of horseradish peroxidase has been investigated extensively. The action of peroxidase (frequently referred to as IAA oxidase) results in the production of either decarboxylated oxindoles or decarboxylated indoles (Fig. 17.45). Although decarboxylated oxindoles have not been observed in vivo, the decarboxylated indoles—indole-3-methanol, indole-3-aldehyde, and indole-3-carboxylic acid—have been isolated from higher plants and appear to be endogenous compounds. Reports indicate that indole-3-methanol and indole-3-carboxylic acid can form glycosylated conjugates.

For many years, decarboxylative catabolism was thought to represent the major IAA degradation pathway in plant tissues. However, evidence obtained with *Z. mays*, tomato, and pea indicates that peroxidases have, at best, only a minor role in the regulation of endogenous IAA pools. IAA–amino acid conjugates that were once perceived as storage products have been identified as intermediates in nondecarboxylative catabolic pathways that deactivate IAA irreversibly.

The principal IAA deactivation pathway in green tomato fruits converts IAA to *N*-(indole-3-acetyl)-L-aspartic acid (IAA-aspartate). The indole ring of IAA-aspartate is oxidized to form *N*-(oxindole-3-acetyl)-L-aspartic acid (OxIAA-aspartate), which is subjected to successive glycosylations at the

indole nitrogen (Fig. 17.46). An IAA to OxIAA-aspartate deactivation pathway also operates in *Dalbergia,* but instead of *N*-glycosylation, the indole ring of OxIAA-aspartate is hydroxylated at C-3 or C-4. A parallel pathway in *Dalbergia* forms similar products that are conjugated to glutamic acid rather than aspartate. Additional nondecarboxylative IAA degradation pathways that involve conjugation reactions have been identified in *Vicia faba* and *Z. mays* seedlings (Fig. 17.47).

17.4.5 IAA ester conjugates serve as storage products in maize seeds.

The conjugation reactions discussed above, including aspartylation or glutamylation of the 1′ carboxyl, *N*-glycosylation of the indole ring, and glycosylation of either the 3 or 7

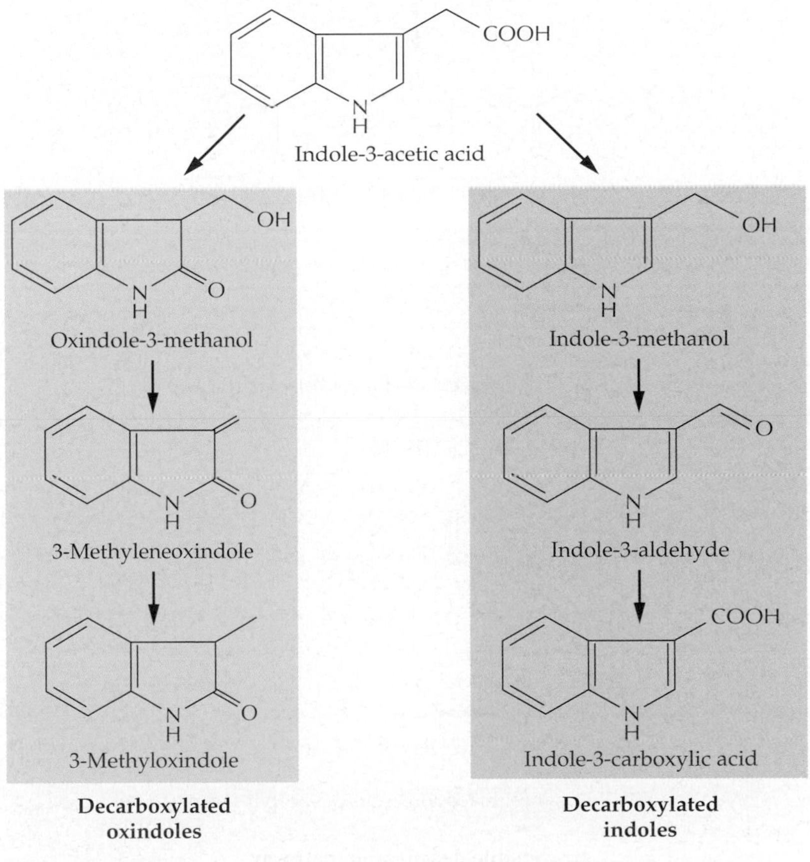

Figure 17.45
Peroxidase-catalyzed decarboxylative catabolism of IAA removes the 1′ carboxyl group and generates two classes of molecules in vitro. Decarboxylated oxindoles (e.g., oxindole-3-methanol, 3-methyleneoxindole, and 3-methyloxindole) are oxidized at the 2 position of the indole ring. In contrast, the indole rings of decarboxylated indoles (e.g., indole-3-methanol, indole-3-aldehyde, and indole-3-carboxylic acid) are not modified during catabolism.

Lycopersicon esculentum

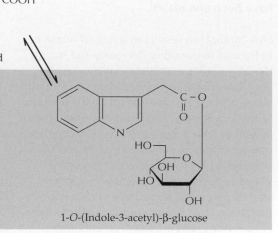

Indole-3-acetic acid

N-(Indole-3-acetyl)-L-aspartic acid

N-(Oxindole-3-acetyl)-L-aspartic acid

N-(1-β-Glucosyloxindole-3-acetyl)-L-aspartic acid

N-[1-(4-*O*-β-Glucosyl-β-glucosyl)oxindole-3-acetyl]-L-aspartic acid

Irreversible deactivation pathway

1-*O*-(Indole-3-acetyl)-β-glucose

Reversible deactivation pathway

Figure 17.46
Nondecarboxylative catabolism and conjugation of IAA in tomato pericarp discs. *N*-(1-β-Glucosyloxindole-3-acetyl)-L-aspartic acid and *N*-[1-(4-*O*-β-glucosyl-β-glucosyl)oxindole-3-acetyl]-L-aspartic acid are permanently deactivated IAA conjugates formed in both green and red tomato fruits, whereas the 1-*O*-(indole-3-acetyl)-β-glucose formed by red pericarp tissues can be converted back to IAA.

hydroxyl groups, appear to permanently deactivate IAA. However, *O*-glycosylation of the 1' carboxyl is typically reversible, so IAA–ester conjugates may function as storage products.

In *Z. mays*, IAA–ester conjugates are formed primarily in the liquid endosperm of developing seeds. IAA is first converted to 1-*O*-(indole-3-acetyl)-β-glucose (1-*O*-IAA-glucose) in a reaction catalyzed by IAA-glucose synthase. Isomers of 1-*O*-IAA-glucose can be formed by nonenzymatic reactions. Two of these isomers, 4-*O*-IAA-glucose and 6-*O*-IAA-glucose, can be cleaved by 6-*O*-IAA-glucose-hydrolase to release active IAA. A second enzyme that liberates active hormone, 1-*O*-IAA-glucose hydrolase, also can transfer IAA from 1-*O*-IAA-glucose to glycerol. However, the major pathway from 1-*O*-IAA-glucose involves its conversion to 2-*O*-(indole-3-acetyl)-*myo*-inositol, which is further conjugated to 2-*O*-(indole-3-acetyl)-*myo*-inositol arabinoside and 2-*O*-(indole-3-acetyl)-*myo*-inositol galactoside (Fig. 17.48). During the first days of germination, the *Z. mays* embryo derives most of its IAA from these three conjugates. However, the supply of hydrolyzable IAA conjugates declines as the seedling grows, and the young plant rapidly develops the capacity to synthesize IAA by way of a tryptophan-independent pathway. This contrasts with the liquid endosperm of developing *Z. mays* seed, which in vitro studies indicate synthesizes IAA primarily from L-tryptophan.

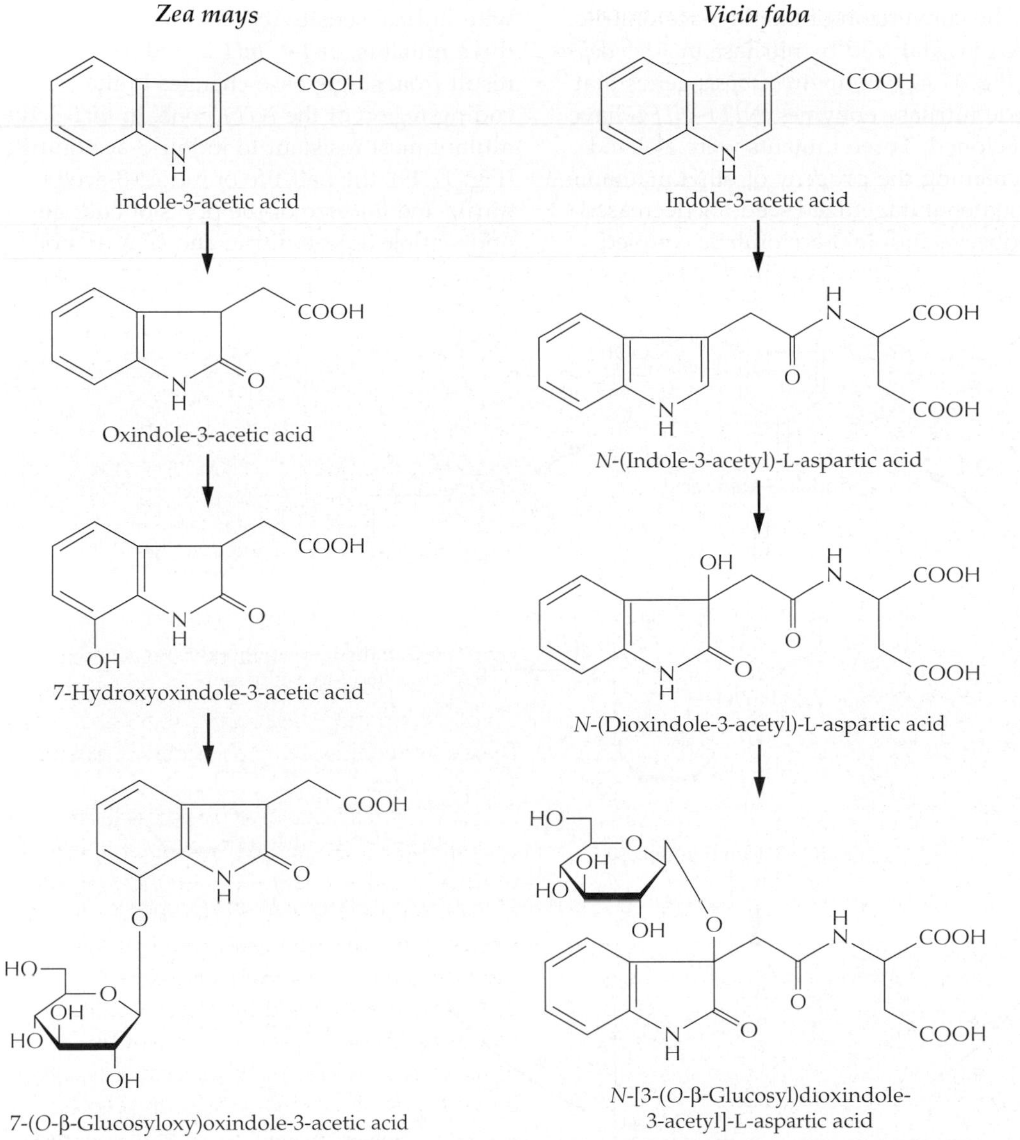

Zea mays

Indole-3-acetic acid

↓

Oxindole-3-acetic acid

↓

7-Hydroxyoxindole-3-acetic acid

↓

7-(O-β-Glucosyloxy)oxindole-3-acetic acid

Vicia faba

Indole-3-acetic acid

↓

N-(Indole-3-acetyl)-L-aspartic acid

↓

N-(Dioxindole-3-acetyl)-L-aspartic acid

↓

N-[3-(O-β-Glucosyl)dioxindole-3-acetyl]-L-aspartic acid

Figure 17.47
Nondecarboxylative catabolism and conjugation of IAA in seedlings of *Zea mays* and *Vicia faba*.

17.4.6 Numerous *Arabidopsis* mutants with an altered IAA content have been isolated and studied.

In addition to *trp2* and *trp3* (see Section 17.4.3), several *Arabidopsis* mutants are beginning to yield important information about the regulation of IAA homeostasis. These include the *trp5* mutant of the *Arabidopsis ASA1* gene. *ASA1* encodes anthranilate synthase, which converts chorismate to anthranilate (see Fig. 17.44). The activity of this enzyme is subject to allosteric feedback inhibition by L-tryptophan. The *trp5* mutant synthesizes feedback-insensitive anthranilate synthase, which increases flux through the tryptophan pathway (see Chapter 8). The phenotype and free IAA content of the *trp5-1* mutant, which was referred to as the *amt* mutant in some of the earlier literature, is relatively normal. However, these plants contain increased concentrations of IAA conjugates and free and conjugated indole-3-butyric acid, suggesting that the mutants compensate for increased rates of IAA biosynthesis by up-regulating IAA conjugation and indole-3-butyric acid production.

The conversion of indole-3-acetonitrile to IAA is catalyzed by nitrilase in *Arabidopsis* (see Fig. 17.42). Four *Arabidopsis* genes that encode nitrilase enzymes, *NIT1–NIT4*, have been cloned. Three mutants were isolated by screening the progeny of ethyl methane-sulfonate–mutagenized seed for decreased response to indole-3-acetonitrile coupled with normal sensitivity to IAA. Each of the three mutants, *nit1-1*, *nit1-2*, and *nit1-3*, result from single base changes in the coding region of the *NIT1* gene. In *nit1-3*, the mutant most resistant to indole-3-acetonitrile (Fig. 17.49), the half-life of indole-3-acetonitrile and the size of the pools of endogenous indole-3-acetonitrile and IAA do not

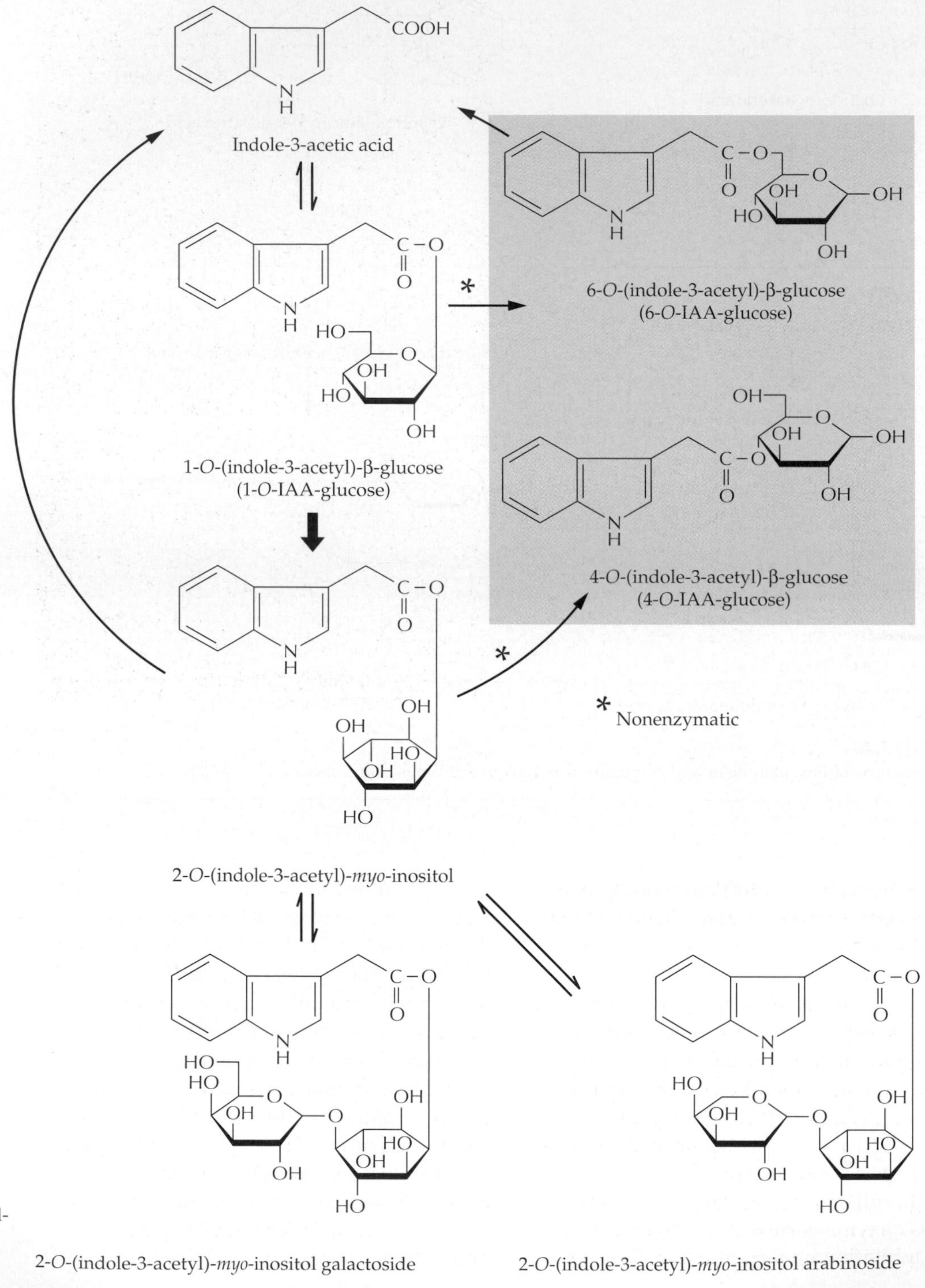

Indole-3-acetic acid

1-*O*-(indole-3-acetyl)-β-glucose
(1-*O*-IAA-glucose)

6-*O*-(indole-3-acetyl)-β-glucose
(6-*O*-IAA-glucose)

4-*O*-(indole-3-acetyl)-β-glucose
(4-*O*-IAA-glucose)

* Nonenzymatic

2-*O*-(indole-3-acetyl)-*myo*-inositol

2-*O*-(indole-3-acetyl)-*myo*-inositol galactoside

2-*O*-(indole-3-acetyl)-*myo*-inositol arabinoside

Figure 17.48
Biosynthesis and hydrolysis of IAA ester conjugates in *Zea mays*.

differ markedly from those of wild-type *Arabidopsis*. The normal morphological phenotype of all three *nit1* mutants is probably a consequence of the synthesis of endogenous IAA by an alternative route. *Arabidopsis* seedlings are known to produce IAA by both the tryptophan-dependent and the tryptophan-independent pathways (see Sections 17.4.1 and 17.4.3).

17.4.7 Some bacterial pathogens encode novel IAA synthesis and conjugation pathways.

As described in Section 17.3.2, some bacterial enzymes catalyze the production of plant hormones. The enhanced synthesis of IAA in *A. tumefaciens*–induced galls and tumors (see Fig. 17.33) results from expression of two bacterial genes that are transferred to the plant when the T-DNA integrates into the host genome. These genes are associated with a unique two-step tryptophan-dependent pathway to IAA. The *iaaM* gene encodes tryptophan monooxygenase, which converts L-tryptophan to indole-3-acetamide. The product of the *iaaH* gene, indole-acetamide hydrolase, catalyzes the conversion of indole-3-acetamide to IAA. Cognate genes with similar functions occur in the plant pathogen *Ps. savastanoi*. A third *Ps. savastanoi* gene, *iaaL*, encodes IAA-lysine synthase. When expressed in the cells of the plant host, this enzyme conjugates IAA and L-lysine to form ε-*N*-(indole-3-acetyl)-L-lysine (IAA-lysine), which is metabolized further to α-*N*-acetyl-ε-*N*-(indole-3-acetyl)-L-lysine (Fig. 17.50).

17.4.8 Transgenic plants expressing IAA biosynthesis genes have been used to study the effects of excess amounts of endogenous IAA.

Transgenic plants expressing the *A. tumefaciens* IAA biosynthesis genes have been produced in several laboratories. Weak coexpression of the *iaaH* and *iaaM* genes in transgenic tobacco (*N. tabacum* cv. Petit Havana SR1) results in a marginal increase in endogenous IAA. Depending on the tissue, there is a two- to threefold increase in the amount of IAA conjugates—primarily IAA-

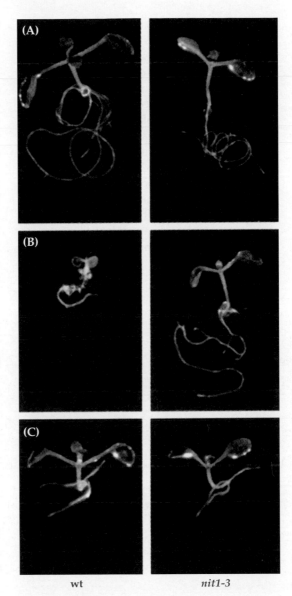

Figure 17.49
Effects of indole-3-acetonitrile and IAA on wild-type (wt) and *nit1-3* mutant seedlings of *Arabidopsis*. Eight-day-old seedlings grown in the presence and absence of 30 μM indole-3-acetonitrile (B) or 1 μM IAA (C). (Control is shown [A].) Note that wild-type plants show a typical auxin-like response to both IAA and indole-3-acetonitrile. The *nit1-3* mutant responds to IAA but does not exhibit an auxin-like response to indole-3-acetonitrile: It lacks nitrilase and cannot convert indole-3-acetonitrile to IAA.

aspartate and IAA-glutamate, with smaller amounts of 1-*O*-IAA-glucose. The transgenic plants exhibited no obvious changes in vegetative phenotype, but their flowers were heterostylous and pollen production was impaired, resulting in infertility. Thus, when IAA production is enhanced, conjugation appears to play a key role in regulating the size of the endogenous IAA pool and maintaining an almost normal phenotype. However, strong *iaaH/iaaM* expression driven by the CaMV 35S promoter increased the concentration of free and conjugated IAA in stems by 10- and 20-fold, respectively, whereas foliar IAA and IAA conjugates increased 3- to 5-fold. This increased IAA content was accompanied by major phenotypic effects, including pronounced apical dominance, dwarfism, excess adventitious root formation, increased phloem and xylem formation, excess lignification, leaf epinasty, and abnormal flower production (Fig. 17.51).

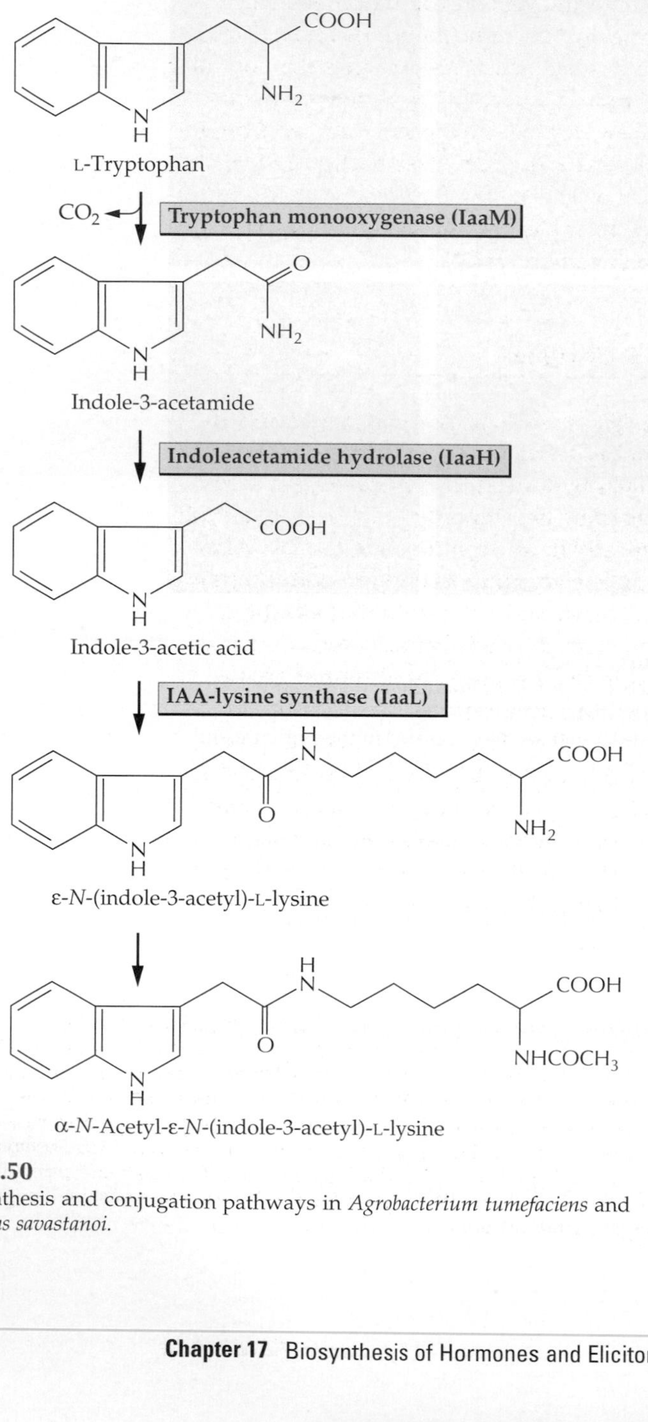

Similar phenotypes have been observed in several transformed plants that overproduce IAA, including petunia and *Arabidopsis*.

High concentrations of IAA are typically accompanied by an increased rate of ethylene biosynthesis, and it was not initially possible to determine whether the phenotypic effects of IAA overproduction were a direct result of IAA or a consequence of increased ethylene levels. To investigate these possibilities, investigators crossed IAA-overproducing transgenic *N. tabacum* cv. Samsun plants with plants expressing a bacterial ACC deaminase that catalyzes the breakdown of the ethylene precursor ACC and thereby reduces ethylene concentrations (see Section 17.5.6). With cv. Samsun (Fig. 17.52) the phenotypic effects of IAA-overproduction are not as severe as with the Petit Havana SR1 transformant (see Fig. 17.51), although inhibition of stem growth, leaf epinasty, and reduced apical dominance are still evident. The phenotype of the Samsun double-transformants, in which IAA overproduction is not accompanied by increased ethylene biosynthesis, shows that apical dominance and leaf epinasty are controlled primarily by IAA, whereas reduced stem elongation is an indirect consequence of high ethylene concentrations (Fig. 17.52).

17.4.9 Gibberellins increase IAA pools, whereas cytokinins may down-regulate IAA synthesis and turnover.

Application of GA_3 to Little Marvel dwarf pea seedlings enhances shoot growth with a concomitant eightfold increase in the IAA content of the elongating tissues. Conversely, the size of the endogenous IAA pool is reduced in the Alaska pea seedlings dwarfed by treatment with uniconazole, the GA biosynthesis inhibitor that blocks the oxidation of *ent*-kaurene (see Section 17.1.5). The effects of uniconazole on both internode growth and IAA content are counteracted by GA_3 treatment.

Surprisingly little is known about the mechanism by which GA_3 increases endogenous IAA concentrations in peas. One proposal postulates that D-tryptophan is converted to IAA more effectively than its L-stereoisomer and acts as an intermediate between L-tryptophan and indole-3-pyruvic

Figure 17.50

IAA biosynthesis and conjugation pathways in *Agrobacterium tumefaciens* and *Pseudomonas savastanoi*.

Figure 17.51

Eight-week-old tobacco plants, *Nicotiana tabacum* cv. Petit Havana SR1: wild-type plant (left); IAA-over-producing plant expressing *Agrobacterium tumefaciens iaaH* and *iaaM* genes under the control of the CaMV 35S promoter (right). Note the severe stunting associated with production of IAA at about 500% of wild-type concentrations.

and [9R-5'P]Z, not only reduce the size of the endogenous IAA and IAA conjugate pools but also appear to lower the rate of IAA turnover.

Just as cytokinin overexpression antagonizes IAA synthesis, IAA overexpression appears to down-regulate cytokinin production. IAA-overproducing tobacco plants that express *iaaH* and *iaaM* (see Fig. 17.51) contain less cytokinin oxidase activity and lower concentrations of endogenous Z and related cytokinins than wild-type plants do. Diminished cytokinin oxidase activity would be expected to reduce cytokinin degradation and to increase the size of cytokinin pools. But this did not occur in the IAA-overproducing transgenic tobacco plants; instead, both cytokinin concentrations and oxidase activity declined. This provides support, albeit indirectly, for the view that cytokinin oxidase degrades exogenous rather than endogenous cytokinins (see Section 17.3.6). As yet, the mechanism for the IAA-mediated reduction in the endogenous cytokinin pool remains unknown.

acid in the IAA biosynthesis pathway; moreover, GA₃ treatment supposedly enhances the activity of the racemase that regulates the conversion of L- to D-tryptophan, thereby increasing the rate of IAA biosynthesis.

In contrast to GA₃, cytokinins reduce the size of endogenous IAA pools. Transgenic tobacco plants that express the *A. tumefaciens ipt* gene overproduce cytokinins (see Fig. 17.34). Compared with wild-type plants, these transgenics contain significantly lower concentrations of free IAA and, in most cases, IAA conjugates. Rates of IAA biosynthesis also are reduced, as determined by ²H₂O incorporation studies, and exogenous [¹³C₆]IAA is degraded more slowly in the *ipt*-transformed plants. Thus, elevated amounts of cytokinins, primarily Z, [9R]Z,

17.5 Ethylene

In 1886, while a 17-year-old graduate student in St. Petersburg, Dimitry Nikolayevich Neljubow noticed that etiolated pea seedlings grew horizontally in laboratory air but vertically in air from outside the lab. After an extensive study to exclude cultural practices, light, and temperature as causative agents, he showed that **ethylene,** a component in the gas used for illumination, induced this abnormal growth. Many of the physiological effects of ethylene on plant growth and development, including its impact on seed germination, root and shoot growth, flower development, senescence and abscission of flowers and leaves, and ripening of fruit, were discovered prior to 1940.

Figure 17.52

Uncoupling of auxin and ethylene effects in eight-week-old transgenic tobacco plants, *Nicotiana tabacum* cv. Samsun. (Left) An ethylene-deficient plant expressing a *Pseudomonas* ACC deaminase gene under the control of the figwort mosaic virus 19S promoter. The phenotype is indistinguishable from that of wild-type plants. (Middle) A double-transformant with increased IAA content and decreased ethylene production. (Right) An IAA-overproducer expressing the *Agrobacterium tumefaciens iaaM* gene under the control of the CaMV 35S promoter. The phenotype of these plants indicates that apical dominance and leaf epinasty are primarily controlled by IAA, whereas ethylene is partially responsible for the inhibition of stem elongation observed in IAA-overproducing plants.

Subsequent work has revealed that ethylene also participates in the modulation of plant responses to a range of biotic and abiotic stresses.

When investigators in the 1930s first proposed that ethylene was both an endogenous plant growth regulator and a fruit-ripening hormone, their hypothesis was met with skepticism by their contemporaries, many of whom considered ethylene an unimportant product of the paramount plant hormone, auxin. Studies were hampered by this prejudice, and ethylene research remained practically dormant for more than a quarter of a century while attention was focused on other plant hormones, initially auxin, and later GAs and cytokinins.

Although research on ethylene may have been unfashionable, the primary re-

strictions on its investigation during this period were technical. The procedures available for measuring ethylene were cumbersome and insensitive, and physiologically relevant concentrations of ethylene were too low to analyze. Until the early 1960s, ethylene was quantified with manometric (pressure-based) procedures, crude colorimetric techniques, and bioassays based on leaf epinasty or the growth of etiolated pea and mung bean seedlings (Fig. 17.53). All this changed with the commercial availability of the gas chromatograph and the development of the flame ionization detector, which lowered the limits of ethylene detection by a factor of 10^6. This major development led to rapid advances in the field of ethylene research, and ethylene was quickly established as an endogenous plant growth regulator produced by almost all higher plants.

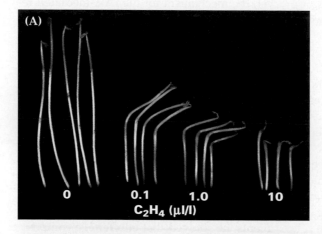

Figure 17.53
The triple response to ethylene of six-day-old etiolated pea seedlings and four-day-old etiolated mung bean seedlings. (A) Untreated control pea seedlings (0) and pea seedlings grown for two days in air supplemented with ethylene at 0.1, 1.0, and 10 µl/ml. Note the concentration-dependent effects of ethylene on diageotropism, inhibition of epicotyl elongation, and lateral enlargement of the epicotyl. (B) Control mung bean seedlings (0) and mung bean seedlings grown for two days in air supplemented with 1 and 10 µl/ml ethylene, which induces a concentration-dependent inhibition of hypocotyl elongation, lateral enlargement of the hypocotyl, and extreme bending of the apical hook. (C) Magnification of ethylene-treated etiolated mung bean seedlings.

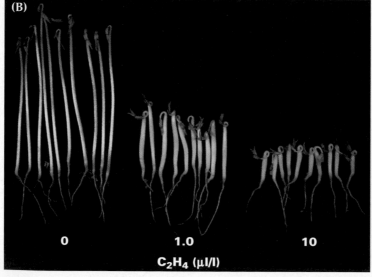

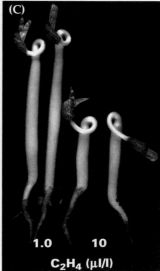

Chapter 17 Biosynthesis of Hormones and Elicitor Molecules

17.5.1 Ethylene is synthesized from S-adenosyl-L-methionine (SAM) via the intermediate 1-aminocyclopropane-1-carboxylic acid (ACC).

The synthesis of ethylene from its immediate precursor ACC is catalyzed by ACC oxidase, whereas ACC is produced from SAM in a reaction catalyzed by ACC synthase as part of the methionine cycle (also called the Yang cycle, after S. F. Yang, who carried out much of the early work in the elucidation of this pathway) (Fig. 17.54). In addition to its role in ethylene production, SAM is involved in

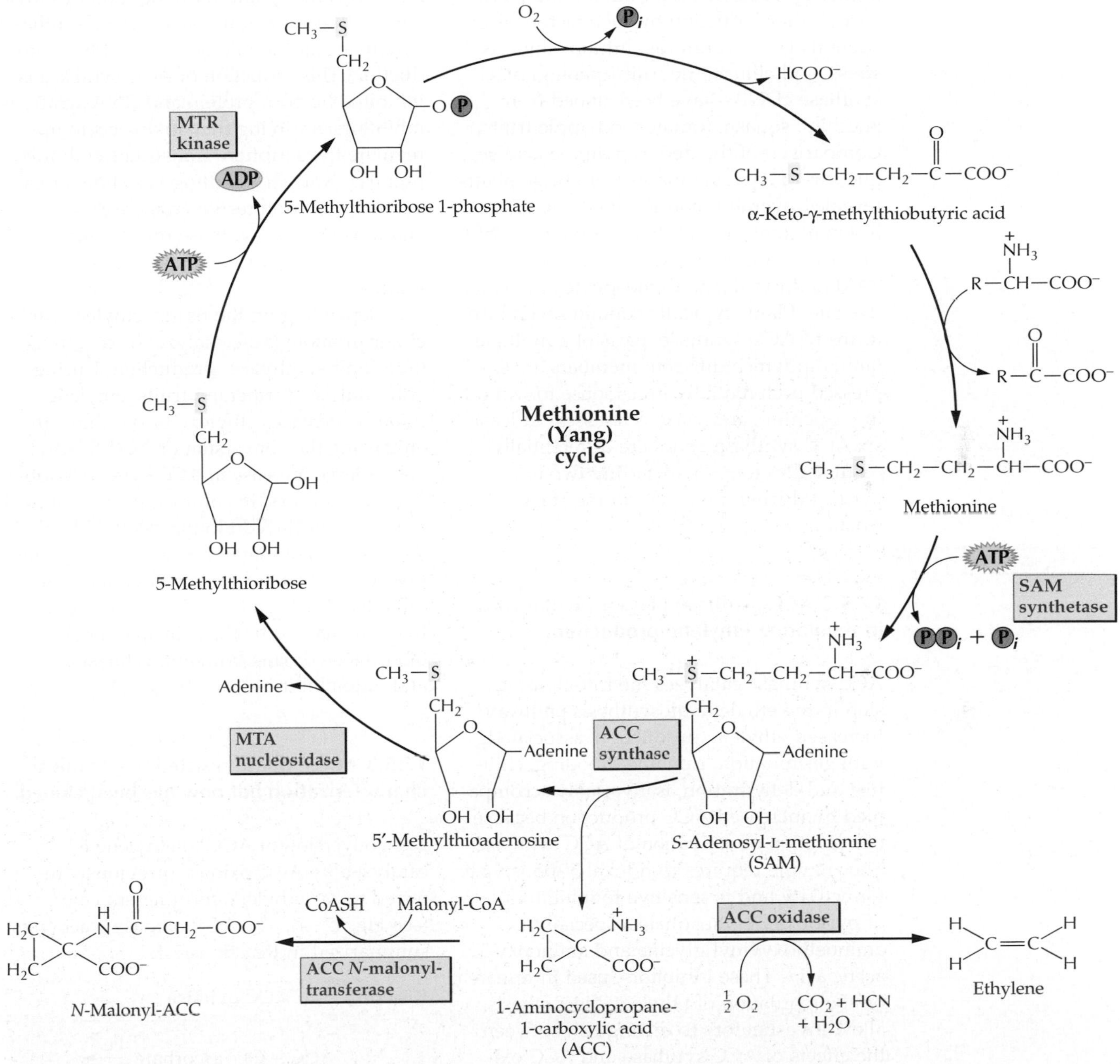

Figure 17.54

The methionine cycle and ethylene biosynthesis. Ethylene is synthesized from methionine by way of SAM and ACC. The enzymes that catalyze these three steps are ATP:methionine S-adenosyl-transferase (SAM synthase), S-adenosyl-L-methionine methyl-thioadenosine-lyase (ACC synthase), and ACC oxidase. 5'-Methyl-thioadenosine, a product of the ACC oxidase reaction, is salvaged for the resynthesis of methionine through the methionine cycle (see Chapter 8). If the methylthio-group from SAM were not recycled, methionine availability and ethylene biosynthesis would probably be restricted by sulfur availability. By converting ACC to N-malonyl-ACC instead of to ethylene, plants can deplete the ACC pool and thereby reduce the rate of ethylene production.

the biosynthesis of polyamines (see Section 17.7) and participates in a wide range of methylation reactions.

ACC synthase, which cleaves SAM to form the ethylene precursor ACC and 5'-methylthioadenosine (Fig. 17.54), was first characterized in a semipurified preparation from tomato pericarp. The enzyme has subsequently been isolated from several plant tissues after induction by such factors as exogenous IAA, wounding, lithium chloride stress, and climacteric fruit ripening. ACC synthase cDNAs have been cloned from zucchini, squash, tomato, and apple fruits. Comparison of the deduced amino acid sequences of ACC synthases from these plants revealed overall identities of 48% to 97%. Seven regions are highly conserved; the fifth includes a lysine residue that reacts with SAM and pyridoxal 5'-phosphate at the active site. Plants typically contain several isoforms of ACC synthase, parts of a multigene family in which different members are expressed preferentially in response to wounding, ripening, and various stresses. At least six ACC synthase genes are differentially regulated in tomato, including two expressed during the usual course of fruit ripening.

17.5.2 ACC synthase plays a pivotal role in regulating ethylene production.

ACC synthase catalyzes the rate-limiting step in the ethylene biosynthesis pathway. Increased ethylene production, associated with germination, ripening, flooding, chilling, and dehydration, is invariably accompanied by increased ACC production because of induction or activation of ACC synthase. The enzyme requires pyridoxal 5'-phosphate for activity and is sensitive to inhibitors of pyridoxal 5'-phosphate, especially aminoethoxyvinyl glycine and aminooxyacetic acid. These inhibitors, used in research on the regulation of ethylene biosynthesis, allow investigators to distinguish between the effects of ACC synthase and ACC oxidase. The naturally occurring isomer of SAM, (−)-S-adenosyl-L-methionine, is the preferred substrate for ACC synthase, whereas (+)-SAM is an effective inhibitor. However, incubating the enzyme with high concentrations of (−)-SAM can irreversibly

modify and inhibit ACC synthase. This "suicide inactivation" involves covalent linkage of a fragment of the SAM molecule to the active site of the enzyme. This substrate-dependent inactivation may be at least partially responsible for the rapid turnover of ACC synthase in plant tissues.

Ethylene biosynthesis rates are influenced by other plant hormones and by ethylene itself. Auxins promote ethylene synthesis by enhancing the rate of ACC synthase production. This induction of ACC synthase is inhibited by both protein and RNA synthesis inhibitors, implying that auxin acts to upregulate transcription. Consistent with this premise, Northern blotting has shown that auxin application results in increased amounts of ACC synthase mRNA but only a few kinds of ACC synthase isoforms are induced.

Depending on the tissue, ethylene can either promote (autocatalyze) or constrain (autoinhibit) ethylene production. During autocatalysis in ripening fruits, ethylene initially increases ethylene biosynthesis by enhancing the conversion of ACC to ethylene. A large increase in ACC synthesis subsequently occurs. In contrast, prior treatment of grapefruit flavedo (outer peel) with ethylene results in a marked decrease in the rate of ethylene production, a consequence of inhibited ACC synthesis. Thus, ACC synthase appears to be the principal target of ethylene during both autocatalysis and autoinhibition.

17.5.3 ACC oxidase resisted biochemical characterization but now has been cloned.

The conversion of ACC to ethylene is catalyzed by ACC oxidase, previously referred to as "ethylene-forming enzyme" (see Fig. 17.54). The ACC oxidase reaction is summarized in Reaction 17.1.

Reaction 17.1: ACC oxidase

$$\text{ACC} + \text{O}_2 + \text{ascorbate} \xrightarrow{\text{Fe}^{2+}}$$
$$\text{C}_2\text{H}_4 + \text{CO}_2 + \text{dehydroascorbate}$$
$$+ \text{HCN} + 2\,\text{H}_2\text{O}$$

ACC oxidase is activated by one of its products, carbon dioxide. The cyanide gas generated by the reaction is detoxified by conversion to β-cyanoalanine, which is further metabolized to asparagine or γ-glutamyl-β-cyanoalanine.

The characteristics of the in vivo conversion of ACC to ethylene, such as substrate specificity, stereoselectivity, and competitive inhibition of activity, provide a basis for evaluating putative ACC oxidases. Substituting an ethyl group for one of the ring hydrogens of ACC generates four stereoisomers of 1-amino-2-ethylcyclopropane-1-carboxylic acid (AEC) (Fig. 17.55). In apple fruit, cantaloupe fruit, and etiolated mung bean hypocotyls, ACC oxidase preferentially utilizes one of the stereoisomers, $(1R, 2S)$-AEC, for synthesis of the ethylene analog, 1-butene. In addition to stereoselectivity among AEC isomers, authentic ACC oxidase displays a high affinity for ACC as its in vivo substrate. Other important characteristics of ACC oxidase include requirements for Fe^{2+}, ascorbate, and O_2, and inhibition by 10 to 100 μM Co^{2+} applied to tissues.

ACC oxidase, which is extremely unstable, proved difficult to purify by conventional techniques. The enzyme was eventually identified by molecular cloning. In the course of investigating gene expression in ripening tomato fruit, a cDNA (pTOM13) was isolated through use of differential screening techniques. The pTOM13 clone then hybridized to an mRNA that accumulated at the onset of fruit ripening and after the wounding of both unripe tomato fruit and leaves, processes associated with a rapid increase in the rate of ethylene production. The link between the pTOM13 gene product and ethylene formation received additional support when ethylene biosynthesis and ACC oxidase activity were shown to be greatly decreased in transgenic tomato

Figure 17.56
Effect of antisense ACC-oxidase genes on the ripening and spoilage of Ailsa Craig tomatoes in fruits picked three weeks after the onset of ripening and stored at room temperature for three weeks. (Left) Fruits from the descendents of the original TOM13-antisense plants, which generate about 5% of the normal amount of ethylene. They ripen fully but do not overripen and deteriorate. (Right) Fruits from wild-type plants grown and stored under identical conditions. They produce normal amounts of ethylene and consequently exhibit severe signs of overripening.

plants that expressed a pTOM13 antisense gene (Fig. 17.56). Confirmation of the function of the pTOM13 product was obtained when pTOM13 cDNA was expressed in the sense orientation in yeast. Cultures of the transformed yeast were able to convert ACC to ethylene, whereas control cultures did not catalyze the reaction. The recombinant 35-kDa protein met all the criteria for authentic ACC oxidase activity, including inhibition by Co^{2+}.

The activity of ACC oxidase was traditionally studied in vivo by applying ACC to plant tissues and measuring subsequent ethylene production. Detailed in vitro research proved difficult because ACC oxidase rapidly lost activity after extraction. The failure of earlier attempts to demonstrate

	(1R,2S)-AEC	(1S,2R)-AEC	(1S,2S)-AEC	(1R,2R)-AEC
Relative efficiency of conversion to 1-butene	100	1.2	0.5	0.5

Figure 17.55
Stereoisomers of 1-amino-2-ethylcyclopropane-1-carboxylic acid (AEC) and the relative efficacy (from 0 to 100) with which they are utilized for in vivo synthesis of the ethylene analog 1-butene by apple fruit, cantaloupe fruit, and etiolated mung bean hypocotyls.

significant ACC oxidase activity in vitro is now attributed to the loss of cofactors during extraction. ACC oxidase from apple fruits has been purified to near homogeneity. The enzyme is a monomer and has a molecular mass of 35 kDa.

All plant tissues appear to contain ACC oxidase, as measured by the rate of ethylene evolution in the presence of a saturating concentration of exogenous ACC. Under stress conditions, in response to ethylene and at selected stages of development (e.g., fruit ripening), ACC oxidase activity increases markedly. Both senescence- and ripening-induced increases in ACC oxidase have been shown to result from increased transcription.

17.5.4 When the supply of available SAM is low, the ethylene and polyamine biosynthetic pathways may compete for this shared substrate.

Biosynthesis of both ACC (see Fig. 17.54) and polyamines (see Fig. 17.68) involves incorporation of the aminopropyl group from SAM. Under certain conditions, competition for SAM may restrict rates of ethylene or polyamine production. Inhibition of ACC synthesis by aminooxyacetic acid results in increased polyamine production. Conversely, inhibition of polyamine biosynthesis leads to increased concentrations of ACC and ethylene. This implies that one SAM-dependent pathway is stimulated when the other is blocked. When competition for the available SAM is circumvented by low demand for polyamines or ACC or is compensated for by up-regulation of the 5′-methylthioribose-recycling enzymes (see Fig.17.54), ethylene and polyamine production do not interact. Whether ACC/polyamine interactions represent a widespread means for controlling ethylene production remains an open question.

17.5.5 Most hormones must be catabolized, but volatile ethylene can be released as a gas.

Before 1975, ethylene metabolism by plants was considered an artifact, caused by bacterial contamination. However, there is now evidence from many plants grown under

aseptic conditions that [^{14}C]ethylene is oxidized to [^{14}C]CO$_2$ or converted to [^{14}C]ethylene oxide and [^{14}C]ethylene glycol. Ethylene metabolism exhibits a very high K_m, indicative of a chemical reaction rather than a controlled physiological process. In peas, the concentration of ethylene yielding a half-maximal rate of ethylene metabolism is about 1000 times the concentration required for half-maximal response in the pea growth test illustrated in Figure 17.53A. Ethylene metabolism is probably largely a consequence of treating plants with artificially high amounts of exogenous ethylene. The major route by which plant tissues lose ethylene is probably diffusion to the surrounding atmosphere.

17.5.6 Repression of ethylene biosynthesis can delay overripening in fruit and is an important field of biotechnological research.

There is a great deal of interest in down-regulating ethylene biosynthesis, given that high concentrations can trigger the ripening and subsequent overripening of climacteric fruits such as bananas, apples, and tomatoes. Two different biotechnological strategies have been used to generate transgenic tomato fruit that resist overripening. One, the overexpression of a *Pseudomonas* gene encoding ACC deaminase, reduces ethylene content in fruits by catalyzing the conversion of ACC to α-ketobutyric acid and NH$_3$. The second approach to limiting ethylene biosynthesis involves the use of antisense gene constructs against either ACC oxidase or ACC synthase. Analysis of transgenic fruit produced by these procedures has demonstrated a direct correlation between the extent of ethylene inhibition and the rate of fruit ripening.

The phenotype of transgenic tomatoes that express antisense pTOM13 RNA has been studied in most detail. In these fruit, ethylene production is inhibited by about 95% during ripening. The ACC oxidase antisense fruit grow normally and begin to change color, losing chlorophyll and accumulating lycopene, at the same stage of development as nontransformed fruit. However, the transgenic fruit exhibit less reddening and an increased resistance to overripening and shriveling when stored at room temperature for prolonged periods (Fig. 17.56). The transgenic fruit do not soften as readily and can be left on the plant longer to ripen more fully. Once purchased by the consumer, these tomatoes can be kept at room temperature for several weeks while remaining fresh and edible. In the future, ethylene production will probably be manipulated in other climacteric fruit, such as melons, pears, kiwi fruit, apples, nectarines, avocados, and a range of exotic and tasty tropical fruit that at present are rarely seen outside their country of origin because they overripen and spoil so readily. The same technology can also be used to delay the senescence of flowers and to enhance the longevity of cut blossoms.

17.6 Brassinosteroids

In the early 1960s, researchers hypothesized that the rapid germination and growth of pollen grains might be associated with the presence of a growth promoter. A crude extract of pollen from *Brassica napus* (rape) induced a rapid elongation of pinto bean internodes that was distinct from GA-mediated stem elongation. This early work led to the isolation and identification of the first steroidal plant growth regulator, brassinolide, in 1979 (Fig. 17.57). For a time, brassinolide was studied intensively by a small group of plant physiologists and chemists, mainly at the US Department of Agriculture but also in Japan, where interest in brassinolide had its origins in a 1968 paper, a report on three relatively pure compounds isolated from leaves of the isunoki plant *(Distylium racemosum)* that exhibited significant biological activity in the rice lamina inclination assay (Fig. 17.58). The structure of these compounds had not been determined because their availability was limited and analytical techniques of the day (e.g., mass spectrometry and NMR; see Chapter 2, Boxes 2.1 and 2.2) were relatively insensitive. Japanese researchers realized that these active components were very similar to brassinolide and might be important growth regulators. A second plant steroid, castasterone (see Fig. 17.57), was isolated in 1982 from insect galls of chestnut *(Castanea crenata)*. Since these reports, a number of related steroidal compounds have been isolated from various plant sources and are now referred to collectively as brassinosteroids (BRs).

BRs occur in algae, ferns, gymnosperms, and angiosperms but have not been detected in microorganisms. More than 40 BRs have been identified; structurally, they are C_{27}-, C_{28}-, and C_{29}-steroids with different substituents on the A and B rings and side chain (Fig. 17.59). Brassinolide, a C_{28}-BR, elicits the highest biological activity of the BRs and is distributed widely throughout the plant kingdom, along with biosynthetically related compounds. The biosynthesis of brassinolide from campesterol has been studied most extensively in normal and transformed cells of *Ca. roseus*. The biosynthetic pathways that generate other BRs with different carbon skeletons have not yet been elucidated.

17.6.1 BRs affect a range of morphological characteristics.

Application of BRs induces a broad spectrum of responses, including an increased rate of stem elongation, pollen tube growth, unrolling of grass leaves, bending of grass leaves at the sheath/blade joints, proton

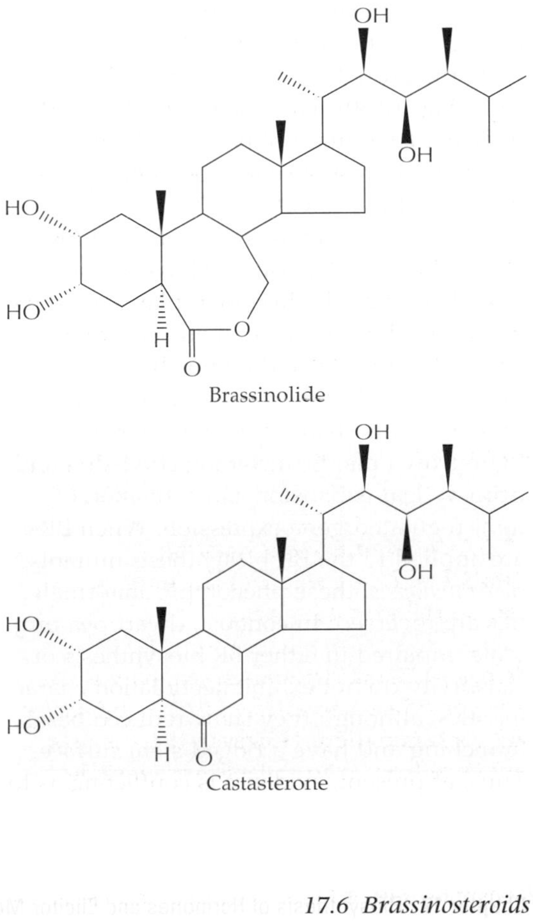

Brassinolide

Castasterone

Figure 17.57
Structures of brassinolide and castasterone.

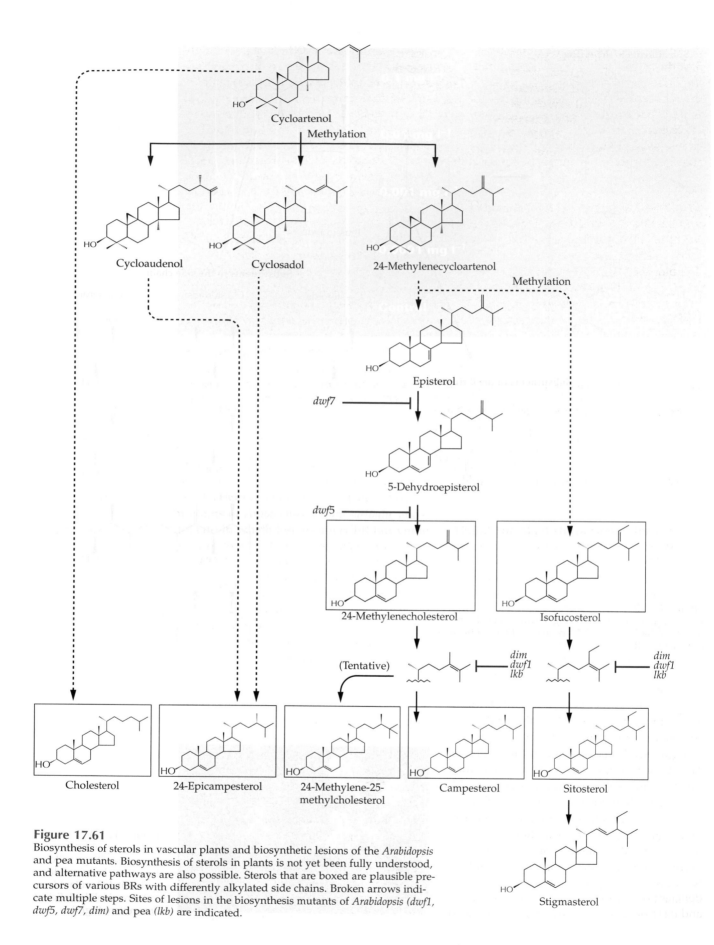

Figure 17.61

Biosynthesis of sterols in vascular plants and biosynthetic lesions of the *Arabidopsis* and pea mutants. Biosynthesis of sterols in plants is not yet been fully understood, and alternative pathways are also possible. Sterols that are boxed are plausible precursors of various BRs with differently alkylated side chains. Broken arrows indicate multiple steps. Sites of lesions in the biosynthesis mutants of *Arabidopsis* (*dwf1*, *dwf5*, *dwf7*, *dim*) and pea (*lkb*) are indicated.

24-methylenecholesterol to campesterol. The dwarf *Arabidopsis* mutants *dwf5* and *dwf7* also are unable to synthesize campesterol because of impaired synthesis of 5-dehydroepisterol and 24-methylenecholesterol, respectively (Fig. 17.61). Thus, campesterol production is apparently a prerequisite for the biosynthesis of brassinolide.

17.6.3 Two routes, the early C-6 oxidation and the late C-6 oxidation pathways, link 5α-campestanol and brassinolide.

The biosynthetic pathway leading to the production of brassinolide was established in cultured cells of *Ca. roseus* (Fig. 17.62). The first step is conversion of campesterol to 5α-campestanol, which is then converted to castasterone via either the early or the late C-6 oxidation pathway. Finally, oxidation of castasterone produces brassinolide.

Campesterol to 5α-campestanol

This reaction was originally detected in *Catharanthus* cell cultures; in subsequent studies with *Arabidopsis*, it was found to proceed by way of three intermediates (Fig. 17.62). Reduction of the Δ^4 bond in the conversion of (24R)-24-methylcholest-4-en-3-one to (24R)-24-methyl-5α-cholestan-3-one is impaired in the *det2* dwarf mutant of *Arabidopsis* and the *lk* dwarf mutant of pea. This reaction is analogous to the conversion of testosterone to dihydrotestosterone in animals. The amino acid sequence deduced from the *DET2* gene shares 40% homology with a mammalian steroid 5α-reductase, which catalyzes the NADPH-dependent conversion of testosterone to dihydrotestosterone.

5α-Campestanol to castasterone through the early C-6 oxidation pathway

The early C-6 oxidation pathway begins with the C-6 hydroxylation of 5α-campestanol to form 6α-hydroxycampestanol (Fig. 17.62). 6α-Hydroxycampestanol is oxidized further to yield 6-oxocampestanol. Although no direct evidence has been obtained for C-22 hydroxylation of 6-oxocampestanol to form cathasterone, this reaction seems likely to occur in *Catharanthus* cells, which contain both compounds as endogenous constituents. Failure to detect the conversion

of 6-oxocampestanol to cathasterone might reflect the 500:1 ratio of putative precursor to product. Cathasterone is hydroxylated at C-23 to form teasterone. Hydroxylations at C-22 and C-23 may be blocked in the respective *dwf4* and *cpd* dwarf mutants of *Arabidopsis*. The *DWF4* and *CPD* genes encode cytochrome P450 oxidases and share homologous domains with animal steroid hydroxylases. Interestingly, expression of the *CPD* gene is suppressed by biologically active BRs, suggesting that, as with GAs (see Section 17.1.11), the biosynthesis of BRs is controlled by a feedback mechanism.

Teasterone is metabolized to typhasterol by way of 3-dehydroteasterone. These reactions involve successive oxidation and reduction that epimerize the 3β-hydroxyl to a 3α-hydroxyl group (Fig. 17.63). A similar sequence of reactions occurs in the biosynthesis of bile acids, ecdysteroids (steroidal insect-molting hormones), and cardenolides (see Chapter 24). Feeding teasterone to *Catharanthus* cell cultures results in its conversion to typhasterol with no 3-dehydroteasterone being detected, presumably because it is turned over too rapidly to accumulate. These two steps in the pathway also undergo partial reversion with some typhasterol being converted back to teasterone by way of 3-dehydroteasterone. Most typhasterol, however, undergoes C-2 oxidation to produce castasterone, which has two sets of vicinal hydroxyl groups in the A ring and side chain. The conversion of teasterone to castasterone is found in seedlings of *Ca. roseus*, tobacco, and rice, indicating that the early C-6 oxidation pathway is also operative in intact plants.

5α-Campestanol to castasterone through the late C-6 oxidation pathway

6-Deoxobrassinosteroids (6-deoxo-BRs), which have no oxygen in the B ring, usually accumulate in plant tissues in relatively high concentrations. These compounds exhibit much less biological activity in the rice lamina inclination assay (see Fig. 17.58) than do the 6-oxobrassinosteroids (6-oxo-BRs) and have been proposed as end-pathway deactivation products. However, 6-deoxocastasterone is converted to castasterone and brassinolide in *Catharanthus* cells, which also synthesize 6-deoxocastasterone from 6-deoxoteasterone by way of

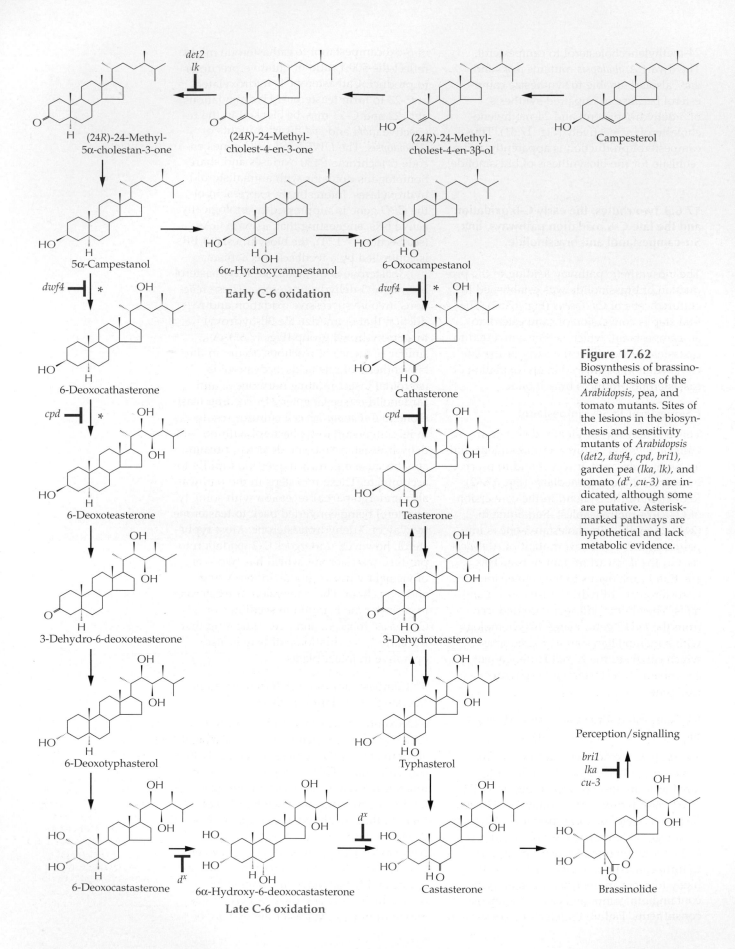

Figure 17.62
Biosynthesis of brassinolide and lesions of the *Arabidopsis*, pea, and tomato mutants. Sites of the lesions in the biosynthesis and sensitivity mutants of *Arabidopsis* (*det2, dwf4, cpd, bri1*), garden pea (*lka, lk*), and tomato (*d^x, cu-3*) are indicated, although some are putative. Asterisk-marked pathways are hypothetical and lack metabolic evidence.

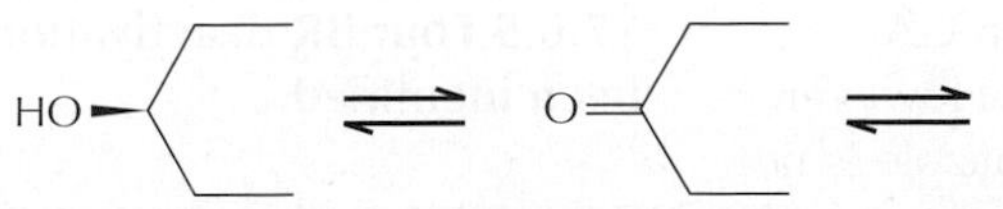

Figure 17.63
Mechanism of epimerization of hydroxyls in BRs. In steroids, $\alpha \to \beta$ and $\beta \to \alpha$ inversions of hydroxyls proceed via ketonic intermediates. This type of reaction is observed in the synthesis of typhasterol from teasterone and in the deactivative metabolism of BRs.

3-dehydro-6-deoxocastasterone (see Fig. 17.62). Tobacco and rice seedlings can also convert 6-deoxocastasterone to castasterone. This route from 5α-campestanol to castasterone is called the late C-6 oxidation pathway, although our knowledge of the steps from 5α-campestanol to 6-deoxoteasterone is tentative. The existence of this alternative pathway to brassinolide is supported indirectly by the observation that several 6-deoxo-BRs exhibit activity in bioassays that use BR-deficient mutants of *Arabidopsis (det2)* and pea *(lkb)*. In the *lkb* mutants of pea, in which biosynthesis of campesterol is blocked (see Fig. 17.61), the amount of 6-deoxocastasterone is reduced 10-fold, an indication that 6-deoxocastasterone and related 6-deoxo-BRs originate from campesterol. The C-6 oxidation in the conversion of 6-deoxocastasterone to castasterone is blocked in the d^x mutant of tomato. The corresponding *DWARF* gene encodes a cytochrome P450 oxidase with a dual role of hydroxylating the C-6 position and then further oxidizing the resulting 6α-hydroxyl group to a keto moiety.

Castasterone to brassinolide

Finally, the 6-oxo group of castasterone is oxidized to a lactone (6-oxo-7-oxa), yielding brassinolide. This reaction has been observed in both cell cultures and seedlings of *Catharanthus* but not in other plant tissues in which castasterone is biologically active. It remains to be determined whether conversion of castasterone to brassinolide is a prerequisite reaction for eliciting biological activity.

Regulation of brassinosteroid biosynthesis

Both the early and late C-6 oxidation BR biosynthesis pathways are operative in *Arabidopsis* and pea; in tomato, only the late C-6 pathway occurs. As yet, little information is available on how the BR biosynthetic pathways are regulated. However, biologically active BRs are responsible for feedback down-regulation of *CPD* gene transcription. Furthermore, BR-insensitive mutants such as pea *lka*, *Arabidopsis bri1*, and tomato *cu-3* (see Fig. 17.62) accumulate biologically active BRs, indicating that perception and signaling also control BR biosynthesis.

17.6.4 Chemical inhibitors have been used to dissect pathways of BR biosynthesis.

The use of inhibitors of plant hormone biosynthesis complements studies with biosynthetic mutants in helping to elucidate metabolic pathways as well as dissecting potential hormonal roles. Triazoles such as uniconazole and paclobutrazol (Fig. 17.64) inhibit *ent*-kaurene synthase, a cytochrome

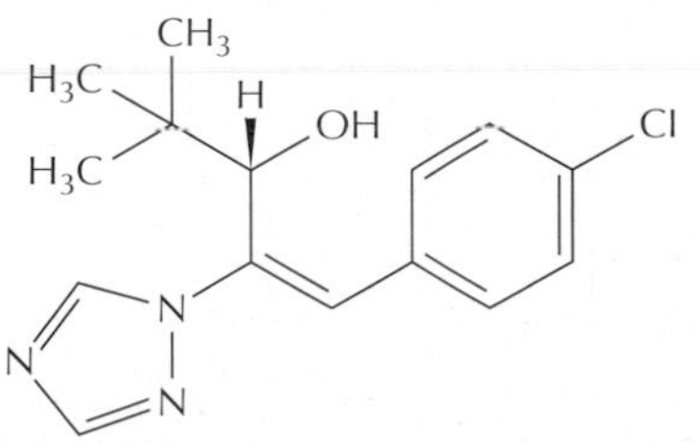

Uniconazole

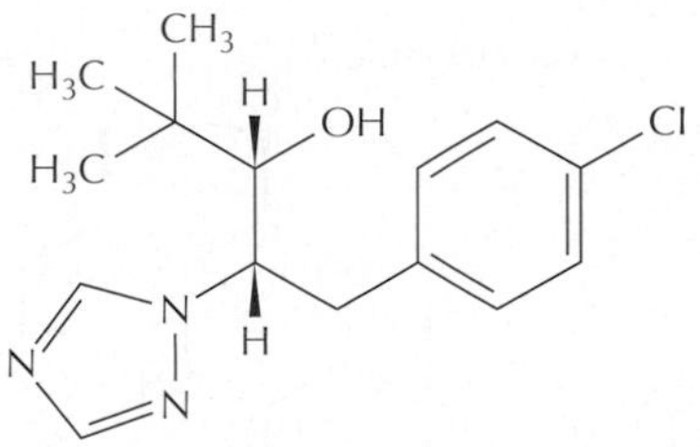

Paclobutrazol

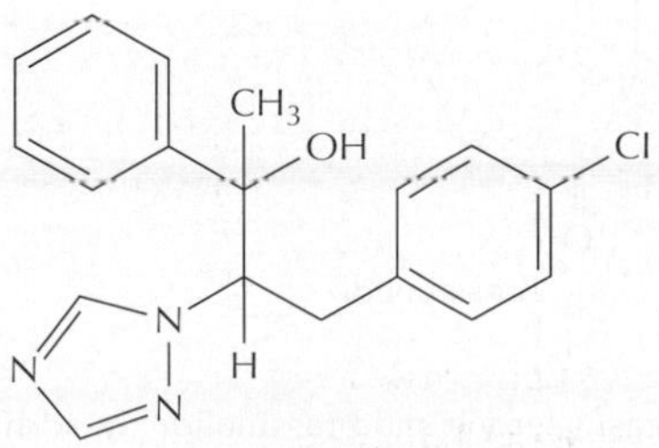

Brassinazole

Figure 17.64
Structures of uniconazole, paclobutrazol, and brassinazole. Uniconazole and paclobutrazol are triazoles that inhibit the GA biosynthesis enzyme *ent*-kaurene oxidase (see Fig 17.12). However, the effects of uniconazole are not specific but also suppress BR biosynthesis. The structurally related compound, brassinazole, is a strong BR inhibitor, blocking at least one cytochrome P450–mediated oxidation step in the BR biosynthesis pathway (see Fig. 17.62). Its effect on GA biosynthesis is undetermined.

P450 monooxygenase involved in GA biosynthesis (see Section 17.1.6). However, new evidence shows that uniconazole is not entirely specific because it also suppresses BR biosynthesis in pea shoots and cultured cells of *Zinnia elegans*. In *Zinnia* cells, BRs are involved in the final stage of tracheid element differentiation, leading to cell death. This process is suppressed by uniconazole. Recent chemical synthesis in which the *tert*-butyl group of uniconazole was replaced by a phenyl group has produced a strong inhibitor of BR biosynthesis, brassinazole (Fig. 17.64). Treatment of *Arabidopsis* seedlings with brassinazole results in dwarfism as well as a deetiolated phenotype in the dark, typical of BR-deficient *Arabidopsis* mutants such as *det2* and *cpd*. Similar effects also have been observed with cress, tomato, and tobacco seedlings. Some cytochrome P450 oxidases are involved in the BR biosynthetic pathway (see Section 17.6.3), and the available evidence suggests that brassinazole blocks at least one of these oxidation steps.

17.6.5 Four BR deactivation routes have been identified.

Research into the metabolism of castasterone and brassinolide has revealed the deactivation steps illustrated in Figure 17.65, although many water-soluble metabolites with as yet undetermined structures also are formed. Metabolism of 24-epicastasterone and 24-epibrassinolide, though rare in plants, has been investigated in the most detail (Figs. 17.66 and 17.67). From these studies, four basic reaction sequences have been identified. Some of these pathways constitute shared events in the metabolism of BRs, ecdysteroids, and animal steroids.

Epimerization of the 2- and 3-hydroxyls, followed by glucosylation or esterification

The α-hydroxyls on the A ring can epimerize to form β-hydroxyls. 3-Epicastasterone is found as a minor metabolite of castasterone in tobacco and rice seedlings (see Fig. 17.65). 3-Epimerization also has been observed after application of 24-epicastasterone and 24-epibrassinolide to cell cultures of tomato (see Fig. 17.66) and serradella *(Ornithopus sativus)*

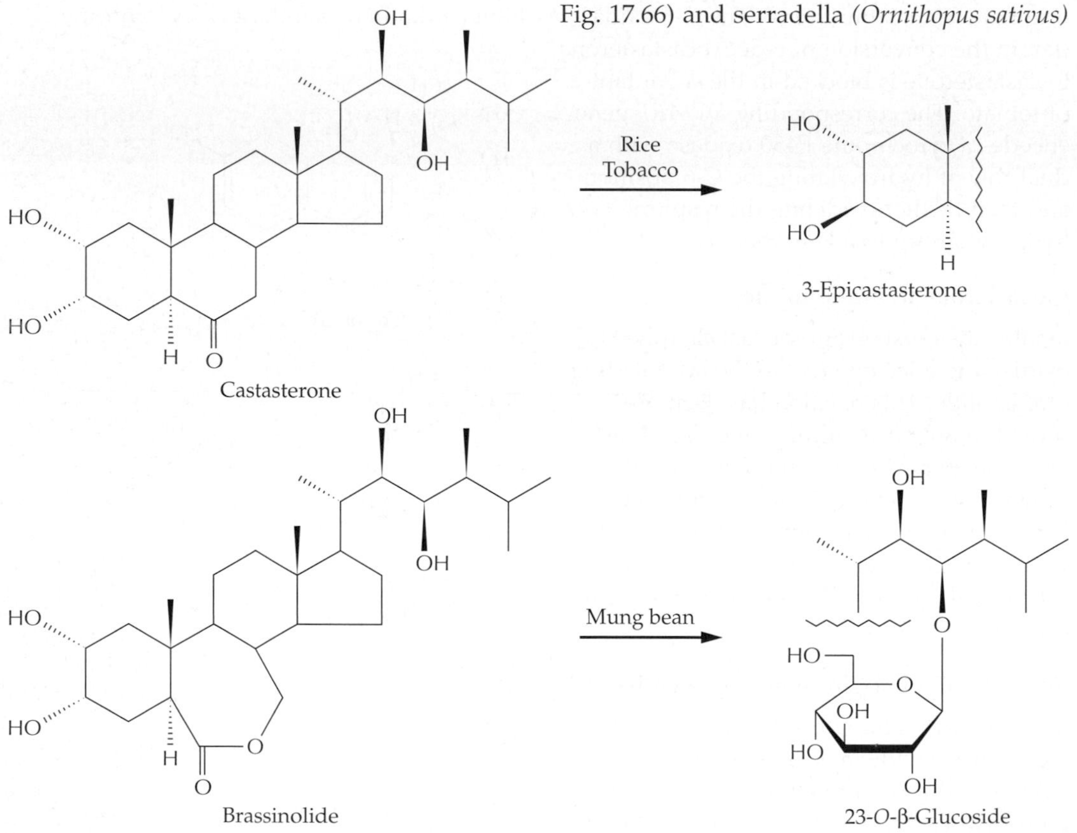

Figure 17.65

Metabolism of castasterone and brassinolide. In addition to the metabolites indicated, water-soluble metabolites are also formed from castasterone and brassinolide, but their structures have not yet been determined.

Chapter 17 Biosynthesis of Hormones and Elicitor Molecules

(see Figs. 17.66 and 17.67). 2-Epimerization is observed when cucumber seedlings metabolize 24-epibrassinolide (Fig. 17.67). Because 2-epicastasterone, 3-epicastasterone, and 2,3-diepicastasterone are major endogenous BRs in *P. vulgaris* seeds, and all are less active biologically than castasterone, epimerization of the 2- and 3-hydroxyls appears to be a general deactivation reaction.

After 3-epimerization, the resulting 3β-hydroxyl group is esterified with lauric acid, myristic acid, or palmitic acid in

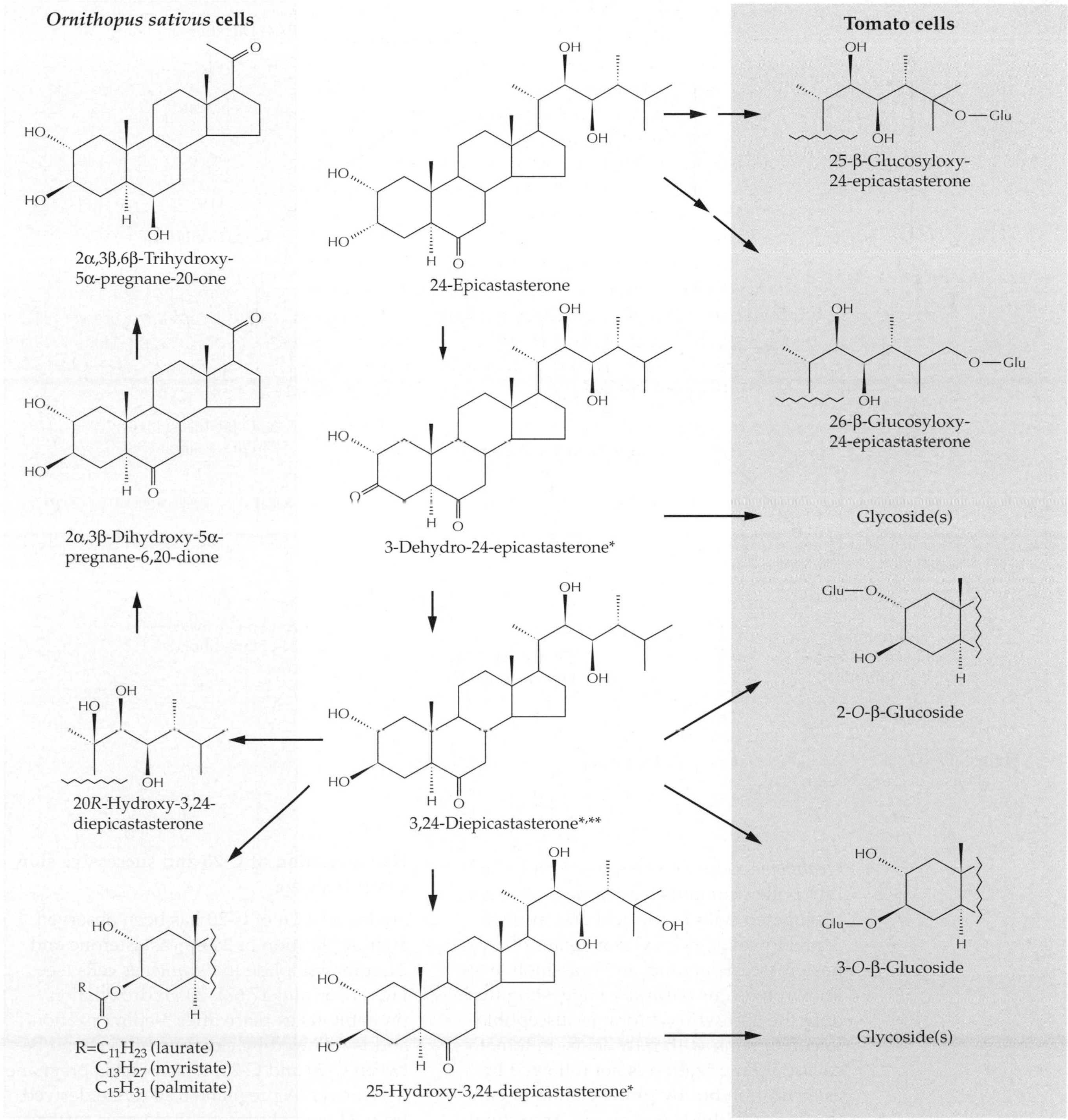

Figure 17.66
Metabolism of 24-epicastasterone. *Present as glycoside(s) in tomato cells. **Identified as the free form in *Ornithopus* cells.

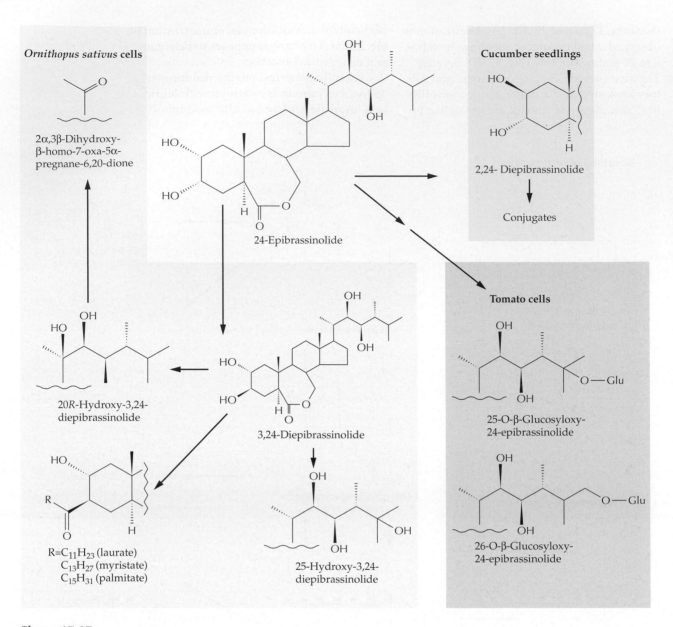

Figure 17.67
Metabolism of 24-epibrassinolide.

Ornithopus cells (see Figs. 17.66 and 17.67). Lily pollen contains teasterone-3-*O*-esters conjugated with lauric acid and myristic acid. However, no acyl conjugates of typhasterol, castasterone, and brassinolide are known to occur naturally, suggesting that only the 3β-hydroxyl form is susceptible to conjugation with fatty acids. In tomato cells, 3-epimerization is not followed by esterification but by glucosylation of either the 2α- or 3β-hydroxyl group. Accordingly, 3-epimerization to form a 3β-hydroxyl group precedes further conjugation.

Hydroxylation of C-20 and successive side chain cleavage

Hydroxylation of C-20 has been observed after application of 24-epicastasterone and 24-epibrassinolide to *Ornithopus* cells (see Figs. 17.66 and 17.67). 20-Hydroxylation probably takes place after 3-epimerization and is followed by cleavage of the bond between C-20 and C-22, which yields pregnane derivatives. A pregnane-6,20-dione derived from 24-epicastasterone undergoes further reduction of the 6-keto group to yield a 6β-hydroxyl moiety.

Glucosylation of C-23 hydroxyl group

In mung bean cuttings, brassinolide is converted almost exclusively to its 23-O-glucoside, whereas castasterone is converted to nonglycosidic metabolites (see Fig. 17.65). 23-O-Glucosides of 25-methyldolichosterone and its 2-epimer are endogenous components in immature seeds of *P. vulgaris*. 23-O-Glucosylation, therefore, may be an important deactivation step at least in some legumes.

Hydroxylation of C-25 and C-26 and subsequent glucosylation

In cultured cells of tomato, 24-epimers of castasterone and brassinolide are not metabolized to 23-O-glucosides. Instead, the C-25 or C-26 positions are subjected to hydroxylation, followed by glucosylation (see Figs. 17.66 and 17.67). Thus, modification of the BR side chain may be determined by whether a 24α- or 24β-methyl group is present. Other possibilities cannot be ruled out, however, because metabolism of the α- and β-isomers has not been examined in the same plant species. 25-Hydroxy-3,24-diepicastasterone is derived from 24-epicastasterone in tomato cells, whereas 25-hydroxy-3,24-diepibrassinolide is produced from 24-epibrassinolide in *Ornithopus* cells.

In the rice lamina inclination assay (see Fig. 17.58), 25-hydroxylation reportedly increases the biological activity of 24-epibrassinolide 10-fold, whereas 26-hydroxylation reduces activity. When applied to BR-deficient mutants of *Arabidopsis*, however, 25-hydroxy compounds are inactive. 25-Hydroxybrassinolide has not been detected in plants and, among naturally occurring BRs, brassinolide exhibits the most biological activity.

17.7 Polyamines

Polyamines are found in all organisms and have been studied in animals and bacteria for more than 50 years. However, their significance in plants has been recognized only recently. Their concentrations correlate with cell division frequency, and polyamines stimulate many reactions involved in the synthesis of DNA, RNA, and proteins. Because of their polycationic nature, polyamines have a high affinity for anionic constituents such as DNA, RNA, phospholipids, and acidic proteins as well as anionic groups in membranes and cell walls. Polyamines are essential for the growth and development of both prokaryotic and eukaryotic cells: Without the ability to synthesize polyamines, living cells would not survive. In plants, polyamines elicit diverse physiological responses, including cell division, tuber formation, root initiation, embryogenesis, flower development, and fruit ripening. Polyamines are much more abundant in plants than are hormones such as GAs and cytokinins, and millimolar amounts of polyamines are required to induce a biological response. Furthermore, the polycationic character of polyamines appears to limit their translocation. On balance, polyamines do not appear to have a truly hormonal role in plants. Instead, as in animals, they appear to participate, directly or indirectly, in several key metabolic pathways essential for efficient functioning at the cellular level.

The polyamines found most frequently in plants and other organisms are putrescine, spermidine, and spermine. Although cadaverine (1,5-diaminopentane) is present much less than the diamine putrescine, it is a common constituent of legumes. Plant polyamines occur both as free amines and as amide conjugates of hydroxycinnamic acids such as *p*-coumaric acid, ferulic acid, and caffeic acid (see Chapter 24). Not only do these conjugates represent a significant portion of the total polyamine pool, they also play important roles in flower, seed, and fruit development, and the hypersensitive response to microbial infection.

17.7.1 Amino acid decarboxylases are active in polyamine biosynthesis.

The same polyamine biosynthetic pathway occurs in plants, microorganisms, and mammals (Fig. 17.68). Putrescine is synthesized from L-arginine by two routes, one involving L-ornithine, the other by way of agmatine and N-carbamoylputrescine. Arginine decarboxylase, which converts L-arginine to agmatine, and ornithine decarboxylase, which converts L-ornithine to putrescine, have been studied extensively with biosynthes mutants. Researchers have also used the enzyme

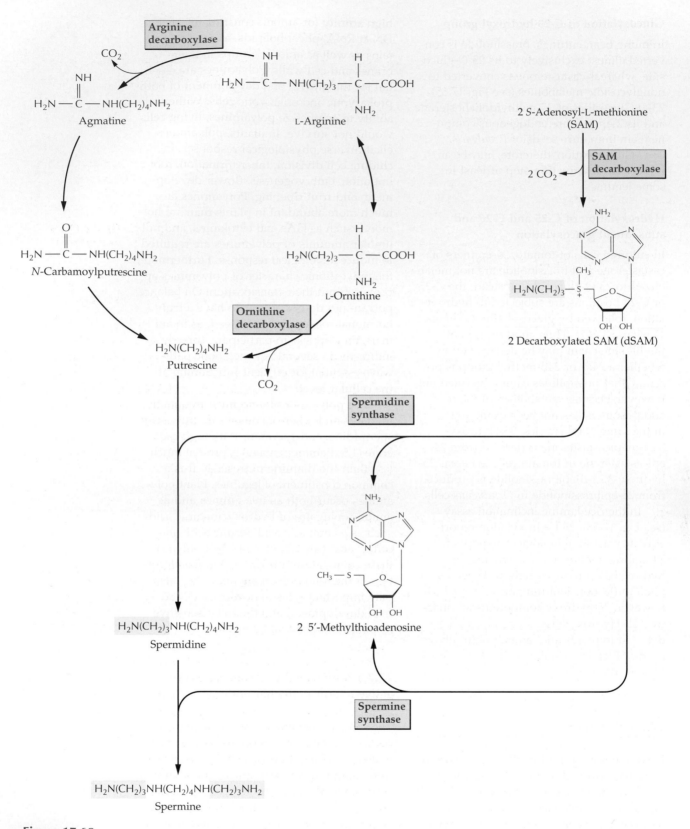

Figure 17.68
Biosynthesis of putrescine, spermidine, and spermine. Arginine decarboxylase is inhibited by α-difluoromethylarginine, and ornithine decarboxylase is inhibited by α-difluoromethylornithine. In most plants arginine decarboxylase is more active than ornithine decarboxylase.

inhibitors α-difluoromethylarginine and α-difluoromethylornithine, which inhibit arginine decarboxylase and ornithine decarboxylase, respectively.

In plants, ornithine decarboxylase is usually less active than arginine decarboxylase. Despite some conflicting evidence, ornithine decarboxylase enzymes from barley, jute, and tomato generally exhibit end-product inhibition in the presence of putrescine, spermidine, and spermine. However, arginine decarboxylase enzymes from oat, rice, cucumber, and *Lathyrus sativus* are inhibited by agmatine but are relatively insensitive to polyamines. In most tissues, arginine decarboxylase activity is modulated by light and hormones. Environmental stresses increase arginine decarboxylase activity, up-regulating putrescine production. Putrescine toxicity is presumably one of the causes of stress symptoms. There is usually a positive correlation between ornithine decarboxylase activity and cell division.

Cadaverine synthesis from L-lysine is catalyzed by lysine decarboxylase (Fig. 17.69). Lysine decarboxylase activity is low in most plant tissues; in some leguminous species, however, lysine decarboxylase activity increases with the accumulation of quinolizidine alkaloids, which are synthesized from cadaverine. In *L. sativus*, cadaverine is produced either from lysine or from L-homoarginine by way of homoagmatine. Both conversions are catalyzed by a single decarboxylase that is distinct from arginine decarboxylase and lysine decarboxylase (Fig. 17.69).

Spermidine and spermine appear to be required for the G_1 to S transition in cell division, and the conversion of putrescine to spermidine is thought to be important in controlling the rate of cell division. Spermidine synthase catalyzes the conversion of putrescine to spermidine, and spermine synthase regulates the conversion of spermidine to spermine (see Fig. 17.68). Both enzymes transfer an aminopropyl group from decarboxylated *S*-adenosyl-L-methionine (dSAM) to their respective substrates. These enzymes exhibit narrow substrate specificities; e.g., spermidine synthase from maize exhibits no spermine synthase activity. However, the aminopropyltransferases associated with the synthesis of some of the less common polyamines may exhibit broader substrate specificities.

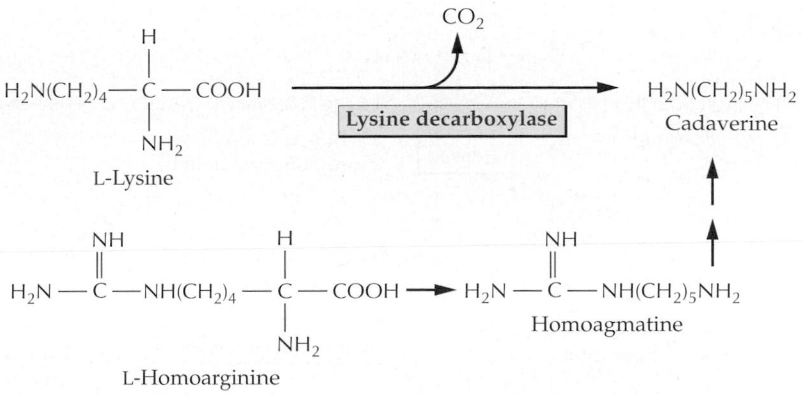

Figure 17.69
Biosynthesis of cadaverine.

S-Adenosyl-L-methionine decarboxylase (SAM decarboxylase) converts SAM to dSAM, which is the rate-limiting step in the polyamine biosynthetic pathway. The synthesis of dSAM is inhibited by spermidine but increases in response to increasing putrescine. Both ethylene and polyamines are derived from SAM, so competition for SAM is possible. Because polyamines and ethylene inhibit each other's biosynthesis, control of the fate of SAM may significantly influence plant growth (see Section 17.5.4).

17.7.2 Two classes of oxidative enzymes participate in polyamine catabolism.

During the catabolism of polyamines, the terminal (primary) amino groups are subjected to oxidative deamination by copper-containing diamine oxidases (Fig. 17.70). Aminoaldehydes are produced from putrescine, cadaverine, and spermidine with the coproduction of hydrogen peroxide and ammonia by these reactions. The amino-aldehydes derived from putrescine and cadaverine cyclize spontaneously to yield 1-pyrroline and 1-piperideine, respectively, whereas the aminoaldehyde from spermidine forms an equilibrium mixture of the cyclic compounds 1-(3-aminopropyl) pyrrolinium and 1,5-diazabicyclononane. Both the terminal amino groups of spermine are oxidized to a dialdehyde (Fig. 17.70). In general, plant diamine oxidases are most active with putrescine and cadaverine as substrates.

A second class of plant enzymes, polyamine oxidases, catabolize both spermidine

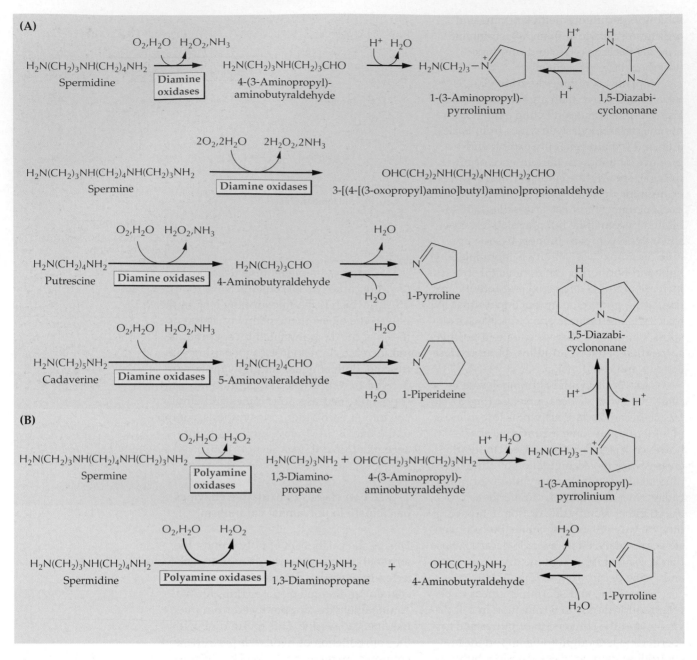

Figure 17.70
(A) Catabolism of polyamines by diamine oxidases. (B) Catabolism of polyamines by polyamine oxidases.

and spermine but not other amines. These flavin enzymes oxidize secondary amino groups, forming hydrogen peroxide, 1,3-diaminopropane, and aminoaldehydes. The aminoaldehydes derived from spermidine and spermine are 4-aminobutyraldehyde and 4-(3-aminopropyl)aminobutyraldehyde. These compounds are also derived from putrescine and spermidine, respectively, in reactions catalyzed by diamine oxidases (Fig. 17.70A).

In addition to regulating polyamine concentrations, catabolic oxidation of polyamine amino groups may play other significant roles; for example, the hydrogen peroxide generated by these oxidases could be used in lignifying cell walls. Polyamines are precursors of various important alkaloids, and diamine oxidases participate in the biosynthesis of alkaloids containing a heterocyclic ring (see Chapter 24). Genes encoding diamine oxidases have been isolated from lentil.

17.7.3 Transgenic plants that express antisense polyamine biosynthesis genes have pleiotropic phenotypes.

Genes encoding arginine decarboxylase, ornithine decarboxylase, SAM decarboxylase, and SAM synthase have been cloned from plant and animal sources and may provide the means to manipulate polyamine biosynthesis by using sense and antisense transgenic approaches. A mouse ornithine decarboxylase gene has been expressed in carrot and, under in vitro culture conditions, promotes an increase in putrescine production, resulting in a high degree of somatic embryogenesis. Antisense constructs containing potato SAM decarboxylase cDNA under the control of the CaMV 35S promoter have been used to create transgenic potatoes with reduced SAM decarboxylase activity, diminished amounts of putrescine, spermine, and spermidine, and enhanced rates of ethylene biosynthesis (Table 17.2). The antisense SAM decarboxylase plants have a stunted phenotype with highly branched stems, short internodes, small chlorotic (yellowed) leaves, and inhibited root growth. In addition, the plants do not flower and the tubers they produce are small and elongated (Fig. 17.71). Further study with transgenic plants in which ethylene biosynthesis also is blocked is needed to determine which of these effects result from reduced polyamine content and which are a consequence of increase concentrations of ethylene.

Figure 17.71
Phenotypic comparison between wild-type and SAM decarboxylase antisense transgenic potato plants. Phenotypes of (A) a normal potato plant grown for five weeks in a greenhouse; (B) 35S SAM decarboxylase antisense potato plants (line A9) grown under the same conditions as the control plants and showing stunted growth characteristics with highly branched stems, small leaves, and early senescence; (C) normal potato tubers; and (D) tubers of antisense A9 plants, which are small and elongated compared with those of control plants.

17.8 Jasmonic acid

Jasmonic acid (JA) and related compounds, initially isolated as inhibitors of growth, have now been implicated in various physiological processes, including defense responses. JA is similar structurally to prostaglandins, autacoidal hormones that have a variety of physiological activities in mammals. Both JA and prostaglandins are derived from fatty acids. Methyl jasmonate and the structurally related compound, *cis*-jasmone, are well known in the perfume industry as fragrant components of the essential oils of jasmine (*Jasminum grandiflorum*).

In 1971, JA was isolated from culture filtrates of a fungus, *Botryodiploidi theobromae*, as a plant growth inhibitor. Cucurbic acid was isolated from immature pumpkin seeds

Table 17.2 Analysis of SAM decarboxylase antisense potato plantlets (line A9) grown in tissue culture

Potato lines	SAM decarboxylase activity (mmol CO_2 per gram fresh weight per hour)	Polyamine content (µg per gram fresh weight)			Ethylene biosynthesis (pmol per gram fresh weight per hour)
		Putrescine	Spermidine	Spermine	
Control	0.116	17.4	33.0	19.1	90.5
A9	0.012	2.8	6.3	6.2	4129.1

All values represent the mean of three independent assays.

in 1977 and identified as a new type of growth inhibitor, different from ABA. In the early 1980s, JA and methyl jasmonate were detected as senescence-promoting or growth-retarding substances in many plant species, including *Artemisia absinthium* (wormwood), *Vicia faba, P. vulgaris, Dolichos lablab,* and *Castanea crenata.* In 1989, the 12-*O*-β-glucoside of tuberonic acid (12-hydroxyjasmonic acid) was characterized as a potato tuber–inducing factor.

(–)-JA and (–)-methyl jasmonate are the major jasmonates in plant tissues. Their naturally occurring stereoisomers, (+)-7-iso-jasmonic acid and methyl (+)-7-isojasmonate, also appear to be active agents in planta because they exhibit higher biological activity in some, but not all, test systems. In vitro, (+)-7-isojasmonic acid is unstable and is isomerized by treatment with acid, base, or heat to form a 9:1 equilibrium mixture of (–)-JA: (+)-7-isojasmonic acid. To what extent this chemical process is involved in the conversion of (+)-7-isojasmonic acid to (–)-JA in plants is unclear. Whether (–)-JA occurs naturally or is an artifact derived from (+)-7-isojasmonic acid has been the subject of some debate. However, identification of the *trans*-alkyl compound, 6-*epi*-7-isocucurbic acid (Fig. 17.72), in plant extracts supports the view that (–)-JA is an endogenous constituent of plant tissues.

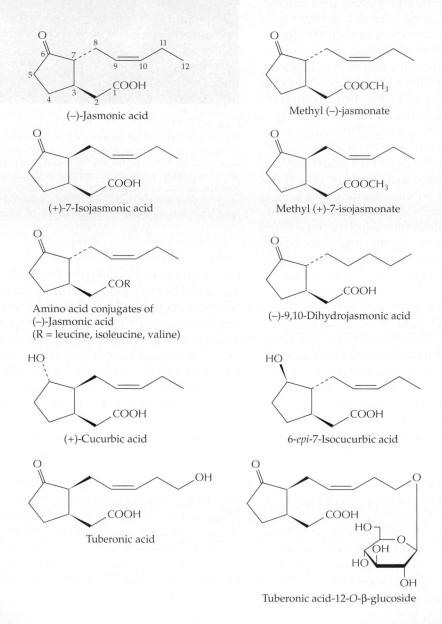

Figure 17.72
Structures of (–)-jasmonic acid and related compounds.

Chapter 17 Biosynthesis of Hormones and Elicitor Molecules

JA and related compounds that have been isolated are classified according the stereochemistry of the alkyl substituents. (+)-7-Isojasmonic acid derivatives have *cis*-alkyl substituents, whereas (–)-JA–related compounds have *trans*-alkyl groups, as shown in Figure 17.72, which illustrates the structures of some of the jasmonates identified as endogenous compounds or metabolites in higher plants and fungi.

17.8.1 (–)-Jasmonic acid inhibits growth processes in many tissues and is active in reproductive development and pathogen resistance.

Exogenous (–)-JA was first reported to inhibit the growth of rice, wheat, and lettuce seedlings; it was subsequently shown to inhibit seed and pollen germination, retard root growth, and promote the curling of tendrils. (–)-JA content is high in vegetative sink tissues, and the compound regulates accumulation of vegetative storage proteins during seed development in soybean and *Arabidopsis*. (–)-JA induces fruit ripening and pigment formation, processes in which ethylene may participate, because ACC oxidase is induced by (–)-JA. The relatively high concentrations of (–)-JA in developing reproductive tissues of plants implies that it plays some role in the formation of flowers, fruit, and seeds. Application of (–)-JA to leaves decreases expression of the nuclear and chloroplast genes involved in photosynthesis.

(–)-JA plays an important role in plant resistance to insects and disease. Plants accumulate (–)-JA when they are wounded or when they are treated with pathogen-derived elicitors (oligosaccharides) or the peptide systemin. (–)-JA activates expression of antifungal proteins such as osmotin and thionin and induces the phytoalexin-related enzymes chalcone synthase, phenylalanine ammonia lyase, and hydroxymethylglutaryl-CoA reductase (see Chapter 20). Induction of protease inhibitors, which appear to be targeted primarily toward certain insects, has been noted in several species. When tomato is treated with (–)-JA, its resistance to *Phytophthora infestans* increases, and mutants of tomato defective in (–)-JA biosynthesis are more susceptible to this pathogen. Just as wounding a plant induces a systemic response that alters the undamaged leaves, exogenous (–)-JA induces proteinase inhibitors in distant, untreated leaves.

In short, (–)-JA modulates the expression of numerous genes involved in both plant development and plant defense. Further analysis of the functions of these genes should reveal the roles of (–)-JA in plants in more detail.

17.8.2 (–)-Jasmonic acid is synthesized from α-linolenic acid.

The starting material for (–)-JA biosynthesis is α-linolenic acid, a C_{18} poly-unsaturated fatty acid synthesized in plants from membranes in reactions catalyzed by enzymes with properties similar to those of lipase-like enzymes involved in fatty acid metabolism in mammals (Fig. 17.73). Several key enzymes in the (–)-JA biosynthetic pathway—lipoxygenase, allene oxide synthase (AOS), and allene oxide cyclase—have transit peptides for chloroplast import, so a chloroplast-localized lipase is probably involved in α-linolenic acid production. α-Linolenic acid is oxidized to 13(S)-hydroperoxylinolenic acid by lipoxygenase.

13(S)-Hydroperoxylinolenic acid is converted to 12,13(S)-epoxylinolenic acid by AOS. 12,13(S)-Epoxylinolenic acid is rapidly cyclized by the enzyme allene oxide cyclase, yielding 12-oxo-*cis*-10,15-phytodienoic acid (12-oxo-PDA), a cyclopentenone derivative. In turn, 12-oxo-PDA is reduced by 12-oxo-PDA reductase, resulting in the formation of 3-oxo-2-(*cis*-2′-pentenyl)cylcopentane-1-octanoic acid, which is converted to (+)-7-isojasmonic acid by three rounds of β-oxidation (see Fig. 17.73 and Chapter 10). (+)-Isojasmonic acid is converted to (–)-JA, but as mentioned earlier, it is unclear to what extent this is an enzymatically regulated step or a spontaneous chemical isomerization. (–)-JA is converted to methyl (–)-jasmonate. Although the free acid predominates in most species, the methyl ester is volatile and contributes to the distinctive aroma of the flowers and fruits of species such as jasmine. Some have proposed that methyl (–)-jasmonate acts as an airborne signal in interplant communication, being released in response to wounding and other stress situations.

17.8.3 AOS is an unusual and important enzyme in biosynthesis of (–)-jasmonic acid.

AOS is a key enzyme in (–)-JA biosynthesis, catalyzing the first step specific to the pathway. AOS, which can be induced by treatment with ethylene and salicylic acid, has been purified and cloned from flax seed, guayule rubber particles, and *Arabidopsis*. This chloroplast-localized cytochrome P450 enzyme has several unusual characteristics: It does not require oxygen and is therefore not an oxidase, nor does it require NADPH or P450 reductase; it also has a low affinity for carbon monoxide. Moreover, the heme-binding domain of AOS has a sequence that differs from the consensus sequence associated with most cytochrome P450 enzymes: The conserved threonine that is thought to mediate O_2 binding is lost. The amino acid sequences deduced from the flax seed and *Arabidopsis* cDNAs include an N-terminal transit peptide that is lacking in the guayule enzyme. That AOS catalyzes the rate-limiting step in (–)-JA biosynthesis is shown by the fact that overexpression of the flax enzyme in transgenic potato plants increases the concentrations of (–)-JA.

The promoter region of the *AOS* gene has been identified in *Arabidopsis,* and a chimeric gene fusion containing the *E. coli* β-glucuronidase (GUS) reporter gene, *uid*A, under the control of the *AOS* promoter, has been introduced into tobacco and *Arabidopsis*. Histochemical studies of GUS activity in these transgenic plants indicate that the *AOS* promoter is subject to developmental control. GUS activity has been detected in older leaves, at the base of petioles and stipules, during the early stages of carpel development, and in maturing pollen grains. After the shedding of floral organs, GUS activity is also present in abscision zone scars. These observations suggest a role for (–)-JA in pollen maturation and floral organ abscision.

Wounding a leaf leads to an increased accumulation of (–)-JA as well as increases in AOS transcript levels and AOS activity. Wounding a single leaf of the transgenic plants that express GUS under the control of the *AOS* promoter results in both local and systemic expression of the *AOS* promoter. In tobacco, 2 hours after wounding, a diffuse zone of GUS activity has developed

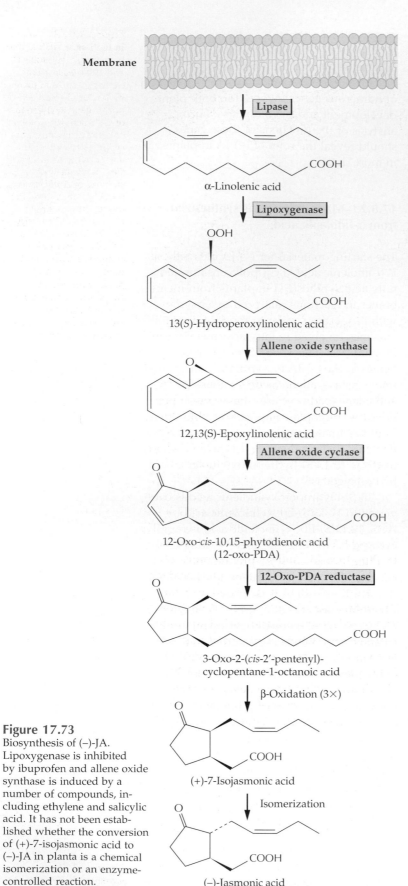

Figure 17.73
Biosynthesis of (–)-JA. Lipoxygenase is inhibited by ibuprofen and allene oxide synthase is induced by a number of compounds, including ethylene and salicylic acid. It has not been established whether the conversion of (+)-7-isojasmonic acid to (–)-JA in planta is a chemical isomerization or an enzyme-controlled reaction.

(A)

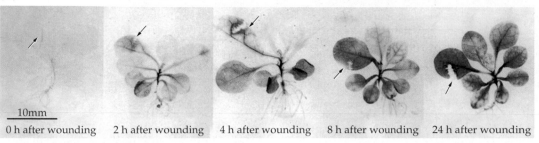

| 0 h after wounding | 2 h after wounding | 4 h after wounding | 8 h after wounding | 24 h after wounding |

10mm

(B)

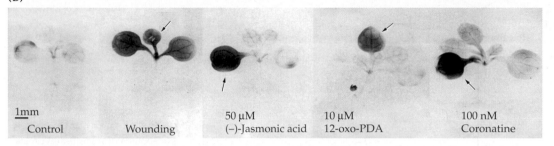

| Control | Wounding | 50 μM (–)-Jasmonic acid | 10 μM 12-oxo-PDA | 100 nM Coronatine |

1mm

Figure 17.74
Induction of GUS activity in transgenic tobacco and *Arabidopsis* plants carrying the AOS promoter fused to a GUS reporter gene. (A) Kinetics of GUS activity in tobacco after wounding of a single leaf (indicated by an arrow) with a hemostat. (B) Pattern of GUS staining of *Arabidopsis* seedlings 16 hours after treatment of a single leaf with 50 μM (–)-JA, 10 μM 12-oxo-PDA, or 100 nM coronatine, compared with an untreated control and a plant wounded at a single leaf with a hemostat 16 hours earlier. Treated leaves are indicated with arrows.

around the wound site, and after 24 hours the AOS promoter is activated throughout the shoot (Fig 17.74A).

In *Arabidopsis*, AOS is induced by 12-oxo-PDA and (–)-JA as well as coronatine (Fig. 17.75), an octadecanoid analog. These compounds also activate the *AOS* promoter when applied locally to a leaf, but in contrast to the wound stimulus, only a local response of the treated leaf is observed (Fig. 17.74B), showing that none of the compounds is transported from the site of application. Thus, although 12-oxo-PDA and (–)-JA are involved in the local response to wounding, they are not the signal molecules responsible for the induction of systemic defense responses.

17.8.4 (–)-Jasmonic acid and (–)-9,10-dihydrojasmonic acid are metabolized to hydroxylated and conjugated products.

Feeding experiments have been carried out with radiolabeled (–)-JA and its dihydro derivative, (–)-9,10-dihydrojasmonic acid. (–)-JA applied to potato leaves undergoes hydroxylation and glucosylation, being converted to tuberonic acid-12-O-β-glucoside (see Fig. 17.72), which is thought to be a tuber-forming substance. Subjecting barley

leaves to osmotic stress results in increases in the endogenous pools of not only (–)-JA but also its conjugates *N*-(–)-jasmonyl-*(S)*-valine, *N*-(–)-jasmonyl-*(S)*-leucine, and *N*-(–)-jasmonyl-*(S)*-isoleucine (see Fig. 17.72). This suggests that both enhanced synthesis of (–)-JA and its conversion to amino acid conjugates occur readily under stress conditions.

Excised barley shoots converted radiolabeled (–)-9,10-dihydrojasmonic acid to (–)-11-hydroxy-9,10-dihydrojasmonic acid and its 11-O-β-glucoside and smaller amounts of (–)-12-hydroxy-9,10-dihydrojasmoic acid via the routes indicated in Figure 17.76. Minor amounts of various amino acid conjugates of 9,10-dihydrojasmonic acid and 11-hydroxy-9,10-dihydrojasmonic acid, but of undetermined stereochemistry, also were formed. In the presence of UDP-glucose, (–)-9,10-dihydrojasmonic acid is converted to its β-glucosyl ester in cell-free preparations from tomato cell suspension cultures (Fig. 17.76).

Coronatine

Figure 17.75
Structure of coronatine.

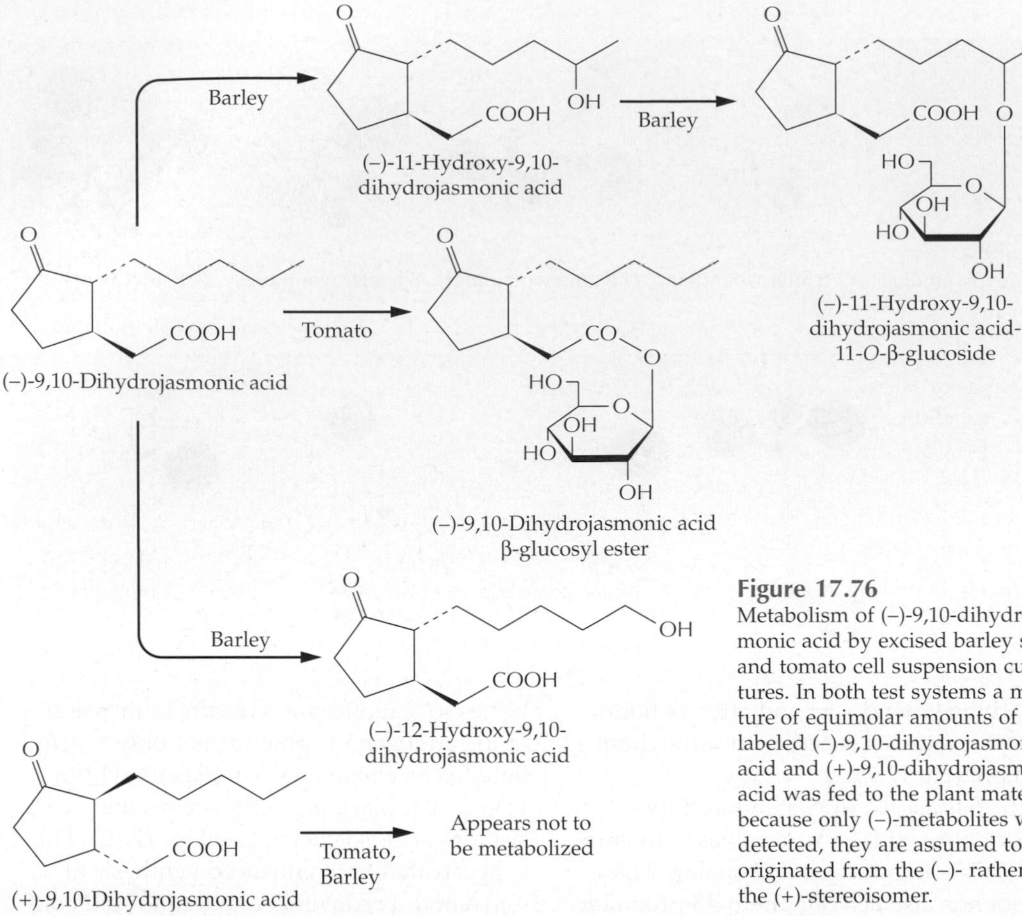

Figure 17.76
Metabolism of (–)-9,10-dihydrojasmonic acid by excised barley shoots and tomato cell suspension cultures. In both test systems a mixture of equimolar amounts of radio-labeled (–)-9,10-dihydrojasmonic acid and (+)-9,10-dihydrojasmonic acid was fed to the plant material; because only (–)-metabolites were detected, they are assumed to have originated from the (–)- rather than the (+)-stereoisomer.

17.9 Salicylic acid

Salicylic acid (SA) (Fig. 17.77) is far better known for its medicinal properties (Box 17.1) than for its regulatory roles in plants. Although many physiological effects of SA application have been reported, only a few have been shown to be physiologically important in plants, for example, the regulatory role of SA in thermogenesis in lilies and in the response of plants to pathogens.

17.9.1 Salicylic acid can retard senescence of petals and induce flowering.

Soluble aspirin tablets disintegrate rapidly in water, and the acetylsalicylic acid is converted to SA. Adding aspirin to a vase of water containing cut flowers will retard senescence of the petals and make the flowers last longer. This is probably a consequence of reduced rates of ethylene biosynthesis, in light of reports that SA blocks the conversion of ACC to ethylene. SA also induces flowering in the duckweeds *Lemna gibba*, *Spirodela polyrrhiza*, and *Wolffia microscopica* when these long-day plants are grown under a noninductive short-day photoperiod (see Chapter 19). This effect is not specific, however, and related phenolics also can stimulate flowering in these and other species. Furthermore, because the SA concentrations are similar in vegetative and induced tissues, the compound is unlikely to be an endogenous flower-inducing signal.

17.9.2 Salicylic acid regulates thermogenesis in voodoo lily.

Clear evidence linking endogenous SA with a regulatory role in plants was first obtained during studies on thermogenesis in flowers of voodoo lily *(Sauromatum guttatum)*. During thermogenesis, much of the electron flow is diverted from the cytochrome respiration pathway to the cyanide-insensitive

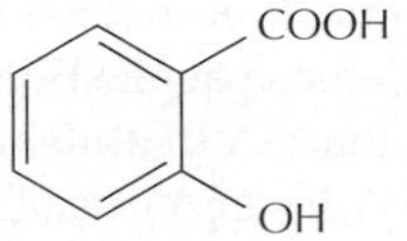

COOH

OH

Salicylic acid

Figure 17.77
Structure of salicylic acid.

nonphosphorylating electron transport pathway that is unique to plant mitochondria (see Chapter 14). Also, the glycolytic pathways and Krebs cycle enzymes, which provide substrates for the alternative oxidase, are activated. The energy released by electron flow through this alternative respiratory pathway is not conserved as chemical energy but is released as heat. The flower of the voodoo lily develops from a corm and can reach more than 70 cm high (see Box 14.2). Early on the day of anthesis, the spathe unfolds; shortly afterwards, the upper spadix begins to generate heat, volatilizing pungent amines and indoles that attract insect pollinators. The temperature of the spadix can increase to 14°C above ambient by early after-

noon but falls to normal values by evening. A second burst of thermogenicity begins in the lower spadix in the late evening and ends early the next morning after an increase in temperature of more than 10°C (Fig. 17.78). In his Ph.D. thesis in 1937, Adriaan van Herk proposed that thermogenesis in the voodoo lily was initiated by what he called "calorigen," a water-soluble substance produced in the staminate flower primordia positioned between the upper and lower spadix. In 1987, calorigen was identified as

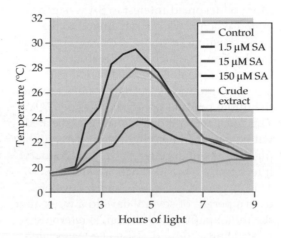

Figure 17.78
The response of the *Sauromatum guttatum* appendix (upper spadix) to application of SA or a crude extract of male flowers of the same species. Both exogenous treatments result in a marked increase in appendix temperature.

Box 17.1

Salicylic acid has been used as an analgesic for more than 2000 years.

Since the 4th century BC, plants such as willow, myrtle, poplar, and meadowsweet have been used to relieve pain caused by a variety of conditions, including diseases of the eye, rheumatism, childbirth, and fever. The active component remained unknown until the 19th century, when SA and related compounds, such as methyl salicylate, saligenin, and their glycosides, were isolated from willow and other plant sources. Oil of wintergreen, extracted from the American plant *Gaultheria procumbens,* was an especially rich source of methyl salicylate and was a widely used analgesic during the mid-19th century.

In 1858, SA was chemically synthesized in Germany, and its commercial production began in 1874 at one-tenth the cost of making wintergreen oil. However, the sharp bitter taste and gastric irritation associated with salicylic acid made long-term use difficult. This problem was overcome in the 1890s, when clinical tests carried out by the German pharma-

ceutical company Frederick Bayer and Co. showed that these side effects could be avoided by replacing SA with its acetylated derivative. In 1898, acetylsalicylic acid was given the trade name aspirin. Within one year of its launch, aspirin began to be manufactured in tablet form, rather than as a powder, and sales soon swelled to amazing proportions.

More than 100 years later, aspirin tablets are still sold widely as a ready remedy for many conditions, including the common cold, headache, fevers, and rheumatoid arthritis. Aspirin is also increasingly being used to prevent and treat heart attacks and cerebral thrombosis. The beneficial effects are ascribed to acetylsalicylic acid retarding the production of prostaglandins, which promote blood clotting. Moreover, a growing number of discerning airline passengers are taking a precautionary aspirin tablet prior to departure to lessen the risks, albeit relatively small, of blood clots and coronary problems associated with long flights. The

current world production is many thousands of tons per year, and average consumption in developed countries is about 100 tablets per person per annum.

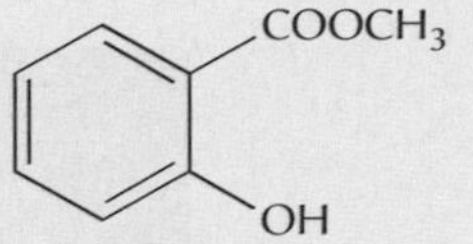

COOCH₃

OH

Methyl salicylate

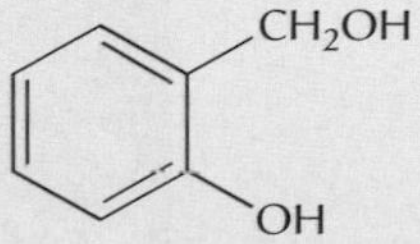

CH₂OH

OH

Saligenin

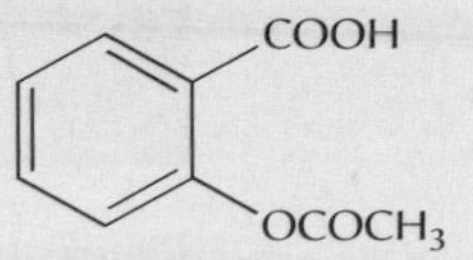

COOH

OCOCH₃

Acetylsalicylic acid

salicylic acid when it was shown that treating immature spadix sections with as little as 0.12 µg of SA per gram fresh weight caused their temperatures to rise by as much as 12°C. Large increases in endogenous concentrations of SA occur immediately before the onset of thermogenicity in both the upper and lower spadix.

17.9.3 Salicylic acid production is associated with disease resistance.

More widespread interest in SA was generated when it was closely linked to the hypersensitive response, a disease-resistance mechanism in which plants restrict the spread of fungal, bacterial, or viral pathogens by producing necrotic lesions around the initial point of penetration (Fig. 17.79). This localized cell death often is associated with changes in healthy, distant parts of the plants that enhance resistance to secondary infection by a broad range of pathogens. This increased resistance, which develops over a period of several days to a week after the initial pathogen invasion, is referred to as systemic acquired resistance. Associated with the hypersensitive response and systemic acquired resistance is the synthesis of five or more families of serologically distinct, pathogenesis-related (PR) proteins. Two of these families encode the hydrolytic β-1,3-glucanases and chitinases. Although the precise function of the other families of PR proteins is not well understood, their expression is associated with resistance to a large number of viral, fungal, and bacterial pathogens.

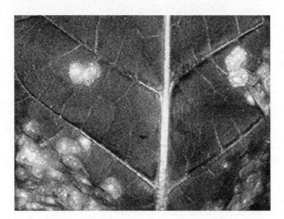

Figure 17.79
Necrotic lesions on the leaf of resistant tobacco formed in response to infection with TMV.

The initial indication that SA was involved in plant defense against disease came with the finding that SA treatment of the leaves of Tobacco Mosaic Virus (TMV)-susceptible tobacco (*N. tabacum* cv. Xanthi-*n*) induced the accumulation of PR proteins and enhanced resistance to TMV infection. It has now been shown that SA treatment also induces acquired resistance against many pathogens in a variety of plants. In the TMV-resistant tobacco cultivar Xanthi-*nc,* but not the susceptible Xanthi-*n,* the endogenous SA pool increases about 40-fold in TMV-inoculated leaves and about 10-fold in uninfected leaves of the same plant. The steep increase in SA content parallels or precedes the induction of PR protein gene expression in both inoculated and uninoculated leaves of TMV-resistant plants. Endogenous SA, therefore, appears to play a key role in the signal transduction pathway that leads to activation of genes encoding PR proteins and in the establishment of the hypersensitive response and systemic acquired resistance. Studies with cucumber, potato, and other species indicate that the involvement of SA in these responses is not unique to tobacco.

Despite arguments that SA acts as the primary long-distance signal that is translocated in the phloem from the site of pathogen invasion to uninfected leaves, there initiating the development of the systemic acquired resistance response, grafting studies with wild-type and TMV-susceptible, transgenic Xanthi-*nc* tobacco plants have refuted this hypothesis. The transgenic plants express the *Ps. putida nahG* gene, the product of which catalyzes the conversion of salicylic acid to catechol and thereby prevents accumulation of SA (Fig. 17.80). This research indicates that local infection leads to the production of an unidentified mobile signal that is transported to distal tissues, where it initiates the accumulation of SA required for induction of systemic acquired resistance. In Xanthi-*nc* tobacco, this signal has recently been proposed to be methyl salicylate, which is produced from SA in TMV-infected leaves. Unlike SA, methyl salicylate is volatile, which, it was suggested, could thus act as an airborne signal to induce PR protein accumulation and disease resistance in both neighboring plants and the healthy tissues of the infected plant. Fascinating as this proposal is, it does not explain the results of

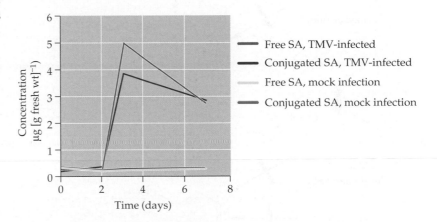

Figure 17.80
The *Pseudomonas putida nahG* gene product catalyzes the metabolism of SA to catechol.

grafting experiments, which have shown that TMV-inoculated transgenic *nahG* tobacco leaves that are deficient in salicylic acid, and therefore in methyl salicylate, produce a long-distance signal that can pass through a graft union and establish systemic acquired disease resistance and PR protein gene expression in leaves of an uninfected wild-type Xanthi-*nc* scion.

Most investigations have used Xanthi-*nc* tobacco at 24°C, in which, after inoculation, SA concentrations increase accompanied by a similar increase in conjugated SA (Fig. 17.81). Incubation of TMV-inoculated tobacco plants at 32°C blocks the development of a hypersensitive response and leads to systemic infection without accumulation of SA or PR proteins. Infected Xanthi-*nc* tobacco plants transferred from 32°C to 24°C undergo systemic necrosis, with a large and rapid increase in both free and conjugated SA (Fig. 17.81).

17.9.4 The salicylic acid biosynthesis pathway in plants is not fully defined.

Despite the wealth of elegant studies demonstrating the key role of SA in disease resistance, the biochemical events that facilitate the large increases in free and conjugated SA that follow TMV inoculation are not yet fully understood. Until recently, studies of SA biosynthesis were fairly rudimentary and compared poorly with equivalent longstanding investigations on GAs, IAA, BRs, cytokinins, and ABA, which have utilized mutants and GC-MS to analyze endogenous compounds and a range of stable isotope- and radioisotope-labeled precursors in metabolic studies.

In 1993, salicylic acid biosynthesis in tobacco was proposed to involve a side branch of the phenylpropanoid pathway, which converts *trans*-cinnamic acid to benzoic acid, the latter then undergoing 2-hydroxylation to form SA (Fig. 17.82). The alternative route from *trans*-cinnamic acid

to SA by way of *o*-coumaric acid appeared not to be functional. It was suggested that after TMV-inoculation, an extraordinarily large pool of benzoic conjugate (100 µg/g fresh weight) is hydrolyzed, releasing substantial amounts of benzoic acid, which activates a putative NADPH-dependent cytochrome P450 monooxygenase, benzoic acid 2-hydroxylase, responsible for catalyzing the conversion of benzoic acid to SA. According to this hypothesis, this SA is further converted to salicylic acid-2-*O*-β-glucoside, the

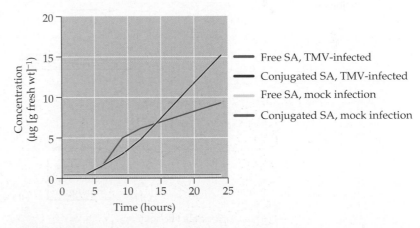

Figure 17.81
Concentrations of free and conjugated SA in leaves of resistant tobacco after mock and TMV inoculations at 24°C (upper figure). In the lower figure the inoculations were carried out at 32°C, but 10 hours later the plants were transferred to 24°C, which induces a large and rapid increase in both free and conjugated SA. Note the different time scales on the two figures.

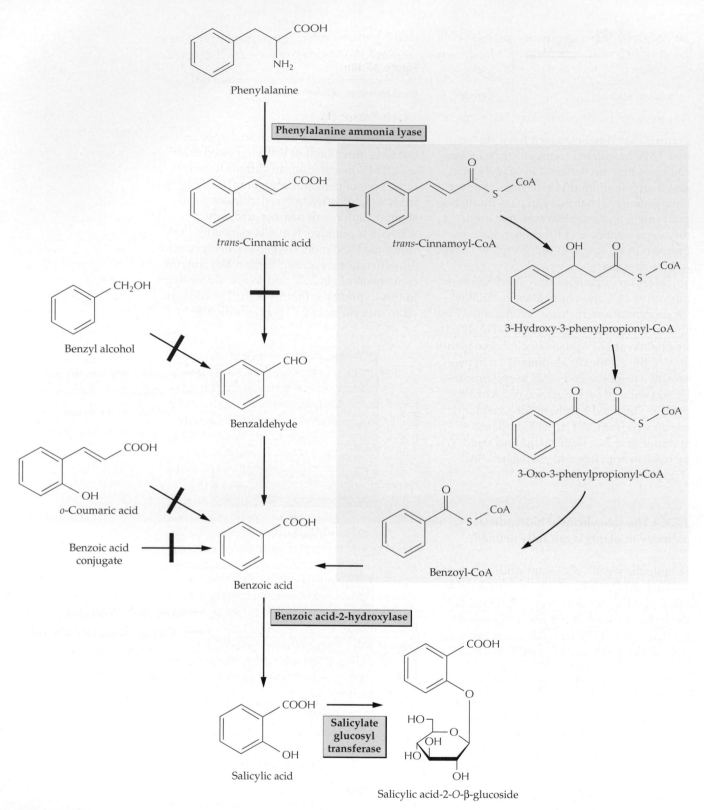

Figure 17.82

Proposed SA biosynthesis pathways in TMV-infected resistant tobacco. Red bars on arrows indicate conversions that do not appear to play a role in the large increase in endogenous salicylic acid that occurs in resistant tobacco following infection by TMV. The current evidence suggests that *trans*-cinnamic acid is converted to benzoic acid and SA by way of the β-oxidation pathway (highlighted in yellow) and not through *o*-coumaric acid. Although exogenous benzyl alcohol and benzaldehyde are both converted efficiently to benzoic acid and SA, these compounds are unlikely to be intermediates in the production of endogenous SA. Other data indicate that some SA may be synthesized from *trans*-cinnamic acid through a route that does not involve benzoic acid.

SA conjugate that accumulates after TMV-infection of Xanthi-*nc* tobacco leaves, by way of the action of SA-inducible UDP-glucose: SA glucosyltransferase (Fig. 17.82).

More recent publications have suggested that the rate-limiting step in SA biosynthesis is the conversion of *trans*-cinnamic acid to benzoic acid, and that this involves a β-oxidation pathway in which *trans*-cinnamoyl-CoA is the first of four intermediates (Fig. 17.82). Although exogenous benzyl alcohol and benzaldehyde are both converted to benzoic acid and SA, they appear not to be precursors of endogenous SA. Indirect evidence supports the involvement of *trans*-cinnamic acid as a key intermediate in the SA biosynthesis pathway: TMV-inoculation of the leaves of transgenic Xanthi-*nc* tobacco plants, in which phenylalanine ammonia lyase activity is suppressed, exhibit a four-fold lower increase in SA than do inoculated leaves of wild-type plants. However, evidence is increasing for an alternative, as yet undefined, route to SA that does not involve benzoic acid. In mycobacteria, SA is synthesized from chorismic acid by way of isochorismic acid (Fig. 17.83); whether this pathway operates in higher plants remains to be determined.

Recent studies, in which [^{14}C]benzoic acid and [^{14}C]SA were applied to mock- and TMV-inoculated Xanthi-*nc* tobacco leaves and their resulting metabolites were identified by GC-MS, have shown that most of the benzoic acid is metabolized to benzoic acid β-glucosyl ester and 3-hydroxybenzoic acid-3-*O*-β-glucoside. Only a relatively small amount is converted to SA, in keeping with the view that not all SA is synthesized by way of benzoic acid. [^{14}C]SA was metabolized extensively, not only to SA-2-*O*-β-glucoside but also to 2,5-dihydroxybenzoic

acid β-glucosyl ester. The probable pathways involved in these conversions are illustrated in Figure 17.84.

17.10 Prospects

Sequencing of plant genomes and the associated genomic technologies will have a great impact on the knowledge and understanding of the genes and enzymes involved in the regulation of plant hormone homeostasis. The first plant genome to be sequenced will be that of *Arabidopsis,* and the sequencing will reveal candidate genes involved in plant hormone biosynthesis. Mutations in these genes will be identified using reverse genetics. Determining the relative importance of the genes in question will require detailed analysis of the pattern of gene expression, localization of the enzyme, and quantification of hormones in both wild-type and mutant plants. It seems likely, therefore, that future research will not focus on single mutants in a pathway but instead on many, especially when several homologs of the same biosynthetic enzyme are present in the genome.

The genomic revolution will provide the necessary material for investigations that will lead towards understanding how the many interactions of biosynthetic pathways lead to control of the sizes of pools of bioactive hormones. Discerning how the different hormones interact and how their quantities are affected by environmental factors such as photoperiod, temperature, and water and nutrient supply, will also be important. Having established such advances in model systems, it will be necessary to transfer this understanding to crop plants to facilitate improvement of traits of agricultural and economic importance.

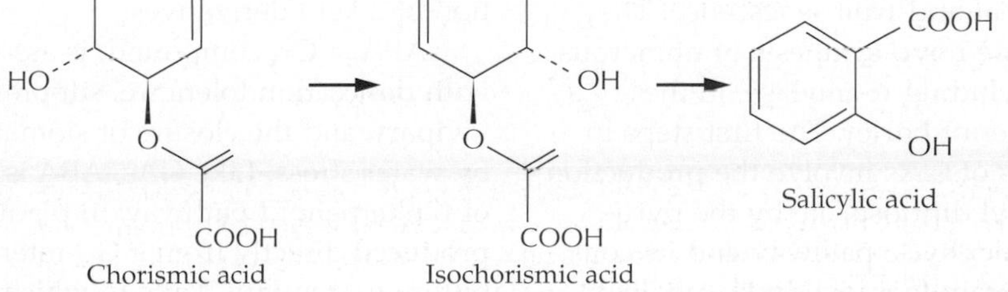

Figure 17.83
Bacterial biosynthesis of salicylic acid from chorismic acid.

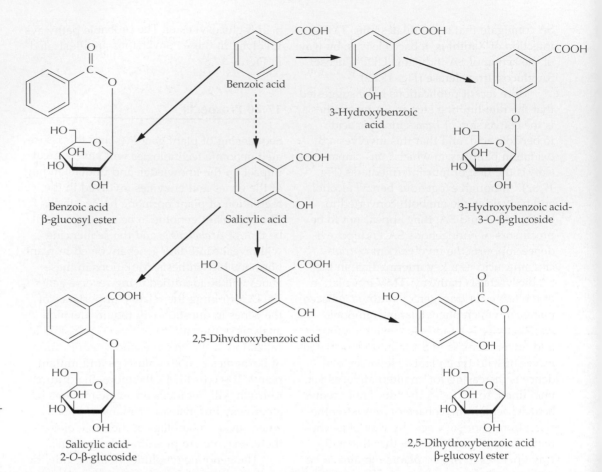

Figure 17.84
Metabolism of benzoic acid and SA in TMV-infected resistant tobacco. Bold arrows represent main pathway, dotted arrow indicates minor route.

Benzoic acid
β-glucosyl ester

Benzoic acid

3-Hydroxybenzoic
acid

3-Hydroxybenzoic acid-
3-*O*-β-glucoside

Salicylic acid

2,5-Dihydroxybenzoic acid

Salicylic acid-
2-*O*-β-glucoside

2,5-Dihydroxybenzoic acid
β-glucosyl ester

Summary

A battery of endogenous hormones regulate plant growth and responses to environmental stimuli, including gibberellins, abscisic acid, cytokinins, the auxin indole-3-acetic acid, ethylene, brassinosteroids, polyamines, jasmonic acid, and salicylic acid. Most of these compounds are present in plant tissues in low concentrations (nanograms per gram fresh weight) and pool sizes, and are tightly controlled by various biosynthetic, catabolic, and conjugation pathways.

GAs promote seed germination, stem elongation, flowering, and cone production and retard leaf and fruit senescence. They also induce de novo synthesis of numerous enzymes, including α-amylase, in the aleurone layer of barley. The first steps in the synthesis of GAs involve the production of isopentenyl diphosphate by the pyruvate/glyceraldehyde pathway and its conversion to geranylgeranyl diphosphate by the terpenoid pathway. The enzymes catalyzing the synthesis of geranylgeranyl diphosphate and its conversion to *ent*-kaurene, which is the first committed step in GA biosynthesis, are plastid-localized terpene cyclases. *ent*-Kaurene is oxidized to bioactive C_{19}-GAs by way of GA_{12}-aldehyde in a series of steps catalyzed by cytochrome-P450 monooxygenases located on the ER and by cytosolic α-keto-glutarate-dependent dioxygenases. The molecular cloning of several of the genes encoding GA biosynthesis enzymes has provided information on feedback regulation of the bioactive C_{19}-GA pools. The bioactive C_{19}-GAs can be deactivated by various reactions, including 2β-hydroxylation, glycosylation, and the formation of 2-keto derivatives.

ABA, a C_{15} compound, is associated with desiccation tolerance, suppression of vivipary, and the closure of stomata induced by water stress. Like GAs, ABA is a product of the terpenoid pathway. In plants it is not produced directly from a C_{15} intermediate but by a circuitous route in which 9'-*cis*-C_{40} compounds undergo oxidative cleavage to yield a C_{15} intermediate, xanthoxin, that is converted to ABA by way of ABA-aldehyde. ABA is metabolized to phaseic acid, dihydrophaseic acid, and dihydrophaseic acid glucoside. ABA-deficient mutants typically exhibit a wilty or viviparous phenotype. Although many such mutants are known, many of the genes encoding ABA biosynthetic enzymes have yet to be cloned.

Cytokinins, in conjugation with auxin, promote cell division and determine cell differentiation. They also are associated with the senescence of plant organs, apical dominance, and stomata opening. The first step in the synthesis of cytokinins is the isopentenylation of 5'-AMP. The resulting [9R-5'P]iP is modified by the trans-hydroxylation of the isopentenyl chain, dephosphorylation, or deribosylation, singly or in combination, to form such cytokinins as [9R]iP, iP, [9R]Z, and Z (see Table 17.1 for full names), of which Z exhibits the highest biological activity. Cytokinins are metabolized by the stereospecific hydrogenation of the side chain, removal of the isopentenyl side chain (catalyzed by cytokinin oxidase), and conjugation reactions. Conjugation steps include glycosylation of the side chain hydroxyl group and N-glucosylation or N-alanylation of the purine ring.

IAA affects apical dominance, tropisms, shoot elongation, the induction of cambial cell division, and root initiation. Synthesized from L-tryptophan (by way of indole-3-pyruvic acid and indole-3-acetaldehyde), IAA can be released by hydrolysis of IAA glucosyl conjugates. In some species, including Arabidopsis, L-tryptophan is converted to IAA by way of indole-3-acetaldoxime and indole-3-acetonitrile; a tryptophan-independent route to IAA also exists. IAA is deactivated by addition of a 2-keto group and formation of aspartyl and N-glucosyl conjugates. Several Arabidopsis IAA homeostasis mutants, displaying a variety of phenotypes, have been isolated. Transgenic tobacco plants expressing bacterial IAA biosynthesis genes contain increased concentrations of free IAA and IAA conjugates; the abnormal phenotype shows pronounced apical dominance, dwarfism, increased adventitious root formation, excess lignification, leaf epinasty, and abnormal flower production. The dwarfism is an indirect consequence of the greater rates of ethylene synthesis in the tissues that overproduce IAA.

Ethylene causes abnormal growth of etiolated seedlings and has an impact on shoot and root growth, flower development, senescence and abscission of flowers and leaves, and ripening of fruit. In planta, ethylene is synthesized from SAM by way of ACC. A gene encoding ACC oxidase has been cloned and expressed in antisense orientation in tomato. The transgenic fruits have 95% less ethylene production, are resistant to overripening, and can be stored at room temperature for long periods while remaining edible.

BRs are essential factors for cell and stem elongation, unrolling of grass leaves, bending of grass leaves at the sheath/blade joints, xylogenesis, and ethylene production. BR biosynthesis and sensitivity mutants show dwarfism and, when grown in the dark, share some characteristics with light-grown plants. Brassinolide, the most biologically active and ubiquitous BR, is synthesized from campesterol. First, campesterol is hydrogenated to campestanol, which is converted to castasterone by the repeated oxidation/hydroxylation of the side chain and A and B rings. Further B ring oxidation of castasterone yields BL. Several genes involved in BR/sterol synthesis have been cloned. BR deactivation reactions can include epimerization of A ring hydroxyls, glucosylation, esterification, modification, cleavage of the side chain, and glucosylation of the 23-OH group.

Polyamines stimulate many reactions involved in the syntheses of DNA, RNA, and protein. The diverse physiological responses elicited by polyamines include cell division, tuber formation, root initiation, embryogenesis, flower development, and fruit ripening. Putrescine, spermidine, and spermine are synthesized from L-arginine and L-ornithine. Synthesis of spermidine and spermine requires an aminopropyl group derived from SAM, and there may be competition between the ethylene and polyamine biosynthesis pathways when concentrations of SAM are limited. The primary (terminal) amines of polyamines are oxidized by diamine oxidases, the secondary amines by polyamine oxidases. Polyamines occur both as free amines and as amide conjugates of hydroxycinnamates such as p-coumaric acid, ferulic acid, and caffeic acid.

(–)-JA is associated with disease resistance, inhibits seed and pollen germination and seedling growth, and induces fruit ripening and abscision of flowers. It is synthesized from α-linolenic acid, a membrane-derived C_{18} polyunsaturated fatty acid. The first specific step in the (–)-JA biosynthesis pathway, the conversion of 13(S)-hydroperoxylinolenic acid to 12,13(S)-epoxylinolenic acid, is catalyzed by allene oxide synthase. Wounding a leaf results in increased AOS activity and accumulation of (–)-JA. (–)-JA is metabolized to hydroxylated products and amino acid and glycosylated conjugates.

SA, synthesized from trans-cinnamic acid by way of a side branch of the phenylpropanoid pathway, is involved in thermogenesis in lilies and pathogen resistance in tobacco and other species. Until recently, benzoic acid was thought to be the immediate precursor of SA, but increasing evidence is supporting an alternative, as yet undefined, route to SA that does not involve benzoic acid. SA is metabolized to SA glucoside and 2,5-dihydroxybenzoic acid glucosyl ester.

Further Reading

General

Hooykaas, P.J.J., Hall, M. A., Libbenga, K. R., eds. (1999) Biochemistry and Molecular Biology of Plant Hormones. Elsevier, Amsterdam.

Gibberellins

Hedden, P., Kamiya, Y. (1997) Gibberellin biosynthesis: enzymes, genes and their regulation. Annu. Rev. Plant Physiol. Plant Mol. Biol. 48: 431–460.

Lange, T. (1998) The molecular biology of gibberellin synthesis. Planta 204: 409–419.

MacMillan, J. (1997) Biosynthesis of the gibberellin plant hormones. Nat. Prod. Rep. 1: 221–243.

Ross, J. J., Murfet, I. C., Reid, J. B. (1997) Gibberellin mutants. Physiol. Plant. 100: 550–560.

Abscisic acid

Cutler, A. J., Krochko, J. E. (1999) Formation and breakdown of ABA. Trends Plant Sci. 12: 472–478.

Walton, D. C., Li, Y. (1995) Abscisic acid biosynthesis and metabolism. In Plant Hormones: Physiology, Biochemistry, and Molecular Biology, P. J. Davies, ed. Kluwer Academic Publishers, Dordrecht, pp. 140–157.

Cytokinins

McGaw, B. A., Burch, L. R. (1995) Cytokinin biosynthesis and metabolism. In Plant Hormones: Physiology, Biochemistry, and Molecular Biology, P. J. Davies, ed. Kluwer Academic Publishers, Dordrecht, pp. 98–117.

Mok, D.W.S., Mok, M. C., eds. (1994) Cytokinins: Chemistry, Action and Function. CRC Press, Boca Raton, FL.

Indole-3-acetic acid

Bartel, B. (1997) Auxin biosynthesis. Annu. Rev. Plant Physiol. Plant Mol. Biol. 48: 51–66.

Normanly, J. (1997) Auxin metabolism. Physiol. Plant. 100: 431–442.

Ethylene

Kanellis, A. K., Chang, C., Kende, H., Grierson, D., eds. (1997) Biology and Biotechnology of the Plant Hormone Ethylene. NATO ASI Series, Kluwer Academic Publishers, Dordrecht.

Kende, H. (1993) Ethylene biosynthesis. Annu. Rev. Plant Physiol. Plant Mol. Biol. 44: 283–307.

Brassinosteroids

Altmann, T. (1998) Recent advances in brassinosteroid molecular genetics. Curr. Opin. Plant Biol. 1: 378–383.

Clouse, S. D., Sasse, J. M. (1998) Brassinosteroids: essential regulators of plant growth and development. Annu. Rev. Plant Physiol. Plant Mol. Biol. 49: 427–451.

Sakurai, A., Yokota, T., Clouse, S. D., eds. (1999) Brassinosteroids. Steroidal Plant Hormones. Springer-Verlag, Tokyo.

Yokota, T. (1997) The structure, biosynthesis and function of brassinosteroids. Trends Plant Sci. 2: 137–143.

Polyamines

Kumar, A., Altabella, T., Taylor, M. A., Tiburcio, A. F. (1997). Recent advances in polyamine research. Trends Plant Sci. 2: 124–130.

Tiburcio, A. F., Altabella, T., Borrell, A., Masgrau, C. (1997) Polyamine metabolism and its regulation. *Physiol. Plant.* 100: 664–674.

Jasmonic acid

Creelman, A., Mullet, J. E. (1997) Biosynthesis and action of jasmonates in plants. *Annu. Rev. Plant Physiol. Plant Mol. Biol.* 48, 355–381.

Mueller, M. J. (1997) Enzymes involved in jasmonic acid biosynthesis. *Physiol. Plant.* 100: 653–663.

Salicylic acid

Durner, J., Shah, J., Klessig, D. F. (1997) Salicylic acid and disease resistance in plants. *Trends Plant Sci.* 7: 266–274.

Pierpoint, W. S. (1994) Salicylic acid. *Adv. Bot. Res.* 20: 163–235.

Raskin, I. (1995) Salicylic acid. In *Plant Hormones: Physiology, Biochemistry, and Molecular Biology*, P. J. Davies, ed. Kluwer Academic Publishers, Dordrecht, pp. 188–205.

Biochemistry & Molecular Biology of Plants, B. Buchanan, W. Gruissem, R. Jones, Eds.
© 2000, American Society of Plant Physiologists

CHAPTER 18

Signal Perception and Transduction

Anthony Trewavas

Introduction

Plant cells are constantly bombarded with information to which they must react. **Signal transduction,** the means whereby cells construct responses to a signal, is a recently defined focus of research in plant biology. The application of biochemical and molecular genetic techniques has resulted in major advances in elucidating the mechanisms that regulate gene expression and in identifying components of many signal transduction pathways in diverse physiological systems. Today, signal transduction research contributes to all aspects of plant science, linking many fields of study in much the same way that signal transduction pathways link myriad cellular processes.

18.1 Overview of signal transduction

18.1.1 The stream of signals to which plant cells react is continuous and complex.

Throughout their life cycle, plants and plant cells continually respond to signals that they use to alter their physiology, morphology, and development. Among the stimuli—both external (Fig. 18.1) and internal (Fig. 18.2)—that convey information to plants are light, mineral nutrients, organic metabolites, gravity, water status, turgor, soil quality, mechanical tensions, wind, heat, cold, freezing, growth regulators and hormones, pH, gases (CO_2, O_2, C_2H_4), wounding and disease, and electrical flux. Signals can vary in quality and quantity from minute to minute. Some of the signals are carried by xylem and phloem, the circulatory system, which can accommodate very large and rapid fluxes.

Plant responses to stimulus are modulated by developmental age, previous environmental experience, and internal clocks that specify the time of year and the time of day. For mature plant cells, the response can be physiological and biochemical; for growing cells, it can be morphological and developmental. Integration of various forms of signaling information is usually crucial to determining the final response. In a seed, for example, the decision to germinate can be irreversible and, if timed inappropriately, fatal. The capacity of seeds to react successfully to many physical, chemical, and temporal variables reflects the presence of a complex system for signal recognition and transduction in the living cells of all plants.

18.1.2 Signal transduction uses a network of interactions within cells, among cells, and throughout the plant.

Two of the principal elements in the signal transduction pathways of plant cells are intracellular Ca^{2+} ($[Ca^{2+}]_i$), and protein kinases, enzymes that phosphorylate and thereby alter the activity of target proteins (Fig. 18.3). The term **second messenger** is often used to describe a readily diffusible molecule involved in conveying information from an extracellular source to the principal target enzymes within the cell. In plants, $[Ca^{2+}]_i$ transduces many signals and is a prominent second messenger; it therefore must be maintained in the cytoplasm at concentrations many orders of magnitude lower than the $[Ca^{2+}]$ in the cell wall. During signaling, **Ca^{2+} transients** (brief increases in $[Ca^{2+}]_i$) are often

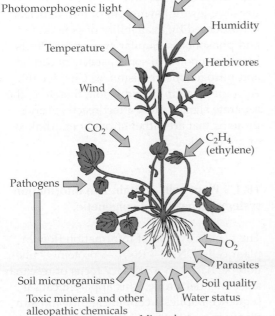

Figure 18.1
External signals that affect plant growth and development include many aspects of the plant's physical, chemical, and biological environments. Some external signals come from other plants. Apart from gravitropic signals, all other signals vary in intensity, often from minute to minute.

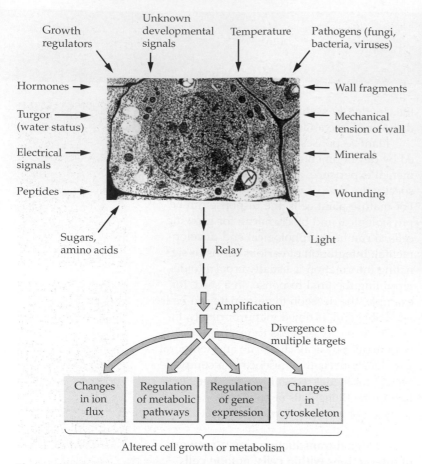

Figure 18.2
A variety of internal signals modify plant cell metabolism, growth, and development. The ability of cells to respond to these signals is not confined to cells that are still growing and developing. Mature cells, too, can initiate metabolic responses and can even reinitiate growth and division in response to signal information.

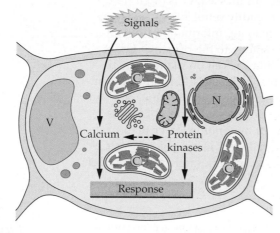

Figure 18.3
Two of the major signal transduction pathways found in plant cells involve cytosolic calcium and protein kinases. Calcium-dependent protein kinases, which are very prevalent in plant cells, connect the two transduction pathways (dashed arrow). V, vacuole; C, chloroplast; N, nucleus.

associated with initiation of responses. The many signal transduction pathways on which $[Ca^{2+}]_i$ acts involve hundreds of different proteins, as well as other second messengers in the cytoplasm and the plasma membrane. Protein kinases are similarly ubiquitous: Genes encoding these enzymes represent an estimated 3% to 4% of the genome; at any one time, cells will be using hundreds of different protein kinases. Together, the signaling pathways that utilize protein kinases and $[Ca^{2+}]_i$ constitute a network of great complexity.

Changes in $[Ca^{2+}]_i$ can initiate diverse responses that vary according to cellular structure and are sensitive to interaction between specific components of signaling systems. For example, increases in $[Ca^{2+}]_i$ can initiate closure of the stomatal aperture in guard cells, reorientation of growth in pollen tubes, or wall thickening in young tobacco seedlings in response to wind. Likewise, a single protein kinase can have many target proteins, but the targets differ among distinct cell types and developmental stages.

Many signals interact cooperatively and synergistically with each other to produce the final response. Signal combinations that induce such complex plant responses include red and blue light (Fig. 18.4), gravity and light, nitrate availability and light (Fig. 18.5), growth regulators (see Chapter 17), and mineral nutrients (see Chapter 23). As illustrated for tobacco in Table 18.1, the chlorosis (yellowing) symptom of iron deficiency is influenced by availability of potassium and phosphorus. Similar synergistic effects have been noted in some aspects of carbon and nitrogen metabolism (see Chapter 16). At some stage in transduction, therefore, the separate signals must affect reactions and proteins that are either strongly interlinked or are identical.

18.1.3 Plant cells contain two information systems: genetic and epigenetic.

The genetic system of information flow in plant cells, DNA → RNA → protein → phenotype, has been a primary focus of research since Mendel's time. This apparent causal sequence implies a certain rigidity and simplicity in result. Some of the genes studied by Mendel, such as those that specify flower

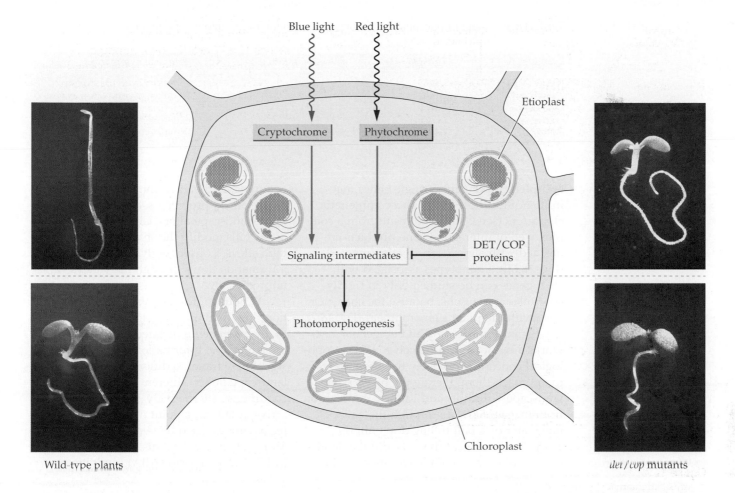

Blue light   Red light

Cryptochrome   Phytochrome

Etioplast

Signaling intermediates   DET/COP proteins

Photomorphogenesis

Chloroplast

Wild-type plants   *det/cop* mutants

Figure 18.4
Blue light and red light often interact and overlap in their effects on plant development. Two sets of proteins called DET (deetiolated) and COP (constituitively photomorphogenic) ordinarily ensure that the etiolated program is maintained in darkness. The effect of light is to switch off the activity of DET and COP, allowing photomorphogenesis to occur. Repression of DET and COP activity is dependent on the blue light receptor (cryptochrome) and the red light receptor (phytochrome) and the presence of photomorphogenically active light. The effects of red light and blue light on deetiolation indicate that signaling intermediates in the two light-reception pathways form an interactive network.

color or seed morphology (see Chapter 7), are invariant in expression under many different conditions of growth and development (Fig. 18.6A). Geneticists concentrate on the expression of such genes because analysis of their inheritance is relatively simple.

On the other hand, many important phenotypic characteristics are strongly modified by the environment in which the plant grows, including production of biomass, duration of growth, branching, partitioning of photosynthate between reproductive and vegetative structures, and responses to stress. In these cases the phenotypic character is constructed from **epistatic genes,** the products of which alter the expression of other genes that were inherited independently,

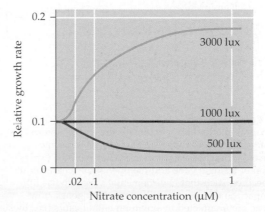

Figure 18.5
Interaction between light and nitrate concentration on relative growth rate of plants. Note that nitrate may inhibit as well as promote growth, depending on the light intensity.

Table 18.1 Effect of phosphate and potassium nutrition on symptoms of mineral deficiency in tobacco

Iron (Fe)	Phosphate (P_i)	Potassium (K)	Apparent symptoms
Low	High	Low	K deficiency in old leaves
High	High	Low	Fe deficiency in young leaves
Low	Low	Low	K deficiency delayed
Low/high	Low	Low	Chlorotic; no K deficiency
High	Low	Low	P_i deficiency

and **pleiotropic genes,** single genes that influence multiple traits. These **epigenetic** characters (see Chapter 7) result from a complex web of interacting gene products enmeshed with signal transduction networks (Fig. 18.6B). Phenotypes associated with such genes can be studied only in rigidly controlled conditions, because the characters vary with the plant's environment (Fig. 18.7).

Signal transduction and epigenetic networks are not easy to conceive; using a topological concept is probably the simplest way to approach them. Picture the network as a landscape with hills and valleys: Information from the signal flows through the valleys like water or a ball rolling downhill (Fig. 18.8). The heights of the hills and the depths of the valleys are initially specified by the genotype; as the environment changes and development progresses, however, the shape of the landscape alters. Changes in the size of the hills modify the shape and position of the valleys. Consequently, the path by which information flows through the valleys varies unpredictably.

Various routes are available for information flow from any one signal, the route actually taken being dictated by the shape of the landscape. Information from the same signal may travel to different regions of the landscape under different environmental conditions. Alternatively, information might arrive at the same point of the landscape after having traveled by very different routes. Thus in *Commelina* plants grown at 10°C to 17°C, abscisic acid (ABA) can initiate

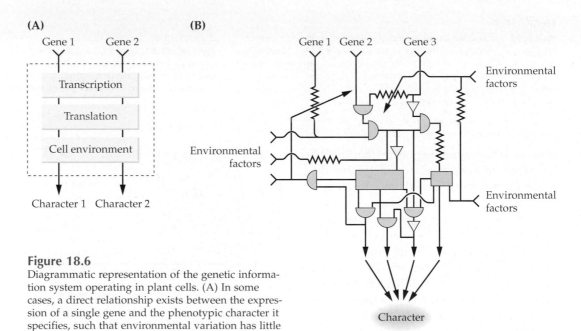

Figure 18.6
Diagrammatic representation of the genetic information system operating in plant cells. (A) In some cases, a direct relationship exists between the expression of a single gene and the phenotypic character it specifies, such that environmental variation has little impact on expression of the gene. Few genes, however, are unaffected by environmental factors (i.e., by signaling events). (B) In the epigenetic information system, a phenotypic character results from complex interactions involving one or more genes and environmental influences that impact signal transduction networks. The computer circuit in the diagram illustrates how difficult it may be to elucidate a precise transduction sequence and predict the influence of any signal. One would expect the actual distribution of information flow through the various branches of the epigenetic network to vary according to environmental cues.

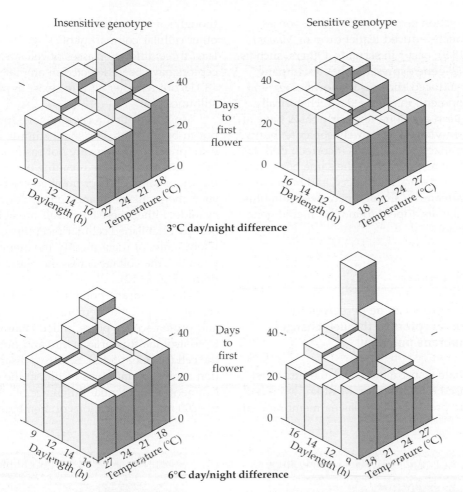

Insensitive genotype   Sensitive genotype

Days to first flower

3°C day/night difference

Days to first flower

6°C day/night difference

Figure 18.7

Phenotypic characters can result from very complex interactions between genotypes and the environment. Flowering time in two genotypes of beans—one early-maturing and photoperiod-insensitive, the other late-maturing and photoperiod-sensitive—is shown to vary with length of daylight and growth temperature. The tops of the columns can be used to construct an epigenetic surface that best describes the characteristics of development illustrated in Figure 18.8. Inclusion of other variables (e.g., nitrate or light intensity) creates a much more complicated surface.

stomatal closure in the absence of changes in $[Ca^{2+}]_i$. For plants grown at temperatures above 25°C, however, an ABA transduction route involving an increase in $[Ca^{2+}]_i$ seems to be essential.

18.1.4 Different signals affect the transduction network in different ways and at different places, but most modify gene expression.

The signals listed in Figures 18.1 and 18.2 impact the cell at different sites and are perceived by different receptors. However, downstream reactions may meet at what can be termed **nodal points,** that is, at proteins or enzymes that are involved in many

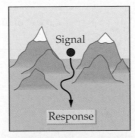

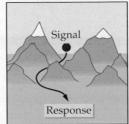

Figure 18.8

The movement of a signal through a transduction network can be thought of as a ball rolling through hills and valleys to reach the final response. Both the environment and development shift the hills and valleys. Achieving similar responses under different environmental conditions involves variations in the activity of different transduction pathways.

transduction sequences. Some responses, e.g., touch-induced leaflet drop of *Mimosa* (Fig. 18.9), occur in seconds. Others, such as shifts in gene expression that accompany touch-induced changes in morphology and development, may take days. Historically, short-term and long-term responses have often been viewed as entirely separate, but we now realize this notion is incorrect. Both fast and slow responses use the same basic transduction network machinery, and both are downstream results of a perceived stimulus. Most signals appear to induce altered gene expression.

18.2 Receptors

18.2.1 Signals can be perceived by protein receptors or through changes in membrane potential.

To initiate transduction, a signal must first be sensed by a **receptor**. Most known receptors are present in the plasma membrane, al-

though some are located in the cytosol or other cellular compartments (Fig. 18.10). At least three different classes of cell surface receptors have been detected in animals (Fig. 18.11), but whether all three exist in plants is still uncertain.

Most identified receptors have turned out to be proteins. For some stimuli, however, protein receptors are not easily identified—for example, the breaking of the dormancy of some buds or imbibed seeds by such chemicals as ethanol, ether, azide, or cyanide. Either these chemicals are able to occupy established cellular receptors or, more likely, many of them modify the membrane potential, the voltage across the plasma membrane (Fig. 18.12).

The membrane potential can act as a receptor. The plasma membrane uses pumps and proteinaceous pores, called **channels,** to control the flux of ions into and out of the cell (see Chapter 3). Selective discrimination against certain ions results in the establishment of a potential difference of –80 to –200 mV. Modifications of membrane

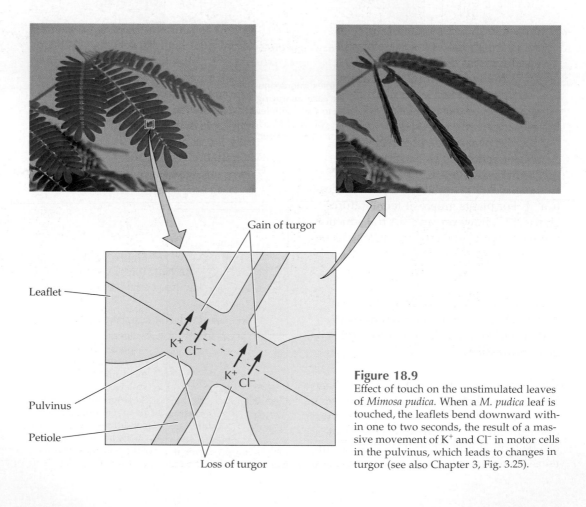

Figure 18.9
Effect of touch on the unstimulated leaves of *Mimosa pudica*. When a *M. pudica* leaf is touched, the leaflets bend downward within one to two seconds, the result of a massive movement of K⁺ and Cl⁻ in motor cells in the pulvinus, which leads to changes in turgor (see also Chapter 3, Fig. 3.25).

Figure 18.10

Extracellular signals bind either to receptors on the cell surface or to receptors inside the cytoplasm or nucleus. Many hydrophilic molecules, such as peptides and carbohydrates, and osmotic signals cannot easily pass through the plasma membrane and therefore are perceived on the cell surface (e.g., ligand 1). Amphiphilic and hydrophobic molecules, such as growth regulators, can pass through the plasma membrane and may be perceived either by surface receptors or inside the cell (e.g., ligand 2). V, vacuole; N, nucleus; C, chloroplast.

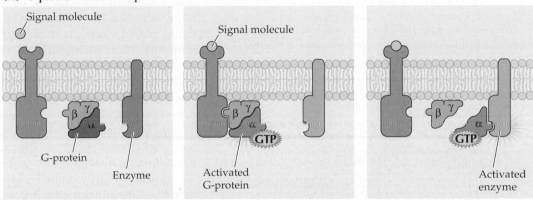

(A) G-protein–linked receptor

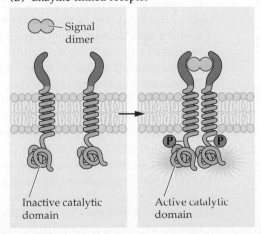

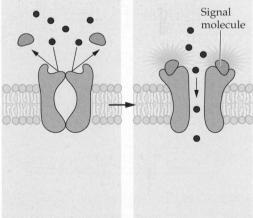

(B) Enzyme-linked receptor

(C) Ion channel–linked receptor

Figure 18.11

Three classes of plasma membrane–located receptors identified in animal cells. (A) When activated, G-protein–linked receptors convey information to a protein that binds GTP as the first stage in transduction. The G-protein α-subunit/GTP complex is usually released from the β/γ-subunits into the cytoplasm, where it can activate other enzymes. (B) Enzyme-linked receptors are commonly protein kinases. Binding of the ligand (signal) causes the receptor to dimerize, leading to intermolecular phosphorylation with activation of the receptor. (C) Ion channel–linked receptors may be coupled directly to important cell surface channels that open when the receptor is occupied. Some ion channel receptors are located on internal membranes as well (see Fig. 18.15 for an example).

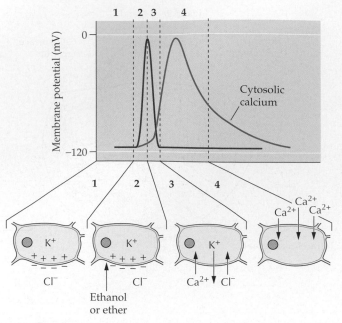

Figure 18.12

Possible sequence of events in transduction of signals such as ethanol, ether, or cyanide. The plant cell membrane potential (in red) is established because the plasma membrane is not equally permeable to K^+ and Cl^- ions (1). Ethanol, ether, or cyanide causes a temporary disruption to the membrane potential (2), which is followed by a brief influx of Ca^{2+} (3). Voltage-sensitive Ca^{2+} channels subsequently open for a longer period, signaling the cell (4). Organic chemicals, minerals, and other depolarizing treatments can also be used to break the dormancy of seeds or buds, increase root formation, and modify flowering.

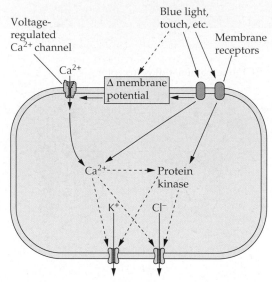

Figure 18.13

Many signals modify the plasma membrane potential and activate voltage-gated channels, permitting entry of Ca^{2+}. When plant cells are exposed to signals such as blue light or touch, detectable changes in membrane potential occur. Voltage-gated channels are opened and concentrations of cytosolic Ca^{2+} are increased. Activation of this transduction pathway can result in opening of the potassium and chloride channels, which leads to loss of turgor (see Fig. 18.9). The separate transduction pathways are linked through Ca^{2+}/calmodulin–regulated protein kinases and other proteins. Blue light and touch may also mobilize intracellular Ca^{2+} by other routes.

potential open a group of voltage-gated channels that allow Ca^{2+} to enter and thus activate a transduction sequence (Figs. 18.12 and 18.13). Ca^{2+} influx, in turn, can cause the subsequent opening of many potassium and chloride channels, resulting in very rapid changes in turgor, such as occur in the pulvini of *Mimosa* (see Fig. 18.9). Many signals are known to modify membrane potential, including red and blue light, fungal elicitors, and many growth regulators.

18.2.2 Many receptors share similar structural attributes and catalytic activities.

The sequences of many receptors have been determined. Although there is often little sequence conservation among receptors from different organisms that bind the same ligand, a unifying characteristic has emerged. Many receptors have seven hydrophobic domains placed strategically throughout the molecule. These hydrophobic domains are thought to represent regions of the receptor that span the plasma membrane (Fig. 18.14). The ligand-binding site may be located in one of these domains or on the extracellular region. Frequently the N terminus of such receptors is located outside the cell and the C terminus is inside.

In some transmembrane protein receptors, the C-terminal region is phosphorylated by protein kinases. Two possible families of protein kinases are distinguished on the basis of the amino acids they phosphorylate on their substrate proteins: serine/threonine residues or tyrosine residues. A few protein kinases phosphorylate all three amino acids. Protein kinases that phosphorylate tyrosine residues exclusively are rare in plant cells.

Another class of receptors is the so-called **receptor-like protein kinases** (RLKs; see Section 18.7.2). The RLKs of plants typically consist of a large extracytoplasmic domain, a single membrane-spanning segment, and a cytoplasmic domain containing the active site of a protein kinase. Binding of the ligand is thought to cause dimerization of the receptor, bringing the cytoplasmic domains of the protein kinases into close proximity. Intermolecular or intramolecular phosphorylation can then stabilize the activated receptor complex. This phosphorylation can maintain the protein kinase activity of some

RLKs in the absence of ligand. Intermolecular phosphorylation usually modifies serine or threonine residues.

In some cases, the activated complex can phosphorylate and activate other downstream proteins. Alternatively, the active RLK complex interacts with membrane-bound or soluble transduction proteins (such as Ras or other G-proteins in animals) to perpetuate the signal transduction sequence in a different direction. Some RLKs form a membrane complex with a protein phosphatase, which dephosphorylates the activated RLK. The inactive RLK is then free to reassociate with other RLKs in further signaling if its ligand is still available.

Numerous RLKs have been identified in plant cells, including protein kinases with seven membrane-spanning domains. Such RLKs have been detected in the male reproductive tissues of plants, where they are implicated in incompatibility reactions that prevent fertilization (see Section 18.7.2).

18.2.3 Intracellular receptors can act as ion channels.

Other receptors are located in intracellular membranes and can act as Ca^{2+} channels. The most well-known receptor in this class binds the second messenger inositol 1,4,5-triphosphate (IP_3; Fig. 18.15). Channels for another second messenger, cyclic ADP-ribose (cADPR), have been reported recently. Both these signaling molecules are synthesized by enzymes in the plasma membrane and then translocate to the vacuole and the ER, where their receptors are located. Occupation of the receptor (which may be composed of four subunits) leads to the opening of Ca^{2+} channels and an influx of Ca^{2+} into the cytoplasm from the vacuole and the ER, each of which contains Ca^{2+} many orders of magnitude greater than the cytosolic concentrations (see Section 18.6.1). In contrast to plasma membrane–bound receptors, these protein subunits each have four membrane-spanning domains. Other membrane-spanning proteins also may have important functions in signal transduction. For example, an auxin-resistant mutant, *aux-1*, results from an alteration in a protein with 10 membrane-spanning domains, which is thought to be a permease.

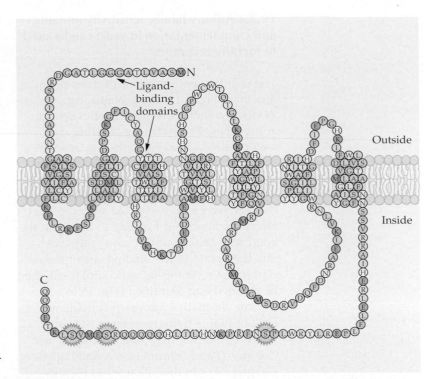

Figure 18.14
Sequence of GCR1, a putative receptor that affects cytokinin sensitivity. The protein structure shown here, with seven membrane-spanning domains and highlighted phosphorylation sites in the C terminus, was deduced from hydropathy plots (see Chapter 3, Box 3.2).

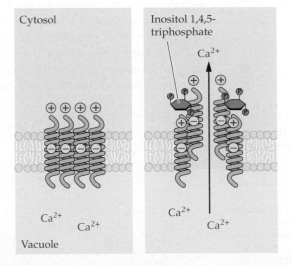

Figure 18.15
The receptor for inositol 1,4,5-triphosphate (IP_3) is located on the tonoplast and ER membranes. Conformational changes in this receptor transduce subsequent signaling. Certain ion channel receptors, including the IP_3 receptor, are composed of four subunits. Each subunit is thought to contain four membrane-spanning domains (not shown). When IP_3 binds to the receptor, conformational changes result in movement of two of the subunits. The distribution of positive and negative charges stabilizes the open conformation of the channel and allows the entry of Ca^{2+} into the cytoplasm.

18.2.4 Affinity labels, sensitivity mutants, and complementation in yeast can be used to identify receptors.

The cellular concentrations of receptors are less than those of other proteins (e.g., enzymes), so their detection requires special approaches. **Photoaffinity labeling** uses isotopically labeled reagents, e.g., reactive carbenes of nitrenes that undergo bond rearrangement on exposure to UV light. When radioactive or fluorescent ligands containing the reactive group are mixed with the receptor preparation and exposed to UV light, the bond rearrangement can covalently cross-link the ligand to the binding site of the receptor; the receptor–ligand complex can then be purified and identified (Fig. 18.16).

An alternative approach is to identify mutant plants that are insensitive to the signal. Molecular mapping techniques (see Chapter 7) and "chromosome walking" (see Chapter 21) can then be used to identify the putative receptor mutation. If the mutation is induced by insertion of characterized fragments of DNA—i.e., a transposon (see Chapter 7) or a T-DNA (see Chapters 6 and 21)—then the gene encoding the protein responsible for the insensitivity can be identified more easily. These methods have identified the receptors for blue light (Fig. 18.17A), for ethylene, and perhaps for ABA but have proven less successful for investigating auxin.

Various other procedures rely on correlations between physiological response and receptor behavior. Identification of **phytochrome** as the red light receptor has relied on the finding that red light effects are frequently counteracted by the immediate exposure of the affected plant to far-red illumination. Purified phytochromes likewise exist in two forms that have different absorption spectra. One form, P_{FR}, primarily absorbs far-red light; the other, P_R, absorbs red light. **Calmodulin** has been functionally identified as a primary Ca^{2+} receptor because the Ca^{2+}–calmodulin complex can activate many other enzymes. Complementation in yeast is potentially an important technology for identifying many plant receptors. For example, responses to osmotic pressure are shared by plants and yeast. Isolation of a yeast mutant that was unresponsive to osmotic pressure led to identification of the yeast receptor involved in osmosensing. This yeast mutant can now be exploited to find plant cDNAs that restore osmotic sensing; such cDNAs might encode plant receptors that signal changes in osmotic pressure (Fig. 18.17B). At present, a plant receptor that senses osmotic status has not yet been cloned.

18.2.5 Receptor–ligand binding is reversible and exhibits saturation kinetics.

Identification of a molecule as a receptor is a difficult research objective. The following set of criteria—all of which are fulfilled by the ethylene receptor—can be used to help in the identification when a functional assay is not available. Ligands should bind to specific sites on their receptors. Binding should be of sufficient strength (and thus occupy the receptor for a sufficient length of time) that the associated downstream processes (which usually require direct interactions with other molecules) can be activated.

- Ligand binding should be of relatively high affinity.
- Ligand binding should be reversible, allowing the system to respond to changes in ligand concentration, because receptors are present in limited abundance.
- The binding of the ligand to its receptor should saturate at a certain ligand concentration.
- The receptor should be selective for biologically active molecules, and binding specificity should mimic in vivo physiological activity.

Figure 18.16
A simple way to detect a receptor is to attach a radioactive affinity label (e.g., ^{3}H-nitrene or ^{3}H-carbene) to a ligand. The receptor preparation is mixed with the radioactive ligand to permit binding, frozen to very low temperatures, and then exposed to UV light. On activation of the reactive nitrene or carbene group (R) by UV light, the labeled compound binds irreversibly to the receptor, which can then be purified and identified.

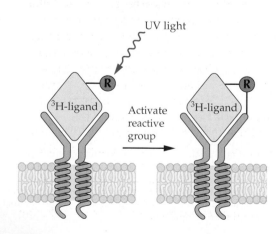

UV light

^{3}H-ligand

Activate reactive group

^{3}H-ligand

(A) Chemical mutagenesis vs. T-DNA mutagenesis

(B) Complementation of mutant yeast

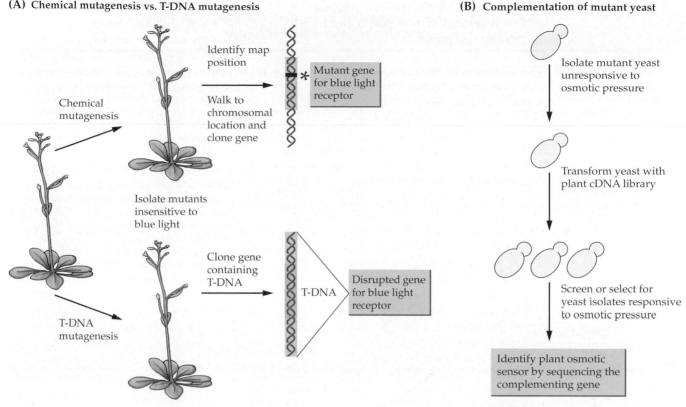

Figure 18.17

(A) Two approaches for identifying plant genes by generating mutant plants with altered phenotypes. Populations of mutants are obtained by chemical mutagenesis and the desired phenotype is identified. Chromosomal regions are identified through selection and mapping. Chromosomal "walks" are conducted to identify the mutant sequence, which is then used to isolate the wild-type sequence and identify the gene of interest. Alternatively, genes can be tagged with T-DNA from the bacterium that causes crown gall disease, *Agrobacterium tumefaciens* (see Chapters 6 and 21). T-DNA is bordered by sequences that facilitate its insertion into the plant genome. Transformation is carried out on seeds or plants, and the required phenotype is identified by screening a population of transformants.

The insertion site can be identified by using the known sequence of T-DNA as a tag. Sequencing around the insertion site enables identification of the gene responsible for the phenotypic character. (B) Identifying a plant gene by functional complementation. In this approach, a plant gene restores a wild-type phenotype in a mutant strain of yeast. Complementation requires isolation of a yeast mutant that is deficient in the plant character being investigated. The yeast is transformed with a plant cDNA library, a complemented yeast clone is isolated, and the plant transgene is sequenced. In the hypothetical example shown here, complementation is used to identify a plant osmotic sensor.

- The affinity constant for binding (K_d, Box 18.1) should be similar to the ligand concentration that is active in vivo. (Variations in receptor concentration bring important caveats to this criterion; see Section 18.2.8.)

18.2.6 Specific receptors for many signals have not yet been identified.

The many signals to which plants specifically respond must necessarily interact with specific receptors that couple them to transduction pathways, but most of these receptors have not yet been identified. For example, plant cells can clearly sense their water status and respond in a variety of ways, including osmotic adjustment through accumulation of compatible solutes such as proline or glycine betaine (see Chapter 22), accumulation of ABA (see Chapter 17), and changes in gene expression, development, and morphology. Bacteria sense their osmotic status through a protein receptor, the conformation of which is determined by the amount of bound water; activation of this receptor leads to changes in gene expression. The bacterial system might serve as a model for an equivalent receptor in plants, although

An important technique for characterizing the interaction of a ligand with its receptor is to measure the affinity constant for binding. Usually such measurements require immobilization of the receptor on a support to expose the receptor to different concentrations of ligand and determine the corresponding amount of binding. According to how the data are treated, important constants characterizing the binding can be determined.

The most common way of dealing with binding data is to treat it on the basis of **occupancy theory,** which was first postulated in the 1930s. This theory requires that ligand [L] and receptor [R] bind together in a simple chemical equilibrium. Thus, $L + R \rightleftharpoons LR$. The velocity of the forward reaction is $K_1[L][R]$, and that of the reverse reaction is $K_{-1}[LR]$. At equilibrium, the rates of these two reactions are equal. A binding (or dissociation) constant can be defined as $K_d = K_{-1}/K_1$, where K_d is equivalent to the concentration of ligand at which half of the binding sites of the receptor are occupied. By manipulating this definition of K_d, the Scatchard equation can be derived:

$$B/F = R_T/K_d - B/K_d$$

where B is the concentration of bound ligand (i.e., [LR]), R_T is the total concentration of receptor binding site, and F is the concentration of free, unbound ligand ([L]). For a single class of receptive site, graphing B/F against B yields a linear plot in which the slope $= -1/K_d$ and the intercept on the B-axis equals the maximum density or concentration of receptor sites (see figure).

Scatchard plot

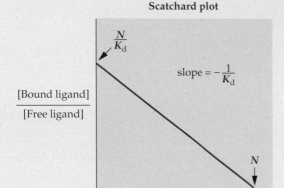

$$\text{slope} = -\frac{1}{K_d}$$

$\dfrac{N}{K_d}$

$\dfrac{[\text{Bound ligand}]}{[\text{Free ligand}]}$

N

[Bound ligand]

$N =$ Number of binding sites per unit protein

$K_d =$ Equilibrium constant for binding

$K_d = \dfrac{[\text{Receptor} \bullet \text{ligand complex}]}{[\text{Free receptor}]\,[\text{Free ligand}]}$ at equilibrium

---

no such putative plant protein has been identifed to date (see Section 18.2.4). However, we have no comparable models for the sensing of carbon dioxide, temperature, and nitrate, even though particular changes in these variables can cause very specific changes in development.

18.2.7 The bacterial two-component system, in which a receptor and an effector interact through phosphorylation of histidine and aspartate residues, may also be present in plants.

The sensory and transduction systems evolved by bacteria enable them to survive and adapt to various different environmental conditions. Bacterial transduction systems facilitate secretion, motility, sporulation, membrane transport, competence, virulence, and metabolic changes in response to a variety of signals. The simplicity of bacterial genetics and transformation technologies has enabled the characterization of many well-defined systems, all of which have some basic elements in common. At least 18 such systems have been identified in *Escherichia coli*, and at least 50 are probably present in prokaryotic cells. These systems, called **two-component systems,** may also be present in plant and fungal cells (Fig. 18.18A).

The first component of a two-component system is usually a receptor protein. The receptor contains a periplasmic domain that binds ligands, includes a variable number of transmembrane domains, and has a C terminal extension. When activated by binding a ligand, a kinase activity located in the C terminus autophosphorylates the receptor, transferring orthophosphate from ATP to a histidine residue. The active receptor is a homodimer in which each monomer phosphorylates the other.

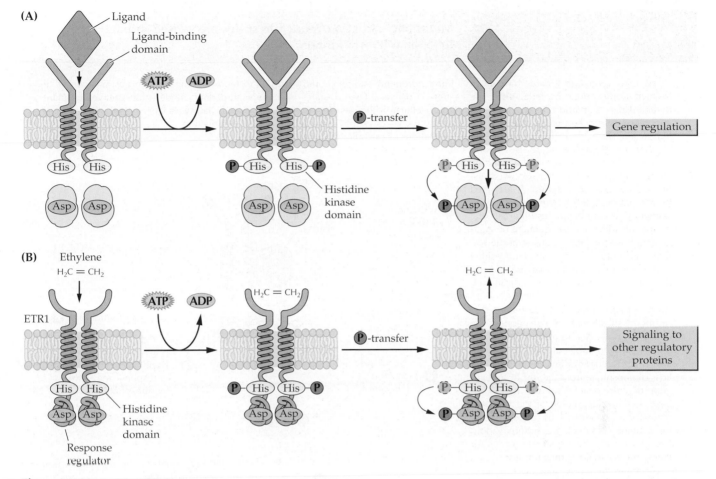

Figure 18.18
The two-component paradigm and the hybrid kinase modification. (A) The two-component system consists of a histidine kinase receptor that, on activation by ligand binding or other processes, autophosphorylates a histidine residue. This phosphate is then transferred to a conserved aspartyl residue on a response regulator, which is often a DNA-binding protein. Gene expression is then altered.

(B) In the hybrid kinase system, the histidine kinase is fused with part of a response regulator. An additional response regulator is typically required to control gene expression. The hybrid system may self-regulate the duration of the response by autoinhibition. The *Arabidopsis* ethylene receptor ETR1 is shown here as an example.

Reaction 18.1: Autophosphorylation of receptor

$$ATP + \text{Histidine kinase} \rightarrow$$
$$ADP + \text{Histidine kinase} \sim P$$

The second component, a **response regulator protein,** is activated when the receptor kinase transfers phosphate from its histidine residue to a conserved aspartate residue on the regulator. Removal of the phosphate group inactivates the response regulator.

Reaction 18.2: Phosphorylation of response regulator

$$\text{Histidine kinase} \sim P + \text{response regulator} \rightarrow$$
$$\text{Histidine kinase} + \text{response regulator} \sim P$$

Reaction 18.3: Dephosphorylation of response regulator

$$\text{Response regulator} \sim P \rightarrow$$
$$\text{Response regulator} + P_i$$

In the phosphorylated form, the response regulator controls a wide range of cellular activities, including motility and gene expression. The extent of phosphorylation of the response regulator is controlled by both the instability of the aspartyl phosphate bond and the protein phosphatase activity. The number of phosphorylated histidyl and aspartyl residues represents a balance between the activating signal and the phosphatase activity.

Box 18.2

Metabolic oxidation/reduction states are sensed in bacteria through a hybrid kinase.

At least 10 hybrid kinases have been isolated from various bacteria. Virulence mechanisms, starvation responses, and adaptation to different redox states all use hybrid kinases as part of the transduction mechanism. ArcB (see figure) is a hybrid kinase created by the fusion of a histidine kinase and a response regulator. ArcB histidine kinase, a redox sensor, is activated by interaction with electron carriers. Subsequently, a conserved histidine residue in the histidine kinase domain of ArcB is autophosphorylated. Transfer of the histidine phosphate to an aspartate residue in the response regulator domain activates the protein. Phosphate is continually transferred from the histidine kinase to ArcA, a transcription factor. This last step results in substantial amplification of the initial signal. Accumulation of many phosphorylated molecules of ArcA initiates specific changes in gene expression. Loss of phosphate from the response regulator domain deactivates the hybrid kinase. Dephosphorylation of the response regulator is controlled by phosphatases, the activities of which are, in turn, modulated by metabolites. A simple means for gating the response is thus present.

Changes in redox state inform plant cells of hazardous situations such as anaerobic conditions, high intensities of light, drought, salinity, and nutrient stress. Changes in redox states may impact cellular responses directly; alternatively, redox states may manipulate concentrations of cytosolic calcium. Bacterial redox-sensing mechanisms may constitute important models for unraveling plant cell pathways for redox sensing and transduction.

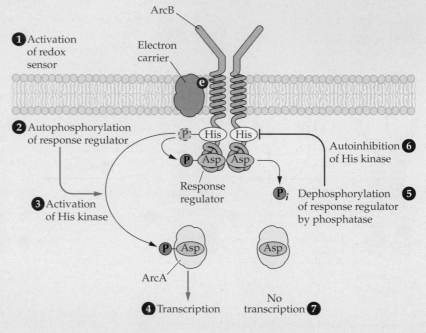

In several two-component systems, the histidine kinase and the ligand-binding domains are located on separate proteins, thus enabling many ligands to activate the same kinase. Another prominent variation is a hybrid kinase system, in which the response regulator is fused to the histidine kinase. The sensing of metabolic redox states in bacteria uses a hybrid kinase transduction system (Box 18.2). ETR-1, the ethylene receptor, is a prominent example of a hybrid kinase in plant cells (see Fig. 18.18B and Section 18.3.1).

18.2.8. Regulation of receptor concentrations can change the sensitivity of cells to signals.

Not all tissues or cell types are able to respond to all signals. For example, fruit tissues become sensitive to ethylene at a certain stage of ripening, whereas guard cells are totally insensitive to high concentrations of the gaseous hormone. Different responses by different tissues to the same signal can in part be explained by families of receptors. Auxin, for example, can induce pericycle cells to form adventitous or lateral roots, but in coleoptile cells it promotes elongation. Different receptors are probably involved in each response. However, divergent downstream elements of the signal transduction pathway may also distinguish the developmental responses to auxin exhibited by different cell types. Tissue-specific signal transduction pathways are thus defined not only by the presence or absence of receptors but also by the presence or absence of downstream apparatus required to transduce the responses.

Tissues can adapt or desensitize themselves to continuous signals, and receptor concentrations can change during development. For example, when etiolated seedlings are exposed to red or white light, the cellular concentrations of phytochromes decrease rapidly. The precise cellular mechanism of phytochrome degradation involves ubiquitin-mediated proteolysis (see Chapter 9), phosphorylation, or some other form of sequestration inside the cell. Phytochrome concentrations in green plants may be 100-fold less than those in etiolated seedlings. Consequently, the sensitivity of the green tissues to the greater light fluxes is adjusted so that phytochrome signals can still be sensed and used as a sensitive regulator of development. This behavior is predicted by the **Strickland–Loeb model of receptor action** (see below).

Modification of receptor concentrations can alter the **dose–response** relationship that links specific concentrations of ligand and receptor to physiological responses. Figure 18.19 illustrates the relationship between calcium and calmodulin concentrations and the activation of a calcium/calmodulin–dependent enzyme, cAMP phosphodiesterase. The relationship between calcium, calmodulin, and target enzymes results from unusual binding affinities among the three constituents. The K_d for calcium and calmodulin is approximately 10 µM; those for the calcium/calmodulin complex and the target protein can be as low as 1 nM. Signal transduction pathways take advantage of this relationship. For example, statocytes, the gravity-sensing cells in the root, contain concentrations of calmodulin an order of magnitude greater than those found in other root meristem cells. Statocytes may therefore respond to much smaller increases in cytosolic

calcium than other root cells do. Likewise, the mechanical signals that cause transient increases of cytosolic calcium in seedlings, i.e., wind and touch, also substantially increase calmodulin synthesis and accumulation. Thus, stimulation can render seedlings much more sensitive to further mechanical signals.

S. Strickland and J. N. Loeb demonstrated mathematically how variations in hormone receptor concentrations, hormone concentrations, and physiological effects of hormones such as protein kinase activation were interrelated. These workers stipulated certain criteria for the analysis:

- The hormone induces the synthesis or release of a second messenger such as cyclic 3′,5′-AMP (cAMP; discussed further in Section 18.5.1).
- The second messenger interacts with a downstream enzyme or protein such as cAMP-dependent protein kinase A (PKA; see Section 18.5.1).
- The affinity of the hormone for its receptor is weaker than the affinity of the second messenger for its target enzyme. That is, the K_d for the second messenger and its target enzyme is greater than the K_d for the hormone and its receptor.

Two important consequences follow if these criteria are fulfilled. First, hormone responses are activated fully when only a small proportion of the receptors are occupied; most receptors can therefore be described

Figure 18.19
Activation of cAMP phosphodiesterase by varying the concentrations of Ca^{2+} and calmodulin. The K_d for Ca^{2+} and calmodulin is about 1 to 10 µM, whereas that for the Ca^{2+}/calmodulin complex and phosphodiesterase is about 1 nM. This binding constant disparity allows full activation of the target enzyme when only a small amount of Ca^{2+}/calmodulin is formed. Activation depends directly on the concentrations of both Ca^{2+} and calmodulin. The x-axis variable is the pCa, the negative logarithm of the Ca^{2+} concentration (M). Illustrated here is cAMP phosphodiesterase activation as a function of calmodulin concentration (the molarity values given next to the individual activation curves).

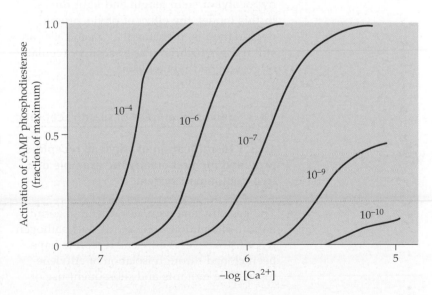

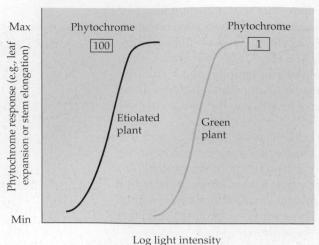

Figure 18.20
Phytochrome concentrations decrease by as much as 100-fold during deetiolation, enabling the plant to retain sensitivity to light over the wide range of light intensities to which dark-grown and green tissues are exposed. Phytochrome concentrations (in boxes) are shown in relative units.

as "spare." Second, changes in the dose–response relationship between hormone concentration and the physiological effects can be produced simply by manipulating receptor concentrations. These proposals suggest a situation analogous to the calcium/calmodulin target enzyme relationship illustrated in Figure 18.19.

If the Strickland–Loeb criteria apply to the deetiolation program, the rapid light-induced decrease in cellular phytochrome adjusts the sensitivity of the tissue to increased light flux. Figure 18.20 indicates a possible relationship between control of hypocotyl or stem length and light fluxes. In this model, the effect of phytochrome degradation is to ensure that green cells can still use phytochrome as a sensitive regulator of growth and metabolism.

18.3 Specific examples of plant receptors

18.3.1 Identification of ethylene receptors provided the first eukaryotic example of a two-component system.

The gas ethylene regulates ripening, germination, elongation, senescence, and pathogen responses. Several ethylene receptors have been cloned through isolation of ethylene-insensitive mutants and subsequent use of

molecular technology to identify the mutant gene. ETR1, a 79-kDa protein with a transmembrane domain, was the first receptor cloned from *Arabidopsis.* The C terminus of ETR1 is homologous to a bacterial two-component system hybrid kinase (see Section 18.2.7). ETR1 exists as a dimer in the plasma membrane. Ethylene joins the two monomers together and permits intermolecular phosphorylation (Fig. 18.21).

Mutations in *ETR1* (designated *etr1*) lead to loss of physiological sensitivity to ethylene. Figure 18.22 illustrates some of the physiological effects of expressing an *etr1-1* transgene. ETR1 has also been expressed in yeast to demonstrate that the protein binds ethylene with high affinity. Competitive ethylene antagonists inhibit this binding. Expression of *etr1* in yeast leads to loss of ethylene binding, confirming that ETR1 is thus a true ethylene receptor.

Genes encoding other ethylene receptors have also been identified, including *ERS* (ethylene response sensor), *Nr* (never ripe, a developmentally regulated gene from tomato; see Fig. 18.22A), and *LeTAE1* (a tomato *ETR1* homolog expressed during flower and fruit senescence).

18.3.2 Many auxin-binding proteins have been detected, but whether they represent receptors for different auxin-mediated processes is still uncertain.

Indole 3-acetic acid (IAA, referred to here as auxin) is a growth regulator with a wide variety of functions in cell division and expansion (see Chapter 17). Auxin has been studied

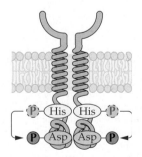

Figure 18.21
The *Arabidopsis* ETR1 protein has been identified as a receptor for ethylene. On binding ethylene, the receptor protein dimerizes and initiates signaling by autophosphorylation and phosphate transfer.

intensively for the past 50 years and, not surprisingly, receptors for the auxin signal have been actively sought. Conventional pharmacological techniques have uncovered one well-characterized **auxin-binding protein** (ABP1). The possible receptor function of this protein was controversial for many years (Box 18.3) but has recently been established.

Use of various affinity-binding techniques and immunoaffinity-labeling approaches has detected several other auxin-binding proteins, including a glutathione *S*-transferase, a (1→3)β-glucanase, and a cytokinin glucosidase. The significance of these proteins to auxin action is unclear at present; perhaps they are expressed by the cell in response to the stressful conditions required for labeling the proteins in vivo. Detecting receptors by isolation of auxin-resistant mutants has yielded some recent success. *AXR1* encodes a protein with similarity to a ubiquitin-conjugating enzyme and may participate in detoxification of auxin (see Section 18.8.3); *AUX1* is a member of a family of closely related genes and encodes an auxin permease with 10 membrane-spanning domains. The roots of auxin-resistant mutants *axr1* and *aux1* have weak or nonexistent

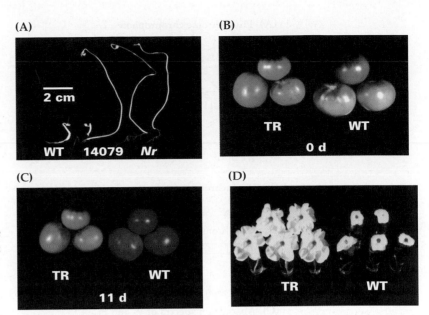

Figure 18.22

Ethylene-insensitive phenotypes displayed by transgenic (TR) plants expressing variants of the *etr1-1* gene. (A) Wild type (WT), TR line containing the *etr1-1* gene (14079), and Never ripe (*Nr*) tomato seedlings germinated in the dark on medium containing the ethylene precursor 1-aminocyclopropane-1-carboxylic acid (ACC). (B, C) Development of three postmature green TR fruits and three mature green WT fruits after storage for 0 (B) and 11 (C) days. (D) Detached TR and WT petunia flowers 16 hours after treatment with ethylene.

Box 18.3 The role of ABP1 is currently under investigation.

ABP1 is a small family of 23-kDa proteins that bind indole 3-acetic acid (IAA) and naphthalene-1-acetic acid (NAA) as well as other molecules with auxin activity. Despite strong circumstantial evidence for its receptor function, critical evidence of the kind obtained for the ethylene receptor is still lacking; this may soon be resolved, however.

Experiments with tobacco protoplasts provide the best early evidence for the role of ABP1 as an auxin receptor. Adding auxin to tobacco protoplasts causes hyperpolarization of the membrane potential; adding ABP1 to the protoplasts greatly increases their sensitivity to auxin (see figure). Adding antibodies against ABP1 to the protoplasts increases the concentration of auxin required to elicit an equivalent response in an untreated control. Moreover, a decapeptide synthesized from the C terminus of ABP1 has auxin-like activities on *Commelina* guard cells. ABP1 contains a C-terminal HDEL putative ER-

retention domain, although small amounts of the protein are found in the plasma membrane as well.

Overexpression of *ABP1* has been achieved in tobacco. Although the phenotype of the transgenic plant remained unaltered and the leaf area unchanged, the size of the average leaf cell increased

more than threefold, most probably because of fewer cell divisions. The results indicate a role for ABP1 in controlling aspects of cell expansion but also suggest the presence of other auxin receptors regulating overall leaf growth in coordination with ABP1, which is responsible for just one aspect: cell elongation.

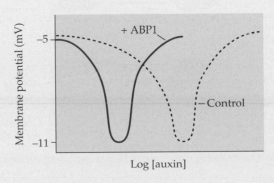

(A) Phytochrome chromophore

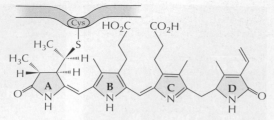

(B) Phytochrome protein

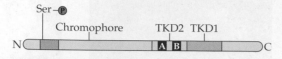

(C) Absorption spectra of P_R and P_FR

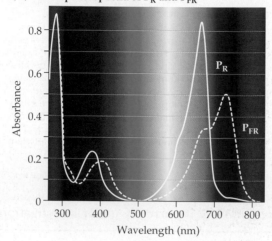

Figure 18.23
Structure and function of phytochromes.
(A) The phytochrome chromophore is a tetrapyrrole that binds to the phytochrome protein.
(B) The C termini of plant phytochromes may contain two distinct transmitter kinase domains, TKD2 and TKD1.
(C) Absorption spectra of P_R and P_{FR} show peaks for red (660 nm) and far-red (730 nm) light, respectively.
(D) Phytochromes can exist as either of two interconvertible forms. Isolated phytochromes in a concentrated solution can undergo reversible changes in absorbance induced by illumination with red or far-red light. These light-induced structural changes are coincident with physiological and developmental changes induced by red light or blocked by far-red light.

(D) Phytochrome activities

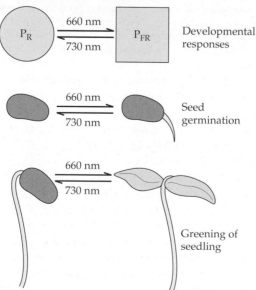

gravitropic responses, but their shoot responses are unaffected, implying that the gravity signal is transduced by different mechanisms in roots and shoots. *AUX1* is a member of a family of closely related genes. Transgenic expression of *AUX1* complements the agravitropic characteristic of *aux1*; expression of *AUX1* is high in root epidermis, where it possibly regulates root growth rates. Some evidence suggests an association of auxin transport with the perception of auxin. More recent investigations have established that AtPIN1, a 67-kDa protein with similarity to bacterial and eukaryotic carrier proteins, is a transmembrane component of the auxin efflux carrier. Mutations in the gene encoding this protein inhibit polar auxin transport and modify production of lateral organs and formation of vascular tissue. In the root, AtPIN2, another member of this auxin efflux family, may control the movement of auxin in the elongation zone. Mutations in the gene for AtPIN2 render roots agravitropic.

18.3.3 Phytochrome, a clearly identified receptor for red light, has protein kinase activity in cyanobacteria.

Red light controls leaf expansion, shade-avoidance reactions, germination, and response to photoperiod. A particular characteristic of many red light–induced effects is their reversibility when an inductive flash of red light (660 nm) is followed by an immediate flash of far-red light (730 nm) (Fig. 18.23). This characteristic was used to identify and purify the red-light receptor phytochrome in 1965, one of the primary achievements of 20th-century plant science. The purified receptor protein undergoes reversible changes in structure in vitro when exposed to red or far-red light.

Phytochromes form a family of 120-kDa proteins (Box 18.4). The photoreactive moiety (chromophore) of these proteins is an open-chain tetrapyrrole. Two forms of phytochrome, A and B, can each form dimers in solution, and physiological evidence suggests that both may dimerize in vivo. The far-red–absorbing form, P_{FR}, is considered to be the active form of phytochrome, although most photomorphogenic

Physiological evidence first revealed that the stability of phytochromes differed in etiolated and deetiolated tissues (see Section 18.2.8). Along with other evidence, this indicated the existence of two pools of P_{FR}, a labile pool and a stable pool—pools that have since been found to represent different phytochromes.

Evidence of distinct types of phytochromes has been provided by the cloning and sequencing of five genes from *Arabidopsis* (designated *PHYA* through *PHYE*) that encode phytochromes A through E. The amino acid sequences of PHYA, PHYB, and PHYC are equally divergent and demonstrate about 50% homology to each other. Mutant analysis has revealed some of the separate functions performed by the various phytochromes. The cucumber *lh* mutant and the *Arabidopsis hy3* mutant, which grow taller than wild-type plants in white light, are deficient in PHYB. The *aurea* mutant of tomato is deficient in PHYA and has been used to dissect phytochrome responses (see Fig. 18.42). A recent mutant screen has identified *PHYC* mutants in *Arabidopsis*, and phenotypic analysis of a *PHYD* mutant in one *Arabidopsis* ecotype indicates that PHYD regulates several of the same responses as PHYB.

Different action spectra have been deduced for PHYA and PHYB. The figure illustrates the capacity of various wavelengths of light to induce seed germination. The action spectrum of PHYA is typically determined on seed imbibed in water for 2 days in darkness; surprisingly, it does not show photoreversibility. The action spectrum for PHYB is determined on seed imbibed in water for very short periods and is reversible by exposure to far-red light.

Several forms of phytochrome response are classified as the high irradiance reaction (HIR) and the induction reaction, based on the fluence and timing of irradiation. The induction reaction is further subdivided into a low fluence red/far-red reversible reaction (LFR) and a very low fluence response (VLFR). LFR and VLFR can be activated by short pulses of light. LFR is the classical low fluence red/far-red reversible control of light-dependent seed germination and is typically detected after only a few minutes of seed imbibition. Mutational studies have implicated PHYB as the prime regulator of LFR. VLFR, a more sensitive response, can be detected in seeds after imbibation for several days in darkness or in many etiolated tissues likewise kept in darkness. Surprisingly, VLFR is not reversible by far-red light, but mutational investigations on seedlings deficient in phytochromes clearly implicate PHYA as the regulator.

HIR requires continuous or prolonged irradiation with high-intensity light. The response in this case is proportional to the irradiance received by the plant; again, photoreversibility is absent. Typical HIR responses are anthocyanin synthesis or inhibition of hypocotyl elongation. Although phytochromes clearly are involved in these responses, evidence indicates that other photoreceptors that absorb UV or blue light contribute to this control.

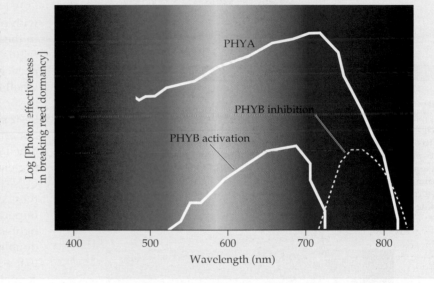

reactions are thought to result from the cellular ratio of P_{FR} to P_R, the form that absorbs red light.

An early hypothesis suggested that P_{FR} is a protein kinase. Plant phytochromes often copurify with a protein kinase activity, but highly purified or cloned phytochromes, and phytochromes expressed in bacteria, do not appear to have conventional protein kinase activity. A phytochrome cloned from the moss *Ceratodon*, however, contains an additional sequence showing homology to protein kinase catalytic domains. The cyanobacterium *Synechocystis* also contains a phytochrome (Cph1). The N terminus of the prokaryotic protein is similar in sequence to that of plant phytochrome, whereas the C terminus contains consensus sequences that define a two-component histidine kinase (Fig. 18.24). A critical histidine has been shown to be autophosphorylated, and the phosphate can be transferred to a conserved aspartyl residue in Rcp1, a separate protein. Only the P_R form of Cph1 displayed substantial kinase activity. Down-regulation of histidine kinase activity may result from transfer of the phosphate on the histidine to a serine residue in the N-terminal region.

18.3.4 The blue light receptor is a DNA lyase-like protein or a protein kinase.

The most prominent blue light responses are phototropism, inhibition of hypocotyl elongation, stomatal opening, anthocyanin production, and expression of blue light–regulated genes. Blue light initiates changes in plant cell plasma membrane potential (Fig. 18.25), participates in redox reactions and electron transport, and induces at the plasma membrane a change in light absorbance involving a flavin-mediated photoreduction of cytochromes.

Cryptochrome, the blue light receptor encoded by the *Arabidopsis CRY1* gene, has been isolated by screening T-DNA insertion lines for blue light–insensitive mutants (see Fig. 18.17A). Genetic evidence suggests that CRY1 is one member of a family of receptors. Cryptochrome has sequence similarity to DNA photolyases, a rare class of flavoproteins that catalyze blue light–dependent electron transfer reactions in bacteria. However, the cryptochrome encoded by *CRY1* does not have detectable DNA-lyase activity.

The *Arabidopsis* mutant *nph1* (nonphototrophic hypocotyl 1) fails to show a phototropic response in blue light. Sequence analysis and biochemical studies have established that NPH1 has protein kinase activity. Specific amino acid domains in NPH1, termed LOV1 and LOV2, share sequence similarity with proteins from bacteria and eukaryotes. In these organisms, light, oxygen, or voltage regulates the activity of LOV proteins, which participate in redox sensing. It is still unclear whether NPH1 is a primary blue light receptor or a downstream component of a blue light signal transduction pathway. Irradiation of isolated plasma membranes with blue light results in detectable changes in protein phosphorylation. The identification of NPH1 as a protein kinase might explain these early observations. NPH1 is clearly an important transduction component in blue light–controlled processes. Formation of double mutants of *cry1* and *nph1* has shown that a separate blue light sensor regulates stomatal aperture.

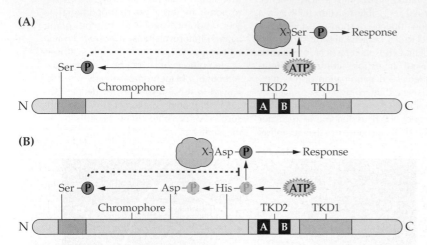

Figure 18.24
Two hypotheses for how kinase activity may mediate phytochrome action in *Synechocystis*. (A) The first model assumes that phytochrome is a serine kinase able to carry out both phosphorylation of a substrate to initiate signaling and autophosphorylation to down-regulate phytochrome activity. (B) The second mechanism assumes phytochrome to be a type of hybrid kinase in which a conserved histidine residue is phosphorylated as a result of light absorption. Transfer of this phosphate to an aspartyl residue and then to a serine residue would down-regulate phytochrome activity.

18.3.5 Is the ABA receptor a protein phosphatase?

The plant growth regulator ABA controls aspects of plant adaptation to variations in the amounts of soil water and responses to water stress. Probably different ABA receptors are used for processes such as seed germination, stomatal closure, or altered gene expression. Work with caged ABA has provided evidence for the presence of ABA receptors inside cells. (Use of caged molecules is discussed further in Box 18.5.) When caged ABA was loaded into individual guard cells and released, the

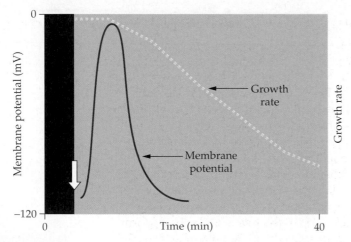

Figure 18.25
Effect of blue light on membrane potential and hypocotyl growth. Etiolated hypocotyls irradiated with blue light (white arrow) undergo a pronounced depolarization of the membrane potential of individual hypocotyl cells, followed by a marked decline in the rate of growth.

Box 18.5

Caged molecules can be used to demonstrate the activity of signaling pathways in plant cells.

One of the primary advances in cell biology technology has been the development and use of photoreactive or "caged" probes. Reaction with a caging group renders the signaling molecule inactive. A flash of UV light can then be used to release the caging moiety. By loading the caged, inactive molecule into cells, the spatial and temporal release of the active molecule inside a cell can be controlled experimentally. A variety of caged molecules are available for study, including hormones (see Fig. 18.26), nucleotides, chelators, and second messengers such as Ca^{2+} and inositol 1,4,5-triphosphate (IP_3). Recent methods have detailed the construction of caged peptides and even caged proteins by modifying critical amino acids. Importantly, loading and photolysis of caged molecules mimic the responses normally induced by ABA, red light, incompatible S-proteins, and treatments that reorient pollen tubes.

Panel A of the figure illustrates an experiment in which caged IP_3 (see structure) is loaded into a guard cell and photolyzed. Brief pulses of UV light release IP_3 into the cell, which is followed by a transient increase of cytosolic Ca^{2+} over several minutes. If a threshold of about 500 nM for $[Ca^{2+}]_i$ is exceeded, the guard cells close (see graph). A similar link between photolysis of caged IP_3 and increased $[Ca^{2+}]_i$ has been determined experimentally in pollen tubes and red light–sensitive protoplasts, indicating the presence of IP_3-sensitive channels in these cell types. Thus, as in animals, IP_3 couples signals to the release of Ca^{2+} from stores within plant cells.

Caged Ca^{2+} (i.e., complexed with diphenyl EGTA) has been released into different regions of the pollen tube to demonstrate that $[Ca^{2+}]_i$ in the pollen tube tip controls orientation of the tube. Pollen tube growth is confined to the apical dome and results from secretion of vesicles containing cell wall material into the wall of the dome. Using focused microbeams of UV light, the cage can be photolyzed in different regions of the pollen tube. A Ca^{2+}-sensitive dye is loaded into the pollen tube and images are generated in which the approximate Ca^{2+} concentrations are color-coded (blue corresponding to low Ca^{2+}, red to high Ca^{2+}), so that the consequences of Ca^{2+} release can be followed. In the fluorescence micrographs shown in panel B of the figure, photolysis was carried out between 30 and 100 seconds after recording started. Images were collected with a confocal microscope (bright-field images of the same pollen tube are included as insets).

When Ca^{2+} was specifically released in one side of the pollen tube tip (indicated by arrow), the pollen tube polarity was altered and the direction of growth was reoriented toward the side having the higher apical concentration of Ca^{2+}.

(A)

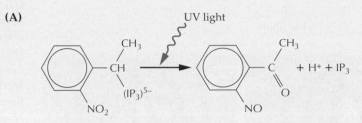

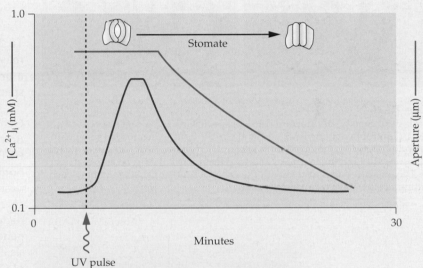

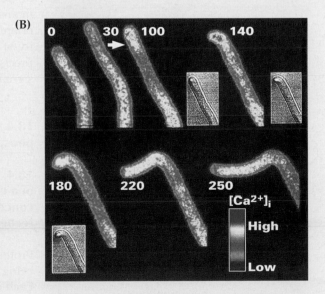

(A)

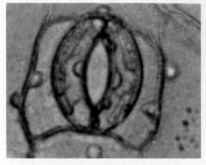
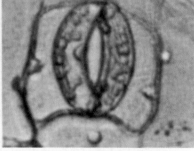
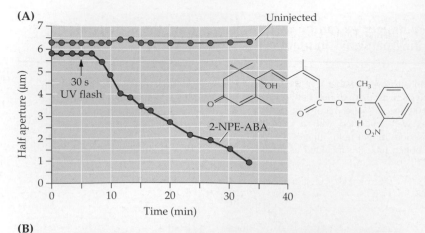

(B)

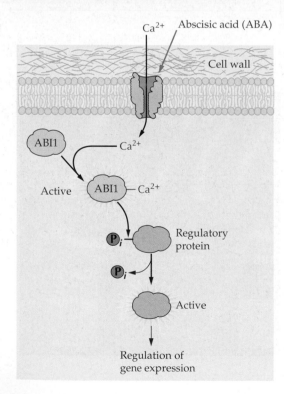

Figure 18.26
Photolysis of caged ABA in guard cells causes closure. (A) A caged species of ABA, 1-(2-nitrophenyl)ethyl-ABA, is labile to UV light. When loaded into guard cells and photolyzed, the caged form is converted to free ABA and causes guard cell closure. (B) Light micrographs showing the impact of ABA on stomatal aperture. The guard cell was loaded with the caged ABA (left panel) and photolyzed, resulting in stomatal closure (right panel).

Figure 18.27
Possible transduction route for ABA. ABA is thought to bind to a receptor (not shown) that allows direct entry of Ca^{2+} at the plasma membrane and activates a Ca^{2+}-sensitive protein phosphatase (ABI1).

guard cell closed (Fig. 18.26). Other evidence suggests the existence of external ABA receptors as well.

At least 10 *abi* (ABA–insensitive) mutants have been identified, indicating that a complex signaling network operates in seed dormancy and other processes involving ABA. Many of the mutants exhibit a wilted phenotype, but the *abi1* mutation interferes with the widest range of responses. ABI1 has sequence similarity to a serine/threonine protein phosphatase that binds Ca^{2+} (see Section 18.8.4). A protein kinase cascade is probably activated by ABI1-catalyzed dephosphorylation (Fig. 18.27). ABI1 does not bind ABA, however, and therefore is not activated directly by ABA. Perhaps ABI1 specifically binds to the ABA receptor or to a receptor complex that also involves a protein kinase.

18.3.6 Cytokinin sensing and transduction may use a two-component system and a protein with seven membrane-spanning domains.

Cytokinin-independent mutants such as *cki1* do not require cytokinin for cell division. CKI1, a 125-kDa protein, has weak sequence homology to a hybrid two-component kinase (Fig. 18.28). CKI1 has two membrane-spanning domains, and its N terminus is predicted to reside outside the cell. A second gene, *GCR1*, has been isolated from *Arabidopsis* cDNA libraries. The deduced structure of GCR1 has seven membrane-spanning domains. Expression of *GCR1* in the antisense orientation specifically reduces the sensitivity of the transformants to cytokinins. The precise function of both proteins in cytokinin signaling, however, remains to be clarified.

18.4 G-proteins and phospholipid signaling

18.4.1 G-proteins, a special subset of a GTPase superfamily, may all be concerned with aspects of accuracy of recognition or interaction.

Proteins that bind and hydrolyze GTP are being discovered at a rapidly increasing rate. Each of these many **GTPases** acts as a molecular switch in which the "on" and "off" states are triggered by binding and hydrolysis of GTP. Conserved structure and mechanism in different types of GTP-binding proteins suggest they are all derived from a single primordial protein that has been repeatedly modified during evolution to perform a large variety of functions.

Figure 18.29 illustrates the basic cycle that occurs in all GTPase proteins. The empty state combines with a GTP to form the active state. Hydrolysis of GTP to GDP inactivates the protein, and subsequent dissociation of GDP returns the protein to the empty state. Two protein families control aspects of this cycle. Members of one family, the GTPase-activating proteins (GAPs), accelerate the rate of GTP hydrolysis. The other set, called guanine nucleotide release proteins (GNRPs) or GDP exchange factors (GEFs), controls release of GDP from the inactive GTPase. Two broad classes of GTPases can be distinguished in mammalian cells. **Heterotrimeric G-proteins,** composed of three subunits (α, β, and γ), are specifically involved in recognition of activated receptor states. The other class, **monomeric GTPases,** includes Ras (Fig. 18.30), Ypt, Rab, Rap, Sec, Rho, and EFTu.

Although GTPases act as switches, their primary function may be to improve the accuracy of recognition during signaling, secretion, and protein synthesis. A specific function for a GTPase was first identified in studies of bacterial protein synthesis. In this process the complex of mRNA, ribosomes, and aminoacyl tRNA is joined by a GTPase (see Chapter 9). To ensure accurate codon: anticodon interactions, the stability of this complex is measured against the rate of hydrolysis of GTP. The amino acid is not added to the growing peptide chain until GTP is hydrolyzed. If the recognition is inaccurate, the aminoacyl tRNA/mRNA codon interaction is less stable and dissociates before GTP hydrolysis has occurred. This form of proofreading during translation improves the accuracy of protein synthesis 100-fold.

The requirement for accurate recognition may also explain the involvement of monomeric GTPases in the docking of vesicles with target membranes during secretion as well as G-protein functioning in ligand–receptor interactions. During signal transduction, enhanced accuracy of ligand–receptor recognition ensures amplification of the correct signal.

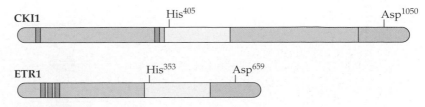

Figure 18.28
Comparison of the structures of CKI1, a putative cytokinin receptor, and ETR1, the ethylene receptor. CKI1 contains two transmembrane domains (shown in green) and conserved histidine and aspartyl residues, which have been recognized by sequence alignment with various two-component hybrid kinases. ETR1 contains three transmembrane domains and conserved histidine and aspartyl residues.

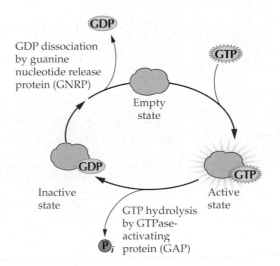

Figure 18.29
The fundamental cycle associated with all signaling GTPases. GTPases are converted from an empty state to an active state by association with GTP. GTP hydrolysis, catalyzed by GTPase-activating protein (GAP), inactivates the GTP-binding proteins. GDP release, catalyzed by guanine nucleotide release proteins (GNRP), returns the GTP-binding protein to the empty state.

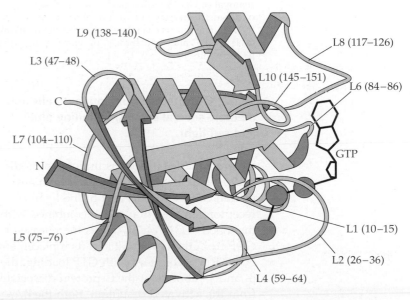

Figure 18.30
Cartoon of the three-dimensional structure of the guanine nucleotide domain of human Ras bound to GTP; the structure was determined by using a nonhydrolyzable GTP analog. The 10 loops, which connect six strands of β sheet and five α helices, are labeled. The structure of Ras serves as a model for other small GTPases found in animals, plants, and fungi.

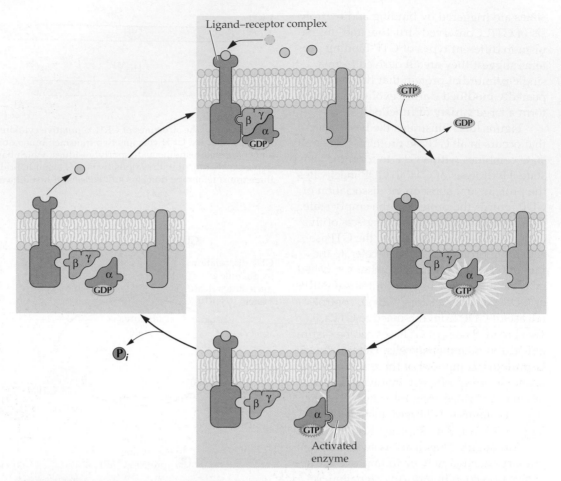

Ligand–receptor complex

GTP

GDP

GTP

Activated
enzyme

P$_i$

GDP

Figure 18.31

The activity cycle of heterotrimeric G-proteins. The ligand–receptor complex acts as the GNRP to catalyze removal of GDP. Binding of GTP to the α-subunit results in its dissociation from the β/γ-subunits. Activation of downstream enzymes such as phospholipase C follows binding by the G-protein α-subunit still bound to GTP. The intrinsic GTPase activity of the α-subunit (stimulated by GAP; not shown) hydrolyzes the GTP to GDP, after which the G-protein returns to the inactive state and the α-subunit reassociates with the β/γ-subunits.

18.4.2 G-proteins found in plant cells may mediate several signals, including blue and red light.

Heterotrimeric G-proteins undergo a modified cycle with GTP (Fig. 18.31). In mammalian cells, GNRP is part of the ligand–receptor complex that, when combined with the inactive G-protein, causes the release of GDP from the α-subunit and its replacement by GTP. Concomitant with GTP loading, the β- and γ-subunits of the G-protein dissociate from the activated α-subunit. Both the β/γ-complex and α-subunit can then interact with and activate other proteins. In plants, the function of heterotrimeric G-proteins and the identities of the downstream target proteins are still uncertain, but phospholipase C

(PLC; see Section 18.4.3) and Ca^{2+} channels are prominent downstream candidates. Others may include potassium channels, phospholipase A$_2$ (PLA$_2$), and perhaps cGMP phosphodiesterase. The α-subunit has intrinsic GTPase activity and, on hydrolysis of GTP, recombines with the plasma membrane by way of the β- and γ-subunits to complete the cycle. Signal amplification is inherent in the G-protein cycle because one activated G-protein can interact with and activate numerous target proteins.

The involvement of G-proteins in signaling can be investigated by using GTP analogs that are not easily hydrolyzed by GTPase, such as GTPγS or Gpp(NH)p (Fig. 18.32A). When microinjected into cells, these molecules can induce responses in the

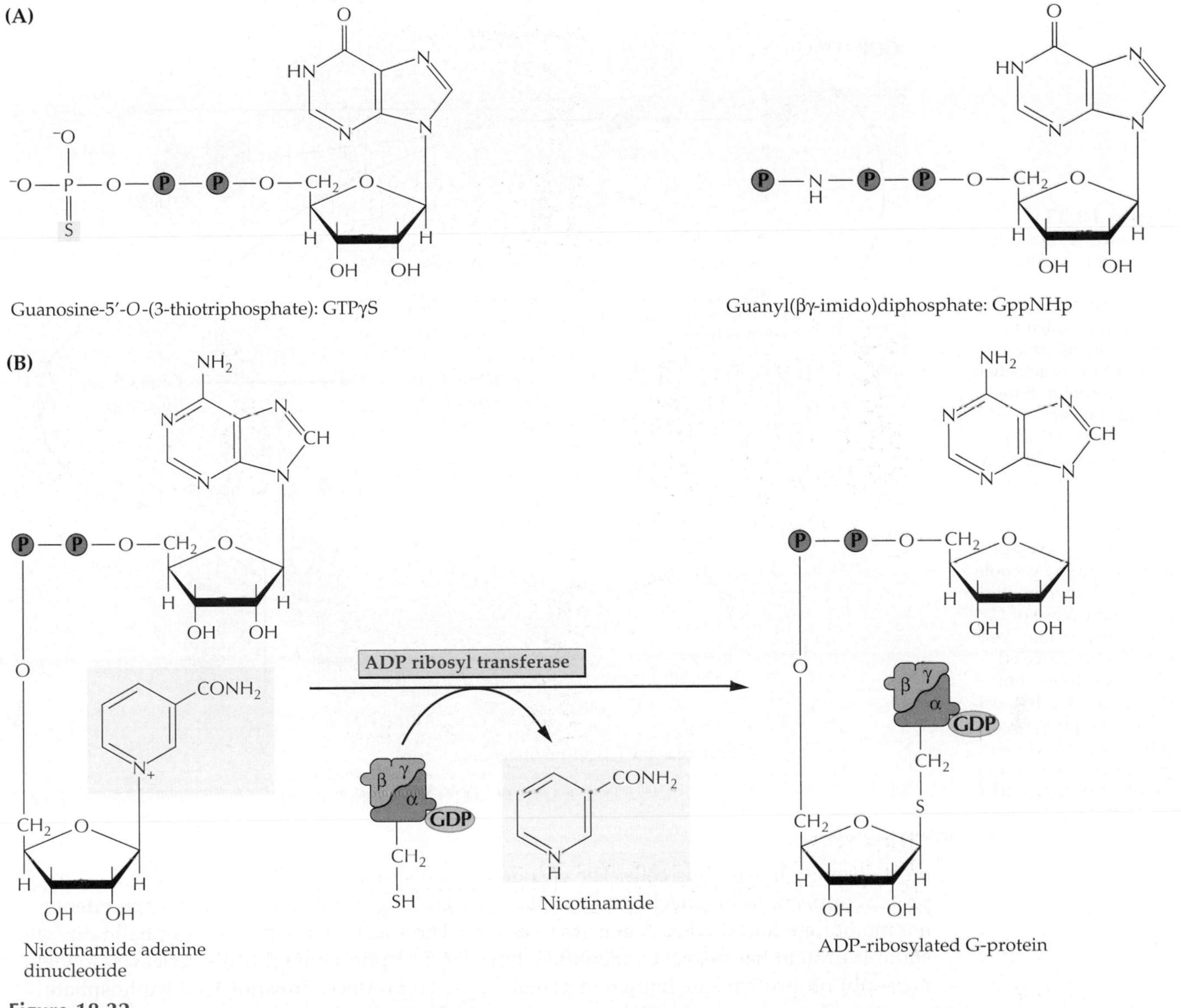

(A)

Guanosine-5'-O-(3-thiotriphosphate): GTPγS

Guanyl(βγ-imido)diphosphate: GppNHp

(B)

ADP ribosyl transferase

Nicotinamide adenine dinucleotide

Nicotinamide

ADP-ribosylated G-protein

Figure 18.32
Structure and action of compounds that inhibit the G-protein cycle. (A) Structures of two nonhydrolyzable analogs of GTP: GTPγS and GppNHp. (B) NAD$^+$ serves as a substrate for cholera and pertussis toxins, which transfer ADP-ribose to proteins. The ADP-ribosyl transferase activity of pertussis toxin can modify and thereby inactivate the trimeric G-protein but does not affect the free α-subunit bound to GTP.

absence of a signal if G-proteins participate in a specific transduction pathway. However, these analogs will not discriminate between heterotrimeric G-proteins and monomeric GTPases. More useful and specific is cholera toxin, which uses NAD$^+$ to transfer an ADP-ribosyl group to the α-subunit of a heterotrimeric G-protein, inhibiting its GTPase activity so that the activated (GTP-bound) state lasts much longer (Fig. 18.32B). Pertussis toxin also catalyzes transfer of ADP ribose from NAD$^+$ to the GDP-bound form of the G-protein α-subunit, which results in irreversible inactivation of the G-protein. The involvement of G-proteins in signaling can therefore be assessed by detecting increased or decreased G-protein activity.

The approaches discussed above have revealed the involvement of G-proteins in the transduction of blue light, red light, auxin, and gibberellin signals and in stomatal aperture regulation. For example, GTPγS was found to reduce inward K$^+$ current in guard cells. Analysis of the expression of the gene for the only G-protein α-subunit known in plants suggests that it is expressed

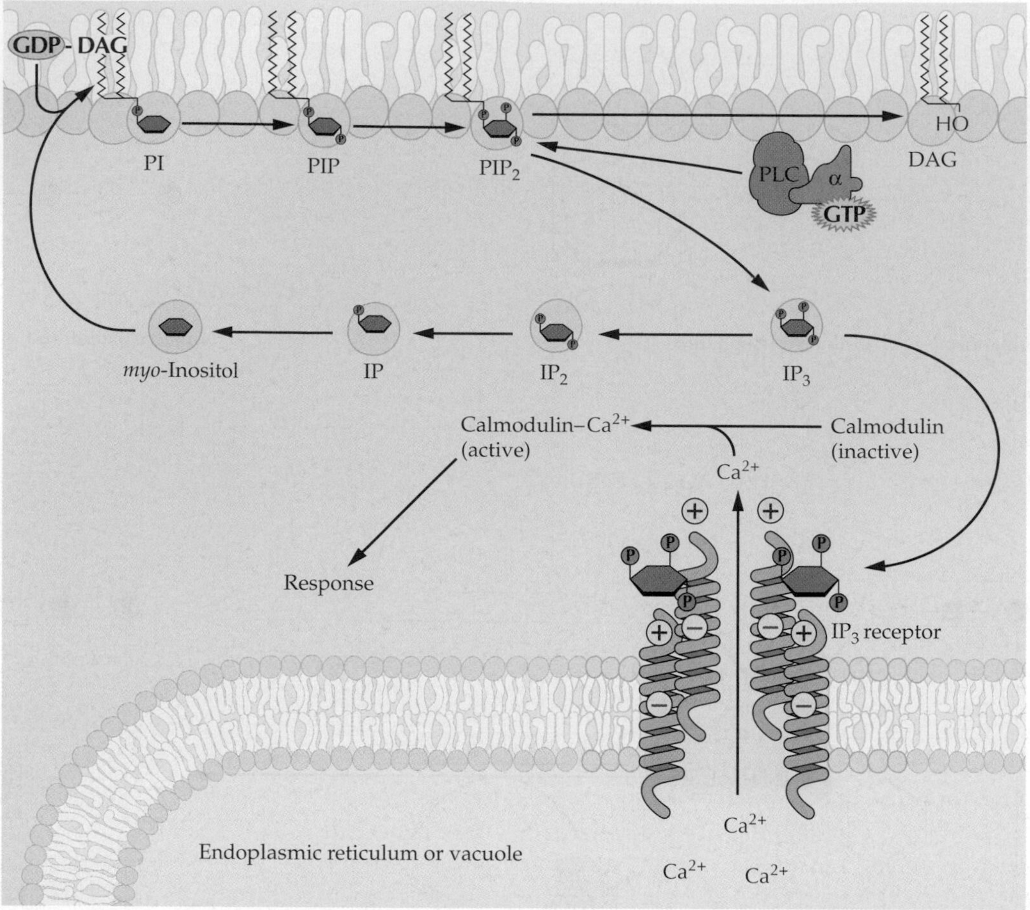

Figure 18.33
Sequence of transduction events leading from activation of phospholipase C (PLC) to increase of cytosolic calcium. Plasma membrane–bound PLC is activated by a G-protein. Phosphatidylinositol 4,5-bisphosphate (PIP_2) is hydrolyzed by PLC to produce the second messengers IP_3 and diacylglycerol (DAG). IP_3 activates the IP_3 receptor attached to the vacuole or the ER, thereby initiating the release of Ca^{2+}. The phospholipid components are recycled phosphatidylinositol (PI), phosphatidylinositol 4-monophosphate (PIP), inositol 1,4-bisphosphate (IP_2), and inositol 1-phosphate (IP).

most strongly in dividing cells. The encoded protein appears to be attached to the plasma membrane and the ER. A gene for a β-subunit protein has also been identified, but no γ-subunit proteins are known in plants.

Rho-like monomeric GTPases have been isolated from several plants. Present indications suggest they have specific functions in pollen tube growth and in cytoplasmic streaming. Rho-like proteins may also be involved in the regulation of actin/myosin–dependent reorganizations of the cytoskeleton during cell division (see Chapter 5).

18.4.3 Phospholipases in the plasma membrane can be activated by G-protein–coupled receptors.

PLC participates with several kinases and phosphatases in an important cycle of inositol phospholipid synthesis and breakdown. PLC, which has been purified from several plants, appears to be a family of pro-

teins that respond to different signals, including one PLC that is Ca^{2+}-dependent. The substrate for PLC—phosphatidylinositol 4,5-bisphosphate (PIP_2)—is cleaved to yield two products: **inositol 1,4,5-triphosphate** (IP_3), a soluble second messenger, and **diacylglycerol** (DAG), which remains membrane-bound (Fig. 18.33). PIP_2 formation is catalyzed by two kinases that successively convert phosphatidylinositol (PI) to phosphatidylinositol 4-phosphate (PIP) and then to PIP_2. A current model suggests that occupation of appropriate receptors by ligands in animals or activation of receptors by red or blue light in plants results in exchange of bound GDP for GTP on the G-protein. The α-subunit bound to GTP is released and activates PLC (Fig. 18.33). IP_3 diffuses freely in the cytoplasm and binds to specific Ca^{2+} channels in the vacuole and the rough ER. Opening these channels releases Ca^{2+} into the cytoplasm.

Inositol phospholipids can also interact with cytoskeletal proteins and thus signal important changes in cellular structure. PIP_2 has been shown to bind to two important proteins involved in microfilament organization. Profilin (see Chapter 5, Box 5.2) is able to bind actin monomers. When phosphorylated, profilin releases actin, which can then participate in the formation of microfilaments. A regulatory domain that binds PIP_2 has been detected in profilin. Gelsolin, a Ca^{2+}-activated F-actin–fragmenting protein, also possesses a PIP_2-binding domain, and PIP_2 is known to modulate gelsolin activity.

18.4.4 The IP_3 signal is constrained by active phosphatases that are sensitive to lithium.

IP_3 is able to mobilize the stores of Ca^{2+} inside the vacuole and rough ER by binding to specific receptors that also act as calcium channels (Fig. 18.33). Both these organelles can accumulate substantial amounts of Ca^{2+}. Associated with the ER are proteins such as calreticulin and calsequestrin, which have many low-affinity binding sites for Ca^{2+}. The IP_3-binding protein has been detected in plant cells and, as in animal cells, its binding is inhibited by heparin. The IP_3 signal is truncated by the activities of phosphatases that successively remove the phosphates to yield sequentially IP_2, IP, and inositol, this last then being reused for synthesis of PI (Fig. 18.33). The phosphatase that degrades IP to inositol has been cloned and, like the animal enzyme, is inhibited by Li^+ ions. Li^+ can thus be used as a general inhibitor of the IP_3 signaling pathway.

18.4.5 Do diacyl lipids play any function in plant cell signaling?

The hydrolysis of phosphatidylinositols generates IP_3 and DAG. In animal cells both products are used to activate downstream signaling components. Some signals activate phospholipases that hydrolyze PIP and PI as well as PIP_2, giving rise to a massive increase in DAG. DAG then activates protein kinase C (PKC) by altering its sensitivity to Ca^{2+} ions. PKC is involved in many signaling events because its target proteins function in the signaling pathways involving cell growth and division (Fig. 18.34). Although typically PKC is activated by a combination of DAG

and Ca^{2+}, enzyme variants have emerged that are DAG-activated but Ca^{2+}-insensitive. All PKCs require phosphatidylserine for activity, and some use unsaturated fatty acids or lysophospholipid as cofactors. PKC is permanently activated by phorbol esters, plant products suspected as carcinogens in mammalian cells (see Chapter 24).

A true plant PKC—i.e., a protein with PKC activity, cofactor requirements, and sequence homology to PKCs in other systems—has not yet been detected, but plant protein kinases activated by phosphatidylserine and Ca^{2+} and activated by phosphatidylserine, Ca^{2+}, and phorbol esters have been reported. Other plant protein kinases activated by free fatty acids or lysophospholipids have been identified, some of which require Ca^{2+}. These proteins may represent functional equivalents to PKC. The absence of a signaling role for DAG in plants seems unlikely, and some evidence points to its direct involvement in the growth of pollen tubes and regulation of stomatal aperture.

18.4.6 Phospholipases A and D can generate other signaling molecules and may be regulated through G-proteins.

In animals, phospholipase A (PLA) catalyzes the hydrolysis of phospholipids to yield lysophospholipids and free fatty acids (Fig. 18.35). The most common free fatty acid

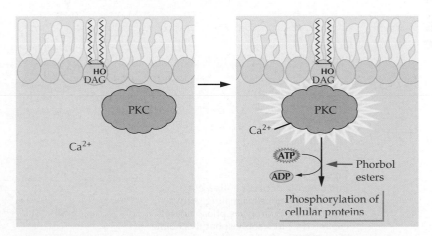

Figure 18.34

Production of DAG by PLC is required for other downstream transduction processes. DAG acting with Ca^{2+} activates protein kinase C (PKC), to which many important functions have been ascribed, including control of cell division and cell growth. A true PKC has yet to be isolated from plant cells but functional protein kinase equivalents have been described. Because PKC is the phorbol ester receptor, its activation can be mimicked by treatment with phorbol esters.

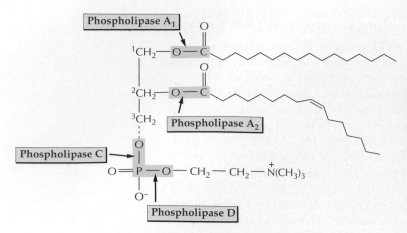

Enzyme	Products of phosphatidylcholine cleavage
PLA	Free fatty acid and lysophospholipid
PLC	DAG and phosphocholine
PLD	Phosphatidic acid and choline

Figure 18.35
Specific phospholipases degrade membrane phospholipids at defined cleavage sites.

released in mammalian cells is arachidonic acid. Lysophospholipids can also serve as a precursor to platelet aggregating factor (PAF), which functions in blood clotting. Two forms of PLA, PLA$_1$ and PLA$_2$, can be distinguished, but only PLA$_2$ can be regulated by G-proteins, proteins kinases, and Ca^{2+} in animal cells.

In plants, free fatty acids and lysophospholipids increase the activity of the plasma membrane H$^+$-ATPase. Auxin treatment may also increase the rate of formation of lysophospholipids. Unsaturated fatty acids trigger stomatal opening, which might reflect

direct effects on potassium channels. A plant type of PLA$_2$ may also mediate the action of fungal elicitors and the associated oxidative burst responses. Several enzymes such as (1→3)β-glucanase are up-regulated as the result of PLA$_2$ activation. One product of PLA$_2$ action, linolenic acid, is a precursor in the octadecanoid pathway that synthesizes jasmonic acid (JA), an important signal generated in response to wounding (Fig. 18.36). PAF, a lysophospholipid-like derivative of phosphatidylcholine, may also influence proton transport through the plasma membrane.

Figure 18.36
Putative role of phospholipase A$_2$ (PLA$_2$) in plant cell signaling. Activation of PLA$_2$ hydrolyzes phosphatidylcholine (PC) to lysophosphatidylcholine (LysoPC). The free fatty acid released can be used for synthesis of jasmonic acid, which mediates induction of gene expression. Although the exact relationships that link all the constituents have not been fully defined, this pathway is undoubtedly important in the transduction of stress signals.

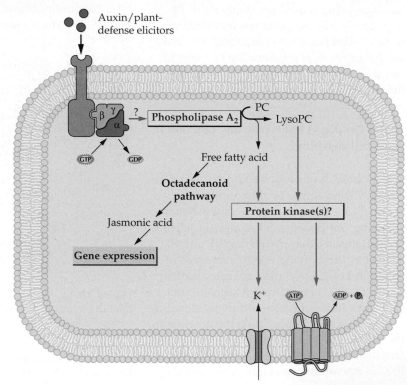

Phospholipase D (PLD) hydrolyzes phosphatidylcholine to release phosphatidic acid (PtdOH; see Fig. 18.35). In plants, PLD has been implicated in senescence (see Chapter 20), fruit ripening, stress (see Chapter 22), and wounding. PtdOH is thought to act as a Ca^{2+} ionophore, permitting the free movement of Ca^{2+} across the plasma membrane (Fig. 18.37). The cellular activities and events regulated by PtdOH include protein kinases, GTP-binding proteins, actin assembly, vesicle trafficking, secretion, and the oxidative burst response to pathogens. When plant cells are challenged by pathogens, PLD is translocated to the plasma membrane and activated in a G-protein–dependent process and by Ca^{2+}. Several PLD isoforms have been identified in plants and are expressed at different times and in different cellular locations. The proposed complex interrelations of PLD and PLC activation are diagrammed in Figure 18.37.

Figure 18.37

Speculative view of possible interactions between phospholipase C (PLC) and phospholipase D (PLD) in plant cell signaling. The release of phosphatidic acid (PtdOH) from membrane phospholipids by PLD may induce several signaling changes involving Ca^{2+} and other targets (e.g., protein kinases, actin assembly and secretion). Various factors, including channel opening, agonist-stimulated activation of heterotrimeric G-proteins, and membrane damage by heat, cold, or parasites, can activate PLD through a calmodulin-binding domain (CalB) and free Ca^{2+}. Mastoparan (a peptide from bee venom), cholera toxin, and alcohol might directly mimic G-protein activation. Activated PLD is recruited to the plasma membrane, where it hydrolyzes phosphatidylcholine (PC). PtdOH can act as an ionophore, enabling more Ca^{2+} to enter the cytoplasm. PtdOH can also be synthesised from diacylglycerol (DAG) in a reaction catalyzed by DAG kinase. Activation of PLC with associated synthesis of DAG will further increase DAG kinase activity. PtdOH can be inactivated through conversion to diacylglycerol pyrophosphate (DGPP) by PtdOH kinase. Any increase in PtdOH can amplify the PLC signaling cascade by activating phosphatidylinositol-4-phosphate-5-kinase (PIPKIN), which converts PIP to the PLC substrate PIP_2.

18.4.7 A signal transduction pathway involving phosphatidylinositide 3-kinases has recently emerged.

Phosphatidylinositide 3-kinase (PI3K) phosphorylates PI, PIP, and PIP_2 at the 3' position (Fig. 18.38). The 3'-phospholipid products are thought to act as second messengers. To date, an enzyme that releases free inositol 3'-phosphates from phospholipids has not been discovered in plants. However, PKC in animal cells can be activated by inositol 3,4,5-P_3.

PI3K appears capable of interacting directly with several proteins to initiate cell division, cell death, and vesicle movement. *Arabidopsis* PI3K, which can complement yeast *PI3K* mutants, contains a Ca^{2+}-dependent lipid-binding domain. Expression of *Arabidopsis PI3K* in the antisense orientation produces severely deformed and stunted plants, probably the result of impaired vesicle transport to the vacuole.

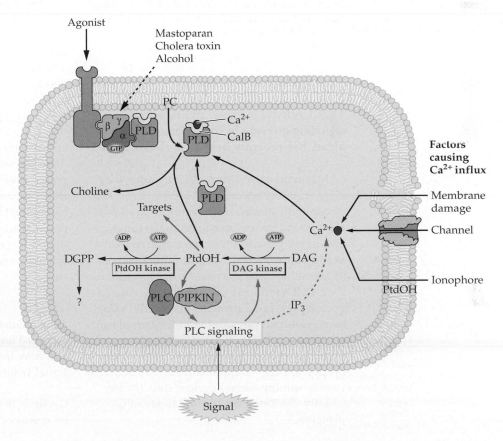

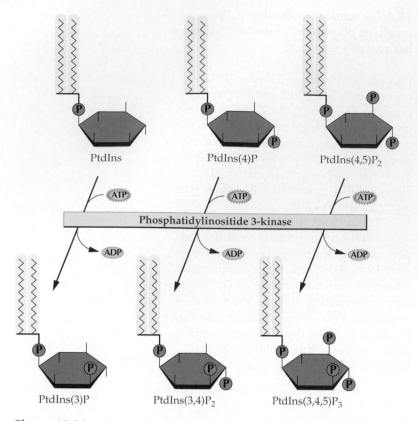

Figure 18.38

The enzymatic action of phosphatidylinositide 3-kinase on phosphatidylinositol phosphates. PtdIns, phosphatidylinositol; PtdIns(4)P, phosphatidylinositol 4-phosphate; PtdIns(4,5)P₂, phosphatidylinositol 4,5-bisphosphate; PtdIns(3)P, phosphatidylinositol 3-phosphate; PtdIns(3,4)P₂, phosphatidylinositol 3,4-bisphosphate; PtdIns(3,4,5)P₃, phosphatidylinositol 3,4,5-trisphosphate.

18.5 Cyclic nucleotides

18.5.1 Adenyl cyclase, an important signaling enzyme in bacteria and motile algae, has recently been detected in plants.

cAMP, synthesized from ATP by adenyl cyclase and degraded to 5'-AMP by cyclic-AMP phosphodiesterase (Fig. 18.39), is an important second messenger in animals, fungi, and many prokaryotes but its role in plants is still controversial. Several different types of adenyl cyclase enzymes are responsible for cAMP synthesis in animal cells. Adenyl cyclases are large proteins (about 120 kDa); some isoforms are soluble in the cytoplasm, whereas others are located in the plasma membrane (Fig. 18.40). Some forms of adenyl cyclase are Ca²⁺-dependent. Signals that activate or inhibit adenyl cyclase directly are usually mediated by specific types of G-proteins.

In mammalian cells, cAMP concentrations are sensed by PKA, which contains both a catalytic subunit and an inhibitory subunit that binds cAMP (Fig. 18.41). In the presence of cAMP, the catalytic subunit is released and can phosphorylate enzymes that control glycogen metabolism and many other metabolic reactions. PKA also regulates transcription factors such as the cAMP response element–binding proteins (CREBs). The activated transcription factors can then bind to cAMP response elements (CREs) in the specific promoters, thereby modifying the transcription of cAMP-regulated genes.

The function of cAMP in higher plant cells was disputed for many years. Detection of plant proteins that synthesize cAMP and have sequence similarity to adenyl cyclase has recently been reported. DNA sequence analysis has also detected putative CREBs and CREs in plants, and several plant and viral promoters contain CRE sequences. Expression of these promoters in transgenic yeast appears to be regulated by cAMP-binding proteins. The nucleotide sequence of the *ETR1* promoter also has a DNA sequence motif that suggests regulation of the gene by cAMP.

Involvement of cAMP has been indicated in many plant-specific processes. For example, signals that initiate stomatal closure may be transduced through adenyl cyclase. The activity of guard cell channels is modified by cAMP-dependent phosphorylation, and cloned plant potassium channel proteins also have a cAMP-binding region. Pollen tube growth is reportedly regulated by cAMP, and adenyl cyclase has been proposed to mediate incompatibility between stigma and pollen tube. Both the plant cell cycle and the rhizobial interactions with root hairs are accompanied by changes in cAMP concentrations. Much more information on the function of this multifunctional second messenger in plants can be expected in the near future.

18.5.2 Guanyl cyclase in plant cells may be more important than adenyl cyclase and may mediate aspects of light signal transduction.

Cyclic 3',5'-GMP (cGMP) is synthesized by guanyl cyclase from GTP. In certain

Figure 18.39
Synthesis and degradation of cyclic 3',5'-AMP by adenyl cyclase and cyclic AMP phosphodiesterase. cAMP is shown as a structural formula and a space-filling model.

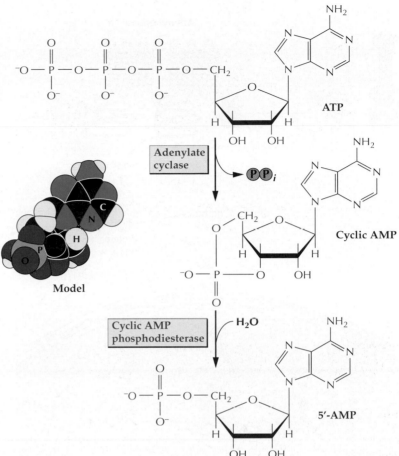

specialized animal cells, most notably rod cells, cGMP helps transduce the visual image. In plant cells, cGMP may participate in transducing signals of fungal invasion, red light signals mediated by phytochrome, and gibberellin signals that regulate synthesis of aleurone amylase.

The tomato *aurea* mutant, which lacks PHYA (see Box 18.4), has been used to demonstrate the involvement of cGMP in phytochrome transduction. Individual etiolated cells injected with cloned and reconstituted phytochrome become sensitive to red light (Fig. 18.42). The transformed cells can then form chloroplasts when irradiated with red light. To elucidate the transduction events that control these processes, researchers microinjected signal transduction components into *aurea* cells and, to monitor the effects of the microinjected transduction component, coinjected DNA constructs containing phytochrome-regulated promoters linked to a reporter gene. Cellular responses observed by microscopy indicated that several transduction pathways were involved in transducing the PHYA signals leading to chloroplast formation. Transduction of red light–activated PHYA signals by G-proteins was observed by using cholera toxin and GTPγS (see Section 18.4.2). Injection of Ca^{2+} alone induced partial chloroplast formation. The expression of genes such as those encoding the chlorophyll *a/b*–binding protein were also induced by microinjection of Ca^{2+}. However, complete differentiation of chloroplasts required concomitant injections of cGMP and Ca^{2+} (Fig. 18.42). Injection of cGMP on its own induced anthocyanin formation. Advanced imaging techniques have demonstrated that brief red light irradiation of etiolated wheat leaf protoplasts induces transient increases in cytosolic Ca^{2+}. Whether these second messengers are active under all cellular circumstances in which phytochrome regulates plant growth and development is currently unknown. However, the molecular events mediating deetiolation seem to involve cGMP and Ca^{2+}.

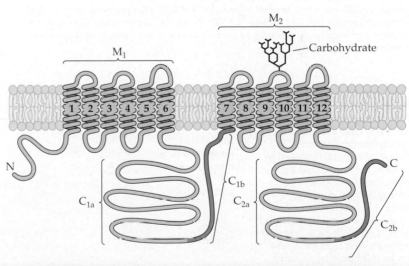

Figure 18.40
Structure of animal adenyl cyclase. This complex enzyme may contain 12 membrane-spanning domains in two groups (M_1 and M_2) and two large intracellular domains ($C_{1a} + C_{1b}$ and $C_{2a} + C_{2b}$). The extracellular domains may be glycosylated. The amino acid sequences in C_{1a} and C_{2a} are highly similar in all known membrane-bound adenyl cyclases.

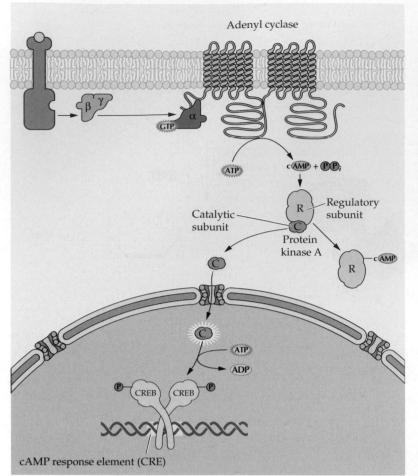

Figure 18.41
The interrelations of heterotrimeric G-proteins, cAMP, and the regulation of gene expression in animal cells. After synthesis by a plasma membrane–bound adenyl cyclase, cAMP can regulate gene expression through phosphorylation of a cAMP response element–binding protein (CREB) by an activated protein kinase A. The catalytic and regulatory subunits of protein kinase A are thought to form a heterotetramer (not shown).

18.6 Calcium

Cytosolic calcium, $[Ca^{2+}]_i$, occupies a pivotal position in plant cell signal transduction. The plant signals thought to be transduced through $[Ca^{2+}]_i$ include touch, wind, temperature shock, fungal elicitors, wounding, oxidative stress, red light, blue light, anaerobiosis, ABA, applied electrical fields, osmotic stresses, and mineral nutrition.

Two practical criteria define the dependence of signaling processes on $[Ca^{2+}]_i$. First, the signal must stimulate observable changes either in $[Ca^{2+}]_i$ or in Ca^{2+} flux across membranes, which in turn must precede physiological responses. (Increased fluxes across membranes or in their vicinity can be crucial transducing events but are often difficult to detect.) Second, the physiological responses associated with changes in $[Ca^{2+}]_i$ must be duplicated artificially by mimicking the observed $[Ca^{2+}]_i$ transient (e.g., with caged Ca^{2+}; see Box 18.5).

18.6.1 Calcium signaling involves a separation of different concentrations of Ca^{2+} by membranes, and the signals must be regulated.

Ca^{2+} signaling depends on the transmembrane electrochemical gradients of Ca^{2+} across the plasma membrane and intracellular membranes. Cells maintain very low resting concentrations of cytosolic Ca^{2+} (100 to 200 nM) to facilitate signaling processes (Fig. 18.43). The vacuole and rough ER constitute

Figure 18.42
Cells of the tomato *aurea* mutant lack PHYA and are impaired in chloroplast development. The mutant cells can be induced to perform the normal red light–controlled development of chloroplasts by microinjection of purified phytochrome followed by exposure to red light (Experiment 1). The effects of this phytochrome/red light treatment can be mimicked by injecting mixtures of cyclic 3′,5′-GMP (cGMP) and Ca^{2+} (Experiment 2). Injection of either second messenger by itself may induce partial chloroplast development. Injection of only cGMP may also induce anthocyanin formation.

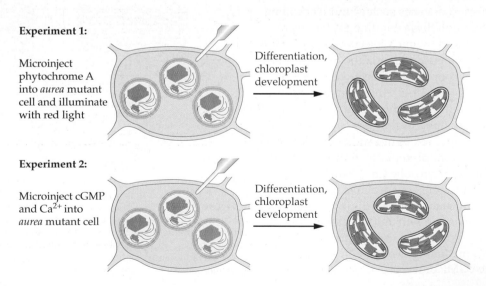

Experiment 1:

Microinject phytochrome A into *aurea* mutant cell and illuminate with red light

Differentiation, chloroplast development

Experiment 2:

Microinject cGMP and Ca^{2+} into *aurea* mutant cell

Differentiation, chloroplast development

large intracellular stores of Ca^{2+} (typically about 1 mM), which can be mobilized by IP_3 and other signals synthesized by the plasma membrane. In the cell wall, where Ca^{2+} is used as a structural molecule, the concentration of Ca^{2+} is estimated to be about 0.5 to 1 mM. Mitochondria, chloroplasts, and even the nucleus may act as Ca^{2+} stores as well. These organelles can also contain other elements of Ca^{2+} signal transduction such as calmodulin, a ubiquitous Ca^{2+} receptor. The nuclear membrane also contains the essential elements of an IP_3-generating system.

When cells receive signals, Ca^{2+} channels are transiently opened and $[Ca^{2+}]_i$ increases rapidly. Numerous Ca^{2+}-binding proteins are thus activated, including calmodulin and Ca^{2+}-dependent calmodulin-like domain protein kinases (CDPKs). Once formed, the Ca^{2+}/calmodulin complex transduces the Ca^{2+} signal by binding to and activating target proteins. Plant cells probably contain several hundred Ca^{2+}- or Ca^{2+}/calmodulin-binding proteins.

$[Ca^{2+}]_i$ signals are truncated by the activity of ATPases located in the plasma membrane, the tonoplast, and ER membranes. These pumps restore and maintain low concentrations of cytoplasmic Ca^{2+} (Fig. 18.43). Just as the H^+-ATPases of the plasma

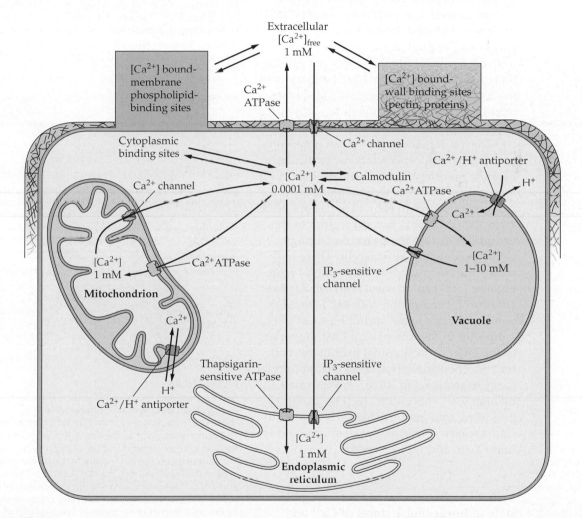

Figure 18.43

Interactions of intracellular and extracellular Ca^{2+} in cell signaling. The relationship of Ca^{2+} stores in plant cells are known to be complex, concentrations of Ca^{2+} being high in organelles and in the cell wall and low in the cytoplasm. When the cell is signaled, channels are opened in various organelles or in the plasma membrane, allowing Ca^{2+} to enter the cytoplasm by diffusing down its electrochemical gradient. Ca^{2+} ATPases and perhaps Ca^{2+}/H^+ antiporters return the cytoplasmic concentration to its resting value. Where known, subcellular concentrations of Ca^{2+} are indicated (quoted values for the ER vary from 0.1 to 1 mM). The concentration of cytoplasmic binding sites has been measured at about 0.5 to 1 mM. Free cytoplasmic Ca^{2+} is in equilibrium with these binding sites. Increases of cytosolic calcium, $[Ca^{2+}]_i$, activate calmodulin, thereby initiating subsequent downstream events. IP_3, inositol 1,4,5-trisphosphate.

membrane and tonoplast remove protons from the cytosol, Ca^{2+}-ATPases use the free energy released by ATP hydrolysis to translocate Ca^{2+} into extracytosolic compartments against its electrochemical gradient. Some of the Ca^{2+}-ATPases are Ca^{2+}/calmodulin-dependent. Inhibitors of calmodulin binding therefore increase [Ca^{2+}]$_i$. A plasma membrane–localized Ca^{2+}-ATPase from plants has membrane-spanning domain sequences similar to those of the well-characterized Ca^{2+}-ATPase from red blood cells. The active sites of both enzymes require a phosphorylated aspartyl residue as an intermediate in the use of ATP and appear to be controlled by phosphorylation/dephosphorylation reactions as well. Ca^{2+}-ATPases in the ER and nuclear membrane are inhibited by thapsigargin and cyclopiazonic acid. Application of these two inhibitors to plants can substantially increase [Ca^{2+}]$_i$. Ca^{2+}/H$^+$ antiporters in the tonoplast and inner mitochondrial membrane may also help maintain low [Ca^{2+}]$_i$ (Fig. 18.43).

During signaling, [Ca^{2+}]$_i$ can transiently reach very high concentrations, particularly near the mouths of open channels (Fig. 18.44). Special luminescence technology using microinjected aequorin (Box 18.6) has detected 10 to 100 μM [Ca^{2+}]$_i$ in local regions of the cytoplasm during the passage of a single action potential in animal cells. These high concentrations cannot be tolerated for long, because Ca^{2+} can interfere with cellular metabolism by competing with Mg^{2+} for ATP. Some carrier proteins, including a Ca^{2+}/H$^+$ antiporter in the tonoplast, a Ca^{2+} uniporter in the inner mitochondrial membrane, and the mitochondrial phosphate/OH$^-$ antiporter (see Chapter 14, Fig. 14.38), are activated by high Ca^{2+} concentrations (Fig. 18.45), providing transport mechanisms to control Ca^{2+} toxicity. The [Ca^{2+}]$_i$ system is poised for immediate and rapid response.

18.6.2 Mechanisms exist for sensing the state of intracellular stores of Ca^{2+} and replenishing them as needed.

A [Ca^{2+}]$_i$ transient induced by red light irradiation of an etiolated leaf protoplast is illustrated in Figures 18.46 and 18.47. When plant cells are signaled and channels are opened, there is an immediate influx of Ca^{2+}

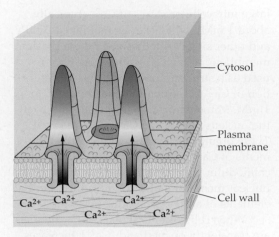

Figure 18.44
Distribution of Ca^{2+} in the cytoplasm within the vicinity of open Ca^{2+} channels. The channels are assumed to be voltage-regulated and open only for about two milliseconds. The color range indicates the concentration of Ca^{2+}, with red being the highest and blue the lowest.

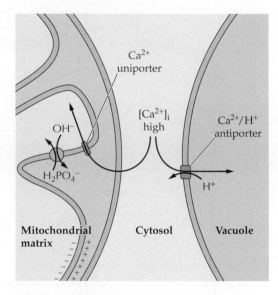

Figure 18.45
Several transport mechanisms activated by high [Ca^{2+}]$_i$. The function of uniporters, such as that shown here on the inner mitochondrial membrane, and antiporters, such as the Ca^{2+}/H$^+$ transporter on the vacuolar membrane and the phosphate/hydroxide antiporter on the inner mitochondrial membrane, is described in detail in Chapters 3 and 14. The transport of Ca^{2+} into the matrix via the mitochondrial uniporter is electrogenic (i.e., not directly compensated by coupled import of an anion or coupled export of a cation). Ca^{2+} uniporter activity is energized by membrane potential and pH gradients across the inner mitochondrial membrane. Under experimental conditions in which isolated mitochondria take up large amounts of Ca^{2+}, the inflow is accompanied by movement of an anion (usually H$_2$PO$_4^-$) from the intermembrane space to the matrix. Movement of the H$_2$PO$_4^-$ is catalyzed by the phosphate/hydroxide exchanger.

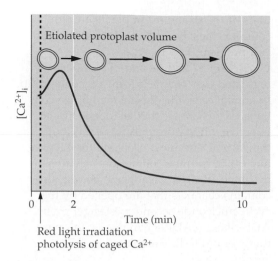

Figure 18.46
Red light induces changes in the concentration of cytosolic Ca^{2+} and the volume of etiolated wheat leaf protoplasts. Red light irradiation of etiolated wheat leaf protoplasts leads to transient increases in cytosolic Ca^{2+} lasting less than two minutes and a subsequent increase of protoplast volume of much longer duration. Photolysis of loaded caged Ca^{2+} or caged IP_3 results in similar changes in volume with a similar 2-minute lag.

down its electrochemical gradient. A single open Ca^{2+} channel can conduct an estimated 10^6 ions per second. In some cases, a counterbalancing activation of Ca^{2+}-ATPases immediately follows the increase of $[Ca^{2+}]_i$. Because calcium released from organelles can be expelled to the wall by plasma membrane–localized Ca^{2+}-ATPases, continuous signaling can rapidly exhaust intracellular stores, which must then be refilled before signal transduction or sensing can continue.

Intracellular stores can communicate their state of emptiness to the plasma membrane. When the stores are empty, special channels are opened in the plasma membrane. These channels have flux rates that are orders of magnitude less than those of the plasma membrane Ca^{2+} channels involved in signaling and remain open until the stores are filled. The intracellular store thus acts rather like an electrical capacitor, and this type of Ca^{2+}-uptake behavior is referred to as **capacitative Ca^{2+} signaling.** The channels involved are called I_{CRAC}, for the current *(I)* carried by a release-activated Ca^{2+} channel. The precise means whereby the store communicates its state of emptiness to the plasma membrane is unknown, but both chemical and structural signals controlling interactions between ER and plasma membrane have been implicated.

18.6.3 Ca^{2+} diffuses slowly in the cytoplasm.

The rate at which free Ca^{2+} diffuses in the cytoplasm is much slower than that in free solution. When isotopes ^{45}Ca and ^{42}K first became available in the late 1950s, they were injected into squid axons and their distributions were measured after several hours. Most surprisingly, ^{45}Ca showed very limited diffusion and remained at the site of injection, whereas ^{42}K diffused readily. Recent measurements have confirmed that the diffusion constant of Ca^{2+} in the cytosol is at least two orders of magnitude less than that

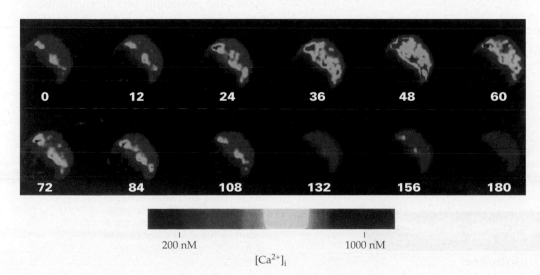

Figure 18.47
Red light–induced changes in cytosolic Ca^{2+} in single wheat leaf protoplasts loaded with a Ca^{2+}-sensitive fluorescent dye, fluo-3, as measured by confocal microscopy imaging. The scale used represents high Ca^{2+} concentration as red color and low Ca^{2+} as blue color. The times (in seconds after dye loading) at which the images were taken are indicated above each protoplast. Only half of the protoplast is visible because the fluorescent dye does not enter the vacuole.

in free solution. Impediments to Ca^{2+} diffusion include uptake into organelles (e.g., chloroplasts, mitochondria, ER, and vacuole) and binding to proteins that are free in the cytosol or are attached to the cytoskeleton.

The low rate of diffusion is an important part of $[Ca^{2+}]_i$ signaling. Because $[Ca^{2+}]_i$ does not disperse quickly in the cytoplasm, standing gradients of Ca^{2+} can form, as in growing pollen tubes (Fig. 18.48). Maintenance of the standing $[Ca^{2+}]_i$ gradient is essential for vesicle fusion and continued growth. In general, the spatial segregation of $[Ca^{2+}]_i$ at defined sites in the cytoplasm can promote signaling specificity (see Section 18.6.7).

18.6.4 Ca^{2+} channel activity can be detected by patch clamp technologies.

Three families of Ca^{2+} channels have been detected in plants by using patch clamp technology (see Chapter 3, Box 3.7). Ca^{2+} channels have been found in the plasma membrane, rough ER, and tonoplast; they may also be present in mitochondria.

Voltage-gated channels have their opening probability determined by a particular value of the membrane potential. The vacuole and the plasma membrane contain a considerable number of different members of the Ca^{2+} channel families that form subfamilies, each of which can be distinguished by the membrane potential required for activation and by the kinetics of opening. In animal cells, the voltage-gated Ca^{2+} channel contains at least four or more separate sub-

units; if the structures in plants are similarly complex, their isolation will present a difficult problem (Fig. 18.49).

Receptor- and second messenger–regulated Ca^{2+} channels form a second group. In animal cells, the plasma membrane contains Ca^{2+} channels that are opened by interaction with G-proteins. In plants, both the vacuole and most probably the ER and nuclear membrane contain Ca^{2+} channels that bind IP_3. The vacuole membrane also contains channels that are opened by the cyclic nucleotide second messenger cyclic ADP-ribose (cADP-R). In plants, the synthesis of cADP-R is regulated by ABA (see Section 18.8.4).

A third group of Ca^{2+} channels is found in both the vacuole and plasma membrane. These "stretch" channels sense tension in the membrane and are opened when the tension is altered. Mechanical signals and turgor status may be mediated through stretch channel activity. Mechanical signals (e.g., touch, wind) modify the interrelation of the plasma membrane and the wall and thus promote a change in tension. Water stress or altered activity of water channels modifies the turgor pressure and hence the pressure exerted between the wall and the plasma membrane. Stretch channels can be prominent in regions of active growth (e.g., the tip of a pollen tube).

Channel activity is not a binary function. Channels may exist in a closed, open, or inactivated condition (see Chapter 3 and Fig. 18.50). Although channels are often described as opening or closing in response to a stimulus, this behavior is not uniform for a population of channels. More accurately stated, the probability of a given channel being open or closed is influenced by the stimulus involved. Phosphorylation of the channel protein can also regulate the probability of opening.

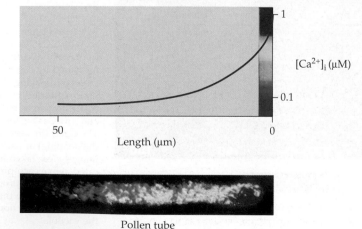

Pollen tube

Figure 18.48
Pollen tubes maintain a standing gradient of cytosolic Ca^{2+} in their tip region. The standing gradient of $[Ca^{2+}]_i$ is essential for growth and results from a tip-associated cluster of Ca^{2+} channels. Pollen tubes can be loaded with Ca^{2+}-sensitive fluorescent ratio imaging dyes such as indo-1 or fura-2 for quantification of free Ca^{2+} by fluorescence microscopy.

Box 18.6

A luminescent Ca²⁺-sensitive protein can help measure [Ca²⁺]ᵢ directly.

(A)

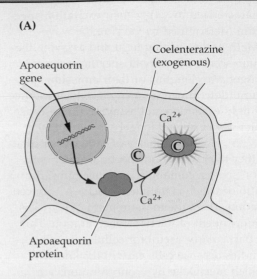

Apoaequorin gene

Coelenterazine (exogenous)

Ca^{2+}

Ca^{2+}

Apoaequorin protein

(B)

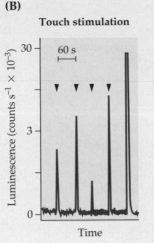

Touch stimulation

60 s

Luminescence (counts s⁻¹ × 10⁻³)

Time

(C)

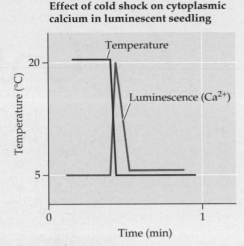

Effect of cold shock on cytoplasmic calcium in luminescent seedling

Temperature

Temperature (°C)

Luminescence (Ca^{2+})

Time (min)

The jellyfish *Aequorea victorea* contains a calcium-sensitive luminescent protein, aequorin. Aequorin consists of two constituents, an apoprotein (apoaequorin) and coelenterazine, a hydrophobic luminophore. When reconstituted, aequorin binds Ca^{2+} atoms with low affinity but high specificity. The Ca^{2+}–aequorin complex undergoes a conformational change that results in oxidation of the bound coelenterazine and an accompanying emission of luminescent light. Transformation of plants with the cDNA for apoaequorin and reconstitution with coelenterazine (shown in panel A above as circled C) generates luminous plants, the luminescence of which directly reports [Ca²⁺]ᵢ (panel A). This method for measuring [Ca²⁺]ᵢ is very simple. Many coelenterazines are available that yield aequorins with different properties, including some useful for ratio measurements. Aequorin can be targeted to different cell compartments and attached to cell membranes. Luminescence can be measured continuously for many weeks. This novel method

offers a broad scope for obtaining very precise and significant information on [Ca²⁺]ᵢ in plant cells.

Transgenic plants containing reconstituted aequorin have been used to detect the effects of touch, wind, cold, oxidative stress, hyperosmotic stress, auxin, blue light, and anaerobiosis—among many other signals. When touched, for example, transgenic seedlings exhibit rapid luminescence spikes (panel B, arrowheads). If the temperature is lowered rapidly, the resulting Ca^{2+} spike induces changes in gene expression (panel C).

Special cameras, originally developed for astronomers, can be used to image luminescent light. Panel D shows tissues of transgenic aequorin-containing seedlings induced to luminesce by cold-shock treatment. The images include a whole seedling (magnification × 2) and a whole cotyledon (magnification × 20; note spottiness in response).

(D)

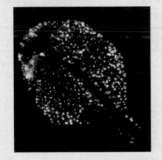

18.6.5 Advanced fluorescence and luminescence technologies allow the imaging of free calcium concentrations inside living cells.

[Ca²⁺]ᵢ can be measured and imaged by the use of fluorescent dyes. In addition, cells can be transformed to express the protein component of a luminescent compound from jelly-

fish. When exposed to a second, light-reactive component, the protein luminesces and can be used to monitor Ca^{2+} concentrations for extended periods in vivo (see Box 18.6).

Two types of fluorescent dyes can be distinguished (Fig. 18.51). **Ratio dyes** undergo shifts in wavelength of the fluorescence spectrum when binding Ca^{2+}. In contrast, Ca^{2+} binding increases the fluorescence

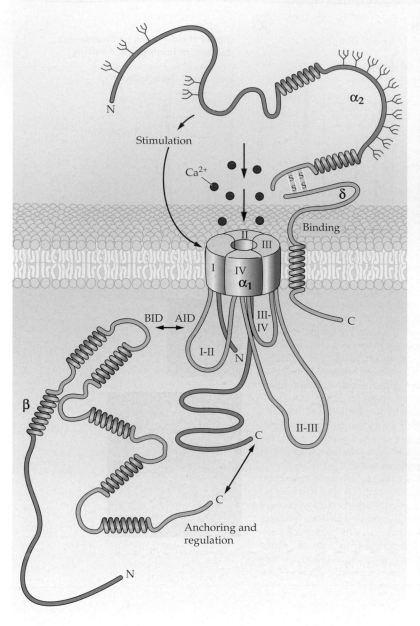

Figure 18.49
Structural organization of voltage-gated calcium channels in animals. The calcium channel regulated by membrane potential shares common features among many different organisms. The basic channel is composed of four different proteins (α_1, α_2, β, and δ), of which the sites of interaction are indicated by double-headed arrows. The major membrane-spanning subunit (α_1) consists of four homologous domains (I–IV), each composed of six transmembrane segments (not shown). Cytoplasmic loops are labeled according to the domains they connect. The α_2 protein is linked by disulfide bridges to δ, which in turn interacts with α_1. Inside the cytoplasm, the β-subunit (through BID [beta-subunit interacting domain]) interacts with the I–II domain of α_1 (AID [alpha-subunit interacting domain]). The extracellular regions of the channel may be glycosylated and the internal regions phosphorylated by protein kinases. Anchorage of the calcium channel to the cytoskeleton and further regulation take place through the C termini of β and α_1. Why such a large structure is necessary to transmit Ca^{2+} is not understood.

intensity of **single-wavelength dyes** but does not induce a spectral shift. Dyes are characterized by either their **excitation spectrum** (determined by varying the wavelength of the exciting light and assaying the fluorescence intensity at specific, defined longer wavelengths) or their **emission spectrum** (determined by using a constant source of light excitation and assaying fluorescence intensity at several longer wavelengths).

A valuable property of ratio dyes is that either their excitation spectra or their emission spectra contain two wavelengths, the ratio of which is unique to each Ca^{2+} concentration. Thus, the measurement of $[Ca^{2+}]_i$ is independent of the dye concentration. This is particularly useful for cellular imaging studies because cells vary in thickness, which means the dye concentrations are rarely uniform. Two common ratio dyes used to monitor Ca^{2+} in fluorescence-ratio imaging experiments are fura-2 and indo-1. Figure 18.52 illustrates the use of fluorescence-ratio imaging to demonstrate the involvement of Ca^{2+} in stomatal closure and, in particular, the role of the vacuole.

The common single-wavelength dyes are calcium green and fluo-3, which are excited by visible light and can be used in conjunction with the confocal microscope. Single-wavelength dyes are usually much brighter than ratio dyes; consequently, the irradiance needed to excite single-wavelength dyes is less, thus reducing photodamage to the cell and photobleaching of the dye. Comparison of successive fluorescence images of a cell during signaling can ameliorate the lack of exact quantitation with single-wavelength dyes. Moreover, dyes can be coupled to dextran to prevent their accumulation in the vacuole or other organelles.

Imaging the distribution of Ca^{2+} in single living cells has been a major achievement of research, requiring fluorescence microscopes coupled to powerful computers. Use of these technologies has enabled the demonstration of permanent gradients of $[Ca^{2+}]_i$ in growing pollen tubes and root hairs, the production of waves and oscillations in guard cells and pollen tubes, and the release of Ca^{2+} from vacuoles inside guard cells. They have also been used to demonstrate that specific modification of the $[Ca^{2+}]_i$ gradient in the pollen tube tip initiates reorientation (see Box 18.5).

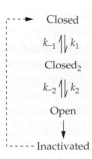

Figure 18.50
At any one instant, a Ca^{2+} channel may exist in various closed, open, or inactivated states. Like all channels, Ca^{2+} channels are only open for very brief periods of time and the behavior of individual channels is stochastic. Thus the channels are better described in terms of the probability of their opening or closing rather than their being open or closed. Opening conditions merely ensure the channel has a high probability that it will be found in the open state; the converse is true for the closed state. The variety of states ensures the presence of a diverse set of channel activities, rather than a uniform population, all open or all closed.

18.6.6 Signaling through $[Ca^{2+}]_i$ may involve waves, cascades, oscillations, capacitative calcium entry, and pacemaker cells.

Calcium signaling can take place through simple transients or through more complex patterns. In some cells (e.g., guard cells, pollen tubes), oscillations of $[Ca^{2+}]_i$ can be observed—apparently resulting from a sequential filling and emptying of the Ca^{2+} stores in the ER and perhaps the vacuole. The emptying and filling are thought to be regulated by a capacitative Ca^{2+} signaling mechanism involving Ca^{2+}-induced Ca^{2+}-release (CICR), probably by way of Ca^{2+}-dependent calcium channels (Fig. 18.53; see also Chapter 3). Oscillations may also enable cells to distinguish genuine $[Ca^{2+}]_i$ signals from noise. Organelles such as mitochondria might initiate metabolic changes only in

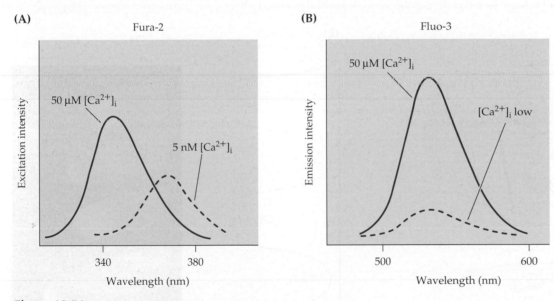

Figure 18.51
Fluorescence spectra for fura-2, a ratio dye, and fluo-3, a single-wavelength dye. All fluorescent dyes respond to wavelengths of exciting light by emitting fluorescent light at longer wavelengths. Dyes can be characterized by subjecting them to different wavelengths of exciting light to generate an excitation spectrum, with fluorescence measured at defined longer wavelengths. Alternatively, a spectrum of emitted fluorescent light can be constructed by using an invariant source of exciting light. The Ca^{2+}-sensitive fluorescent ratio dye, fura-2 (A), exhibits a shift in its excitation spectrum when it binds Ca^{2+}. The ratio of fluorescence emitted after excitation at two wavelengths, 340 and 380 nm, is unique for each Ca^{2+} concentration. The value of ratio imaging is that it obviates variations in dye concentration, dye bleaching, and cell thickness. Single-wavelength dyes, e.g., fluo-3 (B), do not exhibit a spectral shift but are frequently used because they are much brighter than dual-wavelength dyes. Further, single-wavelength dyes, which require less time for resolution can be useful for detecting changes in cytosolic calcium by comparing images taken close together in time.

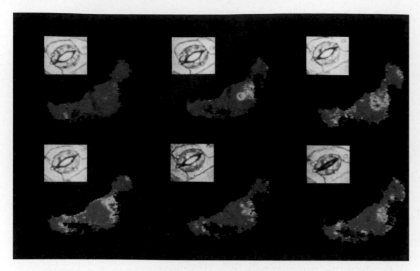

Figure 18.52
Fluorescence-ratio imaging of $[Ca^{2+}]_i$ in single guard cells of *Commelina communis* after changing the concentration of Ca^{2+} in the bathing medium from 20 μM to 1 mM at time zero. Individual guard cells were loaded with the ratio dye, indo-1. Ratio images were taken (left to right) 0, 2, and 5 (top row) and 10, 20, and 30 (bottom row) minutes after the medium change. Bright field images are shown as insets to indicate guard cell apertures.

response to oscillations rather than to sustained or transient $[Ca^{2+}]_i$ changes.

IP$_3$, which may also participate in the oscillations discussed above, produces Ca^{2+} waves that regulate the growth and orientation of pollen tubes. IP$_3$ is thought to function as a relay between intracellular Ca^{2+} stores (Fig. 18.54) and sites on the plasma membrane and thus overcome the constraint on diffusion of Ca^{2+} described in Section 18.6.3. Both waves and oscillations can pass between cells in a form of intercellular communication (Fig. 18.55) and may give rise to more complex forms of Ca^{2+} signaling such as circadian oscillations.

18.6.7 Wherein lies the specificity of calcium signaling?

Few signals do not involve changes in $[Ca^{2+}]_i$, raising the question of how such a simple molecule can give rise to so many

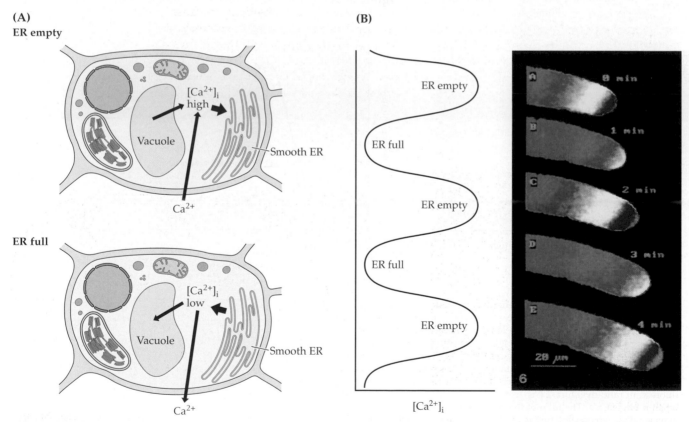

(A)
ER empty
ER full

(B)

Figure 18.53
(A) Oscillations in cytosolic Ca^{2+} may arise from a filling and emptying of the Ca^{2+} stores in the endoplasmic reticulum (ER). The size of the arrows indicates flux rates. When the ER store is empty, Ca^{2+} is sequestered into the store; when the store is full, Ca^{2+} is released. The basis of the mechanism may operate through alternation of the activities of channels and ATPases in the organelles involved. (B) Oscillations in $[Ca^{2+}]_i$ accompany pollen tube growth. Fluorescence-ratio images taken at 1-minute intervals reveal oscillations in $[Ca^{2+}]_i$ in the tip.

different physiological responses. One aspect of this specificity involves the duration of the signaling itself. Some transient increases in Ca^{2+} concentration last longer than others. The longer-lasting influxes can be expected to penetrate farther into the cytoplasm and therefore encounter more centrally located Ca^{2+}-dependent enzymes (Fig. 18.56). Spatial specificity can also play a role in determining the physiological response to a given Ca^{2+} signal. Just as Ca^{2+} channels are concentrated in the tips of growing pollen tubes, so can receptors be clustered such that only certain parts of the cytoplasm receive signals. Cells may also "read" the frequency of oscillations or the speed of waves.

A useful analogy for the specificity associated with ubiquitous signaling molecules such as calcium is electrical wiring. In any house one switch will turn on a light, another will activate a TV. Identical switches can electrify any appliance. The factor determining the result of toggling a switch is the wiring that links switches to components. The cytoskeleton/scaffold is analogously hard-wired with Ca^{2+}-sensitive proteins. All are subject to control by the same switch—Ca^{2+}—but spatial differentiation of the receptors ensures that only some of the circuitry is activated in response.

18.6.8 The eukaryotic Ca^{2+}-based signaling systems may have evolved as detoxification mechanisms.

The use of $[Ca^{2+}]_i$ as a signaling molecule seems to be largely limited to eukaryotic cells, raising questions about the evolution of signal transduction. One suggestion is that eukaryotic cells experienced a Ca^{2+} catastrophe during early evolution, which resulted in the elaboration of detoxification mechanisms to remove Ca^{2+} and thereby avoid its toxic effects on ATP metabolism: These detoxification mechanisms later developed into signaling pathways. The precedent here is the hypothesis that sees the evolution of aerobic respiration as resulting from an early attempt to detoxify O_2 produced during photosynthesis. An alternative view sees $[Ca^{2+}]_i$ signaling as evolving naturally from the very limited signaling systems in bacteria that regulate aspects of chemotaxis by utilizing $[Ca^{2+}]_i$. According to this hypothesis, eukaryotes learned to recognize perturbations in the plasma membrane because those gave rise to transient increases in $[Ca^{2+}]_i$. With the evident evolutionary need to sense more signals, a primitive $[Ca^{2+}]_i$-based system was elaborated and various divergent mechanisms evolved independently,

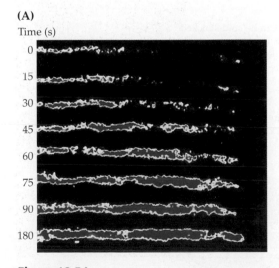

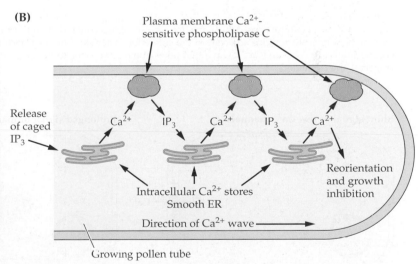

Figure 18.54

Calcium waves in growing pollen tubes. (A) The color image of the Ca^{2+}-sensitive fluorescent dye, fluo-3, in poppy pollen tubes is shown after photolysis of loaded caged inositol 1,4,5-trisphosphate (IP$_3$) at time zero (at top of figure). An induced calcium wave is initiated in the vicinity of the nucleus (the middle of the cytoplasmic region of the pollen tube). The wave reaches the pollen tube apex in about one minute and may oscillate during its progress. Images are taken at defined time intervals after photolytic release. (B) Deduced mechanism of Ca^{2+} wave observed in pollen tubes after photolysis of caged IP$_3$. The wave is most likely propagated by a Ca^{2+}-dependent phospholipase C in the plasma membrane, because the synthesis of IP$_3$ mobilizes Ca^{2+} from intracellular stores. When the wave reaches the tip, growth is inhibited; on recovery, growth is reoriented.

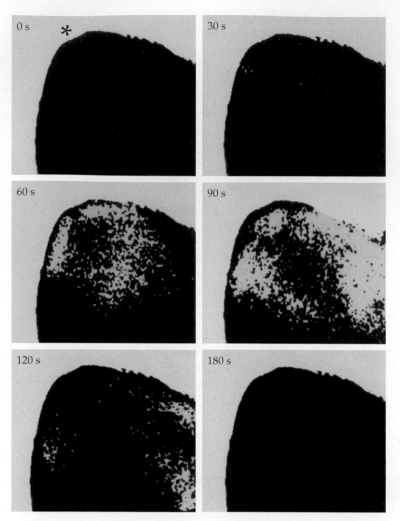

acquiring different signal specificities. Elaboration of the downstream signaling coupled to spatially separate $[Ca^{2+}]_i$ changes would generate a chain of signaling and transduction components specific to each signal, much as seems to occur in present-day yeast.

18.6.9 Calmodulin is the primary calcium receptor, and there are many calmodulin-binding proteins.

Calmodulin, a small (15 to 17 kDa), highly conserved, Ca^{2+}-binding protein, is the primary calcium receptor in both plant and animal cells. Sequences of plant and animal calmodulins differ by as few as 12 to 13 amino acids among about 150. The molecule has four Ca^{2+}-binding regions or loops (Fig. 18.57A), which contain 12 amino acids each and are rich in aspartate and glutamate. Two helices found on either side of the loops have given rise to the terminology of helix–loop–helix proteins, because such Ca^{2+}-binding structures may be found in many other proteins. Sometimes the loop is called an EF hand, in reference to the prominence of glutamate and phenylalanine in the beginning and end of the loop.

Calmodulin is found in both the cytoplasmic and nuclear compartments and can

Figure 18.55
A tissue Ca^{2+} wave in tobacco. Transgenic tobacco seedlings containing aequorin can be cold-shocked by placing a tiny block of ice adjacent to the cotyledon (asterisk). Luminescence is imaged here at 30-second intervals. A tissue $[Ca^{2+}]_i$ wave that takes about 30 seconds to develop is clearly visible as it traverses the cotyledon.

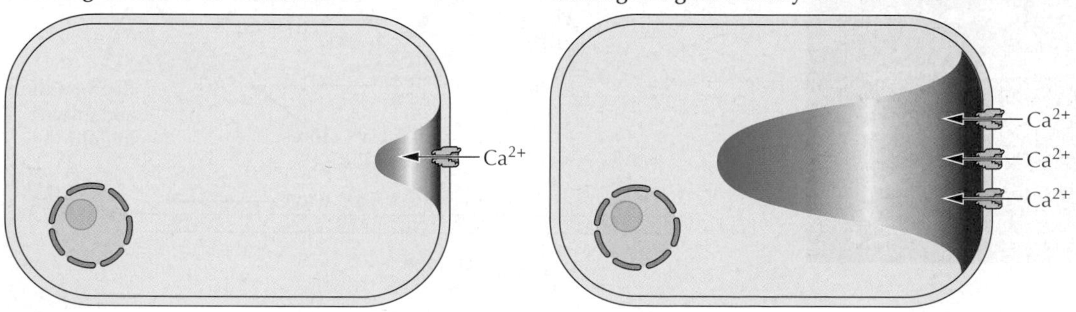

Figure 18.56
The duration of a transient influx of Ca^{2+} determines the distance that a Ca^{2+} signal penetrates from the plasma membrane to the cytoplasm because Ca^{2+} diffuses slowly in cytoplasm. Clustering of Ca^{2+} channels ensures a greater distance of penetration of the signal by limiting Ca^{2+} buffering by the cytoplasm. The slow diffusion of the Ca^{2+} signal is therefore partially offset by the numbers of channels involved in responding to the signal.

be attached to the plasma membrane. The cellular concentration of calmodulin varies greatly among plant cell types and developmental stages: As mentioned in Section 18.2.8, root statocytes contain calmodulin concentrations an order of magnitude greater than that in meristematic cells and thus are probably more sensitive to small changes in $[Ca^{2+}]_i$. In polarized cells (e.g., the polarizing *Fucus* zygote), calmodulin may be concentrated in regions of active growth and metabolism.

On binding Ca^{2+}, calmodulin undergoes a substantial change in tertiary structure, exposing a very hydrophobic patch rich in methionine, leucine, and phenylalanine. Specific regions on target proteins recognize this patch and combine with the calmodulin, resulting in activation of the target proteins. Unlike calmodulin itself, the sequences of the target peptides are not conserved. The prominence of methionine in the calmodulin-binding region is thought to ensure flexibility in the interaction.

Calmodulin-binding proteins or peptides (e.g., M13; Fig. 18.57B) may be detected in several ways. Isotopically labeled calmodulin can be used to probe expression libraries or even separated proteins on gel electrophoresis. The characterized calmodulin-binding proteins in plants are myosin V, kinesin, NAD$^+$ kinase, glutamate decarboxylase, protein kinases, and Ca^{2+}-ATPases. Plant calmodulin is also important in formation of actin filaments and cytoplasmic streaming (see Chapter 5), polarized growth, and the cell division cycle (at both S phase and M phase; see Chapter 11). Many more calmodulin-binding proteins have been identified and await characterization.

18.7 Protein kinases: primary elements in signal transduction

18.7.1 Protein kinases are ubiquitous enzymes, and many are signal specific.

Members of the protein kinase superfamily catalyze the reversible transfer of the γ-phosphate from ATP to serine, threonine, or tyrosine amino acid side chains on target proteins. Protein kinase activity is counterbalanced by the action of specific protein phosphatases that remove the phosphate

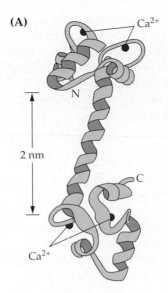

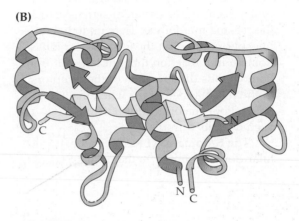

Figure 18.57
Structures of calmodulin. (A) The uncomplexed, calcium-loaded form is a dumbbell-shaped structure with two Ca^{2+}-binding domains at each end. Binding to Ca^{2+} exposes hydrophobic patches on each end that are rich in methionine, leucine, and phenylalanine residues. These hydrophobic interactions play a key role in the binding of calmodulin to target proteins. (B) The Ca^{2+}/calmodulin complex bound to M13, a calmodulin-binding peptide (Ca^{2+} ions not shown). Calmodulin is shown in purple, the M13 peptide in green.

from proteins. In most cases phosphorylation modifies target protein activity. One protein kinase molecule can phosphorylate many hundreds of target proteins, thereby greatly amplifying weak signals.

Activation of protein kinase has been implicated in responses to light, pathogen attack, growth regulators, temperature stress, and nutrient deprivation. Several important protein kinases are concerned with regulation of metabolic pathways. Hundreds of plant genes for different protein kinases have been identified but at least a thousand must exist. Figure 18.58 illustrates the various groups of protein kinases that have been identified in plant cells.

18.7.2 RLKs represent a complex family of protein kinases with diverse functions in signaling.

The mechanisms used by plants to transmit extracellular signals into the cytoplasm require receptors located in the plasma membrane. RLKs (see Section 18.2.2) are an important group of protein kinases with direct functions in transmission of signals across the plasma membrane. They undergo autophosphorylation on the intracellular kinase domain in a reaction thought to result from homologous dimerization of the receptors in the plasma membrane when the ligand binds. Four major groups of RLKs have been described in plants.

The **S-domain RLKs** have an extracellular domain similar to that of the S-locus glycoproteins concerned with *Brassica* sporophytic incompatibility (see Chapter 19). Incompatibility occurs as the result of a cell-to-cell interaction between pollen tube and stigma (Fig. 18.59). This intercellular communication results in a rapid cessation of pollen tube growth. The proteins involved in this communication are the S-locus glycoproteins and an S-locus glycoprotein receptor kinase located in the stigma tissue. Even though its members are expressed in many different tissues and in different species, signaling that leads to incompatibility reactions might be the primary function of this group.

The **leucine rich-repeat (LRR) RLKs** contain a recognition core of leucine and asparagine, which is thought to engage in protein–protein interactions. LRR RLKs are found in many different tissues, including the shoot apex. One example is RLK5, which is found in association with a protein phosphatase (Fig. 18.60). The *Cf-9* gene encodes a putative transmembrane protein with an extracellular domain of 28 LRRs and a short cytoplasmic tail. The extracellular domain exhibits considerable similarity to the LRR class of RLK. *Cf-9* confers resistance against tomato mold (see Chapter 21).

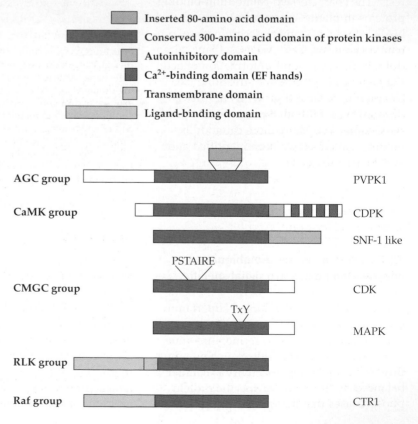

Figure 18.58
The various classes of protein kinase (PK) found in plant cells. The catalytic domain of most PKs consists of a 300-amino acid region. This conserved catalytic domain has various insertions and modifications that enable the classification of kinases into distinct groups. The AGC group is represented by cyclic nucleotide PKs (e.g., PKA; see Section 18.5.1) and calcium- and phospholipid-dependent PK (e.g., PKC; see Section 18.4.5). A plant protein member of this group, PVPK1, has an insert of approximately 80 amino acids within the kinase catalytic domain. The CaMK group contains PKs that are activated by or dependent on calcium and calmodulin. CDPK, a prominent member of this group in plant cells, has a C-terminal calmodulin-like region containing four (or fewer) calcium-binding EF hands. This enzyme also contains an autoinhibitory region involved in the regulation of CDPK activity. The CMGC group includes the CDK (cyclin-dependent PK), MAPK (mitogen-activated PK), GSK-3 (glycogen synthase kinase), and CKII (CaM kinase II) families. CDK has a conserved PSTAIRE region. MAPK has a conserved Thr-X-Tyr (TxY) motif. RLKs (receptor-like kinases) have a ligand/signal-binding domain joined to a conserved catalytic sequence by a membrane-spanning domain; a generic RLK structure is shown here. CTR1, a Raf-like PK active in ethylene transduction, also has a putative ligand-binding domain.

The **epidermal growth factor-like (EGF-like) RLKs** contain EGF-like repeats in the extracellular domain. A particular protein kinase in this group has been found attached to the cell wall (Fig. 18.61A). The binding domain located in the cell wall is linked through a single membrane-spanning domain to an internal kinase catalytic site. This kinase may transduce mechanical signals, which are sensed at the wall/plasma membrane junction.

Only one **lectin receptor kinase** has been identified in plant cells (Fig. 18.61B). The large extracellular domain of this protein kinase might bind important carbohydrate signals such as oligosaccharides and other wall fragments. Because many lectins are glycoproteins, there may be some similarity with the S-locus class of RLKs.

Identification of downstream elements of these RLKs has begun. If RLKs are similar to the well-characterized growth receptor-like tyrosine kinases, a cascade of protein kinases should link signal reception to regulation of gene expression. As noted above, an important aspect to regulation by phosphorylation is the large-scale amplification built into the mechanism.

18.7.3 Plants have an unusual Ca²⁺-dependent protein kinase with a calmodulin-like domain.

The calmodulin kinase (CaMK) group of protein kinases includes the Ca^{2+}/calmodulin–dependent protein kinases and SNF1/AMP–activated protein kinase families. CDPK (see Section 18.6.1) was originally purified from soybean but has now been found in many plant cells. The N-terminal half shares homology with other members of the CaMK group, whereas the C-terminal region shows homology to calmodulin with four helix–loop–helix Ca^{2+}-binding sites (Fig. 18.62). A junction that joins the kinase and calmodulin-like domains may function as an autoinhibitory site. In the absence of Ca^{2+}, the junction covers the active site; in the presence of Ca^{2+}, the site is exposed and catalytic activity can commence.

CDPKs have been found attached to the cytoskeleton or the plasma membrane and in the cytoplasm. These proteins may phosphorylate the plasma membrane H⁺-ATPase and

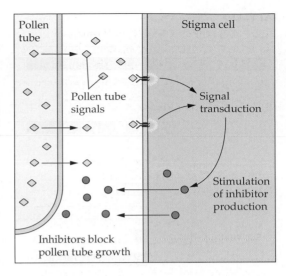

Stigma cell

Pollen tube

Pollen tube signals

Signal transduction

Stimulation of inhibitor production

Inhibitors block pollen tube growth

Figure 18.59
General model for pollen tube incompatibility as a result of two-way signaling between pollen tube and stigma in the Brassicaceae. Signals are derived from the growing pollen tubes of genetically incompatible pollen grains and induce a response in the stigma cells, leading to the secretion of growth inhibitors. S-domain receptor-like kinases are located in the stigma cells, suggesting that proteins secreted by pollen tubes might represent the pollen tube signal (see Chapter 19).

membrane transporters. CpCK1 is a CDPK variant that contains a consensus myristylation site at its N terminus. This type of CDPK is thought to use a Ca^{2+}/myristylation switch to associate reversibly with the plasma membrane. Ca^{2+} binding alters the conformation of the protein, allowing myristylation to occur and thus activating the kinase domain. Several other Ca^{2+}/calmodulin–dependent protein kinases that most likely function in Ca^{2+}-activated signal transduction pathways have also been identified in plant cells.

SNF1-like (sucrose-nonfermenting) kinases are required in yeast for catabolite repression (see Chapter 8); equivalent enzymes from plants have been identified by complementation of yeast *snf1* mutants. These kinases may act as metabolic sensors of the ATP/AMP ratio and may control the metabolic flux between anabolism and catabolism by regulating the transcription of genes encoding enzymes that function in carbohydrate metabolism. In plants, some

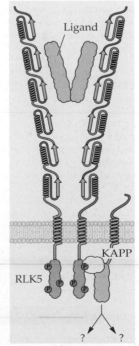

Ligand

RLK5

KAPP

Figure 18.60
Model of a homodimeric receptor-like protein kinase, RLK5, binding its ligand and undergoing autophosphorylation. RLKs are located in the plasma membrane. The RLK shown here, a member of the LRR (leucine rich-repeat) class, contains a recognition core of leucine and asparagine residues. The extracellular region of such a kinase has extended numbers of LRRs, suggesting that the ligand may be another protein. On binding of the ligand, the receptor dimerizes, undergoing intramolecular phosphorylation and activation. A kinase-associated protein phosphatase (KAPP) controls the duration of signaling, thereby regulating the degree of signal amplification.

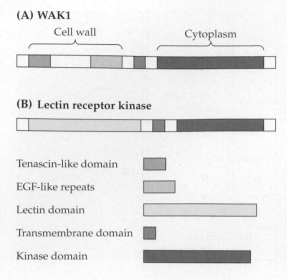

(A) WAK1

Cell wall   Cytoplasm

(B) Lectin receptor kinase

Tenascin-like domain

EGF-like repeats

Lectin domain

Transmembrane domain

Kinase domain

Figure 18.61
Potential signaling molecules at the plasma membrane–cell wall continuum. (A) WAK1, a wall-associated kinase, spans the plasma membrane so that its extracellular domain can interact with the wall and its cytoplasmic kinase domain can relay this interaction to proteins within the cell. Some wall-associated kinases have sequences that suggest similarity with signaling molecules in animals, such as epidermal growth factor (EGF), collagen, and neurexin, or with plant extensins (see Chapter 2). (B) A lectin-like receptor kinase traverses the plasma membrane. Lectins may interact with carbohydrate molecules or glycoproteins.

members of this group of kinases are synthesized when cells are water-stressed or exposed to high concentrations of ABA.

18.7.4 Growth factor kinases and mitogen-activated protein kinases are critical elements in the transduction of numerous signals, many of which affect growth.

Members of the CMGC (i.e., CDK/MAPK/GSK-3/CKII) group include some of the kinases most important for growth and development. Mitogen-activated protein (MAP) kinase (MAPK), responsible for the direct regulation of transcription factors, is in turn activated by a protein kinase cascade consisting of MAPKK (MAPK kinase) and MAPKKK (MAPKK kinase, sometimes called Raf) (Fig. 18.63). MAPKKK can be activated in animal cells by an important small GTP-binding protein called Ras or by other GTP-binding proteins that participate in parallel cascades (e.g., Rho or Rac). These cascades are controlled by growth factors and may mediate the action of growth regulators.

In plants, ABA may regulate MAPK activity by removing an inhibitory phosphate on a Raf-like kinase by way of the putative

ABA receptor/phosphatase; auxin may regulate MAPKK. MAPK cascades are implicated in the transduction of mechanical signals such as touch or wind and in signal transduction cascades triggered by gibberellin, ethylene, osmotic stress, wounding, and fungal elicitors. In pollen, a MAPK is synthesized within 1 or 2 minutes after imbibition. These enzymes will phosphorylate serine, threonine, and tyrosine residues in target proteins.

Members of one important class of kinases in the CMGC group regulate cell division and are activated by cyclins, proteins that control progression through the cell cycle (see Chapter 11). These cyclin-dependent protein kinases (CDKs) are themselves regulated by phosphorylation.

Proteins belonging to another important subgroup of CMGC kinases are homologous to members of the GSK-3 (glycogen synthase kinase) group. This subgroup is referred to as shaggy/zeste white 3, named for gene products that specify cell fate and polarity in *Drosophila*. GSK-3s are encoded by a family of genes in plant cells, different members of

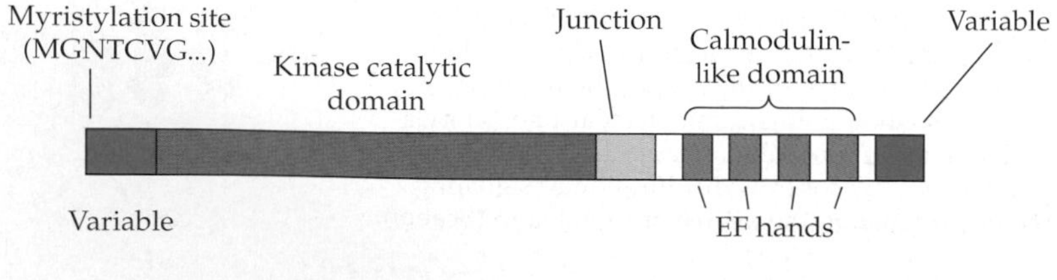

Myristylation site
(MGNTCVG...)

Kinase catalytic
domain

Junction

Calmodulin-
like domain

Variable

Variable

EF hands

Figure 18.62
Calmodulin-like domain protein kinases (CDPKs) are ubiquitous protein kinases in plant cells. All the plant CDPKs cloned thus far contain a calmodulin-like sequence that binds Ca^{2+} at four (or fewer) EF hands. A junction region contains an autoinhibitory sequence that binds to the active site in the absence of Ca^{2+}. Some forms of CDPK may bind to the plasma membrane by way of a myristyl moiety that modifies the N terminus. Perhaps a combined Ca^{2+} and myristylation signal is responsible for the movement of soluble CDPK to the plasma membrane.

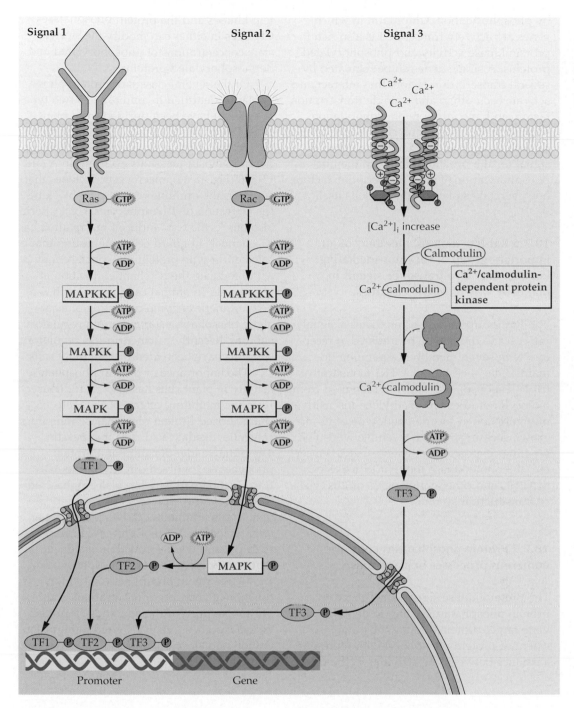

Figure 18.63
Many signals are transduced by protein kinase cascades that regulate gene expression. One such transduction sequence is believed to underlie regulation of gene expression involving a small GTPase (Ras- or Rac-like protein) and mitogen-activated protein kinase (MAPK) cascade leading to transcription factor phosphorylation. Transduction chains from different signals using a MAPK cascade (e.g., signals 1 and 2) or through alteration of $[Ca^{2+}]_i$ (e.g., signal 3) can all affect the same gene through the phosphorylation of different transcription factors (TFs). The TFs may move through the nuclear membrane when they are phosphorylated (signals 1 and 3); alternatively, MAPK may move into the nucleus after phosphorylation by MAPKK (signal 2).

which are differentially expressed, indicating their significance in the regulation of development. In *Arabidopsis*, at least five genes encode homologs of GSK-3 that can autophosphorylate serine, threonine, and tyrosine residues. In animals, these protein kinases are important for specifying cell fate and polarity; they may have a similar function in plants.

CKII (CaM kinase II) is discussed in Section 18.9.3.

18.7.5 Kinases can regulate transcription through phosphorylation of transcription factors.

Expression of a gene depends on the binding of transcription factors to the promoter region of the gene. Phosphorylation (e.g., by MAPK or CKII) represents a primary means for regulating transcription. Many transcription factors are unable to bind DNA unless phosphorylated, although some are inhibited by phosphorylation. Chromatin in which genes are actively transcribed is also rich in protein kinase activity and phosphorylated proteins. A single gene can be regulated by several transcription factors that interact and activate each other, and a single transcription factor can have multiple phosphorylation sites. In some cases, transcription factors are phosphorylated in the cytoplasm, which is a signal for the protein to move to the nucleus; phosphorylation of most transcription factors, however, takes place in the nucleus itself.

18.7.6 Raf-like kinases, members of an important group of receptor-mediating kinases, may also transduce signals in plants by way of MAPK cascades.

The Raf group of kinases mediates a signaling cascade thought to be initiated at receptors with seven membrane-spanning domains. One such kinase, CTR1 (constitutive triple response), was the first identified from molecular studies on an ethylene-insensitive mutant. Named for the phenotype of its air-grown seedlings, which resemble wild-type seedlings exposed to ethylene, CTR1 is a serine/threonine kinase thought to activate MAPKK and other components of this type of transduction system.

18.7.7 Protein phosphatases control numerous processes in plant cells.

The protein kinase signal is regulated by protein phosphatases acting to dephosphorylate target proteins. At any one time, the extent of protein phosphorylation represents a balance between the activities of the protein kinases and the protein phosphatases. Changes in either can modify the steady-state concentrations of phosphorylated and dephosphorylated proteins.

Of the several classes of protein phosphatases identified in animal cells, two types (1 and 2A) have been purified from plant cells. Protein phosphatase 2B (calcineurin) is Ca^{2+}/calmodulin–activated and inhibited by the immunosuppressants cyclosporin A and FK506 (Fig. 18.64), macrocyclic lactones that suppress the immune response by blocking the activation of T-lymphocytes. Cyclosporin also blocks the Ca^{2+}-induced inactivation of K^+ channels in guard cells, suggesting that a calcineurin-type protein phosphatase may act to regulate the K^+ channel function in guard cells. In animal cells, calcineurin is a heterodimeric enzyme consisting of a catalytic phosphatase subunit and a regulatory subunit. Recently, calcineurin-like regulatory subunits have been identified in plant cells.

The importance of protein phosphatase activity in plant cells has also been deduced from the inhibitory effects of the well-characterized protein phosphatase inhibitors calyculin, okadaic acid, and microcystin. The putative identification of a protein phosphatase involved directly in the ABA transduction pathway has also emphasized the importance of protein dephosphorylation. These inhibitors can either prolong an apparent stimulation of protein kinase activity or prevent the transduction of signals that require target protein phosphorylation turnover. Protein phosphatase inhibitors function at very low concentrations and are able to prevent the operation of protein kinase cascades initiated by red light, proliferation, growth regulators, and pollen tube signaling.

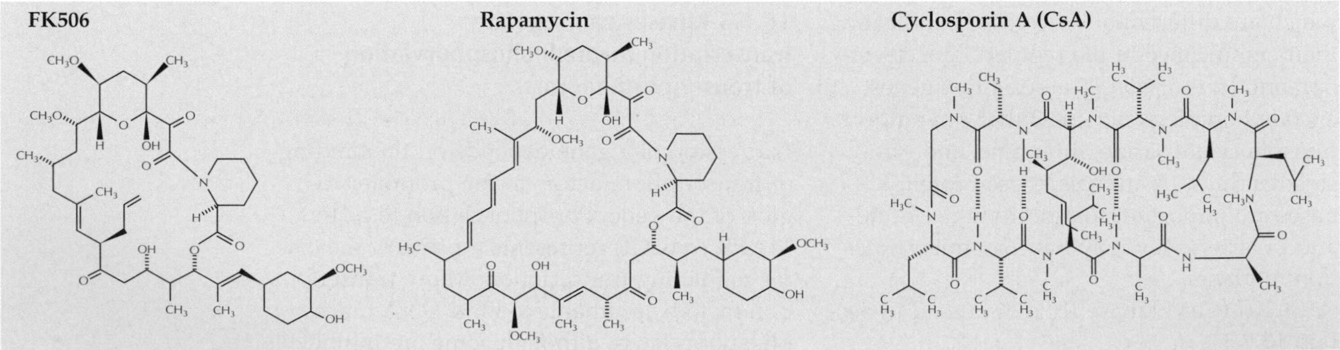

Figure 18.64
Structures of FK506, rapamycin, and cyclosporin—compounds used medicinally to suppress the activity of the immune system, which they do by inhibiting lymphocyte signal sensitivity. All three are thought to inhibit protein phosphatase activity. Remarkably, they all block Ca^{2+}-induced inactivation of specific K^+ channels in guard cells.

FK506   Rapamycin   Cyclosporin A (CsA)

Chapter 18 Signal Perception and Transduction

18.8 Particular pathways of signal transduction associated with plant growth regulators

18.8.1 Ethylene transduction uses protein kinase cascades.

Several classes of mutants obtained by using the seedling bioassay as a screen (Fig. 18.65A) affect a broad range of ethylene responses throughout the life cycle of the plant. These observations indicate that the responses to ethylene share a primary signal transduction pathway. Multiple mutant alleles at the *ETR1* and *EIN* (ethylene-insensitive) loci confer global insensitivity to ethylene. *CTR1* encodes another transduction protein that, when mutated, constitutively activates three ethylene-regulated processes in seedlings: hook development, hypocotyl elongation, and root growth (Fig. 18.65B). CTR1 is therefore a negative regulator of the ethylene response. Double-mutant analysis involving *etr1*, *ctr1*, and *ein2* indicates that CTR1 is downstream from ETR1 and upstream from EIN2. CTR1 has sequence similarity to the Raf group of protein kinases (i.e., MAPKKK; Fig. 18.66). CTR1 activation might specifically regulate ethylene-dependent processes exclusively, especially if it is linked to downstream elements through direct structural interaction.

18.8.2 Gibberellin signal–response pathways indicate the involvement of transcription factors.

Numerous mutants defective in gibberellin (gibberellic acid; GA) signaling have been identified (Fig. 18.67). Most mutants fall into either of two classes: those that resemble GA-deficient plants but do not respond to GA, and those that resemble plants in which a GA response pathway has been constitutively activated. The *rht* (reduced height) mutants are in the first class. Mutants in GA biosynthesis have been exploited by breeders to create "green revolution" cereals, including short-length wheat varieties commonly grown worldwide.

In *Arabidopsis*, SPY (spindly) is a negative regulator of GA signal transduction. *spy* plants are tall but capable of some increased height response to applied gibberellin.

(A)

(B)

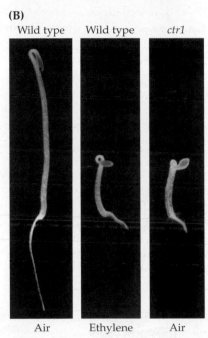

Wild type	Wild type	*ctr1*
Air	Ethylene	Air

Figure 18.65
Ethylene-signaling mutants have striking phenotypes. (A) Mutation in the *ETR1* gene prevents ethylene-dependent hypocotyl shortening in early germination. (B) The constitutive triple response mutation (*ctr1*, right panel) has an air-grown seedling grown in the presence of 10 ppm ethylene (middle panel). Wild-type CTR1 (left panel) functions as a negative regulator of the ethylene response.

Sequence analysis of *SPY* suggests that it encodes an *O*-glucosyl-*N*-acetyltransferase. *gai* (gibberellic acid–insensitive) was identified during characterization of mutant plants with a GA-deficient phenotype. *gai* plants are unresponsive to applied GA but mimic all other aspects of GA deficiency. *GRS* (GAI-related sequence) and *GAI* have been sequenced and most likely encode transcription factors.

Figure 18.66
Comparison of ethylene-signaling pathway and animal Ras-signaling pathway. CTR1 has distinct sequence similarity to Raf (MAPKKK). A protein kinase cascade is indicated for transduction of the ethylene signal, but the physiological function of many of the proteins in the ethylene pathway is currently unknown.

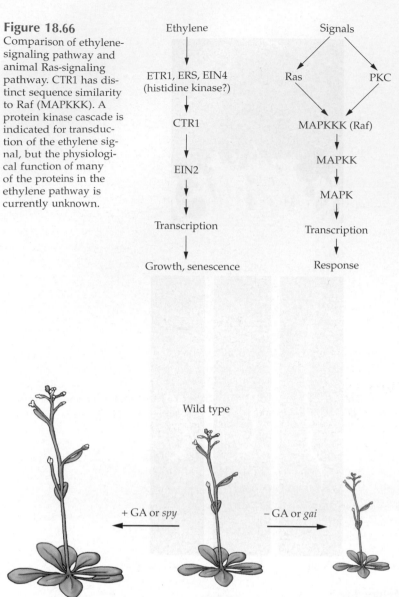

Finally, RGA (repressor of GAI3) does not suppress as many aspects of the GA-deficient phenotype as *spy* does. Double-mutants of *spy* and *rga* exhibit an additive phenotype, which suggests these two genes are in separate GA signal transduction pathways. Figure 18.68 suggests one possible outline of GA signal transduction.

GA-dependent transduction pathways in barley cells involve cytosolic Ca^{2+} and cGMP. These aleurone cells secrete α-amylase in response to gibberellin. MAPK pathways have also been implicated in GA signal transduction.

18.8.3 Auxin signal transduction may involve a protein kinase cascade, 14-3-3 proteins, and ubiquitin-degradation pathways.

Auxin is often regarded as a "master" hormone because cell division, growth, maturation, and differentiation are all associated with auxin regulation. At least some auxin-mediated processes (e.g., acidification of the apoplast during cell wall expansion; see below) appear to be mediated by 14-3-3 proteins, a recently discovered group of proteins. 14-3-3 proteins facilitate phosphorylation of other molecules and may discriminate between phosphorylated and nonphosphorylated target proteins. Therefore, 14-3-3 proteins act as intracellular messengers that can cross-link signal transduction chains.

The plasma membrane–localized H^+-ATPase (P-type ATPase) is an important target for regulation of growth by auxin, because the enzyme is able to increase wall extensibility by acidifying the apoplast. In addition, the establishment of a proton gradient between the plasma membrane and the wall provides the necessary energy for active uptake of the potassium ion, which is required to maintain turgor pressure as the cell expands. The toxin fusicoccin (see Chapter 3, Fig. 3.8) binds to and increases the activity of the electrogenic ATPase. Fusicoccin induces stomatal opening, enhances growth rates, and breaks seed dormancy. Because the fusicoccin receptor is a 14-3-3 protein, it is likely that auxin modifies ATPase activity through a 14-3-3 protein, probably through activation of a protein kinase.

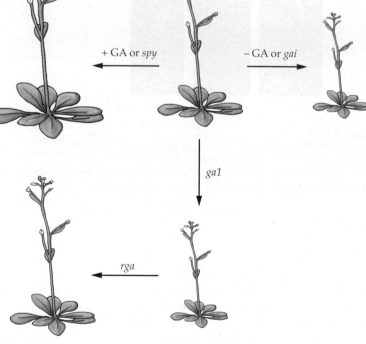

Figure 18.67
Cartoon of the various types of *Arabidopsis* mutants with altered responses to gibberellic acid (GA). The *spy* mutation phenocopies the effect of treating wild-type plants with GA. The *gai* mutation phenocopies the effect of treating plants with a GA biosynthesis inhibitor. The *ga1* mutation disrupts GA biosynthesis. The *rga* mutation suppresses the GA-deficient phenotypes associated with the *ga1* mutation.

Several auxin-resistant mutants have been isolated and analyzed. Sequence analysis of *AXR1* indicates that it encodes a protein with homology to a ubiquitin-activating enzyme, which prepares proteins for ubiquitin-mediated degradation (see Chapter 9). The auxin-resistant mutant *tir1* is defective in synthesis of a protein having a conserved amino acid domain similar to that found in a ubiquitin protein ligase. These observations suggest an unusual mechanism to explain auxin action (Fig. 18.69).

Auxin is generally abundant in shoot tips, which contain cells that progress through a defined developmental program: division → expansion → maturation → differentiation. The continued presence of auxin is necessary for these processes to occur. Each stage of development is associated with a unique set of proteins, some of which must be degraded by coupling to ubiquitin before the subsequent stage can be initiated. The critical proteins of one stage are proposed to repress by feedback the transcription of genes necessary for the next stage. Only if the critical proteins are degraded will the next stage of the developmental program be enabled. Regulation of the ubiquitin-conjugating pathway by auxin is thus indicated, but the mechanism for this is not yet clear. Direct interaction with enzymes in this pathway might explain, for example, the rapidity with which growth diminishes when auxin is removed. Another interesting possibility is that progress through development might be accelerated to some extent if the pathway is enhanced (Fig. 18.69).

18.8.4 ABA transduction involves Ca²⁺-dependent pathways and protein kinases.

The control of guard cell aperture has provided substantial information on the transduction of ABA signals. Both Ca^{2+}-dependent and Ca^{2+}-independent pathways are involved, the latter pathway being known to require changes in intracellular pH, i.e., uses H^+ ions as a second messenger. The discovery of mutants that are ABA-insensitive has enabled identification of ABI1, a type-2C protein phosphatase. A transduction sequence using phosphorylation and Ca^{2+} is thus indicated. Patch clamp investigations

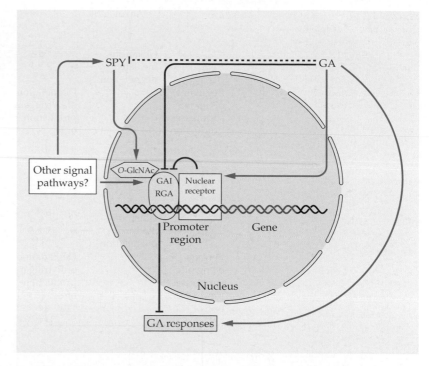

Figure 18.68

Model of how SPY, GAI, and RGA may act in GA signal transduction. SPY is predicted to be a *N*-acetylglucosamine (*O*-GlcNAc) transferase, whereas GAI and RGA are predicted to be transcription factors.

and the use of immunosuppressors that inhibit protein phosphatase activity (e.g., cyclosporin) indicate that the potassium channels that open to permit turgor loss during stomatal closure are regulated by phosphorylation.

The mobilization of Ca^{2+} by ABA may involve the second messenger cADP-R (Fig. 18.70), which is synthesized from NAD^+. cADP-R can mobilize Ca^{2+} from intracellular stores. Microinjection of cADP-R into tomato cells mimics the effect of ABA on gene expression. Reporter constructs containing the promoter *LTI78*, an ABA-dependent gene that also responds separately to water stress and low-temperature signals, have shown that coinjection of cADP-R and protein phosphatase inhibitors into individual cells induces expression of *LTI78*, thus circumventing the normal requirement for ABA induction. Coinjection of cADP-R and either of two kinase inhibitors, staurosporine or K252a, blocks *LTI78* expression. Again, a transduction pathway involving Ca^{2+} and protein kinase is implicated.

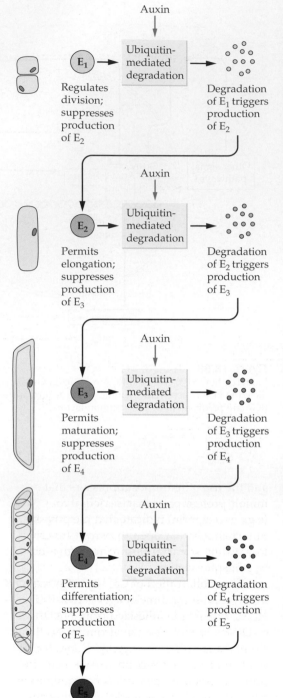

18.9 The future of plant cell signal transduction research

At present, knowledge of plant cell signal transduction is in its infancy. There is considerable room for filling in details of the transduction map, and there are many uncharted directions in which the field will soon expand. Only some can be predicted.

18.9.1 The main signaling pathways cross-talk with each other.

Until recently, the molecular switches that set signaling into action were neatly divided into a few discrete groups, each comprising a distinct set of protein families that receive a particular set of extracellular stimuli and mediate distinct cell responses. One such chain uses G-proteins and $[Ca^{2+}]_i$ as its primary transduction pathway; the other uses protein kinase and leads to growth and proliferation. We now know that such division is simplistic, because the two pathways are linked by many connections, generally referred to as **cross talk.**

We can expect the number of signaling pathways that have been identified to proliferate. In addition, each pathway bifurcates, presumably providing partial or full redundancy in each pathway. For example, guard cells can be closed by an increase in $[Ca^{2+}]_i$ but do not necessarily require changes in $[Ca^{2+}]_i$ for closure to occur. With environmental fluctuation, the flux through each pathway can be expected to vary substantially. Complex integration processes lead to the physiological response. Various different Ca^{2+}-regulated protein kinases clearly cross-talk along the two established pathways. In animals, protein kinases can be activated by small GTPases of the Ras family, which can

Figure 18.69

Isolation and sequencing of auxin-resistant mutants suggest that auxin might regulate protein degradation through ubiquitination. The development of a tracheid is used as an example. It has been suggested that the progress of cells through their developmental schedule requires degradation of the specific proteins associated with one developmental phase before the next stage can occur. Thus, the presence of specific proteins concerned with cell division represses the genes encoding specific proteins required for cell growth and so on. At each stage ubiquitination of these critical proteins may be required for degradation (see Chapter 9). Auxin acts to maintain the activity of this developmental pathway. Removal of auxin disrupts the developmental schedule, arresting development at whatever stage has previously been reached. Restoring auxin concentrations enables completion of the entire program.

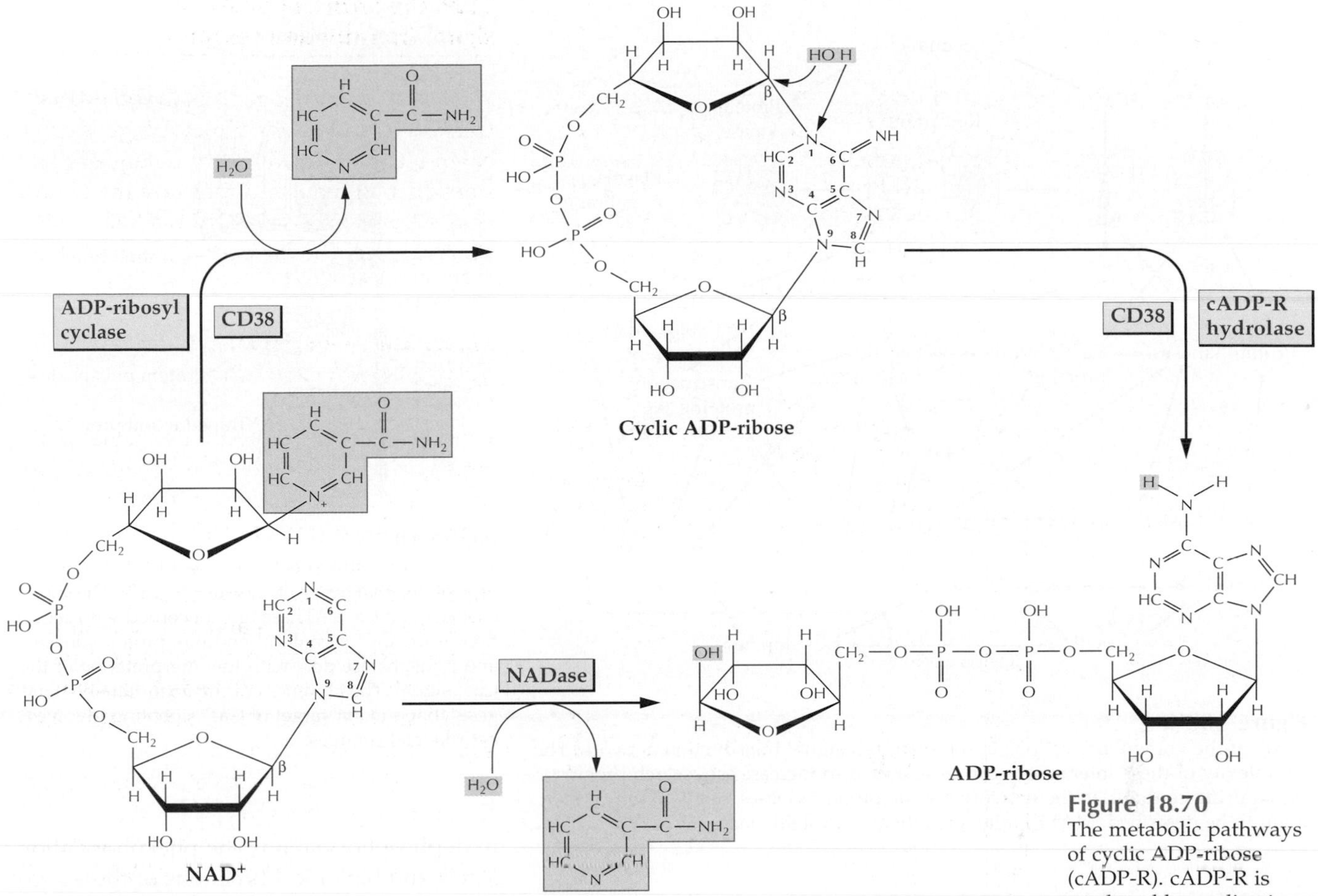

Cyclic ADP-ribose

NAD+

ADP-ribose

ADP-ribosyl cyclase | CD38

CD38 | cADP-R hydrolase

NADase

H_2O

Figure 18.70
The metabolic pathways of cyclic ADP-ribose (cADP-R). cADP-R is produced by cyclization of NAD+, a reaction catalyzed by ADP-ribosyl cyclase, and is degraded by cADP-R hydrolase. Enzymes such as CD38 (from animal sources) can catalyze both reactions. NADase catalyzes the breakdown of NAD+ to ADP-ribose and is involved in cADP-R metabolism.

also increase $[Ca^{2+}]_i$ and various other adapter proteins, thereby establishing cross-talk between the two pathways. Ras can also modify $[Ca^{2+}]_i$, but the β/γ-subunits from G-proteins can activate Ras. Clearly, our expanding base of knowledge indicates that this system is very complex. Figure 18.71 maps some of the known connections that form the basis of a complex transduction network.

18.9.2 Specificity in signaling arises from the specific spatial locations of the signaling elements.

Nature has generated a remarkable array of amino acid sequence motifs to ensure that the right enzymes are in the right place at the right time. Precise spatial and temporal activation of signal transduction pathway components is particularly critical because biochemical "noise" that might obscure signaling must be kept to a minimum. As the second messengers relay signals through cellular compartments that contain many different kinases and phosphatases, it is critical to activate only the appropriate signal transducers; uncontrolled activation would be fatal to the cell. A mechanism for targeting involves anchoring specific components of a signaling cascade to scaffold proteins, thereby producing a large protein complex termed a **transducon** (Fig. 18.72). Several well-characterized protein kinases (e.g., PKC and PKA, MAPK) and phosphatases are coupled to the cytoskeletal scaffold by means of specific tethering subunits. Calmodulin binds to specific structures during cell division and attaches to putative growth poles during polarization; CDPK binds to actin microfilaments. Clustering of Ca^{2+} channels is in some way directed by microfilaments. Crucial research goals include understanding the spatial distribution of transducons, the possibility of spatial overlapping, the sharing of individual constituents, and the importance of mobile second messengers in

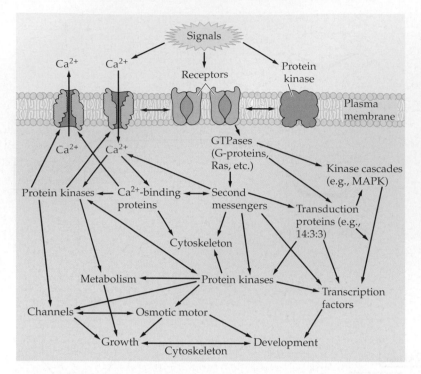

Figure 18.71
Some of the known interactions in the plant cell signal transduction network. The complexity of these interactions can be expected to increase as research progresses. Figuring centrally in the network are the protein kinases, many of which remain to be described. MAPK, mitogen-activated protein kinase.

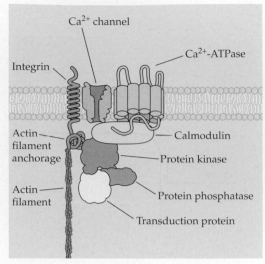

Figure 18.72
Suggested constituents for a transducon, a complex of proteins that acts to transduce signals. The Ca^{2+} channel and Ca^{2+}-ATPase are concerned with regulation of the Ca^{2+} signal; calmodulin, protein kinase, and phosphatase deal with the interpretation of the Ca^{2+} signal. The closeness of the actin filament suggests that a major target of Ca^{2+} signaling involves cytoskeletal changes.

integrating the activities of limited numbers of transducons. Signaling complexes have already been discovered in yeast.

18.9.3 Integration of many signals might result from multiple phosphorylation sites or from integrating enzymes such as CKII.

CKII is a calcium/calmodulin–activated protein kinase originally detected in brain. An equivalent form appears to be present in plant cells. Part of its cycle of activation by Ca^{2+} leads to autophosphorylation and, once a threshold is exceeded, this activation is irreversible and becomes independent of Ca^{2+}. Moreover, the enzyme depends for activation not only on the size and number of Ca^{2+} spikes it perceives but also on their frequency (Fig. 18.73). This enables integration of all the signals that contribute to the $[Ca^{2+}]_i$ signal.

Integration may occur in other ways. Many nuclear proteins have multiple phosphorylation sites that may be phosphorylated by several separate protein kinases. Multiple phosphatases may also participate in dephosphorylation. The phosphorylation states and biological functions of these proteins therefore represent the integrated activity of various different signals transduced through different kinases. Protein kinase activation may also signal the cell without additional phosphorylation. For example, a receptor protein kinase can dimerize in the plasma membrane and undergo autophosphorylation. The two subunits can then separate and their phosphorylation status can be perceived by other proteins, which are then activated and continue transduction of this active state.

18.9.4 Adapter proteins may help cross-link signaling pathways or act as phosphorylation receptors.

Members of the highly conserved and abundant eukaryotic family of 14-3-3 proteins interact with many important signaling proteins in animal cells (e.g., PKC). In plant cells, they interact with and regulate nitrate reductase (see Chapter 16) and act as the receptor for fusicoccin, a fungal toxin that activates the plasma membrane H⁺-ATPase (see Chapter 3 and Section 18.8.3). There are at least 10 isoforms of 14-3-3 in *Arabidopsis*, which have been proposed as potential adapters between pairs of different signaling components, thereby controlling signal transduction. One of the important 14-3-3–binding molecules in animal cells is Raf.

Phosphorylation of the target protein seems to be an essential signal for 14-3-3 binding, and a peptide motif containing a serine phosphate is shared by most 14-3-3–binding proteins. Thus, the 14-3-3 proteins may act as receptors for the phosphorylated state of proteins and activate downstream constituents as a consequence. Plants probably contain many proteins that bind 14-3-3 proteins but their identification and characterization are difficult because they must be phosphorylated to bind to a 14-3-3 protein. Proteins involved in cell cycle induction and cell cycle checkpoints may also be likely targets for 14-3-3–binding proteins.

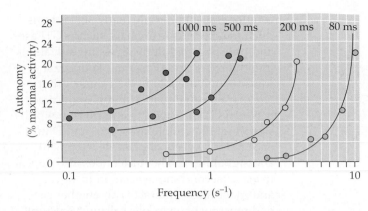

Figure 18.73

Modulation of the frequency response of CKII, a complex Calcium/calmodulin (CaM)–activated protein kinase found in substantial quantities in brain tissue. This unusual enzyme responds to activation by Ca^{2+}/CaM by becoming autonomous of these two activators, using autophosphorylation to provide a stable activated state. Remarkably, CKII can interpret oscillations of Ca^{2+}. Autonomy, as a percentage of maximum activity (i.e., the extent to which the enzyme is activated and no longer needs Ca^{2+}/CaM), has been measured against the frequency with which particular activating pulses are applied. Short pulses (80 milliseconds) must be applied with greater frequency than long pulses (1000 milliseconds) to achieve comparable effects.

18.9.5 Cells communicate by means of integrins and plasmodesmata.

The identity of each cell is specified by signals from neighboring cells. Cells may respond to mechanical signals from the cell wall, to chemical signals diffusing from other cells, or to signals that arrive through the plasmodesmata. Mechanical signals are probably important in the specification of cell form and may be determined by the tensions and compressions resulting from wall- and turgor-based interactions with the surrounding cells. Cellular receptors in the wall/extracellular matrix probably bind to **integrins** that span the plasma membrane and then connect to the cytoskeleton through other proteins (see Chapter 5). Animal integrins include vinculin, actinin, and talin; equivalent receptors and integrins are probably present in plant cells as well. Interaction with integrins can be inhibited by peptides such as RGD (Arg-Gly-Asp). Transduction through integrins is initiated by integrin clustering and results in activation of the $[Ca^{2+}]_i$ pathway, probably to ensure microfilament rearrangements in response to signaling. Protein kinases are almost certainly involved as well.

The passage of signals through the plasmodesmata (see Chapter 15) is thought to be regulated directly by $[Ca^{2+}]_i$, which can cause plasmodesmatal closure. Although it might be difficult for Ca^{2+} waves to move directly from one cell to another, Ca^{2+} could generate action potentials as an alternative form of communication capable of moving from cell to cell, increasing $[Ca^{2+}]_i$ sequentially. IP_3 might also carry a Ca^{2+} wave from cell to cell through the plasmodesmata. Recent studies on plasmodesmata have indicated that this intercellular communication route may also permit the direct exchange of transcription factors, thereby allowing the formation of a supracellular network of interactions controlled, in turn, by signal transduction.

Summary

Signal transduction is an actively expanding topic of research in plant biology. Signals, which include a wide array of external and internal stimuli, are amplified and communicated by complex signal transduction networks, most of which initiate with the activation of receptor proteins. Bacterial receptor and transduction systems provide models for plant receptors, including proteins that sense ethylene and phytochrome. Among the various plant signal transduction pathways that have been identified are other components common to many signal transduction networks in animals, such as GTPases and phospholipid derivatives. Investigations into the roles of GTPases in plant signal transduction are still in their infancy, but already a strong relationship is implicated between GTPase activity and phospholipid signaling. Phospholipases A, C, and D influence many aspects of plant development and signaling. Cyclic nucleotides also appear to act as second messengers in plant cells and most likely interact with another second messenger, cytosolic calcium. Calcium channels and other calcium transporters form the basis of a complex Ca^{2+} signaling network in plants. Protein kinases are the most common transduction components interpreting signals in plant cells. Various classes of protein kinase act in concert with protein phosphatases to mediate plant cell signaling and control metabolism. Plant hormones are important elements in controlling plant growth and development, and progress is being made in understanding how cells transduce these signals. Advances in signal transduction research are rapidly expanding our understanding of how plant cells communicate and cooperate.

Further Reading

General

Trewavas, A. J., Malho, R. (1997) Signal perception and transduction: the origin of the phenotype. *Plant Cell* 9: 1181–1195.

Receptors

Bleecker, A. B., Schaller, G. E. (1996) The mechanism of ethylene perception. *Plant Physiol.* 111: 653–660.

Furuya, M., Schafer, E. (1996) Photoperception and signaling of induction reactions by different phytochromes. *Trends Plant Sci.* 1: 301–307.

Khurana, J. P., Kochhar, A., Tyagi, A. K. (1998) Photosensory perception and signal transduction in higher plants: molecular genetic analysis. *Crit. Rev. Plant Sci.* 17: 465–539.

Kieber, J. J. (1997) The ethylene response pathway in *Arabidopsis*. *Annu. Rev. Plant Physiol. Plant Mol. Biol.* 48: 277–296.

Leung, J. (1998) Abscisic acid signal transduction. *Annu. Rev. Plant Physiol. Plant Mol. Biol.* 49: 199–222.

G-proteins

Ma, H. (1994) GTP-binding proteins in plants: new members of an old family. *Plant Mol. Biol.* 26: 1611–1636.

Phospholipid signaling

Munnik, T., Irvine, R. F., Musgrave, A. (1998) Phospholipid signaling in plants. *Biochim. Biophys. Acta* 1389: 222–272.

Cyclic nucleotides

Assmann, S. M. (1995) Cyclic AMP as a second messenger in higher plants: status and future prospects. *Plant Physiol.* 108: 885–889.

Neuhaus, G., Bowler, C., Hiratsuka, K., Yamagata, H., Chua, N.-H. (1997) Phytochrome-regulated repression of gene expression requires calcium and cGMP. *EMBO J.* 16: 2554–2564.

Calcium

Gilroy, S. (1997) Fluorescence microscopy of living plant cells. *Annu. Rev. Plant Physiol. Plant Mol. Biol.* 48: 165–190.

Malho, R., Moutinho, A., Van Der Luit, A., Trewavas, A. J. (1998) Spatial characteristics of calcium signaling: the calcium wave as a basic unit in plant cell calcium signaling. *Philos. Trans. R. Soc. Lond. Ser. B* 353: 1463–1473.

Sanders, D., Brownlee, C., Harper, J. F. (1999) Communicating with calcium. *Plant Cell* 11: 691–706.

Zielinski, R. E. (1998) Calmodulin and calmodulin-binding proteins in plants. *Annu. Rev. Plant Physiol. Plant Mol. Biol.* 49: 697–727.

Growth substances

Kende, H., Zeevart, J.A.D. (1997) The five "classical" plant hormones. *Plant Cell* 9: 1197–1210.

Protein kinase

Mirinov, V., De Veylder, L., Van Montagu, M., Inze, D. (1999) Cyclin-dependent kinases and cell division in plants: the nexus. *Plant Cell* 11: 509–521.

Smith, R. D. (1996) Plant protein phosphatases. *Annu. Rev. Plant Physiol. Plant Mol. Biol.* 47: 101–125.

Stone, J. M., Walker, J. C. (1995) Plant protein kinase families and signal transduction. *Plant Physiol.* 108: 451–457.

Newer approaches

Faux, M. C., Scott, J. D. (1996) More on target with protein phosphorylation: conferring specificity by location. *Trends Biochem. Sci.* 21: 312–315.

Ferl, R. J. (1996) 14-3-3 proteins and signal transduction. *Annu. Rev. Plant Physiol. Plant Mol. Biol.* 47: 49–73.

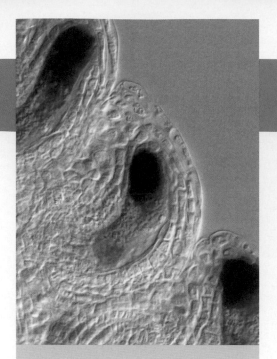

Biochemistry & Molecular Biology of Plants, B. Buchanan, W. Gruissem, R. Jones, Eds.
© 2000, American Society of Plant Physiologists

CHAPTER *19*

Reproductive Development

J. Derek Bewley
Frederick D. Hempel
Sheila McCormick
Patricia Zambryski

CHAPTER OUTLINE

Introduction

Throughout history, humans have celebrated the beauty and fertility of flowering plants. In addition to their aesthetic appeal, flowers contain the reproductive organs of the plant and are therefore essential for sexual propagation of plant life. Our dependence on flowering is illustrated by the dietary importance of fruits and seed crops. Advances in molecular biology and biochemistry are now revealing the cellular mechanisms that underlie the development and symmetry of flowers and the nutritive value of seeds.

Figure 19.1 outlines the life cycle of a flowering plant, highlighting the reproductive organs that we will discuss at length in this chapter. After first describing how flowering is induced and how flower primordia are patterned, we will consider the production and union of male and female gametes. Finally, we will illustrate the formation of seeds, in which plant embryos are packaged and remain quiescent until induced to germinate. Throughout the chapter we will highlight recent developments in genetics, molecular biology, and biochemistry that are being utilized to unravel the details of reproductive mechanisms in plants.

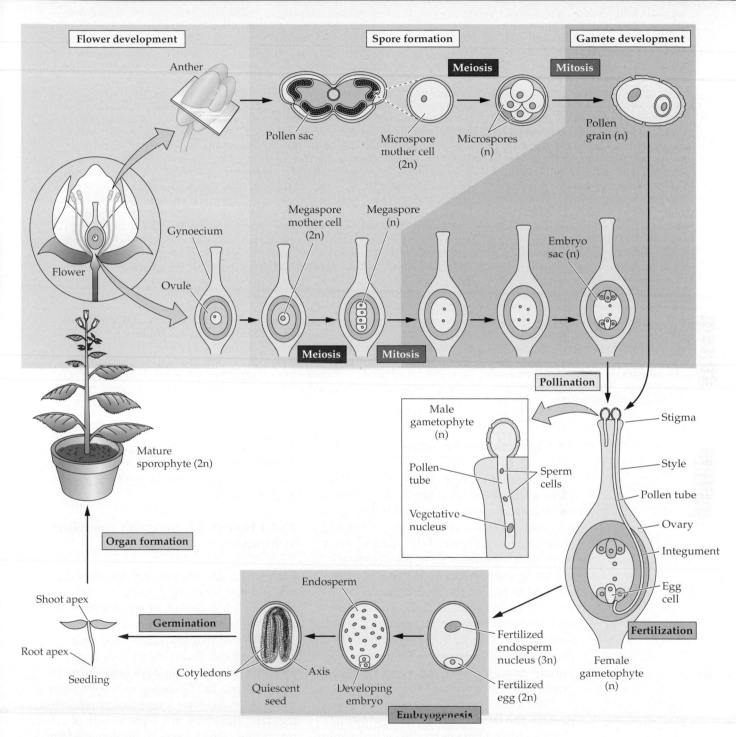

Figure 19.1
The life cycle of a flowering plant. Note the alternation of multicellular haploid (gametophytic) and diploid (sporophytic) generations.

Although seed plants such as angiosperms and higher animals such as mammals both produce large diploid individuals, these groups of organisms use distinctly different developmental and reproductive strategies. One of the most fundamental differences is mobility: Both animals and the cells within them can move; plants and their cells generally do not. The ability to move dictates how the organism responds to the environment, nourishes itself, and reproduces. Animals commonly respond to changes in the environment behaviorally as well as physiologically, but plants are limited predominantly to physiological responses because they cannot move to avoid adverse conditions. For example, an animal might attempt to hide from or outrun a predator, whereas a plant might up-regulate its metabolic production of foul-tasting compounds to discourage herbivores.

Angiosperms and higher animals are shaped by very different developmental strategies. The development of higher animals is generally viewed as a series of complex pathways that lead from undifferentiated cells to progressively more differentiated cell types and lineages. Animal development, therefore, tends to restrict the developmental potential of individual cells: Once an entire animal is formed, the cells and organs within it are often terminally differentiated. In contrast, plant cells often remain totipotent or pluripotent, a characteristic of various plant tissues that has been widely exploited to develop tissue culture methods. The ability to regenerate entire plants from single plant cells is especially useful for genetic engineering strategies. Notably, the major plant organ systems, the shoot and the root, retain the ability to differentiate new organs in the fully mature plant. The new organs generally are derived from meristems, which are regions of active cell division that allow the mature plant to produce new roots, shoots, and leaves.

In animal development, the gametes and the reproductive organs are set aside early during embryonic development, and these germ line cells do not contribute to the somatic cells of the animal. In contrast, plants mature before producing reproductive organs. The mature plant forms new organs vegetatively from the shoot and root meristems. Then, in response to developmental and environmental signals, the shoot apex of the plant switches from a vegetative mode of differentiation to a reproductive mode. This shift results in the production of flowers and gametes. Thus, the cells that give rise to the plant germ line are derived from meristematic regions that previously gave rise to the vegetative (somatic) parts of the plant. Consequently, somatic mutations in meristem cells have the potential to alter the genotype of the gametes.

The early sequestration of the germ line in animal development ensures that the reproductive cells are not exposed to the developmental signals that subsequently induce formation of the major animal organ systems. The plant reproductive system, however, is derived from cells of the mature plant that have been through many cell divisions as well as exposure to various environmental changes throughout the life of the plant. Nonetheless, plants have an efficient mechanism for selecting against somatic mutations that occur before gametogenesis. After meiosis has occurred in either pollen (male) or ovule (female) progenitor cells, a series of mitotic divisions generates essential accessory cells. These divisions constitute a **haplosufficiency** test for critical cellular functions: Haploid gametophytes with mutations in the genes required for cell survival are unlikely to reproduce successfully.

19.1 Induction of flowering

The molecular aspects of flowering interest plant scientists because flower development represents an important switch in the developmental program of the shoot. During vegetative development, cell divisions in shoot meristems repeatedly give rise to indeterminate secondary shoots. When a plant is induced to flower, the meristems are reprogrammed to produce flowers, that is, determinate shoots from which additional shoots do not arise. This shift from somatic to reproductive development highlights an important developmental difference between plants and animals (Box 19.1).

The induction of flowering represents perhaps the most dramatic developmental change in the angiosperm life cycle. In **annuals,** short-lived plants that flower only once, the induction of flowering represents not only the onset of reproduction but also the start of senescence (see Chapter 20). Thus, plants must regulate flowering precisely, to ensure that it will occur at a favorable time for complete seed development and reproductive success.

19.1.1 Flowering is commonly controlled by photoperiod.

Early in the 20th century, investigators discovered that flowering routinely is controlled by photoperiod, or the length of exposure to light. Experiments at that time demonstrated that short days (and long nights) induced flowering in hops and cannabis, whereas long days (and short nights) promoted flowering in *Sempervivum funkii.* Scientists later found that photoperiod regulates flowering in a wide variety of plants and coined the term **photoperiodism,** defined as a response to the timing of light and darkness. Plants in which flowering is promoted when they are grown under long-night/short-day conditions are classified as **short-day plants,** and plants in which flowering is promoted in short-night/long-day conditions are classified as **long-day plants** (Fig. 19.2). Flowering in a third class of plants, known as **day-neutral plants,** is not affected by photoperiod. Further variations include plants that flower in response to long days followed by short days and plants that flower in response to short days followed by long days.

The discovery of photoperiodism greatly stimulated and shaped experimental research on flowering in the decades that followed, because in many plants the photoperiodic control of flowering provided a simple means for precisely controlling the developmental state of the plant and the onset of flowering. Photoinduction experiments also revealed that the rapid changes in the flowering status of many plants are mediated by the leaves, which detect photoperiodic stimuli.

Experiments with directed light in the late 1940s provided the first evidence that leaves discern photoperiods. For example, spinach was induced to flower when only the leaves but no other parts of the plant were exposed to inductive (short-day) photoperiods. However, when only the shoot apex was exposed to short-day photoperiods, no flowering occurred. Grafting experiments supported the conclusion that leaves perceive photoperiodic signals. For example, when a leaf from a photoperiodically induced *Perilla* plant was grafted successively onto the shoot apices of several vegetative plants, a single photoinduced leaf was able to induce flowering in seven vegetative plants (Fig. 19.3). The stable commitment of *Perilla* leaves to the promotion of flowering suggests that, at least in some plants, the leaves control flowering status.

Although photoperiod regulates flowering in most species, flowering clearly also is modulated by other factors, including temperature and light quality. Some plants have an absolute photoperiodic requirement for flowering, but almost all will flower eventually, even in photoperiods that do not promote flowering. Two plants that illustrate this principle are the model systems *Arabidopsis* and *Antirrhinum.* Both are facultative long-day plants that flower eventually in short days but rapidly are induced to flower by long days. The sensitivity of both plants to floral signals increases with plant age.

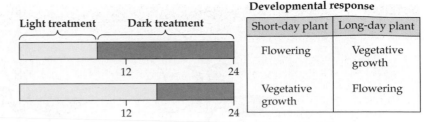

Figure 19.2
The relationship between photoperiod and flowering response is different in short-day and long-day plants.

19.1.2 Phytochromes are the primary photoreceptors involved in sensing photoperiod and light quality.

The molecular mechanisms by which plants perceive light quality involve photoresponsive pigments, particularly the **phytochromes,** which are found in the cytoplasm of green plants and algae. Phytochromes absorb both red and far-red light and are involved in several timing processes, including flowering and germination (see Chapter 18). Photoperiodic studies have implicated phytochromes both in photoperiodism and in sensing the light quality, although the connection between phytochromes and photoperiodism still is incompletely understood.

Arabidopsis has five phytochrome genes (*PHYA* through *PHYE*), each of which encodes a distinct phytochrome protein (see Chapter 18, Box 18.4). *PHYA* and *PHYB* are involved in regulating the time to flowering, although neither is essential for the repression or induction of flowering. The phenotype of the *phyB* mutant, which flowers early, indicates that *PHYB* delays flowering time (Fig. 19.4). In contrast, *PHYA* accelerates the time to flowering in response to flower-promoting light signals. For example, although wild-type plants flower early when exposed to one-hour light periods in the middle of the dark period, *phyA* mutants do not. However, the *phyA phyB* double mutant will flower even earlier than the *phyB* mutant, an indication that the effect of the two phytochromes on flowering is complex. One of the major challenges in photobiology is to identify the specific effects of the various phytochromes on the developmental processes, including flowering.

Although phytochromes are the primary photoreceptors involved in the control of

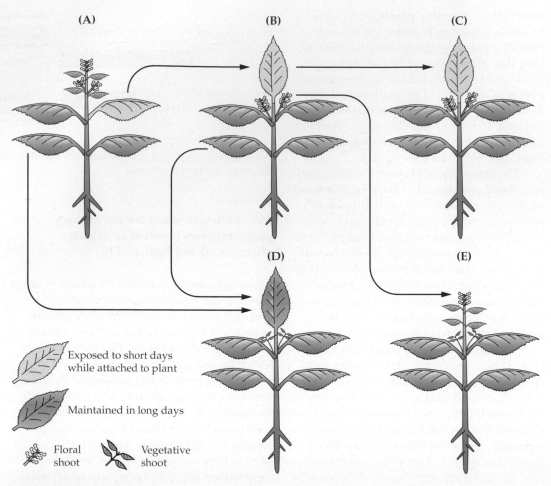

Figure 19.3
Permanent photoinduction of *Perilla* leaves. (A) A single *Perilla* leaf exposed to numerous short days (light green) can induce flowering of an attached plant that has been exposed only to noninductive long days (dark green). (B) The same short-day leaf, when grafted onto a second vegetative plant grown in noninductive long days, can induce flowering on the second plant. (C) If the same leaf is again removed and is placed on a third plant, it again induces flowering. (D) Noninduced (long-day) leaves from plants in flower do not induce flowering when grafted onto a noninduced plant. (E) A flowering apex grafted onto a noninduced plant does not induce flowering in the host plant. Taken together, these grafting experiments suggest that the leaves of *Perilla* are permanently induced by short-day photoperiods to produce flower-promoting signals.

flowering, they are not the only photoreceptors that influence flowering. **Cryptochromes,** which perceive blue light (see Chapter 18), also affect photomorphogenesis—although their influence on flowering is yet to be fully determined. One complicating factor is that phytochromes, too, perceive blue light, and separating phytochrome-mediated responses from cryptochrome-mediated responses is often difficult. Photosynthetic pigments (see Chapter 12) also are involved in the regulation of flowering, although their effects on flowering are assumed to be largely indirect. For example, increased photosynthesis may affect the timing of flowering by shortening the time required for a plant to mature and become competent to flower.

19.1.3 Flowering is under the control of a multifactorial system.

Classic grafting experiments (see Fig. 19.3) suggest that a transmissible signal moves from "photoinduced" leaves to the meristems, where flowers are produced. This suggestion of a transmissible signal has prompted the search for a universal flowering hormone, or **florigen.** Although no single flower-inducing substance has yet been found,

Figure 19.4
Early flowering in the *phyB* mutant of *Arabidopsis*. The two plants shown were grown together in long days for the same length of time. On the right is a wild-type plant that has not yet initiated a macroscopic inflorescence shoot. On the left is a *phyB* mutant with flowers already evident at the tip of the shoot.

these studies have successfully identified several molecules with important flower-promoting effects. For example, gibberellic acid (GA; see Chapter 17) promotes flowering in numerous species, particularly in rosette-type plants such as *Arabidopsis*.

Many molecules promote flowering, but none is sufficient to trigger flowering universally. It is logical to conclude, therefore, that control of flowering is a multifactorial process. One model for the multifactorial control of flowering has been developed on the basis of experiments utilizing the long-day plant *Sinapis alba* (white mustard). This model postulates that flowering is regulated by several molecules, including plant hormones and other common metabolites. Those researchers and others have identified flower-promoting roles for GA, cytokinins, sucrose, and polyamines in annual mustards such as *S. alba* and *Arabidopsis*.

The multifactorial-control model also suggests that different factors will become limiting for flowering in different plants or under different growth conditions—a result consistent with observations on flowering in many species, including *Arabidopsis*. For example, when the GA-deficient *Arabidopsis* mutant *ga1-3* is grown in long days, flowering occurs later in the mutant than in the wild type. Furthermore, when plants are grown in noninductive short days, flowering does not occur at all in the mutant, unless exogenous GA is applied. These results indicate that GA is an important component in the multifactorial control of flowering in *Arabidopsis* and is a limiting factor for flowering under short-day conditions.

Little is known about the mechanisms by which flowering signals are transmitted from the leaves to the meristem during the induction of flowering. Analyses of phloem exudates, before and after the photoinduction of flowering, indicate that phloem transport is involved in the transport of known flower-promoting signals, including sucrose and cytokinin. Photoinduction treatments lead to increased fluxes of sucrose and cytokinin toward the shoot apex. However, many aspects of sucrose and cytokinin synthesis and movement during floral induction remain undefined. Although the symplasmic cell-to-cell movement of signaling molecules by way of the plasmodesmata may explain the transport of many floral signaling molecules, other types of signaling pathways undoubtedly are involved as well. Additional work is required to define the details of multifactorial signaling during floral induction.

19.1.4 Many genes involved in the production, transmission, or perception of signals that induce flowering have been identified genetically.

Mutants that flower earlier or later than wild type have been identified in *Arabidopsis*, indicating that flowering also is genetically multifactorial. In general, mutants that flower early are indicative of genes that ordinarily repress flowering, and mutants that flower late are indicative of genes that promote flowering. Some of these flowering-time mutants should be useful in identifying genes involved in the production of floral signals in the leaves; others probably will serve to identify genes involved in the perception of floral signals within meristems.

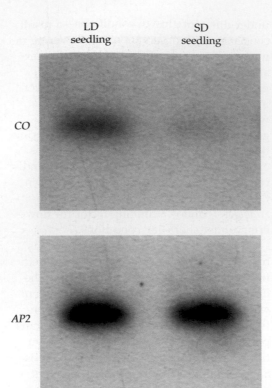

LD seedling   SD seedling

CO

AP2

Figure 19.5
CONSTANS (*CO*) expression increases in long days. RNA gel blot analysis indicates that seedlings grown in long days (LD) show increased expression of the flower-promoting gene, *CO*, which is expressed both in shoots and in the leaves. The expression of *APETALA2* (*AP2*), however, is similar in long and short days (SD). The role of *AP2* in flowering is discussed in Sections 19.1.5 and 19.3.

(A)   **(B)**

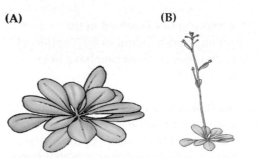

Wild-type control (ecotype Landsberg erecta)   *35S::CO* transformant

Figure 19.6
CO expression induces flowering in *Arabidopsis*, even in short days. The cartoons represent two plants grown under short-day conditions. (A) The wild-type plant is not flowering and will remain vegetative for at least three more weeks. (B) A plant expressing a *35S::CO* transgene constitutively. This transgenic plant has flowered precociously in short days.

Arabidopsis is well suited to the molecular analysis of flowering-time mutants. Two late-flowering phenotypes correspond to mutant alleles of the genes *CONSTANS* (*CO*) and *LUMINIDEPENDENS* (*LD*). Both of these genes have been cloned and are expressed in the young leaves of vegetative plants. The mechanism by which *LD* affects flowering is unclear at present. *CO*, however, is a putative transcription factor with similarity to zinc-finger proteins (see Chapter 7) and is up-regulated in flower-promoting long days (Fig. 19.5). Furthermore, the ectopic expression of *CO* rapidly induces the expression of flowering genes and promotes flowering (Fig. 19.6). This suggests that *CO* is involved in activation of flowering genes. Whether *CO* is important for the synthesis or transport of flower-promoting signals, or is important for the perception of signals, is not known. Further characterization of *CO* and other time-to-flowering genes will greatly facilitate our understanding of floral induction processes.

Analysis of another late-flowering gene in *Arabidopsis*, *FRIGIDA* (*FRI*), indicates that prolonged cold treatments also can regulate flowering at the shoot apex. The dominant *FRI* alleles, which confer a late-flowering phenotype, have been isolated from naturally occurring populations. Their late-flowering phenotype, however, can be reversed if a prolonged treatment with cold (**vernalization** treatment) is given (Fig. 19.7). *FRI*, in combination with a second gene, *FLOWERING LOCUS C*, modulates a requirement for vernalization. Evidently, the role of *FRI* under natural conditions is to regulate flowering by sensing cold temperatures.

Among other effects, vernalization promotes cytosine demethylation. In the crucifer *Thlaspi arvense*, the correlation between demethylation and the transcriptional activation of *KAH*, a gene that encodes the GA biosynthesis enzyme *ent*-kaurenoic acid hydroxylase (see Chapter 17), indicates that vernalization may activate expression of *KAH* and possibly other genes involved in the promotion of flowering. Vernalization effects can be mimicked by treatment with the demethylating agent 5-azacytidine, thus supporting the connection between vernalization and methylation in controlling time to flowering. These results suggest that the demethylation of specific genes is correlated

(A)

(B)

0 days of
vernalization at 4°C

100 days of
vernalization at 4°C

Figure 19.7
FRIGIDA (*FRI*) plants respond to cold treatments
(vernalization). Both plants shown are *FRI* plants in
the Columbia ecotype of *Arabidopsis*. The dominant
allele of *FRI*, which is common in some natural
ecotypes of *Arabidopsis*, confers late flowering. Note
that the nonvernalized plant (A) has elaborated numerous leaves during its extended vegetative development. The late-flowering phenotype, however, can be reversed by a long cold treatment (B).
Both plants are shown at the time their first flowers opened.

(A)

(B)

Figure 19.8
LEAFY (*LFY*), a gene for floral meristem identity, is
necessary and sufficient to convert indeterminate
shoots into flowers in *Arabidopsis*. (A) In *lfy* mutants,
shoots that would normally develop as flowers are
converted to indeterminate shoots. All of the secondary shoots shown here have emerged in the positions in which flowers would develop on a wild-type
inflorescence. (B) In a *35S::LFY* plant that expresses
LFY constitutively, secondary shoots are converted
into flowers and the primary shoot terminates early
in a flower or cluster of flowers.

with maturation and flowering. Interestingly,
recent results suggest that methylation also
is involved in controlling the gene *SUPER-
MAN* (*SUP*), which defines the identity of
various floral organs (see Section 19.3.2).

19.1.5 Expression of most genes that define the identity of floral meristems is up-regulated by floral induction signals.

The ultimate targets of the signals that promote flowering are the genes that define the
identity of floral meristems, particularly
LEAFY (*LFY*), APETALA1 (*AP1*), and *CAU-
LIFLOWER* (*CAL*). Whether the products of
genes for floral meristem identity interact
with transmissible flower-promoting signals
directly or indirectly is not yet known.

The *lfy* mutant of *Arabidopsis* produces
more inflorescence branches than does wild
type. *lfy* flowers are green and consist of
whorls of sepal-like and carpel-like organs
(Fig. 19.8A). Conversely, the ectopic expression of *LFY* promotes early flowering and
the conversion of shoots to flowers (Fig.
19.8B). Thus, the *LFY* gene functions both in

defining floral meristem identity and in affecting the time to flowering.

Another gene that much affects the determinacy of *Arabidopsis* meristems is *TER-
MINAL FLOWER 1* (*TFL1*). *tfl1* mutants
flower early, and the primary inflorescence
branches produce terminal flowers. *tfl1*
plants also lack lateral branches and exhibit
greatly reduced flower numbers (Fig. 19.9A).
TFL1 protein is expressed in cells in the shoot
meristem but not in young flower primordia
(Fig. 19.9B). The generally opposite phenotypes of *tfl1* (conversion of indeterminate

(A)

WT   *tfl1*

(B)

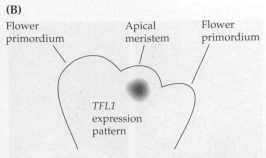

Flower primordium   Apical meristem   Flower primordium

TFL1 expression pattern

Figure 19.9

TFL1 regulates flowering and indeterminacy. (A) *tfl1* inflorescence phenotype (right and enlargement) compared with wild type (left). Mutant plants terminate early in a terminal flower or a terminal cluster of flowers. (B) *TFL1* is expressed in the center of the inflorescence meristem.

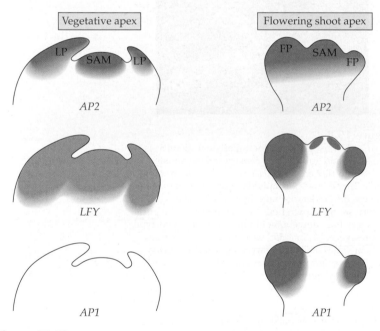

Vegetative apex

LP   SAM   LP

AP2

LFY

AP1

Flowering shoot apex

FP   SAM   FP

AP2

LFY

AP1

Figure 19.10

Expression of genes for floral meristem identity in apices of vegetative and flowering shoots. These genes tend to be expressed strongly in flower meristems. *LFY*, *AP1*, and *AP2* are all expressed early in primordium ontogeny when the identity of a flower primordium (FP) is established. Expression of *LFY* and *AP1* is not strictly limited to flower primordia, but their normal function probably is. *LFY* expression is low within the apices of vegetative plants and increases greatly in response to stimuli that induce flowering; this indicates a potential role for *LFY* in the sensing and integration of flower-promoting signals during the induction of flowering. *AP2* expression also is evident in the vegetative shoot apex, although the significance of this expression is yet to be determined. *AP1* is expressed only in plants that have been induced to flower, although its expression is not strictly limited to young flower primordia. The expression of another MADS box transcription factor, encoded by the *CAL* gene, is very similar to *AP1* (not shown). LP, leaf primordium; SAM, shoot apical meristem.

shoots to determinate shoots) and *lfy* (conversion of determinate shoots to indeterminate shoots) suggest that their gene products may act antagonistically.

Three other genes influence the determinacy of the floral meristem: *AP1*, *APETALA2* (*AP2*), and *CAL*. For example, the *ap1* and *ap2* mutations enhance the *lfy* phenotype. Although *cal* mutants have no visible phenotype, *cal ap1* mutants are more highly indeterminate and markedly overproduce floral meristems, resulting in cauliflower-like inflorescences. The sequences and expression patterns of the proteins encoded by *AP1* and *CAL* are very similar, suggesting that their functions may be largely redundant.

Expression patterns of *LFY*, *AP1*, and *TFL* support their proposed early roles in floral meristem specification (Fig. 19.10). The floral meristem identity genes *LFY* and *AP1* are expressed in floral meristems before the organ primordia develop. *LFY* is expressed in low amounts during the vegetative phase of the plant life cycle. Expression is up-regulated rapidly on floral induction by photoperiod treatment or in response to GA. *LFY* expression is increased greatly after the induction of ectopic *CO* expression, which suggests that one function of *CO* may be to up-regulate *LFY*. *LFY* is already expressed during the switch to floral development; *AP1*, however, is not generally expressed in flower primordia until one to two days after floral induction begins. *AP2*, although primarily required for proper floral meristem specification, is expressed throughout the life cycle of the plant (Fig. 19.10). *TFL1*, like *AP2*, is expressed throughout development in cells that lie just below the apical dome of the primary shoot meristem (see Fig. 19.9B).

LFY, *AP1*, *AP2*, and *CAL* have distinct functions that overlap at several stages of early flower development, possibly acting to enhance each other's activity. One reason for this mutual stimulation might be that all four genes probably encode transcription factors. Molecular studies indicate that *AP1* and *CAL* encode transcription factors containing a MADS box region (see Section 19.3.2). *AP2* encodes a putative transcription factor, and *LFY* encodes a protein with both proline-rich and acidic domains, which may represent a new type of transcription factor. Only *TFL1* does not appear to encode a transcription factor; instead, its coding sequence is homologous to those of the phosphatidylethanolamine-binding proteins that associate with membrane-localized protein complexes in animals.

19.2 Flower development

The genetic and molecular dissection of flower development in research laboratories has mainly utilized a few flowering plants: *Arabidopsis*, maize, petunia, snapdragon, and tobacco. Because the most extensive information available is for *Arabidopsis*, however, our discussion of flowering will focus on this model plant system.

19.2.1 *Arabidopsis* flowers contain four sets of organs arranged in four whorls.

Once flowering has been induced in *Arabidopsis*, the apical meristem produces flowers. The main flower-bearing stem, which produces flowers indeterminately, is called the **primary inflorescence,** for it is the first inflorescence to form (Fig. 19.11A). Additional **secondary inflorescences,** which bear flowers along their length, arise from the axillary buds of cauline leaves on the primary shoot. Close ultrastructural examination of the shoot apical meristem during the induction of flowering shows that the primary shoot meristem starts to initiate flowers shortly before the secondary inflorescences are initiated. Flower primordia form as bulges on the shoot apical meristem. Organ primordia form as bulges on a flower primordium.

The mature *Arabidopsis* flower, which is relatively typical of flowers produced by advanced angiosperms, contains four different organ types, organized in four concentric rings called whorls (Fig. 19.11B). The outermost whorl consists of four **sepals** in a

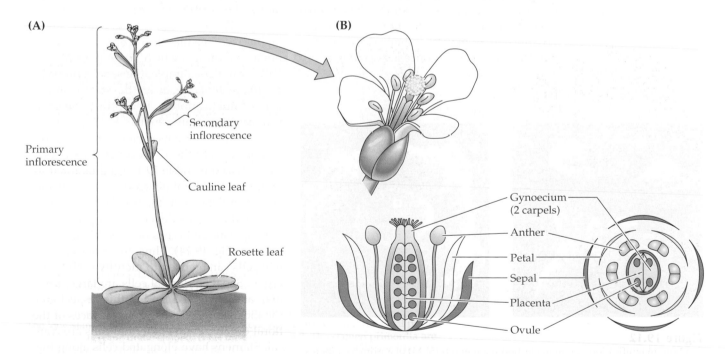

(A)

Primary inflorescence

Secondary inflorescence

Cauline leaf

Rosette leaf

(B)

Gynoecium (2 carpels)

Anther

Petal

Sepal

Placenta

Ovule

Figure 19.11
(A) Diagram of an *Arabidopsis* plant, showing rosette leaves produced before flowering, cauline leaves produced on the primary inflorescence stem, several secondary inflorescence stems, and flowers. (B) Diagrammatic representation of longitudinal and transverse sections through an *Arabidopsis* flower. The outermost whorl contains four sepals, the second whorl four petals. The third whorl contains six stamens, each with a filament and a pollen-bearing anther. The fourth (innermost) whorl contains the central gynoecium, which houses the ovules.

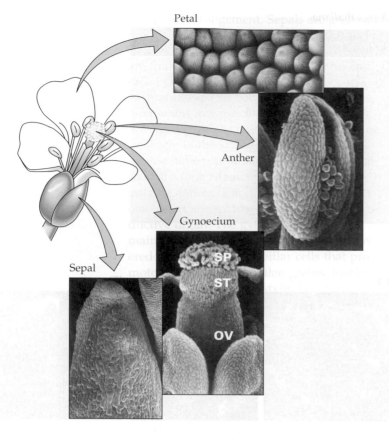

Petal

Anther

Gynoecium

Sepal

SP
ST
OV

Figure 19.14
Surface morphology of mature and developing floral organs. The epidermal cells of the floral organs have very different morphologies. The epidermal cells of petal are conical. Those of the sepal and style (ST) are elongated. The stigma is covered with papillate cells (SP), and the epidermal cells of anthers are irregular. OV, ovary.

surface of the ovary; the less regular, elongated cells on the surface of the style; and the papillar cells at the top of the stigma.

19.3 Genetic and molecular analysis of flower development

Analysis of the genetic mutations that alter the specification of floral development provides a means to dissect these fundamental processes at the molecular level. In this section we will review what is known about the genes for floral organ identity and the establishment of organ identity in *Arabidopsis*.

19.3.1 The ABC model describes the specification of floral organs.

Genes that affect the development of the floral organs can be identified by mutations that result in reduced, altered, or misplaced floral organs. Mutations in four genes, *AP2*, *AGAMOUS* (*AG*), *PISTILLATA* (*PI*), and *APETALA3* (*AP3*), cause misspecification of organ types (Fig. 19.15). Generally, mutations in these genes affect development in two adjacent whorls of the flower. Mutations in *AP2* affect development in the first and second whorls. Two of the effects of *ap2* mutations are the conversion of whorls 1 and 2, such that carpels replace sepals and stamens often replace petals. However, *ap2* mutations also lead to loss of organs, a result of the formation of fewer organ primordia. This loss is variable but often all organs are missing from whorl 2 and only a few organs form in whorl 3; when organs form in both whorls, they are all stamens. Defective alleles of *AG* affect flower development in the third and fourth whorls: Stamens are replaced by petals, and the gynoecium is replaced by a second *ag* flower, so that the pattern sepals–petals–petals is reiterated several times. The indeterminate development of *ag* flowers causes them to look like many-petaled cultivated roses. Phenotypes of *pi* and *ap3* are similar; mutations in each gene affect the second and third whorls of the flower, such that sepals replace petals and carpel-like organs replace stamen. By analogy to genes in other organisms that affect development by causing transformation of organs in positions where they would not normally arise, the *AG*, *AP2*, *PI*, and *AP3* genes have been classified as **homeotic.** Although many homeotic genes encode homeodomain proteins (see Chapter 7), the homeotic genes involved in floral development do not.

In its simplest form, the flower primordium can be viewed as three concentric and overlapping fields of gene activity, designated A, B, and C (Fig. 19.16A and B). In the early wild-type flower primordium, the three developmental fields require the activity of *AP2* for A (whorls 1 and 2), *PI/AP3* for B (whorls 2 and 3), and *AG* for C (whorls 3 and 4). Thus, each gene or gene pair controls the identity of two adjacent whorls of organs. This model assumes that (*a*) the combination of homeotic gene products present in each of the four flower whorls is responsible for specifying the developmental fate of the organs that will arise at that position and (*b*) the A and C functions act antagonistically (Fig. 19.16B). Thus, *AP2* activity in the first whorl is required for sepal production; *AP2*

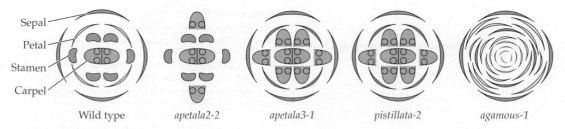

Sepal
Petal
Stamen
Carpel

Wild type   apetala2-2   apetala3-1   pistillata-2   agamous-1

Figure 19.15
The four original ABC mutants: *ap2*, *ap3*, *pi*, and *ag*. Each mutant demonstrates a homeotic transformation, that is, a transformation in which one organ type replaces another. The organs are color-coded: sepals, green; petals, yellow; stamens, red; carpels, blue.

activity in combination with *PI/AP3* expression in whorl 2 is required for petal production; *PI/AP3* in combination with *AG* in whorl 3 is required for stamen specification; and, finally, *AG* expression in whorl 4 is required for gynoecium development (Fig. 19.16C).

This simple ABC model can be used to predict the patterns of organ formation in *ag*, *pi*, *ap3*, and *ap2* mutants, as illustrated in Figure 19.17. In a null *ag* mutant, the domain

of *AP2* function would be expected to expand into whorls 3 and 4, replacing stamens with petals and carpels with sepals. In a null *ap2* mutant plant, the domain of expression of *AG* should extend to whorls 1 and 2, resulting in the production of carpels in whorl 1 and stamens in whorl 2. In the absence of *PI/AP3* function, no gene activity is combined with *AP2* in the second whorl or with *AG* in the third whorl; thus, in a *pi* or *ap3*

(A) Floral whorls

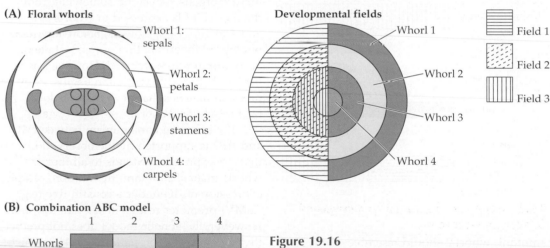

Whorl 1: sepals
Whorl 2: petals
Whorl 3: stamens
Whorl 4: carpels

Developmental fields

Whorl 1
Whorl 2
Whorl 3
Whorl 4

Field 1
Field 2
Field 3

(B) Combination ABC model

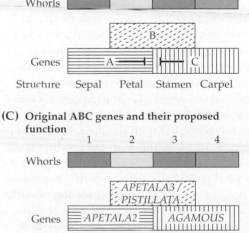

Whorls   1   2   3   4

Genes   A   B   C

Structure   Sepal   Petal   Stamen   Carpel

(C) Original ABC genes and their proposed function

Whorls   1   2   3   4

Genes   APETALA2   APETALA3 / PISTILLATA   AGAMOUS

Structure   Sepal   Petal   Stamen   Carpel

Figure 19.16
The ABC model for floral organ specification in *Arabidopsis*. (A) The wild-type *Arabidopsis* flower consists of four nested whorls, shown on the left: sepals (green; whorl 1), petals (yellow; whorl 2), stamens (red; whorl 3), and carpels (blue; whorl 4). This general pattern is common among flowering plants. The four whorls schematically correspond to three overlapping developmental fields, shown on the right. (B) The ABC model for floral organ specification in *Arabidopsis* is a combinatorial model; it posits that the four whorls of a flower are specified by three "functions" corresponding to the developmental fields defined in (A). Each box represents a defined function acting in a developmental field that corresponds to half of a floral meristem. The function of A alone specifies sepals. The A function in combination with the B function specifies petals. The B function in combination with the C function specifies stamens. The function of C alone specifies carpels. The model suggests that the A and C functions are antagonistic; that is, the C function prevents the A function, and vice versa. (C) Development of the ABC model was based on the phenotypes of a few carefully chosen mutations. Because *ap2* mutants showed evidence of increased C function, *AP2* was proposed to be an A function gene. Likewise, *ag* mutants suggested an increase in the A function, so *AG* was proposed to be a C function gene. The *ap3* and *pi* phenotypes suggest a loss of B function and so *AP3* and *PI* were proposed to be the B function genes.

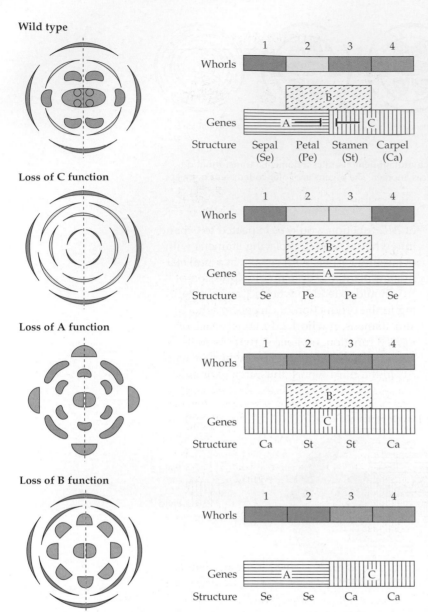

Wild type

Loss of C function

Loss of A function

Loss of B function

	1	2	3	4
Whorls				
Genes		B		
	A		C	
Structure	Sepal (Se)	Petal (Pe)	Stamen (St)	Carpel (Ca)

	1	2	3	4
Whorls				
Genes		B		
	A			
Structure	Se	Pe	Pe	Se

	1	2	3	4
Whorls				
Genes		B		
	C			
Structure	Ca	St	St	Ca

	1	2	3	4
Whorls				
Genes	A		C	
Structure	Se	Se	Ca	Ca

Figure 19.17

The proposed antagonism of the A and C functions leads to the following predictions: If the C function is removed, the A function will occur throughout the floral meristem. Conversely, if the A function is removed, the C function will occur throughout the floral meristem. Loss of the B function prevents modification of the A and C functions in the central whorls. Shown are the predicted phenotypes resulting from these altered functions. Note that the observed phenotypes of the *ag* and *ap2* (see Fig. 19.15) null mutants are not exactly equivalent to the phenotypes predicted by the ABC model.

used to monitor the temporal and spatial expression patterns displayed by these genes throughout flower development (Fig. 19.18A). As predicted by the ABC model, *AG* is expressed in the center of the flower meristem and continues to be expressed in this central region until the formation of stamens and the gynoecium is complete. Also consistent with the model, *AP3* is expressed between the inner and outer regions of the floral meristem and in this region in the sepal and stamen primordia as well as in the mature organs. Deviating somewhat from the model, *PI* is expressed in whorls 2, 3, and 4 in very early flowers, before any morphological determination of petal, stamen, or gynoecium has occurred. However, once morphological differentiation begins, *PI* expression matches that of *AP3* and is localized to whorls 2 and 3.

The presumed antagonism between *AG* and *AP2* is supported by the observation that *AG* expression extends to all four whorls in an *ap2* mutant. In addition, ectopic expression of *AG* under a constitutive (e.g., CaMV) promoter results in *ap2*-like flowers. However, the genetic model does not predict the actual expression patterns of *AP2*, which is detected in all four whorls throughout flower development (Fig. 19.18A). This last result implies that *AG* does not suppress *AP2* expression in whorls 3 and 4 and, further, that *AP2* expression in these whorls does not suppress *AG* gene expression the way it does in whorls 1 and 2. These data can be reconciled with the model if *AP2* suppresses *AG* by acting with the product of an unknown gene, "X," expressed in whorls 1 and 2 (Fig. 19.18B). Although *AP2* expression is not limited by AG, the expression of *AP1*, a gene for floral meristem identity, is suppressed by *AG*. Given that *AP1* also has a role in the specification of whorls 1 and 2,

mutant, whorl 2 should resemble whorl 1, and whorl 3 should resemble whorl 4. These predictions correspond well, but not perfectly, to the phenotypes of the homeotic mutants (see Fig. 19.15).

19.3.2 Cloning of homeotic genes has led to subsequent refinement of the ABC model.

Molecular analysis of the homeotic gene products has supported most premises of the ABC model. In situ hybridization with labeled probes homologous to mRNA transcripts of *AP2, PI, AP3,* and *AG* has been

(A)

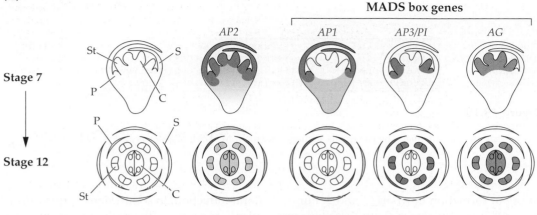

(B)

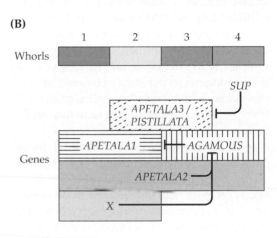

it has been included in revised ABC models (Fig. 19.18B).

Another revision to the ABC model is the inclusion of *SUP*, a negative regulator of the B function genes *PI* and *AP3*. *sup* mutant plants expand the pattern of expression of *PI* and *AP3* to include the center of the floral meristem, thus extending stamen production into the fourth whorl. The *SUP* gene is subject to epigenetic regulation: Hypermethylated *SUP* alleles (called the *clark kent* alleles) have defects resembling those of the *sup* mutations. *SUP* has been proposed to regulate communication between whorls, acting at the junction between whorls 3 and 4.

Molecular cloning of the homeotic genes has allowed researchers to predict the sequences of their protein products, providing insight into how these genes might control flower organogenesis. The proteins encoded by *AG*, *PI*, *AP1*, and *AP3* are thought to belong to an evolutionarily conserved family of transcription factors. Each of these gene products contains a highly conserved DNA-binding domain called the MADS box (Fig.

Figure 19.18
The ABC model has been revised to reflect advances in our understanding of floral development. (A) The expression of *AP2* and the floral homeotic MADS box genes in the developing flower. AP2 is expressed throughout flower development in all four whorls. The MADS box genes, however, are each limited to a pair of whorls, as predicted by the ABC model. S, sepal; P, petal; St, stamen; C, carpel (gynoecium). (B) Cloning of all of the genes shown (except for the unknown gene X) has led to minor revisions of the ABC model. The revised model indicates both gene function and gene expression. The areas within the boxes generally indicate the limits of gene expression. Note that *AP2* is expressed throughout the meristem. The prediction that *AP2* is necessary to suppress *AG* in whorls 1 and 2 has been validated, but apparently at least one other gene (*X*) also may be necessary to act with *AP2*. Although *AG* does not suppress *AP2* expression in whorls 3 and 4, it does suppress the expression of another C function gene, *AP1*. *SUP* contributes to the repression of B genes in whorl 4.

19.19). Because these flower organ identity genes are transcription factors, they are likely to control the expression of other genes that ultimately specify the floral organs. *SUP* encodes a protein with several domains commonly found in transcription factors (see Chapter 7), including a zinc finger, a basic region, a serine/proline–rich region, and a leucine zipper. The predicted product of *AP2* has no homology to any known gene products; however, its serine-rich acidic domain may function in DNA binding.

Until recently, floral meristem identity genes were assumed to be distinct from floral organ identity (ABC) genes. Now, however, the genes are recognized as overlapping in time and specificity of function. *AP2* has roles both in determinacy (see Section 19.1.5) and in organ identity within whorls 1 and 2. *AP1* also has an early role in the floral meristem (see Section 19.1.5) and a later role

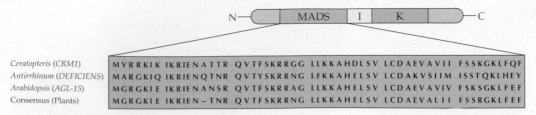

Figure 19.19

Comparison of MADS-domain sequences from *Ceratopteris* (a fern) and from the two model systems *Arabidopsis* and *Antirrhinum*. The consensus sequence for plant MADS-box genes is also shown.

in organ formation in whorls 1 and 2. Further, *AG* is a regulator of floral determinacy: *ag* flowers do not terminate but produce more than 12 whorls (see Fig. 19.15).

The genetic and molecular analyses of the five genes, *AP1, AP2, PI, AP3,* and *AG,* highlight some basic principles for how flower development is controlled. First, the genes that encode transcription factors probably control a complex array of target genes, an array that ultimately specifies the fates of the flower primordia and their differentiation into mature organs. Second, these regulatory genes act in different combinations to control developmental fates. Third, the products of some of these identity genes, such as AP2 and AG, control each other's activity. This highly regulative mode of development, involving the control of gene activity, is not unexpected; indeed, it has been used successfully throughout evolution in unicellular and multicellular organisms. The importance of homeotic proteins as transcriptional regulators is underscored by the marked alterations of developmental pathways observed in mutants that are defective in homeotic genes. The simple ABC model for combinatorial gene action to specify the floral organ types is highly consistent with having the organ identity genes encode various transcription factors.

19.3.3 Many of the gene products associated with flower development appear to be negative regulators of gene expression.

The model presented in Figure 19.18B suggests that certain interactions between floral meristem and organ identity genes are negative. There are also many indications that

floral induction, to some degree, represents a release from normally repressive signals. The *CURLY LEAF* (*CLF*) gene may represent just such a negative regulator of floral development because in *clf* mutants, *AG* is expressed ectopically in vegetative leaf tissues. *CLF* is homologous to the *Polycomb* genes of *Drosophila,* which encode proteins that strongly influence chromatin structure and thereby affect gene expression. Thus, *CLF* may be an essential repressor of floral development that is itself repressed by floral induction. The product of the *EMBRYONIC FLOWER* gene may also repress floral development: *emf* mutants produce very little vegetative tissue that is intermixed with floral cell types, and both *AP1* and *AG* are ectopically expressed in *emf* mutants (Fig. 19.20). *TFL* is another candidate as a repressor of floral development, given that *tfl* mutants flower early. Because ectopic expression of *LFY* and *AP1* produce early flowers that mimic the *tfl* phenotype, *LFY* and *AP1* are probably positive regulators of the floral program.

19.3.4 Numerous mutations affecting floral development have been described, and many more probably will be discovered.

Many other mutations that affect floral organ number, organ shape, and regional differentiation without changing the identity of either the floral shoot or its component organs have been described. For example, mutations in *CLAVATA1* (encoding a cell surface receptor kinase), *CLAVATA3* (encoding a putative peptide ligand for the CLAVATA1 protein), and *ETTIN* (encoding a putative auxin response factor) increase the number of organs or alter organ shape. In contrast,

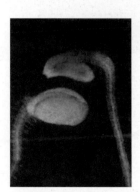

Figure 19.20

Ectopic expression of *AP1,* the gene for flower meristem identity, in *embryonic flower* (*emf*) mutants. In an *emf1-2* background, *AP1::GUS* activity is seen throughout the shoot in four-day-old seedlings. In wild-type plants (not shown), *AP1::GUS* is present only after the commitment to flowering. Early expression of *AP1::GUS* indicates that its product is involved in the repression of flowering.

Box 19.2

Similar genes participate in the formation of very different flowers.

Although *Arabidopsis* floral development has been the focus of the accompanying discussion, genetic and molecular analyses of flower development in *Antirrhinum majus* (snapdragon) have revealed the involvement of genes nearly identical to those defined in *Arabidopsis*. In fact, the *Arabidopsis* genes *LFY* and *TFL* were cloned by using *Antirrhinum* sequences for *FLORICAULA* and *CENTRORADIALIS*. Comparison of *Arabidopsis* and snapdragon reveals that the *AP1* gene is homologous to *SQUAMOSA*, the *LFY* gene to *FLORICAULA*, the *AP3* gene to *DEFICIENS*, the *PI* gene to *GLOBOSA*, and the *AG* gene to *PLENA*. Both the mutant phenotypes and the encoded gene products are highly conserved in these two different plant families.

Although the genes that dictate meristem and organ identity in *Ara-* *bidopsis* and snapdragon are homologous, the morphologies of resulting flowers are highly divergent (compare panels A and E, B and F, C and G, D and H in figure below). Indeed, floral development among all angiosperms may utilize highly conserved regulatory genes to establish whorl identities. Certainly the functions of these genes and the arrangement of their structural domains are highly conserved. However, not all flowering plants have ABC homologs that pattern floral whorls in precisely the same manner. The dicots *Arabidopsis* and snapdragon, for example, may have only minor differences in the mechanisms for floral whorl patterning, but more substantial differences are apparent between *Arabidopsis* and the monocot maize.

Of particular evolutionary interest is the observation that homologs of the "flowering genes" discussed in this chapter are found in nonflowering plants. For example, a *LFY/FLORI-CAULA* homolog, *NEEDLY*, has been isolated from *Pinus radiata*. Interestingly, the expression of *NEEDLY* in *Arabidopsis* rescues *LFY* mutants, and its overexpression in transgenic *Arabidopsis* plants yields a phenotype strikingly similar to that of transgenic *Arabidopsis* plants that overexpress *LFY*. In *P. radiata*, *NEEDLY* is expressed both in vegetative and reproductive meristems. MADS box genes have been isolated from both gymnosperms and ferns, which suggests that these genes arose before the evolution of the flower; apparently, some of them were later recruited to function as flowering genes in angiosperms.

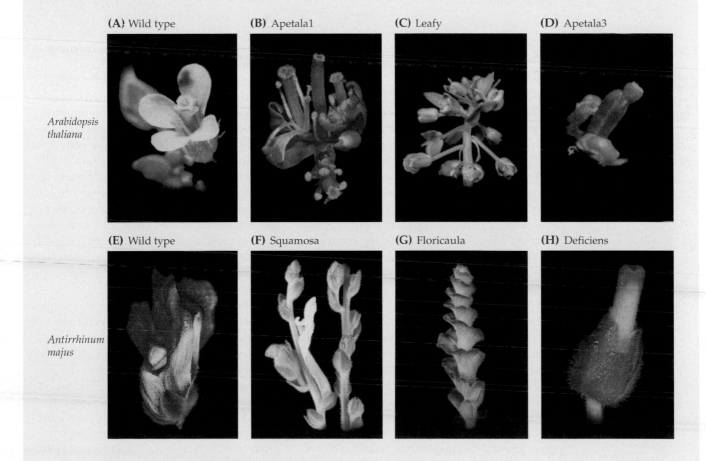

(A) Wild type   **(B)** Apetala1   **(C)** Leafy   **(D)** Apetala3

Arabidopsis thaliana

(E) Wild type   **(F)** Squamosa   **(G)** Floricaula   **(H)** Deficiens

Antirrhinum majus

mutations in *TOUSLED* (encoding a protein kinase) decrease organ number and organ size as well as alter the shapes of organs. These genes probably act to pattern growth and cell division before or during the acquisition of identity within the developing flower.

Many more genes regulating flower development remain to be uncovered. A possible approach to identifying these genes is to search for additional genes containing domains that are conserved in known regulatory proteins. For example, by using the region of *AG* encoding the MADS box as a probe, researchers have identified more than 20 additional MADS box genes. Many of these genes are expressed in the flowers, and some have been found to regulate novel developmental processes. For example, the *FRUIT-FULL* MADS box gene regulates formation of specific regions of the gynoecium associated with fruit dehiscence. The fact that such genes continue to be identified indicates that many additional regulatory genes remain undiscovered.

19.4 Formation of gametes

Once a plant has formed flowers, the next step in its reproductive cycle is **gametogenesis.** In plants, gametes are formed by meiotic division of a diploid **sporophytic** cell, followed by mitotic divisions of one or more haploid cells to generate a multicellular **gametophyte.** As we will discuss in the following sections, much more is known about **microgametogenesis,** the formation of **pollen** (the male gametophyte), than about **megagametogenesis,** the formation of the **embryo sac** (the female gametophyte). This is the case at least partly because pollen is produced in copious amounts and is easy to collect. For example, a single pollen grain of maize is only 100 μm in diameter but a single plant can produce more than 25 million pollen grains. The female gametophyte of maize is larger but less abundant: A single ear of maize produces only about 300 embryo sacs. Furthermore, the embryo sac is formed and remains buried within the ovule tissue, making it very difficult to isolate for experimental manipulations or molecular biology experiments. Over the years, plant breeders have accumulated numerous male-sterile mutations, but only recently have scientists undertaken large-scale mutagenesis screens to identify female-sterile mutations. We will discuss first male gametophyte development, then female gametophyte development, and finally the interactions between the male and female that lead to successful fertilization.

19.4.1 The male gametophyte is formed in the anther.

Figure 19.1 illustrated some of the important features of gametogenesis in the context of the plant life cycle. Figure 19.21 shows in greater detail the processes that give rise to the three-celled male gametophyte, or pollen grain. The formation of the male gametophyte occurs in the anther, starting when a sporophytic cell divides to give a tapetal initial cell (see below) and a sporogenous initial cell. Each sporogenous initial cell (pollen mother cell) undergoes meiosis, giving rise to a tetrad of haploid cells, which are held together by a wall of **callose,** a $(1{\rightarrow}3)\beta$-glucan (see Chapter 2). The cells in the tetrad are released as free microspores by the action of callase, an enzyme produced by the tapetal cells. After release from the callose wall, the microspores undergo two mitotic divisions. The first division is asymmetric and yields a large cell, the **vegetative cell,** and a smaller cell, the **generative cell.** A second mitotic division of the generative cell yields two **sperm cells.** In some plants, such as *Arabidopsis* and maize, the second mitotic division of the generative cell can occur before the pollen is shed from the plant. However, in most angiosperms, the pollen grain is released from the anther at the two-celled stage, and the second mitotic division occurs during pollen tube growth.

The **tapetum,** a nutritive tissue that lines the locule of the anther, is derived from subsequent divisions of the tapetal initial cells. The first important role for the tapetum is to produce the callase that releases the microspores from the callose wall. The tapetal cells produce and deposit numerous compounds, including structural polymers and pigments, in the outer wall of the pollen grain (see Section 19.4.4). The tapetal cells eventually disintegrate, and some of the cytoplasmic contents, including proteins and lipids, are deposited on the pollen coat. The

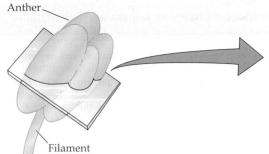

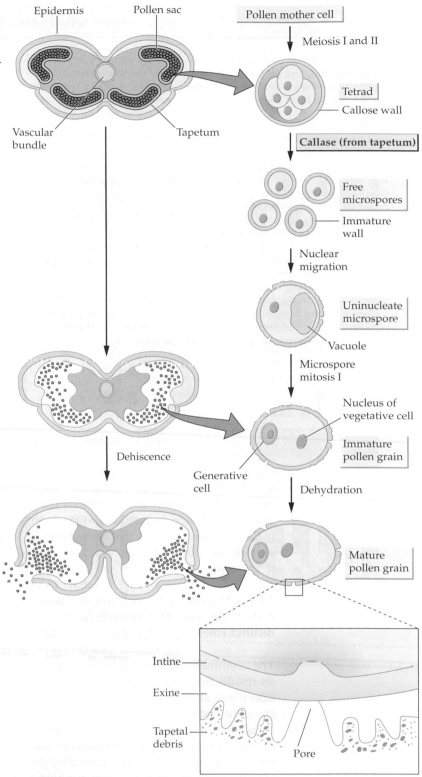

pollen coat plays an important role in interactions with the female tissue (see Sections 19.6 and 19.7).

Cytological and molecular evidence has demonstrated extreme conservation of anther and pollen development throughout the angiosperms, although some minor details of this process may differ among plant families. This degree of conservation means that each aspect of anther and pollen development can be studied in a model plant amenable to the particular experimental approach being used, and the information gained in such research will be generally applicable to other plants.

19.4.2 Many genes are specifically expressed in the male gametophyte.

Estimates based on RNA/DNA hybridization studies suggest that 60% to 90% of the genes that are expressed in the gametophyte also are expressed in sporophytic tissues. Much research has focused on the isolation of pollen-specific genes to determine their roles during pollen development and to understand the mechanisms of pollen-specific gene expression.

A recent calculation suggests that approximately 350 genes are expressed exclusively in the *Arabidopsis* anther. Some of the proteins encoded by these genes are unique to pollen (see Box 19.3 for some examples of pollen-specific proteins). Others include pollen-specific variants of proteins that are found in other parts of the plant. For example, multigene families encode many cytoskeletal proteins (e.g., actin and tubulin; see Chapter 5). Certain members of these gene families are expressed only in reproductive cells. Analysis of several pollen-specific promoters has shown that their specificity

Figure 19.21
Diagram illustrating the pollen development pathway typical of most angiosperms. Meiosis occurs in the anther and yields a tetrad of microspores. Free microspores are released after callase degrades the cell wall that has been holding the tetrad together. The first mitotic division is asymmetric, forming a large vegetative cell and a small generative cell. At dehiscence, partially desiccated pollen is released from the stoma of the anther.

Box 19.4

Microspore culture can revert the gametophytic developmental program to a sporophytic program, yielding haploid plants.

Cultured anthers or microspores can be used to generate haploid plants. This technique is of interest to plant breeders because doubling the chromosome number of the resulting haploid plants, either spontaneously or by chemical treatment, can generate homozygous diploids.

Flower buds at the appropriate developmental stage are collected and microspores isolated and cultured as shown in the figure. For some plants, microspore culture (left pathway) is not efficient, and the anthers must be cultured instead (right pathway). The sporophytic anther tissue is thought to provide factors required for the development of microspore-derived embryos. Formation of embryos is typically induced by a stress treatment. In some cases, whole plants are grown under stress conditions (such as low temperatures or nitrogen starvation). In other cases, anthers are excised or microspores isolated from a nonstressed plant; after the isolated tissues or cells are subjected to a stress treatment (e.g., heat shock or starvation for nitrogen and sucrose), they are transferred to a medium containing sucrose and nitrogen to promote further development of microspore embryos.

After stress treatment, microspore embryos arise by one of two developmental pathways. In the first pathway (shown at the left), the vegetative cell of a binucleate pollen grain continues to proliferate, forming a multinucleate pollen grain and eventually an embryo. In the second, the uninucleate microspore undergoes a symmetrical cell division, giving rise to two equal-sized cells that both continue to divide (not shown in the figure). Both paths involve a release of the cell cycle arrest that the vegetative cell undergoes during normal pollen development.

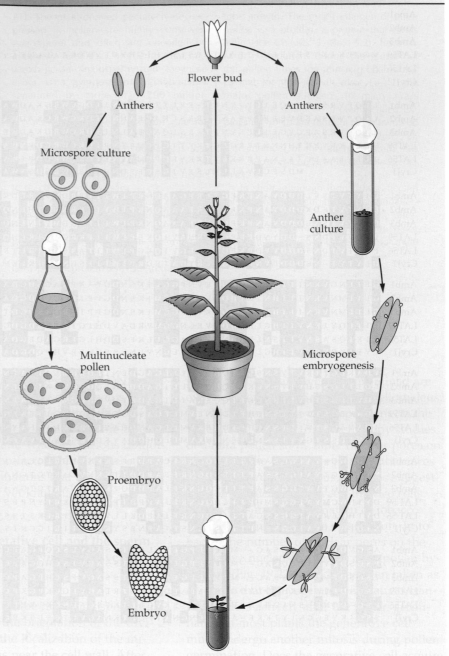

histones are thought to play a role in chromatin condensation. Another gene, *LGC1* (lily generative cell), shows generative cell–specific expression. Interestingly, an antibody raised against LGC1 labels the sperm surface, suggesting that this protein might be important for interactions with the egg cell.

19.4.4 The male gametophyte is enclosed in a complex cell wall.

A mature pollen grain has a two-layered cell wall, an inner or **intine** wall, and an outer, or **exine** wall (see Fig. 19.21). The intine wall is not covered by exine at the position of the

pollen apertures, or pores. The pollen tube forms at one of the pollen apertures and elongates by tip growth (see Chapter 2), making the wall of the pollen tube an extension of the intine cell wall. Like other plant cell walls, the intine wall has a cellulose backbone. However, the intine wall differs from those of most other plant cells by having as its principal component $(1{\rightarrow}3)\beta$-glucan callose, although lesser amounts of cellulose and arabinans are also present (see Chapter 2).

The exine wall of pollen grains is frequently decorated with complex patterns of spines and ridges (Fig. 19.23). These patterns differ among plant species and can help identify the species that produced the pollen. The main structural component of the exine wall is **sporopollenin.** The composition of sporopollenin has been debated for many years but is now believed to be a polymer of phenolics. Sporopollenin is extremely resistant to decay. To extract it, the investigator treats the pollen with hot acetic acid anhydride and concentrated (9.9 M) sulfuric acid—a treatment that hydrolyzes, rather than extracts, other organic molecules. The durability of sporopollenin accounts for the presence of pollen in the fossil record. Pollen exine patterns in fossils provide clues as to the types of plants that lived millions of years ago.

How do different plants produce different exine patterns? When species that are closely related but differ in pollen grain morphology are crossed, the pollen grains of the resulting F1 plants appear identical. This result suggests that the patterns are established by the sporophyte, either before meiosis or contributed from the tapetum. If the patterns were determined gametophytically, the F1 plants would be predicted to generate both morphologies of pollen in a 1:1 ratio. However, pollen of the F1 plants can have morphological characteristics from both parents (e.g., the spine density of one parent and the spine length of the other), which suggests that different and unlinked genes control pollen grain morphology. To determine what types of proteins are encoded by such genes is an interesting question for future research. One approach would be to search for mutants that exhibit an altered exine pattern in their pollen; however, a screen for the mutants that affect exine patterns would be time-consuming with current

technology—perhaps requiring scanning electron microscopic examination of millions of grains of pollen from individual plants of mutagenized populations.

19.4.5 Female gametophyte development involves one meiotic and several mitotic divisions.

Female gametogenesis occurs within the ovule and, as in the male, begins with meiosis. Although many variations are possible in the female cell division patterns that give rise to the embryo sac, more than 70% of flowering plant families demonstrate *Polygonum*-type embryo sac development, in which three of the products of meiosis undergo cell death and the fourth forms the megaspore (Fig. 19.24). The surviving megaspore undergoes three mitotic divisions to produce the female gametophyte, or embryo sac. The typical mature embryo sac thus has seven cells: three **antipodal cells,** two **synergids,** a **central cell** (containing two of the haploid nuclei), and the **egg cell.**

As in the search for pollen-specific genes (see Section 19.4.2), researchers have performed differential screens to identify genes expressed specifically in the female gametophyte. Early experiments were largely unsuccessful, probably because the ovule is embedded within sporophytic tissue, which makes it difficult to obtain large quantities of uncontaminated ovules for isolating mRNA and preparing a cDNA library. The clones that were identified from pistil cDNA libraries often were expressed in the sporophytic cells of the pistil, rather than in the ovule or embryo sac. More recently, progress in the search for female gametophyte-specific transcripts has relied on orchid as a model system (Fig. 19.25).

Orchid has two advantages for the isolation of ovule-specific genes. Because ovule development in orchid does not begin until pollination occurs, the development of all ovules can be synchronized. Furthermore, this ovule development occurs over an 11-week period, providing ample opportunity to isolate ovule tissues at particular stages of development. The mRNAs extracted from ovules at each developmental stage have been used to construct cDNA libraries. Ovule-specific cDNAs from orchid can be

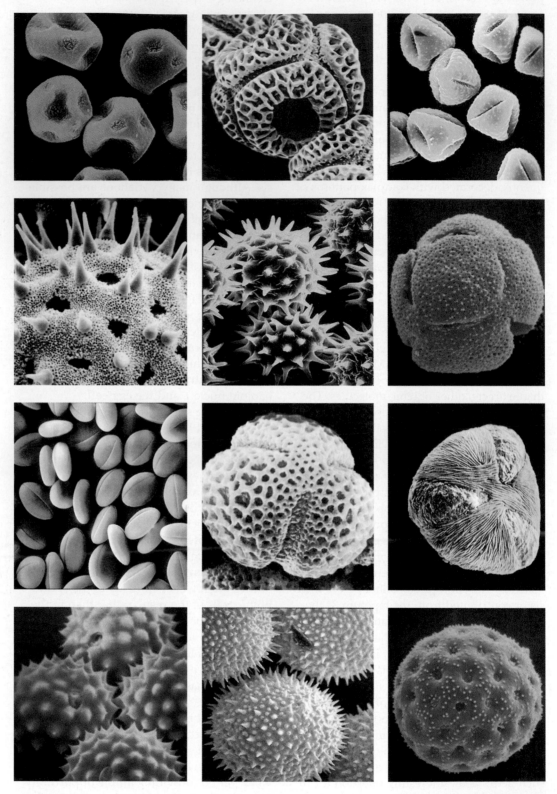

Figure 19.23
Pollen from diverse families, showing attractive sculpturing patterns in the exine wall. Left to right, top row: plantain, winter's bark, buttercup. Second row: morning glory, ox-eye daisy, rose cactus. Third row: nettle, poppy, sycamore. Fourth row: ragweed, hollyhock, aheaahea. Coloration was used for visual enhancement.

1) 1500 - 200 word summary of the current Project Proposal.

2) Limitations and Potential Problems of the current Project Proposal.

3) Contingency Plans and alternative approaches to deal with anticipated technical Problems.

4) Additional objectives to enhance the proposal.

5) Estimated Costing of extra Plans.

PLANT BIOTECHNOLOGY

SEEN QUESTION
SECTION C

The funding available to your company has increased by 50% over the whole time frame of your project. Describe how would you use this extra funding to create a better solution to your Biotechnology Problem. And of increase the likelihood of achieving your aim. Structure your response around the following issues:-

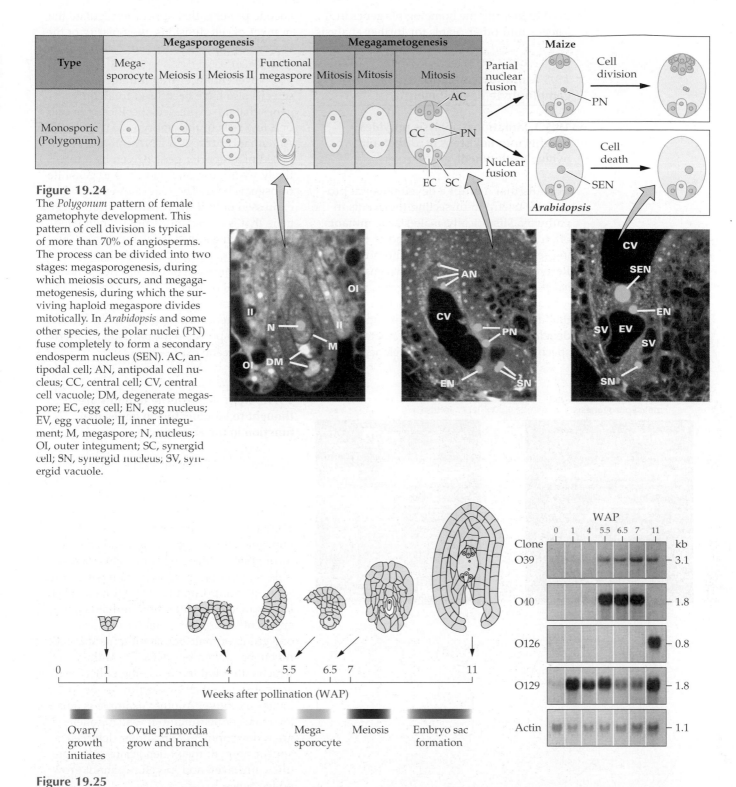

Figure 19.24

The *Polygonum* pattern of female gametophyte development. This pattern of cell division is typical of more than 70% of angiosperms. The process can be divided into two stages: megasporogenesis, during which meiosis occurs, and megagametogenesis, during which the surviving haploid megaspore divides mitotically. In *Arabidopsis* and some other species, the polar nuclei (PN) fuse completely to form a secondary endosperm nucleus (SEN). AC, antipodal cell; AN, antipodal cell nucleus; CC, central cell; CV, central cell vacuole; DM, degenerate megaspore; EC, egg cell; EN, egg nucleus; EV, egg vacuole; II, inner integument; M, megaspore; N, nucleus; OI, outer integument; SC, synergid cell; SN, synergid nucleus; SV, synergid vacuole.

Figure 19.25

Ovule development in orchid is initiated by pollination and takes 11 weeks to complete. Key stages are indicated below the timeline. Pollen tubes first enter the ovary around week 2. Fertilization occurs at week 11. cDNA libraries were prepared from ovules at different developmental stages. The ovule primordia library includes genes expressed 5.5 weeks after pollination (WAP). The megasporocyte library includes genes expressed at 6.5 to 7 WAP. The female gametophyte library includes genes expressed at 11 WAP.

These libraries were subjected to differential screens to identify cDNAs expressed in the ovule tissue during particular stages of development. Stage-specific expression of selected genes represented by these cDNAs are shown in the RNA gel blots (right panels; the cDNA names are indicated to the left of the panels). O39 and O40 were from the megasporocyte library; O126 and O129 from the female gametophyte library. The actin-loading control shows that RNA is present in all lanes of the blot.

used to identify the homologous genes in *Arabidopsis* and other plants for further analysis.

19.5 Mutations affecting gametophyte development

19.5.1 Mutational analysis provides insights into key steps of anther and ovule development.

Mutants that disrupt a developmental process are useful in dissecting the events in the pathway. Historically, male-sterile mutants often were identified at the end of the growing season as the plants that did not set fruit. Recently, large-scale mutant screens in *Arabidopsis* have identified both genes required for anther development or function and genes that act in the sporophyte to disrupt ovule development. Of the female gametophyte-specific genes that have been cloned, several encode proteins that appear to regulate the number of cell divisions, the polarity of the embryo sac, or the process of cellularization (Fig. 19.26).

Novel enhancer-trap mutagenesis screens also have been used to identify genes that are expressed specifically in one or a few cells of the embryo sac (Fig. 19.27). An **enhancer trap** is a DNA construct composed of a reporter gene driven by a relatively weak promoter and used to generate transgenic lines. The reporter gene will be expressed only if the construct inserts near a gene that is transcriptionally up-regulated by a cis-acting enhancer element (see Chapter 7). The transgenic lines are then examined for expression of the reporter gene in specific cells or tissues, such as the gametophytes. These transgenic lines allow the cloning of genes that contribute the enhancer elements responsible for particular patterns of reporter gene expression.

19.5.2 Most male-sterile mutants are thought to be defective in genes that function in the sporophyte.

In most cases, the male-sterile phenotype results from a homozygous recessive genotype (*ms/ms*). Plants that are heterozygous (+/*ms*) for a male-sterile mutation typically have 100% normal pollen. Some male-sterile mutants are defective in meiosis and cannot form pollen. Others have defective tapetal cells, suggesting that the mutant genes affect some step important for tapetal metabolism. However, other male-sterile mutants have apparently normal tapetal tissues, as does one group of male-sterile mutants that has functional pollen but defective anthers, which either fail to release the pollen or release it so late that female flowers are no longer receptive. Another example is a barley male-sterile mutant in which pollen grain development and anther dehiscence appear normal; the pollen grains lack apertures, however, and therefore cannot grow pollen tubes.

Recently, a few male-sterile genes have been cloned, but the proteins encoded by these genes have not revealed any clues to their function. For example, one male-sterile gene from maize encodes a protein with limited sequence similarity to strictocidine

fem2: Never progresses beyond megasporogenesis

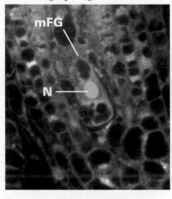

gfa2: Polar nuclei fail to fuse

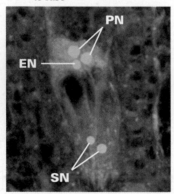

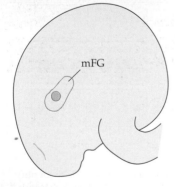

Figure 19.26
Confocal laser scanning micrographs of *Arabidopsis* mutants with defects in megagametogenesis. The *fem2* mutant stops development at the one-cell stage. In the *gfa* mutant the polar nuclei fail to fuse. mFG, mutant female gametophyte; N, nucleus; SN, synergid nucleus; EN, egg nucleus; other abbreviations as in Figure 19.24.

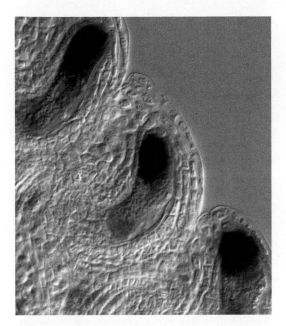

Figure 19.27
Enhancer trap in female gametophyte. The GUS transgene is expressed when it inserts into the genome at a position where transcription is up-regulated by a megagametophyte-specific cis-element.

synthase, an enzyme that acts in indole synthesis (see Chapter 24). If this protein is indeed involved in indole biosynthesis, then perhaps the male-sterile mutant lacks an indole compound that is critical at some point in pollen development. However, the identity of such an indole compound and its possible function are completely unknown. The first male-sterile gene cloned from *Arabidopsis* encodes a protein with no known function. A short amino acid sequence domain of the protein is similar to that of a mitochondrion-encoded protein, but the function of the latter protein also is unknown.

19.5.3 Cytoplasmic male sterility and nuclear restorer genes illustrate the importance of mitochondrial function to pollen development.

Male sterility that is transmitted through the female, called **cytoplasmic male sterility** (CMS), has been found in more than 150 different plant species and has been intensively studied in maize, sunflower, rice, petunia, and bean (*Phaseolus vulgaris*). Geneticists have deduced that either the chloroplast or the mitochondrial genome could be responsible for the CMS trait. In all cases studied

so far, the sterility is caused by the expression of an abnormal protein in the mitochondria of the anthers. Different CMS types express different abnormal proteins. Interestingly, although the mutation is present in all the mitochondria of the plant, only the anthers show a phenotype. It is not yet understood how the abnormal proteins disrupt mitochondrial function; however, these results suggest that efficient mitochondrial function is particularly important during pollen development.

If the expression of the abnormal mitochondrial protein is reduced, fertility is restored. Almost all known CMS systems have nuclear restorer genes that can suppress, by unknown mechanisms, the expression of the abnormal mitochondrial protein. For example, in the CMS-T system in maize, restoration of fertility requires the expression of two nuclear-encoded genes, *Rf1* and *Rf2* (Fig. 19.28). The abnormal mitochondrial protein that is important in CMS-T is called URF13. If a plant has only *Rf1* (in other words, if its genotype is *Rf1 rf2*), fertility is not restored, although the amount of URF13 is greatly reduced. This result suggests that the *Rf2* gene product must provide an additional important

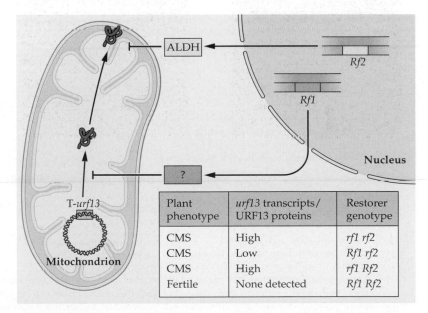

Plant phenotype	*urf13* transcripts / URF13 proteins	Restorer genotype
CMS	High	*rf1 rf2*
CMS	Low	*Rf1 rf2*
CMS	High	*rf1 Rf2*
Fertile	None detected	*Rf1 Rf2*

Figure 19.28
The CMS-T system in maize: A schematic illustration of the roles played by the mitochondrial-encoded toxic protein URF13 and the nuclear-encoded restorer proteins Rf1 and Rf2. Both *Rf1* and *Rf2* are required to restore fertility. *Rf1* greatly reduces the accumulation of *urf13* transcripts, although how this occurs is not yet known. *Rf2* encodes an aldehyde dehydrogenase (ALDH) but has no effect on the abundance of *urf13* transcripts. The ALDH may remove a toxic compound produced by the URF13 protein.

function that restores fertility, for example, complete suppression of URF13 expression. However, *rf1 Rf2* plants have high amounts of the URF13 protein, thus excluding that possibility. Cloning of the *Rf2* gene has demonstrated that it encodes an aldehyde dehydrogenase. One model suggests that URF13 may cause the production of aldehydes that are toxic to the mitochondria of tapetal cells, but that *Rf2* can detoxify these aldehydes or otherwise prevent their accumulation.

Cytoplasmic male sterility is commercially useful for hybrid production in many crops, because it does not require laborious emasculation of the intended female parent. To make F1 hybrids, plant breeders use CMS plants as the female parent and plants with the nuclear restorer genes as the male, re-

sulting in a male-fertile F1 hybrid (see also Box 19.5).

19.6 Germination of pollen

19.6.1 Pollen hydration is the first step in pollen tube germination.

Pollen is released from the anther in a partially dehydrated state. When it lands on the stigma, it must hydrate and grow a pollen tube toward the ovule. What controls the hydration of pollen and the initiation of pollen tube growth? An early cue for hydration seems to be lipid molecules in the pollen coat. For example, the pollen coat in plants that contact so-called "dry" stigmas contain lipids. Mutants that are deficient in this lipid coat, such as the *cer* mutants of *Arabidopsis*

Box 19.5   **Hybrid seed production can be facilitated by using genetically engineered male sterility.**

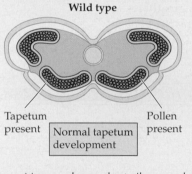

Wild type

Tapetum present — Pollen present

Normal tapetum development

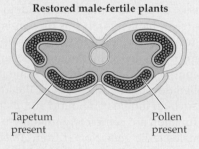

Male-sterile plants

No tapetum — No pollen

Restored male-fertile plants

Tapetum present — Pollen present

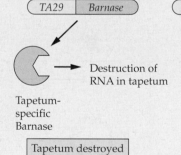

TA29 | Barnase

Tapetum-specific Barnase → Destruction of RNA in tapetum

Tapetum destroyed

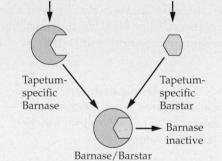

TA29 | Barnase   TA29 | Barstar

Tapetum-specific Barnase   Tapetum-specific Barstar

Barnase/Barstar complex → Barnase inactive

Normal tapetum development

Many nuclear male-sterile genes have been described in crop plants; interestingly, however, they are not widely used for hybrid seed production. Although seed producers prefer to use male-sterile plants for easier hybrid seed production, farmers need male-fertile plants to be able to generate seed for sale. A clever solution to this problem uses a tapetum-specific promoter from tobacco, *TA29*, fused to a gene that codes for Barnase, a cytotoxic protein from *Bacillus amyloliquefaciens*. This system, when transferred into plants, produces male-sterile plants that can be easily pollinated with the desired male parent to generate hybrid seed (see figure).

However, this modification still fails to meet farmers' needs. Because male sterility induced by the Barnase-encoding gene is a dominant trait, the seed sold to the farmer will germinate to form male-sterile

plants, which are not useful in seed production. This male sterility can be corrected, however, by expressing a specific inhibitor protein, called Barstar, to block the cytotoxicity of Barnase. If the plant breeder crosses a female parent that expresses *TA29::Barnase* to a male parent that carries a *TA29::Barstar* transgene, the hybrid seed sold to the farmer will be male-fertile (see figure).

(see Chapter 10), are defective in hydration and are thus male-sterile under dry conditions. If relative humidity is high, however, the *cer* pollen grains can hydrate.

In plants with so-called "wet" stigmas, such as tobacco, the lipids of the pollen coat have been thought to be less important to pollen function. However, certain classes of lipids can replace the requirement for stigma exudate, as demonstrated by applying lipids to tobacco stigmas that have been ablated (see also Box 19.6). More recently, researchers have suggested that the initial guidance cue for the emerging pollen tube is actually water and that the lipids in the pollen coat or stigma exudate establish a concentration gradient of water that directs growth of the pollen tubes.

Secondary metabolites such as flavonols can also facilitate pollen tube growth in some plants. Flavonol pigments in pollen are produced by the tapetum and deposited on the pollen grain. Maize and petunia plants that are deficient in chalcone synthase, a key enzyme in the biosynthesis of flavonols (see Chapter 24), are self-sterile. However, if the pollen grains are treated with exogenous flavonols, the pollen can germinate and self-fertilize successfully. Furthermore, the pollen grains deficient in chalcone synthase can successfully pollinate a wild-type plant—which suggests that the style tissue of the female can also provide flavonols to the flavonol-deficient pollen grains.

19.6.2 The growing pollen tube is directed toward the embryo sac by signaling mechanisms that have not yet been characterized.

The vegetative cell of the male gametophyte grows a pollen tube through the female stigma and style and delivers its passengers, the two sperm cells, to the embryo sac. Pollen tubes grow by tip extension of the intine wall, which first extrudes through one of the pollen apertures. Pollen tubes can achieve very rapid growth rates, as fast as 10 μm min^{-1}. As pollen tubes grow, the cytoplasm is concentrated near the tip, and the tube closest to the grain is blocked off by deposition of callose plugs (Fig. 19.29). Calcium chelators have been used to show that a calcium gradient is required for pollen tube growth. The direction of pollen tube growth can be changed if the calcium concentrations at the tip are manipulated (see Chapter 18, Box 18.5).

Pollens of many plants can hydrate and grow pollen tubes in a solution containing boric acid, calcium, and sucrose (which acts as both a source of carbon and an osmoticum). Although pollen tubes can grow in a simple medium, the rates of tube growth in vitro are not as fast as those measured in vivo, implying that other factors, most likely from the female, also are required for optimal growth. It also is possible that pollen tubes require adherence to the female tissue

---

Box 19.6

The *fiddlehead* mutant of *Arabidopsis* allows promiscuous pollen germination on nonreproductive parts of the plant.

Usually pollen grains hydrate and germinate when they contact the stigma of the flower. *Arabidopsis* plants that carry the *fiddlehead* (*fdh*) mutation have a modified epidermis: As shown in the photographs, pollen grains can hydrate and germinate on leaf surfaces of the *fdh* mutant (panel A, see arrows) but not on wild-type leaves (panel B). Interestingly, some aspects of pollen recognition are maintained even on the leaf epidermis, because only pollen from *Arabidopsis* or from closely related species can hydrate on *fdh* plants.

Why can pollen germinate on *fdh* leaf cells? One possibility is that the leaf cells in the *fdh* mutant have acquired pistil characteristics. This hypothesis appears unlikely because *fdh* mutant leaves do not express the *AGAMOUS* gene, which is expressed early in pistil development (see Section 19.3.2). However, experiments have shown that the permeability of the epidermal cells is altered in the *fdh* mutant, and that the high-molecular-mass lipids in *fdh* leaf cells differ from those in wild-type leaves. These results implicate a further role for lipids in pollen–pistil interactions.

(A)   (B)

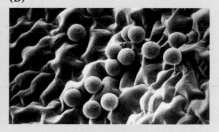

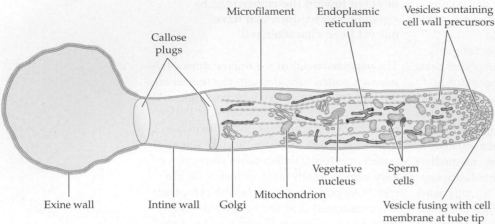

Figure 19.29
(A) Schematic illustration of a pollen tube growing by tip extension. Callose plugs limit the volume occupied by the cytoplasm to regions close to the growing tip. (B) Fluorescence micrograph of pollen tubes growing through a pistil. The callose plugs that block off the pollen tube at regular intervals are visualized by staining with aniline blue.

to maintain rapid growth. Animal cells adhere to each other by way of extracellular proteins or other components in the extracellular matrix; they will also adhere to tissue culture plates that have been coated with protein purified from the extracellular matrix of the animal cells.

If pollen tube adhesion occurs, what components of the pollen tube and of the style exudate mediate the interaction? In many plants, this would be a difficult question to resolve, because the pollen tubes grow through the small intercellular spaces of solid stylar tissue. Easter lily is a convenient experimental model for such studies, however, because its style is hollow, allowing access to the pollen tubes growing through the style. An adhesion assay can be used to purify the components of lily style exudate that are required for pollen tube adhesion (Fig. 19.30).

Further evidence for guidance cues from the embryo sac has been found in the plant *Torenia fournieri,* in which the embryo sac is "naked" rather than enclosed by a typical cell wall. In this plant, pollen tube growth toward embryo sacs is much more precise and directed after the pollen tubes have traversed the style (Fig. 19.31). Probably what provides the guidance cue are the synergids of the embryo sacs: First, the pollen tubes enter the embryo sac at the synergid end; second, synergids are secretory cells and

contain high concentrations of calcium; and third, the pollen tubes in *Torenia* are not attracted to embryo sacs that have already been fertilized, to embryo sacs with disrupted synergids, or to heat-killed embryo sacs.

19.7 Self-incompatibility

19.7.1 Self-incompatibility mechanisms disrupt normal pollen–pistil interactions and prevent inbreeding.

The structure of most flowers favors pollen landing on the stigma of the same flower. Nonetheless, many physical mechanisms can prevent a successful pollination. For example, if pollen is produced either too early or too late in relation to when the female tissue is receptive to pollen tube growth, pollination will fail. Perhaps the most interesting examples of failed pollination involve the phenomenon termed **self-incompatibility,** in which pollen of an individual plant fails on the female organs of that plant but pollen from other plants of the same species is successful. Darwin realized that self-incompatibility would increase outcrossing and prevent inbreeding. Self-incompatibility has evolved in many plant families, but different plant families use different proteins to accomplish the same goal.

The genetic control of self-incompatibility is now well understood. Self-incompatibility is usually controlled by a single genetic locus with multiple alleles. To account for

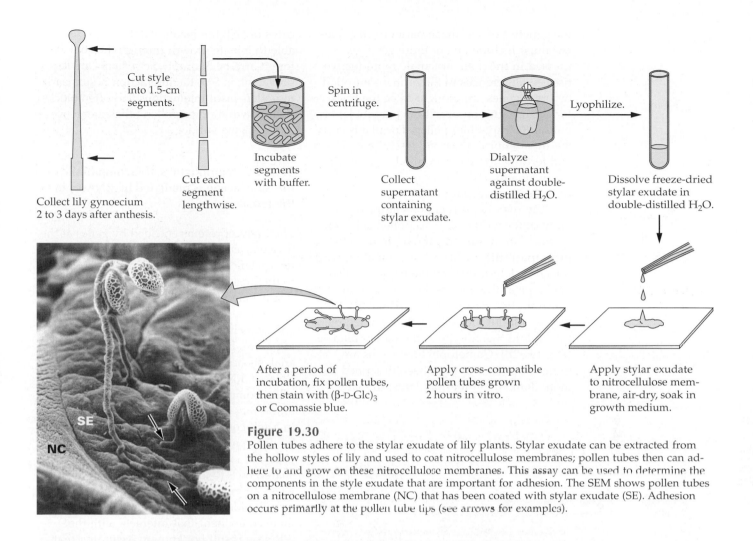

Cut style into 1.5-cm segments.

Collect lily gynoecium 2 to 3 days after anthesis.

Cut each segment lengthwise.

Incubate segments with buffer.

Spin in centrifuge.

Collect supernatant containing stylar exudate.

Dialyze supernatant against double-distilled H$_2$O.

Lyophilize.

Dissolve freeze-dried stylar exudate in double-distilled H$_2$O.

After a period of incubation, fix pollen tubes, then stain with (β-D-Glc)$_3$ or Coomassie blue.

Apply cross-compatible pollen tubes grown 2 hours in vitro.

Apply stylar exudate to nitrocellulose membrane, air-dry, soak in growth medium.

Figure 19.30
Pollen tubes adhere to the stylar exudate of lily plants. Stylar exudate can be extracted from the hollow styles of lily and used to coat nitrocellulose membranes; pollen tubes then can adhere to and grow on these nitrocellulose membranes. This assay can be used to determine the components in the style exudate that are important for adhesion. The SEM shows pollen tubes on a nitrocellulose membrane (NC) that has been coated with stylar exudate (SE). Adhesion occurs primarily at the pollen tube tips (see arrows for examples).

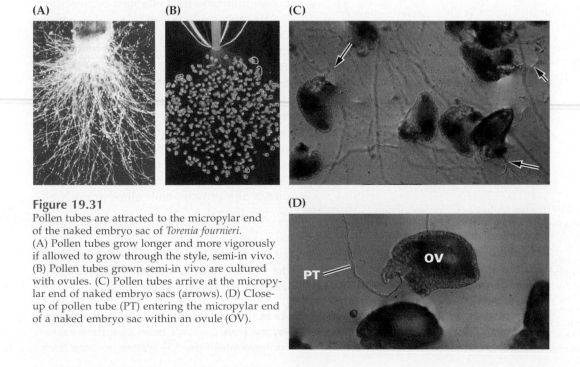

(A) **(B)** **(C)**

(D)

Figure 19.31
Pollen tubes are attracted to the micropylar end of the naked embryo sac of *Torenia fournieri.*
(A) Pollen tubes grow longer and more vigorously if allowed to grow through the style, semi-in vivo. (B) Pollen tubes grown semi-in vivo are cultured with ovules. (C) Pollen tubes arrive at the micropylar end of naked embryo sacs (arrows). (D) Close-up of pollen tube (PT) entering the micropylar end of a naked embryo sac within an ovule (OV).

the genetics of self-incompatibility, the S locus must include one or more genes expressed in the male or female reproductive tissues. Differences in the proteins encoded by these genes are thought to be the basis for the recognition of self (incompatible) or nonself (compatible) pollen. For plants carrying the identical allele, successful pollination and fertilization are prevented, whereas if the male and female carry different alleles, the cross is successful.

The two major types of self-incompatibility differ with respect to pollen behavior on the pistil tissue. In **gametophytic self-incompatibility** (GSI), the incompatibility of pollen is determined by the haploid pollen genotype at the S locus (Fig. 19.32A). In **sporophytic self-incompatibility** (SSI), the incompatibility of pollen is determined by the diploid S genotype of the parent plant (Fig. 19.32B). Gametophytic systems are more common and have been described in more than 60 families of plants. In many

cases of GSI, the incompatible pollen tube is able to initiate growth through the style before it arrests. Sporophytic systems are less prevalent; in SSI, the interaction occurs early in pollen–pistil interactions and often blocks pollen hydration or pollen tube emergence at the stigma surface.

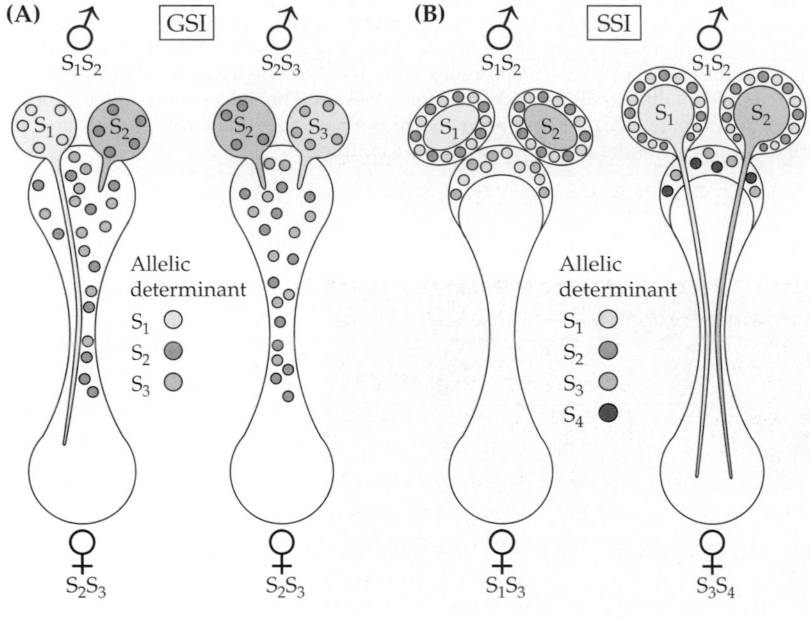

19.7.2 Gametophytic self-incompatibility in the Solanaceae is mediated by RNAses in the female.

The S-glycoproteins encoded by genes at the S locus were identified in stylar extracts as the specific abundant proteins that correlated to the presence of particular S alleles. The distinct electrophoretic mobilities of the different S proteins allowed their purification and sequencing; as a result, their corresponding genes have been cloned. Subsequent analysis has revealed that the S proteins are RNases. RNase activity is necessary for the self-incompatibility response because a mutation that inactivates RNase activity results in a self-compatible plant. S-RNases encoded by different alleles can be highly polymorphic. For example, the amino acid identities shared by S-RNases encoded by different alleles can range from 38% to 98%.

The identity of the pollen-expressed male component that interacts with the S-RNase is still not known. Assuming that pollen tube growth is inhibited because the S-RNases degrade pollen tube RNAs, at least two models could explain the interaction of the male component with the stylar S-RNases. The pollen S gene product could be a receptor that allows the S-RNase of the same allelic specificity to enter the pollen tube and there inhibit pollen tube growth by degrading pollen mRNA. Alternatively, all S-RNases could freely enter the pollen cytoplasm, but only the S-RNase of the same allelic specificity would remain active and able to inhibit pollen tube growth while all other S-RNases are inactivated.

19.7.3 Sporophytic self-incompatibility in the Brassicaceae is mediated by receptor-like protein kinases in the female.

As in the gametophytic system, abundant stigma proteins have been identified as components of the SSI system because their

Figure 19.32
Genetics of self-incompatibility. In gametophytic self-incompatibility (GSI), pollen is successful only when the genotype of the pollen does not match the genotype of the female. As shown in (A), both the S_2 and S_3 pollen tubes are arrested after some growth through an S_2S_3 style. Only the S_1 pollen tube can continue to grow through the style. In sporophytic self-incompatibility (SSI), the pollen grains can germinate and grow tubes only when they do not carry determinants that match the genotype of the female. As shown in (B), even though the S_2 pollen on the style on the left is genotypically different from the S_1S_3 style, it still cannot grow a pollen tube. Presumably, it has acquired determinants on its pollen coat that match the S_1 style, leading to an incompatible response. In contrast, neither the S_1 nor the S_2 pollen on the style on the right has any determinants that match the S_3S_4 style, so both can grow pollen tubes.

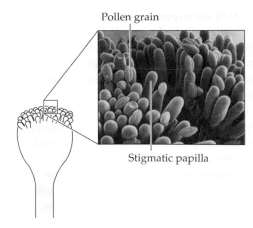

Pollen grain

Stigmatic papilla

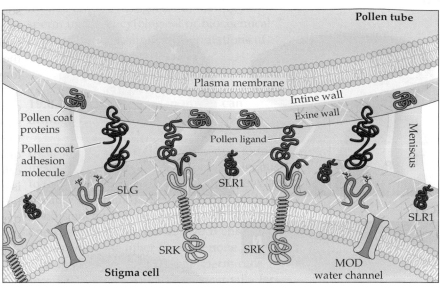

Figure 19.33
The scanning electron micrograph at left shows the surface of a stigma. Stigmatic papillae and pollen grains are visible. The simplified schematic representation at right illustrates the contact zone between a *Brassica* stigmatic papillar cell and an incompatible pollen grain. Some of the proteins thought to be involved in the interaction are shown (the shapes and sizes of the proteins in this model are hypothetical). S locus receptor kinase (SRK) may bind the pollen ligand, and in some cases this binding might be facilitated by interaction with S locus glycoprotein (SLG). The pollen ligand probably is a component of the pollen coat. SLG-like receptor 1 (SLR1) and other components of the pollen coat probably are involved in pollen adhesion to the papillar cell. MOD, modifier of self-incompatibility.

presence correlates with particular S alleles. Over the years, the number of known SSI proteins has increased and the system is now quite complex (Fig. 19.33). Two S locus genes have been studied extensively: The *SLG* gene encodes a glycoprotein that is secreted into the cell wall of the stigmatic papillar cells, whereas the *SRK* gene encodes a **receptor-like kinase** that contains an extracellular domain, a transmembrane domain that anchors the protein in the plasma membrane, and a cytoplasmic domain that has serine/threonine kinase activity (see Chapter 18). The extracellular domain of SRK frequently shares significant amino acid identity with the corresponding SLG protein. This sequence conservation is striking, considering that the amino acid sequences of SLGs (or SRKs) from different S alleles can vary substantially. The amino acid conservation between these two components of any one S allele, the SRK extracellular domain and SLG, suggests that both SRK and SLG are

required for recognition of the pollen component that participates in self-incompatibility.

Are both SLG and SRK required for the SSI response in the female? Plants with defects in the expression of *SRK* are self-compatible, thus establishing that SRK is important. The requirement for SLG, however, is still uncertain. Some plants show only very low *SLG* expression but are still self-incompatible, whereas other plants express SLG strongly but are self-compatible.

19.7.4 The search for the male components of self-incompatibility continues.

In most well-studied self-incompatibility systems, the female component has been identified but the male component has not. A successful candidate for this role must meet several criteria. The male component must be encoded by a gene that is linked to the S locus and must show S allelic specificity. Most likely, the male component also interacts directly with the female component that is encoded by the S locus (i.e., with pistil-localized S-RNases in the gametophytic system typical of the Solanaceae, or with pistil-localized SRK or SLG in the sporophytic system typical of the Brassicaceae). Because both the male and female components must be linked to the S locus, many researchers have tried to identify the pollen component by looking for linked

19.9 Seed formation

During its development, the angiosperm seed is composed of several tissues, including the **embryo,** the **endosperm,** and the **testa.** The embryo comprises the axis, which contains the root and shoot meristems that ultimately grow to form the vegetative plant, and the cotyledons (or scutellum in cereals), which are terminally differentiated organs that accumulate stored food reserves to be used during early seedling establishment. The endosperm is a nutritive storage tissue; it may be reabsorbed during seed development as a food source, or it may persist in the mature seed, as in cereals and endospermic legumes. The testa (seed coat) is derived from the integuments. In some species, the testa is rudimentary, and the layer surrounding the seed tissue is the pericarp (fruit coat), which originates from the ovary wall.

Many studies of embryo development have been conducted on dicots. Crucifers such as *Arabidopsis* have frequently been used as models of embryo development be-cause they show regular and predictable patterns of cell division and include a variety of developmental mutants. Development of the endosperm, another process for which mutants have been identified, has been documented for many species, particularly cereals.

19.9.1 Mutants defective in various stages of embryo development have been identified.

A typical pattern of dicot embryo development is shown in Figure 19.36. Embryogenesis occurs in three overlapping phases: differentiation of tissues, cell enlargement, and maturation (drying). During histodifferentiation, the fertilized egg undergoes multiple cellular divisions and differentiates to form the

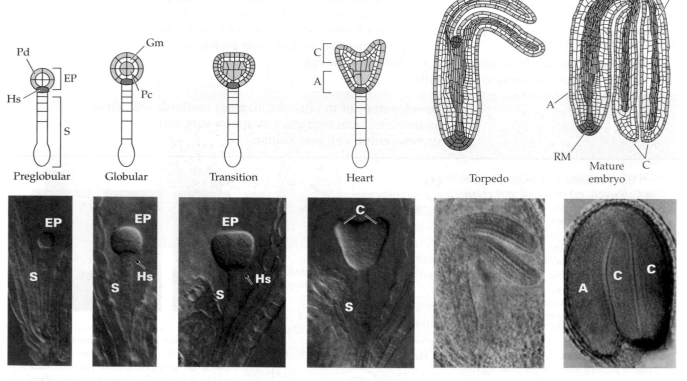

Figure 19.36
Stages of development of a wild-type *Arabidopsis* embryo. Initially the fertilized zygote cell divides to form an apical and a basal cell (not shown). The former divides to become the embryo proper (EP) region of the preglobular stage, and the latter becomes the suspensor (S). The protoderm (Pd) develops into the epidermal (Ed) layer in the mature embryo; the ground meristem (Gm) develops into the storage parenchyma (P); the procambium (Pc) develops into the vascular tissue (V); and the hypophysis (Hs) develops into root and shoot (axial) meristems (RM, SM). A, axis; C, cotyledons. The testa and the single cell-layer of endosperm that surrounds the embryo are not shown in the diagram.

embryonic tissue and organ systems (globular to heart stages). A suspensor forms and facilitates embryo development until mid-maturation (the torpedo stage), when it is usually occluded by the growing embryo. Subsequent cell enlargement accommodates the deposition of stored reserves during the torpedo stage. During maturation drying, developmental processes are terminated as the embryo prepares for and is subjected to desiccation.

Many genes are involved in the regulation of development. Genetic analysis coupled with molecular characterization is an important part of the identification of these genes and their functions. Defects or mutations of the genes involved in the synthesis of stored reserves have been invaluable in elucidating biosynthetic pathways and their control points. Identifying genes that control morphological development presents a greater challenge, but some progress has been made by using chemical and T-DNA mutagenesis techniques. Embryo-defective mutants of *Arabidopsis* have been identified that are blocked at various stages of devel-

opment, from as early as the preglobular and globular stages to late maturation. Although most specific developmental genes and their products remain to be identified, the pleiotropic effects of the mutations reveal interesting regulatory patterns. *Raspberry* mutants, for example, exhibit morphological arrest at the globular stage (Fig. 19.37), yet their procambial, ground meristem, and protoderm tissues complete cell differentiation up to the terminal stages of embryogenesis.

LEAFY COTYLEDON (*LEC*) and *FUSCA* (*FUS*) genes control late embryogenesis, after the heart stage, and encode proteins that might function as critical components in the transduction of environmental and cellular signals during late development (Fig. 19.37). Mutations that eliminate the function of these genes have such defects as precocious germination, desiccation intolerance, failure to synthesize some storage proteins and lipids, and cotyledons that develop leaf-like traits. Other developmental mutants are known, and during the next few years more will probably be identified and ascribed specific gene functions. This should eventually

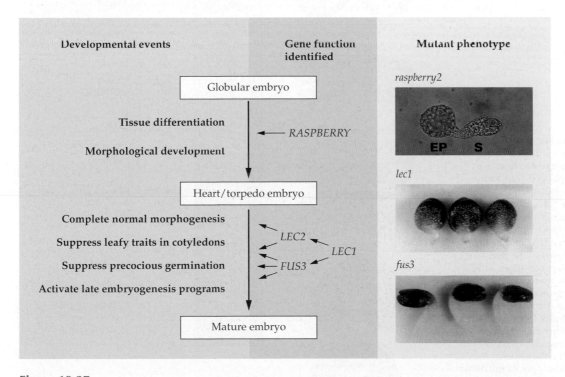

Figure 19.37
Control points in the regulation of late embryogenesis in *Arabidopsis,* as suggested by studies on the *raspberry* mutation and on the embryo-defective *leafy cotyledon* (*lec*) and *fusca* (*fus*) mutants. Loss-of-function mutant phenotypes indicate that successful completion of specific events requires wild-type gene products (arrows). EP, embryo proper; S, suspensor.

unravel the molecular mechanisms responsible for specifying different cell and tissue types during embryogenesis.

19.9.2 Endosperm development follows different pathways.

In many angiosperms, the endosperm is an ephemeral tissue that stores reserves temporarily until the developing embryo absorbs them into its own cotyledon storage tissues. In some seeds where the endosperm persists (e.g., lettuce, tomato), it may be only one to a few cells thick and may play a minor storage role compared with that of the embryo it surrounds. In other seeds, the endosperm may act as the major storage tissue, with the embryo containing few or different reserves. In castor bean, for example, the rel-atively massive endosperm consists of living cells that store oil and protein and produce the enzymes necessary for the mobilization of these reserves. In the endosperms of cereals and some endospermic legumes, most of the endosperm cells are nonliving at maturity; the storage reserves they contain are hydro-lyzed by enzymes synthesized in and released from a surrounding, thin layer of living aleurone (Fig. 19.38; see also Chapter 20).

Two major types of endosperm development have been observed. In nuclear endosperm development, the endosperm undergoes several free nuclear divisions before the cell wall is formed. This is the most common pattern, and it is found in developing cereal grains. In seeds that demonstrate cellular endosperm development (e.g., *Impatiens, Lobelia*), there is no free nuclear stage.

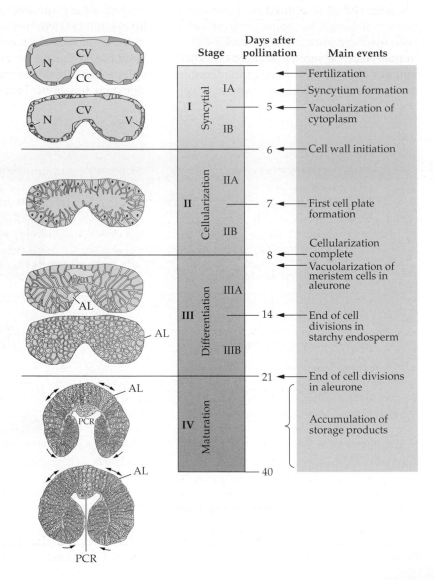

Figure 19.38
Endosperm development in barley, showing the stages and main events. (Stage IA) Development in the midregion of the embryo sac results in nuclear (N) divisions without cell wall formation (the syncytial nuclear endosperm) surrounding a central vacuole (CV). The central crease (CC) is present. (Stage IB) Nuclei undergo further mitotic divisions and vacuoles (V) appear in the cytoplasm. (Stage IIA) Endosperm cell walls start to be laid down in the periphery, and (Stage IIB) cellularization continues until approximately 100,000 cells are formed. (Stage IIIA) The first cells of the aleurone layer (AL) form over the crease. (Stage IIIB) Aleurone layer differentiation spreads laterally from the central crease to join with other localized regions of formation. (Stage IV) The endosperm expands over the crease as a result of cell enlargement; reserve deposition commences. The supply of nutrients through the phloem to the endosperm runs in the placento-chalazal region (PCR), which is eventually occluded as the endosperm expands to fill the crease area. Endosperm expansion and reserve deposition are completed.

Many mutants with perturbations in the biosynthesis of stored reserves have been isolated. Members of another class of mutants demonstrate altered endosperm morphology, including *de* and *dek* mutants in maize and *dex* and *sex* mutants in barley. Several *sex* mutants are defective in cell wall formation, and their development is arrested at syncytial stage I, resulting in an empty endosperm phenotype. Other mutants are developmentally arrested at stage II and demonstrate incomplete cellularization, although the starchy endosperm and aleurone layer cells differentiate. As with the embryo mutants, the specific regulatory genes and their products involved in endosperm development remain to be identified.

19.10 Deposition of stored reserves during seed development

19.10.1 Carbon and nitrogen are transferred from parental tissues to the endosperm or embryo for synthesis of reserve compounds.

There is no direct vascular connection between the parent plant and the embryo or endosperm (see Chapter 15). The vascular tissue ends in the seed coat or fruit coat of dicots or in the placenta-chalazal region of cereals. Assimilates are transported from the parental seed coat cells to the extracellular space (apoplast) by a passive, facilitated membrane-transport process and are subsequently taken up by the filial tissues of the endosperm or embryo.

During early development, sucrose is generated by the parent plant (as a product of photosynthesis or mobilization of stored starch reserves), released from the phloem into the apoplast, and hydrolyzed by invertases (see Chapter 13) in the cell wall. The hexose products are loaded into the filial storage tissues. The amount of hexose made available to these tissues before synthesis of the reserves can influence embryogenesis. In *Vicia faba* seeds, for example, highly active seed coat invertase and the high concentrations of hexose it promotes are associated with extended mitotic activity of the developing cotyledons. The resulting cotyledons are larger than in seeds with low seed coat invertase activity and contain more cells in

which reserves can later be deposited. Invertase also influences the development of cereal seeds. *miniature-1*, a maize mutant lacking invertase activity, demonstrates aberrant development of both the endosperm and the area of phloem unloading (pedicel). These examples illustrate how metabolite availability can influence both morphological and biosynthetic aspects of seed development.

During the early stages of development, sucrose may be resynthesized from the hexoses imported into the storage tissues by the sucrose-phosphate synthase pathway (Fig. 19.39A; see also Chapter 13). At later stages of seed development, during deposition of reserves, the ratio of sucrose to hexose entering the storage tissues increases, and hydrolysis of sucrose is not an obligatory step for uptake. In *V. faba* cotyledons, sucrose-phosphate synthase activity declines in the later stages of development, when sucrose itself is imported (Fig. 19.39B).

The major forms of nitrogen translocated into the seed in the phloem are asparagine and glutamine, although in some legumes (e.g., soybean and cowpea) about 10% to 15% of nitrogen translocated into seeds is ureide-N (e.g., allantoin and allantoic acid). These compounds are converted to asparagine and glutamine by enzymes in the seed coat. The composition of the amino acids reaching the developing storage tissues often differs from that entering the phloem of the parental tissues, so enzymes that interconvert amino acids must be present in the seed coat. These enzymes may vary in activity with changes in the supply of nitrogen, the environment, and the stage of development. In the seed coats of developing peas, maximum activities of asparaginase and aspartate aminotransferase (see Chapter 8) are attained before synthesis of storage proteins commences. The products of seed coat asparaginase and aspartate aminotransferase are imported into the embryo or endosperm and serve as substrates for synthesis and interconversion of amino acids. Later during development, when reserve protein synthesis is high, asparaginase activity in the seed coat is very low, and asparagine is taken up directly into the cotyledons, where it is deaminated by a cotyledon-localized isoenzyme of asparaginase, the activity of which increases during cotyledon development. Glutamine synthetase and glutamate

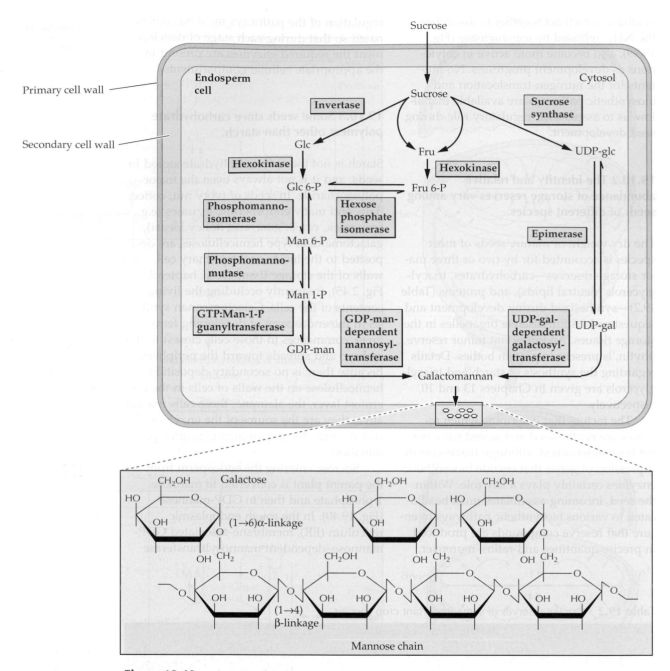

Figure 19.40

Proposed pathway for the synthesis of galactomannan in the endosperm of developing endospermic legumes. It has not been determined whether the cleavage of sucrose by invertase occurs outside or inside the cytoplasm of the endosperm cells. The secondary cell wall is composed almost exclusively of galactomannan; a brief segment of its structure is shown in the lower panel. Glc, glucose; Glc 6-P, glucose 6-phosphate; Fru, fructose; Fru 6-P, fructose 6-phosphate; Man, mannose; Man 6-P, mannose 6-phosphate; Man 1-P, mannose 1-phosphate; GDP-man, guanosine diphosphate mannose; UDP-glc, uridine diphosphate glucose; UDP-gal, uridine diphosphate galactose.

then transfers mannose resudue to the nonreducing end of a linear (1→4)β-linked mannose primer to form the growing backbone chain of the galactomannan polymer. Simultaneously, UDP-galactose–dependent galactosyltransferase, another membrane-

associated enzyme, transfers a galactose residue to a mannose at or near the nonreducing end of the growing mannan chain, forming a (1→6)α-linkage; however, galactose cannot be transferred to preformed mannose chains. The activities of the two transferase enzymes increase in parallel during galactomannan synthesis, with the result that the mannose:galactose ratio in the polymer remains constant.

19.10.4 Historically, seed storage proteins have been grouped according to solubility.

About a century ago, T. B. Osborne pioneered the systematic study of plant proteins by categorizing them on the basis of their solubility characteristics. Proteins were extracted sequentially in a series of solvents, as follows: First, water solubilizes the **albumins;** next, extraction in dilute salt solutions solvates the **globulins;** subsequent extraction in alcohol–water mixtures yields the **prolamins;** and finally, treatment with dilute acids or alkali releases the **glutelins.** Although these names remain in use, the classification scheme on which they are based is no longer current, because of cases in which structurally similar proteins belong to different groups. For example, prolamins that lack interchain disulfide bonds are soluble in aqueous alcohol, as expected, but those that have interchain bonds are alcohol-insoluble and are classically, but incorrectly, assigned to the glutelin group. This inconsistency has been eliminated by redefining glutelins as one class of prolamins.

Prolamins are unique in that they occur only in the endosperms of cereals and other grasses (Poaceae), where they are usually the major storage protein; exceptions are rice and oats, in which glutelin-type prolamins and globulins, respectively, predominate. Globulins are the major storage proteins in most dicot seeds, especially legumes, although a low-molecular-mass sulfur-rich albumin (one of the 2S albumins, named for their sedimentation coefficient) also is present in these and other species.

19.10.5 Most prolamins, storage proteins unique to cereal endosperms, appear to have evolved from a single ancestral gene by duplications, insertions, and deletions.

The prolamins of three cereals grown in temperate climates—barley, rye, and wheat—are highly polymorphic mixtures of polypeptides with molecular masses of 30 to 90 kDa.

Prolamins fall into one of three groups, based on their amino acid sequence: sulfur-rich, sulfur-poor, or HMW (high-molecular-mass) (Fig. 19.41). The S-rich group accounts for as much as 90% of the prolamin proteins known and includes both polymeric proteins linked by interchain disulfide bonds and monomeric proteins with intrachain disulfide bonds. Their amino acid sequences consist of an N-terminal domain of repeated proline and glutamine sequences and a nonrepetitive C-terminal domain that contains the sulfur-bearing cysteine residues. The S-poor prolamins, not surprisingly, have a truncated C-terminal region and a larger N-terminal domain of repeated sequences. The HMW prolamins have extensive repeated

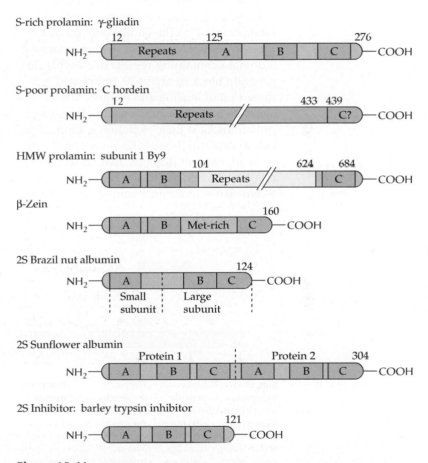

Figure 19.41
The prolamin superfamily, illustrating relationships among the sulfur-rich (S-rich), sulfur-poor (S-poor), and HMW (high-molecular-mass) prolamins of cereals and the 2S albumins of cereals and dicots. Three conserved domains (A, B, C) are present in all but the S-poor prolamins (e.g., C hordein). The 2S albumin from Brazil nut is cleaved proteolytically at a site between A and B, and the resulting polypeptides are joined by a disulfide bond. The barley trypsin inhibitor and the sunflower albumins do not contain any internal sites for proteolysis, but the latter is synthesized as two similar confluent proteins that are then cleaved posttranslationally. Note that S in the expression 2S refers to Svedbergs (a unit of sedimentation), not sulfur. Met-rich, methionine-rich.

sequences flanked by nonrepetitive C- and N-terminal domains that are rich in glycine and glutamine. All three prolamin types probably evolved from a single ancestral gene (Fig. 19.42). The ancestral protein may have been about 90 amino acids long with domains A, B, and C. During evolution, S-rich and HMW prolamins acquired insertions of repeat sequences in domains A, B, C, and the flanking regions. In the S-poor prolamins, the repeated sequences have been amplified, A and B have been lost, and only part of C remains.

The prolamins of other cereals illustrate variations on this evolutionary theme. For example, oat prolamins (e.g., avenin) contain domains A, B, and C and two blocks of proline- and glutamine-rich repeats. Maize α-zeins, which account for about 80% of the total prolamin in the endosperm, consist of unique C- and N-terminal domains flanking a domain containing repeats of a highly degenerate block of about 20 amino acid residues, rich in leucine and alanine. These zeins show no sequence similarity to other cereal prolamins, and their evolution is more difficult to explain. β-Zein and γ-zein, on the other hand, contain domains that clearly are related to A, B, and C (see Fig. 19.41), with γ-zein also containing a unique repeat region.

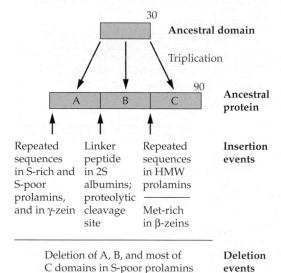

Figure 19.42
Model for the possible sequence of events in the evolution of prolamins and 2S albumins, suggested by the sequences of the extant proteins. Abbreviations as in the legend for Figure 19.41.

19.10.6 2S albumins are present in seeds of both dicots and monocots.

2S albumins are widely distributed in dicot seeds, including those of the legumes, members of the Brassicaceae (in rapeseed, the proteins are called napins), the Euphorbiaceae (e.g., castor bean), and the Asteraceae (e.g., sunflower). They also are present in cereals, where they have the additional property of being inhibitors of trypsin, α-amylase, or both. Some legume 2S albumins (e.g., soybean α-conglycinin) likewise function as enzyme inhibitors. The 2S albumins are compact globular proteins with conserved cysteine residues. Frequently, they are heterodimers that contain two polypeptides of different molecular masses linked by interchain disulfide bonds (e.g., napins contain two proteins, one of 9 kDa and the other 4 kDa). 2S albumins usually are synthesized by cleavage from a single precursor protein, accompanied by the loss of a linker peptide and short peptides from both termini.

Evolutionary homologies between the 2S albumins and prolamins include that both contain conserved regions related to A, B, and C (see Figs. 19.41 and 19.42). The albumins lack the repeated sequences but usually contain a proteolytic cleavage site between A and B that, when cut, yields the two polypeptide chains.

19.10.7 Globulins are subcategorized according to their sedimentation coefficient.

Globulins, the most widespread group of storage proteins, are divided into two groups based on their sedimentation coefficient—the 7S vicilins and the 11S legumins. The 7S vicilins, trimeric proteins of 150 to 200 kDa, lack cysteine residues and hence do not form disulfide bonds. They are initially synthesized as groups of 45- to 50-kDa polypeptides. In *Pisum sativum* (pea; Fig. 19.43) and *V. faba* (broad bean), 7S vicilins are cleaved into smaller polypeptides and glycosylated to produce a heterogeneous set of proteins, most of which range in size from 12 to 35 kDa. Convicilin in pea is not cleaved but is glycosylated extensively to produce a polypeptide of 70 to 75 kDa. Likewise, in soybean and *Ph. vulgaris* (french bean),

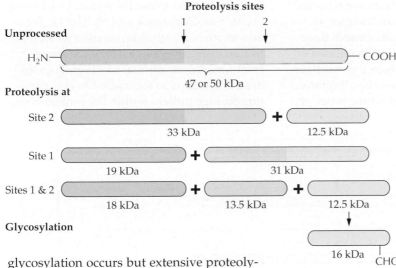

Proteolysis sites

1   2

Unprocessed

H₂N—[   ]— COOH

47 or 50 kDa

Proteolysis at

Site 2   [   ] + [   ]
         33 kDa   12.5 kDa

Site 1   [   ] + [   ]
         19 kDa   31 kDa

Sites 1 & 2 [   ] + [   ] + [   ]
             18 kDa   13.5 kDa   12.5 kDa

Glycosylation

[   ]
16 kDa   CHO

Figure 19.43
Origin of the various-sized polypeptides of the pea
7S vicilin. Polypeptides are generated by cleavage
at one or two proteolytic sites. Glycosylation of the
cleaved and uncleaved polypeptides in different
legumes generates 7S vicilin polypeptides that range
widely in size (12 to 75 kDa).

glycosylation occurs but extensive proteolysis does not: The mature polypeptides in
these plants range from 42 to 57 kDa. Despite the great diversity in polypeptide size,
most 7S globulins have highly homologous
primary structures and probably have homologous tertiary structures as well. These
proteins may have originated from a common ancestral gene, which in pea has given
rise to at least 24 genes at six or more genetic loci.

The mature 11S legumins have molecular masses of 350 to 400 kDa and consist of
six subunit pairs, each pair consisting of an
acidic 40-kDa polypeptide and a basic 20-
kDa polypeptide (Fig. 19.44). The subunit
pair is synthesized as a single precursor protein, which is proteolytically cleaved after
disulfide bond formation (Fig. 19.45). Unlike
the vicilins, the 11S legumins are very rarely
glycosylated. Recent comparative studies on
the structures and sequence homologies of

the 11S and 7S globulins have led to the suggestion that they evolved from a common
ancestral protein.

The 7S globulins are not major storage
proteins in any cereal and are located only
in the embryo and aleurone layer; in dicots
they are frequently less abundant than the
11S legumins. The globulin-rich oat grains
contain the 11S storage protein within the
endosperm region.

19.10.8 Storage protein synthesis and deposition are highly regulated processes.

All of the major storage proteins are synthesized on the rough ER (Fig. 19.46). They possess signal peptides that direct the newly
translated polypeptides into the lumen of the

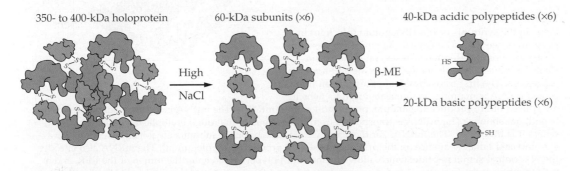

350- to 400-kDa holoprotein   60-kDa subunits (×6)   40-kDa acidic polypeptides (×6)

                    High   β-ME
                    ────→   ────→   HS—

                    NaCl

                                                            20-kDa basic polypeptides (×6)

                                                                    —SH

Figure 19.44
An 11S legumin protein is disassembled to release its acidic (red) and basic (blue) component polypeptides by
sequential treatment with high-salt extraction buffer and β-mercaptoethanol (β-ME). The holoprotein itself
(comprising six pairs of acidic and basic subunits) is dissociated into its individual subunits by the high concentrations of salt through disruption of the noncovalent, electrostatic bonds holding the subunits together.
β-ME, a reducing agent, breaks the disulfide bonds that link the subunit pairs.

ER, where the signal peptides are removed by proteolytic cleavage (see Chapter 4). The protein is folded into its correct three-dimensional structure within the lumen of the ER, the same place where the disulfide bonds in legumins and in some albumins and prolamins are formed. Three types of proteins located in the ER lumen assist these events (see Chapters 4 and 9). The 11S legumin of soybean (glycinin) is used as an illustration in Figure 19.45 to show the stages from initial transcription of the gene in soybean cotyledons to the assembly of the mature storage protein within the protein body.

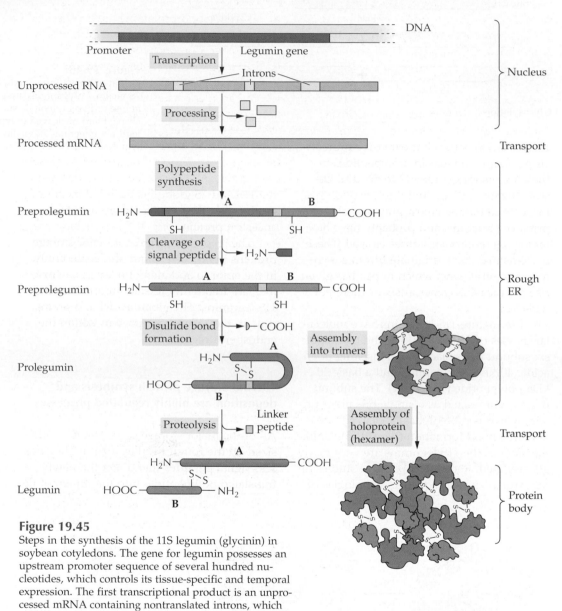

Figure 19.45
Steps in the synthesis of the 11S legumin (glycinin) in soybean cotyledons. The gene for legumin possesses an upstream promoter sequence of several hundred nucleotides, which controls its tissue-specific and temporal expression. The first transcriptional product is an unprocessed mRNA containing nontranslated introns, which are spliced out to produce the mature, processed legumin mRNA. (The mRNA is also polyadenylated at the 3′ end; not shown.) The mRNA is transported from the nucleus to the rough endoplasmic reticulum (RER) where it is translated. The mRNAs for the acidic (A) and the basic (B) subunits are joined by codons for a 4-amino acid linker sequence on the primary translation product (preprolegumin). The mRNA also encodes the N-terminal signal peptide, which allows the nascent polypeptide to enter the lumen of the RER, and a pentapeptide at the C-terminal end. The signal peptide and the pentapeptide are cleaved from the primary protein product (prolegumin) to yield the A and B subunits joined by the linker sequence and disulfide bonds. In the final step of processing, a thiol-containing protease removes the linker sequence, forming a mature protein in which the acidic and basic subunits are joined only by disulfide bonds. The mature acidic subunit contains 278 amino acids (about 40 kDa) and the basic subunit 180 amino acids (about 20 kDa). Within the RER, prolegumin assembles into trimers that enter the protein bodies. The mature hexameric legumin holoprotein assembles after the proteolytic separation of the acidic and basic subunits.

Some seed storage proteins are glycoproteins (e.g., 7S vicilins and some albumins), and the initial glycosylation of storage protein polypeptides occurs as a cotranslational event. Two major types of oligosaccharide side chains are attached to storage proteins: (a) simple, or high-mannose, oligosaccharides containing only mannose (Man) and N-acetylglucosamine (Glc-NAc) residues, usually in a ratio of 5:2 to 9:2; and (b) complex, or modified, oligosaccharides that in addition contain xylose (Xyl) or fucose (Fuc) or both. The formation of both types of oligosaccharides is described in Chapter 4.

Protein body formation occurs in two different ways in developing cereal grain endosperms, in that the protein body forms from either the vacuole or the rough ER. The major storage proteins in rice and oat endosperms are 11S globulins; after being transported from the lumen of the ER to the vacuole by way of the Golgi apparatus (see Chapter 4), the protein bodies form by fragmentation of the vacuole. A similar pathway operates in the aleurone layer of cereals, which contains globulins and albumins but never prolamins or glutelins. In contrast, the prolamins of rice and maize are sequestered

in protein bodies that are formed as localized distentions as the lumen of the rough ER becomes filled. The rice endosperm thus contains two populations of protein bodies, those of vacuolar origins, which contain globulins (80% of the storage protein), and those of ER origin, which contain prolamins (5% of the storage protein). Interestingly, the globulins and prolamins are synthesized in different regions of the rough ER, which contains two domains, the cisternal ER (C-ER), a network composed of stacks of rough ER, and the protein body ER (PB-ER), which has a less ordered structure. Globulins are preferentially synthesized on polysomes associated with the C-ER, from where they are transported to the Golgi and targeted to the vacuole. In contrast, the prolamin mRNAs are translated on ribosomes associated with the PB-ER, where the prolamins then enter the lumen and form protein bodies (Fig. 19.46; see also Chapter 4). The mechanisms that segregate the two types of mRNAs to different regions of the rough ER remain to be elucidated. Within the PB-ER of the maize endosperm are sequestered four different types of zein proteins: α-, β-, γ-, and δ-zein. Initially, the developing protein bodies

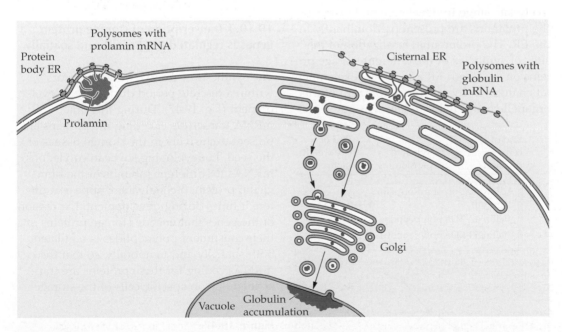

Figure 19.46
The rough ER of developing rice endosperms is composed of two distinct domains: the cisternal ER, where the polysomes translate mRNA for a globulin-like protein (glutelin), and the protein body ER, which contains polysomes translating mRNA for prolamin (oryzenin). The prolamin is sequestered in the ER of the lumen and directly distends to form a protein body, whereas the globulin passes through the Golgi apparatus to the vacuoles (see Chapter 4, Fig. 4.22 for details). The storage protein content of a mature rice grain is 5% oryzenin and 80% glutelin.

Table 19.3 Location of events involved in processing the storage proteins deposited within vacuolar-derived protein bodies[a]

Rough endoplasmic reticulum	Golgi	Protein body
Removal of N-terminal signal peptide and C-terminal sequence	Glycan modification	Proteolytic cleaving/trimming
N-Glycosylation		Glycan modification/removal
Polypeptide folding		Oligomer assembly
Disulfide bond formation		
Oligomer assembly		

[a]Note that these events do not occur for all storage proteins (e.g., legumins are not N-glycosylated).

consist only of the cysteine-rich β- and γ-zeins. As these bodies enlarge and distend the ER, the β- and γ-zeins remain at the periphery as a highly cross-linked network, and the most abundant type, α-zein, along with the minor δ-zein component, somehow penetrate this network to occupy the central region. An extensive array of microtubules associates with the developing maize protein bodies. This cytoskeletal network may play a role in regulating the deposition of storage proteins, perhaps by localizing mRNA to the appropriate region on the rough ER (a mechanism that might also operate in rice to segregate the two mRNA types). Prolamins in developing wheat, barley, and rye endosperms appear to be present in both ER-derived and vacuolar protein bodies; at a relatively late stage in development, however, the prolamins are present predominantly in the ER. The preferential localization of the sites of synthesis of the different storage proteins on the rough ER as well as developmental changes in the microenvironment of this organelle could help explain why some are targeted to the vacuole and others are retained with the lumen of the ER. The mechanisms involved in this differential targeting and retention currently are not well understood.

The 2S albumins and the 7S and 11S globulins of legume and other dicot seeds are transported to the vacuole through the Golgi (Fig. 19.46; see also Chapter 4). As storage proteins are deposited within the vacuole, the vacuole is thought to fragment to form many smaller protein bodies. Assembly of the storage proteins in the protein bodies is a highly ordered event, with different steps occurring within the ER, Golgi, and protein body itself (Table 19.3; see also Fig. 19.45).

19.10.9 Transcription of storage protein genes is regulated temporally and spatially.

The synthesis of reserve proteins occurs within a discrete period during seed development (Fig. 19.47). During this time, mRNA transcripts encoding the proteins are present exclusively in the storage tissues of the seed. In developing soybean cotyledons, mRNAs for different subunits of the same vicilin protein, β-conglycinin, appear at different times during development. Expression of the genes that encode storage proteins or their constituent polypeptides is regulated both spatially and temporally, so that the mRNAs coding for these proteins are transcribed only in specific cells of the storage

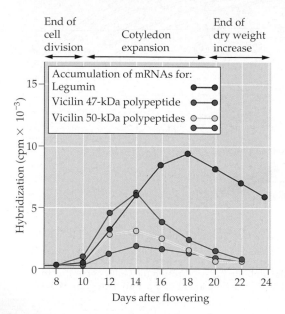

Figure 19.47
Graph illustrating relative abundance of messenger RNAs for one 11S legumin and three 7S vicilin storage polypeptides in developing pea seed cotyledons. RNA from cotyledons at various stages of development was subjected to RNA blot hybridization with appropriate radiolabeled cDNA probes.

tissues (e.g., those for prolamins are transcribed in the starchy endosperm of cereals but not in the surrounding aleurone layer, which is also part of the endosperm).

Our understanding of the regulation of storage protein genes has resulted largely from their expression in seeds of transgenic plants. The transgene constructs used include fusions between (a) the upstream regulatory regions (promoters) of storage protein genes and reporter genes, (b) the coding regions of storage protein genes and other regulatory sequences, and (c) promoter deletion mutants and reporter genes (see Chapter 7). Generally, storage protein genes driven by their own promoters are expressed only in the appropriate storage tissue of transgenic seeds; for example, legume storage protein genes transferred into tobacco are expressed mostly in the cotyledons and at the correct time during development. Thus the regulatory regions of different storage protein genes all carry the same information and can all be recognized by similar transcription factors in

the nuclei of widely different species. Sequences that control the temporal regulation of expression and determine seed and tissue specificity have been identified for a few storage proteins. This research has already revealed that more than one promoter element is required for correct temporal and spatial gene expression. Part of the promoter operates to prevent expression in nonseed tissues.

Detailed analysis of the promoter of the gene that encodes β-phaseolin (a vicilin from the french bean, *Ph. vulgaris*) has revealed several target sequences (cis-elements) to which specific DNA-binding proteins (transacting factors) bind to activate or repress gene expression (Fig. 19.48). Two regions appear to affect temporal expression. A seed-specific enhancer up-regulates expression late in development, whereas a minimal promoter drives low expression early in development. Some elements involved in temporal expression can be deleted without altering spatial expression. The major regulatory elements

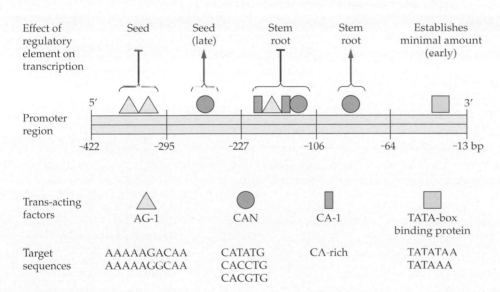

Figure 19.48
Temporal and spatial regulatory elements in the 5′ upstream promoter region of the β-phaseolin gene. Expression of the β-phaseolin gene requires activation of elements in the proximal (–295 to +20 bp) region, although distal elements extending as far as –1470 bp upstream can modulate this expression. Within the proximal region, distinct elements can be recognized. One upstream element (–422 to –295 bp) is involved in down-regulation of expression in the seed. A 68-bp seed-specific enhancer (SSE; –295 to –227 bp) strongly up-regulates expression in the seed late in development. A middle region (–227 to –106 bp) down-regulates gene expression in the stem and root of the vegetative plant. Further downstream is a basal β-phaseolin promoter (*phas*; –106 to +20 bp) that includes two cis-elements. One element (–106 to –64 bp) moderately up-regulates expression in the stem and root, whereas the other, a minimal promoter immediately upstream from the coding sequence (–64 to –13 bp), establishes a small amount of gene expression during early development. The middle region and SSE interact synergistically to activate gene expression; in combination, they promote about 300 times more *GUS* gene expression than either does alone. The SSE and *phas* interact to determine within which tissues the seed gene expression occurs. Each positive or negative regulatory element binds one or more putative transcription factors at defined nucleotide sequences.

identified in the upstream regions of storage protein genes are detailed in Table 19.4. Certain trans-acting factors can bind to the DNA in several places because the regulatory region contains repeats of the target sequence they recognize. The concomitant activation of different elements within the promoter region may be necessary to obtain full expression of a particular storage protein gene.

In developing seeds of some species, the synthesis of several proteins, including storage proteins, is positively influenced by the growth regulator abscisic acid (ABA). For example, mRNAs for 2S napin and 11S cruciferin from rapeseed and the β-subunit of 7S β-conglycinin from soybean increase when seeds are treated with ABA. In cereals, ABA tends to regulate protein synthesis within the embryo, rather than storage protein synthesis in the endosperm. In some dicots, ABA induces synthesis of proteins associated with seed maturation, e.g., late embryogenesis abundant (LEA) proteins, rather than seed storage proteins. An ABA-responsive element (ABRE; see Chapter 22) has been identified in the regulatory region of several storage protein genes, which frequently includes G-box motifs (Table 19.4) that bind bZIP transcription factors (see Chapter 7). These motifs have been identified in other, non–ABA-regulated genes, however, and probably other elements outside of the core G-box also are involved in regulation by ABA.

19.10.10 Expression of storage proteins is also subject to posttranscriptional controls.

Seed storage protein synthesis is controlled mainly at the level of transcription, but posttranscriptional regulation can also occur. A particularly striking example of this is seen in developing oat grains: The predominant endosperm storage protein, a globulin,

Table 19.4 Sequences and possible functions of some conserved regulatory elements associated with genes that encode seed storage proteins

Regulatory element	Structural gene	Functions
CATGCATG (RY element)	Seed-specific genes Storage protein (SP) genes in legumes and some cereals	Essential for expression of some 11S and 7S SP genes, alone or in combination with other elements Suppresses SP expression in vegetative tissues Trans-acting factor not identified
TGTAAAG (–300 box)	Prolamin and legumin genes	Not essential for SP expression May affect expression quantitatively Trans-acting factor known
CACA (CA-rich box)	Legume and cereal SP genes Seed and vegetative tissue genes	General regulatory function Not essential for all SP genes Trans-acting factor is CA-1 nuclear protein (see Fig. 19.48)
GCCAC(C/T)TC (Octanucleotide box)	7S, 11S, and lectin SP genes	Confers seed-specific expression and suppresses transcription in stems and roots May affect temporal expression Putative trans-acting factor identified
AA/G/CCCA	7S soybean SP gene	Role in seed SP gene expression not confirmed Trans-acting factor known
T/CACGTG/A/C	Seed-specific and vegetative genes Zein SP genes	Essential for gene expression Binds transcription factors O2 (bZIP) in maize and CAN in bean (see Fig. 19.48)
Vicilin box I	Vicilin genes of legumes only	May determine temporal or cell-specific patterns of expression
Vicilin box II	Most vicilin genes of legumes	Not essential for SP gene expression Binds TEMP-1 trans-acting factor
Legumin box	Legumin genes of legumes only	CATGCATG motif in legumin box important for seed-specific expression

accounts for as much as 80% of the total protein, but its mRNA is less abundant than that for the minor prolamin storage protein. The predominance of the globular protein might indicate that (*a*) its mRNA is translated more efficiently, (*b*) the protein itself is processed more efficiently, (*c*) the protein is more stable, or (*d*) some combination of these factors is operating. Timing of the appearance of storage proteins in developing seeds can also be modulated posttranscriptionally. Accumulation of the mRNA for the α'/α-subunit of conglycinin in soybean cotyledons commences before accumulation of the β-subunit mRNA, yet transcription of both mRNAs occurs at similar rates, beginning and stopping simultaneously. Differential stability of the mRNA may account for the differences, resulting in later accumulation and longer persistence of the more stable mRNA for the β-subunit. Regulation of storage protein turnover after translation is another potential posttranscriptional control point.

Developing seeds of plants grown in an S-deficient environment respond by altering the ratio of the S-rich and S-poor proteins they synthesize. In legumes, the S-rich 2S albumin and 11S legumin proteins decrease relative to the S-poor 7S vicilins, which increase to maintain seed protein content. A similar change in the proportions of S-rich and S-poor proteins occurs in cereals and nonlegume dicots and involves both transcriptional and posttranscriptional events. Two small S-rich albumins are synthesized in the cotyledons of developing pea seeds during their development. Both are encoded by a single gene (*PA1*), the product of which is cleaved posttranslationally to yield two albumins (PA1a and PA1b). In S-sufficient conditions, expression of *PA1* is under transcriptional control, and the amount of PA1a and PA1b produced is related to the amount of *PA1* mRNA transcribed. Under S-deficient conditions, accumulation of PA1a and PA1b is down-regulated by a decrease in the abundance of *PA1* mRNA, a consequence of reduced posttranscriptional stability of the mRNA rather than an altered rate of *PA1* gene transcription. In contrast, the increase in synthesis of S-poor 7S vicilin in developing pea seeds in S-deficient conditions reflects increased transcription of its gene. During recovery from S deficiency, transcription of the vicilin gene decreases substantially.

19.10.11 Phytate synthesis is not well understood.

Mineral nutrients are stored in all seeds, most commonly as **phytate,** a mixed cation salt of *myo*-inositolhexaphosphoric acid (phytic acid). Making up about 1% to 8% of the dry weight of mature seeds, phytate accounts for as much as 90% of their total phosphorus content. Associated with the phosphate groups of phytic acid are Mg^{2+} and K^+ and, to a lesser extent, Ca^{2+}, Mn^{2+}, Ba^{2+}, and Fe^{2+}. Phytate is stored within protein bodies, dispersed throughout the proteinaceous matrix in some plant species and as discrete electron-dense aggregates called globoids or globoid crystals in others.

The biosynthetic pathway for phytic acid is not known. Three models have been proposed, albeit without any substantive supporting evidence (Fig. 19.49). The site of phytate synthesis also is unknown, although light and electron microscopy studies of the developing castor bean endosperm suggest that phytate is synthesized in association with the rough ER and accumulates initially in the lumen. Phytate may then be transported by Golgi-derived vesicles to the vacuoles/developing protein bodies in the same manner as storage proteins are transported (Fig. 19.50).

19.11 Embryo maturation and desiccation

19.11.1 ABA-induced protein synthesis is thought to play a role in establishing desiccation tolerance.

Besides storage proteins, other proteins are synthesized in developing seeds at specific times of development. Of particular interest are the ABA-induced genes associated with seed maturation and acquisition of desiccation tolerance. Some of these genes encode small heat-shock proteins and LEA proteins (see Chapter 22). Synthesis of these proteins increases as seed development proceeds beyond the torpedo stage, when synthesis of storage proteins is declining. *LEA*-related genes contain ABA-regulatory elements (ABREs) with a core sequence of CACGTG (G-box) to which bZIP-type DNA proteins bind specifically. LEA-type proteins

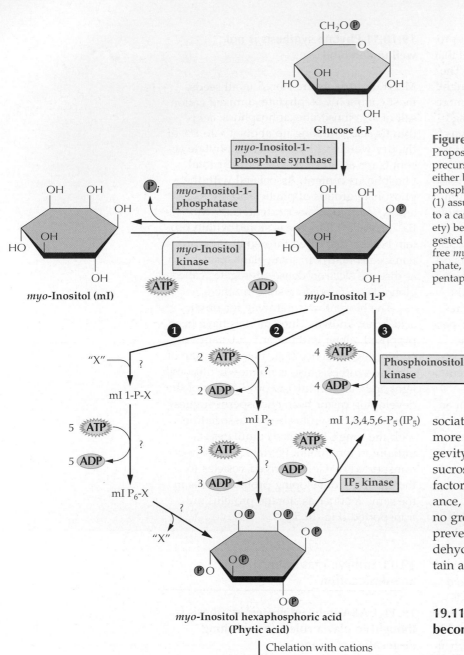

Figure 19.49
Proposed pathways for phytate biosynthesis. The precursor, *myo*-inositol 1-phosphate, is synthesized either by cyclization of glucose 6-phosphate or by phosphorylation of free *myo*-inositol. One pathway (1) assumes that *myo*-inositol 1-phosphate is bound to a carrier molecule (perhaps a membrane lipid moiety) before phosphorylation. In the other two suggested pathways, ATP-dependent phosphorylation of free *myo*-inositol 1-phosphate yields the hexaphosphate, possibly with a stable triphosphorylated (2) or pentaphosphorylated (3) intermediate.

sociated with desiccation tolerance, although more probably, it is important for the longevity of the mature dry seed. The ratio of sucrose to raffinose might be a determining factor: For maintenance of desiccation tolerance, the ratio in the embryo cells must be no greater than 20:1. Raffinose appears to prevent sucrose crystallization during cell dehydration, allowing the cytoplasm to retain a stable, glassy state.

19.11.2 During desiccation, seed tissues become less sensitive to ABA.

ABA is important for maintaining some of the developmental and maturation processes in seeds, e.g., synthesis of storage and LEA proteins. A decrease in both ABA content and sensitivity to ABA occurs as seeds dry out, for reasons still to be elucidated. On rehydration, mature dry seeds, or even seeds desiccated prematurely during development, switch from a developmental program to a germination program. Development-related mRNAs are no longer transcribed, but those associated with germination and postgermination mobilization of reserves are activated (Fig. 19.51).

Such a complete switch in gene expression as a consequence of maturation drying

frequently are expressed in nonseed tissues in response to environmental stress (see Chapter 22); their importance in seeds might be for the protection of other proteins and membranes. LEA proteins might solvate these cellular components by forming amorphous coils, protecting cell contents from disruption or damage in the near-dry state.

An increase in soluble nonreducing oligosaccharides also occurs in seeds during late development. In particular, accumulation of raffinose (see Chapter 15) may be as-

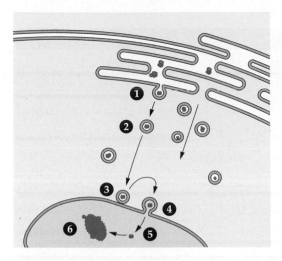

Figure 19.50
Schematic representation of phytin synthesis and deposition of phytin in protein bodies of developing endosperm cells of castor bean. Phytin is released in vesicles (1, 2) from the ER and is targeted to the vacuole/protein body where it fuses with the membrane (3, 4). Phytin is released into the vacuole/protein body (5), where it coalesces to form a globoid (6).

is probably brought about in several ways, but evidence suggests that the upstream promoter region of some genes that encode storage proteins itself becomes insensitive to ABA after drying. Chimeric reporter genes under the control of a promoter region from a storage protein gene are expressed in the seeds of transgenic tobacco plants during their development but are repressed when the seeds are subjected to premature or maturation desiccation. Chimeric constructs that comprise a viral promoter with no known sensitivity to desiccation or ABA and a storage protein gene are not suppressed by drying in the transgenic seed.

If mature seeds of alfalfa are permitted to complete germination without maturation drying, they remain responsive to ABA, and their cotyledons will even recommence synthesizing storage proteins in seedlings treated with ABA at a time when storage proteins ordinarily are mobilized, not synthesized.

19.12 Germination

By definition, germination commences when the quiescent dry seed begins to take up water (imbibition) and is completed when the embryonic axis elongates. The visible sign that germination is completed is usually the penetration by the radicle of structures surrounding the embryo. Once germination is complete, the reserves within the storage tissues of the seed are mobilized to support seedling growth. Thus, mobilization of storage reserves is a postgermination event (Fig. 19.52).

As a result of the rapid influx of water during imbibition, solutes and low-molecular-mass metabolites leak from the cells. This is symptomatic of the unstable configuration of the cell membranes, particularly the plasma membrane, being perhaps in a gel phase or a liquid–crystalline phase (see Chapters 1 and 10) associated with protective sugar or proteins. Shortly after rehydration, the stable liquid–crystalline configuration is resumed, and leakage ceases.

The imbibing seed resumes metabolic activity within minutes of the water entering its cells, initially using structural and enzymatic components synthesized during development and conserved in the dry state. The mitochondria responsible for respiratory ATP synthesis are initially few in number and poorly differentiated in structure, a consequence of cell desiccation and rehydration, although they contain sufficient citric cycle enzymes and terminal oxidases

Hormonal level
• Decline in ABA production
• Increase in ABA destruction/sequestration
• Decline in ABA sensitivity

Genomic level
Storage protein, *Lea* genes, etc.:
• Transcriptional inactivation
• Loss of promoter activity
• Loss of trans-acting factors
Germination/postgermination genes:
• Transcriptional activation
• Activation of promoters
• Synthesis of trans-acting factors

Cytoplasmic level
• Destruction of residual developmental mRNAs
• Translation of stored and newly synthesized germination-related mRNAs
• Expression of postgermination genes

Figure 19.51
Changes occurring to seed protein synthesis as a consequence of both premature and maturation drying, which result in a switch from a developmental to a germinative mode of metabolism.

Figure 19.52
Time course of some important events associated with germination and subsequent postgermination growth. During phase I, water is taken up by the seed (imbibition). Water uptake reaches a plateau (phase II), during which metabolism proceeds and repair is completed to cellular components damaged in the processes of maturation drying and rehydration. After germination, the fresh weight of the seed increases as it grows (phase III) and develops into a seedling, during which time the major storage reserves deposited during development are mobilized.

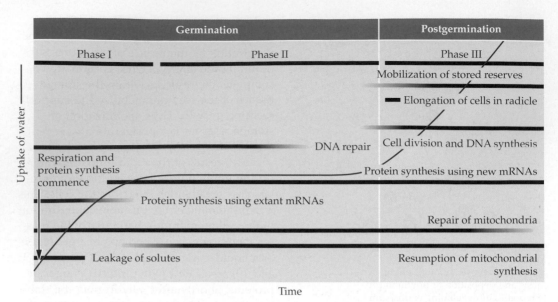

(see Chapter 14) to support the requirements for early germination. In some seeds, preexisting mitochondria are repaired and activated; in others, biogenesis of new organelles occurs, involving nuclear and organellar genes. Resumption of protein synthesis is another very early event during imbibition, utilizing machinery that was conserved in the dry seed (e.g., ribosomes, mRNAs, and cytoplasmic initiation and elongation factors; see Chapter 9).

As imbibition time elapses, extant RNAs are replaced by de novo transcription and turnover. New mRNAs are produced as germination proceeds, as are transcripts encoding proteins that support basic metabolism. Although some proteins are unique to seed development (e.g., storage proteins and enzymes associated with the synthesis of the other storage reserves), none has been identified that is unique to, or essential for, germination. Thus, the fundamental bases of this very important event in the life cycle of the plant are unknown. Moreover, the processes required for cell growth and wall expansion in the embryonic axis during the completion of germination remain to be elucidated.

Summary

The onset of flowering in plants represents a major developmental shift in the plant life cycle. This shift is controlled by environmen-

tal cues, physiological changes, and alterations in gene expression. Many of the genes involved in the induction of flowering have been identified in model organisms. At the shoot apex, a small number of floral meristem identity genes control the positioning and identity of flowers. In *Arabidopsis*, flowers are organized into four whorls. The organs on each of these whorls are specified by combinations of transcription factors, which are encoded by MADS box genes. These genes originally were identified by studying homeotic mutant phenotypes in which whorl identity is altered. The MADS box genes that control the patterning of floral organ identity evidently are conserved among angiosperms.

Gametes must be formed to complete the plant life cycle. In the anthers, sporogenous cells undergo meiosis and each resulting haploid cell eventually develops into a male gametophyte (pollen grain). In the ovules, sporogenous cells undergo meiosis and eventually form the female gametophyte (embryo sac). Many genes are expressed only during gametophyte development. When pollen grains contact the stigmatic tissues of the gynoecium, they hydrate and then extrude a pollen tube, growing though stylar tissue to deliver the two sperm cells to the embryo sac. Typically, one sperm cell nucleus fuses with the egg cell nucleus and the other fuses with the nuclei of the central cell. Genetic analysis of mutants and molecular analysis of gene expression have contributed to our understanding of gametophyte development and pollen–pistil interactions.

Seed development proceeds from the single-celled fertilized egg to a multicellular structure comprising an embryo, which is the next generation of the plant; the seed coat (testa); and a nutritive endosperm that persists in the mature dry seed of some species. Many genes are involved in the regulation of seed development. The roles of some of these genes have been elucidated by using developmental mutants. An important aspect of seed development is the deposition of the stored reserves—carbohydrates, proteins, and oils—that are utilized after germination to support early seedling growth. The terminal phase of seed development is desiccation, which results in a mature dry seed. On subsequent imbibition of water, metabolism is renewed as the seed commences germination, which is completed with the emergence of the radicle.

Further Reading

Bewley, J. D. (1997) Seed germination and dormancy. *Plant Cell* 9: 1055–1066.

Bewley, J. D., Black, M. (1994) *Seeds: Physiology of Development and Germination.* Plenum, New York.

Bowman, J. L., Smyth, D. R., Meyerowitz, E. M. (1991) Genetic interactions among floral homeotic genes of *Arabidopsis. Development* 112: 1–20.

Bradley, D., Ratcliffe, O., Vincent, C., Carpenter, R., Coen, E. (1997) Inflorescence commitment and architecture in *Arabidopsis. Science* 275: 80–83.

Drews, G. N., Lee, D., Christensen, C. (1998) Genetic analysis of female gametophyte development. *Plant Cell* 10: 5–17.

Gasser, C. S., Broadhvest, J., Hauser, B. A. (1998) Genetic analysis of ovule development. *Annu. Rev. Plant Physiol. Plant Mol. Biol.* 49: 1–24.

Goldberg, R. B., de Paiva, G., Yadegari, R. (1994) Plant embryogenesis: zygote to seed. *Science* 266: 605–614.

Kigel, J., Galili, G., eds. (1995) *Seed Development and Germination.* Marcel Dekker, New York.

Lang, A. (1965) Physiology of flower initiation. In *Handbuch der Pflanzenphysiologie (Encyclopedia of Plant Physiology),* W. Ruhland, ed., Springer-Verlag, Berlin, pp. 1379–1536.

McCormick, S. (1998) Self-incompatibility and other pollen–pistil interactions. *Curr. Opin. Plant Biol.* 1: 18–25.

Meyerowitz, E. M., Bowman, J. L., Brockman, L. L., Drews, G. N., Jack, T., Sieburth, L. E., Weigel, D. (1991) A genetic and molecular model for flower development in *Arabidopsis thaliana. J. Cell. Biochem.* (Suppl. 15).

Okamuro, J. K., den Boer, B.G.W., Jofuku, K. D. (1993) Regulation of *Arabidopsis* flower development. *Plant Cell* 5: 1183–1193.

Plant Cell. (1993) Vol. 5, No. 10. Special issue on plant reproduction.

Schneitz, K., Balasubramanian, S., Schiefthaler, U. (1998) Organogenesis in plants: the molecular and genetic control of ovule development. *Trends Plant Sci.* 3: 468–472.

Scott, R. J. (1994) Pollen exine—the sporopollenin enigma and the physics of pattern. *Soc. Exp. Biol. Sem. Ser.* 55: 49–81.

Sexual Plant Reproduction. (1996) Vol. 9, No. 6. Special issue.

Smyth, D. R., Bowman, J. L., Meyerowitz, E. M. (1990) Early flower development in *Arabidopsis. Plant Cell* 2: 755–767.

Taylor, L. P., Hepler, P. K. (1997) Pollen germination and tube growth. *Annu. Rev. Plant Physiol. Plant Mol. Biol.* 48: 461–491.

Weber, H., Borisjuk, L., Wobus, U. (1997) Sugar import and metabolism during seed development. *Trends Plant Sci.* 2: 169–174.

Yanofsky, M. F. (1995) Floral meristems to floral organs: genes controlling early events in *Arabidopsis* development. *Annu. Rev. Plant Physiol. Plant Mol. Biol.* 46: 167–188.

Biochemistry & Molecular Biology of Plants, B. Buchanan, W. Gruissem, R. Jones, Eds.
© 2000, American Society of Plant Physiologists

CHAPTER 20

Senescence and Programmed Cell Death

Jeffery L. Dangl
Robert A. Dietrich
Howard Thomas

CHAPTER OUTLINE

Introduction

Paradoxically, the death of specific sets of cells is an essential part of the growth and development of many eukaryotic organisms, including plants and animals. In addition to its role in development, cell death can be one component of the response to biotic and abiotic stresses. Because the organism controls the initiation and execution of the cell death process, these types of cell death are referred to as **programmed cell death** (PCD). This broad definition of PCD, however, implies nothing about the mechanisms involved in the execution of cell death.

Two examples of PCD in plants, **senescence** and the cell death associated with the **hypersensitive response** (HR), demonstrate the range of forms that PCD can take in plants. Senescence is the relatively slow cell death of tissues at the end of their life span. Senescence involves the ordered disassembly of cellular components in the senescing tissue and allows for maximum recovery of nutrients from the senescing tissues for recycling to the parts of the plant that survive. The PCD seen in the HR, the cell death that is triggered in plant cells in and around the point of attempted infection by some pathogens, is quite different. Because one function of the localized cell death in HR might be to block the further spread of infection, the emphasis is on rapid execution, to kill the host tissue before the pathogen gets established, rather than maximum recovery of nutrients. These are just two examples of PCD that occur in plants, but they demonstrate how diverse PCD processes in plants can be.

In this chapter, the current state of knowledge regarding PCD in plants will be discussed to provide an understanding of how death, at the appropriate time and place, can in fact be an essential part of life.

20.1 Types of cell death observed in animals and plants

20.1.1 Apoptotic cell death contributes to the morphological and physiological development of animals.

The recent upsurge in the study of cell death in eukaryotes was sparked in the early 1970s by the research of John Kerr, Andrew Wyllie, and their colleagues, who were investigating the way animal cells died. They identified two major types of animal cell death: **apoptosis,** a highly regulated, energy-dependent process, and **necrosis,** which results from trauma. Apoptosis in animals is recognized as a ubiquitous phenomenon with distinct morphological hallmarks (Fig. 20.1), including the formation and pinching off (blebbing) of vesicles from the outer surface of the plasma membrane and the nuclear envelope, chromatin condensation, cleavage of DNA into characteristic 180-bp internucleosomal fragments, and formation of **apoptotic bodies** (Box 20.1), which are engulfed and degraded by neighboring cells. In animals, cells that have been physically damaged die by necrosis, a process that results in rupture of the plasma membrane and endomembranes, causing hydrolytic enzymes and other mate-

rials to be released rapidly (Fig. 20.1). Release of enzymes after a trauma results in inflammation, whereas cells that die by apoptosis do not rupture and therefore do not cause inflammation.

More than two types of cell death occur in animals and plants. Cells that die by processes that require energy and are regulated by distinct sets of genes are said to undergo PCD. Among the instances of PCD known to occur in animal embryology and physiology are the following:

- Death of interdigital tissues during embryonic development of the mammalian hand, which originates as a webbed structure. Cells are removed from the webbing by apoptosis to give rise to fingers and toes. In birds, this process is what distinguishes a chicken's foot from the webbed foot of a duck.
- Liquidation, preceding reassembly, of insect muscle cells during metamorphosis
- Resorption of tadpole tail in amphibian development
- Normal development of the nematode *Caenorhabditis elegans*, during which 131 specific cells of the 1090 that make up the mature animal die according to a precise schedule

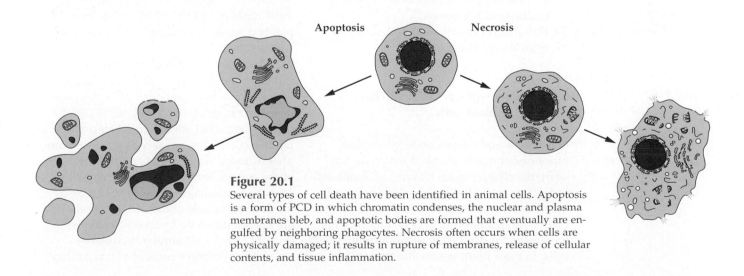

Apoptosis   **Necrosis**

Figure 20.1
Several types of cell death have been identified in animal cells. Apoptosis is a form of PCD in which chromatin condenses, the nuclear and plasma membranes bleb, and apoptotic bodies are formed that eventually are engulfed by neighboring phagocytes. Necrosis often occurs when cells are physically damaged; it results in rupture of membranes, release of cellular contents, and tissue inflammation.

Box 20.1   **Apoptosis is one type of PCD.**

Apoptosis is a form of PCD that has been described in animals ranging from nematodes to mammals. Although it is not yet clear whether apoptotic cell death also occurs in plants, the word apoptosis ironically has its origins in botany. Pronounced a-po-to′-sis, this is the Greek word used to describe the dropping of petals from flowers and leaves from trees.

In animals, apoptotic cell death, an ordered process characterized by a distinct set of morphological features, is seen during normal development as well as in response to certain pathogens. Cells undergoing apoptosis shrink and the contents of the cytoplasm condense, although the integrity of the plasma membrane and the organelles is retained. The nuclear DNA also condenses and is cleaved into fragments of approximately 50 kb each. Nuclear DNA often, but not always, is digested further by endogenous Ca^{2+}-dependent endonucleases to form oligonucleosomal fragments. These fragments, which are diagnostic for some, but not all, forms of apoptosis in animal cells, can be visualized on a gel and appear as a ladder of DNA with fragment sizes in multiples of approximately 180 bp. The photograph at right illustrates DNA degradation in thymocytes (lanes, 1, 3, and 5) and lymph node nuclei (lanes 2, 4,

and 6) as catalyzed by Ca^{2+}- and Mg^{2+}-dependent enzymes (lanes 1 and 2). The presence of the Ca^{2+} chelator EGTA (lanes 3 and 4) or the Zn^{2+} (lanes 5 and 6) results in no DNA laddering. The fragmentation of the nuclear DNA results in the breakdown of the chromatin structure. At the same time, the nucleus breaks up and the cell itself is converted into apoptotic bodies, small membrane-bound structures containing the debris of the animal cell nucleus. These bodies migrate to the margin of the cell and are taken up by adjacent cells.

In contrast to apoptosis, necrotic cell death in animals is characterized in the early stages by chromatin clumping and organelles swelling. Finally, the plasma membrane breaks down and the cell disintegrates. Most cellular debris is engulfed by professional phagocytes.

Caspases, a type of cysteine protease involved in apoptotic cell death in animals, participate both in initiation of the apoptosis program and in disassembly of cellular components. Synthesized as inactive precursors, caspases are activated by proteolytic cleavage—primarily autocatalytic, triggered by interactions with cofactors or removal of inhibitors. In some cases, caspase activation involves cleavage by other caspases, such that a caspase

cleavage cascade participates in apoptotic signaling. Caspases have very high substrate specificity, cleaving on the C-terminal side of aspartate residues only; specific secondary and tertiary structural elements also are required. No caspase-type cysteine protease genes or proteins have been identified in plants to date.

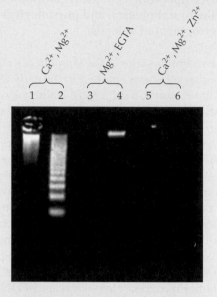

- Normal embryonic development of brain and eyes as well as other organ systems in mice
- Physiological regulation of tissues such as the placenta
- Pathological responses to mutagens, toxins, and hormones
- Elimination of self-reactive leukocytes from the immune system

20.1.2 Many types of cell death have been observed in plant cells.

Plants cells and tissues undergo various types of cell death, many of which do not share the characteristic features of apoptosis. Indeed, as we will discuss below, apoptosis may be rare in plants. Apoptosis and phagocytosis by neighboring cells cannot provide the means for degrading dead or dying plant cells. In most plant tissues the presence of

cell walls precludes the absorption of apoptotic bodies; even if cell walls were absent, however, plant cells do not contain phagocytes. Furthermore, chromatin condensation and DNA laddering, both characteristic features of apoptosis, do not accompany most types of PCD in plants. How then do plants dispose of unwanted cytoplasm or whole cells?

One way in which cultured plant cells can degrade their contents is **autophagy,** which in plants is similar to what has been described in detail in baker's yeast (*Saccharomyces cerevisiae*) after nutrient starvation. During autophagy, yeast cells produce **autophagosomes,** vesicles that engulf portions of the cytosol, including intact organelles. Then the autophagosomes are taken up by the central vacuole of the cell, where they are broken down by hydrolytic enzymes (Fig. 20.2A). A very similar picture has emerged in certain tissues of plants as they

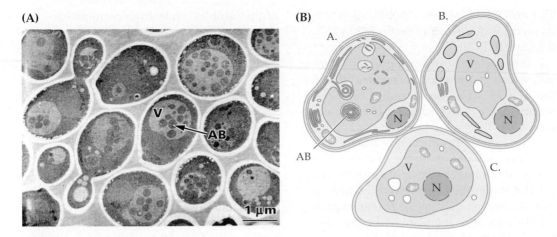

(A)

(B)

Figure 20.2
(A) *Saccharomyces cerevisiae* undergoes autophagy when starved of nutrients. The yeast shown in this figure was nitrogen-starved, bringing about autophagy, in which autophagic bodies (AB) engulf the cytoplasm and transport it to the vacuole (V), where proteases and other lytic enzymes release nitrogen to compensate for the lack of nitrogen in the medium. (B) Autophagy occurs in dying plant cells as in the senescing *Ipomoea tricolor* corolla cells diagrammed here. Autophagic bodies engulf a portion of cytoplasm, including the nucleus (N), to the vacuole(s), where breakdown of the cell's contents occurs. The three cells shown represent different stages of PCD, A being the earliest, and B and C later stages of autophagy.

undergo senescence and in cultured plant cells subject to nutrient starvation. Cells of the senescing corolla of Japanese morning glory (*Ipomoea tricolor*) contain autophagosomes, which have been found to fuse with the tonoplast and release their contents into the vacuole, where they are broken down by vacuolar enzymes (Fig. 20.2B). Another form of autophagy occurs in the cereal aleurone, in which the number of organelles is dramatically reduced as the central vacuole increases to occupy almost the entire cellular volume before cell death (Fig. 20.3). How the cellular organelles are disposed of during this type of autophagy has not been determined; autophagic vacuoles do not appear to participate.

A third way in which the plant cell protoplast is disposed of has been observed in tracheid differentiation. During the formation of tracheary elements (TE), which are dead at functional maturity, the degradation of organelles and other cellular contents occurs when rupture of the tonoplast releases hydrolases (e.g., proteases, nucleases, and phosphatases). This process is described in detail in Sections 20.12.1 through 20.12.4, illustrated with cultured *Zinnia* cells as a model.

Some types of PCD may be unique to plants. The formation of the cereal endosperm is an example. Early in grain formation, the cereal endosperm differentiates into two tissues: the starchy endosperm and the surrounding aleurone layer. Starchy endo-

sperm cells accumulate starch and storage proteins during grain maturation. When the grain reaches harvest ripeness, endosperm cells die, whereas the surrounding aleurone cells remain alive. The unique aspect of starchy endosperm PCD is that the dead cell retains all of its contents, including nuclei and organelles. Endosperm cell **corpses** become **mummified** and do not undergo degradation until germination and seedling growth are initiated, often months or years later, when enzymes secreted from the aleurone layer attack the

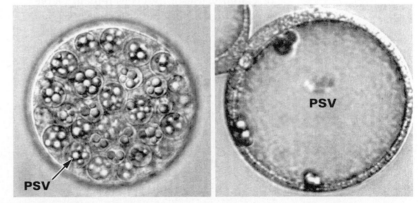

Figure 20.3
PCD in the cereal aleurone is accompanied by autophagy and extensive vacuolation of the cytoplasm without rupture of membranes or the formation of autophagic vacuoles. In the barley aleurone cell (left) small protein storage vacuoles (PSV) lose their stored protein and fuse to form one large central vacuole (right). Membrane integrity is maintained until death ensues.

mummified endosperm. Mobilization of endosperm is discussed in Sections 20.12.5 through 20.12.10.

It is unclear how death proceeds during HR, which is discussed in Sections 20.13.4 through 20.13.16. The currently understood sequence of events includes pathogen recognition, ion fluxes and influx of calcium ions into the cytosol, production of reactive oxygen intermediates and nitric oxide, and activation of a mitogen-activating protein kinase (MAPK)-controlled signal transduction cascade. However, the specific events required for HR vary with the plant and pathogen species involved. Some evidence suggests the involvement of caspases (see Box 20.1) but this is far from clear. The morphological hallmarks of animal apoptosis are found in some but not all instances of HR.

20.1.3 Plant cells remain viable during most of the developmental program that leads to cell death.

An understanding of PCD requires making a clear distinction between the processes that lead to death, during which time the cells are viable, and the final act of death. Experiments in tobacco indicate that the phase preceding death can be arrested or even reversed (see Section 20.10.2). This contrasts sharply with animal systems, in which the cell death program cannot be stopped once it is initiated.

Senescence of the green (mesophyll) cells of leaves illustrates this prolonged period of viability. Mesophyll cells remain viable until almost all of the leaf's resources have been exported to other parts of the plant. During salvage of the plastids and other organelles, the senescing leaves remain turgid, indicating the continued integrity of membranes, cell compartments, and water relations. Wilting, which results from loss of cellular turgor followed by cell death, is a late event in the leaf's senescence program. In comparison with the time-scale of the senescence phase, the execution process itself is probably very rapid.

20.2 PCD in the life cycle of plants

Cell death in plants permits developmental and biochemical plasticity. Almost all phases of the plant life cycle, from germination through vegetative and reproductive development, are influenced by PCD (Fig. 20.4). In addition, responses to pathogens and abiotic stresses involve the controlled death of cells.

20.2.1 PCD is essential for normal reproductive development.

Flower development is radically affected by PCD of selected cells or groups of cells. In most plants that have unisexual flowers, the developing flower initially contains primordia for both male and female organs. At early stages, male and female flowers are indistinguishable. During flower formation, at a developmental stage that varies with the species of plant, either the male or female parts cease growing and are eliminated via a cell death program. In maize, for example, the male inflorescence (the tassel) is spatially separate from the female inflorescence (the ear). The young flowers in the tassel contain primordia for both stamens and gynoecium, but as the flower develops, gynoecial cells stop growing and dividing, and the organelles, including the nucleus, break down. In the *tasselseed2* mutant (Fig. 20.5), however, the arrest and degeneration of the gynoecia do not occur, and female flowers are produced in the tassel. From this, we can conclude that the *TASSELSEED2* gene is required for the death of the developing female organs in the tassel. The *TASSELSEED2* gene is expressed in gynoecial cells in the tassel just before the start of gynoecial degeneration. The product encoded by the gene is similar to hydroxysteroid dehydrogenases, which raises the possibility that the protein may regulate cell death by generating a steroid-like molecule that acts as a signal in the cell death pathway.

Haploid tissues of many plants are also influenced by cell death programs. During megagametogenesis in angiosperms, three of the four megaspores that form after meiosis of the megaspore mother cell undergo PCD, leaving one megaspore to give rise to the egg and other components of the embryo sac (see Fig. 20.4, diagram 1). Likewise, PCD plays a role in microsporogenesis, in which the tapetum that surrounds the microsporocytes dies and disintegrates.

After fertilization in most angiosperms, the first mitotic division of the zygote gives rise to two cells: One produces the embryo, the other the suspensor (see Chapter 19). The suspensor may undergo a few rounds of mitosis, but eventually suspensor cells undergo PCD (see Fig. 20.4, diagram 2).

20.2.2 Many aspects of vegetative development are PCD-dependent.

During germination of seeds of monocots and dicots, the endosperm undergoes PCD. As already mentioned, the starchy endo-sperm of cereals forms a unique mummified tissue; in other seeds, however, the endosperm dies after its storage or protective function has been served. In seeds as diverse as those of lettuce and *Arabidopsis*, the endosperm is a structure of one or two cell layers that envelops the embryo. This tissue functions to store reserves of protein, lipid, and carbohydrate, but because of the unique structure of the cell wall, it also acts to restrict growth of the embryo. The weakening of the endosperm during the first few hours of imbibition allows the embryo to expand.

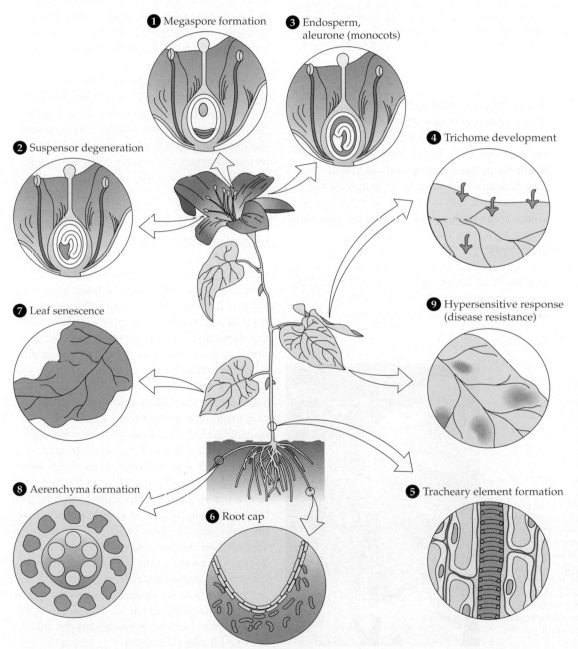

Figure 20.4
Cell death occurs in almost all plant cells and tissues. PCD is involved in numerous processes, including the following illustrated in this figure: gamete formation, including megaspore formation (1); embryo development (2); degeneration of tissues in the seed and fruit (3); tissue and organ development (4 through 6); senescence (7); and responses to environmental signals and pathogens (8 and 9).

❶ Megaspore formation
❸ Endosperm, aleurone (monocots)
❷ Suspensor degeneration
❹ Trichome development
❼ Leaf senescence
❾ Hypersensitive response (disease resistance)
❽ Aerenchyma formation
❻ Root cap
❺ Tracheary element formation

(A)

(B)

Figure 20.5
The inflorescences of maize contain flowers that are initially bisexual, but PCD results in the death of male or female tissues to give rise to female inflorescence (ear) or male inflorescence (tassel), respectively. In the *tasselseed2* mutant (A), female tissues in the tassel do not undergo PCD, and the resulting tassel flowers are mostly pistillate. A wild-type tassel is included for comparison (B).

Eventually, PCD leads to complete breakdown of the endosperm by autolysis.

PCD plays an essential role in the early stages of tissue differentiation, particularly in the development of xylem TE. The formation of mature TE requires that the cytoplasm of the TE initials be removed so that a clear path can be formed for efficient water movement through either tracheids or vessels.

The overall form of many plants also relies on PCD occurring at various times during organ formation. One example of PCD that brings about dramatic changes in form

occurs during development of the leaves of *Monstera*, the Swiss cheese plant (Fig. 20.6). Other aspects of vegetative development are also affected by selective cell death. Many trichomes, prickles, thorns, or spines on the surfaces of leaves and stems are dead at maturity. An extreme example of this is the cacti, in which the green stem functionally replaces leaves and the leaves are reduced to spines.

Various glands on the aerial surfaces of plants arise as a result of cell death. For instance, oil glands on the surface of citrus fruit develop because a group of subepidermal cells undergoing PCD formed a cavity that has filled with essential oil (Fig. 20.7). Botanists use the term **lysigeny** to refer to the disintegration of cells that occurs when new structures (such as oil glands) are differentiating. Lysigeny, with or without cell separation (**schizogeny**), is responsible for the differentiation of secretory ducts, cavities, or canals in many species. For example, the mucilaginous canals found in bud scales of the lime tree *Tilia cordata* arise as a result of lysigenous cell death.

20.2.3 PCD is a component of some plant responses to stress.

One of the more interesting examples of lysigeny occurs when aerenchyma are formed in roots of plants subject to stresses such as hypoxia (see Chapter 22). Aerenchyma formation recently has been studied in detail in corn roots, the details of which will be discussed in Sections 20.13.1 through 20.13.3. One of the unique features of aerenchyma formation is that the cell wall and protoplast are removed from mature roots (Fig. 20.8), forming channels through which air can diffuse from shoot to root.

Another example of PCD in plants as a response to stress is the HR that characterizes a plant's response to attempted infection by a pathogen. If a plant is able to recognize the pathogen, the host cells in the immediate area of the infection site undergo rapid cell death. The death of the host tissue thus may kill the pathogen by depriving it of the living plant tissue it requires for growth. Alternatively, the death of the host tissue may be a consequence of the mechanisms the plant uses to kill the pathogen.

Figure 20.6
One visible example of PCD in plants is seen in the ornamental plant *Monstera deliciosa*. The leaves of this plant exhibit deep indentations and holes in the lamina, which result from the programmed death of specific regions of tissue in the developing leaf primordia. As the leaf expands, these areas are not replaced, and the resulting leaf lamina has the characteristic pattern that inspires the common name, "Swiss cheese plant."

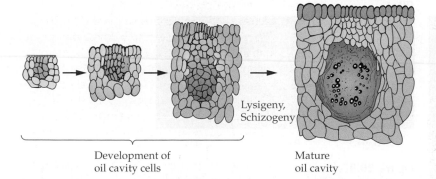

Figure 20.7
Formation of an oil gland in citrus peel occurs by lysigeny, the death and dissolution of the protoplast and cell wall. Sometimes lysigeny is accompanied by schizogeny, wall separation to produce an intercellular space. In the case of oil glands, the combination of lysigeny and schizogeny results in formation of a cavity that becomes filled with essential oils secreted from cells surrounding the cavity.

Lysigeny, Schizogeny

Development of oil cavity cells

Mature oil cavity

20.2.4 Senescence is a PCD-like process associated with the terminal stage of development in both vegetative and reproductive organs.

Senescence and subsequent death are terminal phases in the development of all organs of the plant, including leaves, stems, roots, and flowers (Fig. 20.9). Senescence remodels the form of the plant by disposing of unwanted or inappropriate cells and tissues while simultaneously reclaiming valuable nutrients, especially nitrogen and phosphorus. Senescence generally follows organ maturity and occurs without growth or morphogenesis; it can, however, be dramatically influenced by environmental and endogenous (e.g., hormonal) perturbations. Sometimes senescence can be very rapid: The flower petals of *I. tricolor* senesce after the flower has been open for only one day (Fig. 20.10). In other cases, senescence can take months, even years, to complete. For example, *Pinus longaeva* leaves reportedly have a life span of 45 years, and the single large leaf of the desert plant *Welwitschia mirabilis* survives for decades.

We do not know how the various cellular senescence programs are integrated during the development and life history of organs or whole plants. Perhaps a "die now" signal is constantly present, and a particular cell, tissue, or organ becomes competent to respond to the signal at a time dictated by its own individual developmental program. Alternatively, certain cell-specific "die now" stimuli may invoke senescence or death programs in their targets. The phenotypes of some genetic variants, including certain *"stay-green"* mutants (Fig. 20.11) as well as necrotic hybrids and disease lesion mimics (see Sections 20.13.6 through 20.13.10), ap-

pear to reflect mutations in the genes that presumably regulate the timing or location of normal senescence and PCD. Analyzing these genes should give important insights into how selective cell death is deployed in plant differentiation. In the remainder of this chapter we will discuss in detail both the various forms of PCD found in plants and the mechanisms that control this complex process.

20.3 Overview of senescence

20.3.1 Senescing cells undergo internal reorganization and are metabolically active.

Senescence is a highly regulated process, during which new metabolic pathways are activated and others are turned off. A scheme

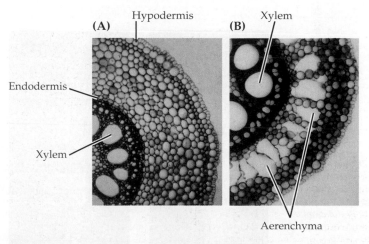

(A) Hypodermis Endodermis Xylem

(B) Xylem Aerenchyma

Figure 20.8
Aerenchyma formation in maize roots in response to hypoxia. Roots were grown under aerobic (A) and hypoxic (B) conditions. Under low-oxygen conditions, cortical cells between the endodermis and hypodermis undergo lysigeny to form air spaces that are continuous throughout the root, thereby allowing submerged roots access to atmospheric gases obtained by above-ground tissues.

(A)

(B)

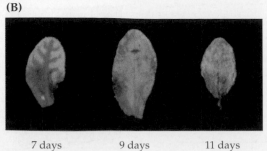

7 days   9 days   11 days

Figure 20.9
Senescence in *Arabidopsis thaliana*. (A) Development of *Arabidopsis* plants is shown at various times after germination. Photographs show plants at 14, 21, 37, and 53 days after germination (left to right). Note the yellowing of shoots of the 53-day-old plant. (B) Age-related changes in rosette leaves of *Arabidopsis* 7, 9, and 11 days after leaf expansion had ceased. Note the progressive yellowing of leaves, beginning farthest from the main veins.

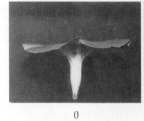

0   1   2   3   4

Figure 20.10
Senescence in the *Ipomoea tricolor* corolla is very rapid. Flowers begin to open at around 0500 to 0600 hours (stage 0) and remain open until 1300 to 1400 (1) the same day. At that time, the corolla begins to curl (2) and the flower changes color from blue to purple. Curling continues during the day until reaching the final stage (4), where the corolla has completely curled inward.

Figure 20.11
Senescence in a wild-type plant and a *stay-green* mutant of the hybrid grass *Festulolium*. Leaves of the mutant (right) remain green, whether senescing attached to the plant or (as here) excised and incubated in darkness, whereas wild-type leaves (left) turn yellow.

outlining the likely steps in the pathway leading to leaf senescence is shown in Figure 20.12. In this scheme, signal transduction cascades linked to various triggers such as hormonal or environmental stimuli lead to an initiation phase involving the activation and inactivation of many genes, followed by the transition to a phase of controlled redifferentiation of cell structures and remobilizing of materials. Finally, a termination phase ensues, in which many of the features of generalized PCD can be recognized. The integrity of subcellular membranes and the compartmentation of biochemical pathways are preserved until late in the terminal phase.

Leaf and fruit senescence are characterized by dramatic changes in the major organelles, particularly the plastids of mesophyll cells and parenchyma of fruit pericarp tissue. Far from undergoing deterioration in structure and function, the chloroplasts of leaf mesophyll cells and green immature fruits redifferentiate into gerontoplasts (Fig. 20.13) and chromoplasts (see Chapter 1, Fig. 1.52), respectively. Subcellular reorganization

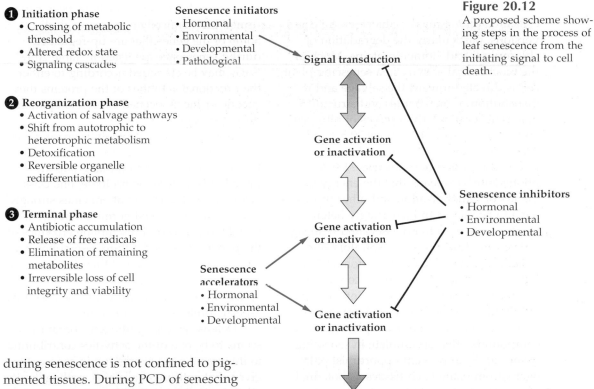

1 Initiation phase
- Crossing of metabolic threshold
- Altered redox state
- Signaling cascades

2 Reorganization phase
- Activation of salvage pathways
- Shift from autotrophic to heterotrophic metabolism
- Detoxification
- Reversible organelle redifferentiation

3 Terminal phase
- Antibiotic accumulation
- Release of free radicals
- Elimination of remaining metabolites
- Irreversible loss of cell integrity and viability

Senescence initiators
- Hormonal
- Environmental
- Developmental
- Pathological

Signal transduction

Gene activation or inactivation

Gene activation or inactivation

Senescence accelerators
- Hormonal
- Environmental
- Developmental

Gene activation or inactivation

Senescence inhibitors
- Hormonal
- Environmental
- Developmental

Cell death

Figure 20.12
A proposed scheme showing steps in the process of leaf senescence from the initiating signal to cell death.

during senescence is not confined to pigmented tissues. During PCD of senescing storage cotyledons and the endosperm of many dicot seeds, and in the aleurone layer of cereal grains, vacuoles change from protein storage organelles to large central vacuoles that maintain the integrity of the tonoplast (see Fig. 20.3). Cotyledons and endosperm tissue of dicots and monocots also store lipids in specialized organelles, called **oleosomes,** which are lost during senescence. Lipid metabolism in storage tissues also is associated with the formation of a new organelle, the **glyoxysome,** which plays an important role in **gluconeogenesis.** In senescing photosynthetic tissues, peroxisomes are converted into glyoxysomes. These changes in cellular compartmentation distinguish differentiation from deterioration and emphasize that senescence is truly a programmed process, not simply a form of necrosis.

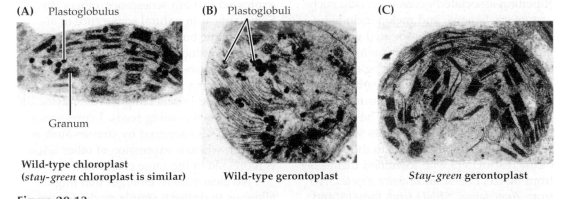

(A) Plastoglobulus

Granum

Wild-type chloroplast
(*stay-green* chloroplast is similar)

(B) Plastoglobuli

Wild-type gerontoplast

(C)

Stay-green **gerontoplast**

Figure 20.13
Impact of senescence on plastid ultrastructure in leaves of wild-type and a *stay-green* mutant of the C3 grass *X. Festulolium.* (A) Prior to senescence, the chloroplasts of a wild-type and mutant plants contain numerous grana, stacks of appressed thylakoid membranes. (B) These internal membrane structures are lost during senescence of a wild-type mesophyll cell, and electron-dense lipid droplets known as plastoglobuli accumulate. (C) Retention of intrinsic thylakoid membrane proteins, pigments, and other hydrophobic components gives the gerontoplasts of mutant tissues a distinctive appearance, with persistent grana stacks and few plastoglobuli.

Ethylene-responsive sequences in the regulatory regions of genes expressed during flower senescence and fruit ripening have been defined by promoter deletion analysis and techniques such as DNase I footprinting to identify trans-acting factors (see Chapter 7, Box 7.5). The promoter of *SEN1*, an *Arabidopsis SAG*, is responsive to dark and to abscisic acid (ABA). Induction of this gene by darkness is suppressed by physiological concentrations of sucrose, glucose, or fructose but not by mannitol or 3-*O*-methyl-D-glucose, carbohydrates that are not metabolized by *Arabidopsis*. The promoters of the cucumber genes encoding the gluconeogenic enzymes malate synthase and isocitrate lyase also have sugar-responsive elements that are distinct from the regions conferring germination specificity.

Promoters of other senescence-enhanced genes also are likely to include regions that are sensitive to the status of other metabolites. For example, expression of GS1, the cytosolic isoenzyme of glutamine synthetase (see Section 20.5.3), is highly responsive to the cellular ratio of glutamine to glutamate. Carbon "starvation" and amide production are characteristic of senescing tissues and are clearly important factors in the coordinated expression of several *SAGs*.

20.3.3 Senescence mutants and variants include plants that are defective in specific senescence-associated enzyme activities as well as plants in which the timing of senescence is altered.

Surveying the genetic variants that modify senescence reveals two broad classes. Members of the first class encode the individual components of the senescence syndrome,

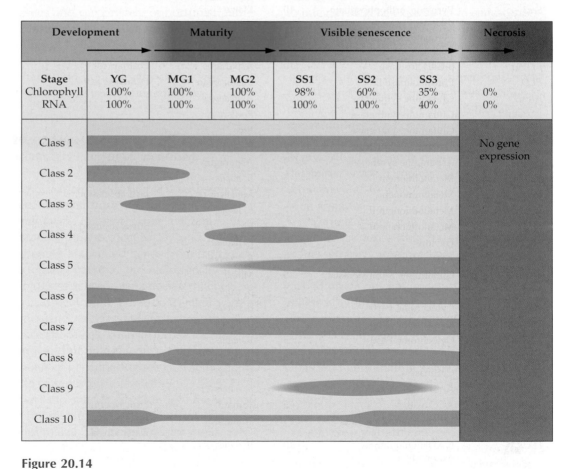

Figure 20.14

Stages in *Brassica napus* leaf senescence. Senescence-associated genes (*SAGs*) are divided into 10 classes according to their temporal pattern of expression. Concentrations of chlorophyll and RNA are quantified as percentages of the concentrations of chlorophyll and RNA present in a mature expanded leaf. YG, fully expanded leaf; MG1, leaves from flowering plants; MG2, leaves from plants with developing siliques; SS1, SS2, and SS3, leaves from plants having 98%, 60%, and 35% of green leaf chlorophyll, respectively.

such as the enzymes of metabolic pathways activated in senescence. Genes in the second class regulate the initiation of the senescence program or its rate of progress. Each class of senescence gene is biologically and practically important in its own way.

Although senescence per se is a relatively recent subject of plant genetic analysis, we realize in retrospect that cotyledon senescence was one of the seven traits investigated by the first geneticist, Gregor Mendel. Pea cotyledons are leaf homologs, and the recessive allele of the gene that Mendel called *B* (now *I*), which specifies green cotyledons, is also expressed as delayed yellowing of leaves during senescence. Plants homozygous for the recessive allele *i* are deficient in the chlorophyll-degrading enzyme pheophorbide *a* oxygenase (PaO; see Section 20.4.1). Other legumes have similar, possibly homologous, genes. For example, the "flageolet" variety of french bean, *Phaseolus vulgaris*, also has green cotyledons, nonyellowing leaves, and a blockage in chlorophyll catabolism. In soybean, as many as 10 separate genetic loci with direct, indirect, or interactive effects on senescence have been described. One of them, *cytG*, is a cytoplasmically inherited gene. Many fruit-ripening mutants are known in numerous species, and some (*greenflesh* in tomato, for example) have pleiotropic effects on foliage.

Genetic variants with delayed leaf senescence are often called "stay-green" mutants (see Figs. 20.11 and 20.13). *Stay-green* cereals and other grasses have economically important agricultural and amenity uses. For example, the record yield of maize grain, almost 24,000 kg ha^{-1}, was achieved in 1985 on a farm in Illinois with the variety FS854, a *stay-green* maize line. Although the factors that contributed to this extraordinary productivity (more than 70% of the theoretical maximum) are complex, the *stay-green* phenotype was an important component. The maize cultivar FS854 and other *stay-green* plants are most likely genetically variant with regard to the overall timing of senescence than to deficiencies in specific enzyme activities.

In crops of the semiarid tropics, the *stay-green* character is of vital importance in the struggle to feed populations living under harsh economic and environmental circumstances. Disease and insect resistance, wide environmental adaptability, appropriate time to maturity, good tillering, and reasonably good grain yield as well as juiciness, palatability, and digestibility of crop residues are the major criteria for improving the sorghum and millet crops of the semiarid zone. Sorghum and millet lines exhibiting the *stay-green* phenotype often show enhancement of other desirable characteristics as well.

A *stay-green* mutation of the locus *sid* in grass species of the genera *Lolium* and *Festuca* has been an important tool for establishing the biochemical pathway of chlorophyll degradation. By conferring unlimited greenness, this allele is a valuable gene for improving turf grasses. The gene also is useful in forages because foliar tissues harvested from plants expressing *sid* retain their chlorophyll, which acts to stabilize high-quality leaf protein (see Section 20.5.1).

Greenness is a readily quantified character and highly convenient for genetic mapping by procedures of quantitative trait analysis. This approach, which has identified a few genetic loci that control senescence in millet and sorghum, has opened the way to breeding for rapid improvement in these crops by exploiting molecular markers that map near the senescence genes. Three independently inherited *stay-green* mutations of *Arabidopsis* also have been described. Once these mutant loci are classified and mapped, the powerful tools of molecular genetics available for this species can be used to assign molecular and physiological functions to the corresponding genes.

20.4 Pigment metabolism during senescence

20.4.1 Chlorophyll is degraded by way of a complex enzymatic pathway involving several subcellular compartments.

The clearest symptom of senescence in leaves is the visible loss of green color with time (see Fig. 20.9). The first step in chlorophyll catabolism is conversion of chlorophyll to chlorophyllide in a reaction catalyzed by chlorophyllase (Fig. 20.15). The other product of chlorophyllase action is phytol, which usually accumulates in the lipid globules of gerontoplasts (see Fig. 20.13), mostly

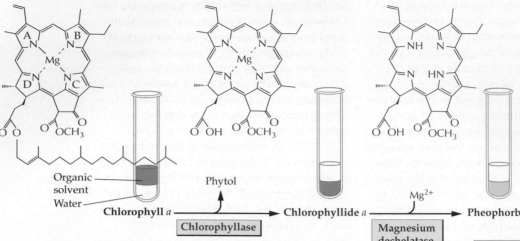

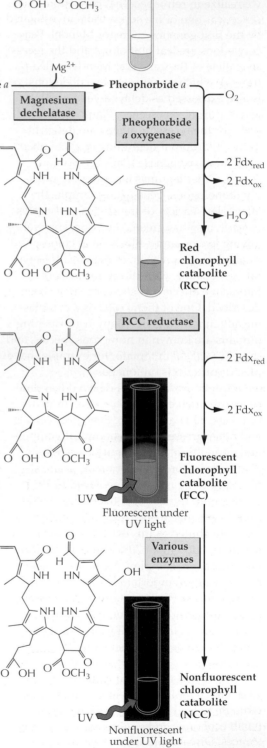

Figure 20.15

Chlorophyll *a* and its breakdown products. Removal of the long phytol side chain from the chlorophyll *a* molecule to yield chlorophyllide has little effect on color but greatly increases the water-solubility of the pigment. Pheophorbide *a* oxidase (PaO) opens the pheophorbide ring, introducing one oxygen atom from O_2 and one from water across the methine bridge that links pyrrole groups A and B of pheophorbide *a*. The resulting red intermediate, red chlorophyll catabolite (RCC), is converted by RCC reductase to fluorescent chlorophyll catabolite (FCC), a colorless product that fluoresces blue when excited with ultraviolet light. Chlorophyll breakdown ends with accumulation of one or more nonfluorescent chlorophyll catabolites (NCCs) similar to the structure shown here. The B ring of chlorophyll *b* has an aldehyde group instead of the methyl group characteristic of chlorophyll *a*. No *b*-type NCCs have been found in plants, suggesting that chlorophyll *b* is converted to chlorophyll *a* or a chlorophyll *a* catabolite before it is degraded.

in the form of esters. An enzyme called Mg dechelatase is required to remove Mg from chlorophyllide, yielding pheophorbide. Both chlorophyllide and pheophorbide retain their intact porphyrin ring structures and are green. A critical step in the degradation pathway opens the ring to generate a colorless straight-chain tetrapyrrole. Two enzymes are responsible for this, PaO and red chlorophyll catabolite (RCC) reductase. The PaO reaction requires O_2 and involves Fe, which operates in a redox cycle driven by reduced ferredoxin. PaO uses pheophorbide *a* but not pheophorbide *b* as a substrate, so chlorophyll *b* must be converted to chlorophyllide *a* or pheophorbide *a* before it can be catabolized. The bright red bilin compound produced by the PaO reaction is similar to pigments excreted by certain single-celled algae when they are starved of nitrogen or are transferred to heterotrophic conditions. This catabolite (RCC) does not accumulate in plants but is immediately metabolized further by RCC reductase, which catalyzes the ferredoxin-dependent reduction of a double bond in the pyrrole system of RCC to produce an almost colorless tetrapyrrole with a strong blue fluorescence (fluorescent chlorophyll catabolite; FCC). Studies of the subcellular location of the enzymes of chlorophyll catabolism have established that chlorophyllase, PaO, and probably Mg dechelatase activities are associated with the gerontoplast envelope, whereas RCC reductase is a soluble plastid protein (Fig. 20.16). Chlorophyll is thought to be conveyed from the thylakoid membrane to the envelope by a carrier protein. FCC export from the gerontoplast is ATP dependent. The final destination of catabolites is the vacuole, which they enter through ATP-binding casette (ABC) transporters (see Chapter 3). En route to the vacuole, FCC may be conjugated—malonyl and β-glucosyl derivatives have been identified—or otherwise modified. The end products of these modifications are various nonfluorescent catabolites (NCCs), which vary greatly in number and type, depending on the species.

20.4.2 Chlorophyll catabolism unmasks foliar carotenoids, which diminish or accumulate during senescence, depending on the plant species.

In the leaves of some species, chlorophyll loss is accompanied by decreases in carotenoids. However, the foliage of deciduous trees and many other plants develops vivid colors before being shed. Fruits also become brightly colored during ripening. In such cases, the loss of chlorophyll unmasks underlying carotenoids, which provide a yellow or orange background against which new pigments accumulate.

New carotenoids are synthesized by the isoprenoid pathway (Fig. 20.17). Genes for three of the key enzymes have been cloned from several plant sources. Geranylgeranyl diphosphate (GGPP) synthase makes the basic C_{20} unit from which all carotenoids are constructed. Phytoene synthase condenses two molecules of GGPP into the C_{40} carotenoid phytoene. Phytoene desaturase (PDS) activity leads to formation of ζ-carotene,

which is desaturated further to generate lycopene, the red pigment of tomato, bell pepper, and similar fruits. The activities of GGPP synthase and PDS increase markedly during ripening of pepper, and the expression of phytoene synthase in tomato is enhanced by treatment with the ripening hormone ethylene (see Section 20.10.1).

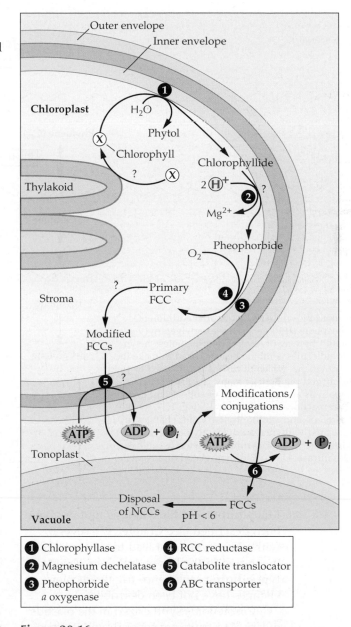

❶ Chlorophyllase	❹ RCC reductase
❷ Magnesium dechelatase	❺ Catabolite translocator
❸ Pheophorbide *a* oxygenase	❻ ABC transporter

Figure 20.16
Subcellular compartmentation of the pheophorbide *a* pathway of chlorophyll catabolism in leaf mesophyll cells. X represents a hypothetical part of the catabolic system thought to be responsible for dismantling the thylakoid pigment–protein complexes and the transport of the resulting chlorophyll molecules to the inner envelope membrane.

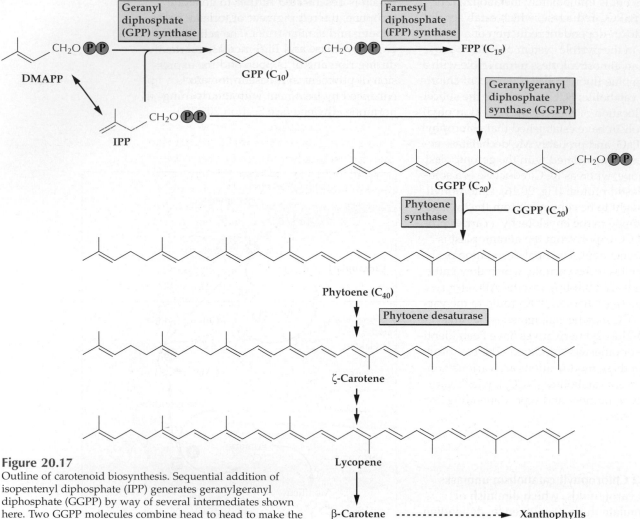

Figure 20.17

Outline of carotenoid biosynthesis. Sequential addition of isopentenyl diphosphate (IPP) generates geranylgeranyl diphosphate (GGPP) by way of several intermediates shown here. Two GGPP molecules combine head to head to make the symmetrical C_{40} phytoene, which in turn is desaturated to yield lycopene, the bright red pigment of many ripe fruit such as tomatoes. Xanthophylls are the products of subsequent cyclization and oxidation reactions (see Chapter 12, Fig. 12.7, for the structures of β-carotene and xanthophylls). Color changes in ripening are associated with increased activity of enzymes in the pathway, particularly GGPP synthase, phytoene synthase, and phytoene desaturase. DMAPP, dimethylallyl diphosphate.

Many carotenoid biosynthesis mutants have been described (see Chapter 17), several with phenotypes that lead to abnormal pigmentation in the leaves, but their phenotypes during senescence-triggered color changes have not been described.

Carotenoids synthesized in the plastids of ripening fruits are concentrated in structures variously described as fibrils, crystals, or globules, which become increasingly numerous as chloroplasts redifferentiate into chromoplasts. **Fibrillins,** specific proteins associated with the carotenoid bodies of ripening fruit tissues, have also been detected on the surface of plastoglobuli, the lipid droplets that accumulate in the plastids of senescing leaves. *CHRC,* a cDNA encoding a carotenoid-associated protein with marked sequence similarity to fibrillin, has been cloned from several plant species. These proteins may have a role in storing lipids, transporting newly synthesized or modified carotenoids, or stabilizing plastid structure in fruits, leaves, and other colored organs.

20.4.3 Phenylpropanoid metabolism also can be altered during senescence.

Other pigments and secondary compounds that accumulate in senescing tissues are products of phenylpropanoid metabolism, which produces a diverse group of phytochemicals, including phenolics, tannins, flavonoids, and lignins (see Chapter 24). Wounding, ethylene treatment, exposure to ozone, and other stimuli that invoke senescence- or PCD-like pathological responses also commonly increase the rate and alter the pattern of phenolic product synthesis. Among the important products of phenylpropanoid metabolism is salicylic acid, an important factor in the regulation of disease resistance and associated PCD events (see Sections 20.13.4 through 20.13.16; see also Chapters 17 and 21).

Phenylpropanoid pathways are complex, branched metabolic sequences with several control points (Fig. 20.18). Phenylalanine ammonia lyase (PAL) is a key early enzyme in phenylpropanoid metabolism. PAL activity increases as leaves age. Transgenic tobacco plants overexpressing PAL have an altered balance of phenylpropanoid products, an in-dication that changes in enzyme concentration during senescence are associated with an altered flux into different branches of secondary metabolism. Individual regulatory genes, such as the *Lc* (leaf color) transcription factor of maize, have been shown to up-regulate regions of the flavonoid pathway in a coordinated manner.

Many of the striking red, purple, and yellow pigments of autumnal foliage are anthocyanins, betacyanins, and flavanoids—water-soluble phenylpropanoid derivatives that accumulate in cell vacuoles. The pigments of ripe strawberry fruit likewise include anthocyanins and proanthocyanins. Expression of the branch-point enzyme dihydroflavonol 4-reductase (DFR; Fig. 20.18) varies during the development of strawberry fruit. Active in early development as condensed tannins accumulate, DFR is subsequently down-regulated and then is

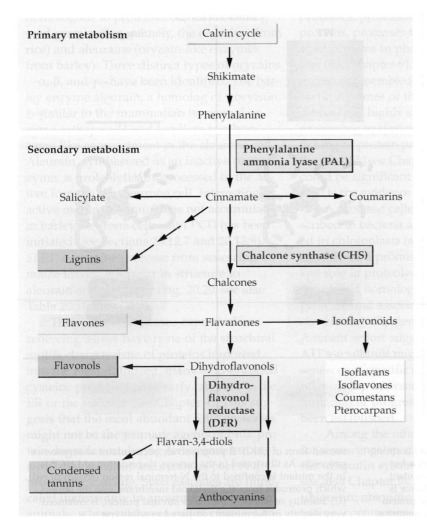

Figure 20.18

Phenylpropanoid metabolism. Phenylalanine, a product of the shikimate pathway (see Chapter 8), enters secondary metabolism by way of the reaction catalyzed by phenylalanine ammonia lyase (PAL). Among the phenylpropanoid end products important in senescence and PCD are salicylate, which is implicated in PCD-like resistance to pathogen infection (see Chapters 17 and 21); lignins and tannins, which accumulate as tissues age (see Chapter 24); isoflavonoid phytoalexins, which have antipathogen properties (see Chapters 21 and 24); and anthocyanins (see Chapter 24), which are responsible for some of the vibrant pigments of ripening fruits and senescing leaves. Anthocyanin production depends on dihydroflavonol 4-reductase (DFR), a developmentally regulated enzyme that is active during the ripening of strawberry fruit.

the ubiquitin cycle; and components of the proteasome have been cloned from various plant sources. Some of these genes have turned up among the collections of clones isolated from senescing tissues, and the mRNAs for polyubiquitin and conjugating enzyme are abundant in senescing tobacco leaves. Experiments with promoter–reporter gene fusions indicate that the polyubiquitin promoter is more active in senescent leaves than in young ones. Nevertheless, biochemical evidence for a specific role of the ubiquitin system in protein remobilization during senescence is meager; the functioning of the pathway in stress responses is much more convincingly established. Perhaps the ubiquitin system is not a causative agent in senescence but rather is up-regulated in response to the stress-like conditions that develop then.

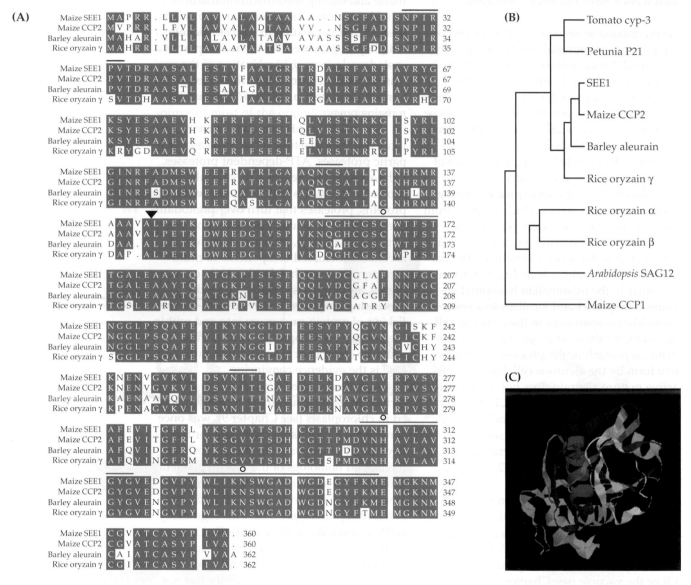

Figure 20.20

Comparison of the foliar senescence-associated protease *SEE1* from maize with other cysteine endopeptidases. (A) The amino acid sequence inferred from *See1* cDNA aligned with the sequences of other proteases from the same family. Conserved structural features include a putative vacuole sorting motif (residues 29 through 34; red line), potential glycosylation sites (125 through 127, 256 through 258; blue lines), and the likely cleavage point that removes the prodomain (between 142 and 143; arrowhead). Open circles indicate the cysteine–histidine–asparagine triad of the active site, and the green lines identify the surrounding conserved regions. (B) Dendrogram locating *SEE1* in the same subfamily as the cereal germination proteases aleurain and oryzain. Note that *SAG12* of *Arabidopsis* is comparatively distantly related to *SEE1*. (C). The degree of conservation of cysteine protease structural motifs in *SEE1* allows a three-dimensional model to be devised.

20.5.3 Organic nitrogen and organic sulfur are exported from senescing leaves.

The proportions of various amino acids in the phloem connecting senescing leaves with young developing tissues differ from the amino acid profile of total leaf protein. Although evidence suggests that amino acids can be modified during transport, the products of proteolysis are metabolized extensively before being loaded into the vascular system.

The amides glutamine and asparagine typically are prominent among the organic nitrogen exports derived from the breakdown of protein during leaf senescence (Fig. 20.22). Assimilatory and catabolic formation of glutamine involves addition of a second amino group to a preexisting glutamate; this ATP-dependent reaction between glutamate and NH_4^+ is catalyzed by glutamine synthetase (GS). The other major amide, asparagine, probably is not produced by a synthetase reaction analogous to GS but rather arises by transamidation from glutamine to aspartate. In senescence, a major source of ammonium ion for amide synthesis is thought to be the catabolic deamination reaction

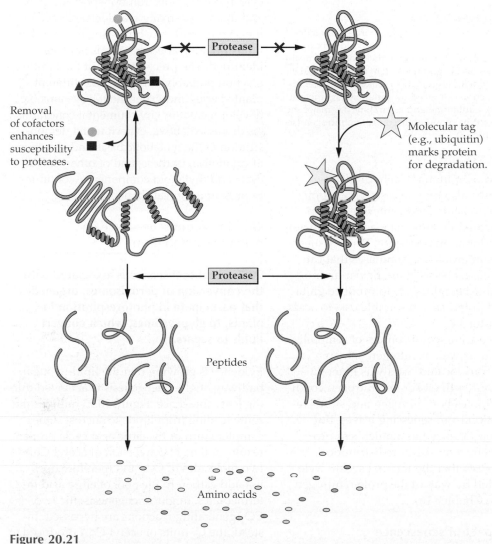

Removal of cofactors enhances susceptibility to proteases.

Molecular tag (e.g., ubiquitin) marks protein for degradation.

Peptides

Amino acids

Figure 20.21

Senescence may be associated with increasing susceptibility of proteins to breakdown. Two possible mechanisms are illustrated. In one case (left side of figure), occupation of binding sites by cofactors (e.g., enzyme substrates) or allosteric effectors locks the protein into a conformation resistant to proteolysis, and the equilibrium between this structure and the susceptible conformation is thus determined by the abundance of these ligands. In the other case (right side), the protein is marked for degradation by the addition of a tag such as ubiquitin and is thereby made available for proteolysis. (Both the ubiquitin system and the impact of chlorophyll degradation on the stability of some chloroplast proteins are discussed in Chapter 9.)

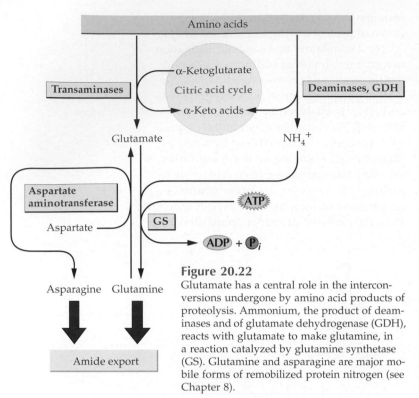

Figure 20.22

Glutamate has a central role in the interconversions undergone by amino acid products of proteolysis. Ammonium, the product of deaminases and of glutamate dehydrogenase (GDH), reacts with glutamate to make glutamine, in a reaction catalyzed by glutamine synthetase (GS). Glutamine and asparagine are major mobile forms of remobilized protein nitrogen (see Chapter 8).

promoted by glutamate dehydrogenase. Deaminases also may release NH_4^+ from other amino acids. A key intermediate in amide formation, glutamate acts as both ammonium donor and acceptor. Plant tissues contain numerous transaminases that catalyze reactions between particular amino acids and α-ketoglutarate to produce glutamate and potentially respirable α-keto acids (see Chapter 14).

Proteins are major sources of remobilizable sulfur as well as nitrogen. Organic sulfur is translocated mainly as homoglutathione. Growth under low-nitrogen conditions promotes both nitrogen and sulfur remobilization from senescing leaves, but sulfur insufficiency has a much less pronounced effect on sulfur redistribution. This suggests that the internal sulfur cycle is regulated by way of the protein nitrogen cycle (see Chapter 16).

20.6 Impact of senescence on photosynthesis

Diminishing rates of photosynthesis in leaves and other green tissues such as pericarp often are cited as the clearest symptom of physiological decline in senescence. Esti-

mates of gas exchange (CO_2 fixation measured by infrared detection, O_2 emission by polarography) show that photosynthetic capacity (expressed, for example, as the rate at light saturation) diminishes roughly in parallel with other indices such as chlorophyll or protein content. In general, the efficiency with which the surviving chlorophyll gathers light energy does not deteriorate significantly, whereas the energy-using reactions of photosynthesis (e.g., the electron transfer reactions, ATP synthesis, and the carbon-linked reactions) decline progressively. This decline in photosynthetic capacity involves, among other components, Photosystem II (PSII), the cytochrome $b_6 f$ electron transport complex, and the carbon-fixing reaction catalyzed by the enzyme Rubisco, but a single factor that limits photosynthetic rates has not been identified. The precise nature of the limiting reaction probably varies among different plant species and experimental systems (i.e., developmental or environmental conditions). As discussed above, experimental demonstration of the metabolic basis and regulation of catabolism of these, and of other Calvin cycle and thylakoid components, continues to present technical challenges.

20.7 Impact of senescence on oxidative metabolism

20.7.1 Leaf senescence is associated with the conversion of peroxisomes, organelles that participate in photorespiration in C_3 plants, to glyoxysomes, which convert lipids to sugars.

Evidence is growing that the lipid-to-sugar pathway of gluconeogenesis is activated during leaf senescence. Figure 20.23 outlines the route leading from lipids to sucrose. One complete turn of the citric acid cycle proper results in the net conversion of acetyl-CoA to two molecules of CO_2 and therefore does not add carbon to the pool of citric acid intermediates. In gluconeogenesis, the two CO_2-generating reactions are bypassed. Instead, the C_2-units of acetyl-CoA produced by β-oxidation of fatty acids contribute to a net increase in organic acids.

The glyoxylate bypass is characterized by two distinctive enzymes, isocitrate lyase and malate synthase. These enzymes become activated as senescence proceeds, and the

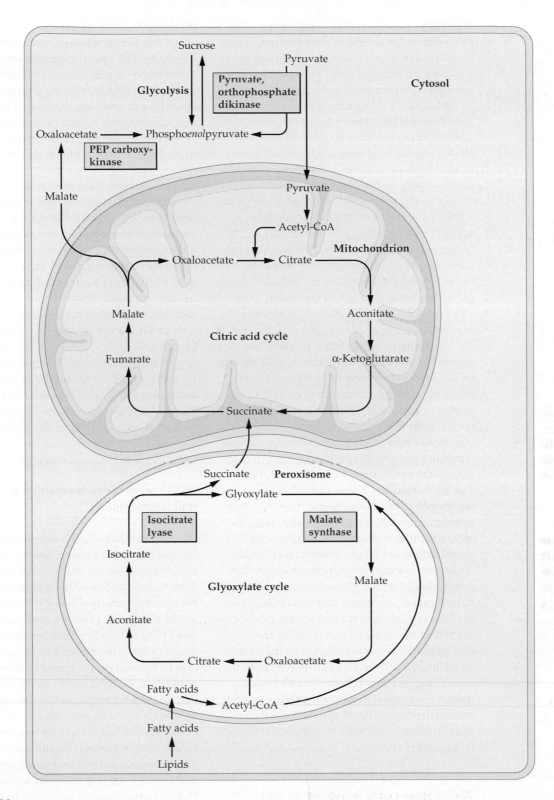

Figure 20.23

Gluconeogenesis in senescing leaf cells. The glyoxylate bypass is activated during senescence and allows the carbon (acetyl-CoA) from lipolysis and β-oxidation to be routed around the two CO_2-liberating steps of the citric acid cycle. Malate synthase and isocitrate lyase are the key activities in this sequence. The carbon salvaged from lipid can then be converted into transportable sucrose by the reversible steps of glycolysis, provided the irreversible pyruvate kinase reaction can be bypassed. This may be achieved by converting an organic acid to a glycolytic intermediate upstream from pyruvate (e.g., decarboxylation of oxaloacetate to phospho*enol*pyruvate [PEP] by PEP carboxykinase). Another type of pyruvate kinase bypass involves using an alternative route from pyruvate to PEP, such as the reaction catalyzed by pyruvate, orthophosphate dikinase, the gene for which is activated during maize leaf senescence.

regulatory elements that control expression of isocitrate lyase and malate synthase genes are stimulated by factors that promote senescence (expression of these genes is also up-regulated during seed germination).

Malate, oxaloacetate, and pyruvate are readily interconverted and represent entry points for carbon into the glycolytic pathway. Reversing the reactions of glycolysis leads to synthesis of hexose (and subsequently sucrose). There is, however, one essentially irreversible step in glycolysis, namely, the pyruvate kinase reaction. If gluconeogenesis is an important pathway in mobilizing leaf reserves, one or more of the enzymes that bypass the pyruvate kinase reaction would be expected to be active, or even induced, during senescence. The mRNA encoding pyruvate, orthophosphate dikinase is abundant in senescing maize leaves; thus, all the necessary elements are in place to allow the senescing leaf to salvage a substantial amount of the carbon invested in lipids, particularly those of chloroplast membranes, and to export it to sinks elsewhere in the plant as water-soluble transport sugars.

Activation of gluconeogenesis in senescence is associated with functional transformation of peroxisomes into glyoxysomes. The transition has been studied in detail in cucumber cotyledons. As seedling development progresses, the glyoxysomes responsible for lipid mobilization during seed germination are reconverted in turn to peroxisomes. Elegant immunomicroscopy has established that the glyoxysome marker enzymes malate synthase and isocitrate lyase and the peroxisome marker glycolate oxidase coexist in the same particle during the changeover from one function to the other (see Chapter 1, Box 1.1). Consistent with the reworking of protein complement inside the glyoxysome/peroxisome is the presence within the organelle of specific endo- and exopeptidases, some of which are particularly active in senescence.

20.7.2 Senescence is associated with carbon-limiting conditions in which the organic acid skeletons of amino acids are respired to generate energy.

Senescence is an energy-demanding process, and respiratory poisons such as the uncoupling agent dinitrophenol are very effective at halting the progress of senescence. Among the many ATP-consuming reactions that become active in senescence are chlorophyll catabolism, amide synthesis, transport and loading activities, the reversal of glycolysis in gluconeogenesis, and de novo protein synthesis. Declining photosynthesis and continued active export of sugars combine to make senescing tissues increasingly carbon-starved, so new sources of respirable energy are required to meet metabolic and transport demands. Amino acids derived from proteolysis are an important source of carbon skeletons. One hypothesis for why amides are the most common form of organic nitrogen transported out of senescing tissues is that carbon starvation favors a carbon skeleton carrying two amino moieties (such as asparagine or glutamine) rather than just one (see Chapter 8).

In some but not all fruits, ripening is associated with the **climacteric,** a characteristic burst of respiration. This burst is usually associated with an increase in ethylene production, as discussed further in Section 20.10.1.

20.7.3 Senescence is sensitive to cellular redox conditions.

Oxidative metabolism is a central feature of senescence at the cellular level. Some of the signaling pathways that regulate senescence may include redox-sensing components. For example, expression of the genes encoding fibrillins, the proteins associated with lipid-pigment bodies in chromoplasts and gerontoplasts, is sensitive to oxygen. Induction of fibrillin biosynthesis by drought, wounding, or ABA involves a signaling pathway in which production of the superoxide anion (O_2^-) is a key step. Redox regulation of transcriptional activators in animal cells, which occurs through highly conserved cysteine residues in the DNA-binding domains, is thought to be important in some kinds of PCD. Similar events may occur in plant cells.

There are many sources of reactive oxygen species in senescing cells. H_2O_2, and the superoxide anion from which it is generated by disproportionation, are products of normal enzymatic reactions in peroxisomes, glyoxysomes, chloroplasts, and other cell compartments. Defenses against accumulation of harmful reactive oxygen species include catalases and superoxide dismutases. The important antioxidants ascorbate and glutathione participate in a cycle driven by four enzymes: ascorbate peroxidase, monodehydroascorbate reductase, dehydroascorbate reductase, and glutathione reductase (Fig. 20.24; see also Chapter 22).

A plausible model for regulation of the senescence program is that the balance is shifted between the production of reactive oxygen species and their removal by antioxidant systems, and the resulting change in signals consonant with the cellular redox state alters gene expression. In this context, the redifferentiation of peroxisomes into glyoxysomes described above could be of special significance in senescence, because the transition clearly is associated with an altered profile of superoxide-producing and antioxidant enzymes.

We emphasize that reactive oxygen species are considered here to act as part of a subtle signaling mechanism in coherent, viable cells. In excess, however, they are capable of rapidly destroying the cell through the propagation of free radical cascades, a process that certainly leads to death in some pathological conditions (see Chapters 21 and 22). Such an unregulated, catastrophic mechanism is likely to contribute to only the most terminal phase of normal senescence.

20.8 Degradation of nucleic acids during senescence

20.8.1 Catabolism of nucleic acids releases inorganic phosphate.

Much of the organic phosphorus in leaves is in the form of nucleic acids, which, like proteins, seem to have the dual function of carrying genetic information and acting as salvageable reserve substances of high molecular mass. A fundamental chemical difference between phosphorus and nitrogen in organic combination within cells is that the former does not bond directly to carbon but instead exists as ester-linked phosphate, which has a high, negative free energy of hydrolysis. Phosphatases are ubiquitous and especially active during senescence, such that conditions during senescence are

both energetically and enzymatically favorable for the release of phosphate from organic compounds. Phosphorus is then readily translocated through the vascular system in the form of inorganic phosphate. The nucleoside products of nuclease and phosphatase action presumably are cleaved into sugars, purines, and pyrimidines.

The pathways leading to the catabolism of guanosine are shown in Figure 20.25. Xanthine dehydrogenase and uricase, glyoxysomal enzymes in the pathway that catabolizes adenine and guanine eventually into ammonia and CO_2, are reported to be activated strongly during leaf senescence. Further biochemical and molecular investigations of these almost unexplored regions of nucleic acid catabolism in senescence should be fruitful.

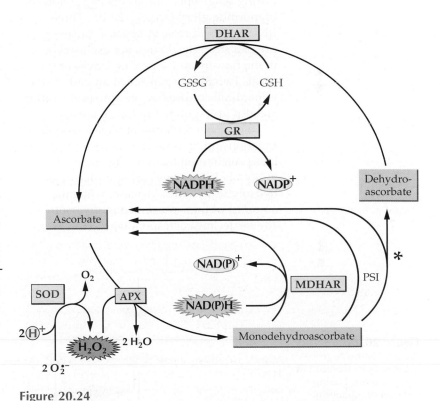

Figure 20.24
The ascorbate–glutathione redox cycle in senescence. Ascorbate is regenerated from oxidized forms by a cycle catalyzed by ascorbate peroxidase (APX), monodehydroascorbate reductase (MDHAR), dehydroascorbate reductase (DHAR), and glutathione reductase (GR). Superoxide dismutase (SOD) is one of the enzymatic defenses against the buildup of reactive oxygen species. In mature leaves, photosynthetic electron transfer generates NADPH, which donates electrons to ascorbate by way of reduced glutathione (GSH). In senescing leaves, NADPH production decreases, resulting in increased ratios of oxidized glutathione (GSSG) to GSH and of dehydroascorbate to ascorbate. These relatively subtle changes can result in substantial changes in cellular redox conditions and signal new patterns of gene expression. PSI, Photosystem I; *, nonenzymatic reaction.

20.8.2 The activity of some ribonucleases increases during senescence.

Quantitatively, the largest nucleic acid fraction contributing to phosphorus remobilization is the RNA of plastid ribosomes. Ribonuclease (RNase) cleavage rapidly breaks RNA down into low-molecular-mass products with little accumulation of large intermediate fragments. All plant RNases for which gene sequences are known belong to the so-called S/T2 superfamily, members of which share several common structural features, including two conserved histidines in the catalytic center. Some RNases, particularly those associated with self-incompatibility (see Chapter 19), are extracellular in location, but others contain motifs that target them to the ER or vacuole (or both) (see Chapter 4). Vacuolar RNase is synthesized de novo during senescence of the ephemeral flowers of morning glory (see Fig. 20.10). Three RNases cloned from *Arabidopsis* are expressed strongly, though not exclusively, in the flowers and senescing leaves of this plant. Two RNase genes that are induced in tomato cell cultures by phosphate starvation are also up-regulated in leaf senescence.

The general picture of nucleic acid degradation and its control in senescence resembles protein mobilization: The genes encoding catabolic enzymes are stimulated and hydrolytic activities increase, while the amounts of their presumed in vivo substrates decline correspondingly. However, the locations of the known enzymes do not appear to correspond to the principal sites of degradation. Researchers are seeking new insights to resolve this anomaly. It is also important to differentiate (although so far this has been done rarely) between the nutrient recycling role of RNases in senescence and the enzymes' other functions, including RNA turnover and processing, pathogen defense, and (of particular significance in the present discussion) PCD. Often the increased expression of RNase occurs rather late in the course of senescence, which might mean it is connected to the point of irreversibility, that is, when senescence inevitably leads to necrosis and death.

20.9 Regulation of metabolic activity in senescing cells

20.9.1 Differential susceptibility of enzymes to proteolysis can be mediated by several mechanisms.

Much of the primary metabolic machinery of senescing cells is present before senescence begins. We are only beginning to understand how senescing cells retain structural and functional integrity of critical enzymes in a cellular environment where proteolysis and other lytic processes predominate. How do enzymes of gluconeogenesis or amino acid metabolism survive to carry out their metabolic roles when other enzymes, such as those of photosynthetic carbon fixation, are being degraded?

Differential stability may be ensured in several ways. If enzymes are more or less equally exposed to degradation, differential activation of gene expression can ensure that

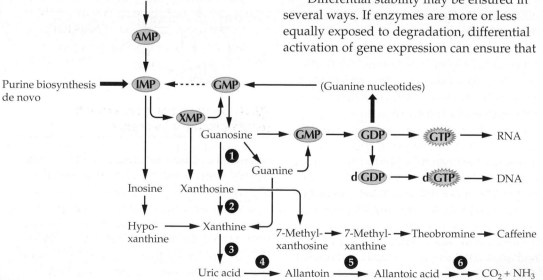

Figure 20.25
Possible pathway for guanosine metabolism in higher plants. During senescence, guanosine and other nucleotides are broken down to yield CO_2 and NH_3. Among the enzymes that participate in guanosine catabolism are (1) guanosine deaminase, (2) nucleosidase, (3) xanthine dehydrogenase, (4) uricase, (5) allantoinase, and (6) allantoicase.

some proteins are replenished continuously, whereas others are removed and not replaced. Another mechanism is compartmentation, discussed in the next section. A further interesting possibility is suggested by experiments on the lability of isolated enzymes (for example, GS1) in the presence and absence of substrates, cofactors, and allosteric effectors. In general, if the catalytic and allosteric sites of an enzyme are occupied by the appropriate balance of metabolic effectors and substrates, the enzyme is comparatively resistant to attack by proteases (see Fig. 20.21, left side). Thus, an enzyme that is gainfully employed in an active metabolic pathway is apparently less likely to be proteolyzed than if it is operating below capacity. If the supply of energy for photosynthesis declines and the flux through the Calvin cycle decreases, as happens in senescence, the Calvin cycle enzymes might be the ones preferentially degraded—compared with, for example, the amino acid–metabolizing enzymes that are fully occupied in processing the products of protein breakdown.

The notion that control of degradation of a particular protein might be exercised by a cofactor through its influence on conformational state could explain other aspects of senescence. For example, the chlorophyll–protein complexes of the thylakoid membrane are highly resistant to attack by proteases; when pigments are not present in the correct stoichiometric ratios, however, the chlorophyll-binding proteins cannot fold properly and then are susceptible to proteolytic attack (see Chapter 9). Thus removal of chlorophyll by the degradation mechanism involving PaO is necessary before proteins of the thylakoid complexes can be broken down in senescence. Mutants that lack PaO activity not only retain chlorophyll but also have much less breakdown of light-harvesting and other pigment-binding thylakoid proteins (see Fig. 20.19).

20.9.2 Many cytosolic enzymes are unaffected or up-regulated during senescence while the organellar isoforms of the same protein are being degraded.

Although their organelles undergo structural changes, the cells of senescing tissues retain a high degree of internal organization and compartmentation until late in PCD. Chlorophyll degradation illustrates the way in which controlled traffic between organelles is necessary for the orderly progress of senescence. The pathway includes highly localized actions in the thylakoid membranes, the envelope, and the stroma of the plastid, the cytosol (possibly including ER-associated cytochrome P450–mediated oxidations), the tonoplast membrane, and the vacuolar sap, where terminal catabolites accumulate. It may seem strange to consider chlorophyll as a toxin, but free chlorophyll is a potent photosensitizer, capable of killing any cell that accumulates it. To gain access to a major source of salvageable amino acids in the pigment–protein complexes of the chloroplast, the plant must remove the chlorophyll and rapidly metabolize it by the general route used for other dangerous compounds.

Many enzymes exist as isoforms, each of which is located in a different organelle. For example, one form of aspartate aminotransferase is present in chloroplasts, another in mitochondria, another in peroxisomes, and yet another in the cytosol. During senescence, the total activities of the aminotransferases and glutamine synthetases decline, but the activities of the cytosolic isoforms remain unaffected or even increase. Loss of the plastid isoforms of several of the enzymes of primary carbon and nitrogen metabolism during senescence strongly suggests that proteolysis is compartmentalized. However, the plastid isoforms of some enzymes—alanine aminotransferase, for example—are intrinsically less resistant to proteolytic attack than are the cytosolic forms. Thus, the highly precise mechanisms that direct the senescence program include biochemical regulation in addition to control of gene expression.

20.10 Endogenous plant growth regulators and senescence

Senescence caused by internal and external factors is mediated by complex interactions of several factors, including hormones. In discussing the role of plant growth regulators in senescence, one must emphasize two points. First, the ultimate determination to initiate senescence in a plant depends on the interplay of various signals—including, but certainly not limited to, endogenous growth regulators. Second, the response to specific growth regulators with respect to senescence is not the same in all plants. For example, ethylene can induce leaf senescence in most plants but has no apparent effect in others. Despite these caveats, growth regulators can play a prominent role in regulating senescence.

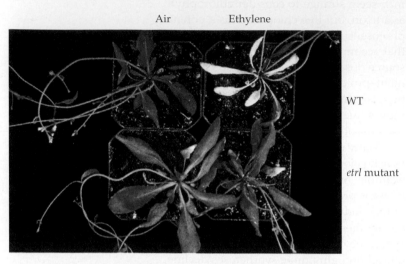

Air   Ethylene

WT

etrl mutant

Figure 20.26
Ethylene promotes leaf senescence in wild-type (WT) *Arabidopsis*, but not in the ethylene-insensitive *etr1* mutant.

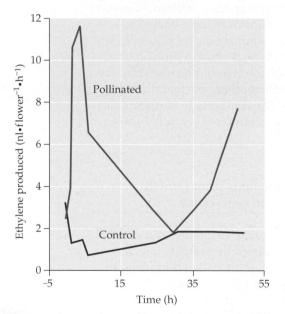

Figure 20.27
Ethylene production by unpollinated control and pollinated *Petunia* flowers. Note the burst of ethylene synthesis that rapidly follows application of pollen.

20.10.1 Ethylene acts primarily as a promoter of senescence.

Among the known plant hormones, ethylene is typically the strongest promoter of senescence. In many plants, treatment with exogenous ethylene can induce senescence in leaves and flowers and ripening of fruit. For example, treatment of *Arabidopsis* leaves with ethylene causes changes paralleling those seen in naturally senescing leaves (Fig. 20.26). The leaves begin to yellow, accompanied by a corresponding decrease in the expression of photosynthesis-related genes and an increase in the expression of *SAGs*. The concentrations of endogenous ethylene correlate with leaf senescence: As leaves get older, or are induced to senesce by dark-treatment, they produce more ethylene. In some flowers, senescence of specific organs such as petals follows pollination. Increases in ethylene concentrations have been measured in these flowers before petal senescence. In petunias, ethylene production can be detected within 20 minutes of pollination and precedes penetration of the stigma by the pollen tube (Fig. 20.27). Thus, recognition of the pollen by the stigma may be sufficient to induce ethylene production.

Because of its economic importance, much of the work on ethylene regulation of senescence has been done in the context of fruit ripening. During normal development of climacteric fruit such as tomato, a developmentally regulated burst of respiration and ethylene production is followed by an increase in the expression of specific genes and ripening-related activities; these result in the changes in fruit color, texture, and flavor that characterize ripening (Fig. 20.28). The effect of exogenous ethylene as a ripening agent has been known for more than a century, and controlled exposure to ethylene still is a major factor in the successful marketing of many fruits. Thus, an increase in ethylene concentrations coincides with senescence in many plants, and exogenous treatment with ethylene can trigger senescence.

Further evidence for the critical role of ethylene in senescence comes from experiments designed to test whether ethylene perception is required for senescence. The *Arabidopsis etr1* mutant (see Fig. 20.26) and the tomato *never ripe* mutant are insensitive to ethylene, the result of mutations in an ethylene receptor protein. These plants show delayed leaf senescence, and in the case of tomato, the normal senescence of floral organs is delayed as well. The tomato fruits never fully ripen, giving the mutant its name (Fig. 20.29). Another indication of a critical role for ethylene in senescence has been the finding that an *Arabidopsis* mutant causing delayed leaf senescence is in fact an allele of the previously identified ethylene-insensitive *ein2* mutant.

Experimental reduction of ethylene production in plants has provided more evidence for the role of ethylene in senescence. Chemical inhibitors of ethylene biosynthesis or action have been shown to inhibit fruit ripening and leaf senescence. Similar results have been obtained with plants engineered to produce less ethylene. 1-Aminocyclopropane-1-carboxylic acid (ACC) synthase and ACC oxidase are two enzymes that catalyze the last steps in ethylene biosynthesis in plants. The development of tomato plants that express these genes in the antisense orientation has resulted in individuals with greatly reduced ethylene concentrations

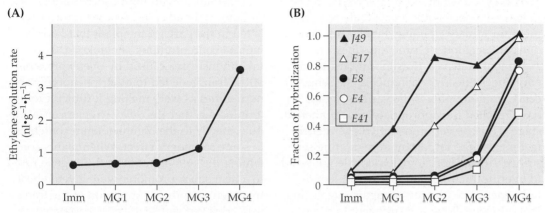

Figure 20.28
Ethylene production (A) and the expression of ethylene-induced genes (B) in ripening tomato fruit. Genes *J49* (black triangles), *E17* (white triangles), *E8* (black circles), *E4* (white circles), and *E41* (white squares) were monitored by dot blotting and reported as a fraction of maximum hybridization. MG1 through MG4 refer to sequential stages in tomato fruit development. MG, mature green; Imm, immature.

Figure 20.29
Exogenous treatment with ethylene induces senescence in many plants. The tomato inflorescence on the left in (A) was treated with ethylene, causing the flowers to senesce and abscise. The inflorescence in (B) is from a plant carrying the *never ripe* mutation, which is insensitive to ethylene because of a mutation in an ethylene receptor. As a result, ethylene does not trigger senescence and abscission in these flowers. In untreated wild-type flowers, floral senescence follows pollination (C), whereas in the ethylene-insensitive mutants, the flowers do not senesce after pollination but instead remain intact, even as the fruit begin to develop (D). Thus, ethylene is a key signal in inducing floral senescence after pollination in tomato. As the name implies, in the *never ripe* mutant, the fruit do not ripen. The two tomato fruits in (E) are of approximately the same age and were grown under the same conditions. However, the fruit on the left is from a *never ripe* mutant, whereas the fruit on the right is from the wild-type plant, thus emphasizing the importance of ethylene signaling in the induction of ripening.

(Fig. 20.30; see also Chapter 17). When these plants are grown in air, their fruit never completely ripens; when treated with exogenous ethylene, however, they show normal senescence and ripening (Fig. 20.31). Two final points can be made concerning ethylene and the regulation of senescence. The role of ethylene can differ somewhat in different organs, even in the same plant. In the ethylene-insensitive mutants of *Arabidopsis* and tomato, leaf senescence is delayed, but the leaves eventually senesce even in the absence of their perception of ethylene. However, the fruits of the ethylene-insensitive tomato mutants never fully ripen. Therefore, ethylene promotes but is not essential for senescence in leaves, whereas it is an absolute requirement for tomato ripening. Finally, ethylene does not act alone to promote senescence. Treatment of leaves, flowers, and fruits with ethylene is effective in inducing the senescence program only if the various organs have reached the appropriate developmental stage. Thus, as yet unidentified age-related factors also are important in regulating senescence. In addition, tomato plants engineered to overproduce ethylene show no altered leaf senescence phenotype, again suggesting that ethylene alone is insufficient to trigger senescence.

20.10.2 Cytokinins act as senescence antagonists.

Cytokinins also appear to play a major role in regulating senescence but with an effect opposite to that of ethylene: Cytokinins inhibit the senescence process. Two lines of evidence initially suggested that cytokinins might function in blocking senescence. First, the concentrations of endogenous cytokinins decrease in most senescing tissues; second, exogenous application of cytokinins can cause a delay in senescence in most tissues. As with ethylene, the effect of cytokinins varies with the age and type of tissue treated and also from species to species. The ability of cytokinins to delay senescence has been exploited by some plant pathogens that infect leaves. Increased cytokinin content in a zone surrounding the infection site delays senescence in this tissue, producing "green islands," areas of green tissue in a background of senescing, yellowing leaf (Fig.

20.32). Whether this cytokinin is generated directly by the pathogen or the pathogen induces the plant to produce it is not clear.

Molecular techniques now are being used to examine further the role of cytokinins at various stages of plant development, including senescence. The *ipt* gene from *Agrobacterium tumefaciens* encodes the enzyme isopentenyl transferase, which catalyzes a limiting step in cytokinin biosynthesis (see Chapter 17). This bacterial gene has been used to generate transgenic plants that overproduce cytokinins. The *ipt* gene has been fused to various plant gene promoters to promote tissue-specific expression and localized areas with high concentrations of cytokinins. In all cases, senescence was delayed in the parts of the plant that had the highest concentrations of cytokinin. The cytokinin increases caused by overexpression of the transgene led to other phenotypic abnormalities as well, making it difficult to determine whether the delay in senescence was due directly to the high cytokinin content or to some secondary effect. When plants containing the cytokinin biosynthesis gene fused to a heat shock promoter were given a heat treatment to induce expression of the transgene, delayed leaf senescence correlated with increased concentrations of cytokinins. These results are complicated somewhat by the fact that the heat treatment needed to

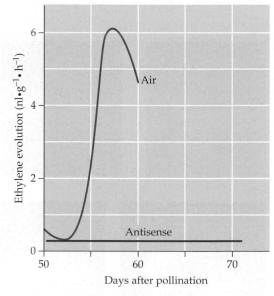

Figure 20.30
Ethylene synthesis by leaves of control (blue) or antisense (red) tomato plants expressing ACC synthase.

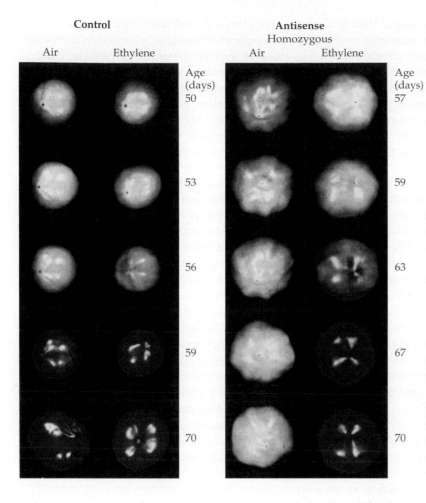

Control   Antisense
                                 Homozygous
Air   Ethylene   Air   Ethylene

Age   Age
(days)   (days)
50   57

53   59

56   63

59   67

70   70

Figure 20.31
Tomato fruits require both ethylene and unknown age-dependent factors for normal ripening. Fruits from wild-type control plants (left panel) maintained in air or in air supplemented with ethylene ripen at roughly the same time. Tomato plants expressing the ACC synthase gene in the antisense orientation (right panel) produce markedly less ethylene than do wild-type plants, and their fruits do not ripen when maintained in air but do ripen normally when kept in air with added ethylene. These results demonstrate two points. First, ethylene is necessary for fruit ripening. In the plants that produce little ethylene, the fruit do not ripen unless they are treated with exogenous ethylene. Second, ethylene alone is insufficient to induce ripening. For example, ethylene does not induce ripening in fruits from wild-type plants that are less than 59 days old. Thus both age-dependent factors and ethylene are required for ripening to occur.

induce the expression of the transgene is in itself a form of stress that might affect the senescence program.

In an elegant refinement of these experiments on cytokinin overproduction, the cytokinin biosynthesis gene was fused to the promoter of *SAG12*, a gene encoding a cysteine endopeptidase specifically expressed in senescing *Arabidopsis* leaf tissue. Tobacco plants transformed with the *SAG12* promoter–*ipt* fusion produced cytokinin in an autoregulated fashion (Fig. 20.33). As the tissue started to senesce, the transgene was induced and cytokinins were produced. The cytokinins inhibited senescence, which in turn decreased expression of the transgene. Thus, cytokinin was produced only in senescing tissue and only in the amount needed to block senescence. The plants had no significant morphological differences from the untransformed plants (height, leaf number, flower number, or flowering time). They did, however, have a significant delay in leaf senescence, with older leaves retaining

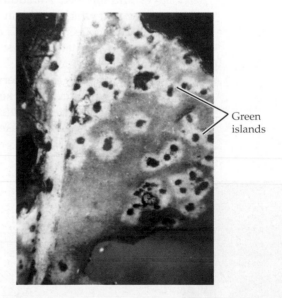

Green islands

Figure 20.32
A dandelion leaf infected with a fungal pathogen exhibits green islands in the regions around the points of fungal infection. The pathogen causes an increase in cytokinin concentrations around the infection site, resulting in delayed senescence in these localized areas, and allowing longer survival of the pathogen, which is an obligate biotroph (see Chapter 21).

photosynthetic activity longer than leaves of control plants lacking the transgene (Fig. 20.33). These experiments clearly demonstrate that the production of cytokinins in leaves can delay senescence. One consequence of the longer photosynthetic life span of leaves on these plants is a concurrent increase in the number of seed produced. The practical implications for agriculture are obvious.

One possible explanation for the senescence-delaying effects of cytokinins is their enhancement of sink activity in plants. Tissues with the greatest cytokinin content are the strongest metabolic sinks, and thus the majority of the nutrients are directed to them. Cytokinins are produced ordinarily in the roots and transported to the leaves. One hypothesis to explain leaf senescence after flowering suggests that cytokinins formed in the root are redirected into the developing seed instead of into the leaf. The seed, therefore, becomes a stronger sink, diverting nutrients from leaves into the seed and triggering leaf senescence. This hypothesis is not consistent with results in soybean, however. Although foliar concentrations of cytokinin decrease during seed development, this appears to reflect an overall decrease in cytokinin concentrations, including in the seed pod, rather than a redirection of cytokinin from leaves into the seed. An additional or alternative way in which cytokinin may influence the course of senescence is as part of a signaling pathway that represses the expression of key SAGs.

Among a group of genes observed to be up-regulated when *Petunia* callus cultures are transferred to low-cytokinin medium are a gene encoding a cysteine protease and one

encoding a peroxidase. Expression of these two genes is strongly enhanced in senescing leaves. Up-regulation of the protease gene *See1* in maize also has been shown to be suppressed by cytokinin treatment. Cytokinins, therefore, probably act at two levels—at a distance, by promoting differentiation and strong sink activity, and locally, in senescing cells themselves, by repressing the launch of the senescence program.

The dual action of cytokinins is seen most dramatically in regreening (Fig. 20.34). At flowering, the lowest, oldest leaves of a mature plant of *Nicotiana rustica* are almost completely yellow. If the shoot is cut off just above the lowest node and the plant is kept in dim light, the leaves will gradually regain their green color. This process is greatly accelerated if the leaf is treated with a cytokinin solution. During regreening, expression of SAGs is suppressed, genes for plastid assembly are turned on, the gerontoplasts of the yellow leaf redifferentiate into chloroplasts, and photosynthetic activity returns. Not only does this demonstrate the potency of cytokinin as an antisenescence factor; it also shows that senescence is potentially reversible at an advanced stage and thus is fundamentally distinct from other PCD processes.

20.10.3 The impact on senescence of growth regulators other than ethylene and cytokinins is less understood.

The role of plant growth regulators other than cytokinins and ethylene in controlling senescence is much less clear or consistent, and we mention them here only briefly. ABA acts in most situations as a promoter

Wild type   Transgenic plant

Figure 20.33

Autoregulated expression of a cytokinin biosynthesis gene in tobacco results in delayed leaf senescence in the transgenic plant (right) in comparison with the wild-type plant (left). The *Agrobacterium tumefaciens* gene encoding isopentenyl transferase (*ipt1*), an enzyme that catalyzes the synthesis of cytokinin, was fused to the promoter of a gene specifically expressed in senescing tissue. The construct was used to transform tobacco, resulting in a plant with very delayed leaf senescence. When the leaf begins to senesce, expression of the fusion gene is induced. The resulting increase in cytokinin content blocks senescence.

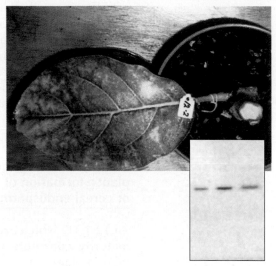

Detop and treat
with cytokinin
(benzylaminopurine)

Protein from
yellow leaf

Protein from
regreened leaf

Figure 20.34

Regreening in tobacco. Nothing illustrates the fundamental distinction between senescence and many other kinds of PCD better than
reversibility. The most convincing model system in this regard is
Nicotiana. At left, the most basal (oldest) leaf on a flowering shoot
has lost virtually all measurable chlorophyll. However, as shown at
right, the leaf can recover full greenness if the stem is removed at
the node, particularly if the leaf is treated with the cytokinin benzylaminopurine. Even terminally senescent tobacco leaf cells can redifferentiate their gerontoplasts into functional chloroplasts. Proteins
characteristic of chloroplast assembly in very young cells—such as
the chlorophyll synthesis enzyme protochlorophyllide oxidoreductase—are reexpressed.

of senescence. Although leaf abscission clearly is related to senescence, and exogenous
ABA often can cause yellowing in leaves,
there is no consistent correlation between the
concentrations of endogenous ABA and tissue senescence. ABA most likely affects
senescence through its action with other
growth regulators. Gibberellins (GAs)
usually retard senescence, but this is by no
means a universal response. The effects of
auxin on senescence are similar to those of
GA; in most cases, treatment with exogenous
auxins delays senescence, but in some species, exogenous auxins promote senescence.
Attempts to correlate the concentrations of
endogenous auxin or GA with senescence
have been inconclusive.

Other growth regulators such as the
jasmonates, brassinosteroids, and polyamines also may have roles to play in senescence. Polyamine concentrations in plants
are strongly correlated with senescence in
many tissues. In tomato ovaries, for example, fruit growth is associated with a dramatic decline in free and conjugated polyamines, whereas in unpollinated ovaries
undergoing senescence, the polyamine
content remains high.

20.11 Environmental influences on senescence

Senescence is as responsive as any other
physiological activity to sub- or supra-
optimal environmental conditions. Seasonal
or otherwise predictable environmental cues
can trigger senescence as part of an adaptive
strategy. Thus trees prepare for winter by
mass leaf senescence in the fall, which is in
turn initiated by the declining length of day
after midsummer. Senescence is also a tactic
deployed when an unpredictable stress is experienced. Not surprisingly, therefore, both
gene promoters and mechanisms that control
metabolism at the posttranscriptional or epigenetic level contain regulatory elements responsive to a diversity of internal and external signals (see Chapter 18). Another complex feature of the environmental sensitivities of senescence is the way in which a
given factor can determine initiation of the
entire senescence syndrome but also often
can have a very different effect once senescence is under way. For instance, drought
invokes premature foliar senescence in
many species, but water limitation during
senescence actually slows the development

of yellowing and other symptoms. Frequently the speed and severity of a developing stress overrun the capacity of the tissue to invoke, coordinate, and express the senescence program; as a result, cells are diverted more or less directly to the PCD pathways described in the remaining sections of this chapter.

20.12 Examples of developmental PCD in plants: formation of TE and mobilization of cereal endosperm

20.12.1 TE, which are dead at functional maturity, constitute a model system for PCD in plants.

Xylogenesis, an interesting example of developmentally regulated PCD in plants, has been developed into a useful model system for explaining plant PCD. Initiated during embryogenesis, xylogenesis, or development of the xylem, continues throughout the life of the plant as new TE are formed from cells of the procambium and cambium. Xylem is primarily composed of TE, the vessels and tracheids of the water-conducting system (Fig. 20.35; see also Chapter 15). During the final stages of differentiation, TE undergo secondary thickening of the cell wall, followed by cell death and autolysis of the contents of the cell. All that remains of the cell

(A)　　　　**(B)**

50 μm　　　50 μm

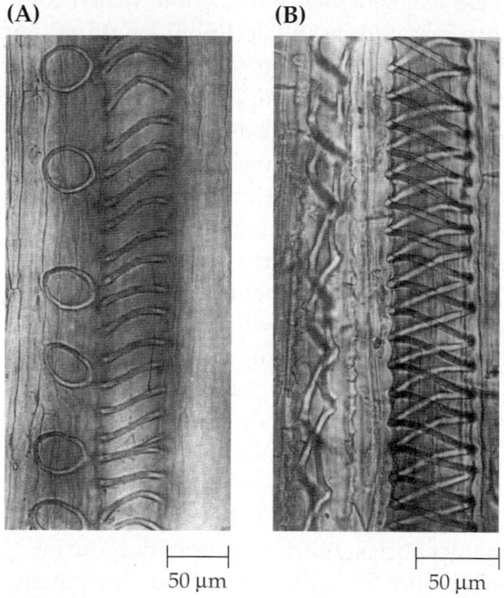

Figure 20.35
TE in the primary xylem of castor bean showing annular (A) and double helical (B) thickenings in the cell walls of vessel elements.

in its final differentiated form is the cell wall with its characteristic bands of secondary thickening.

20.12.2 *Zinnia* leaf mesophyll cells can be induced to form TE in culture.

TE ordinarily derive from the procambium and cambium, although the process can be reproduced in vitro by using cultured *Zinnia* mesophyll cells. These cells can be isolated from intact *Zinnia* leaves by grinding leaf pieces with a mortar and pestle. When mechanically isolated cells are incubated in a culture medium containing auxin and cytokinin, as many as 40% to 60% dedifferentiate and then redifferentiate into TE, all in the absence of cell division (Fig. 20.36). Differentiating TE develop the banded secondary thickening of the cell walls characteristic of those found during differentiation in vivo. Shortly after the secondary wall thickenings become visible, the tonoplast ruptures. Perhaps this is the committed step to death, because it is followed by breakdown of the remaining organelles and ultimately by loss of the cell's contents. The breakdown of cellular contents is paralleled by an increase in the activity of various degradative enzymes, including DNases, RNases, and proteases.

This in vitro induction of tracheary differentiation provides an ideal system for studying the differentiation and cell death processes involved in the development of the cells. A large population of cells can be induced to undergo the process more or less synchronously, which allows biochemical and molecular analyses of the events accompanying differentiation. In vivo, in contrast, only a few cells differentiate at a time, and these are embedded within other tissues. Study of the in vitro *Zinnia* system has allowed investigators to divide the process of TE formation into three distinct stages, which take place over four days.

The expression patterns of many genes change during stages 1 to 3 of TE differentiation in *Zinnia*, as shown in Figure 20.37. Genes expressed during the dedifferentiation phase (stage 1) include those involved in the wound response in plants, such as protease inhibitor genes *ZePI1* and *ZePI2*, and those involved in protein synthesis, such as ribosomal protein (*RP*) and elongation factor (*EF*) genes. Interestingly, genes involved in phenylpropanoid metabolism are expressed both early and late in TE differentiation in *Zinnia* cells. Whereas expression of the genes encoding PAL and cinnamic acid-4-hydroxylase (C4H) during formation of the secondary wall thickenings (TE-specific development, stage 3) might be expected, the finding that these two genes also are

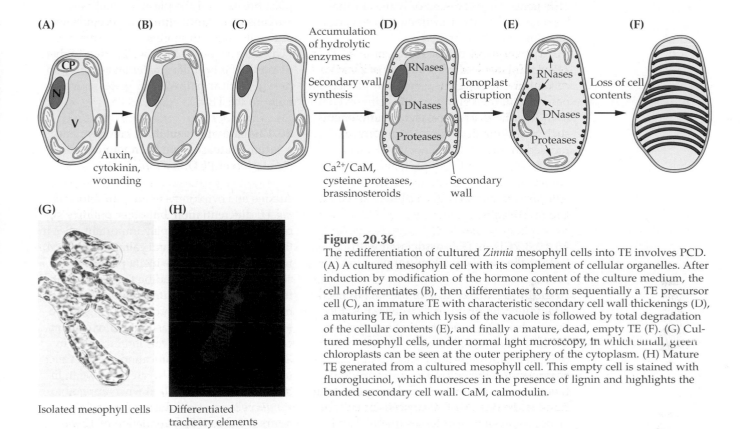

Figure 20.36

The redifferentiation of cultured *Zinnia* mesophyll cells into TE involves PCD. (A) A cultured mesophyll cell with its complement of cellular organelles. After induction by modification of the hormone content of the culture medium, the cell dedifferentiates (B), then differentiates to form sequentially a TE precursor cell (C), an immature TE with characteristic secondary cell wall thickenings (D), a maturing TE, in which lysis of the vacuole is followed by total degradation of the cellular contents (E), and finally a mature, dead, empty TE (F). (G) Cultured mesophyll cells, under normal light microscopy, in which small, green chloroplasts can be seen at the outer periphery of the cytoplasm. (H) Mature TE generated from a cultured mesophyll cell. This empty cell is stained with fluoroglucinol, which fluoresces in the presence of lignin and highlights the banded secondary cell wall. CaM, calmodulin.

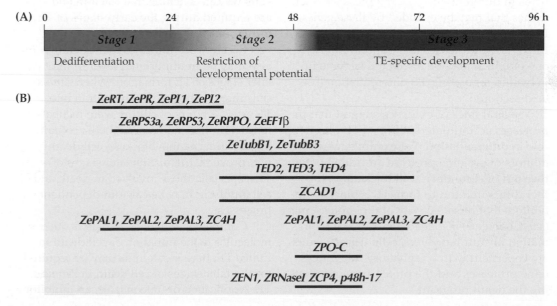

Figure 20.37

Stages of gene expression during *Zinnia* mesophyll redifferentiation. TE differentiation has been divided into three stages totaling about 96 hours. The identities of most of the genes expressed during these stages are discussed in Sections 20.12.2 and 20.12.3 of the text. Not mentioned in the text are the stage 3 genes *ZRNaseI* and *ZPO-C* (a peroxidase gene).

expressed in stage 1 is surprising. The onset of stage 2, during which the developmental potential of the cultured cells narrows to TE formation, is characterized by the appearance of transcripts for *ZCAD1* (encoding cinnamoyl alcohol dehydrogenase) as well as TE differentiation–related (*TED*) transcripts. *TED* genes are expressed only in cells that become TE, but the function of these genes remains to be determined.

The specialized thickenings of mature TE are laid down during stage 3 of *Zinnia* cell culture. Among the genes that participate in wall thickening are those involved in the synthesis of cytoskeleton components such as tubulin (TUB), which are first expressed in stage 1. Immunocytochemistry has shown that cell wall proteins such as arabinogalactan-like and extensin-like proteins as well as novel glycoproteins accumulate in TE walls.

20.12.3 PCD in TE is associated with lysis of the vacuolar membrane.

After the secondary wall of the TE has been laid down, the protoplast degenerates. Electron microscopy indicates that autolysis of the cell's contents is brought about when the tonoplast ruptures and releases lytic vacuolar enzymes into the cytosol. Genes encoding lytic enzymes are expressed during late stage 2 and early stage 3 of *Zinnia* cells in culture. Genes for cysteine proteases (*p48h-17* and *ZCP4*) and an S1-type nuclease (*ZEN1*) have been cloned, and biochemical evidence confirms in differentiating TE the presence of enzymes that may be encoded by these genes. In addition to cysteine proteases, at least three serine proteases are found in *Zinnia* TE, one of which may be a secreted enzyme involved in regulating TE differentiation (see next section).

Several lines of evidence suggest that the proteases accumulated in *Zinnia* TE play a role in differentiation. For example, when inhibitors of cysteine protease are added before the start of secondary wall thickening, further differentiation to form TE is blocked. Inhibitors that block activity of the proteasome (see Chapter 9) also lead to arrest of TE formation in cultured *Zinnia* cells. These results are consistent with a regulatory role for cysteine proteases and the proteasome in initiating the death program.

PCD in *Zinnia* lacks the hallmark cellular events associated with apoptosis and may represent a unique form of cell death. Rupture of the tonoplast is accompanied by changes in the organization of cellular organelles and the cell wall (Fig. 20.38). Lignification of the cell wall occurs after the tonoplast breaks, and the plasma membrane remains intact until almost all recognizable organelles have disappeared. This form of PCD may be particularly well suited to the disposal of all cellular materials required to provide a clear pathway for water movement in the TE.

20.12.4 Growth regulators, calcium, and a serine protease have been implicated in regulation of PCD in *Zinnia* TE.

Auxins and cytokinins are required for *Zinnia* TE differentiation, but other putative signaling molecules also play important roles in this process. Calcium and calmodulin (CaM) have been shown to be likely players in the signaling pathways leading to TE differentiation. Removal of calcium from the culture medium reduces TE formation from 50% to less than 10%, and measurement of cellular calcium indicates that this ion increases in *Zinnia* cells before lignification begins. Calcium channel blockers also are effective in inhibiting TE differentiation when applied to *Zinnia* cells any time during the first 48 hours of culture. Differentiation of TE can also be blocked by applying CaM antagonists to the culture medium. From these results we can conclude that calcium and CaM are required during the early stages of TE formation.

Recent work indicates that calcium also may be an important player in the later stages of TE formation. When *Zinnia* cells that had begun differentiation into TE after 72 hours of culture were incubated in calcium-free media or in media containing calcium channel blockers, cell death was prevented. Thus the later stages of TE, especially those involved in death and cell autolysis, may be calcium-dependent processes.

Compelling evidence also links other molecules to the initiation of cell death in *Zinnia* TE. Brassinosteroids may be required for the later stages of cell death in *Zinnia* TE. Application of uniconazole, an inhibitor

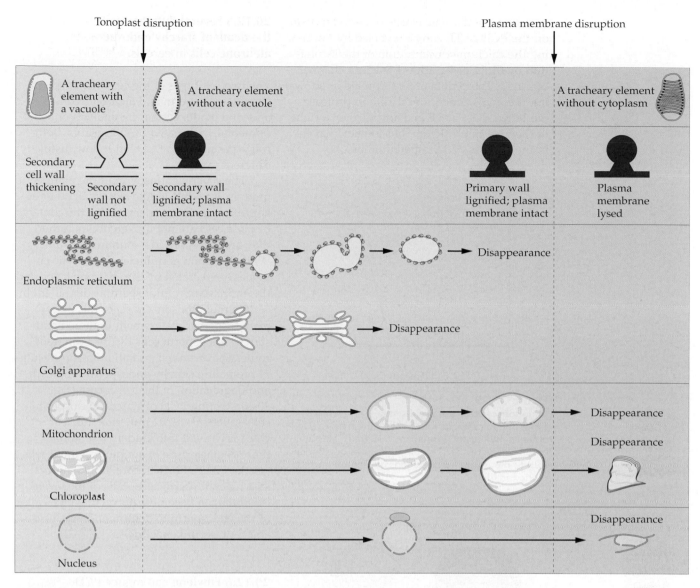

Figure 20.38
Structural changes in the organelles of differentiating *Zinnia* TE. The sequence of organellar changes is shown in relation to the status of the tonoplast and plasma membrane.

of brassinosteroid synthesis (see Chapter 17), prevents TE formation (Fig. 20.39) and specifically represses genes that are expressed late in differentiation (see Fig. 20.37). In one experiment, for example, uniconazole had no effect on the accumulation of *TED* transcripts but down-regulated the transcripts encoding PAL and C4H. The effects of uniconazole on TE differentiation and on gene expression could be reversed completely by the application of brassinolide, indicating that uniconazole indeed was affecting TE development through an effect on brassinosteroid biosynthesis.

A secreted serine protease that accumulates in the incubation medium of cultured *Zinnia* also has been shown to affect the PCD of TE. Several types of experiments show that this serine protease may play a role in PCD of cultured *Zinnia* cells. First, addition of soybean trypsin inhibitor to TE culture media blocks cell death and arrests TE differentiation. The soybean protein also blocks the activity of a 40-kDa serine protease that accumulates in the medium coincident with cell death. Second, proteolytic enzymes such as trypsin or papain dramatically accelerate cell death of TE when added to culture media. The effects of added trypsin on the PCD of TE can be reversed by lowering the calcium concentration of the incubation medium or by adding calcium channel blockers. These data are interpreted as showing the involvement of calcium in protease-mediated PCD in TE and add further weight to the idea that calcium ions are key players in the process of TE differentiation.

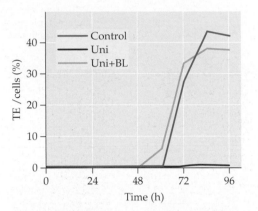

Figure 20.39
Inhibitors of brassinosteroid synthesis prevent TE formation in cultured *Zinnia* cells. Uniconazole (Uni) at 5 μM inhibits the formation of TE (red). This inhibition can be reversed by the addition of brassinolide (BL) to the culture medium (green). Control data are plotted in blue.

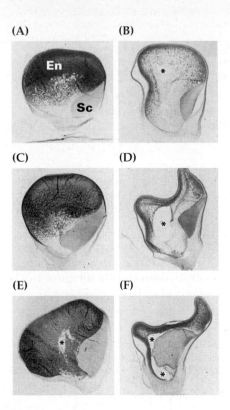

Figure 20.40
Endosperm (En) development in wild-type and *shrunken2* maize genotypes. The *shrunken2* mutant undergoes premature degradation of the starchy endosperm to yield cavities (*). Shown are kernels of wild type (left panels) and *shrunken2* (right panels) at 28 days (A, B), 32 days (C, D), and 40 days (E, F) after pollination. The starch is stained with iodine–potassium iodide. Sc, scutellum/embryo.

20.12.5 Several different pathways lead to the death of starchy endosperm and aleurone cells in cereals.

The two cell types of the cereal endosperm, starchy endosperm and aleurone, constitute a second model system for analysis of developmentally regulated PCD in plants. Both cell types undergo PCD but follow distinctly different routes.

The starchy endosperm of the mature grain is dead, but unlike almost all other eukaryotic cells that undergo PCD, the contents of these cells are not broken down; rather, they are preserved in a mummified state. When the grain germinates, hydrolytic enzymes are secreted by the aleurone layer, and the entire starchy endosperm is degraded. In certain mutations in *Zea mays*, this pattern of endosperm development is altered and starchy endosperm cells die by a different route. The *shrunken2* mutation in corn results in premature starchy endosperm cell death and degradation of the cell corpse (Fig. 20.40). During PCD of starchy endosperm cells in the *shrunken2* mutation, nuclear DNA is cleaved into ladders resembling those characteristic of apoptotic animal cells (see Box 20.1). Furthermore, unlike normal starchy endosperm cells, cells of the *shrunken2* mutation become autolyzed, which causes the endosperm to deform and produces aberrant, shrunken kernels.

20.12.6 Ethylene can induce PCD in starchy endosperm cells.

An interesting aspect of cell death in the starchy endosperm of corn concerns the role of ethylene. Much more ethylene is produced in kernels of the *shrunken2* mutant than in the wild type. Moreover, treatment of developing wild-type kernels with ethylene increases the number of autolytic cell deaths and produces aberrant kernels (Fig. 20.41). Additional evidence that ethylene might be an important player in regulating PCD in corn kernels comes from experiments with the ethylene inhibitor aminoethoxyvinylglycine (AVG). AVG reduced the extent of DNA fragmentation in the *shrunken2* mutant and decreased the size of the central cavity formed in the aberrant kernel (see asterisks in Fig. 20.40).

20.12.7 GA and ABA regulate cell death in cereal aleurone.

Unlike the cells of the starchy endosperm, aleurone cells remain alive until germination and endosperm mobilization are complete. Death of the cereal aleurone was first described more than 100 years ago by Gottlieb Haberlandt, who worked with maize, oat, rye, and wheat grains. Haberlandt characterized the changes taking place within the cereal aleurone during germination, including progressive vacuolation, followed by the death and collapse of the protoplast. Aleurone cells undergo autolysis and die after the storage reserves of the starchy endosperm have been mobilized (see Fig. 20.3). GA and ABA tightly regulate this process. Whereas GA stimulates the onset of PCD in the aleurone layer of barley and wheat, ABA postpones PCD (Fig. 20.42). The effects of ABA on cell death in barley aleurone are dramatic. ABA-treated aleurone protoplasts can be kept alive for more than six months, whereas after GA treatment of protoplasts most cells die in five to eight days.

20.12.8 Aleurone storage vacuoles become lytic organelles after treatment with GA.

The protein storage vacuoles (PSVs) of aleurone cells become acidic, lytic compartments within the first few hours after treatment with GA. Aleurone cells contain PSVs that have a luminal pH of around 7, but after three to four hours of GA treatment, the lumen of these vacuoles reaches pH 5.5. The vacuoles of GA-treated cells also accumulate a spectrum of acid hydrolases, including several aspartic and cysteine proteases, and nuclease activities. Vacuoles in ABA-treated cells that do not undergo PCD lack these enzyme activities and maintain a pH near neutrality.

Although PSVs of GA-treated aleurone cells are lytic organelles, their role in PCD is not understood. Unlike cell death in *Zinnia* TE, there is no evidence that the tonoplast of this cell ruptures and allows hydrolases access to the cell's contents. Furthermore, microscopy does not show the presence of any autophagosomes that could deliver the cytosol to the vacuole for degradation, as were seen in yeast (see Fig. 20.2A). What has be-

come clear from studying PCD in aleurone cells microscopically is that death occurs along a novel route. Aleurone cell death does not follow the apoptotic route, and autophagosomes do not appear to participate in autolysis. Aleurone cells undergoing PCD are dismantled in an orderly way that preserves the compartmentation of the cell's organelles. The aleurone degradation pathway may involve ubiquitin and the proteasome.

(A)   **(B)**

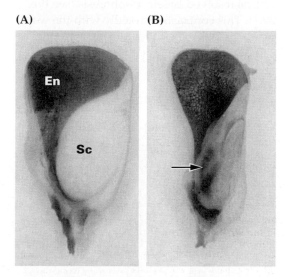

Figure 20.41
Ethylene can cause kernel deformation in wild-type maize kernels. Shown are a control kernel of wild-type maize (A) and an ethylene-treated kernel (B) 32 days after pollination. The arrow points to dead cells in the scutellum. The starch is stained with iodine–potassium iodide. En, endosperm; Sc, scutellum/embryo.

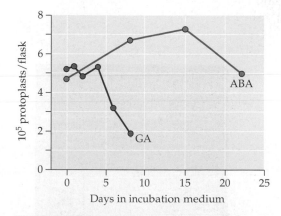

Figure 20.42
Gibberellic acid (GA) promotes PCD and abscisic acid (ABA) delays PCD in barley aleurone cells. Aleurone cells were incubated in a medium containing 5 μM GA or 25 μM ABA; cell death was monitored by counting the number of live cells.

20.12.9 DNA degradation in aleurone cells does not result in the formation of apoptotic ladders.

Degradation of DNA in the aleurone cell provides further evidence that PCD does not follow the apoptotic route. During apoptosis, DNA undergoes internucleosomal cleavage, generating fragments of approximately 180 bp. These fragments appear as characteristic ladders when DNA from dying cells is resolved by electrophoresis (see Box 20.1). This contrasts markedly with the way in which DNA is cleaved during aleurone cell death. DNA degradation in aleurone cells is a late event in PCD (Fig. 20.43A) and begins only after almost the entire contents of the cell have been autolyzed (see Fig. 20.3). In addition, DNA in aleurone cells is degraded into small fragments that are not resolvable into the 180-bp fragments seen during apoptosis (Fig. 20.43B).

20.12.10 Many signal transduction pathways participate in PCD of cereal aleurone.

The signal transduction cascades leading to the GA-induced synthesis and secretion of hydrolases in cereal aleurone have been studied extensively; cytosolic free Ca^{2+}, cytosolic pH, cGMP, and CaM as well as protein kinases and protein phosphatases have been identified as elements of these signaling pathways. Much less is known about the signaling components that promote GA-induced death in cereal aleurone cells. The protein phosphatase inhibitor okadaic acid was found to block GA-induced increases in cytosolic Ca^{2+} and prevented cell death in wheat aleurone layers. It is therefore likely that a protein phosphatase is part of the signaling cascade in this cell type. Microinjection into barley aleurone protoplasts of syntide-2, a synthetic substrate for Ca^{2+}- and CaM-dependent protein kinases, selectively inhibited α-amylase secretion, the expression of the GUS-reporter gene under the control of the α-amylase promoter, and the development of large vacuoles characteristic of the GA-response, but the inhibitor did not affect changes in cytosolic Ca^{2+}. This experiment indicates that protein phosphorylation is also involved in the signaling cascade leading to barley aleurone PCD, because only protoplasts that reach the highly vacuolated stage are destined to die.

Blocking a cGMP-linked GA-signaling pathway with the guanyl cyclase inhibitor LY83583 reduced expression of genes encoding α-amylase and the transcription factor GAMyb and diminished α-amylase secretion. LY83583 also prevented GA-induced increases in nuclease activities as well as DNA degradation and cell death. This makes cGMP a likely candidate for participation in the GA signal transduction pathway that leads to cell death in barley aleurone cells.

(A)

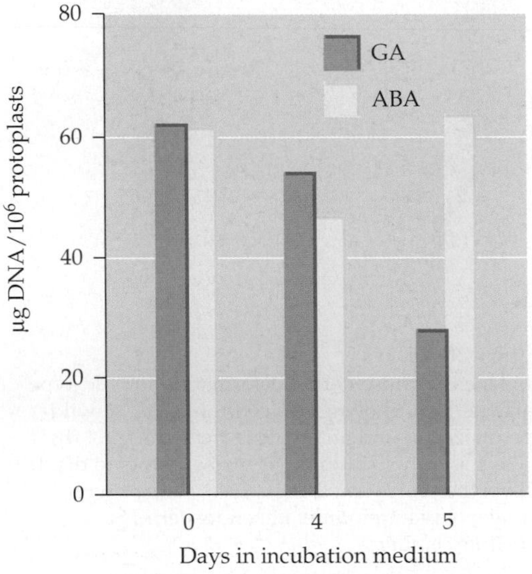

(B)

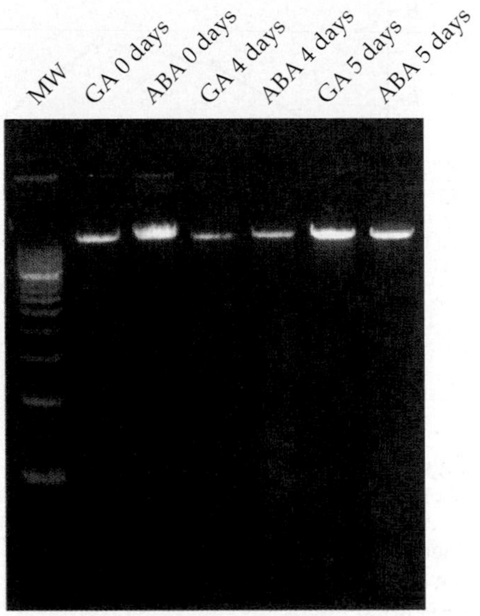

Figure 20.43
DNA degradation is a late event in aleurone PCD. (A) Aleurone cells begin to lose DNA only after they become committed to die. (B) DNA breakdown is not accompanied by formation of the 180-bp ladders characteristic of apoptosis (see figure in Box 20.1 for comparison). MW, molecular mass markers.

20.13 Examples of PCD as a plant response to stress: formation of aerenchyma and the HR

20.13.1 Aerenchyma formation is a response to flooding, which limits the availability of oxygen to roots.

The roots of many plants that are not adapted to wetland conditions often form aerenchyma in response to oxygen deficiency (see Fig. 20.8). Aerenchyma formation has been extensively studied in the roots of Z. *mays* in response to oxygen-limiting conditions as well as other environmental stresses, including the depletion of nitrogen and phosphorus from soil solution. Aerenchyma results from PCD of a highly specific group of root cortical cells located between the endodermis and hypodermis. The response to oxygen stress can be very rapid and results in the removal of the entire cortical cell including the cell wall, forming a space that facilitates the movement of oxygen to the roots.

20.13.2 Ethylene mediates aerenchyma formation.

Although aerenchyma formation in roots was discovered initially as a response to hypoxic conditions, PCD in the root cortex is now understood to be caused by increased ethylene concentrations in the root. Maize roots grown under low-oxygen conditions contain higher amounts of both ethylene and its precursor, ACC. Inhibitors of ethylene synthesis, such as AVG, or ethylene action, such as silver ions, can prevent the formation of aerenchyma under oxygen-limiting conditions. Finally, exogenously supplied ethylene can bring about aerenchyma formation under aerobic conditions, providing strong support for the idea that ethylene regulates aerenchyma formation.

Treatment of corn roots with low concentration of oxygen or with ethylene brings about an increase in the quantities of several cell wall–modifying enzymes, including cellulase and xyloglucan endotransglycosylase (XET; see Chapter 2). Cellulase activity, measured by the depolymerization of carboxymethylcellulose, increases dramatically in the apical region of corn roots grown in 4% oxygen, relative to that in roots grown in

21% oxygen as controls (Fig. 20.44A). This experiment found a strong correlation between cellulase accumulation and activity of ACC synthase, a key enzyme in ethylene biosynthesis (Fig. 20.44B). Hypoxia and ethylene also bring about an increase in the abundance of a transcript that encodes XET, an increase in gene expression that is prevented by the ethylene biosynthesis inhibitor aminooxoacetic acid. Other enzymes likely to be involved in PCD during aerenchyma formation include cysteine proteases, the activities of which also increase in conditions of oxygen starvation.

20.13.3 Calcium mediates the signaling that links hypoxia sensing to ethylene production during aerenchyma formation.

Cytosolic calcium is part of a signal transduction pathway linking both hypoxia and ethylene to PCD in maize cells. Cultured maize cells that respond to hypoxia by

(A)

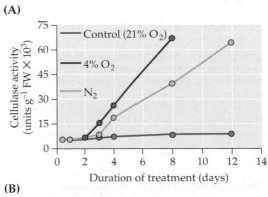

(B)

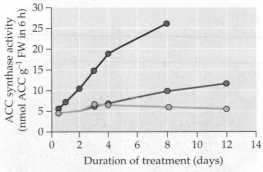

Figure 20.44
Increases in the activities of cellulase (A) and of a rate-limiting enzyme in ethylene biosynthesis, ACC synthase (B), accompany aerenchyma formation in anoxic maize roots. Both enzymes increase with the duration of hypoxia in maize roots.

undergoing PCD show reversible changes in their concentrations of cytosolic free calcium (Fig. 20.45). Cytosolic calcium increases from a resting value of about 50 to 100 nM in cells incubated under aerobic conditions to 250 to 300 nM under hypoxia. These changes in cytosolic calcium are relatively rapid and can be detected within minutes of imposing hypoxia. Changes in cytosolic calcium in cultured maize cells occur independently of the presence of calcium in the incubation medium. Even when incubated in calcium-free media, the cells respond to hypoxia with an increase in cytosolic calcium. Internal stores of calcium in mitochondria are the probable source of calcium that contributes to the cytosolic calcium signal during anaerobiosis in cultured maize cells.

Experiments with intact maize root cells support the idea that cytosolic calcium is involved in signaling aerenchyma formation. Compounds that are known to increase cytosolic calcium enhance aerenchyma formation in roots under aerobic conditions. In contrast to experiments with cultured maize cells, using the calcium chelator EGTA to remove calcium from the medium surrounding the roots prevented aerenchyma formation (Fig. 20.46A) and inhibited the increase in cellulase activity under hypoxia, suggesting that at least in intact roots, extracellular calcium is important for cellular calcium homeostasis. CaM also may be a component of the signal transduction pathway linking hypoxia to aerenchyma formation in corn roots; the CaM antagonist W7 inhibited the production of cellulase and aerenchyma formation under hypoxia.

Specific inhibitors have been used to demonstrate the probable involvement of G-proteins, protein kinases, and protein phosphatases in regulating aerenchyma formation in corn roots in response to both oxygen and ethylene. GTPγS, which locks G-proteins into the active state, and okadaic acid, which inhibits protein phosphatase activity, each promote the formation of cellulase and aerenchyma in roots under aerobic conditions (Figs. 20.46B and C). On the other hand, an inhibitor of protein kinase activity prevented the formation of aerenchyma in roots under hypoxia. Taken together, these data implicate cytosolic calcium and CaM in a signaling cascade that could activate protein kinases and protein phosphatases to promote modification of proteins involved in PCD. As with the differentiating *Zinnia* TE and cereal aleurone cells, the root aerenchyma promises to be an excellent model system for unraveling the mechanisms of PCD in plants.

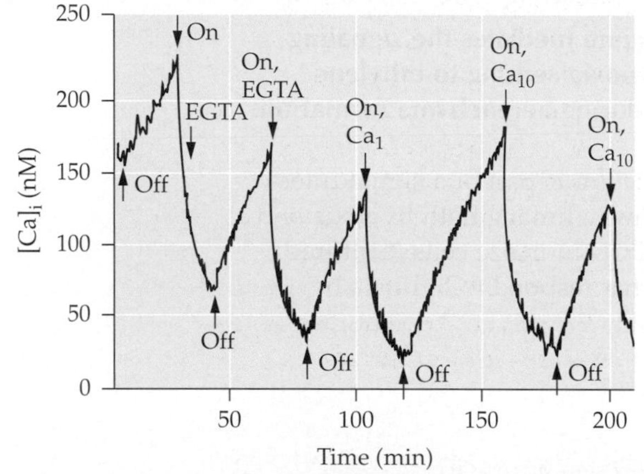

Figure 20.45
Cytosolic calcium [Ca]$_i$ changes rapidly in response to hypoxia in cultured maize cells. Cells were initially subjected to hypoxia by turning off the aerator that circulated buffer around the cells (off). The buffer was reaerated (on), and the cells were then perfused with the calcium chelator EGTA (0.2 mM). The effects of adding EGTA, 1 mM CaCl$_2$ (Ca$_1$), or 10 mM CaCl$_2$ (Ca$_{10}$) to the bathing medium were investigated during subsequent cycles of hypoxia and reaeration. Note that changes in cytosolic calcium were relatively unaffected by changes in external calcium.

20.13.4 Cell death is a common response to attack by pathogens.

One of the best-studied examples of PCD in plants is the cell death that occurs in response to attempted invasion by pathogens.

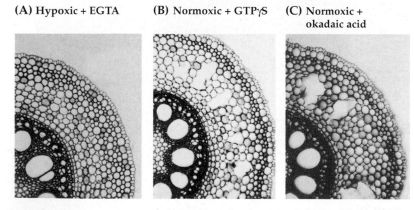

(A) Hypoxic + EGTA **(B) Normoxic + GTPγS** **(C) Normoxic + okadaic acid**

Figure 20.46
Aerenchyma formation is influenced by agonists and antagonists of signal transduction pathways. EGTA prevents aerenchyma in maize roots under hypoxic conditions (A), whereas GTPγS (B) and okadaic acid (C) induce aerenchyma formation under aerobic conditions.

Host-cell death is a nearly ubiquitous feature of one type of resistance response in plant–pathogen interactions. In its most recognizable form, the HR, this host-cell death is manifest as the rapid collapse of tissue (Fig. 20.47). As detailed below, the HR is programmed genetically in the plant and is a consequence of new host transcription and translation. It is also a correlative feature of many, but not all, incompatible interactions between disease-resistant plants and avirulent pathogens. These incompatible interactions are controlled by disease resistance (R) genes that enable the plant to recognize and respond to pathogens carrying specific avirulence (Avr) genes (see Chapter 21). In the absence of either the R allele in the host or the Avr allele in the pathogen, a so-called compatible reaction results, in which the plant does not recognize the pathogen, and disease ensues. Whether the cell death that constitutes the HR actually supports disease resistance by depriving the incoming pathogen of nutrients or by releasing microbiocidal compounds from dying cells is unclear. Alternatively, the HR could be the consequence of a mechanism that kills both host and microbe cells. Subsequent biosynthesis of protective secondary metabolites and cell wall buttressing around the HR site is thought to contribute to overall pathogen containment. Cell death, however, also can be a feature of disease symptoms during compatible interactions. This form of cell death may not be programmed by the plant but rather may be the result of killing by the pathogen, often through the action of a pathogen-derived toxin.

20.13.5 Is cell death during the HR suicide or murder?

Recent investigations have been aimed at genetic and biochemical dissection of the mechanisms of the HR and associated cell death. A key question is whether the cell death that occurs in the HR is programmed genetically by plant cells (suicide) or is entirely a consequence of pathogen-derived products killing plant cells (murder). Several lines of evidence suggest that the cell death associated with the HR in response to pathogens may, in fact, be programmed by the host rather than result from direct killing by the pathogen.

First, as mentioned above, the HR is an active process, requiring transcription and translation on the part of the host. Second, purified preparations of elicitors can induce the full HR response when applied to a resistant host.

Elicitors are molecules produced either by the pathogen or as a result of the host–pathogen interaction. Known active elicitors include peptides and oligosaccharides. When applied to a plant containing a resistance gene that can recognize the pathogen, the purified elicitor induces the biochemical and cellular responses characteristic of an HR, including cell death. When applied to a susceptible host, however, it causes no visible response, indicating that the cell death seen on the resistant host is not the result of direct toxic effects of the elicitor (Fig. 20.48). This suggests that the resistance-gene–mediated recognition of the elicitor by the host is sufficient to trigger an endogenous cell death pathway. Finally, as discussed in Section 20.13.7, genetic evidence points to the cell death accompanying the HR as being completely programmed by the plant.

20.13.6 Lesion-mimic mutants form lesions in the absence of pathogens.

Mutants have been identified in many plant species that spontaneously form localized areas of dead tissue resembling those seen

Figure 20.47
The HR, a common form of plant resistance to pathogens, involves localized PCD. This tobacco leaf is from a plant with a gene that encodes resistance to tobacco mosaic virus (TMV). After inoculation with TMV, the plant responds to the attempted infection by inducing localized cell death at the points of infection. This HR cell death is seen as small areas of dead, brown tissue across the leaf.

MM   Cf9

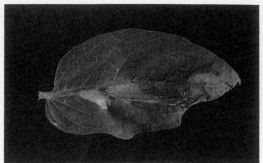

Figure 20.48
An elicitor preparation from the fungal pathogen *Cladosporium fulvum* induces cell death in resistant but not susceptible plants. The two tomato lines MM and Cf9 are genetically very similar, but differ in that Cf9 contains a gene that encodes resistance to the pathogen, whereas MM lacks the resistance gene. Resistance in Cf9 is manifested as a typical HR with the usual cell death at the point of infection. When the resistant Cf9 plant is treated with an elicitor preparation from the pathogen, cell death results. When leaves from the MM plant, which lacks the resistance gene, are treated with the same elicitor, there are no symptoms. Thus, the elicitor itself is not directly toxic to the leaf, but rather the recognition of the elicitor mediated by the product of the resistance gene triggers a cell death program in the resistant host.

in the HR. In these mutants, referred to as lesion-mimic mutants, HR-type lesions form in the absence of a pathogen (Fig. 20.49). The phenotypes of these mutants have led to their description as "paranoid plants" because they behave as if constantly under pathogen attack. Again, the fact that the entire defense response, including cell death, can occur without any input from a pathogen supports the hypothesis that HR involves an endogenous cell death program. Lesion-mimic mutations are found in a wide variety of plants, including maize and *Arabidopsis*.

A key question with mutants of this sort is whether they define genes for which the wild-type functions are involved in disease resistance. The mutants' cell death phenotypes also might result from perturbed metabolism or impaired signal exchange with neighboring cells. Such factors are sufficient to initiate PCD in animals, in which one or more central monitors interpret intra- and extracellular signals; when these signals are absent or altered, the monitor or monitors shunt the cell into a default suicide pathway. Thus, for an animal cell to survive, proliferate, and differentiate, it continuously must be reassured that its position and activities are appropriate. This notion of "social control" of cell death in animal cells also may operate in plant cells during cell death that is either developmentally programmed or the result of the HR.

20.13.7 *Arabidopsis* lesion-mimic mutations fall into two groups: those that trigger defense responses, and those that do not.

In some mutants, the spontaneous lesions exhibit the histochemical, molecular, and physiological markers associated with actual HR lesions. Significantly, other mutants that display dead cells, or other treatments that kill cells, do not trigger the defense–response pathways. It has therefore been possible to set criteria separating the broad class of cell death mutants into those defining genes for which the wild-type function can play a role in disease-resistance pathways and those that do not. The fact that mutations in single genes can result in lesions that resemble HR lesions suggests that the complete machinery necessary for carrying out the cell death indicative of HR must be intrinsically programmed in the plant. Collectively, these mutants offer the opportunity to decipher the signal transduction pathway of HR.

20.13.8 Lesion-mimic mutations in defense pathways result in improper initiation of HR or failure to limit the spread of HR.

Bona fide HR lesions form specifically at the site of pathogen infection and terminate at a discrete boundary. Genetic evidence indicates two separate functions are involved in

Figure 20.49
Lesion-mimic mutants spontaneously develop areas of dead tissue, indicating a misregulation of the functions that control cell death. Lesion mimics have been identified in a wide range of plants, including maize (e.g., *lls1*, left panels) and *Arabidopsis* (e.g., *lsd* mutants, right panels). (A) Expression of the *lls1* phenotype is developmentally regulated. Older leaves on the lower part of the plant are more severely affected, while younger leaves at the top are much less lesioned. (B) *lls1* lesions can be triggered by mechanical wounding, such as pinpricks. *lls1* is a propagation type of lesion mimic, which means that, once initiated, the lesion continues to spread across the leaf. The amount of cell death seen on this leaf exceeds that directly caused by the mechanical damage. (C) Revertant sectors remain green in this *lls1* plant, showing that the *Lls1* function is cell-autonomous. *lsd1*, a conditional lesion-mimicking mutant, remains healthy when grown under permissive conditions (D), but when shifted to nonpermissive conditions, the plant develops areas of dead tissue (E). Like *lls1*, *lsd1* is a propagation type of lesion-mimicking mutant; the tissue death spreads out from the point of initiation until the whole leaf is dead. The mutant *lsd3* is also conditional, as shown under permissive (F) and nonpermissive (G) conditions. *lsd3* is an initiation type of lesion mimic, which means the lesions are determinate in size; once formed, they do not enlarge.

the development of these HR lesions: An initiation function directs the killing of cells at the point of attempted infection, and a second function limits the spread of this cell death once it has been initiated. These two functions are represented by two classes of lesion-mimic mutants that activate disease-resistance pathways. The first, initiation-class mutants, are characterized by discrete areas of necrosis that do not spread. Thus, these mutations may result from misregulation of lesion initiation functions, either in the signal receptor itself (possibly the product of a resistance gene) or in an early step in signal transduction. Supporting this idea are two examples of lesion-mimic mutations that map to known disease-resistance loci (e.g., *RP1* in maize and *mlo* in barley; see Section 20.13.9). Because some cloned resistance genes encode products containing well-known signaling motifs such as kinase domains or domains postulated to act as signal switches by binding nucleotide triphosphates, mutations in these domains could conceivably lead to a constitutive signaling phenotype. Thus, constitutive expression of resistance genes could result in a lesion-mimic phenotype. Alternatively, and by analogy to animal cells as mentioned above, the default pathway of a metabolically injured plant cell may be a form of PCD. If so, then some initiation-class mutants may lack certain negative regulators of the death program that ordinarily prevent activation of PCD pathways in cells receiving proper positional and developmental cues. Without the negative regulators, the cells undergo PCD despite receiving the signals that ordinarily prevent death.

The second class of lesion-mimic mutants, propagation-class mutants, are defective in their ability to limit lesion spread after it is initiated. This mutant phenotype suggests that during the formation of a wild-type HR resulting from an incompatible interaction, signals emanate outward to neighboring cells that can trigger their death. Once the HR site involves a particular number of cells, a second control system apparently inhibits its further spread. If this inhibiting mechanism is absent, any cell death accompanying the HR expands and can eventually consume the entire leaf.

Lesion-mimic mutants can be found naturally, isolated after mutagenesis, or generated by constitutive transgene expression. For example, transgenic tobacco plants ectopically expressing a bacterial proton pump gene develop spontaneous lesions on their leaves (Fig. 20.50). Many of the lesion-mimic mutants are developmentally regulated; consequently, their phenotypes can be exacerbated by manipulation of growth regulators, light regimen, or temperature.

Of some 32 different lesion mimics known in maize, 23 are dominant gain-of-function mutations, and most of these are initiation-class mutants. This is the largest class of gain-of-function mutants found in maize, and their prevalence suggests two divergent hypotheses: Either many dominant mutations can lead to a metabolic perturbation that drives cells down a default PCD pathway, or many of these mutations are in response networks that are ordinarily recruited to fend off pathogens. If the latter is true, then presumably a large number of input points can be activated in the disease resistance response pathway. This could, in turn, reflect a large number of resistance genes for which gain-of-function phenotypes could give rise to lesion mimics.

Some evidence suggests that mutations in disease resistance genes can, in fact, result in a lesion-mimic phenotype. The *Rp1* gene in maize encodes resistance to the rust fungus *Puccinia sorghi* and exhibits a high frequency of unequal crossing-over and gene conversion events. Several alleles responsible for lesion mimics of both the initiation and propagation classes map to the *Rp1* locus. These phenotypes lend further support to the idea that at least some lesion mimics may result from "resistance genes gone bad."

20.13.9 Initiation-class mutants can be associated with disease resistance genes.

As mentioned earlier, mutations at the *Rp1* locus of maize, which encodes resistance to a fungal pathogen, can result in plants with a lesion-mimic phenotype, including both propagation and initiation classes of lesion mimics. Another example of an initiation-type of lesion mimic associated with disease resistance is the barley *mlo* gene (Fig. 20.51). The *mlo* gene is unique among barley disease resistance genes in that it confers resistance to all known races of *Erysiphe graminis*, the fungus responsible for powdery mildew, but remains susceptible to non-*Erysiphe* fungi. The function of the wild-type *MLO* allele is unknown, but mutant *mlo* alleles result in thickening and apposition of cell wall material at the location of attempted penetration by fungal hyphae (Fig. 20.52). Thus, wild-type *MLO* presumably acts to down-regulate genes directing apposition formation. Cell-type–specific spontaneous cell wall appositions can occur in *mlo* plants grown in the absence of pathogen. These spontaneous cell wall thickenings are similar in chemical composition, structure, and tissue localization to appositions triggered by infection and may act as a physical barrier to pathogen attack.

Although apposition formation in response to fungal hyphae does not result in

Figure 20.50
Lesion-mimic phenotypes are sometimes seen in transgenic plants expressing foreign genes. The leaf shown is from a transgenic tobacco plant expressing a bacterial proton pump gene. The plant has not been inoculated with any pathogens but spontaneously develops lesions that resemble HR lesions (compare with Fig. 20.47).

cell death, spontaneous lesions can appear on *mlo* plants at low temperatures in the absence of a pathogen. The size and severity of these lesions depends on the genetic background and the *mlo* allele, suggesting that interplay between the *MLO* gene product and a variety of factors usually determines the extent of apposition formation. Whether there is a connection between disease resistance and cell death in the *mlo* plants is not clear, because resistance can occur in the absence of any cell death. To separate disease resistance from spontaneous cell death, one can select for suppressors of the *mlo* phenotype, but in both the original *mlo* allelic series and in the suppressors identified thus far, the amount of spontaneous lesion formation and the degree of resistance to *Erysiphe* are consistently correlated. The sequence of the *MLO* gene indicates it encodes a novel protein, probably one integrated into a membrane.

20.13.10 Propagation-class lesion-mimic mutants are defective in a function that blocks the spread of cell death once it has been initiated.

Propagation-class mutants, which negatively regulate the spread of at least some forms of cell death, are found in a wide variety of plants. Interestingly, light affects the development of lesions in propagation-class mutants in maize, *Arabidopsis*, and tomato. The light effect may be mediated by reactive oxygen intermediates, which are generated during photorespiration. In one of these mutants, the *Arabidopsis lsd1* mutant, the superoxide radical is required for lesion formation.

The maize mutant lethal leaf spot 1 (*lls1*), a recessive propagation-class mutant, causes developmentally programmed and environmentally influenced lesions that mimic those formed on a plant infected by the fungal

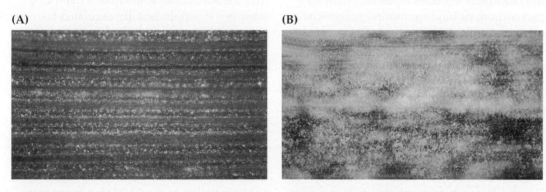

Figure 20.51
mlo genes confer resistance to powdery mildew in barley. Leaves of barley that carry *mlo* demonstrate resistance when inoculated with powdery mildew (A), whereas *MLO* plants are susceptible (B). The lesions on the resistant phenotype of barley are limited in comparison with those on the susceptible cultivar.

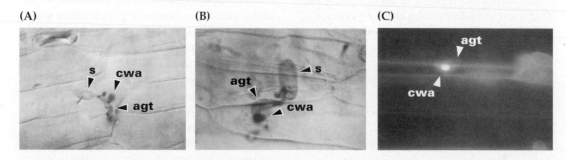

Figure 20.52
Histochemical analysis of leaf cells 48 hours after inoculation with *Erysiphe graminis* f.sp. *hordei*. The *mlo* mutant of barley forms cell wall appositions (cwa) in response to penetration by the fungal appressorial germ tube (agt). Staining with resorcinol blue (A) and lacmoid (B) allows visualization of the cwa, agt, and fungal spore (s). Callose in the cwa fluoresces when stained with analine blue (C).

pathogen *Cochliobolus carbonum.* Mechanical wounding of the plant tissue, such as puncturing it with a pin, will also induce lesion formation (see Fig. 20.49). *LLS1* function is cell-autonomous because revertant sectors do not die, despite being surrounded by dead tissue. Lesion formation in *lls1* mutants requires light, suggesting that reactive oxygen intermediates resulting from photorespiration may play a role in the spread of cell death. These observations imply that a threshold level of tolerance to reactive oxygen intermediates is exceeded in mutant plants, especially when the tissue is actively photosynthetic, and that an excess of reactive oxygen intermediates may be the trigger initiating cell death. The *LLS1* gene product must remove these cell death–promoting factors or inhibit their action as a signal for PCD, although preliminary experiments involving quenchers of free radicals have not stopped lesion formation.

DNA sequence analysis indicates that the *LLS1* gene encodes a protein containing two consensus binding motifs characteristic of phenolic dioxygenases, which suggests that the protein may be involved in modification of a phenolic signal that mediates cell death. A highly homologous cDNA exists in *Arabidopsis.* Thus *LLS1* encodes a cell death–controlling factor that may be conserved among dicots and monocots.

Two propagation-class mutants have been identified in *Arabidopsis.* The recessive *lsd1* mutant forms propagation-class lesions. Its wild-type product apparently is necessary for limiting lesion size during the HR. *lsd1* plants form lesions in response to inoculation with small amounts of a wide variety of pathogens. The same response follows treatment with salicylic acid or other chemicals that induce **systemic acquired resistance** (SAR), an enhanced resistance that occurs throughout a plant after infection by a necrotizing pathogen. Lesions do not form, however, in response to structural analogs of salicylic acid that do not trigger SAR, to heat-killed bacteria, or to mechanical wounding (thus differentiating the *lsd1* phenotype from maize *lls1*). Recently, the superoxide radical was identified as a key signal molecule for both initiation and propagation of *lsd1* lesions, again implicating reactive oxygen intermediates in the spread of cell death. The *LSD1* gene encodes a zinc finger

protein of a class known in animals to act as transcription factors. *LSD1* also may be a transcription factor, possibly acting either as a repressor of genes required for cell death or, alternatively, as an activator of genes that block cell death. The *Arabidopsis* accelerated cell death 1 (*acd1*) mutant also forms propagative lesions similar to *lsd1*. However, like the maize *lls1* mutant but unlike *lsd1*, *acd1* plants also form lesions in response to abiotic stresses, such as mechanical stress. Because *acd1* and *lsd1* are nonallelic, at least two nonredundant functions in *Arabidopsis* are required for halting HR-related cell death by way of negative regulation.

Propagation-class mutants by no means are confined to *Arabidopsis* and maize. One of the earliest reported lesion mimics, *ne,* arose from attempts to breed resistance to tomato leaf mold, *Cladosporium fulvum,* into a commercial tomato line. Small brownish lesions formed near the tips and midribs of the lower leaves subsequently enlarge to engulf the leaves and sometimes extend to the stem in a developmentally regulated fashion. Exposure to intense light enhances the extent of lesion formation, again raising the possibility of a role for reactive oxygen intermediates in the cell death process. Initial experiments suggested that necrosis seemed to be a result of the interplay between the *ne* locus and the *C. fulvum* resistance gene *Cf-2.* Recent results, however, suggest that *ne* is a dominant gene linked to *Cf-2* but that *Cf-2* itself does not interact with *ne* to cause lesions. Isolation of the *Cf-2* locus may shed light not only on how *NE* functions but also on how its function is tied to that of *Cf-2* in the generation of cell death during a normal resistance response.

The Sekiguchi lesion (*sl*) mutant is a naturally occurring recessive propagation mutant found in rice. The lesions on this mutant resemble responses to the ordinarily avirulent blast fungus *Bipolaris oryzae.* They are also triggered by avirulent isolates of *Pyricularia oryzae* but are not seen in response to virulent fungal isolates, supporting the idea that the *sl* mutant is also impaired in the pathway that normally limits a successful HR to cells at or near the infection site. *sl* lesions form in response to applied chemicals such as sodium hypochlorite, pentachlorophenol, and organophosphates. Perhaps these *sl*-inducing chemicals mimic

Chapter 20 Senescence and Programmed Cell Death

endogenous messengers of HR signal transduction in rice. Alternatively, these compounds may simply kill plant cells and thereby trigger propagative cell death. In either scenario, the *sl* mutation should be characterized further in the context of the maize and *Arabidopsis* mutants discussed above. Mapping *sl* should be interesting to determine whether, like maize *Rp1* alleles, it represents a mutation in a specific blast resistance gene.

20.13.11 Cell death phenotypes are not limited to foliar tissues.

The *rn* (*root necrosis*) mutant of soybean exhibits lesion-positive roots (Fig. 20.53) but is partially resistant to an isolate of fungus that is pathogenic to the parental genotype. Additionally, defense gene transcription and the subsequent accumulation of antimicrobial compounds known as phytoalexins are observed in this mutant. No phenotype is seen in the leaves of these plants, a difference from the *lsd1* and *lls1* mutations, which influence the phenotypes of a variety of tissues, including floral organs. Thus, the diversity of foliar cell death phenotypes and the various developmental and environmental controls that define their onset and severity may reflect a set of pathways for control of cell death. Whether this variety represents interplay between developmentally cued and pathogen-invoked PCD remains to be seen.

20.13.12 Lesion-mimic mutants primarily affect cell death in the context of the pathogen defense response but apparently have no effect on other developmental cell death events.

The fact that many mutants affecting pathogen-related cell death are known, whereas far fewer mutants affecting developmental cell death have been identified, probably reflects two factors. First, because of the economic importance of plant disease, the plant defense response has been the subject of intensive research. Second, the mutants affecting cell death events that occur during development are more likely to result in lethality. If, for example, TE death

does not occur as normally programmed, a functional xylem will not develop and the plant will not survive long. On the other hand, mutations affecting defense response are often conditional, because the pathway is required only if the plant is attacked by a pathogen. In the absence of a pathogen, many plants can survive and reproduce despite having defects in defensive pathways.

20.13.13 Overexpression of foreign proteins in plants can result in a lesion-mimic phenotype.

Lesion-mimic phenotypes occasionally are seen in transgenic plants engineered to overexpress various proteins (see Fig. 20.50). Among the introduced genes that have resulted in lesion-mimic phenotypes are a variant of a polyubiquitin gene, a yeast invertase gene, the gene for cholera toxin (which inhibits GTPases), and a bacterial

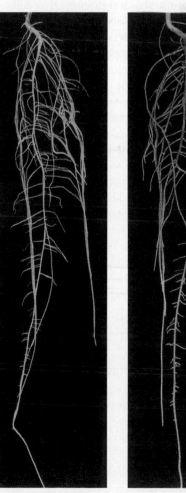

Figure 20.53
The phenotype of the soybean lesion-mimic mutant *rn* (*root necrosis*) is sensitive to photoperiod, suggesting that shoot-derived factors (possibly involving reactive oxygen species generated during photosynthesis) may modulate the induction of root lesions. In *rn* plants grown under a 24-hour photoperiod (right), extensive root necrosis results. This phenotype is not present in wild-type plants grown under the same photoperiod (left).

proton pump gene. Although all the transgenes that result in a lesion phenotype possibly impinge on cell death pathways, it is perhaps more likely that in many cases, overproduction of a particular protein in a cell results in metabolic perturbation and this is what causes the cell to commit suicide. If so, that implies a system existing in the cell that senses this deviation from homeostasis and as a result triggers an endogenous cell death program.

20.13.14 Reactive oxygen intermediates are a key trigger in PCD accompanying the HR.

The production of reactive oxygen intermediates within hours of pathogen inoculation is known as the oxidative burst. In both compatible and incompatible interactions, an oxidative burst occurs within the first hour after inoculation. Incompatible interactions are characterized by a second, stronger, and more prolonged oxidative burst, and it is in these incompatible interactions that cell death occurs. The oxidative burst first was reported to result in accumulation of superoxide, which is neither particularly stable nor particularly toxic and cannot easily cross cell membranes. Superoxide can be enzymatically converted into the more toxic hydrogen peroxide, which can cross membranes and can be converted into even more noxious reactive oxygen intermediates (Fig. 20.54).

The best evidence to date indicates that cell death follows the production of reactive oxygen intermediates, leading to the notion that reactive oxygen intermediates cause cell

death, either directly or indirectly. In addition, inhibiting the generation of superoxide in a plant has been reported to reduce the amount of pathogen-induced cell death. The quantities of reactive oxygen intermediates produced in an incompatible interaction may be sufficient to kill plant cells directly. If reactive oxygen intermediates mediate the cell death response seen in the HR, treatment with similar reactive oxygen intermediates should cause death that mimics the death caused by an incompatible pathogen. Some evidence indicates that this may, in fact, occur in at least one host–pathogen system. However, to mimic the response triggered by a pathogen, the reactive oxygen intermediates must be applied at the right concentration for the right length of time and in the appropriate location in the cell. These experiments are very difficult, given the highly reactive and unstable nature of reactive oxygen intermediates.

In addition to a possible role in directly killing host or pathogen cells (or both), reactive oxygen intermediates may act as signal molecules to trigger an independent cell death pathway or transcription of defense genes (Fig. 20.55). The different roles of reactive oxygen intermediates in the HR may be concentration dependent. High concentrations may cause death, whereas lower concentrations may be involved in signaling. Hydrogen peroxide produced during the HR has been demonstrated to trigger transcription of genes encoding antioxidant proteins in adjacent tissue. This suggests that small amounts of reactive oxygen intermediates might also signal the host to trigger endogenous cellular protectants possibly to limit the extent of cell death. Other reactive oxygen intermediates also may have signaling roles in cell death pathways. For example, superoxide is a key regulator of induction of the spreading cell death associated with the *lsd1* mutation, whereas hydrogen peroxide has no effect.

20.13.15 Does PCD in host–pathogen interactions occur by apoptosis?

To establish precisely how cells die, recent studies have examined plant cells undergoing PCD during host–pathogen interactions, in senescence, and during developmentally cued PCD. As already discussed, plant cells

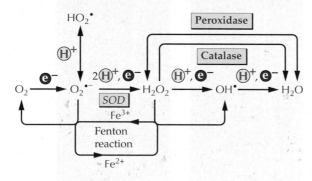

Figure 20.54
Scheme showing the interconversions of reactive oxygen species that can occur within cells. SOD, superoxide dismutase; $O_2^{\bullet-}$, superoxide anion; $HO_2^{\bullet}$, hydroperoxide radical; $OH^{\bullet}$, hydroxyl radical.

follow several routes to death—some that may be unique to plants, and others that resemble apoptosis. Growing evidence suggests that some plant responses to pathogen cells may follow the apoptotic route. The criteria generally applied when defining apoptosis in animal cells include formation of apoptotic bodies, blebbing of the nuclear membrane, chromatin condensation, and cleavage of nuclear DNA into internucleosomal fragments (see Box 20.1).

The ordered cleavage of DNA into fragments of about 50 kb has been demonstrated in two different host–pathogen systems. In one case, however, a similar result was obtained in cells put through a freeze–thaw cycle, which presumably causes rapid traumatic cell death rather than PCD. Thus, although the appearance of large DNA fragments is consistent with the appearance of fragments of a similar size in apoptotic animal cells, their presence may be simply a more general feature of cell death in plants. Ca^{2+}-dependent endonucleases that are activated during HR-associated cell death may be responsible for the DNA fragmentation seen in some instances of cell death.

Although oligonucleosomal DNA fragments were not detected in the above systems, they have been reported in other host–pathogen systems, the fragmentation being detected by the appearance on an agarose gel of a "ladder" of bands, one for each multiple of 180 bp. DNA ladders were detected during the onset of disease symptoms in a compatible interaction involving a fungal pathogen (*Alternaria alternata* f.sp. *lycopersici*) that uses a toxin (AAL toxin) to kill cells in the host plant (tomato). DNA ladders also were detected in tomato cells treated with arachidonic acid, an elicitor of the HR, suggesting that oligonucleosomal DNA laddering can occur in an incompatible plant–pathogen interaction. Again, such fragmentation may not be specific to HR, because a similar result was seen in tomato protoplasts and leaflets after heat shock or treatment with KCN.

Oligonucleosomal DNA fragmentation has been shown more conclusively in leaves of bean cultivars treated with avirulent (i.e., HR-inducing) isolates of bean rust (Fig. 20.56). Virulent isolates of the pathogen, which cause disease but do not induce HR,

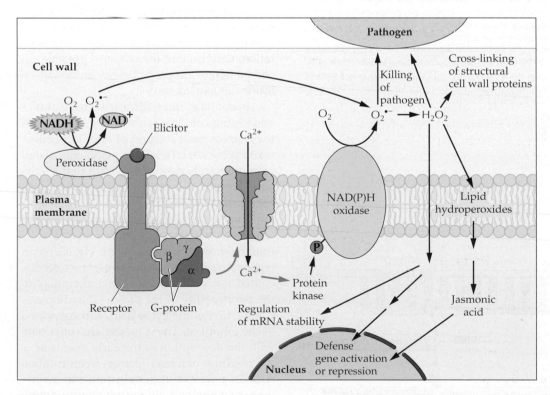

Figure 20.55
Model postulating the relationship between pathogen attack on a host cell and the signal transduction pathway that induces production of reactive oxygen species.

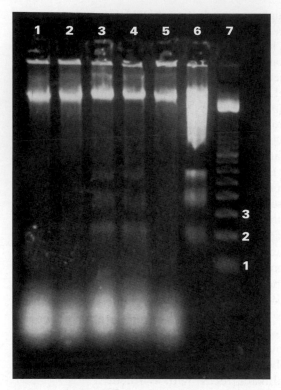

Figure 20.56
DNA cleavage into oligonucleosome-size fragments, a characteristic of some apoptotic cell death events in animal cells, is detected in some plant tissues undergoing HR. Lanes 1 and 2 contain DNA from a bean cultivar susceptible to a fungal pathogen; lane 1 is from uninoculated tissue and lane 2 from inoculated tissue. Lanes 3 and 4 contain DNA from leaves of resistant cultivars that have been inoculated with the pathogen and are undergoing HR. Lane 5 contains DNA from the fungal spores alone, and lane 6 contains DNA from mouse cells that have been induced to undergo apoptosis as a positive control for DNA laddering. DNA laddering of the type seen in the apoptotic mouse cells (lane 6) is seen also in the DNA from leaves undergoing the HR cell death (lanes 3 and 4), but not in leaves from the compatible interaction (lane 2). Lane 7 contains molecular mass standards, with bands 1, 2, and 3 indicating 100-, 200-, and 300-bp fragments, respectively.

Figure 20.57
In the TUNEL assay, terminal deoxynucleotidyl transferase is used to estimate the extent of DNA cleavage, based on the labeling of free 3′-OH ends of DNA with fluorescein-labeled deoxyuridine or some other fluorescently labeled nucleotide triphosphate. Note that the TUNEL method estimates only the extent of DNA cleavage and does not provide information on the size of the cleaved product.

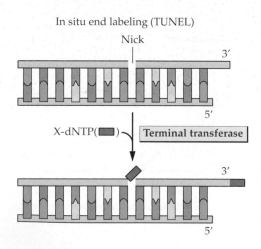

In situ end labeling (TUNEL)

Nick

3′

5′

X-dNTP(▬) → Terminal transferase

3′

5′

did not induce detectable DNA laddering. In the TUNEL assay (Fig. 20.57), which measures DNA degradation by detecting free 3′-OH groups of cleaved DNA, labeling of the DNA fragments (which were presumed to be oligonucleosomal in size) was limited to the host cells containing fungal haustoria and was not seen in adjacent uninfected cells. Several other treatments (including freezing, KCN treatment, and H_2O_2 application) induced non-HR cell death, and no DNA laddering was observed.

Other morphological features characteristic of apoptosis have been reported in various host–pathogen systems. Along with the DNA ladders, structures reminiscent of apoptotic bodies were detected in susceptible tomato protoplasts treated with the AAL toxin. These bodies, which contain fragmented DNA, migrated to the periphery of the cells and eventually were released from the cells into the medium (Fig. 20.58). Interestingly, the toxin from the tomato pathogen used in these studies also can induce apoptosis in mammalian cells. In other experimental systems, including soybean cells and leaves treated with avirulent but not virulent bacteria, and in tobacco cells treated with the fungal peptide elicitor cryptogein, some features of apoptotic cell death were detected: condensation of the nucleus, DNA fragmentation, shrinkage of the cell, and possibly production of apoptotic bodies in the HR lesions of treated leaves.

In contrast, in an ultrastructural study addressing cell death during the HR, none of the characteristic features of apoptotic cell death were seen. In lettuce plants inoculated with an avirulent bacterial strain that resulted in a rapid HR, microscopic examination of the inoculation sites revealed localized alterations in the walls of cells adjacent to bacteria, both with an avirulent strain and a mutant strain unable to trigger HR. Cellular responses specific to the avirulent isolate included mitochondrial swelling, alteration in the permeability of the plasma membrane, vacuolation of the cytoplasm, and collapse of the tonoplast. These results are consistent with the description of the early events in the response of tobacco leaves to inoculation with an avirulent bacterial strain and have more in common with morphological definitions of necrotic cell death in animals than with apoptotic cell death.

Less is known about apoptotic-like cell death in some of the other examples of PCD in plants, and, as with HR, the results are somewhat contradictory. Root cortical cells in soybean *rn* mutants (see Fig. 20.53) also exhibit some of the morphological features of apoptosis at the ultrastructural level, including nuclear blebbing and chromatin condensation preceding phytoalexin accumulation and the induction of some defense response–related genes.

20.13.16 Does apoptosis have molecular parallels in plant cells?

Although evidence is mounting for a type of PCD in some plants with at least some functional parallels to apoptosis, evidence for common features at the molecular level is more elusive. Several positive (e.g., *ced-3* and *ced-4*) and negative (e.g., *ced-9, bcl-2*) regulators of cell death in animal systems have been identified, but (with one exception) homologous genes in plants have not been found. This could indicate that the corre-sponding plant sequences have diverged to the extent that the sequences no longer are detected as homologous, although the function may be conserved, or it may indicate that plants have evolved other means of controlling these pathways. The three plant genes cloned (*LLS1, LSD1,* and *MLO1,* described earlier) that regulate at least some forms of cell death likewise have no known counterparts in animal systems. Also, expression in tobacco of *BCL-X$_L$*, a negative regulator of cell death in animals, did not affect the cell death produced in response to tobacco mosaic virus (TMV) or avirulent bacteria. Normal HR lesions were seen in these plants. Similarly, *BCL-2* did not protect cultured maize cells from PCD induced by a calcium ionophore. However, a recent study has demonstrated that a positive regulator of animal apoptosis, the BAX protein, can induce cell death in plant cells. These sets of data are consistent with plants having evolved some steps for regulating the processes of cell death that differ from those in animals and having conserved some steps that are seen in animals also.

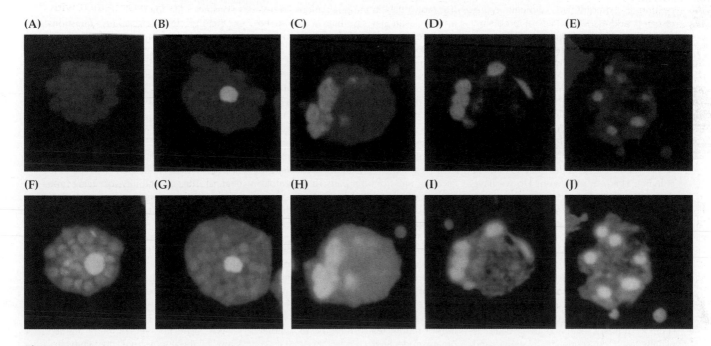

Figure 20.58

Death of susceptible tomato protoplasts induced by a fungal toxin has some morphological features of apoptosis, as shown in a five-step time course following the treatment of tomato protoplasts with AAL toxin. The top panels show the cells labeled with a probe to detect DNA fragmentation. The bottom panels show the same cells counterstained with a DNA dye. In normal, healthy cells, no DNA fragmentation is detected (A), and the nucleus clearly stains for DNA (F). In the early stages of cell death, DNA fragmentation can be detected (B), but the nucleus remains intact (G). In (C) and (H), the nucleus begins to disperse. In late stages—(D) and (E), (I) and (J)—the progressively fragmented DNA is localized to small distinct bodies within the cells, in a process resembling the formation of apoptotic bodies in animal cells undergoing apoptosis.

20.14 Further questions and future directions for PCD research

20.14.1 How similar at the mechanistic level are the various forms of PCD in plants?

In each of the various forms of PCD perhaps some of the same machinery is used in the actual execution phase. However, clearly there are differences in the various cell death programs. Senescence is relatively slow and ordered, maximizes recovery of metabolites from the tissue, and may be reversible until a late stage. Cell death associated with the HR, on the other hand, is relatively rapid, minimizing the pathogen's opportunities to gain a foothold. Studies examining the expression of genes in various forms of PCD may shed light on whether common pathways are used in different types of the process. For example, are any of the genes that are expressed during senescence also expressed in any of the other forms of PCD? Genes expressed in several different death programs could have a more general death function.

A genetic approach also may be useful in identifying which genes are required in regulating cell death. Along with traditional mapping and cloning techniques to identify the genes and proteins responsible for cell death control, seed from lesion-mimic plants can be mutagenized and examined for offspring

---

Box 20.2	Are programmed cellular processes decisive for the aging of the whole organism?

Intuitively, one would think that the progressive accumulation of PCD events must be related to aging, that is, to the time-dependent deteriorative changes leading to the ultimate death of the whole organism. In practice, making such a connection is extraordinarily difficult. Aging research addresses a range of problems, some of which are beginning to yield to modern analytical approaches as well as others that may never be answered and will remain subjects for semantic or philosophical disputation. Consider the issue of definitions, for example. An organism transitions from viability to death by any of a number of routes, in many of which aging either is not involved at all or is a secondary or peripheral influence. By what criteria do we know that something is living, aged, or otherwise? Do we have definitions that allow us to say an organism is really dead and not in diapause or some such suspended state, for example? The comparative biology of aging has revealed a range of lifestyles. Some individuals such as lobsters appear not to age at all, some organisms age but usually die by accident (for example, wild mice, which almost invariably get eaten before they have a chance to grow old), some species such as humans age and their death is usually as a consequence of aging, and some organisms can be rejuvenated (some invertebrates, for example). But what do we mean by an individual? Is it a coral, or a sponge, or a slime-mold, or a tree?—a single organism, or a population? There are special difficulties in trying to translate aging behavior from one level of biological organization to another. One problem of "scaling down" from general to specific cases is this: What is the relationship (if any) between actuarial or demographic definitions of aging and the behavior of the individuals in the population? For example, life tables say that, in developed countries, women live on average three years longer than men. Does this mean that women begin to age three years later, or do women age at a slower rate than men, or is something else going on? Turning to the nature of aging itself, we see that all sorts of biological processes fail or decline with age, but which are causes that dictate the progress and nature of aging, and which are merely symptoms? Here researchers encounter a problem in "scaling up" from the parts to the whole: Is there a relationship between the aging of component systems and that of the entire organism? Eyesight, blood pressure, joints, and mental processes all degenerate with age in humans, but what has this to do with dying? What is the relationship between cell/tissue/organ death and survival of the whole organism? Is it meaningful to think in terms of a "master" reaction, a specific component for which deterioration leads directly to whole-organism decline?

Of particular significance for plants is the question of resource capture and allocation in relation to aging: Is aging a kind of starvation or neglect process, and what meaning do these terms have for autotrophic organisms, in which raw materials and energy generally are not limiting? What is the contribution of nonoptimal (stressful) environments to aging? Does aging literally result from being weather-beaten? How is the integration of an aging period into the full life cycle related to organism life span?

Finally, there are many questions about the mechanisms of aging. Is aging a failure of processes that normally defend against deterioration? What are the cost–benefit tradeoffs of repair, maintenance, and durable construction? How is this related to the distinction between germ line and soma in some organisms (although not most plants)? Can aging be avoided? Is aging a failure to escape from influences that invoke the aging response? Have organisms been able to channel the inevitability of aging into processes that benefit their ecological and evolutionary fitness and, if so, what influence has this had on the programs for cell death and senescence? How can natural selection act to evolve genes with specific functions in aging? Can "aging genes" be mutated, mapped, and isolated? What would be their environmental sensitivities? Can we ultimately disrupt aging by tinkering with such genes?

We have not yet properly framed many of these questions, let alone answered them, but aging research has become one of the most exciting areas of modern biology. Research into aging will both influence and benefit from progress in understanding senescence and PCD.

that revert to wild-type phenotype. The absence of lesions is an easily screened marker, especially for propagation-class mutants such as *lsd1* in *Arabidopsis* and *lls1* in maize and for early-onset initiation-class mutants such as *lsd5,* all of which can be conditionally lethal. Such a screen can uncover extragenic suppressors of the original mutation, thus revealing other parts of the signal transduction pathway for cell death. Also intriguing is the possibility that suppressors of one cell death phenotype could suppress other mutations that affect control of cell death or exhibit altered interactions with pathogens. Loci that suppress multiple cell death–control mutations may define key steps in common pathways leading to cell death.

Summary

Selective death of cells, tissues, and organs is an essential feature of plant development and survival. The term programmed cell death refers to any process by which protoplasm, with or without the cell wall that encloses it, is eliminated as part of a developmental or adaptive event in the life cycle of the plant. Plants dispose of unwanted cytoplasm or whole cells by several mechanisms, including self-ingestion (autophagy), lysis (lysigeny), or a kind of mummification (as seen in endosperm formation). PCD is essential for normal reproductive and vegetative development and for responses to environmental stresses. During senescence (a PCD-like process associated with the terminal stages of organ development), specific genes are expressed, some of which have been cloned and their promoters analyzed. *Stay-green* mutants are genetic variants in which expression of senescence-associated genes is impaired.

Senescing cells are metabolically active. The chlorophyll degradation pathway is turned on, accompanied by the unmasking or accumulation of carotenoids and other pigments. Proteins are broken down, and the mobilized organic nitrogen and organic sulfur are exported from senescing leaves. Catabolism of nucleic acids releases inorganic phosphate. Photosynthesis declines, and peroxisomes are redifferentiated into glyoxysomes, which convert lipids to sugars. Metabolic regulation during senescence involves responses to cellular redox conditions, compartmentation, and differential susceptibility of enzymes to proteolysis. Senescence is sensitive to growth regulators, particularly to the senescence-promoter ethylene and to cytokinins, which act as senescence antagonists.

Senescence is used both as part of adaptive strategies that ensure seasonal survival and as a tactic deployed when an unpredictable stress is experienced. Formation of TE and mobilization of cereal endosperm are examples of developmental PCD. *Zinnia* leaf mesophyll cells induced to form TE in culture are a valuable model of cell death. Several different pathways lead to death of starchy endosperm and aleurone cells in cereals. Ethylene can induce PCD in starchy endosperm cells, and GA and ABA regulate cell death in cereal aleurone. An example of PCD under an abiotic environmental stress is aerenchyma formation, which is a response to limited oxygen availability, such as when roots are flooded. Ethylene mediates aerenchyma formation, and signaling between ethylene production and hypoxia sensing is mediated by calcium.

The hypersensitive response is an example of PCD related to biotic stress and has its own effect on the programmed nature of cell death. Lesion-mimic mutants are useful models for analyzing HR. Such mutations in *Arabidopsis* fall into two groups: those that trigger defense responses, and those that do not. Lesion-mimic mutants primarily affect cell death in the context of the defense response to pathogens but apparently have no effect on other events in developmental cell death. Reactive oxygen intermediates are a key trigger in PCD accompanying the HR. HR can be compared with apoptosis, a type of PCD in animals that has been intensively researched and has been an influential general model for PCD and its regulation. Future directions for PCD research include the relationship between programmed cellular processes and whole-organism aging.

Further Reading

Bethke, P. C., Lonsdale, J. E., Fath, A., Jones, R. L. (1999) Hormonally regulated programmed cell death in barley aleurone cells. *Plant Cell* 11: 1033–1045.

Biochemistry & Molecular Biology of Plants, B. Buchanan, W. Gruissem, R. Jones, Eds.
© 2000, American Society of Plant Physiologists

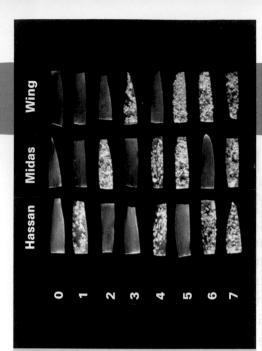

CHAPTER *21*

Responses to Plant Pathogens

Kim Hammond-Kosack
Jonathan D.G. Jones

CHAPTER OUTLINE

Introduction

Plants must continuously defend themselves against attack from bacteria, viruses, fungi, invertebrates, and even other plants. Because their immobility precludes escape, each plant cell possesses both a preformed and an inducible defense capacity. This is in striking contrast to the vertebrate immune system, in which specialized cells devoted to defense are rapidly mobilized to the infection site, where they either kill the invading organism or limit its spread. The noncirculatory defense strategy of the plant nevertheless minimizes infections. In wild plant populations, most plants are healthy most of the time; if disease does occur, it is usually restricted to only a few plants and affects only a small amount of tissue (Fig. 21.1). **Disease,** the outcome of a successful infection, rarely kills a plant. Natural selection probably acts to curtail fatal pathogen toxicity to plants; after all, diseases that keep a host alive longer may permit more reproduction of pathogen.

Why study the interactions between plants and pathogenic organisms? There are three main reasons. First, a detailed study of plant–microbe interactions should provide sustainable practical solutions for the control of plant disease in agricultural crops. Growing monocultures of genetically uniform crop species over vast tracts of land is a practice that frequently leads to severe outbreaks of disease; such **epidemics** lower both crop yield and quality and can diminish the safety of the end product (Fig. 21.2). In addition, the use of agrochemicals to control plant disease can cause serious pollution and increase the costs of production. Second, such studies should help elucidate the signaling mechanisms by which plant cells cope with a stress situation. For example, do plant responses provoked by the invasion of a pathogenic organism differ from those provoked by mechanical wounding or by the stresses of low temperature, high salinity, or desiccation? Third, study of plant–pathogen interactions can lead us to discover how organisms from different kingdoms communicate with one another. What type of messages do they exchange and how are appropriate responses evoked?

In this chapter we examine the biochemical and molecular mechanisms by which plants defend themselves against attack from microbial pathogens and invertebrate pests. To begin, we will look at the numerous strategies utilized by pathogenic organisms for a successful invasion of plant tissue. A **plant pathogen** is defined as an organism that, to complete a part or all of its life cycle, grows inside the plant and in so doing has a detrimental effect on the plant.

21.1 Ways in which plant pathogens cause disease

Roots and shoots of all plants come into intimate contact with plant pathogens. Each pathogen has evolved a specific way to invade plants (Fig. 21.3). Some species directly penetrate surface layers by using mechanical pressure or enzymatic attack. Others pass through natural openings (e.g., stomata or lenticels). A third group invades only tissue that has been previously wounded. Once inside the plant, one of three main attack strategies is deployed to utilize the host plant as a substrate: **necrotrophy,** in which the plant cells are killed; **biotrophy,** in which the plant cells remain alive; and **hemibiotrophy,** in which the pathogen initially keeps cells alive but kills them at later stages of the infection (Table 21.1). This process of infection, colonization, and pathogen reproduction is known as **pathogenesis.** A pathogen strain that causes disease is termed **virulent.** We shall now briefly consider how different types of pathogens cause disease.

21.1.1 Virulent pathogens have unique survival-promoting characteristics.

The success of certain widespread plant pathogens can be attributed to several main factors:

- An extremely effective mode of pathogen esis that permits a rapid and high rate of

Figure 21.1
The leaf mold fungus *Cladosporium fulvum* sporulating through the lower surface of tomato leaves, 11 days after infection. This biotrophic pathogen is restricted to attacking only a few plant species of the genus *Lycopersicon*.

(A)

(B)

Figure 21.2
Sugar beet–root nematode interaction. (A) The central rows of sugar beet show severe damage from the endoparasitic cyst nematode *Heterodera schachtii*. (B) Mature female nematode bodies filled with eggs, attached to the sugar beet roots at the end of the pathogen's life cycle.

reproduction during the main growing season for plants.
- A very efficient dispersal mechanism by wind, water (rain splash), or vector organisms such as insects.
- A different type of reproduction (often sexual) toward the end of each plant growing season to produce a second type of structure (e.g., spore, propagule) that has long-term survival capacity (some times as long as 30 years).
- A high capacity to generate genetic diversity. In many pathogens, haploidy during the phase of greatest reproduction permits mutations of functional importance to give an immediate selective advantage within the pathogen population. Subsequent sexual reproduction creates a novel pool of recombinant genotypes from which new epidemics can arise.
- Monoculture of crop plants and well-adapted pathogen genotypes.

21.1.2 Fungal plant pathogens use a wide range of pathogenesis strategies.

Less than 2% of the approximately 100,000 known fungal species are able to colonize plants and cause disease. Necrotrophic species that produce cell wall–degrading enzymes tend to attack a broad range of plant species. For example, the fungi *Pythium* and *Botrytis* are each known to attack more than 1000 plant species (Fig. 21.4). Some necrotrophs produce host-selective toxins that are active in only a few plant species. Each toxin has a highly specific mode of action, inactivating just a single plant enzyme. For example, the **HC-toxin** produced by the maize fungal pathogen *Cochliobolus carbonum* inhibits the activity of histone deacetylase (Fig. 21.5), an enzyme that is required for the activation of plant defense genes. *Alternaria alternata* f.sp. *lycopersici* produces the AAL toxin, which appears to activate a plant cell death program in tomato plants (see Chapter 20). Other fungi produce non–host-selective toxins. For example, *Fusicoccum amygdali* produces the toxin **fusicoccin**, which targets the plasma membrane–localized H^+-ATPase in many plant species (see Chapter 3). The action of this toxin leads to the irreversible opening of stomata and plant wilting, which is followed by cell death and necrotrophic colonization.

Biotrophic fungi keep host cells alive and usually exhibit a high degree of specialization for individual plant species. To utilize the living plant cells as a food substrate, these fungi often penetrate the rigid cell wall by forming first a penetration peg and then a **haustorium,** which causes invagination of the plasma membrane. This specialized feeding structure increases surface contact between the two organisms, thus maximizing nutrient and water flow to favor fungal growth (Fig. 21.6). Many downy and powdery mildew species utilize haustorial associations to attack plants. Noting the frequent formation of "green islands" on senescing leaves surrounding biotrophic infection sites led to the discovery that biotrophic pathogens can alter the hormonal balance at sites of infection to ensure that host cells are kept alive throughout these intimate associations. A few biotrophic fungi— for example, the tomato leaf mold pathogen *Cladosporium fulvum* (see Fig. 21.1)—do not

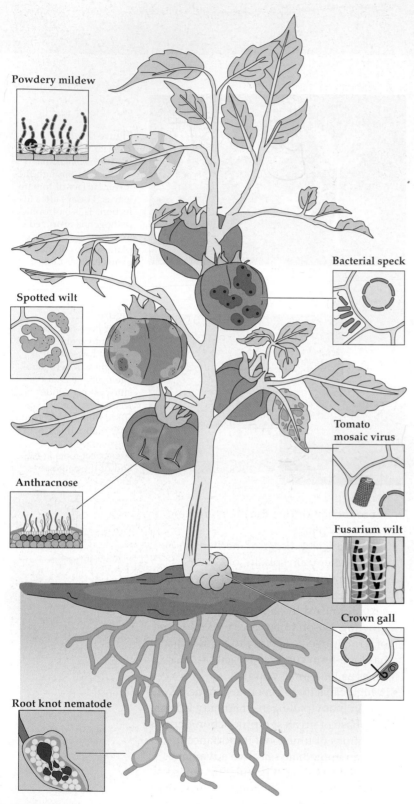

Figure 21.3

Most microbes attack only a specific part of the plant and produce characteristic disease symptoms, such as a mosaic, necrosis, spotting, wilting, or enlarged roots. Tomato plants are attacked by more than 100 different pathogenic microorganisms.

Table 21.1 Strategies utilized by plant pathogens

	Necrotrophic	Biotrophic	Hemibiotrophic
Attack strategy	Secreted cell wall degrading–enzymes, host toxins, or both	Intimate intracellular contact with plant cells	Initial biotrophic phase, then necrotrophic phase
Specific features of interaction	Plant tissue killed and then colonized by the pathogen. Extensive tissue maceration.	Plant cells remain alive throughout the infection. Minimal plant cell damage.	Plant cells alive only in the initial stages of the infection. Extensive plant tissue damage at late stages.
Host range	Broad	Narrow; often only a single species of plant is attacked.	Intermediate
Examples	Rotting bacteria (e.g., *Erwinia* spp.); rotting fungi (e.g., *Botrytis cinerea*)	Fungal mildews and rusts; viruses and endoparasitic nematodes; *Pseudomonas* spp. bacteria	*Phytophthora infestans* (causal agent of potato late blight disease)

Figure 21.4
Botrytis cinerea, the gray mold fungus, sporulating on grapes. This necrotroph secretes large numbers of cell wall–degrading enzymes and thereby destroys plant tissue in advance of the colonizing hyphae.

form haustoria but instead grow exclusively outside the plant cell wall within the intercellular air space (apoplast), subsisting on leaked nutrients. *Hemibiotrophic*

Hemibiotrophic fungi sequentially deploy a biotrophic and then a necrotrophic mode of nutrition. The switch is usually triggered by increasing nutritional demands as the fungal biomass increases. Some of the world's most devastating phytopathogenic species fall into this category. For example, the organism *Phytophthora infestans*, which causes late blight disease of potato (Fig. 21.7), was responsible for the devastating blight disease epidemic in Ireland in 1846 and 1847, resulting in the Irish famine and the emigration of more than one million

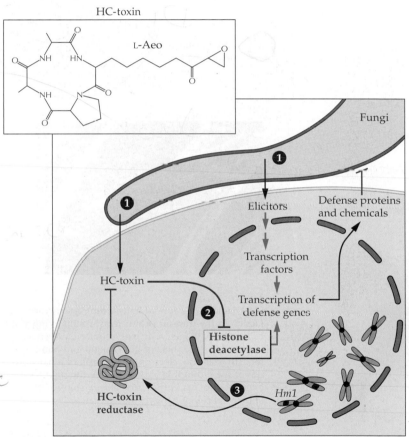

Figure 21.5
(1) The maize pathogen *Cochliobolus carbonum* secretes HC-toxin and numerous nonspecific fungal elicitors that trigger plant defense responses. (2) HC-toxin is hypothesized to enter the responding plant cells, where it inhibits histone deacetylase enzyme activity. Histone deacetylase is responsible for the reversible deacetylation of core histones (H3 and H4) assembled in the chromatin. Inhibition of histone deacetylase activity is believed to interfere with transcription of maize defense genes and thus favors fungal growth and disease development. (3) Hm1-resistant maize plants produce an HC-toxin reductase (HCTR), which detoxifies HC-toxin by reducing the carbonyl group of the side chain of L-Aeo (2-amino-9,10-epoxy-8-oxo-decanoic acid), an unusual epoxidated fatty amino acid.

(A)

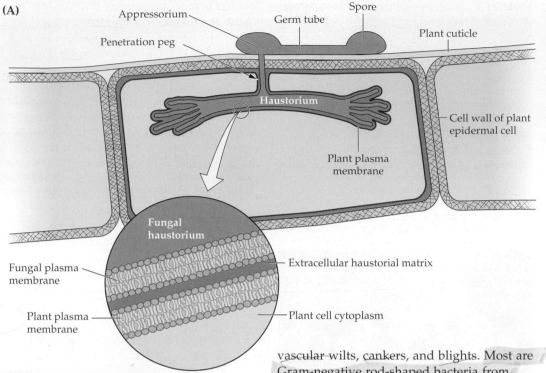

(B)

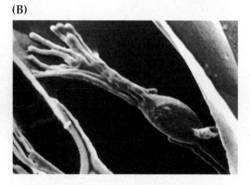

Figure 21.6
(A) Diagram of a fungal haustorium, a structure that facilitates the transfer of nutrients from a living plant cell to the pathogen. The extracellular haustorial matrix contains products of both fungal and plant origin. (B) Scanning electron micrograph of a haustorium of the biotrophic barley mildew fungus *Erysiphe graminis* f.sp. *hordei* (*Blumeria graminis*) inside the epidermal cell of barley.

people to the United States and other countries. Today this fungus still causes large losses in annual yields.

21.1.3 Bacterial pathogens of plants and animals appear to use similar molecular mechanisms for invading host tissues.

Phytopathogenic bacteria specialize in colonizing the apoplast to cause rots, spots,

vascular wilts, cankers, and blights. Most are Gram-negative rod-shaped bacteria from the genera *Pseudomonas*, *Xanthomonas*, and *Erwinia*. Two features characterize bacteria–plant relationships. First, during their parasitic life, most bacteria reside within the intercellular spaces of the various plant organs or in the xylem (Fig. 21.8). Second, many cause considerable plant tissue damage by secreting either toxins, extracellular polysaccharides (EPSs), or cell wall–degrading enzymes at some stage during pathogenesis. Bacterial toxins contribute to the production of symptoms on certain plant species, but targeted mutagenesis has shown that the toxins are not essential for pathogenesis. The secreted EPSs, which entirely surround the growing bacterial colony, may aid bacterial virulence—for example, by saturating intercellular spaces with water or by blocking the xylem to produce wilt symptoms; again, however, they are not absolutely required for pathogenesis. Other potential roles for these bacterial toxins and factors associated with disease are receiving renewed scrutiny. Bacteria that deploy pectic enzymes during pathogenesis, such as *Erwinia*, have a broader host range and cause extensive cell death and tissue maceration. These pectic enzymes cleave plant cell wall polymers either by hydrolysis (polygalacturonases) or through β-eliminations (pectate or pectin lyases). The multigenic nature of the gene

Figure 21.7
The pathogen responsible for late blight, *Phytophthora infestans*, is the most destructive microorganism that affects potato crops, attacking them worldwide. The hemibiotrophic lifestyle of this pathogen facilitates its progress from leaf infection to sporulation in a mere three days. If moist, cool conditions prevail, the entire foliage of a potato field can be destroyed within two weeks.

(A)   **(B)**

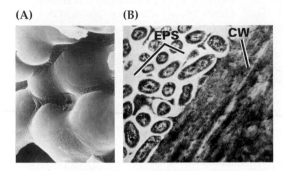

Figure 21.8
(A) Scanning electron micrograph of *Xanthomonas campestris* bacteria colonizing the intercellular air spaces of a *Brassica* leaf. (B) Ordinarily, the bacteria are surrounded by an extracellular polysaccharide material (EPS) and proliferate in close contact with the plant cell walls (CW).

families involved has made it difficult to establish the precise role of individual enzymes in bacterial pathogenesis. However, decreased pectin enzyme activity frequently results in less bacterial virulence.

Several bacterial genes, referred to collectively as the **hypersensitive response and pathogenicity cluster,** are absolutely required for bacterial pathogenesis. Many *hrp* gene sequences from plant bacteria are very similar to the genes required for pathogenesis in bacteria that infect animals, which suggests that these distinct pathogens utilize similar virulence strategies. One known strain of *Pseudomonas aeruginosa* is capable of causing disease in both *Arabidopsis* and mice (Fig. 21.9). Mutation of specific genes destroys the ability of this strain to colonize both organisms. This discovery reinforces the hypothesis that during the evolution of bacterial colonization of animals and plants, certain common mechanisms may have been retained.

21.1.4 Plant pathogenic viruses move by way of plasmodesmata and phloem.

More than 40 families of DNA and RNA plant viruses exist, but the overwhelming majority are single-strand (ss) positive-sense RNA viruses (Fig. 21.10). Symptoms of viral infection include tissue yellowing (chlorosis) or browning (necrosis), mosaic patterns, and plant stunting. Plant viruses are biotrophs and all face the same three basic challenges: how to replicate in the cell initially infected; how to move into adjacent cells and the

(A)   *Pseudomonas aeruginosa* bacterium

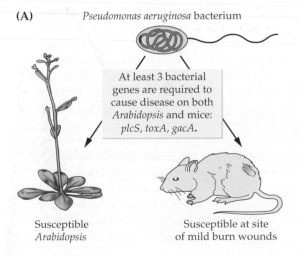

At least 3 bacterial genes are required to cause disease on both *Arabidopsis* and mice: *plcS, toxA, gacA.*

Susceptible *Arabidopsis*

Susceptible at site of mild burn wounds

(B)

P. aeruginosa genotype	*Arabidopsis* (bacterial titer/cm^2)	Mouse (% lethality)
Wild type	1×10^7 (full disease)	77
plcS mutant	1×10^5 (less severe)	40
toxA mutant	1×10^6 (less severe)	40
gacA mutant	5×10^3 (attenuated symptoms)	0

plcS   encodes a phospholipase S enzyme that degrades phospholipids of eukaryotic cells.

toxA   encodes exotoxin A, which inhibits protein synthesis by ribosylating elongation factor 2.

gacA   encodes a transcriptional regulator of several *hrp* genes.

Figure 21.9
(A) The bacterial pathogen *Pseudomonas aeruginosa* has a broad host range, its *hrp* genes encoding products that are required for virulence in both *Arabidopsis* and mice. (B) Bacterial strains with mutant *hrp* genes have diminished virulence for both mouse and *Arabidopsis*.

Box 21.1

Agrobacterium-mediated T-DNA transfer, a mechanism used in pathogenesis, has become a tool for transforming plant genomes.

Agrobacterium tumefaciens, a soil bacterium closely related to *Rhizobium* species (see Chapter 16), causes crown-gall disease in thousands of dicotyledonous plant species. In the only known natural example of DNA transport between kingdoms, *Agrobacterium* genetically transforms plant cells with a DNA fragment called the T-DNA (transferred DNA), which is part of the large tumor-inducing (Ti) plasmid. The T-DNA is transferred from the bacterium into the plant cell nucleus, where it is stably integrated into the plant genome. Expression of the genes of a wild-type T-DNA produces remarkable modifications in plant cell metabolism, for example, the synthesis of plant growth hormones (auxins and cytokinins), which lead to neoplastic growth (tumor formation). Plant cells removed from tumors can be grown indefinitely in sterile tissue culture in the absence of both hormones and *Agrobacterium*. Second, within tumor tissue, various amino compounds called opines are synthesized, which the bacteria use as their main sources of carbon and nitrogen. Indeed, classification of *Agrobacterium* strains is based on the types of opines specified by the bacterial T-DNA, for example, octopine and nopaline. Opine catabolism within bacterial cells requires specialized enzymes, which are encoded by the *Agrobacterium* Ti plasmid. Because neither the plant nor any other soil-dwelling microorganism can metabolize opines, this creates a biological niche for *Agrobacterium* within the tumor tissue.

Molecular genetic studies have been the key to elucidating many of the properties of the *Agrobacterium*–plant cell interaction. The bacterium predominantly infects plants at a wound site, where the damaged plant cells synthesize and secrete unusual phenolic compounds such as acetosyringone (see step 1 in the accompanying illustration). A set of virulence genes (*Vir* region) located on the *Agrobacterium* chromosome are activated by acetosyringone (step 2). Some of the products of the virulence genes are involved in bacterial chemotaxis toward and attachment to the wounded plant cells. The subsequent binding of the bacteria to specific plant cell wall components initiates a complex process in which the T-DNA is synthesized (step 3), transferred to the plant (step 4), targeted to the plant nucleus, and inserted into the plant genome (step 5), ultimately leading to plant expression of bacterial genes that encode proteins for the synthesis of opines and plant hormones (step 6). Overproduction of auxin and cytokinin causes tumor formation (step 7; see also Chapter 11), and the opines synthesized in tumor tissues constitute a food source for *Agrobacterium* (step 8).

The ability of *Agrobacterium* T-DNA to integrate stably into the plant genome has led to its widespread exploitation as a vector for the genetic transformation of cells of higher plants with recombinant DNA molecules. To regenerate plants with a normal appearance, the *Agrobacterium* genes that encode proteins required for tumor formation are eliminated from these vectors and replaced by selectable markers.

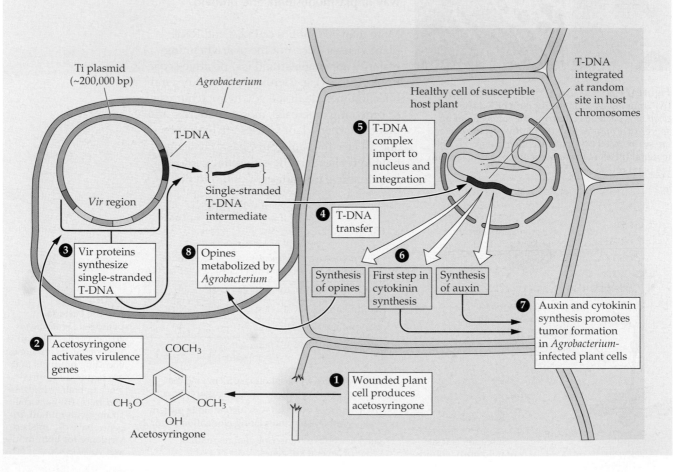

(A)

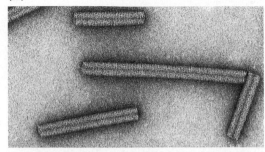

(B)

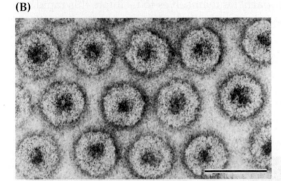

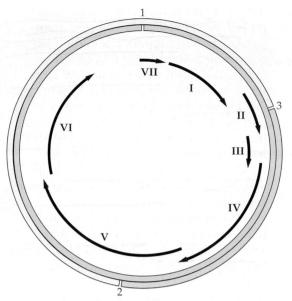

Figure 21.10
Transmission electron micrographs of (A) an encapsidated, ssRNA virus, tobacco mosaic virus, and (B) a double-stranded (ds) DNA virus, cauliflower mosaic virus. Bar = 100 nm.

ORF	Predicted protein MW	Inferred protein function
I	38 kDa	Cell-to-cell movement, plasmodesmata-associated
II	18 kDa	Aphid transmission
III	15 kDa	Non-sequence specific DNA binding protein
IV	57 kDa	Coat protein
V	79 kDa	Reverse transcriptase
VI	61 kDa	Translational transactivator
VII	11 kDa	No known protein function or protein detected

Figure 21.11
Genomic organization of the double-stranded cauliflower mosaic virus (CaMV). The 8-kbp CaMV DNA has three single-stranded interruptions—one in the coding strand, the others in the noncoding strand. The genome encodes seven potential open reading frames (ORF); indicated are the inferred functions of the proteins encoded (where known) and their predicted molecular masses (MW).

vascular system; and how to suppress host defenses and thereby colonize the entire plant. Over the past 25 years great progress has been made in defining the functions of virus-encoded genes. Each gene is now known to have one or more specific functions. For example, the genomes of cauliflower mosaic virus (CaMV; Fig. 21.11) and tobacco mosaic virus (TMV) contain seven and five open reading frames (ORFs), respectively, which function in the replication and movement of the viral DNA, symptom development, and encapsidation.

Genome replication for positive-strand RNA viruses occurs in the cytoplasm, apparently utilizing the translation apparatus of the host. In contrast, genome amplification of ssDNA geminiviruses—and some negative-strand ssRNA viruses—occurs in the nucleus and involves components of the host's DNA replication machinery (see Chapter 6). Subsequent transport of the virus particles occurs by way of intracellular (symplastic) movement through channels between plant cells, the **plasmodesmata** (see Chapter 15). In contrast to animal viruses, plant viruses never

cross the plasma membrane of the infected cell. Plant virus movement proteins (MPs), in association with various components of the cytoskeleton of the host cell (see Chapter 5), facilitate transport of nucleoprotein complexes or virus particles into adjacent cells by way of modified plasmodesmata. Two strategies are deployed for viral movement. Positive-strand ssRNA genomes of Tobamoviruses (e.g., TMV) transiently increase by 10-fold the size-exclusion limits of plasmodesmata to permit the trafficking of the large nucleo-protein complex (Fig. 21.12, upper three panels; see also Chapter 15). In contrast, some double-stranded (ds) DNA viruses (e.g., CaMV) direct the formation of large tubular structures composed of MPs to facilitate the movement of encapsidated virus particles through enlarged plasmodesmata (Fig. 21.12, lower panels). Many viral MPs have only minimal sequence similarity, which suggests that plant viruses initially may have acquired MP function from plant genomes by way of recombination.

The processes controlling long-distance transport of virus particles or viral nucleic acids within the phloem are probably distinct from those controlling movement between mesophyll cells. Exactly how viruses enter or leave this tissue compartment, however, is not understood. Once inside the phloem, movement of virus particles has been documented to reach speeds of 1 cm/h. For some viruses (e.g., TMV), the coat protein (CP) is necessary for this process to work efficiently; for other viruses, however, the CP may not be involved. One of the future challenges is to identify the host components in the phloem that interact with either the CP or the virus particles themselves to facilitate this rapid process; it will also be important to identify those components that facilitate the subsequent reentry of the virus into the mesophyll and the root cortical cells once a new organ is reached.

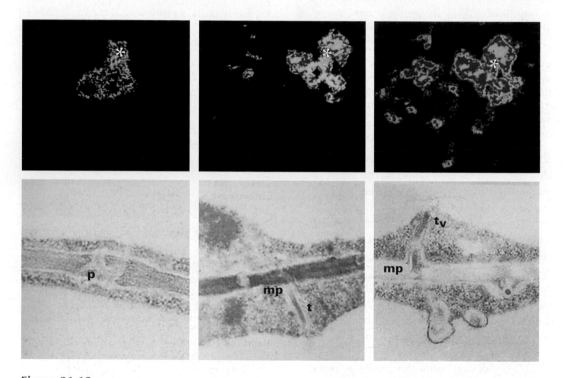

Figure 21.12

Plant pathogenic viruses move between adjacent cells by way of modified plasmodesmata. Two distinct strategies are utilized. Upper three panels: ssRNA viruses transiently increase by 10-fold the size-exclusion limit of the plasmodesmata, allowing the large viral nucleoprotein complex to move through. Using tobacco plants constitutively expressing the 30-kDa movement protein (MP) in tobacco mosaic virus provided an elegant demonstration that the MP controls the plasmodesmata pore size. When fluorescently labeled F-dextrans of various molecular masses were microinfected into individual leaf cells, cells of the transgenic lines could rapidly traffic F-dextran molecules as large as 9.4 kDa. The middle and right panels show UV images taken 15 and 60 seconds after injection, respectively. In the control (nontransformed) tobacco, no trafficking was evident, even after 20 minutes (left panel). The * in each panel indicates the point of microinjection. The size-exclusion limit for control tobacco tissue is 700 to 800 Da. Lower three panels: The dsDNA virus cauliflower mosaic virus (CaMV) is transported as encapsidated virus particles through a large tubular structure that forms and spans enlarged plasmodesmata. Left panel: healthy turnip leaves; middle and right panels: CaMV-infected turnip leaves. p, plasmodesmata; mp, modified plasmodesmata showing extended tubular structures (t), some of which contain visible virus particles (t_v).

21.1.5 Some plant pathogenic nematodes modify the metabolism of root cells, inducing the plant to form specialized feeding structures.

More than 20 genera of nematodes cause plant diseases. Infections by these tiny, round worms (which are about 1 mm long) are nearly always confined to the plant root system. These root infections can, however, dramatically alter the entire metabolism of the plant and often produce substantial modifications in the root architecture. All parasitic nematode species are obligate biotrophs, and all possess a hollow feeding stylet capable of penetrating plant cell walls (Fig. 21.13A). Ectoparasitic nematodes feed exclusively at the root surface, whereas endoparasitic species invade root tissues and spend a large portion of their life cycle in intimate association with root cells. By far the most damaging nematodes in the world are the two groups of sedentary endoparasitic species from the family *Heteroderidae:* the **cyst nematodes** (genera *Heterodera* and *Globodera*) and the root-knot **nematodes** (genus *Meloidogyne*).

The life cycle of the endoparasitic nematodes commences when dormant eggs perceive an as yet uncharacterized chemical signal released by plant roots; receipt of this signal induces the eggs to hatch, releasing juvenile nematodes. The motile second-stage juveniles then penetrate the roots and migrate to the vascular tissue. Once feeding is initiated, the nematode loses motility and becomes sedentary. To commence feeding, the juvenile cyst nematodes push their stylets through the plant cell wall, but not the plasma membrane, and release glandular secretions into the selected cell(s). Molecules in the secreted fluid induce rapid modifications to the cytoplasm of the plant cell, and the metabolic activity of the plant cell increases markedly. In addition, the nematode triggers partial dissolution of the plant cell walls, such that the symplasmic connections of the modified cell to its neighbors become more extensive, until finally protoplast fusion occurs. Eventually, as many as 200 plant cells may be recruited by the nematode to form the **syncytial feeding structure.** In contrast, the feeding of root-knot nematode juveniles induces both mitosis uncoupled from cytokinesis (see Chapters 5 and 11) and DNA endoreduplication (i.e., nuclei with increased DNA content). These result in abnormal cortical

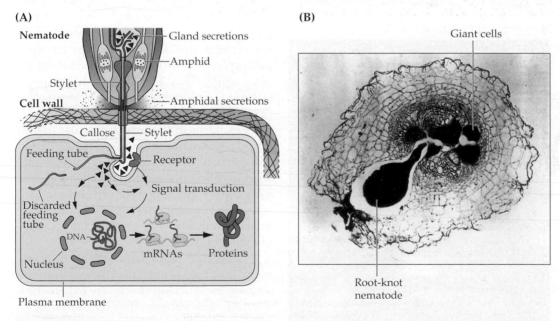

(A)

Nematode
Gland secretions
Amphid
Stylet
Cell wall
Amphidal secretions
Callose — Stylet
Feeding tube
Receptor
Discarded feeding tube
Signal transduction
DNA
Nucleus
mRNAs — Proteins
Plasma membrane

(B)

Giant cells

Root-knot nematode

Figure 21.13
Plant root–endoparasitic nematode interaction. (A) Diagram of the anterior of a nematode, which uses a stylet to enter a plant cell and to pump secretions from the amphid glands into the selected plant cell. Subsequently, inside the invaded plant cell a new feeding tube is synthesized, which the nematode uses to extract nutrients specifically from the modified cell cytoplasm. In each plant cell selected, hundreds of discarded feeding tubes can accumulate. (B) Transverse section of tomato root, showing a root-knot nematode and several giant cells that have formed at the feeding site of this now sedentary feeding nematode.

cell growth and the formation of a series of **giant cells** (Fig. 21.13B). In other respects, the morphology of the syncytial and giant cells appears to be very similar; both become closely associated with the phloem by way of the transfer cell connections and ensure that the entire plant is tapped for nutrients. In effect, the developing female root-knot and cyst nematodes become powerful alternative sinks for photosynthates and thereby lower the crop yields from the infected plants.

One key unresolved issue in plant–sedentary nematode associations is the nature of the biochemical signals within the glandular secretions of the nematodes that cause such dramatic changes in root cell architecture. Detailed microscopic analyses have shown the presence of a **feeding tube** structure, which is now recognized to be exclusively associated with the stylet. Unlike the stylet, however, the feeding tube is located within the plant cell cytoplasm (see Fig. 21.13A). Surprisingly, a new feeding tube is formed every time the nematode feeds. Thus, by the time the pathogenesis is completed, hundreds of feeding tubes are present in each giant or syncytial cell. Microinjection experiments using fluorescently labeled dextrans of various molecular masses have determined that the size-exclusion limit

for molecules passing through the feeding tube is between 20 and 40 kDa. This suggests that the signals from the nematode that modify the plant cell cytoplasm and increase plant cell metabolism will be relatively small molecules. Identification of the bioactive gland secretion components is currently an area of intense research.

21.1.6 Feeding arthropods not only damage plants directly but also facilitate colonization by viral, bacterial, and fungal pathogens.

Myriad insect species feed, reproduce, and shelter on plants. Two broad categories of herbivorous insects are recognized: those that chew and those that suck sap. Chewing insects cause the more spectacular plant tissue damage. For example, Colorado potato beetles (Fig. 21.14) and locust species may defoliate an entire crop covering several acres within a day or so. For other chewing species, the extent of damage frequently depends on the developmental stage of both the pest and the plant. European corn borer larvae attack the leaves of young maize plants; as both plants and pests mature, however, the insects become devastating stalk borers. Other chewing insects feed exclusively on roots or seeds. In contrast, most sap-sucking insects, such as adult leafhoppers, aphids, or thrips, cause minimal direct tissue destruction. These insects use a specialized mouth part, the stylet, to locate, penetrate, and drain sap from the phloem sieve elements of the plant's vascular tissue (Fig. 21.15; see also Box 15.2, Chapter 15, for a discussion of how sap-sucking insects are used to collect phloem sap for chemical analysis). Heavy infestations of sap-sucking insects cause chronic shortages of photosynthates and thus severely reduce the growth potential of the plant.

In addition to damaging plants directly, many insect pests transmit viruses while feeding. Sap-sucking species are extremely effective virus vectors: They deliver virus particles directly into the vascular tissue, which rapidly disseminates the virus throughout the plant. Some virus species also can replicate and persist inside insect vectors. During subsequent feedings on

Figure 21.14
The Colorado potato beetle (*Leptinotarsa decemlineata*) causes devastation to potato crops by chewing and eating green leaf and stems.

plants, therefore, infected insects continue to spread viruses. Chewing insects rarely transmit viruses, but the tissue damage they cause frequently permits attack by necrotrophic fungal and bacterial species.

21.2 Plant defense systems

21.2.1 Complex responses frequently save plants from destruction by pathogens.

Only a very small proportion of pathogen infections are likely to result in a diseased plant. Four main reasons account for most failures of pathogens to infect plants successfully.

- The plant species attacked is unable to support the life-strategy requirements of the particular pathogen and thus is considered a nonhost.
- The plant possesses preformed structural barriers or toxic compounds that confine successful infection to specialized pathogen species (nonhost resistance).
- On recognition of the attacking pathogen, defense mechanisms are activated such that the invasion remains localized.
- Environmental conditions change and the pathogen perishes before the infection process has reached the point at which it is no longer influenced by adverse external stresses.

The first three interactions are said to represent **genetic incompatibility,** but only the third mechanism of **resistance** depends exclusively on induced defense responses to limit pathogen attack. However, in some nonhost plant–pathogen interactions, defense responses also are activated. Successful pathogen infection and disease **(compatibility)** occur only if environmental conditions are favorable, if the preformed plant defenses are inadequate, and if either the plant fails to detect the pathogen or the activated defense responses are ineffective.

The primary objective of this section is to highlight the essential prerequisites for pathogen recognition and the induction of localized defense responses. The difference between induced and preformed defenses also will be outlined. Later, in Section 21.5,

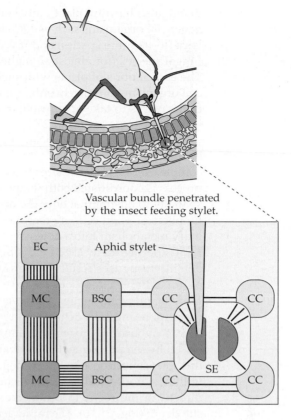

Vascular bundle penetrated by the insect feeding stylet.

Figure 21.15

Insects can be vectors for the transmission of viruses into plant cells. An aphid feeding stylet delivers viruses directly into the phloem sieve elements (SE). The virus particles then move by way of the translocation stream to other regions of the plant, where they can exit into the surrounding cells by way of the SE and the companion cells (CC) through the plasmodesmatal connections. BSC, bundle sheath cell; MC, mesophyll cell; EC, epidermal cell. The relative frequency of plasmodesmatal connections is indicated by the number of lines between cells.

we examine in detail the actual molecular and biochemical events that occur during the induced defense response.

21.2.2 Activated defenses can result in the hypersensitive response, a localized cell death program that prevents pathogen spread.

To activate its defense responses rapidly at the site of pathogen invasion, every plant cell needs a sophisticated surveillance system. This system must be fully functional in healthy plants and capable of distinguishing between self-generated signals and those emitted from pathogens. Moreover, when nonself signals are detected, the plant system must have enough discriminatory capacity to distinguish harmful plant pathogens from beneficial organisms such as *Rhizobium* bacteria (for legume species) and mycorrhizal fungal species (for almost all higher plants). The associations of plants with such beneficial organisms ultimately enhance the ability of the plants to take up nutrients from poor soils (see Chapters 16 and 23).

Recognition of a genetically incompatible pathogen results in the activation of a complex series of defense responses. The process is coordinated both temporally and spatially to ensure that only the necessary numbers of plant cells are recruited from primary metabolism into a defensive role. This rapid and highly localized induction of plant defense responses results in the creation of unfavorable conditions for pathogen growth and reproduction; at the same time, the responding cells detoxify and impair the spread of harmful enzymes and toxins produced by the pathogen. Full activation of this intense response to the pathogen occurs within 24 hours and invariably leads either directly or indirectly to localized cell and tissue death (Fig. 21.16). The rapid activation of defense reactions in association with host cell death is frequently called the **hypersensitive response** (HR). Because the dead cells contain high concentrations of molecules with antimicrobial properties, they are not subsequently attacked by opportunist necrotrophic organisms. However, many well-documented examples of non-HR–mediated resistance also exist (see Section 21.5).

21.2.3 Preformed defenses involve numerous secondary metabolites.

Most healthy plants already possess many different secondary metabolites with antimicrobial properties (see Chapter 24). These compounds may be present in their biologically active forms or may be stored as inactive precursors that are converted to their active forms by host enzymes in response to pathogen attack or tissue damage. In general, these preformed inhibitors are sequestered in vacuoles or organelles in the outer cell layers of plant tissues. Therefore, the type of attack strategy the pathogen pursues will greatly influence the concentrations of inhibitors it encounters. Necrotrophs invariably cause the release of high concentrations of inhibitors, whereas haustorium-forming biotrophic fungi may never encounter those defenses.

Two well-characterized classes of preformed inhibitors are the saponins and the glucosinolates. Saponins are glycosylated compounds, classified as either triterpenoids, steroids, or steroidal glycoalkaloids. A biologically active triterpenoid saponin found in the roots of oat plants, **avenacin A-1**, is highly effective against *Gaeumannomyces graminis* var. *tritici,* a major pathogen of cereal roots that is extremely sensitive to avenacins. Consequently, the widespread disease caused by *G. g.* var. *tritici* affects wheat and barley but never oat plants (Fig. 21.17A). *G. g.* var. *avenae*, a form of the pathogen that is specialized for oats, produces the saponin-detoxifying enzyme avenacinase, which is required for pathogenesis on oats.

Glucosinolates are sulfur-containing glucosides produced by members of the Brassicaceae, including the model experimental plant species *Arabidopsis thaliana.* Glucosinolates are divided into three classes, according to the nature of their side chains, which are derived from aliphatic, indolyl, or aralkyl α-amino acids. In contrast to saponins, glucosinolates become biologically active only in response to tissue damage by the activity of the enzyme myrosinase, a thioglucosidase. In healthy plant cells, subcellular compartmentalization separates myrosinase from the glucosinolate substrate. The unstable aglycone intermediate generated by myrosinase enzyme activity is converted into various products, including volatile isothiocyanates (for example, mustard oils)

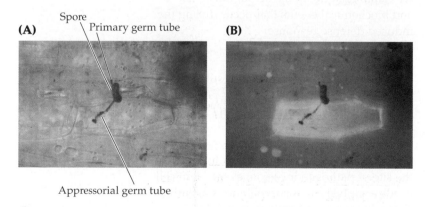

(A) Spore
Primary germ tube
Appressorial germ tube

(B)

Figure 21.16
The hypersensitive cell death response (HR) of barley epidermal cells to attack by germinating spores of the barley powdery mildew fungus. (A) The stained extracellular fungal structures—primary germ tube, appressorial germ tube, and spore—viewed with phase-contrast light microscopy. (B) The corresponding whole cell autofluorescence of the same barley epidermal cells as seen under UV light excitation (wavelength 310 nm); only the single cell undergoing the HR exhibits autofluorescence.

Figure 21.17

Two classes of preformed defense compounds. (A) Avenacin A-1, a saponin, is produced in the roots of oat plants but not in wheat or barley roots. The peripheral location of avenacin in healthy oat roots is revealed when root sections are viewed under UV fluorescence. The root-infecting take-all fungus *Gaeumannomyces graminis* var. *tritici* is extremely sensitive to avenacin A-1, whereas a closely related species, *G. g.* var. *avenae*, carries a saponin-detoxifying enzyme that removes the 1→2- and 1→4-linked terminal D-glucose (glu) sugar molecules, rendering the avenacin A-1 nontoxic. (B) When tissue is damaged, hydrolysis of stored, preformed glucosinolates leads to the formation and release of numerous bioactive toxic compounds. Wounding not only activates myrosinase enzyme activity but also causes cellular decompartmentalization, which brings the activated myrosinase enzyme into contact with the glucosinolate substrates.

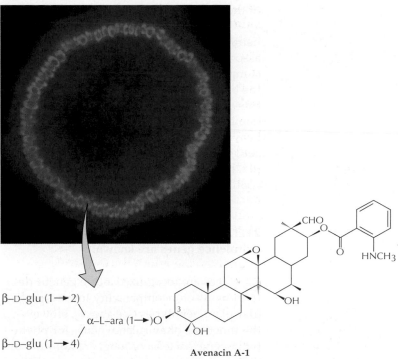

β–D–glu (1→2)

α–L–ara (1→)O

β–D–glu (1→4)

Avenacin A-1

(Fig. 21.17B; see also Chapter 16). Genetic variation for the glucosinolate profiles has been demonstrated in oilseed rape plants. Increasing the proportion of butenyl glucosinolates by eliminating the pentenyl ones decreases the palatability of leaf tissues to unspecialized *Brassica* pests such as rabbits, pigeons, and slugs but increases the susceptibility of the plants to specialized insect pests such as adult flea beetles (*Psylliodes chrysocephala*).

21.3 Genetic basis of plant–pathogen interactions

21.3.1 Disease resistance is usually mediated by dominant genes, but some recessive resistance genes also exist.

Plant collectors in the 19th century frequently noted differences in the disease susceptibility of various accessions of plants collected from the wild. After the rediscovery of Mendel's work, plant breeders in the early 1900s recognized that resistance to plant pathogens was often inherited as a single dominant or semidominant trait. However, not until the 1940s, and the seminal genetic studies of Harold H. Flor on flax and the flax rust pathogen, was the inheritance of not only plant resistance but also pathogen virulence finally elucidated and the **gene-for-gene model** proposed (Fig. 21.18). This model predicts that plant resistance will occur only when a plant possesses a dominant **resistance gene** (*R*) and the pathogen

(B)

S–β-D-glu

Preformed glucosinolates   R—C

NOSO₃⁻

H₂O → Myrosinase ← Wounding increases myrosinase activity

SH

Unstable aglycone intermediate   R—C   + D-glucose

NOSO₃⁻

R—N=C=S   R—N≡C   R—S—C≡N

Isothiocyanate   Nitrile   Thiocyanate

Various bioactive toxic compounds released

expresses the complementary dominant **avirulence gene** (*Avr*). The model holds true for most biotrophic plant–pathogen interactions.

For pathogens that deploy host-selective toxins for successful pathogenesis, a different model is proposed. Pathogen virulence must be dominant because a functional toxin or enzyme (or both) must be produced to cause disease (Fig. 21.19). Plant resistance, in turn, is predominantly inherited as a dominant trait and is achieved through enzymatic detoxification or through loss or alteration of the toxin in the pathogen.

The term **disease tolerance** describes a neutral outcome observed in plant–pathogen associations. The interaction is genetically compatible, but the plant somehow restricts the biochemical processes required for symptom development. As a consequence, tissue damage is kept down even if the plants are heavily infected. Disease-tolerant plants can act as important reservoirs of pathogen inocula, however, which may go on to infect susceptible species.

21.3.2 Properties of only a few pathogen avirulence genes are known.

Avr genes are recognized as the genetic determinants of incompatibility toward particular plant genotypes (see above). However, the functions of avirulence genes for phytopathogenic bacteria or fungi are not fully understood. Plant viruses provide the only conclusive exception. In different incompatible interactions, the viral CP, replicase, and MP are recognized as avirulence factors. Changes in amino acids that do not substantially compromise the primary function of the protein

in pathogenesis can still alter their *avr* specificity. For bacteria and fungi, a mutation from avirulence (*Avr*) to virulence (*avr*) is often associated with a loss of fitness for growth on plants that lack the *R* gene. This suggests that the *Avr* gene products have important roles for microbial pathogenicity either during specific stages of growth and reproduction in plants, for symptom development, or for transfer to other plants. Very few of these suspected additional functions have been rigorously tested by mutation analysis, however.

In 1984 the first avirulence gene was isolated from a soybean-infecting *Pseudomonas* bacteria by a technique known as shotgun cloning (Fig. 21.20). An identical strategy has since been used to isolate more than 30 *avr* genes from various pathogenic Gram-negative *Pseudomonas* and *Xanthomonas* species. The bacterial *avr* gene sequences cloned from the organisms code for soluble, hydrophilic proteins. Avr proteins share little homology with other known proteins, but there is substantial

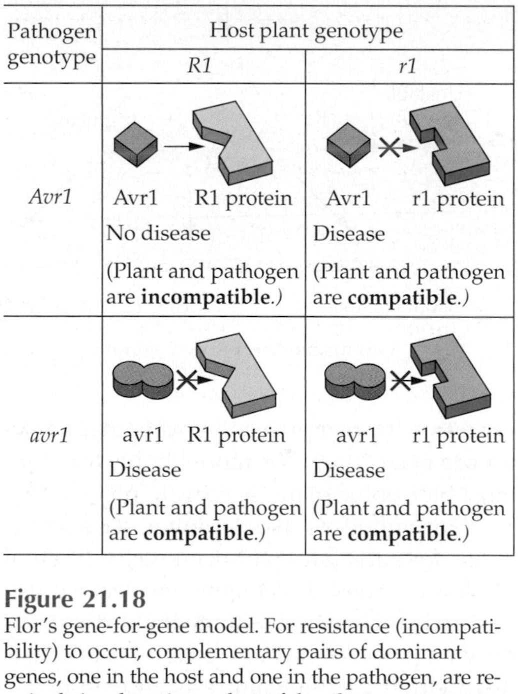

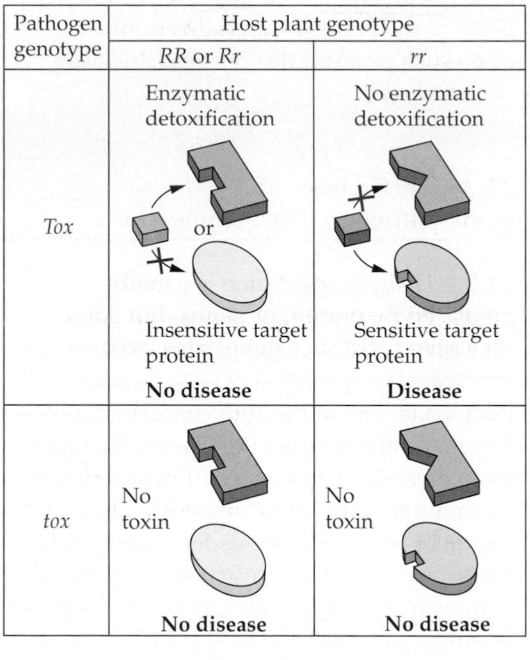

Figure 21.19
Interactions involved in toxin-dependent compatibility. The wild-type pathogen *Tox* gene is required for synthesis of a toxin that is crucial for pathogenesis; *tox* is the corresponding recessive, nonfunctional allele. The dominant allele of the host *R* gene is required for detoxification-based resistance. Resistance can also occur when the host expresses a toxin-insensitive form of the toxin target. Disease results only when the plant cannot detoxify the toxin produced by the pathogen.

Figure 21.18
Flor's gene-for-gene model. For resistance (incompatibility) to occur, complementary pairs of dominant genes, one in the host and one in the pathogen, are required. An alteration or loss of the plant resistance gene (*R* changing to *r*) or of the pathogen avirulence gene (*Avr* changing to *avr*) leads to disease (compatibility).

Box 21.2

Resistance genes can be bred into crop plants to control disease but this approach has only limited success.

After the realization that plant disease resistance is frequently inherited as a dominant trait, 20th-century plant breeders developed breeding programs to identify resistant germplasm in wild relatives of crop plants and then introgressed the corresponding resistance (*R*) genes for agricultural benefit. The discovery that plants have geographical centers of origin, locations where the greatest genetic diversity for both the plant and its pathogens resides, greatly assisted in the collection of novel germplasm for breeding purposes. Plant breeders achieved considerable early success in disease control through using these new resistant cultivars. However, they often found that within a few years of the introduction of an *R* gene into commercial production, the gene ceased to protect the crop from disease, and severe epidemics occurred. This **boom and bust cycle** of disease resistance results because effective disease control (the boom years) occurs only when the pathogen population consists entirely of races that express the corresponding functional *Avr* gene. Once an *avr* mutant race of pathogens appears, disease control fails (the bust years). Introgression of another *R* gene allele typically initiates another boom and bust cy-

cle, as illustrated by the resistance rating curves for *R2* and *R3* (see figure).

Some *R* genes have provided excellent disease control in large-scale commercial production for more than 15 years. For example, the tomato *Cf-9* gene-mediated resistance to the fungus *Cladosporium fulvum* has remained effective since the early 1980s. Why some *R* genes provide more **durable resistance** than others is still poorly understood.

One of the consequences of the cyclic nature of the disease control achieved with many *R* gene alleles has been the development of agrochemicals to provide an alternative solution.

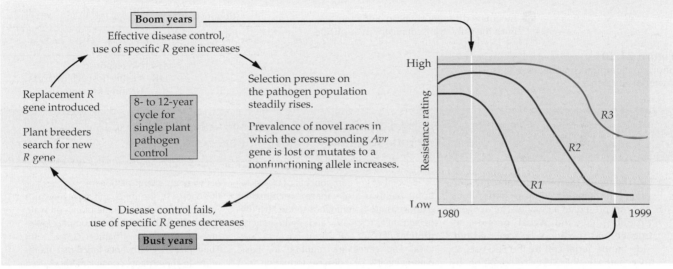

homology between some Avr proteins. Two types of distinct Avr-generated signals are recognized. Exported syringolides (C-glucosides with a novel tricyclic ring; Fig. 21.21) are produced by enzymes encoded by the *avrD* locus of *P. syringae* pv. *glycinea,* and these syringolides trigger a resistance response in soybean plants carrying the corresponding *R* gene, *Rpg4*. For other bacterial species, the Avr protein itself is now thought to be the signal (Box 21.3).

The *Xanthomonas avrBs3* family of *avr* genes is distinct from the *Pseudomonas avr* genes. Different *avrBs3* members encode proteins that contain a reiterated internal motif 34 amino acids long. For example, the *AvrBs3* gene product has 17.5 nearly identical repeats of this motif. By deleting some of these repeats, the specificity of this *Avr* gene can be altered so that it is no longer

recognized by the corresponding *Bs3* resistance gene in pepper plants. Other internal deletions in *AvrBs3* lead to a gain-of-function phenotype, resulting in a resistance response in plants carrying the recessive *bs3* allele (Fig. 21.22). Delivery of AvrBs3 into the plant cell cytoplasm by way of the type III secretion system (see Box 21.3) appears to be necessary for the function of *Xanthomonas Avr* gene products. In addition, mutational studies of the two functional nuclear localization signals in the C terminus of the AvrBs3 protein reveal that the Avr protein must be targeted to the plant cell nucleus for *Bs3*-mediated resistance to operate.

For pathogenic fungi, only a few *Avr* genes are known. For fungal species that colonize only the intercellular airspaces of plants, a biochemical approach can isolate small secreted peptides capable of eliciting

① Grow *Pseudomonas syringae* pv. *glycinea* (race 6) containing *Avr* gene of interest (shown in red). Genome size is approximately 3000 kb.

② Prepare chromosomal DNA, partially digest with restriction enzymes and purify DNA fragments approximately 25 kb in length.

③ Clone DNA fragments into a cosmid vector conferring tetracycline resistance and transform into *Escherichia coli*.

Donor *E. coli* strain carrying a cosmid with *Psg6* DNA fragment.

Helper *E. coli* strain carrying a plasmid to assist bacterial conjugation.

Recipient *Psg5* strain that is tetracycline sensitive and lacks the *Avr* gene

④ Separate triparental matings with 680 selected *E. coli* Tet^R colonies.

⑤ Grow individual Tet^R *Psg5* transconjugants in small liquid cultures to prepare the inoculum for plant disease assays.

Wild-type *Psg6*

Inocula of each *Psg5* transconjugant pressure-inoculated into tissue at 10^7 to 10^9 cfu/ml

Wild-type *Psg5*

⑥ Inoculate different soybean cultivars with the same transconjugant in duplicate.

⑦ Assess pathogenicity.

Response of soybean cultivars

P.s. pv. *glycinea* race	Harosoy	Peking	Flambeau
Psg5 (recipient strain)	—	—	—
Psg6 (donor strain)	HR	HR	HR
Psg5 + cloned *Avr* gene	HR	HR	—

— No response
HR Visible host cell necrosis within 24 hours

Figure 21.20
Shotgun cloning technique used to isolate bacterial *Avr* genes.

Box 21.3 — Bacterial pathogens of plants and animals utilize similar secretion pathways during pathogenesis.

The recent discoveries that some avirulence proteins function as intracellular elicitors presented a puzzle. The *avr* gene products have no signal peptide sequence and so cannot be exported outside the bacterial cell by the SecA secretory pathway. However, recognition by the corresponding plant *R* gene product is likely to occur within the plant cell cytoplasm (see Section 21.4.2). How do *Avr* products get into the plant cell?

Studies of Gram-negative bacterial pathogens of mammals provided the clue. *Yersinia*, *Salmonella*, and *Shigella* species produce specific virulence proteins that are secreted across both the inner and outer bacterial membranes, even though each secreted protein lacks a signal peptide at the N terminus. The mechanism utilized, termed a **type III secretion pathway**, involves a protein complex that spans both the inner and outer bacterial membranes. Specific protein subcomponents of the complex are responsible for translocating each protein to be exported, first across the inner bacterial membrane and then across the outer one. Finally, the protein is delivered into a projecting pilus to complete its transfer into the host cell at the bacterium–host interface (see figure).

Several of the components of the secretion pathway of these mammalian pathogens are homologous to the **Hrp** (hypersensitive response and pathogenicity) **proteins** of plant pathogenic bacteria. Bacterial *Hrp* genes are required for basic pathogenicity, i.e., disease formation in a susceptible plant host, as well as for the *avr* gene product to cause a hypersensitive response in a resistant plant host (see figure 21.16). Nine of the Hrp proteins of pathogenic plant bacteria have a close similarity to components of the type III secretion pathway of animal pathogenic bacteria. Consequently, members of this subset of *Hrp* genes are now known as *Hrc* genes (hypersensitive response and conserved). The genomic organization of the *Hrp* gene cluster of *Pseudomonas syringae*, which is required for type III secretion, is shown in the upper left corner of the figure. The defense response to *Pseudomonas*, mediated by the resistance genes *Pto*, *RPS2*, and *RPM1*, is now known to be induced only if the corresponding Avr proteins are delivered directly into the plant cell cytoplasm. Once Avr proteins are inside the plant cell, the Hrp proteins are not needed for induction of the HR.

A critical, still unresolved question regarding the *Hrp* gene type III secretion pathway in plant pathogenic bacteria is the nature of the mechanism that permits Avr proteins to traverse the plant cell wall. Plant cell walls are highly complex polymer matrices (see Chapters 2 and 24). Their presence prevents the direct membrane-to-membrane contact method used by mammalian pathogenic bacteria to facilitate establishment of the type III secretion pathway. A possible mechanism in plants may involve *Hrp*-dependent fibrillar structures, known as appendages, or pili, which have been observed on the surface of *Pseudomonas* bacteria growing in the intercellular spaces of plants. These appendages, which contain Hrp proteins, are only 6 to 8 nm in diameter and so might be thin enough to traverse the plant cell wall matrix and reach the plant plasma membrane. Cell attachment might then induce transfer of protein through the appendage. Alternatively, these appendages may serve to anchor the bacterial cell close to the plant plasma membrane, allowing protein released from the bacterial pore to be taken up by endocytosis into the plant cell. The *hrpA* mutant of *P. syringae* pv. *tomato*, which is incapable of forming a pilus, was found to be unable to elicit the HR or to cause disease in plants.

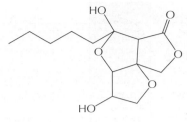

Figure 21.21
Chemical structure of the bacterial elicitor syringolide, produced by enzymes encoded by *avrD* of *Pseudomonas syringae* pv. *glycinea*. This syringolide triggers the resistance response in soybean plants carrying the resistance gene *Rpg4*.

R gene–dependent plant defense responses in the absence of the pathogen (e.g., Avr9; see Fig. 21.23). Each peptide appears to function as the direct elicitor of the R protein–mediated plant signal perception process

(see Section 21.4.2), but the full role of these peptides in fungal pathogenicity remains elusive.

21.4 *R* genes and *R* gene–mediated disease resistance

21.4.1 Most plant R proteins have structural similarities.

In the late 1980s, several laboratories initiated experiments to isolate different *R* genes. Because the biochemical role of the *R* gene product was not known, only two isolation strategies could be deployed. Both relied predominantly on (*a*) locating the *R* gene on the chromosome by using plant populations that segregate for resistant and susceptible individuals, and then (*b*) identifying the correct sequence either by inserting a mobile genetic element, called a **transposon** (see

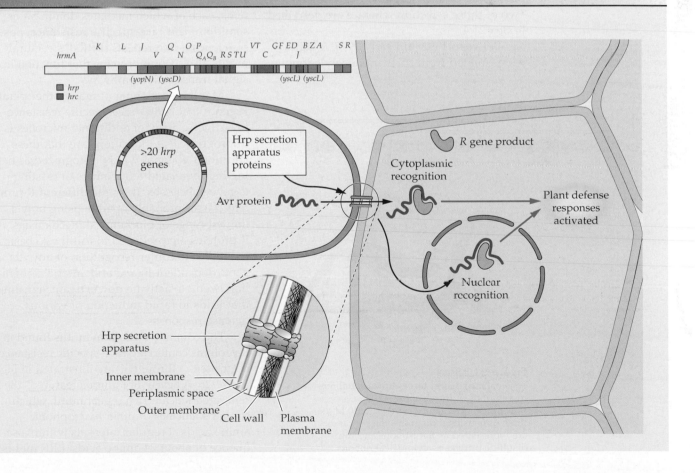

Predicted protein structure

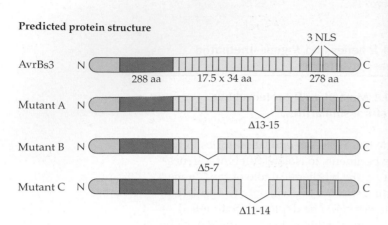

| | Pepper | | Tomato | |
	Susceptible cultivar	Resistant cultivar	Bonny Best susceptible cultivar	Remarks
AvrBs3	S	R	S	Wild-type reaction
Mutant A	S	R	S	Specificity unchanged
Mutant B	S	S	S	Biological activity destroyed
Mutant C	R	S	R	Novel Avr function generated

Figure 21.22

Overall structure of the AvrBs3 protein of the pathogenic bacterium *Xanthomonas campestris* pv. *vesicatoria*. The protein is divided into three major domains, with the central domain consisting of 17.5 nearly identical repeats of 34 amino acids each and the C-terminal domain containing three nuclear localization sequences (NLS). Some internal deletions of the repeat units lead to a loss of AvrBs3 function (mutant B), whereas other deletions lead to the gain of a novel function (mutant C). Adjacent to the structures of the mutants is a diagram of their respective effects when bacteria carrying the mutant genes are inoculated on various plants.

Chapter 7), to destroy biological activity (Fig. 21.24), or by using binary cosmid complementation to confer the resistance phenotype on a susceptible plant (Fig. 21.25). Two of those selection schemes are described in Box 21.4.

Between 1992 and early 1999, *R* genes were isolated from three monocotyledonous and five dicotyledonous plant species. These genes provide resistance to a range of taxonomically unrelated pathogens (Table 21.2). Four of the six classes of predicted R proteins, each of which mediates dominant or semidominant **race-specific resistance**, possess leucine-rich repeats (LRRs), structural motifs also seen in proteins that function in signal transduction pathways.

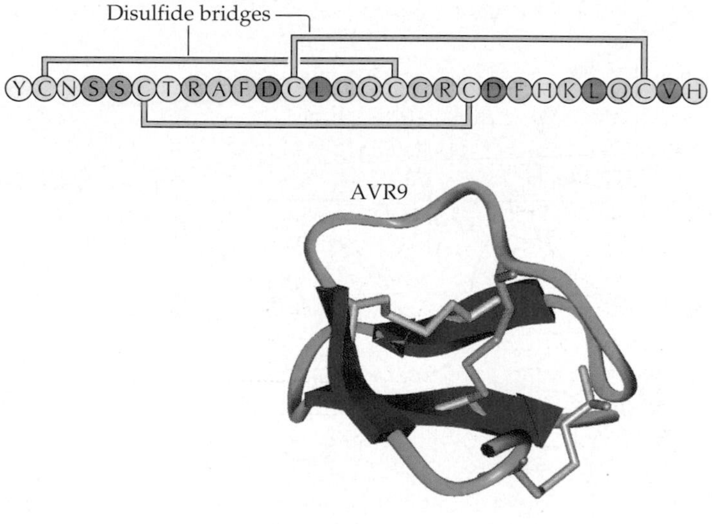

Map-based cloning of many other plant *R* genes that confer race-specific resistance to a diverse array of additional microbes is in progress. The indications are that these additional *R* genes will be recognized as being highly related to members of existing *R* gene subclasses. If so, the different R protein classes probably operate using only a limited range of biochemical mechanisms. R proteins are predicted to fulfill two basic functions: to confer recognition of any *Avr* gene–dependent ligand and, after a recognition event, to activate downstream signaling that leads to rapid induction of various defense responses.

The common structural motifs found in R proteins conferring race-specific resistance, i.e., classes 1 through 4, are illustrated in Figure 21.26. A nearly ubiquitous feature is the reiterated leucine-rich repeat motif, which contains leucines or other hydrophobic amino acids at regular intervals within a sequence of about 24 amino acids. LRR motifs

Disulfide bridges

Y C N S S C T R A F D C L G Q C G R C D F H K L Q C V H

AVR9

Figure 21.23

Schematic two- and three-dimensional representations of the structure of the *Cladosporium fulvum* avirulence protein Avr9. In the three-dimensional diagram, the β-strands are represented as flat ribbons and the disulfide bridges as cylindrical ribbons.

Identification of an *R* gene by transposon inactivation

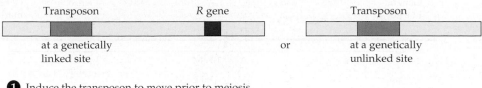

Transposon   *R* gene   Transposon

at a genetically   or   at a genetically
linked site   unlinked site

1 Induce the transposon to move prior to meiosis.

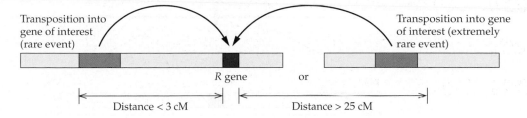

Transposition into
gene of interest
(rare event)

Transposition into gene
of interest (extremely
rare event)

R gene   or

Distance < 3 cM   Distance > 25 cM

2 Screen or apply a selection to a large population of progeny to identify disease-susceptible plants.

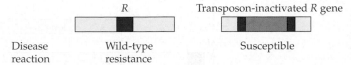

R   Transposon-inactivated *R* gene

Disease   Wild-type   Susceptible
reaction   resistance

3 To confirm the tagging experiments, look for resistant revertants in the next generation.

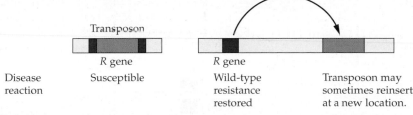

Transposon

R gene   *R* gene

Disease   Susceptible   Wild-type   Transposon may
reaction   resistance   sometimes reinsert
                              restored   at a new location.

4 Sequence the DNA flanking the transposons identified at step 2 to find the *R* gene.

Figure 21.24
Identification of an *R*
gene by inactivation
with transposons.

have been shown to mediate protein–protein and receptor–ligand interactions in many different kinds of organisms. Genetic evidence indicates that the β-strand/β-turn of the LRR (with the consensus sequence xxLxLxx) is a key region in the R protein and appears to determine its specificity. Given the crystal structure determined for porcine ribonuclease inhibitor protein, the conserved leucines (L) in the plant R proteins within this consensus are predicted to occupy the hydrophobic protein core, whereas the other residues (x) form a solvent-exposed surface that can participate in binding other proteins (Fig. 21.27). *R* gene sequence comparisons reveal that the x residues in this region are hypervariable and have high ratios of nonsynonymous to synonymous nucleotide substitutions. These data suggest the xxLxLxx region creates a surface that has evolved to detect variations in the multitude of pathogen-derived ligands. Parts of the LRR motif in plant R proteins may also participate in relaying downstream signaling through interactions with effector proteins. The large size of the LRR domain in most R proteins could even permit both the recognition and the effector functions to be accommodated by different binding specificities within different LRR subdomains or by interactions with more than one pathogen-derived ligand. Both the RPM1 and Mi proteins confer dual resistance specificity (Table 21.2). The binding of an *Avr* gene product or a plant protein to an LRR region has yet to be reported for any R protein so, in fact, *R* gene products might be detecting the binding of Avr products to other plant targets. Consistent

Identification of an *R* gene by map-based cloning

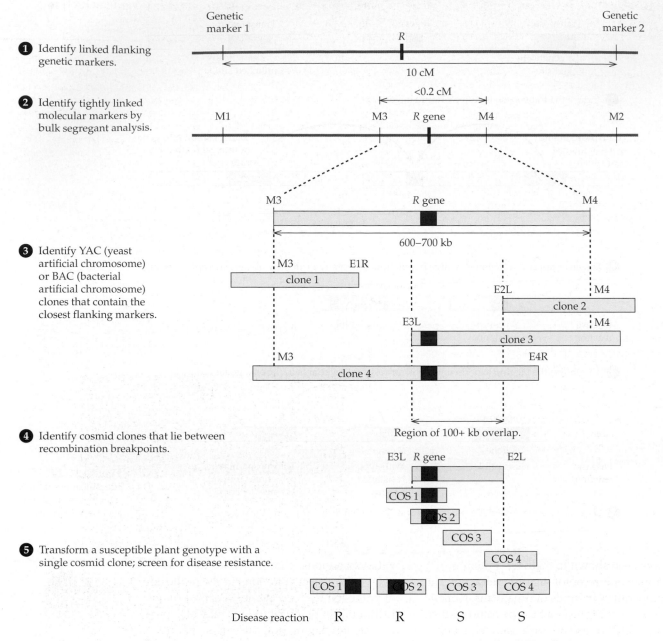

1 Identify linked flanking genetic markers.

Genetic marker 1

Genetic marker 2

R

10 cM

2 Identify tightly linked molecular markers by bulk segregant analysis.

<0.2 cM

M1   M3   *R* gene   M4   M2

3 Identify YAC (yeast artificial chromosome) or BAC (bacterial artificial chromosome) clones that contain the closest flanking markers.

M3   *R* gene   M4

600–700 kb

M3   E1R
clone 1

E2L   M4
clone 2

E3L   M4
clone 3

M3   E4R
clone 4

Region of 100+ kb overlap.

4 Identify cosmid clones that lie between recombination breakpoints.

E3L *R* gene   E2L

COS 1

COS 2

COS 3

COS 4

5 Transform a susceptible plant genotype with a single cosmid clone; screen for disease resistance.

COS 1   COS 2   COS 3   COS 4

Disease reaction   R   R   S   S

6 Sequence the region of overlap between clones 1 and 2 to identify the candidate *R* gene open reading frame.

Figure 21.25
Identification of an *R* gene by map-based cloning. R, resistant; S, susceptible.

with this idea, the Avr9 peptide has been found to bind with high affinity to a plasma membrane protein that is present even in membranes of plants that lack the *Cf-9* gene.

Most LRR R proteins also possess a central nucleotide-binding site (NBS) that contains several conserved motifs, the functions of which are not known. Although these R proteins do not appear to have intrinsic kinase activity, they could bind ATP or GTP and then activate the defense response. Mutations that alter key residues within the proposed NBS destroy R protein function. Some current models view the central NBS as an adaptor region, linking the C-terminal LRR recognition domain to various N-terminal effectors.

Some NBS-LRR R proteins possess a putative leucine zipper (LZ) or coiled-coil

Two of the *R* gene isolation experiments involving mobile transposable elements were coupled with a genetic selection that rapidly identified the rare plant genotypes in which the transposon had been inserted into the *R* gene sequence and thereby destroyed biological function. For the tobacco *N* gene, which confers resistance to tobacco mosaic virus (TMV), investigators took advantage of the fact that the *N* gene functions only at temperatures cooler than 25°C (see pair of photographs below left). A positive selection scheme for *N* gene transposon inactivation was used to rapidly recover tobacco plants susceptible to TMV. A plant population carrying both *N* and the transposon was inoculated with TMV at 30°C to permit systemic viral spread. After three days the growing temperature was lowered to 21°C, resulting in activation of a lethal whole-plant HR in all the plants except those in which the transposon was now residing within the *N* gene sequence. Plants of the latter genotype remained green.

A combined transposon tagging/genetic selection strategy was deployed to isolate the fungal resistance gene *Cf-9* from tomato plants (see panel A). Tomato plants lacking the *Cf-9* gene were genetically en-

gineered to produce fungal Avr9 peptide targeted to the apoplastic space. The *Avr9* plants were crossed with a second tagging line containing a *Dissociator (Ds)* element from maize closely linked in cis relation to the target *Cf-9* gene. This *Ds* element was mobilized to jump by a genetically unlinked transposase source (*Activator [sAc]*, also from maize). In the resulting F₁ generation, all plants that had inherited func-

tional copies of *Cf-9* and the *Avr9* transgene died shortly after seed germination, whereas the mutant plants that had inherited a transposon-tagged *Cf-9* survived (panel B). In plants of the *Cf-9^Ds*–tagged genotypes that also inherited the transposase source, variegated necrosis was evident (panel C), because somatic excision of the transposon from the *Cf-9* gene restored the Cf-9 function.

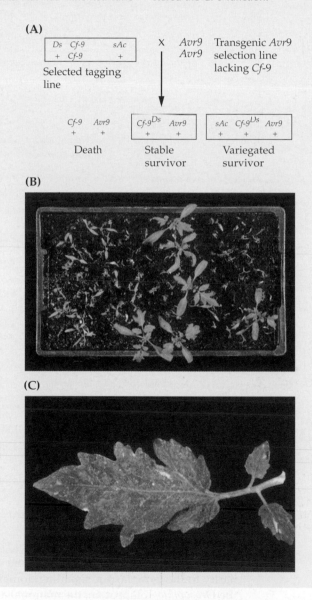

(A)

Ds	Cf-9	sAc
+	Cf-9	+

Selected tagging line

X *Avr9* / *Avr9* Transgenic *Avr9* selection line lacking *Cf-9*

Cf-9	Avr9
+	+

Death

Cf-9^Ds	Avr9
+	+

Stable survivor

sAc	Cf-9^Ds	Avr9
+	+	+

Variegated survivor

(B)

(C)

30°C → 21°C TMV

nn

Nn

sequence between the N terminus and the NBS domains. LZs are well known for their roles in homo- and hetero-dimerization of eukaryotic transcription factors (see Chapter 7) as well as facilitating interactions between proteins with other functions. Other NBS-LRR R proteins contain a large N-terminal domain called the **Toll/interleukin-l/resistance** (TIR)

Table 21.2 The six classes of plant resistance genes

Class	R gene	Plant	Pathogen	Pathogen type	Avr gene	Structure[a]	Cellular location	Size (aa)
1	RPS2	Arabidopsis	Pseudomonas syringae pv. tomato	Bacteria	avrRpt2	LZ-NBS-LRR	Cytoplasmic	909
	RPS5				avrPphB	LZ-NBS-LRR	Cytoplasmic	889
	RPS4				avrRps4	TIR-NBS-LRR	Cytoplasmic	1217
	RPM1	Arabidopsis	P. syringae pv. maculicola	Bacteria	avrRpm1, avrB	LZ-NBS-LRR	Cytoplasmic	926
	Rx	Potato	Potato virus X	Virus	Coat protein	LZ-NBS-LRR	Cytoplasmic	937
	Gpa2	Potato	Globodera pallida	Nematode	Unknown	LZ-NBS-LRR	Cytoplasmic	912
	N	Tobacco	Tobacco mosaic virus	Virus	TMV Replicase	TIR-NBS-LRR	Cytoplasmic	1144
	L[6]	Flax	Melampsora lini	Fungus	AL[6]	TIR-NBS-LRR	Cytoplasmic	1294
	M	Flax	M. lini	Fungus	AM	TIR-NBS-LRR	Cytoplasmic	1305
	RPP5	Arabidopsis	Peronospora parasitica	Fungus	avrRPP5	TIR-NBS-LRR	Cytoplasmic	1361
	RPP1				avrRPP1	TIR-NBS-LRR	Cytoplasmic	1189, 1221 1217
	RPP8				avrRPP8	LZ-NBS-LRR	Cytoplasmic	906
	I2C-1	Tomato	Fusarium oxysporum	Fungus	Unknown	NBS-LRR	Cytoplasmic	1220
	Mi	Tomato	Meloidogyne incognita	Nematode and aphid	Unknown	NBS-LRR	Cytoplasmic	1257
	Rp1-D	Maize	Puccinia sorghi	Fungus	avrRp1D	NBS-LRR	Cytoplasmic	1292
	Xa1	Rice	Xanthomonas oryzae pv. oryzae	Bacteria	avrXa1	NBS-LRR with unique LRR repeats 93 aa long	Cytoplasmic	1802
	Prf	Tomato	P. syringae pv. tomato	Bacteria	avrPto	NBS-LRR	Cytoplasmic	1824
2	Pto	Tomato	P. syringae pv. tomato	Bacteria	avrPto	Protein kinase	Cytoplasmic	321
3	Cf-9	Tomato	Cladosporium fulvum	Fungus	Avr9	eLRR-TM	Transmembrane	863
	Cf-4				Avr4	eLRR-TM		806
	Cf-2				Avr2	eLRR-TM		1112
	Cf-5				Avr5	eLRR-TM		968
4	Xa21	Rice	X. oryzae pv. oryzae	Bacteria	Unknown	LRR, protein kinase	Transmembrane	1025
5	Hml	Maize	Cochliobolus carbonum, race 1	Fungus	None	Toxin reductase	Cytoplasmic	357
6	mlo	Barley	Erysiphe graminis f.sp. hordei	Fungus	Unknown	G-protein–coupled receptor	Transmembrane	533
?	Hs1[pro-1]	Sugar beet	Heterodera schachtii	Nematode	Unknown	Controversial	Unknown	282

[a]Abbreviations: aa, amino acids; e, extracellular; G, GTP-binding; LRR, leucine-rich repeat; LZ, leucine zipper; NBS, nucleotide-binding site; TIR, Toll/interleukin-l/resistance domain; TM, transmembrane.

domain (see Fig. 21.26), which has some similarity to the cytoplasmic signaling domain of the *Drosophila* Toll protein, the mammalian in-terleukin receptor (IL-1R), and a family of mammalian Toll-like receptors, one of which participates in recognition and response to lipopolysaccharides (LPS). Toll, IL-1R, and the mammalian Toll homolog all contribute to

the immune response. The presence of the TIR domain in several R plant proteins suggests a role for this domain in signaling but not in ligand binding.

The serine/threonine protein kinase domain found in Pto and Xa21 proteins suggests that protein kinase–mediated signal transduction plays a crucial role in

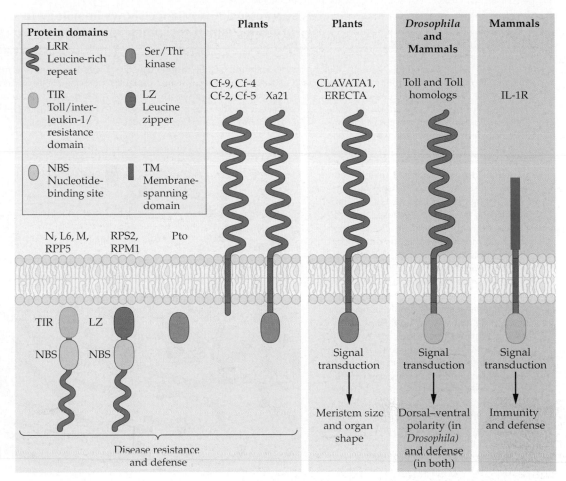

Figure 21.26
Schematic diagram illustrating several plant resistance proteins and various plant LRR-containing proteins synthesized during the defense response. For comparison are included structurally related proteins that are involved in various aspects of plant development as well as other eukaryotic proteins that coordinate development and the induction of the immune response in animals.

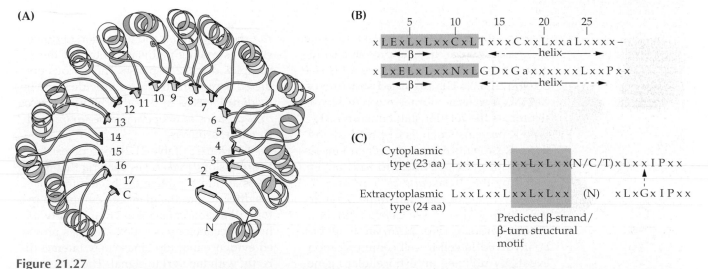

Figure 21.27
Structure of LRR (leucine-rich repeat)-containing proteins. (A) Ribbon diagram of the crystal structure of the porcine ribonuclease inhibitor (RI) protein. (B) Aligned consensus sequences of the LRRs of porcine RI. The one-letter amino acid code is used, including x to indicate any amino acid and a to denote an aliphatic amino acid. The part of the repeat that is strongly conserved in all LRR proteins is enclosed in the boxes. Below the sequences, the core regions of the β-sheet and helix are marked with solid lines; dots denote extensions of helix in different repeats. (C) LRRs of cytoplasmic and extracytoplasmic plant R proteins. The boxed area indicates the

leucine residues within the β-strand/β-turn structure (comparable with residues 3 through 9 in porcine RI, panel B), which are predicted to project into the hydrophobic core of the protein while the side chains of the flanking x amino acids are exposed to solvent. R gene sequence comparisons indicate that the x amino acids in the xxLxLxx region are hypervariable and have ratios of nonsynonymous to synonymous substitutions significantly greater than one. This suggests there is an advantage to high amino acid diversity in this region, which may have evolved to detect variations in interacting ligand molecules, possibly pathogen-derived ligands.

Box 21.5

Not all *R* gene products contain the functional domains described in Figure 21.26. One example is the recessive barley *mlo* mutation, which confers broad resistance against all known isolates of the barley powdery mildew fungus *Erysiphe graminis* f.sp. *hordei* (*Blumeria graminis*). The deduced 60-kDa barley gene product is predicted to be anchored in the membrane by seven membrane-spanning helices (see figure) and may well be a G-protein–linked receptor. Because the Mlo protein does not have any of the common structural motifs present in the race-specific R proteins (see Table 21.2), the wild-type Mlo protein is thought to attenuate resistance, whereas mutants are derepressed for resistance mechanisms (see Section 21.5).

The *mlo* gene is the only plant resistance gene that has been selected for following the mutagenesis of a susceptible wild-type (*Mlo*) variety (undertaken in the 1950s). This mutation has also arisen spontaneously in more than 150 barley varieties and has been detected in some wild barley accessions as well.

As another example, *Hm1* from maize encodes a unique R protein that confers resistance to the leaf spot fungus *Cochliobolus carbonum*. This fungus utilizes a race-specific toxin, HC-toxin, to elicit disease.

The HC-toxin inhibits the histone deacetylase activity of the plant, leaving it unable to activate defense responses. This HC toxicity affects a broad range of plant species. The maize *Hm1* resistance gene encodes a reductase enzyme that specifically detoxifies the HC-toxin (see Fig. 21.5 and Table 21.2). Many plant species are known to have genes that confer this same detoxification function.

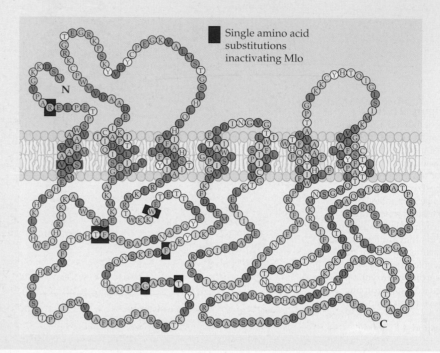

■ Single amino acid substitutions inactivating Mlo

---

race-specific resistance. However, to confer disease resistance, the Pto protein also requires the LRR-containing protein Prf (see Section 21.4.2). The predicted structure of Xa21 as a transmembrane receptor kinase reinforces the relationship between LRR-type R proteins (such as Cf proteins) and those R proteins encoding protein kinases. Two plant proteins similar to Xa21 are encoded by the *Arabidopsis ERECTA* and *CLAVATA1* genes that determine floral organ shape and size (see Chapter 19). Both of these proteins (Fig. 21.26) are thought to be involved in cell-to-cell communication events by utilizing an extracellular ligand.

21.4.2 How *R* and *Avr* gene products activate plant defense responses is not understood.

Each *R* gene product is thought to possess two functions: recognition of the corresponding

Avr-derived signal, and activation of downstream signaling pathways to trigger the complex defense response. The various predicted R protein structures provide some immediate clues as to how the different classes of R proteins may operate as receptors and signal transducers.

Pto kinase (Table 21.2) confers resistance to strains of *Pseudomonas* bacteria that express avrPto, a protein the bacterium delivers directly into the plant cell cytoplasm by type III secretion (see Box 21.3). Yeast two-hybrid technology (see Box 21.6) has provided evidence that Pto kinase may interact directly with the avrPto signal. The *Pto* gene belongs to a linked multigene family and shows 87% sequence similarity to a second kinase gene called *Fen*, the product of which confers sensitivity to the insecticide Fenthion, resulting in an HR type of cell death (Fig. 21.28). Analysis of chimeric Pto-Fen proteins indicates that subdomain VIII of the Pto kinase, which is involved in the protein

Table 21.3 Loci required for disease resistance

Plant species	RDR locus	Resistance	Pathogen affected	Function loss	Structu...
Tomato	Rcr-1, Rcr-2, Rcr-3	Cf-9/Cf-2	Cladosporium fulvum	Partial or complete	?
	Prf	Pto/Fen	Pseudomonas syringae pv. tomato	Complete	LZ-NE...
					Chord...
Barley	Rar1, Rar2	Mla-12	Erysiphe graminis f.sp. hordei	Almost complete	?
	Ror1, Ror2	mlo	E. graminis f.sp. hordei	Almost complete	?
Arabidopsis	Ndr1	RPM1	P. syringae (avrB)	Complete	
		RPS2	P. syringae (avrRpt2)	Complete	
		Some RPPs	Peronospora parasitica	Complete	
	Npr1	Some RPPs	Pe. parasitica	Partial	An...
		RPSs	P. syringae	Partial	An...
	Eds1, Pad4	Some RPPs	Pe. parasitica	Complete	Li...

Barley lines containing either the race-specific R gene *Mla-12* or the non–race-specific R gene *mlo* have been mutagenized, and various mutant genotypes sensitive to barley powdery mildew have been recovered (Table 21.3). Interestingly, the two *rar* mutants that interfere with *Mla-12* gene function do not affect resistance mediated by *mlo* or by a second race-specific R gene, *Mlg*. The two other mutant loci identified, *ror1* and *ror2*, abolish resistance mediated by *mlo* but not race-specific resistance mediated by either *Mla-12* or *Mlg*. Thus the three different R genes may operate by dissimilar mechanisms.

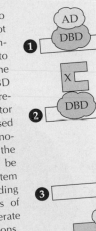
(A) Genomic organization of the gene family in *Pto* tomato

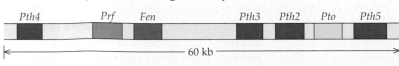

Pth4   Prf   Fen   Pth3   Pth2   Pto   Pth5

← 60 kb →

(B) Lines lacking *Pto* are susceptible to *P. syringae* pv. *tomato* (*avrPto*)

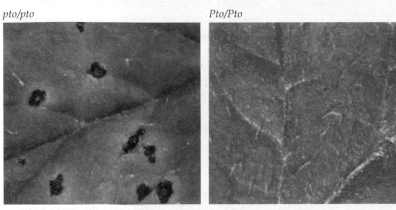

pto/pto   Pto/Pto

(C) The tomato *prf* mutant is susceptible to *P. syringae* pv. *tomato* (*avrPto*) but resistant to Fenthion.

Prf/Pfr; Pto/Pto; Fen/Fen   prf/prf; Pto/Pto; Fen/Fen

Infected with *P. syringae*

Treated with Fenthion

Figure 21.28
Genomic organization of the *Pto* gene family in tomato (A) and the biological phenotypes conferred by three of the genes in the cluster (B, C). The *Pto* gene confers resistance to *Pseudomonas syringae* pv. *tomato* bacteria expressing the avirulence gene *avrPto* (B). The *Prf* gene is required for both the Fen- and Pto-mediated responses; the *Fen* gene confers sensitivity to the insecticide Fenthion (C).

kinase activity, is required for interaction with avrPto. This result was unexpected for two reasons: First, cytoplasmically localized protein kinases were not previously known to function as receptors in any organism. Second, both the recognition and the signaling capacities of the enzyme apparently reside within the same Pto kinase domain because, thus far, it has been impossible to mutate one function without losing the other.

Several plant proteins that can interact with the Pto kinase have been identified with use of the yeast two-hybrid system, including proteins with homology to transcription factors as well as another protein kinase, Pti1. Three of the possible transcription factors—Pti4, Pti5, and Pti6—possess a highly conserved DNA-binding domain that recognizes a core hexanucleotide sequence, a sequence found in the promoters of several genes that encode ethylene-induced defense-related proteins (such as the pathogenesis-related [PR] proteins) (see Section 21.5). Overexpression of Pti kinase in plants carrying the Pto resistance gene enhances the avrPto-mediated HR. Thus the Pto protein appears to activate several distinct signaling pathways simultaneously (Fig. 21.29). The avrPto/Pto pathway, however, is still incomplete. The *Prf* gene, which is located

within the *Pto* gene cluster (see Fig. 21.28), is required for both *Pto*- and *Fen*-specified responses. *Prf* encodes also a LZ-NBS-LRR protein (see Fig. 21.26), for which the molecular function in disease resistance is still unknown. Conceivably, Prf "guards" Pto and recognizes when *Pto* interacts with Avr-Pto to activate defense.

The cytoplasmic TIR-NBS-LRR class of R proteins can confer resistance to biotrophic fungi, bacteria, and viruses. How might these proteins confer pathogen recognition? Direct interaction between the tobacco N protein and the viral Avr determinant—the replicase protein—is plausible because TMV replication is exclusively intracellular. The existence of NBS-LRR R gene products that confer resistance to different haustorial fungi suggests Avr products enter the plant cell cytoplasm where Avr-mediated recognition events occur. The presence of the conserved TIR domain suggests that the immune response in plants, mammals, and invertebrates may utilize an evolutionarily conserved mechanism (see Fig. 21.26). Such a mechanism would involve either direct or indirect activation of a serine/threonine protein kinase by the TIR domain, leading through a series of steps to the activation of transcription factor(s).

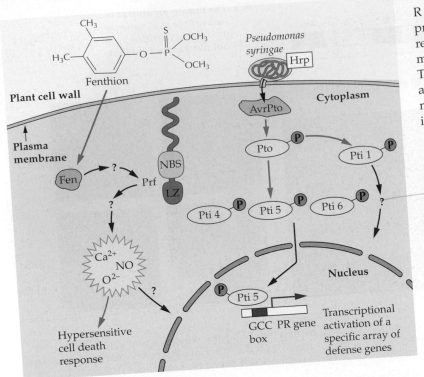

Figure 21.29
Model of the potential Pto-mediated resistance signaling pathway in tomato. Yeast two-hybrid analysis has shown that although Pto kinase interacts directly with bacterial AvrPto protein, this interaction is not required for interaction with the Pti proteins or for in vitro phosphorylation of Pti 1. The NBS-LRR protein Prf perhaps evolved to "guard" Pto and recognize the AvrPto–Pto complex, in order to initiate defense responses in addition to those activated by the transcription factors Pti 4, Pti 5, and Pti 6 and the phosphorylated Pti 1 protein. Similarly, sensitivity to the insecticide Fenthion, leading to induced cell death, may arise from Prf recognizing the activated Fenthion–Fen complex. Hrp, contact-dependent bacterial (type III) secretion system.

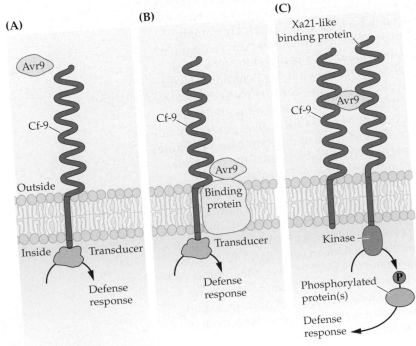

The transmembrane-lo[cated] R proteins, exemplified by [Cf] proteins, exhibit features in [a] receptor-like function but p[rovide] minimal clues about signa[ling] Three plausible molecular [...] account for activation of t[he...] mediated defense respons[es] in Figure 21.30.

21.4.3 Additional plant [genes] for R gene action.

Several plant genes requ[ired for] resistance (RDR genes; [...]) identified. These loci n[...] genetic tools to investig[...] tance phenotype and s[...] components. Two exam[ples of] useful disease-sensitiv[e...] been in increasing our [...] ease resistance.

The *Arabidopsis* m[utant...] resistance to *Pseudomo*[nas...] diated by RPS2 and [...] *Peronospora parasitica* [...] tain RPP genes (Fig. [...]) *Arabidopsis* mutant e[...] not all) RPP gene–m[ediated...] does not interfere w[ith...] tion. Therefore, the [...] provide evidence fo[r...] ways downstream f[rom...] these initially distin[ct...] later appear to conv[erge...] quirement for salic[ylic acid...] of the resistance re[sponse...] Mutagenesis of the [...] and subsequent sc[reening...] restore resistance [...] should permit the [...] regulators in the [...]

Figure 21.30
Three possible expl[anations for how] Avr9 peptide and th[e...] vate plant defense [...] The fungal Avr9 pr[otein...] with the host Cf-9 [...] through a host-enc[oded...] tein (B). Alternativ[ely...] a protein kinase ca[...] dimerization of C[f-9...]

Pseudomonas syringae   *Peronospora parasitica*

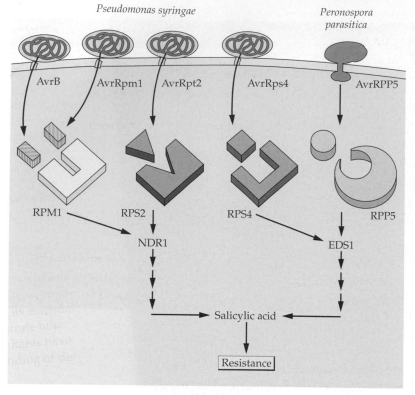

21.4.4 Extensive allelic variation exists at some *R* gene loci.

Many plant pathogens can easily mutate from avirulence to virulence, enabling them to overcome the resistance mediated by a specific *R* gene. If every plant in the population carries this particular *R* gene, such mutations favor the multiplication of novel virulent pathogen races. Moreover, mutations in *Avr* genes and pathogen evolution to virulence are likely to be much more frequent than the creation of corresponding novel plant *R* genes. How then do plants respond to the rapid evolution of novel pathogen races? In wild plant populations, polymorphism at the many known *R* gene loci can

Figure 21.31
The *Ndr1* and *Eds1* gene products participate in parallel signaling events in *R-Avr* gene–mediated plant disease resistance pathways in *Arabidopsis*. The products of the LZ-NBS-LRR genes *RPS2* and *RPM1* are dependent on Ndr1; the products of the TIR-NBS-LRR genes *RPS4* and *RPP5* are dependent on Eds1.

Box 21.7   Some *R* genes function in heterologous plant species.

Several isolated *R* genes introduced into other plant species by way of *Agrobacterium*-mediated transformation often retain biological activity. For example, an inserted tobacco *N* gene is still active in tomato (see panel A), where it restricts tobacco mosaic virus (TMV) infections to localized necrotic lesions, in a manner identical to that in tobacco (panel B). For comparison, untransformed tomato mock inoculates (panel C, left) and the extreme plant stunting and yellow mosaic symptoms resulting from a susceptible TMV infection (panel C, right) are also shown. Interestingly, the *N* resistance response still displays its temperature sensitivity (see Box 21.4) in tomato.

The functionality of R proteins in heterologous plant species indicates that Avr-dependent protein signaling cascades are conserved across some plant taxa. However, overexpression of some *R* genes in heterologous plants has resulted in the development of necrotic symptoms in the absence of pathogen attack and has yielded plants with altered development traits. These results illustrate the finely tuned relationship between R proteins and signaling partners in their native species. They also suggest that *R* gene evolution is constrained not only by selection *for* pathogen recognition but also by selection *against* recognition of related endogenous plant proteins that control plant development (see Fig. 21.47).

vastly exceed polymorphism at other gene loci. Virulent pathogen races are therefore rarely able to cause epidemics, because each plant can have a different *R* genotype. In addition, mutations in *Avr* genes that can overcome the resistance mediated by a specific plant *R* gene often reduce the fitness of the pathogen in a mixed population. Novel virulent pathogen races will therefore be successful only when a specific *R* gene is rare in the plant population and most pathogen races carry the corresponding novel *Avr* gene.

We now know from detailed genetic and molecular analysis of various *R* loci that different organizational types of *R* loci exist (Fig. 21.32). Most *R* genes are members of multigene families that are arranged in clusters to form a **complex locus.** Molecular analysis of complex loci for which the *R* gene sequence is already known has revealed that numerous homologous genes reside within a close physical distance of one another (e.g., at the tomato *Cf-9* and *Pto* loci shown in Fig. 21.32; see also Fig. 21.28). Another type of *R* locus, a single gene with an array of distinct alleles, is called a **simple locus.** The flax *L* locus conforms to this type (Fig. 21.32). Some simple *R* loci contain only a single *R* gene sequence that shows no or very limited allelic diversity among genotypes of the same plant species. In other cases, the single *R* gene sequence is either present or absent in each genotype (e.g., *RPM1*; Fig 21.32).

How then do novel *R* variants arise? Mutation followed by selection on solvent exposed amino acids in the LRR (Fig. 21.27) clearly plays an important role. However, in complex *R* gene loci with multiple repeats, recombination between LRR sequences from different homologs also could create new surfaces with novel recognition capacities. Such events could arise through mispairing, intra- and intergenic recombination, and gene conversion (Fig. 21.33). Comparison of homologs from *R* gene families provides evidence that sequences have been exchanged during the evolution of family members. The relative significance of recombination and point mutation in generating novelty is still unknown and may differ at different loci. Heterozygosity at *R* gene haplotypes might increase the frequency of gene conversion and unequal cross-overs; homozygosity will generate novel events less frequently because the homologous haplotypes will exhibit regular meiotic pairing.

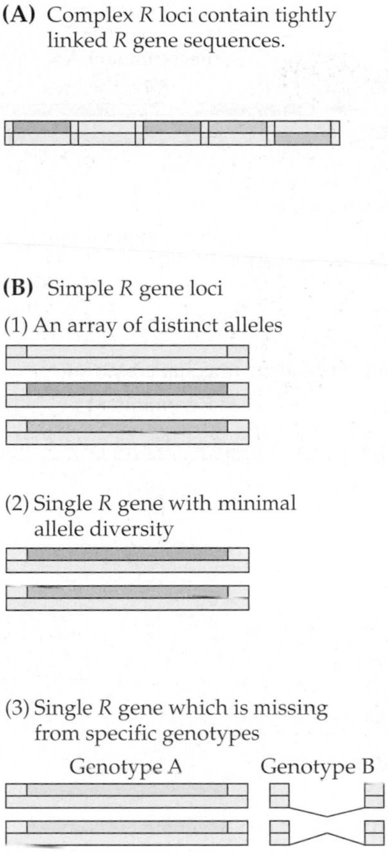

(A) Complex *R* loci contain tightly linked *R* gene sequences.

Examples		Number of *R* homologs at locus
Tomato	*Cf-9*	5
Tomato	*Cf-2*	3
Tomato	*Pto*	5
Arabidopsis	*RPP5*	8–10
Flax	*M*	15

(B) Simple *R* gene loci

(1) An array of distinct alleles

Flax	*L*	13 alleles

(2) Single *R* gene with minimal allele diversity

Arabidopsis	*RPS2*	2 alleles

(3) Single *R* gene which is missing from specific genotypes

Genotype A   Genotype B

Arabidopsis	*RPM1*	2 alleles

Figure 21.32
The four main types of plant resistance (*R*) locus organization: (A) complex *R* loci; (B) three types of simple *R* loci.

21.5 Biochemistry of plant defense reactions

21.5.1 Multiple types of defense reactions in each cell are activated by pathogen attack.

Plant resistance is correlated with the activation of a diverse set of defense mechanisms. The response involves transcriptional activation of numerous defense-related genes, opening of ion channels, modifications of protein phosphorylation status, and activation of preformed enzymes to undertake specific modifications to primary and secondary metabolism (Fig. 21.34). In addition, a range of secondary signaling molecules are generated to ensure coordination of the defense response both temporally and spatially, resulting in rapid containment of the pathogen. Overall, plants appear to

Complex R locus

(A) *R* gene reassortment in a heterozygote

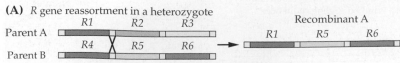

(B) Creating a novel variant *R* gene in a heterozygote

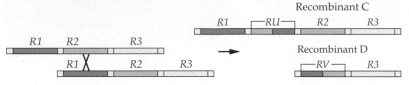

(C) Creating a novel variant *R* gene in a homozygote by unequal crossover

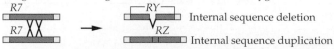

Simple R locus

(D) Intragenic recombination in a heterozygote

(E) Misalignment and intragenic recombination in a homozygote

Internal sequence deletion

Internal sequence duplication

Immediate responses of invaded cells

Generation of reactive oxygen species
Nitric oxide synthesis
Opening of ion channels
Protein phosphorylation/dephosphorylation
Cytoskeletal rearrangements
Hypersensitive cell death (HR)
Gene induction

Local responses and gene activation

Alterations in secondary metabolic pathways
Cessation of cell cycle
Synthesis of pathogenesis-related (PR) proteins
Accumulation of benzoic and salicylic acid
Production of ethylene and jasmonic acid
Fortification of cell walls (lignin, PGIPs, HRGPs)

Systemic responses and gene activation

$(1{\rightarrow}3)\beta$-Glucanases
Chitinases
Peroxidases
Synthesis of other PR proteins

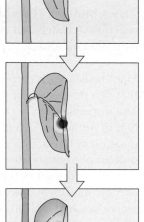

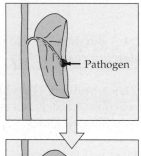

Pathogen

Figure 21.33

New *R* alleles can be generated by several different mechanisms, but the likelihood of their occurrence depends on whether the plant species is maintained as an outbreeding (cross-pollinating) population (A, B) or an inbreeding (self-pollinating) population (C, D, E). The evolution of resistance in plants appears to occur primarily at the single-gene level so that novel specificities arising by way of intergenic recombination between similar genes (C) are rare. The variant *R* alleles eventually selected encode proteins with increased effectiveness or confer a novel recognition capacity. *RT, RU, RV, RW, RX, RY,* and *RZ* represent different recombination products.

elicit a broad-spectrum defense response to most pathogens. Detailed studies on plant defense mechanisms have involved adding either the pathogen or pathogen-derived elicitors to intact plants or to plant cell suspension cultures. The most frequently encountered defense and signaling mechanisms are described below.

21.5.2 The HR results in rapid, localized cell death.

Necrotic flecks resulting from dead plant cells often form at sites of attempted pathogen attack (Fig. 21.35; see also Fig. 21.16). This rapid response is thought to play a causal role in resistance to some pathogens by depriving biotrophs of access to further nutrients. For other pathogen types, cellular decompartmentalization is thought to result in the release of preformed inhibitory substances (see Section 21.2.3). HR is not, however, an obligatory component of plant defense; many *R* genes confer non-HR– mediated resistance. For example, tomato *Cf-9* gene–mediated resistance against *C. fulvum* is not always associated with plant cell death. Potentially, two mechanisms underlie HR formation. Either the attacked cell initiates a regulated cell death program or the responding cells are rapidly poisoned by the toxic compounds and free radicals they have synthesized and thus die as a result of necrosis. Recent evidence suggests that both types of cell death probably occur during plant defense (see Chapter 20).

Figure 21.34

Schematic representation of the temporal order of activation of plant defense responses, both locally and systemically, from the initial site of pathogen invasion.

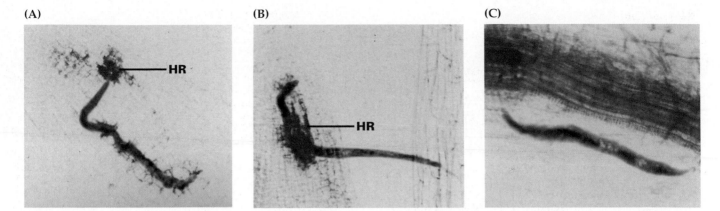

(A) HR

(B) HR

(C)

Figure 21.35

The invasion of resistant tomato roots by the soybean root-knot nematode causes the rapid formation of a highly localized hypersensitive response, surrounding the migrating nematode (A) and at the initial feeding site (B). In comparison, when the nematode feeds on the root of a susceptible soybean cultivar, there is no hypersensitive response and the nematode body rapidly increases in size and development (C).

ROS ⅀ HR

21.5.3 Reactive oxygen species are often produced during the early stages of a plant resistance response.

In many incompatible interactions, the production of reactive oxygen species (ROS) is often the first response detected, occurring within less than five minutes (see Fig. 21.34). The typical ROS species detected are superoxide ($O_2^{\bullet-}$) and hydrogen peroxide (H_2O_2). The mechanism plants have for producing superoxide from molecular oxygen probably involves a plasma membrane–associated **NADPH oxidase,** similar but not identical to that used in mammalian neutrophils for defense (Fig. 21.36). The superoxide anions produced outside the plant cells usually are rapidly converted to H_2O_2, a molecule that can cross the plasma membrane and enter plant cells. H_2O_2 is eventually removed from cells by conversion to water through the action of the enzymes catalase, ascorbate peroxidase, or glutathione peroxidase (see Chapters 20 and 22).

Several roles in plant defense have been proposed for ROS (Fig. 21.37). For example, H_2O_2 may be directly toxic to pathogens. In the presence of iron (Fe), H_2O_2 gives rise to the extremely reactive hydroxyl radical ($OH^{\bullet}$). Alternatively, it may contribute to the structural reinforcement of plant cell walls, either by cross-linking various hydroxyproline and proline-rich glycoproteins to the polysaccharide matrix or by increasing the rate of lignin polymer formation by way of peroxidase enzyme activity (see

Chapter 24)—both of which would make the plant cell wall more resistant to microbial penetration and enzymatic degradation. A signaling role for some ROS is also likely. H_2O_2 induces benzoic acid 2-hydroxylase (BA 2-H) enzyme activity, which is required for biosynthesis of SA (see Section 21.5.7). H_2O_2 is also known to induce genes for proteins involved in certain cell protection mechanisms, for example, glutathione

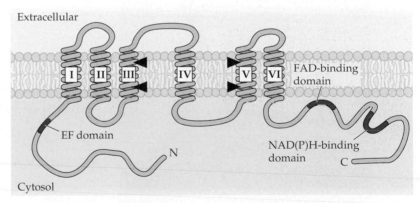

Figure 21.36

Proposed membrane protein topology of respiratory-burst oxidase (RBO) proteins encoded by genes isolated from *Arabidopsis*. Indicated are the six transmembrane domains (roman numerals), the putative positions of the two heme molecules (filled triangles), and the FAD and NAD(P)H binding sites. In the cytoplasmic N-terminal region, a proven calcium-binding motif, the EF hand, is present. The extra N-terminal extension and EF hand in the plant RBO proteins are not present in the homologous gp91 component of the NADPH oxidase complex that is responsible for generating the oxidative burst in mammalian macrophage cells. The presence of a cytoplasmically localized EF hand in the plant protein suggests the possibility of direct regulation of oxidase activity through the sensing of subcellular changes in calcium concentrations, rather than through the assembly of additional cytoplasmic oxidase components, as occurs in the mammalian system.

(A)

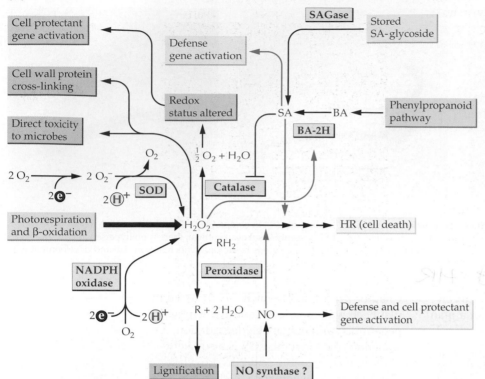

Figure 21.37
(A) The interconnecting roles of H_2O_2, nitric oxide (NO), and salicylic acid (SA) for the activation and coordination of multiple plant defense reactions. SOD, superoxide dismutase; SAGase, SA glycosyltransferase, BA-2H, benzoic acid 2-hydroxylase. (B) Localization of H_2O_2 accumulation, shown as electron-dense deposits of cesium perhydroxide (arrows) in cells adjacent to wild-type avirulent *Pseudomonas syringae* pv. *phaseolicola*, five hours after bacterial inoculation into the intercellular airspace of lettuce leaves (a nonhost species).

(B)

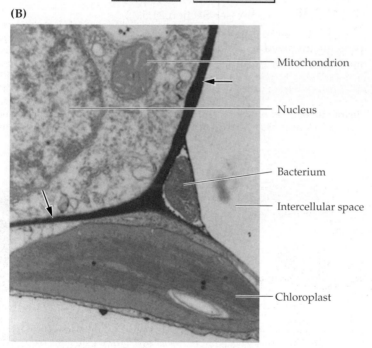

— Mitochondrion

— Nucleus

— Bacterium

— Intercellular space

— Chloroplast

21.5.4 Production of nitric oxide, a signaling molecule in mammals, is induced during incompatible interactions in plants.

Nitric oxide (NO) is a signal molecule used by mammals to regulate various biological processes of the immune, nervous, and vascular systems. In plants, rapid de novo synthesis of NO accompanies the recognition of avirulent pathogenic bacteria. Although a localized HR is a consistent feature of genetically incompatible interactions, the rapid burst of ROS production is insufficient to induce plant cell death but may be able to inhibit pathogen growth. Interestingly, NO has the capacity to potentiate induction of plant cell death by ROS (Fig. 21.38). NO is known to bind heme and thereby could inhibit catalase and ascorbate peroxidase, which detoxify H_2O_2. However, the NO molecule appears to have other roles during plant defense as well, because adding a NO-generating compound to plant cell suspension cultures and leaves leads to the accumulation of mRNAs from several genes involved in defense and cell protection. Furthermore, in the presence of inhibitors of NO production, the HR diminishes, disease symptoms become more severe,

S-transferase. Production of ROS may also substantially alter the redox balance in the responding cells. Like many mammalian transcription factors that are known to be redox-regulated, the activity of specific plant transcription factors may also be regulated by changes in the redox status of cells.

and bacterial growth is increased. These findings indicate that NO and ROS play an important synergistic role in the rapid activation of a wide repertoire of defense responses after pathogen attack.

21.5.5 Cell wall fortifications and extracellular activities contribute to plant disease resistance responses.

Minute papillae often form directly beneath the sites at which biotrophic fungi attempt to penetrate the plant cell wall (Fig. 21.39). These papillae, which are primarily composed of **callose** [a $(1\rightarrow3)\beta$–glucan polymer; see Chapter 2] and **lignin** (a highly complex phenolic polymer; see Chapters 2 and 24), are thought to act as a physical barrier, blocking fungal penetration into plant cells. Induced callose deposition within plasmodesmata is also likely to block virus cell-to-cell movement.

Extracellular basic **hydroxyproline-rich glycoproteins** (HRGPs) contribute to fortifying the cell wall in two ways. Preformed HRGPs cross-link rapidly to the wall matrix by way of tyrosine (in the motif PPPPY) reacting with induced H_2O_2. Later, de novo HRGP synthesis initiates additional lignin polymerization to further reinforce cell walls (see Chapter 24).

Another class of defense-related extracellular proteins are **polygalacturonase-inhibiting proteins** (PGIPs), which carry a

Psm (avrRpm1)

−Inhibitor   +L-NNA   +PBITU

1 d

2 d

−Inhibitor   +PBITU

Figure 21.38
NO is required for the hypersensitive response (HR) to avirulent pathogens. Infiltration of avirulent *P. syringae* bacteria expressing the avirulence gene *avrRpm1* into leaves of an *Arabidopsis* ecotype possessing a functional *Rpm1* resistance gene rapidly induces HR (upper panels, left). Two different NO inhibitors, nitro-L-arginine (L-NNA) and *S,S′*-1,3-phenylene-bis(1,2-ethanediyl)- bis-isothiourea (PBITU), block the early HR and promote a spreading chlorotic reaction at the site of the bacterial infiltration (upper panels, middle and right). This chlorotic response ordinarily is observed only in a compatible interaction with the isogenic virulent bacteria lacking *avrRpm1*. White arrows, HR; black arrows, disease symptoms.

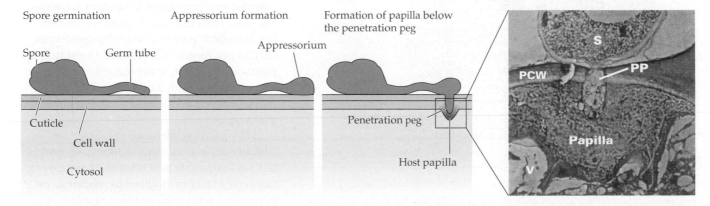

Spore germination   Appressorium formation   Formation of papilla below the penetration peg

Spore   Germ tube   Appressorium

Cuticle

Cell wall

Cytosol

Penetration peg

Host papilla

S

PCW   PP

Papilla

V

Figure 21.39
A mildew fungus spore (S) germinates and, after a short distance of growth over the leaf surface, forms an appressorium structure for attachment. Subsequently, a fungal penetration peg (PP) is produced that passes through the plant cuticle and epidermal cell wall. Papillae frequently form on the inner surface of the cell wall, directly beneath the invasion site, to prevent further ingress of hyphae. In some pathosystems, however, this localized reinforcement of the cell wall is thought to anchor an invading hypha by forming a tight seal between the neck of the haustorium and the plasma membrane of the invaded epidermal cell. PCW, plant cell wall.

LRR motif (see Section 21.4.1) and inhibit a specific subclass of necrotrophic pathogen cell wall–degrading enzymes, called polygalacturonases (PGs). PGIPs possibly retard PG function, resulting in an increased abundance of **oligogalacturonides** with chains more than eight units long; this, in turn, may trigger additional defense responses (Fig. 21.40).

21.5.6 Benzoic acid and SA may participate in numerous plant defense responses.

Free acids and glucoside conjugates of the phenolic compounds benzoic acid (BA) and SA accumulate to high concentrations in the immediate vicinity of incompatible infection sites. Both SA and BA are derived from the phenylpropanoid pathway (see Chapter 24) and have many roles in plant defense responses (Fig. 21.41; see also Chapter 17). An absolute requirement for SA in some incompatible interactions has been demonstrat-

ed in transgenic plants engineered to constitutively express a bacterial *nahG* gene encoding salicylate hydroxylase, the enzyme that converts SA to catechol. These transgenic plants contain much less SA. The lack of SA accumulation in these *nahG*-expressing plants correlates with a weakening of several *R* gene–mediated resistance phenotypes—for example, *N* gene–mediated resistance to TMV (Fig. 21.41) or *Arabidopsis RPP*-mediated resistance to the downy mildew fungus. *nahG* expression by transgenic plants also abolishes the induction of various defense genes.

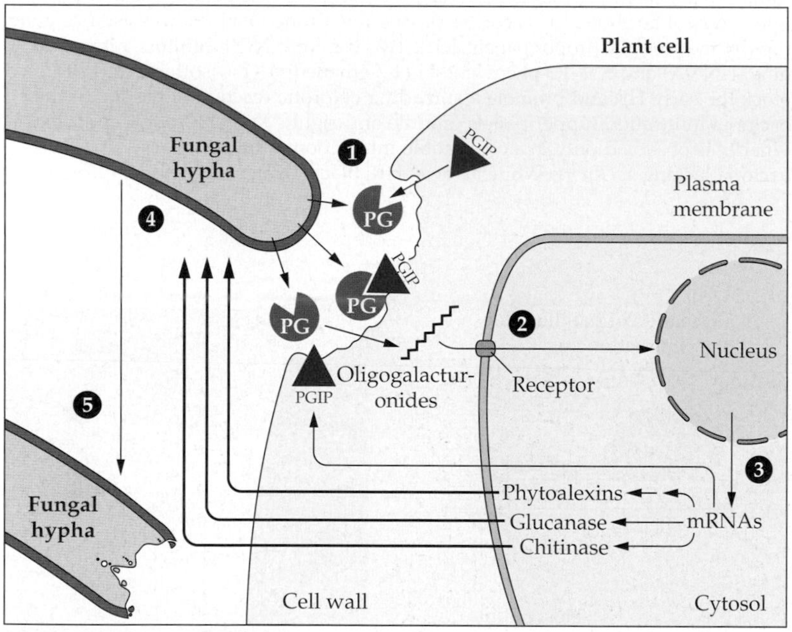

21.5.7 Jasmonic acid and ethylene, which are required for defense against necrotrophic fungi and for induction of some plant defense genes, may exacerbate disease symptoms.

Jasmonic acid (JA) is an oxylipin-like hormone derived from oxygenated linolenic acid (see Chapter 17). Increases in JA in response to pathogen/insect attack occur both locally and systemically. An *Arabidopsis* mutant *coi1* is impaired in the JA response pathway; it shows enhanced susceptibility to various necrotrophic fungal pathogens, including *Alternaria brassicicola, Botrytis cinerea,* and *Pythium mastophorum,* but the gene is not required for resistance to biotrophic fungi, for example, the downy mildew *Pe. parasitica.* Spraying methyl-JA onto plants increases their resistance to some (but not all) necrotrophic fungi, for example, *Al. brassicicola,* but not to biotrophic fungi or bacteria. However, this methyl-JA–induced protection is evident only in *Arabidopsis* genotypes that have a functional JA signal transduction pathway. A subset of the inducible plant defense genes from *Arabidopsis* require a JA-dependent, SA-independent signaling pathway.

The gaseous hormone ethylene (see Chapter 17) is frequently synthesized during both incompatible and compatible interactions. By blocking ethylene biosynthesis or perception with various plant mutants or transgenes, investigators have found that ethylene is apparently not required for several *R-Avr* gene–mediated resistance responses, such as the resistance to *Pseudomonas* bacteria conferred by the *Arabidopsis RPM1* and *RPS2* genes and the tobacco *N* gene–mediated

Figure 21.40
PG–PGIP model. (1) Polygalacturonases (PGs) secreted by the fungal hyphae interact with secreted polygalacturonase-inhibiting proteins (PGIPs) present in the plant cell wall. (2) PGs release from the plant cell wall various oligogalacturonides that interact with the putative oligogalacturonide receptor of the plant. (3) Transduction of the oligogalacturonide signal results in the expression of plant defense proteins, including PGIPs. (4) Chitinases, glucanases, and phytoalexins are secreted by the plant and (5) damage the fungal hyphae.

induction of gene expression. These trans-acting DNA-binding proteins appear to be regulated by either a rapid increase in transcript concentrations at steady state or changes in their phosphorylation status. Cascades of transcription factors, a common theme in gene regulation, may serve to amplify the input signal or to fine-tune the regulation of specific aspects of the complex plant defense response. For example, ethylene-mediated signaling clearly involves transcription factor cascades (see Chapter 18). Some transcription factors activating defenses appear to be unique to plants, but others have mammalian counterparts.

21.5.9 Phytoalexins include both organic and inorganic secondary metabolites.

Phytoalexins are low-molecular-mass, lipophilic antimicrobial compounds that accumulate rapidly at sites of incompatible pathogen infection. Biosynthesis of phytoalexins occur only after primary metabolic precursors are diverted into a novel secondary metabolic pathway. For example, phenylalanine is diverted into the synthesis of various flavonoid phytoalexins by the de novo synthesis of phenylalanine ammonia lyase (PAL), an enzyme that controls a key branchpoint in the phenylpropanoid biosynthetic pathway (see Chapter 24). However, the synthesis of most phytoalexins requires the activities of numerous biosynthetic enzymes and therefore necessitates highly coordinated signal transduction events. One way plants appear to have achieved this coordination is through the use of a common cis-acting DNA sequence element within the promoter of each gene that encodes an enzyme required for phytoalexin synthesis. For example, the multiple *PAL* and *4CL* genes and the single *C4H* gene that encode for the core reactions of phenylpropanoid metabolism are regulated at both the mRNA and protein levels to form the strictly regulated pathway required for biosynthesis of flavonoid phytoalexins.

Although initially discovered in the 1940s, the exact role for most phytoalexins in plant defense has yet to be determined—except for the *Arabidopsis* phytoalexin camalexin (see Chapter 16, Fig. 16.51) and the grapevine phytoalexin resveratrol (see Chapter 24, Fig. 24.77).

(A)

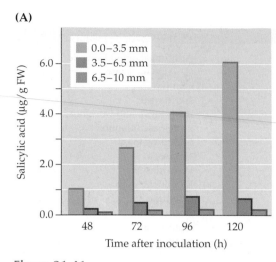

(B)

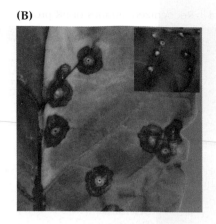

Figure 21.41
(A) Detailed analysis of TMV necrotic lesions forming on *N* gene–expressing resistant tobacco leaves reveals that total salicylic acid (SA) contents are greatest in the necrotic lesion center and rapidly diminish with distance from the center. The size ranges refer to the diameter of the concentric rings of plant tissue obtained from individual tobacco mosaic virus (TMV) necrotic lesions. (B) Constitutive expression in tobacco plants of the *nahG* gene of *Pseudomonas putida* (which encodes the enzyme salicylate hydroxylase) leads to the continuous removal of induced SA. This causes a substantial increase in TMV multiplication in the tobacco leaves, visible as an enlargement of *N* gene–mediated local lesion formation. Shown in the inset at an identical magnification are local lesions on wild-type plants.

resistance to TMV. In contrast, when the ethylene signal is eliminated during compatible interactions, the severity of chlorotic and necrotic symptoms frequently is greatly reduced and the plants appear to be more disease tolerant (Fig. 21.42).

Ethylene is required to mediate both resistance against necrotrophic fungal pathogens and nonhost resistance against soilborne fungal species that are not ordinarily plant pathogens. Engineered ethylene-insensitive (*etr1-1*) tobacco plants are susceptible to several *Pythium* species, and *Arabidopsis ein2* mutants are susceptible to *B. cinerea*. Another proven role for ethylene in defense is in combination with the signal molecule JA, both of which are required for activation of proteinase inhibitor (PI) genes (see Sections 21.5.10 and 21.6.2) and certain PR and chitinase genes (see next section).

(A)

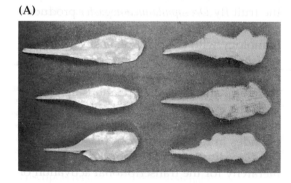

(B)

Figure 21.42
Plants insensitive to the stress-induced gas ethylene exhibit reduced symptoms (tolerance) to bacteria infection but increased susceptibility to soil fungi that are not normally considered plant pathogens. (A) Leaves from (left) wild-type *Arabidopsis* and (right) the ethylene-insensitive (*ein2*) mutant (Columbia ecotype) 4 days after inoculation with a virulent strain of the bacterium *Pseudomonas syringae* pv. *tomato*. All leaves contain a similar titer of bacteria (about $10^7/cm^2$), but the *ein2* leaves retain 50% more chlorophyll. (B) Wild-type (left) and transformed tobacco plants expressing the *Arabidopsis etr1-1* gene, shown 11 days after root inoculation with the necrotrophic fungus *Pythium sylvaticum*. The ethylene-insensitive transgenic line (right) has severe wilting symptoms because the fungal infection has caused degeneration of the vascular tissue at the stem base.

Table 21.4 Recog...

Family
PR-1
PR-2
PR-3
PR-4
PR-5
PR-6
PR-7
PR-8
PR-9
PR-10
PR-11
PR-12
PR-13
PR-14

21.5.8 PR proteins related proteins in wall–degrading en: polypeptides, and transduction casca

The transcripts of n related genes accu: to hours of pathoge treatment (Table 21 genes are also indu actions but much n Some PR proteins a canases, enzymes th polysaccharides of probably reduce fu mediated signal tra regulate the transcr many PR genes. Als been shown to act s enhancing the expr 21.43). Other PR def lipoxygenase, may generating secondar

Air

SA(mM)   0   0.1   1.0   5.0

PR-1
mRNA

Figure 21.43
The cooperative effect o molecules, gaseous ethy on the accumulation of mRNA. Application of g bacco leaves is not suffi

21.5.12 Parallel signaling pathways coordinate the complex, highly localized plant defense responses.

A simplified overview of the signal transduction pathways required to activate and coordinate the localized induction of plant defense responses is given in Figure 21.46. The most salient features of defense activation are as follows:

- Only minutes are required for each plant cell to switch from normal primary metabolism to a multitude of secondary metabolism defense pathways and to activate novel defense enzymes and genes.
- Every cellular compartment is recruited into defense.
- The cellular regulators are varied, including ion channels, phosphorylation/dephosphorylation events, and de novo synthesis of numerous signaling molecules, including H_2O_2, NO, ethylene, and JA.
- Synergies, antagonisms, and positive and negative feedback loops exist both within and between signaling pathways and metabolic pathways to create a complex network that ensures tight coordination of the eventual defense response.
- Similar mechanisms are activated during non-host–induced resistance, in R-Avr–mediated defense, and in response to pathogen-derived elicitors.
- The repertoire of defense responses activated is not usually microbe-species specific.
- Specific cellular protection mechanisms are activated simultaneously and accompany the defense response to minimize the damage to host cells.
- The plant defense signal transduction network can cross-talk with other plant stress-response pathways.
- Many of the key regulators of defense in each plant cell show homology to those used to coordinate defense in the dedicated cell population recruited to the site of attack in vertebrates.

21.6 Systemic plant defense responses

Within minutes of pathogen, insect, or nematode attack, plant defense responses are activated locally. Within hours, defense responses

are also sometimes elaborated in tissues far from the invasion site and even in neighboring plants. The type of systemic response induced, however, is determined by the identity of the attacking organism. As shown in Figure 21.47, induced systemic responses to fungi, bacteria, and viruses are distinct from the responses to insects. Nematodes appear to induce a mixture of the two, and yet another type of responses are those induced by root-colonizing nonpathogenic bacteria. The adaptive significance to plants of inducing and coordinating these different systemic responses is great. Their variety ensures that the plant is primed and can therefore respond more effectively to subsequent pathogen attacks. Some of the specific facets of each type of systemic response will now be considered.

21.6.1 Numerous pathogens can elicit systemic acquired resistance.

Fungi, bacteria, and viruses activate systemically a specific subset of PR-type genes by a mechanism known as **systemic acquired resistance** (SAR). For SAR to occur, the initial infection must result in formation of necrotic lesions, either as part of the HR or as a symptom of disease. SAR activation leads to a marked reduction in disease symptoms after subsequent infection by many different pathogen species. In tobacco, for example, *N* gene–mediated resistance against TMV protects the plant against later infection by the identical TMV strain and by most other tobacco pathogens tested (Fig. 21.48). Thus, SAR converts otherwise genetically compatible plant–pathogen interactions into incompatible ones. Once the molecular and biochemical mechanisms of SAR have been clarified (Box 21.9), it could be possible to engineer broad-spectrum disease control into agriculturally important crops (see Section 21.7).

Various synthetic chemicals induce SAR. Two of the most potent are 2,6-dichloroisonicotinic acid (INA) and benzo-(1,2,3)-thiodiazole-7-carbothionic acid *S*-methyl ester (BTH) (Fig. 21.49, see p. 1145). Neither INA nor BTH treatment causes SA concentrations to increase, and both compounds activate SAR when applied to *nahG*-expressing plants. These findings suggest that both INA and BTH act independently or downstream of SA in SAR signaling

SAR 354

good

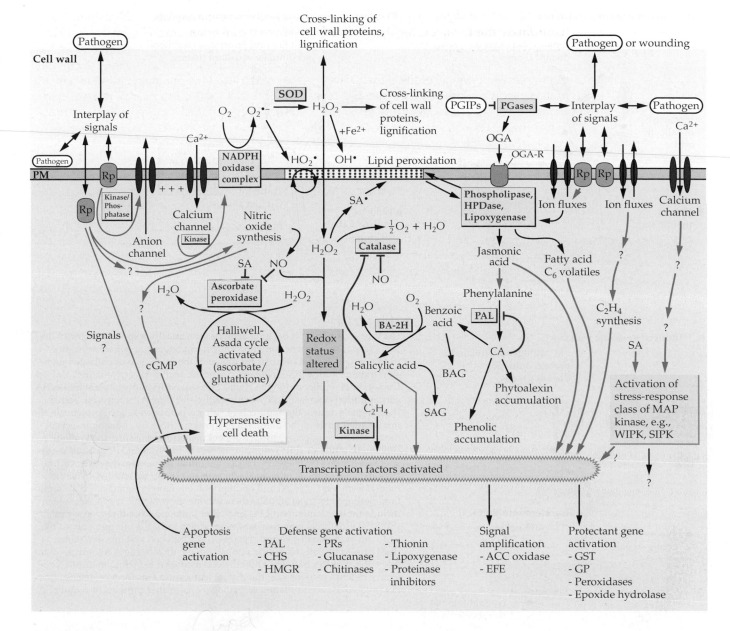

Figure 21.46

Overview of signal transduction pathways activating and coordinating plant defense responses. Plant receptor proteins (Rp) intercept pathogen-derived or interaction-dependent signals. These signals include the direct or indirect products of *Avr* genes, physical contact, and general components of each organism, such as chitin, enzymes, and plant cell wall fragments. Plant receptor proteins may or may not be the products of *R* genes. The immediate downstream signaling events are not known but involve kinases, phosphatases, G proteins, and ion fluxes. Several distinct and rapidly activated outcomes are recognized, including the production of reactive oxygen species (ROS) and nitric oxide, and direct induction of defense gene transcription (or possibly apoptosis genes). Amplification of the initial defense response occurs through the generation of additional signal molecules, that is, other ROS, lipid peroxides, benzoic acid (BA), salicylic acid (SA), jasmonic acid (JA), and ethylene. These, in turn induce other defense-related genes and modify defense proteins and enzymes. Concomitant alterations to cellular redox status or cellular damage will activate preformed mechanisms

for cell protection (e.g., the Halliwell–Asada cycle and plastid-localized superoxide dismutase [SOD] and catalase enzymes) and induce genes that encode various cell protectants. Defense-related stress may also induce cell death. Cross-talk between the various induced pathways appears to coordinate the responses. A green arrow indicates the positive interactions, and a red block the negative ones. ACC, 1-aminocyclopropane-1-carboxylic acid; BAG, benzoic acid glucoside; BA-2H, benzoic acid-2 hydroxylase; CA, cinnamic acid; cGMP, cyclic guanosine 5'-monophosphate; CHS, chalcone synthase; EFE, ethylene-forming enzyme; GP, glutathione peroxidase; GST, glutathione *S*-transferase; HMGR, 3'-hydroxy-3-methylglutaryl-CoA reductase; $HO_2^{\bullet}$, hydroperoxyl radical; HPDase, hydroperoxide dehydrase; MAP, mitogen-activating protein; NO, nitric oxide; $OH^{\bullet}$, hydroxyl radical; OGA and OGA-R, oligogalacturonide fragments and receptor; PAL, phenylalanine ammonialyase; PGases, polygalacturonases; PM, plasma membrane; $SA^{\bullet}$, salicylic acid radical; SAG, salicylic acid glucoside; SIPK, salicylic acid–induced protein kinase; WIPK, wound-induced protein kinase.

(A) Systemic acquired resistance

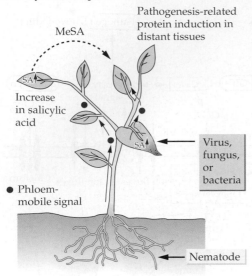

Pathogenesis-related protein induction in distant tissues

MeSA

SA

Increase in salicylic acid

SA

• Phloem-mobile signal

Virus, fungus, or bacteria

Nematode

(B) Systemic proteinase inhibitor/wound response

Systemic wound response proteins and proteinase inhibitors accumulate in distant tissues.

MeJA

JA

Chewing insects or mechanical wounds

Transient C_2H_4 synthesis

• Phloem-mobile signal

Electrical signal

Nematode

(C) Induced systemic resistance

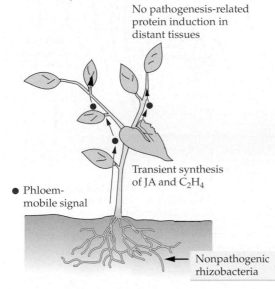

No pathogenesis-related protein induction in distant tissues

Transient synthesis of JA and C_2H_4

• Phloem-mobile signal

Nonpathogenic rhizobacteria

Figure 21.47

(A) Viruses, fungi, and bacteria activate systemically a specific subset of defense responses in a phenomenon known as systemic acquired resistance (SAR), in which local necrosis formation at the initial site of pathogen invasion triggers both a local increase in salicylic acid (SA) accumulation and the formation of a phloem-mobile signal. Subsequently, in distal plant tissue, SA concentrations increase and volatile methyl-SA (MeSA) is released. Together, these signals induce the synthesis of various pathogen-related proteins in the noninvaded parts of the plant. (B) In contrast, attack by chewing insects or mechanical wounds activates a different protective response, the systemic proteinase inhibitor (PI)/wound response. In the systemic PI/wound response activated by chewing insects, the initial tissue damage causes a transient increase in the synthesis of ethylene and jasmonic acid (JA). Volatile methyl jasmonate (MeJA) and another phloem-mobile signal called systemin (and perhaps electrical signals) then activate the systemic responses, which include the accumulation of PIs and other systemic wound response proteins (SWRPs). Root-attacking nematodes appear to induce a mixture of both the SAR and systemic PI/wound responses. (C) Induced systemic resistance (ISR) caused by soil-inhabiting nonpathogenic rhizobacteria colonizing plant roots. ISR requires both JA- and ethylene-mediated signaling to induce protective defense responses in the distant leaf tissue. This form of defense does not involve the accumulation of pathogenesis-related proteins or require SA.

Figure 21.48

Development of systemic acquired resistance (SAR). *N* gene–expressing tobacco leaves are able to restrict the spread of tobacco mosaic virus (TMV) to a small zone of tissue around the point of entry, where a necrotic lesion later appears (right leaf). This resistance phenotype, the hypersensitive response (HR), is subsequently accompanied by the induction of SAR throughout the plant. Consequently, a secondary infection with the virus, occurring several days after the initial infection, results in HR lesions (left leaf) much smaller than those induced by the primary infection. Both leaves are shown four days after infection.

Box 21.9   **Is SA the mobile signal that activates SAR?**

An absolute requirement for salicylic acid (SA) in systemic acquired resistance (SAR) activation was demonstrated by using transgenic tobacco and *Arabidopsis* plants expressing the bacterial *nahG* gene (see Section 21.5.6). These plants did not accumulate free SA and were incapable of activating SAR. Is SA therefore the translocated signal? Studies of in vivo ^{14}C-labeled SA in tobacco have shown that as much as 70% of the increase in SA concentrations observed in uninfected tissue after tobacco mosaic virus (TMV) infection is the result of SA translocation from the infected leaves. However, a series of grafting experiments between wild-type and transgenic *nahG*-expressing plants suggests that for induction of the SAR phenomenon, SA need only be present in the distal plant organs.

Reciprocal grafts and control grafts were generated by using two types of tobacco plants, which expressed either the *N* resistance gene alone (*Xanthi*) or the *N* gene in combination with the *nahG* transgene to remove inducible SA (*nahG*). The four types of grafted plants were inoculated with TMV on the lower rootstock leaves; seven days later, the same TMV isolate was inoculated onto the scion leaves. The photographs show the infec-

tion types on the scion leaves five days after the second TMV inoculation. *nahG* scions grafted onto *Xanthi* rootstocks (*nahG/Xanthi*) were unable to mount an SAR response. In contrast, *Xanthi* scions grafted onto *nahG* rootstocks (*Xanthi/nahG*) demonstrated SAR responses similar to those of control *Xanthi/Xanthi* grafts

not expressing the *nahG* transgene. *nahG/nahG* grafts lacking SA were unable to mount an SAR response.

Thus, although SA is mobile, it is probably not the mobile signal that activates SAR. The mechanisms by which increases in free SA induce SAR in distal plant tissues are not yet known.

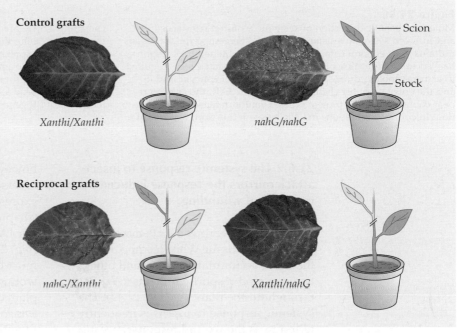

Control grafts

Xanthi/Xanthi   *nahG/nahG*

Reciprocal grafts

nahG/Xanthi   *Xanthi/nahG*

Scion — Stock

NR2 gen

(Fig. 21.50). Two mutants, *nim1* and *npr1* (now known to be alleles of the same gene), fail to induce PR genes when treated with chemical inducers such as SA and INA; these plants fail to develop SAR. However, both mutants do induce the expected HR and SA accumulation in response to genetically incompatible pathogens, an indication that the biochemical lesion is downstream from SA accumulation (Fig. 21.50). The gene *NPR1* encodes a novel protein containing specific amino acid repeats (ankyrin repeats) that are also present in a few other eukaryotic proteins involved in protein–protein interactions. NPR1 is most similar to mammalian IκB, an inhibitor of the NF-κB transcription factor (see Fig. 21.26). This homology thus suggests that NPR1 is involved in the transcriptional regulator of PR gene expression. Several other mutants have been identified that constitutively exhibit high concentrations

of SA, increased PR gene expression, and enhanced resistance to various virulent pathogens (e.g., *cpr1* mutants). Most of these mutants also show spontaneous formation of necrotic lesions and are discussed in detail in Chapter 20.

Salicylic acid
(SA)

Dichloroisonicotinic acid
(INA)

Benzo-(1,2,3)-thiodiazole-
7-carbothionic acid
S-methylester
(BTH)

Figure 21.49
Chemical structures of three systemic acquired resistance (SAR)–inducing compounds.

(A)

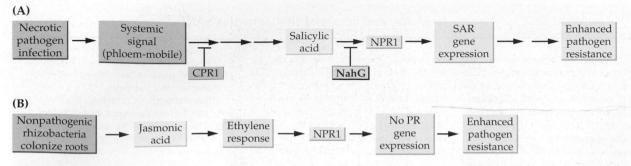

(B)

Figure 21.50

Molecular genetic dissection of systemic acquired resistance (SAR) and induced systemic resistance (ISR). (A) Screening for *Arabidopsis* mutants with either a compromised or a heightened SAR response has led to identification of numerous interconnected steps required to activate enhanced resistance. Downstream of salicylic acid (SA) and the entry point for chemical inducers of SAR, the ankyrin repeat-containing NPR1 protein is essential for defense gene activation. In contrast, mutations in CPR1 protein function, which lead to constitutive SAR, are NahG suppressible and therefore operate through increasing the concentrations of SA. *cpr1* mutants constitutively express PR-1 under normal growing conditions. (B) Similar genetic analyses of the ISR response in *Arabidopsis* has determined a requirement for JA- and ethylene-mediated signaling as well as the need for functional NPR1 protein, even though no induction of PR genes occurs.

21.6.2 The systemic response to insect attack mirrors the response induced by mechanical wounding.

Chewing herbivorous insects mechanically wound plant tissue while feeding, inducing the rapid accumulation of PIs and other systemic wound response proteins (SWRPs) throughout the plant (see Fig. 21.47). The systemic response to insects is frequently referred to as the **wound response,** because identical molecular and biochemical events are triggered after plant tissue is mechanically wounded. Although high concentrations of oligogalacturonides released from damaged plant cell walls induce PI and SWRP genes locally, these do not move systemically. Instead, an 18-amino acid polypeptide called **systemin** (see Chapter 9, Fig. 9.26), which induces PI and SWRP genes at only 10^{-6} of the concentration required for induction by oligogalacturonides, is released from the damaged cells and is transported by the phloem to the upper (unwounded) leaves within 60 to 90 minutes.

Once systemin reaches the target tissue, it activates a lipid-based signaling cascade that produces JA (see Chapter 17). JA then induces the transcriptional activation of PI and other SWRP genes. However, JA induction of PI genes requires a third signal molecule, ethylene. Ethylene transiently accumulates within 30 to 120 minutes of the addition of systemin to the translocation stream. When ethylene synthesis is blocked,

however, whether pharmacologically or by reverse genetics (e.g., using transgenic plants expressing an antisense ACC oxidase; see Chapter 17), neither wounding, systemin, nor JA alone can induce PI gene expression (Fig. 21.51). Therefore, the ethylene signal must be acting downstream of JA in the wound transduction pathway. Certain plants, for example, the tomato mutant *defenseless 1* (*def1*), are no longer able to induce PI genes after wounding and are more susceptible to feeding by insect larvae (Fig. 21.52). The SA inhibition of the wound response may also explain why PI and SWRP genes are not induced during SAR to pathogens, where very high amounts of SA accumulate (see Section 21.6.1).

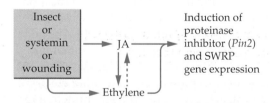

Figure 21.51

Concomitant increases in the concentrations of both JA and ethylene are required for the induction of systemic wound response protein (SWRP) genes in response to insect attack, mechanical wounding, or application of the plant peptide systemin. The PI gene *Pin2*, a typical SWRP gene, is induced strongly in leaves within a couple of hours after the inductive stimulus.

(A)

(B)

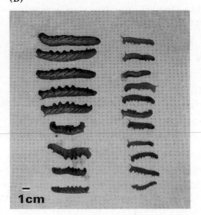

1cm

Figure 21.52
The tomato mutant *defenseless* (*def1*; A, left plant) has a compromised systemic wound response compared with that of the wild-type plant (A, right plant). When larvae of the tobacco hornworm insect (*Manduca sexta*) feed on *defenseless* plants (B, left column), their growth rate is faster than when feeding on wild-type tomato plants, which can systemically synthesize proteinase inhibitor and other systematic wound response proteins in response to larvae feeding (B, right column).

21.6.3 Nonpathogenic root-colonizing rhizobacteria cause induced systemic resistance.

Rhizobacteria that promote specific plant growth, for example, strains of *P. fluorescens*, induce from colonized root tissue a systemic resistance response that does not depend on either SA or PR protein accumulation. Instead, induced systemic resistance (ISR) (see Fig. 21.47C) requires both JA and ethylene signaling and, curiously, also the SAR regulatory protein NPR1 (see Fig. 21.50). This requirement for the NPR1 protein indicates that pathogen-induced SAR and rhizobacteria-induced ISR converge during the latter parts of the signaling pathway. In *Arabidopsis*, ISR pro-tection is usually slightly less than that conferred by SAR. These two types of resistance apparently do not activate the same spectrum of plant defense responses because spraying SA onto ISR plants can heighten plant protection still further.

21.7 Control of plant pathogens by genetic engineering

Interesting opportunities now exist to use the isolated plant *R* genes, pathogen *Avr* genes, and additional plant genes required for disease resistance—both to improve existing plant breeding methods and to devise novel strategies for disease control. In addition, pathogen processes recognized as absolutely required for pathogenesis can now become the targets for novel plant-specific inhibitory mechanisms.

21.7.1 Transforming susceptible plants with cloned *R* genes may provide novel forms of pathogen resistance.

To control diseases in elite commercial cultivars, plant breeders traditionally have used lengthy breeding programs to introgress new *R* genes from wild relatives of crop species (see Box 21.2). Now, the availability of cloned *R* genes for genetic transformations is opening the possibility of direct transfer into elite lines within a single generation (see Chapter 6). Plant transformation also offers the immediate possibility of introducing simultaneously several different *R* gene alleles that are effective against a single pathogen species (Fig. 21.53). In theory, this should slow the process of microbe evolution, because the various *R* genes should be overcome only if all the corresponding *Avr* gene products mutate simultaneously within a single pathogen isolate. The introduction of *R* genes by plant transformation also removes the barriers presented when interspecies infertility prevents gene introduction by conventional plant breeding; as a result, the number of plant species from which useful novel resistance sources can be recovered will increase. *R* genes providing resistance to all or almost all races of a particular pathogen species, for example, rice *Xa21* (see Box 21.5), are attractive candidates for interspecies transfer because they plausibly could confer the same broad-spectrum resistance as they do in their "native" genomes (see Box 21.7). The high degree of sequence conservation exhibited by many race-specific *R* genes will also permit candidate *R* genes to be isolated rapidly from heterologous plant species.

One of the major drawbacks of most *R* genes is their extreme specificity of action toward a single *avr* gene of one specific microbial species. This particular deficit might be overcome by creating new *R* genes in the laboratory. Once the domains in R proteins that contribute to recognition specificity have been identified, perhaps their amino acid sequences can be altered to recognize a wider array of microbial products, potentially providing a broader spectrum of disease control. Durable resistance could result, for example, from targeting the protein known to be essential for the pathogen to complete its life cycle; the loss of such a gene product would disable the pathogen (Fig. 21.54). Alternatively, the R protein structure could be altered to assume a constitutive "on" configuration; such a modified R protein would switch on defense mechanisms even in the absence of Avr ligand perception. However, these genetically engineered constitutive defenses would need to be maintained at a low level of activity to avoid a sizeable crop yield penalty. Natural variants arising from the maize *Rp1* resistance locus, for example,

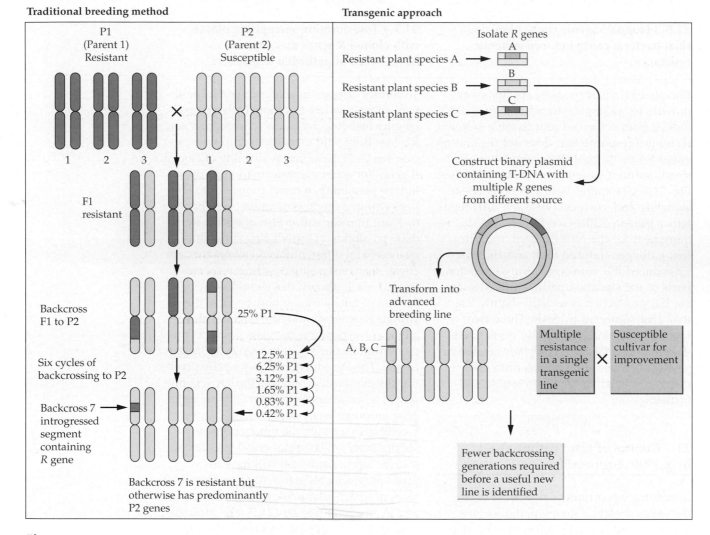

Figure 21.53

Traditional breeding compared with transformation technology approaches to introgressing desirable genes into commercial crop plants. In a traditional breeding program, as much as 0.4% of the genome complement from each donor parent can reside in the seventh backcross generation along with the *R* gene of interest (originally from parent 1). By the transgene transformation approach, multiple *R* genes from several initial sources are first assembled into a single Ti plasmid. After T-DNA integration into the plant genome, these *R* genes cosegregate in all subsequent breeding steps—greatly simplifying the subsequent backcrossing program for introducing multiple new traits into a cultivar. When the transgenic transformation approach is used, the entire sequence of the introduced DNA is known, whereas in the traditional breeding program, neither the total extent of the DNA introgressed nor its sequence identity is known.

are known to be accompanied by deleterious side-effects, including spontaneous necrotic lesion formation (see Chapter 20) and reduced yield.

21.7.2 Cloned *R* and *Avr* genes can be used in combination to promote acquired resistance in transgenic plants.

The rapidly activated and localized defense response that frequently culminates in HR is one of the most prevalent and effective mechanisms deployed by plants to minimize pathogen attack. Through the combined expression of both an *R* gene and the complementary *Avr* gene in a single plant genotype, an engineered "trigger" for HR is possible. However, if both components are expressed continuously in a single transgenic plant, the HR induced is devastating, destroying not only the pathogen but also the entire plant (see Box 21.4). Therefore, the expression of either one component or both must be tightly regulated. The desired resistance phenotype may be obtained by either using a pathogen-inducible promoter (a two-component system) or generating a limited restoration of *R* gene function through the somatic excision of a transposable element from an *R* gene, in combination with constitutive *Avr* expression; this latter is known as the four-component genetically engineered acquired resistance (GEAR) system (Fig. 21.55). These two approaches have the advantage that the entire multifactorial defense response would be activated, thereby potentially achieving broad-spectrum pathogen control.

21.7.3 Constitutive overexpression and silencing of plant genes can enhance plant defenses.

Through the molecular genetic dissection of *R-Avr* gene–mediated resistance and the SAR response, several plant loci have been isolated that are absolutely required for local or systemic induced resistance, for example, the tomato *Prf* and *Arabidopsis NPR1* genes (see Sections 21.4.2 and 21.6.1). Constitutive overexpression of these genes under control of a strong promoter (i.e., 35S from CaMV) enhances resistance to several pathogens, suggesting that in wild-type plants these gene products are ordinarily present at limiting concentrations. Increasing the amounts of Prf or NPR1 proteins in the transgenic plants activates either the constitutive defense mechanisms or the broad, more efficient defense mechanisms against infection by a wider array of pathogens. These experiments and those described in the preceding section reinforce the view that plant defense responses result from biochemical pathways and molecular mechanisms that are not entirely pathogen-specific.

Successful disease control has been achieved by the overexpression of a few specific defense genes (Fig. 21.56A,B). For example, engineering the production of a grapevine phytoalexin in tobacco has enhanced the resistance of the tobacco plant to the necrotrophic fungus *B. cinerea*. The constitutive coexpression of a basic chitinase and a glucanase has provided enhanced control of several fungal diseases, for example, *Cercospora nicotianae* in tobacco.

(A)

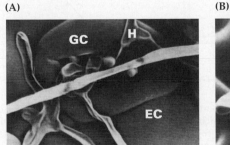

(B)

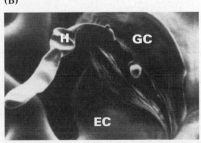

Figure 21.54
Pathogenicity genes are essential for a microorganism to complete its life cycle on a plant. Unlike wild-type *Cladosporium fulvum* (A), *ecp2* mutants (B) grow poorly in tomato leaves and fail to sporulate through stomata. When *ecp2* mutants are growing on synthetic media in Petri dishes, however, the growth rate is identical to that of the wild-type strain. The *ecp2* gene encodes a secreted protein of unknown function. Engineering the plant defense responses to be activated after the plant cell perceives an essential pathogenicity gene should result in durable resistance to this pathogen species. EC, epidermal cell; GC, guard cell; H, fungal hypha.

PI overexpression, either throughout the entire plant (for insect control) or targeted to the specialized nematode-feeding site, has also proved effective. However, several inherent problems are associated with this type of strategy. For example, once the transgenic plants are generated, the laborious search must then begin to identify which pathogens are potentially controlled. Secondly, the disease control obtained relies at best on just a few components of the general defense reaction repertoire. This type of engineered resistance is therefore more prone to circumvention by pathogen mutation than are the other strategies outlined above.

A modified approach could involve engineering plants that constitutively produce a key defense signal molecule to activate a broader array of mechanisms simultaneously, thereby controlling more pathogen species. Increasing hydrogen peroxide production in potato plants through apoplastic expression of a heterologous glucose oxidase, for example, enhanced the resistance of the plants to *Ph. infestans* and the soft rotting bacterium *Erwinia carotovora* subsp. *carotovora* (Fig. 21.56C). Alternatively, by coupling

the promoter of a response gene to that of a signaling gene—for example, a transcription factor that activates various defense genes—a positive feedback loop could be established to increase the speed of the response in an ordinarily susceptible genotype.

Sometimes successful disease control has been achieved by activating a gene-silencing mechanism to eliminate a protein product absolutely required for microbial pathogenesis. TMV ordinarily induces a specific tobacco $(1\rightarrow3)\beta$-glucanase to dissolve the callose plugs that form to block plasmodesmatal cell-to-cell connections. The absence of this plant enzyme in gene-silenced tobacco plants effectively prohibits TMV spread (Fig. 21.56D).

21.7.4 Expression of a pathogen-derived gene product can provide resistance to some viral and bacterial diseases.

Pathogen-derived resistance (PDR) strategies have been applied primarily to the control of diseases caused by viral and bacterial species. The concept is based on transgenic

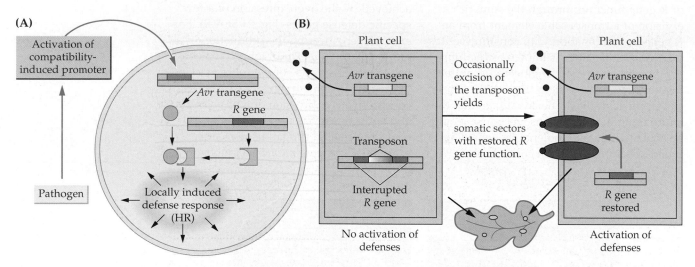

Figure 21.55
Two plant biotechnology approaches have been devised that enable plants to activate the entire multifactorial defense response and thereby achieve broad-spectrum pathogen control. (A) The two-component sensor system includes in one plant the sensor, an *Avr* gene under the control of a pathogen-inducible promoter, and an effector, an *R* gene. The promoter (red box) fused to the *Avr* gene (yellow box) is activated by nonspecific elicitors from the attacking pathogen. The *Avr* gene product (blue circle) then interacts with the resistance gene product, leading to activation of the defense response (HR). (B) Genetically engineered acquired resistance (GEAR) operates through limited restoration of the *R* gene function, in combination with constitutive *Avr* expression. *R* gene function is regulated by inserting a transposable element in the *R* gene coding sequence, which results in low frequency of somatic excision of this transposon coincident with plant cell division. In the plant cells where R protein function is restored, recognition of the cognate Avr protein occurs, which triggers activation of the plant defense responses. Subsequent defense signals emanating from the cell responding to R-Avr induce resistance responses in the surrounding plant cells, in which the *R* gene is still nonfunctional. The GEAR technology creates a plant that is a genetic mosaic for cells with and without restored R protein function; in most cells, however, defense responses are active and give the plant improved protection against pathogen attack.

expression of intact pathogen-derived genes or genomic fragments (only in virus control) that interfere with the normal pathogenicity process of the microbe. The underlying mechanisms operate at either the nucleic acid (Box 21.10) or the protein level. Examples of direct protein interference-mediated resistance include the expression in planta of bacteria-derived toxin-inactivating enzymes to control specific *Pseudomonas* bacteria as

Figure 21.56

Successful disease control has been achieved through the constitutive overexpression of individual/dual defense proteins and signal molecules. However, many of these biotechnology approaches have been accompanied by compromised normal plant growth and development and therefore have not been commercially exploitable. (A) Overexpression of stilbene synthase enzyme in tobacco (lower panel) leads to synthesis of the phytoalexin resveratrol, which confers resistance to attack by the necrotrophic fungus *Botrytis cinerea* (see Fig. 21.4). The upper panel shows a control plant exhibiting disease symptoms. (B) Coexpression of glucanase and chitinase genes in tobacco gives the best protection against the fungus *Cercospora nicotianae*, diminishing lesion size markedly. 2G, homozygous glucanase gene only; 2C, homozygous chitinase gene only; 1G/1C, heterozygous for both defense genes; 2G/2G, homozygous for both defense genes. (C) Potato plants engineered to express the hydrogen peroxide–generating glucose oxidase gene from *Aspergillus niger* are resistant to the fungal pathogen *Verticillium dahliae*, which attacks roots and vasculature (the two rows on the right). The Russet Burbank control lines (shown on the left), whether nontransformed (far left) or transformed with a vector lacking the glucose oxidase gene (second from left), exhibit severe disease symptoms. The plants were photographed 40 days after inoculation. (D) Antisense glucanase tobacco plants exhibit enhanced resistance to infection by tobacco mosaic virus. Smaller local lesions develop on the antisense lines (upper half of the leaf) than on the control lines (lower half of the leaf), as shown in the photograph taken seven days after leaf inoculation.

Box 21.10

Some mechanisms of pathogen-derived resistance are based on nucleic acid sequences.

Legitimate concerns have been raised about using overexpression of functional viral proteins to engineer disease control. Transgenic CPs can transencapsidate heterologous viral RNAs and promote aphid transmission of viruses that are not ordinarily transmitted by this vector. Also, the transgene or its RNA product theoretically could recombine with an infecting virus to generate a variant with novel biological properties. Accordingly, widespread interest is now turning to the alternative strategy of nucleic acid sequence–based expression mechanisms.

Transgenic nucleic acids (either RNA or DNA) could be designed to act as decoy molecules that compete with the infecting viral genome, promoting interactions between host or virally encoded proteins and various inappropriate partners that do not allow viral replication and spreading within the plant. This type of competitive inhibition, posttranscriptional gene silencing (PTGS), has already been demonstrated in cases where the transgene specifies a defective (and therefore interfering) RNA or DNA molecule.

The figure illustrates a model for PTGS to prevent the spread of geminiviruses.

The production of aberrant RNAs (aRNAs) functions as a signal to activate the silencing mechanism (panel A). Viral entry into a cell releases copies of the viral genome, circular ssDNA (1). Direct pairing between viral DNA and the transgene decoy may occur (2), resulting in the formation of aberrant ssRNAs after transcription (3). Alternatively, antisense transcripts from an antisense transgene, or from a sense transgene inserted next to a plant promoter that is transcribed in the opposite orientation (black triangle), may hybridize with viral mRNA (sense transcripts), resulting in the formation of aberrant dsRNAs. If these aRNAs accumulate above a critical threshold value (4), the silencing mechanism will be activated (panel B). aRNA molecules may activate methylases (5) that affect transgene expression (6), or they may activate host-encoded RNA-dependent RNA polymerases (RDRPs) (7). During maintenance of the pathogen-derived resistance (PDR) mechanism (panel C), RDRPs use aRNA molecules as templates to produce copy-RNAs (cRNAs) (8), which are complementary to viral transcripts and so bind the viral (and transgene) transcripts, resulting in the accumu-

lation of 25 nucleotide RNAs that hybridize to the silenced sequences. This shuts down production of the viral gene products, even if the transcription rates remain high (9). cRNAs, or dsRNAs, may also move systemically through the plant to induce the silencing mechanism in all plant tissues (10). Some viruses, however, produce antisilencing proteins, for example, the HC-Pro protein of tobacco etch potyvirus, which enable the virus to evade the gene-silencing surveillance mechanism (see Fig. 21.45).

Exploitation of a PTGS mechanism appears to be an effective way to engineer plant protection against tobamo-, potex- and geminiviruses. To be capable of activating PTGS, the introduced transcribed transgene requires only 300 bp of its sequence to be homologous to any part of the viral genome. Many of the early success stories on providing virus control by introducing specific single viral genes are now thought to have been operating either partially or entirely through activation of PTGS mechanisms.

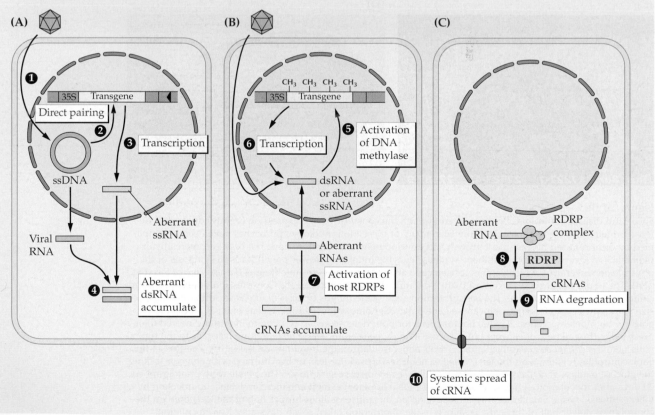

well as the constitutive expression of CPs or MPs to protect against plant viruses (Fig. 21.57). CP-mediated resistance is thought to operate by way of high CP concentrations acting to inhibit virus disassembly in the

(A)

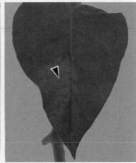

(B)

(C)

initially infected cells. Overexpression of dysfunctional MP probably leads to competition for plasmodesmatal binding sites between the mutant MP and the wild-type MP of the infecting virus. CP-mediated virus resistance was one of the early success stories of plant genetic engineering and has already led to the development of virus-resistant squash cultivars for commercial exploitation (Fig. 21.57). MP-mediated resistance is attractive because it appears to confer broad-spectrum efficacy.

21.7.5 Expression of *Bacillus thuringiensis* toxins can control damage by some insects.

Since the 1930s, numerous related toxins produced by different *Bacillus thuringiensis* (BT) subspecies have been sprayed on commercial crops to control coleopteran and lepidopteran insect species (Fig. 21.58). When ingested by an insect, these BT toxins create holes in the membranes of the cells in its digestive tract and death rapidly ensues. The BT toxins have been attractive for plant genetic engineering because each protein affects only a very few insect species, and the toxin has already been proved to provide stable insect control in crops. High in vivo production of BT proteins has been achieved by designing BT genes that utilize the codon preferences of the plant rather than those of the bacterium. The level of insect control achievable by BT expression transgenically is impressive (Fig. 21.59). In 1996, the first BT cotton and maize plants were introduced into commercial production in the United States, followed by BT potatoes in 1997. The introduction of these BT crops has already seen a substantial decrease in the use of chemical insecticides and a concomitant increase in the diversity of the native insect

Figure 21.57
Plant resistance engineered through the overexpression of specific pathogen genes. (A) Control of infection of bean by *Pseudomonas syringae* pv. *phaseolicola* bacteria is achieved by constitutive overexpression of a bacteria-derived toxin-inactivation enzyme, ornithine carbamyl-transferase (OCTase). This prevents bacterial phaseolo-toxin accumulation; consequently, chlorosis (arrowheads) is minimal and the bacteria titer remains low in the transgenic plant (right) compared with the nontransformed plant (left). (B) Constitutive overexpression of the coat protein (CP) gene from the tobacco mosaic virus (TMV) provides control against TMV infection in tobacco. The plant on the left is nontransformed and shows the characteristic symptoms of systemic infection by TMV: light and dark green areas and deformation of leaf margins. The transformed plant on the right, which contains and expresses the gene for the TMV CP has not developed symptoms, as shown in the photograph taken 14 days after inoculation. (C) Constitutive overexpression of dysfunctional viral movement protein (MP) mediates resistance to wild-type TMV. The untransformed plant on the left exhibits the characteristic systemic symptoms (arrows) by 11 days after inoculation, whereas the plant expressing the dysfunctional TMV MP (right) has minimal symptoms.

Figure 21.58

Figure 21.58
Three-dimensional structure of insecticidal toxins produced by *Bacillus thuringiensis*. (A) The CryIA(a) toxin is active against lepidopteran insects. Domains II (cyan) and III (yellow) pack on helix α7 (magenta) of domain I (green). (B) The CryIIIA toxin is active against coleopteran insects. It has three domains: a bundle of seven α-helices, a group of three β-sheets (lower right), and a β-sandwich (upper right).

(A)

(B)

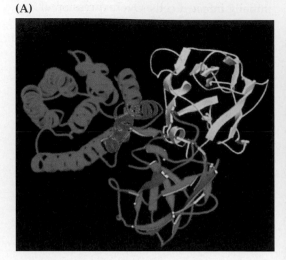
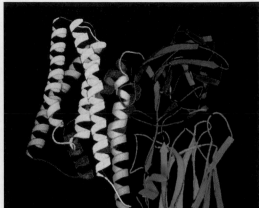

population found within predominantly agricultural areas. However, the BT transgenic approach to insect control clearly must be interwoven with deployment of additional insect-controlling transgenes having their own unique modes of action and with integrated pest-management approaches, including the provision of sensitive host plants as insect refuges to minimize the probability of insects developing resistance to BT toxin.

21.7.6 Disease resistance can be engineered through the expression of novel gene sequences.

A completely new development in the engineering of viral resistance has come through the overexpression of non-plant– and non-pathogen–derived genes. For example, a transgene coding for a single-chain antibody that recognizes a structurally important epitope of a viral CP has provided excellent

Figure 21.59
Use of *Bacillus thuringiensis* (BT) protein in plants to control insects. Constitutive overexpression of the BT CryIA(b) protein in maize plants effectively controls attack by larvae of the European corn borer (*Ostrinia nubilalis*) in both stems and cobs (not shown). The CryIA(b)-expressing plant is shown at left, a damaged control plant at right.

protection against viral attack (Fig. 21.60). In the future, expression of antibodies may become an important general mechanism for neutralizing the function of essential host and pathogen proteins required for disease initiation.

21.7.7 Genetically engineered disease control is in its infancy.

Transgenic disease control will always face stiff competition from more-conventional approaches to disease control, such as safer pesticides and molecular marker–assisted breeding. The successful transfer to the farm of the technologies described above will also depend on many external factors, including government policy, company/university licensing agreements, and consumer acceptance. Moreover, we must remember that farmers cultivate genomes, not genes. For any of these novel transgenic disease control strategies to succeed, they must be bred into elite crop plant germplasms.

Summary

Plants are resistant to most plant pathogens. Every plant cell can defend itself from

(A) Plant T-DNA expression vector

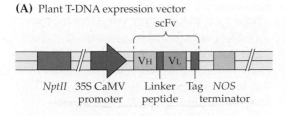

NptII 35S CaMV   Linker   Tag *NOS*
        promoter   peptide   terminator

(B)

Figure 21.60
Plant disease resistance to viruses provided by transgenic mammalian antibodies. Tobacco plants were engineered to express a single-chain Fv antibody (scFv) directed against a structurally important epitope of the viral coat protein of artichoke mottle crinkle (AMC) virus. (A) The plant expression vector used to join the variable heavy and light (V$_H$ and V$_L$) chains together by a linker peptide and to attach a 3′ tag to produce the scFv antibody. This open reading frame was placed under the control of the 35S cauliflower mosaic virus (CaMV) promoter and *NOS* termination sequences. (B) By 14 days after inoculation of untransformed (right) and transgenic (left) plants expressing the scFv antibody with the AMC virus, both the viral symptoms and the virus particle titer in the antibody-expressing plants had decreased.

attacking pathogenic microorganisms and invertebrates. Some defenses, such as antimicrobial secondary metabolites, are constitutive, being located in specific cellular compartments and ready to be released on cell damage. Other defense responses, such as those induced by pathogen invasion, require detection of the pathogen by the plant. Defense activation is correlated with rapid activation of defense-related genes and often culminates in the HR, localized cell death, to impair pathogen spread.

Plant resistance to pathogens can be mediated by dominant resistance (*R*) genes in plants that are complementary to avirulence (*Avr*) genes in pathogens. Avr proteins exhibit extensive sequence diversity, and their function in the pathogen is poorly under-

stood. In contrast, plant R proteins are strikingly similar in structure, sharing such motifs as LRRs, a central NBS, and a serine/threonine protein kinase domain, among others. R proteins both detect pathogens and initiate signal transduction to activate defense mechanisms. In addition, *R* loci/genes can evolve new *R* gene specificities to keep pace with the evolution of virulence in pathogen populations.

Plant defense reactions involve complex biochemical pathways and multiple signal molecules, including ROS, NO, SA, JA, and ethylene, to provoke the induction of antifungal proteins, secondary metabolites, and cell wall fortification reactions—both at the infection site and systemically throughout the attacked plant. Specialist defenses against plant viruses include PTGS; those against insects involve PI proteins. Many aspects of induced plant defense appear to be conserved in other eukaryotes, perhaps indicating the existence of an ancient defense strategy against microbial attack.

The genetic engineering of plants has started to achieve broad-spectrum and durable pest and pathogen control in crops. Still needed are a better understanding of the factors/processes involved and better ways to manipulate resistance mechanisms so as to reduce crop losses.

Further Reading

Agrios, G. N. (1997) *Plant Pathology,* 4th ed. Academic Press, San Diego.

Baker, B., Zambryski, P., Staskawicz, B., Dinesh-Kumar, S. P. (1997) Signaling in plant–microbe interactions. *Science* 276: 726–733.

Bergey, D. R., Howe, G. A., Ryan, C. A. (1996) Polypeptide signaling for plant defensive genes exhibits analogies to defense signaling in animals. *Proc. Natl. Acad. Sci. USA* 93: 12053–12058.

Dangl, J. L. (1995) *Bacterial Pathogenesis of Plants and Animals, Molecular and Cellular Mechanisms.* Springer-Verlag, Berlin, pp. 99–118.

Day, P. (1974) *The Genetics of Host–Pathogen Inter-Relationships.* John Wiley & Sons, San Francisco.

Delledonne, M., Xia, Y., Dixon, R. A., Lamb,

C. (1998) Nitric oxide signal functions in plant disease resistance. *Nature* 394: 585–588.

De Wit, P.J.G.M. (1992) Molecular characterization of gene-for-gene systems in plant–fungal interactions and the application of avirulence genes in control of plant pathogens. *Annu. Rev. Phytopathol.* 30: 391–418.

Hammond-Kosack, K. E., Jones, J.D.G. (1997) Plant disease resistance genes. *Annu. Rev. Plant Physiol. Plant Mol. Biol.* 48: 575–607.

Hartleb, H., Heitefuss, R., Hoppe, H. H. (1997) *Resistance of Crop Plants Against Fungi.* Gustav Fischer, Jena, Germany.

Matthews, R.E.F. (1992) *Fundamentals of Plant Virology.* Academic Press, San Diego.

Rahme, L. G., Stevens, E. J., Wolfort, S. F., Shao, J., Tompkins, R. G., Ausubel, F. M. (1995) Common virulence factors for bacterial pathogenicity in plants and animals. *Science* 268: 1899–1902.

Rushton, P. J., Somssich, I. E. (1998) Transcriptional control of plant genes responsive to pathogens. *Curr. Opin. Plant Biol.* 1: 311–315.

Staskawicz, B. J., Ausubel, F. M., Baker, B. J., Ellis, J. G., Jones, J.D.G. (1995) Molecular genetics of plant disease resistance. *Science* 268: 661–667.

Plant–microbe interactions (1996) *Plant Cell* 8: 1651–1913 (Special Issue).

Plant biotechnology R&D. (1995) *Trends Biotechnol.* [13] (Sept): 313–409 (Special Issue).

Useful Websites

Plant pathology Internet guidebook: http://www.ifgb.uni-hannover.de/extern/ppigb/ppigb.htm

Plant pathology introductory course: http://www.ianr.unl.edu/ianr/plntpath/peartree/homer/public.html

Cereal diseases: http://www.crl.unm.edu

Plant parasite nematodes: http://www.ianr.unl.edu/ianr/plntpath/nematode/wormhome.htm

Table 22.1 Average yields and record yields of eight major crops

Crop	Record yield (kg per hectare)	Average yield (kg per hectare)	Average losses (kg per hectare) Biotic[a]	Average losses (kg per hectare) Abiotic[b]	Abiotic losses as a percentage of record yield
Corn	19,300	4,600	1,952	12,700	65.8
Wheat	14,500	1,880	726	11,900	82.1
Soybean	7,390	1,610	666	5,120	69.3
Sorghum	20,100	2,830	1,051	16,200	80.6
Oat	10,600	1,720	924	7,960	75.1
Barley	11,400	2,050	765	8,590	75.4
Potato	94,100	28,300	17,775	50,900	54.1
Sugar beet	121,000	42,600	17,100	61,300	50.7

[a]Biotic factors include diseases, insects, and weeds.
[b]Abiotic environmental factors include, but are not limited to, drought, salinity, flooding, and low and high temperature.

groundwater, so they can survive long periods without rain. In contrast, desert ephemerals escape drought by germinating and completing their life cycles while adequate water is available. Sunken stomata, light-reflective spines, and deep roots are among the constitutive, genotypically determined traits for stress resistance that are expressed whether the plants are stressed or not. They constitute **adaptations,** evolutionary improvements that enhance the environmental fitness of a population of organisms. Other resistance mechanisms are achieved through **acclimation,** the adjustment of individual organisms in response to changing environmental factors. During acclimation,

an organism alters its **homeostasis,** its steady-state physiology, to accommodate shifts in its external environment. A period of acclimation before stress is encountered may confer resistance to an otherwise vulnerable plant (Fig. 22.2). For example, during the summer, trees in the northern latitudes cannot withstand freezing but many of them can acclimate and eventually withstand temperatures as low as −196°C (−328.8°F) in winter. Whether based on acclimation or genotypically determined traits, successful mechanisms for stress resistance support survival in otherwise lethal conditions or maintain productivity under circumstances that can impair crop yields.

Figure 22.2
Stress resistance can involve tolerating the stressful condition or avoiding it. Some resistance mechanisms are constitutive and are active before exposure to stress. In other cases, plants exposed to stress alter their physiology in response, thereby acclimating themselves to an unfavorable environment. Examples of constitutive mechanisms of drought resistance include the succulent, photosynthetic stem of the saguaro catus (*Cereus giganteus*), a drought-tolerant species; the deep roots of the honey mesquite (*Prosopis glandulosa* var. *glandulosa*), a drought-avoiding species; and the wet-season life cycle of the Mohave desert star (*Monoptilon bellioides*). Examples of acclimation mechanisms include osmotic adjustment (see Section 22.3) in plants such as spinach and freezing tolerance (see Section 22.6) in cold-hardy trees such as black spruce (*Picea mariana* Mill.).

(A)

Abiotic stress

Acclimation

Resistance
• Stress avoidance
• Stress tolerance

(B)
Saguaro
Honey mesquite
Spinach
Mohave desert star
Black spruce

22.1.3 Gene expression patterns often change in response to stress.

Stress-induced changes in metabolism and development can often be attributed to altered patterns of gene expression. A stress response is initiated when a plant recognizes a stress at the cellular level. Stress recognition activates signal transduction pathways that transmit information within individual cells and throughout the plant (Fig. 22.3). Ultimately, changes in gene expression, which occur at the cellular level, are integrated into a response by the whole plant that may modify growth and development and even influence reproductive capabilities. The duration and severity of the stress dictate the scale and the timing of the response.

Little is known about how plants recognize stresses. Our best clues come from yeast and bacterial proteins that initiate signal transduction in response to abiotic stresses such as low osmotic potential. Plants probably contain similar proteins, but comparable functions have not yet been demonstrated. The intricate signaling pathways that are assumed to participate in alterations of plant gene expression in response to stress are yet to be elucidated. However, considerable evidence indicates that the regulation of plant stress responses involves hormones—especially abscisic acid (ABA), jasmonic acid, and ethylene—and secondary messengers, such as Ca^{2+} (see Chapters 17 and 18).

In response to stress, some genes are expressed more strongly, whereas others are repressed. The protein products of stress-induced genes often accumulate in response to unfavorable conditions. The functions of these proteins and the mechanisms that regulate their expression are currently a central topic of research in stress physiology. Although most studies have focused on transcriptional activation of gene expression, growing evidence suggests that the accumulation of gene products is also influenced by posttranscriptional regulatory mechanisms that increase the amounts of specific mRNAs, enhance translation, stabilize proteins, alter protein activity, or some combination of these. Using molecular genetic techniques, researchers have begun to dissect plant responses associated with exposure to specific stresses. The stresses addressed in this chapter include water deficit, low and high temperatures, oxygen deficit, and environmental oxidants. Abiotic stresses described elsewhere in this text include responses to some

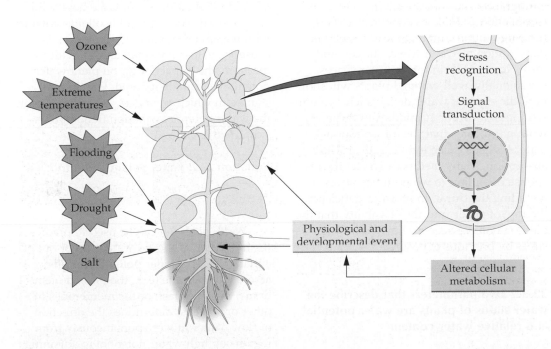

Figure 22.3
Plants respond to stress both as collections of cells and as whole organisms. Stresses constitute environmental signals that are received and recognized by the plant. After stress recognition, the signal is communicated within cells and throughout the plant. Transduction of environmental signals typically results in altered gene expression at the cellular level, which in turn can influence metabolism and development of the whole plant.

nutrient deficiencies (see Chapter 23) and to toxic concentrations of aluminum (see Chapter 23) or cadmium (see Chapter 16).

22.2 Stresses involving water deficit

22.2.1 Water deficit can be induced by many environmental conditions.

Water-related stresses can affect plants if the environment contains excess water or if the quantity or quality of water available is insufficient to meet basic needs. The impact of flooding on terrestrial plants is discussed in Section 22.7. In this section we will review plant responses to inadequate water supplies, termed water deficit.

Many environmental conditions can lead to water deficit in plants. Periods of little or no rainfall can lead to a meteorological condition called drought. Transient or prolonged drought conditions reduce the amount of water available for plant growth. However, water deficit also can occur in environments in which water is not limiting. In saline habitats, the presence of high salt concentrations makes it more difficult for plant roots to extract water from the environment. Low temperatures can also result in water stress (see Section 22.6). For example, exposure to freezing temperatures can lead to cellular dehydration as water leaves the cell and forms ice crystals in intercellular spaces. Occasionally, well-watered plants will show periodic signs of water deficit such as a transient loss of turgor at midday. In this case, wilting indicates that the transpirational water loss has exceeded the rate of water absorption. Many factors can affect the response of a plant to water-deficit stresses, including the duration of water deficiency, the rate of onset, and the possibility that the plant may have been acclimated to water stress by previous exposure.

22.2.2 Two parameters that describe the water status of plants are water potential and relative water content.

Like all molecules, water can be described thermodynamically in terms of its free energy content, also known as its chemical potential. Plant physiologists use a related param-

eter, **water potential** (ψ_w; see Chapter 15, Box 15.1, for a more quantitative discussion of this topic). Such measurements can be used to evaluate the extent to which a cell, organ, or whole plant is "hydrated."

Equation 22.1: Water potential

$$\psi_w = \psi_s + \psi_p + \psi_g + \psi_m$$

The ψ_w of a plant equals the sum of various component potentials. **Solute potential, ψ_s,** is dictated by the number of particles dissolved in water. Water potential decreases as solute concentration increases. The **pressure potential, ψ_p,** reflects physical forces exerted on water by its environment. When water is subjected to negative pressure (tension), ψ_p is less than 0 MPa (megapascals) and ψ_w is diminished. (Note that water potential is typically defined in units of pressure rather than energy.) In contrast, water potential is increased by positive pressure (turgor, ψ_p greater than 0 MPa). **Gravitational potential,** ψ_g, can have a substantial effect when water is transported over vertical distances greater than 5 to 10 m, but this term can be omitted when describing transport between cells or within small plants. A fourth factor, the **matric potential (ψ_m)**, accounts for how solid surfaces (e.g., cell walls and colloids) interact with water and depress ψ_w. However, because ψ_m values are small and difficult to measure, the impact of ψ_m on plant water potential is usually ignored. For circumstances in which ψ_g and ψ_m are insignificant, the water potential equation frequently is simplified as follows:

Equation 22.2: Water potential (simplified)

$$\psi_w = \psi_s + \psi_p$$

Water potential can be used to predict the movement of liquid water into or out of a plant cell. The water potential difference across a membrane (e.g., the plasma membrane, the tonoplast, or the membranes of other organelles) determines the direction of flow. Water moves spontaneously from regions of high water potential to adjoining regions of low water potential. Consider a plant cell placed in a beaker of pure water. The difference in water potential will drive water uptake until the cell is swollen and the plasma membrane presses against the cell wall, exerting positive pressure (Fig. 22.4, upper cell). This turgor pressure increases until water potentials are equal on both sides of the membrane, and the net water transport is zero.

If the same plant cell is placed in a concentrated salt solution, water will flow out of the cell until turgor is lost and the plasmolyzed protoplast pulls away from the cell wall. At the same time, cellular solutes that cannot be quickly metabolized become more concentrated. These factors combine to lower the water potential of the cell (Fig. 22.4, lower cell). However, if the solute concentration within the plasmolyzed (i.e., nonturgid) cell becomes greater than in the outside medium, the water potential gradient across the plasma membrane will help drive water into the cell. As a result, cellular solutes are diluted

Turgid cell: $\psi_{w\ external} = 0$ MPa

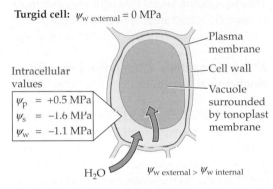

Intracellular values

ψ_p	=	+0.5 MPa
ψ_s	=	−1.6 MPa
ψ_w	=	−1.1 MPa

Plasma membrane

Cell wall

Vacuole surrounded by tonoplast membrane

H_2O $\psi_{w\ external} > \psi_{w\ internal}$

Plasmolyzed cell: $\psi_{w\ external} = -2.5$ MPa

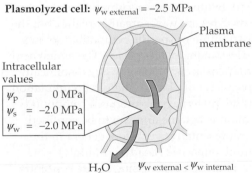

Intracellular values

ψ_p	=	0 MPa
ψ_s	=	−2.0 MPa
ψ_w	=	−2.0 MPa

Plasma membrane

H_2O $\psi_{w\ external} < \psi_{w\ internal}$

Figure 22.4
Movement of water into and out of a cell depends on the water potential gradient across the plasma membrane. Water moves spontaneously down its chemical potential gradient, which in plant cells is defined by pressure and concentration parameters. The water potential outside the cell must be greater than the water potential inside the cell for water to enter the cell from outside (upper cell). The extent to which the lower cell has been plasmolyzed would probably be lethal. Note that the cell will dehydrate further, until it comes to equilibrium with the external solution ($\psi_{w\ cell} = \psi_{w\ external} = -2.5$ MPa).

(ψ_s increases) and the turgor pressure is restored as the protoplast expands. When ψ_w inside the cell is in equilibrium with ψ_w outside, the net inward flow of water will stop.

The water potential of plant organs such as roots or leaves can be measured by using a pressure chamber (see Chapter 15) or a thermocouple psychrometer. Although the validity of using ψ_w as a sole indicator of the water status of the whole plant has been debated among plant physiologists, this parameter is thought to describe accurately the osmotically driven flow across cell membranes and between intracellular compartments.

The physiological or metabolic changes detected in water-stressed plants do not always correlate with changes in plant ψ_w measurements. To address these concerns, a second parameter frequently used to assess water status—**relative water content** (RWC)—often is reported in conjunction with plant ψ_w measurements. Turgid weight is determined by floating tissues on water in an enclosed chamber, preferably under light, until the weight of the tissues stays constant.

Equation 22.3: Relative water content

$$RWC = [(\text{fresh wt.} - \text{dry wt.})/(\text{turgid wt.} - \text{dry wt.})] \times 100$$

When water uptake by roots closely matches water loss by leaves, the RWC of transpiring leaves typically ranges from 85% to 95%. If the RWC for a given organ drops below a critical value, tissue death follows. The value for critical RWC varies among species and tissue types but frequently is less than 50%. In general, when the water potential of a plant drops, the RWC also declines. One noteworthy exception to this involves plants that can maintain a high RWC despite a low or declining ψ_w, a trait associated with plants that can adjust osmotically (see next section).

22.3 Osmotic adjustment and its role in tolerance to drought and salinity

22.3.1 Osmotic adjustment is a biochemical mechanism that helps plants acclimate to dry or saline soil.

A plant cannot extract water from the soil unless the water potential in the root is less than the water potential in the surrounding soil. The plant root must establish a water potential gradient so that water flows toward the root surface from the soil. Some plants are highly sensitive to water stress and wilt (lose turgor) when soil water potential becomes too low. Other plants, however, can endure dry or saline conditions without evident loss of turgor. Many drought-tolerant plants can regulate their solute potentials to compensate for transient or extended periods of water stress. This process, called **osmotic adjustment,** results from a net increase in the number of solute particles present in the plant cell. The osmolalities achieved through osmotic adjustment exceed those that result when solutes are passively concentrated by dehydration. Through decreasing the plant ψ_s, osmotic adjustment can drive root ψ_w to values lower than soil ψ_w, thus allowing water to move from soil to plant down a potential gradient (Fig. 22.5). Osmotic adjustment is believed to play a critical role in helping plants acclimate to drought or saline conditions. Its effect can be difficult to measure, however, because changes in solute composition or concentration that promote drought tolerance are not always easy to distinguish from the solute effects that are symptomatic of a metabolism perturbed by stress.

Because ψ_s is a function of the total number of solute particles in a given volume of water, a remarkable array of organic compounds and inorganic ions contribute to the solute potentials of plant cells and tissues. In principle, osmotic adjustment could involve metabolic changes that alter rates of ion uptake, decrease assimilation of low-molecular-mass organic compounds, or enhance their synthesis. However, comparative biochemical studies have shown that bacterial, animal, and plant cells accumulate relatively few solutes to concentrations that actively contribute to osmotic adjustment (see next section).

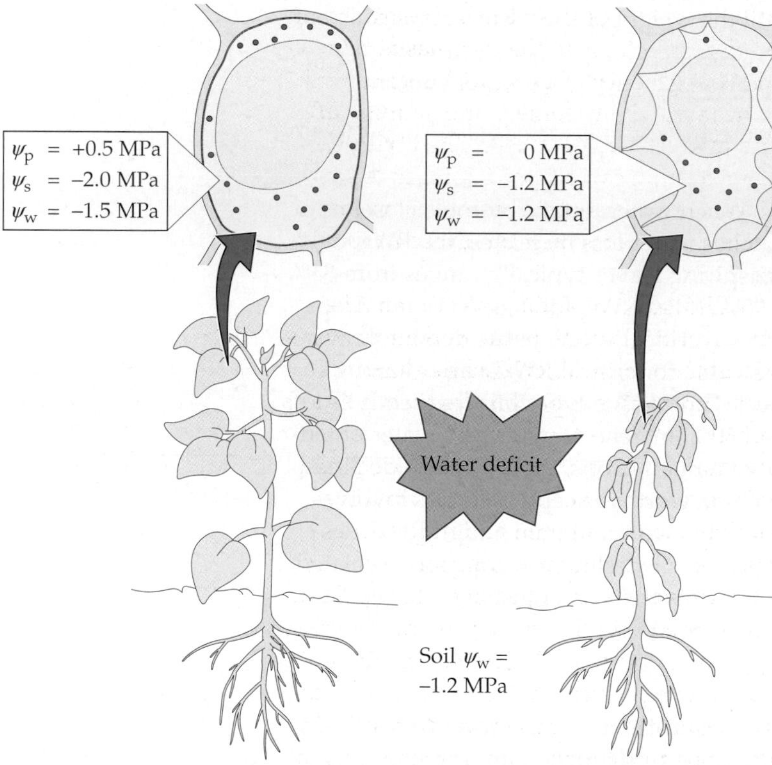

22.3.2 Compatible solutes share specific biochemical attributes.

Compatible solutes, also known as compatible osmolytes, are a small group of chemically diverse organic compounds that are highly soluble and do not interfere with cellular metabolism, even at high concentrations (Fig. 22.6). Synthesis and accumulation of organic osmolytes are widespread in plants, but the distribution of specific compatible solutes varies among plant species. The amino acid proline is accumulated by a taxonomically diverse set of plants, whereas accumulation of the quaternary ammonium compound β-alanine betaine appears to be confined to representatives of a few genera in the Plumbaginaceae (Leadwort family). One mechanism for increasing solute concentrations is the irreversible synthesis of compounds such as glycine betaine (see Section 22.3.5). Concentrations of other compatible solutes (e.g., proline) are maintained through a combination of synthesis and catabolism. Monomeric sugars (e.g., glucose and fructose) can be released from polymeric forms (starch and fructans, respectively) in response to stress. Once the stress is removed, these monomers can be repolymerized to facilitate rapid and reversible osmotic adjustment.

$\psi_p = +0.5$ MPa
$\psi_s = -2.0$ MPa
$\psi_w = -1.5$ MPa

$\psi_p = 0$ MPa
$\psi_s = -1.2$ MPa
$\psi_w = -1.2$ MPa

Water deficit

Soil $\psi_w = -1.2$ MPa

Osmotic adjustment

No osmotic adjustment

Figure 22.5
Osmotic adjustment occurs when the concentrations of solutes within a plant cell increase to maintain a positive turgor pressure within the cell. The cell actively accumulates solutes and, as a result, ψ_s drops, promoting the flow of water into the cell. In cells that fail to adjust osmotically, solutes are concentrated passively but turgor is lost.

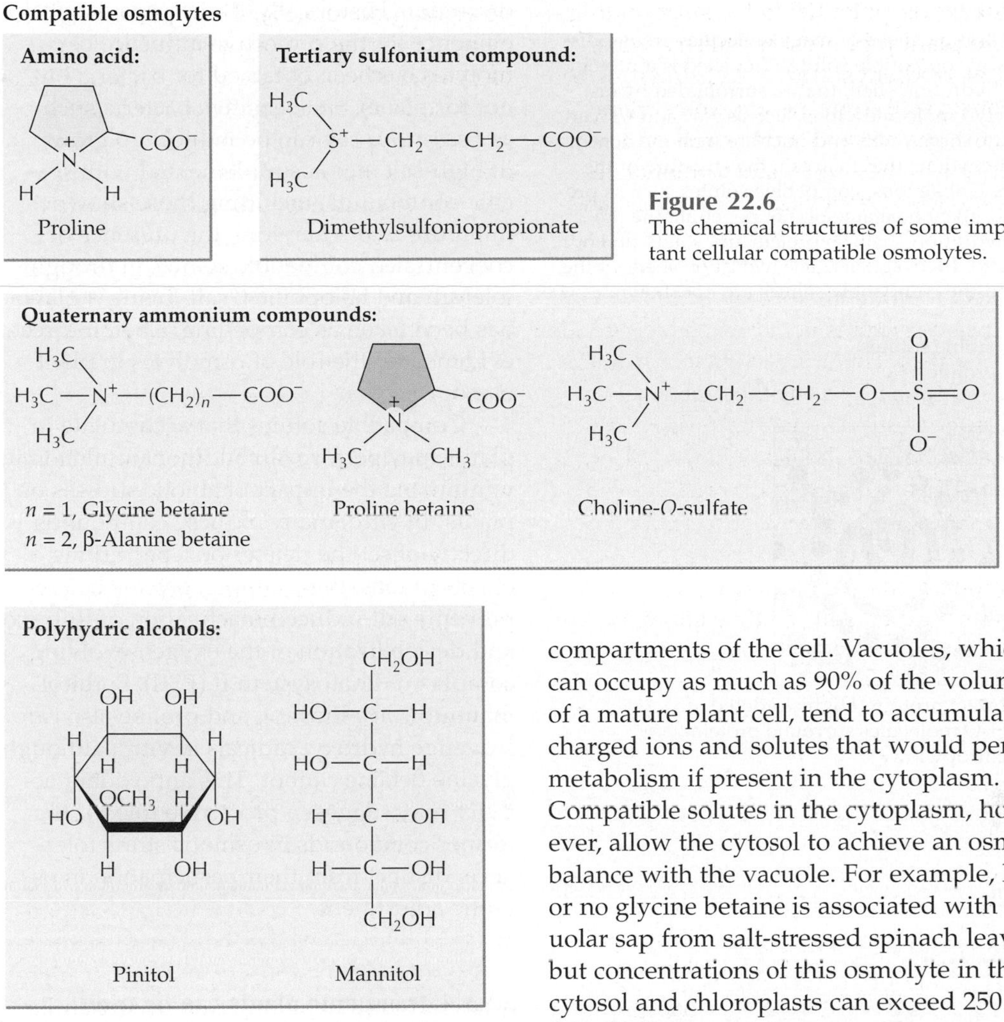

Compatible osmolytes

Amino acid:

Proline

Tertiary sulfonium compound:

Dimethylsulfoniopropionate

Figure 22.6
The chemical structures of some important cellular compatible osmolytes.

Quaternary ammonium compounds:

$n = 1$, Glycine betaine
$n = 2$, β-Alanine betaine

Proline betaine

Choline-O-sulfate

Polyhydric alcohols:

Pinitol

Mannitol

The organic character of osmolyte molecules probably reflects the potential toxicity of concentrated inorganic solutes. Many ions found in cells adversely affect metabolic processes when present at high concentrations, possibly by binding to and altering the properties of cofactors, substrates, membranes, and enzymes. Furthermore, many ions can enter the hydration shells of a protein and promote its denaturation. In contrast, compatible solutes tend to be neutrally charged at physiological pH, either nonionic or **zwitterionic** (dipolar, with spatially separated positive and negative charges), and are excluded from the hydration shells of macromolecules (Fig. 22.7).

In addition to stereotypical charge characteristics, compounds active in osmotic adjustment demonstrate distribution patterns that support water potential (osmotic) equilibria among the various membrane-bound compartments of the cell. Vacuoles, which can occupy as much as 90% of the volume of a mature plant cell, tend to accumulate charged ions and solutes that would perturb metabolism if present in the cytoplasm. Compatible solutes in the cytoplasm, however, allow the cytosol to achieve an osmotic balance with the vacuole. For example, little or no glycine betaine is associated with vacuolar sap from salt-stressed spinach leaves, but concentrations of this osmolyte in the cytosol and chloroplasts can exceed 250 mM (Fig. 22.8).

Although compatible solutes must be compartmentalized to perform their physiological roles, the mechanisms that generate and maintain this compartmentation have received little attention to date. Membrane-associated carriers or transporters are probably involved in differentially distributing osmolytes within the cell and may also participate in regulating the transport and distribution of these solutes throughout the plant. Some evidence indicates that the distribution of proline in osmotically stressed plants involves a transporter. For example, two *Arabidopsis* cDNAs encoding proline transporters were cloned by functional complementation of a yeast amino acid permease-targeting mutation and identified by amino acid transport assays. *ProT2* transcripts, which encode one of the proline transporters, were found in all tissues types examined and showed a marked increase in expression when plants were subjected to water deficit or salt stress.

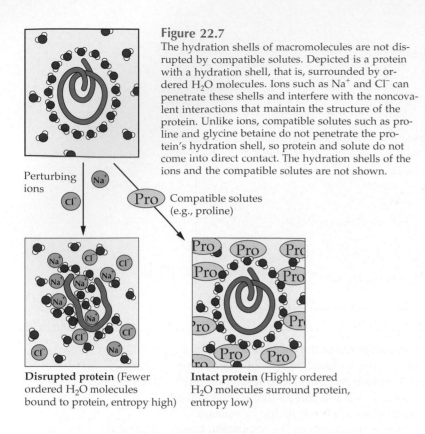

Figure 22.7
The hydration shells of macromolecules are not disrupted by compatible solutes. Depicted is a protein with a hydration shell, that is, surrounded by ordered H_2O molecules. Ions such as Na^+ and Cl^- can penetrate these shells and interfere with the noncovalent interactions that maintain the structure of the protein. Unlike ions, compatible solutes such as proline and glycine betaine do not penetrate the protein's hydration shell, so protein and solute do not come into direct contact. The hydration shells of the ions and the compatible solutes are not shown.

Perturbing ions

Na^+

Cl^-

Pro — Compatible solutes (e.g., proline)

Disrupted protein (Fewer ordered H_2O molecules bound to protein, entropy high)

Intact protein (Highly ordered H_2O molecules surround protein, entropy low)

Salt-stressed spinach leaf cell

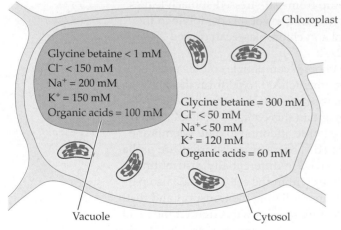

Chloroplast

Glycine betaine < 1 mM
Cl^- < 150 mM
Na^+ = 200 mM
K^+ = 150 mM
Organic acids = 100 mM

Glycine betaine = 300 mM
Cl^- < 50 mM
Na^+ < 50 mM
K^+ = 120 mM
Organic acids = 60 mM

Vacuole

Cytosol

Figure 22.8
Osmotic adjustment in a mesophyll cell of a salt-stressed spinach leaf. Sodium and chloride ions, which can disrupt cytosolic metabolism, are concentrated in the vacuole. In contrast, concentrations of glycine betaine are high in the chloroplasts (data not shown) and cytoplasm but low in the vacuole.

22.3.3 Some compatible solutes may serve protective functions in addition to osmotic adjustment.

The role that compatible solutes play in osmotically stressed plants often has been defined as "osmoprotection." That is, solute accumulation is assumed to have a putative protective function against water deficit. However, this term should be applied with

discretion. Historically, direct physiological evidence for the protective influence of osmolytes has been obtained for bacteria but not for plants. Salt-sensitive bacteria such as *Escherichia coli* can be induced to grow in high-salt media supplemented with specific compounds, including those shown in Figure 22.6. Therefore, the presence of concentrated compatible solutes in drought-tolerant and halophilic ("salt-loving") plants has been taken as compelling, albeit indirect, evidence for the role of osmolytes in plant osmoprotection.

Compatible solutes that accumulate in plants may perform an additional function in minimizing the impact of abiotic stresses on plants. In vitro, many of these compounds directly offset the deleterious, perturbing effects of ions. For example, glycine betaine prevents salt-induced inactivation of Rubisco and destabilization of the oxygen-evolving complex of Photosystem II (PSII). Sorbitol, mannitol, *myo*-inositol, and proline also can scavenge hydroxyl radicals in vitro, although glycine betaine cannot. This antioxidant activity may suggest a protective role for the former compounds in osmotic stress tolerance, distinct from their participation in osmotic adjustment.

22.3.4 Transgenic plants can be used to test the acclimative functions of specific osmolytes.

Genetic engineering of plants offers an opportunity to test the adaptive significance of compatible solutes directly. Drought- or salt-sensitive plants can be transformed with genes encoding enzymes that are critical for the synthesis of a putative osmoprotectant. The transformed plants then can be assayed for enhanced accumulation of compatible solutes and for the ability to adjust osmotically in response to drought and salinity. Ongoing experiments hold promise for the future development of stress-resistant cultivars.

Biotechnological approaches to enhanced drought tolerance typically involve dissection of the biosynthesis pathways of specific osmolytes and subsequent manipulation of the activities of the enzymes operating in the pathways. Among the compounds studied to date are the amino acid proline, the quaternary ammonium glycine betaine,

Betaine synthesis pathway

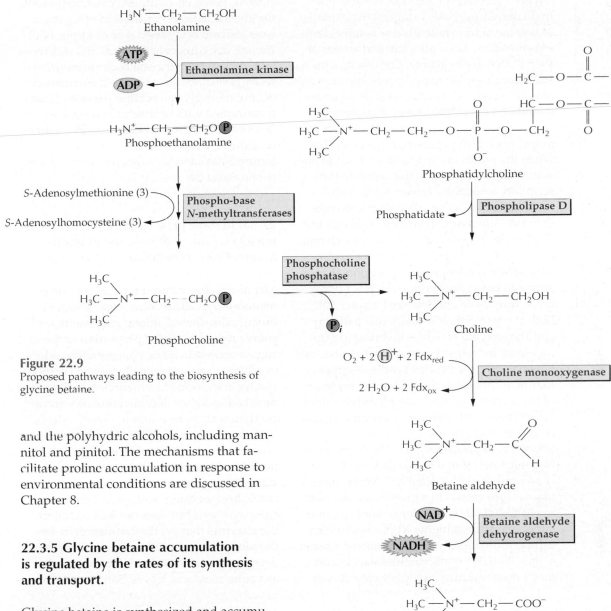

Figure 22.9
Proposed pathways leading to the biosynthesis of glycine betaine.

and the polyhydric alcohols, including mannitol and pinitol. The mechanisms that facilitate proline accumulation in response to environmental conditions are discussed in Chapter 8.

22.3.5 Glycine betaine accumulation is regulated by the rates of its synthesis and transport.

Glycine betaine is synthesized and accumulated by many algae and higher plants. Its distribution among plants is sporadic; in the Chenopodiaceae (Goosefoot family), all members examined to date accumulate glycine betaine, but in most other families this trait is found in only some of the species.

Radiotracer evidence indicates that glycine betaine accumulation in osmotically stressed plants results from increased rates of synthesis. Unlike proline, for which synthesis and catabolism appear to be coordinately regulated in response to water status (see Chapter 8), glycine betaine does not appear to be broken down by plants. In the ab-

sence of glycine betaine catabolism, researchers believe that the amounts of this osmolyte present are dictated by rates of biosynthesis and by transport, through the phloem, to growing tissues.

Glycine betaine is synthesized from choline in a two-step pathway (Fig. 22.9). Evidence from subcellular localization studies with spinach places this pathway in the chloroplast. The first enzyme, choline monooxygenase, catalyzes the oxidation of choline to betaine aldehyde, using photosynthetically reduced ferredoxin and molecular oxygen. The second enzyme, betaine aldehyde dehydrogenase, catalyzes the oxidation of betaine aldehyde to glycine betaine. Both enzymes have been purified, and clones of their cDNA are available. Consistent with results from labeling experiments, the activities of both enzymes increase severalfold under conditions of osmotic stress. This increased enzyme activity is accompanied by increases in the amounts of transcript, which decline when the plants are irrigated with nonsaline water. The mechanisms that regulate transcript accumulation are not yet known.

Plants that accumulate glycine betaine may use different pathways to generate the choline precursor (Fig. 22.9). In spinach and sugar beet, choline synthesis proceeds by way of the methylation of phosphoethanolamine to phosphocholine, which is hydrolyzed directly to choline (see Chapter 10). Both the rate of flux through this pathway and the enzyme activities involved are upregulated by salt stress in spinach. In barley, however, glycine betaine synthesis appears to utilize choline derived from the turnover of the membrane lipid phosphatidylcholine.

Some genetic evidence indicates that accumulation of glycine betaine promotes salt tolerance. Researchers have developed near-isogenic lines of maize that differ in the ability to synthesize this osmolyte. When these plants were grown in a greenhouse and subjected to salt stress, glycine betaine accumulators maintained higher RWC, assimilated carbon more rapidly, and maintained greater turgor than did nonaccumulators. Because many economically important crop species

do not accumulate glycine betaine, biotechnological attempts to introduce this trait are underway. In addition to the available plant cDNAs, genes encoding enzymes responsible for glycine betaine synthesis in other organisms also are being used. For example, both choline dehydrogenase from *E. coli* and choline oxidase from the soil bacterium *Arthrobacter globiformis* can catalyze the oxidation of choline to glycine betaine. Tobacco plants transformed with choline dehydrogenase demonstrate improved growth under saline conditions, as do cultures of the cyanobacterium *Synechococcus* when transformed with choline oxidase.

22.3.6 In some plant species, salt stress inhibits sucrose synthesis and promotes accumulation of mannitol.

Mannitol is the reduced form of the sugar mannose (Fig. 22.10). This sugar alcohol is broadly distributed among plants and accounts for a substantial proportion of the sugars present in some plants such as celery, in which photosynthetically reduced carbon is often translocated as mannitol. In vivo radiolabeling shows that mannitol concentrations increase in response to osmotic stress. In sharp contrast to glycine betaine, mannitol accumulation appears to be regulated by inhibition of competing pathways and by decreased rates of mannitol consumption and catabolism. In celery, salt stress inhibits sucrose synthesis but does not seem to affect the enzymes that synthesize mannitol. At the same time, rates of mannitol utilization decrease, particularly in sink tissues such as young roots and leaves. Salt stress also

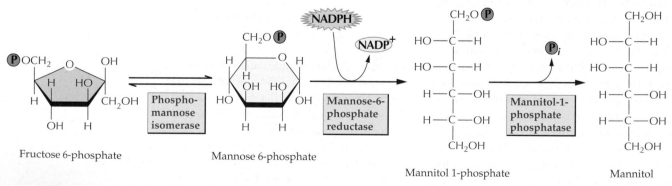

Figure 22.10
The biosynthetic pathway for mannitol.

down-regulates NAD$^+$-dependent mannitol dehydrogenase, the enzyme that oxidizes mannitol in celery. Enzyme activity, protein concentration, and mRNA abundance all decrease, bolstering mannitol accumulation.

Evidence for mannitol as a compatible solute comes from work with cell cultures. Celery cells cultured under conditions that result in mannitol accumulation can withstand salt stress better than can cells grown in a medium that promotes production of sucrose but not mannitol. The improved survival rates for mannitol-synthesizing cells cannot be attributed to the cellular solute potentials, because those are the same with either treatment. Additional evidence has been gained by using molecular genetic approaches. Tobacco and *Arabidopsis*, both salt-sensitive plants, have been genetically engineered to synthesize mannitol, a solute these plants do not usually contain. Transgenic plants expressing the *E. coli* gene for NAD$^+$-dependent mannitol-1-phosphate dehydrogenase, which converts fructose 6-phosphate to mannitol 1-phosphate, accumulated mannitol, albeit at relatively low concentrations. The salt tolerance of the transgenic mannitol-producing tobacco was improved relative to that of the nonaccumulating controls. Seeds of mannitol-accumulating transgenic *Arabidopsis* were able to germinate in the presence of salt.

22.3.7 Taxonomically diverse plants accumulate pinitol in response to salt stress.

D-Pinitol, a cyclic sugar alcohol, is a major solute in members of the Pinaceae (Pine family), Fabaceae (Bean family), and Caryophyllaceae (Pink family). In general, its concentrations are higher among halophytic species and those adapted to drought. Pinitol accumulation occurs in salt-tolerant legumes (e.g., *Sesbania* spp.) and in a facultative halophyte, *Mesembryanthemum crystallinum*, when irrigated with solutions containing sodium chloride. Pinitol can constitute more than 70% of the soluble carbohydrates present in *M. crystallinum* plants treated with 400 mM NaCl, but only 5% in nonsalinized controls. In leaves, pinitol is localized to the chloroplasts and cytosol but is not detected in vacuoles.

The biosynthesis of pinitol is believed to proceed by the *O*-methylation of *myo*-inositol, which forms ononitol in angiosperms (Fig. 22.11) and the ononitol isomer, sequoyitol, in gymnosperms. Both intermediates undergo epimerization to pinitol. A cDNA encoding an *S*-adenosyl-L-methionine-dependent *myo*-inositol 6-*O*-methyltransferase has

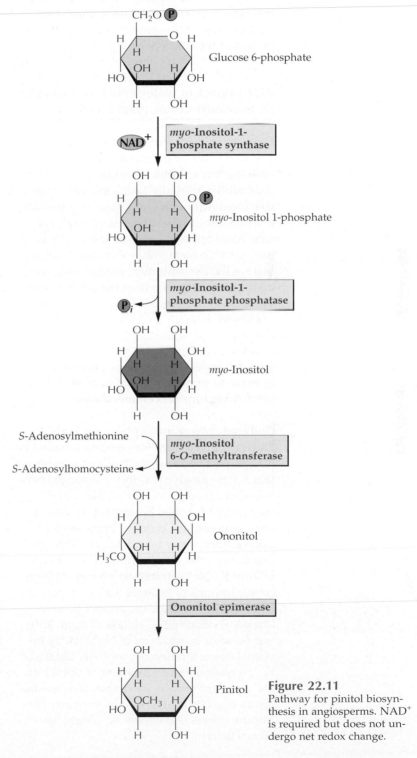

Figure 22.11
Pathway for pinitol biosynthesis in angiosperms. NAD$^+$ is required but does not undergo net redox change.

been isolated from salinized *M. crystallinum.* Increases in the concentration of pinitol in salt-exposed *M. crystallinum* are accompanied by an almost 60-fold induction in the activity of *myo*-inositol 6-*O*-methyltransferase. This increase in enzyme activity is regulated at the transcriptional level and is associated with increased mRNA concentrations. *E. coli* and tobacco transformed with this gene accumulate ononitol, consistent with the role of this enzyme in the pathway proposed for angiosperms.

22.4 Impact of water deficit and salinity on transport across plant membranes

Drought and salt stress both entail acclimation to low water potentials; however, plants growing under saline conditions must also cope with potentially toxic amounts of specific ions. Although ion uptake may provide a means for osmotic adjustment, high cytosolic concentrations of some ions, such as Na^+, perturb metabolism. Consequently, regulating the concentration, composition, and distribution of ions within the cell can be viewed as an essential feature of tolerance to osmotic stress.

22.4.1 Carriers, pumps, and channels operate to minimize the impact of perturbing ions on cell metabolism.

Cultured cells or intact plants subjected to high NaCl concentrations tend to accumulate Na^+, in part because of a membrane potential that favors passive Na^+ influx through channels and carriers. Preventing Na^+ uptake may not be feasible. Functional studies in yeast expressing a putative high-affinity K^+ uptake transporter from wheat indicate that this transporter (HKT1) functions as a high-affinity K^+-Na^+ cotransport system. At high concentrations of external Na^+, transport of K^+ is blocked and the transporter behaves like a low-affinity Na^+ uptake system. Thus, one possible mechanism of Na^+ toxicity involves direct interference with the uptake of K^+, an essential macronutrient. If uptake of Na^+ by plant roots is inevitable, then mechanisms must be available to prevent any perturbing effects that excess Na^+ might have on cell activities.

Na^+ accumulation in the cytoplasm may be prevented in part by the active transport of cytosolic Na^+ across the plasma membrane and out of the cell. In addition, cytosolic Na^+ is transported across the tonoplast and sequestered in the vacuole, where it can accumulate to concentrations that have a marked effect on the osmotic balance of the plant cell. Salt treatments have been shown to increase the activity of vacuolar Na^+/H^+ antiporters in several plant systems, including sugar beet suspension cultures and barley roots. The operation of this carrier requires an electrochemical potential gradient across the tonoplast, a principal component of which arises from pH gradients generated by the action of H^+ pumps. Among the pumps that can energize Na^+/H^+ antiports are the plasma membrane H^+-ATPase, the vacuolar H^+-ATPase, and the H^+-pyrophosphatase (H^+-PPase) of the tonoplast membrane (see Chapter 3). For *M. crystallinum,* a salt-induced increase in tonoplast Na^+/H^+ antiporter activity correlates with increased activity of the vacuolar H^+-ATPase but not the H^+-PPase. This suggests that both pumps may not contribute equally to the membrane potential that energizes transport of excess Na^+ into vacuoles.

Recent work reinforces the importance of the Na^+/H^+ antiporter in salt tolerance. Overexpression of the antiporter allowed *Arabidopsis* plants to grow in 200 mM NaCl, about half the concentration present in sea water. Research of this type could influence the reclamation for agricultural use of land made saline by irrigation.

22.4.2 Synthesis and activity of aquaporin may be up-regulated in response to drought.

The hydrophobic nature of the lipid bilayer presents a considerable barrier to the free movement of water into the cell and between intracellular compartments. However, plasma membranes and tonoplasts can be rendered more permeable to water by proteinaceous transmembrane water channels called aquaporins (see Chapter 3). Water movement through aquaporins can be modulated rapidly. Evidence suggests that these channels may facilitate water movement in drought-stressed tissues and promote the rapid recovery of turgor on watering. In *Arabidopsis*, water deficit strongly induces expression of the *Rd28* gene, which encodes a member of the MIP (major intrinsic protein) family. The aquaporin RD28 is located in the plasma membrane. Genes encoding MIP-related proteins also have been identified in *M. crystallinum*. The abundance of mRNA transcripts of these genes correlates with the turgor changes in leaves of plants subjected to a 400 mM NaCl shock treatment (Fig. 22.12). The amount of transcript first decreases after the initial shock, as does turgor. Transcript concentrations then increase as turgor is restored. Greater transcript concentrations, enhanced translation, and activation of existing proteins each may constitute a mechanism for regulating aquaporin abundance and activity in response to water stress. For example, the plant aquaporin α-TIP (tonoplast intrinsic protein) can undergo phosphorylation (see Chapter 3). Expression studies in *Xenopus* oocytes link α-TIP phosphorylation with increased permeability of the cell membrane to water. The presence of multiple mechanisms for regulating water channel activity may be advantageous to plants coping with water deficit.

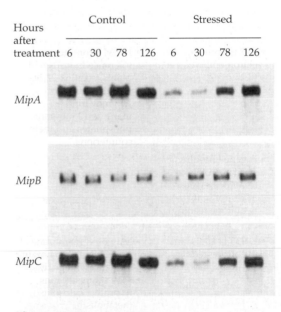

Figure 22.12
RNA gel blot analysis illustrating expression in roots of *Mesembryanthemum crystallinum* of MIP-related genes. Declining transcript abundance in response to osmotic stress correlates with loss of turgor. Transcript concentrations increase as turgor is restored.

22.5 Additional genes induced by water stress

Researchers have assumed that characterizing the genes expressed during water stress will shed light on the metabolic events that protect plants from drought. Differential cDNA screening techniques have been used to identify genes that are up-regulated during periods of water deficit. These screens have enabled researchers to isolate many genes of unknown function; studies are underway to determine how the proteins encoded by these genes act during stress.

When water is limiting, plants can become more susceptible to other stresses. When in light, plants dissipate heat mainly through transpiration, a process inhibited by drought and salinity. Accordingly, heat stress often accompanies water deficit. Water stress can also enhance the production of reactive oxygen species (see Section 22.8) and increase susceptibility to pathogens (see Chapter 21). Some gene products induced by water deficit may function primarily to minimize other stresses, and several appear to have multiple stress-related roles.

22.5.1 Some seed proteins may protect vegetative tissues from stress.

The *Lea* (late embryogenesis abundant) genes were first identified as genes induced in seeds during maturation and desiccation (see Chapter 19). The amounts of some *Lea* gene products are now known to increase in the vegetative tissues of plants exposed to stresses that include a water-deficit component. Most LEA proteins are overwhelmingly hydrophilic, a trait consistent with their cytoplasmic location. In addition, many share a biased amino acid composition, rich in alanine and glycine and lacking cysteine and tryptophan. At present, the importance of these proteins is inferred from their abundance and expression patterns; the in vivo activities of most LEA proteins remain unknown. Overexpression of transgenic LEA proteins in rice and yeast has been shown to enhance resistance to specific water-deficit stresses (Box 22.1). Additional physiological, biochemical, and biophysical studies are needed to determine how LEA proteins function during seed development and stress.

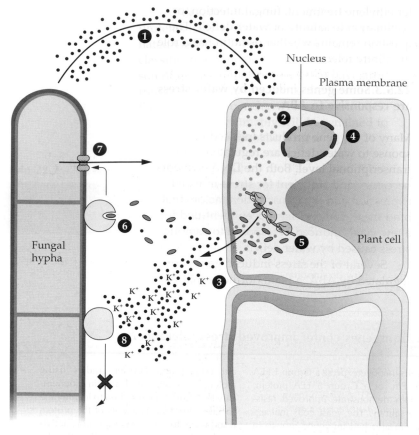

Figure 22.13
Model for antifungal action of osmotin, a plant protein that accumulates in response to many abiotic and biotic stresses. The fungal hypha releases fungal toxins (1), which disrupt the plant membrane (2), causing it to leak nutrients that the fungus utilizes (3). The plant cell loses turgor (4), which promotes the accumulation of osmotin (5). Osmotin from the leaking cell comes into contact with a fungal membrane receptor (6) and facilitates the formation of pores in the fungal membranes (7), rendering them permeable. This loss of membrane integrity inhibits hyphal growth and can kill the fungal pathogen. Cationic interactions, however, limit the effects of osmotin (8).

increased in response to water deficit and low temperature. ABA plays a role in several responses to water stress, most notably stomatal closure and induction of gene expression (see Chapters 17 and 18). ABA biosynthesis mutants have been used to demonstrate that *Arabidopsis*, maize, and tomato require increased concentrations of ABA to express several water-deficit–induced genes. Four genes in *flacca*, an ABA-deficient mutant of tomato (see Chapter 17), were found not to be expressed unless exogenous ABA is applied. The products of these genes are two LEA proteins (members of Groups 2 and 5), a lipid-transfer protein, and a stress-induced isotype of histone H1.

Of the genes that participate in ABA-induced gene expression, several have been identified as participating in the phosphorylation and dephosphorylation events involved in transduction of the ABA signal. The *ABI1* and *ABI2* genes, found in ABA-insensitive mutants of *Arabidopsis* (see Chapter 18), are thought to encode protein phosphatases and can complement a yeast mutant lacking PCT1, a type-2C protein phosphatase. ABI1 functions in stomatal closure and the induction of specific genes. Also active in regulating ABA-responsive gene expression is a tyrosine kinase, similar to MAPKs (mitogen-activated protein kinases; see Chapters 11 and 18). In addition, ABA and experimental conditions that result in cellular water deficits induce many other kinases, including members of the MAPK pathway. However, the function of these kinases and the associated signal transduction pathways have not been delineated.

Because ABA does not regulate all stress-induced genes, other signals must also be involved in responses to water deficit. Moreover, several genes induced by water deficit do not require ABA for expression and are probably regulated by distinct or interacting signal transduction pathways.

22.5.4 Specific cis-elements and trans-acting factors promote transcription in response to ABA and water deficit.

In the wheat *Em* gene, a DNA element containing CACGTG is required for ABA-induced expression of transgenic reporter genes in a transient expression system that uses rice protoplasts. When fused to a minimal CaMV 35S promoter, multiple copies of this ABA-responsive element (ABRE) confer ABA responsiveness on a reporter gene. Interestingly, the ACGT core is found in many other genes regulated by environmental conditions, though not necessarily by ABA (e.g., the G-box; see Section 22.7.3). A bZIP-type transcription factor (see Chapter 7), EmBP1, binds the ABRE (Fig. 22.14). bZIP proteins are characterized by a basic domain adjacent to a leucine heptad repeat region.

Additional promoter sequences called coupling elements (CEs) function with the ABRE to control the expression of ABA-responsive genes (Fig. 22.14). The ABA-regulated barley genes *HVA22* and *HVA1* contain the elements CE1 and CE3,

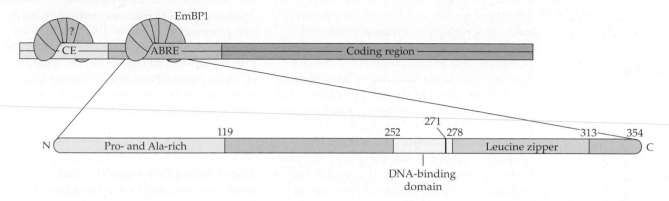

Figure 22.14

EmBP1, a bZIP transcription factor, binds the ABA-responsive element (ABRE) of an ABA-induced gene. The basic DNA-binding domain of EmBP1 is adjacent to a leucine zipper region involved in dimer formation. The function of the proline- and alanine-rich region has not been determined, although it may play a role in protein–protein interactions. ABREs may function in concert with additional promoter sequences called coupling elements (CEs). The DNA-binding proteins that interact with these elements have not been identified.

respectively. These CEs are thought to bind different transcription factors, thereby allowing the plant to activate transcription of each ABA-responsive gene independently. CEs may provide some insight into tissue-specific and temporally distinct patterns of ABA-induced gene expression.

Another family of transcription factors, *ABI3* identified in *Arabidopsis* and *Vp1* identified in maize, are required for ABA-regulated expression of specific genes. The gene products ABI3 and VP1 are ordinarily found in developing seeds. The promoter of the *Craterostigma plantagineum* gene *cDet-27-45* (Group 4 Lea) requires *Abi3* to drive transcription; the *cDet-6-19* (Group 2 Lea) promoter does not.

Not all water-deficit–responsive genes are induced by ABA. The drought-responsive DNA element (DRE; also called the C-repeat) functions in the transcription of *Arabidopsis* genes that are regulated by water deficiency and low temperature (see Section 22.6.4) but not by ABA.

22.6 Freezing stress

22.6.1 Some plants can acclimate to subfreezing temperatures.

Cold temperatures can cause a type of water-deficit stress. Freezing greatly affects cellular water relations. At a given temperature, the chemical potential of ice is less than that of liquid water. Also, the vapor pressure of extracellular ice is less than that of the water in the cytoplasm or vacuole. As ice formation is initiated in the intercellular spaces, cellular water moves down the water potential gradient, across the plasma membrane, and toward the extracellular ice. Therefore, a water deficit develops within the cell in response to freezing (Fig. 22.15). Some freezing-tolerant plants promote the formation of extracellular ice and thereby prevent the formation of damaging ice crystals within the cytoplasm. A different mechanism, accumulation of antifreezing proteins in the apoplast (see Section 22.6.3), slows the formation of ice and so may retard cellular dehydration.

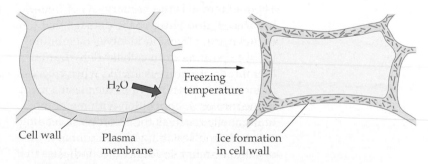

Figure 22.15

The exposure of plants to freezing temperatures results in a cellular water deficit as water travels down its potential gradient, crossing the plasma membrane into the cell wall and intercellular spaces. The rate of freezing shown here was sufficiently slow to prevent formation of ice crystals in the cytoplasm, which is thought to result in cell death. Instead, the cell dehydrates, and freezing occurs in the apoplast.

The ability to survive temperatures below freezing is genotype-specific. Among the genotypes unable to withstand temperatures below freezing are many important crop plants, including corn, tomato, and rice. Other plants are able to survive temperatures below freezing; some can survive temperatures lower than –40°C. However, those plants are not tolerant to freezing throughout the growing season. Freezing tolerance develops in a process known as **cold acclimation,** a response to low but nonfreezing temperatures that occur before freezing. The mechanism of acclimation to temperatures below zero is the subject of intense research study. *Arabidopsis,* which is able to acclimate to cold, has become one of the model plants for these studies. After exposure to temperatures in the range of 1°C to 5°C for one to five days, *Arabidopsis* can survive temperatures of –8°C to –12°C, which makes it a good model for studying the development of freezing tolerance in crop plants. Investigations of plants that are able to survive the subzero temperatures of the northern latitudes are performed with other model species, such as spruce.

The details of the mechanisms that permit freezing tolerance are not well understood. Several processes that can occur in the development of freezing tolerance are (*a*) stabilization of membranes; (*b*) accumulation of sugars, other osmolytes, and antifreeze proteins; and (*c*) multiple changes in gene expression.

22.6.2 A primary function of freeze-tolerance mechanisms is membrane stabilization.

The membrane is the primary site of freeze-induced injury. Much of the damage that occurs during freezing involves membrane lesions resulting from cellular dehydration. Membrane structure is altered when freeze-induced dehydration brings the plasma membrane into close apposition with membranes of organelles such as the chloroplast, leading to membrane destabilization, a primary cause of cellular injury. Evidence also indicates that oxidative stress as well as protein denaturation may occur in response to freezing. The rate of freezing, the temperature at which ice crystals form, and the subcellular location of freezing can each strongly affect the extent of damage to the cell.

Multiple mechanisms are involved in the development of freezing tolerance. One mechanism involves changes in membrane lipid composition (see Chapter 10), including enhanced fatty acid desaturation in membrane phospholipids and changes in the abundance of various membrane sterols and cerebrosides. In addition, the roles of proteins that alter membrane properties are beginning to be determined.

22.6.3 Roles of the osmolytes and antifreeze proteins that accumulate in promoting freezing tolerance remain poorly understood.

Sucrose and other simple sugars are known to accumulate in conjunction with development of freezing tolerance, for example, in *Arabidopsis.* Other osmolytes, such as proline, accumulate after freezing tolerance has already developed and are therefore not considered primary determinants. *eskimo1,* an *Arabidopsis* mutant that is constitutively freeze-tolerant, overexpresses sugars and proline. Although the sugars are proposed to protect membranes, the accumulation of sugars alone is not sufficient for the development of freezing tolerance. Several single-gene mutants of *Arabidopsis* that are defective in freezing tolerance nonetheless accumulate sugars normally in response to the stress of low temperatures.

During periods of cold acclimation, some plants accumulate apoplastic PR proteins that can retard the growth of ice crystals. In winter rye, these antifreezing proteins include endochitinases, $(1{\rightarrow}3)\beta$-endoglucanases, and osmotin (thaumatin)-like proteins (see Section 22.5.2). Interestingly, in frost-sensitive tobacco, PR proteins that have the same catalytic activities but are induced by pathogens instead of by chilling stress do not have the ability to prevent ice crystal formation. Cold-induced PR proteins, which retain their expected enzyme activities, appear to have evolved structural adaptations for protecting cells from freezing stress. These antifreezing proteins have been shown to form oligomeric complexes. The large surfaces of the complexes interact with ice, inhibiting its growth and recrystallization more effectively than the individual polypeptides can. Thus, these proteins may play a dual role, protecting against pathogens while also preventing damage associated with the formation of ice crystals.

22.6.4 Freezing tolerance involves changes in gene expression.

Since the mid-1980s, when changes in gene expression in response to freezing temperatures were first observed in spinach, numerous genes responsive to low-temperature treatments have been identified. Unfortunately, no clear functional role has been determined for many of these genes. Several of the genes induced by low temperature are also induced by water deficit alone or by ABA. Gene products that are hydrophilic and soluble when boiled, including Group 2 and Group 5 LEA proteins and the products of other unique genes, have been found to accumulate during freezing stress. One such unique gene product is COR15a, a small, hydrophilic protein that is targeted to the chloroplast and encoded in the nuclear genome. When imported into the chloroplast, the protein is processed to form the mature protein, COR15am, a 9.4-kDa polypeptide. Protoplasts of nonacclimated transgenic *Arabidopsis* plants that constitutively express the *COR15a* gene have increased freezing tolerance. Researchers have proposed that COR15am alters the intrinsic curvature of the chloroplast membrane.

Many of the cold-regulated genes contain the DRE. This DNA element is bound by the transcriptional activator CBF1 (C-repeat binding factor; also called DREBP, or DRE-binding protein). CBF1 functions as a transcriptional activator in yeast and contains a 60–amino acid DNA-binding domain found in other plant transcription factors, including APETALA2, AINTEGUMENTA, and TINY (Fig. 22.16; see also Chapter 19). The *CBF1* gene has been cloned. Constitutive *CBF1* expression in *Arabidopsis* resulted in increased abundance of all *COR* gene transcripts in nonacclimated plants, which demonstrated increased freezing tolerance. Overexpression of the transcriptional activator *CBF1* enhanced low-temperature tolerance more than did overexpression of *COR15a* alone. These results further implicate the role of cold-regulated genes in the development of tolerance to low temperature and open the door to further research on the overexpression of transcription factors to protect crop plants against freezing.

22.7 Flooding and oxygen deficit

Plants can be damaged not only by the absence of water but also by too much water, which blocks entry of O_2 into the soil so that roots and other organs cannot carry out respiration. Like most eukaryotic organisms, plants are obligate aerobes: Oxygen is the terminal electron acceptor in the mitochondrial electron transfer chain. Under normal aerobic conditions, plants can oxidize 1 mol of hexose sugar through glycolysis, the citric acid cycle, and oxidative phosphorylation to yield 30 to 32 mol of ATP (see Chapters 13 and 14). In the absence of oxygen, ATP production is greatly diminished because glycolysis can produce only 2 mol of ATP per mole of hexose sugar. Intracellular ratios of ATP and ADP decline as mitochondrial ATP production is inhibited. Ironically, oxygen deficit associated with flooding also can prevent plants from obtaining adequate water from the soil.

Water-deficit or cold-induced gene not responsive to ABA

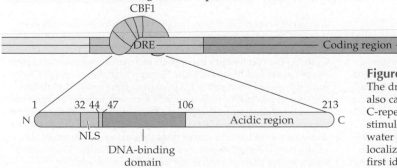

Figure 22.16
The drought-responsive, ABA-independent element (DRE), also called the C-repeat, is bound by CBF1 protein. The C-repeat/DRE is a cis-acting DNA regulatory element that stimulates transcription in response to low temperature and water deficit. The product of the *CBF1* gene contains a nuclear localization signal (NLS), a DNA-binding domain that was first identified in *Apetala2* (AP2 domain), and an acidic region.

Table 22.3   Impact of oxygen deprivation on respiratory metabolism

Oxygen status	Effect on metabolism
Normoxic (aerobic)	Aerobic respiration proceeds normally. Almost all ATP production results from oxidative phosphorylation.
Hypoxic	The partial pressure of O_2 limits ATP production by oxidative phosphorylation. Glycolysis accounts for a larger percentage of ATP yield than under normoxic conditions. Metabolic and developmental changes are stimulated that result in adaptation to a low-oxygen environment.
Anoxic (anaerobic)	ATP is produced only by way of glycolysis. Cells exhibit low ATP contents, diminished protein synthesis, and impaired division and elongation. If anoxic conditions persist, many plant cells die.

Plant or cellular oxygen status can be defined as **normoxic, hypoxic,** or **anoxic** (Table 22.3). To survive short-term flooding, plants must generate sufficient ATP, regenerate $NADP^+$ and NAD^+, and avoid accumulation of toxic metabolites. Periods of oxygen deficit can trigger developmental responses that promote acclimation to hypoxic or anoxic conditions.

22.7.1 Plants vary in ability to tolerate flooding.

Plant species generally can be classified as wetland, flood-tolerant, or flood-sensitive, according to their ability to withstand periods of oxygen deficit (Table 22.4). Wetland plants possess diverse anatomical, morphological, and physiological features that permit survival in aquatic environments or waterlogged soils. Aquatic environments present challenges that are met with unusual adaptations (Fig. 22.17). Growth in a wetland environment promotes formation of a thickened root hypodermis to reduce O_2 loss to the anaerobic soil. To facilitate transport of O_2 from aerial structures to submerged roots and thereby maintain aerobic metabolism and growth, some plants develop specific structures: **aerenchyma** (continuous, columnar intracellular spaces formed in root corti-

Oxygen concentrations in well-drained, porous soil are nearly equal to atmospheric concentrations (20.6% oxygen, 20.6 kPa), but the diffusion coefficient of oxygen in water is four orders of magnitude lower than that in air. Irrigation and precipitation displace soil gases with water and thereby decrease the rate of gas exchange between soil and air. The supply of oxygen to root cells is influenced by several factors, including soil porosity, water content, temperature, root density, and the presence of competing algae and aerobic microorganisms. Oxygen concentrations in root tissues also vary according to root depth, root thickness, the volume of intercellular gaseous spaces, and cellular metabolic activity.

(A)

(B)

Table 22.4   Plant species categorized by sensitivity to oxygen deprivation

Wetland plants	Flood-tolerant plants	Flood-sensitive plants
Acorus calamus (sweet flag)	*Arabidopsis thaliana*	*Glycine max* (soybean)
Echinochloa crus-galli (rice grass)	*Ec. crus-pavonis* (barnyard grass)	*Lycopersicon esculentum* (tomato)
Ec. phyllopogon (barnyard grass)	*Hordeum vulgare* (oat)	*Pisum sativum* (pea)
Erythina caffra (coral tree)	*Solanum tuberosum* (potato)	
Rumex maritimus (golden dock)	*Triticum aestivum* (wheat)	
Oryza sativa (rice)	*Zea mays* (corn)	
Zizania aquatica (wild rice)		

Figure 22.17
Survival strategies of a freshwater weed. (A) *Hydrilla verticillata* (L.f.) Royle, a monocotyledous native of Asia, is an invasive submerged weed in freshwater ecosystems in the southern United States. The aggressive growth of *Hydrilla* results from numerous adaptations to submersion under conditions of minimal dissolved CO_2, high concentrations of dissolved O_2, and variable solar irradiance. The leaves lack stomata and are composed of two cell layers. *Hydrilla* performs C_3 photosynthetic gas exchange. Under low CO_2 conditions that induce photorespiration in C_3 plants, *Hydrilla* induces a C_4-like malate production and maintains photosynthesis, despite the lack of Kranz anatomy. C_4-like photosynthesis occurs when the external pH is high by the conversion of dissolved bicarbonate to CO_2 at the abaxial leaf surface. C_4-like photosynthesis also occurs when the external pH is not alkaline by the concentration of CO_2 in the chloroplast. This occurs by conversion of malate to pyruvate and CO_2 by a NADP-malic enzyme. The shift from C_3 to C_4-like metabolism induces an increase in cytosolic phospho*enol*pyruvate and plastidic malic enzyme that allows *Hydrilla* to reduce the loss of precious CO_2 via photorespiration and maintain photosynthesis under high internal O_2 concentrations. Oxygen produced by photosynthesis moves to the nonphotosynthetic roots by way of extensive intercellular air spaces. (B) *Hydrilla* develops as a dense mat of stems and leaves near the surface of the water. These plants now threaten waterways in Florida and other Southeastern states, where their growth is not limited by cold weather.

(A)

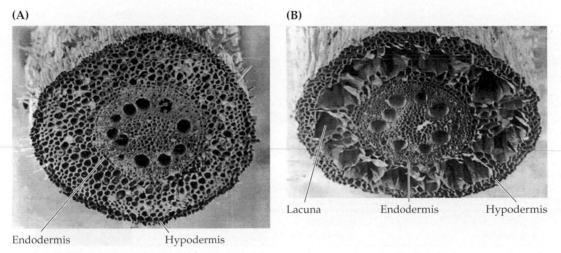

(B)

Endodermis   Hypodermis

Lacuna   Endodermis   Hypodermis

Figure 22.18
Aerenchyma development in the root cortex of maize (*Zea mays* L.) after oxygen deprivation. Photomicrographs of transverse sections of maize roots maintained under aerobic conditions (A) or under 72 hours of hypoxia (B) demonstrate the formation of cortical aerenchyma in hypoxic roots. The hypodermis and endodermis remain intact, and the lacunae created by the death of the central cortical cells form columnar gas-conducting chambers.

cal tissues [Fig. 22.18]; see also Section 22.7.4 and Chapter 20), adventitious roots from the hypocotyl or stem (Fig. 22.19), **lenticels** (openings in the periderm that allow gas exchange [Fig. 22.19]), shallow roots, and **pneumatophores** (shallow roots that grow with negative geotrophy out of the aquatic environment [Fig. 22.20]).

When inundated by water, wetland species alter cellular metabolism to increase survival. For example, rice responds to tran-

sient oxygen deficit with increased ethanolic fermentation. Aerenchyma formation in rice is constitutive and is further promoted by flooding. Submergence of deepwater rice seedlings promotes formation of adventitious roots and accelerates the internodal stem elongation, which enables stems and leaves to be established above the aquatic

Figure 22.19
Adventitious roots and prominent (hypertrophied) lenticels on the stem of young ash (*Fraxinus pennsylvanica* Marshall) after flooding. The black arrow indicates the water depth during flooding.

Figure 22.20
Pneumatophores of mangrove (*Avicennia nitida*) develop from roots submerged in estuarine mud.

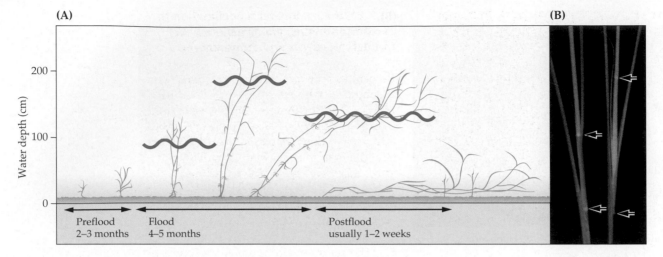

(A)

(B)

Figure 22.21
(A) Diagram illustrating growth responses of seedlings of deep-water rice (*Oryza sativa* L. var. Indica) to flooding. Seedlings are established before the annual flooding. Submergence promotes rapid internodal elongation and development of adventitious roots. Once the flood waters recede, the adventitious roots grow into the soil and aerial portions of the plant grow upward. (B) Photographs comparing internode elongation in aerobic (left) and submerged (right) plants. Arrows indicate positions of nodes.

environment (Fig. 22.21). Mitochondrial morphology is altered, but functional electron transport chain complexes of the inner membrane and enzymes of the citric acid cycle in the matrix are maintained (Fig. 22.22). The ability of the rice weed, *Echinochloa crusgalli*, to grow when submerged is attributed to active glycolysis and ATP production by way of a partially functioning citric acid cycle.

In contrast, other wetland plants, such as the marshland monocot *Acorus calamus*, respond to hypoxic conditions by down-regulating their metabolism, which allows them to maintain a nearly quiescent state for several months while utilizing the starch reserves stored in their rhizomes. Remarkably, the seedlings of many wetland species can withstand long-term anoxia, and the seeds of some, such as rice and *Ec. crus-galli*, even can germinate partially in an anaerobic environment, although their subsequent development requires oxygen.

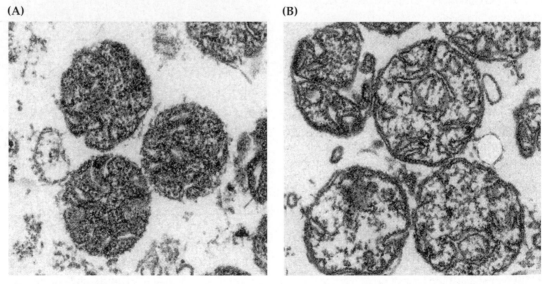

(A)

(B)

Figure 22.22
Photomicrographs comparing mitochondria from rice (*Oryza sativa* L.) seedlings germinated aerobically and exposed to aerobic (A) or anaerobic (B) treatments for 48 hours. Mitochondria from anoxic seedlings have well-developed cristae but a less dense matrix.

Figure 22.23
A flooded maize field. Flooding in the US Midwest in 1993 resulted in an estimated 33% reduction in yield compared with 1992.

Flood-tolerant plants can endure anoxia temporarily but not for prolonged periods. Like wetland species, these plants generate ATP through anaerobic metabolism during short-term flooding. In most cases, root elongation is inhibited, overall rates of protein synthesis diminish, and patterns of gene expression are markedly altered. Seedlings of flood-tolerant species (e.g., maize) can survive anoxia for three to five days, depending on genotype and developmental age (Fig. 22.23). Flood tolerance is increased if young plants become hypoxic before experiencing anoxia. Hypoxia promotes formation of adventitious and nodal roots with cortical aerenchyma as an acclimation to soils with low oxygen content (see Fig. 22.18).

Flood-sensitive plants exhibit an injury response to anoxia. As in flood-tolerant species, ethanol production is enhanced. However, these plants rapidly succumb to flooding because the cytoplasm of their cells acidifies. When deprived of oxygen, flood-sensitive species exhibit greatly diminished protein synthesis, mitochondrial degradation, inhibited cell division and elongation, disrupted ion transport, and cell death within root meristems. These plants do not develop root aerenchyma and typically cannot survive more than 24 hours of anoxia.

22.7.2 During short-term acclimation to anoxic conditions, plants generate ATP through glycolysis and fermentation.

Researchers have characterized extensively the rapid metabolic responses to flooding that occur in seedlings of three species: tomato (flood-sensitive), maize (flood-tolerant), and rice (wetland). In these plants, flooding stimulates an increase in glycolytic flux known as the **Pasteur effect.** Sucrose or glucose from the phloem is directed toward glycolysis in flooded organs. Flooded rice coleoptiles also hydrolyze stored starch (see Chapter 13) to obtain additional sugars.

The energy-yielding steps of glycolysis generate low amounts of ATP while reducing NAD^+ to NADH (see Chapter 13). To support ongoing glycolysis in the absence of mitochondrial respiration, the glycolytic substrate NAD^+ must be regenerated through fermentative reactions. The principal end products of glycolysis in oxygen-deprived plant tissues are lactate and ethanol (Fig. 22.24); alanine, succinate, and γ-aminobutyrate may also be formed (Fig. 22.25). The

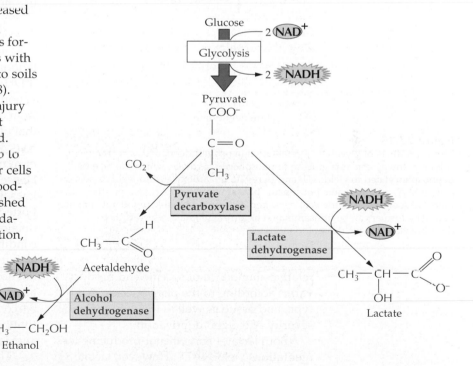

Figure 22.24
Major fermentation products of carbohydrate metabolism in flooded roots. The consumption of sugars and stored starch (in seeds) under oxygen deprivation has been studied in detail in several crop plants, including maize, rice, and barley. The two most important products of glycolytic fermentation are lactate, generated by the reduction of pyruvate, and ethanol, generated by the decarboxylation of pyruvate and reduction of the resulting acetaldehyde. Note that both the lactic and ethanolic fermentations regenerate NAD^+, which is required for glycolysis.

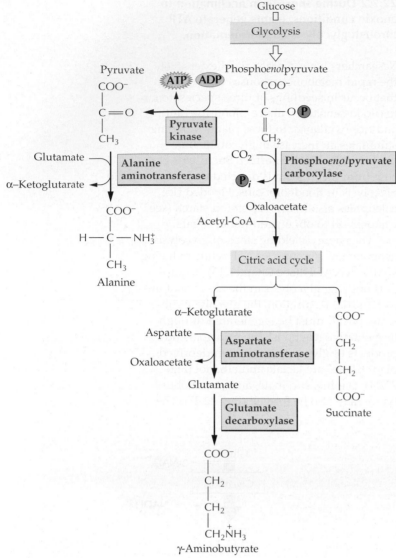

Figure 22.25
Minor products of glycolytic fermentation in oxygen-deprived seeds. The glycolytic pathway converts glucose to phospho*enol*pyruvate, which can be either dephosphorylated to yield ATP and pyruvate or carboxylated to yield oxaloacetate (OAA) and inorganic phosphate. Transamination of pyruvate produces alanine. OAA enters the citric acid cycle and is converted to α-ketoglutarate or succinate. The former can be transaminated to form glutamate, which is then decarboxylated to yield γ-aminobutyrate.

relative abundance of specific end products varies according to the plant species, genotype, and tissue as well as the duration and severity of oxygen deprivation.

Both lactate- and ethanol-producing fermentations yield NAD^+. However, lactate lowers cytosolic pH, whereas ethanol does not. According to the Davies–Roberts lactate dehydrogenase (LDH)/pyruvate decarboxylase (PDC) pH-stat hypothesis, anaerobic metabolism is regulated by the activities of pH-sensitive enzymes (Fig. 22.26). According to this model, the pyruvate produced initially by glycolysis is converted to lactate in a reaction catalyzed by LDH, an enzyme with an optimum at physiological pH. Lactate production reoxidizes NADH but also lowers the cytoplasmic pH. As the cytosol acidifies, LDH is progressively inhibited and a second enzyme, PDC, is activated for which the pH optimum is lower than the normal cytoplasmic pH; therefore, the accumulation of lactate ultimately stimulates the conversion of pyruvate to acetaldehyde. Alcohol dehydrogenase (ADH) subsequently reduces the acetaldehyde to ethanol while oxidizing NADH to NAD^+. Unlike lactate, ethanol is an uncharged molecule at cellular pH and can diffuse across plasma membranes (see below). As a result, the switch to ethanol production stabilizes cytoplasmic pH at a slightly acidic value. In some species, cytosolic pH may also be maintained by the conversion of malate (2 H^+ available) to lactate (1 H^+ available) or ethanol (no H^+ available) and the production of the zwitterionic species alanine and γ-aminobutyrate from the monoproton compound pyruvate.

A different hypothesis predicts that concentrations of pyruvate, rather than cytoplasmic pH, regulate ethanolic fermentation. The PDC/pyruvate dehydrogenase (PDH)-stat hypothesis is based on the observation that the K_m of cytosolic PDC ranges from 0.25 mM to 1.0 mM pyruvate, whereas that of mitochondrial PDH ranges from 50 μM to 75 μM. Cellular concentrations of pyruvate are usually less than 0.4 mM. Hence, an anoxic increase in pyruvate stimulates the activity of PDC. This model is supported by several observations. Ethanol production can be stimulated under aerobic conditions that inhibit mitochondrial PDH activity. In addition, lactate production does not accompany ethanolic fermentation in certain species. Finally, roots of transgenic tobacco plants that have twice the normal activity of PDC produce more ethanol under anoxia than do wild-type roots.

The LDH/PDC pH-stat and PDC/PDH-stat hypotheses are not mutually exclusive; the regulation of the switch to ethanolic fermentation may vary according to the plant species and the conditions of low-oxygen

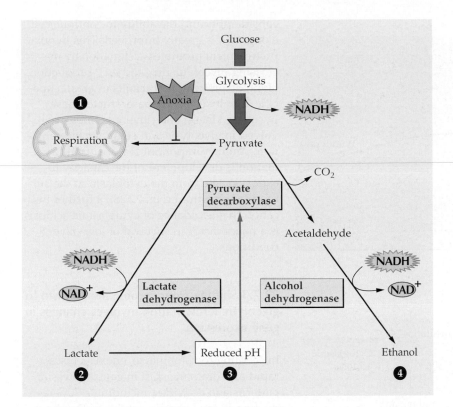

Figure 22.26
The Davies–Roberts LDH/PDC pH-stat hypothesis. The onset of oxygen deprivation results in inhibition of ATP production by way of the citric acid cycle and the
mitochondrial electron transfer chain (1). This stimulates a transient burst in the
conversion of pyruvate to lactate (2), resulting in a decrease in cytosolic pH (3).
Cytoplasmic pH also may be lowered as a result of proton leakage from the vacuole. At the lower cytosolic pH lactate dehydrogenase (LDH) is inhibited but
pyruvate decarboxylase (PDC) activity is stimulated, which promotes the production of ethanol (4), a neutral fermentation end product. The production of ethanol
and the elimination of lactate by way of an efflux mechanism prevent the detrimental acidification of the cytoplasm. An alternative proposal, the PDC/PDH
(pyruvate dehydrogenase) stat hypothesis (not shown), predicts that concentrations of pyruvate regulate ethanolic fermentation (see Section 22.7.2).

stress. Interestingly, the transgenic tobacco
plants with increased PDC activity also show
increased sensitivity to flooding, perhaps
because of loss of energy reserves resulting
from the faster consumption of the available
carbohydrates.

Plants that are flood-tolerant are able to
stimulate ethanolic fermentation and avoid
cytoplasmic acidosis during short-term inun-
dation. The role of ethanol production in
avoidance of cytoplasmic acidosis has been
demonstrated for maize, a flood-tolerant
species. Maize mutants deficient in ADH1,
an enzyme not required for growth under
aerobic conditions, are more sensitive to
flooding than are isogenic wild-type geno-

types. The mutant exhibits acetaldehyde accumulation and a more rapid and sustained
drop in cytoplasmic pH than does the wild
type (Fig. 22.27). However, not all plants that
produce ethanol can endure even short-term
oxygen deprivation. Flood-sensitive species
such as pea rapidly and dramatically can up-
regulate ADH activity but still succumb to
flooding as a result of cytoplasmic acidosis,
the results of lactate accumulation and failure
to sequester excess protons in the vacuole.

In nature, hypoxia frequently precedes
anoxia in flooded roots. If maize or rice seedlings are transferred to hypoxic conditions
(3 kPa oxygen) before transfer to anoxia
(0 kPa oxygen), their survival rates increase

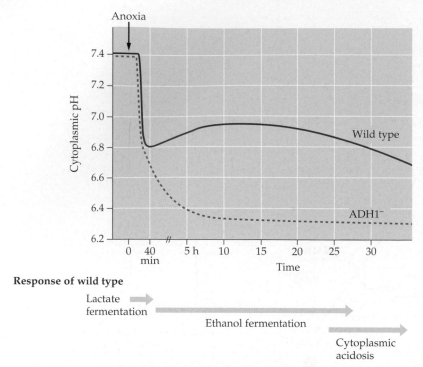

Figure 22.27
Response of root tips of wild-type maize (solid red) and mutants deficient in ADH1 (dotted blue) to anoxia. Wild-type root tips experience a rapid decrease in cytoplasmic pH, followed by a partial recovery; however, the wild-type root cells ultimately succumb to cytoplasmic acidosis. Root tips of ADH1-deficient seedlings show rapid cytoplasmic acidosis and cell death.

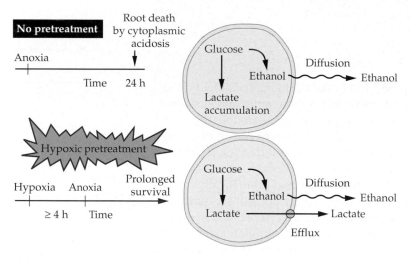

Figure 22.28
Effects of hypoxic pretreatment and acclimation on survival of anoxia: avoidance of cytoplasmic acidosis by lactate efflux. Flooding typically results in hypoxia, followed by anoxia. Lactate and ethanolic fermentation is stimulated in both anoxic and hypoxic cells. Exposure of maize seedlings to hypoxia for several hours before being transferred to anoxia increases the capacity of roots for lactate efflux and prolongs survival.

considerably and the ability to continue cell elongation is greatly improved. This hypoxic pretreatment promotes acclimation by increasing glycolytic flux and ATP production. Hypoxic pretreatment results in greater specific activity of hexokinase, fructokinase, pyruvate kinase, and other enzymes involved in glycolysis and ethanolic fermentation. Another important feature of this acclimation is development of the capacity to export lactate from the cytoplasm to the surrounding medium (Fig. 22.28), a further indication that avoidance of cytoplasmic acidosis is a major factor in survival of low-oxygen conditions.

22.7.3 Shifting from aerobic metabolism to glycolytic fermentation involves changes in gene expression.

In seedling roots of maize, anoxia causes a rapid and dramatic shift in gene expression patterns and activities, including a 70% reduction in total protein synthesis. Protein synthesis is also reduced and altered under hypoxia but to a lesser extent. Modifications in the pattern of protein synthesis observed in anoxic tissues result from transcriptional and posttranscriptional regulation of gene expression. Although most gene expression is repressed in response to oxygen deprivation, an important subset of genes is upregulated. Most of the known proteins synthesized in high amounts in anoxic roots are enzymes involved in sucrose and starch degradation, glycolysis, and ethanol fermentation (Figs. 22.29 and 22.30). Several genes that encode isoforms (e.g., ADH and glyceraldehyde-3-phosphate dehydrogenase) of the same enzyme are expressed differentially in aerobic and anoxic cells because of sequence differences in their promoters.

The *Adh* gene promoters of maize and *Arabidopsis* have been characterized (Fig. 22.31). In both species these promoters possess functional cis-acting elements that are required for expression in hypoxic and anoxic cells. These elements include the anaerobic response element (ARE), found in the promoter region of many genes that are transcriptionally induced by low oxygen in monocots and dicots, and G-box type motifs. Two transcription factors have been identified that interact with the G-box of

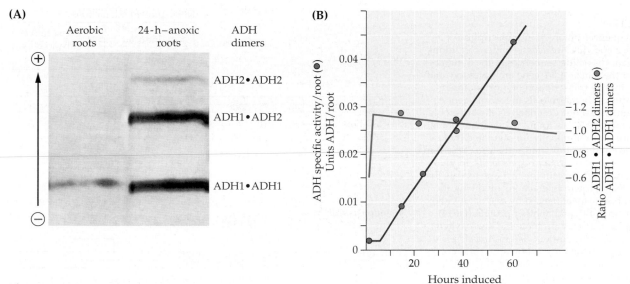

Figure 22.29
Differential expression of dimeric ADH isoenzymes in response to anoxia. Starch gel electrophoresis and activity staining demonstrate that ADH1 is present at low concentrations in aerobic roots of maize. (A) Roots of seedlings submerged for 24 hours contain increased concentrations of ADH1 and ADH2. (B). The specific activity of ADH increases dramatically after anoxic induction (red circles). Quantifying the ratio of ADH1•ADH2 heterodimers to ADH1•ADH1 homodimers demonstrates that the increase in ADH2 is more transient than that of ADH1 (blue circles).

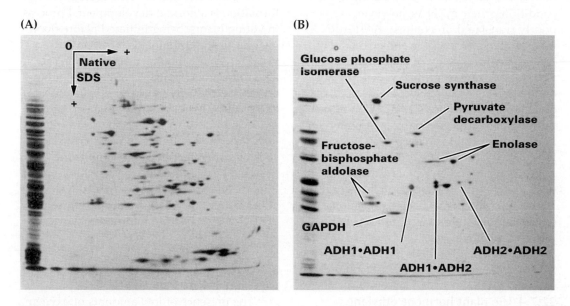

Figure 22.30
Anoxia markedly alters the pattern of protein synthesis in maize root tips, as shown in autoradiographs of native–sodium dodecyl (SDS) sulfate two-dimensional polyacrylamide gels of soluble root proteins. Aerobic roots were labeled with ^{3}H-leucine for 5 hours (A). Anoxic roots were labeled with ^{3}H-leucine during the last 5 hours of a 17-hour stress treatment. The anaerobic polypeptides that have been identified are labeled. ADH, alcohol dehydrogenase; GAPDH, cytosolic glyceraldehyde-3-phosphate dehydrogenase.

Arabidopsis Adh1. Dimethyl sulfate footprinting and DNase I hypersensitivity analyses of *Adh* promoters have demonstrated both constitutive and dynamically modified DNA–protein interactions that correlate with transcriptional activity and have provided information on the higher-order structure of the chromatin of *Adh* genes.

Changes in gene expression cannot be attributed solely to transcriptional controls.

Figure 22.31

Cis-acting sequences involved in transcription of *Adh1* in maize and *Adh Arabidopsis* under conditions of low oxygen. Cis-acting sequences in the 5' flanking region of the maize *Adh1* gene are necessary and sufficient to induce transcription of a promoter–reporter gene fusion in maize cells in low oxygen. A promoter sequence motif described as the anaerobic response element (ARE) is found in some but not all genes that are transcriptionally up-regulated by low oxygen. G-box and ½ G-box motifs are also present in *Adh* promoters from maize and *Arabidopsis*. In vivo footprinting of DNA–protein interactions reveals that only some DNA–protein interactions involving the AREs and G-boxes are enhanced by anoxia. A G-box–binding protein (GBP) and another protein (GF14) form a complex that may promote the low-oxygen induction of *Adh* transcription. *Adh* in *Arabidopsis* also is transcriptionally induced by application of ABA, dehydration, and cold stress. Distinct DNA–protein interactions involving specific promoter elements probably modulate responses to these stimuli.

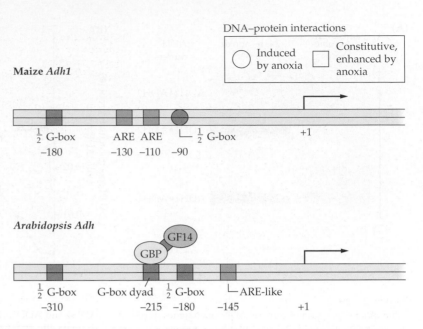

Many of the genes expressed in aerobic roots are transcribed during anoxia in approximately the same amounts as during aerobic conditions; their mRNAs, however, are very poorly translated. In contrast, *Adh1* mRNA is translated efficiently in anoxic cells (Fig. 22.32). The translation of maize *Adh1* mRNA under low-oxygen conditions is dependent on the presence of specific sequences in the 5' region and 3' untranslated region of the mRNA. The reduced synthesis of many normal cellular proteins reflects the failure of their mRNAs to initiate translation. Changes in the translational machinery, including phosphorylation of translation factors and ribosomal proteins, may allow for more-efficient translation of a subset of cellular mRNAs in anoxic cells.

22.7.4 The plant hormone ethylene promotes long-term acclimative responses, including formation of aerenchyma and stem elongation, in wetland and flood-tolerant species.

Aerenchyma (see Fig. 22.18) provide a conduit for gas diffusion between roots and aerial organs; they can form by cell death and dissolution (**lysigeny**), by separation of cells without collapse (**schizogeny**), or by a combination of lysigeny and schizogeny (**schizolysigeny**). The aerenchyma of rice and maize are lysigenous and most likely result from programmed cell death (see Chapter 20).

In many wetland species, aerenchyma formation is a normal developmental process in young plants, being initiated before flooding and further promoted in response to flooding. Plants that have aerenchyma are able to maintain synthesis of high amounts of ATP in root cells submerged in an anoxic environment. The formation of aerenchyma reduces the severity of hypoxia and facilitates growth of submerged roots in species such as rice. When flooding inhibits the growth of existing roots or causes their death in nonwetland species, new roots develop at the upper portion of the primary root, on the hypocotyl, or from internodes above the air–water/air–soil interface. These new roots form aerenchyma that facilitate the transport of oxygen from aerial portions of the plant to oxygen-deprived root tissues.

The presence of low amounts of oxygen (less than 12.5 kPa to 3 kPa) stimulates production of the hormone ethylene, which promotes the formation of aerenchyma in the central portion of the root cortex. Anoxic roots develop fewer aerenchyma than hypoxic roots because oxygen is required for ethylene synthesis (Fig. 22.33A; see also Chapter 17). The abundance of 1-aminocyclopropane-1-carboxylic acid (ACC) synthase and ACC oxidase, enzymes in the ethylene biosynthesis pathway, increases considerably in response to hypoxia in maize root tips. The

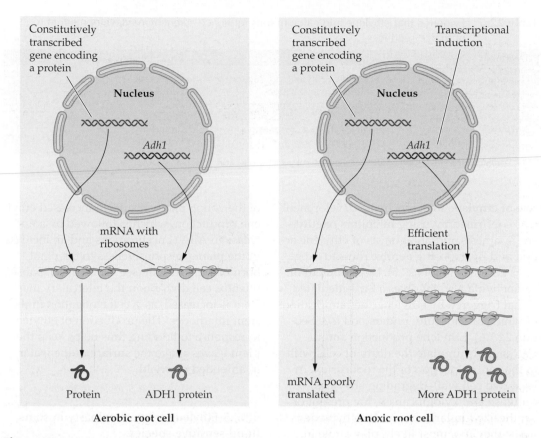

Figure 22.32
Transcriptional and translational mechanisms control gene expression in anoxic root cells of maize. Nuclear run-on experiments have shown that genes encoding many normal cellular proteins are constitutively but poorly transcribed under anoxia. These mRNAs associate with ribosomes inefficiently in anoxic cells. *Adh1* mRNA increases in response to transcriptional induction and competes efficiently for the translation machinery in anoxic cells.

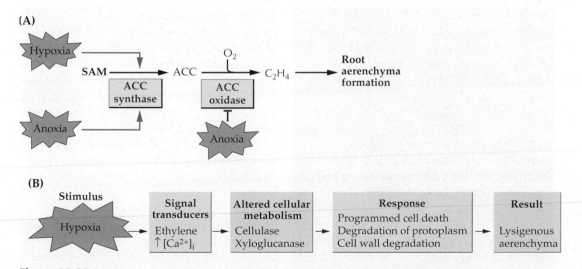

Figure 22.33
(A) The formation of aerenchyma is stimulated by hypoxia but not by anoxia. The conversion of ACC to ethylene requires oxygen. Ethylene production leads to lysigenous aerenchyma in the root cortex. (B) Transduction of the signal for low oxygen in maize leads to programmed cell death during the formation of lysigenous aerenchyma (see Chapter 20). Chemicals that increase the cytosolic Ca^{2+} concentration under aerobic conditions promote aerenchyma formation, whereas compounds that block Ca^{2+} movement under anoxia inhibit aerenchyma formation.

Table 22.5 Evidence that ethylene and calcium ions influence aerenchyma development in roots of maize seedlings

Treatment	Action	Effect on aerenchyma development
Hypoxia	Stimulates ethylene production	Promoted
Hypoxia + Ag$^+$	Inhibits ethylene action	Reduced
Hypoxia + AVG	Inhibits ethylene synthesis	Reduced
Hypoxia + EGTA	Chelates Ca^{2+}	Reduced
Normoxia + ethylene	Induces ethylene response	Promoted
Normoxia + caffeine	Increases cytosolic Ca^{2+}	Promoted

AVG, 1-aminoethoxyvinylglycine; EGTA, ethylene glycol-bis tetraacetic acid

role of ethylene in aerenchyma development was confirmed by using inhibitors of ethylene synthesis and antagonists of ethylene action and by exposing aerobic roots to exogenous ethylene (Table 22.5). Development of aerenchyma also involves a Ca^{2+}-mediated signal triggered by ethylene, the ramifications of which are not fully understood (see Section 22.7.6). Ethylene production and the Ca^{2+} signal stimulate the death of cells within the central portion of the root cortex. At least two cell wall–degrading enzymes, cellulase and xyloglucanase, are present or synthesized in large amounts in hypoxic roots; they too, most likely, play a role in aerenchyma formation (Fig. 22.33B).

Ethylene also is involved in the flooding response of deepwater rice varieties that can grow in water as deep as 4 m. Submergence

of the young plants results in increased ethylene production, which is followed by a decrease in ABA concentrations and an increase in the plants' responsiveness to the plant hormone gibberellin (see Chapter 17). Subsequently, cell division in the intercalary meristem is increased, as is cell elongation in the stem internodes. The production of ethylene in response to flooding thus helps keep the plant leaves above the surface of the water in a flooded rice field.

22.7.5 Ethylene triggers epinasty in some flood-sensitive species.

Ethylene is responsible for transient changes in the morphology of aerial tissues in flood-sensitive plants. Leaf epinasty (Fig. 22.34A),

Figure 22.34
(A) Leaf epinasty (curvature) in tomato. (B) Ethylene production in roots by way of the ACC pathway results in aerenchyma formation if oxygen is available. In the absence of oxygen, the ACC is transported to aerial tissues, where ethylene is formed, resulting in leaf epinasty.

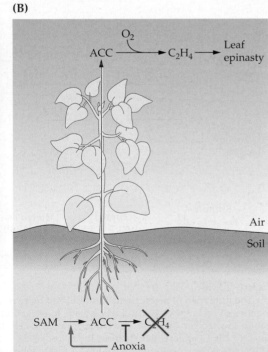

curvature caused by cell expansion of the adaxial cells of the petiole, is a common response to waterlogging associated with flood-sensitive plants, such as tomato. Epinasty reduces the foliar absorption of light, thereby slowing transpirational water loss in plants for which water absorption by roots is limited by anoxia. In tomato, severe flooding stimulates transcription of the ACC synthase gene, which results in increased ACC synthesis in the root. In the absence of oxygen, this metabolite cannot be converted to ethylene by ACC oxidase. Some of the ACC is therefore transported to leaves, where ACC oxidase concentrations subsequently increase and ethylene is produced (Fig. 22.34B). This ethylene synthesis results in leaf epinasty. Leaves of tomato plants transformed with an antisense gene complementary to ACC oxidase mRNA demonstrated less foliar ACC oxidase activity, less ethylene production, and less epinasty in response to flooding. These data raise interesting questions regarding the signals transmitted between root and leaf in response to flooding.

22.7.6 How do plants sense oxygen deprivation?

Plant responses to flooding include transient alterations in gene expression and metabolism, as well as long-term developmental responses. How are these responses triggered by low concentrations of available oxygen? The lack of oxygen rapidly results in less ATP and increased NADH as well as decreased cytosolic pH. Any or all of these factors could participate in signal transduction processes. Plant hormones such as ethylene and ABA also may be involved in transducing the low-oxygen signal. In addition, plants possess hemoglobin-like proteins (e.g., leghemoglobin; see Chapter 16), but whether these are involved in sensing oxygen deprivation remains unclear. Constitutive expression of an antisense barley hemoglobin gene in cultured maize cells resulted in decreased hemoglobin content and decreased ability to maintain ATP concentrations, suggesting that hemoglobin may play some role in acclimation to low-oxygen conditions.

Evidence is increasing that Ca^{2+} may be an important second messenger in transducing the low-oxygen signal, altering gene expression, and promoting aerenchyma formation. Anoxia stimulates a rapid increase in cytosolic Ca^{2+} in maize protoplasts. This flux of Ca^{2+}, coming at least in part from mitochondria, appears to be necessary for the increase in *Adh1* transcripts. Use of aequorin, a protein from jellyfish, for fluorescence reporting of cytosolic Ca^{2+} concentrations has provided evidence that a biphasic flux of Ca^{2+} in the response to anoxia in *Arabidopsis* shoots and cotyledons but not in roots. Ca^{2+} has been implicated as a second messenger in the response to heat and cold stress and many other stimuli in plants, and its role in signal transduction is currently a focus of research (see Chapter 18).

22.8 Oxidative stress

Oxidative stress results from conditions promoting the formation of active oxygen species that damage or kill cells. Environmental factors that cause oxidative stress (Fig. 22.35) include air pollution (increased amounts of ozone or sulfur dioxide), oxidant-forming herbicides such as paraquat dichloride (methyl viologen, 1,1′-dimethyl-4,4′-bipyridinium), heavy metals, drought, heat and cold stress, wounding, UV light, and highly intense light conditions that stimulate **photoinhibition** (see Chapters 9 and 12). Oxidative stress also occurs in response to pathogen infection (see Chapter 21) and during senescence (see Chapter 20).

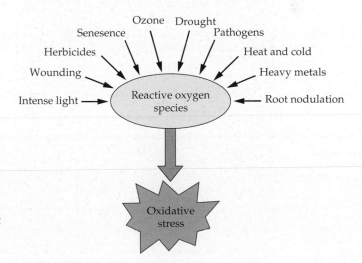

Figure 22.35
Environmental factors that increase the concentrations of reactive oxygen species in plant cells.

Reactive oxygen species (ROS) (Fig. 22.36) are formed during certain redox reactions and during incomplete reduction of oxygen or oxidation of water by the mitochondrial or chloroplast electron transfer chains. Formation of singlet oxygen ($^1O_2$) subsequently stimulates production of other ROS such as hydrogen peroxide (H_2O_2), superoxide anion ($O_2^{\bullet-}$), and hydroxyl ($HO^\bullet$) and perhydroxyl ($O_2H^\bullet$) radicals. Superoxide anions also are produced in the chloroplast when electrons are transferred directly from Photosystem I (PSI) to oxygen. These reactive molecules, especially $HO^\bullet$, are highly destructive to lipids, nucleic acids, and proteins. Nevertheless, reactive oxygen species such as $O_2^{\bullet-}$ and H_2O_2 are required for **lignification** (see Chapters 2, 20, and 24) and function as signals in the defense response to pathogen infection (see Chapters 20 and 21). Plants scavenge and dispose of these reactive molecules by use of antioxidant defense systems present in several subcellular compartments. When these defenses fail to halt the self-propagating autooxidation reactions associated with ROS, cell death ultimately results.

The antioxidant defense systems include nonenzymatic and enzymatic antioxidants (Fig. 22.37). These compounds and enzymes are not distributed uniformly, so defense systems vary among specific subcellular compartments (Tables 22.6 and 22.7). The

Compound	Shorthand notation(s)	Structural representation(s)	Sources
Molecular oxygen (triplet ground state)	O_2; $^3\Sigma$	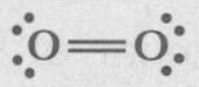 $\ddot{O}=\ddot{O}$ $1s^22s^2(\sigma_s)^2(\sigma_s*)^2(\sigma_x)^2(\pi_y)^2(\pi_z)^2(\pi_y*)^1(\pi_z*)^1$	Most common form of dioxygen gas
Singlet oxygen (first excited singlet state)	$^1O_2$; $^1\Delta$	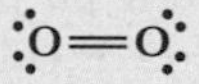 $\ddot{O}=\ddot{O}$ $1s^22s^2(\sigma_s)^2(\sigma_s*)^2(\sigma_x)^2(\pi_y)^2(\pi_z)^2(\pi_y*)^2$	UV irradiation, photoinhibition, photosystem II e$^-$ transfer reactions (chloroplasts)
Superoxide anion	$O_2^{\bullet-}$	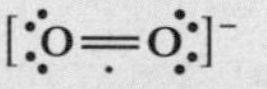 $[\ddot{O}=\ddot{O}]^-$	Mitochondrial e$^-$ transfer reactions, Mehler reaction in chloroplasts (reduction of O_2 by iron–sulfur center F_X of Photosystem I), glyoxysomal photorespiration, peroxisome activity, plasma membrane, oxidation of paraquat, nitrogen fixation, defense against pathogens, reaction of O_3 and OH^- in apoplastic space
Hydrogen peroxide	H_2O_2	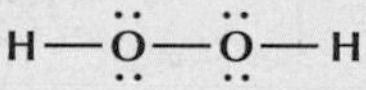 $H-\ddot{O}-\ddot{O}-H$	Photorespiration, β-oxidation, proton-induced decomposition of $O_2^{\bullet-}$, defense against pathogens
Hydroxyl radical	$OH^\bullet$	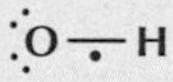 $\ddot{O}{-\hspace{-0.3em}\bullet}\, H$	Decomposition of O_3 in presence of protons in apoplastic space, defense against pathogens
Perhydroxyl radical	$O_2H^\bullet$	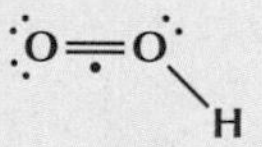 $\ddot{O}=\dot{O}\diagdown_H$	Reaction of O_3 and OH^- in apoplastic space
Ozone	O_3	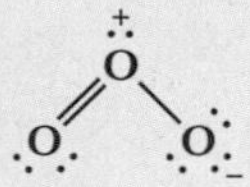 $\overset{+}{O}$ with $\ddot{O}$ and $\ddot{O}^-$	Electrical discharge or UV radiation in stratosphere, reactions involving combustion products of fossil fuels and UV radiation in troposphere

Figure 22.36

Molecular structure of reactive oxygen species active in plants: singlet oxygen, hydrogen peroxide, superoxide anion, hydoxyl radical, and perhydroxyl radical.

major antioxidant species in plants are ascorbate (vitamin C), reduced glutathione (GSH), α-tocopherol (vitamin E), and carotenoids; polyamines and flavonoids also may provide some protection from free radical injury. The ascorbate–glutathione cycle is the major antioxidant pathway in plastids, where ROS are generated during normal biochemical processes that include photosynthetic transfer of electrons. The photosynthetic apparatus receives additional protection from oxidative damage by the exothermic production of the xanthophyll zeaxanthin (see Chapter 12). ROS are produced in root nodules of nitrogen-fixing plants and are scavenged by enzymatic antioxidants. Regulation of the concentrations of antioxidants and antioxidant enzymes constitutes an important mechanism for avoiding oxidative stress.

22.8.1 Tropospheric ozone is linked to oxidative stress in plants.

One of the best characterized causes of oxidative stress is exposure to high concentrations of ozone. Anthropogenic hydrocarbons and oxides of nitrogen (NO, NO_2) and sulfur (SO_x) react with solar UV radiation to generate ozone (O_3). Stratospheric ozone is beneficial because it shields the earth from UV irradiation, but tropospheric ozone is harmful to

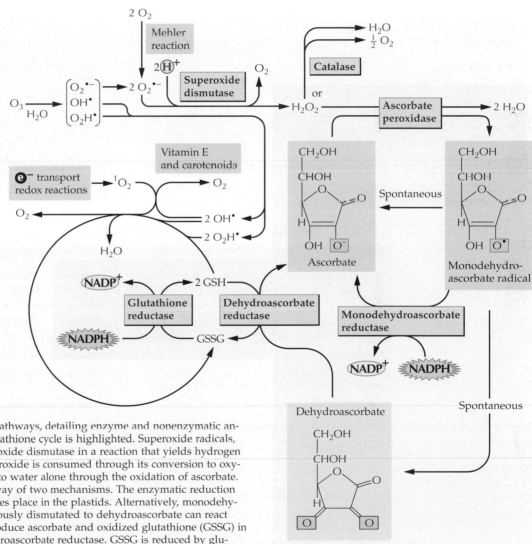

Figure 22.37
Antioxidant defense system pathways, detailing enzyme and nonenzymatic antioxidants. The ascorbate–glutathione cycle is highlighted. Superoxide radicals, $O_2^{\bullet-}$, are eliminated by superoxide dismutase in a reaction that yields hydrogen peroxide, H_2O_2. Hydrogen peroxide is consumed through its conversion to oxygen and water by catalase or to water alone through the oxidation of ascorbate. Ascorbate is regenerated by way of two mechanisms. The enzymatic reduction of monodehydroascorbate takes place in the plastids. Alternatively, monodehydroascorbate that is spontaneously dismutated to dehydroascorbate can react with glutathione (GSH) to produce ascorbate and oxidized glutathione (GSSG) in a reaction catalyzed by dehydroascorbate reductase. GSSG is reduced by glutathione reductase, requiring the consumption of NADPH. Singlet oxygen and hydroxyl ions are eliminated in the glutathione pathway. Damage by singlet oxygen and hydroxyl ions is also diminished by the nonenzymatic antioxidants, vitamin E and carotenoids.

Table 22.6 Subcellular locations of antioxidants

Antioxidant	Structure	Subcellular location
Ascorbate (vitamin C)	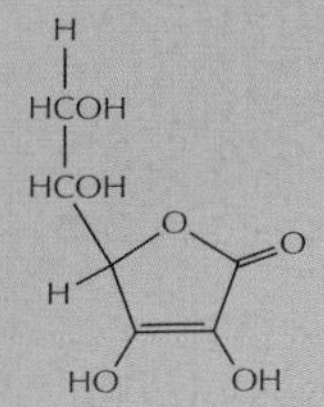	Apoplast, cytosol, plastid, vacuole
β-Carotene	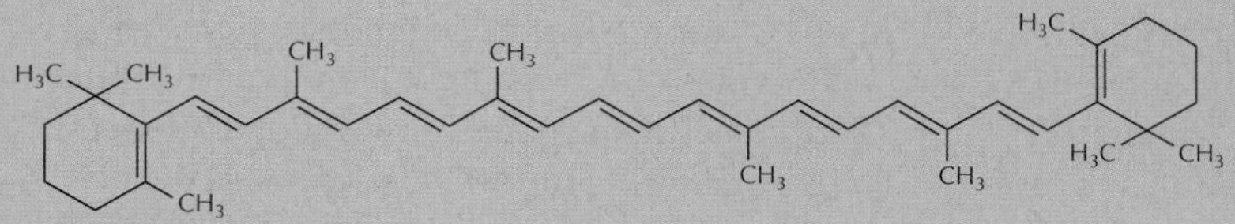	Plastid
Glutathione, reduced (GSH)	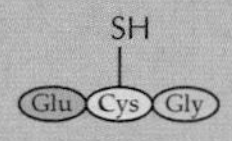	Cytosol, mitochondrion, plastid
Polyamines (e.g., putrescene, shown here)	$H_2N(CH_2)_4NH_2$	Cytosol, mitochondrion, nucleus, plastid
α-Tocopherol (vitamin E)	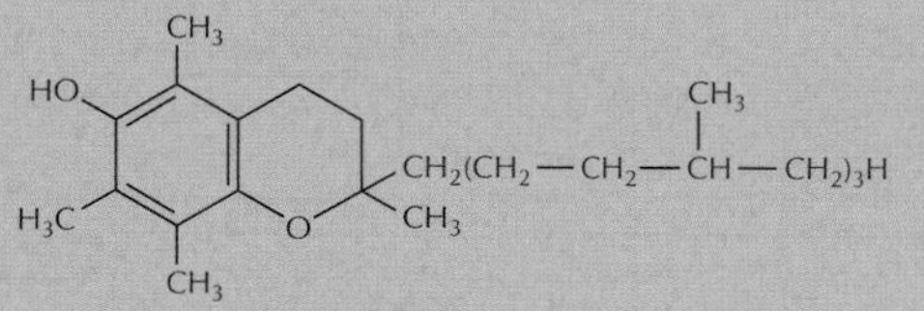	Cell membranes (including membranes of plastid)
Zeaxanthin	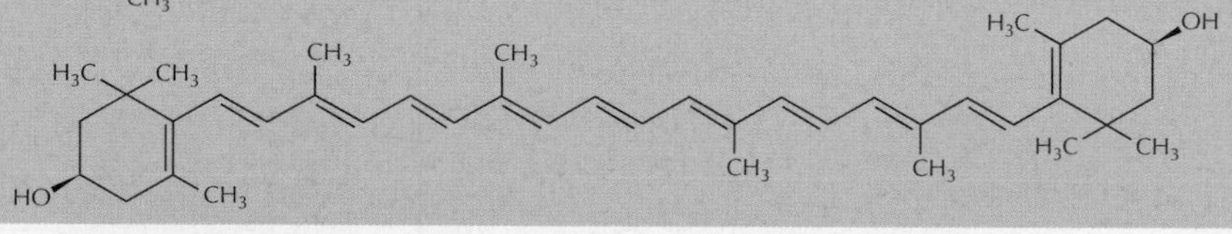	Chloroplast

Table 22.7 Subcellular locations of antioxidant enzymes

Antioxidant enzyme	Abbreviation	Subcellular location
Ascorbate peroxidase	APX	Cytosol, plastid stroma, plastid membrane, root nodules
Catalase	CAT	Cytosol, glyoxysome, peroxisome
Dehydroascorbate reductase	DHAR	Cytosol, plastid stroma, root nodules
Glutathione reductase	GR	Cytosol, mitochondrion, plastid stroma, root nodules
Monodehydroascorbate reductase	MDHAR	Plastid stroma, root nodules
Superoxide dismutase (grouped by metal cofactor)	Cu/ZnSOD	Cytosol, peroxisome, plastid, root nodules
	MnSOD	Mitochondrion
	FeSOD	Plastid

life because it is a highly reactive oxidant (Fig. 22.38).

The negative effects of ozone on plants include decreased rates of photosynthesis, leaf injury (Fig. 22.39), reduced growth of shoots and roots, accelerated senescence, and reduced crop yield. Despite continued efforts to reduce ozone production, this pollutant continues to be a major cause of crop loss. Concentrations of tropospheric ozone range from 0.02 µl l^{-1} to 0.05 µl l^{-1} in pristine locations to as much as 0.40 µl l^{-1} in polluted urban environments. Plants vary greatly in their ability to survive in high-ozone

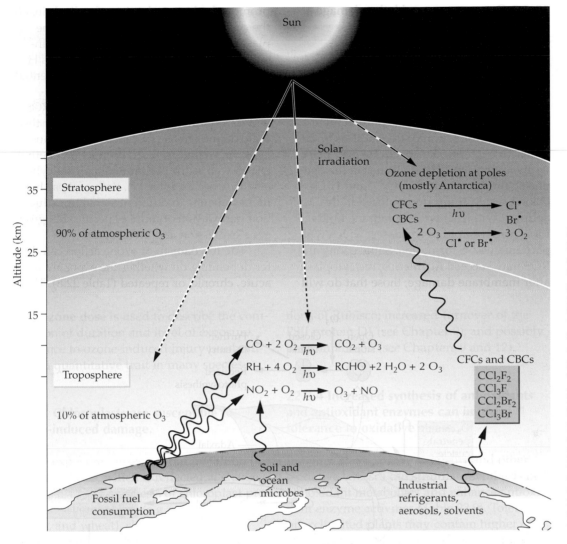

Solar
irradiation

Ozone depletion at poles
(mostly Antarctica)

Stratosphere

35

25

Altitude (km)

90% of atmospheric O_3

CFCs $\longrightarrow$ $Cl^{\bullet}$
CBCs $\xrightarrow{h\upsilon}$ $Br^{\bullet}$
$2 O_3 \xrightarrow[Cl^{\bullet} \text{ or } Br^{\bullet}]{} 3 O_2$

15

$CO + 2 O_2 \xrightarrow{h\upsilon} CO_2 + O_3$

$RH + 4 O_2 \xrightarrow{h\upsilon} RCHO + 2 H_2O + 2 O_3$

$NO_2 + O_2 \xrightarrow{h\upsilon} O_3 + NO$

Troposphere

10% of atmospheric O_3

CFCs and CBCs

CCl_2F_2
CCl_3F
CCl_2Br_2
CCl_3Br

Fossil fuel
consumption

Soil and
ocean
microbes

Industrial
refrigerants,
aerosols, solvents

Figure 22.38
Diagram illustrating the stratospheric consumption and atmospheric generation of O_3. Ninety percent of atmospheric O_3 is in the stratosphere, 10% in the troposphere. Stratospheric ozone beneficially decreases the penetration of solar radiation to the troposphere. Anthropomorphic release of chlorofluorocarbons (CFCs) leads to depletion of stratospheric ozone, especially over polar regions. The burning of fossil fuels increases the production of carbon compounds that react in sunlight with oxygen to form ozone in the troposphere.

environments. Their resistance to ozone utilizes either avoidance or tolerance mechanisms. Avoidance involves physically excluding the pollutant by closing the stomata, the principal sites at which ozone enters a plant. Tolerance can result from biochemical responses that induce or activate the antioxidant defense system and possibly also various repair mechanisms. Resistance also is influenced by environmental factors that can cause oxidative stress (e.g., water status, temperature, and light intensity) as well as by the developmental age of the plant or plant organ. Ozone damage is ameliorated in plants grown in high concentrations of CO_2, probably as a result of enhanced antioxidant defense mechanisms. Although ozone-induced injury can increase the susceptibility of plants to pathogens, ozone exposure para-

doxically can induce pathogen resistance by up-regulating the antioxidant enzymes that are induced by the hypersensitive response (HR; see Chapters 20 and 21) and systemic acquired resistance (SAR; see Chapter 21).

Figure 22.39
Ozone-damaged oat (*Avena sativa* L.) leaves. Chlorotic lesions develop in the middle of the leaves. Leaf tips (oldest leaf cells) and leaf bases (youngest leaf cells) show the least damage.

Table 22.9 Stress conditions that stimulate increased levels or activities of antioxidants and antioxidant enzymes

Antioxidant or antioxidant enzyme[a]	Stress conditions
Anionic peroxidases	Chilling, high CO_2
Ascorbate peroxidase	Drought, high CO_2, high light intensity, ozone, paraquat
Catalase	Chilling
Glutathione	Chilling, drought, γ-irradiation, heat stress, high CO_2, ozone, SO_2
Glutathione reductase	Chilling, drought, high CO_2, ozone, paraquat
Polyamines	Deficiency of K, P, Ca, Mg, Fe, Mn, S, or B; drought; heat; ozone
Superoxide dismutase	Chilling, high CO_2, high light, increased O_2, ozone, paraquat, SO_2

[a]Changes in the amounts of these compounds and enzyme activities are dependent on the developmental stage of the plant and the experimental parameters.

Table 22.10 Phenotypes associated with overexpression of specific antioxidant enzymes and antioxidant biosynthesis enzymes

Protein overexpressed (source of transgene)[a]	Location of protein (transformed species)	Phenotype, compared with that of untransformed control[a]
Glutathione reductase, GR (*Escherichia coli*)	Chloroplasts (tobacco)	3-fold increase in foliar GR concentrations Increased tolerance to paraquat and SO_2 No effect on ozone tolerance
	Cytosol (tobacco)	1- to 3.5-fold increase in GR Increased tolerance to paraquat No effect on ozone tolerance
	Chloroplasts (poplar)	1000-fold increase in GR 2-fold increase in glutathione Increased ascorbate in stressed leaves Increased tolerance to chilling-induced photoinhibition No effect on paraquat tolerance
	Cytosol (poplar)	5- to 10-fold increase in GR Glutathione not increased Tolerance to chilling-induced photoinhibition not increased No effect on paraquat tolerance
Glutathione synthetase, GS (*E. coli*)	Cytosol (poplar)	100-fold increase in GS Glutathione not increased Tolerance to chilling-induced photoinhibition not increased No effect on paraquat tolerance
Chloroplast Cu/ZnSOD (pea)	Chloroplasts (tobacco)	3- to 15-fold increase in SOD Concomitant increase in APX, GR, DHAR Increased tolerance to high light, chilling, paraquat, photoinhibition No increase in tolerance to acute ozone
Cytosolic Cu/ZnSOD (pea)	Cytosol (tobacco)	1.5- to 6-fold increase in SOD Reduced injury from acute ozone exposure
Mitochondrial MnSOD (tobacco)	Mitochondria (tobacco)	8-fold increase in SOD Little to no effect on ozone tolerance
Mitochondrial MnSOD (tobacco)	Chloroplasts (tobacco)	2- to 4-fold increase in SOD Increased tolerance to chronic ozone exposure Decreased sensitivity to paraquat Increased amounts of antioxidants and antioxidant enzymes
Mitochondrial MnSOD (tobacco)	Chloroplasts (alfalfa)	2-fold increase in SOD Increased tolerance to freezing stress Increased field tolerance to drought

[a]See Table 22.7 for enzyme abbreviations.

oxidant-scavenging mechanisms are limiting or if overexpression results in increased production or reduced scavenging of ROS. In certain cases, however, the overexpression of one or more antioxidant enzymes may provide protection from oxidative stress.

Oxidative stress tolerance probably depends on the successful induction of functional detoxification systems in the specific subcellular compartments that are exposed to high concentrations of free radicals. Some evidence indicates that overexpression of antioxidant enzymes can result in tolerance to multiple causes of oxidative stress, including pathogens, paraquat, and osmotic stresses (e.g., chilling, salinity, and drought). Genetic engineering of antioxidant defense systems will most probably be successful in improving crop tolerance to adverse growth conditions.

22.8.5 Oxidative stress or ozone can interact with plant hormones such as SA and ethylene to produce plant responses.

The initial signal in the response to ROS is probably the reactive molecules themselves. An increase in cytosolic concentrations of Ca^{2+} can serve as a second messenger, with plant hormones modulating the response (see Chapter 18). Ozone exposure results in increased amounts of H_2O_2, which stimulate the production of SA. Interestingly, this results in a transient increase in the number of transcripts that encode defense-related secondary metabolites (e.g., phytoalexins; see Chapter 21), cellular barrier molecules (e.g., lignins, callose, and extensins; see Chapters 2 and 24), and PR proteins [e.g., $(1\rightarrow3)\beta$-glucanase, chitinase, glutathione S-transferase, and phenylalanine ammonia lyase; see Chapter 21]. In this way, ozone can increase the resistance of plants to pathogens by overlap with the HR and SAR. Ozone fumigation can also result in ethylene production by inducing increases in ACC synthase and ACC oxidase gene transcription. Overproduction of ethylene, moreover, may be responsible for localized cell death in ozone-treated plants. It is not clear which signal stimulates the increases in antioxidant enzymes that are the key defense to oxidative stress. Both SA and methyl jasmonate (see Chapter 17) have been implicated.

22.9 Heat stress

22.9.1 Heat stress alters cellular function.

Plants exposed to excess heat exhibit a characteristic set of cellular and metabolic responses, many of which are conserved in all organisms. The signature response to heat stress is a decrease in the synthesis of normal proteins, accompanied by an accelerated transcription and translation of a new set of proteins known as **heat shock proteins** (HSPs). This response is observed when plants are exposed to temperatures at least 5°C above their optimal growing conditions. HSPs can be visualized easily on two-dimensional electrophoretic gels (Fig. 22.41).

In addition to altering patterns of gene expression, heat also damages cellular structures, including organelles and the cytoskeleton, and impairs membrane function. Organisms amenable to genetic manipulation, especially *E. coli*, yeast, and *Arabidopsis*, have been used to demonstrate the cellular and metabolic changes required to survive high temperatures.

22.9.2 Plants can acclimate to heat stress.

Heat shock may arise under numerous temporal and developmental circumstances, with results ranging from retarded growth to damaged organs to plant death. In the field, heat shock may arise in leaves, when transpiration is insufficient (i.e., when water is limiting and temperature is high) or when stomata

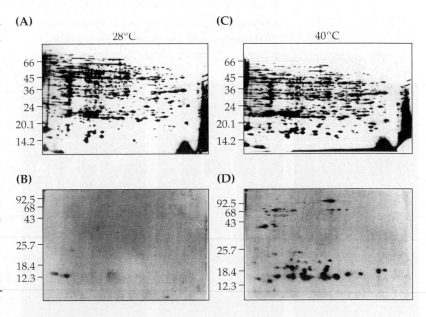

Figure 22.41
HSPs, particularly those of low molecular mass (smHSPs), accumulate in soybean seedlings in response to high temperatures. From seedlings that were incubated in the presence of ^{3}H-leucine at 28°C (A and B) or 40°C (C and D) for three hours, the total protein was extracted and resolved by two-dimensional polyacrylamide gel electrophoresis. Proteins were visualized by silver staining (A and C) or by fluorography (B and D).

are partially or fully closed and irradiance is high; in germinating seedlings, when the soil is warmed by the sun; in organs with reduced capacity for transpiration (e.g., fruit); and overall, from high ambient temperatures. Duration and severity of the stress, susceptibility of different cell types, and stage of development all influence the ability of a particular genotype to survive heat stress.

Plants can acquire thermotolerance if subjected to a nonlethal (permissive) high temperature for a few hours before encountering heat shock conditions. An acclimated plant can survive exposure to a temperature that would otherwise be lethal (Fig. 22.42). The acclimation process is thought to involve new proteins, synthesized in response to the permissive high temperature, that confer thermotolerance to the organism. Despite this capacity for acclimation, there is, of course, a limit to how much heat a plant can withstand.

22.9.3 HSPs are conserved among different organisms.

Some major HSPs are conserved among all living organisms, both prokaryotic and eukaryotic. Many of these proteins function as chaperones and are involved in refolding

proteins denatured by heat. Some HSPs are expressed throughout development. Indeed, some proteins designated as HSPs are not induced by heat at all, but rather are defined as HSPs according to sequence homology and possible functional similarities. In most cases, the functions of plant HSPs have not been demonstrated in vivo. Much of the research on HSP function and significance has been completed in yeast and *Drosophila*, where genetic mutants can be obtained easily. The roles played by plant proteins may be inferred from complementation studies in yeast. For example, an HSP104 deletion mutant of yeast, which is not thermotolerant, can be partially rescued by an *Arabidopsis* member of the HSP100 family. However, the plant protein affords the yeast mutant less thermotolerance than that of the native yeast counterpart.

22.9.4 Five classes of HSPs are defined according to size.

HSPs are grouped into five distinct classes named for their approximate molecular masses (Table 22.11). Members of the HSP100 family, which includes proteins from bacteria, yeast, trypanosomes, mammals, and plants, contain two conserved ATP-binding domains. Three subfamilies can be distinguished on the basis of sequence homology of a spacer domain flanking the ATP-binding domains. One subfamily (ClpB) is inducible by heat. Members of this family are required for thermotolerance in both plants and yeast. Some evidence indicates that HSP104 and HSP70 work together in the same pathway or in parallel, interacting pathways. The HSP100 family of chaperones may function in disaggregation rather than to prevent protein aggregation and misfolding during exposure to high temperature.

Proteins of the HSP90 family are found in bacteria and in the cytosolic, nuclear, and endoplasmic reticulum (ER) compartments of eukaryotic cells, where they may function as molecular chaperones. However, unlike typical chaperones, these proteins may have a specific and long-lived interaction with particular target proteins. In yeast, HSP90 is essential for growth at all temperatures.

HSP70 proteins are essential for normal cell function. Some members of this family

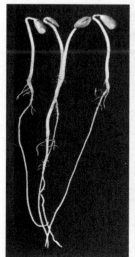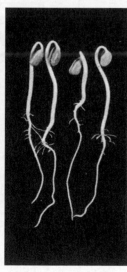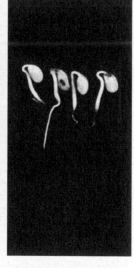

Figure 22.42
Thermotolerance can be induced in seedlings. The soybean seedlings shown were grown at 28°C. The seedlings in the left panel were not subjected to additional heat. The seedlings in the center panel were subjected to a two-hour pretreatment at 40°C before being transferred to 45°C for two hours. The seedlings in the right panel were transferred to the high temperature without any pretreatment.

are expressed constitutively; others are induced by heat or cold. Present in multiple compartments (e.g., cytoplasm, ER, mitochondria, and plastids), HSP70 proteins are ATP-dependent molecular chaperones that may interact with many different proteins, given their involvement in protein folding, unfolding, assembly, and disassembly (see Chapters 4 and 9). Localized to the nucleolus during heat stress, HSP70 is redistributed to the cytoplasm during stress recovery. The most highly conserved feature shared among HSP70 proteins from different organisms is the N-terminal ATP-binding domain; the C terminus, however, varies and has been proposed to determine substrate specificity. In *E. coli*, additional proteins such as DnaJ and GrpE may facilitate interactions involving HSP70 proteins (e.g., DnaK) and their targets. In *E. coli* and yeast, HSP70 is required for survival at moderately high temperatures but not for surviving extreme temperatures. HSP70 expression in plants generally is not as strong as in other organisms.

Members of the HSP60 protein family also are thought to function as molecular chaperones. These proteins are present in the bacterial cytosol (GroEL), the mitochondrial matrix, and the chloroplast stroma. HSP60 proteins are abundant even at normal temperatures; their major role is thought to involve protein assembly (see Chapters 4 and 9). HSP60 proteins share a common oligomeric structure of two seven-membered rings, and require ATP. In plants, the most studied HSP60 protein is chaperonin 60, a nuclear-encoded chloroplast protein that is involved in Rubisco assembly but does not increase in response to heat stress. In vitro, HSP60 proteins prevent other proteins from aggregating at physiologically relevant temperatures and are important in protein refolding as temperatures increase.

One unique aspect of the heat shock response in plants is the abundance of low-molecular-mass HSPs (smHSPs). Plants contain five or perhaps six classes of smHSPs, whereas other eukaryotes have only a single class of smHSP, which is encoded by one to four genes. One explanation for this difference is that these proteins are distributed in different compartments in plants: two in the cytosol, one in the chloroplast, one in the ER, one in the mitochondrion, and possibly another in a membrane compartment that has not been defined. The C termini of all classes of the smHSPs share homology with α-crystallins, proteins of the vertebrate eye lens. Like α-crystallins, smHSPs form complexes ranging in size from 200 kDa to 800 kDa. Although a role for these proteins has not been demonstrated definitively, evidence from mammalian and *Drosophila* studies indicates that the smHSPs may participate in establishing thermotolerance; there is no evidence that they are required for normal cellular function. smHSPs do not require ATP. HSP18.1 from pea has been demonstrated to prevent protein aggregation in response to high temperature in vitro.

Table 22.11 HSPs, examples of proteins in each class, and their characteristics

Protein class	Name	Size (kDa)	Location
HSP100	hsp104 (yeast) ClpB (*Escherichia coli*) ClpA (*E. coli*)	100–114	Cytoplasm
HSP90	hsp90 GRP94	80–94	Cytoplasm ER
HSP70	DnaK (*E. coli*) BiP and GRP SSA1-4 (yeast) SSB1-2 (yeast) SSC1 (yeast) KAR2 (yeast)	69–71	ER Cytoplasm Cytoplasm Mitochondria ER
HSP60	chaperonin 60 GroEL (*E. coli*) GroES (*E. coli*)	60 57 10	Chloroplasts and mitochondria
smHSP		15–30	Cytoplasm Chloroplast ER Mitochondria

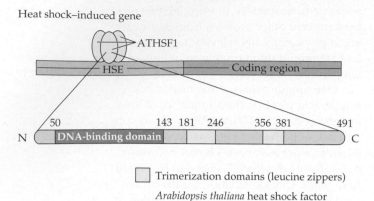

Heat shock–induced gene

Figure 22.43
An *Arabidopsis* HSF, ATHSF1, includes domains conserved in HSFs from other organisms: a DNA-binding domain and two leucine zipper trimerization domains.

22.9.5 Expression of many HSPs is controlled by a transcription factor that recognizes a conserved promoter sequence.

The heat shock transcription factor (HSF) is expressed constitutively but must be activated during heat stress to recognize its DNA target, the heat shock element (HSE). The HSE is made up of 5-bp repeats in alternating orientations with the consensus nGAAn. An HSF-regulated promoter may contain five to seven of these repeats close to the TATA box. Many HSEs contain the DNA element 5'-CTnGAAnnTTCnAG-3'.

Like other HSFs, the HSF of *Arabidopsis* (ATHSF1; Fig. 22.43) only can bind DNA as trimers; heat stress is required for trimerization (Fig. 22.44). The oligomerization and DNA-binding domains of HSF are conserved among different organisms. Trimerization depends on the presence of a leucine zipper configuration of hydrophobic heptad repeats

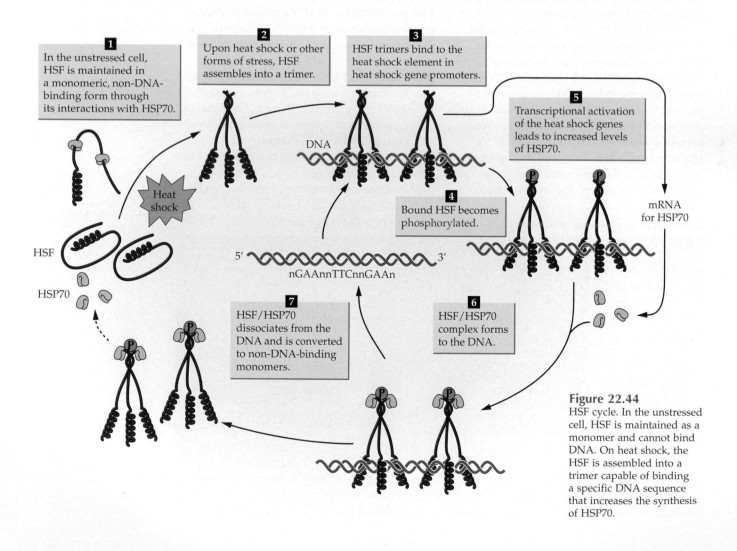

Figure 22.44
HSF cycle. In the unstressed cell, HSF is maintained as a monomer and cannot bind DNA. On heat shock, the HSF is assembled into a trimer capable of binding a specific DNA sequence that increases the synthesis of HSP70.

are expressed constitutively; others are induced by heat or cold. Present in multiple compartments (e.g., cytoplasm, ER, mitochondria, and plastids), HSP70 proteins are ATP-dependent molecular chaperones that may interact with many different proteins, given their involvement in protein folding, unfolding, assembly, and disassembly (see Chapters 4 and 9). Localized to the nucleolus during heat stress, HSP70 is redistributed to the cytoplasm during stress recovery. The most highly conserved feature shared among HSP70 proteins from different organisms is the N-terminal ATP-binding domain; the C terminus, however, varies and has been proposed to determine substrate specificity. In *E. coli*, additional proteins such as DnaJ and GrpE may facilitate interactions involving HSP70 proteins (e.g., DnaK) and their targets. In *E. coli* and yeast, HSP70 is required for survival at moderately high temperatures but not for surviving extreme temperatures. HSP70 expression in plants generally is not as strong as in other organisms.

Members of the HSP60 protein family also are thought to function as molecular chaperones. These proteins are present in the bacterial cytosol (GroEL), the mitochondrial matrix, and the chloroplast stroma. HSP60 proteins are abundant even at normal temperatures; their major role is thought to involve protein assembly (see Chapters 4 and 9). HSP60 proteins share a common oligomeric structure of two seven-membered rings, and require ATP. In plants, the most studied HSP60 protein is chaperonin 60, a nuclear-encoded chloroplast protein that is involved in Rubisco assembly but does not increase in response to heat stress. In vitro, HSP60 proteins prevent other proteins from aggregating at physiologically relevant temperatures and are important in protein refolding as temperatures increase.

One unique aspect of the heat shock response in plants is the abundance of low-molecular-mass HSPs (smHSPs). Plants contain five or perhaps six classes of smHSPs, whereas other eukaryotes have only a single class of smHSP, which is encoded by one to four genes. One explanation for this difference is that these proteins are distributed in different compartments in plants: two in the cytosol, one in the chloroplast, one in the ER, one in the mitochondrion, and possibly another in a membrane compartment that has not been defined. The C termini of all classes of the smHSPs share homology with α-crystallins, proteins of the vertebrate eye lens. Like α-crystallins, smHSPs form complexes ranging in size from 200 kDa to 800 kDa. Although a role for these proteins has not been demonstrated definitively, evidence from mammalian and *Drosophila* studies indicates that the smHSPs may participate in establishing thermotolerance; there is no evidence that they are required for normal cellular function. smHSPs do not require ATP. HSP18.1 from pea has been demonstrated to prevent protein aggregation in response to high temperature in vitro.

Table 22.11   HSPs, examples of proteins in each class, and their characteristics

Protein class	Name	Size (kDa)	Location
HSP100	hsp104 (yeast) ClpB (*Escherichia coli*) ClpA (*E. coli*)	100–114	Cytoplasm
HSP90	hsp90 GRP94	80–94	Cytoplasm ER
HSP70	DnaK (*E. coli*) BiP and GRP SSA1-4 (yeast) SSB1-2 (yeast) SSC1 (yeast) KAR2 (yeast)	69–71	 ER Cytoplasm Cytoplasm Mitochondria ER
HSP60	chaperonin 60 GroEL (*E. coli*) GroES (*E. coli*)	60 57 10	Chloroplasts and mitochondria
smHSP		15–30	Cytoplasm Chloroplast ER Mitochondria

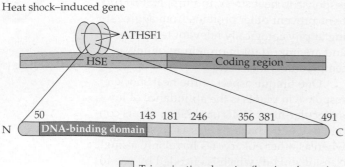

Heat shock–induced gene

ATHSF1

HSE | Coding region

N | 50 | DNA-binding domain | 143 181 | 246 | 356 381 | 491 | C

☐ Trimerization domains (leucine zippers)

Arabidopsis thaliana heat shock factor

Figure 22.43
An *Arabidopsis* HSF, ATHSF1, includes domains conserved in HSFs from other organisms: a DNA-binding domain and two leucine zipper trimerization domains.

22.9.5 Expression of many HSPs is controlled by a transcription factor that recognizes a conserved promoter sequence.

The heat shock transcription factor (HSF) is expressed constitutively but must be activated during heat stress to recognize its DNA target, the heat shock element (HSE). The HSE is made up of 5-bp repeats in alternating orientations with the consensus nGAAn. An HSF-regulated promoter may contain five to seven of these repeats close to the TATA box. Many HSEs contain the DNA element 5'-CTnGAAnnTTCnAG-3'.

Like other HSFs, the HSF of *Arabidopsis* (ATHSF1; Fig. 22.43) only can bind DNA as trimers; heat stress is required for trimerization (Fig. 22.44). The oligomerization and DNA-binding domains of HSF are conserved among different organisms. Trimerization depends on the presence of a leucine zipper configuration of hydrophobic heptad repeats

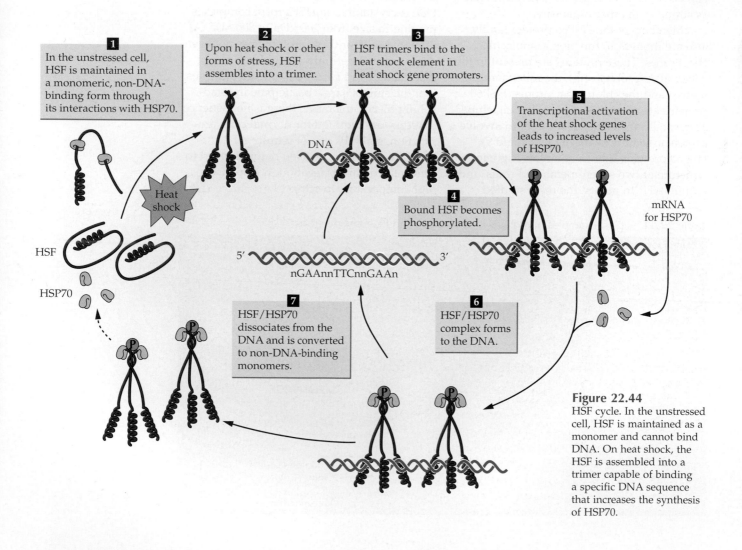

1 In the unstressed cell, HSF is maintained in a monomeric, non-DNA-binding form through its interactions with HSP70.

2 Upon heat shock or other forms of stress, HSF assembles into a trimer.

3 HSF trimers bind to the heat shock element in heat shock gene promoters.

5 Transcriptional activation of the heat shock genes leads to increased levels of HSP70.

4 Bound HSF becomes phosphorylated.

7 HSF/HSP70 dissociates from the DNA and is converted to non-DNA-binding monomers.

6 HSF/HSP70 complex forms to the DNA.

DNA

HSF

HSP70

Heat shock

5' nGAAnnTTCnnGAAn 3'

mRNA for HSP70

Figure 22.44
HSF cycle. In the unstressed cell, HSF is maintained as a monomer and cannot bind DNA. On heat shock, the HSF is assembled into a trimer capable of binding a specific DNA sequence that increases the synthesis of HSP70.

Box 22.2

Plants that express a heat shock transcription factor demonstrate improved heat tolerance in the absence of heat pretreatment.

A molecular-genetic approach has been used to generate *Arabidopsis* plants in which the heat shock transcription factor (HSF) is constitutively active. These plants express a transgenic protein in which ATHSF1 is fused to the N or C terminus of β-glucuronidase, a fusion resulting in an HSF that trimerizes in the absence of heat. The transgenic plants express HSPs constitutively and demonstrate increased heat tolerance in the absence of heat pretreatment. Under conditions that did not promote heat stress, the transgenic plants produced approximately 20% of the maximum heat-inducible amount of HSP18 mRNA and protein. Without heat pretreatment, the transgenic plants survived at 50°C for one hour, whereas wild-type plants survived for one hour at a lower temperature (46°C) only. Interestingly, the transgenic plants did not display an increased acquired thermotolerance in response to a two-hour pretreatment at 35°C: Both wild-type and transgenic plants survived at 54°C after pretreatment.

Overexpression of transcription factors may prove an efficient way to improve stress tolerance, because a single transgenic protein can induce an entire battery of coordinately regulated genes. At present, interest is considerable in determining whether crop plants that overexpress this HSF construct will demonstrate improved heat tolerance in the field.

located adjacent to the DNA-binding domain. The mechanism that controls trimerization is poorly understood but recent evidence indicates that trimerization, DNA binding, and transcriptional activity are repressed in the absence of heat stress. When ATHSF1 is overexpressed in *Arabidopsis,* the transcription factor is not active; overexpression of ATHSF1 in heterologous systems, however, such as *E. coli, Drosophila,* and HeLa cells, results in constitutively active HSF. This indicates that the repression mechanism is selective for the homologous HSF and not the heterologous HSF. ATHSF1 can be derepressed when it is tagged with a reporter gene and then overexpressed (Box 22.2).

Summary

Abiotic stresses are prevalent in nature and can substantially diminish plant yields. Plant responses to stressful environmental factors can be part of the mechanisms that permit the plant to withstand the stress. Alternatively, such responses may be a manifestation of injury that has occurred in response to the stress. The response depends on the severity and duration of the stress, the developmental stage of the affected plant, the tissue type, and the interactions of multiple stresses. Mechanisms that permit stress survival are termed resistance mechanisms and can allow an organism to avoid or tolerate stress. Acclimation, a process that improves stress resistance, may occur in response to a mild nonlethal stress. Changes in gene expression may be involved in the mechanism of stress resistance or may be a result of injury. Described in this chapter are the abiotic stresses arising from drought, salinity, low temperature, flooding, air pollution, and high temperature.

Stresses involving water deficit may arise from drought conditions, saline soils, or low temperature. To quantify the effect of this stress on the plant, one can determine the water status of the plant by using either ψ_w or RWC. Measuring the water status of the plant is important for determining the impact of the environmental condition. Decreases in plant water potential may be brought about by osmotic adjustment, the accumulation of compatible solutes that promote acclimation to dry or saline soils. Compatible solutes, such as glycine betaine, mannitol, pinitol, and proline, do not disrupt cellular function when accumulated to high concentrations in the cytoplasm. In addition to osmotic adjustment, some compatible solutes may serve other protective functions. Effects of water deficit and perturbing ions on the membrane may be minimized by the action of carriers, pumps, and channels. Amelioration of plant stress may also arise from the function of a set of genes discovered during investigation of the desiccation stages of seed development. Five groups of *Lea* genes have been identified, based on homology among different species. The majority of these proteins are hydrophilic and soluble when boiled; however, not all groups have these characteristics. Several LEA-encoding genes have been shown to function in stress resistance by using overexpression technology in transgenic plants or yeast. Various other types of genes also are induced by water deficit, including those that may protect the plant from secondary biotic stresses. Osmotin, a tobacco protein with antifungal activity, accumulates during water deficit. The mechanisms of gene induction are regulated by specific DNA elements: Two classes of elements, ABRE and DRE, have been found in many water-deficit–induced genes.

Flooding causes an oxygen deficit in the cell that impacts respiratory metabolism. The ability to tolerate flooding varies greatly among species and can be altered by acclimation processes involving exposure to hypoxic conditions (3 kPa oxygen). During short-term acclimation to anoxic conditions, plants generate ATP through glycolysis and fermentation; this shift from aerobic metabolism to glycolytic fermentation involves changes in gene expression. The plant hormone ethylene promotes long-term acclimative responses, including formation of aerenchyma and stem elongation. Some wetland genotypes are adapted to long-term flooding.

Oxidative stress may arise from any abiotic or biotic stress that causes the formation of a ROS, such as hydrogen peroxide (H_2O_2), superoxide anion ($O_2^{\bullet-}$), and hydroxyl radical ($HO^{\bullet}$), or perhydroxyl radical ($HO_2^{\bullet}$). Plants scavenge and eliminate these reactive molecules by using antioxidant defense systems—antioxidants and antioxidant enzymes—that are present in various subcellular compartments. Ozone exposure to plants can be used as a model system to determine how ROS cause oxidative damage to biomolecules. Studies in which antioxidant enzymes are overexpressed in transgenic plants have emphasized the important role of subcellular compartmentation in detoxification mechanisms; that is, overexpression of antioxidant enzymes in one compartment may not improve stress tolerance if oxidant-scavenging mechanisms are limiting in other cellular compartments.

Heat stress responses are widely conserved among different organisms. Thermotolerance can be developed as plants acclimate to a nonlethal high temperature. During heat stress in plants, as in other organisms, gene expression patterns, including transcription and translation, are altered to promote the accumulation of HSPs. The five major classes of HSPs, defined according to size, are conserved among different organisms. In general, the HSPs function as chaperones to promote proper folding of proteins. Expression of HSPs is controlled by a transcription factor that recognizes a conserved DNA element, 5′-nGAAn-3′, present in multiple copies in the promoter. The transcription factor is active as a trimer and must be derepressed to activate gene expression.

Progress in understanding plant responses to stress has been impressive. Nonetheless, numerous important questions remain. None of the mechanisms by which higher plants perceive abiotic stresses has been elucidated. Progress in this crucial area will advance substantially our knowledge of stress-initiated signal transduction events. Stress-related signals are propagated by several different agents. In some cases, these signal transduction events involve at least one of the five best-studied plant hormones: ABA, auxins, cytokinins, ethylene, and gibberellins. Calcium is implicated as a second messenger in many stress responses. However, perhaps signaling molecules not yet identified also participate in controlling plant responses to the environment. As plant genomes are analyzed, it has become apparent that many genes associated with mammalian signal transduction cascades, including peptide hormones and membrane receptor kinases, also are present in plants.

Further Reading

Alscher, R. G., Hess, J. L. (1993) *Antioxidants in Higher Plants.* CRC Press, Boca Raton, FL.

Close, T. J. (1996) Dehydrins: emergence of a biochemical role of a family of plant dehydration proteins. *Physiol. Plant* 97: 795–803.

Drew, M. (1997) Oxygen deficiency and root metabolism: injury and acclimation under hypoxia and anoxia. *Annu. Rev. Plant Physiol. Plant Mol. Biol.* 48: 223–250.

Foyer, C. H., Decourvieres, P., Kunert, K. J. (1994) Protection against oxygen radicals: an important defence mechanism studied in transgenic plants. *Plant Cell Environ.* 17: 507–523.

Imai, R., Chang, L., Bray, E. A., Takagi, M. (1996) A lea-class gene of tomato confers salt and freezing tolerance when overexpressed in *Saccharomyces cerevisiae. Gene* 170: 243–248.

Ingram, J., Bartels, D. (1996) The molecular basis of dehydration tolerance in plants. *Annu. Rev. Plant Physiol. Plant Mol. Biol.* 47: 377–403.

Kennedy, R. A., Rumpho, M. E., Fox, T. C. (1992) Anaerobic metabolism in plants. *Plant Physiol.* 11: 1–6.

Lee, J. H., Hübel, A., Schöffl, F. (1995) Derepression of the activity of genetically engineered heat shock factor causes constitutive synthesis of heat shock protein and increased thermotolerance in transgenic *Arabidopsis. Plant J.* 8: 603–612.

McCue, K., Hanson, A. (1990) Drought and salt tolerance: towards understanding and application. *Trends Biotechnol.* 8: 358–362.

Nilsen, E. T., Orcutt, D. M. (1996) *Physiology of Plants under Stress: Abiotic Factors.* John Wiley & Sons, New York, p. 689.

Perata, P., Alpi, A. (1993) Plant responses to anaerobiosis. *Plant Sci.* 93: 1–17.

Richard, B., Couée, I., Raymond, P., Saglio, P. II., Saint-Ges, V., Pradet, A. (1994) Plant metabolism under hypoxia and anoxia. *Plant Physiol. Biochem.* 32: 1–10.

Sachs, M. M., Subbaiah, C. C., Saab, I. N. (1996) Anaerobic gene expression and flooding tolerance in maize. *J. Exp. Bot.* 47: 1–15.

Schrimer, E. C., Lindquist, S., Vierling, E. (1994) An *Arabidopsis* heat shock protein complements a thermotolerance defect in yeast. *Plant Cell* 6: 1899–1909.

Tarczynski, M. C., Jensen, R. G., Bonhert, H. J. (1993) Stress protection of transgenic tobacco by production of the osmolyte mannitol. *Science* 259: 508–510.

Thomashow, M. F. (1999) Plant cold acclimation: freezing tolerance genes and regulatory mechanisms. *Annu. Rev. Plant Physiol. Plant Mol. Biol.* 50: 571–599.

Xu, D., Duan, X., Wang, B., Hong, B., Ho, T.-H.D., Wu, R. (1996). Expression of a late embryogenesis abundant protein gene, *HVA1*, from barley confers tolerance to water deficit and salt stress in transgenic rice. *Plant Physiol.* 110: 249–257.

Yancey, P. H., Clark, M. E., Hand, S. C., Bowlus, R. D., Somero, G. N. (1982) Living with water stress: evolution of osmolyte systems. *Science* 217: 1214–1222.

Biochemistry & Molecular Biology of Plants, B. Buchanan, W. Gruissem, R. Jones, Eds.
© 2000, American Society of Plant Physiologists

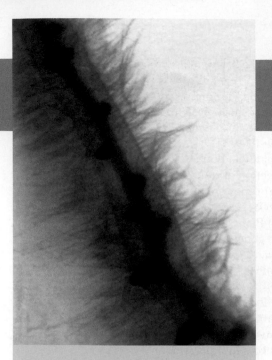

C H A P T E R *23*

Molecular Physiology of Mineral Nutrient Acquisition, Transport, and Utilization

Leon V. Kochian

Introduction

The importance of mineral nutrients to crop production has been recognized for more than 2000 years. Plant mineral nutrition is unique because green plants, the only multicellular autotrophic organisms, can mine inorganic elements from the environment without having to rely on high-energy compounds synthesized by other organisms. Earlier this century, classical physiological research into the processes involved in mineral nutrient acquisition was a major research focus in the field of plant biology. Recently, there has been a great upsurge in research and a renewed interest in the field of plant mineral nutrition now that contemporary experimental approaches in membrane biophysics, molecular biology, and plant physiology are being used to dissect the mechanisms underlying transport and utilization of mineral ion nutrients. We currently are witnessing an exciting period of research into plant mineral ion transport. Many new molecular approaches have allowed the cloning of families of mineral ion transporters, and microelectrode technologies have allowed us to study the functions of these individual transporters, both when expressed in heterologous systems and when studied in planta. The challenge now at hand is to fit these individual pieces back together to begin to understand the molecular physiology of the intact plant, in terms of the mechanisms and regulation of mineral nutrient acquisition and utilization.

This chapter focuses on recent findings concerning the molecular physiology of macronutrient and micronutrient transport as well as the strategies plants use to tolerate stressful soil environments such as toxic concentrations of metals in the soil. Emphasis will be on those essential mineral elements, such as potassium (K) and iron (Fe), for which researchers are beginning to understand the molecular mechanisms of plant acquisition and transport. Recent experimental and technological advances that have enabled researchers to clone the genes encoding mineral ion transporters will also be described.

Many topics related to mineral nutrition have been addressed in other chapters of this text. Plant uptake and assimilation of nitrogen and sulfur, for example, are discussed in Chapters 8 and 16, respectively. Anabolic and catabolic reactions involving the most abundant mineral nutrients—carbon, hydrogen, and oxygen—appear throughout the text and constitute the primary focus of Chapters 12 through 14. Long-distance transport of selected mineral nutrients is discussed in Chapter 15.

23.1 Overview of essential mineral elements

Of the 17 nutrient elements determined to be essential in plants, 14 are presented in Table 23.1—the last mineral element listed, nickel, having been determined to be essential only within the last 10 years. The three elements most abundant in plant tissues—C, H, and O—are not included in the table. Also not listed are Na, Si, and Co, which some researchers view as beneficial but not essential. The symptoms of some mineral deficiencies in sugar beet are illustrated in Figure 23.1.

Plant mineral nutritionists define **essential minerals** more rigorously than do their colleague animal or human nutritionists. A mineral element is essential to plants if

(*a*) without it, the plant cannot complete its life cycle and (*b*) it is a component of an essential plant metabolite or constituent. The 17 essential nutrients can be classified in several different ways—for example, according to their biochemical role in the plant or their chemical form in the soil or plant. The most common classification is based on the relative concentrations of each nutrient in the plant when the nutrient is present in adequate concentrations for normal plant function. The **macronutrients** are C, H, O, N, K, Ca, Mg, P, and S. Except for the more abundant O and C, the concentrations of macronutrients generally found in plants range from 1000 to 15,000 µg per gram of dry weight. In contrast, the concentrations of **micronutrients** (Cl, B, Fe, Mn, Zn, Cu, Mo, and Ni) usually found in plants are 100- to 10,000-fold less than those of the macronutrients.

Macronutrients can be divided into those that maintain their identity as ions within the plant (e.g., K^+, Ca^{2+}, and Mg^{2+}) and those that are assimilated into organic compounds (e.g., N, S, P). By far, our most complete understanding of mineral ion transport comes from studies of K^+ uptake and movement within the plant; meanwhile, advances in molecular biology recently have

Table 23.1 Concentrations of essential mineral nutrients considered to be in the adequate range for plants

Element	Chemical symbol	Concentration in dry material ($\mu g\ g^{-1}$)	Concentration in fresh tissue[a]
Macronutrients			
Nitrogen	N	15,000	71.4 mM
Potassium	K	10,000	17 mM
Calcium	Ca	5,000	8.3 mM
Magnesium	Mg	2,000	5.5 mM
Phosphorus	P	2,000	4.3 mM
Sulfur	S	1,000	2.1 mM
Micronutrients			
Chlorine	Cl	100	188 µM
Boron	B	20	123 µM
Iron	Fe	100	120 µM
Manganese	Mn	50	61 µM
Zinc	Zn	20	20.4 µM
Copper	Cu	6	6.2 µM
Molybdenum	Mo	0.1	0.07 µM
Nickel	Ni	0.005	0.006 µM

[a]Fresh weight concentrations were calculated by assuming a 15:1 fresh weight:dry weight ratio.

Mineral-sufficient
(control)

Potassium-deficient
(−K)

Phosphorous-deficient
(−P)

Iron-deficient
(−Fe)

Zinc-deficient
(−Zn)

Calcium-deficient
(−Ca)

Magnesium-deficient
(−Mg)

Copper-deficient
(−Cu)

Manganese-deficient
(−Mn)

Figure 23.1
Symptoms observed in leaves of strawberry
plants deficient in K, P, Fe, Zn, Ca, Mg, Cu, or
Mn. A leaf from a mineral-sufficient control
plant is also shown.

opened up new avenues of research into P transport and nutrition.

23.2 Mechanisms and regulation of plant K⁺ transport

23.2.1 Potassium transport in plants has been studied extensively.

Potassium is the most abundant cellular cation, with cytoplasmic concentrations regulated between 80 and 200 mM, and total tissue concentrations of approximately 20 mM. It plays a role in myriad cellular and whole-plant functions—serving as an osmoticum for cellular growth and stomatal function, balancing the charges of diffusible and nondiffusable anions, activating more than 50 plant enzymes, and participating in numerous metabolic processes, including photosynthesis, oxidative metabolism, and protein synthesis. The transport of K^{+}, which is mobile within cells, tissues, and whole plants, has been studied intensively for the last 50 years. As a result, we know more about the transport of K^+ by plants than about any other mineral or organic solute. Many of these studies have focused on K^+ transport in roots because roots are the primary mineral-absorbing organs and are amenable to studies of ion transport. Although the importance of controlling and facilitating K^+ entry into the plant by way of transport across the plasma membrane of the root cell is unquestioned, other K^+ transport sites within the plant are of equal importance, particularly when K^+ transport and translocation at the whole-plant level are considered. Potassium recirculation within the plant is well documented, as are complex patterns of K^+ translocation between tissues and organs (see Chapter 15). In white lupine, about one-half of the absorbed K^+ translocated to the shoot by the xylem is returned to the root by the phloem, and nearly 75% of the phloem-borne K^+ returning to the root reenters the xylem stream and again is transported to the shoot. Some sites of K^+ transport are diagrammatically outlined in Figure 23.2.

K^+ enters the root symplasm by way of transport across the root cell plasma membrane. From there, it can travel through the symplasm to the vascular tissues, where it is unloaded from xylem parenchyma into xylem vessels for long-distance transport to the leaves. K^+ is reabsorbed from the xylem into leaf cells. After efflux from fully expanded source leaves, it can be loaded into phloem cells for translocation to actively growing sink tissues (e.g., shoot and root apices), where it can be unloaded by way of symplasmic or apoplastic pathways. K^+ can also cross the tonoplast membrane for storage in vacuoles of both root and shoot cells. The integration and regulation of K^+ transport systems at different sites along the long-distance pathway allow the plant to direct the partitioning and circulation of K^+. Such an integrated system of K^+ transporters probably plays a central role in plant growth and development and in the allocation of mineral nutrients in response to changes in nutrient availability. Note that this model is equally relevant for all essential mineral nutrients, because a range of different transporters localized to specific cell types must function to move each essential element throughout the intact plant.

23.2.2 Early physiological and biochemical studies indicated the existence of multiple K⁺ transporters.

Most of the early studies on plant K^+ transport focused on the use of excised roots and can be traced to the classical studies of Denis Hoagland and Ted Broyer. Before the discovery of radioisotopes, physiologists lacked sensitive methods by which to quantify mineral ion fluxes in roots and other plant tissues or organs. Hoagland and Broyer worked with barley seedlings grown in dilute hydroponic media, typically dilute Ca-salt solutions (e.g., $CaCl_2$ or $CaSO_4$) because Ca was needed to maintain normal membrane function. They found that the roots of these seedlings exhibited extremely high initial rates of mineral ion absorption and thus were useful experimental material. "Low-salt roots," which contain low concentrations of mineral salts and high concentrations of sugar (a replacement osmoticum for K salts), have a large transport capacity and still are used widely, even after the introduction of radioisotopes.

When radioisotopes became available for biological research, new avenues of research in plant mineral nutrition opened.

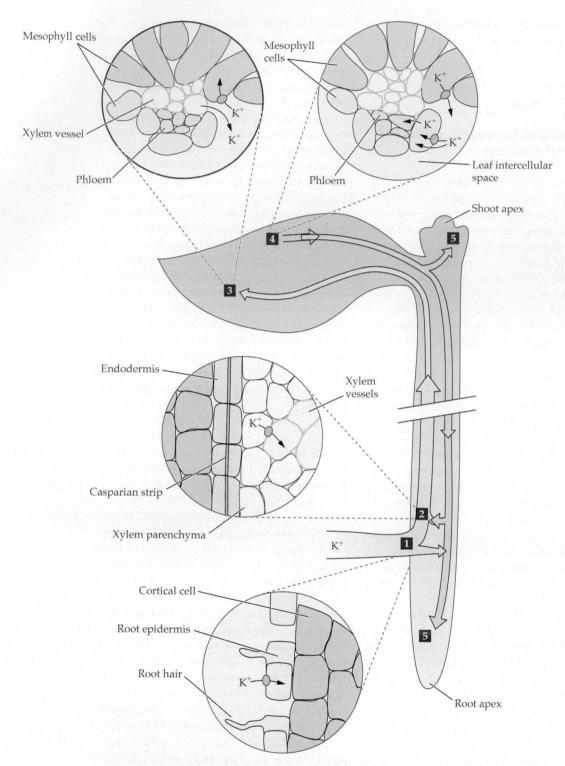

Figure 23.2

Diagrammatic depiction of K⁺ transport into and within the plant. K⁺ is transported within the xylem (red arrows) and phloem (blue arrows). The numbers represent important transport sites along the long-distance K⁺-transport pathway. For four of the five numbered sites, enlargements depict K⁺ transport at the cellular level. (1) K⁺ is absorbed across the root cell plasma membrane (longitudinal view). (2) K⁺ is transported into nonliving, thick-walled xylem vessels by way of efflux across the xylem parenchyma plasma membrane (cross-sectional view). (3) K⁺ is transported by the xylem to the shoot (leaf), moves from the xylem vessel to the apoplast surrounding the neighboring leaf cells, and is absorbed into the leaf parenchyma cells (cross-sectional view). (4) K⁺ is loaded into the phloem of a fully expanded source leaf after K⁺ efflux from leaf cells. Transport into the sieve tube–companion cell complex can occur by a combination of apoplastic and symplasmic routes (cross-sectional view). (5) K⁺ moves through the phloem to the shoot and root apices, where it is unloaded for subsequent use.

Investigators could then label a solution with a radioisotope or a radioactive analog of the mineral ion of interest (e.g., $^{42}K^+$ or $^{86}Rb^+$ for K^+) and then quantify the uptake, translocation, or efflux of the radioactivity administered. However, the use of radioisotopes to quantify ion fluxes properly is not a trivial undertaking; one must take care to measure unidirectional ion influx rather than a combination of cell wall binding and net ion flux (Box 23.1). The pioneering work of Emanuel Epstein in the use of radiotracers to measure mineral ion fluxes advanced the field of ion transport for all researchers. Epstein and coworkers were the first to treat mineral ion transporters as enzymes and to apply analysis from enzyme kinetics to studies of concentration-dependent root K^+ absorption (Box 23.2). When low-salt barley roots were exposed to solutions containing low concentrations (0 to 200 µM) of the K^+ analog $^{86}Rb^+$, the resulting high-affinity uptake demonstrated saturating (Michaelis–Menten) kinetics. The high-affinity K^+ transporter, which Epstein designated Mechanism I, was subsequently shown to transport K^+ (and Rb^+) preferentially over other alkali cations. Over a broader concentration range for K^+, the uptake kinetics of K^+ had a complex appearance that could be more clearly represented by two separate saturating curves (Fig. 23.3). Epstein proposed that roots also contain a low-affinity uptake system, Mechanism II, which is active at higher external concentrations of K^+. This pattern of K^+ up-take was described as the **dual isotherm** and was hypothesized to reflect the activities of two families of transporters. High-affinity K^+ transport was thought to dominate at low concentrations of K^+ in soil, whereas at higher K^+ concentrations, a low-affinity K^+ transporter operated that was less specific for K^+ over other alkali cations (such as Na^+).

By the late 1970s, when the dual isotherm hypothesis was challenged, studies had shown that model organisms in which the ion transporters were amenable to biochemical characterization (such as bacteria) contain multiple transporters for the same solute. The concept that mineral uptake involves distinct transporters with different substrate affinities was further suggested by thermodynamic evidence. At low external concentrations of K^+ (less than 100 µM), root cells often take up K^+ against the electrochemical potential gradient for the cation (see Chapter 3). Under these circumstances, uptake cannot occur by diffusion and instead must be mediated by a mechanism of active transport, either an ATP-driven ion pump or a secondarily active transporter that is coupled to the energy stored in a transmembrane gradient of H^+ or Na^+. Therefore, although controversy regarding Epstein's hypothesis was widespread during the 1970s and 1980s, available information about root transport of K^+ promoted speculation that high-affinity K^+ transport was an active process, whereas low-affinity transport involved passive K^+ uptake—which we now know probably involves K^+ channels.

Physiological studies provided further evidence for separate high- and low-affinity root transporters of K^+. For example, Figure 23.4 illustrates the kinetics of K^+ influx in maize roots, investigated by studying the K^+ analog Rb^+. The transport kinetics are complex and appear to result from the functioning of a saturable high-affinity transporter and a low-affinity K^+ transporter that exhibits first-order kinetics. With the use of specific inhibitors such as N-ethylmaleimide (NEM), the high-affinity uptake could be selectively inhibited without impairing the linear transport component. In other experiments, the application of a K^+-channel blocker such as tetraethylammonium (TEA) ion selectively inhibited the linear, low-affinity transport component. Further physiological studies showed that the high-affinity K^+

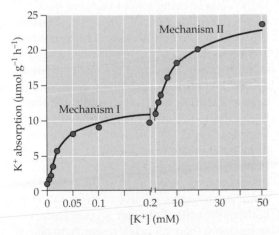

Figure 23.3
Dual isotherm depicting the rate of K^+ absorption in barley roots over a wide range of K^+ concentrations in the absorption solution. Note change in the concentration scale after 0.2 mM K^+.

Box 23.1

Radioisotopes are used to measure ion fluxes in cells.

Radioisotopes have been used to monitor mineral ion transport processes in plants for almost 50 years. It is relatively easy to immerse plant roots (or other tissues such as excised leaves) for a prescribed time in an uptake solution labeled with radioisotope, wash the tissue to remove adhering radiolabeled solution, and quantify the radioisotope associated with the tissue by detecting either β- or γ-radiation. Using radioisotopes to measure ion flux accurately, however, is not easy. The difficulty arises from the anatomical characteristics of plant cells. As shown in the accompanying figure, a plant cell contains three major compartments: cell wall, cytoplasm, and vacuole. When the cell is placed in an uptake solution containing $^{42}K^+$ or $^{86}Rb^+$, cation-binding sites and pores in the cell wall are the first sites to bind or accumulate the radiotracer. As the concentration of radiotracer within the cell wall near the plasma membrane surface increases, the K^+ that is transported across the plasma membrane becomes a mixture of unlabeled ions and the radioisotope analog ion. Given longer uptake times, the radioisotopic fraction of the K^+ pool in the cytoplasm will increase. When the amount of radioisotope in the cytoplasm becomes great enough, a substantial radioisotope efflux out of the cell and influx into the vacuole occur. This combination of efflux and influx confound attempts to measure a unidirectional influx.

Under steady-state conditions, the entry of mineral ions into cells is a net flux consisting of unidirectional influx and efflux components. Because researchers are usually interested in the unidirectional influx of an ion, the duration of the uptake period is crucial: Enough time must elapse for quantifiable concentrations of radioisotope to enter the symplasm. However, an overly long uptake period provides time for a substantial efflux of absorbed radioisotope out of the cell and can create conditions in which the entry of ions into the cell is limited by the rate of transport into the vacuole.

Because the cell wall can bind and accumulate a large amount of cationic radioisotope, a desorption regimen must be developed that effectively removes the radiolabel from the cell wall. This desorption period also must be timed carefully,

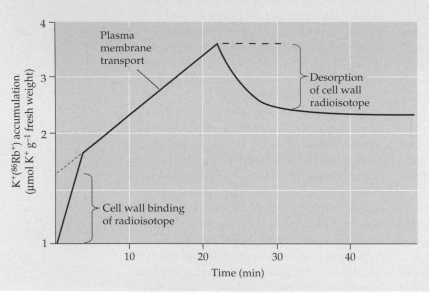

uptake system exhibited several properties, including increased gene expression or transport activity induced by K^+ starvation, a very high affinity for K^+ (K_m = 5 to 30 μM), and a strong selectivity for K^+ and Rb^+ over other alkali cations. In contrast, the low-affinity K^+ transporter was found to be less selective for K^+ and Rb^+ over Na^+ and less influenced by changes in the K^+ status of the plant. However, more-definitive evidence supporting the existence of multiple K^+ transporters has been derived in the molecular and biophysical studies detailed below.

23.2.3 Biophysical investigations provide a mechanistic basis for high- and low-affinity K^+ transport.

Microelectrode-based technologies, such as the patch clamp technique (see Chapter 3) and ion-selective microelectrodes, allow study of the electrical properties of cells, ion transport processes, and ionic relations in single cells and membrane patches. These tools are now being used to resolve questions concerning plant K^+ transport. Electrophysiological experiments in which root cells are impaled with a single salt-filled glass microelectrode to measure the voltage gradient across the plasma membrane (i.e., **membrane potential,** or E_m) have indicated that high-affinity K^+ uptake is **electrogenic** (see Chapters 3 and 14). For example, exposure of low-salt maize roots to low K^+ concentrations (2 to 20 μM) elicited a strong (70 mV) depolarization of the E_m (i.e., the cytosol became less negative relative to the solution bathing the cell). At higher K^+ concentrations (more than 0.5 mM), researchers observed depolarizations of lesser magnitude, responses consistent with the operation of a low-affinity transporter such as the K^+-selective channel.

so that a large fraction of the cell wall label is removed but not the radioisotope that has been transported into the cytoplasm. Generally, for a monovalent cation such as K^+, unlabeled uptake solution is an effective desorption solution. Alternatively, solutions that contain high concentrations of K^+ and Ca^{2+} can be used to optimize desorption of $^{42}K^+$ or $^{86}Rb^+$ from negatively charged cell wall binding sites.

The graph shows a time course for radioisotope accumulation and subsequent desorption in plant roots. After an initial rapid phase of isotope accumulation as label equilibrates with and binds within the wall, there is a slower phase of radioisotope accumulation, resulting from unidirectional cation influx into the cytoplasm. The effect of terminating radiotracer uptake by transferring the radiolabeled roots into desorption solution is also indicated. Under ideal conditions, the desorption phase rapidly removes most of the radiolabel in the cell wall without affecting the radiolabel accumulated in the symplasm. The difference between the amount of desorbed radiotracer and the total amount of radiotracer present in the plant tissue before desorption represents a measure of unidirectional influx into the cytoplasm.

To obtain a reasonable measure of unidirectional K^+ influx into root cells,

one effective uptake–desorption regimen for robust plant roots such as maize or barley, or for smaller roots such as those of *Arabidopsis*, utilizes a radiotracer uptake period of 10 to 20 minutes, followed by a desorption period of 10 to 20 minutes. If uptake is continued for longer periods (several hours), a third, slower phase of accumulation, caused by radiotracer efflux out of the cell or significant flux into the vacuole, often is seen. This process tends to be slower than the ion influx through the plasma membrane.

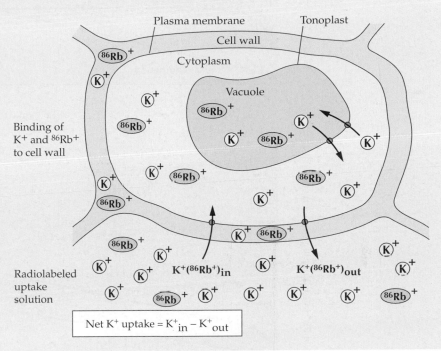

Net K^+ uptake = $K^+_{in} - K^+_{out}$

A combination of several different electrophysiological approaches has been used to investigate the energetics of high-affinity K^+ uptake and the underlying transport mechanisms. Double-barreled K^+–E_m microelectrodes, which measure cytoplasmic K^+ and E_m simultaneously, can be used along with a separate extracellular K^+ microelectrode to measure the net K^+ uptake from a 10 μM K^+ solution. In this approach, all of the variables required to calculate the electrochemical potential gradient for K^+ in single root cells involved in high-affinity K^+ uptake can be determined. The measured values of cytoplasmic K^+ (83 mM), E_m (−150 mV), and net K^+ uptake (0.1 μmol g^{-1} h^{-1}) in *Arabidopsis* root cells lead to the calculation of a strong K^+ electrochemical gradient directed out of the cells—clear evidence that high-affinity K^+ uptake must be mediated by an active transport mechanism. Incidentally, microelectrode measurements of E_m in root

cells underestimate E_m by 20 to 30 mV, a result of the large difference between the electrical resistance of the whole-cell membrane (at least 20 GΩ [gigaohms]) and that of the measuring microelectrode (less than 0.5 GΩ). These differences in electrical resistance do not invalidate the analysis but merely lower from 1 mM to about 0.4 mM the external K^+ concentration at which the transmembrane electrochemical gradient for K^+ changes from outward to inward.

As mentioned in Section 23.2.2, if high-affinity K^+ uptake is an active process, the possible transport mechanism must be an ion pump (e.g., ATPase) or a secondarily active cotransporter (a symporter or antiporter). Root K^+ uptake has long been linked to a stimulation in H^+ efflux, and high-affinity K^+ uptake has often been suggested to involve some type of K^+–H^+ exchanger. When plant plasma membrane ATPases were first characterized, the observation of K^+

(A)

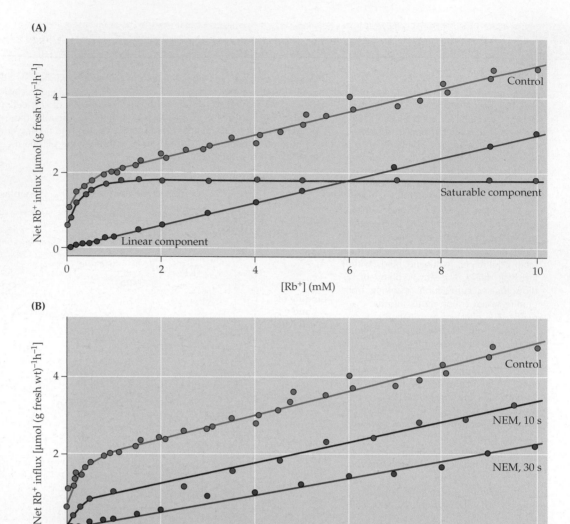

Figure 23.4
Kinetics of $^{86}Rb^+$ influx into maize roots grown in high salt conditions. (A) The control curve (green) has been separated into its saturable high-affinity (red) and linear low-affinity (blue) transport components. (B) Influence of the sulfhydryl modifier N-ethylmaleimide (NEM) on $^{86}Rb^+$ influx in maize roots grown in high salt conditions. Root segments were pretreated with 0.3 mM NEM for 10 seconds (red) or 30 seconds (blue), then washed for 10 minutes in 1 mM dithioerythritol to bind and inactivate unreacted NEM before $^{86}Rb^+$ uptake. NEM inhibits the high-affinity transporter but does not affect the low-affinity transporter.

stimulation of ATPase activity provoked speculation that a plasma membrane K^+-ATPase was involved in active K^+ uptake. Subsequently, this K^+-stimulated enzyme activity was found to be associated with the ubiquitous plasma membrane H^+-ATPase that energizes plant solute transport, which led some researchers to postulate that this ATPase was a K^+/H^+-exchange ATPase, mediating both active H^+ efflux and active K^+ uptake. However, the linkage between K^+ influx and H^+ efflux is more likely to be indirect. For example, use of extracellular K^+ and H^+ microelectrodes to quantify the K^+ and H^+ fluxes simultaneously in a single maize root epidermal cell did not provide stoichiometric data typical of direct coupling between K^+ influx and H^+ efflux. In that work, $K^+:H^+$ flux ratios ranged from 6:1 to 1:6. H^+ efflux and K^+ influx now generally are accepted as being coupled indirectly. K^+-induced depolarization of E_m can result in transient stimulation of H^+-ATPase, which hyperpolarizes the E_m (i.e., makes the cytosol

Box 23.2

The carrier-kinetic approach to plant ion transport was once a highly controversial area of plant research.

When Emanuel Epstein and coworkers first applied the Michaelis–Menten enzyme kinetic analysis to concentration-dependent studies of root K^+ uptake, they had a tremendous impact on the field of ion transport. This approach gave researchers a new way to look at and analyze ion transport processes, both in plants and in a wide range of other organisms. Transport of a wide range of substrates by many plant and animal tissues was observed to involve complex kinetics. Often, ion uptake curves contained several different putative phases or components. During the 1960s and 1970s the field of plant ion transport was filled with contention. Conflicting hypotheses were presented concerning the number of transport "phases," the cellular location of each transport phase, and the physiological significance of each transport component.

One of the most interesting and long-running debates concerning the mechanistic nature of complex ion transport kinetics (such as those presented in Figs. 23.3 and 23.4) arose when Per Nissen presented an alternative interpretation to Epstein's hypothesis that the observed kinetics indicated the function of multiple K^+ transporters. Nissen's approach was to reanalyze the kinetic transport data of other researchers by using a "best fit" statistical program. Nissen's analysis was performed on primary transport data as well as on Lineweaver–Burke kinetic transformations, in which the inverse of the initial transport rate is plotted as a function of the inverse of the substrate concentration and yields a straight line for a transporter that exhibits simple Michaelis–Menten kinetics. In all cases, Nissen found the data were best fitted by a complex, multiphasic transport curve (see figure, where arrows mark the transitions between the four separate transport "phases"). He suggested that each concentration-dependent kinetic curve was discontinuous and consisted of discrete phases; that each phase exhibited Michaelis–Menten kinetics; and that the multiphasic uptake curves for K^+, SO_4^{2-}, HPO_4^{2-}, Na^+, and other ions were the result of a single complex transporter for each ion. Nissen argued that no examples had been found of multiple or redundant transporters for a particular ion and that a complex, multiple-subunit transport protein yielded the complex multiphasic kinetics.

Recent work on ion transport indicates that Epstein and Nissen were both correct, to a certain degree. From the results of cloning and characterization of different plant K^+ transporters, including several different root K^+ transporters, we now know that Epstein was correct in hypothesizing the existence of multiple K^+ transporters operating over a range of K^+ concentrations; however, the identity and function of high- and low-affinity transporters in the root remain confusing and controversial topics. Apparently Nissen was also correct: The characterization of K^+ transport by some cloned transporters has shown that a single K^+ transporter can yield complex kinetics, at least when expressed in a heterologous system. For example, K^+ transport by cells of *Arabidopsis* and of yeast expressing the AtKUP protein (see Section 23.2.6) and by yeast and *Xenopus* oocytes expressing the KAT1 protein demonstrates K^+ uptake kinetics similar to those presented in Figure 23.3 for K^+ uptake by roots of intact plants.

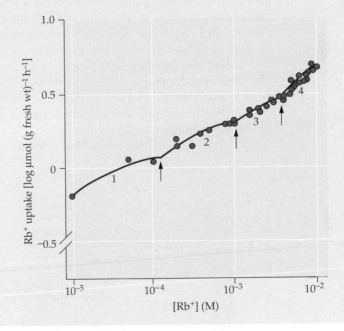

more negative in relation to the apoplast) and thereby increases the activity of low-affinity K^+ channels and the driving force for low-affinity K^+ influx.

Recent biophysical investigations of root K^+ transport have provided evidence for a high-affinity K^+ transport mechanism that is coupled to protons in a different manner. Study of a high-affinity K^+ transporter in *Arabidopsis* roots as the transmembrane gradients for Ca^{2+}, Na^+, and H^+ were varied showed that the transporter responded only to changes in the H^+ gradient and did so in a way that conformed to the operation of a K^+–H^+ symporter with a 1:1 transport stoichiometry. Evidence for a similar transporter was found in a fungus, *Neurospora*, that shares many features of membrane transport with plants, including a P-type plasma membrane H^+-ATPase and a high-affinity

K$^+$ transporter. The K$^+$ concentration–dependent kinetics for high-affinity ^{42}K$^+$ influx and the K$^+$-associated inward current in *Neurospora* shared the same K_m (14.9 µM), whereas the V_{max} current measured electrically was twice the rate of K$^+$ influx that was measured with radioisotopes (Fig. 23.5). That is, for every K$^+$ that entered the cell, an additional positive charge also entered. The investigators determined that the second positive charge entering the cell was a H$^+$, and the transport cycle for the high-affinity K$^+$ transporter and the H$^+$-ATPase indicated that for every two H$^+$ extruded by the ATPase, one H$^+$ returned to the cell by way of the K$^+$–H$^+$ symporter.

Patch clamp studies on root protoplasts have shown that one major component of low-affinity K$^+$ uptake involves inward-rectifying (K$^+_{in}$) channels (see Chapter 3). Patch clamp analysis of protoplasts derived from various types of cells (e.g., root cells, guard cells, leaf cells, and aleurone cells) revealed the presence of two types of K$^+$ channels: inward-rectifying channels, which open on hyperpolarization of E_m and facilitate K$^+$ uptake, and outward-rectifying (K$^+_{out}$) channels, which open on depolarization of E_m and transport K$^+$ out of the cell. The outward-rectifying channels are involved in osmotic adjustment, stomatal function, regulation of E_m, and the unloading of salts from xylem parenchyma into xylem vessels for long-distance transport to the shoot. Plant K$^+_{in}$ channels differ from their animal counterparts by not inactivating readily and by remaining open for long periods. These properties are important if they are to function as a major transport pathway for acquisition of K$^+$ from soil. Kinetic attributes, selectivity, and conductance data indicate that K$^+_{in}$ channels in the root plasma membrane correspond to the low-affinity K$^+$ transporters initially identified in radiotracer K$^+$ transport kinetic studies.

When all of the biophysical studies of plant K$^+$ transport are considered, the preponderance of evidence supports the hypothesis that several distinct K$^+$ transporters operate at different sites along the transport pathway in plants. In terms of K$^+$ absorption by roots, probably one or more high-affinity K$^+$ transporters, acting as K$^+$–H$^+$ symporters energized by H$^+$-ATPase, function in parallel with one or more types of K$^+_{in}$ channels that mediate low-affinity uptake and are important for K$^+$ acquisition at higher concentrations of soil K$^+$.

23.2.4 Molecular investigations have identified many plant genes that encode K$^+$ transporters.

The physiological and biophysical studies described above have been useful in increasing our understanding of plant K$^+$ transport and nutrition. However, these approaches have limitations and have not fully elucidated the mechanisms of K$^+$ transport in complex organs such as roots. The recent cloning and characterization of genes encoding K$^+$ channels or their component subunits, as well as genes for high-affinity transporters, have opened new avenues of study relating to plant K$^+$ nutrition. As researchers gain new information about the range of different K$^+$ transporters and the relationship between transporter structure, function, and regulation, this knowledge can be integrated with the wealth of physiological information concerning K$^+$ transport to better understand the roles K$^+$ can play in the functioning of the intact plant.

Biochemical approaches primarily have been successful for characterizing H$^+$-ATPases, transport proteins that are relatively

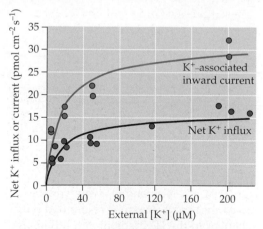

Figure 23.5
Stoichiometry of the proposed K$^+$–H$^+$ cotransport system in *Neurospora crassa*. The smooth curves are Michaelis–Menten functions fitted to either the K$^+$ concentration dependence for net K$^+$ uptake (red) or the K$^+$-associated inward current (green). The K_m for both curves is 14.9 µM; the V_{max} is 15.3 ± 1.0 pmol cm^{-2} s^{-1} for the K$^+$ flux and 30.1 ± 1.6 pmol cm^{-2} s^{-1} for the measured current.

abundant in the plasma membrane and tonoplast. The apparently low abundance of plant K⁺ transporters in plant cell membranes has confounded attempts to purify them by biochemical techniques. Until recently, screening cDNA libraries with heterologous probes that encode K⁺ transporters from other organisms has had limited success. A more successful method for cloning plant K⁺ transport genes as well as a range of other plant solute transporters has involved functional expression cloning in the yeast, *Saccharomyces cerevisiae*. This approach, in which the expression of a heterologous gene restores function to a known mutant of yeast, has been used to clone a diverse range of plant transporters. In addition to *KAT1* and *AKT1* (see below), this technique has been used to identify plant genes that encode transporters of sugars, amino acids, NH_4^+, and SO_4^{2-}.

The first K⁺ transport genes to be cloned in plants, *KAT1* and *AKT1*, were cloned by complementation of the yeast K⁺ transport–defective mutant *trk1trk2* with an *Arabidopsis* cDNA library (Fig. 23.6). *KAT1* and *AKT1*, which share extensive homology, were found to encode members of the Shaker superfamily of K⁺ channels (see Chapter 3). Until *KAT1* and *AKT1* were identified, the Shaker gene superfamily was thought to encode only voltage-dependent, K^+_{out} channels. *KAT1* and *AKT1*, however, encode K^+_{in} channels—identified on the basis of their current–voltage properties characterized in voltage clamp studies in three different het-erologous systems: yeast, *Xenopus* oocytes, and insect cells. Figure 23.7, which depicts a typical current–voltage relationship for *KAT1* expressed in *Xenopus* oocytes, shows that the channel is activated by the shift of E_m to more negative voltages and then mediates a strong inward current (K⁺ influx). The transport characteristics of *KAT1*- and *AKT1*-encoded proteins possess the hallmark properties of K^+_{in} channels studied by patch-clamping plant cells, including their voltage dependence, selectivity for K⁺ over other cations, time-dependent kinetics, lack of inactivation, and responses to the K⁺ channel blockers TEA and Ba^{2+}.

KAT1 and AKT1 share structural features with Shaker-type channels, the specific structural motifs of which are illustrated in Figure 23.8. Hydropathy analysis predicts six membrane-spanning hydrophobic domains near the N terminus (S1–S6), a conserved voltage-sensing region within S4, and an amphipathic P-domain (also called H5) between S5 and S6. The P-domain is highly conserved in all Shaker-type channels and has been proposed to form the channel pore that governs ion permeation and selectivity. The C terminus of each channel includes a putative nucleotide-binding domain that might be involved in transport regulation. In addition, AKT1 contains a C-terminal series of ankyrin-like repeats that has been proposed to link channel proteins to other proteins such as the cytoskeleton. According to present understanding of Shaker-type channels, functional plant K^+_{in} channels

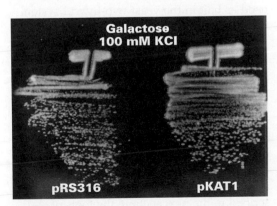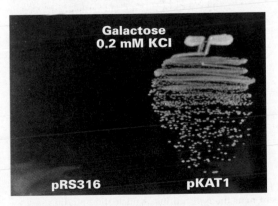

Figure 23.6

Growth of *trk1trk2* K⁺ transport–defective mutant of *Saccharomyces cerevisiae* after transformation with either the empty pRS316 vector (control) or pKAT1. Note that the control and the mutant expressing *KAT1* grow equally well on high-K⁺ medium (100 mM; left panel), whereas at a low concentration of K⁺ (0.2 mM), only the cells expressing *KAT1* can grow (right panel).

probably consist of four subunits of the peptide depicted in Figure 23.8. The S4 transmembrane domains contain positively charged basic amino acids at every third or fourth position. The placement of these charged amino acids within the hydrophobic

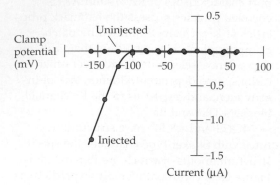

Figure 23.7
Current–voltage relationship for uninjected *Xenopus* oocytes (blue) and oocytes injected with *KAT1* mRNA four days before this experiment (red). *KAT1* mRNA leads to development of inward current in response to hyperpolarizing voltages more negative than –100 mV, which is typical of an inward-rectifying K⁺ channel.

environment of the membrane is consistent with the premise that this region senses voltage and causes conformational changes in response to shifts in E_m that lead to opening or closing of the channel.

The cloning of *KAT1* and *AKT1* led to the identification of additional related genes in *Arabidopsis* and other plant species, indicating the existence of a multigene family encoding K⁺ transporters. As more K⁺ transporter–encoding genes were cloned from plants and other organisms, the increased availability of sequence information and the development of new cloning approaches made it possible to use techniques other than functional complementation to clone K⁺ transport genes, such as DNA hybridization approaches based on sequence homology, computer-assisted searches of expressed sequence tags (ESTs) and other sequence databases, and yeast two-hybrid screens (see Chapter 21). The growing number of known *KAT1*- and *AKT1*-like genes is making it clear that different members of the family are localized to different cell types or plant tissues. *KAT1* and its ortholog in potato, *KST1*, are localized to guard cells and presumably encode K⁺$_{in}$ channels involved in stomatal function. *AKT1*, on the other hand, is localized to the root epidermis and cortex; its presumed role in plant nutrition is discussed below. *AKT2* (also sometimes designated *AKT3*), which shares 60% identity with *AKT1*, is expressed in leaves and not roots, suggesting that it encodes the same K⁺ transport function in leaves as *AKT1* mediates in roots. The C-terminal domain of the potato K⁺ channel, *KST1*, has been used as bait in the yeast two-hybrid system to identify proteins that associate with KST1. This approach, expected to identify proteins that regulate the channel or anchor it in the membrane, instead has led to the cloning of two new K⁺ channel genes, *SKT1* and *SKT2*, which exhibit substantial similarity to *Arabidopsis AKT2*. Subsequent investigations have determined that the C termini of all K⁺$_{in}$ channels contain conserved sequences that are apparently involved in the association of K⁺ channel subunits. As described in the next section, the tools of molecular biology can be used not only to identify genes but also to acquire additional information regarding the function of their products.

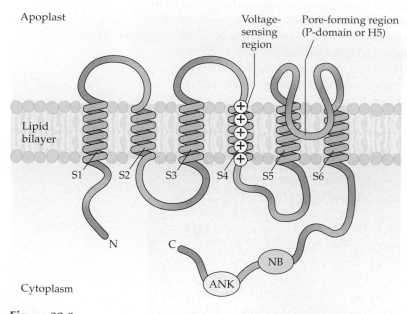

Figure 23.8
Proposed structural model for AKT1, a plant K⁺$_{in}$ channel. Six membrane-spanning domains are present near the N terminus. A nucleotide-binding sequence (NB) and ankyrin-like domains (ANK) are found near the C terminus. The S4 region, believed to be the voltage sensor, contains several positively charged amino acids within the membrane-spanning region. The P-domain, also referred to as H5, is thought to form the mouth of the pore and to play a critical role in channel selectivity.

23.2.5 Molecular techniques have also been used to study how the structure of transporters influences their function.

In addition to its utility in cloning, the expression of plant ion transport genes in heterologous systems is a powerful tool for investigating the mechanisms of transport and elucidating the relationships between protein structure and function. Several such approaches have been used in recent studies of K^+ transporter activity. The expression of plant K^+ transporters in yeast mutants that lack endogenous plasma membrane K^+ transporters generates an experimental system ready-made for functional characterization. Populations of yeast cells are amenable to physiological investigation of transport using radioisotopes, and single yeast cells can be studied by using patch clamp techniques to assess biophysical properties of transport. Expression of plant transporters after injection of cRNA into *Xenopus* oocytes yields a large single-cell system amenable to current–voltage analysis with two microelectrodes. A more recently developed heterologous system involves expressing plant K^+ transporters in cultured insect cells by using recombinant baculoviruses, followed by patch clamp studies with the insect cells. For study of transporters that are not easily expressed in *Xenopus* oocytes, the baculovirus/insect cell system may provide a useful alternative. For example, no laboratory has been successful in studying *AKT1* in oocytes. However, biophysical characterization of AKT1 activity has been carried out by patch-clamping insect cells that express *AKT1*.

The coupled use of heterologous systems and site-directed mutagenesis allows researchers to elucidate the relationship between structure and function of K^+ channels. To date, this approach has focused primarily on the highly conserved pore region (P-domain or H5) of K^+ channels. The ability of plant K^+_{in} channel genes to complement K^+ transport–defective yeast mutants means the use of a genetic approach to identifying specific amino acids involved in channel K^+ selectivity is suitable. Using this approach, several groups have demonstrated the importance of a highly conserved tripeptide (Gly-Tyr-Gly) found in the P-domains of all highly selective K^+ channels. Mutations in this tripeptide, or in a threonine residue proposed to interact with the peptide, causes major changes in the selectivity of KAT1 for K^+ over other cations. In principle, this approach should also be useful for crop improvement by engineering changes in the selectivity of specific mineral ion transporters, which could enhance nutrient efficiency or prevent the entry of toxic cation analogs (e.g., Na^+) into cells.

Molecular information concerning K^+_{out} channels is now accumulating. The first K^+_{out} channel was cloned from *Arabidopsis* by searching an EST database for a Thr-X-Gly-Tyr-Gly-Asp motif that is highly conserved in the P-domain of K^+ channels. This K^+_{out} channel gene, *KCO1*, belongs to a new class of two-pore K^+ channels identified in yeast and humans. The deduced amino acid structure of the protein consists of four membrane-spanning domains and two P-type pore regions (see Chapter 3, Fig. 3.38). K^+ transport by way of the *Arabidopsis KCO1*-encoded pathway was studied by patch-clamping insect cells that expressed this gene. On depolarization of E_m, KCO1 was found to mediate outward-rectifying K^+ currents. Interestingly, KCO1-activated K^+ currents were influenced strongly by changes in cytoplasmic $[Ca^{2+}]$, requiring Ca^{2+} concentrations greater than 150 nM for activation. This finding, along with the observation that *KCO1* contains two EF-hand Ca^{2+}-binding motifs, indicates a link between Ca^{2+}-mediated signaling processes and K^+ transport by way of this channel. *KCO1* is expressed in low amounts in *Arabidopsis* and is not localized to any particular tissue, being found in whole seedlings, leaves, and flowers.

Although the K^+ transport characteristics of KCO1 and its regulation by cytosolic Ca^{2+} show similarities to some K^+_{out} channels studied in plant cells, no specific physiological function has been ascribed to this channel. K^+_{out} channels are thought to affect plant K^+ nutrition by way of xylem parenchyma cells, which load mineral ions into xylem vessels. To isolate xylem parenchyma protoplasts from barley roots for patch-clamping, the stele is dissected from the root, after which the cell walls are digested enzymatically. As shown in Figure 23.9, enzymatic digestion releases protoplasts specifically from the xylem parenchyma bordering the early metaxylem vessels. Patch

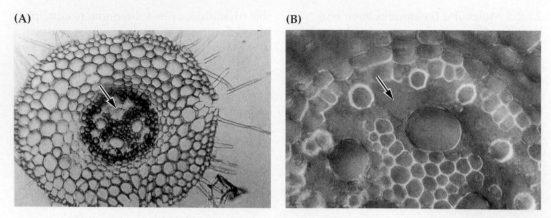

Figure 23.9

Light micrographs of barley root cross-sections, depicting the differential cell wall hydrolysis after a two-hour exposure to the cell wall–digesting enzymes used to isolate xylem parenchyma protoplasts. (A) Effect on the entire root; (B) selective digestion of the walls (highlighted at higher magnification) surrounding the xylem parenchyma adjacent to early metaxylem vessels, with release of xylem parenchyma protoplasts. Arrows indicate the region of selective digestion and protoplast release.

clamp studies with these cells have identified three types of ion channels in the xylem parenchyma plasma membrane: K^+_{in} channel and two channels—one selective for K^+ and the other a nonselective cation channel. None of these channels fits the characteristics exhibited by KCO1 in insect cells,

and a combined molecular–biophysical analysis of xylem transport awaits future investigation.

23.2.6 The cloning of high-affinity K^+ transporters has yielded controversy.

The gene for the first putative plant high-affinity K^+ transporter, *HKT1*, was cloned by complementation of the yeast *trk1trk2* K^+ transport mutant with a cDNA library isolated from K^+-starved wheat roots. The deduced amino acid sequence of HKT1 predicted the protein would have a mass of 59 kDa and contain 12 membrane-spanning domains. Although the transport function of HKT1 could not be predicted from homology to other cation transporters, radiotracer ($^{86}Rb^+$) flux studies in yeast cells confirmed that it mediated high-affinity K^+ uptake (K_m = 29 µM). Localization of *HKT1* mRNA was not consistent with that expected for a root transporter involved in uptake of K^+ from the soil, in that it was localized to the inner root cortex and stele, as well as to the vasculature of leaves and stems (Fig. 23.10). Expression of *HKT1* in *Xenopus* oocytes gave rise to large inward currents in the presence of K^+. Although the transport mechanism was not rigorously investigated, current–voltage analysis performed while varying the external pH or K^+ concentration indicated that the shifts in the zero current potential (E_{rev}, or reversal potential; see Chapter 3)

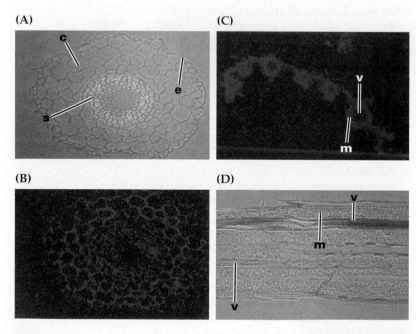

Figure 23.10

In situ localization of *HKT1* mRNA in root and leaf tissues. Note that *HKT1* expression localizes to the root cortex (c) and to a lesser extent the stele (s), but not to the root epidermis (e). In leaves, *HKT1* mRNA is found in the mesophyll cells (m) that surround the vasculature (v). (A) Bright-field image of root in cross-section. (B, C) Dark-field images of root (B) and leaf (C) in cross-section. The hybridization pattern is seen in red. (D) Bright-field image of leaf in longitudinal section.

were consistent with the operation of a K^+–H^+ cotransporter having a K^+:H^+–uptake stoichiometry of 1:1.

Subsequent research suggested that HKT1 is a Na^+–K^+ cotransporter. As shown in Figure 23.11, high-affinity $^{86}Rb^+$ uptake by yeast cells expressing *HKT1* is stimulated by increasing the external concentration of Na^+, whereas $^{22}Na^+$ influx is stimulated by increasing the external concentration of K^+. Electrophysiological investigations of K^+ and Na^+ currents mediated by HKT1 in oocytes demonstrate that the protein is much more selective for K^+ and Na^+ than for Rb^+ and the other alkali cations, suggesting that the transporter might be an important route for the entry of toxic Na^+ into plant roots. These findings, which led to the controversial hypothesis that the putative root high-affinity K^+ transporter is a Na^+–K^+ cotransporter, raised the possibility that altering this transporter might enhance the salt tolerance of crop plants. Precedents do exist in the plant kingdom for sodium-energized ion transport. Charophytic algae, which grow in brackish waters containing considerable Na^+, possess a Na^+-coupled high-affinity K^+ uptake system. Moreover, a Na^+-coupled K^+ transport system operates in the plasma membranes of leaf cells of the aquatic plant, *Egeria*. However, an exhaustive survey of high-affinity K^+ uptake in roots of wheat, *Arabidopsis*, and barley, based on electrophysiological analysis and $^{86}Rb^+$ flux studies, provided no evidence for such a system in the roots of any of these terrestrial plant species.

The hypothesis that a Na^+–K^+ cotransporter mediates high-affinity K^+ transport in roots has been disputed for several reasons. First, it is difficult to envisage a transport system in roots through which absorption of an essential element (K^+) is energized by the influx of a toxic cation such as Na^+. Second, although most soils contain low concentrations of Na^+ (0.1 to 1 mM), Na^+ is not essential or beneficial to the growth of most terrestrial plants. The high concentrations of Na^+ present in saline soils (as much as 200 mM) are toxic to most plant species. Finally, some of the cation transport properties observed for HKT1 in yeast and oocytes do not resemble the K^+ transport characteristics previously shown for high-affinity K^+ transport in plant roots. Physiological experiments on the activity of the high-affinity K^+ uptake system in roots of barley or maize indicate that K^+ influx is not affected by exposing the roots to external Na^+ concentrations as great as 50 mM. Furthermore, the high-affinity K^+ transporters in barley and maize transport K^+ and Rb^+ equally well, whereas HKT1 expressed in yeast and oocytes has a preference for K^+ (and Na^+) over Rb^+.

The controversy surrounding HKT1 is unresolved. The apparent disparity between the activity of HKT1 expressed in yeast cells or oocytes and the characteristics of the high-affinity K^+ uptake system studied in

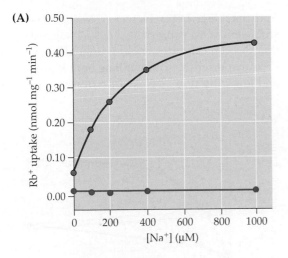

(A)

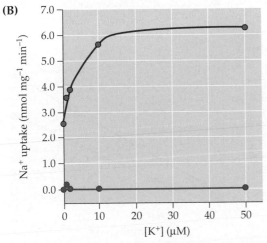

(B)

Figure 23.11
Radiotracer $^{86}Rb^+$ and ^{22}Na flux measurements in yeast cells expressing *HKT1* (red) and in control yeast cells deficient in K^+ uptake (blue). The expression of *HKT1* in yeast yielded Na^+-stimulated Rb^+ uptake (A) and K^+-stimulated Na^+ uptake (B). Based on findings such as these, and results from electrophysiological experiments in oocytes expressing *HKT1*, HKT1 was proposed to be a K^+–Na^+ cotransporter that mediates high-affinity K^+ uptake.

Extracellular space
47
65
186
198
244
270
325
342
413
419
470
476
29
82
166
216
224
289
306
361
395
438
452
496
N
C
Cytoplasm

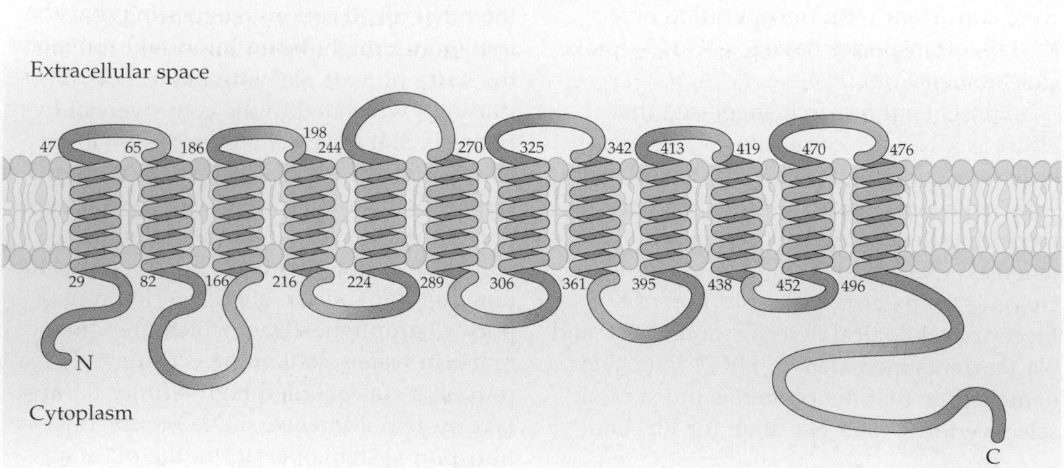

Figure 23.12
Proposed structural model for a plant high-affinity K^+ transporter, AtKUP1, which is predicted to contain 12 membrane-spanning domains.

roots may reflect the difficulties associated with extrapolating results from heterologous systems to intact plants. Although cloning the genes encoding plant transporters and expressing the genes in heterologous systems are powerful tools for gaining insight into transport mechanisms, researchers must take into account that the activity of plant transporters can be affected by the foreign environment of a yeast, an oocyte, or an insect cell. An alternative explanation may be that HKT1 is not the same as the high-affinity transporter studied in intact roots. This possibility is supported by the observation that *HKT1* mRNA is not expressed in the root epidermis and outer cortex, which should be the primary absorption site for high-affinity uptake of K^+ from the soil.

Members of another family of genes encoding putative high-affinity K^+ transporters, cloned in *Arabidopsis* by three different laboratories, are designated *AtKUP1* and *AtKUP2*. This transporter gene family was cloned on the basis of sequence similarity to the *Kup*-encoded K^+ transporter from *Escherichia coli* and to the *HAK1*-encoded K^+ transporter from *Schwanniomyces occidentalis* yeast. *AtKUP1* (Fig. 23.12) and *AtKUP2* are members of a new family of genes encoding K^+ transporters for which the encoded protein has a predicted mass of 79 kDa and is predicted to contain 12 membrane-spanning domains. These transporters appear to be expressed in roots, leaves, and flowers. Based on $^{86}Rb^+$ flux studies in transgenic *Arabidopsis* suspension cells expressing *AtKUP1*, the gene appears to encode primarily a high-affinity K^+ transporter with the ability to mediate a small amount of low-affinity K^+ transport (Fig. 23.13). However, transport studies in yeast expressing *AtKUP1* suggest that the gene encodes either a dual-affinity K^+ transporter that can mediate both high- and low-affinity K^+ uptake or a transporter

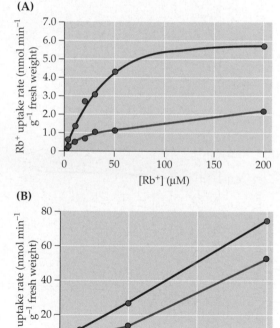

(A)
7.0
6.0
5.0
4.0
3.0
2.0
1.0
0
Rb+ uptake rate (nmol min−1 g−1 fresh weight)
0
50
100
150
200
[Rb+] (μM)
(B)
80
60
40
20
0
Rb+ uptake rate (nmol min−1 g−1 fresh weight)
0
1.0
2.0
3.0
[Rb+] (mM)

Figure 23.13
Concentration-dependent kinetics of Rb^+ (a K^+ analog) uptake in transgenic *Arabidopsis* suspension cells expressing *AtKUP1*. (A) Kinetics of Rb^+ uptake from a micromolar solution by cells transformed with *35S:AtKUP1* (red) or an empty vector (blue). Here, AtKUP1 is shown to mediate high-affinity K^+ uptake. (B) Kinetics of Rb^+ uptake from a millimolar solution reveal that AtKUP1 can also facilitate low-affinity K^+ uptake.

with only a low affinity for K^+ (Fig. 23.14). This area of research will require further study and examination to clarify the nature of the AtKUP transport mechanism.

To summarize, numerous genes that encode K^+ channels or high-affinity K^+ transporters have been cloned from plants (Table 23.2). These experimental results indicate the existence of a multigene family encoding K^+ transporters.

(A)

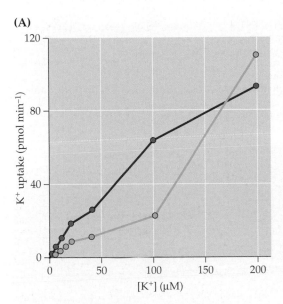

(B)

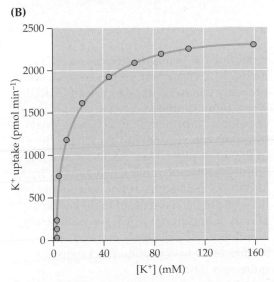

Figure 23.14

(A) Concentration-dependent kinetics for K^+ uptake by yeast expressing the *Arabidopsis* gene *AtKUP1* (green) and the *Saccharomyces* high-affinity K^+ transporter gene *TRK1* (red) from a micromolar solution. (B) Yeast expressing *AtKUP1* take up K^+ from millimolar solutions. These data reinforce the hypothesis that a single K^+ transporter can mediate both high- and low-affinity uptake.

23.2.7 Information gained from molecular and biophysical studies has furthered our understanding of whole-plant K^+ nutrition.

For many of the cloned K^+ transporters, we now have some understanding of their transport mechanism, primarily from studies in heterologous systems. What is lacking, however, is an understanding of how these transporters function and interact to regulate the key aspects of plant mineral nutrition. The task facing plant biologists is to integrate the information gained from molecular and biophysical investigations with our understanding of the physiology of K^+ transport in plants. One approach to studying K^+ transporters in plants is the use of reverse genetics. Promising reverse genetics strategies have been developed that allow a mutational analysis approach to studying plant ion transporters. The transferred DNA (T-DNA) of *Agrobacterium tumefaciens* is used to generate insertional mutagens in the plant genome. Integration of the T-DNA serves two purposes: knocking out the function of a specific gene and simultaneously tagging the gene for subsequent cloning. Because the T-DNA sequence is known, PCR primers for T-DNA and for identified plant transport genes of interest can be designed to detect the presence of the T-DNA within a particular gene. This approach has been used to screen 9100 T-DNA-transformed *Arabidopsis* lines for mutations in genes that are involved in signal transduction and ion transport.

Many activities essential to plant function, including transport of K^+, appear to be controlled by multigene families. The functional redundancy implied by the presence of multiple genes for regulating a specific process such as K^+ transport complicates the use of gene knockout reverse genetics. Nonetheless, it should be possible to knock out each member of a gene family. If knocking out a single, specific gene does not result in an altered phenotype, genetic crosses can be used to create double or multiple mutants until a transport-impaired phenotype is observed. Using this approach, investigators should be able to elucidate how specific K^+ transport gene products act individually and in combination to regulate plant K^+ transport.

This approach has been used to obtain an *Arabidopsis* mutant in which the *AKT1* gene is disrupted. Electrophysiological and

Table 23.2 Summary of cloned plant genes encoding K⁺ transporters

Gene	Species	Cloning approach	Localization
KAT1	Arabidopsis thaliana	Yeast complementation	Guard cells and vascular tissue
KST1	Solanum tuberosum	DNA hybridization using KAT1 sequence	Guard cells and flower buds
AKT1	A. thaliana	Yeast complementation	Epidermal and cortical root tissues
AKT2/13	A. thaliana	DNA hybridization using KAT1 or Shaker sequences	Leaves
SKT1	So. tuberosum	DNA hybridization using AKT1 sequence	Roots, epidermal tissues of leaves
SKT2	So. tuberosum	Yeast two-hybrid using KST1 as bait	?
SKT3	So. tuberosum	Yeast two-hybrid using KST1 as bait	?
KCO1	A. thaliana	Computer search of EST database for P-domain	Flowers, whole seedlings, leaves
SKOR	A. thaliana	DNA hybridization using pore domain of KAT/AKT channel family members	Root stellar tissues
HKT1	Triticum aestivuum	Yeast complementation	Inner cortex of roots, root vasculature
AtKUP1	A. thaliana	Yeast complementation, computer search of EST database for sequences homologous to kup (Escherichia coli) and HAK1 (Schwanniomyces occidentalis)	Roots, leaves, flowers
AtKUP2	A. thaliana	Yeast complementation, computer search of EST database for sequences homologous to kup (E. coli) and HAK1 (Sc. occidentalis)	Roots, leaves, flowers
HvHAK1	Hordeum vulgare	RT-PCR using sequence similarity to fungal HAK and E. coli Kup K⁺ transporters	Roots

RT-PCR, reverse transcriptase–polymerase chain reaction.

radiotracer ($^{86}Rb^+$) flux studies indicate that the roots of the mutant plant lack K^+_{in} channel activity and exhibit reduced Rb^+ uptake even at concentrations as low as 10 to 100 μM. Growth of this mutant was less robust than that of wild-type plants when the K^+ concentration in the growth media was 100 μM or less. These findings indicate *AKT1* is important for obtaining K^+ from the soil even at low K^+ concentrations. Incidentally, this phenotype (reduced Rb^+ uptake and poor growth in low K^+) was observed only when high concentrations (2 to 4 mM) of ammonium were included in the uptake or growth media. Whole-plant studies have already shown that high-affinity K^+ uptake is inhibited by NH_4^+; perhaps the bona fide high-affinity transporter mediates both K^+ and NH_4^+ uptake. Another possibility is that when K^+ concentrations are low, high concentrations of NH_4^+ inhibit another transport pathway that provides K^+ to the plant, thus rendering the plants dependent on AKT1-mediated K^+ transport. In this context, NH_4^+ probably blocks K^+ uptake through AtKUP1. These findings underscore the complexity of plant K^+ nutrition, in which individual transport components function interactively in the intact plant. Does AKT1 act as a high-affinity transporter only when another high-affinity K^+ transporter is not functioning? Or does our understanding of high- and low-affinity transporters, which has been strongly influenced by the early carrier-kinetic studies of K^+ transport in intact roots, require revision?

23.3 Phosphorus nutrition and transport

Phosphorus nutrition is unique in that P availability is one of the major constraints to growth for plants in natural ecosystems, and P is probably the most important fertilizer input for crop plants (see Fig. 23.1). Unlike K^+, P can exist in plants as both inorganic phosphate anions (e.g., P_i), and organophosphate compounds. Unlike nitrate and sulfate, phosphate is not reduced in plants during assimilation but instead remains in its oxidized state, forming phosphate esters in a wide range of organic compounds. P_i constitutes an important structural component of nucleic acids and phospholipids, plays a critical role in energy conversion in the form of high-energy phosphoester and diphosphate bonds, is important both as a substrate and regulatory factor in photosynthesis and oxidative metabolism, participates in signal transduction, and regulates the activities of a diverse array of proteins by way of covalent phosphorylation/dephosphorylation reactions.

Because of its low solubility and high sorption capacity in soil, phosphate is relatively unavailable to plant roots; hence, P supply is one of the major constraints to plant growth (Fig. 23.15). Soil phosphate concentrations are often 1 µM or less, and P is rapidly depleted from the rhizosphere by roots. Thus plants have evolved several strategies for obtaining P from the soil. Roots exhibit an impressive plasticity in response to low availability of P and can modify their structure and function at the cellular, organ, and organ system levels. In response to low P, some plants can directly modify the rhizosphere to obtain soil P that would be otherwise unavailable.

23.3.1 Physiological investigations indicate phosphate is transported into roots by an active, high-affinity mechanism.

Research into the physiology of root P_i absorption by using radiolabeled $^{32}P_i$ to measure unidirectional influx, along with investigations monitoring P_i depletion to quantify net P_i uptake, demonstrate that $H_2PO_4^-$ is the primary form of P transported into root cells. Because root cells must transport this anion against a sizeable negative E_m inside the cell, and because P_i concentration in the soil solution adjacent to roots is 1 µM or less whereas cytoplasmic P_i is in the millimolar range, transport of P_i into root cells is an active process. Usually 25 to 40 kJ is required to transport 1 mol of P_i into root cells, or about the same amount of energy derived from the hydrolysis of 1 mol of ATP. Thus, the transport mechanism for active P_i influx into cells of the root must either be an ATPase, or a secondary active transporter indirectly coupled to the transmembrane electrochemical H^+ gradient generated by the plasma membrane H^+-ATPase. On the basis of physiological studies conducted during the 1970s and 1980s, some investigators have proposed that root P_i absorption as well as the uptake of the other important mineral anions (e.g., NO_3^- and SO_4^{2-}) are mediated by H^+-coupled cotransporters. The evidence in support of this proposal is indirect and can be summarized as follows:

- Root P_i uptake is often accompanied by an alkalinization of the external media, which translates into a $H^+:H_2PO_4^-$ flux stoichiometry of between two and four protons for every P_i.
- After accounting for the effects of changes in pH on P_i speciation, root P_i uptake is enhanced when researchers lower the pH of the uptake solution.

(A)

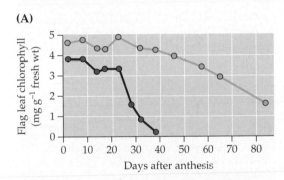

(B)

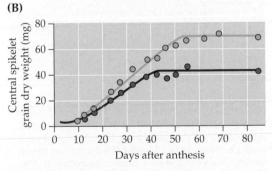

Figure 23.15
P nutrition influences leaf chlorophyll content and accumulation of spikelet biomass in wheat. P deficiency (red) has a strong negative effect on both chlorophyll content (A) and dry weight accumulation (B). Data from P-sufficient control plants are plotted in green.

- Electrophysiological measurements of root P_i influx show that the process is accompanied by a transient depolarization of E_m, followed by a repolarization (and often a hyperpolarization) of E_m. This is consistent with the operation of a H^+–P_i cotransporter to transport net positive charge into the cell (two or more H^+ for every $H_2PO_4^-$). The subsequent acidification of the cytoplasm should stimulate the H^+-ATPase, which would explain the subsequent repolarization.
- Studies using pH microelectrodes and pH-sensitive fluorescent dyes have shown that P_i absorption acidifies the cytoplasm.
- Protonophores such as CCCP, a protonophore that can collapse transmembrane H^+ gradients, abolish root P_i absorption.

The concentration-dependent kinetics of P_i uptake into plant cells have not been studied in the detail described for K^+. However, several studies using roots, leaves, and cell suspensions from different plant species and physiologically relevant concentrations of P_i (low micromolar range) have demonstrated the existence of a high-affinity P_i transporter (K_m = 1 to 5 µM). This high-affinity transporter appears to be highly regulated with regard to plant P status. For example, if P is withheld from roots of tomato and barley for several days, high-affinity P_i influx is markedly stimulated in these plants. When P_i is resupplied to the roots, the influx of P_i declines within one to two hours. Probably an internal pool of P_i exerts a regulatory influence on the expression of P_i transporter genes and the activity of existing P_i-transport proteins, the latter possibly involving allosteric interactions between cytoplasmic P_i and protein structure. The inverse relationship between root P_i influx and plant P status can be seen in electrophysiological traces. Addition of P_i to roots of P-replete plants results in little or no depolarization of E_m, which correlates with the magnitude of P_i influx (Fig. 23.16A and B). However, in plants starved of P for 7 (Fig. 23.16C) or 14 days (Fig. 23.16D), depolarization increases in proportion to the severity of P deficiency. The relationship of root P_i uptake to P nutrition will be revisited when discussing molecular investigations of root P_i transport.

23.3.2 Roots utilize a range of strategies to increase the bioavailability of P in the rhizosphere.

Because of the high sorption of phosphate to soil particles, P is the macronutrient least available to plant roots. Phosphate ions can bond tightly to the surface of soil clay minerals, forming strong chemical bonds with oxides of Fe and Al in the soil. In calcareous soils, similar reactions occur between P_i and calcium carbonate. Additionally, soil microbes can effectively convert P_i to organic forms of P that are not readily utilized by plant roots. Plants have developed a fascinating array of strategies to mine otherwise unavailable soil reserves of P. In response to P stress, the structure and function of root systems can be altered to increase the solubility of soil P and to enhance the

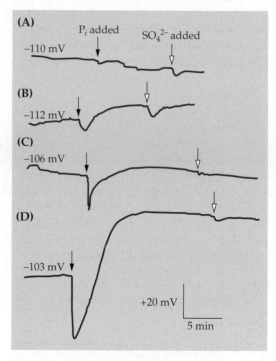

Figure 23.16

Phosphate-induced transients in transmembrane electrical potential in roots of *Trifolium repens*. The solid arrows indicate the addition of 125 µM phosphate to media buffered at pH 5 with MES buffer and the resulting change in the root membrane potential. The open arrows indicate the addition of 63 µM sulfate as a control for the general effect of mineral anions on the membrane potential. (A) Plant grown on P_i for the entire 29-day period; (B) P_i withheld for final day of growth; (C) P_i withheld for last 7 days of growth; (D) P_i withheld for last 14 days of growth.

exploration of soils by root systems. Root-based responses to low P availability include the following:

- **mycorrhizal associations** between roots and some soil fungi to acquire P in non-rhizosphere soil
- alterations in root architecture and branching to explore the soil more effectively
- increases in root hair density and length to increase the absorptive area of roots and reduce the length of the diffusive pathway for P to reach the root surface
- release of organic acids and H^+ to solubilize inorganic P
- exudation of phosphatases to release organically bound P from the soil
- up-regulation of high-affinity P_i transporters in the root cell plasma membrane

Expression of these traits varies among genotypes, not only among different plant species but also among cultivars within species. Thus, both traditional breeding and molecular approaches can be used to increase the ability of crops to utilize soil P more efficiently.

The beneficial effects of mycorrhizae on plant growth are well documented; probably their most important benefit is to increase the uptake of immobile nutrients from the soil, notably P_i. The predominant mycorrhizal associations that influence root P absorption involve the **endomycorrhizae,** also known as **vesicular-arbuscular mycorrhizae** (VAM; Fig. 23.17). VAM are the most abundant mycorrhizae and colonize roots of a wide range of plant species in almost all types of soils. They form extensive hyphal networks in the soil and within root cortical tissues, forming intimate associations with the plasma membrane but not penetrating the cell. VAM are characterized by the formation of branched haustoria (arbuscules) within the tissues of the root cortex that are the sites of solute exchange between the fungus and the host plant. The fungal mycelium extends into the soil. VAM enhance P_i absorption by root systems two- to sixfold over nonmycorrhizal roots. The stimulation of root P_i absorption by VAM is a result of a more thorough exploration of soil P reserves. The majority of P_i in the rhizosphere moves to the root by diffusion, which is a slow process in biological terms. Extensive hyphal growth, similar to increased root hair growth, reduces the distance that P must diffuse to reach the sites of absorption. The symbiotic relationship between fungus and plant involves the provision of P to the plant in exchange for fixed carbon. It is not clear how the absorbed P_i is transferred from fungi to the root or how the fixed C from the plant moves to the fungi. Perhaps P_i transfer from fungi to the plant occurs at the interfaces between fungal arbuscules and plant cortical cell and may involve reabsorption of fungal-delivered P_i by plant P_i transporters across the plasma membrane of the root cortical cells.

The regulation of P nutrition in plants is influenced substantially by mycorrhizal interactions with roots. In nonmycorrhizal

(A)

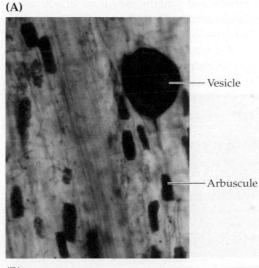

— Vesicle

— Arbuscule

(B)

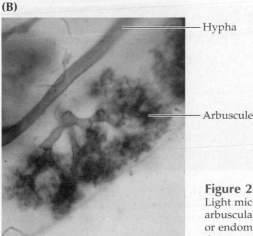

— Hypha

— Arbuscule

Figure 23.17
Light micrographs of vesicular-arbuscular mycorrhizae (VAM, or endomycorrhizae) showing the structure of vesicles (A) and arbuscles (B).

roots, P stress up-regulates the activity of high-affinity P_i transporters at the root cell plasma membrane. However, when roots are infected with VAM that are actively absorbing P from the soil, the endogenous plant P_i transporters become almost inactive. The molecular regulation of this response will be discussed in the next section.

An interesting model system for investigations of P changes in root proliferation and organic acid exudation caused by P starvation involves the induction of proteoid or cluster roots in white lupine (*Lupinus albus*) and several other nonmycorrhizal species. P deficiency triggers a proliferation of tertiary lateral roots from a secondary lateral root so that the lateral roots form a tight cluster; this enables the plant more effectively to extract immobile nutrients such as P from a limited soil volume (Figs. 23.18 and 23.19). The proteoid roots release organic acids, in particular citrate and malate, which are effective chelators of Al and Fe in the soil. The chela-

tion of soil Fe and Al releases P_i from Al–P and Fe–P complexes, making it available for root absorption. Lupine can release as much as 25% of its total fixed C as organic acids into the soil if necessary to solubilize P for absorption. The high price for P acquisition indicates the priority that plants in natural ecosystems place on obtaining this precious nutrient.

Detailed studies of the biochemistry of carbon metabolism in proteoid roots show changes in citric cycle enzyme activities as well as in the abundance of mRNA and protein for key enzymes associated with the citric acid cycle (see Chapter 14). Phospho*enol*-pyruvate carboxylase (PEPC) is a key enzyme in the P-stress response in proteoid roots. When proteoid roots emerge from secondary laterals, PEPC mRNA, PEPC protein, and

(A)

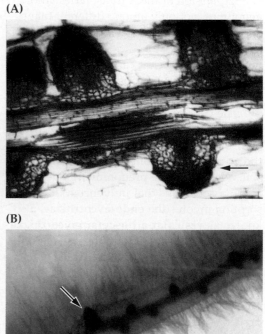

(B)

Days after emergence
5 6 7 8 10 12 14   22

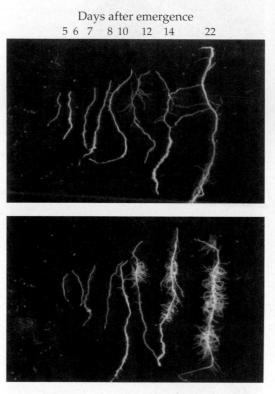

Figure 23.18

Intact lateral roots from *Lupinus albus* plants grown with 1 mM P (top panel) or without P (bottom panel). Top panel depicts normal lateral roots, whereas bottom panel shows typical proteoid roots from P-deficient plants. Numbers at top are number of days after emergence when roots were harvested.

Figure 23.19

(A) Longitudinal section of proteoid root from P-deficient *Lupinus albus* plant showing the dense cluster of emerging lateral roots. Solid arrow shows a developing tertiary root meristem. (B) Secondary lateral root excised from *L. albus* plant grown without P. Root was cleared with sodium hypochlorite and stained with methylene blue. By eight days after emergence, tertiary lateral roots are beginning to emerge through the epidermis (solid arrow). Bar = 1 mm.

specific activity are up-regulated, resulting in an increase in citrate and malate synthesis and enhanced exudation of these organic acids. Activities of citrate synthase and malate dehydrogenase also are increased, supporting the hypothesis that P deficiency triggers the coordinated regulation of proteoid root development, transcriptional regulation of PEPC, and subsequent increased synthesis of organic acids, which leads to release of the acids into the rhizosphere.

Another response to P stress is the release of phosphatases from roots. Plants synthesize a wide range of phosphatases, which are classified as alkaline or acid phosphatases based on their pH optima. The enzymes excreted from roots in response to P deficiency are acid phosphatases, which can hydrolyze a wide range of P-containing substrates. The roles of these secreted phosphatases remain to be established, but it is frequently speculated that they play a role in enhancing P_i availability in the rhizosphere by releasing P_i from organic sources. During the onset of P deficiency, the influx of P_i into the roots and the release of acid phosphatase from the roots are increased, although no direct link between the two responses has been established nor is there any firm evidence for the involvement of the phosphatases in acquisition. The role of secreted phosphatases in P acquisition from the soil remains an attractive, albeit speculative, hypothesis.

23.3.3 Molecular approaches provide insights into the complex regulation of P acquisition and nutrition in plants.

Recently, plant biologists have begun to dissect the molecular regulation of P_i nutrition by cloning genes that play a critical role in P_i acquisition. Much of the research in plants has benefited by earlier discoveries in yeast. Work with *S. cerevisiae* has shown that P_i acquisition is controlled by the *PHO* regulon. In response to P deficiency, several of the genes encoding P_i transporters and phosphatases are transcriptionally activated, including genes encoding both positive and negative regulatory factors in the cellular response to P. *PHO84*, a gene encoding a yeast high-affinity P transporter, has been cloned and characterized. In addition to *PHO84*, several other genes apparently are involved

in high-affinity P_i uptake in yeast, suggesting that P_i uptake could involve a multiple-subunit complex. High-affinity P_i transporters have been cloned recently from plants. An *Arabidopsis* EST has been identified that shows similarity to *PHO84*, and use of this EST to screen an *Arabidopsis* cDNA library has led to the isolation of two cDNAs that encode plant P_i transporters, *AtPT1* and *AtPT2*. Genes encoding similar P_i transporters have been cloned from other plant species. These transporters exhibit a high degree of similarity to PHO84 in yeast and to two other previously cloned fungal transporters from *Neurospora* and the mycorrhizal fungus, *Glomus versiforme* (Fig. 23.20).

All P_i transporters share important structural similarities (Fig. 23.21). The predicted secondary structure is characterized by six N-terminal transmembrane domains and six C-terminal transmembrane domains separated by a central hydrophilic region. This

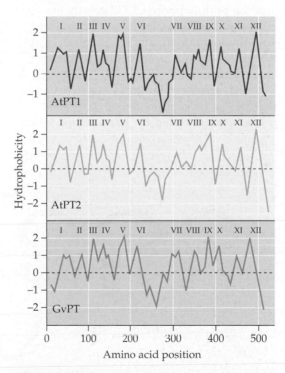

Figure 23.20
Hydrophobicity plots depicting the predicted protein structural similarities between two *Arabidopsis* P transporters (AtPT1 and AtPT2) and a P transporter cloned from the mycorrhizal fungus *Glomus versiforme* (GvPT). All three transporters share structural similarities, including six predicted N-terminal transmembrane domains (I–VI) and six C-terminal transmembrane domains (VII–XII) separated by a central hydrophilic region (six-loop-six structure).

structure, termed the six-loop-six structure, also is found in sugar and amino acid cotransporters from eukaryotes and prokaryotes. Additional structural motifs that are conserved in these P_i transporters include sites for protein kinase C– and casein kinase II–mediated phosphorylation and for N-linked glycosylation. Based on Southern DNA blot analysis, P_i transporters in plants apparently are encoded by members of a small gene family comprising two or three members in each of the plant species studied to date (*Arabidopsis*, potato, tomato, and alfalfa).

As was described in Section 23.3.1, P_i transporters in plants are accepted widely as utilizing H^+–P_i cotransport as the underlying mechanism. The findings from physiological studies of root uptake of P_i can be used to build a circumstantial case for P_i transport by way of H^+–P_i cotransporters. However, the molecular evidence for this type of transport mechanism is weak. When plant P_i transporters were expressed in yeast, P_i uptake increased as the external pH was decreased, and the protonophore CCCP greatly inhibited P_i uptake. These findings are consistent with the functioning of a H^+–P_i cotransporter, but more rigorous proof awaits biophysical characterization.

A P_i-uptake mutant of yeast deficient in PHO84 can be complemented with two P_i transporters encoded by genes cloned from potato (*StPT1* and *StPT2*). Radiotracer ($^{32}P_i$) flux experiments indicate that StPT1 and StPT2 have a lower affinity for P_i when expressed in yeast (K_m values of 280 and 130 µM for StPT1 and StPT2, respectively) than was previously observed in roots (K_m = 1 to 10 µM). This difference in K_m may result from posttranslational modification in plants; putative phosphorylation sites have been identified in *StPT1* and *StPT2*. Alternatively, the proper functioning of P_i transporters in yeast may require several different gene products in addition to PHO84 to form a functioning unit.

The cloning of plant genes encoding P_i transporters has opened up new avenues of research into the molecular regulation of plant P_i acquisition and nutrition. A detailed study on the regulation of P_i transport in tomato indicated that in P-sufficient plants, mRNA transcripts for two transporters (*LePT1* and *LePT2*) are present only in roots. However, P starvation rapidly (within 24 hours) increases the abundance of both mRNAs in roots and *LePT1* mRNA in leaves (Fig. 23.22). Provision of P to previously P-starved plants decreases the amount of LePT1 and LePT2 transcripts to that seen in P-sufficient plants. An antibody specific to LePT1 was used to show that the P deficiency brought about an increase in transporter abundance in roots. Additionally, in situ

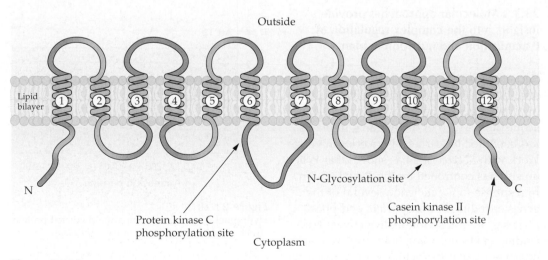

Figure 23.21
Proposed structural model for plant plasma membrane high-affinity P_i transporters showing the six-loop-six motif of 12 membrane-spanning domains. The predicted sites for phosphorylation by protein kinase C and casein kinase II and for N-glycosylation are shown.

mRNA localization of *LePT1* and *LePT2* in P-starved plants demonstrated that the genes are expressed in the root epidermis, and immunocytochemical studies with antibodies to LePT1 indicated that the protein is localized to the root cell plasma membrane, confirming a role for these transporters in P acquisition from the soil.

An important question regarding regulation of plant P status concerns the nature of the signal to which genes encoding P transporters as well as other proteins involved in P acquisition respond. Yeast cells have been suggested to respond to external concentrations of P to regulate the expression of the P_i transporters. Because of the low availability of P in soils, it is more logical for plants to use some indication of internal concentrations of P as a regulatory signal. This question has been addressed by examining the expression of *LePT1* and *LePT2* in tomato plants in which half of the root system was exposed to P-sufficient conditions while the other half was deprived of P in the growth solution. No difference was found in the expression of *LePT1* and *LePT2* in the +P and −P roots, and the amount of expression in plants with split roots was intermediate between plants for which the entire root system either was provided sufficient P or was P-starved. These findings suggest that the signal regulating the transcription of these genes comes from the shoot and may involve the concentrations of P_i in the shoot.

The expression of *LePT1* and *LePT2* in the presence and absence of P correlates well with physiological studies of root P_i influx conducted during the 1980s. Those studies used a split-root experimental system in which a single potato root was provided P and the rest of the root system was deprived of P. Although the experimental P-treated root contained normal concentrations of P, the rate of P_i influx into that root was three times greater than the influx rate observed for the roots of control plants, in which the entire root system had been fed P. The rate of P_i influx into the single P-exposed root was comparable with that observed when P-deprived roots of the plant were exposed briefly to a ^{32}P-labeled phosphate solution. What differentiated the P-limited split-root seedlings from the P-sufficient control seedlings was the P concentration in the shoot, which was considerably higher in the control plants. These findings suggest that shoot concentrations of P play an important role in the molecular and physiological regulation of P_i absorption by the root.

The interaction between mycorrhizal fungi and host-plant roots also influences the regulation of plant P_i transporters. In a high-affinity P_i transporter cloned from a mycorrhizal fungus (*G. versiforme*) and the roots of the host plant (*Medicago truncatula*), expression of the fungal P_i transporter, *GvPT*, was localized to the external hyphae during colonization of the *M. truncatula* roots. During this colonization, the transcript amounts of *MtPT1* and *MtPT2*, the genes encoding the alfalfa root P_i transporters, were much less than in the control, nonmycorrhizal roots. Expression in *M. truncatula* of another P-starvation–inducible gene (an acid phosphatase homolog from *Arabidopsis*) also was down-regulated during root colonization with *G. versiforme*. These findings are consistent with previous studies on the impact of mycorrhizal colonization on the physiology of root P_i absorption, in that root P_i uptake is down-regulated during mycorrhizal symbiosis, presumably because the fungus is providing the bulk of the plant's P requirements.

Investigations into the molecular regulation of plant P acquisition and nutrition are in the early stages. The complex responses that plants exhibit to acquire P from the soil

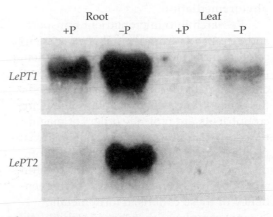

Figure 23.22
Northern blot showing expression of the P_i transporter genes *LePT1* and *LePT2* in roots and leaves of tomato plants grown under P-sufficient (+P) and P-deficient (−P) conditions. P deficiency results in the increased abundance of mRNA transcripts in roots and the appearance of *LePT1* expression in leaves.

will require much more study to elucidate all of the regulatory mechanisms involved.

23.4 The molecular physiology of micronutrient acquisition

Until recently, research on plant mineral nutrient transport and acquisition emphasized macronutrient transport. The body of knowledge on macronutrient transport is much larger than for any micronutrients, except possibly Fe. However, awareness of the importance of micronutrients such as Fe, Zn, and Cu to agriculture has been growing because in many soils, micronutrient availability limits crop production (see Fig. 23.1). Currently, interest is increasing in how the micronutrient content of foods might benefit human health and nutrition. Investigators also are intrigued by possible detrimental links between micronutrient transport and human health, given that some essential micronutrients also can be classified as toxic heavy metals (e.g., Zn, Cu, Ni) or are analogs of heavy metals (e.g., Zn, which is an analog of Cd and Hg). Environmental aspects of agriculture have received considerable attention recently, some of which has focused on plants as a primary site for the entry of heavy metals into the food chain. Moreover, scientists have realized recently that terrestrial plants might constitute an inexpensive mechanism for cleaning soils contaminated with heavy metals, a technology termed **phytoremediation.**

Research into micronutrient transport and acquisition from the soil is more complicated than the acquisition of macronutrients for several reasons. First, most micronutrients are relatively unavailable in soil, and the concentrations of available (soluble) micronutrients are typically orders of magnitude less than those of most of the macronutrients. Second, research into transport of micronutrient cations (Fe^{3+}, Cu^{2+}, Zn^{2+}, Mn^{2+}, and Ni^{2+}) has been complicated by the tendency of these cations to form metalloorganic complexes that vary in stability, size, and charge. The influence of metal–chelate chemistry on micronutrient cation transport at the membrane, organ, and whole-plant levels requires that plant scientists working in this area of research must address complex problems—for example, the presence of metal chelates in the rhizosphere, the possible splitting of chelates at the plasma membrane of the root cell, the forms in which the micronutrient cations are transported into plant cells, and the nature of the metal–chelate complexes present within cells or being transported long distances by the xylem and phloem.

Progress and interest in this area has been stimulated by molecular advances that have identified some of the genes encoding micronutrient transporters. The present section will focus on what is known about the physiology and molecular biology of Fe acquisition and transport, given that more is understood about the transport of this one essential element than about all of the other micronutrients combined. Recent progress in the field of plant Zn transport will be addressed, as will the relevance of these findings to heavy metal transport and the emerging field of phytoremediation.

23.4.1 How do plants acquire Fe from the soil?

Although Fe is abundant in soil, accounting for approximately 5% of the earth's crust, its availability in well-aerated soils in low. In aerobic systems, the solubility of inorganic Fe depends on ferric oxides in the soil (Rx. 23.1).

Reaction 23.1

$$Fe(OH)_3 + 3H^+ \rightleftharpoons Fe^{3+} + 3H_2O$$

Hence, as solution pH is reduced from pH 8 to pH 4, the Fe^{3+} in solution increases from low concentrations (about 10^{-20} M) to about 10^{-9} M (1 nM). Soil redox potential has a large influence on Fe solubility, because reduced, ferrous iron (Fe^{2+}) is considerably more soluble than Fe^{3+}. However, for typical soil pH values of 6.0 to 8.0, the Fe^{3+} species dominates because the Fe^{2+} is oxidized readily. Therefore, in most well-aerated soils, the minimum solubility of total inorganic Fe is fixed at 0.01 to 1 nM. The actual concentration of soluble Fe that has been measured in soil solution is usually higher, in the range of 10 to 1000 nM, because of the chelation of soil Fe by soluble organic ligands. These organic compounds result from the degradation of organic matter or the microbial synthesis of **siderophores** (Greek: "iron bearers"), Fe-chelating compounds released by soil bacteria and fungi.

Chelated ferric ion is the dominant form of Fe available for absorption by the plant; the concentration of chelated Fe required for optimal plant growth is around 0.1 to 10 μM, a concentration one order of magnitude greater than that found in soil solution. Plants have evolved complex responses to increase the bioavailability of Fe in the soil and to maintain substantial root Fe fluxes at low soil Fe concentrations. The processes that plants utilize to acquire Fe from relatively unavailable soil reserves have been the focus of much of the research in Fe nutrition over the past 20 years. Plants can use two distinct strategies to solubilize and absorb Fe from the soil. Some plants reduce ferric chelates at the outer surface of the root cell plasma membrane and absorb the ferrous ions produced, whereas others excrete **phytosiderophores,** specific low-molecular-mass organic compounds with a high affinity for Fe^{3+}, which solubilize ferric ions and make them available for absorption. The phytosiderophore-based mechanism is restricted to the grass family (Poaceae); the Fe reduction mechanism is used by the dicots and by nongrass monocots.

23.4.2 Plants use a strategy of chemical reduction or chelation to acquire Fe.

During the 1970s and 1980s, physiological investigations into Fe uptake and nutrition demonstrated why grasses, such as corn, cannot effectively obtain Fe from nutrient solutions containing Fe that is complexed to strong ferric chelators, whereas dicots can grow well by using strongly complexed Fe(III) as their the sole Fe source. When strong Fe chelators were applied in molar excess to Fe in hydroponic media, dicots could readily obtain Fe but the grasses became Fe-deficient. Also, Fe-deficient dicots could split Fe from Fe(III) chelates, absorbing the Fe while leaving the synthetic chelate in solution. The mechanism of Fe absorption by Fe-deficient soybean roots was found to involve an obligatory reduction of chelated Fe^{3+}, which releases Fe^{2+} ion from the chelate and subsequently absorbs Fe^{2+} into the root. This hypothesis was supported by the observation that application of an Fe^{2+} chelator, bathophenanthrolinedisulfonate (BPDS), to the nutrient solution in excess of Fe(III)-bound ethylenediamine-di(o-hydroxyphenylacetic acid) (EDDHA), resulted in 99% inhibition of Fe uptake (measured as translocation of ^{59}Fe to the shoots). Presumably, in the reduction of ferric to ferrous ion at the root surface, BPDS acted as a ferrous trap, binding the ion at the root surface and thereby preventing its absorption.

Subsequent research on the role of Fe reduction in Fe absorption has indicated that dicots and nongraminaceous monocots respond to Fe deficiency by inducing several different physiological and biochemical responses. These alterations appear to be part of an integrated response that increases the ability of the plant to solubilize and absorb Fe from the environment. These responses to Fe deficiency include the following:

- induction in root epidermal cells of a plasma-membrane ferric reductase, which reduces extracellular Fe(III) chelates and drives the accumulation of Fe^{2+} ions at the outer surface of the plasma membrane
- induction of a root cell plasma membrane Fe^{2+} transporter that facilitates the uptake of Fe into the cell
- enhancement of H^+ efflux at the root in a way that appears to involve induction of the root cell plasma membrane H^+-ATPase (this response helps acidify the rhizosphere, increasing Fe solubility)
- accumulation of citrate and malate in the roots and subsequent exudation of the organic acids, these effective Fe(III) chelators being thought to increase the concentration of soluble Fe(III) in the rhizosphere

The important dicot responses to Fe deficiency are depicted in the model in Figure 23.23. In this model, Fe absorption from the rhizosphere is a two-step process. The first step involves induction by Fe deficiency of ferric reductase, which reduces extracellular Fe(III) chelates, releasing free Fe^{2+} ions. The second step is the transport of Fe^{2+} into the cytoplasm by way of a specific Fe^{2+} transporter or a less specific divalent cation transport system. In addition, induction of the

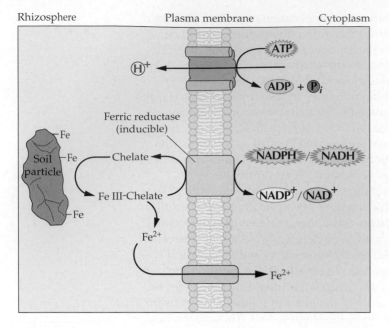

Figure 23.23
Model of Fe uptake by dicots and nongrass monocots, depicting the key plasma membrane–localized components that are induced in roots by Fe deficiency. These include an inducible ferric reductase that reduces Fe(III) chelates in the rhizosphere, releasing Fe^{2+} for transport across the plasma membrane by a high-affinity Fe^{2+} transporter. Also shown is the plasma membrane H^+-ATPase, which is induced by Fe deficiency and acidifies the rhizosphere, increasing Fe solubility.

plasma membrane H^+-ATPase increases Fe availability by acidifying the rhizosphere.

Yeast has been a useful model organism for studying the reductive mechanism of Fe acquisition, because its molecular regulation of Fe uptake has been well characterized. Two genes encoding plasma membrane ferric reductases, *FRE1* and *FRE2*, have been cloned in yeast and have been shown to be transcriptionally up-regulated by Fe deficiency. High- and low-affinity Fe^{2+} transporters also have been cloned from yeast. The high-affinity transporter (K_m = 150 nM) is encoded by the *FTR1* gene. Proper function of this transporter requires another gene product, FET3, an oxidase that contains several Cu ions. Hence, Cu is required for normal Fe absorption by yeast; moreover, the responses induced by Fe deficiency also are induced by Cu deficiency. A low-affinity Fe^{2+} transporter encoded by the *FET4* gene is much less specific for Fe (K_m = 40 μM). Expression of the ferric reductase and the high- and low-affinity Fe^{2+} transporters is induced by Fe deficiency. Among the several genes that are involved in the transcriptional regulation of yeast Fe-transport genes and have

been characterized are *AFT1*—which encodes a transcription factor that activates the transcription of *FRE1*, *FRE2*, *FET3*, and *FTR1*—and the transcription factor *MAC1*, which regulates the transcription of the ferric reductase gene, *FRE1*.

Much of the research on the regulation of Fe absorption by dicot roots has been physiological in nature. An interesting spatial regulation of ferric reductase along the root differs among plant species. In tomato and *Arabidopsis*, for example, Fe deficiency–induced ferric reductase activity is localized to a short zone behind the root apex. However, for legumes such as peas, the reductase is induced along most of the root surface except for the terminal 1 to 2 cm (Fig. 23.24). Deficiency of Cu or Fe induces the ferric reductase, but other mineral deficiencies do not. Fe deficiency induces not only Fe(III) reduction but also Cu(II) reduction with the same spatial localization pattern (Fig. 23.24). The physiological relevance of Cu(II) reduction has not been established.

23.4.3 Cloning plant Fe-acquisition genes has provided insights into the molecular basis for Fe nutrition as well as possible routes for plant acquisition of heavy metals.

Major advances in the field of plant mineral nutrition have come from the cloning of genes encoding the key components in root Fe acquisition in dicots, the ferric reductase and the ferrous transporter. The root ferric reductase gene (*FRO2*) from *Arabidopsis* was cloned recently by taking advantage of its sequence similarity to the genes encoding the FAD- and NADPH-binding site of the yeast ferric reductase and the human phagocyte NADPH oxidase. Sequence analysis indicates that FRO2 (Fig. 23.25) is a member of a superfamily of flavocytochromes that transport electrons across membranes, and functional analysis suggests it is the root ferric reductase. Transformation of *Arabidopsis* mutants with *FRO2* restores root ferric reductase activity in reductase-defective mutants. Also, *FRO2* is transcriptionally activated by Fe deficiency in *Arabidopsis* seedlings.

A second advance in the field of Fe uptake involved the cloning of a putative root Fe^{2+} transporter from *Arabidopsis*. The cloning of high- and low-affinity Fe^{2+} transporters in

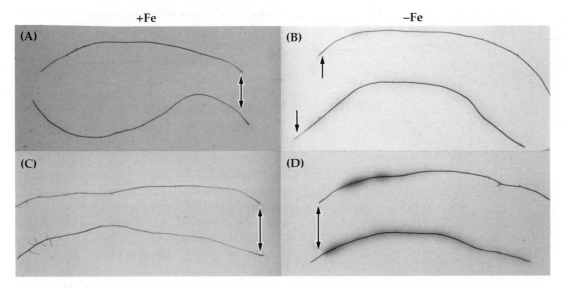

Figure 23.24

Visualization of Fe(III) and Cu(II) reduction along primary lateral roots of 14-day-old control (+Fe) and Fe-deficient (−Fe) pea seedlings. Reduction of Cu(II) to Cu(I) by roots of control (A) and Fe-deficient (B) seedlings is visualized by formation of the yellow complex Cu-BCDS (2,9-dimethyl-4,7-diphenyl-1,10-phenanthrolinedisulfonic acid). Reduction of Fe(III) to Fe(II) by roots of control (C) and Fe-deficient (D) seedlings is visualized by formation of the red complex Fe-BPDS (bathophenanthrolinedisulfonic acid). Note that Fe-sufficient roots do not appear to reduce Cu or Fe. Arrows indicate position of root tips.

S. cerevisiae allowed the generation of yeast Fe transport mutants defective in Fe^{2+} uptake. Complementation of the yeast Fe^{2+} transport double mutant *fet3fet1* with an *Arabidopsis* yeast expression cDNA library allowed the isolation of *IRT1* (iron-regulated transporter). Transformation with *IRT1* enhanced iron accumulation by the *fet3fet4* double mutant (Fig. 23.26) and restored its ability to grow on low Fe. RNA gel blot analysis of *IRT1* expression in *Arabidopsis* showed that it is expressed only in roots and that exposure to Fe deficiency greatly increased the concentrations of *IRT1* mRNA in roots. IRT1 subsequently has been shown to have broad substrate specificity and to transport ions other than Fe (see below). Isolation of *IRT1* led to the discovery of a new family of micronutrient transporters (Table 23.3). Sequence similarities between *IRT1* and previously identified open reading frames with unknown functions in yeast were used to clone two members of this family in yeast, *ZRT1* and *ZRT2*. These genes do not encode Fe transporters but instead encode high- and low-affinity Zn^{2+} transporters, respectively (see Section 23.4.5). Identification of *ZRT1* and *ZRT2* facilitated the subsequent cloning of Zn^{2+} transporters

from *Arabidopsis* (see Section 23.4.5). All of the ZRT/IRT-related proteins encoded by the *ZIP* gene family encoding micronutrient transporters are predicted to have eight membrane-spanning regions, and a region

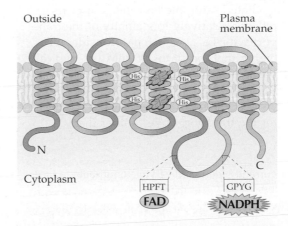

Figure 23.25

Predicted protein structure for FRO2, the root plasma membrane (PM) ferric reductase in *Arabidopsis*. FRO2 structure is predicted to include six membrane-spanning domains within the C-terminal region and two membrane-spanning domains within the N-terminal region. Four histidine residues are predicted to coordinate two heme groups localized to two of the internal membrane-spanning domains, forming a complex that may play a role in transmembrane electron transport. Near the C terminus is a cytoplasmic loop that includes putative FAD- and NADPH-binding domains.

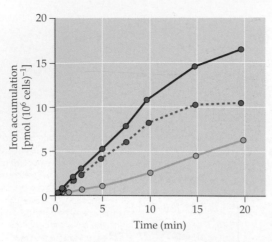

Figure 23.26

The *Arabidopsis* Fe transporter IRT1 increases accumulation of ^{55}Fe by the yeast Fe-uptake mutant *fet3fet4*. The red plot indicates the time course of ^{55}Fe accumulation at 30°C by *fet3fet4* mutant strain DEY1453 transformed with pIRT1. The green plot shows accumulation by an untransformed DEY1453 control. The dotted blue line defines *IRT1*-dependent accumulation at each time interval (i.e., ^{55}Fe accumulation by transformants minus ^{55}Fe accumulation by control).

between the third and fourth transmembrane domains that contains a histidine-rich motif may act as a metal-binding site in micronutrient transport (Fig. 23.27).

The concentration-dependent kinetics of Fe accumulation by yeast expressing *IRT1* suggest that this gene does not encode a high-affinity Fe transporter. Indeed, IRT1-mediated saturable Fe^{2+} influx appears to reflect a relatively low affinity of the transporter for Fe^{2+} (K_m = 6 μM, considerably higher that the 10 to 1000 nM Fe^{2+} typically present in soil solution). In comparison, the K_m of the high-affinity Fe^{2+} transporter in yeast is 150 nM. IRT1 appears to be specific for the transport of Fe^{2+} rather than Cu^{2+},

Mn^{2+}, Zn^{2+}, and Ni^{2+}. However, the presence of Cd in the uptake solution at 10-fold excess inhibits Fe^{2+} influx, which suggests that this transporter might facilitate the uptake of Cd in contaminated soils.

The metal ion transport properties of an *IRT1* homolog isolated from peas, designated *RIT1* (for root iron transporter), have also been characterized. In peas, Fe deficiency induces not only root Fe^{2+} uptake but also saturable Zn^{2+} and Cd^{2+} uptake, with a K_m of about 5 μM. Like *IRT1*, *RIT* is expressed in roots and its expression is up-regulated by the imposition of Fe deficiency. However, when expressed in yeast, RIT1 catalyzes high-affinity Fe^{2+} transport (K_m = 54 nM; Fig. 23.28A). This transporter also mediates a lower-affinity saturable Zn^{2+} influx (K_m = 4 μM; Fig. 23.28B) and Cd^{2+} influx. RIT1 appears to function primarily in the acquisition of Fe, but it also can provide a transport pathway for Zn and Cd, particularly in contaminated soils.

23.4.4 Grasses facilitate Fe uptake through the release of Fe-binding phytosiderophores.

Grasses long have been recognized to differ from other plants by their inability to utilize Fe supplied as stable Fe(III) chelates such as EDDHA. Because of this, and the absence of Fe deficiency–induced responses ordinarily observed in dicots (increased H^+ excretion and reductase activity), grasses have been characterized as "Fe-inefficient". However, when grasses and dicots are grown in calcareous soils containing low amounts of available Fe, grasses often are more effective than dicots in resisting Fe-deficiency chlorosis. This observation led to the hypothesis that grasses acquire Fe from the soil by a different mechanism than that used by dicots.

Washings from rice roots contain a compound that can solubilize ferric iron; moreover, in response to Fe deficiency, both oat and rice roots exhibit a stimulated release of Fe^{3+}-chelating compounds. This response of grasses to Fe deficiency is specific to grasses; the roots of dicots do not release Fe-chelating compounds under Fe deficiency. Subsequent work has identified these Fe-chelating compounds as nonprotein amino acids, and two of the chelating compounds, mugineic and

Table 23.3 Summary of currently identified ZIP micronutrient transport genes

Gene	Species	Speculative transport function	Localization
IRT1	Arabidopsis thaliana	Fe^{2+} uptake	Roots
RIT1	Pisum sativum	$Fe^{2+}/Zn^{2+}/Cd^{2+}$ uptake	Roots
ZIP1	A. thaliana	Zn^{2+} uptake	Roots
ZIP2	A. thaliana	Zn^{2+} uptake	?
ZIP3	A. thaliana	Zn^{2+} uptake	Roots and shoots
ZIP4	A. thaliana	?	Roots and shoots
ZNT1	Thlaspi caerulescens	Zn^{2+}/Cd^{2+} uptake	Roots and shoots

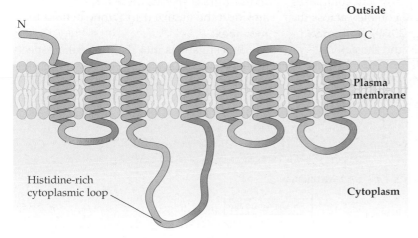

Figure 23.27
Predicted protein structure of the members of the ZIP family of micronutrient transporters. The notable structural features include eight membrane-spanning domains and a cytoplasmic loop of variable length situated between the third and fourth transmembrane helices. This cytoplasmic domain contains histidine repeats that may function in metal-binding.

avenic acid, have been characterized (Fig. 23.29). The pathways for their biosynthesis are still not clearly delineated, but they are known to be synthesized from methionine by way of nicotianamine.

Because mugineic and avenic acids are effective ferric chelators, they have been named phytosiderophores, after the well-characterized iron uptake system in microorganisms that utilizes the release of specific Fe-chelating compounds called siderophores. *E. coli* contains a beautifully orchestrated iron operon of three genes that encode the siderophore, a gene encoding the membrane receptor, and a nearby regulatory sequence consisting of repressor, promoter, and operator regions. In response to Fe deficiency, the siderophore and the membrane receptor are synthesized. The released siderophore, which is specific for Fe^{3+}, binds ferric ion outside the cell. The Fe(III)–siderophore complex binds to the specific receptor at the outer membrane, after which the complex is transported into the cell. When the Fe requirement for the bacterium has been met, the excess Fe binds to a repressor protein, and the Fe–protein complex binds to the operator region of the operon and turns off the system.

Much less is known concerning the mechanism of Fe–phytosiderophore uptake in grasses. Fe-uptake studies in Fe-deficient barley roots using dual-labeled ^{59}Fe(III)–[^{14}C]phytosiderophores, and transport studies in rice investigating Fe uptake from Fe(III) complexes with the D- and L-enantiomeric forms of phytosiderophores, indicate the presence of an Fe transporter in the root cell plasma membrane in grasses that recognizes

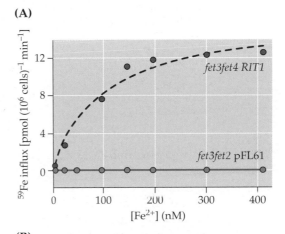

(A)

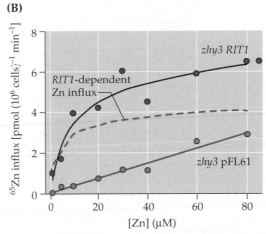

(B)

Figure 23.28
RIT1-mediated Fe^{2+} and Zn^{2+} uptake in yeast. (A) Concentration-dependent kinetics of $^{59}Fe^{2+}$ influx in the *fet3fet4* yeast double-mutant, which is defective in Fe transport. The yeast express the pea root micronutrient transporter, *RIT1* (red) or the empty-vector pFL61 plasmid (control, green). The K_m for *RIT1*-dependent Fe^{2+} influx was 54 nM, and the V_{max} was 5.5 pmol min^{-1} (10^6 cells)$^{-1}$. (B) Concentration-dependent kinetics of $^{65}Zn^{2+}$ influx in a Zn transport–defective yeast mutant, *zhy3*, expressing *RIT1* (red) or the empty-vector pFL61 plasmid (control, green). In this case, the Zn^{2+} influx associated with the control yeast was subtracted from the Zn^{2+} influx associated with +*RIT1* yeast to yield the *RIT1*-dependent Zn^{2+} influx (dashed blue line). The K_m for *RIT1*-dependent Zn^{2+} uptake was 4 µM, and the V_{max} was 4.4 pmol min^{-1} (10^6 cells)$^{-1}$.

specific Fe(III)–phytosiderophore complexes and transports the entire complex across the plasma membrane (Fig. 23.30). Both phytosiderophore synthesis and the activity of the Fe(III)–phytosiderophore transport system are stimulated by Fe deficiency.

In most grass species these responses appear to be localized to a zone behind the root apex.

Investigations into the molecular aspects of this transport system in plants are in the early stages and have focused thus far on

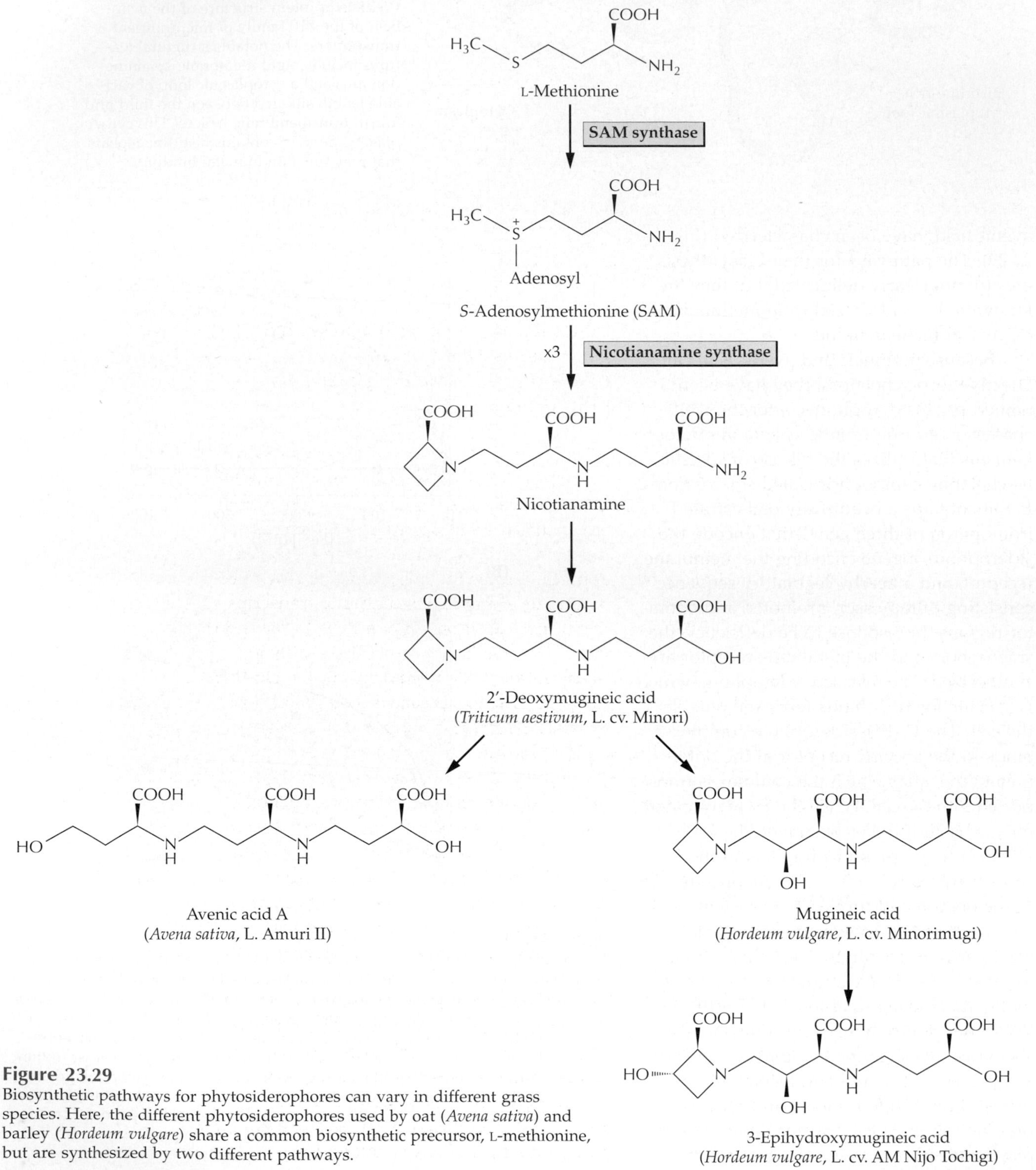

Figure 23.29
Biosynthetic pathways for phytosiderophores can vary in different grass species. Here, the different phytosiderophores used by oat (*Avena sativa*) and barley (*Hordeum vulgare*) share a common biosynthetic precursor, L-methionine, but are synthesized by two different pathways.

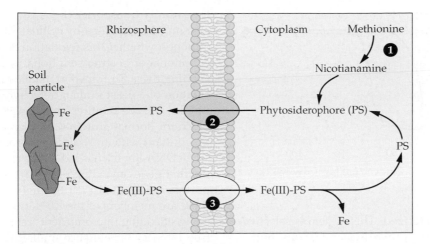

Figure 23.30

Model of Fe uptake by graminaceous plants. Phytosiderophores (PS) are synthesized in the cytoplasm from methionine by way of nicotianamine (1). An Fe deficiency–inducible plasma membrane (PM) transporter is involved in release of the PS into the rhizosphere (2). Another Fe deficiency–inducible transporter in the PM recognizes the Fe–PS complex and transports it intact into the cytoplasm.

the biosynthetic pathway for phytosiderophore production. Nicotianamine synthase, purified from Fe-deficient barley roots, catalyzes the key first step in phytosiderophore synthesis, the conversion of three molecules of *S*-adenosylmethionine to nicotianamine. The enzyme exists as a homodimer with a subunit molecular mass of 30 kDa. Nicotianamine synthase activity, which is stimulated by Fe deficiency, is 100-fold greater in roots than in leaves. Efforts are underway by several groups to identify the genes encoding this enzyme. A subtractive hybridization approach has been used to clone two iron-deficiency specific barley genes, *IDS2* and *IDS3*, that are expressed in barley roots in response to Fe deficiency. The predicted peptides encoded by these genes share similarities with α-ketoglutarate dioxygenases—enzymes involved in hydroxylation reactions similar to those required for the synthesis of certain phytosiderophores (e.g., conversion of deoxymugineic acid to mugineic acid)—which suggests that *IDS2* and *IDS3* may encode important enzymes in the phytosiderophore biosynthetic pathway.

Further advances concerning our understanding of the underlying mechanisms and regulation of the Fe acquisition strategy in grasses depend on obtaining more biochemical and molecular information about each step in phytosiderophore synthesis, phytosiderophore release, and absorption of the Fe(III) complex.

23.4.5 Recent advances in elucidating molecular aspects of Fe acquisition have furthered our understanding of plant Zn uptake.

Until recently, the basic mechanisms of Zn uptake from the soil and Zn transport within the plant were not studied actively by plant scientists. A recent upsurge in research in this area has been spurred by an increased appreciation for the importance of zinc to biology. For example, research into transcriptional regulation has shown that Zn plays an important role in key structural motifs in transcriptional regulatory proteins, including Zn finger, Zn cluster, and RING finger domains. These motifs are found in a large number of different proteins; for example, 2% of the yeast gene products identified during the sequencing of the yeast genome contain Zn-binding domains. Awareness of the importance of Zn nutrition to agriculture is also increasing (Fig. 23.31; see also Fig. 23.1). Zinc deficiency in soils has been recognized recently as an important problem worldwide. A global study conducted by the Food and Agriculture Organization, an agency of the United Nations, found that 30% of the world's cultivated soils are Zn deficient.

Earlier work on absorption by roots focused on the physiology of Zn acquisition. When uptake solutions containing 0 to 10 μM Zn^{2+} were used, a Zn^{2+}-uptake system with a K_m in the low micromolar range

Figure 23.31

Zn, a component of numerous transcription factors, has an important influence on protein synthesis. Zn-deficient rice shoots demonstrate greatly reduced protein content and limited abundance of 80S ribosomes, organelles essential for translation of mRNA.

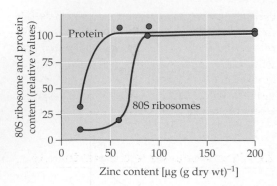

was characterized. This system is selective for Zn^{2+} over other divalent cations, but physiological investigations indicate it also might transport Cd^{2+} and Cu^{2+}. The affinity of the root Zn^{2+} transporter for its putative substrate does not match the physiologically relevant concentrations of Zn^{2+} found in soil solution, which are generally 1 nM to 1 µM—at least two orders of magnitude lower than the K_m determined for the transporter from radiotracer flux experiments in solution culture. Technical limitations have hindered uptake experiments with Zn^{2+} in the nanomolar range, but recent techniques using synthetic chelates to "buffer" the free Zn^{2+} activity in the nanomolar range may make it possible to determine whether the concentration-dependent kinetics for Zn^{2+} uptake indicate the existence of a high-affinity Zn^{2+} transporter in plant roots.

Another complication in studying Zn uptake from the soil concerns the actual substrate for the Zn transporter. Much of the Zn in soil solution can be chelated by various low-molecular-mass organic ligands. Can Zn complexed to organic ligands be transported into root cells? This is particularly important for grasses, as the phytosiderophores released by roots of grass species not only can bind Fe^{3+} but also effectively can chelate Zn^{2+} (as well as Cu^{2+}, Mn^{2+}, Ni^{2+}, and Co^{2+}). Because phytosiderophores chelate other metals, some investigators argue that they should be termed phytometallophores.

As described in Section 23.4.3, information gained from cloning the *IRT1*-encoded Fe transporter from *Arabidopsis* facilitated the cloning of two Zn^{2+} transporters from *S. cerevisiae*. *ZRT1* encodes a high-affinity Zn^{2+} transporter (K_m = 0.5 to 1 µM) and is transcriptionally up-regulated by low concentrations of cellular Zn. *ZRT2* encodes a

low-affinity transporter (K_m = 10 µM) that also may be regulated by cellular Zn status, although whether this regulation is transcriptional or posttranscriptional is not clear. Cloning these Zn transporters allowed the creation of a yeast mutant defective in Zn^{2+} uptake. *zrt1zrt2* requires high concentrations of Zn for normal growth. Complementing this mutant with an *Arabidopsis* yeast expression cDNA library has led to the cloning of the first genes encoding plant Zn^{2+} transporters. Designated *ZIP1* through *ZIP3*, these are members of the *ZIP* family of genes encoding micronutrient transporters (see Table 23.3). Sequence similarities among the *ZIP* clones were used to identify a fourth Zn^{2+} transport gene, *ZIP4*, from an *Arabidopsis* EST database. *ZIP1* and *ZIP3* are expressed in roots in response to Zn deficiency, suggesting they are involved in absorption of Zn from the soil. *ZIP2* mRNA was not detected in *Arabidopsis*, whereas *ZIP4* was expressed in both roots and shoots of the plant. Radiotracer (^{65}Zn) flux studies with yeast expressing *ZIP1/2/3* indicated that these genes encode Zn^{2+}-uptake systems with K_m values between 2 and 14 µM. The affinity of these transporters for Zn^{2+} is similar to that determined from radiotracer studies with roots and appears to be considerably lower than the Zn^{2+} activities typically present in soil solution. Competition studies with other divalent cations suggest indirectly that these transporters might also mediate Cd and Cu uptake. However, the yeast *zrt1zrt2* mutant could not be complemented with *ZIP4*, leading to speculation that *ZIP4* encodes an intracellular Zn transporter that is not localized to the plasma membrane.

Much of the recent research on plant Zn transport has arisen from interest in the use of plants to remediate soils contaminated with heavy metals. Interest in phytoremediation has been triggered by assessment of heavy metal–hyperaccumulating plant species, which are endemic to soils that are naturally metalliferous. These hyperaccumulating plants also have been found on mine spoils and other soils contaminated with heavy metals. One of the best known of hyperaccumulators is *Thlaspi caerulescens*, a member of the Brassicaceae (Box 23.3). This species has been shown to accumulate high concentrations of Zn and Cd in shoots, reportedly as much as 40,000 µg of Zn per gram dry weight. By comparison, the usual concentrations of Zn in plant tissue range from 100 to 200 µg g^{-1}, with 500 µg g^{-1} usually considered toxic. This plant uses effective methods to transport Zn and to detoxify the Zn accumulated in the symplasm. Studies of the molecular physiology of Zn hyperaccumulation in *T. caerulescens* show that Zn transport systems in the plasma membrane are stimulated by Zn. In roots, a saturable Zn^{2+}-uptake system characterized with a K_m of 6 to 8 µM exhibited a V_{max} four- to fivefold greater than the V_{max} observed for a comparable Zn^{2+}-transport system in a related nonaccumulating plant, *T. arvense* (Fig. 23.32). Leaf Zn^{2+} absorption was also stimulated in *T. caerulescens* compared with *T. arvense*, which is consistent with the hyperaccumulation of Zn in leaf cells. Complementation of the Zn transport-defective yeast mutant *zrt1zrt2* with a *T. caerulescens* cDNA library led to the cloning of a *Thlaspi* Zn^{2+} transporter designated *ZNT1* (for zinc transporter). This transporter is encoded by a member of the *ZIP* gene family (see Table 23.3), and sequence comparison between *ZNT1* and the *ZIP*-encoded transporters from *Arabidopsis* indicates the former is a homolog of *ZIP4*. Unlike *ZIP4*, *ZNT1* complements the Zn-transport defects in the yeast *zrt1zrt2* mutant, and radiotracer (^{65}Zn and ^{109}Cd) flux studies have shown that it encodes a saturable Zn transporter with a K_m of 7 µM and also mediates Cd^{2+} uptake (but with a lower affinity). RNA gel blot analysis using *ZNT1* to probe *T. caerulescens* mRNA and a *ZNT1* homolog from *T. arvense* to probe mRNA from that species showed that *ZNT1* is expressed in roots and shoots and that Zn hyperaccumulation involves an increase in the expression of *ZNT1* in *T. caerulescens*. Thus, *ZNT1* is expressed to high amounts in roots and shoots of *T. caerulescens* grown under Zn-deficient and Zn-sufficient conditions, whereas its expression in *T. arvense* is low under Zn-sufficient conditions and is enhanced by Zn deficiency. Apparently *T. caerulescens* behaves as though it is Zn deficient, even while it is accumulating high amounts of Zn. The mechanistic basis for this response is being studied.

Again, as with other mineral nutrient transporters, plants contain a multiplicity of Zn transporters. In general, multiple members of each mineral ion transport family en-

code similar transporters that are expressed in different tissues or cell types. Probably the different members of a specific ion transport gene family respond differently to specific environmental stresses, which would provide the plants with the flexibility to respond to a wide range of environmental stresses.

23.5 Plant responses to mineral toxicity

23.5.1 Aluminum toxicity is a major limitation to crop production on acid soils.

Because aluminum is the most abundant metal and third most abundant element in the earth's crust, plants have evolved in a soil environment where the roots are potentially exposed to high concentrations of aluminum. Fortunately, phytotoxic forms of Al are relatively insoluble at alkaline, neutral, or mildly acidic soil pH values. However, at soil pH 5 or lower, toxic forms of Al are solubilized and accumulate to concentrations that inhibit root growth and function.

Al toxicity is one of the major factors limiting crop production on acid soils. Acid soils diminish agricultural productivity in many regions of the world, including substantial areas of the United States. About 30% of the world's total land area consists

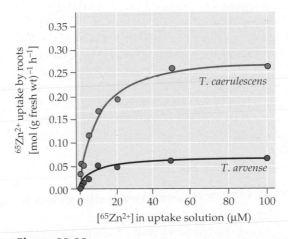

Figure 23.32
Concentration-dependent kinetics of ^{65}Zn^{2+} influx into roots of the Zn/Cd hyperaccumulator, *Thlaspi caerulescens* (blue) and the nonaccumulator, *T. arvense* (red). Roots were immersed in an uptake solution containing ^{65}Zn^{2+} at the concentrations shown. After 20 minutes, roots were desorbed, excised, blotted, and weighed, and their γ-radioactivity was measured. The kinetic parameters for root Zn^{2+} uptake were *T. caerulescens*, $K_m = 8$ µM, $V_{max} = 0.27$ µmol g^{-1}h^{-1}; *T. arvense*, $K_m = 6$ µM; $V_{max} = 0.06$ µmol g^{-1}h^{-1}.

of acid soils; of land currently devoted to the production of crop species, approximately 12% is composed of acid soils. Worldwide, 20% of maize acreage, 13% of rice acreage, and 5% of wheat acreage is grown on such soils. A large proportion of acid soil occurs in the developing countries of the tropics and subtropics: The humid tropics account for an estimated 60% of the acid soils in the world. In developed countries such as the United States, high-input farming practices, such as the extensive use of ammonia

Box 23.3

The conceptual basis for phytoremediation, the use of plants to clean contaminated soils, came from identifying plants that hyperaccumulate metals.

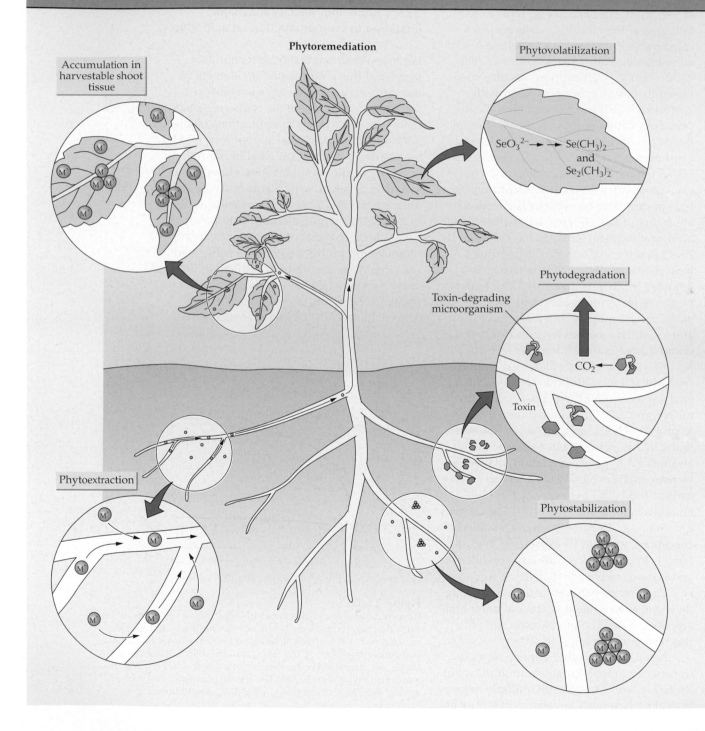

Chapter 23 Molecular Physiology of Mineral Nutrient Acquisition, Transport, and Utilization

fertilizers, are causing further acidification of agricultural soils. Although liming acid soils ameliorates problems associated with soil acidity, this is neither an economic option for poor farmers nor an effective strategy for alleviating subsoil acidity.

Research programs representing a diverse array of scientific disciplines are addressing acid soil–related limitations to crop productivity. Included in this effort is research by plant physiologists, molecular biologists, and geneticists to determine the basis of

Contamination of the environment with heavy metals and radionuclides is a serious problem both for human health and agriculture. In many locations throughout the world, sizeable land areas contain surface soils contaminated with hazardous amounts of toxic heavy metals such as Cd, Zn, Pb, Cu, Cr, Hg, Ni, and U. The engineering-based technologies currently in use to remediate surface-contaminated soils (e.g., removal of topsoil for storage in landfills) can be prohibitively expensive and often disturb the landscape. Recently, planners have taken a great deal of interest in the use of plants as an inexpensive alternative method for remediation of soils polluted with heavy metals. Some plant species can grow in soils that are heavily contaminated with metals and even accumulate these metals to high concentrations in shoots. The existence of these metal hyperaccumulator species suggests that plants could be used to **bioremediate** metal-contaminated soils (i.e., to detoxify them by means of biological activity). Phytoremediation (plant-based bioremediation) can involve **phytoextraction** (removal of toxins from soil), **phytostabilization** (complexation and immobilization of toxins within the soil), and **phytodegradation** of organic contaminants in the rhizosphere (see figure). Phytoextraction involves the use of terrestrial plants to absorb heavy metals from the soil and transport them to the shoots. Phytoextracted metals may accumulate in plant shoot tissues (e.g., Pb, Ni) or may be released as volatile species (e.g., Se). For metals that become concentrated within plant tissues, remediation involves harvesting the shoot biomass, which is reduced in volume (e.g., ashed) and then disposed of in a final repository. Moreover, for Se, the toxicity of the volatile species dimethylselenide and dimethyldiselenide is much lower than that of the soluble oxyanions, such as selenite (SeO_3^{2-}), present in soil solution.

Metal-hyperaccumulating plant species are endemic to metalliferous soils and can tolerate and accumulate high amounts of heavy metals in their shoots. Hyperaccumulators were first identified more than 100 years ago; this fascinating collection of different plants not only can grow in soils contaminated with high concentrations of a specific heavy metal, but also often have a higher tolerance for that metal. To date, hyperaccumulating plant species have been identified in nearly 400 taxa representing about 45 plant families. There is strong evidence for hyperaccumulators of Zn/Cd, Ni, and Se and some indications of plants that accumulate Co, Cu, and possibly Pb.

One well-known heavy metal hyperaccumulator is *Thlaspi caerulescens,* a member of the Brassicaceae that accumulates Zn and Cd (see photograph). Certain ecotypes of *T. caerulescens* have been shown to accumulate and tolerate as much as 40,000 µg of Zn per gram dry weight in their shoots (foliar [Zn] in hydroponically grown plants is ordinarily around 100 µg to 200 µg g^{-1}; 30 µg g^{-1} is considered sufficient). This species also accumulates high concentrations of Cd and has been suggested to accumulate high amounts of other heavy metals. Several studies conducted with *T. caerulescens* to remediate soils contaminated with Zn and Cd have had moderate success. One drawback to using *T. caerulescens* (or most other hyperaccumulators) for phytoremediation is that the species is slow-growing and does not produce substantial shoot biomass. To maximize phytoextraction, a species should have high shoot biomass and should accumulate metals to high concentrations in their shoots. Nonetheless, the unique physiology of heavy metal transport and tolerance in *T. caerulescens* makes it an interesting experimental plant for basic research aimed at elucidating the mechanism of heavy metal hyperaccumulation, especially with the goal of transferring hyperaccumulating traits to species with a higher biomass.

plant resistance to Al. Our understanding of the biology of Al resistance in plants is in its infancy, and broad questions remain regarding Al resistance mechanisms and their underlying genetic and molecular determinants. The results of this research are expected to be of immediate practical use in breeding Al-resistant crops and to provide a framework for future research aimed at isolating genes for use in transgenic manipulation of Al resistance.

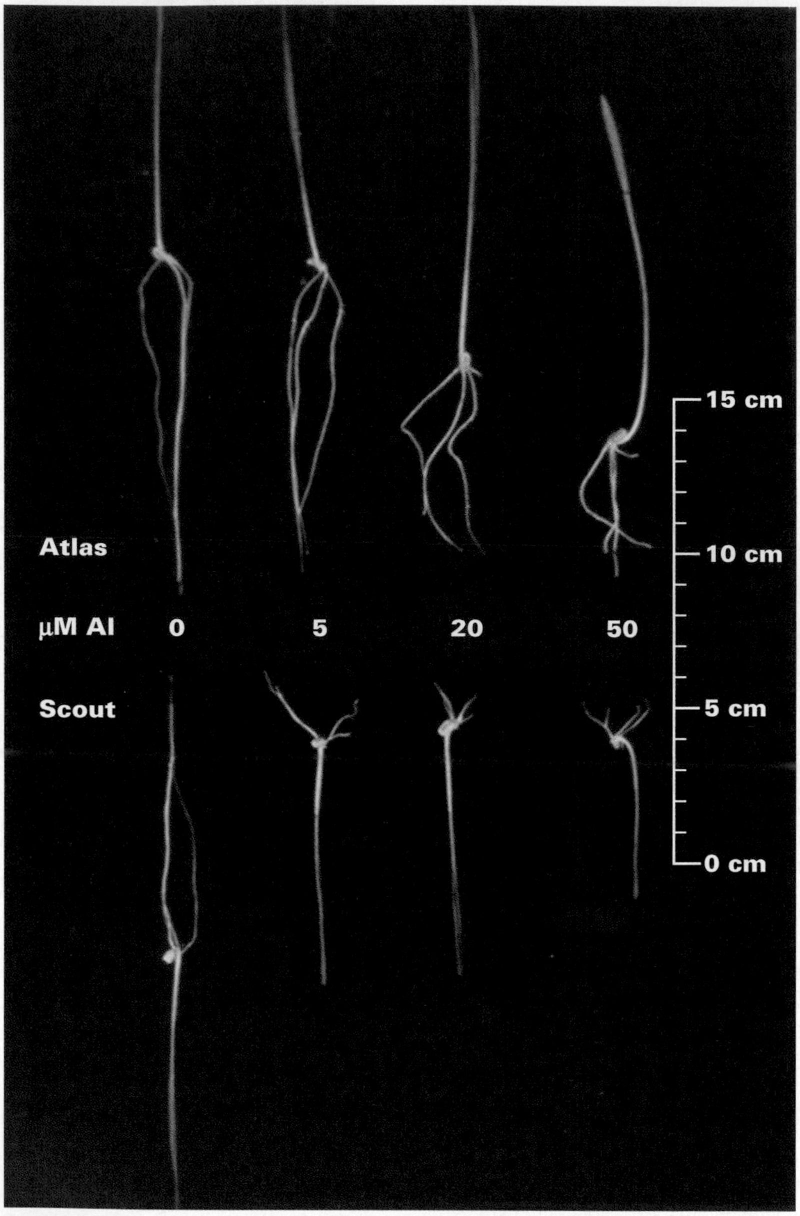

Figure 23.33
Effects of Al exposure on root growth of Al-resistant Atlas 66 and Al-sensitive Scout 66 cultivars of wheat. Seedlings were grown for four days in 0.6 mM $CaSO_4$ solutions containing 0, 5, 20, or 50 µM $AlCl_3$ (pH 4.5).

The initial and most evident symptom of Al toxicity is a rapid inhibition of root elongation, which can occur within minutes of exposure of roots to Al. Figure 23.33 shows Al-tolerant (Atlas 66) and Al-sensitive (Scout 66) cultivars of wheat grown in an acidic $CaSO_4$ solution containing defined concentrations of Al. Al toxicity greatly inhibits root growth; shoot growth is much less affected. Severe Al phytotoxicity reduces and damages the root system, which in turn causes the plant to be susceptible to drought stress and mineral nutrient deficiencies. Effects of Al exposure, such as blockage of Ca^{2+} and K^+ channels, can cause mineral nutrient deficiencies in the long term.

Al-induced inhibition of root growth is recognized as the most important component of Al phytotoxicity, and recent research has focused on this topic. A crucial finding is that this inhibition is caused solely by Al interactions within the root apex. The significance of this finding is that Al resistance mechanisms must act in the root apex to protect this region from the toxic effects of Al.

The speciation of Al in solution is complex, and only recently has it been demonstrated that for cereal species, Al^{3+} is the rhizotoxic Al species. Figure 23.34, which diagrams the various monomeric hydrolysis products of Al in aqueous solution, shows that as the solution pH decreases below 5, phytotoxic Al^{3+} dominates. Monomeric Al also forms low-molecular-mass complexes with various ligands, such as carboxylate, sulfate, and phosphate groups. Thus, Al^{3+} forms complexes with organic acids, inorganic phosphate, and sulfate and will also bind to these groups in macromolecules such as proteins and nucleotides. Because Al^{3+} and other monomeric forms of Al are potentially reactive with biological ligands, researchers have speculated that Al toxicity (root growth inhibition) might result from Al interactions with several different sites within the root, including sites within the cell wall, the plasma membrane, and the symplasm. Unfortunately, a paucity of data is available to confirm or confound such speculation. Research with plants and animals has suggested that Al might interact with signal transduction pathways such as the phosphoinositide pathway in the plasma membrane, which would result in concomitant disruptions of cytoplasmic Ca^{2+} homeostasis, the phospholipid bilayer of the plasma membrane, and components of the cytoskeleton.

23.5.2 Recent research has increased our understanding of the physiological basis of Al resistance.

Because of the widely held view that Al toxicity arises from Al interactions with several different processes within the root, Al resistance is assumed to be equally complex, with plants possessing a variety of different Al resistance mechanisms. Two classes of Al resistance mechanisms have been proposed: those that allow the plant to tolerate Al accumulation in the symplasm (symplasmic tolerance), and those that exclude Al from the root apex (Al exclusion). Possible mechanisms of symplasmic tolerance include chelation of symplasmic Al by low-molecular-mass ligands or Al-binding proteins in the cytoplasm, and sequestration of Al within an internal compartment (e.g., the vacuole). Mechanisms that might mediate Al exclusion are the release of low-molecular-mass, Al-chelating ligands into the rhizosphere; root-induced increases in rhizosphere pH; binding of Al within the cell wall; decreased permeability of the plasma membrane to Al influx; and binding of Al within the mucigel associated with the root apex. To date, evidence supporting the existence of most of these mechanisms is limited.

Recent evidence from work with wheat indicates that Al tolerance is associated with reduced accumulation of Al at the root apex but not in the mature root. Di- and tricarboxylic organic acids (e.g., malate and citrate) long have been known to be effective ligands for chelation of Al^{3+}. Numerous studies have attempted to identify differences in the organic acid content of roots as a potential mechanism of Al detoxification. However, they have not been able to associate Al tolerance with changes in root organic acid content. Indeed, Al exposure elicits similar increases in organic acid content in both Al-resistant and Al-sensitive varieties. However, although internal concentrations of Al-chelating organic acids do not correlate with Al tolerance, the application of malate or citrate to the solution bathing the roots allevi-

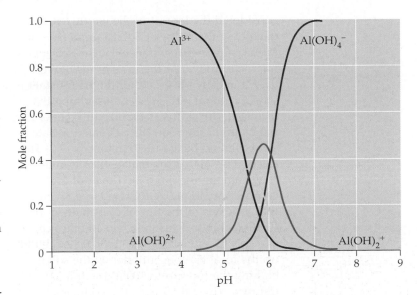

ates Al phytotoxicity. Al complexes formed with citrate and malate are not readily membrane-permeable and are not easily absorbed by roots.

Information available for several years has been used to develop a model whereby Al is excluded from roots by root-mediated increases in organic acids in the rhizosphere. A key recent finding supporting this model was the demonstration that plants can release Al-chelating organic acids from the root apex. The first evidence implicating the release of these compounds as a potential mechanism aiding Al tolerance came from work showing that Al exposure induced the release of 10-fold more citrate from the roots of an Al-resistant snap bean cultivar than from an Al-sensitive cultivar. More compelling evidence corroborating organic acid release as a mechanism aiding Al tolerance came from work with near-isogenic lines of wheat that differed in Al tolerance as encoded by a single dominant gene (*Alt1*). Within minutes of exposure to Al, malate was released from root apices of the tolerant isoline but not the sensitive isoline (Fig. 23.35A). The amount of malate released by the tolerant isoline increased markedly with the severity of Al exposure (Fig. 23.35B). Moreover, as based on measurements of root growth, Al tolerance conferred by *Alt1* cosegregates with lesser amounts of apical Al accumulation and Al-induced malate release.

Is the malate flux out of the wheat root tip a large ion efflux? The fluxes of K^+ and H^+ into and out of roots, which are

Figure 23.34

Distribution of mononuclear aluminum ion species in aqueous solution, showing the mole fraction of each species as a function of solution pH. Note the predominance of phytotoxic Al^{3+} at pH 5 or lower.

considered the largest of the ion fluxes, are usually in the range of 5 to 25 pmol cm^{-2} s^{-1} as measured with extracellular ion-selective microelectrodes. At the rates of malate release typically observed for the Al-resistant wheat cultivar Atlas 66, and assuming that in wheat, this efflux is localized to the terminal 2 mm of the root, the estimated malate efflux should be about 10 pmol cm^{-2} s^{-1}. This calculation places the malate flux at mid-range among the large ion fluxes of roots. In comparison, in maize roots, a large K$^+$ efflux that localizes to the tips of roots bathed in CaSO$_4$ solution is usually 5 to 10 pmol cm^{-2} s^{-1}. This K$^+$ efflux from the root tip can build up the local K$^+$ concentration in the solution bathing the root tip to 100 µM within 30 minutes. Thus, the rates of Al-inducible malate release from root tips of Al-resistant

wheat are likely to be high enough to contribute markedly to Al resistance; development of a malate-selective microelectrode ought to provide the definitive proof needed to resolve this issue.

One very interesting area of research concerns the underlying mechanisms and regulation of the malate excretion response to Al. Organic acids such as malate and citrate exist in the cytoplasm at pH 7.0 as anions, and their release is most likely mediated by anion channels. Thus, the physiological evidence suggests the existence of an interesting signal transduction pathway that involves the binding of Al to an unknown receptor in the plasma membrane, to the anion channel itself, or to some cytoplasmic factor that triggers the opening of the anion channel (Fig. 23.36). Good evidence is now available that supports the existence of an Al-gated anion channel in cells of the root apex in Al-resistant wheat. First, several laboratories have shown that known blockers of plant anion channels block the Al-induced malate release in wheat roots. More definitive evidence for this system comes from a patch clamp study of protoplasts isolated from the root apex of an Al-resistant wheat isoline. The study identified an anion channel that is activated by 20 to 50 µM AlCl$_3$ but not by La^{3+}. Thus, an Al-inducible mechanism at the cellular level involves specific anion channels and could play a role in Al resistance.

We need to determine whether the chelation-based Al exclusion mechanism postulated for wheat is a general response for other valuable crop species. In studies with maize, Al was found to induce the release of citrate but not malate from the root apex. This response was triggered rapidly by Al, was localized only to the root apex, and was found only in Al-resistant cultivars and lines. Given recent studies on the comparative mapping of grass genomes that indicate a considerable synteny (see Chapter 7) between maize and wheat, it will be interesting to ascertain whether a gene identical or similar to *Alt1* is expressed in maize.

Although the *Alt1* gene appears to confer a major component of Al tolerance in wheat, some evidence from physiological investigations suggests a second mechanism of Al exclusion, possibly controlled by a different gene. In a study in which the Al tolerance of 36 wheat cultivars was compared

(A)

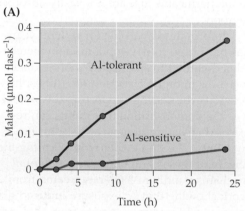

(B)

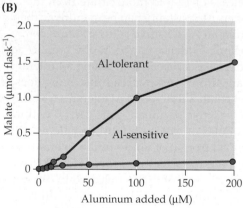

Figure 23.35
Aluminum-inducible release of malate from roots of Al-tolerant and Al-sensitive isolines of wheat. (A) Time course of Al-induced malate release. This response is exhibited only in the Al-tolerant isoline. (B) Dependency of malate release induction on Al concentration. In the Al-tolerant isoline, malate release increases with Al concentration in an almost linear relationship.

with rates of malate release from roots, a good overall correlation was reported between Al tolerance and malate release for two different growth-medium concentrations of Al (3 and 10 μM). Root growth and malate exudation also correlated significantly (P < 0.05) for the seven most Al-resistant cultivars grown in 3 μM Al. However, when the same seven cultivars were grown in 10 μM Al, the correlation between root growth and malate release was not significant. This observation suggests that malate exudation may account for only part of the tolerance to high concentrations of Al exhibited by strongly Al-resistant wheat varieties and may indicate that additional resistance mechanisms protect these plants.

Evidence for this possibility comes from a study in which relative Al resistance (measured as Al inhibition of root growth) was compared among very Al-resistant (Atlas 66) and Al-sensitive (Scout 66) wheat cultivars and the near-isogenic wheat lines used to characterize the *Alt1* locus (ET3 and ES3). Atlas 66 is one of the seven Al-resistant wheat cultivars for which resistance to high concentrations of Al was not found to correlate with malate release. Atlas 66 is considerably more Al-resistant than ET3, which exhibits intermediate Al resistance. Both Scout 66 and ES3 are very Al sensitive. Atlas 66 and ET3 exhibit similar rates of Al-induced malate release, presumably controlled by *Alt1*. In addition, Atlas 66 shows constitutive phosphate release, which also is localized to the root apex. Phosphate is another effective Al-chelating compound, and addition of phosphate to Al-containing growth solution alleviates Al toxicity in Al-sensitive wheat cultivars. These findings provide physiological evidence for the existence of multiple Al resistance mechanisms that might mediate Al exclusion by highly Al-resistant wheat lines and suggest that such resistance is controlled by two or more genes.

23.5.3 Investigations into the genetic basis for Al resistance in crops suggest it is a simple trait.

Genetic variability for Al tolerance has been reported among a large number of important crop species. Wheat (*Triticum aestivum*), perhaps the most comprehensively studied crop

with respect to Al tolerance, serves as a useful model for investigating the genetic complexity of this characteristic. Most studies of Al-tolerance inheritance in wheat have found that this trait can be controlled by one or a few major genes. However, care must be used when inferring the total number of different Al-tolerance genes represented in these and other wheat cultivars. Potentially many genes may contribute to Al tolerance, based upon analyses performed with Chinese Spring wheat ditelosomic lines in which genes on at least seven different chromosomes can influence Al tolerance. (A ditelosomic line is one in which a specific chromosome arm has been lost through cytogenetic-based techniques. In ditelosomic hexaploid Spring wheat, ditelosomic individuals have two copies of one chromosome arm and lack that arm on the third chromosome.) The evidence confirming that such a large number of genes conferring Al tolerance exist among different Al-tolerant wheat lines is at best modest. This evidence includes work with chromosome

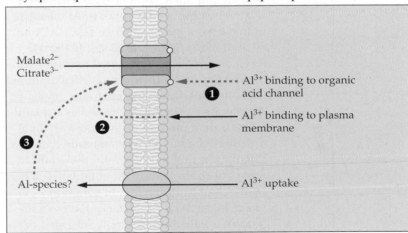

Figure 23.36
Model of a proposed mechanism of Al tolerance: Al^{3+}-induced efflux of organic acids by way of an Al-gated anion channel in the plasma membrane (PM) of cells of the root apex. In this model, a plasma membrane anion channel is opened on exposure of root apex cells to Al, allowing organic acids to flow out of the cell and into the rhizosphere. Shown are three possible mechanisms by which Al might activate the anion channel: (1) Al^{3+} binds directly to the channel protein; (2) Al^{3+} binds to the plasma membrane surface, activating the channel by means of a PM-localized set of events that may involve, singly or in combination, alterations to the lipid bilayer, blocking of ion channels, or binding to an unknown plasma membrane receptor; (3) Al is transported across the plasma membrane and activates the channel from the cytoplasmic side.

substitution lines of wheat, where it was shown that Al-tolerance genes are located on chromosomes 2D, 4D, and 5D. Also, molecular mapping has confirmed that an Al-tolerance gene in the highly Al-tolerant wheat cultivar BH1146 is found on chromosome arm 4DL. Finally, there are a few instances where Al tolerance genes have been differentiated by crossing Al-tolerant parents and following the segregation of the Al-tolerance trait in populations derived from these crosses. This approach was used to suggest that the wheat cultivars Atlas 66 and BH1146 possess different Al-tolerance genes. Thus, although the potential exists for the presence of a relatively large number of different Al-tolerance genes in wheat, evidence supporting this contention is fragmentary, and the issue has yet to be explored in a comprehensive manner.

Compared to wheat, the genetic basis of Al tolerance in other crops has not received as much attention. The genetic control of Al tolerance in barley (Hordeum vulgare) has been attributed to the action of a single gene, with evidence that variation in the degree of Al tolerance among cultivars is conditioned by different alleles of that gene. Likewise, relatively simple inheritance of Al tolerance in maize (Zea mays) has been reported, and as with barley, multiple alleles at a single locus also have been reported to confer various extents of Al tolerance to different inbred lines. In maize breeding programs, however, Al tolerance generally is treated as a quantitative trait, a phenotype that can vary in a quantitative manner when measured among different individuals. Quantitative traits often are controlled by the cumulative action of alleles at multiple loci. In sorghum (Sorghum bicolor), Al tolerance has been reported as being either simply inherited or polygenic. Rye (Secale cereale) Al-tolerance genes have been localized to chromosomes 3R, 4R, and 6R by the use of wheat/rye chromosome addition and substitution lines. Finally, although variability for Al tolerance has been reported in rice (Oryza sativa), no formal Al-tolerance inheritance studies in this species have been carried out. In relation to the global importance of grain crops, our understanding of the genetics of Al tolerance is limited, which is particularly surprising considering that large acreages of these crops are grown on acid soils.

23.5.4 Molecular investigations into Al resistance are in their infancy.

Work is in progress to clone genes that are involved in Al resistance in plants. Using differential hybridization screening of a root tip cDNA library from an Al-sensitive wheat genotype, investigators cloned seven cDNAs that were Al induced. The proteins encoded by these cDNAs show homology to stress-related proteins in plants, including metal-lothionein-like proteins, phenylalanine-ammonia lyase (see Chapters 20 and 24), proteinase inhibitors (see Chapter 21), and asparagine synthetase (see Chapter 20). These genes were induced in both Al-resistant and Al-sensitive genotypes after 24 to 96 hours of exposure to Al. The timing of the induction of these genes, which comes well after an Al-tolerance response is initiated, suggests they are involved in Al-related stress responses and not resistance. Al-resistant (alr) mutants of Arabidopsis have been isolated in a molecular genetic approach to this problem. alr mutants were identified on the basis of enhanced root growth in the presence of Al concentrations that strongly inhibit root growth in wild-type seedlings. Genetic analysis of the alr mutants showed that Al resistance was semidominant, and chromosome mapping of the mutants with microsatellite and RAPD (randomly amplified polymorphic DNA) markers indicated that the mutants mapped to two sites in the Arabidopsis genome: a locus on chromosome 1, and one on chromosome 4. The alr mutants accumulated lower concentrations of Al in the root tips than did the wild-type plants. The mutants that mapped on chromosome 1 released greater amounts of citrate or malate (as well as pyruvate) than did the wild type, suggesting that Al exclusion from roots of these alr mutants results from enhanced organic acid exudation. Roots of alr-104, on the other hand, did not demonstrate increased release of malate or citrate but did exhibit an Al-induced increase in rhizosphere pH localized to the root tip. The increase in rhizosphere pH was sufficient to decrease Al^{3+} activity to less toxic values.

Work with transgenic lines of tobacco and papaya that overexpress the bacterial citrate synthase gene also provides strong conceptual support that enhanced organic acid release confers Al resistance. These lines synthesized much higher concentrations of citrate in the roots, and excreted a considerable amount of citrate into the soil. These citrate-overproducing lines are considerably more Al-resistant than is the wild type. This approach is exciting, but whether it will be of practical value to agriculture remains to be determined.

For further progress in this field, the genes that confer Al resistance must be isolated in those Al-tolerant cereal genotypes that have been characterized physiologically and genetically. Gene discovery in species such as wheat is not a trivial undertaking and may require the development or application of new approaches. Perhaps complementation of yeast with a cDNA library from Al-resistant plants or other gene discovery approaches will be suitable for investigation of the large and complex genomes of some crop species.

Summary

Plants use a wide range of mechanisms and responses to acquire essential mineral nutrients from the soil and to tolerate toxic soil environments. Some of these approaches are exceptionally complex. Diverse, nutrient-specific strategies allow terrestrial plants to increase the availability of sparingly soluble essential minerals in the soil, transport them into the root, and translocate them to critical sites throughout the plant. Plants also must regulate the acquisition of potentially toxic but essential elements such as Fe and Zn to prevent both nutrient deficiency and metal toxicity.

The field of plant mineral nutrition, rooted in physiological investigations, has recently moved into a new era of discovery, in which molecular research approaches are being applied to questions of mineral nutrition. A current challenge faced by plant scientists is to integrate the information gained from molecular dissections of these intricate processes with previous and current research into the physiology of mineral nutrition at the levels of cell, organ, and whole plant. The ultimate goal of this work is to further the scientific understanding of how intact plants grow, whether in a farmer's field or in a native ecosystem.

Further Reading

Anderson, J. A., Huprikar, S. S., Kochian, L. V., Lucas, W. J., Gaber, R. F. (1992) Functional expression of a probable *Arabidopsis thaliana* potassium channel in *Saccharomyces cerevisiae. Proc. Natl. Acad. Sci. USA* 89: 3736–3740.

Aniol, A., Gustafson, J. P. (1984) Chromosome locations of genes controlling aluminum tolerance in wheat, rye, and triticale. *Can. J. Genet. Cytol.* 26: 701–705.

Bertl, A., Anderson, J. A., Slayman, C. L., Gaber, R. F. (1995) Use of *Saccharomyces cerevisiae* for patch-clamp analysis of heterologous membrane proteins: characterization of Kat1, an inward-rectifying K^+ channel from *Arabidopsis thaliana*, and comparison with endogenous yeast channels and carriers. *Proc. Natl. Acad. Sci. USA* 92: 2701–2705.

Briat, J.-F., Lobreaux, S. (1997) Iron transport and storage. *Trends Plant Sci.* 2: 187–193.

Bun-Ya, M., Nishimura, M., Harashima, S., Oshima, Y. (1991) The *PHO84* gene of *Saccharomyces cerevisiae* encodes an inorganic phosphate transporter. *Mol. Cell. Biol.* 11: 3229–3238.

Carver, B. F., Ownby, J. D. (1995) Acid soil tolerance in wheat. *Adv. Agron.* 54: 117–173.

Chilcott, T. C., Frost-Shartzer, S., Iverson, M. W., Garvin, D. F., Kochian, L. V., Lucas, W. J. (1995) Potassium transport kinetics of *KAT1* expressed in *Xenopus* oocytes: a proposed molecular structure and field effect mechanism for membrane transport. *C. R. Acad. Sci. Paris Life Sci.* 318: 761–771.

Czempinski, K., Zimmermann, S., Ehrhardt, T., Muller-Rober, B. (1997) New structure and function in plant K^+ channels: KCO1, an outward rectifier with a steep Ca^{2+}-dependency. *EMBO J.* 16: 2565–2575.

Degenhardt, J., Larsen, P. B., Howell, S. E., Kochian, L. V. (1998) Aluminum resistance in the *Arabidopsis* mutant *alr-104* is caused by an aluminum-induced increase in rhizosphere pH. *Plant Physiol.* 117: 19–27.

de la Fuente, J. M., Ramirez-Rodriguez, V., Cabrera-Ponce, J. L., Herrera-Estrella, L. (1997) Aluminum tolerance in transgenic plants by alteration of citrate synthesis. *Science* 276: 1566–1568.

Delhaize, E., Craig, S., Beaton, C. D., Bennet, R. J., Jagadish, V. C., Randall, P. J. (1993) Aluminum tolerance in wheat (*Triticum* aestivum L.): I. Uptake and distribution of aluminum in root apices. *Plant Physiol.* 103: 685–693.

Delhaize, E., Ryan, P. R., Randall, P. J. (1993) Aluminum tolerance in wheat (*Triticum aestivum* L.): II. Aluminum-stimulated excretion of malic acid from root apices. *Plant Physiol.* 103: 695–702.

Dinkelaker, B., Romheld, V., Marschner, H. (1989) Citric acid excretion and precipitation of calcium citrate in the rhizosphere of white lupin (*Lupinus albus*). *Plant Cell Envir.* 12: 285–292.

Eide, D., Broderius, M., Fett, J., Guerinot, M. L. (1996) A novel iron-regulated metal transporter from plants identified by functional expression in yeast. *Proc. Natl. Acad. Sci. USA* 93: 5624–5628.

Epstein, E. (1972) *Mineral Nutrition of Plants: Principles and Perspectives.* John Wiley & Sons, New York.

Epstein, E., Rains, D. W., Elzam, O. E. (1963) Resolution of dual mechanisms of potassium absorption by barley roots. *Proc. Natl. Acad. Sci. USA* 49: 684–692.

Fox, T. C., Guerinot, M. L. (1998) Molecular biology of cation transport in plants. *Annu. Rev. Plant Physiol. Plant Mol. Biol.* 49: 669–696.

Fu, H.-H., Luan, S. (1998) AtKUP1: a dual-affinity K^+ transporter from *Arabidopsis*. *Plant Cell* 10: 63–73.

Gassmann, W., Rubio, F., Schroeder, J. I. (1996) Alkali cation selectivity of the wheat root high-affinity potassium transporter HKT1. *Plant J.* 10: 869–882.

Gaymard, F., Cerutti, M., Horeau, C., Lemaille, G., Urbach, S., Ravallec, M., Devauchelle, G., Sentenac, H., Thibaud, J.-P. (1996) The baculovirus/insect cell system as an alternative to *Xenopus* oocytes. First characterization of the AKT1 K^+ channel from *Arabidopsis thaliana*. *J. Biol. Chem.* 271: 22863–22870.

Grotz, N., Fox, T., Connolly, E., Park, W., Guerinot, M. L., Eide, D. (1998) Identification of a family of zinc transporter genes from *Arabidopsis* that respond to zinc deficiency. *Proc. Natl. Acad. Sci. USA* 95: 7220–7224.

Hirsch, R. E., Lewis, B. D., Spalding, E. P., Sussman, M. R. (1998) A role for the AKT1 potassium channel in plant nutrition. *Science* 280: 918–921.

Hoagland, D. R., Broyer, T. C. (1936) General nature of the process of salt accumulation by roots with description of experimental methods. *Plant Physiol.* 11: 471–507.

Johnson, J. F., Allen, D. L., Vance, C. P. (1994) Phosphorus stress-induced proteoid roots show altered metabolism in *Lupinus albus*. *Plant Physiol.* 104: 657–665.

Johnson, J. F., Vance, C. P., Allen, D. L. (1996) Phosphorus deficiency in *Lupinus albus*. Altered lateral root development and enhanced expression of phospho*enol*pyruvate carboxylase. *Plant Physiol.* 112: 31–41.

Kim, E. J., Kwak, J. M., Uozumi, N., Schroeder, J. I. (1998) AtKUP1: an *Arabidopsis* gene encoding high-affinity potassium transport activity. *Plant Cell* 10: 51–62.

Kochian, L. V. (1993) Zinc absorption from hydroponic solutions by plant roots. In *Zinc in Soils and Plants*, A. D. Robson, ed. Kluwer Academic Publishers, Dordrecht, The Netherlands, pp. 45–57.

Kochian, L. V. (1995) Cellular mechanisms of aluminum toxicity and resistance in plants. *Annu. Rev. Plant Physiol. Plant Mol. Biol.* 46: 237–260.

Kochian, L. V., Jones, D. L. (1997) Aluminum toxicity and resistance in plants. In *Research Issues in Aluminum Toxicity*, R. Yokel and M. S. Golub, eds. Taylor and Francis Publishers, Washington, D.C., pp. 69–89.

Kochian, L. V., Lucas, W. J. (1988) Potassium transport in roots. *Adv. Bot. Res.* 15: 93–178.

Korshunova, Y. O., Eide, D., Clark, W. G., Guerinot, M. L., Pakrasi, H. B. (1999) The IRT1 protein from *Arabidopsis thaliana* is a metal transporter with a broad substrate range. *Plant Mol. Biol.* 40: 37–44.

Larsen, P. B., Degenhardt, J., Tai, C.-Y., Stenzler, L., Howell, S. H., Kochian, L. V. (1998) Aluminum-resistant *Arabidopsis* mutants that exhibit altered patterns of Al accumulation and organic acid release from roots. *Plant Physiol.* 117: 9–18.

Lasat, M. M., Baker, A.J.M., Kochian, L. V. (1996) Physiological characterization of root Zn^{2+} absorption and translocation to shoots in Zn hyperaccumulator and nonaccumulator species of *Thlaspi*. *Plant Physiol.* 112: 1715–1722.

Lasat, M. M., Baker, A.J.M., Kochian, L. V. (1998) Altered zinc compartmentation in the root symplasm and stimulated Zn^{2+} absorption into the leaf as mechanisms involved in zinc hyperaccumulation in *Thlaspi caerulescens*. *Plant Physiol.* 118: 875–883.

Liu, C., Muchhal, U. S., Uthappa, M., Kononowicz, A. K., Raghothama, K. G. (1998) Tomato phosphate transporter genes are differentially regulated in plant tissues by phosphorus. *Plant Physiol.* 116: 91–99.

Maathuis, F.J.M., Sanders, D. (1993) Energization of potassium uptake in *Arabidopsis thaliana*. *Planta* 191: 302–307.

Maathuis, F.J.M., Sanders, D. (1994) Mechanism of high-affinity potassium uptake in roots of *Arabidopsis thaliana*. *Proc. Natl. Acad. Sci. USA* 91: 9272–9276.

Maathuis, F.J.M., Sanders, D. (1996) Mechanisms of potassium absorption by higher plant roots. *Physiol. Plant* 96: 158–168.

Maathuis, F.J.M., Ichida, A. M., Sanders, D., Schroeder, J. I. (1997) Update on channels: roles of higher plant K$^+$ channels. *Plant Physiol.* 114: 1141–1149.

Maathuis, F.J.M., Verlin, D., Smith, F. A., Sanders, D., Fernandez, J. A., Walker, N. A. (1996) The physiological relevance of Na$^+$-coupled K$^+$ transport. *Plant Physiol.* 112: 1609–1616.

Marschner, H. (1995) *Mineral Nutrition of Higher Plants*, 2nd ed. Academic Press, London.

Minella, E., Sorrells, M. E. (1992) Aluminum tolerance in barley: genetic relationships among genotypes of diverse origin. *Crop Sci.* 32: 593–598.

Muchhal, U. S., Pardo, J. M., Raghothama, K. G. (1996) Phosphate transporters from the higher plant *Arabidopsis thaliana*. *Proc. Natl. Acad. Sci. USA* 93: 10519–10523.

Nissen, P. (1974) Uptake mechanisms: inorganic and organic. *Annu. Rev. Plant Physiol.* 25: 53–79.

Quintero, F. J., Blatt, M. R. (1997) A new family of K$^+$ transporters from *Arabidopsis* that are conserved across phyla. *FEBS Lett.* 415: 206–211.

Raghothama, K. G. (1999) Phosphate acquisition. *Annu. Rev. Plant Physiol. Plant Mol. Biol.* 50: 665–693.

Riede, C. R., Anderson, J. A. (1996) Linkage of RFLP markers to an aluminum tolerance gene in wheat. *Crop Sci.* 36: 905–909.

Robinson, N. J., Procter, C. M., Connolly, E. L., Guerinot, M. L. (1999) A ferric-chelate reductase for iron uptake from soils. *Nature* 397: 694–697.

Rodriguez-Navarro, A., Blatt, M. R., Slayman, C. L. (1986) A potassium–proton symport in *Neurospora crassa*. *J. Gen. Physiol.* 87: 649–674.

Rubio, F., Gassmann, W., Schroeder, J. I. (1995) Sodium-driven potassium uptake by the plant potassium transporter HKT1 and mutations conferring salt tolerance. *Science* 270: 1660–1663.

Salt, D. E., Smith, R. D., Raskin, I. (1998) Phytoremediation. *Annu. Rev. Plant Physiol. Plant Mol. Biol.* 49: 643–668.

Schachtman, D. P., Schroeder, J. I. (1994) Structure and transport mechanism of a high-affinity potassium uptake transporter from higher plants. *Nature* 370: 655–658.

Schroeder, J. I., Ward, J. M., Gassmann, W. (1994) Perspectives on the physiology and structure of inward-rectifying K$^+$ channels in higher plants—biophysical implications for K$^+$ uptake. *Annu. Rev. Biophys. Biomol. Struct.* 23: 441–471.

Sentenac, H., Bonneaud, N., Minet, M., Lacroute, F., Salmon, J.-M., Gaymard, F., Grignon, C. (1992) Cloning and expression in yeast of a plant potassium ion transport system. *Science* 256: 663–665.

von Uexküll, H. R., Mutert, E. (1995) Global extent, development and economic impact of acid soils. In *Plant–Soil Interactions at Low pH: Principles and Management*, R. A. Date, N. J. Grundon, G. E. Raymet, and M. E. Probert, eds. Kluwer Academic Publishers, Dordrecht, The Netherlands, pp. 5–19.

Wegner, L. H., Raschke, K. (1994) Ion channels in the xylem parenchyma of barley roots. A procedure to isolate protoplasts from this tissue and a patch-clamp exploration of salt passageways into xylem vessels. *Plant Physiol.* 105: 799–813.

Welch, R. M. (1995) Micronutrient nutrition of plants. *Crit. Rev. Plant Sci.* 14: 49–82.

Biochemistry & Molecular Biology of Plants, B. Buchanan, W. Gruissem, R. Jones, Eds.
© 2000, American Society of Plant Physiologists

CHAPTER 24

Natural Products (Secondary Metabolites)

Rodney Croteau
Toni M. Kutchan
Norman G. Lewis

CHAPTER OUTLINE

Introduction

Natural products have primary ecological functions.

Plants produce a vast and diverse assortment of organic compounds, the great majority of which do not appear to participate directly in growth and development. These substances, traditionally referred to as **secondary metabolites,** often are differentially distributed among limited taxonomic groups within the plant kingdom. Their functions, many of which remain unknown, are being elucidated with increasing frequency. The **primary metabolites,** in contrast, such as phytosterols, acyl lipids, nucleotides, amino acids, and organic acids, are found in all plants and perform metabolic roles that are essential and usually evident.

Although noted for the complexity of their chemical structures and biosynthetic pathways, **natural products** have been widely perceived as biologically insignificant and have historically received little attention from most plant biologists. Organic chemists, however, have long been interested in these novel phytochemicals and have investigated their chemical properties extensively since the 1850s. Studies of natural products stimulated development of the separation techniques, spectroscopic approaches to structure elucidation, and synthetic methodologies that now constitute the foundation of contemporary organic chemistry. Interest in natural products was not purely academic but rather was prompted by their great utility as dyes, polymers, fibers, glues, oils, waxes, flavoring agents, perfumes, and drugs. Recognition of the biological properties of myriad natural products has fueled the current focus of this field, namely, the search for new drugs, antibiotics, insecticides, and herbicides. Importantly, this growing appreciation of the highly diverse biological effects produced by natural products has prompted a reevaluation of the possible roles these compounds play in plants, especially in the context of ecological interactions. As illustrated in this chapter, many of these compounds now have been shown to have important adaptive significance in protection against herbivory and microbial infection, as attractants for pollinators and seed-dispersing animals, and as allelopathic agents (allelochemicals that influence competition among plant species). These ecological functions affect plant survival profoundly, and we think it reasonable to adopt the less pejorative term "plant natural products" to describe secondary plant metabolites that act primarily on other species.

The boundary between primary and secondary metabolism is blurred.

Based on their biosynthetic origins, plant natural products can be divided into three major groups: the terpenoids, the alkaloids, and the phenylpropanoids and allied phenolic compounds. All terpenoids, including both primary metabolites and more than 25,000 secondary compounds, are derived from the five-carbon precursor isopentenyl diphosphate (IPP). The 12,000 or so known alkaloids, which contain one or more nitrogen atoms, are biosynthesized principally from amino acids. The 8000 or so phenolic compounds are formed by way of either the shikimic acid pathway or the malonate/acetate pathway.

Primary and secondary metabolites cannot readily be distinguished on the basis of precursor molecules, chemical structures, or biosynthetic origins. For example, both primary and secondary metabolites are found among the diterpenes (C_{20}) and triterpenes (C_{30}). In the diterpene series, both kaurenoic acid and abietic acid are formed by a very similar sequence of related enzymatic reactions (Fig. 24.1); the former is an essential intermediate in the synthesis of gibberellins, i.e., growth hormones found in all plants (see Chapter 17), whereas the latter is a resin component largely restricted to members of the Fabaceae and Pinaceae. Similarly, the essential amino acid proline is classified as a primary metabolite, whereas the C_6 analog pipecolic acid is considered an alkaloid and

thus a natural product (Fig. 24.1). Even lignin, the essential structural polymer of wood and second only to cellulose as the most abundant organic substance in plants, is considered a natural product rather than a primary metabolite.

In the absence of a valid distinction based on either structure or biochemistry, we return to a functional definition, with primary products participating in nutrition and essential metabolic processes inside the plant, and natural (secondary) products influencing ecological interactions between the plant and its environment. In this chapter, we provide an overview of the biosynthesis of the major classes of plant natural products, emphasizing the origins of their structural diversity, as well as their physiological functions, human uses, and potential biotechnological applications.

24.1 Terpenoids

Terpenoids perhaps are the most structurally varied class of plant natural products. The name terpenoid, or terpene, derives from the fact that the first members of the class were

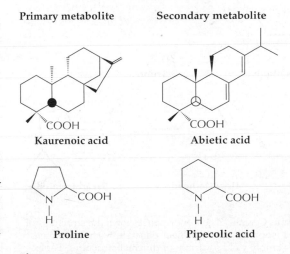

Primary metabolite

Kaurenoic acid

Proline

Secondary metabolite

Abietic acid

Pipecolic acid

Figure 24.1
Kaurenoic acid and proline are primary metabolites, whereas the closely related compounds abietic acid and pipecolic acid are considered secondary metabolites.

isolated from turpentine ("terpentin" in German). All terpenoids are derived by repetitive fusion of branched five-carbon units based on isopentane skeleton. These monomers generally are referred to as isoprene units because thermal decomposition of many terpenoid substances yields the alkene gas isoprene as a product (Fig. 24.2, upper panel) and because suitable chemical conditions can induce isoprene to polymerize in multiples of five carbons, generating numerous terpenoid skeletons. For these reasons, the terpenoids are often called isoprenoids, although researchers have known for well over 100 years that isoprene itself is not the biological precursor of this family of metabolites.

24.1.1 Terpenoids are classified by the number of five-carbon units they contain.

The five-carbon (isoprene) units that make up the terpenoids are often joined in a "head-to-tail" fashion, but head-to-head fusions are also common, and some products are formed by head-to-middle fusions (Fig. 24.2, lower panel). Accordingly, and because extensive structural modifications with carbon–carbon bond rearrangements can occur, tracing the original pattern of isoprene units is sometimes difficult.

The smallest terpenes contain a single isoprene unit; as a group, they are named **hemiterpenes** (half-terpenes). The best

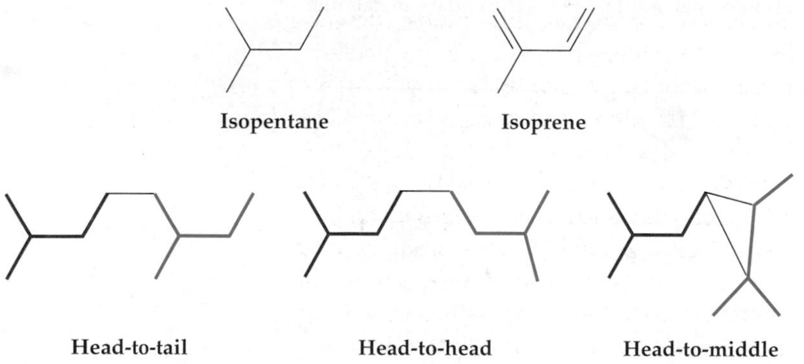

known hemiterpene is isoprene itself, a volatile product released from photosynthetically active tissues. The enzyme isoprene synthase is present in the leaf plastids of numerous C_3 plant species, but the metabolic rationale for the light-dependent production of isoprene is unknown (acclimation to high temperatures has been suggested). Estimated annual foliar emissions of isoprene are quite substantial (5×10^8 metric tons of carbon), and the gas is a principal reactant in the NOx radical–induced formation of tropospheric ozone (see Chapter 22, Fig. 22.37).

C_{10} terpenoids, although they consist of two isoprene units, are called **monoterpenes;** as the first terpenoids isolated from turpentine in the 1850s, they were considered to be the base unit from which the subsequent nomenclature is derived. The monoterpenes are best known as components of the volatile essences of flowers and of the essential oils of herbs and spices, in which they make up as much as 5% of plant dry weight. Monoterpenes are isolated by either distillation or extraction and find considerable industrial use in flavors and perfumes.

The terpenoids that derive from three isoprene units contain 15 carbon atoms and are known as **sesquiterpenes** (i.e., one and one-half terpenes). Like monoterpenes, many sesquiterpenes are found in essential oils. In addition, numerous sesquiterpenoids act as phytoalexins, antibiotic compounds produced by plants in response to microbial challenge, and as antifeedants that discourage opportunistic herbivory. Although the plant hormone abscisic acid is structurally a sesquiterpene, its C_{15} precursor, xanthoxin, is not synthesized directly from three isoprene units but rather is produced by asymmetric cleavage of a C_{40} carotenoid (see Chapter 17).

The **diterpenes,** which contain 20 carbons (four C_5 units), include phytol (the hydrophobic side chain of chlorophyll), the gibberellin hormones, the resin acids of conifer and legume species, phytoalexins, and a host of pharmacologically important metabolites, including taxol, an anticancer agent found at very low concentrations (0.01% dry weight) in yew bark, and forskolin, a compound used to treat glaucoma. Some gibberellins have only 19 carbon atoms and are considered norditerpenoids since they have lost 1 carbon through a metabolic cleavage reaction (see Chapter 17).

Isopentane   Isoprene

Head-to-tail   Head-to-head   Head-to-middle

Figure 24.2

Terpenes are synthesized from C_5 units. The upper panel shows the structures of the isopentane skeleton and isoprene gas. The lower panel shows how different patterns of isoprene unit assembly yield a variety of different structures. For example, the triterpene squalene is formed by head-to-head fusion of two molecules of farnesyl diphosphate (FPP), which itself is the product of the head-to-tail fusion of isopentenyl diphosphate (IPP) and geranyl diphosphate (GPP) (see Fig. 24.7). The monoterpene pyrethrin I (see Fig. 24.10) results from a head-to-middle fusion of two C_5 units.

The **triterpenes,** which contain 30 carbon atoms, are generated by the head-to-head joining of two C_{15} chains, each of which constitutes three isoprene units joined head-to-tail. This large class of molecules includes the brassinosteroids (see Chapter 17), the phytosterol membrane components (see Chapter 1), certain phytoalexins, various toxins and feeding deterrents, and components of surface waxes, such as oleanolic acid of grapes.

The most prevalent **tetraterpenes** (40 carbons, eight isoprene units) are the carotenoid accessory pigments which perform essential functions in photosynthesis (see Chapter 12). The **polyterpenes,** those containing more than eight isoprene units, include the prenylated quinone electron carriers (plastoquinone and ubiquinone; see Chapters 12 and 14), long-chain polyprenols involved in sugar transfer reactions (e.g., dolichol; see Chapters 1 and 4), and enormously long polymers such as rubber (average molecular mass greater than 10^6 Da), often found in latex.

Natural products of mixed biosynthetic origins that are partially derived from terpenoids are often called **meroterpenes.** For example, both cytokinins (see Chapter 17) and numerous phenylpropanoid compounds contain C_5 isoprenoid side chains. Certain alkaloids, including the anticancer drugs vincristine and vinblastine, contain terpenoid fragments in their structures (see Fig. 24.34). Additionally, some modified proteins include a 15- or 20-carbon terpenoid side chain that anchors the protein in a membrane (see Chapter 1).

24.1.2 A diverse array of terpenoid compounds is synthesized by various conserved reaction mechanisms.

At the turn of the 20th century, structural investigations of many terpenoids led Otto Wallach to formulate the "isoprene rule," which postulated that most terpenoids could be constructed hypothetically by repetitively joining isoprene units. This principle provided the first conceptual framework for a common structural relationship among terpenoid natural products (Box 24.1). Wallach's idea was refined in the 1930s, when Leopold Ruzicka formulated the "biogenetic isoprene rule," emphasizing mechanistic considerations of terpenoid synthesis in terms of electrophilic elongations, cyclizations, and rearrangements. This hypothesis ignores the precise character of the biological precursors and assumes only that they are "isoprenoid" in structure. As a working model for terpenoid biosynthesis, the biogenetic isoprene rule has proved essentially correct.

Despite great diversity in form and function, the terpenoids are unified in their common biosynthetic origin. The biosynthesis of all terpenoids from simple, primary metabolites can be divided into four overall steps: (*a*) synthesis of the fundamental precursor IPP; (*b*) repetitive additions of IPP to form a series of prenyl diphosphate homologs, which serve as the immediate precursors of the different classes of terpenoids; (*c*) elaboration of these allylic prenyl diphosphates by specific terpenoid synthases to yield terpenoid skeletons; and (*d*) secondary enzymatic modifications to the skeletons (largely redox reactions) to give rise to the functional properties and great chemical diversity of this family of natural products.

24.2 Synthesis of IPP

24.2.1 Biosynthesis of terpenoids is compartmentalized, as is production of the terpenoid precursor IPP.

Although terpenoid biosynthesis in plants, animals, and microorganisms involves similar classes of enzymes, important differences exist among these processes. In particular, plants produce a much wider variety of terpenoids than do either animals or microbes, a difference reflected in the complex organization of plant terpenoid biosynthesis at the tissue, cellular, subcellular, and genetic levels. The production of large quantities of terpenoid natural products as well as their subsequent accumulation, emission, or secretion is almost always associated with the presence of anatomically highly specialized structures. The glandular trichomes (Fig. 24.3A, B) and secretory cavities of leaves (Fig. 24.3C) and the glandular epiderms of flower petals generate and store or emit terpenoid essential oils that are important because they encourage pollination by insects. The resin ducts and blisters of conifer species

Box 24.1

Early investigators formulated rules for identifying and naming isoprenoid structures.

In the late 1800s, chemists struggled to define the structures of the monoterpenes. The mixed results achieved by these efforts are illustrated by the numerous structures proposed for camphor ($C_{10}H_{16}O$; see structures at left of figure, which include the names of the proposers and the dates proposed). Chromatographic purification techniques and spectroscopic methods for structure elucidation were not available to these early chemists, who relied on the preparation of crystalline derivatives to assess purity and on chemical degradation studies to determine structures. Systematic study of the monoterpenes led the German chemist Otto Wallach to recognize that many terpenoid compounds might be constructed by joining isoprene units, generally in a repetitive head-to-tail fashion, as in Bredt's correct proposed structure for camphor (see figure). This concept, known as the **isoprene rule,** earned Wallach the Nobel Prize in Chemistry in 1910.

By the 1930s, faced with a bewildering array of terpenoid substances, Leopold Ruzicka and his contemporaries sought to develop a unifying principle that could rationalize the natural occurrence of all of the known terpenoids, even those that did not strictly fit Wallach's isoprene rule. Ruzicka's ingenious solution to the problem was to focus on reaction mechanisms and ignore the precise character of the biological precursor, assuming only that it had a terpenoid structure during reaction. He hypothesized the involvement of electrophilic reactions that generated carbocationic intermediates, which underwent subsequent C_5 addition, cyclization, and in some cases skeletal rearrangement before elimination of a proton or capture by a nucleophile to yield the observed terpenoid products. This proposal, which Ruzicka called the **biogenetic isoprene rule,** can be stated simply: A compound is "isoprenoid" if it is derived biologically from an "isoprenoid" precursor, with or without rearrangements. Ruzicka's concept differs from Wallach's in its emphasis on biochemical origin rather than structure. The great strength of the biogenetic isoprene rule lay in its use of mechanistic considerations to classify the bulk of known terpenoids, including structures that did not strictly follow Wallach's isoprene rule. Application of the biogenetic isoprene rule to the origin of several of the common monoterpene skeletons is illustrated in the right panel of the figure (note the bornane skeleton from which camphor is derived). Ruzicka was awarded the Nobel Prize in Chemistry in 1939.

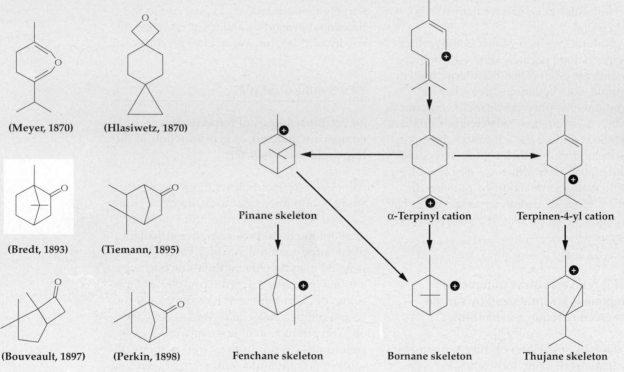

(Meyer, 1870)   **(Hlasiwetz, 1870)**

(Bredt, 1893)   **(Tiemann, 1895)**

(Bouveault, 1897)   **(Perkin, 1898)**

Pinane skeleton   **α-Terpinyl cation**   **Terpinen-4-yl cation**

Fenchane skeleton   **Bornane skeleton**   **Thujane skeleton**

(Fig. 24.3D) produce and accumulate a defensive resin consisting of turpentine (monoterpene olefins) and rosin (diterpenoid resin acids). Triterpenoid surface waxes are formed and excreted from specialized epidermis, and laticifers produce certain triterpenes and polyterpenes such as rubber. These specialized structures sequester natural products away from sensitive metabolic processes and thereby prevent autotoxicity. Most structures of this type are nonphotosynthetic and must therefore rely on adjacent cells to supply the carbon and energy needed to drive terpenoid biosynthesis.

A more fundamental, and perhaps universal, feature of the organization of terpenoid metabolism exists at the subcellular level. The sesquiterpenes (C_{15}), triterpenes (C_{30}), and polyterpenes appear to be produced in the cytosolic and endoplasmic reticulum (ER) compartments, whereas isoprene, the monoterpenes (C_{10}), diterpenes (C_{20}), tetraterpenes (C_{40}), and certain prenylated quinones originate largely, if not exclusively, in the plastids. The evidence now indicates that the biosynthetic pathways for the formation of the fundamental precursor IPP differ markedly in these compartments, with the classical acetate/mevalonate pathway being active in the cytosol and ER and the glyceraldehyde phosphate/pyruvate pathway operating in the plastids. Regulation of these dual pathways may be difficult to assess, given that plastids may supply

IPP to the cytosol for use in biosynthesis, and vice versa. Mitochondria, a third compartment, may generate the ubiquinone prenyl group by the acetate/mevalonate pathway, although little is known about the capability of these organelles for terpenoid biosynthesis.

24.2.2 Hydroxymethylglutaryl-CoA reductase, an enzyme in the acetate/mevalonate pathway, is highly regulated.

The basic enzymology of IPP biosynthesis by way of the acetate/mevalonate pathway is widely accepted (Fig. 24.4). This cytosolic IPP pathway involves the two-step condensation of three molecules of acetyl-CoA catalyzed by thiolase and hydroxymethylglutaryl-CoA synthase. The resulting product, 3-hydroxy-3-methylglutaryl-CoA (HMG-CoA), is subsequently reduced by

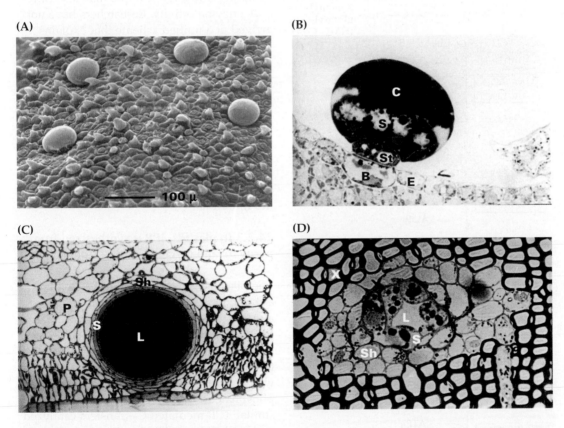

(A)

(B)

(C)

(D)

Figure 24.3
(A) Scanning electron micrograph of the leaf surface of thyme. The round structures are peltate glandular trichomes, in which monoterpenes and sesquiterpenes are synthesized. (B) Light micrograph of a glandular trichome from spearmint, shown in longitudinal section. C, subcuticular space; S, secretory cells; St, stalk; B, basal cell; E, epidermal cell. (C) Light micrograph of a secretory cavity in a lemon leaf, shown in cross-section. L, lumen; Sh, sheath cells; P, parenchyma cell. (D) Light micrograph of a resin duct in wood of Jeffrey pine, shown in cross-section. X, secondary xylem.

HMG-CoA reductase in two coupled reactions that form mevalonic acid. Two sequential ATP-dependent phosphorylations of mevalonic acid and a subsequent phosphorylation/elimination-assisted decarboxylation yield IPP.

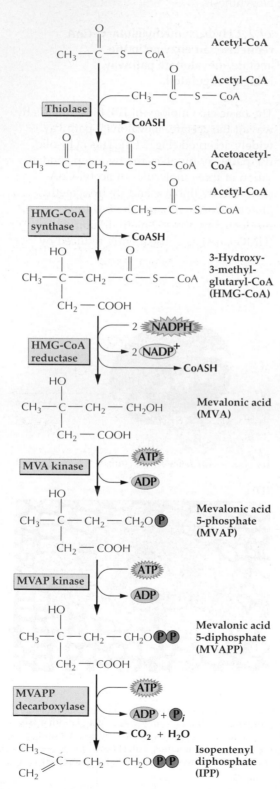

Figure 24.4
The acetate/mevalonate pathway for the formation of IPP, the basic five-carbon unit of terpenoid biosynthesis. Synthesis of each IPP unit requires three molecules of acetyl-CoA.

HMG-CoA reductase is one of the most highly regulated enzymes in animals, being largely responsible for the control of cholesterol biosynthesis. Accumulated evidence indicates that the plant enzyme, which is located in the ER membrane, is also highly regulated. In many cases, small gene families, each containing multiple members, encode this reductase. These gene families are expressed in complex patterns, with individual genes exhibiting constitutive, tissue- or development-specific, or hormone-inducible expression. Specific HMG-CoA reductase genes can be induced by wounding or pathogen infection. The activity of HMG-CoA reductase may be subject to posttranslational regulation, for example, by a protein kinase cascade that phosphorylates and thereby inactivates the enzyme. Allosteric modulation probably also plays a regulatory role. Proteolytic degradation of HMG-CoA reductase protein and the rate of turnover of the corresponding mRNA transcripts may also influence enzyme activity. Researchers have not arrived at a unified scheme that explains how the various mechanisms that regulate HMG-CoA reductase facilitate the production of different terpenoid families. The precise biochemical controls that influence activity have been difficult to assess in vitro because the enzyme is associated with the ER membrane. A model proposed to rationalize the selective participation of HMG-CoA reductase in the biosynthesis of different mevalonate-derived terpenoids is shown in Figure 24.5.

24.2.3 In plastids, IPP is synthesized from pyruvate and glyceraldehyde 3-phosphate.

The plastid-localized route to IPP involves a different pathway, demonstrated in green algae and many eubacteria as well as plants. In this pathway, pyruvate reacts with thiamine pyrophosphate (TPP) to yield a two-carbon fragment, hydroxyethyl-TPP, which condenses with glyceraldehyde 3-phosphate (see Chapter 12, Fig. 12.41, for similar TPP-mediated C_2 transfers catalyzed by transketolase). TPP is released to form a five-carbon intermediate, 1-deoxy-D-xylulose 5-phosphate, which is rearranged and reduced to form 2-C-methyl-D-erythritol 4-phosphate and subsequently transformed to yield IPP (Fig. 24.6, upper pathway). The

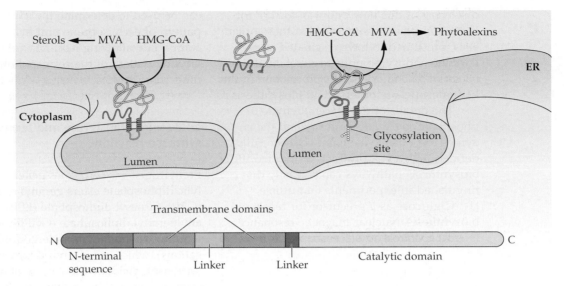

Figure 24.5

Model for the membrane topology of HMG-CoA reductase (HMGR). The protein includes a highly variable hydrophilic N-terminal sequence (blue), a conserved membrane anchor (orange), a highly variable linker sequence (green and purple), and a highly conserved, cytosol-exposed, C-terminal catalytic domain (yellow). Isoforms of HMGR that are associated with elicitor-induced synthe-sis of sesquiterpenoid phytoalexins contain an N-linked glycosyla-tion site exposed to the ER lumen. Differences in N-terminal se-quences and extent of glycosylation may affect targeting of HMGR to various ER domains and to other organelles of the endomem-brane system (see Chapters 1 and 4). ER, endoplasmic reticulum; MVA, mevalonic acid.

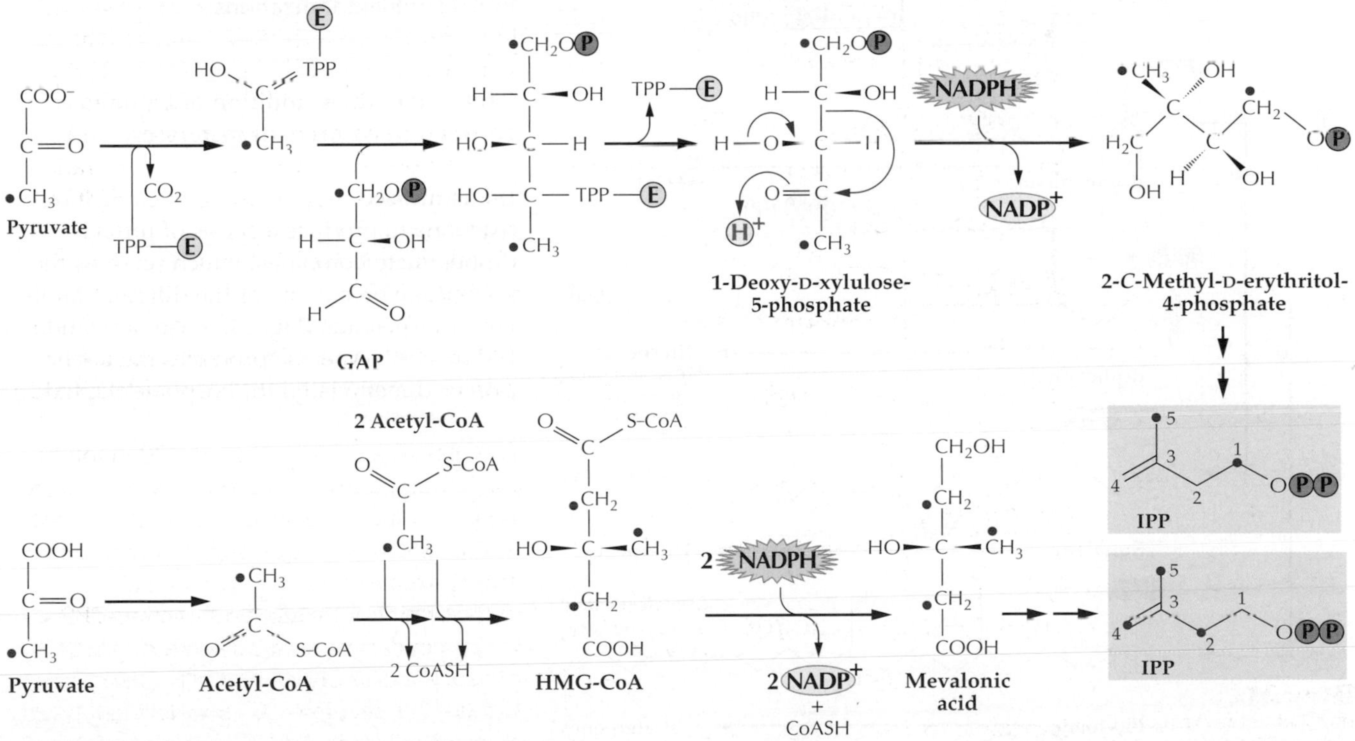

Figure 24.6

Feeding studies distinguish two pathways of isoprenoid biosynthe-sis. When glucose isotopically labeled at C-1 is transformed by gly-colytic enzymes and pyruvate dehydrogenase, the label subse-quently appears in the methyl groups of pyruvate and acetyl-CoA and in C-3 of glyceraldehyde 3-phosphate (GAP). IPP synthezised from labeled pyruvate and GAP by the plastid-localized pathway will be labeled at C-1 and C-5 (upper panel), whereas IPP formed from labeled acetyl-CoA by way of the cytosolic acetate/meval-onate pathway will be labeled at C-2, C-4, and C-5 (lower panel).

discovery of this new pathway for IPP formation in plastids suggests that these organelles, presumed to have originated as prokaryotic endosymbionts, have retained the bacterial machinery for the production of this key intermediate of terpenoid biosynthesis.

The details of the glyceraldehyde 3-phosphate/pyruvate pathway and the enzymes responsible have not yet been fully defined. However, products of the two IPP biosynthesis pathways can be easily distinguished in experiments that utilize [1-^{13}C]glucose as a precursor for terpenoid biosynthesis. Nuclear magnetic resonance (NMR) spectroscopy (see Chapter 2, Box 2.2)

can be used to determine the ^{13}C-labeling pattern of each isoprene unit in a terpenoid compound, allowing researchers to infer the labeling pattern of the corresponding IPP units (Fig. 24.6).

24.3 Prenyltransferase and terpene synthase reactions

Prenyltransferase enzymes generate the allylic diphosphate esters geranyl diphosphate (GPP), farnesyl diphosphate (FPP), and geranylgeranyl diphosphate (GGPP). Reactions that these compounds undergo (often cyclizations), which are catalyzed by terpene synthases, yield a wide variety of terpenoid compounds. Both prenyltransferases and terpene synthases utilize electrophilic reaction mechanisms involving carbocationic intermediates, a feature of terpenoid biochemistry. Enzymes in both groups share similar properties and contain conserved sequence elements, such as an aspartate-rich DDxxD motif involved in substrate binding, which may participate in initiating divalent metal ion–dependent ionizations.

24.3.1 Repetitive addition of C_5 units is carried out by prenyltransferases.

IPP is utilized in a sequence of elongation reactions to produce a series of prenyl diphosphate homologs, which serve as the immediate precursors of the different families of terpenoids (Fig. 24.7). Isomerization of IPP by IPP isomerase produces the allylic isomer dimethylallyl diphosphate (DMAPP),

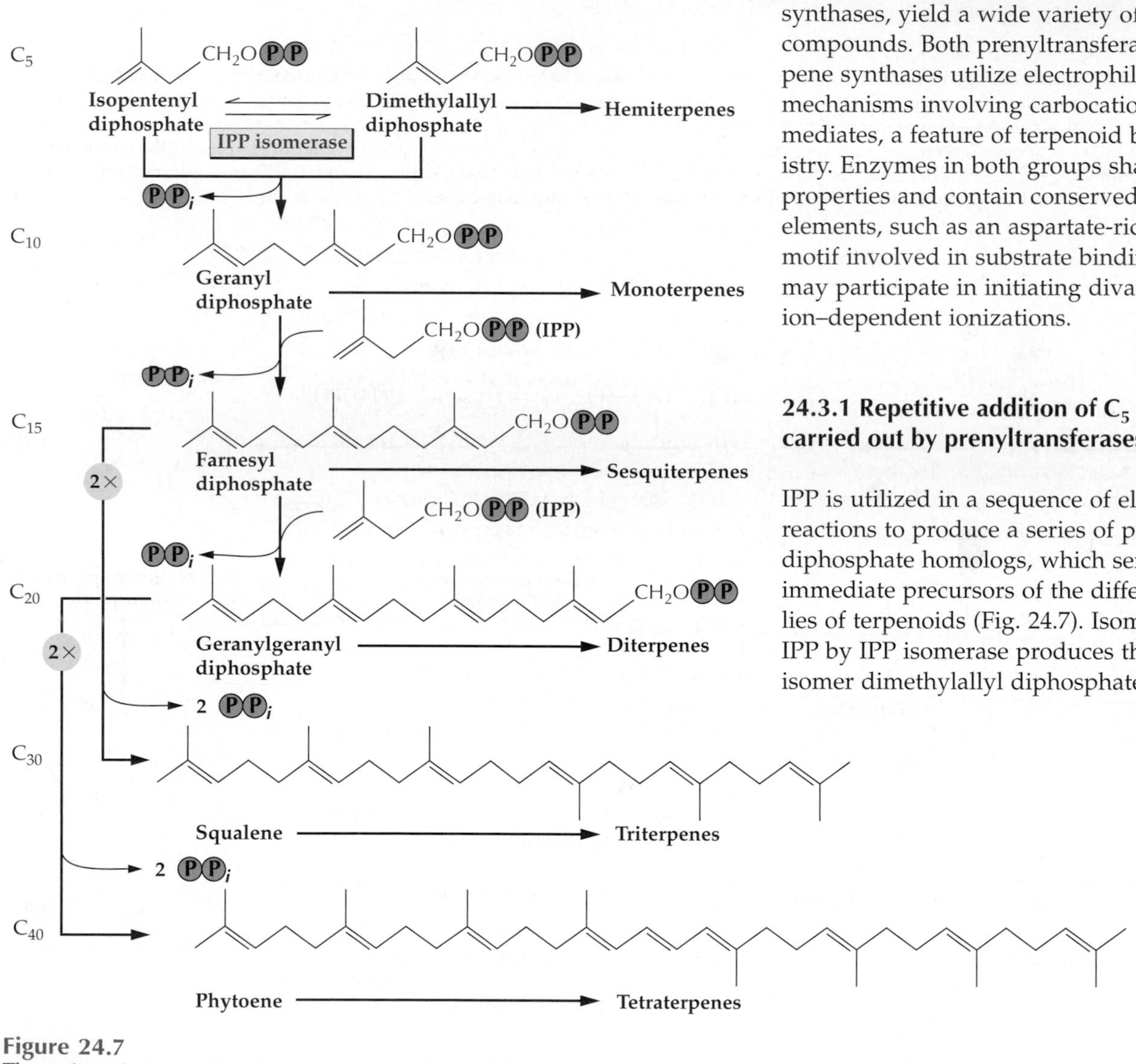

Figure 24.7
The major subclasses of terpenoids are biosynthesized from the basic five-carbon unit, IPP, and from the initial prenyl (allylic) diphosphate, dimethylallyl diphosphate, which is formed by isomerization of IPP. In reactions catalyzed by prenyltransferases, monoterpenes (C_{10}), sesquiterpenes (C_{15}), and diterpenes (C_{20}) are derived from the corresponding intermediates by sequential head-to-tail addition of C_5 units. Triterpenes (C_{30}) are formed from two C_{15} (farnesyl) units joined head-to-head, and tetraterpenes (C_{40}) are formed from two C_{20} (geranylgeranyl) units joined head-to-head.

which is considered the first prenyl diphosphate. Because DMAPP and related prenyl diphosphates contain an allylic double bond, these compounds can be ionized to generate resonance-stabilized carbocations. Once formed, a carbocation intermediate of n carbons can react with IPP to yield a prenyl diphosphate homolog containing $n + 5$ carbons. Thus, the reactive primer DMAPP undergoes condensation with IPP to yield the C_{10} intermediate GPP. Repetition of the reaction cycle by addition of one or two molecules of IPP provides FPP (C_{15}) or GGPP (C_{20}), respectively. Each prenyl homolog in the series arises as an allylic diphosphate ester that can ionize to form a resonance-stabilized carbocation and condense with IPP in another round of elongation (Fig. 24.8).

The electrophilic elongation reactions that yield C_{10}, C_{15}, and C_{20} prenyl diphosphates are catalyzed by enzymes known collectively as prenyltransferases. GPP, FPP, and GGPP are each formed by specific prenyltransferases named for their products (e.g., farnesyl diphosphate synthase). The new allylic double bond introduced in the course of the prenyltransferase reaction is commonly in the trans geometry, although this is not always the case: The transferase responsible for rubber biosynthesis introduces cis-double bonds, which are responsible for the elasticity of that polymer. Prenylation reactions are not limited to elongations involving IPP; the same basic carbocationic mechanism permits the attachment of prenyl side chains to atoms of carbon, oxygen, nitrogen, or sulfur in a wide range of nonterpenoid compounds, including proteins.

The most extensively studied prenyltransferase, farnesyl diphosphate synthase, plays an important role in cholesterol biosynthesis in humans. Farnesyl diphosphate synthases from microbes, plants, and animals exhibit high sequence conservation. The first enzyme of the terpenoid pathway to be structurally defined is recombinant avian farnesyl diphosphate synthase, the crystal structure of which has been determined.

24.3.2 The enzyme limonene synthase is a model for monoterpene synthase action.

The families of enzymes responsible for the formation of terpenoids from GPP, FPP, and GGPP are known as monoterpene, sesquiterpene, and diterpene synthases, respectively. These synthases use the corresponding prenyl diphosphates as substrates to form the enormous diversity of carbon skeletons characteristic of terpenoids. Most terpenoids are cyclic, and many contain multiple ring systems, the basic structures of which are determined by the highly specific synthases. Terpenoid synthases that produce cyclic products are also referred to as "cyclases," although examples of synthases producing acyclic products are also known.

A diverse array of monoterpene synthases has been isolated from essential oil-producing angiosperm species and resin-producing gymnosperms. These enzymes use a common mechanism in which ionization

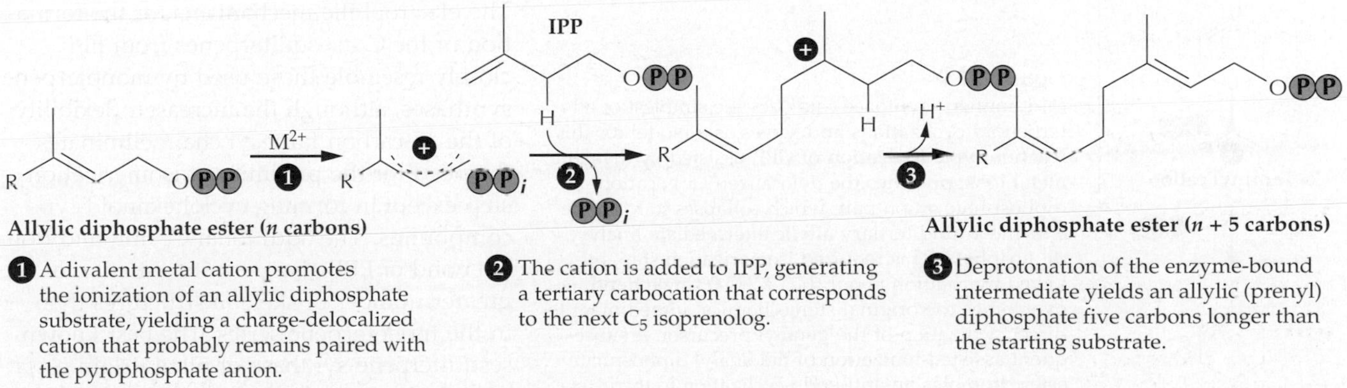

Allylic diphosphate ester (*n* carbons)

Allylic diphosphate ester (*n* + 5 carbons)

1 A divalent metal cation promotes the ionization of an allylic diphosphate substrate, yielding a charge-delocalized cation that probably remains paired with the pyrophosphate anion.

2 The cation is added to IPP, generating a tertiary carbocation that corresponds to the next C_5 isoprenolog.

3 Deprotonation of the enzyme-bound intermediate yields an allylic (prenyl) diphosphate five carbons longer than the starting substrate.

Figure 24.8
The prenyltransferase reaction.

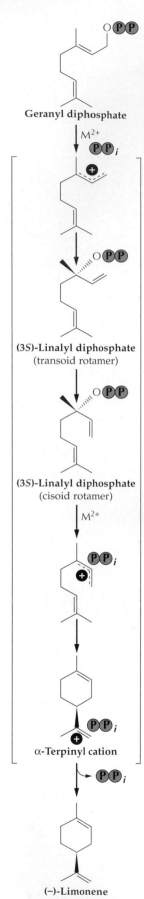

Geranyl diphosphate

M^{2+}

(3S)-Linalyl diphosphate
(transoid rotamer)

(3S)-Linalyl diphosphate
(cisoid rotamer)

M^{2+}

α-Terpinyl cation

(−)-Limonene

Figure 24.9
(−)-Limonene synthase catalyzes the simplest of all terpenoid cyclizations and serves as a model for this reaction type. Ionization of GPP, assisted by divalent metal ions, provides the delocalized carbocation–diphosphate anion pair, which collapses to form the enzyme-bound tertiary allylic intermediate linalyl diphosphate. This required isomerization step, followed by rotation about the C-2–C-3 single bond, overcomes the original stereochemical impediment to direct cyclization of the geranyl precursor. A subsequent assisted ionization of the linalyl diphosphate ester promotes an anti-endo-cyclization to the α-terpinyl cation, which undergoes deprotonation to form limonene, a compound now thought to be an important cancer preventive in humans.

of GPP leads initially to the tertiary allylic isomer linalyl diphosphate (LPP; Fig. 24.9). This isomerization step is required because GPP cannot cyclize directly, given the presence of the trans-double bond. Ionization of the enzyme-bound LPP intermediate promotes cyclization to a six-membered ring carbocation (the α-terpinyl cation), which may undergo additional electrophilic cyclizations, hydride shifts, or other rearrangements before the reaction is terminated by deprotonation of the carbocation or capture by a nucleophile (e.g., water). Variations on this simple mechanistic scheme, involving subsequent reactions of the α-terpinyl carbocation, are responsible for the enzymatic formation of most monoterpene skeletons (see Box 24.1).

The simplest monoterpene synthase reaction is catalyzed by limonene synthase, a useful model for all terpenoid cyclizations (Fig. 24.9). The electrophilic mechanism of action used by limonene synthase can be viewed as an intramolecular equivalent of the prenyltransferase reaction (see Fig. 24.8). Synthases that produce acyclic olefin products (e.g., myrcene) and bicyclic products (α- and β-pinene) from GPP are also known, as are enzymes that transform GPP to oxygenated derivatives such as 1,8-cineole and bornyl diphosphate (Fig. 24.10), the precursor of camphor (see Box 24.1).

An interesting feature of the monoterpene synthases is the ability of these enzymes to produce more than one product; for example, pinene synthase from several plant sources produces both α- and β-pinene. The pinenes are among the most common monoterpenes produced by plants and are principal components of turpentine of the pines, spruces, and firs. The compounds are toxic to bark beetles and their pathogenic fungal symbionts, which cause serious damage to conifer species worldwide. Many conifers respond to bark beetle infestation by up-regulating synthesis of monoterpenes, a process analogous to the production of antimicrobial phytoalexins, when under pathogen attack (Fig. 24.11). Other monoterpenes have quite different functions. Thus, linalool (see Fig. 24.10) and 1,8-cineole emitted by flowers serve as attractants for pollinators, including bees, moths, and bats. 1,8-Cineole and camphor act as foliar feeding deterrents to large herbivores such as hares and deer and also may provide a competitive advantage to several angiosperm species as allelopathic agents that inhibit germination of the seeds of other species.

Exceptions to the general pattern of head-to-tail joining of isoprene units seen in limonene, the pinenes, and most other monoterpenes derived from GPP are the "irregular" monoterpenes. An example of this type is the family of insecticidal monoterpene esters called pyrethrins, found in *Chrysanthemum* and *Tanacetum* species. These monoterpenoids, which exhibit a head-to-middle joining of C_5 units, have gained wide use as commercial insecticides because of their negligible toxicity to mammals and their limited persistence in the environment (see Fig. 24.10).

24.3.3 Sesquiterpene synthases generate several compounds that function in plant defense.

The electrophilic mechanisms for the formation of the C_{15} sesquiterpenes from FPP closely resemble those used by monoterpene synthases, although the increased flexibility of the 15-carbon farnesyl chain eliminates the need for the preliminary isomerization step except in forming cyclohexanoid-type compounds. The additional C_5 unit and double bond of FPP also permit formation of a greater number of skeletal structures than in the monoterpene series. The best known sesquiterpene synthase of plant origin is *epi*-aristolochene synthase from tobacco, the crystal structure of which has been determined (Fig. 24.12). This enzyme cyclizes FPP

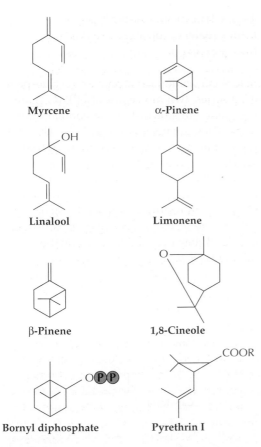

Myrcene

α-Pinene

Linalool

Limonene

β-Pinene

1,8-Cineole

Bornyl diphosphate

Pyrethrin I

Figure 24.10
Structures of monoterpenes, including insecticidal compounds (α- and β-pinene, pyrethrin), pollinator attractants (linalool and 1,8-cineole), and antiherbivory agents (1,8-cineole).

Figure 24.11
Mass attack by mountain pine beetles on a lodgepole pine (*Pinus contorta*) bole. Each white spot on the trunk represents a beetle entry point at which resin has been secreted. This tree has survived the attack because turpentine production was sufficient to kill all of the bark beetles, which have been "pitched out" by resin outflow. On evaporation of the turpentine and exposure to air, the diterpenoid resin acids form a solid plug that seals the wound.

and catalyzes a methyl migration to yield the olefin precursor of the phytoalexin capsidiol, which is elicited by pathogen attack. Vetispiradiene synthase from potato provides the olefin precursor of the phytoalexin lubimin in this species, whereas δ-cadinene synthase from cotton yields the olefin precursor of the important defense compound gossypol, the latter being currently studied as a possible male contraceptive (Fig. 24.13). Some sesquiterpene synthases involved in the production of conifer resin are capable of individually producing more than 25 different olefins.

24.3.4 Diterpene synthases catalyze two distinct types of cyclization reactions.

Two fundamentally different types of enzymatic cyclization reactions occur in the transformation of GGPP to diterpenes (Fig. 24.14). The first resembles the reactions catalyzed by monoterpene and sesquiterpene synthases, in which the cyclization involves ionization of the diphosphate ester and attack of the resulting carbocation on an interior double bond of the geranylgeranyl substrate. An example of this type is casbene synthase, which is responsible for production of the phytoalexin casbene in castor bean. Taxadiene synthase from yew species uses a mechanistically similar, but more complex, cyclization to produce the tricyclic olefin precursor of taxol.

Abietadiene synthase from grand fir exemplifies the second type of cyclization, in which protonation of the terminal double bond to generate a carbonium ion initiates the first cyclization to a bicyclic intermediate (labdadienyl diphosphate, also known as copalyl diphosphate). Ionization of the diphosphate ester promotes the second cyclization step to give the tricyclic olefin product, abietadiene; a single enzyme catalyzes both cyclization steps. Oxidation of a methyl group yields abietic acid (see Fig. 24.1), one of the most common diterpenoid resin acids of conifers and important for wound sealing in these species. Fossilization of this resin produces amber.

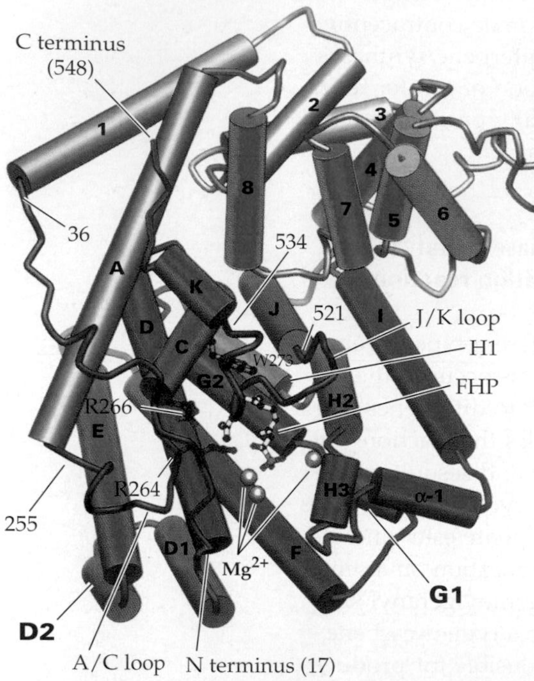

24.3.5 Triterpene synthesis proceeds from squalene, tetraterpene synthesis from phytoene.

Before cyclization can occur in the triterpene (C_{30}) series, two molecules of FPP (C_{15}) are first joined in a head-to-head condensation to produce squalene (see Fig. 24.7). The catalyst, squalene synthase, is a prenyltransferase that catalyzes a complex series of cationic rearrangements to accomplish the chemically difficult chore of joining the C-1 carbons of two farnesyl residues. Squalene is usually oxidized to form the 2,3-epoxide, oxidosqualene, and then cyclized in a protonation-initiated reaction to produce, for example,

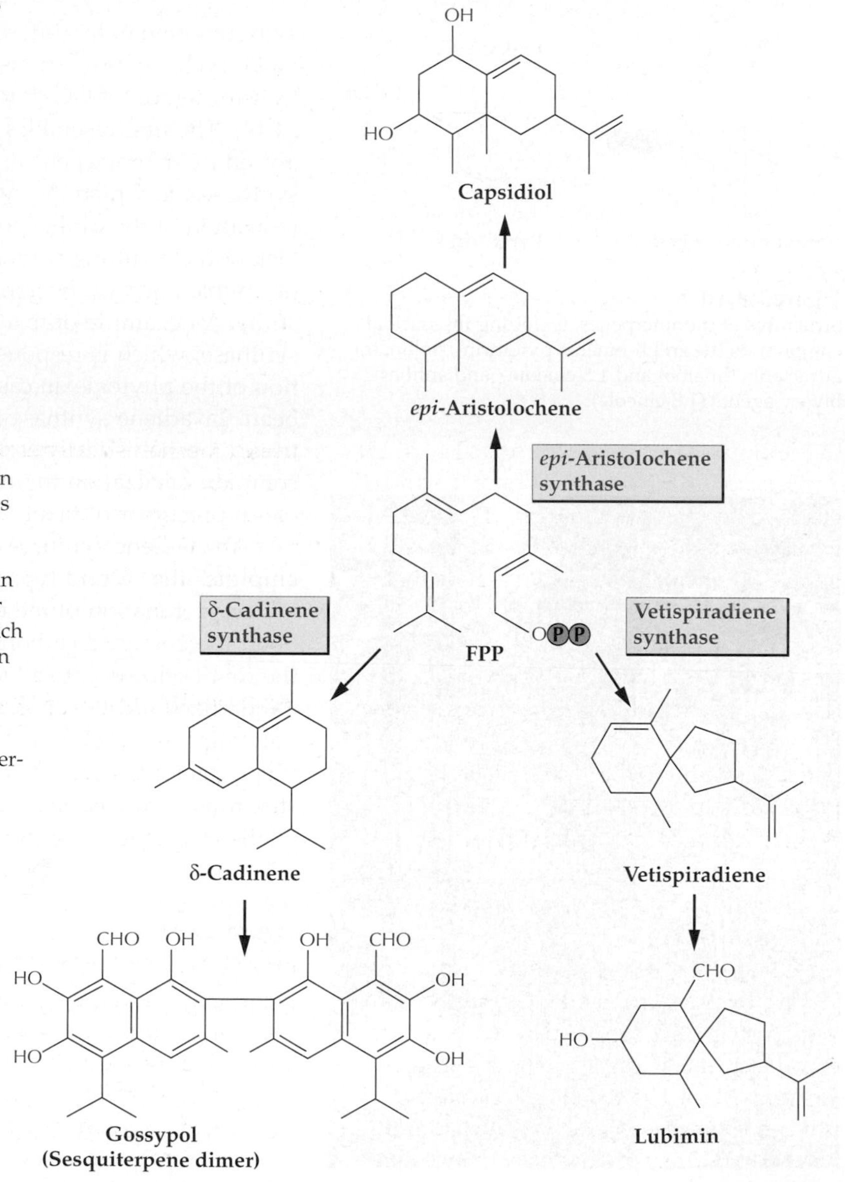

Figure 24.12
Schematic view of *epi*-aristolochene synthase complexed with the substrate analog farnesyl hydroxyphosphonate (FHP). Blue rods represent α-helices in the N-terminal domain; red rods represent α-helices in the C-terminal domain. Loop regions shown in purple are disordered in the native enzyme. Three Mg^{2+} ions and arginines 264 and 266 are involved in the initial steps of the reaction and are labeled near the entrance to the active site. Tryptophan 273, which serves as the general base in the final deprotonation step, is shown within the hydrophobic active-site pocket. The substrate analog FHP is shown in ball-and-stick representation, highlighted with yellow carbon–carbon bonds. Naming of helices in the C-terminal domain corresponds to the convention used for FPP synthase.

Figure 24.13
Structures of sesquiterpenes biosynthetically derived from FPP. The end products function in plant defense.

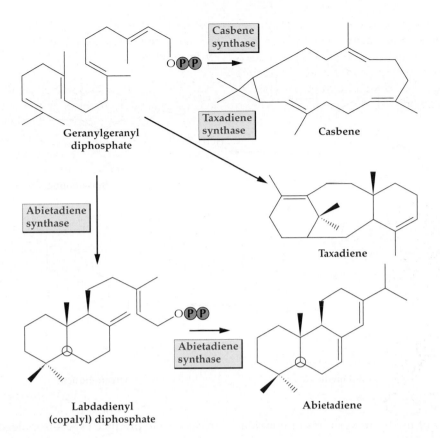

Figure 24.14
Cyclization of GGPP to form the diterpenes casbene, taxadiene, and abietadiene. Cyclization can proceed by one of two distinct mechanims, only one of which yields the intermediate labdadienyl (copalyl) diphosphate.

the common sterol cycloartenol (Fig. 24.15), a precursor of many other phytosterols and brassinosteroids (see Chapter 17). Several alternative modes of cyclization in the triterpene series are also known, such as that leading to the pentacyclic compound β-amyrin, the precursor of oleanolic acid found in the surface wax of several fruits (Fig. 24.15). Preliminary evidence suggests that sesquiterpene biosynthesis and triterpene biosynthesis (both of which utilize cytosolic FPP as a precursor) are reciprocally regulated during the induced defense responses, such that production of C_{15} defensive compounds is enhanced and C_{30} synthesis is repressed.

The tetraterpenes (C_{40}) are produced by joining two molecules of GGPP in head-to-head fashion to produce phytoene, in a manner analogous to the formation of squalene (see Fig. 24.7). The reaction is catalyzed by phytoene synthase, which deploys a mechanism very similar to that of squalene syn-

thase. A series of desaturation steps precedes cyclization in the tetraterpene (carotenoid) series, usually involving formation of six-membered (ionone) rings at the chain termini to produce, for example, β-carotene from lycopene (see Chapter 12, Fig. 12.7).

24.4 Modification of terpenoid skeletons

Subsequent modifications of the basic parent skeletons produced by the terpenoid synthases are responsible for generating the myriad different terpenoids produced by plants. These secondary transformations most commonly involve oxidation, reduction, isomerization, and conjugation reactions, which impart functional properties to the terpenoid molecules. Several oxygenated derivatives of parent terpenoids have already been described in this chapter, including capsidiol, lubimin, gossypol, abietic acid, and oleanolic acid.

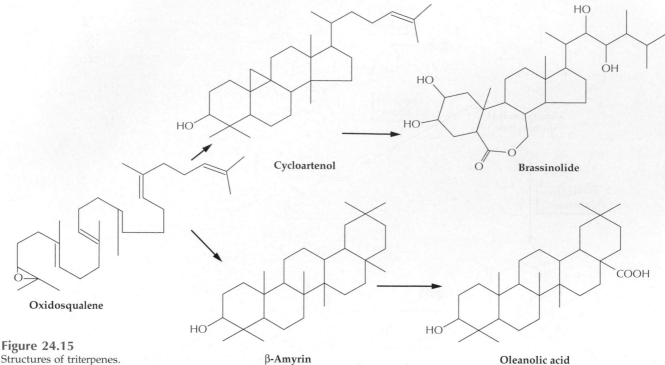

Figure 24.15
Structures of triterpenes. This class of squalene-derived products includes brassinosteroid regulators of plant growth and surface wax components.

Many of the hydroxylations or epoxidations involved in introducing oxygen atoms into the terpenoid skeletons are performed by cytochrome P450 mixed-function oxidases. Because these reactions are not unique to terpenoid biosynthesis, this section will not focus on specific enzyme types but rather on the general role of secondary transformations as the wellspring of diversity in terpenoid structure and function.

24.4.1 The conversion of (–)-limonene to (–)-menthol in peppermint and carvone in spearmint illustrates the biochemistry of terpenoid modification.

The principal and characteristic essential oil components of peppermint (*Mentha piperita*) and spearmint (*M. spicata*) are produced by secondary enzymatic transformations of (–)-limonene (Fig. 24.16). In peppermint, a microsomal cytochrome P450 limonene 3-hydroxylase introduces an oxygen atom at an allylic position to produce (–)-*trans*-isopiperitenol. A soluble NADP⁺-dependent dehydrogenase oxidizes the alcohol to a ketone, (–)-isopiperitenone, thereby activating the adjacent double bond for reduction by a soluble, NADPH-dependent, regiospecific

reductase to produce (+)-*cis*-isopulegone. An isomerase next moves the remaining double bond into conjugation with the carbonyl group, yielding (+)-pulegone. One regiospecific, NADPH-dependent, stereoselective reductase converts (–)-pulegone to either (+)-isomenthone or the predominant species, (–)-menthone. Similar reductases produce the menthol isomers from these ketones. (–)-Menthol greatly predominates among the menthol isomers (constituting as much as 40% of the essential oil) and is the component primarily responsible for the characteristic flavor and cooling sensation of peppermint. The menthol isomers are often found as acetate esters, formed by the action of an acetyl CoA-dependent acetyltransferase. The menthol and menthyl acetate content of peppermint oil glands increases with leaf maturity. Environmental factors greatly influence oil composition. Water stress and warm night growth conditions both promote the accumulation of the more-oxidized pathway intermediates such as (+)-pulegone.

The pathway in spearmint is much shorter. In this instance, a cytochrome P450 limonene 6-hydroxylase specifically introduces oxygen at the alternative allylic position to produce (–)-*trans*-carveol, which is oxidized to (–)-carvone by the soluble

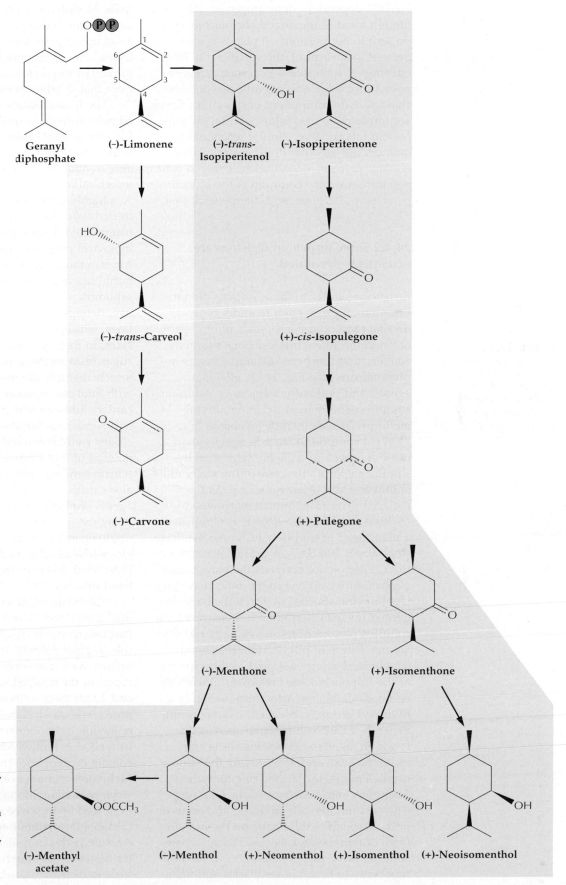

Figure 24.16

Essential oil synthesis in peppermint and spearmint. In peppermint, (–)-limonene is converted to (–)-isopiperitenone, which is modified to form (–)-menthol and related compounds. In spearmint, (–)-limonene is converted to (–)-carvone by a two-step pathway.

Geranyl diphosphate

(–)-Limonene

(–)-*trans*-Isopiperitenol

(–)-Isopiperitenone

(–)-*trans*-Carveol

(+)-*cis*-Isopulegone

(–)-Carvone

(+)-Pulegone

(–)-Menthone

(+)-Isomenthone

(–)-Menthyl acetate

(–)-Menthol

(+)-Neomenthol

(+)-Isomenthol

(+)-Neoisomenthol

NADP⁺-dependent dehydrogenase. Although most of the enzymatic machinery present in peppermint oil glands is also present in spearmint, the specificity of these enzymes is such that (–)-carvone is a very poor substrate. Consequently, carvone, the characteristic component of spearmint flavor, accumulates as the major essential oil component (about 70%). Similar reaction sequences initiated by allylic hydroxylations and subsequent redox metabolism and conjugations are very common in the monoterpene, sesquiterpene, and diterpene classes.

24.4.2 Some terpenoid skeletons are extensively decorated.

Reactions similar to those responsible for essential oil production in mints generate myriad terpenoid compounds of biological or pharmaceutical interest. Such reactions convert sesquiterpene olefin precursors to phytoalexins (see Fig. 24.13), allelopathic agents, and pollinator attractants. Additional sesquiterpenes generated by modifying olefin precursors include juvabione (Fig. 24.17), a compound from fir species that exhibits insect juvenile hormone activity; sirenin, a sperm attractant of the water mold allomyces; and artemisinin, a potent antimalarial drug from annual wormwood (*Artemisia annua*, also known as Qinghaosu, a plant used in traditional Chinese medicine since about 200 B.C.). A related enzymatic reaction sequence converts the parent diterpene olefin taxadiene to the anticancer drug taxol in yew species, in which the basic terpenoid nucleus is modified extensively by a complex pattern of hydroxylations and acylations. Esters of phorbol (another highly oxygenated diterpene) produced by species of the Euphorbiaceae are powerful irritants and cocarcinogens. After introduction of a hydroxyl group, subsequent oxidation can generate a carboxyl function such as that found in abietic acid (see Fig. 24.1) and oleanolic acid, and also provide the structural elements for lactone ring formation. Sesquiterpenes bearing such lactone rings, e.g., costunolide, are produced and accumulated in the glandular hairs on the leaf surfaces of members of the Asteraceae, where some of these compounds serve as feeding repellents to herbivorous insects and mammals. Monoterpene lactones include nepetalactone (the active principle of catnip as well as an aphid pheromone), a member of the iridoid family of monoterpenes, which are formed by a cyclization reaction quite different from that of other monoterpenes (Fig. 24.17).

The limonoids are a family of oxygenated nortriterpene antiherbivore compounds. Like the sesquiterpene lactones, these substances taste very bitter to humans and probably to other mammals as well. A powerful insect antifeedant compound is azadirachtin A, a highly modified limonoid from the neem tree (*Azadirachta indica*). Other oxygenated triterpenoid natural products with unusual biological properties include the phytoecdysones, a family of plant steroids that act as hormones and stimulate insect molting; the saponins, so named because of their soap-like, detergent properties; and the cardenolides, which, like the saponins, are glycosides, in that they bear one or more attached sugar residues. Ingestion of α-ecdysone by insects disrupts the molting cycle, usually with fatal consequences. The saponins and cardenolides are toxic to many vertebrate herbivores; this family of compounds includes well-known fish poisons and snail poisons of significance in the control of schistosomiasis. Many of these products are also cardioactive and anticholesterolemic agents of pharmacological significance. Digitoxin, the glycone (glycosylated form) of digitoxigenin (Fig. 24.17) extracted from foxglove (*Digitalis*), is used widely in carefully prescribed doses for treatment of congestive heart disease.

The broad range of insect and higher animal toxins and deterrents among the modified triterpenes leaves little doubt as to their role in plant defense. Interestingly, some herbivores have developed the means to circumvent the toxic effects of these terpenoids and adapt them to their own defense purposes. The classical example of this phenomenon is the monarch butterfly, a specialist feeder on milkweeds (*Asclepias*) which contain cardenolides that are toxic to most herbivores and are even associated with livestock poisoning. Monarch caterpillars, however, feed on milkweeds and accumulate the cardenolides without apparent ill effects. As a result, both caterpillars and the adult butterflies contain enough cardenolides to be toxic to their own predators such as birds.

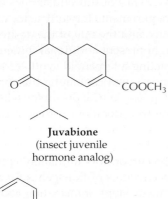

Juvabione
(insect juvenile
hormone analog)

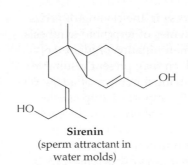

Sirenin
(sperm attractant in
water molds)

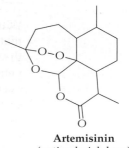

Artemisinin
(antimalarial drug)

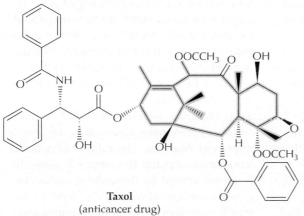

Taxol
(anticancer drug)

Phorbol
(irritant and cocarcinogen)

Costunolide
(insect repellent,
mammal antifeedant)

Nepetalactone
(active principle of catnip)

Azadirachtin A
(insect antifeedant)

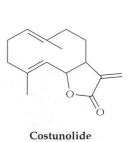

α-Ecdysone
(disrupts insect molting cycle)

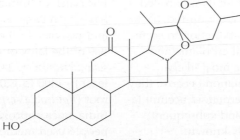

Figure 24.17
Terpenoids formed by
secondary transforma-
tions of parent cyclic
compounds. The yellow
highlighting delineates
the terpenoid portion of
the molecule taxol.

Hecogenin,
the aglycone of a saponin
(a detergent)

Digitoxigenin,
the aglycone of digitoxin, a cardenolide
(treatment of congestive heart disease)

24.5 Toward transgenic terpenoid production

With recent success in the cloning of genes that encode enzymes of terpenoid synthesis, the transgenic manipulation of plant terpenoid metabolism may present a suitable avenue for achieving a number of goals. Several agriculturally important crop species have been bred selectively to produce relatively low amounts of unpalatable terpenoid defense compounds; in the process, these cultivars have lost not only defense capabilities but also, in some cases, quality attributes such as flavor and color. The selective reintroduction of terpenoid-based defense chemistry is certainly conceivable, as is the engineering of pathways into fruits and vegetables to impart desirable flavor properties. The aroma profiles of ornamental plant species might be modified by similar approaches. Likewise, transgene expression might accelerate the rate of slow biosynthetic steps and thereby increase the yields of essential oils used in flavors and perfumes, phytopharmaceuticals (e.g., artemisinin and taxol), insecticides (e.g., pyrethrins and azadirachtin), and a wide range of industrial intermediates that are economically inaccessible by traditional chemical synthesis.

The genetic engineering of terpenoid-based insect defenses is particularly appealing, given the array of available monoterpene, sesquiterpene, diterpene, and triterpene compounds that are toxic to insects not adapted to them. Attracting predators and parasitoids of the target insect or modifying host attractants, oviposition stimulators, and pheromone precursors offers even more sophisticated strategies for pest control. For effective transgenic manipulation of such terpenoid biosynthetic pathways, promoters for tissue-specific, developmentally controlled, and inducible expression are required, as are promoters for targeting production to secretory structures of essential oil plants and conifers. The latter are the most likely species for initial manipulation because they already are adapted for terpenoid accumulation, and the antecedent and subsequent metabolic steps are largely known.

The engineering of terpenoid biosynthetic pathways into plant species that do not ordinarily accumulate these natural products presents a greater opportunity but an even greater challenge, given that little metabolic context exists in these cases. In such species, issues of subcellular sites of synthesis, requirements for sufficiency of precursor flux, and the fate of the desired product might present additional difficulties. Clearly, targeting a terpenoid synthase to the cellular compartment containing the appropriate C_{10}, C_{15}, C_{20}, or C_{30} precursor will be an important consideration. Sufficient flux of IPP at the production site to drive the pathway also will be essential. Because constraints in precursor flow ultimately will limit the effectiveness of transgenes for subsequent pathway steps, information about the flux controls on IPP biosynthesis in both cytosol and plastid, and about the interactions of these controls, is sorely needed.

Very few published examples of the genetic engineering of terpenoid metabolism are currently available, although two notable successes have been achieved in the area of terpenoid vitamins. The ratio of beneficial tocopherol (vitamin E) isomers in oilseeds has been altered by this means, and an increased concentration of β-carotene (a vitamin A precursor) in both rice kernels and rapeseed has been obtained by manipulating the carotenoid pathway. In another, cautionary example, however, overexpression in a transgenic tomato of the enzyme that diverts GGPP to carotenoids resulted in a dwarf phenotype, an unintended consequence of depleting the precursor of the gibberellin plant hormones.

24.6 Alkaloids

24.6.1 Alkaloids have a 3000-year history of human use.

For much of human history, plant extracts have been used as ingredients in potions and poisons. In the eastern Mediterranean, use of the latex of the opium poppy (*Papaver somniferum*; Fig. 24.18) can be traced back at least to 1400 to 1200 B.C. The Sarpagandha root (*Rauwolfia serpentina*) has been used in India since approximately 1000 B.C. Ancient people used medicinal plant extracts as purgatives, antitussives, sedatives, and treatments for a wide range of ailments, including snakebite, fever, and insanity. As the use of medicinal plants spread westward across

(A)   **(B)**

Figure 24.18
(A) Maturing capsule of the opium poppy *Papaver somniferum*. When the capsule is wounded, a white, milky latex is exuded. Poppy latex contains morphine and related alkaloids such as codeine. When the exuded latex is allowed to dry, a hard, brown substance called opium is formed. (B) Statuette from Gazi of a goddess of sleep crowned with capsules of the opium poppy (1250–1200 B.C.).

Arabia and Europe, new infusions and decoctions played a role in famous events. During his execution in 399 B.C., the philosopher Socrates drank an extract of coniine-containing hemlock (*Conium maculatum;* Fig. 24.19). In the last century B.C., Queen Cleopatra used extracts of henbane (*Hyoscyamus*), which contains atropine (Fig. 24.20), to dilate her pupils and appear more alluring to her male political rivals.

Over the centuries, the king of all medicinals has been opium, which was widely consumed in the form of Theriak, a concoction consisting mainly of opium, dried snake meat, and wine (Box 24.2). Analysis of the individual components of opium led to the identification of morphine (Fig. 24.21A), named for Morpheus, the god of dreams in Greek mythology. The isolation of morphine in 1806 by German pharmacist Friedrich Sertürner gave rise to the study of **alkaloids.**

The term alkaloid, coined in 1819 in Halle, Germany, by another pharmacist, Carl Meissner, finds its origin in the Arabic name *al-qali*, the plant from which soda was first isolated. Alkaloids were originally defined as pharmacologically active, nitrogen-containing basic compounds of plant origin.

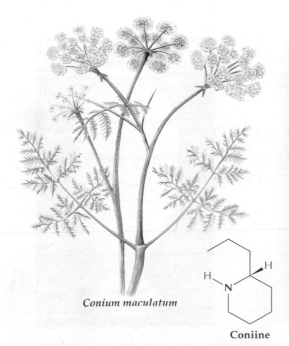

Conium maculatum

Coniine

(B)

Figure 24.19
(A). The piperidine alkaloid coniine, the first alkaloid to be synthesized, is extremely toxic, causing paralysis of motor nerve endings. (B) In 399 B.C., the philosopher Socrates was executed by consuming an extract of coniine-containing poisonous hemlock. This depiction of the event, "The Death of Socrates," was painted by Jacques-Louis David in 1787.

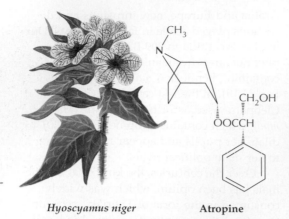

Figure 24.20
Stucture of the anti-cholinergic tropane alkaloid atropine from *Hyoscyamus niger*.

Hyoscyamus niger   **Atropine**

After 190 years of alkaloid research, this definition as such is no longer comprehensive enough to encompass the alkaloid field, but in many cases it is still appropriate. Alkaloids are not unique to plants. They have also been isolated from numerous animal sources (Fig. 24.21B and Box 24.3), although still to be determined is whether biosynthe-

sis de novo occurs in each organism. Many of the alkaloids that have been discovered are not pharmacologically active in mammals and some are neutral rather than basic in character, despite the presence of a nitrogen atom in the molecule.

Alkaloid-containing plants were mankind's original "materia medica." Many are still in use today as prescription drugs (Table 24.1). One of the best-known prescription alkaloids is the antitussive and analgesic codeine from the opium poppy (Fig. 24.21A). Plant alkaloids have also served as models for modern synthetic drugs, such as the tropane alkaloid atropine for tropicamide used to dilate the pupil during eye examinations and the indole-derived antimalarial alkaloid quinine for chloroquine (Fig. 24.22).

In addition to having a major impact on modern medicine, alkaloids have also influenced world geopolitics. Notorious examples include the Opium Wars between China and Britain (1839–1859) and the efforts currently underway in various countries to eradicate

---

Box 24.2   **Theriak, an ancient antipoisoning nostrum containing opium, wine, and snake meat, is still used today in rare instances.**

One of the oldest and most long-lived medications in the history of mankind is Theriak. Originating in Greco-Roman culture, Theriak consists of mainly opium and wine with a variety of plant, animal, and mineral constituents. Panel A of the figure shows a recipe for Theriak from the French Pharmacopée Royale in 1676.

Theriak was developed as an antidote against poisoning, snake bites, spider bites, and scorpion stings. History has it that the Roman Emperor Nero contracted the Greek physician Andromachus to discover a medicine that was effective against all diseases and poisons. Andromachus improved the then-existing recipe to include, in addition to opium, five other plant poisons and 64 plant drugs. Another crucial component was dried snake meat, believed to act against snake bites by neutralizing the venom.

Today, Theriak is still prescribed in rare cases in Europe for pain and other ailments. Panel B shows a valuable Theriak-holding vessel made of Nymphenburg porcelain (in about 1820), which is on display in the Residenz Pharmacy in Munich, Germany.

(A)

(B)

(A)

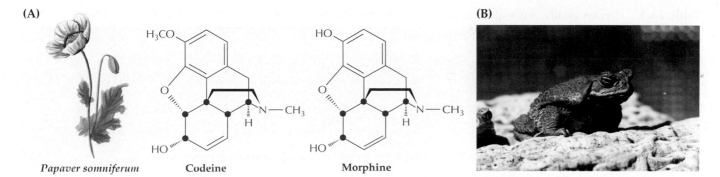

Papaver somniferum   Codeine   Morphine

(B)

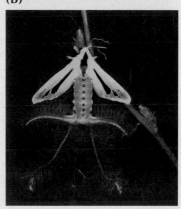

Figure 24.21
(A) Structures of the alkaloids codeine and morphine from the opium poppy *Papaver somniferum.* Asymmetric (chiral) carbons are highlighted with red dots. (B) The frog *Bufo marinus* accumulates a considerable amount of morphine in its skin.

illicit production of heroin, a semisynthetic compound derived by acetylation of morphine (Fig. 24.23), and cocaine, a naturally occurring alkaloid of the coca plant (Fig. 24.24). Because of their various pharmacological activities, alkaloids have influenced

---

Box 24.3	**Some butterflies and moths use alkaloids for sexual signaling or for protection against predators.**

Alkaloid-bearing species have been found in nearly all classes of organisms, including frogs, ants, butterflies, bacteria, sponges, fungi, spiders, beetles, and mammals. Alkaloids of various structures have been isolated from a variety of marine creatures. Some animals, such as amphibians, produce an array of either toxic or noxious alkaloids in the skin or the secretory glands. Others, such as the insects described below, use plant alkaloids as a source of attractants, pheromones, and defense substances.

Some butterflies gather alkaloidal precursors from plants that are not their food sources and convert these compounds into pheromones and defense compounds. Larvae of the cinnabar moth, *Tyria jacobaea*, continuously graze their plant host *Senecio jacobaea* until the plant is completely defoliated (see panel A). The alkaloids thus obtained by the larvae are retained throughout metamorphosis. Male Asian and American arctiid moths incorporate pyrrolizidine alkaloids into their reproductive biology by sequestering these alkaloids in abdominal scent organs called coremata, which are everted in the final stages of their courtship to release the pheromones necessary to gain acceptance by a female. The coremata of a male Asian arctiid moth (*Creatonotos transiens*) are directly proportional to the pyrrolizidine alkaloid content of its diet during the larval stage (see panel B). The

courtship success of these male butterflies therefore depends on their ingesting alkaloids from higher plants.

The larvae of a second insect group, the Ithomiine butterflies, feed on solanaceous plants and sequester the plant toxins, including tropane alkaloids and steroidal glycoalkaloids. However, the adult Ithomiinae do not contain these Solanaceae alkaloids but prefer to ingest plants that produce pyrrolizidine alkaloids, sequestering these bitter substances as N-oxides and monoesters. The pyrrolizidine alkaloid derivatives protect Ithomiinae butterflies from an abundant predator, the giant tropical orb spider. The spider will release a field-caught butterfly from its web but will readily eat a freshly emerged adult that has not yet had an opportunity to feed on the preferred host plant. When palatable butterflies were painted externally with a solution of pyrrolizidine alkaloids, the spider released them from its web. In contrast, palatable butterflies treated the same way with Solanaceae alkaloids were devoured. In general, mostly male butterflies are found feeding on the pyrrolizidine alkaloid-accumulating plants; however, as much as 50% of the pyrrolizidine alkaloids present in these males is sequestered in the spermatophores and transferred to females at mating. In some butterfly species, the protective alkaloids are then transferred to the eggs.

(A)

(B)

Table 24.1 Physiologically active alkaloids used in modern medicine

Alkaloid	Plant source	Use
Ajmaline	*Rauwolfia serpentina*	Antiarrythmic that functions by inhibiting glucose uptake by heart tissue mitochondria
Atropine, -(±)-hyoscyamine	*Hyoscyamus niger*	Anticholinergic, antidote to nerve gas poisoning
Caffeine	*Coffea arabica*	Widely used central nervous system stimulant
Camptothecin	*Camptotheca acuminata*	Potent anticancer agent
Cocaine	*Erythroxylon coca*	Topical anaesthetic, potent central nervous system stimulant, and adrenergic blocking agent; drug of abuse
Codeine	*Papaver somniferum*	Relatively nonaddictive analgesic and antitussive
Coniine	*Conium maculatum*	First alkaloid to be synthesized; extremely toxic, causes paralysis of motor nerve endings, used in homeopathy in small doses
Emetine	*Uragoga ipecacuanha*	Orally active emetic, amoebicide
Morphine	*P. somniferum*	Powerful narcotic analgesic, addictive drug of abuse
Nicotine	*Nicotiana tabacum*	Highly toxic, causes respiratory paralysis, horticultural insecticide; drug of abuse
Pilocarpine	*Pilocarpus jaborandi*	Peripheral stimulant of the parasympathetic system, used to treat glaucoma
Quinine	*Cinchona officinalis*	Traditional antimalarial, important in treating *Plasmodium falciparum* strains that are resistant to other antimalarials
Sanguinarine	*Eschscholzia californica*	Antibacterial showing antiplaque activity, used in toothpastes and oral rinses
Scopolamine	*H. niger*	Powerful narcotic, used as a sedative for motion sickness
Strychnine	*Strychnos nux-vomica*	Violent tetanic poison, rat poison, used in homeopathy
(+)-Tubocurarine	*Chondrodendron tomentosm*	Nondepolarizing muscle relaxant producing paralysis, used as an adjuvant to anaesthesia
Vinblastine	*Catharanthus roseus*	Antineoplastic used to treat Hodgkin's disease and other lymphomas.

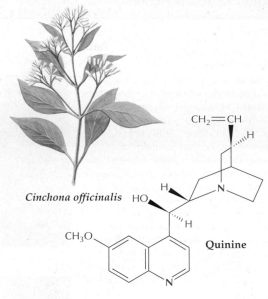

Cinchona officinalis

Quinine

Figure 24.22
Structure of the monoterpenoid indole alkaloid-derived quinine from *Cinchona officinalis*. An antimalarial quinine-containing tonic prepared from the bark of *C. officinalis* greatly facilitated European exploration and inhabitation of the tropics during the past two centuries.

24.6.2 Physiologically active alkaloids participate in plant chemical defenses.

More than 12,000 alkaloids have been isolated since the discovery of morphine. About 20% of the species of flowering plants produce alkaloids, and each of these species accumulates the alkaloids in a unique, defined pattern. Some plants, such as the periwinkle (*Catharanthus roseus*) contain more than 100 different monoterpenoid indole alkaloids. Why should a plant invest so much nitrogen into synthesizing such a large number of alkaloids of such diverse structure? The role of alkaloids in plants has been a longstanding question, but a picture has begun to emerge that supports an ecochemical function for these compounds.

The role of chemical defense for alkaloids in plants is supported by their wide range of physiological effects on animals and by the antibiotic activities many alkaloids possess. Various alkaloids also are toxic to insects or function as feeding deterrents. For example, nicotine, found in tobacco, was one of the first insecticides used by humans and remains one of the most effective (Fig. 24.25). Herbivory has been found to stimulate nicotine biosynthesis in wild tobacco plants. Another effective insect toxin is caffeine, found in seeds and leaves of cocoa,

human history profoundly, both for good and ill. Of interest to plant biologists, however, is the evolutionary selection process in plants that has caused alkaloids to evolve into such a large number of complex structures and to remain effective over the millennia.

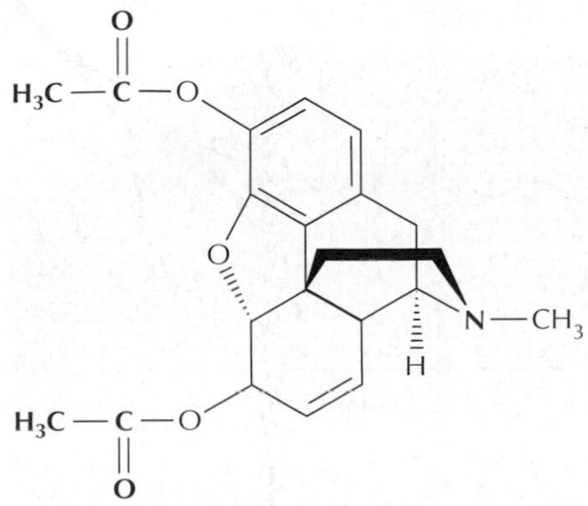

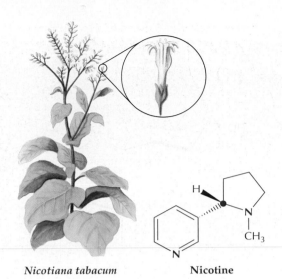

Heroin

Figure 24.23
Structure of diacetyl morphine, commonly known as heroin.

Nicotiana tabacum   **Nicotine**

Figure 24.25
Structure of nicotine from *Nicotiana tabacum*. The asymmetric chiralcarbon is highlghted with a red dot.

coffee, cola, maté, and tea (Fig. 24.26). At a dietary concentration well below that found in fresh coffee beans or tea leaves, caffeine kills nearly all larvae of the tobacco hornworm (*Manduca sexta*) within 24 hours—primarily by inhibiting the phosphodiesterase that hydrolyzes cAMP. The steroid alkaloid α-solanine, a cholinesterase inhibitor found in potato tuber (Fig. 24.27), is the trace toxic constituent thought to be responsible for the teratogenicity of sprouting potatoes.

Two groups of alkaloids that have been well studied with respect to ecochemical function are the pyrrolizidine and quinolizidine alkaloids. The pyrrolizidine alkaloids, frequently found in members of the tribe Senecioneae (Asteraceae) and in the Boraginaceae, render most of these plants toxic to mammals. In *Senecio* species (Fig. 24.28), senecionine N-oxide is synthesized in the roots and translocated throughout the plant. In species such as *Senecio vulgaris* and

S. vernalis, 60% to 80% of the pyrrolizidine alkaloids is found to accumulate in the inflorescences. Members of the *Senecio* genus are responsible for livestock poisonings and also represent a potential health hazard for humans. Naturally occurring pyrrolizidine alkaloids are harmless but become highly toxic when transformed by cytochrome P450 monooxygenases in the liver. On the other hand, several insect species have adapted to the pyrrolizidine alkaloids that accumulate in plants and have evolved mechanisms for using these alkaloids to their own benefit. Some insects can feed on pyrrolizidine alkaloid-producing plants and effectively and

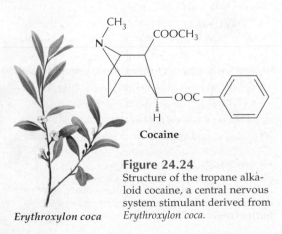

Cocaine

Erythroxylon coca

Figure 24.24
Structure of the tropane alkaloid cocaine, a central nervous system stimulant derived from *Erythroxylon coca.*

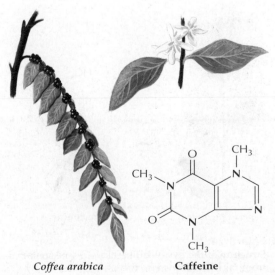

Coffea arabica   **Caffeine**

Figure 24.26
Structure of the purine alkaloid caffeine from *Coffea arabica.*

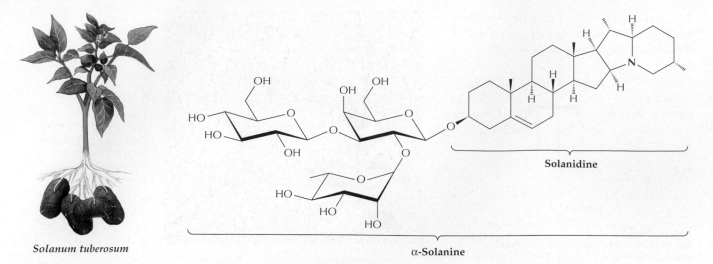

Solanum tuberosum

Solanidine

α-Solanine

Figure 24.27
Structure of the steroid alkaloid glycoside α-solanine from *Solanum tuberosum* (potato). The aglycone solanidine is derived from cholesterol.

efficiently eliminate the alkaloids after enzymatic modification, such as formation of N-oxide derivatives. Other insects not only feed on these plants, but also store the pyrrolizidine alkaloids for their own defense or convert the ingested pyrrolizidine alkaloids to pheromones that attract prospective mates (Box 24.3).

The quinolizidine alkaloids occur primarily in the genus *Lupinus* and are frequently referred to as lupine alkaloids (Fig. 24.29); they are toxic to grazing animals, particularly to sheep. The highest incidence of livestock losses attributable to lupine alkaloid poisoning occurs in autumn during the seed-bearing stage of the plant life cycle— the seeds being the plant parts that accumulate the greatest quantities of these alkaloids. Because of their bitter taste, lupine alkaloids can also function as feeding deterrents. Given a mixed population of sweet and bitter lupines, rabbits and hares will readily eat the alkaloid-free sweet variety and avoid the lupine alkaloid-accumulating bitter variety, indicating that lupine alkaloids in plants serve to reduce herbivory by functioning both as bitter-tasting deterrents and toxins. Given this collection of examples, alkaloids can be viewed as a part of the chemical defense system of the plant that evolved under the selection pressure of predation.

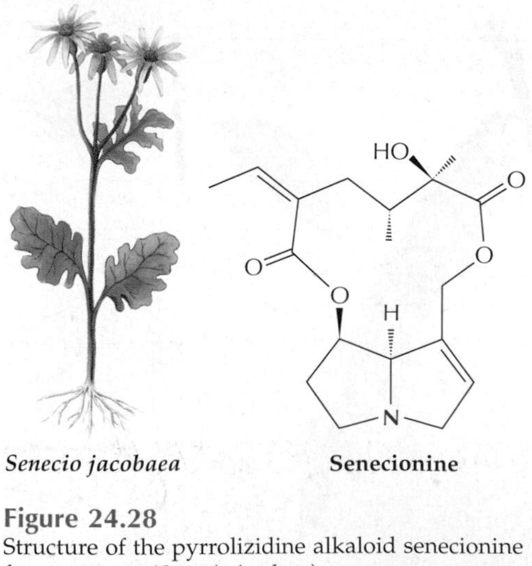

Senecio jacobaea   **Senecionine**

Figure 24.28
Structure of the pyrrolizidine alkaloid senecionine from ragwort (*Senecio jacobaea*).

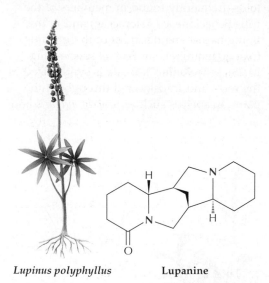

Lupinus polyphyllus   **Lupanine**

Figure 24.29
Structure of the quinolizidine alkaloid lupanine from the bitter lupine *Lupinus polyphyllus.* Lupanine is a bitter compound that functions as a feeding deterrent.

24.6.3 Alkaloid biosynthesis research has been greatly aided by the development of techniques for culturing plant cells.

Many alkaloids have complex chemical structures and contain multiple asymmetric centers, complicating structure elucidation and making study of the biosynthesis of alkaloids quite difficult until relatively recently. For example, although nicotine (one asymmetric center; see Fig. 24.25) was discovered in 1828, its structure was not known until it was synthesized in 1904, and the structure of morphine (five asymmetric centers; see Fig. 24.21) was not unequivocally elucidated until 1952, almost 150 years after its isolation. Almost all of the enzymes involved in the biosynthesis of these two alkaloids have been identified, but 190 years after morphine was first isolated, its biosynthetic pathway remains incomplete.

Why has it been so difficult to elucidate alkaloid biosynthetic pathways? Plants synthesize natural products at a relatively sluggish rate, so steady-state concentrations of the alkaloid biosynthetic enzymes are low. In addition, the large amounts of tannins and other phenolics that accumulate in plants interfere with the extraction of active enzymes. Even when plants are treated with radiolabeled precursors and the resulting radioactive alkaloids are chemically degraded to identify the position of the label, the low rate of natural product metabolism can prevent the high rates of incorporation that yield clearly interpretable results. The use of polyvinylpyrrolidone and Dowex-1 in preparing protein extracts from plant tissues has helped overcome the enzyme inactivation by phenolic compounds, but isolation of the enzymes involved in natural product synthesis has had only limited success because of their very low concentrations in the plant.

Not until the 1970s were suspension cultures of plant cells established that were capable of producing high concentrations of alkaloids (Fig. 24.30). As an experimental system, cell culture provides several advan-

Figure 24.30
Callus cultures established from plants can be optimized to produce high concentrations of a wide variety of natural products. In some of the examples shown, metabolite pigments give the calli distinctive colors.

tages over whole-plant studies, including the year-round availability of plant material; the undifferentiated, relatively uniform state of development of the cells; the absence of interfering microorganisms; and most importantly, the compressed vegetative cycle. Plant cell cultures can synthesize large amounts of secondary products within a two-week cultivation period. This is very favorable in comparison with in planta production, for which the time frame for alkaloid accumulation may vary from one season for annual plants to several years for some perennial species. In plant cell culture, the rate of alkaloid biosynthesis can be increased, greatly facilitating its study (Table 24.2). Moreover, the greater metabolic rates associated with cell cultures promote the incorporation of labeled precursors during alkaloid biosynthesis. Hormones regulate the accumulation of alkaloids in culture, and in many cases, alkaloid biosynthesis can be induced by the addition of abiotic and biotic elicitor substances to the culture. These advances have provided a powerful system with which to analyze the regulation of alkaloid biosynthesis. Since the advent of alkaloid production in culture, more than 80 new enzymes that catalyze steps in the biosynthesis of indole, isoquinoline, tropane, pyrrolizidine, acridone, and purine classes of alkaloids have been discovered and partially characterized.

Table 24.2 Production of selected alkaloids in plant cell culture

Secondary metabolite	Species	Yield (g/l)	% Dry weight
Berberine	*Coptis japonica*	7.0	12
Jatrorrhizine	*Berberis wilsoniae*	3.0	12
Raucaffricine	*Rauwolfia serpentina*	1.6	3

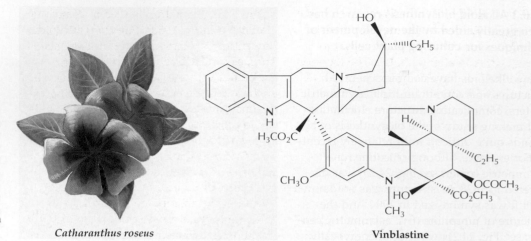

Figure 24.31
Structure of the monoterpenoid indole alkaloid vinblastine from *Catharanthus roseus*.

Catharanthus roseus

Vinblastine

At one time, plant cell suspension cultures were considered an alternative source of industrially significant secondary metabolites, particularly alkaloids of pharmaceutical importance. However, many important compounds such as vincristine, vinblastine (Fig. 24.31), pilocarpine (Fig. 24.32), morphine, and codeine, among many others, are not synthesized to any appreciable extent in cell culture. The reason for this is thought to be tissue-specific expression of alkaloid biosynthesis genes, because in some cases plants regenerated from nonproducing callus cells contained the same alkaloid profile as the parent plant. Although not currently used for commercial alkaloid production, plant cell culture continues to provide biochemists with a rich source of certain alkaloid biosynthesis enzymes and a convenient system with which to study enzyme regulation.

24.6.4 Although typically considered constitutive defense compounds, some alkaloids are synthesized in response to plant tissue damage.

Alkaloids are thought to be part of the constitutive chemical defense system of many plants. The ultimate test of this hypothesis may be future research into molecular genetic suppression of alkaloid biosynthesis. The phenotypes of mutants lacking specific gene products in an alkaloid biosynthesis pathway may provide a direct demonstration of the role of noninducible alkaloids produced constitutively in plants. Near-isogenic species of alkaloid-producing and nonproducing plants might then be subjected to experimental conditions to test their relative resistance.

In a few cases, such as that of nicotine in tobacco, convincing evidence has been presented that an alkaloid is involved in induced chemical defense. Wild species of tobacco have been found to be highly toxic to the hornworm, a tobacco-adapted species that is insensitive to nicotine but susceptible to N-acyl nicotines found in the tobacco leaf. The N-acyl derivatives are not found in unwounded *Nicotiana repanda* but their formation is induced by methyl jasmonate treatment. In response to leaf wounding, tobacco plants increase the alkaloid content of leaves that have not been subjected to wounding. *N*-Acetylnicotine accumulates very rapidly (within 10 hours). The alkaloid content increases and then returns to basal concentrations over a 14-day period. Recent isotope labeling experiments indicate that this

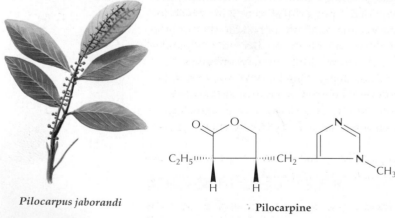

Pilocarpus jaborandi

Pilocarpine

Figure 24.32
The imidazole alkaloid pilocarpine from *Pilocarpus jaborandi*.

derivative is formed from a preexisting pool of nicotine. De novo nicotine biosynthesis occurs in roots, followed by transport to leaves, but only after 36 hours. The increase in nicotine biosynthesis results in a 10-fold increase of the alkaloid in the xylem fluid.

Freshly hatched hornworm larvae fed wounded leaves achieve only half the weight gain obtained by counterparts fed leaf material from unwounded plants. Recent studies demonstrate that, given the choice, hornworms will abandon a wounded plant. Hornworms not permitted to leave a wounded plant exhibit much higher mortality rates and much lower growth rates than those fed on unwounded plants.

Inducible synthesis of nicotine and other alkaloids appears to involve methyl jasmonate, a volatile plant growth regulator (see Chapter 17). Endogenous jasmonate pools increase rapidly when plant cells are treated with an elicitor prepared from yeast cell walls. In turn, jasmonates are known to induce accumulation of secondary metabolites in cell culture. More than 140 different cultured plant species respond to the addition of methyl jasmonate by increasing their production of natural products. Although studies of this type with intact plants are not as extensive as with cell suspension cultures, clear examples have been demonstrated with tobacco plants, in which leaf wounding produces an increase in endogenous jasmonic acid pools in shoots and roots. Moreover, the application of methyl jasmonate to tobacco leaves increases both endogenous jasmonic acid in roots and de novo nicotine biosynthesis. These results imply that jasmonate may play a role in regulating the defense responses of alkaloid-producing plants.

24.7 Alkaloid biosynthesis

24.7.1 Plants biosynthesize alkaloids from simple precursors, using many unique enzymes.

Until the mid-20th century, our view of how alkaloids are synthesized in plants was based on biogenic hypotheses. Pathways suggested by illustrious natural product chemists such as Sir Robert Robinson, Clemens Schöpf, Ernst Winterstein, and Georg Trier were based on projections considered feasible within the realm of organic chemistry. In the 1950s, however, alkaloid biosynthesis became an experimental science, as radioactively labeled organic molecules became available for testing hypotheses. These early precursor-feeding experiments clearly established that alkaloids are in most cases formed from L-amino acids (e.g., tryptophan, tyrosine, phenylalanine, lysine, and arginine), either alone or in combination with a steroidal, secoiridoid (e.g., secologanin), or other terpenoid-type moiety. One or two transformations can convert these ubiquitous amino acids from primary metabolites to substrates for highly species-specific alkaloid metabolism. Although we do not thoroughly understand how most of the 12,000 known alkaloids are made by plants, several well-investigated systems can serve as examples of types of building blocks and enzymatic transformations that have evolved in alkaloid biosynthesis.

The L-tryptophan–derived monoterpenoid indole alkaloid ajmalicine was the first alkaloid for which biosynthesis was clarified at the enzyme level (Fig. 24.33); in that study plant cell suspension cultures of the Madagascar periwinkle C. roseus (see Fig. 24.31) were used. In plants, the biosynthesis of ajmalicine and more than 1800 other monoterpenoid indole alkaloids begins with the decarboxylation of the amino acid L-tryptophan by tryptophan decarboxylase to form tryptamine. Then tryptamine, by action of strictosidine synthase, is stereospecifically condensed with the secoiridoid secologanin (derived in multiple enzymatic steps from geraniol) to form 3α-strictosidine. Strictosidine can then be enzymatically permutated in a species-specific manner to form a multitude of diverse structures (Fig. 24.34). The elucidation of the enzymatic formation of ajmalicine by using plant cell cultures laid the groundwork for analysis of more-complex biosynthetic pathways, such as those leading to two other L-tryptophan–derived monoterpenoid indole alkaloids, ajmaline (Fig. 24.35) and vindoline.

24.7.2 The berberine synthesis pathway has been defined completely.

The first alkaloid for which each biosynthetic enzyme has been identified, isolated, and characterized from the primary metabolite

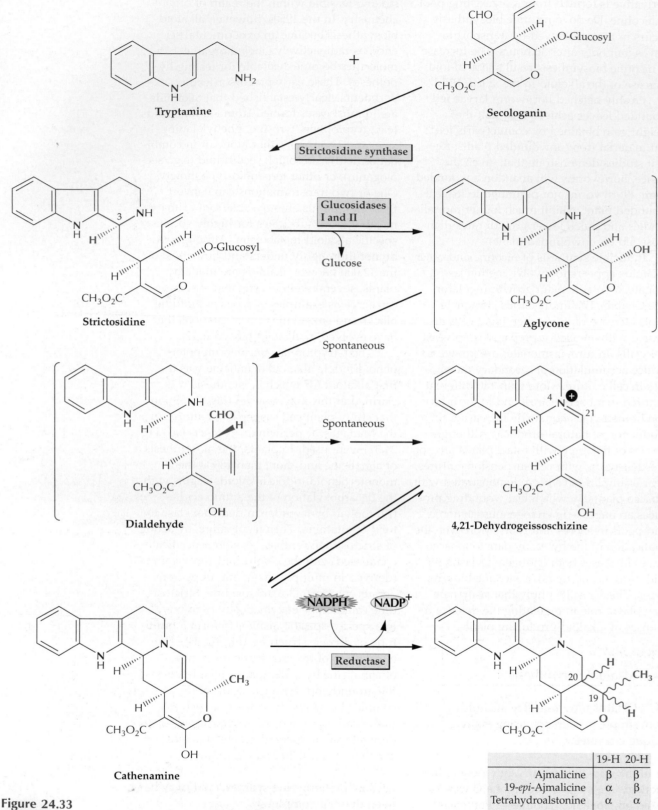

Tryptamine

Secologanin

Strictosidine synthase

Strictosidine

Glucosidases I and II

Glucose

Aglycone

Spontaneous

Spontaneous

Dialdehyde

4,21-Dehydrogeissoschizine

NADPH   NADP⁺

Reductase

Cathenamine

	19-H	20-H
Ajmalicine	β	β
19-*epi*-Ajmalicine	α	β
Tetrahydroalstonine	α	α

Figure 24.33
Biosynthesis of the monoterpenoid indole alkaloid ajmalicine and related compounds in *Catharanthus roseus*. Tryptamine is derived from L-tryptophan by decarboxylation through the action of tryptophan decarboxylase, and the secoiridoid secologanin is derived in multiple steps from the monoterpene geraniol.

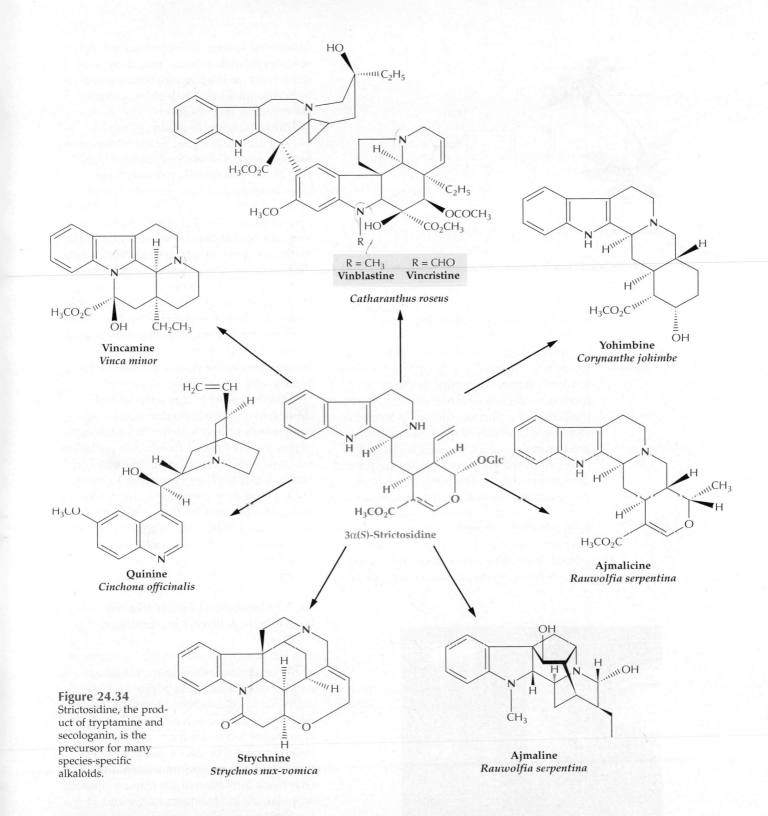

Figure 24.34
Strictosidine, the product of tryptamine and secologanin, is the precursor for many species-specific alkaloids.

R = CH₃ → R = CHO
Vinblastine   Vincristine

Catharanthus roseus

Vincamine
Vinca minor

Yohimbine
Corynanthe johimbe

Quinine
Cinchona officinalis

3α(S)-Strictosidine

Ajmalicine
Rauwolfia serpentina

Strychnine
Strychnos nux-vomica

Ajmaline
Rauwolfia serpentina

precursor through to the end product alkaloid is the antimicrobial tetrahydrobenzylisoquinoline alkaloid, berberine, in *Berberis* (barberry) cell suspension cultures (Fig. 24.36). This pathway will be described in detail because it exemplifies the role of highly substrate-specific enzymes and of compartmentalization in alkaloid biosynthesis.

The biosynthesis of tetrahydrobenzylisoquinoline alkaloids in plants begins in the cytosol with a matrix of reactions that generates the first tetrahydrobenzylisoquinoline

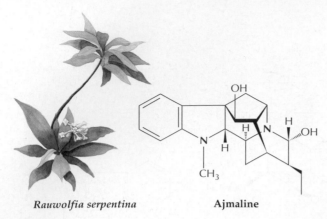

Rauwolfia serpentina   **Ajmaline**

Figure 24.35
Structure of the monoterpenoid indole alkaloid ajmaline from *Rauwolfia serpentina*.

alkaloid, (S)-norcoclaurine (Fig. 24.37). The pathway proceeds from two molecules of L-tyrosine. One is decarboxylated to form tyramine or is acted on by a phenol oxidase to form L-dopa. Dopamine can then be formed by decarboxylation of L-dopa or by the action of a phenol oxidase on tyramine. Determining which of these two pathways is predominant in a given plant is difficult because all of the enzyme activities are present in protein extracts. The benzyl moiety of (S)-norcoclaurine is formed by transamination of the second L-tyrosine molecule to form *p*-hydroxyphenylpyruvate, which is next decarboxylated to *p*-hydroxyphenylacetaldehyde. Dopamine and *p*-hydroxyphenylacetaldehyde are then stereoselectively condensed to form (S)-norcoclaurine. A series of methylation and oxidation reactions yield the branchpoint intermediate of benzylisoquinoline alkaloid biosynthesis, (S)-reticuline (Fig. 24.38).

In *Berberis,* the N-methyl group of (S)-reticuline is oxidized to the berberine bridge carbon C-8 of (S)-scoulerine (see Fig. 24.37). The specific pathway from (S)-scoulerine that leads to berberine proceeds with O-methylation to (S)-tetrahydrocolumbamine. The 3-O-methyl moiety of tetrahydrocolumbamine is converted to the methylenedioxy bridge of canadine by canadine synthase, a microsomal cytochrome P450–dependent oxidase. The final step in the biosynthesis of berberine is catalyzed by (S)-tetrahydroprotoberberine oxidase, an enzyme shown to contain a covalently bound flavin. The end product alkaloid berberine accumulates in the central vacuole of the *Berberis* cell.

The berberine bridge enzyme and (S)-tetrahydroprotoberberine oxidase are compartmentalized together in vesicles apparently derived from the smooth endoplasmic reticulum. Each of these enzymes consumes 1 mol of O_2 and produces 1 mol of H_2O_2 per mole of berberine formed. Overall, the course of reactions from 2 mol of L-tyrosine to 1 mol of berberine consumes 4 mol of S-adenosylmethionine and 2 mol of NADPH.

24.7.3 Elucidation of other alkaloid biosynthetic pathways is progressing.

The enzymes that catalyze the biosynthesis of the benzophenanthridine alkaloid macarpine in the California poppy *Eschscholzia californica* have also been identified, isolated, and characterized, as have nearly all of the enzymes of morphine biosynthesis in the opium poppy (Fig. 24.39). Good progress has been made toward understanding the enzymatic formation of the tropane alkaloid scopolamine in *Hyoscyamus niger* and of the acridone alkaloid furofoline-I in *Ruta graveolens.*

Studies have revealed that the chemical transformations required for alkaloid biosynthesis are catalyzed by highly stereo-, regio-, and substrate-specific enzymes that are present only in specific species. These enzymes

Figure 24.36
The *Berberis wilsoniae* plant (left) and cell suspension culture (right). The cell suspension culture derives its color from optimized production of the highly oxidized benzylisoquinoline alkaloid berberine. Plant cell cultures (like this one) that produce large quantities of alkaloids have led to the complete elucidation of several alkaloid biosynthetic pathways.

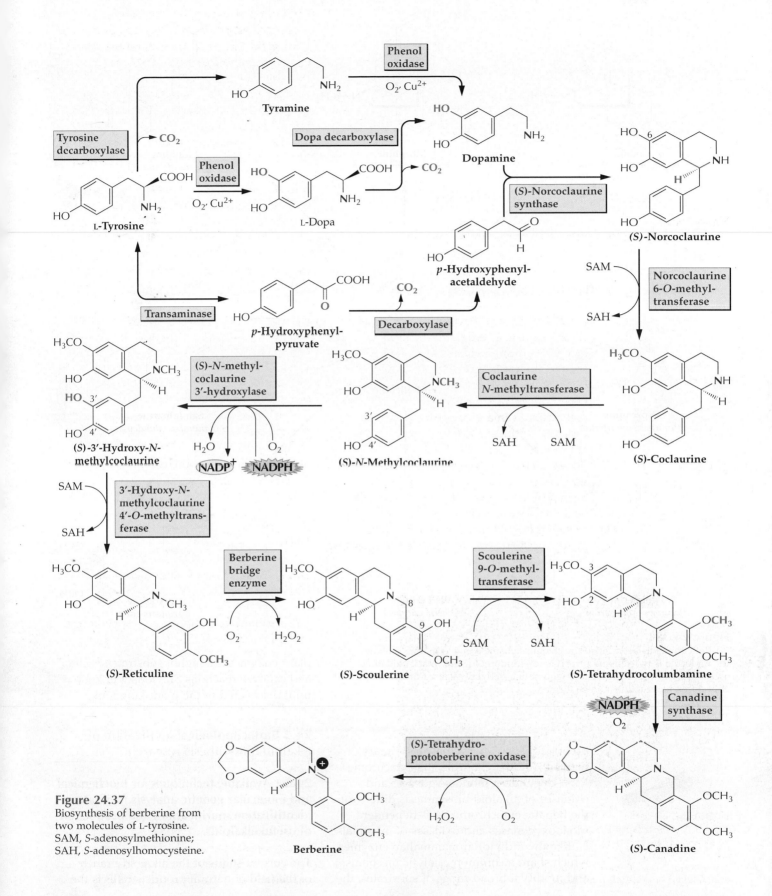

Figure 24.37
Biosynthesis of berberine from two molecules of L-tyrosine. SAM, S-adenosylmethionine; SAH, S-adenosylhomocysteine.

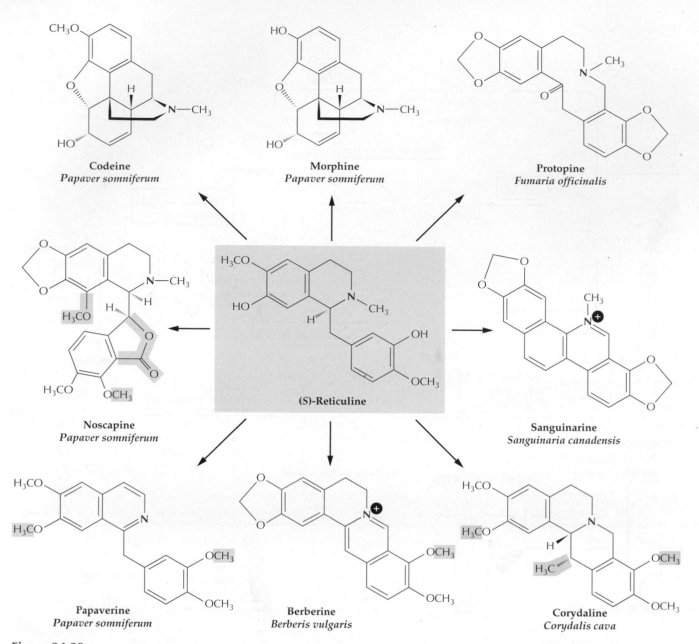

Figure 24.38

(*S*)-Reticuline has been called the chemical chameleon. Depending on how the molecule is twisted and turned before undergoing enzymatic oxidation, a vast array of tetrahydrobenzylisoquinoline-derived alkaloids of remarkably different structures can be formed.

do not appear to participate in primary metabolism. For example, the cytochrome P450–dependent monooxygenases and oxidases of alkaloid biosynthesis differ from the hepatic cytochrome P450–dependent monooxygenases and oxidases of mammals. Unlike the individual mammalian enzymes, which share a common catalytic mechanism and modify a broad range of substrates, the

plant enzymes are highly substrate-specific and catalyze reactions previously unknown until discovered in the plant kingdom.

24.8 Biotechnological application of alkaloid biosynthesis research

24.8.1 Available techniques for biochemical and molecular genetic analysis facilitate identification, purification, and production of useful alkaloids.

The current status of the alkaloid branch of the field of natural products reflects the

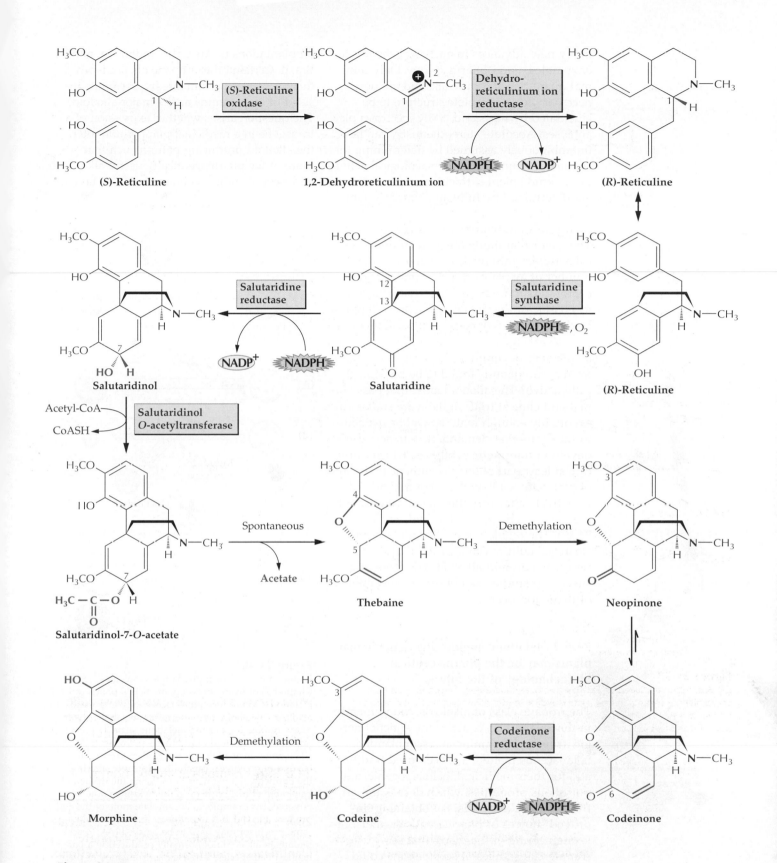

Figure 24.39
Isolation and characterization of all the enzymes of morphine biosynthesis in opium poppy are nearly complete 190 years after discovery of that alkaloid. Many of the equivalent morphine biosynthetic enzymes have been discovered in mammalian liver. Demonstration that the mammalian liver biosynthesizes morphine de novo would have tremendous implications concerning evolutionary development of the opiate receptor in humans.

many new advances in analytical chemistry, enzymology, and pharmacology. Only minimal quantities of a pure alkaloid are now necessary for a complete structure to be elucidated by mass and NMR spectroscopic analyses. Absolute stereochemistry can be unambiguously assigned by determining the crystal structure. The pharmacological activities of crude plant extracts or pure substances are determined by fully automated systems, such that millions of data points are collected each year in industrial screening programs. The factor that limits the number of biological activities for which we can test is the number of available target enzymes and receptors. As more of the underlying biochemical bases for diseases continue to be discovered, the number of test systems will increase.

What happens when a small quantity of an alkaloid of complex chemical structure from a rare plant is found to be physiologically active? The alkaloid must first pass animal and clinical trials; if these are successful, eventually enough material will be needed to satisfy market demand. Researchers can develop **biomimetic** syntheses, which duplicate at least part of the biosynthetic pathways of plants to yield synthetic compounds; alternatively, they can alter the metabolism of the plant to change the alkaloid profile (Fig. 24.40). The regulation of alkaloid biosynthesis in cell culture can also be influenced to produce a desired alkaloid. The following successful studies demonstrate the viability of these approaches.

24.8.2 Metabolic engineering of medicinal plants may be the pharmaceutical biotechnology of the future.

The tropane class of alkaloids, found mainly in the Solanaceae, contains the anticholinergic drugs hyoscyamine and scopolamine. Solanaceous plants have been used traditionally for their medicinal, hallucinogenic, and poisonous properties, which derive, in part, from tropane alkaloids. For obtaining improved sources of pharmaceuticals, metabolic engineering of the plants that serve as commercial sources of scopolamine could augment classical breeding in the effort to develop plants with an optimal alkaloid pattern. The current commercial source of scopolamine is *Duboisia*, which is cultivated

on plantations in Australia, Indonesia, and Brazil. Certain other tropane alkaloid–producing species accumulate hyoscyamine instead of scopolamine as the major alkaloid. The question arises whether expression of a transgene in a medicinal plant would alter the alkaloid-producing pattern such that more of the pharmaceutically useful alkaloid, scopolamine, is obtained. To this end, a

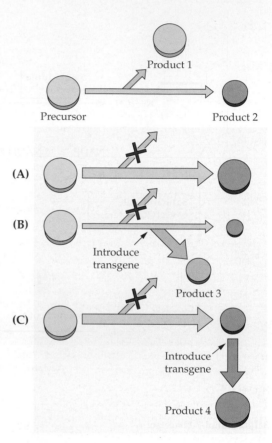

Figure 24.40
Using antisense/cosuppression technologies (see Chapter 7) or overexpression, medicinal plants can be tailored to produce pharmaceutically important alkaloids by eliminating interfering metabolic steps or by introducing desired metabolic steps. Expressing an entire alkaloid biosynthesis pathway of 20 to 30 enzymes in a single microorganism is currently beyond our technical capability. However, altering the pathway in a plant and producing the desired alkaloid either in culture or in the field may now be possible. For example, to accumulate more of the end product alkaloid, a side pathway that also uses the same precursor may have to be blocked (A). To accumulate an alkaloid not normally produced in a particular plant species, a transgene (from another plant or a microorganism) may be introduced (B). If the end product alkaloid would be more useful as a particular derivative, for example, as a more soluble glycoside, a gene that encodes a glycosyl transferase could be introduced (C).

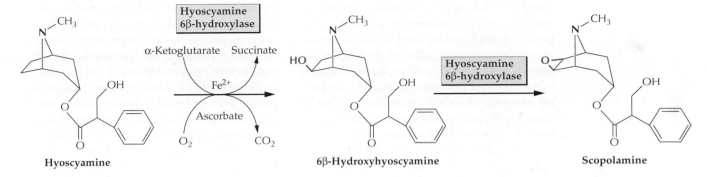

Figure 24.41

The reaction catalyzed by hyoscyamine 6β-hydroxylase along the biosynthetic pathway leading to the tropane alkaloid scopolamine in species of the genera *Hyoscyamus, Duboisia,* and *Datura.* This one enzyme catalyzes two consecutive steps, hydroxylation of hyoscyamine followed by epoxide formation to produce scopolamine.

cDNA encoding hyoscyamine 6β-hydroxylase from *H. niger* (black henbane) has been introduced into *Atropa belladonna* (deadly nightshade) by using *Agrobacterium tumefaciens*– and *A. rhizogenes*–mediated transformation (Fig. 24.41). The resulting transgenic plants and hairy roots each contained greater concentrations of scopolamine than did the wild-type plants. These transgenic *Atropa* plants provided the first example of how medicinal plants could be successfully altered by using molecular genetic techniques to produce increased quantities of a medicinally important alkaloid.

Designing meaningful transformation experiments requires a thorough knowledge of alkaloid biosynthetic pathways. Such studies are also limited by our ability to transform and regenerate medicinal plants. To date, expertise in this important area lags well behind that for tobacco, petunia, and cereal crops. For example, in the area of tropane alkaloids, transformation and regeneration of *Duboisia,* a plant for which plantation, harvesting, and purification techniques have already been established commercially, will have to be developed before any potential commercialization can be considered. Genetic manipulation of plant cell cultures may increase the concentrations of rate-limiting enzymes or may result in expression of gene products not normally induced in cultured cells. If so, alkaloid production in plant cell or tissue culture may become a viable industrial approach (Fig. 24.42).

Another successful example of how metabolic engineering can alter natural products synthesis has been provided by the

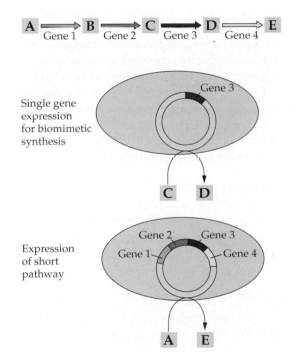

Figure 24.42

One alternative approach to the use of metabolically engineered plants is to use microbes to produce alkaloids. Short biosynthetic pathways can be expressed in either yeast or bacteria and used as a source of alkaloid. Plant alkaloid genes can be functionally expressed in microorganisms to produce either single biotransformation steps or short biosynthetic pathways. For example, a known plant biosynthetic pathway contains an enzyme encoded by gene 3, which catalyzes a transformation step that is difficult to achieve by chemical synthesis. After heterologous expression of gene 3 in a microorganism, the protein product can be used in a biomimetic synthesis of alkaloid D. Likewise, expression of the biosynthetic genes 1 through 4 in a microorganism might produce alkaloid E directly from precursor A.

transformation of *Brassica napus* (canola) with the cDNA encoding the *C. roseus* tryptophan decarboxylase used in biosynthesis of monoterpenoid indole alkaloids. Usefulness of seed from this oil-producing crop as animal feed has been limited in part by the presence of indole glucosinolates (see Chapters 8 and 16), sulfur-containing compounds that make the protein meal less palatable. The tryptophan decarboxylase transgene in canola redirects tryptophan pools away from indole glucosinolate biosynthesis and into tryptamine (Fig. 24.43). The mature seed of the transgenic canola plants contains less of the indole glucosinates and does not accumulate tryptamine, making it more suitable for use as animal feed and achieving a potentially economically useful product.

To date, the elucidation of enzymatic syntheses of at least eight alkaloids is either complete or nearly complete: ajmaline, vindoline, berberine, corydaline, macarpine, morphine, berbamunine, and scopolamine. Of these alkaloids, those in current industrial use, such as morphine and scoploamine, are still being isolated from the plants that produce them rather than synthesized. The future for research on these alkaloids lies in the development of alternative systems of production, such as plant cell or microbial cultures, and in the development of plants with an improved spectrum of alkaloids for a more efficient production of the pharmaceuticals currently isolated from field-grown plants. The design of these alternative systems and optimized plants requires molecular manipulation, which in turn requires knowledge of alkaloid biosynthetic pathways at the enzyme level. Much progress has been made with select alkaloids, but much remains to be discovered about the enzymatic synthesis of pharmaceutically important alkaloids such as camptothecin, quinine, and emetine, to name only a few examples. cDNAs have now been isolated for approximately 20 enzymes of alkaloid biosynthesis, and the rate at which new clones are identified is certain to increase in the coming years. As genes are isolated, we can anticipate that heterologous expression systems will be developed in bacterial, yeast, and insect cell culture systems to allow production of single enzymes, and perhaps even short pathways, for biomimetic syntheses of alkaloids. Our understanding of how the expression of alkaloid biosynthesis genes is regulated by elicitors or in specific tissues will also improve as the promoters of alkaloid biosynthetic genes are analyzed. The future will almost certainly bring genetically engineered microorganisms and eukaryotic cell cultures that produce alkaloids, metabolically engineered medicinal plants with tailored alkaloid spectra, pharmaceutically important alkaloids in plant cell culture, and even enzymatic synthesis of as yet unknown alkaloids through combinatorial biochemistry.

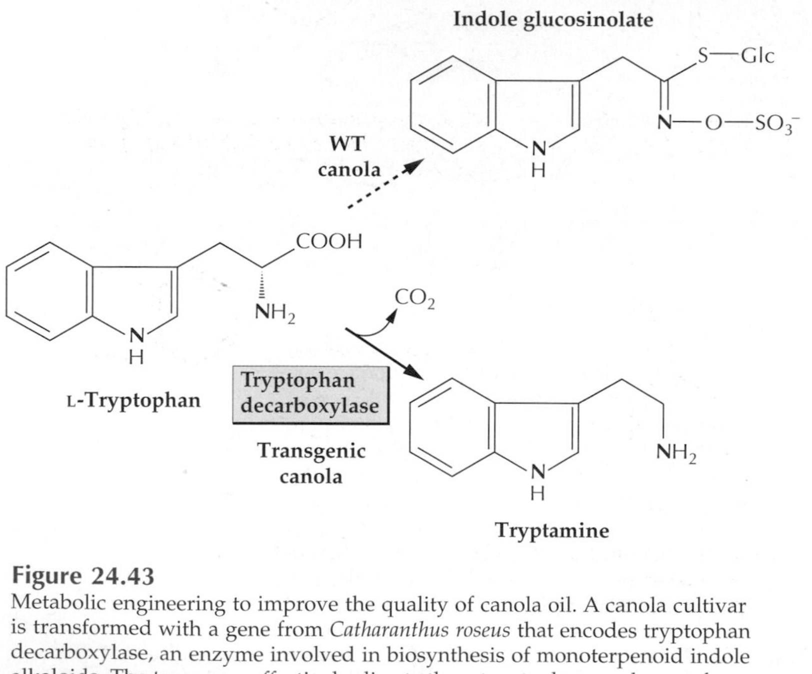

24.9 Phenylpropanoid and phenylpropanoid-acetate pathway metabolites

24.9.1 Plants contain a remarkably diverse array of phenolic compounds.

Plants originated in an aquatic environment. Their successful evolutionary adaptation to land was achieved largely by massive formation of "plant phenolic" compounds. Although the bulk of these substances assumed cell wall structural roles, a vast array of nonstructural constituents was

Figure 24.43
Metabolic engineering to improve the quality of canola oil. A canola cultivar is transformed with a gene from *Catharanthus roseus* that encodes tryptophan decarboxylase, an enzyme involved in biosynthesis of monoterpenoid indole alkaloids. The transgene effectively directs the L-tryptophan pool away from use in biosynthesis of the bitter indole glucosinolate and into the production of tryptamine. WT, wild type.

also formed, having such various roles as defending plants, determining certain distinguishing features of different woods and barks (e.g., durability), establishing flower color, and contributing substantially to certain flavors (tastes and odors). These functions and others performed by plant phenolics are essential for the continued survival of all types of vascular plants. Accounting for about 40% of organic carbon circulating in the biosphere, these phenolic compounds are primarily derived from **phenylpropanoid, phenylpropanoid-acetate,** and related biochemical pathways such as those leading to "hydrolyzable" tannins. Furthermore, it is their reassimilation back to carbon dioxide during biodegradation (mineralization) that presents the rate-limiting step in recycling biological carbon.

Plant phenolics are generally characterized as aromatic metabolites that possess, or formerly possessed, one or more "acidic" hydroxyl groups attached to the aromatic arene (phenyl) ring (Fig. 24.44). These compounds plagued plant scientists for years by interfering with experimental methods. For example, when exposed to air, plant phenolics readily oxidize and turn brown, generating products that form complexes with proteins and inhibit enzyme activity. Many protocols now used to isolate plant proteins and nucleic acids include special precautions designed to minimize interference by phenolic compounds. Cultured plant tissues can also release phenolics that inhibit growth of callus and regeneration of plantlets. At the same time, phenolic compounds are increasingly being recognized for their profound impact on plant growth, development, reproduction, and defense; indeed, scientists have come to appreciate their significance more fully, particularly over the past few decades.

The discussion of plant phenolic substances is a discussion of plant diversity itself. Characteristics unique to each of the roughly 250,000 species of vascular plants arise, at least in part, through differential deposition of highly specialized phenylpropanoid and phenylpropanoid-acetate derivatives. No single species can be used to illustrate the extraordinary diversity of "secondary" metabolites that exists within the plant kingdom, because many branches of the pathways are found or are amplified

only in specific plant families. Placing undue emphasis on any single plant species can obscure the extremely broad variation in biosynthetic capabilities that has yielded this spectrum of different plant types.

24.9.2 Most, but not all, plant phenolic compounds are products of phenylpropanoid metabolism.

Most plant phenolics are derived from the phenylpropanoid and phenylpropanoid-acetate pathways (Fig. 24.45) and fulfill a very broad range of physiological roles in planta. In ferns, fern allies, and seed plants, polymeric **lignins** reinforce specialized cell walls, enabling them to support their massive weights on land and to transport water and minerals from roots to leaves. Closely related to lignins, the **lignans** can vary from dimers to higher oligomers. Widespread throughout the plant kingdom, lignans can, for example, either help defend

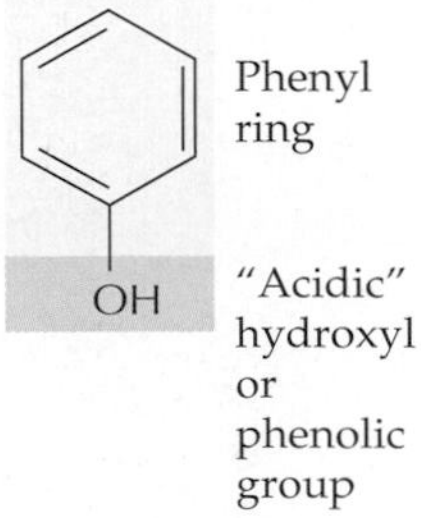

Figure 24.44
Structure of phenol.

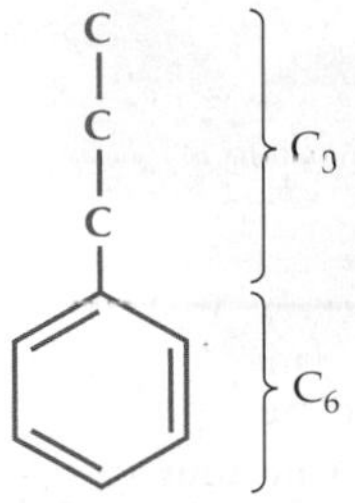

Phenylpropanoid skeleton (C_6C_3)

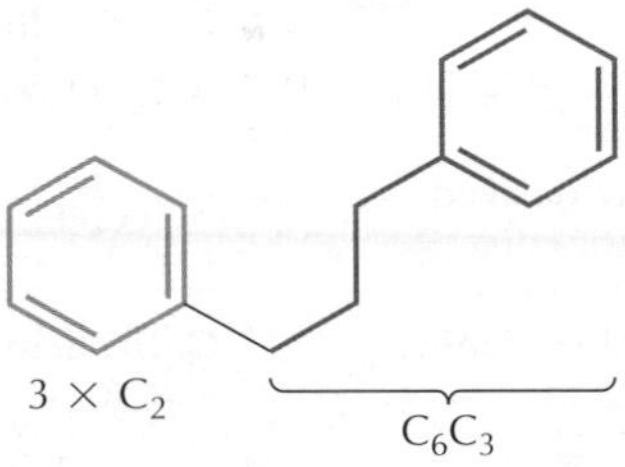

Phenylpropanoid–acetate skeleton (C_6C_3–C_6), with phenylpropanoid-derived (C_6C_3) and acetate-derived ($3 \times C_2$) rings

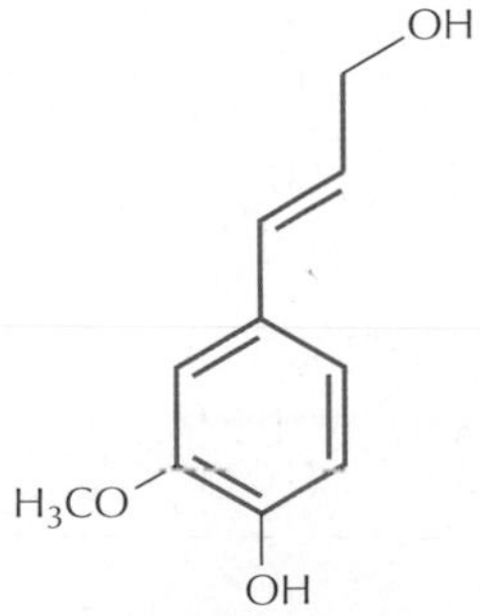

Coniferyl alcohol, a component of lignins and many lignans

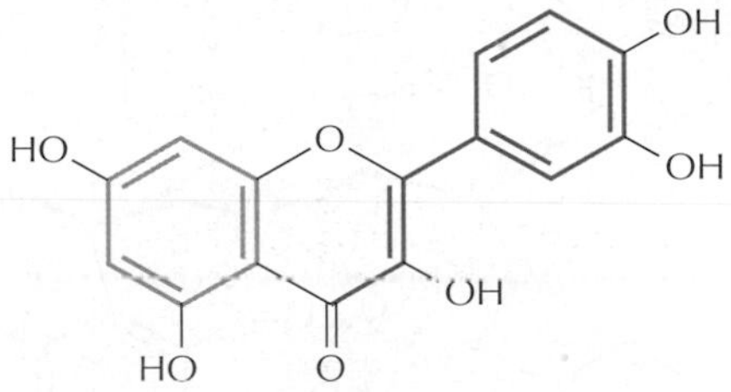

Quercetin, a flavonoid (C_6C_3–C_6)

Figure 24.45
Phenylpropanoid and phenylpropanoid-acetate skeletons and representative plant compounds based on those structures.

—— Phenylpropanoid skeleton
—— Acetate-derived rings

against various pathogens or act as antioxidants in flowers, seeds, seed coats, stems, nuts, bark, leaves, and roots. **Suberized** tissues contain alternating layers of hydrophobic (aliphatic) and hydrophilic (phenolic) structural substances. Present in cork, bark, roots, and certain periderm tissues (e.g.,

potato skin), suberized tissues function, for example, by providing a protective barrier, thereby limiting the effects of desiccation from the atmosphere and pathogen attack. The **flavonoids** comprise an astonishingly diverse group of more than 4500 compounds. Among their subclasses are the anthocyanins (pigments), proanthocyanidins or condensed tannins (feeding deterrents and wood protectants), and isoflavonoids (defensive products and signaling molecules). The **coumarins, furanocoumarins,** and **stilbenes** protect against bacterial and fungal pathogens, discourage herbivory, and inhibit seed germination. Numerous miscellaneous phenolics also play defensive roles or impart characteristic tastes and odors to plant material.

Although most plant phenolics are products of phenylpropanoid metabolism, with the phenylpropanoids, in turn, being derived from phenylalanine and tyrosine (Fig. 24.46), some phenolic compounds are generated through alternative pathways. For example, hydrolyzable tannins, a group of mostly polymeric substances that appear to act in plant defense, are typically copolymers of carbohydrates and the shikimate-derived gallic and ellagic acids (Fig. 24.47). Found in the leaves, fruits, pods, and galls of some woody dicots, hydrolyzable tannins have not yet been identified in monocots. "Condensed tannins," on the other hand, are widespread and occur in practically all trees and shrubs. Known as proanthocyanidins, these compounds are synthesized by the phenylpropanoid-acetate pathway. Some others, such as the phenolic psychoactive compounds of cannabis, the tetrahydrocannabinoids, are polyketide (acetate) or terpenoid derivatives (Fig. 24.48).

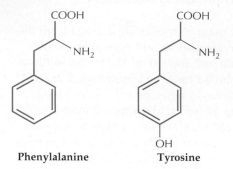

Figure 24.46
The aromatic amino acids phenylalanine and tyrosine are derivatives of the shikimic–chorismic acid pathway (see Chapter 8).

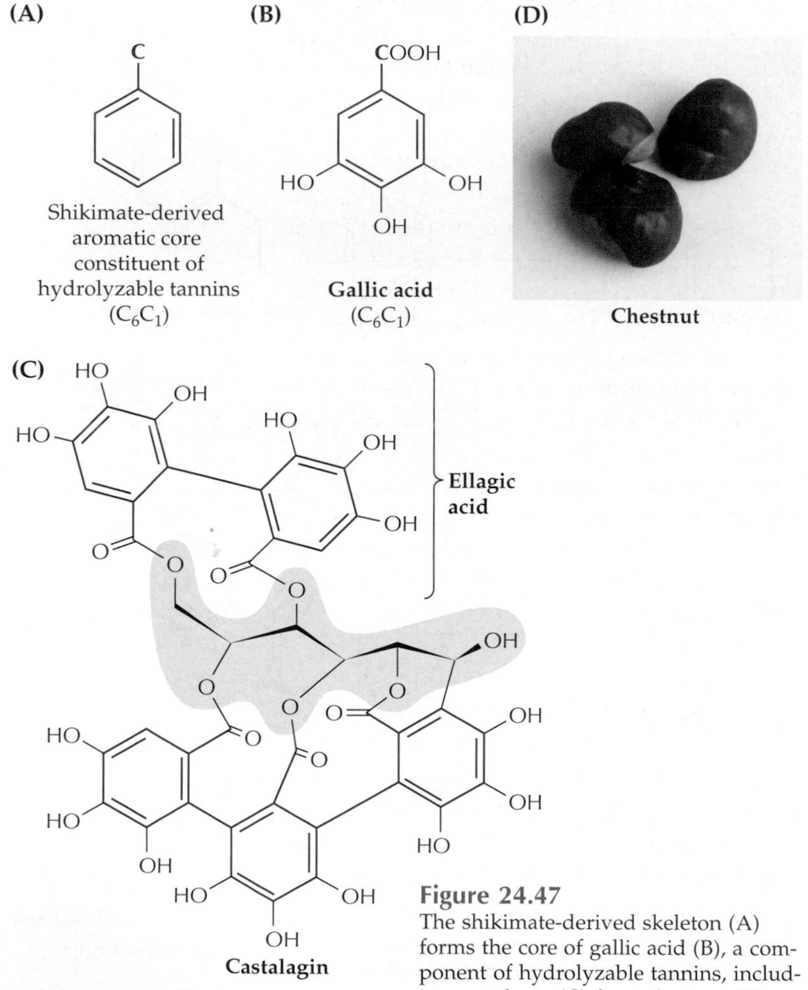

Phenylalanine

Tyrosine

(A)

Shikimate-derived aromatic core constituent of hydrolyzable tannins (C_6C_1)

(B)

Gallic acid
(C_6C_1)

(D)

Chestnut

(C)

Ellagic acid

Castalagin

Figure 24.47
The shikimate-derived skeleton (A) forms the core of gallic acid (B), a component of hydrolyzable tannins, including castalagin (C) from chestnut (D).

24.10 Phenylpropanoid and phenylpropanoid-acetate biosynthesis

24.10.1 Phenylalanine (tyrosine) ammonia-lyase is a central enzyme in phenylpropanoid metabolism.

One enzyme directs carbon from aromatic amino acids to the synthesis of phenylpropanoid metabolites. This enzyme converts phenylalanine (PAL) to cinnamic acid and tyrosine (TAL) to *p*-coumaric acid (Fig. 24.49, reactions 1 and 2). Interestingly, in

(A)

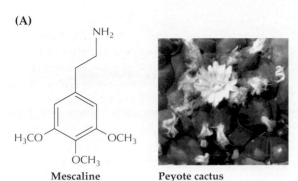

Mescaline   Peyote cactus

(B)

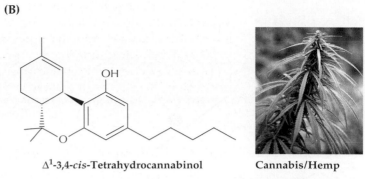

Δ^1-3,4-*cis*-Tetrahydrocannabinol   Cannabis/Hemp

Figure 24.48

Not all plant phenolic compounds are derived from phenyl-propanoid substrates. Whereas mescaline, the psychoactive compo-nent of peyote, is a phenylpropanoid derivative (A), the phenolic compound Δ^1-3,4-*cis*-tetrahydrocannabinol (B), a psychoactive com-ponent of cannabis, is a product of polyketide synthesis, the repeat-ed condensation of acetyl-CoA units derived from malonyl-CoA.

most vascular plants, Phe is the highly pre-ferred substrate, but the monocot enzyme can utilize both Phe and Tyr. PAL has been detected in a few aquatic plants, where it probably functions in formation of simple flavonoids, such as the C-glucosyl–linked lucenin and vicenin of *Nitella* species (Fig. 24.50). Lignins, however, are not present in aquatic plants. Thus, this PAL (TAL) enzy-matic step and the products of the various phenylpropanoid and phenylpropanoid-ac-etate pathways appear to have been key to the plant colonization of land.

PAL is the most extensively studied en-zyme in the phenylpropanoid pathway, per-haps in all secondary metabolism. In some plants, PAL appears to be encoded by a sin-gle gene, whereas in others it is the product of a multigene family. The enzyme requires no cofactor for activity. The ammonium ion liberated by the PAL reaction is recycled by way of glutamine synthetase and glutamate synthetase (GS-GOGAT; see Chapter 8). Once assimilated into glutamate, the amino group can be donated to prephenate, form-ing arogenate, a precursor of both phenylala-nine and tyrosine (Fig. 24.51). This nitrogen-cycling process ensures a steady supply of the aromatic amino acids from which plant phenolics are derived.

24.10.2 Biochemical pathways to distinct phenolic classes share many common features.

During the 1960s and early 1970s, impressive progress was made in defining many of the salient features of the pathway that converts

cinnamic acid to the monolignols (see Fig. 24.49). This pathway essentially comprises four types of enzymatic reactions: aromatic hydroxylations, O-methylations, CoA liga-tions, and NADPH-dependent reductions. More recently, the precise enzymology in-volved in earlier parts of the pathway has come under renewed attention, focusing par-ticularly on aromatic hydroxylations and on whether the O-methylation steps utilize the free acids or CoA esters.

Aromatic ring hydroxylation involves three distinct hydroxylation conversions, all of which are believed to be microsomal. The best studied of these enzymes, cinnamate-4-hydroxylase, is an oxygen-requiring, NADPH-dependent, cytochrome P450 en-zyme that catalyzes the regiospecific hydrox-ylation at the *para*-position of cinnamic acid to give *p*-coumaric acid (see Fig. 24.49, reac-tion 3). The other two hydroxylases original-ly were thought to introduce hydroxyl groups into the free acids *p*-coumarate or ferulate (or their CoA ester forms), yielding the diphenol (catechol) products caffeic acid or 5-hydroxyferulic acid (or their CoA derivatives), respectively (see Fig. 24.49, re-action 4). At this time, however, substantial confusion remains as to how the caffeoyl moiety of caffeic acid or caffeoyl-CoA is formed. Whether this biosynthesis involves a nonspecific phenolase-catalyzed conversion or whether some other enzymatic step (e.g., one involving an NADPH-dependent cy-tochrome P450) is used is still not known. Additionally, although ferulate-5-hydroxy-lase has been established as an NADPH-dependent cytochrome P450 enzyme, there

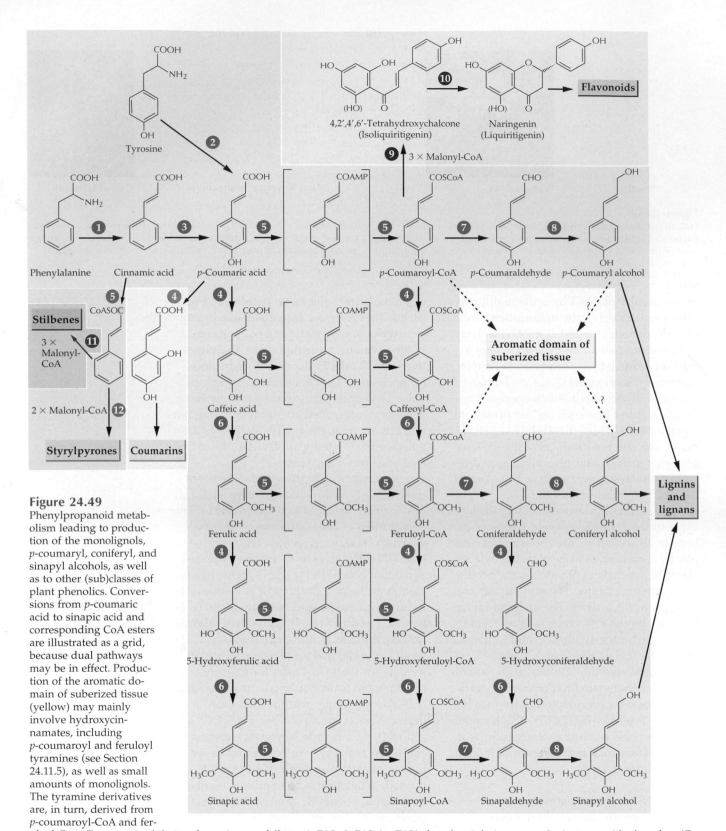

Figure 24.49
Phenylpropanoid metabolism leading to production of the monolignols, *p*-coumaryl, coniferyl, and sinapyl alcohols, as well as to other (sub)classes of plant phenolics. Conversions from *p*-coumaric acid to sinapic acid and corresponding CoA esters are illustrated as a grid, because dual pathways may be in effect. Production of the aromatic domain of suberized tissue (yellow) may mainly involve hydroxycinnamates, including *p*-coumaroyl and feruloyl tyramines (see Section 24.11.5), as well as small amounts of monolignols. The tyramine derivatives are, in turn, derived from *p*-coumaroyl-CoA and feruloyl-CoA. Enzymes (and their cofactors) are as follows: 1. PAL; 2. PAL (or TAL), found mainly in grasses; 3. cinnamate-4-hydroxylase (O_2, cytochrome P450, NADPH); 4. hydroxylases (O_2, cyt. P450, NADPH); 5. CoA ligases that participate in ligation of AMP and CoA (CoASH, ATP); 6. *O*-methyltransferases (SAM); 7. cinnamoyl-CoA:NADPH oxidoreductases (NADPH); 8. cinnamoyl alcohol dehydrogenases (NADPH); 9. chalcone synthase; 10. chalcone isomerase; 11. stilbene synthase; 12. styrylpyrone synthase. Products in parentheses refer to less common pathways. [Note: Sequence of intermediates in the pathways leading to sinapyl alcohol awaits experimental confirmation at the time of writing. The reader is encouraged to read the pertinent literature on developments in this area.]

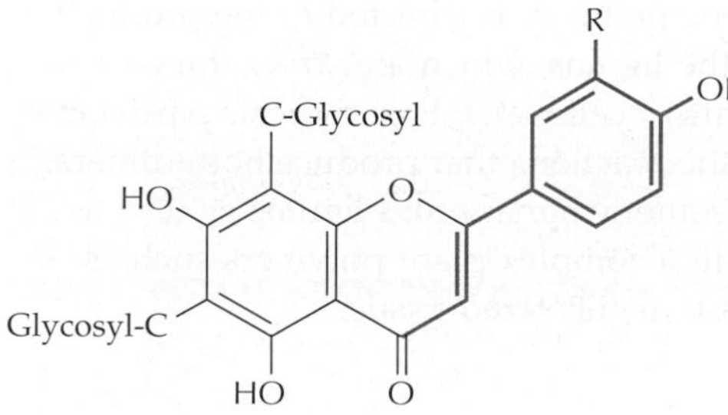

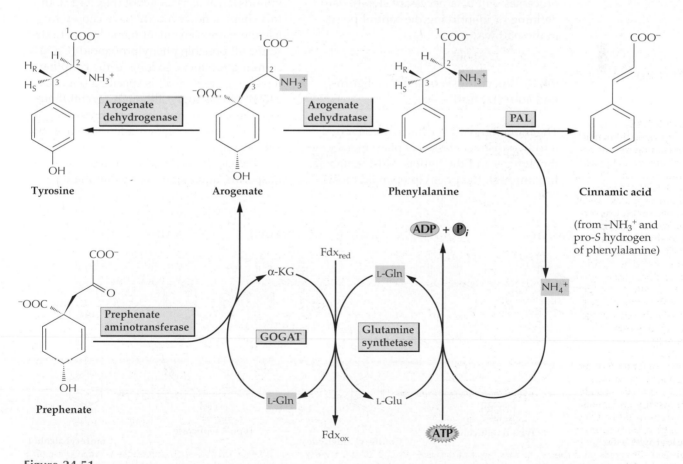

Figure 24.50
C-Glycosyl flavonoid types reported to be present in a green alga, *Nitella hookeri* (Charophyceae).

R = OH   **Lucenin**
R = H   **Vicenin**

is still some uncertainty as to whether it is ferulic acid, feruloyl-CoA, coniferaldehyde, or coniferyl alcohol that serves as the physiological substrate.

Researchers have not yet determined whether in some instances the O-methylation steps precede CoA-ligation, or whether both routes are possible (see Fig. 24.49, reaction 6). In any case, O-methyltransferases, whether acting on free acids or CoA esters, introduce methyl groups in a highly regiospecific manner, methylating the *meta*-hydroxyl group but not the group at the *para*-position. The enzyme catalyzing this transformation uses *S*-adenosylmethionine (SAM) as a cofactor, whereas CoA ligation requires ATP and CoASH. This two-step ligation first generates the AMP derivative, then converts it into the corresponding CoA ester.

After the CoA ester is formed, two sequential NADPH-dependent reductions produce the monolignols, completing the general phenylpropanoid pathway (see Fig. 24.49, reactions 7 and 8). The first of these enzymes, cinnamoyl-CoA reductase, catalyzes formation of *p*-coumaraldehyde (*p*-hydroxycinnamaldehyde), coniferaldehyde, and possibly sinapaldehyde. This type B reductase

Figure 24.51
During active phenylpropanoid metabolism, nitrogen from phenylalanine is recycled. Although TAL activity has been reported in certain plant species, no report has yet established a comparable nitrogen-recycling system for tyrosine. GOGAT, glutamine: α-keto-glutarate aminotransferase; L-Gln, glutamine; L-Glu, glutamate; α-KG, α-ketoglutarate; Fdx$_{red}$, reduced ferredoxin; Fdx$_{ox}$, oxidized ferredoxin.

abstracts the pro-*S* hydride (Fig. 24.52) from behind the nicotinamide plane of NADPH during reduction. The second enzyme, cinnamyl alcohol dehydrogenase, is a type A reductase that abstracts the pro-*R* hydride from in front of the nicotinamide plane to yield the monolignols *p*-coumaryl, coniferyl, and sinapyl alcohols (Fig. 24.52).

The above description is a brief account of the overall biochemical steps that culminate in monolignol formation. However, the pathway shown in Figure 24.49 is deceptive. Not all cells, tissues, or species of plants utilize the entire pathway. In many instances, plants utilize only a small pathway segment that directs substrates to one or more of the main metabolic branchpoints; moreover, they may express that truncated pathway only in specific tissues. Researchers do not yet fully understand metabolic flux and compartmentalization in the phenylpropanoid pathway. Elucidation of these processes will be a necessary step toward defining or identifying the control points in the pathway.

24.11 Biosynthesis of lignans, lignins, and suberization

The monolignols are primarily converted into two distinct classes of plant metabolites: the lignans and the lignins. Most metabolic flux through the phenylpropanoid biosyn-

thetic pathway is directed to the production of the lignins, which are structural components of cell walls. Free radicals participate in the reactions that produce both dimeric/oligomeric lignans and lignins as well as related complex plant polymers such as those in suberized tissue.

24.11.1 Dimeric and oligomeric lignans are formed primarily from coniferyl alcohol.

The term lignan was initially coined by Robert Downs Haworth in 1936 to describe a class of dimeric phenylpropanoid (C_6C_3) metabolites linked by way of their 8–8' bonds. More recently, another term, **neolignan,** was used to define all of the other types of linkages (e.g., 8–1'-linked dimers), but has since been modified to encompass substances derived from allylphenol compounds, such as isoeugenol (Fig. 24.53). In this chapter, however, we have chosen to use the more convenient name lignan to describe all possible phenylpropanoid (C_6C_3) coupling products, so long as the coupling mode (e.g., 8–8', 8–5') is specified (Fig. 24.54). Interestingly, although several thousand lignans are now known in nature, relatively few coupling modes have been encountered.

Lignan dimers are found in ferns, gymnosperms, and angiosperms, but higher

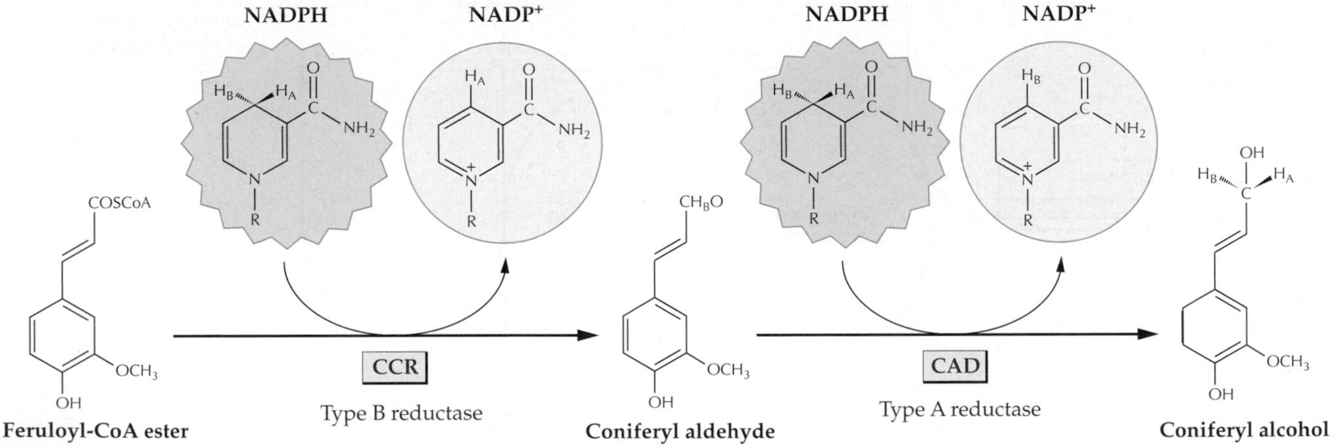

Figure 24.52
The stereospecificity of a type B reductase, NADPH-dependent cinnamoyl-CoA reductase (CCR), and a type A reductase, cinnamyl alcohol dehydrogenase (CAD). H_A, pro-*R* (the hydrogen projecting upward from the A-face of the nicotinamide ring, i.e., out from the plane of the page); H_B, pro-*S* (the hydrogen projecting upward from the B-face of the nicotinamide ring, i.e., behind the plane of the page). R, adenine nucleotide diphosphate.

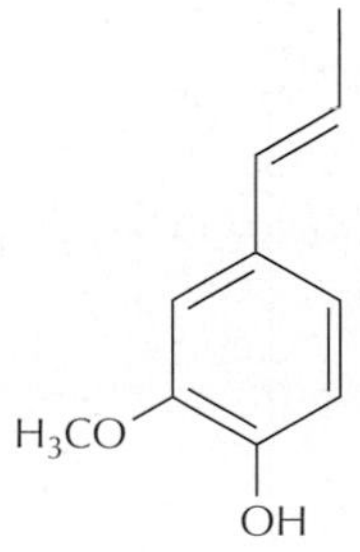

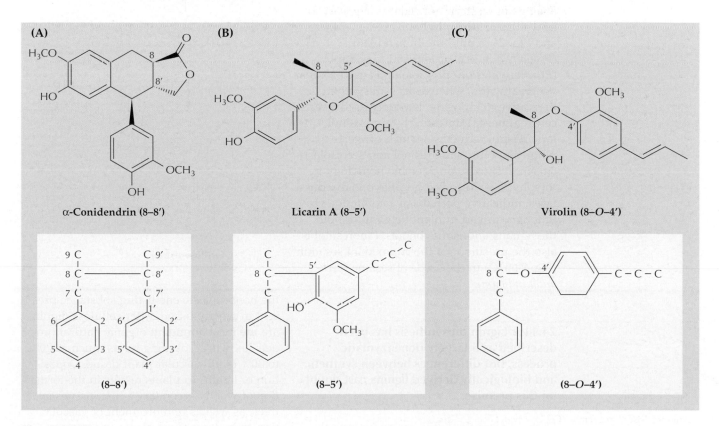

Figure 24.53
Isoeugenol, an allylphenol.

oligomeric forms also occur. Lignan formation utilizes coniferyl alcohol predominantly, along with other monolignols, allylphenols, and phenylpropanoid monomers to a lesser extent. Most lignans are optically active, although the particular antipode (enantiomer) can vary with the plant source.

The biochemistry of lignan formation has only very recently begun to be delineated. To date, work has focused mainly on generation of the most common 8–8′-linked lignans. This class of natural products is formed by a strict stereoselective coupling of two coniferyl alcohol molecules. The first demonstrated example of stereoselective control of phenolic coupling was the in vitro synthesis of (+)-pinoresinol (Fig. 24.55). This overall reaction, discovered in *Forsythia* species, is as follows: A laccase or laccase-like enzyme catalyzes a one-electron oxidation that forms the corresponding free radicals, and a **dirigent protein** (Latin: *dirigere*, to guide or align) orients the putative free radical substrates in such a way that random coupling cannot occur; only formation of the 8–8′-coupled intermediate, (+)-pinoresinol, is permitted. The particular antipode (optical form) of pinoresinol formed also varies with the plant species in question; for example, flax seeds accumulate (–)-pinoresinol. Once formed, pinoresinol can then undergo a variety of conversions, depending on the plant species.

The gene encoding the *Forsythia* dirigent protein has been cloned and the functional recombinant protein expressed. It is not homologous to any other protein. Given the existence of lignans linked by way of other distinct bonding modes and the increasing number of homologous genes and expressed sequence tags found in this and other species,

(A)

α-Conidendrin (8–8′)

(8–8′)

(B)

Licarin A (8–5′)

(8–5′)

(C)

Virolin (8–*O*–4′)

(8–*O*–4′)

Figure 24.54
Examples of lignans derived by distinct coupling modes, e.g., 8–8′, 8–5′, and 8–*O*–4′.

Forsythia intermedia

Figure 24.55
Proposed biochemical mechanism accounting for stereoselective control (regio- and stereochemistry) of *E*-coniferyl alcohol coupling in *Forsythia* species. The particular enantiomer of pinoresinol formed can vary with plant species.

E-Coniferyl alcohol

Dirigent protein
+
Oxidase
or oxidant

(+)-Pinoresinol

we can easily assume that the dirigent protein represents a new class of proteins. Additionally, the mode of action of this protein is of particular interest and may provide new and definitive insight into the macromolecular assembly processes that lead to lignins and suberins (see Sections 24.11.3 and 24.11.5).

Pinoresinol can be enantiospecifically converted into lariciresinol and secoisolariciresinol, followed by dehydrogenation to give matairesinol (Fig. 24.56 and Box 24.4). This last is the presumed precursor of plicatic acid (Figs. 24.56 and 24.57A) and its analogs in western red cedar (*Thuja plicata*), as well as of podophyllotoxin (Figs. 24.56 and Fig. 24.57B) in the Indian plant (*Podophyllum hexandrum*) and may apple (*P. peltatum*). Podophyllotoxin is used to treat venereal warts, whereas its semisynthetic derivative, teniposide, is widely used in cancer treatment. Interestingly, pinoresinol/lariciresinol reductase, which converts pinoresinol into lariciresinol and secoisolariciresinol, shows considerable homology to the phytoalexin-forming isoflavonoid reductases, indicative perhaps of a common evolutionary thread in plant defense for both the lignans and isoflavonoids. Pinoresinol is also the precursor of the antioxidant sesamin (Fig. 24.57C) in the seeds of sesame (*Sesamum indicum*).

24.11.2 Lignin biosynthesis has been described as a largely nonenzymatic process, but differences between synthetic and biologically derived lignins cast doubt on this premise.

Derived from the Latin *lignum* (wood), the term lignin initially was coined to describe

the noncellulosic encrusting substance present in woody tissue. After cellulose, lignins are the most abundant organic natural products known, accounting for as much as 20% to 30% of all vascular plant tissue. Deposition of lignins in plants results in the formation of woody secondary xylem tissues in trees, as well as reinforcement of vascular tissues in herbaceous plants and grasses. There are still no methods available for

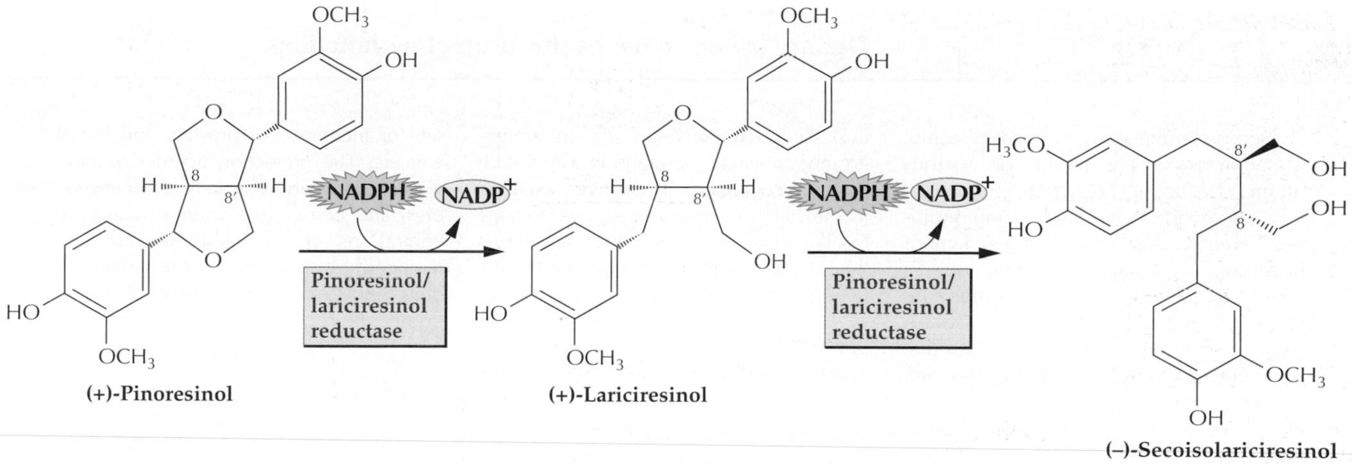

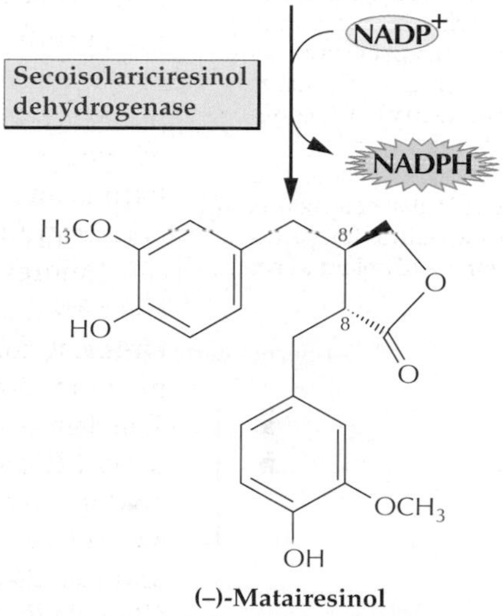

Figure 24.56
Proposed biochemical pathway for interconversions in the various 8–8'-linked lignan classes in *Forsythia*, western red cedar (*Thuja plicata*), and *Podophyllum* species. The pathway from pinoresinal to matairesinol is common to all three plants.

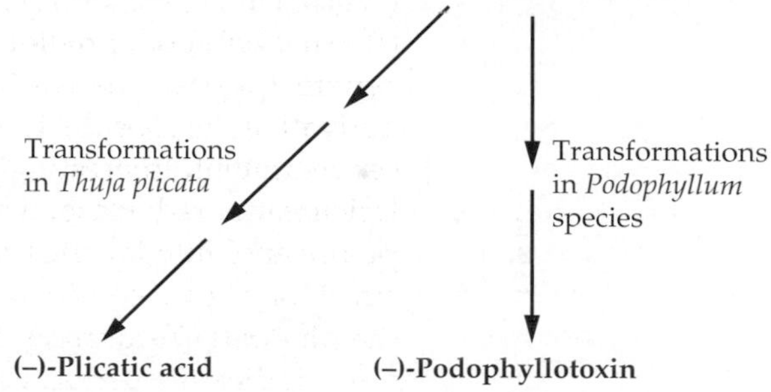

isolating lignins in their native state that do not markedly alter the original structure of the biopolymers during dissolution. In contrast to many of the lignans, lignins are thought to be racemic (optically inactive). Gymnosperm lignins are primarily derived from coniferyl alcohol, and to a lesser extent, *p*-coumaryl alcohol, whereas angiosperms contain coniferyl and sinapyl alcohols in roughly equal proportions (see Fig. 24.49).

For decades, the perceived formation of lignins in vivo has been biochemically incongruous. Investigators originally proposed that monolignols were transported into the cell walls and that the only subsequent enzymatic requirement for biopolymer formation was the one-electron oxidation of the monolignols to give the corresponding free radical intermediates, as shown with coniferyl alcohol (Fig. 24.58). Even today, there is no full agreement on the oxidative enzymes responsible for free radical generation (monolignol oxidation) in lignin biosynthesis. Five or six candidate proteins are still under consideration, although peroxidase remains the most favored.

The free radical intermediates formed by oxidation were initially believed to couple together in a manner requiring no further enzymatic control or input. These nonenzymatic free radical coupling reactions were thought to generate dimeric lignan structures that underwent further reoxidation and coupling to yield the lignin biopolymer (Fig. 24.58). In other words, the random reactions of monolignol-derived free radical intermediates in a test tube were considered to give preparations identical to the lignins formed in vivo. According to this model, nature's second most abundant substance is the only natural product for which its formation is not under enzymatic control.

However, although it is rarely recognized, natural and synthetic lignins differ in terms of bonding frequency, bonding type, and macromolecular size. For example, for lignins in vivo, the 8–*O*–4' interunit linkage predominates (more than 50%), with the 8–5' substructure found in much lower amounts (about 9–12%) (Fig. 24.59). In contrast, in synthetic in vitro preparations, the 8–*O*–4' substructure is present to only a very small extent, and the 8–5' and 8–8' linkages

Box 24.4   **Dietary lignans have health-protecting functions.**

Secoisolariciresinol and matairesinol are common constituents of various plants, including *Forsythia intermedia*, flax, and certain vegetables and grains (e.g., green beans and rye). These lignans have important nutritional functions in health protection. During digestion, intestinal bacteria convert secoisolariciresinol and matairesinol to enterodiol and enterolactone, respectively (see figure). These "mammalian" lignans undergo enterohepatic circulation, in which they are conjugated in the liver, excreted in the bile, deconjugated in the intestine by bacterial enzymes, absorbed across the intestinal mucosa, and returned to the liver in the portal circulation (see figure).

Enterodiol and enterolactone are believed to be responsible for preventing the onset of and substantially reducing the rate of incidence of prostate and breast cancers. The protection accrues to individuals on diets rich in grains, vegetables, and berries that contain high concentrations of secoisolariciresinol and matairesinol. In contrast, typical Western diets tend to be poor in these foods and do not afford comparable protection.

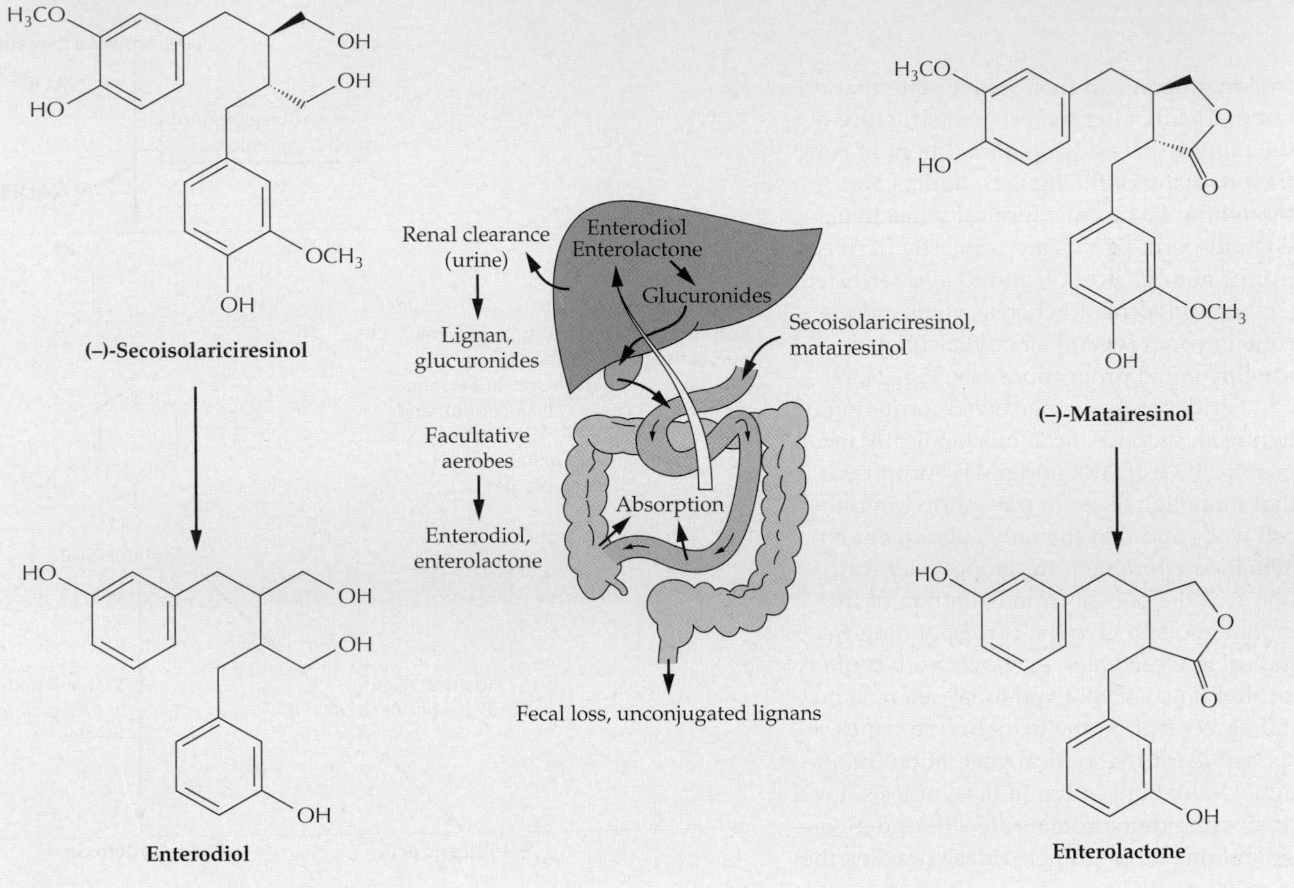

(–)-Secoisolariciresinol

Enterodiol

(–)-Matairesinol

Enterolactone

Renal clearance (urine)

Enterodiol Enterolactone

Glucuronides

Secoisolariciresinol, matairesinol

Lignan, glucuronides

Facultative aerobes

Enterodiol, enterolactone

Absorption

Fecal loss, unconjugated lignans

predominate. This disparity suggests that within woody tissues some mechanism in the lignifying cell wall regulates or mandates the interunit linkage pattern within the native biopolymer.

As is becoming increasingly clear, the lignification process in situ is under very tight biochemical control as part of a cell-specific programmed process. In the following section, we describe known elements that control lignification in vivo. Undoubted-

ly, more details will emerge as this important process is investigated systematically.

24.11.3 Lignin biosynthesis is controlled spatially and temporally and may involve a proteinaceous template.

Before lignin biosynthesis is initiated, the cells destined to form secondary xylem (i.e., wood; Fig. 24.60) undergo specific

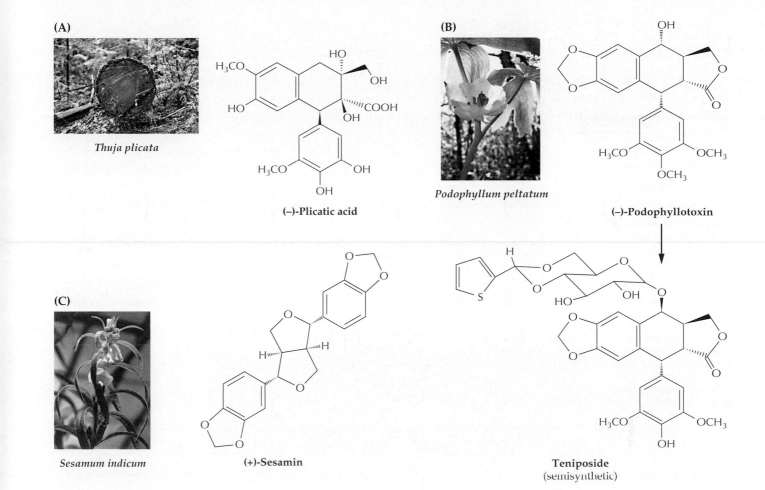

(A)

Thuja plicata

(–)-Plicatic acid

(B)

Podophyllum peltatum

(–)-Podophyllotoxin

(C)

Sesamum indicum

(+)-Sesamin

Teniposide
(semisynthetic)

Figure 24.57

Examples of 8–8′-linked lignans. (A) Plicatic acid (and its oligomeric congeners, not shown) are deposited en masse during heartwood formation in western red cedar. The congeners contribute substantially to the color, quality, and durability of this heartwood and are major components of the biochemical protection that enables such species to survive for more than 3000 years.

(B) Podophyllotoxin, from the may apple. The etoposide or teniposide derivatives of this compound are also used in treatment of certain cancers. (C) Sesamin, from the sesame seed, has in vitro antioxidant properties that stabilize sesame oil against turning rancid during commercial storage.

irreversible changes that ultimately lead to cell death and the formation of conducting elements (e.g., tracheids, vessels) and structural supporting tissues, such as fibers. These cells experience a programmed expansion of their primary walls, followed by so-called secondary thickenings, which involve ordered deposition of cellulose, hemicellulose, pectin, and structural proteins. Thus, the overall architecture of the plant cell wall is established largely before lignification takes place.

At the start of lignin biosynthesis, monolignols are transported from the cytosol into the cell wall during a specific stage of wall development. Electron microscopy investigations have shown that lignin bio-

synthesis is initiated at defined sites in the cell corners and middle lamella, i.e., at the locations farthest from the cytosol and plasma membrane. These loci in the cell walls then form distinctive domains that extend inward through the various cell wall layers, toward the plasma membrane. The domains ultimately coalesce.

UV-microscopy and radiochemical labeling indicate that individual monolignols are deposited differentially. For example, in conifers, *p*-coumaryl alcohol is primarily laid down at the early stages of lignin biosynthesis in the cell corners and middle lamella, whereas coniferyl alcohol is deposited predominantly in the secondary wall (Fig. 24.60A). This controlled deposition of specific

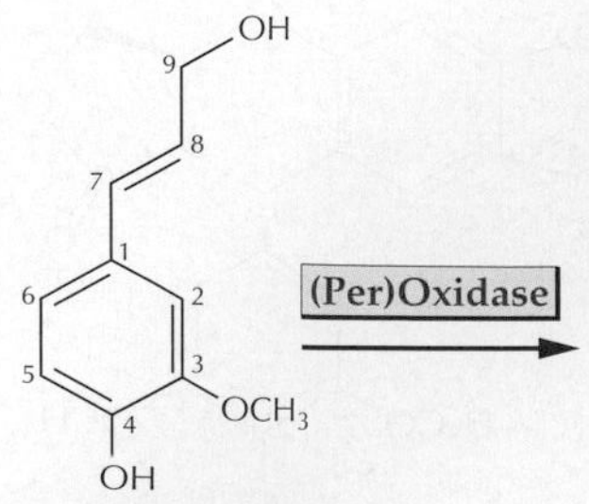

Coniferyl alcohol

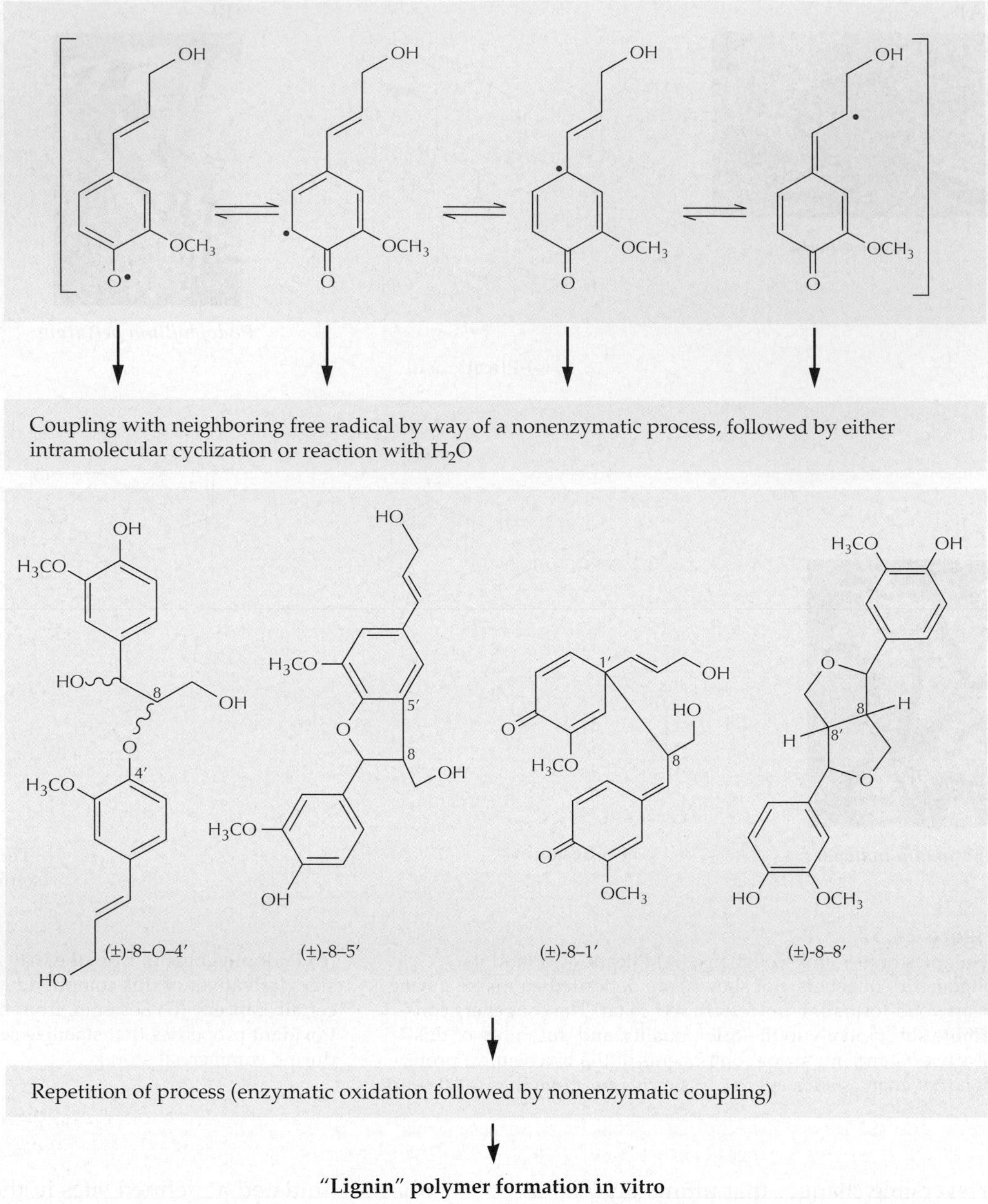

Figure 24.58
The random coupling hypothesis for "lignin" formation in vitro. Free radical intermediates are putatively generated by peroxidase or laccase. The free radicals then couple nonenzymatically to generate (±)-racemic dimers. Repetition of this process, involving further enzymatic oxidation of the dimeric phenols, was originally considered to continue until "lignin" was formed.

Coupling with neighboring free radical by way of a nonenzymatic process, followed by either intramolecular cyclization or reaction with H_2O

(±)-8–O–4'   (±)-8–5'   (±)-8–1'   (±)-8–8'

Repetition of process (enzymatic oxidation followed by nonenzymatic coupling)

"Lignin" polymer formation in vitro

monolignols creates domains with distinct structural configurations.

Perhaps most interesting of all, immunochemical studies demonstrate that initiation of lignin biosynthesis is both temporally and spatially associated with the secretion of distinct proteins from the Golgi apparatus and their deposition into the cell wall, including some that are proline-rich. These or related polypeptides, including some proline-rich proteins, may participate in lignification and may be related to the dirigents identified in lignan biosynthesis. Indeed, dirigent sites have been detected in regions where lignification is initiated. Thus, lignin biopolymer assembly may be under the control of a proteinaceous template.

Taken together, this evidence suggests that lignin assembly in vivo is subject to biochemical regulation, whereby the appropriate monomers are linked in a specific manner to yield a limited number of coupling modes in characteristic proportions. This model assumes that elongation of the primary lignin chain occurs by end-wise polymerization and is guided by an array of proteinaceous sites that stipulate or control linkage type and configuration. Moreover, in this way, the cytosol predetermines the outcome of phenoxy radical coupling. Lignin chain replication is thus envisaged to involve primitive self-replicating polymerization templates, and even the presumed lack of optical activity in lignins might result from, for example, the self-replication process involving generation of mirror-image polymeric assemblies. How lignification is ultimately achieved and what is the precise nature and mechanism of the putative proteinaceous templates now await full clarification at the biochemical level. As we establish the salient details of how lignin biopolymer assembly is controlled, plants are beginning to yield some of the long-hidden secrets involved in cell wall formation.

24.11.4 Variations on lignin deposition can be observed in the formation of reaction wood and in lignification in nonwoody plants.

A programmed plasticity of sorts is built into the overall macromolecular assembly of lignified cell walls. Perhaps the best example of this is seen in the formation of so-called **reaction wood.** When the woody stem becomes misaligned from its vertical axis, reaction wood forms to buttress the growing stem and gradually realign the photosynthetic canopy (Fig. 24.61). In this region, some of the cells originally fated to form ordinary xylem (see Fig. 24.60B) are reprogrammed to generate reaction wood instead. These cells then undergo massive changes in the macromolecular assembly of their cell walls (Fig. 24.61). In conifers, the cell walls of reaction wood, called **compression wood,** become thicker and rounder, the cellulose content is reduced relative to normal wood, and the cellulose microfibril angle is increased; the quantity of lignin also increases, primarily through an increase in the *p*-coumaryl alcohol content. In contrast, the reaction wood formed in angiosperms is known as **tension wood,** because the affected tissue is placed under tension rather than compression. Tension wood forms on the

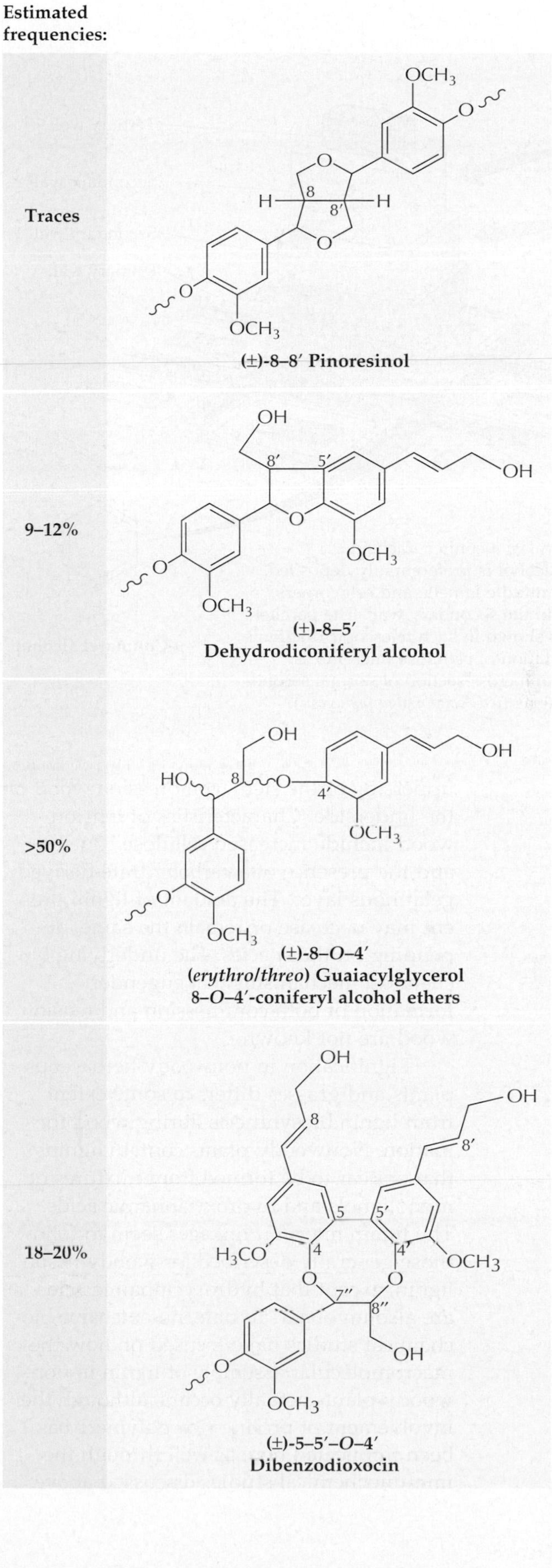

Estimated frequencies:

Figure 24.59
Prevalence of selected interunit linkages in native lignin biopolymers from the gymnosperm Norway spruce (*Picea abies*).

(A)

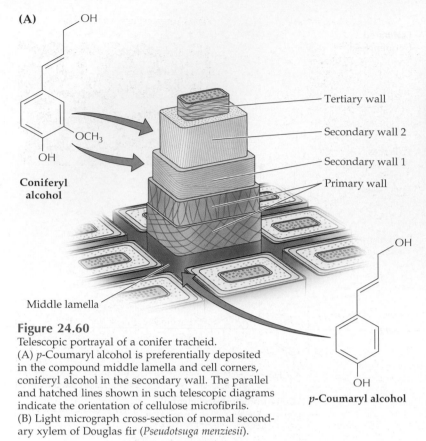

Coniferyl
alcohol

Tertiary wall

Secondary wall 2

Secondary wall 1

Primary wall

Middle lamella

p-Coumaryl alcohol

Figure 24.60
Telescopic portrayal of a conifer tracheid.
(A) *p*-Coumaryl alcohol is preferentially deposited
in the compound middle lamella and cell corners,
coniferyl alcohol in the secondary wall. The parallel
and hatched lines shown in such telescopic diagrams
indicate the orientation of cellulose microfibrils.
(B) Light micrograph cross-section of normal secondary xylem of Douglas fir (*Pseudotsuga menziesii*).

(B) Normal secondary xylem

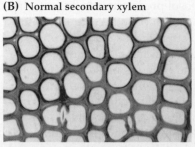

upperside of the stem, compression wood on
the underside. Characteristics of tension
wood include increased cellulose content
and the presence of a carbohydrate-derived
gelatinous layer. The amount of lignin present may decrease or remain the same, depending on the species. The underlying biochemical mechanisms that engender
formation of both compression and tension
wood are not known.

Lignification in nonwoody herbaceous
plants and grasses differs to some extent
from lignin biosynthesis during wood formation. Nonwoody plants contain lignins
that appear to be formed from mixtures of
monolignols and hydroxycinnamic acids.
The lignin interunit linkages seem to follow
those generally described for woody tissue
lignin, except that hydroxycinnamic acids
are also involved. To date, no extensive biochemical studies have focused on how the
macromolecular assembly of lignin in nonwoody plants actually occurs, although the
involvement of proline-rich polymers has
been implicated here as well through the
immunochemical studies discussed above.

24.11.5 Suberization protects tissues from water loss and pathogen invasion.

Suberized tissues are found in various un-
derground organs (e.g., roots, stolons, tubers) as well as in periderm layers (e.g.,
cork, bark). They are also formed as part of
the wound- and pathogen-induced defenses
of specific organs and cell types, perhaps the
most familiar example being the browning
and subsequent encrustation of sliced potato
tubers. Suberized tissues are formed as mul-
tilamellar domains consisting of alternating
polyaliphatic and polyaromatic layers (Fig.
24.62), as shown in the wound-healing layers in potato. These layers contribute to cell
wall strength and provide a means to limit
uncontrolled water loss by the intact organism by forming impenetrable barriers. From
an evolutionary perspective, suberization
was of utmost importance in plant adaptation to living on land and may even have
preceded lignification.

As with lignin, no methods are yet
available to obtain either of the two domains
of suberin in a native or unaltered condition.
The aliphatic component is located between
the primary wall and the plasmalemma.
Suberin aliphatics are generally long-chain
(more than 20 carbons) lipid substances;
they also include α,ω-fatty dioic acids, such
as C_{16}- or C_{18}-alkan-α,ω-dioic acids, which
are considered diagnostic of suberized tissue
(Fig. 24.63). Interestingly, the polyaromatic
domain located in the cell wall is apparently
formed before the aliphatics, primarily from
distinctive monomeric building blocks that
contain hydroxycinnamate-derived substances (Fig. 24.64). Thus, the formation of
suberized tissue is very distinct from the lignification of secondary xylem (where deposition is the last biochemical act of the xylem-
forming cells before cell death).

(A)

Compression wood

Figure 24.61
Compression wood (reaction wood). (A) Gymnosperm showing region of compression wood tissue. (B) Cross-section of *Sequoia sempervirens* showing pith and compression wood. (C) Light micrograph cross-section of compression wood xylem of Douglas fir (*Pseudotsuga menziesii*). (D) Telescopic portrayal of a tracheid in compression wood.

(B) *Sequoia sempervirens* stem   **(C)** Compression wood xylem

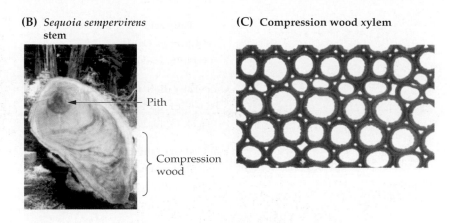

Pith

Compression wood

(D) Compression (reaction) wood tracheid

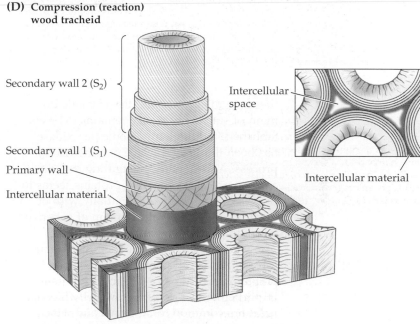

Secondary wall 2 (S$_2$)

Secondary wall 1 (S$_1$)

Primary wall

Intercellular material

Intercellular space

Intercellular material

A further complication to the study of the aromatic domain in suberization is the presence of related phenolic substances. For example, in wound-healing suberizing potato periderm tissues, chlorogenic acid and miscellaneous other phenolics are also present (Fig. 24.65). These compounds do not appear to function in suberization per se but rather may provide a means for topical disinfection of the exposed cell surfaces, thereby preventing or limiting infection/contamination. Some evidence also suggests the presence of low amounts of monolignols in suberized tissues, but these may be from small amounts of lignin.

Although the polymeric suberin phenolic constituents are predominantly derived from hydroxycinnamate, how this aromatic domain is assembled is unknown. Recent studies demonstrated that potato tuber wound-healing suberizing tissues contain two hydroxycinnamoyl-CoA transferases, which catalyze formation of various alkyl ferulates and (*p*-coumaroyl) feruloyl tyramine derivatives, respectively. How, or if,

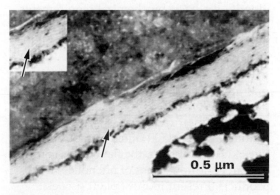

0.5 μm

Figure 24.62
Suberized tissue consists of layered polyaliphatic and polyaromatic domains, as shown for wound-healing potato tuber slices exposed to air. A suberized layer (see arrows) forms five days after exposure of potato tuber slices to air.

Fatty acid

H₃C(CH₂)ₙCOOH

$n = 14, 16, 18, 20, 22$

α,ω-Fatty dioic acids

HOOC(CH₂)ₙCOOH

$n = 14, 16, 18, 20, 22$

α,ω-Hydroxy fatty acids

HO(CH₂)ₙCOOH

$n = 15, 17, 19, 21, 23, 25$

Figure 24.63
Aliphatic components of suberized tissue. Found in combination, these compounds are considered diagnostic for suberin. An α,ω-dioic acid has carboxyl groups on both of the end carbons. An α,ω-hydroxy acid has a hydroxyl on one of the end carbons.

these are integrated into the aromatic domain of suberin of potato remains to be established, although an anionic peroxidase has been implicated in the polymerization process. Additionally, the finding that the appearance of proline-rich proteins seems to correlate temporally and spatially with deposition of the aromatic domain of suberized tissue may be important.

Much remains to be understood about formation of both the polyaromatic and the polyaliphatic domains of suberized tissue. In particular, we do not know yet which features are common to all plants and which are species-specific. For example, the suberized tissues seen in various root, periderm, and woody bark tissues are not identical to one another. This underscores the need to identify the basic biochemical requirements for suberization and to determine how these differ with regard to tissue-specific addition of particular phenolic substances.

24.12 Flavonoids

With more than 4500 different representatives known thus far, the flavonoids constitute an enormous class of phenolic natural products. Present in most plant tissues, often in vacuoles, flavonoids can occur as monomers, dimers, and higher oligomers. They are also found as mixtures of colored oligomeric/polymeric components in various heartwoods and barks.

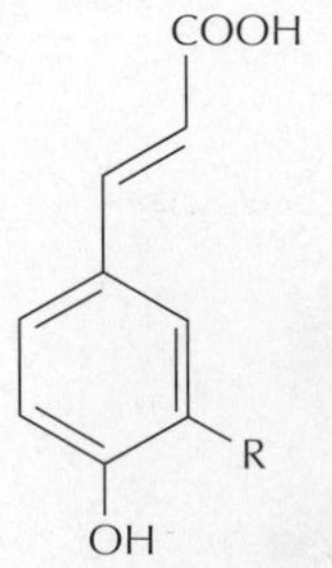

R = H   *p*-Coumaric acid
R = OCH₃   Ferulic acid

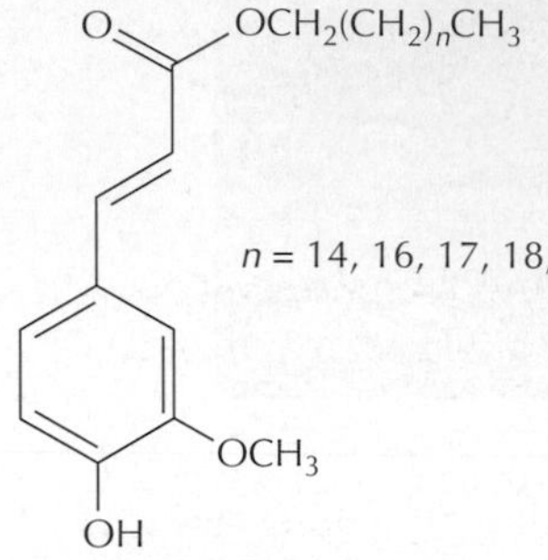

$n = 14, 16, 17, 18, 19, 20, 22, 24, 26$

Alkyl ferulates

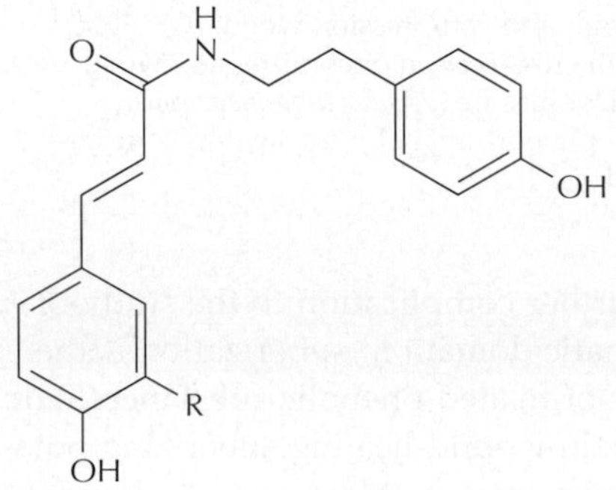

R = H   *p*-Coumaroyl tyramine
R = OCH₃   Feruloyl tyramine

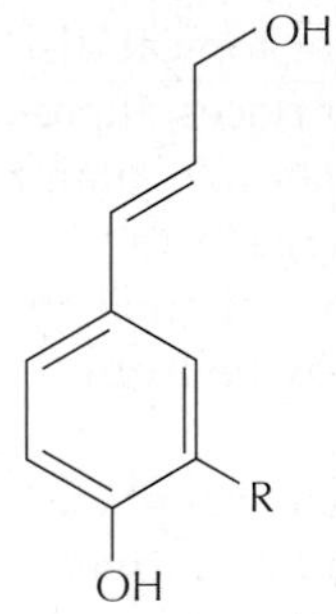

R = H   *p*-Coumaryl alcohol
R = OCH₃   Coniferyl alcohol

Figure 24.64
Aromatic components of suberized tissue, derived primarily from hydroxycinnamates, including alkyl ferulates and *p*-coumaroyl and feruloyl tyramines. Suberized tissue may also contain small amounts of monolignols.

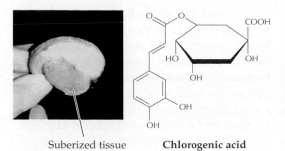

Suberized tissue   **Chlorogenic acid**

Figure 24.65
Suberin deposition has been studied in wounded potato tubers. In these tissues, suberin formation is accompanied by the production of an unrelated phenolic compound, chlorogenic acid.

24.12.1 Flavonoids comprise a diverse set of compounds and perform a wide range of functions.

Many plant–animal interactions are influenced by flavonoids. The colors of flowers and fruits, which often function to attract pollinators and seed dispersers, result primarily from vacuolar anthocyanins (Fig. 24.66) such as the pelargonidins (orange, salmon, pink, and red), the cyanidins (magenta and crimson), and the delphinidins (purple, mauve, and blue). Related flavonoids, such as flavonols, flavones, chalcones, and aurones, also contribute to color definition. Manipulating flower color by targeting various enzymatic steps and genes in flavonoid biosynthesis has been quite successful, particularly in petunia.

Specific flavonoids can also function to protect plants against UV-B irradiation, a role sometimes ascribed to kaempferol (Fig. 24.67). Others can act as insect feeding attractants, such as isoquercetin in mulberry, a factor involved in silkworm recognition of its host species. In contrast, condensed tannins such as the proanthocyanidins add a distinct bitterness or astringency to the taste of certain plant tissues and function as antifeedants (Fig. 24.68). The flavonoids apigenin and luteolin serve as signal molecules in legume–rhizobium bacteria interactions, facilitating nitrogen fixation (Fig. 24.69). In a related function, isoflavonoids are involved in inducible defense against fungal attack in alfalfa (e.g., medicarpin; Fig. 24.69) and other plant species. Perhaps the most poorly studied and least understood classes of the flavonoids are the oligomeric and polymeric substances associated with formation of certain heartwood and bark tissues. These

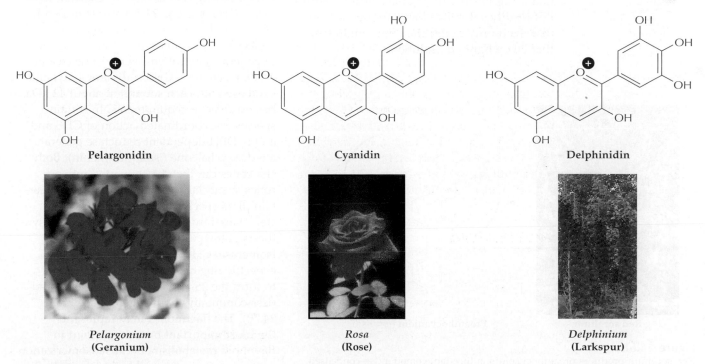

Pelargonidin

Pelargonium
(Geranium)

Cyanidin

Rosa
(Rose)

Delphinidin

Delphinium
(Larkspur)

Figure 24.66
Selected anthocyanin pigments: pelargonidin, cyanidin, and delphinidin from geranium, rose, and larkspur, respectively.

Soybean

Figure 24.67
Kaempferol, a UV-B protectant, is present in many plants such as soybean (*Glycine max*).

Kaempferol

Various flavonoids have also been studied extensively from the perspectives of health protection and pharmacological utility, for which mammalian enzyme systems have been used to assess flavonoid activity. Flavonoids have been analyzed as modulators of immune and inflammatory responses, for their impact on smooth muscle function, and as anticancer, antiviral, antitoxic, and hepatoprotective agents. There is considerable current interest in the use of isoflavonoids in cancer prevention. Dietary consumption of the isoflavonoids daidzein and genistein (Fig. 24.70), which are present in soybeans, is thought to reduce substantially the incidence of breast and prostate cancers in humans.

24.12.2 The flavonoid biosynthesis pathway has several important branchpoints.

The flavonoids consist of various groups of plant metabolites, which include chalcones, aurones, flavonones, isoflavonoids, flavones, flavonols, leucoanthocyanidins (flavan-3,4-diols), catechins, and anthocyanins (Figs. 24.70 and 24.71).

The first committed step of the flavonoid pathway is catalyzed by chalcone synthase (CHS; see Fig. 24.70). Three molecules of acetate-derived malonyl-CoA and one molecule of *p*-coumaryl-CoA are condensed to generate a tetrahydroxychalcone (see Fig. 24.49, reaction 9). CHS, a dimeric polyketide synthase with each subunit at about 42 kDa, has no cofactor requirements. In certain species, the coordinated action of CHS and an NADPH-dependent reductase generates a 6-deoxychalcone (isoliquiritigenin). Both chalcones can then be converted into aurones, a subclass of flavonoids found in certain plant species. Beyond CHS, the next step shared by most of the flavonoid biosynthesis pathways is catalyzed by chalcone isomerase (CHI), which catalyzes a stereospecific ring closure isomerization step to form the 2S-flavanones, naringenin, and (less commonly) liquiritigenin (see Fig. 24.70). The flavanones may represent the most important branching point in flavonoid metabolism, because isomerization of these compounds yields the phytoalexin isoflavonoids (Fig. 24.70), introduction of a C-2–C-3 double bond affords flavones and

compounds include proanthocyanidins and their congeners in woody gymnosperms and isoflavonoids in woody legumes from the tropics. In both cases, their massive deposition during heartwood formation contributes significantly and characteristically to the overall color, quality, and rot resistance of wood. These metabolites can be misidentified as lignins because some constituents are not readily solubilized and are frequently dissolved only under the same conditions that effect lignin dissolution.

Red sorghum

Proanthocyanidin (*n* = 1–30)

Figure 24.68
Red sorghum produces proanthocyanidin antifeedant compounds—condensed tannins, which deter birds from feeding on the seed. White sorghum, which is deficient in these compounds, is rapidly consumed by birds. Similar compounds are present in the heartwood of Douglas fir (not shown).

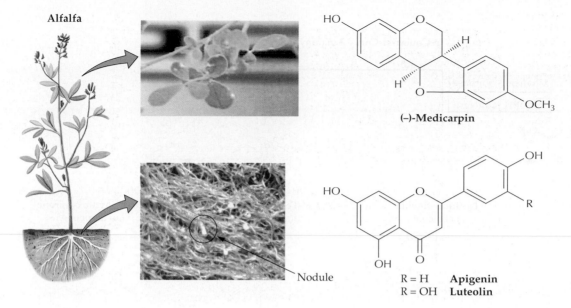

Figure 24.69

Flavonoids perform diverse functions in alfalfa (*Medicago sativa*). The flavonoids apigenin and luteolin function as signaling molecules that induce *Nod* gene expression in compatible *Rhizobium* bacteria, facilitating the development of nitrogen-fixing root nodules. The phytoalexin isoflavonoid medicarpin participates in inducible plant defense.

flavonols (Fig. 24.71), and hydroxylation of the 3-position generates dihydroflavonols (Fig. 24.71).

Entry into the isoflavonoid branchpoint occurs by way of two enzymes (see Fig. 24.70). The first, isoflavone synthase (IFS), catalyzes an unusual C-2 to C-3 aryl migration and hydroxylation to give the 2-hydroxyisoflavanones and has recently been shown to be an NADPH-dependent cytochrome P450 enzyme. Dehydration of the 2-hydroxyisoflavanones, catalyzed by 2-hydroxyisoflavanone dehydratase (IFD), forms the isoflavonoids genistein and daidzein. The isoflavonoids can be further metabolized, primarily in the Fabaceae, to yield phytoalexins (e.g., medicarpin in alfalfa; see Fig. 24.70) or to generate isoflavonoid-derived substances known as rotenoids in tropical legumes (e.g., 9-demethylmunduserone from *Amorpha fruticosa*; see Fig. 24.70). The rotenoids, which are isolated mainly from *Derris elliptica* and related species, are used extensively as insecticidal agents but have other applications as well. For example, rotenone is used as a rat poison and an inhibitor of NADH dehydrogenase. Interestingly, the NADPH-dependent isoflavone reductase (IFR) step involved in isoflavonoid formation shows considerable homology to

pinoresinol/lariciresinol reductase (see Fig. 24.56 and Section 24.11.1), suggesting a phylogenetic link between both lignan and isoflavonoid pathways for plant defense.

The second branching point in general flavonoid metabolism involves that of dehydration of naringenin at the C-2/C-3 positions to give such abundant flavones as apigenin (Fig. 24.71). This conversion is catalyzed by flavone synthase (FNS), which varies in enzymatic type depending on the plant species. For example, in parsley cell cultures, flavone formation is catalyzed by an α-ketoglutarate–dependent dioxygenase (FNS I in Fig. 24.71), whereas an NADPH-dependent microsomal preparation engenders this reaction in *Antirrhinum* flowers (FNS II in Fig. 24.71).

The third major branchpoint in flavonoid metabolism is stereospecific 3-hydroxylation of naringenin (or its 3′-hydroxylated analog) to give dihydroflavonols (Fig. 24.71) such as dihydrokaempferol (or dihydroquercetin). The enzyme involved, flavanone 3-hydroxylase, is an Fe^{2+}-requiring, α-ketoglutarate–dependent dioxygenase. Specific hydroxylation involving an NADPH-dependent cytochrome P450 monooxygenase of naringenin can also directly give dihydroquercetin, which can be converted to quercetin (a flavanol) by

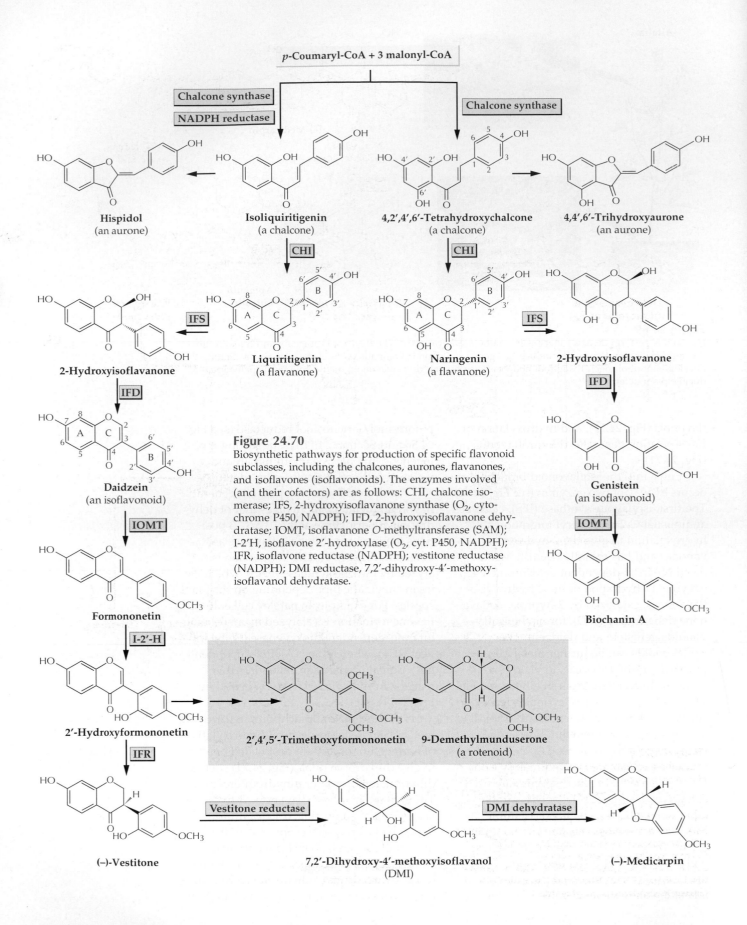

Figure 24.70

Biosynthetic pathways for production of specific flavonoid subclasses, including the chalcones, aurones, flavanones, and isoflavones (isoflavonoids). The enzymes involved (and their cofactors) are as follows: CHI, chalcone isomerase; IFS, 2-hydroxyisoflavanone synthase (O₂, cytochrome P450, NADPH); IFD, 2-hydroxyisoflavanone dehydratase; IOMT, isoflavanone O-methyltransferase (SAM); I-2'H, isoflavone 2'-hydroxylase (O₂, cyt. P450, NADPH); IFR, isoflavone reductase (NADPH); vestitone reductase (NADPH); DMI reductase, 7,2'-dihydroxy-4'-methoxy-isoflavanol dehydratase.

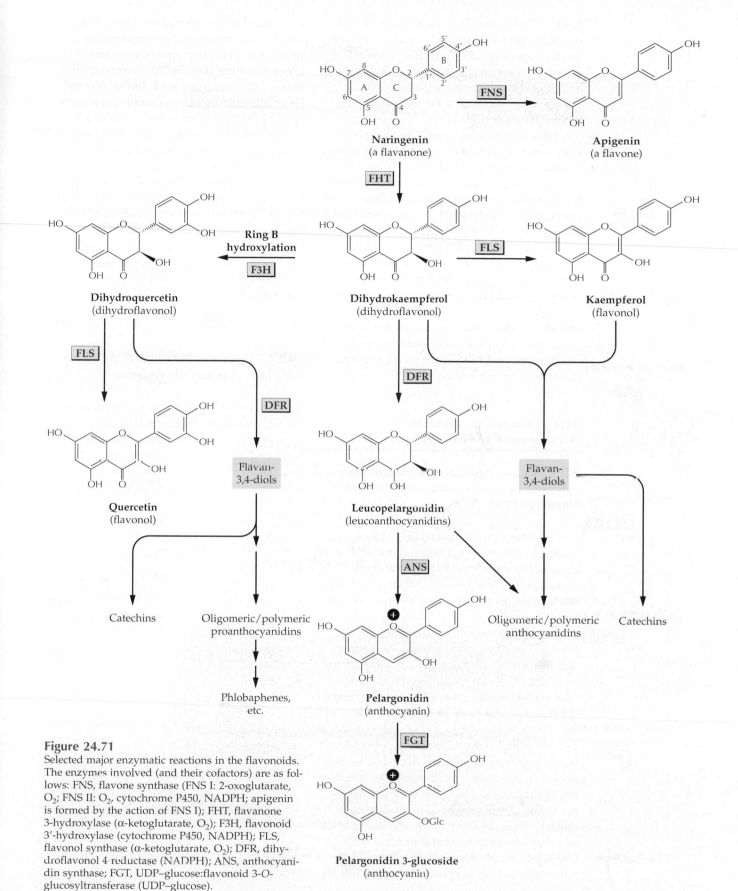

Figure 24.71
Selected major enzymatic reactions in the flavonoids. The enzymes involved (and their cofactors) are as follows: FNS, flavone synthase (FNS I: 2-oxoglutarate, O_2; FNS II: O_2, cytochrome P450, NADPH; apigenin is formed by the action of FNS I); FHT, flavanone 3-hydroxylase (α-ketoglutarate, O_2); F3H, flavonoid 3'-hydroxylase (cytochrome P450, NADPH); FLS, flavonol synthase (α-ketoglutarate, O_2); DFR, dihydroflavonol 4-reductase (NADPH); ANS, anthocyanidin synthase; FGT, UDP–glucose:flavonoid 3-O-glucosyltransferase (UDP–glucose).

flavonol synthase (FLS)–catalyzed C-2–C-3 double bond formation; FLS is an α-ketoglutarate–dependent dioxygenase. Alternatively, dihydroquercetin can be reduced by an NADPH-dependent dihydroflavonol reductase (DFR) to give the corresponding flavan-3,4-diols (Fig. 24.71).

Subsequent species- and tissue-specific enzymatic conversions can create vast arrays of structurally diverse groups of flavonoids. For example, in flower petals, the leucoanthocyanidins (e.g., leucopelargonidin) can be converted to the colored anthocyanins (e.g., pelargonidin) through the action of a dehydratase, anthocyanidin synthase (ANS), which is thought to be an α-ketoglutarate–dependent dioxygenase (Fig. 24.71). Leucoanthocyanidins can also serve as precursors of the (epi-)catechins and condensed tannins. The enzymology associated with those coupling processes, chain extension mechanisms, and oxidative modifications, however, is not yet established.

24.13 Coumarins, stilbenes, styrylpyrones, and arylpyrones

24.13.1 Some coumarins, a class of plant defense compounds, can cause internal bleeding or dermatitis.

Coumarins (e.g., coumarin; Fig. 24.72A) belong to a widespread family of plant metabolites called the benzopyranones, with more than 1500 representatives in more than 800 species. In plants, these compounds can occur in seed coats, fruits, flowers, roots, leaves, and stems, although in general the greatest concentrations are found in fruits and flowers. Their roles in plants appear to be mainly defense-related, given their antimicrobial, antifeedant, UV-screening, and germination inhibitor properties.

The best known properties of coumarins indirectly highlight their roles in plant defense. Ingesting coumarins from plants such as clover can cause massive internal bleeding in mammals. This discovery ultimately led to the development of the rodenticide Warfarin (Fig. 24.72B) and to the use of related compounds to treat and prevent stroke. Likewise, the photosensitizing compound 8-methoxypsoralen, present in leaf tissue

of *Heracleum mantegazzianum* (giant hogweed), can cause photophytodermititis on skin contact and subsequent exposure to UV-A radiation (Fig. 24.73). A comparable form of coumarin-induced dermatitis can also occur during celery handling. Psoralen (Fig. 24.74), however, is now successfully used to treat various skin disorders (eczema, psoriasis) by means of a combination of oral ingestion and UV-A treatment.

The structure of a representative simple coumarin, 7-hydroxycoumarin, is shown in Figure 24.75. Additional families of plant coumarins (see Fig. 24.74) include linear furanocoumarins (e.g., psoralen), angular furanocoumarins (e.g., angelicin), pyranocoumarins (e.g., seselin), and pyrone-substituted coumarins (e.g., 4-hydroxycoumarin).

24.13.2 Coumarin biosynthesis pathways have not yet been fully elucidated.

The biosynthetic pathways to the coumarins are only partially determined at this point; they mainly involve aromatic hydroxylations and additional reactions catalyzed by *trans/cis*-hydroxycinnamic acid isomerases, dimethylallyltransferases, various P450/NADPH/O_2–dependent synthases and *O*-methyltransferases (Fig. 24.75). The simplest

(A)

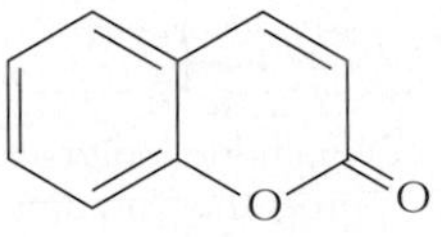

Coumarin

(B)

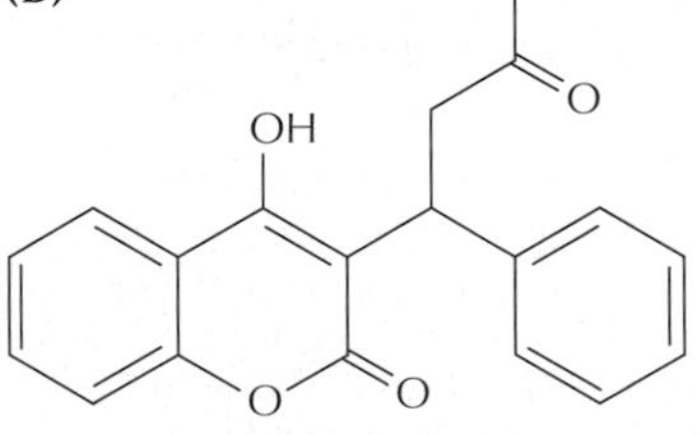

Warfarin
(synthetic coumarin)

Figure 24.72
Structures of (A) coumarin (from clover), and (B) a synthetic coumarin, the rodenticide Warfarin.

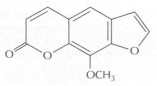

OCH₃

Heracleum

8-Methoxypsoralen
(a furanocoumarin)

Figure 24.73
A linear furanocoumarin, 8-methoxypsoralen, sensitizes human skin to UV-A light. This compound, present in external tissues of *Heracleum* species, causes severe blistering on skin contact followed by exposure to UV-irradiation.

examples, coumarin and 7-hydroxycoumarin (umbelliferone), are believed to be formed by O-hydroxylation of cinnamic and *p*-coumaric acids, respectively, followed by trans/cis-isomerization and ring closure. However, neither the enzymes nor their encoding genes have yet been obtained.

On the other hand, much more is known about the biosynthesis of both linear and angular furanocoumarins. These involve regiospecific prenylation through the action of the corresponding tranferases to yield demethylsuberosin and osthenol, respectively. The subsequent transformations are believed to involve various NADPH-dependent, cytochrome P450 oxidase–catalyzed conversions and O-methylations (Fig. 24.75).

Various fungi and yeasts also biosynthesize coumarins, e.g., the toxic aflatoxins. However, these metabolites are polyketide derivatives and hence are biochemically distinct from their plant analogs.

24.13.3 Stilbenes, styrylpyrones, and arylpyrones constitute another class of chemical defense compounds.

In addition to the products of the flavonoid pathway, cinnamoyl-CoA and malonyl-CoA (acetate-derived) pathways can in certain plant species also undergo condensation reactions to yield the corresponding stilbenes, styrylpyrones, and arylpyrones (Fig. 24.76). Comparison of gene sequences for each entry-point enzyme (CHS and stilbene synthase) reveals significant homology, as would be expected for similar enzymatic

systems. Beyond the initial synthases, however, little has yet been described about subsequent transformations.

Stilbenes are present in bryophytes, pteridophytes, gymnosperms, and angiosperms, with more than 300 different stilbenoids known today. The stilbenes play important roles in plants, particularly in heartwood protection, and also have significance in pharmacology and human health. In plants, they can function as both constitutive and inducible defense mechanisms. Stilbenes display weak antibacterial properties

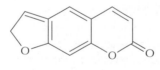

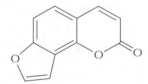

Psoralen
(a linear furanocoumarin)

Angelicin
(an angular furanocoumarin)

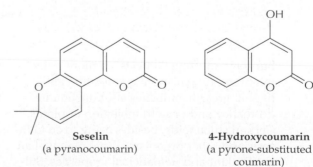

Seselin
(a pyranocoumarin)

4-Hydroxycoumarin
(a pyrone-substituted coumarin)

Figure 24.74
Structures of the linear furanocoumarin psoralen, the angular furanocoumarin angelicin, the pyranocoumarin seselin, and the pyrone-substituted coumarin 4-hydroxycoumarin.

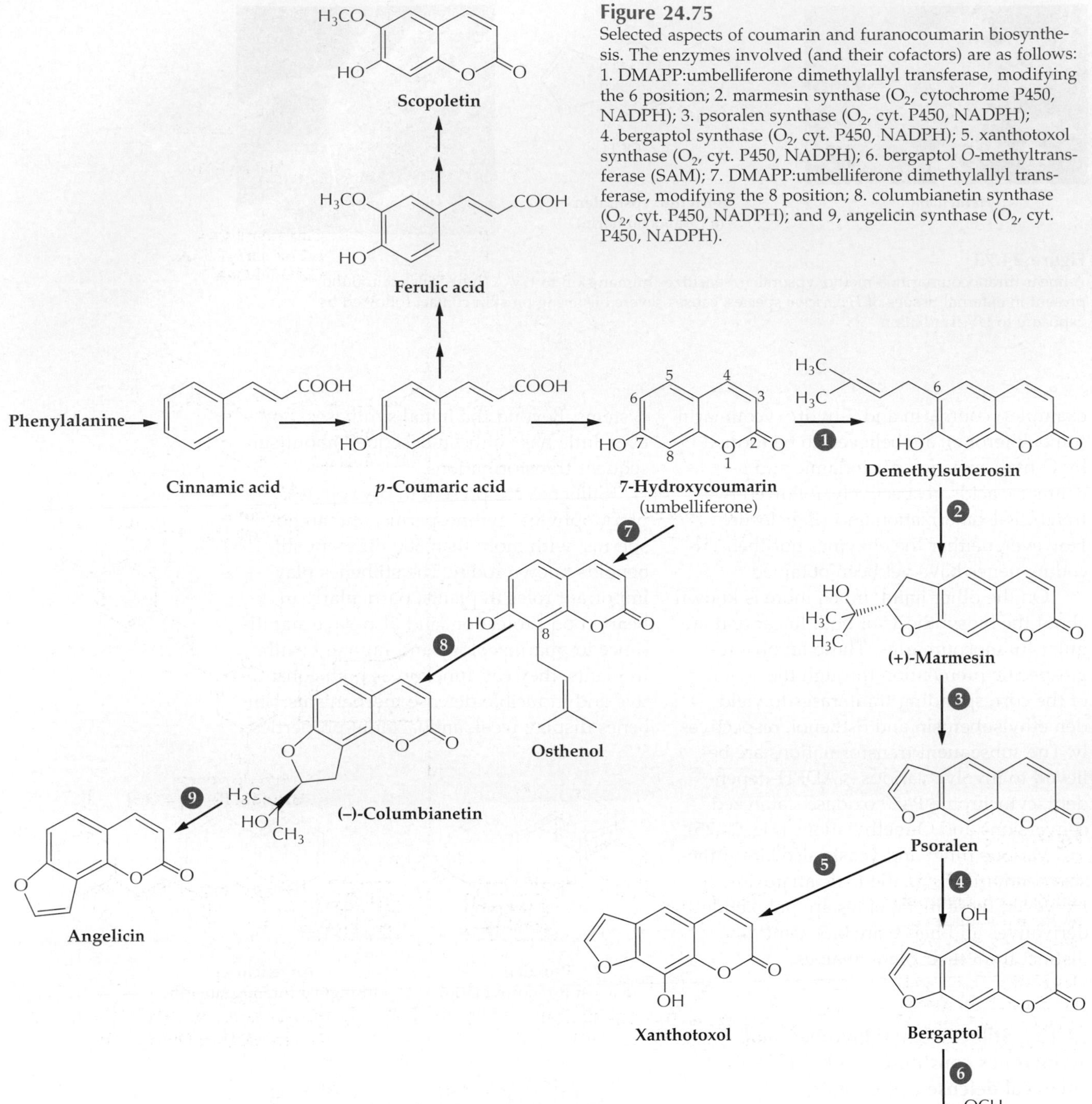

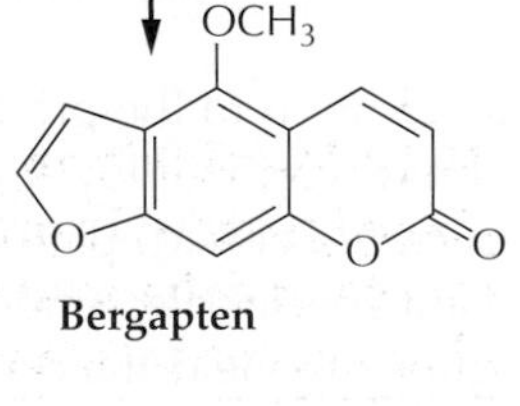

Figure 24.75
Selected aspects of coumarin and furanocoumarin biosynthesis. The enzymes involved (and their cofactors) are as follows: 1. DMAPP:umbelliferone dimethylallyl transferase, modifying the 6 position; 2. marmesin synthase (O_2, cytochrome P450, NADPH); 3. psoralen synthase (O_2, cyt. P450, NADPH); 4. bergaptol synthase (O_2, cyt. P450, NADPH); 5. xanthotoxol synthase (O_2, cyt. P450, NADPH); 6. bergaptol O-methyltransferase (SAM); 7. DMAPP:umbelliferone dimethylallyl transferase, modifying the 8 position; 8. columbianetin synthase (O_2, cyt. P450, NADPH); and 9, angelicin synthase (O_2, cyt. P450, NADPH).

but their antifungal effects are more potent, inhibiting fungal spore germination and hyphal growth; stilbenes also function in dormancy and growth inhibition of plants. Certain stilbenoids, besides being toxic to insects and other organisms, have mammalian antifeedant and nematicidal properties. Stilbenoid formation can be induced by insect attack, as illustrated by the colored deposits formed in radiata pine sapwood when attacked by the *Sarix* wasp (see photographs in Box 24.5).

From a pharmacological perspective, the stilbene combretastatin has important antineoplastic activities, and resveratrol, present in red wine, helps suppress tumor formation (Fig. 24.77).

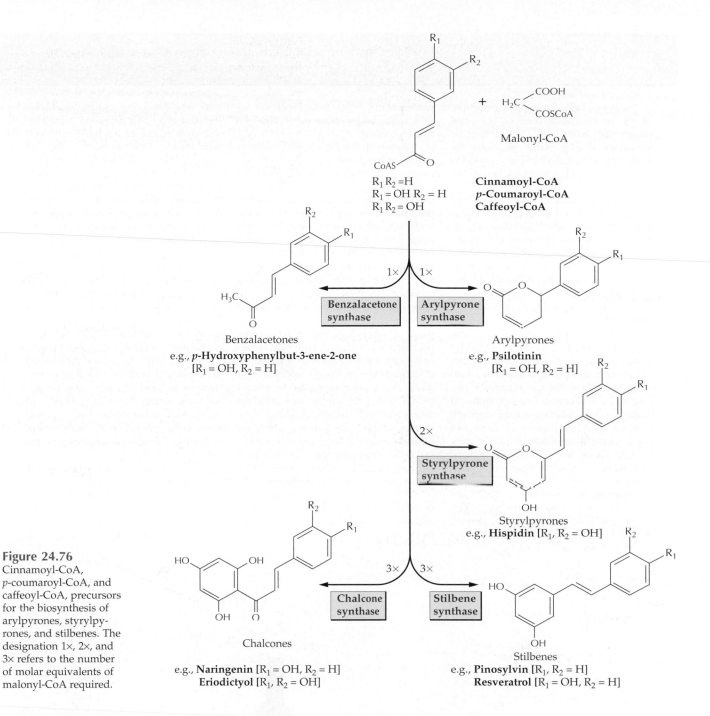

Figure 24.76
Cinnamoyl-CoA,
p-coumaroyl-CoA, and
caffeoyl-CoA, precursors
for the biosynthesis of
arylpyrones, styrylpy-
rones, and stilbenes. The
designation 1×, 2×, and
3× refers to the number
of molar equivalents of
malonyl-CoA required.

Malonyl-CoA

R_1 R_2 =H   **Cinnamoyl-CoA**
R_1 = OH R_2 = H   *p*-**Coumaroyl-CoA**
R_1 R_2 = OH   **Caffeoyl-CoA**

Benzalacetones
e.g., *p*-**Hydroxyphenylbut-3-ene-2-one**
$[R_1 = OH, R_2 = H]$

Benzalacetone synthase

Arylpyrone synthase

Arylpyrones
e.g., **Psilotinin**
$[R_1 = OH, R_2 = H]$

Styrylpyrone synthase

Styrylpyrones
e.g., **Hispidin** $[R_1, R_2 = OH]$

Chalcones
e.g., **Naringenin** $[R_1 = OH, R_2 = H]$
Eriodictyol $[R_1, R_2 = OH]$

Chalcone synthase

Stilbene synthase

Stilbenes
e.g., **Pinosylvin** $[R_1, R_2 = H]$
Resveratrol $[R_1 = OH, R_2 = H]$

24.14 Metabolic engineering of phenylpropanoid production: a possible source of enhanced fibers, pigments, pharmaceuticals, and flavoring agents

The biochemical, chemical, and molecular characterization of how plants produce vari-ous metabolic substances is essential to un-derstanding the very basis of the biodiversi-ty and life of plants. This pursuit also has economic implications, affording new oppor-tunities for systematic modification of com-mercially important plants to engineer or specify particular traits that can benefit hu-manity.

Many biotechnological possibilities await our manipulation of plant phenolic metabolism: plants with increased resistance to pathogens; improvements in the quality of wood and fiber products; new or improved

Box 24.5

Postlignification metabolism and heartwood formation require nonstructural plant phenolic compounds.

Heartwood represents more than 95% of the merchantable bole of harvested wood. The heartwood of commercially important woody plants accounts for more than 60% of the revenues generated from harvesting plant materials before further factory processing. Heartwood serves as the main source of raw material for lumber, solid wood products, fine furniture, paper, and many miscellaneous applications. Despite its economic significance, however, the general mechanism responsible for its formation is one of the most poorly studied and poorly understood areas of plant science today.

Heartwood is formed by the species-specific deposition of distinct and varied metabolites that frequently alter the color, durability, texture, and odor of particular woods relative to that of sapwood. Heartwoods contain strikingly distinctive colored metabolites that can readily be observed by inspecting cross-sections of woody stems of plants, such as tamarack (see panel A of figure), western red cedar (reddish- or pinkish-brown to dark brown), ebony (jet black), and southern pine (yellow-orange). In contrast, spruce wood, highly valued for pulp and paper manufacture, contains less of the highly colored heartwood metabolites and hence has a pale whitish-yellow color. Indeed, the pale color and the very high lignin content of this wood (about 28%) indicate that lignin biopolymers themselves are nearly colorless.

Heartwood production is a postsecondary xylem-forming process, whereby nonstructural highly colored phenolics (primarily lignan, stilbene, and flavonoid-derived compounds) and other characteristic substances (e.g., terpenes or alkaloids) are infused into wood that has already been lignified. Substances similar (if not identical, in some cases) to those in heartwood can also be formed in regions where insects or pathogens have attacked sapwood, but these are manifested as a more-localized containment response. For example, panel B of the figure shows sapwood of radiata pine into which a *Sirex noctilio* wasp has bored, forming two tunnels, one for the wasp's eggs and one for a fungus, *Amylostereum noctilio,* that serves as a foodstuff for the larvae. The attacked plant responds by increasing the deposition of various phenolic substances, in this case stilbenes, which are primarily localized in the affected regions, making them appear lighter-colored than the background in the stained wood section shown in panel C of the figure.

Constitutive heartwood formation, on the other hand, follows several years or decades of sapwood growth and development. According to biochemical details only now becoming known, heartwood metabolites are first deposited in the central (pith) region of the lignified woody stem tissues, which primarily consist of dead, lignified cells. Over years of subsequent growth, heartwood formation gradually extends radially, until almost all of the woody xylem tissue is encompassed. A transition zone sometimes visible between the heartwood and sapwood is presumed to be involved in the final stages of biosynthesis of heartwood metabolites preceding cell death. The composition of heartwood metabolites varies extensively among species. For example, Douglas fir accumulates flavonoids and lignans in its heartwood, whereas yellow poplar deposits lignans, terpenoids, and alkaloids.

Given that wood is composed largely of dead cells, how are heartwood metabolites deposited? Investigators recognized 50 years ago that ray parenchyma cells remain living in lignified sapwood. As their last function before death, these cells accumulate or biosynthesize substances (often in complex species-specific mixtures) that are then infused into lignified woody secondary xylem by way of pit apertures (see figure, panel D). This infusion process may explain why many of the heartwood substances also occur at much lower concentrations in sapwood, where the ray parenchyma cells

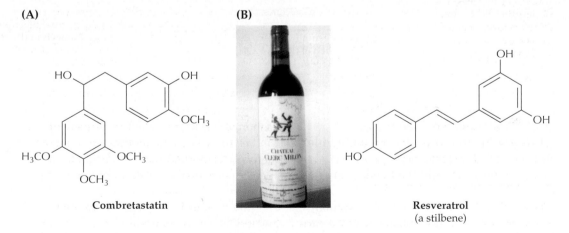

(A)   **(B)**

Combretastatin   Resveratrol
(a stilbene)

Figure 24.77
(A) The stilbene, combretastatin, has antineoplastic activity. (B) Another stilbene, resveratrol, present in red grapes and red wine, has potent antitumor properties.

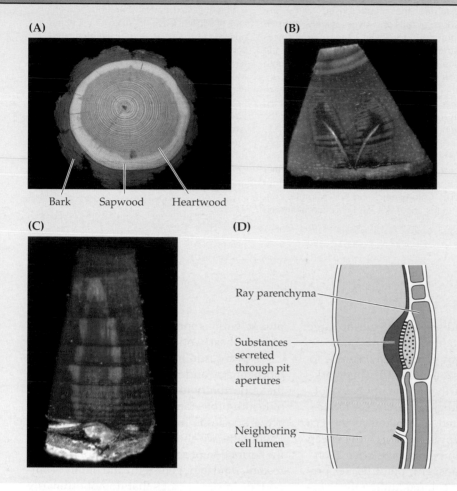

(A)

Bark   Sapwood   Heartwood

(B)

are located. At some point during heartwood formation, the infused phenolic substances apparently undergo oxidation to form nonstructural oligomeric and polymeric components, some of which can be removed only under the conditions typically used for lignin extraction. With the further sealing of bordered pits, heartwood ceases to function in conducting nutrients and water and essentially becomes a protective structural tissue, highly durable and resistant to rot.

(C)

(D)

Ray parenchyma

Substances secreted through pit apertures

Neighboring cell lumen

sources of pharmaceuticals, nutriceuticals, pigments, flavors, and fragrances; and selective adjustments to the taste and odor of selected plant species. Indeed, a biotechnological revolution is now being witnessed in the plant sciences. The combined use of molecular genetic techniques and conventional plant breeding approaches is expected to produce a new generation of plants that are even further optimized for human use.

By far the largest and economically most significant deployment of plant materials is as a fiber source, whether for pulp/paper, lumber for housing and shelter, wood for furniture, or other applications. Accordingly, many biotechnological strategies are directed toward improving fiber and wood properties by manipulating the biochemical processes responsible for cell wall biosynthesis and as-

sociated metabolic functions. This approach could involve modification of lignin, either to render it more susceptible to removal, or to increase its content, thereby increasing the strength and rigidity of certain fragile crops. Modifying heartwood metabolite formation may allow researchers to tailor traits such as rot resistance, texture, color, and durability in various commercially important woody plant heartwoods. These goals require further study of the fundamental mechanisms controlling both macromolecular assembly patterns involved in the biosynthesis of plant cell wall polymers and exploration of how and where the heartwood-forming metabolites are generated.

To this point, most of the biotechnological emphasis placed on attempting to engineer lignin content and composition has involved

Box 24.6   Phenolics flavor our world.

Phenylpropanoid-derived plant phenolics contribute significantly to imparting specific fragrances/odors, flavors, and tastes to various plants widely utilized in the food and beverage industries today (see figure). Although the biochemistry of their formation is scarcely addressed in this brief chapter, their importance cannot be discounted.

The capsaicinoids, such as capsaicin, are responsible for the pungent properties of the red peppers, whereas the piperinoids flavor black pepper. The delightful tastes of cinnamon and ginger are imparted by various cinnamate and gingerol derivatives, respectively; and allylphenols establish the characteristic tastes and odors of oil of cloves, widely used in toothache treatment, and of the spices nutmeg and mace. Vanillin, from the vanilla bean, is used extensively in both baking and confectionery. In most instances, precise biochemical pathways to these compounds are not yet established at the levels of either the enzymes or the genes.

Plant phenolics are important components of the characteristic aromas, flavors, and colors of many beverages, whether for alcoholic or nonalcoholic consumption. Chlorogenic acid, for example, constitutes about 4% of the coffee bean and is thus ingested daily by millions. Green and black tea leaves contain other plant phenolics, such as (epi-)catechins and various other tannins that impart characteristic tastes to these popular beverages. Most drinks consumed today would be watery indeed if not for various phenolics, such as vanillin, ferulic acid, certain flavonoids, tannins, and others. An important endeavor of the flavor and fragrance industry is to define or identify the mixtures of various phenolic substances that create pleasing flavors ranging from maple syrup to whisky.

using antisense and sense strategies to target the genes that encode various enzymatic steps in the pathway from phenylalanine to the monolignols (see Fig. 24.49). This work has focused primarily on cinnamyl alcohol dehydrogenase and cinnamoyl-CoA reductase and has targeted such plants as tobacco, poplar, and eucalyptus. Although the effects on lignin formation per se have often been quite small, the transgenic plant tissues produced were highly colored, unlike the original wild-type plants. Whether these transgenic plants will have any beneficial properties, for example, greater ease of lignin removal for pulp/paper applications, is unclear. The pigmentation effects observed were not anticipated by the researchers involved and point to the fact that attempts to alter lignin-forming processes must also take into account the related biochemical pathways that utilize the same substrates.

The finding that lignin formation proper is somehow temporally and spatially associated with various presumedly proline-rich proteins and dirigent sites holds much promise. Full details of the influence of these proteins on lignin structure may result in the design of new strategies for modifying both lignin deposition and structure. The discovery of dirigent proteins, pinoresinol/lariciresinol reductases, and their corresponding genes also affords the opportunity to pursue various interesting questions, including how heartwood is formed.

Advances in lignan and (iso)flavonoid biochemistry and molecular biology offer the opportunity to modify concentrations of health protectants and pharmacologically active species in particular plants of choice. Eventually, we should be able to engineer the formation of secoisolariciresinol, matairesinol, daidzein, genistein, and similar compounds in staple crops that do not ordinarily produce them in significant quantities. The corresponding transgenic plants thus would provide long-term health benefits as sources of cancer preventives. A similar target for enhanced production might be podophyllotoxin, one of a handful of plant anticancer compounds already in use today.

The potential being unleashed is perhaps most vividly demonstrated by the impressive advances in plant metabolic engineering seen in the manipulation of flower color by application of sense/antisense technologies. Several laboratories in Europe and New Zealand have successfully transformed various plants such as petunia to alter petal color.

Lastly, knowledge of these pathways will eventually lead to the systematic modification and improvement of plant flavors and fragrances, the properties of which define the very essence of many of our foodstuffs, such as pepper, ginger, and vanilla.

Coffee beans

Chlorogenic acid

Cloves

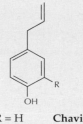

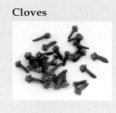

R = H   Chavicol
R = OCH₃ Eugenol

Cinnamon bark

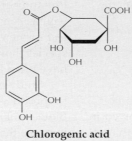

Cinnamaldehyde

Nutmeg

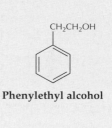

R = H   Safrole
R = OCH₃ Myristicin

Ginger rhizome

n=4,6,8

Gingerols

Orchid

Phenylethyl alcohol

Red and black peppers

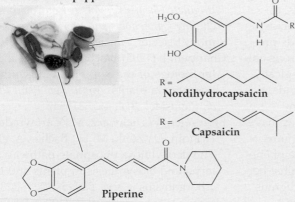

R =
Nordihydrocapsaicin

R =
Capsaicin

Piperine

Vanilla

Vanillin

Green tea

R = H   (−)-Epicatechin
R = OH   (−)-Epigallocatechin

These modifications will ultimately impact the quality of many of our alcoholic and nonalcoholic beverages, which in turn are often largely determined by their aromatic phenolic constituents (Box 24.6).

Summary

Plants produce a great variety of organic compounds that are not directly involved in primary metabolic processes of growth and development. The roles these natural products or secondary metabolites play in plants have only recently come to be appreciated in an analytical context. Natural products appear to function primarily in defense against predators and pathogens and in providing reproductive advantage as attractants of pollinators and seed dispersers. They may also act to create competitive advantage as poisons of rival species.

Most natural products can be classified into three major groups: terpenoids, alkaloids, and phenolic compounds (mostly phenylpropanoids). Terpenoids are composed of five-carbon units synthesized by way of the acetate/mevalonate pathway or the glyceraldehyde 3-phosphate/pyruvate pathway. Many plant terpenoids are toxins and feeding deterrents to herbivores or are attractants of various sorts. Alkaloids are synthesized principally from amino acids. These nitrogen-containing compounds protect plants from a variety of herbivorous animals, and many possess pharmacologically important activity. Phenolic compounds, which are synthesized primarily from products of the shikimic acid pathway, have several important roles in plants. Tannins, lignans, flavonoids, and some simple phenolic compounds serve as defenses against herbivores and pathogens. In addition, lignins strengthen cell walls mechanically, and many flavonoid pigments are important attractants for pollinators and seed dispersers. Some phenolic compounds have allelopathic activity and may adversely influence the growth of neighboring plants.

Throughout the course of evolution, plants have developed defenses against herbivory and microbial attack and produced other natural products to aid competitiveness. The better-defended, more-competitive plants have generated more progeny, and so

the capacity to produce and safely store such ecologically useful metabolites has become widely established in the plant kingdom. Pressures from herbivores and pathogens, as well as constant competition, continue to select for new natural products. In cultivated species, however, such chemical defenses have often been artificially selected against.

Study of the biochemistry of plant natural products has many practical applications. Biotechnological approaches can selectively increase the amounts of defense compounds in crop plants, thereby reducing the need for costly and potentially toxic pesticides. Similarly, genetic engineering can be utilized to increase the yields of pharmaceuticals, flavor and perfumery materials, insecticides, fungicides, and other natural products of commercial value. Although many natural products and their functions have been described in this chapter, the metabolism of natural products in most plant species remains to be elucidated. A great deal of fascinating biochemistry remains to be discovered.

Further Reading

Terpenoids

Cane, D. E., ed. (1999) *Comprehensive Natural Products Chemistry, Vol. 2, Isoprenoids Including Carotenoids and Steroids.* Pergamon/Elsevier, Amsterdam.

Chappell, J. (1995) Biochemistry and molecular biology of the isoprenoid biosynthetic pathway in plants. *Annu. Rev. Plant Physiol. Plant Mol. Biol.* 46: 521–547.

Eisenreich, W., Schwarz, M., Cartayrade, A., Arigoin, D., Zenk, M. H., Bacher, A. (1998) The deoxylulose phosphate pathway of terpenoid biosynthesis in plants and microorganisms. *Chem. Biol.* 5: R221–R233.

Gershenzon, J., Croteau, R. (1993) Terpenoid biosynthesis: the basic pathway and formation of monoterpenes, sesquiterpenes and diterpenes. In *Lipid Metabolism in Plants,* T. S. Moore, Jr., ed. CRC Press, Boca Raton, FL, pp. 339–388.

Harborne, J. B., and Tomas-Barberan, F. A., eds. (1991) *Ecological Chemistry and Biochemistry of Plant Terpenoids.* Clarendon Press, Oxford, UK.

Langenheim, J. H. (1994) Higher plant terpenoids: a phytocentric overview of their ecological roles. *J. Chem. Ecol.* 20: 1223–1280.

Lichtenthaler, H. K. (1999) The 1-deoxy-D-xylulose-5-phosphate pathway of isoprenoid biosynthesis in plants. *Annu. Rev. Plant Physiol. Plant Mol. Biol.* 50: 47–65.

McGarvey, D., Croteau, R. (1995) Terpenoid biosynthesis. *Plant Cell* 7: 1015–1026.

Alkaloids

Cordell, G., ed. (1997) *The Alkaloids*, Vol. 50. Academic Press, San Diego.

Rosenthal, G. A., and Berenbaum, M. R., eds. (1991) *Herbivores: Their Interactions with Secondary Plant Metabolites, Vol. 1: The Chemical Participants*, 2nd ed. Academic Press, San Diego.

Southon, I. W., and Buckingham, J., eds. (1989) *Dictionary of Alkaloids.* Chapman and Hall, London.

Suberin

Bernards, M. A., Lewis, N. G. (1998) The macromolecular aromatic domain in suberized tissues: a changing paradigm. *Phytochemistry* 47: 583–591.

Lignins and lignans

Gang, D. R., Costa, M. A., Fujita, M., Dinkova-Kostova, A. T., Wang, H.-B., Burlat, V., Martin, W., Sarkanen, S., Davin, L. B., Lewis, N. G. (1999) Regiochemical control of monolignol radical coupling: a new paradigm for lignin and lignan biosynthesis. *Chem. Biol.* 6: 143–151.

Lewis, N. G., Davin, L. B. (1999) Lignans: biosynthesis and function. In *Comprehensive Natural Products Chemistry, Vol. 1, Polyketides and Other Secondary Metabolites Including Fatty Acids and Their Derivatives*, D.H.R. Barton, K. Nakanishi, and O. Meth-Cohn, eds.-in-chief. Elsevier, Amsterdam, pp. 639–712.

Lewis, N. G., Davin, L. B., Sarkanen, S. (1999) The nature and functions of lignins. In *Comprehensive Natural Products Chemistry, Vol. 3, Carbohydrates and Their Derivatives Including Tannins, Cellulose and Related Lignins*, D.H.R. Barton, K. Nakanishi, and O. Meth-Cohn, eds.-in-chief. Elsevier, Amsterdam, pp. 617–745.

Sarkanen, S., Lewis, N. G. (1998) Biosynthesis of lignins and lignans. *ACS Symp. Ser.* 697: 1–421.

Timell, T. E. (1986) *Compression Wood in Gymnosperms,* Vols. 1–3. Springer-Verlag, Berlin.

Flavonoids

Dixon, R. A. (1999) Isoflavonoids: biochemistry, molecular biology and biological functions. In *Comprehensive Natural Products Chemistry, Vol. 1, Polyketides and Other Secondary Metabolites Including Fatty Acids and Their Derivatives*, D.H.R. Barton, K. Nakanishi, and O. Meth-Cohn, eds.-in-chief. Elsevier, Amsterdam, pp. 773–824.

Forkmann, G. (1991) Flavonoids as flower pigments: the formation of the natural spectrum and its extension by genetic engineering. *Plant Breed.* 106: 1–26.

Forkmann, G., Heller, W. (1999) Biosynthesis of flavonoids. In *Comprehensive Natural Products Chemistry, Vol. 1, Polyketides and Other Secondary Metabolites Including Fatty Acids and Their Derivatives*, D.H.R. Barton, K. Nakanishi, and O. Meth-Cohn, eds.-in-chief Elsevier, Amsterdam, pp. 713–748.

Harborne, J. B., ed. (1994) *The Flavonoids: Advances in Research Since 1986.* Chapman and Hall, London.

Coumarins and furanocoumarins

Berenbaum, M. R., and Zangerl, A. R. (1996) Phytochemical diversity: adaptation or random variation. *Rec. Adv. Phytochem.* 30: 1–24.

Keating, G. J., O'Kennedy, R. (1997) The chemistry and occurrence of coumarins. In *Coumarins: Biology, Application and Mode of Action*, R. O'Kennedy and R. D. Thornes, eds. John Wiley & Sons, Chichester, pp. 23–66.

Matern, U., Lüer, P., Kreusch, D. (1999) Biosynthesis of coumarins. In *Comprehensive Natural Products Chemistry, Vol. 1, Polyketides and Other Secondary Metabolites Including Fatty Acids and Their Derivatives*, D.H.R. Barton, K. Nakanishi, and O. Meth-Cohn, eds.-in-chief. Elsevier, Amsterdam, pp. 623–638.

Matern, U., Strasser, H., Wendorff, H., Hamerski, D. (1988) Coumarins and furanocoumarins. In *Cell Culture and Somatic Cell Genetics of Plants*, Vol. 5, I. K. Vasil, ed. Academic Press, Orlando, FL, pp. 3–21.

Zobel, A. M. (1997) Coumarins in fruits and vegetables. *Proc. Phytochem. Soc. Eur.* 41: 173–203.

Stilbenes, styrylpyrones, and arylpyrones

Berkert, C., Horn, C., Schnitzler, J.-P., Lehning, A., Heller, W., Veit, M. (1997) Styrylpyrone biosynthesis in *Equisetum arvense. Phytochemistry* 44: 275–283.

Gorham, J., Tori, M., Asakawa, Y. (1995) *The Biochemistry of the Stilbenoids, Vol. 1, Biochemistry of Natural Products Series*, J. B. Harborne, ed. Chapman and Hall, London.

Schröder, J. (1999) The chalcone/stilbene synthase-type family of condensing enzymes. In *Comprehensive Natural Products Chemistry, Vol. 1, Polyketides and Other Secondary Metabolites Including Fatty Acids and Their Derivatives*, D.H.R. Barton, K. Nakanishi, and O. Meth-Cohn, eds.-in-chief. Elsevier, Amsterdam, pp. 749–772.

Sources and Credits

The publisher gratefully acknowledges the generous authors and publishers who have granted permission to reprint or adapt their works for inclusion in this volume. For readers' ease in locating specific materials, the following sources and credits are listed by chapter, with text figures cited first, followed by boxed figures and then tables.

Part 1 opening photograph:
Galen Rowell/Mountain Light Photography Emeryville, CA.

Chapter 1

Figure 1.1B: Staehelin, L. A., Giddings, T. H. Jr., Kiss, J. Z., Sack, F. D. (1990) Macromolecular differentiation of Golgi stacks in root tips of *Arabidopsis* and *Nicotiana* seedlings as visualized in high-pressure frozen and freeze-substituted samples. *Protoplasma* 157: 75–91. Copyright © 1990 Springer-Verlag.

Figure 1.11: A. L. Samuels, University of British Columbia, Vancouver, Canada; previously unpublished.

Figure 1.12: Reproduced from Tilney, L., Cooke, T. J., Connelly, P. S., Tilney, M. S. (1991) The structure of plasmodesmata as revealed by plasmolysis, detergent extraction, and protease digestion. *J. Cell Biol.* 112: 739–748. By copyright permission of the Rockefeller University Press.

Figure 1.13A: Oparka, K., Prior, D.A.M., Crawford, J. W. (1994) Behaviour of plasma membrane, cortical ER and plasmodesmata during plasmolysis of onion epidermal cells. *Plant Cell Environ.* 17: 163–171. Used with permission of Blackwell Science Ltd.

Figure 1.16A,B: Knebel, W., Quader, H., Schnepf, E. (1990) Mobile and immobile endoplasmic reticulum in onion bulb epidermis cells: short-term and long-term observations with a confocal laser scanning microscope. *Eur. J. Cell Biol.* 52: 328–340. Used with permission from Urban & Fischer Verlag, Niederlassung Jena, Germany.

Figure 1.17A,B: Reproduced from Bonnett, H. T., Newcomb, E. H. (1965) Polyribosomes and cisternal accumulations in root cells of radish. *J. Cell Biol.* 27: 423–432. By copyright permission of the Rockefeller University Press.

Figure 1.18: G. W. Turner, Washington State University, Pullman; previously unpublished.

Figure 1.19: Lichtscheidl, I. K., Lancelle, S. A., Hepler, P. K. (1990) Actin–endoplasmic reticulum complexes in *Drosera*: their structural relationship with the plasmalemma, nucleus, and organelles in cells prepared by high pressure freezing. *Protoplasma* 155: 116–126. Copyright © 1990 Springer-Verlag.

Figure 1.20B: Fernandez, D., Staehelin, L. A. (1987) Does gibberellic acid induce the transfer of lipase from protein bodies to lipid bodies in barley aleurone cells? *Plant Phys.* 85: 487–496. Copyright © 1988 Gustav Fischer Verlag.

Figure 1.21B: B. A. Larkins, University of Arizona, Tucson; previously unpublished.

Figure 1.22: Craig, S., Staehelin, L. A. (1988) High-pressure freezing of intact plant tissues: evaluation and characterization of novel features of the endoplasmic reticulum and associated membrane systems. *Eur. J. Cell Biol.* 46: 81–93. Used with permission from Urban & Fischer Verlag, Niederlassung Jena, Germany.

Figure 1.23B: Staehelin, L. A., Giddings, T. H. Jr., Kiss, J. Z., Sack, F. D. (1990) Macromolecular differentiation of Golgi stacks in root tips of *Arabidopsis* and *Nicotiana* seedlings as visualized in high-pressure frozen and freeze-substituted samples. *Protoplasma* 157: 75–91. Copyright © 1990 Springer-Verlag.

Figure 1.24: Craig, S., Staehelin, L. A. (1988) High-pressure freezing of intact plant tissues: evaluation and characterization of novel features of the endoplasmic reticulum and associated membrane systems. *Eur. J. Cell Biol.* 46: 81–93. Used with permission from Urban & Fischer Verlag, Niederlassung Jena, Germany.

Figure 1.25: A. Nebenfuehr, University of Colorado, Boulder; previously unpublished.

Figure 1.26: Craig, S., Staehelin, L. A. (1988) High-pressure freezing of intact plant tissues: evaluation and characterization of novel features of the endoplasmic reticulum and associated membrane systems. *Eur. J. Cell Biol.* 46: 81–93. Used with permission from Urban & Fischer Verlag, Niederlassung Jena, Germany.

Figure 1.27: Staehelin, L. A., Giddings, T. H. Jr., Kiss, J. Z., Sack, F. D. (1990) Macromolecular differentiation of Golgi stacks in root tips of *Arabidopsis* and *Nicotiana* seedlings as visualized in high-pressure frozen and freeze-substituted samples. *Protoplasma* 157: 75–91. Copyright © 1990 Springer-Verlag.

Figure 1.29A: Staehelin, L. A., Giddings, T. H. Jr., Kiss, J. Z., Sack, F. D. (1990) Macromolecular differentiation of Golgi stacks in root tips of *Arabidopsis* and *Nicotiana* seedlings as visualized in high-pressure frozen and freeze-substituted samples. *Protoplasma* 157: 75–91. Copyright © 1990 Springer-Verlag.

Figure 1.29B: Y. Mineyuki, Hiroshima University, Japan; previously unpublished.

Figure 1.30: Craig, S., Staehelin, L. A. (1988) High-pressure freezing of intact plant tissues: evaluation and characterization of novel features of the endoplasmic reticulum and associated membrane systems. *Eur. J. Cell Biol.* 46: 81–93. Used with permission from Urban & Fischer Verlag, Niederlassung Jena, Germany.

Figure 1.31A,B: Fowke, L. C., Tanchak, M. A., Galway, M. E. (1991) Ultrastructural cytology of the endocytotic pathway in plants. In *Endocytosis, Exocytosis and Vesicle Traffic in Plants*, C. R. Hawes, J.O.D. Coleman, and D. E. Evans, eds. Cambridge University Press, Cambridge, UK, pp. 15–40. Reprinted with the permission of Cambridge University Press.

Figure 1.32: L. A. Staehelin, University of Colorado, Boulder; previously unpublished.

Figure 1.33: L. A. Staehelin, University of Colorado, Boulder; previously unpublished.

Figure 1.34: P. C. Bethke, University of California, Berkeley; previously unpublished.

Figure 1.35: Micrograph: E. H. Newcomb, University of Wisconsin, Madison; previously unpublished.

Figure 1.36: L. A. Staehelin, University of Colorado, Boulder; previously unpublished.

Figure 1.37B: T. Murata, University of Tokyo, Japan; previously unpublished.

Figure 1.38: E. H. Newcomb, University of Wisconsin, Madison; previously unpublished.

Figure 1.39A,B: S. E. Frederick, Mount Holyoke College, South Hadley, MA; previously unpublished.

Figure 1.40A: Frederick, S. E., Newcomb, E. H. (1975) Plant microbodies. *Protoplasma* 84: 1–29. Copyright © 1975 Springer-Verlag.

Figure 1.41B: P. J. Gruber, Mt. Holyoke College, South Hadley, MA; previously unpublished.

Figure 1.43: R. H. Köhler, Cornell University, Ithaca, NY; previously unpublished.

Figure 1.45: E. H. Newcomb, University of Wisconsin, Madison; previously unpublished.

Figure 1.46: W. P. Wergin, Agricultural Research Service, Beltsville, MD; previously unpublished.

Figure 1.47: Source: Smith, J. D., Todd, P., Staehelin, L. A. (1997) Modulation of statolith mass and grouping in white clover *Trifolium repens* growth in 1-g, microgravity and on the Clinostat. *Plant J.* 12: 1361–1373. Used with permission from Blackwell Science Ltd.

Figure 1.48: G. W. Turner, Washington State University, Pullman; previously unpublished.

Figure 1.49A: S. E. Frederick, Mount Holyoke College, South Hadley, MA; previously unpublished.

Figure 1.50: L. A. Staehelin, University of Colorado, Boulder; previously unpublished.

Figure 1.51: W. P. Wergin, Agricultural Research Service, Beltsville, MD; previously unpublished.

Figure 1.52: W. P. Wergin, Agricultural Research Service, Beltsville, MD; previously unpublished.

Figure 1.53: W. P. Wergin, Agricultural Research Service, Beltsville, MD; previously unpublished.

Figure 1.54: L. A. Staehelin, University of Colorado, Boulder; previously unpublished.

Figure 1.56: L. A. Staehelin, University of Colorado, Boulder; previously unpublished.

Figure 1.57: P. J. Gruber, Mt. Holyoke College, South Hadley, MA; previously unpublished.

Figure 1.58: E. H. Newcomb, University of Wisconsin, Madison; previously unpublished.

Figure 1.61: M. E. Doohan, Texas Tech University, Lubbock; previously unpublished.

Figure 1.62: M. E. Doohan, Texas Tech University, Lubbock; previously unpublished.

Chapter 2

Figure 2.1: N. J. Stacey, John Innes Centre, Norwich, UK; previously unpublished.

Figure 2.2A–C: K. Findlay, John Innes Centre, Norwich, UK; previously unpublished.

Figure 2.2D: P. Linstead, John Innes Centre, Norwich, UK; previously unpublished.

Figure 2.3: M. Bush, Duke University, Durham, NC; previously unpublished.

Figure 2.4B: Ledbetter, M. C., Porter, K. R. (1970) *Introduction to the Fine Structure of Plant Cells.* Springer-Verlag, New York. Copyright © 1970 Springer-Verlag.

Figure 2.5A,B: Knox, J. P., Day, S., Roberts, K. (1989) A set of cell surface glycoproteins forms an early marker of cell position, but not cell type, in the root apical meristem of *Daucus carota* L. *Development* 106: 47–56. Used with permission from Company of Biologists Ltd.

Figure 2.5C: P. Linstead, John Innes Centre, Norwich, UK; previously unpublished.

Figure 2.6: R. L. Nicholson, Purdue University, West Lafayette, IN; previously unpublished.

Figure 2.7: R. Cotter, Plant Gene Expression Center, Albany, CA; previously unpublished.

Figure 2.11B: Reproduced from Mueller, S. C., Brown, R. M. Jr. (1980) Evidence for an intramembrane component associated with a cellulose microfibril–synthesizing complex in higher plants. *J. Cell Biol.* 84: 315–326. By copyright permission of the Rockefeller University Press.

Figure 2.11E: M. Jarvis, Glasgow University, Scotland; previously unpublished.

Figure 2.13: The relationship of the orders is adapted from Dahlgren, G. (1989) An updated angiosperm classification. *Bot. J. Linn. Soc.* 100: 197–204. Reprinted by permission of Academic Press. Copyright © 1989.

Figure 2.18B: Stafstrom, J. P., Staehelin, L. A. (1986) The role of carbohydrate in maintaining extensin in an extended conformation. *Plant Physiol.* 81: 242–246.

Figure 2.18C: Reprinted by permission from Condit, C. M., Meagher, R. B. (1986) A gene encoding a novel glycine-rich structural protein of petunia. *Nature* 323: 178–181. Copyright © 1986 Macmillan Magazines Ltd.

Figure 2.19A–D: Ye, Z.-H., Varner, J. E. (1991) Tissue-specific expression of cell wall proteins in developing soybean tissues. *Plant Cell* 3: 23–38.

Figure 2.20D: Gane, A. M., Clarke, A. E., Bacic, A. (1995) Localisation and expression of arabinogalactan-proteins in the ovaries of *Nicotiana alata* Link and Otto. *Sex. Plant Reprod.* 8: 278–282. Copyright © 1995 Springer-Verlag.

Figure 2.22: Adapted from Iiyama, K., Lam, T.B.-T., Stone, B. A. (1994) Covalent cross-links in the cell wall. *Plant Physiol.* 104: 315–320.

Figure 2.23A: N. Carpita, Purdue University, West Lafayette, IN; previously unpublished.

Figure 2.26B: Gunning, B.E.S., Steer, M. W. (1996) *Plant Cell Biology: Structure and Function.* Jones and Bartlett, Sudbury, MA. Copyright © 1996 and used by permission from Gustav Fischer Verlag.

Figure 2.26C: Zhang, G. F., Staehelin, L. A. (1992) Functional compartmentation of the Golgi apparatus of plant cells: immunocytochemical analysis of high-pressure frozen and freeze-substituted sycamore maple suspension culture cells. *Plant Physiol.* 99: 1070–1083.

Figure 2.30: Adapted from Gibeaut, D. M., Carpita, N. C. (1993) Synthesis of (1→3), (1→4)β-D-glucan in the Golgi apparatus of maize coleoptiles. *Proc. Natl. Acad. Sci. USA* 90: 3850–3854. Copyright © 1993 National Academy of Sciences, U.S.A.

Figure 2.31A–C: Reproduced from Giddings, T. H. Jr., Brower, D. L., Staehelin, L. A. (1980) Visualization of particle complexes in the plasma membrane of *Micrasterias denticulata* associated with the formation of cellulose fibrils in primary and secondary cell walls. *J. Cell Biol.* 84: 327–339. By copyright permission of the Rockefeller University Press.

Figure 2.31D: Herth, W. (1983) Arrays of plasma-membrane "rosettes" involved in cellulose microfibril formation of *Spirogyra*. *Planta* 157: 347–356. Copyright © 1983 Springer-Verlag.

Figure 2.31E,F: Montezinos, D., Brown, M. Jr. (1976) Surface architecture of the plant cell: biogenesis of the cell wall, with special emphasis on the role of the plasma membrane in cellulose biosynthesis. *J. Supramol. Struct.* 5: 277–290. Reprinted by permission of Wiley-Liss, Inc., a subsidiary of John Wiley & Sons, Inc.

Figure 2.31G,H: Grimson, M. J., Haigler, C. H., Blanton, R. L. (1996) Cellulose microfibrils, cell motility, and plasma membrane protein organization change in parallel during culmination in *Dictyostelium discoideum*. *J. Cell Sci.* 109: 3079–3087. Used with permission from Company of Biologists Ltd.

Figure 2.31I–K: Herth, W. (1985) Plasma-membrane rosettes involved in localized wall thickening during xylem vessel formation of *Lepidium sativum*. *Planta* 164: 12–21. Copyright © 1985 Springer-Verlag.

Figure 2.36: Adapted from Cosgrove, D. J. (1993) How do plant cell walls extend? *Plant Physiol.* 102: 1–6.

Figure 2.37: Adapted from McQueen-Mason, S., Durachko, D. M., Cosgrove, D. J. (1992) Two endogenous proteins that induce cell wall extension in plants. *Plant Cell* 4: 1425–1433.

Figure 2.39A,B: M. Bush, Duke University, Durham, NC; previously unpublished.

Figure 2.39C,D: Stacy, N. J., Roberts, K., Carpita, N. C., Wells, B., McCann, M. C. (1995) Dynamic changes in cell surface molecules are very early events in the differentiation of mesophyll cells from *Zinnia elegans* into tracheary elements. *Plant J.* 8: 891–906. Used with permission from Blackwell Science Ltd.

Figure 2.40A–C: Bowes, B. G. (1996) *A Colour Atlas of Plant Structure.* Manson Publishers, London. Copyright © 1996 Manson Publishers Ltd.

Figure 2.41A: Beasley, C. A. (1975) Developmental morphology of cotton flowers and seed as seen with the scanning electron microscope. *Am. J. Bot.* 62: 584–592. Used with permission from the Botanical Society of America, Inc. Copyright © 1975.

Figure 2.41B–D: Ledbetter, M. C., Porter, K. R. (1970) *Introduction to the Fine Structure of Plant Cells.* Springer-Verlag, New York. Copyright © 1970 Springer-Verlag.

Figure 2.42: Micrographs reprinted from Falconer, M. M., Seagull, R. W. (1985) Immunofluorescent and Calcoofluor white staining of developing tracheary elements in *Zinnia elegans* suspension cultures. *Protoplasma* 125: 190–198. Copyright © 1985 Springer-Verlag.

Figure 2.43: Bowes, B. G. (1996) *A Colour Atlas of Plant Structure.* Manson Publishers, London. Copyright © 1996 Manson Publishers Ltd.

Figure 2.44A: Dharmawardhana, D P., Ellis, B. E., Carlson, J. E. (1992) Characterization of vascular lignification in *Arabidopsis-thaliana*. *Can. J. Bot.* 70: 2238–2244. Used with permission from NRC Research Press. Copyright © 1992.

Figure 2.44B: Bowes, B. G. (1996) *A Colour Atlas of Plant Structure.* Manson Publishers, London. Copyright © 1996 Manson Publishers Ltd.

Figure 2.44C: Zhong, R., Taylor, J. J., Ye, Z.-H. (1997) Disruption of interfascicular fiber differentiation in an *Arabidopsis* mutant. *Plant Cell* 9: 2159–2170.

Figure 2.44D,E: Chapple, C.C.S., Vogt, T., Ellis, B. E., Somerville, C. R. (1992) An *Arabidopsis* mutant defective in the general phenylpropanoid pathway. *Plant Cell* 4: 1413–1424.

Figure 2.45A: D. DeMason, University of California, Riverside; previously unpublished.

Figure 2.45B: M. Buckeridge, Instituto de Botanica, Secao de Fisiologia e Bioquimica Plantas, Sao Paulo, SP, Brasil; previously unpublished.

Figure 2.46A,B: McCann, M. C., Stacey, N. J., Wilson, R., Roberts, K. (1993) Orientation of macromolecules in the walls of elongating carrot cells. *J. Cell Sci.* 106: 1347–1356. Used with permission from Company of Biologists Ltd.

Figure 2.46C: J. Labavitch, University of California, Davis; previously unpublished.

Figure 2.46D: D. Delmer, University of California, Davis; previously unpublished.

Figure 2.46E: C. Haigler, Texas Tech University, Lubbock; previously unpublished.

Box 2.1B: N. Carpita, Purdue University, West Lafayette, IN; previously unpublished.

Box 2.1D: Carpita, N. C., Shea, E. M. (1989) Linkage structure of carbohydrates by gas chromatography–mass spectrometry (GC-MS) of partially methylated alditol acetates. In *Analysis of Carbohydrates by GLC and MS*, C. Biermann and G. D. McGinnis, eds. CRC Press, Boca Raton, FL, pp. 157–216. Copyright © 1989 CRC Press.

Box 2.2D,E: W. York, University of Georgia, Athens; previously unpublished.

Box 2.3: N. Carpita, Purdue University, West Lafayette, IN; previously unpublished.

Box 2.4A: Hooke, R. (1664) *Micrographia*. The Council of the Royal Society of London for Improving Natural Knowledge, London. Reprinted by permission of the President and Council of the Royal Society.

Box 2.4B: McCann, M. C., Wells, B., Roberts, K. (1990) Direct visualization of cross-links in the primary plant cell wall. *J. Cell Sci.* 96: 323–334. Used with permission from Company of Biologists Ltd.

Box 2.4C: McCann, M. C., Roberts, K., Wilson, R. H., Gidley, M. J., Gibeaut, D. M., Kim, J.-B., Carpita, N. C. (1995) Old and new ways to probe plant cell-wall architecture. *Can. J. Bot.* 73 (Suppl. 1): S103–S113. Used with permission from NRC Research Press. Copyright © 1995.

Box 2.4D: Ledbetter, M. C., Porter, K. R. (1970) *Introduction to the Fine Structure of Plant Cells*. Springer-Verlag, New York. Copyright © 1970 Springer-Verlag.

Box 2.4E: B. Wells, John Innes Center, Norwich, UK; previously unpublished.

Box 2.4F: V. Morris, Institute of Food Research, Norwich, UK; previously unpublished.

Box 2.5: Table adapted from Knox, J. P. (1989) The use of antibodies to study the architecture and developmental regulation of plant cell walls. *Int. Rev. Cytol.* 171: 79–120. Reprinted by permission of Academic Press. Copyright © 1989.

Box 2.5A: Steele, N. M., McCann, M. C., Roberts, K. (1997) Pectin modification in cell walls of ripening tomatoes occurs in distinct domains. *Plant Physiol.* 114: 373–381.

Box 2.5B: McCann, M. C., Roberts, K., Wilson, R. H., Gidley, M. J., Gibeaut, D. M., Kim, J.-B., Carpita, N. C. (1995) Old and new ways to probe plant cell-wall architecture. *Can. J. Bot.* 73 (Suppl.): S103–S113. Used with permission from NRC Research Press. Copyright © 1995.

Box 2.5C: Zhang, G. F., Staehelin, L. A. (1992) Functional compartmentation of the Golgi apparatus of plant cells: immunocytochemical analysis of high-pressure frozen and freeze-substituted sycamore maple suspension culture cells. *Plant Physiol.* 99: 1070–1083.

Box 2.5D,E: M. C. McCann, John Innes Center, Norwich, UK; previously unpublished.

Chapter 3

Figure 3.2: Johansson, F., Sommarin, M., Larsson, C. (1994) Rapid purification of the plasma membrane H$^+$-ATPase in its non-activated form using FPLC. *Physiol. Plant.* 92: 389–396. Copyright © 1994 Physiologium Plantarum and published by Munksgaard International Publishers, Ltd., Copenhagen, Denmark.

Figure 3.3: Reprinted from Junge, W., Lill, H., Engelbrecht, S. (1997) ATP synthase: an electrochemical transducer with rotatory mechanics. *Trends Biochem. Sci.* 22: 420–423. Copyright © 1997 with permission from Elsevier Science.

Figure 3.4: Adapted from (1997) ATP generation by photosynthesis. In *Plant Biochemistry and Molecular Biology*, H.-W. Heldt. Oxford University Press, London. Used with permission from Oxford University Press.

Figure 3.7A: DeWitt, N. D., Sussman, M. R. (1995) Immunocytological localization of an epitope-tagged plasma membrane proton pump (H$^+$-ATPase) in phloem companion cells. *Plant Cell* 7: 2053–2067.

Figure 3.7B,C: Harper, J. F., Manney, L., Sussman, M. R. (1994) The plasma membrane H$^+$-ATPase gene family in *Arabidopsis*: genomic sequence of AHA10 which is expressed primarily in developing seeds. *Mol. Gen. Genet.* 244: 572–587. Copyright © 1994 Springer-Verlag.

Figure 3.11: Adapted with permission from Finbow, M. E., Harrison, M. A. (1997) The vacuolar H$^+$-ATPase: a universal proton pump of eukaryotes. *Biochem. J.* 324: 697–712. Copyright © 1997 The Biochemical Society.

Figure 3.14A,B: Data adapted from Lu, Y.-P., Li, Z.-S., Drozdowicz, Y. M., Hoertensteiner, S., Martinoia, E., Rea, P. A. (1998) AtMRP2, an *Arabidopsis* ATP binding cassette transporter able to transport gluathione *S*-conjugates and chlorophyll catabolites: functional comparisons with AtMRP1. *Plant Cell* 10: 267–282.

Figure 3.17: Graphs: D. Sanders, University of York, UK; previously unpublished.

Figure 3.19A,B: Stadler, R., Sauer, N. (1996) The *Arabidopsis thaliana* AtSUC2 gene is specifically expressed in companion cells. *Bot. Acta* 109: 299–306. Copyright © 1996 Georg Thieme Verlag, Stuttgart–New York.

Figure 3.20A,B: Reprinted with permission from Kühn, C., Franceschi, V. R., Schulz, A., Lemoine, R., Frommer, W. B. (1997) Macromolecular trafficking indicated by localization and turnover of sucrose transporters in enucleate sieve elements. *Science* 275: 1298–1302. Copyright © 1997 American Association for the Advancement of Science.

Figure 3.22: Kim, E. J., Kwak, J. M., Uozumi, N., Schroeder, J. I. (1998) AtKUP1: an *Arabidopsis* gene encoding high-affinity potassium transport activity. *Plant Cell* 10: 51–62.

Figure 3.25A: Roblin, G. (1979) *Mimosa pudica*: a model for the study of the excitability in plants. *Biol. Rev.* 54: 135–153. Reprinted with the permission of Cambridge University Press.

Figure 3.25B: Abe, T., Oda, K. (1976) Resting and action potentials of excitable cells in the main pulvinus of *Mimosa pudica*. *Plant Cell Physiol.* 17: 1343–1346. Reprinted with permission from the Japanese Society of Plant Physiologists.

Figure 3.28: Adapted from Pantoja, O., Dainty, J., Blumwald, E. (1989) Ion channels in vacuoles from halophytes and glycophytes *FEBS Lett.* 255: 92–96. Copyright © 1989 with permission from Elsevier Science.

Figure 3.29: Adapted from Pantoja, O., Dainty, J., Blumwald, E. (1989) Ion channels in vacuoles from halophytes and glycophytes *FEBS Lett.* 255: 92–96. Copyright © 1989 with permission from Elsevier Science.

Figure 3.32: Roberts, S. K., Tester, M. (1997) A patch clamp study of Na$^+$ transport in maize roots. *J. Exp. Bot.* 48 (Special Issue): 431–440. Used with permission from Oxford University Press.

Figure 3.33A: Nakamura, R. L., McKendree, W. L. Jr., Hirsch, R. E., Sedbrook, J. C., Gaber, R. F., Sussman, M. R. (1995) Expression of an *Arabidopsis* potassium channel gene in guard cells. *Plant Physiol.* 109: 371–374.

Figure 3.33B: Lagarde, D., Basset, M., Lepetit, M., Conejero, G., Gaymard, R., Astruc, S., Grignon, C. (1996) Tissue-specific expression of *Arabidopsis* AKT1 gene is consistent with a role in K$^+$ nutrition. *Plant J.* 9: 195–203. Used with permission from Blackwell Science Ltd.

Figure 3.35: Adapted and used with permission from Hirsch, R. E., Lewis, B. D., Spalding, E. P., Sussman, M. R. (1998) A role for the AKT1 potassium channel in plant nutrition. *Science* 280: 918–921. Copyright © 1998 American Association for the Advancement of Science.

Figure 3.37: Reprinted by permission from Kreusch, A., Pfaffinger, P. J., Stevens, C. F., Choe, S. (1998) Crystal structure of the tetramerization domain of the *Shaker* potassium channel. *Nature* 392: 945–948. Copyright © 1998 Macmillan Magazines Ltd.

Figure 3.39: Czempinski, K., Zimmerman, S., Ehrhardt, T., Müller-Röber, B. (1997) New structure and function in plant K$^+$ channels: KCO$_1$, an outward rectifier with a steep Ca^{2+} dependency. *EMBO J.* 16: 2565–2575. Used with permission from Oxford University Press.

Figure 3.40: Roberts, S. K., Tester, M. (1997) A patch clamp study of Na$^+$ transport in maize roots. *J. Exp. Bot.* 48 (Special Issue): 431–440. Used with permission from Oxford University Press.

Figure 3.41A,B: Allen, G. J., Amtmann, A., Sanders, D. (1998) Calcium-dependent and calcium-independent K$^+$ mobilization channels in *Vicia faba* guard cell vacuoles. *J. Exp. Bot.* 49 (Special Issue): 305–318. Used with permission from Oxford University Press.

Figure 3.42: Adapted from Bethke, P. G., Jones, G., Jones, R. L. (1995) Calcium and plant hormone action (Fig. 1). In *Plant Hormones—Physiology, Biochemistry, and Molecular Biology*, P. J. Davies, ed. Kluwer Academic Publishers, Dordrecht, The Netherlands, pp. 298–317. Used with kind permission from Kluwer Academic Publishers.

Figure 3.43: Adapted from Piñeros, M., Tester, M. (1995) Characterization of a voltage-dependent Ca^{2+}-selective channel from wheat roots. *Planta* 195: 478–488. Copyright © 1995 Springer-Verlag.

Figure 3.45: Bethke, P., Jones, R. L. (1994) Ca^{2+}-calmodulin modulates ion channel activity in storage protein vacuoles of barley aleurone cells. *Plant Cell* 6: 277–285.

Figure 3.46: Schroeder, J. I., Keller, B. U. (1992) Two types of anion channel currents in guard cells with distinct voltage regulation. *Proc. Natl. Acad. Sci. USA* 89: 5052–5029. Used with permission.

Figure 3.47: Adapted from Cerana, R., Giromini, L., Colombo, R. (1995) Malate-regulated channels permeable to anions in vacuoles of *Arabidopsis thaliana*. *Aust. J. Plant Physiol.* 22: 115–121. Used with permission from Commonwealth Scientific and Industrial Research Organization, Melbourne.

Figure 3.53: Reprinted by permission from Walz, T., Hirai, T., Murata, K., Heymann, J. B., Mitsuoka, K., Fuijyoshi, Y., Smith, B. L., Agre, P., Engel, A. (1997) The three-dimensional structure of aquaporin-1. *Nature* 387: 624–627. Copyright © 1997 Macmillan Magazines Ltd.

Box 3.1: D. Sanders, University of York, UK; previously unpublished.

Box 3.2A–C: Adapted from Uozumi, N., Tatsunosuke, N., Schroeder, J. I., Muto, S. (1998) Determination of transmembrane topology of an inward-rectifying potassium channel from *Arabidopsis thaliana* based on functional expression in *Escherichia coli*. *Proc. Natl. Acad. Sci. USA* 95: 9773–9778. Copyright © 1998 National Academy of Sciences, U.S.A.

Box 3.6: From product literature for the GeneClamp 500. Axon Instruments Inc., Foster City, CA.

Box 3.7B,C: Patch clamp tower: Used with permission from Adams & List Associates Ltd., Great Neck, NY.

Box 3.8A,B: Roberts, S. K., Tester, M. (1997) Permeation of Ca^{2+} and monovalent cations through an outwardly rectifying channel in maize root stelar cells. *J. Exp. Bot.* 48: 839–846. Used with permission from Oxford University Press.

Chapter 4

Figure 4.5B: K. Keegstra, Michigan State University, East Lansing; previously unpublished.

Figure 4.8A–C: M. Lee, R. Trelease, Arizona State University, Tempe; previously unpublished.

Figure 4.10A: A. Heese, N. Raikhel, Michigan State University, East Lansing; previously unpublished.

Figure 4.11A,B: Varagona, M. J., Schmidt, R. J., Raikhel, N. V. (1992) Nuclear localization signal(s) required for nuclear targeting of the maize regulatory protein Opaque-2. *Plant Cell* 4: 1213–1227.

Figure 4.11C,D: Hicks, G. R., Smith, H. M., Lobreaux, S., Raikhel, N. V. (1996) Nuclear import in permeabilized protoplasts from higher plants has unique features. *Plant Cell* 8:1337–1352.

Figure 4.23A,B: Reproduced from Levanony, J., Rubin, R., Altschuler, Y., Galili, G. (1992) Evidence for a novel route of wheat storage proteins to vacuoles. *J. Cell Biol.* 119: 1117–1128. By copyright permission of the Rockefeller University Press.

Figure 4.25: J. Dombrowski, N. Raikhel, Michigan State University, East Lansing; previously unpublished.

Figure 4.26: Bassham, D. C., Gal, S., Concaiçao, A. D., Raikhel, N. V. (1995) An *Arabidopsis* syntaxin homologue isolated by functional complementation of a yeast pep12 mutant. *Proc. Natl. Acad. Sci. USA* 92: 7262–7266. Copyright © 1995 National Academy of Sciences, U.S.A.

Figure 4.31: D. Robinson, University of Goettingen, Germany; previously unpublished.

Box 4.1: Photomicrograph from Herman, E. M., Tague, B. W., Hoffman, L. M., Kjemtrup, S. E., Chrispeels, M. J. (1990) Retention of phytohemagglutinin with carboxy-terminal tetrapeptide KDEL in the nuclear envelope and the endoplasmic reticulum. *Planta* 182: 305–312. Copyright © 1990 Springer-Verlag.

Box 4.4A–D: D. Robinson, University of Goettingen, Germany; previously unpublished.

Chapter 5

Figure 5.1: T. Baskin, University of Missouri, Columbia; previously unpublished.

Figure 5.2: Xu, P., Lloyd, C. W., Staiger, C. J., Drøbak, B. K. (1992) Association of phosphatidylinositol 4-kinase with the plant cytoskeleton. *Plant Cell* 4: 941–951.

Figure 5.6A: Reprinted by permission from Aebi, U., Cohn, J., Buhle, L., Gerace, L. (1986) The nuclear lamina is a meshwork of intermediate-type filaments. *Nature* 323: 560–564. Copyright © 1986 Macmillan Magazines Ltd.

Figure 5.7: Mizuno, K. (1995) A cytoskeletal 50kDa protein in higher plants that forms intermediate-sized filaments and stabilizes microtubules. *Protoplasma* 186: 99–112. Copyright © 1995 Springer-Verlag.

Figure 5.9A–F: Ross, J.H.E., Hutchings, A., Butcher, G. W., Lane, E. B., Lloyd, C. W. (1991) The intermediate filament-related system of higher plant cells shares an epitope with cytokeratin 8. *J. Cell Sci.* 99: 91–98. Used with permission from Company of Biologists Ltd.

Figure 5.10: Adapted from Linder, S., Schliwa, M., Kube-Granderath, E. (1997) Sequence analysis and immunofluorescence study of alpha- and beta-tubulins in *Reticulomyxa filosa*: implications of the high degree of beta-2-tubulin divergence. *Cell Motil. Cytoskeleton* 36: 164–178. Reprinted by permission of Wiley-Liss, Inc., a subsidiary of John Wiley & Sons, Inc.

Figure 5.11A: Adapted from Villemur, R., Joyce, C. M., Haas, N. A., Goddard, R. H., Kopczak, S. D., Hussey, P. J., Snustad, D. P., Silflow, C. D. (1992) α-Tubulin gene family of maize (*Zea mays* L.): evidence for two ancient α-tubulin genes in plants. *J. Mol. Biol.* 227: 81–96. Reprinted by permission of Academic Press. Copyright © 1992.

Figure 5.11B: Adapted from Villemur, R., Haas, N. A., Joyce, C. M., Snustad, D. P., Silflow, C. D. (1994) Characterization of four new beta-tubulin genes and their expression during male flower development in maize (*Zea mays* L.). *Plant Mol. Biol.* 24: 295–315. Used with kind permission from Kluwer Academic Publishers.

Figure 5.12A,B: Reproduced from Hoyle, H. D., Raff, E. C. (1990) Two *Drosophila* beta tubulin isoforms are not functionally equivalent. *J. Cell Biol.* 111: 1009–1026. By copyright permission of the Rockefeller University Press.

Figure 5.13A,B: Carpenter, J. L., Ploense, S. E., Snustad, D. P., Silflow, C. D. (1992) Preferential expression of an α tubulin gene of *Arabidopsis* in pollen. *Plant Cell* 4: 557–571.

Figure 5.16A: Reprinted by permission from Kabsch, W., Mannherz, H. G., Suck, D., Pai, E. F., Holmes, K. C. (1990) Atomic structure of the actin DNase I complex. *Nature* 347: 37–44. Copyright © 1990 Macmillan Magazines Ltd.

Figure 5.16C: Reprinted by permission from Holmes, K. C., Popp, D., Gebhard, W., Kabsch, W. (1990) Atomic model of the actin filament. *Nature* 347: 44–49. Copyright © 1990 Macmillan Magazines Ltd.

Figure 5.16D: Reprinted by permission from Milligan, R. A., Whittaker, M., Safer, D. (1990) Molecular structure of F-actin and location of surface binding sites. *Nature* 348: 217–221. Copyright © 1990 Macmillan Magazines Ltd.

Figure 5.17A: Reprinted by permission from Nogales, E., Wolf, S. G., Downing, K. H. (1998) Structure of the alpha-beta tubulin dimer by electron crystallography. *Nature* 391: 199–203. Copyright © 1998 Macmillan Magazines Ltd.

Figure 5.17C: Juniper, B. E., Lawton, J. R. (1979) The effect of caffeine, different fixation regimes and low temperature on microtubules in the cells of higher plants. *Planta* 145: 411–416. Copyright © 1979 Springer-Verlag.

Figure 5.17D: D. Chrètian, EMBO Laboratory, Heidelberg, Germany; previously unpublished.

Figure 5.19B,D: Reproduced from Mandelkow, E.-M., Mandelkow, E., Milligan, R. A. (1991) Microtubule dynamics and microtubule caps: a time-resolved cryo-electron microscopy study. *J. Cell Biol.* 114: 977–992. By copyright permission of the Rockefeller University Press.

Figure 5.20A–D: Reproduced by permission from Hollenbeck, P. J. (1989) Dissecting a molecular motor. *Nature* 338: 294–295. Copyright © 1989 Macmillan Magazines Ltd.

Figure 5.24B,C: Williamson, R. E. (1975) Cytoplasmic streaming in *Chara*: a cell model activated by ATP and inhibited by cytochalasin B. *J. Cell Sci.* 17: 655–668. Used with permission from Company of Biologists Ltd.

Figure 5.24D: G. Wasteneys, Australian National University, Canberra; previously unpublished.

Figure 5.25A,B: Trojan, A., Gabryś, H. (1996) Chloroplast distribution in *Arabidopsis thaliana* (L.) depends on light conditions during growth. *Plant Physiol.* 111: 419–425.

Figure 5.26B: Mineyuki, Y., Kataoka, H., Masuda, Y., Nagai, R. (1995) Dynamic changes in the actin cytoskeleton during the high-fluence rate response of the *Mougeotia* chloroplast. *Protoplasma* 185: 222–229. Copyright © 1995 Springer-Verlag.

Figure 5.27A,B: Lancelle, S. A., Cresti, M., Hepler, P. K. (1996) Growth inhibition and recovery in freeze-substituted *Lilium longiflorum* pollen tubes: structural effects of caffeine. *Protoplasma* 196: 21–33. Copyright © 1996 Springer-Verlag.

Figure 5.28A: Baskin, T. I., Miller, D. D., Vos, J. W., Wilson, J. E., Hepler, P. K. (1996) Cryofixing single cells and multicellular specimens enhances structure and immunocytochemistry for light microscopy. *J. Microsc.* 182: 149–161. Used with permission from Blackwell Science Ltd.

Figure 5.28B: Reprinted from Gunning, B.E.S., Hardham, A. R. (1979) Microtubules and morphogenesis in plants. *Endeavor* 3:112–117. Copyright © 1979 with permission from Elsevier Science.

Figure 5.29: Reproduced from Euteneuer, U., Jackson, W. T., McIntosh, J. R. (1982) Polarity of spindle microtubules in *Haemanthus* endosperm. *J. Cell Biol.* 94: 644–653. By copyright permission of the Rockefeller University Press.

Figure 5.30A: McDonald, A. R., Liu, B., Joshi, H. C., Palevitz, B. A. (1993) Gamma tubulin is associated with a cortical microtubule-organizing zone in the developing guard cells of *Allium cepa* L. *Planta* 191: 357–361. Copyright © 1993 Springer-Verlag.

Figure 5.30B,C: Liu, B., Joshi, H. C., Palevitz, B. A. (1995) Experimental manipulation of gamma-tubulin distribution in *Arabidopsis* using anti-microtubule drugs. *Cell Motil. Cytoskeleton* 31: 113–129. Reprinted by permission of Wiley-Liss, Inc., a subsidiary of John Wiley & Sons, Inc.

Figure 5.31A,B: Baskin, T. I., Wilson, J. E. (1997) Inhibitors of protein kinases and phosphatases alter root morphology and disorganize cortical microtubules. *Plant Physiol.* 113: 493–502.

Figure 5.32: Vesk, P. A., Vesk, M., Gunning, B.E.S. (1996) Field emission scanning electron microscopy of microtubule arrays in higher plant cells. *Protoplasma* 195: 168–182. Copyright © 1996 Springer-Verlag.

Figure 5.34: Zhang, D., Wadsworth, P., Hepler, P. K. (1990) Microtubule dynamics in living dividing plant cells: confocal imaging of microinjected fluorescent brain tubulin. *Proc. Natl. Acad. Sci. USA* 87: 8820–8824. Used with permission.

Figure 5.35A: Lloyd, C. W., Shaw, P. J., Warn, R. M., Yuan, M. (1996) Gibberellic-acid induced reorientation of cortical microtubules in living plant cells. *J. Microsc.* 181: 140–144. Used with permission from Blackwell Science Ltd.

Figure 5.35B: Cleary, A.L. (1995) F-actin redistributions at the division site in living *Tradescantia* stomatal complexes as revealed by microinjection of rhodamine-phalloidin. *Protoplasma* 185: 152–165. Copyright © 1995 Springer-Verlag.

Figure 5.36: R. Cyr, Pennsylvania State University, Abington; previously unpublished.

Figure 5.37: Adapted from Williamson, R. E. (1990) Alignment of cortical microtubules by anisotropic wall stresses. *Aust. J. Plant Physiol.* 17: 601–614 Used with permission from Commonwealth Scientific and Industrial Research Organization, Melbourne.

Figure 5.38: Vesk, P. A., Vesk, M., Gunning, B.E.S. (1996) Field emission scanning electron microscopy of microtubule arrays in higher plant cells. *Protoplasma* 195: 168–182. Copyright © 1996 Springer-Verlag.

Figure 5.39C: Reproduced from Waters, J. C., Chen, R.-H., Murray, A. W., Salmon, E. D. (1998) Localization of Mad2 to kinetochores depends on microtubule attachment, not tension. *J. Cell Biol.* 141: 1181–1191. By copyright permission of the Rockefeller University Press.

Figure 5.39D: Reproduced from Khodjakov, A., Cole, R. W., Bajer, A. S., Rieder, C. L. (1996) The force for poleward chromosome motion in *Haemanthus* cells acts along the length of the chromosome during metaphase but only at the kinetochore during anaphase. *J. Cell Biol.* 132: 1093–1104. By copyright permission of the Rockefeller University Press.

Figure 5.40C: Reproduced from Moritz, M., Braunfeld, M. B., Fung, J. C., Sedat, J. W., Alberts, B. M., Agard, D. A. (1995) Three-dimensional structural characterization of centrosomes from early *Drosophila* embryos. *J. Cell Biol.* 130: 1149–1159. By copyright permission of the Rockefeller University Press.

Figure 5.40D: H. Schatten, University of Missouri, Columbia, previously unpublished.

Figure 5.43A–K: Gunning, B.E.S., Steer, M. W. (1996) *Plant Cell Biology: Structure and Function.* Jones and Bartlett, Sudbury, MA. Copyright © 1996 and used by permission from Gustav Fischer Verlag.

Figure 5.44A: R. Martin, Gerhard Wanner, University of Munich, Germany; previously unpublished.

Figure 5.44C: Van Hooser, A., Brinkley, B. R. (1999) Methods for in situ localization of proteins and DNA in the centromere–kinetochone complex. *Methods Cell Biol.* 61: 57–80. Reprinted by permission of Academic Press. Copyright © 1999.

Figure 5.44D: K. McDonald, University of California, Berkeley; previously unpublished.

Figure 5.52A: Galatis, B., Apostolakos, P., Katsaros, C. (1984) Experimental studies on the function of the cortical cytoplasmic zone of the preprophase microtubule band. *Protoplasma* 122: 11–26. Copyright © 1984 Springer-Verlag.

Figure 5.52B,C: Kakimoto, T., Shibaoka, H. (1987) Actin filaments and microtubules in the preprophase band and phragmoplast of tobacco cells. *Protoplasma* 140: 151–156. Copyright © 1987 Springer-Verlag.

Figure 5.53A–F: Cleary, A. L., Gunning, B.E.S., Wasteneys, G. O., Hepler, P. K. (1992) Microtubule and F actin dynamics at the division site in living *Tradescantia* stamen hair cells. *J. Cell Sci.* 103: 977–988. Used with permission from Company of Biologists Ltd.

Figure 5.54: Mineyuki, Y., Gunning, B.E.S. (1990) A role for preprophase bands of microtubules in maturation of new cell walls, and a general proposal on the function of preprophase band sites in cell division in higher plants. *J. Cell Sci.* 97: 527–538. Used with permission from Company of Biologists Ltd.

Figure 5.56A–D: Gunning, B.E.S., Steer, M. W. (1996) *Plant Cell Biology: Structure and Function* Jones and Bartlett, Sudbury, MA. Copyright © 1996 and used by permission from Gustav Fischer Verlag.

Figure 5.56E,F: Smirnova, E. A., Cox, D. L., Bajer, A. S. (1995) Antibody against phosphorylated proteins (MPM-2) recognizes mitotic microtubules in endosperm cells of higher plant *Haemanthus. Cell Motil. Cytoskeleton* 31: 34–44. Reprinted with permission from Wiley-Liss, Inc., a subsidiary of John Wiley & Sons, Inc.

Figure 5.57A–V: Zhang, D., Wadsworth, P., Hepler, P. K. (1993) Dynamics of microfilaments are similar but distinct from microtubules during cytokinesis in living dividing plant cells. *Cell Motil. Cytoskeleton* 24: 151–155. Reprinted with permission from Wiley-Liss, Inc., a subsidiary of John Wiley & Sons, Inc.

Figure 5.58: Kakimoto, T., Shibaoka, H. (1988) Cytoskeletal ultrastructure of phragmoplast-nuclei complexes isolated from cultured tobacco cells. *Protoplasma* (Suppl. 2): 95–103. Copyright © 1988 Springer-Verlag.

Figure 5.59A–C: Venverloo, C. J., Libbenga, K. R. (1981) Cell division in *Nautilocalyx* explants. II Duration of cytokinesis and velocity of cell-plate growth in large highly vacuolate cells. *Z. Pflanzenphysiol.* 102: 389–395. Used with permission from Urban & Fischer Verlag, Niederlassung Jena, Germany.

Figure 5.59D: Asada, T., Shibaoka, H. (1994) Isolation of polypeptides with microtubule-translocating activity from phragmoplasts of tobacco BY-2 cells. *J. Cell Sci.* 107: 2249–2257. Used with permission from Company of Biologists Ltd.

Figure 5.60A: Asada, T., Shibaoka, H. (1994) Isolation of polypeptides with microtubule-translocating activity from phragmoplasts of tobacco BY-2 cells. *J. Cell Sci.* 107: 2249–2257. Used with permission from Company of Biologists Ltd.

Figure 5.60B. Asada, Osaka University, Japan, previously unpublished.

Figure 5.61A: Gunning, B.E.S., Steer, M. W. (1996) *Plant Cell Biology: Structure and Function.* Jones and Bartlett, Sudbury, MA. Copyright © 1996 Gustav Fischer Verlag.

Figure 5.61B: Kakimoto, T., Shibaoka, H. (1992) Synthesis of polysaccharides in phragmoplasts isolated from tobacco BY-2 cells. *Plant Cell Physiol.* 33: 353–361. Used with permission from the Japanese Society of Plant Physiology.

Box 5.1A: Portrait used by permission of the Linnean Society of London.

Box 5.1B,C: B. J. Ford, Rothay House, Cambridgeshire, UK; previously unpublished.

Box 5.2: Ribbon diagrams adapted from Federov, A. A., Ball, T., Mahoney, N. M., Valenta, R., Almo, S. C. (1997) Molecular basis for allergen cross-reactivity: crystal structure and IgE-epitope mapping of birch pollen profilin. *Structure* 5: 33–45. Used with permission from Current Biology Ltd.

Box 5.2A,B: Valster, A. H., Pierson, E. S., Valenta, R., Hepler, P. K., Emons, A.M.C. (1997) Probing the plant actin cytoskeleton during cytokinesis and interphase by profilin microinjection. *Plant Cell* 9: 815–1824.

Box 5.2C: Upper panel: Miller, D. D., Lancelle, S. A., Hepler, P. K. (1996) Actin microfilaments do not form a dense meshwork in *Lilium longiflorum* pollen tubes. *Protoplasma* 195:123–132. Copyright © 1996 Springer-Verlag. Middle and lower panels: Vidali, L., Hepler, P. K. (1997) Characterization and localization of profilin pollen grains and tubes of *Lilium longiflorum. Cell Motil. & Cytoskeleton* 36:323–338. Reprinted with permission from Wiley-Liss, Inc., a subsidiary of John Wiley & Sons, Inc.

Box 5.3A: Inouè, S. (1953) Polarization optical studies of the mitotic spindle. I The demonstration of spindle fibers in living cells. *Chromosoma* 5: 487–506. Copyright © 1953 Springer-Verlag.

Box 5.3B: Inouè, S., Bajer, A. (1961) Birefringence in endosperm mitosis. *Chromosoma* 12: 48–63. Copyright © 1961 Springer-Verlag.

Box 5.3C: Oldenbourg, R., Mei, G. (1995) New polarized light microscope. *J. Microsc.* 180: 140–147. Used by permission from Blackwell Science Ltd.

Box 5.4A–D: Kiss, J. Z., Giddings, T. H. Jr., Staehelin, L. A., Sack, F. D. (1990) Comparison of the ultrastructure of conventionally fixed and high-pressure frozen/freeze-substituted root tips of *Nicotiana* and *Arabidopsis. Protoplasma* 157: 64–74. Copyright © 1990 Springer-Verlag.

Box 5.4E,F: Reproduced (E) and adapted (F) from Samuels, A. L., Giddings, T. H., Staehelin, L. A. (1995) Cytokinesis in tobacco BY-2 and root tip cells: a new model of cell plate formation in higher plants. *J. Cell Biol.* 130: 1345–1357. By copyright permission of the Rockefeller University Press.

Part 2 opening photograph: Galen Rowell/Mountain Light Photography, Emeryville, CA.

Chapter 6

Chapter 6 opening photograph: D. Richardson, Northampton, MA; published in Purves, W., Orians, G., Heller, H. C., Sadava, D. (1998) *Life: The Science of Biology*, 5th edition. Sinauer/Freeman, Sunderland, MA, p. 65. Used with permission from the photographer.

Figure 6.10B: Reprinted from Shippen, D. I., McKnight, T. (1998) Telomeres, telomerase, and plant development. *Trends Plant Sci.* 3: 126–129. Copyright © 1998 with permission from Elsevier Science.

Figures 6.12–6.16: Adapted from Cooper, G. M. (1997) *The Cell: A Molecular Approach.* Co-published by ASM Press, Washington, DC, and Sinauer Associates, Sunderland, MA.

Figure 6.21: Adapted from Cooper, G. M. (1997) *The Cell: A Molecular Approach.* Co-published by ASM Press, Washington, DC, and Sinauer Associates, Sunderland, MA.

Figure 6.22: Photomicrographs: Madigan, M. T., Martinko, J. M., Parker, J. (1997) *Brock Biology of Microorganisms*, 8th ed. Prentice Hall, Upper Saddle River, NJ.

Figure 6.23 top: Hiratsuka, J., Shimada, H., Whittier, R., Ishibashi, T., Sakamoto, M., Mori, M., Kondo, C., Honji, Y., Sun, C.-R., Meng, B.-Y., Li, Y.-Q., Kanno, A., Nishizawa, Y., Hirai, A., Shinozaki, K., Sugiura, M. (1989) The complete sequence of the rice *Oryza sativa* chloroplast genome: intermolecular recombination between distinct transfer RNA genes accounts for a major plastid DNA inversion during the evolution of the cereals. *Mol. Gen. Genet.* 217: 185–194. Copyright © 1989 Springer-Verlag.

Figure 6.23 bottom: Adapted from Maier, L.R.M., Neckermann, K., Igloi, G. L., Kössel, H. (1995) Complete sequence of the maize chloroplast genome: gene content, hotspots of divergence and fine tuning of genetic information by transcription editing. *J. Mol. Biol.* 251: 614–628. Adapted by permission of Academic Press. Copyright © 1995.

Figure 6.25: Adapted from Kornberg, A., Baker, T. A. (1992) *DNA Replication,* 2nd ed. W. H. Freeman, New York. Used with permission.

Figure 6.26: Fauron, C., Casper, M., Gao, Y., Moore, B. (1995) The maize mitochondrial genome: dynamic, yet functional. *Trends Genet.* 11: 228–235. Copyright © 1995 with permission from Elsevier Science.

Figure 6.27A–C: Fauron, C., Casper, M., Gao, Y., Moore, B. (1995) The maize mitochondrial genome: dynamic, yet functional. *Trends Genet.* 11: 228–235. Copyright © 1995 with permission from Elsevier Science.

Figure 6.28: DNA diagram adapted from Boer, P. H., Gray, M. W. (1991) Short dispersed repeats localized in spacer regions of *Chlamydomonas reinhardtii* mitochondrial DNA. *Curr. Genet.* 19: 309–312.

Figure 6.29: Cover of *Plant Mol. Biol. Rep.* 11 (June), accompanying article by Hirai, A., Nakasono, M. (1993). Used with permission from the International Society for Plant Molecular Biology.

Figure 6.30: Bacterial structure adapted from Cooper, G. M. (1997) *The Cell: A Molecular Approach.* Copublished by ASM Press, Washington, DC, and Sinauer Associates, Sunderland, MA.

Figure 6.33: Graph: M. Sugiura, Center for Gene Research, Nagoya University, Japan; previously unpublished.

Figure 6.35: Adapted from Harris, E. H., Boynton, J. E., Gillham, N. W. (1994) Chloroplast ribosomes and protein synthesis. *Microbiol. Rev.* 58: 700–754. Used with permission from the American Society for Microbiology.

Figure 6.36: Adapted from Dudock, B. S., Katz, G., Taylor, E. K., Holley, R. (1969) Primary structure of wheat germ phenylalanine transfer RNA. *Proc. Natl. Acad. Sci. USA* 62: 941–945. Used with permission.

Figure 6.37: Adapted from Canaday, J., Guillemaut, P., Weil, J. H. (1980) The nucleotide sequences of the initiator transfer RNAs from bean cytoplasm and chloroplasts. *Nucleic Acids Res.* 8: 999–1008. Copyright © 1980 Springer-Verlag.

Figure 6.42: Adapted from Westaway, S. K. (1995) Splicing of tRNA precursors. In *tRNA: Structure, Biosynthesis, and Function,* D. Soell and U. L. RajBhandary, eds. ASM Press, Washington, DC, pp. 79–92. Used with permission from the American Society for Microbiology.

Figure 6.43: Adapted from Cooper, G. M. (1997) *The Cell: A Molecular Approach.* Copublished by ASM Press, Washington, DC, and Sinauer Associates, Sunderland, MA.

Figure 6.46: Adapted from Westhoff, P., Herrmann, R. G. (1988) Complex RNA maturation in chloroplasts: the Psb-B operon from spinach. *Eur. J. Biochem.* 171: 551–564. Used by permission from Blackwell Science Ltd.

Figure 6.47: Adapted from Russell, P. (1998) *Genetics,* 5th ed. Addison Wesley Longman Publishers, Menlo Park, CA. Reprinted by permission of Addison Wesley Educational Publishers, Inc.

Box 6.2: Photograph: M. Sugiura, Center for Gene Research, Nagoya University, Japan; previously unpublished.

Box 6.4: W. Gruissem, University of California, Berkeley; previously unpublished.

Chapter 7

Figure 7.1: Rattner, J. B. 1991. The structure of the mammalian centromere. *BioEssays* 13: 51–56. Used with permission from Wiley-Liss, Inc., a subsidiary of John Wiley & Sons, Inc.

Figure 7.9: L. Jesaitis, University of California, Berkeley; previously unpublished.

Figure 7.10A,B: Rhoades, M. M. (1950) Meiosis in maize. *J. Hered.* 41: 58–67. Used with permission from OxfordUniversity Press.

Figure 7.11: Photographs from F. W. Goro, published in Federov, N. V. (1984) Transposable genetic elements in maize. *Sci. Am.* 250(6): 84–98. Used with permission of the Estate of F. W. Goro.

Figure 7.12: A. Brousseau, St. Mary's College, Marga, CA. Copyright © 1995 Brother Eric Vogel, St. Mary's College.

Figure 7.20: M. P. Tolson, University of California, Davis; previously unpublished.

Figure 7.22A: Meyne, J. (1993) In *Chromosomes: A Synthesis,* R. P. Wagner, M. P. Maguire, R. L. Stallings, eds. Wiley-Liss, New York. Used with permission from Wiley-Liss, Inc., a subsidiary of John Wiley & Sons, Inc.

Figure 7.22B: Reprinted from Willard, H. F. (1990) Centromeres of mammalian chromosomes. Reprinted from *Trends Genet.* 6: 410–416. Copyright © 1990 with permission from Elsevier Science.

Figure 7.22C: S. Ruzin, University of California, Berkeley; previously unpublished.

Figure 7.23: Adapted with permission from San Miguel, P., Tikhonov, A., Jin, Y. K., Motchoulskaia, N., Zakharov, D., Melake Berhan, A., Springer, P. S., Edwards, K. J., Lee, M., Avramova, Z., et al. (1996) Nested retrotransposons in the intergenic regions of the maize genome. *Science* 274: 765–768. Copyright © 1996 American Association for the Advancement of Science.

Figure 7.25: O. L. Miller, B. R. Beatty, D. W. Fawcett/Visuals Unlimited, Swanzey, NH.

Figure 7.35: Figure adapted from McClintock, B. (1952) Chromosome organization and genic expression. *Cold Spring Harbor Symp. Quant. Biol.* 16: 13–47. Used with permission from Cold Spring Harbor Laboratory, Cold Spring Harbor, NY. Legend from Keller, E. F. (1983) *A Feeling for the Organism: The Life and Work of Barbara McClintock.* W. H. Freeman, New York, p. 82. Copyright © 1983. Used with permission.

Figure 7.36: Photographs from F. W. Goro, published in Federov, N. V. (1984) Transposable genetic elements in maize. *Sci. Am.* 250(6): 84–98. Used with permission of the Estate of F. W. Goro.

Figure 7.37: Keller, J., Lim, E., James, D. W. Jr., Dooner, H. K. (1992) Germinal and somatic activity of the maize element activator (*Ac*) in *Arabidopsis. Genetics* 131: 449–459. Used with permission from the Genetic Society of America.

Figure 7.38: Neuffer, M. G., Coe, E. H., Wessler, S. R. (1997) *Mutants of Maize.* Cold Spring Harbor Laboratory Press, Cold Spring Harbor, NY.

Figure 7.39: Chuck, G., Lincoln, C., Hake, S. (1996) *KNAT1* induces lobed leaves with ectopic meristems when overexpressed in *Arabidopsis. Plant Cell* 8: 1277–1289.

Figure 7.40: Taylor, J. E., Renwick, K. F., Webb, A. A., McAinsh, M. R., Furini, A., Bartels, C., Quatrano, R. S., Marcotte, W. R. Jr., Hetherington, A. M. (1995) ABA-regulated promoter activity in stomatal guard cells. *Plant J.* 7: 129–134. Used with permission from Blackwell Science Ltd.

Figure 7.47A: Reprinted by permission from Luger, K., Mäder, A. W., Richmond, R. K., Sargent, D. F., Richmond, T. J. (1997) Crystal structure of the nucleosome core particle at 2.8 Å resolution. *Nature* 389: 251–260. Copyright © 1997 Macmillan Magazines Ltd.

Figure 7.47B,C: Karp, G. (1996) *Cell and Molecular Biology: Concepts and Experiments.* John Wiley & Sons, New York. Used with permission from John Wiley & Sons. Copyright © 1996.

Figure 7.48: Paul, A.-L., Vasil, V., Vasil, I. K., Ferl, R. J. (1987) Constitutive and anaerobically induced DNase-I-hypersensitive sites in the 5' region of the maize *ADH1* gene. *Proc. Natl. Acad. Sci. USA* 84: 799–803. Copyright © 1987 National Academy of Sciences, U.S.A.

Figure 7.50: Electronmicrograph from Paulson, J. R., Laemmli, U. K. (1977) The structure of histone-depleted metaphase chromosomes. *Cell* 12: 817–828. Copyright © 1977 Cell Press.

Figure 7.51: Kermicle, J. L. (1996) Epigenetic silencing and activation of a maize *r* gene. In *Cold Spring Harbor Monograph Series, 32. Epigenetic mechanisms of gene regulation,* V.E.A. Russo, R. A. Martienssen, and A. D. Riggs, eds. Cold Spring Harbor Laboratory Press, Cold Spring Harbor, NY, pp. 267–287.

Figure 7.52: Hollick, J. B., Dorweiler, J. E., Chandler, V. L. (1997) Paramutation and related allelic interactions. Reprinted from *Trends Genet.* 13: 302–308, Copyright © 1997 with permission from Elsevier Science.

Figure 7.53: Meyer, P. (1996) DNA methylation of flower color transgenes in *Petunia hybrida.* In *Cold Spring Harbor Monograph Series, 32. Epigenetic mechanisms of gene regulation,* V.E.A. Russo, R. A. Martienssen, and A. D. Riggs, eds. Cold Spring Harbor Laboratory Press, Cold Spring Harbor, NY, pp. 305–317.

Figure 7.54A: Jorgenson, R. (1993) The germinal inheritance of epigenetic information in plants. *Philos. Trans. R. Soc. Lond. B* 339: 173–181. Reprinted by permission of the President and Council of the Royal Society.

Figure 7.54B: Jorgenson, R. A., Napoli, C. A. (1996) A responsive regulatory system is revealed by sense suppression of pigment genes in *Petunia* flowers. In *Genomes: Proceedings of the 22nd Stadler Genetics Symposium*, J. P. Gustafson and R. B. Flavell, eds. Plenum Press, New York, pp. 159–176.

Figure 7.54C: Napoli, C., Lemieux, C., Jorgenson, R. (1990) Introduction of a chimeric chalcone synthase gene into petunia results in reversible co-suppression of homologous genes in trans. *Plant Cell* 2: 279–289.

Figure 7.54D: Jorgenson, R. (1994) Developmental significance of epigenetic impositions on the plant genome: a paragenetic function for chromosomes. *Dev. Genet.* 15: 523–532. Used with permission from John Wiley & Sons, Inc. Copyright © 1994.

Figure 7.56: Reprinted with permission from Ronemus, M. J., Galbiati, M., Ticknor, C., Chen, J., Dellaporta, S. L. (1996) Demethylation-induced developmental pleiotropy in *Arabidopsis*. *Science* 273: 654–657. Copyright © 1996 American Association for the Advancement of Science.

Box 7.5 (left page): Paul, A.-L., Ferl, R. L. (1991) In vivo footprinting reveals unique cis-elements and different modes of hypoxic induction in maize Adh1 and Adh2. *Plant Cell* 3: 159–168.

Box 7.5A,C: Paul, A.-L., Ferl, R. L. (1994) In vivo footprinting identifies an activating element of the maize Adh2 promoter specific for root and vascular tissues. *Plant J.* 5: 523–533. Used with permission from Blackwell Science Ltd.

Chapter 8

Figure 8.3: Adapted from Lam, H.-M., Coschigano, K., Schultz, C., Melo-Oliveira, R., Tjaden, G., Oliveira, I., Ngai, N., Hsieh, M.-H., Coruzzi, G. (1995) Use of *Arabidopsis* mutants and genes to study amide amino acid biosynthesis. *Plant Cell* 7: 887–898.

Figure 8.8: Adapted from Wallgrove, R. M., Turner, J. C., Hall, N. P., Kendall, A. C., Bright, S.W.J. (1987) Barley mutants lacking chloroplast glutamine synthetase—biochemical and genetic analysis. *Plant Physiol.* 83: 155–158.

Figure 8.9: Edwards, J. W., Walker, E. L., Coruzzi, G. M. (1990) Cell-specific expression in transgenic plants reveals nonoverlapping roles for chloroplast and cytosolic glutamine synthetase. *Proc. Natl. Acad. Sci. USA* 87: 3459–3463. Used with permission.

Figure 8.11: Photographs: Coschigano, K. T., Melo-Oliveira, R., Lim, J., Coruzzi, G. M. (1998) *Arabidopsis gls* mutants and distinct Fd-GOGAT genes: implications for photorespiration and primary nitrogen assimilation. *Plant Cell* 10: 741–752.

Figure 8.12: Electrophoretogram: Coschigano, K. T., Melo-Oliveira, R., Lim, J., Coruzzi, G. M. (1998) *Arabidopsis gls* mutants and distinct Fd-GOGAT genes: implications for photorespiration and primary nitrogen assimilation. *Plant Cell* 10: 741–752.

Figure 8.15: Melo-Oliveira, R., Oliveira, I. C., Coruzzi, G. M. (1996) *Arabidopsis* mutant analysis and gene regulation define a nonredundant role for glutamate dehydrogenase in nitrogen assimilation. *Proc. Natl. Acad. Sci. USA* 93: 4718–4723. Copyright © 1996 National Academy of Sciences, U.S.A.

Figure 8.18A,B: Schultz, C. J., Hsu, M., Miesak, B., Coruzzi, G. M. (1998) *Arabidopsis* mutants define an in vivo role for isoenzymes of aspartate aminotransferase in plant nitrogen assimilation. *Genetics* 149: 491–499. Copyright © 1998 Genetic Society of America.

Figure 8.22: Eckes, P., Schmitt, P., Winfried, D., Wegenmeyer, F. (1989) Overproduction of alfalfa glutamine synthetase in transgenic tobacco plants. *Mol. Gen. Genet.* 217: 263–268. Copyright © 1989 Springer-Verlag.

Figure 8.23A,B: Adapted (A) and reprinted (B) by permission from Kozaki, A., Takeba, G. (1996) Photorespiration protects C_3 plants from photooxidation. *Nature* 384: 557–560. Copyright © 1996 Macmillan Magazines Ltd.

Figure 8.26: Adapted from Dyer, W. E., Weaver, L. M., Zhao, J., Kuhn, D. N., Weller, S. C., Herrmann, K. M. (1990) A complementary DNA encoding 3-deoxy-D-arabino-heptulosonate-7-phosphate synthase from *Solanum tuberosum* L. *J. Biol. Chem.* 265: 1608–1614. Copyright © 1990 American Society for Biochemistry and Molecular Biology.

Figure 8.37: R. Last, Cereon Genomics LLC, Cambridge, MA; previously unpublished.

Figure 8.41A: Hyde, C. C., Ahmed, S. A., Padlan, E. A., Miles, E. W., Davies, D. R. (1998) Three-dimensional structure of the tryptophan synthase $\alpha_2\beta_2$ multienzyme complex from *Salmonella typhimurium*. *J. Biol. Chem.* 263: 17857–17871. Copyright © 1988 American Society for Biochemistry and Molecular Biology.

Figure 8.41B: Adapted from Miles, E. W. (1995) Tryptophan synthase structure, function, and protein engineering. In *Subcellular Biochemistry; Proteins: Structure, Function, and Engineering*, B. B. Biswas and S. Roy, eds. Plenum Press, New York, pp. 207–254.

Figure 8.43A: Graph adapted from Dyer, W. E., Henstrand, J. M., Handa, A. K., Herrmann, K. M. (1989) Wounding induces the first enzyme of the shikimate pathway in Solanaceae. *Proc. Natl. Acad. Sci. USA* 86: 7370–7373. Used with permission.

Figure 8.43B: Immunoblot adapted from Zhao, J., Last, R. L. (1996) Coordinate regulation of the tryptophan biosynthetic pathway and idolic phytoalexin accumulation in *Arabidopsis*. *Plant Cell* 8: 2235–2244.

Figure 8.55: Biou, V., Dumas, R., Cohen-Addad, C., Douce, R., Job, D., Pebay-Peroula, E. (1997) The crystal structure of plant acetohydroxy acid isomer or reductase complexed with NADPH, two magnesium ions, and a herbicidal transition site analog determined at 1.645 Å resolution. *EMBO J.* 16: 3405–3415. Used with permission from Oxford University Press.

Box 8.3: Dutch National Asparagus and Mushroom Museum, Horst-Melderslo, The Netherlands.

Box 8.4: Padgette, S. R., Re., D. B., Barry, G. F., Eicholtz, D. E., Delannay, X., Fuchs, R. C., Kishore, G. M., Fraley, R. T. (1996) New weed control opportunities: development of soybeans with a Roundup Ready gene. In *Herbicide-Resistant Crops: Agricultural, Environmental, Economic, Regulatory, and Technical Aspects*, S. O. Duke, ed. CRC Press, Boca Raton, FL, pp. 53–84. Copyright © 1996 CRC Press.

Box 8.5B: R. Last, Cereon Genomics LLC, Cambridge, MA, previously unpublished.

Box 8.6B: Bohlmann, J., De Luca, V., Eilert, U., Martin, W. (1995) Purification and cDNA cloning of anthranilate synthase from *Ruta graveolens*: modes of expression and properties of native and recombinant enzymes. *Plant J.* 7: 491–501. Used with permission from Blackwell Science Ltd.

Box 8.10A: Frankard, V., Ghislam, M., Jacobs, M. (1992) Two feedback-insensitive enzymes of the aspartate pathway in *Nicotiana sylvestris*. *Plant Physiol.* 99: 1285–1293.

Table 8.1: *Arabidopsis* data from Somerville, C. R., Ogren, W. L. (1980) Inhibition of photosynthesis in *Arabidopsis* mutants lacking leaf glutamate synthase activity. *Nature* 286: 257–259. Copyright © 1980 Macmillan Magazines Ltd. Barley data from Kendall, A. C., Wallsgrove, R. M., Hall, N. P., Turner, J. C., Lea, P. J. (1986) Carbon and nitrogen metabolism in barley (*Hordeum vulgare*) mutants lacking ferredoxin-dependent glutamate synthase. *Planta* 168: 316–323. Copyright © 1986 Springer-Verlag.

Chapter 9

Figure 9.6: Sussman, J. L., Holbrook, S. R., Warrant, R. W., Church, G. M., Kim, S.-H. (1978) Crystal structure of yeast phenylalanine t-RNA. I. Crystallographic refinement. *J. Mol. Biol.* 123: 607. Reprinted by permission of Academic Press. Copyright © 1978.

Figure 9.7: Reprinted by permission from Perona, J. J., Rould, M. A., Steitz, T. A. (1991) Structural basis of antecodon loop recognition by glutaminyl transfer RNA synthetase. *Nature* 352: 213–218. Copyright © 1991 Macmillan Magazines Ltd.

Figure 9.9: Hill, W. E. (1990) Probing ribosome structure and function by using short complementary DNA oligomers. In *The Ribosome: Structure, Function, and Evolution*, W. E. Hill, J. Weller, T. Gluick, C. Merryman, R. T. Marconi, A. Tassanahajohn, and W E. Tepprich, eds. ASM Press, Washington, DC, pp. 253–262. Used with permission from the American Society for Microbiology.

Figure 9.32B: Baneyx, F., Bertsch, U., Kalbach, C. E., van Der Vies, S. M., Soll, J., Gatenby, A. A. (1995) Spinach chloroplast cpn 21 co-chaperonin possesses two functional domains fused together in a toroidal structure and exhibits nucleotide-dependent binding to plastid chaperonin 60. *J. Biol. Chem.* 270: 10695–10702. Copyright © 1995 American Society for Biochemistry and Molecular Biology.

Figure 9.33: Bukau, B., Horwich, A. L. (1998) The Hsp70 and Hsp60 chaperone machines. *Cell* 92: 351–366. Copyright © 1998 Cell Press.

Figure 9.38: I. Andersson, Swedish University of Agricultural Sciences, Uppsala; previously unpublished.

Figure 9.43B: Vierstra, R. D. (1996) Proteolysis in plants—mechanisms and functions. *Plant Mol. Biol.* 32: 275–302. Used with kind permission from Kluwer Academic Publishers.

Box 9.3: Mitsuko Williams, University of Illinois at Urbana-Champaign; previously unpublished.

Box 9.4: Electrophoretogram from Danon, A., Mayfield, S., Light regulated translational activators: identification of chloroplast gene-specific mRNA binding proteins. (1991) *EMBO J.* 10: 3993–4001. Used with permission from Oxford University Press.

Chapter 10

Figure 10.2: Adapted from Browse, J., Somerville, C. R. (1994) Glycerolipids. In *Arabidopsis*, E. Meyerowitz and C. R. Somerville, eds. Cold Spring Harbor Laboratory, Cold Spring Harbor, NY, pp. 881–912. Used with permission from Cold Spring Harbor Laboratory Press.

Figure 10.4: Reprinted with permission from Post-Beittenmiller, D. (1996) Biochemistry and molecular biology of wax production in plants. *Annu. Rev. Plant Physiol. Plant Mol. Biol.* 47: 405–430. Copyright © 1996 Annual Reviews. www.AnnualReviews.org

Figure 10.6: Reprinted with permission from Venable, R. M., Zhang, Y., Hardy, B. J., Pastor, R. W. (1993) Molecular dynamics simulation of a lipid bilayer and of hexadecane: an investigation of membrane fluidity. *Science* 262: 223–226. Copyright © 1993 American Association for the Advancement of Science.

Figure 10.17: Barton, P. G., Gunstone, F. D. (1975) Hydrocarbon chain packing and molecular motion in phospholipid bilayers formed from unsaturated lecithins. Synthesis and properties of 16 positional isomers of 1,2-dioctadecenoyl-*sn*-glycero-3-phosphorylcholine. *J. Biol. Chem.* 250: 4470–4476. Copyright © 1975 American Society for Biochemistry and Molecular Biology.

Figure 10.19A: J. Shanklin, Brookhaven National Laboratory, Upton, NY; previously unpublished.

Figure 10.19B,C: Lindqvist, Y., Huang, W., Schneider, G., Shanklin, J. (1996) Crystal structure of Δ-9 stearoyl-acyl carrier protein desaturase from castor seed and its relationship to other di-iron proteins. *EMBO J.* 15: 4081–4092. Used with permission from Oxford University Press.

Figure 10.21: Adapted from Browse, J., Somerville, C. R. (1994) Glycerolipids. In *Arabidopsis*, E. Meyerowitz and C. R. Somerville, eds. Cold Spring Harbor Laboratory, Cold Spring Harbor Laboratory Press, NY, pp. 881–912. Used with permission from Cold Spring Harbor Laboratory Press.

Figure 10.23: Vigh, L., Los, D. A., Horvath, I., Murata, N. (1993) The primary signal in the biological perception of temperature: Pd-catalyzed hydrogenation of membrane lipids stimulated the expression of the *desA* gene in *Synechocystis* PCC6803. *Proc. Natl. Acad. Sci. USA* 90: 9090–9094. Copyright © 1993 National Academy of Sciences, U.S.A.

Figure 10.28: Data from Sauer, F. D., Kramer, J.K.G. (1983) The problems associated with the feeding of high erucic acid rapeseed oils and some fish oils to experimental animals. In *High and Low Erucic Acid Rapeseed Oils: Production, Usage, Chemistry, and Toxicological Evaluation*, J.K.G. Kramer, F. D. Sauer, and W. J. Pidgen, eds. Academic Press, New York, pp. 253–292. Used with permission from Academic Press. Copyright © 1983.

Figure 10.31: Browse, J., Somerville, C. R. (1994) Glycerolipids. In *Arabidopsis*, E. Meyerowitz and C. R. Somerville, eds. Cold Spring Harbor Laboratory Press, Cold Spring Harbor, NY, pp. 881–912. Used with permission from Cold Spring Harbor Laboratory Press.

Figure 10.33: Kunst, L., Browse, J., Somerville, S. R. (1988) Altered regulation of lipid biosynthesis in a mutant of *Arabidopsis* deficient in chloroplast glycerol phosphate acyltransferase activity. *Proc. Natl. Acad. Sci. USA* 85: 4143–4147. Used with permission.

Figure 10.42: C. Somerville, Carnegie Institute of Washington, Stanford, CA; previously unpublished.

Figure 10.43A–E: Lightner, J., Sames, D. W. Jr., Donner, H. K., Browse, J. (1994) Altered body morphology is caused by increased stearate levels in a mutant of *Arabidopsis*. *Plant J.* 6: 401–412. Reprinted with permission from Blackwell Science Ltd.

Figure 10.44A,B: McConn, M., Browse, J. (1996) The critical requirement for linolenic acid is pollen development, not photosynthesis, in an *Arabidopsis* mutant. *Plant Cell* 8: 403–416.

Figure 10.44C: McConn, M., Browse, J. (1998) Polyunsaturated membranes are required for photosynthetic competence in a mutant of *Arabidopsis*. *Plant J.* 15: 521–530. Reprinted with permission from Blackwell Science Ltd.

Figure 10.45: C. Somerville, Carnegie Institute of Washington, Stanford, CA; previously unpublished results.

Figure 10.46: Adapted from Janiak, M. J., Small, D. M., Shipley, G. G. (1976) Nature of the thermal pretransition of synthetic phospholoids: dimyristoyl- and dipalmitoyl-lecithin, *Biochemistry* 15: 4574–4580. Copyright © 1976 American Chemical Society.

Figure 10.47A: Hugly, S., Somerville, C. R. (1992) A role for membrane lipid polyunsaturation in biogenesis at low temperature. *Plant Physiol.* 99: 197–202.

Figure 10.47B: Miquel, M., James, D. Jr., Dooner, H., Browse, J. (1993) *Arabidopsis* requires polyunsaturated lipids for low temperature survival. *Proc. Natl. Acad. Sci. USA* 90: 6208–6212. Copyright © 1993 National Academy of Sciences, U.S.A.

Figure 10.47C,D: Wu, J., Lightner, J., Warwick, N., Browse, J. (1997) Low-temperature damage and subsequent recovery of *fab1* mutant *Arabidopsis* exposed to 2°C. *Plant Physiol.* 113: 347–356.

Figure 10.48: Mike Thomashow, Michigan State University, East Lansing; previously unpublished.

Figure 10.53: McConn, M., Creelman, R. A., Bell, E., Mullet, J. E., Browse, J. (1997) Jasmonate is essential for insect defense in *Arabidopsis*. *Proc. Natl. Acad. Sci. USA* 94: 5473–5477. Copyright © 1997 National Academy of Sciences, U.S.A.

Figure 10.54A: Riederer, M., Schonherr, J. (1988) Development of plant cuticles: fine structure and cutin composition of *Clivia miniatra* Reg. leaves. *Planta* 174: 127–138. Copyright © 1988 Springer-Verlag.

Figure 10.54B: Lendzian, K. J., Schonherr, J. (1983) In vivo study of cutin synthesis in leaves of *Clivia miniata* Reg. *Planta* 158: 70–75. Copyright © 1983 Springer-Verlag.

Figure 10.55: Jeffree, C. E. (1996) Structure and ontogeny of plant cuticles. In *Plant Cuticles: An Integrated Functional Approach*, G. Kerstiens, ed. BIOS Scientific Publishers, Oxford, UK, pp. 33–82. Used with permission of BIOS Scientific Publishers Ltd.

Figure 10.56: Riederer, M., Schonherr, J. (1988) Development of plant cuticles: fine structure and cutin composition of *Clivia miniatra* Reg. leaves. *Planta* 174: 127–138. Copyright © 1988 Springer-Verlag.

Figure 10.59: Sutter, E. (1984) Chemical composition of epicuticular wax in cabbage plants grown in vitro. *Can. J. Bot.* 62: 74–77. Reprinted with permission from NRC Research Press.

Figure 10.60: Jenks, M. A., Rich, P. J., Ashworth, E. N. (1994) Involvement of cork cells in the secretion of epicuticular wax filaments on *Sorghum bicolor* (L.) Moench. *Int. J. Plant Sci.* 155: 506–518. Reprinted with permission from the University of Chicago Press.

Figure 10.62: D. Preuss, University of Chicago, Illinois; previously unpublished.

Figure 10.63: Huang, A.H.C. (1987) Lipases. In *The Biochemistry of Plants: A Comprehensive Treatise. Vol. 9, Lipids: Structure and Function*, P. K. Stumpf, ed. Academic Press, New York, pp. 91–119. Reprinted by permission of Academic Press. Copyright © 1987.

Figure 10.65A: Fernandez, D. E., Staehelin, L. A. (1987) Does gibberellic acid induce the transfer of lipase from protein bodies to lipid bodies in barley aleurone cells? *Plant Physiol.* 85: 487–496.

Figure 10.65B: Reprinted with permission from Huang, A.H.C. (1992) Oil bodies and oleosins in seeds. *Annu. Rev. Plant Physiol. Plant Mol. Biol.* 43: 177–200. Copyright © 1992 Annual Reviews. www.Annual Reviews.org

Figure 10.65C: Qu, R., Huang, A.H.C. (1990) Oleosin kD18 on the surface of oil bodies in maize: genomic and complementary DNA sequences and the deduced protein structure. *J. Biol. Chem.* 6: 2238–2243. Copyright © 1990 American Society for Biochemistry and Molecular Biology.

Figure 10.66: Adapted with permission from Huang, A.H.C. (1992) Oil bodies and oleosins in seeds. *Annu. Rev. Plant Physiol. Plant Mol. Biol.* 43: 177–200. Copyright © 1992 Annual Reviews. www.Annual Reviews.org

Figure 10.69: Ackman, R. G. (1983) Chemical composition of rapeseed oil. In *High and Low Erucic Acid Rapeseed Oils*, J.K.G. Kramer, F. D. Sauer, and W. J. Pidgen, eds. Academic Press, Toronto, pp. 85–129. Used with permission from Academic Press. Copyright © 1983.

Figure 10.69 Inset: Illustration by B. E. Nicholson. Reprinted from *The Oxford Book of Food Plants*. (1969) by permission of Oxford University Press.

Figure 10.70: Based on data from Roesler, K. R., Shintani, D., Savage, L., Boddupalli, S., Ohlrogge, J. B. (1997) Targeting of the *Arabidopsis* homomeric acetyl-CoA carboxylase to *Brassica napus* seed plastids. *Plant Physiol.* 13: 75–81.

Figure 10.71: J. Ohlrogge, Michigan State University, East Lansing; previously unpublished.

Figure 10.73A: B. Page, University of Alberta, Edmonton, Canada; previously unpublished.

Figure 10.75: Nawrath, C., Poirier, Y., Somerville, C. R. (1994) Targeting of the polyhydroxybutyrate biosynthetic pathway to the plastids of *Arabidopsis thaliana* results in high levels of polymer accumulation. *Proc. Natl. Acad. Sci. USA* 91: 12760–12764. Copyright © 1994 National Academy of Sciences, U.S.A.

Box 10.1: J. Ohlrogge, Michigan State University, East Lansing; previously unpublished.

Box 10.4: C. Somerville, Carnegie Institute of Washington, Stanford, CA; previously unpublished.

Table 10.5: Data from Battey, J. F., Schmid, K. M., Ohlrogge, J. B. (1989) Genetic engineering for plant oils: potential and limitations. *Trends Biotechnol.* 7: 122–126. Copyright © 1989 Elsevier Science.

Chapter 11

Chapter 11 opening photograph: P. Heslop-Harrison, John Innes Centre, Norwich, UK; previously unpublished.

Figure 11.1: Photographs: Russel, P., Nurse, P. (1987) Negative regulation of mitosis by *Wee1*[+], a gene encoding a protein kinase homolog. *Cell* 49: 559–567. Copyright © 1987 Cell Press.

Figure 11.5B: Minshull, J. (1989) The role of cyclin synthesis modification and destruction in the control of cell division. In *The Cell Cycle*, R. Brooks et al., eds. *J. Cell Sci. Suppl.* 12: 77–98. Used with permission from Company of Biologists Ltd.

Figure 11.6B: Szabados, I., Dudits, D. (1980) Fusion between interphase and mitotic plant protoplasts. Induction of premature chromosome condensation. *Exp. Cell Res.* 127: 441–446.

Figure 11.9: Adapted from Reichheld, J. P., Gigot, C., Chaubet-Gigot, N. (1998) Multilevel regulation of histone gene expression during the cell cycle in tobacco cells. *Nucleic Acids Res.* 26: 3255–3262. Used by permission of Oxford University Press.

Figure 11.10: Photographs: Sitte, P., Ziegler, H., Ehrendorfer, F., Bresinsky, A., eds. (1998) *Strasburger Lehrbuch der Botanik*, 34th ed. Gustav Fischer Verlag. Copyright © 1998 Spektrum Akademischer Verlag, Heidelberg.

Figure 11.11: Ribbon structures: CDK—Reprinted by permission from de Bondt, H. L., Rosenblatt, J., Jancarik, J., Jones, H. D., Morgan, D. O., Kim, S.-H. (1993) Crystal structure of cyclin-dependent kinase 2. *Nature* 363: 595–602. Copyright © 1993 Macmillan Magazines Ltd. CDK2–cyclin A complex—Reprinted by permission from Jeffrey, P. D., Russo, A. A., Polyak, K., Gibbs, E., Hurwitz, J., Massague, J., Pavletich, N. P. (1995) Mechanism of CDK activation revealed by the structure of a cyclinA–CDK2 complex. *Nature* 376: 313–320. Copyright © 1995 Macmillan Magazines Ltd. T160-phosphorylated CDK2–cyclinA complex—Reprinted by permission from Russo, A. A., Jeffrey, P. D., Pavletich, N. P. (1996) Structural basis of cyclin-dependent kinase activation by phosphorylation. *Nat. Struct. Biol.* 3: 696–700. Copyright © 1996 Nature Publishing Group, New York. CDK2–cycA–CKI complex—Reprinted by permission from Russo, A. A., Jeffrey, P. D., Patten, A. K., Massague, J., Pavletich, N. P. (1996) Crystal structure of the p27[Kip1] cyclin-dependent kinase inhibitor bound to the cyclin A–Cdk2 complex. *Nature* 382: 325–331. Copyright © 1996 Macmillan Magazines Ltd.

Figure 11.18B,C: Ferreria, P.C.G., Hemerly, A. S., Engler, J.D.A., Van Montagu, M., Inzè, D., Engler, G. (1994) Developmental expression of the *Arabidopsis* cyclin gene *cyc1AT*. *Plant Cell* 6: 1763–1774.

Figure 11.19A: Lyndon, R. F. (1998) *The Shoot Apical Meristem: Its Growth and Development*. Cambridge University Press, Cambridge, UK. Reprinted with permission from the Cambridge University Press.

Figure 11.20A: Colón-Carmona, A., You, R., Haimovich-Gal, T., Doerner, P. (1999) Spatio-temporal analysis of mitotic activity with a labile cyclin–GUS fusion protein. *Plant J.* 20: 503–508. Used with permission from Blackwell Science Ltd.

Figure 11.20B: Clowes, originally published in O'Brien, T. P., McCully, M. E. (1969) *Plant Structure and Development—A Pictoral and Physiological Approach*. Collier-Macmillan, Ltd. London.

Figure 11.22A: Reprinted from Yokota, T. (1997) The structure, biosynthesis and function of brassinosteroids. *Trends Plant Sci.* 2: 137–143. Copyright © 1997 with permission from Elsevier Science.

Figure 11.22B: Boerjan, W., Cervera, M.-T., Delarue, M., Beeckman, T., Van Montagu, M., Inzè, D. (1995) *superroot*, a recessive mutation in *Arabidopsis*, confers auxin overproduction. *Plant Cell* 7: 1405–1419.

Figure 11.22C: McCarty, D. R., Hattori, T., Carson, C. B., Vasil, V., Lazar, M., Vasil, I. K. (1991) The *Viviparous-1* developmental gene of maize encodes a novel transcription activator. *Cell* 66: 895–905. Copyright © 1991 Cell Press.

Figure 11.24: Top: Reprinted by permission from Doerner, P. D., Jorgensen, J.-E., You, R., Steppuhn, J., Lamb, C. (1996) Control of root growth and development by cyclin expression. *Nature* 380: 520–523. Copyright © 1996 Macmillan Magazines Ltd. Bottom: P. Doerner, University of Edinburgh, Scotland; previously unpublished.

Figure 11.25A: Reprinted by permission from Doerner, P. D., Jorgensen, J.-E., You, R., Steppuhn, J., Lamb, C. (1996) Control of root growth and development by cyclin expression. *Nature* 380: 520–523. Copyright © 1996 Macmillan Magazines Ltd.

Figure 11.25B: Hemerly, A., Engler, J.D.A., Bergounioux, C., Van Montagu, M., Engler, G., Inzè, D., Ferreira, P. (1995) Dominant negative mutants of the Cdc2 kinase uncouple cell division from iterative plant development. *EMBO J.* 14: 3925–3936.

Figure 11.26A: S. Smith, University of Edinburgh, Scotland; previously unpublished.

Figure 11.26B: Raven, P. H., Evert, R. F., Eichhorn, S. E., eds. (1992) *Biology of Plants*, 5th ed. Worth Publishers, New York. Used with permission.

Figure 11.27: Adapted from Galbraith, D. W., Harkins, K. R., Knapp, S. (1991) Systemic endopolyploidy in *Arabidopsis thaliana*. *Plant Physiol.* 96: 985–989.

Box 11.3: Reprinted with permission from Koshland, D.E.J., Goldbeter, S., Stock, J. B. (1984) Amplification and adaptation in regulatory and sensory systems. *Science* 217: 220–225. Copyright © 1984 American Association for the Advancement of Science.

Part 3 opening photograph:
Galen Rowell/Mountain Light Photography, Emeryville, CA.

Chapter 12

Figure 12.1B: Staehelin, L. A., van der Staay, G.W.M. (1996) Structure, composition, functional organization and dynamic properties of thylakoid membranes. In *Oxygenic Photosynthesis: The Light Reactions*, D. R. Ort and C. F. Yocum, eds. Kluwer Academic Publishers, Dordrecht, The Netherlands, pp. 11–30. Copyright © 1996. Used with kind permission of Kluwer Academic Publishers.

Figure 12.8: Cunningham, F. X. Jr., Gantt, E. (1998) Genes and enzymes of carotenoid biosynthesis in plants. *Annu. Rev. Plant Physiol. Plant Mol. Biol.* 49: 557–583. Copyright © 1998 Annual Reviews. www.AnnualReviews.org

Figure 12.10A,B: Lancaster, C.P.D., Michel, H. (1996) Three-dimensional structures of photosynthetic reaction centers. *Photosynth. Res.* 48: 65–740. Copyright © 1996. Used with kind permission of Kluwer Academic Publishers.

Figure 12.13: Adapted from He, W. Z. (1998) Photosystems I and II. In *Photosynthesis: A Comprehensive Treatise*, A. S. Reghavendra, ed. Cambridge University Press, New York and Cambridge, pp. 29–43. Reprinted with the permission of Cambridge University Press.

Figure 12.14: Schubert, W. D., Klukas, O., Krauss, N., Saenger, W., Fromme, P., Witt, H. T. (1997) Photosystem I of *Synechococcus elongatus* at 4 Å resolution: comprehensive structure analysis. *J. Mol. Biol.* 272: 741–769. Reprinted by permission of Academic Press. Copyright © 1997.

Figure 12.15: Schubert, W. D., Klukas, O., Krauss, N., Saenger, W., Fromme, P., Witt, H. T. (1997) Photosystem I of *Synechococcus elongatus* at 4 Å resolution: comprehensive structure analysis. *J. Mol. Biol.* 272: 741–769. Reprinted by permission of Academic Press. Copyright © 1997.

Figure 12.16A: Reprinted by permission from Kühlbrandt, W., Wang, D. N., Fujiyoshi, Y. (1994) Atomic model of plant light harvesting complex by electron crystallography. *Nature* 367: 614–621. Copyright © 1994 Macmillan Magazines Ltd.

Figure 12.16B: Reprinted from Kühlbrandt, W. (1994) Structure and function of the plant light harvesting complex, LHC-II. *Curr. Opin. Struct. Biol.* 4: 519–528. Copyright © 1994 with permission from Elsevier Science.

Figure 12.17: Adapted from Jansson, S. (1994) The light harvesting chlorophyll *a/b* binding proteins. *Biochim. Biophys. Acta* 1184: 1–19. Copyright © 1994 with permission from Elsevier Science.

Figure 12.23: Adapted from He, W. Z. (1998) Photosystems I and II. In *Photosynthesis: A Comprehensive Treatise*, A. S. Reghavendra, ed. Cambridge University Press, New York and Cambridge, pp. 29–43. Reprinted with the permission of Cambridge University Press.

Figure 12.26: J. Fernandez-Velasco, University of California, Berkeley, and E. Berry, Ernest O. Lawrence Berkeley National Laboratory; previously unpublished.

Figure 12.27: Carrell, C. J., Zhang, H., Cramer, W. A., Smith, J. L. (1997) Biological identity and diversity in photosynthesis and respiration: structure of the lumen-side domain of the chloroplast Rieske protein. *Structure* 5: 1613–1625. Used with permission from Current Biology Ltd.

Figure 12.28: Adapted from He, W. Z. (1998) Photosystems I and II. In *Photosynthesis: A Comprehensive Treatise*, A. S. Reghavendra, ed. Cambridge University Press, New York and Cambridge, pp. 29–43. Reprinted with the permission of Cambridge University Press.

Figure 12.29: Babcock, G. T. (1973) Kinetics and intermediates in photosynthetic oxygen evolution. Ph.D. thesis, University of California, Berkeley.

Figure 12.31: Reprinted with permission from Yachandra, V. K., DeRose, V. J., Latimer, M. J., Mukerji, I., Sauer, K., Klein, M. P. (1993) Where plants make oxygen: a structural model for the photosynthetic oxygen-evolving manganese cluster. *Science* 260: 675–679. Copyright © 1993 American Association for the Advancement of Science.

Figure 12.35: Reprinted from Junge, W., Lill, H., Engelbrecht, S. (1997) ATP synthase: an electrochemical transducer with rotatory mechanics. *Trends Biochem. Sci.* 22: 420–423. Copyright © 1997 with permission from Elsevier Science.

Figure 12.36: Adapted from Cross, R. L., Duncan, T. M. (1996) Subunit rotator in F_0F_1-ATP synthases as a means of coupling proton transport through F_0 to the binding changes in F_1. *J. Bioenerg. Biomembr.* 28: 403–408. Copyright © 1996. Used with kind permission of Kluwer Academic Publishers.

Figure 12.37: Ernest O. Lawrence Berkeley National Laboratory, Photography and Digital Imaging Services; previously unpublished.

Figure 12.39: I. Andersson, Swedish University of Agricultural Sciences, Uppsala; previously unpublished.

Figure 12.45: E. H. Newcomb, University of Wisconsin, Madison; previously unpublished.

Figure 12.47: E. H. Newcomb, University of Wisconsin, Madison; previously unpublished.

Figure 12.48: Taiz, L., Zeiger, E. (1991) CO_2 Concentrating mechanisms II: The C_4 photosynthetic carbon assimilation (PCA) cycle. In *Plant Physiology* Benjamin/Cummings, Redwood City, CA. Copyright © 1991 The Benjamin/Cummings Publishing Co., Inc. Reprinted by permission of Addison Wesley Longman.

Box 12.2A: Melis, A. (1990) Regulation of photosystem stoichiometry in oxygenic photosynthesis. *Bot. Mag. Tokyo* 2 (Special Issue): 9–28. Copyright © 1990 Botanical Society of Japan.

Box 12.2B: R. Malkin, University of California, Berkeley; previously unpublished.

Box 12.3: The molecular models are from S. Dai and P. Schürmann, Universitè de Neuchatel, Neuchatel, Switzerland; previously unpublished.

Chapter 13

Chapter 13 opening photograph: Agricultural Research Service, US Department of Agriculture.

Box 13.3: Cover photograph for *Cell* 60 (1), by Bhattacharyya. Copyright © 1990 Cell Press.

Chapter 14

Figure 14.1B: R. Wiser, University of Wisconsin, Oshkosh; previously unpublished.

Figure 14.22B: Reprinted with permission from Iwata, S., Lee, J. W., Okada, K., Lee, J. K., Iwata, M., Rasmussen, B., Link, T. A., Ramaswamy, S., Jap, B. K. (1998) Complete structure of the 11-subunit bovine mitochondrial cytochrome bc_1 complex. *Science* 281: 64–71. Copyright © 1998 American Association for the Advancement of Science.

Figure 14.23B: Reprinted with permission from Tsukihara, T., Aoyama, H., Yamashita, E., Tomizaki, T., Yamaguchi, H., Shinzawa-Itoh, K., Nakashima, R., Yaono, R., Yoshikawa, S. (1996) The whole structure of the 13-subunit oxidized cytochrome-c oxidase at 2.8 Å. *Science* 272: 1136–1144. Copyright © 1996 American Association for the Advancement of Science.

Figure 14.44: Berry, J. A., Downton, J. S. (1982) Environmental regulation of photosynthesis. In *Photosynthesis, Development, Carbon Metabolism and Plant Productivity*, Vol. 2, Govindjee, ed. Academic Press, New York, pp. 263–343. Reprinted by permission of Academic Press. Copyright © 1982.

Figure 14.49: Berry, J. A., Downton, J. S. (1982) Environmental regulation of photosynthesis. In *Photosynthesis, Development, Carbon Metabolism and Plant Productivity*, Vol. 2, Govindjee, ed. Academic Press, New York, pp. 263–343. Reprinted by permission of Academic Press. Copyright © 1982.

Figure 14.50: Berry, J., Bjorkman, O. (1980) Photosynthetic response and adaptation to temperature in higher plants. *Annu. Rev. Plant Physiol.* 31: 491–543.

Box 14.2: M. Benstead, Fife, Scotland, UK; previously unpublished.

Part 4 opening photograph:
Galen Rowell/Mountain Light Photography, Emeryville, CA.

Chapter 15
Figure 15.1A–J, photograph: D. B. Fisher, Washington State University, Pullman; previously unpublished.

Figure 15.7: Protein data adapted from le Maire, M., Aggerbeck, L. P., Monteilhet, C., Andersen, J. P., Moller, J. V. (1986) The use of high-performance liquid chromatography for the determination of size and molecular weight of proteins: a caution and a list of membrane proteins suitable as standards. *Anal. Biochem.* 154: 525–535. Copyright © Academic Press. Dextran data adapted from Jorgensen, D. E., Moller, J. V. (1979) Use of flexible polymers as probes of glomerular pore size. *Am. J. Physiol.* 236: F103–F111. Copyright © American Physiological Society.

Figure 15.10A–D: Koontz, H., Biddulph, O. (1957) Factors regulating absorption and translocation of foliar applied phosphorus. *Plant Physiol.* 32: 463–470.

Figure 15.13A,B: Gunning, B.E.S., Pate, J. S., Green, L. W. (1970) Transfer cells in the vascular system of stems: taxonomy, associations with nodes, and structure. *Protoplasma* 71: 147–171. Copyright © 1970 Springer-Verlag.

Figure 15.16A,B: Sperry, J. S., Donnelly, J. R., Tyree, M. T. (1988) A method for measuring hydraulic conductivity and embolism in xylem. *Plant Cell Environ.* 11: 35–40. Used with permission from Blackwell Science Ltd.

Figure 15.18: Adapted by permission from Sperry, J. S. (1995) Limitations on stem water transport and their consequences. In *Plant Stems. Physiological and Functional Morphology*, B. L. Gartner, ed. Academic Press, San Diego, pp. 105–125. Copyright © 1995 Academic Press.

Figure 15.19: D. B. Fisher, Washington State University, Pullman; previously unpublished.

Figure 15.20A: Longitudinal section: D. B. Fisher, Washington State University, Pullman; previously unpublished. Cross-section: Overall, R. L., Wolfe, J., Gunning, B.E.S. (1982) Intercellular communication in *Azolla* roots: I. Ultrastructure of plasmodesmata. *Protoplasma* 111: 134–150. Copyright © 1982 Springer-Verlag.

Figure 15.21: Radford, J. E., Vesk, M., Overall, R. L. (1998) Callose deposition in plasmodesmata. *Protoplasma* 201: 30–37. Copyright © 1998 Springer-Verlag.

Figure 15.22: Radford, J. E., Vesk, M., Overall, R. L. (1998) Callose deposition in plasmodesmata. *Protoplasma* 201: 30–37. Copyright © 1998 Springer-Verlag.

Figure 15.23: Hepler, P. K. (1982) Endoplasmic reticulum in the formation of the cell plate and plasmodesmata. *Protoplasma* 111: 121–133. Copyright © 1982 Springer-Verlag.

Figure 15.25: Adapted from van Bell, A.J.E., Oparka, K. J. (1995) On the validity of plasmodesmograms. *Bot. Acta* 108: 174–182. Copyright © 1995 Georg Thieme Verlag, Stuttgart–New York.

Figure 15.26A–F: Adapted from Kollmann, R., Glockmann, C. (1991) Studies on graft unions. III. On the mechanism of secondary formation of plasmodesmata at the graft interface. *Protoplasma* 165: 71–85. Copyright © 1991 Springer-Verlag.

Figure 15.29: Adapted from Overall, R. L., Gunning, B.E.S. (1982) Intercellular communication in *Azolla* roots: II. Electrical coupling. *Protoplasma* 111: 151–160. Copyright © 1982 Springer-Verlag.

Figure 15.30A–D: Terry, B. R., Robards, A. W. (1987) Hydrodynamic radius alone governs the mobility of molecules through plasmodesmata. *Planta* 171: 145–157. Copyright © 1987 Springer-Verlag.

Figure 15.31: Duckett, C. M., Oparka, K. J., Prior, D.A.M., Dolan, L., Roberts, K. (1994) Dye-coupling in the root epidermis of *Arabidopsis* is progressively reduced during development. *Development* 120: 3247–3255. Used with permission from Company of Biologists Ltd.

Figure 15.32A,B: Wright, K. M., Oparka, K. J. (1997) Metabolic inhibitors induce symplastic movement of solutes from the transport phloem of *Arabidopsis* roots. *J. Exp. Bot.* 48: 1807–1814. Used with permission from Oxford University Press.

Figure 15.33A,B: Adapted from data of Tyree, M. T., Tammes, P.M.L. (1972) Translocation of uranin in the symplasm of staminal hairs of *Tradescantia*. *Can. J. Bot.* 53: 2038–2046. Reprinted with permission from NRC Research Press.

Figure 15.34A,B: Yahalom, A., Warmbrodt, R. D., Laird, D. W., Traub, O., Revel, J. P., Willecke, K., Epel, B. L. (1991) Maize mesocotyl plasmodesmata proteins cross-react with connexin gap junction protein antibodies. *Plant Cell* 3: 407–417.

Figure 15.35A–C: Blackman, L. M., Boevink, P., Santa Cruz, S., Palukaitis, P., Oparka, K. J. (1998) The movement protein of cucumber mosaic virus traffics into sieve elements in minor veins of *Nicotiana clevelandi*. *Plant Cell* 10: 525–537.

Figure 15.36A,B: Fisher, D. B. (1975) Structure of functional soybean sieve elements. *Plant Physiol.* 56: 555–569.

Figure 15.37: Knoblauch, M., van Bel, A.J.E. (1998) Sieve tubes in action. *Plant Cell* 10: 35–50.

Figure 15.38: Adapted with permission from Watson, B. T. (1975) The influence of low temperature on the rate of translocation in the phloem of *Salix viminalis* L. *Ann. Bot.* 39: 889–900. Copyright © 1975 Academic Press.

Figure 15.39: Adapted with permission from Hocking, P. J. (1980) The composition of phloem exudate and xylem sap from tree tobacco (*Nicotiana glauca* Grah.). *Ann. Bot.* 45: 633–643. Copyright © 1980 Academic Press.

Figure 15.40: Esau, K., Thorsch, J. (1985) Sieve plate pores and plasmodesmata, the communication channels of the symplast: ultrastructural aspects and developmental relations. *Am. J. Bot.* 72: 1641–1653. Used with permission from Botanical Society of America, Inc. Copyright © 1985.

Figure 15.41(left): Turgeon, R., Beebe, D. U., Gowan, E. (1993) The intermediary cell: minor-vein anatomy and raffinose oligosaccharide synthesis in the Scrophulariaceae. *Planta* 191: 446–456. Copyright © 1993 Springer-Verlag.

Figure 15.41(right): Fisher, D. B., Housley, T. L., Christy A. L. (1978) Source pool kinetics for ^{14}C-photosynthate translocation in morning glory and soybean. *Plant Physiol.* 61: 291–295.

Figure 15.42A,B: Geiger, D. R., Sovonick, S. A., Shock, T. L., Fellows, R. J. (1974) Role of free space in translocation in sugar beet. *Plant Physiol.* 54: 892–898.

Figure 15.43A,B: Giaquinta, R. (1976) Evidence for phloem unloading from the apoplast. Chemical modification of membrane sulfhydryl groups. *Plant Physiol.* 57: 872–875.

Figure 15.44: Adapted from Wright, J. P., Fisher, D. B. (1981) Measurement of the sieve tube membrane potential. *Plant Physiol.* 67: 845–848.

Figure 15.45A,B: Turgeon, R., Wimmers, L. E. (1988) Different patterns of vein loading of exogenous [^{14}C]sucrose in leaves of *Pisum sativum* and *Coleus blumei*. *Plant Physiol.* 87: 179–182.

Figure 15.46: Adapted from Russin, W. A., Evert, R. F. (1985) Studies on the leaf of *Populus deltoides* (Salicaceae): ultrastructure, plasmodesmatal frequency, and solute concentrations. *Am. J. Bot.* 72: 1232–1247. Used with permission from Botanical Society of America, Inc. Copyright © 1985.

Figure 15.48: Adapted from Schmalstig, J., Cosgrove, D. (1990) Coupling of solute transport and cell expansion in pea stems. *Plant Physiol.* 94: 1625–1633.

Figure 15.49: Oparka, K. J., Duckett, C. M., Prior, D.A.M., Fisher, D. B. (1994) Real-time imaging of phloem unloading in the root tip of *Arabidopsis*. *Plant J.* 6: 759–766. Reprinted with permission from Blackwell Science Ltd.

Figure 15.51A–C: D. B. Fisher, Washington State University, Pullman; previously unpublished.

Figure 15.52A,B: Fisher, D. B., Wang, N. (1993) A kinetic and microautoradiographic analysis of [^{14}C]sucrose import by developing wheat grains. *Plant Physiol.* 101: 391–398.

Figure 15.54A: Johnson, R. W., Dixon, M. A., Lee, D. R. (1992) Water relations of the tomato during fruit growth. *Plant Cell Environ.* 15: 947–953. Reprinted with permission from Blackwell Science Ltd.

Figure 15.54B: Gandar, P. W., Turner, C. B. (1976) Potato leaf and tuber water potential measurements with a pressure chamber. *Am. Potato J.* 53: 1–14. Used with permission from the Potato Association of America.

Figure 15.55: Wang, N., Fisher, D. B. (1994) The use of fluorescent tracers to characterize the post-phloem transport pathway in maternal tissues of developing wheat grains. *Plant Physiol.* 104: 17–27.

Figure 15.56: Lang, A. (1990) Xylem, phloem and transpiration flows in developing apple fruits. *J. Exp. Bot.* 41: 645–651. Used with permission from Oxford University Press.

Figure 15.57: Pritchard, J., Tomos, A. D. (1993) Correlating biophysical and biochemical control of root cell expansion. In *Water Deficits: Plant Responses from Cell to Community*, J.A.C. Smith and H. Griffiths, eds. BIOS Scientific Publishers, Oxford, UK, pp. 53–72. Used with permission of BIOS Scientific Publishers Ltd.

Figure 15.58: Fisher, D. B., Wu, Y., Ku, M.S.B. (1992) Turnover of soluble proteins in the wheat sieve tube. *Plant Physiol.* 100: 1433–1441.

Figure 15.59A,B: Oparka, K. J., Roberts, A. G., Boevink, P., Santa Cruz, S., Roberts, I., Pradel, K. S., Imlau, A., Kotlizky, G., Saure, N., Epel, B. (1999) Simple, but not branched, plasmodesmata allow the nonspecific trafficking of proteins in developing tobacco leaves. *Cell* 97: 743–754. Copyright © 1999 Cell Press.

Figure 15.59C,D: Imlau, A., Truernit, E., Sauer, N. (1999) Cell-to-cell and long-distance trafficking of the green fluorescent protein in the phloem and symplastic unloading of the protein into sink tissues. *Plant Cell* 11: 309–322.

Figure 15.60: Fisher, D. B., Wu, Y., Ku, M.S.B. (1992) Turnover of soluble proteins in the wheat sieve tube. *Plant Physiol.* 100: 1433–1441.

Figure 15.61A,B: Reprinted with permission from Xoconostle-Cazares, B., Xiang, Y., Ruiz-Medrano, R., Wang, H.-L., Monzer, J., Yoo, B. C., McFarland, K. C., Franceschi, V. R., Lucas, W. J. (1999) Plant paralog to viral movement protein that potentiates transport of mRNA into the phloem. *Science*

283: 94–98. Copyright © 1999 American Association for the Advancement of Science.

Figure 15.63A,B: Reprinted with permission from Lucas, W. J., Bouche-Pillon, S., Jackson, D. P., Nguyen, L., Baker, L., Ding, B., Hake, S. (1995) Selective trafficking of KNOTTED-1 homeodomain protein and its mRNA through plasmodesmata. *Science* 270: 1980–1983. Copyright © 1995 American Association for the Advancement of Science.

Figure 15.64: Ding, B. (1998) Intercellular protein trafficking through plasmodesmata. *Plant Mol. Biol.* 38: 279–310. With kind permission from Kluwer Academic Publishers.

Figure 15.65: Adapted from Kragler, F., Monzer, J., Shash, K., Xoconostle-Cazares, B., Lucas, W. J. (1998) Cell-to-cell transport of proteins: requirement for unfolding and characterization of binding to a putative plasmodesmal receptor. *Plant J.* 15: 367–381. Reprinted with permission from Blackwell Science Ltd.

Figure 15.66: Ding, B., Kwon, M.-O., Hammond, R., Owens, R. (1997) Cell-to-cell movement of potato spindle tuber viroid. *Plant J.* 12: 931–936. Reprinted with permission from Blackwell Science Ltd.

Box 15.2C–E: D. B. Fisher, Washington State University, Pullman; previously unpublished.

Box 15.3: Graph data from Holbrook, N. M. (1995) Negative xylem pressures in plants: a test of the balancing pressure technique. *Science* 270: 1193–1194. Copyright © 1995 American Association for the Advancement of Science.

Box 15.4A: Eleftheriou, E. P., Tsekos, I. (1982) Development of protophloem in roots of *Aegilops comosa* var. *thessalica*. II. Sieve-element differentiation. *Protoplasma* 113: 221–233. Copyright © 1982 Springer-Verlag.

Box 15.4B,C: Eleftheriou, E. P. (1993) Differentiation of abnormal sieve elements in roots of wheat (*Tritium aestivum* L.) affected by colchicine. *New Phytol.* 125: 813–827. Reprinted with permission of Academic Press. Copyright © 1993.

Table 15.3: Gamalei, Y. (1991) Phloem loading and its development related to plant evolution from trees to herbs. *Trees* 5: 50–64. Copyright © 1991 Springer-Verlag.

Chapter 16

Figure 16.3: Ulrich, A., Hills, F. J. (1969) *Sugar Beet Nutrient Deficiency Symptoms. A Color Atlas and Chemical Guide*. University of California, Division of Agriculture and Natural Resources, Davis. Copyright © 1969 University of California Board of Regents.

Figure 16.5: Adapted from Stacey, G., Burris, R. H., Evans, H. J., eds. (1992) *Biological Nitrogen Fixation*. Chapman & Hall, New York. Used with kind permission from Kluwer Academic Publishers.

Figure 16.6B: B. Newton, Virginia Polytechnic Institute and State University, Blacksburg, Virginia, unpublished; based on data from Kim, J., Rees, D. C. (1994) Ribbon structure of nitrogenase complex including Fe and MoFe proteins. Nitrogenase and biological nitrogen fixation. *Biochemistry* 33: 389–397.

Figure 16.7: Adapted by permission from Schindelin, H., Kisker, C., Schlessman, J. L., Howard, J. B., Rees, D. C. (1987) Structure of ADP•AlF$_4$–stabilized nitrogenase complex and its implications for signal transduction. *Nature* 387: 370–376. Copyright © 1987 Macmillan Magazines Ltd.

Figure 16.8: Peters, J. W., Stowell, M.H.B., Soltis, S. M., Finnegan, M. G., Johnson, M. K., Rees, D. C. (1997) Redox-dependent structural changes in the nitrogenase P-cluster. *Biochemistry* 36: 1181–1187. Copyright © 1997 American Chemical Society.

Figure 16.10: Reprinted by permission from Schindelin, H., Kisker, C., Schlessman, J. L., Howard, J. B., Rees, D. C. (1987) Structure of ADP•AlF$_4$–stabilized nitrogenase complex and its implications for signal transduction. *Nature* 387: 370–376. Copyright © 1987 Macmillan Magazines Ltd.

Figure 16.11: Lowe, D. J. (1992) A role of the 'P' centres of *Klebsiella pneumoniae* MoFe protein in reducing dinitrogen. In *Ninth International Congress on Nitrogen Fixation*, R. Palacios, J. Mora, and W. E. Newton, eds. Kluwer Academic Publishers, Dordrecht, The Netherlands. Used with kind permission from Kluwer Academic Publishers.

Figure 16.12: L. Feldman, University of California, Berkeley; previously unpublished.

Figure 16.13: Adapted from Soltis, D. E., Soltis, P. S., Morgan, D. R., Swensen, S. M., Mullin, B. C., Dowd, J. M., Martin, P. G. (1995) Chloroplast gene sequence data suggest a single origin of the predisposition for symbiotic nitrogen fixation in angiosperms. *Proc. Natl. Acad. Sci. USA* 92: 2647–2651. Copyright © 1995 National Academy of Sciences, U.S.A.

Figure 16.14: S. R. Long, Stanford University, Palo Alto, CA; previously unpublished.

Figure 16.15: Adapted from van Berkum, P., Eardly, B. D. (1998) Molecular evolutionary systematics of the Rhizobiaceae. In *The Rhizobiaceae*, H. P. Spaink, A. Kondorosi, and P.J.J. Hooykaas, eds. Kluwer Academic Publishers, Dordrecht, The Netherlands, pp. 1–24. Used with kind permission from Kluwer Academic Publishers.

Figure 16.17A,B: Photographs: D. Ehrhardt and S. R. Long, Stanford University, Palo Alto, CA; previously unpublished.

Figure 16.17C: Photographs: VandenBosch, K. A., Newcomb, E. H. (1986) Immunogold localization of nodule-specific uricase in developing soybean root nodules. *Planta*

167: 425–436. Copyright © 1987 Springer-Verlag.

Figure 16.21A: Maps for *R. leuminosarum* bv. *viciae, S. meliloti,* and *B. japonicum* adapted from Schlaman, H.R.M., Phillips, D. A., Kondorosi, E. (1998) Genetic organization and transcriptional regulation of rhizobial nodulation genes. In *The Rhizobiaceae,* H. P. Spaink, A. Kondorosi, and P.J.J. Hooykaas, eds. Kluwer Academic Publishers, Dordrecht, The Netherlands, pp. 361–386. Used with kind permission from Kluwer Academic Publishers.

Figure 16.21B: Map for pNGR34 adapted by permission from Freiberg, C., Fellay, R., Bairoch, A., Broughton, W. J., Rosenthal, A., Perret, X. (1997) Molecular basis of symbiosis between *Rhizobium* and legumes. *Nature* 387: 394–401. Copyright © 1997 Macmillan Magazines Ltd.

Figure 16.25: Adapted from Becker, A., Pühler, A. (1999) Production of exopolysaccharides. In *The Rhizobiaceae,* H. P. Spaink, A. Kondorosi, and P.J.J. Hooykaas, eds. Kluwer Academic Publishers, Dordrecht, The Netherlands, p. 110. Used with kind permission from Kluwer Academic Publishers.

Figure 16.26: Felle, H. H., Kondorosi, E., Kondorosi, A., Schultze, M. (1998) The role of ion fluxes in Nod factor signalling in *Mendicago sativa. Plant J.* 13: 455–463. Reprinted with permission from Blackwell Science Ltd.

Figure 16.27: Ehrhardt, D. W., Wais, R., Long, S. R. (1996) Calcium spiking in plant root hairs responding to *Rhizobium* nodulation signals. *Cell* 85: 673–681. Copyright © 1996 Cell Press.

Figure 16.28: Scheres, B., Van De Wiel, C., Zalensky, A., Horvath, B., Spaink, H., Van Eck, H., Zwartkruis, F., Wolters, A.-M., Gloudemans, T., Van Kammen, A., Bisseling, T. (1990) The ENOD12 gene product is involved in the infection process during the pea–*Rhizobium* interaction. *Cell* 60: 281–294. Copyright © 1990 Cell Press.

Figure 16.29: Heidstra, R., Yang, W. C., Yalcin, Y., Peck, S., Emons, A., Van Kammen, A., Bisseing, T. (1997) Ethylene provides positional information on cortical cell division but is not involved in Nod factor-induced root hair tip growth in *Rhizobium*–legume interaction. *Development* 124: 1781–1787. Used with permission from Company of Biologists Ltd.

Figure 16.35A,B: Adapted from Touraine, B., Glass, A.D.M. (1997) NO$_3^-$ and ClO$_3^-$ fluxes in the *chl1-5* mutant of *Arabidopsis thaliana*—does the *CHL1-5* gene encode a low-affinity NO$_3^-$ transporter? *Plant Physiol.* 114: 137–144.

Figure 16.32B: Dennis, D. T., Turpin, D. H., Lefebvre, D. D., Layzell, D. B., eds. (1997) *Plant Metabolism,* 2nd ed. Longman Publishing, Addison Wesley Longman Ltd., Singapore, pp. 450–477.

Figure 16.36: N. M. Crawford, University of California–San Diego, La Jolla, CA, previously unpublished.

Figure 16.37: Adapted from Wang, R., Crawford, N. M. (1996) Genetic identification of a gene involved in constitutive, high affinity, nitrate transport in *Arabidopsis thaliana. Proc. Natl. Acad. Sci. USA* 93: 9297–9301. Copyright © 1996 National Academy of Sciences, U.S.A.

Figure 16.38B: W. H. Campbell, Michigan Technical Institute, Houghton; previously unpublished.

Figure 16.39: Campbell, W. H. (1999) Nitrate reductase structure, function and regulation: bridging the gap between biochemistry and physiology. *Annu. Rev. Plant Physiol. Plant Mol. Biol.* 50: 277–303. Copyright © 1999 Annual Reviews. www.AnnualReviews.org.

Figure 16.40: Reprinted from Guoguang, L., Campbell, W. H., Schneider, G., Lindquist, Y. (1994) Crystal structure of the FAD-containing fragment of corn nitrate reductase at 2.5 Å resolution: relationship to other flavoprotein reductases. *Structure* 2: 809–821. Copyright © 1994 with permission from Elsevier Science.

Figure 16.41A: Adapted from Melzer, J. M., Kleinhofs, A., Warner, R. L. (1989) Nitrate reductase regulation: effects of nitrate and light on nitrate reductase mRNA accumulation. *Mol. Gen. Genet.* 217: 341–346. Copyright © 1989 Springer-Verlag.

Figure 16.41B: Adapted from Pilgrim, M. L., Caspar, T., Quail, P. H., McClung, C. R. (1993) Circadian and light-regulated expression of nitrate reductase in *Arabidopsis. Plant Mol. Biol.* 23: 349–364. Used with kind permission from Kluwer Academic Publishers.

Figure 16.42: Adapted from Huber, S. C., Bachmann, M., Huber, J. L. (1996) Post-translational regulation of nitrate reductase activity: a role for Ca^{2+} and 14-3-3 proteins. *Trends Plant Sci.* 1: 432–438. Copyright © 1996 with permission from Elsevier Science.

Figure 16.52: Schnug, E., Haneklaus, S. (1994) Symptomatology of S deficiency in *Brassica* species. *Landbauforsch. Voelkenrode Sonderh.* 144: 14–21. Reprinted with permission.

Figure 16.53: Byers, M., Franklin, J., Smith, S. J. (1987) The nitrogen and sulphur nutrition of wheat and its effect on the composition and baking quality of the grain. *Aspects Appl. Biol.* 15: 337–344. Used by permission from the Association of Applied Biologists.

Figure 16.57: Adapted from Nissen, P. (1971) Uptake of sulfate by roots and leaf slices of barley mediated by single, multiphasic mechanisms. *Physiol. Plant.* 24: 315–324. Copyright © 1971 Physiologium Plantarum and published by Munksgaard International Publishers Ltd., Copenhagen, Denmark.

Figure 16.58: Hawkesford, M. J., Smith, F. W. (1997) Molecular biology of higher plant sulphate transporters. In *Sulphur Metabolism*

in Higher Plants, W. J. Cram, L. J. DeKok, I. Stulen, C. Brunold, and H. Rennenberg, eds. Backhuys Publishers, Leiden, The Netherlands, pp. 13–26. Used with permission.

Figure 16.63A,B: Sangman, L. (1999) Molecular analysis of sulfate assimilation in higher plants: effect of cysteine, sulfur, and nitrogen nutrients, heavy metal stress, and genomic DNA cloning. Ph.D. thesis. Rutgers, The State University of New Jersey, New Brunswick, NJ, 155 pp. Used with permission of the author.

Figure 16.69B: Adapted from Rauser, W. (1995) Phytochelatins and related peptides. *Plant Physiol.* 109: 1141–1149.

Box 16.1: Chinese characters: Ho, P.-T. (1997) Indigenous origins of Chinese agriculture. In *Origins of Agriculture,* C. A. Reed, ed. Mouton Publishers, The Hague. Flowchart: National Fertilizer Development Center, Tennessee Valley Authority, Chattanooga, TN.

Table 16.2: Vitousek, P. M., Aber, J. D., Howart, R. W., Likens, G. E., Matson, P. A., Schindler, D. W., Schlesinger, W. H., Tilman, D. G. (1997) Human alteration of the global nitrogen cycle: sources and consequences. *Ecol. Appl.* 7: 737–750. Reprinted with permission from Ecological Society of America.

Chapter 17

Figure 17.5: A. Crozier, University of Glasgow, UK; previously unpublished.

Figure 17.6: Lang, A. (1957) The effect of gibberellin upon flower formation. *Proc. Natl. Acad. Sci. USA* 43: 709–717.

Figure 17.7: A. Crozier, University of Glasgow, UK; previously unpublished.

Figure 17.11A–G: Silverstone, A. L., Chang, C.-W., Krol, E., Sun, T.-P. (1997) Developmental regulation of the gibberellin biosynthetic gene *GA1* in *Arabidopsis thaliana. Plant J.* 12: 9–119. Used with permission from Blackwell Science Ltd.

Figure 17.14: Phillips, A. L., Ward, D. A., Ukness, S., Appleford, N.E.J., Lange, T., Huttly, A. K., Gaskin, P., Graebe, J. E., Hedden, P. (1995) Isolation and expression of three gibberellin 20-oxidase cDNA clones from *Arabidopsis. Plant Physiol.* 108: 1049–1057.

Figure 17.18: Phillips, A. L., Ward, D. A., Ukness, S., Appleford, N.E.J., Lange, T., Huttly, A. K., Gaskin, P., Graebe, J. E., Hedden, P. (1995) Isolation and expression of three gibberellin 20-oxidase cDNA clones from *Arabidopsis. Plant Physiol.* 108: 1049–1057.

Figure 17.19: Adapted from Gilmour, S. J., Zeevaart, J.A.D., Schwenen, L., Graebe, J. E. (1986) Gibberellin metabolism in cell-free extracts from spinach (*Spinacia oleracea*) leaves in relation to photoperiod. *Plant Physiol.* 82: 190–195.

Figure 17.20: Electrophoretograms: Toyomasu, T., Kawaide, H., Mitsuhashi, W., Inoue, Y., Kamiya, Y. (1998) Phytochrome regulates gibberellin biosynthesis during germination of photoblastic lettuce seeds. *Plant Physiol.* 118: 1517–1523.

Figure 17.22: S. McCormick, University of California, Berkeley; previously unpublished.

Figure 17.23: J. Weyers, University of Dundee, UK; previously unpublished.

Figure 17.31: T. Kakimoto, Osaka University, Japan; previously unpublished.

Figure 17.33: I. Scott, University of Wales, Aberystwyth, UK; previously unpublished.

Figure 17.34: G. Sandberg, Swedish University of Agricultural Sciences, Umeå; previously unpublished.

Figure 17.35A,B: Faiss, M., Zalubilova, J., Strand, M., Schmülling, T. (1997) Conditional transgenic expression of the *ipt* gene indicates a function for cytokinins in paracrine signaling in whole tobacco plants. *Plant J.* 12: 401–415. Used with permission from Blackwell Science Ltd.

Figure 17.36: Martineau, B., Houck, C. M., Sheehy, R. E., Hiatt, W. R. (1994) Fruit-specific expression of the *A. tumefaciens* isopentenyl transferase gene in tomato: effects on fruit ripening and defense-related gene expression in leaves. *Plant J.* 5: 11–19. Used with permission from Blackwell Science Ltd.

Figure 17.43: J. Cohen, U.S. Department of Agriculture, Beltsville, MD, and A. Wright, University of Missouri, Columbia; previously unpublished.

Figure 17.49: Adapted from Normanly, J., Grisafi, P., Fink, G. R., Bartel, B. (1997) *Arabidopsis* mutants resistant to the auxin effects of indole-3-acetonitrile are defective in the nitrilase encoded by the *NIT1* gene. *Plant Cell* 9: 1781–1790.

Figure 17.51: Adapted from Nilsson, O., Crozier, A., Schmulling, T., Sandberg, G., Olsson, O. (1993) Indole-3-acetic acid homeostasis in transgenic tobacco plants expressing *Agrobacterium rhizogenes rolB* gene. *Plant J.* 3: 681–689. Used with permission from Blackwell Science Ltd.

Figure 17.52: Romano, C. P., Cooper, M. L., Klee, H. J. (1993) Uncoupling auxin and ethylene effects in transgenic tobacco and *Arabidopsis* plants. *Plant Cell* 5: 181–189.

Figure 17.53A–C: H. Mori, Nagoya University, Japan; previously unpublished.

Figure 17.56: D. Grierson, University of Nottingham, UK; previously unpublished.

Figure 17.58: T. Yokota, Teikyo University Utsunomiya, Japan; previously unpublished.

Figure 17.60: T. Yokota, Teikyo University, Utsunomiya, Japan; previously unpublished.

Figure 17.71: Adapted from Kumar, A., Altabella, T., Taylor, M. A., Tiburcio, A. F. (1997) Recent advances in polyamine research. *Trends Plant Sci.* 2: 124–130. Copyright © 1997 with permission from Elsevier Science.

Figure 17.74 A,B: Kubigsteltig, I., Laudert, D., Weiler, E. W. (1999) Structure and regulation of the *Arabidopsis thaliana* allene oxide synthase gene. *Planta* 208: 463–471. Copyright © 1999 Springer-Verlag.

Figure 17.78: Reprinted with permission from Raskin, I., Ehmann, A., Melander, W. R., Meeuse, B.J.D. (1987) Salicylic acid—a natural inducer of heat production in Arum lilies. *Science* 237: 1545–1556. Copyright © 1987 American Association for the Advancement of Science.

Figure 17.79: J. Draper, University of Wales, Aberystwyth, UK; previously unpublished.

Figure 17.81: J. Draper, University of Wales, Aberystwyth, UK; previously unpublished.

Table 17.2: Kumar, A., Taylor, M. A., Arif, S.A.M., Davies, H. V. (1996) Potato plants expressing antisense and sense S-adenosylmethionine decarboxylase (SAMDC) transgenes sho altered levels of polyamines and ethylene: antisense plants display abnormal phenotypes. *Plant J.* 9: 147–158. Used with permission from Blackwell Science Ltd.

Chapter 18

Figure 18.2: Staehelin, L. A., Giddings, T. H. Jr., Kiss, J. Z., Sack, F. D. (1990) Macromolecular differentiation of Golgi stacks in root tips of *Arabidopsis* and *Nicotiana* seedlings as visualized in high-pressure frozen and freeze-substituted samples. *Protoplasma* 157: 75–91. Copyright © 1990 Springer-Verlag.

Figure 18.4: X.-W. Deng, Yale University, New Haven, CT; previously unpublished.

Figure 18.7: Adapted from Wallace, D.H., Enriquez, G.A. (1980) Daylength and temperature effects on days to flowering of early and late maturing beans (*Phaseolus vulgaris* L.). *J. Am. Hort. Sci.* 105: 583–591.

Figure 18.9: W. Gruissem, University of California, Berkeley; previously unpublished.

Figure 18.14: Adapted from Plakidou-Dymock, S., Dymock, D., Hooley, R. (1998) A higher plant seven-transmembrane receptor that influences sensitivity to cytokinins. *Curr. Biol.* 8: 315–324. Copyright © 1998 with permission from Elsevier Science.

Figure 18.22: Wilkinson, J. Q., Lanahan, M. B., Clark, D. G., Bleecker, A. B., Chang, C., Meyerowitz, E. M., Klee, H. J. (1997) A dominant mutant receptor from *Arabidopsis* confers ethylene insensitivity in heterologous plants. *Nat. Biotechnol.* 15: 444–447.

Figure 18.25: A. J. Trewavas, Institute of Cell and Molecular Biology, University of Edinburgh, UK; previously unpublished.

Figure 18.26A,B: Data from Allan, A. C., Fricker, M. D., Ward, J. L., Beale, M. H., Trewavas, A. J. (1994) Two transduction pathways mediate rapid effects of abscisic acid

in *Commelina* guard cells. *Plant Cell* 6: 1319–1328.

Figure 18.30: Adapted from Wittinghofer, A., Pai, E. F. (1991) The structure of Ras protein: a model for a universal molecular switch. *Trends Biochem. Sci.* 16: 382–387. Copyright © 1991 with permission from Elsevier Science.

Figure 18.36: Adapted from Munnik, T., Irvine, R. F., and Musgrave, A. (1998) Phospholipid signalling in plants. *Biochim. Biophys. Acta* 1389: 222–272.

Figure 18.37: Adapted from Munnik, T., Irvine, R. F., and Musgrave, A. (1998) Phospholipid signalling in plants. *Biochim. Biophys. Acta* 1389: 222–272.

Figure 18.46: A. J. Trewavas, Institute of Cell and Molecular Biology, University of Edinburgh, Scotland; previously unpublished.

Figure 18.47: Reprinted by permission from Shacklock, P. S., Read, N. D., Trewavas, A. J. (1992) Cytosolic free calcium mediates red light-induced photomorphogenesis. *Nature* 358: 753–755. Copyright © 1992 Macmillan Magazines Ltd.

Figure 18.48: Mahlo, R., Read, N. D., Pais, M. S., Trewavas, A. J. (1994) Role of cytosolic free calcium in the reorientation of pollen tube growth. *Plant J.* 5: 331–341. Used with permission from Blackwell Science Ltd.

Figure 18.51: A. J. Trewavas, Institute of Cell and Molecular Biology, University of Edinburgh, Scotland; previously unpublished.

Figure 18.52: Gilroy, S., Fricker, M. D., Read, N. D., Trewavas, A. J. (1991) Role of calcium in signal transduction of *Commelina* guard cells. *Plant Cell* 3: 333–341.

Figure 18.53B: Pierson, E. S., Miller, D. D., Callahan, D. A., Van Aken, J., Hackett, G., Hepler, P. K. (1996) Tip-localized calcium entry fluctuates during pollen tube growth. *Dev. Biol.* 174: 160–173. Reprinted by permission of Academic Press. Copyright © 1996.

Figure 18.54A: Franklin-Tong, V. E., Drobak, B. K., Allan, A. C., Watkins, P.A.C., Trewavas, A. J. (1996) Growth of pollen tubes of *Papaver rhoeas* is regulated by a slow-moving calcium wave propagated by inositol 1,4,5-trisphosphate. *Plant Cell* 8: 1305–1321.

Figure 18.55: Reproduced from Knight, M. R., Read, N. D., Campbell, A. K., Trewavas, A. J. (1993) Imaging calcium dynamics in living plants using semi-synthetic recombinant aequorins. *J. Cell Biol.* 121: 83–90. By copyright permission from the Rockefeller University Press.

Figure 18.57: Adapted with permission from Ikura, M., Clore, G. M., Gronenborn, A. M., Zhu, G., Klee, C. B., Bax, A. (1992) Solution structure of a calmodulin–target peptide complex by multidimensional

NMR. *Science* 256: 632–638. Published by American Association for the Advancement of Science.

Figure 18.65A: K. Stepnitz, Michigan State University, East Lansing; published as the cover of *Science,* 26 August 1988, accompanying the paper by Bleecker, A. B., Estelle, M. A., Somerville, C., Kende, H. (1988) Insensitivity to ethylene conferred by a dominant mutation in *Arabidopsis thaliana. Science* 241: 1086–1089. Used with permission of the photographer.

Figure 18.65B: J. Keiber, University of Chicago; previously unpublished.

Fig. 18.73: Adapted with permission from DeKoninck, P., Schulman, H. (1998) Sensitivity of CaM kinase II to the frequency of Ca^{2+} oscillations. *Science* 279: 227–230. Copyright © 1998 American Association for the Advancement of Science.

Box 18.4: Adapted from Furuya, M., Schafer, E. (1996) Photo perception and signalling of induction reactions by different phytochromes. *Trends Plant Sci.* 1: 301–307. Copyright © 1996 by permission from Elsevier Science.

Box 18.5A,B: Mahlo, R., Trewavas, A. J. (1996) Localized apical increases of cytosolic free calcium control pollen tube orientation. *Plant Cell* 8: 1935–1949.

Box 18.6B,C: Data from Trewavas, A. J., Malho, R. (1998) Ca^{2+} signalling in plant cells: the big network! *Curr. Opin. Plant Biol.* 1: 428–434.

Box 18.6D: H. Page, N. D. Read, and A. J. Trewavas, Institute of Cell and Molecular Biology, University of Edinburgh, Scotland; previously unpublished.

Table 18.1: Data from Hewitt, E. J. (1963) Inorganic nutrition. In *Plant Physiology; a Treatise,* Vol. 3, F. C. Steward, ed. Academic Press, New York, pp. 137–361.

Chapter 19

Figure 19.1: Adapted from Goldberg, R. B. (1988) Plants' novel developmental processes. *Science* 240: 1460–1467.

Figure 19.3: Adapted from Lang, A. (1965) Physiology of flower initiation. In *Handbuch der Pflanzenphysiologie (Encyclopedia of Plant Physiology),* W. Ruhland, ed. Springer-Verlag, Berlin, pp. 1379–1536. Copyright © 1965 Springer-Verlag.

Figure 19.4: F. Hempel, University of California, Berkeley; previously unpublished.

Figure 19.5: G. Coupland, John Innes Centre, Norwich, UK; previously unpublished.

Figure 19.7A,B: R. Amasino, University of Wisconsin, Madison; previously unpublished.

Figure 19.8: D. Weigel, Salk Institute, San Diego, CA; previously unpublished.

Figure 19.9A: Photograph: Shannon, S., Meeks-Wagner, D. R. (1991) A mutation in the *Arabidopsis TFL1* gene affects inflorescence meristem development. *Plant Cell* 3: 877–892.

Figure 19.9B: Adapted with permission from Bradley, D., Ratcliffe, O., Vincent, C., Carpenter, R., Coen, E. (1997) Inflorescence commitment and architecture in *Arabidopsis. Science* 275: 80–83. Copyright © 1997 American Association for the Advancement of Science.

Figure 19.12A,B: Smyth, D. R., Bowman, J. L., Meyerowitz, E. M. (1990) Early flower development in *Arabidopsis. Plant Cell* 2: 755–767.

Figure 19.13: Smyth, D. R., Bowman, J. L., Meyerowitz, E. M. (1990) Early flower development in *Arabidopsis. Plant Cell* 2: 755–767.

Figure 19.14: Smyth, D. R., Bowman, J. L., Meyerowitz, E. M. (1990) Early flower development in *Arabidopsis. Plant Cell* 2: 755–767.

Figure 19.15: Meyerowitz, E. M., Bowman, J. L., Brockman, L. L., Drews, G. N., Jack, T., Sieburth, L. E., Weigel, D. (1991) A genetic and molecular model for flower development in *Arabidopsis thaliana. Development* (Suppl. 1): 157–167. Used with permission from Company of Biologists Ltd.

Figure 19.18A: Adapted from Okamuro, J. K., den Boer, B.G.W., Jofuku, K. D. (1993) Regulation of *Arabidopsis* flower development. *Plant Cell* 5: 1183–1193.

Figure 19.19: Münster, T., Pahnke, J., Di Rosa, A., Kim, J. T., Martin, W., Saedler, H., Theissen, G. (1997) Floral homeotic genes were recruited from homologous MADS-box genes preexisting in the common ancestor of ferns and seed plants. *Proc. Natl. Acad. Sci. USA* 94: 2415–2420. Copyright © 1997 National Academy of Sciences, U.S.A.

Figure 19.20: Chen, L., Cheng, J.-C., Castle, L., Sung, Z. R. (1997) *EMF* genes regulate *Arabidopsis* inflorescence development. *Plant Cell* 9: 2011–2024.

Figure 19.22: Wagner, V. T., Cresti, M., Salvatice, P., Tiezzi, A. (1990) Changes in volume, surface area, and frequency of nuclear pores on the vegetative nucleus of tobacco pollen in fresh, hydrated, and activated conditions. *Planta* 181: 304–309. Copyright © 1990 Springer-Verlag.

Figure 19.23: Plantain, buttercup, ox-eye daisy, nettle: M. P. Kage, Okapia, Frankfurt, Germany; published in Attenborough, D. (1995) *The Private Lives of Plants. A Natural History of Plant Behaviour.* Princeton University Press, Princeton, NJ. Used with permission from Okapia. Winter's bark, morning glory, rose cactus, aheaahea: Gifford, E. M., and Foster, A. S. (1987) *Morphology and Evolution of Vascular Plants,* 3rd ed. W. H. Freeman and Co., New York. Poppy: Stern, K. R. (1991) *Introductory Plant Biology,* 5th ed. W. C. Brown Publishers, Dubuque, IA. Used with permission of E. M. Gifford. Ragweed: Howlett, B. J., Knox, R. B., Heslop-Harrison,

J. (1973) Pollen-wall proteins: release of the allergen antigen E from intine and exine sites in pollen grains of ragweed and *Cosmos. J. Cell Sci.* 13: 603–619. Used with permission from Company of Biologists Ltd. Sycamore and hollyhock: Andrew & Polly Syred, Microscopix, Powys, UK. Used with permission from Microscopix.

Figure 19.24: Photograph: Drews, G. N., Lee, D., Christensen, C. (1998) Genetic analysis of female gametophyte development. *Plant Cell* 10: 5–17.

Figure 19.25: Adapted from Nadeau, J. A., Zhang, X. S., Li, J., O'Neill, S. D.(1996) Ovule development: identification of state-specific and tissue-specific cDNAs. *Plant Cell* 8: 213–239.

Figure 19.26: Drews, G. N., Lee, D., Christen- sen, C. (1998) Genetic analysis of female gametophyte development. *Plant Cell* 10: 5–17.

Figure 19.27: U. Grossniklaus, Basel, Switzerland; published as the cover of the Cold Spring Harbor Laboratory 1998 Annual Report, Cold Spring Harbor Laboratory Press, Cold Spring Harbor, NY.

Figure 19.28: Adapted from Wise, R. P., Brownson, C., Schnable, P. S., Horner, H. T. (1998) T-cytoplasmic male sterility of maize. *Adv. Agron.* 65: 79–130. Reprinted by permission of Academic Press. Copyright © 1998.

Figure 19.29A: Adapted from Mascarenjas, J. P. (1993) Molecular mechanisms of pollen tube growth and differentiation. *Plant Cell* 5: 1303–1314.

Figure 19.29B: Photograph: Newbigin, E., Anderson, M. A., Clarke, A. E. (1993) Gametophytic self-incompatibility systems. *Plant Cell* 5: 1315–1324.

Figure 19.30: Jauh, G. Y., Eckhard, K. J., Nothnagal, E. A., Lord, E. M. (1997) Adhesion of lily pollen tubes on an artificial matrix. *Sex. Plant Reprod.* 10: 173–180. Copyright © 1997 Springer-Verlag.

Figure 19.31A–D: Higashiyama, T., Kuroiwa, H., Kawano, S., Kuroiwa, T. (1998) Guidance in vitro of the pollen tube to the naked embryo sac of *Torenia fournieri. Plant Cell* 10: 2019–2031.

Figure 19.32: Adapted from McCormick, S. (1998) Self-incompatibility and other pollen–pistil interactions. *Curr. Opin. Plant Biol.* 1: 18–25.

Figure 19.33: Photograph: Nasrallah, J. B., Nasrallah, M. E. (1993) Pollen-stigma signaling in the sporophytic self-incompatibility response. *Plant Cell* 5: 1325–1335.

Figure 19.34A–D: Photograph: Dickinson, H. G., Elleman, C. J. (1994) Pollen hydrodynamics and self-incompatibility in *Brassica oleracea.* In *Pollen–Pistil Interactions and Pollen Tube Growth (Current Topics in Plant Physiology,* Vol. 12), A. G. Stephenson and T.-H. Kao, eds. American Society of Plant Physiologists, Rockville, MD, pp. 45–61.

Figure 19.36: Drawings and pre-globular through heart photos: Yadegari, R., de Paiva, G. R., Laux, T., Koltunow, A.M., Apuya, N., Zimmerman, J. L., Fischer, R. L., Harada, J. J., Goldberg, R. B. (1994) Cell differentiation and morphogenesis are uncoupled in *Arabidopsis raspberry* embryos. *Plant Cell* 6: 1713–1729. Torpedo and mature embryo photos: West, M.A.L., Harada, J. J. (1993) Embryogenesis in higher plants: An overview. *Plant Cell* 5: 1361–1369.

Figure 19.37: *lec1* and *fus3*: West, M.A.L., Yee, K. M., Danao, J., Zimmerman, J. L., Fischer, R. L., Goldberg, R. B., Harada, J. J. (1994) *LEAFY COTYLEDON1* is an essential regulator of late embryogenesis and cotyledon identity in *Arabidopsis*. Plant Cell 6: 1731–1745. *raspberry*: Yadegari, R., de Paiva, G. R., Laux, T., Koltunow, A.M., Apuya, N., Zimmerman, J. L., Fischer, R. L., Harada, J. J., Goldberg, R. B. (1994) Cell differentiation and morphogenesis are uncoupled in *Arabidopsis raspberry* embryos. *Plant Cell* 6: 1713–1729. Torpedo and mature embryo photos: West, M.A.L., Harada, J. J. (1993) Embryogenesis in higher plants: An overview. *Plant Cell* 5: 1361–1369.

Figure 19.38: Lopes, M. A., Larkins, B. A. (1993) Endosperm origin, development, and function. *Plant Cell* 5: 1383–1399.

Figure 19.45: Adapted from Krochko, J. E., Bewley, J. D. (1989) Use of electrophoretic techniques in determining the composition of seed storage proteins in alfalfa. *Electrophoresis* 9: 751–763. Copyright © 1989 Springer-Verlag.

Figure 19.47: Data from Gatehouse, J. A., Evans, I. M., Croy, R.R.D., Boulter, D. (1986) Differential expression of genes during legume seed development. *Philos. Trans. R. Soc. Lond. B Biol. Sci.* 314: 367–384; and Boulter, D., Evans, I. M., Ellis, J. R., Shirsat, A., Gatehouse, J. A., Croy, R.R.D. (1987) Differential gene expression in the development of *Pisum-sativum*. *Plant Physiol. Biochem. (Paris)* 25: 283–290. Copyright Elsevier Science.

Figure 19.48: Adapted from Kawagoe, Y., Murai, N. (1992) Four distinct nuclear proteins recognize in vitro the proximal promoter of the bean seed storage protein β-phaseolin conferring spatial and temporal control. *Plant J.* 2: 927–936; and data from Burow, M. D., Sen, P., Chlan, C. A., Murai, N. (1992) Developmental control of the β-phaseolin gene requires positive, negative, and temporal seed-specific transcriptional regulatory elements and a negative element for stem and root expression. *Plant J.* 2: 537–548.

Figure 19.52: Adapted from Bewley, J. D. (1997) Seed germination and dormancy. *Plant Cell* 9: 1055–1066.

Box 19.2A,B: Gustafson-Brown, C., Savidge, B., Yanofsky, M. F. (1994) Regulation of the *Arabidopsis* floral homeotic gene *APETA-*

LA1. *Cell* 76: 131–143. Copyright © 1994 Cell Press.

Box 19.2C: D. Weigel, Salk Institute, San Diego, CA; previously unpublished.

Box 19.2D: Meyerowitz, E. M., Bowman, J. L., Brockman, L. L., Drews, G. N., Jack, T., Sieburth, L. E., Weigel, D. (1991) A genetic and molecular model for flower development in *Arabidopsis thaliana*. *Development* (Suppl. 1): 157–167. Used with permission from Company of Biologists Ltd.

Box 19.2E,H: Zachgo, S., Silva, E. de A., Motte, P., Tröbner, W., Saedler, H., Schwarz-Sommer, Z. (1995) Functional analysis of the *Antirrhinum* floral homeotic *DEFICIENS* gene in vivo and in vitro by using a temperature-sensitive mutant. *Development* 121: 2861–2875. Used with permission from Company of Biologists Ltd.

Box 19.2F: Huijser, P., Klein, J., Lönnig, W.-E., Meijer, H., Saedler, H., Sommer, H. (1992) Bracteomania, an inflorescence anomaly, is caused by the loss of function of the MADS-box gene *squamosa* in *Antirrhinum majus*. *EMBO J.* 11: 1239–1249. Used with permission from Oxford University Press.

Box 19.2G: Carpenter, R., Copsey, L., Vincent, C., Doyle, S., Magrath, R., Coen, E. (1995) Control of flower development and phyllotaxy by meristem identity genes in *Antirrhinum*. *Plant Cell* 7: 2001–2011.

Box 19.3: Adapted from McCormick, S. (1991) Molecular analysis of male gametogenesis in plants. *Trends Genet.* 7: 298–303. Copyright © 1991 with permission from Elsevier Science.

Box 19.6A,B: Lolle, S. J., Cheung, A. Y. (1993) Promiscuous germination and growth of wildtype pollen from *Arabidopsis* and related species on the shoot of the *Arabidopsis* mutant, *fiddlehead*. *Dev. Biol.* 155: 250–258. Reprinted by permission of Academic Press. Copyright © 1993.

Box 19.7: Kranz, E., Bautor, J., Lörz, H. (1994) In vitro fertilization of single, isolated gametes of maize mediated by electrofusion. *Sex. Plant Reprod.* 4: 12–16.

Table 19.1: Smyth, D. R., Bowman, J. L., Meyerowitz, E. M. (1990) Early flower development in *Arabidopsis*. *Plant Cell* 2: 755–767.

Table 19.2: Data from Bewley, J. D., and Black, M. (1994) *Seeds. Physiology of Development and Germination*. Plenum Press, New York.

Chapter 20

Figure 20.2A: Reproduced from Takeshige, K., Baba, M., Tsuboi, S., Noda, T., Ohsumi, Y. (1992) Autophagy in yeast demonstrated with proteinase-deficient plants and conditions for its induction. *J. Cell Biol.* 119: 301–311. By copyright permission of the Rockefeller University Press.

Figure 20.2B: Adapted from Matile, P., Winkenbach, F. (1971) Function of lysosomes and lysosomal enzymes in the senescing corolla of the morning glory (*Ipomoea purpurea* [sic]). *J. Exp. Bot.* 22: 759–771. Used with permission from Oxford University Press.

Figure 20.3: P. Bethke, R. Jones, University of California, Berkeley; previously unpublished.

Figure 20.5A: Neuffer, M. G., Coe, E. H., Wessler, S. R. (1997) *Mutants of Maize*. Cold Spring Harbor Laboratory Press, Cold Spring Harbor, NY. Copyright © 1997 Cold Spring Harbor Laboratory Press.

Figure 20.5B: L. Jesaitis, University of California, Berkeley; previously unpublished.

Figure 20.6: R. A. Dietrich, Novartis Crop Protection, Inc., Research Triangle Park, NC; previously unpublished.

Figure 20.8A,B: He, C.-J., Morgan, P. W., Drew, M. C. (1996) Transduction of an ethylene signal is required for cell death and lysis in the root cortex of maize during aerenchyma formation induced by hypoxia. *Plant Physiol.* 112: 463–472.

Figure 20.9: Bleecker, A. B., Patterson, S. E. (1997) Last exit: senescence, abscission, and meristem arrest in *Arabidopsis*. *Plant Cell* 9: 1169–1179.

Figure 20.10: Kende, H., Baumgartner, B. (1974) Regulation of aging in flower of *Ipomoea tricolor* by ethylene. *Planta* 116: 279–289. Copyright © 1974 Springer-Verlag.

Figure 20.11: H. Thomas, IGER, Aberystwyth, UK; previously unpublished.

Figure 20.12: Adapted from Noodèn, L. D., Guiamet, J. J., John, I. (1997) Senescence mechanisms. *Physiol. Plant.* 101: 746–753. Copyright © 1997 Physiologium Plantarum and published by Munksgaard International Publishers Ltd., Copenhagen, Denmark.

Figure 20.13A–C: Thomas, H. (1977) Ultrastructure, polypeptide composition and photochemical activity of chloroplasts during foliar senescence of a non-yellowing mutant genotype of *Festuca pratensis*. *Planta* 137: 53–60. Copyright © 1977 Springer-Verlag.

Figure 20.14: Adapted from Buchanan-Wollaston, V. (1997) The molecular biology of leaf senescence. *J. Exp. Bot.* 48: 181–199. Reprinted with permission from Oxford University Press.

Figure 20.16: Adapted with permission from Matile, P., Hörtensteiner, S., Thomas, H. (1999) Chlorophyll degradation. *Annu. Rev. Plant Physiol. Plant Mol. Biol.* 50: 67–95. Copyright © 1999 Annual Reviews. www.AnnualReviews.org

Figure 20.19A–C: PaO data: Vicentini, F., Hörtensteiner, S., Schellenberg, M., Thomas, H., Matile, P. (1995) Chlorophyll breakdown in senescent leaves: identification of the biochemical lesion in a *stay-green* genotype of *Festuca pratensis* Huds. *New Phytol.* 129: 247–252. Reprinted by permission of Academic Press. Copyright © 1995. Protein data: Adapted from Hilditch, P., Thomas, H., Thomas, B. J., Rogers, L. J. (1989) Leaf senescence in a non-yellowing mutant of *Festuca pratensis:* proteins of Photosystem II. *Planta* 177: 265–272. Copyright © 1997 Springer-Verlag.

Figure 20.20A,B: Griffiths, C. M., Hosken, S. E., Oliver, D., Chojecki, J., Thomas, H. (1997) Sequencing, expression pattern and RFLP mapping of a senescence-enhanced cDNA from *Zea mays* with high homology to oryzain γ and aleurain. *Plant Mol. Biol.* 34: 815–821. Used with kind permission of Kluwer Academic Publishers.

Figure 20.20C: I. Donnison, IGER, Aberystwyth, UK, previously unpublished.

Figure 20.23: Adapted from Buchanan-Wollaston, V. (1997) The molecular biology of leaf senescence. *J. Exp. Bot.* 48: 181–199. Reprinted with permission from Oxford University Press.

Figure 20.25: Adapted from Ashihara, H., Takasawa, Y., Suzuki, T. (1997) Metabolic fate of guanosine in higher plants. *Physiol. Plant.* 100: 909–916. Copyright © 1997 Physiologium Plantarum and published by Munksgaard International Publishers Ltd., Copenhagen, Denmark.

Figure 20.26: A. Bleeker, University of Wisconsin, Madison; previously unpublished.

Figure 20.27: Whitehead, C. S., Halevy, A. H., Reid, M. S. (1984) Roles of ethylene and 1-aminocyclopropane-1-carboxylic acid in pollination and wound-induced senescence of *Petunia hybrida* flowers. *Physiol. Plant.* 61: 643–648. Copyright © 1984 Physiologium Plantarum and published by Munksgaard International Publishers Ltd., Copenhagen, Denmark.

Figure 20.28A,B: Lincoln, J. E., Cordes, S., Read, E., Fischer, R. L. (1987) Regulation of gene expression by ethylene during *Lycopersicon esculentum* (tomato) fruit development. *Proc. Natl. Acad. Sci. USA* 84: 2793–2797. Used with permission.

Figure 20.29A–E: Lanahan, M. B., Yen, H. C., Giovannoni, J. J., Klee, H. J. (1994) The *never ripe* mutation blocks ethylene perception in tomato. *Plant Cell* 6: 521–530.

Figure 20.30: Adapted with permission from Oeller, P. W., Min-Wong, L., Taylor, L. P., Pike, D. A., Theologis, A. (1991) Reversible inhibition of tomato fruit senescence by antisense RNA. *Science* 254: 437–439. Copyright © 1991 American Association for the Advancement of Science.

Figure 20.31: Theologis, A., Oeller, P. W., Wong, L. M., Rottmann, W. H., Gantz, D. M. (1993) Use of a tomato mutant constructed with reverse genetics to study fruit ripening, a complex developmental process. *Dev. Genet.* 14: 282–295. Reprinted by permission of Wiley-Liss, Inc., a subsidiary of John Wiley & Sons, Inc.

Figure 20.32: Johal, G. S., Hulbert, S., and Briggs, S. P. (1995) Disease lesion mimics in maize: a model for cell death in plants. *BioEssays* 17: 685–692. Reprinted by permission of Wiley-Liss, Inc., a subsidiary of John Wiley & Sons, Inc.

Figure 20.33: Photograph: Gan, S., Amasino, R. M. (1997) Making sense of senescence. *Plant Physiol.* 113: 313–319.

Figure 20.34: Adapted from Thomas, H., Ougham, H., Hörtensteiner, S. (1998) Ring in the new, ring out the old. *IGER Innovations* 2: 5–9.

Figure 20.35: Raven, P. H. (1999) Cells and tissues of the plant body. In *Biology of Plants,* P. H. Raven, R. F. Evert, and S. E. Eichorn, eds. Worth Publishers, New York, pp. 570–588. Copyright © 1971, 1976, 1981, 1992, and 1999 by Worth Publishers. Used with permission.

Figure 20.36A–F: Adapted from Fukuda, H. (1997) Tracheary element differentiation. *Plant Cell* 9: 1147–1156.

Figure 20.36G,H: A. Groover, University of North Carolina, Chapel Hill; previously unpublished.

Figure 20.37A,B: Adapted from Fukuda, H. (1997) Tracheary element differentiation. *Plant Cell* 9: 1147–1156.

Figure 20.38: Watanabe, Y., Fukada, H. (1995) *Plant Cell Physiol.* (Suppl.): 87. Reprinted with permission from the Japanese Society of Plant Physiologists.

Figure 20.39: Yamamoto, R., Demura, T., Fukuda, H. (1997) Brassinosteroids induce entry into the final stage of tracheary element differentiation in cultured Zinnia cells. *Plant Cell Physiol.* 38: 980–983. Reprinted with permission from the Japanese Society of Plant Physiologists.

Figure 20.40A–F: Young, T. E., Gallie, D. R., DeMason, D. A. (1997) Ethylene-mediated programmed cell death during maize endosperm development of wild-type and *shrunken2* genotypes. *Plant Physiol.* 115: 737–751.

Figure 20.41A,B: Young, T. E., Gallie, D. R., DeMason, D. A. (1997) Ethylene-mediated programmed cell death during maize endosperm development of wild-type and *shrunken2* genotypes. *Plant Physiol.* 115: 737–751.

Figure 20.42: Bethke, P. C., Lonsdale, J. E., Fath, A., Jones, R. L. (1999) Hormonally regulated programmed cell death in barley aleurone cells. *Plant Cell* 11: 1033–1045.

Figure 20.43A,B: Bethke, P. C., Lonsdale, J. E., Fath, A., Jones, R. L. (1999) Hormonally regulated programmed cell death in barley aleurone cells. *Plant Cell* 11: 033–1045.

Figure 20.44A,B: He, C.-J., Drew, M. C., Morgan, P. W. (1994) Induction of enzymes associated with lysigenous aerenchyma formation in roots of *Zea mays* during hypoxia or nitrogen starvation. *Plant Physiol.* 105: 861–865.

Figure 20.45: Subbaiah, C. C., Bush, D. S., Sachs, M. M. (1994) Elevation of cytosolic calcium precedes anoxic gene expression in maize suspension-cultured cells. *Plant Cell* 6: 1747–1762.

Figure 20.46A–C: He, C.-J., Morgan, P. W., Drew, M. C. (1996) Transduction of an ethylene signal is required for cell death and lysis in the root cortex of maize during aerenchyma formation induced by hypoxia. *Plant Physiol.* 112: 463–472.

Figure 20.47: Ward, E. R., Uknes, S. J., Williams, S. C., Dincher, S. S., Wiederhold, D. L., Alexander, D. C., Ahl-Goy, P., Metraux, J.-P., Ryals, J. A. (1991) Coordinate gene activity in response to agents that induce systemic acquired resistance. *Plant Cell* 3: 1085–1094.

Figure 20.48: Van den Ackerveken, G.F.J.M., Van Kan, J.A.L., De Wit, P.J.G.M. (1992) Molecular analysis of the avirulence gene *avr9* of the fungal tomato pathogen *Cladosporium fulvum* fully supports the gene-for-gene hypothesis. *Plant J.* 2: 359–366. Used with permission from Blackwell Science Ltd.

Figure 20.49A–G: Dangl, J. L., Dietrich, R. A., Richberg, M. H. (1996) Death don't have no mercy: cell death programs in plant–microbe interactions. *Plant Cell* 8: 1793–1807.

Figure 20.50: Mittler, R., Shulaev, V., Lam, E. (1995) Coordinated activation of programmed cell death and defense mechanisms in transgenic tobacco plants expressing a bacterial proton pump. *Plant Cell* 7: 29–42.

Figure 20.51A,B: Freialdenhoven, A., Peterhänsel, C., Kurth, J., Kreuzaler, F., Schulze-Lefert, P. (1996) Identification of genes required for the function of non-race-specific *mlo* resistance to powdery mildew in barley. *Plant Cell* 8: 5–14.

Figure 20.52: Wolter, M., Hollricher, K., Salamini, F., Schulze-Lefert, P. (1993) The *mlo* resistance alleles to powdery mildew infection in barley trigger a developmentally controlled defence mimic phenotype. *Mol. Gen. Genet.* 239: 122–128. Copyright © 1993 Springer-Verlag.

Figure 20.53: Kosslak, R. M., Chamberlin, M. A., Palmer, R. G., Bowen, B. A. (1997) Programmed cell death in the root cortex of soybean *root necrosis* mutants. *Plant J.* 11: 729–745. Used with permission from Blackwell Science Ltd.

Figure 20.54: Adapted from Hammond-Kosack, K. E., Jones, J.D.G. (1996) Resistance gene-dependent plant defense responses. *Plant Cell* 8: 1773–1791.

Figure 20.55: Adapted from Mehdy, M. (1994) Active oxygen species in plant defense against pathogens. *Plant Physiol.* 105: 457–472.

Figure 20.56: Ryerson, D. E., Heath, M. C. (1996) Cleavage of nuclear DNA into oligonucleosomal fragments during cell death induced by fungal infection or by abiotic treatments. *Plant Cell* 8: 393–402.

Figure 20.58: Wang, H., Li, J., Bostock, R. M., Gilchrist, D. G. (1996) Apoptosis: a functional paradigm for programmed plant cell death induced by a host-selective phytotoxin and invoked during development. *Plant Cell* 8: 375–391.

Box 20.1: Gel photograph: Peitsch, M. C., Polzar, B., Stephan, H., Crompton, T., Mac-Donald, H. R., Mannherz, H. G., Tschopp, J. (1993) Characterization of the endogenous deoxyribonuclease involved in nuclear DNA degradation during apoptosis (programmed cell death). *EMBO J.* 12: 371–377. Reprinted with permission from Oxford University Press.

Table 20.1: Data from Buchanan-Wollaston, V. (1997) The molecular biology of leaf senescence. *J. Exp. Bot.* 48: 181–199. Copyright Oxford University Press.

Part 5 opening photograph:
Galen Rowell/Mountain Light Photography, Emeryville, CA.

Chapter 21

Figure 21.1: K. Hammond-Kosack, Monsanto Co., St. Louis, MO; previously unpublished.

Figure 21.2: E. P. Caswell-Chen, University of California, Davis; published in Moffat, A. S. (1997) First nematode-resistance gene found. *Science* 275: 757.

Figure 21.3: Adapted from Jackson, A. O., Taylor, C. B. (1996) Plant–microbe interactions: life and death at the interface. *Plant Cell* 8: 1651–1668.

Figure 21.4: Reprinted with the permission of Simon & Schuster, from *The Simon & Schuster's Beginner's Guide to Understanding Wine* by M. Schuster, p. 53. Text copyright © 1990 by M. Schuster.

Figure 21.6B: Jones, D. G., Clifford, B. C. (1983) Nature of pathogenicity. In *Cereal Diseases: Their Pathology and Control*, 2nd ed. John Wiley & Sons, New York, pp. 16–26. Reprinted by permission of John Wiley & Sons, Inc.

Figure 21.7: K. Hammond-Kosack, The Sainesbury Laboratory, Norwich, UK; previously unpublished.

Figure 21.8A,B: M. Dow, The Sainsbury Laboratory, Norwich UK; previously unpublished

Figure 21.10A: Franki, R.I.B., Milne, R. G., Hatta, T. (1985) *Atlas of Plant Viruses*, Vol. 2, CRC Press, Boca Raton, FL, p. 138. Copyright © 1985 CRC Press.

Figure 21.10B: Franki, R.I.B., Milne, R. G., Hatta, T. (1985) *Atlas of Plant Viruses*, Vol. 1, CRC Press, Boca Raton, FL, p. 20. Copyright © 1985 CRC Press.

Figure 21.12: Upper panels: Reprinted with permission from Wolf, S., Deom, C. M., Beachy, R. N., Lucas, W. J. (1989) Movement protein of tobacco mosaic virus modifies plasmodesmatal size exclusion limit. *Science* 246: 377–379. Copyright © 1989 American Association for the Advancement of Science. Lower panels: Linstead, P. J., Hills, G. J., Plaskitt, K. A., Willson, I. G., Harker, C. I., Maule, A. J. (1988) The subcellular location of the gene 1 product of cauliflower mosaic virus is consistent with a function associated with virus spread. *J. Gen. Virol.* 69: 1809–1818. Reprinted with permission of the Society of General Microbiology.

Figure 21.13A: Williamson, V. M., and Hussey, R. S. (1996) Nematode pathogenesis and resistance in plants. *Plant Cell* 8: 1735–1745.

Figure 21.14: Monsanto Co., St. Louis, MO; previously unpublished.

Figure 21.15: Adapted with permission from Lucas, W. J. (1994) Plasmodesmata in relation to viral movement within leaf tissues. *Annu. Rev. Phytopathol.* 32: 387–411. Copyright © 1994 Annual Reviews. www.AnnualReviews.org

Figure 21.16: Gorg, R., Hollricher, K., Schulze-Lefert, P. (1993) Functional analysis and RFLP-mediated mapping of the *Mlg* resistance locus in barley. *Plant J.* 3: 857–866. Used with permission from Blackwell Science Ltd.

Figure 21.17A: Osbourn, A. E. (1999) Antimicrobial phytoprotectants and fungal pathogens: a commentary. *Fungal Genet. Biol.* 26: 163–168. Reprinted by permission of Academic Press. Copyright © 1999.

Figure 21.18B: J. Brown, John Innes Centre, Norwalk, UK; previously unpublished.

Figure 21.22: Adapted by permission from Herbers, K., Conrads-Strauch, J., Bonas, U. (1992) Race-specificity of plant resistance to bacterial spot disease determined by repetitive motifs in a bacterial avirulence protein. *Nature* 356: 172–174. Copyright © 1992 Macmillan Magazines Ltd.

Figure 21.23: Three-dimensional model adapted from Vervoort, J., Van Den Hooven, H. W., Berg, A., Vossen, P., Vogelsand, R., Joosten, M.H.A.J., De Wit, P.G.M. (1997) The race-specific elicitor AVR9 of the tomato pathogen *Cladosporium fulvum*: a cystine knot protein. *FEBS Lett.* 404: 153–158. Copyright © 1997 with permission from Elsevier Science.

Figure 21.27A,B: Reprinted from Kobe, B., Deisenhofer, J. (1995) Proteins with leucine-rich repeats. *Curr. Opin. Struct. Biol.* 5: 409–416. Copyright © 1995 with permission from Elsevier Science.

Figure 21.28B: Carland, F. M., Staskawicz, B. J. (1993) Genetic characterization of the *Pto* locus of tomato: semi-dominance and cosegregation of resistance to *Pseudomonas syringae* pathovar tomato and sensitivity to the insecticide Fenthion. *Mol. Gen. Genet.* 239: 17–27. Copyright © 1993 Springer-Verlag.

Figure 21.28C: Salmeron, J. M., Oldroyd, G.E.D., Rommons, C.M.T., Scofield, S. R., Kim, H.-S., Lavelle, D. T., Dahlbeck, D., Staskawicz, B. J. (1996) Tomato *Prf* is a member of the leucine-rich repeat class of plant disease resistance genes and lies embedded with the *Pto* kinase gene cluster. *Cell* 86: 123–133. Copyright © 1996 Cell Press.

Figure 21.35: Ho, J.-Y., Weide, R., Ma, H. M., van Wordragen, M. F., Lambert, K. N., Koomneef, M., Zabel, P., Williamson, V. M. (1992) The root-know nematode resistance gene (*Mi*) in tomato: construction of a molecular linkage map and identification of dominant cDNA markers in resistant genotypes. *Plant J.* 2: 971–973. Used with permission from Elsevier Science.

Figure 21.37B: Bestwick, C. S., Brown, I. R., Bennett, M.H.R., Mansfield, J. W. (1997) Localization of hydrogen peroxide accumulation during the hypersensitive reaction of lettuce cells to *Pseudomonas syringae* pv. *phaseolicola*. *Plant Cell* 9: 209–221.

Figure 21.38: Reprinted by permission from Delledonne, M., Xia, Y., Dixon, R. A., Lamb, C. (1998) Nitric oxide functions as a signal in plant disease resistance. *Nature* 394: 585–588. Copyright © 1998 Macmillan Magazines Ltd.

Figure 21.39: Photograph: Hammond-Kosack, K. E., Jones, J.D.G. (1996) Resistance gene dependent plant defense responses. *Plant Cell* 8: 1773–1791.

Figure 21.41A: Adapted from Enyedi, A.J., Yalpani, N., Silverman, P., Raskin, I. (1992) Localization, conjugation, and function of salicylic acid in tobacco during the hypersensitive reactants to tobacco mosaic virus. *Proc. Natl. Acad. Sci. USA* 89: 2480–2484.

Figure 21.41B: Hammond-Kosack, K. E., Jones, J.D.G. (1996) Resistance gene–dependent plant defense responses. *Plant Cell* 8: 1773–1791.

Figure 21.42A: Bent, A. F., Innes, R. W., Ecker, J. R., Staskawicz, B. J. (1997) Disease development in ethylene-insensitive *Arabidopsis thaliana* infected with virulent and avirulent *Pseudomonas* and *Xanthomonas* pathogens. *Mol. Plant Microbe Interact.* 5: 372–378. Published by the American Phytopathological Society.

Figure 21.42B: Knoester, M., van Loon, L. C., van den Heuvel, J., Hennig, J., Bol, J. F., Linthorst, H.U.B. (1998) Ethylene-insensitive tobacco lacks nonhost resistance against soil-borne fungi. *Proc. Natl. Acad. Sci. USA* 95: 1933–1937. Copyright © 1998 National Academy of Sciences, U.S.A.

Figure 21.43: Lawton, K. A., Potter, S. L., Uknes, S., Ryals, J. (1994) Acquired resistance signal transduction in *Arabidopsis* is ethylene independent. *Plant Cell* 6: 581–588.

Figure 21.44A,B: Terras, F.R.G., Eggermont, K., Kovaleva, V., Raikhel, N. V., Osborn, R. W., Kester, A., Torrekens, S., Van Leuven, F., Vanderleyden, J., Cammue, V.P.A., Broek-aert, W. F. (1995) Small cyseine-rich antifun-gal proteins from radish: their role in host defense. *Plant Cell* 7: 573–588.

Figure 21.46: Hammond-Kosack, K. E., Jones, J.D.G. (1996) Resistance gene–dependent plant defense responses. *Plant Cell* 8: 1773–1791.

Figure 21.48: Reprinted from Durner, J., Shah, J., Klessig, D. F. (1997) Salicylic acid and disease resistance in plants. *Trends Plant Sci.* 2: 266–274. Copyright © 1997 with permission from Elsevier Science.

Figure 21.52: Howe, G. A., Lightner, J., Browse, J., Ryan, C. A. (1996) An octadec-anoid pathway mutant (JL5) of tomato is compromised in signaling for defense against insect attack. *Plant Cell* 8: 2067–2077.

Figure 21.54A,B: Lauge, R., Joosten, M.H.A.J., Van den Ackerveken, G.F.J.M., Van den Broek, H.W.J., De Wit, P.J.G.M. (1997) The in planta–produced extracellular proteins ECP1 and ECP2 of *Cladosporium fulvum* are virulence factors. *Mol. Plant Mi-crobe Interact.* 10: 725–734. Reprinted with permission from the American Phytopatho-logical Society.

Figure 21.55A: Adapted with permission from De Wit, P.J.G.M. (1992) molecular characterization of gene-for-gene systems in plant–fungus interactions and the appli-cation of avirulence genes in control of plant pathogens. *Annu. Rev. Phytopathol.* 30: 391–418. Copyright © 1992 Annual Re-views. www.AnnualReviews.org

Figure 21.56A: Reprinted by permission from Hain, R., Reif, H.-J., Krause, E., Langebartels, R., Kindl, H., Vornam, B., Wiese, W., Schmelzer, E., Schreier, P. H., Stöcker, R. H., Stenzel, K. (1993) Disease re-sistance results from foreign phytoalexin expression in a novel plant. *Nature* 361: 153–156. Copyright © 1993 Macmillan Mag-azines Ltd.

Figure 21.56B: Zhu, Q., Maher, E. A., Masoud, S., Dixon, R A., Lamb, C. J. (1994) Enhanced protection against fungal attack by constitutive co-expression of chitinase and glucanase genes in gransgenic tobacco. *io/Technology* 12: 807–812. Used with per-sion of Nature Publishing Group, York.

?1.56C: Reprinted with permission h, D. M., Rommens, C.M.T., N. (1995) Resistance to diseases transgenic plants: progress ns to agriculture. *Trends* 3–368. Copyright © 1995 om Elsevier Science.

Figure 21.56D: Beffa, R. S., Hofer, R.-M., Thomas, M., Meins, F. Jr. (1996) Decreased susceptibility to viral disease of β-1,3-glu-canase–deficient plants generated by anti-sense transformation. *Plant Cell* 8: 1001–1011.

Figure 21.57A: de la Fuenta-Martínez, J. M., Mosqueda-Cana, G., Alvarez-Morales, A., Herrera-Estrella, L. (1992) Expression of a bacterial phaseolotoxin-resistant ornithyl thranscarbamylase in transgenic tobacco confers resistance to *Pseudomonal syringae* pv. *phaseolicola*. *Bio/Technology* 10: 905–909. Used with permission of Nature Publishing Group, New York.

Figure 21.57B: Reprinted with permission from Abel, P. P., Nelson, R. S., De, B., Hoff-mann, N., Rogers, S. G., Fraley, R. T., Beachy, R. N. (1986) Delay of disease devel-opment in transgenic plants that express the tobacco mosaic virus coat protein gene. *Science* 232: 738–743. Copyright © 1986 American Association for the Advancement of Science.

Figure 21.57C: Lapidot, M., Gafny, R., Ding, B., Wolf, S., Lucas, W. J., Beachy, R. N. (1993) A dysfunctional movement protein of tobacco mosaic virus that partially modi-fies the plasmodesmata and limits virus spread in transgenic plants. *Plant J.* 4: 959–970. Used with permission from Black-well Science Ltd.

Figure 21.58A: Grochulski, P., Masson, L., Borisova, S., Pusztai-Carey, M., Schwartz, J. L., Brousseau, R., Cygler, M. (1995) *Bacillus thuringiensis* CryIA(a) insecticidal toxin: crystal structure and channel formation. *J. Mol. Biol.* 254: 447–464. Reprinted by permis-sion of Academic Press. Copyright © 1999.

Figure 21.58B: Reprinted by permission from Li, J., Carroll, J., Ellar, D. J. (1991) Crystal structure of insecticidal δ-endotoxin from *Bacillus thuringiensis* at 2.5 Å resolu-tion. *Nature* 353: 815–821. Copyright © 1991 Macmillan Magazines Ltd.

Figure 21.59: Photograph: Reprinted with permission from Shah, D. M., Rommens, C.M.T., Beachy, R. N. (1995) Resistance to diseases and insects in transgenic plants: progress and applications to agriculture. *Trends Biotechnol.* 13: 362–368. Copyright © 1995 with permission from Elsevier Science. Graph: Adapted from Munkvold, G. P., Helmuh, R. L., Showers, W. B. (1997) Re-duced fusarium ear rot and symptomless infection in kernels of maize genetically en-gineered for European corn borer resis-tance. *Phytopathology* 87: 1071–1077. Pub-lished by the American Phytopathological Society.

Figure 21.60A,B: Reprinted with permission from Tavladoraki, P., Benvenuto, E., Trinca, S., De Martinis, D., Cattaneo, A., Galeffi, P. (1993) Transgenic plants expressing a func-tional single-chain Fv antibody are specifi-

cally protected from virus attack. *Nature* 366: 469–472. Copyright © 1993 Macmillan Magazines Ltd.

Box 21.2: J. Brown, John Innes Centre, Nor-walk, UK; previously unpublished.

Box 21.3: Adapted from Alfano, J. R., Coll-mer, A. (1997) The Type III (Hrp) secretion pathway of plant pathogenic bacteria: traf-ficking harpins, Avr proteins, and death. *J. Bacteriol.* 179: 5655–5662. Used with permis-sion from the American Society for Micro-biology.

Box 21.4: Pair of photographs at right: Dinesh-Kumar, S. P., Whitham, S., Choi, D., Hehl, R., Corr, C., Baker, B. (1995) Transpo-son tagging of tobacco mosaic virus resis-tance gene *N*: its possible role in the TMV–*N*-mediated signal transduction pathway. *Proc. Natl. Acad. Sci. USA* 92: 4175–4180. Copyright © 1995 National Academy of Sciences, U.S.A.

Box 21.4A–C: Reprinted with permission from Jones, D. A., Thomas, C. M., Ham-mond-Kosack, K. E., Balint-Kurti, P. J., Jones, J.D.G. (1994) Isolation of the tomato *Cf-9* gene for resistance to *Cladosporium ful-vum* by transposon tagging. *Science* 266: 789–793. Copyright © 1994 American Asso-ciation for the Advancement of Science.

Box 21.5: Adapted from Devoto, A., Pif-fanelli, P., Nilsson, I., Wallin, E., Panstruga, R., Von Heijne, G., Schulze-Lefert, P. (1999) Topology, subcellular localization, and se-quence diversity of the Mlo family in plants. *J. Biol. Chem.* 274: 34993–35004. Copyright © 1999 American Society for Bio-chemistry and Molecular Biology.

Box 21.7: Whitham, S., McCormick, S., Bak-er, B. (1996) The *N* gene of tobacco confers resistance to tobacco mosaic virus in trans-genic tomato. *Proc. Natl. Acad. Sci. USA* 93: 8776–8781. Copyright © 1996 National Academy of Sciences, U.S.A.

Box 21.8: Jabs, T., Tschoepe, M., Colling, C., Hahlbrock, K., Scheel, D. (1997) Elicitor-stimulated ion fluxes and O_2^- from the ox-idative burst are essential components in triggering defense gene activation and phy-toalexin synthesis in parsley. *Proc. Natl. Acad. Sci. USA* 94: 4800–4805. Copyright © 1997 National Academy of Sciences, U.S.A.

Box 21.9: Photographs: Vernooij, B., Friedrich, L., Morse, A., Reist, R., Kolditz-Jawhar, R., Ward, E., Uknes, S., Kessman, H., Ryals, J. (1994) Salicylic acid is not the translocated signal responsible for inducing systemic acquired resistance but is required in signal transduction. *Plant Cell* 6: 959–965.

Chapter 22

Figure 22.2: L. E. Epple, in Epple, A. O. (1995) *A Field Guide to the Plants of Arizona*. Falcon Press Publishing, Helena, MT.

Figure 22.12: Yamada, S., Katsuhara, M., Kelly, W. B., Michalowski, C. B., Bohnert, H. J. (1995) A family of transcripts encoding water channel proteins: tissue-specific expression in the common ice plant. *Plant Cell* 7: 1129–1142.

Figure 22.13: Adapted from Kononowicz, A. K., Ragothama, K. G., Casa, A. M., Reuveni, M., Watad, A.-E. A., Liu, D., Bressan, R. A., Hasegawa, P. M. (1993) Osmotin: regulation of gene expression and function. In *Plant Response to Cellular Dehydration During Environmental Stress,* T. J. Close and E. A. Bray, eds. American Society of Plant Physiologists, Rockville, MD, pp. 144–158.

Figure 22.17A: V. Ramey, University of Florida Center for Aquatic and Invasive Plants, http://aquat1.ifas.ufl.edu/hyvepic.htm. Copyright © 1999 University of Florida.

Figure 22.17B: B. Nelson, http://aquat1.ifas.ufl.edu/hyvepic.htm. Copyright © 1997 Southwest Florida Water Management District.

Figure 22.18: He, C.-J., Morgan, P. W., Drew, M. C. (1996) Transduction of an ethylene signal is required for cell death and lysis in the root cortex of maize during aerenchyma formation induced by hypoxia. *Plant Physiol.* 112: 463–472.

Figure 22.19: Kozlowski, T. T. (1984) *Physiological Ecology.* Academic Press, New York, p. 147. Reprinted by permission of Academic Press. Copyright © 1984.

Figure 22.20: Bowes, B. G. (1996) *A Colour Atlas of Plant Structures.* Manson Publishers, p. 136. Copyright © 1996 Manson Publishers Ltd.

Figure 22.21A: Adapted from Catling, D. (1992) *Rice in Deep Water* (Macmillan, London) and published with permission of the International Rice Research Institute in Kende, H., van der Knaap, E., Cho, H.-T. (1998) Deepwater rice: a model plant to study stem elongation. *Plant Physiol.* 118: 1105–1110.

Figure 22.21B: Kende, H., van der Knaap, E., Cho, H.-T. (1998) Deepwater rice: a model plant to study stem elongation. *Plant Physiol.* 118:1105–1110.

Figure 22.22: Couèe, I., Defontaine, S., Carde, J.-P., Pradet, A. (1992) Effects of anoxia on mitochondrial biogenesis in rice shoots: modification of *in organello* translation characteristics. *Plant Physiol.* 98: 411–421.

Figure 22.23: M. S. Martin, University of Illinois at Urbana-Champaign; previously unpublished.

Figure 22.27: Adapted from Roberts, J.K.M., Callis, J., Wemmer, D., Walbot, V., Jardetzky, O. (1984) Mechanism of cytoplasmic pH regulation in hypoxic maize root tips

and its role in survival under hypoxia. *Proc. Natl. Acad. Sci. USA* 81: 3379–3383. Used with permission.

Figure 22.29A,B: Adapted from Freeling, M. (1973) Simultaneous induction by anaerobiosis or 2,4-D of multiple enzymes specified by two unlinked genes: differential *Adh1–Adh2* expression in maize. *Mol. Gen. Genet.* 127: 215–227. Copyright © 1973 Springer-Verlag.

Figure 22.30: Adapted from Sachs, M. M., Freeling, M., Okimoto, R. (1980) The anaerobic proteins of maize. *Cell* 20: 761–767. Copyright © 1980 Cell Press.

Figure 22.34A: Reprinted from Jackson, M. (1997) Hormones from roots as signals for the shoots of stressed plants. *Trends Plant Sci.* 2: 22–28. Copyright © 1997 with permission from Elsevier Science.

Figure 22.39: Jacobson, J. S., Hill, A. C., eds. (1970) *Recognition of Air Pollution Injury to Vegetation: A Pictorial Atlas.* Air Pollution Control Association, Pittsburgh, PA.

Figure 22.41A–D: Mansfield, M. A., Key, J. L. (1987) Synthesis of the low molecular weight heat shock proteins in plants. *Plant Physiol.* 84: 1007–1017.

Figure 22.42: Key, J. L., Lin, C. Y., Ceglarz, E., Schoffl, F. (1983) The heat shock response in soybean seedlings. (In *Structure and Function of Plant Genomes,* O. Ciferri and L. Dure, eds.) *NATO ASI Ser. A* 63: 25–36.

Figure 22.44: Adapted from Morimoto, R. I., Jurivich, D. A., Kroeger, P. E., Mathur, S. K., Murphy, S. P., Nakai, A., Sarge, K., Abravaya, K., Sistonen, L. T. (1994) Regulation of heat shock gene transcription by a family of heat shock factors. In *The Biology of Heat Shock Proteins and Molecular Chaperones,* R. I. Morimoto, A. Tissières, and C. Georgopoulos, eds. Cold Spring Harbor Laboratory Press, Cold Spring Harbor, NY, pp. 417–455. Copyright © 1994 Cold Spring Harbor Laboratory Press.

Box 22.1: Photograph: Xu, D., Duan, X., Wang, B., Hong, B., Ho, T.-H.D., Wu, R. (1996) Expression of a late embryogenesis abundant protein gene, *HVA1,* from barley confers tolerance to water deficit and salt stress in transgenic rice. *Plant Physiol.* 110: 249–257.

Table 22.1: Reprinted with permission from Boyer, J. S. (1982) Plant productivity and environment. *Science* 218: 443–448. Copyright © 1982 American Association for the Advancement of Science.

Chapter 23

Figure 23.1: Ulrich, M.A.E., Allen, W. W. (1992) Strawberry deficiency symptoms: a visual and plant analysis guide to fertilization. Agri. Exp. Station Bull. 1917. Copyright © University of California Board of

Regents, Division of Agriculture and Natural Resources, Davis.

Figure 23.3: Adapted from Epstein, E., Rains, D. W., Elzam, O. E. (1962) Resolution of dual mechanisms of potassium absorption by barley roots. *Proc. Natl. Acad. Sci. USA* 49: 684–692. Used with permission.

Figure 23.4: Data from Kochian, L. V., Lucas, W. J. (1982) Potassium transport in corn roots. *Plant Physiol.* 70: 1723–1731.

Figure 23.5: Adapted from Rodriguez-Navarro, A., Blatt, M. R., Slayman, C. L. (1986) A potassium–proton symport in *Neurospora crassa. J. Gen. Physiol.* 87: 649–674. By copyright permission from the Rockefeller University Press.

Figure 23.6: Anderson, J. A., Huprikar, S. S., Kochian, L. V., Lucas, W. J., Gaber, R. F. (1992) Functional expression of a probable *Arabidopsis thaliana* potassium channel in *Saccharomyces cerevisiae. Proc. Natl. Acad. Sci. USA* 89: 3736–3740. Used with permission.

Figure 23.7: L. Kochian, Cornell University, Ithaca, NY, and W. Lucas, University of California, Davis; previously unpublished.

Figure 23.9: Wegner, L. H., Raschke, K. (1994) Ion channels in the xylem parenchyma of barley roots. A procedure to isolate protoplasts from this tissue and a patch-clamp exploration of salt passageways into xylem vessels. *Plant Physiol.* 105: 799–813.

Figure 23.10: Reprinted with permission from Schachtman, D. P., Schroeder, J. I. (1994) Structure and transport mechanism of a high-affinity potassium uptake transporter from higher plants. *Nature* 370: 655–658. Copyright © 1994 Macmillan Magazines Ltd.

Figure 23.11: Adapted with permission from Rubio, F., Gassmann, W., Schroeder, J. I. (1995) Sodium-driven potassium uptake by the plant potassium transporter HKT1 and mutations conferring salt tolerance. *Science* 270: 1660–1663. Copyright © 1995 American Association for the Advancement of Science.

Figure 23.12: Adapted from Kim, E. J., Kwak, J. M., Uozumi, N., Schroeder, J. I. (1998) *AtKUP1:* an *Arabidopsis* gene encoding high-affinity potassium transport activity. *Plant Cell* 10: 51–62.

Figure 23.13: Adapted from Kim, E. J., Kwak, J. M., Uozumi, N., Schroeder, J. I. (1998) *AtKUP1:* an *Arabidopsis* gene encoding high-affinity potassium transport activity. *Plant Cell* 10: 51–62.

Figure 23.14: Adapted from Fu, H.-H., Luan, S. (1998) AtKUP1: a dual-affinity K^+ transporter from *Arabidopsis. Plant Cell* 10: 63–73.

Figure 23.15: Adapted from Batten, G. D., Wardlaw, I. F. (1987) Senescence and grain development in wheat plants grown with contrasting phosphorus regimes. *Aust. J. Plant Physiol.* 14: 253–268. Reprinted with copyright permission from Commonwealth Scientific and Industrial Research Organization, Melbourne.

Figure 23.16: Dunlop, J., Gardiner, S. (1993) Phosphate uptake, proton extrusion and membrane electropotentials of phosphorus-deficient *Trifolium repens* L. *J. Exp. Bot.* 44: 1801–1808. Used with permission from the Oxford University Press.

Figure 23.17A: D. Redecke, University of Marburg, Germany; previously unpublished.

Figure 23.17B: K. Wex, University of Marburg, Germany; previously unpublished.

Figure 23.18: Johnson, J. F., Vance, C. P., Allen, D. L. (1996) Phosphorus deficiency in *Lupinus albus*. Altered lateral root development and enhanced expression of phospho*enol*pyruvate carboxylase. *Plant Physiol.* 112: 31–41.

Figure 23.19A,B: Johnson, J. F., Vance, C. P., Allen, D. L. (1996) Phosphorus deficiency in *Lupinus albus*. Altered lateral root development and enhanced expression of phospho*enol*pyruvate carboxylase. *Plant Physiol.* 112: 31–41.

Figure 23.20: Adapted from Muchhal, U. S., Pardo, J. M., Raghothama, K. G. (1996) Phosphate transporters from the higher plant *Arabidopsis thaliana*. *Proc. Natl. Acad. Sci. USA* 93: 10519–10523. Copyright © 1996 National Academy of Sciences, U.S.A.

Figure 23.22: Liu, C., Muchhal, U. S., Uthappa, M., Kononowicz, A. K., Raghothama, K. G. (1998) Tomato phosphate transporter genes are differentially regulated in plant tissues by phosphorus. *Plant Physiol.* 116: 91–99.

Figure 23.24: Welch, R. M., Norvell, W. A., Schaefer, S. C. (1993) Induction of iron(III) and copper(II) reduction in pea roots by Fe and Cu status: does the root-cell plasmalemma Fe(II)-chelate reductase perform a general role in regulating cation uptake? *Planta* 190: 555–561. Copyright © 1993 Springer-Verlag.

Figure 23.25: Reprinted with permission from Robinson, N. J., Procter, C. M., Connolly, E. L., Guerinot, M. L. (1999) A ferric-chelate reductase for iron uptake from soils. *Nature* 397: 694–697. Copyright © 1999 Macmillan Magazines Ltd.

Figure 23.26: Adapted from Eide, D., Broderius, M., Fett, J., Gurinot, M. L. (1996) A novel iron-regulated metal transporter from plants identified by functional expression in yeast. *Proc. Natl Acad. Sci. USA* 93: 5624–5628. Copyright © 1996 National Academy of Sciences, U.S.A.

Figure 23.28: C. K. Cohen and L. V. Kochian, Cornell University, Ithaca, NY; previously unpublished.

Figure 23.29: Ma, J. F., Nomoto, K. (1993) Two related biosynthetic pathways of mugineic acids in gramineous plants. *Plant Physiol.* 102: 373–378.

Figure 23.31: Adapted from Delhaize, E., Ryan, P. R., Randall, P. J. (1993) Aluminum tolerance in wheat (*Triticum aestivum* L.). *Plant Physiol.* 103: 695–702.

Figure 23.32: Lasat, M. M., Baker, A.J.M., Kochian, L. V. (1996) Physiological characterization of root Zn^{2+} absorption and translocation to shoots in Zn hyperaccumulator and nonaccumulator species of *Thlaspi*. *Plant Physiol.* 112: 1715–1722.

Figure 23.33: I. Raskin, Rutgers University, NJ; previously unpublished.

Figure 23.34: Martin, R. B. (1988) Bioinorganic chemistry of aluminum. In *Metal Ions in Biological Systems: Aluminum and Its Role in Biology*, Vol. 24, H. Sigel and A. Sigel, eds. Marcel Dekker, New York, pp. 2–57. Reprinted by courtesy of Marcel Dekker, Inc.

Figure 23.35: Adapted from Delhaize, E., Ryan, P. R., Randall, P. J. (1993) Aluminum tolerance in wheat (*Triticum aestivum* L.). *Plant Physiol.* 103: 695–702.

Box 23.2: Adapted from Nissen, P. (1989) Multiphasic uptake of potassium by corn roots. *Plant Physiol.* 89: 231–237.

Box 23.3: Photograph: L. V. Kochian, Cornell University, Ithaca, NY; previously unpublished.

Table 23.1: Epstein, E. (1972) *Mineral Nutrition of Plants: Principles and Perspectives*. John Wiley & Sons, New York. Reprinted by permission of John Wiley & Sons, Inc.

Chapter 24

Figure 24.3A: R. Croteau, Washington State University, Pullman; previously unpublished.

Figure 24.3B–D: Wise, M. L., Croteau, R. (1999) Biosynthesis of monoterpenes. In *Comprehensive Natural Products Chemistry, Vol. 2, Isoprenoids Including Carotenoids and Steroids*, D. E. Cane, ed. Pergamon/Elsevier, Oxford, UK, pp. 97–153. Copyright © 1999 with permission from Elsevier Science.

Figure 24.5: Adapted from McCaskill, D., Croteau, R. (1998) Some caveats for bioengineering terpenoid metabolism in plants. *Trends Biotechnol.* 16: 349–354. Copyright © 1998 with permission from Elsevier Science.

Figure 24.11: Reprinted with permission from Gijzen, M., Lewinsohn, E., Savage, T. J., Croteau, R. B. (1993) Conifer monoterpenes: biochemistry and bark beetle chemical ecology. *ACS Symp. Ser.* 525: 8–22. Copyright © 1993 American Chemical Society.

Figure 24.12: Reprinted with permission from Starks, C. M., Back, K., Chappell, J., Noel, J. P. (1997) Structural basis for cyclic terpene biosynthesis by 5-*epi*-aristolochene synthase from tobacco. *Science* 277: 1815–1820. Copyright © 1997 American Association for the Advancement of Science.

Figure 24.18A: T. M. Kutchan, Liebniz Institut für Pflanzenbiochemie, Halle, Germany; previously unpublished.

Figure 24.18B: Photograph courtesy of the Ministry of Culture Archaeological Receipts Fund, Athens, Greece.

Figure 24.19B: The Metropolitan Museum of Art, New York, NY.

Figure 24.21B: V. Rickl, Universität München, Germany; previously unpublished.

Figure 24.30: M. H. Zenk, Universität München, Germany; previously unpublished.

Figure 24.36: M. H. Zenk, Universität München, Germany; previously unpublished.

Figure 24.47D: L. B. Davin, Washington State University, Pullman; previously unpublished.

Figure 24.48A: L. B. Davin, Washington State University, Pullman; previously unpublished.

Figure 24.48B: Hemp-Museum, Berlin, Germany; previously unpublished.

Figure 24.55: Photograph: L. B. Davin, Washington State University, Pullman; previously unpublished.

Figure 24.57A,C: L. B. Davin, Washington State University, Pullman; previously unpublished.

Figure 24.57B: Photograph: G.H.N. Towers, R. A. Norton, University of British Columbia, Vancouver, Canada; previously unpublished.

Figure 24.59: Data from Adler, E. (1977) Lignin chemistry: past, present and future. *Wood Sci. Technol.* 11: 169–218, and Brunow, G., Kilpeläinen, I., Sipilä, J. Syrjänen, K., Karhunen, P., Setälä, H., Rummakko, P. (1998) Oxidative coupling of phenols and the biosynthesis of lignin. *ACS Symp. Ser.* 697: 131–147.

Figure 24.60B: L. B. Davin, L. H. Levine, Washington State University, Pullman; previously unpublished.

Figure 24.61A–C: L. B. Davin, L. H. Levine, Washington State University, Pullman; previously unpublished.

Figure 24.62: L. H. Levine, Washington State University, Pullman; previously unpublished.

Figure 24.65: Photograph: M. A. Bernards, University of Western Ontario, London, Canada; previously unpublished

Figure 24.67: Photograph: L. B. Davin, Washington State University, Pullman; previously unpublished.

Figure 24.68: Photograph: T. Lumpkin, US Department of Agriculture, Washington State University, Pullman; previously unpublished.

Figure 24.69: Photographs: L. B. Davin (alfalfa plant), M. L. Kahn (nodules), Washington State University, Pullman; previously unpublished.

Figure 24.73: Photographs: G.H.N. Towers, University of British Columbia, Vancouver, Canada; previously unpublished.

Figure 24.77: Photograph: L. B. Davin, Washington State University, Pullman; previously unpublished.

Box 24.2A: French Pharmacopèe Royale; photograph: T. Kutchan, Leibniz Institut für Pflanzenbiochemie, Halle, Germany; previously unpublished.

Box 24.2B: Residenz Pharmacy, Munich, Germany; photograph: T. Kutchan, Leibniz Institut für Pflanzenbiochemie, Halle, Germany; previously unpublished.

Box 24.3A,B: T. Kutchan, Leibniz Institut für Pflanzenbiochemie, Halle, Germany; previously unpublished.

Box 24.4: Diagram adapted from Setchell, K.C.R., Lawson, A. M., Borriello, S. P., Adlercreutz, H., Axelson, M. (1982) Formation of lignan by intestinal microflora. *Falk Symp.* 31: 93–97. Used with kind permission from Kluwer Academic Publishers.

Box 24.5A: Gang, D. R., Fujita, M., Davin, L. B., Lewis, N. G. (1998) The "abnormal lignins": mapping heartwood formation through the lignan biosynthetic pathway. In *Lignan and Lignan Biosynthesis (ACS Symp. Ser.)*, N. G. Lewis, and S. Sarkanen, eds., Washington, DC. Copyright © 1998 American Chemical Society.

Box 24.5B,C: L. Shain, W. E. Hillis, Commonwealth Scientific and Industrial Research Organization, Clayton South MDC, Victoria, Australia; previously unpublished.

Box 24.5D: Adapted from Chattaway, M. M. (1949) The development of tyloses and secretion of gum in heartwood formation. *Aust. J. Sci. Res.* 227–240.

Box 24.6: L. B. Davin, Washington State University, Pullman; previously unpublished.

Index

Page numbers in *italics* indicate illustrations; those followed by t indicate tables; and those followed by b indicate boxed material.

DNA ligase in, 269
DNA polymerase in, 269
DNA topoisomerases in, 270b
elongation in, *268,* 268–269
enzymes in, *268,* 269, 270b
errors in, 269
initiation of, 267–268
lagging strand in, 269
leading strand in, 269
in mitochondria, 563–564
origins of, 267–268, *268*
in plastids, *287,* 287–288
primer in, 269
proofreading in, 269
proteins in, 268, *268*
replication fork in, 268, *268*
semiconservative, *268,* 269
semidiscontinuous, 269
stages of, 267
telomeres in, 271, *271, 274*
termination of, 271
timing of, 271–272
translesion, 277
in viruses, 271
DNase I, 348, *349*
DNase I hypersensitivity, 348, *349*
DNA topoisomerases, 270b
DNA transcription. *See* Transcription
Dominant mutation, 338
Dominant traits, *315,* 315–316, *316*
Donnan potential, 117b
Double fertilization, 1022–1023, *1023*
Double helix, 266, *267*
Double-strand break repair model, 279, *280*
DRE, in freezing tolerance, 1177, *1177*
Drought. *See* Water deficit
Drought-responsive ABA-independent element (DRE)
  in freezing tolerance, 1177, *1177*
  in water deficit, 1175
Ds element, 334–335, *336*
D-type cyclins, 545
Dye coupling experiments, 753
Dyes
  ratio, 967, *968*
  single-wavelength, 967
Dynamic instability, microtubular, 216–219, *217*
Dynein, *219,* 219–221, *220,* 241

E2F transcription factors, 553
Early nodulins, 808
αEcdysone, 1266, *1267*
Edible oils. *See* Oils
Egg cell, development of, 1011–1014, *1013*
18:3 plants, 490
Electrical coupling, plasmodesmata and, 752–753, *753, 754*
Electrochemical gradient, in membrane transport, *10, 11,* 11–13
Electrochemical potential, 116b
Electrochemical potential difference, 116b
Electrogenic proton pump, 118
Electrogenic transport, 118, 708, 1210
Electron distribution model, 703, *703*
Electroneutral transport, 708
Electron-impact mass spectrometry, 61b
Electron overflow model, 703, *703*
Electron transfer. *See also* Oxidative phosphorylation
  mitochondrial, *687–690,* 687–697, 702–705
    alternative oxidase in, 696–697, *698,* 702–705, *703–705*

inhibitors of, 692, *692*
nonphosphorylating bypasses in, 705–707, 706b
Electron transfer chain, photosynthetic, 571, 602–605, *603, 605*
Elongases, fatty acid, 481–483, *482*
Elongation, in polymerization, 213, *213*
Embryo, 1024
Embryogenesis, 561, *1024,* 1024–1026, *1025*
  mutations in, 1024–1025
Embryo sac, formation of, 1006, 1011–1014
Enantiomers, 56
Endocytic pathway, 178
Endocytosis, 23, 24–25, *25, 26,* 199–200, *200*
Endomembrane system, 14, *16*
Endomycorrhizae, 1225–1226
Endoplasmic reticulum (ER), 14–18
  appressed, 748, *749*
  chaperones in, 184, 186, 187
  cisternae of, 14–17, *16, 17*
  cortical, *16,* 17
  endomembrane system and, 14, *16*
  export site, 18
  functional domains of, 14, *15*
  functions of, 14
  lipid movement to/from, *488,* 488–489, 490–491, *491*
  movement of, 222
  oil body formation in, 17–18, *18*
  plasmodesmata and, 748–750, *749, 750*
  protein body formation in, 17–18, *18*
  protein degradation in, 184, 185, 186
  protein modification in, 184, *185*
  protein oligomerization in, 185–187
  in protein sorting and transport, 18, *19,* 175–190
  protein synthesis in, 175–177, *176*
  protein targeting in, 162
  protein targeting to, 162, 179–182, *181, 442,* 442–443
  protein translocation into, 179–182, *181*
  resident proteins in, 187
  ribosome attachment to, 175
  rough, 14, *16*
  smooth, 14, *17*
  structural rearrangements in, 14–17, *16, 17*
  transport vesicles of, 18, *19*
  triacylglyceride synthesis in, 17–18, *18*
Endopolyploidy, 562
Endosomes, 178
Endosperm, 1024
  cell death in, 1049–1050, 1082
  development of, *1026,* 1026–1027
  mummification in, 1047, 1072
Endosymbiont hypothesis, 283, *283*
Endosymbionts, 788
Energy transduction, membrane transport in, 114
Energy transfer, 574
Energy yield, from fatty acid vs. carbohydrate metabolism, 457–458, *458*
Enhancers, 340, *340*
Enhancer trap, 341, 1014, *1015*
ENOD sequences, 808
Enolase, 670
Enoyl-ACP reductase, 475, *475*
Environmental stress, respiratory bypass mechanisms and, 696, *696,* 705–707
Enzyme(s). *See also specific enzymes*
  in cell wall expansion, 95–96, *96*
  in ripening fruit, 99, *102*
  committing, 383–384
  proteolytic. *See* Protein degradation

Infection thread, 798, 800, *802*
Inflorescence, 997, *997*
Inheritance
  chromosome theory of, 317
  Mendelian, *315*, 315–316, 317b
Inositol 1,4,5-triphosphate (IP$_3$), in signaling, 956–957
Insects. *See also* Plant pathogens
  Bacillus thuringiensis toxins for, 1153–1154
  pathogenetic strategies of, *1112*, 1112–1113, *1113*
  systemic response to, 1146
Integral membrane proteins, 8–9, *9*, 111, 177, 179
  identification of, 113b
  orientation of, 182–184, *183*
  in translocation complex, *181*, 182
  type I, 182
  type II, 182
Integrins, 985
  in signaling, 985
Intercisternal elements, 21
Intermediate filaments, 205–208, *206–209*
Intine wall, *1007*, 1010
Intron(s), 323, *324*, 301t, 301–304, *302–304*, 305b
  gene products and, 301b
  as mobile genetic elements, 303–304, *304*
  self-splicing, 303–304, *304*
  splicing of, 301, 302–304, *302–304*, 305b
  types of, 301t, 301–302
Intron homing, 303–304
Invertase, 646–648, *647, 648*
In vitro fertilization, 1023b
In vivo footprinting, 344b–345b
Inward rectifiers, 142–145, *143–146*
Ion channels, 12
  AKT1, 143–144, *144*, 145
  anion, *150*, 150–151
  Ca^{2+}. *See* Ca^{2+} ion channels
  cation, 146–147
  chloride, 151
  concentrations of, 119
  current in, 137–139
    direction of, 137–139
    energy source for, 137–139, *139, 140*
    time-dependent, 142, *143*
  current-voltage relationship for, 137, *139*
  functioning of, 142–152
  FV, 146, *148*
  integrated activity of, 152
  ionic selectivity of, 139–140, 140b–141b
  K$^+$, 142–145, *145–147*
    inward rectifying, 142–145, *143–146*
    outward rectifying, 145, *146*
    two-pore, 145
  KAT1, 113b, 143–144, *144*
  KCO1, 145, *146, 147*
  locations of, 136
  Na$^+$, 146, *147*
  patch clamp studies of, 136–137, 138b, *139*
  permeability of, 137
  properties of, 135–142
  receptors as, 939
  rectifying, 142–145, *143–147*
  Shaker, 143–145, *145, 146*
  in signal transduction, 147–149, *147–150*, *148, 149*, 152, *153*, 936–938, *938, 939*, *939*
  vacuolar, 146, *148*
    fast, 146, *148*
    K$^+$, 146, *148*

  malate, 151, *151*
    slowly activating, 149
  VK, 146, *148*
  VMAL, 151, *151*, 152
  voltage clamp analysis of, 136–137, *139*
  voltage-gated, 140–142, *141, 142*
  voltage sensors for, 144, *145*
Ion fluxes, measurement of, 1210b–1211b
Ion pumps. *See also* Ion channels; Proton pumps
  Ca^{2+}, 125–126, *126*
  H$^+$, 119–128. *See also* H$^+$ pumps
  Na$^+$, 135
Iontophoresis, 753
Ion transport
  carrier-kinetic approach to, 1213b
  iron, 1232–1234
  phosphate, 1222–1230
  potassium, 1205–1222
  zinc, 1237–1239
Iron deficiency, *1206*, 1231–1232
Iron nutrition, 1232–1234
Iron transporters, 1232–1234
Iron uptake, 1230–1239
  chelation in, 1231
  phytosiderophores in, 1231, 1234–1237, *1236, 1237*
  reduction in, 1231
Isoenzymes, 359
Isoleucine, synthesis of, 396, *404*, 404–406, *405*
Isopentenyl diphosphate (IPP), 1255
  synthesis of, 1253–1258
Isoprene rule, 1254b
Isoprenoids, identification and naming of, 1254b
Isotype, 209

Jasmonate, 504–507
Jasmonic acid, 459, 915–919, *916, 918–920*
  discovery of, 915–916
  effects of, 917
  metabolism of, 919, *920*
  plant pathogens and, 1136–1137, 1146, *1146*
  structure of, *851, 916*, 916–917
  synthesis of, 917–918, *918–920*
Junction zones, 76, *82*
Juvabione, *1266, 1267*

KAT1 ion channel, 113b, 143–144, *144*
ent-Kaurene, in gibberellin synthesis, 855–858
KCO1 ion channel, 145, *146, 147*
KDEL receptor, 190
KDEL tail, 187
Keratins, 206, *206*
3-Ketoacyl-ACP reductase, 474
3-Ketoacyl-ACP synthase, 472–474, *473*
α-Ketoglutarate
  in amino acid synthesis, 359, *360*, 710
  in ammonia assimilation, 359, *360*, 710
  synthesis of, 372, *372*
α-Ketoglutarate carrier, *709*, 709–710
α-Ketoglutarate dehydrogenase complex, 683
Ketoses, 55
Kinesin, *219*, 219–221, *220*
Kinetochore, 235, *236, 238*, 238–239, *239*, *541*, 542
  chromosome movement and, 241–245, *242–245*
  microtubule binding by, 238–240, *240*, *541*, 542

K$^+$ ion channels, 142–145, *145–147*. *See also* Ion channels
  inward-rectifying, 142–145, *143–146*
  outward-rectifying, 145, *146*
  two-pore, 145
Knotted1 (Kn1), *338*, *338, 339*, 347, *347*
Kranz anatomy, 620, *620*
Krebs cycle. *See* Citric acid cycle

Lactones, 1266, *1267*
Lamellae
  middle, 53, *54*
  of proplastid, 38
Lamina, nuclear, 206, *206*, 207
Lateral heterogeneity, 42, 591
LATS, 816
Leaf epinasty, *1188*, 1188–1189
Leaf form, programmed cell death and, 1050, *1050*
Leaky scanning, 420b
LEA proteins, 1171–1172, 1172t
Lectin receptor kinase, 975, *976*
Leghemoglobin, 810
Legume-rhizobial symbiotic nitrogen fixation, 796–815. *See also* Nitrogen fixation, legume-rhizobial symbiosis in
Lenticels, 1178, *1179*
Lesch-Nyhan syndrome, 265b
Lesion-mimic mutants, *1087–1091*, 1087–1094, *1093*
Leucine, synthesis of, 396, 404, *404, 405*, 406
Leucine-rich repeat proteins, 1120–1126, 1124t, *1125*
Leucine rich-repeat receptor-like protein kinases, 974
Leucine zipper, 343, *343*
Leucoplasts, 39, *40*. *See also* Plastid(s)
Levorotatory rotation, 56
Lianyl diphsophate, *1260, 1260*
Licensing factor, 538
Light, properties of, 572
Light-harvesting complex II, 588–589, 589t, *590*
  phosphorylation of, 591–592, *593*
Lignans, 1287–1288
  dietary, 1296b
  synthesis of, 1292–1294, *1293*
Lignification, 1300
Lignin, 75, *79*, 102–103, *105*, 1136, 1287
  synthesis of, 1294–1299
  in woody plants, 1299–1300
LIM15, in homologous recombination, 281, *281*
Limonene synthase, in terpenoid synthesis, 1259–1260, *1260*
Limonoids, 1266, *1267*
Linkage group, 317–318, *319*, 320
Linkage mapping, 320–321, *320–322*
Linkage structures, of sugars, 57–58, *58*, 59b–61b
Linoleic acid, 460, 461t
  genetic engineering of, 523
  in jasmonic acid synthesis, 917, *918*
Linolenic acid, 461t
Lipid(s)
  amphipathic, 5–6, 463
  definition of, 457
  fatty acid composition of, 460, 461t. *See also* Fatty acids
  functions of, 457–459, 459t
  genetic engineering of, 518–525
  gluconeogenesis from, 710, *714*
  melting temperature of, 476, *476*

O-linked glycans
  modification of
    in endoplasmic reticulum, 186
    in Golgi apparatus, 196, *197*, 198
  synthesis of, 22, *23*
One-sided invasion model, 279, *280*,
  280–281, *281*
Open reading frame, 420b
Operons, 284
Opium, 1269, *1269*, 1270b
Orchid, female gametophyte development
  in, 1011, *1013*
Organelle genes, 312–313
Organelles(s)
  isolation of, sucrose density gradients in,
    178b–179b
  pleiomophoric, 45
Organogenesis, 561
Origin recognition complex (ORC),
  536–537, *538*, 539t
Ornithine aminotransferase, in proline syn-
    thesis, 407–408, *408*
Ornithine decarboxylase, 911–913, 915
Orotate, synthesis of, 261–262, *263*
Orotic acid pathway, 261, *263*
Oryzains, 1063
Osmolytes, compatible. *See* Compatible
    solutes
Osmosis, transcellular, 154b
Osmotic adjustment, 1163–1170
  compatible solutes and, 1164–1170
  definition of, 1164
  glycine betaine and, *1167*, 1167–1168
  mannitol and, *1168*, 1168–1169
  pinitol and, 1169–1170, *1170*
Osmotically generated pressure flow
  (OGPF), 731–735, 758–760
Osmotic potential, 734b
Osmotic pressure, 734b
Osmotin, 1172–1173, *1174*
Outward rectifiers, 145–146, *147*
Ovary, 998
Ovule
  development of, 1011–1014, *1013*
    mutations and, 1014
  gametogenesis in, 1011–1014
Oxaloacetate (OAA)
  in amino acid synthesis, 359, *360*
  in citric acid cycle, 682, *682*, 683–685
  in gluconeogenesis, 710
  OAA shuttle and, *709*, 710
Oxidation-reduction process, photosynthe-
    sis as, 569
Oxidative metabolism, senescence and,
  1066–1069
Oxidative phosphorylation, 679. *See also*
    Electron transfer
  mechanism of, 679–680, *681*
  respiration and, 678, 679
  substrates and products of, 680t, *682*
Oxidative photosynthetic carbon cycle,
  719–724, *721*
Oxidative stress, 1189–1196
  causes of, 1189, *1189*
  defenses for, 1190–1191, *1191*, 1195–1196,
    1196t
  hormones and, 1197
  ozone and, 1191–1195
Oxygenase, Rubisco as, 34, *612*, 617–618
Oxygenation. *See also* Respiration
  Rubisco and, 719, *719*
Oxygen deficit stress
  aerenchyma formation and, *1085*,

1085–1086, *1086*, 1178, *1179*,
  1186–1188, *1187*, 1188t
ATP synthesis and, 1181–1184
epinasty and, *1188*, 1188–1189
flooding and, 1177–1189
gene expression and, 1184–1186
glycolytic fermentation and, 1181–1186
levels of, 1178t
long-term acclimation to, 1186–1188
metabolic response to, 1178t, 1181–1189
short-term acclimation to, 1181–1184
signaling and, 1189
Oxygenic photosynthesis, 569
Ozone, oxidative stress and, 1191–1195
Ozone dose, 1195

P_i carrier, *701*, 701–702
PAPS reductase, 836
Paramutation, 351–352, *352*
Particle rosettes, 88, *90*
Pasteur effect, 1181
Patch clamp studies, 136–137, 138b, *139*
Pathogen-derived resistance, 1150–1153,
  1152b
Pathogenesis, 1103
Pathogenesis-related proteins, 1138t,
  1138–1139
Pathogens. *See* Plant pathogens
P-clusters, 793, *793*
PDC/PDH-stat hypothesis, 1182–1183, *1183*
Pdslp, 539t, *541*, 542
  ubiquitination of, 542
Pea (*Pisum sativum*), Mendel's experiments
  with, 315–316, *316*, 317b
Pectin methylesterase, 76
Pectins, in cell wall, 65–69, *72*, *73*, 76–78, *79*,
  81, *82*
  in expansion, 96–97
  of ripening fruit, 99, *102*
pep12, in protein sorting, 194, *195*
PEP carboxykinase pathway, 712, *715*
Peppermint oil, 1264–1266
Peptide bonds, protein folding and, 444,
  444
Peptidyl prolyl isomerases, in protein folding, 444, *444*
Perhydroxyl radicals, as reactive oxygen
  species, 1190
Perinuclear space, 29, *29*
Peroxidase, in indole-3-acetic acid
  catabolism, 889, *889*
Peroxidatic reaction, 33
Peroxisomes, *32*, 32–35, *33*, 35–37
  catalase in, 32–34, *33*
  definition of, 32
  discovery of, 32
  fatty acid oxidation in, 517–518, *519*
  functions of, 32–37, 170
  in glyoxylate cycle, 34–35, *35*, 36b
  in nitrogen conversion, 35–36
  photorespiration and, *33*, 34, 720, *721*, 723
  production of, 36–37
  protein targeting to, 170, *170*
  structure of, 32
Petals, 998
PFK, 657
PGase, 99
Phalloidin, in fluorescence microscopy, 229,
  233
Phaseic acid, 872, *873*
Phase transition, 7t, 7–8, *8*
Phenolic compounds, 1287–1292,
  1314b–1315b

in heartwood formation, 1302b
synthesis of, 1288–1292, *1290–1292*
Phenotype, genotype and, 315, *315*
Phenylalanine, synthesis of, 383–385, *384*,
  385, 1288
Phenylalanine (tyrosine) ammonia-lyase,
  1288–1289
Phenylpropanoid-acetate, 1287–1292
  synthesis of, 1288–1292
Phenylpropanoids, 78, *83*, 1287–1292
  biotechnologic applications of, 1312–1314
  in senescence, 1060–1062, *1061*
  synthesis of, 1288–1292
Pheophorbide, in chlorophyll catabolism,
  1058–1059, *1059*
Phloem, *732*
  exudates from
    collection of, 740b–741b
    properties of, 738, 738t, 739t
  virus movement in, 1110
Phloem loading, 761–763, *762*, 766
Phloem transport, 738–743, 758–776. *See also*
    Transport
  apoplastic loading in, 751, 761–765
  inhibition of, 740, *742*
  minor vein configuration and, 761–763,
    762
  of nonviral RNA, 778–779
  osmotically generated pressure flow in,
    731–735, 758–760
  plasmodesmata in, 748–758
  pore–plasmodesma complex in, 760–761,
    761
  SE/CC complex in, 760–771, 761–771. *See
    also* SE/CC complex
  sieve element unloading in, 768–771,
    770b
  symplasmic loading in, 761, 765–768
  vs. xylem transport, 738–743
Phloem-xylem solute transfer, 743, *743*
Phorbol, 1266, *1267*
Phosphatases, 532b
Phosphate. *See also* Phosphorus
  conservation of, 499b
  root release of, 1227
  root uptake of, 1223–1227
  transport of, 1222–1230
Phosphate transport, 1222–1230
Phosphate transporters, 1227–1230
Phosphatidate, 492–493
Phosphatidic acid, in lipid synthesis, 488, *488*
Phosphatidylcholine
  structure of, 6
  synthesis of, 488, 492, 494, *496*
  in triacylglycerol synthesis, 513, *514*
Phosphatidylethanolamine, synthesis of,
  492, 494, *495*, 496
Phosphatidylglycerol, synthesis of, 493–494,
  495
Phosphatidylinositide 3-kinase, in signaling, 958, *960*
Phosphatidylinositol
  in signaling, 504, *505*
  synthesis of, 494, *495*
Phosphatidylinositol-anchored membrane
  proteins, 8–9, *9*
Phosphatidylserine, synthesis of, 494, *495*,
  497
Phosphoenolpyruvate (PEP)
  in amino acid synthesis, 359, *360*
  in nitrogen fixation, 812
  in photosynthesis, 621–624, 625, 712, *715*
  in respiration, 682, *682*

Porins, 678
Positive selection, 359
Postillumination burst, 718
Postphloem transport, 771–776
Post-replication repair, 277–279, *278*
Post-transcriptional gene silencing, 1141, *1141*
Potassium. *See also under* K⁺
  deficiency of, *1206*
  functions of, 1207
  vacuolar release of, 146, *148*
Potassium ion channels, 12–13, 142–145, *145–147*. *See also* Ion channels
  inward-rectifying, 142–145, *143–146*
  outward-rectifying, 145, *146*
  two-pore, 145
Potassium nutrition, whole-plant, 1221–1222
Potassium transport, 1207–1221
  high-affinity, 1209–1214
  low-affinity, 1209–1214
  multiple transporters in, 1207–1210
  overview of, *1208*
  regulation of, 1214–1216
Potassium transporters
  cloning of, 1218–1221, 1222t
  structure of, functional implications of, 1217–1218
P-proteins, 777
pRb, 553
Precession, 62
Precursor RNA (pre-RNA), 304–306, *306, 307*
Prenyl group-linked membrane proteins, 8–9, *9*
Prenyltransferase, in terpenoid synthesis, 1258–1259
Prephenate aminotransferase, in aromatic amino acid synthesis, 385
Preprophase band, *247*, 247–248, *248*
Prereplication complex, 536–537, *538*, 539–540
Presequence, mitochondrial, 169
Pressure bomb, 745b
Pressure potential, 734b, 1162
Pressure probe, 154b
Primary assimilation, 362. *See also* Nitrogen assimilation
Primary inflorescence, 997, *997*
Primary metabolites, 1250–1251, *1251*
Proanthyocyanidins, 1288
Profilin, 221, 222b–223b
Programmed cell death, 1044–1099
  aging and, 1098b
  in aleurone, 1063, *1083*, 1083–1084, *1084*
  apoptotic, 1044–1046, *1045*, 1046b, 1094–1097
  autophagy in, 1046–1047, *1047*
  cell viability during, 1048
  definition of, 1044
  developmental, 1048–1051, *1049*
  in flower development, 1048–1049
  hypersensitive response and, 1044, 1048, 1050, 1087–1094, 1114, *1114*
    lesion-specific mutants and, *1087–1091*, 1087–1094, *1093*
  lysigeny and, 1050
  necrotic, 1044
  oil gland formation and, 1050, *1051*
  pathogen attacks and, 1086–1087
    apoptosis and, 1095–1097
  plant form and, 1050, *1050*
  in plant life cycle, 1048–1051, *1049*
  schizogeny and, 1050

senescence and, 1044. *See also* Senescence
  signaling in, 1052–1053, *1053*, 1084
  as stress response, 1050, *1051*, 1077–1078, 1085–1097
  in tissue differentiation, 1050
  tracheary elements in, 1047, *1078*, 1078–1082, *1079, 1081, 1082*
  types of, 1044–1048, 1098–1099
  vegetative development and, 1049–1050
Prokaryotic pathway, *480, 486*, 486–490
Prolamellar body, 39, *40*
Prolamins, 17–18, *18*, 180b, 191, 1031, *1031*, 1031–1032, 1036
  classification of, *1031*, 1031–1032
  seed storage of, 1031–1032
Proline, synthesis of, 407–409, *408*
Proline dehydrogenase, 409
Proline-rich glycoproteins, 69
Proline-rich proteins, 69, 97–98, 102
Prometaphase, *531*
Promiscuous DNA, 290–291, *291*
Promoters, 292–294, *294, 340*, 340, 341–342, *342*
  identification of, 344b–345b
Proofreading, 269
Propeptides
  C-terminal, 192, *193*, 194
  N-terminal, 192, *193*, 194
Prophase, *242–243*, 531
Prophase spindle, 236, *238, 239*
Proplastids, 37–38, *38*, 43
n-Propylgallate, 697, *697*
Proteases
  in cell cycle regulation, 549–552
  in chromosome separation, 542
  in protein degradation, *448*, 448–449, *449*
Proteasome, 449–450
Protein(s). *See also* Gene(s) *and specific types*
  cargo, 177, *177*, 177–178
  endoplasmic reticulum-resident, 187
  life cycle of, *413*
  membrane. *See* Membrane proteins
  mitochondrial, 48
  motor, *219*, 219–221, *220*
  oligomerization of, *185*, 185–187
  postranslational modification of, 418, 434–445
  proteolytic processing of, 434–435, *436*
  rod-like, 207–208, *208*
  stability of, amino acid residues and, 450–453, *452, 453*
  storage, 178, 180b, 191, *191*, 1027–1039. *See also* Storage proteins
  structural, in cell wall, 69–75, *74, 78*
Protein disulfide isomerase, in chloroplast 432, *433*
Proteinase inhibitors, 1141
Protein bodies, 18, *18*, 191, 448
Protein degradation, 447–453
  in cell cycle regulation, 546–548, *547*
  in chloroplast, *448*, 449, *449*
  in cytoplasm, 448–449
  destabilizing residues in, 452–453
  enzyme susceptibility in, 1070–1071
  functions of, 447–448
  N-acetylation in, 450–452
  proteases in, *448*, 448–449, *449*
  in senescence, 1062–1064, *1064, 1065*, 1070–1071
  ubiquitin in, 449–450, *450, 451*, 546–548, *547*, 1063–1064
    auxin and, *981*, 982
  in vacuoles, *448*, 448–449

Protein disulfide-isomerase, in protein folding, 184, *185*, 443, 443–444
Protein folding, 184, *185*, 435–444
  chaperones in, 161–162, *162*, 184, 186, 437–440, *439–441*
  chaperonins in, 437, 438t, 438–440, *440, 441*
  in cytoplasm, *442*, 442–443
  N-linked glycans in, *186*, 186–187
  peptide bonds and, 444, *444*
  protein disulfide-isomerase in, 184, *185*, 443, 443–444
  protein targeting and, *442*, 442–443
  protein transport and, 444
  signal recognition particle in, *442*, 442–443
  steps in, *437, 441*, 442–443, *443, 444*
Protein kinase C, in signaling, 957, *957*
Protein kinase cascade, 543b
Protein kinases, 532b
  activation of, 973–974
  calmodulin, 975–978, *976–978*
  classification of, *974*, 974–975
  functions of, 973
  growth factor, 976
  mitogen-activated, 543, 559–560, 976
  Raf-like, 978
  receptor-like, 938–939, 974–975, *975, 976*
  as receptors, 932, 938–939
  in signaling, 932, 938–939, 973–978
    in auxins, 981
  in transcription regulation, 977–978
Protein sorting. *See also* Protein targeting; Protein transport
  chaperones in, 161–162, *162*
  endoplasmic reticulum in, 18, *19*, 175–190, *176*
  Golgi complex in, *176*, 177
  machinery of, 161–164
  membrane transport in, 161–162, *162*
  signaling in, 162, 163b, 170–171, 179–182, *181*, 181t, *182*, 192–194, *193*
  targeting domains in, 161, *161*, 161t, 162
  in yeast, 193–194, *195*
Protein storage vacuoles, 178, 180b, 191, *191*
Protein synthesis, 412–453
  amino acid activation in, 416–417
  in chloroplasts, 426–434
  in cytoplasm, 413–426
  elongation in, 423, *424*
  in endoplasmic reticulum, 175–177, *176*
  in eukaryotes, 418–426
  initiation of, 418–423, *419, 421, 423*
    regulation of, 422–423
    steps in, *421*
  light-activated, 429–432, *431, 433, 434*
  messenger RNA in, 413–417, *414–416*
  mitochondrial, 48, 413, *414*
  of photosynthetic proteins, 429–432, *430–434*
  plastids in, 43, 413, *414*, 426–434
  ribosomes in, 417–418, *418*
  ricin inhibition of, 427b
  signal recognition particle in, 181, *182*
  sites of, 413, *414*
  stages of, 418, *418*
  termination of, 423–426, *425*
  in thylakoid membrane, 429, *429*
  transfer RNA in, 415–417, *416, 417*
  translation in, 413–418. *See also* Translation
Protein targeting. *See also* Protein sorting; Protein transport
  to chloroplasts, 164–168, *165–167*